International Cosmetic Ingredient Dictionary and Handbook

Tenth Edition
2004

Editors
Tara E. Gottschalck
Gerald N. McEwen, Jr., Ph.D., J.D.

Volume 2

Published by
The Cosmetic, Toiletry, and Fragrance Association
1101 17th Street, NW, Suite 300
Washington, D.C. 20036-4702
www.ctfa.org

Contents

Volume 2 of 4

- L -

LACCAIC ACID

CAS No.: 60687-93-6

JPN Translation:
ラッカイン酸

Empirical Formula:
$C_{26}H_{19}NO_{12}$

Definition: Laccaic Acid is the pigment found in the lac resin produced by the insect *Coccus laccae* on certain tress in India. It consists chiefly of :

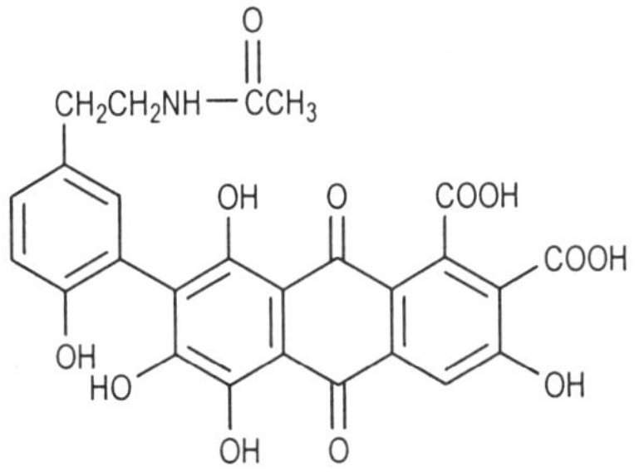

See "Regulatory and Ingredient Use Information," for Colorants in Volume 1, Introduction, Part A. This ingredient is not an approved colorant for the US.

Information Sources: JCIC, JCLS, JSQI, MI-13(5345)

Chemical Classes: Amides; Carboxylic Acids; Phenols

Function: Not Reported

Technical/Other Name:
Natural Red 25

LACCASE

CAS No.	**EINECS No.**
80498-15-3	420-150-4

Definition: Laccase is an enzyme produced by fermentation of *Aspergillus oryzae* expressing the gene encoding a laccase from *Myceliophthora thermophila*.

Chemical Class: Proteins

Function: Oxidizing Agent

Technical/Other Names:
p-Diphenol Oxidase
Urushiol Oxidase

Trade Name Mixture:
Novozym 809 (Novozymes)

LACTAMIDE

CAS No.	**EINECS No.**
2043-43-8	218-049-3

Empirical Formula:
$C_3H_7NO_2$

Definition: Lactamide is the organic compound that conforms to the formula:

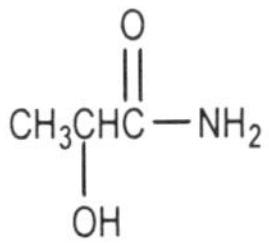

Chemical Class: Amides

Function: Skin-Conditioning Agent - Humectant

Technical/Other Names:
Lactic Acid Amide
Propanamide, 2-hydroxy-

Trade Name:
(-)-Lactamide (Merck KGaA)

LACTAMIDE DEA

Empirical Formula:
$C_7H_{13}NO_4$

Definition: Lactamide DEA is the amide that conforms to the formula:

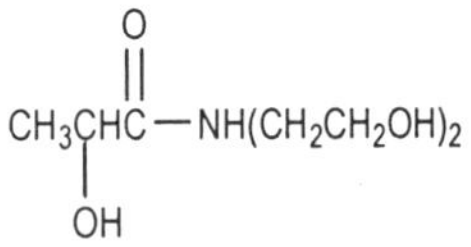

Chemical Class: Alkoxylated Amides

Function: Skin-Conditioning Agent - Humectant

Technical/Other Names:
Diethylene Glycol Lactamide
PEG-2 Lactamide

Trade Name:
Naetex-L (Lanaetex)

LACTAMIDE MEA

CAS No.	**EINECS No.**
5422-34-4	226-546-1

Empirical Formula:
$C_5H_{11}NO_3$

Definition: Lactamide MEA is a mixture of ethanolamides of lactic acid. It conforms generally to the formula:

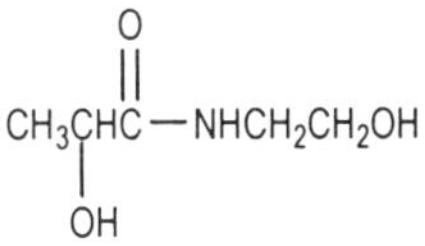

Chemical Class: Alkanolamides

Functions: Hair Conditioning Agent; Skin-Conditioning Agent - Humectant; Surfactant - Foam Booster; Viscosity Increasing Agent - Aqueous

Reported Product Categories: Hair Preparations (Non-coloring), Misc.; Tonics, Dressings, and Other Hair Grooming Aids; Moisturizing Preparations; Hair Conditioners; Shampoos (Non-coloring)

Technical/Other Names:
2-Hydroxy-N-(2-Hydroxyethyl)Propanamide
Lactamide, N-(2-hydroxyethy)-
Lactic Acid Monoethanolamide
Monoethanolamine Lactic Acid Amide
Propanamide, 2-Hydroxy-N-(2-Hydroxyethyl)-

Trade Names:
Foamid LAM (Alzo)
Incromectant LMEA (Croda, Inc.)
820752 Lactamide (Merck Schuchardt OHG)
Lipamide LMEA (Lipo)
Mackamide LME (McIntyre)
Parapel LAM-1000 (Bernel)
Schercomid LME (Scher)

Trade Name Mixtures:
Incromectant LAMEA (Croda Chemicals)
Incromectant LAMEA (Croda, Inc.)
Lipomectant AL (Lipo)

LACTAMIDOPROPYL TRIMONIUM CHLORIDE

CAS No.: 93507-51-8

Empirical Formula:
$C_9H_{21}N_2O_2 \cdot Cl$

Definition: Lactamidopropyl Trimonium Chloride is the quaternary ammonium salt that conforms to the formula:

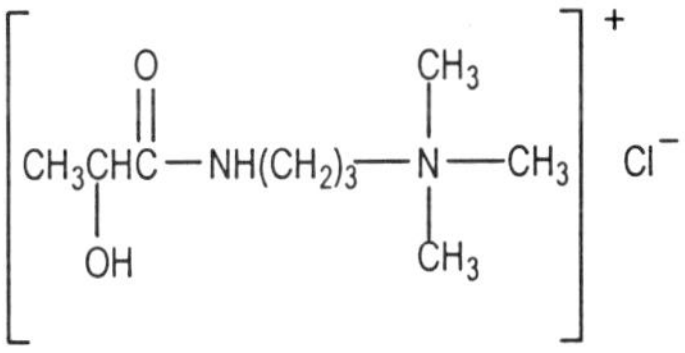

Chemical Class: Quaternary Ammonium Compounds

Function: Antistatic Agent

Reported Product Category: Hair Conditioners

Technical/Other Names:
(3-Lactamidopropyl)Trimethylammonium Chloride
1-Propanaminium, 3-((2-hydroxy-1-oxopropyl)amino)-N,N,N-trimethyl-, chloride

Trade Names:
AEC Lactamidopropyl Trimonium Chloride (A & E Connock)
Incromectant LQ (Croda, Inc.)

The inclusion of any compound in the *Dictionary and Handbook* does not indicate that use of that substance as a cosmetic ingredient complies with the laws and regulations governing such use in the United States or any other country.

LACTATE DEHYDROGENASE

CAS No.	EINECS No.
9001-60-9	232-617-8

Definition: Lactate Dehydrogenase is an enzyme which dehydrogenates lactic acid.

Information Source: MI-13(5349)

Chemical Class: Proteins

Function: Skin-Conditioning Agent - Miscellaneous

Trade Name:
L-Lactic Dehydrogenase (I.R.A.)

LACTIC ACID

CAS Nos.	EINECS Nos.
50-21-5	200-018-0
79-33-4	201-196-2

JPN Translation:
乳酸

Empirical Formula:
$C_3H_6O_3$

Definition: Lactic Acid is the organic acid that conforms to the formula:

$$CH_3CHCOOH$$
$$|$$
$$OH$$

Information Sources: ARG, AUS, BP, BPC, BRA, 21CFR133, 21CFR150.141, 21CFR150.161, 21CFR172.814, 21CFR178.1010, 21CFR184.1061, CIR: [SQ] IJT-17(Suppl. 1)1998, CZE, DA, DDR, EGY, FCC, FIN, HP, HUN, IND, ITA, JAN, JCLS, JSCI, MAR, MEX, MI-13(5350), MI-13(5352), PF, PN, POR, RIFM, ROM, TSCA, USAN, USD, USP XXIV, WHO, YUG

Chemical Class: Carboxylic Acids

Functions: Exfoliant; Fragrance Ingredient; Humectant; pH Adjuster; Skin-Conditioning Agent - Humectant; Skin-Conditioning Agent - Miscellaneous

Reported Product Categories: Shampoos (Non-coloring); Hair Conditioners; Hair Dyes and Colors (All Types Requiring Caution Statements and Patch Tests); Tonics, Dressings, and Other Hair Grooming Aids; Bath Oils, Tablets, and Salts; Cleansing Products (Cold Creams, Cleansing Lotions, Liquids and Pads); Moisturizing Preparations; Skin Care Preparations, Misc.; Body and Hand Preparations (Excluding Shaving Preparations); Bath Preparations, Misc.; Bath Capsules; Face and Neck Preparations (Excluding Shaving Preparations); Night Skin Care Preparations; Permanent Waves; Hair Preparations (Non-coloring), Misc.; Paste Masks (Mud Packs); Hair Sprays (Aerosol Fixatives); Eyeliners; Aftershave Lotions; Baby Shampoos; Skin Fresheners; Indoor Tanning Preparations; Hair Rinses (Non-coloring); Hair Coloring Preparations, Misc.; Hair Wave Sets; Hair Straighteners

Technical/Other Names:
α-Hydroxypropanoic Acid
2-Hydroxypropanoic Acid
2-Hydroxypropionic Acid
Lactic acid (RIFM)
Lactic acid (RIFM)
Propanoic Acid, 2-Hydroxy-

Trade Names:
AEC Lactic Acid (A & E Connock)
Lexalt L (Inolex)
PURAC HiPure 90 (Purac)
PURAC PF 90 (Purac)
Unichem LACA (Universal Preserv-A-Chem)

Trade Name Mixtures:
Acifructol Complex P 63 (Gattefosse s.a.)
Acifructol Tomato P 62 (Gattefosse s.a.)
A.H.A. 40 (Ennagram)
A.H.A. Extracts (Ennagram)
A.H.A. Extracts (Phytochim)
Alfhac 5 (Fabriquimica)
Amisil-L (Alban Muller)
Crodarom Hygroderm (Croda, Inc.)
Curasan (CLR)
Dragocare W 2/032700 (Symrise)
DS-CERIX (Doosan)
Extrapone Arnica Special 2/034591 (Symrise)
Extrapone Balm Mint 2/033121 (Symrise)
Extrapone Bamboo 2/032635 (Symrise)
Extrapone Coco-Nut Special 2/033055 (Symrise)
Extrapone Hops Special 2/032971 (Symrise)
Extrapone Macadamia Nut 2/032160 (Symrise)
Extrapone #4 Herbs 2/032495 (Symrise)
Extrapone #7 Herbs 2/032535 (Symrise)
Extrapone #1 Special 2/032451 (Symrise)
Extrapone #3 Special 2/034481 (Symrise)
Extrapone #5 Special 2/032501 (Symrise)
Extrapone Orris 2/033460 (Symrise)
Extrapone Poppy Flower 2/395321 (Symrise)
Extrapone 3 Special New 2/034484 (Symrise)
Extrapone Spruce Needle-Special (2/034831) (Symrise)
Facteur Hydratant PH (Prod'Hyg)
Fruitedone (UCIB (Solabia))
Haircare Complex CLR (CLR)
Hydeoviton 5,5 N 2/059359 (Symrise)
Hydrofacteur LC (LCW)
Hydrolyzed NMF (Proalan)
Hydroveg VV (Variati)
Hydroviton 2/059353 (Symrise)
Hydroviton 24 (2/059351) (Symrise)
Lactil (Degussa Care Specialties)
Lipoid Liposome 0040 (Lipoid)
Milk "AHA" (Cosmetochem) (Cosmetochem International Ltd.)
Moisturising Factor Hydrogerm (Crodarom)
Moisturizing Factor Hygro-Complex ARO 5272 (Crodarom)
Molecularsource LA (C.I.T.)
PURAC BF/P (Purac)
PURAC BF/S (Purac)
Tegodeo HY 77 (Degussa Care Specialties)

LACTIC ACID/HYDROXYSTEARIC ACID COPOLYMER

Definition: Lactic Acid/Hydroxystearic Acid Copolymer is a copolymer of lactic acid and hydroxystearic acid monomers.

Chemical Class: Synthetic Polymers

Function: Skin-Conditioning Agent - Miscellaneous

Trade Name:
Ethox PLPHS (Ethox)

LACTIC YEASTS

Definition: Lactic Yeasts is a Yeast (q.v.) obtained from milk. *See "Regulatory and Ingredient Use Information," regarding use of EU Trivial names in Volume 1, Introduction, Part A.*

Chemical Class: Biological Products

Function: Not Reported

Trade Name Mixture:
Lactic Yeast Extract 2 (Nonogawa)

LACTITOL

CAS No.	EINECS No.
585-86-4	209-566-5

Empirical Formula:
$C_{12}H_{24}O_{11}$

Definition: Lactitol is a disaccharide polyol obtained by the controlled hydrogenation of lactose. It conforms to the formula:

Information Sources: BAN, INN, MI-13 (5354)

Chemical Classes: Carbohydrates; Polyols

Functions: Flavoring Agent; Humectant; Skin-Conditioning Agent - Humectant

Reported Product Category: Shampoos (Non-coloring)

Technical/Other Name:
4-O-(β-galactopyranosyl)-D-glucitol

Trade Name:
Lacty (Purac)

Trade Name Mixture:
Ecodermine (Sederma)

LACTOBACCILLUS/PHOENIX DACTYLI-FERA (DATE) FRUIT FERMENT EXTRACT

Definition: Lactobaccillus/Phoenix Dactylifera (Date) Fruit Ferment Extract is an extract of the fermentation of the fruit of *Phoenix dactylifera* by the organism *Lactobaccilus*.

Chemical Class: Biological Products

Functions: Hair Conditioning Agent; Humectant; Skin-Conditioning Agent - Humectant

Trade Name:
ACB Date Palm Extract (Active Concepts)

LACTOBACILLUS/ACEROLA CHERRY FERMENT

Definition: Lactobacillus/Acerola Cherry Ferment is the product obtained by the fermentation of acerola cherry by the organism *Lactobacillus*.

Chemical Class: Biological Products

Function: Not Reported

LACTOBACILLUS/ALGAE EXTRACT FERMENT

Definition: Lactobacillus/Algae Extract Ferment is the product obtained by the fermentation of algae extract by the organism *Lactobacillus*.

Chemical Class: Biological Products

Function: Not Reported

LACTOBACILLUS/BRASSICA NIGRA SEED FERMENT EXTRACT

Definition: Lactobacillus/Brassica Nigra Seed Ferment Extract is an extract of the fermentation of *Brassica nigra* seeds by the organism, *Lactobacillus*.

Chemical Class: Biological Products

Function: Skin-Conditioning Agent - Miscellaneous

Trade Name:
ACB Mustard Bioferment (Active Concepts)

LACTOBACILLUS/CURCURBITA PEPO FRUIT FERMENT EXTRACT

Definition: Lactobacillus/Curcurbita Pepo Fruit Ferment Extract is an extract of the product obtained by the fermentation of the fruit of *Curcurbita pepo* by the organism, *Lactobacillus*.

Chemical Class: Biological Products

Function: Skin-Conditioning Agent - Miscellaneous

Trade Names:
ACB Pumpkin Enzyme (Active Concepts)
ACB Pumpkin Enzyme EF (Active Concepts)

LACTOBACILLUS/ERIODICTYON CALI-FORNICUM FERMENT EXTRACT

Definition: Lactobacillus/Eriodictyon Californicum Ferment Extract is an extract of the product obtained by the fermentation of *Eriodictyon californicum* by the organism, *Lactobacillus*.

Chemical Class: Biological Products

Function: Skin-Conditioning Agent - Miscellaneous

Trade Name:
ACB Yerba Santa Glycoprotein (Active Concepts)

LACTOBACILLUS FERMENT

Definition: Lactobacillus Ferment is an extract of the product resulting from the fermentation of *Lactobacillus*.

Chemical Class: Biological Products

Function: Not Reported

Trade Name Mixture:
Phosphovital (Sederma)

LACTOBACILLUS FERMENT LYSATE

Definition: Lactobacillus Ferment Lysate is a lysate of the product obtained by the fermentation of *Lactobacillus*.

Chemical Class: Biological Products

Function: Skin-Conditioning Agent - Miscellaneous

LACTOBACILLUS/GLYCERIN/ HYDROLYZED CASEIN/LACTOSE/ CATHARANTHUS ROSEUS SEED/YEAST EXTRACT FERMENT

Definition: Lactobacillus/Glycerin/ Hydrolyzed Casein/Lactose/Catharanthus Roseus Seed/Yeast Extract Ferment is the product obtained from the fermentation of glycerin, hydrolyzed casein, lactose, crushed periwinkle seeds from the plant *Catharanthus roseus*, and yeast extract by the organism, Lactobacillus.

Chemical Class: Biological Products

Function: Not Reported

LACTOBACILLUS/GLYCERIN/ HYDROLYZED SOY PROTEIN/ PERIWINKLE SEED/YEAST EXTRACT FERMENT

Definition: Lactobacillus/Glycerin/ Hydrolyzed Soy Protein/Periwinkle Seed/ Yeast Extract Ferment is the fermentation of Glycerin (q.v.), Hydrolyzed Soy Protein (q.v.), crushed seeds of *Catharanthus roseus*, and Yeast Extract (q.v.) by the organism, *Lactobacillus*.

Chemical Class: Biological Products

Function: Skin-Conditioning Agent - Miscellaneous

Technical/Other Name:
Lactobacillus/Glycerin/Hydrolyzed Soy Protein/Periwinkle (Catharanthus Roseus) Seed/Yeast Extract Ferment

LACTOBACILLUS-MILK/CALCIUM/ PHOSPHORUS/MAGNESIUM/ZINC FERMENT

Definition: Lactobacillus-Milk/Calcium/ Phosphorus/Magnesium/Zinc Ferment is an extract of a fermentation product of lactobacillus in the presence of calcium, phosphorus, magnesium and zinc ions.

Chemical Class: Biological Products

Function: Not Reported

Trade Name Mixture:
Biomin Yoghurt MAGZCaP (Arch Personal Care Products)

LACTOBACILLUS/MILK FERMENT FILTRATE

Definition: Lactobacillus/Milk Ferment Filtrate is a filtrate of the fermentation of milk by the microorganism *Lactogacillus*.

Chemical Class: Biological Products

Function: Skin-Conditioning Agent - Miscellaneous

Trade Name Mixtures:
FM Extract LA (Ichimaru Pharcos)
FM Extract LA-B (Ichimaru Pharcos)

LACTOBACILLUS-MILK/MANGANESE/ZINC FERMENT LYSATE

Definition: Lactobacillus-Milk/Manganese/Zinc Ferment Lysate is a lysate of the fermentation product of Milk (q.v.) by the organism *Lactobacillus* in the presence of manganese and zinc ions.

Chemical Class: Biological Products

Function: Not Reported

Trade Name:
Biotide (Pacific)

LACTOBACILLUS/MILK SOLIDS/GLYCINE SOJA (SOYBEAN) OIL FERMENT

Definition: Lactobacillus/Milk Solids/Glycine Soja (Soybean) Oil Ferment is the product obtained by the fermentation of milk solids and soybean oil by the organism, *Lactobacillus*.

Chemical Class: Biological Products

Function: Skin-Conditioning Agent - Humectant

Trade Name Mixture:
Lactopro CLP (Arch Personal Care Products)

LACTOBACILLUS/OLEA EUROPAEA (OLIVE) LEAF FERMENT EXTRACT

Definition: Lactobacillus/Olea Europaea (Olive) Leaf Ferment Extract is an extract of the product obtained by the fermentation of the leaves of *Olea europaea* by the organism *Lactobacillus*.

Chemical Class: Biological Products

Functions: Hair Conditioning Agent; Skin-Conditioning Agent - Humectant

Trade Name:
ACB Olive Leaf Extract (Active Concepts)

LACTOBACILLUS/ORYZA SATIVA (RICE) FERMENT

Definition: Lactobacillus/Oryza Sativa (Rice) Ferment is the product obtained by the fermentation of Oryza Sativa (Rice) (q.v.) by the organism, *Lactobacillus*.

Chemical Class: Biological Products

Functions: Hair Conditioning Agent; Skin-Conditioning Agent - Miscellaneous

Trade Name:
Lafrin-TN (Technoble)

LACTOBACILLUS/PORPHYRIDIUM FERMENT

Definition: Lactobacillus/Porphyridium Ferment is the product obtained by the fermentation of *Porphyridium* by the organism *Lactobacillus*.

Chemical Class: Biological Products

Function: Not Reported

Trade Name Mixture:
Antoline (Sederma)

LACTOBACILLUS/PUNICA GRANATUM FRUIT FERMENT EXTRACT

Definition: Lactobacillus/Punica Granatum Fruit Ferment Extract is an extract of the product obtained by the fermentation of the fruit of *Punica granatum* by the organism *Lactobacillus*.

Chemical Class: Biological Products

Functions: Hair Conditioning Agent; Skin-Conditioning Agent - Humectant

Trade Name:
ACB Pomegranate Enzyme (Active Concepts)

LACTOBACILLUS/RICE BRAN/SACCHAROMYCES/CAMELLIA SINENSIS LEAF EXTRACT FERMENT

Definition: Lactobacillus/Rice Bran/Saccharomyces/Camellia Sinensis Leaf Extract Ferment is the product obtained by the fermentation of Oryza Sativa (Rice) Bran (q.v.) by the microorganism *Lactobacillus*, and the fermentation of Camellia Sinensis Leaf Extract (q.v.) by the microorganism *Saccharomyces*. *See Reported Ingredient Functions-The Cosmetic Drug Distinction, in Regulatory and Ingredient Use Information, Volume I, Part A.*

Chemical Class: Biological Products

Functions: Antidandruff Agent; Antioxidant; Skin Bleaching Agent; Skin-Conditioning Agent - Miscellaneous

Trade Name:
aod.CHK (EM Kankyo Jyoka Giken)

LACTOBACILLUS/RYE FLOUR FERMENT

Definition: Lactobacillus/Rye Flour Ferment is the product obtained by the fermentation of rye flour by the microorganism *Lactobacillus*.

Chemical Class: Biological Products

Functions: Hair Conditioning Agent; Skin-Conditioning Agent - Humectant

Trade Names:
Woresan Rye Fluid (Woresan GmbH)
Woresan Rye Gel (Woresan GmbH)

LACTOBACILLUS/RYE FLOUR FERMENT FILTRATE

Definition: Lactobacillus/Rye Flour Ferment Filtrate is a filtrate of the product obtained from the fermentaion of rye flour by the microorganism *Lactobacillus*.

Chemical Class: Biological Products

Functions: Hair Conditioning Agent; Skin-Conditioning Agent - Humectant

Trade Name:
Woresan Rye Serum (Woresan GmbH)

LACTOBACILLUS/SKELETONEMA FERMENT

Definition: Lactobacillus/Skeletonema Ferment is the product obtained by the fermentation of *Skeletonema* by the organism *Lactobacillus*.

Chemical Class: Biological Products

Function: Not Reported

Trade Name Mixture:
Antoxine (Sederma)

LACTOBACILLUS/SOYBEAN FERMENT EXTRACT

Definition: Lactobacillus/Soybean Ferment Extract is an extract of the fermentation of soybeans by the microorganism *Lactobacillus*.

Chemical Class: Biological Products

Function: Skin-Conditioning Agent - Miscellaneous

Trade Name Mixture:
SoyAct-w (Kikkoman Corporation)

LACTOBACILLUS/WASABIA JAPONICA ROOT FERMENT EXTRACT

Definition: Lactobacillus/Wasabia Japonica Root Ferment Extract is an extract of the product of the fermentation of the roots of *Wasabia japonica* by the microorganism, *Lactobacillus.*

Chemical Class: Biological Products

Function: Antioxidant

Trade Name:
ACB Wasabi Extract (Active Concepts)

Trade Name Mixture:
AC Colorplex (Active Concepts)

LACTOBACILLUS/WATER HYACINTH FERMENT

Definition: Lactobacillus/Water Hyacinth Ferment is the product obtained from the fermentation of water hyacinth, *Eichornia crassipes* by the organism, *Lactobacillus.*

Chemical Class: Biological Products

Function: Not Reported

Trade Name Mixture:
Biacinth (Sederma)

LACTOBACILLUS/WHEY FERMENT

Definition: Lactobacillus/Whey Ferment is the product obtained by the fermentation of whey by the organism *Lactobacillus.*

Chemical Class: Biological Products

Function: Not Reported

Trade Names:
Biogen Active (Invekta)
Biogen Peptid Complex (Invekta)

Trade Name Mixtures:
FM Extract LA (Ichimaru Pharcos)
FM Extract LA-B (Ichimaru Pharcos)

LACTOBIONIC ACID

CAS No. 96-82-2 **EINECS No.** 202-538-3

Empirical Formula:
$C_{12}H_{22}O_{12}$

Definition: Lactobionic Acid is the organic acid that conforms to the formula:

Information Source: MI-13(5356)

Chemical Classes: Carbohydrates; Carboxylic Acids; Ethers

Function: pH Adjuster

Technical/Other Names:
4-O-β-Galactopyranosyl-D-Gluconic Acid
D-Lactobionic Acid

LACTOCOCCUS FERMENT

Definition: Lactococcus Ferment is an extract of the bacterial culture derived from *Lactococcus.*

Chemical Class: Biological Products

Function: Not Reported

Trade Name Mixtures:
Amaryl Hydro (Laboratoires Serobiologiques)
Elespher Vitaplex Hydro (Laboratoires Serobiologiques)
Neo Placenta LS 8407 (Laboratoires Serobiologiques)
Technobion BL (Laboratoires Serobiologiques)

LACTOCOCCUS FERMENT EXTRACT

Definition: Lactococcus Ferment Extract is an extract of the fermentation product of *Lactococcus.*

Chemical Class: Biological Products

Function: Not Reported

Technical/Other Name:
Extract of Lactococcus Ferment

LACTOCOCCUS FERMENT LYSATE

Definition: Lactococcus Ferment Lysate is a lysate of the fermentation product of *Lactococcus.*

Chemical Class: Biological Products

Function: Not Reported

Trade Name:
Protectan (CLR)

LACTOCOCCUS/MILK FERMENT LYSATE

Definition: Lactococcus/Milk Ferment Lysate is the product obtained by the fermentation of milk by the microorganism *Lactococcus lactis* with subsequent lysing of the microorganism's cells.

Chemical Class: Biological Products

Function: Not Reported

LACTOFERRIN

JPN Translation:
ラクトフェリン

Definition: Lactoferrin is the iron-binding glycoprotein component of mammalian milk.

Information Sources: JCIC, JCLS, MI-13 (9647)

Chemical Class: Proteins

Functions: Hair Conditioning Agent; Skin-Conditioning Agent - Miscellaneous

Reported Product Category: Moisturizing Preparations

Technical/Other Names:
Lactoferrin Solution
Lactotransferrin

Trade Name Mixtures:
Capigen CG (Sederma)
Enzyami 3 (Alban Muller)
Enzyami 5 (Alban Muller)
Enzyami 6 (Alban Muller)
FRS-Diffuser Microreservoir (Sederma)
Iniferine (Sederma)
Lactoferrin-S (Ichimaru Pharcos)
Lactoferrin S Free (Ichimaru Pharcos)
SB-12 (Sederma)

LACTOFLAVIN

CAS No. 83-88-5 **EINECS No.** 201-507-1

Empirical Formula:
$C_{17}H_{20}N_4O_6$

Definition: Lactoflavin is the organic compound that conforms to the formula:

See "Regulatory and Ingredient Use Information," for Colorants in Volume 1, Introduction, Part A. This ingredient is not an approved colorant for the US. To identify

---

The inclusion of any compound in the *Dictionary and Handbook* does not indicate that use of that substance as a cosmetic ingredient complies with the laws and regulations governing such use in the United States or any other country.

the colorant allowed for use in the European Union (EU), the INCI Name *Lactoflavin* must be used, except for hair dye products.

Information Sources: BAN, EEC(IV/1), INN, JAN, MI-13(8284), TSCA, USAN, USP XXIV

Chemical Class: Color Additives - Approved in the EU

Function: Colorant

Technical/Other Names:
Beflavin
6,7-Dimethyl-9-ribitylisoalloxazine
Lactoflavine
Riboflavin
Riboflavine
Vitamin B2
Vitamin G

LACTOGLOBULIN

Definition: Lactoglobulin is a globular protein isolated from milk.

Chemical Class: Proteins

Functions: Hair Conditioning Agent; Skin-Conditioning Agent - Miscellaneous

Trade Name:
Lactobiol (I.D. bio)

Trade Name Mixtures:
Lacteclat (I.D. bio)
Lacteclat Gel (I.D. bio)
Moisturizing Factor L (Sederma)

LACTOPEROXIDASE

CAS No.
9003-99-0

EINECS No.
232-668-6

Definition: Lactoperoxidase is an enzyme obtained from milk.

Chemical Class: Proteins

Function: Skin-Conditioning Agent - Miscellaneous

Reported Product Category: Moisturizing Preparations

Technical/Other Name:
Peroxidase

Trade Name Mixtures:
Capigen CG (Sederma)
Enzyami 1 (Alban Muller)
Enzyami 3 (Alban Muller)
Enzyami 5 (Alban Muller)
Enzyami 6 (Alban Muller)
Myavert C (Boots)
SB-12 (Sederma)

LACTOSE

CAS No.
63-42-3

EINECS No.
200-559-2

JPN Translation:
乳糖

Empirical Formula:
$C_{12}H_{22}O_{11}$

Definition: Lactose is a disaccharide conforming to the formula:

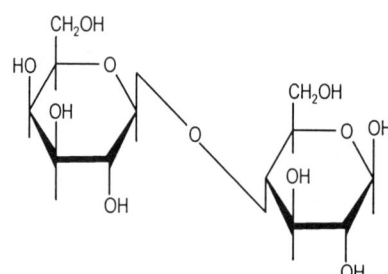

Information Sources: ARG, AUS, BP, BPC, BRA, 21CFR73.85, 21CFR133.124, 21CFR133.178, 21CFR133.179, 21CFR168.122, 21CFR169.179, 21CFR169.182, 21CFR184.1979, 21CFR184.1979a, 21CFR184.1979b, 21CFR184.1979c, 21CFR310.545, 21CFR460.6, DA, DDR, EGY, EP, FIN, HUN, IND, ITA, JAN, JCLS, JSCI, MAR, MEX, MI-13(5358), NF XIX, PF, PN, POR, ROM, TSCA, USAN, USD, USP XXIV, YUG

Chemical Classes: Carbohydrates; Polyols

Function: Skin-Conditioning Agent - Humectant

Reported Product Category: Skin Fresheners

Technical/Other Names:
4-O-β-D-Galactopyranosyl-D-Glucose
D-Glucose, 4-O-β-D-Galactopyranosyl-
Milk Sugar
Saccharum Lactin

Trade Names:
Milveta-Lactose (Milei)
Unisweet L (Universal Preserv-A-Chem)

Trade Name Mixtures:
Dry Vitamin A Palmitate, Type 250-SD (Roche)
Follicusan (CLR)
Granule AA (Ichimaru Pharcos)
Happyness Complex II (Greentech)
Hydratherm CGI Lyophilized (Universal Flavors)
Hydromilk ENL-SD (Arch Personal Care Products)
Lactofil Debacterise (Gattefosse s.a.)
Lactofil Foam (Gattefosse s.a.)
Lactofil Moist (Gattefosse s.a.)
Lactofil Sensitive (Gattefosse s.a.)
Lipobead Blue-AE (Lipo)

Lipobead Blue-AEL (Lipo)
Lipobead Blue-T (Lipo)
Lipobead Green-E (Lipo)
Lipobead Green-EL (Lipo)
Lipobead Pink-AEC (Lipo)
Lipobead Red-E (Lipo)
Lipobead Red-EL (Lipo)
Lipobead Yellow-EL (Lipo)
Milk Hydrolysate (Provital/Centerchem)
Natuchrom Super Green (Quest International)
Unicerin C-30 (Induchem)
Unispheres Series (Induchem)

LACTOYL METHYLSILANOL ELASTINATE

JPN Translation:
ラクトイルメチルシラノールエラスチネート

Definition: Lactoyl Methylsilanol Elastinate is the ester of Lactic Acid (q.v.) and Methylsilanol Elastinate (q.v.).

Chemical Classes: Protein Derivatives; Siloxanes and Silanes

Functions: Hair Conditioning Agent; Skin-Conditioning Agent - Miscellaneous

Reported Product Categories: Night Skin Care Preparations; Skin Care Preparations, Misc.

Trade Name:
Lasilium C (Exsymol)

LACTOYL PHYTOSPHINGOSINE

CAS No.
100403-19-8

EINECS No.
309-560-3

Definition: Lactoyl Phytosphingosine is a synthetic N-acylated sphingoid that conforms generally to the formula:

$$CH_3(CH_2)_nCH_2CHCHCHCH_2OH$$

where n has a value of 10 to 20.

Chemical Classes: Alcohols; Amides

Functions: Hair Conditioning Agent; Skin-Conditioning Agent - Miscellaneous

Reported Product Categories: Moisturizing Preparations; Face and Neck Preparations (Excluding Shaving Preparations); Cleansing Products (Cold Creams, Cleansing Lotions,

Liquids and Pads); Eye Makeup Preparations, Misc.; Hair Conditioners; Hair Preparations (Non-coloring), Misc.

Technical/Other Names:
Hydroxypropanoyl-C18-Phytosphingosine
Hydroxypropanoyl-4-Hydroxysphinganine
1,3,4-Octadecanetriol, 2-(2-Hydroxy) Propanamide

Trade Name:
Ceramide VID (C3:0) (Degussa Care Specialties)

LACTUCA SCARIOLA SATIVA (LETTUCE) LEAF EXTRACT

CAS No.   **EINECS No.**
84776-66-9   283-995-6

JPN Translation:
レタスエキス

Definition: Lactuca Scariola Sativa (Lettuce) Leaf Extract is an extract of the leaf of the lettuce, *Lactuca scariola sativa.* *See "Regulatory and Ingredient Use Information," regarding the labeling names for botanical ingredients in Volume 1, Introduction, Part A.*

Information Sources: JCIC, JCLS, JSQI

Chemical Class: Biological Products

Function: Skin-Conditioning Agent - Miscellaneous

Reported Product Category: Bath Soaps and Detergents

Technical/Other Names:
Extract of Lactuca Scariola Sativa
Extract of Lettuce
Lactuca Extract
Lettuce Extract
Lettuce Extract (1)
Lettuce Extract (2)
Lettuce (Lactuca Scariola Sativa) Extract

Trade Name Mixtures:
Actiphyte of Lettuce BG50 (Active Organics)
Actiphyte of Lettuce GL50 (Active Organics)
Actiphyte of Lettuce Lipo S (Active Organics)
Actiphyte of Lettuce PG50 (Active Organics)
Glycolysat of Lettuce (CEP (Solabia))
Lactuca Scariola Sativa (Lettuce) Extract ies (IES LABO)
Lettuce Extract (Maruzen Pharmaceuticals Co., Ltd.)
Lettuce Extract BG (Maruzen Pharmaceuticals Co., Ltd.)
Lettuce Extract HG (Provital/Centerchem)
Lettuce Extract HS 2575 G (Grau)
Lettuce HS (Alban Muller)
Lettuce Leaf Extract (Libiol)
Lettuce Liquid B (Ichimaru Pharcos)
Phytami Lettuce (Alban Muller)
VT-217 Extract of Lettuce (Vege-Tech)

LACTUCA SCARIOLA SATIVA (LETTUCE) LEAF JUICE

JPN Translation:
レタス液汁

Definition: Lactuca Scariola Sativa (Lettuce) Leaf Juice is the juice derived from the leaf of *Lactuca scariola sativa.* *See "Regulatory and Ingredient Use Information," regarding the labeling names for botanical ingredients in Volume 1, Introduction, Part A.*

Information Sources: JCIC, JCLS

Chemical Class: Biological Products

Function: Not Reported

Technical/Other Names:
Lactuca Scariola Sativa Juice
Lettuce Juice
Lettuce Leaf Juice

LACTUCA VIROSA LEAF EXTRACT

Definition: Lactuca Virosa Leaf Extract is an extract of the leaves of *Lactuca virosa.* *See "Regulatory and Ingredient Use Information," regarding the labeling names for botanical ingredients in Volume 1, Introduction, Part A.*

Chemical Class: Biological Products

Function: Skin-Conditioning Agent - Miscellaneous

Technical/Other Name:
Extract of Lactuca Virosa Leaf

Trade Name Mixture:
Wild Lettuce Leaf Extract (Libiol)

LACTULOSE

CAS No.   **EINECS No.**
4618-18-2   225-027-7

Empirical Formula:
$C_{12}H_{22}O_{11}$

Definition: Lactulose is the disaccharide that conforms to the formula:

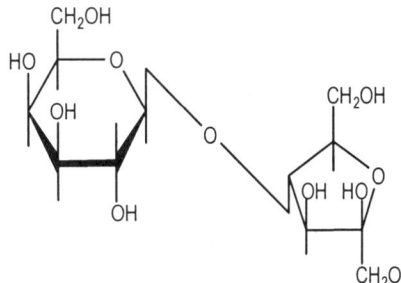

Information Source: MI-13(5361)

Chemical Class: Carbohydrates

Function: Skin-Conditioning Agent - Humectant

Technical/Other Names:
D-Fructose, 4-O-βδ-Galactopyranosyl-Isolactose
D-Lactulose

LAGENARIA SICERARIA FRUIT EXTRACT

Definition: Lagenaria Siceraria Fruit Extract is an extract of the fruit of *Lagenaria vulgaris.* *See "Regulatory and Ingredient Use Information," regarding the labeling names for botanical ingredients in Volume 1, Introduction, Part A.*

Chemical Class: Biological Products

Function: Skin-Conditioning Agent - Humectant

Technical/Other Names:
Extract of Lagenaria Siceraria Fruit
Lagenaria Vulgaris Fruit Extract

Trade Name Mixtures:
Bottle Gourd Fruit Extract (Ajinomoto)
Yuugao Extract (Nonogawa)
Yuugao Extract 2 (Nonogawa)

LAGERSTROEMIA INDICA EXTRACT

Definition: Lagerstroemia Indica Extract is an extract of the whole plant *Lagerstroemia indica.* *See "Regulatory and Ingredient Use Information," regarding the labeling names for botanical ingredients in Volume 1, Introduction, Part A.*

Chemical Class: Biological Products

Function: Skin-Conditioning Agent - Miscellaneous

Technical/Other Name:
Extract of Lagerstroemia Indica

Trade Name Mixture:
VT-612 Extract of Crape Myrtle (Vege-Tech)

LAGERSTROEMIA INDICA FLOWER EXTRACT

Definition: Lagerstroemia Indica Flower Extract is an extract of the flowers of *Lagerstroemia indica.* *See "Regulatory and Ingredient Use Information," regarding the labeling names for botanical ingredients in Volume 1, Introduction, Part A.*

Chemical Class: Biological Products

Function: Skin-Conditioning Agent - Miscellaneous

Technical/Other Name:
Extract of Lagerstroemia Indica Flower

Trade Name Mixtures:
Crape Myrtle Extract (Carrubba)
Crape Myrtle Extract and BG (Carrubba)

LAGERSTROEMIA SPECIOSA EXTRACT

JPN Translation:
オオバナサルスベリエキス

Definition: Lagerstroemia Speciosa Extract is an extract of the leaves of *Lagerstroemia speciosa*. See *"Regulatory and Ingredient Use Information," regarding the labeling names for botanical ingredients in Volume 1, Introduction, Part A.*

Chemical Class: Biological Products

Function: Skin-Conditioning Agent - Miscellaneous

Technical/Other Name:
Extract of Lagerstroemia Speciosa

Trade Name:
Lagerstroemia Speciosa Extract MP
(Maruzen Pharmaceuticals Co., Ltd.)

Trade Name Mixtures:
Banaba Glycolic Extract (Vinyals)
Banaba Liquid B (Ichimaru Pharcos)
Banaba Liquid E (Ichimaru Pharcos)
Lagerstroemia Speciosa Extract
(Yamakawa)

LAMINARIA CLOUSTONI EXTRACT

CAS Nos.   **EINECS No.**
90046-11-0   289-979-5
92128-82-0

JPN Translation:
ラミナリアクロウストニエキス

Definition: Laminaria Cloustoni Extract is an extract of the alga, *Laminaria cloustoni*. See *"Regulatory and Ingredient Use Information," regarding the labeling names for botanical ingredients in Volume 1, Introduction, Part A.*

Information Source: RIFM

Chemical Class: Biological Products

Functions: Fragrance Ingredient; Not Reported

Technical/Other Name:
Kelp (Laminaria & Mereocystis spp.)
(RIFM)

Trade Name:
Phyconnexine (SECMA)

Trade Name Mixture:
ACP 941 (Bottger)

LAMINARIA DIGITATA EXTRACT

CAS Nos.   **EINECS No.**
90046-12-1   289-980-0
92128-82-0

JPN Translation:
ラミナリアディギタータエキス

Definition: Laminaria Digitata Extract is an extract of the alga, *Laminaria digitata*. See *"Regulatory and Ingredient Use Information," regarding the labeling names for botanical ingredients in Volume 1, Introduction, Part A.*

Information Source: RIFM

Chemical Class: Biological Products

Functions: Fragrance Ingredient; Skin-Conditioning Agent - Miscellaneous

Reported Product Category: Shampoos (Non-coloring)

Technical/Other Names:
Extract of Laminaria Digitata
Kelp (Laminaria & Mereocystis spp.)
(RIFM)

Trade Name Mixtures:
Actiphyte of Laminaria (Active Organics)
Algenextrakt-Konzentrat Spezial
(Labopharma)
Aquaphycol LD (SECMA)
Bio-Energizer (SECMA)
Bioenergizer BG (SECMA)
Campo Bo Hai Cai Extract (Campo)
Complexe Algomarin (Codif)
Devil's Apron HS (Alban Muller)
Oleaphycol LD (SECMA)
Pheofiltrat Laminaria HG (Codif)
PHYCARINE (Goemar)
Phycojuvenine (Codif)
Phycol LD (SECMA)
Phycol LD BG (SECMA)
Phyco R75 (Codif)
Polyplant Slimming (Prival/Centerchem)
Prodhy Extract Laminaire (Prod'Hyg)
Pronalen Slimming (Provital/Centerchem)
Sea Extract Laminaria (E.U.K)
Seanergilium (Coletica SA)
Seanergilium BG (Coletica SA)

LAMINARIA DIGITATA POWDER

Definition: Laminaria Digitata Powder is the powder derived from the dried, crushed thallus of *Laminaria digitata*. See *"Regulatory and Ingredient Use*

Information," regarding the labeling names for botanical ingredients in Volume 1, Introduction, Part A.

Chemical Class: Biological Products

Function: Skin-Conditioning Agent - Miscellaneous

Technical/Other Name:
Powder, Laminaria Digitata

Trade Names:
AEC Oarweed Powdered (A & E Connock)
Devil's Apron Powder (Alban Muller)

LAMINARIA HYPERBOREA EXTRACT

CAS Nos.   **EINECS No.**
90046-13-2   289-981-6
92128-82-0

JPN Translation:
ラミナリアヒベルボレアエキス

Definition: Laminaria Hyperborea Extract is an extract of the alga, *Laminaria hyperborea*. See *"Regulatory and Ingredient Use Information," regarding the labeling names for botanical ingredients in Volume 1, Introduction, Part A.*

Information Source: RIFM

Chemical Class: Biological Products

Functions: Fragrance Ingredient; Not Reported

Technical/Other Names:
Extract of Laminaria Hyperborea
Kelp (Laminaria & Mereocystis spp.)
(RIFM)

Trade Name Mixtures:
Laminaria Gel BG (Crodarom)
Phycoboreane (SECMA)

LAMINARIA JAPONICA EXTRACT

CAS No.: 92128-82-0

JPN Translation:
マコンブエキス

Definition: Laminaria Japonica Extract is an extract of the seaweed, *Laminaria japonica*. See *"Regulatory and Ingredient Use Information," regarding the labeling names for botanical ingredients in Volume 1, Introduction, Part A.*

Information Source: RIFM

Chemical Class: Biological Products

Functions: Fragrance Ingredient; Not Reported

Technical/Other Names:
Extract of Laminaria Japonica

Kelp (Laminaria & Mereocystis spp.)
(RIFM)

Trade Name Mixtures:
Campo Long Xu Cai Extract (Campo)
China Extract Lung Xu Cai (E.U.K)
Chine Extract Laminaria Japonica
(Ennagram)

LAMINARIA OCHROLEUCA EXTRACT

CAS No.   **EINECS No.**
92128-82-0   295-780-4

Definition: Laminaria Ochroleuca Extract
is an extract of the alga, *Laminaria
ochroleuca. See "Regulatory and Ingredient Use Information," regarding the labeling
names for botanical ingredients in Volume
1, Introduction, Part A.*

Information Source: RIFM

Chemical Class: Biological Products

Functions: Fragrance Ingredient; Skin-
Conditioning Agent - Miscellaneous

Technical/Other Names:
Extract of Laminaria Ochroleuca
Kelp (Laminaria & Mereocystis spp.)
(RIFM)

Trade Name:
Laminaine (SECMA)

Trade Name Mixture:
Antileukine 6 (SECMA)

LAMINARIA SACCHARINA EXTRACT

CAS No.: 92128-82-0

JPN Translation:
カラフトコンブエキス

Definition: Laminaria Saccharina Extract
is an extract of the thallus of the alga,
*Laminaria saccharina. See "Regulatory and
Ingredient Use Information," regarding the
labeling names for botanical ingredients in
Volume 1, Introduction, Part A.*

Information Source: RIFM

Chemical Class: Biological Products

Functions: Fragrance Ingredient; Not
Reported

Technical/Other Names:
Extract of Laminaria Saccharina
Kelp (Laminaria & Mereocystis spp.)
(RIFM)

Trade Name Mixtures:
Phlorogine (SECMA)
Phlorogine BG (SECMA)
Recelderm - 503 (Bioland)

LAMIUM ALBUM EXTRACT

Definition: Lamium Album Extract is an
extract of the whole plant of *Lamium
album. See "Regulatory and Ingredient Use
Information," regarding the labeling names
for botanical ingredients in Volume 1, Introduction, Part A.*

Chemical Class: Biological Products

Function: Skin-Conditioning Agent - Miscellaneous

Technical/Other Names:
Extract of Lamium Album
Extract of White Nettle
White Nettle Extract

Trade Name Mixtures:
Pharcolex BX32 (Ichimaru Pharcos)
White Nettle Liquid B (Ichimaru Pharcos)

LAMIUM ALBUM FLOWER EXTRACT

CAS No.   **EINECS No.**
84012-23-7   281-669-8

JPN Translation:
オドリコソウ花エキス

Definition: Lamium Album Flower Extract
is an extract of the flowers of the white
nettle, *Lamium album. See "Regulatory and
Ingredient Use Information," regarding the
labeling names for botanical ingredients in
Volume 1, Introduction, Part A.*

Information Sources: JCIC, JCLS, JSQI

Chemical Class: Biological Products

Functions: Cosmetic Astringent; Skin-
Conditioning Agent - Miscellaneous

Technical/Other Names:
Extract of Lamium Album
Extract of White Nettle
Odorikosou Ekisu (JPN)
White Nettle Extract
White Nettle (Lamium Album) Extract

Trade Names:
Phytelene of Ortie Blanche EN 103 powder
(Indena SA)
Phytelene of White Nettle EN 103 powder
(Indena SA)
Phytogreen of White Nettle EP 487 Powder
(Phytochim)

Trade Name Mixtures:
Actiphyte of White Nettle (Active Organics)
Actiphyte of White Nettle BG50 (Active
Organics)
Actiphyte of White Nettle GL50 (Active
Organics)
Actiphyte of White Nettle Lipo S (Active
Organics)
Actiphyte of White Nettle PG50 (Active
Organics)

Dead Nettle Extract HS 2430 G (Grau)
Extrait D'Ortie MPE 100 (Yves Rocher)
Extrapone White Nettle 2/034017 (Symrise)
Fruitapone Gooseberry G 2/036710
(Symrise)
Glycolysat of White Nettle (CEP (Solabia))
Hair Treatment Complex 260 (Ennagram)
Hair Treatment Phytogreen Complex GXH
260 (Phytochim)
HAIR VOLUME (Greentech S.A)
Herbasol-Extract Dwarf Pine
(Cosmetochem)
Herbasol-Extract White/Dead Nettle
(Cosmetochem)
Phytelene Complex EGX 232 (Indena SA)
Phytelene of Ortie Blanche EG 157 liquid
(Indena SA)
Phytelene of White Nettle EG 157 liquid
(Indena SA)
Phytogreen 55 of White Nettle EXH 631
Liquid (Phytochim)
Prodhy Extract Ortie Blanche (Prod'Hyg)
Super Extrapone White Nettle 2/500016
(Symrise)
Vegebios of White Nettle (CEP (Solabia))
Vegetol White Nettle MCF 796 Hydro
(Gattefosse s.a.)
Vegetol White Nettle MCF 1233 Oily
(Gattefosse s.a.)
VT-259 Extract of White Nettle (Vege-
Tech)
White Dead Nettle HS (Alban Muller)
White Nettle Extract HG (Provital/
Centerchem)
White Nettle Phytexcell (Crodarom)

LAMIUM ALBUM LEAF EXTRACT

Definition: Lamium Album Leaf Extract is
an extract of the leaves of *Lamium album.
See "Regulatory and Ingredient Use Information," regarding the labeling names for
botanical ingredients in Volume 1, Introduction, Part A.*

Chemical Class: Biological Products

Function: Skin-Conditioning Agent - Miscellaneous

Technical/Other Names:
Extract of Lamium Album Leaf
Odorikosou Ekisu (JPN)

Trade Name Mixture:
Herbasol Dead Nettle Extract (Leaf)
(Cosmetochem) (Cosmetochem
International Ltd.)

LANDOLPHIA OWARIENSIS EXTRACT

Definition: Landolphia Owariensis Extract
is an extract of the plant, *Landolphia*

---

owariensis. See "Regulatory and Ingredient Use Information," regarding the labeling names for botanical ingredients in Volume 1, Introduction, Part A.

Chemical Class: Biological Products

Function: Not Reported

Technical/Other Name:
Extract of Landolphia Owariensis

Trade Name Mixture:
Apocynaceae (COL-AR)

LANETH-5

CAS No.: 61791-20-6 (Generic)

JPN Translation:
ラネス - 5

Definition: Laneth-5 is the polyethylene glycol ether of Lanolin Alcohol (q.v.) with an average ethoxylation value of 5.

Information Sources: CIR: [S] JACT-1(4)-1982, CTFA D, JCLS, JSCI, MI-13(7659), TSCA

Chemical Classes: Alkoxylated Alcohols; Lanolin and Lanolin Derivatives

Functions: Skin-Conditioning Agent - Miscellaneous; Surfactant - Emulsifying Agent

Reported Product Categories: Hair Dyes and Colors (All Types Requiring Caution Statements and Patch Tests); Hair Tints; Hair Straighteners

Technical/Other Names:
PEG-5 Lanolin Ether
Polyethylene Glycol (5) Lanolin Ether
Polyoxyethylene (5) Lanolin Alcohol
Polyoxyethylene (5) Lanolin Ether

Trade Names:
Nikkol BWA-5 (Nikko)
Polychol 5 (Croda Chemicals)
Polychol 5 (Croda, Inc.)

LANETH-10

CAS No.: 61791-20-6 (Generic)

JPN Translation:
ラネス - 10

Definition: Laneth-10 is the polyethylene glycol ether of Lanolin Alcohol (q.v.) with an average ethoxylation value of 10.

Information Sources: JCLS, JSCI, MI-13 (7659), TSCA

Chemical Classes: Alkoxylated Alcohols; Lanolin and Lanolin Derivatives

Function: Surfactant - Emulsifying Agent

Technical/Other Names:
PEG-10 Lanolin Ether
Polyethylene Glycol 500 Lanolin Ether
Polyoxyethylene (10) Lanolin Alcohol
Polyoxyethylene (10) Lanolin Ether

Trade Names:
Nikkol BWA-10 (Nikko)
Polychol 10 (Croda Chemicals)

LANETH-15

CAS No.: 61791-20-6 (Generic)

JPN Translation:
ラネス - 15

Definition: Laneth-15 is the polyethylene glycol ether of Lanolin Alcohol (q.v.) with an average ethoxylation value of 15.

Information Sources: CTFA D, JCLS, JSCI, MI-13(7659), TSCA

Chemical Classes: Alkoxylated Alcohols; Lanolin and Lanolin Derivatives

Function: Surfactant - Emulsifying Agent

Reported Product Categories: Hair Conditioners; Hair Straighteners

Technical/Other Names:
PEG-15 Lanolin Ether
Polyethylene Glycol (15) Lanolin Ether
Polyoxyethylene (15) Lanolin Alcohol
Polyoxyethylene (15) Lanolin Ether
Polyoxyethylene (15) Wool Wax Alcohol Ethers

Trade Names:
Fancol LA-15 (Fanning)
Ivarlan 3442 (Arch Personal Care Products)
Polychol 15 (Croda Chemicals)
Polychol 15 (Croda, Inc.)

Trade Name Mixtures:
Relaxer Concentrate #2 (Arch Personal Care Products)
Relaxer Concentrate #3 (Arch Personal Care Products)
Relaxer Concentrate #4 (Arch Personal Care Products)

LANETH-16

CAS No.: 61791-20-6 (Generic)

JPN Translation:
ラネス - 16

Definition: Laneth-16 is the polyethylene glycol ether of Lanolin Alcohol (q.v.) with an average ethoxylation value of 16.

Information Sources: CIR: [S] JACT-1(4)-1982, JSQI, MI-13(7659), TSCA

Chemical Classes: Alkoxylated Alcohols; Lanolin and Lanolin Derivatives

Function: Surfactant - Emulsifying Agent

Reported Product Categories: Makeup Bases; Moisturizing Preparations; Mascara; Tonics, Dressings, and Other Hair Grooming Aids; Shampoos (Non-coloring)

Technical/Other Names:
PEG-16 Lanolin Ether
Polyethylene Glycol (16) Lanolin Ether
Polyoxyethylene (16) Lanolin Ether

Trade Name Mixture:
Solulan 16 (Amerchol)

LANETH-20

CAS No.: 61791-20-6 (Generic)

JPN Translation:
ラネス - 20

Definition: Laneth-20 is the polyethylene glycol ether of Lanolin Alcohol (q.v.) with an average ethoxylation value of 20.

Information Sources: CTFA D, JCLS, JSCI, MI-13(7659), TSCA

Chemical Classes: Alkoxylated Alcohols; Lanolin and Lanolin Derivatives

Functions: Surfactant - Cleansing Agent; Surfactant - Solubilizing Agent

Technical/Other Names:
PEG-20 Lanolin Ether
Polyethylene Glycol 1000 Lanolin Ether
Polyoxyethylene (20) Lanolin Alcohol
Polyoxyethylene (20) Lanolin Ether

Trade Names:
AEC Laneth-20 (A & E Connock)
Nikkol BWA-20 (Nikko)
Polychol 20 (Croda Chemicals)

LANETH-25

CAS No.: 61791-20-6 (Generic)

JPN Translation:
ラネス - 25

Definition: Laneth-25 is the polyethylene glycol ether of Lanolin Alcohol (q.v.) with an average ethoxylation value of 25.

Information Sources: CIR: [S] JACT-1(4)-1982, CTFA D, JCLS, JSCI, MI-13(7659), TSCA

Chemical Classes: Alkoxylated Alcohols; Lanolin and Lanolin Derivatives

Functions: Surfactant - Cleansing Agent; Surfactant - Solubilizing Agent

Technical/Other Names:
PEG-25 Lanolin Ether

Polyethylene Glycol (25) Lanolin Ether
Polyoxyethylene (25) Lanolin Alcohol
Polyoxyethylene (25) Lanolin Ether

LANETH-40

CAS No.: 61791-20-6 (Generic)

JPN Translation:
ラネス - 40

Definition: Laneth-40 is the polyethylene glycol ether of Lanolin Alcohol (q.v.) with an average ethoxylation value of 40.

Information Sources: CTFA D, JCLS, JSCI, MI-13(7659), TSCA

Chemical Classes: Alkoxylated Alcohols; Lanolin and Lanolin Derivatives

Functions: Surfactant - Cleansing Agent; Surfactant - Solubilizing Agent

Reported Product Category: Foundations

Technical/Other Names:
PEG-40 Lanolin Ether
Polyethylene Glycol 2000 Lanolin Ether
Polyoxyethylene (40) Lanolin Alcohol
Polyoxyethylene (40) Lanolin Ether

Trade Names:
Nikkol BWA-40 (Nikko)
Polychol 40 (Croda Chemicals)

LANETH-50

CAS No.: 61791-20-6 (Generic)

JPN Translation:
ラネス - 50

Definition: Laneth-50 is the polyethylene glycol ether of Lanolin Alcohol (q.v.) with an average ethoxylation value of 50.

Information Sources: MI-13(7659), TSCA

Chemical Classes: Alkoxylated Alcohols; Lanolin and Lanolin Derivatives

Functions: Surfactant - Cleansing Agent; Surfactant - Solubilizing Agent

Technical/Other Names:
PEG-50 Lanolin Ether
Polyethylene Glycol (50) Lanolin Ether
Polyoxyethylene (50) Lanolin Ether

LANETH-60

CAS No.: 61791-20-6 (Generic)

JPN Translation:
ラネス - 60

Definition: Laneth-60 is the polyethylene glycol ether of Lanolin Alcohol (q.v.) with an average ethoxylation value of 60.

Information Source: MI-13(7659)

Chemical Classes: Alkoxylated Alcohols; Lanolin and Lanolin Derivatives

Functions: Surfactant - Cleansing Agent; Surfactant - Solubilizing Agent

Technical/Other Names:
PEG-60 Lanolin Ether
Polyethylene Glycol 3000 Lanolin Ether
Polyoxyethylene (60) Lanolin Ether

Trade Name Mixture:
Relaxer Concentrate #4 (Arch Personal Care Products)

LANETH-75

CAS No.: 61791-20-6 (Generic)

JPN Translation:
ラネス - 75

Definition: Laneth-75 is the polyethylene glycol ether of Lanolin Alcohol (q.v.) with an average ethoxylation value of 75.

Information Sources: JSQI, MI-13(7659), TSCA

Chemical Classes: Alkoxylated Alcohols; Lanolin and Lanolin Derivatives

Functions: Surfactant - Cleansing Agent; Surfactant - Solubilizing Agent

Technical/Other Names:
PEG-75 Lanolin Ether
Polyethylene Glycol 4000 Lanolin Ether
Polyoxyethylene (75) Lanolin Ether

LANETH-9 ACETATE

JPN Translation:
酢酸ラネス - 9

Definition: Laneth-9 Acetate is the acetylated ester of an ethoxylated ether of Lanolin Alcohol (q.v.), with an average ethoxylation value of 9.

Information Sources: CIR: [S] JACT-1(4)-1982, JSQI

Chemical Classes: Esters; Lanolin and Lanolin Derivatives

Functions: Hair Conditioning Agent; Skin-Conditioning Agent - Emollient

Technical/Other Names:
Acetylated Polyoxyethylene (9) Lanolin Alcohol
PEG-9 Lanolin Ether, Acetylated
Polyethylene Glycol 450 Lanolin Ether Acetate
Polyoxyethylene (9) Lanolin Ether, Acetylated

LANETH-10 ACETATE

CAS No.: 65071-98-9 (Generic)

JPN Translation:
酢酸ラネス - 10

Definition: Laneth-10 Acetate is the acetylated ester of an ethoxylated ether of Lanolin Alcohol (q.v.) with an average ethoxylation value of 10.

Information Sources: CIR: [S] JACT-1(4)-1982, CTFA S

Chemical Classes: Esters; Lanolin and Lanolin Derivatives

Functions: Hair Conditioning Agent; Skin-Conditioning Agent - Emollient

Reported Product Categories: Bath Preparations, Misc.; Body and Hand Preparations (Excluding Shaving Preparations); Bath Soaps and Detergents; Hair Sprays (Aerosol Fixatives); Personal Cleanliness Products, Misc.; Aftershave Lotions; Baby Shampoos; Bath Oils, Tablets, and Salts; Cleansing Products (Cold Creams, Cleansing Lotions, Liquids and Pads); Foundations; Moisturizing Preparations; Night Skin Care Preparations; Shampoos (Non-coloring); Tonics, Dressings, and Other Hair Grooming Aids

Technical/Other Names:
Acetylated Polyoxyethylene (10) Lanolin Alcohol
PEG-10 Lanolin Ether, Acetylated
Polyethylene Glycol 500 Lanolin Ether Acetate
Polyoxyethylene (10) Lanolin Ether, Acetylated

LANETH-4 PHOSPHATE

Definition: Laneth-4 Phosphate is a complex mixture of phosphoric acid and an ethoxylated ether of Lanolin Alcohol (q.v.), with an average ethoxylation value of 4.

Chemical Classes: Esters; Lanolin and Lanolin Derivatives; Phosphorus Compounds

Function: Surfactant - Emulsifying Agent

Technical/Other Names:
PEG-4 Lanolin Ether Phosphate
Polyethylene Glycol 200 Lanolin Ether Phosphate
Polyoxyethylene (4) Lanolin Ether Phosphate

LANNEA COROMANDELICA BARK EXTRACT

Definition: Lannea Coromandelica Bark Extract is an extract of the bark of *Lannea*

coromandelica. See "Regulatory and Ingredient Use Information," regarding the labeling names for botanical ingredients in Volume 1, Introduction, Part A.

Chemical Class: Biological Products

Function: Skin-Conditioning Agent - Miscellaneous

Technical/Other Name:
Extract of Lannea Coromandelica Bark

Trade Name:
P.A Lannea LS (Laboratoires Sero-biologiques)

LANOLIN

CAS No.	EINECS No.
8006-54-0 (Anhydrous)	232-348-6

JPN Translation:
ラノリン

Definition: Lanolin is a refined derivative of the unctuous fat-like sebaceous secretion of sheep. It consists of a highly complex mixture of esters of high molecular weight aliphatic, steroid or triterpenoid alcohols and fatty acids. In the United States, Lanolin may be used as an active ingredient in OTC drug products. When used as an active drug ingredient, the established name is Lanolin. See "Regulatory and Ingredient Use Information," regarding use of EU Trivial names in Volume 1, Introduction, Part A. See "Regulatory and Ingredient Use Information," regarding the labeling names for U.S. OTC Drug Ingredients in Volume 1, Introduction, Part A.

Information Sources: ARG, AUS, BEL, BP, BPC, BRA, 21CFR172.615, 21CFR175.300, 21CFR176.170, 21CFR176.210, 21CFR177.1200, 21CFR177.2600, 21CFR178.3910, 21CFR310.545, 21CFR346.14, 21CFR349.14, CIR: [S] JEPT-4(4)1980, CTFA S, CZE, DA, DDR, EGY, FCC, FI, FIN, HUN, ITA, JAN, JCIC, JCLS, JP, JSCI, JSQI, MAR, MEX, MI-13(5373), OTC-I-AR, OTC-I-OP, OTC-I-SK, PF, PN, POL, ROM, SNPF, TSCA, USAN, USD, USP XXIV, YUG

Chemical Class: Lanolin and Lanolin Derivatives

Functions: Emulsion Stabilizer; Hair Conditioning Agent; Skin-Conditioning Agent - Emollient; Skin Protectant; Surfactant - Emulsifying Agent

Reported Product Categories: Blushers (All types); Bath Preparations, Misc.; Body and Hand Preparations (Excluding Shaving Preparations); Tonics, Dressings, and Other Hair Grooming Aids; Foundations; Moisturizing Preparations; Bath Oils, Tablets, and Salts; Cleansing Products (Cold Creams, Cleansing Lotions, Liquids and Pads); Permanent Waves; Eye Shadows; Skin Care Preparations, Misc.; Makeup Preparations (Not eye), Misc.; Paste Masks (Mud Packs); Bath Soaps and Detergents; Face Powders; Night Skin Care Preparations; Sachets; Shaving Cream (Aerosol, Brushless and Lather); Bath Capsules; Eye Makeup Preparations, Misc.; Face and Neck Preparations (Excluding Shaving Preparations); Hair Straighteners; Lipsticks; Cuticle Softeners; Hair Coloring Preparations, Misc.; Personal Cleanliness Products, Misc.; Suntan Preparations, Misc.; Eyeliners; Indoor Tanning Preparations; Makeup Bases; Rouges; Shampoos (Non-coloring); Aftershave Lotions; Baby Lotions, Oils, Powders and Creams; Deodorants (Underarm); Eyebrow Pencils; Fragrance Preparations, Misc.; Hair Conditioners; Hair Preparations (Non-coloring), Misc.; Mascara; Suntan Gels, Creams, and Liquids; Basecoats and Undercoats; Eye Lotions; Hair Sprays (Aerosol Fixatives); Hair Wave Sets; Manicuring Preparations, Misc.; Nail Polish and Enamels

Technical/Other Names:
Adsorption Refined Lanolin
Anhydrous Lanolin
Hard Lanolin
Hydrous Lanolin
Lanolin, Anhydrous
Lanolin Anhydrous USP
Purified Lanolin
Wool Fat
Wool Wax

Trade Names:
AEC Lanolin (A & E Connock)
Anhydrous Lanolin USP (Protameen)
Corona PNL (Croda, Inc.)
Cosmetic Lanolin (Croda, Inc.)
Cosmetic Lanolin (Fanning)
Cosmetic Lanolin (RITA)
Emery 1650 (Cognis Care Chemicals/NJ)
Emery 1650 (Cognis Care Chemicals/PA)
Emery 1660 (Cognis Care Chemicals/NJ)
Emery 1660 (Cognis Care Chemicals/PA)
Ivarlan 3000 (Arch Personal Care Products)
Ivarlan 3001 (Arch Personal Care Products)
Ivarlan 3006 Light (Arch Personal Care Products)
Lanolin, Anhydrous (Lanolines de la Tossee)
Lanolin Anhydrous USP (ChemMark)
Lanolin Anhydrous USP (Lanolines de la Tossee)
Lanolin Anhydrous USP Modified (Lanolines de la Tossee)
Lanolin Technical Grade (Lanolines de la Tossee)
Lanolin Technical Grade (RITA)
Lanolin USP AAA (Amerchol)
Lanolin USP Odorless (Lanolines de la Tossee)
Lanolin X-tra Deodorized (RITA)
Medilan (Croda Chemicals)
Pharmaceutical Lanolin (Croda, Inc.)
Pharmaceutical Lanolin (Fanning)
Pharmaceutical Lanolin (RITA)
Pharmalan (Croda Chemicals)
Prime Cosmetic Lanolin (Lanaetex)
Superfine Lanolin (Croda, Inc.)
Superfine Lanolin (Fanning)
Ultra Fine Lanolin (Fanning)
Yofco Lanolin (Nippon Chemical)
Yofco Lanolin SS (Nippon Chemical)

Trade Name Mixtures:
Amerchol C (Amerchol)
Aquaphil K (Croda Chemicals)
Argobase L2 (Croda Chemicals)
Argobase S1 (Croda Chemicals)
Base 323 MS (LCW)
Base RAL W 323 T (LCW)
Base Rouge A Levres 323 TAL (LCW)
Covashine (LCW)
Emery 1740 (Cognis Care Chemicals/NJ)
Emery 1740 (Cognis Care Chemicals/PA)
Emulgator Apicerol 2/014081 (Symrise)
Fancol C (Fanning)
Forlan 300 (RITA)
Forlan 500 (RITA)
Forlan L (RITA)
Homulgator 1330 G (Grau)
Isocreme CB 0279 (Croda Chemicals)
Ivarbase T (Arch Personal Care Products)
Lanaetex CLC (Lanaetex)
Lanaetex FB (Lanaetex)
Lanaetex-H (Lanaetex)
Lanaetex L-15 (Lanaetex)
Lanola 90 (Lanaetex)
Lanosil (Lanaetex)
Lanosoluble A (Prod'Hyg)
Lanosoluble M (Prod'Hyg)
Neo-PCL w/o s.e. 2/066255 (Symrise)
PCL SE w/o 2/066255 (Symrise)
Ritaderm (RITA)
Sabowax FL 84 (Sabo)

LANOLIN ACID

CAS No.	EINECS No.
68424-43-1	270-302-7

JPN Translation:
ラノリン脂肪酸

Definition: Lanolin Acid is a mixture of organic acids obtained from the hydrolysis of Lanolin (q.v.).

Information Sources: CIR: [S] JEPT-4(4)-1980, CTFA S, JCLS, JSCI, TSCA

Chemical Classes: Carboxylic Acids; Lanolin and Lanolin Derivatives

Function: Surfactant - Cleansing Agent

Surfactant-Cleansing Agent is included as a function for the soap form of Lanolin Acid.

Reported Product Categories: Mascara; Lipsticks; Eye Shadows; Foundations; Hair Conditioners; Moisturizing Preparations; Eyeliners; Hair Straighteners; Shaving Cream (Aerosol, Brushless and Lather); Paste Masks (Mud Packs)

Technical/Other Names:
Acids, Lanolin
Fatty Acids, Lanolin
Hard Lanolin Fatty Acid
Lanolic Acids
Lanolin Fatty Acid
Lanolin Fatty Acids
Soft Lanolin Fatty Acid

Trade Names:
Amerlate LFA, Low Odor (Amerchol)
Lanolic Acid (Croda Chemicals)
Lanolin Fatty Acid A (Nippon Chemical)
Lanolin Fatty Acid Hard (Nippon Chemical)
Lanolin Fatty Acid LIV (Nippon Chemical)
Lanolin Fatty Acids (Lanolines de la Tossee)
Ritalafa (RITA)

LANOLIN ALCOHOL

CAS No.   **EINECS No.**
8027-33-6   232-430-1

JPN Translation:
ラノリンアルコール

Definition: Lanolin Alcohol is a mixture of organic alcohols obtained from the hydrolysis of Lanolin (q.v.). *See "Regulatory and Ingredient Use Information," regarding use of EU Trivial names in Volume 1, Introduction, Part A.*

Information Sources: AUS, BP, BPC, CIR: [S] JEPT-4(4)1980, CTFA S, CZE, DA, DDR, HUN, IND, JCLS, JSCI, TSCA

Chemical Classes: Alcohols; Lanolin and Lanolin Derivatives; Sterols

Functions: Binder; Emulsion Stabilizer; Hair Conditioning Agent; Viscosity Increasing Agent - Nonaqueous

Reported Product Categories: Bath Preparations, Misc.; Body and Hand Preparations (Excluding Shaving Preparations); Tonics, Dressings, and Other Hair Grooming Aids; Moisturizing Preparations; Foundations; Makeup Bases; Night Skin Care Preparations; Lipsticks; Blushers (All types); Eyebrow Pencils; Suntan Gels, Creams, and Liquids; Bath Oils, Tablets, and Salts; Cleansing Products (Cold Creams, Cleansing Lotions, Liquids and Pads); Hair Conditioners; Skin Care Preparations, Misc.; Bath Capsules; Face and Neck Preparations (Excluding Shaving Preparations); Sachets; Eye Makeup Preparations, Misc.; Face Powders; Hair Straighteners; Makeup Preparations (Not eye), Misc.; Paste Masks (Mud Packs); Cuticle Softeners; Eyeliners; Personal Cleanliness Products, Misc.; Shaving Cream (Aerosol, Brushless and Lather); Shaving Preparations, Misc.; Fragrance Preparations, Misc.; Indoor Tanning Preparations; Shampoos (Non-coloring); Makeup Fixatives; Rouges; Aftershave Lotions; Baby Lotions, Oils, Powders and Creams; Baby Shampoos; Bath Soaps and Detergents; Eye Makeup Removers; Eye Shadows; Hair Preparations (Non-coloring), Misc.; Mascara; Suntan Preparations, Misc.; Hair Dyes and Colors (All Types Requiring Caution Statements and Patch Tests)

Technical/Other Names:
Alcohols, Lanolin
Wool Wax Alcohol

Trade Names:
AEC Lanolin Alcohol (A & E Connock)
Anatol (Lanaetex)
Argowax Cosmetic Super (Croda Chemicals)
Argowax Standard (Croda Chemicals)
Ceralan (Amerchol)
Emery 1780 (Cognis Care Chemicals/NJ)
Emery 1780 (Cognis Care Chemicals/PA)
Fancol LA (Fanning)
Hartolan (Croda Chemicals)
Hartolan (Croda, Inc.)
Ivarlan 3310 (Arch Personal Care Products)
Jeelan Alcohol (Jeen)
Lanalol Distilled (Maybrook)
Lanalol Standard (Maybrook)
Lanolin Alcohol A (Nippon Chemical)
Lanolin Alcohol B-2E (Nippon Chemical)
Ritawax (RITA)
Super Hartolan (Croda Chemicals)
Super Hartolan (Croda, Inc.)

Trade Name Mixtures:
Almolan Lis (Alma Chimica)
Amerchol C (Amerchol)
Amerchol CAB (Amerchol)
Amerchol L-101 (Amerchol)
Amphocerin K (Cognis Deutschland)
Aquaphil K (Croda Chemicals)
Argobase 125 (Croda Chemicals)
Argobase EU (Croda Chemicals)
Argobase L2 (Croda Chemicals)
Argobase S1 (Croda Chemicals)
Celanol A.S.-L (Lanaetex)
Crosterol SFA (Croda Chemicals)
Emery 1732 (Cognis Care Chemicals/PA)
Emery 1740 (Cognis Care Chemicals/NJ)
Emery 1740 (Cognis Care Chemicals/PA)
Fancol C (Fanning)
Fancol CAB (Fanning)
Fancol LAO (Fanning)
Forlan 200 (RITA)
Forlan 300 (RITA)
Forlan 500 (RITA)
Isocreme CB 0279 (Croda Chemicals)
Ivarbase 101 (Arch Personal Care Products)
Ivarbase T (Arch Personal Care Products)
Jeelan M-16 (Jeen)
Jeelan M-26 (Jeen)
Lanaetex CLC (Lanaetex)
Lanaetex FB (Lanaetex)
Lanaetex-H (Lanaetex)
Lanaetex L-15 (Lanaetex)
Lanalene ABS (Maybrook)
Lexate PX (Inolex)
Liquid Absorption Base Type T (Croda, Inc.)
Megabase L101 (Megachem)
Multilan A (Fabriquimica)
Oleo-Coll LP (Arch Personal Care Products)
Pionier MAA (Hansen & Rosenthal)
Protalan M-16 (Protameen)
Protalan M-26 (Protameen)
Protegin V (Degussa Care Specialties)
Protegin XV (Degussa Care Specialties)
Proto-Lan 8 (Maybrook)
Ritachol (RITA)
Steralchol (Lanaetex)
Unieucerin (Chemyunion)
Uniliquid (Chemyunion)

LANOLINAMIDE DEA

JPN Translation:
ラノリン脂肪酸アミド DEA

Definition: Lanolinamide DEA is a mixture of ethanolamides of Lanolin Acid (q.v.). It conforms generally to the formula:

$$RC \overset{\overset{\displaystyle O}{\|}}{} - N(CH_2CH_2OH)_2$$

where RCO- represents the fatty acids derived from lanolin.

Information Sources: EEC(III/1-60), JCIC, JCLS, JSQI

Chemical Classes: Alkanolamides; Lanolin and Lanolin Derivatives

Functions: Surfactant - Foam Booster; Viscosity Increasing Agent - Aqueous

Technical/Other Names:
N,N-Bis(2-Hydroxyethyl)Lanolin Acid Amide
Diethanolamine Lanolin Acid Amide
Lanolin Acid Amide, N,N-Bis-2-Hydroxyethyl-
Lanolin Diethanolamide
Lanolin Fatty Acid Diethanolamide

LANOLIN LINOLEATE
JPN Translation:
リノール酸ラノリル

Definition: Lanolin Linoleate is the ester of Lanolin Alcohol (q.v.) and linoleic acid.

Information Sources: JCIC, JCLS

Chemical Class: Lanolin and Lanolin Derivatives

Functions: Hair Conditioning Agent; Skin-Conditioning Agent - Occlusive

Technical/Other Name:
9,12-Octadienoic Acid, Lanolin Alcohol Ester

LANOLIN OIL

CAS Nos.
8038-43-5
70321-63-0

EINECS No.
274-559-6

JPN Translation:
液状ラノリン

Definition: Lanolin Oil is the liquid fraction of lanolin obtained by physical means from whole lanolin.

Information Sources: CIR: [S] JEPT-4(4)-1980, CTFA S, JCLS

Chemical Class: Lanolin and Lanolin Derivatives

Functions: Hair Conditioning Agent; Skin-Conditioning Agent - Emollient

Reported Product Categories: Lipsticks; Moisturizing Preparations; Bath Preparations, Misc.; Body and Hand Preparations (Excluding Shaving Preparations); Skin Care Preparations, Misc.; Bath Oils, Tablets, and Salts; Blushers (All types); Cleansing Products (Cold Creams, Cleansing Lotions, Liquids and Pads); Face Powders; Permanent Waves; Tonics, Dressings, and Other Hair Grooming Aids; Foundations; Makeup Bases; Bath Soaps and Detergents; Eyeliners; Makeup Preparations (Not eye), Misc.; Eyebrow Pencils; Night Skin Care Preparations; Suntan Gels, Creams, and Liquids; Eye Makeup Preparations, Misc.; Hair Conditioners; Hair Dyes and Colors (All Types Requiring Caution Statements and Patch Tests); Rouges; Bath Capsules; Eye Shadows; Face and Neck Preparations (Excluding Shaving Preparations); Fragrance Preparations, Misc.; Shampoos (Non-coloring); Shaving Cream (Aerosol, Brushless and Lather); Suntan Preparations, Misc.; Indoor Tanning Preparations; Baby Lotions, Oils, Powders and Creams; Cuticle Softeners; Hair Sprays (Aerosol Fixatives); Mascara; Nail Creams and Lotions; Nail Polish and Enamels; Paste Masks (Mud Packs)

Technical/Other Names:
Dewaxed Lanolin
Oils, Lanolin

Trade Names:
Fluilan (Croda Chemicals)
Fluilan (Croda, Inc.)
Ivarlan 3100 (Arch Personal Care Products)
Jeelan Oil (Jeen)
Lanogene (Amerchol)
Lanoil (Lanaetex)
Lanor Crystal (Lanolines de la Tossee)
Lantrol 1673 (Cognis Care Chemicals/NJ)
Lantrol 1673 (Cognis Care Chemicals/PA)
Lantrol 1674 (Cognis Care Chemicals/NJ)
Lantrol 1674 (Cognis Care Chemicals/PA)
Lipolan R (Lipo)
Liquid Lanolin (Nippon Chemical)
Protalan Oil (Protameen)
Ritalan (RITA)
Vigilan (Fanning)
Yofco Liquid Lanolin (Nippon Chemical)
Yofco Liquid Lanolin SS (Nippon Chemical)

Trade Name Mixtures:
Aloe Extract #103 (Florida Food Products)
Aqualose SLT (Croda Chemicals)
Bentone Gel LOI (ELE)
Crestalan CB3910 (Croda Chemicals)
Dermoil (Lanaetex)
Fancol ISO (Fanning)
Ivarbase 3240 (Arch Personal Care Products)
Ivarbase 3250 (Arch Personal Care Products)
Linolan (Variati)
Microlan (Croda Chemicals)
Proto-Lan 20 (Maybrook)
Pulvi-Lan (Lanaetex)
Ritalan C (RITA)
Tixogel LAN (Sud-Chemie, United Catalysts)

LANOLIN RICINOLEATE

Definition: Lanolin Ricinoleate is the ester of Lanolin Alcohol (q.v.) and ricinoleic acid.

Chemical Class: Lanolin and Lanolin Derivatives

Functions: Hair Conditioning Agent; Skin-Conditioning Agent - Occlusive

Technical/Other Names:
12-Hydroxy-9-Octadecenoic Acid, Lanolin Alcohol Ester
9-Octadecenoic Acid, 12-Hydroxy-, Lanolin Alcohol Ester

LANOLIN WAX

CAS No.
68201-49-0

EINECS No.
269-220-4

JPN Translation:
ラノリンロウ

Definition: Lanolin Wax is the semisolid fraction of lanolin obtained by physical means from whole lanolin. *See "Regulatory and Ingredient Use Information," regarding use of EU Trivial names in Volume 1, Introduction, Part A.*

Information Sources: CIR: [S] JEPT-4(4)-1980, CTFA S, TSCA

Chemical Classes: Lanolin and Lanolin Derivatives; Waxes

Functions: Binder; Hair Conditioning Agent; Skin-Conditioning Agent - Emollient; Viscosity Increasing Agent - Nonaqueous

Reported Product Categories: Lipsticks; Mascara; Moisturizing Preparations; Bath Oils, Tablets, and Salts; Cleansing Products (Cold Creams, Cleansing Lotions, Liquids and Pads); Eyebrow Pencils; Eyeliners; Skin Care Preparations, Misc.; Tonics, Dressings, and Other Hair Grooming Aids; Bath Preparations, Misc.; Body and Hand Preparations (Excluding Shaving Preparations); Blushers (All types); Eye Shadows; Foundations; Makeup Preparations (Not eye), Misc.; Rouges

Technical/Other Name:
De-Oiled Lanolin

Trade Names:
Crodalan SWL (Croda Japan)
Fancor Lanwax (Fanning)
Hard Lanolin (Nippon Chemical)
Ivarwax (Arch Personal Care Products)
Lanfrax 1776 (Cognis Care Chemicals/PA)
Lanfrax 1779 (Cognis Care Chemicals/PA)
Lanocerin (Amerchol)
Lanolin Wax H (Nippon Chemical)
Protalan Wax (Protameen)

Trade Name Mixture:
Argobase L2 (Croda Chemicals)

LANOSTEROL

CAS No.
79-63-0

EINECS No.
201-214-9

JPN Translation:
ラノステロール

Empirical Formula:
$C_{30}H_{50}O$

Definition: Lanosterol is a sterol obtained from lanolin. It conforms generally to the formula:

Information Sources: JCIC, JCLS, JSQI, MI-13(5375)

Chemical Classes: Lanolin and Lanolin Derivatives; Sterols

Functions: Hair Conditioning Agent; Skin-Conditioning Agent - Miscellaneous

Reported Product Category: Moisturizing Preparations

Technical/Other Names:
Lanosta-8,24-Dien-3-ol, (3β)-
Lanosterin
4m4m14α-Trimethylcholesta-8,24-dien-3β-ol

Trade Names:
AEC Lanosterol (A & E Connock)
(Croda) Lanosterol (Croda Chemicals)

LANSIUM DOMESTICUM EXTRACT

JPN Translation:
デュークエキス

Definition: Lansium Domesticum Extract is an extract of the pericarp of *Ransium domesticum*. See "Regulatory and Ingredient Use Information," regarding the labeling names for botanical ingredients in Volume 1, Introduction, Part A.

Information Source: JCIC

Chemical Class: Biological Products

Function: Skin-Conditioning Agent - Miscellaneous

Technical/Other Names:
Duku Extract
Extract of Lansium Domesticum

LANTANA CAMARA LEAF EXTRACT

Definition: Lantana Camara Leaf Extract is an extract of dried leaves of *Lantana camara*. See "Regulatory and Ingredient Use Information," regarding the labeling names for botanical ingredients in Volume 1, Introduction, Part A.

Chemical Class: Biological Products

Function: Not Reported

Technical/Other Name:
Extract of Lantana Camara

Trade Name Mixture:
Lantanine (CEP (Solabia))

LANTANA CAMARA LEAF WATER

Definition: Lantana Camara Leaf Water is an aqueous solution of the steam distillate obtained from the dried leaves of *Lantana camara*. See "Regulatory and Ingredient Use Information," regarding the labeling names for botanical ingredients in Volume 1, Introduction, Part A.

Chemical Class: Biological Products

Function: Fragrance Ingredient

Trade Name:
Vegebios of Golden Alyssum (CEP (Solabia))

LANTANA CAMARA ROOT EXTRACT

Definition: Lantana Camara Root Extract is an extract of the roots of Lantana camara. See "Regulatory and Ingredient Use Information," regarding the labeling names for botanical ingredients in Volume 1, Introduction, Part A.

Chemical Class: Biological Products

Function: Skin-Conditioning Agent - Humectant

Technical/Other Name:
Extract of Lantana Camara Root

Trade Name Mixture:
Lantana Camara Root Extract BG (Maruzen Pharmaceuticals Co., Ltd.)

LANTHANUM CHLORIDE

CAS No.	EINECS No.
10099-58-8	233-237-5

Empirical Formula:
Cl$_3$La

Definition: Lanthanum Chloride is the inorganic salt that conforms to the formula:

$$LaCl_3$$

Information Source: TSCA

Chemical Class: Inorganic Salts

Function: Not Reported

Technical/Other Name:
Lanthanum Truchloride

LAPYRIUM CHLORIDE

CAS No.	EINECS No.
6272-74-8	228-464-1

JPN Translation:
ラピリウムクロリド

Empirical Formula:
C$_{21}$H$_{35}$N$_2$O$_3$ • Cl

Definition: Lapyrium Chloride is the quaternary ammonium salt that conforms to the formula:

Information Sources: CIR: [S] JACT-10(1)-1991, INN, JCIC, JCLS, MI-13(5385), TSCA, USAN

Chemical Class: Quaternary Ammonium Compounds

Functions: Antistatic Agent; Cosmetic Biocide

Reported Product Categories: Deodorants (Underarm); Personal Cleanliness Products, Misc.

Technical/Other Names:
1-(2-Hydroxyethyl)Carbamoyl Methyl Pyridinium Chloride Laurate
Lapirium Chloride
N-(Lauroyl Colamino Formyl Methyl)-Pyridinium Chloride
N-(Lauryl Colamino Formyl Methyl) Pyridinium Chloride
1-[2-Oxo-2-[[2-[(1-Oxododecyl)Oxy]Ethyl]-Amino]Ethyl]Pyridinium Chloride
Pyridinium, 1-[2-Oxo-2-[[2-[(1-Oxododecyl)Oxy]Ethyl]Amino]Ethyl]-, Chloride

LARD

CAS No.	EINECS No.
61789-99-9	263-100-5

Definition: Lard is the purified fat obtained from the abdomen of the hog. See "Regulatory and Ingredient Use Information," regarding use of EU Trivial names in Volume 1, Introduction, Part A.

Information Sources: BP 1963, 21CFR176.210, 21CFR182.70, CIR: [SQ] IJT-20(SUPPL. 2)2001, CZE, DA, FIN, JAN, MAR, MI-13(5386), POL, ROM, TSCA

Chemical Class: Fats and Oils

Function: Skin-Conditioning Agent - Occlusive

Technical/Other Name:
Fats and Glycerdic Oils, Lard

LARD GLYCERIDE

CAS No.	EINECS No.
61789-10-4	263-032-6

The inclusion of any compound in the *Dictionary and Handbook* does not indicate that use of that substance as a cosmetic ingredient complies with the laws and regulations governing such use in the United States or any other country.

Definition: Lard Glyceride is the mono-glyceride derived from Lard (q.v.).

Information Sources: CIR: [SQ] IJT-20 (SUPPL. 2)2001, TSCA

Chemical Class: Glyceryl Esters and Derivatives

Functions: Emulsion Stabilizer; Skin-Conditioning Agent - Emollient; Viscosity Increasing Agent - Nonaqueous

Reported Product Category: Body and Hand Preparations (Excluding Shaving Preparations)

Technical/Other Names:
 Glycerides, Lard Mono-
 Glycerol Monester of Lard Acids
 Lard Monglycerides

LARD GLYCERIDES

CAS No. 91744-46-6
EINECS No. 294-609-0

Definition: Lard Glycerides is a mixture of mono, di and triglycerides derived from Lard (q.v.).

Information Source: CIR: [SQ] IJT-20 (SUPPL. 2) 2001

Chemical Class: Glyceryl Esters and Derivatives

Functions: Skin-Conditioning Agent - Emollient; Viscosity Increasing Agent - Nonaqueous

Technical/Other Name:
 Glycerides, Lard Mono-, Di- and Tri-

LARIX EUROPAEA WOOD EXTRACT

Definition: Larix Europaea Wood Extract is an extract of the wood of *Larix europaea*. See *"Regulatory and Ingredient Use Information,"* regarding the labeling names for botanical ingredients in Volume 1, Introduction, Part A.

Chemical Class: Biological Products

Function: Skin-Conditioning Agent - Humectant

Technical/Other Name:
 Extract of Larix Europaea

Trade Name:
 LARCH (Larix Europaea) EXTRACT LS 8978 (Laboratoires Serobiologiques)

Trade Name Mixture:
 Cryocytol (Laboratoires Serobiologiques)

LARREA DIVARICATA EXTRACT

CAS No. 91722-67-7
EINECS No. 294-468-5

Definition: Larrea Divaricata Extract is an extract of the chaparral, *Larrea divaricata*. See *"Regulatory and Ingredient Use Information,"* regarding the labeling names for botanical ingredients in Volume 1, Introduction, Part A.

Chemical Class: Biological Products

Function: Not Reported

Technical/Other Names:
 Chaparral Extract
 Chaparral (Larrea Divaricata) Extract
 Extract of Chaparral
 Extract of Larrea Divaricata

Trade Name Mixtures:
 Actiphyte of Chaparral BG50 (Active Organics)
 Actiphyte of Chaparral GL50 (Active Organics)
 Actiphyte of Chaparral Lipo S (Active Organics)
 Actiphyte of Chaparral PG50 (Active Organics)
 Chapparal BG (Alban Muller)
 Cosflor Creosote Bush HGS (A & E Connock)
 Floraceutical Chaparral Extract-Standardized (Bio-Botanica)

LARREA MEXICANA EXTRACT

CAS No. 84603-72-5
EINECS No. 283-270-4

Definition: Larrea Mexicana Extract is an extract of the chaparral, *Larrea mexicana*. See *"Regulatory and Ingredient Use Information,"* regarding the labeling names for botanical ingredients in Volume 1, Introduction, Part A.

Chemical Class: Biological Products

Function: Not Reported

Technical/Other Names:
 Chaparral Extract
 Chaparral (Larrea Mexicana) Extract
 Extract of Chaparral
 Extract of Larrea Mexicana
 Zygophyllum Tridentatum

Trade Name Mixtures:
 Bio-Dandra Plex (Bio-Botanica)
 Nutriplant (Bio-Botanica)

LAURALDEHYDE

CAS No. 112-54-9
EINECS No. 203-983-6

Empirical Formula:
 $C_{12}H_{24}O$

Definition: Lauraldehyde is the organic compound that conforms to the formula:

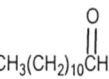

Information Source: RIFM

Chemical Class: Aldehydes

Functions: Flavoring Agent; Fragrance Ingredient

Technical/Other Names:
 1-Dodecanal
 Lauric aldehyde (RIFM)
 Lauryl Aldehyde

Trade Name:
 C-12 (LC United)

LAURALKONIUM BROMIDE

CAS No. 7281-04-1
EINECS No. 230-698-4

Empirical Formula:
 $C_{21}H_{38}N \cdot Br$

Definition: Lauralkonium Bromide is the quaternary ammonium salt that conforms to the formula:

$$\left[CH_3(CH_2)_{11} - \underset{\underset{CH_3}{|}}{\overset{\overset{CH_3}{|}}{N}} - CH_3 \right]^+ \quad Br^-$$

Information Source: TSCA

Chemical Class: Quaternary Ammonium Compounds

Functions: Antistatic Agent; Cosmetic Biocide

Technical/Other Names:
 Ammonium, Benzyldodecyldimethyl-, Bromide
 Benzenemethanaminium, N-Dodecyl-N,N-Dimethyl-, Bromide
 Benzyldodecyldimethylammonium Bromide
 Benzyllauryldimethylammonium Bromide
 N-Dodecyl-N,N-DimethylBenzene-methanaminium Bromide
 Lauryldimethylbenzylammonium Bromide

LAURALKONIUM CHLORIDE

CAS No. 139-07-1
EINECS No. 205-351-5

Empirical Formula:
 $C_{21}H_{38}N \cdot Cl$

Definition: Lauralkonium Chloride is a quaternary ammonium salt that conforms to the formula:

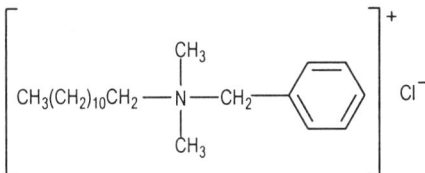

Information Sources: 21CFR172.165, 21CFR173.320, 21CFR175.105, CZE, INN, TSCA

Chemical Class: Quaternary Ammonium Compounds

Functions: Antistatic Agent; Cosmetic Biocide

Technical/Other Names:
Ammonium, Benzyldodecyldiemthyl-, Chloride
Benzenemethanaminium, N,N-Dimethyl-N-Dodecyl-, Chloride
Benzyldodecyldiemthylammonium Cloride
N,N-Dimethyl-N-Dodecylbenzene-methanaminium Chloride
Lauryl Benzalkonium Chloride
Lauryl Dimethyl Benzyl Ammonium Chloride

Trade Name Mixture:
Catinal CB-50 (Toho)

LAURAMIDE

CAS No.   **EINECS No.**
1120-16-7   214-298-7

Empirical Formula:
$C_{12}H_{25}NO$

Definition: Lauramide is the aliphatic amide that conforms generally to the formula:

$$CH_3(CH_2)_{10}C \overset{O}{\underset{||}{\;}} -NH_2$$

Information Sources: 21CFR178.3860, TSCA

Chemical Class: Amides

Function: Not Reported

Technical/Other Names:
Dodecanamide
Dodecylamide
Lauric Acid Amide
Lauryl Amide

LAURAMIDE DEA

CAS No.   **EINECS No.**
120-40-1   204-393-1

JPN Translation:
ラウラミド DEA

Empirical Formula:
$C_{16}H_{33}NO_3$

Definition: Lauramide DEA is a mixture of ethanolamides of lauric acid. It conforms generally to the formula:

$$CH_3(CH_2)_{10}C \overset{O}{\underset{||}{\;}} -N(CH_2CH_2OH)_2$$

Information Sources: 21CFR172.710, 21CFR173.315, 21CFR175.105, 21CFR176.180, 21CFR176.210, 21CFR177.2260, 21CFR177.2800, 21CFR178.3130, CIR: [SQ] JACT-5(5)1986, CTFA S, EEC(III/1-60), JCLS, JSCI, TSCA

Chemical Class: Alkanolamides

Functions: Surfactant - Foam Booster; Viscosity Increasing Agent - Aqueous

Reported Product Categories: Hair Dyes and Colors (All Types Requiring Caution Statements and Patch Tests); Shampoos (Non-coloring); Bubble Baths; Bath Soaps and Detergents; Bath Preparations, Misc.; Bath Oils, Tablets, and Salts; Cleansing Products (Cold Creams, Cleansing Lotions, Liquids and Pads); Personal Cleanliness Products, Misc.; Hair Sprays (Aerosol Fixatives); Hair Bleaches; Hair Preparations (Non-coloring), Misc.; Tonics, Dressings, and Other Hair Grooming Aids; Hair Shampoos (Coloring); Shaving Cream (Aerosol, Brushless and Lather); Hair Conditioners; Shaving Preparations, Misc.; Hair Tints; Hair Wave Sets; Skin Care Preparations, Misc.; Baby Products, Misc.; Baby Shampoos; Deodorants (Underarm); Feminine Hygiene Deodorants; Foot Powders and Sprays; Fragrance Preparations, Misc.; Leg and Body Paints; Oral Hygiene Products, Misc.; Permanent Waves

Technical/Other Names:
N,N-Bis(2-Hydroxyethyl)Dodecanamide
N,N-Bis(2-Hydroxyethyl)Lauramide
Diethanolamine Lauric Acid Amide
N,N-Diethanoldodecanamide
Dodecanamide, N,N-Bis(2-Hydroxyethyl)-
Lauric Acid Diethanolamide
Lauric Diethanolamide
Lauroyl Diethanolamide

Trade Names:
AEC Lauramide DEA (A & E Connock)
Alkamide LE (Rhodia)
Alkamide DL 203/S (Rhodia)
Alkamide DL 207/S (Rhodia)
Amidex LD (Chemron)
Amidex LSM (Chemron)
Aminol LM-30C (Finetex)
Aminol LM-30-C Special (Finetex)
Calamide LL (Pilot)
Colamid 0071 (Colonial Chemical Inc)
Colamid AL (Colonial Chemical Inc)
Colamid GDD (Colonial Chemical Inc)
Colamid KDO (Colonial Chemical Inc)
Colamid LA (Colonial Chemical Inc)
Colamid·LA-31 (Colonial Chemical Inc)
Colamid 150-LW (Colonial Chemical Inc)
Colamid 3462 MFA (Colonial Chemical Inc)
DeMIDE LA-100 (DeForest)
DeMIDE LLA-100 (DeForest)
DeMIDE ML-100 (DeForest)
DeMIDE MLY-100 (DeForest)
Empilan LDE (Albright & Wilson UK)
Foamid SL-Extra (Alzo)
Hetamide ML (Heterene)
Hetamide MOC (Heterene)
Incromide L-90 (Croda, Inc.)
Incromide LR (Croda, Inc.)
Jeemide 1224 (Jeen)
Jeemide L-90 (Jeen)
Jeemide LM-73 (Jeen)
Jeemide L-80-M (Jeen)
Jeemide LMAV (Jeen)
Lauramina (Vevy)
Mackamide L-5 (McIntyre)
Mackamide L-10 (McIntyre)
Mackamide L-95 (McIntyre)
Mackamide LLM (McIntyre)
Mackamide LMD (McIntyre)
Monamid 31 (Uniqema)
Monamid 716 (Uniqema)
Monamid 150-GLT (Uniqema)
Monamid 150 LMWC (Uniqema)
Monamid 150 LWA (Uniqema)
NINOL L-9 (Stepan)
NINOL 30-LL (Stepan)
NINOL 50-LL (Stepan)
NINOL 55-LL (Stepan)
NINOL 70-SL (Stepan)
NINOL 96-SL (Stepan)
Norfox DLSA (Norman, Fox & Co.)
Norfox 55-LL (Norman, Fox & Co.)
Protamide 1224 (Protameen)
Protamide L90 (Protameen)
Protamide 1224-LD (Protameen)
Protamide LM 73 (Protameen)
Protamide LMAV (Protameen)
Protamide LM-73-L (Protameen)
Protamide L80M-LD (Protameen)
Protamide LM-73 PG (Protameen)
Protamide L-80MA (Protameen)
Protamide L-80M (Protameen)
Schercomid SL Extra (Scher)
Schercomid SLMC-75 (Scher)
Schercomid SL-ML (Scher)
Schercomid SLM-LC (Scher)
Schercomid SLM-S (Scher)
Schercomid SL-Special (Scher)
Standamid KDS (Cognis Care Chemicals/NJ)
Standamid KDS (Cognis Care Chemicals/PA)
Standamid LAC (Cognis Care Chemicals/NJ)
Standamid LAC (Cognis Care Chemicals/PA)

Standamid LD (Cognis Care Chemicals/NJ)
Standamid LD (Cognis Care Chemicals/ PA)
Standamid LDO (Cognis Care Chemicals/ NJ)
Standamid LDO (Cognis Care Chemicals/ PA)
Standamid LDS (Cognis Care Chemicals/ NJ)
Standamid LDS (Cognis Care Chemicals/ PA)
Upamide LD (Universal Preserv-A-Chem)
Upamide LDS (Universal Preserv-A-Chem)
Upamide LM-20 (Universal Preserv-A-Chem)
Upamide LS-173 (Universal Preserv-A-Chem)
Upamide LS-196 (Universal Preserv-A-Chem)

Trade Name Mixtures:
BIO-SOFT LD-95 (Stepan)
BIO-TERGE 804 (Stepan)
DeMIDE LMSB-100 (DeForest)
Foamid 24 (Alzo)
Miracare AXB/X (Rhodia)
Miracare XEK/A (Rhodia)
Monamid 1007 (Uniqema)
Protamide LM-73-LD (Protameen)
Schercomid 1214 (Scher)
Standamid LD 80/20 (Cognis Care Chemicals/PA)
STEPANOL 360 (Stepan)

LAURAMIDE MEA

CAS No. 142-78-9
EINECS No. 205-560-1

JPN Translation:
ラウラミド MEA

Empirical Formula:
$C_{14}H_{29}NO_2$

Definition: Lauramide MEA is a mixture of ethanolamides of lauric acid. It conforms generally to the formula:

$$CH_3(CH_2)_{10}\overset{\displaystyle O}{\overset{\|}{C}}—NHCH_2CH_2OH$$

Information Sources: JCIC, JCLS, JSQI, TSCA

Chemical Class: Alkanolamides

Functions: Surfactant - Foam Booster; Viscosity Increasing Agent - Aqueous

Reported Product Categories: Bubble Baths; Bath Preparations, Misc.; Bath Oils, Tablets, and Salts; Shampoos (Non-coloring); Hair Rinses (Non-coloring); Personal Cleanliness Products, Misc.

Technical/Other Names:
Dodecanamide, N-(2-Hydroxyethyl)-

N-(2-Hydroxyethyl)Dodecanamide
N-(2-Hydroxyethyl)Lauramide
Lauric Acid Monoethanolamide
Lauroyl Monoethanolamide
Laurylethanoalamide
Laurylethanolamide
Monoethanolamine Lauric Acid Amide

Trade Names:
Alkamide L203 (Rhodia)
Amidex LMEA (Chemron)
Empilan LME/A (Albright & Wilson UK)
Hetamide MML (Heterene)
Jeemide LME (Jeen)
Mackamide LMA (McIntyre)
Mackamide LMM (McIntyre)
Monamid LM-MA (Uniqema)
NINOL C12LMP (Stepan)
NINOL LMP (Stepan)
Protamide LME (Protameen)

Trade Name Mixture:
Cosmacol P-50 (Sasol Italy)

LAURAMIDE MIPA

CAS No. 142-54-1
EINECS No. 205-541-8

JPN Translation:
ラウラミド MIPA

Empirical Formula:
$C_{15}H_{31}NO_2$

Definition: Lauramide MIPA is a mixture of isopropanolamides of lauric acid. It conforms generally to the formula:

$$CH_3(CH_2)_{10}\overset{\displaystyle O}{\overset{\|}{C}}—NHCH_2\underset{\underset{\displaystyle CH_3}{|}}{C}HOH$$

Information Sources: JCIC, JCLS, JSQI, TSCA

Chemical Class: Alkanolamides

Functions: Surfactant - Foam Booster; Viscosity Increasing Agent - Aqueous

Reported Product Categories: Bath Soaps and Detergents; Bath Preparations, Misc.; Personal Cleanliness Products, Misc.; Shampoos (Non-coloring)

Technical/Other Names:
Dodecanamide, N-(2-Hydroxypropyl)-
N-(2-Hydroxypropyl)Dodecanamide
2-Hydroxypropyllauramide
Lauroyl Isopropanolamide
N-Lauroyl Isopropanolamine
Lauryl Monoisopropaanolamide
LAuryl Monoisopropanolamide
Monoisopropanolamine Lauric Acid Amide

Trade Names:
Amidex LIPA (Chemron)

Empilan LIS/B (Albright & Wilson UK)
Mackamide LPA (McIntyre)
Monamid LIPA (Uniqema)
Nidaba 3 (Vevy)

Trade Name Mixture:
Texapon EVR K 400 (Cognis Deutschland)

LAURAMIDE/MYRISTAMIDE DEA

JPN Translation:
(ラウラミド / ミリスタミド) DEA

Definition: Lauramide/Myristamide DEA is a mixture of ethanolamides of a blend of lauric and myristic acids. It conforms generally to the formula:

$$R\overset{\displaystyle O}{\overset{\|}{C}}—N(CH_2CH_2OH)_2$$

where RCO- represents the lauric/myristic acid radical.

Information Sources: EEC(III/1-60), JCIC, JCLS, JSQI

Chemical Class: Alkanolamides

Functions: Surfactant - Foam Booster; Viscosity Increasing Agent - Aqueous

Technical/Other Name:
Diethanolamide Laurate Myristate

Trade Name:
Colamid LM-73 (Colonial Chemical Inc)

LAURAMIDOBUTYL GUANIDINE ACETATE

CAS No.: 161865-39-0

Definition: Lauramidobutyl Guanidine Acetate is the organic compound that conforms to the formula:

$$CH_3(CH_2)_{10}\overset{\displaystyle O}{\overset{\|}{C}}—NH(CH_2)_4N=\underset{\underset{\displaystyle NH_2}{|}}{\overset{\overset{\displaystyle NH_2}{|}}{C}} \cdot CH_3COOH$$

Chemical Classes: Amides; Amines

Functions: Antistatic Agent; Hair Conditioning Agent; Humectant

Trade Names:
B-C12A4G(32%) (Lion Corporation)
C12A4G (Lion Corporation)
C112A4G(50%) (Lion Corporation)

Trade Name Mixtures:
C12A4G (32%) (Lion Corporation)
C12A4G (50%) (Lion Corporation)

LAURAMIDOBUTYL GUANIDINE HCl

Empirical Formula:
$C_{17}H_{36}N_4O \cdot ClH$

Definition: Lauramidobutyl Guanidine HCl is the organic comopund that conforms to the formula:

$$CH_3(CH_2)_{10}\overset{\overset{O}{\|}}{C}-NH(CH_2)_4N=\overset{NH_2}{\underset{NH_2}{C}} \cdot HCl$$

Chemical Classes: Amides; Amines

Functions: Hair Conditioning Agent; Humectant; Skin-Conditioning Agent - Miscellaneous

Trade Name:
LAG Hydrochloride (Ajinomoto)

LAURAMIDOPROPYL ACETAMIDO-DIMONIUM CHLORIDE

CAS No.
68259-01-8

EINECS No.
269-504-8

Empirical Formula:
$C_{19}H_{40}N_3O_2 \cdot Cl$

Definition: Lauramidopropyl Acetamidodimonium Chloride is the quaternary ammonium salt that conforms to the formula:

$$\left[CH_3(CH_2)_{10}\overset{\overset{O}{\|}}{C}-NH(CH_2)_3\overset{\overset{CH_3}{|}}{\underset{CH_3}{N}}-CH_2\overset{\overset{O}{\|}}{C}-NH_2\right]^+ Cl^-$$

Chemical Class: Quaternary Ammonium Compounds

Function: Antistatic Agent

Technical/Other Names:
N-(2-Amino-2-Oxoethyl)-N,N-Dimethyl-3-[(1-Oxododecyl)Amino]-1-Propanaminium Chloride
Lauraminopropyl Acetamidodimonium Chloride
1-Propanaminium, N-(2-Amino-2-Oxoethyl)-N,N-Dimethyl-3-[(1-Oxododecyl)Amino]-, Chloride

LAURAMIDOPROPYLAMINE OXIDE

CAS No.
61792-31-2

EINECS No.
263-218-7

Empirical Formula:
$C_{17}H_{36}N_2O_2$

Definition: Lauramidopropylamine Oxide is the aliphatic amine oxide that conforms generally to the formula:

$$CH_3(CH_2)_{10}\overset{\overset{O}{\|}}{C}-NH(CH_2)_3-\overset{\overset{CH_3}{|}}{\underset{CH_3}{N}}\longrightarrow O$$

Information Source: TSCA

Chemical Class: Amine Oxides

Functions: Hair Conditioning Agent; Surfactant - Cleansing Agent; Surfactant - Foam Booster; Surfactant - Hydrotrope

Technical/Other Names:
Amides, Lauric, N-[3-(Dimethylamino)-Propyl], N-Oxide
N-[3-(Dimethylamino)Propyl]Dodecanamide-N-Oxide
Dodecanamide, N-[3-(Dimethylamino)-Propyl]-, N-Oxide
Dodecylamidopropyldimethylamine Oxide
3-Laurammidopropyl-N,N-dimethylamine Oxide

Trade Names:
Ammonyx LMDO (Stepan)
Amphotensid Cox (Zschimmer & Schwarz Italiana)
Mackamine LAO (McIntyre)

Trade Name Mixture:
Afron-N (Vevy)

LAURAMIDOPROPYL BETAINE

CAS Nos.
4292-10-8
86438-78-0

EINECS No.
224-292-6

JPN Translation:
ラウラミドプロピルベタイン

Empirical Formula:
$C_{19}H_{38}N_2O_3$

Definition: Lauramidopropyl Betaine is the zwitterion (inner salt) that conforms generally to the formula:

$$CH_3(CH_2)_{10}\overset{\overset{O}{\|}}{C}-NH(CH_2)_3-\overset{\overset{CH_3}{|}}{\underset{CH_3}{N^+}}-CH_2COO^-$$

Information Sources: JCIC, JCLS, JSQI, TSCA

Chemical Class: Betaines

Functions: Antistatic Agent; Hair Conditioning Agent; Skin-Conditioning Agent - Miscellaneous; Surfactant - Cleansing Agent; Surfactant - Foam Booster; Viscosity Increasing Agent - Aqueous

Reported Product Categories: Bath Preparations, Misc.; Shampoos (Non-coloring); Bath Oils, Tablets, and Salts; Bubble Baths; Skin Care Preparations, Misc.; Cleansing Products (Cold Creams, Cleansing Lotions, Liquids and Pads); Bath Soaps and Detergents

Technical/Other Names:
Ammonium, (carboxymethyl)(3-lauramidopropyl)diemthyl-, Hydroxide, Inner Salt
N-(Carboxymethyl)-N,N-Dimethyl-3-[(1-Oxodecyl)Amino]-1-Propanaminium Hydroxide, Inner Salt
N-(Dodecylamidopropyl)-N,N-diemthylammonium Betaine
Glycine, (3-lauramidopropyl)-Diemthylbetaine
Lauroyl Amide Propyldimethyl Glycine Solution
1-Propanaminium, N-(Carboxymethyl)-N,N-Dimethyl-3-[(1-Oxodecyl)Amino]-, Hydroxide, Inner Salt

Trade Names:
Amido Betaine L (Zohar)
AMPHOSOL LB (Stepan)
Amphotensid B4 (Zschimmer & Schwarz Italiana)
Colateric LMB (Colonial Chemical Inc)
Empigen BR (Albright & Wilson UK)
Mackam LMB (McIntyre)
Mirataine BB (Rhodia)
Monateric LMAB (Uniqema)
Monateric LMAB-B (Uniqema)
Softazoline LPB (Kawaken)

LAURAMIDOPROPYL DIMETHYLAMINE

CAS No.
3179-80-4

EINECS No.
221-661-3

Empirical Formula:
$C_{17}H_{36}N_2O$

Definition: Lauramidopropyl Dimethylamine is the amidoamine which conforms to the formula:

$$CH_3(CH_2)_{10}\overset{\overset{O}{\|}}{C}-NH(CH_2)_3-N\overset{CH_3}{\underset{CH_3}{\diagdown}}$$

Information Source: TSCA

Chemical Class: Amines

Function: Antistatic Agent

Technical/Other Names:
N-[3-(Dimethylamino)Propyl]Dodecanamide
Dimethylaminopropyl Lauramide
Dodecanamide, N-[3-(Dimethylamino)-Propyl]-
3-Dodecanamidopropyldimethylamine

Trade Names:
Mackine 801 (McIntyre)
Schercodine L (Scher)
Zohar Amido Amine L (Zohar)

LAURAMIDOPROPYL DIMETHYLAMINE PROPIONATE

Empirical Formula:
$C_{17}H_{36}N_2O \cdot C_3H_6O_2$

Definition: Lauramidopropyl Dimethylamine Propionate is the propionic acid salt of Lauramidopropyl Dimethylamine (q.v.). It conforms to the formula:

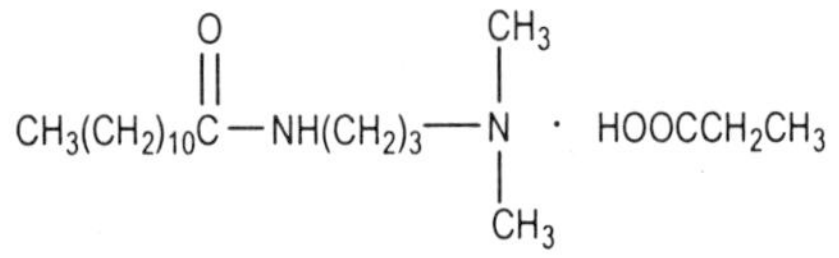

Chemical Class: Amines

Function: Antistatic Agent

Technical/Other Name:
Dimethylaminopropyl Lauramide Propionate

Trade Name:
Mackalene LAP (McIntyre)

LAURAMIDOPROPYL HYDROXYSULTAINE

Definition: Lauramidopropyl Hydroxysultaine is the zwitter ion (inner salt) that conforms to the formula:

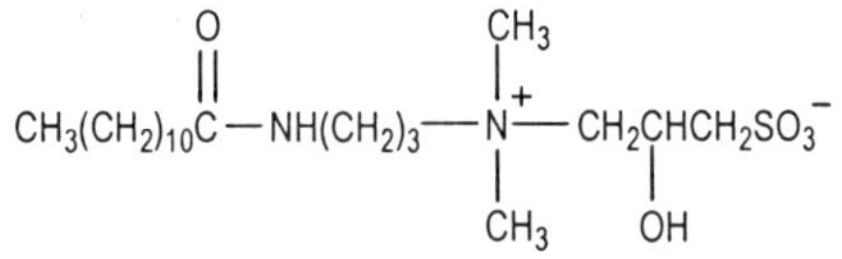

Chemical Class: Betaines

Functions: Antistatic Agent; Hair Conditioning Agent; Skin-Conditioning Agent - Miscellaneous; Surfactant - Cleansing Agent; Surfactant - Foam Booster; Viscosity Increasing Agent - Aqueous

Trade Name:
Softazoline LSB (Kawaken)

LAURAMIDOPROPYL PG-DIMONIUM CHLORIDE

Empirical Formula:
$C_{20}H_{43}N_2O_3 \cdot Cl$

Definition: Lauramidopropyl PG-Dimonium Chloride is the quaternary ammonium salt that conforms to the formula:

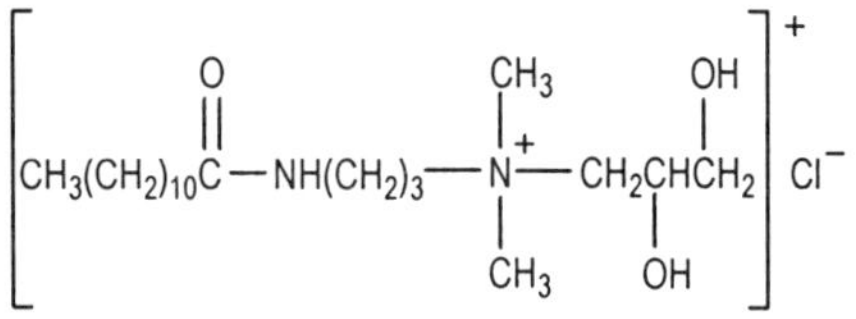

Chemical Class: Quaternary Ammonium Compounds

Functions: Antistatic Agent; Hair Conditioning Agent

LAURAMINE

CAS No.	EINECS No.
124-22-1	204-690-6

Empirical Formula:
$C_{12}H_{27}N$

Definition: Lauramine is the aliphatic amine that conforms to the formula:

$$CH_3(CH_2)_{11}NH_2$$

Information Sources: CIR: [I] JACT-14(3)-1995, TSCA

Chemical Class: Amines

Function: Antistatic Agent

Technical/Other Names:
1-Aminododecane
1-Dodecanamine
Dodecylamine
Lauryl Amine

Trade Name:
Armeen 12D (Akzo Nobel)

LAURAMINE OXIDE

CAS No.	EINECS No.
1643-20-5	216-700-6

JPN Translation:
ラウラミンオキシド

Empirical Formula:
$C_{14}H_{31}NO$

Definition: Lauramine Oxide is the tertiary amine oxide that conforms generally to the formula:

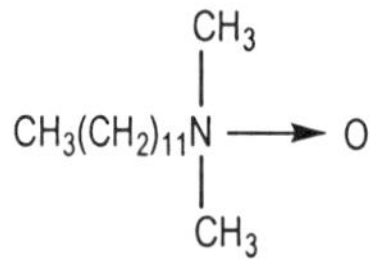

Information Sources: CIR: [SQ] JACT-13 (3)1994, CTFA S, JCIC, JCLS, JSQI, RIFM, TSCA

Chemical Class: Amine Oxides

Functions: Fragrance Ingredient; Hair Conditioning Agent; Surfactant - Cleansing Agent; Surfactant - Foam Booster; Surfactant - Hydrotrope

Reported Product Categories: Bath Soaps and Detergents; Tonics, Dressings, and Other Hair Grooming Aids; Shampoos (Noncoloring); Bath Oils, Tablets, and Salts; Cleansing Products (Cold Creams, Cleansing Lotions, Liquids and Pads); Personal Cleanliness Products, Misc.; Body and Hand Preparations (Excluding Shaving Preparations); Face and Neck Preparations (Excluding Shaving Preparations); Foundations; Hair Conditioners; Hair Preparations (Noncoloring), Misc.; Moisturizing Preparations; Skin Care Preparations, Misc.

Technical/Other Names:
N,N-Dimethyl-1-Dodecanamine-N-Oxide
1-Dodecanamine, N,N-Dimethyl-, N-Oxide
Dodecyldimethylamine oxide (RIFM)
Dodecyldimethylamine Oxide
Laurylamine Oxide
Lauryl Dimethyl Amine Oxide
Lauryl Dimethylamine Oxide Solution

Trade Names:
AEC Lauramine Oxide (A & E Connock)
Ammonyx DMCD-40 (Stepan)
AMMONYX LO (Stepan)
Aromox DMMC-W (Akzo Nobel)
Chemoxide LM-30 (Chemron)
Colalux LO (Colonial Chemical Inc)
DeMOX LAO (DeForest)
Empigen OB (Albright & Wilson UK)
Empigen OB/EB (Albright & Wilson UK)
Empigen OC/B (Albright & Wilson UK)
Euroxide LO (EOC Surfactants)
Incromine Oxide L (Croda, Inc.)
Jeechem LO (Jeen)
Mackamine LO (McIntyre)
Mackamine LO-SP (McIntyre)
Rhodamox LO (Rhodia)
Schercamox DML (Scher)
Softamin L (Toho)
Unimox LO (Universal Preserv-A-Chem)
Zoramox LO (Zohar)

LAURAMINOPROPIONIC ACID

CAS No.	EINECS No.
1462-54-0	215-968-1

JPN Translation:
ラウラミノプロピオン酸

Empirical Formula:
$C_{15}H_{31}NO_2$

Definition: Lauraminopropionic Acid is the substituted propionic acid that conforms generally to the formula:

$$CH_3(CH_2)_{11}NH(CH_2)_2COOH$$

Information Sources: JCIC, JCLS, JSQI, TSCA

Chemical Class: Alkyl-Substituted Amino Acids

Functions: Hair Conditioning Agent; Surfactant - Cleansing Agent

Technical/Other Names:
β-Alanine, N-Dodecyl-
N-Dodecyl-β-Alanine
N-Lauryl β-Aminopropionic Acid Solution
N-Lauryl, Myristyl Beta-Aminopropionic Acid

Trade Names:
Mackam 151 L (McIntyre)
Unitex 710-L (Universal Preserv-A-Chem)

LAURAMINOPROPYLAMINE

CAS No.
5538-95-4

EINECS No.
226-902-6

Empirical Formula:
$C_{15}H_{34}N_2$

Definition: Lauraminopropylamine is the substituted amine that conforms to the formula:

$$CH_3(CH_2)_{11}NH(CH_2)_3NH_2$$

Information Source: TSCA

Chemical Class: Amines

Function: Antistatic Agent

Technical/Other Names:
N-Dodecyl-1,3-Diaminopropane
1,3-Propanediamine, N-Dodecyl-

Trade Names:
Genamin LAP 100 D (Clariant)
Genamin LAP 100 D (Clariant GmbH, Personal Care)

LAURDIMONIUM HYDROXYPROPYL HYDROLYZED JOJOBA PROTEIN

CAS No.: 333338-09-3

Definition: Laurdimonium Hydroxypropyl Hydrolyzed Jojoba Protein is the quaternary ammonium salt that conforms generally to the formula:

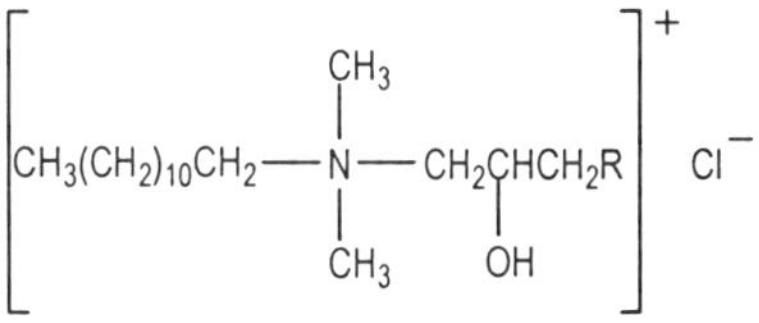

where R represents the Hydrolyzed Jojoba Protein (q.v.) moiety.

Chemical Classes: Protein Derivatives; Quaternary Ammonium Compounds

Functions: Antistatic Agent; Hair Conditioning Agent

Technical/Other Name:
Protein Hydrolyzates, Jojoba, [3-(Dodecyldimethylammonio)-2-Hydroxypropyl], Chlorides

Trade Name:
Jojoba Quat - LH (Desert Whale)

LAURDIMONIUM HYDROXYPROPYL HYDROLYZED WHEAT PROTEIN

CAS No.: 130381-06-5

Definition: Laurdimonium Hydroxypropyl Hydrolyzed Wheat Protein is the quaternary ammonium chloride that conforms generally to the formula:

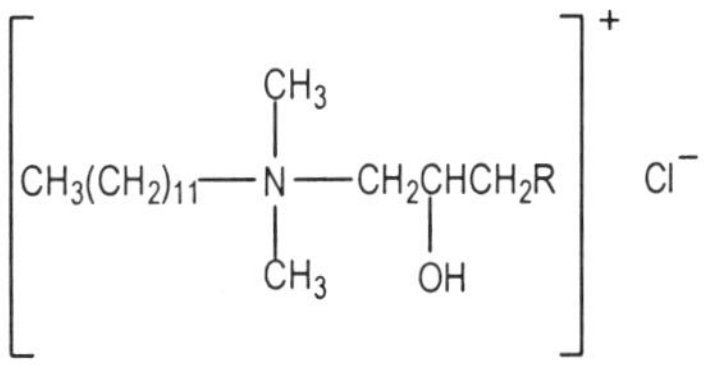

where R represents the hydrolyzed wheat protein moiety.

Chemical Classes: Protein Derivatives; Quaternary Ammonium Compounds

Functions: Antistatic Agent; Hair Conditioning Agent

Reported Product Categories: Hair Dyes and Colors (All Types Requiring Caution Statements and Patch Tests); Hair Conditioners

Technical/Other Names:
Protein Hydrolysates, Wheat Germ, [3-(Dodecyl)dimethylammonio)-2-Hydroxypropyl], Chlorides
Protein Hydrolysates, Wheat Germ, [3-(Dodecyldimethylammonio)-2-Hydroxypropyl], Chlorides

Trade Names:
Aqua Pro II QWL (MGP)
Aqua Pro II VPL (MGP)
Gluadin WQ (Cognis Care Chemicals/NJ)
Gluadin WQ (Cognis Care Chemicals/PA)
Gluadin WQ (Cognis Deutschland)
Hydrotriticum QL (Croda Chemicals)
Hydrotriticum QL (Croda, Inc.)

Trade Name Mixtures:
Aqua Pro II QWOL (MGP)
Aqua Pro WAPC (MGP)
Cropeptide QL (Croda, Inc.)

LAURDIMONIUM HYDROXYPROPYL HYDROLYZED WHEAT PROTEIN/ SILOXYSILICATE

Definition: Laurdimonium Hydroxypropyl Hydrolyzed Wheat Protein/Siloxysilicate is the quaternary ammonium salt that conforms to the formula:

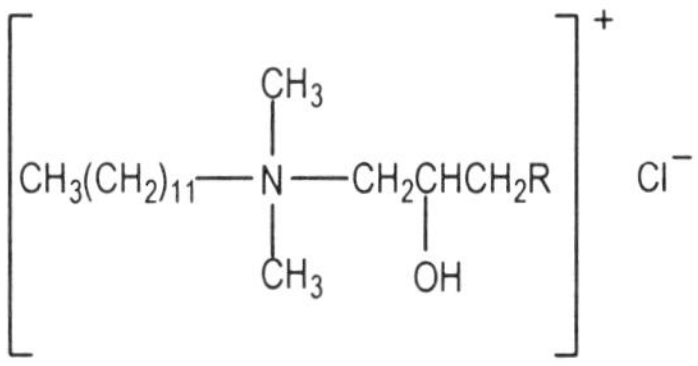

where R represents the hydrolyzed wheat protein siloxysilicate moiety.

Chemical Class: Quaternary Ammonium Compounds

Functions: Antistatic Agent; Hair Conditioning Agent

LAURDIMONIUM HYDROXYPROPYL HYDROLYZED WHEAT STARCH

Definition: Laurdimonium Hydroxypropyl Hydrolyzed Wheat Starch is the quaternary ammonium chloride that conforms generally to the formula:

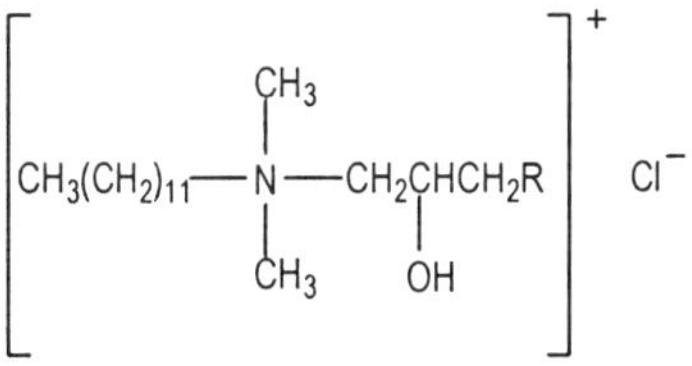

where R represents the hydrolyzed wheat starch moiety.

Chemical Class: Quaternary Ammonium Compounds

Functions: Antistatic Agent; Hair Conditioning Agent

Trade Name Mixtures:
Aqua Pro II QWOL (MGP)
Cropeptide QL (Croda, Inc.)

LAURDIMONIUM HYDROXYPROPYL WHEAT AMINO ACIDS

Definition: Laurdimonium Hydroxypropyl Wheat Amino Acids is the quaternary ammonium salt that conforms generally to the formula:

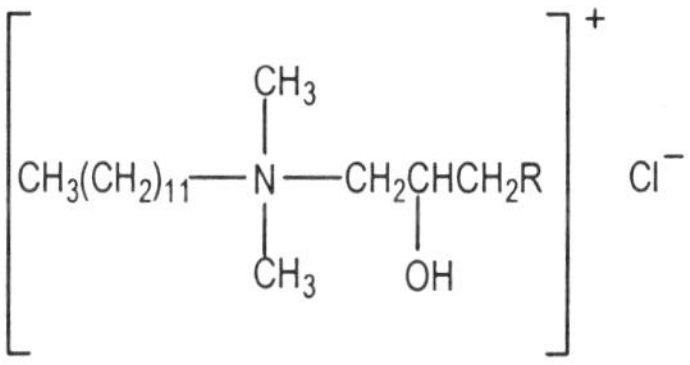

where R represents the wheat amino acids grouping.

Chemical Class: Quaternary Ammonium Compounds

Functions: Antistatic Agent; Hair Conditioning Agent

Trade Name:
Aqua Pro II QWAAL (MGP)

LAURETH-1

CAS No.
4536-30-5

EINECS No.
224-886-5

JPN Translation:
ラウレス - 1

Definition: Laureth-1 is the ethylene glycol ether of Lauryl Alcohol (q.v.) that conforms to the formula:

$$CH_3(CH_2)_{11}OCH_2CH_2OH$$

Information Sources: CTFA D, JCLS, MI-13(7659), TSCA

Chemical Class: Alkoxylated Alcohols

Function: Surfactant - Emulsifying Agent

Technical/Other Names:
2-(Dodecyloxy)Ethanol
Ethanol,2-(Dodecyloxy)-
Ethylene Glycol Monolauryl Ether
2-Hydroxyethyl Lauryl Ether
PEG-1 Lauryl Ether

Trade Name:
Hetoxol L-1 (Heterene)

LAURETH-2

CAS Nos.
3055-93-4
9002-92-0 (Generic)

EINECS No.
221-279-7

JPN Translation:
ラウレス - 2

Definition: Laureth-2 is the polyethylene glycol ether of Lauryl Alcohol (q.v.) that conforms to the formula:

$$CH_3(CH_2)_{11}(OCH_2CH_2)_nOH$$

where n has an average value of 2.

Information Sources: JCLS, MI-13(7659), TSCA

Chemical Class: Alkoxylated Alcohols

Function: Surfactant - Emulsifying Agent

Reported Product Category: Shampoos (Non-coloring)

Technical/Other Names:
Diethylene Glycol Didecyl Ether
Diethyleneglycol Lauryl Ether
2-[2-(Dodecyloxy)Ethoxy]Ethanol
Ethanol, 2-[2-(Dodecyloxy)Ethoxy]-
Ethanol, 2-[2-(Dodecyloxy)Ethoxy]-
2-[2-(Dodecyloxy)Ethoxy]Ethanol
PEG-2 Lauryl Ether
Polyethylene Glycol 100 Lauryl Ether
Polyoxyethylene (2) Lauryl Ether

Trade Names:
Akyporox RLM 22 (Kao GmbH)
Alfonic 1214GC-2.0 Ethoxylate (Sasol North America)
Arlypon F (Cognis Deutschland)
Arlypon F-T (Cognis Deutschland)
Empilan KB2 (Albright & Wilson UK)
Empilan KB 2/ZA (Albright & Wilson Asia)
Hetoxol L-2 (Heterene)
Jeecol LA-2 (Jeen)
Marlipal 24/20 (Sasol GmbH - Marl)
Mergital LM 2 (Cognis France)
Nikkol BL-2 (Nikko)
Novel II 1214GC-2.0 Ethoxylate (Sasol North America)
Oxetal VD 20 (Zschimmer & Schwarz)
Procol LA-2 (Protameen)
Sabowax LM 2 (Sabo)
Sympatens ALM-020 (Kolb)
Unihydol LS-2 (Universal Preserv-A-Chem)

Trade Name Mixtures:
Oxetal VD 92 (Zschimmer & Schwarz)
Saboamid RIT (Sabo)

LAURETH-3

CAS No.
3055-94-5

EINECS No.
221-280-2

JPN Translation:
ラウレス - 3

Definition: Laureth-3 is the polyethylene glycol ether of Lauryl Alcohol (q.v.) that conforms to the formula:

$$CH_3(CH_2)_{11}(OCH_2CH_2)_nOH$$

where n has an average value of 3.

Information Sources: JCLS, MI-13(7659), TSCA

Chemical Class: Alkoxylated Alcohols

Function: Surfactant - Emulsifying Agent

Reported Product Categories: Hair Preparations (Non-coloring), Misc.; Shampoos (Non-coloring)

Technical/Other Names:
2-[2-[2-(Dodecyloxy)Ethoxy]Ethoxy]Ethanol
Ethal 326
Ethanol, 2-[2-[2-(Dodecyloxy)Ethoxy]-Ethoxy]-
Lauryl Triglycol Ether
PEG-3 Lauryl Ether
Polyethylene Glycol (3) Lauryl Ether
Polyoxyethylene (3) Lauryl Ether
Triethylene Glycol Dodecyl Ether

Trade Names:
AE-1214/3 (Procter & Gamble)
AEC Laureth-3 (A & E Connock)
Alfonic 1412-3.0 Ethoxylate (Sasol North America)
Dehydol LS DEO-N (Cognis Deutschland)
Emalex 703 (Nihon Emulsion)
Empilan KB3 (Albright & Wilson UK)
Empilan KB 3/ZA (Albright & Wilson Asia)
Genapol L-3 (Clariant)
Genapol L-3 (Clariant GmbH, Personal Care)
Genapol 26-L-3 (Clariant)
Genapol 26-L-3 (Clariant GmbH, Personal Care)
Glycolene (Vevy)
Hetoxol L-3N (Heterene)
Jeecol LA-3 (Jeen)
Marlipal 24/30 (Sasol GmbH - Marl)
Mergital LM 3 (Cognis France)
Sabowax LM 3 (Sabo)
Sympatens-ALM/030 (Kolb)
Unihydol LS-3 (Universal Preserv-A-Chem)

Trade Name Mixtures:
Amilan GST 40 (Degussa Care Specialties)
Destressine 2000 (Sederma)
Euromix MEA (EOC Surfactants)
Hostapon SCI-40 L (Clariant GmbH, Personal Care)
Isoxal 5 (Vevy)
Oxocap (LCW)
Sabowax FL 1 (Sabo)
Standapol CS Paste (Cognis Care Chemicals/NJ)
Standapol CS Paste (Cognis Care Chemicals/PA)
Tegodeo CW 90 (Degussa Care Specialties)
Unipol CS-50 (Universal Preserv-A-Chem)
Unipol CS PASTE (Universal Preserv-A-Chem)

LAURETH-4

CAS Nos.
5274-68-0
68002-97-1 (Generic)
68439-50-9 (Generic)

EINECS No.
226-097-1

JPN Translation:
ラウレス - 4

Definition: Laureth-4 is the polyethylene glycol ether of Lauryl Alcohol (q.v.) that conforms to the formula:

$$CH_3(CH_2)_{11}(OCH_2CH_2)_nOH$$

where n has an average value of 4.

Information Sources: 21CFR178.3520, CIR: [S] JACT-2(7)1983, CTFA S, JCLS, MI-13(7659), SNPF, TSCA, USAN

Chemical Class: Alkoxylated Alcohols

Function: Surfactant - Emulsifying Agent

Reported Product Categories: Hair Dyes and Colors (All Types Requiring Caution Statements and Patch Tests); Blushers (All types); Hair Bleaches; Eye Shadows; Eyeliners; Shampoos (Non-coloring); Hair Conditioners; Personal Cleanliness Products, Misc.; Deodorants (Underarm); Permanent Waves; Bath Preparations, Misc.; Hair Rinses (Non-coloring); Hair Preparations (Non-coloring), Misc.; Skin Care Preparations, Misc.; Aftershave Lotions; Bath Oils, Tablets, and Salts; Eye Makeup Preparations, Misc.; Face Powders; Moisturizing Preparations

Technical/Other Names:
PEG-4 Lauryl Ether
Polyethylene Glycol 200 Lauryl Ether
Polyoxyethylene (4) Lauryl Ether
Tetraethylene Glycol Dodecyl Ether

Trade Names:
Akyporox RLM 40 (Kao GmbH)
Brij 30 (Uniqema Americas)
Chemonic L-4 (Chemron)
DeTHOX LA-4 (DeForest)
Empilan KBE4 (Albright & Wilson UK)
Ethal LA-4 (Ethox)
Ethosperse LA-4 (Lonza Inc./Lonza Ltd.)
Hetoxol L-4 (Heterene)
Jeecol LA-4 (Jeen)
Lanycol-30 (Lanaetex)
Lipocol L-4 (Lipo)
Lumulse L-4 (Lambent)
Marlipal 24/40 (Sasol GmbH - Marl)
Marlowet BL (Sasol GmbH - Marl)
Mergital LM 4 L (Cognis France)
Mulsifan CPA (Zschimmer & Schwarz)
Nikkol BL-4.2 (Nikko)
Pegnol L-4 (Toho)
Procol LA-4 (Protameen)
Rhodasurf L-4 (Rhodia)
Sabowax LM 4 (Sabo)
Simulsol P 4 (SEPPIC)
Sipol LAL-4 (Specialty Industrial)
Sympatens-ALM/040 (Kolb)
Tego Alkanol L 4 (Degussa Care Specialties)
Unicol LA-4 (Universal Preserv-A-Chem)
Unihydol LS-4 (Universal Preserv-A-Chem)

Trade Name Mixtures:
AEC Dimethicone (&) Laureth-4 (&) Laureth-23 (A & E Connock)
Atlas G-1823 (Uniqema Americas)
Dow Corning 1664 Emulsion (Dow Corning)
Euperlan PK-3000 (Cognis Care Chemicals/NJ)
Euperlan PK-3000 (Cognis Care Chemicals/PA)
Euperlan PK-4000 (Cognis Deutschland)
Euperlan PK 3000 AM (Cognis Deutschland)
Euperlan PK 3000 OK (Cognis Care Chemicals/NJ)
Euperlan PK 3000 OK (Cognis Deutschland)
Eur-Amid N (EOC Surfactants)
EuroNac AMF (EOC Surfactants)
Genapol PDB (Clariant)
Genapol PDB (Clariant GmbH, Personal Care)
Genapol PDC (Clariant)
Genapol PDC (Clariant GmbH, Personal Care)
KM-901 (Shin-Etsu Chemical Co.)
KM-902 (Shin-Etsu Chemical Co.)
KM-903 (Shin-Etsu Chemical Co.)
KM-904 (Shin-Etsu Chemical Co.)
KM-905 (Shin-Etsu Chemical Co.)
KM-910 (Shin-Etsu Chemical Co.)
KM-902C (Shin-Etsu Chemical Co.)
Lowenol Conditioner 288 (Lowenstein)
Lowenol Emulsion 79 (Lowenstein)
Sabowax GT (Sabo)
SM2169 (GE Silicones)
Tego Pearl N 300 (Degussa Care Specialties)
Texapon WW100 (Cognis Care Chemicals/PA)
Zetesol 100 (Zschimmer & Schwarz)

LAURETH-5

CAS No. 3055-95-6 **EINECS No.** 221-281-8

JPN Translation:
ラウレス - 5

Definition: Laureth-5 is the polyethylene glycol ether of Lauryl Alcohol (q.v.) that conforms to the formula:

$$CH_3(CH_2)_{11}(OCH_2CH_2)_nOH$$

where n has an average value of 5.

Information Sources: JCLS, MI-13(7659), TSCA

Chemical Class: Alkoxylated Alcohols

Function: Surfactant - Emulsifying Agent

Technical/Other Names:
PEG-5 Lauryl Ether
Pentaethylene Glycol Lauryl Ether
Polyethylene Glycol (5) Lauryl Ether
Polyoxyethylene (5) Lauryl Ether

Trade Name:
Intrasol FA 12/18/5 (Stockhausen GmbH)

Trade Name Mixture:
Isoxal 12 (Vevy)

LAURETH-6

CAS Nos. 3055-96-7 68002-97-1 (Generic) **EINECS No.** 221-282-3

JPN Translation:
ラウレス - 6

Definition: Laureth-6 is the polyethylene glycol ether of Lauryl Alcohol (q.v.) that conforms to the formula:

$$CH_3(CH_2)_{11}(OCH_2CH_2)_nOH$$

where n has an average value of 6.

Information Sources: JCLS, MI-13(7659), TSCA

Chemical Class: Alkoxylated Alcohols

Function: Surfactant - Emulsifying Agent

Technical/Other Names:
Hexaethylene Glycol Dodecyl Ether
PEG-6 Lauryl Ether
Polyethylene Glycol 300 Lauryl Ether
Polyoxyethylene (6) Lauryl Ether

Trade Names:
Empilan KB6 (Albright & Wilson UK)
Marlipal 24/60 (Sasol GmbH - Marl)
Pegnol L-6 (Toho)

Trade Name Mixture:
Rewopol HM 80 (Degussa Care Specialties)

LAURETH-7

CAS No. 3055-97-8 **EINECS No.** 221-283-9

JPN Translation:
ラウレス - 7

Definition: Laureth-7 is the polyethylene glycol ether of Lauryl Alcohol (q.v.) that conforms to the formula:

$$CH_3(CH_2)_{11}(OCH_2CH_2)_nOH$$

where n has an average value of 7.

Information Sources: JCLS, MI-13(7659), TSCA

Chemical Class: Alkoxylated Alcohols

Function: Surfactant - Emulsifying Agent

Reported Product Categories: Moisturizing Preparations; Bath Preparations, Misc.; Body and Hand Preparations (Excluding Shaving Preparations); Foundations; Bath Capsules; Face and Neck Preparations (Excluding Shaving Preparations); Hair Shampoos (Coloring); Night Skin Care Preparations; Bath Soaps and Detergents; Makeup Preparations (Not eye), Misc.; Paste Masks (Mud Packs); Skin Care Preparations, Misc.; Eyeliners; Indoor Tanning Preparations

Technical/Other Names:
Hyptaethylene Glycol Dodecyl Ether
PEG-7 Lauryl Ether
Polyethylene Glycol (7) Lauryl Ether
Polyoxyethylene (7) Lauryl Ether

Trade Names:
Alfonic 1412-7.0 Ethoxylate (Sasol North America)
Bio-Soft EC-600 (Stepan)
Chemonic L-7 (Chemron)
Empilan KB 7/ZA (Albright & Wilson Asia)
Empilan KB7/ZA (Albright & Wilson Asia)
Ethal LA-7 (Ethox)
Jeecol LA-7 (Jeen)
Marlipal 24/70 (Sasol GmbH - Marl)
Marlipal 24/79 (Sasol GmbH - Marl)

Marlipal MG (Sasol GmbH - Marl)
Novel II 1412-7.0 Ethoxylate (Sasol North America)
Procol LA-7 (Protameen)
Rhodasurf L-7 90 (Rhodia)
Sabowax LM 7 (Sabo)
Sipol LAL-7 (Specialty Industrial)

Trade Name Mixtures:
Anise Softcream (CEP (Solabia))
Apricot Softcream (CEP (Solabia))
Avocado Softcream (CEP (Solabia))
Bitter Orange Softcream (CEP (Solabia))
Cardamom Softcream (CEP (Solabia))
Cinnamon Softcream (CEP (Solabia))
CREAGEL C 13-14 (Creations Couleurs)
Creagel EZ 7 (C.I.T.)
CREAGEL EZ PFC (Creations Couleurs)
Erase (Degussa Care Specialties)
Eucalyptus Softcream (CEP (Solabia))
Ginger Softcream (CEP (Solabia))
Grapefruit Softcream (CEP (Solabia))
Grape Seed Softcream (CEP (Solabia))
Lavender Softcream (CEP (Solabia))
Lime Softcream (CEP (Solabia))
Mirasheen 207 (Rhodia)
Mirasheen CP-920 (Rhodia)
Mirasheen CP-820/G (Rhodia)
Peppermint Softcream (CEP (Solabia))
Pistachio Softcream (CEP (Solabia))
RheoThik 110 (International Additive)
Rice Bran Softcream (CEP (Solabia))
Rosemary Softcream (CEP (Solabia))
Saboamid RIT (Sabo)
Saboperl 300 (Sabo)
Sepigel 305 (SEPPIC)
Sweet Almond Softcream (CEP (Solabia))
White Tea Softcream (CEP (Solabia))

LAURETH-8

CAS Nos.: 3055-98-9; 9002-92-0 (Generic); 68002-97-1 (Generic)

JPN Translation:
ラウレス - 8

Definition: Laureth-8 is the polyethylene glycol ether of Lauryl Alcohol (q.v.) that conforms to the formula:

$$CH_3(CH_2)_{11}(OCH_2CH_2)_nOH$$

where n has an average value of 8.

Information Sources: 21CFR177.2800, JCLS, MAR, MHLW-331/2, MI-13(7659), TSCA

Chemical Class: Alkoxylated Alcohols

Function: Surfactant - Emulsifying Agent

Technical/Other Names:
Octaethylene Glycol Dodecyl Ether
PEG-8 Lauryl Ether
Polyethylene Glycol 400 Lauryl Ether
Polyoxyethylene (8) Lauryl Ether

Trade Names:
AEC Laureth-8 (A & E Connock)
Akyporox RLM 80 (Kao GmbH)
Elfapur LM 75 S (Akzo Nobel Surface AB)
Empilan KB8 (Albright & Wilson UK)
Pegnol TH-8 (Toho)
Sabowax LM8 (Sabo)
Sympatens-ALM/080 (Kolb)

Trade Name Mixtures:
Mirasil DME-2 (Rhodia)
Mirasil DME-30 (Rhodia)
Mirasil DME-40A (Rhodia)
Oxocap (LCW)
Sabowax GT (Sabo)
Solvariane (LCW)

LAURETH-9

CAS Nos.	**EINECS No.**
3055-99-0	221-284-4
9002-92-0 (Generic)	

JPN Translation:
ラウレス - 9

Definition: Laureth-9 is the polyethylene glycol ether of Lauryl Alcohol (q.v.) that conforms to the formula:

$$CH_3(CH_2)_{11}(OCH_2CH_2)_nOH$$

where n has an average value of 9.

Information Sources: 21CFR177.2800, 21CFR178.3130, INN, JAN, JCLS, MHLW-331/2, MI-13(7659), TSCA, USAN

Chemical Class: Alkoxylated Alcohols

Function: Surfactant - Emulsifying Agent

Reported Product Categories: Bath Oils, Tablets, and Salts; Bubble Baths; Cleansing Products (Cold Creams, Cleansing Lotions, Liquids and Pads); Shampoos (Non-coloring)

Technical/Other Names:
Nonaoxyethylene Monododecyl Ether
PEG-9 Lauryl Ether
Polyethylene Glycol 450 Lauryl Ether
Polyoxyethylene (9) Lauryl Ether

Trade Names:
Alfonic 1412-9.0 Ethoxylate (Sasol North America)
Atlas G-4829 (Uniqema Americas)
Ethal 926 (Ethox)
Hetoxol L-9 (Heterene)
Hetoxol LS-9 (Heterene)
Jeecol LA-9 (Jeen)
Marlipal 24/90 (Sasol GmbH - Marl)
Marlipal 24/99 (Sasol GmbH - Marl)
Nikkol BL-9EX (Nikko)
Pegnol L-9A (Toho)
Procol LA-9 (Protameen)
Sympatens-AL/090 (Kolb)
Unicol LA-9 (Universal Preserv-A-Chem)

Trade Name Mixtures:
Botanisil ME-12 (Botanigenics)
Lipacide SHCO90 (SEPPIC)
Lipacide SHK90 (SEPPIC)
Lipacide SHU90 (SEPPIC)
Lubrasil DS (Guardian)
Mirasil DPDM-E (Rhodia HPCII)
SME253 (GE Silicones)
Standapol Pearl Conc. 7130 (Cognis Care Chemicals/NJ)
Standapol Pearl Conc. 7130 (Cognis Care Chemicals/PA)

LAURETH-10

CAS Nos.: 6540-99-4; 9002-92-0 (Generic); 68002-97-1 (Generic)

JPN Translation:
ラウレス - 10

Definition: Laureth-10 is the polyethylene glycol ether of Lauryl Alcohol (q.v.) that conforms to the formula:

$$CH_3(CH_2)_{11}(OCH_2CH_2)_nOH$$

where n has an average value of 10.

Information Sources: 21CFR177.2800, JCLS, MHLW-331/2, MI-13(7659), TSCA, USAN

Chemical Class: Alkoxylated Alcohols

Function: Surfactant - Emulsifying Agent

Reported Product Categories: Shampoos (Non-coloring); Hair Dyes and Colors (All Types Requiring Caution Statements and Patch Tests); Bubble Baths; Bath Soaps and Detergents

Technical/Other Names:
PEG-10 Lauryl Ether
Polyethylene Glycol 500 Lauryl Ether
Polyoxyethylene (10) Lauryl Ether

Trade Names:
Empilan KB10 (Albright & Wilson UK)
Intrasol FA 12/18/10 (Stockhausen GmbH)
Jeecol LA-10 (Jeen)
Pegnol L-10 (Toho)
Procol LA-10 (Protameen)
Sabowax LMT 10 (Sabo)
Unihydol 100 (Universal Preserv-A-Chem)

Trade Name Mixtures:
AEC Glycol Distearate (&) Sodium Laureth Sulfate (&) Cocamide MEA (&) Laureth-10 (A & E Connock)
Akypo Soft 100 BVC (Kao GmbH)
Euperlan PK-771 (Cognis Care Chemicals/NJ)
Euperlan PK-771 (Cognis Care Chemicals/PA)
Euperlan PK-810 (Cognis Care Chemicals/NJ)

Euperlan PK-810 (Cognis Care Chemicals/ PA)
EuroNac AN10 (EOC Surfactants)
Isoxal 11 (Vevy)
Texapon CS Paste (Cognis Deutschland)
Texapon EVR K 400 (Cognis Deutschland)

LAURETH-11

CAS No.: 9002-92-0 (Generic)

JPN Translation:
ラウレス - 11

Definition: Laureth-11 is the polyethylene glycol ether of Lauryl Alcohol (q.v.) that conforms to the formula:

$$CH_3(CH_2)_{11}(OCH_2CH_2)_nOH$$

where n has an average value of 11.

Information Sources: 21CFR177.2800, JCLS, MI-13(7659), TSCA

Chemical Class: Alkoxylated Alcohols

Function: Surfactant - Emulsifying Agent

Reported Product Category: Permanent Waves

Technical/Other Names:
PEG-11 Lauryl Ether
Polyethylene Glycol (11) Lauryl Ether
Polyoxyethylene (11) Lauryl Ether

Trade Name:
Mergital LM 11 (Cognis France)

Trade Name Mixture:
Mirasil DPDM-E (Rhodia HPCII)

LAURETH-12

CAS No.	EINECS No.
9002-92-0 (Generic)	221-286-5

JPN Translation:
ラウレス - 12

Definition: Laureth-12 is the polyethylene glycol ether of Lauryl Alcohol (q.v.) that conforms to the formula:

$$CH_3(CH_2)_{11}(OCH_2CH_2)_nOH$$

where n has an average value of 12.

Information Sources: 21CFR177.2800, CTFA D, JCLS, MI-13(7659), SNPF, TSCA

Chemical Class: Alkoxylated Alcohols

Function: Surfactant - Emulsifying Agent

Reported Product Categories: Hair Dyes and Colors (All Types Requiring Caution Statements and Patch Tests); Shampoos (Non-coloring); Bath Soaps and Detergents; Cleansing Products (Cold Creams, Cleansing Lotions, Liquids and Pads)

Technical/Other Names:
PEG-12 Lauryl Ether
Polyethylene Glycol 600 Lauryl Ether
Polyoxyethylene (12) Lauryl Ether

Trade Names:
Chemonic L-12 (Chemron)
Empilan KB12 (Albright & Wilson UK)
Ethosperse LA-12 (Lonza Inc./Lonza Ltd.)
Jeecol LA-12 (Jeen)
Lipocol L-12 (Lipo)
Pegnol L-12S (Toho)
Procol LA-12 (Protameen)
Rhodasurf L-12 (Rhodia)
Sipol LAL-12 (Specialty Industrial)

Trade Name Mixtures:
Silderm Powder Gel (Active Concepts)
STEPAN PEARL 4 (Stepan)

LAURETH-13

CAS No.: 9002-92-0 (Generic)

JPN Translation:
ラウレス - 13

Definition: Laureth-13 is the polyethylene glycol ether of Lauryl Alcohol (q.v.) that conforms to the formula:

$$CH_3(CH_2)_{11}(OCH_2CH_2)_nOH$$

where n has an average value of 13.

Information Sources: 21CFR177.2800, JCLS, MI-13(7659), TSCA

Chemical Class: Alkoxylated Alcohols

Function: Surfactant - Emulsifying Agent

Technical/Other Names:
PEG-13 Lauryl Ether
Polyethylene Glycol (13) Lauryl Ether
Polyoxyethylene (13) Lauryl Ether

LAURETH-14

CAS No.: 9002-92-0 (Generic)

JPN Translation:
ラウレス - 14

Definition: Laureth-14 is the polyethylene glycol ether of Lauryl Alcohol (q.v.) that conforms to the formula:

$$CH_3(CH_2)_{11}(OCH_2CH_2)_nOH$$

where n has an average value of 14.

Information Sources: 21CFR177.2800, JCLS, MI-13(7659)

Chemical Class: Alkoxylated Alcohols

Function: Surfactant - Emulsifying Agent

Technical/Other Names:
PEG-14 Lauryl Ether

Polyethylene Glycol (14) Lauryl Ether
Polyoxyethylene (14) Lauryl Ether

LAURETH-15

CAS No.: 9002-92-0 (Generic)

JPN Translation:
ラウレス - 15

Definition: Laureth-15 is the polyethylene glycol ether of Lauryl Alcohol (q.v.) that conforms to the formula:

$$CH_3(CH_2)_{11}(OCH_2CH_2)_nOH$$

where n has an average value of 15.

Information Sources: 21CFR177.2800, JCLS, MI-13(7659), TSCA

Chemical Class: Alkoxylated Alcohols

Function: Surfactant - Emulsifying Agent

Technical/Other Names:
PEG-15 Lauryl Ether
Polyethylene Glycol (15) Lauryl Ether
Polyoxyethylene (15) Lauryl Ether

Trade Names:
Jeecol LA-15 (Jeen)
Procol LA-15 (Protameen)

LAURETH-16

CAS No.: 9002-92-0 (Generic)

JPN Translation:
ラウレス - 16

Definition: Laureth-16 is the polyethylene glycol ether of Lauryl Alcohol (q.v.) that conforms to the formula:

$$CH_3(CH_2)_{11}(OCH_2CH_2)_nOH$$

where n has an average value of 16.

Information Sources: 21CFR177.2800, JCLS, MI-13(7659)

Chemical Class: Alkoxylated Alcohols

Functions: Surfactant - Cleansing Agent; Surfactant - Emulsifying Agent

Technical/Other Names:
PEG-16 Lauryl Ether
Polyethylene Glycol (16) Lauryl Ether
Polyoxyethylene (16) Lauryl Ether

Trade Name:
Akyporox RLM 160 (Kao GmbH)

LAURETH-20

CAS No.: 9002-92-0 (Generic)

JPN Translation:
ラウレス - 20

Definition: Laureth-20 is the polyethylene glycol ether of Lauryl Alcohol (q.v.) that conforms to the formula:

$$CH_3(CH_2)_{11}(OCH_2CH_2)_nOH$$

where n has an average value of 20.

Information Sources: 21CFR177.2800, JCLS, MI-13(7659), TSCA

Chemical Class: Alkoxylated Alcohols

Functions: Surfactant - Cleansing Agent; Surfactant - Solubilizing Agent

Technical/Other Names:
PEG-20 Lauryl Ether
Polyethylene Glycol 1000 Lauryl Ether
Polyoxyethylene (20) Lauryl Ether

Trade Names:
Emalex 720 (Nihon Emulsion)
Pegnol L-20S (Toho)
Procol LA-20 (Protameen)

LAURETH-21

Definition: Laureth-21 is the polyethylene glycol ether of lauryl alcohol that conforms generally to the formula:

$$CH_3(CH_2)_{11}(OCH_2CH_2)_nOH$$

where n has an average value of 21.

Information Source: JCLS

Chemical Class: Alkoxylated Alcohols

Functions: Surfactant - Cleansing Agent; Surfactant - Solubilizing Agent

Technical/Other Names:
Polyethylene Glycol (21) Lauryl Ether
Polyoxyethylene (21) Lauryl Ether

Trade Name:
Nikkol BL-21 (Nikko)

LAURETH-23

CAS No.: 9002-92-0 (Generic)

JPN Translation:
ラウレス - 23

Definition: Laureth-23 is the polyethylene glycol ether of Lauryl Alcohol (q.v.) that conforms to the formula:

$$CH_3(CH_2)_{11}(OCH_2CH_2)_nOH$$

where n has an average value of 23.

Information Sources: 21CFR177.2800, CIR: [S] JACT-2(7)1983, CTFA S, JCLS, MI-13(7659), SNPF, TSCA

Chemical Class: Alkoxylated Alcohols

Functions: Surfactant - Cleansing Agent; Surfactant - Solubilizing Agent

Reported Product Categories: Hair Dyes and Colors (All Types Requiring Caution Statements and Patch Tests); Permanent Waves; Shaving Cream (Aerosol, Brushless and Lather); Moisturizing Preparations; Bath Preparations, Misc.; Body and Hand Preparations (Excluding Shaving Preparations); Tonics, Dressings, and Other Hair Grooming Aids; Skin Care Preparations, Misc.; Hair Bleaches; Deodorants (Underarm); Hair Conditioners; Personal Cleanliness Products, Misc.; Hair Rinses (Non-coloring); Bath Capsules; Bath Oils, Tablets, and Salts; Cleansing Products (Cold Creams, Cleansing Lotions, Liquids and Pads); Face and Neck Preparations (Excluding Shaving Preparations); Shampoos (Non-coloring); Colognes and Toilet Waters; Hair Preparations (Non-coloring), Misc.; Hair Wave Sets

Technical/Other Names:
PEG-23 Lauryl Ether
Polyethylene Glycol (23) Lauryl Ether
Polyoxyethylene (23) Lauryl Ether

Trade Names:
Brij 35 (Uniqema Americas)
Brij 35 Liquid (Uniqema Americas)
Brij 35SP (Uniqema Americas)
Chemonic L-23 (Chemron)
DeTHOX LA-23 (DeForest)
Ethal LA-23 (Ethox)
Ethosperse LA-23 (Lonza Inc./Lonza Ltd.)
Eumulgin LM 23 (Cognis France)
Genapol 26-L-23 (Clariant)
Genapol 26-L-23 (Clariant GmbH, Personal Care)
Hetoxol L-23 (Heterene)
Jeecol LA-23 (Jeen)
Lanycol-35 (Lanaetex)
Lipocol L-23 (Lipo)
Lumulse L-23 (Lambent)
Procol LA-23 (Protameen)
Rhodasurf L-25 (Rhodia)
Ritox 35 (RITA)
Sabowax LM 23 (Sabo)
Simulsol P 23 (SEPPIC)
Sipol LAL-23 (Specialty Industrial)
Sympatens-ALM/230 (Kolb)
Tego Alkanol L 23 P (Degussa Care Specialties)
Unicol LA-23 (Universal Preserv-A-Chem)

Trade Name Mixtures:
AEC Dimethicone (&) Laureth-4 (&) Laureth-23 (A & E Connock)
Atlas G-1875 (Uniqema Americas)
Cerasynt 945 (International Specialty Products)
CosmoTurb SB (CosmoCare)
Dow Corning 1664 Emulsion (Dow Corning)
Extrapone Rosemary 2/783630 (Symrise)
KM-901 (Shin-Etsu Chemical Co.)
KM-902 (Shin-Etsu Chemical Co.)
KM-903 (Shin-Etsu Chemical Co.)
KM-904 (Shin-Etsu Chemical Co.)
KM-905 (Shin-Etsu Chemical Co.)
KM-910 (Shin-Etsu Chemical Co.)
KM-902C (Shin-Etsu Chemical Co.)
SM2169 (GE Silicones)

LAURETH-25

CAS No.: 9002-92-0 (Generic)

JPN Translation:
ラウレス - 25

Definition: Laureth-25 is the polyethylene glycol ether of Lauryl Alcohol (q.v.) that conforms to the formula:

$$CH_3(CH_2)_{11}(OCH_2CH_2)_nOH$$

where n has an average value of 25.

Information Sources: 21CFR177.2800, JCLS, MI-13(7659), TSCA

Chemical Class: Alkoxylated Alcohols

Functions: Surfactant - Cleansing Agent; Surfactant - Solubilizing Agent

Technical/Other Names:
PEG-25 Lauryl Ether
Polyethylene Glycol (25) Lauryl Ether
Polyoxyethylene (25) Lauryl Ether

Trade Name:
Nikkol BL-25 (Nikko)

LAURETH-30

CAS No.: 9002-92-0 (Generic)

JPN Translation:
ラウレス - 30

Definition: Laureth-30 is the polyethylene glycol ether of Lauryl Alcohol (q.v.) that conforms to the formula:

$$CH_3(CH_2)_{11}(OCH_2CH_2)_nOH$$

where n has an average value of 30.

Information Sources: 21CFR177.2800, JCLS, MI-13(7659), TSCA

Chemical Class: Alkoxylated Alcohols

Functions: Surfactant - Cleansing Agent; Surfactant - Solubilizing Agent

Technical/Other Names:
PEG-30 Lauryl Ether
Polyethylene Glycol (30) Lauryl Ether
Polyoxyethylene (30) Lauryl Ether

Trade Names:
Jeecol LA-30 (Jeen)
Procol LA-30 (Protameen)

LAURETH-38

CAS No.: 9002-92-0 (Generic)

Definition: Laureth-38 is the polyethylene glycol ether of lauryl alcohol that conforms generally to the formula:

$$CH_3(CH_2)_{10}CH_2(OCH_2CH_2)_nOH$$

where n has an average value of 38.

Chemical Class: Alkoxylated Alcohols

Functions: Surfactant - Cleansing Agent; Surfactant - Solubilizing Agent

Technical/Other Names:
Polyethylene Glycol (38) Lauryl Ether
Polyoxyethylene (38) Lauryl Ether

Trade Name:
Emulmin L-380 (Sanyo Chemical)

LAURETH-40

CAS No.: 9002-92-0 (Generic)

JPN Translation:
ラウレス - 40

Definition: Laureth-40 is the polyethylene glycol ether of Lauryl Alcohol (q.v.) that conforms to the formula:

$$CH_3(CH_2)_{11}(OCH_2CH_2)_nOH$$

where n has an average value of 40.

Information Sources: 21CFR177.2800, JCLS, MI-13(7659), TSCA

Chemical Class: Alkoxylated Alcohols

Functions: Surfactant - Cleansing Agent; Surfactant - Solubilizing Agent

Technical/Other Names:
PEG-40 Lauryl Ether
Polyethylene Glycol 2000 Lauryl Ether
Polyoxyethylene (40) Lauryl Ether

LAURETH-50

Definition: Laureth-50 is the polyethylene glycol ether of lauryl alcohol that conforms generally to the formula:

$$CH_3(CH_2)_{11}(OCH_2CH_2)_nOH$$

where n has an average value of 50.

Chemical Class: Alkoxylated Alcohols

Function: Surfactant - Emulsifying Agent

Trade Names:
Emalex 750 (Nihon Emulsion)
Ethal LA-50 (Ethox)

LAURETH-2 ACETATE

Empirical Formula:
$C_{18}H_{36}O_4$

Definition: Laureth-2 Acetate is the ester of Laureth-2 (q.v.) and acetic acid. It conforms generally to the formula:

$$CH_3\overset{\displaystyle O}{\overset{\|}{C}} - (OCH_2CH_2)_nO(CH_2)_{11}CH_3$$

where n has an average value of 2.

Information Source: JCLS

Chemical Class: Esters

Function: Skin-Conditioning Agent - Emollient

Technical/Other Names:
PEG-2 Lauryl Ether Acetate
Polyethylene Glycol 100 Lauryl Ether Acetate
Polyoxyethylene (2) Lauryl Ether Acetate

Trade Names:
Estalan 42 (Lanaetex)
Pelemol L2A (Phoenix)

Trade Name Mixture:
Lancol (Lanaetex)

LAURETH-2 BENZOATE

Empirical Formula:
$C_{23}H_{38}O_3$

Definition: Laureth-2 Benzoate is the ester of Benzoic Acid (q.v.) and Laureth-2 (q.v.) that conforms generally to the formula:

where n has an average value of 2.

Chemical Class: Esters

Function: Skin-Conditioning Agent - Emollient

Technical/Other Name:
Benzoic Acid, Laureth-2 Ester

Trade Name:
Bernel Ester 126 (Bernel)

LAURETH-5 BUTYL ETHER

Definition: Laureth-5 Butyl Ether is the organic compound that conforms generally to the formula:

$$CH_3(CH_2)_{11}(OCH_2CH_2)_nO\overset{\displaystyle CH_3}{\underset{\displaystyle CH_3}{C}}CH_3$$

where n has an average value of 5.

Chemical Classes: Alkoxylated Alcohols; Ethers

Functions: Antifoaming Agent; Surfactant - Emulsifying Agent

Technical/Other Name:
PEG-5 Lauryl t-Butyl Ether

Trade Name:
Marlox B 24/50 (Sasol GmbH - Marl)

LAURETH-3 CARBOXYLIC ACID

CAS Nos.: 20858-24-6; 27306-90-7

JPN Translation:
ラウレス- 3 カルボン酸

Empirical Formula:
$C_{18}H_{36}O_5$

Definition: Laureth-3 Carboxylic Acid is the organic acid that conforms generally to the formula:

$$CH_3(CH_2)_{11}(OCH_2CH_2)_nOCH_2COOH$$

where n has an average value of 2.

Chemical Class: Carboxylic Acids

Functions: Surfactant - Cleansing Agent; Surfactant - Emulsifying Agent

Technical/Other Names:
Acetic Acid, [2-[2-(Dodecyloxy)Ethoxy]-Ethoxy]-
[2-[2-(Dodecyloxy)Ethoxy]Ethoxy]Acetic Acid
PEG-3 Lauryl Ether Carboxylic Acid
Polyethylene Glycol (3) Lauryl Ether Carboxylic Acid
Polyoxyethylene (3) Lauryl Ether Carboxylic Acid

Trade Name:
Empicol CBB (Albright & Wilson UK)

LAURETH-4 CARBOXYLIC ACID

CAS Nos.: 20858-25-7; 27306-90-7

JPN Translation:
ラウレス- 4 カルボン酸

Empirical Formula:
$C_{20}H_{40}O_6$

Definition: Laureth-4 Carboxylic Acid is the organic acid that conforms generally to the formula:

$$CH_3(CH_2)_{11}(OCH_2CH_2)_nOCH_2COOH$$

where n has an average value of 3.

Chemical Class: Carboxylic Acids

Function: Surfactant - Cleansing Agent

Technical/Other Names:
PEG-4 Lauryl Ether Carboxylic Acid
Polyethylene Glycol 200 Lauryl Ether
Carboxylic Acid
Polyoxyethylene (4) Lauryl Ether
Carboxylic Acid
3,6,9,12-Tetraoxatetracosanoic Acid

Trade Names:
Akypo RLM 25 (Kao GmbH)
Empicol CBC (Albright & Wilson UK)

LAURETH-5 CARBOXYLIC ACID

CAS Nos.: 21127-45-7; 27306-90-7

JPN Translation:
ラウレス-5カルボン酸

Empirical Formula:
$C_{22}H_{44}O_7$

Definition: Laureth-5 Carboxylic Acid is the organic acid that conforms generally to the formula:

$$CH_3(CH_2)_{11}(OCH_2CH_2)_nOCH_2COOH$$

where n has an average value of 4.

Chemical Class: Carboxylic Acids

Function: Surfactant - Cleansing Agent

Technical/Other Names:
PEG-5 Lauryl Ether Carboxylic Acid
3,6,9,12,15-Pentaoxaheptacosanoic Acid
Polyethylene Glycol (5) Lauryl Ether
Carboxylic Acid
Polyoxyethylene (5) Lauryl Ether
Carboxylic Acid

Trade Names:
Akypo RLM 45 (Kao GmbH)
Akypo RLMQ 38 (Kao GmbH)
Empicol CED 5 (Albright & Wilson UK)
Marlowet 1072 (Sasol GmbH - Marl)
Sandopan LA-8 HCM Liquid (Clariant)
Sandopan LA-8 HCM Liquid (Clariant
GmbH, Personal Care)

LAURETH-6 CARBOXYLIC ACID

CAS Nos.: 20260-64-4; 27306-90-7

JPN Translation:
ラウレス-6カルボン酸

Empirical Formula:
$C_{24}H_{48}O_8$

Definition: Laureth-6 Carboxylic Acid is the organic acid that conforms generally to the formula:

$$CH_3(CH_2)_{11}(OCH_2CH_2)_nOCH_2COOH$$

where n has an average value of 5.

Chemical Class: Carboxylic Acids

Function: Surfactant - Cleansing Agent

Technical/Other Names:
3,6,9,12,15,18-Hexaoxatriacontanoic Acid
PEG-6 Lauryl Ether Carboxylic Acid
Polyethylene Glycol 300 Lauryl Ether
Carboxylic Acid
Polyoxyethylene (6) Lauryl Ether
Carboxylic Acid

LAURETH-8 CARBOXYLIC ACID

Definition: Laureth-8 Carboxylic Acid is the organic acid that conforms generally to the formula:

$$CH_3(CH_2)_{11}(OCH_2CH_2)_nOCH_2COOH$$

where n has an average value of 7.

Chemical Class: Carboxylic Acids

Function: Surfactant - Cleansing Agent

Technical/Other Names:
PEG-8 Lauryl Ether Carboxylic Acid
Polyethylene Glycol (8) Lauryl Ether
Carboxylic Acid
Polyoxyethylene (8) Lauryl Ether
Carboxylic Acid

Trade Names:
Sandosan LNA (Clariant)
Sandosan LNA (Clariant GmbH, Personal
Care)

LAURETH-10 CARBOXYLIC ACID

CAS No.: 27306-90-7

JPN Translation:
ラウレス-10カルボン酸

Empirical Formula:
$C_{32}H_{64}O_{12}$

Definition: Laureth-10 Carboxylic Acid is the organic acid that conforms generally to the formula:

$$CH_3(CH_2)_{11}(OCH_2CH_2)_nOCH_2COOH$$

where n has an average value of 9.

Chemical Class: Carboxylic Acids

Function: Surfactant - Cleansing Agent

Technical/Other Names:
PEG-10 Lauryl Ether Carboxylic Acid
Polyethylene Glycol 500 Lauryl Ether
Carboxylic Acid
Polyoxyethylene (10) Lauryl Ether
Carboxylic Acid

LAURETH-11 CARBOXYLIC ACID

CAS No.: 27306-90-7

Definition: Laureth-11 Carboxylic Acid is the organic acid that conforms generally to the formula:

$$CH_3(CH_2)_{11}(OCH_2CH_2)_nOCH_2COOH$$

where n has an average value of 10.

Chemical Class: Carboxylic Acids

Function: Surfactant - Cleansing Agent

Technical/Other Names:
PEG-11 Lauryl Ether Carboxylic Acid
Polyethylene Glycol (11) Lauryl Ether
Carboxylic Acid
Polyoxyethylene (11) Lauryl Ether
Carboxylic Acid

Trade Names:
Akypo RLM 100 (Kao GmbH)
Empicol CBJ (Albright & Wilson UK)

LAURETH-12 CARBOXYLIC ACID

Definition: Laureth-12 Carboxylic Acid is the organic acid that conforms generally to the formula:

$$CH_3(CH_2)_{11}(OCH_2CH_2)_nOCH_2COOH$$

where n has an average value of 11.

Chemical Class: Carboxylic Acids

Function: Surfactant - Cleansing Agent

Technical/Other Names:
PEG-12 Lauryl Ether Carboxylic Acid
Polyethylene Glycol 600 Lauryl Ether
Carboxylic Acid
Polyoxyethylene (12) Carboxylic Acid

Trade Names:
Sandosan LNCA (Clariant)
Sandosan LNCA (Clariant GmbH, Personal
Care)

LAURETH-13 CARBOXYLIC ACID

CAS No.: 27306-90-7

Definition: Laureth-13 Carboxylic Acid is the organic acid that conforms generally to the formula:

$$CH_3(CH_2)_{11}(OCH_2CH_2)_nOCH_2COOH$$

where n has an average value of 12.

Chemical Class: Carboxylic Acids

Function: Surfactant - Cleansing Agent

Technical/Other Names:
PEG-13 Lauryl Ether Carboxylic Acid
Polyethylene Glycol (13) Lauryl Ether
Carboxylic Acid
Polyoxyethylene (13) Lauryl Ether
Carboxylic Acid

Trade Name:
    SURFINE WLL-A (Finetex)

LAURETH-14 CARBOXYLIC ACID

CAS No.: 27306-90-7

Definition: Laureth-14 Carboxylic Acid is the organic acid that conforms generally to the formula:

$$CH_3(CH_2)_{11}(OCH_2CH_2)_nOCH_2COOH$$

where n has an average value of 13.

Chemical Class: Carboxylic Acids

Function: Surfactant - Cleansing Agent

Technical/Other Names:
    PEG-14 Lauryl Ether Carboxylic Acid
    Polyethylene Glycol (14) Lauryl Ether
      Carboxylic Acid
    Polyoxyethylene (14) Lauryl Ether
      Carboxylic Acid

LAURETH-17 CARBOXYLIC ACID

CAS No.: 27306-90-7

Definition: Laureth-17 Carboxylic Acid is the organic acid that conforms generally to the formula:

$$CH_3(CH_2)_{11}(OCH_2CH_2)_nOCH_2COOH$$

where n has an average value of 16.

Chemical Class: Carboxylic Acids

Function: Surfactant - Cleansing Agent

Technical/Other Names:
    PEG-17 Lauryl Ether Carboxylic Acid
    Polyethylene Glycol (17) Lauryl Ether
      Carboxylic Acid
    Polyoxyethylene (17) Lauryl Ether
      Carboxylic Acid

Trade Name:
    Akypo RLM 160 (Kao GmbH)

LAURETH-6 CITRATE

CAS No.: 161756-30-5

Empirical Formula:
    $C_{30}H_{56}O_3$

Definition: Laureth-6 Citrate is the ester of citric acid and Laureth-6 (q.v.).

Chemical Class: Esters

Function: Surfactant - Cleansing Agent

Technical/Other Names:
    PEG-6 Lauryl Ether Citrate
    Polyethylene Glycol 300 Lauryl Ether
      Citrate
    Polyoxyethylene (6) Lauryl Ether Citrate

LAURETH-7 CITRATE

Empirical Formula:
    $C_{32}H_{60}O_{14}$

Definition: Laureth-7 Citrate is the ester of Laureth-7 (q.v.) and Citric Acid (q.v.).

Chemical Classes: Carboxylic Acids; Esters

Function: Surfactant - Cleansing Agent

Technical/Other Names:
    Polyethylene Glycol (7) Lauryl Ether Citrate
    Polyoxyethylene (7) Lauryl Ether Citrate

LAURETH-2 ETHYLHEXANOATE

CAS No.: 125804-14-0 (Generic)

Empirical Formula:
    $C_{24}H_{48}O_4$

Definition: Laureth-2 Ethylhexanoate is the ester of Laureth-2 (q.v.) and 2-ethylhexanoic acid. It conforms generally to the formula:

$$CH_3(CH_2)_3\overset{\displaystyle |}{\underset{\displaystyle CH_2CH_3}{CHC}}\overset{\displaystyle O}{\overset{\displaystyle ||}{}}{-}(OCH_2CH_2)_nO(CH_2)_{11}CH_3$$

where n has an average value of 2.

Chemical Class: Esters

Function: Skin-Conditioning Agent - Emollient

Reported Product Categories: Manicuring Preparations, Misc.; Nail Creams and Lotions

Technical/Other Names:
    Laureth-2 Octanoate
    PEG-2 Lauryl Ether Octanoate
    Polyethylene Glycol 100 Lauryl Ether
      Octanoate
    Polyoxyethylene (2) Lauryl Ether Octanoate

Trade Names:
    Dermol 1012 (Alzo)
    Estalan 12 (Lanaetex)
    Pelemol L20 (Phoenix)

LAURETH-1 PHOSPHATE

Definition: Laureth-1 Phosphate is a complex mixture of esters of Laureth-1 (q.v.) and phosphoric acid.

Chemical Classes: Alkoxylated Alcohols; Phosphorus Compounds

Function: Surfactant - Cleansing Agent

Technical/Other Names:
    Polyethylene Glycol (1) Lauryl Ether Phos-
      phate
    Polyoxyethylene (1) Lauryl Ether Phos-
      phate

Trade Names:
    Dermalcare MAP L-210 (Rhodia)
    Nikkol Phosten HLP-1 (Nikko)

LAURETH-2 PHOSPHATE

Definition: Laureth-2 Phosphate is a complex mixture of esters of phosphoric acid and Laureth-2 (q.v.).

Information Source: JCLS

Chemical Class: Phosphorus Compounds

Function: Surfactant - Cleansing Agent

Technical/Other Names:
    PEG-2 Lauryl Ether Phosphate
    Polyethylene Glycol (2) Lauryl Ether Phos-
      phate
    Polyoxyethylene (2) Lauryl Ether Phos-
      phate

Trade Name:
    Phosphanol ML-220 (Toho)

LAURETH-3 PHOSPHATE

CAS Nos.: 25852-45-3; 39464-66-9 (Generic)

JPN Translation:
    ラウレス - 3 リン酸

Definition: Laureth-3 Phosphate is a complex mixture of esters of phosphoric acid and Laureth-3 (q.v.).

Information Sources: JCLS, TSCA

Chemical Class: Phosphorus Compounds

Function: Surfactant - Cleansing Agent

Technical/Other Names:
    PEG-3 Lauryl Ether Phosphate
    Polyethylene Glycol (3) Lauryl Ether Phos-
      phate
    Poly(Oxy-1,2-Ethanediyl), α-Phosphono-
      ω-, (Dodecyloxy)-
    Polyoxyethylene (3) Lauryl Ether Phos-
      phate

Trade Names:
    Servoxyl VPAZ 3/100 (Sasol Servo)
    Surfagene FDD 402 (Kao GmbH)

LAURETH-4 PHOSPHATE

CAS No.: 39464-66-9 (Generic)

The inclusion of any compound in the *Dictionary and Handbook* does not indicate that use of that substance as a cosmetic ingredient complies with the laws and regulations governing such use in the United States or any other country.

JPN Translation:
ラウレス - 4 リン酸

Definition: Laureth-4 Phosphate is a complex mixture of esters of phosphoric acid and Laureth-4 (q.v.).

Information Source: JCLS

Chemical Class: Phosphorus Compounds

Function: Surfactant - Cleansing Agent

Technical/Other Names:
PEG-4 Lauryl Ether Phosphate
Polyethylene Glycol 200 Lauryl Ether Phosphate
Polyoxyethylene (4) Laury Ether Phosphate

Trade Names:
Ethfac 142W (Ethox)
Phosphanol RD-510Y (Toho)

Trade Name Mixtures:
Hostaphat KML (Clariant)
Hostaphat KML (Clariant GmbH, Personal Care)

LAURETH-7 PHOSPHATE

CAS No.: 39464-66-9 (Generic)

JPN Translation:
ラウレス - 7 リン酸

Definition: Laureth-7 Phosphate is a complex mixture of esters of phosphoric acid and Laureth-7 (q.v.).

Information Source: JCLS

Chemical Class: Phosphorus Compounds

Function: Surfactant - Cleansing Agent

Technical/Other Names:
PEG-7 Lauryl Ether Phosphate
Polyethylene Glycol (7) Lauryl Ether Phosphate
Polyoxyethylene (7) Lauryl Ether Phosphate

Trade Name:
Akypomine MW 05 (Kao GmbH)

LAURETH-8 PHOSPHATE

CAS No.: 39464-66-9 (Generic)

JPN Translation:
ラウレス - 8 リン酸

Definition: Laureth-8 Phosphate is a complex mixture of esters of phosphoric acid and Laureth-8 (q.v.).

Information Sources: JCLS, TSCA

Chemical Class: Phosphorus Compounds

Function: Surfactant - Cleansing Agent

Technical/Other Names:
PEG-8 Lauryl Ether Phosphate
Polyethylene Glycol 400 Lauryl Ether Phosphate
Polyoxyethylene (8) Lauryl Ether Phosphate

LAURETH-12 SUCCINATE

Definition: Laureth-12 Succinate is the monoester of Laureth-12 (q.v.) and Succinic Acid (q.v.).

Chemical Class: Esters

Functions: Hair Conditioning Agent; Skin-Conditioning Agent - Miscellaneous

Technical/Other Names:
Polyethylene Glycol (12) Lauryl Ether Succinate
Polyoxyethylene (12) Lauryl Ether Succinate

Trade Name Mixture:
Biosil Basics Cetylsil NS (Biosil Technologies, Inc.)

LAURETH-7 TARTRATE

Empirical Formula:
$C_{30}H_{58}O_{13}$

Definition: Laureth-7 Tartrate is the ester of Laureth-7 (q.v.) and Tartaric Acid (q.v.).

Chemical Classes: Carboxylic Acids; Esters

Function: Surfactant - Cleansing Agent

Technical/Other Names:
Polyethylene Glycol (7) Lauryl Ether Tartrate
Polyoxyethylene (7) Lauryl Ether Tartrate

LAURIC ACID

CAS No.	EINECS No.
143-07-7	205-582-1

JPN Translation:
ラウリン酸

Empirical Formula:
$C_{12}H_{24}O_2$

Definition: Lauric Acid is the fatty acid that conforms generally to the formula:

$$CH_3(CH_2)_{10}COOH$$

Information Sources: 21CFR172.210, 21CFR172.860, 21CFR173.340, 21CFR175.105, 21CFR175.320, 21CFR176.170, 21CFR176.200, 21CFR176.210, 21CFR177.1010, 21CFR177.1200, 21CFR177.2260, 21CFR177.2600, 21CFR177.2800, 21CFR178.3570, 21CFR178.3910, CIR: [S] JACT-6(3)1987, CTFA S, FCC, JCLS, JSCI, MI-13(5400), RIFM, SNPF, TSCA

Chemical Class: Fatty Acids

Functions: Fragrance Ingredient; Surfactant - Cleansing Agent

Surfactant-Cleansing Agent is included as a function for the soap form of Lauric Acid.

Reported Product Categories: Hair Dyes and Colors (All Types Requiring Caution Statements and Patch Tests); Bath Oils, Tablets, and Salts; Cleansing Products (Cold Creams, Cleansing Lotions, Liquids and Pads); Bath Preparations, Misc.; Bath Soaps and Detergents; Tonics, Dressings, and Other Hair Grooming Aids; Deodorants (Underarm); Personal Cleanliness Products, Misc.; Shampoos (Non-coloring); Shaving Cream (Aerosol, Brushless and Lather)

Technical/Other Names:
n-Dodecanoic Acid
Dodecylic Acid
Lauric acid (RIFM)

Trade Names:
AEC Lauric Acid (A & E Connock)
Kortacid 1295 (Akzo Nobel Surface AB)
Kortacid 1299 (Akzo Nobel)
Kortacid 1299 (Akzo Nobel Surface AB)
Lauric Acid PC (Protameen)
PRIFRAC 2920 (Uniqema Europe)
PRIFRAC 2922 (Uniqema Europe)

Trade Name Mixtures:
Lipokel 12E (Giovanni Bozzetto SpA)
Lipokel 12 G (Bozzetto)

LAURIC/PALMITIC/OLEIC TRIGLYCERIDE

Definition: Lauric/Palmitic/Oleic Triglyceride is a mixed triester of glycerin with lauric, palmitic and oleic acids.

Chemical Class: Fats and Oils

Function: Skin-Conditioning Agent - Occlusive

Trade Name:
Turtleoil (Vevy)

LAURIMINO BISPROPANEDIOL

CAS No.	EINECS No.
817-01-6	212-441-8

Empirical Formula:
$C_{18}H_{39}NO_4$

Definition: Laurimino Bispropanediol is the organic compound that conforms to the formula:

$$CH_3(CH_2)_{11}N \underset{CH_2CHCH_2OH}{\overset{CH_2CHCH_2OH}{}}$$

with OH groups attached

Chemical Classes: Alkyl-Substituted Amino Acids; Polyols

Functions: Surfactant - Cleansing Agent; Surfactant - Foam Booster; Viscosity Increasing Agent - Aqueous

Technical/Other Name:
1,2-Propanediol, 3,3'(Dodecylimino)Bis-

Trade Name:
Nissan Nymeen LG-302 (NOF)

LAURIMINODIPROPIONIC ACID

Definition: Lauriminodipropionic Acid is the organic compound that conforms to the formula:

$$CH_3(CH_2)_{11}N \underset{CH_2CH_2COOH}{\overset{CH_2CH_2COOH}{}}$$

Chemical Classes: Alkyl-Substituted Amino Acids; Amines

Functions: Hair Conditioning Agent; Surfactant - Cleansing Agent

Trade Name Mixture:
Vitol IEP (Vital Personal Care Specialties, Inc)

LAUR/MYRIST/PALMITAMIDOBUTYL GUANIDINE ACETATE

Definition: Laur/Myrist/Palmitamidobutyl Guanidine Acetate is the organic compound that conforms to the formula:

$$RC \overset{O}{\underset{\|}{-}} NH(CH_2)_4N = C \underset{NH_2}{\overset{NH_2}{}} \cdot CH_3COOH$$

where RCO- represents a mixture of lauroyl, myristoyl and palmitoyl radicals.

Chemical Classes: Amides; Amines

Functions: Antistatic Agent; Hair Conditioning Agent; Humectant; Surfactant - Emulsifying Agent

Trade Name:
C1246A4G (Lion Corporation)

LAUROAMPHODIPROPIONIC ACID

CAS No.: 64265-41-4

Empirical Formula:
$C_{22}H_{42}N_2O_6$

Definition: Lauroamphodipropionic Acid is the amphoteric organic compound that conforms generally to the formula:

$$CH_3(CH_2)_{10}\overset{O}{\underset{\|}{C}} - NH(CH_2)_2N(CH_2)_2COOH \text{ with } (CH_2)_2O(CH_2)_2COOH$$

Chemical Class: Alkylamido Alkylamines

Functions: Hair Conditioning Agent; Surfactant - Cleansing Agent; Surfactant - Foam Booster; Surfactant - Hydrotrope

Technical/Other Names:
β-Alanine, N-[2-(2-Carboxyethoxy)Ethyl]-N-[2-[(1-Oxododecyl)Amino]Ethyl]-Lauroamphocarboxypropionic Acid

LAUROYL BETA-ALANINE

CAS No.	EINECS No.
21539-56-0	410-320-6

Empirical Formula:
$C_{15}H_{29}NO_3$

Definition: Lauroyl Beta-Alanine is the substituted amino acid that conforms to the formula:

$$CH_3(CH_2)_{10}\overset{O}{\underset{\|}{C}} - NH(CH_2)_2COOH$$

Chemical Class: Alkyl-Substituted Amino Acids

Function: Surfactant - Cleansing Agent

LAUROYL ARGININE

CAS No.	EINECS No.
42492-22-8	255-851-2

Empirical Formula:
$C_{18}H_{36}N_4O_3$

Definition: Lauroyl Arginine is the substituted amino acid that conforms to the formula:

$$CH_3(CH_2)_{10}\overset{O}{\underset{\|}{C}} - NHCH \underset{COOH}{\overset{}{|}}(CH_2)_3NHC \underset{NH_2}{\overset{}{\|}} = NH$$

Chemical Class: Amino Acids

Functions: Hair Conditioning Agent; Skin-Conditioning Agent - Emollient

Technical/Other Names:
L-Arginine, N2-(1-Oxododecyl)-N-Lauroyl-L-Arginine

Trade Name:
Amisafe AL-01 (Ajinomoto)

LAUROYL COLLAGEN AMINO ACIDS

CAS No.: 68920-59-2

Definition: Lauroyl Collagen Amino Acids is the product obtained by the condensation of lauric acid chloride with Collagen Amino Acids (q.v.).

Chemical Class: Amino Acids

Functions: Hair Conditioning Agent; Surfactant - Cleansing Agent

LAUROYL ETHYLENEDIAMINE TRIACETIC ACID

CAS No.: 148124-42-9

Empirical Formula:
$C_{20}H_{36}O_7N_2$

Definition: Lauroyl Ethylenediamine Triacetic Acid is the substituted diamine that conforms to the formula:

$$CH_3(CH_2)_{10}\overset{O}{\underset{\|}{C}} - NCH_2CH_2N \underset{CH_2COOH}{\overset{CH_2COOH}{}} - CH_2COOH$$

Information Source: TSCA

Chemical Class: Amino Acids

Functions: Chelating Agent; Corrosion Inhibitor; Surfactant - Hydrotrope; Surfactant - Solubilizing Agent; Surfactant - Suspending Agent

Technical/Other Name:
Glycine, N-[2-[Bis(Carboxymethyl)Amino]-Ethyl]-N-(1-Oxododecyl)-N-Lauroyl-N,N',N '-Ethylenediamine Triacetic Acid

Trade Name:
Hampshire LED3A Acid (Amerchol)

LAUROYL ETHYL GLUCOSIDE

Definition: Lauroyl Ethyl Glucoside is the ester of Lauric Acid (q.v.) and ethyl glucoside.

Chemical Classes: Carbohydrates; Esters

Function: Surfactant - Emulsifying Agent

LAUROYL GLUTAMIC ACID

CAS No.	EINECS No.
3397-65-7	222-261-1

JPN Translation:
ラウロイルグルタミン酸

Empirical Formula:
$C_{17}H_{31}NO_5$

Definition: Lauroyl Glutamic Acid is the substituted amino acid that conforms to the formula:

HOOCCH$_2$CH$_2$CHCOOH
HN — C(CH$_2$)$_{10}$CH$_3$
O

Information Sources: JCIC, JCLS

Chemical Classes: Amides; Amino Acids

Function: Skin-Conditioning Agent - Miscellaneous

Technical/Other Names:
L-Glutamic Acid, N-(1-oxododecyl)-
N-Lauroyl-L-Glutamic Acid

Trade Name:
Amisoft LA (Ajinomoto)

LAUROYL HYDROLYZED COLLAGEN

CAS No.: 68920-59-2

Definition: Lauroyl Hydrolyzed Collagen is the condensation product of lauric acid chloride and Hydrolyzed Collagen (q.v.).

Information Source: TSCA

Chemical Class: Protein Derivatives

Functions: Hair Conditioning Agent; Skin-Conditioning Agent - Miscellaneous; Surfactant - Cleansing Agent

Reported Product Category: Shampoos (Non-coloring)

Technical/Other Names:
Collagens, Lauroyl Derivs.
Lauroyl Hydrolyzed Animal Protein
Proteins, Hydrolysates, Reaction Products
with Lauroyl Chloride

Trade Names:
Collodex (Dextran)
Pro-Tein SA-20 (Maybrook)

Trade Name Mixture:
Lipo-Peptide AME 30 (Maybrook)

LAUROYL HYDROLYZED ELASTIN

Definition: Lauroyl Hydrolyzed Elastin is the condensation product of lauric acid chloride and Hydrolyzed Elastin (q.v.).

Chemical Class: Protein Derivatives

Functions: Hair Conditioning Agent; Skin-Conditioning Agent - Miscellaneous; Surfactant - Cleansing Agent

Trade Name:
Crolastin AS (Croda Chemicals)

LAUROYL LACTYLIC ACID

Empirical Formula:
$C_{18}H_{32}O_6$

Definition: Lauroyl Lactylic Acid is the organic compound that conforms to the formula:

CH$_3$(CH$_2$)$_{10}$C — OCHC — OCHCOOH
CH$_3$ CH$_3$

Chemical Classes: Carboxylic Acids; Esters

Function: Surfactant - Emulsifying Agent

Trade Name Mixture:
PRIAZUL 2131 (Uniqema Europe)

LAUROYL LYSINE

CAS No.
52315-75-0

EINECS No.
257-843-4

JPN Translation:
ラウロイルリシン

Empirical Formula:
$C_{18}H_{36}N_2O_3$

Definition: Lauroyl Lysine is the lauroyl derivative of Lysine (q.v.) that conforms to the formula:

CH$_3$(CH$_2$)$_{10}$C — NH(CH$_2$)$_4$CHCOOH
NH$_2$

Information Sources: JCIC, JCLS, JSQI

Chemical Class: Amino Acids

Functions: Hair Conditioning Agent; Skin-Conditioning Agent - Miscellaneous

Reported Product Categories: Eye Shadows; Face Powders; Foundations; Lipsticks; Blushers (All types); Mascara; Powders (Dusting and Talcum, Excluding Aftershave Talcs); Makeup Bases; Makeup Preparations (Not eye), Misc.; Rouges

Technical/Other Name:
N-ε-Lauroyl-L-lysine

Trade Name:
Amihope LL (Ajinomoto)

Trade Name Mixtures:
Amilon (Ikeda)
Covamat (LCW)
Dry- Flo Elite LL (National Starch)

Essential Vital Elements - S (Dipta)
Liponyl 20-LL (Lipo)
Mearlcite SRA (Engelhard Corp.)
Mearlmica SVA (Engelhard Corp.)
Mearltalc TCA (Engelhard Corp.)
NYLONPOLY WL 10 LL (C.I.T.)
Sericite WL (Ikeda)
TALCPOLY LL (Creations Couleurs)

LAUROYL METHYL BETA-ALANINE

JPN Translation:
ラウロイルメチルアラニン

Empirical Formula:
$C_{16}H_{31}NO_3$

Definition: Lauroyl Methyl Beta-Alanine is the substituted amino acid that conforms to the formula:

CH$_3$(CH$_2$)$_{10}$C — NCH$_2$CH$_2$COOH
CH$_3$

Information Sources: JCIC, JCLS

Chemical Classes: Amides; Amino Acids

Function: Skin-Conditioning Agent - Miscellaneous

Technical/Other Names:
Lauroyl Methyl Alanine
N-Lauroyl-N-Methyl-β-Alanine

Trade Name:
Alanon ALA (Kawaken)

LAUROYL METHYL GLUCAMIDE

Empirical Formula:
$C_{19}H_{39}NO_6$

Definition: Lauroyl Methyl Glucamide is the organic compound that conforms to the formula:

CH$_3$(CH$_2$)$_{10}$C — NCH$_2$CHCHCHCHCH$_2$OH
CH$_3$ OH   OH

Chemical Classes: Amides; Polyols

Functions: Skin-Conditioning Agent - Miscellaneous; Surfactant - Cleansing Agent

LAUROYL PG-TRIMONIUM CHLORIDE

Empirical Formula:
$C_{18}H_{38}NO_3 \cdot Cl$

Definition: Lauroyl PG-Trimonium Chloride is the quaternary ammonium salt that conforms to the formula:

$$\left[CH_3(CH_2)_{10}C(=O)-OCH_2CHCH_2-N(CH_3)_2-CH_3 \right]^+ \quad Cl^-$$

Chemical Class: Quaternary Ammonium Compounds

Functions: Antistatic Agent; Hair Conditioning Agent

Trade Name:
Akypoquat 132 (Kao GmbH)

LAUROYL SARCOSINE

CAS No. 97-78-9
EINECS No. 202-608-3

JPN Translation:
ラウロイルサルコシン

Empirical Formula:
$C_{15}H_{29}NO_3$

Definition: Lauroyl Sarcosine is the N-lauroyl derivative of N-methylglycine that conforms generally to the formula:

$$CH_3(CH_2)_{10}C(=O)-NCH_2COOH \quad (CH_3)$$

Information Sources: 21CFR177.1200, 21CFR178.3130, CIR: [SQ] IJT-20(SUPPL. 1)2001, CTFA D, JCIC, JCLS, JSQI, TSCA

Chemical Class: Sarcosinates and Sarcosine Derivatives

Functions: Hair Conditioning Agent; Surfactant - Cleansing Agent

Reported Product Category: Shampoos (Non-coloring)

Technical/Other Names:
N-Dodecanoylsarcosine
N-Dodeccanoyl-N-methylglycine
Glycine, N-Methyl-N-(1-Oxododecyl)-
N-Lauroyl-N-nethylaminoacetic Acid
N-Methyl-N-(1-Oxododecyl)Glycine

Trade Names:
Crodasinic L (Croda Chemicals)
Hamposyl L (Amerchol)
Lowenol "L" Acid (Lowenstein)
Nikkol Sarcosinate LH (Nikko)
Vanseal LS (Vanderbilt)

LAUROYL SILK AMINO ACIDS

Definition: Lauroyl Silk Amino Acids is the product obtained by the condensation of lauric acid chloride and Silk Amino Acids (q.v.).

Chemical Class: Amino Acids

Functions: Hair Conditioning Agent; Surfactant - Cleansing Agent

Trade Name:
Crosilk AS (Croda Chemicals)

LAURTRIMONIUM BROMIDE

CAS No. 1119-94-4
EINECS No. 214-290-3

JPN Translation:
ラウルトリモニウムブロミド

Empirical Formula:
$C_{15}H_{34}N \cdot Br$

Definition: Laurtrimonium Bromide is the quaternary ammonium salt that conforms to the formula:

$$\left[CH_3(CH_2)_{11}-N(CH_3)_2-CH_3 \right]^+ \quad Br^-$$

Information Sources: JCIC, JCLS, JSQI, TSCA

Chemical Class: Quaternary Ammonium Compounds

Functions: Cosmetic Biocide; Hair Conditioning Agent

Technical/Other Names:
Ammonium, Dodecyltrimethyl-,Bromide
1-Dodecanaminium, N,N,N-Trimethyl-, Bromide
Dodecyltrimethylammonium Bromide
Lauryltrimethylammonium Bromide
N,N,N-Trimethyl-1-Dodecanaminiuim Bromide

LAURTRIMONIUM CHLORIDE

CAS No. 112-00-5
EINECS No. 203-927-0

JPN Translation:
ラウリルトリモニウムクロリド

Empirical Formula:
$C_{15}H_{34}N \cdot Cl$

Definition: Laurtrimonium Chloride is the quaternary ammonium salt that conforms generally to the formula:

$$\left[CH_3(CH_2)_{11}-N(CH_3)_2-CH_3 \right]^+ \quad Cl^-$$

Information Sources: JCLS, JSCI, TSCA

Chemical Class: Quaternary Ammonium Compounds

Functions: Antistatic Agent; Cosmetic Biocide; Surfactant - Emulsifying Agent

Reported Product Category: Hair Conditioners

Technical/Other Names:
Ammonium, Dodecyltrimethyl-, Chloride
1-Dodecanaminium, N,N,N-Trimethyl-, Chloride
Dodecyltrimethylammonium Chloride
Lauryl Trimethyl Ammonium Chloride
N,N,N-Trimethyl-1-Dodecanaminium Chloride

Trade Names:
Arquad 12-37W (Akzo Nobel)
Chemquat 12-33 (Chemax)
Chemquat 12-50 (Chemax)
Empigen 5089 (Albright & Wilson UK)
Laurene (Vevy)
Nikkol CA-2150 (Nikko)

Trade Name Mixture:
Arquad 12-50 (Akzo Nobel)

LAURTRIMONIUM TRICHLOROPHEN-OXIDE

Empirical Formula:
$C_{15}H_{35}N \cdot C_6H_2Cl_3O$

Definition: Laurtrimonium Trichlorophenoxide is the quaternary ammonium compound that conforms generally to the formula:

$$\left[CH_3(CH_2)_{11}-N(CH_3)_2-CH_3 \right]^+ \quad \left[\text{2,4,5-trichlorophenoxide} \right]^-$$

Chemical Classes: Halogen Compounds; Quaternary Ammonium Compounds

Function: Cosmetic Biocide

Technical/Other Name:
Lauryl Trimethyl Ammonium 2,4,5-Trichloro Phenoxide

LAURUS NOBILIS

Definition: *See "Regulatory and Ingredient Use Information," regarding EU labeling names for botanical ingredients in Volume 1, Introduction, Part A.*

Chemical Class: Biological Products

Technical/Other Names:
Laurus Nobilis Leaf (U.S.)
Laurus Nobilis Leaf Extract (U.S.)
Laurus Nobilis Oil (U.S.)

LAURUS NOBILIS LEAF

Definition: Laurus Nobilis Leaf is the plant material derived from the dried, crushed leaves of *Laurus nobilis*. See "Regulatory and Ingredient Use Information," regarding the labeling names for botanical ingredients in Volume 1, Introduction, Part A.

Chemical Class: Biological Products

Function: Fragrance Ingredient

Technical/Other Name:
Laurus Nobilis (EU)

Trade Names:
Laurel (Laurus Nobilis) Leaves (Lebermuth)
Laurel Leaf - KTS (Sun Trade)

LAURUS NOBILIS LEAF EXTRACT

CAS No.	EINECS No.
84603-73-6	283-272-5

Definition: Laurus Nobilis Leaf Extract is an extract of the leaves of the laurel, *Laurus nobilis*. See "Regulatory and Ingredient Use Information," regarding the labeling names for botanical ingredients in Volume 1, Introduction, Part A.

Information Sources: 21CFR182.20, RIFM

Chemical Class: Biological Products

Functions: Fragrance Ingredient; Skin-Conditioning Agent - Miscellaneous; Skin-Conditioning Agent - Occlusive

Technical/Other Names:
Bay Laurel Extract
Extract of Grecian Laurel
Extract of Laurel
Extract of Laurus Nobilis
Extract of Sweet Bay
Grecian Laurel Extract
Laurel Extract
Laurel (Laurus Nobilis) Extract
Laurel leaves extract (Laurus nobilis L.) (RIFM)
Laurus Nobilis (EU)
Sweet Bay Extract

Trade Name:
Hydroessential Laurus (Vevy)

Trade Name Mixtures:
Actiphyte of Bay Laurel BG50 (Active Organics)
Actiphyte of Bay Laurel GL50 (Active Organics)
Actiphyte of Bay Laurel Lipo S (Active Organics)
Actiphyte of Bay Laurel PG50 (Active Organics)
Aromaphyte of Bay Laurel (Active Organics)
Herbasol Complex "Herbes de Provence" (Cosmetochem) (Cosmetochem International Ltd.)
Herbasol Extract Laurel (Cosmetochem) (Cosmetochem International Ltd.)
Herbasol Extract Oil Soluble Laurel (Cosmetochem) (Cosmetochem International Ltd.)
Hydroplastidine Laurus (Vevy)
Laurel Extract HS 2385 G (Grau)
Laurel HS (Alban Muller)
LAUREL LEAF EXTRACT (Libiol)
Laurus Nobilis Leaf Extract ies (IES LABO)
278 Relaxant HS (Alban Muller)
678 Relaxant LS (Alban Muller)
270 Solarium HS (Alban Muller)
670 Solarium LS (Alban Muller)
Spicypone Laurel 2/035500 (Symrise)
VT-096 Extract of Bay Leaves (Vege-Tech)

LAURUS NOBILIS OIL

CAS Nos.: 8002-41-3; 8007-48-5

Definition: Laurus Nobilis Oil is the volatile oil obtained from *Laurus nobilis*. See "Regulatory and Ingredient Use Information," regarding the labeling names for botanical ingredients in Volume 1, Introduction, Part A.

Information Sources: MI-13(5398), RIFM

Chemical Class: Essential Oils

Function: Fragrance Ingredient

Technical/Other Names:
Bay oil, sweet (Laurus nobilis L.) (RIFM)
Bay, sweet (Laurus nobilis L.) (RIFM)
Laurel berries (Laurus nobilis L.) (RIFM)
Laurel (Laurus Nobilis) Oil
Laurel Oil
Laurus Nobilis (EU)
Oil of Laurel
Sweet Bay Oil

Trade Name:
AEC Laurel Leaf Oil (A & E Connock)

Trade Name Mixtures:
Aromaphyte of Bay Laurel (Active Organics)
Essentiaderm N.20 (Universal Flavors)
Essentiaderm N.21 (Universal Flavors)

LAURYL ACRYLATE CROSSPOLYMER

Definition: Lauryl Acrylate Crosspolymer is a polymer of lauryl acrylate crosslinked with divinylbenzene.

Chemical Class: Synthetic Polymers

Function: Hair Fixative

LAURYL ACRYLATE/VA COPOLYMER

Definition: Lauryl Acrylate/VA Copolymer is a copolymer of lauryl acrylate and vinyl acetate monomers.

Information Sources: CIR: [SQ] IJT 21 (SUPPL. 3) 2002, JCLS

Chemical Class: Synthetic Polymers

Function: Film Former

LAURYL ACRYLATE/VA CROSSPOLYMER

Definition: Lauryl Acrylate/VA Crosspolymer is a copolymer of lauryl acrylate and vinyl acetate crosslinked with divinylbenzene.

Chemical Class: Synthetic Polymers

Function: Abrasive

Trade Name:
Polymer EB (Kao Corp.)

LAURYL ALCOHOL

CAS No.	EINECS No.
112-53-8	203-982-0

JPN Translation:
ラウリルアルコール

Empirical Formula:
$C_{12}H_{26}O$

Definition: Lauryl Alcohol is the fatty alcohol that conforms generally to the formula:

$$CH_3(CH_2)_{11}OH$$

Information Sources: 21CFR172.515, 21CFR172.864, 21CFR175.105, 21CFR175.300, 21CFR177.1010, 21CFR177.1200, 21CFR177.2800, 21CFR178.3480, 21CFR178.3910, CTFA D, FCC, JCLS, JSCI, MI-13(3439), RIFM, TSCA

Chemical Class: Fatty Alcohols

Functions: Emulsion Stabilizer; Fragrance Ingredient; Skin-Conditioning Agent - Emollient; Surfactant - Foam Booster; Viscosity Increasing Agent - Aqueous; Viscosity Increasing Agent - Nonaqueous

Reported Product Categories: Hair Dyes and Colors (All Types Requiring Caution Statements and Patch Tests); Hair Conditioners; Shampoos (Non-coloring)

Technical/Other Names:
Didecyl Alcohol

1-Dodecanol
1-Hydroxydodecane
Lauryl alcohol (RIFM)

Trade Names:
Alfol 1216 CO Alcohol (Sasol North America)
Cachalot L-90 (Michel)
CO-1214 (Procter & Gamble)
Laurex L1 (Albright & Wilson UK)
Laurex NC (Albright & Wilson UK)
Lipocol L (Lipo)
Nacol 12-99 Alcohol (Sasol GmbH - Hamburg)

Trade Name Mixtures:
ProLipid 141 (International Specialty Products)
Sabonal C12 14 (Sabo)
Tewax TC 1 (Cesalpinia)

LAURYL ALCOHOL DIPHOSPHONIC ACID

CAS No.: 16610-63-2

Empirical Formula:
$C_{12}H_{28}O_7P_2$

Definition: Lauryl Alcohol Diphosphonic Acid is the organic compound that conforms to the formula:

$$CH_3(CH_2)_{10}\overset{\overset{\displaystyle PO_3H_2}{|}}{\underset{\underset{\displaystyle PO_3H_2}{|}}{C}}\!-\!OH$$

Chemical Classes: Alcohols; Phosphorus Compounds

Function: Emulsion Stabilizer

Technical/Other Names:
1-Hydroxydodecane-1,1-diphosphonic Acid
Phosphonic Acid, (1-Hydroxydodecylidene)Bis-

Trade Name Mixtures:
Lipokel 12E (Giovanni Bozzetto SpA)
Lipokel 12 G (Bozzetto)

LAURYLAMINE DIPROPYLENEDIAMINE

CAS No.   **EINECS No.**
2372-82-9   219-145-8

Empirical Formula:
$C_{18}H_{41}N_3$

Definition: Laurylamine Dipropylenediamine is the organic compound that conforms to the formula:

$$CH_3(CH_2)_{11}NH(CH_2)_3NH(CH_2)_3NH_2$$

Chemical Class: Amines

Function: Hair Conditioning Agent

Technical/Other Names:
N,N-Bis(3-Aminopropyl)Dodecylamine
Dodecylamine, N,N-Bis(3-Aminopropyl)-
1,3-Propanediamine, N-(3-Aminopropyl)-N-Dodecyl

Trade Name:
Lonzabac-12.100 (Lonza Ltd.)

LAURYL AMINOPROPYLGLYCINE

CAS No.   **EINECS No.**
34395-72-7   251-993-4

Empirical Formula:
$C_{17}H_{36}N_2O_2$

Definition: Lauryl Aminopropylglycine is the substituted amino acid that conforms to the formula:

$$CH_3(CH_2)_{11}NH(CH_2)_3NHCH_2COOH$$

Information Source: TSCA

Chemical Classes: Alkyl-Substituted Amino Acids; Amines

Functions: Hair Conditioning Agent; Skin-Conditioning Agent - Miscellaneous

Reported Product Category: Permanent Waves

Technical/Other Names:
2-(3-(Dodecylamino)propylamino)acetic Acid
N-[3-(Dodecylamino)Propyl]Glycine
Glycine, N-[3-(Dodecylamino)Propyl]-

Trade Name Mixtures:
Facteur Hydratant PH (Prod'Hyg)
Hydroviton 2/059353 (Symrise)

LAURYL BEHENATE

CAS No.: 42233-07-8

Empirical Formula:
$C_{34}H_{68}O_2$

Definition: Lauryl Behenate is the ester of lauryl alcohol and behenic acid. It conforms to the formula:

$$CH_3(CH_2)_{20}\overset{\overset{\displaystyle O}{\|}}{C}\!-\!O(CH_2)_{11}CH_3$$

Chemical Class: Esters

Function: Skin-Conditioning Agent - Occlusive

Technical/Other Names:
Docosanoic Acid, Dodecyl Ester
Dodecyl Behenate
Dodecyl Docosanoate

Trade Name:
Pelemol LB (Phoenix)

LAURYL BETAINE

CAS No.   **EINECS No.**
683-10-3   211-669-5

JPN Translation:
ラウリルベタイン

Empirical Formula:
$C_{16}H_{33}NO_2$

Definition: Lauryl Betaine is the zwitterion (inner salt) that conforms generally to the formula:

$$CH_3(CH_2)_{11}\!-\!\overset{\overset{\displaystyle CH_3}{|}}{\underset{\underset{\displaystyle CH_3}{|}}{N^+}}\!-\!CH_2COO^-$$

Information Sources: JCLS, JSCI, TSCA

Chemical Class: Betaines

Functions: Antistatic Agent; Hair Conditioning Agent; Skin-Conditioning Agent - Miscellaneous; Surfactant - Cleansing Agent; Surfactant - Foam Booster; Viscosity Increasing Agent - Aqueous

Reported Product Categories: Shampoos (Non-coloring); Bath Preparations, Misc.; Bath Oils, Tablets, and Salts; Bubble Baths; Cleansing Products (Cold Creams, Cleansing Lotions, Liquids and Pads); Hair Conditioners; Hair Tints

Technical/Other Names:
1-Dodecanaminium, N-(Carboxymethyl)-N, N-Dimethyl-, Hydroxide, Inner Salt
Dodecylbetaine
Dodecyldimethylbetaine
Lauryl Dimethylaminoacetic Acid Betaine
Lauryl Dimethyl Glycine
Lauryl-N-methylsarcosine

Trade Names:
Amphoteen 24 (Akzo Nobel Surface AB)
Empigen BB (Albright & Wilson UK)
Euroquat CF (EOC Surfactants)
LMB (Vevy)
Mackam LB (McIntyre)
Mackam LB-35 (McIntyre)
Nikkol AM-301 (Nikko)
Obazolin LB (Toho)
Product DDN (DuPont de Nemours)
Unibetaine LB (Universal Preserv-A-Chem)
Zohartaine AB (Zohar)

Trade Name Mixture:
Afron-N (Vevy)

LAURYL COCOATE

Definition: Lauryl Cocoate is the ester of lauryl alcohol and the fatty acids derived from coconut oil that conforms to the formula:

$$RC\overset{\overset{\displaystyle O}{\|}}{}\!-\!O(CH_2)_{11}CH_3$$

where RCO- represents the fatty acids derived from coconut oil.

Chemical Class: Esters

Functions: Skin-Conditioning Agent - Emollient; Skin-Conditioning Agent - Occlusive

Trade Name:
Cetinol 1212 (Fabriquimica)

LAURYL p-CRESOL KETOXIME

CAS No.: 50652-76-1

Empirical Formula:
$C_{19}H_{31}NO_2$

Definition: Lauryl p-Cresol Ketoxime is the organic compound that conforms to the formula:

Chemical Class: Phenols

Functions: Buffering Agent; Skin-Conditioning Agent - Emollient

Technical/Other Name:
1-Dodecanone, 1-(2-Hydroxy-5-Methyl-phenyl)-, Oxime

Trade Name:
RonaCare LPO (Merck KGaA)

N-LAURYL DIETHANOLAMINE

JPN Translation:
ラウリル DEA

Definition: *See "Regulatory and Ingredient Use Information," regarding use of Japan Trivial names in Volume 1, Introduction, Part A.*

Information Source: JCIC

Chemical Class: Alkanolamines

Function: Not Reported

LAURYL DIETHYLENEDIAMINOGLYCINE

CAS No.	EINECS No.
6843-97-6	229-930-7

Empirical Formula:
$C_{18}H_{39}N_3O_2$

Definition: Lauryl Diethylenediaminoglycine is the substituted amino acid that conforms to the formula:

$$CH_3(CH_2)_{11}NH(CH_2)_2NH(CH_2)_2NHCH_2COOH$$

Information Sources: BAN, JCLS, TSCA

Chemical Classes: Alkyl-Substituted Amino Acids; Amines

Functions: Hair Conditioning Agent; Skin-Conditioning Agent - Miscellaneous

Reported Product Category: Permanent Waves

Technical/Other Names:
N-[2-[[-(Dodecylamino)Ethyl]Amino]Ethyl]-Glycine
Glycine, N-[2-[[2-(Dodecylamino)Ethyl]-Amino]Ethyl]-
N-Lauryldiethylenetriaminoacetic Acid

Trade Name Mixtures:
Facteur Hydratant PH (Prod'Hyg)
Hydeoviton 5,5 N 2/059359 (Symrise)
Hydroviton 2/059353 (Symrise)

LAURYL DIETHYLENEDIAMINOGLYCINE HCl

Definition: Lauryl Diethylenediaminoglycine HCl is the hydrochloric acid salt of Lauryl Diethylenediaminoglycine (q.v.).

Information Source: MHLW-331/3

Chemical Classes: Alkyl-Substituted Amino Acids; Amines

Function: Preservative

LAURYL DIMETHICONE

Definition: Lauryl Dimethicone is the siloxane polymer that conforms generally to the formula:

Chemical Class: Siloxanes and Silanes

Function: Skin-Conditioning Agent - Occlusive

Trade Name:
Wacker - Belsil LDM 3107 VP (Wacker-Chemie)

LAURYL DIMETHICONE PEG-15 CROSS-POLYMER

Definition: Lauryl Dimethicone PEG-15 Crosspolymer is a crosslinked copolymer formed from PEG-15 and Lauryl Dimethicone (q.v.).

Chemical Class: Siloxanes and Silanes

Functions: Surfactant - Emulsifying Agent; Surfactant - Suspending Agent; Viscosity Controlling Agent

Trade Name:
KSG-30 (Shin Etsu)

LAURYL DIMETHICONE/POLYGLYCERIN-3 CROSSPOLYMER

Definition: Lauryl Dimethicone/Polyglycerin-3 Crosspolymer is a polymer of Lauryl Dimethicone (q.v.) crosslinked with diallyl polyglycerin-3.

Chemical Classes: Siloxanes and Silanes; Synthetic Polymers

Functions: Skin-Conditioning Agent - Miscellaneous; Surfactant - Cleansing Agent; Surfactant - Emulsifying Agent; Surfactant - Solubilizing Agent; Viscosity Increasing Agent - Nonaqueous

Trade Name Mixtures:
KSG-830 (Shin-Etsu Chemical Co.)
KSG-840 (Shin-Etsu Chemical Co.)

LAURYL DIMETHYLAMINE CYCLOCARBOXYPROPYLOLEATE

Empirical Formula:
$C_{35}H_{67}NO_4$

Definition: Lauryl Dimethylamine Cyclocarboxypropyloleate is the amine salt that conforms generally to the formula:

Chemical Class: Amines

Function: Hair Conditioning Agent

Technical/Other Names:
Cyclocarboxypropyloleic Acid, Lauryl Dimethylamine Salt
Lauroyl Dimethylamine Acrylinoleate
Lauryl Dimethylamine Acrylinoleate
Lauryl Dimethylamine C21-Dicarboxylate

LAURYLDIMONIUM HYDROXYPROPYL HYDROLYZED CASEIN

Definition: Lauryldimonium Hydroxypropyl Hydrolyzed Casein is the quaternary ammonium chloride that conforms generally to the formula:

$$CH_3(CH_2)_{11} - \overset{\overset{\displaystyle CH_3}{|}}{\underset{\underset{\displaystyle CH_3}{|}}{N}} - CH_2\overset{\overset{}{}}{\underset{\underset{\displaystyle OH}{|}}{C}}H CH_2 - R \Big]^+ \quad Cl^-$$

where R represents the hydrolyzed casein moiety.

Chemical Classes: Protein Derivatives; Quaternary Ammonium Compounds

Functions: Antistatic Agent; Hair Conditioning Agent; Skin-Conditioning Agent - Miscellaneous

Trade Name:
Promois Milk-LAQ (Seiwa Kasei)

LAURYLDIMONIUM HYDROXYPROPYL HYDROLYZED COLLAGEN

JPN Translation:
ラウリルジモニウムヒドロキシプロピル加水分解コラーゲン

Definition: Lauryldimonium Hydroxypropyl Hydrolyzed Collagen is the quaternary ammonium chloride that conforms generally to the formula:

$$CH_3(CH_2)_{11} - \overset{\overset{\displaystyle CH_3}{|}}{\underset{\underset{\displaystyle CH_3}{|}}{N}} - CH_2\overset{\overset{}{}}{\underset{\underset{\displaystyle OH}{|}}{C}}H CH_2 - R \Big]^+ \quad Cl^-$$

where R represents the hydrolyzed collagen moiety.

Information Sources: JCIC, JCLS

Chemical Classes: Protein Derivatives; Quaternary Ammonium Compounds

Functions: Antistatic Agent; Hair Conditioning Agent; Skin-Conditioning Agent - Miscellaneous

Reported Product Categories: Hair Dyes and Colors (All Types Requiring Caution Statements and Patch Tests); Permanent Waves

Technical/Other Names:
N-[2-Hydroxy-3-(lauryldimethylammonio)-propyl] Hydrolyzed Collagen Chloride
Lauryldimonium Hydroxypropyl Hydrolyzed Animal Protein

Trade Names:
Croquat L (Croda Chemicals)
Croquat LK (Croda Chemicals)
Promois W-32LAQ (Seiwa Kasei)

Promois W-42LAQ (Seiwa Kasei)
Promois W-32RHLAQ (Seiwa Kasei)
Quat-Coll LDMA-40 (Arch Personal Care Products)

LAURYLDIMONIUM HYDROXYPROPYL HYDROLYZED KERATIN

JPN Translation:
ラウリルジモニウムヒドロキシプロピル加水分解ケラチン

Definition: Lauryldimonium Hydroxypropyl Hydrolyzed Keratin is the quaternary ammonium chloride that conforms generally to the formula:

$$CH_3(CH_2)_{11} - \overset{\overset{\displaystyle CH_3}{|}}{\underset{\underset{\displaystyle CH_3}{|}}{N}} - CH_2\overset{\overset{}{}}{\underset{\underset{\displaystyle OH}{|}}{C}}H CH_2 - R \Big]^+ \quad Cl^-$$

where R represents the hydrolyzed keratin moiety.

Information Sources: JCIC, JCLS

Chemical Classes: Protein Derivatives; Quaternary Ammonium Compounds

Functions: Antistatic Agent; Hair Conditioning Agent; Skin-Conditioning Agent - Miscellaneous

Technical/Other Name:
N-[2-Hydroxy-3-(lauryldimethylammonio)-propyl] Hydrolyzed Keratin Chloride

Trade Names:
Croquat K (Croda Chemicals)
Promois WK-HLAQ (Seiwa Kasei)
Proticute C Gamma 12 (Ichimaru Pharcos)

LAURYLDIMONIUM HYDROXYPROPYL HYDROLYZED SILK

JPN Translation:
ラウリルジモニウムヒドロキシプロピル加水分解シルク

Definition: Lauryldimonium Hydroxypropyl Hydrolyzed Silk is the quaternary ammonium chloride that conforms generally to the formula:

$$CH_3(CH_2)_{11} - \overset{\overset{\displaystyle CH_3}{|}}{\underset{\underset{\displaystyle CH_3}{|}}{N}} - CH_2\overset{\overset{}{}}{\underset{\underset{\displaystyle OH}{|}}{C}}H CH_2 - R \Big]^+ \quad Cl^-$$

where R represents the hydrolyzed silk moiety.

Information Sources: JCIC, JCLS

Chemical Classes: Protein Derivatives; Quaternary Ammonium Compounds

Functions: Antistatic Agent; Hair Conditioning Agent; Skin-Conditioning Agent - Miscellaneous

Technical/Other Name:
N-[2-Hydroxy-3-(lauryldimethylammonio)-propyl] Hydrolyzed Silk Chloride

Trade Name:
Promois Silk-LAQ (Seiwa Kasei)

LAURYLDIMONIUM HYDROXYPROPYL HYDROLYZED SOY PROTEIN

Definition: Lauryldimonium Hydroxypropyl Hydrolyzed Soy Protein is the quaternary ammonium chloride that conforms generally to the formula:

$$CH_3(CH_2)_{11} - \overset{\overset{\displaystyle CH_3}{|}}{\underset{\underset{\displaystyle CH_3}{|}}{N}} - CH_2\overset{\overset{}{}}{\underset{\underset{\displaystyle OH}{|}}{C}}H CH_2 - R \Big]^+ \quad Cl^-$$

where R represents the hydrolyzed soy protein moiety.

Chemical Classes: Protein Derivatives; Quaternary Ammonium Compounds

Functions: Antistatic Agent; Hair Conditioning Agent; Skin-Conditioning Agent - Miscellaneous

Trade Names:
AC Quaternized Soy LD (Active Concepts)
Aqua Pro II QSL (MGP)
Croquat Soya (Croda Chemicals)
Quat-Soy LDMA-25 (Arch Personal Care Products)

LAURYL ETHYLHEXANOATE

CAS No.	EINECS No.
56078-38-7	259-982-6

Empirical Formula:
$C_{20}H_{40}O_2$

Definition: Lauryl Ethylhexanoate is the ester of lauryl alcohol and 2-ethylhexanoic acid. It conforms to the formula:

$$CH_3(CH_2)_3\overset{\overset{}{}}{\underset{\underset{\displaystyle CH_2CH_3}{|}}{C}}H \overset{\overset{\displaystyle O}{\parallel}}{C} - O(CH_2)_{11}CH_3$$

Chemical Class: Esters

Function: Skin-Conditioning Agent - Emollient

Technical/Other Names:
2-Ethylhexanoic Acid, Lauryl Ester
Hexanoic Acid, 2-ethyl-, Dodecyl Ester
Lauryl 2-Ethylhexanoate
Lauryl Octanoate

Trade Name Mixture:
DUB Liquide 1214 (Stearinerie Dubois Fils)

LAURYLGLUCONAMIDE PALMITATES

Definition: Laurylgluconamide Palmitates is the organic compound that conforms generally to the formula:

$$CH_3(CH_2)_{11}NH-\overset{\overset{O}{\|}}{C}-\overset{\overset{OR}{|}}{\underset{\underset{OR}{|}}{CH}}\overset{\overset{OR}{|}}{CH}\overset{}{\underset{\underset{OR}{|}}{CH}}CHCH_2OR$$

where R represents H or the palmitoyl grouping.

Chemical Classes: Amides; Esters

Functions: Hair Conditioning Agent; Skin-Conditioning Agent - Emollient

Trade Name:
LGA-16,16 (Nippon Chemical)

LAURYL GLUCOSIDE

CAS Nos.	EINECS No.
27836-64-2	248-685-7
110615-47-9	

JPN Translation:
ラウリルグルコシド

Definition: Lauryl Glucoside is the product obtained by the condensation of lauryl alcohol with a glucose polymer.

Chemical Class: Carbohydrates

Function: Surfactant - Cleansing Agent

Reported Product Categories: Hair Dyes and Colors (All Types Requiring Caution Statements and Patch Tests); Bubble Baths; Shampoos (Non-coloring)

Technical/Other Names:
Dodecyl D-glucoside
D-Glucopyranoside, Dodecyl
Lauryl D-glucopyranoside

Trade Names:
AEC Lauryl Glucoside (A & E Connock)
DeSULF GOS-P-60WCG (DeForest)
Plantaren 1200 (Cognis Care Chemicals/NJ)
Plantaren 1200 (Cognis Care Chemicals/PA)
Plantaren 1300 (Cognis Care Chemicals/NJ)
Plantaren 1300 (Cognis Care Chemicals/PA)

Trade Name Mixtures:
Cerasperse H (Degussa Care Specialties)
Eumulgin VL 75 (Cognis Deutschland)
Euperlan PL-1000 (Cognis Care Chemicals/PA)
Lamesoft T 120 (Cognis Care Chemicals/NJ)
Plantapon CL 30 (Cognis Care Chemicals/NJ)
Plantapon LGC (Cognis Deutschland)
Plantaren LSC (Cognis Care Chemicals/NJ)
Plantaren LSC (Cognis Care Chemicals/PA)
Plantaren PS 10 (Cognis Deutschland)
Plantaren PS-200 (Cognis Care Chemicals/NJ)
Plantaren PS-200 (Cognis Care Chemicals/PA)
Plantaren PS-400 (Cognis Care Chemicals/NJ)
Plantaren PS-400 (Cognis Care Chemicals/PA)

LAURYL GLYCOL

CAS No.	EINECS No.
1119-87-5	214-289-8

Empirical Formula:
$C_{12}H_{26}O_2$

Definition: Lauryl Glycol is the diol that conforms to the formula:

$$CH_3(CH_2)_9\underset{\underset{OH}{|}}{CH}CH_2OH$$

Chemical Class: Alcohols

Functions: Hair Conditioning Agent; Skin-Conditioning Agent - Emollient

Technical/Other Names:
1,2-Dihydroxydodecane
1,2-Dodecanediol
1,2-Dodecylene Glycol

Trade Name:
Mexanyl GU (Chimex)

LAURYL GLYCOL HYDROXYPROPYL ETHER

Empirical Formula:
$C_{15}H_{32}O_3$

Definition: Lauryl Glycol Hydroxypropyl Ether is the organic comound that conforms to the formula:

$$CH_3(CH_2)_9\underset{\underset{CH_2OH}{|}}{CH}O\underset{\underset{}{}}{CH_2}\overset{\overset{OH}{|}}{CH}CH_3$$

Chemical Class: Ethers

Functions: Surfactant - Cleansing Agent; Surfactant - Emulsifying Agent; Surfactant - Foam Booster

Technical/Other Name:
2-(2-Hydroxypropoxy)-1-Dodecanol

Trade Name:
Viscosafe LPE (Kawaken)

LAURYL HYDROXYETHYL IMIDAZOLINE

CAS No.	EINECS No.
136-99-2	205-271-0

Empirical Formula:
$C_{16}H_{32}N_2O$

Definition: Lauryl Hydroxyethyl Imidazoline is the heterocyclic compound that conforms to the formula:

Information Source: TSCA

Chemical Class: Imidazoline Compounds

Functions: Antistatic Agent; Hair Conditioning Agent

Technical/Other Name:
1H-Imidazole-1-Ethanol, 4,5-Dihydro-2-Undecyl-

Trade Names:
Mackazoline L (McIntyre)
Schercozoline L (Scher)

LAURYL HYDROXYSULTAINE

CAS No.	EINECS No.
13197-76-7	236-164-7

JPN Translation:
ラウリルヒドロキシスルタイン

Empirical Formula:
$C_{17}H_{37}NO_4S$

Definition: Lauryl Hydroxysultaine is the zwitterion (inner salt) that conforms to the formula:

$$CH_3(CH_2)_{11}-\overset{\overset{CH_3}{|}}{\underset{\underset{CH_3}{|}}{N^+}}-CH_2\overset{\overset{}{}}{\underset{\underset{OH}{|}}{CH}}CH_2SO_3^-$$

Information Sources: JCIC, JCLS

Chemical Class: Betaines

Functions: Antistatic Agent; Hair Conditioning Agent; Skin-Conditioning Agent - Miscel-

---

laneous; Surfactant - Cleansing Agent; Surfactant - Foam Booster; Viscosity Increasing Agent - Aqueous

Reported Product Category: Shampoos (Non-coloring)

Technical/Other Names:
Ammonium, dodecyl(2-hydroxy-3-sulfopropyl)dimethyl-, hydroxide, Inner Salt
1-Dodecanaminium, N-(2-hydroxy-3-sulfopropyl)-N,N-dimethyl-, Inner Salt
Lauryl Hydroxy Sulfobetaine Solution

Trade Names:
Mackam CD-117 (McIntyre)
Mackam LHS (McIntyre)
Obazolin AHS-103 (Toho)

LAURYL ISOQUINOLINIUM BROMIDE

CAS No.	EINECS No.
93-23-2	202-230-9

JPN Translation:
ラウリルイソキノリニウムブロミド

Empirical Formula:
$C_{21}H_{32}N \cdot Br$

Definition: Lauryl Isoquinolinium Bromide is the quaternary ammonium compound that conforms generally to the formula:

Information Sources: CTFA D, JCLS, JSCI, MHLW-331/3, TSCA, USAN

Chemical Classes: Heterocyclic Compounds; Quaternary Ammonium Compounds

Functions: Antistatic Agent; Cosmetic Biocide; Deodorant Agent

Technical/Other Names:
Alkylisoquinolinium Bromide Solution
2-Dodecylisoquinolinium Bromide

LAURYL ISOQUINOLINIUM SACCHARINATE

JPN Translation:
ラウリルイソキノリニウムサッカリン

Empirical Formula:
$C_{21}H_{32}N \cdot C_7H_4NO_3S$

Definition: Lauryl Isoquinolinium Saccharinate is the quaternary ammonium salt that conforms to the formula:

Information Sources: JCIC, JCLS

Chemical Classes: Heterocyclic Compounds; Organic Salts; Quaternary Ammonium Compounds

Function: Cosmetic Biocide

Technical/Other Names:
Isoquinolinium, 2-dodecyl-, Salt with 1,2-benzisothiazol-3(2H)-one 1,1-dioxide (1:1)
Laurylisoquinolinium Benzosulfimide

LAURYL ISOSTEARATE

CAS No.	EINECS No.
93803-85-1	298-359-3

Empirical Formula:
$C_{30}H_{60}O_2$

Definition: Lauryl Isostearate is the ester of lauryl alcohol and Isostearic Acid (q.v.). It conforms generally to the formula:

Chemical Class: Esters

Function: Skin-Conditioning Agent - Emollient

Technical/Other Names:
Dodecyl Isooctadecanoate
Isooctadecanoic Acid, Dodecyl Ester
Isostearic Acid, Lauryl Ester

Trade Names:
AEC Lauryl Isostearate (A & E Connock)
Isostearene L (Vevy)

LAURYL LACTATE

CAS No.	EINECS No.
6283-92-7	228-504-8

JPN Translation:
乳酸ラウリル

Empirical Formula:
$C_{15}H_{30}O_3$

Definition: Lauryl Lactate is the ester of lauryl alcohol and lactic acid. It conforms generally to the formula:

Information Sources: CIR: [SQ] IJT-17 (Suppl. 1)1998, JCIC, JCLS, JSQI, RIFM, TSCA

Chemical Class: Esters

Functions: Fragrance Ingredient; Skin-Conditioning Agent - Emollient

Reported Product Category: Powders (Dusting and Talcum, Excluding Aftershave Talcs)

Technical/Other Names:
Dodecyl 2-Hydroxypropanoate
Dodecyl lactate (RIFM)
Dodecyl Lactate
2-Hydroxypropanoic Acid, Dodecyl Ester
Lactic Acid, Dodecyl Ester
Propanoic Acid, 2-Hydroxy-, Dodecyl Ester

Trade Names:
AEC Lauryl Lactate (A & E Connock)
Ceraphyl 31 (International Specialty Products)
Cetinol LL (Fabriquimica)
Crodamol LL (Croda Chemicals)
Dermol LL (Alzo)
DUB LL (Stearinerie Dubois Fils)
Pelemol LL (Phoenix)
Schercemol LL (Scher)

Trade Name Mixture:
DUB Synersol (Stearinerie Dubois Fils)

LAURYL LAURATE

CAS No.: 13945-76-1

Empirical Formula:
$C_{24}H_{48}O_2$

Definition: Lauryl Laurate is the ester of Lauryl Alcohol (q.v.) and Lauric Acid (q.v.).

Chemical Class: Esters

Functions: Binder; Emulsion Stabilizer; Hair Conditioning Agent; Opacifying Agent; Skin-Conditioning Agent - Miscellaneous

Technical/Other Name:
Dodecanoic Acid, Dodecyl Ester

Trade Names:
ESP LL-24 (Earth Supplied Products)
Lauryl Laurate (Procter & Gamble)
Purester 24 (Strahl & Pitsch)

Trade Name Mixtures:
ESP Dry feel-Olive (Earth Supplied Products)
ESP Dry Oil-MO (Earth Supplied Products)
ESP Dry Wax hi vis (Earth Supplied Products)
ESP Dry Wax low vis (Earth Supplied Products)
ESP Dry Wax med vis (Earth Supplied Products)

LAURYL MALAMIDE

Definition: Lauryl Malamide is the organic compound that conforms to the formula:

$$HOOCCH_2CHC-NH(CH_2)_{11}CH_3$$

with O double-bonded to C and OH below the first CH.

Chemical Classes: Alkanolamides; Carboxylic Acids

Functions: Humectant; Skin-Conditioning Agent - Miscellaneous; Surfactant - Cleansing Agent

Trade Name:
MAA (Lion Corporation)

LAURYL METHACRYLATE

CAS Nos.	EINECS Nos.
142-90-5	205-570-6
93804-49-0	298-425-1

Empirical Formula:
$C_{16}H_{30}O_2$

Definition: Lauryl Methacrylate is the ester of lauryl alcohol and methacrylic acid. It conforms generally to the formula:

$$CH_2=CC-O(CH_2)_{11}CH_3$$

with O double-bonded and CH_3 below.

Information Sources: CIR: [SQ], TSCA

Chemical Class: Esters

Function: Artificial Nail Builder

Technical/Other Names:
Dodecyl Methacrylate
Dodecyl 2-Methyl-2-Propenoate
Methacrylic Acid, Dodecyl Ester
2-Methyl-2-Propenoic Acid, Dodecyl Ester
2-Propenoic Acid, 2-Methyl-, Dodecyl Ester

LAURYL METHACRYLATE/GLYCOL DIMETHACRYLATE CROSSPOLYMER

JPN Translations:
(メタクリル酸ラウリル / ジメタクリル酸
エチレングリコール) クロスポリマー
アクリル樹脂

Definition: Lauryl Methacrylate/Glycol Dimethacrylate Crosspolymer is a crosspolymer of lauryl methacrylate and ethylene glycol dimethacrylate monomers.

Information Sources: JCIC, JCLS

Chemical Class: Synthetic Polymers

Functions: Film Former; Hair Fixative

Technical/Other Names:
Lauryl Methacrylate•Ethylene Glycol Dimethacrylate Polymer

Lauryl Methacrylate/Glycol Dimethacrylate Copolymer

Trade Name:
Polytrap 6603 Adsorber (EDT, Inc.)

Trade Name Mixtures:
Polytrap 6035 Cyclomethicone (EDT, Inc.)
Polytrap 7100 Dimethicone Macrobeads (EDT, Inc.)
Polytrap 6500 Dimethicone/Petrolatum Powder (EDT, Inc.)
Polytrap 6038 Mineral Oil Macrobeads (EDT, Inc.)

LAURYL METHICONE

CAS No.: 139614-44-1

Definition: Lauryl Methicone is the siloxane polymer that conforms to the formula:

$$(CH_3)_3SiO-\left[\begin{array}{c}CH_3\\|\\SiO\\|\\(CH_2)_{11}\\|\\CH_3\end{array}\right]_x-Si(CH_3)_3$$

Chemical Class: Siloxanes and Silanes

Function: Skin-Conditioning Agent - Occlusive

Technical/Other Name:
3-Dodecyl-1,1,1,3,5,5,5-Heptamethyl-trisiloxane

Trade Names:
SilCare Silicone 41M20 Lauryl Methicone (Clariant)
SilCare Silicone 41M20 Lauryl Methicone (Clariant GmbH, Personal Care)

LAURYL METHYL GLUCETH-10 HYDROXYPROPYLDIMONIUM CHLORIDE

Definition: Lauryl Methyl Gluceth-10 Hydroxypropyldimonium Chloride is the quaternary ammonium salt prepared by the reaction of Methyl Gluceth-10 (q.v.) with a dimethyl dodecylammonium substituted epoxide.

Chemical Classes: Carbohydrates; Polyols; Quaternary Ammonium Compounds

Functions: Antistatic Agent; Hair Conditioning Agent

Reported Product Category: Tonics, Dressings, and Other Hair Grooming Aids

Trade Name:
Glucquat 125 (Amerchol)

LAURYL MYRISTATE

CAS No.	EINECS No.
2040-64-4	218-039-9

Empirical Formula:
$C_{26}H_{52}O_2$

Definition: Lauryl Myristate is the ester of lauryl alcohol and myristic acid. It conforms to the formula:

$$CH_3(CH_2)_{12}C-O(CH_2)_{11}CH_3$$

with O double-bonded to C.

Chemical Class: Esters

Functions: Hair Conditioning Agent; Skin-Conditioning Agent - Occlusive

Reported Product Category: Shampoos (Non-coloring)

Technical/Other Names:
Dodecyl Tetradecanoate
Myristic Acid, Dodecyl Ester
Tetradecanoic Acid, Dodecyl Ester

Trade Name:
AEC Lauryl Myristate (A & E Connock)

LAURYL OLEATE

CAS No.: 36078-10-1

Empirical Formula:
$C_{30}H_{58}O_2$

Definition: Lauryl Oleate is ester of lauryl alcohol and oleic acid that conforms to formula:

$$CH_3(CH_2)_7CH=CH(CH_2)_7C-O(CH_2)_{11}CH_3$$

with O double-bonded to C.

Chemical Class: Esters

Function: Skin-Conditioning Agent - Occlusive

Technical/Other Names:
Dodecyl Oleate
9-Octadecenoic Acid, Dodecyl Ester
Oleic Acid, Dodecyl Ester
Oleic Acid, Lauryl Ester

Trade Name:
Dermol CV (Fabriquimica)

LAURYL PALMITATE

CAS No.	EINECS No.
42232-29-1	255-725-7

Empirical Formula:
$C_{28}H_{56}O_2$

Definition: Lauryl Palmitate is the ester of lauryl alcohol and palmitic acid. It conforms to the formula:

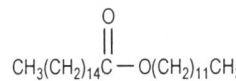

Chemical Class: Esters

Function: Skin-Conditioning Agent - Occlusive

Technical/Other Names:
Dodecyl Palmitate
Hexadecanoic Acid, Dodecyl Ester

Trade Names:
AEC Lauryl Palmitate (A & E Connock)
Palmitate De Lauryle (Prod'Hyg)

LAURYL PCA

CAS No.   **EINECS No.**
22794-26-9   245-224-1

JPN Translation:
PCA ラウリル

Empirical Formula:
$C_{17}H_{31}NO_3$

Definition: Lauryl PCA is the ester of lauryl alcohol and PCA (q.v.) that conforms to the formula:

Chemical Classes: Heterocyclic Compounds; Organic Salts

Function: Skin-Conditioning Agent - Miscellaneous

Reported Product Category: Lipsticks

Technical/Other Names:
Lauryl Pyrrolidonecarboxylate
Pyrrolidone Carboxylic Acid, Lauryl Ester

Trade Names:
AEC Lauryl PCA (A & E Connock)
Laurydone (UCIB (Solabia))

Trade Name Mixtures:
Cryolidone (UCIB (Solabia))
Waxidone (UCIB (Solabia))

LAURYL PEG/PPG-18/18 METHICONE

Definition: Lauryl PEG/PPG-18/18 Methicone is an alkoxylated derivative of Lauryl Methicone (q.v.) containing an average of 18 moles of ethylene oxide and 18 moles of propylene oxide.

Chemical Class: Siloxanes and Silanes

Functions: Skin-Conditioning Agent - Miscellaneous; Surfactant - Emulsifying Agent

Trade Name:
Dow Corning 5200 Formulation Aid (Dow Corning)

LAURYL PHOSPHATE

CAS No.   **EINECS No.**
12751-23-4   235-798-1

JPN Translation:
ラウリルリン酸

Definition: Lauryl Phosphate is the monolauryl ester of phosphoric acid.

Information Sources: JCIC, JCLS, JSQI

Chemical Class: Phosphorus Compounds

Function: Surfactant - Emulsifying Agent

Technical/Other Names:
Lauryl Monophosphate
Phosphoric Acid, Dodecyl Ester

Trade Names:
Ethfac 102 (Ethox)
Nikkol Phosten HLP (Nikko)
Phosphanol ML-200 (Toho)

LAURYL POLYGLYCERYL-6 CETEARYL GLYCOL ETHER

Definition: Lauryl Polyglyceryl-6 Cetearyl Glycol Ether is the ether of polyglycerine-6 and a mixture of cetearyl glycol and lauryl alcohol.

Chemical Classes: Ethers; Glyceryl Esters and Derivatives

Functions: Skin-Conditioning Agent - Emollient; Surfactant - Emulsifying Agent

Trade Name:
Chimexane NS (Chimex)

LAURYLPYRIDINIUM CHLORIDE

CAS No.   **EINECS No.**
104-74-5   203-232-2

JPN Translation:
ラウリルピリジニウムクロリド

Empirical Formula:
$C_{17}H_{30}N \cdot Cl$

Definition: Laurylpyridinium Chloride is the quaternary ammonium compound that conforms generally to the formula:

Information Sources: JCIC, JCLS, JSCI, JSQI, TSCA

Chemical Classes: Heterocyclic Compounds; Quaternary Ammonium Compounds

Functions: Antistatic Agent; Cosmetic Biocide; Deodorant Agent

Reported Product Categories: Hair Conditioners; Hair Rinses (Non-coloring); Skin Care Preparations, Misc.

Technical/Other Names:
1-Dodecylpyridinium Chloride
Lauryl Pyridinium Chloride Solution
Lauryl Pyridium Chloride
Pyridinium, 1-Dodecyl-, Chloride

Trade Name Mixture:
Zeinquat (Variati)

LAURYL PYRROLIDONE

CAS No.: 2687-96-9

Empirical Formula:
$C_{16}H_{31}NO$

Definition: Lauryl Pyrrolidone is the substituted heterocyclic organic compound that conforms to the formula:

Chemical Class: Heterocyclic Compounds

Functions: Hair Conditioning Agent; Surfactant - Cleansing Agent

Reported Product Categories: Shampoos (Non-coloring); Hair Conditioners

Technical/Other Names:
1-Dodecyl-2-Pyrrolidinone
2-Pyrrolidinone, 1-Dodecyl-

Trade Name:
Surfadone LP-300 (International Specialty Products)

LAURYL STEARATE

CAS No.   **EINECS No.**
5303-25-3   226-150-9

Empirical Formula:
$C_{30}H_{60}O_2$

Definition: Lauryl Stearate is the ester of lauryl alcohol and stearic acid. It conforms to the formula:

Information Source: TSCA

Chemical Class: Esters

Function: Skin-Conditioning Agent - Occlusive

Technical/Other Names:
Dodecyldimethyl(3-sulfopropyl)ammonium hydroxide Inner Salt
Dodecyl Octadecanoate
Octadecanoic Acid, Dodecyl Ester
Stearic Acid, Dodecyl Ester

Trade Name:
AEC Lauryl Stearate (A & E Connock)

LAURYL SULTAINE

CAS Nos.
14933-08-5
52667-78-4

EINECS No.
239-002-3

Empirical Formula:
$C_{17}H_{37}NO_3S$

Definition: Lauryl Sultaine is the zwitterion (inner salt) that conforms generally to the formula:

$$CH_3(CH_2)_{11}\overset{\overset{\displaystyle CH_3}{|}}{\underset{\underset{\displaystyle CH_3}{|}}{N^+}}(CH_2)_3SO_3^-$$

Chemical Class: Betaines

Functions: Antistatic Agent; Hair Conditioning Agent; Skin-Conditioning Agent - Miscellaneous; Surfactant - Cleansing Agent; Surfactant - Foam Booster; Viscosity Increasing Agent - Aqueous

Technical/Other Names:
N,N-Dimethyl-N-(3-Sulfopropyl)-1-Decanaminium Hydroxide, Inner Salt
N,N-Dimethyl-N-(3-Sulfopropyl)-1-Dodecanaminium Hydroxide, Inner Salt
1-Dodecanaminium, N,N-Dimethyl-N- (3-Sulfopropyl)-, Hydroxide, Inner Salt
Lauryldimethylsulfobetain
Lauryl Sulfobetaine

LAVANDULA ANGUSTIFOLIA (LAVENDER) EXTRACT

CAS Nos.
84776-65-8
90063-37-9

EINECS Nos.
283-994-0
289-995-2

JPN Translation:
ラベンダーエキス

Definition: Lavandula Angustifolia (Lavender) Extract is an extract of the aerial parts of *Lavandula angustifolia*. See *"Regulatory and Ingredient Use Information,"* regarding the labeling names for botanical ingredients in Volume 1, Introduction, Part A.

Information Sources: 21CFR182.20, JCIC, JCLS, RIFM

Chemical Class: Biological Products

Functions: Fragrance Ingredient; Skin-Conditioning Agent - Miscellaneous; Skin-Conditioning Agent - Occlusive

Reported Product Categories: Bath Soaps and Detergents; Shampoos (Non-coloring); Tonics, Dressings, and Other Hair Grooming Aids; Bath Preparations, Misc.; Hair Conditioners

Technical/Other Names:
Extract of Lavandula Angustifolia
Extract of Lavender
Lavandula Angustifolia Extract
Lavandula Officinalis Extract
Lavandula Vera Extract
Lavender Extract
Lavender Extract (1)
Lavender Extract (2)
Lavender, Lavandula angustifolia angustifolia, ext. (RIFM)
Lavender, Lavandula angustifolia, ext. (RIFM)

Trade Names:
Hydroessential Lavandula (Vevy)
Lavendin Extract (Robertet, Inc.)

Trade Name Mixtures:
Actiphyte of Lavender BG50 (Active Organics)
Actiphyte of Lavender GL50 (Active Organics)
Actiphyte of Lavender Lipo S (Active Organics)
Actiphyte of Lavender PG50 (Active Organics)
Aromaphyte of Lavender (Active Organics)
BBC Relaxing Complex (Bio-Botanica)
Bio-Chelated Derma Plex II (Bio-Botanica)
Bio-Chelated Sauna-Derm II (Bio-Botanica)
Cosflor Spike Lavender HGS (A & E Connock)
Extrait De Lavande PP PG40 (Yves Rocher)
Extrait Hydroglycolique de Lavande (Greentech)
Extrapone Lavender 2/033080 (Symrise)
Herbaliquid Lavender Special (Crodarom)
Herbal Vinegar (Provital/Centerchem)
Herbasec Lavender (Cosmetochem) (Cosmetochem International Ltd.)
Herbasol Complex "Herbes de Provence" (Cosmetochem) (Cosmetochem International Ltd.)
Herbasol-Extract Lavender (Cosmetochem)
Lavender Extract (Maruzen Pharmaceuticals Co., Ltd.)
Lavender Extract BG (Maruzen Pharmaceuticals Co., Ltd.)

Lavender Extract BG-J (Maruzen Pharmaceuticals Co., Ltd.)
Lavender Extract HG (Provital/Centerchem)
Lavender Extract LA (Maruzen Pharmaceuticals Co., Ltd.)
Lavender Flowers Extract HS 2565 G (Grau)
Lavender HS (Alban Muller)
Lavender LS (Alban Muller)
Lavender Milk J (Alban Muller)
Lavender Tincture (Rahn)
Nutriplant (Bio-Botanica)
Phytelene of Lavender EG 503 Liquid (Indena SA)
Phytoderm P/25 Hydroalcoholic (Universal Flavors)
Phytogreen 55 of Lavender EXH 706 Liquid (Phytochim)
Prodhy Extract Lavande (Prod'Hyg)
Spike Lavender Extract (Cosmetic Developments)
280 Stimulant HS (Alban Muller)
680 Stimulant LS (Alban Muller)
Vegebios of Lavender (CEP (Solabia))
2350 Vege-Plex Body Complex (Vege-Tech)
Vege Plex VP#1334 (Vege-Tech)
Vegetol Lavender MCF 1484 Hydro (Gattefosse s.a.)
VT-028 Extract of Lavender (Vege-Tech)

LAVANDULA ANGUSTIFOLIA (LAVENDER) FLOWER EXTRACT

Definition: Lavandula Angustifolia (Lavender) Flower Extract is an extract of the flower of the lavender, *Lavandula angustifolia*. See *"Regulatory and Ingredient Use Information,"* regarding the labeling names for botanical ingredients in Volume 1, Introduction, Part A.

Chemical Class: Biological Products

Function: Fragrance Ingredient

Technical/Other Name:
Lavender Flower Extract

Trade Names:
AEC Lavender Tincture (A & E Connock)
Lavender Flower (Kansai Koso)

Trade Name Mixtures:
Aroma Lavender B (Ichimaru Pharcos)
Biopein (Bio-Botanica)
Lavandula Angustifolia Essential Water (Bioland)
Lavender Extract BG-15 (Maruzen Pharmaceuticals Co., Ltd.)
Lavender Extract SQ (Maruzen Pharmaceuticals Co., Ltd.)
Lavender Liquid B (Ichimaru Pharcos)
Premier Lavender Flower 10% Extract (Premier Specialties)

LAVANDULA ANGUSTIFOLIA (LAVENDER) FLOWER POWDER

Definition: Lavandula Angustifolia (Lavender) Flower Powder is the crushed, dried flowers, *Lavandula angustifolia.* See *"Regulatory and Ingredient Use Information,"* regarding the labeling names for botanical ingredients in Volume 1, Introduction, Part A.

Information Source: JCLS

Chemical Class: Biological Products

Function: Fragrance Ingredient

LAVANDULA ANGUSTIFOLIA (LAVENDER) FLOWER WATER

JPN Translation:
ラベンダー水

Definition: Lavandula Angustifolia (Lavender) Flower Water is an aqueous solution of the steam distillate obtained from the flowers of *Lavandula angustifolia.* See *"Regulatory and Ingredient Use Information,"* regarding the labeling names for botanical ingredients in Volume 1, Introduction, Part A.

Information Sources: JCIC, JCLS

Chemical Class: Biological Products

Functions: Fragrance Ingredient; Skin-Conditioning Agent - Miscellaneous

Technical/Other Names:
Lavandula Vera Water
Lavender Flower Water
Lavender Water

Trade Names:
AEC Lavender Water (A & E Connock)
Extrait Orig=inel Lavande (Gattefosse s.a.)
Lavender Ecoconcentrate Natural (Robertet S.A.)
Lavender Water (Alban Muller)
Lavender Water (Maruzen Pharmaceuticals Co., Ltd.)

Trade Name Mixtures:
Aroma Lavender B (Ichimaru Pharcos)
Eau de Aroma Lavender (Ichimaru Pharcos)
Lavender Water K-A (Koei Perfumery)

LAVANDULA ANGUSTIFOLIA (LAVENDER) FLOWER WAX

Definition: Lavandula Angustifolia (Lavender) Flower Wax is a wax obtained from the flower of *Lavandula angustifolia.* See *"Regulatory and Ingredient Use Information,"* regarding the labeling names for botanical ingredients in Volume 1, Intro-duction, Part A. See Reported Ingredient Functions-The Cosmetic Drug Distinction, in Regulatory and Ingredient Use Information, Volume I, Part A.

Chemical Class: Waxes

Functions: Skin-Conditioning Agent - Emollient; Skin Protectant

Technical/Other Names:
Lavender Flower Wax
Lavender Wax

Trade Names:
AEC Cire Essentielle De Fleurs De Lavande (A & E Connock)
AEC Lavande Cire Essentielle (A & E Connock)
Cire Essentielle de fleurs de Lavande (Bertin)

LAVANDULA ANGUSTIFOLIA (LAVENDER) OIL

CAS No.: 8000-28-0

JPN Translation:
ラベンダー油

Definition: Lavandula Angustifolia (Lavender) Oil is the volatile oil obtained from *Lavendula officinalis.* See *"Regulatory and Ingredient Use Information,"* regarding the labeling names for botanical ingredients in Volume 1, Introduction, Part A.

Information Sources: AUS, BEL, BPC, BRA, 21CFR182.20, 27CFR21.65, 27CFR21.151, CZE, DA, EGY, FCC, FIN, HUN, JAN, JCIC, JCLS, MAR, MI-13(6848), NF XV, PF, PN, POR, RIFM, ROM, SNPF, TSCA, USD, USSR, YUG

Chemical Class: Essential Oils

Functions: Fragrance Ingredient; Skin-Conditioning Agent - Miscellaneous

Reported Product Categories: Skin Care Preparations, Misc.; Bath Preparations, Misc.; Body and Hand Preparations (Excluding Shaving Preparations); Bath Capsules; Moisturizing Preparations; Bath Oils, Tablets, and Salts; Cleansing Products (Cold Creams, Cleansing Lotions, Liquids and Pads); Face and Neck Preparations (Excluding Shaving Preparations); Paste Masks (Mud Packs); Shampoos (Noncoloring); Skin Fresheners; Bath Soaps and Detergents; Hair Conditioners; Suntan Gels, Creams, and Liquids; Tonics, Dressings, and Other Hair Grooming Aids; Foot Powders and Sprays; Night Skin Care Preparations; Suntan Preparations, Misc.

Technical/Other Names:
Lavandula Officinalis Oil

Lavender absolute (Lavandula officinalis Chaix) (RIFM)
Lavender Flowers Oil
Lavender (Lavandula officinalis Chaix) (RIFM)
Lavender Oil
Lavender oil (Lavandula officinalis Chaix) (RIFM)
Oil of Lavender

Trade Names:
AEC Lavender Oil (A & E Connock)
Custosense LAV (lavender oil) (Custom Ingredients)

Trade Name Mixtures:
Aromaphyte of Lavender (Active Organics)
Essentiaderm n.1 (Universal Flavors)
Essentiaderm n.2 (Universal Flavors)
Essentiaderm n.3 (Universal Flavors)
Essentiaderm n.5 (Universal Flavors)
Essentiaderm n.6 (Universal Flavors)
Essentiaderm n.7 (Universal Flavors)
Essentiaderm n.8 (Universal Flavors)
Essentiaderm n.9 (Universal Flavors)
Essentiaderm N.14 (Universal Flavors)
Essentiaderm N.19 (Universal Flavors)
Essentiaderm N.20 (Universal Flavors)
Essentiaderm N.21 (Universal Flavors)
Lavender Milk J (Alban Muller)
Lavender Softcream (CEP (Solabia))
Relaxing Phytospa (Alban Muller)
Relaxing Phytospa NaCl (Alban Muller)

LAVANDULA ANGUSTIFOLIA (LAVENDER) WATER

Definition: Lavandula Angustifolia (Lavender) Water is an aqueous solution of the steam distillate obtained from *Lavandula angustifolia.* See *"Regulatory and Ingredient Use Information,"* regarding the labeling names for botanical ingredients in Volume 1, Introduction, Part A.

Chemical Class: Biological Products

Function: Fragrance Ingredient

Trade Name:
Lavender Hydroflorate (Bayliss Ranch)

LAVANDULA HYBRIDA EXTRACT

Definition: Lavandula Hybrida Extract is an extract of *Lavandula hybrida.* See *"Regulatory and Ingredient Use Information,"* regarding the labeling names for botanical ingredients in Volume 1, Introduction, Part A.

Information Source: RIFM

Chemical Class: Biological Products

Function: Fragrance Ingredient

Technical/Other Names:
Extract of Lavandin
Extract of Lavandula Hybrida
Lavandin (Lavandula Hybrida) Extract
Lavandula Latifolia Extract
Lavandula Spica Extract
Lavender, Lavandula spica, ext. (RIFM)

Trade Names:
Lavandin Ecoconcentrate (Robertet, Inc.)
Lavandin Ecoconcentrate Natural (Robertet S.A.)

LAVANDULA HYBRIDA OIL

CAS No.: 8022-15-9

Definition: Lavandula Hybrida Oil is the essential oil obtained from *Lavandula hybrida.* See *"Regulatory and Ingredient Use Information,"* regarding the labeling names for botanical ingredients in Volume 1, Introduction, Part A.

Information Source: RIFM

Chemical Class: Essential Oils

Function: Fragrance Ingredient

Technical/Other Names:
AEC Lavandin Oil
Lavandin (Lavandula Hybrida) Oil
Lavandin Oil
Lavandin oil (Lavandula hybrida) (RIFM)
Oil of Lavandin
Oils, Lavandin

Trade Names:
AEC Spike Lavender Oil (A & E Connock)
Lavandin abrialis Oil (Charabot)

Trade Name Mixtures:
Covazen Detox (LCW)
Covazen Relax (LCW)
V-Tonic (Gattefosse s.a.)

LAVANDULA SPICA (LAVENDER) EXTRACT

Definition: Lavandula Spica (Lavender) Extract is an extract of the aerial parts of *Lavandula spica.* See *"Regulatory and Ingredient Use Information,"* regarding the labeling names for botanical ingredients in Volume 1, Introduction, Part A.

Chemical Class: Biological Products

Functions: Fragrance Ingredient; Skin-Conditioning Agent - Miscellaneous

Reported Product Categories: Bath Soaps and Detergents; Shampoos (Non-coloring); Tonics, Dressings, and Other Hair Grooming Aids; Bath Preparations, Misc.; Hair Conditioners

Technical/Other Name:
Lavender Extract

LAVANDULA STOECHAS EXTRACT

Definition: Lavandula Stoechas Extract is an extract of *Lavandula stoechas.* See *"Regulatory and Ingredient Use Information,"* regarding the labeling names for botanical ingredients in Volume 1, Introduction, Part A.

Chemical Class: Biological Products

Function: Fragrance Ingredient

Technical/Other Name:
Extract of Lavandula Stoechas

Trade Name Mixtures:
Areaumat Lavanda (Codif)
Areaumat Lavanda Glycerine (Codif)
Aroleat Lavanda (Codif)
Lavandula stoechas HG (Codif)

LAWSONE

CAS No.	EINECS No.
83-72-7	201-496-3

Empirical Formula:
$C_{10}H_6O_3$

Definition: Lawsone is the substituted naphthoquinone that conforms to the formula:

See Reported Ingredient Functions-The Cosmetic Drug Distinction, in Regulatory and Ingredient Use Information, Volume I, Part A.

Information Sources: MI-13(5410), OTC-I-SU, TSCA

Chemical Classes: Color Additives - Hair; Phenols

Functions: Colorant; Sunscreen Agent

Technical/Other Names:
2-Hydroxy-1,4-Naphthalenedione
2-Hydroxy-1,4-naphthoquinone
2-Hydroxy-1,4-Napthoquinone
1,4-Naphthalenedione, 2-Hydroxy-

Trade Name:
Imexine OG (Chimex)

LAWSONIA INERMIS (HENNA) EXTRACT

CAS No.	EINECS No.
84929-30-6	284-514-2

Definition: Lawsonia Inermis (Henna) Extract is an extract of the dried flowers, fruit and leaves of the henna, *Lawsonia inermis.* See *"Regulatory and Ingredient Use Information,"* regarding the labeling names for botanical ingredients in Volume 1, Introduction, Part A.

Chemical Class: Biological Products

Function: Skin-Conditioning Agent - Miscellaneous

Reported Product Categories: Shampoos (Non-coloring); Hair Conditioners; Hair Sprays (Aerosol Fixatives); Tonics, Dressings, and Other Hair Grooming Aids

Technical/Other Names:
Colorless Henna
Extract of Henna
Extract of Lawsonia Inermis
Henna Extract
Lawsonia Alba Extract
Lawsonia Inermis Extract
Neutral Henna

Trade Names:
Bio-Chelated Neutral Henna Extract (Bio-Botanica)
Phytelene of Henna Leaf EN 413 powder (Indena SA)

Trade Name Mixtures:
Actiphyte of Black Henna BG50 (Active Organics)
Actiphyte of Black Henna GL50 (Active Organics)
Actiphyte of Black Henna Lipo S (Active Organics)
Actiphyte of Black Henna PG50 (Active Organics)
Actiphyte of Neutral Henna BG50 (Active Organics)
Actiphyte of Neutral Henna GL50 (Active Organics)
Actiphyte of Neutral Henna Lipo S (Active Organics)
Actiphyte of Neutral Henna PG50 (Active Organics)
Actiphyte of Red Henna BG50 (Active Organics)
Actiphyte of Red Henna GL50 (Active Organics)
Actiphyte of Red Henna Lipo S (Active Organics)
Actiphyte of Red Henna PG50 (Active Organics)
Bio-Chelated Neutral Henna Plus I (Bio-Botanica)
Bio-Chelated Neutral Henna Plus II (Bio-Botanica)
Bio-Chelated Nutra Plant Complex I (Bio-Botanica)
Bio-Chelated Nutra Plant Complex II (Bio-Botanica)
Bio-Dandra Plex (Bio-Botanica)

Complex Henna (Fabriquimica)
Cremogen Henna neutral (730399)
(Haarmann & Reimer GmbH)
Extrait de Henne PE 100 (Yves Rocher)
Extrapone Henna Special 2/032930
(Symrise)
Fitopur B (Sederma)
Glycolysat of Natural Henna (CEP
(Solabia))
Henna Concentrate Black Coloring HS
2687 G (Grau)
Henna Concentrate Red Coloring HS 2686
G (Grau)
Henna Extract (Maruzen Pharmaceuticals
Co., Ltd.)
Henna Extract AL (Maruzen
Pharmaceuticals Co., Ltd.)
Henna Extract BG (Maruzen
Pharmaceuticals Co., Ltd.)
HENNA EXTRACT BG01 (Maruzen
Pharmaceuticals Co., Ltd.)
Henna Extract BGC (Maruzen
Pharmaceuticals Co., Ltd.)
Henna Extract ET-BG (Maruzen
Pharmaceuticals Co., Ltd.)
Henna Extract EV (Maruzen
Pharmaceuticals Co., Ltd.)
Henna Extract HG (Provital/Centerchem)
Henna Extract HS 2448 G Red Coloring
(Grau)
Henna Extract LA (Maruzen
Pharmaceuticals Co., Ltd.)
Henna HS (Alban Muller)
Herbasec Henna (Cosmetochem)
(Cosmetochem International Ltd.)
Herbasol Distillate Henna (Cosmetochem)
(Cosmetochem International Ltd.)
Herbasol-Extract Henna (Cosmetochem)
Herbasol Extract Oil Soluble Henna
(Cosmetochem) (Cosmetochem
International Ltd.)
Hydroplastidine Lawsonia (Vevy)
Phytelene of Henna Leaf EG 405 liquid
(Indena SA)
Phytogreen 55 of Henna Leaf EXH 666
Liquid (Phytochim)
RED-BROWN HAIR (Greentech S.A)
Tensoplex Henna (Fabriquimica)
Vegebios of Natural Henna (CEP (Solabia))
2100 Vege-Plex Hair Complex (Vege-Tech)
2210 Vege-Plex Hair Complex (Vege-Tech)
2235 Vege-Plex Hair Complex (Vege-Tech)
2230 Vege-Plex Hair Complex Conditioner
(Vege-Tech)
2240 Vege-Plex Hair Complex Conditioner
(Vege-Tech)
2110 Vege-Plex Hair Complex Shampoo
(Vege-Tech)
2120 Vege-Plex Hair Complex Shampoo
(Vege-Tech)
Vegetol Henna MCF 1232 Hydro
(Gattefosse s.a.)
VT-083 Extract of Black Henna (Vege-
Tech)
VT-021 Extract of Henna (Vege-Tech)

LAWSONIA INERMIS WAX

Definition: Lawsonia Inermis Wax is the wax obtained from the leaf of the henna, *Lawsonia inermis. See "Regulatory and Ingredient Use Information," regarding the labeling names for botanical ingredients in Volume 1, Introduction, Part A.*

Chemical Class: Waxes

Function: Hair Conditioning Agent

Technical/Other Name:
Waxes, Lawsonia Inermis

LEAD ACETATE

CAS Nos.	EINECS Nos.
301-04-2	206-104-4
15347-57-6	239-379-4

Empirical Formula:
$C_2H_4O_2 \cdot \frac{1}{2}Pb$

Definition: Lead Acetate is the inorganic salt that conforms to the formula:

$$\left[CH_3COO \right]_2 Pb$$

See "Regulatory and Ingredient Use Information," for Colorants in Volume 1, Introduction, Part A. To identify the colorant meeting the requirements for labeling purposes in the US, the INCI Name Lead Acetate must be used.

Information Sources: ARG, AUS, BPC, BRA, 21CFR73.2396, 21CFR184.1333, CZE, DDR, EEC(III/1-55), EGY, FI, FIN, IND, ITA, MAR, MI-13(5415), PF, POR, ROM, TSCA, YUG

Chemical Classes: Color Additives - Exempt from Batch Certification by the U.S. Food and Drug Administration; Color Additives - Hair; Organic Salts

Function: Hair Colorant

Reported Product Category: Hair Coloring Preparations, Misc.

Technical/Other Names:
Acetic Acid, Lead Salt
Lead Diacetate
Plumbous Acetate

Trade Name:
Unichem PBA (Universal Preserv-A-Chem)

LECITHIN

CAS Nos.	EINECS Nos.
8002-43-5	232-307-2
8030-76-0	
93685-90-6	297-639-2

JPN Translation:
レシチン

Definition: Lecithin is a naturally occuring mixture of the diglycerides of stearic, palmitic and oleic acids, linked to the choline ester of phosphoric acid. It is found in living plants and animals.

Information Sources: ARG, AUS, 21CFR133.169, 21CFR133.173, 21CFR133.179, 21CFR136.110, 21CFR136.115, 21CFR136.130, 21CFR136.160, 21CFR136.180, 21CFR163.123, 21CFR163.130, 21CFR163.135, 21CFR163.140, 21CFR163.145, 21CFR163.150, 21CFR163.155, 21CFR166.40, 21CFR166.110, 21CFR169.115, 21CFR169.140, 21CFR169.150, 21CFR175.300, 21CFR176.170, 21CFR176.200, 21CFR184.1063, 21CFR184.1400, 21CFR310.545, 21CFR582.1400, CIR: [SQ] IJT-20(SUPPL. 1)2001, CTFA D, FCC, JCIC, JCLS, JSCI, JSQI, MAR, MI-13(5447), NF XIX, POR, TSCA, USAN

Chemical Classes: Glyceryl Esters and Derivatives; Phosphorus Compounds

Functions: Skin-Conditioning Agent - Miscellaneous; Surfactant - Emulsifying Agent

Reported Product Categories: Foundations; Eye Shadows; Moisturizing Preparations; Makeup Bases; Lipsticks; Hair Conditioners; Shampoos (Non-coloring); Skin Care Preparations, Misc.; Bath Preparations, Misc.; Body and Hand Preparations (Excluding Shaving Preparations); Night Skin Care Preparations; Mascara; Blushers (All types); Makeup Preparations (Not eye), Misc.; Face Powders; Tonics, Dressings, and Other Hair Grooming Aids; Bath Capsules; Hair Sprays (Aerosol Fixatives); Face and Neck Preparations (Excluding Shaving Preparations); Paste Masks (Mud Packs); Eye Makeup Preparations, Misc.; Bath Oils, Tablets, and Salts; Cleansing Products (Cold Creams, Cleansing Lotions, Liquids and Pads); Eyeliners; Hair Preparations (Non-coloring), Misc.; Permanent Waves; Eye Lotions; Rouges; Shaving Preparations, Misc.; Skin Fresheners; Eyebrow Pencils; Fragrance Preparations, Misc.; Makeup Fixatives

Technical/Other Names:
Egg Yolk Lecithin
Lecithins, Egg Yolk
Lecithin, Soybean
Soybean Phospholipid

Trade Names:
Actiflo (Central Soya)
AEC Lecithin Liquid (A & E Connock)
AEC Lecithin Powder (A & E Connock)

Alcolec (American Lecithin)
Alcolec BS (American Lecithin)
Alcolec F-100 (American Lecithin)
Alcolec Granules (American Lecithin)
Alcolec PG (American Lecithin)
Alcolec S (American Lecithin)
Amisol 329 (Lucas Meyer GmbH)
Augon 1000 (Vevy)
Basis LP-20 (Nisshin OilliO)
Basis LS-60 (Nisshin OilliO)
Centrocap (Central Soya)
Centrol (Central Soya)
Centrolex (Central Soya)
Centromix (Central Soya)
Centrophase (Central Soya)
Centrophil (Central Soya)
Egg Yolk Lecithin Pl-100E (Q.P.)
Emulmetik 100 (Lucas Meyer GmbH)
Emulmetik 300 (Lucas Meyer GmbH)
Emulmetik 900 (Lucas Meyer GmbH)
Emulmetik 970 (Lucas Meyer GmbH)
Lecithin-Extract KOSMAFLOR, Water-Dispersible (Crodarom)
Lecsoy (Fabriquimica)
Lexin K (American Lecithin)
Lipoid E 80 (Lipoid)
Lipoid S 20 (Lipoid)
Lipoid S 40 (Lipoid)
Lipoid S 75 (Lipoid)
Phospholipon 25G (Phospholipid)
Phospholipon 25P (Phospholipid)
Precept 8140 (Central Soya)
RPF Complex (Greentech)
Unilec DS (Universal Preserv-A-Chem)
Unilec S (Universal Preserv-A-Chem)
Unilec SH (Universal Preserv-A-Chem)
Unilec-WD (Universal Preserv-A-Chem)

Trade Name Mixtures:
Aceromine (Greentech)
Actiprime 100 (Active Organics)
A.F.R./Veg. (Laboratoires Serobiologiques)
Amisol 110 (Lucas Meyer)
Amisol 634 (Lucas Meyer GmbH)
Amisol 638 (Lucas Meyer GmbH)
Amisol 688 (Lucas Meyer GmbH)
Amisol 4135 (Lucas Meyer GmbH)
Amisol 641-A (Lucas Meyer GmbH)
Amisol HS-2 (Lucas Meyer GmbH)
Amisol HS-3 (Lucas Meyer GmbH)
Amisol HS-6 (Lucas Meyer GmbH)
Amisol HS3US (Lucas Meyer)
Amisol MS-10 (Lucas Meyer GmbH)
Amisol 406-N (Lucas Meyer GmbH)
Antioxidant G-2 (Provital/Centerchem)
Arbusomes (Alban Muller)
Azelosome (Coletica SA)
Bee's Milk (Koster Keunen)
Bellsilk TL-LT (Ichimaru Pharcos)
Biophilic S (Lucas Meyer GmbH)
Black Iron Oxide PP (Cardre)
Brookosome ACEBC Concentrate (Arch Personal Care Products)
Brookosome SOD (Arch Personal Care Products)
Champagene Complex (Greentech)

Champagne Truffle Complex (Greentech)
Clarisome (Coletica SA)
Collagen Stimulation Factor MAP (Cosmetochem) (Cosmetochem International Ltd.)
Complex 3C (Vincience)
Controx VP (Cognis Deutschland)
Crodasome A/E (Croda, Inc.)
Crodasome UV-A/B (Croda, Inc.)
Depigmentation Factor 2U (Cosmetochem) (Cosmetochem International Ltd.)
Dermaceride (Sederma)
Dermasome-A (ChemMark)
Dermasome-E (ChemMark)
Dermasome-P (ChemMark)
Dermasome-S (ChemMark)
Dermasome-U (ChemMark)
Dermasome-V (ChemMark)
E-15 Aescufos (Va Ma Farmacosmetica S.R.L)
Emulmetik 110 (Lucas Meyer GmbH)
Emulmetik 310 (Lucas Meyer GmbH)
Emulmetik 910 (Lucas Meyer GmbH)
Emulmetik 920 (Lucas Meyer GmbH)
FRS-Diffuser Microreservoir (Sederma)
Glycosphere-GT (Kobo)
Glycosphere-PCO (Kobo SA)
Greenosome Ace (Greentech)
Greenosome Smart M (Greentech)
Guardian GP (Earth Supplied Products)
Happyness Complex I (Greentech)
Hydro-Diffuser Microreservoir (Sederma)
Hydro MLK (Vama Farmacosmetica)
Isocell Care (Lucas Meyer)
Isocell Citrus (Lucas Meyer)
Isocell Life (Lucas Meyer)
Isocell OFT (Lucas Meyer)
Isocell Slim (Lucas Meyer)
Kalixide Grassa (Vevy)
Lactohydrol (I.D. bio)
Lanapene (Lanaetex)
Lecithin water-dispersible CLR (CLR)
Lecsoy S (Fabriquimica)
Lifidrem COSTLJVE (Coletica SA)
Lipobelle Soyaglycone (Mibelle AG)
Lipoderma - Shield LT (Lipo)
Lipodermol Veg (Laboratoires Sero-biologiques)
Lipoid Liposome 0003 (Lipoid)
Lipoid Liposome 0040 (Lipoid)
Lipoid Liposomes 0001 (Lipoid)
Lipopepso (I.D. bio)
Lipophos (Vevy)
Liposome Concentrate (Cosmetochem) (Cosmetochem International Ltd.)
Liposome DTO (Sederma)
Liposome/Oxylho3 (Sederma)
Liposomes Anti-Age Veg (Laboratoires Serobiologiques)
Liposomes Trichogen Veg (Laboratoires Serobiologiques)
Lipotrophyne-A (Vevy)
Lipotrophyne-M (Vevy)
Lyc-O-Mato 2-4%SG (LycoRed USA)

Magilyne (Greentech)
Medullenna (Ennagram)
Melaclear (Sederma)
Membranol PC-30 (Formula One Sciences)
Membranol PC-35 (Formula One Sciences)
Merospheres (Barnet)
Mica PP (Cardre)
Microreservoir Capigen (Sederma)
Nanocos A-73 (Induchem)
Nanocos E-72 (Induchem)
Nikkol Nikkolipid 81S (Nikko)
N.S.L.E. (Sederma)
Nucleolys (Greentech)
Nutri-Diffuser Microreservoir (Sederma)
Oleo-Coll LP (Arch Personal Care Products)
Oleo-Coll LP/LF (Arch Personal Care Products)
Oryza Ceramide-LC (Ikeda)
Oryza Gamma Milky (Ichimaru Pharcos)
Oryza Tocotrienol-L (Oryza Oil)
Oxynet AP (Merck KGaA/EMD Chemicals Inc.)
Oxynex LM (Merck KGaA/EMD Chemicals Inc.)
Oxysomes (Barnet)
Papaya Complex (Greentech)
Phosal 53MCT (Phospholipid)
Phosal 50PG (Phospholipid)
Phosal 75 SA (Phospholipid)
Phosal 35SB (Phospholipid)
Photosomes (Barnet)
Phytomelanin (Dr. Carlo Ghisalberti)
Phyto Sphingoceramid (New Standard)
Phytotal AI (Phybiotex/Sederma)
Phytotal AW (Phybiotex/Sederma)
Phytotal FM (Phybiotex/Sederma)
Phytotal MS (Phybiotex/Sederma)
Phytotal OS (Phybiotex/Sederma)
Phytotal RS (Phybiotex/Sederma)
Phytotal SL (Phybiotex/Sederma)
Phytotal VT (Phybiotex/Sederma)
PL-Troxe (Va Ma Farmacosmetica S.R.L)
Polyphenols de cacao (Greentech)
Pongamia Complex (Greentech)
ProLipid 141 (International Specialty Products)
Protachem 35A (Protameen)
Protelin L (Greentech)
Proto-Lan 8 (Maybrook)
Red Iron Oxide PP (Cardre)
Retimine II (Greentech)
Retimine III (Greentech)
RonaCare ASC III (Merck KGaA)
RonaCare ASC III (Merck KGaA/EMD Chemicals Inc.)
RonaCare VTA (Merck KGaA)
Rosamine (Greentech)
Rovisome AA (Rovi)
Rovisome ACE (Rovi)
Rovisome H A (Rovi)
Rovisome AHA - Citric Acid (Rovi)
Rovisome AHA - Glycolic Acid (Rovi)
Rovisome AHA - Lactic Acid (Rovi)

Rovisome AHA - Malic Acid (Rovi)
Rovisome C (Rovi)
Rovisome Caffeine (Rovi)
Rovisome DHA (Rovi)
Rovisome Melanin (Rovi)
Rovisome Retinol Moist (Rovi)
Rovisome Whitening (Rovi)
Rovisome H Whitening (Rovi)
Salicysome (Coletica SA)
Seborilys (Greentech)
Self Tanning Complex (Greentech)
Self-Tanning Liposomes Concentrate
   (Crodarom)
Sericite PP (Cardre)
Slimisome Caffeine (Coletica SA)
Slimisome Carnitine (Coletica SA)
Slimisome Esculoside (Coletica SA)
Slimming Factor "T" (Cosmetochem)
   (Cosmetochem International Ltd.)
Spherobiol (I.D. bio)
Sphingosome AL/VEG (Laboratoires Sero-
   biologiques)
Stabolec C (Lucas Meyer GmbH)
Supra Molecular Bio Vector (SMBV) (Kobo
   SA)
Sveltine (Coletica SA)
Talc PP (Cardre)
Tensami 1/05 (Alban Muller)
Titanium Dioxide PP (Cardre)
Toshiki SP-565-41 Red (Nikko)
Transomes Anti-Age EGX 246-TR (Indena
   SA)
Transomes of Factor of MIcro-circulation
   No 3 (Indena SA)
Transomes of Factor of Micro-circulation
   No 5 (Indena SA)
Ultracolor Blue 1 (Ultra Chemical)
Ultracolor I.O. Black (Ultra Chemical)
Ultracolor I.O. Red (Ultra Chemical)
Ultracolor I.O. Yellow (Ultra Chemical)
Ultracolor Red 40 (Ultra Chemical)
Ultracolor Soft Focus (Ultra Chemical)
Ultracolor Yellow 5 (Ultra Chemical)
Ultrafine Titanium Dioxide PP (Cardre)
Ultramarine Blue PP (Cardre)
Ultrasomes (Barnet)
Ultraspheres-5012 (Lipoid)
Ultraspheres-5014 (Lipoid)
Ultraspheres-5409 (Lipoid)
Ultraspheres-6100 (Lipoid)
Ultraspheres-7900 (Lipoid)
Ultraspheres-8009 (Lipoid)
Ultraspheres 8017 (Lipoid)
Ultraspheres-8401 (Lipoid)
Unipherol U-14 (Induchem)
Ursolisome (Coletica SA)
The Vert Encapsule (Greentech)
Vexel (Sederma)
VitAine (AGI Dermatics)
Vitamin A Palmitate Lipomicron (Sederma)
Vitamin E Acetate Lipomicron (Sederma)
Yellow Iron Oxide PP (Cardre)

LECITHINAMIDE DEA

Definition: Lecithinamide DEA is the mixture of reaction products of Lecithin (q.v.) and diethanolamine.

Information Source: EEC(III/1-60)

Chemical Classes: Alkanolamides; Phosphorus Compounds

Functions: Hair Conditioning Agent; Surfactant - Foam Booster; Viscosity Increasing Agent - Aqueous

Technical/Other Names:
   Amides, Lecithin, N,N-Bis(Hydroxyethyl)-
   N,N-Bis(Hydroxyethyl)Lecithin Amides
   Lecithin Amides, N,N-Bis(Hydroxyethyl)-

LEDEBOURIELLA DIVARICATA ROOT EXTRACT

Definition: Ledebouriella Divaricata Root Extract is an extract of the roots of *Saposhnikovia divaricata*. See "Regulatory and Ingredient Use Information," regarding the labeling names for botanical ingredients in Volume 1, Introduction, Part A.

Chemical Class: Biological Products

Function: Skin-Conditioning Agent - Miscellaneous

Technical/Other Name:
   Extract of Ledebouriella Divaricata Root

Trade Name Mixture:
   Siler Extract (Korea Kolmar)

LEDUM GROENLANDICUM

Definition: *See "Regulatory and Ingredient Use Information," regarding EU labeling names for botanical ingredients in Volume 1, Introduction, Part A.*

Chemical Class: Biological Products

Technical/Other Name:
   Ledum Groenlandicum Extract (U.S.)

LEDUM GROENLANDICUM EXTRACT

Definition: Ledum Groenlandicum Extract is an extract of the dried flowering plant of the Labrador tea, *Ledum groenlandicum*. *See "Regulatory and Ingredient Use Information," regarding the labeling names for botanical ingredients in Volume 1, Introduction, Part A.*

Chemical Class: Biological Products

Function: Not Reported

Technical/Other Names:
   Extract of Labrador Tea
   Extract of Ledum Groenlandicum
   Labrador Tea Extract
   Labrador Tea (Ledum Groenlandicum)
      Extract
   Ledum Groenlandicum (EU)

LEDUM PALUSTRE EXTRACT

CAS No.	EINECS No.
90063-39-1	289-997-3

Definition: Ledum Palustre Extract is an extract of the dried flowering plant of the Labrador tea, *Ledum palustre*. See "Regulatory and Ingredient Use Information," regarding the labeling names for botanical ingredients in Volume 1, Introduction, Part A.

Chemical Class: Biological Products

Function: Not Reported

Technical/Other Names:
   Extract of Labrador Tea
   Extract of Ledum Palustre
   Labrador Tea Extract
   Labrador Tea (Ledum Palustre) Extract

Trade Name Mixture:
   Herbasol-Extract Laborador/Marsu Tea
      (Cosmetochem)

LEMON EKISU

Definition: Lemon Ekisu is an extract of the fruit or of the fruit juice of *Citrus medica limonum*. *See "Regulatory and Ingredient Use Information," regarding use of Japan Trivial names in Volume 1, Introduction, Part A.*

Chemical Class: Biological Products

Functions: Fragrance Ingredient; Skin-Conditioning Agent - Miscellaneous

Technical/Other Names:
   Citrus Medica Limonum (Lemon) Fruit
      Extract (U.S.)
   Citrus Medica Limonum (Lemon) Juice
      Extract (U.S.)

LENS CULINARIS (LENTIL) SYMBIOSOME EXTRACT

Definition: Lens Culinaris (Lentil) Symbiosome Extract is an extract of the symbiosome (root nodule) of the lentil, *Lens culinaris*. *See "Regulatory and Ingredient Use Information," regarding the labeling names for botanical ingredients in Volume 1, Introduction, Part A.*

Chemical Class: Biological Products

Functions: Antioxidant; Skin-Conditioning Agent - Miscellaneous

Technical/Other Name:
Extract of Lens Culinaris (Lentil) Symbiosome

Trade Name Mixture:
Lentil Zymbiozome Fermentum (Arch Personal Care Products)

LENS ESCULENTA (LENTIL) FRUIT EXTRACT

CAS No.	EINECS No.
90063-40-4	289-998-9

Definition: Lens Esculenta (Lentil) Fruit Extract is an extract of the fruit of the lentil, *Lens esculenta. See "Regulatory and Ingredient Use Information," regarding the labeling names for botanical ingredients in Volume 1, Introduction, Part A.*

Chemical Class: Biological Products

Function: Skin-Conditioning Agent - Miscellaneous

Technical/Other Names:
Extract of Lens Esculenta
Extract of Lentil
Lens Esculenta Extract
Lentil Extract
Lentil Fruit Extract

Trade Name Mixture:
Lentil HS (Alban Muller)

LENTINUS EDODES EXTRACT

Definition: Lentinus Edodes Extract is an extract of the shiitake mushroom, *Lentinus edodes. See "Regulatory and Ingredient Use Information," regarding the labeling names for botanical ingredients in Volume 1, Introduction, Part A.*

Information Source: JAN

Chemical Class: Biological Products

Function: Skin-Conditioning Agent - Miscellaneous

Technical/Other Name:
Shiitake Mushroom Extract

Trade Name:
Shiitake Spray Dried Extract (Alban Muller)

Trade Name Mixtures:
ABS Mushroom Extract SM (Active Concepts)
Actiphyte of Shiitake Mushroom (Active Organics)
BIOMODULINE (Greentech)
Biomoduline D (Greentech)
Campo Citisu (Campo)
Far East Extract Shii-Take (E.U.K)
Fermiskin (Silab)
Lentisterol (Bioland)
NAB Mushroom Extract (Arch Personal Care Products)
Shiitake HG (Alban Muller)
Shiitake Liquid B (Ichimaru Pharcos)
Shiitake Liquid Concentrate Extract (Alban Muller)
Shiitake Liquid E (Ichimaru Pharcos)

LENTINUS EDODES MYCELIUM EXTRACT

Definition: Lentinus Edodes Mycelium Extract is an extract of the mycelium of *Lentinus edodes.*

Chemical Class: Biological Products

Function: Skin-Conditioning Agent - Miscellaneous

Technical/Other Name:
Extract of Lentinus Edodes Mycelium

Trade Name:
AHCC (Amino Up)

LEONTOPODIUM ALPINUM EXTRACT

Definition: Leontopodium Alpinum Extract is an extract of the flowers and leaves of *Leontopodium alpinum. See "Regulatory and Ingredient Use Information," regarding the labeling names for botanical ingredients in Volume 1, Introduction, Part A.*

Chemical Class: Biological Products

Function: Not Reported

Technical/Other Name:
Extract of Leontopodium Alpinum

Trade Names:
Edelweiss Floral Water (Alban Muller)
Edelweiss purified Spray Dried Extract (Alban Muller)

Trade Name Mixtures:
ABS Plant Sil Blend (REU) (Active Concepts)
Edelweiss BG (Alban Muller)
Edelweiss Extract (Alpaflor)
Edelweiss HBG (Alban Muller)
Edelweiss HPG Titrated (Alban Muller)
Edelweiss HS (Alban Muller)
Extrait Hydroglycolique d'Edelweiss (Greentech)
Herbasec Edelweiss (Cosmetochem) (Cosmetochem International Ltd.)
Herbasol Extract Edelweiss (Cosmetochem) (Cosmetochem International Ltd.)

LEONURUS SIBIRICUS EXTRACT

Definition: Leonurus Sibiricus Extract is an extract of the aerial parts of *Leonurus sibiricus. See "Regulatory and Ingredient Use Information," regarding the labeling names for botanical ingredients in Volume 1, Introduction, Part A.*

Chemical Class: Biological Products

Function: Not Reported

Technical/Other Name:
Extract of Leonurus Sibiricus

Trade Name Mixtures:
Motherwort Extract (Matuura Yakugyo)
Yakumosou Liquid E (Ichimaru Pharcos)

LEPIDIUM MEYENII ROOT EXTRACT

Definition: Lepidium Meyenii Root Extract is an extract of the roots of *lepidium meyenii. See "Regulatory and Ingredient Use Information," regarding the labeling names for botanical ingredients in Volume 1, Introduction, Part A.*

Chemical Class: Biological Products

Function: Skin-Conditioning Agent - Humectant

Trade Name Mixture:
Maca Kefir LS (Laboratoires Serobiologiques)

LEPTOSPERMUM PETERSONII OIL

Definition: Leptospermum Petersonii Oil is the volatile oil obtained from the leaves of *Leptospermum petersonii. See "Regulatory and Ingredient Use Information," regarding the labeling names for botanical ingredients in Volume 1, Introduction, Part A.*

Chemical Class: Essential Oils

Functions: Fragrance Ingredient; Pesticide

Technical/Other Name:
Oils, Leptospermum Petersonii

Trade Name:
Lemon Tea Tree Oil (Southern Cross Botanicals)

LEPTOSPERMUM SCOPARIUM OIL

Definition: Leptospermum Scoparium Oil is the volatile oil obtained from the leaves and branches of *Leptospermum scoparium. See "Regulatory and Ingredient Use Infor-*

mation," regarding the labeling names for botanical ingredients in Volume 1, Introduction, Part A.

Chemical Class: Essential Oils

Function: Fragrance Ingredient

Trade Names:
AEC Manuka Oil (A & E Connock)
AEC Tairawhiti Manuka Oil (A & E Connock)
East Cape Manuka (Tairawhiti)
Manex Manuka (Tairawhiti)
Manukaoil (Düllberg Konzentra)
Tairawhiti Manuka (Tairawhiti)

Trade Name Mixture:
LEMA (Southern Cross Botanicals)

LESPEDEZA BICOLOR BARK EXTRACT

Definition: Lespedeza Bicolor Bark Extract is an extract of the bark of *Lespedeza bicolor*. See "Regulatory and Ingredient Use Information," regarding the labeling names for botanical ingredients in Volume 1, Introduction, Part A.

Chemical Class: Biological Products

Function: Skin-Conditioning Agent - Humectant

Technical/Other Name:
Extract of Lespedeza Bicolor Bark

Trade Name Mixture:
Lespedeza Extract (Korea Kolmar)

LESPEDEZA CAPITATA EXTRACT

CAS No. 84837-05-8

EINECS No. 284-291-1

Definition: Lespedeza Capitata Extract is an extract of the leaves and stems of *Lespedeza capitata*. See "Regulatory and Ingredient Use Information," regarding the labeling names for botanical ingredients in Volume 1, Introduction, Part A.

Chemical Class: Biological Products

Function: Not Reported

Technical/Other Name:
Extract of Lespedeza Capitata

Trade Name Mixture:
Sveltonyl (Laboratoires Serobiologiques)

LESQUERELLA FENDLERI

Definition: See "Regulatory and Ingredient Use Information," regarding EU labeling

names for botanical ingredients in Volume 1, Introduction, Part A.

Chemical Class: Biological Products

Technical/Other Name:
Lesquerella Fendleri Seed Oil (U.S.)

LESQUERELLA FENDLERI SEED OIL

Definition: Lesquerella Fendleri Seed Oil is the oil expressed from the seed of *Lesquerella fendleri*. See "Regulatory and Ingredient Use Information," regarding the labeling names for botanical ingredients in Volume 1, Introduction, Part A.

Chemical Class: Biological Products

Functions: Hair Conditioning Agent; Skin-Conditioning Agent - Emollient; Skin-Conditioning Agent - Miscellaneous

Technical/Other Names:
Lesquerella Fendleri (EU)
Lesquerella Oil
Oil of Lesquerella

Trade Name:
AEC Lesquerella Oil (A & E Connock)

LEUCINE

CAS Nos.
61-90-5 (L-Form)
328-39-2 (dl-alpha)

EINECS No.
200-522-0

JPN Translation:
ロイシン

Empirical Formula:
$C_6H_{13}NO_2$

Definition: Leucine is the amino acid that conforms to the formula:

$$CH_3CHCH_2CHCOOH$$

with a CH_3 group on the third carbon and an NH_2 group on the first carbon.

Information Sources: 21CFR172.320, 21CFR310.545, 21CFR582.5406, JAN, JCLS, JP, MI-13(5470), RIFM, TSCA, USAN, USP XXIV

Chemical Class: Amino Acids

Functions: Fragrance Ingredient; Hair Conditioning Agent; Skin-Conditioning Agent - Miscellaneous

Reported Product Categories: Hair Conditioners; Permanent Waves

Technical/Other Names:
L-α-Aminoisocaproic Acid
2-Amino-4-Methylpentanoic Acid
2-Amino-4-Methylvaleric Acid

L-Leucine (RIFM)
L-Norvaline, 4-Methyl-

Trade Name:
AEC Leucine (A & E Connock)

Trade Name Mixtures:
Activated Aminos SI 1010 (Norjin)
Amiderm Phytoamine Biocomplex (Alban Muller)
Anti-Inflammatory Phytoamine Biocomplex (Alban Muller)
Anti-Seborrhoeic Phytoamine Biocomplex (Alban Muller)
Anti-Stress Phytoamine Biocomplex (Alban Muller)
Anti-Stress Phytoamine Biocomplex (Alban Muller)
Anti-Stress Phytoamine Biocomplex In Glycerin (Alban Muller)
Omega-CHS-Activator (GfN)
Essential Vital Elements (Dipta)
Essential Vital Elements - S (Dipta)
Free Radical Scavenger Phytoamine Biocomplex (Alban Muller)
Moisturizing Phytoamine Biocomplex (Alban Muller)
Regenerative Phytoamine Biocomplex (Alban Muller)
Sel-Smooth (Seltzer)

LEUKOCYTE EXTRACT

Definition: Leukocyte Extract is an extract of white blood cells.

Chemical Class: Biological Products

Function: Not Reported

Technical/Other Name:
Extract of Leukocyte

Trade Name Mixture:
Melawhite (Pentapharm/Centerchem)

LEVISTICUM OFFICINALE

Definition: See "Regulatory and Ingredient Use Information," regarding EU labeling names for botanical ingredients in Volume 1, Introduction, Part A.

Chemical Class: Biological Products

Technical/Other Names:
Levisticum Officinale Leaf Extract (U.S.)
Levisticum Officinale Oil (U.S.)
Levisticum Officinale Root Extract (U.S.)

LEVISTICUM OFFICINALE LEAF EXTRACT

Definition: Levisticum Officinale Leaf Extract is an extract of the leaves of

Levisticum officinale. See "Regulatory and Ingredient Use Information," regarding the labeling names for botanical ingredients in Volume 1, Introduction, Part A.

Chemical Class: Biological Products

Function: Not Reported

Technical/Other Name:
Levisticum Officinale (EU)

Trade Name Mixture:
Herbasol Extract Lovage (Leaf) (Cosmetochem International Ltd.)

LEVISTICUM OFFICINALE OIL

CAS No.: 8016-31-7

Definition: Levisticum Officinale Oil is the volatile oil distilled from *Levisticum officinale.* See "Regulatory and Ingredient Use Information," regarding the labeling names for botanical ingredients in Volume 1, Introduction, Part A.

Information Sources: 21CFR172.510, FCC, RIFM, TSCA

Chemical Class: Essential Oils

Function: Fragrance Ingredient

Technical/Other Names:
Levisiticum Officinale Oil
Levisticum Officinale (EU)
Lovage extract (Levisticum officinale Koch) (RIFM)
Lovage (Levisticum officinale Koch) (RIFM)
Lovage (Levisticum Officinale) Oil
Lovage Oil
Lovage oil (Levisticum officinale Koch) (RIFM)
Oil of Lovage

LEVISTICUM OFFICINALE ROOT EXTRACT

Definition: Levisticum Officinale Root Extract is an extract of the roots of *Levisticum officinale.* See "Regulatory and Ingredient Use Information," regarding the labeling names for botanical ingredients in Volume 1, Introduction, Part A.

Chemical Class: Biological Products

Function: Not Reported

Technical/Other Names:
Extract of Levisticum Officinale
Extract of Lovage
Extract of Lovage Root
Levisticum Officinale (EU)
Lovage Extract
Lovage (Levisticum Officinale) Extract
Lovage Root Extract

Trade Name Mixtures:
Actiphyte of Ligusticum BG50 (Active Organics)
Actiphyte of Ligusticum GL50 (Active Organics)
Actiphyte of Ligusticum Lipo S (Active Organics)
Actiphyte of Ligusticum PG50 (Active Organics)
Actiphyte of Lovage BG50 (Active Organics)
Actiphyte of Lovage GL50 (Active Organics)
Actiphyte of Lovage Lipo S (Active Organics)
Actiphyte of Lovage PG50 (Active Organics)
Phytotal MS (Phybiotex/Sederma)
VT-180 Extract of Lingusticum (Vege-Tech)

LEVULINIC ACID

CAS No. 123-76-2   **EINECS No.** 204-649-2

JPN Translation:
レブリン酸

Empirical Formula:
$C_5H_8O_3$

Definition: Levulinic Acid is the organic acid that conforms to the formula:

$$CH_3\overset{O}{\overset{\|}{C}}CH_2CH_2COOH$$

Information Sources: JCIC, JCLS, JSQI, MI-13(5492), RIFM, TSCA

Chemical Classes: Carboxylic Acids; Ketones

Functions: Fragrance Ingredient; Skin-Conditioning Agent - Miscellaneous

Technical/Other Names:
3-Acetylpropionic Acid
4-Ketovaleric Acid
Levulinic acid (RIFM)
4-Oxopentanoic Acid
4-Oxovaleric Acid
Pentanoic Acid, 4-Oxo

LIATRIS ODORATISSIMA LEAF EXTRACT

CAS No. 68602-86-8   **EINECS No.** 271-627-7

Definition: Liatris Odoratissima Leaf Extract is an extract of the leaves of *Liatris odoratissima.* See "Regulatory and Ingredient Use Information," regarding the labeling names for botanical ingredients in Volume 1, Introduction, Part A.

Chemical Class: Biological Products

Function: Not Reported

Technical/Other Name:
Extract of Liatris Odoratissima

Trade Name Mixture:
Abs. Liatrix GHG 822 (Charabot)

LIDOCAINE HCl

CAS No.: 6108-05-0

Empirical Formula:
$C_{14}H_{22}N_2O \cdot ClH$

Definition: Lidocaine HCl is the organic compound that conforms to the formula:

Information Sources: 21CFR310.545, 21CFR346.10, 21CFR348.10, EEC(II-399), MI-13(5503)

Chemical Classes: Amides; Amines

Function: Not Reported

Technical/Other Names:
Acetamide, 2-(Diethylamino)-N-(2,6-Dimethylphenyl)-, Monohydrochloride, Monohydrate
Lidocaine Monohydrate Monohydrochloride

LIGNOCERYL ERUCATE

Empirical Formula:
$C_{46}H_{90}O_2$

Definition: Lignoceryl Erucate is the ester of lignoceryl alcohol and erucic acid. It conforms to the formula:

$$CH_3(CH_2)_7CH=CH(CH_2)_{11}\overset{O}{\overset{\|}{C}}-O(CH_2)_{23}CH_3$$

Chemical Class: Esters

Function: Skin-Conditioning Agent - Emollient

Technical/Other Names:
13-Docosenoic Acid, Lignoceryl Ester
Erucic Acid, Lignoceryl Ester

Trade Name:
Crodamol LGE (Croda, Inc.)

LIGUSTICUM CHUANXIONG EXTRACT

Definition: Ligusticum Chuanxiong Extract is an extract of the whole plant, *Ligusticum* *chuanxiong. See "Regulatory and Ingredient Use Information," regarding the labeling names for botanical ingredients in Volume 1, Introduction, Part A.*

Chemical Class: Biological Products

Function: Skin-Conditioning Agent - Miscellaneous

Technical/Other Name:
Extract of Ligusticum Chuanxiong

Trade Name Mixture:
Actiphyte Szechuan Lovage Root (Active Organics)

LIGUSTICUM JEHOLENSE EXTRACT

Definition: Ligusticum Jeholense Extract is an extract of the flowers and roots of *Ligusticum jeholense. See "Regulatory and Ingredient Use Information," regarding the labeling names for botanical ingredients in Volume 1, Introduction, Part A.*

Chemical Class: Biological Products

Function: Not Reported

Technical/Other Name:
Extract of Ligusticum Jeholense

Trade Name Mixture:
Campo Gao Ben Hua Extract (Campo)

LIGUSTICUM SINENSE ROOT EXTRACT

Definition: Ligusticum Sinense Root Extract is an extract of the roots and rhizomes of *Ligusticum sinense. See "Regulatory and Ingredient Use Information," regarding the labeling names for botanical ingredients in Volume 1, Introduction, Part A.*

Chemical Class: Biological Products

Function: Not Reported

Technical/Other Name:
Extract of Ligusticum Sinense

Trade Name Mixture:
Kouhon Liquid E (Ichimaru Pharcos)

LIGUSTICUM STRIATUM ROOT EXTRACT

Definition: Ligusticum Striatum Root Extract is an extract of the roots of *Ligusticum striatum. See "Regulatory and Ingredient Use Information," regarding the labeling names for botanical ingredients in Volume 1, Introduction, Part A.*

Chemical Class: Biological Products

Function: Antioxidant

Technical/Other Name:
Extract of Ligusticum Striatum Root

Trade Name Mixture:
Draco Ligusticum Striatum Full Spectrum Standardized Extract (Draco)

LIGUSTRUM JAPONICUM FRUIT EXTRACT

Definition: Ligustrum Japonicum Fruit Extract is an extract of the fruit of *Ligustrum japonicum. See "Regulatory and Ingredient Use Information," regarding the labeling names for botanical ingredients in Volume 1, Introduction, Part A.*

Chemical Class: Biological Products

Function: Not Reported

Technical/Other Name:
Extract of Ligustrum Japonicum

Trade Name Mixture:
Nyotei Liquid E (Ichimaru Pharcos)

LIGUSTRUM LUCIDUM EXTRACT

Definition: Ligustrum Lucidum Extract is an extract of the berries of *Ligustrum lucidum.*

Chemical Class: Biological Products

Function: Not Reported

Technical/Other Name:
Extract of Ligustrum Lucidum

Trade Name Mixtures:
Campo Nu Chen Extract (Campo)
China Extract Nu Chen (E.U.K)
Chine Extract Lugustrum Lucidum (Ennagram)

LILIUM CANDIDUM BULB EXTRACT

CAS No.
84776-67-0

EINECS No.
283-996-1

JPN Translation:
ユリエキス

Definition: Lilium Candidum Bulb Extract is an extract of the bulbs of the white lily, *Lilium candidum. See "Regulatory and Ingredient Use Information," regarding the labeling names for botanical ingredients in Volume 1, Introduction, Part A.*

Chemical Class: Biological Products

Function: Skin-Conditioning Agent - Miscellaneous

Reported Product Categories: Skin Care Preparations, Misc.; Moisturizing Preparations; Bath Oils, Tablets, and Salts; Cleansing Products (Cold Creams, Cleansing Lotions, Liquids and Pads); Paste Masks (Mud Packs)

Technical/Other Names:
Extract of Lilium Candidum
Extract of White Lily
Lilium Album Extract
White Lily Extract
White Lily (Lilium Candidum) Extract

Trade Name Mixtures:
Actiphyte of White Lily BG50 (Active Organics)
Actiphyte of White Lily GL50 (Active Organics)
Actiphyte of White Lily Lipo S (Active Organics)
Actiphyte of White Lily PG50 (Active Organics)
Extrait de Bulbes de Lys ME 100 (Yves Rocher)
Glycolysat of White Lily Bulb (CEP (Solabia))
Lily Extract (Maruzen Pharmaceuticals Co., Ltd.)
Lily Extract BG (Maruzen Pharmaceuticals Co., Ltd.)
Lily Extract LA (Maruzen Pharmaceuticals Co., Ltd.)
Oleat of White Lily Bulb (CEP (Solabia))
Oleat T of White Lily Bulb (CEP (Solabia))
Phytelene of Lily EG 162 Liquid (Indena SA)
Phytogreen 55 of Lily EXH 632 Liquid (Phytochim)
Prodhy Extract LYS (Prod'Hyg)
Vegebios of White Lily Bulb (CEP (Solabia))
Vegetol Lily GR 464 Hydro (Gattefosse s.a.)
Vegetol Lily MCF 1968 Hydro (Gattefosse s.a.)
VT-055 Extract of White Lily (Vege-Tech)
White Lily Bulb HPG Titrated (Alban Muller)
White Lily Bulb HS (Alban Muller)
White Lily Extract (Maruzen Pharmaceuticals Co., Ltd.)
White Lily Extract BG (Maruzen Pharmaceuticals Co., Ltd.)
White Lily Extract LA (Maruzen Pharmaceuticals Co., Ltd.)

LILIUM CANDIDUM FLOWER EXTRACT

Definition: Lilium Candidum Flower Extract is an extract of the flowers of the white lily, *Lilium candidum. See "Regulatory and Ingredient Use Information," regarding the labeling names for botanical ingredients in Volume 1, Introduction, Part A.*

Chemical Class: Biological Products

Function: Skin-Conditioning Agent - Miscellaneous

Technical/Other Names:
Extract of Lilium Candidum Flower
Extract of White Lily (Lilium Candidum)
Flower
White Lily (Lilium Candidum) Flower
Extract

Trade Name:
White Lily Flower Spray Dried Extract
(Alban Muller)

Trade Name Mixtures:
Amidroxy 4 Flowers (Alban Muller)
Amidroxy 4 Flowers (Alban Muller)
Lilium Candidum Flower Extract ies (IES
LABO)
Phytelene (R) of Lily (Indena SA)
Vege Plex VP#1341 (Vege-Tech)
White Lily Flower HS (Alban Muller)

LILIUM CANDIDUM FLOWER WATER

Definition: Lilium Candidum Flower Water
is an aqueous solution of the steam
distillate obtained from *Lilium candidum.*
*See "Regulatory and Ingredient Use Information," regarding the labeling names for
botanical ingredients in Volume 1, Introduction, Part A.*

Chemical Class: Biological Products

Function: Skin-Conditioning Agent -
Emollient

Trade Name:
Essential Lily Nectar (Libiol)

LILIUM SPECIOSUM RUBRUM BULB
EXTRACT

Definition: Lilium Speciosum Rubrum
Bulb Extract is an extract of the bulbs of
*Lilium speciosum rubrum. See "Regulatory
and Ingredient Use Information," regarding
the labeling names for botanical ingredients
in Volume 1, Introduction, Part A.*

Chemical Class: Biological Products

Function: Not Reported

Technical/Other Name:
Extract of Lilium Speciosum Rubrum Bulb

Trade Name Mixture:
Extrait huileux de Lis (Greentech)

LILIUM TIGRINUM EXTRACT

Definition: Lilium Tigrinum Extract is an
extract of the aerial parts of *Lilium tigrinum.*

*See "Regulatory and Ingredient Use Information," regarding the labeling names for
botanical ingredients in Volume 1, Introduction, Part A.*

Chemical Class: Biological Products

Function: Not Reported

Technical/Other Names:
Extract of Lilium Tigrinum
Extract of Tiger Lily
Tiger Lily Extract

Trade Name Mixtures:
Actiphyte of Tiger Lily GL (Active Organics)
Cosflor Tiger Lily HGS (A & E Connock)
Tiger Lily Extract (Cosmetic Developments)

LIMNANTHES ALBA (MEADOWFOAM)
SEED OIL

JPN Translation:
メドウフォーム油

Definition: Limnanthes Alba
(Meadowfoam) Seed Oil is the oil extracted
from the seeds of the meadowfoam plant,
*Limnanthes alba. See "Regulatory and
Ingredient Use Information," regarding the
labeling names for botanical ingredients in
Volume 1, Introduction, Part A.*

Information Sources: JCIC, JCLS

Chemical Class: Fats and Oils

Function: Skin-Conditioning Agent -
Occlusive

Reported Product Categories: Lipsticks;
Hair Conditioners; Shampoos (Non-coloring);
Tonics, Dressings, and Other Hair Grooming
Aids; Bath Capsules; Bath Preparations,
Misc.; Body and Hand Preparations
(Excluding Shaving Preparations); Face and
Neck Preparations (Excluding Shaving Preparations); Moisturizing Preparations

Technical/Other Names:
Meadowfoam Seed Oil
Oils, Meadowfoam Seed

Trade Names:
AEC Meadowfoam Oil (A & E Connock)
Botanol MO (Botanigenics)
Cropure Meadowfoam (Croda Chemicals)
Huile de Perle (Bertin)
Nikkol Meadowfoam Oil (Nikko)

Trade Name Mixtures:
CREALBA AQUASOL (Creations Couleurs)
Creanatural LAB (Creations Couleurs)
Fancol Meadowlan (Fanning)
Fancol VB (Fanning)

Empirical Formula:
$C_{10}H_{15}$

Definition: Limonene is a terpene that
contains one or more of the following
stereoisomers: d-limonene, l-limonene, or
(recemic) dl-limonene. It conforms to the
formula:

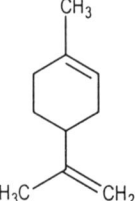

Information Sources: 21CFR175.105,
21CFR175.300, 21CFR177.2600, JAN, MI-
13(5515), RIFM, TSCA

Chemical Class: Hydrocarbons

Functions: Fragrance Ingredient; Solvent

Technical/Other Names:
Cyclohexene, 1-Methyl-4-(1-
Methylethenyl)-
d-Limonene
DL-Limonene
d,l-Limonene (isomer unspecified) (RIFM)
l-Limonene
dl-Limonene (racemic) (RIFM)
1-Methyl-4-Isopropenylcyclohexene
1-Methyl-4-(1-Methylethenyl)Cyclohexene

Trade Name:
Dipentene No. 122 (Hercules)

LIMONIUM VULGARE EXTRACT

Definition: Limonium Vulgare Extract is an
extract of the aerial parts of *Limonium
vulgare. See "Regulatory and Ingredient
Use Information," regarding the labeling
names for botanical ingredients in Volume
1, Introduction, Part A.*

Chemical Class: Biological Products

Function: Not Reported

Technical/Other Name:
Extract of Limonium Vulgare

Trade Name Mixtures:
Cosflor Sea Lavender HGS (A & E
Connock)
Sea Lavender Extract (Cosmetic Developments)

LIMONENE

CAS Nos.	EINECS No.
138-86-3 (dl-alpha)	205-341-0
5989-27-5 (d-alpha)	

LINALOOL

CAS No.	EINECS No.
78-70-6	201-134-4

JPN Translation:
リナロール

Empirical Formula:
$C_{10}H_{18}O$

Definition: Linalool is the terpene that conforms to the formula:

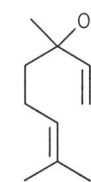

Information Sources: 21CFR172.515, 21CFR182.60, 21CFR582.60, EEC(III/1-84), JCLS, JSCI, MI-13(5517), RIFM

Chemical Class: Alcohols

Function: Fragrance Ingredient

Technical/Other Names:
2,6-Dimethyl-2,7-Octadien-6-ol
3,7-Dimethyl-1,6-Octadien-3-ol
Linalool (RIFM)
Linalyl Alcohol
1,6-Octadien-3-ol, 3,7-Dimethyl-
2,7-Octadien-6-ol, 2,6-Dimethyl-

Trade Name Mixtures:
Farnesol KSN 2/060060 (Symrise)
Hexatrate (Vevy)
Hexatrate Al-Free (Vevy)
Lipofresh (Lipo)
Unistab S-69 (Induchem)

LINALYL ACETATE

CAS No. 115-95-7

EINECS No. 204-116-4

JPN Translation:
酢酸リナリル

Empirical Formula:
$C_{12}H_{20}O_2$

Definition: Linalyl Acetate is the ester of Linalool (q.v.) and acetic acid. It conforms to the formula:

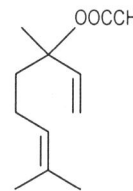

Information Sources: JCLS, JSCI, MI-13 (5518), RIFM, TSCA

Chemical Class: Esters

Function: Fragrance Ingredient

Technical/Other Names:
3,7-Dimethyl-1,6-ctadien-3-ol Acetate

Linalool Acetate
Linalyl acetate (RIFM)
1,6-Octadien-3-ol, 3,7-Dimethyl-, Acetate

LINARIA CYMBALARIA EXTRACT

Definition: Linaria Cymbalaria Extract is the extract obtained from the aerial parts of *Linaria cymbalaria. See "Regulatory and Ingredient Use Information," regarding the labeling names for botanical ingredients in Volume 1, Introduction, Part A.*

Chemical Class: Biological Products

Function: Not Reported

Technical/Other Name:
Extract of Linaria Cymbalaria

Trade Name Mixture:
Cosflor Pennywort HGS (A & E Connock)

LINDERA STRYCHNIFOLIA ROOT EXTRACT

Definition: Lindera Strychnifolia Root Extract is an extract of the roots of *Lindera strychnifolia. See "Regulatory and Ingredient Use Information," regarding the labeling names for botanical ingredients in Volume 1, Introduction, Part A.*

Chemical Class: Biological Products

Function: Not Reported

Technical/Other Name:
Extract of Lindera Strychnifolia

Trade Name Mixture:
Uyaku Liquid E (Ichimaru Pharcos)

LINOLEAMIDE

CAS Nos.
3072-13-7
3999-01-7

EINECS No.
223-644-6

Empirical Formula:
$C_{18}H_{33}NO$

Definition: Linoleamide is the aliphatic amide of linoleic acid. It conforms generally to the formula:

$$CH_3(CH_2)_4CH = CHCH_2CH = CH(CH_2)_7 \overset{O}{\overset{\|}{C}} - NH_2$$

Information Sources: 21CFR175.105, 21CFR175.300, 21CFR178.3910, 21CFR179.45, 21CFR181.22, 21CFR181.28, TSCA

Chemical Class: Amides

Functions: Hair Conditioning Agent; Opacifying Agent; Viscosity Increasing Agent - Nonaqueous

Technical/Other Names:
Linoleic Acid Amide
9,12-Octadecadienamide

LINOLEAMIDE DEA

CAS Nos.
56863-02-6
94094-35-6

EINECS No.
260-410-2

JPN Translation:
リノレアミド DEA

Empirical Formula:
$C_{22}H_{41}NO_3$

Definition: Linoleamide DEA is a mixture of ethanolamides of linoleic acid. It conforms generally to the formula:

$$CHCH_2CH = CH(CH_2)_7 \overset{O}{\overset{\|}{C}} - N(CH_2CH_2OH)_2$$
(with $CH(CH_2)_4CH_3$ above)

Information Sources: CIR: [SQ] JACT-5 (5)1986, CTFA D, EEC(III/1-60), JCIC, JCLS, JSQI, TSCA

Chemical Class: Alkanolamides

Functions: Hair Conditioning Agent; Surfactant - Foam Booster; Viscosity Increasing Agent - Aqueous

Reported Product Categories: Hair Dyes and Colors (All Types Requiring Caution Statements and Patch Tests); Bath Preparations, Misc.; Bath Soaps and Detergents; Shampoos (Non-coloring); Bubble Baths; Bath Oils, Tablets, and Salts; Cleansing Products (Cold Creams, Cleansing Lotions, Liquids and Pads); Feminine Hygiene Deodorants; Shaving Cream (Aerosol, Brushless and Lather); Tonics, Dressings, and Other Hair Grooming Aids

Technical/Other Names:
N,N-Bis(2-Hydroxyethyl)Linoleamide
N,N-Bis(2-Hydroxyethyl)-9,12-Octadecadienamide
Diethanolamine Linoleic Acid Amide
Linoleic Acid Diethanolamide
Linoleoyl Diethanolamide
9,12-Octadecadienamide, N,N-Bis(2-Hydroxyethyl)-

Trade Names:
Alkamide DIN 295/S (Rhodia)
Colamid 3458 MFA (Colonial Chemical Inc)
DeMIDE SBA-100 (DeForest)
Hetamide LL (Heterene)
Hetamide LN (Heterene)
Jeemide LNO (Jeen)
Jeemide 15-W (Jeen)

Mackamide LOL (McIntyre)
Mackamide LOL-5 (McIntyre)
Monamid 150-ADY (Uniqema)
Protamide LNO (Protameen)
Protamide 15W (Protameen)
Schercomid SLE (Scher)
Schercomid SLE-Special (Scher)

Trade Name Mixtures:
Calamide F (Pilot)
Condipon (Cosmetochem)
Condisoap (Cosmetochem) (Cosmetochem
International Ltd.)
DeCONC SC 10-2-1 (DeForest)
DeMIDE LMSB-100 (DeForest)
Extrapone Rosemary 2/783630 (Symrise)
Monamid 1007 (Uniqema)
Monamine R8-26 (Uniqema)

LINOLEAMIDE MEA

CAS Nos.   **EINECS No.**
10015-67-5
68171-52-8   269-029-6

Empirical Formula:
$C_{20}H_{37}NO_2$

Definition: Linoleamide MEA is a mixture of
ethanolamides of linoleic acid. It conforms
generally to the formula:

$$CH(CH_2)_4CH_3$$
$$CHCH_2CH = CH(CH_2)_7C-NHCH_2CH_2OH$$

Information Source: TSCA

Chemical Class: Alkanolamides

Functions: Hair Conditioning Agent;
Surfactant - Foam Booster; Viscosity
Increasing Agent - Aqueous

Reported Product Category: Hair Dyes
and Colors (All Types Requiring Caution
Statements and Patch Tests)

Technical/Other Names:
N-(2-Hydroxyethyl)Linoleamide
N-(2-Hydroxyethyl)-9,12-
Octadecadienamide
Linoleoyl Monoethanolamide
Monoethanolamine Linoleic Acid Amide
9,12-Octadecadienamide, N-(2-Hydroxy-
ethyl)-

Trade Names:
AMINOL LNO-2 (Finetex)
Hetamide MLL (Heterene)
Mackamide LOMA (McIntyre)

LINOLEAMIDE MIPA

Empirical Formula:
$C_{21}H_{39}NO_2$

Definition: Linoleamide MIPA is a mixture of
isopropanolamides of linoleic acid. It
conforms generally to the formula:

$$CH(CH_2)_4CH_3 \qquad O$$
$$CHCH_2CH = CH(CH_2)_7C-NHCH_2CHOH$$
$$CH_3$$

Chemical Class: Alkanolamides

Functions: Hair Conditioning Agent;
Surfactant - Foam Booster; Viscosity
Increasing Agent - Aqueous

Technical/Other Names:
N-(2-Hydroxypropyl)-9,12-
Octadecadienamide
Linoleoyl Monoisopropanolamide
Monoisopropanolamine Linoleic Acid
Amide
9,12-Octadecadienamide, N-(2-Hydroxy-
propyl)-

Trade Name:
Schercomid DMI (Scher)

LINOLEAMIDOPROPALKONIUM CHLORIDE

Empirical Formula:
$C_{30}H_{51}N_2O \cdot Cl$

Definition: Linoleamidopropalkonium
Chloride is the quaternary ammonium salt
that conforms to the formula:

$$CH(CH_2)_4CH_3$$
$$CHCH_2CH \qquad O \qquad CH_3$$
$$CH(CH_2)_7C-NH(CH_2)_3-N^+-CH_3 \quad Cl^-$$
$$CH_2$$

Chemical Class: Quaternary Ammonium
Compounds

Functions: Antistatic Agent; Hair Conditioning Agent

LINOLEAMIDOPROPYL DIMETHYLAMINE

CAS No.: 81613-56-1

Empirical Formula:
$C_{23}H_{44}N_2O$

Definition: Linoleamidopropyl Dimethylamine is the amidoamine that conforms
generally to the formula:

$$CH(CH_2)_4CH_3$$
$$CHCH_2CH \qquad O$$
$$CH(CH_2)_7C-NH(CH_2)_3-N \begin{array}{c} CH_3 \\ CH_3 \end{array}$$

Chemical Class: Amines

Function: Antistatic Agent

Technical/Other Names:
Dimethylaminopropyl Linoleamide
N-[3-(Dimethylamino)Propyl]-9,12-
Octadecadienamide
9,12-Octadecadienamide, N-[3-(Dimethyl-
amino)Propyl]-

Trade Name:
Foamine O-80 (Alzo)

LINOLEAMIDOPROPYL DIMETHYLAMINE DIMER DILINOLEATE

CAS No.: 125804-10-6

Empirical Formula:
$(C_{18}H_{32}O_2)_2 \cdot C_{23}H_{44}N_2O$

Definition: Linoleamidopropyl Dimethylamine Dimer Dilinoleate is the salt of
Linoleamidopropyl Dimethylamine (q.v.) and
a dimer of linoleic acid.

Chemical Class: Amines

Functions: Hair Conditioning Agent; Skin-
Conditioning Agent - Miscellaneous

Reported Product Categories: Hair Dyes
and Colors (All Types Requiring Caution
Statements and Patch Tests); Hair Conditioners

Technical/Other Name:
9,12-Octadecadienoic Acid, Dimer, Compd.
with N-[3-(Dimethylamino)Propyl]-9,12-
Octadecadienamide (1:1)

Trade Name:
Necon LO-80 (Alzo)

LINOLEAMIDOPROPYL DIMONIUM LACTATE

Empirical Formula:
$C_{23}H_{44}N_2O \cdot C_3H_6O_3$

Definition: Linoleamidopropyl Dimonium
Lactate is the amidoamine salt that conforms
generally to the formula:

$$CH(CH_2)_4CH_3$$
$$CHCH_2CH \qquad O \qquad CH_3 \qquad CH_3$$
$$CH(CH_2)_7C-NH(CH_2)_3-N \cdot HOOCCH$$
$$CH_3 \qquad OH$$

Chemical Class: Amines

Function: Antistatic Agent

Technical/Other Name:
Dimethylaminopropyl Linoleamide Lactate

Trade Name:
Linoquat VG (Sinerga)

LINOLEAMIDOPROPYL ETHYLDIMONIUM ETHOSULFATE

CAS No.: 99542-23-1

Empirical Formula:
$C_{25}H_{49}N_2O \cdot C_2H_5O_4S$

Definition: Linoleamidopropyl Ethyldimonium Ethosulfate is the quaternary ammonium salt that conforms to the formula:

Chemical Class: Quaternary Ammonium Compounds

Functions: Antistatic Agent; Hair Conditioning Agent

Reported Product Category: Hair Conditioners

Technical/Other Name:
1-Propanaminium, N-Ethyl-N,N-Dimethyl-3-[(1-Oxo-9,12-Octadecadienyl)Amino]-, Ethosulfate

Trade Names:
Foamquat SAQ-90 (Alzo)
Naetex Q (Lanaetex)

Trade Name Mixture:
Parapel HC (Bernel)

LINOLEAMIDOPROPYL PG-DIMONIUM CHLORIDE PHOSPHATE

Definition: Linoleamidopropyl PG-Dimonium Chloride Phosphate is the quaternary ammonium salt that conforms to the formula:

where RCO- represents the linoleic acid radical.

Chemical Classes: Phosphorus Compounds; Quaternary Ammonium Compounds

Function: Antistatic Agent

Reported Product Categories: Personal Cleanliness Products, Misc.; Hair Conditioners; Shampoos (Non-coloring); Hair Preparations (Non-coloring), Misc.; Bubble Baths

Trade Names:
Colalipid SAFL (Colonial Chemical Inc)
Phospholipid EFA (Uniqema)

LINOLEAMIDOPROPYL PG-DIMONIUM CHLORIDE PHOSPHATE DIMETHICONE

CAS No.: 179005-04-0

Definition: Linoleamidopropyl PG-Dimonium Chloride Phosphate Dimethicone is the product obtained by the reaction of Linoleamidopropyl PG-Dimonium Chloride Phosphate (q.v.) and Dimethicone (q.v.).

Chemical Classes: Quaternary Ammonium Compounds; Siloxanes and Silanes

Function: Hair Conditioning Agent

Technical/Other Name:
Siloxanes and Silicones, 3-[4-[[[3-[(2,3-Dihydroxypropyl)Dimethylammonio]-Propyl]Amino]Carbonyl]-2-Oxo-1-Acylamino)Propyl]Dimethylammonio]-2-Hydroxypropyl Phosphates], Chlorides, Sodium Salts

Trade Name:
Monasil PLN (Uniqema)

LINOLEIC ACID

CAS No. 60-33-3   **EINECS No.** 200-470-9

JPN Translation:
リノール酸

Empirical Formula:
$C_{18}H_{32}O_2$

Definition: Linoleic Acid is the unsaturated fatty acid that conforms generally to the formula:

$$CH_3(CH_2)_4CH = CHCH_2CH = CH(CH_2)_7COOH$$

Information Sources: 21CFR175.105, 21CFR184.1065, JCIC, JCLS, JSQI, MI-13 (5527), RIFM, TSCA

Chemical Class: Fatty Acids

Functions: Fragrance Ingredient; Hair Conditioning Agent; Skin-Conditioning Agent - Miscellaneous; Surfactant - Cleansing Agent

Surfactant-Cleansing Agent is included as a function for the soap form of Linoleic Acid.

Reported Product Categories: Hair Dyes and Colors (All Types Requiring Caution Statements and Patch Tests); Moisturizing Preparations; Hair Conditioners; Shampoos (Non-coloring); Eyebrow Pencils; Suntan Gels, Creams, and Liquids; Bath Capsules; Bath Oils, Tablets, and Salts; Face and Neck Preparations (Excluding Shaving Preparations); Skin Care Preparations, Misc.; Cleansing Products (Cold Creams, Cleansing Lotions, Liquids and Pads); Eye Shadows; Night Skin Care Preparations; Suntan Preparations, Misc.; Body and Hand Preparations (Excluding Shaving Preparations); Eyeliners; Indoor Tanning Preparations; Lipsticks; Tonics, Dressings, and Other Hair Grooming Aids

Technical/Other Names:
Linoleic acid (RIFM)
9,12-Octadecadienoic Acid

Trade Name:
Emersol 315 (Cognis Corp.)

Trade Name Mixtures:
AC EFA Liposome (Active Concepts)
Biosil Basics HMC - Hair Moisture Complex (Biosil Technologies, Inc.)
Biosil Basics HMC-1 Hair Moisture Complex (Biosil Technologies, Inc.)
Biosil Basics HMV- Hair Moisture Complex (Biosil Technologies, Inc.)
Biosil Basics HMW - Hair Moisture Complex (Biosil Technologies, Inc.)
Botaniceutical BR-1 (Botanigenics)
Botaniceutical BR-2 (Botanigenics)
Botaniceutical BR-T (Botanigenics)
Brookosome EFA (Arch Personal Care Products)
Brookosome ELL (Arch Personal Care Products)
Cutavit Richter (CLR)
DUB VFA (Stearinerie Dubois Fils)
Efaderma (Vevy)
Efadermasterolo (Vevy)
EFA-Plex (Arch Personal Care Products)
EFA-Plexsol (Arch Personal Care Products)
Germinol (Dr. Gerhard Steidl)
Glycosan VIT A, E, F-12 (Chemyunion)
Liant TW 876 (LCW)
Lipotrophyne-A (Vevy)
N.S.L.E. (Sederma)
RonaCare ASC III (Merck KGaA)
RonaCare ASC III (Merck KGaA/EMD Chemicals Inc.)
Soluvit Richter (CLR)
Tagravit F1 (Tagra)
Thiolin (General Topics)
Vitacap (LCW)

Vitamin Concentrate "O" (Cosmetochem) (Cosmetochem International Ltd.)
Vitamin Extract AEFH' Water Soluble (Crodarom)
Vitamin Extract AEF, Oil Soluble (Crodarom)
Vitaminextract VC, Oil Soluble (Crodarom)
Vitaminextract VC, Water Soluble (Crodarom)
Vitamin F (Tagra)
Vitamin F Forte CLR (CLR)
Vitamin F free acid (Crodarom)
Vitamin F Oilsoluble (Cosmetochem) (Cosmetochem International Ltd.)
Vitamin F, Water Soluble (Crodarom)
Vitamin F water-soluble CLR (CLR)
Vitamin F Watersoluble (Cosmetochem) (Cosmetochem International Ltd.)
Vitasol, Vitamin-Horsechestnut-Complex (Crodarom)

LINOLENIC ACID

CAS No. 463-40-1

EINECS No. 207-334-8

JPN Translation:
リノレン酸

Empirical Formula:
$C_{18}H_{30}O_2$

Definition: Linolenic Acid is the unsaturated fatty acid that conforms generally to the formula:

CHCH₂CH₃
‖
CHCH₂CH══CHCH₂CH══CH(CH₂)₇COOH

Information Sources: MI-13(5528), MI-13 (5529), RIFM, TSCA

Chemical Class: Fatty Acids

Functions: Fragrance Ingredient; Hair Conditioning Agent; Skin-Conditioning Agent - Miscellaneous; Surfactant - Cleansing Agent

Surfactant-Cleansing Agent is included as a function for the soap form of Linolenic Acid.

Reported Product Categories: Hair Conditioners; Shampoos (Non-coloring); Eyebrow Pencils; Suntan Gels, Creams, and Liquids; Eye Shadows; Moisturizing Preparations; Suntan Preparations, Misc.; Bath Capsules; Face and Neck Preparations (Excluding Shaving Preparations); Skin Care Preparations, Misc.; Tonics, Dressings, and Other Hair Grooming Aids

Technical/Other Names:
Linolenic acid (RIFM)
9,12,15-Octadecatrienoic Acid

Trade Name Mixtures:
Biosil Basics HMC - Hair Moisture Complex (Biosil Technologies, Inc.)

Biosil Basics HMC-1 Hair Moisture Complex (Biosil Technologies, Inc.)
Biosil Basics HMV- Hair Moisture Complex (Biosil Technologies, Inc.)
Biosil Basics HMW - Hair Moisture Complex (Biosil Technologies, Inc.)
Cutavit Richter (CLR)
DUB VFA (Stearinerie Dubois Fils)
Efaderma (Vevy)
Efadermasterolo (Vevy)
EFA-Plex (Arch Personal Care Products)
EFA-Plexsol (Arch Personal Care Products)
Liant TW 876 (LCW)
Lipotrophyne-A (Vevy)
Tagravit F1 (Tagra)
Vitamin Concentrate "O" (Cosmetochem) (Cosmetochem International Ltd.)
Vitamin F (Tagra)
Vitamin F Forte CLR (CLR)
Vitamin F free acid (Crodarom)
Vitamin F Oilsoluble (Cosmetochem) (Cosmetochem International Ltd.)
Vitamin F water-soluble CLR (CLR)
Vitamin F Watersoluble (Cosmetochem) (Cosmetochem International Ltd.)

LINOLEYL LACTATE

Empirical Formula:
$C_{21}H_{38}O_3$

Definition: Linoleyl Lactate is the ester of linoleyl alcohol and lactic acid. It conforms to the formula:

O
‖
CH₃CHC─OCH₂(CH₂)₇CH══CHCH₂CH══CH
|
OH
(CH₂)₄CH₃

Chemical Class: Esters

Function: Skin-Conditioning Agent - Emollient

Technical/Other Name:
2-Hydroxypropanoic Acid, Linoleyl Ester

Trade Name:
Pelemol LIL (Phoenix)

LINSEED ACID

CAS No. 68424-45-3

EINECS No. 270-304-8

Definition: Linseed Acid is the mixture of fatty acids derived from Linum Usitatissimum (Linseed) Oil (q.v.).

Information Sources: 21CFR175.105, 21CFR176.200, 21CFR176.210, TSCA

Chemical Class: Fatty Acids

Function: Surfactant - Cleansing Agent

Surfactant-Cleansing Agent is included as a function for the soap form of Linseed Acid.

Technical/Other Names:
Acids, Linseed
Fatty Acids, Linseed Oil
Linseed Oil Fatty Acid
Linum Usitatissimum (Linseed) Acid

LINUM USITATISSIMUM FLOWER EXTRACT

Definition: Linum Usitatissimum Flower Extract is an extract of the flowers of *Linum Usitatissimum*. See "Regulatory and Ingredient Use Information," regarding the labeling names for botanical ingredients in Volume 1, Introduction, Part A.

Chemical Class: Biological Products

Function: Not Reported

Technical/Other Name:
Extract of Linum Usitatissimum Flower

Trade Name Mixture:
Huile de Fleurs de Lin - GT10 (Greentech)

LINUM USITATISSIMUM (LINSEED) SEED EXTRACT

Definition: Linum Usitatissimum (Linseed) Seed Extract is an extract of the seeds of the linseed, *Linum usitatissimum*. See "Regulatory and Ingredient Use Information," regarding the labeling names for botanical ingredients in Volume 1, Introduction, Part A.

Chemical Class: Biological Products

Function: Skin-Conditioning Agent - Miscellaneous

Reported Product Categories: Bath Capsules; Bath Oils, Tablets, and Salts; Bath Preparations, Misc.; Body and Hand Preparations (Excluding Shaving Preparations); Cleansing Products (Cold Creams, Cleansing Lotions, Liquids and Pads); Face and Neck Preparations (Excluding Shaving Preparations); Fragrance Preparations, Misc.

Technical/Other Names:
Extract of Flaxseed
Extract of Linseed
Extract of Linum Usitatissimum
Flaxseed Extract
Linseed Extract
Linseed Seed Extract
Linum Usitatissimum Extract

Trade Name Mixtures:
Actiphyte of Flaxseed BG50 (Active Organics)

Actiphyte of Flaxseed GL50 (Active
Organics)
Actiphyte of Flaxseed Lipo S (Active
Organics)
Actiphyte of Flaxseed PG50 (Active
Organics)
Almondermin LS (Laboratoires Sero-
biologiques)
Elespher Almondermin (Laboratoires Sero-
biologiques)
Extrait de Lin MHC 100 (Yves Rocher)
Flaxseed HS (Alban Muller)
Flaxseed LS (Alban Muller)
Herbasol Extract Linseed (Cosmetochem)
(Cosmetochem International Ltd.)
Highcareen LS (Cognis Deutschland)
Natunola Flax Extract 130 (Natunola)
Natunola Flax Extract 160 (Natunola)
Natunola Flax Extract 1621 (Natunola)
VT-111 Extract of Flax Seed (Vege-Tech)

LINUM USITATISSIMUM (LINSEED) SEED OIL

CAS No.	EINECS No.
8001-26-1	232-278-6

Definition: Linum Usitatissimum (Linseed)
Seed Oil is the expressed oil from the dried
ripe seed of *Linum usitatissimum. See
"Regulatory and Ingredient Use
Information," regarding the labeling names
for botanical ingredients in Volume 1, Introduction, Part A.*

Information Sources: AUS, BP 1963,
21CFR175.105, 21CFR175.300,
21CFR176.200, 21CFR176.210,
21CFR181.22, 21CFR181.26, CZE, DA,
DDR, EGY, FI, FIN, HUN, IND, MAR, MI-12
(5533), PN, POL, POR, RIFM, ROM, TSCA,
YUG

Chemical Class: Fats and Oils

Functions: Fragrance Ingredient; Skin-
Conditioning Agent - Miscellaneous; Skin-
Conditioning Agent - Occlusive

Reported Product Categories: Bath Oils,
Tablets, and Salts; Cleansing Products (Cold
Creams, Cleansing Lotions, Liquids and
Pads); Skin Care Preparations, Misc.

Technical/Other Names:
Linseed Oil
Linseed oil absolute (RIFM)
Linseed Seed Oil
Oils, Linseed

Trade Names:
AC Flax Seed Oil (Active Concepts)
AEC Linseed Oil (A & E Connock)
Certified Organic Flax Seed Oil (Formula
One Sciences)
Flaxseed Oil (Alban Muller)
Linola Oil (Unilever (Czech))

Linum Usitatissmum Seed Oil ies (IES
LABO)

Trade Name Mixtures:
Extrapone Cereals GW 2/031300 (Symrise)
Extrapone Linseed GW 2/031305
(Symrise)
Rapeseed Oil Special (Henry Lamotte)

LIPASE

CAS No.	EINECS No.
9001-62-1	232-619-9

JPN Translation:
リパーゼ

Definition: Lipase is an enzyme that
hydrolyzes triglycerides.

Information Sources: 21CFR184.1415,
21CFR310.545, JCIC, JCLS, MI-13(5533)

Chemical Class: Proteins

Functions: Lytic Agent; Skin-Conditioning
Agent - Miscellaneous

Technical/Other Names:
Lipase (1)
Lipase (2)

Trade Names:
Lipase-MY (Meito Sangyo)
Lipase-OF (Meito Sangyo)
Lipase SNS (Degussa Care Specialties)

Trade Name Mixtures:
Bellsilk EZ-L (Ichimaru Pharcos)
Cyclolipase (Sederma)
Lipase MC (Suzuki Yushi)

LIPPIA CITRIODORA FLOWER EXTRACT

CAS No.	EINECS No.
85116-63-8	285-515-0

Definition: Lippia Citriodora Flower
Extract is an extract of the flowering ends
of the lemon verbena, *Lippia citriodora.
See "Regulatory and Ingredient Use Information," regarding the labeling names for
botanical ingredients in Volume 1, Introduction, Part A.*

Information Source: 21CFR172.510

Chemical Class: Biological Products

Function: Skin-Conditioning Agent - Miscellaneous

Technical/Other Names:
Extract of Lemon Verbena
Extract of Lippia Citriodora
Lemon Verbena Extract
Lemon Verbena (Lippia Citriodora) Extract

Trade Name Mixtures:
Actiphyte of Lemon Verbena BG50 (Active
Organics)

Actiphyte of Lemon Verbena GL50 (Active
Organics)
Actiphyte of Lemon Verbena Lipo S (Active
Organics)
Actiphyte of Lemon Verbena PG50 (Active
Organics)
Extrait de Verveine MP PG 40 (Yves
Rocher)
Lemon Verbena HS (Alban Muller)

LIPPIA CITRIODORA FLOWER WATER

Definition: Lippia Citriodora Flower Water
is an aqueous solution of the steam
distillate obtained from the flowers of *Lippia
citriodora. See "Regulatory and Ingredient
Use Information," regarding the labeling
names for botanical ingredients in Volume
1, Introduction, Part A.*

Chemical Class: Biological Products

Function: Fragrance Ingredient

Technical/Other Name:
Water, Lippia Citriodora Flower

Trade Name:
Lemon Verbena Water (Alban Muller)

LIPPIA CITRIODORA LEAF EXTRACT

Definition: Lippia Citriodora Leaf Extract
is an extract of the leaves of *Lippia
citriodora. See "Regulatory and Ingredient
Use Information," regarding the labeling
names for botanical ingredients in Volume
1, Introduction, Part A.*

Chemical Class: Biological Products

Function: Cosmetic Astringent

Technical/Other Name:
Extract of Lippia Citriodora Leaves

Trade Name Mixtures:
Actiphyte Lemon Verbena Lipo AK (Active
Organics)
Glycolysat of Lemon Verbena (CEP
(Solabia))
Herbasol-Extract Vervain (Cosmetochem)

LIPPIA CITRIODORA WATER

Definition: Lippia Citriodora Water is an
aqueous solution of the steam distillate
obtained from *Lippia citriodora. See
"Regulatory and Ingredient Use
Information," regarding the labeling names
for botanical ingredients in Volume 1, Introduction, Part A.*

Chemical Class: Biological Products

Function: Fragrance Ingredient

Trade Name:
Lemon Verbena Hydroflorate (Bayliss Ranch)

LIQUIDAMBAR STYRACIFLUA OIL

Definition: Liquidambar Styraciflua Oil is the volatile oil obtained from the exudate of *Liquidambar styraciflua. See "Regulatory and Ingredient Use Information," regarding the labeling names for botanical ingredients in Volume 1, Introduction, Part A.*

Chemical Class: Essential Oils

Function: Fragrance Ingredient

Technical/Other Name:
Oils, Liquidambar Styraciflua

Trade Name:
Styrax Oil (Chauvet)

LIRIOSMA OVATA EXTRACT

Definition: Liriosma Ovata Extract is an extract of the wood of *Liriosma ovata See "Regulatory and Ingredient Use Information," regarding the labeling names for botanical ingredients in Volume 1, Introduction, Part A.*

Chemical Class: Biological Products

Function: Not Reported

Trade Name Mixture:
Herbasol Extract Muira-Puama (Cosmetochem International Ltd.)

LITCHI CHINENSIS FRUIT EXTRACT

Definition: Litchi Chinensis Fruit Extract is an extract of the fruit of *Litchi chinensis. See "Regulatory and Ingredient Use Information," regarding the labeling names for botanical ingredients in Volume 1, Introduction, Part A.*

Chemical Class: Biological Products

Function: Skin-Conditioning Agent - Miscellaneous

Technical/Other Name:
Extract of Litchi Chinensis

Trade Name Mixtures:
Cosflor Lychee HGS (A & E Connock)
Extrait Hydroglycolique de Litchis-GT10 (Greentech)
Litchee HS (Alban Muller)
Litchi Chinensis Fruit Extract ies (IES LABO)
Lychee Extract (Cosmetic Developments)

LITCHI CHINENSIS SEED EXTRACT

Definition: Litchi Chinensis Seed Extract is an extract of the seeds of *Litchi chinensis. See "Regulatory and Ingredient Use Information," regarding the labeling names for botanical ingredients in Volume 1, Introduction, Part A.*

Chemical Class: Biological Products

Function: Skin-Conditioning Agent - Miscellaneous

Technical/Other Name:
Extract of Litchi Chinensis Seed

Trade Name Mixture:
Litchi Liquid B (Ichimaru Pharcos)

LITCHI CHINENSIS WATER

Definition: Litchi Chinensis Water is an aqueous solution of the steam distillate obtained from the fruit of *Litchi chinensis. See "Regulatory and Ingredient Use Information," regarding the labeling names for botanical ingredients in Volume 1, Introduction, Part A.*

Chemical Class: Biological Products

Function: Skin-Conditioning Agent - Miscellaneous

Technical/Other Name:
Water, Litchi Chinensis

Trade Name:
Extrait Originel Litchi (Gattefosse s.a.)

LITHIUM FLUORIDE

CAS No. 7789-24-4   **EINECS No.** 232-152-0

Definition: Lithium Fluoride is the inorganic salt that conforms to the formula:

$$LiF$$

Information Source: MI-13(5553)

Chemical Class: Inorganic Salts

Functions: Buffering Agent; Oral Care Agent

LITHIUM GLUCONATE

CAS No. 60816-70-8   **EINECS No.** 262-443-8

Empirical Formula:
$C_6H_{11}O_7 \cdot Li$

Definition: Lithium Gluconate is the lithium salt of Gluconic Acid (q.v.). It conforms to the formula:

$$[CH_2OH(CHOH)_4COO^-] \quad Li^+$$

Chemical Classes: Organic Salts; Polyols

Function: Not Reported

Technical/Other Name:
Gluconic Acid, Lithium Salt

Trade Name:
Givobio GLi (SEPPIC)

LITHIUM GUANOSINE TRIPHOSPHATE

CAS No. 85737-04-8   **EINECS No.** 288-514-3

Empirical Formula:
$C_{10}H_{16}N_5O_{14}P \cdot 3Li$

Definition: Lithium Guanosine Triphosphate is the heterocyclic compound that conforms to the formula:

Chemical Classes: Amines; Carbohydrates; Heterocyclic Compounds; Phosphorus Compounds

Function: Skin-Conditioning Agent - Miscellaneous

Technical/Other Name:
Guanosine 5'-(Tetrahydrogen Triphosphate), Lithium Salt

LITHIUM HYDROXIDE

CAS No. 1310-65-2   **EINECS No.** 215-183-4

Empirical Formula:
HLiO

Definition: Lithium Hydroxide is the inorganic base that conforms to the formula:

$$LiOH$$

Information Sources: EEC(III/1-15b), MI-13(5556), TSCA, USAN, USP XXIV

Chemical Class: Inorganic Bases

Function: pH Adjuster

Reported Product Category: Hair Straighteners

Technical/Other Name:
Lithium Hydrate

LITHIUM MAGNESIUM SILICATE

CAS No. 37220-90-9
EINECS No. 253-408-8

Definition: Lithium Magnesium Silicate is a synthetic silicate clay consisting mainly of lithium and magnesium silicates.

Information Source: CIR: [S] IJT-22 (SUPPL. 1)2003

Chemical Classes: Inorganic Salts; Inorganics

Functions: Binder; Bulking Agent; Viscosity Increasing Agent - Aqueous

Reported Product Category: Lipsticks

Technical/Other Name:
Silicic Acid, Lithium Magnesium Salt

Trade Name:
LUCENTITE SAN (Co-Op/Kobo)

LITHIUM MAGNESIUM SODIUM SILICATE

CAS No. 53320-86-8
EINECS No. 258-476-2

Definition: Lithium Magnesium Sodium Silicate is a synthetic silicate clay consisting mainly of lithium, magnesium and sodium silicates.

Information Sources: CIR: [S] IJT-22 (SUPPL. 1)2003, TSCA

Chemical Class: Inorganics

Functions: Bulking Agent; Viscosity Increasing Agent - Aqueous

Technical/Other Names:
Magnesium Lithium Sodium Silicate
Silicic Acid, Lithium, Magnesium, Sodium Salt
Sodium Lithium Magnesium Silicate

Trade Names:
LUCENTITE-SWN (Co-Op/Kobo)
Smectite SWN (Nikko)

LITHIUM OXIDIZED POLYETHYLENE

Definition: Lithium Oxidized Polyethylene is the lithium salt of Oxidized Polyethylene (q.v.).

Chemical Classes: Organic Salts; Synthetic Polymers

Function: Viscosity Increasing Agent - Nonaqueous

Trade Name:
Petrolite C-400 Polymer (Baker Petrolite)

LITHIUM STEARATE

CAS No. 4485-12-5
EINECS No. 224-772-5

Empirical Formula:
$C_{18}H_{36}O_2 \cdot Li$

Definition: Lithium Stearate is the lithium salt of stearic acid. It conforms generally to the formula:

$$\left[CH_3(CH_2)_{16}COO^- \right] \; Li^+$$

Information Sources: 21CFR175.300, CIR: [S] JACT-1(2)1982, CTFA D, TSCA

Chemical Class: Soaps

Functions: Anticaking Agent; Binder; Opacifying Agent; Slip Modifier; Viscosity Increasing Agent - Nonaqueous

Reported Product Categories: Eye Shadows; Foundations; Blushers (All types); Makeup Bases

Technical/Other Names:
Lithium Octadecanoate
Octadecanoic Acid, Lithium Salt
Stearic Acid, Lithium Salt

LITHIUM SULFIDE

CAS No. 12136-58-2
EINECS No. 235-228-1

Empirical Formula:
Li_2S

Definition: Lithium Sulfide is the inorganic salt that conforms to the formula:

$$Li_2S$$

Information Sources: EEC(III/1-23), TSCA

Chemical Class: Inorganic Salts

Function: Depilating Agent

Technical/Other Name:
Dilithium Sulfide

LITHOSPERMUM ERYTHRORHIZON ROOT

JPN Translation:
シコン

Definition: Lithospermum Erythrorhizon Root is the powdered root of *Lithospermum erythrorhizon*. See *"Regulatory and Ingredient Use Information,"* regarding the labeling names for botanical ingredients in *Volume 1, Introduction, Part A.*

Information Source: JCIC

Chemical Class: Biological Products

Function: Skin-Conditioning Agent - Miscellaneous

Technical/Other Name:
Root of Lithospermum Erythrorhizon

LITHOSPERMUM ERYTHRORHIZON ROOT EXTRACT

JPN Translation:
シコンエキス

Definition: Lithospermum Erythrorhizon Root Extract is an extract of the dried roots of *Lithospermum erythrorhizon*. See *"Regulatory and Ingredient Use Information,"* regarding the labeling names for botanical ingredients in Volume 1, Introduction, Part A.

Information Source: JCLS

Chemical Class: Biological Products

Function: Not Reported

Technical/Other Name:
Extract of Lithospermum Erythrorhizon

Trade Name:
Shikonin (Maruzen Pharmaceuticals Co., Ltd.)

Trade Name Mixtures:
Extract LE (Sino Lion)
Shikon Extract TH (T.HASEGAWA)
Shikon Extract TH-L (T.HASEGAWA)

LITHOSPERMUM OFFICINALE EXTRACT

CAS No. 90063-58-4
EINECS No. 290-017-1

Definition: Lithospermum Officinale Extract is an extract of *Lithospermum officinale*. See *"Regulatory and Ingredient Use Information,"* regarding the labeling names for botanical ingredients in Volume 1, Introduction, Part A.

Chemical Class: Biological Products

Function: Skin-Conditioning Agent - Miscellaneous

Technical/Other Name:
Extract of Lithospermum Officinale

Trade Name Mixture:
Gromwell HS (Alban Muller)

LITHOSPERMUM OFFICINALE ROOT EXTRACT

JPN Translation:
シコンエキス

Definition: Lithospermum Officinale Root Extract is an extract of the roots of *Lithospermum officinale*. See *"Regulatory and Ingredient Use Information,"* regarding the labeling names for botanical ingredients in Volume 1, Introduction, Part A.

Information Sources: JCIC, JCLS

Chemical Class: Biological Products

Function: Not Reported

Technical/Other Names:
Extract of Lithospermum
Extract of Lithospermum Officinale
Lithospermum Extract
Lithospermum Root Extract
Radix Lithospermi Extract

Trade Name Mixtures:
Crodarom Zi Cao (Croda, Inc.)
NOVAPUR Zi Cao oil soluble (Crodarom)
NOVAPUR Zi Cao water soluble (Crodarom)
Shiconix Liquid AB (N) (Ichimaru Pharcos)
Shiconix Liquid BG (Ichimaru Pharcos)
Shiconix Liquid MD (Ichimaru Pharcos)

LITHOSPERMUM ROOT EXTRACT SERUM ALBUMIN SUCCINATE

Definition: Lithospermum Root Extract Serum Albumin Succinate is the product obtained by the reaction of Lithospermum Erythrorhizone Root Extract (q.v.) with succinylated Serum Albumin (q.v.).

Chemical Class: Biological Products

Function: Skin-Conditioning Agent - Miscellaneous

Trade Name Mixture:
Shiconix Liquid AB (Ichimaru Pharcos)

LITHOTHAMNIUM CALCARUM EXTRACT

Definition: Lithothamnium Calcarum Extract is an extract of the red alga, *Lithothamnium calcarum. See "Regulatory and Ingredient Use Information," regarding the labeling names for botanical ingredients in Volume 1, Introduction, Part A.*

Chemical Class: Biological Products

Function: Not Reported

Technical/Other Names:
Extract of Lithotamnium Calcarum
Extract of Lithotamniun Calcarum

Trade Name Mixtures:
Maerl MP PG 40 (Yves Rocher)
Sea Extract Lithothamnium (E.U.K)

LITHOTHAMNIUM CALCARUM POWDER

Definition: Lithothamnium Calcarum Powder is a powder of the finely ground red alga, *Lithothamnium calcarum. See "Regulatory and Ingredient Use*

Information," regarding the labeling names for botanical ingredients in Volume 1, Introduction, Part A.*

Chemical Class: Biological Products

Function: Abrasive

Technical/Other Name:
Powdered Lithothamnium Calcarum

Trade Name:
Exfosea Lithothamnium (CEP (Solabia))

Trade Name Mixtures:
AquaMin C-F (MG FORCE)
Odyceane (Setalg)

LITHOTHAMNIUM CORALLIOIDES POWDER

Definition: Lithothamnium Corallioides Powder is a powder of the finely ground alga, *Lithothamnium corralioides. See "Regulatory and Ingredient Use Information," regarding the labeling names for botanical ingredients in Volume 1, Introduction, Part A.*

Chemical Class: Biological Products

Function: Abrasive

Trade Name Mixture:
AquaMin C-F (MG FORCE)

LITSEA CUBEBA FRUIT OIL

CAS No.: 68855-99-2

Definition: Litsea Cubeba Fruit Oil is the volatile oil obtained from the berries of *Litsea cubeba. See "Regulatory and Ingredient Use Information," regarding the labeling names for botanical ingredients in Volume 1, Introduction, Part A.*

Information Sources: MI-13(6830), RIFM

Chemical Class: Essential Oils

Function: Fragrance Ingredient

Technical/Other Names:
Cubeba Oil
Litsea cubeba oil (RIFM)

Trade Names:
AEC Litsea Cubeba Oil (A & E Connock)
Litsea Cubeba Essential Oil (Robertet S.A.)

LITSEA GLUTINOSA BARK EXTRACT

Definition: Litsea Glutinosa Bark Extract is an extract of the bark of *Litsea glutinosa. See "Regulatory and Ingredient Use Information," regarding the labeling names for*

botanical ingredients in Volume 1, Introduction, Part A.*

Chemical Class: Biological Products

Function: Skin-Conditioning Agent - Miscellaneous

Technical/Other Name:
Extract of Litsea Glutinosa Bark

Trade Name Mixture:
Litsea Glutinosa Extract (Libiol)

LIVER EXTRACT

CAS No.	EINECS No.
8002-47-9	232-309-3

JPN Translation:
ウシ肝臓エキス

Definition: Liver Extract is an extract of bovine livers. *See "Regulatory and Ingredient Use Information," regarding use of EU Trivial names in Volume 1, Introduction, Part A.*

Information Sources: BAN, MI-12(5575)

Chemical Class: Biological Products

Function: Not Reported

LIVER HYDROLYSATE

Definition: Liver Hydrolysate is the hydrolysate of animal liver tissue derived by acid, enzyme or other method of hydrolysis.

Chemical Class: Biological Products

Function: Not Reported

LOCUST BEAN HYDROXYPROPYL-TRIMONIUM CHLORIDE

Definition: Locust Bean Hydroxypropyltrimonium Chloride is the quaternary ammonium chloride formed by the reaction of hydroxypropyl trimethylamine and Ceratonia siliqua (locust bean) gum. It conforms generally to the formula:

$$\left[RCH_2CHCH_2-\overset{\overset{\displaystyle CH_3}{|}}{\underset{\underset{\displaystyle CH_3}{|}}{N}}-CH_3 \atop \underset{\displaystyle OH}{|} \right]^+ \quad Cl^-$$

where R represents the locust bean gum moiety.

Chemical Classes: Gums, Hydrophilic Colloids and Derivatives; Quaternary Ammonium Compounds

Functions: Antistatic Agent; Hair Conditioning Agent; Skin-Conditioning Agent - Miscellaneous

Trade Name:
Catinal CL-100 (Toho)

LOESS

Definition: Loess is a loose mineral sediment formed by wind action during the Ice Age.

Chemical Class: Inorganics

Functions: Abrasive; Absorbent; Bulking Agent

Trade Name:
Luvos (Heilerde)

LONGIFOLENE

CAS No. 475-20-7   **EINECS No.** 207-491-2

Empirical Formula:
$C_{15}H_{24}$

Definition: Longifolene is the organic compound that conforms to the formula:

$$CH_3CHCHCCH_2O—\overset{\overset{CH_3}{|}}{\underset{|}{|}}$$

Information Sources: MI-13(5588), RIFM

Chemical Class: Hydrocarbons

Functions: Fragrance Ingredient; Skin-Conditioning Agent - Miscellaneous

Technical/Other Names:
Junipene
Longifolene (RIFM)
1,4-Methanoazulene, Decahydro-4,8,8-Trimethyl-9-Methylene-, (1S,3aR,4S,8aS)-

LONICERA CAERULEA FRUIT JUICE

Definition: Lonicera Caerulea Fruit Juice is the juice expressed from the fruit of Haskaap berries, *Lonicera caerulea*. See *"Regulatory and Ingredient Use Information,"* regarding the labeling names for botanical ingredients in Volume 1, Introduction, Part A.

Chemical Class: Biological Products

Function: Humectant

Technical/Other Name:
Lonicera Caerulea (Haskaap) Fruit Juice

Trade Name Mixture:
Haskaap Juice BG (Nippon Shinyaku)

LONICERA CAERULEA FRUIT WATER

Definition: Lonicera Caerulea Fruit Water is an aqueous solution of the steam distillate obtained from the fruit of *Lonicera caerulea*. See *"Regulatory and Ingredient Use Information,"* regarding the labeling names for botanical ingredients in Volume 1, Introduction, Part A.

Chemical Class: Biological Products

Functions: Flavoring Agent; Humectant

Trade Name Mixture:
Haskaap Fruit Water FN (Nippon Shinyaku)

LONICERA CAPRIFOLIUM (HONEY-SUCKLE) EXTRACT

Definition: Lonicera Caprifolium (Honeysuckle) Extract is an extract of the whole plant of *Lonicera caprifolium*. See *"Regulatory and Ingredient Use Information,"* regarding the labeling names for botanical ingredients in Volume 1, Introduction, Part A.

Chemical Class: Biological Products

Function: Skin-Conditioning Agent - Miscellaneous

Technical/Other Names:
Honeysuckle Plant Extract
Lonicera caprifolium Extract

Trade Name Mixture:
Vegetol Honeysuckle GR 115 Hydro (Gattefosse s.a.)

LONICERA CAPRIFOLIUM (HONEY-SUCKLE) FLOWER EXTRACT

CAS No. 84603-62-3   **EINECS No.** 283-263-6

JPN Translation:
ハニーサックル花エキス

Definition: Lonicera Caprifolium (Honeysuckle) Flower Extract is an extract of the flowers of the honeysuckle, *Lonicera caprifolium*. See *"Regulatory and Ingredient Use Information,"* regarding the labeling names for botanical ingredients in Volume 1, Introduction, Part A.

Information Sources: JCIC, JCLS, JSQI

Chemical Class: Biological Products

Function: Not Reported

Reported Product Category: Shampoos (Non-coloring)

Technical/Other Names:
Extract of Honeysuckle
Extract of Lonicera Caprifolium
Honeysuckle Extract
Honeysuckle Flower Extract
Lonicera caprifolium Extract

Trade Name Mixtures:
Actiphyte of Honeysuckle BG50 (Active Organics)
Actiphyte of Honeysuckle GL50 (Active Organics)
Actiphyte of Honeysuckle Lipo S (Active Organics)
Actiphyte of Honeysuckle PG50 (Active Organics)
Extrait de Chevrefeuille PPE PG 20 (Yves Rocher)
Honey Suckle Extract HS 2551 G (Grau)
Lipoplastidine Lonicera (Vevy)
Lonicera Caprifolium (Honey-Suckle) Flower Extract ies (IES LABO)
VT-306 Extract of Honeysuckle (Vege-Tech)

LONICERA JAPONICA (HONEYSUCKLE) FLOWER EXTRACT

JPN Translation:
スイカズラ花エキス

Definition: Lonicera Japonica (Honeysuckle) Flower Extract is an extract of the flowers of *Lonicera japonica*. See *"Regulatory and Ingredient Use Information,"* regarding the labeling names for botanical ingredients in Volume 1, Introduction, Part A.

Chemical Class: Biological Products

Function: Skin-Conditioning Agent - Miscellaneous

Reported Product Category: Shampoos (Non-coloring)

Technical/Other Names:
Extract of Honeysuckle
Extract of Lonicera Japonica
Honeysuckle Extract
Honeysuckle Flower Extract
Kinginka Ekisu (JPN)
Lonicera Japonica Extract
Suikazura Ekisu (JPN)

Trade Name:
Honeysuckle Flower Extract Powder (Maruzen Pharmaceuticals Co., Ltd.)

Trade Name Mixtures:
Bathgranue Suikazura (Ichimaru Pharcos)
Honeysuckle Flower Extract (Maruzen Pharmaceuticals Co., Ltd.)
Honeysuckle Flower Extract BG (Maruzen Pharmaceuticals Co., Ltd.)

Honeysuckle Flower Extract-J (Maruzen Pharmaceuticals Co., Ltd.)
Kinginka Liquid B (Ichimaru Pharcos)
Kinginka Liquid E (Ichimaru Pharcos)
Sinopurete (I.D. bio)
YSK Magic 11 (Phyto-Technologies)

LONICERA JAPONICA (HONEYSUCKLE) LEAF EXTRACT

JPN Translation:
スイカズラ葉エキス

Definition: Lonicera Japonica (Honeysuckle) Leaf Extract is an extract of the leaves of the honeysuckle, *Lonicera japonica*. See "Regulatory and Ingredient Use Information," regarding the labeling names for botanical ingredients in Volume 1, Introduction, Part A.

Chemical Class: Biological Products

Function: Not Reported

Technical/Other Names:
Extract of Honeysuckle Leaf
Extract of Lonicera Japonica Leaf
Honeysuckle Leaf Extract
Lonicera Japonica Leaf Extract
Nindou Ekisu (JPN)
Suikazura Ekisu (JPN)

Trade Name Mixtures:
Honeysuckle Extract (Maruzen Pharmaceuticals Co., Ltd.)
Honeysuckle Extract BG (Maruzen Pharmaceuticals Co., Ltd.)
Honeysuckle Extract-J (Maruzen Pharmaceuticals Co., Ltd.)
Honeysuckle Extract Powder-S (Maruzen Pharmaceuticals Co., Ltd.)
Suikazura Liquid B (Ichimaru Pharcos)
Suikazura Liquid E (Ichimaru Pharcos)

LOTUS CORNICULATUS FLOWER EXTRACT

CAS No. 84696-24-2
EINECS No. 283-643-1

Definition: Lotus Corniculatus Flower Extract is an extract of the flowers of *Lotus corniculatus*. See "Regulatory and Ingredient Use Information," regarding the labeling names for botanical ingredients in Volume 1, Introduction, Part A.

Chemical Class: Biological Products

Function: Not Reported

Technical/Other Names:
Bird's Foot Trefoil Extract
Extract of Lotus Corniculatus

Trade Name Mixtures:
Actiphyte of Lotus Blossom BG50 (Active Organics)
Actiphyte of Lotus Blossom GL50 (Active Organics)
Actiphyte of Lotus Blossom Lipo S (Active Organics)
Actiphyte of Lotus Blossom PG50 (Active Organics)
Lotus Corniculatus Extract HS 3130 G (Grau)

LOTUS CORNICULATUS SEED EXTRACT

Definition: Lotus Corniculatus Seed Extract is an extract of the seeds of *Lotus corniculatus*. See "Regulatory and Ingredient Use Information," regarding the labeling names for botanical ingredients in Volume 1, Introduction, Part A.

Chemical Class: Biological Products

Function: Skin-Conditioning Agent - Miscellaneous

Technical/Other Name:
Extract of Lotus Corniculatus Seed

Trade Name Mixture:
Lotus Extract (IES LABO)

LUFFA CYLINDRICA EXTRACT

JPN Translation:
ヘチマエキス

Definition: Luffa Cylindrica Extract is an extract of the fruit and aerial parts of the luffa, *Luffa cylindrica*. See "Regulatory and Ingredient Use Information," regarding the labeling names for botanical ingredients in Volume 1, Introduction, Part A.

Information Sources: JCIC, JCLS, JSQI

Chemical Class: Biological Products

Function: Not Reported

Technical/Other Names:
Extract of Loofah
Extract of Luffa Cylindrica
Luffa Extract
Sponge Gourd Extract (1)
Sponge Gourd Extract (2)

Trade Name:
Loofah Extract (Koshiro)

Trade Name Mixtures:
Bathgranue Hetima (Ichimaru Pharcos)
Hetima Liquid (BG) (Ichimaru Pharcos)
Luffa Extract (Maruzen Pharmaceuticals Co., Ltd.)
Luffa Extract BG (Maruzen Pharmaceuticals Co., Ltd.)
Luffa Extract BG-J (Maruzen Pharmaceuticals Co., Ltd.)
Luffa Extract BG-JC (Maruzen Pharmaceuticals Co., Ltd.)
Luffa Extract-J (Maruzen Pharmaceuticals Co., Ltd.)
Luffa Extract-JC (Maruzen Pharmaceuticals Co., Ltd.)
Luffa Extract LA (Maruzen Pharmaceuticals Co., Ltd.)
Luffa Extract LA-EV (Maruzen Pharmaceuticals Co., Ltd.)
Luffa Extract LA-J (Maruzen Pharmaceuticals Co., Ltd.)
Luffa Extract LA-JC (Maruzen Pharmaceuticals Co., Ltd.)
Luffa Extract Powder-S (Maruzen Pharmaceuticals Co., Ltd.)

LUFFA CYLINDRICA FRUIT

JPN Translation:
ヘチマ

Definition: Luffa Cylindrica Fruit is a plant material obtained from the fruit of the sponge *Luffa cylindrica*. See "Regulatory and Ingredient Use Information," regarding the labeling names for botanical ingredients in Volume 1, Introduction, Part A.

Information Sources: JCIC, JCLS, JSQI

Chemical Class: Biological Products

Function: Abrasive

Technical/Other Names:
Loofah
Luffa
Sponge Gourd Powder

Trade Names:
AEC Loofah Ground (A & E Connock)
Lipo Lufa 30/100 (Lipo)

Trade Name Mixtures:
Lipo Lufa Blue (Lipo)
Lipo Lufa Burgandy (Lipo)
Lipo Lufa Green (Lipo)
Lipo Lufa Violet (Lipo)

LUFFA CYLINDRICA LEAF EXTRACT

Definition: Luffa Cylindrica Leaf Extract is an extract of the leaf of *Luffa cylindrica*. See "Regulatory and Ingredient Use Information," regarding the labeling names for botanical ingredients in Volume 1, Introduction, Part A.

Chemical Class: Biological Products

Function: Skin-Conditioning Agent - Miscellaneous

Technical/Other Names:
Extract of Luffa Cylindrica
Loofah Leaf Extract

LUFFA CYLINDRICA SEED OIL

Definition: Luffa Cylindrica Seed Oil is the fixed oil expressed from the seeds of *Luffa* *cylindrica. See "Regulatory and Ingredient Use Information," regarding the labeling names for botanical ingredients in Volume 1, Introduction, Part A.*

Chemical Class: Fats and Oils

Function: Skin-Conditioning Agent - Occlusive

Trade Name:
Sponge Gourd Oil (Sederma)

LUFFA CYLINDRICA STEM SAP

Definition: Luffa Cylindrica Stem Sap is the sap obtained from the stems of *Luffa cylindrica. See "Regulatory and Ingredient Use Information," regarding the labeling names for botanical ingredients in Volume 1, Introduction, Part A.*

Chemical Class: Biological Products

Function: Skin-Conditioning Agent - Miscellaneous

Technical/Other Name:
Juice, Luffa Cylindrica Stem

Trade Name Mixture:
Hetimasui 96 (Sunstar)

LUFFA OPERCULATA EXTRACT

CAS No.	EINECS No.
90063-68-6	290-028-1

Definition: Luffa Operculata Extract is an extract of the plant, *Luffa operculata. See "Regulatory and Ingredient Use Information," regarding the labeling names for botanical ingredients in Volume 1, Introduction, Part A.*

Chemical Class: Biological Products

Function: Skin-Conditioning Agent - Miscellaneous

Technical/Other Name:
Extract of Luffa Operculata

Trade Name Mixture:
Luffa Extract 3516 G (Grau)

LUPINE AMINO ACIDS

Definition: Lupine Amino Acids is a mixture of amino acids derived from the complete hydrolysis of Lupine Protein (q.v.).

Chemical Class: Amino Acids

Functions: Hair Conditioning Agent; Skin-Conditioning Agent - Humectant

Trade Names:
AC Lupein Amino Acids (Active Concepts)
Hydrolupin AA (Croda, Inc.)

LUPINUS ALBUS

Definition: *See "Regulatory and Ingredient Use Information," regarding EU labeling names for botanical ingredients in Volume 1, Introduction, Part A.*

Chemical Class: Biological Products

Technical/Other Names:
Lupinus Albus Oil Unsaponifiables (U.S.)
Lupinus Albus Protein (U.S.)
Lupinus Albus Seed Extract (U.S.)
Lupinus Albus Seed Oil (U.S.)

LUPINUS ALBUS OIL UNSAPONIFIABLES

Definition: Lupinus Albus Oil Unsaponifiables is the fraction of Lupinus Albus Oil (q.v.) which is not saponified during the refining of Lupinus Albus Oil (q.v.). *See "Regulatory and Ingredient Use Information," regarding the labeling names for botanical ingredients in Volume 1, Introduction, Part A.*

Chemical Class: Unsaponifiables

Functions: Hair Conditioning Agent; Skin-Conditioning Agent - Miscellaneous

Technical/Other Names:
Lupin (Lupinus Albus) Oil Unsaponifiables
Lupinus Albus (EU)

Trade Name:
Lupin Oil Unsaponifiable (Expanscience)

LUPINUS ALBUS PROTEIN

Definition: Lupinus Albus Protein is the protein derived from the seeds of *Lupinus albus. See "Regulatory and Ingredient Use Information," regarding the labeling names for botanical ingredients in Volume 1, Introduction, Part A.*

Chemical Class: Proteins

Functions: Skin-Conditioning Agent - Emollient; Skin-Conditioning Agent - Miscellaneous

Technical/Other Names:
Lupine Protein
Lupinus Albus (EU)

LUPINUS ALBUS SEED EXTRACT

CAS No.	EINECS No.
84082-55-3	282-001-8

Definition: Lupinus Albus Seed Extract is an extract of the seeds of the lupin, *Lupinus albus. See "Regulatory and Ingredient Use Information," regarding the labeling names for botanical ingredients in Volume 1, Introduction, Part A.*

Information Source: MI-13(5629)

Chemical Class: Biological Products

Function: Skin-Conditioning Agent - Miscellaneous

Technical/Other Names:
Extract of Lupin Seeds
Extract of Lupinus Albus Seeds
Lupin Extract
Lupin (Lupinus Albus) Extract
Lupin Seed Extract
Lupinus Albus (EU)

Trade Name Mixtures:
Actiphyte of Lupine BG50 (Active Organics)
Actiphyte of Lupine GL50 (Active Organics)
Actiphyte of Lupine Lipo S (Active Organics)
Actiphyte of Lupine PG50 (Active Organics)
Gatuline Lifting (Gattefosse s.a.)
Lupeol (Expanscience)
Lupilift (Silab)
Lupilift ST (Silab)
Lupine HS (Alban Muller)
Lupine LS (Alban Muller)
Lupine Phytolait (Alban Muller)
Structurine AVP (Silab)
White Lupin Milk (CEP (Solabia))

LUPINUS ALBUS SEED OIL

Definition: Lupinus Albus Seed Oil is the fixed oil expressed from the seeds of the lupin, *Lupinus albus. See "Regulatory and Ingredient Use Information," regarding the labeling names for botanical ingredients in Volume 1, Introduction, Part A.*

Chemical Class: Fats and Oils

Functions: Skin-Conditioning Agent - Emollient; Skin-Conditioning Agent - Occlusive

Technical/Other Names:
Lupin (Lupinus Albus) Oil
Lupinus Albus (EU)
Oils, Lupin

Trade Name:
Lupin Oil (Expanscience)

Trade Name Mixture:
Alpha-Lupaline (Expanscience)

LUPINUS LUTEUS

Definition: *See "Regulatory and Ingredient Use Information," regarding EU labeling*

The inclusion of any compound in the *Dictionary and Handbook* does not indicate that use of that substance as a cosmetic ingredient complies with the laws and regulations governing such use in the United States or any other country.

names for botanical ingredients in Volume 1, Introduction, Part A.

Chemical Class: Biological Products

Technical/Other Name:
Lupinus Luteus Seed Extract (U.S.)

LUPINUS LUTEUS SEED EXTRACT

Definition: Lupinus Luteus Seed Extract is an extract of the seeds of the lupin, *Lupinus luteus*. See "Regulatory and Ingredient Use Information," regarding the labeling names for botanical ingredients in Volume 1, Introduction, Part A.

Chemical Class: Biological Products

Function: Not Reported

Technical/Other Names:
Extract of Lupin Seeds
Extract of Lupinus Luteus Seeds
Lupin Extract
Lupin (Lupinus Luteus) Extract
Lupin Seed Extract
Lupinus Luteus (EU)

Trade Name Mixture:
Lupine Extract HS 3007 G (Grau)

LUPINUS SUBCARNOSUS SYMBIOSOME EXTRACT

Definition: Lupinus Subcarnosus Symbiosome Extract is an extract of the symbiosome (root nodule) of *Lupinus subcarnosus*. See "Regulatory and Ingredient Use Information," regarding the labeling names for botanical ingredients in Volume 1, Introduction, Part A.

Chemical Class: Biological Products

Functions: Antioxidant; Skin-Conditioning Agent - Miscellaneous

Technical/Other Name:
Extract of Lupinus Subcarnosus Symbiosome

Trade Name Mixture:
Lupine Zymbiozome Fermentum (Arch Personal Care Products)

LYCIUM CHINENSE FRUIT EXTRACT

Definition: Lycium Chinense Fruit Extract is an extract of the fruit of *Lycium chinense*. See "Regulatory and Ingredient Use Information," regarding the labeling names for botanical ingredients in Volume 1, Introduction, Part A.

Chemical Class: Biological Products
Function: Antioxidant
Technical/Other Name:
Extract of Lycium Chinense Fruit
Trade Name Mixture:
Draco Lycium Chinensis Fructus Full Spectrum Standardized Extract (Draco)

LYCIUM CHINENSE ROOT EXTRACT

Definition: Lycium Chinense Root Extract is an extract of the root bark of *Lycium chinense*. See "Regulatory and Ingredient Use Information," regarding the labeling names for botanical ingredients in Volume 1, Introduction, Part A.

Chemical Class: Biological Products
Function: Not Reported
Technical/Other Name:
Extract of Lycium Chinense
Trade Name Mixtures:
Draco Lycium Chinensis Cortex Full Spectrum Standardized Extract (Draco)
Jikoppi Liquid E (Ichimaru Pharcos)

LYCOPENE
CAS No.　　　**EINECS No.**
502-65-8　　　207-949-1
Empirical Formula:
$C_{40}H_{56}$
Definition: Lycopene is the organic compound that conforms to the formula:

Chemical Class: Hydrocarbons

Function: Antioxidant

Technical/Other Names:
(all-E)-2,6,10,14,19,23,27,31-Octamethyl-2, 6,8,10,12,14,16,18,20,22,24,26,30-Dotriacontatridecaene
CI 75125
trans-Lycopene

Trade Name Mixtures:
Lyc-O-Mato 2-4%SG NG(LycoRed USA)
Lyc-O-Mato 2-4%SG (LycoRed USA)

LYCOPODIUM CLAVATUM EXTRACT

Definition: Lycopodium Clavatum Extract is an extract of the whole plant, *Lycopodium clavatum*. See "Regulatory and Ingredient Use Information," regarding the labeling names for botanical ingredients in Volume 1, Introduction, Part A.

Chemical Class: Biological Products

Function: Skin-Conditioning Agent - Humectant

Technical/Other Name:
Extract of Lycopodium Clavatum

Trade Name Mixture:
Shinkinsou Ekisu (Noevir)

LYCOPSIS ARVENSIS EXTRACT

Definition: Lycopsis Arvensis Extract is an extract of the fruit, leaves, roots and stems of the bugloss, *Lycopsis arvensis*. See "Regulatory and Ingredient Use Information," regarding the labeling names for botanical ingredients in Volume 1, Introduction, Part A.

Chemical Class: Biological Products

Function: Skin-Conditioning Agent - Miscellaneous

Technical/Other Names:
Bugloss Extract
Bugloss (Lycopsis Arvensis) Extract
Extract of Bugloss
Extract of Lycopsis Arvensis

Trade Name Mixture:
Bugloss HS (Alban Muller)

LYSIMACHIA FOENUM-GRAECUM EXTRACT

Definition: Lysimachia Foenum-graecum Extract is an extract of the herb, *Lysimachia foenum-graecum*. See "Regulatory and Ingredient Use

---

Information," regarding the labeling names for botanical ingredients in Volume 1, Introduction, Part A.

Chemical Class: Biological Products

Function: Not Reported

Technical/Other Name:
Extract of Lysimachia Foenum-graecum

Trade Name Mixture:
Campo Ling Ling Xiang Extract (Campo)

LYSINE

CAS Nos.	EINECS No.
56-87-1 (L-Form)	200-294-2
70-54-2 (dl-alpha)	

JPN Translation:
リシン

Empirical Formula:
$C_6H_{14}N_2O_2$

Definition: Lysine is the amino acid that conforms to the formula:

$$NH_2(CH_2)_4CHCOOH$$
$$|$$
$$NH_2$$

Information Sources: 21CFR172.320, 21CFR310.545, 21CFR431.53, 21CFR440.10, 21CFR582.5411, INN, JCIC, JCLS, MI-13(5656), RIFM, TSCA, USAN

Chemical Class: Amino Acids

Functions: Fragrance Ingredient; Hair Conditioning Agent; Skin-Conditioning Agent - Miscellaneous

Reported Product Categories: Hair Conditioners; Shampoos (Non-coloring); Tonics, Dressings, and Other Hair Grooming Aids

Technical/Other Names:
2,6-Diaminohexanoic Acid
Lysine (RIFM)
L-Lysine (RIFM)
L-Lysine Solution
L-Norleucine, 6-Amino-

Trade Name:
AEC Lysine (A & E Connock)

Trade Name Mixtures:
Anti-Inflammatory Phytoamine Biocomplex (Alban Muller)
Anti-Seborrhoeic Phytoamine Biocomplex (Alban Muller)
Anti-Stress Phytoamine Biocomplex (Alban Muller)
Anti-Stress Phytoamine Biocomplex (Alban Muller)
Anti-Stress Phytoamine Biocomplex In Glycerin (Alban Muller)
Omega-CHS-Activator (GfN)
Free Radical Scavenger Phytoamine Biocomplex (Alban Muller)
Hydroveg VV (Variati)
Lysofat OL (CEP (Solabia))
Moisturizing Liposomes (Collaborative Labs)
Prodew 400 (Ajinomoto)
Regenerative Phytoamine Biocomplex (Alban Muller)
Tegodeo LYS (Degussa Care Specialties)

LYSINE ASPARTATE

Empirical Formula:
$C_6H_{14}N_2O_2$ • $C_4H_7NO_4$

Definition: Lysine Aspartate is the lysine salt of Aspartic Acid (q.v.).

Chemical Class: Amino Acids

Functions: Hair Conditioning Agent; Skin-Conditioning Agent - Miscellaneous

Trade Name:
Asparlyne (CEP (Solabia))

LYSINE CARBOXYMETHYL CYSTEINATE

Definition: Lysine Carboxymethyl Cysteinate is the lysine salt of carboxymethyl cysteine.

Chemical Classes: Amino Acids; Thio Compounds

Function: Skin-Conditioning Agent - Miscellaneous

Reported Product Category: Bath Oils, Tablets, and Salts

Technical/Other Name:
Lysine Carbocysteinate

Trade Name:
Elastocell (Sinerga)

Trade Name Mixture:
Tiolisina Complex 30 (Sinerga)

LYSINE COCOATE

JPN Translation:
ヤシ脂肪酸リシン

Definition: Lysine Cocoate is the salt of Lysine (q.v.) and Coconut Acid (q.v.).

Chemical Classes: Amines; Soaps

Function: Surfactant - Cleansing Agent

Trade Names:
Aminosoap LYC-12 (Ajinomoto)
Aminosoap LYC-12S (Ajinomoto)

LYSINE DNA

Definition: Lysine DNA is the Lysine (q.v.) salt of DNA (q.v.).

Chemical Classes: Biological Polymers and their Derivatives; Organic Salts; Phosphorus Compounds

Function: Skin-Conditioning Agent - Miscellaneous

Technical/Other Name:
DNA, Lysine Salt

Trade Name:
HPDR: Highly Polymerized Desoxyribonucleic Acid (Lysine Salt) (Javenech)

LYSINE GLUTAMATE

CAS No.	EINECS No.
5408-52-6	226-474-0

JPN Translation:
グルタミン酸リシン

Definition: Lysine Glutamate is a mixture of salts of Lysine (q.v.) and Glutamic Acid (q.v.).

Information Source: MI-13(5658)

Chemical Classes: Amino Acids; Organic Salts

Function: Skin-Conditioning Agent - Miscellaneous

Technical/Other Names:
L-Glutamic Acid mono-L-lysine Salt
L-Lysine L-Glutamate

LYSINE HCl

CAS Nos.	EINECS No.
657-27-2	211-519-9
10098-89-2 (L-Form)	
22834-80-6 (dl-alpha)	

JPN Translation:
リシン HCl

Empirical Formula:
$C_6H_{14}N_2O_2$ • ClH

Definition: Lysine HCl is the amine salt that conforms to the formula:

$$NH_2(CH_2)_4CHCOOH \quad \cdot \quad HCl$$
$$|$$
$$NH_2$$

Information Sources: 21CFR172.320, JAN, JCLS, JSCI, TSCA, USAN, USP XXIV

Chemical Class: Amino Acids

Function: Skin-Conditioning Agent - Miscellaneous

Reported Product Categories: Permanent Waves; Hair Conditioners; Hair Wave Sets

Technical/Other Names:
Lysine Hydrochloride
L-Lysine Monohydrochloride

Trade Name Mixtures:
Amino Acid Microspheres (Coletica SA)
Cytokinol (Laboratoires Serobiologiques)
Osmhydran (Laboratoires Serobiologiques)
Osmhydran LS 8453 (Laboratoires Sero-
biologiques)
Sel-Smooth (Seltzer)

LYSINE LAUROYL METHIONATE

Definition: Lysine Lauroyl Methionate is the
lysine salt of the product formed by the
condensation of lauroyl fatty acid chloride
with methionine.

Chemical Classes: Amino Acids; Thio
Compounds

Function: Skin-Conditioning Agent - Miscellaneous

Trade Name:
Lipacide LML (SEPPIC)

LYSINE PCA

CAS Nos.	EINECS Nos.
30657-38-6	250-275-8
97635-56-8	307-418-5

Empirical Formula:
$C_6H_{14}N_2O_2 \cdot C_5H_7NO_3$

Definition: Lysine PCA is the lysine salt of
PCA (q.v.).

Chemical Classes: Amides; Amino Acids;
Heterocyclic Compounds

Function: Skin-Conditioning Agent -
Humectant

Reported Product Categories: Bath Capsules; Face and Neck Preparations
(Excluding Shaving Preparations)

Technical/Other Names:
L-Lysine, Compd. with 5-Oxo-L-Proline
(1:1)
5-Oxoproline, Compd. with Lysine (1:1)
Proline, 5-Oxo-, Compd. with Lysine (1:1)

Trade Name:
Lysidone (UCIB (Solabia))

Trade Name Mixture:
Fruitedone (UCIB (Solabia))

LYSINE THIAZOLIDINE CARBOXYLATE

Empirical Formula:
$C_6H_{14}N_2O_2 \cdot C_4H_6NO_2S$

Definition: Lysine Thiazolidine Carboxylate
is the lysine salt of thiazolidine carboxylic
acid that conforms to the formula:

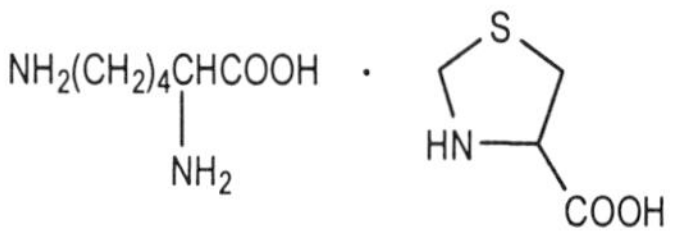

$NH_2(CH_2)_4CHCOOH \cdot$ [thiazolidine carboxylic acid structure with S, HN, COOH groups]
NH_2

Chemical Classes: Amino Acids; Heterocyclic Compounds; Thio Compounds

Function: Skin-Conditioning Agent - Miscellaneous

Reported Product Category: Bath Oils,
Tablets, and Salts

Trade Name Mixture:
Tiolisina Complex 30 (Sinerga)

LYSOLECITHIN

JPN Translation:
リゾレシチン

Definition: Lysolecithin is the product
obtained from acid, enzyme or other method
of hydrolysis of lecithin.

Chemical Classes: Glyceryl Esters and
Derivatives; Phosphorus Compounds

Function: Surfactant - Emulsifying Agent

Reported Product Category: Moisturizing
Preparations

Technical/Other Name:
Hydrolyzed Lecithin

Trade Names:
Blendmax (Central Soya)
Egg Yolk Lecithin LPC-1 (Q.P.)
Emulmetik 120 (Lucas Meyer GmbH)
Precept 8160 (Central Soya)

Trade Name Mixture:
Laminactine (Advanced Beauty)

LYSOPHOSPHATIDYLETHANOLAMINE

Definition: Lysophosphatidylethanolamine
is the hydrolysate of phosphatidylethanol-
amine obtained by acid, enzyme or other
method of hydrolysis. *See Reported Ingredient Functions-The Cosmetic Drug Distinction, in Regulatory and Ingredient Use
Information, Volume I, Part A.*

Chemical Class: Phosphorus Compounds

Functions: Humectant; Skin Bleaching
Agent; Skin-Conditioning Agent - Miscellaneous; Skin Protectant

Trade Name:
Natulox (Doosan)

LYSOZYME

CAS No.	EINECS No.
9001-63-2	232-622-5

JPN Translation:
塩化リゾチーム

Definition: Lysozyme is an enzyme isolated
from egg white.

Information Sources: JCLS, MI-13(5660)

Chemical Class: Proteins

Function: Skin-Conditioning Agent - Miscellaneous

Trade Name:
Lysozyme Chloride (Q.P.)

Trade Name Mixtures:
Bellsilk EZ-M (Ichimaru Pharcos)
Enzyami 3 (Alban Muller)
Stazyme-LSM (Pacific)

LYSOZYME BETA-GLUCAN

Definition: Lysozyme Beta-Glucan is the
product obtained by the reaction of Lysozyme
(q.v.) with Beta-Glucan (q.v.).

Chemical Classes: Carbohydrates; Proteins

Functions: Hair Conditioning Agent; Skin-
Conditioning Agent - Miscellaneous

Trade Name Mixture:
Stazyme-LS (Pacific)

LYTHRUM SALICARIA EXTRACT

CAS No.	EINECS No.
84603-74-7	283-273-0

Definition: Lythrum Salicaria Extract is an
extract of the flowering herb of the purple
loosestrife, *Lythrum salicaria. See
"Regulatory and Ingredient Use
Information," regarding the labeling names
for botanical ingredients in Volume 1, Introduction, Part A.*

Chemical Class: Biological Products

Function: Skin-Conditioning Agent - Miscellaneous

Technical/Other Names:
Extract of Lythrum Salicaria
Extract of Purple Loosestrife
Extract of Salicaria
Purple Loosestrife Extract
Salicaria Extract
Salicaria (Lythrum Salicaria) Extract

Trade Name Mixtures:
Grass Polly HS (Alban Muller)
Lythrum Salicaria Extract ies (IES LABO)

- M -

MACADAMIA INTEGRIFOLIA SEED OIL

Definition: Macadamia Integrifolia Seed Oil is the fixed oil obtained from the nut of *Macadamia integrifolia*. See "Regulatory and Ingredient Use Information," regarding the labeling names for botanical ingredients in Volume 1, Introduction, Part A.

Chemical Class: Fats and Oils

Function: Skin-Conditioning Agent - Occlusive

Technical/Other Names:
Macadamia Integrifolia Nut Oil
Oils, Macadamia Nut

Trade Name:
Floramac Macadamia Nut Oil, Hawaiian (Floratech)

MACADAMIA TERNIFOLIA SEED EXTRACT

Definition: Macadamia Ternifolia Seed Extract is an extract of the nuts of the macadamia tree, *Macadamia ternifolia*. See "Regulatory and Ingredient Use Information," regarding the labeling names for botanical ingredients in Volume 1, Introduction, Part A.

Chemical Class: Biological Products

Function: Not Reported

Technical/Other Names:
Extract of Macadamia Nut
Extract of Macadamia Ternifolia Nut
Macadamia Integrifolia Nut Extract
Macadamia Nut Extract

Trade Name Mixtures:
Actiphyte of Macadamia Nut BG50 (Active Organics)
Actiphyte of Macadamia Nut GL50 (Active Organics)
Actiphyte of Macadamia Nut Lipo S (Active Organics)
Actiphyte of Macadamia Nut PG50 (Active Organics)
VT-256 Extract of Macadamia Nut (Vege-Tech)

MACADAMIA TERNIFOLIA SEED OIL

CAS Nos.: 128497-20-1; 129811-19-4

JPN Translation:
マカデミアナッツ油

Definition: Macadamia Ternifolia Seed Oil is the fixed oil obtained from the nuts of *Macadamia ternifolia*. See "Regulatory and Ingredient Use Information," regarding the labeling names for botanical ingredients in Volume 1, Introduction, Part A.

Information Sources: JCIC, JCLS, JSQI

Chemical Class: Fats and Oils

Function: Skin-Conditioning Agent - Occlusive

Reported Product Categories: Lipsticks; Face Powders; Eye Makeup Preparations, Misc.; Nail Polish and Enamels; Foundations; Bath Capsules; Eye Shadows; Makeup Preparations (Not eye), Misc.; Moisturizing Preparations; Face and Neck Preparations (Excluding Shaving Preparations); Makeup Bases; Bath Oils, Tablets, and Salts; Blushers (All types); Cleansing Products (Cold Creams, Cleansing Lotions, Liquids and Pads); Skin Care Preparations, Misc.; Fragrance Preparations, Misc.; Night Skin Care Preparations; Rouges; Mascara; Powders (Dusting and Talcum, Excluding Aftershave Talcs); Eyebrow Pencils; Hair Conditioners; Paste Masks (Mud Packs); Suntan Gels, Creams, and Liquids

Technical/Other Names:
Macadamia (Macadamia Ternifolia) Nut Oil
Macadamia Nut Oil
Macadamia Ternifolia Oil
Oils, Macadamia
Oils, Macadamia Nut
Oils, Macadamia Tetraphylla

Trade Names:
AEC Macadamia Nut Oil (A & E Connock)
Cropure Macadamian (Croda Chemicals)
Huile de Macadamia (LCW)
Huile de Macadamia Vierge (Bertin)
Jeen Macadamia Nut Oil (Jeen)
Lipovol MAC (Lipo)
Macadamia Nut Oil (Desert Whale)
Macadamia Nut Oil (Nestle World Trade)
Macadamia Nut Oil (NOF)
Macadamia Nut Oil EX (Ikeda)
Macadamia Nut Oil YZ (Ikeda)
Macadamia Oil (Dekker)
Macadamia Ternifolia Seed Oil ies (IES LABO)
Nikkol Macadamia Nut Oil (Nikko)
Oils of Aloha Macadamia Nut Oil (Oils of Aloha)
Phytol MAC (Custom Ingredients)
Sozio Macadamia Nut Oil (Sozio)
TZ Oil (NOF)

Trade Name Mixtures:
AEC Macadamia Nut Wax (A & E Connock)
AEC Mac-Kui Nut Wax (A & E Connock)
Covacrem MK (LCW)
Covawax 501 (LCW)
Dragobotania 2/H00005 (Symrise)
DS-SBS (Doosan)
Extrapone Macadamia Nut 2/032160 (Symrise)
Extrapone Macadamia Nut 2/032160 (Symrise)
Extrapone Macadamia Nut Milk 2/033848 (Symrise)
Gelhyperm (Macadamia nut oil) (Novoselect)
Macadamia Milk (Cosmetochem) (Cosmetochem International Ltd.)
Macamat Wax (LCW)
Macarose (LCW)
Melange Huile Vierge de Macadamia / Huile de Graine de Kiwi (Bertin)
Micromac Wax (LCW)
Nikkol Aquasome AE Conc (Nikko)
Toshiki BINS-4 (Nikko)
Ultraspheres-5012 (Lipoid)
Ultraspheres-5014 (Lipoid)
Vitacap (LCW)

MACROCYSTIS PYRIFERA EXTRACT

JPN Translation:
マクロシスティスピリフェラエキス

Definition: Macrocystis Pyrifera Extract is an extract of the kelp, *Macrocystis pyrifera*. See "Regulatory and Ingredient Use Information," regarding the labeling names for botanical ingredients in Volume 1, Introduction, Part A.

Chemical Class: Biological Products

Function: Not Reported

Reported Product Categories: Hair Conditioners; Bath Oils, Tablets, and Salts; Bath Preparations, Misc.; Makeup Preparations (Not eye), Misc.; Shampoos (Non-coloring); Skin Care Preparations, Misc.; Skin Fresheners

Technical/Other Names:
Extract of Kelp
Extract of Macrocystis Pyrifera
Extract of Macrocystis Pyriferae
Kelp Extract
Kelp (Macrocystis Pyrifera) Extract
Macrocystis Pyriferae Extract

Trade Name:
Kelpadelie (SECMA)

Trade Name Mixtures:
Actiphyte of Sea Kelp BG50 (Active Organics)
Actiphyte of Sea Kelp GL50 (Active Organics)
Actiphyte of Sea Kelp Lipo S (Active Organics)
Actiphyte of Sea Kelp PG50 (Active Organics)
Kelp Extract (Draco)

Marine Plasma Extract III (Arch Personal
Care Products)
Pacific Sea Kelp Extract (Bell Flavors)
Pacific Sea Kelp Oil Extract (Bell Flavors)
2325 Vege-Plex Body Complex (Vege-
Tech)
2600 Vege-Plex Skin Complex (Vege-
Tech)
VT-027 Extract of Kelp (Vege-Tech)

MACROCYSTIS PYRIFERA (KELP)

Definition: Macrocystis Pyrifera (Kelp) is
a marine product recovered from the giant
Pacific kelp, *Macrocystis pyriferae. See
"Regulatory and Ingredient Use
Information," regarding the labeling names
for botanical ingredients in Volume 1, Introduction, Part A.*

Information Sources: 21CFR172.365,
21CFR184.1120, 21CFR184.1121,
21CFR310.545, 21CFR582.40

Chemical Class: Biological Products

Function: Viscosity Increasing Agent -
Aqueous

Reported Product Categories: Bath Preparations, Misc.; Bath Soaps and Detergents;
Body and Hand Preparations (Excluding
Shaving Preparations); Moisturizing Preparations; Paste Masks (Mud Packs)

Technical/Other Name:
Kelp Protein J

Trade Name:
Pacific Kelp (Meer)

MACROCYSTIS PYRIFERA (KELP) PROTEIN

Definition: Macrocystis Pyrifera (Kelp)
Protein is the protein derived from the kelp,
Macrocystis pyrifera.

Chemical Class: Proteins

Function: Skin-Conditioning Agent - Miscellaneous

MACROTOMIA EUCHROMA ROOT EXTRACT

Definition: Macrotomia Euchroma Root
Extract is an extract of the roots of
*Macrotomia euchroma. See "Regulatory
and Ingredient Use Information," regarding
the labeling names for botanical ingredients
in Volume 1, Introduction, Part A.*

Chemical Class: Biological Products

Function: Skin-Conditioning Agent -
Humectant

Technical/Other Name:
Extract of Macrotomia Euchroma Root

Trade Name Mixture:
Shikon Extract TH-M (T.HASEGAWA)

MADECASSIC ACID

CAS No.: 18449-41-7

Empirical Formula:
$C_{30}H_{48}O_6$

Definition: Madecassic Acid is the organic
compound that conforms to the formula:

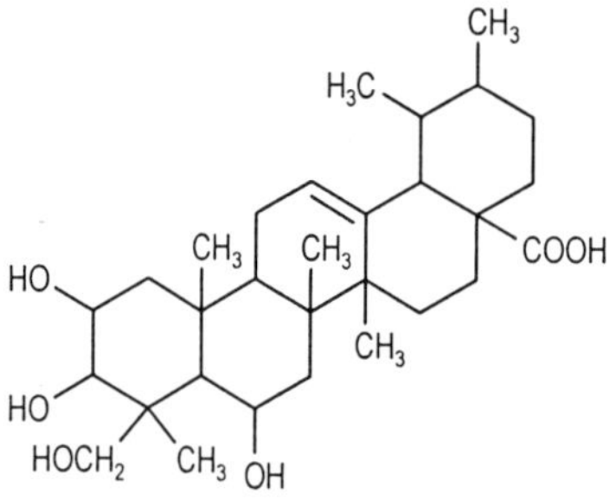

Chemical Classes: Carboxylic Acids;
Sterols

Function: Skin-Conditioning Agent - Miscellaneous

Technical/Other Names:
Brahmic Acid
6β-Hydroxyasiatic Acid
Urs-12-en-28-oic Acid, 2,3,6,23-
tetrahydroxy-, (2α,3β,4α,6β)-

Trade Name Mixtures:
Centella Asiatica Purified Triterpenes
(Vinyals)
Plantactiv Centella (Cognis Deutschland)
ViaPure Centella (Actives International)

MADECASSICOSIDE

Empirical Formula:
$C_{48}H_{78}O_{19}$

Definition: Madecassicoside is the organic
compound that conforms to the formula:

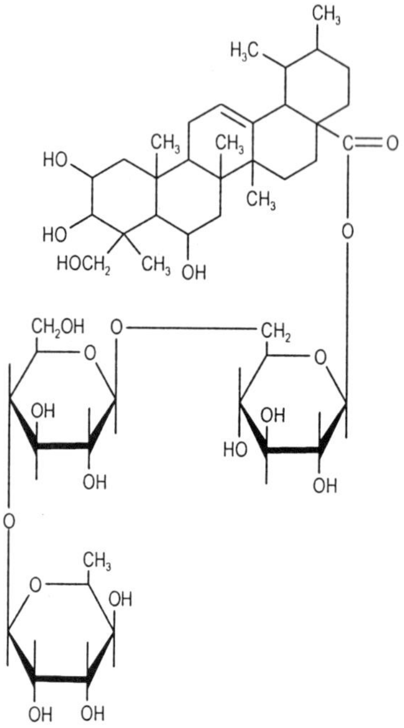

Information Source: MI-13(839)

Chemical Classes: Carbohydrates; Esters;
Sterols

Functions: Antioxidant; Skin-Conditioning
Agent - Miscellaneous

Trade Name:
Madesid (Indena SpA)

MAGNESIUM ACETATE

CAS No.	EINECS No.
142-72-3	205-554-9

Empirical Formula:
$C_4H_8O_4 \cdot Mg$

Definition: Magnesium Acetate is the
magnesium salt of acetic acid that conforms
to the formula:

$$\left[CH_3COO^- \right]_2 Mg^{-2}$$

Information Source: MI-13(5676)

Chemical Class: Organic Salts

Function: Buffering Agent

Technical/Other Names:
Acetic Acid, Magnesium Salt
Magnesium Diacetate

MAGNESIUM ACETYLMETHIONATE

CAS No.: 105883-49-6

Definition: Magnesium Acetylmethionate is
the magnesium salt of N-acetylmethionine.

Chemical Class: Amino Acids

Function: Not Reported

Trade Name:
Exsymagnesium (Exsymol)

MAGNESIUM ALGINATE

CAS No.: 37251-44-8

Definition: Magnesium Alginate is the magnesium salt of Alginic Acid (q.v.).

Chemical Class: Gums, Hydrophilic Colloids and Derivatives

Functions: Binder; Emulsion Stabilizer; Viscosity Increasing Agent - Aqueous

Technical/Other Name:
Alginic Acid, Magnesium Salt

Trade Name:
Protanal LFMg 5/60 (Pronova Biopolymer Inc.)

MAGNESIUM/ALUMINUM/HYDROXIDE/ CARBONATE

CAS Nos.	EINECS Nos.
11097-59-9	234-319-3
85585-93-9	287-796-5

Empirical Formula:
$Al_2Mg_6O_{19}$

Definition: Magnesium/Aluminum/ Hydroxide/Carbonate is the inorganic carbonate that conforms generally to the formula:

$$Mg_6Al_2(OH)_{16}CO_3 \cdot H_2O$$

Information Source: TSCA

Chemical Class: Inorganics

Function: Viscosity Increasing Agent - Aqueous

Technical/Other Names:
[Carbonato(2-)]Hexadecahydroxybis (Aluminum)Hexamagnesium
Magnaesium, Carbonate Hydroxy Aluminum Complexes
Magnaesium, [Carbonato(2-)]- Hexadecahydroxybis(Aluminum)Hexa-

Trade Name:
Hydrotalcite (Giulini/Giulini Chemie)

MAGNESIUM ALUMINUM SILICATE

CAS Nos.	EINECS Nos.
12199-37-0	235-374-6
12511-31-8	235-682-0

JPN Translation:
ケイ酸 (Al / Mg)

Definition: Magnesium Aluminum Silicate is a complex silicate refined from naturally occurring minerals.

Information Sources: BPC, CIR: [S] IJT-22(SUPPL. 1)2003, CTFA S, DDR, JAN, JCLS, JSCI, MAR, MI-13(349), NF XIX, OTC-I-AA, TSCA, USAN

Chemical Class: Inorganic Salts

Functions: Absorbent; Anticaking Agent; Opacifying Agent; Slip Modifier; Viscosity Increasing Agent - Aqueous

Reported Product Categories: Foundations; Moisturizing Preparations; Makeup Bases; Bath Preparations, Misc.; Body and Hand Preparations (Excluding Shaving Preparations); Mascara; Bath Oils, Tablets, and Salts; Cleansing Products (Cold Creams, Cleansing Lotions, Liquids and Pads); Paste Masks (Mud Packs); Skin Care Preparations, Misc.; Makeup Preparations (Not eye), Misc.; Eyeliners; Indoor Tanning Preparations; Bath Capsules; Face and Neck Preparations (Excluding Shaving Preparations); Personal Cleanliness Products, Misc.; Eye Makeup Preparations, Misc.; Night Skin Care Preparations; Aftershave Lotions; Baby Shampoos; Deodorants (Underarm); Eyebrow Pencils; Suntan Gels, Creams, and Liquids; Suntan Preparations, Misc.; Eye Shadows; Foot Powders and Sprays; Hair Preparations (Non-coloring), Misc.; Hair Straighteners; Lipsticks; Makeup Fixatives; Shampoos (Non-coloring); Shaving Preparations, Misc.; Bath Soaps and Detergents; Dentifrices (Aerosol, Liquid, Pastes and Powders); Eye Lotions; Hair Dyes and Colors (All Types Requiring Caution Statements and Patch Tests)

Technical/Other Names:
Aluminum Magnesium Silicate
Aluminum Magnesium Silicon Oxide
Silicic Acid, Aluminum Magnesium Salt

Trade Names:
AEC Magnesium Aluminium Silicate (A & E Connock)
Gelwhite MAS-H (Southern Clay)
Gelwhite MAS-L (Southern Clay)
Magnabrite F (American Colloid, Consumer Specialties)
Magnabrite FS (American Colloid, Consumer Specialties)
Magnabrite HS (American Colloid, Consumer Specialties)
Magnabrite HV (American Colloid, Consumer Specialties)
Magnabrite K (American Colloid, Consumer Specialties)
Magnabrite S (American Colloid, Consumer Specialties)

Neusilin (Fuji Chemical)
Sebumase (US Cosmetics)
Veegum (Vanderbilt)
Veegum D (Vanderbilt)
Veegum F (Vanderbilt)
Veegum HS (Vanderbilt)
Veegum HV (Vanderbilt)
Veegum K (Vanderbilt)
Veegum Ultra (Vanderbilt)

Trade Name Mixtures:
Integrahair RE (Chemyunion)
Matipure (Advanced Beauty)
Veegum Plus (Vanderbilt)
WHP-CL-1 (US Cosmetics)

MAGNESIUM/ALUMINUM/ZINC/ HYDROXIDE/CARBONATE

Definition: Magnesium/Aluminum/Zinc/ Hydroxide/Carbonate is the inorganic compound that conforms generally to the formula:

$$Mg_3ZnAl_2(OH)_{12}CO_3 \cdot H_2O$$

See Reported Ingredient Functions-The Cosmetic Drug Distinction, in Regulatory and Ingredient Use Information, Volume I, Part A.

Chemical Class: Inorganics

Functions: Antidandruff Agent; Deodorant Agent

Trade Name:
Alcamizer P93 (Kyowa Chemical)

MAGNESIUM ASCORBATE

CAS No.	EINECS No.
15431-40-0	239-442-6

Empirical Formula:
$C_6H_8O_6 \cdot \frac{1}{2}Mg$

Definition: Magnesium Ascorbate is the magnesium salt of Ascorbic Acid (q.v.).

Information Source: CIR: [S]

Chemical Class: Organic Salts

Functions: Antioxidant; Skin-Conditioning Agent - Miscellaneous

Technical/Other Name:
Ascorbic Acid, Magnesium Salt

MAGNESIUM ASCORBATE/PCA

Definition: Magnesium Ascorbate/PCA is the Magnesium salt of Ascorbic Acid (q.v.) and PCA (q.v.).

Chemical Class: Organic Salts

Function: Antioxidant

Trade Name:
Vitacedone (UCIB (Solabia))

MAGNESIUM ASCORBYL PHOSPHATE

CAS Nos.: 113170-55-1; 114040-31-2

JPN Translation:
リン酸アスコルビル Mg

Empirical Formula:
$C_6H_8O_9P \cdot {}^3/_2Mg$

Definition: Magnesium Ascorbyl Phosphate is the organic compound that conforms to the formula:

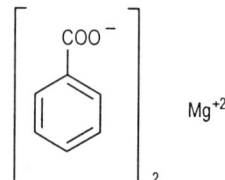

Information Sources: CIR: [S], JCIC, JCLS, JSQI

Chemical Classes: Heterocyclic Compounds; Organic Salts; Phosphorus Compounds

Function: Antioxidant

Reported Product Categories: Bath Capsules; Moisturizing Preparations; Face and Neck Preparations (Excluding Shaving Preparations); Suntan Gels, Creams, and Liquids; Bath Oils, Tablets, and Salts; Cleansing Products (Cold Creams, Cleansing Lotions, Liquids and Pads); Suntan Preparations, Misc.; Body and Hand Preparations (Excluding Shaving Preparations); Cuticle Softeners; Eye Lotions; Face Powders; Foundations; Hair Conditioners; Makeup Bases; Makeup Fixatives; Mascara; Night Skin Care Preparations; Paste Masks (Mud Packs); Shampoos (Non-coloring); Skin Fresheners; Tonics, Dressings, and Other Hair Grooming Aids

Technical/Other Name:
Magnesium L-Ascorbyl-2-phosphate

Trade Names:
Ascorbyl Phosphate Magnesium (Showa Denko)
Ascorbyl PM (Presperse)
C-Mate (Takeda Vitamin & Food)
MAP SL (Sino Lion)
Nikkol VC-PMG (Nikko)
Rona Care MAP (Merck KGaA)
Rona Care MAP (Merck KGaA/EMD Chemicals Inc.)

Trade Name Mixtures:
Collagen Stimulation Factor MAP (Cosmetochem) (Cosmetochem International Ltd.)
Color Marine Vitamine CPMG Spheres (Coletica SA)
Deperoxidium Marin (Coletica SA)
Deperoxidium Vegetal (Coletica SA)
Elespher Vitaplex Lipo (Laboratoires Sero-biologiques)
Greenosome Ace (Greentech)
Induxin (Laboratoires Serobiologiques)
Isocell MAP (Lucas Meyer)
Melfade J (Pentapharm/Centerchem)
Nikkol Aquasome EC-5 (Nikko)
NIKKOL Aquasome EC-30 (Nikko)
Oxygen Complex LS (Laboratoires Sero-biologiques)
Oxysomes (Barnet)
RonaCare VTA (Merck KGaA)
Rovisome AA (Rovi)
Rovisome ACE (Rovi)
Rovisome C (Rovi)
Toshiki Nylon IVCF-1 (Nikko)

MAGNESIUM ASPARTATE

CAS Nos.	EINECS Nos.
1187-91-3 (dl-alpha)	214-702-1
2068-80-6 (L-Form)	
18962-61-3 (L-Form)	242-703-7
52101-01-6 (dl-alpha)	

JPN Translation:
アスパラギン酸 Mg

Empirical Formula:
$C_4H_7NO_4 \cdot xMg$

Definition: Magnesium Aspartate is the magnesium salt of Aspartic Acid (q.v.) that conforms to the formula:

$$\left[\begin{array}{c} HOOCCHCH_2COO^- \\ | \\ NH_2 \end{array} \right]_2 Mg^{+2}$$

Information Sources: JCIC, JCLS, TSCA

Chemical Class: Amino Acids

Function: Skin-Conditioning Agent - Miscellaneous

Technical/Other Names:
Aspartic Acid, Magnesium Salt
DL-Aspartic Acid, Magnesium Salt (2:1)
L-Aspartic Acid, Magnesium Salt (2:1)
Magnesium L-Aspartate

Trade Names:
Givobio AMgDL (SEPPIC)
Givobio AMgL (SEPPIC)
Oligoidyne Magnesium (Vevy)

Trade Name Mixtures:
Afron 22 (Vevy)
Afron-A (Vevy)
Afron-N (Vevy)
Carbossalina (Vevy)
Givobio AMgKL (SEPPIC)
Lipoderma AA (Lipo)
Oligoidyne-1-Complex (Vevy)
Sel-Smooth (Seltzer)
Sepicalm S (SEPPIC)
Sepitonic M3 (SEPPIC)
Sepitonic M4 (SEPPIC)
Tensioplastidina Avena (Vevy)
Unimoist U-125 (Induchem)

MAGNESIUM BENZOATE

CAS No.	EINECS No.
553-70-8	209-045-2

Empirical Formula:
$C_7H_6O_2 \cdot \frac{1}{2}Mg$

Definition: Magnesium Benzoate is the magnesium salt of Benzoic Acid (q.v.). It conforms to the formula:

$$\left[\underset{}{\overset{COO^-}{\bigcirc}} \right]_2 Mg^{+2}$$

Information Sources: EEC(VI/1-1), MHLW-331/3, MI-13(5679), TSCA

Chemical Class: Organic Salts

Function: Preservative

Technical/Other Names:
Benzoic Acid, Magnesium Salt
Magnesium Dibenzoate
Magnesium Dibromide

MAGNESIUM BROMIDE

CAS No.	EINECS No.
7789-48-2	232-170-9

Empirical Formula:
Br_2Mg

Definition: Magnesium Bromide is the inorganic salt that conforms to the formula:

$$MgBr_2$$

Information Source: MI-13(5681)

Chemical Class: Inorganic Salts

Function: Not Reported

MAGNESIUM CARBONATE

CAS Nos.	EINECS No.
546-93-0	208-915-9
7757-69-9	

JPN Translation:
炭酸 Mg

Empirical Formula:
$CH_2O_3 \cdot Mg$

Definition: Magnesium Carbonate is a basic dehydrated magnesium carbonate or a normal hydrated magnesium carbonate. *This ingredient is not an approved colorant for the US. To identify the colorant allowed for use in the European Union (EU), the INCI Name CI 77713 must be used, except for hair dye products.*

Information Sources: ARG, AUS, BP, BPC, BRA, 21CFR133.102, 21CFR133.106, 21CFR133.111, 21CFR133.141, 21CFR133.165, 21CFR133.181, 21CFR133.183, 21CFR133.195, 21CFR137.105, 21CFR137.155, 21CFR137.160, 21CFR137.165, 21CFR137.170, 21CFR137.175, 21CFR137.180, 21CFR137.185, 21CFR163.110, 21CFR177.2600, 21CFR184.1425, CI 77713, CTFA S, DA, EP, FCC, FIN, HUN, IND, ITA, JAN, JCLS, JSCI, MAR, OTC-I-AA, OTC-I-OR, PF, PN, POR, ROM, TSCA, USAN, USD, USP XXIV, USSR

Chemical Class: Inorganic Salts

Functions: Absorbent; Bulking Agent; Opacifying Agent; pH Adjuster

Reported Product Categories: Powders (Dusting and Talcum, Excluding Aftershave Talcs); Eye Shadows; Face Powders; Blushers (All types); Mascara; Foundations; Sachets; Hair Bleaches; Baby Products, Misc.; Bath Preparations, Misc.; Body and Hand Preparations (Excluding Shaving Preparations); Lipsticks; Men's Talcum; Tonics, Dressings, and Other Hair Grooming Aids

Technical/Other Names:
Carbonic Acid, Magnesium Salt
CI 77713
Heavy Magnesium Carbonate
Light Magnesium Carbonate

Trade Names:
Magnesium Carbonate, Heavy USP 320 - S (Whittaker, Clark & Daniels)
Magnesium Carbonate, Light USP 309 -S (Whittaker, Clark & Daniels)
Unichem MC (Universal Preserv-A-Chem)

Trade Name Mixtures:
Natrasorb HFB (National Starch)
Pad-1 (Vevy)

MAGNESIUM CARBONATE HYDROXIDE

CAS Nos.	EINECS Nos.
7760-50-1	231-851-8
12125-28-9	235-192-7

Empirical Formula:
$C_4H_2Mg_5O_{14}$

Definition: Magnesium Carbonate Hydroxide is an inorganic basic carbonate that conforms generally to the formula:

$$(MgCO_3)_4 \cdot Mg(OH)_2 \cdot 5H_2O$$

Information Sources: DDR, EGY, HP, MI-13(5682), PF, TSCA, YUG

Chemical Class: Inorganic Salts

Functions: Bulking Agent; pH Adjuster

Technical/Other Names:
Basic magnesium Carbonate
Carbonic Acid, Magnesium Complex
Magnesium, Tetrakis[Carbonato(2-)]Dihydroxypenta-

Trade Name:
Magnesium hydroxide carbonate light (Merck KGaA)

MAGNESIUM CHLORIDE

CAS No.	EINECS No.
7786-30-3	232-094-6

JPN Translation:
塩化 Mg

Empirical Formula:
Cl_2Mg

Definition: Magnesium Chloride is the inorganic salt that conforms to the formula:

$$MgCl_2$$

Information Sources: 21CFR172.560, 21CFR177.1650, 21CFR184.1426, JCIC, JCLS, JSQI, MI-13(5684), USAN, USP XXIV

Chemical Class: Inorganic Salts

Function: Not Reported

Reported Product Categories: Bath Capsules; Shampoos (Non-coloring); Bath Oils, Tablets, and Salts; Hair Conditioners; Skin Care Preparations, Misc.; Moisturizing Preparations; Cleansing Products (Cold Creams, Cleansing Lotions, Liquids and Pads)

Technical/Other Name:
Magnesium Dichloride

Trade Name Mixtures:
Dacriosalt (Vevy)
Essential Vital Elements (Dipta)
Essential Vital Elements - S (Dipta)
Liposerve MM (Lipo)

MAGNESIUM CITRATE

CAS Nos.	EINECS Nos.
144-23-0	
6150-79-4	222-093-9
7779-25-1	231-923-9

Empirical Formula:
$C_6H_6O_7 \cdot Mg$

Definition: Magnesium Citrate is the magnesium salt of Citric Acid (q.v.).

Information Sources: MI-13(5685), MI-13 (5686), TSCA, USP XXIV

Chemical Class: Organic Salts

Function: Skin-Conditioning Agent - Miscellaneous

Reported Product Category: Hair Conditioners

Technical/Other Names:
Citric Acid, Magnesium Salt
1,2,3-Propanetricarboxylic Acid, 2-Hydroxy-, Magnesium Salt

Trade Names:
Magnesium Citrate Extra Pure (Merck KGaA)
Tri-Magnesium Dicitrate 14-Hydrate (Merck KGaA)

MAGNESIUM COCETH SULFATE

Definition: Magnesium Coceth Sulfate is the magnesium salt of sulfated, ethoxylated coconut alcohol. It conforms generally to the formula:

$$R(OCH_2CH_2)_nOSO_3Na$$

where R represents the the alkyl groups derived from coconut oil and n has an average value of 3.

Chemical Class: Alkyl Ether Sulfates

Functions: Surfactant - Cleansing Agent; Surfactant - Emulsifying Agent

Trade Name:
Zetesol MG/C (Zschimmer & Schwarz Italiana)

MAGNESIUM COCOATE

Definition: Magnesium Cocoate is the magnesium salt of Coconut Acid (q.v.).

Information Sources: 21CFR175.105, 21CFR175.320, 21CFR176.170, 21CFR176.200, 21CFR176.210, 21CFR177.1200, 21CFR177.2260, 21CFR178.3910

Chemical Class: Soaps

Functions: Anticaking Agent; Slip Modifier; Viscosity Increasing Agent - Nonaqueous

Reported Product Category: Bath Soaps and Detergents

Technical/Other Names:
Coconut Fatty Acids, Magnesium Salts
Fatty Acids, Coconut Oil, Magnesium Salt

MAGNESIUM COCO-SULFATE

Definition: Magnesium Coco-Sulfate is the magnesium salt of sulfated coconut alcohol that conforms generally to the formula:

$$(ROSO_3)_2Mg$$

where R represents the alkyl groups derived from coconut oil.

Chemical Class: Alkyl Sulfates

Function: Surfactant - Cleansing Agent

Technical/Other Name:
Sulfuric Acid, Cocoyl Ester, Magnesium Salt

MAGNESIUM DNA

Definition: Magnesium DNA is the magnesium salt of DNA (q.v.).

Chemical Classes: Biological Polymers and their Derivatives; Organic Salts; Phosphorus Compounds

Function: Skin-Conditioning Agent - Miscellaneous

Technical/Other Names:
DNA, Magnesium Salt
Magnesium Deoxyribonucleic Acid

Trade Name Mixture:
HPDR: Highly Polymerized Desoxyribonucleic Acid (Mixed Sodium, Calcium, and Magnesium Salt) (Javenech)

MAGNESIUM FLUORIDE

CAS No.	EINECS No.
7783-40-6	231-995-1

Empirical Formula:
F_2Mg

Definition: Magnesium Fluoride is the inorganic salt that conforms to the formula:

$$MgF_2$$

Information Sources: EEC(III/1-56), MI-13 (5688), TSCA

Chemical Classes: Halogen Compounds; Inorganic Salts

Function: Oral Care Agent

Technical/Other Name:
Magnesium Difluoride

Trade Name Mixtures:
ChromaFlair Light Interference Pigment (Flex Products)
Spectraflair Pigment (Flex Products)

MAGNESIUM FLUOROSILICATE

CAS No.	EINECS No.
16949-65-8	241-022-2

Empirical Formula:
$F_6H_2Si \cdot Mg$

Definition: Magnesium Fluorosilicate is the inorganic salt that conforms to the formula:

$$MgF_6Si$$

Information Sources: EEC(III/1-43), MI-13 (5691), TSCA

Chemical Class: Inorganic Salts

Function: Oral Care Agent

Technical/Other Names:
Fluosilicic Acid Magnesium Salt
Magnesium Hexafluorosilicate
Silicate, Hexafluoro-, Magnesium
Silicon Fluoride Magnesium Salt

MAGNESIUM GLUCOHEPTONATE

Definition: Magnesium Glucoheptonate is a mixture of magnesium salts of α- and β-glucoheptonic acid. It conforms generally to the formula:

$$\left[CH_2OH(CHOH)_5COO^- \right]_2 Mg^{+2}$$

Chemical Classes: Carboxylic Acids; Organic Salts; Polyols

Function: Skin-Conditioning Agent - Miscellaneous

Trade Name:
Givobio GHMg (SEPPIC)

Trade Name Mixture:
Givobio GGHMg (SEPPIC)

MAGNESIUM GLUCONATE

CAS No.	EINECS No.
3632-91-5	222-848-2

Empirical Formula:
$C_{12}H_{24}O_{14} \cdot Mg$

Definition: Magnesium Gluconate is the magnesium salt of gluconic acid that conforms to the formula:

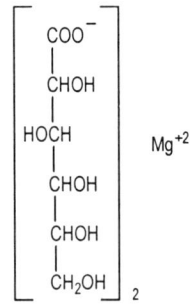

Information Sources: MI-13(4469), TSCA, USAN, USP XXIV

Chemical Class: Organic Salts

Function: Not Reported

Reported Product Categories: Bath Oils, Tablets, and Salts; Cleansing Products (Cold Creams, Cleansing Lotions, Liquids and Pads); Bath Capsules; Face and Neck Preparations (Excluding Shaving Preparations); Shampoos (Non-coloring); Skin Care Preparations, Misc.; Skin Fresheners

Technical/Other Name:
D-Gluconic Acid, Magnesium Salt (2:1)

Trade Names:
Givobio GMg (SEPPIC)
Gluconal MG (Glucona America)

Trade Name Mixtures:
Givobio GGHMg (SEPPIC)
Givobio GMgK (SEPPIC)

MAGNESIUM GLYCEROPHOSPHATE

Empirical Formula:
$C_3H_7O_6P \cdot Mg$

Definition: Magnesium Glycerophosphate is the organic compound that conforms generally to the formula:

$$\left[\begin{array}{c} CH_2OPO_3 \\ | \\ CHOH \\ | \\ CH_2OH \end{array} \right]^{-2} Mg^{+2}$$

Chemical Classes: Glyceryl Esters and Derivatives; Organic Salts; Phosphorus Compounds

Function: Oral Care Agent

Technical/Other Name:
Magnesium Glycerol Phosphate (1:1:1)

Trade Name:
Givobio GPMg (SEPPIC)

MAGNESIUM HYDROGEN PHOSPHATE

CAS No.: 7782-75-4

Definition: Magnesium Hydrogen Phosphate is the inorganic salt that conforms to the formula:

$$MgHPO_4 \cdot 3H_2O$$

Information Source: MI-13(5705)

Chemical Class: Inorganic Salts

Function: Anticaking Agent

Technical/Other Names:
Magnesium Hydrogen Phosphate Trihydrate
Magnesium Phosphate, Dibasic
1,2,3-Propanetriol, 1-(dihydrogen phosphate), Magnesium Salt (1:1)

Trade Name:
Magnesium Hydrogen Phosphate Trihydrate (Merck KGaA)

MAGNESIUM HYDROXIDE

CAS No.   **EINECS No.**
1309-42-8   215-170-3

JPN Translation:
水酸化 Mg

Empirical Formula:
H_2MgO_2

Definition: Magnesium Hydroxide is an inorganic base that conforms to the formula:

$$Mg(OH)_2$$

Information Sources: ARG, BPC, 21CFR184.1428, FCC, JAN, JCIC, JCLS, KOR, MAR, MI-13(5693), NFJ, OTC-I-AA, PF, TSCA, USAN, USD, USP XXIV

Chemical Class: Inorganic Bases

Functions: Absorbent; pH Adjuster

Trade Name:
Magnesium Hydroxide, Powder USP 370-S (Whittaker, Clark & Daniels)

MAGNESIUM LANOLATE

Definition: Magnesium Lanolate is the magnesium salt of Lanolin Acid (q.v.).

Chemical Classes: Lanolin and Lanolin Derivatives; Soaps

Functions: Anticaking Agent; Skin-Conditioning Agent - Miscellaneous; Viscosity Increasing Agent - Nonaqueous

MAGNESIUM LAURETH-11 CARBOXYLATE

Definition: Magnesium Laureth-11 Carboxylate is the magnesium salt of the carboxylic acid derived from Laureth-11 (q.v.). It conforms generally to the formula:

$$\left[CH_3(CH_2)_{11}(OCH_2CH_2)_{10}OCOO^- \right]_2 Mg^{+2}$$

Chemical Class: Organic Salts

Function: Surfactant - Cleansing Agent

Technical/Other Names:
Magnesium Polyethylene Glycol (11) Lauryl Ether Carboxylate
Magnesium Polyoxyethylene (11) Lauryl Ether Carboxylate
PEG-11 Lauryl Ether Carboxylic Acid, Magnesium Salt
Polyethylene Glycol (11) Lauryl Ether Carboxylic Acid, Magnesium Salt
Polyoxyethylene (11) Lauryl Ether Carboxylic Acid, Magnesium Salt

Trade Name:
Akypo Soft 100 MgV (Kao GmbH)

MAGNESIUM LAURETH SULFATE

CAS No.: 62755-21-9

Definition: Magnesium Laureth Sulfate is the magnesium salt of ethoxylated lauryl sulfate that conforms generally to the formula:

$$\left[CH_3(CH_2)_{11}(OCH_2CH_2)_nOSO_3^- \right]_2 Mg^{+2}$$

where n has a value between 1 and 4.

Chemical Class: Alkyl Ether Sulfates

Function: Surfactant - Cleansing Agent

Reported Product Categories: Shampoos (Non-coloring); Bath Oils, Tablets, and Salts; Eye Makeup Removers; Cleansing Products (Cold Creams, Cleansing Lotions, Liquids and Pads); Baby Shampoos

Technical/Other Name:
Magnesium Lauryl Ether Sulfate

Trade Names:
AEC Magnesium Laureth Sulphate (A & E Connock)
Empicol EGB (Albright & Wilson UK)
Empicol EGC (Albright & Wilson UK)
Empicol EGC 70 (Albright & Wilson UK)
Zoharpon MGES (Zohar)

Trade Name Mixtures:
Empicol BSD (Albright & Wilson UK)
Empicol BSD 52 (Albright & Wilson UK)
LP110 (Phytocos)
Rhodapex BSD-FL/A2 (Rhodia)
Texapon ASV (Cognis Deutschland)
Texapon ASV-50 (Cognis Care Chemicals/ NJ)
Texapon ASV-50 (Cognis Care Chemicals/ PA)
Texapon ASV-50 (Cognis Deutschland)

MAGNESIUM LAURETH-5 SULFATE

CAS No.: 62755-21-9 (Generic)

Empirical Formula:
$C_{22}H_{45}O_9S \cdot \frac{1}{2}Mg$

Definition: Magnesium Laureth-5 Sulfate is the magnesium salt of the sulfuric acid ester of Laureth-5 (q.v.). It conforms generally to the formula:

$$\left[CH_3(CH_2)_{11}(OCH_2CH_2)_nOSO_3^- \right]_2 Mg^{+2}$$

where n has an average value of 5.

Chemical Class: Alkyl Ether Sulfates

Function: Surfactant - Cleansing Agent

Reported Product Categories: Shampoos (Non-coloring); Bath Oils, Tablets, and Salts; Eye Makeup Removers; Cleansing Products (Cold Creams, Cleansing Lotions, Liquids and Pads); Baby Shampoos

Technical/Other Names:
Magnesium Polyethylene Glycol (5) Lauryl Ether Sulfate
Magnesium Polyoxyethylene (5) Lauryl Ether Sulfate

MAGNESIUM LAURETH-8 SULFATE

CAS No.: 62755-21-9 (Generic)

Definition: Magnesium Laureth-8 Sulfate is the magnesium salt of the sulfuric acid ester of Laureth-8 (q.v.). It conforms generally to the formula:

$$\left[CH_3(CH_2)_{11}(OCH_2CH_2)_nOSO_3^- \right]_2 Mg^{+2}$$

where n has an average value of 8.

Chemical Class: Alkyl Ether Sulfates

Function: Surfactant - Cleansing Agent

Reported Product Categories: Eye Makeup Removers; Bath Oils, Tablets, and Salts; Cleansing Products (Cold Creams, Cleansing Lotions, Liquids and Pads); Shampoos (Non-coloring); Baby Shampoos

Technical/Other Name:
Magnesium Polyethylene Glycol 400 Lauryl Ether Sulfate

Trade Name Mixtures:
Texapon ASV (Cognis Deutschland)
Texapon ASV-50 (Cognis Care Chemicals/ NJ)
Texapon ASV-50 (Cognis Care Chemicals/ PA)
Texapon ASV-50 (Cognis Deutschland)

MAGNESIUM LAURETH-16 SULFATE

CAS No.: 62755-21-9 (Generic)

Definition: Magnesium Laureth-16 Sulfate is the magnesium salt of the sulfuric acid ester of Laureth-16 (q.v.). It conforms generally to the formula:

$$\left[CH_3(CH_2)_{11}(OCH_2CH_2)_nOSO_3^- \right]_2 Mg^{+2}$$

where n has an average value of 16.

Chemical Class: Alkyl Ether Sulfates

Function: Surfactant - Cleansing Agent

Reported Product Categories: Shampoos (Non-coloring); Bath Oils, Tablets, and Salts; Eye Makeup Removers; Cleansing Products (Cold Creams, Cleansing Lotions, Liquids and Pads); Baby Shampoos

Technical/Other Name:
Magnesium Polyoxyethylene (16) Lauryl Ether Sulfate

MAGNESIUM LAURYL HYDROXYPROPYL SULFONATE

Empirical Formula:
$C_{15}H_{32}O_5S \cdot \frac{1}{2}Mg$

Definition: Magnesium Lauryl Hydroxypropyl Sulfonate is the organic compound that conforms to the formula:

$$\left[\begin{array}{c} CH_3(CH_2)_{11}OCH_2CHCH_2SO_3^- \\ | \\ OH \end{array} \right]_2 Mg^{+2}$$

Chemical Class: Sulfonic Acids

Function: Surfactant - Cleansing Agent

Trade Name:
Ages 2006/Mg (Solvay GmbH)

MAGNESIUM LAURYL SULFATE

CAS No.	EINECS No.
3097-08-3	221-450-6

JPN Translation:
ラウリル硫酸 Mg

Empirical Formula:
$C_{12}H_{26}O_4S \cdot \frac{1}{2}Mg$

Definition: Magnesium Lauryl Sulfate is the magnesium salt of lauryl sulfate that conforms generally to the formula:

$$\left[CH_3(CH_2)_{11}OSO_3^- \right]_2 Mg^{+2}$$

Information Sources: 21CFR175.105, 21CFR176.170, 21CFR177.1200, CTFA D, JCIC, JCLS, SNPF, TSCA

Chemical Class: Alkyl Sulfates

Function: Surfactant - Cleansing Agent

Technical/Other Names:
Magnesium Lauryl Sulfate Solution

Magnesium Monododecyl Sulfate
Sulfuric Acid, Monododecyl Ester, Magnesium Salt

Trade Names:
Colonial LM (Colonial Chemical Inc)
Desulf MLS-30 (DeForest)
Empicol ML (Albright & Wilson UK)
Norfox MLS (Norman, Fox & Co.)
Rhodapon LM (Rhodia)
STEPANOL MG (Stepan)
Sulfochem MG (Chemron)
Zoharpon MgS (Zohar)

MAGNESIUM METHYL COCOYL TAURATE

JPN Translation:
ココイルメチルタウリン Mg

Definition: Magnesium Methyl Cocoyl Taurate is the magnesium salt of the coconut fatty acid amide of N-methyltaurine. It conforms generally to the formula:

$$\left[\begin{array}{c} O \\ || \\ RC-NCH_2CH_2SO_3^- \\ | \\ CH_3 \end{array} \right]_2 Mg^{+2}$$

where RCO- represents the coconut acid radical.

Information Sources: JCIC, JCLS

Chemical Class: Sulfonic Acids

Function: Surfactant - Cleansing Agent

Technical/Other Names:
Amides, Coconut Oil, with N-Methyltaurine, Magnesium Salts
Magnesium Cocoyl Methyl Taurate Solution

MAGNESIUM MYRETH SULFATE

Definition: Magnesium Myreth Sulfate is the magnesium salt of the sulfated ethoxylated myristyl alcohol that conforms generally to the formula:

$$\left[CH_3(CH_2)_{13}(OCH_2CH_2)_nOSO_3^- \right]_2 Mg^{+2}$$

where n has an average value of 1 to 4.

Chemical Class: Alkyl Ether Sulfates

Function: Surfactant - Cleansing Agent

Technical/Other Names:
Magnesium Polyethylene Glycol (1-4) Myristyl Ether Sulfate
Magnesium Polyoxyethylene (1-4) Myristyl Ether Sulfate

MAGNESIUM MYRISTATE

CAS No.	EINECS No.
4086-70-8	223-817-6

JPN Translation:
ミリスチン酸 Mg

Empirical Formula:
$C_{14}H_{28}O_2 \cdot \frac{1}{2}Mg$

Definition: Magnesium Myristate is the magnesium salt of myristic acid. It conforms generally to the formula:

$$\left[CH_3(CH_2)_{12}COO^- \right]_2 Mg^{+2}$$

Information Sources: 21CFR172.863, 21CFR175.105, 21CFR175.320, 21CFR176.170, 21CFR176.200, 21CFR176.210, 21CFR177.1200, 21CFR177.2260, 21CFR178.3910, JCLS, JSCI, TSCA

Chemical Class: Soaps

Functions: Anticaking Agent; Slip Modifier; Viscosity Increasing Agent - Nonaqueous

Reported Product Categories: Eye Shadows; Face Powders; Blushers (All types)

Technical/Other Name:
Tetradecanoic Acid, Magnesium Salt

Trade Name Mixtures:
ASO/MM3 (Kobo)
BBO/MM3 (Kobo)
BRO/MM3 (Kobo)
BTD/MM3 (Kobo)
BYO/MM3 (Kobo)
Mica S/MM3 (Kobo)

MAGNESIUM NITRATE

CAS No.	EINECS No.
10377-60-3	233-826-3

Empirical Formula:
$HNO_3 \cdot \frac{1}{2}Mg$

Definition: Magnesium Nitrate is the inorganic salt that conforms to the formula:

$$Mg(NO_3)_2$$

Information Sources: MI-13(5697), TSCA

Chemical Class: Inorganics

Function: Not Reported

Reported Product Categories: Hair Conditioners; Shampoos (Non-coloring)

Technical/Other Name:
Nitric Acid, Magnesium Salt

Trade Name Mixture:
Liposerve MM (Lipo)

MAGNESIUM OLETH SULFATE

CAS No.: 87569-97-9

Definition: Magnesium Oleth Sulfate is the magnesium salt of sulfated, ethoxylated oleyl alcohol that conforms generally to the formula:

$$\left[\begin{array}{c} CH_3(CH_2)_7CH \\ \| \\ CH(CH_2)_8(OCH_2CH_2)_nOSO_3 \end{array} \right]_2^- Mg^{+2}$$

where n has an average value between 1 and 4.

Chemical Class: Alkyl Ether Sulfates

Function: Surfactant - Cleansing Agent

Reported Product Categories: Eye Makeup Removers; Bath Oils, Tablets, and Salts; Cleansing Products (Cold Creams, Cleansing Lotions, Liquids and Pads); Shampoos (Noncoloring)

Technical/Other Names:
Magnesium Polyethylene Glycol (1-4) Oleyl Ether Sulfate
Magnesium Polyoxyethylene (1-4) Oleyl Ether Sulfate

Trade Name Mixtures:
LP110 (Phytocos)
Texapon ASV (Cognis Deutschland)
Texapon ASV-50 (Cognis Care Chemicals/ NJ)
Texapon ASV-50 (Cognis Care Chemicals/ PA)
Texapon ASV-50 (Cognis Deutschland)

MAGNESIUM OXIDE

CAS No.	EINECS No.
1309-48-4	215-171-9

JPN Translation:
酸化 Mg

Empirical Formula:
MgO

Definition: Magnesium Oxide is the inorganic oxide that conforms generally to the formula:

MgO

Information Sources: AUS, BP, BPC, BRA, 21CFR163.110, 21CFR175.300, 21CFR177.1460, 21CFR177.2400, 21CFR177.2600, 21CFR178.1010, 21CFR184.1431, 21CFR558.311, CI 77711, CTFA S, CZE, DA, DDR, EGY, EP, FCC, FIN, HP, HUN, IND, ITA, JAN, JCLS, JSCI, MAR, MEX, MI-13(5700), OTC-I-AA, PF, PN, POR, ROM, TSCA, USAN, USD, USP XXIV, USSR, YUG

Chemical Class: Inorganics

Functions: Absorbent; Opacifying Agent; pH Adjuster

Reported Product Categories: Bath Oils, Tablets, and Salts; Hair Bleaches

Technical/Other Name:
CI 77711

Trade Names:
Magnesium Oxide, Heavy USP 310-S (Whittaker, Clark & Daniels)
Magnesium Oxide, Light USP 311-S (Whittaker, Clark & Daniels)

Trade Name Mixtures:
Kalixide CT (Vevy)
Thermo Power (LG Cosmetic)
Vulcanyl CG (ARCO)
White Charcoal MB-15153 (Ikeda)

MAGNESIUM PALMITATE

CAS No.	EINECS No.
2601-98-1	220-010-0

Empirical Formula:
$C_{16}H_{32}O_2 \cdot \frac{1}{2}Mg$

Definition: Magnesium Palmitate is the magnesium salt of palmitic acid. It conforms generally to the formula:

$$\left[CH_3(CH_2)_{14}COO^- \right]_2 Mg^{+2}$$

Information Sources: 21CFR172.863, 21CFR175.105, 21CFR175.300, 21CFR175.320, 21CFR176.170, 21CFR176.200, 21CFR176.210, 21CFR177.1200, 21CFR177.2260, 21CFR178.3910, TSCA

Chemical Class: Soaps

Functions: Anticaking Agent; Slip Modifier; Viscosity Increasing Agent - Nonaqueous

Technical/Other Names:
Hexadecanoic Acid, Magnesium Salt
Magnesium Hexadecanoate
Palmitic Acid, Magnesium Salt

MAGNESIUM PALMITOYL GLUTAMATE

CAS No.	EINECS No.
57539-47-6	260-800-2

Definition: Magnesium Palmitoyl Glutamate is the substituted amino acid that conforms to the formula:

$$\left[\begin{array}{c} {}^-OOCCH_2CH_2CHCOO^- \\ | \\ HN-C(CH_2)_{14}CH_3 \\ \| \\ O \end{array} \right] Mg^{+2}$$

Chemical Class: Amino Acids

Function: Skin-Conditioning Agent - Miscellaneous

Technical/Other Names:
Dihydrogen Bis[N-Palmitoyl-L-Glutamato (2-)-N,O]Magnesate(2-)
L-Glutamic Acid, N-(1-Oxohexadecyl)-, Magnesium Complex
Magnesate(2-), Bis[N-(1-Oxohexadecyl)-L-Glutamato(2-)-N,O1], Dihydrogen, (T-4)-

Trade Name Mixture:
Sepifeel One (SEPPIC)

MAGNESIUM PCA

Empirical Formula:
$C_5H_7NO_3 \cdot \frac{1}{2}Mg$

Definition: Magnesium PCA is the magnesium salt of PCA (q.v). It conforms to the formula:

$$\left[\begin{array}{c} O=\overset{NH}{\diagdown}\diagup COO^- \end{array} \right]_2 Mg^{+2}$$

Chemical Classes: Heterocyclic Compounds; Organic Salts

Function: Skin-Conditioning Agent - Humectant

Trade Names:
Magnesium Oligoelement Mg Complex (L.C.S.)
Magnolidone (UCIB (Solabia))

Trade Name Mixtures:
Caorange (CEP (Solabia))
Physiogenyl (UCIB (Solabia))

MAGNESIUM PEG-3 COCAMIDE SULFATE

Definition: Magnesium PEG-3 Cocamide Sulfate is the magnesium salt of a mixture of sulfated esters of PEG-3 Cocamide (q.v.).

Chemical Class: Alkoxylated Amides

Function: Surfactant - Emulsifying Agent

Technical/Other Names:
PEG-3 Cocamide Ether Sulfuric Acid, Magnesium Salt
Polyethylene Glycol (3) Cocamide Ether Sulfuric Acid, Magnesium Salt
Polyoxyethylene (3) Cocamide Ether Sulfuric Acid, Magnesium Salt

Trade Names:
Genapol AMG (Clariant)
Genapol AMG (Clariant GmbH, Personal Care)

MAGNESIUM PEROXIDE

CAS Nos.	EINECS Nos.
1335-26-8	215-627-7
14452-57-4	238-438-1

Empirical Formula:
MgO_2

Definition: Magnesium Peroxide is the inorganic oxide that conforms to the formula:

$$MgO_2$$

Information Sources: EEC(III/1-12), MI-13 (5704), TSCA

Chemical Class: Inorganics

Function: Oxidizing Agent

Technical/Other Name:
Magnesium Dioxide

Trade Name:
Magnesium Peroxide, 25% (Lohmann)

MAGNESIUM PHOSPHATE

CAS No.	EINECS No.
10043-83-1	233-142-9

Definition: Magnesium Phosphate is the inorganic salt that conforms to the formula:

$$Mg_3(PO_4)_2$$

Chemical Classes: Inorganic Salts; Phosphorus Compounds

Function: Suspending Agent - Nonsurfactant

Technical/Other Name:
Phosphoric Acid, Magnesium Salt

Trade Name Mixture:
Ionpure Type H (US Cosmetics)

MAGNESIUM POTASSIUM FLUORO-SILICATE

Definition: Magnesium Potassium Fluoro-silicate is the inorganic salt that conforms generally to the formula:

$$KMg_{2.5}Si_4O_{10}F_2$$

Chemical Class: Inorganics

Function: Abrasive

Trade Names:
Micro Mica (Co-Op/Kobo)
Soft Sericite T-6 (Dainihon Kasei)

Trade Name Mixture:
Micro Mica (Kobo)

MAGNESIUM/POTASSIUM/SILICON/FLUORIDE/HYDROXIDE/OXIDE

Definition: Magnesium/Potassium/Silicon/Fluoride/Hydroxide/Oxide is the product obtained by heating talc with potassium silicofluoride.

Chemical Class: Inorganics

Function: Bulking Agent

Trade Name:
Micro Mica (Co-Op/Kobo)

Trade Name Mixture:
Micro Mica (Kobo)

MAGNESIUM PROPIONATE

CAS No.	EINECS No.
557-27-7	209-166-0

Empirical Formula:
$C_3H_6O_2 \cdot \frac{1}{2}Mg$

Definition: Magnesium Propionate is the magnesium salt of propionic acid that conforms to the formula:

$$\left[CH_3CH_2COO^- \right]_2 Mg^{+2}$$

Information Source: EEC(VI/1-2)

Chemical Class: Organic Salts

Function: Preservative

Technical/Other Names:
Magnesium Dipropionate
Propanoic Acid, Magnesium Salt
Propionic Acid, Magnesium Salt

MAGNESIUM SALICYLATE

CAS No.	EINECS No.
18917-89-0	242-669-3

Empirical Formula:
$C_7H_6O_3 \cdot \frac{1}{2}Mg$

Definition: Magnesium Salicylate is the magnesium salt of Salicylic Acid (q.v.) that conforms to the formula:

Information Sources: CIR: [SQ], EEC(VI/1-3), MHLW-331/3, MI-13(5709), OTC-I-IA, OTC-I-MD, TSCA, USAN, USP XXIV

Chemical Classes: Organic Salts; Phenols

Function: Preservative

Technical/Other Names:
Benzoic Acid, 2-Hydroxy-, Magnesium Salt
Magnesium 2-Hydroxybenzoate
Salicylic Acid, Magnesium Salt

MAGNESIUM SILICATE

CAS No.	EINECS No.
1343-88-0	215-681-1

Definition: Magnesium Silicate is an inorganic salt of variable composition which consists mainly of:

$$MgO \cdot SiO_2 \cdot xH_2O$$

Information Sources: 21CFR169.179, 21CFR169.182, 21CFR182.2437, CIR: [S] IJT-22(SUPPL. 1)2003, CTFA D, FCC, JAN, JCLS, JSCI, MI-13(5711), NF XIX, PN, TSCA, USAN

Chemical Class: Inorganic Salts

Functions: Absorbent; Anticaking Agent; Bulking Agent; Opacifying Agent; Slip Modifier; Viscosity Increasing Agent - Aqueous

Reported Product Category: Paste Masks (Mud Packs)

Technical/Other Name:
Silicic Acid, Magnesium Salt (1:1)

Trade Names:
Celkate T-21 (Celite)
Magnesol (Dallas Group)
3M Cosmetic Microspheres CM-111 (3M)
Silprec SM (PQ Corporation)

Trade Name Mixture:
Kalixide LT (Vevy)

MAGNESIUM SODIUM FLUOROSILICATE

Definition: Magnesium Sodium Fluoro-silicate is the inorganic salt that conforms generally to the formula:

$$NaMg_{2.5}Si_4O_{10}F_2$$

Chemical Class: Inorganics

Function: Abrasive

Trade Names:
SOMASIF (Co-Op/Kobo)
Somasif ME-100 (Co-Op/Kobo)

MAGNESIUM STEARATE

CAS No.	EINECS No.
557-04-0	209-150-3

JPN Translation:
ステアリン酸 Mg

Empirical Formula:
$C_{18}H_{36}O_2 \cdot \frac{1}{2}Mg$

Definition: Magnesium Stearate is the magnesium salt of stearic acid. It conforms generally to the formula:

$$\left[CH_3(CH_2)_{16}COO^- \right]_2 Mg^{+2}$$

See "Regulatory and Ingredient Use Information," for Colorants in Volume 1, Introduction, Part A. This ingredient is not an approved colorant for the US. To identify the colorant allowed for use in the European Union (EU), the INCI Name Magnesium Stearate must be used, except for hair dye products.

Information Sources: AUS, BP, BPC, BRA, 21CFR172.863, 21CFR173.340, 21CFR175.105, 21CFR175.300, 21CFR175.320, 21CFR176.170, 21CFR176.200, 21CFR176.210, 21CFR177.1200, 21CFR177.2260, 21CFR178.3910, 21CFR179.45, 21CFR181.22, 21CFR181.29, 21CFR184.1440, CIR: [S] JACT-1(2)1982, CTFA S, CZE, EEC(IV/1), FCC, HUN, ITA, JAN, JCLS, JSCI, MAR, MI-13(5714), NF XIX, PN, POR, SNPF, TSCA, USAN, USD

Chemical Classes: Color Additives - Approved in the EU; Soaps

Functions: Anticaking Agent; Bulking Agent; Colorant; Viscosity Increasing Agent - Nonaqueous

Technical/Other Names:
Magnesium Octadecanoate
Octadecanoic Acid, Magnesium Salt
Stearic Acid, Magnesium Salt

Trade Names:
Radiastar 1100 (Oleon NV)
Unichem MS (Universal Preserv-A-Chem)

Trade Name Mixtures:
Base W/O 126 (LCW)
Hostacerin WO (Clariant)
Hostacerin WO (Clariant GmbH, Personal Care)
Mascawax 012 (LCW)

MAGNESIUM SULFATE

CAS No.   EINECS No.
7487-88-9   231-298-2

JPN Translation:
硫酸 Mg

Empirical Formula:
$H_2O_4S \cdot Mg$

Definition: Magnesium Sulfate is the inorganic salt that conforms to the formula:

$$MgSO_4$$

Information Sources: ARG, AUS, BP, BPC, BRA, 21CFR184.1443, 21CFR558.311, CZE, DA, DDR, EGY, EP, FCC, FIN, HP, HUN, IND, ITA, JAN, JCIC, JCLS, JSCI, MAR, MEX, MI-13(5715), OTC-I-LX, PF, PN, POR, ROM, TSCA, USAN, USD, USP XXIV, USSR, YUG

Chemical Class: Inorganic Salts

Function: Bulking Agent

Reported Product Categories: Bath Preparations, Misc.; Body and Hand Preparations (Excluding Shaving Preparations); Bath Soaps and Detergents; Night Skin Care Preparations; Skin Care Preparations, Misc.; Foundations; Moisturizing Preparations; Cleansing Products (Cold Creams, Cleansing Lotions, Liquids and Pads); Makeup Preparations (Not eye), Misc.; Bath Oils, Tablets, and Salts; Permanent Waves; Bath Capsules; Face and Neck Preparations (Excluding Shaving Preparations); Hair Conditioners; Shampoos (Non-coloring); Suntan Gels, Creams, and Liquids

Technical/Other Names:
Anhydrous Magnesium Sulfate
Epsom Salt
Sulfuric Acid, Magnesium Salt (1:1)

Trade Name Mixtures:
Activator Omega MO Type B (Derma-Search)
OMADINE MDS (Arch Chemical)
Phycosaccharide AG (Codif)

MAGNESIUM SULFIDE

CAS No.   EINECS No.
12032-36-9   234-771-1

Empirical Formula:
MgS

Definition: Magnesium Sulfide is the inorganic salt that conforms to the formula:

$$MgS$$

Information Sources: EEC(III/1-23), TSCA

Chemical Class: Inorganic Salts

Function: Depilating Agent

Technical/Other Name:
Magnesium Monosulfide

MAGNESIUM TALLOWATE

Definition: Magnesium Tallowate is the magnesium salt of Tallow Acid (q.v.).

Information Sources: 21CFR175.105, 21CFR176.170, 21CFR178.3910

Chemical Class: Soaps

Functions: Anticaking Agent; Bulking Agent; Viscosity Increasing Agent - Nonaqueous

Technical/Other Name:
Fatty Acids, Tallow, Magnesium Salt

MAGNESIUM/TEA-COCO-SULFATE

JPN Translation:
ココアルキル硫酸（ Mg / TEA ）

Definition: Magnesium/TEA-Coco-Sulfate is the mixed magnesium and triethanolamine salt of Coco-Sulfate (q.v.).

Information Sources: JCIC, JCLS

Chemical Class: Alkyl Sulfates

Function: Surfactant - Cleansing Agent

Technical/Other Name:
Magnesium•Triethanolamine Coco Sulfate

MAGNESIUM THIOGLYCOLATE

CAS No.   EINECS No.
63592-16-5   264-358-1

Empirical Formula:
$C_2H_4O_2S \cdot \frac{1}{2}Mg$

Definition: Magnesium Thioglycolate is the magnesium salt of thioglycolic acid. It conforms to the formula:

$$\left[HSCH_2COO^- \right]_2 Mg^{+2}$$

Information Source: EEC(III/1-2a)

Chemical Classes: Organic Salts; Thio Compounds

Functions: Depilating Agent; Hair-Waving/Straightening Agent; Reducing Agent

Technical/Other Names:
Acetic Acid, Mercapto-, Magnesium Salt
Magnesium Mercaptoacetate
Magnesium Thioglycollate
Thiooglycolic Acid, Magnesium Salt

MAGNESIUM TRISILICATE

CAS No.   EINECS No.
14987-04-3   239-076-7

JPN Translation:
ケイ酸 Mg

Empirical Formula:
$H_4O_8Si_3 \cdot 2Mg$

Definition: Magnesium Trisilicate is the inorganic compound that conforms generally to the formula:

$$2MgO \cdot 3SiO_2 \cdot xH_2O$$

Information Sources: ARG, AUS, BP, BPC, BRA, CIR: [S] IJT-22(SUPPL. 1)2003, EGY, HUN, IND, MAR, MEX, MI-13(5711), OTC-I-AA, OTC-I-OR, PN, TSCA, USAN, USD, USP XXIV, YUG

Chemical Class: Inorganic Salts

Functions: Abrasive; Absorbent; Anticaking Agent; Bulking Agent; Opacifying Agent; Slip

Modifier; Viscosity Increasing Agent - Aqueous

Reported Product Category: Powders (Dusting and Talcum, Excluding Aftershave Talcs)

Technical/Other Names:
Magnesium Silicon Oxide
Silicic Acid, Magnesium Salt (1:2)

MAGNOLIA ACUMINATA BARK EXTRACT

Definition: Magnolia Acuminata Bark Extract is an extract of the bark of the magnolia, *Magnolia acuminata. See "Regulatory and Ingredient Use Information," regarding the labeling names for botanical ingredients in Volume 1, Introduction, Part A.*

Chemical Class: Biological Products

Function: Not Reported

Technical/Other Names:
Cucumber Tree Bark Extract
Extract of Cucumber Tree Bark
Extract of Magnolia Acuminata Bark
Extract of Magnolia Bark
Magnolia Bark Extract

Trade Name Mixtures:
Actiphyte of Magnolia Bark BG50 (Active Organics)
Actiphyte of Magnolia Bark GL50 (Active Organics)
Actiphyte of Magnolia Bark Lipo S (Active Organics)
Actiphyte of Magnolia Bark PG50 (Active Organics)

MAGNOLIA ACUMINATA FLOWER EXTRACT

Definition: Magnolia Acuminata Flower Extract is an extract of the flower of *Magnolia acuminata. See "Regulatory and Ingredient Use Information," regarding the labeling names for botanical ingredients in Volume 1, Introduction, Part A.*

Chemical Class: Biological Products

Function: Skin-Conditioning Agent - Miscellaneous

Technical/Other Name:
Extract of Magnolia Acuminata Flower

Trade Name Mixture:
Magnolia HG (Alban Muller)

MAGNOLIA BIONDII BARK EXTRACT

Definition: Magnolia Biondii Bark Extract is an extract of the bark of the Chinese magnolia, *Magnolia biondii. See "Regulatory and Ingredient Use Information," regarding the labeling names for botanical ingredients in Volume 1, Introduction, Part A.*

Chemical Class: Biological Products

Function: Not Reported

Technical/Other Names:
Chinese Magnolia (Magnolia Biondii) Bark Extract
Extract of Magnolia Biondii Bark

Trade Name Mixture:
VT-264 Extract of Chinese Magnolia (Vege-Tech)

MAGNOLIA BIONDII FLOWER EXTRACT

Definition: Magnolia Biondii Flower Extract is an extract of the flowers and buds of the Chinese magnolia, *Magnolia biondii. See "Regulatory and Ingredient Use Information," regarding the labeling names for botanical ingredients in Volume 1, Introduction, Part A.*

Chemical Class: Biological Products

Function: Skin-Conditioning Agent - Miscellaneous

Technical/Other Names:
Chinese Magnolia Extract
Chinese Magnolia (Magnolia Biondii) Extract
Extract of Chinese Magnolia
Extract of Magnolia Biondii

Trade Name Mixtures:
Chinese Magnolia Flower Extract HS (Tri-K)
Phytotal AI (Phybiotex/Sederma)
YSK Magic 11 (Phyto-Technologies)

MAGNOLIA GRANDIFLORA LEAF EXTRACT

Definition: Magnolia Grandiflora Leaf Extract is an extract of the leaf of *Magnolia grandiflora. See "Regulatory and Ingredient Use Information," regarding the labeling names for botanical ingredients in Volume 1, Introduction, Part A.*

Chemical Class: Biological Products

Function: Not Reported

Technical/Other Name:
Extract of Magnolia Grandiflora

MAGNOLIA KOBUS BARK EXTRACT

Definition: Magnolia Kobus Bark Extract is an extract of the bark of *Magnolia kobus.* See *"Regulatory and Ingredient Use Information," regarding the labeling names for botanical ingredients in Volume 1, Introduction, Part A.*

Chemical Class: Biological Products

Function: Skin-Conditioning Agent - Humectant

Technical/Other Name:
Extract of Magnolia Kobus

Trade Name:
Kobushi Extract Powder (Ichimaru Pharcos)

Trade Name Mixture:
Kobushi Liquid E (Ichimaru Pharcos)

MAGNOLIA LILIFLORA EXTRACT

Definition: Magnolia Liliflora Extract is an extract of the buds of the flower, *Magnolia liliflora. See "Regulatory and Ingredient Use Information," regarding the labeling names for botanical ingredients in Volume 1, Introduction, Part A.*

Chemical Class: Biological Products

Function: Not Reported

Technical/Other Name:
Extract of Magnolia Liliflora

Trade Name Mixtures:
China Extract Xin Yi (E.U.K)
Chine Extract Magnolia Liliflora (Ennagram)
Extrait Hydroglycolique De Magnolia (Greentech)

MAGNOLIA OBOVATA BARK EXTRACT

Definition: Magnolia Obovata Bark Extract is an extract of the bark of *Magnolia obovata. See "Regulatory and Ingredient Use Information," regarding the labeling names for botanical ingredients in Volume 1, Introduction, Part A.*

Chemical Class: Biological Products

Function: Not Reported

Technical/Other Name:
Extract of Magnolia Obovata

Trade Name:
Kouboku Extract Powder (Ichimaru Pharcos)

Trade Name Mixtures:
Honoki Liquid (Sato Forestry Co.)
Kouboku Liquid B (Ichimaru Pharcos)
Kouboku Liquid E (Ichimaru Pharcos)

MAGNOLIA OFFICINALIS BARK EXTRACT

Definition: Magnolia Officinalis Bark Extract is an extract of the bark of *Magnolia* *officinalis.* See *"Regulatory and Ingredient Use Information,"* regarding the labeling names for botanical ingredients in Volume 1, Introduction, Part A.

Chemical Class: Biological Products

Function: Skin-Conditioning Agent - Miscellaneous

Technical/Other Name:
Extract of Magnolia Officinalis Bark

Trade Name:
Premier Magnolia 100% Extract (Premier Specialties)

MAHONIA AQUIFOLIUM ROOT EXTRACT

Definition: Mahonia Aquifolium Root Extract is an extract of the roots of *Mahonia aquifolium.* See *"Regulatory and Ingredient Use Information,"* regarding the labeling names for botanical ingredients in Volume 1, Introduction, Part A.

Chemical Class: Biological Products

Function: Cosmetic Astringent

Technical/Other Name:
Extract of Mahonia Aquifolium Root

Trade Name Mixtures:
Actiphyte Oregon Grape (Active Organics)
Actiphyte Oregon Grape BG50P (Active Organics)

MALACHITE

CAS No.: 1319-53-5

Empirical Formula:
$CH_2Cu_2O_5$

Definition: Malachite is the mineral consisting chiefly of copper carbonate hydroxide. It conforms generally to the formula:

$$Cu_2CO_3(OH)_2$$

Information Source: MI-13(2658)

Chemical Class: Inorganics

Function: Skin-Conditioning Agent - Miscellaneous

Technical/Other Names:
Basic Copper Carbonate
Copper Carbonate Hydroxide

Trade Name:
Malachite KJ (Aveda)

MALACHITE EXTRACT

Definition: Malachite Extract is an extract of Malachite (q.v.) obtained by grinding the stone, with subsequent solubilization by hydrochloric acid and neutralization by sodium bicarbonate.

Chemical Class: Inorganics

Function: Antioxidant

Trade Name Mixture:
Mala' Kite (Libiol)

MALEATED SOYBEAN OIL

CAS No. 68648-66-8   **EINECS No.** 272-000-0

Definition: Maleated Soybean Oil is a modified soybean oil in which some of the unsaturation has been converted to a cyclic dicarboxylic acid generally conforming to the formula:

where R represents the unreacted residual soybean oil.

Information Source: TSCA

Chemical Class: Fats and Oils

Function: Skin-Conditioning Agent - Miscellaneous

Reported Product Categories: Lipsticks; Eyebrow Pencils; Suntan Gels, Creams, and Liquids; Bath Soaps and Detergents

Technical/Other Names:
Glyceridacid
Oils, Soybean, Maleated
Soybean Oil, Maleated

MALEIC ACID

CAS Nos. 110-16-7   **EINECS No.** 203-742-5
6915-18-0

Empirical Formula:
$C_4H_4O_4$

Definition: Maleic Acid is the cis unsaturated organic acid that conforms to the formula:

$$HOOCCH = CHCOOH$$

Information Sources: BP, BPC, 21CFR175.105, 21CFR177.1200, CTFA D, EGY, IND, MAR, MI-13(5726), RIFM, TSCA

Chemical Class: Carboxylic Acids

Functions: Fragrance Ingredient; pH Adjuster

Technical/Other Names:
2-Butenedioic Acid
cis-1,2-Ethylenedicarboxylic Acid
Maleic acid (RIFM)

MALEIC ANHYDRIDE

CAS No. 108-31-6   **EINECS No.** 203-571-6

Empirical Formula:
$C_4H_2O_3$

Definition: Maleic Anhydride is the organic compound that conforms to the formula:

Information Source: MI-13(5727)

Chemical Class: Heterocyclic Compounds

Function: Artificial Nail Builder

Technical/Other Name:
2,5-Furandione

MALIC ACID

CAS Nos. 97-67-6 (L-Form)   **EINECS Nos.** 202-601-5
617-48-1 (dl-alpha)
636-61-3 (D-Form)
6915-15-7   230-022-8

JPN Translation:
リンゴ酸

Empirical Formula:
$C_4H_6O_5$

Definition: Malic Acid is the organic acid that conforms to the formula:

$$\underset{\underset{OH}{|}}{HOOCCHCH_2COOH}$$

Information Sources: 21CFR150, 21CFR150.161, 21CFR169.115, 21CFR169.140, 21CFR169.150, 21CFR184.1069, CIR: [SQ] IJT-20(SUPPL. 1)2001, CTFA D, FCC, IND, JCIC, JCLS, JSQI, MAR, MI-13(5730), NF XVIII, RIFM, TSCA, USAN

Chemical Class: Carboxylic Acids

Functions: Fragrance Ingredient; pH Adjuster

Reported Product Categories: Nail Polish and Enamels; Basecoats and Undercoats;

Hair Conditioners; Shampoos (Non-coloring); Moisturizing Preparations; Manicuring Preparations, Misc.; Night Skin Care Preparations; Bath Capsules; Face and Neck Preparations (Excluding Shaving Preparations)

Technical/Other Names:
Butanedioic Acid, Hydroxy-
Deoxytetraric Acid
Hydroxybutanedioic Acid
2-Hydroxyethane-1,2-dicarboxylic Acid
Hydroxysuccinnic Acid
DL-Malic Acid
l-Malic acid (RIFM)

Trade Names:
AC Malic Acid (Active Concepts)
Keresine (Vevy)

Trade Name Mixtures:
Acifructol Complex P 63 (Gattefosse s.a.)
A.H.A. 40 (Ennagram)
Alfhac 5 (Fabriquimica)
Amidroxy 4 Flowers (Alban Muller)
Apple "AHA" (Cosmetochem)
  (Cosmetochem International Ltd.)
Apricot "AHA" (Cosmetochem)
  (Cosmetochem International Ltd.)
Elesponge AHA LS 8911 B (Laboratoires Serobiologiques)
Fruitedone (UCIB (Solabia))
Glycacid CMMA (Coletica SA)
Glycacid INMA (Coletica SA)
Green Apple Complex (Greentech)
Silesil (Chemyunion)
Vegeles AHA 8702 (Laboratoires Sero-biologiques)
Vegeles AHA LS 8763 (Laboratoires Sero-biologiques)

MALLOTUS JAPONICUS BARK EXTRACT

Definition: Mallotus Japonicus Bark Extract is an extract of the bark of *Mallotus japonicus*. See "Regulatory and Ingredient Use Information," regarding the labeling names for botanical ingredients in Volume 1, Introduction, Part A.

Chemical Class: Biological Products

Function: Skin-Conditioning Agent - Miscellaneous

Technical/Other Name:
Extract of Mallotus Japonicus

Trade Name Mixtures:
Malloti Extract (Maruzen Pharmaceuticals Co., Ltd.)
Malloti Extract BG (Maruzen Pharmaceuticals Co., Ltd.)

MALONIC ACID

CAS No.	EINECS No.
141-82-2	205-503-0

Empirical Formula:
$C_3H_4O_4$

Definition: Malonic Acid is the organic compound that conforms to the formula:

$$HOOCCH_2COOH$$

Information Sources: MI-13(5732), RIFM

Chemical Class: Carboxylic Acids

Functions: Fragrance Ingredient; pH Adjuster

Technical/Other Names:
Carboxyacetic Acid
Dicarboxymethane
Malonic acid (RIFM)
Propanedioic Acid
1,3-Propanedioic Acid

MALPIGHIA EMARGINATA (ACEROLA) FRUIT EXTRACT

Definition: Malpighia Emarginata (Acerola) Fruit Extract is an extract of the ripe fruit of *Malpighia emarginata*. See "Regulatory and Ingredient Use Information," regarding the labeling names for botanical ingredients in Volume 1, Introduction, Part A.

Chemical Class: Biological Products

Functions: Hair Conditioning Agent; Skin-Conditioning Agent - Miscellaneous

Technical/Other Names:
Acerola (Malpighia Emarginata) Extract
Extract of Malpighia Emarginata
Malpighia Emarginata Extract

Trade Name Mixtures:
Acerola Extract WB (Nichirei) (Nichirei)
Acerola Extract W (Nichirei) (Nichirei)

MALPIGHIA EMARGINATA (ACEROLA) SEED EXTRACT

Definition: Malpighia Emarginata (Acerola) Seed Extract is an extract of the seeds of *Malpighia emarginata*. See "Regulatory and Ingredient Use Information," regarding the labeling names for botanical ingredients in Volume 1, Introduction, Part A.

Chemical Class: Biological Products

Functions: Hair Conditioning Agent; Humectant; Skin-Conditioning Agent - Emollient

Technical/Other Name:
Extract of Malpighia Emarginata (Acerola) Seed

Trade Name Mixture:
Acerola Seed Extract B30 (Nichirei) (Nichirei)

MALPIGHIA GLABRA (ACEROLA) FRUIT EXTRACT

Definition: Malpighia Glabra (Acerola) Fruit Extract is an extract of the fruit of *Malpighia glabra*. See "Regulatory and Ingredient Use Information," regarding the labeling names for botanical ingredients in Volume 1, Introduction, Part A.

Chemical Class: Biological Products

Function: Not Reported

Technical/Other Names:
Acerola Extract
Acerola Extract BG-25
Acerola (Malpighia Glabra) Extract
Barbados Cherry Extract
Extract of Acerola
Extract of Malpighia glabra

Trade Name:
Frulix TF Acerola (Assessa-Industria)

Trade Name Mixtures:
Acerola Dry Extract 15% Natural Vitamin C (Centroflora)
Acerola Secrets (Gattefosse s.a.)
Cytoflavin-C (Chemyunion)
Ecofloral Acerola (Centroflora)
Vegetol Acerola ME 219 Hydro (Gattefosse s.a.)

MALPIGHIA GLABRA FRUIT

Definition: Malpighia Glabra Fruit is the fruit of *Malpighia glabra*. See "Regulatory and Ingredient Use Information," regarding the labeling names for botanical ingredients in Volume 1, Introduction, Part A.

Chemical Class: Biological Products

Function: Skin-Conditioning Agent - Emollient

Trade Name Mixture:
Unioil Fruits (Chemyunion)

MALPIGHIA PUNICIFOLIA (ACEROLA) FRUIT

Definition: Malpighia Punicifolia (Acerola) Fruit is a plant material derived from the ripe fruit of the acerola, *Malpighia punicifolia*. See "Regulatory and Ingredient Use Information," regarding the labeling names for botanical ingredients in Volume 1, Introduction, Part A.

Information Sources: 21CFR101.30, MAR, MI-13(36)

Chemical Class: Biological Products

Function: Not Reported

Technical/Other Names:
Acerola
Acerola B Concentrate
Acerola (Malpighia Punicifolia)

MALPIGHIA PUNICIFOLIA (ACEROLA) FRUIT EXTRACT

Definition: Malpighia Punicifolia (Acerola) Fruit Extract is the extract obtained from the fruit of *Malpighia punicifolia. See "Regulatory and Ingredient Use Information," regarding the labeling names for botanical ingredients in Volume 1, Intro-duction, Part A.*

Chemical Class: Biological Products

Function: Antioxidant

Technical/Other Names:
Acerola Extract
Acerola (Malpighia Punicifolia) Extract
Extract of Acerola (Malpighia Puncifolia)

Trade Name Mixtures:
Acerola Extract HS 3629 G (Grau)
Acerola HS (Alban Muller)
Acerola HSA (Alban Muller)
Aceromine (Greentech)
Actiphyte of Acerola (Active Organics)
Actiphyte of Acerola GL (Active Organics)
Fruitapone Acerola GT 2/037010 (Symrise)
Rosamine (Greentech)

MALT EXTRACT

CAS No. 8002-48-0   **EINECS No.** 232-310-9

JPN Translation:
バクガエキス

Definition: Malt Extract is the dark syrup obtained by evaporating an aqueous extract of partially germinated and dried barley seeds. *See Reported Ingredient Functions-The Cosmetic Drug Distinction, in Regulatory and Ingredient Use Informa-tion, Volume I, Part A.*

Information Sources: BPC, BRA, 21CFR133.178, 21CFR184.1445, EGY, IND, JAN, JCIC, JCLS, JSQI, MAR, NED, NFJ

Chemical Class: Biological Products

Functions: Skin-Conditioning Agent - Mis-cellaneous; Skin Protectant

Technical/Other Names:
Extract of Malt
Malt Root Extract

Trade Name Mixtures:
Hops Malt Extract HS 2518 G (Grau)
Pronalen Sunlife (Provital/Centerchem)
Questex Malt Wine Concentrate (Quest International)

MALTITOL

CAS No. 585-88-6 (D-Form)   **EINECS No.** 209-567-0

JPN Translation:
マルチトール

Empirical Formula:
$C_{12}H_{24}O_{11}$

Definition: Maltitol is a disaccharide polyol obtained by hydrogenation of maltose. It conforms to the formula:

Information Sources: JCIC, JCLS, JSQI, NF XIX, USAN

Chemical Classes: Carbohydrates; Polyols

Functions: Flavoring Agent; Humectant; Skin-Conditioning Agent - Humectant

Reported Product Categories: Moisturizing Preparations; Skin Care Preparations, Misc.; Bath Preparations, Misc.; Body and Hand Preparations (Excluding Shaving Prepara-tions); Bath Oils, Tablets, and Salts; Cleans-ing Products (Cold Creams, Cleansing Lotions, Liquids and Pads)

Technical/Other Names:
D-Glucitol, 4-O-α-D-glucopyranosyl-Maltitol Solution

Trade Names:
Crystalline Maltisorb (Roquette)
Malbit (Hayashibara)

Trade Name Mixture:
Nikkol Aquasome LA (Nikko)

MALTITOL LAURATE

CAS No.: 75765-49-0

JPN Translation:
ラウリン酸マルチトール

Empirical Formula:
$C_{24}H_{45}O_{12}$

Definition: Maltitol Laurate is the ester of Maltitol (q.v) and lauric acid that conforms to the formula:

Chemical Classes: Carbohydrates; Esters; Polyols

Functions: Emulsion Stabilizer; Skin-Conditioning Agent - Miscellaneous

Technical/Other Names:
D-Glucitol, 4-O-α-D-Glucopyranosyl-, Monododecanoate
Maltitol Monolaurate

Trade Name:
Maltel SML (Croda Japan)

MALT JUICE

JPN Translation:
バクガ液汁

Definition: Malt Juice is the liquid expressed from partially germinated and dried barley seeds.

Information Source: JCLS

Chemical Class: Biological Products

Function: Skin-Conditioning Agent - Mis-cellaneous

MALTODEXTRIN

CAS No. 9050-36-6   **EINECS No.** 232-940-4

Definition: Maltodextrin is the saccharide material obtained by hydrolysis of starch.

Information Sources: 21CFR184.1444, 21CFR310.545, NF XIX, TSCA, USAN

Chemical Class: Gums, Hydrophilic Colloids and Derivatives

Functions: Absorbent; Binder; Emulsion Stabilizer; Film Former; Hair Conditioning

Agent; Skin-Conditioning Agent - Miscellaneous; Suspending Agent - Nonsurfactant

Reported Product Categories: Face Powders; Foundations; Baby Products, Misc.; Bath Oils, Tablets, and Salts; Bath Preparations, Misc.; Eye Makeup Preparations, Misc.

Trade Names:
Glucidex (Roquette)
Maltrin (Grain Processing)

Trade Name Mixtures:
Acerola Dry Extract 15% Natural Vitamin C (Centroflora)
Acerola Secrets (Gattefosse s.a.)
Activa TI (Ajinomoto)
AEC Coconut Cream Powder (A & E Connock)
AEC Honey Powder (A & E Connock)
Albacan (Bio-Botanica)
Aloe Vera Gel Spray Dried Powder 100:1 SP2l1OO (Aloestar)
Blaukraut P-AC-6 (Heidelberger)
Brennessel P-AC-9 (Heidelberger)
Carmine Passion (Quest International)
Carosol 3% (Carotech)
Champagene Complex (Greentech)
Champagne Truffle Complex (Greentech)
C-Med-100 (AF Nutraceutical)
Dried Truffle Extract (Greentech)
Dry Extract of Champagne (Greentech)
Dry Extract of Red Wine (Greentech)
Dry Extract of White Tea (Greentech)
Eclaicissant G (Greentech)
Ecophylane of Cocoa (CEP (Solabia))
Ecophylane of Coriander (CEP (Solabia))
Ecophylane of Cotton (CEP (Solabia))
Ecophylane of Green Tea (CEP (Solabia))
Ecophylane of Kiwi (CEP (Solabia))
Enzyami 3 (Alban Muller)
Extrait sec de Figue (Greentech)
Extrapone Calendula 2/B04042 (Symrise)
Farberdistel P-AC-10 (Heidelberger)
Farnesol Powder 2/027038 (Symrise)
Figeline (Greentech)
Gardenia P-AC-2 (Heidelberger)
Gardenia Grun P-AC-8 (Heidelberger)
Genotexin (Bioland)
Gravinol (Kikkoman Corporation)
Green Tea with 6% Caffeine (Bio-Botanica)
Herbasec Arnica (Cosmetochem) (Cosmetochem International Ltd.)
Herbasec Balm Mint (Cosmetochem) (Cosmetochem International Ltd.)
Herbasec Barberry Fruit (Cosmetochem International Ltd.)
Herbasec Bearberry (Cosmetochem) (Cosmetochem International Ltd.)
Herbasec Borage (Cosmetochem) (Cosmetochem International Ltd.)
Herbasec Burdock Root (Cosmetochem) (Cosmetochem International Ltd.)

Herbasec Centella Asiatica (Cosmetochem) (Cosmetochem International Ltd.)
Herbasec Chamomile (Cosmetochem) (Cosmetochem International Ltd.)
Herbasec Comfrey (Cosmetochem) (Cosmetochem International Ltd.)
Herbasec Cornflower (Cosmetochem) (Cosmetochem International Ltd.)
Herbasec Echinacea Purpurea (Cosmetochem) (Cosmetochem International Ltd.)
Herbasec Edelweiss (Cosmetochem) (Cosmetochem International Ltd.)
Herbasec Eucalyptus (Cosmetochem) (Cosmetochem International Ltd.)
Herbasec Eyebright (Cosmetochem) (Cosmetochem International Ltd.)
Herbasec Ginseng (Cosmetochem) (Cosmetochem International Ltd.)
Herbasec Green Tea (Cosmetochem) (Cosmetochem International Ltd.)
Herbasec Hawthorn (Cosmetochem International Ltd.)
Herbasec Hawthorn (Cosmetochem) (Cosmetochem International Ltd.)
Herbasec Henna (Cosmetochem) (Cosmetochem International Ltd.)
Herbasec Jasmine Flowers (Cosmetochem) (Cosmetochem International Ltd.)
Herbasec Lavender (Cosmetochem) (Cosmetochem International Ltd.)
Herbasec Mallow (Cosmetochem) (Cosmetochem International Ltd.)
Herbasec Mallow Leaves (Cosmetochem) (Cosmetochem International Ltd.)
Herbasec Marigold/Calendula (Cosmetochem) (Cosmetochem International Ltd.)
Herbasec Marsh Mallow (Cosmetochem) (Cosmetochem International Ltd.)
Herbasec Milk Thistle (St Mary's Thistle) (Cosmetochem) (Cosmetochem International Ltd.)
Herbasec Mimosa Tenuiflora (Cosmetochem) (Cosmetochem International Ltd.)
Herbasec MPE Deo (Cosmetochem) (Cosmetochem International Ltd.)
Herbasec MPE Rooibos (Cosmetochem) (Cosmetochem International Ltd.)
Herbasec MPE Sebostat (Cosmetochem) (Cosmetochem International Ltd.)
Herbasec Olive (Leaf) (Cosmetochem) (Cosmetochem International Ltd.)
Herbasec Orange Flowers (Cosmetochem) (Cosmetochem International Ltd.)
Herbasec Orris (Iris) (Cosmetochem) (Cosmetochem International Ltd.)
Herbasec Peppermint (Cosmetochem) (Cosmetochem International Ltd.)

Herbasec Pineapple (Fruit) (Cosmetochem) (Cosmetochem International Ltd.)
Herbasec Pond Lily (Cosmetochem) (Cosmetochem International Ltd.)
Herbasec Rooibos (Cosmetochem) (Cosmetochem International Ltd.)
Herbasec Rose Bay Willow-Herb (Cosmetochem) (Cosmetochem International Ltd.)
Herbasec Rose Flower (Cosmetochem) (Cosmetochem International Ltd.)
Herbasec Sage (Cosmetochem) (Cosmetochem International Ltd.)
Herbasec Sea Weed (Cosmetochem) (Cosmetochem International Ltd.)
Herbasec Stinging Nettle (Cosmetochem) (Cosmetochem International Ltd.)
Herbasec St John's Wort (Cosmetochem) (Cosmetochem International Ltd.)
Herbasec Sudanese Tea/Hibiscus (Cosmetochem) (Cosmetochem International Ltd.)
Herbasec Tomato KBA (Cosmetochem) (Cosmetochem International Ltd.)
Herbasec Violet (Cosmetochem) (Cosmetochem International Ltd.)
Herbasec White Tea (Cosmetochem) (Cosmetochem International Ltd.)
Herbasec Witch Hazel (Cosmetochem) (Cosmetochem International Ltd.)
Herbasol Complex Stinging Nettle/Horsetail (Cosmetochem) (Cosmetochem International Ltd.)
Holunder P-AC-5 (Heidelberger)
Indische Dattel P-AC-11 (Heidelberger)
Japanische Blaualge P-AC-7 (Heidelberger)
Kikkoman Cranberry Powder (Kikkoman Corporation)
Menthol Starchosomes 2/012690 (Symrise)
Molecularsource Arbutin (C.I.T.)
Molecularsource LA (C.I.T.)
Molecularsource LPC (C.I.T.)
Molecularsource SA (C.I.T.)
Molecularsource UQ (C.I.T.)
MT-250 (MAFCO)
Natuchrom Beetrage (Quest International)
Natuchrom Mulberry Vine (Quest International)
Neo Extrapone Chamomille 2/060350 (Symrise)
Oleavine LS (Laboratoires Serobiologiques)
Papaya Secrets (Gattefosse s.a.)
Paprika P-AC-3 (Heidelberger)
Pelabath (C.I.T.)
Poudre de Citron Ananas (Greentech)
Pronalen Ruscus SPE (Provital/Centerchem)
Raspa de Jua Soluble Dry Extract (Centroflora)
Red Tea (Greentech)
REGU-SLIM (Pentapharm/Centerchem)

Rote Beete P-AC-4 (Heidelberger)
Sophorine (Solabia)
SoyAct-w (Kikkoman Corporation)
Terra-Dry - Freeze Dried Aloe Vera Powder (Terry)
Terra-Spray - Spray Dried Aloe Vera Powder (Terry)
Vegepone Asparagus 2/031308 (Symrise)
White Tea Extract (Greentech)
Wilder Safran P-AC-1 (Heidelberger)

MALTOSE

CAS Nos.
69-79-4
16984-36-4

EINECS No.
200-716-5

JPN Translation:
マルトース

Empirical Formula:
$C_{12}H_{22}O_{11}$

Definition: Maltose is the sugar that conforms to the formula:

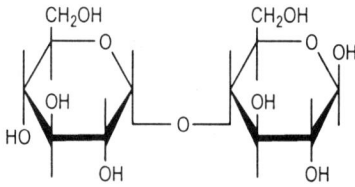

Information Sources: JAN, JCIC, JCLS, JSQI, MI-13(5736)

Chemical Classes: Carbohydrates; Polyols

Functions: Flavoring Agent; Humectant; Skin-Conditioning Agent - Humectant

Technical/Other Names:
Anhydrous Maltose
4-O-α-D-Glucopyranosyl-D-Glucose
D-Glucose, 4-O-α-D-Glucopyranosyl-
Maltobiose
Maltodiose
Malt Sugar

Trade Name Mixture:
Spirulina Extract Powder (MicroAlgae)

MALTOSYL CYCLODEXTRIN

CAS No.: 104723-60-6

Definition: Maltosyl Cyclodextrin is the product obtained by the reaction of Cyclodextrin (q.v.) with Maltose (q.v.).

Chemical Class: Carbohydrates

Function: Skin-Conditioning Agent - Miscellaneous

Trade Name Mixtures:
Hinokitiol AQ25 (Nikko)
Spirulina Extract Powder (MicroAlgae)

MALVA MOSCHATA LEAF EXTRACT

Definition: Malva Moschata Leaf Extract is an extract of the leaf of *Malva moschata*. *See "Regulatory and Ingredient Use Information," regarding the labeling names for botanical ingredients in Volume 1, Introduction, Part A.*

Chemical Class: Biological Products

Function: Not Reported

Technical/Other Name:
Extract of Malva Moschata Leaf

Trade Name Mixture:
Cosflor Musk Mallow HGS (A & E Connock)

MALVA MOSCHATA ROOT EXTRACT

Definition: Malva Moschata Root Extract is an extract of the roots of *Malva moschata. See "Regulatory and Ingredient Use Information," regarding the labeling names for botanical ingredients in Volume 1, Introduction, Part A.*

Chemical Class: Biological Products

Function: Not Reported

Technical/Other Names:
Extract of Malva Moschata
Extract of Musk Mallow

Trade Name Mixtures:
Cosflor Musk Mallow HGS (A & E Connock)
Musk Mallow Extract (Cosmetic Developments)

MALVA ROTUNDAFOLIA FLOWER

Definition: Malva Rotundafolia Flower is the whole dried flower of *Malva rotundafolia. See "Regulatory and Ingredient Use Information," regarding the labeling names for botanical ingredients in Volume 1, Introduction, Part A.*

Chemical Class: Biological Products

Function: Not Reported

Technical/Other Names:
Black Malva
Black Malva (Malva Rotundafolia)

MALVA SYLVESTRIS (MALLOW) EXTRACT

CAS No.
84082-57-5

EINECS No.
282-003-9

JPN Translation:
ゼニアオイエキス

Definition: Malva Sylvestris (Mallow) Extract is an extract of the flowers, leaves or stems of *Malva sylvestris. See "Regulatory and Ingredient Use Information," regarding the labeling names for botanical ingredients in Volume 1, Introduction, Part A.*

Information Sources: ARG, JCIC, JCLS, JSQI

Chemical Class: Biological Products

Function: Skin-Conditioning Agent - Miscellaneous

Reported Product Categories: Bath Oils, Tablets, and Salts; Moisturizing Preparations; Cleansing Products (Cold Creams, Cleansing Lotions, Liquids and Pads); Bath Preparations, Misc.; Body and Hand Preparations (Excluding Shaving Preparations); Eye Makeup Preparations, Misc.; Skin Care Preparations, Misc.; Skin Fresheners; Paste Masks (Mud Packs); Baby Lotions, Oils, Powders and Creams; Eyeliners; Indoor Tanning Preparations; Night Skin Care Preparations; Shampoos (Non-coloring); Suntan Preparations, Misc.

Technical/Other Names:
Extract of Malva Silvestris
Mallow Blossom Extract
Mallow Extract
Malva Silvestris Extract

Trade Name Mixtures:
ABS Mallow Extract (Active Concepts)
Actiphyte of Blue Malva BG50 (Active Organics)
Actiphyte of Blue Malva GL50 (Active Organics)
Actiphyte of Blue Malva Lipo S (Active Organics)
Actiphyte of Blue Malva PG50 (Active Organics)
342 Babyderme HS (Alban Muller)
642 Babyderme LS (Alban Muller)
250 Blend for Greasy Skin HS (Alban Muller)
650 Blend for Greasy Skin LS (Alban Muller)
658 Demaquillant LS (Alban Muller)
235 Emollient HS (Alban Muller)
635 Emollient LS (Alban Muller)
Extrait De Mauve PP PG40 (Yves Rocher)
Extrapone Mallow 2/030112 (Symrise)
Extrapone Mallow-Special (2/033111) (Symrise)
Gigawhite (Cosmetic Ingredient Resources/Centerchem)
Glycolysat of Mallow (CEP (Solabia))
Glycolysat of Soothing Plants (CEP (Solabia))
Herbasec Mallow (Cosmetochem) (Cosmetochem International Ltd.)

Herbasec Mallow Leaves (Cosmetochem)
(Cosmetochem International Ltd.)
Herbasol Distillate Mallow (Cosmetochem)
(Cosmetochem International Ltd.)
Herbasol-Extract Mallow Flowers
(Cosmetochem)
Mallow Extract (Maruzen Pharmaceuticals
Co., Ltd.)
Mallow Extract BG (Maruzen
Pharmaceuticals Co., Ltd.)
Mallow Extract HG (Provital/Centerchem)
Mallow Extract HS 2387G (Grau)
Mallow Extract LA (Maruzen
Pharmaceuticals Co., Ltd.)
Mallow Extract PG (Rahn)
Mallow Extract SQ (Maruzen
Pharmaceuticals Co., Ltd.)
Mallow HS (Alban Muller)
Mallow LS (Alban Muller)
Mallow Phytexcell (Crodarom)
Mallow Tincture (Rahn)
Moisturizing Complex 266 (Ennagram)
Moisturizing Phytogreen Complex GXH
266 (Phytochim)
Novaplant Mallow Extract (Crodarom)
Oral Mucous Protection Complex MU 3776
(Greentech S.A)
Phystrogene (Coletica SA)
Phytelene Complex EGX 251 (Indena SA)
Phytelene of Mauve EG 216 liquid (Indena
SA)
Phytelene of Wood Mallow EG 216 liquid
(Indena SA)
Phytogreen 55 of Mallow EXH 640 Liquid
(Phytochim)
Phytogreen 55 of Wood Mallow EXH 640
Liquid (Phytochim)
Polyplant Anti-Inflammation (Provital/
Centerchem)
Polyplant Descongestant (Provital/
Centerchem)
Polyplant Emollient (Provital/Centerchem)
Polyplant Moisturizing (Provital/
Centerchem)
Prodhy Extract Mauve (Prod'Hyg)
PROTECTIVE MOISTURIZER (Greentech
S.A)
Sederma Mallow (Sederma)
SUN PROTECTION (Greentech S.A)
Toothpaste Complex MU 3319 (Greentech
S.A)
Unioil Herbs (Chemyunion)
Vegebios of Mallow (CEP (Solabia))
2560 Vege-Plex Skin Complex (Vege-
Tech)
Vegetol Flowers MCF 1161 Hydro
(Gattefosse s.a.)
Vegetol Mallow MCF 784 Hydro
(Gattefosse s.a.)
Vegetol Mallow 4142 Oily (Gattefosse s.a.)
Vitactyl (Silab)

MALVA SYLVESTRIS (MALLOW) FLOWER EXTRACT

CAS No.: 84082-57-5

Definition: Malva Sylvestris (Mallow) Flower Extract is an extract of the flowers of *Malva sylvestris*. See *"Regulatory and Ingredient Use Information,"* regarding the labeling names for botanical ingredients in Volume 1, Introduction, Part A.

Chemical Class: Biological Products

Function: Skin-Conditioning Agent - Miscellaneous

Reported Product Categories: Bath Oils, Tablets, and Salts; Bath Preparations, Misc.; Eye Makeup Preparations, Misc.; Baby Lotions, Oils, Powders and Creams; Eyeliners

Technical/Other Names:
Extract of Mallow Flower
Extract of Malva Sylvestris (Mallow) Flower
Mallow Flower Extract

Trade Name Mixtures:
Mallow Liquid B (Ichimaru Pharcos)
Mallow Liquid E (Ichimaru Pharcos)
Pharcolex BX51 (Ichimaru Pharcos)

MALVA SYLVESTRIS (MALLOW) FLOWER POWDER

Definition: Malva Sylvestris (Mallow) Flower Powder is the crushed, dried flowers of *Malva sylvestris*. See *"Regulatory and Ingredient Use Information,"* regarding the labeling names for botanical ingredients in Volume 1, Introduction, Part A.

Chemical Class: Biological Products

Function: Not Reported

Technical/Other Names:
Mallow Flower Powder
Mallow Powder
Malva Sylvestris Powder
Powdered Mallow (Malva Sylvestris)

MALVA SYLVESTRIS (MALLOW) LEAF POWDER

JPN Translation:
ゼニアオイ

Definition: Malva Sylvestris (Mallow) Leaf Powder is the powder obtained from the leaf of *Malva sylvestris*. See *"Regulatory and Ingredient Use Information,"* regarding the labeling names for botanical ingredients in Volume 1, Introduction, Part A.

Information Sources: JCIC, JCLS

Chemical Class: Biological Products

Function: Not Reported

Technical/Other Names:
Mallow Leaf Powder
Mallow Powder

MAMMARIAN HYDROLYSATE

Definition: Mammarian Hydrolysate is the hydrolysate of animal mammarian tissue derived by acid, enzyme or other method of hydrolysis.

Chemical Class: Biological Products

Function: Not Reported

MAMMARY EXTRACT

CAS No.	EINECS No.
85883-82-5	288-732-9

Definition: Mammary Extract is an extract of bovine mammary tissue.

Chemical Class: Biological Products

Function: Not Reported

Technical/Other Name:
Mammarian Extract

MAMUSHI OIL

JPN Translation:
マムシ油

Definition: Mamushi Oil is the oil obtained from the Japanese Mamushi snake, *Agkistrodon halys blomhoffi*.

Chemical Class: Fats and Oils

Function: Skin-Conditioning Agent - Emollient

Trade Name:
Mamushi Oil (Isehan Co., Ltd)

MANDELIC ACID

CAS No.	EINECS No.
90-64-2	202-007-6

Empirical Formula:
$C_8H_8O_3$

Definition: Mandelic Acid is the organic acid that conforms to the formula:

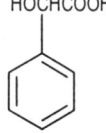

HOCHCOOH

Information Source: MI-13(5739)

Chemical Class: Carboxylic Acids

Function: Not Reported

Technical/Other Names:
Almond Acid
Amygdalic Acid
Benzeneacetic Acid, α-Hydroxy-
α-Hydroxybenzeneacetic Acid
α-Hydroxyphenyl Acetic Acid
2-Phenylglycolic Acid

Trade Name Mixtures:
A.H.A. Extracts (Ennagram)
A.H.A. Extracts (Phytochim)
Extrapone Peach 2/033319 (Symrise)

MANDRAGORA OFFICINARUM

Definition: *See "Regulatory and Ingredient Use Information," regarding EU labeling names for botanical ingredients in Volume 1, Introduction, Part A.*

Chemical Class: Biological Products

Technical/Other Name:
Mandragora Officinarum Root Extract (U.S.)

MANDRAGORA OFFICINARUM ROOT EXTRACT

Definition: Mandragora Officinarum Root Extract is an extract of the roots of *Mandragora officinarum*. See "Regulatory and Ingredient Use Information," regarding the labeling names for botanical ingredients in Volume 1, Introduction, Part A.

Chemical Class: Biological Products

Function: Not Reported

Technical/Other Names:
Extract of Mandragora Officinarum Root
Mandragora Officinarum (EU)

Trade Name Mixture:
Mandragora Extract HS 3131 G (Grau)

MANGANESE ACETYLMETHIONATE

CAS No.: 105883-50-9

Definition: Manganese Acetylmethionate is the manganese salt of N-acetylmethionine.

Chemical Class: Thio Compounds

Function: Not Reported

Trade Name:
Exsymanganese (Exsymol)

MANGANESE ASPARTATE

Empirical Formula:
$C_4H_7NO_4 \cdot \frac{1}{2}Mn$

Definition: Manganese Aspartate is the manganese salt of Aspartic Acid (q.v.).

Chemical Class: Amino Acids

Function: Skin-Conditioning Agent - Occlusive

Technical/Other Name:
Aspartic Acid, Manganese Salt

Trade Name:
Oligoidyne Manganese (Vevy)

Trade Name Mixtures:
Oligoidyne-1-Complex (Vevy)
Oligoidyne-2-Complex (Vevy)

MANGANESE CHLORIDE

CAS Nos.   **EINECS No.**
7773-01-5   231-869-6
11132-78-8

Empirical Formula:
Cl_2Mn

Definition: Manganese Chloride is the inorganic salt that conforms to the formula:

$$MnCl_2$$

Information Sources: 21CFR184.1446, MI-13(5751), USAN, USP XXIV

Chemical Class: Inorganic Salts

Function: Proprietary

Technical/Other Names:
Manganese Dichloride
Manganous Chloride

Trade Name:
Manganese (II) Chloride Tetrahydrate (Merck KGaA)

MANGANESE DIOXIDE

CAS No.   **EINECS No.**
1313-13-9   215-202-6

Definition: Manganese Dioxide is the inorganic oxide that conforms to the formula:

$$MnO_2$$

Information Source: MI-13(5753)

Chemical Class: Inorganics

Function: Antioxidant

Trade Name:
Manganese Oxide MBK (MBK)

MANGANESE GLUCONATE

CAS No.   **EINECS No.**
6485-39-8   229-350-4

Empirical Formula:
$C_{12}H_{24}O_{14} \cdot Mn$

Definition: Manganese Gluconate is the manganese salt of gluconic acid that conforms to the formula:

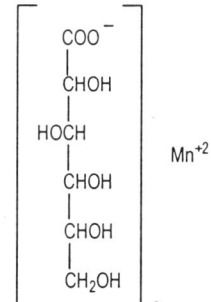

Information Sources: USAN, USP XXIV

Chemical Class: Organic Salts

Function: Not Reported

Reported Product Categories: Bath Oils, Tablets, and Salts; Cleansing Products (Cold Creams, Cleansing Lotions, Liquids and Pads); Bath Capsules; Face and Neck Preparations (Excluding Shaving Preparations); Skin Care Preparations, Misc.; Skin Fresheners

Technical/Other Name:
Gluconic Acid, Manganese Salt (2:1)

Trade Names:
Givobio GMn (SEPPIC)
Gluconal MN (Glucona America)

MANGANESE GLYCEROPHOSPHATE

CAS Nos.   **EINECS No.**
1320-46-3   215-301-4
1335-36-0

Empirical Formula:
$C_3H_7O_6P \cdot Mn$

Definition: Manganese Glycerophosphate is the organic compound that conforms to the formula:

$$\left[\begin{array}{c} CH_2OPO_3 \\ | \\ CHOH \\ | \\ CH_2OH \end{array} \right]^{-2}_{} Mn^{+2}$$

Chemical Classes: Glyceryl Esters and Derivatives; Organic Salts; Phosphorus Compounds

Function: Oral Care Agent

Technical/Other Names:
Glycerol, Phosphate, Maganese Salt

1,2,3-Propanetriol, Dihydrogen Phosphate, Maganese Salt

Trade Name:
Givobio GPMn (SEPPIC)

MANGANESE PCA

Empirical Formula:
$C_5H_7NO_3 \cdot \frac{1}{2}Mn$

Definition: Manganese PCA is the manganese salt of PCA (q.v.). It conforms to the formula:

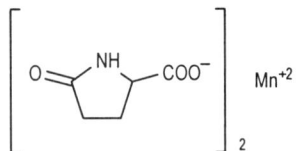

Chemical Classes: Heterocyclic Compounds; Organic Salts

Function: Skin-Conditioning Agent - Humectant

Trade Names:
Mangalidone (UCIB (Solabia))
Manganese Oligoelement Mn Complex (L.C.S.)
Manganese Pidolate (UCIB (Solabia))

Trade Name Mixture:
Physiogenyl (UCIB (Solabia))

MANGANESE SULFATE

CAS Nos.	**EINECS Nos.**
7785-87-7	232-089-9
10124-55-7	233-342-6

Definition: Manganese Sulfate is the inorganic salt that conforms to the formula:

$$MnSO_4$$

Information Sources: MI-13(5763), USP XXIV

Chemical Class: Inorganic Salts

Function: Skin-Conditioning Agent - Miscellaneous

Technical/Other Names:
Manganese Monosulfate
Sulfuric Acid, Manganese Salt (1:1)

Trade Name Mixture:
Phycosaccharide AG (Codif)

MANGANESE VIOLET

CAS No.	**EINECS No.**
10101-66-3	233-257-4

JPN Translation:
マンガンバイオレット

Empirical Formula:
$H_4O_7P_2 \cdot H_3N \cdot Mn$

Definition: Manganese Violet is the inorganic salt that conforms generally to the formula:

$$MnNH_4P_2O_7$$

See "Regulatory and Ingredient Use Information," for Colorants in Volume 1, Introduction, Part A. To identify the colorant meeting the requirements for labeling purposes in the US, the INCI Name Manganese Violet must be used. To identify the colorant allowed for use in the European Union (EU), the INCI Name CI 77742 must be used, except for hair dye products.

Information Sources: 21CFR73.2775, CI 77742, JCIC, JCLS, JSQI, M3, TSCA

Chemical Classes: Color Additives - Exempt from Batch Certification by the U.S. Food and Drug Administration; Inorganic Salts

Function: Colorant

Reported Product Categories: Eye Shadows; Lipsticks; Eyeliners; Blushers (All types); Face Powders; Eyebrow Pencils; Makeup Preparations (Not eye), Misc.; Mascara; Eye Makeup Preparations, Misc.; Bath Oils, Tablets, and Salts; Makeup Bases; Rouges

Technical/Other Names:
Ammonium Manganese Pyrophosphate CI 77742
Diphosphoric Acid, Ammonium Manganese (3+) Salt (1:1:1)
Manganese Ammonium Pyrophosphate
Mango Violet
Mineral Violet
Pigment Violet 16

Trade Names:
A303 Tudor Violet (Kingfisher Colours)
C43-001 Mango Violet (Sun Pigments)
MANGANESE VIOLET 10-34-PC-3512 (Noveon Hilton Davis)
Rose De Manganese W 4801 (LCW)
Unipure Pink LC 583 (LCW)
Unipure Violet LC 581 (LCW)
Violet De Manganese W 632 (LCW)
Violet De Manganese W 5802 (LCW)
Violet De Manganese W 5803 (LCW)
Violet de Manganese W 5807 (LCW)

Trade Name Mixtures:
A303.30 Tudor Violet Hydrophobic (Kingfisher Colours)
Cardre Manganese Violet RFHC (Cardre)
Chroma-Lite Violet 4507 (Engelhard Corp.)

Hydrophobic Manganese Violet C9401 (LCW)
LipoCrystals LIP Blue (Lipo)
LipoCrystals LIP Pearl (Lipo)
LT-MV-43001 (US Cosmetics)
Manganese Violet SI 2 (Cardre)
Micapoly WOE Manganese Violet (Creations Couleurs)
Micapoly WOE Violet (Creations Couleurs)
Tefpoly Violet (C.I.T.)

MANGIFERA INDICA LEAF EXTRACT

Definition: Mangifera Indica Leaf Extract is an extract of the leaves of *Mangifera indica*. See "Regulatory and Ingredient Use Information," regarding the labeling names for botanical ingredients in Volume 1, Introduction, Part A.

Chemical Class: Biological Products

Function: Skin-Conditioning Agent - Miscellaneous

Technical/Other Name:
Extract of Mangifera Indica Leaf

Trade Name:
ViaPure Mango (Actives International)

MANGIFERA INDICA (MANGO) FRUIT

Definition: Mangifera Indica (Mango) Fruit is the fruit of the mango, *Mangifera indica*. See "Regulatory and Ingredient Use Information," regarding the labeling names for botanical ingredients in Volume 1, Introduction, Part A.

Chemical Class: Biological Products

Function: Cosmetic Astringent

Technical/Other Names:
Mango
Mango Fruit

Trade Name:
AEC Mango Puree (A & E Connock)

MANGIFERA INDICA (MANGO) FRUIT EXTRACT

CAS No.	**EINECS No.**
90063-86-8	290-045-4

Definition: Mangifera Indica (Mango) Fruit Extract is an extract of the fruit of the mango, *Mangifera indica*. See "Regulatory and Ingredient Use Information," regarding the labeling names for botanical ingredients in Volume 1, Introduction, Part A.

Chemical Class: Biological Products

Function: Skin-Conditioning Agent - Miscellaneous

Reported Product Categories: Hair Conditioners; Shampoos (Non-coloring); Tonics, Dressings, and Other Hair Grooming Aids; Bath Soaps and Detergents; Paste Masks (Mud Packs)

Technical/Other Names:
Extract of Mangifera Indica
Extract of Mango
Mangifera Indica Extract
Mango Extract
Mango Fruit Extract

Trade Name:
Frulix TF Manga (Assessa-Industria)

Trade Name Mixtures:
ABS Pap-Ango Enzyme (Active Concepts)
Actiphyte of Mango BG50 (Active Organics)
Actiphyte of Mango GL50 (Active Organics)
Actiphyte of Mango Lipo S (Active Organics)
Actiphyte of Mango PG50 (Active Organics)
A.H.A. Extracts (Ennagram)
A.H.A. Extracts (Phytochim)
Biogreen Mangue (Greentech)
Extrapone Mango 2/033123 (Symrise)
FRESH CELLS MANGO (Libiol)
Glycolysat of Mango (CEP (Solabia))
Herbasol Extract Mango (Cosmetochem) (Cosmetochem International Ltd.)
Mangifera Indica (Mango) Fruit Extract ies (IES LABO)
Mango Extract (Bio-Botanica)
Mango Extract (Jeen)
Mango Extract HG (Provital/Centerchem)
Mango Extract HG 597 PB (Ennagram)
Mango Extract HS 2756 G (Grau)
Mango HPG Titrated (Alban Muller)
Mango HS (Alban Muller)
Mango Milk (Cosmetochem) (Cosmetochem International Ltd.)
Phytelene of Mango BG 740 (Indena SA)
Phytelene of Mango EG 740 (Indena SA)
Six Fruit Concentrate 3519 G (Grau)
VT-137 Extract of Mango (Vege-Tech)

MANGIFERA INDICA (MANGO) JUICE

Definition: Mangifera Indica (Mango) Juice is the juice of the mango, *Mangifera indica. See "Regulatory and Ingredient Use Information," regarding the labeling names for botanical ingredients in Volume 1, Introduction, Part A.*

Chemical Class: Biological Products

Function: Skin-Conditioning Agent - Miscellaneous

Technical/Other Names:
Juice, Mangifera Inidica
Juice, Mango
Mangifera Indica Juice
Mango Juice

Trade Name:
AEC Mango Conc (A & E Connock)

Trade Name Mixtures:
Fruitapone Mango B 2/036350 (Symrise)
Fruitapone Mango GT 2/037350 (Symrise)

MANGIFERA INDICA (MANGO) SEED

Definition: Mangifera Indica (Mango) Seed is the seed derived from *Mangifera indica. See "Regulatory and Ingredient Use Information," regarding the labeling names for botanical ingredients in Volume 1, Introduction, Part A.*

Chemical Class: Biological Products

Function: Abrasive

Technical/Other Name:
Mango (Mangifera Indica) Seed

Trade Name:
Actiscrub M (Active Organics)

MANGIFERA INDICA (MANGO) SEED BUTTER

Definition: Mangifera Indica (Mango) Seed Butter is the fat obtained from the seeds of *Mangifera indica. See "Regulatory and Ingredient Use Information," regarding the labeling names for botanical ingredients in Volume 1, Introduction, Part A.*

Chemical Class: Fats and Oils

Function: Skin-Conditioning Agent - Occlusive

Technical/Other Names:
Mango Butter
Mango Seed Butter

Trade Names:
ESP Mango Butter (Earth Supplied Products)
Mango Butter - Refined ASMB001 (Aloestar)
Mango Butter - Ultra Refined (Biochemicals Int'l)
Trivent Mango Butter (Trivent)

MANGIFERA INDICA (MANGO) SEED OIL

Definition: Mangifera Indica (Mango) Seed Oil is the fixed oil expressed from the kernels of *Mangifera indica. See "Regulatory and Ingredient Use*

Information," regarding the labeling names for botanical ingredients in Volume 1, Introduction, Part A.

Chemical Class: Fats and Oils

Function: Skin-Conditioning Agent - Occlusive

Technical/Other Names:
Mango Seed Oil
Oil of Mango Seed
Oils, Mango Seed

Trade Names:
AEC Mango Seed Oil (A & E Connock)
Lipex 203 (Karlshamns AB)

Trade Name Mixture:
Aloe Butter - Mango Butter ASMBOO2 (Aloestar)

MANGO SEED OIL PEG-70 ESTERS

Definition: Mango Seed Oil PEG-70 Esters is a complex mixture formed from the transesterification of Mangifera Indica (Mango) Seed Oil (q.v.) and PEG-70.

Chemical Class: Glyceryl Esters and Derivatives

Functions: Skin-Conditioning Agent - Emollient; Surfactant - Emulsifying Agent

MANIHOT UTILISSIMA LEAF EXTRACT

CAS No.	EINECS No.
92456-72-9	296-251-0

Definition: Manihot Utilissima Leaf Extract is an extract of the leaves of *Manihot utilissima. See "Regulatory and Ingredient Use Information," regarding the labeling names for botanical ingredients in Volume 1, Introduction, Part A.*

Chemical Class: Biological Products

Function: Not Reported

Technical/Other Names:
Cassava Extract
Extract of Cassava
Extract of Manihot Utilissima

Trade Name Mixture:
Cassava II Extract (Sederma)

MANNAN

CAS Nos.: 9036-88-8; 51395-96-1

Definition: Mannan is a natural polysaccharide consisting of glucose, mannose, and acetyl mannose monomers. It conforms generally to the formula:

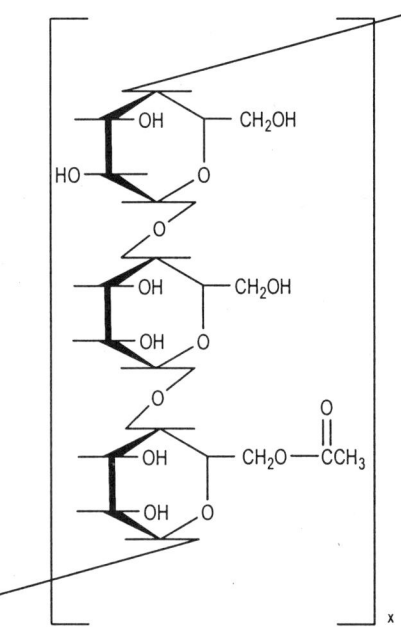

Chemical Classes: Carbohydrates; Polyols

Function: Film Former

Technical/Other Name:
D-Mannosan

Trade Names:
Konjac-Mannan 1.0 (GfN)
Propol ISLB (Shimizu)
Propol Mannan (Shimizu)

MANNITAN LAURATE

CAS No.: 120200-78-4

Empirical Formula:
$C_{18}H_{34}O_6$

Definition: Mannitan Laurate is the monoester of lauric acid and anhydrides derived from Mannitol (q.v.). It conforms generally to the formula:

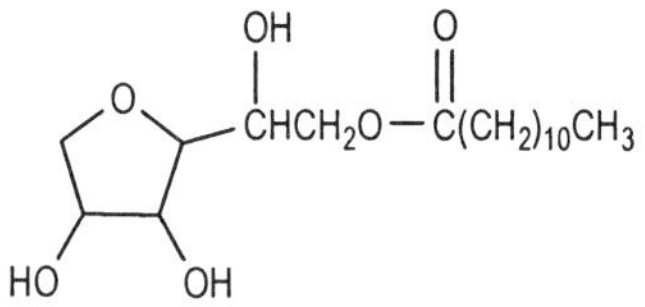

OH   O
| ||
CHCH₂O — C(CH₂)₁₀CH₃

Chemical Classes: Esters; Polyols

Function: Surfactant - Emulsifying Agent

Technical/Other Name:
D-Mannitol, 1,4-Anhydro-, 6-Dodecanoate

MANNITAN OLEATE

Empirical Formula:
$C_{24}H_{44}O_6$

Definition: Mannitan Oleate is the monoester of oleic acid and anhydrides derived from Mannitol (q.v.). It conforms to the formula:

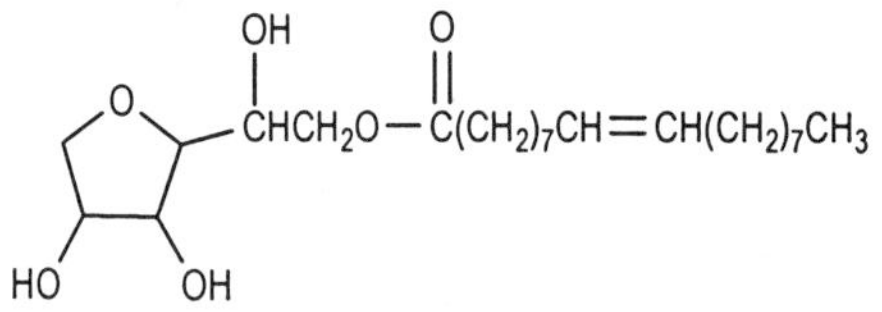

OH   O
| ||
CHCH₂O — C(CH₂)₇CH=CH(CH₂)₇CH₃

Chemical Classes: Esters; Heterocyclic Compounds; Polyols

Function: Surfactant - Emulsifying Agent

Trade Name:
Arlacel A (Uniqema Americas)

MANNITOL

CAS Nos.
69-65-8
87-78-5

EINECS Nos.
200-711-8
201-770-2

JPN Translation:
マンニトール

Empirical Formula:
$C_6H_{14}O_6$

Definition: Mannitol is the hexahydric alcohol that conforms to the formula:

CH₂OH
|
HOCH
|
HOCH
|
HCOH
|
HCOH
|
CH₂OH

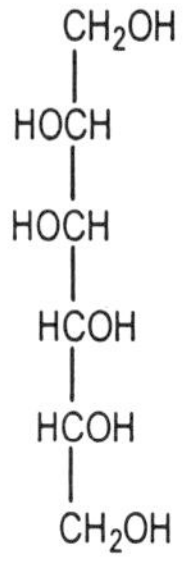

Information Sources: BP, BPC, BRA, 21CFR175.300, 21CFR175.320, 21CFR177.2420, 21CFR180.25, 21CFR250.102, 21CFR310.545, 21CFR450.240, 21CFR582.5470, FCC, ITA, JAN, JCLS, JSCI, KOR, MAR, MI-13 (5769), POR, TSCA, USAN, USD, USP XXIV

Chemical Class: Polyols

Functions: Binder; Flavoring Agent; Humectant; Skin-Conditioning Agent - Humectant

Reported Product Categories: Bath Capsules; Moisturizing Preparations; Face and Neck Preparations (Excluding Shaving Preparations); Rouges; Bath Preparations, Misc.; Body and Hand Preparations (Excluding Shaving Preparations); Eye Makeup Preparations, Misc.; Foundations; Skin Care Preparations, Misc.

Technical/Other Names:
Manna Sugar
D-Mannitol

Trade Name Mixtures:
A.F.R./Veg. (Laboratoires Serobiologiques)
Aglycal (Laboratoires Serobiologiques)
Antistressine (Coletica SA)
A.P.S. LS 8425 (Laboratoires Sero-biologiques)
Biacinth (Sederma)
Omega-CHS-Activator (GfN)
Connexan LS (Laboratoires Sero-biologiques)
Connexan LS (Laboratoires Sero-biologiques)
Dermawhite HS (Laboratoires Sero-biologiques)
Dermawhite NF (Laboratoires Sero-biologiques)
Dermosaccharides SEA (Laboratoires Serobiologiques)
Induxin (Laboratoires Serobiologiques)
Osmhydran (Laboratoires Serobiologiques)
Osmhydran LS 8453 (Laboratoires Sero-biologiques)
Oxygen Complex LS (Laboratoires Sero-biologiques)
P.A. Reviviscence PW (Laboratoires Sero-biologiques)
Photonyl (Laboratoires Serobiologiques)
Polyglucomannan KM (Polygon)
Polysol FS (Polygon)
Pronalen Ginseng SPE (Provital/Centerchem)
RonaCare ASC III (Merck KGaA)
RonaCare ASC III (Merck KGaA/EMD Chemicals Inc.)
RonaCare VTA (Merck KGaA)
Seanamine BD (Laboratoires Sero-biologiques)
Sunactyl LS 9610 (Laboratoires Sero-biologiques)
Trichodyn LS (Laboratoires Sero-biologiques)
Unispheres NT-Series (Induchem)
Vegeles Phytofiltre AD LS (Laboratoires Serobiologiques)
Vegetens OR (LCW)
Vitacell Powder (Laboratoires Sero-biologiques)

MANNOSE

CAS No.
3458-28-4

EINECS No.
222-392-4

Empirical Formula:
$C_6H_{12}O_6$

Definition: Mannose is the sugar that conforms to the formula:

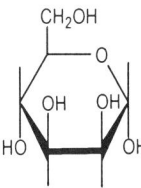

Information Source: MI-13(5772)

Chemical Classes: Carbohydrates; Polyols

Functions: Humectant; Skin-Conditioning Agent - Humectant

Technical/Other Names:
Carubinose
α-D-Mannose
Seminose

Trade Name:
D-Mannose (Senn Chemicals)

MARANTA ARUNDINACEA ROOT EXTRACT

Definition: Maranta Arundinacea Root Extract is an extract of the rhizomes of the arrowroot, *Maranta arundinacea. See "Regulatory and Ingredient Use Information," regarding the labeling names for botanical ingredients in Volume 1, Introduction, Part A.*

Chemical Class: Biological Products

Function: Not Reported

Technical/Other Names:
Arrowroot Extract
Arrowroot (Maranta Arundicacea) Extract
Extract of Arrowroot
Extract of Maranta Arundinacea

Trade Name Mixtures:
Actiphyte of Arrowroot BG50 (Active Organics)
Actiphyte of Arrowroot GL50 (Active Organics)
Actiphyte of Arrowroot Lipo S (Active Organics)
Actiphyte of Arrowroot PG50 (Active Organics)
VT-197 Extract of Arrowroot (Vege-Tech)

MARE MILK

Definition: Mare Milk is whole milk from female horses.

Chemical Class: Biological Products

Function: Skin-Conditioning Agent - Miscellaneous

Trade Name:
Mare's Milk (Gestuet Sickinger Hoehe)

MARMOT OIL

Definition: Marmot Oil is the oil obtained from the fatty tissues of the marmot.

Chemical Class: Fats and Oils

Functions: Hair Conditioning Agent; Skin-Conditioning Agent - Occlusive

Technical/Other Name:
Oils, Marmot

Trade Names:
Axungia Muris Alpinae (Josef Mack)
Marmot Oil, Murmeltierol, Murmeltierfett (B. Mueller KG)

MARROW EXTRACT

JPN Translation:
ウシ骨髄脂

Definition: Marrow Extract is the extract of bovine bone marrow.

Chemical Class: Biological Products

Function: Skin-Conditioning Agent - Humectant

Trade Names:
Marrow Fat A (Q.P.)
Snow Fat (Snowden)

Trade Name Mixture:
Herbasol-Extract Marrow (Cosmetochem)

MARROW LIPIDS

Definition: Marrow Lipids are the lipids derived from bone marrow.

Information Sources: JCIC, JCLS

Chemical Class: Fats and Oils

Function: Skin-Conditioning Agent - Emollient

Technical/Other Name:
Bovine Bone Marrow Fat

Trade Name:
Ukokei Fowl Bone Marrow Fat (Chugoku)

MARRUBIUM VULGARE EXTRACT

CAS No.
84696-20-8

EINECS No.
283-638-4

Definition: Marrubium Vulgare Extract is an extract of the herb of the horehound, *Marrubium vulgare. See "Regulatory and Ingredient Use Information," regarding the labeling names for botanical ingredients in Volume 1, Introduction, Part A.*

Information Source: RIFM

Chemical Class: Biological Products

Functions: Fragrance Ingredient; Not Reported

Technical/Other Names:
Extract of Horehound
Extract of Marrubium Vulgare
Horehound Extract
Horehound (hoarhound) extract (Marrubium vulgare L.) (RIFM)
Horehound (Marrubium Vulgare) Extract

Trade Name Mixtures:
Actiphyte of Horehound BG50 (Active Organics)
Actiphyte of Horehound GL50 (Active Organics)
Actiphyte of Horehound Lipo S (Active Organics)
Actiphyte of Horehound PG50 (Active Organics)
White Hore Hound Wort Extract HS 3350 G (Grau)

MARSDENIA CONDURANGO BARK EXTRACT

CAS No.
84787-66-6

EINECS No.
284-108-5

Definition: Marsdenia Condurango Bark Extract is an extract of the bark of the condurango, *Marsdenia condurango. See "Regulatory and Ingredient Use Information," regarding the labeling names for botanical ingredients in Volume 1, Introduction, Part A.*

Chemical Class: Biological Products

Function: Skin-Conditioning Agent - Miscellaneous

Technical/Other Names:
Condurango Extract
Condurango (Marsdenia Condurango) Extract
Extract of Condurango
Extract of Marsdenia Condurango
Gonolobus Condurango Extract

Trade Name Mixtures:
Actiphyte of Condurango BG50 (Active Organics)
Actiphyte of Condurango GL50 (Active Organics)
Actiphyte of Condurango Lipo S (Active Organics)
Actiphyte of Condurango PG50 (Active Organics)
Condurango Bark Extract HS 2663 G (Grau)
Condurango HPG Titrated (Alban Muller)
Condurango HS (Alban Muller)
Condurango Milk (CEP (Solabia))
Herbasol-Extract Condurango (Cosmetochem)

Vegebios of Condurango (CEP (Solabia))
VT-198 Extract of Condurango (Vege-Tech)

MARSDENIA CONDURANGO ROOT EXTRACT

CAS No.
84787-66-6

EINECS No.
284-108-5

Definition: Marsdenia Condurango Root Extract is an extract of the roots of *Marsdenia condurango*. See "Regulatory and Ingredient Use Information," regarding the labeling names for botanical ingredients in Volume 1, Introduction, Part A.

Chemical Class: Biological Products

Function: Skin-Conditioning Agent - Miscellaneous

Technical/Other Name:
Extract of Marsdenia Condurango Root

Trade Name Mixture:
Glycolysat of Condurango (CEP (Solabia))

MARSILEA MINUTA EXTRACT

Definition: Marsilea Minuta Extract is an extract of the whole plant of *Marsilea minuta*. See "Regulatory and Ingredient Use Information," regarding the labeling names for botanical ingredients in Volume 1, Introduction, Part A.

Chemical Class: Biological Products

Function: Not Reported

Technical/Other Name:
Extract of Marsilea Minuta

Trade Name Mixture:
Campo Sunisannaka (Campo)

MASSOY BARK OIL

Definition: Massoy Bark Oil is the volatile oil distilled from the bark of *Cryptocario massoia*. See "Regulatory and Ingredient Use Information," regarding the labeling names for botanical ingredients in Volume 1, Introduction, Part A.

Chemical Class: Essential Oils

Function: Fragrance Ingredient

Trade Name:
Essential Oil of Massoia Bark (Argeville)

MAURITIA FLEXUOSA EXTRACT

Definition: Mauritia Flexuosa Extract is an extract of the flowers and nuts of *Mauritia flexuosa*. See "Regulatory and Ingredient Use Information," regarding the labeling names for botanical ingredients in Volume 1, Introduction, Part A.

Chemical Class: Biological Products

Function: Not Reported

Technical/Other Name:
Extract of Mauritia Flexuosa

Trade Name Mixtures:
Campo Muruity-Muruity (Campo)
Campo Muruity-Muruity Oil (Campo)

MAURITIA FLEXUOSA FRUIT OIL

Definition: Mauritia Flexuosa Fruit Oil is the volatile oil obtained from the fruit of *Mauritia flexuosa*. See "Regulatory and Ingredient Use Information," regarding the labeling names for botanical ingredients in Volume 1, Introduction, Part A.

Chemical Class: Essential Oils

Function: Skin-Conditioning Agent - Miscellaneous

Technical/Other Name:
Oils, Mauritia Flexuosa Fruit

Trade Names:
Buriti Oil (Beraca industria E Comercio LTDA)
Buriti Oil (Provital/Centerchem)
Buriti Oil HR (Croda Brasil)
Melscreen Buriti (Chemyunion)

Trade Name Mixtures:
NS Ingredients II (Chemyunion)
Unioil Fruits (Chemyunion)

MAYTENUS SENEGALENSIS LEAF EXTRACT

Definition: Maytenus Senegalensis Leaf Extract is an extract of the leaves of *Maytenus senegalensis*. See "Regulatory and Ingredient Use Information," regarding the labeling names for botanical ingredients in Volume 1, Introduction, Part A.

Chemical Class: Biological Products

Function: Skin-Conditioning Agent - Miscellaneous

MDM HYDANTOIN

CAS Nos.
116-25-6
16228-00-5
27636-82-4

EINECS Nos.
204-132-1
240-352-4

Empirical Formula:
$C_6H_{10}N_2O_3$

Definition: MDM Hydantoin is the organic compound that conforms to one of the following formulas:

Information Sources: CTFA D, MI-13 (4856), TSCA

Chemical Classes: Amides; Heterocyclic Compounds

Function: Preservative

Technical/Other Names:
1-(Hydroxymethyl)-5,5-Dimethyl Hydantoin
3-Hydroxymethyl-5,5-dimethylhydantoin
1-(Hydroxymethyl)-5,5-Dimethyl-2,4-Imidazolidinedione
2,4-Imidazolidinedione, 1-(Hydroxymethyl)-5,5-Dimethyl-
2,4-Imidazolidinedione, 3-(hydroxymethyl)-5,5-dimethyl
MDMH
1-Methylol 5,5-Dimethylhydantoin
3-Methylol 5,5-Dimethylhydantoin
MMDMH
Monomethylol Dimethyl Hydantoin

MEA-BENZOATE

CAS Nos.
4337-66-0
33545-23-2

EINECS No.
224-287-2

Empirical Formula:
$C_7H_6O_2 \cdot C_2H_7NO$

Definition: MEA-Benzoate is the salt of monoethanolamine and Benzoic Acid (q.v.). It conforms to the formula:

Information Source: TSCA

Chemical Classes: Amines; Organic Salts

Function: Preservative

Technical/Other Names:
Benzoic Acid, Compd. with 2-Aminoethanol
Ehtanol, 2-Amino-, Benzoate
Ethanolamine Benzoate
Monoethanolamine Benzoate

MEA-BIOTINATE

Empirical Formula:
$C_{12}H_{23}N_3O_4S$

The inclusion of any compound in the *Dictionary and Handbook* does not indicate that use of that substance as a cosmetic ingredient complies with the laws and regulations governing such use in the United States or any other country.

Definition: MEA-Biotinate is the monoethanolamine salt of Biotin (q.v.) that conforms to the formula:

Chemical Classes: Heterocyclic Compounds; Organic Salts

Function: Skin-Conditioning Agent - Miscellaneous

MEA-BORATE

CAS Nos.	EINECS No.
10377-81-8	233-829-3
68130-12-1	

Empirical Formula:
$C_2H_8BNO_3$

Definition: MEA-Borate is the ester of Ethanolamine (q.v.) and Boric Acid (q.v.).

Information Source: EEC(III/1-1)

Chemical Class: Amines

Function: Buffering Agent

Reported Product Category: Hair Sprays (Aerosol Fixatives)

Technical/Other Names:
Boric Acid, 2-Aminoethyl Ester
Ethanolamine Borate
Ethanol, 2-Amino-, Monoester with Boric Acid
Monoethanolamine Borate

Trade Name Mixture:
Monacor BE (Uniqema)

MEA-COCOATE

CAS No.	EINECS No.
66071-80-5	266-105-0

Definition: MEA-Cocoate is the monoethanolamine salt of Coconut Acid (q.v.).

Chemical Class: Organic Salts

Functions: Surfactant - Cleansing Agent; Surfactant - Foam Booster

Technical/Other Names:
Fatty Acids, Coco, Compds. with Ethanolamine
Monoethanolamine Cocoate

MEA-DICETEARYL PHOSPHATE

JPN Translation:
ジセテアリルリン酸 MEA

Definition: MEA-Dicetearyl Phosphate is the monoethanolamine salt of a complex mixture of diesters of Cetearyl Alcohol (q.v.) and phosphoric acid.

Information Sources: JCIC, JCLS, JSQI

Chemical Classes: Organic Salts; Phosphorus Compounds

Function: Surfactant - Emulsifying Agent

Technical/Other Name:
Monoethanolamine Dicetostearyl Phosphate

MEADOWFOAM ESTOLIDE

CAS No.	EINECS No.
68937-92-8	273-100-7

Definition: Meadowfoam Estolide is the oligomeric ester derived from meadowfoam fatty acids. It conforms generally to the formula:

$$CH_3(CH_2)_{13}CH = CH(CH_2)_3C - OCH(CH_2)_3COOH$$
$$(CH_2)_{14}CH_3$$

Chemical Class: Esters

Functions: Hair Conditioning Agent; Skin-Conditioning Agent - Emollient

Trade Name:
Meadowestolide (Fanning)

MEADOWFOAM DELTA-LACTONE

Definition: Meadowfoam Delta-Lactone is the lactone formed from meadowfoam seed oil fatty acids. It conforms generally to the formula:

Chemical Class: Ketones

Functions: Hair Conditioning Agent; Skin-Conditioning Agent - Emollient

Trade Name:
Meadowfoam Delta Lactone (Fanning)

MEA-HYDROLYZED COLLAGEN

Definition: MEA-Hydrolyzed Collagen is the monoethanolamine salt of Hydrolyzed Collagen (q.v.).

Chemical Class: Protein Derivatives

Functions: Hair Conditioning Agent; Skin-Conditioning Agent - Miscellaneous

Technical/Other Name:
MEA-Hydrolyzed Animal Protein

MEA-HYDROLYZED SILK

Definition: MEA-Hydrolyzed Silk is the monoethanolamine salt of Hydrolyzed Silk (q.v.).

Chemical Class: Protein Derivatives

Functions: Hair Conditioning Agent; Skin-Conditioning Agent - Miscellaneous

Trade Name Mixture:
Aqua-Tein S (Maybrook)

MEA-IODINE

Empirical Formula:
$C_2H_7NO \cdot 2I$

Definition: MEA-Iodine is the organic compound that conforms to the formula:

$$HOCH_2CH_2NH_2 \quad \cdot \quad I_2$$

Chemical Class: Amines

Function: Hair Conditioning Agent

Trade Name:
Iodamicid (Vevy)

MEA-LAURETH-6 CARBOXYLATE

Empirical Formula:
$C_{24}H_{48}O_8 \cdot C_2H_7NO$

Definition: MEA-Laureth-6 Carboxylate is the monoethanolamine salt of laureth-6 carboxylic acid. It conforms generally to the formula:

$$CH_3(CH_2)_{11}(OCH_2CH_2)_nOCH_2COOH \cdot NH_2CH_2CH_2OH$$

where n has an average value of 5.

Chemical Class: Organic Salts

Function: Surfactant - Cleansing Agent

Technical/Other Names:
PEG-6 Lauryl Ether Carboxylic Acid, Monoethanolamine Salt
Polyethylene Glycol 300 Lauryl Ether Carboxylic Acid, Monoethanolamine Salt
Polyoxyethylene (6) Lauryl Ether Carboxylic Acid, Monoethanolamine Salt

MEA-LAURETH SULFATE

CAS No.: 68184-04-3

Definition: MEA-Laureth Sulfate is the monoethanolamine salt of an ethoxylated lauryl sulfate that conforms generally to the formula:

$$CH_3(CH_2)_{11}(OCH_2CH_2)_nOSO_3H \cdot NH_2CH_2CH_2OH$$

where n has a value between 1 and 4.

Chemical Class: Alkyl Ether Sulfates

Function: Surfactant - Cleansing Agent

Reported Product Category: Hair Dyes and Colors (All Types Requiring Caution Statements and Patch Tests)

Technical/Other Name:
Monoethanolamine Lauryl Ether Sulfate

Trade Names:
Akyposal MLES 35 (Kao GmbH)
Sabosol EMM (Sabo)

Trade Name Mixtures:
Sabosol CST (Sabo)
Sabosol EM (Sabo)

MEA-LAURYL SULFATE

CAS No.	EINECS No.
4722-98-9	225-214-3

JPN Translation:
ラウリル硫酸 MEA

Empirical Formula:
$C_{12}H_{26}O_4S \cdot C_2H_7NO$

Definition: MEA-Lauryl Sulfate is the monoethanolamine salt of sulfated lauryl alcohol. It conforms generally to the formula:

$$CH_3(CH_2)_{11}OSO_3H \cdot NH_2CH_2CH_2OH$$

Information Sources: JCIC, JCLS, JSQI, SNPF, TSCA

Chemical Class: Alkyl Sulfates

Function: Surfactant - Cleansing Agent

Technical/Other Names:
Dodecyl Sulfate Ethanolamine Salt
MEA-dodecyl Sulfate
Monoethanolamine Lauryl Sulfate
Sulfuric Acid, Monododecyl Ester, Compd. with 2-Aminoethanol (1:1)

Trade Names:
Akyposal MLS 30 (Kao GmbH)
Empicol LQ33/F (Albright & Wilson UK)
Manro ML30 (Manro)
Sabosol MEA (Sabo)
Sulfetal MF (Zschimmer & Schwarz)
Texapon MLS (Cognis Deutschland)
Zoharpon LAM (Zohar)

Trade Name Mixtures:
Afron 22 (Vevy)
Afron-LS (Vevy)
Afron-N (Vevy)
Tensioplastidina Avena (Vevy)

MEA o-PHENYLPHENATE

CAS No.	EINECS No.
84145-04-0	282-227-7

Empirical Formula:
$C_{12}H_{10}O \cdot C_2H_7NO$

Definition: MEA o-Phenylphenate is the monoethanolamine salt of o-phenylphenol that conforms to the formula:

Information Source: EEC(VI/1-7)

Chemical Class: Organic Salts

Function: Preservative

Technical/Other Names:
(1,1'-Bisphenyl)-2-ol, Monoethanolamine Salt
Ethanolamine o-Phenylphenol
Monoethanolamine o-Phenylphenol
o-Phenylphenol, Monoethanolamine Salt

MEA PPG-6 LAURETH-7 CARBOXYLATE

Definition: MEA PPG-6 Laureth-7 Carboxylate is the monoethanolamine salt of the carboxylic acid which conforms generally to the formula:

$$CH_3(CH_2)_{11}(OCHCH_2)_x(OCH_2CH_2)_y \cdot NH_2CH_2CH_2OH$$
$$\qquad\qquad CH_3 \qquad\qquad OCH_2COOH$$

where x has an average value of 6 and y has an average value of 6.

Chemical Class: Organic Salts

Function: Surfactant - Cleansing Agent

Technical/Other Names:
PEG-6-PPG-6 Lauryl Ether Carboxylic Acid, Monoethanolamine Salt
Polyoxyethylene (6) Polyoxypropylene (6) Lauryl Ether Carboxylic Acid, Monoethanolamine Salt
Polyoxypropylene (6) Polyoxyethylene (6) Lauryl Ether Carboxylic Acid, Monoethanolamine Salt

Trade Name Mixture:
Akypogene SO (Kao GmbH)

MEA-PPG-8-STEARETH-7 CARBOXYLATE

Definition: MEA-PPG-8-Steareth-7 Carboxylate is the monoethanolamine salt of

the carboxylic acid which conforms generally to the formula:

$$CH_3(CH_2)_{17}(OCHCH_2)_x(OCH_2CH_2)_y \cdot NH_2CH_2CH_2OH$$
$$\qquad\qquad CH_3 \qquad\qquad OCH_2COOH$$

where x has an average value of 8 and y has an average value of 6.

Chemical Class: Organic Salts

Function: Surfactant - Cleansing Agent

Trade Name:
Product LC 378 (Kao GmbH)

MEA-SALICYLATE

CAS No.	EINECS No.
59866-70-5	261-963-2

Empirical Formula:
$C_7H_6O_3 \cdot C_2H_7NO$

Definition: MEA-Salicylate is the monoethanolamine salt of Salicylic Acid (q.v.) that conforms to the formula:

Information Sources: CIR: [SQ], EEC(VI/1-3), MHLW-331/3

Chemical Classes: Organic Salts; Phenols

Function: Preservative

Technical/Other Names:
Benzoic Acid, 2-Hydroxy-, Monoethanol-amine Salt
Ethanolamine Salicylate
Monoethanolamine 2-Hydroxybenzoate
Salicylic Acid, Monoethanolamine Salt

MEA-SULFITE

CAS No.	EINECS No.
13427-63-9	236-546-3

Empirical Formula:
$C_2H_7NO \cdot H_2O_3S$

Definition: MEA-Sulfite is the organic compound that conforms to the formula:

$$HOCH_2CH_2NH_2 \cdot H_2SO_3$$

Chemical Class: Alkanolamines

Function: Hair Fixative

Technical/Other Names:
Ethanol, 2-Amino-, Sulfite (1:1) Salt
2-Hydroxyethylammonium Hydrogen Sulphite
Monoethanolamine Sulfite

Trade Name:
M - ESN (Ohshika Perfumery)

MEA-THIOLACTATE

CAS No.   **EINECS No.**
54266-38-5   259-050-9

Empirical Formula:
$C_2H_7NO \cdot C_3H_6O_2S$

Definition: MEA-Thiolactate is the monoethanolamine salt of thiolactic acid. It conforms to the formula:

$$HOOCCHCH_3 \cdot NH_2CH_2CH_2OH$$
$$|$$
$$SH$$

Chemical Classes: Organic Salts; Thio Compounds

Function: Hair-Waving/Straightening Agent

Technical/Other Names:
(2-Hydroxyethyl)Ammonium 2-Mercapto-propionate
Propanoic Acid, 2-Mercapto-, Compd. with 2-Aminoethanol

MEA-UNDECYLENATE

CAS No.   **EINECS No.**
56532-40-2   260-247-7

Empirical Formula:
$C_{11}H_{20}O_2 \cdot C_2H_7NO$

Definition: MEA-Undecylenate is the organic salt that conforms to the formula:

$$CH_2 = CH(CH_2)_8COOH \cdot NH_2CH_2CH_2OH$$

Information Source: EEC(VI/1-18)

Chemical Classes: Organic Salts; Soaps

Function: Surfactant - Cleansing Agent

Technical/Other Names:
Monoethanolamine 10-Undecenoate
10-Undecenoic Acid, Monoethanolamine Salt

MEDICAGO SATIVA (ALFALFA) EXTRACT

CAS No.   **EINECS No.**
84082-36-0   281-984-0

Definition: Medicago Sativa (Alfalfa) Extract is an extract of the alfalfa, *Medicago sativa. See "Regulatory and Ingredient Use Information," regarding the labeling names for botanical ingredients in Volume 1, Introduction, Part A.*

Information Sources: 21CFR182.20, RIFM

Chemical Class: Biological Products

Functions: Fragrance Ingredient; Not Reported

Reported Product Categories: Eyebrow Pencils; Suntan Gels, Creams, and Liquids; Hair Conditioners; Tonics, Dressings, and Other Hair Grooming Aids

Technical/Other Names:
Alfalfa Extract
Alfalfa extract (Medicago sativa L.) (RIFM)
Extract of Alfalfa
Extract of Medicago Sativa
Lucerne Extract
Medicago Sativa Extract
Purple Medick Extract

Trade Name Mixtures:
Actiphyte of Alfalfa BG50 (Active Organics)
Actiphyte of Alfalfa GL50 (Active Organics)
Actiphyte of Alfalfa Lipo S (Active Organics)
Actiphyte of Alfalfa PG50 (Active Organics)
Activated Botanicals Estroherb Complex AB 106 (Norjin)
Alfalfa Herb Extract HS 2967 G (Grau)
BBC Mineral Complex (Bio-Botanica)
Extrait Huileux de Luzerne (Yves Rocher)
Herbasol Extract Alfalfa (Cosmetochem) (Cosmetochem International Ltd.)
Hydroplastidine Medicago (Vevy)
Lutexan 10 (Burlington Bio-Medical)
Lutexan 50 (Burlington Bio-Medical)
Medicago Sativa (Alfalfa) Extract ies (IES LABO)
Phytosterol Complex Concentrate SI 1006 (Norjin)
Pronalen Refirming BG (Provital/Centerchem)
Pronalen Refirming HSC (Provital/Centerchem)
2300 Vege-Plex Body Complex (Vege-Tech)
VT-001 Extract of Alfalfa (Vege-Tech)

MEDICAGO SATIVA (ALFALFA) LEAF POWDER

Definition: Medicago Sativa (Alfalfa) Leaf Powder is the powder derived from the ground leaves of *Medicago sativa. See "Regulatory and Ingredient Use Information," regarding the labeling names for botanical ingredients in Volume 1, Introduction, Part A.*

Chemical Class: Biological Products

Function: Skin-Conditioning Agent - Miscellaneous

Trade Name:
Alfalfa Leaf Powder (Aveda)

MEDICAGO SATIVA (ALFALFA) OIL UNSAPONIFIABLES

Definition: Medicago Sativa (Alfalfa) Oil Unsaponifiables is the fraction of alfalfa oil which is not saponified in the recovery of alfalfa oil fatty acids. *See "Regulatory and Ingredient Use Information," regarding the labeling names for botanical ingredients in Volume 1, Introduction, Part A.*

Chemical Class: Unsaponifiables

Functions: Hair Conditioning Agent; Skin-Conditioning Agent - Miscellaneous

Technical/Other Name:
Alfalfa Oil Unsaponifiables

Trade Names:
AEC Alfalfa Oil (A & E Connock)
Extrait RNG "L" (Bertin)

MEDICAGO SATIVA (ALFALFA) SEED POWDER

Definition: Medicago Sativa (Alfalfa) Seed Powder is the powder derived from crushed alfalfa seeds, *Medicago sativa. See "Regulatory and Ingredient Use Information," regarding the labeling names for botanical ingredients in Volume 1, Introduction, Part A.*

Chemical Class: Biological Products

Function: Skin-Conditioning Agent - Miscellaneous

Technical/Other Name:
Alfalfa Seed Powder

MEDICAGO SATIVA (ALFALFA) SYMBIOSOME EXTRACT

Definition: Medicago Sativa (Alfalfa) Symbiosome Extract is an extract of the symbiosome (root nodule) of *Medicago sativa. See "Regulatory and Ingredient Use Information," regarding the labeling names for botanical ingredients in Volume 1, Introduction, Part A.*

Chemical Class: Biological Products

Functions: Antioxidant; Skin-Conditioning Agent - Miscellaneous

Technical/Other Name:
Extract of Medicago Sativa (Alfalfa) Symbiosome

Trade Name Mixture:
Alfalfa Zymbiozome Fermentum (Arch Personal Care Products)

MEK

CAS No.   **EINECS No.**
78-93-3   201-159-0

JPN Translation:
MEK

Empirical Formula:
C_4H_8O

Definition: MEK is the aliphatic ketone that conforms to the formula:

$$CH_3\overset{\displaystyle O}{\overset{\|}{C}}CH_2CH_3$$

Information Sources: 21CFR172.515, 21CFR172.859, 21CFR175.105, 21CFR175.320, 21CFR177.1200, 21CFR1310.02, 21CFR1310.04, 21CFR1313.15, 21CFR1313.24, CTFA D, FCC, JCLS, JSCI, MI-13(6097), RIFM, TSCA

Chemical Class: Ketones

Functions: Fragrance Ingredient; Solvent

Reported Product Category: Nail Polish and Enamel Removers

Technical/Other Names:
2-Butanone (RIFM)
2-Butanone
Methyl Ethyl Ketone

MELALEUCA ALTERNIFOLIA LEAF POWDER

Definition: Melaleuca Alternifolia Leaf Powder is the dried powder obtained from the leaves of *Melaleuca alternifolia. See "Regulatory and Ingredient Use Information," regarding the labeling names for botanical ingredients in Volume 1, Introduction, Part A.*

Chemical Class: Biological Products

Function: Abrasive

Trade Name:
Melafresh EXFOL (Southern Cross Botanicals)

MELALEUCA ALTERNIFOLIA (TEA TREE) EXTRACT

Definition: Melaleuca Alternifolia (Tea Tree) Extract is an extract of the leaves, flowers, and twigs of *Melaleuca alternifolia. See "Regulatory and Ingredient Use Information," regarding the labeling names for botanical ingredients in Volume 1, Introduction, Part A.*

Chemical Class: Biological Products

Function: Not Reported

Technical/Other Names:
Extract of Melaleuca Alternifolia

Extract of Tea Tree (Melaleuca Alternifolia)
Melaleuca Alternifolia Extract
Tea Tree Extract

Trade Name Mixtures:
Actiphyte Melaleuca (Active Organics)
Actiphyte Melaleuca BG50P (Active Organics)
Actiphyte Melaleuca Lipo M (Active Organics)
Actiphyte Melaleuca Lipo S (Active Organics)
Melaleuca Alternifolia (Tea Tree) Extract ies (IES LABO)
Tea Tree Extract (Bell Flavors)
Tea Tree Extract (Carrubba)

MELALEUCA ALTERNIFOLIA (TEA TREE) LEAF EXTRACT

Definition: Melaleuca Alternifolia (Tea Tree) Leaf Extract is an extract of the leaves of the tea tree, *Melaleuca alternifolia. See "Regulatory and Ingredient Use Information," regarding the labeling names for botanical ingredients in Volume 1, Introduction, Part A.*

Chemical Class: Biological Products

Function: Skin-Conditioning Agent - Miscellaneous

Technical/Other Name:
Extract of Melaleuca Alternifolia (Tea Tree) Leaf

MELALEUCA ALTERNIFOLIA (TEA TREE) LEAF OIL

CAS No.: 68647-73-4

Definition: Melaleuca Alternifolia (Tea Tree) Leaf Oil is the oil distilled from the leaves of the *Melaleuca alternifolia. See "Regulatory and Ingredient Use Information," regarding the labeling names for botanical ingredients in Volume 1, Introduction, Part A.*

Information Sources: ARG, MAR, MI-13 (9175), RIFM, TSCA

Chemical Class: Fats and Oils

Functions: Antioxidant; Fragrance Ingredient

Reported Product Categories: Shampoos (Non-coloring); Bath Capsules; Face and Neck Preparations (Excluding Shaving Preparations); Paste Masks (Mud Packs); Hair Conditioners

Technical/Other Names:
Oils, Tea Tree

Tea resinoid (RIFM)
Tea Tree Leaf Oil
Tea tree oil (RIFM)
Tea Tree Oil

Trade Names:
AEC Tea Tree Oil (A & E Connock)
Botanileuca TT (Botanigenics)
EmCon Tea Tree (Fanning)
Epicutin - TT (CLR)
Melaleucol (SNP Natural)
Teatree Oil (Megachem)
Tea Tree Oil - Pharmaceutical Grade (Desert Whale)

Trade Name Mixtures:
Essentiaderm Capillare N.18 (Universal Flavors)
Herbasol Extract Tea Tree Oil (Cosmetochem) (Cosmetochem International Ltd.)
LEMA (Southern Cross Botanicals)
Melafresh SLR (Southern Cross Botanicals)
Microfolia (LCW)
VT-293 Extract of Tea Tree (Vege-Tech)
Water-Soluble Tea Tree Oil (Southern Cross Botanicals)

MELALEUCA BRACTEATA LEAF EXTRACT

Definition: Melaleuca Bracteata Leaf Extract is an extract of the leaves of *Melaleuca bracteata. See "Regulatory and Ingredient Use Information," regarding the labeling names for botanical ingredients in Volume 1, Introduction, Part A.*

Chemical Class: Biological Products

Function: Not Reported

Technical/Other Name:
Extract of Melaleuca Bracteata

Trade Name Mixture:
Campo Black Tea Tree (Campo)

MELALEUCA ERICIFOLIA LEAF OIL

Definition: Melaleuca Ericifolia Leaf Oil is the volatile oil distilled from the leaves of *Melaleuca ericifolia. See "Regulatory and Ingredient Use Information," regarding the labeling names for botanical ingredients in Volume 1, Introduction, Part A.*

Chemical Class: Essential Oils

Function: Fragrance Ingredient

Trade Name:
Rosalina Oil (Provital/Centerchem)

MELALEUCA LEUCADENDRON (CAJAPUT) FRUIT EXTRACT

Definition: Melaleuca Leucadendron (Cajaput) Fruit Extract is an extract of the fruit of *Melaleuca leucadendron cajaputi*. See "Regulatory and Ingredient Use Information," regarding the labeling names for botanical ingredients in Volume 1, Introduction, Part A.

Chemical Class: Biological Products

Function: Not Reported

Technical/Other Name:
Extract of Melaleuca Leucadendron Cajaput

Trade Name Mixtures:
Cajuput Extract (Cosmetic Developments)
VT-285 Extract of Cajeput (Vege-Tech)

MELALEUCA LEUCADENDRON CAJAPUT OIL

CAS No.	EINECS No.
85480-37-1	287-316-4

Definition: Melaleuca Leucadendron Cajaput Oil is the volatile oil obtained from *Melaleuca leucadendron cajaput*. See "Regulatory and Ingredient Use Information," regarding the labeling names for botanical ingredients in Volume 1, Introduction, Part A.

Chemical Class: Essential Oils

Function: Fragrance Ingredient

Technical/Other Name:
Oils, Melaleuca Leucadendron Cajaputi

Trade Names:
AEC Cajuput Oil (A & E Connock)
Cajeput Oil (JPM)
Minyak Kayu Putih Cap Lang (Eagle Brand Cajuut Oil) (PT Eagle Indo Pharma)
Tien Cajuput Oil (Tien Yuan Chemical Pte)

MELALEUCA LEUCADENDRON VIRIDIFLORA EXTRACT

Definition: Melaleuca Leucadendron Viridiflora Extract is the extract obtained from the leaves and flowers of *Melaleuca leucadendron viridiflora*. See "Regulatory and Ingredient Use Information," regarding the labeling names for botanical ingredients in Volume 1, Introduction, Part A.

Chemical Class: Biological Products

Function: Not Reported

Technical/Other Name:
Extract of Melaleuca Leucadendron Viridiflora

Trade Name Mixtures:
Campo Broad Leafed Tea Tree (Campo)
Cosflor Cajuput HGS (A & E Connock)

MELALEUCA SYMPHYOCARP EXTRACT

Definition: Melaleuca Symphyocarp Extract is an extract of the leaves and flowers of *Melaleuca symphyocarp*. See "Regulatory and Ingredient Use Information," regarding the labeling names for botanical ingredients in Volume 1, Introduction, Part A.

Chemical Class: Biological Products

Function: Not Reported

Technical/Other Name:
Extract of Melaleuca Symphyocarp

MELALEUCA UNCINATA EXTRACT

Definition: Melaleuca Uncinata Extract is an extract of the leaf and flower of *Melaleuca uncinata*. See "Regulatory and Ingredient Use Information," regarding the labeling names for botanical ingredients in Volume 1, Introduction, Part A.

Chemical Class: Biological Products

Function: Not Reported

Technical/Other Name:
Extract of Melaleuca Uncinata

Trade Name Mixtures:
Campo Broom Brush Ti-Tri (Campo)
Campo Liniment Ti-Tri (Campo)

MELAMINE PEROXIDE

Empirical Formula:
$C_3H_6N_6 \cdot H_2O_2$

Definition: Melamine Peroxide is the heterocyclic organic compound that conforms to the formula:

Information Source: EEC(III/1-12)

Chemical Classes: Amines; Heterocyclic Compounds

Function: Oxidizing Agent

MELANIN

CAS Nos.: 8049-97-6; 77465-45-3

JPN Translation:
イカスミ

Definition: Melanin is the pigment responsible for the color of animal skin, hair, feathers and fur. See "Regulatory and Ingredient Use Information," for Colorants in Volume 1, Introduction, Part A. This ingredient is not an approved colorant for the US.

Information Source: MI-13(5835)

Chemical Class: Biological Polymers and their Derivatives

Function: Skin-Conditioning Agent - Miscellaneous

Reported Product Categories: Foundations; Eyebrow Pencils; Suntan Gels, Creams, and Liquids; Mascara

Technical/Other Name:
Cuttlefish Ink Powder

Trade Names:
Marine Melanin (Vincience)
MelaneZe 2/060040 (Symrise)
Melanin FCG (Gaskin)
MelanInk (MeL-Co)
Melanin 10% Solution (Lipo)
Synthetic Melanin (Lipotec/Centerchem)

Trade Name Mixtures:
AC Colorplex (Active Concepts)
AC Melanin Liposome (Active Concepts)
Melanosponge (EDT, Inc.)
Phytomelanin (Dr. Carlo Ghisalberti)
Rovisome Melanin (Rovi)

MELATONIN

CAS No.	EINECS No.
73-31-4	200-797-7

Empirical Formula:
$C_{13}H_{16}N_2O_2$

Definition: Melatonin is the organic compound that conforms to the formula:

Information Source: MI-13(5838)

Chemical Classes: Amides; Ethers; Heterocyclic Compounds

Function: Antioxidant

Technical/Other Names:
N-Acetyl-5-Methoxytryptamine
N-(2-(5-Methoxyindol-3-yl)ethyl)Acetamide

Trade Name:
Genzyme MelaPure melatonin (Genzyme)

MELIA AZADIRACHTA BARK EXTRACT

Definition: Melia Azadirachta Bark Extract is an extract of the bark of *Melia*

azadirachta. See "Regulatory and Ingredient Use Information," regarding the labeling names for botanical ingredients in Volume 1, Introduction, Part A. See Reported Ingredient Functions-The Cosmetic Drug Distinction, in Regulatory and Ingredient Use Information, Volume I, Part A.

Information Source: MI-13(6465)

Chemical Class: Biological Products

Functions: Antiacne Agent; Antifungal Agent; Antimicrobial Agent; Skin Protectant

Technical/Other Name:
Extract of Melia Azadirachta Bark

Trade Name Mixture:
Neem (Heritage Bio-Natural)

MELIA AZADIRACHTA CONDITIONED MEDIA/CULTURE

Definition: Melia Azadirachta Conditioned Media/Culture is a uniform suspension prepared from the growth media of the cultured cells of *Melia Azadirachta* and the disrupted *Melia Azadirachta* cells. *See Reported Ingredient Functions-The Cosmetic Drug Distinction, in Regulatory and Ingredient Use Information, Volume I, Part A.*

Chemical Class: Biological Products

Functions: Antioxidant; Film Former; Skin-Conditioning Agent - Miscellaneous; Skin Protectant

Trade Name:
PCF Enhanced Neem Culture (Phyton)

MELIA AZADIRACHTA EXTRACT

Definition: Melia Azadirachta Extract is an extract of the neem tree, *Melia azadirachta*. *See "Regulatory and Ingredient Use Information," regarding the labeling names for botanical ingredients in Volume 1, Introduction, Part A.*

Chemical Class: Biological Products

Function: Not Reported

Technical/Other Names:
Extract of Melia Azadirachta
Extract of Neem
Neem (Melia Azadirachta) Extract

Trade Name:
Silvose TPS (SynPharma)

MELIA AZADIRACHTA FLOWER EXTRACT

Definition: Melia Azadirachta Flower Extract is an extract of the flowers of *Melia*

azadirachta. See "Regulatory and Ingredient Use Information," regarding the labeling names for botanical ingredients in Volume 1, Introduction, Part A.

Chemical Class: Biological Products

Function: Not Reported

Technical/Other Names:
Extract of Melia Azadirachta Flower
Extract of Neem (Melia Azadirachta) Flower
Neem (Melia Azadirachta) Flower Extract

Trade Name Mixtures:
Campo Mahanimba (Campo)
Campo Siddha Neer Kattari (Campo)
Campo Siddha Neer Pazchai (Campo)
Campo Siddha Neer Sikkappu (Campo)
Campo Siddha Pazchai Yanai (Campo)
Campo Siddha Vepuvillai Nahi (Campo)

MELIA AZADIRACHTA LEAF EXTRACT

CAS No.	EINECS No.
90063-92-6	290-052-2

Definition: Melia Azadirachta Leaf Extract is an extract of the leaves of *Melia azadirachta*. *See "Regulatory and Ingredient Use Information," regarding the labeling names for botanical ingredients in Volume 1, Introduction, Part A.*

Chemical Class: Biological Products

Function: Not Reported

Technical/Other Names:
Extract of Melia Azadirachta Leaf
Extract of Neem Leaf
Neem Leaf Extract
Neem (Melia Azadirachta) Leaf Extract

Trade Name Mixtures:
Actiphyte Neem (Active Organics)
Actiphyte Neem AQ (Active Organics)
Actiphyte Neem GL (Active Organics)
Actiphyte Neem Lipo S (Active Organics)
Campo Australian Neem Tree (Campo)
Campo Vaipillai (Campo)
Neem Tree Leaf HS (Alban Muller)
Phytotal OS (Phybiotex/Sederma)

MELIA AZADIRACHTA SEED OIL

Definition: Melia Azadirachta Seed Oil is the oil expressed from the seeds of the neem tree, *Melia azadirachta*. *See "Regulatory and Ingredient Use Information," regarding the labeling names for botanical ingredients in Volume 1, Introduction, Part A.*

Information Source: MI-13(6466)

Chemical Class: Fats and Oils

Function: Skin-Conditioning Agent - Occlusive

Technical/Other Names:
Neem (Melia Azadirachta) Seed Oil
Neem Seed Oil

Trade Names:
AEC Neem Oil (A & E Connock)
Melia Azadirachta Seed Oil ies (IES LABO)
Midecol CF (Microbial Systems)
Neem Oil (Aldivia)

Trade Name Mixture:
Vedacalm (Libiol)

MELIBIOSE

CAS Nos.	EINECS No.
585-99-9	209-568-6
5340-95-4	

Empirical Formula:
$C_{12}H_{22}O_{11}$

Definition: Melibiose is the carbohydrate that conforms to the formula:

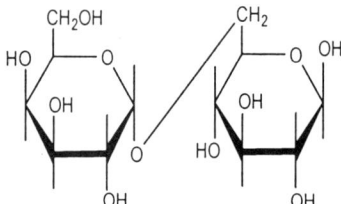

Information Source: MI-13(5842)

Chemical Class: Carbohydrates

Function: Skin-Conditioning Agent - Humectant

Technical/Other Names:
D-Glucopyranose, 6-O-α-D-galactopyrano-syl-
D-Glucose, 6-O-α-D-Galactopyranosyl-
D-Melibiose

Trade Name:
Melibiose monohydrate (RoC)

MELILOTUS OFFICINALIS EXTRACT

CAS No.	EINECS No.
84082-81-5	282-028-5

JPN Translation:
メリロートエキス

Definition: Melilotus Officinalis Extract is an extract of the sweet clover, *Melilotus officinalis*. *See "Regulatory and Ingredient Use Information," regarding the labeling names for botanical ingredients in Volume 1, Introduction, Part A.*

Information Sources: JCIC, JCLS, RIFM

Chemical Class: Biological Products

Functions: Fragrance Ingredient; Skin-Conditioning Agent - Miscellaneous

Reported Product Categories: Bath Preparations, Misc.; Body and Hand Preparations (Excluding Shaving Preparations)

Technical/Other Names:
Extract of Melilotus Officinalis
Extract of Sweet Clover
Melilotus Extract
Sweet Clover Extract
Sweet clover, Melilotus officinalis, ext. (RIFM)
Sweet Clover (Melilotus Officinalis) Extract

Trade Name Mixtures:
Coobato Meliloto (Universal Flavors)
Extrait De Melilot PPE PG40 (Yves Rocher)
Glycolysat of Sweet Clover (CEP (Solabia))
Integrals of Sweet Clover (Solabia)
Melilot Extract HS 2376 G (Grau)
Melilot HS (Alban Muller)
Melilot Liquid B (Ichimaru Pharcos)
Melilot Liquid E (Ichimaru Pharcos)
Melilot LS (Alban Muller)
Melilotus Officinalis Extract ies (IES LABO)
Natupure Sweet Clover (E.U.K)
Prodhy Extract Melilot (Prod'Hyg)
Vegebios of Sweet Clover (CEP (Solabia))
Vegetol Sweet Clover GR 436 Hydro (Gattefosse s.a.)

MELISSA OFFICINALIS EXTRACT

Definition: Melissa Officinalis Extract is an extract of the whole herb of *Melissa officinalis*. See *"Regulatory and Ingredient Use Information,"* regarding the labeling names for botanical ingredients in Volume 1, Introduction, Part A.

Chemical Class: Biological Products

Function: Not Reported

Reported Product Categories: Skin Fresheners; Bath Preparations, Misc.; Body and Hand Preparations (Excluding Shaving Preparations); Moisturizing Preparations; Bath Capsules; Bath Oils, Tablets, and Salts; Face and Neck Preparations (Excluding Shaving Preparations); Hair Sprays (Aerosol Fixatives)

Technical/Other Names:
Balm Mint Extract
Extract of Balm Mint
Extract of Melissa Officinalis (Balm Mint)
Melissa Officinalis (Balm Mint) Extract

Trade Names:
AEC Melissa Extract Powder (A & E Connock)
Melissin (Sabinsa)

Trade Name Mixtures:
ACB Lemon Balm Extract (Active Concepts)
Actiphyte Balm Mint AQ (Active Organics)
Actiphyte Balm Mint Lipo M (Active Organics)
Herbal Extract Glycolic - Article 251172 (Plantextrakt)

MELISSA OFFICINALIS LEAF EXTRACT

CAS No. 84082-61-1   **EINECS No.** 282-007-0

JPN Translation:
メリッサエキス

Definition: Melissa Officinalis Leaf Extract is an extract of the leaves and tops of the balm mint, *Melissa officinalis. See "Regulatory and Ingredient Use Information," regarding the labeling names for botanical ingredients in Volume 1, Introduction, Part A.*

Information Sources: 21CFR182.20, JCIC, JCLS, JSQI, RIFM

Chemical Class: Biological Products

Functions: Fragrance Ingredient; Skin-Conditioning Agent - Miscellaneous; Skin-Conditioning Agent - Occlusive

Reported Product Categories: Skin Fresheners; Bath Preparations, Misc.; Body and Hand Preparations (Excluding Shaving Preparations); Moisturizing Preparations; Personal Cleanliness Products, Misc.; Bath Capsules; Bath Oils, Tablets, and Salts; Cleansing Products (Cold Creams, Cleansing Lotions, Liquids and Pads); Face and Neck Preparations (Excluding Shaving Preparations); Hair Conditioners; Hair Sprays (Aerosol Fixatives); Skin Care Preparations, Misc.

Technical/Other Names:
Balm leaves extract (Melissa officinalis L.) (RIFM)
Balm Mint Extract
Balm Mint Leaf Extract
Extract of Balm Mint
Extract of Lemon Balm
Extract of Melissa Officinalis
Lemon Balm Extract
Melissa Officinalis Extract

Trade Names:
Flavex Balm Leaf CO2-to extract, Type 078.001 (Flavex)
Herbalia Balm Mint (Cognis Deutschland)
Hydroessential Melissa (Vevy)

Trade Name Mixtures:
Actiphyte of Balm Mint BG50 (Active Organics)
Actiphyte of Balm Mint GL50 (Active Organics)
Actiphyte of Balm Mint Lipo S (Active Organics)
Actiphyte of Balm Mint PG50 (Active Organics)
Balm Mint Extract (Maruzen Pharmaceuticals Co., Ltd.)
Balm Mint Extract BG (Maruzen Pharmaceuticals Co., Ltd.)
Balm Mint Extract HG (Provital/ Centerchem)
Balm Mint Extract LA (Maruzen Pharmaceuticals Co., Ltd.)
Balm Mint Extract PG (Rahn)
Balm Mint Extract Powder-S (Maruzen Pharmaceuticals Co., Ltd.)
Balm Mint Extract W (Maruzen Pharmaceuticals Co., Ltd.)
Balm Mint HS (Alban Muller)
Balm Mint Oil infus. (Crodarom)
Balm Mint Phytexcell (Crodarom)
Balm Mint Tincture (Rahn)
Coobato Melissa (Universal Flavors)
Cremogen AF (PN 736567) (Haarmann & Reimer GmbH)
Cremogen Melissa Balm Mint (PN 739 013) (Haarmann & Reimer GmbH)
Cremogen MZ/N (PN 755 321) (Haarmann & Reimer GmbH)
Extrapone Alpine Herbs GW 2/031560 (Symrise)
Extrapone Alpine Herbs Special 2/032561 (Symrise)
Extrapone Balm Mint 2/033121 (Symrise)
Extrapone Cooling Complex 2/B16500 (Symrise)
Extrapone Cooling Complex N 2/B16501 (Symrise)
Extrapone #7 Herbs 2/032535 (Symrise)
Extrapone #3 Special 2/034481 (Symrise)
Extrapone #3 Special 2/789490 (Symrise)
Extrapone 3 Special 2/789490 (Symrise)
Extrapone 3 Special New 2/034484 (Symrise)
Gigawhite (Cosmetic Ingredient Resources/ Centerchem)
Herbalcomplex 3 Special (Crodarom)
Herbaliquid Balm Mint Special (Crodarom)
Herbasec Balm Mint (Cosmetochem) (Cosmetochem International Ltd.)
Herbasol Complex "Sedative/Relaxing" (Cosmetochem) (Cosmetochem International Ltd.)
Herbasol-Extract Balm Mint (Cosmetochem)
Hexaflor-Complex (Crodarom)
Hexaplant Richter (CLR)
InCyte Lemon Balm MG (Collaborative Labs)
Melissa Cl 2/033128 (Symrise)
Melissa Extract HS 2453 G (Grau)
Melissa Liquid B (Ichimaru Pharcos)
Melissa Liquid E (Ichimaru Pharcos)
Novaplant Balm Mint Extract (Crodarom)

Oleat of Balm Mint (CEP (Solabia))
Phytelene of Balm Mint EG 456 liquid (Indena SA)
Phytogreen 55 of Balm EXH 680 Liquid (Phytochim)
Phytotal AI (Phybiotex/Sederma)
Polyplant Base (Provital/Centerchem)
Polyplant Base S.E. (Provital/Centerchem)
Polyplant Sedative (Provital/Centerchem)
Polyplant Skin Purifying (Provital/Centerchem)
Prodhy Extract Melisse (Prod'Hyg)
Sedaflor-Complex (Crodarom)
Sedaplant Richter (CLR)
Sederma Balm Mint (Sederma)
Solarcat (Collaborative Labs)
Solarease II (Collaborative Labs)
Toothpaste Complex MU 3319 (Greentech S.A)
Vegebios of Balm Mint (CEP (Solabia))
2300 Vege-Plex Body Complex (Vege-Tech)
2303 Vege-Plex Body Complex (Vege-Tech)
2330 Vege-Plex Body Complex (Vege-Tech)
2270 Vege-Plex Hair Complex (Vege-Tech)
2290 Vege-Plex Hair Complex (Vege-Tech)
Vege Plex VP#1341 (Vege-Tech)
VT-004 Extract of Balm Mint (Vege-Tech)
VT-271 Extract of Citronella (Vege-Tech)

MELISSA OFFICINALIS LEAF OIL

CAS No.: 8014-71-9

Definition: Melissa Officinalis Leaf Oil is a volatile oil obtained from the leaves and tops of *Melissa officinalis. See "Regulatory and Ingredient Use Information," regarding the labeling names for botanical ingredients in Volume 1, Introduction, Part A.*

Information Sources: 21CFR182.20, MAR, MI-13(6807), RIFM, TSCA, YUG

Chemical Class: Essential Oils

Function: Fragrance Ingredient

Reported Product Category: Skin Care Preparations, Misc.

Technical/Other Names:
Balm (Melissa officinalis L.) (RIFM)
Balm Mint Leaf Oil
Balm Mint Oil
Balm oil (Melissa officinalis L.) (RIFM)
Lemon Balm
Melissa Officinalis Oil
Oil, Essential, Balm Mint
Oil of Balm
Oil of Melissa
Oils, Melissa Officinalis

Trade Name:
AEC Melissa Oil (A & E Connock)

Trade Name Mixtures:
Cremogen MZ (PN 739 032) (Haarmann & Reimer GmbH)
Cremogen M-82 (PN 730 337) (Haarmann & Reimer GmbH)
Essentiaderm n.4 (Universal Flavors)
Essentiaderm n.6 (Universal Flavors)
Essentiaderm n.8 (Universal Flavors)

MELISSA OFFICINALIS LEAF POWDER

Definition: Melissa Officinalis Leaf Powder is the powder derived from the ground leaves of *Melissa officinalis. See "Regulatory and Ingredient Use Information," regarding the labeling names for botanical ingredients in Volume 1, Introduction, Part A.*

Chemical Class: Biological Products

Function: Skin-Conditioning Agent - Miscellaneous

Trade Name:
Lemon Balm Powder (Aveda)

MELISSA OFFICINALIS SEED OIL

Definition: Melissa Officinalis Seed Oil is the oil expressed from the seeds of *Melissa officinalis. See "Regulatory and Ingredient Use Information," regarding the labeling names for botanical ingredients in Volume 1, Introduction, Part A.*

Chemical Class: Fats and Oils

Function: Skin-Conditioning Agent - Occlusive

Technical/Other Names:
Balm Mint Seed Oil
Oils, Balm Mint Seed
Oils, Melissa Officinalis

Trade Name:
Huile Vierge de Graines de Melisse (Bertin)

MELISSA OFFICINALIS WATER

Definition: Melissa Officinalis Water is an aqueous solution containing volatile oils obtained by the distillation of the aerial parts of *Melissa officinalis. See "Regulatory and Ingredient Use Information," regarding the labeling names for botanical ingredients in Volume 1, Introduction, Part A.*

Chemical Class: Biological Products

Function: Not Reported

Technical/Other Names:
Balm Mint Distillate
Melissa Officinalis Distillate

Trade Names:
Lemon Balm Hydroflorate (Bayliss Ranch)
Melissa Ecoconcentrate Natural (Robertet S.A.)

MELLISIC ACID

CAS Nos.	EINECS No.
506-50-3	208-042-3
38232-01-8	

Empirical Formula:
$C_{30}H_{60}O_2$

Definition: Mellisic Acid is the aliphatic acid that conforms to the formula:

$$CH_3(CH_2)_{29}COOH$$

Chemical Class: Carboxylic Acids

Function: Viscosity Increasing Agent - Nonaqueous

Technical/Other Names:
Hentriacontanoic Acid
Melissic Acid
Triacontanoic Acid

Trade Name Mixture:
Ross Bayberry Wax Substitute 1641 (Ross)

MELOTHRIA HETEROPHYLLA EXTRACT

Definition: Melothria Heterophylla Extract is an extract of the roots of *Melothria heterophylla. See "Regulatory and Ingredient Use Information," regarding the labeling names for botanical ingredients in Volume 1, Introduction, Part A.*

Chemical Class: Biological Products

Function: Not Reported

Technical/Other Name:
Extract of Melothria Heterophylla

Trade Name Mixture:
Melothria Extract (Suntory)

MENADIONE

CAS No.	EINECS No.
58-27-5	200-372-6

Empirical Formula:
$C_{11}H_8O_2$

Definition: Menadione is the polycyclic aromatic compound that conforms to the formula:

The inclusion of any compound in the *Dictionary and Handbook* does not indicate that use of that substance as a cosmetic ingredient complies with the laws and regulations governing such use in the United States or any other country.

Information Sources: BAN, BRA, 21CFR573.620, 21CFR573.625, EP, ITA, MEX, MI-13(5853), NF XV, PF, TSCA, USAN, USD, USP XXIV

Chemical Class: Ketones

Function: Not Reported

Technical/Other Names:
1,4-Dihydro-1,4-dioxo-2-methylnaphthalene
2-Methyl-1,4-Naphthalenedione
1,4-Naphthalenedione, 2-Methyl-
1,4-Naphthoquinone, 2-methyl-
Vitamin K3

MENHADEN OIL

CAS No.	EINECS No.
8002-50-4	232-311-4

Definition: Menhaden Oil is the oil obtained from the small North Atlantic fish, *Brevoortia tyrannus*. See *"Regulatory and Ingredient Use Information," regarding use of EU Trivial names in Volume 1, Introduction, Part A.*

Information Sources: 21CFR175.300, 21CFR176.200, 21CFR176.210, 21CFR177.2800, MI-13(5859), TSCA

Chemical Class: Fats and Oils

Function: Skin-Conditioning Agent - Occlusive

Technical/Other Names:
Brevoortia Tyrannus Oil
Menhaden Fish Oil
Mossbunker Oil
Oils, Menhaden

MENTHA AQUATICA LEAF EXTRACT

CAS No.: 90063-96-0

Definition: Mentha Aquatica Leaf Extract is an extract of the leaves of *Mentha aquatica*. See *"Regulatory and Ingredient Use Information," regarding the labeling names for botanical ingredients in Volume 1, Introduction, Part A.*

Chemical Class: Biological Products

Function: Skin-Conditioning Agent - Miscellaneous

Technical/Other Name:
Extract of Mentha Aquatica

Trade Name Mixtures:
Herbasol Extract Water Mint (Cosmetochem) (Cosmetochem International Ltd.)
Watermint Extract (Cosmetic Developments)
Watermint HS (Alban Muller)

MENTHA AQUATICA WATER

Definition: Mentha Aquatica Water is an aqueous solution of the steam distillate obtained from *Mentha aquatica*. See *"Regulatory and Ingredient Use Information," regarding the labeling names for botanical ingredients in Volume 1, Introduction, Part A.*

Chemical Class: Biological Products

Functions: Cosmetic Astringent; Deodorant Agent; Fragrance Ingredient

Trade Name:
Chocolate Mint Hydroflorate (Bayliss Ranch)

MENTHA ARVENSIS EXTRACT

Definition: Mentha Arvensis Extract is an extract of the aerial parts of *Mentha Arvensis*. See *"Regulatory and Ingredient Use Information," regarding the labeling names for botanical ingredients in Volume 1, Introduction, Part A.*

Chemical Class: Biological Products

Function: Fragrance Ingredient

Technical/Other Names:
Extract of Mentha Arvensis
Field Mint Extract
Wild Mint Extract

Trade Name Mixture:
Hakka Liquid B (Ichimaru Pharcos)

MENTHA ARVENSIS LEAF EXTRACT

CAS No.	EINECS No.
90063-97-1	290-058-5

Definition: Mentha Arvensis Leaf Extract is an extract of the leaves and twigs of the wild mint, *Mentha arvensis*. See *"Regulatory and Ingredient Use Information," regarding the labeling names for botanical ingredients in Volume 1, Introduction, Part A.*

Chemical Class: Biological Products

Function: Not Reported

Technical/Other Names:
Extract of Field Mint
Extract of Mentha Arvensis
Extract of Wild Mint
Field Mint Extract
Wild Mint Extract
Wild Mint (Mentha Arvensis) Extract

Trade Name Mixtures:
Actiphyte of Wild Mint (Active Organics)
Herbasol-Extract Wild Mint (Cosmetochem)

MENTHA ARVENSIS LEAF OIL

CAS No.: 68917-18-0

Definition: Mentha Arvensis Leaf Oil is the oil derived from the leaves of *Mentha arvensis*. See *"Regulatory and Ingredient Use Information," regarding the labeling names for botanical ingredients in Volume 1, Introduction, Part A.*

Information Sources: RIFM, TSCA

Chemical Class: Essential Oils

Function: Fragrance Ingredient

Technical/Other Names:
Hakka Yu (JPN)
Mentha arvensis oil (RIFM)
Oil of Cornmint
Oil of Wild Mint
Wild Mint (Mentha Arvensis) Oil
Wild Mint Oil

Trade Names:
Cornmint Oil (Ungerer)
Mentha Arvensis oil (Mane)

Trade Name Mixture:
V-Tonic (Gattefosse s.a.)

MENTHA ARVENSIS POWDER

Definition: Mentha Arvensis Powder is the powder derived from *Mentha arvensis*. See *"Regulatory and Ingredient Use Information," regarding the labeling names for botanical ingredients in Volume 1, Introduction, Part A.*

Information Sources: JCIC, JCLS

Chemical Class: Biological Products

Function: Fragrance Ingredient

Technical/Other Names:
Mentha Herb Powder
Wild Mint (Mentha Arvensis) Powder

MENTHANEDIOL

CAS No.	EINECS No.
42822-86-6	255-953-7

Empirical Formula:
$C_{10}H_{20}O_2$

Definition: Menthanediol is the organic compound that conforms to the formula:

Chemical Class: Alcohols

Functions: Hair Conditioning Agent; Oral Care Agent; Skin-Conditioning Agent - Miscellaneous

Technical/Other Names:
2-Hydroxy-α,α,4-Trimethylcyclohexane-methanol
p-Menthane-3,8-diol

Trade Name:
Coolact 38D (Takasago)

p-MENTHAN-7-OL

Empirical Formula:
$C_{10}H_{20}O$

Definition: p-Menthan-7-ol is the organic compound that conforms to the formula:

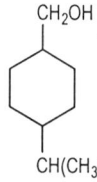

Chemical Class: Alcohols

Function: Fragrance Ingredient

Technical/Other Name:
Cyclohexanemethanol, 4-(1-methylethyl)-

Trade Name:
Mayol (Firmenich)

MENTHA PIPERITA (PEPPERMINT) EXTRACT

Definition: Mentha Piperita (Peppermint) Extract is an extract of the whole plant, *Mentha piperita. See "Regulatory and Ingredient Use Information," regarding the labeling names for botanical ingredients in Volume 1, Introduction, Part A.*

Chemical Class: Biological Products

Function: Skin-Conditioning Agent - Miscellaneous

Trade Name Mixture:
Calmiskin (Silab)

MENTHA PIPERITA (PEPPERMINT) LEAF

Definition: Mentha Piperita (Peppermint) Leaf are the dried leaves and tops of the peppermint, *Mentha piperita. See "Regulatory and Ingredient Use Information," regarding the labeling names for botanical ingredients in Volume 1, Introduction, Part A.*

Information Sources: BEL, 21CFR182.10, CIR: [SQ] IJT-20(SUPPL. 3)2001, CZE, DDR, EGY, HP, HUN, MAR, MI-13(7224), NED, POL, POR, RIFM, ROM, USD, USSR, YUG

Chemical Class: Biological Products

Function: Fragrance Ingredient

Technical/Other Names:
Leaf, Mentha Piperita
Leaf, Peppermint
Mentha Piperita Leaves
Pepperment leaves (Mentha piperita L.) (RIFM)
Peppermint Leaf
Peppermint Leaves

Trade Name:
AEC Peppermint Leaf Cut (A & E Connock)

MENTHA PIPERITA (PEPPERMINT) LEAF EXTRACT

CAS No.   **EINECS No.**
84082-70-2   282-015-4

JPN Translation:
セイヨウハッカエキス

Definition: Mentha Piperita (Peppermint) Leaf Extract is an extract of the leaves of the peppermint, *Mentha piperita. See "Regulatory and Ingredient Use Information," regarding the labeling names for botanical ingredients in Volume 1, Introduction, Part A.*

Information Sources: 21CFR182.20, CIR: [SQ] IJT-20(SUPPL. 3)2001, HP, JCIC, JCLS, JSQI, RIFM

Chemical Class: Biological Products

Functions: Fragrance Ingredient; Skin-Conditioning Agent - Miscellaneous; Skin-Conditioning Agent - Occlusive

Reported Product Categories: Hair Conditioners; Moisturizing Preparations; Paste Masks (Mud Packs); Bath Preparations, Misc.; Body and Hand Preparations (Excluding Shaving Preparations); Bath Capsules; Personal Cleanliness Products, Misc.; Shampoos (Non-coloring); Skin Care Preparations, Misc.

Technical/Other Names:
Extract of Mentha Piperita
Extract of Peppermint
Extract of Peppermint Leaves
Mentha Piperita Extract
Pepperment leaves (Mentha piperita L.) (RIFM)
Peppermint Extract
Peppermint Leaf Extract

Trade Names:
Flavex Peppermint Leaf CO2-se extract, Type 036.001 (Flavex)

Hydroessential Mentha (Vevy)
Menthe Ecoconcentrate (Robertet, Inc.)
Mint Ecoconcentrate Natural (Robertet S.A.)

Trade Name Mixtures:
Actiphyte of Peppermint BG50 (Active Organics)
Actiphyte of Peppermint GL50 (Active Organics)
Actiphyte of Peppermint Lipo S (Active Organics)
Actiphyte of Peppermint PG50 (Active Organics)
Aromaphyte of Peppermint (Active Organics)
Bathgranue Peppermint (Ichimaru Pharcos)
BBC Moisture Trol (Bio-Botanica)
BBC Relaxing Complex (Bio-Botanica)
Caomint (CEP (Solabia))
Cremogen Peppermint (PN 774 201) (Haarmann & Reimer GmbH)
Extrait Hydroglycolique de menthe de la baie saint John's-GT110 (Greentech)
Extrapone Cooling Complex 2/B16500 (Symrise)
Extrapone Cooling Complex N 2/B16501 (Symrise)
Extrapone #4 Herbs 2/032495 (Symrise)
Extrapone Peppermint-Special (2/033171) (Symrise)
Folistim-2 (Bio-Botanica)
Gigawhite (Cosmetic Ingredient Resources/ Centerchem)
Glycolysat of Mint (CEP (Solabia))
Herbaliquid Peppermint Special (Crodarom)
Herbasec Peppermint (Cosmetochem) (Cosmetochem International Ltd.)
Herbasol Distillate Peppermint (Cosmetochem) (Cosmetochem International Ltd.)
Herbasol-Extract Peppermint (Cosmetochem)
Keshastim (Bio-Botanica)
Keshastim-A (Bio-Botanica)
Keshastim-D (Bio-Botanica)
Mentha Piperita (Peppermint) Leaf Extract IES (IES LABO)
Mint Extract HG (Provital/Centerchem)
Natupure Peppermint (E.U.K)
Peppermint CL 2/033173 (Symrise)
Peppermint Extract (Maruzen Pharmaceuticals Co., Ltd.)
Peppermint Extract BG (Maruzen Pharmaceuticals Co., Ltd.)
Peppermint Extract HS 2365 G (Grau)
Peppermint Extract LA (Maruzen Pharmaceuticals Co., Ltd.)
Peppermint Extract Powder-S (Maruzen Pharmaceuticals Co., Ltd.)
Peppermint HS (Alban Muller)
Peppermint Liquid B (Ichimaru Pharcos)
Peppermint Liquid E (Ichimaru Pharcos)

Peppermint Liquid PO (Ichimaru Pharcos)
Peppermint LS (Alban Muller)
Phytelene of Pepper Mint EG 339 liquid
(Indena SA)
Phytogreen 55 of Pepper Mint EXH 656
Liquid (Phytochim)
Polyplant Descongestant (Provital/
Centerchem)
Prodhy Extract Menthe (Prod'Hyg)
Vegebios of Mint (CEP (Solabia))
Vegetol Peppermint GR 105 Hydro
(Gattefosse s.a.)
VT-050 Extract of Peppermint (Vege-Tech)

MENTHA PIPERITA (PEPPERMINT) LEAF JUICE

Definition: Mentha Piperita (Peppermint) Leaf Juice is the juice expressed from the leaves of *Mentha piperita*. *See "Regulatory and Ingredient Use Information," regarding the labeling names for botanical ingredients in Volume 1, Introduction, Part A.*

Chemical Class: Biological Products

Function: Skin-Conditioning Agent - Miscellaneous

Technical/Other Name:
Peppermint Leaf Juice

Trade Name:
Authenticals of Peppermint (CEP (Solabia))

MENTHA PIPERITA (PEPPERMINT) LEAF WATER

JPN Translation:
ハッカ水

Definition: Mentha Piperita (Peppermint) Leaf Water is an aqueous solution of the steam distillate obtained from the leaves of *Mentha piperita*. *See "Regulatory and Ingredient Use Information," regarding the labeling names for botanical ingredients in Volume 1, Introduction, Part A.*

Information Sources: CIR: [SQ] IJT-20 (SUPPL. 3)2001, JCLS

Chemical Class: Biological Products

Functions: Flavoring Agent; Fragrance Ingredient; Skin-Conditioning Agent - Miscellaneous

Technical/Other Names:
Pepermint Water
Peppermint Leaf Water
Peppermint Water
Water, Peppermint

Trade Names:
Extrait Originel Menthe (Gattefosse s.a.)

Peppermint Distillate (Maruzen
Pharmaceuticals Co., Ltd.)
Peppermint Water (Alban Muller)

MENTHA PIPERITA (PEPPERMINT) OIL

CAS No.: 8006-90-4

JPN Translation:
セイヨウハッカ油

Definition: Mentha Piperita (Peppermint) Oil is a volatile oil obtained from the plant *Mentha piperita*. *See "Regulatory and Ingredient Use Information," regarding the labeling names for botanical ingredients in Volume 1, Introduction, Part A.*

Information Sources: BEL, BP, BPC, BRA, 21CFR182.20, 27CFR21.65, 27CFR21.151, CIR: [SQ] IJT-20(SUPPL. 3)-2001, CZE, DA, DDR, EGY, FCC, FI, FIN, HUN, ITA, JCIC, JCLS, JP, JSQI, MAR, MEX, MI-13(6866), NF XIX, OTC-I-CT, PF, PN, POL, RIFM, ROM, TSCA, USAN, USD, USSR, YUG

Chemical Class: Essential Oils

Functions: Fragrance Ingredient; Skin-Conditioning Agent - Miscellaneous

Reported Product Categories: Skin Care Preparations, Misc.; Bath Oils, Tablets, and Salts; Cleansing Products (Cold Creams, Cleansing Lotions, Liquids and Pads); Dentifrices (Aerosol, Liquid, Pastes and Powders); Mouthwashes and Breath Fresheners (Liquids and Sprays); Skin Fresheners; Bath Preparations, Misc.; Body and Hand Preparations (Excluding Shaving Preparations); Paste Masks (Mud Packs); Foot Powders and Sprays; Bath Capsules; Face and Neck Preparations (Excluding Shaving Preparations); Lipsticks; Moisturizing Preparations

Technical/Other Names:
Hakka Yu (JPN)
Mentha Oil
Mentha Piperita Oil
Oil of Peppermint
Peppermint oil (RIFM)
Peppermint Oil

Trade Names:
AEC Peppermint Oil (A & E Connock)
Custosense PM (peppermint oil) (Custom Ingredients)

Trade Name Mixtures:
Aromaphyte of Peppermint (Active Organics)
Cooling Mols (Chemyunion)
Essentiaderm n.1 (Universal Flavors)
Essentiaderm n.6 (Universal Flavors)
Essentiaderm n.8 (Universal Flavors)

Essentiaderm N.14 (Universal Flavors)
Essentiaderm N.19 (Universal Flavors)
Germinol (Dr. Gerhard Steidl)
Peppermint Cl Forte 2/J33172 (Symrise)
Peppermint Softcream (CEP (Solabia))
Stimulating Phytospa (Alban Muller)
Stimulating Phytospa NaCl (Alban Muller)
Tonic Phytospa (Alban Muller)

MENTHA PULEGIUM

Definition: *See "Regulatory and Ingredient Use Information," regarding EU labeling names for botanical ingredients in Volume 1, Introduction, Part A.*

Chemical Class: Biological Products

Technical/Other Names:
Mentha Pulegium Extract (U.S.)
Mentha Pulegium Oil (U.S.)

MENTHA PULEGIUM EXTRACT

CAS No.	EINECS No.
90064-00-9	290-061-1

Definition: Mentha Pulegium Extract is an extract of the flowering herb of the pennyroyal, *Mentha pulegium*. *See "Regulatory and Ingredient Use Information," regarding the labeling names for botanical ingredients in Volume 1, Introduction, Part A.*

Information Source: 21CFR172.510

Chemical Class: Biological Products

Function: Skin-Conditioning Agent - Miscellaneous

Technical/Other Names:
Extract of Mentha Pulegium
Extract of Pennyroyal
Mentha Pulegium (EU)
Pennyroyal Extract
Pennyroyal (Mentha Pulegium) Extract

Trade Name Mixtures:
Actiphyte of Pennyroyal BG50 (Active Organics)
Actiphyte of Pennyroyal GL50 (Active Organics)
Actiphyte of Pennyroyal Lipo S (Active Organics)
Actiphyte of Pennyroyal PG50 (Active Organics)
Aromaphyte of Pennyroyal (Active Organics)
Herbasol Extract Pennyroyal Mint (Cosmetochem) (Cosmetochem International Ltd.)
Pennyroyal Extract HS 2390 G (Grau)
Pennyroyal Fluid Extract (Alban Muller)
Pennyroyal HS (Alban Muller)

Pennyroyal LS (Alban Muller)
Toothpaste Complex MU 3319 (Greentech S.A)
VT-097 Extract of Pennyroyal (Vege-Tech)

MENTHA PULEGIUM OIL

CAS Nos.: 8007-44-1; 8013-99-8

Definition: Mentha Pulegium Oil is the volatile oil obtained from *Mentha pulegium*. *See "Regulatory and Ingredient Use Information," regarding the labeling names for botanical ingredients in Volume 1, Introduction, Part A.*

Information Sources: MI-13(6864), RIFM, TSCA

Chemical Class: Essential Oils

Function: Fragrance Ingredient

Technical/Other Names:
Mentha Pulegium (EU)
Oil of Pennyroyal-European
Pennyroyal (Mentha Pulegium) Oil
Pennyroyal Oil
Pennyroyal oil, American (Hedeoma pulegioides) (RIFM)
Pennyroyal oil (Mentha pulegium L.) (RIFM)

Trade Name:
AEC Pennyroyal Oil (A & E Connock)

Trade Name Mixture:
Aromaphyte of Pennyroyal (Active Organics)

MENTHA ROTUNDIFOLIA LEAF EXTRACT

Definition: Mentha Rotundifolia Leaf Extract is an extract of the leaves of *Mentha rotundifolia*. *See "Regulatory and Ingredient Use Information," regarding the labeling names for botanical ingredients in Volume 1, Introduction, Part A.*

Chemical Class: Biological Products

Function: Not Reported

Technical/Other Name:
Extract of Mentha Rotundifolia

Trade Name Mixture:
Applemint Extract (Cosmetic Developments)

MENTHA SUAVEOLENS LEAF EXTRACT

Definition: Mentha Suaveolens Leaf Extract is an extract of the leaves of *Mentha suaveolens*. *See "Regulatory and Ingredient Use Information," regarding the*

labeling names for botanical ingredients in Volume 1, Introduction, Part A.

Chemical Class: Biological Products

Functions: Cosmetic Astringent; Skin-Conditioning Agent - Miscellaneous

Technical/Other Name:
Extract of Mentha Suaveolens Leaf

Trade Name Mixture:
Actiphyte of Apple Mint (Active Organics)

MENTHA VIRIDIS (SPEARMINT) EXTRACT

CAS No.: 90064-01-0

Definition: Mentha Viridis (Spearmint) Extract is an extract of the spearmint, *Mentha viridis*. *See "Regulatory and Ingredient Use Information," regarding the labeling names for botanical ingredients in Volume 1, Introduction, Part A.*

Information Source: RIFM

Chemical Class: Biological Products

Function: Fragrance Ingredient

Technical/Other Names:
Extract of Mentha Viridis
Extract of Spearmint
Mentha Spicata Extract
Spearmint Extract
Spearmint extract (Mentha spicata L.) (RIFM)

Trade Name:
Spearmint Ecoconcentrate Natural (Robertet S.A.)

Trade Name Mixtures:
Actiphyte of Spearmint BG50 (Active Organics)
Actiphyte of Spearmint GL50 (Active Organics)
Actiphyte of Spearmint Lipo S (Active Organics)
Actiphyte of Spearmint PG50 (Active Organics)
Aromaphyte of Spearmint (Active Organics)
VT-129 Extract of Spearmint Leaves (Vege-Tech)

MENTHA VIRIDIS (SPEARMINT) LEAF JUICE

Definition: Mentha Viridis (Spearmint) Leaf Juice is the juice expressed from the leaves of *Mentha viridis*. *See "Regulatory and Ingredient Use Information," regarding the labeling names for botanical ingredients in Volume 1, Introduction, Part A.*

Chemical Class: Biological Products

Function: Skin-Conditioning Agent - Miscellaneous

MENTHA VIRIDIS (SPEARMINT) LEAF OIL

CAS No.: 8008-79-5

JPN Translation:
スペアミント油

Definition: Mentha Viridis (Spearmint) Leaf Oil is the volatile oil obtained from the dried tops and leaves of *Mentha viridis*. It consists largely of carvone. *See "Regulatory and Ingredient Use Information," regarding the labeling names for botanical ingredients in Volume 1, Introduction, Part A.*

Information Sources: BPC, 21CFR182.20, 27CFR21.65, 27CFR21.128, 27CFR21.151, FCC, JCIC, JCLS, MAR, MI-13(6876), NF XV, NFJ, RIFM, TSCA, USD

Chemical Class: Essential Oils

Function: Fragrance Ingredient

Reported Product Categories: Dentifrices (Aerosol, Liquid, Pastes and Powders); Mouthwashes and Breath Fresheners (Liquids and Sprays); Bath Preparations, Misc.; Shampoos (Non-coloring)

Technical/Other Names:
Mentha Spicata Oil
Oil of Spearmint
Oils, Spearmint
Spearment oil, Scotch (RIFM)
Spearmint Leaf Oil
Spearmint oil (RIFM)
Spearmint Oil

Trade Name:
AEC Spearmint Oil (A & E Connock)

Trade Name Mixture:
Aromaphyte of Spearmint (Active Organics)

MENTHOL

CAS Nos.	EINECS Nos.
89-78-1	201-939-0
1490-04-6	216-074-4

JPN Translation:
メントール

Empirical Formula:
$C_{10}H_{20}O$

Definition: Menthol is the diterpene that conforms to the formula:

In the United States, Menthol may be used as an active ingredient in OTC drug products. When used as an active drug ingredient, the established name is *Menthol. See "Regulatory and Ingredient Use Information,"* regarding the labeling names for U.S. OTC Drug Ingredients in Volume 1, Introduction, Part A.

Information Sources: ARG, AUS, BAN, BP, BPC, BRA, 21CFR172.515, 21CFR182.20, 21CFR310.545, 21CFR341.14, 21CFR346.16, 21CFR582.20, 27CFR21.65, 27CFR21.151, CTFA S, CZE, DA, DDR, EGY, FCC, FIN, HUN, IND, INN, ITA, JAN, JCLS, JSCI, MAR, MEX, MI-13(5861), NED, OTC-I-AR, OTC-I-CT, OTC-I-EA, OTC-I-OH, PF, PN, POR, RIFM, ROM, TSCA, USAN, USD, USP XXIV, USSR, YUG

Chemical Class: Alcohols

Functions: Denaturant; External Analgesic; Flavoring Agent; Fragrance Ingredient; Oral Health Care Drug

Reported Product Categories: Skin Care Preparations, Misc.; Bath Oils, Tablets, and Salts; Cleansing Products (Cold Creams, Cleansing Lotions, Liquids and Pads); After-shave Lotions; Baby Shampoos; Bath Preparations, Misc.; Body and Hand Preparations (Excluding Shaving Preparations); Paste Masks (Mud Packs); Skin Fresheners; Shaving Cream (Aerosol, Brushless and Lather); Moisturizing Preparations; Mouth-washes and Breath Fresheners (Liquids and Sprays); Tonics, Dressings, and Other Hair Grooming Aids; Foot Powders and Sprays; Personal Cleanliness Products, Misc.; Deodorants (Underarm); Eyebrow Pencils; Shampoos (Non-coloring); Suntan Gels, Creams, and Liquids; Hair Straighteners; Powders (Dusting and Talcum, Excluding Aftershave Talcs); Bath Soaps and Detergents; Colognes and Toilet Waters; Dentifrices (Aerosol, Liquid, Pastes and Powders); Face Powders; Foundations; Hair Conditioners; Preshave Lotions (All types); Shaving Preparations, Misc.; Bath Capsules; Blushers (All types); Face and Neck Preparations (Excluding Shaving Preparations); Fragrance Preparations, Misc.

Technical/Other Names:
Cyclohexanol, 5-Methyl-2-(1-Methylethyl)-
3-Hydroxy-p-menthane
2-Isopropyl-5-methylcyclohexanol
Menthol (RIFM)
d-Menthol (RIFM)
DL-Menthol
d,l-Menthol (isomer unspecified) (RIFM)
l-Menthol (RIFM)
l-Menthol
Menthol racemic (RIFM)
p-Methan-3-ol

5-Methyl-2-(1-Methylethyl)Cyclohexanol
Racemic Menthol

Trade Names:
AEC Menthol Crystals BP (A & E Connock)
AEC Menthol Liquid Synthetic (A & E Connock)
Custosense Menthol (menthol) (Custom Ingredients)
Jeen Menthol Racemic USP (Jeen)
Menthol Crystals (Dekker)
OriStar MC (Orient Stars)
Unichem MENT (Universal Preserv-A-Chem)

Trade Name Mixtures:
Cooling Mols LE (Chemyunion)
Extrapone Cooling Complex 2/B16500 (Symrise)
Extrapone Cooling Complex N 2/B16501 (Symrise)
Lipo CD-Menthol (Lipo)
Menthol Starchosomes 2/012690 (Symrise)
Pilinhib Veg (Laboratoires Serobiologiques)
Questice Plus (Quest International)

MENTHONE GLYCERIN ACETAL

CAS No.: 63187-91-7

Empirical Formula:
$C_{13}H_{24}O_3$

Definition: Menthone Glycerin Acetal is the organic compound that conforms to the formula:

Information Sources: RIFM, TSCA

Chemical Classes: Alcohols; Ethers; Heterocyclic Compounds

Functions: Flavoring Agent; Fragrance Ingredient

Technical/Other Names:
1,4-Dioxaspiro (4,5)Decane-2-Methanol 9-Methyl-6-(1-Methylethyl)-
(6-Isopropyl-9-Methyl-1,4-Dioxa-Spiro(4,5)Decy-2-yl)-Methanol
d,l-Menthone 1,2-glycerol ketal (RIFM)
l-Menthone 1,2-glycerol ketal (RIFM)

Trade Name:
Frescolat, Type MGA (PN 600 165) (Haarmann & Reimer GmbH)

MENTHOXYPROPANEDIOL

CAS No.	EINECS No.
87061-04-9	289-296-2

JPN Translation:
メントキシプロパンジオール

Empirical Formula:
$C_{13}H_{26}O_3$

Definition: Menthoxypropanediol is the organic compound that conforms to the formula:

Information Sources: JCIC, JCLS, RIFM

Chemical Class: Alcohols

Functions: Flavoring Agent; Fragrance Ingredient

Reported Product Categories: Fragrance Preparations, Misc.; Paste Masks (Mud Packs); Bath Oils, Tablets, and Salts; Cleansing Products (Cold Creams, Cleansing Lotions, Liquids and Pads)

Technical/Other Names:
3-l-Menthoxypropane-1,2-diol (RIFM)
L-Menthylglyceryl Ether
3-[[5-Methyl-2-(1-Methylethyl)Cyclohexyl]-Oxy]-1,2-Propanediol
1,2-Propanediol, 3-[[5-Methyl-2-(1-Methyl-ethyl)Cyclohexyl]Oxy]-

Trade Name:
Cooling Agent No.10 (Takasago)

MENTHYL ACETATE

CAS Nos.	EINECS Nos.
89-48-5	201-911-8
16409-45-3	240-459-6

Empirical Formula:
$C_{12}H_{22}O_2$

Definition: Menthyl Acetate is the ester of menthol and acetic acid. It conforms to the formula:

Information Sources: 21CFR172.515, 21CFR182.20, MI-13(5863), RIFM, TSCA

Chemical Class: Esters

Functions: Flavoring Agent; Fragrance Ingredient

Technical/Other Names:
Cyclohexanol, 5-Methyl-2-(1-Methylethyl)-, Acetate
Menthol Acetate
Menthyl acetate (1alpha,2beta,5alpha) (RIFM)
dl-Menthyl acetate (RIFM)
Menthyl acetate (isomer unspecified) (RIFM)
5-Methyl-2-(1-Methylethyl)Cyclohexanol Acetate

MENTHYL ANTHRANILATE

CAS No.
134-09-8

EINECS No.
205-129-8

Empirical Formula:
$C_{17}H_{25}NO_2$

Definition: Menthyl Anthranilate is the ester of menthol and o-anthranilic acid. It conforms to the formula:

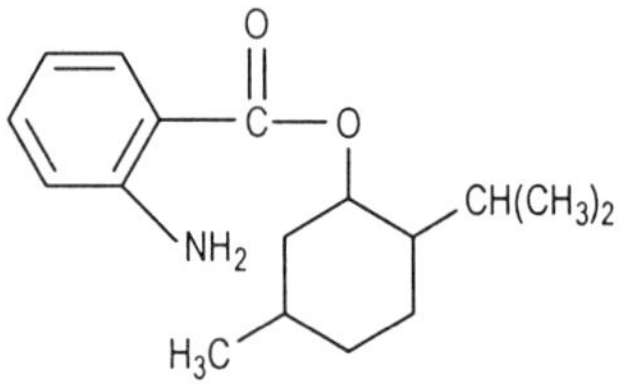

In the United States, Menthyl Anthranilate may be used as an active ingredient in OTC drug products. When used as an active drug ingredient, the established name is *Meradimate. See "Regulatory and Ingredient Use Information," regarding the labeling names for U.S. OTC Drug Ingredients in Volume 1, Introduction, Part A.*

Information Sources: 21CFR1310.02, 21CFR1310.04, OTC-I-SU, TSCA, USAN, USP XXIV

Chemical Class: Esters

Functions: Sunscreen Agent; Ultraviolet Light Absorber

Reported Product Categories: Eyebrow Pencils; Suntan Gels, Creams, and Liquids; Suntan Preparations, Misc.

Technical/Other Names:
Anthranilic Acid, p-menth-3-yl Ester
Cyclohexanol, 5-Methyl-2-(1-Methylethyl)-, 2-Aminobenzoate
Menthol, anthranilate
Menthyl o-Aminobenzoate
Meradimate
5-Methyl-2-(1-Methylethyl)Cyclohexanol-2-Aminobenzoate

Trade Names:
Dermoblock MA (Alzo)
Neo Heliopan, Type MA (Haarmann & Reimer GmbH)

MENTHYL LACTATE

CAS No.
59259-38-0

EINECS No.
261-678-3

Empirical Formula:
$C_{13}H_{24}O_3$

Definition: Menthyl Lactate is the ester of menthol and lactic acid. It conforms to the formula:

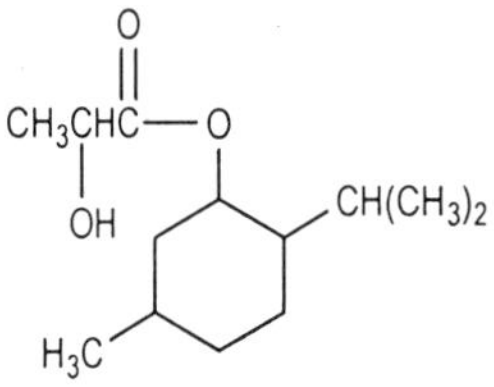

Information Source: RIFM

Chemical Class: Esters

Functions: Flavoring Agent; Fragrance Ingredient

Reported Product Categories: Aftershave Lotions; Baby Shampoos; Bath Capsules; Bath Oils, Tablets, and Salts; Body and Hand Preparations (Excluding Shaving Preparations); Cleansing Products (Cold Creams, Cleansing Lotions, Liquids and Pads); Skin Care Preparations, Misc.; Face and Neck Preparations (Excluding Shaving Preparations); Skin Fresheners

Technical/Other Names:
2-Hydroxypropanoic Acid, 5-Methyl-2-(1-Methylethyl)Cyclohexyl Ester
Lactic Acid, p-menth-3-yl Ester
l-Menthyl lactate (RIFM)
Propanoic Acid, 2-Hydroxy-, 5-Methyl-2-(1-Methylethyl)Cyclohexyl Ester

Trade Names:
AEC Menthyl Lactate (A & E Connock)
Covafresh LM (LCW)
Frescolat ML cryst (Symrise)
Frigydil (Prod'Hyg)
Koko ML (Sino Lion)

Trade Name Mixtures:
AEC Menthyl Lactate (Water Soluble) (A & E Connock)
Covafresh (LCW)
Covafresh II (LCW)

MENTHYL PCA

Empirical Formula:
$C_{15}H_{26}NO_3$

Definition: Menthyl PCA is the ester of menthol and PCA (q.v.). It conforms to the formula:

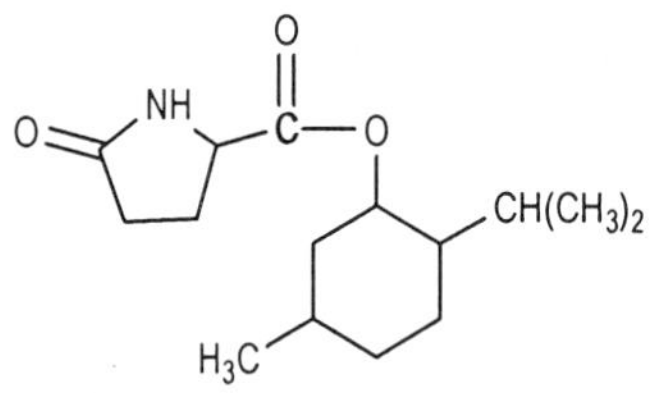

Chemical Classes: Esters; Heterocyclic Compounds

Function: Skin-Conditioning Agent - Miscellaneous

Trade Names:
Questice (Quest International)
Questice L (Quest International)

Trade Name Mixtures:
Cryolidone (UCIB (Solabia))
Questice Plus (Quest International)

MENTHYL SALICYLATE

CAS Nos.
89-46-3
109423-22-5

EINECS No.
201-909-7

Empirical Formula:
$C_{17}H_{24}O_3$

Definition: Menthyl Salicylate is the ester of menthol and salicylic acid. It conforms to the formula:

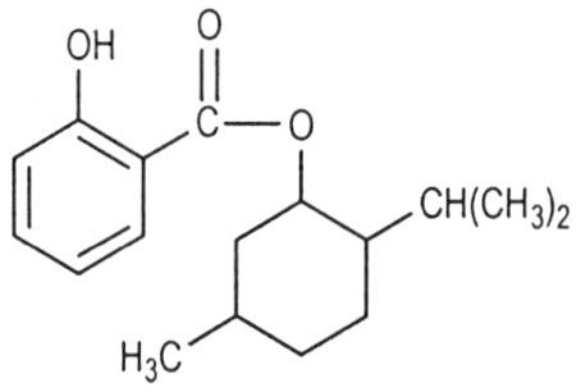

Information Sources: MAR, MI-13(5865)

Chemical Classes: Esters; Phenols

Functions: Flavoring Agent; Fragrance Ingredient; Ultraviolet Light Absorber

Technical/Other Names:
Benzoic Acid, 2-Hydroxy-, 5-Methyl-2-(1-Methylethyl)Cyclohexyl Ester
2-Hydroxybenzoic Acid, 5-Methyl-2-(1-Methylethyl)Cyclohexyl Ester
Methanol, Salicylate
5-Methyl-2-(1-Methylethyl)Cyclohexyl 2-Hydroxybenzoate
Salicylic Acid, p-menth-3-yl Ester

MENYANTHES TRIFOLIATA LEAF EXTRACT

CAS No.
84082-63-3

EINECS No.
282-009-1

Definition: Menyanthes Trifoliata Leaf Extract is an extract of the dried leaves of the buckbean, *Menyanthes trifoliata. See "Regulatory and Ingredient Use Information," regarding the labeling names for botanical ingredients in Volume 1, Introduction, Part A.*

Information Source: 21CFR172.510

Chemical Class: Biological Products

Function: Not Reported

Technical/Other Names:
Buckbean Extract
Buckbean (Menyanthes Trifoliata) Extract
Extract of Buckbean
Extract of Menyanthes Trifoliata

Trade Name Mixtures:
Buckbean Extract HG (Provital/
Centerchem)
Herbasol-Extract Buckbean
(Cosmetochem)

MERCAPTOPROPIONIC ACID

CAS Nos.	**EINECS No.**
107-96-0	203-537-0
30232-12-3	

Empirical Formula:
$C_3H_6O_2S$

Definition: Mercaptopropionic Acid is the organic acid that conforms to the formula:

$$HSCH_2CH_2COOH$$

Information Source: TSCA

Chemical Classes: Carboxylic Acids; Thio Compounds

Functions: Depilating Agent; Hair-Waving/ Straightening Agent; Reducing Agent

Reported Product Category: Hair Dyes and Colors (All Types Requiring Caution Statements and Patch Tests)

Technical/Other Names:
β-Mercaptopropionic Acid
3-Mercaptopropionic Acid
Propanoic Acid, 3-Mercapto-
3-Thiopropanoic Acid
3-Thiopropionic Acid

Trade Name Mixture:
Nonychosine E (Exsymol)

MERCURIALIS ANNUA

Definition: *See "Regulatory and Ingredient Use Information," regarding EU labeling names for botanical ingredients in Volume 1, Introduction, Part A.*

Chemical Class: Biological Products

Technical/Other Name:
Mercurialis Annua Extract (U.S.)

MERCURIALIS ANNUA EXTRACT

CAS No.	**EINECS No.**
90064-02-1	290-062-7

Definition: Mercurialis Annua Extract is an extract of the herb of the mercurialis, *Mercurialis annua. See "Regulatory and Ingredient Use Information," regarding the labeling names for botanical ingredients in Volume 1, Introduction, Part A.*

Chemical Class: Biological Products

Function: Not Reported

Technical/Other Names:
Extract of Mercurialis Annua
Mercurialis Annua (EU)
Mercurialis Extract

MERCURIALIS PERENNIS

Definition: *See "Regulatory and Ingredient Use Information," regarding EU labeling names for botanical ingredients in Volume 1, Introduction, Part A.*

Chemical Class: Biological Products

Technical/Other Name:
Mercurialis Perennis Extract (U.S.)

MERCURIALIS PERENNIS EXTRACT

CAS No.	**EINECS No.**
90064-03-2	290-063-2

Definition: Mercurialis Perennis Extract is an extract of the herb of the mercurialis, *Mercurialis perennis. See "Regulatory and Ingredient Use Information," regarding the labeling names for botanical ingredients in Volume 1, Introduction, Part A.*

Chemical Class: Biological Products

Function: Not Reported

Technical/Other Names:
Extract of Mercurialis Perennis
Mercurialis Extract
Mercurialis Perennis (EU)

Trade Name Mixture:
Herbasol-Extract Mercury (Cosmetochem)

MERCURIC OXIDE

CAS No.	**EINECS No.**
21908-53-2	244-654-7

Empirical Formula:
HgO

Definition: Mercuric Oxide is the inorganic oxide that conforms to the formula:

$$HgO$$

Information Sources: ARG, BPC, BRA, CI 77760, CZE, DDR, EEC(II-221), EGY, HUN, IND, ITA, KOR, MAR, MEX, MHLW-331/1, MI-13(5908), MI-13(5909), NF XIII, POR, TSCA, USSR, WHO, YUG

Chemical Class: Inorganics

Function: Cosmetic Biocide

Technical/Other Names:
Mercury Oxide
Red Mercuric Oxide
Yellow Mercuric Oxide

MEROXAPOL 105

CAS No.: 9003-11-6 (Generic)

Definition: Meroxapol 105 is the polyoxypropylene, polyoxyethylene block polymer that conforms generally to the formula:

$$HO(CHCH_2O)_x(CH_2CH_2O)_y(CH_2CHO)_zH$$
$$\quad\quad |\quad\quad\quad\quad\quad\quad\quad\quad\quad\quad |$$
$$\quad\quad CH_3\quad\quad\quad\quad\quad\quad\quad\quad CH_3$$

in which the average values of x, y and z are respectively 7, 22 and 7.

Information Sources: 21CFR172.808, 21CFR173.340, 21CFR177.1680, 21CFR178.1010, MI-13(7644), TSCA

Chemical Class: Polymeric Ethers

Functions: Surfactant - Emulsifying Agent; Surfactant - Solubilizing Agent

Trade Name:
Pluronic 10 R5 (BASF)

MEROXAPOL 108

CAS No.: 9003-11-6 (Generic)

Definition: Meroxapol 108 is the polyoxypropylene, polyoxyethylene block polymer that conforms generally to the formula:

$$HO(CHCH_2O)_x(CH_2CH_2O)_y(CH_2CHO)_zH$$
$$\quad\quad |\quad\quad\quad\quad\quad\quad\quad\quad\quad\quad |$$
$$\quad\quad CH_3\quad\quad\quad\quad\quad\quad\quad\quad CH_3$$

in which the average values of x, y and z are respectively 7, 91 and 7.

Information Source: TSCA

Chemical Class: Polymeric Ethers

Functions: Surfactant - Cleansing Agent; Surfactant - Emulsifying Agent; Surfactant - Solubilizing Agent

Technical/Other Names:
Ethylene Glycol-propylene Glycol Block
Copolymer
Ethylene Oxide-propylene Oxide Block
Polymer
Methyloxirane-oxirane Block Copolymer
Oxyethylene-oxypropylene Block
Copolymer
Polyoxyethylene-polyoxypropylene Block

MEROXAPOL 171

CAS No.: 9003-11-6 (Generic)

Definition: Meroxapol 171 is the polyoxypropylene, polyoxyethylene block polymer that conforms generally to the formula:

$$HO(CHCH_2O)_x(CH_2CH_2O)_y(CH_2CHO)_zH$$
$$\qquad\quad CH_3 \qquad\qquad\qquad\qquad CH_3$$

in which the average values of x, y and z are respectively 12, 4 and 12.

Information Source: TSCA

Chemical Class: Polymeric Ethers

Function: Surfactant - Solubilizing Agent

MEROXAPOL 172

CAS No.: 9003-11-6 (Generic)

Definition: Meroxapol 172 is the polyoxypropylene, polyoxyethylene block polymer that conforms generally to the formula:

$$HO(CHCH_2O)_x(CH_2CH_2O)_y(CH_2CHO)_zH$$
$$\qquad\quad CH_3 \qquad\qquad\qquad\qquad CH_3$$

in which the average values of x, y and z are respectively 12, 9 and 12.

Information Source: TSCA

Chemical Class: Polymeric Ethers

Function: Surfactant - Solubilizing Agent

Trade Name:
Pluronic 17 R2 (BASF)

MEROXAPOL 174

CAS No.: 9003-11-6 (Generic)

Definition: Meroxapol 174 is the polyoxypropylene, polyoxyethylene block polymer that conforms generally to the formula:

$$HO(CHCH_2O)_x(CH_2CH_2O)_y(CH_2CHO)_zH$$
$$\qquad\quad CH_3 \qquad\qquad\qquad\qquad CH_3$$

in which the average values of x, y and z are respectively 12, 23 and 12.

Information Source: TSCA

Chemical Class: Polymeric Ethers

Functions: Surfactant - Cleansing Agent; Surfactant - Emulsifying Agent; Surfactant - Solubilizing Agent

Trade Name:
Pluronic 17 R4 (BASF)

MEROXAPOL 178

CAS No.: 9003-11-6 (Generic)

Definition: Meroxapol 178 is the polyoxypropylene, polyoxyethylene block polymer that conforms generally to the formula:

$$HO(CHCH_2O)_x(CH_2CH_2O)_y(CH_2CHO)_zH$$
$$\qquad\quad CH_3 \qquad\qquad\qquad\qquad CH_3$$

in which the average values of x, y and z are respectively 12, 136 and 12.

Information Source: TSCA

Chemical Class: Polymeric Ethers

Functions: Surfactant - Cleansing Agent; Surfactant - Solubilizing Agent

MEROXAPOL 251

CAS No.: 9003-11-6 (Generic)

Definition: Meroxapol 251 is the polyoxypropylene, polyoxyethylene block polymer that conforms generally to the formula:

$$HO(CHCH_2O)_x(CH_2CH_2O)_y(CH_2CHO)_zH$$
$$\qquad\quad CH_3 \qquad\qquad\qquad\qquad CH_3$$

in which the average values of x, y and z are respectively 18, 6 and 18.

Information Source: TSCA

Chemical Class: Polymeric Ethers

Functions: Surfactant - Emulsifying Agent; Surfactant - Solubilizing Agent

MEROXAPOL 252

CAS Nos.: 9003-11-6 (Generic); 106392-12-5

Definition: Meroxapol 252 is the polyoxypropylene, polyoxyethylene block polymer that conforms generally to the formula:

$$HO(CHCH_2O)_x(CH_2CH_2O)_y(CH_2CHO)_zH$$
$$\qquad\quad CH_3 \qquad\qquad\qquad\qquad CH_3$$

in which the average values of x, y and z are respectively 18, 14 and 18.

Information Source: TSCA

Chemical Class: Polymeric Ethers

Functions: Surfactant - Emulsifying Agent; Surfactant - Solubilizing Agent

Technical/Other Names:
Ethylene Oxide-propylene Oxide Block Polymer
Methyloxirane-oxirane Block Copolymer
Polyoxyethylene-polyoxypropylene Block Copolymer

Trade Name:
Pluronic 25 R2 (BASF)

MEROXAPOL 254

CAS No.: 9003-11-6 (Generic)

Definition: Meroxapol 254 is the polyoxypropylene, polyoxyethylene block polymer that conforms generally to the formula:

$$HO(CHCH_2O)_x(CH_2CH_2O)_y(CH_2CHO)_zH$$
$$\qquad\quad CH_3 \qquad\qquad\qquad\qquad CH_3$$

in which the average values of x, y and z are respectively 18, 34 and 18.

Information Source: TSCA

Chemical Class: Polymeric Ethers

Functions: Surfactant - Cleansing Agent; Surfactant - Solubilizing Agent

Trade Name:
Pluronic 25 R4 (BASF)

MEROXAPOL 255

CAS No.: 9003-11-6 (Generic)

Definition: Meroxapol 255 is the polyoxypropylene, polyoxyethylene block polymer that conforms generally to the formula:

$$HO(CHCH_2O)_x(CH_2CH_2O)_y(CH_2CHO)_zH$$
$$\qquad\quad CH_3 \qquad\qquad\qquad\qquad CH_3$$

in which the average values of x, y and z are respectively 18, 51 and 18.

Information Source: TSCA

Chemical Class: Polymeric Ethers

Functions: Surfactant - Cleansing Agent; Surfactant - Solubilizing Agent

MEROXAPOL 258

CAS Nos.: 9003-11-6 (Generic); 106392-12-5

Definition: Meroxapol 258 is the polyoxypropylene, polyoxyethylene block polymer that conforms generally to the formula:

$$HO(CHCH_2O)_x(CH_2CH_2O)_y(CH_2CHO)_zH$$
$$\qquad\quad CH_3 \qquad\qquad\qquad\qquad CH_3$$

in which the average values of x, y and z are respectively 18, 163 and 18.

Information Source: TSCA

Chemical Class: Polymeric Ethers

Functions: Surfactant - Cleansing Agent; Surfactant - Solubilizing Agent

Technical/Other Names:
Ethylene Oxide-propylene Oxide Block Polymer

Methyloxirane-oxirane Block Copolymer
Oxyethylene-oxypropylene Block Polymer
Polyoxyethylebe-polyoxypropylene Block
Copolymer

MEROXAPOL 311

CAS No.: 9003-11-6 (Generic)

Definition: Meroxapol 311 is the polyoxy-propylene, polyoxyethylene block polymer that conforms generally to the formula:

$$HO(CHCH_2O)_x(CH_2CH_2O)_y(CH_2CHO)_zH$$
$$\quad\quad CH_3 \quad\quad\quad\quad\quad\quad\quad CH_3$$

in which the average values of x, y and z are respectively 21, 7 and 21.

Information Source: TSCA

Chemical Class: Polymeric Ethers

Functions: Surfactant - Emulsifying Agent; Surfactant - Solubilizing Agent

Trade Names:
Ethox 31-R-1 (Ethox)
Pluronic 31 R1 (BASF)

MEROXAPOL 312

CAS No.: 9003-11-6 (Generic)

Definition: Meroxapol 312 is the polyoxy-propylene, polyoxyethylene block polymer that conforms generally to the formula:

$$HO(CHCH_2O)_x(CH_2CH_2O)_y(CH_2CHO)_zH$$
$$\quad\quad CH_3 \quad\quad\quad\quad\quad\quad\quad CH_3$$

in which the average values of x, y and z are respectively 21, 15, and 21.

Information Source: TSCA

Chemical Class: Polymeric Ethers

Functions: Surfactant - Emulsifying Agent; Surfactant - Solubilizing Agent

MEROXAPOL 314

CAS No.: 9003-11-6 (Generic)

Definition: Meroxapol 314 is the polyoxy-propylene, polyoxyethylene block polymer that conforms generally to the formula:

$$HO(CHCH_2O)_x(CH_2CH_2O)_y(CH_2CHO)_zH$$
$$\quad\quad CH_3 \quad\quad\quad\quad\quad\quad\quad CH_3$$

in which the average values of x, y and z are respectively 21, 39 and 21.

Information Source: TSCA

Chemical Class: Polymeric Ethers

Functions: Surfactant - Cleansing Agent; Surfactant - Solubilizing Agent

MESOPHYLLUM LICHENOIDES EXTRACT

Definition: Mesophyllum Lichenoides Extract is an extract of the alga, *Mesophyllum lichenoides. See "Regulatory and Ingredient Use Information," regarding the labeling names for botanical ingredients in Volume 1, Introduction, Part A.*

Chemical Class: Biological Products

Function: Skin-Conditioning Agent - Miscellaneous

Technical/Other Name:
Extract of Mesophyllm Lichenoides

Trade Name Mixture:
Extrait de Mesophyllum (Codif)

MESUA FERREA SEED EXTRACT

Definition: Mesua Ferrea Seed Extract is an extract of the seeds of *Mesua ferrea. See "Regulatory and Ingredient Use Information," regarding the labeling names for botanical ingredients in Volume 1, Introduction, Part A.*

Chemical Class: Biological Products

Function: Not Reported

Technical/Other Name:
Extract of Mesua Ferrea Seed

Trade Name Mixture:
Nagasari Seed Extract (Yamakawa)

METAPHOSPHORIC ACID

CAS Nos.	EINECS Nos.
10343-62-1	233-750-4
37267-86-0	253-433-4

Definition: Metaphosphoric Acid is the inorganic acid that conforms to the formula:

$$(HPO_3)_n$$

Information Source: MI-13(7431)

Chemical Classes: Inorganic Acids; Phosphorus Compounds

Function: pH Adjuster

Technical/Other Names:
Phosphenic Acid
Phosphoric Acid, Meta

METHACRYLIC ACID

CAS No.	EINECS No.
79-41-4	201-204-4

Empirical Formula:
$C_4H_6O_2$

Definition: Methacrylic Acid is the organic compound that conforms to the formula:

Information Sources: CIR: [SQ], MI-13 (5967)

Chemical Class: Carboxylic Acids

Function: Artificial Nail Builder

Technical/Other Name:
2-Propenoic Acid, 2-Methyl-

METHACRYLIC ACID/SODIUM ACRYLAMIDOMETHYL PROPANE SULFONATE COPOLYMER

Definition: Methacrylic Acid/Sodium Acrylamidomethyl Propane Sulfonate Copolymer is a copolymer of methacrylic acid and sodium acrylamidomethyl propane sulfonate monomers.

Chemical Class: Synthetic Polymers

Functions: Film Former; Hair Fixative; Hair-Waving/Straightening Agent

Trade Names:
Fixomer A-30 (ONDEO Nalco)
Fixomer N-28 (ONDEO Nalco)

METHACRYLOYL ETHYL BETAINE/ ACRYLATES COPOLYMER

JPN Translation:
（ メタクリロイルオキシエチルカルボキシ
ベタイン / メタクリル酸アルキル ）コポ
リマー

Definition: Methacryloyl Ethyl Betaine/ Acrylates Copolymer is a polymer of methacryloyl ethyl betaine and two or more monomers of methacrylic acid or its simple esters.

Information Source: CIR: [SQ] IJT 21 (SUPPL. 3) 2002

Chemical Class: Synthetic Polymers

Functions: Film Former; Hair Fixative; Suspending Agent - Nonsurfactant

Technical/Other Name:
Methacryloyl Ethyl Betaine/Methacrylates Copolymer

Trade Names:
Diaformer Z-301 (Clariant)

The inclusion of any compound in the Dictionary and Handbook does not indicate that use of that substance as a cosmetic ingredient complies with the laws and regulations governing such use in the United States or any other country.

Diaformer Z-301 (Clariant GmbH, Personal Care)
Diaformer Z-400 (Clariant)
Diaformer Z-400 (Clariant GmbH, Personal Care)
Diaformer Z-AT (Clariant)
Diaformer Z-AT (Clariant GmbH, Personal Care)
Diaformer Z-SM (Clariant)
Diaformer Z-SM (Clariant GmbH, Personal Care)
Diaformer Z-W (Clariant)
Diaformer Z-W (Clariant GmbH, Personal Care)
Yukaformer 202 (Mitsubishi Petrochemical)
Yukaformer 204 (Mitsubishi Petrochemical)
Yukaformer 206 (Mitsubishi Petrochemical)
Yukaformer 301 (Mitsubishi Petrochemical)
Yukaformer Amphoset (Mitsubishi Petro-chemical)
Yukaformer 104-D (Mitsubishi Petro-chemical)
Yukaformer 105-D (Mitsubishi Petro-chemical)
Yukaformer FH (Mitsubishi Petrochemical)
Yukaformer M-75 (Mitsubishi Petro-chemical)
Yukaformer R 102 (Mitsubishi Petro-chemical)
Yukaformer R205 (Mitsubishi Petro-chemical)
Yukaformer R402 (Mitsubishi Petro-chemical)
Yukaformer RFN (Mitsubishi Petrochemical)
Yukaformer R205S (Mitsubishi Petro-chemical)
Yukaformer 202S (Mitsubishi Petro-chemical)
Yukaformer SM (Mitsubishi Petrochemical)
Yukaformer SMP (Mitsubishi Petrochemical)
Yukaformer 204WL (Mitsubishi Petro-chemical)
Yukaformer 204WL2 (Mitsubishi Petro-chemical)
Yukaformer WPS (Mitsubishi Petro-chemical)
Yukaformer WPS-D (Mitsubishi Petro-chemical)
Yukaformer WPS-DD (Mitsubishi Petro-chemical)

METHACRYLOYL PROPYLTRIMETHOXY-SILANE

CAS No.	EINECS No.
2530-85-0	219-785-8

Empirical Formula:
$C_{10}H_{20}O_5Si$

Definition: Methacryloyl Propyltrimethoxy-silane is the organic compound that conforms to the formula:

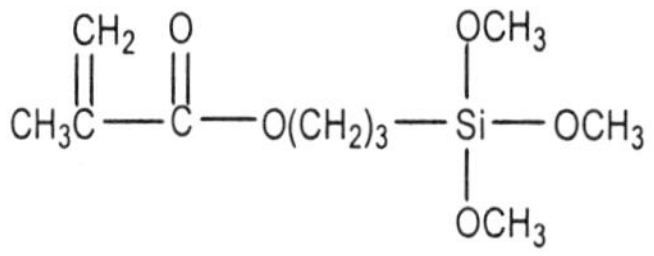

Chemical Classes: Esters; Siloxanes and Silanes

Functions: Binder; Film Former; Surface Modifier

Technical/Other Names:
2-Propenoic Acid, 2-Methyl-, 3-(Trimethoxysilyl)Propyl Ester
3-(Trimethoxysilyl)Propyl Methacrylate

Trade Name Mixture:
COSMO S-40SB (Catalysts & Chemicals)

METHENAMINE

CAS No.	EINECS No.
100-97-0	202-905-8

Empirical Formula:
$C_6H_{12}N_4$

Definition: Methenamine is the organic amine that conforms to the formula:

Information Sources: AUS, BRA, 21CFR175.105, 21CFR176.180, 21CFR177.2410, 21CFR181.30, 21CFR310.545, CIR: [SQ] JACT-11(4)1992, CZE, DA, DDR, EEC(VI/1-30), EGY, FIN, HUN, INN, MI-13(5994), PF, POR, TSCA, USAN, USD, USP XXIV, YUG

Chemical Class: Amines

Function: Cosmetic Biocide

Reported Product Categories: Shampoos (Non-coloring); Hair Conditioners

Technical/Other Names:
Aminoform
Formamine
Hexamethyleneamine
Hexamethylene Tetramine
Methenamide
Methenamin
1,3,5,7-Tetraazatricyclo[3,3,1,1]Decane

Trade Name Mixture:
Elestab 48 (Laboratoires Serobiologiques)

METHENAMMONIUM CHLORIDE

Empirical Formula:
$C_7H_{15}N_4 \cdot Cl$

Definition: Methenammonium Chloride is the heterocyclic compound which conforms to the formula:

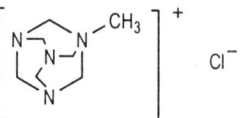

Chemical Class: Amines

Function: Cosmetic Biocide

Trade Name:
Busan 1500 (Buckman Labs)

METHICONE

CAS Nos.: 9004-73-3; 63148-57-2

JPN Translation:
メチコン

Empirical Formula:
$(CH_4OSi)_x$

Definition: Methicone is a linear monomethyl polysiloxane. It conforms generally to the formula:

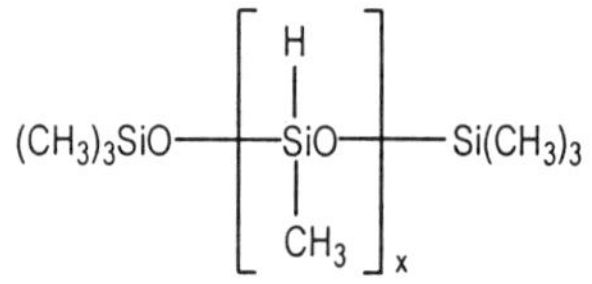

Information Sources: 21CFR175.105, 21CFR175.300, 21CFR176.130, 21CFR177.1200, CIR: [S], JCIC, JCLS, JSQI, TSCA

Chemical Class: Siloxanes and Silanes

Functions: Skin-Conditioning Agent - Occlusive; Surface Modifier

Reported Product Categories: Lipsticks; Eye Shadows; Foundations; Face Powders; Makeup Preparations (Not eye), Misc.; Blushers (All types); Eyeliners; Rouges; Mascara; Makeup Fixatives; Eyebrow Pencils; Makeup Bases; Moisturizing Preparations; Eye Makeup Preparations, Misc.; Powders (Dusting and Talcum, Excluding Aftershave Talcs); Eye Makeup Removers; Hair Coloring Preparations, Misc.; Nail Polish and Enamels

Technical/Other Names:
Hydrogen Methyl Polysiloxane
Methyl Hydrogen Polysiloxane
Poly[Oxy(Methylsilylene)]
Siloxanes and Silicones, Me Hydrogen

Trade Names:
Dow Corning 1107 Fluid (Dow Corning)
F-9W-9 (Shin Etsu)

Trade Name Mixtures:
A302.30 Tudor Harebell Hydrophobic (Kingfisher Colours)
A303.30 Tudor Violet Hydrophobic (Kingfisher Colours)
A304.30 Tudor Lavender Hydrophobic (Kingfisher Colours)

A305.30 Tudor Mallow Hydrophobic
  (Kingfisher Colours)
A306.30 Tudor Mint Hydrophobic
  (Kingfisher Colours)
A307.30 Tudor Sage Hydrophobic
  (Kingfisher Colours)
A310.30 Tudor Aspen Hydrophobic
  (Kingfisher Colours)
A310.35 Tudor Aspen (Kingfisher Colours)
A311.30 Tudor Ash Hydrophobic
  (Kingfisher Colours)
A401.30 Tudor Ebony Hydrophobic
  (Kingfisher Colours)
A402.30 Tudor Rosewood Hydrophobic
  (Kingfisher Colours)
A403.30 Tudor Oak Hydrophobic
  (Kingfisher Colours)
A405.30 Tudor Chestnut Hydrophobic
  (Kingfisher Colours)
A406.30 Tudor Mahogany Hydrophobic
  (Kingfisher Colours)
A407.30 Tudor Willow Hydrophobic
  (Kingfisher Colours)
A408.30 Tudor Walnut Hydrophobic
  (Kingfisher Colours)
Black Iron Oxide SI 2 (Cardre)
Chromium Hydroxide SI 2 (Cardre)
Chromium Oxide Green SI 2 (Cardre)
COULOURMAT BLACK SIL 45%
  (Creations Couleurs)
COULOURMAT RED SIL 45% (Creations
  Couleurs)
COULOURMAT WHITE R SIL 45%
  (Creations Couleurs)
COULOURMAT YELLOW SIL 45%
  (Creations Couleurs)
Hydrophobic Black Oxide C9333 (LCW)
Hydrophobic Brown Oxide C9458 (LCW)
Hydrophobic Chromium Oxide C9409
  (LCW)
Hydrophobic Kaolin C9400 (LCW)
Hydrophobic Manganese Violet C9401
  (LCW)
Hydrophobic Red Oxide C9454 (LCW)
Hydrophobic Talc C9441 (LCW)
Hydrophobic TiO2 C9428 (LCW)
Hydrophobic Ultramarine Blue C9404
  (LCW)
Hydrophobic Ultra Violet C9402 (LCW)
Hydrophobic Yellow C9455 (LCW)
Manganese Violet SI 2 (Cardre)
Mica SI 2 (Cardre)
Micro-Ace P-2-030 (Presperse)
Red Iron Oxide SI 2 (Cardre)
SAS-TTO-S-3 (16%) (US Cosmetics)
SAS-TTO-S-3/D5 (50%) (US Cosmetics)
Sericite SI 2 (Cardre)
Sericite SL-012 (Presperse)
Sericite SLZ-012P (Presperse)
SI-TTO-S-1 (LHC) (US Cosmetics)
SNI-PT-46 (US Cosmetics)
Talc SI 2 (Cardre)
Titanium Dioxide SI 2 (Cardre)
Toshiki Black AS-61D (Nikko)
Toshiki Mica GS-61D (Nikko)

Toshiki Red AS-61D (Nikko)
Toshiki Sericite JSF-25-3 (Nikko)
Toshiki Sericite OS-61D (Nikko)
Toshiki Talc LS-61D (Nikko)
Toshiki Talc LSF-25-3 (Nikko)
Toshiki TiO2 AS-61D (Nikko)
Toshiki TiO2 ASF-25-3 (Nikko)
Toshiki Yellow BS-61D (Nikko)
TZ-Powder Type 1 (EU) (US Cosmetics)
TZ-Powder Type 2 (US Cosmetics)
Ultracolor Soft Focus (Ultra Chemical)
Ultrafine Titanium Dioxide SI 2 (Cardre)
Ultramarine Blue SI 2 (Cardre)
Yellow Iron Oxide SI 2 (Cardre)

METHIONINE

CAS Nos.
59-51-8
63-68-3

EINECS Nos.
200-432-1
200-562-9

JPN Translation:
メチオニン

Empirical Formula:
$C_5H_{11}NO_2S$

Definition: Methionine is the amino acid that conforms to the formula:

$$CH_3SCH_2CH_2\underset{\underset{NH_2}{|}}{C}HCOOH$$

Information Sources: AUS, BRA, 21CFR172.320, 21CFR310.545, 21CFR573.870, 21CFR573.940, 21CFR582.5475, 21CFR582.5477, CZE, DA, DDR, FCC, INN, ITA, JAN, JCLS, JSCI, MAR, MI-13(6004), RIFM, TSCA, USAN, USD, USP XXIV

Chemical Classes: Amino Acids; Thio Compounds

Functions: Fragrance Ingredient; Hair Conditioning Agent; Skin-Conditioning Agent - Miscellaneous

Reported Product Categories: Hair Conditioners; Permanent Waves; Shampoos (Non-coloring); Tonics, Dressings, and Other Hair Grooming Aids; Skin Care Preparations, Misc.; Body and Hand Preparations (Excluding Shaving Preparations)

Technical/Other Names:
α-Amino-γ-methylmercaptobutyric Acid
2-Amino-4-(methylthio)butyric Acid
L-Homocysterine, S-methyl-
D,L-Methionine (RIFM)
DL-Methionine
L-Methionine

Trade Name:
AEC Methionine (A & E Connock)

Trade Name Mixtures:
Aminodermin CLR (CLR)

Erase (Degussa Care Specialties)
Hair Care Phytoamine Biocomplex (Alban
  Muller)
Hair Complex 20/70 n (CLR)
Nonychosine V (Exsymol)
Phosphovital (Sederma)

METHOXY AMODIMETHICONE/ SILSESQUIOXANE COPOLYMER

Definition: Methoxy Amodimethicone/ Silsesquioxane Copolymer is a copolymer of methyl trimethoxysilane and methoxy amodimethicone monomers.

Chemical Classes: Siloxanes and Silanes; Synthetic Polymers

Functions: Film Former; Hair Conditioning Agent; Hair Fixative; Skin-Conditioning Agent - Emollient

Trade Name:
SF1706 (GE Silicones)

3-METHOXYBUTANOL

CAS No.
2517-43-3

EINECS No.
219-741-8

Empirical Formula:
$C_5H_{12}O_2$

Definition: 3-Methoxybutanol is the organic compound that conforms to the formula:

$$HOCH_2CH_2\underset{\underset{OCH_3}{|}}{C}HCH_3$$

Information Source: TSCA

Chemical Classes: Alcohols; Ethers

Function: Solvent

Technical/Other Name:
1-Butanol, 3-Methoxy-

Trade Name:
Methoxybutanol (Schwarzkopf GmbH)

METHOXYCINNAMIDOPROPYL HYDROXYSULTAINE

Empirical Formula:
$C_{18}H_{28}O_6N_2S$

Definition: Methoxycinnamidopropyl Hydroxysultaine is the zwitterion (inner salt) that conforms to the formula:

Chemical Class: Betaines

Functions: Hair Conditioning Agent; Ultraviolet Light Absorber

Trade Name:
Galaxy - SunBeat (Galaxy Surfactants)

METHOXYCINNAMIDOPROPYL LAURDIMONIUM TOSYLATE

Empirical Formula:
$C_{34}H_{54}N_2O_5S$

Definition: Methoxycinnamidopropyl Laurdimonium Tosylate is the quaternary ammonium salt that conforms to the formula:

Chemical Class: Quaternary Ammonium Compounds

Functions: Hair Conditioning Agent; Ultraviolet Light Absorber

Technical/Other Name:
p-Methoxy Cinnamidopropyldimethyl-laurylammonium Tosylate

Trade Name:
Galaxy - TosyQuat (Galaxy Surfactants)

METHOXYDIGLYCOL

CAS No.   **EINECS No.**
111-77-3   203-906-6

Empirical Formula:
$C_5H_{12}O_3$

Definition: Methoxydiglycol is the aliphatic ether alcohol that conforms to the formula:

$$CH_3OCH_2CH_2OCH_2CH_2OH$$

Information Sources: 21CFR175.105, CTFA D, MI-13(6064), RIFM, TSCA

Chemical Class: Ethers

Functions: Fragrance Ingredient; Solvent; Viscosity Decreasing Agent

Technical/Other Names:
Diethylene glycol monomethyl ether (RIFM)
Diethylene Glycol Monomethyl Ether
Diglycol Monomethyl Ether
3,6-Dioxa-1-heptanol
Ethanol, 2-(2-Methoxyethoxy)-
2-(2-Methoxyethoxy)Ethanol
Methyl Dioxitol

Trade Names:
DOWANOL DM (Dow Chemical)
Eastman DM Solvent (Eastman Chemical)
Hisolve DM (Toho)
Methyl CARBITOL Solvent (Dow Chemical)

Trade Name Mixture:
Lowenol Solvent 402 (Lowenstein)

METHOXYDIGLYCOL METHACRYLATE

CAS No.   **EINECS No.**
45103-58-0   256-190-2

Empirical Formula:
$C_9H_{16}O_4$

Definition: Methoxydiglycol Methacrylate is the organic compound that conforms to the formula:

Information Source: CIR: [SQ]

Chemical Class: Esters

Function: Artificial Nail Builder

Technical/Other Name:
2-Propenoic Acid, 2-Methyl-, 2-(2-Methoxyethoxy)Ethyl Ester

METHOXYDIMETHICONE/TITANATE CROSSPOLYMER

Definition: Methoxydimethicone/Titanate Crosspolymer is the crosslinked polymer formed by the reaction of titanium tetraisopropoxide and methoxy dimethicone.

Chemical Class: Synthetic Polymers

Function: Bulking Agent

Trade Name:
TS Hybrid Powder (KOSE)

METHOXYETHANOL

CAS Nos.   **EINECS Nos.**
109-86-4   203-713-7
32718-54-0   251-174-1

JPN Translation:
メトキシエタノール

Empirical Formula:
$C_3H_8O_2$

Definition: Methoxyethanol is the aliphatic ether alcohol that conforms to the formula:

$$CH_3OCH_2CH_2OH$$

Information Sources: 21CFR175.105, 21CFR449.550b, CTFA D, JCIC, JCLS, JSQI, MI-13(6066), RIFM

Chemical Classes: Alcohols; Ethers

Functions: Fragrance Ingredient; Solvent; Viscosity Decreasing Agent

Technical/Other Names:
Ethanol, 2-Methoxy-
Ethylene glycol methyl ether (RIFM)
Ethylene Glycol Monomethyl Ether
2-Methoxyethanol
Methoxyethylene Glycol
Methoxyhydroxyethane
Methyl Glycol
Methyl Oxitol
Monomethyl Ethylene Glycol Ether

Trade Names:
Hisolve MC (Toho)
Methylglykol (Clariant)
Methylglykol (Clariant GmbH, Personal Care)

METHOXYETHANOL ACETATE

CAS No.   **EINECS No.**
110-49-6   203-772-9

Empirical Formula:
$C_5H_{10}O_3$

Definition: Methoxyethanol Acetate is the ester of Methoxyethanol (q.v.) and Acetic Acid (q.v.). It conforms to the formula:

Information Sources: MI-13(6067), RIFM, TSCA

Chemical Classes: Esters; Ethers

Functions: Fragrance Ingredient; Solvent; Viscosity Decreasing Agent

Technical/Other Names:
Ethanol, 2-Methoxy, Acetate
Ethylene glycol methyl ether acetate (RIFM)
Ethylene Glycol Monomethyl Ether Acetate
2-Methoxyethyl Acetate
Methyl Glycol Acetate

Trade Name:
Hisolve MC Acetate (Toho)

N-METHOXYETHYL-p-PHENYLENEDIAMINE HCl

Empirical Formula:
$C_9H_{14}N_2O \cdot ClH$

Definition: N-Methoxyethyl-p-Phenylene-diamine HCl is the substituted aromatic amine salt that conforms to the formula:

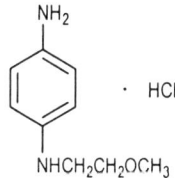

See "Regulatory and Ingredient Use Information," for Colorants in Volume 1, Introduction, Part A.

Information Source: EEC(III/1-8)

Chemical Classes: Amines; Color Additives - Hair

Function: Hair Colorant

Technical/Other Name:
1,4-Benzenediamine, N-Methoxyethyl, Hydrochloride

Trade Name:
Imexine OAH (Chimex)

METHOXYINDANE

CAS No.	EINECS No.
1006-27-5	213-743-2

Empirical Formula:
$C_{10}H_{12}O$

Definition: Methoxyindane is the aromatic ether that conforms to the formula:

Information Source: RIFM

Chemical Class: Ethers

Function: Fragrance Ingredient

Technical/Other Names:
1H-Indene, 2,3-dihydro-1-methoxy-
Indan, 1-Methoxy-
1-Methoxyindan (RIFM)
1-Methoxyindan

Trade Name:
Phloralid (Hercules/PFW)

METHOXYISOPROPANOL

CAS No.	EINECS No.
107-98-2	203-539-1

Empirical Formula:
$C_4H_{10}O_2$

Definition: Methoxyisopropanol is the aliphatic ether alcohol that conforms to the formula:

$$HOCHCH_2OCH_3$$
$$|$$
$$CH_3$$

Information Sources: 21CFR175.105, 21CFR181.22, 21CFR181.30, RIFM, TSCA

Chemical Classes: Alcohols; Ethers

Functions: Fragrance Ingredient; Solvent; Viscosity Decreasing Agent

Reported Product Categories: Nail Polish and Enamels; Nail Polish and Enamel Removers

Technical/Other Names:
1-Methoxy-2-hydroxypropane
1-Methoxypropan-2-ol (RIFM)
1-Methoxy-2-Propanol
2-Propanol, 1-Methoxy-
Propylene Glycol Monomethyl Ether

Trade Names:
DOWANOL PM (Dow Chemical)
Eastman PM Solvent (Eastman Chemical)
Hisolve MP (Toho)

METHOXYISOPROPYL ACETATE

CAS Nos.	EINECS Nos.
108-65-6	203-603-9
84540-57-8	283-152-2

Empirical Formula:
$C_6H_{12}O_3$

Definition: Methoxyisopropyl Acetate is the organic compound that conforms to the formula:

$$\overset{O}{\overset{||}{CH_3C}}-OCHCH_2OCH_3$$
$$|$$
$$CH_3$$

Chemical Classes: Esters; Ethers

Function: Solvent

Technical/Other Names:
2-Methoxy-1-Methylethyl Acetate
Propylene Glycol Monomethyl Ether Acetate

Trade Names:
Arcosolv PMA (Lyondell Chemical)
DOWANOL PMA (Dow Chemical)
Eastman PM Acetate (Eastman Chemical)
Methyl Proxitol Acetate (SHELL UK)

6-METHOXY-2-METHYLAMINO-3-AMINOPYRIDINE HCl

CAS No.: 90817-34-8

Empirical Formula:
$C_7H_{11}N_3O \cdot 2ClH$

Definition: 6-Methoxy-2-methylamino-3-aminopyridine HCL is the hair color that conforms to the formula:

See "Regulatory and Ingredient Use Information," for Colorants in Volume 1, Introduction, Part A.

Chemical Class: Color Additives - Hair

Function: Hair Colorant

Technical/Other Names:
HC Blue No. 7
6-Methoxy-N2-Methyl-2,3-Pyridinediamine
2,3-Pyridinediamine, 6-Methoxy-N2-Methyl-

2-METHOXYMETHYL-p-AMINOPHENOL HCl

CAS No.: 135043-65-1

Empirical Formula:
$C_8H_{11}NO_2 \cdot ClH$

Definition: 2-Methoxymethyl-p-Aminophenol HCl is the substituted aromatic amine that conforms to the formula:

See "Regulatory and Ingredient Use Information," for Colorants in Volume 1, Introduction, Part A.

Chemical Classes: Amines; Color Additives - Hair

Function: Hair Colorant

Technical/Other Names:
4-Amino-2-Methoxymethylphenol Hydrochloride
Phenol, 4-Amino-2-Methoxymethyl-, Hydrochloride

METHOXYMETHYLBUTANOL

CAS No.	EINECS No.
56539-66-3	260-252-4

Empirical Formula:
$C_6H_{14}O_2$

Definition: Methoxymethylbutanol is the organic compound that conforms to the formula:

$$CH_3$$
$$|$$
$$CH_3CCH_2CH_2OH$$
$$|$$
$$OCH_3$$

Information Source: TSCA

Chemical Classes: Alcohols; Ethers

Function: Solvent

Technical/Other Names:
1-Butanol, 3-Methoxy-3-Methyl-
3-Methoxy-3-Methyl-1-Butanol

METHOXY PEG-7

CAS No.: 9004-74-4 (Generic)

Definition: Methoxy PEG-7 is the ether that conforms generally to the formula:

$$H(OCH_2CH_2)_nOCH_3$$

where n has an average value of 7.

Chemical Class: Alkoxylated Alcohols

Functions: Humectant; Solvent

Technical/Other Name:
Polyethylene Glycol (7) Monomethyl Ether

Trade Names:
Polyglycol M 350 (Clariant)
Polyglycol M 350 (Clariant GmbH, Personal Care)

METHOXY PEG-10

CAS No.: 9004-74-4 (Generic)

Empirical Formula:
$C_{21}H_{44}O_{11}$

Definition: Methoxy PEG-10 is the ether that conforms generally to the formula:

$$H(OCH_2CH_2)_nOCH_3$$

where n has an average value of 10.

Information Sources: NF XVIII, USAN

Chemical Class: Alkoxylated Alcohols

Functions: Humectant; Solvent

Technical/Other Names:
Polyethylene Glycol 500 Monomethyl Ether
Polyoxyethylene (10) Monomethyl Ether

Trade Names:
CARBOWAX MPEG 550 (Dow Chemical)
Polyglycol M 500 (Clariant)
Polyglycol M 500 (Clariant GmbH, Personal Care)

METHOXY PEG-16

CAS No.: 9004-74-4 (Generic)

Definition: Methoxy PEG-16 is the ether that conforms generally to the formula:

$$H(OCH_2CH_2)_nOCH_3$$

where n has an average value of 16.

Information Sources: NF XVIII, USAN

Chemical Class: Alkoxylated Alcohols

Functions: Humectant; Solvent

Technical/Other Names:
Polyethylene Glycol (16) Monomethyl Ether
Polyoxyethylene (16) Monomethyl Ether

Trade Names:
CARBOWAX MPEG 750 (Dow Chemical)
Polyglycol M 750 (Clariant)
Polyglycol M 750 (Clariant GmbH, Personal Care)

METHOXY PEG-25

CAS No.: 9004-74-4 (Generic)

Definition: Methoxy PEG-25 is the ether that conforms generally to the formula:

$$H(OCH_2CH_2)_nOCH_3$$

where n has an average value of 25.

Chemical Class: Alkoxylated Alcohols

Functions: Humectant; Solvent

Technical/Other Name:
Polyethylene Glycol (25) Monomethyl Ether

Trade Names:
Polyglycol M 1100 (Clariant)
Polyglycol M 1100 (Clariant GmbH, Personal Care)

METHOXY PEG-40

CAS No.: 9004-74-4 (Generic)

Definition: Methoxy PEG-40 is the ether that conforms generally to the formula:

$$H(OCH_2CH_2)_nOCH_3$$

where n has an average value of 40.

Information Sources: NF XVIII, USAN

Chemical Class: Alkoxylated Alcohols

Functions: Humectant; Solvent

Technical/Other Names:
Polyethylene Glycol 2000 Monomethyl Ether
Polyoxyethylene (40) Monomethyl Ether

Trade Names:
CARBOWAX MPEG 2000 (Dow Chemical)
Polyglycol M 2000 (Clariant)
Polyglycol M 2000 (Clariant GmbH, Personal Care)

METHOXY PEG-100

CAS No.: 9004-74-4 (Generic)

Definition: Methoxy PEG-100 is the ether that conforms generally to the formula:

$$H(OCH_2CH_2)_nOCH_3$$

where n has an average value of 100.

Information Sources: NF XVIII, USAN

Chemical Class: Alkoxylated Alcohols

Functions: Humectant; Solvent

Technical/Other Names:
Polyethylene Glycol (100) Monomethyl Ether
Polyoxyethylene (100) Monomethyl Ether

Trade Names:
CARBOWAX MPEG 5000 (Dow Chemical)
Polyglycol M 5000 (Clariant)
Polyglycol M 5000 (Clariant GmbH, Personal Care)

METHOXY PEG-7 ASCORBIC ACID

Empirical Formula:
$C_{21}H_{38}O_{13}$

Definition: Methoxy PEG-7 Ascorbic Acid is the organic compound that conforms generally to the formula:

where n has an average value of 6.

Chemical Classes: Ethers; Heterocyclic Compounds; Polyols

Function: Antioxidant

Trade Name:
Medimin C (LG Cosmetic)

METHOXY PEG-17/DODECYL GLYCOL COPOLYMER

Definition: Methoxy PEG-17/Dodecyl Glycol Copolymer is the polymer that conforms generally to the formula:

$$CH_3O(CH_2CH_2O)_x(CH_2CHO)_yH$$
$$|$$
$$C_{10}H_{21}$$

where x has an average value of 17 and y has an average value of 1.

Chemical Classes: Alkoxylated Alcohols; Synthetic Polymers

Functions: Emulsion Stabilizer; Skin-Conditioning Agent - Miscellaneous;

Suspending Agent - Nonsurfactant; Viscosity Increasing Agent - Nonaqueous

Trade Names:
AEC Methoxy PEG-17/Dodecyl Glycol Copolymer (A & E Connock)
Elfacos OW-100 (Akzo Nobel)
Elfacos OW-100 (Akzo Nobel Surface AB)

METHOXY PEG-22/DODECYL GLYCOL COPOLYMER

Definition: Methoxy PEG-22/Dodecyl Glycol Copolymer is the polymer that conforms generally to the formula:

$$CH_3O(CH_2CH_2O)_x(CH_2CHO)_yH$$
$$|$$
$$C_{10}H_{21}$$

where x has an average value of 22 and y has an average value of 7.

Chemical Classes: Alkoxylated Alcohols; Synthetic Polymers

Functions: Emulsion Stabilizer; Suspending Agent - Nonsurfactant; Viscosity Increasing Agent - Aqueous; Viscosity Increasing Agent - Nonaqueous

Reported Product Category: Night Skin Care Preparations

Trade Names:
AEC Methoxy PEG-22/Dodecyl Glycol Copolymer (A & E Connock)
Elfacos E 200 (Akzo Nobel)
Elfacos E 200 (Akzo Nobel Surface AB)

METHOXY PEG/PPG-7/3 AMINOPROPYL DIMETHICONE

CAS No.: 298211-68-4

Definition: Methoxy PEG/PPG-7/3 Aminopropyl Dimethicone is the silicone polymer that conforms generally to the formula:

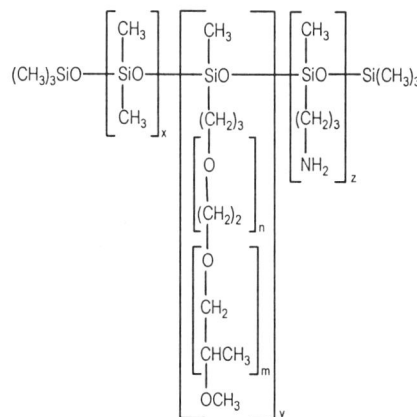

where n has an average value of 7 and m has an average value of 3.

Chemical Classes: Siloxanes and Silanes; Synthetic Polymers

Function: Not Reported

Technical/Other Name:
Siloxanes and Silicones, 3-Aminopropyl Me, Di-Me, 3-Hydroxypropyl Me, Ethers with Polyethylene-Polypropylene Glycol Mono-Me Ether

Trade Name:
Abil Soft AF 100 (Degussa Care Specialties)

METHOXY PEG-12 RETINAMIDE

Definition: Methoxy PEG-12 Retinamide is the organic compound that conforms to the formula:

where n has an average value of 11.

Chemical Class: Alkoxylated Amides

Function: Skin-Conditioning Agent - Miscellaneous

Trade Name:
Medimin A (LG Cosmetic)

METHOXY-PEG-7 RUTINYL SUCCINATE

Definition: Methoxy-PEG-7 Rutinyl Succinate is the ester of methoxy PEG-7 and rutinyl succinate.

Chemical Class: Esters

Function: Antioxidant

Trade Name:
Biorusol (Euphar)

2-METHOXY-p-PHENYLENEDIAMINE SULFATE

CAS Nos.	EINECS Nos.
42909-29-5	255-999-8
66671-82-7	266-443-9

Empirical Formula:
$C_7H_{10}N_2O \cdot H_2O_4S$

Definition: 2-Methoxy-p-Phenylenediamine Sulfate is the substituted aromatic amine salt that conforms to the formula:

See "Regulatory and Ingredient Use Information," for Colorants in Volume 1, Introduction, Part A.

Information Source: EEC(II-377)

Chemical Classes: Amines; Color Additives - Hair

Function: Hair Colorant

Technical/Other Names:
1,4-Benzenediamine, 2-Methoxy-, Sulfate
2,5-Diaminoanisol Sulfate
2-Methoxy-1,4-Benzenediamine Sulfate
1-Methoxy-2,5-Diaminobenzene Sulfate
Methoxy-p-phenylenediamine Sulfate

METHOXYPOLYOXYMETHYLENE MELAMINE

CAS No.: 68002-20-0

Definition: Methoxypolyoxymethylene Melamine is the reaction product of methanol and Polyoxymethylene Melamine (q.v.).

Information Sources: 21CFR177.1630, TSCA

Chemical Class: Synthetic Polymers

Functions: Binder; Film Former

Technical/Other Name:
1,3,5-Triazine-2,4,6-Triamine, Polymer with Formaldehyde, Methylated

Trade Name:
Cymel 385 (Cytec Specialty Resins)

METHOXYPROPYLGLUCONAMIDE

Empirical Formula:
$C_{10}H_{21}NO_7$

Definition: Methoxypropylgluconamide is the organic compound that conforms to the formula:

Chemical Classes: Amides; Ethers

Function: Skin-Conditioning Agent - Humectant

Reported Product Categories: Moisturizing Preparations; Foundations; Hair Conditioners; Shampoos (Non-coloring)

6-METHOXY-2,3-PYRIDINEDIAMINE HCI

CAS No.	EINECS No.
94166-62-8	303-358-9

Empirical Formula:
$C_6H_9N_3O \cdot 2HCl$

Definition: 6-Methoxy-2,3-Pyridinediamine HCl is the heterocyclic organic compound that conforms to the formula:

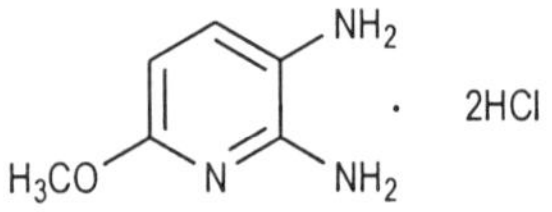

See "Regulatory and Ingredient Use Information," for Colorants in Volume 1, Introduction, Part A.

Chemical Classes: Amines; Color Additives - Hair

Function: Hair Colorant

Technical/Other Name:
2,3-Pyridinediamine, 6-Methoxy-, Dihydrochloride

Trade Name:
2,3-Diamino-6-Methoxypyridine 2HCl (Rutgers Organics)

4-METHOXYTOLUENE-2,5-DIAMINE HCI

CAS No.: 56496-88-9

Empirical Formula:
$C_8H_{12}N_2O \cdot ClH$

Definition: 4-Methoxytoluene-2,5-Diamine HCl is the aromatic amine salt that conforms to the formula:

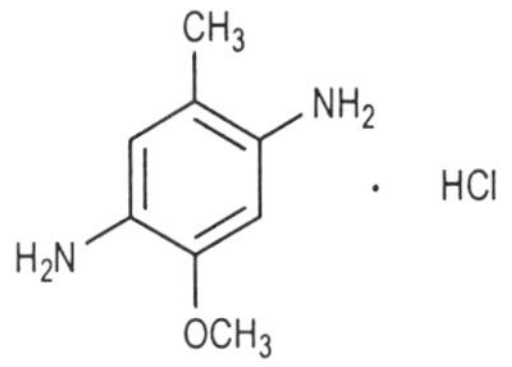

See "Regulatory and Ingredient Use Information," for Colorants in Volume 1, Introduction, Part A.

Chemical Classes: Amines; Color Additives - Hair

Function: Hair Colorant

METHOXYTRIMETHYLHEPTANOL

CAS No.	EINECS No.
41890-92-0	225-574-7

Empirical Formula:
$C_{11}H_{24}O_2$

Definition: Methoxytrimethylheptanol is the organic compound that conforms to the formula:

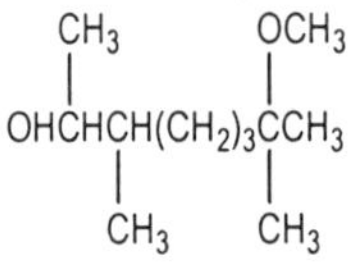

Chemical Classes: Alcohols; Ethers

Function: Fragrance Ingredient

Technical/Other Name:
2-Octanol, 7-Methoxy-3,7-Dimethyl-

Trade Name:
Osyrol (International Flavors)

METHOXYTRIMETHYLPHENYL DIHY-DROXYPHENYL PROPANOL

Empirical Formula:
$C_{19}H_{24}O_5$

Definition: Methoxytrimethylphenyl Dihydroxyphenyl Propanol is the organic compound that conforms to the formula:

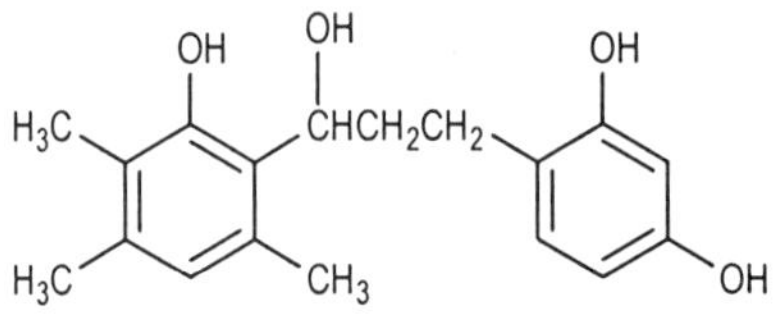

Chemical Classes: Alcohols; Ethers; Phenols

Functions: Antioxidant; Skin-Conditioning Agent - Miscellaneous

Technical/Other Name:
Methoxytrimethylphenyl Hydroxyphenyl Propanol

Trade Name:
LG-106W (LG Cosmetic)

METHYL ACETAMIDE

CAS No.	EINECS No.
79-16-3	201-182-6

Empirical Formula:
C_3H_7NO

Definition: Methyl Acetamide is the substituted amide of acetic acid that conforms to the formula:

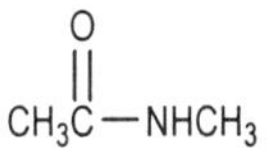

Information Source: TSCA

Chemical Class: Amides

Function: Not Reported

Technical/Other Names:
Acetamide, N-Methyl-
N-Methyl Acetamide

METHYL ACETATE

CAS No.	EINECS No.
79-20-9	201-185-2

Empirical Formula:
$C_3H_6O_2$

Definition: Methyl Acetate is the ester of methyl alcohol and acetic acid. It conforms to the formula:

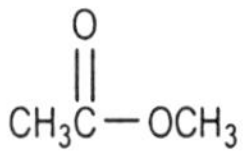

Information Sources: 21CFR172.515, 21CFR175.105, MI-13(6038), RIFM, TSCA

Chemical Class: Esters

Functions: Fragrance Ingredient; Solvent

Technical/Other Names:
Acetic Acid, Methyl Ester
Methyl acetate (RIFM)
Methyl Ethanoate

Trade Name:
Eastman Methyl Acetate (Eastman Chemical)

p-METHYL ACETOPHENONE

CAS No.	EINECS No.
122-00-9	204-514-8

JPN Translation:
メチルアセトフェノン

Empirical Formula:
$C_9H_{10}O$

Definition: p-Methyl Acetophenone is the organic compound that conforms to the formula:

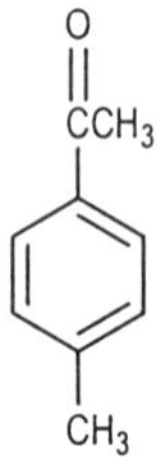

Information Sources: JCLS, JSCI, RIFM, TSCA

Chemical Class: Ketones

Function: Fragrance Ingredient

Technical/Other Names:
1-Acetyl-4-methylbenzene
4-Acetyltoluene
Ethanone, 1-(4-MethylPhenyl)-
4'-Methylacetophenone (RIFM)
Methyl 4-methylphenyl Ketone
1-(4-MethylPhenyl)Ethanone
p-Tolyl Methyl Ketone

METHYL ACETYL RICINOLEATE

CAS No.
140-03-4

EINECS No.
205-392-9

Empirical Formula:
$C_{21}H_{38}O_4$

Definition: Methyl Acetyl Ricinoleate is the methyl ester of acetyl ricinoleic acid. It conforms generally to the formula:

$$CH_3(CH_2)_5CHCH_2CH = CH(CH_2)_7C - OCH_3$$

(with $O - CCH_3$ and O substituent)

Information Sources: 21CFR175.105, TSCA

Chemical Class: Esters

Function: Skin-Conditioning Agent - Emollient

Reported Product Category: Suntan Gels, Creams, and Liquids

Technical/Other Names:
12-(Acetyloxy)-9-Octadecenoic Acid, Methyl Ester
Methyl Ricinoleate Acetate
9-Octadecenoic Acid, 12-(Acetyloxy)-, Methyl Ester
9-Octadecenoic Acid, 12-Hydroxy-, Methyl Ester, Acetate
Ricinoleic Acid, Methyl Ester, Acetate

Trade Names:
Naturechem MAR (CasChem)
Pelemol MAR (Phoenix)

METHYLAL

CAS No.
109-87-5

EINECS No.
203-713-7

Empirical Formula:
$C_3H_8O_2$

Definition: Methylal is the ether that conforms to the formula:

$$CH_3OCH_2OCH_3$$

Information Sources: 21CFR100.2, 21CFR172.515, MI-13(6042), RIFM, TSCA

Chemical Class: Ethers

Functions: Fragrance Ingredient; Solvent

Technical/Other Names:
Dimethoxymethane
2,4-Dioxapentane
Formal
Formaldehyde dimethyl acetal (RIFM)
Methane, Dimethoxy-
Methylene Dimethyl Ether

METHYL ALCOHOL

CAS No.
67-56-1

EINECS No.
200-659-6

Empirical Formula:
CH_4O

Definition: Methyl Alcohol is the aliphatic alcohol that conforms to the formula:

$$CH_3OH$$

Information Sources: 21CFR172.560, 21CFR172.859, 21CFR172.867, 21CFR173.250, 21CFR173.385, 21CFR175.105, 21CFR176.180, 21CFR176.200, 21CFR176.210, 21CFR177.1010, 21CFR177.2420, 21CFR177.2460, 21CFR436.201, 21CFR436.306, 21CFR436.334, 21CFR436.339, 21CFR446.20, 21CFR460.11, 21CFR556.760, 27CFR21.115, CIR: [S] IJT-20(SUPPL. 1)-2001, EEC(III/1-52), EGY, FCC, MAR, MHLW-331/1, MI-13(5984), MI-13(6465), NF XIX, RIFM, TSCA, USAN

Chemical Class: Alcohols

Functions: Denaturant; Fragrance Ingredient; Solvent

Reported Product Category: Bath Preparations, Misc.

Technical/Other Names:
Carbinol
Methanol (RIFM)
Methyl Hydroxide
Methylol
Wood Alcohol

Trade Name:
Eastman Methyl Alcohol (Eastman Chemical)

Trade Name Mixtures:
Baheda (Heritage Bio-Natural)
Bhuiamla (Heritage Bio-Natural)
Manjishta (Heritage Bio-Natural)
Neem (Heritage Bio-Natural)
Parsol-SLX (Roche)
Shatavari (Heritage Bio-Natural)

METHYL ALOESINYL CINNAMATE

CAS No.: 175413-23-7

Empirical Formula:
$C_{28}H_{30}O_{10}$

Definition: Methyl Aloesinyl Cinnamate is the organic compound that conforms to the formula:

Chemical Classes: Carbohydrates; Esters

Function: Not Reported

Technical/Other Name:
4H-1-Benzopyran-4-one, 2-[(2R)-2-Hydroxypropyl]-7-Methoxy-5-Methyl-8-[2-O-[(2E)-1-Oxo-3-Pheny l-2-Propenyl]-β-D-Glucopyranosyl]-

Trade Name:
UP540 (Univera Pharmaceuticals)

3-METHYLAMINO-4-NITROPHENOXY-ETHANOL

CAS No.
59820-63-2

EINECS No.
261-940-7

Empirical Formula:
$C_9H_{12}N_2O_4$

Definition: 3-Methylamino-4-Nitrophenoxyethanol is the substituted aromatic compound that conforms to the formula:

See "Regulatory and Ingredient Use Information," for Colorants in Volume 1, Introduction, Part A.

Information Source: EEC(III/2-28)

Chemical Classes: Alcohols; Amines; Color Additives - Hair

Function: Hair Colorant

The inclusion of any compound in the Dictionary and Handbook does not indicate that use of that substance as a cosmetic ingredient complies with the laws and regulations governing such use in the United States or any other country.

Reported Product Category: Hair Dyes and Colors (All Types Requiring Caution Statements and Patch Tests)

Technical/Other Names:
Ethanol, 2-[3-(Methylamino)-4-Nitrophenoxy]-
2-[3-(Methylamino)-4-Nitrophenoxy] Ethanol

Trade Name:
Imexine FR (Chimex)

p-METHYLAMINOPHENOL

CAS No.	EINECS No.
150-75-4	205-768-2

Empirical Formula:
C_7H_9NO

Definition: p-Methylaminophenol is the substituted phenol that conforms to the formula:

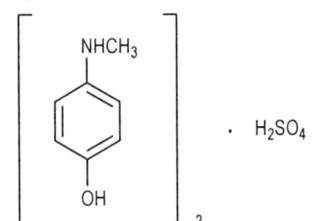

See "Regulatory and Ingredient Use Information," for Colorants in Volume 1, Introduction, Part A.

Information Source: EEC(III/2-12)

Chemical Classes: Amines; Color Additives - Hair; Phenols

Function: Hair Colorant

Technical/Other Names:
4-Hydroxy-N-Methylaniline
4-(Methylamino)Phenol
N-Methyl-p-Hydroxyaniline
Paramethylaminophenol
Phenol, 4-(Methylamino)-

p-METHYLAMINOPHENOL SULFATE

CAS Nos.	EINECS Nos.
55-55-0	200-237-1
1936-57-8	217-706-1

Empirical Formula:
$C_7H_9NO \cdot \frac{1}{2}H_2O_4S$

Definition: p-Methylaminophenol Sulfate is the substituted phenol that conforms to the formula:

See "Regulatory and Ingredient Use Information," for Colorants in Volume 1, Introduction, Part A.

Information Sources: CIR: [S] JACT-10(1)-1991, EEC(III/2-12), MI-13(6046), TSCA

Chemical Classes: Amines; Color Additives - Hair; Phenols

Function: Hair Colorant

Reported Product Category: Hair Dyes and Colors (All Types Requiring Caution Statements and Patch Tests)

Technical/Other Names:
4-(Methylamino)Phenol Sulfate
Metol
Paramethylaminophenol Sulfate
Phenol, 4-(Methylamino)-, Sulfate Salt (2:1)

Trade Name:
Rodol PM (Lowenstein)

METHYL ANTHRANILATE

CAS No.	EINECS No.
134-20-3	205-132-4

Empirical Formula:
$C_8H_9NO_2$

Definition: Methyl Anthranilate is the ester of methyl alcohol and 2-aminobenzoic acid. It conforms to the formula:

Information Sources: 21CFR1310.02, 21CFR1310.04, MI-13(6049), RIFM, TSCA

Chemical Classes: Amines; Esters

Functions: Flavoring Agent; Fragrance Ingredient

Technical/Other Names:
2-Aminobenzoic Acid, Methyl Ester
Benzoic Acid, 2-Amino-, Methyl Ester
Methyl 2-Aminobenzoate
Methyl anthranilate (RIFM)

METHYL ASPARTIC ACID

CAS Nos.: 4226-18-0; 6384-92-5; 17833-53-3

Empirical Formula:
$C_5H_9NO_4$

Definition: Methyl Aspartic Acid is the organic compound that conforms to the formula:

$$HOOCCH_2CHCOOH$$
$$|$$
$$NHCH_3$$

Information Source: MI-13(6696)

Chemical Class: Alkyl-Substituted Amino Acids

Functions: Hair Conditioning Agent; Skin-Conditioning Agent - Miscellaneous

Technical/Other Names:
Aspartic Acid, N-Methyl-
N-Methyl Aspartic Acid
NMDA

Trade Name:
MASPA (Penn-Squire)

METHYL BEHENATE

CAS No.	EINECS No.
929-77-1	213-207-8

Empirical Formula:
$C_{23}H_{46}O_2$

Definition: Methyl Behenate is the ester of methyl alcohol and Behenic Acid (q.v.). It conforms to the formula:

$$CH_3(CH_2)_{20}C-OCH_3$$

Chemical Class: Esters

Function: Skin-Conditioning Agent - Emollient

Technical/Other Names:
Behenic Acid, Methyl Ester
Docosanoic Acid, Methyl Ester
Methyl Docosanoate

4-METHYLBENZALDEHYDE

CAS No.	EINECS No.
104-87-0	203-246-9

Empirical Formula:
C_8H_8O

Definition: 4-Methylbenzaldehyde is the organic compound that conforms to the formula:

Information Source: RIFM

Chemical Class: Aldehydes

Functions: Flavoring Agent; Fragrance Ingredient

Technical/Other Names:
Benzaldehyde, 4-Methyl-
p-Methylbenzaldehyde
p-Tolualdehyde (RIFM)

Trade Name:
PMB (LC United)

METHYLBENZETHONIUM CHLORIDE

CAS No.
25155-18-4

EINECS No.
246-675-7

JPN Translation:
メチルベンゼトニウムクロリド

Empirical Formula:
$C_{28}H_{45}NO_2 \cdot Cl$

Definition: Methylbenzethonium Chloride is the quaternary ammonium salt that conforms generally to the formula:

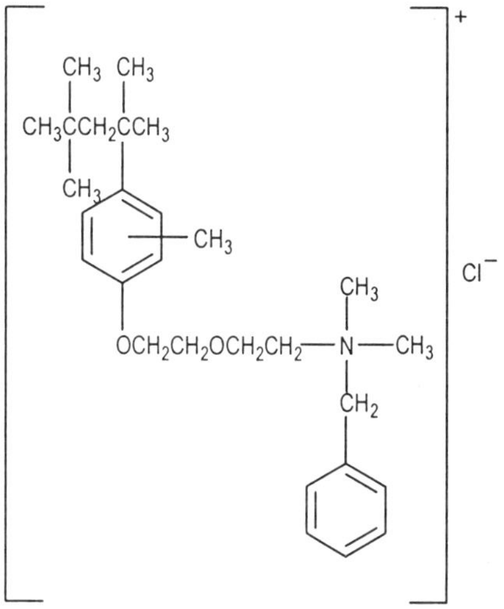

In the United States, Methylbenzethonium Chloride may be used as an active ingredient in OTC drug products. When used as an active drug ingredient, the established name is *Methylbenzethonium Chloride. See "Regulatory and Ingredient Use Information," regarding the labeling names for U.S. OTC Drug Ingredients in Volume 1, Introduction, Part A.*

Information Sources: BAN, CIR: [SQ] JACT-4(5)1985, CTFA D, INN, JCLS, JSCI, KOR, MAR, MI-13(6051), OTC-I-AM, USAN, USD, USP XXIV

Chemical Class: Quaternary Ammonium Compounds

Functions: Antimicrobial Agent; Antistatic Agent; Cosmetic Biocide; Deodorant Agent

Technical/Other Names:
Ammonium, benzyldimethyl(2-(2-((4-(1,1,3, 3-tetramethylbutyl)tolyl)oxy)ethoxy)ethyl)-Chloride

Benzenemethanaminium, N,N-Dimethyl-N-[2-[2-[Methyl-4-(1,1,3,3-Tetramethyl-butyl)Phenoxy]Ethoxy]Ethyl]-, Chloride
Diisobutyl Cresoxy Ethoxy Ethyl Dimethyl Benzyl Ammonium Chloride
N,N-Dimethyl-N-[2-[2-[Methyl-4-(1,1,3,3-Tetramethylbutyl)Phenoxy]Ethoxy]-Ethyl]Benzenemethanaminium Chloride

METHYL BENZOATE

CAS No.
93-58-3

EINECS No.
202-259-7

Empirical Formula:
$C_8H_8O_2$

Definition: Methyl Benzoate is the ester of methyl alcohol and benzoic acid that conforms to the formula:

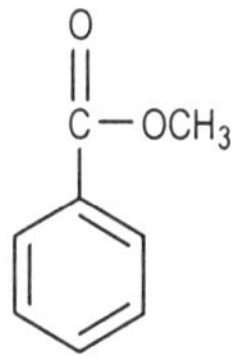

Information Sources: 21CFR172.515, EEC(VI/1-1), MI-13(6052), RIFM

Chemical Class: Esters

Functions: Fragrance Ingredient; Skin-Conditioning Agent - Emollient; Solvent

Technical/Other Names:
Benzoic Acid, Methyl Ester
Methyl Benzenecarboxylate
Methyl benzoate (RIFM)
Oil of Niobe

Trade Name:
Morflex Methyl Benzoate (Morflex)

METHYL BENZODIOXEPINONE

CAS No.
28940-11-6

EINECS No.
249-320-4

Empirical Formula:
$C_{10}H_{10}O_3$

Definition: Methyl Benzodioxepinone is the heterocyclic compound that conforms to the formula:

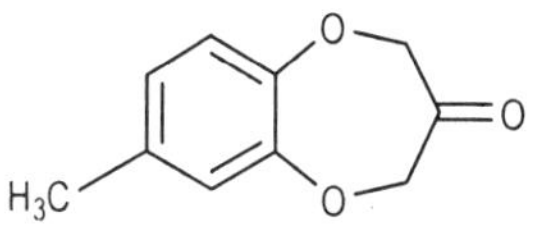

Information Source: RIFM

Chemical Class: Heterocyclic Compounds

Function: Fragrance Ingredient

Technical/Other Names:
2H-1,5-Benzodioxepin-3(4H)-one, 7-Methyl-
7-Methyl-2H-benzo-1,5-dioxepin-3(4H)-one (RIFM)

Trade Name:
Calone (Symrise)

METHYLBENZYL ACETATE

CAS Nos.
93-92-5
29759-11-3

EINECS No.
202-288-5

Empirical Formula:
$C_{10}H_{12}O_2$

Definition: Methylbenzyl Acetate is the organic compound that conforms to the formula:

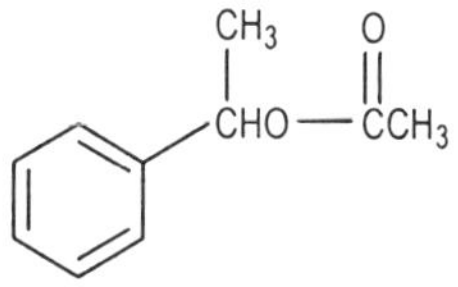

Chemical Class: Esters

Function: Fragrance Ingredient

Technical/Other Names:
Benzenemethanol, α-Methyl-, Acetate
Gardenol
sec-Phenethyl Acetate
Styrallyl Acetate

4-METHYLBENZYL 4,5-DIAMINO PYRAZOLE SULFATE

CAS No.: 173994-77-9

Empirical Formula:
$C_{11}H_{14}N_4 \cdot \frac{1}{2}H_2O_4S$

Definition: 4-Methylbenzyl 4,5-Diamino Pyrazole Sulfate is the heterocyclic compound that conforms to the formula:

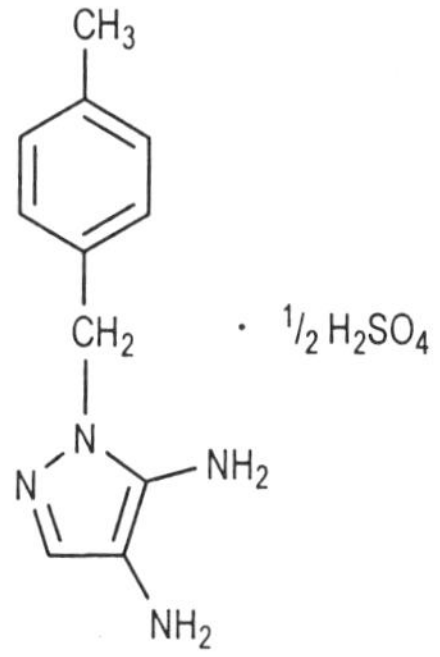

Chemical Classes: Amines; Color Additives - Hair; Heterocyclic Compounds

Function: Hair Colorant

The inclusion of any compound in the *Dictionary and Handbook* does not indicate that use of that substance as a cosmetic ingredient complies with the laws and regulations governing such use in the United States or any other country.

Technical/Other Name:
1H-Pyrazole-4,5-Diamine, 1-[(4-Methyl-phenyl)Methyl]-, Sulfate (2:1)

4-METHYLBENZYLIDENE CAMPHOR

CAS Nos.
36861-47-9
38102-62-4

EINECS No.
253-242-6

Empirical Formula:
$C_{18}H_{22}O$

Definition: 4-Methylbenzylidene Camphor is the aromatic organic compound that conforms to the formula:

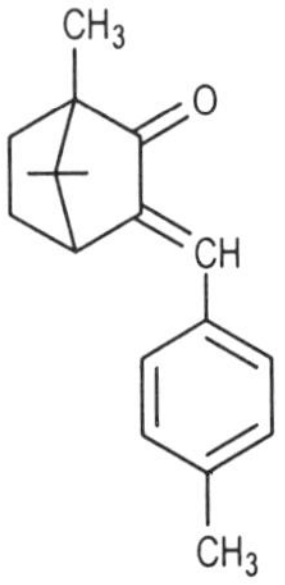

Information Sources: EEC(VII/1-18), USAN

Chemical Class: Ketones

Function: Ultraviolet Light Absorber

Reported Product Categories: Eyebrow Pencils; Suntan Gels, Creams, and Liquids

Technical/Other Names:
Bicyclo[2.2.1]Heptan-2-One, 1,7,7-Trimethyl-3-[(4-Methylphenyl)Methylene]-Enacamene
3-(4-Methylbenzylidene)-dl-Camphor
1,7,7-Trimethyl-3-[(4-Methylphenyl)-Methylene]Bicyclo[2.2.1]Heptan-2-One

Trade Names:
Eusolex 6300 (Merck KGaA/EMD Chemicals Inc.)
Neo Heliopan, Type MBC (PN600266) (Haarmann & Reimer GmbH)
Uvinul MBC 95 (BASF)
UVSOB MBC (LC United)

Trade Name Mixtures:
Novaprotex-37 (Crodarom)
Unifilter B-42 (Induchem)

METHYLBUTENE/PIPERYLENE COPOLYMER

CAS No.: 79586-89-3

Definition: Methylbutene/Piperylene Copolymer is a copolymer of 2-methyl-2-butene and 1,3-pentadiene monomers.

Chemical Class: Synthetic Polymers

Function: Plasticizer

Technical/Other Name:
1,3-Pentadiene, Polymer with 2-Methyl-2-Butene

Trade Name:
STA - TAC (Coloplast Consumer)

METHYLBUTENES

CAS No.
513-35-9

EINECS No.
208-156-3

Definition: Methylbutenes is a mixture containing 2-methyl-2-butene and 2-methyl-1-butene.

Information Source: MI-13(612)

Chemical Class: Hydrocarbons

Function: Solvent

Technical/Other Names:
Amylene
Trimethylene
Trimethylethene
1,1,2-Trimethylethylene

Trade Name:
Bayer Isoamylene (Bayer)

METHYL 4-t-BUTYLBENZOATE

CAS No.
26537-19-9

EINECS No.
247-768-5

Empirical Formula:
$C_{12}H_{16}O_2$

Definition: Methyl 4-t-Butylbenzoate is the organic compound that conforms to the formula:

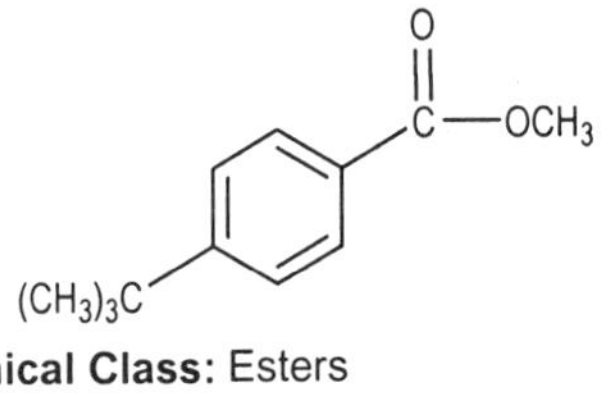

Chemical Class: Esters

Functions: Flavoring Agent; Fragrance Ingredient

Technical/Other Names:
Benzoic Acid, 4-(1,1-Dimethylethyl)-, Methyl Ester
Methyl p-tert-Butylbenzoate

Trade Name:
MBB (LC United)

METHYLBUTYLPHENYL DECYLOXY-BENZOATE

CAS No.
69777-63-5

EINECS No.
274-114-6

Empirical Formula:
$C_{28}H_{40}O_3$

Definition: Methylbutylphenyl Decyloxy-benzoate is the organic compound that conforms to the formula:

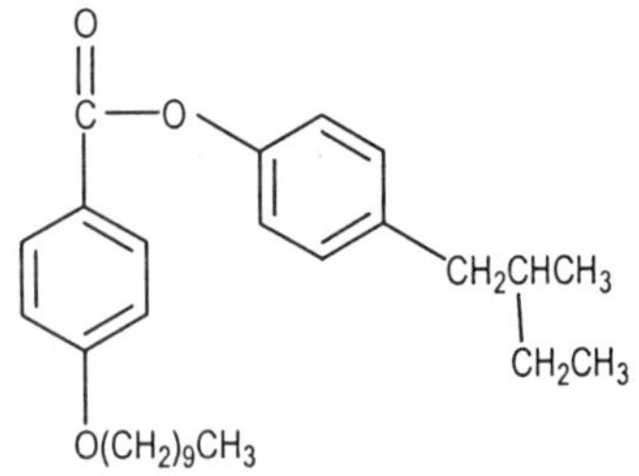

Chemical Class: Esters

Functions: Skin-Conditioning Agent - Miscellaneous; Skin-Conditioning Agent - Occlusive

Technical/Other Name:
Benzoic Acid, 4-(Decyloxy)-, 4-[(2S)-2-Methylbutyl]Phenyl Ester

Trade Name Mixtures:
Liquid Crystal BN 533 (Hallcrest Limited)
Liquid Crystal BN 1001 (Hallcrest Limited)

METHYLBUTYLPHENYL DODECYLOXY-BENZOATE

CAS No.
83846-95-1

EINECS No.
281-075-9

Empirical Formula:
$C_{30}H_{44}O_3$

Definition: Methylbutylphenyl Dodecyloxy-benzoate is the organic compound that conforms to the formula:

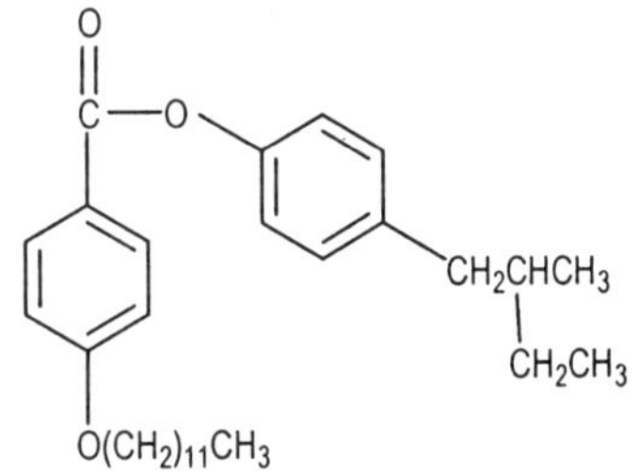

Chemical Class: Esters

Functions: Skin-Conditioning Agent - Miscellaneous; Skin-Conditioning Agent - Occlusive

Technical/Other Name:
Benzoic Acid, 4-(Dodecyloxy)- 4-[(2s)-2-Methylbutyl)Phenyl Ester

Trade Name Mixture:
Liquid Crystal BN 533 (Hallcrest Limited)

METHYLBUTYLPHENYL HEPTYL-BIPHENYLCARBOXYLATE

CAS No.
69777-71-5

EINECS No.
274-116-1

Empirical Formula:
$C_{31}H_{38}O_2$

Definition: Methylbutylphenyl Heptyl-biphenylcarboxylate is the organic compound that conforms to the formula:

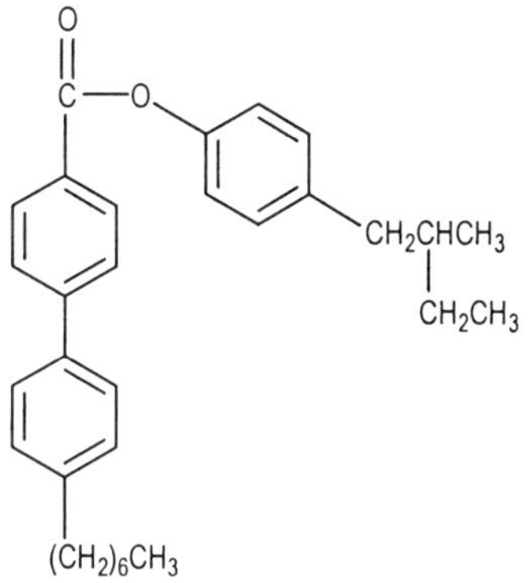

Chemical Class: Esters

Functions: Skin-Conditioning Agent - Miscellaneous; Skin-Conditioning Agent - Occlusive

Technical/Other Name:
[1,1'-Biphenyl]-4-Carboxylic Acid, 4'-Heptyl-, 4-[(2S)-2-Methylbutyl]Phenyl Ester

Trade Name Mixtures:
Liquid Crystal BN 533 (Hallcrest Limited)
Liquid Crystal BN 600 (Hallcrest Limited)
Liquid Crystal BN 823 (Hallcrest Limited)
Liquid Crystal BN 825 (Hallcrest Limited)
Liquid Crystal BN 826 (Hallcrest Limited)

METHYLBUTYLPHENYL HEXYLOXY-BENZOATE

CAS Nos.	EINECS Nos.
69777-59-9	274-112-5
84620-33-7	283-397-5

Empirical Formula:
$C_{24}H_{32}O_3$

Definition: Methylbutylphenyl Hexyloxy-benzoate is the organic compound that conforms to the formula:

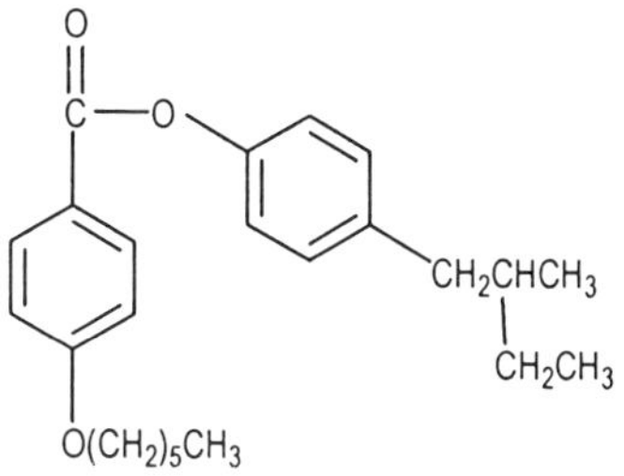

Chemical Class: Esters

Functions: Skin-Conditioning Agent - Miscellaneous; Skin-Conditioning Agent - Occlusive

Technical/Other Names:
Benzoic Acid, 4-(Hexyloxy)-, 4-(2-Methyl-butyl)Phenyl Ester
Benzoic Acid, 4-(Hexyloxy)-, [(2S)-2-Meth-ylbutyl]Phenyl Ester

Trade Name Mixtures:
Liquid Crystal BN 533 (Hallcrest Limited)
Liquid Crystal BN 600 (Hallcrest Limited)
Liquid Crystal BN 823 (Hallcrest Limited)
Liquid Crystal BN 825 (Hallcrest Limited)
Liquid Crystal BN 826 (Hallcrest Limited)
Liquid Crystal BN 1001 (Hallcrest Limited)

METHYLBUTYLPHENYL OCTYLOXY-BENZOATE

CAS Nos.	EINECS Nos.
69777-61-3	274-113-0
84236-44-2	282-476-1

Empirical Formula:
$C_{26}H_{36}O_3$

Definition: Methylbutylphenyl Octyloxy-benzoate is the organic compound that conforms to the formula:

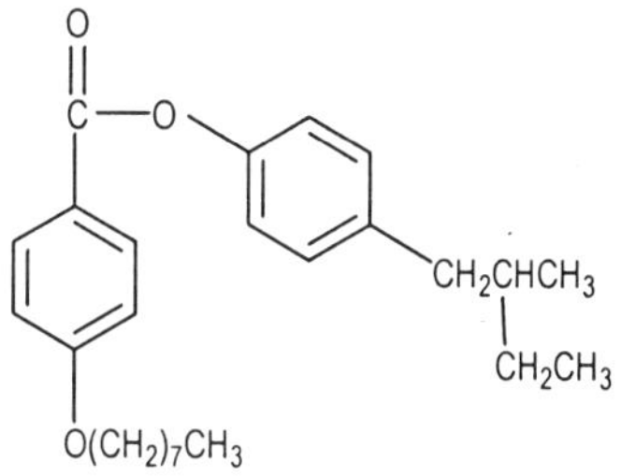

Chemical Classes: Esters; Phenols

Functions: Skin-Conditioning Agent - Miscellaneous; Skin-Conditioning Agent - Occlusive

Technical/Other Names:
Benzoic Acid, 4-(Octyloxy)-, 4-(2-Methyl-butyl)Phenyl Ester
Benzoic Acid, 4-(Octyloxy)-, 4-[(2S)-2-Methylbutyl]Phenyl Ester

Trade Name Mixtures:
Liquid Crystal BN 533 (Hallcrest Limited)
Liquid Crystal BN 600 (Hallcrest Limited)
Liquid Crystal BN 823 (Hallcrest Limited)
Liquid Crystal BN 825 (Hallcrest Limited)
Liquid Crystal BN 826 (Hallcrest Limited)
Liquid Crystal BN 1001 (Hallcrest Limited)

METHYLBUTYLPHENYL PENTYL-BENZOATE

CAS No.	EINECS No.
69777-64-6	274-115-1

Empirical Formula:
$C_{23}H_{30}O_2$

Definition: Methylbutylphenyl Pentyl-benzoate is the organic compound that conforms to the formula:

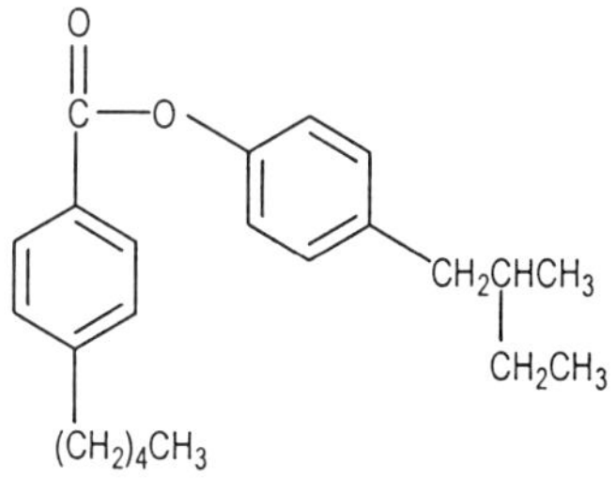

Chemical Class: Esters

Functions: Skin-Conditioning Agent - Miscellaneous; Skin-Conditioning Agent - Occlusive

Technical/Other Name:
Benzoic Acid, 4-Pentyl-, 4-[(2S)-2-Methyl-butyl]Phenyl Ester

Trade Name Mixture:
Liquid Crystal BN 533 (Hallcrest Limited)

METHYLBUTYLPHENYL PROPYL-BENZOATE

CAS No.	EINECS No.
94442-17-8	305-343-2

Empirical Formula:
$C_{21}H_{26}O_2$

Definition: Methylbutylphenyl Propyl-benzoate is the organic compound that conforms to the formula:

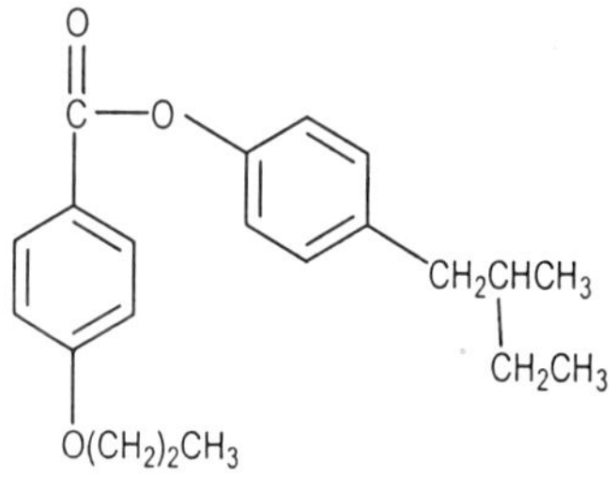

Chemical Class: Esters

Functions: Skin-Conditioning Agent - Miscellaneous; Skin-Conditioning Agent - Occlusive

Technical/Other Name:
Benzoic Acid, 4-Propyl-, 4-[(2S)-2-Methyl-butyl]Phenyl Ester

Trade Name Mixtures:
Liquid Crystal BN 533 (Hallcrest Limited)
Liquid Crystal BN 825 (Hallcrest Limited)
Liquid Crystal BN 826 (Hallcrest Limited)

METHYL CAPROATE

CAS No.	EINECS No.
106-70-7	203-425-1

Empirical Formula:
$C_7H_{14}O_2$

Definition: Methyl Caproate is the ester of methyl alcohol and caproic acid. It conforms generally to the formula:

$$CH_3(CH_2)_4C \overset{\overset{\textstyle O}{\|}}{{}} - OCH_3$$

Information Sources: 21CFR172.515, RIFM, TSCA

Chemical Class: Esters

Functions: Fragrance Ingredient; Skin-Conditioning Agent - Emollient

Technical/Other Names:
Hexanoic Acid, Methyl Ester
Methyl hexanoate (RIFM)
Methyl Hexanoate

METHYL CAPRYLATE

CAS No.   **EINECS No.**
111-11-5   203-835-0

Empirical Formula:
$C_9H_{18}O_2$

Definition: Methyl Caprylate is the ester of methyl alcohol and caprylic acid. It conforms generally to the formula:

$$CH_3(CH_2)_6C \overset{\overset{\textstyle O}{\|}}{{}} - OCH_3$$

Information Sources: 21CFR172.225, 21CFR172.515, 21CFR176.200, 21CFR176.210, 21CFR177.2800, RIFM, TSCA

Chemical Class: Esters

Functions: Fragrance Ingredient; Skin-Conditioning Agent - Emollient

Technical/Other Names:
Caprylioc Acid Methyl Ester
Methyl octanoate (RIFM)
Methyl Octanoate
Octanoic Acid, Methyl Ester

Trade Name:
AEC Methyl Caprylate (A & E Connock)

METHYL CAPRYLATE/CAPRATE

Definition: Methyl Caprylate/Caprate is the mixture of esters of methyl alcohol and caprylic and capric acids.

Information Sources: 21CFR172.225, 21CFR176.200, 21CFR176.210, 21CFR177.2800

Chemical Class: Esters

Function: Skin-Conditioning Agent - Emollient

Trade Name:
AEC Methyl Caprylate/Caprate (A & E Connock)

METHYLCELLULOSE

CAS No.: 9004-67-5

JPN Translation:
メチルセルロース

Definition: Methylcellulose is the methyl ether of cellulose.

Information Sources: AUS, BP, BPC, BRA, 21CFR150.141, 21CFR150.161, 21CFR175.105, 21CFR175.210, 21CFR175.300, 21CFR176.200, 21CFR182.1480, 21CFR310.545, 21CFR349.12, 21CFR582.1480, CIR: [S] JACT-5(3)1986, CTFA S, FCC, HUN, INN, ITA, JAN, JCLS, JSCI, MAR, MI-13(6068), NFJ, OTC-I-LX, OTC-I-OP, RIFM, ROM, TSCA, USAN, USD, USP XXIV

Chemical Class: Gums, Hydrophilic Colloids and Derivatives

Functions: Binder; Emulsion Stabilizer; Fragrance Ingredient; Viscosity Increasing Agent - Aqueous

Reported Product Categories: Bath Preparations, Misc.; Body and Hand Preparations (Excluding Shaving Preparations); Shampoos (Non-coloring); Paste Masks (Mud Packs); Eyeliners; Skin Care Preparations, Misc.

Technical/Other Names:
Cellulose, Methyl Ether
Methyl cellulose (RIFM)

Trade Names:
Benecel Methylcellulose (Hercules/ Aqualon)
Methocel A (Amerchol)

Trade Name Mixtures:
Flamenco Pearl SEC (Engelhard Corp.)
Mearlmaid AA (Engelhard Corp.)
Mearlmaid PLN (Engelhard Corp.)
Mearlmaid TR (Engelhard Corp.)
Mykon ATC (WarwickIntl)

METHYLCHLOROISOTHIAZOLINONE

CAS No.   **EINECS No.**
26172-55-4   247-500-7

JPN Translation:
メチルクロロイソチアゾリノン

Empirical Formula:
C_4H_4ClNOS

Definition: Methylchloroisothiazolinone is the heterocyclic organic compound that conforms to the formula:

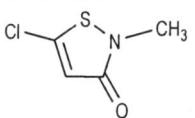

Information Sources: 21CFR175.105, 21CFR176.170, CIR: [SQ] JACT-11(1)1992, EEC(VI/1-39), JCIC, JCLS, MHLW-331/3

Chemical Class: Heterocyclic Compounds

Function: Preservative

Reported Product Categories: Shampoos (Non-coloring); Hair Conditioners; Hair Tints; Hair Preparations (Non-coloring), Misc.; Bath Oils, Tablets, and Salts; Cleansing Products (Cold Creams, Cleansing Lotions, Liquids and Pads); Skin Care Preparations, Misc.; Bath Preparations, Misc.; Body and Hand Preparations (Excluding Shaving Preparations); Bath Soaps and Detergents; Bubble Baths; Moisturizing Preparations; Tonics, Dressings, and Other Hair Grooming Aids; Bath Capsules; Hair Shampoos (Coloring); Face and Neck Preparations (Excluding Shaving Preparations); Personal Cleanliness Products, Misc.; Skin Fresheners; Baby Products, Misc.; Eyebrow Pencils; Suntan Gels, Creams, and Liquids; Hair Rinses (Coloring); Hair Rinses (Non-coloring); Shaving Preparations, Misc.; Baby Lotions, Oils, Powders and Creams; Baby Shampoos; Eye Makeup Removers; Foundations; Hair Wave Sets; Night Skin Care Preparations; Paste Masks (Mud Packs); Permanent Waves; Suntan Preparations, Misc.

Technical/Other Names:
5-Chloro-2-Methyl-4-Isothiazolin-3-one
4-Isothiazolin-3-one, 5-Chloro-2-Methyl-

Trade Name Mixtures:
Conservateur GD500 (Phytocos)
Conservateur GD700 (Phytocos)
Dekaben IT (Dekker)
Euxyl K 100 (Schulke & Mayr)
Euxyl K 727 (Schulke & Mayr)
Kathon CG (Rohm and Haas)
Kathon CG II Biocide (Rohm and Haas)
Kathon CG III (Rohm and Haas)
Liposerve MB (Lipo)
Liposerve MM (Lipo)
Microcare CB (Acti-Chem)
Microcare CBA (Acti-Chem)
Microcare ITA (Acti-Chem)
Microcare ITL (Acti-Chem)
Nipaguard CMB (Clariant)
Nipaguard CMB (Clariant GmbH, Personal Care)

METHYL COCOATE

CAS No.   **EINECS No.**
61788-59-8   262-988-1

Definition: Methyl Cocoate is the ester of methyl alcohol and coconut fatty acids. It conforms generally to the formula:

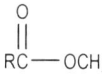

where RCO- represents the fatty acids derived from coconut oil.

Information Sources: 21CFR172.225, 21CFR175.105, 21CFR176.200, 21CFR176.210, 21CFR177.2260, 21CFR177.2800, 21CFR178.3910, TSCA

Chemical Class: Esters

Function: Skin-Conditioning Agent - Emollient

Technical/Other Name:
Fatty Acids, Coco, Methyl Esters

Trade Names:
AEC Methyl Cocoate (A & E Connock)
CE-618 (Procter & Gamble)

6-METHYL COUMARIN

CAS No.	EINECS No.
92-48-8	202-158-8

Empirical Formula:
$C_{10}H_8O_2$

Definition: 6-Methyl Coumarin is the heterocyclic compound that conforms to the formula:

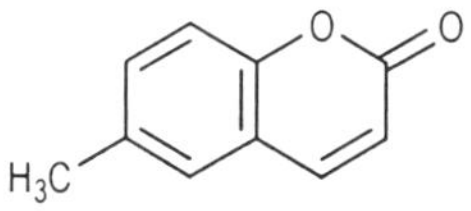

Information Sources: EEC(III/1-46), RIFM, TSCA

Chemical Classes: Esters; Heterocyclic Compounds

Function: Fragrance Ingredient

Technical/Other Names:
2H-1-Benzopyran-2-one, 6-Methyl-
6-Methylbenzopyrone
6-Methylcoumarin (RIFM)
6-Methyl-2H-1-Benzopyran-2-one

METHYL CYCLODEXTRIN

CAS No.: 128446-36-6

JPN Translation:
メトキシシクロデキストリン

Definition: Methyl Cyclodextrin is the product obtained by the methylation of Cyclodextrin (q.v.).

Chemical Class: Carbohydrates
Function: Chelating Agent
Trade Name:
Cavasol W7 M (Wacker-Chemie)

METHYLCYCLOHEXENYL BUTANOL

CAS No.	EINECS No.
15760-18-6	239-845-7

Empirical Formula:
$C_{11}H_{20}O$

Definition: Methylcyclohexenyl Butanol is the organic compound that conforms to the formula:

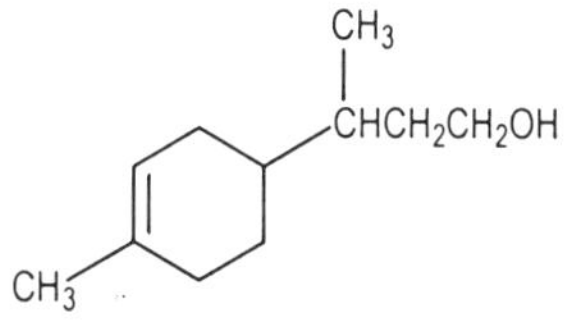

Chemical Class: Alcohols
Function: Fragrance Ingredient
Technical/Other Name:
3-Cyclohexene-1-Propanol, γ, 4-Dimethyl-

METHYLCYCLOPENTADECENONE

CAS No.	EINECS No.
82356-51-2	429-900-5

Empirical Formula:
$C_{16}H_{28}O$

Definition: Methylcyclopentadecenone is the organic compound that conforms to the formula:

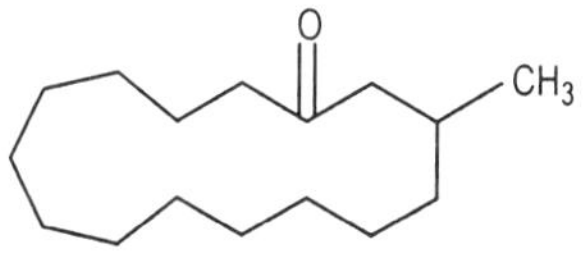

Chemical Class: Ketones
Function: Fragrance Ingredient
Technical/Other Name:
Cyclopentadecenone, 3-Methyl-

METHYL DEHYDROABIETATE

CAS No.: 1235-74-1
Empirical Formula:
$C_{21}H_{30}O_2$

Definition: Methyl Dehydroabietate is the ester that conforms to the formula:

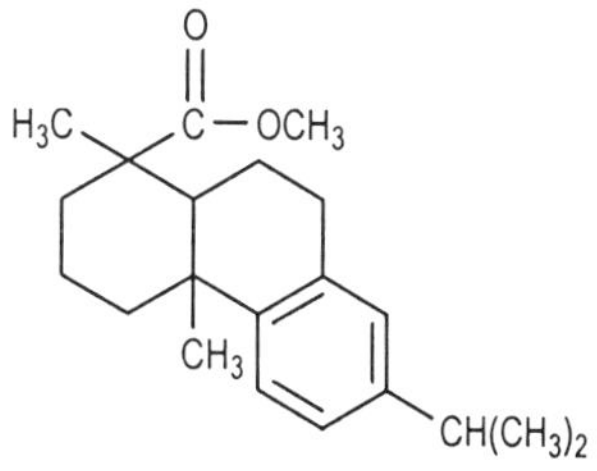

Chemical Class: Esters

Functions: Skin-Conditioning Agent - Emollient; Viscosity Increasing Agent - Nonaqueous

Technical/Other Names:
Dehydroabietic Acid Methyl Ester
1-Phenanthrenecarboxylic Acid, 1,2,3,4,4a, 9,10,10a-Octahydro-1, 4a-dimethyl-7-(1-methylethyl)-,methyl ester

METHYLDIBROMO GLUTARONITRILE

CAS No.	EINECS No.
35691-65-7	252-681-0

Empirical Formula:
$C_6H_6Br_2N_2$

Definition: Methyldibromo Glutaronitrile is the brominated methylene glutaronitrile that conforms to the formula:

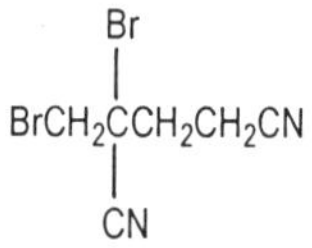

Information Sources: 21CFR176.210, CIR: [SQ] JACT-15(2)1996, EEC(VI/1-36), MI-13(3045)

Chemical Class: Halogen Compounds

Function: Preservative

Reported Product Categories: Shampoos (Non-coloring); Hair Conditioners; Bubble Baths; Face Powders; Hair Preparations (Non-coloring), Misc.; Tonics, Dressings, and Other Hair Grooming Aids; Bath Capsules; Blushers (All types); Eyeliners; Face and Neck Preparations (Excluding Shaving Preparations); Indoor Tanning Preparations; Permanent Waves

Technical/Other Names:
2-Bromo-2-(Bromomethyl) Glutaronitrile
2-Bromo-2-(Bromomethyl) Pentanedinitrile
1,2-Dibromo-2,4-Dicyanobutane
Glutaronitrile, 2-Bromo-2-(Bromomethyl)-
Pentanedinitrile, 2-Bromo-2-(Bromomethyl)-

Trade Names:
Merguard 1105 (ONDEO Nalco)
Merguard 1105 (ONDEO Nalco Europe)

Trade Name Mixtures:
AEC Methyldibromo Glutaronitrile (&) Dipropylene Glycol (A & E Connock)
Dekaben GN (Dekker)
Dekaben IGN (Dekker)
Euxyl K 135 (Schulke & Mayr)
Euxyl K400 (Schulke & Mayr)
Euxyl K 446 (Schulke & Mayr)
Euxyl K 727 (Schulke & Mayr)

The inclusion of any compound in the *Dictionary and Handbook* does not indicate that use of that substance as a cosmetic ingredient complies with the laws and regulations governing such use in the United States or any other country.

Merguard 1190 (ONDEO Nalco)
Merguard 1190 (ONDEO Nalco Europe)
Merguard 1200 (ONDEO Nalco)
Merguard 1200 (ONDEO Nalco Europe)
Merguard X-18 (ONDEO Nalco)
Microcare MG (Acti-Chem)
Microcare MGI (Acti-Chem)
Nipaguard DCB (Clariant)
Nipaguard DCB (Clariant GmbH, Personal Care)

METHYL DICOCAMINE

CAS No. 61788-62-3

EINECS No. 262-990-2

Definition: Methyl Dicocamine is the aliphatic amine that conforms generally to the formula:

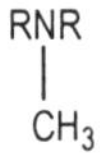

$$RNR$$
$$|$$
$$CH_3$$

where R represents the alkyl groups derived from coconut oil.

Chemical Class: Amines

Function: Antistatic Agent

Technical/Other Names:
Amines, Dicoco Alkylmethyl
N-Methylbis(coconut oil alkyl)amine

Trade Name:
Armeen M2C (Akzo Nobel)

METHYL DIHYDROABIETATE

CAS Nos.: 30968-45-7; 33892-18-1

Empirical Formula:
$C_{21}H_{34}O_2$

Definition: Methyl Dihydroabietate is the organic compound that conforms to the formula:

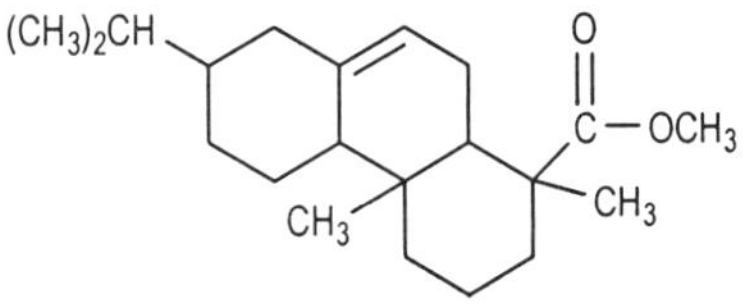

Chemical Class: Esters

Function: Viscosity Increasing Agent - Nonaqueous

Technical/Other Names:
Abietic Acid, dihydro-, Methyl Ester
1-Phenanthrenecarboxylic Acid,
Dodecahydro-1,4a-Dimethyl-7-(-Methyl-ethyl)-, Methyl Ester
Podocarp-13-en-15-oic Acid, 13-Isopropyl-, Methyl Ester

METHYLDIHYDROJASMONATE

CAS Nos.
2630-39-9
24851-98-7

EINECS No.
220-112-5

Empirical Formula:
$C_{13}H_{22}O_3$

Definition: Methyldihydrojasmonate is the organic compound that conforms to the formula:

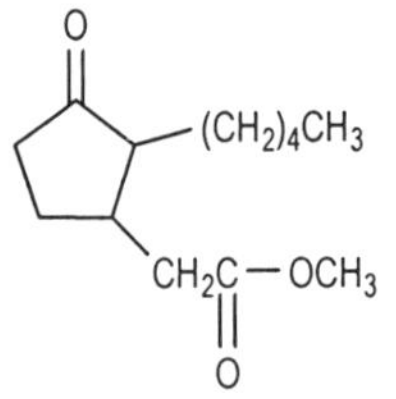

Information Sources: RIFM, TSCA

Chemical Classes: Esters; Ketones

Function: Fragrance Ingredient

Reported Product Categories: Foundations; Moisturizing Preparations; Skin Care Preparations, Misc.

Technical/Other Names:
2-Amylcyclopentanoneacetic acid, Methyl Ester
Cyclopentaneacetic Acid, 3-Oxo-2-Pentyl-, Methyl Ester
Dihyrojasmonic Acid Methyl Ester
Hedione
Methyl (2-Amyl-3-Oxocyclopentyl)Acetate
Methyl dihydrojasmonate (RIFM)
3-Oxo-2-Pentylcyclopentaneacetic Acid, Methyl Ester

Trade Name:
Hedione (Firmenich)

Trade Name Mixture:
Laura 0/719511 (Symrise)

METHYL DIHYDROXYBENZOATE

CAS No. 2150-46-1

EINECS No. 218-427-8

Empirical Formula:
$C_8H_8O_4$

Definition: Methyl Dihydroxybenzoate is the organic compound that conforms to the formula:

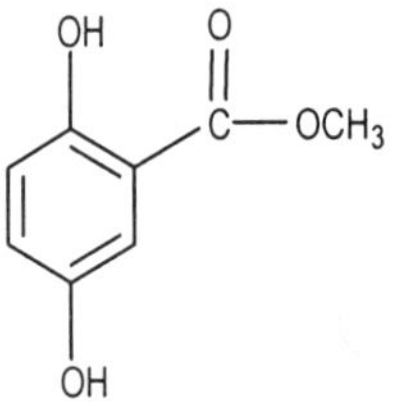

Chemical Classes: Esters; Phenols

Function: Chelating Agent

Technical/Other Names:
Benzoic Acid, 2,5-Dihydroxy-, Methyl Ester
Methyl 2,5-Dihydroxybenzoate
Methyl Gentisate

Trade Name:
Metil Gentisato (I.C.I.M.)

METHYL DIISOPROPYL PROPIONAMIDE

CAS No. 51115-67-4

EINECS No. 256-974-4

Empirical Formula:
$C_{10}H_{21}NO$

Definition: Methyl Diisopropyl Propionamide is the organic compound that conforms to the formula:

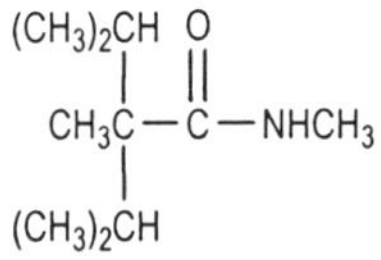

Information Sources: MI-13(9788), RIFM

Chemical Class: Amides

Function: Fragrance Ingredient

Technical/Other Names:
Butanamide, N,2,3-Trimethyl-2-(1-Methyl-ethyl)-
2-isopropyl-N,2,3-trimethylbutanamide
2-Isopropyl-N,2,3-trimethylbutyramide (RIFM)
N,2,3-Trimethyl-2-(1-Methylethyl)-Butanamide

Trade Name:
WS23 (Rhodia ChiRex)

METHYLEICOSAMIDOPROPYL ETHYLDIMONIUM ETHOSULFATE

Definition: Methyleicosamidopropyl Ethyldimonium Ethosulfate is the quaternary ammonium salt that conforms to the formula:

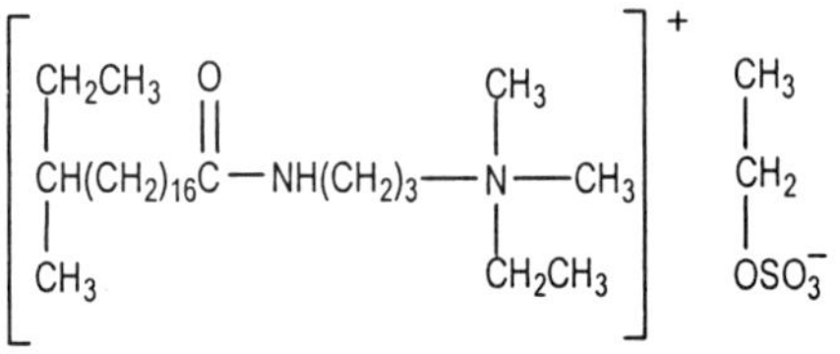

Chemical Class: Quaternary Ammonium Compounds

Functions: Antistatic Agent; Hair Conditioning Agent

2,2'-METHYLENEBIS 4-AMINOPHENOL

CAS No.: 63969-46-0

Empirical Formula:
$C_{13}H_{14}N_2O_2$

Definition: 2,2'-Methylenebis 4-Amino-phenol is the organic compound that conforms to the formula:

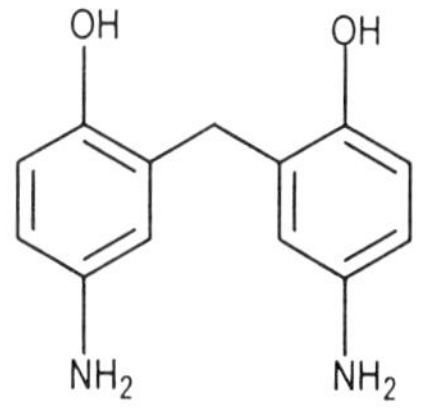

Chemical Classes: Amines; Phenols

Function: Hair Colorant

Technical/Other Name:
Phenol, 2,2'-Methylenebis[4-Amino-]

Trade Name:
Bis-(5-amino-2-hydroxyphenyl)-methan (Henkel KgaA)

2,2'-METHYLENEBIS-4-AMINOPHENOL HCl

CAS No.: 27311-52-0

Empirical Formula:
$C_{13}H_{14}N_2O_2 \cdot 2HCl$

Definition: 2,2'-Methylenebis-4-Amino-phenol HCl is the organic compound that conforms to the formula:

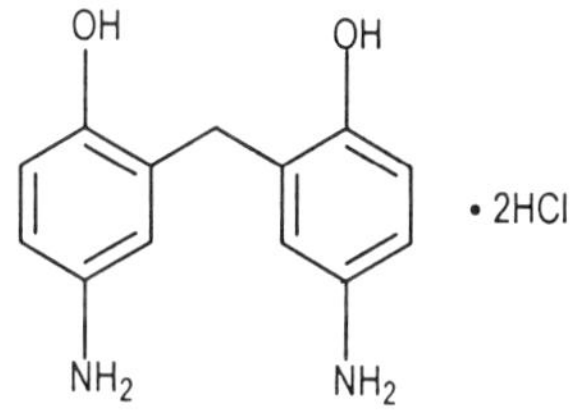

Chemical Classes: Amines; Phenols

Function: Hair Colorant

Technical/Other Name:
Phenol, 2,2'-Methylenebis[4-Amino, Dihydrochloride]

METHYLENE BIS-BENZOTRIAZOLYL TETRAMETHYLBUTYLPHENOL

CAS No.: 103597-45-1

Empirical Formula:
$C_{41}H_{50}N_6O_2$

Definition: Methylene Bis-Benzotriazolyl Tetramethylbutylphenol is the heterocyclic compound that conforms to the formula:

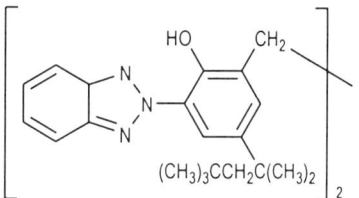

Information Source: EEC(VII/1-23)

Chemical Classes: Heterocyclic Compounds; Phenols

Function: Ultraviolet Light Absorber

Technical/Other Name:
2,2'-Methylenebis[4-(1,1,3,3-Tetramethyl-butyl)-6-(2H-Benzotriazol-2-yl)Phenol]

Trade Name:
Tinosorb M (Ciba Specialty Chemicals)

METHYLENEBIS TALLOW ACETAMIDO-DIMONIUM CHLORIDE

Definition: Methylenebis Tallow Acetamido-dimonium Chloride is the quaternary ammonium salt that conforms generally to the formula:

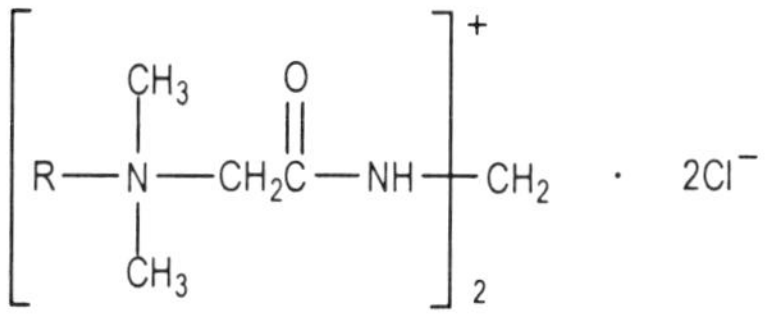

where R represents the alkyl groups derived from tallow.

Chemical Class: Quaternary Ammonium Compounds

Function: Antistatic Agent

METHYLENE DI-t-BUTYLCRESOL

CAS No.	EINECS No.
119-47-1	204-327-1

Empirical Formula:
$C_{23}H_{33}O_2$

Definition: Methylene Di-t-Butylcresol is the organic compound that conforms to the formula:

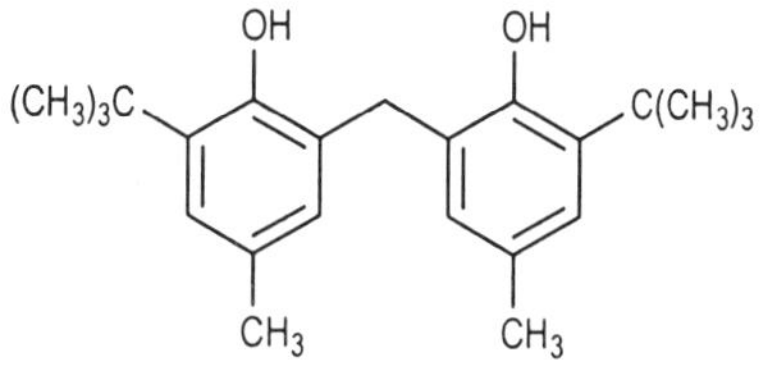

Chemical Class: Phenols

Function: Antioxidant

Technical/Other Names:
3,3'-Di-tert-butyl-2,2'-dihydroxy-5-5'-dimethyldiphenylmethane
6,6'-Di-tert-Butyl-2,2'-Methylenedi-p-Cresol
2,2'-Methylenebis(6-tert-Butyl-4-Methyl-phenol)
Phenol, 2,2'-Methylenebis[6-(1,1-Dimethylethyl)-4-Methyl-

Trade Name:
Lowinox 22M46 (GLM(G))

3,4-METHYLENEDIOXYANILINE

CAS No.	EINECS No.
14268-66-7	238-161-6

Empirical Formula:
$C_7H_7NO_2$

Definition: 3,4-Methylenedioxyaniline is the aromatic amine that conforms to the formula:

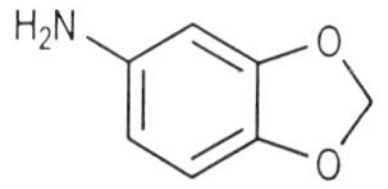

See "Regulatory and Ingredient Use Information," for Colorants in Volume 1, Introduction, Part A.

Chemical Classes: Amines; Color Additives - Hair; Heterocyclic Compounds

Function: Hair Colorant

Technical/Other Names:
5-Amino-1,3-benzodioxole
1,3-Benzodioxol-5-Amine
1,2-(Methylenedioxy)-4-aminobenzene

3,4-METHYLENEDIOXYPHENOL

CAS No.	EINECS No.
533-31-3	208-561-5

Empirical Formula:
$C_7H_6O_3$

Definition: 3,4-Methylenedioxyphenol is substituted phenol that conforms to the formula:

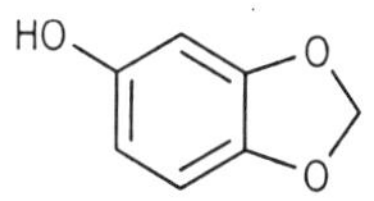

See "Regulatory and Ingredient Use Information," for Colorants in Volume 1, Introduction, Part A.

Chemical Classes: Color Additives - Hair; Heterocyclic Compounds

Function: Hair Colorant

Technical/Other Names:
1,3-Benzodioxol-5-ol

4-Hydroxy-1,2-methylenedioxybenzene
Sesamol

METHYLETHANOLAMINE

CAS No.	EINECS No.
109-83-1	203-710-0

Empirical Formula:
C_3H_9NO

Definition: Methylethanolamine is the organic compound that conforms to the formula:

$$CH_3NHCH_2CH_2OH$$

Information Source: MI-13(6045)

Chemical Class: Alkanolamines

Function: pH Adjuster

Reported Product Category: Hair Dyes and Colors (All Types Requiring Caution Statements and Patch Tests)

Technical/Other Names:
Ethanol, 2-(Methylamino)-
N-(2-Hydroxyethyl)-N-Methylamine
2-(Methylamino)Ethanol

Trade Name:
N-Methylethanolamine (Dow Chemical)

METHYL ETHYLCELLULOSE

CAS No.: 9004-59-5

Definition: Methyl Ethylcellulose is the methyl ether of Ethylcellulose (q.v.).

Chemical Classes: Carbohydrates; Ethers

Functions: Binder; Film Former; Viscosity Increasing Agent - Aqueous

Technical/Other Names:
Cellulose, Ethyl Methyl Ether
Ethyl Methyl Cellulose

Trade Name:
Celacol CTP 991 (Courtaulde)

METHYL EUGENOL

CAS No.	EINECS No.
93-15-2	202-223-0

Empirical Formula:
$C_{11}H_{14}O_2$

Definition: Methyl Eugenol is the aromatic compound that conforms to the formula:

Information Sources: 21CFR172.515, EEC(II-451), FCC, RIFM, TSCA

Chemical Class: Ethers

Function: Fragrance Ingredient

Technical/Other Names:
Allylveratrole
Benzene, 1,2-Dimethoxy-4-(2-Propenyl)-
3,4-Dimethoxyallylbenzene
1,2-Dimethoxy-4-(2-Propenyl)Benzene
Eugenol Methyl Ether
Eugenyl methyl ether (RIFM)
Eugenyl Methyl Ether
Veratrole Methyl Ether

METHYLEUGENYL PEG-8 DIMETHICONE

CAS No.: 200443-93-2

Definition: Methyleugenyl PEG-8 Dimethicone is a copolymer of methyleugenyl dimethicone and PEG-8 Dimethicone (q.v.).

Chemical Class: Siloxanes and Silanes

Functions: Hair Conditioning Agent; Skin-Conditioning Agent - Miscellaneous

Technical/Other Names:
Siloxanes and Silicones, Di-Me, 3-Hydroxypropyl Me, Ethoxylated
Siloxanes, di-Me, 3-hydroxypropyl Me, Polymers with Ethylene Oxide

Trade Name:
Silsoft Shine (OSi Specialties)

METHYLGLUCAMINE

CAS No.	EINECS No.
6284-40-8	228-506-9

Empirical Formula:
$C_7H_{17}NO_5$

Definition: Methylglucamine is the organic compound that conforms to the formula:

Information Sources: MI-13(6102), TSCA

Chemical Classes: Amines; Polyols

Function: pH Adjuster

Technical/Other Names:
1-Deoxy-1-(Methylamino)-D-Glucitol
D-Glucitol, 1-Deoxy-1-(Methylamino)-
N-Methylsorbitylamine
Sorbitol, 1-deoxy-1-methylamino

Trade Name:
D(-)-N-Methylglucamine (Merck KGaA)

METHYL GLUCETH-10

JPN Translation:
メチルグルセス - 10

Empirical Formula:
$C_{27}H_{54}O_{16}$

Definition: Methyl Gluceth-10 is the polyethylene glycol ether of methyl glucose that conforms generally to the formula:

$$CH_3C_6H_{10}O_5(OCH_2CH_2)_nOH$$

where n has an average value of 10.

Information Sources: JSQI, TSCA

Chemical Classes: Alkoxylated Alcohols; Carbohydrates; Polyols

Function: Skin-Conditioning Agent - Humectant

Reported Product Categories: Bath Capsules; Moisturizing Preparations; Bath Preparations, Misc.; Body and Hand Preparations (Excluding Shaving Preparations); Hair Conditioners; Bath Oils, Tablets, and Salts; Cleansing Products (Cold Creams, Cleansing Lotions, Liquids and Pads); Face and Neck Preparations (Excluding Shaving Preparations); Skin Care Preparations, Misc.

Trade Names:
Glucam E-10 (Amerchol)
Ritacam E-10 (RITA)

Trade Name Mixture:
Milk Albumin (Cosmetochem)
(Cosmetochem International Ltd.)

METHYL GLUCETH-20

CAS No.: 68239-42-9 (Generic)

JPN Translation:
メチルグルセス - 20

Definition: Methyl Gluceth-20 is the polyethylene glycol ether of methyl glucose that conforms generally to the formula:

$$CH_3C_6H_{10}O_5(OCH_2CH_2)_nOH$$

where n has an average value of 20.

Information Sources: JSQI, TSCA

Chemical Classes: Alkoxylated Alcohols; Carbohydrates; Polyols

Function: Skin-Conditioning Agent - Humectant

Reported Product Categories: Moisturizing Preparations; Bath Soaps and Detergents; Skin Care Preparations, Misc.; Night Skin Care Preparations; Bath Preparations, Misc.; Body and Hand Preparations (Excluding Shaving Preparations); Bath Capsules; Bath Oils, Tablets, and Salts; Cleansing Products (Cold Creams, Cleansing Lotions, Liquids and Pads); Face and Neck Preparations (Excluding Shaving Preparations); Fragrance Preparations, Misc.; Hair Wave Sets; Manicuring Preparations, Misc.; Paste Masks (Mud Packs)

Trade Names:
 Aloxe MG-20 (Alzo)
 Glucam E-20 (Amerchol)
 Hetoxide MG-20 (Heterene)
 Ritacam E-20 (RITA)

Trade Name Mixtures:
 Dermaseal Allergen-Blocker 100-1
  (Hydromer)
 Dermaseal Allergen-Blocker E-100
  (Hydromer)

METHYL GLUCETH-20 BENZOATE

Definition: Methyl Gluceth-20 Benzoate is the ester of Methyl Gluceth-20 (q.v.) and benzoic acid.

Chemical Classes: Alkoxylated Alcohols; Esters

Functions: Skin-Conditioning Agent - Emollient; Solvent

Trade Name:
 Finsolv EMG-20 (Finetex)

METHYL GLUCOSE CAPRYLATE/ CAPRATE

JPN Translation:
 モノ（カプリル／カプリン酸）メチルグルコース

Definition: Methyl Glucose Caprylate/ Caprate is the ester of methyl glucoside and a mixture of caprylic and capric acids.

Information Source: JCIC

Chemical Classes: Carbohydrates; Esters

Function: Skin-Conditioning Agent - Emollient

Technical/Other Name:
 Methyl Glucoside Mono (Caprylate/ Caprate) Solution

METHYL GLUCOSE DIOLEATE

Empirical Formula:
 $C_{43}H_{78}O_8$

Definition: Methyl Glucose Dioleate is the diester of a methyl glucoside and oleic acid.

Chemical Classes: Carbohydrates; Esters; Polyols

Function: Skin-Conditioning Agent - Emollient

Trade Name:
 Glucate DO (Amerchol)

METHYLGLUCOSE DIOLEATE/ HYDROXYSTEARATE

Definition: Methylglucose Dioleate/ Hydroxystearate is the mixed ester of methylglucose and oleic and hydroxystearic acids.

Chemical Class: Carbohydrates

Function: Surfactant - Emulsifying Agent

METHYL GLUCOSE ISOSTEARATE

Empirical Formula:
 $C_{25}H_{48}O_7$

Definition: Methyl Glucose Isostearate is the ester of methyl glucoside and Isostearic Acid (q.v.).

Chemical Classes: Carbohydrates; Esters

Function: Skin-Conditioning Agent - Emollient

Trade Name:
 Isolan IS (Degussa Care Specialties)

METHYL GLUCOSE LAURATE

Empirical Formula:
 $C_{19}H_{34}O_7$

Definition: Methyl Glucose Laurate is the ester of methyl glucoside and lauric acid.

Chemical Classes: Carbohydrates; Esters

Function: Skin-Conditioning Agent - Emollient

METHYL GLUCOSE SESQUICAPRYLATE/ SESQUICAPRATE

Definition: Methyl Glucose Sesquicaprylate/ Sesquicaprate is a mixture of mono- and diesters of a methyl glucoside and caprylic and capric acids.

Chemical Classes: Carbohydrates; Esters; Polyols

Function: Skin-Conditioning Agent - Emollient

METHYL GLUCOSE SESQUICOCOATE

Definition: Methyl Glucose Sesquicocoate is a mixture of mono- and diesters of a methyl glucoside and Coconut Acid (q.v.).

Chemical Classes: Carbohydrates; Esters; Polyols

Function: Skin-Conditioning Agent - Emollient

METHYL GLUCOSE SESQUI- ISOSTEARATE

Definition: Methyl Glucose Sesqui- isostearate is a mixture of mono- and diesters of a methyl glucoside and isostearic acid.

Chemical Classes: Carbohydrates; Esters; Polyols

Function: Skin-Conditioning Agent - Emollient

METHYL GLUCOSE SESQUILAURATE

Definition: Methyl Glucose Sesquilaurate is a mixture of mono- and diesters of methyl glucoside and lauric acid.

Chemical Classes: Carbohydrates; Esters; Polyols

Function: Skin-Conditioning Agent - Emollient

METHYL GLUCOSE SESQUIOLEATE

Definition: Methyl Glucose Sesquioleate is a mixture of mono- and diesters of a methyl glucoside and oleic acid.

Chemical Classes: Carbohydrates; Esters

Function: Skin-Conditioning Agent - Emollient

METHYL GLUCOSE SESQUISTEARATE

CAS No. 68936-95-8   **EINECS No.** 273-049-0

JPN Translation:
 セスキステアリン酸メチルグルコース

Definition: Methyl Glucose Sesquistearate is a mixture of mono- and diesters of a methyl glucoside and stearic acid.

Information Sources: JCIC, JCLS, JSQI, TSCA

Chemical Classes: Carbohydrates; Esters; Polyols

Function: Skin-Conditioning Agent - Emollient

Reported Product Categories: Lipsticks; Bath Oils, Tablets, and Salts; Cleansing Products (Cold Creams, Cleansing Lotions, Liquids and Pads); Foundations; Bath Preparations, Misc.; Body and Hand Preparations (Excluding Shaving Preparations); Makeup Bases; Makeup Preparations (Not eye), Misc.; Moisturizing Preparations; Aftershave Lotions; Baby Shampoos; Bath Capsules; Eye Makeup Preparations, Misc.; Face and Neck Preparations (Excluding Shaving Preparations)

Technical/Other Name:
D-Glucopyranoside, Methyl, Octadecanoate (2:3)

Trade Names:
Glucate SS (Amerchol)
Tego Care PS (Degussa Care Specialties)

METHYL GLUTAMIC ACID

CAS Nos.: 6753-62-4; 35989-16-3

Empirical Formula:
$C_6H_{11}NO_4$

Definition: Methyl Glutamic Acid is the organic compound that conforms to the formula:

$$HOOCCH_2CH_2CHCOOH$$
$$|$$
$$NHCH_3$$

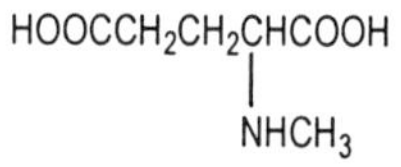

Chemical Class: Alkyl-Substituted Amino Acids

Functions: Hair Conditioning Agent; Skin-Conditioning Agent - Miscellaneous

Technical/Other Names:
Glutamic Acid, N-Methyl-
N-Methyl Glutamic Acid

Trade Name:
N-Methyl Glutamic Acid (Penn-Squire)

METHYL GLYCYRRHIZATE

CAS No.: 104191-95-9

Empirical Formula:
$C_{43}H_{64}O_{16}$

Definition: Methyl Glycyrrhizate is the ester of methyl alcohol and Glycyrrhizic Acid (q.v.).

Chemical Classes: Alcohols; Esters; Ketones

Function: Flavoring Agent

Technical/Other Name:
Glycyrrhizic Acid, Methyl Ester

METHYLHEPTYL ISOSTEARATE

CAS No.: 209802-43-7

Empirical Formula:
$C_{26}H_{52}O_2$

Definition: Methylheptyl Isostearate is the ester of caprylic alcohol and isostearic acid. It conforms to the formula:

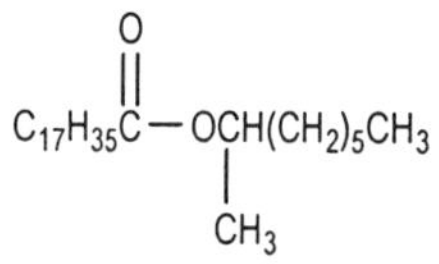

Chemical Class: Esters

Function: Skin-Conditioning Agent - Occlusive

Technical/Other Name:
Isooctadecanoic Acid, Octyl Ester

Trade Name:
Beantree (Bernel)

METHYL HESPERIDIN

CAS No.: 11013-97-1

JPN Translation:
メチルヘスペリジン

Empirical Formula:
$C_{29}H_{36}O_{15}$

Definition: Methyl Hesperidin is the organic compound that conforms to the formula:

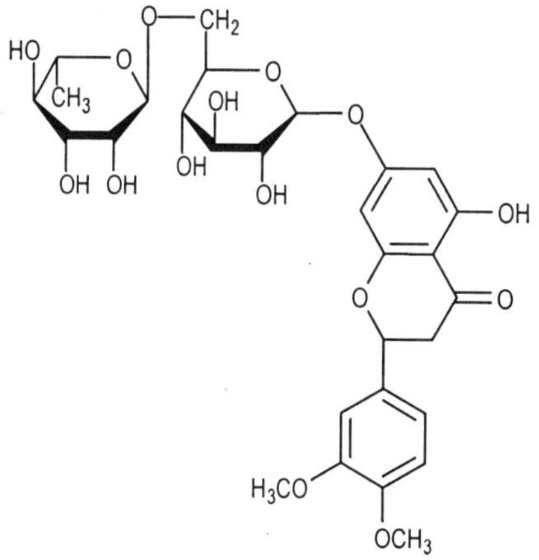

Information Source: JCLS

Chemical Classes: Carbohydrates; Ethers; Heterocyclic Compounds; Ketones; Phenols

Function: Skin-Conditioning Agent - Miscellaneous

Technical/Other Name:
4H-1-Benzopyran-4-one, 7-[[6-O-(6Deoxy-α-L-Mannopyranosyl)-β-D-Glucopyranosyl]Oxy]-2,3-Dihydro-5-Hydroxy-2-(3-Hydroxy-4-Methoxyphenyl)-, Monomethyl Ether, (2S)-

Trade Name:
Methyl Hesperidin (Iwaki & CO.)

METHYL HEXYL ETHER

CAS No.	EINECS No.
4747-07-3	225-263-0

Empirical Formula:
$C_7H_{16}O$

Definition: Methyl Hexyl Ether is the aliphatic ether that conforms to the formula:

$$CH_3O(CH_2)_5CH_3$$

Information Sources: RIFM, TSCA

Chemical Class: Ethers

Functions: Fragrance Ingredient; Solvent; Viscosity Decreasing Agent

Technical/Other Names:
Hexane, 1-Methoxy-
1-Methoxyhexane (RIFM)
1-Methoxyhexane

1-METHYLHYDANTOIN-2-IMIDE

CAS No.	EINECS No.
60-27-5	200-466-7

JPN Translation:
クレアチニン

Empirical Formula:
$C_4H_7N_3O$

Definition: 1-Methylhydantoin-2-Imide is the organic compound that conforms to the formula:

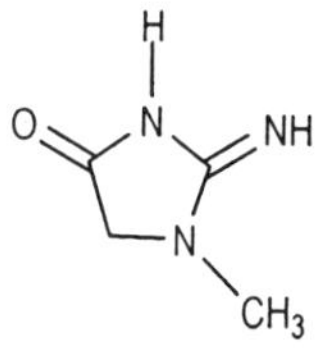

Information Source: MI-13(2597)

Chemical Class: Heterocyclic Compounds

Function: Skin-Conditioning Agent - Miscellaneous

Technical/Other Names:
Creatinine
4H-Imidazoool-4-one, 2-Amino-1,5-Dihydro-1-Methyl-

Trade Names:
Cosmocair C 250 (Degussa Care Specialties)
Creatinine (Yuki Gosei Kogyo)

METHYL HYDROGENATED ROSINATE

CAS No.: 8050-15-5

Definition: Methyl Hydrogenated Rosinate is the ester of methyl alcohol and the hydrogenated alicyclic acids derived from rosin.

Information Sources: RIFM, TSCA

Chemical Class: Esters

Functions: Fragrance Ingredient; Skin-Conditioning Agent - Emollient; Viscosity Increasing Agent - Nonaqueous

Reported Product Category: Permanent Waves

Technical/Other Names:
Methyl ester of rosin (partially hydrogenated) (RIFM)
Resin Acids and Rosin Acids, Hydrogenated, Methyl Esters

Trade Names:
Brilloline (Laserson)
Foralyn 5020-F (Eastman Chemical)
Foralyn 5020-F Ester of Hydrocarbon Resin (Eastman Chemical)

METHYL HYDROXYCETYL GLUCAMINIUM LACTATE

Empirical Formula:
$C_{23}H_{50}NO_6 \cdot C_3H_6O_3$

Definition: Methyl Hydroxycetyl Glucaminium Lactate is the organic salt that conforms to the formula:

Chemical Class: Organic Salts

Functions: Hair Conditioning Agent; Skin-Conditioning Agent - Miscellaneous

Technical/Other Name:
N-Methyl, N-(2-Hydroxycetyl) Glucaminium Lactate

2-METHYL-5-HYDROXYETHYLAMINO-PHENOL

CAS No.	EINECS No.
55302-96-0	259-583-7

Empirical Formula:
$C_9H_{13}NO_2$

Definition: 2-Methyl-5-Hydroxyethylaminophenol is the substituted aromatic amine that conforms to the formula:

See "Regulatory and Ingredient Use Information," for Colorants in Volume 1, Introduction, Part A.

Information Sources: CIR: [S] JACT-9(2)-1990, EEC(III/2-21)

Chemical Classes: Amines; Color Additives - Hair; Phenols

Function: Hair Colorant

Reported Product Categories: Hair Dyes and Colors (All Types Requiring Caution Statements and Patch Tests); Hair Color Sprays (Aerosol)

Technical/Other Names:
5-[(2-Hydroxyethyl)Amino]-2-Methylphenol
Phenol, 5-[(2-Hydroxyethyl)Amino]-2-Methyl-

Trade Names:
Colorex PAOX (Chemical Compounds, Inc.)
Imexine OAG (Chimex)
JAROCOL 2M5HEAP (Robinson)

METHYL HYDROXYETHYLCELLULOSE

CAS No.: 9032-42-2

Definition: Methyl Hydroxyethylcellulose is the methyl ether of Hydroxyethylcellulose (q.v.).

Chemical Class: Gums, Hydrophilic Colloids and Derivatives

Functions: Adhesive; Emulsion Stabilizer; Viscosity Increasing Agent - Aqueous

Technical/Other Name:
Cellulose, 2-hydroxyethyl Methyl Ether

Trade Names:
Tylopur MH (Clariant)
Tylopur MH (Clariant GmbH, Personal Care)
Tylopur MHB (Clariant)
Tylopur MHB (Clariant GmbH, Personal Care)
Tylose MB (Clariant)
Tylose MB (Clariant GmbH, Personal Care)
Tylose MH (Clariant)
Tylose MH (Clariant GmbH, Personal Care)
Tylose MHB (Clariant)
Tylose MHB (Clariant GmbH, Personal Care)

METHYL HYDROXYMETHYL OLEYL OXAZOLINE

CAS No.	EINECS No.
14408-42-5	238-387-5

Empirical Formula:
$C_{22}H_{41}NO_2$

Definition: Methyl Hydroxymethyl Oleyl Oxazoline is the substituted heterocyclic compound that conforms generally to the formula:

Information Sources: 21CFR176.210, CTFA D, TSCA

Chemical Class: Heterocyclic Compounds

Function: Hair Conditioning Agent

Technical/Other Names:
2-(8-Heptadecenyl)-4-Methyl-2-Oxazoline-4-Methanol
4-Oxazolemthanol, 2-(8-heptadecenyl)-4,5-dihydro-4-Methyl-
2-Oxazoline-4-Methanol, 2-(8-Heptadecenyl)-4-Methyl-

2-METHYL-4-HYDROXYPYRROLIDINE

Empirical Formula:
$C_5H_{11}NO$

Definition: 2-Methyl-4-Hydroxypyrrolidine is the substituted heterocyclic organic compound that conforms to the formula:

Chemical Classes: Alcohols; Heterocyclic Compounds

Function: Skin-Conditioning Agent - Miscellaneous

Trade Name:
Tiazamina (Vevy)

METHYL HYDROXYSTEARATE

CAS Nos.	EINECS No.
141-23-1	205-471-8
1331-93-7	

Empirical Formula:
$C_{19}H_{38}O_3$

Definition: Methyl Hydroxystearate is the ester of methyl alcohol and Hydroxystearic Acid (q.v.). It conforms to the formula:

Information Sources: 21CFR176.210, CTFA D, TSCA

Chemical Class: Esters

Function: Skin-Conditioning Agent - Emollient

Technical/Other Names:
12-Hydroxyoctadecanoic Acid, Methyl Ester
Methyl 12-Hydroxyoctadecanoate
Methyl 12-Hydroxystearate
Octadecanoic Acid, 12-Hydroxy-, Methyl Ester

Trade Name:
AEC Methyl Hydroxystearate (A & E Connock)

METHYL ISOSTEARATE

CAS No.: 68517-10-2

Empirical Formula:
$C_{19}H_{38}O_2$

Definition: Methyl Isostearate is the ester of methyl alcohol and isostearic acid. It conforms to the formula:

$$C_{17}H_{35}\overset{\overset{O}{\|}}{C}-OCH_3$$

Information Source: TSCA

Chemical Class: Esters

Function: Skin-Conditioning Agent - Emollient

Technical/Other Names:
Isooctadecanoic Acid, Methyl Ester
Methyl Isooctadecanoate

Trade Name:
PRISORINE 3760 (Uniqema Europe)

METHYLISOTHIAZOLINONE

CAS No.	EINECS No.
2682-20-4	220-239-6

JPN Translation:
メチルイソチアゾリノン

Empirical Formula:
C_4H_5NOS

Definition: Methylisothiazolinone is the heterocyclic organic compound that conforms to the formula:

Information Sources: 21CFR175.105, 21CFR176.170, CIR: [SQ] JACT-11(1)1992, EEC(VI/1-39), JCIC, JCLS, MHLW-331/3

Chemical Class: Heterocyclic Compounds

Function: Preservative

Reported Product Categories: Shampoos (Non-coloring); Hair Conditioners; Hair Tints; Hair Preparations (Non-coloring), Misc.; Bath Oils, Tablets, and Salts; Cleansing Products (Cold Creams, Cleansing Lotions, Liquids and Pads); Bath Preparations, Misc.; Body and Hand Preparations (Excluding Shaving Preparations); Skin Care Preparations, Misc.; Bubble Baths; Moisturizing Preparations; Bath Soaps and Detergents; Tonics, Dressings, and Other Hair Grooming Aids; Bath Capsules; Face and Neck Preparations (Excluding Shaving Preparations); Hair Shampoos (Coloring); Personal Cleanliness Products, Misc.; Skin Fresheners; Baby Products, Misc.; Eyebrow Pencils; Suntan Gels, Creams, and Liquids; Hair Rinses (Coloring); Hair Rinses (Non-coloring); Shaving Preparations, Misc.; Baby Lotions, Oils, Powders and Creams; Baby Shampoos; Eye Makeup Removers; Foundations; Hair Wave Sets; Night Skin Care Preparations; Paste Masks (Mud Packs); Permanent Waves; Suntan Preparations, Misc.

Technical/Other Names:
3(2H)-Isothiazolone, 2-Methyl-
Methylchloroisothiazolinone•Methyliso-thiazolinone Solution
2-Methyl-3(2H)-Isothiazolone
2-Methyl-4-Isothiazolin-3-one

Trade Names:
Microcare MT (Acti-Chem)
Neolone 950 Preservative (Rohm and Haas)
Neolone 5000 preservative (Rohm and Haas)

Trade Name Mixtures:
Conservateur GD500 (Phytocos)
Conservateur GD700 (Phytocos)
Dekaben IT (Dekker)
Euxyl K 100 (Schulke & Mayr)
Euxyl K 727 (Schulke & Mayr)
Kathon CG (Rohm and Haas)
Kathon CG II Biocide (Rohm and Haas)
Kathon CG III (Rohm and Haas)
Liposerve MB (Lipo)
Liposerve MM (Lipo)
Microcare CB (Acti-Chem)
Microcare CBA (Acti-Chem)
Microcare ITA (Acti-Chem)
Microcare ITL (Acti-Chem)
Microcare SI (Acti-Chem)
Nipaguard CMB (Clariant)
Nipaguard CMB (Clariant GmbH, Personal Care)

METHYL LACTATE

CAS Nos.	EINECS Nos.
547-64-8	208-930-0
27871-49-4	248-704-9

Empirical Formula:
$C_4H_8O_3$

Definition: Methyl Lactate is the ester of methyl alcohol and Lactic Acid (q.v.). It conforms to the formula:

$$CH_3\overset{}{\underset{\underset{OH}{|}}{C}}H-\overset{\overset{O}{\|}}{C}-OCH_3$$

Information Sources: CIR: [SQ] IJT-17 (Suppl. 1)1998, MI-13(6116), RIFM

Chemical Classes: Alcohols; Esters

Functions: Flavoring Agent; Fragrance Ingredient; Solvent

Technical/Other Names:
Lactic Acid, Methyl Ester
Methyl 2-hydroxypropanoate
Methyl 2-Hydroxypropionate
Methyl lactate (RIFM)
Methyl (S)-(-)-Lactate
Propanoic Acid, 2-hydroxy-, Methyl Ester

Trade Name:
(S)-(-)-Methyl Lactate (Merck KGaA)

METHYL LACTIC ACID

CAS No.	EINECS No.
594-61-6	209-848-8

Empirical Formula:
$C_4H_8O_3$

Definition: Methyl Lactic Acid is the organic compound that conforms to the formula:

$$CH_3\overset{}{\underset{\underset{CH_3}{|}}{\overset{\overset{OH}{|}}{C}}}COOH$$

Information Source: TSCA

Chemical Classes: Alcohols; Carboxylic Acids

Function: Fragrance Ingredient

Technical/Other Names:
Acetonic Acid
Hydroxydimethylactetic Acid
α-Hydroxyisobutyric Acid
2-Hydroxy-2-Methylpropionic Acid
Lactic Acid, 2-Methyl
Propanoic Acid, 2-Hydroxy-2-Methyl-

METHYL LAURATE

CAS No.	EINECS No.
111-82-0	203-911-3

Empirical Formula:
$C_{13}H_{26}O_2$

Definition: Methyl Laurate is the ester of methyl alcohol and lauric acid. It conforms to the formula:

$$CH_3(CH_2)_{10}\overset{\displaystyle O}{\overset{\displaystyle \|}{C}}-OCH_3$$

Information Sources: 21CFR172.225, 21CFR172.515, 21CFR176.200, 21CFR176.210, 21CFR177.2260, 21CFR177.2800, RIFM, TSCA

Chemical Class: Esters

Functions: Fragrance Ingredient; Skin-Conditioning Agent - Emollient

Technical/Other Names:
Dodecanoic Acid, Methyl Ester
Lauric Acid, Methyl Ester
Methyl Dodecanoate
Methyl laurate (RIFM)

Trade Names:
AEC Methyl Laurate (A & E Connock)
CE-1218 (Procter & Gamble)
CE-1270 (Procter & Gamble)
CE-1295 (Procter & Gamble)
DUB Lame (Stearinerie Dubois Fils)
ESTOL 1502 (Uniqema Europe)
ESTOL 1507 (Uniqema Europe)

Trade Name Mixture:
ESTOL 1519 (Uniqema Europe)

METHYL LINOLEATE

CAS Nos.	EINECS Nos.
112-63-0	203-993-0
2462-85-3	219-560-4

Empirical Formula:
$C_{19}H_{34}O_2$

Definition: Methyl Linoleate is the ester of methyl alcohol and linoleic acid. It conforms to the formula:

$$CH_3(CH_2)_4CH=CHCH_2CH=CH(CH_2)_7\overset{\displaystyle O}{\overset{\displaystyle \|}{C}}-OCH_3$$

Information Sources: 21CFR172.225, MI-13(6117), RIFM, TSCA

Chemical Class: Esters

Functions: Fragrance Ingredient; Skin-Conditioning Agent - Emollient

Technical/Other Names:
Linoleic Acid, Methyl Ester
Methyl linoleate (RIFM)
Methyl 9,12-octadecadienoate
9,12-Octadecadienoic Acid, Methyl Ester

Trade Names:
AEC Methyl Linoleate (A & E Connock)
DUB LIM (Stearinerie Dubois Fils)

METHYL METHACRYLATE/ACRYLO-NITRILE COPOLYMER

CAS No.: 30396-85-1

Definition: Methyl Methacrylate/Acrylonitrile Copolymer is a copolymer of methyl methcrylate and acrylonitrile monomers.

Chemical Class: Synthetic Polymers

Function: Film Former

Technical/Other Names:
Methacrylic Acid Methyl Ester, Polymer with Acrylontrile
2-Propenoic Acid, 2-methyl-, Methyl Ester, Polymer with 2-propenenitrile

Trade Name:
Micropearl F-50ED (Matsumoto Yushi-Seiyaku)

METHYL METHACRYLATE CROSS-POLYMER

CAS No.: 25777-71-3

Empirical Formula:
$(C_{10}H_{14}O_4 \cdot C_5H_8O_2)_x$

Definition: Methyl Methacrylate Cross-polymer is a copolymer of methyl methacrylate crosslinked with glycol dimethacrylate.

Information Source: TSCA

Chemical Class: Synthetic Polymers

Functions: Film Former; Viscosity Increasing Agent - Nonaqueous

Reported Product Category: Makeup Preparations (Not eye), Misc.

Technical/Other Names:
Methyl 2-Methyl-2-Propenoate, Polymer with 2-Methyl-2-Propenoic Acid, 1,2-Ethanediyl Ester
2-Methyl-2-Propenoic Acid, 1,2-Ethanediyl Ester, Polymer with Methyl 2-Methyl-2-Propenoate
2-Propenoic Acid, 2-Methyl, 1,2-Ethanediyl Ester, Polymer with Methyl 2-Methyl-2-Propenoate

Trade Names:
Covabead LH 85 (LCW)
Ganzpearl GMP-0800 (Ganz Chemical)
Ganzpearl GMX-0610 (Presperse)
Jurymer MB-1P (Nihon Junyaku)
Jurymer MB-1XJ (Nihon Junyaku)
Micropearl M305 (SEPPIC)
Microsphere M-305 (Tomen America)
Techpolymer LMX-C Series (Sekisui)
TECHPOLYMER MBP series (Sekisui)
TECHPOLYMER MBX-C series (Sekisui)

METHYL METHACRYLATE/GLYCOL DIMETHACRYLATE CROSSPOLYMER

Definition: Methyl Methacrylate/Glycol Dimethacrylate Crosspolymer is a cross-polymer of methyl methacrylate and ethylene glycol dimethacrylate monomers.

Chemical Class: Synthetic Polymers

Function: Film Former

Trade Name:
Microsponge (EDT, Inc.)

Trade Name Mixtures:
Matsumoto Microsphere S-100 (Tomen America)
Matsumoto Microsphere S-101 (Tomen America)
Matsumoto Microsphere S-102 (Tomen America)
Microsponge 5700 Dimethicone (EDT, Inc.)
Microsponge 5647 Glycerin (EDT, Inc.)

METHYL 3-METHYLRESORCYLATE

CAS No.	EINECS No.
33662-58-7	251-618-4

Empirical Formula:
$C_9H_{10}O_4$

Definition: Methyl 3-Methylresorcylate is the organic compound that conforms to the formula:

Information Source: RIFM

Chemical Classes: Esters; Phenols

Function: Fragrance Ingredient

Technical/Other Names:
Benzoic Acid, 2,4-Dihydroxy-3-Methyl-, Methyl Ester
2,4-Dihydroxy-3-Methylbenzoic Acid, Methyl Ester
Methyl 2,4-dihydroxy-m-toluate (RIFM)
Methyl 2,4-dihydroxy-m-toluate
Methyl 2,4-diydroxy-3-methylbenzoate

Trade Name:
Seamoss (Hercules/PFW)

METHYL MORPHOLINE OXIDE

CAS No.	EINECS No.
7529-22-8	231-391-8

Empirical Formula:
$C_5H_{11}NO_2$

Definition: Methyl Morpholine Oxide is the substituted cyclic tertiary amine oxide that conforms to the formula:

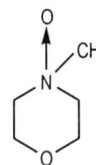

Chemical Classes: Amine Oxides; Heterocyclic Compounds

Functions: Surfactant - Cleansing Agent; Surfactant - Foam Booster; Surfactant - Hydrotrope

Technical/Other Names:
4-Methylmorpholine-4-Oxide
Morpholine, 4-Methyl-, 4-Oxide

METHYL MYRISTATE

CAS No.
124-10-7

EINECS No.
204-680-1

Empirical Formula:
$C_{15}H_{30}O_2$

Definition: Methyl Myristate is the ester of methyl alcohol and myristic acid. It conforms generally to the formula:

$$CH_3(CH_2)_{12}\overset{O}{\overset{\|}{C}}-OCH_3$$

Information Sources: 21CFR172.225, 21CFR172.515, 21CFR176.200, 21CFR176.210, 21CFR177.2260, 21CFR177.2800, RIFM, TSCA

Chemical Class: Esters

Functions: Fragrance Ingredient; Skin-Conditioning Agent - Emollient

Technical/Other Names:
Methyl myristate (RIFM)
Methyl Tetradecanoate
Myristic Acid, Methyl Ester
Tetradecanoic Acid, Methyl Ester

Trade Names:
AEC Methyl Myristate (A & E Connock)
DUB MME (Stearinerie Dubois Fils)
ESTOL 1520 (Uniqema Europe)

Trade Name Mixture:
ESTOL 1519 (Uniqema Europe)

2-METHYL-1-NAPHTHOL

CAS No.
7469-77-4

EINECS No.
231-265-2

Empirical Formula:
$C_{11}H_{10}O$

Definition: 2-Methyl-1-Naphthol is the hair colorant that conforms to the formula:

Chemical Classes: Color Additives - Hair; Phenols

Function: Hair Colorant

Technical/Other Name:
Naphthalene, 1-Hydroxy, 2-Methyl-

METHYL NICOTINATE

CAS No.
93-60-7

EINECS No.
202-261-8

JPN Translation:
ニコチン酸メチル

Empirical Formula:
$C_7H_7NO_2$

Definition: Methyl Nicotinate is an ester of methyl alcohol and nicotinic acid. It conforms to the formula:

See Reported Ingredient Functions-The Cosmetic Drug Distinction, in Regulatory and Ingredient Use Information, Volume I, Part A.

Information Sources: JCIC, JCLS, MI-13 (6122), OTC-I-EA, RIFM, TSCA, USAN

Chemical Classes: Esters; Heterocyclic Compounds

Functions: External Analgesic; Fragrance Ingredient; Skin-Conditioning Agent - Miscellaneous

Technical/Other Names:
3-(Carbomethoxy)pyridine
3-(Methoxycarbonyl)pyridine
Methyl nicotinate (RIFM)
Methyl 3-Pyridinecarboxylate
Nicotinic Acid, Methyl Ester
3-Pyridinecarboxylic Acid, Methyl Ester

Trade Names:
AEC Methyl Nicotinate (A & E Connock)
Methyl 3-Pyridinecarboxylate (Merck KGaA)

N-METHYL-3-NITRO-p-PHENYLENE-DIAMINE

CAS No.
2973-21-9

EINECS No.
221-014-5

Empirical Formula:
$C_7H_9N_3O_2$

Definition: N-Methyl-3-Nitro-p-Phenylene-diamine is the substituted aromatic amine that conforms to the formula:

See "Regulatory and Ingredient Use Information," for Colorants in Volume 1, Introduction, Part A.

Chemical Classes: Amines; Color Additives - Hair

Function: Hair Colorant

Reported Product Category: Hair Dyes and Colors (All Types Requiring Caution Statements and Patch Tests)

Technical/Other Names:
1,4-Benzenediamine, N-4-Methyl-2-Nitro-
N-4-Methyl-2-Nitro-1,4-Benzenediamine
N-Methyl-2-Nitro-p-Phenylenediamine

Trade Name:
Imexine FB (Chimex)

METHYL 2-OCTYNOATE

CAS No.: 111-12-6

Empirical Formula:
$C_9H_{14}O_2$

Definition: Methyl 2-Octynoate is the organic compound that conforms to the formula:

$$CH_3(CH_2)_4C\equiv C\overset{O}{\overset{\|}{C}}-OCH_3$$

Information Sources: EEC(III/1-89), RIFM

Chemical Class: Esters

Function: Fragrance Ingredient

Technical/Other Names:
Folione
Methyl 2-octynoate (RIFM)
Methyl Pentylacetylenecarboxylate
2-Octynoic Acid, Methyl Ester

METHYL OLEATE

CAS Nos.
112-62-9
2462-84-2

EINECS Nos.
203-992-5
219-559-9

Empirical Formula:
$C_{19}H_{36}O_2$

Definition: Methyl Oleate is the ester of methyl alcohol and oleic acid. It conforms generally to the formula:

$$CH_3(CH_2)_7CH = CH(CH_2)_7C - OCH_3$$

Information Sources: 21CFR172.225, 21CFR175.105, 21CFR176.200, 21CFR176.210, 21CFR177.2260, 21CFR177.2800, MI-13(6898), RIFM, TSCA

Chemical Class: Esters

Functions: Fragrance Ingredient; Skin-Conditioning Agent - Emollient

Technical/Other Names:
Methyl 9-octadecenoate (RIFM)
Methyl 9-Octadecenoate
9-Octadecenoic Acid, Methyl Ester
Oleic Acid, Methyl Ester

Trade Names:
AEC Methyl Oleate (A & E Connock)
STEPAN C-68 (Stepan)

METHYL PALMATE

CAS No. 91051-34-2   **EINECS No.** 293-086-6

Definition: Methyl Palmate is the ester of methyl alcohol and the fatty acids derived from Elaeis Guineensis (Palm) Oil (q.v.). It conforms generally to the formula:

$$RC - OCH_3$$

where RCO- represents the fatty acids derived from palm oil.

Chemical Class: Esters

Function: Skin-Conditioning Agent - Emollient

Technical/Other Name:
Palm Oil Fatty Acids, Methyl Ester

METHYL PALMITATE

CAS No. 112-39-0   **EINECS No.** 203-966-3

Empirical Formula:
$C_{17}H_{34}O_2$

Definition: Methyl Palmitate is the ester of methyl alcohol and palmitic acid. It conforms to the formula:

$$CH_3(CH_2)_{14}C - OCH_3$$

Information Sources: 21CFR172.225, 21CFR175.105, 21CFR176.200, 21CFR176.210, 21CFR177.2260, 21CFR177.2800, 21CFR178.3910, RIFM, TSCA

Chemical Class: Esters

Functions: Fragrance Ingredient; Skin-Conditioning Agent - Emollient

Technical/Other Names:
Hexadecanoic Acid, Methyl Ester
Methyl hexadecanoate (RIFM)
Methyl Hexadecanoate
Palmitic Acid, Methyl Ester

Trade Names:
AEC Methyl Palmitate (A & E Connock)
DUB PM COS (Stearinerie Dubois Fils)
ESTOL 1503 (Uniqema Europe)

METHYLPARABEN

CAS No. 99-76-3   **EINECS No.** 202-785-7

JPN Translation:
メチルパラベン

Empirical Formula:
$C_8H_8O_3$

Definition: Methylparaben is the ester of methyl alcohol and p-hydroxybenzoic acid. It conforms to the formula:

$$C - OCH_3$$

OH

Information Sources: AUS, BP, BPC, BRA, 21CFR150.141, 21CFR150.161, 21CFR172.515, 21CFR181.22, 21CFR181.23, 21CFR184.1490, 21CFR310.545, 21CFR556.390, 21CFR582.3490, CIR: [S] JACT-3(5)1984, CTFA S, CZE, DA, DDR, EEC(VI/1-12), EGY, FCC, FIN, HUN, IND, ITA, JAN, JCLS, JSCI, MAR, MEX, MHLW-331/3, MI-13 (6129), NF XIX, PF, PN, POR, RIFM, ROM, TSCA, USAN, USD, YUG

Chemical Classes: Esters; Phenols

Functions: Fragrance Ingredient; Preservative

Reported Product Categories: Moisturizing Preparations; Bath Preparations, Misc.; Body and Hand Preparations (Excluding Shaving Preparations); Bath Oils, Tablets, and Salts; Cleansing Products (Cold Creams, Cleansing Lotions, Liquids and Pads); Skin Care Preparations, Misc.; Shampoos (Non-coloring); Hair Conditioners; Lipsticks; Foundations; Face Powders; Bath Capsules; Blushers (All types); Face and Neck Preparations (Excluding Shaving Preparations); Tonics, Dressings, and Other Hair Grooming Aids; Mascara; Paste Masks (Mud Packs); Night Skin Care Preparations; Bath Soaps and Detergents; Eye Makeup Preparations, Misc.; Powders (Dusting and Talcum, Excluding Aftershave Talcs); Makeup Bases; Makeup Preparations (Not eye), Misc.; Hair Dyes and Colors (All Types Requiring Caution Statements and Patch Tests); Skin Fresheners; Fragrance Preparations, Misc.; Aftershave Lotions; Baby Shampoos; Eyebrow Pencils; Suntan Gels, Creams, and Liquids; Personal Cleanliness Products, Misc.; Eye Makeup Removers; Hair Preparations (Non-coloring), Misc.; Eyeliners; Indoor Tanning Preparations; Shaving Cream (Aerosol, Brushless and Lather); Bubble Baths; Deodorants (Underarm); Baby Lotions, Oils, Powders and Creams; Eye Shadows; Shaving Preparations, Misc.; Permanent Waves; Suntan Preparations, Misc.; Hair Coloring Preparations, Misc.; Colognes and Toilet Waters; Hair Wave Sets; Eye Lotions; Makeup Fixatives; Sachets; Foot Powders and Sprays; Hair Rinses (Non-coloring); Cuticle Softeners; Perfumes; Hair Shampoos (Coloring); Hair Straighteners; Nail Creams and Lotions; Dentifrices (Aerosol, Liquid, Pastes and Powders); Rouges; Manicuring Preparations, Misc.; Hair Sprays (Aerosol Fixatives); Douches; Hair Rinses (Coloring); Baby Products, Misc.; Men's Talcum; Depilatories; Nail Polish and Enamels; Feminine Hygiene Deodorants; Hair Bleaches; Hair Lighteners with Color; Hair Tints; Leg and Body Paints; Nail Polish and Enamel Removers; Preshave Lotions (All types)

Technical/Other Names:
Benzoic Acid, 4-Hydroxy-, Methyl Ester
p-Carbomryhoxyphenol
4-Hydroxybenzoic Acid, Methyl Ester
p-Methoxycarbonylphenol
Methyl 4-Hydroxybenzoate
Methyl p-hydroxybenzoate (RIFM)
Methyl Parahydroxybenzoate
Parahydroxybenzoate Ester

Trade Names:
Aseptoform (Greeff)
Botanistat MP (Botanigenics)
CoSept M (Costec)
Jeen Methyl Paraben NF (Jeen)
Lexgard M (Inolex)
Methyl 4-Hydroxybenzoate (Merck KGaA)
Methylparaben NF (RITA)
Methylparaben NF-PC (Protameen)
Methyl Parasept (Tenneco)
Nipagin M (Clariant)
Nipagin M (Clariant GmbH, Personal Care)

NS 3550 (Nutri-Shield)
Paridol M (Dekker)
S&M Methylparaben (Schulke & Mayr)
Solbrol M (Bayer AG)
Unisept M (Universal Preserv-A-Chem)

Trade Name Mixtures:
AEC Cosflor Blend 017 Moisture Factor WSS (A & E Connock)
AEC Moisture Factor HV (A & E Connock)
AEC Papaya Extract (Papain 0.25%) (A & E Connock)
AEC Pineapple Extract (Bromelain 0.25%) (A & E Connock)
Bactecar 125S (Phytocos)
Bactiphen 2506 G (Grau)
Chemynol (Chemyunion)
Compositum (Vevy)
Conservateur GD500 (Phytocos)
Conservateur GD700 (Phytocos)
CoSept PEP (RTD Hall Star)
Cosmocil AF (Zeneca)
Dekaben (Dekker)
Dekaben P (Dekker)
Dekacydol (Dekker)
Dermocide L (Fabriquimica)
Dragocid Forte 2/027045 (Symrise)
Elastase Inhibitor-3 (Arval)
Elestab 305 (Laboratoires Serobiologiques)
Elestab 388 (Laboratoires Serobiologiques)
Elestab 4112 (Laboratoires Sero-biologiques)
Elestab FL 15 (Laboratoires Sero-biologiques)
Elestab 50 J (Laboratoires Serobiologiques)
Euxyl K 300 (Schulke & Mayr)
Fenilight (Sinerga)
Fenossiparaben (Sinerga)
Germaben II (Sutton)
Germaben II-E (Sutton)
Germazide MPB (Collaborative Labs)
Gramben II (Sinerga)
Killitol II (Collaborative Labs)
Liposerve DUP (Lipo)
Liposerve PP (Lipo)
Microcare DMP (Acti-Chem)
Microcare IMP (Acti-Chem)
Microcare PM (Acti-Chem)
Microcare PM5 (Acti-Chem)
Mikrokill 300 (Arch Personal Care Products)
Neo Dragocide Powder 2/060100 (Symrise)
Neo Dragocid Liquid 2/060110 (Symrise)
Nipacide A (Clariant)
Nipacide A (Clariant GmbH, Personal Care)
Nipaguard BPX (Clariant)
Nipaguard BPX (Clariant GmbH, Personal Care)
Nipaguard MPA (Clariant)
Nipaguard MPA (Clariant GmbH, Personal Care)
Nipaguard MPS (Clariant)

Nipaguard MPS (Clariant GmbH, Personal Care)
Nipaguard PDU (Clariant)
Nipaguard PDU (Clariant GmbH, Personal Care)
Nipasept (Clariant)
Nipasept (Clariant GmbH, Personal Care)
Nipastat (Clariant)
Nipastat (Clariant GmbH, Personal Care)
Ocean Collagen B-03 (Air Water)
Ocean Collagen B-05 (Air Water)
Paragon (McIntyre)
Paragon II (McIntyre)
Paragon III (McIntyre)
Paragon MEPB (McIntyre)
Paraoxiben (Vevy)
Phenonip (Clariant)
Phenonip (Clariant GmbH, Personal Care)
Phenova (Crodarom)
Pongamia Complex (Greentech)
RonaCare ASC III (Merck KGaA)
RonaCare VTA (Merck KGaA)
Saccaluronate CC (LCW)
Saccaluronate LC (LCW)
Self Tanning Complex (Greentech)
Sepicide HB (SEPPIC)
Sepicide HB2 (SEPPIC)
Sepicide WP1 (SEPPIC)
Talcoseptic C (Vevy)
Undebenzofene C (Vevy)
Uniphen P-23 (Induchem)

METHYL PELARGONATE

CAS No. 1731-84-6   **EINECS No.** 217-052-7

Empirical Formula:
$C_{10}H_{20}O_2$

Definition: Methyl Pelargonate is the ester of methyl alcohol and Pelargonic Acid (q.v.). It conforms to the formula:

$$CH_3(CH_2)_7\overset{\overset{\displaystyle O}{\|}}{C}-OCH_3$$

Information Sources: 21CFR172.515, RIFM, TSCA

Chemical Class: Esters

Functions: Fragrance Ingredient; Skin-Conditioning Agent - Emollient

Technical/Other Names:
Methyl nonanoate (RIFM)
Methyl Nonanoate
Nonanoic Acid, Methyl Ester
Pelargonic Acid Methyl Ester

METHYL PERFLUOROBUTYL ETHER

CAS No.: 163702-08-7

Empirical Formula:
$C_5H_3F_9O$

Definition: Methyl Perfluorobutyl Ether is the organic compound that conforms to the formula:

$$\underset{\underset{\displaystyle CF_3}{|}}{CF_3CFCF_2OCH_3}$$

Chemical Class: Halogen Compounds

Functions: Solvent; Viscosity Decreasing Agent

Technical/Other Name:
Propane, 2-(Difluoromethoxymethyl)-1,1,1,2,3,3,3-Heptafluoro-

Trade Name:
3M Cosmetic Fluid CF-61 (3M)

METHYL PHENYLBUTANOL

CAS No. 103-05-9   **EINECS No.** 203-074-4

Empirical Formula:
$C_{11}H_{16}O$

Definition: Methyl Phenylbutanol is the organic compound that conforms to the formula:

Information Source: RIFM

Chemical Class: Alcohols

Function: Fragrance Ingredient

Technical/Other Names:
Benzyl-tert-butanol
2-Butanol, 2-Methyl-4-Phenyl
Dimethylbenzenepropanol
1,1-Dimethyl-3-phenylpropanol
2-Methyl-4-phenyl-2-butanol (RIFM)
2-Methyl-4-Phenylbutan-2-ol
2-Phenethyl-2-propanol
Phenylethyl Dimethyl Carbinol

Trade Name:
Dimethyl Phenyl Ethyl Carbinol (International Flavors & Fragrances)

METHYLPOLYSILOXANE EMULSION

Definition: *See "Regulatory and Ingredient Use Information," regarding use of Japan Trivial names in Volume 1, Introduction, Part A.*

Information Source: JCLS

Chemical Classes: Alkoxylated Alcohols; Siloxanes and Silanes

Function: Not Reported

METHYLPROPANEDIOL

CAS No.: 2163-42-0

Empirical Formula:
$C_4H_{10}O_2$

Definition: Methylpropanediol is the organic compound that conforms to the formula:

$$HOCH_2 \overset{\overset{\displaystyle CH_3}{|}}{CH} CH_2OH$$

Chemical Class: Polyols

Function: Solvent

Reported Product Category: Deodorants (Underarm)

Technical/Other Names:
β-Hydroxyisobutanol
2-Methyl-1,3-Propanediol

Trade Names:
AEC Methylpropanediol (A & E Connock)
MPDiol glycol (Lyondell Chemical)

Trade Name Mixture:
Emiliania 5 MMPD (Eclosarium)

METHYL PYRROLIDONE

CAS Nos.: 872-50-4; 51013-18-4

JPN Translation:
メチルピロリドン

Empirical Formula:
C_5H_9NO

Definition: Methyl Pyrrolidone is the substituted heterocyclic organic compound that conforms to the formula:

Information Sources: 21CFR176.300, JCIC, JCLS, JSQI, MI-13(6140), TSCA

Chemical Class: Heterocyclic Compounds

Function: Solvent

Technical/Other Names:
1-Methylazacyclopentan-2-one
N-Methyl-γ-butyrolactam
1-Methyl-2-Pyrrolidone
N-Methyl-2-Pyrrolidone
2-Pyrrolidone, 1-Methyl

Trade Name Mixtures:
Flouwet EA 093 (Clariant)
Flouwet EA 093 (Clariant GmbH, Personal Care)
Liposcreen PC4 (Lipo)
Lowenol Solvent 402 (Lowenstein)

2-METHYLRESORCINOL

CAS No.	EINECS No.
608-25-3	210-155-8

Empirical Formula:
$C_7H_8O_2$

Definition: 2-Methylresorcinol is the aromatic compound that conforms to the formula:

See "Regulatory and Ingredient Use Information," for Colorants in Volume 1, Introduction, Part A.

Information Sources: CIR: [S] JACT-5(3)-1986, EEC(III/2-37), TSCA

Chemical Classes: Color Additives - Hair; Phenols

Function: Hair Colorant

Reported Product Categories: Hair Dyes and Colors (All Types Requiring Caution Statements and Patch Tests); Hair Tints

Technical/Other Names:
1,3-Benzenediol, 2-Methyl-
2,6-Dihydroxytoluene
2-Methyl-1,3-Benzenediol
2-Methyl-1,3-dihydroxybenzene
Tolunene-2,6-diol

Trade Names:
Colorex 2MR-HP (Chemical Compounds, Inc.)
Jarocol 2MR (James Robinson)
Rodol MRP (Lowenstein)
Rodol MRS (Lowenstein)

METHYL RICINOLEATE

CAS Nos.	EINECS No.
141-24-2	205-472-3
23224-20-6	

JPN Translation:
ヒマシ脂肪酸メチル

Empirical Formula:
$C_{19}H_{36}O_3$

Definition: Methyl Ricinoleate is the ester of methyl alcohol and ricinoleic acid. It conforms generally to the formula:

$$CH_3(CH_2)_5 \overset{\overset{\displaystyle }{|}}{\underset{\underset{\displaystyle OH}{|}}{CH}} CH_2CH = CH(CH_2)_7 \overset{\overset{\displaystyle O}{\|}}{C} - OCH_3$$

Information Sources: 21CFR175.105, 21CFR176.210, JCIC, JCLS, JSQI, TSCA

Chemical Class: Esters

Function: Skin-Conditioning Agent - Emollient

Reported Product Category: Tonics, Dressings, and Other Hair Grooming Aids

Technical/Other Names:
Castor Oil Acid, Methyl Ester
12-Hydroxy-9-Octadecenoic Acid, Methyl Ester
Methyl Castor Oil Fatty Acid
Methyl 12-Hydroxy-9-Octadecenoate
9-Octadecenoic Acid, 12-Hydroxy-, Methyl Ester
Ricinoleic Acid, Methyl Ester

Trade Name:
DUB RM (Stearinerie Dubois Fils)

METHYL ROSINATE

CAS No.	EINECS No.
68186-14-1	269-035-9

Definition: Methyl Rosinate is the methyl ester of acids recovered from Rosin (q.v.).

Information Sources: 21CFR172.615, 21CFR175.105, 21CFR175.300, 21CFR176.170, 21CFR176.200, 21CFR176.210, 21CFR177.1200, 21CFR177.2600, 21CFR178.3120, 21CFR178.3800, 21CFR178.3870, MI-13 (6036), RIFM, TSCA

Chemical Class: Esters

Functions: Fragrance Ingredient; Skin-Conditioning Agent - Emollient; Viscosity Increasing Agent - Nonaqueous

Technical/Other Names:
Methyl abietate (RIFM)
Rosin Acid, Methyl Ester

Trade Names:
Abalyn (Hercules)
Abalyn E (Eastman Chemical)

METHYL SALICYLATE

CAS No.	EINECS No.
119-36-8	204-317-7

JPN Translation:
サリチル酸メチル

Empirical Formula:
$C_8H_8O_3$

Definition: Methyl Salicylate is the ester of methyl alcohol and salicylic acid. It conforms to the formula:

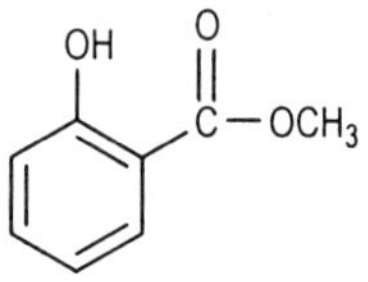

See Reported Ingredient Functions-The Cosmetic Drug Distinction, in Regulatory and Ingredient Use Information, Volume I, Part A.

Information Sources: ARG, AUS, BP, BPC, BRA, 21CFR175.105, 21CFR177.1010, 27CFR21.65, 27CFR21.151, CIR: [SQ], CTFA S, CZE, DDR, EGY, EP, FCC, FIN, HUN, IND, ITA, JAN, JCLS, JSCI, MAR, MEX, MI-13(6143), NF XIX, OTC-I-EA, PF, POR, RIFM, ROM, TSCA, USAN, USD, USSR, YUG

Chemical Classes: Esters; Phenols

Functions: Denaturant; External Analgesic; Flavoring Agent; Fragrance Ingredient

Reported Product Categories: Mouthwashes and Breath Fresheners (Liquids and Sprays); Dentifrices (Aerosol, Liquid, Pastes and Powders); Skin Care Preparations, Misc.; Bath Soaps and Detergents; Body and Hand Preparations (Excluding Shaving Preparations); Deodorants (Underarm); Foot Powders and Sprays; Skin Fresheners; Suntan Gels, Creams, and Liquids

Technical/Other Names:
Benzoic Acid, 2-Hydroxy-, Methyl Ester
Birch Oil, Sweet
Birch, sweet, oil (Betula lenta L.) (RIFM)
2-Carbomethoxyphenol
2-Hydroxybenzoic Acid, Methyl Ester
Methyl 2-Hydroxybenzoate
Methyl salicylate (RIFM)
Oil of Wintergreen
Salicylic Acid, Methyl Ester
Sweet Birch Oil

Trade Names:
AEC Methyl Salicylate (A & E Connock)
Custosense MS (methyl salicylate) (Custom Ingredients)
Unichem METSAL (Universal Preserv-A-Chem)

METHYLSERINE

CAS No.: 2480-26-4

Empirical Formula:
$C_4H_9NO_3$

Definition: Methylserine is the organic compound that conforms to the formula:

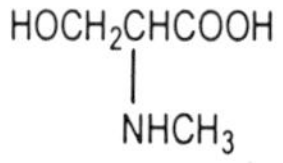

Chemical Class: Amino Acids

Function: Skin-Conditioning Agent - Humectant

Technical/Other Name:
L-Serine, N-Methyl

Trade Name:
Aquaserine (Takasago)

METHYLSILANOL ACETYLMETHIONATE

Empirical Formula:
$C_8H_{17}NO_5SSi$

Definition: Methylsilanol Acetylmethionate is the organic compound that conforms to the formula:

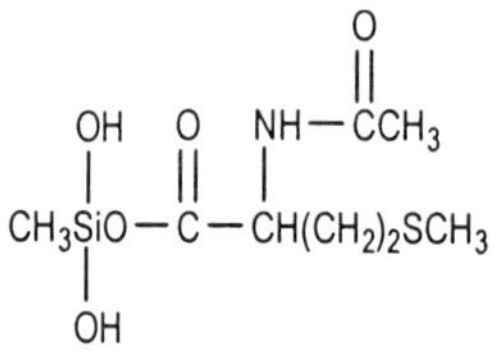

Chemical Classes: Amino Acids; Esters; Siloxanes and Silanes; Thio Compounds

Functions: Hair Conditioning Agent; Skin-Conditioning Agent - Miscellaneous

Trade Name:
Methiosilane (Exsymol)

METHYLSILANOL ACETYLTYROSINE

Empirical Formula:
$C_{12}H_{17}NO_6Si$

Definition: Methylsilanol Acetyltyrosine is the organic compound that conforms to the formula:

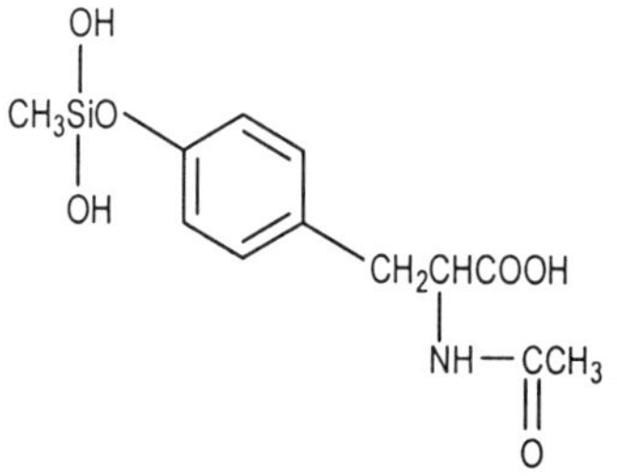

Chemical Classes: Amino Acids; Siloxanes and Silanes

Function: Skin-Conditioning Agent - Miscellaneous

Trade Name:
Tyrosilane (Exsymol)

METHYLSILANOL ASCORBATE

Empirical Formula:
$C_7H_{12}O_8Si$

Definition: Methylsilanol Ascorbate is the organic compound that conforms to the formula:

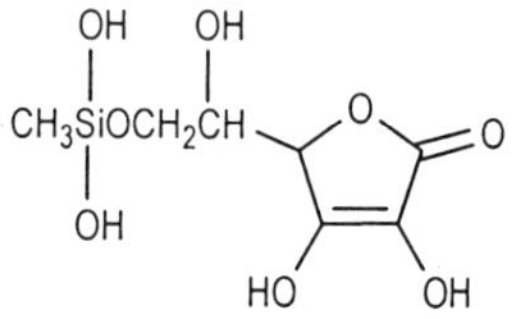

Chemical Classes: Ethers; Heterocyclic Compounds; Polyols; Siloxanes and Silanes

Function: Antioxidant

METHYLSILANOL CARBOXYMETHYL THEOPHYLLINE

Empirical Formula:
$C_{10}H_{14}N_4O_6Si$

Definition: Methylsilanol Carboxymethyl Theophylline is the organic compound that conforms to the formula:

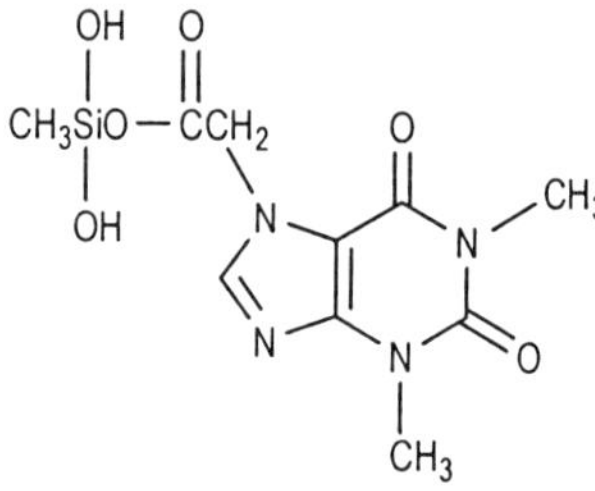

Chemical Classes: Esters; Heterocyclic Compounds; Siloxanes and Silanes

Function: Not Reported

Trade Name:
Theophyllisilane (Exsymol)

METHYLSILANOL CARBOXYMETHYL THEOPHYLLINE ALGINATE

Definition: Methylsilanol Carboxymethyl Theophylline Alginate is the reaction product of Methylsilanol Carboxymethyl Theophylline (q.v.) and Alginic Acid (q.v.).

Chemical Classes: Esters; Gums, Hydrophilic Colloids and Derivatives; Heterocyclic Compounds; Siloxanes and Silanes

Function: Not Reported

Trade Name:
Theophyllisilane C (Exsymol)

METHYLSILANOL ELASTINATE

JPN Translation:
メチルシラノールエラスチネート

Definition: Methylsilanol Elastinate is the reaction product of elastin and methylsilanol.

Chemical Classes: Protein Derivatives; Siloxanes and Silanes

Functions: Hair Conditioning Agent; Skin-Conditioning Agent - Miscellaneous

Reported Product Category: Night Skin Care Preparations

Trade Name:
Proteosilane C (Exsymol)

METHYLSILANOL GLYCYRRHIZINATE

Definition: Methylsilanol Glycyrrhizinate is the product obtained by the hydrolysis of monomethylsilanetriol in the presence of Glycyrrhizic Acid (q.v). It conforms to the formula:

where R represents the glycyrrhizoyl moiety.

Chemical Classes: Alcohols; Carboxylic Acids; Ketones; Siloxanes and Silanes

Function: Skin-Conditioning Agent - Miscellaneous

Trade Name:
Glysinol C (Exsymol)

METHYLSILANOL HYDROXYPROLINE

JPN Translation:
ヒドロキシプロリンメチルシラノール

Empirical Formula:
$C_6H_{13}NO_5Si$

Definition: Methylsilanol Hydroxyproline is the ester of monomethyl silanol and hydroxyproline. It conforms to the formula:

Chemical Classes: Alcohols; Amino Acids; Esters; Heterocyclic Compounds; Siloxanes and Silanes

Function: Skin-Conditioning Agent - Miscellaneous

METHYLSILANOL HYDROXYPROLINE ASPARTATE

Definition: Methylsilanol Hydroxyproline Aspartate is the reaction product of Methylsilanol Hydroxyproline (q.v.) and Aspartic Acid (q.v.).

Information Source: JCLS

Chemical Classes: Amino Acids; Esters; Heterocyclic Compounds; Organic Salts; Siloxanes and Silanes

Function: Skin-Conditioning Agent - Miscellaneous

Trade Name:
Hydroxyprolisilane C (Exsymol)

METHYLSILANOL MANNURONATE

JPN Translation:
アルギン酸メチルシラノール

Definition: Methylsilanol Mannuronate is the ester of monomethylsilanol and oligomeric mannuronic acid. It conforms generally to the formula:

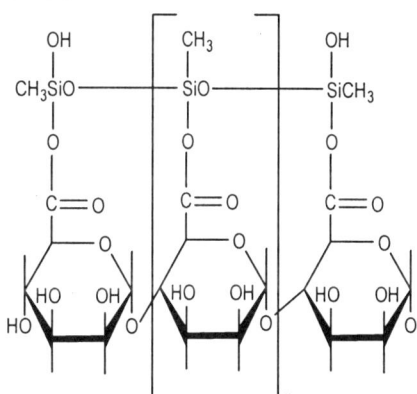

Chemical Classes: Carbohydrates; Esters; Siloxanes and Silanes

Function: Skin-Conditioning Agent - Miscellaneous

Reported Product Categories: Skin Care Preparations, Misc.; Bath Capsules; Face and Neck Preparations (Excluding Shaving Preparations)

Trade Name:
Algisium C (Exsymol)

METHYLSILANOL PCA

Empirical Formula:
$C_6H_{11}NO_5Si$

Definition: Methylsilanol PCA is the ester of monomethylsilanol and PCA (q.v). It conforms to the formula:

Chemical Classes: Amino Acids; Esters; Heterocyclic Compounds; Siloxanes and Silanes

Function: Skin-Conditioning Agent - Miscellaneous

Reported Product Category: Moisturizing Preparations

METHYLSILANOL PEG-7 GLYCERYL COCOATE

Definition: Methylsilanol PEG-7 Glyceryl Cocoate is the ethoxylated ester that conforms to the formula:

$$CH_3Si(OH)_2(OCH_2CH_2)_7OCH_2CHCH_2O - CR$$

where RCO- represents the fatty acid radical derived from coconut oil.

Chemical Classes: Glyceryl Esters and Derivatives; Siloxanes and Silanes

Function: Skin-Conditioning Agent - Emollient

Trade Name:
Monosiliol (Exsymol)

METHYLSILANOL/SILICATE CROSS-POLYMER

CAS No.: 68584-81-6

Definition: Methylsilanol/Silicate Crosspolymer is the crosspolymer formed by the reaction of silica and methylsilanol.

Chemical Class: Siloxanes and Silanes

Function: Not Reported

Trade Name:
TAK-110 (Takemoto Oil)

METHYLSILANOL SPIRULINATE

Definition: Methylsilanol Spirulinate is the reaction product of spirulina protein and methylsilanol.

Chemical Classes: Protein Derivatives; Siloxanes and Silanes

Function: Skin-Conditioning Agent - Miscellaneous

Trade Name:
Protulisilane C (Exsymol)

METHYLSILANOL TRI-PEG-8 GLYCERYL COCOATE

JPN Translation:
メチルシラノール / トリ PEG - 8 ヤシ脂肪酸グリセリル

Definition: Methylsilanol Tri-PEG-8 Glyceryl Cocoate is the ethoxylated ester that conforms to the formula:

$$CH_3Si \left[(OCH_2CH_2)_8OCH_2\underset{OH}{CH}CH_2O - \overset{O}{\underset{\|}{C}}R \right]_3$$

where RCO- represents the fatty acid radical derived from coconut oil.

Chemical Classes: Glyceryl Esters and Derivatives; Siloxanes and Silanes

Function: Skin-Conditioning Agent - Emollient

Trade Name:
Monosiliol C (Exsymol)

METHYL SOYATE

CAS No.	EINECS No.
68919-53-9	272-898-4

Definition: Methyl Soyate is the ester of methyl alcohol and Soy Acid (q.v.). It conforms generally to the formula:

$$R\overset{O}{\underset{\|}{C}} - OCH_3$$

where RCO- represents the fatty acids derived from soybean oil.

Chemical Class: Esters

Function: Skin-Conditioning Agent - Emollient

Technical/Other Name:
Soy Acid, Methyl Ester

Trade Names:
SoyClear 1500 (AG Environmental)
Soygold 1000 (AG Environmental)
SoyGold 1100 (AG Environmental)

METHYL STEARATE

CAS No.	EINECS No.
112-61-8	203-990-4

Empirical Formula:
$C_{19}H_{38}O_2$

Definition: Methyl Stearate is the ester of methyl alcohol and stearic acid. It conforms to the formula:

$$CH_3(CH_2)_{16}\overset{O}{\underset{\|}{C}} - OCH_3$$

Information Sources: 21CFR172.225, 21CFR176.200, 21CFR176.210, 21CFR177.2260, 21CFR177.2800, 21CFR178.3910, MI-13(8882), RIFM, TSCA

Chemical Class: Esters

Functions: Fragrance Ingredient; Skin-Conditioning Agent - Emollient

Technical/Other Names:
Methyl octadecanoate (RIFM)
Octadecanoic Acid, Methyl Ester
Stearic Acid, Methyl Ester

Trade Names:
AEC Methyl Stearate (A & E Connock)
DUB SME (Stearinerie Dubois Fils)

METHYLSTYRENE/VINYLTOLUENE COPOLYMER

CAS No.: 9017-27-0

Empirical Formula:
$(C_9H_{10} \cdot C_9H_{10})_x$

Definition: Methylstyrene/Vinyltoluene Copolymer is the polymer of methylstyrene and vinyltoluene monomers.

Information Sources: 21CFR175.105, 21CFR175.320, 21CFR176.170, 21CFR177.1200, 21CFR177.2600, TSCA

Chemical Class: Synthetic Polymers

Function: Viscosity Increasing Agent - Nonaqueous

Reported Product Category: Lipsticks

Technical/Other Name:
Benzene, Ethenylmethyl-, Polymer with (1-Methylethenyl)Benzene

Trade Name Mixture:
Durawax #1032 (Ross)

METHYL SUNFLOWERSEEDATE

Definition: Methyl Sunflowerseedate is the methyl ester of the fatty acids derived from Helianthus Annuus (Sunflower) Seed Oil (q.v.).

Chemical Class: Esters

Functions: Skin-Conditioning Agent - Emollient; Solvent

METHYLTHIOADENOSINE

CAS No.: 2457-80-9

Empirical Formula:
$C_{11}H_{15}N_5O_3S$

Definition: Methylthioadenosine is the heterocyclic compound that conforms to the formula:

Chemical Classes: Amines; Heterocyclic Compounds; Thio Compounds

Function: Skin-Conditioning Agent - Miscellaneous

Trade Name:
Adenosine thiomethylpentose (APR)

METHYL THIOGLYCOLATE

CAS No.	EINECS No.
2365-48-2	219-121-7

Empirical Formula:
$C_3H_6O_2S$

Definition: Methyl Thioglycolate is the ester of methyl alcohol and Thioglycolic Acid (q.v.). It conforms to the formula:

$$HSCH_2\overset{O}{\underset{\|}{C}} - OCH_3$$

Information Source: EEC(III/1-2b)

Chemical Classes: Esters; Thio Compounds

Function: Hair-Waving/Straightening Agent

Technical/Other Names:
Acetic Acid, Mercapto-, Methyl Ester
Methyl Mercaptoacetate
Methyl Thioglycollate
Thioglycolic Acid, Methyl Ester

METHYL TRIMETHICONE

Definition: Methyl Trimethicone is the organic compound that conforms to the formula:

$$(CH_3)_3SiO - \underset{OSi(CH_3)_3}{\overset{CH_3}{Si}}O - Si(CH_3)_3$$

The inclusion of any compound in the Dictionary and Handbook does not indicate that use of that substance as a cosmetic ingredient complies with the laws and regulations governing such use in the United States or any other country.

Chemical Class: Siloxanes and Silanes

Functions: Hair Conditioning Agent; Skin-Conditioning Agent - Miscellaneous; Solvent

Trade Name:
TMF-1.5 (Shin Etsu)

METHYL TRYPTOPHANATE HCl

CAS No.	EINECS No.
7524-52-9	231-385-5

Empirical Formula:
$C_{12}H_{14}N_2O_2 \cdot ClH$

Definition: Methyl Tryptophanate HCl is the organic compound that conforms to the formula:

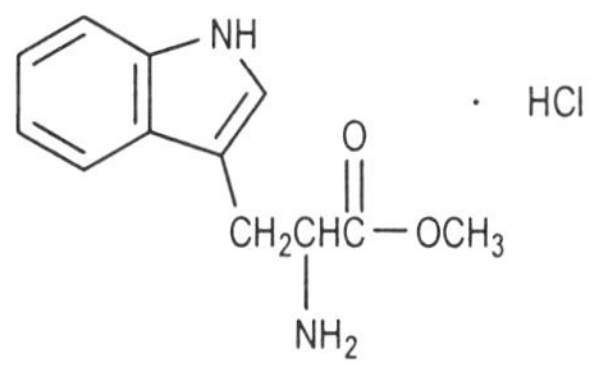

Chemical Classes: Amino Acids; Esters

Function: Hair Conditioning Agent

Technical/Other Names:
Methyl L-Tryptophanate Hydrochloride
L-Tryptophan, Methyl Ester, Monohydro-chloride

Trade Name:
L-Tryptophan Methyl Ester Hydrochloride (Nippon Rika)

METHYL TYROSINATE HCl

CAS No.	EINECS No.
3417-91-2	222-313-3

Empirical Formula:
$C_{10}H_{13}NO_3 \cdot ClH$

Definition: Methyl Tyrosinate HCl is the organic compound that conforms to the formula:

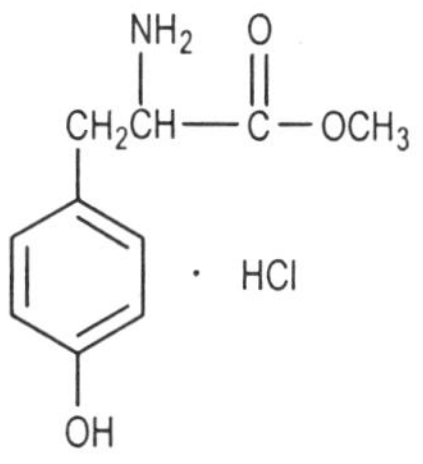

Chemical Classes: Amino Acids; Esters

Function: Hair Conditioning Agent

Technical/Other Names:
Methyl L-Tyrosinate Hydrochloride
L-Tyrosine, Methyl Ester, Hydrochloride

Trade Name:
L-Tyrosine, Methyl Ester, Hydrochloride (Nippon Rika)

MEVALONOLACTONE

CAS Nos.	EINECS No.
674-26-0	211-615-0
19115-49-2	

Empirical Formula:
$C_6H_{10}O_3$

Definition: Mevalonolactone is the organic compound that conforms to the formula:

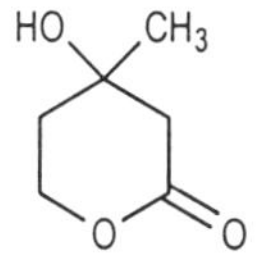

Information Source: MI-13(6185)

Chemical Class: Ketones

Function: Skin-Conditioning Agent - Humectant

Technical/Other Names:
2H-Pyran-2-one, Tetrahydro-4-Hydroxy-4-Methyl-, (R)-
Mevalonic Acid Lactone

Trade Name:
ADEKA Mevalonolactone (Asahi Denka Kogyo)

MIBK

CAS No.	EINECS No.
108-10-1	203-550-1

JPN Translation:
MIBK

Empirical Formula:
$C_6H_{12}O$

Definition: MIBK is the aliphatic ketone that conforms to the formula:

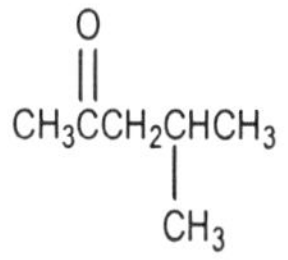

Information Sources: 21CFR172.515, 21CFR175.105, 21CFR177.1650, 21CFR1310.02, 21CFR1310.04, 21CFR1310.08, 27CFR21.116, CIR: [SQ], JCLS, JSCI, MI-13(5227), NF XIX, RIFM, TSCA, USAN

Chemical Class: Ketones

Functions: Denaturant; Fragrance Ingredient; Solvent

Technical/Other Names:
Hexone
Isopropylacetone
Methyl Isobutyl Ketone
4-Methyl-2-Oxopentane
4-Methyl-2-pentanone (RIFM)
4-Methyl-2-Pentanone
2-Pentanone, 4-Methyl-

Trade Name:
Eastman Methyl Isobutyl Ketone (Eastman Chemical)

MICA

CAS No.	EINECS No.
12001-26-2	310-127-6

JPN Translations:
金雲母
マイカ

Definition: Mica is a series of silicate minerals of varying chemical composition but with similar physical properties. Mica has well-defined cleavage, and splits into very thin sheets. *See "Regulatory and Ingredient Use Information," for Colorants in Volume 1, Introduction, Part A. To identify the colorant meeting the requirements for labeling purposes in the US, the INCI Name Mica must be used.*

Information Sources: 21CFR73.1496, 21CFR73.2496, 21CFR175.300, 21CFR177.1460, 21CFR177.2410, 21CFR177.2600, 21CFR178.3297, CI 77019, CTFA S, JCIC, JCLS, JSCI, JSQI, TSCA

Chemical Classes: Color Additives - Exempt from Batch Certification by the U.S. Food and Drug Administration; Inorganics

Function: Colorant

Reported Product Categories: Lipsticks; Eye Shadows; Face Powders; Hair Dyes and Colors (All Types Requiring Caution Statements and Patch Tests); Blushers (All types); Eyeliners; Foundations; Nail Polish and Enamels; Makeup Preparations (Not eye), Misc.; Powders (Dusting and Talcum, Excluding Aftershave Talcs); Eye Makeup Preparations, Misc.; Mascara; Makeup Bases; Rouges; Shampoos (Non-coloring); Moisturizing Preparations; Bath Preparations, Misc.; Body and Hand Preparations (Excluding Shaving Preparations); Makeup Fixatives; Eyebrow Pencils; Suntan Gels, Creams, and Liquids; Fragrance Preparations, Misc.; Skin Care Preparations, Misc.; Bath Oils, Tablets, and Salts; Bath Capsules; Cleansing Products (Cold Creams, Cleansing Lotions, Liquids and Pads); Paste Masks (Mud Packs); Face and Neck Preparations

(Excluding Shaving Preparations); Hair Coloring Preparations, Misc.; Hair Conditioners; Bath Soaps and Detergents; Tonics, Dressings, and Other Hair Grooming Aids; Basecoats and Undercoats; Colognes and Toilet Waters; Hair Bleaches; Hair Preparations (Non-coloring), Misc.; Manicuring Preparations, Misc.

Technical/Other Names:
CI 77019
Golden Mica
Muscovite Mica
Pigment White 20
Sericite
Sericite GMS-C
Sericite GMS-2C
Sericite MK-A
Sericite MK-B

Trade Names:
Argile Verte (Argiletz)
Lipomic 601 (Lipo)
Mearlmica MMCF (Engelhard Corp.)
Mearlmica MMSV (Engelhard Corp.)
Mica, Cosmetic, Bacteria Controlled 280 (Whittaker, Clark & Daniels)
Mica DD 8302 (Engelhard Corp.)
Mica M (Merck KGaA/EMD Chemicals Inc.)
Micro Mica (Chrystal)
Satin Mica (Merck KGaA/EMD Chemicals Inc.)
Serica 5 (LCW)
Sericite DNN (Ikeda)
Sericite FSE (Presperse)
Sericite 300 S (LCW)
Sericite SL (Presperse)
Silk Mica (Merck KGaA/EMD Chemicals Inc.)
Submica E (LCW)
Submica N (LCW)
Ultrapearl Silver White Luster (Ultra Chemical)

Trade Name Mixtures:
AEC Microcapsule White Vitamin E (A & E Connock)
Bi-Lite 20, 1070 (Engelhard Corp.)
Bi-Lite Ultralite 3186 (Engelhard Corp.)
Bi-Lite Ultrapress 1082 (Engelhard Corp.)
Bi-Lite Ultrawhite 1084 (Engelhard Corp.)
Cardre Mica 8 RFHC (Cardre)
Cardre Sericite RFHC (Cardre)
Cashmir K-II (Presperse)
Cellini Blue (Engelhard Corp.)
Cellini Coral (Engelhard Corp.)
Cellini Green (Engelhard Corp.)
Cellini Red (Engelhard Corp.)
Cellini Yellow (Engelhard Corp.)
Chroma-Lite Aqua 4508 (Engelhard Corp.)
Chroma-Lite Black 4498 (Engelhard Corp.)
Chroma-Lite Bronze 4499 (Engelhard Corp.)
Chroma-Lite Brown 4509 (Engelhard Corp.)
Chroma-Lite Dark Blue 4501 (Engelhard Corp.)
Chroma-Lite Gold 4504 (Engelhard Corp.)
Chroma-Lite Green 4503 (Engelhard Corp.)
Chroma-Lite Light Blue 4500 (Engelhard Corp.)
Chroma-Lite Magenta 4505 (Engelhard Corp.)
Chroma-Lite Mauve 4511 (Engelhard Corp.)
Chroma-Lite Purple 4510 (Engelhard Corp.)
Chroma-Lite Red 4506 (Engelhard Corp.)
Chroma-Lite Violet 4507 (Engelhard Corp.)
Chroma-Lite Yellow 4502 (Engelhard Corp.)
Cloisonne Blue (Engelhard Corp.)
Cloisonne Blue Flambe (Engelhard Corp.)
Cloisonne Blue-Green (Engelhard Corp.)
Cloisonne Cerise Flambe (Engelhard Corp.)
Cloisonne Copper (Engelhard Corp.)
Cloisonne Gold (Engelhard Corp.)
Cloisonne Gold CC (Engelhard Corp.)
Cloisonne Golden Bronze (Engelhard Corp.)
Cloisonne Green (Engelhard Corp.)
Cloisonne Imperial Gold (Engelhard Corp.)
Cloisonne Monarch Gold (Engelhard Corp.)
Cloisonne Nu-Antique Blue (Engelhard Corp.)
Cloisonne Nu-Antique Bronze (Engelhard Corp.)
Cloisonne Nu-Antique Copper (Engelhard Corp.)
Cloisonne Nu-Antique Gold (Engelhard Corp.)
Cloisonne Nu-Antique Green (Engelhard Corp.)
Cloisonne Nu-Antique Red (Engelhard Corp.)
Cloisonne Nu-Antique Rouge Flambe (Engelhard Corp.)
Cloisonne Nu-Antique Super Green (Engelhard Corp.)
Cloisonne Orange (Engelhard Corp.)
Cloisonne Red (Engelhard Corp.)
Cloisonne Red CC (Engelhard Corp.)
Cloisonne Regal Gold (Engelhard Corp.)
Cloisonne Rouge Flambe (Engelhard Corp.)
Cloisonne Satin Bronze (Engelhard Corp.)
Cloisonne Satin Copper (Engelhard Corp.)
Cloisonne Satin Gold (Engelhard Corp.)
Cloisonne Satin Rouge (Engelhard Corp.)
Cloisonne sparkle Blue (Engelhard Corp.)
Cloisonne Sparkle Blue Rouge (Engelhard Corp.)
Cloisonne Sparkle Bronze (Engelhard Corp.)
Cloisonne Sparkle Copper (Engelhard Corp.)
Cloisonne Sparkle Gold (Engelhard Corp.)
Cloisonne Sparkle Red (Engelhard Corp.)
Cloisonne Sparkle Rouge (Engelhard Corp.)
Cloisonne Super Blue (Engelhard Corp.)
Cloisonne Super Bronze (Engelhard Corp.)
Cloisonne Super Copper (Engelhard Corp.)
Cloisonne Super Gold (Engelhard Corp.)
Cloisonne Super Green (Engelhard Corp.)
Cloisonne Super Red (Engelhard Corp.)
Cloisonne Super Rouge (Engelhard Corp.)
Cloisonne Violet (Engelhard Corp.)
Colorona Aborigine Amber (Merck KGaA/EMD Chemicals Inc.)
Colorona Aztec Aqua (Merck KGaA/EMD Chemicals Inc.)
Colorona Blackstar Blue (Merck KGaA/EMD Chemicals Inc.)
Colorona Blackstar Gold (Merck KGaA/EMD Chemicals Inc.)
Colorona Blackstar Green (Merck KGaA/EMD Chemicals Inc.)
Colorona Blackstar Red (Merck KGaA/EMD Chemicals Inc.)
Colorona Bordeaux (Merck KGaA/EMD Chemicals Inc.)
Colorona Bright Gold (Merck KGaA/EMD Chemicals Inc.)
Colorona Bronze (Merck KGaA/EMD Chemicals Inc.)
Colorona Bronze Fine (Merck KGaA/EMD Chemicals Inc.)
Colorona Bronze Sparkle (Merck KGaA/EMD Chemicals Inc.)
Colorona Carmine Red (Merck KGaA/EMD Chemicals Inc.)
Colorona Chameleon (Merck KGaA/EMD Chemicals Inc.)
Colorona Copper (Merck KGaA/EMD Chemicals Inc.)
Colorona Copper Fine (Merck KGaA/EMD Chemicals Inc.)
Colorona Copper Sparkle (Merck KGaA/EMD Chemicals Inc.)
Colorona Dark Blue (Merck KGaA/EMD Chemicals Inc.)
Colorona Egyptian Emerald (Merck KGaA/EMD Chemicals Inc.)
Colorona Glitter Bordeaux (Merck KGaA/EMD Chemicals Inc.)
Colorona Glitter Bronze (Merck KGaA/EMD Chemicals Inc.)
Colorona Glitter Chameleon (Merck KGaA/EMD Chemicals Inc.)
Colorona Glitter Copper (Merck KGaA/EMD Chemicals Inc.)
Colorona Glitter Orange (Merck KGaA/EMD Chemicals Inc.)
Colorona Glitter Sienna (Merck KGaA/EMD Chemicals Inc.)
Colorona Imperial Red (Merck KGaA/EMD Chemicals Inc.)
Colorona Light Blue (Merck KGaA/EMD Chemicals Inc.)
Colorona Magenta (Merck KGaA/EMD Chemicals Inc.)

Colorona Majestic Green (Merck KGaA/
EMD Chemicals Inc.)
Colorona Oriental Beige (Merck KGaA/
EMD Chemicals Inc.)
Colorona Passion Orange (Merck KGaA/
EMD Chemicals Inc.)
Colorona Patagonian Purple (Merck KGaA/
EMD Chemicals Inc.)
Colorona Patina Gold (Merck KGaA/EMD
Chemicals Inc.)
Colorona Patina Silver (Merck KGaA/EMD
Chemicals Inc.)
Colorona Red Brown (Merck KGaA/EMD
Chemicals Inc.)
Colorona Red Gold (Merck KGaA/EMD
Chemicals Inc.)
Colorona Russet (Merck KGaA/EMD
Chemicals Inc.)
Colorona Sienna (Merck KGaA/EMD
Chemicals Inc.)
Colorona Sienna Fine (Merck KGaA/EMD
Chemicals Inc.)
Colorona Sienna Sparkle (Merck KGaA/
EMD Chemicals Inc.)
Colorona Tibetan Ochre (Merck KGaA/
EMD Chemicals Inc.)
Cosmica Blue (Engelhard Corp.)
Cosmica Orange (Engelhard Corp.)
Cosmica Red (Engelhard Corp.)
COULOURMAT BLACK SIL 45%
(Creations Couleurs)
COULOURMAT PL BLACK (Creations
Couleurs)
COULOURMAT PL BLUE (Creations
Couleurs)
COULOURMAT PL RED (Creations
Couleurs)
COULOURMAT PL WHITE R (Creations
Couleurs)
COULOURMAT PL YELLOW (Creations
Couleurs)
COULOURMAT RED SIL 45% (Creations
Couleurs)
COULOURMAT WHITE R SIL 45%
(Creations Couleurs)
COULOURMAT WOE BLACK (Creations
Couleurs)
COULOURMAT WOE BLUE (Creations
Couleurs)
COULOURMAT WOE GREEN (Creations
Couleurs)
COULOURMAT WOE GREEN LEAF
(Creations Couleurs)
COULOURMAT WOE RED (Creations
Couleurs)
COULOURMAT WOE WHITE (Creations
Couleurs)
COULOURMAT WOE YELLOW (Creations
Couleurs)
COULOURMAT YELLOW SIL 45%
(Creations Couleurs)
Covamat (LCW)
Covazur GHA (LCW)
Covazur LAC2 (LCW)
Covazur ZO3 (LCW)

Desert Reflections Canyon Sunset
(Engelhard Corp.)
Desert Reflections Midnight Sagebrush
(Engelhard Corp.)
Desert Reflections Painted Desert Plum
(Engelhard Corp.)
Desert Reflections Sunlit Cactus
(Engelhard Corp.)
Dichrona BG (Merck KGaA/EMD
Chemicals Inc.)
Dichrona RB (Merck KGaA/EMD
Chemicals Inc.)
Dichrona RY (Merck KGaA/EMD
Chemicals Inc.)
Dichrona Spendid BY (Merck KGaA/EMD
Chemicals Inc.)
Dichrona Splendid BR (Merck KGaA/EMD
Chemicals Inc.)
Diophase Gold (Nihon Koken Kogyo)
Diophase Silver (Nihon Koken Kogyo)
Duocrome BG (Engelhard Corp.)
Duocrome BR (Engelhard Corp.)
Duocrome BV (Engelhard Corp.)
Duocrome BY (Engelhard Corp.)
Duocrome GY (Engelhard Corp.)
Duocrome RB (Engelhard Corp.)
Duocrome RO (Engelhard Corp.)
Duocrome RV (Engelhard Corp.)
Duocrome RY (Engelhard Corp.)
Duocrome Sparkle BR (Engelhard Corp.)
Duocrome Sparkle BY (Engelhard Corp.)
Duocrome Sparkle RB (Engelhard Corp.)
Duocrome Sparkle RY (Engelhard Corp.)
Duocrome YB (Engelhard Corp.)
Duocrome YG (Engelhard Corp.)
Duocrome YR (Engelhard Corp.)
Extender W (Merck KGaA/EMD Chemicals
Inc.)
Facemat (LCW)
FI-S-100 (US Cosmetics)
Flamenco Blue (Engelhard Corp.)
Flamenco Gold (Engelhard Corp.)
Flamenco Gold CC (Engelhard Corp.)
Flamenco Green (Engelhard Corp.)
Flamenco Orange (Engelhard Corp.)
Flamenco Orange CC (Engelhard Corp.)
Flamenco Pearl (Engelhard Corp.)
Flamenco Pearl CC (Engelhard Corp.)
Flamenco Pearl SEC (Engelhard Corp.)
Flamenco Red (Engelhard Corp.)
Flamenco Satina (Engelhard Corp.)
Flamenco Satin Blue (Engelhard Corp.)
Flamenco Satin Gold (Engelhard Corp.)
Flamenco Satin Green (Engelhard Corp.)
Flamenco Satin Orange (Engelhard Corp.)
Flamenco Satin Pearl (Engelhard Corp.)
Flamenco Satin Red (Engelhard Corp.)
Flamenco Satin Violet (Engelhard Corp.)
Flamenco Sparkle Blue (Engelhard Corp.)
Flamenco Sparkle Gold (Engelhard Corp.)
Flamenco Sparkle Green (Engelhard
Corp.)
Flamenco Sparkle Orange (Engelhard
Corp.)

Flamenco Sparkle Red (Engelhard Corp.)
Flamenco Sparkle Violet (Engelhard Corp.)
Flamenco Summit Blue (Engelhard Corp.)
Flamenco Summit Gold (Engelhard Corp.)
Flamenco Summit Green (Engelhard
Corp.)
Flamenco Summit Red (Engelhard Corp.)
Flamenco Summit Turquoise (Engelhard
Corp.)
Flamenco Super Blue (Engelhard Corp.)
Flamenco Super Gold (Engelhard Corp.)
Flamenco Super Green (Engelhard Corp.)
Flamenco Superpearl (Engelhard Corp.)
Flamenco Superpearl CC (Engelhard
Corp.)
Flamenco Super Red (Engelhard Corp.)
Flamenco Super Red CC (Engelhard
Corp.)
Flamenco Twilight Blue (Engelhard Corp.)
Flamenco Twilight Gold (Engelhard Corp.)
Flamenco Twilight Green (Engelhard
Corp.)
Flamenco Twilight Red (Engelhard Corp.)
Flamenco Twilight Red CC (Engelhard
Corp.)
Flamenco Ultra Fine (Engelhard Corp.)
Flamenco Ultra Silk 2500 (Engelhard
Corp.)
Flamenco Ultra Sparkle (Engelhard Corp.)
Flamenco Velvet (Engelhard Corp.)
Flamenco Violet (Engelhard Corp.)
Flonac ME 10 C (Eckart) (Eckart GmbH)
Flonac MF 10 C (Eckart) (Eckart GmbH)
Flonac MG 10 C (Eckart) (Eckart GmbH)
Flonac MG 30 C (Eckart) (Eckart GmbH)
Flonac MI 10 C (Eckart) (Eckart GmbH)
Flonac MI 33 C (Eckart) (Eckart GmbH)
Flonac ML 10 C (Eckart) (Eckart GmbH)
Flonac MM 10 C (Eckart) (Eckart GmbH)
Flonac MS 10 C (Eckart) (Eckart GmbH)
Flonac MS 20 C (Eckart) (Eckart GmbH)
Flonac MS 30 C (Eckart) (Eckart GmbH)
Flonac MS 33 C (Eckart) (Eckart GmbH)
Flonac MS 40 C (Eckart) (Eckart GmbH)
Flonac MS 50 C (Eckart) (Eckart GmbH)
Flonac MS 60 C (Eckart) (Eckart GmbH)
Flonac MS 70 C (Eckart) (Eckart GmbH)
Flonac MU 10 C (Eckart) (Eckart GmbH)
Flonac MX 10 C (Eckart) (Eckart GmbH)
Flonac MX 30 C (Eckart) (Eckart GmbH)
FN-S100 (US Cosmetics)
Gemtone Amber (Engelhard Corp.)
Gemtone Amethyst (Engelhard Corp.)
Gemtone Emerald (Engelhard Corp.)
Gemtone Garnet (Engelhard Corp.)
Gemtone Goldstone (Engelhard Corp.)
Gemtone Jade (Engelhard Corp.)
Gemtone Mauve Quartz (Engelhard Corp.)
Gemtone Moonstone (Engelhard Corp.)
Gemtone Ruby (Engelhard Corp.)
Gemtone Sapphire (Engelhard Corp.)
Gemtone Sunstone (Engelhard Corp.)
Gemtone Sunstone CC (Engelhard Corp.)
Gemtone Tan Opal (Engelhard Corp.)
Gemtone Tan Opal CC (Engelhard Corp.)

Gemtone Topaz (Engelhard Corp.)
Genapol PDC (Clariant)
Genapol PDC (Clariant GmbH, Personal Care)
Hot Dots - Blue (Charm Girl)
Hot Dots - Green (Charm Girl)
Hot Dots - Red (Charm Girl)
HydroSpers-B-335198 (US Cosmetics)
Lipmat (LCW)
Lipobead Red-E (Lipo)
Lipobead Red-EL (Lipo)
Lipomic 601 BN (Lipo)
LipoSphere Glitter Pink (Lipo)
LI-S-100 (US Cosmetics)
Low Lustre Pigment (Merck KGaA/EMD Chemicals Inc.)
Lumiral (LCW)
MCP-45 (Presperse)
Mearlcite SRA (Engelhard Corp.)
Mearlmica SVA (Engelhard Corp.)
Mibiron N-50 (Merck KGaA/EMD Chemicals Inc.)
Mica Black (Merck KGaA/EMD Chemicals Inc.)
Mica FHC (Cardre)
MICAPOLY 320 A (Creations Couleurs)
MICAPOLY 320 B (Creations Couleurs)
MICAPOLY 320 C (Creations Couleurs)
Micapoly MSL Red (Creations Couleurs)
Micapoly MSL White (Creations Couleurs)
Micapoly MSL Yellow (Creations Couleurs)
Micapoly UV Cristal (C.I.T.)
MICAPOLY UV SHADOW (Creations Couleurs)
MICAPOLY UV ZINC (Creations Couleurs)
MICAPOLY WL 12 PFC (Creations Couleurs)
Micapoly WOE Black (Creations Couleurs)
Micapoly WOE Blue (Creations Couleurs)
Micapoly WOE Brown (Creations Couleurs)
Micapoly WOE Geranium (Creations Couleurs)
Micapoly WOE Green 70% (Creations Couleurs)
Micapoly WOE Manganese Violet (Creations Couleurs)
Micapoly WOE Melon (Creations Couleurs)
Micapoly WOE Pink (Creations Couleurs)
Micapoly WOE Red (Creations Couleurs)
Micapoly WOE Violet (Creations Couleurs)
Micapoly WOE White (Creations Couleurs)
Micapoly WOE Yellow (Creations Couleurs)
Mica PP (Cardre)
Mica SI 2 (Cardre)
Mica S/MM3 (Kobo)
Micro Mica (Kobo)
Microna Matte Black (Merck KGaA/EMD Chemicals Inc.)
Microna Matte Blue (Merck KGaA/EMD Chemicals Inc.)
Microna Matte Orange (Merck KGaA/EMD Chemicals Inc.)

Microna Matte Red (Merck KGaA/EMD Chemicals Inc.)
Microna Matte White (Merck KGaA/EMD Chemicals Inc.)
Microna Matte Yellow (Merck KGaA/EMD Chemicals Inc.)
Micronasphere M (Merck KGaA/EMD Chemicals Inc.)
MI-S-100 (US Cosmetics)
MPU-WS2 (Catalysts & Chemicals)
NAI-S (US Cosmetics)
Naturaleaf Powder (Merck KGaA/EMD Chemicals Inc.)
NFI-MICA (US Cosmetics)
NFI-Sericite (US Cosmetics)
PALI-M-102 (US Cosmetics)
PALI-S-100 (US Cosmetics)
PFI-Powder La Vie (US Cosmetics)
Powder La Vie (US Cosmetics)
Prestige Blue (Eckart) (Eckart America)
Prestige Bright Blue (Eckart) (Eckart America)
Prestige Bright Bronze (Eckart) (Eckart America)
Prestige Bright Copper (Eckart) (Eckart America)
Prestige Bright Fire Red (Eckart) (Eckart America)
Prestige Bright Gold (Eckart) (Eckart America)
Prestige Bright Green (Eckart) (Eckart America)
Prestige Bright Lemon Gold (Eckart) (Eckart America)
Prestige Bright Red (Eckart) (Eckart America)
Prestige Bright Silver (Eckart) (Eckart America)
Prestige Bright Silver Star (Eckart) (Eckart America)
Prestige Bright Sun Gold (Eckart) (Eckart America)
Prestige Bright Violet (Eckart) (Eckart America)
Prestige Bronze (Eckart) (Eckart America)
Prestige Copper (Eckart) (Eckart America)
Prestige Dazzling Red Gold (Eckart) (Eckart America)
Prestige Dazzling Silver (Eckart) (Eckart America)
Prestige Fire Red (Eckart) (Eckart America)
Prestige Gold (Eckart) (Eckart America)
Prestige Green (Eckart) (Eckart America)
Prestige Lemon Gold (Eckart) (Eckart America)
Prestige Orange (Eckart) (Eckart America)
Prestige Red (Eckart) (Eckart America)
Prestige Silk Blue (Eckart) (Eckart America)
Prestige Silk Gold (Eckart) (Eckart America)
Prestige Silk Green (Eckart) (Eckart America)

Prestige Silk Orange (Eckart) (Eckart America)
Prestige Silk Red (Eckart) (Eckart America)
Prestige Silk Silver (Eckart) (Eckart America)
Prestige Silk Silver Star (Eckart) (Eckart America)
Prestige Silk Violet (Eckart) (Eckart America)
Prestige Silver (Eckart) (Eckart America)
Prestige Silver Star (Eckart) (Eckart America)
Prestige Sparkling Lemon Gold (Eckart) (Eckart America)
Prestige Sparkling Silver (Eckart) (Eckart America)
Prestige Sparkling Silver Star (Eckart) (Eckart America)
Prestige Sun Gold (Eckart) (Eckart America)
Prestige Violet (Eckart) (Eckart America)
SA-Excel Mica JP-2 (US Cosmetics)
Sericite FHC (Cardre)
Sericite PP (Cardre)
Sericite SI 2 (Cardre)
Sericite SL-012 (Presperse)
Sericite SLZ-012P (Presperse)
Sericite WL (Ikeda)
Shinju White 100T (Engelhard Corp.)
SilPower (LG Cosmetic)
SilPower Plus (LG Cosmetic)
SM-1000 (Presperse)
SM-2000 (Presperse)
Soloron Silver (Merck KGaA/EMD Chemicals Inc.)
Soloron Silver Fine (Merck KGaA/EMD Chemicals Inc.)
SP-29 UVS (Presperse)
SXI-9 (US Cosmetics)
SXI-9S (US Cosmetics)
Timica Brilliant Gold (Engelhard Corp.)
Timica Copper (Engelhard Corp.)
Timica Extra Bright 1500 (Engelhard Corp.)
Timica Extra Large Sparkle (Engelhard Corp.)
Timica Golden Bronze (Engelhard Corp.)
Timica Gold Sparkle (Engelhard Corp.)
Timica Nu-Antique Bronze (Engelhard Corp.)
Timica Nu-Antique Copper (Engelhard Corp.)
Timica Nu-Antique Gold (Engelhard Corp.)
Timica Nu-Antique Silver (Engelhard Corp.)
Timica Pearlwhite (Engelhard Corp.)
Timica Radiant Gold (Engelhard Corp.)
Timica Silkwhite (Engelhard Corp.)
Timica Silver Sparkle (Engelhard Corp.)
Timica Sparkle (Engelhard Corp.)
Timiron Arctic Silver (Merck KGaA/EMD Chemicals Inc.)
Timiron MP-155 Blue (Merck KGaA/EMD Chemicals Inc.)

Timiron MP-176 Blue Red (Merck KGaA/
EMD Chemicals Inc.)
Timiron MP-149 Diamond Cluster (Merck
KGaA/EMD Chemicals Inc.)
Timiron MP-20 Fine Gold (Merck KGaA/
EMD Chemicals Inc.)
Timiron MP-45 Gleamer Flake (Merck
KGaA/EMD Chemicals Inc.)
Timiron MP-127 Gold (Merck KGaA/EMD
Chemicals Inc.)
Timiron MP-26 Gold Glow (Merck KGaA/
EMD Chemicals Inc.)
Timiron MP-25 Gold Plus (Merck KGaA/
EMD Chemicals Inc.)
Timiron MP-24 Karat Gold (Merck KGaA/
EMD Chemicals Inc.)
Timiron MP-10 Pearl Flake (Merck KGaA/
EMD Chemicals Inc.)
Timiron MP-30 Pearl Sheen (Merck KGaA/
EMD Chemicals Inc.)
Timiron MP-175 Red (Merck KGaA/EMD
Chemicals Inc.)
Timiron MP-1117 Satin (Merck KGaA/EMD
Chemicals Inc.)
Timiron MP-99 Snowflake (Merck KGaA/
EMD Chemicals Inc.)
Timiron MP-47 Sparkle (Merck KGaA/EMD
Chemicals Inc.)
Timiron MP-115 Starluster (Merck KGaA/
EMD Chemicals Inc.)
Timiron MP-29 Sun Gold Sparkle (Merck
KGaA/EMD Chemicals Inc.)
Timiron MP-1001 Supersheen (Merck
KGaA/EMD Chemicals Inc.)
Timiron MP-1005 Super Silk (Merck KGaA/
EMD Chemicals Inc.)
Timiron MP-28 Transgold (Merck KGaA/
EMD Chemicals Inc.)
Timiron MP-18 Transwhite (Merck KGaA/
EMD Chemicals Inc.)
Timiron MP-111 Ultra Luster (Merck KGaA/
EMD Chemicals Inc.)
Timiron Silk Blue (Merck KGaA/EMD
Chemicals Inc.)
Timiron Silk Gold (Merck KGaA/EMD
Chemicals Inc.)
Timiron Silk Green (Merck KGaA/EMD
Chemicals Inc.)
Timiron Silk Red (Merck KGaA/EMD
Chemicals Inc.)
Timiron Splendid Blue (Merck KGaA/EMD
Chemicals Inc.)
Timiron Splendid Copper (Merck KGaA/
EMD Chemicals Inc.)
Timiron Splendid Gold (Merck KGaA/EMD
Chemicals Inc.)
Timiron Splendid Green (Merck KGaA/
EMD Chemicals Inc.)
Timiron Splendid Red (Merck KGaA/EMD
Chemicals Inc.)
Timiron Splendid Violet (Merck KGaA/EMD
Chemicals Inc.)
Timiron Starlight Blue (Merck KGaA/EMD
Chemicals Inc.)

Timiron Starlight Gold (Merck KGaA/EMD
Chemicals Inc.)
Timiron Starlight Green (Merck KGaA/EMD
Chemicals Inc.)
Timiron Starlight Red (Merck KGaA/EMD
Chemicals Inc.)
Timiron Super Blue (Merck KGaA/EMD
Chemicals Inc.)
Timiron Super Copper (Merck KGaA/EMD
Chemicals Inc.)
Timiron Super Gold (Merck KGaA/EMD
Chemicals Inc.)
Timiron Super Green (Merck KGaA/EMD
Chemicals Inc.)
Timiron Super Red (Merck KGaA/EMD
Chemicals Inc.)
Timiron Super Silver (Merck KGaA/EMD
Chemicals Inc.)
Timiron Super Silver Fine (Merck KGaA/
EMD Chemicals Inc.)
Timiron Super Violet (Merck KGaA/EMD
Chemicals Inc.)
Toshiki Mica GS-61D (Nikko)
Toshiki Mica HP-18K (Nikko)
Toshiki Sericite JA-A3 (Nikko)
Toshiki Sericite JF-25-3 (Nikko)
Toshiki Sericite JSF-25-3 (Nikko)
Toshiki Sericite JW-A5 (Nikko)
Toshiki Sericite OS-61D (Nikko)
Toshiki Super Mica A (Nikko)
Toshiki Super Mica D (Nikko)
Velvet Veil 310 (Presperse)
Velvetveil A (Ikeda)
Velvetveil X (Ikeda)
Velvetveil Y (Ikeda)
Velvetveil Z (Ikeda)
WHP-CL-1 (US Cosmetics)

MICHELIA ALBA FLOWER OIL

Definition: Michelia Alba Flower Oils the
volatile oil obtained from the flowers of
*Michelia alba. See "Regulatory and Ingre-
dient Use Information," regarding the
labeling names for botanical ingredients in
Volume 1, Introduction, Part A.*

Chemical Class: Essential Oils

Function: Fragrance Ingredient

Trade Name:
    Magnolia Flowers Oil (International
    Flavors)

MICHELIA ALBA LEAF OIL

Definition: Michelia Alba Leaf Oil is the
volatile oil obtained from the leaves of
*Michelia alba. See "Regulatory and Ingre-
dient Use Information," regarding the
labeling names for botanical ingredients in
Volume 1, Introduction, Part A.*

Chemical Class: Essential Oils

Function: Fragrance Ingredient

Technical/Other Name:
    Oils, Michelia Alba Leaf

Trade Name:
    Magnolia Leaves Oil (International Flavors)

MICHELIA CHAMPACA FLOWER EXTRACT

Definition: Michelia Champaca Flower
Extract is an extract of the flowers of
*Michelia champaca. See "Regulatory and
Ingredient Use Information," regarding the
labeling names for botanical ingredients in
Volume 1, Introduction, Part A.*

Chemical Class: Biological Products

Function: Skin-Conditioning Agent - Miscellaneous

Technical/Other Name:
    Extract of Michelia Champaca Flower

Trade Name Mixture:
    Champaca Absolute (Haldin Pacific)

MICHELIA CHAMPACA OIL

Definition: Michelia Champaca Oil is the
volatile oil obtained from the flowers of
*Michelia champaca. See "Regulatory and
Ingredient Use Information," regarding the
labeling names for botanical ingredients in
Volume 1, Introduction, Part A.*

Chemical Class: Essential Oils

Functions: Deodorant Agent; Fragrance
Ingredient; Skin-Conditioning Agent - Miscellaneous

Technical/Other Name:
    Oils, Michelia Champaca

Trade Name:
    Champaca Oil (Haldin Pacific)

MICROCOCCUS LYSATE

CAS No.: 158765-79-8

Definition: Micrococcus Lysate is the end
product of the controlled lysis of various
species of *Micrococcus.*

Chemical Class: Biological Products

Function: Not Reported

Trade Name:
    Luteus Extract (AGI Dermatics)

Trade Name Mixture:
    Ultrasomes (Barnet)

MICROCRYSTALLINE CELLULOSE

CAS No.: 9004-34-6

JPN Translation:
結晶セルロース

Definition: Microcrystalline Cellulose is the isolated, colloidal crystalline portion of cellulose fibers.

Information Sources: BAN, JCLS, JSCI, MI-13(1977), NF XIX, USAN

Chemical Classes: Biological Polymers and their Derivatives; Carbohydrates

Functions: Abrasive; Absorbent; Anticaking Agent; Bulking Agent; Emulsion Stabilizer; Slip Modifier; Viscosity Increasing Agent - Aqueous

Reported Product Categories: Lipsticks; Powders (Dusting and Talcum, Excluding Aftershave Talcs); Bath Capsules; Foundations; Hair Dyes and Colors (All Types Requiring Caution Statements and Patch Tests); Rouges; Baby Products, Misc.; Bath Preparations, Misc.; Body and Hand Preparations (Excluding Shaving Preparations); Eyeliners; Face and Neck Preparations (Excluding Shaving Preparations); Makeup Preparations (Not eye), Misc.; Paste Masks (Mud Packs); Skin Care Preparations, Misc.; Tonics, Dressings, and Other Hair Grooming Aids

Trade Names:
Acticel 12 (Active Organics)
Avicel (FMC/Pharmaceutical Division)
Avicel PH (FMC/Pharmaceutical Division)
Avicel PH-101, NF (FMC/Pharmaceutical Division)
Avicel PH-102, NF (FMC/Pharmaceutical Division)
Avicel PH-103, NF (FMC/Pharmaceutical Division)
Avicel PH-105, NF (FMC/Pharmaceutical Division)
Avicel PH-112, NF (FMC/Pharmaceutical Division)
Avicel PH-113, NF (FMC/Pharmaceutical Division)
Avicel PH-200, NF (FMC/Pharmaceutical Division)
Avicel PH-301, NF (FMC/Pharmaceutical Division)
Avicel PH-302, NF (FMC/Pharmaceutical Division)
Toshiki SP-White (Nikko)
Vivapur (Rettenmaier)

Trade Name Mixtures:
Avicel CL-611, NF (FMC/Pharmaceutical Division)
Avicel RC (FMC/Pharmaceutical Division)
Avicel RC-581, NF (FMC/Pharmaceutical Division)
Avicel RC-591, NF (FMC/Pharmaceutical Division)
Granule AA (Ichimaru Pharcos)
REGU-SLIM (Pentapharm/Centerchem)
Toshiki SP-565-41 Red (Nikko)
Vitacel MCG 591 (Rettenmaier)

MICROCRYSTALLINE WAX

CAS No.	EINECS No.
63231-60-7	264-038-1

JPN Translation:
マイクロクリスタリンワックス

Definition: Microcrystalline Wax is a wax derived from petroleum and characterized by the fineness of its crystals in contrast to the larger crystals of paraffin wax. It consists of high molecular weight saturated aliphatic hydrocarbons. *See "Regulatory and Ingredient Use Information," regarding use of EU Trivial names in Volume 1, Introduction, Part A.*

Information Sources: 21CFR172.886, 21CFR173.340, 21CFR175.105, 21CFR175.320, 21CFR176.170, 21CFR176.180, 21CFR176.200, 21CFR177.1200, 21CFR177.2600, CIR: [S] JACT-3(3)1984, CTFA S, JCLS, JSCI, NF XIX, SNPF, TSCA, USAN

Chemical Classes: Hydrocarbons; Waxes

Functions: Binder; Bulking Agent; Emulsion Stabilizer; Viscosity Increasing Agent - Nonaqueous

Reported Product Categories: Lipsticks; Eyeliners; Foundations; Eye Makeup Preparations, Misc.; Makeup Preparations (Not eye), Misc.; Mascara; Eye Shadows; Blushers (All types); Eyebrow Pencils; Tonics, Dressings, and Other Hair Grooming Aids; Skin Care Preparations, Misc.; Bath Oils, Tablets, and Salts; Rouges; Bath Preparations, Misc.; Body and Hand Preparations (Excluding Shaving Preparations); Cleansing Products (Cold Creams, Cleansing Lotions, Liquids and Pads); Face Powders; Moisturizing Preparations; Night Skin Care Preparations; Bath Capsules; Hair Conditioners; Perfumes; Personal Cleanliness Products, Misc.; Baby Products, Misc.; Deodorants (Underarm); Depilatories; Eye Lotions; Hair Preparations (Non-coloring), Misc.; Hair Straighteners; Indoor Tanning Preparations; Oral Hygiene Products, Misc.; Shaving Preparations, Misc.; Suntan Gels, Creams, and Liquids; Suntan Preparations, Misc.

Technical/Other Names:
Hydrocarbon Waxes, Microcryst
Petroleum Wax, Microcrystalline

Trade Names:
AEC Microcrystalline Wax (A & E Connock)
AEC Microwax Beads (A & E Connock)
Be Square 175 Amber Wax (Bareco)
Be Square 185 Amber Wax (Bareco)
Be Square 195 Amber Wax (Bareco)
Be Square 195 White Wax (Bareco)
CS-2080W (Huntington)
Mekon White Wax (Bareco)
Metabeads Microwax (Floratech)
Microcrystalline Wax 180 (Ross)
Microcrystalline Wax-Black 170 (Ross)
Microcrystalline Wax SP 16 (Strahl & Pitsch)
Microcrystalline Wax SP 18 (Strahl & Pitsch)
Microcrystalline Wax SP 19 (Strahl & Pitsch)
Microcrystalline Wax SP 26 (Strahl & Pitsch)
Microcrystalline Wax SP 34 (Strahl & Pitsch)
Microcrystalline Wax SP 60 (Strahl & Pitsch)
Microcrystalline Wax SP-89 (Strahl & Pitsch)
Microcrystalline Wax SP 94 (Strahl & Pitsch)
Microcrystalline Wax SP 95 (Strahl & Pitsch)
Microcrystalline Wax SP 96 (Strahl & Pitsch)
Microcrystalline Wax SP 410 (Strahl & Pitsch)
Microcrystalline Wax SP-433 (Strahl & Pitsch)
Microcrystalline Wax SP 617 (Strahl & Pitsch)
Microcrystalline Wax SP 624 (Strahl & Pitsch)
Microcrystalline Wax SP-1204 (Strahl & Pitsch)
Microcrystalline Wax SP-1674 (Strahl & Pitsch)
Microcrystalline Wax SP-16W (Strahl & Pitsch)
Microcrystalline Wax SP-60W (Strahl & Pitsch)
Microcrystalline Wax 1275WH (Ross)
Microcrystalline Wax-White 190 (Ross)
Microcrystalline Wax-White 214 (Ross)
Microcrystalline Wax-White 1251/7 (Ross)
Microcrystalline Wax-White 1365 (Ross)
Microcrystalline Wax-Yellow 165 (Ross)
Microcrystalline Wax-Yellow 170 (Ross)
Microcrystalline Wax-Yellow 916 (Ross)
Microcrystalline Wax-Yellow 1149/4 (Ross)
Microcrystalline Wax-Yellow 1385 (Ross)
Micro. Wax (Ross)
Micro. Wax 2305 (Ross)
Micro. Wax 1135/15W (Ross)
Multiwax 180-M (Crompton Corporation)
Multiwax ML-445 (Crompton Corporation)
Multiwax 180-W (Crompton Corporation)
Multiwax W-445 (Crompton Corporation)
Multiwax W-835 (Crompton Corporation)

Starwax 100 Wax (Bareco)
TeCero-Wachs (Tromm)
Ultraflex Amber Wax (Bareco)
Victory Wax (Bareco)
Waxcerite (Vevy)
White Microcrystalline 1275 (Ross)
White Microcrystalline 1329/1 (Ross)
White Microcrystalline 1275W (Ross)
White Micro. Wax 863 (Ross)
White Micro. Wax 1160/14 (Ross)
Yellow Micrcrystalline 1275 (Ross)
Yellow Microcrystalline 669 (Ross)

Trade Name Mixtures:
Beeswax Substitute 81-1104 (Ross)
B-Wax White (Ross)
Cocoa Butter Fractionated (Browne)
Dehymuls F (Cognis Care Chemicals/NJ)
Dehymuls K (Cognis Deutschland)
Durawax #1032 (Ross)
Emery 1740 (Cognis Care Chemicals/NJ)
Emery 1740 (Cognis Care Chemicals/PA)
Enviro Pure 302 (React Inc)
Enviro Pure 305 (React Inc)
Hairwax 7686 o.E. (Kahl)
Hostacerin WO (Clariant)
Hostacerin WO (Clariant GmbH, Personal
Care)
Japan Wax Substitute No. 473 (Ross)
Japan Wax Substitute No. 525 (Ross)
Lipcare Wax 7782 (Kahl)
Petrolite C-8500 Polymer (Baker Petrolite)
PIONIER 17106 Mineral Oil Free Vaseline
(Hansen & Rosenthal)
PM Wax 82 (Nikko Rica)
Ross Japan Wax Substitute No. 473 (Ross)
Ross Japan Wax Substitute No. 525 (Ross)
Ross Japan Wax Substitute No. 930 (Ross)
Ultrapure HMP (Ultra Chemical)
Ultrapure HMP-S (Ultra Chemical)
Unitina BW (Universal Preserv-A-Chem)
Unitina LM (Universal Preserv-A-Chem)

MICROMERIA CHAMISSONIS EXTRACT

Definition: Micromeria Chamissonis
Extract is the extract obtained from the
aerial parts of *Micromeria chamissonis*.
*See "Regulatory and Ingredient Use Information," regarding the labeling names for
botanical ingredients in Volume 1, Introduction, Part A.*

Chemical Class: Biological Products

Function: Not Reported

Technical/Other Name:
Extract of Micromeria Chamissonis

Trade Name Mixture:
Cosflor Yerba Buena HGS (A & E
Connock)

MIDORI3
CAS No.
2353-45-9
EINECS No.
219-091-5

JPN Translation:
緑 3

Empirical Formula:
$C_{37}H_{36}N_2O_{10}S_3 \cdot 2Na$

Definition: Midori3 is classed chemically as
a triphenylmethane color. It conforms to the
formula:

See "Regulatory and Ingredient Use Information," for Colorants in Volume 1, Introduction, Part A. To identify the colorant
allowed for use in Japan, the INCI name
Midori3 must be used. To identify the
colorant (aluminum lake) allowed for use in
Japan, the INCI name Midori3 must be
used. This INCI name may not be used for
ingredient labeling in the US or the EU. To
identify the colorant allowed for use in the
European Union (EU), the INCI Name CI
42053 must be used except for hair dye
products. To identify the certified colorant
for labeling purposes in the US, the INCI
name Green 3 must be used. To identify
the certified colorant (aluminum lake) for
labeling purposes in the US, the INCI name
Green 3 Lake must be used. The INCI
Name for batches of this colorant that have
not been certified is Fast Green FCF.

Information Sources: CI 42053, M3,
MHLW Ord. No. 30, MI-13(3970), TSCA

Chemical Class: Color Additives - Approved
in Japan

Function: Colorant

Technical/Other Names:
Benzenemethanaminium, N-Ethyl-
N-[4-[[4-[Ethyl[3-Sulfophenyl)Methyl]-
Amino]Phenyl](4-Hydroxy-2-
Sulfophenyl)Methylene]-2,5-Cyclohexa-
dien-1-ylidene]-3-Sulfo-, Hydroxide, Inner
Salt, Disodium Salt
CI 42053
Fast Green FCF
Food Green 3
Japan Green 3

MIDORI201
CAS No.
4403-90-1
EINECS No.
224-546-6

JPN Translation:
緑 201

Empirical Formula:
$C_{28}H_{22}N_2O_8S_2 \cdot 2Na$

Definition: Midori201 is classed chemically
as an anthraquinone color. It conforms to the
formula:

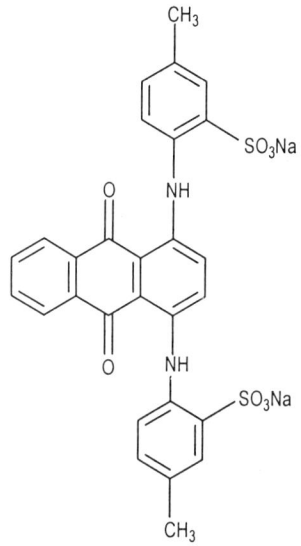

See "Regulatory and Ingredient Use Information," for Colorants in Volume 1, Introduction, Part A. To identify the colorant
allowed for use in Japan, the INCI name
Midori201 must be used. To identify the
colorant (aluminum lake) allowed for use in
Japan, the INCI name Midori201 must be
used. This INCI name may not be used for
ingredient labeling in the US or the EU. To
identify the colorant allowed for use in the
European Union (EU), the INCI Name CI
61570 must be used except for hair dye
products. To identify the certified colorant
for labeling purposes in the US, the INCI
name Green 5 must be used. The INCI
Name for batches of this colorant that have
not been certified is Acid Green 25.

Information Sources: CI 61570, M3,
MHLW Ord. No. 30, MI-13(247), TSCA

Chemical Class: Color Additives - Approved
in Japan

Function: Colorant

Technical/Other Names:
Acid Green 25
Alizarin Cyanine Green
Benzenesulfonic Acid, 2,2'-[(9,10-Dihydro-
9,10-Dioxo-1,4-Anthracenediyl)-
Diimino]Bis(5-Methyl)-, Disodium Salt
CI 61570
2,2'-[(9,10-Dihydro-9,19-Dioxo-1,4-
Anthracenediyl)Diimino]Bis[5-Methyl]-
Benzenesulfonic Acid, Disodium Salt

Japan Green 201
μ-Toluenesulfonic Acid, 6,6'-(1,4-Anthra-quinonylenediimino)di-, Disodium Salt

MIDORI202

CAS No.
128-80-3

EINECS No.
204-909-5

JPN Translation:
緑 202

Empirical Formula:
$C_{28}H_{22}N_2O_2$

Definition: Midori202 is classed chemically as an anthraquinone color. It conforms to the formula:

See "Regulatory and Ingredient Use Information," for Colorants in Volume 1, Introduction, Part A. To identify the colorant allowed for use in Japan, the INCI name Midori202 must be used. This INCI name may not be used for ingredient labeling in the US or the EU. To identify the colorant allowed for use in the European Union (EU), the INCI Name CI 61565 must be used except for hair dye products. To identify the certified colorant for labeling purposes in the US, the INCI name Green 6 must be used. The INCI Name for batches of this colorant that have not been certified is Solvent Green 3.

Information Sources: CI 61565, M3, MHLW Ord. No. 30, MI-13(8157), TSCA

Chemical Class: Color Additives - Approved in Japan

Function: Colorant

Technical/Other Names:
9,10-Anthracenedione, 1,4-Bis[(4-Methyl-phenyl)Amino]-
Anthraquinone, 1,4-Di-ψ-Toluidino-
1,4-bis(4'-Methylanilino)Anthraquinone
1,4-Bis[(4-Methylphenyl)Amino-9,10-Anthracenedione
Ceres Green BB
CI 61565
Quinizarin Green SS
Solvent Green 3

MIDORI204

CAS No.
6358-69-6

EINECS No.
228-783-6

JPN Translation:
緑 204

Empirical Formula:
$C_{16}H_{10}O_{10}S_3$ • 3Na

Definition: Midori204 is classed chemically as a pyrene color. It conforms to the formula:

See "Regulatory and Ingredient Use Information," for Colorants in Volume 1, Introduction, Part A. To identify the colorant allowed for use in Japan, the INCI name Midori204 must be used. To identify the colorant (aluminum lake) allowed for use in Japan, the INCI name Midori204 must be used. This INCI name may not be used for ingredient labeling in the US or the EU. To identify the colorant allowed for use in the European Union (EU), the INCI Name CI 59040 must be used except for hair dye products. To identify the certified colorant for labeling purposes in the US, the INCI name Green 8 must be used. The INCI Name for batches of this colorant that have not been certified is Solvent Green 7.

Information Sources: CI 59040, M3, MHLW Ord. No. 30, TSCA

Chemical Class: Color Additives - Approved in Japan

Function: Colorant

Technical/Other Names:
CI 59040
8-Hydroxy-1,3,6-Pyrenetrisulfonic Acid, Trisodium Salt
Japan Green 204
Pyranine
1,3,6-Pyrenetrisulfonic Acid, 8-Hydroxy-, Trisodium Salt
Solvent Green 7

MIDORI205

CAS No.: 5141-20-8

JPN Translation:
緑 205

Empirical Formula:
$C_{37}H_{34}O_9N_2S_3$ • 2Na

Definition: Midori205 is classed chemically as a triarylmethane color. It conforms to the formula:

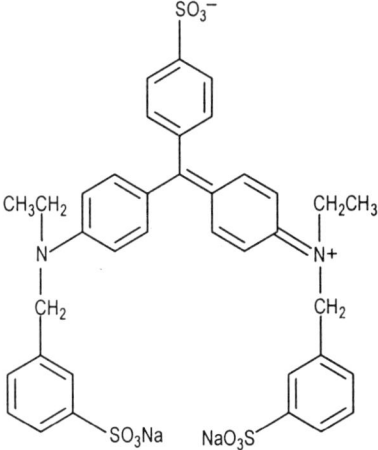

See "Regulatory and Ingredient Use Information," for Colorants in Volume 1, Introduction, Part A. To identify the colorant allowed for use in Japan, the INCI name Midori205 must be used. To identify the colorant (aluminum or zirconium lake) allowed for use in Japan, the INCI name Midori205 must be used. This ingredient is not an approved colorant for the US or the EU. This INCI name may not be used for ingredient labeling in the US or the EU.

Information Sources: CI 42095, MHLW Ord. No. 30, MI-13(5506)

Chemical Class: Color Additives - Approved in Japan

Function: Colorant

Technical/Other Names:
Acid Green 5
Benzenemethanaminium, N-Ethyl-N-[4-[[4-[Ethyl(3-Sulfophenyl)methyl]-amino]phenyl](4-Sulfophenyl)Methylene]-2,5-Cyclohexadien-1-ylidene]-3-Sulfo, Inner Salt, Disodium Salt
CI 42095
Food Green 2
Green No. 205

MIDORI401

CAS No.
19381-50-1

EINECS No.
243-010-2

JPN Translation:
緑 401

Empirical Formula:
$C_{30}H_{15}FeN_3O_{15}S_3$ • 3Na

Definition: Midori401 is classed chemically as a nitroso color. It conforms to the formula:

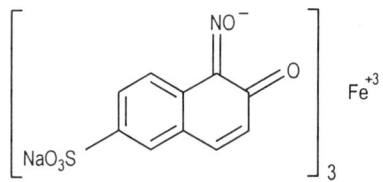

See "Regulatory and Ingredient Use Information," for Colorants in Volume 1, Introduction, Part A. To identify the colorant allowed for use in Japan, the INCI name Midori401 must be used. This INCI name may not be used for ingredient labeling in the US or the EU. To identify the colorant allowed for use in the European Union (EU), the INCI Name CI 10020 must be used except for hair dye products. This ingredient is not an approved colorant for the US.

Information Sources: CI 10020, M3, MHLW Ord. No. 30, TSCA

Chemical Class: Color Additives - Approved in Japan

Function: Colorant

Technical/Other Names:
Acid Green 1
CI 10020
Ferrate(3)-, Tris[5,6-Dihydro-5-(Hydroxyimino)-6-Oxo-2-Naphthalenesulfonato(2-)-N5,O 6, Trisodium
Green No. 401
Green PLX
Japan Green 401
2-Naphthalenesulfonic Acid, 5,6-Dihydro-5-(hydroxyimino)-6-oxo, Iron Complex
Naphthol Green B
Tris[5,6-Dihydro-5-(Hydroxyimino)-6-Oxo-2-Naphthalenesulfonato(2-)-N5,O6]Ferrate(3), Trisodium

MIDORI402

CAS No.: 4680-78-8

JPN Translation:
緑 402

Empirical Formula:
$C_{37}H_{35}O_6O_2S_2 \cdot Na$

Definition: Midori402 is classed chemically as a triarylmethanecolor. It conforms to the formula:

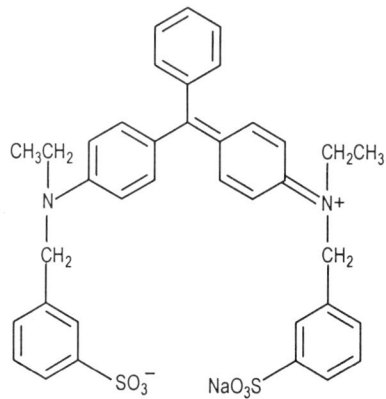

See "Regulatory and Ingredient Use Information," for Colorants in Volume 1, Introduction, Part A. To identify the colorant allowed for use in Japan, the INCI name Midori402 must be used. This ingredient is not an approved colorant for the US or the EU. This INCI name may not be used for ingredient labeling in the US or the EU.

Information Sources: CI 42085, MHLW Ord. No. 30, MI-13(4589)

Chemical Class: Color Additives - Approved in Japan

Function: Colorant

Technical/Other Names:
Acid Green 3
Benzenemethanaminium, N-Ethyl-N-[4-[[4-[Ethyl(3-Sulfophenyl)methyl]-amino]phenyl](Phenylmethylene)-2,5-Cyclohexadien-1-ylidene]-3-Sulfo, Inner Salt, Sodium Salt
CI 42085
Food Green 1
Green No. 402
Guinea Green B

MILK

CAS No.: 8049-98-7

JPN Translation:
牛乳

Definition: Milk is whole milk from cows. See "Regulatory and Ingredient Use Information," regarding use of EU Trivial names in Volume 1, Introduction, Part A.

Information Sources: 21CFR101.4, 21CFR101.12, 21CFR101.13, 21CFR131.110, 21CFR131.200, 21CFR131.203, 21CFR131.206, 21CFR184.1349, 21CFR184.1366, 21CFR184.1950, 21CFR184.1979, 21CFR184.1979a, 21CFR184.1979b, 21CFR184.1979c, 21CFR331.11, 21CFR1210.1, 21CFR1210.3, 21CFR1210.17, 21CFR1210.21, 21CFR1210.22, 21CFR1210.24, 21CFR1210.25, 21CFR1210.26, 21CFR1210.27, 21CFR1210.28, 21CFR1210.31, 21CFR1240.3, 21CFR1250.26, JCIC, JCLS, MI-13(6215)

Chemical Class: Biological Products

Function: Skin-Conditioning Agent - Miscellaneous

Technical/Other Name:
Cow's Milk

Trade Names:
AEC Milk Powder (Full Cream) (A & E Connock)
AEC Milk Powder (Skimmed) (A & E Connock)
Siero Di Latte Disidratato (Lactose Milk Protein) (Rottapharm)

Trade Name Mixture:
Extrapone Soy Milk 2/033858 (Symrise)

MILKAMIDOPROPYL AMINE OXIDE

Definition: Milkamidopropyl Amine Oxide is the tertiary amine oxide that conforms generally to the formula:

$$RC-NH(CH_2)_3-N \rightarrow O$$

where RCO- represents the fatty acids derived from milk.

Chemical Classes: Amides; Amine Oxides

Functions: Hair Conditioning Agent; Surfactant - Cleansing Agent; Surfactant - Foam Booster; Surfactant - Hydrotrope

Trade Name:
Monalac MO (Uniqema)

MILKAMIDOPROPYL BETAINE

Definition: Milkamidopropyl Betaine is the zwitterion (inner salt) that conforms generally to the formula:

$$RC-NH(CH_2)_3-N^+-CH_2COO^-$$

where RCO- represents the fatty acids derived from milk.

Chemical Classes: Amides; Betaines

Functions: Antistatic Agent; Hair Conditioning Agent; Skin-Conditioning Agent - Miscellaneous; Surfactant - Cleansing Agent; Surfactant - Foam Booster; Viscosity Increasing Agent - Aqueous

Trade Name:
Monalac MAB (Uniqema)

MILK AMINO ACIDS

CAS No.: 65072-00-6

Definition: Milk Amino Acids is the mixture of amino acids resulting from the complete hydrolysis of Milk Protein (q.v.).

Chemical Class: Amino Acids

Functions: Hair Conditioning Agent; Skin-Conditioning Agent - Miscellaneous

Reported Product Categories: Shampoos (Non-coloring); Skin Care Preparations, Misc.

Trade Names:
AEC Milk Amino Acids (A & E Connock)
Milkamino 20 (Arch Personal Care Products)

MILK FERMENT

Definition: Milk Ferment is a product obtained by the fermentation of milk (q.v.) by the microorganisms Streptococcus lactis, S. cremoris, S. Thermophilus, Mesophilic lactobacilli, and lactate fermenting yeasts.

Chemical Class: Biological Products

Function: Skin-Conditioning Agent - Miscellaneous

Trade Name:
Miferm (fermented) (Heidelberger)

MILK LIPIDS

JPN Translation:
乳脂

Definition: Milk Lipids is a mixture of lipids derived from milk. *See "Regulatory and Ingredient Use Information," regarding use of EU Trivial names in Volume 1, Introduction, Part A.*

Information Sources: JCIC, JCLS, JSQI

Chemical Class: Fats and Oils

Function: Skin-Conditioning Agent - Emollient

Technical/Other Name:
Milk Fat

Trade Name:
Milk Lipid Complex EL (Pentapharm/Centerchem)

Trade Name Mixture:
Lactomide (Pentapharm/Centerchem)

MILK PROTEIN

CAS No.: 91053-68-8

JPN Translation:
乳タンパク

Definition: Milk Protein is a mixture of proteins obtained from cow's milk. *See "Regulatory and Ingredient Use Information," regarding use of EU Trivial names in Volume 1, Introduction, Part A.*

Information Sources: JCIC, JCLS

Chemical Class: Proteins

Functions: Hair Conditioning Agent; Skin-Conditioning Agent - Miscellaneous

Reported Product Categories: Face and Neck Preparations (Excluding Shaving Preparations); Hair Conditioners; Shampoos (Non-coloring)

Technical/Other Name:
Proteins, Milk

Trade Names:
CMP-I (American Casein)
Complete Milk Protein (American Casein)
Lactobiol (I.D. bio)
Lactokine Fluid (CLR)
Lactokine Powder (CLR)
Milveta-TMP (Milei)

Trade Name Mixtures:
Follicusan (CLR)
Lactensyl PA (Laboratoires Sero-biologiques)
Lactofil Debacterise (Gattefosse s.a.)
Milk Albumin (Cosmetochem) (Cosmetochem International Ltd.)
Tensami 3/06 (Alban Muller)

MILK PROTEIN EXTRACT

Definition: Milk Protein Extract is an extract of Milk Protein (q.v.).

Chemical Class: Biological Products

Function: Not Reported

Trade Name Mixture:
Small Peptide-Fraction G (Arch Personal Care Products)

MILT EXTRACT

Definition: Milt Extract is an extract of fish milt.

Chemical Class: Biological Products

Function: Skin-Conditioning Agent - Miscellaneous

Technical/Other Name:
Extract of Milt

Trade Name Mixture:
SLENDERENNA (Ennagram)

MIMOSA PUDICA LEAF EXTRACT

Definition: Mimosa Pudica Leaf Extract is an extract of the leaves of *Mimosa pudica*. *See "Regulatory and Ingredient Use Information," regarding the labeling names for botanical ingredients in Volume 1, Introduction, Part A.*

Chemical Class: Biological Products

Function: Skin-Conditioning Agent - Miscellaneous

Technical/Other Name:
Extract of Mimosa Pudica Leaf

Trade Name:
Premier Mimosa Pudica 100% Extract (Premier Specialties)

MIMOSA TENUIFLORA BARK EXTRACT

CAS No.	EINECS No.
93685-96-2	297-646-0

Definition: Mimosa Tenuiflora Bark Extract is an extract of the bark of the mimosa, *Mimosa tenuiflora*. *See "Regulatory and Ingredient Use Information," regarding the labeling names for botanical ingredients in Volume 1, Introduction, Part A.*

Information Source: RIFM

Chemical Class: Biological Products

Functions: Fragrance Ingredient; Skin-Conditioning Agent - Miscellaneous

Reported Product Categories: Skin Care Preparations, Misc.; Paste Masks (Mud Packs)

Technical/Other Names:
Extract of Mimosa Bark
Extract of Mimosa Tenuiflora Bark
Mimosa absolute (Acacia decurrens Willd. var. dealbata) (RIFM)
Mimosa Bark Extract

Trade Name Mixtures:
Extrait de Mimosa Tenuiflora MBE BG 30 (Yves Rocher)
Extrapone Mimose 2/0333577 (Symrise)
Glycolysat of Tepescohuite (CEP (Solabia))
Herbasec Mimosa Tenuiflora (Cosmetochem) (Cosmetochem International Ltd.)
Herbasol Extract Mimose Tenuiflora (Cosmetochem) (Cosmetochem International Ltd.)

Mimosa Extract (Plantextrakt)
Mimosa Extract, Glycolic (Plantextrakt)
Mimosa Tenuiflora Extract HS 3386 G
(Grau)
Mimosa Tenuiflora Extract-NOVA
(Crodarom)
Mimosa Tenuiflora 5% Solution (Libiol)
Mimosoie (Alban Muller)
Nail Regenerative Complex (Alban Muller)
Phytelene of Mimosa Tenuiflora EG 363
Liquid (Indena SA)
Regenerative Phytoamine Biocomplex
(Alban Muller)
Skin Tree Extract PG (Rahn)
Tepescohuite Extract HG (Provital/
Centerchem)
Tepescohuite HPG Titrated (Alban Muller)

MIMOSA TENUIFLORA LEAF EXTRACT

CAS No.	EINECS No.
93685-96-2	297-646-0

Definition: Mimosa Tenuiflora Leaf Extract is an extract of the leaves of the mimosa, *Mimosa tenuiflora*. See *"Regulatory and Ingredient Use Information,"* regarding the labeling names for botanical ingredients in Volume 1, Introduction, Part A.

Information Source: RIFM

Chemical Class: Biological Products

Functions: Fragrance Ingredient; Proprietary

Reported Product Categories: Skin Care Preparations, Misc.; Paste Masks (Mud Packs)

Technical/Other Names:
Extract of Mimosa Tenuiflora Leaf
Mimosa absolute (Acacia decurrens Willd. var. dealbata) (RIFM)
Mimosa Leaf Extract

Trade Name Mixtures:
Actiphyte of Mimosa Leaf BG50 (Active Organics)
Actiphyte of Mimosa Leaf GL50 (Active Organics)
Actiphyte of Mimosa Leaf Lipo S (Active Organics)
Actiphyte of Mimosa Leaf PG50 (Active Organics)
VT-223 Extract of Mimosa (Vege-Tech)

MIMOSA TENUIFLORA LEAF POWDER

Definition: Mimosa Tenuiflora Leaf Powder is the powder obtained from the dried leaves of *Mimosa tenuiflora*. See *"Regulatory and Ingredient Use*

Information," regarding the labeling names for botanical ingredients in Volume 1, Introduction, Part A.

Chemical Class: Biological Products

Function: Skin-Conditioning Agent - Miscellaneous

Trade Name:
AEC Tepescohuite PF (A & E Connock)

MINERAL OIL

CAS Nos.	EINECS Nos.
8012-95-1	232-384-2
8020-83-5	
8042-47-5	232-455-8

JPN Translation:
ミネラルオイル

Definition: Mineral Oil is a liquid mixture of hydrocarbons obtained from petroleum. In the United States, Mineral Oil may be used as an active ingredient in OTC drug products. When used as an active drug ingredient, the established name is *Mineral Oil. See "Regulatory and Ingredient Use Information,"* regarding use of EU Trivial names in Volume 1, Introduction, Part A. See *"Regulatory and Ingredient Use Information,"* regarding the labeling names for U.S. OTC Drug Ingredients in Volume 1, Introduction, Part A.

Information Sources: AUS, BEL, BRA, 21CFR172.878, 21CFR173.340, 21CFR175.105, 21CFR175.210, 21CFR175.230, 21CFR175.300, 21CFR176.170, 21CFR176.200, 21CFR176.210, 21CFR177.1200, 21CFR177.2260, 21CFR177.2600, 21CFR177.2800, 21CFR178.3570, 21CFR178.3620, 21CFR178.3740, 21CFR178.3910, 21CFR179.45, 21CFR346.14, 21CFR369.20, 21CFR573.680, CTFA S, CZE, DA, DDR, FCC, FI, FIN, HUN, IND, ITA, JAN, JCLS, JSCI, KOR, MAR, MEX, MI-13(7263), NF XIX, OTC-I-AR, OTC-I-LX, OTC-I-OP, OTC-I-SK, PF, PN, POR, RIFM, ROM, SNPF, TSCA, USAN, USD, USP XXIV

Chemical Class: Hydrocarbons

Functions: Fragrance Ingredient; Hair Conditioning Agent; Skin-Conditioning Agent - Emollient; Skin-Conditioning Agent - Occlusive; Skin Protectant; Solvent

Reported Product Categories: Moisturizing Preparations; Bath Preparations, Misc.; Body and Hand Preparations (Excluding Shaving Preparations); Bath Oils, Tablets, and Salts; Cleansing Products (Cold Creams, Cleansing Lotions, Liquids and Pads); Lipsticks; Skin Care Preparations, Misc.; Bath Capsules; Face and Neck Preparations (Excluding Shaving Preparations); Hair Conditioners; Tonics, Dressings, and Other Hair Grooming Aids; Blushers (All types); Foundations; Makeup Bases; Face Powders; Night Skin Care Preparations; Eyebrow Pencils; Paste Masks (Mud Packs); Suntan Gels, Creams, and Liquids; Makeup Preparations (Not eye), Misc.; Shaving Cream (Aerosol, Brushless and Lather); Baby Lotions, Oils, Powders and Creams; Eye Makeup Removers; Fragrance Preparations, Misc.; Powders (Dusting and Talcum, Excluding Aftershave Talcs); Aftershave Lotions; Baby Shampoos; Eye Makeup Preparations, Misc.; Eyeliners; Hair Bleaches; Hair Straighteners; Indoor Tanning Preparations; Perfumes; Bath Soaps and Detergents; Hair Dyes and Colors (All Types Requiring Caution Statements and Patch Tests); Hair Preparations (Non-coloring), Misc.; Shaving Preparations, Misc.; Eye Shadows; Manicuring Preparations, Misc.; Suntan Preparations, Misc.; Makeup Fixatives; Mascara; Personal Cleanliness Products, Misc.; Hair Coloring Preparations, Misc.; Nail Creams and Lotions; Permanent Waves; Bubble Baths; Cuticle Softeners; Deodorants (Underarm); Depilatories; Feminine Hygiene Deodorants; Hair Rinses (Non-coloring); Rouges; Shampoos (Non-coloring)

Technical/Other Names:
Deobase (RIFM)
Heavy Mineral Oil
Hydrocarbon Oils
Light Mineral Oil
Liquid Paraffin
Liquid Petrolatum
Paraffin Oil
Paraffin oils (RIFM)
Prolatum Oil
White mineral oil, petroleum (RIFM)

Trade Names:
Benol (Crompton Corporation)
Blandol (Crompton Corporation)
Carnation (Crompton Corporation)
Drakeol 5 (Penreco)
Drakeol 6 (Penreco)
Drakeol 7 (Penreco)
Drakeol 8 (Penreco)
Drakeol 9 (Penreco)
Drakeol 10 (Penreco)
Drakeol 13 (Penreco)
Drakeol 15 (Penreco)
Drakeol 19 (Penreco)
Drakeol 21 (Penreco)
Drakeol 32 (Penreco)
Drakeol 34 (Penreco)
Drakeol 35 (Penreco)
Draketex 50 (Penreco)
Ervol (Crompton Corporation)
Gloria (Crompton Corporation)
Hydrobrite (Crompton Corporation)
Jeen Mineral Oil NF (Jeen)

Kaydol (Crompton Corporation)
Klearol (Crompton Corporation)
Marcol 52 (Exxon Mobil L & S)
Marcol 72 (Exxon Mobil L & S)
Marcol 82 (Exxon Mobil L & S)
Marcol 122 (Exxon Mobil L & S)
Marcol 152 (Exxon Mobil L & S)
Merkur White Oil Pharma (MERKUR Vaseline)
MERKUR White Oil Pharma; VARA 200; VARA 600 (MERKUR Vaseline)
Parol 70 (Penreco)
Parol 80 (Penreco)
Parol 100 (Penreco)
Peneteck (Penreco)
PIONIER White Oil/Liquid Paraffin (Hansen & Rosenthal)
Primol 262 (Exxon Mobil L & S)
Primol 352 (Exxon Mobil L & S)
Primol 382 (Exxon Mobil L & S)
Primol 542 (Exxon Mobil L & S)
Protol (Crompton Corporation)
Rudol (Crompton Corporation)
Superla White Oil 5 (Chevron Lubricants)
Superla White Oil 7 (Chevron Lubricants)
Superla White Oil 9 (Chevron Lubricants)
Superla White Oil 10 (Chevron Lubricants)
Superla White Oil 13 (Chevron Lubricants)
Superla White Oil 18 (Chevron Lubricants)
Superla White Oil 21 (Chevron Lubricants)
Superla White Oil 31 (Chevron Lubricants)
Superla White Oil 35 (Chevron Lubricants)
Superla White Oil 38 (Chevron Lubricants)
Superla White Oil 7A (Chevron Lubricants)
Superla White Oil 9A (Chevron Lubricants)
Uniwhite Oil 55 (Universal Preserv-A-Chem)
Uniwhite Oil 70 (Universal Preserv-A-Chem)
Uniwhite Oil 85 (Universal Preserv-A-Chem)
Uniwhite Oil 350 (Universal Preserv-A-Chem)

Trade Name Mixtures:
Actiphyte Indian Hemp Root Lipo M (Active Organics)
Actiphyte Almond Lipo M (Active Organics)
Actiphyte Anise Lipo M (Active Organics)
Actiphyte Balm Mint Lipo M (Active Organics)
Actiphyte Bitter Orange Lipo M (Active Organics)
Actiphyte Evening Primrose Lipo M (Active Organics)
Actiphyte Fennel Seed Lipo M (Active Organics)
Actiphyte Horsetail Lipo M (Active Organics)
Actiphyte Melaleuca Lipo M (Active Organics)
Actiphyte Myrtle Lipo M (Active Organics)
Actiphyte Orange Blossom Lipo M (Active Organics)

Actiphyte Orchid Lipo M (Active Organics)
Actiphyte Passionflower Lipo M (Active Organics)
Actiphyte Pyrethrum Lipo M (Active Organics)
Actiphyte Rose Lipo M (Active Organics)
Activera 106 Lipo M (Active Organics)
AEC Microcapsule White Vitamin E (A & E Connock)
Almolan Lis (Alma Chimica)
Aloe Oil Extract, Mineral (Concentrated Aloe Corp. (CAC))
Aloe Oil Extract, Mineral-Coconut (Concentrated Aloe Corp. (CAC))
Aloe Vera Oil Extract (A0001) (Terry)
Aloe Vera oil - Mineral/Coconut Oil Base AO2WOO2 (Aloestar)
Aloe Vera Oil - Mineral Oil Base AO2WOO1 (Aloestar)
Amerchol L-101 (Amerchol)
Argobase 125 (Croda Chemicals)
Argobase EU (Croda Chemicals)
Argobase L2 (Croda Chemicals)
Argobase S1 (Croda Chemicals)
Base 323 MS (LCW)
Base O/W 097 (LCW)
Base RAL W 323 T (LCW)
Base Rouge A Levres 323 TAL (LCW)
Base RW 135 (LCW)
Base RW 136 (LCW)
Base W/O 126 (LCW)
Bentone Gel MIO (ELE)
Bentone Gel MIO-A40 (ELE)
Botanivera 106 (Botanigenics)
C-Base (Maybrook)
Chamazulene - V (Vincience)
Cocoa Butter Fractionated (Browne)
Covagloss (LCW)
Covalip 94 (LCW)
Covalip 99 (LCW)
Covapencil 07 (LCW)
Covashine (LCW)
CREAGEL CRYSTAL MO (C.I.T.)
Crodabase SQ (Croda Brasil)
Crosterol SFA (Croda Chemicals)
Dehymuls K (Cognis Deutschland)
DP 705-9339 (Ciba Specialty Chemicals)
Dromicine (Cornerstone)
Emery 1732 (Cognis Care Chemicals/PA)
Emery 1740 (Cognis Care Chemicals/NJ)
Emery 1740 (Cognis Care Chemicals/PA)
Emulgator Apicerol 2/014081 (Symrise)
Emulzome (Exsymol)
ESP Dry Oil-MO (Earth Supplied Products)
Extract LE (Sino Lion)
Fancol LAO (Fanning)
Fancorgel A (Fanning)
Flocare ET 30 (SNF)
Flocare ET 58 (SNF)
Flocare ET 75 (SNF)
Gilugel MIN (Giulini/Giulini Chemie)
Guaiazulene 50% 2/012990 (Symrise)
Hostacerin WO (Clariant)

Hostacerin WO (Clariant GmbH, Personal Care)
Hydrobrite 2000 Gel (Crompton Corporation)
Hydrophobic Black Oxide C9333 (LCW)
Hydrophobic Brown Oxide C9458 (LCW)
Hydrophobic Chromium Oxide C9409 (LCW)
Hydrophobic Kaolin C9400 (LCW)
Hydrophobic Manganese Violet C9401 (LCW)
Hydrophobic Red Oxide C9454 (LCW)
Hydrophobic Talc C9441 (LCW)
Hydrophobic TiO2 C9428 (LCW)
Hydrophobic Ultramarine Blue C9404 (LCW)
Hydrophobic Ultra Violet C9402 (LCW)
Hydrophobic Yellow C9455 (LCW)
Isocreme CB 0279 (Croda Chemicals)
Ivarbase 101 (Arch Personal Care Products)
Ivarbase 3230 (Arch Personal Care Products)
Jeelan M-16 (Jeen)
Jeelan M-26 (Jeen)
KeraCoat (Arch Personal Care Products)
KSG-31 (Shin-Etsu Chemical Co.)
KSG-41 (Shin-Etsu Chemical Co.)
KSG-310 (Shin-Etsu Chemical Co.)
Lanaetex FB (Lanaetex)
Lanalene ABS (Maybrook)
Lanosoluble M (Prod'Hyg)
Liant TW 406 (LCW)
Liant TW 729 (LCW)
Liant TW 876 (LCW)
Lipofacteur Vitentiel (LCW)
Lipoplastidine Althaea (Vevy)
Lipoplastidine Calendula (Vevy)
Lipoplastidine Daucus (Vevy)
Lipoplastidine Equisetum (Vevy)
Lipoplastidine Matricaria (Vevy)
Lipoplastidine Mel (Vevy)
Lipoplastidine Oenothera Biennis (Vevy)
Lipoplastidine Ostrea (Vevy)
Lipoplastidine Pappa Regalis (Vevy)
Lipoplastidine Valeriana (Vevy)
Lipshine (LCW)
Liquid Absorption Base Type T (Croda, Inc.)
Mascawax 012 (LCW)
Mearlite GEJ (Engelhard Corp.)
Megabase L101 (Megachem)
Microsponge 5645 Mineral Oil (EDT, Inc.)
Multilan A (Fabriquimica)
Neo PCL SE o/w 2/066280 (Symrise)
Neo-PCL w/o s.e. 2/066255 (Symrise)
Pacific Sea Kelp Oil Extract (Bell Flavors)
Paprika-Extract, Oil Soluble (Crodarom)
PCL SE w/o 2/066255 (Symrise)
Permulgin 3510 (Koster Keunen Holland)
Pionier PLW (Hansen & Rosenthal)
Polymer EX-617 (Noveon)
Polytrap 6038 Mineral Oil Macrobeads (EDT, Inc.)
Prodhycreme (Prod'Hyg)

Prodhycreme 013 (Prod'Hyg)
Prodhyrouge 2000 (Prod'Hyg)
Protalan M-16 (Protameen)
Protalan M-26 (Protameen)
Protegin V (Degussa Care Specialties)
Protegin XV (Degussa Care Specialties)
Rheocare ATC (Cosmetic Rheologies)
Ritachol (RITA)
Saboderm 818 (Sabo)
Salcare SC91 (Ciba Specialty Chemicals)
Salcare SC92 (Ciba Specialty Chemicals)
Salcare SC95 (Ciba Specialty Chemicals)
Sebase (Croda Chemicals)
Sepigel 501 (SEPPIC)
Sericite SLZ-012P (Presperse)
Sexadecyl Alcohol-Cosmetic Grade
  (Lanaetex)
Simagel M (Biophil)
Spectraveil MOTG (Uniqema, Belgium)
Steralchol (Lanaetex)
Tagravit A1 (Tagra)
Tioveil MOTG (Uniqema, Belgium)
Tioveil 50 MOTG (Uniqema, Belgium)
Tixogel MIO (Sud-Chemie, United
  Catalysts)
Tixogel MIO 1584 (Sud-Chemie, United
  Catalysts)
Unieucerin (Chemyunion)
Uniliquid (Chemyunion)
Unitina LM (Universal Preserv-A-Chem)
Vegetol Aloe GR 335 Oily (Gattefosse s.a.)
Vegetol Arnica MCF 1397 Oily (Gattefosse
  s.a.)
Vegetol Burdock 4148 Oily (Gattefosse
  s.a.)
Vegetol Calendula WL 1072 Oily
  (Gattefosse s.a.)
Vegetol Capsicum LC 481 Oily (Gattefosse
  s.a.)
Vegetol Elder 4144 Oily (Gattefosse s.a.)
Vegetol Hawthorn MCF 066 Oily
  (Gattefosse s.a.)
Vegetol Hop 4150 Oily (Gattefosse s.a.)
Vegetol Ivy 4149 Oily (Gattefosse s.a.)
Vegetol Linden 4141 Oily (Gattefosse s.a.)
Vegetol Mallow 4142 Oily (Gattefosse s.a.)
Vegetol Matricaria 4140 Oily (Gattefosse
  s.a.)
Vegetol Rosemary 4145 Oily (Gattefosse
  s.a.)
Vegetol Sage 4138 Oily (Gattefosse s.a.)
Vegetol St. John's Wort MCF 883 Oily
  (Gattefosse s.a.)
Vegetol White Nettle MCF 1233 Oily
  (Gattefosse s.a.)
Vegetol White Willow 4151 Oily (Gattefosse
  s.a.)
Vegetol Wild Roseberry MCF 1837 Oily
  (Gattefosse s.a.)
Versagel M (Penreco)
Vitaceane (Sederma)
Vitaphyle ACE (LCW)

MINERAL SALTS

Definition: Mineral Salts is a mixture of inorganic salts derived from mineral water.

Chemical Class: Inorganic Salts

Function: Skin-Conditioning Agent - Miscellaneous

Technical/Other Name:
  Legenaria Vulgaris

Trade Names:
  Dry mineral salts of La Toja spring water
    (Henkel Iberica)
  Elguea Thermal salts (Natural thermal salts
    obtained from Elguea Thermal Cuba)
    (Natural Beauty s.r.l.)
  EMS Salt (Siemens)
  Luneburge Solecreme (Intrapharm)
  Miyabi Spa Extract 2 (Nonogawa)

MINERAL SPIRITS

CAS Nos.   **EINECS No.**
8032-32-4   232-453-7
64475-85-0

Definition: Mineral Spirits is a mixture of hydrocarbons obtained from petroleum with a distillation range between 318 and 400 degrees F.

Information Sources: 21CFR178.3800, MI-13(5510), MI-13(6223), TSCA

Chemical Class: Hydrocarbons

Function: Solvent

Reported Product Categories: Personal Cleanliness Products, Misc.; Mascara

Technical/Other Names:
  Ligroin
  Ligroine
  Petroleum Spirits

Trade Name Mixture:
  Tixogel OMS (Sud-Chemie, United
    Catalysts)

MINKAMIDE DEA

CAS No.: 124046-27-1

Definition: Minkamide DEA is a mixture of ethanolamides of the fatty acids derived from mink oil. It conforms to the formula:

$$RC \overset{O}{\overset{\|}{-}} N(CH_2CH_2OH)_2$$

where RCO- represents the fatty acids derived from mink oil.

Information Source: EEC(III/1-60)

Chemical Class: Alkanolamides

Functions: Surfactant - Foam Booster; Viscosity Increasing Agent - Aqueous

Technical/Other Names:
  Amides, Mink Oil, N,N-Bis(Hydroxyethyl)-
  N,N-Bis(2-Hydroxyethyl)Mink Fatty Acid
    Amide
  Diethanolamine Mink Fatty Acid
    Condensate
  Mink Amides, N,N-Bis(2-Hydroxyethyl)-
  Mink Fatty Acid Diethanolamide

MINKAMIDOPROPALKONIUM CHLORIDE

CAS No.: 124046-06-6

Definition: Minkamidopropalkonium Chloride is the quaternary ammonium salt that conforms generally to the formula:

$$\left[RCNH(CH_2)_3 - \overset{\overset{CH_3}{|}}{\underset{\underset{CH_3}{|}}{N}} - CH_2 - C_6H_5 \right]^{+} Cl^{-}$$

where RCO- represents the fatty acids derived from mink oil.

Chemical Class: Quaternary Ammonium Compounds

Functions: Antistatic Agent; Hair Conditioning Agent

Technical/Other Name:
  Benzenemethanaminium, N-(3-Amino-
    propyl)-N,N-Dimethyl-, N-Mink Oil Acyl
    Derivs., Chlorides

MINKAMIDOPROPYLAMINE OXIDE

CAS No.: 124046-29-3

Definition: Minkamidopropylamine Oxide is the tertiary amine oxide that conforms generally to the formula:

$$RC \overset{O}{\overset{\|}{-}} NH(CH_2)_3 - \overset{\overset{CH_3}{|}}{\underset{\underset{CH_3}{|}}{N}} \rightarrow O$$

where RCO- represents the fatty acids derived from mink oil.

Chemical Class: Amine Oxides

Functions: Hair Conditioning Agent; Surfactant - Cleansing Agent; Surfactant - Foam Booster; Surfactant - Hydrotrope

Technical/Other Names:
  Amides, Mink, N-[3-(Dimethylamino)-
    Propyl], N-Oxide

N-[3-(Dimethylamino)Propyl]Mink Amides-
N-Oxide

Mink Amides, N-[3-(Dimethylamino)Propyl],
N-Oxide

Dimethylaminopropyl Mink Fatty Acids
Amide

N-[3-(Dimethylamino)Propyl]Mink Oil
Amides

MINKAMIDOPROPYL BETAINE

Definition: Minkamidopropyl Betaine is the zwitterion (inner salt) that conforms generally to the formula:

$$\underset{O}{\overset{\parallel}{RC}}-NH(CH_2)_3-\overset{CH_3}{\underset{CH_3}{\overset{|}{N^+}}}-CH_2COO^-$$

where RCO- represents the fatty acids derived from mink oil.

Chemical Classes: Amides; Betaines

Functions: Antistatic Agent; Hair Conditioning Agent; Skin-Conditioning Agent - Miscellaneous; Surfactant - Cleansing Agent; Surfactant - Foam Booster; Viscosity Increasing Agent - Aqueous

Technical/Other Names:
N-(Carboxymethyl)-N,N-Dimethyl-3-[(1-Oxomink)Amino]-1-Propanaminium Hydroxide, Inner Salt
Mink Amide Propylbetaine
Minkamidopropyl Dimethyl Glycine
1-Propanaminium, N-(Carboxymethyl)-N,N-Dimethyl-3-[(1-Oxomink)Amino]-, Hydroxide, Inner Salt
Quaternary Ammonium Compounds, (Carboxymethyl)(3-Minkamidopropyl) Dimethyl, Hydroxide, Inner Salt

MINKAMIDOPROPYL DIMETHYLAMINE

CAS No.	EINECS No.
68953-11-7	273-187-1

Definition: Minkamidopropyl Dimethylamine is the amidoamine that conforms generally to the formula:

$$\underset{O}{\overset{\parallel}{RC}}-NH(CH_2)_3N\overset{CH_3}{\underset{CH_3}{<}}$$

where RCO- represents the fatty groups derived from mink oil.

Information Source: TSCA

Chemical Class: Amines

Function: Antistatic Agent

Reported Product Category: Hair Conditioners

Technical/Other Names:
Amides, Mink Oil, N-[3-(Dimethylamine)-Propyl]-

MINKAMIDOPROPYL ETHYLDIMONIUM ETHOSULFATE

CAS No.: 115340-79-9

Definition: Minkamidopropyl Ethyldimonium Ethosulfate is the quaternary ammonium salt that conforms generally to the formula:

$$\left[\underset{O}{\overset{\parallel}{RC}}-NH(CH_2)_3-\overset{CH_3}{\underset{CH_3}{\overset{|}{N}}}-CH_2CH_3\right]^+ \quad \underset{OSO_3^-}{\overset{CH_3CH_2}{|}}$$

where RCO- represents the fatty acids derived from mink oil.

Chemical Class: Quaternary Ammonium Compounds

Functions: Antistatic Agent; Hair Conditioning Agent

Technical/Other Name:
1-Propanaminium, 3-Amino-N-Ethyl-N,N-Dimethyl-, N-Mink Oil Acyl Derivs., Ethyl Sulfates

MINK OIL

CAS No.	EINECS No.
8023-74-3	232-423-3

JPN Translation:
ミンク油

Definition: Mink Oil is an oil obtained from the sub-dermal fatty tissues of the mink. *See "Regulatory and Ingredient Use Information," regarding use of EU Trivial names in Volume 1, Introduction, Part A.*

Information Sources: CIR: [I] IJT-17 (SUPPL. 4)1998, CIR: [S], CTFA D, JCLS, JSCI

Chemical Class: Fats and Oils

Functions: Hair Conditioning Agent; Skin-Conditioning Agent - Occlusive

Reported Product Categories: Hair Conditioners; Lipsticks; Bath Preparations, Misc.; Body and Hand Preparations (Excluding Shaving Preparations); Tonics, Dressings, and Other Hair Grooming Aids; Moisturizing Preparations; Bath Oils, Tablets, and Salts; Cleansing Products (Cold Creams, Cleansing Lotions, Liquids and Pads); Skin Care Preparations, Misc.; Night Skin Care Preparations; Hair Preparations (Non-coloring), Misc.; Hair Straighteners; Permanent Waves; Shampoos (Non-coloring); Hair Sprays (Aerosol Fixatives); Shaving Cream (Aerosol, Brushless and Lather); Blushers (All types); Eye Shadows; Face and Neck Preparations (Excluding Shaving Preparations); Face Powders; Fragrance Preparations, Misc.; Hair Rinses (Non-coloring); Makeup Bases; Mascara; Suntan Gels, Creams, and Liquids; Suntan Preparations, Misc.

Technical/Other Names:
Mustele Oil
Oil of Mink
Oils, Mink

Trade Names:
AEC Mink Oil (A & E Connock)
Cropure Mink (Croda Chemicals)
Emulan (Emulan)
Jeen Mink Oil (Jeen)
Mink Oil (U.S.) (Nikko Rica)
Naturol (Lanaetex)
Peacock Mink Oil (Pfau)
Super Mink Oil (Nikko Rica)

Trade Name Mixtures:
Algae Extract MOS (Maruzen Pharmaceuticals Co., Ltd.)
Dermoil (Lanaetex)

MINK OIL PEG-13 ESTERS

Definition: Mink Oil PEG-13 Esters is a complex mixture formed from the transesterification of Mink Oil (q.v.) and PEG-13.

Chemical Class: Glyceryl Esters and Derivatives

Functions: Skin-Conditioning Agent - Emollient; Surfactant - Emulsifying Agent

MINK WAX

JPN Translation:
ミンクロウ

Definition: Mink Wax is the solid fraction derived from Mink Oil (q.v.). *See "Regulatory and Ingredient Use Information," regarding use of EU Trivial names in Volume 1, Introduction, Part A.*

Information Sources: JCIC, JCLS

Chemical Class: Waxes

Functions: Hair Conditioning Agent; Skin-Conditioning Agent - Occlusive

Technical/Other Name:
Waxes, Mink

MIPA-BORATE

CAS No.	EINECS No.
68003-13-4	268-109-8

Empirical Formula:
$C_3H_9NO \cdot BH_3O_3$

Definition: MIPA-Borate is the reaction product of Isopropanolamine (q.v.) and Boric Acid (q.v.).

Information Source: EEC(III/1-1)

Chemical Class: Amines

Function: Buffering Agent

Reported Product Category: Hair Sprays (Aerosol Fixatives)

Technical/Other Names:
Boric Acid, Compd. with 1-Amino-2-Propanol
Isopropanolamine Borate
Monoisopropanolamine Borate

Trade Name Mixture:
Monacor BE (Uniqema)

MIPA C12-15 PARETH SULFATE

Definition: MIPA C12-15 Pareth Sulfate is the monoisopropanolamine salt of a sulfated ethoxylated C12-15 fatty alcohol that conforms generally to the formula:

$$R(OCH_2CH_2)_nOSO_3H \quad \cdot \quad NH_2CH_2CHCH_3$$
$$| $$
$$OH$$

where n has an value between 1 and 4 and R represents the C12-15 fatty alcohol.

Chemical Class: Alkyl Ether Sulfates

Function: Surfactant - Cleansing Agent

Technical/Other Name:
Monoisopropanolamine C12-15 Pareth Sulfate

MIPA-DODECYLBENZENESULFONATE

CAS Nos. **EINECS Nos.**
42504-46-1   255-854-9
54590-52-2   259-249-0

Empirical Formula:
$C_{18}H_{30}O_3S \cdot C_3H_9NO$

Definition: MIPA-Dodecylbenzenesulfonate is the monoisopropanolamine salt of a substituted aromatic compound that conforms generally to the formula:

Information Sources: 21CFR176.210, TSCA

Chemical Class: Alkyl Aryl Sulfonates

Function: Surfactant - Cleansing Agent

Technical/Other Names:
Benzenesulfonic Acid, Dodecyl-, Compd. with 1-Amino-2-Propanol (1:1)
Monoisopropanolamine Dodecylbenzenesulfonate

Trade Name:
Hetsulf IPA (Heterene)

MIPA-LAURETH SULFATE

CAS No.: 83016-76-6

Definition: MIPA-Laureth Sulfate is the monoisopropanolamine salt of sulfated ethoxylated lauryl alcohol that conforms generally to the formula:

$$CH_3(CH_2)_{11}(OCH_2CH_2)_nOSO_3H \quad \cdot \quad NH_2CH_2CHCH_3$$
$$|$$
$$OH$$

where n has a value between 1 and 4.

Chemical Class: Alkyl Ether Sulfates

Function: Surfactant - Cleansing Agent

Technical/Other Names:
Monoisopropanolamine Lauryl Ether Sulfate
Poly(Oxy-1,2-Ethanediyl), α-Sulfo-ω-(Dodecyloxy)-, Compd. with 1-Amino-2-Propanol

Trade Names:
Zetesol 856 (Zschimmer & Schwarz)
Zetesol 2056 (Zschimmer & Schwarz)

Trade Name Mixtures:
Marlinat 242/90M (Sasol GmbH - Marl)
Perlglanzmittel GM 4055 (Zschimmer & Schwarz)
Texapon WW100 (Cognis Care Chemicals/PA)
Zetesol 100 (Zschimmer & Schwarz)
Zetesol 856 T (Zschimmer & Schwarz)

MIPA-LAURYL SULFATE

CAS No. **EINECS No.**
21142-28-9   244-238-5

Empirical Formula:
$C_{12}H_{26}O_4S \cdot C_3H_9NO$

Definition: MIPA-Lauryl Sulfate is the monoisopropanolamine salt of lauryl sulfate that conforms generally to the formula:

$$CH_3(CH_2)_{11}OSO_3H \quad \cdot \quad NH_2CH_2CHCH_3$$
$$|$$
$$OH$$

Information Source: TSCA

Chemical Class: Alkyl Sulfates

Function: Surfactant - Cleansing Agent

Technical/Other Names:
Dodecyl Sulfate, Comp. with 1-Amino-2-Propanol (1:1)
Monoisopropanolamine Lauryl Sulfate
Sulfuric Acid, Monododecyl Ester, Compd. with 1-Amino-2-Propanol (1:1)

Trade Name:
Empicol YL60 (Albright & Wilson UK)

MIPA-MYRISTATE

Definition: MIPA-Myristate is the salt of monoisopropanolamine and Myrustic Acid (q.v.). It conforms to the formula:

$$CH_3(CH_2)_{12}COOH \quad \cdot \quad NH_2CH_2CHCH_3$$
$$|$$
$$OH$$

Information Source: JCLS

Chemical Class: Esters

Functions: Surfactant - Foam Booster; Viscosity Increasing Agent - Aqueous

MIRABILIS JALAPA EXTRACT

CAS No. **EINECS No.**
91722-88-2   294-491-0

Definition: Mirabilis Jalapa Extract is an extract of the aerial parts of *Mirabilis jalapa*. See "Regulatory and Ingredient Use Information," regarding the labeling names for botanical ingredients in Volume 1, Introduction, Part A.

Chemical Class: Biological Products

Function: Skin-Conditioning Agent - Miscellaneous

Technical/Other Name:
Extract of Mirabilis Jalapa

Trade Name Mixture:
Belle De Nuit Extract (Sederma)

MITRACARPUS SCABER EXTRACT

Definition: Mitracarpus Scaber Extract is an extract of *Mitracarpus scaber*. See "Regulatory and Ingredient Use Information," regarding the labeling names for botanical ingredients in Volume 1, Introduction, Part A.

Chemical Class: Biological Products

Function: Not Reported

The inclusion of any compound in the *Dictionary and Handbook* does not indicate that use of that substance as a cosmetic ingredient complies with the laws and regulations governing such use in the United States or any other country.

Technical/Other Name:
Extract of Mitracarpe

Trade Name Mixture:
Etioline (Sederma)

MIXED CRESOLS

CAS No.	EINECS No.
1319-77-3	215-293-2

JPN Translation:
クレゾール

Empirical Formula:
C_7H_8O

Definition: Mixed Cresols is a mixture of o-Cresol (q.v.), m-Cresol (q.v.) and p-Cresol (q.v.).

Information Sources: 21CFR172.515, 21CFR175.300, 21CFR177.2410, 21CFR310.545, CIR: [I], JAN, JCLS, JSCI, MHLW-331/3, NF XVIII, RIFM, TSCA, USAN

Chemical Class: Phenols

Functions: Fragrance Ingredient; Preservative

Technical/Other Names:
Cresol
Cresol (mixed isomers) (RIFM)
Methylphenol, Mixed
Phenol, Methyl-. Mixed

MIXED IONONES

Empirical Formula:
$C_{13}H_{20}O$

Definition: Mixed Ionones is α-ionone, or β-ionone, or a mixture of these isomers. It conforms to the formula:

a -form   b-form

Information Source: MI-13(5072)

Chemical Class: Ketones

Function: Fragrance Ingredient

MIXED ISOPROPANOLAMINES

Definition: Mixed Isopropanolamines is a blend of Isopropanolamine (q.v.), Diiso-propanolamine (q.v.) and Triiso-propanolamine (q.v.).

Information Sources: CIR: [SQ] JACT-6 (1)1987, EEC(II-411)

Chemical Class: Alkanolamines

Function: pH Adjuster

MIXED ISOPROPANOLAMINES LANOLATE

Definition: Mixed Isopropanolamines Lanolate is a mixture of amine salts formed by neutralizing lanolin acids with Mixed Isopropanolamines (q.v.).

Chemical Class: Soaps

Functions: Hair Conditioning Agent; Skin-Conditioning Agent - Miscellaneous; Surfactant - Cleansing Agent

MIXED ISOPROPANOLAMINES LAURYL SULFATE

CAS No.: 68877-25-8

Definition: Mixed Isopropanolamines Lauryl Sulfate is the Mixed Isopropanolamines (q.v.) salt of lauryl sulfate.

Information Source: TSCA

Chemical Class: Alkyl Sulfates

Function: Surfactant - Cleansing Agent

MIXED ISOPROPANOLAMINES MYRISTATE

Definition: Mixed Isopropanolamines Myristate is a mixture of amine salts formed by neutralizing myristic acid with Mixed Isopropanolamines (q.v.).

Information Sources: JCIC, JCLS

Chemical Class: Soaps

Function: Surfactant - Cleansing Agent

Technical/Other Name:
Isopropanolamine Myristate Solution

Trade Name:
Lanamine (Amerchol)

MIXED TERPENES

Definition: Mixed Terpenes is a mixture of hydrocarbons distilled from certain plants and trees such as the pine and citrus. The mixture consists of terpenes, sesquiterpenes, diterpenes and polyterpenes.

Chemical Class: Hydrocarbons

Function: Solvent

Trade Name:
Solvenol No. 2 (Hercules)

MOLASSES EXTRACT

Definition: Molasses Extract is an extract of molasses. *See "Regulatory and Ingredient Use Information," regarding use of EU Trivial names in Volume 1, Introduction, Part A.*

Information Source: JCLS

Chemical Class: Biological Products

Function: Skin-Conditioning Agent - Miscellaneous

Technical/Other Name:
Extract of Molasses

Trade Name:
Actilac M (Active Organics)

Trade Name Mixture:
Molasses Liquid (Straetmans)

MOLYBDENUM ASPARTATE

Definition: Molybdenum Aspartate is the molybdenum salt of Aspartic Acid (q.v.).

Chemical Classes: Amino Acids; Organic Salts

Function: Skin-Conditioning Agent - Miscellaneous

Trade Name:
Oligoidyne Molybdenum (Vevy)

MOMORDICA CHARANTIA FRUIT EXTRACT

Definition: Momordica Charantia Fruit Extract is the extract of the fruit of *Momordica charantia*. *See "Regulatory and Ingredient Use Information," regarding the labeling names for botanical ingredients in Volume 1, Introduction, Part A.*

Chemical Class: Biological Products

Functions: Hair Conditioning Agent; Humectant; Skin-Conditioning Agent - Humectant; Skin-Conditioning Agent - Miscellaneous; Skin-Conditioning Agent - Occlusive

Technical/Other Name:
Extract of Momordica Charantia Fruit

Trade Name Mixtures:
Bitter Melon (Greentech)
Turureishi Extract (Eikodo & Co)

MOMORDICA CHARANTIA FRUIT POWDER

Definition: Momordica Charantia Fruit Powder is the powder obtained from the dried fruit of *Momordica charantia*. *See "Regulatory and Ingredient Use Information," regarding the labeling names for botanical ingredients in Volume 1, Introduction, Part A.*

Chemical Class: Biological Products

Function: Skin-Conditioning Agent - Humectant

Trade Name:
Turureishi Powder (Eikodo & Co)

MOMORDICA GROSVENORI FRUIT JUICE

Definition: Momordica Grosvenori Fruit Juice is the juice expressed from the fruit of *Mormordica grosvenori*. *See "Regulatory and Ingredient Use Information," regarding the labeling names for botanical ingredients in Volume 1, Introduction, Part A.*

Chemical Class: Biological Products

Functions: Skin-Conditioning Agent - Humectant; Solvent

Trade Name Mixture:
Momordica Grosvenori Fruit Water F (Maruzen Pharmaceuticals Co., Ltd.)

MONARDA DIDYMA

Definition: *See "Regulatory and Ingredient Use Information," regarding EU labeling names for botanical ingredients in Volume 1, Introduction, Part A.*

Chemical Class: Biological Products

Technical/Other Names:
Monarda Didyma Leaf Extract (U.S.)
Monarda Didyma Oil (U.S.)

MONARDA DIDYMA LEAF EXTRACT

Definition: Monarda Didyma Leaf Extract is an extract of the leaves of the bee balm, *Monarda didyma*. *See "Regulatory and Ingredient Use Information," regarding the labeling names for botanical ingredients in Volume 1, Introduction, Part A.*

Chemical Class: Biological Products

Function: Not Reported

Technical/Other Names:
Bee Balm Extract
Bee Balm (Monarda Didyma) Extract
Extract of Bee Balm
Extract of Monarda Didyma
Extract of Oswego Tea
Monarda Didyma (EU)
Oswego Tea Extract

Trade Name Mixtures:
Actiphyte of Bee Balm BG50 (Active Organics)
Actiphyte of Bee Balm GL50 (Active Organics)
Actiphyte of Bee Balm Lipo S (Active Organics)
Actiphyte of Bee Balm PG50 (Active Organics)

MONARDA DIDYMA OIL

Definition: Monarda Didyma Oil is the volatile oil obtained from *Monarda didyma*. *See "Regulatory and Ingredient Use Information," regarding the labeling names for botanical ingredients in Volume 1, Introduction, Part A.*

Chemical Class: Essential Oils

Function: Fragrance Ingredient

Technical/Other Names:
Bee Balm (Monarda Didyma) Oil
Bee Balm Oil
Monarda Didyma (EU)
Oil, Essential, Bee Balm
Oil of Bee Balm

MONASCUS EXTRACT

Definition: Monascus Extract is an extract of *Monascus purpureus*.

Chemical Class: Biological Products

Function: Not Reported

Technical/Other Name:
Extract of Monascus

Trade Name Mixture:
Monascus Extract TH (T.HASEGAWA)

MONASCUS/RICE FERMENT

Definition: Monascus/Rice Ferment is the product obtained by the fermentation of rice by the organism, *Monascus purpureus*.

Chemical Class: Biological Products

Function: Antioxidant

Trade Name:
Tri-K Fre (Tri-K)

MONOGLYCERIDES ACETATE

JPN Translation:
酢酸脂肪酸グリセリル

Definition: *See "Regulatory and Ingredient Use Information," regarding use of Japan Trivial names in Volume 1, Introduction, Part A.*

Information Source: JCLS

Chemical Class: Esters

Function: Not Reported

MONOGLYCERIDES CITRATE

JPN Translation:
クエン酸脂肪酸グリセリル

Definition: *See "Regulatory and Ingredient Use Information," regarding use of Japan Trivial names in Volume 1, Introduction, Part A.*

Information Source: JCLS

Chemical Class: Esters

Function: Not Reported

MONOGLYCERIDES/DIACETYL TARTARATE ESTERS

JPN Translation:
レタスエキス

Definition: *See "Regulatory and Ingredient Use Information," regarding use of Japan Trivial names in Volume 1, Introduction, Part A.*

Information Source: JCLS

Chemical Class: Esters

Function: Not Reported

MONOGLYCERIDES LACTATE

JPN Translation:
乳酸脂肪酸グリセリル

Definition: Monoglycerides Lactate is a monoester of lactic acid and fatty acid monoglycerides. *See "Regulatory and Ingredient Use Information," regarding use of Japan Trivial names in Volume 1, Introduction, Part A.*

Information Source: JCLS

Chemical Class: Esters

Function: Not Reported

MONOGLYCERIDES SUCCINATE

JPN Translation:
コハク酸脂肪酸グリセリル

Definition: *See "Regulatory and Ingredient Use Information," regarding use of Japan Trivial names in Volume 1, Introduction, Part A.*

Information Source: JCLS

Chemical Class: Esters

Function: Not Reported

MONOSACCHARIDE LACTATE CONDENSATE

Definition: Monosaccharide Lactate Condensate is a condensation product of sodium lactate and the following monosaccharides: Glucose, fructose, glucosamine, ribose and deoxyribose.

Chemical Classes: Carbohydrates; Esters

Function: Skin-Conditioning Agent - Miscellaneous

MONOSODIUM CITRATE

CAS No.	EINECS No.
18996-35-5	242-734-6

Empirical Formula:
$C_6H_8O_7 \cdot Na$

Definition: Monosodium Citrate is the organic salt that conforms to the formula:

$$
\begin{array}{c}
OH \\
|\\
NaOOCCH_2CCH_2COOH \\
|\\
COOH
\end{array}
$$

Information Source: TSCA

Chemical Class: Organic Salts

Function: pH Adjuster

Technical/Other Names:
Citric Acid, Monsodium Salt
2-Hydroxy-1,2,3-Propanetricarboxylic Acid, Monosodium Salt
1,2,3-Propanetricarboxylic Acid, 2-Hydroxy-, Monosodium Salt

Trade Name:
Jungbunzlauer Monosodium Citrate (Jungbunzlauer)

MONTAN ACID WAX

CAS No.	EINECS No.
68476-03-9	270-664-6

Definition: Montan Acid Wax is the product obtained by the oxidation of Montan Wax (q.v.).

Information Source: TSCA

Chemical Class: Waxes

Functions: Binder; Viscosity Increasing Agent - Nonaqueous

Technical/Other Names:
Fatty Acids, Montan Wax
Waxes, Montan Fatty Acids

Trade Names:
Hoechst Wax LP (Clariant)
Hoechst Wax LP (Clariant GmbH, Personal Care)
Hoechst Wax S (Clariant)
Hoechst Wax S (Clariant GmbH, Personal Care)
Hoechst Wax SW (Clariant)
Hoechst Wax SW (Clariant GmbH, Personal Care)

Trade Name Mixture:
Ross Carnauba Wax Replacement (Ross)

MONTAN WAX

CAS No.	EINECS No.
8002-53-7	232-313-5

JPN Translation:
モンタンロウ

Definition: Montan Wax is a wax obtained by extraction of lignite. *See "Regulatory and Ingredient Use Information," regarding use of EU Trivial names in Volume 1, Introduction, Part A.*

Information Sources: 21CFR175.105, 21CFR176.210, 21CFR177.2600, CIR: [S] JACT-3(3)1984, JCIC, JCLS, MI-13(6280), TSCA

Chemical Class: Waxes

Functions: Binder; Viscosity Increasing Agent - Nonaqueous

Reported Product Categories: Lipsticks; Foundations; Eye Makeup Preparations, Misc.; Eye Shadows; Eyeliners; Makeup Preparations (Not eye), Misc.; Mascara; Nail Polish and Enamels

Technical/Other Names:
Breached Montan Wax
Waxes, Montan

Trade Names:
Bleached Montan Wax (Ross)
Bleached Montan Wax Cosmetic Grade (Ross)
Hansonwax S, Hansonwax E, Hansonwax OP (Hansotech)
Hoechst Wax BJ (Clariant)
Hoechst Wax BJ (Clariant GmbH, Personal Care)
Hoechst Wax F (Clariant)
Hoechst Wax F (Clariant GmbH, Personal Care)
Hoechst Wax W (Clariant)
Hoechst Wax W (Clariant GmbH, Personal Care)
Montan Wax (Megachem)
Montan Wax SP 48 (Strahl & Pitsch)
Montan Wax SP 164 (Strahl & Pitsch)
Montan Wax - STRALPITZ (Strahl & Pitsch)
S&P Montan Wax Refined (Strahl & Pitsch)

Trade Name Mixtures:
7304 Candelilla Substitute (Kahl)
Enviro Pure 308 (React Inc)
2901 Synthetic Carnauba (Kahl)

MONTMORILLONITE

CAS No.	EINECS No.
1318-93-0	215-288-5

JPN Translation:
モンモリロナイト

Definition: Montmorillonite is a complex aluminum/magnesium silicate clay.

Information Sources: CIR: [S] IJT-22 (SUPPL. 1)2003, JCIC, JCLS, MI-13(6283)

Chemical Class: Inorganics

Functions: Abrasive; Absorbent; Bulking Agent; Emulsion Stabilizer; Opacifying Agent; Viscosity Increasing Agent - Aqueous

Reported Product Category: Paste Masks (Mud Packs)

Trade Names:
Gelwhite GP (Southern Clay)
Gelwhite L (Southern Clay)
Gelwhite H, NF (Southern Clay)
Mineral Colloid BP (Southern Clay)
Mineral Colloid MO (Southern Clay)

MORINDA CITRIFOLIA EXTRACT

CAS No.	EINECS No.
84929-68-0	284-551-4

Definition: Morinda Citrifolia Extract is an extract of the whole plant, *Morinda citrifolia*. *See "Regulatory and Ingredient Use Information," regarding the labeling names for botanical ingredients in Volume 1, Introduction, Part A.*

Chemical Class: Biological Products

Function: Cosmetic Astringent

Technical/Other Name:
Extract of Morinda Citrifolia

Trade Name Mixtures:
Actiphyte Noni (Active Organics)
Actiphyte Noni BG50P (Active Organics)

MORINDA CITRIFOLIA FRUIT EXTRACT

CAS No.: 84929-68-0

Definition: Morinda Citrifolia Fruit Extract is an extract of the fruit of *Morinda citrifolia*. See "*Regulatory and Ingredient Use Information," regarding the labeling names for botanical ingredients in Volume 1, Intro-duction, Part A.*

Chemical Class: Biological Products

Function: Not Reported

Trade Names:
Extrait Fluide De Morinda Citrifolia (Plantes et Industrie)
Tahitian Noni Extract (Morinda, Inc)

Trade Name Mixtures:
BIOGREEN NONI (Greentech)
Extrait Fluide De Morinda Citrifolia (Plantes et Industrie)

MORINDA CITRIFOLIA FRUIT JUICE

Definition: Morinda Citrifolia Fruit Juice is the juice expressed from the fruit of *Morinda citrifolia*. See "*Regulatory and Ingredient Use Information," regarding the labeling names for botanical ingredients in Volume 1, Introduction, Part A.*

Chemical Class: Biological Products

Function: Skin-Conditioning Agent - Mis-cellaneous

Trade Names:
Noni Juice (Ennagram)
Tahitian Noni Juice (Morinda, Inc)

MORINDA CITRIFOLIA FRUIT POWDER

Definition: Morinda Citrifolia Fruit Powder is the dried powder obtained from the fruit of *Morinda citrifolia*. See "*Regulatory and Ingredient Use Information," regarding the labeling names for botanical ingredients in Volume 1, Introduction, Part A.*

Chemical Class: Biological Products

Function: Not Reported

MORINDA CITRIFOLIA LEAF EXTRACT

CAS No. 84929-68-0
EINECS No. 284-551-4

Definition: Morinda Citrifolia Leaf Extract is an extract of the leaves of *Morinda citrifolia*. See "*Regulatory and Ingredient Use Information," regarding the labeling names for botanical ingredients in Volume 1, Introduction, Part A.*

Chemical Class: Biological Products

Function: Skin-Conditioning Agent - Mis-cellaneous

Technical/Other Name:
Extract of Morinda Citrifolia Leaf

Trade Name:
Citrifoline (CEP (Solabia))

MORINDA CITRIFOLIA SEED OIL

Definition: Morinda Citrifolia Seed Oil is the oil expressed from the seed of *Morinda citrifolia*. See "*Regulatory and Ingredient Use Information," regarding the labeling names for botanical ingredients in Volume 1, Introduction, Part A.*

Chemical Class: Fats and Oils

Function: Skin-Conditioning Agent - Emollient

Technical/Other Name:
Oil, Morinda Citrifolia Seed

Trade Name:
Tahitian Noni Oil (Morinda, Inc)

MORINGA OIL/HYDROGENATED MORINGA OIL ESTERS

Definition: Moringa Oil/Hydrogenated Moringa Oil Esters is the product obtained by the interesterification of Moringa Pterygosperma Seed Oil (q.v.) and hydrogenated moringa oil.

Chemical Class: Glyceryl Esters and Derivatives

Function: Skin-Conditioning Agent - Emollient

Trade Name:
Moringa Esters (Floratech)

MORINGA PTERYGOSPERMA SEED EXTRACT

Definition: Moringa Pterygosperma Seed Extract is an extract of the seeds of *Moringa pterygosperma*. See "*Regulatory and Ingredient Use Information," regarding the labeling names for botanical ingredients in Volume 1, Introduction, Part A.*

Chemical Class: Biological Products

Function: Skin-Conditioning Agent - Mis-cellaneous

Technical/Other Name:
Extract of Moringa Pterygosperma Seed

Trade Name Mixtures:
Puricare LS (Laboratoires Serobiologiques)
Purisoft (Laboratoires Serobiologiques)

MORINGA PTERYGOSPERMA SEED OIL

Definition: Moringa Pterygosperma Seed Oil is the oil obtained from the seeds of the tropical tree *Moringa pterygosperma*. See "*Regulatory and Ingredient Use Information," regarding the labeling names for botanical ingredients in Volume 1, Introduction, Part A.*

Chemical Class: Fats and Oils

Function: Skin-Conditioning Agent - Occlusive

Technical/Other Names:
Ben Oil
Moringa Oil
Moringa Oleafera Oil
Oil of Moringa

Trade Names:
AEC Moringa Oil (A & E Connock)
Floralipids Moringa (International Flora)
Floralipids Moringa Oil (Floratech)
Lipofructyl MO (Laboratoires Sero-biologiques)
Moringa Oil (Greentech)
Moringa Oil (Sederma)
Moringa Oleifera Huile raffinee (Sedaherb)

MOROCCAN LAVA CLAY

Definition: Moroccan Lava Clay is a type of smectite clay mined in Morocco.

Chemical Class: Inorganics

Functions: Abrasive; Absorbent; Bulking Agent; Emulsion Stabilizer; Opacifying Agent; Viscosity Increasing Agent - Aqueous

Trade Names:
D- Extra Ghassoul (Ghassoul Japan)
Ghassoul (Alban Muller)
Ghassoul (Alva)
Ghassoul " Saponiferous Clay" (Societe du Ghassoul)

MORPHOLINE

CAS No. 110-91-8
EINECS No. 203-815-1

JPN Translation:
モルホリン

Empirical Formula:
C_4H_9NO

Definition: Morpholine is the heterocyclic organic compound that conforms to the formula:

Information Sources: 21CFR172.235, 21CFR173.310, 21CFR175.105, 21CFR176.210, 21CFR178.3300, CIR: [I] JACT-8(4)1989, CTFA S, EEC(II-344), JCIC, JCLS, MI-13(6303), TSCA

Chemical Classes: Amines; Heterocyclic Compounds

Function: pH Adjuster

Technical/Other Names:
Diethylene Oximide
Diethylenimide Oxide
1-Oxa-4-Azacyclohexane
Tetrahydro-1,4-Oxazine

MORPHOLINE OLEATE

CAS No.	EINECS No.
1095-66-5	214-139-1

Empirical Formula:
$C_4H_9NO \cdot C_{18}H_{34}O_2$

Definition: Morpholine Oleate is the soap formed from oleic acid and morpholine that conforms generally to the formula:

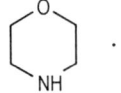 · $CH_3(CH_2)_7CH = CH(CH_2)_7COOH$

Information Source: EEC(II-344)

Chemical Classes: Heterocyclic Compounds; Soaps

Function: Surfactant - Cleansing Agent

Technical/Other Names:
Morpholine 9-Octadecenoate
Morpholinium Oleate
9-Octadecenoic Acid, Compd. with Morpholine (1:1)
Oleic Acid, Morpholine Salt

MORPHOLINE STEARATE

CAS No.	EINECS No.
22707-25-1	245-164-6

Empirical Formula:
$C_{18}H_{36}O_2 \cdot C_4H_9NO$

Definition: Morpholine Stearate is the soap formed from oleic acid and morpholine that conforms generally to the formula:

 · $CH_3(CH_2)_{16}COOH$

Information Sources: 21CFR176.210, 21CFR178.3300, EEC(II-344), TSCA

Chemical Classes: Heterocyclic Compounds; Soaps

Function: Surfactant - Cleansing Agent

Technical/Other Names:
Morpholine, Octadecanoate
Octadecanoic Acid, Compd. with Morpholine (1:1)
Stearic Acid Morpholine Salt

MORTIERELLA OIL

JPN Translation:
モルティエレラ油

Definition: Mortierella Oil is an oil obtained from culturing *Mortierella isabellina. See "Regulatory and Ingredient Use Information," regarding use of EU Trivial names in Volume 1, Introduction, Part A. See "Regulatory and Ingredient Use Information," regarding the labeling names for botanical ingredients in Volume 1, Introduction, Part A.*

Information Sources: JCIC, JCLS

Chemical Class: Fats and Oils

Function: Skin-Conditioning Agent - Occlusive

Reported Product Categories: Bath Capsules; Moisturizing Preparations

Technical/Other Names:
Mortierela Isabellina Oil
Oils, Mortierella

Trade Name:
Bio-EPO (Shiseido Company)

MORUS ALBA BARK EXTRACT

CAS No.	EINECS No.
94167-05-2	303-403-2

Definition: Morus Alba Bark Extract is an extract of the bark of *Morus alba. See "Regulatory and Ingredient Use Information," regarding the labeling names for botanical ingredients in Volume 1, Introduction, Part A.*

Chemical Class: Biological Products

Function: Not Reported

Technical/Other Name:
Extract of Morus Alba Bark

Trade Name Mixtures:
Actiphyte Mulberry Bark (Active Organics)
Actiphyte Mulberry Bark BG50P (Active Organics)
Actiphyte Mulberry Bark GL (Active Organics)
Actiphyte Mulberry Bark Lipo S (Active Organics)
Clerilys (Greentech)

MORUS ALBA LEAF EXTRACT

CAS No.	EINECS No.
94167-05-2	303-403-2

JPN Translation:
クワ葉エキス

Definition: Morus Alba Leaf Extract is an extract of the dried leaves of the white mulberry, *Morus alba. See "Regulatory and Ingredient Use Information," regarding the labeling names for botanical ingredients in Volume 1, Introduction, Part A.*

Chemical Class: Biological Products

Function: Not Reported

Reported Product Categories: Bath Capsules; Face and Neck Preparations (Excluding Shaving Preparations)

Technical/Other Names:
Extract of Morus Alba Leaves
Extract of Mulberry Leaves
Mulberry Extract
Mulberry Leaf Extract
Mulberry (Morus Alba) Extract

Trade Name Mixtures:
Actiphyte Mulberry Leaf (Active Organics)
Actiphyte Mulberry Leaf BG50P (Active Organics)
Actiphyte Mulberry Leaf GL (Active Organics)
Campo Tohasaku (Campo)
Morus Alba Leaf Extract ies (IES LABO)
Mulberry BG Concentrate (CEP (Solabia))
Mulberry Concentrate (CEP (Solabia))
Turn White Complex (Ennagram)

MORUS ALBA ROOT EXTRACT

CAS No.	EINECS No.
94167-05-2	303-403-2

JPN Translation:
マグワ根皮エキス

Definition: Morus Alba Root Extract is an extract of the roots of the white mulberry, *Morus alba. See "Regulatory and Ingredient Use Information," regarding the labeling names for botanical ingredients in Volume 1, Introduction, Part A.*

Information Source: JSQI

Chemical Class: Biological Products

Function: Skin-Conditioning Agent - Miscellaneous

Technical/Other Names:
Extract of Morus Alba Roots
Extract of Mulberry Roots
Mulberry (Morus Alba) Root Extract
Mulberry Root Extract
Souhakuhi Ekisu (JPN)

Trade Names:
AEC White Mulberry Extract Powder (A & E Connock)
Mulberry Extract Powder (Maruzen Pharmaceuticals Co., Ltd.)
Mulberry Extract Powder-J (Maruzen Pharmaceuticals Co., Ltd.)
Mulberry Extract SP (Maruzen Pharmaceuticals Co., Ltd.)
Souhakuhi Ex. Powder K (Ichimaru Pharcos)

Trade Name Mixtures:
Black Mulberry Root HS (Alban Muller)
Morus Root Extract BG-100 (Tri-K)
Mulberry Extract BG100 (Maruzen Pharmaceuticals Co., Ltd.)
Mulberry Extract BG-J (Maruzen Pharmaceuticals Co., Ltd.)
Mulberry Extract-J (Maruzen Pharmaceuticals Co., Ltd.)
Mulberry Extract-JO (Maruzen Pharmaceuticals Co., Ltd.)
Mulberry Extract LA-J (Maruzen Pharmaceuticals Co., Ltd.)
Mulberry Extract Powder-S (Maruzen Pharmaceuticals Co., Ltd.)
Mulberry Extract SQ (Maruzen Pharmaceuticals Co., Ltd.)
Mulberry Root Extract (Bioland)
Pharcolex MSTC (Ichimaru Pharcos)
Prodhy Extract Murier Blanc (Prod'Hyg)
Souhakuhi Liquid (Ichimaru Pharcos)
Souhakuhi Liquid (BG) (Ichimaru Pharcos)
Souhakuhi Liquid E (Ichimaru Pharcos)
VER Mulberry (Technoble)
White Mulberry HS (Alban Muller)

MORUS BOMBYCIS LEAF EXTRACT

Definition: Morus Bombycis Leaf Extract is an extract of the dried leaves of the mulberry, *Morus bombycis. See "Regulatory and Ingredient Use Information," regarding the labeling names for botanical ingredients in Volume 1, Introduction, Part A.*

Chemical Class: Biological Products

Function: Not Reported

Reported Product Categories: Bath Capsules; Face and Neck Preparations (Excluding Shaving Preparations)

Technical/Other Names:
Extract of Morus Bombycis Leaves
Extract of Mulberry Leaves
Mulberry Extract
Mulberry Leaf Extract
Mulberry (Morus Bombycis) Extract

MORUS BOMBYCIS ROOT EXTRACT

JPN Translation:
ヤマグワ根皮エキス

Definition: Morus Bombycis Root Extract is an extract of the roots of the mulberry, *Morus bombycis. See "Regulatory and Ingredient Use Information," regarding the labeling names for botanical ingredients in Volume 1, Introduction, Part A.*

Information Source: JSQI

Chemical Class: Biological Products

Function: Not Reported

Reported Product Categories: Bath Capsules; Face and Neck Preparations (Excluding Shaving Preparations)

Technical/Other Names:
Extract of Morus Bombycis Roots
Extract of Mulberry Roots
Mulberry (Morus Bombycis) Root Extract
Mulberry Root Extract

Trade Name:
Mulberry root powder (Bioland)

Trade Name Mixtures:
Biowhite (Coletica SA)
Clarisome (Coletica SA)
Mulberry Extract (Maruzen Pharmaceuticals Co., Ltd.)
Mulberry Extract BG (Maruzen Pharmaceuticals Co., Ltd.)
Mulberry Extract LA (Maruzen Pharmaceuticals Co., Ltd.)
Mulberry Root Extract (Bioland)
Phytoclar (Coletica SA)
Phytoclar II (Coletica SA)
Ultrawhite (Coletica SA)

MORUS NIGRA FRUIT EXTRACT

Definition: Morus Nigra Fruit Extract is an extract of the fruit of the mulberry, *Morus nigra. See "Regulatory and Ingredient Use Information," regarding the labeling names for botanical ingredients in Volume 1, Introduction, Part A.*

Chemical Class: Biological Products

Function: Skin-Conditioning Agent - Miscellaneous

Technical/Other Names:
Extract of Morus Nigra
Morus Nigra Extract
Mulberry Fruit Extract
Mulberry (Morus Nigra) Fruit Extract

MORUS NIGRA LEAF EXTRACT

CAS No.	EINECS No.
90064-11-2	290-072-1

Definition: Morus Nigra Leaf Extract is an extract of the dried leaves of the black mulberry, *Morus nigra. See "Regulatory and Ingredient Use Information," regarding the labeling names for botanical ingredients in Volume 1, Introduction, Part A.*

Chemical Class: Biological Products

Function: Skin-Conditioning Agent - Miscellaneous

Reported Product Categories: Bath Capsules; Face and Neck Preparations (Excluding Shaving Preparations)

Technical/Other Names:
Extract of Morus Nigra Leaves
Extract of Mulberry Leaves
Mulberry Extract
Mulberry Leaf Extract
Mulberry (Morus Nigra) Extract

Trade Name Mixtures:
Black Mulberry Leaf HS (Alban Muller)
Extrait de Murier (Silab)
Herbasol Extract Mulberry (Leaf) (Cosmetochem) (Cosmetochem International Ltd.)

MORUS NIGRA ROOT EXTRACT

JPN Translation:
クロミグワ根皮エキス

Definition: Morus Nigra Root Extract is an extract of the roots of the mulberry, *Morus nigra. See "Regulatory and Ingredient Use Information," regarding the labeling names for botanical ingredients in Volume 1, Introduction, Part A.*

Information Source: JSQI

Chemical Class: Biological Products

Function: Not Reported

Technical/Other Names:
Extract of Morus Nigra Roots
Extract of Mulberry Roots
Mulberry (Morus Nigra) Root Extract
Mulberry Root Extract

Trade Name Mixture:
Morus Nigra Root Extract ies (IES LABO)

MOTHER OF PEARL

Definition: Mother of Pearl is a mineral consisting chiefly of calcium carbonate.

Chemical Class: Inorganics

Function: Abrasive

Trade Name:
Extrait de Nacre (Greentech)

MOTHER OF PEARL EXTRACT

Definition: Mother of Pearl Extract is an extract of Mother of Pearl (q.v.).

The inclusion of any compound in the *Dictionary and Handbook* does not indicate that use of that substance as a cosmetic ingredient complies with the laws and regulations governing such use in the United States or any other country.

Chemical Class: Biological Products

Function: Skin-Conditioning Agent - Emollient

Technical/Other Name:
Extract of Mother of Pearl

Trade Name:
Perlarine (Greentech)

MOURERA FLUVIATILIS EXTRACT

Definition: Mourera Fluviatilis Extract is an extract of the whole plant, *Mourera fluviatilis*. See "Regulatory and Ingredient Use Information," regarding the labeling names for botanical ingredients in Volume 1, Introduction, Part A.

Chemical Class: Biological Products

Function: Skin-Conditioning Agent - Miscellaneous

Technical/Other Name:
Extract of Mourera Fluviatilis

Trade Name Mixture:
Fluxhydran (Laboratoires Serobiologiques)

MUCOR CIRCINELLOIDES OIL

Definition: Mucor Circinelloides Oil is an oil produced by the fungus, *Mucor circinelloides*. See "Regulatory and Ingredient Use Information," regarding the labeling names for botanical ingredients in Volume 1, Introduction, Part A.

Chemical Class: Fats and Oils

Function: Skin-Conditioning Agent - Emollient

Technical/Other Name:
Extract of Mucor Circinelloides

Trade Name:
Glanoil HGC (Idemitsu Technofine Co., Ltd)

MUCOR MEIHEI EXTRACT

Definition: Mucor Meihei Extract is an extract of the protease enzyme, *Mucor meihei*.

Chemical Class: Biological Products

Functions: Cosmetic Astringent; Skin-Conditioning Agent - Miscellaneous

Technical/Other Name:
Extract of Mucor Meihei

Trade Name Mixture:
Actizyme E3M-M (Active Organics)

MUCOR MIEHEI EXTRACT

Definition: Mucor Miehei Extract is an extract of the mycelium of *Mucor miehei*.

Chemical Class: Biological Products

Function: Skin-Conditioning Agent - Miscellaneous

Technical/Other Name:
Extract of Mucor Miehei

Trade Name Mixture:
Actizyme E3M-M (Active Organics)

MUCUNA BIRDWOODIANA STEM EXTRACT

JPN Translation:
ケイケツトウエキス

Definition: Mucuna Birdwoodiana Stem Extract is an extract of the stems of *Mucuna birdwoodiana*. See "Regulatory and Ingredient Use Information," regarding the labeling names for botanical ingredients in Volume 1, Introduction, Part A.

Chemical Class: Biological Products

Function: Skin-Conditioning Agent - Miscellaneous

Technical/Other Name:
Extract of Mucuna Birdwoodiana Stem

Trade Name Mixture:
Birdwoodiana Stalk Extract (Maruzen Pharmaceuticals Co., Ltd.)

MURASAKI201

CAS No. 81-48-1

EINECS No. 201-353-5

JPN Translation:
紫 201

Empirical Formula:
$C_{21}H_{15}NO_3$

Definition: Murasaki201 is classed chemically as an anthraquinone color. It conforms to the formula:

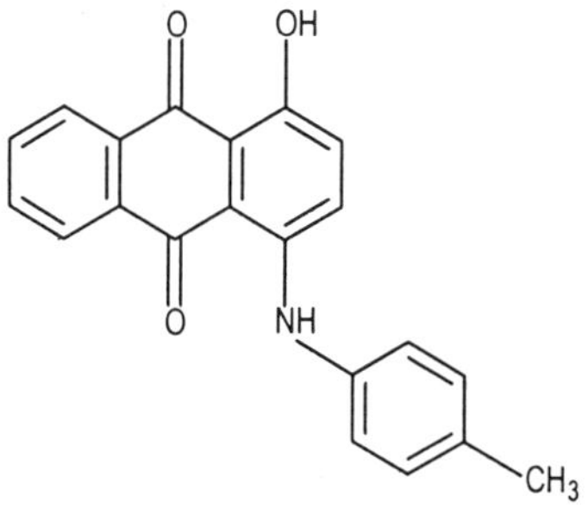

See "Regulatory and Ingredient Use Information," for Colorants in Volume 1, Intro-

duction, Part A. To identify the colorant allowed for use in Japan, the INCI name Murasaki201 must be used. This INCI name may not be used for ingredient labeling in the US or the EU. To identify the colorant allowed for use in the European Union (EU), the INCI Name CI 60725 must be used except for hair dye products. To identify the certified colorant for labeling purposes in the US, the INCI name Violet 2 must be used. The INCI Name for batches of this colorant that have not been certified is Solvent Violet 13.

Information Sources: CI 60725, M3, MHLW Ord. No. 30, TSCA

Chemical Class: Color Additives - Approved in Japan

Function: Colorant

Technical/Other Names:
9,10-Anthracenedione, 1-Hydroxy-4-[(4-Methylphenyl)Amino]-
Anthraquinone, 1-Hydroxy-4-ψ-Toluidino
CI 60725
Disperse Blue 72
1-Hydroxy-4-(4-Methylanilino)Anthraquinone
1-Hydroxy-4-[(4-Methylphenyl)Amino]-9,10-Anthracenedione
Japan Violet 201
Solvent Violet 13

MURASAKI401

CAS No. 4430-18-6

EINECS No. 224-618-7

JPN Translation:
紫 401

Empirical Formula:
$C_{21}H_{15}NO_6S$ • Na

Definition: Murasaki401 is classed chemically as an anthraquinone color. It conforms to the formula:

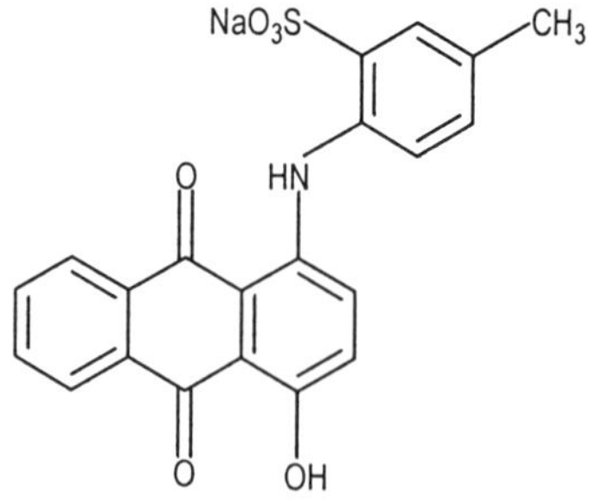

See "Regulatory and Ingredient Use Information," for Colorants in Volume 1, Introduction, Part A. To identify the colorant allowed for use in Japan, the INCI name Murasaki401 must be used. To identify the colorant (aluminum lake) allowed for use in

Japan, the INCI name Murasaki401 must be used. This INCI name may not be used for ingredient labeling in the US or the EU. To identify the colorant allowed for use in the European Union (EU), the INCI Name CI 60730 must be used except for hair dye products. To identify the certified colorant for labeling purposes in the US, the INCI name Ext. Violet 2 must be used. The INCI Name for batches of this colorant that have not been certified is Acid Violet 43.

Information Sources: CI 60730, M3, MHLW Ord. No. 30, TSCA

Chemical Class: Color Additives - Approved in Japan

Function: Colorant

Technical/Other Names:
Acid Violet 43
Benzenesulfonic Acid, 2-[(9,10-Dihydro-4-Hydroxy-9,10-Dioxo-1-Anthracenyl)-Amino]-5-Methyl-, Monosodium Salt
CI 60730
2-[(9,10-Dihydro-4-Hydroxy-9,10-Dioxo-1-Anthracenyl)Amino]-5-Methylbenzene-sulfonic Acid, Monosodium Salt
Japan Violet 401
μ-Toluenesulfonic Acid, 6-((4-Hydroxy-1-Anthraquinonyl)Amino)-, Monosodium Salt

MURRAYA EXOTICA LEAF EXTRACT

Definition: Murraya Exotica Leaf Extract is an extract of the leaves of *Murraya exotica*. See "Regulatory and Ingredient Use Information," regarding the labeling names for botanical ingredients in Volume 1, Introduction, Part A.

Chemical Class: Biological Products

Functions: Antioxidant; Oral Care Agent; Skin-Conditioning Agent - Emollient

Technical/Other Name:
Extract of Murraya Exotica Leaf

Trade Name:
Chinese Myrtle (Haldin Pacific)

MURRAYA KOENIGII EXTRACT

JPN Translation:
ムラヤコエンジーエキス

Definition: Murraya Koenigii Extract is an extract of the stems and twigs of *Murraya koenigii*. See "Regulatory and Ingredient Use Information," regarding the labeling names for botanical ingredients in Volume 1, Introduction, Part A.

Chemical Class: Biological Products

Function: Skin-Conditioning Agent - Miscellaneous

Technical/Other Name:
Extract of Murraya Koenigii

Trade Name Mixture:
Murraya Koenigii Extract (Yamakawa)

MUSA NANA FRUIT EXTRACT

Definition: Musa Nana Fruit Extract is an extract of the fruit of *Musa nana*. See "Regulatory and Ingredient Use Information," regarding the labeling names for botanical ingredients in Volume 1, Introduction, Part A.

Chemical Class: Biological Products

Functions: Hair Conditioning Agent; Skin-Conditioning Agent - Miscellaneous

Technical/Other Names:
Extract of Musa Nana Fruit
Musa Cavendishii Fruit Extract

Trade Name:
Frulix TF Banana (Assessa-Industria)

MUSA PARADISICA (BANANA) FRUIT

Definition: Musa Paradisica (Banana) Fruit is the fruit of the banana, *Musa paradisica*. See "Regulatory and Ingredient Use Information," regarding the labeling names for botanical ingredients in Volume 1, Introduction, Part A.

Chemical Class: Biological Products

Function: Cosmetic Astringent

Technical/Other Names:
Banana
Banana Fruit

Trade Name:
AEC Banana Puree (A & E Connock)

MUSA PARADISICA (BANANA) FRUIT JUICE

Definition: Musa Paradisica (Banana) Fruit Juice is the liquid expressed from the fruit of the banana, *Musa paradisica*. See "Regulatory and Ingredient Use Information," regarding the labeling names for botanical ingredients in Volume 1, Introduction, Part A.

Chemical Class: Biological Products

Function: Skin-Conditioning Agent - Miscellaneous

Technical/Other Names:
Banana Fruit Juice
Banana Juice
Juice, Banana
Juice, Musa Paradisica
Musa Paradisica Juice

Trade Name:
AEC Banana Conc. (A & E Connock)

MUSA SAPIENTUM (BANANA) FLOWER EXTRACT

Definition: Musa Sapientum (Banana) Flower Extract is an extract of the flower of *Musa sapientum*. See "Regulatory and Ingredient Use Information," regarding the labeling names for botanical ingredients in Volume 1, Introduction, Part A.

Chemical Class: Biological Products

Function: Not Reported

Technical/Other Name:
Banana Flower Extract

Trade Name Mixtures:
Banana (Tree) Flowers Milk (CEP (Solabia))
Glycolysat BG of Banana Flowers (CEP (Solabia))
Maturine (CEP (Solabia))

MUSA SAPIENTUM (BANANA) FLOWER WATER

Definition: Musa Sapientum (Banana) Flower Water is an aqueous solution of the steam distillate obtained from the flowers of *Musa sapientum*. See "Regulatory and Ingredient Use Information," regarding the labeling names for botanical ingredients in Volume 1, Introduction, Part A.

Chemical Class: Biological Products

Function: Fragrance Ingredient

Technical/Other Names:
Banana Flower Water
Banana Water
Musa Sapientum Flower Water
Water, Banana

Trade Names:
Essential banana Nectar (Libiol)
Vegebios Of Banana Flowers (CEP (Solabia))

MUSA SAPIENTUM (BANANA) FRUIT EXTRACT

CAS No.	EINECS No.
89957-82-4	289-602-4

Definition: Musa Sapientum (Banana) Fruit Extract is an extract of the fruit of the banana, *Musa sapientum. See "Regulatory and Ingredient Use Information," regarding the labeling names for botanical ingredients in Volume 1, Introduction, Part A.*

Chemical Class: Biological Products

Function: Skin-Conditioning Agent - Miscellaneous

Technical/Other Names:
Banana Extract
Banana Fruit Extract
Extract of Banana
Extract of Musa Sapientum
Musa Paradisiaca Extract
Musa Sapientum Extract

Trade Name Mixtures:
Actiphyte of Banana BG50 (Active Organics)
Actiphyte of Banana GL50 (Active Organics)
Actiphyte of Banana Lipo S (Active Organics)
Actiphyte of Banana PG50 (Active Organics)
Banana Extract HG (Provital/Centerchem)
Banana Extract HS 3381 G (Grau)
Banana HS (Alban Muller)
Herbasol-Extract Banana (Cosmetochem)
Musa Sapientum (Banana) Fruit Extract ies (IES LABO)
VT-202 Extract of Banana Fruit (Vege-Tech)

MUSA SAPIENTUM (BANANA) LEAF EXTRACT

CAS No.	EINECS No.
89957-82-4	289-602-4

Definition: Musa Sapientum (Banana) Leaf Extract is an extract of the leaves of the banana, *Musa sapientum. See "Regulatory and Ingredient Use Information," regarding the labeling names for botanical ingredients in Volume 1, Introduction, Part A.*

Chemical Class: Biological Products

Function: Not Reported

Technical/Other Names:
Banana Leaf Extract
Extract of Banana Leaves
Extract of Musa Sapientum Leaf
Musa Paradisiaca Leaf Extract

Trade Name Mixtures:
Actiphyte of Banana Leaves BG50 (Active Organics)
Actiphyte of Banana Leaves GL50 (Active Organics)
Actiphyte of Banana Leaves Lipo S (Active Organics)
Actiphyte of Banana Leaves PG50 (Active Organics)
VT-1161 Extract of Banana Leaf (Vege-Tech)

MUSA SAPIENTUM (BANANA) WATER

Definition: Musa Sapientum (Banana) Water is an aqueous solution of the steam distillate obtained from *Musa sapientum. See "Regulatory and Ingredient Use Information," regarding the labeling names for botanical ingredients in Volume 1, Introduction, Part A.*

Chemical Class: Biological Products

Function: Skin-Conditioning Agent - Miscellaneous

MUSCLE EXTRACT

Definition: Muscle Extract is an extract of bovine muscle.

Chemical Class: Biological Products

Function: Not Reported

Technical/Other Name:
Extract of Muscles

MUSK KETONE

CAS No.	EINECS No.
81-14-1	201-328-9

Empirical Formula:
$C_{14}H_{18}N_2O_5$

Definition: Musk Ketone is the organic compound that conforms to the formula:

Information Sources: EEC(III/2-61), RIFM, TSCA

Chemical Class: Ketones

Function: Fragrance Ingredient

Technical/Other Names:
Acetonphenone, 4'-tert-butyl-2',6'-dimethyl-3',5'-dinitro-
1-[4-(1,1-Dimethylethyl)-2,6-Dimethyl-3,5-Dinitrophenyl]Ethanone
Ethanone, 1-[4-(1,1-Dimethylethyl)-2,6-Dimethyl-3,5-Dinitrophenyl]-
Musk ketone (RIFM)

MUSSEL EXTRACT

CAS No.	EINECS No.
94465-78-8	305-363-1

Definition: Mussel Extract is an extract of sea mussels.

Chemical Class: Biological Products

Function: Not Reported

Technical/Other Name:
Extract of Mussel

MUSTELIC/PALMITIC TRIGLYCERIDE

JPN Translation:
トリ（ミンク脂肪酸／パルミチン酸）グリセリル

Definition: Mustelic/Palmitic Triglyceride is the product obtained by the reaction of glycerin with mustelic and palmitic acids.

Information Source: JCIC

Chemical Class: Fats and Oils

Functions: Skin-Conditioning Agent - Emollient; Solvent

MYOSOTIS SYLVATICA EXTRACT

Definition: Myosotis Sylvatica Extract is an extract of the aerial parts of *Myosotis sylvatica. See "Regulatory and Ingredient Use Information," regarding the labeling names for botanical ingredients in Volume 1, Introduction, Part A.*

Chemical Class: Biological Products

Function: Not Reported

Technical/Other Name:
Extract of Myosotis Sylvatica

Trade Name Mixtures:
Cosflor Forget-Me-Not HGS (A & E Connock)
Forget-Me-Not Extract (Cosmetic Developments)

MYRCENOL

CAS No.	EINECS No.
543-39-5	208-843-8

Empirical Formula:
$C_{10}H_{18}O$

Definition: Myrcenol is the organic compound that conforms to the formula:

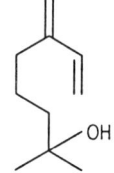

Information Sources: RIFM, TSCA

Chemical Class: Alcohols

Function: Fragrance Ingredient

Technical/Other Names:
7-Hydroxy-7-Methyl-3-Methylene-1-Octene
2-Methyl-6-Methyleneoct-7-en-2-ol
Myrcenol (RIFM)

Trade Name:
Myrcenol 50 (International Flavors)

MYRCIARIA DUBIA FRUIT EXTRACT

Definition: Myrciaria Dubia Fruit Extract is an extract of the fruit of *Myrciaria dubia. See "Regulatory and Ingredient Use Information," regarding the labeling names for botanical ingredients in Volume 1, Introduction, Part A.*

Chemical Class: Biological Products

Function: Skin-Conditioning Agent - Miscellaneous

Technical/Other Names:
Camu Camu Extract
Extract of Myrciaria Dubia Fruit

Trade Name Mixtures:
Camu-Camu Extract B30 (Nichirei) (Nichirei)
Camu-Camu Extract W (Nichirei) (Nichirei)

MYRCIARIA DUBIA SEED EXTRACT

Definition: Myrciaria Dubia Seed Extract is an extract of the seeds of Myrciaria dubia. See "Regulatory and Ingredient Use Information," regarding the labeling names for botanical ingredients in Volume 1, Introduction, Part A.

Chemical Class: Biological Products

Functions: Hair Conditioning Agent; Humectant; Skin-Conditioning Agent - Emollient

Technical/Other Name:
Extract of Myrciaria Dubia Seed

Trade Name Mixture:
Camu-Camu Seed Extract B30 (Nichirei) (Nichirei)

MYRETH-2

CAS No.: 27306-79-2

JPN Translation:
ミレス - 2

Empirical Formula:
$C_{18}H_{38}O_3$

Definition: Myreth-2 is the polyethylene glycol ether of Myristyl Alcohol (q.v.) that conforms to the formula:

$$CH_3(CH_2)_{13}(OCH_2CH_2)_nOH$$

where n has an average value of 2.

Information Sources: JCLS, TSCA

Chemical Class: Alkoxylated Alcohols

Function: Surfactant - Emulsifying Agent

MYRETH-3

CAS Nos.
26826-30-2
27306-79-2 (Generic)

EINECS No.
248-016-9

JPN Translation:
ミレス - 3

Empirical Formula:
$C_{20}H_{42}O_4$

Definition: Myreth-3 is the polyethylene glycol ether of Myristyl Alcohol (q.v.) that conforms generally to the formula:

$$CH_3(CH_2)_{13}(OCH_2CH_2)_nOH$$

where n has an average value of 3.

Information Sources: JCLS, MI-13(7659), TSCA

Chemical Class: Alkoxylated Alcohols

Function: Surfactant - Emulsifying Agent

Technical/Other Names:
Ethanol, 2-[2-[2-(Tetradecyloxy)Ethoxy]-Ethoxy]-
Myristyl Triethoxylate
PEG-3 Myristyl Ether
Polyethylene Glycol (3) Myristyl Ether
Polyoxyethylene (3) Myristyl Ether
2-[2-[2-(Tetradecyloxy)Ethoxy]Ethoxy]-Ethanol
2-[2-[2-(Tetradecyloxyyy)Ethoxy]Ethoxy]-Ethanol
Triethylene Glycoll Tetradecyl Ether

Trade Name:
Hetoxol M-3 (Heterene)

Trade Name Mixture:
Isoxal 5 (Vevy)

MYRETH-4

CAS Nos.: 27306-79-2 (Generic); 39034-24-7

JPN Translation:
ミレス - 4

Empirical Formula:
$C_{22}H_{46}O_5$

Definition: Myreth-4 is the polyethylene glycol ether of Myristyl Alcohol (q.v.) that conforms generally to the formula:

$$CH_3(CH_2)_{13}(OCH_2CH_2)_nOH$$

where n has an average value of 4.

Information Sources: CTFA D, JCLS, MI-13(7659), SNPF, TSCA

Chemical Class: Alkoxylated Alcohols

Function: Surfactant - Emulsifying Agent

Technical/Other Names:
PEG-4 Myristyl Ether
Polyethylene Glycol 200 Myristyl Ether
Polyoxyethylene (4) Myristyl Ether
Tetraethylene Glycol Myristyl Ether
Tetraethylene Glycol Tetradecyl Ether
3,6,9,12-Tetraoxahexacosan-1-ol

Trade Name Mixtures:
Homulgator 920 G (Grau)
Homulgator 1330 G (Grau)

MYRETH-5

CAS Nos.: 27306-79-2 (Generic); 92669-01-7

JPN Translation:
ミレス - 5

Empirical Formula:
$C_{24}H_{50}O_6$

Definition: Myreth-5 is the polyethylene glycol ether of Myristyl Alcohol (q.v.) that conforms generally to the formula:

$$CH_3(CH_2)_{13}(OCH_2CH_2)_nOH$$

where n has average value of 5.

Information Sources: JCLS, MI-13(7659)

Chemical Class: Alkoxylated Alcohols

Function: Surfactant - Emulsifying Agent

Technical/Other Names:
Peg-5 Myristyl Ether
Pentaethylene Glycol Tetradecyl Ether
3,6,9,12,15-Pentaoxanonacosan-1-ol
Pentaoxyethylene Monotetradecyl Ether
Polyethylene Glycol (5) Myristyl Ether
Polyoxyethylene (5) Myristyl Ether

Trade Name Mixture:
Isoxal 12 (Vevy)

MYRETH-10

CAS No.: 27306-79-2 (Generic)

JPN Translation:
ミレス - 10

Empirical Formula:
$C_{34}H_{70}O_{11}$

Definition: Myreth-10 is the polyethylene glycol ether of Myristyl Alcohol (q.v.) that conforms generally to the formula:

$$CH_3(CH_2)_{13}(OCH_2CH_2)_nOH$$

where n has an average value of 10.

Information Sources: 21CFR177.2800, JCLS, MI-13(7659)

Chemical Class: Alkoxylated Alcohols

Function: Surfactant - Emulsifying Agent

Technical/Other Names:
PEG-10 Myristyl Ether
Polyethylene Glycol 500 Myristyl Ether
Polyoxyethylene (10) Myristyl Ether

Trade Name Mixture:
Isoxal 11 (Vevy)

MYRETH-3 CAPRATE

CAS No.: 59599-56-3

Empirical Formula:
$C_{30}H_{60}O_5$

Definition: Myreth-3 Caprate is the ester of Myreth-3 (q.v.) and capric acid. It conforms generally to the formula:

$$CH_3(CH_2)_8\overset{\displaystyle O}{\overset{\displaystyle \|}{C}}\!\!-\!\!(OCH_2CH_2)_3O(CH_2)_{13}CH_3$$

Chemical Class: Esters

Function: Skin-Conditioning Agent - Emollient

Technical/Other Names:
Myristyl Ethoxy Caprate
PEG-3 Myristyl Ether Caprate
Polyethylene Glycol (3) Myristyl Ether Caprate
Polyoxyethylene (3) Myristyl Ether Caprate
Triethylene Glycol Decanoate Tetradecyl Ether

Trade Name:
Unimul-1410 (Universal Preserv-A-Chem)

MYRETH-3 CARBOXYLIC ACID

Empirical Formula:
$C_{20}H_{40}O_5$

Definition: Myreth-3 Carboxylic Acid is the organic acid that conforms generally to the formula:

$$CH_3(CH_2)_{13}(OCH_2CH_2)_nOCH_2COOH$$

where n has an average value of 2.

Chemical Class: Carboxylic Acids

Functions: Surfactant - Cleansing Agent; Surfactant - Emulsifying Agent

Technical/Other Names:
PEG-3 Myristyl Ether Carboxylic Acid
Polyethylene Glycol (3) Myristyl Ether Carboxylic Acid
Polyoxyethylene (3) Myristyl Ether Carboxylic Acid

MYRETH-5 CARBOXYLIC ACID

CAS Nos.: 38720-61-5; 120001-52-7

Empirical Formula:
$C_{24}H_{48}O_7$

Definition: Myreth-5 Carboxylic Acid is the organic acid that conforms generally to the formula:

$$CH_3(CH_2)_{13}(OCH_2CH_2)_nOCH_2COOH$$

where n has an average value of 4.

Chemical Class: Carboxylic Acids

Function: Surfactant - Cleansing Agent

Technical/Other Names:
PEG-5 Myristyl Ether Carboxylic Acid
3,6,9,12,15-Pentaoxanonacosanoic Acid
Polyethylene Glycol (5) Myristyl Ether Carboxylic Acid
Polyoxyethylene (5) Myristyl Ether Carboxylic Acid

Trade Name:
Akypostat MA 35 (Kao GmbH)

MYRETH-3 ETHYLHEXANOATE

Empirical Formula:
$C_{28}H_{56}O_5$

Definition: Myreth-3 Ethylhexanoate is the ester of Myreth-3 (q.v.) and 2-ethylhexanoic acid. It conforms generally to the formula:

$$CH_3(CH_2)_3\underset{\displaystyle CH_2CH_3}{\overset{\displaystyle O}{\overset{\displaystyle \|}{CHC}}}\!\!-\!\!(OCH_2CH_2)_3O(CH_2)_{13}CH_3$$

Chemical Class: Esters

Function: Skin-Conditioning Agent - Emollient

Reported Product Category: Indoor Tanning Preparations

Technical/Other Names:
Myreth-3 Octanoate
PEG-3 Myristyl Ether Octanoate
Polyethylene Glycol (3) Myristyl Ether Octanoate
Polyoxyethylene (3) Myristyl Ether Octanoate

Trade Name:
Trivent OC-143 (Trivent)

MYRETH-3 LAURATE

Empirical Formula:
$C_{32}H_{64}O_5$

Definition: Myreth-3 Laurate is the ester of Myreth-3 (q.v.) and lauric acid. It conforms generally to the formula:

$$CH_3(CH_2)_{10}\overset{\displaystyle O}{\overset{\displaystyle \|}{C}}\!\!-\!\!(OCH_2CH_2)_3O(CH_2)_{13}CH_3$$

Chemical Class: Esters

Function: Skin-Conditioning Agent - Emollient

Technical/Other Names:
PEG-3 Myristyl Ether Laurate
Polyethylene Glycol (3) Myristyl Ether Laurate
Polyoxyethylene (3) Myristyl Ether Laurate

Trade Name:
Schercemol MEL-3 (Scher)

MYRETH-2 MYRISTATE

Empirical Formula:
$C_{32}H_{64}O_4$

Definition: Myreth-2 Myristate is the ester of Myreth-2 (q.v.) and myristic acid. It conforms generally to the formula:

$$CH_3(CH_2)_{12}\overset{\displaystyle O}{\overset{\displaystyle \|}{C}}\!\!-\!\!(OCH_2CH_2)_2O(CH_2)_{13}CH_3$$

Chemical Class: Esters

Function: Skin-Conditioning Agent - Emollient

Technical/Other Names:
PEG-2 Myristyl Ether Myristate
Polyethylene Glycol 100 Myristyl Ether Myristate
Polyoxyethylene (2) Myristyl Ether Myristate

Trade Name:
Atlas G-4964 (Uniqema Americas)

MYRETH-3 MYRISTATE

CAS No.: 59686-68-9

JPN Translation:
ミリスチン酸ミレス - 3

Empirical Formula:
$C_{34}H_{68}O_5$

Definition: Myreth-3 Myristate is the ester of Myreth-3 (q.v.) and myristic acid. It conforms generally to the formula:

$$CH_3(CH_2)_{12}\overset{\overset{\displaystyle O}{\|}}{C}\text{—}(OCH_2CH_2)_3O(CH_2)_{13}CH_3$$

Information Sources: CTFA D, JCIC, JCLS, JSQI

Chemical Class: Esters

Function: Skin-Conditioning Agent - Emollient

Reported Product Categories: Bath Oils, Tablets, and Salts; Cleansing Products (Cold Creams, Cleansing Lotions, Liquids and Pads); Paste Masks (Mud Packs); Bath Capsules; Bath Preparations, Misc.; Body and Hand Preparations (Excluding Shaving Preparations); Face and Neck Preparations (Excluding Shaving Preparations); Makeup Bases

Technical/Other Names:
- PEG-3 Myristyl Ether Myristate
- Polyethylene Glycol (3) Myristyl Ether Myristate
- Polyoxyethylene (3) Mryistyl Ether Myristate
- Polyoxyethylene Myristyl Ether Myristate (3E.O.)

Trade Names:
- Cetiol 1414 E (Cognis Care Chemicals/NJ)
- Cetiol 1414 E (Cognis Care Chemicals/PA)
- Lanol 14 M (SEPPIC)
- Liponate 143-M (Lipo)
- Schercemol MEM-3 (Scher)
- Unimul-1414EW (Universal Preserv-A-Chem)

MYRETH-3 PALMITATE

Empirical Formula:
$C_{36}H_{72}O_5$

Definition: Myreth-3 Palmitate is the ester of Myreth-3 (q.v.) and palmitic acid. It conforms generally to the formula:

$$CH_3(CH_2)_{14}\overset{\overset{\displaystyle O}{\|}}{C}\text{—}(OCH_2CH_2)_3O(CH_2)_{13}CH_3$$

where n has an average value of 3.

Chemical Class: Esters

Function: Skin-Conditioning Agent - Emollient

Technical/Other Names:
- PEG-3 Myristyl Ether Palmitate
- Polyethylene Glycol (3) Myristyl Ether Palmitate
- Polyoxyethylene (3) Myristyl Ether Palmitate

Trade Name:
- Schercemol MEP-3 (Scher)

MYRICA CERIFERA BARK EXTRACT

Definition: Myrica Cerifera Bark Extract is an extract of the bark of *Myrica cerifera. See "Regulatory and Ingredient Use Information," regarding the labeling names for botanical ingredients in Volume 1, Introduction, Part A.*

Chemical Class: Biological Products

Function: Cosmetic Astringent

Trade Name Mixtures:
- Actiphyte Bayberry Bark (Active Organics)
- Actiphyte of Siberian Ginseng GL50 (Active Organics)
- Actiphyte of Siberian Ginseng Lipo S (Active Organics)
- Actiphyte of Siberian Ginseng PG50 (Active Organics)

MYRICA CERIFERA (BAYBERRY) FRUIT EXTRACT

CAS No.	EINECS No.
84929-34-0	284-518-4

Definition: Myrica Cerifera (Bayberry) Fruit Extract is an extract of the fruit of the bayberry, *Myrica cerifera. See "Regulatory and Ingredient Use Information," regarding the labeling names for botanical ingredients in Volume 1, Introduction, Part A.*

Chemical Class: Biological Products

Function: Not Reported

Technical/Other Names:
- Bayberry Extract
- Bayberry Fruit Extract
- Candleberry Extract
- Extract of Bayberry
- Extract of Candleberry
- Extract of Myrica Cerifera
- Extract of Wax Myrtle
- Myrica Cerifera Extract
- Wax Myrtle Extract

Trade Name Mixtures:
- Actiphyte of Bayberry BG50 (Active Organics)
- Actiphyte of Bayberry GL50 (Active Organics)
- Actiphyte of Bayberry Lipo S (Active Organics)
- Actiphyte of Bayberry PG50 (Active Organics)

MYRICA CERIFERA (BAYBERRY) FRUIT WAX

CAS No.: 8038-77-5

Definition: Myrica Cerifera (Bayberry) Fruit Wax is a wax obtained from the covering of the berries of the bayberry, *Myrica cerifera. See "Regulatory and Ingredient Use Information," regarding the labeling names for botanical ingredients in Volume 1, Introduction, Part A.*

Information Sources: CTFA D, HP, MI-13 (1015), TSCA

Chemical Class: Waxes

Functions: Emulsion Stabilizer; Film Former; Skin-Conditioning Agent - Emollient; Skin-Conditioning Agent - Miscellaneous; Skin-Conditioning Agent - Occlusive; Viscosity Increasing Agent - Nonaqueous

Technical/Other Names:
- Bayberry Fruit Wax
- Bayberry Wax
- Myrica Cerifera Wax
- Myrtle Wax
- Waxes, Bayberry
- Waxes, Myrica Cerifera
- Waxes, Myrtle

Trade Names:
- Bayberry Wax (Ross)
- Bayberry Wax - STRALPITZ (Strahl & Pitsch)
- Filtered Bayberry Wax (Alban Muller)
- Filtered Bayberry Wax (Ross)
- Hansonwax JH-5000 (Hansotech)
- Refined Bayberry Wax (Ross)
- Waxes, Bayberry (Ross)

Trade Name Mixtures:
- Enviro Pure 306 (React Inc)
- Enviro Pure 310 (React Inc)
- Enviro Pure 306 PG (React Inc)

MYRICA CERIFERA (BAYBERRY) LEAF EXTRACT

CAS No.	EINECS No.
84929-34-0	284-518-4

Definition: Myrica Cerifera (Bayberry) Leaf Extract is an extract of the leaves of the bayberry, *Myrica cerifera. See "Regulatory and Ingredient Use Information," regarding the labeling names for botanical ingredients in Volume 1, Introduction, Part A.*

Chemical Class: Biological Products

Function: Not Reported

Technical/Other Names:
- Bayberry Leaf Extract
- Candleberry Leaf Extract
- Extract of Bayberry Leaves
- Extract of Candleberry Leaves
- Extract of Myrica Cerifera Leaf

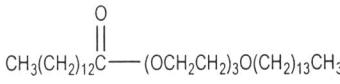

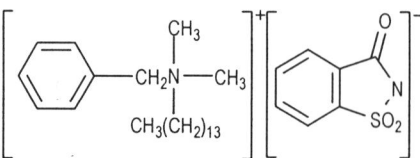

Extract of Wax Myrtle Leaves
Myrica Cerifera Leaf Extract
Wax Myrtle Leaf Extract

MYRICA GALE EXTRACT

Definition: Myrica Gale Extract is an extract of the aerial parts of *Myrica gale. See "Regulatory and Ingredient Use Information," regarding the labeling names for botanical ingredients in Volume 1, Introduction, Part A.*

Chemical Class: Biological Products

Function: Fragrance Ingredient

Technical/Other Names:
Extract of Myrica Gale
Extract of Sweet Gale
Sweet Gale Extract

Trade Name Mixtures:
Cosflor Sweet Gale HGS (A & E Connock)
Sweet Gale Extract (Cosmetic Developments)

MYRICYL ALCOHOL

CAS Nos.	EINECS Nos.
544-86-5	208-882-0
593-50-0	209-794-5

Empirical Formula:
$C_{30}H_{62}O$

Definition: Myricyl Alcohol is the alcohol that conforms to the formula:

$$CH_3(CH_2)_{29}OH$$

Information Source: MI-13(9665)

Chemical Class: Alcohols

Function: Viscosity Increasing Agent - Nonaqueous

Technical/Other Names:
1-Hydrocytriacontane
1-Triacontanol
Triacontyl Alcohol

Trade Name Mixture:
Ross Bayberry Wax Substitute 1641 (Ross)

MYRISTALKONIUM CHLORIDE

CAS No.	EINECS No.
139-08-2	205-352-0

JPN Translation:
ミリスタルコニウムクロリド

Empirical Formula:
$C_{23}H_{42}N \cdot Cl$

Definition: Myristalkonium Chloride is the quaternary ammonium salt that conforms generally to the formula:

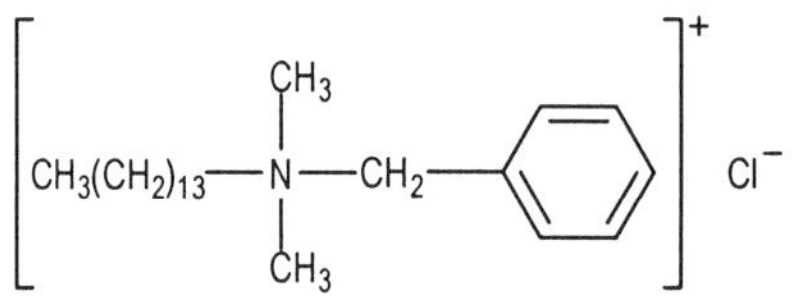

Information Sources: BAN, 21CFR172.165, 21CFR173.320, 21CFR175.105, 21CFR178.1010, CTFA D, INN, JCLS, JSCI, JSQI, TSCA

Chemical Class: Quaternary Ammonium Compounds

Functions: Antistatic Agent; Cosmetic Biocide; Surfactant - Cleansing Agent

Technical/Other Names:
Ammonium, Benzyldimethyltetradecyl-, Chloride
Benzenemethanaminium, N,N-Dimethyl-N-Tetradecyl-, Chloride
Benzyldimethyltetradecylammonium Chloride
N,N-Dimethyl-N-Tetradecylbenzene-methanaminium Chloride
Myristyl Dimethyl Benzyl Ammonium Chloride

Trade Names:
Barquat MX-50 (Lonza Inc./Lonza Ltd.)
Barquat MX-80 (Lonza Inc./Lonza Ltd.)
BTC 824 (Stepan)
FMB 65-15 Quat (Huntington)
FMB 65-28 Quat (Huntington)
FMB 451-5 Quat (Huntington)
FMB 451-8 Quat (Huntington)
FMB 4500-5 Quat (Huntington)
JAQ Powdered Quat (Huntington)
Maquat MC1416-50% (Mason)
Maquat MC1416-80% (Mason)

Trade Name Mixtures:
BTC 2125 (Stepan)
BTC 2125 M (Stepan)
FMB 504-5 Quat (Huntington)
FMB 504-8 Quat (Huntington)
FMB 1210-5 Quat (Huntington)
FMB 1210-8 Quat (Huntington)
FMB 3328-5 Quat (Huntington)
FMB 3328-8 Quat (Huntington)
LOWENOL T-1106-A (Lowenstein)

MYRISTALKONIUM SACCHARINATE

CAS Nos.	EINECS No.
68989-01-5	273-545-7
137951-75-8	

Empirical Formula:
$C_{23}H_{42}N \cdot C_7H_5NO_3S$

Definition: Myristalkonium Saccharinate is the quaternary ammonium salt that conforms generally to the formula:

Information Source: TSCA

Chemical Classes: Heterocyclic Compounds; Quaternary Ammonium Compounds

Function: Cosmetic Biocide

Technical/Other Names:
Ammonium, Benzldimethyltetradecyl-, salt with 1,2-benzisothiazolin-3-one 1,1-dioxide (1:1)
Benzenemethanaminium, N,N-Dimethyl-N-Tetradecyl-, Saccharinate
Benzenemethanaminium, N,N-dimethyl-N-tetradecyl-, Salt with 1,2-benzisothiazol-3 (2H)-one 1,1-dioxide (1:1)
N,N-Dimethyl-N-Tetradecylbenzene-methanaminium Saccharinate
Myristyl Dimethyl Benzyl Ammonium Saccharinate
Quaternium-3

Trade Name:
ONYXIDE 3300 (Stepan)

MYRISTAMIDE DEA

CAS No.	EINECS No.
7545-23-5	231-426-7

JPN Translation:
ミリスタミド DEA

Empirical Formula:
$C_{18}H_{37}NO_3$

Definition: Myristamide DEA is a mixture of ethanolamides of myristic acid. It conforms generally to the formula:

$$CH_3(CH_2)_{12}C(=O)-N(CH_2CH_2OH)_2$$

Information Sources: 21CFR175.105, 21CFR176.180, 21CFR176.210, 21CFR177.2260, 21CFR177.2800, CIR: [SQ], CTFA D, EEC(III/1-60), JCIC, JCLS, JSQI, TSCA

Chemical Class: Alkanolamides

Functions: Surfactant - Foam Booster; Viscosity Increasing Agent - Aqueous

Reported Product Categories: Shampoos (Non-coloring); Cleansing Products (Cold Creams, Cleansing Lotions, Liquids and Pads); Leg and Body Paints

Technical/Other Names:
N,N-Bis(2-Hydroxyethyl)Myristamide

N,N-Bis(2-Hydroxyethyl)Tetradecanamide
Diethanolamine Myristic Acid Condensate
Myristic Diethanolamide
Myristoyl Diethanolamide
Myristoyl Diethanolamine
Tetradecamide, N,N-Bis

Trade Names:
Hetamide M (Heterene)
Jeemide MRCA (Jeen)
Myristamine (Vevy)
Protamide MRCA (Protameen)

Trade Name Mixture:
Protamide LM-73-LD (Protameen)

MYRISTAMIDE MEA

CAS No. 142-58-5
EINECS No. 205-546-5

Empirical Formula:
$C_{16}H_{33}NO_2$

Definition: Myristamide MEA is a mixture of ethanolamides of myristic acid. It conforms generally to the formula:

$$CH_3(CH_2)_{12}C(O)-NHCH_2CH_2OH$$

Information Sources: CIR: [SQ], CTFA D, TSCA

Chemical Class: Alkanolamides

Functions: Surfactant - Foam Booster; Viscosity Increasing Agent - Aqueous

Technical/Other Names:
N-(2-Hydroxyethyl)myristamide
N-(2-Hydroxyethyl)Tetradecanamide
Monoethanolamine Myristic Acid
Condensate
Myristic Monoethanolamide
Myristoyl Monoethanolamide
Tetradecanamide, N-(2-Hydroxyethyl)-

Trade Name:
Hetamide MM (Heterene)

MYRISTAMIDE MIPA

CAS No. 10525-14-1
EINECS No. 234-077-9

Empirical Formula:
$C_{17}H_{35}NO_2$

Definition: Myristamide MIPA is a mixture of isopropanolamides of myristic acid. It conforms generally to the formula:

$$CH_3(CH_2)_{12}C(O)-NHCH_2CHOH$$
$$CH_3$$

Chemical Class: Alkanolamides

Functions: Surfactant - Foam Booster; Viscosity Increasing Agent - Aqueous

Technical/Other Names:
N-(2-Hydroxypropyl)Tetradecanamide
Monoisopropanolamine Myristic Acid
Amide
Myristoyl Isopropanolamide
Tetradecanamide, N-(2-Hydroxypropyl)-

MYRISTAMIDOBUTYL GUANIDINE ACETATE

Definition: Myristamidobutyl Guanidine Acetate is the organic compound that conforms to the formula:

$$CH_3(CH_2)_{12}C(O)-NH(CH_2)_4N=CNH_2 \cdot CH_3COOH$$
$$NH_2$$

Chemical Classes: Amides; Amines

Functions: Antistatic Agent; Hair Conditioning Agent; Humectant; Surfactant - Emulsifying Agent

Trade Name:
C14A4G (Lion Corporation)

MYRISTAMIDOPROPYLAMINE OXIDE

CAS No. 67806-10-4
EINECS No. 267-191-2

Empirical Formula:
$C_{19}H_{40}N_2O_2$

Definition: Myristamidopropylamine Oxide is the aliphatic amine oxide that conforms generally to the formula:

$$CH_3(CH_2)_{12}C(O)-NH(CH_2)_3N(CH_3) \rightarrow O$$

Information Source: TSCA

Chemical Class: Amine Oxides

Functions: Hair Conditioning Agent; Surfactant - Cleansing Agent; Surfactant - Foam Booster; Surfactant - Hydrotrope

Technical/Other Names:
Amides, Myristic, N-[3-(Dimethylamino)-Propyl], N-Oxide
Dimethylaminopropyl Myristamide N-oxide
N-[3-(Dimethylamino)Propyl]-Tetradecanamide-N-Oxide
Tetradecanamide, N-[3-(Dimethylamino)-Propyl]-, N-Oxide
Tetradecylamidopropyldimethylamine Oxide

MYRISTAMIDOPROPYL BETAINE

CAS No. 59272-84-3
EINECS No. 261-684-6

Empirical Formula:
$C_{21}H_{42}N_2O_3$

Definition: Myristamidopropyl Betaine is the zwitterion (inner salt) that conforms generally to the formula:

$$CH_3(CH_2)_{12}C(O)-NH(CH_2)_3-N^+(CH_3)_2-CH_2COO^-$$

Information Source: TSCA

Chemical Class: Betaines

Functions: Antistatic Agent; Hair Conditioning Agent; Skin-Conditioning Agent - Miscellaneous; Surfactant - Cleansing Agent; Surfactant - Foam Booster; Viscosity Increasing Agent - Aqueous

Technical/Other Names:
N-(Carboxymethyl)-N,N-Dimethyl-3-[(1-Oxotetradecyl)Amino]-1-Propanaminium Hydroxide, Inner Salt
Myristamidopropyl Dimethyl Glycine
1-Propanaminium, N-(Carboxymethyl)-N,N-Dimethyl-3-[(1-Oxotetradecyl)Amino]-, Hydroxide, Inner Salt

Trade Name:
Schercotaine MAB (Scher)

MYRISTAMIDOPROPYL DIMETHYLAMINE

CAS No. 45267-19-4
EINECS No. 256-214-1

Empirical Formula:
$C_{19}H_{40}N_2O$

Definition: Myristamidopropyl Dimethylamine is the amidoamine that conforms generally to the formula:

$$CH_3(CH_2)_{12}C(O)-NH(CH_2)_3N(CH_3)_2$$

Information Source: TSCA

Chemical Class: Amines

Function: Antistatic Agent

Technical/Other Names:
Dimethylaminopropyl Myristamide
N-[3-(Dimethylamino)Propyl]-Tetradecanamide
Tetradecanamide, N-[3-(Dimethylamino)-Propyl]-

Trade Name:
Schercodine M (Scher)

Trade Name Mixture:
Categel (Collaborative Labs)

MYRISTAMIDOPROPYL DIMETHYLAMINE PHOSPHATE

CAS No.: 129541-40-8

Definition: Myristamidopropyl Dimethyl-amine Phosphate is a complex mixture of salts of phosphoric acid and Myristamido-propyl Dimethylamine (q.v.).

Chemical Classes: Amines; Phosphorus Compounds

Function: Surfactant - Cleansing Agent

Technical/Other Name:
Tetradecanamide, N-[3-(Dimethylamino)-Propyl]-, Phosphate (1:1)

Trade Name:
Katemul MP-80 (Scher)

MYRISTAMIDOPROPYL HYDROXY-SULTAINE

Empirical Formula:
$C_{22}H_{46}N_2O_5S$

Definition: Myristamidopropyl Hydroxy-sultaine is the zwitter ion (inner salt) that conforms to the formula:

$$CH_3(CH_2)_{12}\overset{\overset{\textstyle O}{\|}}{C}-NH(CH_2)_3-\overset{\overset{\textstyle CH_3}{|}}{\underset{\underset{\textstyle CH_3}{|}}{N^+}}-CH_2\overset{\underset{\textstyle OH}{|}}{C}HCH_2SO_3^-$$

Chemical Class: Betaines

Functions: Antistatic Agent; Hair Condition-ing Agent; Skin-Conditioning Agent - Miscel-laneous; Surfactant - Cleansing Agent; Surfactant - Foam Booster; Viscosity Increasing Agent - Aqueous

Trade Name:
Obazolin MHS-121 (Toho)

MYRISTAMINE OXIDE

CAS No.	EINECS No.
3332-27-2	222-059-3

JPN Translation:
ミリスタミンオキシド

Empirical Formula:
$C_{16}H_{35}NO$

Definition: Myristamine Oxide is the tertiary amine oxide that conforms generally to the formula:

$$CH_3(CH_2)_{13}\overset{\overset{\textstyle CH_3}{|}}{\underset{\underset{\textstyle CH_3}{|}}{N}}\longrightarrow O$$

Information Sources: JCIC, JCLS, TSCA

Chemical Class: Amine Oxides

Functions: Hair Conditioning Agent; Surfactant - Cleansing Agent; Surfactant - Foam Booster; Surfactant - Hydrotrope

Reported Product Categories: Shampoos (Non-coloring); Hair Wave Sets; Douches

Technical/Other Names:
N,N-Dimethyl-1-Tetradecanamine-N-Oxide
Dimethyltetradecylamine
Myristyl Dimethyl Amine Oxide
Myristyl Dimethyl Amine Oxide Solution
Tetradecanamine, N,N-Dimethyl-, N-Oxide
1-Tetradecanamine, N,N-Dimethyl-, N-Oxide

Trade Names:
Ammonyx MCO (Stepan)
AMMONYX MO (Stepan)
Barlox 14 (Lonza Inc./Lonza Ltd.)
Empigen OH25 (Albright & Wilson UK)
Euroxide M25 (EOC Surfactants)
Incromine Oxide M (Croda, Inc.)
Schercamox DMA (Scher)
Schercamox DMM (Scher)

MYRISTAMINOPROPIONIC ACID

CAS No.	EINECS No.
14960-08-8	239-033-2

Empirical Formula:
$C_{17}H_{35}NO_2$

Definition: Myristaminopropionic Acid is the substituted propionic acid that conforms generally to the formula:

$$CH_3(CH_2)_{13}NHCH_2CH_2COOH$$

Information Source: TSCA

Chemical Classes: Alkyl-Substituted Amino Acids; Amines

Functions: Hair Conditioning Agent; Surfactant - Cleansing Agent; Surfactant - Foam Booster

Technical/Other Names:
β-Alanine, N-Tetradecyl-
N-Tetradecyl-β-Alanine

MYRISTIC ACID

CAS No.	EINECS No.
544-63-8	208-875-2

JPN Translation:
ミリスチン酸

Empirical Formula:
$C_{14}H_{28}O_2$

Definition: Myristic Acid is the organic acid that conforms generally to the formula:

$$CH_3(CH_2)_{12}COOH$$

Information Sources: 21CFR172.210, 21CFR172.860, 21CFR173.340, 21CFR175.105, 21CFR175.320, 21CFR176.170, 21CFR176.200, 21CFR176.210, 21CFR177.1010, 21CFR177.1200, 21CFR177.2260, 21CFR177.2600, 21CFR177.2800, 21CFR178.3570, 21CFR178.3910, CIR: [S] JACT-6(3)1987, CTFA S, FCC, JCLS, JSCI, MI-13(6359), RIFM, TSCA

Chemical Class: Fatty Acids

Functions: Fragrance Ingredient; Opacifying Agent; Surfactant - Cleansing Agent

Surfactant-Cleansing Agent is included as a function for the soap form of Myristic Acid.

Reported Product Categories: Bath Oils, Tablets, and Salts; Cleansing Products (Cold Creams, Cleansing Lotions, Liquids and Pads); Shaving Cream (Aerosol, Brushless and Lather); Bath Preparations, Misc.; Bath Soaps and Detergents; Foundations; Sham-poos (Non-coloring); Shaving Preparations, Misc.

Technical/Other Names:
Myristic acid (RIFM)
Tetradecanoic Acid

Trade Names:
AEC Myristic Acid (A & E Connock)
Emery 655 (Cognis Corp.)
Kortacid 1495 (Akzo Nobel Surface AB)
Kortacid 1499 (Akzo Nobel)
Kortacid 1499 (Akzo Nobel Surface AB)
Myristic Acid PC (Protameen)
PRIFRAC 2940 (Uniqema Europe)
PRIFRAC 2942 (Uniqema Europe)

Trade Name Mixtures:
Capispheres (Barnet)
Questamix H (Quest International)
SAMT-UFZO-450 (13%) (US Cosmetics)
SAMT-UFZO-450/D5 (60%) (US Cosmetics)

MYRISTICA FRAGRANS (NUTMEG) EXTRACT

CAS No.	EINECS No.
84082-68-8	282-013-3

Definition: Myristica Fragrans (Nutmeg) Extract is an extract of the nutmeg, *Myristica fragrans. See "Regulatory and Ingredient Use Information," regarding the labeling names for botanical ingredients in Volume 1, Introduction, Part A.*

Information Source: RIFM

Chemical Class: Biological Products

Functions: Fragrance Ingredient; Skin-Conditioning Agent - Miscellaneous

Technical/Other Names:
Extract of Myristica Fragrans
Extract of Nutmeg
Myristica Fragrans Extract
Nutmeg Extract
Nutmeg (Myristica fragrans Houtt.) (RIFM)

Trade Name:
Nutmeg Extract (Quest International)

Trade Name Mixtures:
Aromaphyte of Nutmeg (Active Organics)
Glycolysat of Nutmeg (CEP (Solabia))
Herbasol Extract Nutmeg (Cosmetochem)
(Cosmetochem International Ltd.)
Nutmeg Extract (Matsuura Yakugyo)
Nutmeg Fluid Extract (Alban Muller)
Nutmeg HS (Alban Muller)
Nutmeg LS (Alban Muller)
Vegebios of Nutmeg (CEP (Solabia))
VT-093 Extract of Nutmeg (Vege-Tech)

MYRISTICA FRAGRANS (NUTMEG) KERNEL OIL

CAS Nos.: 8007-12-3; 8008-45-5

Definition: Myristica Fragrans (Nutmeg) Kernel Oil is the oil obtained from the kernel of *Myristica fragrans. See "Regulatory and Ingredient Use Information," regarding the labeling names for botanical ingredients in Volume 1, Introduction, Part A.*

Information Sources: ARG, AUS, BPC, BRA, 21CFR182.20, FCC, FI, IND, MAR, MI-13(6857), NF XV, RIFM, SNPF, TSCA, USSR

Chemical Class: Essential Oils

Functions: Flavoring Agent; Fragrance Ingredient

Technical/Other Names:
Mace (Myristica fragrans Houtt.) (RIFM)
Mace oil (Myristica fragrans Houtt.) (RIFM)
Mace oleoresin (Myristica fragrans Houtt.) (RIFM)
Myristica Fragrans Oil
Myristica Oil
Nutmeg Kernel Oil
Nutmeg oil (RIFM)
Nutmeg Oil
Oil of Myristica Fragrens
Oil of Nutmeg

Trade Name:
AEC Numeg Oil (A & E Connock)

Trade Name Mixture:
Aromaphyte of Nutmeg (Active Organics)

MYRISTIC/PALMITIC/STEARIC/RICINOLEIC/EICOSANEDIOIC GLYCERIDES

Definition: Myristic/Palmitic/Stearic/Ricinoleic/Eicosanedioic Glycerides is a mixture of mono-, di- and triglycerides of myristic, palmitic, stearic, ricinoleic and eicosanedioic acids.

Chemical Class: Glyceryl Esters and Derivatives

Functions: Hair Conditioning Agent; Hair Fixative; Skin-Conditioning Agent - Occlusive

Trade Name Mixture:
Nikkol Woodwax (Nikko)

MYRISTOYL ETHYL GLUCOSIDE

Definition: Myristoyl Ethyl Glucoside is the ester of Myristic Acid (q.v.) and ethyl glucoside.

Chemical Classes: Carbohydrates; Esters

Function: Surfactant - Emulsifying Agent

MYRISTOYL GLUTAMIC ACID

JPN Translation:
ミリストイルグルタミン酸

Empirical Formula:
$C_{19}H_{35}NO_5$

Definition: Myristoyl Glutamic Acid is the substituted amino acid that conforms to the formula:

$$HO-\overset{\overset{O}{\|}}{C}CH_2CH_2\overset{\overset{}{|}}{\underset{\underset{\underset{\underset{O}{\|}}{C(CH_2)_{12}CH_3}}{NH}}{C}}H\overset{\overset{O}{\|}}{C}-OH$$

Information Sources: JCIC, JCLS

Chemical Classes: Amides; Amino Acids

Functions: Hair Conditioning Agent; Skin-Conditioning Agent - Miscellaneous; Surfactant - Cleansing Agent

Technical/Other Name:
N-Myristoyl L-Glutamic Acid

MYRISTOYL GLYCINE/HISTIDINE/LYSINE POLYPEPTIDE

Definition: Myristoyl Glycine/Histidine/Lysine Polypeptide is the product obtained from the reaction of myristic acid with a polypeptide containing glycine, histidine and lysine residues.

Chemical Class: Proteins

Function: Skin-Conditioning Agent - Miscellaneous

Trade Name Mixture:
Myristyl-GHK (Sederma)

MYRISTOYL HEXAPEPTIDE-5

Definition: Myristoyl Hexapeptide-5 is a reaction product of Myristic Acid (q.v.) with a synthetic peptide containing Valine (q.v.), Tyrosine (q.v.)., Glutamic Acid (q.v.), Proline (q.v.) and Isoleucine (q.v.) residues.

Chemical Class: Protein Derivatives

Function: Skin-Conditioning Agent - Miscellaneous

Trade Name:
Collasyn 614 VY (Therapeutic Peptide Inc.)

MYRISTOYL HEXAPEPTIDE-6

Definition: Myristoyl Hexapeptide-6 is the reaction product of Myristic Acid (q.v.) with a synthetic peptide containing Valine (q.v.), Glycine (q.v.), Alanine (q.v.) and Proline (q.v.) residues.

Chemical Class: Protein Derivatives

Function: Skin-Conditioning Agent - Miscellaneous

Trade Name:
Collasyn 614 VG (Therapeutic Peptide Inc.)

MYRISTOYL HYDROLYZED COLLAGEN

CAS No.: 72319-06-3

JPN Translation:
ミリストイル加水分解コラーゲン

Definition: Myristoyl Hydrolyzed Collagen is the condensation product of myristic acid chloride and Hydrolyzed Collagen (q.v.).

Information Sources: JCIC, JCLS, JSQI

Chemical Class: Protein Derivatives

Functions: Hair Conditioning Agent; Skin-Conditioning Agent - Miscellaneous; Surfactant - Cleansing Agent

Reported Product Categories: Hair Preparations (Non-coloring), Misc.; Hair Sprays (Aerosol Fixatives); Tonics, Dressings, and Other Hair Grooming Aids; Mascara

Technical/Other Names:
Myristoyl Hydrolyzed Animal Protein
Myristoyl Hydrolyzed Animal Protein Solution
Proteins, Hydrolysates, Reaction Products with Myristoyl Chloride

Trade Names:
Etha-Coll 210-20 (Arch Personal Care Products)

Promois Resin AM (Seiwa Kasei)
Pro-Tein SM-20 (Maybrook)

Trade Name Mixtures:
Oleo-Coll A240 (Arch Personal Care Products)
Oleo-Coll A240-20 (Arch Personal Care Products)

MYRISTOYL LACTYLIC ACID

Empirical Formula:
$C_{20}H_{36}O_6$

Definition: Myristoyl Lactylic Acid is the organic compound that conforms to the formula:

$$CH_3(CH_2)_{12}C \overset{O}{\overset{\|}{}} - OCHC \overset{O}{\overset{\|}{}} - OCHCOOH$$
$$\underset{CH_3}{|} \quad \underset{CH_3}{|}$$

Chemical Classes: Carboxylic Acids; Esters

Function: Surfactant - Emulsifying Agent

Trade Name Mixture:
PRIAZUL 2131 (Uniqema Europe)

MYRISTOYL METHYL BETA-ALANINE

CAS No.: 21539-71-9

JPN Translation:
ミリストイルメチルアラニン

Empirical Formula:
$C_{18}H_{35}NO_3$

Definition: Myristoyl Methyl Beta-Alanine is the substituted amino acid that conforms to the formula:

$$CH_3(CH_2)_{12}C \overset{O}{\overset{\|}{}} - NCHCOOH$$
$$\underset{CH_3}{|}$$

Chemical Classes: Amides; Amino Acids

Functions: Skin-Conditioning Agent - Miscellaneous; Surfactant - Emulsifying Agent

Technical/Other Names:
β-Alanine, N-Methyl-N-(1-Oxotetradecyl)-
N-Methyl-N-(1-Oxotetradecyl)-β-Alanine
N-Myristoyl-N-Methylalanine

Trade Name:
Alanon AMA (Kawaken)

MYRISTOYL NONAPEPTIDE-2

Definition: Myristoyl Nonapeptide-2 is the reaction product of Myristic Acid (q.v.) with a synthetic peptide containing glutamic acid, glutamine, valine, glycine, serine, aspartic acid, asparagine, and lysine.

Chemical Class: Protein Derivatives

Function: Skin-Conditioning Agent - Miscellaneous

Trade Name:
Prolifersyn 914 VK (Therapeutic Peptide Inc.)

MYRISTOYL/PCA CHITIN

Definition: Myristoyl/PCA Chitin is the product obtained by the reaction of myristic acid and acetic acid with Chitosan PCA (q.v.).

Chemical Class: Biological Polymers and their Derivatives

Functions: Skin-Conditioning Agent - Humectant; Suspending Agent - Nonsurfactant

Trade Name Mixture:
PM-Chitosan (PIAS)

MYRISTOYL PENTAPEPTIDE-3

Definition: Myristoyl Pentapeptide-3 is the reaction product of Myristic Acid (q.v.) with a synthetic peptide containing Threonine (q.v.), Serine (q.v.) and Lysine (q.v.).

Chemical Class: Protein Derivatives

Function: Skin-Conditioning Agent - Miscellaneous

Trade Name:
Collasyn 514KS (Therapeutic Peptide Inc.)

MYRISTOYL SARCOSINE

CAS No.	EINECS No.
52558-73-3	258-007-1

Empirical Formula:
$C_{17}H_{33}NO_3$

Definition: Myristoyl Sarcosine is the N-myristoyl derivative of N-methylglycine that conforms to the formula:

$$CH_3(CH_2)_{12}C \overset{O}{\overset{\|}{}} - NCH_2COOH$$
$$\underset{CH_3}{|}$$

Information Sources: CIR: [SQ] IJT-20 (SUPPL. 1)2001, TSCA

Chemical Class: Sarcosinates and Sarcosine Derivatives

Functions: Hair Conditioning Agent; Surfactant - Cleansing Agent

Reported Product Category: Shaving Preparations, Misc.

Technical/Other Names:
Glycine, N-Methyl-N-(1-Oxotetradecyl)-
N-Methyl-N-(1-Oxotetradecyl)Glycine
Myristoyl N-Methylglycine
N-Tetradecanoylsarcosine

Trade Names:
Hamposyl M (Amerchol)
Vanseal MS (Vanderbilt)

MYRISTOYL SUCCINOYL ATELOCOLLAGEN

Definition: Myristoyl Succinoyl Atelocollagen is the product formed by the reaction of succinic anhydride and the reaction product of myristoyl chloride and Antelocollagen (q.v.).

Information Source: JCLS

Chemical Classes: Amides; Protein Derivatives

Function: Skin-Conditioning Agent - Miscellaneous

MYRISTOYL TETRAPEPTIDE-4

Definition: Myristoyl Tetrapeptide-4 is the reaction product of Myristic Acid (q.v.) with a synthetic peptide containing Glycine (q.v.), Glutaminic Acid (q.v.) and Proline (q.v.) residues.

Chemical Class: Protein Derivatives

Function: Skin-Conditioning Agent - Miscellaneous

Trade Name:
Collasyn 414 GG (Therapeutic Peptide Inc.)

MYRISTOYL TETRAPEPTIDE-5

Definition: Myristoyl Tetrapeptide-5 is the reaction product of Myristic Acid (q.v.) with a synthetic peptide containing Valine (q.v.), Proline (q.v.) and Alanine (q.v.) residues.

Chemical Class: Protein Derivatives

Function: Skin-Conditioning Agent - Miscellaneous

Trade Name:
Collasyn 414 VA (Therapeutic Peptide Inc.)

MYRISTOYL TRIPEPTIDE-3

Definition: Myristoyl Tripeptide-3 is the reaction product of Myristic Acid (q.v.) with a synthetic peptide containing Glycine (q.v.), Histidine (q.v.) and Arginine (q.v.) residues.

Chemical Class: Protein Derivatives

Function: Skin-Conditioning Agent - Miscellaneous

Trade Name:
Collasyn 314 GR (Therapeutic Peptide Inc.)

MYRIST/PALMITAMIDOBUTYL GUANIDINE ACETATE

Definition: Myrist/Palmitamidobutyl Guanidine Acetate is the organic compound that conforms to the formula:

$$RC\overset{O}{\overset{\|}{-}}NH(CH_2)_4N=CNH_2 \cdot CH_3COOH$$
$$\underset{NH_2}{|}$$

where RCO- represents a mixture of the myristoyl and palmitoyl radicals.

Chemical Classes: Amides; Amines

Functions: Antistatic Agent; Hair Conditioning Agent; Humectant; Surfactant - Emulsifying Agent

Trade Name:
C146A4G (Lion Corporation)

MYRISTYL ACETATE

CAS No. 638-59-5   **EINECS No.** 211-344-8

Empirical Formula:
$C_{16}H_{32}O_2$

Definition: Myristyl Acetate is the organic compound that conforms to the formula:

$$CH_3C\overset{O}{\overset{\|}{-}}O(CH_2)_{13}CH_3$$

Chemical Class: Esters

Function: Skin-Conditioning Agent - Emollient

Technical/Other Names:
Acetic Acid, Tetradecyl Ester
Tetradecamol Acetate
Tetradecyl Acetate

Trade Name Mixture:
Placenta Extract, Oil- Soluble (Bottger)

MYRISTYL ALANINATE

Definition: Myristyl Alaninate is the ester of myristic acid and alanine. It conforms to the formula:

$$CH_3CHC\overset{O}{\overset{\|}{-}}O(CH_2)_{13}CH_3$$
$$\underset{NH_2}{|}$$

Information Source: JCIC

Chemical Classes: Amino Acids; Esters

Functions: Hair Conditioning Agent; Skin-Conditioning Agent - Emollient; Surfactant - Emulsifying Agent

MYRISTYL ALCOHOL

CAS No. 112-72-1   **EINECS No.** 204-000-3

JPN Translation:
ミリスチルアルコール

Empirical Formula:
$C_{14}H_{30}O$

Definition: Myristyl Alcohol is the fatty alcohol that conforms generally to the formula:

$$CH_3(CH_2)_{13}OH$$

Information Sources: 21CFR172.864, 21CFR175.105, 21CFR175.300, 21CFR176.200, 21CFR176.210, 21CFR177.1010, 21CFR177.2800, 21CFR178.3480, 21CFR178.3910, CIR: [S] JACT-7(3)1988, CTFA S, JCIC, JCLS, JSQI, MI-13(6361), NF XVIII, RIFM, TSCA, USAN

Chemical Class: Fatty Alcohols

Functions: Emulsion Stabilizer; Fragrance Ingredient; Skin-Conditioning Agent - Emollient; Surfactant - Foam Booster; Viscosity Increasing Agent - Aqueous; Viscosity Increasing Agent - Nonaqueous

Reported Product Categories: Hair Dyes and Colors (All Types Requiring Caution Statements and Patch Tests); Moisturizing Preparations; Shaving Preparations, Misc.; Bath Oils, Tablets, and Salts; Cleansing Products (Cold Creams, Cleansing Lotions, Liquids and Pads)

Technical/Other Names:
1-Hydroxytetradecane
Tetradecanol
1-Tetradecanol (RIFM)
Tetradecyl Alcohol

Trade Names:
AEC Myristyl Alcohol (A & E Connock)
Alfol 14 Alcohol (Sasol North America)
Cachalot myristyl alcohol M-43 (Michel)
CoChem MA (Costec)
Lanette 14 (Cognis Care Chemicals/NJ)
Lanette 14 (Cognis Care Chemicals/PA)
Lanette 14 (Cognis Deutschland)

Nacol 14-99 Alcohol (Sasol GmbH - Hamburg)
Sabonal C14 (Sabo)
Unihydag WAX-14 (Universal Preserv-A-Chem)

Trade Name Mixtures:
Ceraphyl 50 (International Specialty Products)
Homulgator 920 G (Grau)
Homulgator 1330 G (Grau)
Homulgator 910 G Extra (Grau)
Montanov 14 (SEPPIC)
Neo PCL SE o/w 2/066280 (Symrise)
ProLipid 141 (International Specialty Products)
Sabonal C12 14 (Sabo)
Unieucerin (Chemyunion)

MYRISTYLAMIDOPROPYL DIMETHYL-AMINE DIMETHICONE PEG-7 PHOSPHATE

CAS No.: 137145-36-9

Definition: Myristylamidopropyl Dimethylamine Dimethicone PEG-7 Phosphate is the Myristamidopropyl Dimethylamine (q.v.) salt of Dimethicone PEG-7 Phosphate (q.v.).

Chemical Classes: Phosphorus Compounds; Siloxanes and Silanes

Function: Hair Conditioning Agent

Technical/Other Name:
Siloxanes and Silicones, Dimethyl, 3-Hydroxypropylmethyl, Ethers with Polyethylene Glycol Dihydrogen Phosphate, Compds. with N-[3-(Dimethylamino)-Propyl]Tetradecanamide

Trade Name:
Pecosil 14 PS (Phoenix)

MYRISTYL BETAINE

CAS No. 2601-33-4   **EINECS No.** 220-006-9

JPN Translation:
ミリスチルベタイン

Empirical Formula:
$C_{18}H_{37}NO_2$

Definition: Myristyl Betaine is the zwitterion (inner salt) that conforms generally to the formula:

$$CH_3(CH_2)_{13}\overset{CH_3}{\underset{CH_3}{-N^+-}}CH_2COO^-$$

Information Sources: JCIC, JCLS, TSCA

Chemical Class: Betaines

Functions: Abrasive; Antistatic Agent; Hair Conditioning Agent; Skin-Conditioning Agent - Miscellaneous; Surfactant - Cleansing Agent; Surfactant - Foam Booster; Viscosity Increasing Agent - Aqueous

Technical/Other Names:
Ammonium, (carboxymethyl)-diemthyltetradecyl-, Hydroxide, Inner Salt
N-(Carboxymethyl)-N,N-Dimethyl-1-Tetradecanaminium Hydroxide, Inner Salt
Myristyl Betaine Solution
Myristyl Dimethyl Glycine
1-Tetradecanaminium, N-(Carboxymethyl)-N,N-Dimethyl-, Hydroxide, Inner Salt
N-Tetradecyl Betaine

MYRISTYL/CETYL AMINE OXIDE

Definition: Myristyl/Cetyl Amine Oxide is the aliphatic amine oxide that conforms to the formula:

$$R - \overset{\overset{\displaystyle CH_3}{|}}{\underset{\underset{\displaystyle CH_3}{|}}{N}} \longrightarrow O$$

where R represents a mixture of myristyl and cetyl alkyl groups.

Chemical Class: Amine Oxides

Functions: Hair Conditioning Agent; Surfactant - Cleansing Agent; Surfactant - Foam Booster; Surfactant - Hydrotrope

Technical/Other Names:
N,N-Dimethyl-1-Myristylamine/Cetylamine-N-Oxide
1-Myristylamine/Cetylamine, N,N-Dimethyl-, N-Oxide

MYRISTYL ETHYLHEXANOATE

Empirical Formula:
$C_{22}H_{44}O_2$

Definition: Myristyl Ethylhexanoate is ester of myristyl alcohol and 2-ethylhexanoic acid that conforms to the formula:

$$CH_3(CH_2)_3\overset{\overset{\displaystyle O}{\|}}{C}HC - O(CH_2)_{13}CH_3$$
$$\underset{\underset{\displaystyle CH_2CH_3}{|}}{}$$

Chemical Class: Esters

Function: Skin-Conditioning Agent - Emollient

Technical/Other Names:
2-Ethylhexanoic Acid, Myristyl Ester

Myristyl Caprylate
Myristyl Octanoate
Octanoic Acod, Tetradecyl Ester
Tetradecyl Caprylate
Tetradecyl Octanoate

Trade Name Mixture:
DUB Liquide 1214 (Stearinerie Dubois Fils)

MYRISTYL GLUCOSIDE

Definition: Myristyl Glucoside is the product obtained by the condensation of myristyl alcohol with a glucose polymer.

Chemical Class: Carbohydrates

Function: Surfactant - Cleansing Agent

Technical/Other Names:
Myristyl D-glucopyranoside
Tetradecyl D-glucopyranoside
Tetradecyl D-Glucoside

Trade Name Mixture:
Montanov 14 (SEPPIC)

MYRISTYL HYDROXYETHYL IMIDAZOLINE

CAS No.	EINECS No.
6942-02-5	230-097-7

Empirical Formula:
$C_{18}H_{36}N_2O$

Definition: Myristyl Hydroxyethyl Imidazoline is the heterocyclic compound that conforms to the formula:

Information Source: TSCA

Chemical Class: Imidazoline Compounds

Functions: Antistatic Agent; Hair Conditioning Agent

Technical/Other Names:
4,5-Dihydro-2-Tridecyl-1H-Imidazole-1-Ethanol
1H-Imidazole-1-Ethanol, 4,5-Dihydro-2-Tridecyl
1-(2-Hydroxyethyl)-2-Tridecylimidazoline
Myristyl Imidazoline

MYRISTYL ISOSTEARATE

CAS No.	EINECS No.
94247-26-4	304-203-8

Empirical Formula:
$C_{32}H_{64}O_2$

Definition: Myristyl Isostearate is the ester of myristyl alcohol and isostearic acid. It conforms generally to the formula:

$$C_{17}H_{35}\overset{\overset{\displaystyle O}{\|}}{C} - O(CH_2)_{13}CH_3$$

Chemical Class: Esters

Function: Skin-Conditioning Agent - Emollient

Technical/Other Names:
Isooctadecanoic Acid, Tetradecyl Ester
Tetradecyl Isooctadecanoate

Trade Name:
AEC Myristyl Isostearate (A & E Connock)

MYRISTYL LACTATE

CAS No.	EINECS No.
1323-03-1	215-350-1

JPN Translation:
乳酸ミリスチル

Empirical Formula:
$C_{17}H_{34}O_3$

Definition: Myristyl Lactate is the ester of myristyl alcohol and lactic acid. It conforms generally to the formula:

$$CH_3\overset{\overset{\displaystyle O}{\|}}{C}HC - O(CH_2)_{13}CH_3$$
$$\underset{\underset{\displaystyle OH}{|}}{}$$

Information Sources: CIR: [S] JACT-1(2)-1982, CIR: [SQ] IJT-17(Suppl. 1)1998, CTFA S, JCLS, JSCI, TSCA

Chemical Class: Esters

Function: Skin-Conditioning Agent - Emollient

Reported Product Categories: Lipsticks; Eye Shadows; Eyebrow Pencils; Moisturizing Preparations; Eyeliners; Indoor Tanning Preparations; Bath Preparations, Misc.; Body and Hand Preparations (Excluding Shaving Preparations); Eye Makeup Preparations, Misc.; Skin Care Preparations, Misc.; Blushers (All types); Foundations; Personal Cleanliness Products, Misc.; Aftershave Lotions; Paste Masks (Mud Packs)

Technical/Other Names:
2-Hydroxypropanoic Acid, Tetradecyl Ester
Myristic Acid, Tetradecyl Ester
Propanoic Acid, 2-Hydroxy-, Tetradecyl Ester
Tetradecyl 2-Hydroxypropanoate
Tetradecyl Lactate

Trade Names:
AEC Myristyl Lactate (A & E Connock)

Cegesoft C 17 (Cognis Deutschland)
Cetinol LM (Fabriquimica)
Crodamol ML (Croda Chemicals)
Dermol ML (Alzo)
DUB LM (Stearinerie Dubois Fils)
Jeechem ML (Jeen)
Lactabase C14 (Prod'Hyg)
Liponate ML (Lipo)
Nikkol Myristyl Lactate (Nikko)
Uniester LM (Chemyunion)

Trade Name Mixtures:
Base 323 MTC (LCW)
Ceraphyl 50 (International Specialty
  Products)
Covalip 22 (LCW)
Covalip 25 (LCW)
Covalip LL 48 (LCW)

MYRISTYL LIGNOCERATE

CAS No.: 42233-51-2

Empirical Formula:
$C_{38}H_{76}O_2$

Definition: Myristyl Lignocerate is the ester of Myristyl Alcohol (q.v.) and lignoceric acid. It conforms to the formula:

$$CH_3(CH)_{22}\overset{\overset{O}{\|}}{C}-O(CH_2)_{13}CH_3$$

Chemical Class: Esters

Function: Skin-Conditioning Agent - Emollient

Technical/Other Names:
Lignoceric Acid, Myristyl Ether
Tetracosanoic Acid, Tetradecyl Ester
Tetradecyl Tetracosanoate

Trade Name Mixture:
Ross Synthetic Candelilla Wax (Ross)

MYRISTYL MYRISTATE

CAS No.	EINECS No.
3234-85-3	221-787-9

JPN Translation:
ミリスチン酸ミリスチル

Empirical Formula:
$C_{28}H_{56}O_2$

Definition: Myristyl Myristate is the ester of myristyl alcohol and myristic acid. It conforms generally to the formula:

$$CH_3(CH)_{12}\overset{\overset{O}{\|}}{C}-O(CH_2)_{13}CH_3$$

Information Sources: CIR: [S] JACT-1(4)-1982, CTFA S, JCLS, JSCI, TSCA

Chemical Class: Esters

Function: Skin-Conditioning Agent - Occlusive

Reported Product Categories: Moisturizing Preparations; Bath Preparations, Misc.; Body and Hand Preparations (Excluding Shaving Preparations); Eye Makeup Preparations, Misc.; Perfumes; Makeup Preparations (Not eye), Misc.; Skin Care Preparations, Misc.; Aftershave Lotions; Baby Shampoos; Makeup Bases; Bath Capsules; Face and Neck Preparations (Excluding Shaving Preparations); Hair Conditioners; Fragrance Preparations, Misc.; Lipsticks; Shaving Cream (Aerosol, Brushless and Lather); Bath Oils, Tablets, and Salts; Eyebrow Pencils; Foundations; Night Skin Care Preparations; Baby Lotions, Oils, Powders and Creams; Cleansing Products (Cold Creams, Cleansing Lotions, Liquids and Pads); Foot Powders and Sprays; Powders (Dusting and Talcum, Excluding Aftershave Talcs); Blushers (All types); Cuticle Softeners; Eye Shadows; Indoor Tanning Preparations; Manicuring Preparations, Misc.; Paste Masks (Mud Packs); Shampoos (Non-coloring)

Technical/Other Names:
Tetradecanoic Acid, Tetradecyl Ester
Tetradecyl Tetradecanoate

Trade Names:
AEC Myristyl Myristate (A & E Connock)
Ceraphyl 424 (International Specialty
  Products)
Cetinol MM (Fabriquimica)
Cetiol MM (Cognis Care Chemicals/NJ)
Cetiol MM (Cognis Care Chemicals/PA)
Cetiol MM (Cognis Deutschland)
Crodamol MM (Croda Chemicals)
Crodamol MM (Croda, Inc.)
Customate MM (myristyl myristate)
  (Custom Ingredients)
DUB MM (Stearinerie Dubois Fils)
HallStar MM (RTD Hall Star)
Jeechem MM (Jeen)
Liponate MM (Lipo)
Myristate De Myristyle (Prod'Hyg)
Nikkol MM (Nikko)
Pelemol MM (Phoenix)
Protachem MM (Protameen)
Saboderm MM (Sabo)
Schercemol MM (Scher)
Tegosoft MM (Degussa Care Specialties)
Uniester MM (Chemyunion)

Trade Name Mixtures:
EFA-Plex (Arch Personal Care Products)
Oleo Keratin ISO (Arch Personal Care
  Products)
Prodhyrouge 2000 (Prod'Hyg)

MYRISTYL NEOPENTANOATE

CAS No.: 144610-93-5

Empirical Formula:
$C_{19}H_{38}O_2$

Definition: Myristyl Neopentanoate is the ester of myristyl alcohol and neopentanoic acid. It conforms generally to the formula:

$$CH_3\overset{\overset{CH_3}{|}}{\underset{\underset{CH_3}{|}}{C}}-\overset{\overset{O}{\|}}{C}-O(CH_2)_{13}CH_3$$

Chemical Class: Esters

Function: Skin-Conditioning Agent - Emollient

Technical/Other Names:
2,2-Dimethylpropanoic acid, Tetradecyl
  Ester
Propanoic Acid, 2,2-Dimethyl-, Tetradecyl
  Ester
Tetradecyl 2,2-Dimethyl Propanoate

Trade Names:
AEC Myristyl Neopentanoate (A & E
  Connock)
Schercemol 145 (Scher)

MYRISTYL NICOTINATE

CAS No.: 273203-62-6

Empirical Formula:
$C_{20}H_{33}NO_2$

Definition: Myristyl Nicotinate is the organic compound that conforms to the formula:

$$CH_3(CH_2)_{13}O-\overset{\overset{O}{\|}}{C}-\text{(pyridine ring)}$$

Chemical Class: Heterocyclic Compounds

Functions: Hair Conditioning Agent; Skin-Conditioning Agent - Miscellaneous

Technical/Other Names:
3-Pyridinecarboxylic Acid, Tetradecyl Ester
Tetradecyl Nicotinate

Trade Name:
Pro-NAD (Niadyne)

MYRISTYL/PALMITYL OXOSTEARAMIDE/ ARACHAMIDE MEA

Definition: Myristyl/Palmityl Oxostearamide/Arachamide MEA is the organic compound that conforms generally to the formula:

$$CH_3(CH_2)_m\overset{\overset{O}{\|}}{C}CH\overset{\overset{O}{\|}}{\underset{\underset{(CH_2)_nCH_3}{|}}{C}}-NHCH_2CHOH$$

where m has a value of 14 or 16, and n has a value of 13 or 15, respectively.

Chemical Classes: Amides; Ketones

Functions: Hair Conditioning Agent; Skin-Conditioning Agent - Humectant

Trade Names:
PC-9S (Aekyung)
PC-9SP (NeoPharm)
Pseudo-dermal lipid (NeoPharm)

MYRISTYL PCA

CAS No.	EINECS No.
37673-27-1	253-596-6

Empirical Formula:
$C_{19}H_{35}NO_3$

Definition: Myristyl PCA is the ester of myristyl alcohol and PCA (q.v.) that conforms to the fomrula:

Chemical Classes: Esters; Heterocyclic Compounds

Function: Skin-Conditioning Agent - Miscellaneous

Technical/Other Name:
L-Proline, 4-Oxo-, Tetradecyl Ester

Trade Name:
Myristidone (UCIB (Solabia))

MYRISTYL-PG HYDROXYETHYL DECANAMIDE

JPN Translation:
ミリスチル PG ヒドロキシエチルデカナミド

Empirical Formula:
$C_{29}H_{59}NO_4$

Definition: Myristyl-PG Hydroxyethyl Decanamide is the organic compound that conforms to the formula:

Information Sources: JCIC, JCLS

Chemical Classes: Amides; Ethers

Function: Skin-Conditioning Agent - Miscellaneous

Technical/Other Name:
N-(Tetradecyloxyhydroxypropyl)-N-hydroxyethyldecanamide

Trade Name:
Sphingolipid E40 (Kao Corp.)

MYRISTYL PROPIONATE

CAS No.	EINECS No.
6221-95-0	228-300-9

Empirical Formula:
$C_{17}H_{34}O_2$

Definition: Myristyl Propionate is the ester of myristyl alcohol and propionic acid. It conforms to the formula:

Information Source: TSCA

Chemical Class: Esters

Function: Skin-Conditioning Agent - Emollient

Reported Product Categories: Moisturizing Preparations; Preshave Lotions (All types)

Technical/Other Names:
1-Tetradecyl Propanoate
Tetraecyl Propionate

Trade Names:
AEC Myristyl Propionate (A & E Connock)
Lonzest 143-S (Lonza Inc./Lonza Ltd.)
Schercemol MP (Scher)

MYRISTYL SALICYLATE

CAS No.: 19666-17-2

Empirical Formula:
$C_{21}H_{34}O_3$

Definition: Myristyl Salicylate is the ester of myristyl alcohol and salicylic acid. It conforms to the formula:

Information Source: CIR: [SQ]

Chemical Classes: Esters; Phenols

Function: Not Reported

Technical/Other Names:
Benzoic Acid, 2-Hydroxy-, Tetradecyl Ester
2-Hydroxybenzoic Acid, Tetradecyl Ester
Myristyl 2-hydroxybenzoate
Salicylic Acid, Tetradecyl Ester
Tetradecyl 2-Hydroxybenzoate
Tetradecyl Salicylate

MYRISTYL STEARATE

CAS No.	EINECS No.
17661-50-6	241-640-2

Empirical Formula:
$C_{32}H_{64}O_2$

Definition: Myristyl Stearate is the ester of myristyl alcohol and stearic acid. It conforms to the formula:

Information Sources: CIR: [S] JACT-4(5)-1985, TSCA

Chemical Class: Esters

Function: Skin-Conditioning Agent - Occlusive

Reported Product Categories: Body and Hand Preparations (Excluding Shaving Preparations); Cleansing Products (Cold Creams, Cleansing Lotions, Liquids and Pads); Cuticle Softeners; Face and Neck Preparations (Excluding Shaving Preparations); Hair Preparations (Non-coloring), Misc.; Makeup Preparations (Not eye), Misc.; Night Skin Care Preparations; Rouges

Technical/Other Names:
Octadecanoic Acid, Tetradecyl Ester
Stearic Acid, Tetradecyl Ester
Tetradecyl Octadecanoate
Tetradecyl Stearate

Trade Name:
AEC Myristyl Stearate (A & E Connock)

MYROCARPUS FASTIGIATUS OIL

CAS No.	EINECS No.
68188-03-4	294-497-3

Definition: Myrocarpus Fastigiatus Oil is the volatile oil obtained from *Myrocarpus fastigiatus*. See *"Regulatory and Ingredient Use Information," regarding the labeling names for botanical ingredients in Volume 1, Introduction, Part A.*

Information Sources: RIFM, TSCA

Chemical Class: Essential Oils

Function: Fragrance Ingredient

Technical/Other Name:
Cabreuva oil (Myrocarpus frondosus & M. fastigiatus) (RIFM)

Trade Name:
Cabreuva Oil (Kneipp)

The inclusion of any compound in the *Dictionary and Handbook* does not indicate that use of that substance as a cosmetic ingredient complies with the laws and regulations governing such use in the United States or any other country.

MYROTHAMNUS FLABELLIFOLIA EXTRACT

Definition: Myrothamnus Flabellifolia Extract is an extract of the aerial parts of *Myrothamnus flabellifolia. See "Regulatory and Ingredient Use Information," regarding the labeling names for botanical ingredients in Volume 1, Introduction, Part A.*

Chemical Class: Biological Products

Function: Skin-Conditioning Agent - Emollient

Technical/Other Name:
Extract of Myrothamnus Flabellifolia

Trade Name Mixtures:
P.A. Reviviscence (Laboratoires Sero-biologiques)
P.A. Reviviscence Lea PW LS (Laboratoires Serobiologiques)
P.A. Reviviscence PW (Laboratoires Serobiologiques)

MYROXYLON BALSAMUM

Definition: *See "Regulatory and Ingredient Use Information," regarding EU labeling names for botanical ingredients in Volume 1, Introduction, Part A.*

Chemical Class: Biological Products

Technical/Other Name:
Myroxylon Balsamum (Balsam Tolu) Resin (U.S.)

MYROXYLON BALSAMUM (BALSAM TOLU) RESIN

CAS Nos.	EINECS No.
8011-89-0	
9000-64-0	232-550-4

Definition: Myroxylon Balsamum (Balsam Tolu) Resin is an oleoresin obtained from *Myroxylon balsamum. See "Regulatory and Ingredient Use Information," regarding the labeling names for botanical ingredients in Volume 1, Introduction, Part A.*

Information Sources: ARG, AUS, BEL, BP 1971, BPC, BRA, 21CFR172.510, 27CFR21.65, 27CFR21.151, EGY, FI, IND, ITA, MAR, MEX, MI-13(952), PF, POR, RIFM, ROM, TSCA, USAN, USD, USP XXIV, YUG

Chemical Class: Biological Products

Functions: Film Former; Fragrance Ingredient

Technical/Other Names:
Balsams, Myroxylon Balsamum

Balsams, Tolu
Balsam Tolu
Balsam Tolu Resin
Myroxylon Balsamum (EU)
Myroxylon Balsamum Balsam
Resin Tolu
Tolu, balsam, extract (Myroxylon spp.) (RIFM)
Tolu, balsam, gum (Myroxylon spp.) (RIFM)
Toluifera Balsamam Resin
Tolu Resin

Trade Name Mixture:
Cosflor Tree Balsam HGS (A & E Connock)

MYROXYLON PEREIRAE

Definition: *See "Regulatory and Ingredient Use Information," regarding EU labeling names for botanical ingredients in Volume 1, Introduction, Part A.*

Chemical Class: Biological Products

Technical/Other Names:
Myroxylon Pereirae (Balsam Peru) Oil (U.S.)
Myroxylon Pereirae (Balsam Peru) Resin (U.S.)

MYROXYLON PEREIRAE (BALSAM PERU) OIL

CAS No.	EINECS No.
8007-00-9	232-352-8

Definition: Myroxylon Pereirae (Balsam Peru) Oil is the volatile oil obtained from *Myroxylon pereirae. See "Regulatory and Ingredient Use Information," regarding the labeling names for botanical ingredients in Volume 1, Introduction, Part A.*

Information Source: RIFM

Chemical Class: Essential Oils

Function: Fragrance Ingredient

Technical/Other Names:
Balsam oil, Peru (Myroxylon pereirae Klotzsch) (RIFM)
Balsam, Peru (Myroxylon pereirae Klotzsch) (RIFM)
Balsam Peru Oil
Indian Balsam
Myroxylon Pereirae (EU)
Myroxylon Pereirae Oil
Oils, Balsam Peru (Myroxylon Pereirae)
Oils, Myroxylon Pereirae
Peru balsam absolute (RIFM)
Peru balsam anhydrol (RIFM)

Trade Name:
Balsam Peru Oil (JPM)

MYROXYLON PEREIRAE (BALSAM PERU) RESIN

CAS Nos.	EINECS No.
8007-00-9	232-352-8
8016-42-0	

Definition: Myroxylon Pereirae (Balsam Peru) Resin is an oleoresin obtained from *Myroxylon pereirae. See "Regulatory and Ingredient Use Information," regarding the labeling names for botanical ingredients in Volume 1, Introduction, Part A.*

Information Sources: ARG, AUS, BEL, BPC, BRA, 21CFR182.20, 21CFR582.20, CZE, DA, DDR, EGY, FCC, FI, FIN, HP, HUN, MAR, MI-13(951), NF XIII, PF, PN, POL, POR, RIFM, ROM, SNPF, TSCA, USD, YUG

Chemical Class: Biological Products

Functions: Film Former; Fragrance Ingredient

Technical/Other Names:
Balsam fir oil (Abies balsamea (L.) Mill.) (RIFM)
Balsam fir oleoresin (Abies balsamea (L.) Mill.) (RIFM)
Balsam of Peru
Balsam oil, Peru (Myroxylon pereirae Klotzsch) (RIFM)
Balsam Peru
Balsam, Peru (Myroxylon pereirae Klotzsch) (RIFM)
Balsam Peru Resin
Balsams, Myroxylon Pereirae
Indian Balsam
Myroxylon Pereirae (EU)
Myroxylon Pereirae Balsam
Myroxylon Pereirae Oleoresin
Peru balsam absolute (RIFM)
Peru balsam anhydrol (RIFM)
Peruvian Balsam

Trade Name:
AEC Peru Balsam (A & E Connock)

MYRRHIS ODORATA EXTRACT

CAS No.	EINECS No.
90064-24-7	290-085-2

Definition: Myrrhis Odorata Extract is an extract of *Myrrhis odorata. See "Regulatory and Ingredient Use Information," regarding the labeling names for botanical ingredients in Volume 1, Introduction, Part A.*

Chemical Class: Biological Products

Function: Fragrance Ingredient

Technical/Other Names:
Extract of Myrrhis Odorata
Extract of Sweet Cicely
Sweet Cicely Extract

Trade Name Mixture:
Sweet Cicely Extract (Cosmetic Developments)

MYRTRIMONIUM BROMIDE

CAS No.	EINECS No.
1119-97-7	214-291-9

Empirical Formula:
$C_{17}H_{38}N \cdot Br$

Definition: Myrtrimonium Bromide is the quaternary ammonium salt that conforms generally to the formula:

Information Sources: INN, MI-13(6362), TSCA

Chemical Class: Quaternary Ammonium Compounds

Functions: Antistatic Agent; Cosmetic Biocide

Technical/Other Names:
Ammonium, Trimethyltertradecyl-, Bromide
Myristyl Trimethyl Ammonium Bromide
Quaternium-13
1-Tetradecanaminium, N,N,N-Trimethyl-, Bromide
Tetradecytrimethylammonium Bromide
N,N,N-Trimethyl-1-Tetradecanaminium Bromide

Trade Names:
Mytab (Zeeland)
Rhodaquat M-214C/99 (Rhodia)
Sumquat 6110 (Zeeland)

MYRTUS COMMUNIS EXTRACT

CAS No.	EINECS No.
84082-67-7	282-012-8

Definition: Myrtus Communis Extract is an extract of the myrtle, *Myrtus communis*. See "Regulatory and Ingredient Use Information," regarding the labeling names for botanical ingredients in Volume 1, Introduction, Part A.

Information Source: 21CFR172.510

Chemical Class: Biological Products

Function: Skin-Conditioning Agent - Miscellaneous

Technical/Other Names:
Extract of Myrtle
Myrtle Extract
Myrtle (Myrtus Communis) Extract

Trade Name:
Myrtle Ecoconcentrate Natural (Robertet S.A.)

Trade Name Mixtures:
Actiphyte Myrtle (Active Organics)
Actiphyte Myrtle AL (Active Organics)
Actiphyte Myrtle BG50P (Active Organics)
Actiphyte Myrtle Lipo M (Active Organics)
Actiphyte Myrtle Lipo S (Active Organics)
Essentiaderm n.2 (Universal Flavors)
Extrait de Myrte MPE PG 40 (Yves Rocher)
Myrtle Ecoconcentrate (Robertet, Inc.)
Myrtle Extract HG (Provital/Centerchem)
Myrtle HS (Alban Muller)
Myrtus Communis Extract ies (IES LABO)
Phytelene of Myrtle EG 519 liquid (Indena SA)
Phytogreen 55 of Myrtle EXH 718 Liquid (Phytochim)
Vegetol Myrtle GR 288 Hydro (Gattefosse s.a.)
Vegetol Tp GR 052 Hydro (Gattefosse s.a.)
VT-146 Extract of Myrtle (Vege-Tech)

MYRTUS COMMUNIS LEAF EXTRACT

Definition: Myrtus Communis Leaf Extract is an extract of the leaves of *Myrtus communis*. See "Regulatory and Ingredient Use Information," regarding the labeling names for botanical ingredients in Volume 1, Introduction, Part A.

Chemical Class: Biological Products

Function: Skin-Conditioning Agent - Miscellaneous

Technical/Other Name:
Extract of Myrtus Communis Leaf

MYRTUS COMMUNIS LEAF WATER

Definition: Myrtus Communis Leaf Water is an aqueous solution of the steam distillate obtained from the leaves of *Myrtus communis*. See "Regulatory and Ingredient Use Information," regarding the labeling names for botanical ingredients in Volume 1, Introduction, Part A.

Chemical Class: Biological Products

Function: Fragrance Ingredient

Technical/Other Name:
Water, Myrtus Communis Leaf

Trade Name:
Myrtle Water (Alban Muller)

MYRTUS COMMUNIS OIL

CAS No.: 8008-46-6

Definition: Myrtus Communis Oil is a volatile oil obtained from *Myrtus communis*. See "Regulatory and Ingredient Use Information," regarding the labeling names for botanical ingredients in Volume 1, Introduction, Part A.

Information Source: RIFM

Chemical Class: Essential Oils

Functions: Fragrance Ingredient; Skin-Conditioning Agent - Miscellaneous

Technical/Other Names:
Myrtle Oi
Myrtle Oil
Myrtle oil (Myrtus communis L. (Fam. Myrtaceae)) (RIFM)
Oils, Myrtle

Trade Name Mixtures:
Essentiaderm N.20 (Universal Flavors)
Essentiaderm N.21 (Universal Flavors)
Sage Foot Spray (Alban Muller)

International Cosmetic Ingredient Dictionary and Handbook

- N -

NACRE POWDER

Definition: Nacre Powder is the dried powder obtained from Mother of Pearl (q.v.).

Chemical Class: Biological Products

Functions: Abrasive; Bulking Agent

Trade Name Mixtures:
Aragoline (Siera)
Biocrystal (Siera)

1,5-NAPHTHALENEDIOL

CAS No.
83-56-7

EINECS No.
201-487-4

Empirical Formula:
$C_{10}H_8O_2$

Definition: 1,5-Naphthalenediol is the bicyclic phenol that conforms to the formula:

See "Regulatory and Ingredient Use Information," for Colorants in Volume 1, Introduction, Part A.

Information Sources: CI 76625, EEC(III/2-32), TSCA

Chemical Classes: Color Additives - Hair; Phenols

Function: Hair Colorant

Reported Product Category: Hair Dyes and Colors (All Types Requiring Caution Statements and Patch Tests)

Technical/Other Names:
CI 76625
1,5-Dihydroxynaphthalene
1,6-Naphthalenediol

Trade Name:
Rodol 15N (Lowenstein)

Trade Name Mixtures:
Rodol Chestnut Brown 5/42A (Lowenstein)
Rodol Chestnut Brown 5/42B (Lowenstein)
Rodol Light Brown 5/02A (Lowenstein)
Rodol Light Brown 5/02B (Lowenstein)
Rodol Medium Brown 4/0-PW (Lowenstein)

1,7-NAPHTHALENEDIOL

CAS No.
575-38-2

EINECS No.
209-383-0

Empirical Formula:
$C_{10}H_8O_2$

Definition: 1,7-Naphthalenediol is the bicyclic phenol that conforms to the formula:

See "Regulatory and Ingredient Use Information," for Colorants in Volume 1, Introduction, Part A.

Information Sources: EEC(III/2-23), TSCA

Chemical Classes: Color Additives - Hair; Phenols

Function: Hair Colorant

Technical/Other Name:
1,7-Dihydroxynaphthalene

Trade Name:
1,7-Dihydroxynaphthalene (Schwarzkopf GmbH)

2,3-NAPHTHALENEDIOL

CAS No.
92-44-4

EINECS No.
202-156-7

Empirical Formula:
$C_{10}H_8O_2$

Definition: 2,3-Naphthalenediol is the bicyclic phenol that conforms to the formula:

See "Regulatory and Ingredient Use Information," for Colorants in Volume 1, Introduction, Part A.

Information Sources: CIR: [I] JACT-7(3)-1988, TSCA

Chemical Classes: Color Additives - Hair; Phenols

Function: Hair Colorant

Reported Product Category: Hair Dyes and Colors (All Types Requiring Caution Statements and Patch Tests)

Technical/Other Name:
2,3-Dihydroxynaphthalene

Trade Name:
Rodol 23N (Lowenstein)

2,7-NAPHTHALENEDIOL

CAS No.
582-17-2

EINECS No.
209-478-7

Empirical Formula:
$C_{10}H_8O_2$

Definition: 2,7-Naphthalenediol is the bicyclic phenol that conforms to the formula:

See "Regulatory and Ingredient Use Information," for Colorants in Volume 1, Introduction, Part A.

Information Sources: CI 76645, EEC(III/2-4), TSCA

Chemical Classes: Color Additives - Hair; Phenols

Function: Hair Colorant

Reported Product Category: Hair Dyes and Colors (All Types Requiring Caution Statements and Patch Tests)

Technical/Other Names:
CI 76645
2,7-Dihydroxynaphthalene

Trade Name:
Rodol 27N (Lowenstein)

1-NAPHTHOL

CAS No.
90-15-3

EINECS No.
201-969-4

Empirical Formula:
$C_{10}H_8O$

Definition: 1-Naphthol is the bicyclic phenol that conforms to the formula:

See "Regulatory and Ingredient Use Information," for Colorants in Volume 1, Introduction, Part A.

Information Sources: 21CFR74.1707a, CI 76605, CIR: [S] JACT-8(4)1989, EEC(III/1-16), MI-13(6409), TSCA

Chemical Classes: Color Additives - Hair; Phenols

Function: Hair Colorant

Reported Product Categories: Hair Dyes and Colors (All Types Requiring Caution Statements and Patch Tests); Hair Tints

Technical/Other Names:
CI 76605
1-Hydroxynaphthalene
1-Naphthalenol

The inclusion of any compound in the *Dictionary and Handbook* does not indicate that use of that substance as a cosmetic ingredient complies with the laws and regulations governing such use in the United States or any other country.

Alpha-Naphthol
1-Naphthyl Alcohol
Oxidation Base 33

Trade Names:
Colorex 1NAP (Chemical Compounds, Inc.)
Jarocol AN (James Robinson)
Rodol ERN (Lowenstein)

2-NAPHTHOL

CAS No.   **EINECS No.**
135-19-3   205-182-7

Empirical Formula:
$C_{10}H_8O$

Definition: 2-Naphthol is the bicyclic phenol that conforms to the formula:

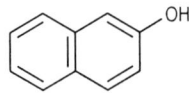

See "Regulatory and Ingredient Use Information," for Colorants in Volume 1, Introduction, Part A.

Information Sources: ARG, BRA, 21CFR74.1254, 21CFR74.1317, 21CFR74.1336, 21CFR176.200, 21CFR176.210, CZE, DDR, EEC(II-241), EGY, HUN, MAR, MI-13(6410), PF, PN, POR, RIFM, TSCA, YUG

Chemical Classes: Color Additives - Hair; Phenols

Functions: Fragrance Ingredient; Hair Colorant

Technical/Other Names:
Azoic Coupling Component 1
2-Hydroxynaphthalene
Isonaphthol
2-Naphthalenol
2-Naphthol (RIFM)
Beta-Naphthol

NARCISSUS JONQUILLA

Definition: *See "Regulatory and Ingredient Use Information," regarding EU labeling names for botanical ingredients in Volume 1, Introduction, Part A.*

Chemical Class: Biological Products

Technical/Other Name:
Narcissus Jonquilla Extract (U.S.)

NARCISSUS JONQUILLA EXTRACT

CAS No.   **EINECS No.**
90064-25-8   290-086-8

Definition: Narcissus Jonquilla Extract is an extract of the jonquil, *Narcissus jonquilla*. See "Regulatory and Ingredient Use Information," regarding the labeling names for botanical ingredients in Volume 1, Introduction, Part A.

Chemical Class: Biological Products

Function: Not Reported

Technical/Other Names:
Extract of Jonquil
Extract of Narcissus Jonquilla
Jonquil Extract
Jonquil (Narcissus Jonquilla) Extract
Narcissus Jonquilla (EU)

Trade Name Mixtures:
Phytelene of Junquil EG 433 liquid (Indena SA)
Phytogreen 55 of Junquil EXH 674 Liquid (Phytochim)

NARCISSUS POETICUS EXTRACT

Definition: Narcissus Poeticus Extract is an extract of the flowers of *Narcissus poeticus*.

Chemical Class: Biological Products

Function: Fragrance Ingredient

NARCISSUS POETICUS FLOWER WAX

Definition: Narcissus Poeticus Flower Wax is a wax obtained from the flower of *Narcissus poeticus*. See "Regulatory and Ingredient Use Information," regarding the labeling names for botanical ingredients in Volume 1, Introduction, Part A.

Chemical Class: Essential Oils

Function: Fragrance Ingredient

Technical/Other Name:
Wax, Narcissus Poeticus

Trade Names:
AEC Cire Essentielle De Fleurs De Narcisse (A & E Connock)
Cire Essentielle de fleurs de Narcisse (Bertin)
Narcissus Wax (Koster Keunen Holland)

Trade Name Mixtures:
AEC Cire Essentielle De NCR (A & E Connock)
Cire Essentielle NCR (Bertin)

NARCISSUS PSEUDO-NARCISSUS

Definition: *See "Regulatory and Ingredient Use Information," regarding EU labeling*

names for botanical ingredients in Volume 1, Introduction, Part A.

Chemical Class: Biological Products

Technical/Other Name:
Narcissus Pseudo-Narcissus (Daffodil) Flower Extract (U.S.)

NARCISSUS PSEUDO-NARCISSUS (DAFFODIL) FLOWER EXTRACT

Definition: Narcissus Pseudo-Narcissus (Daffodil) Flower Extract is an extract of the flowers of the daffodil, *Narcissus pseudo-narcissus*. See "Regulatory and Ingredient Use Information," regarding the labeling names for botanical ingredients in Volume 1, Introduction, Part A.

Chemical Class: Biological Products

Function: Not Reported

Technical/Other Names:
Daffodil Extract
Daffodil Flower Extract
Extract of Daffodil
Extract of Narcissus Pseudonarcissus
Extract of Naricssus Pseudo-Narcissus
Narcissus Pseudo-Narcissus (EU)
Narcissus Pseudonarcissus Extract
Narcissus Pseudo-Narcissus Extract

Trade Name Mixtures:
Daffodil Extract HS 3633 G (Grau)
Extrait De Narcisse ME 100 (Yves Rocher)

NARCISSUS TAZETTA BULB EXTRACT

Definition: Narcissus Tazetta Bulb Extract is an extract of the bulbs of *Narcissus tazetta*. See "Regulatory and Ingredient Use Information," regarding the labeling names for botanical ingredients in Volume 1, Introduction, Part A.

Chemical Class: Biological Products

Function: Not Reported

Technical/Other Name:
Extract of Narcissus Tazetta

Trade Name Mixture:
IBR-dormin (IBR)

NASTURTIUM OFFICINALE EXTRACT

CAS No.   **EINECS No.**
84775-70-2   283-899-4

JPN Translation:
オランダカラシエキス

Definition: Nasturtium Officinale Extract is an extract of the flowers and leaves of the watercress, *Nasturtium officinale. See "Regulatory and Ingredient Use Information," regarding the labeling names for botanical ingredients in Volume 1, Introduction, Part A.*

Information Sources: JCIC, JCLS, JSQI, RIFM

Chemical Class: Biological Products

Functions: Fragrance Ingredient; Skin-Conditioning Agent - Miscellaneous

Reported Product Categories: Tonics, Dressings, and Other Hair Grooming Aids; Hair Conditioners

Technical/Other Names:
Extract of Nasturtium Officinale
Extract of Watercress
Nasturtium officinale, ext. (RIFM)
Nasturtium Officinalis Extract
Watercress Extract
Watercress (Nasturtium Officinale) Extract

Trade Name Mixtures:
Actiphyte of Watercress BG50 (Active Organics)
Actiphyte of Watercress GL50 (Active Organics)
Actiphyte of Watercress Lipo S (Active Organics)
Actiphyte of Watercress PG50 (Active Organics)
Activated Botanicals P8-MIN AB 102 (Norjin)
Anti-Seborrhoeic Phytoamine Biocomplex (Alban Muller)
312 Blend for Greasy Hair HS (Alban Muller)
315 Blend for Greasy Skin Imperfections HS (Alban Muller)
245 Capilotonique HS (Alban Muller)
Cresson Liquid B (Ichimaru Pharcos)
Cresson Liquid E (Ichimaru Pharcos)
238 Emollient HS (Alban Muller)
Extrapone Watercress 2/032122 (Symrise)
GREASY HAIR (Greentech S.A)
Hair Treatment Complex 260 (Ennagram)
Hair Treatment Phytogreen Complex GXH 260 (Phytochim)
Herbaliquid Watercress Special (Crodarom)
Herbasec MPE Sebostat (Cosmetochem) (Cosmetochem International Ltd.)
Herbasol-Extract Watercress (Cosmetochem)
Herbasol-Extract watercress (Nasturtium) (Cosmetochem)
Nasturtium Officinale Extract ies (IES LABO)
210 Nutriderme HS (Alban Muller)
243 Nutriderme HS (Alban Muller)
Odraline (Silab)
OILY SKIN (Greentech S.A)

Oily Skins Complex 264 (Ennagram)
Oily Skins Phytogreen Complex GXH 264 (Phytochim)
Pharcolex BX32 (Ichimaru Pharcos)
Pharcolex BX47 (Ichimaru Pharcos)
Phytelene Complex EGX 232 (Indena SA)
Phytelene Complex EGX 247 (Indena SA)
Phytelene of Cresson EG 224 liquid (Indena SA)
Phytelene of Watercress EG 224 liquid (Indena SA)
Phytogreen 55 of Watercress EXH 643 Liquid (Phytochim)
Polyplant Hair (Provital/Centerchem)
Polyplant Oily Skin (Provital/Centerchem)
Prodhy Extract Cresson (Prod'Hyg)
Sederma Watercress (Sederma)
Vegebois of Watercress (CEP (Solabia))
2305 Vege-Plex Body Complex (Vege-Tech)
2325 Vege-Plex Body Complex (Vege-Tech)
2290 Vege-Plex Hair Complex (Vege-Tech)
Vege Plex VP#1335 (Vege-Tech)
Vegetol Cress LC 422 Hydro (Gattefosse s.a.)
VT-047 Extract of Watercress (Vege-Tech)
Watercress Extract (Maruzen Pharmaceuticals Co., Ltd.)
Watercress Extract BG (Maruzen Pharmaceuticals Co., Ltd.)
Watercress Extract G (Provital/Centerchem)
Water Cress Extract HS 2345 G (Grau)
Watercress Extract LA (Maruzen Pharmaceuticals Co., Ltd.)
Watercress Extract PG (Rahn)
Watercress HS (Alban Muller)
Watercress Tincture (Rahn)

NATTO GUM

CAS No.: 9079-02-1

JPN Translations:
セトキシメチルポリシロキサン
ダイズ発酵エキス

Definition: Natto Gum is a fermentation product of soy protein by *Bacillus natto.*

Information Source: MI-13(8802)

Chemical Class: Biological Polymers and their Derivatives

Function: Viscosity Increasing Agent - Aqueous

Technical/Other Name:
Gum, Natto

Trade Names:
Phyto Collage BD-II (Ichimaru Pharcos)
Soypol EX (Pacific)

Trade Name Mixtures:
FSP Liquid (BG) (Ichimaru Pharcos)

Phyto Collage B (Ichimaru Pharcos)
Phyto Collage E (Ichimaru Pharcos)
Phyto Collage PFE (Ichimaru Pharcos)
SP Liquid (Ichimaru Pharcos)

NATURAL RED 26

CAS No.	EINECS No.
36338-96-2	252-981-1

Empirical Formula:
$C_{43}H_{42}O_{22}$

Definition: Natural Red 26 is a natural pigment derived from the petals of *Carthamus tinctorius* (q.v.). *See "Regulatory and Ingredient Use Information," for Colorants in Volume 1, Introduction, Part A.*

Information Sources: MI-13(1882), MI-13 (1883)

Chemical Class: Color Additives - Miscellaneous

Function: Colorant

Technical/Other Names:
Carthamin
Carthamus
CI Natural Red 26
4-Cyclohexe-1,3-dione, 6-β-d-Glucopyranosyl-2-[[3-β-d-Glucopyranosyl-2,3,4-Trihydroxy-5-[3-(4-Hydroxyphenyl)-1-Oxo-2-Propenyl]-6-Oxo-1,4-Cyclohexadien-1-yl]Methylene]-5,6-Dihydroxy-4-[3-(4-Hydroxyphenyl)-1-Oxo-2-Propenyl]-

Trade Names:
Carthamus Red (Alps)
Liofresh Red CR-7000 (Toyo)

NEATSFOOT OIL

CAS No.	EINECS No.
8002-64-0	232-314-0

Definition: Neatsfoot Oil is a fixed oil obtained from cattle feet. *See "Regulatory and Ingredient Use Information," regarding use of EU Trivial names in Volume 1, Introduction, Part A.*

Information Sources: MI-13(6458), TSCA

Chemical Class: Fats and Oils

Function: Skin-Conditioning Agent - Occlusive

Trade Name:
Megachem NON (Megachem)

NELUMBIUM SPECIOSUM FLOWER EXTRACT

Definition: Nelumbium Speciosum Flower Extract is an extract of the flowers of the lotus, *Nelumbium speciosum*. See "*Regulatory and Ingredient Use Information*," *regarding the labeling names for botanical ingredients in Volume 1, Introduction, Part A.*

Chemical Class: Biological Products

Function: Skin-Conditioning Agent - Miscellaneous

Technical/Other Names:
Extract of Nelumbium Speciosum
Extract of Nelumbo Nucifera
Nelumbo Nucifera Extract

Trade Name:
Vegetol Lotus ME 218 Hydro (Gattefosse s.a.)

Trade Name Mixtures:
Campo Po Zhulin Hua Extract (Campo)
China Extract Po Zhu Lin Hua (E.U.K)
Chine Extract Nelumbium Nelumbo (Ennagram)
Extrait Hydroglycolique De Lotus-GT 10 (Greentech)
Hydralphatine Asiatique (Lanatech)
Lotus HPG Titrated (Alban Muller)
Lotus HS (Alban Muller)
Lotus LS (Alban Muller)
Lotus Phytolait (Alban Muller)

NELUMBIUM SPECIOSUM FLOWER OIL

Definition: Nelumbium Speciosum Flower Oil is the oil expressed from the flowers of *Nelumbium speciosum*. See "*Regulatory and Ingredient Use Information*," *regarding the labeling names for botanical ingredients in Volume 1, Introduction, Part A.*

Chemical Class: Fats and Oils

Function: Skin-Conditioning Agent - Miscellaneous

Technical/Other Name:
Oils, Nelumbium Speciosum

Trade Name Mixture:
Lotus Oil Special (Alban Muller)

NELUMBIUM SPECIOSUM FLOWER WATER

Definition: Nelumbium Speciosum Flower Water is an aqueous solution of the steam distillate obtained from the flowers of *Nelumbium speciosum*. See "*Regulatory and Ingredient Use Information*," *regarding the labeling names for botanical ingredients in Volume 1, Introduction, Part A.*

Chemical Class: Biological Products

Function: Fragrance Ingredient

Technical/Other Name:
Water, Nelumbium Speciosum

NELUMBO NUCIFERA FLOWER EXTRACT

CAS No.	EINECS No.
85085-51-4	285-379-2

Definition: Nelumbo Nucifera Flower Extract is an extract of the flower of *Nelumbo nucifera*. See "*Regulatory and Ingredient Use Information*," *regarding the labeling names for botanical ingredients in Volume 1, Introduction, Part A.*

Chemical Class: Biological Products

Function: Not Reported

Technical/Other Names:
Extract of Nelumbo Nucifera
Nelumbo Nucifera Extract

Trade Name Mixtures:
Actiphyte Sacred Lotus BG50P (Active Organics)
Actiphyte Sacred Lotus (Active Organics)
Actiphyte Sacred Lotus Lipo S (Active Organics)
Actiphyte Sacred Lotus Seed (Active Organics)
Glycolysat BG of Lotus Flowers (CEP (Solabia))
Hasu Liquid B (Ichimaru Pharcos)
Herbasol Extract Lotus (Cosmetochem) (Cosmetochem International Ltd.)
Lotus Flowers Milk (CEP (Solabia))
Oleat of Lotus Flowers (CEP (Solabia))
Renshu Extract A (Nonogawa)
Vegetol Lotus ME 169 Hydro (Gattefosse s.a.)

NELUMBO NUCIFERA FLOWER WATER

Definition: Nelumbo Nucifera Flower Water is an aqueous solution of the steam distillate obtained from the flowers of *Nelumbo nucifera*. See "*Regulatory and Ingredient Use Information*," *regarding the labeling names for botanical ingredients in Volume 1, Introduction, Part A.*

Chemical Class: Biological Products

Function: Fragrance Ingredient

Technical/Other Name:
Water, Nelumbo Nucifera Flower

Trade Name:
Vegebios of Lotus Flowers (CEP (Solabia))

NELUMBO NUCIFERA GERM EXTRACT

Definition: Nelumbo Nucifera Germ Extract is an extract of the germ of

Nelumbo nucifera. See "*Regulatory and Ingredient Use Information*," *regarding the labeling names for botanical ingredients in Volume 1, Introduction, Part A.*

Chemical Class: Biological Products

Function: Skin-Conditioning Agent - Humectant

Technical/Other Name:
Extract of Nelumbo Nucifera Germ

Trade Name Mixture:
Lotus Germ Extract BG (Maruzen Pharmaceuticals Co., Ltd.)

NELUMBO NUCIFERA ROOT WATER

Definition: Nelumbo Nucifera Root Water is an aqueous solution of the steam distillate obtained from the roots of *Nelumbo nucifera*. See "*Regulatory and Ingredient Use Information*," *regarding the labeling names for botanical ingredients in Volume 1, Introduction, Part A.*

Chemical Class: Biological Products

Technical/Other Name:
Extract of Nelumbo Nucifera Root

Trade Name:
Extrait Originel Lotus (Gattefosse s.a.)

NEOHESPERIDIN DIHYDROCHALCONE

CAS No.	EINECS No.
20702-77-6	243-978-6

Empirical Formula:
$C_{28}H_{36}O_{15}$

Definition: Neohesperidin Dihydrochalcone is the organic compound that conforms to the formula:

Information Sources: MI-13(6480), RIFM

Chemical Classes: Carbohydrates; Ethers; Ketones

Functions: Flavoring Agent; Fragrance Ingredient

Technical/Other Names:
Neohesperidin dihydrochalcone (RIFM)
1-Propanone, 1-[4[[2-O-(6-Deoxy-α-L-Mannopyranosyl)-β-D-Glucopyranosyl]-Oxy]-2,6- Dihydroxyphenyl]-3-(3-Hydroxy-4-Methoxyphenyl)-

Trade Name:
Citrosa (Exquim)

NEOPENTYL GLYCOL

CAS No.	EINECS No.
126-30-7	204-781-0

Empirical Formula:
$C_5H_{12}O_2$

Definition: Neopentyl Glycol is organic compound that conforms to the formula:

$$\begin{array}{c} CH_3 \\ | \\ HOCH_2CCH_2OH \\ | \\ CH_3 \end{array}$$

Information Sources: MI-13(6317), MI-13 (6486)

Chemical Class: Alcohols

Functions: Plasticizer; Solvent

Technical/Other Names:
2,2-Dimethyl-1,3-Dihydroxypropane
Dimethylolpropane
2,2-Dimethyltrimethylene Glycol
Neopentanediol
Neopentylene Glycol
1,3-Propanediol, 2,2-Dimethyl-

Trade Name:
Neopentylglycol Ecailles (Jouvance)

NEOPENTYL GLYCOL DICAPRATE

CAS No.	EINECS No.
27841-06-1	248-688-3

JPN Translation:
ジカプリン酸ネオペンチルグリコール

Empirical Formula:
$C_{25}H_{48}O_4$

Definition: Neopentyl Glycol Dicaprate is the diester of neopentyl glycol and decanoic acid that conforms to the formula:

$$\begin{array}{ccc} O & CH_3 & O \\ || & | & || \\ CH_3(CH_2)_8C-OCH_2CCH_2O-C(CH_2)_8CH_3 \\ & | & \\ & CH_3 & \end{array}$$

Information Sources: JCIC, JCLS, JSQI

Chemical Class: Esters

Functions: Skin-Conditioning Agent - Emollient; Viscosity Increasing Agent - Nonaqueous

Reported Product Categories: Lipsticks; Body and Hand Preparations (Excluding Shaving Preparations); Eye Makeup Removers

Technical/Other Names:
Decanoic Acid, 2,2-Dimethyl-1,3-Propane-diyl Diester
Decanoic Acid, 2,2-Dimethyltrimethylene Ester
2,2-Dimethyl Propanediol Dicaprate
2,2-Dimethylpropane-1,3-diol Didecanoate
Neopentyl Glycol Didecanoate

Trade Names:
AEC Neopentyl Glycol Dicaprate (A & E Connock)
Dermol NGDC (Alzo)
Estemol N-01 (Nisshin OilliO)
Schercemol NGDC (Scher)
Technol 210 (Technonet)

NEOPENTYL GLYCOL DICAPRYLATE/ DICAPRATE

CAS No.	EINECS No.
70693-32-2	274-764-0

JPN Translation:
ジ（カプリル / カプリン酸）ネオペンチルグリコール

Definition: Neopentyl Glycol Dicaprylate/Dicaprate is the diester of neopentyl glycol and a blend of caprylic and capric acids.

Chemical Class: Esters

Functions: Skin-Conditioning Agent - Emollient; Viscosity Increasing Agent - Nonaqueous

Reported Product Categories: Face and Neck Preparations (Excluding Shaving Preparations); Foundations; Blushers (All types); Face Powders; Moisturizing Preparations

Technical/Other Names:
Decanoic Acid, Mixed Esters with Neopentyl Glycol and Octanoic Acid
Octanoic Acid, Mixed Esters with Neopentyl Glycol and Decanoic Acid

Trade Names:
DUB 810 NPG (Stearinerie Dubois Fils)
Hatcol 5191 (Hatco)
Liponate NPGC-2 (Lipo)

Trade Name Mixtures:
Lipovol MOS-70 (Lipo)
Lipovol MOS-350 (Lipo)

NEOPENTYL GLYCOL DICAPRYLATE/ DIPELARGONATE/DICAPRATE

Definition: Neopentyl Glycol Dicaprylate/Dipelargonate/Dicaprate is the diester of neopentyl glycol and a blend of caprylic, pelargonic and capric acids.

Chemical Class: Esters

Functions: Skin-Conditioning Agent - Emollient; Viscosity Increasing Agent - Nonaqueous

NEOPENTYL GLYCOL DIETHYLHEXANOATE

CAS No.	EINECS No.
28510-23-8	249-060-1

JPN Translation:
ジオクタン酸ネオペンチルグリコール

Empirical Formula:
$C_{21}H_{40}O_4$

Definition: Neopentyl Glycol Diethylhexanoate is the diester of neopentyl glycol and 2-ethylhexanoic acid. It conforms to the formula:

$$\begin{array}{ccccc} & O & CH_3 & O & \\ & || & | & || & \\ CH_3(CH_2)_3CHC-OCH_2CCH_2O-CCH(CH_2)_3CH_3 \\ & | & | & | & \\ & CH_2CH_3 & CH_3 & CH_2CH_3 & \end{array}$$

Information Sources: JCIC, JCLS, JSQI, TSCA

Chemical Class: Esters

Functions: Skin-Conditioning Agent - Emollient; Viscosity Increasing Agent - Nonaqueous

Reported Product Categories: Cleansing Products (Cold Creams, Cleansing Lotions, Liquids and Pads); Eye Makeup Preparations, Misc.; Eye Shadows; Eyebrow Pencils; Face and Neck Preparations (Excluding Shaving Preparations); Face Powders; Foundations; Hair Conditioners; Hair Preparations (Non-coloring), Misc.; Indoor Tanning Preparations; Lipsticks; Makeup Bases; Makeup Preparations (Not eye), Misc.; Moisturizing Preparations; Night Skin Care Preparations

Technical/Other Names:
2,2-Dimethyl Propanediol Di(2-Ethyl-hexanoate)
2,2-Dimethyl-1,3-Propanediyl 2-Ethyl-hexanoate
Hexanoic Acid, 2-Ethyl-, 2,2-Dimethyl-1,3-Propanediyl Ester
Hexanoic Acid, 2-ethyl-,2,2-dimethyltri-methylene Ester
Neopentyl Glycol Dioctanoate
Octanoic Acid, 2,2-Dimethyl-1,3-Propane-diyl Diester

Trade Names:
AEC Neopentyl Glycol Diethylhexanoate (A & E Connock)

Cosmol 525 (Nisshin OilliO)
DUB DONPG (Stearinerie Dubois Fils)
KAK NDO (Kokyu Alcohol)
Puresyn 2E7 (ExxonMobil/Synthetics)
Salacos 525 (Nisshin OilliO)
Schercemol NGDO (Scher)

Trade Name Mixtures:
Minno 21 (Bernel)
Minno 41 (Bernel)

NEOPENTYL GLYCOL DIHEPTANOATE

Empirical Formula:
$C_{19}H_{36}O_4$

Definition: Neopentyl Glycol Diheptanoate is the diester of neopentyl glycol and heptanoic acid. It conforms to the formula:

$$CH_3(CH_2)_5C - OCH_2CCH_2O - C(CH_2)_5CH_3$$

Chemical Class: Esters

Functions: Skin-Conditioning Agent - Emollient; Viscosity Increasing Agent - Nonaqueous

Technical/Other Names:
Heptanoic Acid, 2,2-Dimethyl-1,3-Propane-diyl Diester
Heptanoic Acid 2,2-Dimethyltrimethylene Ester
1,3-Propanediol, 2,2-dimethyl-, Diheptanoate

Trade Names:
AEC Neopentyl Glycol Diheptanoate (A & E Connock)
DUB DNPG (Stearinerie Dubois Fils)
PureSyn 2E5 (ExxonMobil/Synthetics)

NEOPENTYL GLYCOL DIISOSTEARATE

CAS No.: 109884-54-0

Empirical Formula:
$C_{41}H_{80}O_4$

Definition: Neopentyl Glycol Diisostearate is the diester of neopentyl glycol and isostearic acid. It conforms to the formula:

$$C_{17}H_{35}C - OCH_2CCH_2O - CC_{17}H_{35}$$

Chemical Class: Esters

Functions: Skin-Conditioning Agent - Occlusive; Viscosity Increasing Agent - Nonaqueous

Technical/Other Name:
Isostearic Acid, 2,2-Dimethyl-1,3-Propane-diyl Diester

Trade Name Mixtures:
Minno 21 (Bernel)
Minno 41 (Bernel)

NEOPENTYL GLYCOL DILAURATE

CAS No.	EINECS No.
10525-39-0	234-081-0

Empirical Formula:
$C_{29}H_{56}O_4$

Definition: Neopentyl Glycol Dilaurate is the diester of neopentyl glycol and lauric acid. It conforms to the formula:

$$CH_3(CH_2)_{10}C - OCH_2CCH_2O - C(CH_2)_{10}CH_3$$

Chemical Class: Esters

Functions: Skin-Conditioning Agent - Emollient; Viscosity Increasing Agent - Nonaqueous

Technical/Other Names:
Isoctadecanoic Acid, 2,2-dimethyl-1,3-propanediyl Ester
Lauric Acid, 2,2-Dimethyl-1,3-Propanediyl Diester
Neopentylene Dilaurate

Trade Name:
Schercemol NGDL (Scher)

NEORUSCOGENIN

CAS Nos.	EINECS No.
17676-33-4	241-660-1
35882-30-5	

Empirical Formula:
$C_{27}H_{40}O_4$

Definition: Neoruscogenin is the sterol that conforms to the formula:

Chemical Class: Sterols

Function: Skin-Conditioning Agent - Miscellaneous

Technical/Other Names:
25(27)-Dehydroruscogenin
Spirosta-5,25(27)-Diene-1,3-Diol, (1β,3β)-

Trade Name Mixtures:
Plantactiv Ruscus (Cognis Deutschland)
Ruscogenines Vinyals (Vinyals)
Ruscogenins (Indena SpA)

NEOTAME

CAS No.: 165450-17-9

Empirical Formula:
$C_{20}H_{30}N_2O_5$

Definition: Neotame is the organic compound that conforms to the formula:

$$HOOCCH_2CHC - NHCHC - OCH_3$$
$$(CH_3)_3CCH_2CH_2NH \qquad CH_2$$

Information Source: MI-13(6493)

Chemical Classes: Amides; Esters

Function: Flavoring Agent

Technical/Other Name:
L-Phenylalanine, N--(3,3-Dimethylbutyl)-L-α-Aspartyl-, 2-Methyl Ester

Trade Name:
Neotame (The NutraSweet)

NEPETA CATARIA EXTRACT

CAS No.	EINECS No.
84929-35-1	284-520-5

Definition: Nepeta Cataria Extract is an extract of the herb of the catnip, *Nepeta cataria*. See "Regulatory and Ingredient Use Information," regarding the labeling names for botanical ingredients in Volume 1, Introduction, Part A.

Information Source: RIFM

Chemical Class: Biological Products

Functions: Fragrance Ingredient; Not Reported

Technical/Other Names:
Catnip Extract
Catnip (Nepeta Cataria) Extract
Extract of Catnip
Extract of Nepeta Cataria
Nepeta cataria, ext. (RIFM)

Trade Name Mixtures:
Actiphyte of Catnip BG50 (Active Organics)
Actiphyte of Catnip GL50 (Active Organics)
Actiphyte of Catnip Lipo S (Active Organics)

Actiphyte of Catnip PG50 (Active Organics)
Catnip Herb Extract HS 2805 G (Grau)

NEPHELIUM LAPPACEUM EXTRACT

Definition: Nephelium Lappaceum Extract is an extract of the branches, leaves and fruit of *Nephelium lappaceum. See "Regulatory and Ingredient Use Information," regarding the labeling names for botanical ingredients in Volume 1, Introduction, Part A.*

Chemical Class: Biological Products

Function: Skin-Conditioning Agent - Humectant

Technical/Other Name:
Extract of Nephelium Lappaceum

Trade Name Mixtures:
Rambutan Extract (Nonogawa)
Rambutan Extract 2 (Nonogawa)

NEPHELIUM LAPPACEUM PEEL EXTRACT

Definition: Nephelium Lappaceum Peel Extract is an extract of the peel of *Nephelium lappaceum. See "Regulatory and Ingredient Use Information," regarding the labeling names for botanical ingredients in Volume 1, Introduction, Part A.*

Chemical Class: Biological Products

Function: Not Reported

Technical/Other Name:
Extract of Nephelium Lappaceum

NEPHELIUM LONGANA SEED EXTRACT

Definition: Nephelium Longana Seed Extract is an extract of the seeds of *Nephelium longana. See "Regulatory and Ingredient Use Information," regarding the labeling names for botanical ingredients in Volume 1, Introduction, Part A.*

Chemical Class: Biological Products

Function: Skin-Conditioning Agent - Miscellaneous

Technical/Other Names:
Dimocarpus Longan
Extract of Nephelium Longana Seed

Trade Name Mixture:
Agetect (Katakura)

NEPHRITE POWDER

CAS No.: 12174-03-7

Definition: Nephrite Powder is a fibrous crystalline aggregate consisting chiefly of:

$$CaMg_5(OH)_2(Si_4O_{11})_2$$

Chemical Class: Inorganics

Functions: Anticaking Agent; Bulking Agent; Exfoliant; Opacifying Agent; Slip Modifier

Trade Name:
Jade Powder (Kiosi)

NERIUM OLEANDER

Definition: *See "Regulatory and Ingredient Use Information," regarding EU labeling names for botanical ingredients in Volume 1, Introduction, Part A.*

Chemical Class: Biological Products

Technical/Other Name:
Nerium Oleander (Oleander) Leaf Extract (U.S.)

NERIUM OLEANDER (OLEANDER) LEAF EXTRACT

CAS No. 84929-39-5   **EINECS No.** 284-522-6

Definition: Nerium Oleander (Oleander) Leaf Extract is an extract of the leaves of the oleander, *Nerium oleander. See "Regulatory and Ingredient Use Information," regarding the labeling names for botanical ingredients in Volume 1, Introduction, Part A.*

Chemical Class: Biological Products

Function: Not Reported

Technical/Other Names:
Extract of Nerium Oleander
Extract of Oleander Leaves
Nerium Oleander (EU)
Nerium Oleander Leaf Extract
Oleander Extract
Oleander Leaf Extract

Trade Name Mixture:
Rosebay Leaves Extract HS 2715 G (Grau)

NEURAL EXTRACT

Definition: Neural Extract is an extract of animal neural tissue.

Chemical Class: Biological Products

Function: Not Reported

Technical/Other Name:
Extract of Neural Tissue

NIACIN

CAS No. 59-67-6   **EINECS No.** 200-441-0

JPN Translation:
ナイアシン

Empirical Formula:
$C_6H_5NO_2$

Definition: Niacin is the heterocyclic aromatic compound that conforms to the formula:

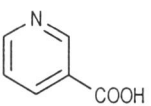

Information Sources: AUS, BAN, BP, BPC, BRA, 21CFR101.9, 21CFR102.23, 21CFR104.20, 21CFR104.47, 21CFR107.10, 21CFR107.100, 21CFR135.115, 21CFR136.115, 21CFR137, 21CFR137.165, 21CFR137.185, 21CFR139, 21CFR184.1530, 21CFR184.1535, 21CFR310.545, 21CFR582.5530, 21CFR582.5535, CIR: [S], DA, DDR, EGY, FCC, FIN, HUN, IND, INN, ITA, JAN, JCIC, JCLS, JP, MAR, MEX, MI-13(6552), NED, PN, POR, ROM, TSCA, USAN, USD, USP XXIV, WHO, YUG

Chemical Classes: Carboxylic Acids; Heterocyclic Compounds

Functions: Hair Conditioning Agent; Skin-Conditioning Agent - Miscellaneous

Technical/Other Names:
3-Carboxypyridine
Nicotinic Acid
3-Pyridinecarboxylic Acid

Trade Name Mixtures:
ANTI-WRINKLE (Greentech S.A)
Fortified Yeast T-6361 (Universal Foods)

NIACINAMIDE

CAS No. 98-92-0   **EINECS No.** 202-713-4

JPN Translation:
ナイアシンアミド

Empirical Formula:
$C_6H_6N_2O$

Definition: Niacinamide is the heterocyclic aromatic amide that conforms to the formula:

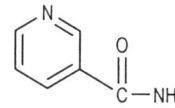

Information Sources: ARG, AUS, BAN, BP, BPC, BRA, 21CFR172.315, 21CFR184.1535, CIR: [S], CZE, DA, DDR, EGY, EP, FCC, FIN, HUN, IND, INN, ITA, JAN, JCLS, JSCI, MAR, MEX, MI-13(6550), PF, PN, POR, ROM, TSCA, USAN, USD, USP XXIV, WHO

Chemical Classes: Amides; Heterocyclic Compounds

Functions: Hair Conditioning Agent; Skin-Conditioning Agent - Miscellaneous

Reported Product Categories: Moisturizing Preparations; Skin Care Preparations, Misc.; Bath Oils, Tablets, and Salts; Night Skin Care Preparations; Tonics, Dressings, and Other Hair Grooming Aids; Cleansing Products (Cold Creams, Cleansing Lotions, Liquids and Pads); Shampoos (Non-coloring); Body and Hand Preparations (Excluding Shaving Preparations); Hair Preparations (Non-coloring), Misc.; Face and Neck Preparations (Excluding Shaving Preparations); Hair Conditioners

Technical/Other Names:
m-(Aminocarbonyl)pyridine
3-Aminopyridine
3-Carbamoylpyridine
Nicotinamide
Nicotninic Acid Amide
3-Pyridinecarboxamide
Vitamin B3

Trade Names:
Niacinamide, USP #0409634 (Roche)
Nicotinamide (Merck KGaA)

Trade Name Mixtures:
Aquaderm (Crodarom)
Ascorbyl Niacinamide (Bioderm Research)
Asebiol LS 2539 BT2 (Laboratoires Sero-
   biologiques)
CYTOBIOL ULMAIRE (Libiol)
Elespher Vitaplex Hydro (Laboratoires
   Serobiologiques)
Hair Complex NOVA (Crodarom)
Lactil (Degussa Care Specialties)
Lipoderma LAC (Lipo)
Liposomes Trichogen Veg (Laboratoires
   Serobiologiques)
Marine Plasma Extract (Arch Personal
   Care Products)
Neo-Derma Vitamin Complex, Water
   Soluble (Crodarom)
Niacin Lipoic Acid (Bioderm Research)
Niacin Salicylic Acid (Bioderm Research)
Sebaryl FL (Laboratoires Serobiologiques)
Sveltonyl (Laboratoires Serobiologiques)
Trichogen Veg (Laboratoires Sero-
   biologiques)
Unilactamin L-17 (Induchem)

NIACINAMIDE GLYCOLATE

Empirical Formula:
$C_6H_6N_2O \cdot C_2H_4O_3$

Definition: Niacinamide Glycolate is the salt of niacinamide and glycolic acid that conforms to the formula:

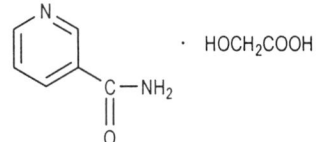

Chemical Classes: Amides; Heterocyclic Compounds

Function: Skin-Conditioning Agent - Miscellaneous

Trade Name:
Niacinamide Glycolate (Bioderm Research)

NIACINAMIDE HYDROXYCITRATE

Empirical Formula:
$C_6H_6N_2O \cdot C_6H_8O_8$

Definition: Niacinamide Hydroxycitrate is the salt of niacinamide and hydroxycitric acid that conforms to the formula:

Chemical Classes: Amides; Heterocyclic Compounds

Function: Skin-Conditioning Agent - Miscellaneous

Trade Name:
Niacinamide Hydroxycitrate (Bioderm Research)

NIACINAMIDE LACTATE

Empirical Formula:
$C_6H_6N_2O \cdot C_3H_6O_3$

Definition: Niacinamide Lactate is the salt of niacinamide and lactic acid that conforms to the formula:

See Reported Ingredient Functions-The Cosmetic Drug Distinction, in Regulatory and Ingredient Use Information, Volume I, Part A.

Chemical Classes: Amides; Heterocyclic Compounds

Functions: Skin Bleaching Agent; Skin-Conditioning Agent - Miscellaneous

Trade Name:
Niacinamide Lactate (Bioderm Research)

NIACINAMIDE MALATE

Empirical Formula:
$C_6H_6N_2O \cdot C_4H_6O_5$

Definition: Niacinamide Malate is the salt of niacinamide and malic acid that conforms to the formula:

Chemical Classes: Amides; Heterocyclic Compounds

Function: Skin-Conditioning Agent - Miscellaneous

Trade Name:
Niacinamide Malate (Bioderm Research)

NIACINAMIDE MANDELATE

Empirical Formula:
$C_6H_6N_2O \cdot C_8H_8O_3$

Definition: Niacinamide Mandelate is the salt of niacinamide and mandelic acid that conforms to the formula:

Chemical Classes: Amides; Heterocyclic Compounds

Function: Skin-Conditioning Agent - Miscellaneous

Trade Name:
Niacinamide Mandelate (Bioderm Research)

NICKEL ACETYLMETHIONATE

Definition: Nickel Acetylmethionate is the nickel salt of acetylmethionine.

Chemical Classes: Amino Acids; Thio Compounds

Function: Not Reported

NICKEL GLUCONATE

Empirical Formula:
$C_{12}H_{22}O_{14} \cdot Ni$

Definition: Nickel Gluconate is the nickel salt of gluconic acid. It conforms to the formula:

$$\left[CH_2OH(CHOH)_4COO^- \right]_2 Ni^{+2}$$

Chemical Classes: Organic Salts; Polyols

Function: Not Reported

Trade Name:
Givobio GNi (SEPPIC)

NICOTIANA TABACUM

Definition: *See "Regulatory and Ingredient Use Information," regarding EU labeling names for botanical ingredients in Volume 1, Introduction, Part A.*

Chemical Class: Biological Products

Technical/Other Name:
Nicotiana Tabacum (Tobacco) Leaf Extract (U.S.)

NICOTIANA TABACUM (TOBACCO) LEAF EXTRACT

CAS No.	EINECS No.
84961-66-0	284-656-5

Definition: Nicotiana Tabacum (Tobacco) Leaf Extract is an extract of the leaves of the tobacco, *Nicotiana tabacum. See "Regulatory and Ingredient Use Information," regarding the labeling names for botanical ingredients in Volume 1, Introduction, Part A.*

Information Source: RIFM

Chemical Class: Biological Products

Functions: Fragrance Ingredient; Not Reported

Technical/Other Names:
Extract of Nicotiana Tabacum Leaf
Extract of Tobacco Leaf
Nicotiana Tabacum (EU)
Nicotiana Tabacum Extract
Tobacco, ext. (RIFM)
Tobacco Extract
Tobacco Leaf Extract

Trade Name Mixtures:
Actiphyte of Tobacco BG50 (Active Organics)
Actiphyte of Tobacco GL50 (Active Organics)
Actiphyte of Tobacco Lipo S (Active Organics)
Actiphyte of Tobacco PG50 (Active Organics)
Nicotiana Tabacum (Tobacco) Leaf Extract ies (IES LABO)

NICOTINAMIDE ADENINE DINUCLEOTIDE

CAS No.: 53-84-9

Definition: Nicotinamide Adenine Dinucleotide is the organic compound that conforms to the formula:

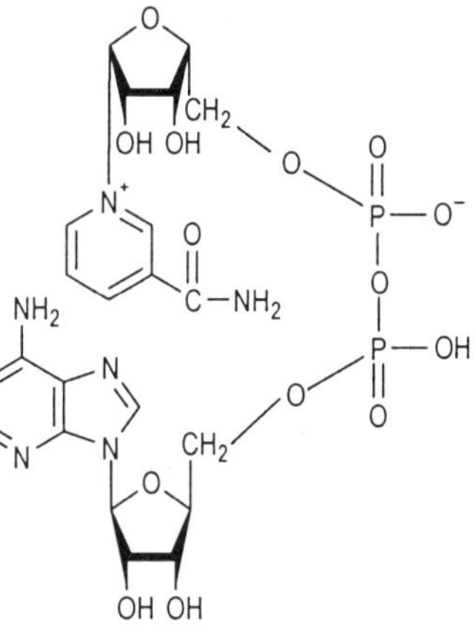

Information Source: MI-13(6370)

Chemical Classes: Amides; Phosphorus Compounds

Function: Skin-Conditioning Agent - Miscellaneous

Technical/Other Name:
NAD

Trade Name Mixture:
Sunactyl LS 9610 (Laboratoires Serobiologiques)

NICOTINYL ALCOHOL

CAS No.	EINECS No.
100-55-0	202-864-6

Empirical Formula:
C_6H_7NO

Definition: Nicotinyl Alcohol is the heterocyclic compound that conforms to the formula:

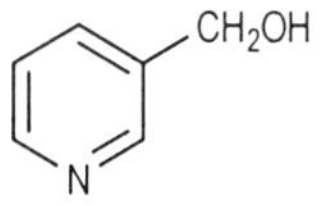

Information Sources: MI-13(6554), TSCA

Chemical Classes: Alcohols; Heterocyclic Compounds

Function: Skin-Conditioning Agent - Miscellaneous

Technical/Other Names:
3-(Hydroxymethyl)pyridine
Nictinic Alcohol
β-Picolyl Alcohol
Pyridine-3-Carbinol
3-Pyridinemethanol

Trade Name:
3-pyridylmethanol (Merck KGaA)

NICOTINYL TARTRATE

CAS No.	EINECS No.
6164-87-0	228-199-1

Empirical Formula:
$C_6H_7NO \cdot C_4H_6O_6$

Definition: Nicotinyl Tartrate is the organic salt that conforms to the formula:

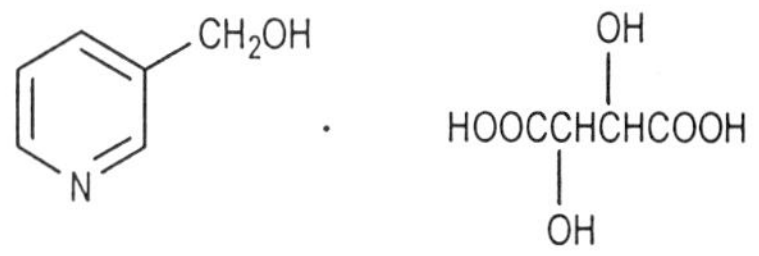

Chemical Classes: Alcohols; Heterocyclic Compounds; Organic Salts

Function: Skin-Conditioning Agent - Miscellaneous

Trade Name:
3-Pyridylmethanol-(RR)-Hydrogentartrate (Merck KGaA)

NIGELLA SATIVA SEED EXTRACT

Definition: Nigella Sativa Seed Extract is an extract of the seeds of *Nigella sativa. See "Regulatory and Ingredient Use Information," regarding the labeling names for botanical ingredients in Volume 1, Introduction, Part A.*

Chemical Class: Biological Products

Function: Skin-Conditioning Agent - Miscellaneous

Technical/Other Name:
Extract of Nigella Sativa Seed

Trade Name Mixture:
Extrait de Nigella Sativa (Greentech)

NIGELLA SATIVA SEED OIL

Definition: Nigella Sativa Seed Oil is the oil expressed from the seeds of *Nigella sativa. See "Regulatory and Ingredient Use Information," regarding the labeling names for botanical ingredients in Volume 1, Introduction, Part A.*

Chemical Class: Fats and Oils

Function: Skin-Conditioning Agent - Occlusive

Technical/Other Name:
Oils, Nigella Sativa

Trade Names:
  AEC Black Cumin Seed Oil (A & E
    Connock)
  Black Caraway Oil 3661 G (Grau)
  Black Cumin Oil, Pressed (Henry Lamotte)

Trade Name Mixture:
  Matipure (Advanced Beauty)

NINDOU EKISU

JPN Translation:
ニンドウエキス

Definition: Nindou Ekisu is an extract of
the leaves and stems of the honeysuckle,
Lonicera japonica or other related species
of the family *Caprifoliacea. See "Regulatory
and Ingredient Use Information," regarding
use of Japan Trivial names in Volume 1,
Introduction, Part A.*

Chemical Class: Biological Products

Functions: Fragrance Ingredient; Skin-
Conditioning Agent - Miscellaneous

Technical/Other Name:
  Lonicera Japonica (Honeysuckle) Leaf
  Extract (U.S.)

NINHYDRIN

CAS No.	EINECS No.
485-47-2	207-618-1

Empirical Formula:
  $C_9H_6O_4$

Definition: Ninhydrin is the organic
compound that conforms to the formula:

*See "Regulatory and Ingredient Use Infor-
mation," for Colorants in Volume 1, Introduction, Part A.*

Information Sources: 21CFR444.20a, MI-
13(6581), MI-13(6582), TSCA

Chemical Classes: Alcohols; Color
Additives - Miscellaneous; Ketones

Function: Colorant

Technical/Other Names:
  2,2-Dihydroxy-1,3-Indandione
  1H-Indene-1,3(2H)-Dione, 2,2-Dihydroxy-
  Indan-1,2,3-Trione

NISIN

CAS No.	EINECS No.
1414-45-5	215-807-5

Definition: Nisin Is a polypeptide
produced by the fermentation of
Lactococcus lactis.

Information Sources: 21CFR133.179,
21CFR184.1538, MI-13(6591), MI-13(6592)

Chemical Class: Proteins

Function: Preservative

Trade Name:
  Ambicin (AMBI)

3-NITRO-4-AMINOPHENOXYETHANOL

CAS No.: 50982-74-6

Empirical Formula:
  $C_8H_{10}N_2O_4$

Definition: 3-Nitro-4-Aminophenoxyethanol
is the substituted aromatic ether alcohol that
conforms to the formula:

*See "Regulatory and Ingredient Use Infor-
mation," for Colorants in Volume 1, Introduction, Part A.*

Chemical Classes: Alcohols; Amines; Color
Additives - Hair

Function: Hair Colorant

Technical/Other Name:
  Ethanol, 2-(4-Amino-3-Nitrophenoxy)-

NITROCELLULOSE

CAS No.: 9004-70-0

JPN Translation:
ニトロセルロース

Definition: Nitrocellulose is a cellulose
derivative that conforms generally to the
formula:

$$C_{12}H_{16}N_4O_{18}$$

Information Sources: 21CFR175.105,
21CFR175.300, 21CFR176.170,
21CFR177.1200, 21CFR179.45,
21CFR181.22, 21CFR181.30, CTFA D,
INN, JAN, JCLS, JSCI, MI-13(8101), TSCA,
USAN, USP XXIII

Chemical Classes: Biological Polymers and
their Derivatives; Carbohydrates

Functions: Film Former; Suspending Agent
- Nonsurfactant

Reported Product Categories: Nail Polish
and Enamels; Basecoats and Undercoats;
Manicuring Preparations, Misc.

Technical/Other Names:
  Cellulose, Nitrate
  Guncotton
  Nitrocellulose Solution

Trade Name:
  Hercules Nitrocellulose (Hercules/Aqualon)

Trade Name Mixtures:
  Biju BNT (Engelhard Corp.)
  Biju BTF-WD (Engelhard Corp.)
  Biju BTF-XD (Engelhard Corp.)
  BIJU BXD (Engelhard Corp.)
  Biju Ultra UNT (Engelhard Corp.)
  Biju Ultra UTF-WD (Engelhard Corp.)
  Biju Ultra UTF-XD (Engelhard Corp.)
  BIJU Ultra UXD (Engelhard Corp.)
  Mearlmaid CKD (Engelhard Corp.)
  Mearlmaid CP (Engelhard Corp.)
  Mearlmaid KN (Engelhard Corp.)
  Mearlmaid KND (Engelhard Corp.)
  Nailsyn C60 (Merck KGaA/EMD Chemicals
    Inc.)
  Nailsyn II C2X (Merck KGaA/EMD
    Chemicals Inc.)
  Nailsyn II Platinum 25 (Merck KGaA/EMD
    Chemicals Inc.)
  Nailsyn Platinum 60 (Merck KGaA/EMD
    Chemicals Inc.)
  Nailsyn Sterling 60 (Merck KGaA/EMD
    Chemicals Inc.)

3-NITRO-p-CRESOL

CAS No.	EINECS No.
2042-14-0	218-044-6

Empirical Formula:
  $C_7H_7NO_3$

Definition: 3-Nitro-p-Cresol is the aromatic
compound that conforms to the formula:

*See "Regulatory and Ingredient Use Infor-
mation," for Colorants in Volume 1, Introduction, Part A.*

Information Sources: EEC(II-250), TSCA

Chemical Classes: Color Additives - Hair;
Phenols

Function: Hair Colorant

Technical/Other Names:
  4-Hydroxy-2-Nitrotoluene

4-Methyl-3-Nitrophenol
Phenol, 4-Methyl-3-Nitro-
Toluene, 4-Hydroxy-2-Nitro

NITROGEN

CAS No.   **EINECS No.**
7727-37-9   231-783-9

JPN Translation:
窒素

Empirical Formula:
N_2

Definition: Nitrogen is the gaseous molecule constituting approximately 78% of the earth's atmosphere.

Information Sources: AUS, BRA, 21CFR73.1125, 21CFR73.3110, 21CFR135.130, 21CFR165.110, 21CFR169.115, 21CFR169.140, 21CFR169.150, 21CFR172.250, 21CFR172.864, 21CFR172.886, 21CFR177.1020, 21CFR177.1030, 21CFR177.1040, 21CFR177.1050, 21CFR177.1480, 21CFR177.1520, 21CFR177.2450, 21CFR178.3620, 21CFR178.3770, 21CFR178.3910, 21CFR184.1540, 21CFR184.1979, 21CFR184.1979a, 21CFR184.1979b, 21CFR184.1979c, 21CFR429.30, 21CFR429.40, 21CFR436.324, 21CFR582.1540, CZE, DDR, HUN, JCLS, JP, MAR, MI-13(6633), MI-13(6634), NF XIX, NFJ, PN, TSCA, USAN

Chemical Class: Inorganics

Function: Propellant

Reported Product Category: Hair Dyes and Colors (All Types Requiring Caution Statements and Patch Tests)

Technical/Other Names:
Diatomic Nitrogen
Dinitrogen
Molecular Nitrogen
Nitrogen Gas

Trade Name Mixtures:
Micropearl PN (C.I.T.)
Nanosource PN (C.I.T.)

2-NITRO-5-GLYCERYL METHYLANILINE

CAS No.   **EINECS No.**
80062-31-3   279-383-3

Empirical Formula:
$C_{10}H_{14}N_2O_5$

Definition: 2-Nitro-5-Glyceryl Methylaniline is the substituted aniline that conforms to the formula:

See "Regulatory and Ingredient Use Information," for Colorants in Volume 1, Introduction, Part A.

Chemical Classes: Amines; Color Additives - Hair

Function: Hair Colorant

Reported Product Category: Hair Dyes and Colors (All Types Requiring Caution Statements and Patch Tests)

Technical/Other Names:
3-[3-(Methylamino)-4-Nitrophenoxy]-1,2-Propanediol
1,2-Propanediol, 3-[3-(Methylamino)-4-Nitrophenoxy]-

Trade Name:
Imexine FT (Chimex)

4-NITROGUAIACOL

CAS No.   **EINECS No.**
3251-56-7   221-839-0

JPN Translation:
ニトログアヤコール

Empirical Formula:
$C_7H_7NO_4$

Definition: 4-Nitroguaiacol is the aromatic compound that conforms to the formula:

Information Sources: JCIC, JCLS, JSQI

Chemical Classes: Color Additives - Hair; Ethers; Phenols

Function: Hair Colorant

Technical/Other Names:
4-Hydroxy-3-methoxynitrobenzene
2-Methoxy-4-Nitrophenol
Mononitro Guaiacol
3-Nitro-6-hydroxyanisole
Phenol, 2-Methoxy-4-Nitro-

3-NITRO-p-HYDROXYETHYLAMINO-PHENOL

CAS No.   **EINECS No.**
65235-31-6   265-648-0

Empirical Formula:
$C_8H_{10}N_2O_4$

Definition: 3-Nitro-p-Hydroxyethylamino-phenol is the substituted aromatic compound that conforms to the formula:

See "Regulatory and Ingredient Use Information," for Colorants in Volume 1, Introduction, Part A.

Information Source: EEC(III/2-48)

Chemical Classes: Amines; Color Additives - Hair; Phenols

Function: Hair Colorant

Reported Product Category: Hair Dyes and Colors (All Types Requiring Caution Statements and Patch Tests)

Technical/Other Names:
4-[(2-Hydroxyethyl)Amino]-3-Nitrophenol
Phenol, 4-[(2-Hydroxyethyl)Amino]-3-Nitro-

Trade Names:
Colorex RED 54 (Chemical Compounds, Inc.)
Imexine FH (Chimex)
Jarocol NHEAP (Robinson)
Rodol HENP (Lowenstein)
Velsol Red 54 (Clariant)
Velsol Red 54 (Clariant GmbH, Personal Care)

2-NITRO-N-HYDROXYETHYL-p-ANISIDINE

CAS No.: 57524-53-5

Empirical Formula:
$C_9H_{12}N_2O_4$

Definition: 2-Nitro-N-Hydroxyethyl-p-Anisidine is the substituted aromatic amine that conforms to the formula:

See "Regulatory and Ingredient Use Information," for Colorants in Volume 1, Introduction, Part A.

Chemical Classes: Amines; Color Additives - Hair

Function: Hair Colorant

Technical/Other Names:
Ethanol, 2-[(4-Methoxy-2-Nitrophenyl)-
Amino]-
2-[(4-Methoxy-2-Nitrophenyl)Amino]Ethanol

Trade Name:
Imexine FE (Chimex)

NITROMETHANE

CAS No.	EINECS No.
75-52-5	200-876-6

Empirical Formula:
CH_3NO_2

Definition: Nitromethane is the organic compound that conforms to the formula:

$$CH_3NO_2$$

Information Sources: 21CFR178.3620, EEC(III/1-18), MI-13(6644)

Chemical Class: Hydrocarbons

Function: Corrosion Inhibitor

Technical/Other Name:
Nitocarbol

NITROPHENOL

CAS Nos.	EINECS Nos.
88-75-5 (ortho)	201-857-5
100-02-7 (para)	202-811-7
554-84-7 (meta)	209-073-5

Empirical Formula:
$C_6H_5NO_3$

Definition: Nitrophenol is the substituted phenol that conforms to the formula:

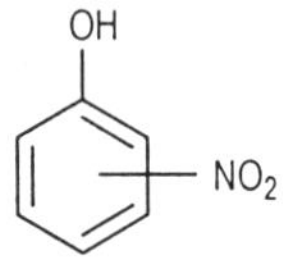

Information Sources: MI-13(6652), MI-13 (6653), MI-13(6654)

Chemical Classes: Color Additives - Hair; Phenols

Function: Hair Colorant

Technical/Other Name:
Mononitrophenol

4-NITROPHENYL AMINOETHYLUREA

CAS No.: 27080-42-8

Empirical Formula:
$C_9H_{12}N_4O_3$

Definition: 4-Nitrophenyl Aminoethylurea is the substituted aromatic amine that conforms to the formla:

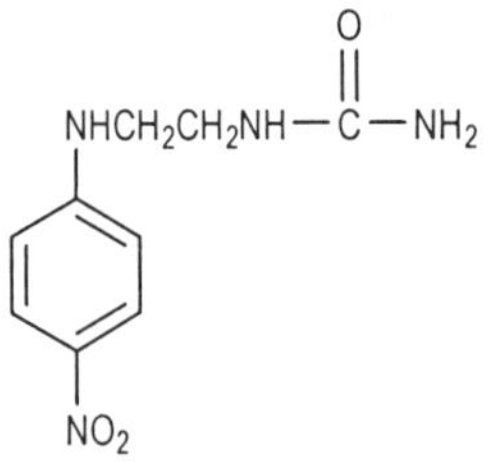

See "Regulatory and Ingredient Use Information," for Colorants in Volume 1, Introduction, Part A.

Information Source: EEC(III/2-49)

Chemical Classes: Amines; Color Additives - Hair

Function: Hair Colorant

Technical/Other Names:
[2-(p-Nitroanilino)Ethyl]Urea
[2-[(4-Nitrophenyl)Amino]Ethyl]Urea
Urea, [2-(p-Nitroanilino)Ethyl]-
Urea, [2-[(4-Nitrophenyl)Amino]Ethyl]-

4-NITRO-o-PHENYLENEDIAMINE DIHYDROCHLORIDE

CAS No.	EINECS No.
6219-77-8	228-293-2

Empirical Formula:
$C_6H_7N_3O_2 \cdot 2ClH$

Definition: 4-Nitro-o-Phenylenediamine Dihydrochloride is the substituted aromatic amine salt that conforms to the formula:

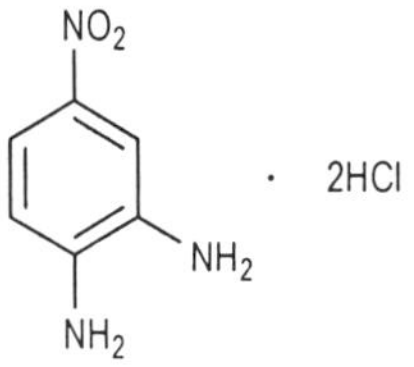

See "Regulatory and Ingredient Use Information," for Colorants in Volume 1, Introduction, Part A.

Information Source: TSCA

Chemical Classes: Amines; Color Additives - Hair

Function: Hair Colorant

Technical/Other Names:
1,2-Benzenediamine, 4-Nitro-, Dihydrochloride
4-Nitro-1,2-Benzenediamine Dihydrochloride
Oxidation Base 9A

2-NITRO-p-PHENYLENEDIAMINE DIHYDROCHLORIDE

Empirical Formula:
$C_6H_7N_3O_2 \cdot 2HCl$

Definition: 2-Nitro-p-Phenylenediamine Dihydrochloride is the hair colorant that conforms to the formula:

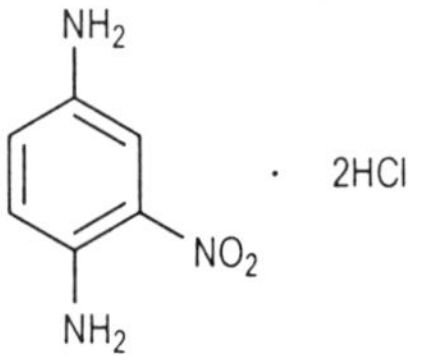

Information Source: EEC(III/2-46)

Chemical Classes: Amines; Color Additives - Hair

Function: Hair Colorant

4-NITRO-o-PHENYLENEDIAMINE HCl

CAS No.	EINECS No.
53209-19-1	258-429-6

Empirical Formula:
$C_6H_7N_3O_2 \cdot ClH$

Definition: 4-Nitro-o-Phenylenediamine HCl is the substituted aromatic amine salt that conforms to the formula:

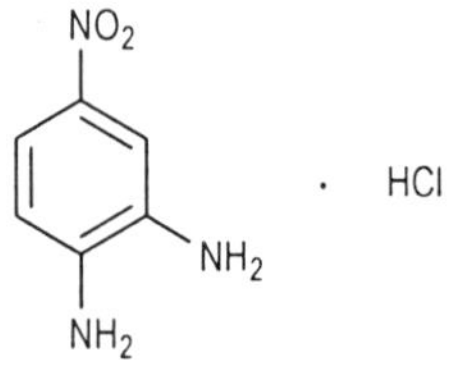

See "Regulatory and Ingredient Use Information," for Colorants in Volume 1, Introduction, Part A.

Information Source: TSCA

Chemical Classes: Amines; Color Additives - Hair

Function: Hair Colorant

Technical/Other Names:
1,2-Benzenediamine, 4-Nitro-, Hydrochloride
4-Nitro-1,2-Benzenediamine Hydrochloride

4-NITRO-m-PHENYLENEDIAMINE

CAS No.	EINECS No.
5131-58-8	225-876-3

Empirical Formula:
$C_6H_7N_3O_2$

Definition: 4-Nitro-m-Phenylenediamine is the substituted aromatic amine that conforms to the formula:

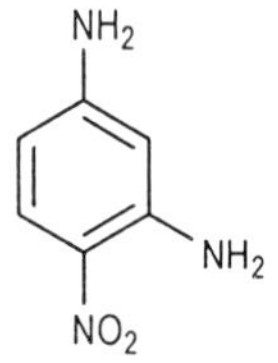

See "Regulatory and Ingredient Use Information," for Colorants in Volume 1, Introduction, Part A.

Information Sources: CI 76030, CIR: [I] JACT-11(4)1992, TSCA

Chemical Classes: Amines; Color Additives - Hair

Function: Hair Colorant

Technical/Other Names:
1,3-Benzenediamine, 4-Nitro-
CI 76030
2,4-Diaminonitrobenzene
4-Nitro-1,3-Diaminobenzene

Trade Names:
Colorex NMPD (Chemical Compounds, Inc.)
Rodol LY (Lowenstein)

4-NITRO-o-PHENYLENEDIAMINE

CAS No.   **EINECS No.**
99-56-9   202-766-3

Empirical Formula:
$C_6H_7N_3O_2$

Definition: 4-Nitro-o-Phenylenediamine is the substituted aromatic amine that conforms to the formula:

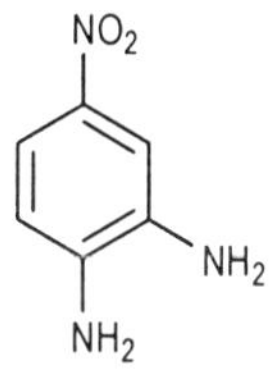

See "Regulatory and Ingredient Use Information," for Colorants in Volume 1, Introduction, Part A.

Information Sources: CI 76020, CIR: [S] JACT-4(3)1985, MI-13(6655), MI-13(6656), TSCA

Chemical Classes: Amines; Color Additives - Hair

Function: Hair Colorant

Reported Product Categories: Hair Dyes and Colors (All Types Requiring Caution Statements and Patch Tests); Hair Tints

Technical/Other Names:
2-Amino-4-nitroaniline
1,2-Benzenediamine, 4-Nitro-
CI 76020
3,4-Diaminonitrobenzene
4-Nitro-1,2-Diaminobenzene

Trade Names:
Colorex NOPD (Chemical Compounds, Inc.)

Jarocol 4NOPD (Robinson)
Rodol 4J (Lowenstein)
Rodol 4JP (Lowenstein)

Trade Name Mixture:
Blonde R-50 (Fusion) (Lowenstein)

2-NITRO-p-PHENYLENEDIAMINE

CAS No.   **EINECS No.**
5307-14-2   226-164-5

Empirical Formula:
$C_6H_7N_3O_2$

Definition: 2-Nitro-p-Phenylenediamine is the substituted aromatic amine that conforms to the formula:

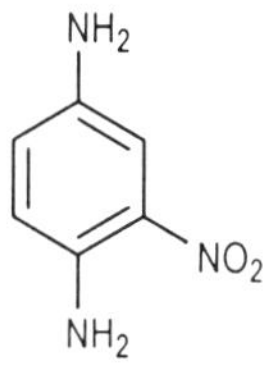

See "Regulatory and Ingredient Use Information," for Colorants in Volume 1, Introduction, Part A.

Information Sources: CI 76070, CIR: [S] JACT-4(3)1985, EEC(III/2-46), TSCA

Chemical Classes: Amines; Color Additives - Hair

Function: Hair Colorant

Reported Product Category: Hair Dyes and Colors (All Types Requiring Caution Statements and Patch Tests)

Technical/Other Names:
4-Amino-2-nitroaniline
1,4-Benzenediamine, 2-Nitro-
CI 76070
2,5-Diaminonitrobenzene
2-Nitro-1,4-Diaminobenzene
o-Nitro-p-Phenylenediamine
Oxidation Base 9A

Trade Names:
Colorex NPD (Chemical Compounds, Inc.)
Covastyle 2NPPD (LCW)
Jarocol 2NPPD (Robinson)
Rodol Brown 2R (Lowenstein)

Trade Name Mixtures:
Blonde 90 (Fusion) (Lowenstein)
Blonde R-50 (Fusion) (Lowenstein)

4-NITRO-m-PHENYLENEDIAMINE SULFATE

Empirical Formula:
$C_6H_7N_3O_2 \cdot \frac{1}{2}H_2SO_4$

Definition: 4-Nitro-m-Phenylenediamine Sulfate is the hair colorant that conforms to the formula:

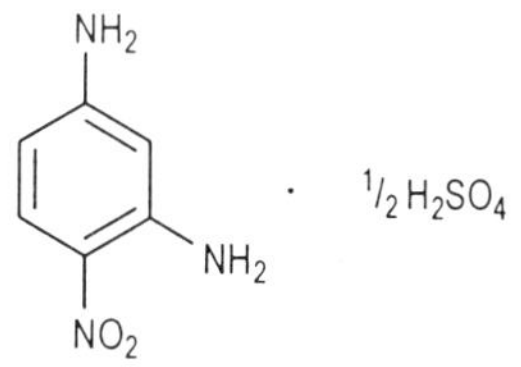

Chemical Classes: Amines; Color Additives - Hair

Function: Hair Colorant

Technical/Other Name:
p-Nitro-m-phenylenediamine sulfate

Trade Name:
Colorex NMPDS (Chemical Compounds, Inc.)

4-NITRO-o-PHENYLENEDIAMINE SULFATE

CAS No.   **EINECS No.**
68239-82-7   269-476-7

Empirical Formula:
$C_6H_7N_3O_2 \cdot H_2SO_4$

Definition: 4-Nitro-o-Phenylenediamine Sulfate is the hair colorant that conforms to the formula:

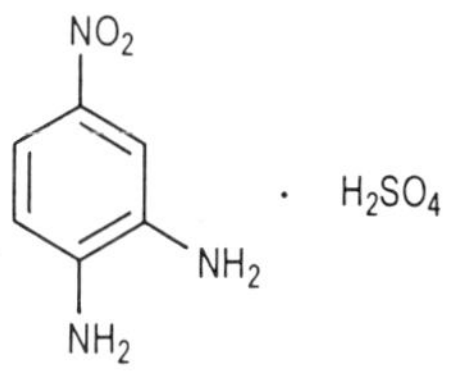

Chemical Classes: Amines; Color Additives - Hair

Function: Hair Colorant

Reported Product Category: Hair Dyes and Colors (All Types Requiring Caution Statements and Patch Tests)

Trade Name:
Rodol 4JS (Lowenstein)

2-NITRO-p-PHENYLENEDIAMINE SULFATE

CAS No.: 68239-83-8

Empirical Formula:
$C_6H_7N_3O_2 \cdot H_2SO_4$

Definition: 2-Nitro-p-Phenylenediamine Sulfate is the aromatic amine salt that conforms to the formula:

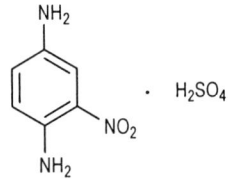

See "Regulatory and Ingredient Use Information," for Colorants in Volume 1, Introduction, Part A.

Information Source: EEC(III/2-46)

Chemical Class: Color Additives - Hair

Function: Hair Colorant

Trade Name:
Rodol Brown 2RS (Lowenstein)

Trade Name Mixture:
Rodol Burgundy 4/5 (Lowenstein)

6-NITRO-2,5-PYRIDINEDIAMINE

CAS No.: 69825-83-8

Empirical Formula:
$C_5H_6N_4O_2$

Definition: 6-Nitro-2,5-Pyridinediamine is the heterocyclic aromatic amine that conforms to the formula:

H_2N —⟨ring⟩— NO_2, NH_2

See "Regulatory and Ingredient Use Information," for Colorants in Volume 1, Introduction, Part A.

Information Source: EEC(III/2-8)

Chemical Classes: Amines; Color Additives - Hair; Heterocyclic Compounds

Function: Hair Colorant

Technical/Other Name:
2,5-Pyridinediamine, 6-Nitro

6-NITRO-o-TOLUIDINE

CAS No.	EINECS No.
570-24-1	209-329-6

Empirical Formula:
$C_7H_8N_2O_2$

Definition: 6-Nitro-o-Toluidine is the aromatic amine that conforms to the formula:

O_2N —⟨ring⟩— CH_3, NH_2

See "Regulatory and Ingredient Use Information," for Colorants in Volume 1, Introduction, Part A.

Chemical Classes: Amines; Color Additives - Hair

Function: Hair Colorant

Technical/Other Names:
1-Amino-2-Methyl-6-Nitrobenzene
2-Amino-3-nitrotoluene
Benzenamine, 2-Methyl-6-Nitro-
2-Methyl-6-nitroaniline
2-Methyl-6-Nitro-Benzenamine
3-Nitro-2-aminotoluene

Trade Name:
Imexine FP (Chimex)

NITROUS OXIDE

CAS No.	EINECS No.
10024-97-2	233-032-0

Empirical Formula:
N_2O

Definition: Nitrous Oxide is the gas that conforms to the formula:

$$N_2O$$

Information Sources: ARG, AUS, BP, BPC, BRA, 21CFR184.1545, DDR, EGY, EP, FIN, HUN, IND, ITA, JAN, MAR, MEX, MI-13(6687), NED, PN, RIFM, TSCA, USAN, USP XXIV, WHO

Chemical Class: Inorganics

Functions: Fragrance Ingredient; Propellant

Technical/Other Names:
Dinitrogen Monoxide
Nitrogen Oxide
Nitrous oxide (RIFM)

GAMMA-NONALACTONE

CAS No.	EINECS No.
104-61-0	203-219-1

JPN Translation:
ノナラクトン

Empirical Formula:
$C_9H_{16}O_2$

Definition: Gamma-Nonalactone is the heterocyclic compound that conforms to the formula:

$CH_3(CH_2)_4$ —⟨ring with O⟩= O

Information Sources: JCLS, JSCI, RIFM

Chemical Classes: Esters; Heterocyclic Compounds

Function: Fragrance Ingredient

Technical/Other Names:
γ-Nonalactone
gamma-Nonalactone (RIFM)

Trade Name:
Aldehyde C 18 soc. coconut (3/010921)
(Symrise)

NONAPEPTIDE-1

Definition: Nonapeptide-1 is the synthetic peptide consisting of Arginine (q.v.), Lysine (q.v.), Methionine (q.v.), Phenylalanine (q.v.), Proline (q.v.), Tryptophan (q,v,) and Valine (q.v.).

Chemical Class: Protein Derivatives

Function: Skin-Conditioning Agent - Miscellaneous

Trade Names:
Melanostatine 5 (I.E.B SA)
White 05 (I.E.B SA)

Trade Name Mixtures:
Melanostatine 5 PP (Peptide Powder) (I.E.B SA)
Melanostatine 5 PS (Peptide Solution) (I.E.B SA)
White 05 PP (Peptide Powder) (I.E.B SA)
White 05 PS (Peptide Solution) (I.E.B SA)

NONETH-8

Empirical Formula:
$C_{25}H_{52}O_9$

Definition: Noneth-8 is the polyethylene glycol ether of nonyl alcohol that conforms generally to the formula:

$$CH_3(CH_2)_8(OCH_2CH_2)_nOH$$

where n has an average value of 8.

Chemical Class: Alkoxylated Alcohols

Function: Surfactant - Emulsifying Agent

NONFAT DRY COLOSTRUM

Definition: Nonfat Dry Colostrum is the solid residue produced by the dehydration of defatted Colostrum (q.v.). See "Regulatory and Ingredient Use Information," regarding use of EU Trivial names in Volume 1, Introduction, Part A.

Chemical Class: Biological Products

Functions: Hair Conditioning Agent; Skin-Conditioning Agent - Miscellaneous

Technical/Other Name:
Colostrum, Nonfat

Trade Name:
 Clar 111 (Clar)

NONFAT DRY MILK

JPN Translation:
 スキムミルク

Definition: Nonfat Dry Milk is the solid residue produced by the dehydration of defatted Milk (q.v.). *See "Regulatory and Ingredient Use Information," regarding use of EU Trivial names in Volume 1, Introduction, Part A.*

Information Sources: 21CFR131.125, 21CFR184.1979, 21CFR184.1979a, 21CFR184.1979b, 21CFR184.1979c, CTFA D, JCIC, JCLS, JSCI

Chemical Class: Biological Products

Functions: Hair Conditioning Agent; Skin-Conditioning Agent - Miscellaneous

Reported Product Categories: Bubble Baths; Moisturizing Preparations; Cleansing Products (Cold Creams, Cleansing Lotions, Liquids and Pads); Body and Hand Preparations (Excluding Shaving Preparations); Paste Masks (Mud Packs); Shampoos (Non-coloring)

Technical/Other Names:
 Milk, Nonfat Dry
 Powdered Skim Milk
 Skimmed Milk Powder
 Skim Milk

Trade Name Mixtures:
 Extrapone Acacia Milk 2/032300 (Symrise)
 Extrapone Acacia Milk 2/033805 (Symrise)
 Extrapone Almond Milk 2/033132 (Symrise)
 Extrapone Almond Milk 2/033895 (Symrise)
 Extrapone Aloe Vera Milk 2/033890 (Symrise)
 Extrapone Caramel Milk 2/033834 (Symrise)
 Extrapone Fig-Mile 2/033875 (Symrise)
 Extrapone Honey Rice Milk P 2/030830 (Symrise)
 Extrapone Macadamia Nut Milk 2/033848 (Symrise)
 Extrapone Soy Milk 2/033858 (Symrise)
 Extrpone Avacado Milk 2/033820 (Symrise)
 Lactofil Sensitive (Gattefosse s.a.)

NONOXYNOL-1

CAS Nos.	**EINECS No.**
26027-38-3 (Generic)	
27986-36-3	248-762-5
37205-87-1 (Generic)	

Empirical Formula:
 $C_{17}H_{28}O_2$

Definition: Nonoxynol-1 is the ethoxylated alkyl phenol that conforms generally to the formula:

$$C_9H_{19}C_6H_4(OCH_2CH_2)_nOH$$

Information Sources: 21CFR175.105, 21CFR176.180, CIR: [SQ] IJT-18(SUPPL. 1)1999, JCLS, JSCI, MI-13(6711), TSCA

Chemical Class: Alkoxylated Alcohols

Function: Surfactant - Emulsifying Agent

Reported Product Categories: Hair Dyes and Colors (All Types Requiring Caution Statements and Patch Tests); Hair Tints

Technical/Other Names:
 Ethanol, 2-(Nonylphenoxy)-
 Ethoxylated Nonylphenol
 Ethylene Glycol Nonyl Phenyl Ether
 2-(Nonylphenoxy)Ethanol
 PEG-1 Nonyl Phenyl Ether
 Polyoxyethylene Nonylphenyl Ether

Trade Names:
 Alkasurf NP-1 (Rhodia)
 Surfonic N-10 (Texaco)

NONOXYNOL-2

CAS Nos.	**EINECS No.**
9016-45-9 (Generic)	
26027-38-3 (Generic)	
27176-93-8 (Generic)	248-291-5
37205-87-1 (Generic)	

JPN Translation:
 ノノキシノール - 2

Empirical Formula:
 $C_{19}H_{32}O_3$

Definition: Nonoxynol-2 is the ethoxylated alkyl phenol that conforms generally to the formula:

$$C_9H_{19}C_6H_4(OCH_2CH_2)_nOH$$

where n has an average value of 2.

Information Sources: 21CFR175.105, 21CFR176.180, 21CFR176.210, CIR: [SQ] IJT-18(SUPPL. 1)1999, CIR: [SQ] JACT-2 (7)1983, MI-13(6711), TSCA

Chemical Class: Alkoxylated Alcohols

Function: Surfactant - Emulsifying Agent

Reported Product Category: Hair Dyes and Colors (All Types Requiring Caution Statements and Patch Tests)

Technical/Other Names:
 Diethylene Glycol Mono(nonylphenyl) Ether
 Ethanol, 2-[2-(Nonylphenoxy)Ethoxy]-
 2-[2-(Nonylphenoxy)Ethoxy]Ethanol
 PEG-2 Nonyl Phenyl Ether
 Polyethylene Glycol 100 Nonyl Phenyl Ether
 Polyoxyethylene (2) Nonyl Phenyl Ether

Trade Names:
 Chemax NP-1,5 (Chemax)
 Empilan NP2 (Albright & Wilson UK)
 Igepal CO-210 (Rhodia)
 Nikkol NP-2 (Nikko)
 Nonal 202 (Toho)
 Unicol NP-2 (Universal Preserv-A-Chem)

Trade Name Mixture:
 Akypogene SO (Kao GmbH)

NONOXYNOL-3

CAS Nos.: 9016-45-9 (Generic); 26027-38-3 (Generic); 27176-95-0 (Generic); 37205-87-1 (Generic); 51437-95-7 (Generic); 84562-92-5 (Generic)

JPN Translation:
 ノノキシノール - 3

Empirical Formula:
 $C_{21}H_{36}O_4$

Definition: Nonoxynol-3 is the ethoxylated alkyl phenol that conforms generally to the formula:

$$C_9H_{19}C_6H_4(OCH_2CH_2)_nOH$$

where n has an average of 3.

Information Sources: 21CFR175.105, 21CFR176.180, 21CFR176.210, CIR: [SQ] IJT-18(SUPPL. 1)1999, MI-13(6711)

Chemical Class: Alkoxylated Alcohols

Function: Surfactant - Emulsifying Agent

Technical/Other Names:
 Ethanol, 2-[2-[2-(Nonylphenoxy)Ethoxy]-
 Ethoxy]-
 2-[2-[2-(Nonylphenoxy)Ethoxy]Ethoxy]-
 Ethanol
 PEG-3 Nonyl Phenyl Ether
 Polyethylene Glycol (3) Nonyl Phenyl Ether
 Polyoxyethylene (3) Nonyl Phenyl Ether
 Triethylene Glycol Nonlphenyl Ether

Trade Names:
 Akyporox NP 30 (Kao GmbH)
 Marlophen NP 3 (Sasol GmbH - Marl)

NONOXYNOL-4

CAS Nos.	**EINECS No.**
7311-27-5	230-770-5
9016-45-9 (Generic)	
26027-38-3 (Generic)	
27176-97-2	
37205-87-1 (Generic)	
68412-54-4	

JPN Translation:
 ノノキシノール - 4

Empirical Formula:
$C_{23}H_{40}O_5$

Definition: Nonoxynol-4 is the ethoxylated alkyl phenol that conforms generally to the formula:

$$C_9H_{19}C_6H_4(OCH_2CH_2)_nOH$$

where n has an average value of 4.

Information Sources: 21CFR175.105, 21CFR176.180, 21CFR176.210, 21CFR178.3400, CIR: [SQ] IJT-18(SUPPL. 1)1999, CIR: [SQ] JACT-2(7)1983, CTFA D, INN, MI-13(6711), TSCA, USAN

Chemical Class: Alkoxylated Alcohols

Function: Surfactant - Emulsifying Agent

Reported Product Categories: Hair Dyes and Colors (All Types Requiring Caution Statements and Patch Tests); Hair Bleaches; Hair Tints; Bath Oils, Tablets, and Salts; Hair Coloring Preparations, Misc.; Eyeliners; Hair Preparations (Non-coloring), Misc.

Technical/Other Names:
Decaethylene Glycol p-nonylphenyl Ether
Ethanol, 2-[2-[2-[2-(Nonylphenoxy)-Ethoxy]Ethoxy]Ethoxy]-
Ethanol, 2-[2-[2-[2-(4-Nonylphenoxy)-Ethoxy]Ethoxy]Ethoxy]-
2-[2-[2-[2-(Nonylphenoxy)Ethoxy]Ethoxy]-Ethoxy]Ethanol
PEG-4 Nonyl Phenyl Ether
Polyethylene Glycol 200 Nonyl Phenyl Ether
Polyoxyethylene (4) Nonyl Phenyl Ether
Tetraethylene Glycol Mono(p-nonylohenyl)ether
Tetraethylene Glycol Nonylphnyl Ether
Tetraoxyethylene Nonylphenyl Ether

Trade Names:
Alkasurf NP-4 (Rhodia)
Arkopal N-040 (Clariant)
Arkopal N-040 (Clariant GmbH, Personal Care)
Chemax NP-4 (Chemax)
T-DET N-4 (Harcros)
Empilan NP4 (Albright & Wilson UK)
Ethal NP-4 (Ethox)
Hetoxide NP-4 (Heterene)
Igepal CO-430 (Rhodia)
Jeechem NP-4 (Jeen)
MAKON 4 (Stepan)
Nonal 204 (Toho)
Norfox NP-4 (Norman, Fox & Co.)
Protachem NP-4 (Protameen)
Sabofen AF 4 (Sabo)
Surfonic N-40 (Texaco)
TERGITOL NP-4 Surfactant (Dow Chemical)
TRITON N-42 Surfactant (Dow Chemical)
Unicol NP-4 (Universal Preserv-A-Chem)
Uniterge NP-4 (Universal Preserv-A-Chem)

Trade Name Mixture:
Emulsifier 227 G (Grau)

NONOXYNOL-5

CAS Nos.	**EINECS No.**
9016-45-9 (Generic)	
20636-48-0	
26027-38-3 (Generic)	
26264-02-8	247-555-7
37205-87-1 (Generic)	

JPN Translation:
ノノキシノール - 5

Empirical Formula:
$C_{25}H_{44}O_6$

Definition: Nonoxynol-5 is the ethoxylated alkyl phenol that conforms generally to the formula:

$$C_9H_{19}C_6H_4(OCH_2CH_2)_nOH$$

where n has an average value of 5.

Information Sources: 21CFR175.105, 21CFR176.180, 21CFR176.210, 21CFR178.3400, CIR: [SQ] IJT-18(SUPPL. 1)1999, MI-13(6711), TSCA

Chemical Class: Alkoxylated Alcohols

Function: Surfactant - Emulsifying Agent

Technical/Other Names:
14-(Nonylphenoxy)-3,6,9,12-Tetraoxatetradecan-1-ol
PEG-5 Nonyl Phenyl Ether
Pentaethylene Glycol Nonylphenyl Ether
Polyethylene Glycol (5) Nonyl Phenyl Ether
Polyoxyethylene (5) Nonyl Phenyl Ether
3,6,9,12-Tetraoxatetradecan-1-ol, 14-(Nonylphenoxy)-

Trade Names:
Igepal CO-520 (Rhodia)
Lowenol PMS (Lowenstein)
Marlophen NP 5 (Sasol GmbH - Marl)
Nikkol NP-5 (Nikko)
Nonal 205 (Toho)
Norfox NP-5 (Norman, Fox & Co.)
TRITON N-57 Surfactant (Dow Chemical)
Unicol NP-5 (Universal Preserv-A-Chem)

NONOXYNOL-6

CAS Nos.: 9016-45-9 (Generic); 26027-38-3 (Generic); 27177-01-1; 37205-87-1 (Generic); 68412-54-4

JPN Translation:
ノノキシノール - 6

Empirical Formula:
$C_{27}H_{48}O_7$

Definition: Nonoxynol-6 is the ethoxylated alkyl phenol that conforms generally to the formula:

$$C_9H_{19}C_6H_4(OCH_2CH_2)_nOH$$

where n has an average value of 6.

Information Sources: 21CFR175.105, 21CFR176.180, 21CFR176.210, 21CFR178.3400, CIR: [SQ] IJT-18(SUPPL. 1)1999, CTFA D, MI-13(6711), TSCA

Chemical Class: Alkoxylated Alcohols

Function: Surfactant - Emulsifying Agent

Reported Product Categories: Hair Bleaches; Hair Coloring Preparations, Misc.; Personal Cleanliness Products, Misc.

Technical/Other Names:
Decaethylene Glycol p-nonylphenyl Ether
Hexaethylene Glycol Nonylphenol Ether
17-(Nonylphenoxy)-3,6,9,12,15-Pentaoxa-heptadecan-1-ol
PEG-6 Nonyl Phenyl Ether
3,6,9,12,15-Pentaoxaheptadecan-1-ol, 17-(Nonylphenoxy)-
Polyethylene Glycol 300 Nonyl Phenyl Ether
Polyethylene Glycol p-nonyphenyl Ether
Polyoxyethylene (6) Nonyl Phenyl Ether
Polyoxyethylene p-nonylphenyl Ether

Trade Names:
Alkasurf NP-6 (Rhodia)
Arkopal N-060 (Clariant)
Arkopal N-060 (Clariant GmbH, Personal Care)
Chemax NP-6 (Chemax)
T-DET N-6 (Harcros)
Empilan NP6 (Albright & Wilson UK)
Ethal NP-6 (Ethox)
Hetoxide NP-6 (Heterene)
Igepal CO-530 (Rhodia)
Jeechem NP-6 (Jeen)
Lowenol CS-1200 (Lowenstein)
MAKON 6 (Stepan)
Marlophen NP 6 (Sasol GmbH - Marl)
Nonal 206 (Toho)
Norfox NP-6 (Norman, Fox & Co.)
Protachem NP-6 (Protameen)
Sabofen AF 6 (Sabo)
Surfonic N-60 (Texaco)
TRITON N-60 Surfactant (Dow Chemical)
Unicol NP-6 (Universal Preserv-A-Chem)

Trade Name Mixture:
Petrolite C-8500 Polymer (Baker Petrolite)

NONOXYNOL-7

CAS Nos.	**EINECS No.**
9016-45-9 (Generic)	
26027-38-3 (Generic)	
27177-03-3	248-292-0
37205-87-1 (Generic)	
68412-54-4	

JPN Translation:
ノノキシノール - 7

Empirical Formula:
$C_{29}H_{52}O_8$

Definition: Nonoxynol-7 is the ethoxylated alkyl phenol that conforms generally to the formula:

$$C_9H_{19}C_6H_4(OCH_2CH_2)_nOH$$

where n has an average value of 7.

Information Sources: 21CFR175.105, 21CFR176.180, 21CFR176.210, 21CFR178.3400, CIR: [SQ] IJT-18(SUPPL. 1)1999, MI-13(6711), TSCA

Chemical Class: Alkoxylated Alcohols

Function: Surfactant - Emulsifying Agent

Reported Product Category: Personal Cleanliness Products, Misc.

Technical/Other Names:
Heptaethylene Glycol Nonylphenyl Ether
3,6,9,12,15,18-Hexaoxaeicosan-1-ol, 20-(Nonylphenoxy)-
20-(Nonylphenoxy)-3,6,9,12,15,18-Hexaoxaeicosan-1-ol
PEG-7 Nonyl Phenyl Ether
Polyethylene Glycol (7) Nonyl Phenyl Ether
Polyoxyethylene (7) Nonyl Phenyl Ether

Trade Names:
Lowenol 2689 (Lowenstein)
Marlophen NP 7 (Sasol GmbH - Marl)
Norfox NP-7 (Norman, Fox & Co.)
Synperonic NP7 (Uniqema Americas)
TERGITOL NP-7 Surfactant (Dow Chemical)
Unicol NP-7 (Universal Preserv-A-Chem)
Uniterge NP-7 (Universal Preserv-A-Chem)

NONOXYNOL-8

CAS Nos.	**EINECS Nos.**
9016-45-9 (Generic)	
26027-38-3 (Generic)	
26571-11-9	247-816-5
27177-05-5	248-293-6
37205-87-1 (Generic)	
68412-54-4	

JPN Translation:
ノノキシノール - 8

Empirical Formula:
$C_{31}H_{56}O_9$

Definition: Nonoxynol-8 is the ethoxylated alkyl phenol that conforms generally to the formula:

$$C_9H_{19}C_6H_4(OCH_2CH_2)_nOH$$

where n has an average value of 8.

Information Sources: 21CFR175.105, 21CFR176.180, 21CFR176.210, 21CFR178.3400, CIR: [SQ] IJT-18(SUPPL. 1)1999, CIR: [SQ] JACT-2(7)1983, MI-13 (6711), TSCA

Chemical Class: Alkoxylated Alcohols

Function: Surfactant - Emulsifying Agent

Technical/Other Names:
3,6,9,12,15,18,21-Heptaoxatricosan-1-ol, 23-(Nonylphenoxy)-
23-(Nonylphenoxy)-3,6,9,12,15,18,21-Heptaoxatricosan-1-ol
Octaethylene Glycol Nonylphenyl Ether
Octaoxyethylenenonylophenyl Ether
PEG-8 Nonyl Phenyl Ether
Polyethylene Glycol 400 Nonyl Phenyl Ether
Polyoxyethylene (8) Nonyl Phenyl Ether

Trade Names:
Alkasurf NP-8 (Rhodia)
Arkopal N-080 (Clariant)
Arkopal N-080 (Clariant GmbH, Personal Care)
T-DET N-8 (Harcros)
Empilan NP8 (Albright & Wilson UK)
Igepal CO-610 (Rhodia)
MAKON 8 (Stepan)
Nikkol NP-7.5 (Nikko)
Sabofen AF 8 (Sabo)
Unicol NP-8 (Universal Preserv-A-Chem)

Trade Name Mixture:
Simulsol 5719 (SEPPIC)

NONOXYNOL-9

CAS Nos.: 9016-45-9 (Generic); 14409-72-4; 26027-38-3 (Generic); 26571-11-9; 37205-87-1 (Generic); 68412-54-4

JPN Translation:
ノノキシノール - 9

Empirical Formula:
$C_{33}H_{60}O_{10}$

Definition: Nonoxynol-9 is the ethoxylated alkyl phenol that conforms generally to the formula:

$$C_9H_{19}C_6H_4(OCH_2CH_2)_nOH$$

where n has an average value of 9.

Information Sources: 21CFR175.105, 21CFR176.180, 21CFR176.210, 21CFR176.300, 21CFR178.3400, CIR: [S] JACT-2(7)1983, CTFA D, INN, MAR, MI-13 (6711), OTC-I-CV, TSCA, USAN, USP XXIV

Chemical Class: Alkoxylated Alcohols

Function: Surfactant - Emulsifying Agent

Reported Product Categories: Hair Dyes and Colors (All Types Requiring Caution Statements and Patch Tests); Hair Bleaches; Hair Preparations (Non-coloring), Misc.; Personal Cleanliness Products, Misc.; Shampoos (Non-coloring); Bath Oils, Tablets, and Salts; Cleansing Products (Cold Creams, Cleansing Lotions, Liquids and Pads); Hair

Conditioners; Tonics, Dressings, and Other Hair Grooming Aids

Technical/Other Names:
Nonaethylene Glycol Nonylphenyl Ether
26-(Nonylphenoxy)-3,6,9,12,15,18,21,24-Octaoxahexacosan-1-ol
3,6,9,12,15,18,21,24-Octaoxahexacosan-1-ol, 26-(Nonylphenoxy)-
PEG-9 Nonyl Phenyl Ether
Polyethylene Glycol 450 Nonyl Phenyl Ether
Polyoxyethylene (9) Nonyl Phenyl Ether

Trade Names:
Alkasurf NP-9 (Rhodia)
Arkopal N-090 (Clariant)
Arkopal N-090 (Clariant GmbH, Personal Care)
Chemax NP-9 (Chemax)
Empilan NP9 (Albright & Wilson UK)
Ethal NP-9 (Ethox)
Hetoxide NP-9 (Heterene)
Igepal CO-630 (Rhodia)
Igepal CO-630 Special (Rhodia)
Jeechem NP-9 (Jeen)
Lipocol NP-9 (Lipo)
Marlophen NP 9 (Sasol GmbH - Marl)
Nonal 209 (Toho)
Norfox NP-9 (Norman, Fox & Co.)
Protachem NP-9 (Protameen)
Sabofen AF 9 (Sabo)
Sympatens-NP/090 (Kolb)
TERGITOL NP-9 Surfactant (Dow Chemical)
Unicol NP-9 (Universal Preserv-A-Chem)

Trade Name Mixtures:
Emulsifier 227 G (Grau)
OPD-140 (Importaciones y Suministros)

NONOXYNOL-10

CAS Nos.	**EINECS No.**
9016-45-9 (Generic)	
26027-38-3 (Generic)	
27177-08-8	248-292-0
27942-26-3	
37205-87-1 (Generic)	
68412-54-4	

JPN Translation:
ノノキシノール - 10

Empirical Formula:
$C_{35}H_{64}O_{11}$

Definition: Nonoxynol-10 is the ethoxylated alkyl phenol that conforms generally to the formula:

$$C_9H_{19}C_6H_4(OCH_2CH_2)_nOH$$

where n has an average value of 10.

Information Sources: 21CFR175.105, 21CFR176.180, 21CFR176.210,

21CFR176.300, 21CFR178.3400, CIR: [S] JACT-2(7)1983, CTFA D, MI-13(6711), NF XIX, TSCA, USAN

Chemical Class: Alkoxylated Alcohols

Function: Surfactant - Emulsifying Agent

Reported Product Categories: Hair Dyes and Colors (All Types Requiring Caution Statements and Patch Tests); Hair Conditioners; Permanent Waves; Tonics, Dressings, and Other Hair Grooming Aids; Hair Preparations (Non-coloring), Misc.; Mascara; Shampoos (Non-coloring); Hair Wave Sets; Cleansing Products (Cold Creams, Cleansing Lotions, Liquids and Pads); Eyeliners; Hair Sprays (Aerosol Fixatives); Hair Tints; Personal Cleanliness Products, Misc.; Hair Bleaches; Hair Straighteners; Aftershave Lotions

Technical/Other Names:
Decaethylene Glycol Nonylohenyl Ether
Decaethylene Glycol p-nonylphenyl Ether
3,6,9,12,15,18,21,24,27-Nonaoxanonacosan-1-ol, 29-(Nonyl-phenoxy)-
3,6,9,12,15,18,21,24,27-Nonaoxanonacosan-1-ol, 29-(4-Nonyl-phenoxy)-
29-(Nonylphenoxy)-3,6,9,12,15,18,21,24,27-Nonaoxanonacosan-1-ol
29-(4-Nonylphenoxy)-3,6,9,12,15,18,21,24,27-Nonaoxanonacosan-1-ol
PEG-10 Nonyl Phenyl Ether
Polyethylene Glycol 500 Nonyl Phenyl Ether
Polyoxyethylene (10) Nonylphenol Ether
Polyoxyethylene (10) Nonyl Phenyl Ether

Trade Names:
Alkasurf NP-10 (Rhodia)
Arkopal N-100 (Clariant)
Arkopal N-100 (Clariant GmbH, Personal Care)
T-DET N-9.5 (Harcros)
Empilan NP10 (Albright & Wilson UK)
Hetoxide NP-10 (Heterene)
Igepal CO-660 (Rhodia)
Igepal CO-710 (Rhodia)
Jeechem NP-10 (Jeen)
MAKON 10 (Stepan)
Nikkol NP-10 (Nikko)
Nonal 210 (Toho)
Norfox NP-10.2 (Norman, Fox & Co.)
Protachem NP-10 (Protameen)
Sabofen AF 10 (Sabo)
Surfonic N-95 (Texaco)
Surfonic N-100 (Texaco)
Surfonic N-102 (Texaco)
TERGITOL NP-10 Surfactant (Dow Chemical)
TRITON N-101 Surfactant (Dow Chemical)
Unicol NP-10 (Universal Preserv-A-Chem)
Uniterge NP-10 (Universal Preserv-A-Chem)

Trade Name Mixtures:
Biofloreol hydrosoluble (Esperis)
Dow Corning 929 Cationic Emulsion (Dow Corning)
Propoli Extract H 5 (Esperis)
TAYLOR TE- AMD - 2 (Taylor Chemical Company)

NONOXYNOL-11

CAS Nos.: 9016-45-9 (Generic); 37205-87-1 (Generic); 68412-54-4

JPN Translation:
ノノキシノール - 11

Definition: Nonoxynol-11 is the ethoxylated alkyl phenol that conforms generally to the formula:

$$C_9H_{19}C_6H_4(OCH_2CH_2)_nOH$$

where n has an average value of 11.

Information Sources: 21CFR175.105, 21CFR176.180, 21CFR176.210, 21CFR176.300, 21CFR178.3400, MI-13 (6711), TSCA

Chemical Class: Alkoxylated Alcohols

Function: Surfactant - Emulsifying Agent

Technical/Other Names:
PEG-11 Nonyl Phenyl Ether
Polyethylene Glycol (11) Nonyl Phenyl Ether
Polyoxyethylene (11) Nonyl Phenyl Ether

Trade Names:
Alkasurf NP-11 (Rhodia)
Arkopal N-110 (Clariant)
Arkopal N-110 (Clariant GmbH, Personal Care)
Chemax NP-10 (Chemax)
T-DET N-10.5 (Harcros)
Jeechem NP-11 (Jeen)
Nonal 211 (Toho)
Norfox NP-11 (Norman, Fox & Co.)
Sabofen AF 11 (Sabo)
TRITON N-111 Surfactant (Dow Chemical)

NONOXYNOL-12

CAS Nos.: 9016-45-9 (Generic); 26027-38-3 (Generic); 37205-87-1 (Generic); 68412-54-4

JPN Translation:
ノノキシノール - 12

Definition: Nonoxynol-12 is the ethoxylated alkyl phenol that conforms generally to the formula:

$$C_9H_{19}C_6H_4(OCH_2CH_2)_nOH$$

where n has an average value of 12.

Information Sources: 21CFR175.105, 21CFR176.180, 21CFR176.210,

21CFR176.300, 21CFR178.3400, CIR: [S] JACT-2(7)1983, MI-13(6711), TSCA

Chemical Class: Alkoxylated Alcohols

Function: Surfactant - Emulsifying Agent

Reported Product Categories: Permanent Waves; Skin Fresheners; Tonics, Dressings, and Other Hair Grooming Aids; Bath Soaps and Detergents; Hair Preparations (Non-coloring), Misc.; Skin Care Preparations, Misc.

Technical/Other Names:
PEG-12 Nonyl Phenyl Ether
Polyethylene Glycol 600 Nonyl Phenyl Ether
Polyoxyethylene (12) Nonyl Phenyl Ether

Trade Names:
Alkasurf NP-12 (Rhodia)
Chemax NP-12 (Chemax)
T-DET N-12 (Harcros)
Empilan NP12 (Albright & Wilson UK)
Hetoxide NP-12 (Heterene)
Igepal CO-720 (Rhodia)
Jeechem NP-12 (Jeen)
MAKON 12 (Stepan)
Norfox NP-12 (Norman, Fox & Co.)
Sabofen AF 12 (Sabo)
Surfonic N-120 (Texaco)

Trade Name Mixture:
Neo PCL Prime 2/789230 (Symrise)

NONOXYNOL-13

CAS Nos.: 9016-45-9 (Generic); 26027-38-3 (Generic); 37205-87-1 (Generic); 68412-54-4 (Generic)

JPN Translation:
ノノキシノール - 13

Definition: Nonoxynol-13 is the ethoxylated alkyl phenol that conforms generally to the formula:

$$C_9H_{19}C_6H_4(OCH_2CH_2)_nOH$$

where n has an average value of 13.

Information Sources: 21CFR175.105, 21CFR176.180, 21CFR176.210, 21CFR178.3400, MI-13(6711), TSCA

Chemical Class: Alkoxylated Alcohols

Function: Surfactant - Emulsifying Agent

Technical/Other Names:
PEG-13 Nonyl Phenyl Ether
Polyethylene Glycol (13) Nonyl Phenyl Ether
Polyoxyethylene (13) Nonyl Phenyl Ether

Trade Names:
Arkopal N-130 (Clariant)
Arkopal N-130 (Clariant GmbH, Personal Care)
Norfox NP-13 (Norman, Fox & Co.)

TERGITOL NP-13 Surfactant (Dow Chemical)
Uniterge NP-13 (Universal Preserv-A-Chem)

NONOXYNOL-14

CAS Nos.: 9016-45-9 (Generic); 26027-38-3 (Generic); 37205-87-1 (Generic); 68412-54-4

JPN Translation:
ノノキシノール - 14

Definition: Nonoxynol-14 is the ethoxylated alkyl phenol that conforms generally to the formula:

$$C_9H_{19}C_6H_4(OCH_2CH_2)_nOH$$

where n has an average value of 14.

Information Sources: 21CFR175.105, 21CFR176.180, 21CFR176.210, 21CFR178.3400, CIR: [S] JACT-2(7)1983, MI-13(6711), TSCA

Chemical Class: Alkoxylated Alcohols

Function: Surfactant - Emulsifying Agent

Reported Product Categories: Skin Care Preparations, Misc.; Bath Oils, Tablets, and Salts; Cleansing Products (Cold Creams, Cleansing Lotions, Liquids and Pads); Bath Soaps and Detergents

Technical/Other Names:
PEG-14 Nonyl Phenyl Ether
Polyethylene Glycol (14) Nonyl Phenyl Ether
Polyoxyethylene (14) Nonyl Phenyl Ether

Trade Name:
MAKON 14 (Stepan)

Trade Name Mixtures:
Novaplant Pine Needle Extract (Crodarom)
Obanol 516 (Toho)
Obazolin 516 (Toho)

NONOXYNOL-15

CAS Nos.: 9016-45-9 (Generic); 26027-38-3 (Generic); 37205-87-1 (Generic); 68412-54-4

JPN Translation:
ノノキシノール - 15

Definition: Nonoxynol-15 is the ethoxylated alkyl phenol that conforms generally to the formula:

$$C_9H_{19}C_6H_4(OCH_2CH_2)_nOH$$

where n has an average value of 15.

Information Sources: 21CFR175.105, 21CFR176.180, 21CFR176.210, CIR: [S] JACT-2(7)1983, CTFA D, INN, MI-13(6711), TSCA, USAN

Chemical Class: Alkoxylated Alcohols

Function: Surfactant - Emulsifying Agent

Reported Product Categories: Permanent Waves; Hair Conditioners; Tonics, Dressings, and Other Hair Grooming Aids; Bath Preparations, Misc.; Eyeliners; Hair Preparations (Non-coloring), Misc.

Technical/Other Names:
PEG-15 Nonyl Phenyl Ether
Polyethylene Glycol (15) Nonyl Phenyl Ether
Polyoxyethylene (15) Nonyl Phenyl Ether

Trade Names:
Alkasurf NP-15 (Rhodia)
Arkopal N-150 (Clariant)
Arkopal N-150 (Clariant GmbH, Personal Care)
Chemax NP-15 (Chemax)
Empilan NP15 (Albright & Wilson UK)
Hetoxide NP-15 - 85% (Heterene)
Igepal CO-730 (Rhodia)
Jeechem NP-15 (Jeen)
Nikkol NP-15 (Nikko)
Norfox NP-15 (Norman, Fox & Co.)
Surfonic N-150 (Texaco)
Sympatens-NP/150 (Kolb)
TERGITOL NP-15 Surfactant (Dow Chemical)
TRITON N-150 Surfactant (Dow Chemical)
Uniterge NP-15 (Universal Preserv-A-Chem)

Trade Name Mixtures:
ALE-75 (OSi Specialties)
Silsoft E-50 (OSi Specialties)

NONOXYNOL-18

CAS Nos.: 9016-45-9 (Generic); 26027-38-3 (Generic); 37205-87-1 (Generic); 68412-54-4

JPN Translation:
ノノキシノール - 18

Definition: Nonoxynol-18 is the ethoxylated alkyl phenol that conforms generally to the formula:

$$C_9H_{19}C_6H_4(OCH_2CH_2)_nOH$$

where n has an average value of 18.

Information Sources: 21CFR175.105, 21CFR176.180, MI-13(6711), TSCA

Chemical Class: Alkoxylated Alcohols

Function: Surfactant - Emulsifying Agent

Technical/Other Names:
PEG-18 Nonyl Phenyl Ether
Polyethylene Glycol (18) Nonyl Phenyl Ether
Polyoxyethylene (18) Nonyl Phenyl Ether

Trade Name:
Nikkol NP-18TX (Nikko)

NONOXYNOL-20

CAS Nos.: 9016-45-9 (Generic); 26027-38-3 (Generic); 37205-87-1 (Generic); 68412-54-4

JPN Translation:
ノノキシノール - 20

Definition: Nonoxynol-20 is the ethoxylated alkyl phenol that conforms generally to the formula:

$$C_9H_{19}C_6H_4(OCH_2CH_2)_nOH$$

where n has an average value of 20.

Information Sources: 21CFR175.105, 21CFR176.180, MI-13(6711), TSCA

Chemical Class: Alkoxylated Alcohols

Functions: Surfactant - Cleansing Agent; Surfactant - Emulsifying Agent; Surfactant - Solubilizing Agent

Reported Product Category: Bubble Baths

Technical/Other Names:
PEG-20 Nonyl Phenyl Ether
Polyethylene Glycol 1000 Nonyl Phenyl Ether
Polyoxyethylene (20) Nonyl Phenyl Ether

Trade Names:
Alkasurf NP-20 (Rhodia)
Chemax NP-20 (Chemax)
Empilan NP20 (Albright & Wilson UK)
Ethal NP-20 (Ethox)
Igepal CO-850 (Rhodia)
Lipocol NP-20 (Lipo)
Nikkol NP-20 (Nikko)
Norfox NP-20 (Norman, Fox & Co.)
Surfonic N-200 (Texaco)
TERGITOL NP-20 Surfactant (Dow Chemical)
Unicol NP-20 (Universal Preserv-A-Chem)
Uniterge NP-20 (Universal Preserv-A-Chem)

Trade Name Mixture:
SM2140 (GE Silicones)

NONOXYNOL-23

CAS Nos.: 9016-45-9 (Generic); 26027-38-3 (Generic); 37205-87-1 (Generic); 68412-54-4

JPN Translation:
ノノキシノール - 23

Definition: Nonoxynol-23 is the ethoxylated alkyl phenol that conforms generally to the formula:

$$C_9H_{19}C_6H_4(OCH_2CH_2)_nOH$$

where n has an average value of 23.

Information Sources: 21CFR175.105, 21CFR176.180, MI-13(6711), TSCA

Chemical Class: Alkoxylated Alcohols

Functions: Surfactant - Cleansing Agent; Surfactant - Solubilizing Agent

Technical/Other Names:
PEG-23 Nonyl Phenyl Ether
Polyethylene Glycol (23) Nonyl Phenyl Ether
Polyoxyethylene (23) Nonyl Phenyl Ether

Trade Names:
Arkopal N-230 (Clariant)
Arkopal N-230 (Clariant GmbH, Personal Care)
Emulsogen ELN (Clariant)
Emulsogen ELN (Clariant GmbH, Personal Care)

NONOXYNOL-25

CAS No.: 9016-45-9 (Generic)

Definition: Nonoxynol-25 is the ethoxylated alkyl phenol that conforms generally to the formula:

$$C_9H_{19}C_6H_4(OCH_2CH_2)_nOH$$

where n has an average value of 25.

Chemical Class: Alkoxylated Alcohols

Functions: Surfactant - Cleansing Agent; Surfactant - Solubilizing Agent

Technical/Other Names:
PEG-25 Nonyl Phenyl Ether
Polyethylene Glycol (25) Nonyl Phenyl Ether
Polyoxyethylene (25) Nonyl Phenyl Ether

Trade Name:
Empilan NP25 (Albright & Wilson UK)

NONOXYNOL-30

CAS Nos.: 9016-45-9 (Generic); 26027-38-3 (Generic); 37205-87-1 (Generic); 68412-54-4

JPN Translation:
ノノキシノール - 30

Definition: Nonoxynol-30 is the ethoxylated alkyl phenol that conforms generally to the formula:

$$C_9H_{19}C_6H_4(OCH_2CH_2)_nOH$$

where n has an average value of 30.

Information Sources: 21CFR175.105, 21CFR176.180, 21CFR178.3400, CIR: [S] JACT-2(7)1983, CTFA D, INN, MI-13(6711), TSCA, USAN

Chemical Class: Alkoxylated Alcohols

Functions: Surfactant - Cleansing Agent; Surfactant - Solubilizing Agent

Technical/Other Names:
PEG-30 Nonyl Phenyl Ether
Polyethylene Glycol (30) Nonyl Phenyl Ether
Polyoxyethylene (30) Nonyl Phenyl Ether

Trade Names:
Arkopal N-300 (Clariant)
Arkopal N-300 (Clariant GmbH, Personal Care)
DelONIC NPE-30 (DeForest)
T-DET N-30 (Harcros)
T-DET N-307 (Harcros)
Empilan NP30 (Albright & Wilson UK)
Ethal NP-307 (Ethox)
Igepal CO-880 (Rhodia)
Jeechem NP-30 (Jeen)
MAKON 30 (Stepan)
Surfonic N-300 (Texaco)
Surfonic NB-5 (Texaco)
Unicol NP-30 (Universal Preserv-A-Chem)

Trade Name Mixture:
TAYLOR TE-600 DM (Taylor Chemical Company)

NONOXYNOL-35

CAS Nos.: 9016-45-9 (Generic); 26027-38-3 (Generic); 37205-87-1 (Generic); 68412-54-4

JPN Translation:
ノノキシノール - 35

Definition: Nonoxynol-35 is the ethoxylated alkyl phenol that conforms generally to the formula:

$$C_9H_{19}C_6H_4(OCH_2CH_2)_nOH$$

where n has an average value of 35.

Information Source: TSCA

Chemical Class: Alkoxylated Alcohols

Functions: Surfactant - Cleansing Agent; Surfactant - Solubilizing Agent

Technical/Other Names:
PEG-35 Nonyl Phenyl Ether
Polyethylene Glycol (35) Nonyl Phenyl Ether
Polyoxyethylene (35) Nonyl Phenyl Ether

Trade Name Mixture:
Lytron 318 (Rohm and Haas)

NONOXYNOL-40

CAS Nos.: 9016-45-9 (Generic); 26027-38-3 (Generic); 37205-87-1 (Generic); 68412-54-4

JPN Translation:
ノノキシノール - 40

Definition: Nonoxynol-40 is the ethoxylated alkyl phenol that conforms generally to the formula:

$$C_9H_{19}C_6H_4(OCH_2CH_2)_nOH$$

where n has an average value of 40.

Information Sources: 21CFR175.105, 21CFR176.180, 21CFR178.3400, CIR: [S] JACT-2(7)1983, MI-13(6711), TSCA

Chemical Class: Alkoxylated Alcohols

Functions: Surfactant - Cleansing Agent; Surfactant - Solubilizing Agent

Reported Product Category: Hair Tints

Technical/Other Names:
PEG-40 Nonyl Phenyl Ether
Polyethylene Glycol 2000 Nonyl Phenyl Ether
Polyoxyethylene (40) Nonyl Phenyl Ether

Trade Names:
Alkasurf NP-40 (Rhodia)
Chemax NP-40 (Chemax)
DelONIC NPE-40 (DeForest)
Ethal NP-407 (Ethox)
Hetoxide NP-40 (Heterene)
Igepal CO-890 (Rhodia)
Jeechem NP-40 (Jeen)
Norfox NP-40 (Norman, Fox & Co.)
Norfox NP-40-70% (Norman, Fox & Co.)
Surfonic N-400 (Texaco)
Surfonic NB-14 (Texaco)
TERGITOL NP-40 Surfactant (Dow Chemical)
TRITON N-401 Surfactant (Dow Chemical)
Unicol NP-40 (Universal Preserv-A-Chem)

Trade Name Mixture:
OPD-140 (Importaciones y Suministros)

NONOXYNOL-44

CAS Nos.: 9016-45-9 (Generic); 26027-38-3 (Generic); 37205-87-1 (Generic); 68412-54-4

JPN Translation:
ノノキシノール - 44

Definition: Nonoxynol-44 is the ethoxylated alkyl phenol that conforms generally to the formula:

$$C_9H_{19}C_6H_4(OCH_2CH_2)_nOH$$

where n has an average value of 44.

Information Sources: 21CFR176.180, 21CFR178.3400, MI-13(6711), TSCA

Chemical Class: Alkoxylated Alcohols

Function: Surfactant - Cleansing Agent

Technical/Other Names:
PEG-44 Nonyl Phenyl Ether
Polyethylene Glycol (44) Nonyl Phenyl Ether
Polyoxyethylene (44) Nonyl Phenyl Ether

Trade Name:
Uniterge NP-44 (Universal Preserv-A-Chem)

NONOXYNOL-50

CAS Nos.: 9016-45-9 (Generic); 26027-38-3 (Generic); 37205-87-1 (Generic); 68412-54-4

JPN Translation:
ノノキシノール - 50

Definition: Nonoxynol-50 is the ethoxylated alkyl phenol that conforms generally to the formula:

$$C_9H_{19}C_6H_4(OCH_2CH_2)_nOH$$

where n has an average value of 50.

Information Sources: 21CFR176.180, 21CFR178.3400, CIR: [S] JACT-2(7)1983, MI-13(6711), TSCA

Chemical Class: Alkoxylated Alcohols

Function: Surfactant - Cleansing Agent

Technical/Other Names:
PEG-50 Nonyl Phenyl Ether
Polyethylene Glycol (50) Nonyl Phenyl Ether
Polyoxyethylene (50) Nonyl Phenyl Ether

Trade Names:
Alkasurf NP-50 (Rhodia)
Chemax NP-50 (Chemax)
T-DET N-50 (Harcros)
Ethal NP-506 (Ethox)
Hetoxide NP-50 (Heterene)
Igepal CO-970 (Rhodia)
Norfox NP-50 (Norman, Fox & Co.)
Synperonic NP50 (Uniqema Americas)
Unicol NP-50 (Universal Preserv-A-Chem)

NONOXYNOL-100

CAS Nos.: 9016-45-9 (Generic); 26027-38-3 (Generic); 37205-87-1 (Generic); 68412-54-4

JPN Translation:
ノノキシノール - 100

Definition: Nonoxynol-100 is the ethoxylated alkyl phenol that conforms generally to the formula:

$$C_9H_{19}C_6H_4(OCH_2CH_2)_nOH$$

where n has an average value of 100.

Information Sources: 21CFR176.180, MI-13(6711), TSCA

Chemical Class: Alkoxylated Alcohols

Function: Surfactant - Cleansing Agent

Technical/Other Names:
PEG-100 Nonyl Phenyl Ether
Polyethylene Glycol 100 Nonyl Phenyl Ether
Polyoxyethylene (100) Nonyl Phenyl Ether

Trade Names:
DelONIC NPE-100 (DeForest)

T-DET N-100 (Harcros)
T-DET N-1007 (Harcros)
Norfox NP-100 (Norman, Fox & Co.)
Norfox NP-977 (Norman, Fox & Co.)
Unicol NP-100 (Universal Preserv-A-Chem)

NONOXYNOL-120

CAS Nos.: 9016-45-9 (Generic); 26027-38-3 (Generic); 37205-87-1 (Generic)

JPN Translation:
ノノキシノール - 120

Definition: Nonoxynol-120 is the ethoxylated alkyl phenol that conforms generally to the formula:

$$C_9H_{19}C_6H_4(OCH_2CH_2)_nOH$$

where n has an average value of 120.

Information Sources: 21CFR176.180, MI-13(6711)

Chemical Class: Alkoxylated Alcohols

Function: Surfactant - Cleansing Agent

Technical/Other Names:
PEG-120 Nonyl Phenyl Ether
Polyethylene Glycol (120) Nonyl Phenyl Ether
Polyoxyethylene (120) Nonyl Phenyl Ether

Trade Name:
Akyporox NP 1200 V (Kao GmbH)

NONOXYNOL-5 CARBOXYLIC ACID

CAS Nos.: 28212-44-4 (Generic); 53610-02-9

Empirical Formula:
$C_{25}H_{42}O_7$

Definition: Nonoxynol-5 Carboxylic Acid is the organic acid that conforms generally to the formula:

$$C_9H_{19}C_6H_4(OCH_2CH_2)_nOCH_2COOH$$

where n has an average value of 4.

Chemical Class: Carboxylic Acids

Function: Surfactant - Cleansing Agent

Technical/Other Names:
PEG-5 Nonyl Phenyl Ether Carboxylic Acid
Polyethylene Glycol (5) Nonyl Phenyl Ether Carboxylic Acid
Polyoxyethylene (5) Nonyl Phenyl Ether Carboxylic Acid

NONOXYNOL-8 CARBOXYLIC ACID

CAS Nos.: 28212-44-4 (Generic); 53610-02-9

Empirical Formula:
$C_{31}H_{54}O_{10}$

Definition: Nonoxynol-8 Carboxylic Acid is the organic acid that conforms generally to the formula:

$$C_9H_{19}C_6H_4(OCH_2CH_2)_nOCH_2COOH$$

where n has an average value of 7.

Chemical Class: Carboxylic Acids

Function: Surfactant - Cleansing Agent

Technical/Other Names:
PEG-8 Nonyl Phenyl Ether Carboxylic Acid
Polyethylene Glycol 400 Nonyl Phenyl Ether Carboxylic Acid
Polyoxyethylene (8) Nonyl Phenyl Ether Carboxylic Acid

Trade Name:
Akypo NP 70 (Kao GmbH)

NONOXYNOL-10 CARBOXYLIC ACID

CAS Nos.: 28212-44-4 (Generic); 53610-02-9

Empirical Formula:
$C_{35}H_{62}O_{12}$

Definition: Nonoxynol-10 Carboxylic Acid is the organic acid that conforms generally to the formula:

$$C_9H_{19}C_6H_4(OCH_2CH_2)_nOCH_2COOH$$

where n has an average value of 9.

Chemical Class: Carboxylic Acids

Function: Surfactant - Cleansing Agent

Technical/Other Names:
PEG-10 Nonyl Phenyl Ether Carboxylic Acid
Polyethylene Glycol 500 Nonyl Phenyl Ether Carboxylic Acid
Polyoxyethylene (10) Nonyl Phenyl Ether Carboxylic Acid

Trade Names:
Sandopan MA-18 (Clariant)
Sandopan MA-18 (Clariant GmbH, Personal Care)
Surfine AZI-A (Finetex)

NONOXYNOL-9 IODINE

CAS No.	EINECS No.
94349-40-3	305-157-1

Empirical Formula:
$C_{33}H_{60}O_{10} \cdot xI_2$

Definition: Nonoxynol-9 Iodine is a complex of Nonoxynol-9 (q.v.) and iodine.

Chemical Class: Alkoxylated Alcohols

Function: Cosmetic Biocide

Technical/Other Names:
26-(Nonylphenoxy)-3,6,9,12,15,18,21,24-Octaoxahexacosan-1-ol, compd. with Iodine

3,6,9,12,15,18,21,24-Octaoxahexacosan-1-ol, 26-(Nonylphenoxy)-, compd. with Iodine
PEG-9 Nonyl Phenyl Ether Iodine Complex
Polyethylene Glycol 450 Nonyl Phenyl Ether Iodine Complex
Polyoxyethylene (9) Nonyl Phenyl Ether Iodine Complex

NONOXYNOL-12 IODINE

Definition: Nonoxynol-12 Iodine is a complex of Nonoxynol-12 (q.v.) and iodine.

Chemical Class: Alkoxylated Alcohols

Function: Cosmetic Biocide

Technical/Other Names:
PEG-12 Nonyl Phenyl Ether Iodine Complex
Polyethylene Glycol 600 Nonyl Phenyl Ether Iodine Complex
Polyoxyethylene (12) Nonyl Phenyl Ether Iodine Complex

Trade Name:
M-2205 20% Iodine Concentrate (Huntington)

NONOXYNOL-3 PHOSPHATE

CAS No.: 51811-79-1 (Generic)

Definition: Nonoxynol-3 Phosphate is a complex mixture of esters of phosphoric acid and Nonoxynol-3 (q.v.).

Information Source: TSCA

Chemical Class: Phosphorus Compounds

Functions: Surfactant - Cleansing Agent; Surfactant - Emulsifying Agent; Surfactant - Hydrotrope

Technical/Other Names:
PEG-3 Nonyl Phenyl Ether Phosphate
Polyethylene Glycol (3) Nonyl Phenyl Ether Phosphate
Polyoxyethylene (3) Nonyl Phenyl Ether Phosphate

Trade Name:
Phosphanol RE-410 (Toho)

NONOXYNOL-4 PHOSPHATE

Definition: Nonoxynol-4 Phosphate is a complex mixture of esters of phosphoric acid and Nonoxynol-4 (q.v.).

Chemical Classes: Alkoxylated Alcohols; Phosphorus Compounds

Functions: Surfactant - Cleansing Agent; Surfactant - Emulsifying Agent; Surfactant - Hydrotrope

Trade Name:
DePHOS PE-481 (DeForest)

NONOXYNOL-6 PHOSPHATE

CAS Nos. | **EINECS No.**
29994-44-3 | 249-992-9
51609-41-7 (Generic)

JPN Translation:
ノノキシノール - 6 リン酸

Definition: Nonoxynol-6 Phosphate is a complex mixture of esters of phosphoric acid and Nonoxynol-6 (q.v.).

Information Sources: 21CFR175.105, 21CFR178.3400

Chemical Class: Phosphorus Compounds

Functions: Surfactant - Cleansing Agent; Surfactant - Emulsifying Agent; Surfactant - Hydrotrope

Technical/Other Names:
17-(Nonylphenoxy)-3,6,9,12,15-Pentaoxa-heptadecan-1-ol, Dihydrogen Phosphate
PEG-6 Nonyl Phenyl Ether Phosphate
3,6,9,12,15-Pentaoxaheptadecan-1-ol, 17-(Nonylphenoxy)-, Dihydrogen Phosphate
Polyethylene Glycol 300 Nonyl Phenyl Ether Phosphate
Polyoxyethylene (6) Nonyl Phenyl Ether Phosphate

Trade Names:
DePHOS PE-786 (DeForest)
Ethfac NP-16 (Ethox)
Monafax 786 (Uniqema)
Phosphanol PE-510 (Toho)

NONOXYNOL-9 PHOSPHATE

CAS Nos. | **EINECS No.**
51609-41-7 (Generic) |
66197-78-2 | 266-231-6

JPN Translation:
ノノキシノール - 9 リン酸

Empirical Formula:
$C_{33}H_{61}O_{13}P$

Definition: Nonoxynol-9 Phosphate is a complex mixture of esters of phosphoric acid and Nonoxynol-9 (q.v.).

Information Sources: 21CFR175.105, 21CFR178.3400

Chemical Class: Phosphorus Compounds

Functions: Surfactant - Cleansing Agent; Surfactant - Emulsifying Agent; Surfactant - Hydrotrope

Technical/Other Names:
26-(Nonylphenoxy)-3,6,9,12,15,18,21,24-Octaoxahexacosan-1-ol, Dihydrogen Phosphate

3,6,9,12,15,18,21,24-Octaoxahexacosan-1-ol, 26-(Nonylphenoxy)-, Dihydrogen Phosphate
PEG-9 Nonyl Phenyl Ether Phosphate
Polyethylene Glycol 450 Nonyl Phenyl Ether Phosphate
Polyoxyethylene (9) Nonyl Phenyl Ether Phosphate

Trade Names:
DePHOS RA-831 (DeForest)
Monafax 785 (Uniqema)
Rhodafac RE-610 (Rhodia)

NONOXYNOL-10 PHOSPHATE

CAS No.: 51609-41-7 (Generic)

JPN Translation:
ノノキシノール - 10 リン酸

Definition: Nonoxynol-10 Phosphate is a complex mixture of esters of phosphoric acid and Nonoxynol-10 (q.v.).

Information Source: 21CFR178.3400

Chemical Class: Phosphorus Compounds

Functions: Surfactant - Cleansing Agent; Surfactant - Emulsifying Agent; Surfactant - Hydrotrope

Technical/Other Names:
PEG-10 Nonyl Phenyl Ether Phosphate
Polyethylene Glycol 500 Nonyl Phenyl Ether Phosphate
Polyoxyethylene (10) Nonyl Phenyl Ether Phosphate

Trade Names:
Ethfac NP-110 (Ethox)
Jeephos P-610 (Jeen)

NONOXYNYL HYDROXYETHYL-CELLULOSE

Definition: Nonoxynyl Hydroxyethylcellulose is the ether formed by the reaction of a nonylphenol-derived epoxide with Hydroxyethylcellulose (q.v.).

Chemical Classes: Ethers; Gums, Hydrophilic Colloids and Derivatives

Function: Viscosity Increasing Agent - Aqueous

NONYL ACETATE

CAS No. | **EINECS No.**
143-13-5 | 205-585-8

Empirical Formula:
$C_{11}H_{22}O_2$

Definition: Nonyl Acetate is the ester of nonyl alcohol and acetic acid. It conforms to the formula:

$$CH_3\overset{\displaystyle O}{\overset{\|}{C}}-O(CH_2)_8CH_3$$

Information Sources: 21CFR172.515, FCC, MI-13(6712), RIFM, TSCA

Chemical Class: Esters

Functions: Fragrance Ingredient; Skin-Conditioning Agent - Emollient

Technical/Other Names:
Acetic Acid, Nonyl Ester
1-Acetoxynonane
Nonyl acetate (RIFM)
n-Nonyl Ethanoate
Pelargonyl Acetate

NONYL NONOXYNOL-5

CAS No.: 9014-93-1 (Generic)

JPN Translation:
ノニルノノキシノール - 5

Empirical Formula:
$C_{24}H_{62}O_6$

Definition: Nonyl Nonoxynol-5 is the ethoxylated alkyl phenol that conforms generally to the formula:

$$(C_9H_{19})_2C_6H_3(OCH_2CH_2)_nOH$$

where n has an average value of 5.

Information Sources: JSQI, TSCA

Chemical Class: Alkoxylated Alcohols

Function: Surfactant - Emulsifying Agent

Technical/Other Names:
PEG-5 Dinonyl Phenyl Ether
Polyethylene Glycol (5) Dinonyl Phenyl Ether
Polyoxyethylene (5) Dinonyl Phenyl Ether

Trade Names:
Emalex DNP-5 (Nihon Emulsion)
Hetoxide DNP-5 (Heterene)

NONYL NONOXYNOL-10

CAS No.: 9014-93-1 (Generic)

JPN Translation:
ノニルノノキシノール - 10

Empirical Formula:
$C_{44}H_{44}O_{11}$

Definition: Nonyl Nonoxynol-10 is the ethoxylated alkyl phenol that conforms generally to the formula:

$$(C_9H_{19})_2C_6H_3(OCH_2CH_2)_nOH$$

where n has an average value of 10.

Information Sources: JSQI, TSCA

Chemical Class: Alkoxylated Alcohols

Function: Surfactant - Emulsifying Agent

Reported Product Category: Hair Bleaches

Technical/Other Names:
PEG-10 Dinonyl Phenyl Ether
Polyethylene Glycol 500 Dinonyl Phenyl Ether
Polyoxyethylene (10) Dinonyl Phenyl Ether

Trade Name:
Hetoxide DNP-10 (Heterene)

NONYL NONOXYNOL-30

CAS No.: 9014-93-1 (Generic)

Definition: Nonyl Nonoxynol-30 is the ethoxylated alkyl phenol that conforms generally to the formula:

$$(C_9H_{19})_2C_6H_3(OCH_2CH_2)_nOH$$

where n has an average value of 30.

Chemical Class: Alkoxylated Alcohols

Functions: Surfactant - Cleansing Agent; Surfactant - Emulsifying Agent

Technical/Other Names:
Polyethylene Glycol (30) Dinonyl Phenyl Ether
Polyoxyethylene (30) Dinonyl Phenyl Ether

Trade Name:
Nonal 530 (Toho)

NONYL NONOXYNOL-49

CAS No.: 9014-93-1 (Generic)

JPN Translation:
ノニルノノキシノール - 49

Definition: Nonyl Nonoxynol-49 is the ethoxylated alkyl phenol that conforms generally to the formula:

$$(C_9H_{19})_2C_6H_3(OCH_2CH_2)_nOH$$

where n has an average value of 49.

Information Sources: JSQI, TSCA

Chemical Class: Alkoxylated Alcohols

Function: Surfactant - Cleansing Agent

Reported Product Categories: Hair Dyes and Colors (All Types Requiring Caution Statements and Patch Tests); Hair Bleaches

Technical/Other Names:
PEG-49 Dinonyl Phenyl Ether
Polyethylene Glycol (49) Dinonyl Phenyl Ether
Polyoxyethylene (49) Dinonyl Phenyl Ether

NONYL NONOXYNOL-100

CAS No.: 9014-93-1 (Generic)

JPN Translation:
ノニルノノキシノール - 100

Definition: Nonyl Nonoxynol-100 is the ethoxylated alkyl phenol that conforms generally to the formula:

$$(C_9H_{19})_2C_6H_3(OCH_2CH_2)_nOH$$

where n has an average value of 100.

Information Sources: JSQI, TSCA

Chemical Class: Alkoxylated Alcohols

Function: Surfactant - Cleansing Agent

Technical/Other Names:
PEG-100 Dinonyl Phenyl Ether
Polyethylene Glycol (100) Dinonyl Phenyl Ether
Polyoxyethylene (100) Dinonyl Phenyl Ether

NONYL NONOXYNOL-150

CAS No.: 9014-93-1 (Generic)

JPN Translation:
ノニルノノキシノール - 150

Definition: Nonyl Nonoxynol-150 is the ethoxylated alkyl phenol that conforms generally to the formula:

$$(C_9H_{19})_2C_6H_3(OCH_2CH_2)_nOH$$

where n has an average value of 150.

Information Sources: JSQI, TSCA

Chemical Class: Alkoxylated Alcohols

Function: Surfactant - Cleansing Agent

Technical/Other Names:
PEG-150 Dinonyl Phenyl Ether
Polyethylene Glycol (150) Dinonyl Phenyl Ether
Polyoxyethylene (150) Dinonyl Phenyl Ether

NONYL NONOXYNOL-7 PHOSPHATE

CAS Nos.	EINECS No.
66172-78-9	266-215-9
66172-83-6	

Empirical Formula:
$C_{38}H_{71}O_{11}P$

Definition: Nonyl Nonoxynol-7 Phosphate is a complex mixture of esters of phosphoric acid and nonyl nonoxynol-7.

Chemical Class: Phosphorus Compounds

Functions: Surfactant - Cleansing Agent; Surfactant - Emulsifying Agent

---

The inclusion of any compound in the *Dictionary and Handbook* does not indicate that use of that substance as a cosmetic ingredient complies with the laws and regulations governing such use in the United States or any other country.

Technical/Other Names:
20-(Dinonylphenoxy)-3,6,5,12,15,18-Hexaoxaeicosan-1-ol, Hydrogen Phosphate
3,6,9,12,15,18-Hexaoxaeicosan-1-ol, 20-(Dinonlphenoxy)-, Hydrogen Phosphate
PEG-7 Dinonyl Phenyl Ether Phosphate
Polyethylene Glycol (7) Dinonyl Phenyl Ether Phosphate
Polyoxyethylene (7) Dinonyl Phenyl Ether Phosphate

NONYL NONOXYNOL-8 PHOSPHATE

CAS No.: 39464-64-7 (Generic)

Definition: Nonyl Nonoxynol-8 Phosphate is a complex mixture of esters of phosphoric acid and nonyl noxynol-8.

Information Source: TSCA

Chemical Class: Phosphorus Compounds

Functions: Surfactant - Cleansing Agent; Surfactant - Emulsifying Agent

Technical/Other Names:
PEG-8 Dinonyl Phenyl Ether Phosphate
Polyethylene Glycol 400 Dinonyl Phenyl Ether Phosphate
Polyoxyethylene (8) Dinonyl Phenyl Ether Phosphate

Trade Name:
Phosphanol RM-410 (Toho)

NONYL NONOXYNOL-9 PHOSPHATE

CAS No.	EINECS No.
66172-82-5	266-218-5

Definition: Nonyl Nonoxynol-9 Phosphate is a complex mixture of esters of phosphoric acid and nonyl nonoxynol-9.

Chemical Class: Phosphorus Compounds

Functions: Surfactant - Cleansing Agent; Surfactant - Emulsifying Agent

Technical/Other Names:
PEG-9 Dinonyl Phenyl Ether Phosphate
Polyethylene Glycol 450 Dinonyl Phenyl Ether Phosphate
Polyoxyethylene (9) Dinonyl Phenyl Ether Phosphate

NONYL NONOXYNOL-10 PHOSPHATE

Definition: Nonyl Nonoxynol-10 Phosphate is a complex mixture of esters of phosphoric acid and Nonyl Nonoxynol-10 (q.v.).

Chemical Class: Phosphorus Compounds

Functions: Surfactant - Cleansing Agent; Surfactant - Emulsifying Agent

Technical/Other Names:
PEG-10 Dinonyl Phenyl Ether Phosphate
Polyethylene Glycol 500 Dinonyl Phenyl Ether Phosphate
Polyoxyethylene (10) Dinonyl Phenyl Ether Phosphate

NONYL NONOXYNOL-11 PHOSPHATE

CAS No.: 39464-64-7 (Generic)

Definition: Nonyl Nonoxynol-11 Phosphate is a complex mixture of esters of phosphoric acid and nonyl nonoxynol-11.

Information Source: TSCA

Chemical Class: Phosphorus Compounds

Functions: Surfactant - Cleansing Agent; Surfactant - Emulsifying Agent

Technical/Other Names:
PEG-11 Dinonyl Phenyl Ether Phosphate
Polyethylene Glycol (11) Dinonyl Phenyl Ether Phosphate
Polyoxyethylene (11) Dinonyl Phenyl Ether Phosphate

Trade Name:
Phosphanol RM-510 (Toho)

NONYL NONOXYNOL-15 PHOSPHATE

Definition: Nonyl Nonoxynol-15 Phosphate is a complex mixture of esters of phosphoric acid and nonyl nonoxynol-15.

Chemical Class: Phosphorus Compounds

Function: Surfactant - Cleansing Agent

Technical/Other Names:
PEG-15 Dinonyl Phenyl Ether Phosphate
Polyethylene Glycol (15) Dinonyl Phenyl Ether Phosphate
Polyoxyethylene (15) Dinonyl Phenyl Ether Phosphate

Trade Name:
Phosphanol RM-710 (Toho)

NONYL NONOXYNOL-24 PHOSPHATE

Definition: Nonyl Nonoxynol-24 Phosphate is a complex mixture of esters of phosphoric acid and nonyl nonoxynol-24.

Chemical Class: Phosphorus Compounds

Function: Surfactant - Cleansing Agent

Technical/Other Names:
PEG-24 Dinonyl Phenyl Ether Phosphate
Polyethylene Glycol (24) Dinonyl Phenyl Ether Phosphate
Polyoxyethylene (24) Dinonyl Phenyl Ether Phosphate

NOPYL ACETATE

CAS No.	EINECS No.
128-51-8	204-891-9

Empirical Formula:
$C_{13}H_{20}O_2$

Definition: Nopyl Acetate is the organic compound that conforms to the formula:

Information Source: RIFM

Chemical Class: Esters

Function: Fragrance Ingredient

Technical/Other Names:
2-(6,6-Dimethylbicyclo[3,1,1]hept-2-en-2yl)ethyl acetate
Nopol Acetate
Nopyl acetate (RIFM)
2-Norpinene-2-ethanol, 6,6-dimethyl-, Acetate

NORDIHYDROGUAIARETIC ACID

CAS No.	EINECS No.
500-38-9	207-903-0

Empirical Formula:
$C_{18}H_{22}O_4$

Definition: Nordihydroguaiaretic Acid is the organic compound that conforms to the formula:

Information Sources: 21CFR175.300, 21CFR181.22, 21CFR181.24, 21CFR189.165, CTFA S, INN, MI-13(6726), TSCA, USAN

Chemical Class: Phenols

Function: Antioxidant

Technical/Other Names:
1,2-Benzenediol, 4,4'-(2,3-Dimethyl-1,4-Butanediyl)Bis-

1,4-Bis(3,4-dihydroxyphenyl)-2,3-
dimethylbutane
Dihydronorguaiaretic Acid
4,4'-(2,3-Dimethyl-1,4-Butanediyl)Bis-1,2-
Benzenediol
Pyrocatechol, 4,4'-(2,3-diemthyltetra-
methylene)di-

Trade Name:
N.D.G.A. (MMP)

NORVALINE

CAS Nos.	EINECS Nos.
760-78-1 (dl-alpha)	212-082-7
2013-12-9 (D-Form)	217-936-2
6600-40-4 (L-Form)	229-543-3

Empirical Formula:
$C_5H_{11}NO_2$

Definition: Norvaline is the amino acid that conforms to the formula:

$$CH_3(CH_2)_2CHCOOH$$
$$|$$
$$NH_2$$

Information Source: MI-13(6750)

Chemical Class: Amino Acids

Functions: Hair Conditioning Agent; Skin-Conditioning Agent - Miscellaneous

Technical/Other Names:
2-Aminopentanoic Acid
2-Aminovaleric Acid

NUPHAR JAPONICUM ROOT EXTRACT

JPN Translation:
コウホネエキス

Definition: Nuphar Japonicum Root Extract is an extract of the roots of *Nuphar japonicum. See "Regulatory and Ingredient Use Information," regarding the labeling names for botanical ingredients in Volume 1, Introduction, Part A.*

Information Source: JCLS

Chemical Class: Biological Products

Function: Skin-Conditioning Agent - Miscellaneous

Technical/Other Names:
Extract of Nuphar Japonicum Root
Nuphar Japonicum Extract

Trade Name Mixture:
Kouhone Liquid B (Ichimaru Pharcos)

NUPHAR LUTEUM ROOT EXTRACT

Definition: Nuphar Luteum Root Extract is an extract of the roots of *Nuphar luteum.*

See "Regulatory and Ingredient Use Information," regarding the labeling names for botanical ingredients in Volume 1, Introduction, Part A.

Chemical Class: Biological Products

Function: Skin-Conditioning Agent - Miscellaneous

Technical/Other Name:
Extract of Nuphar Luteum Root

Trade Name Mixture:
Cosflor Yellow Water Lily HG-1 (A & E Connock)

NYLON-6

CAS No.: 25038-54-4

JPN Translation:
ナイロン - 6

Empirical Formula:
$(C_6H_{11}NO)_n$

Definition: Nylon-6 is the polyamide that conforms to the formula:

$$\left[-C(CH_2)_5NH- \right]_x$$

Information Sources: 21CFR177.1500, 21CFR177.2260, 21CFR177.2470, 21CFR177.2480, INN, MI-13(6768), USAN

Chemical Class: Synthetic Polymers

Functions: Bulking Agent; Opacifying Agent

Technical/Other Names:
Caprolactam Polymer
2H-Azepin-2-one, Hexahydro-, Homo-
polymer
Hexanoic Acid, 6-amino-, Homopolymer
Polycaproamide
Polycaprolactam
Poly[Imino(1-Oxo-1,6 Hexanediyl)]

Trade Names:
Inducos 10 Series (Induchem)
Micropan 777 (Chemopharma)
Orgasol 1002 EXD BL 10 COS 10 Microns
(Atofina)
Orgasol 1002 D Nat Cos (Lipo)
Orgasol 1002 D NAT COS 20 Microns
(Atofina)

Trade Name Mixtures:
Fiberlon Y2 (LCW)
Fiberlon Y10 (LCW)
Morphotex (Kishimoto)
Morphotex Blue (Kishimoto)
Morphotex Red (Kishimoto)
Morphotex Yellow (Kishimoto)
Orgasol 1002D White 10 Cos. (Lipo)

Toshiki Nylon AP-18K (Nikko)
Toshiki Nylon IVCF-1 (Nikko)

NYLON 6/12

JPN Translation:
ナイロン - 6/12

Definition: Nylon 6/12 is a polyamide formed by the reaction of caprolactam and dodecyllactam.

Chemical Class: Synthetic Polymers

Functions: Absorbent; Bulking Agent; Opacifying Agent

Trade Names:
Orgasol 4000 EXD Nat Cos (Lipo)
Orgasol 4000 EXD NAT COS 10 Microns
(Atofina)

Trade Name Mixtures:
Liponyl N30SA (Lipo)
Liponyl Yeast Extract AE (Lipo)

NYLON-11

CAS No.: 25035-04-5

JPN Translation:
ナイロン - 11

Empirical Formula:
$(C_{11}H_{21}NO)_n$

Definition: Nylon-11 is the polyamide that conforms to the formula:

$$\left[-C(CH_2)_{10}NH- \right]_x$$

Information Sources: 21CFR177.1500, 21CFR177.2260

Chemical Class: Synthetic Polymers

Functions: Bulking Agent; Opacifying Agent

Technical/Other Names:
11-Aminoundecanoic Acid Polymer
Poly[Imino(1-Oxo-1,11-Undecanediyl)]
Polyundecaneamide

NYLON-12

CAS No.: 25038-74-8

JPN Translation:
ナイロン - 12

Empirical Formula:
$(C_{12}H_{23}NO)_n$

Definition: Nylon-12 is a polyamide derived from 12-aminododecanoic acid. It conforms generally to the formula:

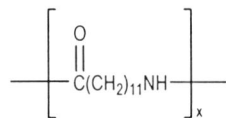

Information Sources: 21CFR175.300, 21CFR177.1500, 21CFR177.2260

Chemical Class: Synthetic Polymers

Functions: Bulking Agent; Opacifying Agent

Reported Product Categories: Foundations; Face Powders; Makeup Preparations (Not eye), Misc.; Eye Shadows; Blushers (All types); Mascara; Bath Capsules; Eyeliners; Makeup Bases; Face and Neck Preparations (Excluding Shaving Preparations); Eye Makeup Preparations, Misc.; Lipsticks; Moisturizing Preparations; Powders (Dusting and Talcum, Excluding Aftershave Talcs); Rouges; Skin Care Preparations, Misc.

Technical/Other Names:
Adipic Acid-Hexamethylenediamine Polymer
Azacyclotridecane-2-One, Homopolymer
1,8-Diazacyclotetradecane-2,7-dione Homopolymer
Dodecalactam Polymer
Hexanedioic Acid, Polymer with 1,6-hexanediamine
Laurolactum Polymer
Polydodecanamide
Poly(hexamethyleneadipamide)
Polylauramide

Trade Names:
Custom Nylon Poly (nylon powder) (Custom Ingredients)
ESP Nylon 12-5 (Earth Supplied Products)
ESP Nylon 12-10 (Earth Supplied Products)
ESP Nylon 12-22 (Earth Supplied Products)
ESP Nylon 12-57 (Earth Supplied Products)
NYLONPOLY WL 5 (C.I.T.)
NYLONPOLY WL 10 (C.I.T.)
NYLONPOLY WL12 (C.I.T.)
Orgasol 2002 EXD Nat Cos (Lipo)
Orgasol 2002 EXD NAT COS 10 Microns (Atofina)
Orgasol 2002 D Ex Nat Cos (Lipo)
Orgasol 2002 D Nat Cos (Lipo)
Orgasol 2002 D NAT COS 20 Microns (Atofina)
Orgasol 2002 UD Nat Cos (Lipo)
Orgasol 2002 UD NAT COS 5 Microns (Atofina)
SP-500 (Kobo)
Vestosint 2158 (Creanova)

Trade Name Mixtures:
Elesponge AHA LS 8911 B (Laboratoires Serobiologiques)
Elesponge Melhydran (Laboratoires Sero-biologiques)
Liponyl 10-BN 6058 (Lipo)
Liponyl 10-BN 6069 (Lipo)
Liponyl 20-LL (Lipo)
NYLONPOLY WL10 AF (C.I.T.)
NYLONPOLY WL 10 LL (C.I.T.)
NYLONPOLY WL 5 PFC (Creations Couleurs)
NYLONPOLY WL10 SIL (C.I.T.)
Sphingoceryl Powder VEG (Laboratoires Serobiologiques)

NYLON-66

CAS No.: 32131-17-2

JPN Translation:
ナイロン - 66

Empirical Formula:
$(C_{12}H_{22}N_2O_2)_x$

Definition: Nylon-66 is a polyamide formed by the reaction of adipic acid with hexylenediamine. It conforms generally to the formula:

$$\left[\begin{array}{c} O \quad\quad O \\ \| \quad\quad \| \\ C(CH_2)_4C-NH(CH_2)_6NH \end{array} \right]_x$$

Information Sources: 21CFR175.300, 21CFR177.1200, 21CFR177.1390, 21CFR177.1395, 21CFR177.1500, 21CFR177.2260, 21CFR177.2470, 21CFR177.2480, 21CFR177.2600, 21CFR178.2010, 21CFR179.45, JCIC, JCLS, MI-13(6767), TSCA

Chemical Class: Synthetic Polymers

Functions: Bulking Agent; Opacifying Agent

Reported Product Categories: Face Powders; Blushers (All types); Makeup Bases; Foundations; Lipsticks; Moisturizing Preparations; Powders (Dusting and Talcum, Excluding Aftershave Talcs); Eye Shadows; Manicuring Preparations, Misc.; Mascara

Technical/Other Names:
Nylon
Nylon Powder
Poly[Imino(1,6-Dioxo-1,6-Hexanediyl)-Imino-1,6-Hexanediyl]

NYLON-611

Definition: Nylon-611 is a polyamide formed by the reaction undecanoic acid with hexylenediamine. It conforms generally to the formula:

$$\left[\begin{array}{c} O \quad\quad O \\ \| \quad\quad \| \\ C(CH_2)_9C-NH(CH_2)_6NH \end{array} \right]_x$$

Chemical Classes: Amides; Synthetic Polymers

Function: Absorbent

NYLON-12/6/66 COPOLYMER

JPN Translation:
ナイロン - 12/6/66 コポリマー

Definition: Nylon-12/6/66 Copolymer is a copolymer formed from the monomers used in the manufacture of Nylon-12 (q.v.), Nylon-6 (q.v.) and Nylon-66 (q.v.).

Chemical Class: Synthetic Polymers

Function: Film Former

Trade Name:
Abifor 501 (Billeter)

NYLON-611/DIMETHICONE COPOLYMER

Definition: Nylon-611/Dimethicone Copolymer is a copolymer of Nylon-611 (q.v.) and Dimethicone (q.v.) monomers.

Chemical Classes: Amides; Siloxanes and Silanes; Synthetic Polymers

Functions: Hair Conditioning Agent; Skin-Conditioning Agent - Miscellaneous; Viscosity Increasing Agent - Nonaqueous

Trade Name Mixture:
Dow Corning 2-8178 Gellant (Dow Corning)

NYMPHAEA ALBA FLOWER EXTRACT

CAS No.	EINECS No.
84696-27-5	283-645-2

Definition: Nymphaea Alba Flower Extract is an extract of the flowers of *Nymphaea alba*. See *"Regulatory and Ingredient Use Information,"* regarding the labeling names for botanical ingredients in Volume 1, Introduction, Part A.

Chemical Class: Biological Products

Function: Skin-Conditioning Agent - Miscellaneous

Technical/Other Name:
Water Lily (Nymphaea Alba) Extract

Trade Name Mixtures:
Aquabios of Water Lily Flowers (CEP (Solabia))
ASTRINGENT EMOLLIENT COMPLEX (Greentech S.A)
Hydralphatine Asiatique (Lanatech)
Nympheline (CEP (Solabia))
Oleat of Water Lily (CEP (Solabia))

Sepicalm VG (SEPPIC)
Vegebios of Water Lily (CEP (Solabia))
Water Lily Flowers Milk (CEP (Solabia))
White Water Lily Flower BG (Alban Muller)
White Water Lily HS (Alban Muller)

NYMPHAEA ALBA ROOT EXTRACT

Definition: Nymphaea Alba Root Extract is an extract of the roots of the water lily, *Nymphaea alba. See "Regulatory and Ingredient Use Information," regarding the labeling names for botanical ingredients in Volume 1, Introduction, Part A.*

Chemical Class: Biological Products

Function: Not Reported

Technical/Other Names:
Extract of Nymphaea Alba Root
Extract of Water Lily
Water Lily Extract
Water Lily (Nymphaea Alba) Root Extract
White Lotus Root Extract

Trade Name Mixtures:
Herbasec Pond Lily (Cosmetochem)
(Cosmetochem International Ltd.)
Herbasol Extract Pond Lily (Cosmetochem)
(Cosmetochem International Ltd.)
Nymphaeae-Alba-Root Extract HS 3136 G
(Grau)
White Water Lily Rhizome HS (Alban Muller)
White Water Lily Rhizome LS (Alban Muller)

NYMPHAEA LOTUS ROOT EXTRACT

Definition: Nymphaea Lotus Root Extract is an extract of the roots of *Nymphaea lotus. See "Regulatory and Ingredient Use Information," regarding the labeling names for botanical ingredients in Volume 1, Introduction, Part A.*

Chemical Class: Biological Products

Function: Cosmetic Astringent

Technical/Other Name:
Extract of Nymphaea Lotus Root

NYMPHAEA ODORATA

Definition: *See "Regulatory and Ingredient Use Information," regarding EU labeling names for botanical ingredients in Volume 1, Introduction, Part A.*

Chemical Class: Biological Products

Technical/Other Name:
Nymphaea Odorata Root Extract (U.S.)

NYMPHAEA ODORATA ROOT EXTRACT

CAS No. 90064-35-0
EINECS No. 290-097-8

Definition: Nymphaea Odorata Root Extract is an extract of the roots of the water lily, *Nymphaea odorata. See "Regulatory and Ingredient Use Information," regarding the labeling names for botanical ingredients in Volume 1, Introduction, Part A.*

Chemical Class: Biological Products

Function: Not Reported

Technical/Other Names:
Extract of Nymphaea Odorata Root
Extract of Water Lily
Nymphaea Odorata (EU)
Water Lily Extract
Water Lily (Nymphaea Odorata) Root Extract
Water Lily Root Extract

Trade Name Mixtures:
Actiphyte of Water Lily (Active Organics)
Actiphyte of White Pond Lily BG50 (Active Organics)
Actiphyte of White Pond Lily GL50 (Active Organics)
Actiphyte of White Pond Lily Lipo S (Active Organics)
Actiphyte of White Pond Lily PG50 (Active Organics)
Actiphyte of White Pond Lily Root (Active Organics)
Cosflor White Pond Lily HGS (A & E Connock)
VT-224 Extract of White Lily Pond (Vege-Tech)

- O -

OAK ROOT EXTRACT

Definition: Oak Root Extract is the extract of the roots of oak tree species *Quercus*.

Information Source: 21CFR172.510

Chemical Class: Biological Products

Function: Not Reported

Reported Product Categories: Skin Care Preparations, Misc.; Moisturizing Preparations; Bath Oils, Tablets, and Salts; Cleansing Products (Cold Creams, Cleansing Lotions, Liquids and Pads); Night Skin Care Preparations

Technical/Other Name:
Extract of Oak Root

Trade Name:
Meristem Extract (Grau)

OATAMIDE MEA

Definition: Oatamide MEA is a mixture of ethanolamides of the fatty acids derived from oat kernel oil, It conforms generally to the formula:

$$RC\overset{\overset{O}{\parallel}}{-}NHCH_2CH_2OH$$

where RCO- represents the fatty acids derived from Avena Sativa (Oat) Kernel Oil (q.v.).

Chemical Class: Alkanolamides

Functions: Surfactant - Foam Booster; Viscosity Increasing Agent - Aqueous

Trade Name:
Natrlfine A (Finetex)

Trade Name Mixture:
Natrlfine T-1 (Finetex)

OATAMIDOPROPYL BETAINE

Definition: Oatamidopropyl Betaine is the zwitterion (inner salt) that conforms generally to the formula:

$$RC\overset{\overset{O}{\parallel}}{-}NH(CH_2)_3\overset{\overset{CH_3}{|}}{\underset{\underset{CH_3}{|}}{N^+}}-CH_2COO^-$$

where RCO- represents the fatty acids derived from Avena Sativa (Oat) Kernel Oil (q.v.).

Chemical Class: Betaines

Functions: Antistatic Agent; Skin-Conditioning Agent - Miscellaneous; Surfactant - Cleansing Agent; Surfactant - Foam Booster; Viscosity Increasing Agent - Aqueous

Trade Name:
Natrlfine AB-40 (Finetex)

OATAMIDOPROPYL DIMETHYLAMINE

Definition: Oatamidopropyl Dimethylamine is the amidoamine that conforms to the formula:

$$RC\overset{\overset{O}{\parallel}}{-}NH(CH_2)_3-N\overset{\nearrow CH_3}{\underset{\searrow CH_3}{}}$$

where RCO- represents the fatty acids derived from Avena Sativa (Oat) Kernel Oil (q.v.).

Chemical Class: Amines

Function: Antistatic Agent

Trade Name:
Natrlfine MD (Finetex)

OAT AMINO ACIDS

Definition: Oat Amino Acids is a mixture of amino acids derived by the complete hydrolysis of oat protein.

Chemical Class: Amino Acids

Functions: Hair Conditioning Agent; Skin-Conditioning Agent - Miscellaneous

Trade Name:
Aqua Pro II OAA (MGP)

OCIMUM BASILICUM

Definition: *See "Regulatory and Ingredient Use Information," regarding EU labeling names for botanical ingredients in Volume 1, Introduction, Part A.*

Chemical Class: Biological Products

Technical/Other Names:
Ocimum Basilicum (Basil) Extract (U.S.)
Ocimum Basilicum (Basil) Leaf Powder (U.S.)
Ocimum Basilicum (Basil) Oil (U.S.)

OCIMUM BASILICUM (BASIL) EXTRACT

CAS No.	EINECS No.
84775-71-3	283-900-8

Definition: Ocimum Basilicum (Basil) Extract is an extract of the flowers and leaves of the basil, *Ocimum basilicum. See "Regulatory and Ingredient Use Information," regarding the labeling names for botanical ingredients in Volume 1, Introduction, Part A.*

Information Sources: 21CFR182.20, RIFM

Chemical Class: Biological Products

Functions: Fragrance Ingredient; Skin-Conditioning Agent - Miscellaneous

Reported Product Category: Shampoos (Non-coloring)

Technical/Other Names:
Basil Extract
Basil oleoresin (Ocimum basilicum L.) (RIFM)
Common Basil Extract
Extract of Basil
Extract of Ocimum Basilicum
Extract of Sweet Basil
Ocimum Basilicum (EU)
Ocimum basilicum Extract
Sweet Basil Extract

Trade Name Mixtures:
Actiphyte of Basil BG50 (Active Organics)
Actiphyte of Basil GL50 (Active Organics)
Actiphyte of Basil Lipo S (Active Organics)
Actiphyte of Basil PG50 (Active Organics)
Aromaphyte of Basil (Active Organics)
Basil Extract (Libiol)
Basil HS (Alban Muller)
Basil LS (Alban Muller)
Campo Vasa Kovil Tulsi (Campo)
Common Basil Herb Extract HS 2580 G (Grau)
Herbasol Extract Basil (Cosmetochem) (Cosmetochem International Ltd.)
Vegebios of Basil (CEP (Solabia))
VT-170 Extract of Basil (Vege-Tech)

OCIMUM BASILICUM (BASIL) LEAF POWDER

Definition: Ocimum Basilicum (Basil) Leaf Powder is the powder obtained from the ground leaves of *Ocimum basilicum. See "Regulatory and Ingredient Use Information," regarding the labeling names for botanical ingredients in Volume 1, Introduction, Part A.*

Chemical Class: Biological Products

Function: Skin-Conditioning Agent - Miscellaneous

Technical/Other Name:
Ocimum Basilicum (EU)

Trade Name:
Basil Powder (Aveda)

OCIMUM BASILICUM (BASIL) OIL

CAS No.: 8015-73-4

Definition: Ocimum Basilicum (Basil) Oil is the volatile oil obtained from *Ocimum basilicum. See "Regulatory and Ingredient Use Information," regarding the labeling names for botanical ingredients in Volume 1, Introduction, Part A.*

Information Sources: MI-13(6808), RIFM

Chemical Class: Essential Oils

Functions: Fragrance Ingredient; Skin-Conditioning Agent - Miscellaneous

Reported Product Category: Skin Care Preparations, Misc.

Technical/Other Names:
AEC Basil Oil
Basil (Ocimum basilicum L.) (RIFM)
Basil Oil
Basil oil (Ocimum basilicum L.) (RIFM)
Ocimum Basilicum (EU)
Ocimum Basilicum Oil
Oil, Essential, Basil
Oil of Basil

Trade Name:
Basil Oil Exotic (Kao Europe)

Trade Name Mixtures:
Aromaphyte of Basil (Active Organics)
Tonic Phytospa (Alban Muller)

OCIMUM SANCTUM LEAF EXTRACT

Definition: Ocimum Sanctum Leaf Extract is an extract of the leaves of *Ocimum sanctum. See "Regulatory and Ingredient Use Information," regarding the labeling names for botanical ingredients in Volume 1, Introduction, Part A.*

Chemical Class: Biological Products

Function: Skin-Conditioning Agent - Miscellaneous

Technical/Other Name:
Extract of Ocimum Sanctum Leaves

Trade Name Mixture:
Pronalen Sensitive Skin (Provital/Centerchem)

OCIMUM TENUIFLORUM EXTRACT

Definition: Ocimum Tenuiflorum Extract is an extract of the plant *Ocimum tenuiflorum. See "Regulatory and Ingredient Use Information," regarding the labeling names for botanical ingredients in Volume 1, Introduction, Part A.*

Chemical Class: Biological Products

Function: Not Reported

Technical/Other Names:
Extract of Holy Basil
Extract of Ocimum Tenuiflorum

Trade Name Mixtures:
Campo Kovil Tulsi (Campo)
Tulsi Extract (Carlisle)

OCIMUM TENUIFLORUM OIL

Definition: Ocimum Tenuiflorum Oil is the wax obtained from the leaf of *Ocimum tenuiflorum. See "Regulatory and Ingredient Use Information," regarding the labeling names for botanical ingredients in Volume 1, Introduction, Part A.*

Chemical Class: Waxes

Functions: Hair Conditioning Agent; Skin-Conditioning Agent - Miscellaneous

Technical/Other Name:
Waxes, Ocimum Tenuiflorum

Trade Name:
Tulsi (Heritage Bio-Natural)

OCTACOSANYL GLYCOL

CAS No. 97338-11-9

EINECS No. 306-603-8

Empirical Formula:
$C_{28}H_{58}O_2$

Definition: Octacosanyl Glycol is the diol that conforms to the formula:

$$CH_3(CH_2)_{25}\underset{|}{C}HCH_2OH$$
$$OH$$

Chemical Class: Alcohols

Functions: Emulsion Stabilizer; Viscosity Increasing Agent - Nonaqueous

Technical/Other Name:
1,2-Octacosanediol

OCTACOSANYL GLYCOL ISOSTEARATE

Empirical Formula:
$C_{46}H_{92}O_3$

Definition: Octacosanyl Glycol Isostearate is the ester of Octacosanyl Glycol (q.v.) and Isostearic Acid (q.v.).

Chemical Class: Esters

Functions: Skin-Conditioning Agent - Occlusive; Viscosity Increasing Agent - Aqueous

OCTACOSATRIMONIUM CHLORIDE

JPN Translation:
アルキル（C28）トリモニウムクロリド

Definition: Octacosatrimonium Chloride is the quaternary ammonium salt that conforms to the formula:

$$\left[R-\overset{\overset{\displaystyle CH_3}{|}}{\underset{\underset{\displaystyle CH_3}{|}}{N}}-CH_3 \right]^{+} \quad Cl^{-}$$

where R represents the octacosanyl (C28) alkyl group.

Information Source: JCIC

Chemical Class: Quaternary Ammonium Compounds

Functions: Antistatic Agent; Hair Conditioning Agent

Technical/Other Name:
Alkyl (28) Trimethyl Ammonium Chloride

OCTADECANE

CAS No. 593-45-3

EINECS No. 209-790-3

Empirical Formula:
$C_{18}H_{38}$

Definition: Octadecane is the saturated straight chain alkane that conforms to the formula:

$$CH_3(CH_2)_{16}CH_3$$

Information Source: RIFM

Chemical Class: Hydrocarbons

Functions: Fragrance Ingredient; Skin-Conditioning Agent - Emollient; Solvent

Technical/Other Names:
C18-n-Alkane
Octadecane (RIFM)
n-Octadecane

OCTADECENE

CAS No. 27070-58-2

EINECS No. 248-205-6

Empirical Formula:
$C_{18}H_{36}$

Definition: Octadecene is the hydrocarbon that conforms to the formula:

$$CH_2=CH(CH_2)_{15}CH_3$$

Chemical Class: Hydrocarbons

Functions: Solvent; Viscosity Decreasing Agent

Trade Name:
GULFTENE 18 (Chevron Chemical)

OCTADECENEDIOIC ACID

CAS No.: 20701-68-2

Empirical Formula:
$C_{18}H_{32}O_4$

Definition: Octadecenedioic Acid is the organic compound that conforms to the formula:

$$HO-\overset{\displaystyle O}{\overset{\|}{C}}(CH_2)_7CH=CH(CH_2)_7\overset{\displaystyle O}{\overset{\|}{C}}-OH$$

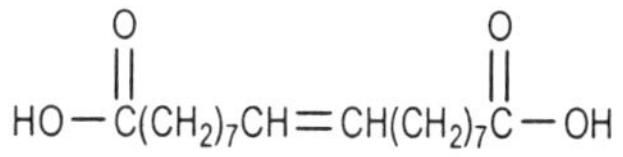

Chemical Class: Carboxylic Acids

Function: Not Reported

Technical/Other Name:
9-Octadecenedioic Acid, (9Z)-

Trade Name:
Pripure 2151 (Uniqema Europe)

OCTADECENE/MA COPOLYMER

CAS No.: 25266-02-8

Empirical Formula:
$(C_{18}H_{36} \cdot C_4H_2O_3)_x$

Definition: Octadecene/MA Copolymer is a polymer of octadecene and maleic anhydride monomers.

Information Source: TSCA

Chemical Class: Synthetic Polymers

Functions: Emulsion Stabilizer; Film Former; Viscosity Increasing Agent - Aqueous; Viscosity Increasing Agent - Nonaqueous

Technical/Other Names:
2,5-Furandione, Polymer with 1-Octadecene
Octadecene/Maleic Anhydride Copolymer
1-Octadecene, Polymer with 2,5-Furandione

Trade Name Mixture:
Stantiv OMA-1 (CasChem)

OCTADECYL DI-t-BUTYL-4-HYDROXY-HYDROCINNAMATE

CAS No. 2082-79-3

EINECS No. 218-216-0

Empirical Formula:
$C_{35}H_{62}O_3$

Definition: Octadecyl Di-t-butyl-4-hydroxy-hydrocinnamate is the organic compound that conforms to the formula:

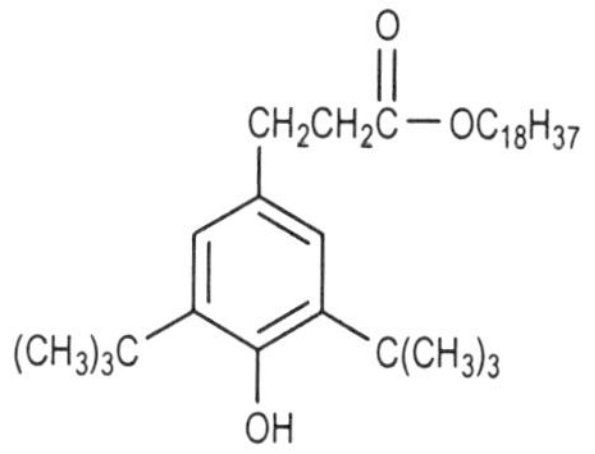

Chemical Classes: Esters; Phenols

Function: Antioxidant

Technical/Other Names:
Benzenepropanoic Acid, 3,5-Bis(1,1-Dimethylethyl)-4-Hydroxy-, Octadecyl Ester
Octadecyl 3-(2,5-Di-Tert-Butyl-4-Hydroxy-phenyl)Propionate

Trade Name:
Tinogard TS (Ciba Specialty Chemicals)

OCTANE

CAS No. 111-65-9

EINECS No. 203-892-1

Empirical Formula:
C_8H_{18}

Definition: Octane is the hydrocarbon that conforms to the formula:

$$CH_3(CH_2)_6CH_3$$

Information Sources: MI-13(6782), TSCA

Chemical Class: Hydrocarbons

Function: Solvent

Technical/Other Name:
n-Octane

Trade Name:
n-Octane (Merck KGaA)

OCTANICOTINOYL EPIGALLOCATECHIN GALLATE

Empirical Formula:
$C_{70}H_{42}N_8O_{19}$

Definition: Octanicotinoyl Epigallocatechin Gallate is the organic compound that conforms to the formula:

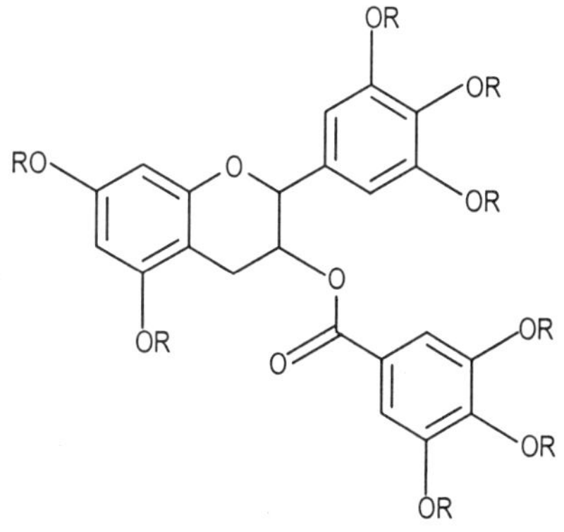

where R represents the nicotinoyl moiety.

Chemical Class: Esters

Function: Antioxidant

Trade Name:
Phyto-SEGNA (Pacific)

OCTENE

CAS No. 111-66-0

EINECS No. 203-893-7

Empirical Formula:
C_8H_{16}

Definition: Octene is the hydrocarbon that conforms to the formula:

$$CH_2=CH(CH_2)_5CH_3$$

Information Source: MI-13(1770)

Chemical Class: Hydrocarbons

Functions: Solvent; Viscosity Decreasing Agent

Trade Name:
GULFTENE 8 (Chevron Chemical)

OCTENIDINE HCl

CAS No. 70775-75-6

EINECS No. 274-861-8

Empirical Formula:
$C_{36}H_{62}N_4 \cdot 2HCl$

Definition: Octenidine HCl is the organic compound that conforms to the formula:

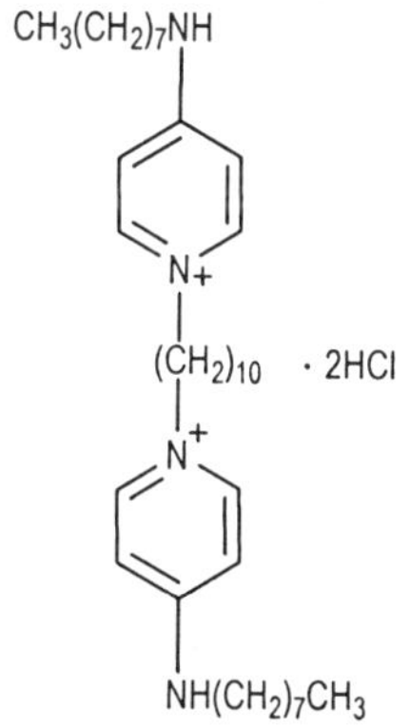

See "Regulatory and Ingredient Use Information," regarding the labeling names for U.S. OTC Drug Ingredients in Volume 1, Introduction, Part A. See Reported Ingredient Functions-The Cosmetic Drug Distinction, in Regulatory and Ingredient Use Information, Volume I, Part A.

Information Source: MI-13(6787)

Chemical Class: Heterocyclic Compounds

Function: Antimicrobial Agent

Technical/Other Name:
1-Octanamine, N,N'-(1,10-Decanediyldi-1 (4H)-Pyridinyl-4-ylidene)Bis-, Dihydrochloride

Trade Name:
Octenidine Dihydrochloride (Schulke & Mayr)

OCTOCRYLENE

CAS No. 6197-30-4
EINECS No. 228-250-8

Empirical Formula:
$C_{24}H_{27}NO_2$

Definition: Octocrylene is the substituted acrylate that conforms to the formula:

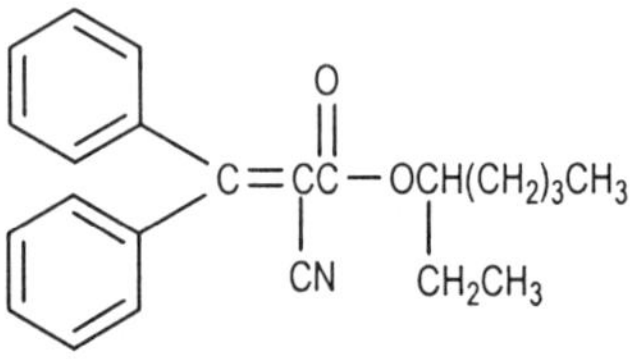

In the United States, Octocrylene may be used as an active ingredient in OTC drug products. When used as an active drug ingredient, the established name is *Octocrylene. See "Regulatory and Ingredient Use Information," regarding the labeling names for U.S. OTC Drug Ingredients in Volume 1, Introduction, Part A.*

Information Sources: CTFA D, EEC(VII/1-10), INN, OTC-I-SU, TSCA, USAN, USP XXIV

Chemical Class: Esters

Functions: Sunscreen Agent; Ultraviolet Light Absorber

Reported Product Categories: Basecoats and Undercoats; Suntan Gels, Creams, and Liquids; Suntan Preparations, Misc.

Technical/Other Names:
2-Cyano-3,3-Diphenyl Acrylic Acid, 2-Ethylhexyl Ester
2-Ethylhexyl 2-Cyano-3,3-Diphenylacrylate
2-Ethylhexyl 2-Cyano-3,3-Diphenyl-2-Propenoate
2-Ethylhexyl 2-Cyano-3-Phenylcinnamate
2-Propenoic Acid, 2-Cyano-3,3-Diphenyl-, 2-Ethylhexyl Ether
UV Absorber-3

Trade Names:
Custoscreen OC (octocrylene) (Custom Ingredients)
Escalol 597 (International Specialty Products)
Eusolex OCR (Merck KGaA/EMD Chemicals Inc.)

Neo Heliopan, Type 303 (Haarmann & Reimer GmbH)
Parsol 340 (Hoffmann-La Roche)
Parsol 340 (Roche)
Uvinul N 539 T (BASF)
UVSOB 320 (LC United)

OCTOPAMINE

CAS No. 104-14-3
EINECS No. 203-179-5

Empirical Formula:
$C_8H_{11}NO_2$

Definition: Octopamine is the organic compound that conforms to the formula:

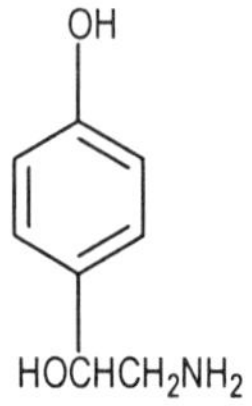

Information Source: MI-13(6791)

Chemical Class: Phenols

Function: Skin-Conditioning Agent - Miscellaneous

Technical/Other Names:
Benzenemethanol, α-(Aminomethyl)-4-Hydroxy-
4-Hydroxyphenylethanolamine

Trade Name:
Para-hydroxy-Phenyl Aminoethanol Pure (Sederma)

OCTOXYNOL-1

CAS Nos.: 2315-67-5; 9002-93-1 (Generic); 9004-87-9 (Generic); 9036-19-5 (Generic)

JPN Translation:
オクトキシノール - 1

Empirical Formula:
$C_{16}H_{26}O_2$

Definition: Octoxynol-1 is the ethoxylated alkyl phenol that conforms generally to the formula:

$$C_8H_{17}C_6H_4OCH_2CH_2OH$$

Information Sources: 21CFR172.710, 21CFR175.105, 21CFR176.180, CIR: [SQ], CTFA D, JCLS, JSCI, MI-13(6793), TSCA

Chemical Class: Alkoxylated Alcohols

Function: Surfactant - Emulsifying Agent

Reported Product Categories: Hair Dyes and Colors (All Types Requiring Caution

Statements and Patch Tests); Hair Rinses (Coloring); Hair Straighteners; Hair Coloring Preparations, Misc.; Hair Conditioners; Hair Lighteners with Color; Permanent Waves

Technical/Other Names:
Ethanol, 2-[p-(1,1,3,3-Tetramethylbutyl)-Phenoxy]-
Ethylene Glycol Octyl Phenyl Ether
PEG-1 Octyl Phenyl Ether
Polyoxyethylene Octylphenyl Ether
2-[p-(1,1,3,3-Tetramethylbutyl)Phenoxy]-Ethanol

Trade Names:
Igepal CA-210 (Rhodia)
TRITON X-15 Surfactant (Dow Chemical)

OCTOXYNOL-3

CAS Nos.: 2315-62-0; 9002-93-1 (Generic); 9004-87-9 (Generic); 9036-19-5 (Generic); 27176-94-9

JPN Translation:
オクトキシノール - 3

Empirical Formula:
$C_{20}H_{34}O_4$

Definition: Octoxynol-3 is the ethoxylated alkyl phenol that conforms generally to the formula:

$$C_8H_{17}C_6H_4(OCH_2CH_2)_nOH$$

where n has an average value of 3.

Information Sources: 21CFR175.105, 21CFR176.180, 21CFR176.210, CIR: [SQ], MI-13(6793), TSCA

Chemical Class: Alkoxylated Alcohols

Function: Surfactant - Emulsifying Agent

Technical/Other Names:
Ethanol, 2-[2-[2-(Octylphenoxy)Ethoxy]-Ethoxy]-
Ethanol, 2-[2-[2-p-(1,1,3,3-Tetramethyl-butyl)Phenoxy]Ethoxy]Ethoxy]-
2-[2-[2-(Octylphenoxy)Ethoxy]Ethoxy]-Ethanol
PEG-3 Octyl Phenyl Ether
Polyethylene Glycol (3) Octyl Phenyl Ether
Polyoxyethylene (3) Octyl Phenyl Ether
2-[2-[2-[p-(1,1,3,3-Tetramethylbutyl)-Phenoxy]Ethoxy]Ethoxy]Ethanol
Triethylene Glycol Octylphenyl Ether

Trade Names:
Igepal CA-420 (Rhodia)
Nikkol OP-3 (Nikko)
Synperonic OP3 (Uniqema Americas)
TRITON X-35 Surfactant (Dow Chemical)

OCTOXYNOL-5

CAS Nos.: 2315-64-2; 9002-93-1 (Generic); 9004-87-9 (Generic); 9036-19-5 (Generic); 27176-99-4

The inclusion of any compound in the *Dictionary and Handbook* does not indicate that use of that substance as a cosmetic ingredient complies with the laws and regulations governing such use in the United States or any other country.

JPN Translation:
オクトキシノール - 5

Empirical Formula:
$C_{24}H_{42}O_6$

Definition: Octoxynol-5 is the ethoxylated alkyl phenol that conforms generally to the formula:

$$C_8H_{17}C_6H_4(OCH_2CH_2)_nOH$$

where n has an average value of 5.

Information Sources: 21CFR172.710, 21CFR175.105, 21CFR176.180, 21CFR176.210, 21CFR178.3400, CIR: [SQ], CTFA S, MI-13(6693), TSCA

Chemical Class: Alkoxylated Alcohols

Function: Surfactant - Emulsifying Agent

Technical/Other Names:
14-(Octylphenoxy)-3,6,9,12-Tetraoxatetradecan-1-ol
PEG-5 Octyl Phenyl Ether
Pentaethylene Glycol p-tert-Octylphhenyl Ether
Polyethylene Glycol (5) Octyl Phenyl Ether
Polyoxyethylene (5) Octyl Phenyl Ether
14-[4-(1,1,3,3-Tetramethylbutyl)Phenoxy]-3,6,9,12-Tetraoxatetradecan-1-ol
3,6,9,12-Tetraoxatetradecan-1-ol, 14-(Octylphenoxy)-
3,6,9,12-Tetraoxatetradecan-1-ol, 14-[4-(1,1,3,3-Tetramethylbutyl)Phenoxy]-

Trade Names:
DelONIC OPE-5 (DeForest)
Igepal CA-520 (Rhodia)
Norfox OP-45 (Norman, Fox & Co.)
TRITON X-45 Surfactant (Dow Chemical)

OCTOXYNOL-6

CAS Nos.: 9002-93-1 (Generic); 9004-87-9 (Generic); 9036-19-5 (Generic)

Definition: Octoxynol-6 is the ethoxylated alkyl phenol that conforms generally to the formula:

$$C_8H_{17}C_6H_4(OCH_2CH_2)_nOH$$

where n has an average value of 6.

Information Source: CIR: [SQ]

Chemical Class: Alkoxylated Alcohols

Function: Surfactant - Emulsifying Agent

Reported Product Category: Paste Masks (Mud Packs)

Technical/Other Names:
Polyethylene Glycol (6) Octyl Phenyl Ether
Polyoxyethylene (6) Octyl Phenyl Ether

Trade Name:
Nonal 106 (Toho)

OCTOXYNOL-7

CAS Nos.: 9002-93-1 (Generic); 9004-87-9 (Generic); 9036-19-5 (Generic); 27177-02-2

JPN Translation:
オクトキシノール - 7

Empirical Formula:
$C_{28}H_{50}O_8$

Definition: Octoxynol-7 is the ethoxylated alkyl phenol that conforms generally to the formula:

$$C_8H_{17}C_6H_4(OCH_2CH_2)_nOH$$

where n has an average value of 7.

Information Sources: 21CFR172.710, 21CFR175.105, 21CFR176.180, 21CFR176.210, 21CFR178.3400, CIR: [SQ], MI-13(6693), TSCA

Chemical Class: Alkoxylated Alcohols

Function: Surfactant - Emulsifying Agent

Technical/Other Names:
Heptaethylene Glycol Octylphenyl Ether
3,6,9,12,15,18-Hexaoxaeicosan-1-ol, 20-(Octylphenoxy)-
20-(Octylphenoxy)-3,6,9,12,15,18-Hexaoxaeicosan-1-ol
PEG-7 Octyl Phenyl Ether
Polyethylene Glycol (7) Octyl Phenyl Ether
Polyoxyethylene (7) Octyl Phenyl Ether

Trade Name:
Igepal CA-620 (Rhodia)

OCTOXYNOL-8

CAS Nos.: 2638-43-9; 3520-90-9; 9002-93-1 (Generic); 9004-87-9 (Generic); 9036-19-5 (Generic)

JPN Translation:
オクトキシノール - 8

Empirical Formula:
$C_{30}H_{54}O_9$

Definition: Octoxynol-8 is the ethoxylated alkyl phenol that comforms generally to the formula:

$$C_8H_{17}C_6H_4(OCH_2CH_2)_nOH$$

where n has an average value of 8.

Information Sources: 21CFR172.710, 21CFR175.105, 21CFR176.180, 21CFR176.210, 21CFR178.3400, CIR: [SQ], MI-13(6693)

Chemical Class: Alkoxylated Alcohols

Function: Surfactant - Emulsifying Agent

Technical/Other Names:
3,6,9,12,15,18,21-Heptaoxatricosan-1-ol, 23-(4-Octylphenoxy)-
Octaethylene Glycol Octylphenyl Ether
23-(4-Octylphenoxy-3,6,9,12,15,18,21-Heptaoxatricosan-1-ol
PEG-8 Octyl Phenyl Ether
Polyethylene Glycol 400 Octyl Phenyl Ether
Polyoxyethylene (8) Octyl Phenyl Ether

Trade Names:
DelONIC OPE-7.5 (DeForest)
Norfox OP-114 (Norman, Fox & Co.)
TRITON X-114 Surfactant (Dow Chemical)

OCTOXYNOL-9

CAS Nos.	**EINECS No.**
9002-93-1 (Generic)	
9004-87-9 (Generic)	
9010-43-9	
9036-19-5 (Generic)	
42173-90-0	255-695-5

JPN Translation:
オクトキシノール - 9

Empirical Formula:
$C_{32}H_{58}O_{10}$

Definition: Octoxynol-9 is the ethoxylated alkyl phenol that conforms generally to the formula:

$$C_8H_{17}C_6H_4(OCH_2CH_2)_nOH$$

where n has an average value of 9.

Information Sources: BAN, 21CFR175.105, 21CFR176.180, 21CFR176.210, 21CFR178.3400, CIR: [S], CTFA S, INN, MI-13(6693), NF XIX, OTC-I-CV, TSCA, USAN

Chemical Class: Alkoxylated Alcohols

Function: Surfactant - Emulsifying Agent

Reported Product Categories: Hair Dyes and Colors (All Types Requiring Caution Statements and Patch Tests); Permanent Waves; Hair Preparations (Non-coloring), Misc.; Hair Conditioners; Skin Fresheners; Tonics, Dressings, and Other Hair Grooming Aids; Bath Oils, Tablets, and Salts; Cleansing Products (Cold Creams, Cleansing Lotions, Liquids and Pads); Paste Masks (Mud Packs); Shampoos (Non-coloring); Colognes and Toilet Waters; Foot Powders and Sprays; Hair Coloring Preparations, Misc.; Hair Straighteners; Perfumes; Personal Cleanliness Products, Misc.; Shaving Cream (Aerosol, Brushless and Lather)

Technical/Other Names:
Nonaethylene Glycol Octylphenyl Ether
3,6,9,12,15,18,21,24-Octaoxahexacosan-1-ol, 26-(Octylphenoxy)-
3,6,9,12,15,18,21,24-Octaoxahexacosan-1-ol, 26-(4-Octylphenoxy)-
26-(Octylphenoxy)-3,6,9,12,15,18,21,24-Octaoxahexacosan-1-ol
26-(4-Octylphenoxy)-3,6,9,12,15,18,21,24-Octaoxahexacosan-1-ol

PEG-9 Octyl Phenyl Ether
Polyethylene Glycol 450 Octyl Phenyl Ether
Polyoxyethylene (9) Octyl Phenyl Ether

Trade Names:
Igepal CA-630 (Rhodia)
Jeechem OP-9 (Jeen)
Nonal 109 (Toho)
Protachem OP-9 (Protameen)
TRITON X-100 Surfactant (Dow Chemical)

Trade Name Mixtures:
Etival 5L (Coletica SA)
Lytron 614 (Rohm and Haas)
Lytron 621 (Rohm and Haas)

OCTOXYNOL-10

CAS Nos.: 2315-66-4; 9002-93-1 (Generic); 9004-87-9 (Generic); 9036-19-5 (Generic); 27177-07-7

JPN Translation:
オクトキシノール - 10

Empirical Formula:
$C_{34}H_{62}O_{11}$

Definition: Octoxynol-10 is the ethoxylated alkyl phenol that conforms generally to the formula:

$$C_8H_{17}C_6H_4(OCH_2CH_2)_nOH$$

where n has an average value of 10.

Information Sources: 21CFR172.710, 21CFR175.105, 21CFR176.180, 21CFR176.210, 21CFR178.3400, CIR: [S], MI-13(6793), TSCA

Chemical Class: Alkoxylated Alcohols

Function: Surfactant - Emulsifying Agent

Reported Product Category: Hair Bleaches

Technical/Other Names:
Decaethylene Glycol Octylphenyl Ether
3,6,9,12,15,18,21,24,27-Nonaoxanonacosan-1-ol; 29-(Octylphenoxy)-
3,6,9,12,15,18,21,24,27-Nonaoxanonacosan-1-ol, 29-[4-(1,1,3,3-Tetramethylbutyl)Phenyl]-
29-(Octylphenoxy)-3,6,9,12,15,18,21,24,27-Nonaoxanonacosan-1-ol
PEG-10 Octyl Phenyl Ether
Polyethylene Glycol 500 Octyl Phenyl Ether
Polyoxyethylene (10) Octyl Phenyl Ether
29-[4-(1,1,3,3-Tetramethylbutyl)Phenoxy]-3,6,9,12,15,18,21,24,27-Nonaoxanonacosan-1-ol

Trade Names:
DelONIC OPE-10 (DeForest)
Jeechem OP-10 (Jeen)
Nikkol OP-10 (Nikko)
Nonal 310 (Toho)
Norfox OP-100 (Norman, Fox & Co.)

Sipol OP 10 CF (Specialty Industrial)
Synperonic OP10 (Uniqema Americas)

OCTOXYNOL-11

CAS Nos.: 9002-93-1 (Generic); 9004-87-9 (Generic); 9036-19-5 (Generic); 108437-62-3

JPN Translation:
オクトキシノール - 11

Empirical Formula:
$C_{36}H_{66}O_{12}$

Definition: Octoxynol-11 is the ethoxylated alkyl phenol that conforms generally to the formula:

$$C_8H_{17}C_6H_4(OCH_2CH_2)_nOH$$

where n has an average value of 11.

Information Sources: 21CFR172.710, 21CFR175.105, 21CFR176.180, 21CFR176.210, 21CFR178.3400, CIR: [S], MI-13(6793)

Chemical Class: Alkoxylated Alcohols

Function: Surfactant - Emulsifying Agent

Reported Product Categories: Cleansing Products (Cold Creams, Cleansing Lotions, Liquids and Pads); Skin Care Preparations, Misc.; Moisturizing Preparations; Shampoos (Non-coloring)

Technical/Other Names:
PEG-11 Octyl Phenyl Ether
Polyethylene Glycol (11) Octyl Phenyl Ether
Polyoxyethylene (11) Octyl Phenyl Ether
32-[4-(1,1,3,3-Tetramethylbutyl)Phenoxy]-3,6,9,12,15,18,21,24,27,30-Decaoxado-triacontan-1-ol

Trade Names:
Oxypol (Gattefosse s.a.)
Synperonic OP11 (Uniqema Americas)

Trade Name Mixtures:
Post-Depilatory (Greentech S.A)
Solubilisant Gamma 2420 (Gattefosse s.a.)
Solubilisant Gamma 2428 (Gattefosse s.a.)

OCTOXYNOL-12

CAS Nos.: 9002-93-1 (Generic); 9004-87-9 (Generic); 9036-19-5 (Generic)

JPN Translation:
オクトキシノール - 12

Definition: Octoxynol-12 is the ethoxylated alkyl phenol that conforms generally to the formula:

$$C_8H_{17}C_6H_4(OCH_2CH_2)_nOH$$

where n has an average value of 12.

Information Sources: 21CFR172.710, 21CFR175.105, 21CFR176.180, 21CFR176.210, 21CFR178.3400, CIR: [S], MI-13(6793)

Chemical Class: Alkoxylated Alcohols

Function: Surfactant - Emulsifying Agent

Technical/Other Names:
PEG-12 Octyl Phenyl Ether
Polyethylene Glycol 600 Octyl Phenyl Ether
Polyoxyethylene (12) Octyl Phenyl Ether

Trade Names:
Akyporox OP 115 SPC (Kao GmbH)
Norfox OP-102 (Norman, Fox & Co.)

OCTOXYNOL-13

CAS Nos.: 9002-93-1 (Generic); 9004-87-9 (Generic); 9036-19-5 (Generic)

JPN Translation:
オクトキシノール - 13

Empirical Formula:
$C_{40}H_{74}O_{14}$

Definition: Octoxynol-13 is the ethoxylated alkyl phenol that conforms generally to the formula:

$$C_8H_{17}C_6H_4(OCH_2CH_2)_nOH$$

where n has an average value of 13.

Information Sources: 21CFR172.710, 21CFR175.105, 21CFR176.180, 21CFR176.210, 21CFR178.3400, CIR: [S], MI-13(6793), TSCA

Chemical Class: Alkoxylated Alcohols

Function: Surfactant - Emulsifying Agent

Reported Product Categories: Tonics, Dressings, and Other Hair Grooming Aids; Hair Wave Sets; Skin Care Preparations, Misc.; Hair Conditioners; Hair Rinses (Non-coloring); Bath Preparations, Misc.; Body and Hand Preparations (Excluding Shaving Preparations); Face Powders; Mascara; Moisturizing Preparations; Shampoos (Non-coloring); Bubble Baths; Eye Makeup Removers; Hair Preparations (Non-coloring), Misc.

Technical/Other Names:
3,6,9,12,15,18,21,24,27,30,33,36-Dodecaoxatriacontan-1-ol, 38-[4-(1,1,3,3,-Tetramethylbutyl)Phenoxy]-
PEG-13 Octyl Phenyl Ether
Polyethylene Glycol (13) Octyl Phenyl Ether
Polyoxyethylene (13) Octyl Phenyl Ether
38-[4-(1,1,3,3-Tetramethylbutyl)Phenoxy]-3,6,9,12,15,18,21,24,27, 30,33,36-Dodecaoxaoctatriacontan-1-ol

Trade Names:
 DelONIC OPE-12 (DeForest)
 Igepal CA-720 (Rhodia)
 Igepal CA-730 (Rhodia)
 Protachem OP-13 (Protameen)
 TRITON X-102 Surfactant (Dow Chemical)

Trade Name Mixtures:
 C8 Soie Hydro (Phytocos)
 LP110 (Phytocos)

OCTOXYNOL-16

CAS Nos.: 9002-93-1 (Generic); 9004-87-9 (Generic); 9036-19-5 (Generic)

JPN Translation:
 オクトキシノール - 16

Definition: Octoxynol-16 is the ethoxylated alkyl phenol that conforms generally to the formula:

$$C_8H_{17}C_6H_4(OCH_2CH_2)_nOH$$

where n has an average value of 16.

Information Sources: 21CFR175.105, 21CFR176.180, CIR: [S], MI-13(6793)

Chemical Class: Alkoxylated Alcohols

Functions: Surfactant - Cleansing Agent; Surfactant - Emulsifying Agent

Technical/Other Names:
 PEG-16 Octyl Phenyl Ether
 Polyethylene Glycol (16) Octyl Phenyl Ether
 Polyoxyethylene (16) Octyl Phenyl Ether

Trade Name:
 TRITON X-165 Surfactant (Dow Chemical)

OCTOXYNOL-20

CAS Nos.: 9002-93-1 (Generic); 9004-87-9 (Generic); 9036-19-5 (Generic)

JPN Translation:
 オクトキシノール - 20

Definition: Octoxynol-20 is the ethoxylated alkyl phenol that conforms generally to the formula:

$$C_8H_{17}C_6H_4(OCH_2CH_2)_nOH$$

where n has an average value of 20.

Information Sources: 21CFR175.105, 21CFR176.180, CIR: [S], MI-13(6793)

Chemical Class: Alkoxylated Alcohols

Functions: Surfactant - Emulsifying Agent; Surfactant - Solubilizing Agent

Technical/Other Names:
 PEG-20 Octyl Phenyl Ether
 Polyethylene Glycol 1000 Octyl Phenyl Ether
 Polyoxyethylene (20) Octyl Phenyl Ether

Trade Name:
 Synperonic OP 20 (Uniqema Americas)

OCTOXYNOL-25

CAS Nos.: 9002-93-1 (Generic); 9004-87-9 (Generic); 9036-19-5 (Generic)

JPN Translation:
 オクトキシノール - 25

Definition: Octoxynol-25 is the ethoxylated alkyl phenol that conforms generally to the formula:

$$C_8H_{17}C_6H_4(OCH_2CH_2)_nOH$$

where n has an average value of 25.

Information Sources: 21CFR175.105, 21CFR176.180, CIR: [S], MI-13(6793)

Chemical Class: Alkoxylated Alcohols

Functions: Surfactant - Cleansing Agent; Surfactant - Solubilizing Agent

Technical/Other Names:
 PEG-25 Octyl Phenyl Ether
 Polyethylene Glycol (25) Octyl Phenyl Ether
 Polyoxyethylene (25) Octyl Phenyl Ether

Trade Name:
 Akyporox OP 250 V (Kao GmbH)

OCTOXYNOL-30

CAS Nos.: 9002-93-1 (Generic); 9004-87-9 (Generic); 9036-19-5 (Generic)

JPN Translation:
 オクトキシノール - 30

Definition: Octoxynol-30 is the ethoxylated alkyl phenol that conforms generally to the formula:

$$C_8H_{17}C_6H_4(OCH_2CH_2)_nOH$$

where n has an average value of 30.

Information Sources: 21CFR172.710, 21CFR175.105, 21CFR176.180, 21CFR178.3400, CIR: [S], MI-13(6793)

Chemical Class: Alkoxylated Alcohols

Functions: Surfactant - Cleansing Agent; Surfactant - Solubilizing Agent

Reported Product Categories: Eyeliners; Mascara

Technical/Other Names:
 PEG-30 Octyl Phenyl Ether
 Polyethylene Glycol (30) Octyl Phenyl Ether
 Polyoxyethylene (30) Octyl Phenyl Ether

Trade Names:
 DelONIC OPE-30 (DeForest)
 Nikkol OP-30 (Nikko)
 TRITON X-305 Surfactant (Dow Chemical)

OCTOXYNOL-33

CAS Nos.: 9002-93-1 (Generic); 9004-87-9 (Generic); 9036-19-5 (Generic)

JPN Translation:
 オクトキシノール - 33

Definition: Octoxynol-33 is the ethoxylated alkyl phenol that conforms generally to the formula:

$$C_8H_{17}C_6H_4(OCH_2CH_2)_nOH$$

where n has an average value of 33.

Information Sources: 21CFR172.710, 21CFR175.105, 21CFR176.180, 21CFR178.3400, CIR: [S], MI-13(6793)

Chemical Class: Alkoxylated Alcohols

Functions: Surfactant - Cleansing Agent; Surfactant - Solubilizing Agent

Technical/Other Names:
 PEG-33 Octyl Phenyl Ether
 Polyethylene Glycol (33) Octyl Phenyl Ether
 Polyoxyethylene (33) Octyl Phenyl Ether

Trade Name Mixture:
 Abex VA 50 (Kao GmbH)

OCTOXYNOL-40

CAS Nos.: 9002-93-1 (Generic); 9004-87-9 (Generic); 9036-19-5 (Generic)

JPN Translation:
 オクトキシノール - 40

Definition: Octoxynol-40 is the ethoxylated alkyl phenol that conforms generally to the formula:

$$C_8H_{17}C_6H_4(OCH_2CH_2)_nOH$$

where n has an average value of 40.

Information Sources: 21CFR172.710, 21CFR175.105, 21CFR176.180, 21CFR178.3400, CIR: [S], MI-13(6793), TSCA

Chemical Class: Alkoxylated Alcohols

Functions: Surfactant - Cleansing Agent; Surfactant - Solubilizing Agent

Reported Product Categories: Hair Bleaches; Hair Conditioners; Hair Wave Sets; Hair Preparations (Non-coloring), Misc.; Hair Dyes and Colors (All Types Requiring Caution Statements and Patch Tests); Shampoos (Non-coloring)

Technical/Other Names:
PEG-40 Octyl Phenyl Ether
Polyethylene Glycol 2000 Octyl Phenyl
Ether
Polyoxyethylene (40) Octyl Phenyl Ether

Trade Names:
Chemax OP-40/70 (Chemax)
DeIONIC OPE-40 (DeForest)
T-DET O-407 (Harcros)
Synperonic OP 40 (Uniqema Americas)
TRITON X-405 Surfactant (Dow Chemical)

Trade Name Mixtures:
Dow Corning 7224 Conditioning Agent
(Dow Corning)
SM2112 (GE Silicones)
SM2115 (GE Silicones)

OCTOXYNOL-70

CAS Nos.: 9002-93-1 (Generic); 9004-87-9
(Generic); 9036-19-5 (Generic)

JPN Translation:
オクトキシノール - 70

Definition: Octoxynol-70 is the ethoxylated
alkyl phenol that conforms generally to the
formula:

$$C_8H_{17}C_6H_4(OCH_2CH_2)_nOH$$

where n has an average value of 70.

Information Sources: 21CFR172.710,
21CFR176.180, CIR: [S], MI-13(6793)

Chemical Class: Alkoxylated Alcohols

Function: Surfactant - Cleansing Agent

Technical/Other Names:
PEG-70 Octyl Phenyl Ether
Polyethylene Glycol (70) Octyl Phenyl
Ether
Polyoxyethylene (70) Octyl Phenyl Ether

Trade Name:
TRITON X-705 Surfactant (Dow Chemical)

OCTOXYNOL-9 CARBOXYLIC ACID

CAS No.: 25338-58-3

Empirical Formula:
$C_{32}H_{56}O_{11}$

Definition: Octoxynol-9 Carboxylic Acid is
the organic acid that conforms generally to
the formula:

$$C_8H_{17}C_6H_4(OCH_2CH_2)_nOCH_2COOH$$

where n has an average value of 8.

Information Source: CIR: [S]

Chemical Class: Carboxylic Acids

Function: Surfactant - Emulsifying Agent

Technical/Other Names:
3,6,9,12,15,18,21,24-Octaoxa-
hexacosanoic Acid, 26-(Octylphenoxy)-
26-(Octylphenoxy)-3,6,9,12,15,18,21,24-
Octaoxahexacosanoic Acid
PEG-9 Octyl Phenyl Ether Carboxylic Acid
Polyethylene Glycol 450 Octyl Phenyl Ether
Carboxylic Acid
Polyoxyethylene (9) Octyl Phenyl Ether
Carboxylic Acid

Trade Name:
Akypo OP 80 (Kao GmbH)

OCTOXYNOL-20 CARBOXYLIC ACID

Definition: Octoxynol-20 Carboxylic Acid is
the organic acid that conforms generally to
the formula:

$$C_8H_{17}C_6H_4(OCH_2CH_2)_nOCH_2COOH$$

where n has an average value of 19.

Information Source: CIR: [S]

Chemical Class: Carboxylic Acids

Function: Surfactant - Cleansing Agent

Technical/Other Names:
PEG-20 Octyl Phenyl Ether Carboxylic Acid
Polyethylene Glycol 1000 Octyl Phenyl
Ether Carboxylic Acid
Polyoxyethylene (20) Octyl Phenyl Ether
Carboxylic Acid

Trade Name:
Akypo OP 190 (Kao GmbH)

OCTRIZOLE

CAS No.	EINECS No.
3147-75-9	221-573-5

Empirical Formula:
$C_{20}H_{25}N_3O$

Definition: Octrizole is the organic
compound that conforms to the formula:

Information Sources: INN, TSCA, USAN

Chemical Class: Heterocyclic Compounds

Function: Ultraviolet Light Absorber

Technical/Other Names:
2-Benzotriazolyl-4-tert-Octylphenol

2-(2H-Benzotriazol-2-yl)-4-(1,1,3,3-Tetra-
methylbutyl)Phenol
2-(2'-Hydroxy-5'-t-Octylphenyl)Benzo-
triazole
Phenol, 2-(2H-Benzotriazol-2-yl)-4-(1,1,3,3-
Tetramethylbutyl)-
UV Absorber-5

Trade Name:
Spectra-Sorb UV-5411 (American
Cyanamid/Fine Chemicals)

OCTYLACRYLAMIDE/ACRYLATES/BUTYLAMINOETHYL METHACRYLATE COPOLYMER

CAS No.: 70801-07-9

JPN Translation:
(オクチルアクリルアミド / アクリル酸ヒ
ドロキシプロピル / メタクリル酸ブチル
アミノエチル) コポリマー

Definition: Octylacrylamide/Acrylates/
Butylaminoethyl Methacrylate Copolymer is
a polymer formed from octylacrylamide, t-
butylaminoethyl methacrylate and two or
more monomers consisting of acrylic acid,
methacrylic acid or any of their simple esters.

Information Sources: CTFA D, JCIC,
JCLS, JSQI

Chemical Class: Synthetic Polymers

Functions: Film Former; Hair Fixative

Reported Product Categories: Hair Sprays
(Aerosol Fixatives); Hair Preparations (Non-
coloring), Misc.; Tonics, Dressings, and
Other Hair Grooming Aids

Technical/Other Name:
2-Propenoic Acid, 2-Methyl-, 2-[(1,1-
Dimethylethyl)Amino]Ethyl Ester, Polymer
with Methyl 2-Methyl-2-Propenoate, 1,2-
Propanediol Mono(2-Methyl-2-
Propenoate), 2-Propenoic Acid and N-(1,
1,3,3-Tetramethylbutyl)-2-Propenamide

Trade Names:
Amphomer (National Starch)
Amphomer LV-71 (National Starch)
Amphomer 30S (National Starch/Specialty
Polymers)
Balance 47 (National Starch)
Hairfix (3V Sigma S.P.A.)

Trade Name Mixture:
Hair Gloss Polymer A (Cosmetochem)
(Cosmetochem International Ltd.)

OCTYLDECANOL

Empirical Formula:
$C_{18}H_{38}O$

Definition: Octyldecanol is the aliphatic alcohol that conforms generally to the formula:

$$CH_3(CH_2)_7CHCH_2OH$$
$$|$$
$$CH_3(CH_2)_6CH_2$$

Information Sources: JCLS, JSCI

Chemical Class: Alcohols

Function: Skin-Conditioning Agent - Emollient

Technical/Other Names:
1-Decanol, 2-Octyl
2-Octyl Decanol

Trade Names:
Exxal 18 (Exxon Chemical)
Isofol 18 E Alcohol (Sasol GmbH - Hamburg)
Isofol 18T Alcohol (Sasol GmbH - Hamburg)

Trade Name Mixture:
Zenibee Cream - Q (Zenitech)

OCTYLDECYL OLEATE

Empirical Formula:
$C_{36}H_{70}O_2$

Definition: Octyldecyl Oleate is the ester of octyldecanol and oleic acid. It conforms to the formula:

$$CH_3(CH_2)_7CH \qquad O$$
$$|| \qquad\qquad ||$$
$$CH(CH_2)_7C - OCH_2CH(CH_2)_7CH_3$$
$$|$$
$$(CH_2)_7CH_3$$

Chemical Class: Esters

Function: Skin-Conditioning Agent - Occlusive

Technical/Other Name:
Oleic Acid, 2-Octyldecyl Ester

OCTYLDECYL PHOSPHATE

CAS No. 97553-81-6

EINECS No. 307-243-4

Definition: Octyldecyl Phosphate is a complex mixture of esters of phosphoric acid and Octyldecanol (q.v.).

Chemical Class: Phosphorus Compounds

Function: Surfactant - Emulsifying Agent

Technical/Other Name:
Phosphoric Acid, Mono- and Bis(Branched and Linear Stearyl) Esters

Trade Names:
Hostaphat CG 120 (Clariant)

Hostaphat CG 120 (Clariant GmbH, Personal Care)

OCTYLDODECANOL

CAS No. 5333-42-6

EINECS No. 226-242-9

JPN Translation:
オクチルドデカノール

Empirical Formula:
$C_{20}H_{42}O$

Definition: Octyldodecanol is an aliphatic alcohol that conforms generally to the formula:

$$CH_3(CH_2)_9CHCH_2OH$$
$$|$$
$$CH_3(CH_2)_7$$

Information Sources: CIR: [S] JACT-4(5)-1985, CTFA S, NF XIX, RIFM, SNPF, TSCA, USAN

Chemical Class: Alcohols

Functions: Fragrance Ingredient; Skin-Conditioning Agent - Emollient

Reported Product Categories: Eyeliners; Lipsticks; Hair Dyes and Colors (All Types Requiring Caution Statements and Patch Tests); Bath Preparations, Misc.; Body and Hand Preparations (Excluding Shaving Preparations); Moisturizing Preparations; Foundations; Makeup Preparations (Not eye), Misc.; Skin Care Preparations, Misc.; Bath Oils, Tablets, and Salts; Cleansing Products (Cold Creams, Cleansing Lotions, Liquids and Pads); Eye Makeup Preparations, Misc.; Bath Capsules; Face and Neck Preparations (Excluding Shaving Preparations); Hair Color Sprays (Aerosol); Night Skin Care Preparations; Hair Conditioners; Blushers (All types); Eyebrow Pencils; Suntan Gels, Creams, and Liquids; Paste Masks (Mud Packs); Face Powders; Makeup Bases; Eye Shadows; Suntan Preparations, Misc.; Deodorants (Underarm); Eye Makeup Removers; Hair Rinses (Non-coloring); Preshave Lotions (All types); Rouges; Shaving Preparations, Misc.

Technical/Other Names:
1-Dodecanol, 2-Octyl-
2-Octyldodecanol
2-Octyl Dodecanol
2-Octyldodecan-1-ol (RIFM)
2-Octyldodecyl Alcohol

Trade Names:
AEC Octyldodecanol (A & E Connock)
Eutanol G (Cognis Care Chemicals/NJ)
Eutanol G (Cognis Care Chemicals/PA)
Eutanol G (Cognis Deutschland)
Exxal 20 (Exxon Chemical)
Isofol 20 Alcohol (Sasol GmbH - Hamburg)
Jarcol I-20 (Jarchem)

Michel XO-150-20 (Michel)
Saboderm G20 (Sabo)
U-Tanol G (Universal Preserv-A-Chem)

Trade Name Mixtures:
Argobase 125 (Croda Chemicals)
Bentone Gel EUG (ELE)
Ceravitin (Laboratoires Serobiologiques)
Cranberry Butter (Zenitech)
Cutina LM Conc (Cognis Deutschland)
Defensil (Rahn)
Gilugel EUG (Giulini/Giulini Chemie)
Irwinol (Laboratoires Serobiologiques)
Lipodermol Veg (Laboratoires Serobiologiques)
Nanospheres 100 Lipophilic (Exsymol)
N.S.L.E. (Sederma)
Paprika-Extract, Oil Soluble (Crodarom)
Raspberry Butter (Zenitech)
Sphingoceryl Powder VEG (Laboratoires Serobiologiques)
Sphingoceryl Veg (Laboratoires Serobiologiques)
Tioveil EUT (Uniqema, Belgium)
Tioveil GCM (Uniqema, Belgium)
Tioveil 50 GCM (Uniqema, Belgium)
Tixogel ODD-1419 (Sud-Chemie, United Catalysts)
Tixogel ODD-1520 (Sud-Chemie, United Catalysts)
Unitina LM (Universal Preserv-A-Chem)
Vetiver Extract 69711 O.S. (Fragrance Oils Int. Ltd.)
Vitamin A Palmitate Lipomicron (Sederma)
Vitamin E Acetate Lipomicron (Sederma)
Zenibee Cream (Zenitech)

OCTYLDODECETH-2

CAS No.: 32128-65-7

JPN Translation:
オクチルドデセス - 2

Empirical Formula:
$C_{24}H_{50}O_3$

Definition: Octyldodeceth-2 is the polyethylene glycol ether of Octyldodecanol (q.v.) that conforms to the formula:

$$CH_3(CH_2)_9CHCH_2(OCH_2CH_2)_nOH$$
$$|$$
$$CH_3(CH_2)_7$$

where n has an average value of 2.

Information Source: JCLS

Chemical Class: Alkoxylated Alcohols

Function: Surfactant - Emulsifying Agent

Technical/Other Names:
PEG-2 Octyldodecyl Ether
Polyethylene Glycol 100 Octyldodecyl Ether
Polyoxyethylene (2) Octyldodecyl Ether

OCTYLDODECETH-5

JPN Translation:
オクチルドデセス - 5

Empirical Formula:
$C_{30}H_{62}O_6$

Definition: Octyldodeceth-5 is the polyethylene glycol ether of Octyldodecanol (q.v.) that conforms to the formula:

$$CH_3(CH_2)_9CHCH_2(OCH_2CH_2)_nOH$$
$$|$$
$$CH_3(CH_2)_7$$

where n has an average value of 5.

Information Source: JCLS

Chemical Class: Alkoxylated Alcohols

Function: Surfactant - Emulsifying Agent

Technical/Other Names:
PEG-5 Octyldodecyl Ether
Polyethylene Glycol (5) Octyldodecyl Ether
Polyoxyethylene (5) Octyldodecyl Ether

OCTYLDODECETH-10

JPN Translation:
オクチルドデセス - 10

Definition: Octyldodeceth-10 is the polyethylene glycol ether of Octyldodecanol (q.v.) that conforms generally to the formula:

$$CH_3(CH_2)_9CHCH_2(OCH_2CH_2)_nOH$$
$$|$$
$$CH_3(CH_2)_7$$

where n has an average value of 10.

Chemical Class: Alkoxylated Alcohols

Function: Surfactant - Emulsifying Agent

Technical/Other Names:
PEG-10 Octyldodecyl Ether
Polyethylene Glycol (10) Octyldodecyl Ether
Polyoxyethylene (10) Octyldodecyl Ether

Trade Name:
Emalex OD-10 (Ikeda)

OCTYLDODECETH-16

JPN Translation:
オクチルドデセス - 16

Definition: Octyldodeceth-16 is the polyethylene glycol ether of Octyldodecanol (q.v.) that conforms to the formula:

$$CH_3(CH_2)_9CHCH_2(OCH_2CH_2)_nOH$$
$$|$$
$$CH_3(CH_2)_7$$

where n has an average value of 16.

Information Source: JCLS

Chemical Class: Alkoxylated Alcohols

Function: Surfactant - Emulsifying Agent

Technical/Other Names:
PEG-16 Octyldodecyl Ether
Polyethylene Glycol (16) Octyldodecyl Ether
Polyoxyethylene (16) Octyldodecyl Ether

Trade Name:
Emalex OD-16 (Nihon Emulsion)

OCTYLDODECETH-20

JPN Translation:
オクチルドデセス - 20

Definition: Octyldodeceth-20 is the polyethylene glycol ether of Octyldodecanol (q.v.) that conforms to the formula:

$$CH_3(CH_2)_9CHCH_2(OCH_2CH_2)_nOH$$
$$|$$
$$CH_3(CH_2)_7$$

where n has an average value of 20.

Information Sources: JCLS, JSQI

Chemical Class: Alkoxylated Alcohols

Functions: Surfactant - Cleansing Agent; Surfactant - Emulsifying Agent; Surfactant - Solubilizing Agent

Reported Product Categories: Bath Preparations, Misc.; Body and Hand Preparations (Excluding Shaving Preparations)

Technical/Other Names:
PEG-20 Octyldodecyl Ether
Polyethylene Glycol 1000 Octyldodecyl Ether
Polyoxyethylene (20) Octyldodecyl Ether

Trade Names:
Emalex OD-20 (Nihon Emulsion)
Emulgen 2020G (Kao Specialties)

OCTYLDODECETH-25

JPN Translation:
オクチルドデセス - 25

Definition: Octyldodeceth-25 is the polyethylene glycol ether of Octyldodecanol (q.v.) that conforms to the formula:

$$CH_3(CH_2)_9CHCH_2(OCH_2CH_2)_nOH$$
$$|$$
$$CH_3(CH_2)_7$$

where n has an average value of 25.

Information Sources: JCLS, JSQI

Chemical Class: Alkoxylated Alcohols

Functions: Surfactant - Cleansing Agent; Surfactant - Solubilizing Agent

Technical/Other Names:
PEG-25 Octyldodecyl Ether
Polyethylene Glycol (25) Octyldodecyl Ether
Polyoxyethylene (25) Octyldodecyl Ether

Trade Name:
Emalex OD-25 (Nihon Emulsion)

OCTYLDODECETH-30

JPN Translation:
オクチルドデセス - 30

Definition: Octyldodeceth-30 is the polyethylene glycol ether of Octyldodecanol (q.v.) that conforms generally to the formula:

$$CH_3(CH_2)_9CHCH_2(OCH_2CH_2)_nOH$$
$$|$$
$$CH_3(CH_2)_7$$

where n has an average value of 30.

Information Source: JCLS

Chemical Class: Alkoxylated Alcohols

Functions: Surfactant - Cleansing Agent; Surfactant - Solubilizing Agent

Technical/Other Names:
Polyethylene Glycol (30) Octyldodecyl Ether
Polyoxyethylene (30) Octyldodecyl Ether

OCTYLDODECYL BEESWAX

Definition: Octyldodecyl Beeswax is the ester of Octyldodecanol (q.v.) and Beeswax Acid (q.v.).

Chemical Class: Esters

Function: Skin-Conditioning Agent - Emollient

Trade Name:
Beesbutter Guerbet Ester (Noveon)

OCTYLDODECYL BEHENATE

CAS No.: 125804-08-2

Empirical Formula:
$C_{42}H_{84}O_2$

Definition: Octyldodecyl Behenate is the ester of Octyldodecanol (q.v.) and behenic acid that conforms to the formula:

$$CH_3(CH_2)_{20}C-OCH_2CH(CH_2)_9CH_3$$

with a carbonyl O above the C, and below:

$$|$$
$$(CH_2)_7CH_3$$

Chemical Class: Esters

Function: Skin-Conditioning Agent - Occlusive

Technical/Other Name:
Docosanoic Acid, 2-Octyldodecyl Ester

Trade Names:
Dermol 2022 (Alzo)
Pelemol 2022 (Phoenix)

OCTYLDODECYL BENZOATE

Empirical Formula:
$C_{27}H_{46}O_2$

Definition: Octyldodecyl Benzoate is the ester of Octyldodecanol (q.v.) and benzoic acid. It conforms to the formula:

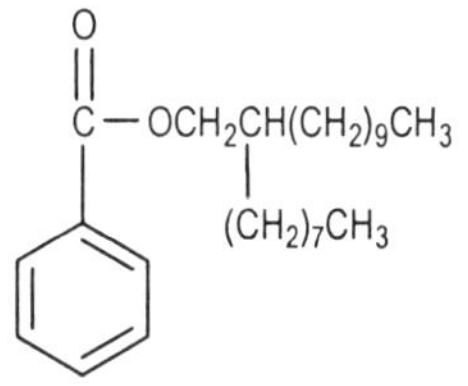

Chemical Class: Esters

Function: Skin-Conditioning Agent - Emollient

Technical/Other Name:
Benzoic Acid, 2-Octyldodecyl Ester

Trade Name:
Finsolv BOD (Finetex)

OCTYLDODECYL ERUCATE

CAS No.: 88103-59-7

JPN Translation:
エルカ酸オクチルドデシル

Empirical Formula:
$C_{42}H_{82}O_2$

Definition: Octyldodecyl Erucate is the ester of octyldodecanol and erucic acid. It conforms to the formula:

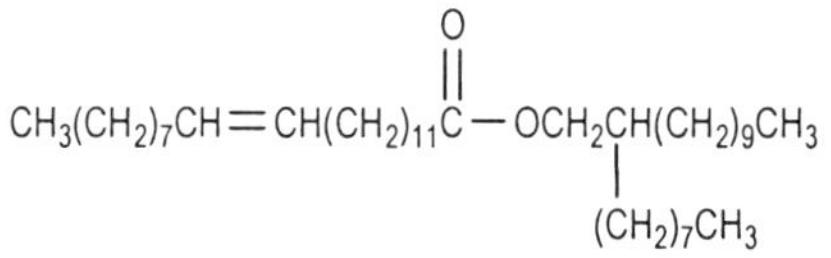

Information Sources: JCIC, JCLS

Chemical Class: Esters

Function: Skin-Conditioning Agent - Occlusive

Technical/Other Names:
13-Docosenoic Acid, 2-Octyldodecyl Ester
Erucic Acid, 2-Octyldodecyl Ester
2-Octyldodecyl 13-Docosenoate
2-Octyldodecyl Erucate

Trade Names:
EOD (Shin-Ei Chemical)
Pelemol EE (Phoenix)

OCTYLDODECYL ETHYLHEXANOATE

CAS No.	EINECS No.
69275-04-3	273-943-0

Empirical Formula:
$C_{28}H_{56}O_2$

Definition: Octyldodecyl Ethylhexanoate is the ester of Octyldodecanol (q.v.) and 2-ethylhexanoic acid. It conforms to the formula:

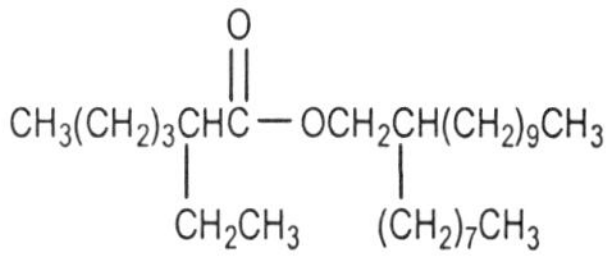

Chemical Class: Esters

Function: Skin-Conditioning Agent - Emollient

Technical/Other Names:
2-Ethylhexanoic Acid, 2-Octyldodecyl Ester
Hexanoic Acid, 2-Ethyl-, 2-Octyldodecyl Ester

OCTYLDODECYL HYDROXYSTEARATE

Empirical Formula:
$C_{38}H_{76}O_3$

Definition: Octyldodecyl Hydroxystearate is the ester of Octyldodecanol (q.v.) and 12-hydroxystearic acid. It conforms to the formula:

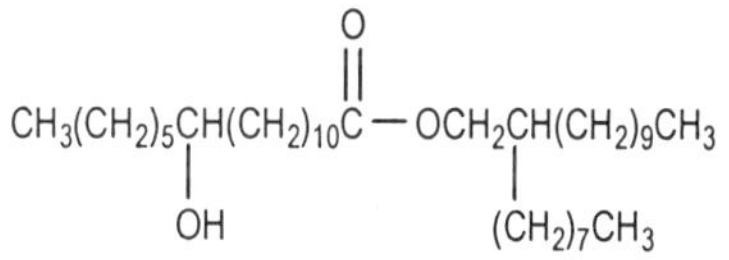

Chemical Class: Esters

Function: Skin-Conditioning Agent - Occlusive

Technical/Other Name:
Hydroxyoctadecanoic Acid, Octadodecyl Ester

Trade Names:
Dermol 20-S (Alzo)
DUB HSCI 20 (Stearinerie Dubois Fils)

OCTYLDODECYL ISOSTEARATE

CAS No.	EINECS No.
93803-87-3	298-361-4

JPN Translation:
イソステアリン酸オクチルドデシル

Empirical Formula:
$C_{38}H_{76}O_2$

Definition: Octyldodecyl Isostearate is the ester of Octyldodecanol (q.v.) and isostearic acid. It conforms to the formula:

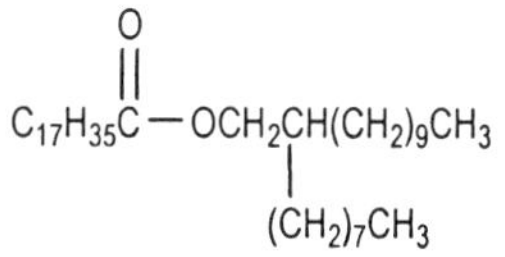

Information Sources: JCIC, JCLS

Chemical Class: Esters

Function: Skin-Conditioning Agent - Occlusive

Technical/Other Names:
Isooctadecanoic Acid, 2-Octyldodecyl Ester
2-Octyldodecyl Isooctadecanoate
2-Octyldodecyl Isostearate

Trade Names:
G-38 Guerbet Ester (Noveon)
Isod (Shin-Ei Chemical)
Isod (Tomen America)

OCTYLDODECYL LACTATE

CAS No.: 57568-20-4

JPN Translation:
乳酸オクチルドデシル

Empirical Formula:
$C_{23}H_{46}O_3$

Definition: Octyldodecyl Lactate is the ester of Octyldodecanol (q.v.) and lactic acid. It conforms to formula:

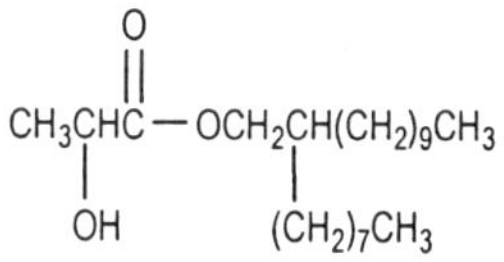

Information Sources: JCIC, JCLS, JSQI

Chemical Class: Esters

Function: Skin-Conditioning Agent - Emollient

Technical/Other Names:
Lactic Acid, 2-Octyldodecyl Ester
Propanoic Acid, 2-Hydroxy-, 2-Octyldodecyl Ester

Trade Names:
AEC Octyldodecyl Lactate (A & E Connock)
Cosmol 13 (Nisshin OilliO)
Salacos 13 (Nisshin OilliO)
Schercemol 203 (Scher)

OCTYLDODECYL LANOLATE

JPN Translation:
ラノリン脂肪酸オクチルドデシル

Definition: Octyldodecyl Lanolate is the ester of Octyldodecanol (q.v.) and Lanolin Acid (q.v.).

Information Sources: JCIC, JCLS

Chemical Class: Lanolin and Lanolin Derivatives

Functions: Hair Conditioning Agent; Skin-Conditioning Agent - Occlusive

Technical/Other Name:
Lanolin Fatty Acid Octyl Dodecyl Ester

Trade Names:
Crodamol ODL (Croda Japan)
Yofco FE-1 (Nippon Chemical)
Yofco FE-101 (Nippon Chemical)
Yofco FE-102 (Nippon Chemical)

OCTYLDODECYL MEADOWFOAMATE

Definition: Octyldodecyl Meadowfoamate is the ester of Octyldodecanol (q.v.) and the fatty acids derived from Limnanthes Alba (Meadowfoam) Seed Oil (q.v.).

Chemical Class: Esters

Function: Skin-Conditioning Agent - Occlusive

Trade Name:
Fancor Meadowester GME (Dare Corp)

OCTYLDODECYL MYRISTATE

CAS Nos.   **EINECS No.**
22766-83-2   245-205-8
83826-43-1

JPN Translation:
ミリスチン酸オクチルドデシル

Empirical Formula:
$C_{34}H_{68}O_2$

Definition: Octyldodecyl Myristate is the ester of Octyldodecanol (q.v.) and myristic acid. It conforms to the formula:

Information Sources: JCLS, JSCI

Chemical Class: Esters

Function: Skin-Conditioning Agent - Occlusive

Reported Product Categories: Moisturizing Preparations; Bath Capsules; Face and Neck Preparations (Excluding Shaving Preparations); Bath Preparations, Misc.; Body and Hand Preparations (Excluding Shaving Preparations); Skin Care Preparations, Misc.

Technical/Other Names:
Myristic Acid, 2-Octyldodecyl Ester
2-Octyldodecyl Myristate
Tetradecanoic Acid, Octyldodecyl Ester
Tetradecanoic Acid, 2-Octyldodecyl Ester

Trade Names:
AEC Octyldodecyl Myristate (A & E Connock)
Dermol 2014 (Alzo)
DUB MOD (Stearinerie Dubois Fils)
MOD (Gattefosse s.a.)
MOD (Shin-Ei Chemical)
Myristol 2-8-12 (Vevy)
Nikkol ODM-100 (Nikko)
ODM (Kokyu Alcohol)
Pelemol 2014 (Phoenix)
Saboderm ODM (Sabo)

Trade Name Mixtures:
HEG/MOD (US Cosmetics)
Kamitsure Liquid MD (Ichimaru Pharcos)
PEGMOD (Shin-Ei Chemical)
Purified Ester Gum/M.O.D. (US Cosmetics)
Shiconix Liquid MD (Ichimaru Pharcos)
Yokuinin Liquid MD (Ichimaru Pharcos)

OCTYLDODECYL NEODECANOATE

JPN Translation:
ネオデカン酸オクチルドデシル

Empirical Formula:
$C_{30}H_{60}O_2$

Definition: Octyldodecyl Neodecanoate is the ester of Octyldodecanol (q.v.) and neodecanoic acid. It conforms generally to the formula:

Information Sources: JCIC, JCLS, JSQI

Chemical Class: Esters

Function: Skin-Conditioning Agent - Emollient

Technical/Other Names:
Neodecanoic Acid, Octyldodecyl Ester
Octyldodecyl Dimethyloctanoate

Trade Names:
AEC Octyldodecyl Neodecanoate (A & E Connock)
Nikkol Neodecanoate 20 (Nikko)

OCTYLDODECYL NEOPENTANOATE

JPN Translation:
ネオペンタン酸オクチルドデシル

Empirical Formula:
$C_{25}H_{50}O_2$

Definition: Octyldodecyl Neopentanoate is the ester of Octyldodecanol (q.v.) and neopentanoic acid. It conforms to the formula:

Chemical Class: Esters

Function: Skin-Conditioning Agent - Emollient

Reported Product Category: Makeup Preparations (Not eye), Misc.

Technical/Other Name:
Neopentanoic Acid, 2-Octyldodecyl Ester

Trade Names:
AEC Octyldodecyl Neopentanoate (A & E Connock)
DUB VCI 20 (Stearinerie Dubois Fils)
Elefac I-205 (Bernel)

Trade Name Mixtures:
NanoGard FE45B EF (Nanophase)
NanoGard FE45BL EF (Nanophase)
NanoGard FE45R EF (Nanophase)

OCTYLDODECYL OCTYLDODECANOATE

Empirical Formula:
$C_{40}H_{80}O_2$

Definition: Octyldodecyl Octyldodecanoate is the ester of Octyldecanol (q.v.) and octyldodecanoic acid. It conforms to the formula:

Chemical Class: Esters

Function: Skin-Conditioning Agent - Occlusive

Technical/Other Name:
2-Octyldodecanoic Acid, 2-Octyldodecyl Ester

Trade Name:
Biosil Basics C-38 (Biosil Technologies, Inc.)

OCTYLDODECYL OLEATE

CAS No.   **EINECS No.**
22801-45-2   245-228-3

JPN Translation:
オレイン酸オクチルドデシル

Empirical Formula:
$C_{38}H_{74}O_2$

Definition: Octyldodecyl Oleate is the ester of Octyldodecanol (q.v.) and oleic acid. It conforms to the formula:

$$CH_3(CH_2)_7CH = CH(CH_2)_7C - OCH_2CH(CH_2)_9CH_3$$
$$(CH_2)_7CH_3$$

Information Sources: JCLS, JSCI

Chemical Class: Esters

Function: Skin-Conditioning Agent - Occlusive

Technical/Other Names:
9-Octadecenoic Acid, 2-Octyldodecyl Ester
2-Octyldodecyl Oleate
Oleic Acid, Octyldodecyl Ester

Trade Name:
OOD (Shin-Ei Chemical)

OCTYLDODECYL OLIVATE

Definition: Octyldodecyl Olivate is the ester of Octyldodecanol (q.v.) and the fatty acids derived from Olea Europaea (Olive) Oil (q.v.).

Chemical Class: Esters

Function: Skin-Conditioning Agent - Occlusive

OCTYLDODECYL PCA

CAS No. 37673-37-3
EINECS No. 253-604-3

Empirical Formula:
$C_{25}H_{47}NO_3$

Definition: Octyldodecyl PCA is the ester of octyldodecanol and PCA (q.v.) that conforms to the formula:

$$O = \quad N - C - OCH_2CH(CH_2)_9CH_3$$
$$(CH_2)_7CH_3$$

Chemical Classes: Esters; Heterocyclic Compounds

Function: Skin-Conditioning Agent - Emollient

Technical/Other Name:
L-Proline, 5-Oxo-, 2-Octyldodecyl Ester

Trade Name:
Ceramidone (UCIB (Solabia))

OCTYLDODECYL RICINOLEATE

CAS Nos.: 79490-62-3; 125093-27-8

JPN Translation:
リシノレイン酸オクチルドデシル

Empirical Formula:
$C_{38}H_{74}O_3$

Definition: Octyldodecyl Ricinoleate is the ester of Octyldodecanol (q.v.) and ricinoleic acid. It conforms to the formula:

$$CH_3(CH_2)_5CHCH_2CH$$
$$OH \quad CH(CH_2)_7C - OCH_2CH(CH_2)_9CH_3$$
$$(CH_2)_7CH_3$$

Information Sources: JCIC, JCLS, JSQI

Chemical Class: Esters

Function: Skin-Conditioning Agent - Occlusive

Technical/Other Names:
12-Hydroxy-9-Octadecenoic Acid, 2-Octyldodecyl Ester
9-Octadecenoic Acid, 12-Hydroxy-, 2-Octyldodecyl Ester
2-Octyldodecyl 12-Hydroxy-9-Octadecenoate
2-Octyldodecyl Ricinoleate

Trade Names:
Pelemol ODR (Phoenix)
ROD (Shin-Ei Chemical)
Ultracas G-20 Guerbet Ester (Noveon)

OCTYLDODECYL STEARATE

CAS No. 22766-82-1
EINECS No. 245-204-2

Empirical Formula:
$C_{38}H_{76}O_2$

Definition: Octyldodecyl Stearate is the ester of Octyldodecanol (q.v.) and stearic acid. It conforms to the formula:

$$CH_3(CH_2)_{16}C - OCH_2CH(CH_2)_9CH_3$$
$$(CH_2)_7CH_3$$

Information Source: TSCA

Chemical Class: Esters

Function: Skin-Conditioning Agent - Occlusive

Technical/Other Names:
Octadecanoic Acid, 2-Octyldodecyl Ester
Stearic Acid, 2-Octyldodecyl Ester

Trade Names:
Cetiol G20S (Cognis Care Chemicals/NJ)
Cetiol G20S (Cognis Care Chemicals/PA)

OCTYLDODECYL STEAROYL STEARATE

CAS No. 90052-75-8
EINECS No. 289-991-0

JPN Translation:
ステアロイルオキシステアリン酸オクチルドデシル

Empirical Formula:
$C_{56}H_{110}O_4$

Definition: Octyldodecyl Stearoyl Stearate is the ester that conforms generally to the formula:

$$CH_3(CH_2)_{16}C - OCH(CH_2)_{10}C - OCH_2CH(CH_2)_9CH_3$$
$$(CH_2)_5CH_3 \quad (CH_2)_7CH_3$$

Information Sources: CIR: [I] IJT-20 (SUPPL. 3)2001, CIR: [S], JCIC, JCLS

Chemical Class: Esters

Functions: Skin-Conditioning Agent - Occlusive; Viscosity Increasing Agent - Nonaqueous

Reported Product Categories: Eye Shadows; Face Powders; Blushers (All types); Foundations; Makeup Bases; Eye Makeup Preparations, Misc.; Rouges; Lipsticks; Makeup Preparations (Not eye), Misc.; Moisturizing Preparations; Cuticle Softeners; Eyebrow Pencils; Eyeliners; Fragrance Preparations, Misc.; Hair Tints; Powders (Dusting and Talcum, Excluding Aftershave Talcs); Shaving Cream (Aerosol, Brushless and Lather)

Technical/Other Names:
Octadecanoic Acid, 12-[(1-Oxooctadecyl)Oxy]-, 2-Octydodecyl Ester
12-[(1-Oxooctadecyl) Oxy] Octadecanoic Acid, 2-Octyldodecyl Ester

Trade Names:
Ceraphyl 847 (International Specialty Products)
Dermol 20-SS (Alzo)
DUB SSOD (Stearinerie Dubois Fils)
Jeechem OSS (Jeen)
Pelemol ODSS (Phoenix)
Sterol OSS (Cesalpinia)
Trivent SS-20 (Trivent)

OCTYLDODECYLTRIMONIUM CHLORIDE

Empirical Formula:
$C_{23}H_{50}N \cdot Cl$

Definition: Octyldodecyltrimonium Chloride is the quaternary ammonium salt that conforms to the formula:

$$\left[CH_3 - N - CH_2CH(CH_2)_9CH_3 \right]^+ \quad Cl^-$$
$$CH_3 \quad (CH_2)_7CH_3$$

Information Source: JCLS

Chemical Class: Quaternary Ammonium Compounds

Functions: Antistatic Agent; Hair Conditioning Agent

OCTYLISOTHIAZOLINONE

CAS No.	EINECS No.
26530-20-1	247-761-7

Empirical Formula:
$C_{11}H_{19}NOS$

Definition: Octylisothiazolinone is the organic compound that conforms to the formula:

Information Sources: MI-13(6788), TSCA

Chemical Classes: Amides; Heterocyclic Compounds; Thio Compounds

Function: Preservative

Technical/Other Names:
3(2H)-Isothiazolone, 2-Octyl
2-Octyl-2H-Isothiazol-3-One

Trade Name:
Koralone 500 Preservative (Rohm and Haas)

ODONTELLA AURITA EXTRACT

Definition: Odontella Aurita Extract is an extract of whole plant *Odontella aurita*. See *"Regulatory and Ingredient Use Information,"* regarding the labeling names for botanical ingredients in Volume 1, Introduction, Part A.

Chemical Class: Biological Products

Function: Skin-Conditioning Agent - Miscellaneous

Technical/Other Name:
Extract of Odontella Aurita

Trade Name Mixture:
Odontella PFST (Innovalg)

ODONTELLA AURITA OIL

Definition: Odontella Aurita Oil is the oil extracted from the alga, *Odontella aurita*. See *"Regulatory and Ingredient Use Information,"* regarding the labeling names for botanical ingredients in Volume 1, Introduction, Part A.

Chemical Class: Fats and Oils

Function: Skin-Conditioning Agent - Emollient

Technical/Other Name:
Oils, Odontella Aurita

Trade Name:
Huile D' Odontella Aurita (Codif)

ODORIKOSOU EKISU

JPN Translation:
オドリコソウエキス

Definition: Odorikosou Ekisu is an extract of the flowers, stems, and leaves of the white nettle, *Lamium album*. See *"Regulatory and Ingredient Use Information,"* regarding use of Japan Trivial names in Volume 1, Introduction, Part A.

Chemical Class: Biological Products

Function: Skin-Conditioning Agent - Miscellaneous

Technical/Other Names:
Lamium Album Flower Extract (U.S.)
Lamium Album Leaf Extract (U.S.)

OENOCARPUS BACABA FRUIT OIL

Definition: Oenocarpus Bacaba Fruit Oil is the oil expressed from the fruit of *Oenocarpus bacaba*. See *"Regulatory and Ingredient Use Information,"* regarding the labeling names for botanical ingredients in Volume 1, Introduction, Part A.

Chemical Class: Fats and Oils

Function: Skin-Conditioning Agent - Occlusive

Technical/Other Name:
Oils, Oenocarpus Bacaba Fruit

Trade Name:
Chemyforest Bacaba CG (Chemyunion)

OENOCARPUS BATAUA FRUIT OIL

Definition: Oenocarpus Bataua Fruit Oil is the oil expressed from the fruit of *Oenocarpus bataua*. See *"Regulatory and Ingredient Use Information,"* regarding the labeling names for botanical ingredients in Volume 1, Introduction, Part A.

Chemical Class: Fats and Oils

Function: Skin-Conditioning Agent - Occlusive

Technical/Other Name:
Oils, Oenocarpus Bataua Fruit

Trade Name:
Chemyforest Pataua CG (Chemyunion)

OENOTHERA BIENNIS (EVENING PRIMROSE) FLOWER EXTRACT

Definition: Oenothera Biennis (Evening Primrose) Flower Extract is an extract of the flowers of *Oenothera biennis*. See *"Regulatory and Ingredient Use Information,"* regarding the labeling names for botanical ingredients in Volume 1, Introduction, Part A.

Chemical Class: Biological Products

Function: Cosmetic Astringent

Technical/Other Names:
Evening Primrose Flower Extract
Extract of Oenothera Biennis Flower

Trade Name Mixtures:
Actiphyte Evening Primrose Lipo J (Active Organics)
Actiphyte Evening Primrose Lipo M (Active Organics)
Actiphyte of Evening Primrose BG50 (Active Organics)
Actiphyte of Evening Primrose GL50 (Active Organics)
Actiphyte of Evening Primrose Lipo S (Active Organics)
Actiphyte of Evening Primrose PG50 (Active Organics)
Herbasol Evening Primrose (Leaf) (Cosmetochem) (Cosmetochem International Ltd.)
Herbasol Extract Oil Soluble Evening Primrose (Cosmetochem) (Cosmetochem International Ltd.)

OENOTHERA BIENNIS (EVENING PRIMROSE) OIL

JPN Translation:
月見草油

Definition: Oenothera Biennis (Evening Primrose) Oil is an oil obtained from *Enethera biennis*. See *"Regulatory and Ingredient Use Information,"* regarding the labeling names for botanical ingredients in Volume 1, Introduction, Part A.

Information Sources: JCIC, JCLS, JSQI, MI-13(3939)

Chemical Class: Fats and Oils

Function: Skin-Conditioning Agent - Miscellaneous

Reported Product Categories: Bath Capsules; Face and Neck Preparations

(Excluding Shaving Preparations); Bath Preparations, Misc.; Body and Hand Preparations (Excluding Shaving Preparations); Hair Conditioners; Moisturizing Preparations; Lipsticks; Night Skin Care Preparations; Skin Care Preparations, Misc.

Technical/Other Names:
Evening Primrose Oil
Oil of Evening Primrose

Trade Names:
AEC Evening Primrose Oil (A & E Connock)
Cropure EPO (Croda, Inc.)
Crossential EPO (Croda Chemicals)
E.P.O. (Nisshin OilliO)
Evening Oil-J (Maruzen Pharmaceuticals Co., Ltd.)
Evening Primrose Oil (Arch Personal Care Products)
Evening Primrose Oil (Dekker)
Evening Primrose Oil (Desert Whale)
Evening Primrose Oil (Freeman)
Evening Primrose Oil (Maruzen Pharmaceuticals Co., Ltd.)
Evening Primrose Oil, CO2 Extracted (Paninkret)
Huile D'Onagre Vierge (Bertin)
Oenothera Biennis (Evening Primrose) Oil ies (IES LABO)
Phytol EPO (Evening Primrose Oil) (Custom Ingredients)
Salacos E.P.O. (Nisshin OilliO)

Trade Name Mixtures:
Actiphyte Siberian Ginseng Lipo E (Active Organics)
Bio-Oil GLA-10 (Arch Personal Care Products)
Brookosome EPO (Arch Personal Care Products)
Chelonine (JUVEX)
Evening Primrose Oil "W" Watersoluble (Cosmetochem) (Cosmetochem International Ltd.)
Melange huileux a base de the vert (Bertin)
Ropufa '10' N-6 Oil (Roche)
Tagrol EPO1 (Tagra)

OENOTHERA BIENNIS (EVENING PRIMROSE) ROOT EXTRACT

CAS No.	EINECS No.
90028-66-3	289-859-2

Definition: Oenothera Biennis (Evening Primrose) Root Extract is an extract of the roots and herb of the evening primrose, *Oenothera biennis. See "Regulatory and Ingredient Use Information," regarding the labeling names for botanical ingredients in Volume 1, Introduction, Part A.*

Chemical Class: Biological Products

Function: Not Reported

Reported Product Categories: Bath Capsules; Face and Neck Preparations (Excluding Shaving Preparations); Bath Preparations, Misc.; Body and Hand Preparations (Excluding Shaving Preparations); Hair Conditioners; Moisturizing Preparations; Night Skin Care Preparations; Skin Care Preparations, Misc.

Technical/Other Names:
Evening Primrose Extract
Evening Primrose Root Extract
Extract of Evening Primrose
Extract of Oenothera Biennis
Oenothera biennis Extract

Trade Name Mixtures:
Helionagre (Bertin)
Lipoplastidine Oenothera Biennis (Vevy)
Oenotherol (Somaig)
Phytelene of Evening Primrose EG 564 Liquid (Indena SA)
VT-194 Extract of Evening Primrose (Vege-Tech)

OENOTHERA BIENNIS (EVENING PRIMROSE) SEED

Definition: Oenothera Biennis (Evening Primrose) Seed is the seed of the evening primrose, *Oenothera biennis. See "Regulatory and Ingredient Use Information," regarding the labeling names for botanical ingredients in Volume 1, Introduction, Part A.*

Chemical Class: Biological Products

Function: Abrasive

Trade Name:
AEC Evening Primrose Seed (A & E Connock)

OENOTHERA BIENNIS SEED EXTRACT

Definition: Oenothera Biennis Seed Extract is an extract of the seeds of *Oenothera biennis. See "Regulatory and Ingredient Use Information," regarding the labeling names for botanical ingredients in Volume 1, Introduction, Part A.*

Chemical Class: Biological Products

Function: Skin-Conditioning Agent - Miscellaneous

Trade Names:
Evening Primrose Extract P (Nikko)
Evening Primrose Extract-WSP (Nikko)
Flavex Evening Primrose Seed CO2-to extract, Type 069.001 (Flavex)
Lunawhite-P (Ichimaru Pharcos)

Trade Name Mixtures:
Evening Primrose Extract-LC (Nikko)
Lunawhite-B (Ichimaru Pharcos)
Lunawhite-E (Ichimaru Pharcos)

OLAFLUR

CAS No.	EINECS No.
6818-37-7	229-891-6

Empirical Formula:
$C_{22}H_{58}N_2O_3 \cdot 2HF$

Definition: Olaflur is the substituted amine salt that conforms to the formula:

$$CH_3(CH_2)_{17}N(CH_2)_3NCH_2CH_2OH \cdot 2HF$$

with CH_2CH_2OH groups

Information Source: EEC(III/1-37)

Chemical Classes: Alkoxylated Amines; Amines

Function: Oral Care Agent

Technical/Other Names:
Ethanol, 2,2'-[[3-[(2-Hydroxyethyl)-Octadecylamino]Propyl]Imino]Bis-, Dihydrofluoride
3-(N-Hexadecy-N-2-hydroxyethylammonio) propyl bis (2-hydroxyethyl) ammonium dihydrofluoride
2,2'-[[3-[(2-Hydroxyethyl)-Octadecylamino]Propyl]Imino]Diethanol Dihydrofluoride
Propane, N,N,N-Tris(2-Hydroxyethyl)-N-Octadecyl-1,3-Diamino-, Dihydrofluoride
Stearyl Trihydroxyethyl Propylenediamine Dihydrofluoride
N,N,N-Tris(2-Hydroxyethyl)-N-Octadecyl-1,3-Diaminopropane Dihydrofluoride

Trade Name:
Unifluorid D-401 (Induchem)

Trade Name Mixture:
RonaCare Olaflur (Merck KGaA)

OLAX DISSITIFLORA ROOT OIL

Definition: Olax Dissitiflora Root Oil is the refined, fixed oil obtained from the roots of *Olax dissitiflora. See "Regulatory and Ingredient Use Information," regarding the labeling names for botanical ingredients in Volume 1, Introduction, Part A.*

Chemical Class: Essential Oils

Function: Fragrance Ingredient

Trade Name:
Phytoselect Olax (Indena SpA)

OLDENLANDIA DIFFUSA EXTRACT

Definition: Oldenlandia Diffusa Extract is an extract of the whole plant *Oldenlandia diffusa.* See *"Regulatory and Ingredient Use Information,"* regarding the labeling names for botanical ingredients in Volume 1, Introduction, Part A.

Chemical Class: Biological Products

Function: Skin-Conditioning Agent - Miscellaneous

Technical/Other Name:
Extract of Oldenlandia Diffusa

Trade Name:
Hedyotis Extract (ACT)

OLEA EUROPAEA (OLIVE) BARK EXTRACT

Definition: Olea Europaea (Olive) Bark Extract is an extract of the bark of the olive tree, *Oleo europaea.* See *"Regulatory and Ingredient Use Information,"* regarding the labeling names for botanical ingredients in Volume 1, Introduction, Part A.

Chemical Class: Biological Products

Function: Skin-Conditioning Agent - Miscellaneous

Technical/Other Name:
Extract of Olea Europea (Olive) Bark

OLEA EUROPAEA (OLIVE) BUD EXTRACT

Definition: Olea Europaea (Olive) Bud Extract is an extract of the buds of the *Olea europaea.* See *"Regulatory and Ingredient Use Information,"* regarding the labeling names for botanical ingredients in Volume 1, Introduction, Part A.

Chemical Class: Biological Products

Functions: Antioxidant; Skin-Conditioning Agent - Emollient

Technical/Other Name:
Extract of Olea Europaea (Olive) Bud

Trade Name Mixture:
Cryo-Bourgeon D'Olivier (Greentech)

OLEA EUROPAEA (OLIVE) FRUIT

Definition: Olea Europaea (Olive) Fruit is the fruit obtained from *Olea europaea.* See *"Regulatory and Ingredient Use Information,"* regarding the labeling names for botanical ingredients in Volume 1, Introduction, Part A.

Chemical Class: Biological Products

Functions: Abrasive; Skin-Conditioning Agent - Miscellaneous

Trade Names:
Olive Smooth 40-60 mesh (American Natural Products)
Olive Smooth 100-200 mesh (American Natural Products)

OLEA EUROPAEA (OLIVE) FRUIT EXTRACT

CAS No.: 84012-27-1

Definition: Olea Europaea (Olive) Fruit Extract is an extract of the fruit of the olive, *Olea europaea.* See *"Regulatory and Ingredient Use Information,"* regarding the labeling names for botanical ingredients in Volume 1, Introduction, Part A.

Chemical Class: Biological Products

Function: Not Reported

Technical/Other Names:
Extract of Olea Europaea
Extract of Olives
Olea Europaea Extract
Olive Extract
Olive Fruit Extract

Trade Name:
Green Olive Fruits Juice (Nippon Olive)

Trade Name Mixtures:
Actiphyte of Italian Olive BG50 (Active Organics)
Actiphyte of Italian Olive GL50 (Active Organics)
Actiphyte of Italian Olive Lipo S (Active Organics)
Actiphyte of Italian Olive PG50 (Active Organics)
Enna 212 Bust Firmness (Ennagram)
Filagrinol-Ext (Vevy)
Greenosome Smart M (Greentech)
Hair Lipid Complex (Natural and Marine Resources)
SCComplex (Natural and Marine Resources)
VT-284 Extract of Olive (Vege-Tech)

OLEA EUROPAEA (OLIVE) FRUIT OIL

CAS No.	EINECS No.
8001-25-0	232-277-0

JPN Translation:
オリーブ油

Definition: Olea Europaea (Olive) Fruit Oil is the fixed oil obtained from the ripe fruit of *Olea europaea.* See *"Regulatory and Ingredient Use Information,"* regarding the labeling names for botanical ingredients in Volume 1, Introduction, Part A.

Information Sources: ARG, AUS, BEL, BP, BPC, BRA, 21CFR175.105, 21CFR176.200, 21CFR176.210, CZE, DA, DDR, EGY, FI, FIN, ITA, JCLS, JSCI, MAR, MEX, MI-13(6905), NF XIX, PF, POR, RIFM, USAN, USD, YUG

Chemical Class: Fats and Oils

Functions: Fragrance Ingredient; Skin-Conditioning Agent - Occlusive

Reported Product Categories: Tonics, Dressings, and Other Hair Grooming Aids; Bath Soaps and Detergents; Moisturizing Preparations; Hair Conditioners; Paste Masks (Mud Packs); Skin Care Preparations, Misc.; Bath Oils, Tablets, and Salts; Bath Capsules; Bath Preparations, Misc.; Body and Hand Preparations (Excluding Shaving Preparations); Cleansing Products (Cold Creams, Cleansing Lotions, Liquids and Pads); Eyebrow Pencils; Shampoos (Non-coloring); Shaving Cream (Aerosol, Brushless and Lather); Suntan Gels, Creams, and Liquids; Suntan Preparations, Misc.

Technical/Other Names:
Oils, Olive
Olea Europaea Oil
Olive Fruit Oil
Olive oil (RIFM)
Olive Oil

Trade Names:
AEC Olive Oil (A & E Connock)
Certified Organic Olive Oil (Formula One Sciences)
Cropure Olive (Croda Chemicals)
Cropure Olive (Croda, Inc.)
EmCon Olive (Fanning)
Fanoliv Oil (Fanning)
Huile D'Olives Raffinee (Bertin)
Huile D'Olives Vierge (Bertin)
Jeen Olive Oil (Jeen)
Lipex Olive (Karlshamns AB)
Lipovol O (Lipo)
Nikkol Olive Oil (Nikko)
Olea-DCG (Vevy)
Olea Europaea Olive (Fruit) Oil ies (IES LABO)
Phytol O Pure (olive oil) (Custom Ingredients)
Tri-OL OLV (Tri-K)

Trade Name Mixtures:
Actiphyte Horsetail Lipo O (Active Organics)
Actiphyte of Arctic Cloudberry Lipo O (Active Organics)
Algae Comp-C (Bio-Botanica)
Chelonine (JUVEX)
Enviro Pure 310 (React Inc)
ESP Dry feel-Olive (Earth Supplied Products)
Extrapone Olive GW 2/033655 (Symrise)

Geldolive (Naturactiva)
Harmonic ASNP (Dr. Gerhard Steidl)
Hypericum Oil CLR (CLR)
Inholive (Naturactiva)
Lipocer (Plantech)
Lipoplastidine Olea Folium (Vevy)
Malvaceae Extract (Plantech)
Melange Huile de Colza / Huile D'Olive /
  Huile de Bourache (Bertin)
Olive Butter (R) (Strahl & Pitsch)
Oliwax (B & T)
Pharconix OV (Ichimaru Pharcos)
Skin Acne Complex MU 3690 (Greentech
  S.A)
St. John's Wort Oil (Cognis Care
  Chemicals/PA)
St. John's Wort Oil NOVAROM (Crodarom)
THIN AND DULL HAIR (Greentech S.A)

OLEA EUROPAEA (OLIVE) FRUIT UNSAPONIFIABLES

Definition: Olea Europaea (Olive) Fruit Unsaponifiables is the fraction of olive fruit remaining after fractional distillation.

Chemical Class: Unsaponifiables

Functions: Antioxidant; Binder; Emulsion Stabilizer; Hair Conditioning Agent; Skin-Conditioning Agent - Emollient

Trade Name:
Vitamolive (Naturactiva)

OLEA EUROPAEA (OLIVE) HUSK OIL

Definition: Olea Europaea (Olive) Husk Oil is the oil obtained from the solvent extraction of the olive husk, *Olea europaea. See "Regulatory and Ingredient Use Information," regarding the labeling names for botanical ingredients in Volume 1, Introduction, Part A.*

Chemical Class: Fats and Oils

Function: Skin-Conditioning Agent - Occlusive

Technical/Other Name:
Olive Husk Oil

Trade Name Mixtures:
Dermoliv (Alma Chimica)
Dermoliv T (Alma Chimica)

OLEA EUROPAEA (OLIVE) HUSK POWDER

Definition: Olea Europaea (Olive) Husk Powder is the powder obtained from the crushed husks of the olive, *Olea europaea. See "Regulatory and Ingredient Use Infor-*

mation," regarding the labeling names for botanical ingredients in Volume 1, Introduction, Part A.

Chemical Class: Biological Products

Function: Abrasive

Technical/Other Name:
Olive Husk Powder

Trade Names:
Olive Stone Powder 200/300 (Alban Muller)
Olive Stone Powder 400/500 (Alban Muller)

OLEA EUROPAEA (OLIVE) LEAF

Definition: Olea Europaea (Olive) Leaf is the leaf and/or stem obtained from the olive tree, *Olea europaea. See "Regulatory and Ingredient Use Information," regarding the labeling names for botanical ingredients in Volume 1, Introduction, Part A.*

Chemical Class: Biological Products

Function: Skin-Conditioning Agent - Miscellaneous

OLEA EUROPAEA (OLIVE) LEAF EXTRACT

CAS No.: 8060-29-5

Definition: Olea Europaea (Olive) Leaf Extract is an extract of the leaves of the olive, *Olea europaea. See "Regulatory and Ingredient Use Information," regarding the labeling names for botanical ingredients in Volume 1, Introduction, Part A.*

Chemical Class: Biological Products

Function: Skin-Conditioning Agent - Miscellaneous

Reported Product Category: Skin Care Preparations, Misc.

Technical/Other Names:
Extract of Olea Europaea Leaf
Extract of Olive Leaves
Olea Europaea Leaf Extract
Olive Leaf Extract

Trade Names:
ACB Olive Leaf Extract Powder (Active
  Concepts)
Exoleaves (Caro'iline)
Herbalia Olive (Cognis Deutschland)
Oleanolic Extract 80% (Sabinsa)
Oleuropein 80% (Sabinsa)

Trade Name Mixtures:
ACB Olive Leaf Extract BG (Active
  Concepts)
Actiphyte of Olive Leaf BG50 (Active
  Organics)
Actiphyte of Olive Leaf GL50 (Active
  Organics)

Actiphyte of Olive Leaf Lipo S (Active
  Organics)
Actiphyte of Olive Leaf PG50 (Active
  Organics)
Eurol BT (B & T)
Herbasec Olive (Leaf) (Cosmetochem)
  (Cosmetochem International Ltd.)
Herbasol-Extract Olive (Leaf)
  (Cosmetochem)
Herbasol Extract Olive (Leaf)
  (Cosmetochem) (Cosmetochem
  International Ltd.)
Hydroplastidine Olea Folium (Vevy)
Lipoplastidine Olea Folium (Vevy)
Macerat Huileux de Feuilles D'Olivier
  (Bertin)
Olea Europaea Olive (Leaf) Extract ies
  (IES LABO)
Oleanoline DPG (Vincience)
Oleavine LS (Laboratoires
  Serobiologiques)
Olive Extract (Maruzen Pharmaceuticals
  Co., Ltd.)
Olive Extract BG (Maruzen
  Pharmaceuticals Co., Ltd.)
Olive Leaf Extract (Maruzen
  Pharmaceuticals Co., Ltd.)
Olive Leaf Extract BG (Maruzen
  Pharmaceuticals Co., Ltd.)
Olive Leaf Extract HG (Provital/
  Centerchem)
Olive Leaf Liquid B (Ichimaru Pharcos)
Olive Leaf Liquid E (Ichimaru Pharcos)
Olive Tree HS (Alban Muller)
Olive Tree Leaf HPG Titrated (Alban
  Muller)
Olive Tree LS (Alban Muller)
Skin Acne Complex MU 3690 (Greentech
  S.A)
Xyleine (Vincience)

OLEA EUROPAEA (OLIVE) OIL UNSAPONIFIABLES

CAS No.: 156798-12-8

Definition: Olea Europaea (Olive) Oil Unsaponifiables is the fraction of olive oil which is not saponified in the refining recovery of olive oil fatty acids. *See "Regulatory and Ingredient Use Information," regarding the labeling names for botanical ingredients in Volume 1, Introduction, Part A.*

Chemical Class: Unsaponifiables

Functions: Hair Conditioning Agent; Skin-Conditioning Agent - Miscellaneous

Technical/Other Names:
Olea Europaea Unsaponifiables
Olive Oil Unsaponifiables
Unsaponifiable Olive Oil

Unsaponifiables, Olea Europaea
Unsaponifiables, Olive Oil

Trade Names:
Fanoliv ActiveE (Fanning)
Planell Oil (Arch Personal Care Products)
Planell Oil MO (Arch Personal Care Products)
Tocolive (Naturactiva)

Trade Name Mixtures:
Dermoliv (Alma Chimica)
Dermoliv T (Alma Chimica)
Filagrinol (Vevy)
Geldolive (Naturactiva)
Inholive (Naturactiva)
Oleawax (Sinerga)
Oliwax (B & T)

OLEA EUROPAEA (OLIVE) SEED POWDER

Definition: Olea Europaea (Olive) Seed Powder is a powder of the crushed pits of *Olea europaea*. See *"Regulatory and Ingredient Use Information," regarding the labeling names for botanical ingredients in Volume 1, Introduction, Part A.*

Chemical Class: Biological Products

Function: Abrasive

Technical/Other Names:
Olive Pit Powder
Powdered Olive Pit

Trade Names:
Olive Pits Abrasive "GBU" Coarse, Medium-fine & Std (Cosmetochem) (Cosmetochem International Ltd.)
Olive Stone Granules (Brightstern)

OLEA EUROPAEA (OLIVE) WOOD EXTRACT

Definition: Olea Europaea (Olive) Wood Extract is an extract of the bark of *Olea europaea*. See *"Regulatory and Ingredient Use Information," regarding the labeling names for botanical ingredients in Volume 1, Introduction, Part A.*

Chemical Class: Biological Products

Function: Not Reported

Trade Name Mixture:
Herbasol Extract Olive Wood (Cosmetochem International Ltd.)

OLEALKONIUM CHLORIDE

CAS Nos.
37139-99-4
80458-20-4

EINECS No.
253-363-4

Empirical Formula:
$C_{27}H_{48}N \cdot Cl$

Definition: Olealkonium Chloride is the quaternary ammonium salt that conforms to the formula:

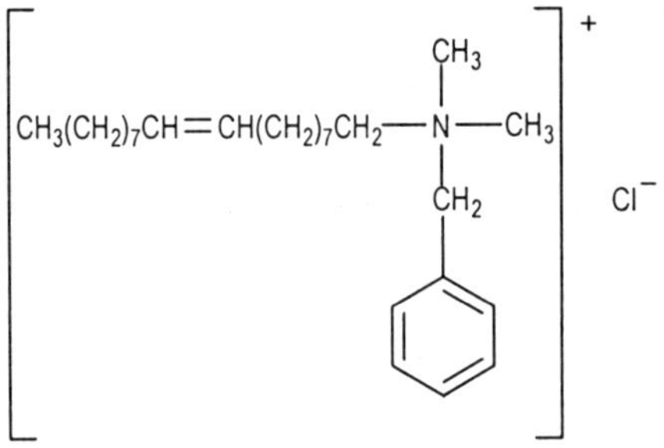

Information Sources: 21CFR175.105, 21CFR178.1010, TSCA

Chemical Class: Quaternary Ammonium Compounds

Functions: Antistatic Agent; Cosmetic Biocide; Hair Conditioning Agent

Reported Product Categories: Hair Dyes and Colors (All Types Requiring Caution Statements and Patch Tests); Tonics, Dressings, and Other Hair Grooming Aids; Hair Conditioners; Hair Sprays (Aerosol Fixatives); Shampoos (Non-coloring)

Technical/Other Names:
Benzenemethanaminium, N,N-Dimethyl-N-9-Octadecenyl-, Chloride
N,N-Dimethyl-N-9-Octadecenylbenzene-methanaminium Chloride
Oleyl Dimethyl Benzyl Ammonium Chloride

Trade Names:
AMMONYX KP (Stepan)
Incroquat O-50 (Croda, Inc.)
Mackernium KP (McIntyre)

OLEAMIDE

CAS Nos.
301-02-0
3322-62-1

EINECS No.
206-103-9

Empirical Formula:
$C_{18}H_{35}NO$

Definition: Oleamide is the aliphatic amide that conforms generally to the formula:

Information Sources: 21CFR175.105, 21CFR175.300, 21CFR178.3860, 21CFR178.3910, 21CFR179.45, 21CFR181.22, 21CFR181.28, TSCA

Chemical Class: Amides

Function: Not Reported

Technical/Other Names:
9-Octadecenamide
Oleic Acid Amide
Oleyl Amide

Trade Name:
Armid O (Akzo Nobel)

Trade Name Mixture:
Amisol 110 (Lucas Meyer)

OLEAMIDE DEA

CAS Nos.
93-83-4
5299-69-4

EINECS No.
202-281-7

JPN Translation:
オレアミド DEA

Empirical Formula:
$C_{22}H_{43}NO_3$

Definition: Oleamide DEA is a mixture of ethanolamides of oleic acid. It conforms generally to the formula:

Information Sources: 21CFR175.105, 21CFR176.210, 21CFR177.2260, 21CFR177.2800, CIR: [SQ] JACT-5(5)1986, EEC(III/1-60), JCIC, JCLS, JSQI, TSCA

Chemical Class: Alkanolamides

Functions: Surfactant - Foam Booster; Viscosity Increasing Agent - Aqueous

Reported Product Categories: Hair Dyes and Colors (All Types Requiring Caution Statements and Patch Tests); Hair Color Sprays (Aerosol); Bubble Baths; Hair Coloring Preparations, Misc.; Shampoos (Non-coloring); Bath Preparations, Misc.; Bath Soaps and Detergents; Cleansing Products (Cold Creams, Cleansing Lotions, Liquids and Pads)

Technical/Other Names:
N,N-Bis(2-Hydroxyethyl)-9-Octadecenamide
N,N-Bis(2-Hydroxyethyl)Oleamide
Diethanolamine Oleic Acid Amide
9-Octadecenamide, N,N-Bis(2-Hydroxy-ethyl)-
Oleic Diethanolamide
Oleylamide DEA

Trade Names:
Active #18 (Blew Chemical)
Amidex O (Chemron)
Calamide O (Pilot)
Colamid OA (Colonial Chemical Inc)
Jeemide OFO (Jeen)

Mackamide MO (McIntyre)
Mackamide NOA (McIntyre)
Mackamide O (McIntyre)
NINOL 201 (Stepan)
Olamida SAV (Fabriquimica)
Protamide OFO-LD (Protameen)
Rewomid DO 280 SE (Degussa Care Specialties)
Saboamid DEO (Sabo)
Schercomid SO-A (Scher)
Schercomid SO-A-Special (Scher)
Upamide O-20 (Universal Preserv-A-Chem)
Upamide OD (Universal Preserv-A-Chem)

Trade Name Mixtures:
Calamide F (Pilot)
CalBlend GEL (Pilot)
Emulmetik 110 (Lucas Meyer GmbH)
Hetamide DO (Heterene)
Naetex O-20 (Lanaetex)
Olamida AV (Fabriquimica)
Protamide OFO (Protameen)
Schercomid ODA (Scher)

OLEAMIDE MEA

CAS Nos.	EINECS No.
111-58-0	203-884-8
7545-20-2	

Empirical Formula:
$C_{20}H_{39}NO_2$

Definition: Oleamide MEA is a mixture of ethanolamides of oleic acid. It conforms generally to the formula:

$$CH_3(CH_2)_7CH = CH(CH_2)_7C - NHCH_2CH_2OH$$
$$\overset{\displaystyle O}{\overset{\displaystyle \|}{}}$$

Information Source: TSCA

Chemical Class: Alkanolamides

Functions: Surfactant - Foam Booster; Viscosity Increasing Agent - Aqueous

Technical/Other Names:
N-(2-Hydroxyethyl)-9-Octadecenamide
N-(2-Hydroxyethyl)Oleamide
Monoethanolamine Oleic Acid Amide
9-Octadecenamide, N-(2-Hydroxyethyl)-
Oleoyl Monoethanolamide

Trade Names:
Hetamide MO (Heterene)
Mackamide OMA (McIntyre)
Schercomid OME (Scher)

OLEAMIDE MIPA

CAS Nos.	EINECS No.
111-05-7	203-828-2
54375-42-7	

Empirical Formula:
$C_{21}H_{41}NO_2$

Definition: Oleamide MIPA is a mixture of isopropanolamides of oleic acid. It conforms generally to the formula:

$$CH_3(CH_2)_7CH = CH(CH_2)_7C - NHCH_2CHOH$$
$$\overset{\displaystyle O}{\overset{\displaystyle \|}{}} \qquad \underset{\displaystyle CH_3}{\overset{\displaystyle |}{}}$$

Information Sources: CTFA D, TSCA

Chemical Class: Alkanolamides

Functions: Surfactant - Foam Booster; Viscosity Increasing Agent - Aqueous

Reported Product Category: Hair Conditioners

Technical/Other Names:
N-(2-Hydroxypropyl)-9-Octadecenamide
Monoisopropanolamine Oleic Acid Amide
9-Octadecenamide, N-(2-Hydroxypropyl)-
Oleamide, N-(2-Hydroxypropyl)-
Oleic Monoisopropanolamide
Oleylmonoisopropanolamide

Trade Names:
Mackamide OP (McIntyre)
Schercomid OMI (Scher)

Trade Name Mixtures:
Custoblend ALO (Amide Free Blend) (Custom Ingredients)
Lowenol 4667 (Lowenstein)
Lowenol T-163A (Lowenstein)
Lowenol T-163 (Lowenstein)

2-OLEAMIDO-1,3-OCTADECANEDIOL

Empirical Formula:
$C_{36}H_{71}NO_3$

Definition: 2-Oleamido-1,3-Octadecanediol is the organic compound that conforms to the formula:

$$CH_3(CH_2)_7CH = CH(CH_2)_7C - NHCHCH(CH_2)_{14}CH_3$$

with O, OH, and CH₂OH substituents.

Chemical Classes: Alcohols; Amides

Functions: Skin-Conditioning Agent - Emollient; Skin-Conditioning Agent - Miscellaneous

Reported Product Category: Moisturizing Preparations

Trade Name:
Mexanyl GZ (Chimex)

OLEAMIDOPROPYLAMINE OXIDE

CAS No.	EINECS No.
25159-40-4	246-684-6

Empirical Formula:
$C_{23}H_{46}N_2O_2$

Definition: Oleamidopropylamine Oxide is the aliphatic amine oxide that conforms generally to the formula:

$$CH_3(CH_2)_7CH = CH(CH_2)_7C - NH(CH_2)_3N \longrightarrow O$$

with O and two CH₃ substituents on N.

Chemical Class: Amine Oxides

Functions: Hair Conditioning Agent; Surfactant - Cleansing Agent; Surfactant - Foam Booster; Surfactant - Hydrotrope

Technical/Other Names:
Amides, Oleic, N-[3-(Dimethylamino)-Propyl], N-Oxide
N-[-3-(Dimethylamino)propyl]-9-Octadecenamide-N-Oxide
Dimethylaminopropyl Oleamide N-Oxide
9-Octadecenamide, N-[3-(Dimethylamino)Propyl]-, N-Oxide
9-Octadecenamide, N-((3-(Dimethyloxido-amino)Propyl)-

Trade Name:
Mackamine OAO (McIntyre)

OLEAMIDOPROPYL BETAINE

CAS No.	EINECS No.
25054-76-6	246-584-2

Empirical Formula:
$C_{25}H_{48}N_2O_3$

Definition: Oleamidopropyl Betaine is the zwitterion (inner salt) that conforms to the formula:

$$CH_3(CH_2)_7CH$$
$$CH(CH_2)_7C - NH(CH_2)_3N^+ - CH_2COO^-$$

with O and CH₃ substituents.

Information Source: TSCA

Chemical Class: Betaines

Functions: Antistatic Agent; Hair Conditioning Agent; Skin-Conditioning Agent - Miscellaneous; Surfactant - Cleansing Agent; Surfactant - Foam Booster; Viscosity Increasing Agent - Aqueous

Technical/Other Names:
Ammonium, (Carboxymethyl)Dimethyl(3-Oleamidopropyl)-, Hydroxide, Inner Salt
N-(Carboxymethyl)-N,N-Dimethyl-3-[(1-Oxooctadecenyl)Amino]-1-Propanaminium Hydroxide, Inner Salt
Oleamidopropyl Dimethyl Glycine
1-Propanaminium, N-(Carboxymethyl)-N,N-Dimethyl-3-[(1-Oxooctadecenyl)Amino]-, Hydroxide, Inner Salt

Trade Names:
Mackam HV (McIntyre)
Mirataine BET O-30 (Rhodia)

OLEAMIDOPROPYL DIMETHYLAMINE

CAS No.   **EINECS No.**
109-28-4   203-661-5

Empirical Formula:
$C_{23}H_{46}N_2O$

Definition: Oleamidopropyl Dimethylamine is the amidoamine that conforms generally to the formula:

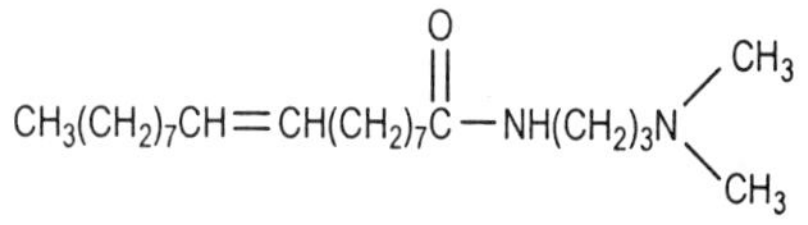

Information Sources: CTFA D, TSCA

Chemical Class: Amines

Function: Antistatic Agent

Reported Product Categories: Hair Dyes and Colors (All Types Requiring Caution Statements and Patch Tests); Hair Coloring Preparations, Misc.; Hair Bleaches; Hair Conditioners; Permanent Waves; Shampoos (Non-coloring); Tonics, Dressings, and Other Hair Grooming Aids

Technical/Other Names:
N-[3-Dimethylamino)Propyl]-9-
   Octadecenamide
Dimethylaminopropyl Oleamide
9-Octadecenamide,
   N-[3-(Dimethylamino)Propyl]-

Trade Names:
Mackine 501 (McIntyre)
Schercodine O (Scher)
Unizeen OA (Universal Preserv-A-Chem)

OLEAMIDOPROPYL DIMETHYLAMINE GLYCOLATE

Empirical Formula:
$C_{23}H_{46}N_2O \cdot C_2H_4O_3$

Definition: Oleamidopropyl Dimethylamine Glycolate is the amidoamine salt that conforms generally to the formula:

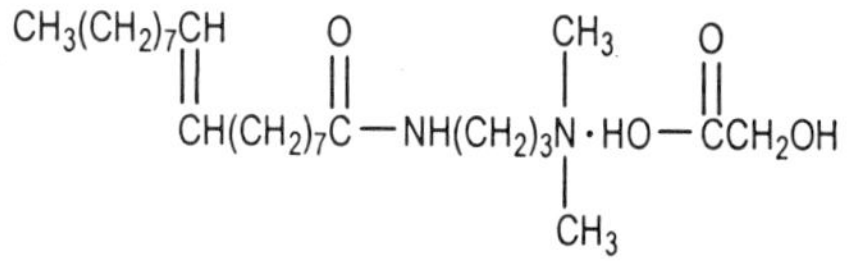

Chemical Class: Amines

Functions: Antistatic Agent; Hair Conditioning Agent

Technical/Other Name:
Dimethylaminopropyl Oleamide Glycolate

OLEAMIDOPROPYL DIMETHYLAMINE HYDROLYZED COLLAGEN

Definition: Oleamidopropyl Dimethylamine Hydrolyzed Collagen is the amine salt of Oleamidopropyl Dimethylamine (q.v.) and Hydrolyzed Collagen (q.v.).

Chemical Class: Protein Derivatives

Functions: Antistatic Agent; Hair Conditioning Agent; Skin-Conditioning Agent - Miscellaneous

Technical/Other Name:
Oleamidopropyl Dimethylamine Hydrolyzed Animal Protein

OLEAMIDOPROPYL DIMETHYLAMINE LACTATE

Empirical Formula:
$C_{23}H_{46}N_2O \cdot C_3H_6O_3$

Definition: Oleamidopropyl Dimethylamine Lactate is the amidoamine salt that conforms generally to the formula:

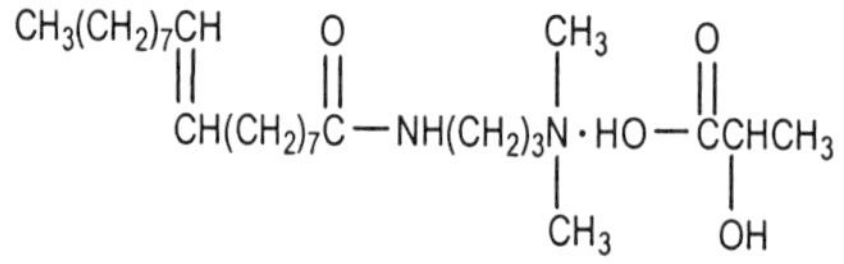

Chemical Class: Amines

Functions: Antistatic Agent; Hair Conditioning Agent

Technical/Other Name:
Dimethylaminopropyl Oleamide Lactate

Trade Name:
Mackalene 516 (McIntyre)

OLEAMIDOPROPYL DIMETHYLAMINE PROPIONATE

Empirical Formula:
$C_{23}H_{46}N_2O \cdot C_3H_6O_2$

Definition: Oleamidopropyl Dimethylamine Propionate is the organic salt that conforms to the formula:

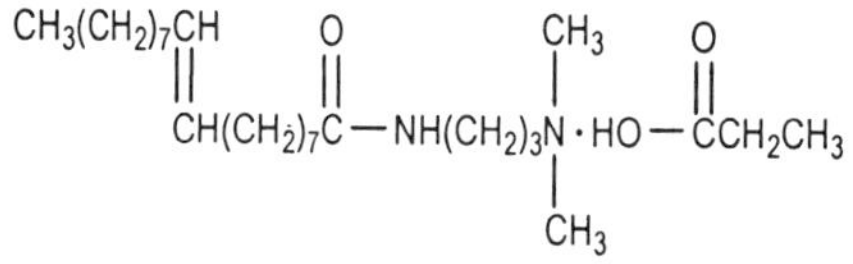

Chemical Class: Amines

Functions: Antistatic Agent; Hair Conditioning Agent

OLEAMIDOPROPYLDIMONIUM HYDROXYPROPYL HYDROLYZED COLLAGEN

Definition: Oleamidopropyldimonium Hydroxypropyl Hydrolyzed Collagen is the quaternary ammonium chloride that conforms generally to the formula:

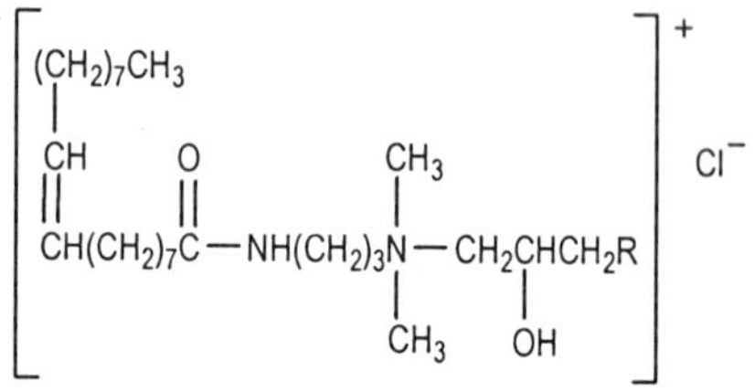

where R represents the hydrolyzed collagen moiety.

Chemical Classes: Protein Derivatives; Quaternary Ammonium Compounds

Functions: Antistatic Agent; Hair Conditioning Agent; Skin-Conditioning Agent - Miscellaneous

OLEAMIDOPROPYL ETHYLDIMONIUM ETHOSULFATE

Empirical Formula:
$C_{25}H_{51}N_2O \cdot C_2H_5O_4S$

Definition: Oleamidopropyl Ethyldimonium Ethosulfate is the quaternary ammonium salt that conforms to the formula:

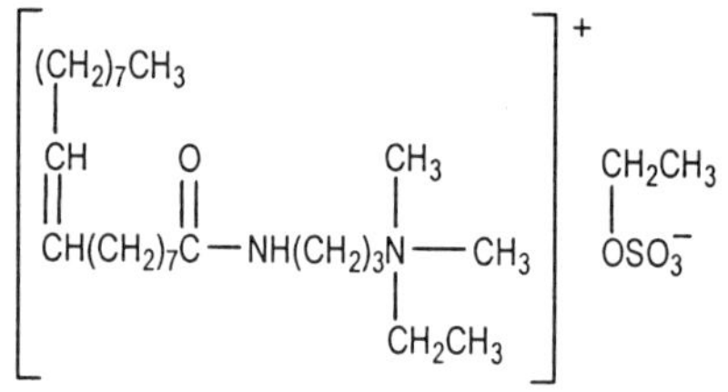

Chemical Class: Quaternary Ammonium Compounds

Function: Antistatic Agent

Trade Names:
Foamquat ODES (Alzo)
Mackernium OAPDES (McIntyre)

OLEAMIDOPROPYL HYDROXYSULTAINE

Empirical Formula:
$C_{26}H_{52}NO_5S$

Definition: Oleamidopropyl Hydroxysultaine is the zwitterion (inner salt) that conforms generally to the formula:

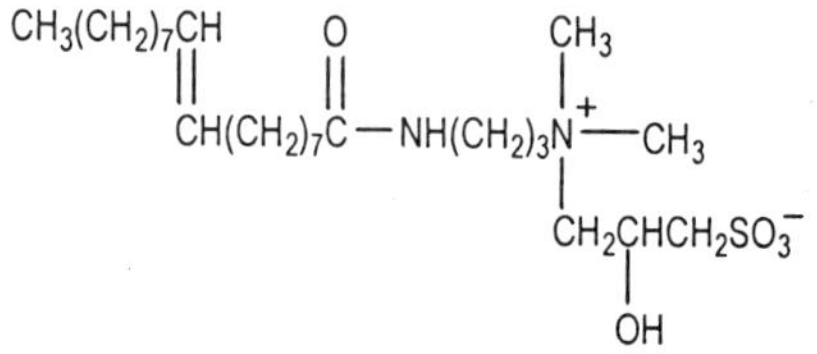

Chemical Class: Betaines

Functions: Antistatic Agent; Hair Conditioning Agent; Skin-Conditioning Agent - Miscellaneous; Surfactant - Cleansing Agent; Surfactant - Foam Booster; Viscosity Increasing Agent - Aqueous

Technical/Other Names:
[3-(9-Octadecenamido)Propyl](2-Hydroxy-3-Sulfopropyl)Dimethyl Quaternary Ammonium Compounds, Hydroxide, Inner Salt
Quaternary Ammonium Compounds, (3-Oleamidopropyl)(2-Hydroxy-3-Sulfopropyl)Dimethyl, Hydroxide, Inner Salt

OLEAMIDOPROPYL PG-DIMONIUM CHLORIDE

Empirical Formula:
$C_{26}H_{52}N_2O_3 \cdot Cl$

Definition: Oleamidopropyl PG-Dimonium Chloride is the quaternary ammonium salt that conforms to the formula:

$$\left[\begin{array}{c} CH_3(CH_2)_7CH \\ CH(CH_2)_7-NH(CH_2)_3\overset{CH_3}{\underset{HOCH_2CHCH_2}{\overset{O}{\parallel}}}NCH_3 \\ OH \end{array} \right]^+ Cl^-$$

Chemical Class: Quaternary Ammonium Compounds

Functions: Antistatic Agent; Hair Conditioning Agent

Trade Name:
Lexquat AMG-O (Inolex)

OLEAMINE

CAS No.	EINECS No.
112-90-3	204-015-5

Empirical Formula:
$C_{18}H_{37}N$

Definition: Oleamine is the primary aliphatic amine that conforms generally to the formula:

$$CH_3(CH_2)_7CH=CH(CH_2)_7CH_2NH_2$$

Information Source: TSCA

Chemical Class: Amines

Function: Antistatic Agent

Technical/Other Names:
9-Octadecen-1-Amine
Oleyl Amine

Trade Names:
Armeen O (Akzo Nobel)
Armeen OD (Akzo Nobel)

OLEAMINE BISHYDROXYPROPYL-TRIMONIUM CHLORIDE

Empirical Formula:
$C_{30}H_{63}N_3O_2 \cdot Cl$

Definition: Oleamine Bishydroxypropyl-trimonium Chloride is the quaternary ammonium compound that conforms to the formula:

$$\begin{array}{c} CH(CH_2)_7CH_3 \\ \parallel \\ CH(CH_2)_7CH_2N \end{array} - \left[CH_2CHCH_2 - \overset{CH_3}{\underset{CH_3}{\overset{|}{N}}} - CH_3 \atop \underset{OH}{} \right]_2^+ 2Cl^-$$

Chemical Class: Quaternary Ammonium Compounds

Functions: Antistatic Agent; Hair Conditioning Agent

Technical/Other Names:
3,3'-(9-Octadecenylimino)Bis[2-Hydroxy-N,N,N-Trimethyl-1-Propanaminium, Chloride
1-Propanaminium, 3,3'-(9-Octadecenylimino)Bis[2-Hydroxy-N,N,N-Trimethyl-, Chloride

Trade Name Mixture:
Akypomine P 191 (Kao GmbH)

OLEAMINE OXIDE

CAS Nos.	EINECS No.
14351-50-9	238-311-0
61792-38-9	

JPN Translation:
オレアミンオキシド

Empirical Formula:
$C_{20}H_{41}NO$

Definition: Oleamine Oxide is the tertiary amine oxide that conforms generally to the formula:

$$CH_3(CH_2)_7CH=CH(CH_2)_7CH_2\overset{CH_3}{\underset{CH_3}{\overset{|}{N}}} \longrightarrow O$$

Information Sources: JCIC, JCLS, JSQI

Chemical Class: Amine Oxides

Functions: Hair Conditioning Agent; Surfactant - Cleansing Agent; Surfactant - Foam Booster; Surfactant - Hydrotrope

Reported Product Categories: Hair Conditioners; Hair Bleaches; Hair Rinses (Coloring)

Technical/Other Names:
N,N-Dimethyl-9-Octadecen-1-Amine-N-Oxide
9-Octadecen-1-Amine, N,N-Dimethyl-, N-Oxide
Oleyl Dimethyl Amine Oxide

Trade Names:
Chemoxide O (Chemron)
Mackamine O2 (McIntyre)
Schercamox DMO (Scher)
Standamox 01 (Cognis Care Chemicals/PA)
Standamox O1 (Cognis Care Chemicals/NJ)
Unimox OL (Universal Preserv-A-Chem)

OLEANOLIC ACID

CAS No.	EINECS No.
508-02-1	208-081-6

Empirical Formula:
$C_{30}H_{48}O_3$

Definition: Oleanolic Acid is the organic compound that conforms to the formula:

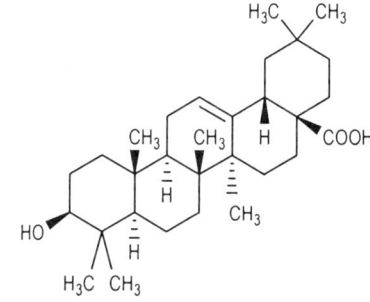

Information Source: MI-13(6897)

Chemical Class: Carboxylic Acids

Function: Skin-Conditioning Agent - Miscellaneous

Technical/Other Name:
Olean-12-en-28-oic Acid, 3-Hydroxy-, (3β)-

Trade Name:
Oleanolic Acid Pure (Sederma)

OLEIC ACID

CAS Nos.	EINECS Nos.
112-80-1	204-007-1
2027-47-6	217-977-6

JPN Translation:
オレイン酸

Empirical Formula:
$C_{18}H_{34}O_2$

Definition: Oleic Acid is the unsaturated fatty acid that conforms generally to the formula:

$$CH_3(CH_2)_7CH = CH(CH_2)_7C - OH$$

with O double bonded at the carboxyl carbon.

Information Sources: ARG, BP, BPC, BRA, 21CFR172.210, 21CFR172.860, 21CFR172.862, 21CFR173.315, 21CFR173.340, 21CFR175.105, 21CFR175.320, 21CFR176.170, 21CFR176.200, 21CFR176.210, 21CFR177.1010, 21CFR177.1200, 21CFR177.2260, 21CFR177.2600, 21CFR177.2800, 21CFR178.3570, 21CFR178.3910, 21CFR182.70, 21CFR182.90, CIR: [S] JACT-6(3)1987, CTFA S, DDR, EGY, FCC, FIN, HUN, JCLS, JSCI, KOR, MAR, MI-13(6898), NED, NF XIX, RIFM, SNPF, TSCA, USAN, USD

Chemical Class: Fatty Acids

Functions: Fragrance Ingredient; Surfactant - Cleansing Agent

Surfactant-Cleansing Agent is included as a function for the soap form of Oleic Acid.

Reported Product Categories: Hair Dyes and Colors (All Types Requiring Caution Statements and Patch Tests); Mascara; Bath Soaps and Detergents; Hair Bleaches; Hair Color Sprays (Aerosol); Eyeliners; Bath Preparations, Misc.; Hair Tints; Foundations; Moisturizing Preparations; Bath Oils, Tablets, and Salts; Cleansing Products (Cold Creams, Cleansing Lotions, Liquids and Pads); Eyebrow Pencils; Fragrance Preparations, Misc.; Lipsticks; Permanent Waves; Shampoos (Non-coloring); Suntan Gels, Creams, and Liquids; Baby Products, Misc.; Body and Hand Preparations (Excluding Shaving Preparations); Hair Coloring Preparations, Misc.; Makeup Bases; Personal Cleanliness Products, Misc.; Makeup Preparations (Not eye), Misc.; Shaving Cream (Aerosol, Brushless and Lather); Skin Care Preparations, Misc.

Technical/Other Names:
9-Octadecenoic Acid
Oleic acid (RIFM)

Trade Names:
AEC Oleic Acid (A & E Connock)
Emersol 210 (Cognis Corp.)
Emersol 213 (Cognis Corp.)
Emersol 221 (Cognis Corp.)
Emersol 6321 (Cognis Corp.)
Emersol 233LL (Cognis Corp.)
Neo-Fat 90-04 (Akzo Nobel)
Neo-Fat 94-04 (Akzo Nobel)
Oleic acid (RIFM) (Lipo)
Pamolyn 100 (Eastman Chemical)
PRIOLENE 6900 (Uniqema Europe)
PRIOLENE 6907 (Uniqema Europe)
PRIOLENE 6928 (Uniqema Europe)
PRIOLENE 6996 (Uniqema Europe)
Radiacid 0212-0216 (Oleon NV)
Unifat 5L (Universal Preserv-A-Chem)

Trade Name Mixtures:
Cerasome (Lipoid)
Ceraspheres 9000 (Lipoid)
Ceraspheres-G 9507 (Lipoid)
DS-CERIX (Doosan)
DUB VFA (Stearinerie Dubois Fils)
EFA-Plex (Arch Personal Care Products)
EFA-Plexsol (Arch Personal Care Products)
Lowenol Conditioner 288 (Lowenstein)
Lowenol Emulsion 79 (Lowenstein)
Lysofat OL (CEP (Solabia))
N.S.L.E. (Sederma)
Phosal 53MCT (Phospholipid)
Protachem 35A (Protameen)
Thiolin (General Topics)
TYR-OL (Sederma)
Vitamin F Oilsoluble (Cosmetochem) (Cosmetochem International Ltd.)
Vitamin F Watersoluble (Cosmetochem) (Cosmetochem International Ltd.)

OLEIC/LINOLEIC/LINOLENIC POLY-GLYCERIDES

Definition: Oleic/Linoleic/Linolenic Poly-glycerides is formed by the polymerization of the mixture of the mono, di, and triglycerides of oleic, linoleic and linolenlic acids.

Chemical Class: Glyceryl Esters and Derivatives

Functions: Binder; Emulsion Stabilizer; Skin-Conditioning Agent - Emollient; Surfactant - Emulsifying Agent; Viscosity Increasing Agent - Nonaqueous

Trade Names:
Viamerine 2500 (Aldivia)
Viamerine 4000 (Aldivia)
Viamerine 10000 (Aldivia)

OLEIC/LINOLEIC TRIGLYCERIDE

Definition: Oleic/Linoleic Triglyceride is the mixed ester of glycerin with oleic and linoleic acids.

Chemical Classes: Fats and Oils; Glyceryl Esters and Derivatives

Function: Skin-Conditioning Agent - Occlusive

OLEIC/PALMITIC/LAURIC/MYRISTIC/LINOLEIC TRIGLYCERIDE

Definition: Oleic/Palmitic/Lauric/Myristic/Linoleic Triglyceride is the mixed triester of glycerin with oleic, palmitic, lauric, myristic and linoleic acids.

Chemical Classes: Fats and Oils; Glyceryl Esters and Derivatives

Function: Skin-Conditioning Agent - Occlusive

Trade Name:
Dermol T (Fabriquimica)

OLEOSTEARINE

JPN Translation:
オレオステアリン

Definition: Oleostearine is a mixture of fatty acid trigylcerides remaining after the physical separation of the low titre oils from beef tallow.

Chemical Class: Fats and Oils

Functions: Binder; Viscosity Increasing Agent - Nonaqueous

Reported Product Category: Eyeliners

Trade Name Mixtures:
Japan Wax Substitute No. 473 (Ross)
Japan Wax Substitute No. 525 (Ross)
Ross Japan Wax Substitute No. 473 (Ross)
Ross Japan Wax Substitute No. 525 (Ross)

OLEOYL EPOXY RESIN

JPN Translation:
エポキシエステル - 2

Definition: Oleoyl Epoxy Resin is the Oleic Acid (q.v.) ester of 4,4'-isopropylidene-diphenol/Epichlorohydrin Copolymer (q.v.), also known as epoxy resin.

Information Source: JCLS

Chemical Class: Synthetic Polymers

Function: Film Former

OLEOYL ETHYL GLUCOSIDE

Definition: Oleoyl Ethyl Glucoside is the ester of Oleic Acid (q.v.) and ethyl glucoside.

Chemical Classes: Carbohydrates; Esters

Function: Surfactant - Emulsifying Agent

OLEOYL HYDROLYZED COLLAGEN

CAS No.: 68458-51-5

JPN Translation:
オレオイル加水分解コラーゲン

Definition: Oleoyl Hydrolyzed Collagen is the condensation product of oleic acid chloride and Hydrolyzed Collagen (q.v.).

Information Sources: JCIC, JCLS, JSQI, TSCA

Chemical Class: Protein Derivatives

Functions: Hair Conditioning Agent; Skin-Conditioning Agent - Miscellaneous; Surfactant - Cleansing Agent

Technical/Other Names:
Oleoyl Hydrolyzed Animal Protein
Proteins, Hydrolysates, Reaction Products with Oleoyl Chloride

Trade Name:
Oleoyl Polypeptid Pulver (Cognis Deutschland)

OLEOYL PG-TRIMONIUM CHLORIDE

Empirical Formula:
$C_{24}H_{48}NO_3 \cdot Cl$

Definition: Oleoyl PG-Trimonium Chloride is the quaternary ammonium salt that conforms to the formula:

$$\left[\begin{array}{c} (CH_2)_7CH_3 \\ | \\ CH \\ || \quad O \\ || \quad || \\ CH(CH_2)_7C-OCH_2CHCH_2-N-CH_3 \\ | \quad\quad | \\ OH \quad\quad CH_3 \end{array} \begin{array}{c} CH_3 \\ | \\ -CH_3 \end{array} \right]^+ Cl^-$$

Chemical Class: Quaternary Ammonium Compounds

Functions: Antistatic Agent; Hair Conditioning Agent

Trade Name Mixture:
Akypoquat 40 (Kao GmbH)

OLEOYL SARCOSINE

CAS No.	EINECS No.
110-25-8	203-749-3

JPN Translation:
オレオイルサルコシン

Empirical Formula:
$C_{21}H_{39}NO_3$

Definition: Oleoyl Sarcosine is the condensation product of oleic acid with N-methylglycine. It conforms generally to the formula:

$$CH_3(CH_2)_7CH=CH(CH_2)_7C-NCH_2COOH$$
$$\underset{CH_3}{|}$$
(with O double bond on carbonyl)

Information Sources: 21CFR178.3130, CIR: [SQ] IJT-20(SUPPL. 1)2001, JCIC, JCLS, JSQI, TSCA

Chemical Class: Sarcosinates and Sarcosine Derivatives

Functions: Hair Conditioning Agent; Surfactant - Cleansing Agent

Technical/Other Names:
Glycine, N-Methyl-N-(1-Oxo-9-Octadecenyl)-
N-Methyl-N-(1-Oxo-9-Octadecenyl)Glycine
Oleic Sarcoside
Oleoyl N-Methylaminoacetic Acid
Oleyl N-Methylglycine

Trade Names:
Crodasinic O (Croda Chemicals)
Hamposyl O (Amerchol)
Nikkol Sarcosinate OH (Nikko)
Oramix O (SEPPIC)
Vanseal OS (Vanderbilt)

Trade Name Mixtures:
Oleo-Coll LP (Arch Personal Care Products)
Oleo-Coll LP/LF (Arch Personal Care Products)
Proto-Lan 8 (Maybrook)

OLEOYL TYROSINE

Empirical Formula:
$C_{27}H_{43}O_4N$

Definition: Oleoyl Tyrosine is the organic compound that conforms to the formula:

$$CH_3(CH_2)_7CH=CH(CH_2)_7C-NH$$

with O carbonyl, then $CH_2CHCOOH$ attached to a benzene ring bearing OH.

Chemical Classes: Amides; Amino Acids

Function: Skin-Conditioning Agent - Miscellaneous

Trade Name Mixture:
TYR-OL (Sederma)

OLETH-2

CAS Nos.: 5274-65-7; 9004-98-2 (Generic); 95287-03-9

JPN Translation:
オレス - 2

Empirical Formula:
$C_{22}H_{44}O_3$

Definition: Oleth-2 is the polyethylene glycol ether of Oleyl Alcohol (q.v.) that conforms generally to the formula:

$$CH_3(CH_2)_7CH=CH(CH_2)_7CH_2(OCH_2CH_2)_nOH$$

where n has an average value of 2.

Information Sources: CIR: [S] IJT-18 (SUPPL. 2)1999, CTFA D, JCLS, MI-13 (7659), RIFM, SNPF, TSCA

Chemical Class: Alkoxylated Alcohols

Functions: Fragrance Ingredient; Surfactant - Emulsifying Agent

Reported Product Categories: Hair Bleaches; Tonics, Dressings, and Other Hair Grooming Aids; Hair Conditioners; Permanent Waves

Technical/Other Names:
Diethylene Glycol Oleyl Ether
Ethanol, 2-[2-[(9-Octadecenyloxy)Ethoxy]-
Ethanol, 2-[2-[(9-Octadecylenyloxy)Ethoxy]-
2-[2-[(9-Octadecenyloxy)Ethoxy]Ethanol
PEG-2 Oleyl Ether
Polyethylene Glycol 100 Oleyl Ether
Poly(oxy-1,2-ethanediyl), .alpha.-9-octadecenyl-.omega.-hydroxy-, (Z)- (RIFM)
Polyoxyethylene (2) Oleyl Ether

Trade Names:
Ameroxol OE-2 (Amerchol)
Brij 92 (Uniqema Americas)
Chemonic O-2 (Chemron)
Eumulgin EP 2 L (Cognis France)
Genapol O 020 (Clariant)
Genapol O 020 (Clariant GmbH, Personal Care)
Hetoxol OL-2 (Heterene)
Jeecol OA-2 (Jeen)
Lanycol-92 (Lanaetex)
Lipocol O-2 (Lipo)
Nikkol BO-2 (Nikko)
Procol OA-2 (Protameen)
Procol OA2-SP (Protameen)
Ritoleth-2 (RITA)
Sabowax OC 2 (Sabo)
Sipol OAL-2 (Specialty Industrial)
Unicol OA-2 (Universal Preserv-A-Chem)
Volpo N2 (Croda Chemicals)

OLETH-3

CAS Nos.: 5274-66-8; 9004-98-2 (Generic); 96459-08-4

JPN Translation:
オレス - 3

Empirical Formula:
$C_{24}H_{48}O_4$

Definition: Oleth-3 is the polyethylene glycol ether of Oleyl Alcohol (q.v.) that conforms generally to the formula:

$$CH_3(CH_2)_7CH=CH(CH_2)_7CH_2(OCH_2CH_2)_nOH$$

where n has an average value of 3.

Information Sources: CIR: [S] IJT-18 (SUPPL. 2)1999, CTFA D, JCLS, MI-13 (7659), RIFM, SNPF, TSCA

Chemical Class: Alkoxylated Alcohols

Functions: Fragrance Ingredient; Surfactant - Emulsifying Agent

Reported Product Categories: Hair Conditioners; Hair Dyes and Colors (All Types Requiring Caution Statements and Patch Tests)

Technical/Other Names:
Ethanol, 2-[2-[2-(9-Octadecenyloxy)-Ethoxy]Ethoxy]-
2-[2-[2-(9-Octadecenyloxy)Ethoxy]-Ethoxy]Ethanol
PEG-3 Oleyl Ether
Polyethylene Glycol (3) Oleyl Ether
Poly(oxy-1,2-ethanediyl), .alpha.-9-octa-decenyl-.omega.-hydroxy-, (Z)- (RIFM)
Polyoxyethylene (3) Oleyl Ether
Triethylene Glycol Oleyl Ether

Trade Names:
Brox OL-3 (Arch Personal Care Products)
Ethoxol 3 (Lanaetex)
Hetoxol OL-3 (Heterene)
Jeecol OA-3 (Jeen)
Lipocol O-3 (Lipo)
Procol OA-3 (Protameen)
Volpo 3 (Croda, Inc.)
Volpo N3 (Croda Chemicals)

OLETH-4

CAS Nos.: 5353-26-4; 9004-98-2 (Generic); 103622-85-1

JPN Translation:
オレス - 4

Empirical Formula:
$C_{26}H_{52}O_5$

Definition: Oleth-4 is the polyethylene glycol ether of Oleyl Alcohol (q.v.) that conforms generally to the formula:

$$CH_3(CH_2)_7CH = CH(CH_2)_7CH_2(OCH_2CH_2)_nOH$$

where n has an average value of 4.

Information Sources: CIR: [S] IJT-18 (SUPPL. 2)1999, JCLS, MI-13(7659), RIFM, TSCA

Chemical Class: Alkoxylated Alcohols

Functions: Fragrance Ingredient; Surfactant - Emulsifying Agent

Technical/Other Names:
PEG-4 Oleyl Ether
Polyethylene Glycol 200 Oleyl Ether
Poly(oxy-1,2-ethanediyl), .alpha.-9-octa-decenyl-.omega.-hydroxy-, (Z)- (RIFM)
Polyoxyethylene (4) Oleyl Ether
3,6,9,12-Tetraoxatriacont-21-en-1-ol

Trade Names:
Chemal OA-4 (Chemax)
Hetoxol OL-4 (Heterene)
Jeecol OA-4 (Jeen)
Lumulse1804 (Lambent)
Procol OA-4 (Protameen)
Unicol OA-4 (Universal Preserv-A-Chem)

OLETH-5

CAS Nos.: 5353-27-5; 9004-98-2 (Generic)

JPN Translation:
オレス - 5

Empirical Formula:
$C_{28}H_{56}O_6$

Definition: Oleth-5 is the polyethylene glycol ether of Oleyl Alcohol (q.v.) that conforms generally to the formula:

$$CH_3(CH_2)_7CH = CH(CH_2)_7CH_2(OCH_2CH_2)_nOH$$

where n has an average value of 5.

Information Sources: 21CFR176.200, CIR: [S] IJT-18(SUPPL. 2)1999, CTFA D, JCLS, MI-13(7659), RIFM, SNPF, TSCA

Chemical Class: Alkoxylated Alcohols

Functions: Fragrance Ingredient; Surfactant - Emulsifying Agent

Reported Product Categories: Hair Conditioners; Tonics, Dressings, and Other Hair Grooming Aids; Skin Care Preparations, Misc.; Fragrance Preparations, Misc.; Moisturizing Preparations

Technical/Other Names:
PEG-5 Oleyl Ether
3,6,9,12,15-Pentaoxatriacont-24-en-1-ol
Polyethylene Glycol (5) Oleyl Ether
Poly(oxy-1,2-ethanediyl), .alpha.-9-octa-decenyl-.omega.-hydroxy-, (Z)- (RIFM)
Polyoxyethylene (5) Oleyl Ether

Trade Names:
Chemonic O-5 (Chemron)
Emulsogen LP (Clariant)
Emulsogen LP (Clariant GmbH, Personal Care)
Ethal OA-5 (Ethox)
Ethoxol-5 (Lanaetex)
Eumulgin EP 5 L (Cognis France)
Eumulgin O 5 (Cognis Deutschland)
Genapol O 050 (Clariant)
Genapol O 050 (Clariant GmbH, Personal Care)
Hetoxol OL-5 (Heterene)
Jeecol OA-5 (Jeen)
Lipocol O-5 (Lipo)
Procol OA-5 (Protameen)
Procol OA-5 SP (Protameen)
Ritoleth-5 (RITA)
Sabowax OC 6 (Sabo)
Sipol OAL-5 (Specialty Industrial)
Sympatens-AO/050 (Kolb)
Sympatens-AOC/050 (Kolb)

Unimul-05 (Universal Preserv-A-Chem)
Volpo 5 (Croda, Inc.)
Volpo N5 (Croda Chemicals)

Trade Name Mixture:
Emulgin M 8 (Cognis Deutschland)

OLETH-6

CAS No.: 9004-98-2 (Generic)

JPN Translation:
オレス - 6

Empirical Formula:
$C_{30}H_{60}O_7$

Definition: Oleth-6 is the polyethylene glycol ether of Oleyl Alcohol (q.v.) that conforms to the formula:

$$CH_3(CH_2)_7CH = CH(CH_2)_7CH_2(OCH_2CH_2)_nOH$$

where n has an average value of 6.

Information Sources: 21CFR176.200, CIR: [S] IJT-18(SUPPL. 2)1999, JCLS, MI-13(7659), RIFM, TSCA

Chemical Class: Alkoxylated Alcohols

Functions: Fragrance Ingredient; Surfactant - Emulsifying Agent

Technical/Other Names:
PEG-6 Oleyl Ether
Polyethylene Glycol 300 Oleyl Ether
Poly(oxy-1,2-ethanediyl), .alpha.-9-octa-decenyl-.omega.-hydroxy-, (Z)- (RIFM)
Polyoxyethylene (6) Oleyl Ether

Trade Names:
Lowenol OP-6 (Lowenstein)
Pegnol O-6A (Toho)

OLETH-7

CAS No.: 9004-98-2 (Generic)

JPN Translation:
オレス - 7

Empirical Formula:
$C_{32}H_{64}O_8$

Definition: Oleth-7 is the polyethylene glycol ether of Oleyl Alcohol (q.v.) that conforms generally to the formula:

$$CH_3(CH_2)_7CH = CH(CH_2)_7CH_2(OCH_2CH_2)_nOH$$

where n has an average value of 7.

Information Sources: 21CFR176.200, CIR: [S] IJT-18(SUPPL. 2)1999, JCLS, MI-13(7659), RIFM, TSCA

Chemical Class: Alkoxylated Alcohols

Functions: Fragrance Ingredient; Surfactant - Emulsifying Agent

Technical/Other Names:
PEG-7 Oleyl Ether
Polyethylene Glycol (7) Oleyl Ether
Poly(oxy-1,2-ethanediyl), .alpha.-9-octa-
decenyl-.omega.-hydroxy-, (Z)- (RIFM)
Polyoxyethylene (7) Oleyl Ether

Trade Names:
Akyporox RTO 70 (Kao GmbH)
Nikkol BO-7 (Nikko)
Pegnol O-107 (Toho)

OLETH-8

CAS Nos.: 9004-98-2 (Generic); 26996-03-2; 27040-03-5

JPN Translation:
オレス - 8

Empirical Formula:
$C_{34}H_{68}O_9$

Definition: Oleth-8 is the polyethylene glycol ether of Oleyl Alcohol (q.v.) that conforms generally to the formula:

$$CH_3(CH_2)_7CH = CH(CH_2)_7CH_2(OCH_2CH_2)_nOH$$

where n has an average value of 8.

Information Sources: 21CFR176.200, 21CFR177.2800, CIR: [S] IJT-18(SUPPL. 2)-1999, JCLS, MI-13(7659), RIFM, TSCA

Chemical Class: Alkoxylated Alcohols

Functions: Fragrance Ingredient; Surfactant - Emulsifying Agent

Reported Product Category: Permanent Waves

Technical/Other Names:
3,6,9,12,15,18,21,24-Octaoxadotetracont-
33-en-1-ol
PEG-8 Oleyl Ether
Polyethylene Glycol 400 Oleyl Ether
Poly(oxy-1,2-ethanediyl), .alpha.-9-octa-
decenyl-.omega.-hydroxy-, (Z)- (RIFM)
Polyoxyethylene (8) Oleyl Ether

Trade Name:
Emalex 508 (Nihon Emulsion)

OLETH-9

CAS No.: 9004-98-2 (Generic)

JPN Translation:
オレス - 9

Empirical Formula:
$C_{36}H_{72}O_{10}$

Definition: Oleth-9 is the polyethylene glycol ether of Oleyl Alcohol (q.v.) that conforms generally to the formula:

$$CH_3(CH_2)_7CH = CH(CH_2)_7CH_2(OCH_2CH_2)_nOH$$

where n has an average value of 9.

Information Sources: 21CFR176.200, 21CFR177.2800, CIR: [S] IJT-18(SUPPL. 2)-1999, JCLS, MI-13(7659), RIFM, TSCA

Chemical Class: Alkoxylated Alcohols

Functions: Fragrance Ingredient; Surfactant - Emulsifying Agent

Technical/Other Names:
PEG-9 Oleyl Ether
Polyethylene Glycol 450 Oleyl Ether
Poly(oxy-1,2-ethanediyl), .alpha.-9-octa-
decenyl-.omega.-hydroxy-, (Z)- (RIFM)
Polyoxyethylene (9) Oleyl Ether

OLETH-10

CAS Nos.: 9004-98-2 (Generic); 24871-34-9; 71976-00-6

JPN Translation:
オレス - 10

Empirical Formula:
$C_{38}H_{76}O_{11}$

Definition: Oleth-10 is the polyethylene glycol ether of Oleyl Alcohol (q.v.) that conforms generally to the formula:

$$CH_3(CH_2)_7CH = CH(CH_2)_7CH_2(OCH_2CH_2)_nOH$$

where n has an average value of 10.

Information Sources: 21CFR176.200, 21CFR177.2800, CIR: [S] IJT-18(SUPPL. 2)-1999, CTFA S, JCLS, MI-13(7659), NF XIX, RIFM, SNPF, TSCA, USAN

Chemical Class: Alkoxylated Alcohols

Functions: Fragrance Ingredient; Surfactant - Emulsifying Agent

Reported Product Categories: Hair Dyes and Colors (All Types Requiring Caution Statements and Patch Tests); Paste Masks (Mud Packs); Bath Oils, Tablets, and Salts; Cleansing Products (Cold Creams, Cleansing Lotions, Liquids and Pads); Colognes and Toilet Waters; Bath Preparations, Misc.; Body and Hand Preparations (Excluding Shaving Preparations); Moisturizing Preparations; Skin Care Preparations, Misc.; Bath Capsules; Face and Neck Preparations (Excluding Shaving Preparations); Hair Bleaches; Skin Fresheners; Hair Preparations (Non-coloring), Misc.; Tonics, Dressings, and Other Hair Grooming Aids

Technical/Other Names:
Decaethylene Glycol Monooleyl Ether
3,6,9,12,15,18,21,24,27,30-
Decaoxaoctatetracont-39-en-1-ol
PEG-10 Oleyl Ether
Polyethylene Glycol 500 Oleyl Ether
Poly(oxy-1,2-ethanediyl), .alpha.-9-octa-
decenyl-.omega.-hydroxy-, (Z)- (RIFM)
Polyoxyethylene (10) Oleyl Ether

Trade Names:
AEC Oleth-10 (A & E Connock)
Ameroxol OE-10 (Amerchol)
Brij 96 (Uniqema Americas)
Brij 97 (Uniqema Americas)
Chemal OA-10 (Chemax)
Chemonic O-10 (Chemron)
Customol O-10 (Oleth-10) (Custom Ingredients)
Ethal OA-10 (Ethox)
Ethoxol 10 (Lanaetex)
Genapol O 100 (Clariant)
Genapol O 100 (Clariant GmbH, Personal Care)
Hetoxol OL-10 (Heterene)
Jeecol OA-10 (Jeen)
Lipocol O-10 (Lipo)
NIKKOL BO-10 (Nikko)
Nikkol BO-10TX (Nikko)
Procol OA-10 (Protameen)
Procol OA-10 SP (Protameen)
Ritoleth-10 (RITA)
Sabowax O 10 (Sabo)
Sabowax OC 9 (Sabo)
Sipol OAL-10 (Specialty Industrial)
Sympatens-AO/100 (Kolb)
Unicol OA-10 (Universal Preserv-A-Chem)
Unimul-10 (Universal Preserv-A-Chem)
Volpo 10 (Croda, Inc.)
Volpo N10 (Croda Chemicals)

Trade Name Mixtures:
Emulgin M 8 (Cognis Deutschland)
Lipowax R-2 (Lipo)
Procol OA-1010 (Protameen)
Relaxer Concentrate No. 1 (Arch Personal Care Products)
Solimate E (Guardian)

OLETH-11

CAS No.: 9004-98-2 (Generic)

JPN Translation:
オレス - 11

Definition: Oleth-11 is the polyethylene glycol ether of Oleyl Alcohol (q.v.) that conforms generally to the formula:

$$CH_3(CH_2)_7CH = CH(CH_2)_7CH_2(OCH_2CH_2)_nOH$$

where n has an average value of 11.

Information Sources: CIR: [S] IJT-18 (SUPPL. 2)1999, JCLS, RIFM

Chemical Class: Alkoxylated Alcohols

Functions: Fragrance Ingredient; Surfactant - Emulsifying Agent

Technical/Other Names:
PEG-11 Oleyl Ether
Polyethylene Glycol (11) Oleyl Ether
Poly(oxy-1,2-ethanediyl), .alpha.-9-octa-
decenyl-.omega.-hydroxy-, (Z)- (RIFM)
Polyoxyethylene (11) Oleyl Ether

OLETH-12

CAS No.: 9004-98-2 (Generic)

JPN Translation:
オレス - 12

Definition: Oleth-12 is the polyethylene glycol ether of Oleyl Alcohol (q.v.) that conforms generally to the formula:

$$CH_3(CH_2)_7CH = CH(CH_2)_7CH_2(OCH_2CH_2)_nOH$$

where n has an average value of 12.

Information Sources: 21CFR176.200, 21CFR177.2800, CIR: [S] IJT-18(SUPPL. 2)-1999, JCLS, MI-13(7659), RIFM, TSCA

Chemical Class: Alkoxylated Alcohols

Functions: Fragrance Ingredient; Surfactant - Emulsifying Agent

Technical/Other Names:
PEG-12 Oleyl Ether
Polyethylene Glycol 600 Oleyl Ether
Poly(oxy-1,2-ethanediyl), .alpha.-9-octa-decenyl-.omega.-hydroxy-, (Z)- (RIFM)
Polyoxyethylene (12) Oleyl Ether

Trade Name:
Mergital OC 12 (Cognis France)

Trade Name Mixtures:
Beeswax Selfemulsifying 8074 (Kahl)
Sinnowax CSNO (Cognis France)

OLETH-15

CAS No.: 9004-98-2 (Generic)

JPN Translation:
オレス - 15

Definition: Oleth-15 is the polyethylene glycol ether of Oleyl Alcohol (q.v.) that conforms generally to the formula:

$$CH_3(CH_2)_7CH = CH(CH_2)_7CH_2(OCH_2CH_2)_nOH$$

where n has an average value of 15.

Information Sources: 21CFR176.200, 21CFR177.2800, CIR: [S] IJT-18(SUPPL. 2)-1999, JCLS, MI-13(7659), RIFM, TSCA

Chemical Class: Alkoxylated Alcohols

Functions: Fragrance Ingredient; Surfactant - Emulsifying Agent

Technical/Other Names:
PEG-15 Oleyl Ether
Polyethylene Glycol (15) Oleyl Ether
Poly(oxy-1,2-ethanediyl), .alpha.-9-octa-decenyl-.omega.-hydroxy-, (Z)- (RIFM)
Polyoxyethylene (15) Oleyl Ether

Trade Names:
Emalex 515 (Nihon Emulsion)
Emalex 515P (Nihon Emulsion)
NIKKOL BO-15 (Nikko)

NIKKOL BO-15TX (Nikko)
Volpo N15 (Croda Chemicals)

OLETH-16

CAS Nos.: 9004-98-2 (Generic); 25190-05-0 (Generic)

JPN Translation:
オレス - 16

Definition: Oleth-16 is the polyethylene glycol ether of Oleyl Alcohol (q.v.) that conforms generally to the formula:

$$CH_3(CH_2)_7CH = CH(CH_2)_7CH_2(OCH_2CH_2)_nOH$$

where n has an average value of 16.

Information Sources: 21CFR176.200, 21CFR177.2800, CIR: [S] IJT-18(SUPPL. 2)-1999, JCLS, MI-13(7659), RIFM

Chemical Class: Alkoxylated Alcohols

Functions: Fragrance Ingredient; Surfactant - Emulsifying Agent

Reported Product Categories: Moisturizing Preparations; Tonics, Dressings, and Other Hair Grooming Aids

Technical/Other Names:
PEG-16 Oleyl Ether
Polyethylene Glycol (16) Oleyl Ether
Poly(oxy-1,2-ethanediyl), .alpha.-9-octa-decenyl-.omega.-hydroxy-, (Z)- (RIFM)
Polyoxyethylene (16) Oleyl Ether

Trade Name:
Pegnol O-16A (Toho)

Trade Name Mixture:
Solulan 16 (Amerchol)

OLETH-20

CAS No.: 9004-98-2 (Generic)

JPN Translation:
オレス - 20

Definition: Oleth-20 is the polyethylene glycol ether of Oleyl Alcohol (q.v.) that conforms generally to the formula:

$$CH_3(CH_2)_7CH = CH(CH_2)_7CH_2(OCH_2CH_2)_nOH$$

where n has an average value of 20.

Information Sources: 21CFR175.105, 21CFR176.180, 21CFR176.200, 21CFR177.1210, 21CFR177.2800, CIR: [S] IJT-18(SUPPL. 2)1999, CTFA S, JCLS, MI-13(7659), RIFM, SNPF, TSCA

Chemical Class: Alkoxylated Alcohols

Functions: Fragrance Ingredient; Surfactant - Cleansing Agent; Surfactant - Emulsifying Agent; Surfactant - Solubilizing Agent

Reported Product Categories: Tonics, Dressings, and Other Hair Grooming Aids; Permanent Waves; Hair Conditioners; Hair Preparations (Non-coloring), Misc.; Skin Care Preparations, Misc.; Bath Oils, Tablets, and Salts; Shaving Preparations, Misc.; Skin Fresheners; Paste Masks (Mud Packs); Aftershave Lotions; Baby Shampoos; Cleansing Products (Cold Creams, Cleansing Lotions, Liquids and Pads); Hair Wave Sets; Bath Preparations, Misc.; Body and Hand Preparations (Excluding Shaving Preparations); Deodorants (Underarm); Hair Dyes and Colors (All Types Requiring Caution Statements and Patch Tests); Hair Straighteners; Shaving Cream (Aerosol, Brushless and Lather); Baby Products, Misc.; Moisturizing Preparations; Personal Cleanliness Products, Misc.; Bubble Baths; Colognes and Toilet Waters; Makeup Bases; Makeup Preparations (Not eye), Misc.

Technical/Other Names:
PEG-20 Oleyl Ether
Polyethylene Glycol 1000 Oleyl Ether
Poly(oxy-1,2-ethanediyl), .alpha.-9-octa-decenyl-.omega.-hydroxy-, (Z)- (RIFM)
Polyoxyethylene (20) Oleyl Ether

Trade Names:
AEC Oleth-20 (A & E Connock)
Ameroxol OE-20 (Amerchol)
Brij 98 (Uniqema Americas)
Chemal OA-20 (Chemax)
Chemonic O-20 (Chemron)
Ethoxol 20 (Lanaetex)
Genapol O 200 (Clariant)
Genapol O 200 (Clariant GmbH, Personal Care)
Hetoxol OL-20 (Heterene)
Jeecol OA-20 (Jeen)
Lanycol-98 (Lanaetex)
Lanycol-99 (Lanaetex)
Lipocol O-20 (Lipo)
Lumulse 1820 (Lambent)
Nikkol BO-20 (Nikko)
Nikkol BO-20TX (Nikko)
Pegnol O-20 (Toho)
Procol OA-20 (Protameen)
Procol OA-20 SP (Protameen)
Rhodasurf ON-870 (Rhodia)
Ritoleth-20 (RITA)
Sabowax O 20 (Sabo)
Sabowax OC 20 (Sabo)
Simulsol 98 (SEPPIC)
Sipol OAL-20 (Specialty Industrial)
Sympatens-AO/200 (Kolb)
Unicol OA-20 (Universal Preserv-A-Chem)
Volpo 20 (Croda, Inc.)
Volpo N20 (Croda Chemicals)

Trade Name Mixture:
Umordant P (Cosmetochem)
(Cosmetochem International Ltd.)

OLETH-23

CAS No.: 9004-98-2 (Generic)

JPN Translation:
オレス - 23

Definition: Oleth-23 is the polyethylene glycol ether of Oleyl Alcohol (q.v.) that conforms generally to the formula:

$$CH_3(CH_2)_7CH = CH(CH_2)_7CH_2(OCH_2CH_2)_nOH$$

where n has an average value of 23.

Information Sources: 21CFR176.180, 21CFR176.200, 21CFR177.2800, CIR: [S] IJT-18(SUPPL. 2)1999, JCLS, MI-13(7659), RIFM, TSCA

Chemical Class: Alkoxylated Alcohols

Functions: Fragrance Ingredient; Surfactant - Cleansing Agent; Surfactant - Solubilizing Agent

Technical/Other Names:
PEG-23 Oleyl Ether
Polyethylene Glycol (23) Oleyl Ether
Poly(oxy-1,2-ethanediyl), .alpha.-9-octa-decenyl-.omega.-hydroxy-, (Z)- (RIFM)
Polyoxyethylene (23) Oleyl Ether

Trade Names:
Chemal OA-23 (Chemax)
Emalex 523 (Nihon Emulsion)
Ethal OA-23 (Ethox)
Hetoxol OL-23 (Heterene)
Jeecol OA-23 (Jeen)
Procol OA-23 (Protameen)

OLETH-24

CAS No.: 9004-98-2 (Generic)

Definition: Oleth-24 is the polyethylene glycol ether of Oleyl Alcohol (q.v.) that conforms generally to the formula:

$$CH_3(CH_2)_7CH = CH(CH_2)_7CH_2(OCH_2CH_2)_nOH$$

where n has an average value of 24.

Information Source: RIFM

Chemical Class: Alkoxylated Alcohols

Functions: Fragrance Ingredient; Surfactant - Cleansing Agent; Surfactant - Solubilizing Agent

Technical/Other Names:
Polyethylene Glycol (24) Oleyl Ether
Poly(oxy-1,2-ethanediyl), .alpha.-9-octa-decenyl-.omega.-hydroxy-, (Z)- (RIFM)
Polyoxyethylene (24) Oleyl Ether

Trade Name:
Pegnol O-24 (Toho)

OLETH-25

CAS No.: 9004-98-2 (Generic)

JPN Translation:
オレス - 25

Definition: Oleth-25 is the polyethylene glycol ether of Oleyl Alcohol (q.v.) that conforms generally to the formula:

$$CH_3(CH_2)_7CH = CH(CH_2)_7CH_2(OCH_2CH_2)_nOH$$

where n has an average value of 25.

Information Sources: 21CFR176.180, 21CFR176.200, 21CFR177.2800, CIR: [S] IJT-18(SUPPL. 2)1999, JCLS, MI-13(7659), RIFM, TSCA

Chemical Class: Alkoxylated Alcohols

Functions: Fragrance Ingredient; Surfactant - Cleansing Agent; Surfactant - Solubilizing Agent

Technical/Other Names:
PEG-25 Oleyl Ether
Polyethylene Glycol (25) Oleyl Ether
Poly(oxy-1,2-ethanediyl), .alpha.-9-octa-decenyl-.omega.-hydroxy-, (Z)- (RIFM)
Polyoxyethylene (25) Oleyl Ether

Trade Names:
Mergital OC 25 (Cognis France)
Procol OA-25 (Protameen)

Trade Name Mixture:
Hydrolactol 70 (Gattefosse s.a.)

OLETH-30

CAS No.: 9004-98-2 (Generic)

JPN Translation:
オレス - 30

Definition: Oleth-30 is the polyethylene glycol ether of oleyl alcohol that conforms generally to the formula:

$$CH_3(CH_2)_7CH = CH(CH_2)_7CH_2(OCH_2CH_2)_nOH$$

where n has an average value of 30.

Information Sources: 21CFR176.180, 21CFR176.200, CIR: [S] IJT-18(SUPPL. 2)-1999, JCLS, MI-13(7659), RIFM

Chemical Class: Alkoxylated Alcohols

Functions: Fragrance Ingredient; Surfactant - Cleansing Agent; Surfactant - Solubilizing Agent

Reported Product Categories: Hair Dyes and Colors (All Types Requiring Caution Statements and Patch Tests); Hair Color Sprays (Aerosol)

Technical/Other Name:
Poly(oxy-1,2-ethanediyl), .alpha.-9-octa-decenyl-.omega.-hydroxy-, (Z)- (RIFM)

Trade Names:
Mergital OC 30 (Cognis France)
Sabowax O 30 (Sabo)
Sabowax OC 30 (Sabo)

OLETH-35

CAS No.: 9004-98-2 (Generic)

Definition: Oleth-35 is the polyethylene glycol ether of Oleyl Alcohol (q.v.) that conforms to the formula:

$$CH_3(CH_2)_7CH = CH(CH_2)_7CH_2(OCH_2CH_2)_nOH$$

where n has an average value of 35.

Information Sources: 21CFR176.180, 21CFR176.200, 21CFR177.2800, JCLS, MI-13(7659), RIFM

Chemical Class: Alkoxylated Alcohols

Functions: Fragrance Ingredient; Surfactant - Cleansing Agent; Surfactant - Solubilizing Agent

Technical/Other Names:
PEG-35 Oleyl Ether
Polyethylene Glycol (35) Oleyl Ether
Poly(oxy-1,2-ethanediyl), .alpha.-9-octa-decenyl-.omega.-hydroxy-, (Z)- (RIFM)
Polyoxyethylene (35) Oleyl Ether

Trade Name:
Ethal OA-35 (Ethox)

OLETH-40

CAS No.: 9004-98-2 (Generic)

JPN Translation:
オレス - 40

Definition: Oleth-40 is the polyethylene glycol ether of Oleyl Alcohol (q.v.) that conforms generally to the formula:

$$CH_3(CH_2)_7CH = CH(CH_2)_7CH_2(OCH_2CH_2)_nOH$$

where n has an average value of 40.

Information Sources: 21CFR176.180, 21CFR176.200, 21CFR177.2800, CIR: [S] IJT-18(SUPPL. 2)1999, JCLS, MI-13(7659), RIFM

Chemical Class: Alkoxylated Alcohols

Functions: Fragrance Ingredient; Surfactant - Cleansing Agent; Surfactant - Solubilizing Agent

Technical/Other Names:
PEG-40 Oleyl Ether
Polyethylene Glycol 2000 Oleyl Ether
Poly(oxy-1,2-ethanediyl), .alpha.-9-octa-decenyl-.omega.-hydroxy-, (Z)- (RIFM)
Polyoxyethylene (40) Oleyl Ether

OLETH-44

CAS No.: 9004-98-2 (Generic)

JPN Translation:
オレス - 44

Definition: Oleth-44 is the polyethylene glycol ether of Oleyl Alcohol (q.v.) that conforms generally to the formula:

$$CH_3(CH_2)_7CH = CH(CH_2)_7CH_2(OCH_2CH_2)_nOH$$

where n has an average value of 44.

Information Sources: 21CFR176.180, 21CFR176.200, 21CFR177.2800, CIR: [S] IJT-18(SUPPL. 2)1999, MI-13(7659), RIFM, TSCA

Chemical Class: Alkoxylated Alcohols

Functions: Fragrance Ingredient; Surfactant - Cleansing Agent

Technical/Other Names:
PEG-44 Oleyl Ether
Polyethylene Glycol (44) Oleyl Ether
Poly(oxy-1,2-ethanediyl), .alpha.-9-octa-decenyl-.omega.-hydroxy-, (Z)- (RIFM)
Polyoxyethylene (44) Oleyl Ether

OLETH-50

CAS No.: 9004-98-2 (Generic)

JPN Translation:
オレス - 50

Definition: Oleth-50 is the polyethylene glycol ether of Oleyl Alcohol (q.v.) that conforms generally to the formula:

$$CH_3(CH_2)_7CH = CH(CH_2)_7CH_2(OCH_2CH_2)_nOH$$

where n has an average value of 50.

Information Sources: 21CFR176.180, 21CFR176.200, 21CFR177.2800, CIR: [S] IJT-18(SUPPL. 2)1999, JCLS, MI-13(7659), RIFM, TSCA

Chemical Class: Alkoxylated Alcohols

Functions: Fragrance Ingredient; Surfactant - Cleansing Agent

Technical/Other Names:
PEG-50 Oleyl Ether
Polyethylene Glycol (50) Oleyl Ether
Poly(oxy-1,2-ethanediyl), .alpha.-9-octa-decenyl-.omega.-hydroxy-, (Z)- (RIFM)
Polyoxyethylene (50) Oleyl Ether

Trade Names:
Emalex 550 (Nihon Emulsion)
Nikkol BO-50 (Nikko)
Sabowax OC 50 (Sabo)

OLETH-82

CAS No.: 9004-98-2 (Generic)

Definition: Oleth-82 is the polyethylene glycol ether of oleyl alcohol that conforms generally to the formula:

$$CH_3(CH_2)_7CH = CH(CH_2)_7CH_2(OCH_2CH_2)_nOH$$

where n has an average value of 82.

Information Source: RIFM

Chemical Class: Alkoxylated Alcohols

Functions: Fragrance Ingredient; Surfactant - Emulsifying Agent; Surfactant - Solubilizing Agent

Technical/Other Names:
Polyethylene Glycol (82) Oleyl Ether
Poly(oxy-1,2-ethanediyl), .alpha.-9-octa-decenyl-.omega.-hydroxy-, (Z)- (RIFM)
Polyoxyethylene (82) Oleyl Ether

Trade Name:
7287A (Nicca Chemical)

OLETH-106

CAS No.: 9004-98-2 (Generic)

Definition: Oleth-106 is the polyethylene glycol ether of Oleyl Alcohol (q.v.) that conforms generally to the formula:

$$CH_3(CH_2)_7CH = CH(CH_2)_7CH_2(OCH_2CH_2)_nOH$$

where n has an average value of 106.

Information Source: RIFM

Chemical Class: Alkoxylated Alcohols

Functions: Fragrance Ingredient; Surfactant - Emulsifying Agent; Surfactant - Solubilizing Agent

Technical/Other Names:
Polyethylene Glycol (106) Oleyl Ether
Poly(oxy-1,2-ethanediyl), .alpha.-9-octa-decenyl-.omega.-hydroxy-, (Z)- (RIFM)
Polyoxyethylene (106) Oleyl Ether

Trade Name:
Nikkol BO-106 (Nikko)

OLETH-2 BENZOATE

Definition: Oleth-2 Benzoate is the ester of Oleth-2 (q.v.) and Benzoic Acid (q.v.).

Chemical Classes: Alkoxylated Alcohols; Esters

Function: Skin-Conditioning Agent - Emollient

Technical/Other Names:
PEG-2 Oleyl Ether Benzoate
Polyethylene Glycol (2) Oleyl Ether Benzoate
Polyoxyethylene (2) Oleyl Ether Benzoate

Trade Name:
Bernel Ester OE-2 (Bernel)

OLETH-3 CARBOXYLIC ACID

CAS No.: 57635-48-0

Empirical Formula:
$C_{24}H_{46}O_5$

Definition: Oleth-3 Carboxylic Acid is the organic acid that conforms generally to the formula:

$$CH_3(CH_2)_7CH = CH(CH_2)_7CH_2(OCH_2CH_2)_nOCH_2COOH$$

where n has an average value of 2.

Chemical Class: Carboxylic Acids

Function: Surfactant - Cleansing Agent

Technical/Other Names:
PEG-3 Oleyl Ether Carboxylic Acid
Polyethylene Glycol (3) Oleyl Ether Carboxylic Acid
Polyoxyethylene (3) Oleyl Ether Carboxylic Acid

Trade Name:
Akypo RO 20 (Kao GmbH)

OLETH-6 CARBOXYLIC ACID

CAS No.: 57635-48-0

Empirical Formula:
$C_{30}H_{58}O_8$

Definition: Oleth-6 Carboxylic Acid is the organic acid that conforms generally to the formula:

$$CH_3(CH_2)_7CH = CH(CH_2)_7CH_2(OCH_2CH_2)_nOCH_2COOH$$

where n has an average value of 5.

Chemical Class: Carboxylic Acids

Function: Surfactant - Cleansing Agent

Technical/Other Names:
PEG-6 Oleyl Ether Carboxylic Acid
Polyethylene Glycol 300 Oleyl Ether Carboxylic Acid
Polyoxyethylene (6) Oleyl Ether Carboxylic Acid

Trade Name:
Akypo RO 50 (Kao GmbH)

OLETH-10 CARBOXYLIC ACID

CAS No.: 57635-48-0

Empirical Formula:
$C_{38}H_{74}O_{12}$

Definition: Oleth-10 Carboxylic Acid is the organic acid that conforms generally to the formula:

$$CH_3(CH_2)_7CH = CH(CH_2)_7CH_2(OCH_2CH_2)_nOCH_2COOH$$

where n has an average value of 9.

Chemical Class: Carboxylic Acids

Function: Surfactant - Cleansing Agent

Technical/Other Names:
PEG-10 Oleyl Ether Carboxylic Acid
Polyethylene Glycol 500 Oleyl Ether
Carboxylic Acid
Polyoxyethylene (10) Oleyl Ether
Carboxylic Acid

Trade Names:
Akypo RO 90 (Kao GmbH)
Empicol CLI (Albright & Wilson UK)

OLETH-2 PHOSPHATE

CAS No.: 39464-69-2 (Generic)

JPN Translation:
オレス - 2 リン酸

Definition: Oleth-2 Phosphate is a complex mixture of esters of phosphoric acid and Oleth-2 (q.v.).

Information Source: JCLS

Chemical Class: Phosphorus Compounds

Function: Surfactant - Emulsifying Agent

Technical/Other Names:
PEG-2 Oleyl Ether Phosphate
Polyethylene Glycol 100 Oleyl Ether Phosphate
Polyoxyethylene (2) Oleyl Ether Phosphate

OLETH-3 PHOSPHATE

CAS No.: 39464-69-2 (Generic)

JPN Translation:
オレス - 3 リン酸

Definition: Oleth-3 Phosphate is a complex mixture of esters of phosphoric acid and Oleth-3 (q.v.).

Information Sources: JCLS, SNPF, TSCA

Chemical Class: Phosphorus Compounds

Function: Surfactant - Emulsifying Agent

Reported Product Categories: Foundations; Shampoos (Non-coloring)

Technical/Other Names:
PEG-3 Oleyl Ether Phosphate
Polyethylene Glycol (3) Oleyl Ether Phosphate
Polyoxyethylene (3) Oleyl Ether Phosphate

Trade Names:
Crodafos N-3 Acid (Croda, Inc.)
Crodafos N3A (Croda Chemicals)
Empiphos O3D (Albright & Wilson UK)
Jeephos OA-3P (Jeen)

OLETH-4 PHOSPHATE

CAS No.: 39464-69-2 (Generic)

JPN Translation:
オレス - 4 リン酸

Definition: Oleth-4 Phosphate is a complex mixture of esters of phosphoric acid and Oleth-4 (q.v.).

Information Sources: JCLS, SNPF, TSCA

Chemical Class: Phosphorus Compounds

Function: Surfactant - Emulsifying Agent

Technical/Other Names:
PEG-4 Oleyl Ether Phosphate
Polyethylene Glycol 200 Oleyl Ether Phosphate
Polyoxyethylene (4) Oleyl Ether Phosphate

Trade Names:
Chemfac PB-184 (Chemax)
Ethfac 140 (Ethox)
Phosphanol RB-410 (Toho)

OLETH-5 PHOSPHATE

CAS No.: 39464-69-2 (Generic)

JPN Translation:
オレス - 5 リン酸

Definition: Oleth-5 Phosphate is a complex mixture of esters of phosphoric acid and Oleth-5 (q.v.).

Information Sources: JCLS, TSCA

Chemical Class: Phosphorus Compounds

Function: Surfactant - Emulsifying Agent

Trade Name:
Crodafos N5A (Croda Chemicals)

OLETH-10 PHOSPHATE

CAS No.: 39464-69-2 (Generic)

JPN Translation:
オレス - 10 リン酸

Definition: Oleth-10 Phosphate is a complex mixture of esters of phosphoric acid and Oleth-10 (q.v.).

Information Sources: CTFA D, JCLS, TSCA

Chemical Class: Phosphorus Compounds

Function: Surfactant - Emulsifying Agent

Technical/Other Names:
PEG-10 Oleyl Ether Phosphate
Polyethylene Glycol 500 Oleyl Ether Phosphate
Polyoxyethylene (10) Oleyl Ether Phosphate

Trade Names:
Crodafos N-10 Acid (Croda, Inc.)
Crodafos N10A (Croda Chemicals)

OLETH-20 PHOSPHATE

CAS No.: 39464-69-2 (Generic)

JPN Translation:
オレス - 20 リン酸

Definition: Oleth-20 Phosphate is a complex mixture of esters of phosphoric acid and Oleth-20 (q.v.).

Information Sources: JCLS, TSCA

Chemical Class: Phosphorus Compounds

Function: Surfactant - Emulsifying Agent

Technical/Other Names:
PEG-20 Oleyl Ether Phosphate
Polyethylene Glycol 1000 Oleyl Ether Phosphate
Polyoxyethylene (20) Oleyl Ether Phosphate

Trade Name:
Gelasol E-50 (Goo Chemical)

OLEYL ACETATE

CAS No.: 693-80-1

Empirical Formula:
$C_{20}H_{38}O_2$

Definition: Oleyl Acetate is the acetyl ester of oleyl alcohol that conforms generally to the formula:

$$CH_3\overset{\overset{O}{\|}}{C}-O(CH_2)_8CH=CH(CH_2)_7CH_3$$

Chemical Class: Esters

Function: Skin-Conditioning Agent - Emollient

Technical/Other Names:
Acetic Acid, 9-Octadecenyl Ester
Acetic Acid, Oleyl Ester
9-Octadecen-1-yl Acetate

Trade Name:
DUB OA (Stearinerie Dubois Fils)

OLEYL ACETYL GLUTAMINATE

Definition: Oleyl Acetyl Glutaminate is the ester of Oleyl Alcohol (q.v.) and Acetyl Glutamine (q.v.) that conforms to the formula:

$$H_2N-\overset{\overset{O}{\|}}{C}CH_2CH_2\underset{\underset{HN-\overset{\overset{O}{\|}}{C}CH_3}{|}}{C}HC-OR$$

where R represents the oleyl moiety.

Chemical Class: Amino Acids

Functions: Hair Conditioning Agent; Skin-Conditioning Agent - Miscellaneous

Trade Name:
NAGO (Kyowa Hakko Kogyo)

OLEYL ALCOHOL

CAS Nos.	EINECS Nos.
143-28-2	205-597-3
593-47-5	209-791-9

JPN Translation:
オレイルアルコール

Empirical Formula:
$C_{18}H_{36}O$

Definition: Oleyl Alcohol is the unsaturated fatty alcohol that conforms generally to the formula:

$$CH_3(CH_2)_7CH = CH(CH_2)_8OH$$

Information Sources: 21CFR176.170, 21CFR176.210, 21CFR177.1010, 21CFR177.1210, 21CFR177.2800, 21CFR178.3910, CIR: [S] JACT-4(5)1985, CTFA S, JCLS, JSCI, MAR, MI-13(6900), NF XIX, RIFM, SNPF, TSCA, USAN

Chemical Class: Fatty Alcohols

Functions: Fragrance Ingredient; Skin-Conditioning Agent - Emollient; Solvent; Viscosity Increasing Agent - Nonaqueous

Reported Product Categories: Hair Dyes and Colors (All Types Requiring Caution Statements and Patch Tests); Lipsticks; Hair Conditioners; Hair Straighteners; Mascara; Moisturizing Preparations; Bath Preparations, Misc.; Body and Hand Preparations (Excluding Shaving Preparations); Skin Care Preparations, Misc.; Tonics, Dressings, and Other Hair Grooming Aids; Eye Shadows; Eyeliners; Foundations; Makeup Preparations (Not eye), Misc.; Bath Capsules; Bath Soaps and Detergents; Blushers (All types); Eyebrow Pencils; Hair Bleaches; Hair Sprays (Aerosol Fixatives); Paste Masks (Mud Packs); Personal Cleanliness Products, Misc.; Suntan Gels, Creams, and Liquids; Suntan Preparations, Misc.

Technical/Other Names:
cis-9-Octadecenyl Alcohol
Hydroxyoctadec-9-ene
9-Octadecen-1-ol
Oleic Alcohol
(Z)-Octadec-9-enol (RIFM)

Trade Names:
AEC Oleyl Alcohol (A & E Connock)
Jeecol O (Jeen)
Lipocol O/95 (Lipo)
Novol (Croda Chemicals)
Novol (Croda, Inc.)

Protachem OA 70/75 (Protameen)
Sabonal 90/95 (Sabo)
U-Tanol HD 80/85 (Universal Preserv-A-Chem)
U-Tanol HD 90/95 (Universal Preserv-A-Chem)
U-Tanol HD CG (Universal Preserv-A-Chem)

Trade Name Mixtures:
Base 323 MS (LCW)
Base RAL W 323 T (LCW)
Base Rouge A Levres 323 TAL (LCW)
Base RW 101 (LCW)
Casto Cetyle (LCW)
Castorcet (Lanaetex)
Cloisonne Gold CC (Engelhard Corp.)
Cloisonne Red CC (Engelhard Corp.)
Fancol ISO (Fanning)
Flamenco Gold CC (Engelhard Corp.)
Flamenco Orange CC (Engelhard Corp.)
Flamenco Pearl CC (Engelhard Corp.)
Flamenco Superpearl CC (Engelhard Corp.)
Flamenco Super Red CC (Engelhard Corp.)
Flamenco Twilight Red CC (Engelhard Corp.)
Gemtone Sunstone CC (Engelhard Corp.)
Gemtone Tan Opal CC (Engelhard Corp.)
Ivarbase 3240 (Arch Personal Care Products)
Liant TW 729 (LCW)
Lipoderma YS (Lipo)
Monamilk (Argeville)
Oxocap (LCW)
Oxowax (LCW)
Prodhyrouge 2000 (Prod'Hyg)
Ricino - Cetyle (Prod'Hyg)
Unisteron Y-50 (Induchem)

OLEYL ARACHIDATE

CAS Nos.	EINECS No.
22393-96-0	244-952-7
156952-79-3	

Empirical Formula:
$C_{38}H_{74}O_2$

Definition: Oleyl Arachidate is the ester of oleyl alcohol and Arachidic Acid (q.v.). It conforms to the formula:

$$CH_3(CH_2)_{18}C-O(CH_2)_8CH = CH(CH_2)_7CH_3$$

Information Source: TSCA

Chemical Class: Esters

Function: Skin-Conditioning Agent - Occlusive

Technical/Other Names:
Eicosanoic Acid, 9-Octadecenyl Ester
9-Octadecenyl Eicosanoate

OLEYL BETAINE

CAS No.	EINECS No.
871-37-4	212-806-1

JPN Translation:
オレイルベタイン

Empirical Formula:
$C_{22}H_{43}NO_2$

Definition: Oleyl Betaine is the zwitterion (inner salt) that conforms generally to the formula:

$$CH_3(CH_2)_7CH = CH(CH_2)_8 - \overset{CH_3}{\underset{CH_3}{\overset{+}{N}}} - CH_2COO^-$$

Information Source: TSCA

Chemical Class: Betaines

Functions: Antistatic Agent; Hair Conditioning Agent; Skin-Conditioning Agent - Miscellaneous; Surfactant - Cleansing Agent; Surfactant - Foam Booster; Viscosity Increasing Agent - Aqueous

Reported Product Categories: Bath Soaps and Detergents; Bath Oils, Tablets, and Salts; Cleansing Products (Cold Creams, Cleansing Lotions, Liquids and Pads)

Technical/Other Names:
N-(Carboxymethyl)-N,N-Dimethyl-9-Octadecen-1-aminium Hydroxide, Inner Salt
9-Octadecen-1-Aminium, N-(Carboxymethyl)-N,N-Dimethyl-, Hydroxide, Inner Salt
Oleyl Dimethyl Glycine

Trade Names:
Chembetaine OL (Chemron)
Mackam OB-30 (McIntyre)
Unibetaine OLB-50 (Universal Preserv-A-Chem)
Velvetex OLB-50 (Cognis Care Chemicals/NJ)
Velvetex OLB-50 (Cognis Care Chemicals/PA)

OLEYL EPOXYPROPYLDIMONIUM CHLORIDE

Empirical Formula:
$C_{23}H_{46}NOCl$

Definition: Oleyl Epoxypropyldimonium Chloride is the quaternary ammonium compound that conforms to the formula:

$$\left[CH_3(CH_2)_7CH = CH(CH_2)_8 \overset{CH_3}{\underset{CH_3}{N}} - CH_2CHCH_2 \right]^+ Cl^-$$

Chemical Class: Quaternary Ammonium Compounds

Function: Hair Conditioning Agent

Trade Name:
Mackernium CG64 (McIntyre)

OLEYL ERUCATE

CAS Nos.   **EINECS No.**
17673-56-2   241-654-9
143485-69-2

JPN Translation:
エルカ酸オレイル

Empirical Formula:
$C_{40}H_{76}O_2$

Definition: Oleyl Erucate is the ester of Oleyl Alcohol (q.v.) and erucic acid. It conforms to the formula:

$$CH_3(CH_2)_7CH \quad O$$
$$\quad\quad\quad \| \quad\quad\quad \|$$
$$CH(CH_2)_{11}C - O(CH_2)_8CH = CH(CH_2)_7CH_3$$

Chemical Class: Esters

Function: Skin-Conditioning Agent - Occlusive

Reported Product Categories: Eyebrow Pencils; Suntan Gels, Creams, and Liquids; Skin Care Preparations, Misc.; Lipsticks

Technical/Other Names:
13-Docosenoic Acid, 9-Octadecenyl Ester
Erucic Acid, Oleyl Ester
9-Octadecenyl 13-Docosenoate

Trade Names:
AEC Oleyl Erucate (A & E Connock)
Cetiol J 600 (Cognis Care Chemicals/NJ)
Cetiol J 600 (Cognis Care Chemicals/PA)
Cetiol J 600 (Cognis Deutschland)
Dynacerin 660 (Sasol GmbH - Witten)
Pelemol OE (Phoenix)

Trade Name Mixture:
Chelonine (JUVEX)

OLEYL ETHYL PHOSPHATE

CAS No.: 10483-96-2

Definition: Oleyl Ethyl Phosphate is a complex mixture of phosphate esters of oleyl and ethyl alcohols.

Chemical Classes: Esters; Phosphorus Compounds

Function: Surfactant - Emulsifying Agent

Technical/Other Name:
Phosphoric Acid, Monoethyl Mono-9-Octadecenyl Ester (Z)-

Trade Name:
Petrolite 110 Compound (Baker Petrolite)

OLEYL GLYCERYL ETHER

JPN Translation:
オレイルグリセリル

Empirical Formula:
$C_{21}H_{42}O_3$

Definition: Oleyl Glyceryl Ether is the organic compound that conforms to the formula:

$$CH_3(CH_2)_7CH = CH(CH_2)_8OCH_2CHCH_2OH$$
$$\quad\quad\quad\quad\quad\quad\quad\quad\quad\quad\quad |$$
$$\quad\quad\quad\quad\quad\quad\quad\quad\quad\quad OH$$

Information Sources: JCIC, JCLS

Chemical Class: Ethers

Function: Skin-Conditioning Agent - Emollient

Technical/Other Name:
Monooleyl Glyceryl Ether

Trade Name:
Nikkol Selachyl Alcohol (Nikko)

OLEYL HYDROXYETHYL IMIDAZOLINE

CAS Nos.   **EINECS Nos.**
95-38-5   202-414-9
21652-27-7   244-501-4
27136-73-8   248-248-0
51023-21-3

Empirical Formula:
$C_{22}H_{42}N_2O$

Definition: Oleyl Hydroxyethyl Imidazoline is the heterocyclic compound that conforms to the formula:

$$CH_3(CH_2)_7CH = CH(CH_2)_7 \quad\quad CH_2CH_2OH$$

Information Sources: 21CFR178.3570, CTFA D, MI-13(407), TSCA

Chemical Class: Imidazoline Compounds

Functions: Antistatic Agent; Hair Conditioning Agent

Technical/Other Names:
2-(8-Heptadecenyl)-4,5-Dihydro-1H-
Imidazole-1-Ethanol
1H-Imidazole-1-Ethanol, 2-(8-
Heptadecenyl)-4,5-Dihydro-
1H-Imidazolol, Ethyldihydro(9-
Octadecenyl)-(Z)-
Oleyl Imidazoline Ethanol

Trade Names:
Ethimid O (Ethox)

Mackazoline O (McIntyre)
Miramine O (Rhodia)
Monazoline O (Uniqema)
Schercozoline O (Scher)

OLEYL LACTATE

CAS No.: 42175-36-0

Empirical Formula:
$C_{21}H_{40}O_3$

Definition: Oleyl Lactate is the ester of oleyl alcohol and lactic acid. It conforms to the formula:

$$\quad\quad\quad\quad O$$
$$\quad\quad\quad\quad \|$$
$$CH_3CHC - O(CH_2)_8CH = CH(CH_2)_7CH_3$$
$$\quad |$$
$$\quad OH$$

Chemical Class: Esters

Function: Skin-Conditioning Agent - Emollient

Technical/Other Name:
Oleyl 2-Hydroxypropionate

Trade Names:
Dermol OL (Alzo)
Pelemol OL (Phoenix)

OLEYL LANOLATE

Definition: Oleyl Lanolate is the ester of oleyl alcohol and lanolin acid.

Chemical Class: Lanolin and Lanolin Derivatives

Functions: Hair Conditioning Agent; Skin-Conditioning Agent - Occlusive

Technical/Other Name:
Lanolin Fatty Acids, 9-Octadecenyl Ester

Trade Name:
AEC Oleyl Lanolate (A & E Connock)

OLEYL LINOLEATE

CAS No.   **EINECS No.**
17673-59-5   241-655-4

Empirical Formula:
$C_{36}H_{66}O_2$

Definition: Oleyl Linoleate is the ester of Oleyl Alcohol (q.v.) and Linoleic Acid (q.v.). It conforms to the formula:

$$CH(CH_2)_4CH_3$$
$$\|$$
$$CHCH_2CH \quad\quad O$$
$$\quad\quad\quad \| \quad\quad\quad \|$$
$$CH(CH_2)_7C - O(CH_2)_8CH = CH(CH_2)_7CH_3$$

Chemical Class: Esters

Functions: Hair Conditioning Agent; Skin-Conditioning Agent - Occlusive

Technical/Other Names:
Linoleic Acid, Oleyl Ester
9,12-Octadecadienoic Acid, 9-Octadecenyl Ester
9-Octadecenyl 9,12-Octadecadienoate

OLEYL MYRISTATE

CAS No.: 22393-93-7

Empirical Formula:
$C_{32}H_{62}O_2$

Definition: Oleyl Myristate is the ester of oleyl alcohol and myristic acid. It conforms to the formula:

$$CH_3(CH_2)_{12}\overset{O}{\overset{\|}{C}}-O(CH_2)_8CH=CH(CH_2)_7CH_3$$

Chemical Class: Esters

Functions: Hair Conditioning Agent; Skin-Conditioning Agent - Occlusive

Technical/Other Names:
9-Octadecenyl Tetradecanoate
Tetradecanoic Acid, 9-Octadecenyl Ester

Trade Name:
AEC Oleyl Myristate (A & E Connock)

OLEYL OLEATE

CAS Nos.   **EINECS No.**
3687-45-4   222-980-0
17363-94-9

JPN Translation:
オレイン酸オレイル

Empirical Formula:
$C_{36}H_{68}O_2$

Definition: Oleyl Oleate is the ester of Oleyl Alcohol (q.v.) and oleic acid. It conforms to the formula:

$$CH_3(CH_2)_7CH$$
$$CH(CH_2)_7\overset{O}{\overset{\|}{C}}-O(CH_2)_8CH=CH(CH_2)_7CH_3$$

Information Sources: HUN, JCIC, JCLS, JSQI, TSCA

Chemical Class: Esters

Functions: Hair Conditioning Agent; Skin-Conditioning Agent - Emollient

Reported Product Categories: Lipsticks; Eye Shadows

Technical/Other Names:
9-Octadecenoic Acid, 9-Octadecenyl Ester
Oleic Acid, Oleyl Ester

Trade Names:
AEC Oleyl Oleate (A & E Connock)
Cetiol (Cognis Deutschland)
Ethox OLO (Ethox)
Schercemol OLO (Scher)
Unitolate (Universal Preserv-A-Chem)

Trade Name Mixture:
Crodasome UV-A/B (Croda, Inc.)

OLEYL PALMITAMIDE

CAS Nos.   **EINECS No.**
16260-09-6   240-367-6
96674-02-1

Empirical Formula:
$C_{34}H_{67}NO$

Definition: Oleyl Palmitamide is the substituted aliphatic amide that conforms to the formula:

$$CH_3(CH_2)_{14}\overset{O}{\overset{\|}{C}}-NH(CH_2)_8CH=CH(CH_2)_7CH_3$$

Information Sources: 21CFR177.1200, 21CFR178.3860

Chemical Class: Amides

Function: Viscosity Increasing Agent - Nonaqueous

Technical/Other Names:
Hexadecanamide, N-9-Octadecenyl-
N-9-Octadecenyl Hexadecanamide

OLEYL PHOSPHATE

CAS No.   **EINECS No.**
37310-83-1   253-455-4

Definition: Oleyl Phosphate is a mixture of mono- and diesters of Oleyl Alcohol (q.v.) and phosphoric acid.

Chemical Classes: Esters; Phosphorus Compounds

Function: Surfactant - Emulsifying Agent

Technical/Other Names:
9-Octadecen-1-ol, Phosphate
Oleyl Alcohol Phosphate

OLEYL STEARATE

CAS Nos.   **EINECS No.**
17673-50-6   241-652-8
33057-39-5

Empirical Formula:
$C_{36}H_{70}O_2$

Definition: Oleyl Stearate is the ester of oleyl alcohol and stearic acid. It conforms to the formula:

$$CH_3(CH_2)_{16}\overset{O}{\overset{\|}{C}}-O(CH_2)_8CH=CH(CH_2)_7CH_3$$

Chemical Class: Esters

Functions: Hair Conditioning Agent; Skin-Conditioning Agent - Occlusive

Technical/Other Names:
Octadecanoic Acid, 9-Octadecenyl Ester
Octadec-9-enyl Octadecanoate

Trade Name:
AEC Oleyl Stearate (A & E Connock)

OLIBANUM

CAS No.   **EINECS No.**
8050-07-5   232-474-1

Definition: Olibanum is a gum resin obtained from *Boswellia carterii.*

Information Sources: 21CFR172.510, 27CFR21.141, MI-13(6901), RIFM

Chemical Classes: Essential Oils; Gums, Hydrophilic Colloids and Derivatives

Functions: Fragrance Ingredient; Skin-Conditioning Agent - Miscellaneous

Technical/Other Names:
Boswellia Carterii Resin
Frankincense
Frankincense gum (RIFM)
Gum Olibanum
Incense
Olibanum absolute (Boswellia spp.) (RIFM)
Olibanum oil (Boswellia spp.) (RIFM)
Olibanum Resin
Olibanum resinoid (Boswellia spp.) (RIFM)
Resin Olibanum

Trade Names:
Encens Oil (Floressence)
Incense EA (Alban Muller)
Incense H (Alban Muller)

OLIGOPEPTIDE-1

Definition: Oligopeptide-1 is a synthetic peptide consisting of Lysine (q.v.), Glycine (q.v.), and Histidine (q.v.).

Chemical Class: Protein Derivatives

Function: Skin-Conditioning Agent - Miscellaneous

Trade Name:
Transtide-S (LG Cosmetic)

OLIGOPEPTIDE-2

Definition: Oligopeptide-2 is a synthetic peptide consisting of Lysine (q.v.), Threonine (q.v.), and Serine (q.v.) .

Chemical Class: Protein Derivatives

Function: Skin-Conditioning Agent - Miscellaneous

Trade Name:
  Transtide-W (LG Cosmetic)

OLIGOPEPTIDE-3

Definition: Oligopeptide-3 is a synthetic 13-amino acid peptide containing phenyl-alanine, alanine, lysine and leucine residues. *See Reported Ingredient Functions-The Cosmetic Drug Distinction, in Regulatory and Ingredient Use Information, Volume I, Part A.*

Chemical Class: Protein Derivatives

Function: Antimicrobial Agent

Trade Name:
  Tridecapeptide HB-55 (Helix BioMedix)

OLIGOPEPTIDE-4

Definition: Oligopeptide-4 is a synthetic 91 amino acid peptide consisting of arginine, hydroxyproline, serine, aspartic acid, glutamic acid, threonine, glycine, alanine, tyrosine, proline, methionine, valine, phenyl-alanine, isoleucine, leucine, histidine and lysine residues.

Chemical Class: Proteins

Function: Skin-Conditioning Agent - Miscellaneous

Trade Name:
  Collagenon (Vevy)

OLIGOPEPTIDE-5

Definition: Oligopeptide-5 is a synthetic 73 amino acid peptide containing arginine, serine, aspartic acid, glutamic acid, threonine, glycine, alanine, tyrosine, proline, methionine, valine, phenylalanine, isoleucine, leucine, histidine and lysine residues.

Chemical Class: Proteins

Function: Skin-Conditioning Agent - Miscellaneous

Trade Name:
  Dermonectin (Vevy)

OLIGOPEPTIDE-6

Definition: Oligopeptide-6 is a synthetic peptide consisting of Alanine (q.v.), Arginine (q.v.), Asparagine (q.v.), Aspartamine (q.v.), Isoleucine (q.v.), Leucine (q.v.), Lysine (q.v.), Phenylalanine (q.v.) and Threonine (q.v.).

Chemical Class: Proteins

Function: Skin-Conditioning Agent - Miscellaneous

Trade Name Mixture:
  Peptide Vinci 01 (Vincience)

OLIVAMIDE DEA

CAS No.: 124046-30-6

Definition: Olivamide DEA is a mixture of ethanolamides of the fatty acids derived from olive oil. It conforms generally to the formula:

$$RC\overset{O}{\underset{||}{}}-N(CH_2CH_2OH)_2$$

where RCO- represents the fatty acids derived from olive oil.

Information Source: EEC(III/1-60)

Chemical Class: Alkanolamides

Functions: Surfactant - Foam Booster; Viscosity Increasing Agent - Aqueous

Technical/Other Names:
  Amides, Olive Oil, N,N-Bis(Hydroxyethyl)-
  N,N-Bis(2-Hydroxyethyl)Olive Fatty Acid Amide
  Diethanolamine Olive Fatty Acid Condensate
  Olive Amides, N,N-Bis(2-Hydroxyethyl)-
  Olive Oil Fatty Acid Diethanolamide

OLIVAMIDOPROPYLAMINE OXIDE

CAS No.: 124046-32-8

Definition: Olivamidopropylamine Oxide is the tertiary amine oxide that conforms generally to the formula:

$$RC\overset{O}{\underset{||}{}}-NH(CH_2)_3-\overset{CH_3}{\underset{CH_3}{N}}\rightarrow O$$

where RCO- represents the fatty acids derived from olive oil.

Chemical Class: Amine Oxides

Functions: Hair Conditioning Agent; Surfactant - Cleansing Agent; Surfactant - Foam Booster; Surfactant - Hydrotrope

Technical/Other Names:
  Amides, Olive, N-[3-(Dimethylamino)-Propyl], N-Oxide
  N-[3-(Dimethylamino)Propyl]Olive Amides-N-Oxide
  Olive Amides, N-[3-(Dimethylamino)Propyl], N-Oxide

OLIVAMIDOPROPYL BETAINE

Definition: Olivamidopropyl Betaine is the zwitterion (inner salt) that conforms to the formula:

$$RC\overset{O}{\underset{||}{}}-NH(CH_2)_3-\overset{CH_3}{\underset{CH_3}{\overset{+}{N}}}-CH_2COO^-$$

where RCO- represents the fatty acids derived from olive oil.

Chemical Class: Betaines

Functions: Antistatic Agent; Hair Conditioning Agent; Skin-Conditioning Agent - Miscellaneous; Surfactant - Cleansing Agent; Surfactant - Foam Booster; Viscosity Increasing Agent - Aqueous

Technical/Other Names:
  N-(Carboxymethyl)-N,N-Dimethyl-3-[(1-Oxoolive)Amino]-1-Propanaminium Hydroxide, Inner Salt
  Olivamidopropyl Dimethyl Glycine
  Olive Amide Propylbetaine
  1-Propanaminium, N-(Carboxymethyl)-N,N-Dimethyl-3-[(1-Oxoolive)Amino]-, Hydroxide, Inner Salt
  Quaternary Ammonium Compounds, (Carboxymethyl)(3-Oliveamidopopyl) Dimethyl, Hydroxide, Inner Salt

OLIVAMIDOPROPYL DIMETHYLAMINE

Definition: Olivamidopropyl Dimethylamine is the amidoamine that conforms generally to the formula:

$$RC\overset{O}{\underset{||}{}}-NH(CH_2)_3N\overset{CH_3}{\underset{CH_3}{}}$$

where RCO- represents the fatty acids derived from olive oil.

Chemical Class: Amines

Function: Antistatic Agent

Technical/Other Names:
  Amides, Olive, N-[3-(Dimethylamino)-Propyl]-
  N-[3-(Dimethylamino)Propyl]Olive Amides

OLIVAMIDOPROPYL DIMETHYLAMINE LACTATE

CAS No.: 124046-31-7

Definition: Olivamidopropyl Dimethylamine Lactate is the lactic acid salt of olivamidopropyl dimethylamine. It conforms generally to the formula:

$$RC-NH(CH_2)_3N \cdot HOOCCHCH_3$$

where RCO- represents the fatty acids derived from olive oil.

Chemical Classes: Amines; Organic Salts

Function: Hair Conditioning Agent

Technical/Other Name:
Amides, Olive Oil, N-[3-(Dimethylamino)-Propyl], Lactates

OLIVAMIDOPROPYLTRIMONIUM CHLORIDE

Definition: Olivamidopropyltrimonium Chloride is the quaternary ammonium salt that conforms generally to the formula:

$$[RC-NH(CH_2)_3-N-CH_3]^+ \quad Cl^-$$

where RCO- represents the fatty acids derived from olive oil.

Chemical Class: Quaternary Ammonium Compounds

Functions: Antistatic Agent; Hair Conditioning Agent; Surfactant - Emulsifying Agent

OLIVE ACID

CAS No.	EINECS No.
92044-96-7	295-376-8

Definition: Olive Acid is a mixture of fatty acids derived from Olea Europaea (Olive) Oil (q.v.).

Chemical Class: Fatty Acids

Function: Surfactant - Cleansing Agent

Technical/Other Names:
Acids, Olive
Olea Europaea (Olive) Acid

Trade Names:
PRIFAC 7986 (Uniqema Europe)
PRIFAC 7987 (Uniqema Europe)

OLIVE ALCOHOL

JPN Translation:
オリーブアルコール

Definition: Olive Alcohol is a mixture of fatty alcohols derived from Olea Europaea (Olive) Fruit Oil (q.v.).

Information Source: JCIC

Chemical Class: Fatty Alcohols

Function: Emulsion Stabilizer

OLIVEAMIDE MEA

Definition: Oliveamide MEA is a mixture of ethanolamides of the fatty acids derived from olive oil. It conforms generally to the formula:

$$RC-NHCH_2CH_2OH$$

where RCO- represents the fatty acids derived from olive oil.

Chemical Class: Alkanolamides

Functions: Hair Conditioning Agent; Surfactant - Foam Booster; Surfactant - Solubilizing Agent; Viscosity Increasing Agent - Aqueous

Technical/Other Names:
Oliveoyl Monoethanolamine
Olivoyl Monoethanolamine

OLIVE OIL PEG-6 ESTERS

CAS No.: 103819-46-1

JPN Translation:
オリーブ油 PEG - 6

Definition: Olive Oil PEG-6 Esters is a complex mixture formed from the transesterification of Olea Europaea (Olive) Oil (q.v.) and PEG-6 (q.v.).

Chemical Class: Glyceryl Esters and Derivatives

Functions: Skin-Conditioning Agent - Emollient; Surfactant - Emulsifying Agent

Trade Name:
Labrafil M 1980 CS (Gattefosse s.a.)

OLIVE OIL PEG-7 ESTERS

Definition: Olive Oil PEG-7 Esters is the complex mixture formed from the transesterification of Olea Europaea (Olive) Oil (q.v.) and PEG-7 (q.v.).

Chemical Class: Glyceryl Esters and Derivatives

Functions: Skin-Conditioning Agent - Emollient; Surfactant - Emulsifying Agent

Trade Name:
Olivem 300 (B & T)

OLIVE OIL.PEG-8 ESTERS

Definition: Olive Oil PEG-8 Esters is a complex mixture formed from the transester-

ification of Olea Europaea (Olive) Oil (q.v.) and PEG-8 (q.v.).

Chemical Class: Glyceryl Esters and Derivatives

Functions: Skin-Conditioning Agent - Emollient; Surfactant - Emulsifying Agent

Trade Names:
Resplanta Olea (Res Pharma)
Viatenza Olive PE8 (Aldivia)

OLIVE OIL PEG-10 ESTERS

CAS No.: 103819-46-1 (Generic)

Definition: Olive Oil PEG-10 Esters is a complex mixture formed from the transesterification of Olea Europaea (Olive) Oil (q.v) and PEG-10 (q.v.).

Chemical Class: Glyceryl Esters and Derivatives

Functions: Skin-Conditioning Agent - Emollient; Surfactant - Emulsifying Agent

Trade Name:
Sympatens-TOL/100 (Kolb)

OLIVE OIL POLYGLYCERYL-6 ESTERS

Definition: Olive Oil Polyglyceryl-6 Esters is the product obtained by the transesterification of Olea Europaea (Olive) Oil (q.v.) and Polyglycerin-6 (q.v.).

Chemical Class: Glyceryl Esters and Derivatives

Functions: Skin-Conditioning Agent - Emollient; Surfactant - Emulsifying Agent

Trade Name:
Viatenza Olive PO6 (Aldivia)

OLIVOYL HYDROLYZED WHEAT PROTEIN

Definition: Olivoyl Hydrolyzed Wheat Protein is the condensation product of olive acid chloride and Hydrolyzed Wheat Protein (q.v.).

Chemical Class: Protein Derivatives

Functions: Hair Conditioning Agent; Skin-Conditioning Agent - Miscellaneous; Surfactant - Cleansing Agent

Trade Name:
Olivoil Glutinate (Keminova Italiana)

OMENTAL LIPIDS

Definition: Omental Lipids are the lipids obtained from bovine omentum.

Chemical Class: Fats and Oils

Function: Skin-Conditioning Agent - Occlusive

ONONIS ARVENSIS ROOT EXTRACT

Definition: Ononis Arvensis Root Extract is an extract of the roots of the restharrow, *Ononis arvensis*. See "Regulatory and Ingredient Use Information," regarding the labeling names for botanical ingredients in Volume 1, Introduction, Part A.

Information Sources: JCIC, JCLS

Chemical Class: Biological Products

Function: Not Reported

Technical/Other Names:
Extract of Ononis Arvensis Roots
Extract of Restharrow Roots
Restharrow Extract
Restharrow (Ononis Arvensis) Extract
Restharrow Root Extract

ONONIS SPINOSA ROOT EXTRACT

CAS No. 84775-89-3

EINECS No. 283-913-9

JPN Translation:
オノニスエキス

Definition: Ononis Spinosa Root Extract is an extract of the roots of the restharrow, *Ononis spinosa*. See "Regulatory and Ingredient Use Information," regarding the labeling names for botanical ingredients in Volume 1, Introduction, Part A.

Chemical Class: Biological Products

Function: Not Reported

Technical/Other Names:
Extract of Ononis Spinosa Roots
Extract of Restharrow Roots
Restharrow Extract
Restharrow (Ononis Spinosa) Extract
Restharrow Root Extract

Trade Name Mixtures:
Actiphyte of Rest Harrow BG50 (Active Organics)
Actiphyte of Rest Harrow GL50 (Active Organics)
Actiphyte of Rest Harrow Lipo S (Active Organics)
Actiphyte of Rest Harrow PG50 (Active Organics)
Extrait de Bugrane (Silab)
Extrait De Bugrane MEG100 (Yves Rocher)
Extrapone #4 GW 2/031040 (Symrise)
Extrapone #3 Special 2/034481 (Symrise)
Extrapone #3 Special 2/789490 (Symrise)

Extrapone Restharrow GW 2/031350 (Symrise)
Extrapone 3 Special 2/789490 (Symrise)
Extrapone 4 Special 2/788400 (Symrise)
Extrapone 3 Special New 2/034484 (Symrise)
Herbalcomplex 3 Special (Crodarom)
Herbal Extract Glycolic - Article 251172 (Plantextrakt)
Herbasol-Extract Restharrow (Cosmetochem)
Hydroplastidine Ononis (Vevy)
VT-263 Extract of Restharrow (Vege-Tech)

ONSEN-SUI

Definition: Onsen-Sui is the water obtained from hot springs, as defined in Japan by the Prefectural Governor under Article 12 of the Hot Spring Law under the jurisdiction of the Ministry of the Environment. See "Regulatory and Ingredient Use Information," regarding use of Japan Trivial names in Volume 1, Introduction, Part A.

Chemical Class: Inorganics

Function: Skin-Conditioning Agent - Miscellaneous

Reported Product Categories: Hair Dyes and Colors (All Types Requiring Caution Statements and Patch Tests); Shampoos (Non-coloring); Bath Preparations, Misc.; Bath Oils, Tablets, and Salts; Hair Conditioners; Colognes and Toilet Waters; Tonics, Dressings, and Other Hair Grooming Aids; Bath Capsules; Hair Preparations (Non-coloring), Misc.; Baby Shampoos; Foundations; Permanent Waves; Mascara; Hair Sprays (Aerosol Fixatives); Fragrance Preparations, Misc.; Perfumes; Makeup Bases; Eyebrow Pencils; Hair Bleaches; Eye Makeup Preparations, Misc.; Lipsticks; Eye Makeup Removers; Eyeliners; Makeup Preparations (Not eye), Misc.; Hair Straighteners; Hair Wave Sets; Hair Tints; Mouthwashes and Breath Fresheners (Liquids and Sprays); Hair Coloring Preparations, Misc.; Hair Rinses (Non-coloring); Dentifrices (Aerosol, Liquid, Pastes and Powders); Baby Lotions, Oils, Powders and Creams; Eye Shadows; Hair Shampoos (Coloring); Baby Products, Misc.; Nail Polish and Enamel Removers; Bubble Baths; Eye Lotions; Hair Rinses (Coloring); Blushers (All types); Sachets; Face Powders; Makeup Fixatives; Cuticle Softeners; Manicuring Preparations, Misc.; Nail Creams and Lotions; Powders (Dusting and Talcum, Excluding Aftershave Talcs); Basecoats and Undercoats; Hair Color Sprays (Aerosol); Hair Lighteners with Color; Nail Polish and Enamels; Rouges; Leg and Body Paints

Technical/Other Name:
Water (U.S.)

OPAL POWDER

CAS No.: 14639-88-4

Definition: Opal Powder is the powder obtained from crushed opals which consists chiefly of silicic acid.

Chemical Class: Inorganics

Function: Skin-Conditioning Agent - Miscellaneous

Trade Name:
Ultrapure Lavender Opal M1 (Ultra Chemical)

OPHIOPOGON EXTRACT STEARATE

JPN Translation:
ステアリン酸バクモンドウエキス

Definition: Ophiopogon Extract Stearate is the product obtained by the reaction of stearic acid with Ophiopogon Japonicus Extract (q.v.).

Information Source: JCLS

Chemical Class: Esters

Function: Surfactant - Cleansing Agent

Trade Name:
Bakumondou Extract Powder ST (Ichimaru Pharcos)

OPHIOPOGON JAPONICUS ROOT EXTRACT

JPN Translation:
ジャノヒゲ根エキス

Definition: Ophiopogon Japonicus Root Extract is an extract of the roots of *Ophiopogon japonicus*. See "Regulatory and Ingredient Use Information," regarding the labeling names for botanical ingredients in Volume 1, Introduction, Part A.

Information Sources: JCIC, JCLS

Chemical Class: Biological Products

Function: Skin-Conditioning Agent - Humectant

Technical/Other Names:
Bakumondou Ekisu (JPN)
Extract of Ophiopogon Japonicus
Ophiopogon Tuber Extract

Trade Name:
Bakumondou Extract Powder (Ichimaru Pharcos)

Trade Name Mixtures:
Bakumondou Extract BG (NOF)
Bakumondou Liquid B (Ichimaru Pharcos)
Boumdan (Bioland)
Phytostan (EUROCOSTECH)

OPOPONAX OIL

CAS No.: 8021-36-1

Definition: Opoponax Oil is the volatile oil obtained from *Commiphora erythraea*.

Information Sources: RIFM, TSCA

Chemical Class: Essential Oils

Function: Fragrance Ingredient

Technical/Other Names:
Commiphora Erythraea Oil
Oils, Opoponax
Opoponax oil (RIFM)

OPUNTIA COCCINELLIFERA

Definition: *See "Regulatory and Ingredient Use Information," regarding EU labeling names for botanical ingredients in Volume 1, Introduction, Part A.*

Chemical Class: Biological Products

Technical/Other Names:
Opuntia Coccinellifera Flower Extract (U.S.)
Opuntia Coccinellifera Fruit Extract (U.S.)

OPUNTIA COCCINELLIFERA FLOWER EXTRACT

CAS No.: 90082-21-6

Definition: Opuntia Coccinellifera Flower Extract is an extract of the dried flowers of *Opuntia coccinellifera. See "Regulatory and Ingredient Use Information," regarding the labeling names for botanical ingredients in Volume 1, Introduction, Part A.*

Chemical Class: Biological Products

Function: Not Reported

Technical/Other Name:
Opuntia Coccinellifera (EU)

Trade Name Mixtures:
Cactus Milk (CEP (Solabia))
Glycolysat of Cactus (CEP (Solabia))
Vegebios of Cactus (CEP (Solabia))

OPUNTIA COCCINELLIFERA FRUIT EXTRACT

Definition: Opuntia Coccinellifera Fruit Extract is an extract of the fruit of *Opuntia coccinellifera. See "Regulatory and Ingredient Use Information," regarding the labeling names for botanical ingredients in Volume 1, Introduction, Part A.*

Chemical Class: Biological Products

Function: Skin-Conditioning Agent - Miscellaneous

Technical/Other Names:
Extract of Opuntia Coccinellifera Fruit
Opuntia Coccinellifera (EU)

Trade Name Mixture:
Prickly Pear Fruit Extract (Libiol)

OPUNTIA STREPTACANTHA LEAF EXTRACT

Definition: Opuntia Streptacantha Leaf Extract is an extract of the leaves of *Opuntia streptacantha. See "Regulatory and Ingredient Use Information," regarding the labeling names for botanical ingredients in Volume 1, Introduction, Part A.*

Chemical Class: Biological Products

Function: Skin-Conditioning Agent - Miscellaneous

Technical/Other Name:
Extract of Opuntia Streptacantha Leaves

Trade Name Mixture:
Extrait Hydroglycolique de Cactus (Greentech)

OPUNTIA STREPTACANTHA STEM EXTRACT

Definition: Opuntia Streptacantha Stem Extract is an extract of the stems of *Opuntia streptacantha. See "Regulatory and Ingredient Use Information," regarding the labeling names for botanical ingredients in Volume 1, Introduction, Part A.*

Chemical Class: Biological Products

Function: Skin-Conditioning Agent - Humectant

Technical/Other Name:
Extract of Opuntia Streptacantha Stem

Trade Name Mixtures:
Saboten Extract OS (C&F Koei Phyto Corp.)
Saboten Extract OS (Koei Perfumery)

OPUNTIA TUNA EXTRACT

Definition: Opuntia Tuna Extract is an extract of the flowers and stems of the prickly pear, *Opuntia tuna. See "Regulatory and Ingredient Use Information," regarding the labeling names for botanical ingredients in Volume 1, Introduction, Part A.*

Chemical Class: Biological Products

Function: Not Reported

Technical/Other Names:
Extract of Opuntia Tuna
Extract of Prickly Pear
Opuntia Vulgaris Extract
Prickly Pear Extract
Prickly Pear (Opuntia Tuna) Extract

Trade Name Mixtures:
Actiphyte Cactus AQ (Active Organics)
Actiphyte of Cactus BG50 (Active Organics)
Actiphyte of Cactus GL50 (Active Organics)
Actiphyte of Cactus Lipo S (Active Organics)
Actiphyte Prickly Pear (Active Organics)
Actiphyte Prickly Pear Cactus (Active Organics)
Complex 2 - Moisturizing (Provital/ Centerchem)
Opuntia Extract HG (Provital/Centerchem)
Pronalen Moisturizing HSC (Provital/ Centerchem)
PRONALEN MOISTURIZING-II (Provital/ Centerchem)

OPUNTIA TUNA FRUIT

Definition: Opuntia Tuna Fruit is the fruit of *Opuntia tuna. See "Regulatory and Ingredient Use Information," regarding the labeling names for botanical ingredients in Volume 1, Introduction, Part A.*

Chemical Class: Biological Products

Function: Not Reported

Technical/Other Names:
Prickly Pear
Prickly Pear (Opuntia Tuna)

Trade Names:
Indian Fig Puree, VT-840 (Vege-Tech)
Indian Fig Puree, VT-853 (Vege-Tech)

OPUNTIA TUNA FRUIT EXTRACT

Definition: Opuntia Tuna Fruit Extract is the extract obtained from the fruit of *Optunia tuna. See "Regulatory and Ingredient Use Information," regarding the labeling names for botanical ingredients in Volume 1, Introduction, Part A.*

Chemical Class: Biological Products

Function: Not Reported

Technical/Other Names:
Extract of Prickly Pear (Opuntia Tuna) Fruit
Prickly Pear Fruit Extract
Prickly Pear (Opuntia Tuna) Fruit Extract

Trade Name Mixtures:
Cosflor Chumba HGS (A & E Connock)
Prickly Pear Fruit Extract (Libiol)

ORANGE 4

CAS No.: 633-96-5

Empirical Formula:
$C_{16}H_{12}N_2O_4S \cdot Na$

Definition: Orange 4 is classed chemically as a monoazo color. It conforms to the formula:

See "Regulatory and Ingredient Use Information," for Colorants in Volume 1, Introduction, Part A. To identify the certified colorant for labeling purposes in the US, the INCI Name Orange 4 must be used. The INCI Name for batches of this colorant that have not been certified is Acid Orange 7. To identify the colorant allowed for use in the European Union (EU), the INCI Name CI 15510 must be used, except for hair dye products. To identify the colorant allowed for use in Japan, the INCI name Daidai205 must be used.

Information Sources: 21CFR74.1254, 21CFR74.2254, 21CFR81.30, 21CFR82.1254, CI 15510, M3, MI-13(6927), TSCA

Chemical Class: Color Additives - Batch Certified by the U.S. Food and Drug Administration

Function: Colorant

Reported Product Categories: Colognes and Toilet Waters; Shampoos (Non-coloring); Perfumes; Hair Conditioners; Bath Soaps and Detergents; Bubble Baths; Aftershave Lotions; Baby Shampoos; Bath Preparations, Misc.; Body and Hand Preparations (Excluding Shaving Preparations); Tonics, Dressings, and Other Hair Grooming Aids; Moisturizing Preparations; Bath Oils, Tablets, and Salts; Deodorants (Underarm); Hair

Rinses (Coloring); Skin Fresheners; Hair Dyes and Colors (All Types Requiring Caution Statements and Patch Tests); Paste Masks (Mud Packs); Skin Care Preparations, Misc.; Blushers (All types); Cleansing Products (Cold Creams, Cleansing Lotions, Liquids and Pads); Face Powders; Hair Bleaches; Hair Straighteners

Technical/Other Name:
D&C Orange No. 4

Trade Names:
D&C Orange 4 10-25-DA-3204 (Noveon Hilton Davis)
D & C Orange 4 W 094 (LCW)

Trade Name Mixtures:
Colourspheres Mandarin Dye HL 25% (Creations Couleurs)
Tefpoly Mandarin Dye (C.I.T.)

ORANGE 4 LAKE

Definition: Orange 4 Lake is the salt of Orange 4 extended on an appropriate substrate in compliance with 21CFR82.1051. See "Regulatory and Ingredient Use Information," for Colorants in Volume 1, Introduction, Part A. To identify the certified colorant for labeling purposes in the US, the INCI Name Orange 4 Lake must be used. To identify the colorant allowed for use in the European Union (EU), the INCI Name CI 15510 must be used, except for hair dye products. To identify the colorant allowed for use in Japan, the INCI name Daidai205 must be used.

Information Sources: 21CFR81.1, 21CFR82.1051

Chemical Class: Color Additives Lakes - Batch Certified by the U.S. Food and Drug Administration

Function: Colorant

Reported Product Categories: Blushers (All types); Face Powders

Technical/Other Name:
D&C Orange No. 4 Aluminum Lake

Trade Names:
D & C Orange #4 Aluminum Lake K7074 (LCW)
K7074 D&C 04 Aluminum Lake (LCW)

ORANGE 5

CAS No.: 596-03-2

Empirical Formula:
$C_{20}H_{10}Br_2O_5$

Definition: Orange 5 is classed chemically as a fluoran color. It conforms to the formula:

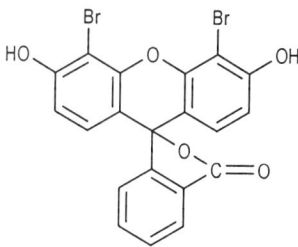

See "Regulatory and Ingredient Use Information," for Colorants in Volume 1, Introduction, Part A. To identify the certified colorant for labeling purposes in the US, the INCI Name Orange 5 must be used. The INCI Name for batches of this colorant that have not been certified is Solvent Red 72. To identify the colorant allowed for use in the European Union (EU), the INCI Name CI 45370 must be used, except for hair dye products. To identify the colorant allowed for use in Japan, the INCI name Daidai201 must be used.

Information Sources: 21CFR74.1255, 21CFR74.2255, 21CFR82.1255, CI 45370:1, M3, MHLW Ord. No. 30, MI-13 (3046), TSCA

Chemical Class: Color Additives - Batch Certified by the U.S. Food and Drug Administration

Function: Colorant

Reported Product Categories: Lipsticks; Makeup Preparations (Not eye), Misc.; Blushers (All types); Rouges

Technical/Other Name:
D&C Orange No. 5

Trade Names:
A514 Tudor Orange (Kingfisher Colours)
C14-033 Dibromo (Sun Pigments)
D & C Orange 5 W 038 (LCW)
D & C Orange #5 K7003 (LCW)

Trade Name Mixtures:
Colourspheres Orange Dye HL 25% (Creations Couleurs)
Tefpoly Orange Dye (C.I.T.)

ORANGE 5 LAKE

Definition: Orange 5 Lake is the salt of Orange 5 extended on an appropriate substrate in compliance with 21CFR82.1051. See "Regulatory and Ingredient Use Information," for Colorants in Volume 1, Introduction, Part A. To identify the certified colorant for labeling purposes in the US, the INCI Name Orange 5 Lake must be used. To identify the colorant allowed for use in the European

Union (EU), the INCI Name CI 45370 must be used, except for hair dye products.

Information Sources: 21CFR74.1255, 21CFR74.2255, 21CFR81.1, 21CFR82.1051

Chemical Classes: Color Additives Lakes - Batch Certified by the U.S. Food and Drug Administration; Halogen Compounds

Function: Colorant

Reported Product Categories: Lipsticks; Makeup Preparations (Not eye), Misc.

Technical/Other Names:
D&C Orange No. 5 Aluminum Lake
D&C Orange No. 5 Aluminum/Zirconium Lake
D&C Orange No. 5 Zirconium Lake

Trade Names:
A517 Tudor Buttercup (Kingfisher Colours)
C14-038 Manchu Orange (Sun Pigments)
D&C Orange No. 5 Aluminum/Zirconium Lake C6905 (LCW)

Trade Name Mixtures:
Colourspheres Orange Lake HL 25% (Creations Couleurs)
Tefpoly Orange Lake (C.I.T.)

ORANGE 10

CAS Nos.: 518-40-1; 38577-97-8

Empirical Formula:
$C_{20}H_{10}I_2O_5$

Definition: Orange 10 is classed chemically as a fluoran color. It conforms to the formula:

See "Regulatory and Ingredient Use Information," for Colorants in Volume 1, Introduction, Part A. To identify the certified colorant for labeling purposes in the US, the INCI Name Orange 10 must be used. To identify the certified colorant (sodium salt) for labeling purposes in the US, the INCI Name Orange 11 must be used. The INCI Name for batches of this colorant that have not been certified is Solvent Red 73. The INCI Name for batches of this colorant (sodium salt) that have not been certified is Acid Red 95. To identify the colorant allowed for use in the European Union (EU), the INCI Name CI 45425 must be used, except for hair dye products. To

identify the colorant allowed for use in Japan, the INCI name Daidai206 must be used.

Information Sources: 21CFR74.1260, 21CFR74.1261, 21CFR74.2260, 21CFR81.10, 21CFR81.30, 21CFR82.1260, CI 45425:1, MI-13(3211), TSCA

Chemical Classes: Color Additives - Batch Certified by the U.S. Food and Drug Administration; Halogen Compounds

Function: Colorant

Technical/Other Name:
D&C Orange No. 10

ORANGE 10 LAKE

Definition: Orange 10 Lake is the salt of Orange 10 extended on an appropriate substrate in compliance with 21CFR82.1051. *See "Regulatory and Ingredient Use Information," for Colorants in Volume 1, Introduction, Part A. To identify the certified colorant for labeling purposes in the US, the INCI Name Orange 10 Lake must be used. To identify the colorant allowed for use in the European Union (EU), the INCI Name CI 45425 must be used, except for hair dye products.*

Information Sources: 21CFR81.1, 21CFR82.1051

Chemical Classes: Color Additives Lakes - Batch Certified by the U.S. Food and Drug Administration; Halogen Compounds

Function: Colorant

Technical/Other Name:
D&C Orange No. 10 Aluminum Lake

ORANGE 11

CAS No.: 33239-19-9

Empirical Formula:
$C_{20}H_{10}I_2O_5 \cdot 2Na$

Definition: Orange 11 is classed chemically as a xanthene color. It conforms to the formula:

See "Regulatory and Ingredient Use Information," for Colorants in Volume 1, Introduction, Part A. The INCI Name for batches of this colorant that have not been certified is Acid Red 95. The INCI Name for batches of this colorant (acid form) that have not been certified is Solvent Red 73. To identify the certified colorant for labeling purposes in the US, the INCI Name Orange 11 must be used. To identify the certified colorant (acid form) for labeling purposes in the US, the INCI Name Orange 10 must be used. To identify the colorant allowed for use in the European Union (EU), the INCI Name CI 45425 must be used, except for hair dye products. To identify the colorant allowed for use in Japan, the INCI name Daidai207 must be used.

Information Sources: 21CFR74.1261, 21CFR74.2260, 21CFR74.2261, 21CFR81.10, 21CFR81.30, 21CFR82.1261, CI 45425, MI-13(3211)

Chemical Classes: Color Additives - Batch Certified by the U.S. Food and Drug Administration; Halogen Compounds

Function: Colorant

Technical/Other Name:
D&C Orange No. 11

ORANGE ROUGHY OIL

JPN Translation:
オレンジラフィー油

Definition: Orange Roughy Oil is the lipid derived from the subcutaneous fat of the deep sea fish *Hoplostethus atlanticus*. See *"Regulatory and Ingredient Use Information," regarding use of EU Trivial names in Volume 1, Introduction, Part A.*

Information Sources: JCIC, JCLS, JSQI

Chemical Class: Fats and Oils

Function: Skin-Conditioning Agent - Occlusive

Technical/Other Name:
Oils, Orange Roughy

Trade Name:
AEC Orange Roughy Oil (A & E Connock)

ORANGE YU

JPN Translation:
オレンジ油

Definition: Orange Yu is the essential oil obtained from the peel of the fruit of *Citrus spp.. See "Regulatory and Ingredient Use Information," regarding use of Japan Trivial names in Volume 1, Introduction, Part A.*

Chemical Class: Essential Oils

Functions: Fragrance Ingredient; Skin-Conditioning Agent - Miscellaneous

Technical/Other Names:
Citrus Aurantium Amara (Bitter Orange) Oil (U.S.)
Citrus Aurantium Dulcis (Orange) Oil (U.S.)
Citrus Grandis (EU)
Citrus Grandis (Grapefruit) Peel Oil (U.S.)

ORBIGNYA COHUNE SEED OIL

Definition: Orbignya Cohune Seed Oil is the fixed oil expressed from the seeds of the cohune palm, *Orbignya cohune*. See *"Regulatory and Ingredient Use Information,"* regarding the labeling names for botanical ingredients in Volume 1, Introduction, Part A.

Chemical Class: Fats and Oils

Functions: Hair Conditioning Agent; Skin-Conditioning Agent - Occlusive

Technical/Other Name:
Oils, Orbignya Cohune

Trade Name:
AEC Cohune Oil (A & E Connock)

ORBIGNYA OLEIFERA

Definition: See *"Regulatory and Ingredient Use Information,"* regarding EU labeling names for botanical ingredients in Volume 1, Introduction, Part A.

Chemical Class: Biological Products

Technical/Other Name:
Orbignya Oleifera Seed Oil (U.S.)

ORBIGNYA OLEIFERA SEED OIL

CAS No.	EINECS No.
91078-92-1	293-376-2

Definition: Orbignya Oleifera Seed Oil is the fixed oil obtained from the nuts of *Orbignya oleifera*. See *"Regulatory and Ingredient Use Information,"* regarding the labeling names for botanical ingredients in Volume 1, Introduction, Part A.

Chemical Class: Fats and Oils

Function: Skin-Conditioning Agent - Occlusive

Reported Product Category: Bath Soaps and Detergents

Technical/Other Names:
Babassu (Orbignya Oleifera) Oil

Oil of Babassu
Oils, Babassu
Orbignya Oleifera (EU)

Trade Names:
AEC Babassu Oil (A & E Connock)
Babassu Oil (Active Concepts)
Babassu Oil (Desert Whale)
Cropure Babassu (Croda Chemicals)
Cropure Babassu (Croda, Inc.)
Polygreen Orbignya T (Polytechno Ind.)

Trade Name Mixture:
Babassu Milk (CEP (Solabia))

ORBIGNYA PHALERATA SEED POWDER

Definition: Orbignya Phalerata Seed Powder is the powder obtained from the crushed seeds of *Orbignya phalerata*. See *"Regulatory and Ingredient Use Information,"* regarding the labeling names for botanical ingredients in Volume 1, Introduction, Part A.

Chemical Class: Biological Products

Function: Skin-Conditioning Agent - Miscellaneous

Trade Name:
Babassu Powder (Nu Skin)

ORBIGNYA SPECIOSA KERNEL OIL

Definition: Orbignya Speciosa Kernel Oil is the oil expressed from the kernels of *Orbignya speciosa*. See *"Regulatory and Ingredient Use Information,"* regarding the labeling names for botanical ingredients in Volume 1, Introduction, Part A.

Chemical Class: Fats and Oils

Function: Skin-Conditioning Agent - Emollient

Technical/Other Name:
Oils, Orbignya Speciosa Kernel

Trade Name:
Chemyforest Babacu CG (Chemyunion)

ORCHID EXTRACT

Definition: Orchid Extract is an extract of various species of the orchid, *Orchis*. See *"Regulatory and Ingredient Use Information,"* regarding the labeling names for botanical ingredients in Volume 1, Introduction, Part A.

Chemical Class: Biological Products

Function: Not Reported

Reported Product Categories: Hair Conditioners; Shampoos (Non-coloring); Tonics, Dressings, and Other Hair Grooming Aids

Trade Name Mixtures:
Orchid Liquid (Ichimaru Pharcos)
Orchid Phalaenopsis Extract HS 3634 G (Grau)

ORCHIS MACULATA FLOWER EXTRACT

Definition: Orchis Maculata Flower Extract is an extract of the flowers of the Orchid, *Orchis maculata*. See *"Regulatory and Ingredient Use Information,"* regarding the labeling names for botanical ingredients in Volume 1, Introduction, Part A.

Chemical Class: Biological Products

Function: Skin-Conditioning Agent - Miscellaneous

Reported Product Categories: Hair Conditioners; Shampoos (Non-coloring); Tonics, Dressings, and Other Hair Grooming Aids

Technical/Other Names:
Extract of Orchid
Extract of Orchis Maculata
Orchid Extract
Orchid (Orchis Maculata) Extract

ORCHIS MASCULA FLOWER EXTRACT

CAS No.	EINECS No.
90082-24-9	290-112-8

Definition: Orchis Mascula Flower Extract is an extract of the flowers of the orchid, *Orchis mascula*. See *"Regulatory and Ingredient Use Information,"* regarding the labeling names for botanical ingredients in Volume 1, Introduction, Part A.

Chemical Class: Biological Products

Function: Skin-Conditioning Agent - Miscellaneous

Reported Product Categories: Hair Conditioners; Shampoos (Non-coloring); Tonics, Dressings, and Other Hair Grooming Aids

Technical/Other Names:
Extract of Orchid
Extract of Orchis Mascula
Orchid Extract
Orchid (Orchis Mascula) Extract

Trade Name Mixtures:
Actiphyte Orchid (Active Organics)
Actiphyte Orchid AL (Active Organics)
Actiphyte Orchid BG50P (Active Organics)
Actiphyte Orchid GL (Active Organics)
Actiphyte Orchid Lipo M (Active Organics)

Actiphyte Orchid Lipo S (Active Organics)
Orchidana (GfN)

ORCHIS MORIO

Definition: *See "Regulatory and Ingredient Use Information," regarding EU labeling names for botanical ingredients in Volume 1, Introduction, Part A.*

Chemical Class: Biological Products

Technical/Other Name:
Orchis Morio Flower Extract (U.S.)

ORCHIS MORIO FLOWER EXTRACT

Definition: Orchis Morio Flower Extract is an extract of the flowers of the orchid, *Orchis morio. See "Regulatory and Ingredient Use Information," regarding the labeling names for botanical ingredients in Volume 1, Introduction, Part A.*

Chemical Class: Biological Products

Function: Not Reported

Reported Product Categories: Hair Conditioners; Shampoos (Non-coloring); Tonics, Dressings, and Other Hair Grooming Aids

Technical/Other Names:
Extract of Orchid
Extract of Orchis Morio
Orchid Extract
Orchid (Orchis Morio) Extract
Orchis Morio (EU)

Trade Name Mixtures:
Phytelene of Catleya Orchid EG 568 Liquid (Indena SA)
Phytelene of Cymbidium Orchid EG 590 Liquid (Indena SA)
Phytogreen 55 of Catleya Orchid EXH 742 Liquid (Phytochim)
VT-296 (Vege-Tech)

ORIGANUM HERACLEOTICUM FLOWER OIL

Definition: Origanum Heracleoticum Flower Oil is the volatile oil obtained from the flowers of *Origanum heracleoticum. See "Regulatory and Ingredient Use Information," regarding the labeling names for botanical ingredients in Volume 1, Introduction, Part A.*

Chemical Class: Essential Oils

Function: Fragrance Ingredient

Technical/Other Name:
Oils, Origanum Heracleoticum Flower

Trade Name Mixture:
C-Protect (Gattefosse s.a.)

ORIGANUM MAJORANA FLOWER OIL

Definition: Origanum Majorana Flower Oil is the volatile oil obtained from the flowers of *Origanum majorana. See "Regulatory and Ingredient Use Information," regarding the labeling names for botanical ingredients in Volume 1, Introduction, Part A.*

Chemical Class: Essential Oils

Function: Fragrance Ingredient

Technical/Other Name:
Oils, Origanum Majorana Flower

Trade Name Mixture:
C-Protect (Gattefosse s.a.)

ORIGANUM MAJORANA LEAF EXTRACT

CAS No.	EINECS No.
84082-58-6	282-004-4

Definition: Origanum Majorana Leaf Extract is an extract of the leaves of the sweet marjoram, *Origanum majorana. See "Regulatory and Ingredient Use Information," regarding the labeling names for botanical ingredients in Volume 1, Introduction, Part A.*

Information Sources: 21CFR182.20, RIFM

Chemical Class: Biological Products

Functions: Fragrance Ingredient; Skin-Conditioning Agent - Miscellaneous

Technical/Other Names:
Extract of Origanum Majorana
Extract of Sweet Marjoram
Marjoram oleoresin (Majorana hortensis Moench-Origanum majorana L.) (RIFM)
Sweet Marjoram Extract
Sweet Marjoram (Origanum Majorana) Extract

Trade Name Mixtures:
Actiphyte of Marjoram BG50 (Active Organics)
Actiphyte of Marjoram GL50 (Active Organics)
Actiphyte of Marjoram Lipo S (Active Organics)
Actiphyte of Marjoram PG50 (Active Organics)
Extrait de Marjolaine PPE PG 100 (Yves Rocher)
Glycolysat of Marjoram (CEP (Solabia))
Marjoram Extact BG (Maruzen Pharmaceuticals Co., Ltd.)

Marjoram Extract HS 2474 G (Grau)
Marjoram HS (Alban Muller)
Marjoram LS (Alban Muller)
Murva Extract (Carlisle)
Phytelene of Marjoram EG 478 Liquid (Indena SA)
Phytoderm Maggiorana Glycolic (Universal Flavors)
Phytogreen 55 of Marjoram EXH 690 Liquid (Phytochim)
Spicypone Sweet Marjoram 2/035600 (Symrise)
Sweet Marjoram Extract HG (Provital/Centerchem)
Vegebois of Marjoram (CEP (Solabia))
VT-107 Extract of Marjoram (Vege-Tech)

ORIGANUM MAJORANA LEAF OIL

CAS No.: 8015-01-8

Definition: Origanum Majorana Leaf Oil is the volatile oil distilled from the leaves of *Origanum majorana*. It consists largely of terpenes. *See "Regulatory and Ingredient Use Information," regarding the labeling names for botanical ingredients in Volume 1, Introduction, Part A.*

Information Sources: 21CFR182.20, RIFM

Chemical Class: Essential Oils

Function: Fragrance Ingredient

Reported Product Categories: Bath Capsules; Face and Neck Preparations (Excluding Shaving Preparations)

Technical/Other Names:
Marjoram oil, sweet (Origanum majorana) (RIFM)
Marjoram seed (Majorana hortensis Moench-Origanum majorana L.) (RIFM)
Marjoram, sweet (Majorana hortensis Moench-Origanum majorana L.) (RIFM)
Oil of Sweet Marjoram
Oils, Marjoram, Sweet
Sweet Marjoram Oil
Sweet Marjoram (Origanum Majorana) Oil

Trade Name:
AEC Sweet Marjoram Oil (A & E Connock)

Trade Name Mixtures:
Essentiaderm n.2 (Universal Flavors)
Essentiaderm n.7 (Universal Flavors)
Essentiaderm n.9 (Universal Flavors)

ORIGANUM VULGARE FLOWER EXTRACT

CAS No.	EINECS No.
84012-24-8	281-670-3

The inclusion of any compound in the *Dictionary and Handbook* does not indicate that use of that substance as a cosmetic ingredient complies with the laws and regulations governing such use in the United States or any other country.

Definition: Origanum Vulgare Flower Extract is an extract of the flowering ends of the wild marjoram, *Origanum vulgare*. See *"Regulatory and Ingredient Use Information,"* regarding the labeling names for botanical ingredients in Volume 1, Introduction, Part A.

Information Sources: 21CFR182.20, RIFM

Chemical Class: Biological Products

Functions: Fragrance Ingredient; Skin-Conditioning Agent - Miscellaneous

Technical/Other Names:
Extract of Origanum Vulgare
Extract of Wild Marjoram
Marjoram, pot (Origanum vulgare L.)
(RIFM)
Wild Marjoram Extract
Wild Marjoram (Origanum Vulgare) Extract

Trade Name Mixtures:
Herbasol-Extract Marjoram
(Cosmetochem)
Oregano HS (Alban Muller)
Origan-Extract HS 2465 G (Grau)
Origanol (Sederma)
Origanum Vulgare Essential Water
(Bioland)
Origanum Vulgare Flower Extract ies (IES LABO)
285 Stimulant HS (Alban Muller)
VT-286 Extract of Oregano (Vege-Tech)

ORIGANUM VULGARE LEAF EXTRACT

Definition: Origanum Vulgare Leaf Extract is an extract of the leaves of *Origanum vulgare*. See *"Regulatory and Ingredient Use Information,"* regarding the labeling names for botanical ingredients in Volume 1, Introduction, Part A.

Chemical Class: Biological Products

Function: Skin-Conditioning Agent - Miscellaneous

Technical/Other Name:
Extract of Origanum Vulgare Leaf

Trade Name Mixtures:
Biopein (Bio-Botanica)
Pronalen Origanum HSC (Provital/
Centerchem)

ORMENIS MULTICAULIS EXTRACT

CAS No.	EINECS No.
92202-02-3	296-034-0

Definition: Ormenis Multicaulis Extract is an extract of *Ormenis multicaulis*. See

"Regulatory and Ingredient Use Information," regarding the labeling names for botanical ingredients in Volume 1, Introduction, Part A.

Information Source: RIFM

Chemical Class: Biological Products

Functions: Fragrance Ingredient; Skin-Conditioning Agent - Emollient

Technical/Other Names:
Extract of Ormenis Multicaulis
Ormenis multicaulis, ext. (RIFM)

ORMENIS MULTICAULIS FLOWER WAX

Definition: Ormenis Multicaulis Flower Wax is the wax obtained from the flowers of *Ormenis multicaulis*. See *"Regulatory and Ingredient Use Information,"* regarding the labeling names for botanical ingredients in Volume 1, Introduction, Part A.

Chemical Class: Biological Products

Function: Not Reported

Technical/Other Names:
Chamomile (Ormenis Multicaulis) Wax
Wax, Chamomile

Trade Names:
AEC Cire Essentielle De Fleurs De
Camomille Sauvage (A & E Connock)
Cire Essentielle de fleurs de Camomille
sauvage (Bertin)

ORMENIS MULTICAULIS OIL

Definition: Ormenis Multicaulis Oil is the volatile oil obtained from *Ormenis multicaulis*. See *"Regulatory and Ingredient Use Information,"* regarding the labeling names for botanical ingredients in Volume 1, Introduction, Part A.

Chemical Class: Essential Oils

Function: Fragrance Ingredient

Technical/Other Name:
Oils, Ormenis Multicaulis

Trade Names:
Chamomile Maroc Oil (JPM)
Chamomille Oil, Morroco (Chauvet)

ORNITHINE

CAS Nos.	EINECS Nos.
70-26-8	200-731-7
616-07-9	210-463-2

Empirical Formula:
$C_5H_{12}N_2O_2$

Definition: Ornithine is the amino acid that conforms to the formula:

$$NH_2(CH_2)_3CHCOOH$$
$$|$$
$$NH_2$$

Information Sources: INN, MI-13(6940), TSCA

Chemical Class: Amino Acids

Function: Skin-Conditioning Agent - Miscellaneous

Technical/Other Names:
2,5-Diaminopentanoic Acid
2,5-Diaminovaleric Acid
Norvaline, 5-Amino-

Trade Name Mixture:
Hydro-Diffuser Microreservoir (Sederma)

ORNITHINE HCl

CAS No.	EINECS No.
3184-13-2	221-678-6

Empirical Formula:
$C_5H_{12}N_2O_2 \cdot ClH$

Definition: Ornithine HCl is the amine salt that conforms to the formula:

$$NH_2(CH_2)_3CHCOOH \cdot HCl$$
$$|$$
$$NH_2$$

Information Source: TSCA

Chemical Class: Amino Acids

Function: Skin-Conditioning Agent - Miscellaneous

Trade Name Mixtures:
Amino Acid Microspheres (Coletica SA)
Liposomes Trichogen Veg (Laboratoires
Serobiologiques)
Phototan (Laboratoires Serobiologiques)
Sel-Smooth (Seltzer)
Trichogen Veg (Laboratoires Sero-
biologiques)

ORNITHOGALUM UMBELLATUM BULB EXTRACT

Definition: Ornithogalum Umbellatum Bulb Extract is an extract of the bulbs of the star of bethlehem, *Ornithogalum umbellatum*. See *"Regulatory and Ingredient Use Information,"* regarding the labeling names for botanical ingredients in Volume 1, Introduction, Part A.

Chemical Class: Biological Products

Function: Not Reported

Technical/Other Names:
Extract of Ornithogalum Umbellatum

Extract of Star of Bethlehem
Star of Bethlehem Extract
Star of Bethlehem (Ornithogalum
Umbellatum) Extract

Trade Name Mixtures:
Actiphyte of Star of Bethlehem BG50
(Active Organics)
Actiphyte of Star of Bethlehem GL50
(Active Organics)
Actiphyte of Star of Bethlehem Lipo S
(Active Organics)
Actiphyte of Star of Bethlehem PG50
(Active Organics)

OROBANCHE RAPUM EXTRACT

Definition: Orobanche Rapum Extract is
an extract of *Orobanche rapum*. See
*"Regulatory and Ingredient Use
Information," regarding the labeling names
for botanical ingredients in Volume 1, Introduction, Part A.*

Chemical Class: Biological Products

Function: Not Reported

Technical/Other Name:
Broomrape Extract

Trade Name Mixture:
Oraposide (Somaig)

OROTIC ACID

CAS No.   **EINECS No.**
65-86-1   200-619-8

JPN Translation:
オロット酸

Empirical Formula:
$C_5H_4O_4N_2$

Definition: Orotic Acid is the heterocyclic
compound that conforms to the formula:

Information Sources: JCIC, MI-13(6942)

Chemical Classes: Carboxylic Acids;
Heterocyclic Compounds

Function: Skin-Conditioning Agent - Miscellaneous

ORTHOSIPHON STAMINEUS EXTRACT

CAS No.   **EINECS No.**
84012-29-3   281-674-5

Definition: Orthosiphon Stamineus Extract
is an extract of the orthosiphon, *Orthosiphon stamineus*. See *"Regulatory and
Ingredient Use Information," regarding the
labeling names for botanical ingredients in
Volume 1, Introduction, Part A.*

Chemical Class: Biological Products

Function: Not Reported

Technical/Other Names:
Extract of Orthosiphon
Extract of Orthosiphon Stamineus
Orthosiphon Extract

Trade Name Mixtures:
Herbasol Extract Java Tea (Cosmetochem)
(Cosmetochem International Ltd.)
Java Tea Extract HS 3739 G (Grau)
Phytelene of Orthosiphon EG 211 liquid
(Indena SA)
Phytogreen 55 of Orthosiphon EXH 637
Liquid (Phytochim)
Vegebios of Java Tea (CEP (Solabia))

ORYZANOL

CAS Nos.: 11042-64-1; 12738-23-7

JPN Translation:
オリザノール

Empirical Formula:
$C_{40}H_{58}O_3$

Definition: Oryzanol is an ester of ferulic
acid and a terpene alcohol that conforms
generally to the formula:

It is derived from rice bran oil.

Information Sources: JAN, JCLS, JSCI,
MI-13(6954)

Chemical Classes: Esters; Phenols

Function: Skin-Conditioning Agent - Miscellaneous

Reported Product Categories: Moisturizing
Preparations; Skin Care Preparations, Misc.;
Bath Capsules; Bath Preparations, Misc.;
Body and Hand Preparations (Excluding
Shaving Preparations); Skin Fresheners

Technical/Other Names:
Orizanol
gamma-Orizanol

Trade Names:
AEC Gamma Oryzanol (A & E Connock)
Cycro Artenyl Ferulate-60 (Oryza Oil)
Cycro Artenyl Ferulate-70 (Oryza Oil)
Gamma Oryzanol (Ikeda)
Oryza Gamma-V (Ichimaru Pharcos)

Trade Name Mixtures:
Oryza Gamma Milky (Ichimaru Pharcos)
Sel Moist GQ (Seltzer)
Sel Moist J (Seltzer)

ORYZA SATIVA (RICE) BRAN

JPN Translation:
コメヌカ

Definition: Oryza Sativa (Rice) Bran is the
broken hulls of rice *Oryza sativa*. See
*"Regulatory and Ingredient Use
Information," regarding the labeling names
for botanical ingredients in Volume 1, Introduction, Part A.*

Information Sources: CIR: [S], JCIC,
JCLS, JSQI

Chemical Class: Biological Products

Functions: Abrasive; Bulking Agent

Technical/Other Names:
Bran, Oryza Sativa
Bran, Rice
Oryza Sativa Bran
Rice Bran

Trade Names:
AEC Rice Bran Ground (A & E Connock)
ESP Rice Flour (Earth Supplied Products)

ORYZA SATIVA (RICE) BRAN EXTRACT

JPN Translation:
コメヌカエキス

Definition: Oryza Sativa (Rice) Bran
Extract is an extract of the bran of rice,
Oryza sativa. See *"Regulatory and Ingre-
dient Use Information," regarding the
labeling names for botanical ingredients in
Volume 1, Introduction, Part A.*

Information Sources: CIR: [S], JCIC,
JCLS

Chemical Class: Biological Products

Function: Not Reported

Technical/Other Names:
Extract of Oryza Sativa Bran
Extract of Rice Bran
Extract of Rice (Oryza Sativa) Bran
Rice Bran Extract

Trade Name:
Rice Bran Extract Powder-S (Maruzen
Pharmaceuticals Co., Ltd.)

Trade Name Mixtures:
Actiphyte of Rice Bran BG50 (Active
Organics)

Actiphyte of Rice Bran GL50 (Active
 Organics)
Actiphyte of Rice Bran Lipo S (Active
 Organics)
Actiphyte of Rice Bran PG50 (Active
 Organics)
Eastman NuTriene Tocotrienol Concentrate
 (Eastman Chemical)
Extrapone Cereals GW 2/031300 (Symrise)
Extrapone Honey/Rice Blend GW 2/031306
 (Symrise)
Extrapone Honey Rice Milk P 2/030830
 (Symrise)
Extrapone Rice GW 2/031228 (Symrise)
Lipoplastidine Oryza Furfur (Vevy)
Megasol-1 (Vevy)
Oryza Ceramide-LC (Ikeda)
Rice Bran Extract BG (Maruzen
 Pharmaceuticals Co., Ltd.)
Rice Bran Extract LA (Maruzen
 Pharmaceuticals Co., Ltd.)
Skin Whitening Complex (Alban Muller)
VT-113 Extract of Rice Bran (Vege-Tech)

ORYZA SATIVA (RICE) BRAN OIL

CAS Nos.   **EINECS No.**
68553-81-1   271-397-8
84696-37-7

JPN Translation:
コメヌカ油

Definition: Oryza Sativa (Rice) Bran Oil is
the oil expressed from rice bran, *Oryza
sativa. See "Regulatory and Ingredient Use
Information," regarding the labeling names
for botanical ingredients in Volume 1, Introduction, Part A.*

Information Sources: 21CFR175.105,
21CFR176.200, 21CFR176.210,
21CFR177.2260, 21CFR177.2800, CIR: [S],
JCIC, JCLS, JSQI, MI-13(8292), TSCA

Chemical Class: Fats and Oils

Function: Skin-Conditioning Agent -
Occlusive

Reported Product Categories: Moisturizing
Preparations; Bath Capsules; Cleansing
Products (Cold Creams, Cleansing Lotions,
Liquids and Pads); Face and Neck Prepara-
tions (Excluding Shaving Preparations); Skin
Care Preparations, Misc.; Hair Conditioners;
Suntan Gels, Creams, and Liquids; Bath Oils,
Tablets, and Salts; Bath Preparations, Misc.;
Bath Soaps and Detergents; Body and Hand
Preparations (Excluding Shaving Prepara-
tions); Eye Makeup Preparations, Misc.;
Eyebrow Pencils; Lipsticks; Makeup Bases;
Night Skin Care Preparations; Paste Masks
(Mud Packs); Shampoos (Non-coloring)

Technical/Other Name:
 Oils, Rice Bran

Trade Names:
 AEC Rice Bran Oil (A & E Connock)
 Black Rice Bran Oil (Nonogawa)
 EmCon Rice Bran (Fanning)
 Lipovol RB (Lipo)
 Oryza Oil S-1 (Ichimaru Pharcos)
 Oryza Tocotrienol 30G (Ichimaru Pharcos)
 Rice Bran Oil (Desert Whale)
 Rice Germ Oil (Ikeda)
 Tri-OL RBO (Tri-K)

Trade Name Mixtures:
 Extrapone Honey/Rice Blend GW 2/031306
  (Symrise)
 Extrapone Honey Rice Milk P 2/030830
  (Symrise)
 Extrapone Rice GW 2/031228 (Symrise)
 Ferulan Proactiv (GfN)
 Natunola Macrice 1501 (Natunola)
 Oryza Ceramide-LC (Ikeda)
 Oryza Ceramide - PC (Ichimaru Pharcos)
 Oryza Tocotrienol (Oryza Oil)
 Oryza Tocotrienol - 40 (Oryza Oil)
 Oryza Tocotrienol - 70 (Oryza Oil)
 Oryza Tocotrienol - 90 (Oryza Oil)
 Oryza Tocotrienol - 30G (Oryza Oil)
 Oryza Tocotrienol-L (Oryza Oil)
 Rice Bran Softcream (CEP (Solabia))
 Rice Milk (CEP (Solabia))

ORYZA SATIVA (RICE) BRAN STEROL

Definition: Oryza Sativa (Rice) Bran
Sterol is a mixture of sterols obtained from
the rice bran, *Oryza sativa. See
"Regulatory and Ingredient Use
Information," regarding the labeling names
for botanical ingredients in Volume 1, Introduction, Part A.*

Chemical Class: Biological Products

Function: Skin-Conditioning Agent -
Emollient

Trade Name:
 Oryza Sterol (Ichimaru Pharcos)

ORYZA SATIVA (RICE) BRAN WATER

Definition: Oryza Sativa (Rice) Bran
Water is an aqueous solution of the steam
distillate obtained from the bran of rice,
*Oryza sativa. See "Regulatory and Ingre-
dient Use Information," regarding the
labeling names for botanical ingredients in
Volume 1, Introduction, Part A.*

Chemical Class: Biological Products

Function: Fragrance Ingredient

Technical/Other Name:
 Oryza Sativa (Rice) Water

Trade Names:
 Extrait Originel Riz (Gattefosse s.a.)
 Komenuka Water K (C&F Koei Phyto
  Corp.)
 Komenuka Water K (Koei Perfumery)

ORYZA SATIVA (RICE) BRAN WAX

CAS No.   **EINECS No.**
8016-60-2   232-409-7

JPN Translation:
コメヌカロウ

Definition: Oryza Sativa (Rice) Bran Wax
is a wax obtained from rice bran, *Oryza
sativa. See "Regulatory and Ingredient Use
Information," regarding the labeling names
for botanical ingredients in Volume 1, Introduction, Part A.*

Information Sources: 21CFR172.615,
21CFR172.890, 21CFR178.3860, CIR: [S],
FCC, JCIC, JCLS, TSCA

Chemical Class: Waxes

Function: Skin-Conditioning Agent -
Occlusive

Reported Product Categories: Bath Capsules; Lipsticks; Eyeliners

Technical/Other Names:
 AEC Rice Wax
 Oryza Sative Bran Wax
 Rice Bran Wax
 Rice Wax
 Waxes, Oryza Sativa
 Waxes, Rice Bran

Trade Names:
 Cerewax (Chemyunion)
 ESP Rice Bran Wax (Earth Supplied
  Products)
 Florabeads RBW (Floratech)
 Oryza Soft "COS" (Cosmetochem)
  (Cosmetochem International Ltd.)
 ORYZA Wax (Ichimaru Pharcos)
 Ricebran Wax SP 8000 (Strahl & Pitsch)
 Rice Wax No. 1 (Ikeda)
 Rice Wax No. 1 (Tri-K)

Trade Name Mixture:
 Nutriene Tocotrienols (Eastman Chemical)

ORYZA SATIVA (RICE) EXTRACT

CAS Nos.: 68553-81-1; 90106-37-9

Definition: Oryza Sativa (Rice) Extract is
an extract of the grains of rice, *Oryza
sativa. See "Regulatory and Ingredient Use
Information," regarding the labeling names
for botanical ingredients in Volume 1, Introduction, Part A.*

Information Source: CIR: [S]

Chemical Class: Biological Products

Functions: Hair Conditioning Agent; Skin-Conditioning Agent - Miscellaneous

Reported Product Categories: Moisturizing Preparations; Bath Capsules; Face and Neck Preparations (Excluding Shaving Preparations); Skin Care Preparations, Misc.

Technical/Other Names:
Extract of Oryza Sativa
Extract of Rice
Orysa Sativa Extract
Rice Extract

Trade Name:
ACB Biomilk Rice (Active Concepts)

Trade Name Mixtures:
4-Cereals Milk (Cosmetochem) (Cosmetochem International Ltd.)
Glycolysat of Rice (CEP (Solabia))
Hair Care Blend (Alban Muller)
Hair Care Phytoamine Biocomplex (Alban Muller)
Hair Care Phytoamine Biocomplex SP Lotion (Alban Muller)
Herbasol Extract Oil Soluble Rice (Cosmetochem) (Cosmetochem International Ltd.)
Herbasol Extract Rice (Cosmetochem) (Cosmetochem International Ltd.)
Oryza Sativa (Rice) Extract ies (IES LABO)
Purple Rice Extract-PC (Nikko)
Rice Extract "COS" (Cosmetochem) (Cosmetochem International Ltd.)
Rice HS (Alban Muller)
Rice Phytolait (Alban Muller)

ORYZA SATIVA (RICE) GERM EXTRACT

Definition: Oryza Sativa (Rice) Germ Extract is an extract of the rice germ, *Oryza sativa. See "Regulatory and Ingredient Use Information," regarding the labeling names for botanical ingredients in Volume 1, Introduction, Part A.*

Chemical Class: Biological Products

Function: Skin-Conditioning Agent - Miscellaneous

Technical/Other Names:
Extract of Oryza Sativa (Rice) Germ
Rice Germ Extract

Trade Name Mixtures:
Oryza Gaba Extract - C (Ichimaru Pharcos)
Oryza Gaba Extract - HC5 (Ichimaru Pharcos)

ORYZA SATIVA (RICE) GERM OIL

JPN Translation:
コメ胚芽油

Definition: Oryza Sativa (Rice) Germ Oil is the oil obtained by the expression of germs of rice, *Oryza sativa. See "Regulatory and Ingredient Use Information," regarding the labeling names for botanical ingredients in Volume 1, Introduction, Part A.*

Information Sources: CIR: [S], JCIC, JCLS, JSQI

Chemical Class: Fats and Oils

Function: Skin-Conditioning Agent - Occlusive

Reported Product Category: Body and Hand Preparations (Excluding Shaving Preparations)

Technical/Other Names:
Oil of Rice Germ
Rice Germ Oil

Trade Name:
PRO-15 (Tsuno)

Trade Name Mixture:
Rice Phytolait (Alban Muller)

ORYZA SATIVA (RICE) GERM POWDER

JPN Translation:
コメ胚芽

Definition: Oryza Sativa (Rice) Germ Powder is the powder derived from the rice germ, *Oryza sativa. See "Regulatory and Ingredient Use Information," regarding the labeling names for botanical ingredients in Volume 1, Introduction, Part A.*

Information Sources: CIR: [S], JCIC, JCLS

Chemical Class: Biological Products

Function: Abrasive

Technical/Other Names:
Rice Germ Powder
Rice Powder

Trade Name:
Exfogreen Rice (CEP (Solabia))

Trade Name Mixture:
Eclaicissant G (Greentech)

ORYZA SATIVA (RICE) HULL EXTRACT

Definition: Oryza Sativa (Rice) Hull Extract is an extract of the hulls of *Oryza sativa. See "Regulatory and Ingredient Use Information," regarding the labeling names for botanical ingredients in Volume 1, Introduction, Part A.*

Chemical Class: Biological Products

Function: Skin-Conditioning Agent - Occlusive

Technical/Other Name:
Extract of Oryza Sativa (Rice) Hull

Trade Name Mixture:
Actiphyte of Black Rice Lipo S (Active Organics)

ORYZA SATIVA (RICE) LEES EXTRACT

Definition: Oryza Sativa (Rice) Lees Extract is an extract of the lees of the rice, *Oryza sativa* obtained during the manufacture of wine.

Chemical Class: Biological Products

Function: Skin-Conditioning Agent - Miscellaneous

Technical/Other Name:
Extract of Oryza Sativa (Rice) Lees

Trade Name Mixture:
Excellent Sake-Kojic Extract (Horus)

ORYZA SATIVA (RICE) POWDER

Definition: Oryza Sativa (Rice) Powder is the powder obtained from the ground rice, *Oryza sativa. See "Regulatory and Ingredient Use Information," regarding the labeling names for botanical ingredients in Volume 1, Introduction, Part A.*

Chemical Class: Biological Products

Function: Bulking Agent

Trade Name:
Oryza Powder (Ichimaru Pharcos)

ORYZA SATIVA (RICE) STARCH

CAS No.	EINECS No.
9005-25-8	232-679-6

JPN Translation:
コメデンプン

Definition: Oryza Sativa (Rice) Starch is a starch obtained from rice, *Oryza sativa. See "Regulatory and Ingredient Use Information," regarding the labeling names for botanical ingredients in Volume 1, Introduction, Part A.*

Information Sources: AUS, BEL, BP, BPC, BRA, 21CFR175.105, 21CFR178.3520, CIR: [S], CTFA D, EGY, EP, FI, IND, ITA, JAN, JCLS, JSCI, MAR, MEX, MI-13(8877), NF XIX, SNPF, TSCA, USAN, USSR, YUG

Chemical Class: Carbohydrates

Functions: Absorbent; Bulking Agent

Reported Product Categories: Paste Masks (Mud Packs); Bath Preparations, Misc.; Body and Hand Preparations (Excluding Shaving Preparations); Face Powders; Skin Care Preparations, Misc.; Bath Capsules; Face and Neck Preparations (Excluding Shaving Preparations); Mascara; Eye Makeup Preparations, Misc.; Eye Shadows; Moisturizing Preparations; Night Skin Care Preparations

Technical/Other Names:
Oryza Sativa Starch
Rice Starch
Starch, Oryza Sativa
Starch, Rice

Trade Names:
AEC Rice Starch (A & E Connock)
D.S.A. 7 (Agrana Zucker Und Starke AG)
Oryzapearl (Ichimaru Pharcos)

OSMANTHUS FRAGRANS FLOWER EXTRACT

CAS No. 92347-21-2 **EINECS No.** 296-209-1

Definition: Osmanthus Fragrans Flower Extract is an extract of the flowers of *Osmanthus fragrans*. *See "Regulatory and Ingredient Use Information," regarding the labeling names for botanical ingredients in Volume 1, Introduction, Part A.*

Chemical Class: Biological Products

Functions: Flavoring Agent; Fragrance Ingredient

Technical/Other Name:
Extract of Osmanthus Fragrans Flower

Trade Name:
Osmanthus Absolute (Robertet S.A.)

Trade Name Mixture:
Osmanthus Absolute 1/510702 (Symrise)

OSTRICH OIL

Definition: Ostrich Oil is the oil rendered from the fatty tissue of the ostrich.

Chemical Class: Fats and Oils

Functions: Hair Conditioning Agent; Skin-Conditioning Agent - Occlusive

Technical/Other Names:
Oil of Ostrich
Oils, Ostrich

Trade Name:
Ashanti Exotic Ostrich Oil (Ashanti)

OTOGIRISOU EKISU

JPN Translation:
オトギリソウエキス

Definition: Otogirisou Ekisu is an extract of the capsules, flowers, leaves, and stem heads of *Hypericum perforatum* or *Hypericum erectum*. *See "Regulatory and Ingredient Use Information," regarding use of Japan Trivial names in Volume 1, Introduction, Part A.*

Chemical Class: Biological Products

Function: Skin-Conditioning Agent - Miscellaneous

Technical/Other Name:
Hypericum Perforatum Extract (U.S.)

OUBAKU

JPN Translation:
オウバク

Definition: Oubaku is the bark of *Phellodendron amurense* or other species of the family *Rutaceae*, from which the periderm has been removed. *See "Regulatory and Ingredient Use Information," regarding use of Japan Trivial names in Volume 1, Introduction, Part A.*

Chemical Class: Biological Products

Functions: Abrasive; Skin-Conditioning Agent - Miscellaneous

Technical/Other Name:
Phellodendron Amurense Bark (U.S.)

OUBAKU EKISU

JPN Translation:
オウバクエキス

Definition: Oubaku Ekisu is an extract of Oubaku (q.v.) *See "Regulatory and Ingredient Use Information," regarding use of Japan Trivial names in Volume 1, Introduction, Part A.*

Chemical Class: Biological Products

Function: Skin-Conditioning Agent - Miscellaneous

Technical/Other Name:
Phellodendron Amurense Bark Extract (U.S.)

OUREN EKISU

JPN Translation:
オウレンエキス

Definition: Ouren Ekisu is an extract of the rhizomes of *Copis japonica* or other species of the family *Ranunculaceae*. *See "Regulatory and Ingredient Use Information," regarding use of Japan Trivial names in Volume 1, Introduction, Part A.*

Chemical Class: Biological Products

Function: Skin-Conditioning Agent - Miscellaneous

Technical/Other Name:
Coptis Japonica Root Extract (U.S.)

OURICURY WAX

CAS No. 68917-70-4 **EINECS No.** 272-847-6

Definition: Ouricury Wax is the wax exuded from the leaves of the Ouricury Palm, *Syagrus coronata*.

Information Sources: MAR, TSCA

Chemical Class: Waxes

Functions: Binder; Viscosity Increasing Agent - Nonaqueous

Technical/Other Name:
Waxes, Ouricury

Trade Names:
Ouricury Wax SP 46 (Strahl & Pitsch)
Ouricury Wax SP 79 (Strahl & Pitsch)
Ouricury Wax - STRALPITZ (Strahl & Pitsch)
S&P Ouricury Wax Refined (Strahl & Pitsch)

OVARIAN EXTRACT

CAS No. 84540-05-6 **EINECS No.** 283-094-8

Definition: Ovarian Extract is an extract derived from bovine ovaries.

Chemical Class: Biological Products

Function: Not Reported

Technical/Other Name:
Extract of Ovaries

OXALIC ACID

CAS No. 144-62-7 **EINECS No.** 205-634-3

Empirical Formula:
$C_2H_2O_4$

Definition: Oxalic Acid is the organic dicarboxylic acid that conforms to the formula:

HOOC — COOH

Information Sources: EEC(III/1-3), MI-13 (6980), TSCA

Chemical Class: Carboxylic Acids

Function: pH Adjuster

Technical/Other Name:
Ethanedioic Acid

.

OXIDIZED BEESWAX

CAS No.: 138724-55-7

Definition: Oxidized Beeswax is the product obtained by the oxidation of Beeswax (q.v.).

Chemical Class: Waxes

Function: Viscosity Increasing Agent - Nonaqueous

Technical/Other Name:
Beeswax, Oxidized

Trade Name:
AO2535 (Koster Keunen)

OXIDIZED CORN OIL

Definition: Oxidized Corn Oil is the product obtained by the oxidation of Zea Mays (Corn) Oil (q.v.).

Chemical Class: Fats and Oils

Function: Skin-Conditioning Agent - Miscellaneous

Technical/Other Name:
Corn Oil, Oxidized

Trade Names:
Oleoxyne m. (GERMANDRE S.A.)
Oxygenated Fatty Acids (Carilene)
Oxygenated Glycerol Triesters-D (Vincience)
Oxylipine 100 (Biophysis)
Ulline 46 (MMP)

OXIDIZED GLUTATHIONE

CAS No.	EINECS No.
27025-41-8	248-107-7

Empirical Formula:
$C_{20}H_{32}N_6O_{12}S_2$

Definition: Oxidized Glutathione is a reaction product of Glutathione (q.v.) and oxygen.

Chemical Classes: Amino Acids; Thio Compounds

Function: Skin-Conditioning Agent - Emollient

Technical/Other Name:
Glycine, L-γ-Glutamyl-L-Cysteinyl-, Bimol. (2.fwdarw. 2')-Disulfide

Trade Name:
L-Glutathione Oxidized (Kyowa Hakko Kogyo)

OXIDIZED HAZEL SEED OIL

Definition: Oxidized Hazel Seed Oil is the product obtained by the oxidation of Corylus Avellana (Hazel) Seed Oil (q.v.).

Chemical Class: Fats and Oils

Function: Skin-Conditioning Agent - Miscellaneous

Technical/Other Name:
Oxidized Hazel Nut Oil

Trade Name:
Oxynut Oil (ISPE)

OXIDIZED KERATIN

Definition: Oxidized Keratin is the material derived chemically from Keratin (q.v.) by oxidation with hydrogen peroxide. This reaction converts some of the sulfur atoms in Cysteine (q.v.) and Cystine (q.v.) residues in keratin to the corresponding sulfonic acid grouping (cysteic acid).

Chemical Classes: Protein Derivatives; Sulfonic Acids

Functions: Hair Conditioning Agent; Skin-Conditioning Agent - Miscellaneous

Technical/Other Name:
Keratin, Oxidized

Trade Name:
Mexoryl SU (Chimex)

OXIDIZED MICROCRYSTALLINE WAX

Definition: Oxidized Microcrystalline Wax is the reaction product of Microcrystalline Wax (q.v.) and oxygen.

Chemical Class: Waxes

Function: Viscosity Increasing Agent - Nonaqueous

Technical/Other Name:
Microcrystalline Wax, Oxidized

Trade Names:
Cardis 314 Oxidized Wax (Baker Petrolite)
Cardis 370 Oxidized Wax (Baker Petrolite)

Trade Name Mixtures:
Cardis 36 Oxidized Wax (Baker Petrolite)
Cardis 320 Oxidized Wax (Baker Petrolite)

OXIDIZED POLYETHYLENE

CAS No.: 68441-17-8

Definition: Oxidized Polyethylene is a reaction product of polyethylene and oxygen.

Information Sources: 21CFR172.260, 21CFR175.105, 21CFR175.125, 21CFR176.170, 21CFR176.200, 21CFR176.210, 21CFR177.1200, 21CFR177.1620, 21CFR177.2800, TSCA

Chemical Class: Synthetic Polymers

Function: Viscosity Increasing Agent - Nonaqueous

Reported Product Categories: Bath Oils, Tablets, and Salts; Cleansing Products (Cold Creams, Cleansing Lotions, Liquids and Pads); Nail Polish and Enamels

Technical/Other Name:
Ethene, Homopolymer, Oxidized

Trade Names:
A-C Polyethylene 316 (Honeywell)
A-C Polyethylene 325 (Honeywell)
A-C Polyethylene 330 (Honeywell)
A-C Polyethylene 392 (Honeywell)
A-C Polyethylene 395 (Honeywell)
A-C Polyethylene 629 (Honeywell)
A-C Polyethylene 680 (Honeywell)
A-C Polyethylene 6702 (Honeywell)
A-C Polyethylene 316A (Honeywell)
A-C Polyethylene 629A (Honeywell)
ACumist A-12 (Honeywell)
ACumist A-18 (Honeywell)
AEC Oxidized Polyethylene (A & E Connock)
Aquascrub 50 (Micro Powders)
Aquascrub 80 (Micro Powders)
Aquascrub 100 (Micro Powders)
Cardis 10 Oxidized Wax (Baker Petrolite)
Epolene Oxidized Polyethylene Wax (Eastman Chemical)
Hoechst Wax PED 121 (Clariant)
Hoechst Wax PED 121 (Clariant GmbH, Personal Care)
Hoechst Wax PED 261 (Clariant)
Hoechst Wax PED 261 (Clariant GmbH, Personal Care)
Hoechst Wax PED 521 (Clariant)
Hoechst Wax PED 521 (Clariant GmbH, Personal Care)
Hoechst Wax PED 522 (Clariant)
Hoechst Wax PED 522 (Clariant GmbH, Personal Care)
Micropoly 210 (Micro Powders)
Petrolite C-7500 Polymer (Baker Petrolite)
Petrolite C-9500 Polymer (Baker Petrolite)
Petrolite E-1040 (Baker Petrolite)
Petrolite E-2020 Polymer (Baker Petrolite)

Trade Name Mixtures:
Cardis 36 Oxidized Wax (Baker Petrolite)
Cardis 320 Oxidized Wax (Baker Petrolite)
Petrolite C-8500 Polymer (Baker Petrolite)

PI-C47051-10 (US Cosmetics)
PT-T-47051 (US Cosmetics)
PT-T-47056 (US Cosmetics)
Ross Carnauba Wax Replacement (Ross)

OXIDIZED POLYPROPYLENE

CAS No.: 68649-58-1

Definition: Oxidized Polypropylene is the reaction product of polypropylene with oxygen.

Chemical Class: Synthetic Polymers

Function: Viscosity Increasing Agent - Nonaqueous

Technical/Other Name:
Polypropylene, Oxidized

Trade Name:
Epolene Oxidized Polypropylene Wax (Eastman Chemical)

OXIDO REDUCTASES

Definition: Oxido Reductases is a mixture of enzymes derived from Yeast (q.v.).

Chemical Class: Proteins

Function: Skin-Conditioning Agent - Miscellaneous

Trade Name Mixture:
Preregen (Pentapharm/Centerchem)

OXOBENZOXAZINYL NAPHTHALENE SULFOANILIDE

CAS No.: 10128-55-9

Empirical Formula:
$C_{24}H_{16}N_2O_4S$

Definition: Oxobenzoxazinyl Naphthalene Sulfoanilide is the fluorescent dye that conforms to the formula:

Chemical Classes: Heterocyclic Compounds; Ketones; Sulfonic Acids

Function: Colorant

Technical/Other Name:
2-Naphthalenesulfonamide, N-[2-(4-Oxo-4H-3,1-Benzoxazin-2-yl)Phenyl]-

Trade Name Mixture:
Covazur GHA (LCW)

OXOTHIAZOLIDINECARBOXYLIC ACID

CAS No.: 19771-63-2

Empirical Formula:
$C_4H_5NO_3S$

Definition: Oxothiazolidinecarboxylic Acid is the heterocyclic organic compound that conforms to the formula:

See Reported Ingredient Functions-The Cosmetic Drug Distinction, in Regulatory and Ingredient Use Information, Volume I, Part A.

Information Source: MI-13(7019)

Chemical Class: Heterocyclic Compounds

Functions: Skin Bleaching Agent; Skin-Conditioning Agent - Miscellaneous

Trade Name:
L-2-Oxothiazolidine-4-carboxylic acid (Nippon Rika)

OXYCOCCUS PALUSTRIS SEED OIL

Definition: Oxycoccus Palustris Seed Oil is the oil expressed from the seeds of *Oxycoccus palustris. See "Regulatory and Ingredient Use Information," regarding the labeling names for botanical ingredients in Volume 1, Introduction, Part A. See Reported Ingredient Functions-The Cosmetic Drug Distinction, in Regulatory and Ingredient Use Information, Volume I, Part A.*

Chemical Class: Fats and Oils

Functions: Antioxidant; Skin-Conditioning Agent - Emollient; Skin Protectant

Technical/Other Name:
Oils, Oxycoccus Palustris Seed

Trade Name:
Red Tocol Arctic Cranberry Seed Oil (CO2 extract) (Aromtech)

OXYGEN

CAS No.
7782-44-7

EINECS No.
231-956-9

Empirical Formula:
O_2

Definition: Oxygen is the diatomic gaseous molecule constituting approximately 21% of the earth's atmosphere.

Information Sources: 21CFR172.723, 21CFR176.210, 21CFR177.2490, 21CFR177.2800, 21CFR556.760, MI-13 (7033), USP XXIV

Chemical Class: Inorganics

Function: Not Reported

Technical/Other Name:
Molecular Oxygen

Trade Name Mixtures:
Micropearl PO (C.I.T.)
Nanosource PO (C.I.T.)

OXYMETHYLENE/MELAMINE COPOLYMER

Definition: Oxymethylene/Melamine Copolymer is the complete reaction product of melamine and formaldehyde.

Chemical Class: Synthetic Polymers

Function: Film Former

Trade Name:
Microsilk (Grantec)

Trade Name Mixtures:
Diamond Piece Co-EP Type, Silver (Daiya Kogyo)
Diamond Piece CO Type, Blue (Daiya Kogyo)
Diamond Piece CO Type, DG Gold (Daiya Kogyo)
Diamond Piece CO Type, Green (Daiya Kogyo)
Diamond Piece CO Type, LG Gold (Daiya Kogyo)
Diamond Piece CO Type, Pink (Daiya Kogyo)
Diamond Piece CO Type, Red (Daiya Kogyo)
Diamond Piece CO Type, Silver (Daiya Kogyo)
Diamond Piece CO Type, Violet (Daiya Kogyo)

OXYQUINOLINE

CAS No.
148-24-3

EINECS No.
205-711-1

Empirical Formula:
C_9H_7NO

Definition: Oxyquinoline is the heterocyclic compound that conforms to the formula:

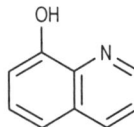

Information Sources: 21CFR310.545, CIR: [I] JACT-11(4)1992, CIR: [SQ], EEC(II-395), EEC(III/1-51), MAR, MI-13(4869), TSCA, USAN

Chemical Classes: Heterocyclic Compounds; Phenols

Functions: Chelating Agent; Cosmetic Biocide

Technical/Other Names:
1-Azanaphthalene-8-ol
8-Hydroxyquinoline
Phenopyridine
Quinolin-8-ol
8-Quinolinol
Quinophenol

OXYQUINOLINE BENZOATE

CAS Nos.	EINECS Nos.
86-75-9	201-697-6
7091-57-8	230-395-7

Empirical Formula:
$C_9H_7NO \cdot C_7H_6O_2$

Definition: Oxyquinoline Benzoate is the salt of oxyquinoline and benzoic acid. It conforms to the formula:

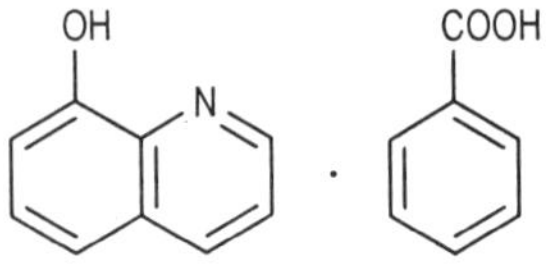

Chemical Classes: Heterocyclic Compounds; Organic Salts; Phenols

Function: Cosmetic Biocide

Technical/Other Names:
Benzoic Acid, Compd. with 8-Quinolinol (1:1)
Dioxyline
8-Hydroxyquinoline Benzoate (Salt)
Oxine Benzoate
8-Quinolinol Benzoate (Salt)
8-Quinolinol Compd. with Benzoic Acid (1:1)
8-Quinolinol Monobenzoate

OXYQUINOLINE SULFATE

CAS No.	EINECS No.
134-31-6	205-137-1

Empirical Formula:
$C_9H_7NO \cdot \frac{1}{2}H_2O_4S$

Definition: Oxyquinoline Sulfate is the salt of oxyquinoline and sulfuric acid. It conforms to the formula:

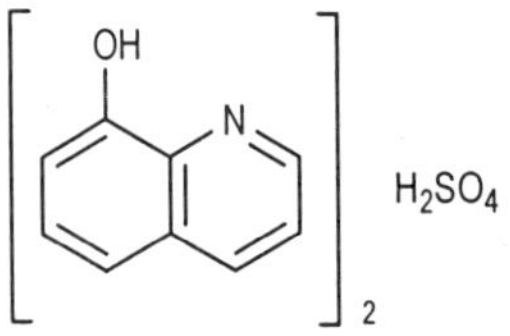

Information Sources: CIR: [I] JACT-11(4)-1992, CIR: [SQ], CTFA D, EEC(II-395), EEC (III/1-51), JSQI, MAR, MI-11(4779), NF XIX, NFJ, TSCA, USAN

Chemical Classes: Heterocyclic Compounds; Organic Salts; Phenols

Functions: Chelating Agent; Cosmetic Biocide

Reported Product Categories: Body and Hand Preparations (Excluding Shaving Preparations); Permanent Waves

Technical/Other Names:
Chinosol
8-Hydroxyquinoline Sulfate
Hydroxy-8-Quinolinium Sulfate
Oxine Sulfate
8-Quinolinol, Sulfate (2:1)(Salt)

Trade Name:
Stabilizer 8HQ (Lowenstein)

OYSTER EXTRACT

CAS No.: 94465-79-9

JPN Translation:
オイスターエキス

Definition: Oyster Extract is an extract of the meat of the oyster.

Information Sources: JCIC, JCLS

Chemical Class: Biological Products

Function: Not Reported

Technical/Other Name:
Extract of Oyster

Trade Name Mixtures:
Extrait d'Huitre MD PG100 (Yves Rocher)
Glycolysat of Oysters (CEP (Solabia))
Glycolysat of (Perles de Caviar) (CEP (Solabia))

OYSTER SHELL EXTRACT

Definition: Oyster Shell Extract is an extract of the shells of oysters. *See "Regulatory and Ingredient Use Information," regarding use of EU Trivial names in Volume 1, Introduction, Part A.*

Chemical Class: Biological Products

Function: Not Reported

Technical/Other Name:
Extract of Oyster Shell

Trade Name Mixtures:
Lipoplastidine Ostrea (Vevy)
Oligoceane (Sederma)

OYSTER SHELL POWDER

JPN Translation:
ボレイ

Definition: Oyster Shell Powder is the powder derived from ground oyster shells.

Information Sources: JCIC, JCLS

Chemical Class: Biological Products

Function: Abrasive

Technical/Other Name:
Powdered Oyster Shell

Trade Names:
Exfosea Mother-of-Pearl (CEP (Solabia))
Exfosea Nacred Oyster Shell (CEP (Solabia))
Shell Splender Dust (pH Solutions)

OZOKERITE

CAS No.: 12198-93-5

JPN Translation:
オゾケライト

Definition: Ozokerite is a hydrocarbon wax derived from mineral or petroleum sources.

Information Sources: CIR: [S] JACT-3(3)-1984, JCIC, JCLS, JSQI

Chemical Class: Waxes

Functions: Binder; Emulsion Stabilizer; Viscosity Increasing Agent - Nonaqueous

Reported Product Categories: Lipsticks; Hair Dyes and Colors (All Types Requiring Caution Statements and Patch Tests); Makeup Preparations (Not eye), Misc.; Foundations; Eye Makeup Preparations, Misc.; Mascara; Moisturizing Preparations; Eyeliners; Makeup Bases; Bath Preparations, Misc.; Blushers (All types); Body and Hand Preparations (Excluding Shaving Preparations); Perfumes; Eyebrow Pencils; Suntan Gels, Creams, and Liquids; Night Skin Care Preparations; Skin Care Preparations, Misc.; Bath Oils, Tablets, and Salts; Cleansing Products (Cold Creams, Cleansing Lotions, Liquids and Pads); Paste Masks (Mud Packs); Eye Makeup Removers; Eye Shadows; Suntan Preparations, Misc.; Tonics, Dressings, and Other Hair Grooming

Aids; Baby Lotions, Oils, Powders and Creams; Bath Soaps and Detergents; Colognes and Toilet Waters; Face Powders; Fragrance Preparations, Misc.; Hair Coloring Preparations, Misc.; Leg and Body Paints; Nail Polish and Enamels; Rouges

Technical/Other Names:
Earth Wax
Fossil Wax
Mineral Wax
Ozocerite
Waxes, Ozokerite

Trade Names:
Botaniwax O-103 (Botanigenics)
Botaniwax O-119 (Botanigenics)
Botaniwax O-128 (Botanigenics)
Hansonwax JH-1174A, 1176,1184,1680 (Hansotech)
Koster Keunen Ozokerite 140/170oF (Koster Keunen)
Ozokerite Wax (Megachem)
Ozokerite Wax (Ross)
Ozokerite Wax (Strahl & Pitsch)
Ozokerite Wax 1899 (Kahl)
Ozokerite Wax 2389 (Kahl)
Ozokerite Wax 6001 (Kahl)
Ozokerite Wax 6095 (Kahl)
Ozokerite Wax Black SP 160 (Strahl & Pitsch)
Ozokerite Wax 170 D (Ross)
Ozokerite Wax 170 D (Strahl & Pitsch)
Ozokerite Wax Green SP 162 (Strahl & Pitsch)
Ozokerite Wax 170 M.F. (Strahl & Pitsch)
Ozokerite Wax 170 M.P. (Ross)
Ozokerite Wax SP 15 (Strahl & Pitsch)
Ozokerite Wax SP 89 (Strahl & Pitsch)
Ozokerite Wax SP 367 (Strahl & Pitsch)
Ozokerite Wax SP 368 (Strahl & Pitsch)
Ozokerite Wax SP 490 (Strahl & Pitsch)
Ozokerite Wax SP 1152 (Strahl & Pitsch)
Ozokerite Wax SP 1155 (Strahl & Pitsch)
Ozokerite Wax White SP 56 (Strahl & Pitsch)
Ozokerite Wax White SP 102 (Strahl & Pitsch)
Ozokerite Wax White SP 109 (Strahl & Pitsch)
Ozokerite Wax White SP 136 (Strahl & Pitsch)
Ozokerite Wax White SP 273 (Strahl & Pitsch)
Ozokerite Wax White SP 681 (Strahl & Pitsch)
Ozokerite Wax White SP 682 (Strahl & Pitsch)
Ozokerite Wax White SP-996 (Strahl & Pitsch)
Ozokerite Wax White SP 1016 (Strahl & Pitsch)
Ozokerite Wax White SP 1017 (Strahl & Pitsch)
Ozokerite Wax White SP 1018 (Strahl & Pitsch)
Ozokerite Wax White SP 1020 (Strahl & Pitsch)
Ozokerite Wax White SP 1021 (Strahl & Pitsch)
Ozokerite Wax White SP 1022 (Strahl & Pitsch)
Ozokerite Wax White SP 1023 (Strahl & Pitsch)
Ozokerite Wax White SP 1024 (Strahl & Pitsch)
Ozokerite Wax White SP 1025 (Strahl & Pitsch)
Ozokerite Wax White SP 1026 (Strahl & Pitsch)
Ozokerite Wax White SP 1027 (Strahl & Pitsch)
Ozokerite Wax White SP 1028 (Strahl & Pitsch)
Ozokerite Wax White SP-1140 (Strahl & Pitsch)
Ozokerite Wax White SP 1190 (Strahl & Pitsch)
Ozokerite Wax White SP 1554 (Strahl & Pitsch)
Ozokerite Wax White SP 1020C (Strahl & Pitsch)
Ozokerite Wax Yellow SP-82 (Strahl & Pitsch)
Ozokerite Wax Yellow SP 603 (Strahl & Pitsch)
Ozokerite Wax Yellow SP-869 (Strahl & Pitsch)
Ozokerite Wax Yellow SP 982 (Strahl & Pitsch)
Ozokerite Wax Yellow SP 984 (Strahl & Pitsch)
Ozokerite Wax Yellow SP 986 (Strahl & Pitsch)
Ross Ozokerite Wax 2690 (Ross)
S&P Ozokerite Wax White (Strahl & Pitsch)
S&P Ozokerite Wax Yellow (Strahl & Pitsch)
White Ozokerite Wax (Strahl & Pitsch)
White Ozokerite Wax 56-1114 (Ross)
White Ozokerite Wax 871 (Ross)
White Ozokerite Wax 1477 (Ross)
White Ozokerite Wax 1543 (Ross)
White Ozokerite Wax 1544 (Ross)
White Ozokerite Wax 1545 (Ross)
White Ozokerite Wax 1556 (Ross)
White Ozokerite Wax 1823 (Ross)
White Ozokerite Wax 1981 (Ross)
White Ozokerite Wax 64W (Ross)
White Ozokerite Wax 71W (Ross)
White Ozokerite Wax 77W (Ross)
Yellow Ozokerite Wax (Strahl & Pitsch)
Yellow Ozokerite Wax 71Y (Ross)
Yellow Ozokerite Wax 77Y (Ross)
Yellow Ozokerite Wax 861 (Ross)
Yellow Ozokerite Wax 863 (Ross)
Yellow Ozokerite Wax 869 (Ross)
Yellow Ozokerite Wax 2318 (Ross)
Yellow Ozokerite Wax S Special (Strahl & Pitsch)

Trade Name Mixtures:
Argobase L2 (Croda Chemicals)
Base 323 MS (LCW)
Base 323 MTC (LCW)
Base RAL W 323 T (LCW)
Base Rouge A Levres 323 TAL (LCW)
Base RW 101 (LCW)
Covacrem LP (LCW)
Covacrem MK (LCW)
Covalip 22 (LCW)
Covalip 25 (LCW)
Covalip 94 (LCW)
Covalip 99 (LCW)
Covalip LL 48 (LCW)
Covapencil 07 (LCW)
Covawax 501 (LCW)
Macamat Wax (LCW)
Mascawax 012 (LCW)
Micromac Wax (LCW)
Protegin V (Degussa Care Specialties)
Protegin W (Degussa Care Specialties)
Protegin WX (Degussa Care Specialties)
Protegin XV (Degussa Care Specialties)
Rosswax 2660 (Ross)

OZONIZED CASTOR OIL

Definition: Ozonized Castor Oil is the end product of the controlled treatment of Ricinus Communis (Castor) Oil (q.v.) with ozone.

Chemical Class: Esters

Function: Skin-Conditioning Agent - Miscellaneous

Trade Name Mixture:
Germinol (Dr. Gerhard Steidl)

OZONIZED JOJOBA OIL

Definition: Ozonized Jojoba Oil is the end product of the controlled treatment of jojoba oil with ozone.

Chemical Class: Esters

Function: Skin-Conditioning Agent - Miscellaneous

Technical/Other Names:
Jojoba Oil, Ozonized
Oils, Jojoba, Ozonized

OZONIZED OLIVE OIL

Definition: Ozonized Olive Oil is the end product of the controlled treatment of Olea Europaea (Olive) Oil (q.v.) with ozone.

Chemical Class: Esters

Function: Skin-Conditioning Agent - Miscellaneous

Trade Name:
OXAKTIV (Pharmoxid)

Trade Name Mixtures:
Germinol (Dr. Gerhard Steidl)
Harmonic ASNP (Dr. Gerhard Steidl)
Harmonic ASP (Dr. Gerhard Steidl)
Jonat AS (Dr. Gerhard Steidl)

OZONIZED SUNFLOWER SEED OIL

Definition: Ozonized Sunflower Seed Oil is the end-product of the controlled treatment of Helianthus Annuus (Sunflower) Seed Oil (q.v.) with ozone.

Chemical Class: Esters

Function: Skin-Conditioning Agent - Miscellaneous

Trade Name:
Peroxoil-G (Ozonoil)

- P -

PABA

CAS No.	EINECS No.
150-13-0	205-753-0

JPN Translation:
 PABA

Empirical Formula:
 $C_7H_7NO_2$

Definition: PABA is the aromatic acid that conforms to the formula:

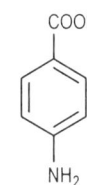

In the United States, PABA may be used as an active ingredient in OTC drug products. When used as an active drug ingredient the established name for PABA is *Aminobenzoic Acid. See "Regulatory and Ingredient Use Information," regarding the labeling names for U.S. OTC Drug Ingredients in Volume 1, Introduction, Part A.*

Information Sources: AUS, BPC, BRA, CZE, DDR, EEC(VII/1-1), JCIC, JCLS, JSQI, MAR, MHLW-331/4, MI-13(422), NFJ, OTC-I-SU, ROM, TSCA, USAN, USD, USP XXIV

Chemical Classes: Amines; Amino Acids; PABA Derivatives

Functions: Sunscreen Agent; Ultraviolet Light Absorber

Reported Product Categories: Hair Conditioners; Hair Preparations (Non-coloring), Misc.; Tonics, Dressings, and Other Hair Grooming Aids; Bath Capsules; Bath Preparations, Misc.; Body and Hand Preparations (Excluding Shaving Preparations); Face and Neck Preparations (Excluding Shaving Preparations); Hair Sprays (Aerosol Fixatives); Moisturizing Preparations

Technical/Other Names:
 Aminobenzoic Acid
 4-Aminobenzoic Acid
 p-Aminobenzoic Acid
 Aniline-4-Carboxylic Acid
 Benzoic Acid, 4-Amino-
 p-Carboxyaniline
 P-Carboxyphenylamine

Trade Name Mixtures:
 Cutavit Richter (CLR)
 Hair Complex 20/70 n (CLR)
 Soluvit Richter (CLR)
 Vitamin Extract AEFH' Water Soluble (Crodarom)
 Vitaminextract VC, Water Soluble (Crodarom)
 810 Yellow 27 (Sterling)

PACHYRRHIZUS EROSUS ROOT EXTRACT

Definition: Pachyrrhizus Erosus Root Extract is an extract of the roots of *Pachyrrhizus erosus. See "Regulatory and Ingredient Use Information," regarding the labeling names for botanical ingredients in Volume 1, Introduction, Part A.*

Chemical Class: Biological Products

Function: Skin-Conditioning Agent - Miscellaneous

Technical/Other Name:
 Extract of Pachyrrhizus Erosus Root

Trade Name:
 Yam Bean (Haldin Pacific)

PADINA PAVONICA EXTRACT

Definition: Padina Pavonica Extract is an extract of the thallus of *Padina pavonica. See "Regulatory and Ingredient Use Information," regarding the labeling names for botanical ingredients in Volume 1, Introduction, Part A.*

Chemical Class: Biological Products

Function: Skin-Conditioning Agent - Miscellaneous

Technical/Other Name:
 Extract of Padina Pavonica

Trade Name Mixtures:
 HPS3 (Alban Muller)
 Protectami (Alban Muller)

PAEONIA ALBIFLORA FLOWER EXTRACT

Definition: Paeonia Albiflora Flower Extract is an extract of the flowers of the peony, *Paeonia albiflora. See "Regulatory and Ingredient Use Information," regarding the labeling names for botanical ingredients in Volume 1, Introduction, Part A.*

Chemical Class: Biological Products

Function: Skin-Conditioning Agent - Miscellaneous

Technical/Other Names:
 Extract of Paeonia Albiflora
 Extract of Peony
 Extract of Peony Flowers
 Peony Extract
 Peony (Paeonia Albiflora) Extract

Trade Name Mixtures:
 Actiphyte of Peony Flower BG50 (Active Organics)
 Actiphyte of Peony Flower GL50 (Active Organics)
 Actiphyte of Peony Flower Lipo S (Active Organics)
 Actiphyte of Peony Flower PG50 (Active Organics)
 Chinese Peony Extract (Alban Muller)
 Hydroplastidine Paeonia (Vevy)
 Paeonia Albiflora Flower Extract ies (IES LABO)
 VT-211 Extract of Peony Flower (Vege-Tech)

PAEONIA ALBIFLORA ROOT EXTRACT

JPN Translation:
 シャクヤク根エキス

Definition: Paeonia Albiflora Root Extract is an extract of the roots of the peony, *Paeonia albiflora. See "Regulatory and Ingredient Use Information," regarding the labeling names for botanical ingredients in Volume 1, Introduction, Part A.*

Information Sources: JCIC, JCLS, JSQI

Chemical Class: Biological Products

Function: Not Reported

Reported Product Categories: Bath Capsules; Bath Oils, Tablets, and Salts; Skin Care Preparations, Misc.

Technical/Other Names:
 Extract of Paeonia Albiflora Root
 Extract of Peony Root
 Paeonia Extract
 Peony (Paeonia Albiflora) Root Extract
 Peony Root Extract

Trade Name:
 Peony Root Extract Powder (Maruzen Pharmaceuticals Co., Ltd.)

Trade Name Mixtures:
 Actiphyte of Peony Root BG50 (Active Organics)
 Actiphyte of Peony Root GL50 (Active Organics)
 Actiphyte of Peony Root Lipo S (Active Organics)
 Actiphyte of Peony Root PG50 (Active Organics)
 Bathgranue Shakuyaku (Ichimaru Pharcos)
 Boumdan (Bioland)
 Herbasol Distillate Paeony (Root) (Cosmetochem) (Cosmetochem International Ltd.)
 Peony Root Extract (Maruzen Pharmaceuticals Co., Ltd.)
 Peony Root Extract BG (Maruzen Pharmaceuticals Co., Ltd.)
 Peony Root Extract BG-JC (Maruzen Pharmaceuticals Co., Ltd.)
 Peony Root Extract-J (Maruzen Pharmaceuticals Co., Ltd.)
 Peony Root Extract LA (Maruzen Pharmaceuticals Co., Ltd.)
 Peony Root Extract Powder-S (Maruzen Pharmaceuticals Co., Ltd.)

Peony Root Extract W (Maruzen
Pharmaceuticals Co., Ltd.)
Phytostan (EUROCOSTECH)
Pivoine R. (Silab)
Shakyaku Liquid (Ichimaru Pharcos)
Vegetol Peony ME 170 Hydro (Gattefosse
s.a.)
VT-195 Extract of Peony Root (Vege-Tech)

PAEONIA ALBIFLORA ROOT POWDER

JPN Translation:
シャクヤク根

Definition: Paeonia Albiflora Root Powder
is a powder of the dried roots of *Paeonia
albiflora.* See "Regulatory and Ingredient
Use Information," regarding the labeling
names for botanical ingredients in Volume
1, Introduction, Part A.

Information Source: JCLS

Chemical Class: Biological Products

Function: Skin-Conditioning Agent - Miscellaneous

Technical/Other Names:
Peony (Paeonia Albiflora) Powder
Peony Powder
Powdered Peony (Paeonia Albiflora)

Trade Names:
Peony Root Powder (Matsuura Yakugyo)
Shakyaku Powder (Ichimaru Pharcos)

PAEONIA LACTIFLORA EXTRACT

Definition: Paeonia Lactiflora Extract is an
extract of the bark and sap of *Paeonia
lactiflora.* See "Regulatory and Ingredient
Use Information," regarding the labeling
names for botanical ingredients in Volume
1, Introduction, Part A.

Chemical Class: Biological Products

Function: Not Reported

Technical/Other Name:
Extract of Paeonia Lactiflora

Trade Name Mixtures:
Campo Shako Nakai (Campo)
Campo Shao Yao Extract (Campo)

PAEONIA LACTIFLORA ROOT

Definition: Paeonia Lactiflora Root is the
root of *Paeonia lactiflora.* See "Regulatory
and Ingredient Use Information," regarding
the labeling names for botanical ingredients
in Volume 1, Introduction, Part A.

Chemical Class: Biological Products

Function: Not Reported

Technical/Other Name:
Shakyaku (JPN)

PAEONIA LACTIFLORA ROOT EXTRACT

Definition: Paeonia Lactiflora Root Extract
is an extract of the root of *Paeonia
lactiflora.* See "Regulatory and Ingredient
Use Information," regarding the labeling
names for botanical ingredients in Volume
1, Introduction, Part A.

Chemical Class: Biological Products

Function: Not Reported

Technical/Other Name:
Shakyaku Ekisu (JPN)

PAEONIA OFFICINALIS FLOWER EXTRACT

Definition: Paeonia Officinalis Flower
Extract is an extract of the flowers of
Paeonia officinalis. See "Regulatory and
Ingredient Use Information," regarding the
labeling names for botanical ingredients in
Volume 1, Introduction, Part A.

Chemical Class: Biological Products

Function: Skin-Conditioning Agent - Miscellaneous

Technical/Other Name:
Extract of Paeonia Officinalis Flower

Trade Name Mixture:
Phytofleur Peony (Crodarom)

PAEONIA OFFICINALIS ROOT EXTRACT

Definition: Paeonia Officinalis Root
Extract is an extract of the roots of *Paeonia
officinalis.* See "Regulatory and Ingredient
Use Information," regarding the labeling
names for botanical ingredients in Volume
1, Introduction, Part A.

Chemical Class: Biological Products

Function: Skin-Conditioning Agent - Miscellaneous

Technical/Other Name:
Extract of Paeonia Officinalis Root

Trade Name Mixtures:
Herbasol Extract Paeony (Root)
(Cosmetochem International Ltd.)
Peony Root HPG Titrated (Alban Muller)
Peony Root HS (Alban Muller)
Peony Root LS (Alban Muller)

PAEONIA SUFFRUTICOSA ROOT EXTRACT

JPN Translation:
ボタンエキス

Definition: Paeonia Suffruticosa Root
Extract is an extract of the roots of the
peony, *Paeonia suffruticosa.* See
"Regulatory and Ingredient Use
Information," regarding the labeling names
for botanical ingredients in Volume 1, Introduction, Part A.

Chemical Class: Biological Products

Function: Not Reported

Technical/Other Names:
Extract of Paeonia Suffruticosa
Extract of Peony
Peony Extract

Trade Name Mixtures:
Alprotector (Ichimaru Pharcos)
Biocleanact Paeonia Extract (MSK-NE 150)
(Micro Science)
Botanpi Liquid B (Ichimaru Pharcos)
Botanpi Liquid E (Ichimaru Pharcos)
Extract of Paeonia Suffruticosa (Arda
Natura)
Extract of Paeonia Suffruticosa I (Arda
Natura)
Hair Growth Complex (Ichimaru Pharcos)
Harmowhite (Ichimaru Pharcos)
Moutan Bark Extract (Maruzen
Pharmaceuticals Co., Ltd.)
Paeonia Extract (Maruzen Pharmaceuticals
Co., Ltd.)
Pharcolex PSP (Ichimaru Pharcos)
Phytoblend TIPS (Ichimaru Pharcos)
Phyto Desensitizer (Ichimaru Pharcos)

PALAU WHITE CLAY EXTRACT

Definition: Palau White Clay Extract is an
extract of the clay mined in Ngeruktabel
Island in Palau.

Chemical Class: Inorganics

Function: Skin-Conditioning Agent -
Humectant

Trade Name:
Palau White Clay (Lotus 21,Co.,Ltd)

PALM ACID

JPN Translation:
パーム脂肪酸

Definition: Palm Acid is a mixture of fatty
acids derived from Elaeis Guineensis (Palm)
Oil (q.v.).

Information Source: JCLS

Chemical Class: Fatty Acids

Functions: Opacifying Agent; Surfactant - Cleansing Agent; Surfactant - Emulsifying Agent

Technical/Other Name:
Elaeis Guineensis (Palm) Acid

Trade Names:
Palm Stearin Fatty Acid (Dial)
PRIFAC 7971 (Uniqema Europe)

PALM ALCOHOL

Definition: Palm Alcohol is a mixture of fatty alcohols derived from Elaeis Guineensis (Palm) Oil (q.v.).

Chemical Class: Fatty Alcohols

Functions: Emulsion Stabilizer; Skin-Conditioning Agent - Emollient; Viscosity Increasing Agent - Nonaqueous

Technical/Other Names:
Alcohols, Palm
Elaeis Guineensis (Palm) Alcohol

PALMAMIDE DEA

Definition: Palmamide DEA is a mixture of ethanolamides of the fatty acids derived from Elaeis Guineensis (Palm) Oil (q.v.). It conforms generally to the formula:

$$\underset{\underset{RC}{\shortparallel}{O}}{}-N(CH_2CH_2OH)_2$$

where RCO- represents the fatty acids derived from palm oil.

Information Sources: 21CFR175.105, 21CFR176.210, 21CFR177.2800, EEC(III/1-60)

Chemical Class: Alkanolamides

Functions: Surfactant - Foam Booster; Viscosity Increasing Agent - Aqueous

Technical/Other Names:
N,N-Bis(2-Hydroxyethyl)Palm Oil Acid Amide
Diethanolamine Palm Oil Acid Amide
Palm Oil Acid Amide, N,N-Bis(2-Hydroxyethyl)-
Palm Oil Acid Diethanolamide

PALMAMIDE MEA

Definition: Palmamide MEA is a mixture of ethanolamides of the fatty acids derived from Elaeis Guineensis (Palm) Oil (q.v.). It conforms generally to the formula:

$$RC{\overset{O}{\shortparallel}}-NHCH_2CH_2OH$$

where RCO- represents the fatty acids derived from palm oil.

Chemical Class: Alkanolamides

Functions: Surfactant - Foam Booster; Viscosity Increasing Agent - Aqueous

Technical/Other Names:
N-(2-Hydroxyethyl) Palm Oil Acid Amide
Monoethanolamine Palm Oil Acid Amide
Palm Oil Acid Amide, N-(2-Hydroxyethyl)-
Palm Oil Acid Monoethanolamide

PALMAMIDE MIPA

Definition: Palmamide MIPA is a mixture of isopropanolamides of the fatty acids derived from Elaeis Guineensis (Palm) Oil (q.v.). It conforms generally to the formula:

$$RC{\overset{O}{\shortparallel}}-NHCH_2\underset{\underset{CH_3}{|}}{C}HOH$$

where RCO- represents the fatty acids derived from palm oil.

Chemical Class: Alkanolamides

Functions: Surfactant - Foam Booster; Viscosity Increasing Agent - Aqueous

Technical/Other Names:
N-(2-Hydroxypropyl)Palm Oil Acid Amide
Monoisopropanolamine Palm Oil Acid Amide
Palm Oil Acid Amide, N-(2-Hydroxypropyl)-
Palm Oil Acid Monoisopropanolamide

PALMAMIDOPROPYL BETAINE

Definition: Palmamidopropyl Betaine is the zwitterion (inner salt) that conforms to the formula:

$$RC{\overset{O}{\shortparallel}}-NH(CH_2)_3-\underset{\underset{CH_3}{|}}{\overset{\overset{CH_3}{|}}{N^+}}-CH_2COO^-$$

where RCO- represents the fatty acids derived from palm oil.

Chemical Class: Betaines

Functions: Antistatic Agent; Hair Conditioning Agent; Skin-Conditioning Agent - Miscellaneous; Surfactant - Cleansing Agent; Surfactant - Foam Booster; Viscosity Increasing Agent - Aqueous

PALMARIA PALMATA EXTRACT

Definition: Palmaria Palmata Extract is an extract of the alga, *Palmaria palmata. See "Regulatory and Ingredient Use Information," regarding the labeling names for botanical ingredients in Volume 1, Introduction, Part A.*

Chemical Class: Biological Products

Function: Skin-Conditioning Agent - Miscellaneous

Reported Product Categories: Bath Preparations, Misc.; Body and Hand Preparations (Excluding Shaving Preparations)

Trade Name Mixtures:
Actiphyte Dulse (Active Organics)
Actiphyte Dulse AQ (Active Organics)
Actiphyte Dulse BG50P (Active Organics)
"Dulse Extract" (Bio-Botanica)
Palmaria Palmata Extract ME 008 (Ennagram) (Ennagram)
Palmaria Palmata HS (Alban Muller)
Rhodofiltrat Palmaria HG (Codif)
Sea Extract Dulse (E.U.K)
Vege Plex VP-1297.050WB Sea Plex in Butylene Glycol (Vege-Tech)
Vege-Tech VT-352 (Vege-Tech)

PALMETH-2 PHOSPHATE

Definition: Palmeth-2 Phosphate is a complex mixture of esters of phosphoric acid and ethoxylated Palm Kernel Alcohol (q.v.) with an average ethoxylation value of 2.

Chemical Classes: Alkoxylated Alcohols; Phosphorus Compounds

Function: Surfactant - Cleansing Agent

Technical/Other Names:
PEG-2 Palm Kernel Ether Phosphate
Polyethylene Glycol (2) Palm Kernel Ether Phosphate
Polyoxyethylene (2) Palm Kernel Ether Phosphate

Trade Name:
Emulpharma 35 (Res Pharma)

PALM GLYCERIDE

CAS No.	EINECS No.
97553-32-7	307-189-1

Definition: Palm Glyceride is the monoglyceride derived from Elaeis Guineensis (Palm) Oil (q.v.).

Chemical Class: Glyceryl Esters and Derivatives

Function: Surfactant - Emulsifying Agent

The inclusion of any compound in the *Dictionary and Handbook* does not indicate that use of that substance as a cosmetic ingredient complies with the laws and regulations governing such use in the United States or any other country.

Technical/Other Name:
Glycerides, Palm Oil Mono-

PALM GLYCERIDES

CAS Nos.	EINECS No.
91744-64-8	294-628-4
129521-59-1	

Definition: Palm Glycerides is a mixture of mono, di and triglycerides derived from Elaeis Guineensis (Palm) Oil (q.v.).

Chemical Class: Glyceryl Esters and Derivatives

Function: Skin-Conditioning Agent - Emollient

Reported Product Category: Body and Hand Preparations (Excluding Shaving Preparations)

Technical/Other Name:
Glycerides, Palm Oil Mono-, Di- and Tri-

Trade Names:
Cremao CE-34 (Aarhus)
Palmotex EE/CT (Aarhus)

PALMITAMIDE

CAS No.	EINECS No.
629-54-9	211-095-5

JPN Translation:
パルミチン酸アミド

Empirical Formula:
$C_{16}H_{33}NO$

Definition: Palmitamide is the organic compound that conforms to the formula:

$$CH_3(CH_2)_{14}C \overset{O}{\underset{\|}{-}} NH_2$$

Information Source: JCLS

Chemical Class: Amides

Function: Skin-Conditioning Agent - Miscellaneous

PALMITAMIDE DEA

CAS No.	EINECS No.
7545-24-6	231-427-2

Empirical Formula:
$C_{20}H_{41}NO_3$

Definition: Palmitamide DEA is a mixture of ethanolamides of palmitic acid. It conforms generally to the formula:

$$CH_3(CH_2)_{14}C \overset{O}{\underset{\|}{-}} N(CH_2CH_2OH)_2$$

Information Sources: 21CFR175.105, 21CFR176.180, 21CFR176.210, 21CFR177.2260, 21CFR177.2800, EEC(III/1-60), TSCA

Chemical Class: Alkanolamides

Functions: Surfactant - Foam Booster; Viscosity Increasing Agent - Aqueous

Technical/Other Names:
N,N-Bis(2-Hydroxyethyl)Hexadecanamide
N,N-Bis(2-Hydroxyethyl)Palmitamide
Diethanolamine Palmitic Acid Amide
N,N-Diethylolpalmitamide
Hexadecanamide, N,N-Bis(2-Hydroxyethyl)-
Palmitic Diethanolamide
Palmityldiethanolamide

PALMITAMIDE MEA

CAS No.	EINECS No.
544-31-0	208-867-9

JPN Translation:
パルミタミド MEA

Empirical Formula:
$C_{18}H_{37}NO_2$

Definition: Palmitamide MEA is a mixture of ethanolamides of palmitic acid. It conforms generally to the formula:

$$CH_3(CH_2)_{14}C \overset{O}{\underset{\|}{-}} NHCH_2CH_2OH$$

Information Sources: INN, JCIC, JCLS, JSQI, MI-13(7063), TSCA

Chemical Class: Alkanolamides

Functions: Surfactant - Foam Booster; Viscosity Increasing Agent - Aqueous

Technical/Other Names:
Hexadecanamide, N-(2-Hydroxyethyl)-
N-(2-Hydroxyethyl)Hexadecanamide
N-(2-Hydroxyethyl)Palmitamide
Monoethanolamine Palmitic Acid Amide
2-Palmitamidoethanol
N-Palmitoylethanolamine

Trade Name:
Nikkol PMEA (Nikko)

PALMITAMIDOBUTYL GUANIDINE ACETATE

Definition: Palmitamidobutyl Guanidine Acetate is the organic compound that conforms to the formula:

$$CH_3(CH_2)_{14}C \overset{O}{\underset{\|}{-}} NH(CH_2)_4N=\overset{}{\underset{NH_2}{C}}NH_2 \cdot CH_3COOH$$

Chemical Classes: Amides; Amines

Functions: Antistatic Agent; Hair Conditioning Agent; Humectant; Surfactant - Emulsifying Agent

Trade Name:
C16A4G (Lion Corporation)

PALMITAMIDOHEXADECANEDIOL

CAS No.: 129426-19-3

Empirical Formula:
$C_{32}H_{65}NO_3$

Definition: Palmitamidohexadecanediol is the organic compound that conforms to the formula:

$$CH_3(CH_2)_{14}C \overset{O}{\underset{\|}{-}} NHCHCH(CH_2)_{12}CH_3$$

(with CH_2OH and OH substituents)

Chemical Classes: Alcohols; Amides

Function: Skin-Conditioning Agent - Miscellaneous

Technical/Other Names:
Hexadecanamide, N-[2-Hydroxy-1-(Hydroxymethyl)Pentadecyl]-
N-[2-Hydroxy-1-(Hydroxymethyl)-Pentadecyl]Hexadecanamide

PALMITAMIDOPROPYLAMINE OXIDE

CAS No.: 67806-12-6

Empirical Formula:
$C_{21}H_{44}N_2O_2$

Definition: Palmitamidopropylamine Oxide is the tertiary amine oxide that conforms to the formula:

$$CH_3(CH_2)_{14}C \overset{O}{\underset{\|}{-}} NH(CH_2)_3N \longrightarrow O$$

(with two CH_3 groups on N)

Information Source: TSCA

Chemical Class: Amine Oxides

Functions: Hair Conditioning Agent; Surfactant - Cleansing Agent; Surfactant - Foam Booster; Surfactant - Hydrotrope

Technical/Other Names:
Amides, Palmitic, N-[3-(Dimethylamino)-Propyl], N-Oxide
N-[3-(Dimethylamino)Propyl]Hexadecanamide-N-Oxide
Hexadecanamide, N-[3-(Dimethylamino)-Propyl]-N-Oxide

Hexadecanamide, N-(3-(Dimethyloxido-amino)Propyl)-

Diethylaminopropyl Palmitamide
Hexadecanamide, N-[3-(Diethylamino)-Propyl]-

PALMITAMIDOPROPYL BETAINE

CAS No. 32954-43-1   **EINECS No.** 251-306-8

Empirical Formula:
$C_{23}H_{46}N_2O_3$

Definition: Palmitamidopropyl Betaine is the zwitterion (inner salt) that conforms to the formula:

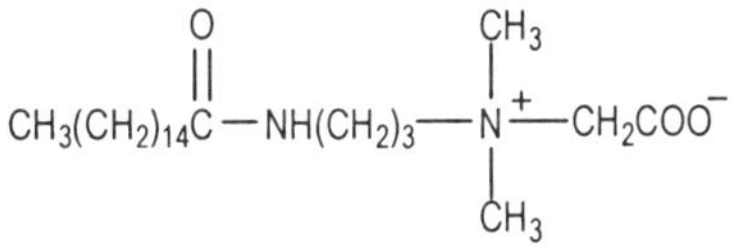

Information Sources: BAN, INN, TSCA

Chemical Class: Betaines

Functions: Antistatic Agent; Hair Conditioning Agent; Skin-Conditioning Agent - Miscellaneous; Surfactant - Cleansing Agent; Surfactant - Foam Booster; Viscosity Increasing Agent - Aqueous

Technical/Other Names:
Ammonium, (Carboxymethyl)Dimethyl(3-Palmitamidopropyl)-, Hydroxide, Inner Salt
N-(Carboxymethyl)-N,N-Dimethyl-3-[(1-Oxohexadecyl)Amino]-1-Propanaminium Hydroxide, Inner Salt
1-Propanaminium, N-(Carboxymethyl)-N,N-Dimethyl-3-[(1-Oxohexadecyl)Amino]-, Hydroxide, Inner Salt

Trade Name:
Schercotaine PAB (Scher)

PALMITAMIDOPROPYL DIETHYLAMINE

CAS No.: 67806-13-7

Empirical Formula:
$C_{23}H_{48}N_2O$

Definition: Palmitamidopropyl Diethylamine is the amidoamine that conforms to the formula:

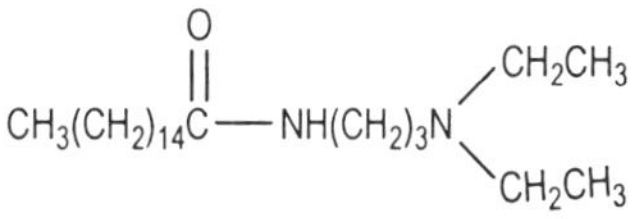

Chemical Class: Amines

Function: Antistatic Agent

Technical/Other Names:
N-[3-(Diethylamino)Propyl]Hexadecanamide

PALMITAMIDOPROPYL DIMETHYLAMINE

CAS No. 39669-97-1   **EINECS No.** 254-585-4

Empirical Formula:
$C_{21}H_{44}N_2O$

Definition: Palmitamidopropyl Dimethylamine is the amidoamine that conforms to the formula:

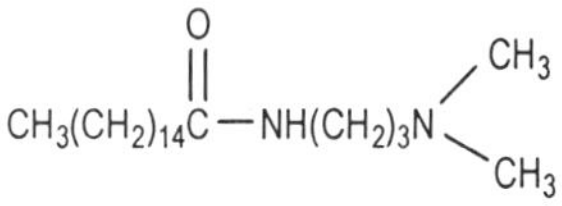

Chemical Class: Amines

Function: Antistatic Agent

Technical/Other Names:
N-[3-(Dimethylamino)Propyl]Hexadecanamide
Dimethylaminopropyl Palmitamide
Hexadecanamide, N-[3-(Dimethylamino)-Propyl]-

Trade Name:
Schercodine P (Scher)

PALMITAMIDOPROPYL DIMETHYLAMINE LACTATE

Empirical Formula:
$C_{21}H_{44}N_2O \cdot C_3H_6O_3$

Definition: Palmitamidopropyl Dimethylamine Lactate is the amidoamine salt that conforms generally to the formula:

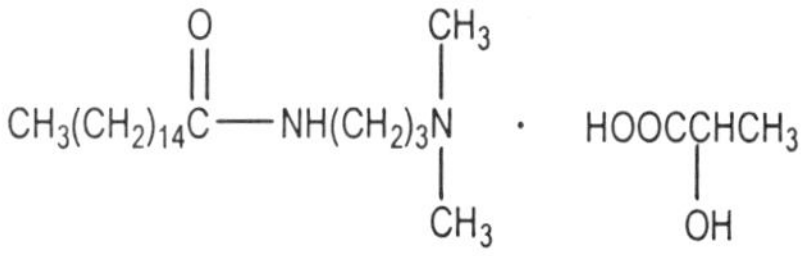

Chemical Class: Amines

Function: Antistatic Agent

Technical/Other Name:
Dimethylaminopropylpalmitamide Lactate

PALMITAMIDOPROPYL DIMETHYLAMINE PROPIONATE

Empirical Formula:
$C_{21}H_{44}N_2O \cdot C_3H_6O_2$

Definition: Palmitamidopropyl Dimethylamine Propionate is the amidoamine salt that conforms to the formula:

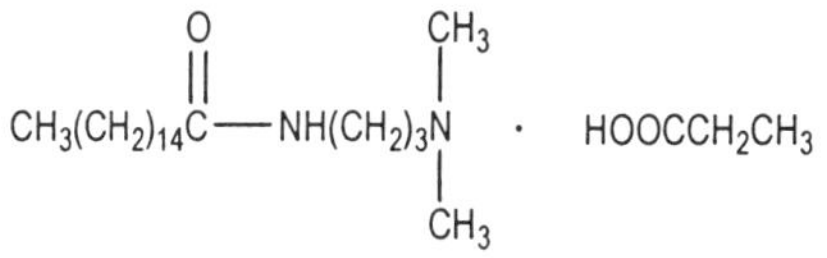

Chemical Class: Amines

Function: Antistatic Agent

Technical/Other Name:
Dimethylaminopropylpalmitamide Propionate

PALMITAMIDOPROPYLTRIMONIUM CHLORIDE

CAS No. 51277-96-4   **EINECS No.** 257-104-6

Empirical Formula:
$C_{22}H_{47}N_2O \cdot Cl$

Definition: Palmitamidopropyltrimonium Chloride is the quaternary ammonium salt that conforms to the formula:

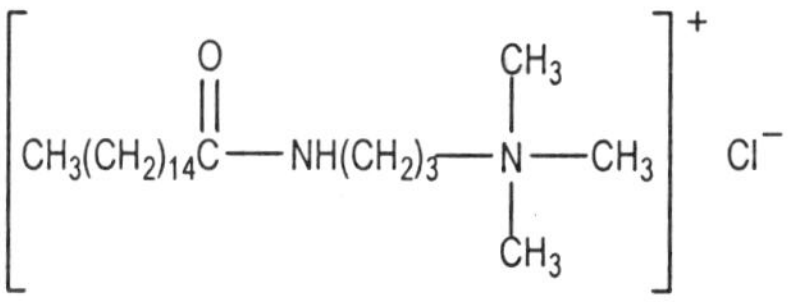

Chemical Class: Quaternary Ammonium Compounds

Functions: Antistatic Agent; Hair Conditioning Agent

Technical/Other Names:
(Hexadecylamidopropyl)trimethylammonium Chloride
1-Propanaminium, N,N,N-Trimethyl-3-((1-oxohexadecyl)amino)-, Chloride
Trimethyl(3-Palmitamidopropyl)Ammonium Chloride

Trade Name:
Varisoft PATC (Degussa Care Specialties)

Trade Name Mixture:
Varisoft Clear (Degussa Care Specialties)

PALMITAMINE

CAS No. 143-27-1   **EINECS No.** 205-596-8

Empirical Formula:
$C_{16}H_{35}N$

Definition: Palmitamine is hexadecylamine. It conforms to the formula:

$$CH_3(CH_2)_{15}NH_2$$

Information Source: TSCA

Chemical Class: Amines

Function: Antistatic Agent

Technical/Other Names:
1-Aminohexadecane
Cetylamine
1-Hexadecanamine
Hexadecylamine
Palmityl Amine

Trade Name:
Armeen 16D (Akzo Nobel)

PALMITAMINE OXIDE

CAS No.	EINECS No.
7128-91-8	230-429-0

Empirical Formula:
$C_{18}H_{39}NO$

Definition: Palmitamine Oxide is the tertiary amine oxide that conforms to the formula:

$$CH_3(CH_2)_{15}N \longrightarrow O$$
with CH_3 and CH_3 groups

Information Source: TSCA

Chemical Class: Amine Oxides

Functions: Hair Conditioning Agent; Surfactant - Cleansing Agent; Surfactant - Foam Booster; Surfactant - Hydrotrope

Reported Product Category: Shampoos (Non-coloring)

Technical/Other Names:
Cetamine Oxide
Cetyl Dimethyl Amine Oxide
N,N-Dimethyl-1-Hexadecanamine-N-Oxide
1-Hexadecanamine, N,N-Dimethyl-, N-Oxide
Palmityl Dimethylamine Oxide

Trade Name:
Aromox DM16W (Akzo Nobel)

Trade Name Mixture:
Aromox DM16 (Akzo Nobel)

PALMITIC ACID

CAS No.	EINECS No.
57-10-3	200-312-9

JPN Translation:
パルミチン酸

Empirical Formula:
$C_{16}H_{32}O_2$

Definition: Palmitic Acid is the fatty acid that conforms generally to the formula:

$$CH_3(CH_2)_{14}COOH$$

Information Sources: 21CFR172.210, 21CFR172.860, 21CFR173.340, 21CFR175.105, 21CFR175.320, 21CFR176.170, 21CFR176.200, 21CFR176.210, 21CFR177.1010, 21CFR177.1200, 21CFR177.2260, 21CFR177.2600, 21CFR177.2800, 21CFR178.3570, 21CFR178.3910, CIR: [S] JACT-6(3)1987, CTFA S, FCC, JCLS, JSCI, MI-13(7064), RIFM, TSCA, USAN

Chemical Class: Fatty Acids

Functions: Fragrance Ingredient; Opacifying Agent; Surfactant - Cleansing Agent; Surfactant - Emulsifying Agent

Surfactant-Cleansing Agent is included as a function for the soap form of Palmitic Acid.

Reported Product Categories: Shampoos (Non-coloring); Shaving Preparations, Misc.; Foundations; Shaving Cream (Aerosol, Brushless and Lather); Bath Soaps and Detergents; Bath Oils, Tablets, and Salts; Makeup Fixatives; Moisturizing Preparations; Cleansing Products (Cold Creams, Cleansing Lotions, Liquids and Pads); Eyebrow Pencils; Suntan Gels, Creams, and Liquids; Skin Care Preparations, Misc.; Bath Preparations, Misc.; Body and Hand Preparations (Excluding Shaving Preparations); Hair Preparations (Non-coloring), Misc.; Mascara

Technical/Other Names:
Cetylic Acid
n-Hexadecanoic Acid
Palmitic acid (RIFM)

Trade Names:
AEC Palmitic Acid (A & E Connock)
Emersol 7043 (Cognis Corp.)
Kortacid 1690 (Akzo Nobel)
Kortacid 1698 (Akzo Nobel Surface AB)
Palmitic Acid PC (Protameen)
PRIFRAC 2960 (Uniqema Europe)
PRIFRAC 2962 (Uniqema Europe)

Trade Name Mixtures:
Biophilic H (Lucas Meyer GmbH)
Biophilic S (Lucas Meyer GmbH)
Capispheres (Barnet)
Cellini Blue (Engelhard Corp.)
Cellini Coral (Engelhard Corp.)
Cellini Green (Engelhard Corp.)
Cellini Red (Engelhard Corp.)
Cellini Yellow (Engelhard Corp.)
Cerasome (Lipoid)
Cleraspheres 9000 (Lipoid)
Cleraspheres-G 9507 (Lipoid)
Chelonine (JUVEX)
Covacrem LP (LCW)
Covacrem MK (LCW)
Cutina FS 45 (Cognis Deutschland)
Macamat Wax (LCW)
Micromac Wax (LCW)
N.S.L.E. (Sederma)
ProLipid 141 (International Specialty Products)
Questamix H (Quest International)
Sabowax FL 81 (Sabo)
Stearina COS (Sabo)
Vitamin F Oilsoluble (Cosmetochem) (Cosmetochem International Ltd.)
Vitamin F Watersoluble (Cosmetochem) (Cosmetochem International Ltd.)

PALMITIC ACID/PENTAERYTHRITOL/ STEARIC ACID/TEREPHTHALIC ACID COPOLYMER

Definition: Palmitic Acid/Pentaerythritol/ Stearic Acid/Terephthalic Acid Copolymer is a copolymer of palmitic acid, pentaerythritol, stearic acid, and terephthalic acid monomers.

Information Source: JCIC

Chemical Class: Synthetic Polymers

Function: Film Former

PALMITOLEAMIDOPROPYL DIMETHYL-AMINE LACTATE

Empirical Formula:
$C_{21}H_{42}N_2O \cdot C_3H_6O_3$

Definition: Palmitoleamidopropyl Dimethyl-amine Lactate is amidoamine salt that conforms generally to the formula:

$$CH_3(CH_2)_7CH=CH(CH_2)_5C(O)-NH(CH_2)_3N(CH_3) \cdot HO-C(O)CHCH_3(OH)$$

Chemical Class: Amines

Function: Antistatic Agent

Technical/Other Name:
Dimethylaminopropylpalmitoleamide Lactate

PALMITOLEAMIDOPROPYL DIMETHYL-AMINE PROPIONATE

Empirical Formula:
$C_{21}H_{42}N_2O \cdot C_3H_6O_2$

Definition: Palmitoleamidopropyl Dimethyl-amine Propionate is the amidoamine salt that conforms to the formula:

$$CH_3(CH_2)_7CH=CH(CH_2)_5C(O)-NH(CH_2)_3N(CH_3) \cdot HO-C(O)CH_2CH_3$$

Chemical Class: Amines

Function: Antistatic Agent

Technical/Other Name:
Dimethylaminopropylpalmitoleamide
Propionate

PALMITOYL ARGININE

CAS No.   **EINECS No.**
58725-47-6   261-412-5

Empirical Formula:
$C_{22}H_{44}N_4O_3$

Definition: Palmitoyl Arginine is the organic compound that conforms to the formula:

$$CH_3(CH_2)_{14}\overset{\overset{O}{\|}}{C}-NHCH(CH_2)_3NH\overset{}{C}=NH$$
$$COOH \qquad NH_2$$

Chemical Class: Amino Acids

Functions: Hair Conditioning Agent; Skin-Conditioning Agent - Emollient

Technical/Other Names:
L-Arginine, N2-(1-Oxohexadecyl)-
N-Palmitoyl-L-Arginine

Trade Name:
Amisafe AP-01 (Ajinomoto)

PALMITOYL CAMELLIA SINENSIS EXTRACT

Definition: Palmitoyl Camellia Sinensis Extract is the condensation product of palmitic acid chloride and Camellia Sinensis Extract (q.v.). *See Reported Ingredient Functions-The Cosmetic Drug Distinction, in Regulatory and Ingredient Use Information, Volume I, Part A.*

Chemical Class: Biological Products

Functions: Antioxidant; Skin Protectant

Trade Name:
Berkemyol The Vert (Berkem)

PALMITOYL CARNITINE

CAS Nos.: 1935-18-8; 2364-67-2

Empirical Formula:
$C_{23}H_{45}NO_4$

Definition: Palmitoyl Carnitine is the ester of Carnitine (q.v.) and palmitic acid that conforms to the formula:

$$CH_3(CH_2)_{14}\overset{\overset{O}{\|}}{C}-OCHCH_2\overset{\overset{CH_2COO^-}{|}}{\underset{\underset{CH_3}{|}}{N^+}}\overset{CH_3}{\underset{}{}}CH_3$$

Chemical Classes: Amino Acids; Betaines; Esters

Function: Skin-Conditioning Agent - Miscellaneous

Technical/Other Names:
Ammonium, (3-Carboxy-2-Hydroxypropyl)trimethyl-, Hydroxide, Inner Salt, Palmitate
Hexadecanoyl Carnitine
Palmitic Acid, Ester with (3-Carboxy-2-Hydroxypropyl)Trimethylammonium Hydroxide Inner Salt
Propanaminium, 3-Carboxy-N,N,N-Trimethyl-2-((1-Oxohexadecyl)Oxy)-, Inner Salt

Trade Name Mixture:
Vexel (Sederma)

PALMITOYL COLLAGEN AMINO ACIDS

Definition: Palmitoyl Collagen Amino Acids is the condensation product of palmitic acid chloride and Collagen Amino Acids (q.v.).

Chemical Class: Amino Acids

Functions: Hair Conditioning Agent; Skin-Conditioning Agent - Miscellaneous; Surfactant - Cleansing Agent

Technical/Other Name:
Palmitoyl Animal Collagen Amino Acids

Trade Name:
Lipacide PCO (SEPPIC)

PALMITOYL GLUTAMIC ACID

CAS No.: 38079-66-2

Empirical Formula:
$C_{21}H_{39}NO_5$

Definition: Palmitoyl Glutamic Acid is the substituted amino acid that conforms to the formula:

$$CH_3(CH_2)_{14}\overset{\overset{O}{\|}}{C}-NHCHCH_2CH_2COOH$$
$$COOH$$

Chemical Classes: Amides; Amino Acids

Function: Skin-Conditioning Agent - Miscellaneous

Technical/Other Names:
L-Glutamic Acid, N-(1-Oxohexadecyl)-
N-(1-Oxohexadecyl)-L-Glutamic Acid
N-Palmitoyl Glutamic Acid

PALMITOYL GLYCINE

Empirical Formula:
$C_{18}H_{35}NO_3$

Definition: Palmitoyl Glycine is the acylation product of glycine with palmitic acid chloride. It conforms to the formula:

$$CH_3(CH_2)_{14}\overset{\overset{O}{\|}}{C}-NHCH_2COOH$$

Chemical Classes: Amides; Amino Acids

Functions: Hair Conditioning Agent; Surfactant - Cleansing Agent

Trade Name:
PALMITOILGLICINA (Keminova Italiana)

PALMITOYL GRAPE SEED EXTRACT

Definition: Palmitoyl Grape Seed Extract is the product obtained by the condensation of palmitic acid chloride and grape seed extract. *See Reported Ingredient Functions-The Cosmetic Drug Distinction, in Regulatory and Ingredient Use Information, Volume I, Part A.*

Chemical Class: Biological Products

Functions: Antioxidant; Skin Protectant

Trade Name:
Berkemyol Pepin de Raisin (Berkem)

PALMITOYL HYDROLYZED COLLAGEN

CAS No.: 68915-45-7

Definition: Palmitoyl Hydrolyzed Collagen is the condensation product of palmitic acid chloride and Hydrolyzed Collagen (q.v.).

Information Source: JCLS

Chemical Class: Protein Derivatives

Functions: Hair Conditioning Agent; Skin-Conditioning Agent - Miscellaneous; Surfactant - Cleansing Agent

Technical/Other Names:
Collagens, Palmitoyl Derivs.
Palmitoyl Hydrolyzed Animal Protein
Proteins, Hydrolysates, Reaction Products with Palmitoyl Chloride

PALMITOYL HYDROLYZED MILK PROTEIN

Definition: Palmitoyl Hydrolyzed Milk Protein is the condensation product of palmitic acid chloride and Hydrolyzed Milk Protein (q.v.).

Chemical Class: Protein Derivatives

Functions: Hair Conditioning Agent; Skin-Conditioning Agent - Miscellaneous; Surfactant - Cleansing Agent

Trade Name:
  Lipacide PCA (SEPPIC)

PALMITOYL HYDROLYZED WHEAT PROTEIN

Definition: Palmitoyl Hydrolyzed Wheat Protein is the condensation product of palmitic acid chloride and Hydrolyzed Wheat Protein (q.v.).

Chemical Class: Protein Derivatives

Functions: Hair Conditioning Agent; Skin-Conditioning Agent - Miscellaneous; Surfactant - Cleansing Agent

Trade Name:
  Lipacide PVB (SEPPIC)

PALMITOYL HYDROXYPROPYL-TRIMONIUM AMYLOPECTIN/GLYCERIN CROSSPOLYMER

Definition: Palmitoyl Hydroxypropyl-trimonium Amylopectin/Glycerin Cross-polymer is the palmitic acid ester of a polymer of the hydroxypropyltrimonium derivative of Amylopectin (q.v.) crosslinked with glycerin.

Chemical Classes: Carbohydrates; Quaternary Ammonium Compounds; Synthetic Polymers

Function: Skin-Conditioning Agent - Miscellaneous

Trade Name Mixtures:
  Glycosphere-GT (Kobo)
  Glycosphere-PCO (Kobo SA)
  Supra Molecular Bio Vector (SMBV) (Kobo SA)

PALMITOYL INULIN

Definition: Palmitoyl Inulin is the condensation product of palmitic acid chloride and the carbohydrate, Inulin (q.v.).

Chemical Classes: Carbohydrates; Esters

Functions: Skin-Conditioning Agent - Emollient; Surfactant - Emulsifying Agent

PALMITOYL KERATIN AMINO ACIDS

Definition: Palmitoyl Keratin Amino Acids is the condensation product of palmitic acid chloride and Keratin Amino Acids (q.v.).

Chemical Class: Amino Acids

Functions: Hair Conditioning Agent; Skin-Conditioning Agent - Miscellaneous; Surfactant - Cleansing Agent

Reported Product Category: Hair Conditioners

Trade Name:
  Lipacide PK (SEPPIC)

PALMITOYL MARE MILK

Definition: Palmitoyl Mare Milk is the condensation product of Mare Milk (q.v.) and palmitic acid chloride.

Chemical Class: Protein Derivatives

Function: Skin-Conditioning Agent - Miscellaneous

Trade Name:
  Equiperm (Novoselect)

PALMITOYL METHOXYTRYPTAMINE

Empirical Formula:
  $C_{27}H_{44}N_2O_2$

Definition: Palmitoyl Methoxytryptamine is the heterocyclic compound that conforms to the formula:

Chemical Classes: Amides; Ethers; Heterocyclic Compounds

Function: Skin-Conditioning Agent - Miscellaneous

Technical/Other Name:
  N-Palmitoyl-5-Methoxytryptamine

Trade Name:
  Promelatonine (Sederma)

PALMITOYL MYRISTYL SERINATE

Definition: Palmitoyl Myristyl Serinate is the product obtained by the condensation of Palmitic Acid (q.v.) with Serine (q.v.), followed by esterification with Myristyl Alcohol (q.v.).

Chemical Classes: Amides; Amino Acids; Esters

Function: Skin-Conditioning Agent - Miscellaneous

Trade Name Mixtures:
  Ceramide A2 (Sederma)
  Dermaceride (Sederma)

PALMITOYL OLIGOPEPTIDE

Definition: Palmitoyl Oligopeptide is the palmitic acid ester of a synthetic peptide consisting of two or more of the following amino acids: alanine, arginine, aspartic acid, glycine, histidine, lysine, proline, serine, or valine.

Chemical Class: Protein Derivatives

Functions: Skin-Conditioning Agent - Miscellaneous; Surfactant - Cleansing Agent

Trade Name Mixtures:
  Bio-Bustyl (Sederma)
  Biopeptide-CL (Sederma)
  Biopeptide-EL (Sederma)
  Biopeptide FN (Sederma)

PALMITOYL PEA AMINO ACIDS

Definition: Palmitoyl Pea Amino Acids is the condensation product of palmitic acid chloride and pea amino acids.

Chemical Class: Amino Acids

Functions: Hair Conditioning Agent; Skin-Conditioning Agent - Miscellaneous

Trade Name:
  C16 Pois Blond (Phytocos)

PALMITOYL PENTAPEPTIDE-2

Definition: Palmitoyl Pentapeptide-2 is the reaction product of Palmitic Acid (q.v.) and a synthetic peptide consisting of Tyrosine (q.v.), Glycine (q.v.), Phenylalanine (q.v.) and Leucine (q.v.) residues.

Chemical Class: Protein Derivatives

Function: Skin-Conditioning Agent - Miscellaneous

Trade Name:
  Lipopentapeptide 2 (Sederma)

PALMITOYL PENTAPEPTIDE-3

Definition: Palmitoyl Pentapeptide-3 is the reaction product of Palmitic Acid (q.v.) and a synthetic peptide consisting of Lysine (q.v.), Threonine (q.v.) and Serine (q.v.) residues.

Chemical Class: Protein Derivatives

Function: Skin-Conditioning Agent - Miscellaneous

Trade Name:
  Lipopentapeptide 3 (Sederma)

PALMITOYL PG-TRIMONIUM CHLORIDE

Empirical Formula:
  $C_{22}H_{46}NO_3 \cdot Cl$

Definition: Palmitoyl PG-Trimonium Chloride is the quaternary ammonium salt that conforms to the formula:

$$CH_3(CH_2)_{14}\overset{\overset{\displaystyle O}{||}}{C}-OCH_2\overset{\overset{\displaystyle }{|}}{\underset{\underset{\displaystyle OH}{|}}{CH}}CH_2-\overset{\overset{\displaystyle CH_3}{|}}{\underset{\underset{\displaystyle CH_3}{|}}{N}}-CH_3 \right]^+ Cl^-$$

Chemical Class: Quaternary Ammonium Compounds

Functions: Antistatic Agent; Hair Conditioning Agent

Trade Name Mixture:
Akypoquat 40 (Kao GmbH)

PALMITOYL PROLINE

CAS No. 59441-32-6

EINECS No. 261-763-5

Empirical Formula:
$C_{21}H_{39}NO_3$

Definition: Palmitoyl Proline is the product obtained by the condensation of palmitic acid chloride with Proline (q.v.).

Chemical Class: Alkyl-Substituted Amino Acids

Function: Not Reported

Technical/Other Names:
1-(1-Oxohexadecyl)-L-Proline
N-Palmitoyl-L-Proline
L-Proline, 1-(1-Oxohexadecyl)-

Trade Name Mixture:
Sepifeel One (SEPPIC)

PALMITOYL QUINOA AMINO ACIDS

Definition: Palmitoyl Quinoa Amino Acids is the condensation product of palmitic acid chloride and quinoa amino acids.

Chemical Class: Amino Acids

Functions: Hair Conditioning Agent; Skin-Conditioning Agent - Miscellaneous

Trade Name:
C16 Quinoa (Phytocos)

PALMITOYL SERINE/SILK AMINO ACIDS METHYL ESTERS

CAS No.: 297157-14-3

Definition: Palmitoyl Serine/Silk Amino Acids Methyl Esters is the condensation

product of palmitic acid chloride and the methyl esters of Silk Amino Acids (q.v.) and Serine (q.v.)

Chemical Class: Amino Acids

Functions: Hair Conditioning Agent; Nail Conditioning Agent; Skin-Conditioning Agent - Miscellaneous

Technical/Other Name:
Amino Acids, Silk, Me Esters, Reaction Products with Palmitoyl Chloride and L-Serine Me Ester

Trade Name:
Lipesters PSS 4060 (Asepta)

PALMITOYL SILK AMINO ACIDS

Definition: Palmitoyl Silk Amino Acids is the condensation product of palmitic acid chloride and Silk Amino Acids (q.v.).

Chemical Class: Amino Acids

Functions: Hair Conditioning Agent; Surfactant - Cleansing Agent

Trade Name:
C16 Soie Acide (Phytocos)

PALMITOYL TETRAPEPTIDE-3

Definition: Palmitoyl Tetrapeptide-3 is the reaction product of Palmitic Acid (q.v.) and a synthetic peptide containing glycine, glutamine, proline and arginine.

Chemical Class: Proteins

Function: Skin-Conditioning Agent - Miscellaneous

Trade Name:
N-Palmitoyl-Rigin (Sederma)

PALMITOYL TRIPEPTIDE-1

Definition: Palmitoyl Tripeptide-1 is the reaction product of palmitic acid and Tripeptide-1 (q.v.).

Chemical Class: Proteins

Function: Skin-Conditioning Agent - Miscellaneous

Trade Name:
Lipo-GKH (Sederma)

PALM KERNEL ACID

JPN Translation:
パーム核脂肪酸

Definition: Palm Kernel Acid is a mixture of fatty acids derived from Elaeis Guineensis (Palm) Kernel Oil (q.v.).

Information Sources: JCIC, JCLS, JSQI

Chemical Class: Fatty Acids

Functions: Opacifying Agent; Surfactant - Cleansing Agent; Surfactant - Emulsifying Agent

Reported Product Category: Bath Soaps and Detergents

Technical/Other Names:
Acids, Palm Kernel
Elaeis Guineensis (Palm) Kernel Acid
Palm Kernel Fatty Acids
Palm Kernel Oil Fatty Acid

Trade Names:
Kortacid PKG (Akzo Nobel Surface AB)
PRIFAC 7905 (Uniqema Europe)

Trade Name Mixture:
PRIFAC 7970 (Uniqema Europe)

PALM KERNEL ALCOHOL

Definition: Palm Kernel Alcohol is the mixture of fatty alcohols derived from Elaeis Guineensis (Palm) Kernel Oil (q.v.).

Information Source: MI-13(1889)

Chemical Class: Fatty Alcohols

Functions: Emulsion Stabilizer; Surfactant - Foam Booster; Viscosity Increasing Agent - Aqueous; Viscosity Increasing Agent - Nonaqueous

Technical/Other Names:
Alcohols, Palm Kernel
Elaeis Guineensis (Palm) Kernel Alcohol

PALM KERNELAMIDE DEA

CAS No. 73807-15-5

EINECS No. 277-612-1

JPN Translation:
パーム核脂肪酸アミド DEA

Definition: Palm Kernelamide DEA is a mixture of ethanolamides of the fatty acids derived from Elaeis Guineensis (Palm) Kernel Oil (q.v.). It conforms generally to the formula:

$$R\overset{\overset{\displaystyle O}{||}}{C}-N(CH_2CH_2OH)_2$$

where RCO- represents the fatty acids derived from palm kernel oil.

Information Sources: 21CFR175.105, EEC(III/1-60), JCIC, JCLS

Chemical Class: Alkanolamides

Functions: Surfactant - Foam Booster; Viscosity Increasing Agent - Aqueous

Technical/Other Names:
Amides, Palm Kernel Oil, N,N-Bis (Hydroxyethyl)-
N,N-Bis(2-Hydroxyethyl)Palm Kernel Oil Acid Amide
Diethanolamine Palm Kernel Oil Acid Amide
Palm Kernel Oil Acid Amide, N,N-Bis(2-Hydroxyethyl)-
Palm Kernel Oil Acid Diethanolamide
Palm Kernel Oil Fatty Acid Diethanolamide (1)

Trade Names:
Accomid 50 (Abitec Corporation)
Accomid PK (Abitec Corporation)
Mackamide PK (McIntyre)
T-Tergamide 1 PD (Harcros)

PALM KERNELAMIDE MEA

Definition: Palm Kernelamide MEA is a mixture of ethanolamides of the fatty acids derived from Elaeis Guineensis (Palm) Kernel Oil (q.v.). It conforms generally to the formula:

$$RC \overset{\overset{\displaystyle O}{\|}}{\quad} NHCH_2CH_2OH$$

where RCO- represents the fatty acids derived from palm kernel oil.

Chemical Class: Alkanolamides

Functions: Surfactant - Foam Booster; Viscosity Increasing Agent - Aqueous

Technical/Other Names:
N-(2-Hydroxyethyl) Palm Kernel Oil Acid Amide
Monoethanolamine Palm Kernel Oil Acid Amide
Palm Kernel Oil Acid Amide, N-(2-Hydroxyethyl)-
Palm Kernel Oil Acid Monoethanolamide

Trade Name:
Mackamide PKM (McIntyre)

PALM KERNELAMIDE MIPA

Definition: Palm Kernelamide MIPA is a mixture of isopropanolamides of the fatty acids derived from Elaeis Guineensis (Palm) Kernel Oil (q.v.). It conforms generally to the formula:

$$RC \overset{\overset{\displaystyle O}{\|}}{\quad} NHCH_2\underset{\underset{\displaystyle CH_3}{|}}{CHOH}$$

where RCO- represents the fatty acids derived from palm kernel oil.

Chemical Class: Alkanolamides

Functions: Surfactant - Foam Booster; Viscosity Increasing Agent - Aqueous

Technical/Other Names:
N-(2-Hydroxypropyl)Palm Kernel Oil Acid Amide
Monoisopropanolamine Palm Kernel Oil Acid Amide
Palm Kernel Oil Acid Amide, N-(2-Hydroxypropyl)-
Palm Kernel Oil Acid Monoiso-propanolamide

PALM KERNELAMIDOPROPYL BETAINE

JPN Translation:
パーム核脂肪酸アミドプロピルベタイン

Definition: Palm Kernelamidopropyl Betaine is the zwitter (inner salt) that conforms generally to the formula:

$$RC \overset{\overset{\displaystyle O}{\|}}{\quad} NH(CH_2)_3 \overset{\overset{\displaystyle CH_3}{|}}{\underset{\underset{\displaystyle CH_3}{|}}{N^+}} CH_2COO^-$$

where RCO- represents the fatty acids derived from palm kernel oil.

Information Sources: JCIC, JCLS

Chemical Class: Betaines

Functions: Antistatic Agent; Hair Conditioning Agent; Skin-Conditioning Agent - Miscellaneous; Surfactant - Cleansing Agent; Surfactant - Foam Booster; Viscosity Increasing Agent - Aqueous

Technical/Other Names:
N-(Carboxymethyl)-N,N-Dimethyl-3-[(1-Oxopalm Kernel)Amino]-1-Propanaminium Hydroxide, Inner Salt
Palm Kernel Amide Propylbetaine
Palm Kernelamidopropyl Dimethyl Glycine
Palm Kernel Oil Amide Propyl Dimethyl Glycine Solution
1-Propanaminium, N-(Carboxymethyl)-N,N-Dimethyl-3-[(1-Oxopalm Kernel)Amino]-, Hydroxide, Inner Salt
Quaternary Ammonium Compounds, (Carboxymethyl)(3-Palm Kernel-amidopropyl) Dimethyl, Hydroxide, Inner Salt

PALM KERNEL GLYCERIDES

Definition: Palm Kernel Glycerides is a mixture of mono, di and triglycerides derived from Elaeis Guineensis (Palm) Kernel Oil (q.v.).

Chemical Class: Glyceryl Esters and Derivatives

Function: Skin-Conditioning Agent - Emollient

Technical/Other Name:
Glycerides, Palm Kernel Mono-, Di- and Tri

Trade Names:
Cremao CS-33 (Aarhus)
Cremao CS-34 (Aarhus)
Rylo MD 12 (Danisco)

PALM KERNEL WAX

Definition: Palm Kernel Wax is the wax fraction of Elaeis Guineensis (Palm) Kernel Oil (q.v.).

Chemical Class: Waxes

Function: Skin-Conditioning Agent - Occlusive

Technical/Other Name:
Waxes, Palm Kernel

PALM OIL PEG-8 ESTERS

Definition: Palm Oil PEG-8 Esters is a complex mixture formed by the transesterification of Elaeis Guineensis (Palm) Oil and PEG-8 (q.v.).

Chemical Class: Glyceryl Esters and Derivatives

Functions: Skin-Conditioning Agent - Emollient; Surfactant - Emulsifying Agent

Trade Name:
Resassol Pseudoplastico (Res Pharma)

PANAX GINSENG ROOT

CAS No.: 50647-08-0

JPN Translation:
オタネニンジン

Definition: Panax Ginseng Root is a plant material derived from the dried roots of the ginseng, *Panax ginseng*. See "Regulatory and Ingredient Use Information," regarding the labeling names for botanical ingredients in Volume 1, Introduction, Part A.

Information Sources: HP, JCIC, JCLS, JSQI, MAR, MI-13(4439), NFJ

Chemical Class: Biological Products

Function: Not Reported

Reported Product Category: Skin Care Preparations, Misc.

Technical/Other Names:
Ginseng
Ginseng (Panax Ginseng)
Ginseng Powder

Trade Names:
AEC Ginseng Root Powder (A & E Connock)
AEC Ginseng Root Whole (A & E Connock)

PANAX GINSENG ROOT EXTRACT

CAS No.	EINECS No.
90045-38-8	289-898-5

JPN Translation:
オタネニンジンエキス

Definition: Panax Ginseng Root Extract is an extract of the roots of the ginseng, *Panax ginseng.* See *"Regulatory and Ingredient Use Information," regarding the labeling names for botanical ingredients in Volume 1, Introduction, Part A.*

Information Sources: JCIC, JCLS, JSQI

Chemical Class: Biological Products

Function: Skin-Conditioning Agent - Miscellaneous

Reported Product Categories: Skin Care Preparations, Misc.; Bath Preparations, Misc.; Body and Hand Preparations (Excluding Shaving Preparations); Hair Dyes and Colors (All Types Requiring Caution Statements and Patch Tests); Moisturizing Preparations; Bath Capsules; Face and Neck Preparations (Excluding Shaving Preparations); Bath Oils, Tablets, and Salts; Cleansing Products (Cold Creams, Cleansing Lotions, Liquids and Pads); Hair Conditioners; Night Skin Care Preparations; Shampoos (Non-coloring); Skin Fresheners; Eye Makeup Preparations, Misc.; Colognes and Toilet Waters; Eye Shadows; Paste Masks (Mud Packs); Tonics, Dressings, and Other Hair Grooming Aids; Fragrance Preparations, Misc.; Hair Wave Sets; Makeup Preparations (Not eye), Misc.; Suntan Preparations, Misc.

Technical/Other Names:
American Ginseng Extract
Extract of Ginseng
Ginseng Extract
Ginseng (Panax Ginseng) Extract
Oriental Ginseng Extract
Panax Quinquefolium Extract

Trade Names:
AEC Ginseng Root Extract Powder (A & E Connock)
Ginseng Dry Extract (Euromed)
Ginseng Extract Powder (Maruzen Pharmaceuticals Co., Ltd.)

Herbalia Ginseng (Cognis Deutschland)
Phytelene of Ginseng EN 359 powder (Indena SA)
Phytogreen of Ginseng EP 510 Powder (Phytochim)
SFE 6 Years Red Ginseng Extract (Green Tek 21)

Trade Name Mixtures:
Actigen O2 GL (Active Organics)
Actiphyte American Ginseng BG50P (Active Organics)
Actiphyte of Ginseng BG50 (Active Organics)
Actiphyte of Ginseng GL50 (Active Organics)
Actiphyte of Ginseng Lipo S (Active Organics)
Actiphyte of Ginseng PG50 (Active Organics)
A.P.S. LS 8425 (Laboratoires Sero-biologiques)
Bio-Chelated Derma-Plex I (Bio-Botanica)
Complex 3 - Anti-Aging (Provital/Centerchem)
Complex 1 - Antipollution (Provital/Centerchem)
Complex Ginseng (Fabriquimica)
Cosmelene of Ginseng (Indena SA)
Cremogen Ginseng (PN 739 020) (Haarmann & Reimer GmbH)
Cremogen M-88(PN 458 469) (Haarmann & Reimer GmbH)
Enna 216 Breast Development Extract (Ennagram)
Enna 212 Bust Firmness (Ennagram)
Enna 231 Eye Contour (Ennagram)
Enzyami 6 (Alban Muller)
Extrait de Ginseng MPE PG 40 (Yves Rocher)
Extrapone Ginseng Special 2/032861 (Symrise)
Floraceutical American Ginseng-Standardized (Bio-Botanica)
Gentle Cleansing Milk (CEP (Solabia))
Ginseng CV (Ichimaru Pharcos)
Ginseng Extract (Maruzen Pharmaceuticals Co., Ltd.)
Ginseng Extract BG (Maruzen Pharmaceuticals Co., Ltd.)
Ginseng Extract BG-100 (Maruzen Pharmaceuticals Co., Ltd.)
Ginseng Extract BG-DS (Maruzen Pharmaceuticals Co., Ltd.)
Ginseng Extract HS 2457 G (Grau)
Ginseng Extract LA (Maruzen Pharmaceuticals Co., Ltd.)
Ginseng Extract LA20 (Maruzen Pharmaceuticals Co., Ltd.)
Ginseng Extract LA-30 (Maruzen Pharmaceuticals Co., Ltd.)
Ginseng Extract LAH (Maruzen Pharmaceuticals Co., Ltd.)
Ginseng Extract LAH-30 (Maruzen Pharmaceuticals Co., Ltd.)

Ginseng Extract LAH2-A (Maruzen Pharmaceuticals Co., Ltd.)
Ginseng Extract LAH-II (Maruzen Pharmaceuticals Co., Ltd.)
Ginseng Extract LAH2-P (Maruzen Pharmaceuticals Co., Ltd.)
Ginseng Extract Powder-S (Maruzen Pharmaceuticals Co., Ltd.)
Ginseng Extract SO (Maruzen Pharmaceuticals Co., Ltd.)
Ginseng Extract SQ (Maruzen Pharmaceuticals Co., Ltd.)
Ginseng HPG Titrated (Alban Muller)
Ginseng HS (Alban Muller)
Ginseng LS (Alban Muller)
Ginseng Phytexcell (Crodarom)
Ginseng Phytosome (Indena SpA)
Glycolysat of Ginseng (CEP (Solabia))
HAIR VOLUME (Greentech S.A)
Herbaliquid Ginseng Special (Crodarom)
Herbasec Ginseng (Cosmetochem) (Cosmetochem International Ltd.)
Herbasol Distillate Ginseng (Cosmetochem) (Cosmetochem International Ltd.)
Herbasol-Extract Ginseng (Cosmetochem)
Korea Ginseng Extract (Bioland)
Korean Ginseng Extract (Ichimaru Pharcos)
Liposomes Trichogen Veg (Laboratoires Serobiologiques)
Mountain Ginseng Extract (Bioland)
Nat Ginseng Extract (Natiris)
Ninjin Liquid B (Ichimaru Pharcos)
Ninjin Liquid E (Ichimaru Pharcos)
Novaplant Ginseng Extract (Crodarom)
NOVAPUR Ren Shen (Crodarom)
Oleat of Ginseng (CEP (Solabia))
Phytelene of Ginseng EG 209 liquid (Indena SA)
Phytoamine Energy (Alban Muller)
Phytoderm Ginseng Glycolic (Universal Flavors)
Phytogreen 55 of Ginseng EXH 636 Liquid (Phytochim)
Phytotec Repair Factor (Sederma)
Polyplant Intimate Hygiene (Provital/Centerchem)
Polyplant Stimulant (Provital/Centerchem)
Prodhy Extract Ginseng (Prod'Hyg)
Pronalen Bio-Protect (Provital/Centerchem)
Pronalen Ginseng HSC (Provital/Centerchem)
Pronalen Ginseng SPE (Provital/Centerchem)
Sansam Extract (EUROCOSTECH)
Stimulating Phytospa (Alban Muller)
Stimulating Phytospa NaCl (Alban Muller)
Tonic Phytospa (Alban Muller)
Trichogen Veg (Laboratoires Serobiologiques)
Vegebios of Ginseng (CEP (Solabia))
Vegetol Ginseng GR 471 Hydro (Gattefosse s.a.)
VT-075 Extract of Ginseng (Vege-Tech)

YSK Magic 3 (Phyto-Technologies)
YSK Magic 9 (Phyto-Technologies)

PANAX GINSENG ROOT PROTOPLASTS

Definition: Panax Ginseng Root Protoplasts are the protoplasts derived from the roots of *Panax ginseng. See "Regulatory and Ingredient Use Information," regarding the labeling names for botanical ingredients in Volume 1, Introduction, Part A.*

Chemical Class: Biological Products

Function: Skin-Conditioning Agent - Humectant

Trade Name Mixture:
   Panax Ginseng CRS (Shiseido Company)

PANAX GINSENG ROOT WATER

Definition: Panax Ginseng Root Water is an aqueous solution of the steam distillate obtained from the roots of *Panax ginseng. See "Regulatory and Ingredient Use Information," regarding the labeling names for botanical ingredients in Volume 1, Introduction, Part A.*

Chemical Class: Biological Products

Function: Fragrance Ingredient

Technical/Other Name:
   Water, Panax Ginseng Root

Trade Name:
   Ginseng Water (Bioland)

PANAX JAPONICUS ROOT EXTRACT

Definition: Panax Japonicus Root Extract is an extract of the rhizomes of *Panax japonicus. See "Regulatory and Ingredient Use Information," regarding the labeling names for botanical ingredients in Volume 1, Introduction, Part A.*

Chemical Class: Biological Products

Function: Not Reported

Technical/Other Name:
   Extract of Panax Japonicus

Trade Name Mixture:
   Tochibaninjin Liquid E (Ichimaru Pharcos)

PANAX NOTOGINSENG EXTRACT

Definition: Panax Notoginseng Extract is an extract of the roots of *Panax noto-*

ginseng. See "Regulatory and Ingredient Use Information," regarding the labeling names for botanical ingredients in Volume 1, Introduction, Part A.

Chemical Class: Biological Products

Function: Skin-Conditioning Agent - Humectant

Technical/Other Name:
   Extract of Panax Notoginseng

Trade Name Mixtures:
   Notoginseng Extract (Cardre)
   Sanchi Ginseng Extract (Nonogawa)

PANAX NOTOGINSENG ROOT POWDER

Definition: Panax Notoginseng Root Powder is the powder obtained from the dried, crushed roots of *Panax notoginseng. See "Regulatory and Ingredient Use Information," regarding the labeling names for botanical ingredients in Volume 1, Introduction, Part A.*

Chemical Class: Biological Products

Function: Skin-Conditioning Agent - Miscellaneous

Trade Name:
   Panax Notoginseng Root Powder
   (Maleave)

PANAX QUINQUEFOLIUM ROOT EXTRACT

Definition: Panax Quinquefolium Root Extract is an extract of the roots of *Panax quinquefolium. See "Regulatory and Ingredient Use Information," regarding the labeling names for botanical ingredients in Volume 1, Introduction, Part A.*

Chemical Class: Biological Products

Function: Cosmetic Astringent

Technical/Other Name:
   Extract of Panax Quinquefolium Roots

PANCREATIN

CAS No.	EINECS No.
8049-47-6	232-468-9

JPN Translation:
パンクレアチン

Definition: Pancreatin is a mixture of enzymes obtained from the fresh pancreas of animals.

Information Sources: BAN, 21CFR184.1583, 21CFR310.543,

21CFR310.545, JAN, JCLS, JSCI, MI-13 (7075), OTC-I-EP, TSCA, USAN, USP XXIV

Chemical Class: Proteins

Functions: Hair Conditioning Agent; Lytic Agent; Skin-Conditioning Agent - Miscellaneous

Trade Name Mixture:
   Complexe DM 50 (Gattefosse s.a.)

PANDANUS AMARYLLIFOLIUS LEAF EXTRACT

Definition: Pandanus Amaryllifolius Leaf Extract is an extract of the leaves of *Pandanus amaryllifolius. See "Regulatory and Ingredient Use Information," regarding the labeling names for botanical ingredients in Volume 1, Introduction, Part A.*

Chemical Class: Biological Products

Functions: Deodorant Agent; Flavoring Agent; Fragrance Ingredient

Technical/Other Name:
   Extract of Pandanus Amaryllifolius Leaf

Trade Name:
   Pandan (Haldin Pacific)

PANICUM MILIACEUM

Definition: *See "Regulatory and Ingredient Use Information," regarding EU labeling names for botanical ingredients in Volume 1, Introduction, Part A.*

Chemical Class: Biological Products

Technical/Other Names:
   Panicum Miliaceum (Millet) Seed Extract (U.S.)
   Panicum Miliaceum (Millet) Seed Flour (U.S.)

PANICUM MILIACEUM (MILLET) SEED EXTRACT

CAS No.	EINECS No.
90082-36-3	290-125-9

Definition: Panicum Miliaceum (Millet) Seed Extract is an extract of the seeds of the millet, *Panicum miliaceum. See "Regulatory and Ingredient Use Information," regarding the labeling names for botanical ingredients in Volume 1, Introduction, Part A.*

Chemical Class: Biological Products

Function: Not Reported

Reported Product Categories: Shampoos (Non-coloring); Moisturizing Preparations

Technical/Other Names:
Extract of Millet
Extract of Panicum Miliaceum
Millet Extract
Millet Seed Extract
Panicum Miliaceum (EU)
Panicum Miliaceum Extract

Trade Name:
Phytoglycolipid (Barnet)

Trade Name Mixtures:
Actiphyte of Millet BG50 (Active Organics)
Actiphyte of Millet GL50 (Active Organics)
Actiphyte of Millet Lipo S (Active Organics)
Actiphyte of Millet PG50 (Active Organics)
Common Millet Extract HS 3460 G (Grau)
Glycolysat of 7 Cereals (CEP (Solabia))
Herbasol-Extract Millet (Cosmetochem)
Hydrocos "P" (Cosmetochem)
   (Cosmetochem International Ltd.)
Lipocer (Plantech)
Millet Extract HS 3460 G (Grau)
Octaprotein-Colloid (Vevy)
Panicum Miliaceum (Millet) Seed Extract
   ies (IES LABO)
VT-273 Extract of Millet (Vege-Tech)

PANICUM MILIACEUM (MILLET) SEED FLOUR

Definition: Panicum Miliaceum (Millet) Seed Flour is the ground seed of millet, *Panicum miliaceum. See "Regulatory and Ingredient Use Information," regarding the labeling names for botanical ingredients in Volume 1, Introduction, Part A.*

Chemical Class: Biological Products

Function: Not Reported

Technical/Other Names:
Flour, Millet (Panicum Miliaceum)
Millet Flour
Millet Seed Flour
Panicum Miliaceum (EU)

PANTETHEINE SULFONATE

Empirical Formula:
$C_{11}H_{21}N_2O_7S_2$

Definition: Pantetheine Sulfonate is the organic compound that conforms to the formula:

$$HOCHC-NH(CH_2)_2C-NH(CH_2)_2S-S-O^-$$

Chemical Classes: Alcohols; Amides; Sulfonic Acids

Function: Skin-Conditioning Agent - Miscellaneous

Technical/Other Name:
Pantetheine-S-Sulfonate

Trade Name Mixture:
REGU-SLIM (Pentapharm/Centerchem)

PANTETHINE

CAS Nos.	EINECS No.
16816-67-4	240-842-8
138148-35-3	

JPN Translation:
パンテチン

Empirical Formula:
$C_{22}H_{42}N_4O_8S_2$

Definition: Pantethine is the organic compound that conforms to the formula:

$$\left[HOCH_2C-CHC-NH(CH_2)_2C-NH(CH_2)_2S- \right]_2$$

Information Sources: JAN, JCIC, JCLS, MI-13(7082)

Chemical Classes: Amides; Esters; Thio Compounds

Function: Hair Conditioning Agent

Reported Product Categories: Hair Conditioners; Shampoos (Non-coloring); Bath Oils, Tablets, and Salts; Moisturizing Preparations; Skin Care Preparations, Misc.; Tonics, Dressings, and Other Hair Grooming Aids; Mascara; Cleansing Products (Cold Creams, Cleansing Lotions, Liquids and Pads); Hair Preparations (Non-coloring), Misc.; Hair Sprays (Aerosol Fixatives)

Technical/Other Names:
Bis(Pantothenamidoethyl) Disulfide
Butanamide, N,N'-[Dithiobis[2,1-Ethanediylimino(3-Oxo-3,1-Propanediyl)]]Bis[2,4-Dihydroxy-3,3-Dimethyl-
N,N'-[Dithiobis[2,1-Ethanediylimino(3-Oxo-3,1-Propanediyl)]]Bis[2,4-Dihydroxy-3,3-Dimethylbutanamide
Pantethine Solution

PANTHENOL

CAS Nos.	EINECS Nos.
81-13-0 (D-Form)	201-327-3
16485-10-2	240-540-6

JPN Translation:
パンテノール

Empirical Formula:
$C_9H_{19}NO_4$

Definition: Panthenol is the alcohol that conforms to the formula:

$$HOCH_2C-CHC-NH(CH_2)_3OH$$

Information Sources: AUS, BAN, 21CFR310.545, CIR: [S] JACT-6(1)1987, DDR, FCC, INN, JCLS, JSCI, JSQI, MAR, MI-13(2964), PN, TSCA, USAN, USP XXIV

Chemical Classes: Alcohols; Amides

Function: Hair Conditioning Agent

Reported Product Categories: Mascara; Hair Dyes and Colors (All Types Requiring Caution Statements and Patch Tests); Foundations; Hair Conditioners; Paste Masks (Mud Packs); Shampoos (Non-coloring); Tonics, Dressings, and Other Hair Grooming Aids; Bath Soaps and Detergents; Eye Makeup Preparations, Misc.; Skin Fresheners; Aftershave Lotions; Baby Shampoos; Fragrance Preparations, Misc.; Hair Wave Sets; Manicuring Preparations, Misc.; Nail Polish and Enamels; Basecoats and Undercoats; Eyebrow Pencils; Suntan Gels, Creams, and Liquids; Eye Makeup Removers; Hair Preparations (Non-coloring), Misc.; Makeup Bases; Lipsticks; Permanent Waves; Personal Cleanliness Products, Misc.; Colognes and Toilet Waters; Hair Coloring Preparations, Misc.; Hair Rinses (Non-coloring); Nail Polish and Enamel Removers; Bath Capsules; Bath Preparations, Misc.; Body and Hand Preparations (Excluding Shaving Preparations); Douches; Eye Lotions; Face and Neck Preparations (Excluding Shaving Preparations); Makeup Preparations (Not eye), Misc.; Skin Care Preparations, Misc.; Baby Lotions, Oils, Powders and Creams; Bath Oils, Tablets, and Salts; Cleansing Products (Cold Creams, Cleansing Lotions, Liquids and Pads); Cuticle Softeners; Deodorants (Underarm); Eye Shadows; Face Powders; Hair Sprays (Aerosol Fixatives); Moisturizing Preparations; Night Skin Care Preparations; Powders (Dusting and Talcum, Excluding Aftershave Talcs); Suntan Preparations, Misc.

Technical/Other Names:
Butanamide, 2,4-Dihydroxy-N-(3-Hydroxypropyl)-3,3-Dimethyl-
Dexpanthenol
2,4-Dihydroxy-N-(3-Hydroxypropyl)-3,3-Dimethylbutanamide

Pantothenol
Pantothenyl Alcohol
D-Pantothenyl Alcohol
DL-Pantothenyl Alcohol
Provitamin B5

Trade Names:
AEC Panthenol (A & E Connock)
Dexpanthenol (Daiichi)
DL Panthenol CG (Jeen)
DL Panthenol (Protameen)
DL-Panthenol 50L (Roche)
DL-Panthenol TK (Tri-K)
D-Panthenol USP (BASF)
D-Panthenol USP, FCC Regular Type
  (Roche)
D-Panthenol 75 W (BASF)
D,L-Panthenol 50 W (BASF)
Ritapan D (RITA)
Ritapan DL (RITA)
TINODERM P (Ciba Specialty Chemicals)

Trade Name Mixtures:
AC Panthenol Liposome (Active Concepts)
Activator Omega MO Type B (Derma-
  Search)
Alpantha (Uniqema Americas)
Ami Nail Bioregenerator (Alban Muller)
Antidandruff Agent Special (Crodarom)
Antiphlogistic "aro" (Crodarom)
Asebiol LS 2539 BT2 (Laboratoires Sero-
  biologiques)
Bio-Keratin Haircomplex 230389
  (Crodarom)
Biosil Basics HMC - Hair Moisture Complex
  (Biosil Technologies, Inc.)
Biosil Basics HMV- Hair Moisture Complex
  (Biosil Technologies, Inc.)
Biosil Basics HMW - Hair Moisture
  Complex (Biosil Technologies, Inc.)
Brookosome P (Arch Personal Care
  Products)
Catezomes P-20 (Collaborative Labs)
Omega-CH-Activator-A (GfN)
Omega-CH-Activator (GfN)
Omega-CHS-Activator (GfN)
Dermasome-P (ChemMark)
Hair Care Phytoamine Biocomplex (Alban
  Muller)
Hair Care Phytoamine Biocomplex SP
  Lotion (Alban Muller)
Hair Complex Aquosum (CLR)
Hair Complex NOVA (Crodarom)
Hydralphatine Asiatique (Lanatech)
Hydroxan (Lanatech)
Hydroxan BG (Lanatech)
Hydroxan CH (Lanatech)
Induxin (Laboratoires Serobiologiques)
Lipofirm LCW (LCW)
Neo-Derma Vitamin Complex, Water
  Soluble (Crodarom)
D-Panthenol 50P (BASF)
Sebaryl FL (Laboratoires Serobiologiques)
Sedaflor-Complex (Crodarom)

PANTHENYL ETHYL ETHER

CAS No.	EINECS No.
667-83-4	211-569-1

JPN Translation:
パンテニルエチル

Empirical Formula:
$C_{11}H_{23}NO_4$

Definition: Panthenyl Ethyl Ether is the ethyl ether of Panthenol (q.v.). It conforms to the formula:

Information Sources: JCIC, JCLS, JSQI

Chemical Classes: Alcohols; Amides; Ethers

Function: Hair Conditioning Agent

Reported Product Categories: Hair Conditioners; Shampoos (Non-coloring); Tonics, Dressings, and Other Hair Grooming Aids; Hair Preparations (Non-coloring), Misc.; Moisturizing Preparations; Permanent Waves; Foundations; Hair Wave Sets; Skin Care Preparations, Misc.; Hair Sprays (Aerosol Fixatives); Bath Preparations, Misc.; Body and Hand Preparations (Excluding Shaving Preparations); Paste Masks (Mud Packs)

Technical/Other Names:
Butanamide, N-(3-Ethoxypropyl)-2,4-
  Dihydroxy-3,3-Dimethyl-
Pantothenyl Ethylether

Trade Names:
Ethyl Panthenol (Roche)
D-Pantothenyl-Ethyl-Ether (Daiichi)

Trade Name Mixtures:
Curasan (CLR)
Follicusan (CLR)
Haircare Complex CLR (CLR)

PANTHENYL ETHYL ETHER ACETATE

JPN Translation:
アセチルパントテニルエチル

Empirical Formula:
$C_{13}H_{25}NO_5$

Definition: Panthenyl Ethyl Ether Acetate is the ester of acetic acid and the ethyl ether of Panthenol (q.v.). It conforms to the formula:

Information Sources: JCLS, JSCI

Chemical Classes: Amides; Esters; Ethers

Function: Hair Conditioning Agent

Technical/Other Name:
Acetyl Pantothenyl Ethyl Ether

PANTHENYL ETHYL ETHER BENZOATE

JPN Translation:
安息香酸パントテニルエチル

Empirical Formula:
$C_{18}H_{27}NO_5$

Definition: Panthenyl Ethyl Ether Benzoate is the ester of benzoic acid and the ethyl ether of panthenol. It conforms to the formula:

Information Sources: JCIC, MHLW-331/3

Chemical Classes: Alcohols; Amides; Esters

Function: Preservative

PANTHENYL HYDROXYPROPYL STEAR-DIMONIUM CHLORIDE

CAS No.: 132467-76-6

Empirical Formula:
$C_{32}H_{67}N_2O_5 \cdot Cl$

Definition: Panthenyl Hydroxypropyl Stear-dimonium Chloride is the quaternary ammonium salt that conforms to the formula:

Chemical Classes: Alcohols; Amides; Ethers; Quaternary Ammonium Compounds

Function: Hair Conditioning Agent

Reported Product Categories: Hair Dyes and Colors (All Types Requiring Caution Statements and Patch Tests); Hair Conditioners; Mascara

---

Technical/Other Names:
N-[2-Hydroxy-3-[3-Hydroxy-4-[(3-Hydroxy-
propyl)Amino]-2,2-Dimethyl-4-
Oxobutoxy]Propyl]-N,N-Dimethyl-1-
Octadecanaminium Chloride
1-Octadecanaminium, N-[2-Hydroxy-3-[3-
Hydroxoy-4--[(3-Hydroxypropyl)Amino]-2,
2-Dimethyl-4- Oxobutoxy]Propyl-N,N-
Dimethyl-, Chloride
1-Octadecanaminium, N-[2-Hydroxy-3-[3-
Hydroxy-4-[(3-Hydroxypropyl)Amino]-2,2-
Dimethyl-4-Oxobutoxy]Propyl-N,N-
Dimethyl-, Chloride
1-Octadecanaminium, N-[2-Hydroxy-[3-
Hydroxy-4-[(3-Hydroxypropyl)Amino]-2,2-
Dimethyl-4-Oxobutoxy)Propyl-N,N-
Dimethyl-, Chloride
Panthenyl Hydroxypropyl Stearyldimonium
Chloride

Trade Names:
Panthequat (Innovachem)
Panthequat (Tri-K)

Trade Name Mixture:
Hair Care Blend (Alban Muller)

PANTHENYL TRIACETATE

CAS Nos. 94089-18-6, 98133-47-2
EINECS No. 302-118-0

Empirical Formula:
$C_{15}H_{25}NO_7$

Definition: Panthenyl Triacetate is the triacetyl ester of Panthenol (q.v.) that conforms to the formula:

$$CH_3C-OCH_2C-CHC-NH(CH_2)_3O-CCH_3$$

Chemical Classes: Amides; Esters

Function: Hair Conditioning Agent

Reported Product Category: Skin Care Preparations, Misc.

Trade Names:
Lipoderma DPT (Lipo)
D-Panthenyltriacetate (Induchem)
D-Panthenyl Triacetate (Lipo)
Ritapan TA (RITA)

Trade Name Mixtures:
Lipoderma P (Lipo)
Lipoderma PF (Lipo)
Lipoderma PF-WS (Lipo)
Unitrienol T-27 (Induchem)
Unitrienol T-272 watersoluble (Induchem)
Unitrienol T-272 WS (Lipo)

PANTOLACTONE

CAS No. 599-04-2
EINECS No. 209-963-3

Empirical Formula:
$C_6H_{10}O_3$

Definition: Pantolactone is the organic compound that conforms to the formula:

Information Source: MI-13(7083)

Chemical Classes: Alcohols; Esters; Heterocyclic Compounds

Function: Skin-Conditioning Agent - Humectant

Technical/Other Names:
Dihydro-3-Hydroxy-4H-Dimethyl-2(3H)-
Furanone
Hydroxy Dimethyl Butyrolactone
Pantothenic Lactone
Pantoyl Lactone

Trade Names:
d-l-lactone (Roche Nicholas)
D-Pantolactone (Sogo)
DL-Pantolactone (Jouvance)

PANTOTHENAMIDE MEA

Empirical Formula:
$C_{11}H_{22}N_2O_5$

Definition: Pantothenamide MEA is the organic compound that conforms to the formula:

$$HOCH_2-C-CHCNH(CH_2)_2CNH(CH_2)_2OH$$

Chemical Class: Alkanolamides

Function: Skin-Conditioning Agent - Miscellaneous

Technical/Other Name:
N-D-Pantothenoyl-2-Aminoethanol

Trade Name:
N-Hydroxyethyl Pantothenamide (Sogo)

PANTOTHENIC ACID

CAS No. 79-83-4
EINECS No. 201-229-0

Empirical Formula:
$C_9H_{17}NO_5$

Definition: Pantothenic Acid is the organic acid that conforms to the formula:

$$HOCH_2C-CHC-NH(CH_2)_2COOH$$

Information Sources: BAN, CIR: [S] JACT-6(1)1987, MI-13(7085)

Chemical Class: Carboxylic Acids

Function: Hair Conditioning Agent

Reported Product Categories: Tonics, Dressings, and Other Hair Grooming Aids; Hair Conditioners; Shampoos (Non-coloring)

Technical/Other Names:
Alanine, N-(2,4-Dihydroxy-3,3-Dimethyl-1-
Oxobutyl)-
Vitamin B5

Trade Name Mixtures:
AC Vitamin A Liposome (Active Concepts)
AC Vitamin B5 Liposome (Active Concepts)
Fortified Yeast T-6361 (Universal Foods)
Integrahair Sphere (Chemyunion)

PANTOTHENIC ACID POLYPEPTIDE

Definition: Pantothenic Acid Polypeptide is the reaction product of Pantothenic Acid (q.v.) and polypeptides.

Chemical Classes: Carboxylic Acids; Protein Derivatives

Functions: Hair Conditioning Agent; Skin-Conditioning Agent - Miscellaneous

Trade Name:
Vitazyme B-5 (Arch Personal Care Products)

Trade Name Mixture:
Brookosome ACEBC Plus (Arch Personal Care Products)

PAPAIN

CAS No. 9001-73-4
EINECS No. 232-627-2

JPN Translation:
パパイン

Definition: Papain is the proteolytic enzyme isolated from the latex of the green fruit and leaves of *Carica papaya*.

Information Sources: 21CFR184.1585, 21CFR310.545, 21CFR582.1585, JCIC, JCLS, JSQI, MI-13(7086), TSCA, USAN, USP XXIV

Chemical Class: Proteins

Functions: Hair Conditioning Agent; Lytic Agent; Skin-Conditioning Agent - Miscellaneous

Technical/Other Names:
Papainase
Papaine

Trade Name Mixtures:
AEC Papaya Extract (Papain 0.25%) (A & E Connock)
EHG Papaye / Papain (I.D. bio)
Linked Papain C-MP6 (Collaborative Labs)
Linked Papain C-MPB (Collaborative Labs)
Stazyme-PNM (Pacific)
2410 Vege-Plex Skin Complex (Vege-Tech)
2510 Vege-Plex Skin Complex (Vege-Tech)

PAPAIN BETA-GLUCAN

Definition: Papain Beta-Glucan is the product obtained by the reaction of Papain (q.v.) with Beta-Glucan (q.v.).

Chemical Classes: Carbohydrates; Proteins

Functions: Hair Conditioning Agent; Skin-Conditioning Agent - Miscellaneous

Trade Name Mixture:
Stazyme-PN (Pacific)

PAPAVER ORIENTALE (POPPY) SEED OIL

Definition: Papaver Orientale (Poppy) Seed Oil is the fixed oil expressed from the seeds of *Papaver orientale. See "Regulatory and Ingredient Use Information," regarding the labeling names for botanical ingredients in Volume 1, Introduction, Part A.*

Information Source: MI-13(7674)

Chemical Class: Fats and Oils

Function: Skin-Conditioning Agent - Occlusive

Technical/Other Names:
Oil of Poppy Seeds
Oils, Papaver Oriental
Poppy Oil
Poppy Seed Oil

Trade Names:
AEC Poppy Seed Oil (A & E Connock)
Poppy Seed Oil NOVAROM (Crodarom)

Trade Name Mixture:
VT- 0934 Poppy Seed Paste (Vege)

PAPAVER RHOEAS EXTRACT

CAS No.	EINECS No.
84696-43-5	283-651-5

Definition: Papaver Rhoeas Extract is an extract of the petals of the corn poppy, *Papaver rhoeas. See "Regulatory and Ingredient Use Information," regarding the labeling names for botanical ingredients in Volume 1, Introduction, Part A.*

Chemical Class: Biological Products

Function: Skin-Conditioning Agent - Miscellaneous

Reported Product Categories: Bath Capsules; Face and Neck Preparations (Excluding Shaving Preparations)

Technical/Other Names:
Corn Poppy Extract
Corn Poppy (Papaver Rhoeas) Extract
Extract of Corn Poppy

Trade Names:
Phytelene of Coquelicot EN 109 powder (Indena SA)
Phytelene of Red Poppy EN 109 powder (Indena SA)
Phytogreen of Red Poppy EP 488 Powder (Phytochim)

Trade Name Mixtures:
Corn Poppy HPG Titrated (Alban Muller)
Corn Poppy HS (Alban Muller)
Extrait de Coquelicot MP PG 40 (Yves Rocher)
Extrapone Poppy Flower 2/395321 (Symrise)
Fieldpoppy Extract HG (Provital/Centerchem)
Herbasol Extract Poppy (Cosmetochem) (Cosmetochem International Ltd.)
Papaver Rhoeas Extract ies (IES LABO)
Phytelene of Red Poppy EG 517 Liquid (Indena SA)
Phytogreen 55 of Red Poppy EXH 716 Liquid (Phytochim)
Prodhy Extract Coquelicot (Prod'Hyg)
Red Poppy Extract HS 2472 G (Grau)
Vegebios of Corn Poppy (CEP) (Solabia))
Vegetol Red Poppy GR 106 Hydro (Gattefosse s.a.)

PAPAVER SOMNIFERUM SEED

Definition: Papaver Somniferum Seed is the seed of *Papaver somniferum. See "Regulatory and Ingredient Use Information," regarding the labeling names for botanical ingredients in Volume 1, Introduction, Part A.*

Chemical Class: Biological Products

Function: Abrasive

Trade Name:
AEC Blue Poppy Seeds (A & E Connock)

PAPAVER SOMNIFERUM SEED EXTRACT

Definition: Papaver Somniferum Seed Extract is an extract of the seeds of *Papaver somniferum. See "Regulatory and Ingredient Use Information," regarding the labeling names for botanical ingredients in Volume 1, Introduction, Part A.*

Chemical Class: Biological Products

Function: Skin-Conditioning Agent - Miscellaneous

Trade Name Mixtures:
Actiphyte Poppy Seed (Active Organics)
Actiphyte Poppy Seed Lipo S (Active Organics)
Blue Poppy Seed HS (Alban Muller)

PAPHIOPEDILUM MAUDIAE (ORCHID) FLOWER EXTRACT

Definition: Paphiopedilum Maudiae (Orchid) Flower Extract is an extract of the flowers of *Paphiopedilum maudiae. See "Regulatory and Ingredient Use Information," regarding the labeling names for botanical ingredients in Volume 1, Introduction, Part A.*

Chemical Class: Biological Products

Function: Skin-Conditioning Agent - Miscellaneous

Technical/Other Names:
Extract of Paphiopedilum Maudiae (Orchid) Flower
Orchid Flower Extract

Trade Name Mixture:
Black Orchid Flower HG (Alban Muller)

PARAFFIN

CAS No.	EINECS No.
8002-74-2	232-315-6

JPN Translation:
パラフィン

Definition: Paraffin is a solid mixture of hydrocarbons obtained from petroleum characterized by relatively large crystals.

Information Sources: ARG, AUS, BEL, BP, BPC, BRA, 21CFR133.181, 21CFR172.615, 21CFR175.105, 21CFR175.210, 21CFR175.250, 21CFR175.300, 21CFR175.320, 21CFR176.170, 21CFR176.200, 21CFR177.1200, 21CFR177.2420, 21CFR177.2600, 21CFR177.2800, 21CFR178.3710, 21CFR178.3800, 21CFR178.3910, 21CFR179.45, 21CFR349.14, CIR: [S] JACT-3(3)1984, CTFA S, DA, DDR, EGY, FI, FIN, HUN, IND, ITA, JAN, JCIC, JCLS, JSCI, MAR, MI-13(7093), NF XIX, OTC-I-OP, PF, PN, POR,

RIFM, ROM, SNPF, TSCA, USAN, USD, YUG

Chemical Classes: Hydrocarbons; Waxes

Functions: Fragrance Ingredient; Skin-Conditioning Agent - Occlusive; Viscosity Increasing Agent - Nonaqueous

Reported Product Categories: Lipsticks; Makeup Bases; Foundations; Mascara; Skin Care Preparations, Misc.; Bath Oils, Tablets, and Salts; Cleansing Products (Cold Creams, Cleansing Lotions, Liquids and Pads); Moisturizing Preparations; Tonics, Dressings, and Other Hair Grooming Aids; Bath Preparations, Misc.; Body and Hand Preparations (Excluding Shaving Preparations); Makeup Preparations (Not eye), Misc.; Personal Cleanliness Products, Misc.; Blushers (All types); Hair Coloring Preparations, Misc.; Eyeliners; Bath Soaps and Detergents; Hair Conditioners; Paste Masks (Mud Packs); Eye Shadows; Night Skin Care Preparations; Nail Polish and Enamels; Bath Capsules; Deodorants (Underarm); Eye Makeup Preparations, Misc.; Face Powders; Colognes and Toilet Waters; Eye Makeup Removers; Eyebrow Pencils; Face and Neck Preparations (Excluding Shaving Preparations); Hair Preparations (Non-coloring), Misc.; Nail Creams and Lotions; Suntan Gels, Creams, and Liquids; Suntan Preparations, Misc.; Baby Lotions, Oils, Powders and Creams; Eye Lotions; Indoor Tanning Preparations; Manicuring Preparations, Misc.; Shaving Cream (Aerosol, Brushless and Lather)

Technical/Other Names:
High Melting Point Paraffin
Low Melting Point Paraffin
Paraffin wax (RIFM)
Petroleum Wax, Crystalline
Waxes, Paraffin

Trade Names:
AEC Paraffin Wax (A & E Connock)
Botaniwax P-110 (Botanigenics)
Chevron Refined Wax 128 (Chevron Lubricants)
Chevron Refined Wax 130 (Chevron Lubricants)
Chevron Refined Wax 135 (Chevron Lubricants)
Chevron Refined Wax 141 (Chevron Lubricants)
Chevron Refined Wax 143 (Chevron Lubricants)
Crude Scale Wax (Ross)
Crude Scale Wax 125 (Ross)
CS-2032 (Huntington)
CS-2037 (Huntington)
CS-2043 (Huntington)
CS-2054 (Huntington)
Hansonwax JH-145, JH-150, JH-160, JH-165 (Hansotech)
Linpar 13-14 (Sasol Italy)

Paraffine Wax Fully Refined 125/130 (Ross)
Paraffine Wax Fully Refined 130 (Ross)
Paraffine Wax Fully Refined 140 (Ross)
Paraffine Wax Fully Refined 145 (Ross)
Paraffine Wax Fully Refined 150 (Ross)
Paraffine Wax Fully Refined 160 (Ross)
Paraffine Wax Fully Refined 165 (Ross)
Paraffin Wax (Ross)
Paraffin Wax Fully Refined 112 (Ross)
Paraffin Wax Fully Refined 112/118 (Ross)
Paraffin Wax Fully Refined 118 (Ross)
Paraffin Wax SP 173 (Strahl & Pitsch)
Paraffin Wax SP-174 (Strahl & Pitsch)
Paraffin Wax SP-192 (Strahl & Pitsch)
Paraffin Wax SP 206 (Strahl & Pitsch)
Paraffin Wax SP 272 (Strahl & Pitsch)
Paraffin Wax SP 324 (Strahl & Pitsch)
Paraffin Wax SP 434 (Strahl & Pitsch)
Paraffin Wax SP 673 (Strahl & Pitsch)
Paraffin Wax SP 674 (Strahl & Pitsch)
Paraffin Wax SP-1158 (Strahl & Pitsch)
Paraffin Wax SP 1275 (Strahl & Pitsch)
Paraffin Wax SP 227B (Strahl & Pitsch)
Paraffin Wax SP 674C (Strahl & Pitsch)
Paraffin Wax - STRALPITZ (Strahl & Pitsch)
Parvan 127 (Exxon)
Parvan 131 (Exxon)
Parvan 137 (Exxon)
Parvan 142 (Exxon)
Parvan 145 (Exxon)
Parvan 152 (Exxon)
Parvan 161 (Exxon)
PIONIER Paraffin Wax (Hansen & Rosenthal)
1-100 S Paraffin Wax (Stevenson-Cooper)

Trade Name Mixtures:
Amphocerin K (Cognis Deutschland)
Argobase EU (Croda Chemicals)
Beeswax Substitute 81-1104 (Ross)
Beeswax Substitute 628/5 (Ross)
7304 Candelilla Substitute (Kahl)
Chamomile Extract LP (Maruzen Pharmaceuticals Co., Ltd.)
Chamomile Extract LP-J (Maruzen Pharmaceuticals Co., Ltd.)
Citrus Unshiu Extract-LP (Maruzen Pharmaceuticals Co., Ltd.)
Covalip 94 (LCW)
Covalip 99 (LCW)
Covapencil 07 (LCW)
HPS3 (Alban Muller)
Isobeeswax SP 154 (Strahl & Pitsch)
Isobeeswax - STRALPITZ (Strahl & Pitsch)
Japanese Angelica Extract-LP (Maruzen Pharmaceuticals Co., Ltd.)
Lanaetex FB (Lanaetex)
Ross Bayberry Wax Substitute 1641 (Ross)
Ross Japan Wax Substitute No. 966 (Ross)
Ross Synthetic Candelilla Wax (Ross)
Solid Vegetable Squalane (Cognis Iberia/Centerchem)
Synthetic Beeswax 6103 (Kahl)

2901 Synthetic Carnauba (Kahl)
Zetesap 5165 (Zschimmer & Schwarz)
Zetesap 5213 (Zschimmer & Schwarz)
Zetesap 813 A (Zschimmer & Schwarz)
Zetesap 813 P (Zschimmer & Schwarz)

PARFUM

Definition: Parfum is a term for ingredient labeling used to identify that a product contains a material or a combination of materials normally added to a cosmetic to produce or to mask a particular odor. The term Parfum shall be used for ingredient labeling in the European Union (EU) instead of listing the individual components of the fragrance under the 6th Amendment to the EC Cosmetics Directive. The INCI name permitted for labeling in the U.S. that corresponds to this EU labeling name is Fragrance. *See "Regulatory and Ingredient Use Information," regarding use of the INCI Names, Fragrance and Parfum in Volume 1, Introduction, Part A.*

PARIETARIA OFFICINALIS EXTRACT

CAS No.
84012-32-8

EINECS No.
281-676-6

JPN Translation:
パリエタリアエキス

Definition: Parietaria Officinalis Extract is an extract of the leaves and stems of the pellitory, *Parietaria officinalis*. See *"Regulatory and Ingredient Use Information," regarding the labeling names for botanical ingredients in Volume 1, Introduction, Part A.*

Information Sources: JCIC, JCLS

Chemical Class: Biological Products

Function: Skin-Conditioning Agent - Miscellaneous

Technical/Other Names:
Extract of Pellitory
Parietary Extract
Pellitory Extract
Pellitory (Parietaria Officinalis) Extract

Trade Name Mixtures:
342 Babyderme HS (Alban Muller)
642 Babyderme LS (Alban Muller)
287 Demaquillant HS (Alban Muller)
687 Demaquillant LS (Alban Muller)
235 Emollient HS (Alban Muller)
635 Emollient LS (Alban Muller)
Extrait de Parietaire PPE PG 40 (Yves Rocher)
Moisturizing Complex 266 (Ennagram)
Moisturizing Phytogreen Complex GXH 266 (Phytochim)

Parietaria Officinalis Extract ies (IES LABO)
Pellitory Liquid B (Ichimaru Pharcos)
Pellitory of the Wall HS (Alban Muller)
Pharcolex BX51 (Ichimaru Pharcos)
Phytelene Complex EGX 251 (Indena SA)
Phytelene of Parietary EG 233 liquid (Indena SA)
Phytelene of Pellitory of Spain EG 481 liquid (Indena SA)
Phytogreen 55 of Parietary EXH 648 Liquid (Phytochim)
Polyplant Moisturizing (Provital/ Centerchem)
Prodhy Extract Parietaire (Prod'Hyg)
PROTECTIVE MOISTURIZER (Greentech S.A)
Wall Pellitory Extract HG (Provital/ Centerchem)

PASSIFLORA ALATA FRUIT EXTRACT

Definition: Passiflora Alata Fruit Extract is an extract of the fruit of *Passiflora alata. See "Regulatory and Ingredient Use Information," regarding the labeling names for botanical ingredients in Volume 1, Introduction, Part A.*

Chemical Class: Biological Products

Function: Hair Conditioning Agent

Technical/Other Name:
Extract of Passiflora Alata Fruit

Trade Name:
Frulix TF Maracuja (Assessa-Industria)

PASSIFLORA EDULIS FLOWER EXTRACT

Definition: Passiflora Edulis Flower Extract is an extract of the flowers of the passionflower, *Passiflora edulis. See "Regulatory and Ingredient Use Information," regarding the labeling names for botanical ingredients in Volume 1, Introduction, Part A.*

Chemical Class: Biological Products

Function: Not Reported

Reported Product Categories: Bath Oils, Tablets, and Salts; Cleansing Products (Cold Creams, Cleansing Lotions, Liquids and Pads); Tonics, Dressings, and Other Hair Grooming Aids; Hair Conditioners; Paste Masks (Mud Packs); Shampoos (Non-coloring)

Technical/Other Names:
Extract of Passiflora Edulis
Extract of Passionflower
Passionflower Extract
Passionflower (Passiflora Edulis) Extract

Trade Name Mixture:
A.H.A. Extracts (Ennagram)

PASSIFLORA EDULIS FRUIT EXTRACT

Definition: Passiflora Edulis Fruit Extract is an extract of the fruit of *Passiflora edulis. See "Regulatory and Ingredient Use Information," regarding the labeling names for botanical ingredients in Volume 1, Introduction, Part A.*

Chemical Class: Biological Products

Function: Not Reported

Technical/Other Names:
Extract of Passiflora Edulis Fruit
Extract of Passionflower Fruit
Passionflower Fruit Extract
Passionflower (Passiflora Edulis) Fruit Extract

Trade Name Mixtures:
Fruitapone Passionfruit B 2/036500 (Symrise)
Fruitapone Passionfruit GT 2/037500 (Symrise)
Glycolysat of Passion Fruit (CEP (Solabia))
Maracuja Extract HS 3349 G (Grau)
Passion Fruit HG (Alban Muller)

PASSIFLORA EDULIS FRUIT JUICE

Definition: Passiflora Edulis Fruit Juice is the juice expressed from the fruit of *Passiflora edulis. See "Regulatory and Ingredient Use Information," regarding the labeling names for botanical ingredients in Volume 1, Introduction, Part A.*

Chemical Class: Biological Products

Function: Skin-Conditioning Agent - Miscellaneous

Trade Name:
Passionfruit Juice Concentrate 9/019702 (Symrise)

PASSIFLORA EDULIS SEED OIL

CAS No.: 97676-26-1

Definition: Passiflora Edulis Seed Oil is the fixed oil expressed from the seeds of the passionfruit, *Passiflora edulis. See "Regulatory and Ingredient Use Information," regarding the labeling names for botanical ingredients in Volume 1, Introduction, Part A.*

Chemical Class: Fats and Oils

Functions: Skin-Conditioning Agent - Emollient; Skin-Conditioning Agent - Occlusive

Technical/Other Names:
Oil of Passiflora Edulis
Oil of Passionflower
Oils, Passiflora Edulis
Oils, Passionflower
Passionflower Oil
Passionflower (Passiflora Edulis) Oil

Trade Names:
AEC Passionflower Oil (A & E Connock)
Passionflower Oil (Dekker)
Passion Fruit Oil (Beraca industria E Comercio LTDA)
Passionfruit Seed Oil (Nestle World Trade)

Trade Name Mixture:
Crodamazon Maracuja (Croda, Inc.)

PASSIFLORA INCARNATA EXTRACT

Definition: Passiflora Incarnata Extract is an extract of the whole plant, *Passiflora incarnata. See "Regulatory and Ingredient Use Information," regarding the labeling names for botanical ingredients in Volume 1, Introduction, Part A.*

Chemical Class: Biological Products

Function: Cosmetic Astringent

Technical/Other Name:
Extract of Passiflora Incarnata

Trade Name:
Herbalia Passiflora (Cognis Deutschland)

Trade Name Mixtures:
Actiphyte Passionflower AL (Active Organics)
Actiphyte Passionflower Lipo M (Active Organics)
Actiphyte Passionflower Lipo Soy (Active Organics)

PASSIFLORA INCARNATA FLOWER EXTRACT

CAS No.: 84012-31-7

Definition: Passiflora Incarnata Flower Extract is an extract of the flowers of the passionflower, *Passiflora incarnata. See "Regulatory and Ingredient Use Information," regarding the labeling names for botanical ingredients in Volume 1, Introduction, Part A.*

Information Source: 21CFR172.510

Chemical Class: Biological Products

Function: Skin-Conditioning Agent - Miscellaneous

Reported Product Categories: Bath Oils, Tablets, and Salts; Cleansing Products (Cold

Creams, Cleansing Lotions, Liquids and Pads); Tonics, Dressings, and Other Hair Grooming Aids; Hair Conditioners; Paste Masks (Mud Packs); Shampoos (Non-coloring)

Technical/Other Names:
Extract of Passiflora Incarnata
Extract of Passionflower
Passionflower Extract
Passionflower (Passiflora Incarnata) Extract

Trade Name Mixtures:
Actiphyte of Passionflower BG50 (Active Organics)
Actiphyte of Passionflower GL50 (Active Organics)
Actiphyte of Passionflower Lipo S (Active Organics)
Actiphyte of Passionflower PG50 (Active Organics)
Actiphyte of Passion Fruit Hull BG50 (Active Organics)
Actiphyte of Passion Fruit Hull GL50 (Active Organics)
Actiphyte of Passion Fruit Hull Lipo S (Active Organics)
Actiphyte of Passion Fruit Hull PG50 (Active Organics)
A.H.A. Extracts (Phytochim)
BBC Relaxing Complex (Bio-Botanica)
Caresse Phytospa (Alban Muller)
Cosmelene (R) of Passionflower (Indena SA)
Extrait De Passiflore MPE100 (Yves Rocher)
Extrapone Passionflower 2/033165 (Symrise)
Flower Passiflora Oil (Greentech)
Glycolysat of Passionflower (CEP (Solabia))
Herbasol-Extract Passion Flower (Cosmetochem)
Maypop Extract HS 2828 G (Grau)
Passion Flower HS (Alban Muller)
Passion Flowers Milk (CEP (Solabia))
Phytelene of Passion Flower EG 389 liquid (Indena SA)
Phytogreen 55 of Passion Flower EXH 661 Liquid (Phytochim)
Prodhy Extract Passiflore (Prod'Hyg)
Radicaptol (Solabia)
Relax Complex RPJ (CEP (Solabia))
Relaxing Extract (CEP (Solabia))
Soothing Phytospa (Alban Muller)
Vegebois of Passionflower (CEP (Solabia))
Vegetol Passionflower MCF 773 Hydro (Gattefosse s.a.)
VT-081 Extract of Passionflower (Vege-Tech)

PASSIFLORA INCARNATA FRUIT EXTRACT

Definition: Passiflora Incarnata Fruit Extract is an extract of the fruit of the passionflower, *Passiflora incarnata.* See *"Regulatory and Ingredient Use Information,"* regarding the labeling names for botanical ingredients in Volume 1, Introduction, Part A.

Chemical Class: Biological Products

Function: Not Reported

Technical/Other Names:
Extract of Passiflora Incarnata Fruit
Extract of Passionfruit
Passionflower Fruit Extract
Passionflower (Passiflora Incarnata) Fruit Extract
Passionfruit Extract

Trade Name Mixtures:
Actiphyte of Passion Fruit BG50 (Active Organics)
Actiphyte of Passion Fruit GL50 (Active Organics)
Actiphyte of Passion Fruit Lipo S (Active Organics)
Actiphyte of Passion Fruit PG50 (Active Organics)
Herbasol Extract Passion Fruit (Cosmetochem) (Cosmetochem International Ltd.)
Natupure Passion Flower (E.U.K)
VT-116 Extract of Passionfruit (Vege-Tech)

PASSIFLORA INCARNATA SEED OIL

CAS No.: 97676-26-1

Definition: Passiflora Incarnata Seed Oil is the oil expressed from the seeds of *Passiflora incarnata.* See *"Regulatory and Ingredient Use Information,"* regarding the labeling names for botanical ingredients in Volume 1, Introduction, Part A.

Information Source: TSCA

Chemical Class: Essential Oils

Function: Skin-Conditioning Agent - Occlusive

Technical/Other Names:
Oil of Passiflora Incarnata
Oil of Passionflower
Oils, Passiflora Incarnata
Oils, Passionflower
Passionflower Oil
Passionflower (Passiflora Incarnata) Oil

Trade Names:
Cegesoft PFO (Cognis Deutschland)
Huile de Graines de Fruits de la Passion (Bertin)

Trade Name Mixture:
Crodamazon Maracuja (Croda, Inc.)

PASSIFLORA LAURIFOLIA FLOWER EXTRACT

Definition: Passiflora Laurifolia Flower Extract is an extract of the flowers of *Passiflora laurifolia.* See *"Regulatory and Ingredient Use Information,"* regarding the labeling names for botanical ingredients in Volume 1, Introduction, Part A.

Chemical Class: Biological Products

Function: Not Reported

Reported Product Categories: Bath Oils, Tablets, and Salts; Cleansing Products (Cold Creams, Cleansing Lotions, Liquids and Pads); Tonics, Dressings, and Other Hair Grooming Aids; Hair Conditioners; Paste Masks (Mud Packs); Shampoos (Non-coloring)

Technical/Other Names:
Extract of Passiflora Laurifolia
Extract of Passionflower
Passionflower (Passiflora Laurifolia) Extract

Trade Name:
Passionflower Extract (Bell Flavors)

PASSIFLORA LAURIFOLIA FRUIT EXTRACT

Definition: Passiflora Laurifolia Fruit Extract is an extract of the fruit of the passionflower, *Passiflora laurifolia.* See *"Regulatory and Ingredient Use Information,"* regarding the labeling names for botanical ingredients in Volume 1, Introduction, Part A.

Chemical Class: Biological Products

Function: Not Reported

Technical/Other Names:
Extract of Passiflora Laurifolia Fruit
Extract of Passionfruit
Passionflower Fruit Extract
Passionflower (Passiflora Laurifolia) Fruit Extract
Passionfruit Extract

PASSIFLORA QUADRANGULARIS FLOWER EXTRACT

Definition: Passiflora Quadrangularis Flower Extract is an extract of the flowers of the passionflower, *Passiflora quadrangularis.* See *"Regulatory and Ingredient Use*

Information," regarding the labeling names for botanical ingredients in Volume 1, Introduction, Part A.

Information Source: 21CFR172.510

Chemical Class: Biological Products

Function: Not Reported

Reported Product Categories: Bath Oils, Tablets, and Salts; Cleansing Products (Cold Creams, Cleansing Lotions, Liquids and Pads); Tonics, Dressings, and Other Hair Grooming Aids; Hair Conditioners; Paste Masks (Mud Packs); Shampoos (Non-coloring)

Technical/Other Names:
Extract of Passiflora Quadrangularis
Extract of Passionflower
Passionflower Extract
Passionflower (Passiflora Quadrangularis) Extract

PASSIFLORA QUADRANGULARIS FRUIT

Definition: Passiflora Quadrangularis Fruit is the fruit of the passionflower, *Passiflora quadrangularis. See "Regulatory and Ingredient Use Information," regarding the labeling names for botanical ingredients in Volume 1, Introduction, Part A.*

Chemical Class: Biological Products

Functions: Cosmetic Astringent; Skin-Conditioning Agent - Miscellaneous

Technical/Other Name:
Passionflower (Passiflora Quadrangularis) Fruit

Trade Name:
AEC Passionfruit Puree (A & E Connock)

PASSIFLORA QUADRANGULARIS FRUIT EXTRACT

Definition: Passiflora Quadrangularis Fruit Extract is an extract of the fruit of the passionflower, *Passiflora quadrangularis. See "Regulatory and Ingredient Use Information," regarding the labeling names for botanical ingredients in Volume 1, Introduction, Part A.*

Chemical Class: Biological Products

Function: Not Reported

Technical/Other Names:
Extract of Passiflora Quadrangularis Fruit
Extract of Passionfruit
Passionflower Fruit Extract
Passionflower (Passiflora Quadrangularis) Fruit Extract
Passionfruit Extract

Trade Name Mixtures:
Complex 2 - Moisturizing (Provital/Centerchem)
Pronalen Fruit Acid AHA-5 (Provital/Centerchem)
Pronalen Fruit Acid AHA-20 (Provital/Centerchem)
Pronalen Fruit Acid AHA-50 (Provital/Centerchem)
PRONALEN MOISTURIZING-II (Provital/Centerchem)

PASSIFLORA QUADRANGULARIS FRUIT JUICE

Definition: Passiflora Quadrangularis Fruit Juice is the liquid expressed from the fruit of the passionflower, *Passiflora quadrangularis. See "Regulatory and Ingredient Use Information," regarding the labeling names for botanical ingredients in Volume 1, Introduction, Part A.*

Chemical Class: Biological Products

Functions: Cosmetic Astringent; Skin-Conditioning Agent - Miscellaneous

Technical/Other Names:
Juice, Passiflora Quadrangularis
Juice, Passionflower
Passiflora Quadrangularis Juice
Passionflower (Passiflora Quadrangularis) Juice

Trade Name:
AEC Passionfruit Conc (A & E Connock)

PASSIFLORA QUADRANGULARIS HULL EXTRACT

Definition: Passiflora Quadrangularis Hull Extract is an extract of the hull of the passionflower, *Passiflora quadrangularis. See "Regulatory and Ingredient Use Information," regarding the labeling names for botanical ingredients in Volume 1, Introduction, Part A.*

Chemical Class: Biological Products

Function: Not Reported

Technical/Other Name:
Passionflower (Passiflora Quadrangularis) Hull Extract

PAULLINIA CUPANA FRUIT EXTRACT

Definition: Paullinia Cupana Fruit Extract is an extract of the fruit of the guarana, *Paullinia cupana. See "Regulatory and Ingredient Use Information," regarding the*

labeling names for botanical ingredients in Volume 1, Introduction, Part A.

Chemical Class: Biological Products

Function: Not Reported

Technical/Other Names:
Extract of Guarana Fruit
Extract of Paullinia Cupana
Guarana (Paullinia Cupana) Fruit Extract
Paullinia Cupana Extract

Trade Name Mixture:
Pronalen Guarana (Provital/Centerchem)

PAULLINIA CUPANA SEED EXTRACT

CAS No.	EINECS No.
84929-28-2	284-512-1

Definition: Paullinia Cupana Seed Extract is an extract of the seeds of the guarana, *Paullinia cupana. See "Regulatory and Ingredient Use Information," regarding the labeling names for botanical ingredients in Volume 1, Introduction, Part A.*

Chemical Class: Biological Products

Functions: Lytic Agent; Skin-Conditioning Agent - Miscellaneous

Technical/Other Names:
Extract of Guarana
Guarana Extract
Guarana (Paullinia Cupana) Extract

Trade Name:
Dried Guarana Extract (Greentech)

Trade Name Mixtures:
Actiphyte of Guarana BG50 (Active Organics)
Actiphyte of Guarana GL50 (Active Organics)
Actiphyte of Guarana Lipo S (Active Organics)
Actiphyte of Guarana PG50 (Active Organics)
Ecofloral Anti-Cellulite Complex (Centroflora)
Ecofloral Guarana (Centroflora)
Extrait huileux de Guarana (Greentech)
Fruitapone Guarana B 2/036200 (Symrise)
Fruitapone Guarana GT 2/037200 (Symrise)
Glycolysat of Guarana (CEP (Solabia)
Guarana Glycolic Extract NTR2 (Centroflora Group)
Guarana HPG Titrated (Alban Muller)
Guarana Phytexcell (Crodarom)
Herbasol Extract Guarana (Cosmetochem) (Cosmetochem International Ltd.)
Liquid Guaranine (Libiol)
Optivegetol Guarana P107 Hydro (Gattefosse s.a.)
Optivegetol Guarana P128 Hydro (Gattefosse s.a.)

Pronalen A/C HSC (Provital/Centerchem)
Pronalen Anti-Cellulite HSC (Provital/
Centerchem)
Pronalen Modeling (Provital/Centerchem)
Pronalen Slimming (Provital/Centerchem)
QuenchT (Cosmetic Ingredient Resources/
Centerchem)
REGU-SLIM (Pentapharm/Centerchem)
Vegecocktail of Caffeine (CEP (Solabia))
VT-261 Extract of Guarana (Vege-Tech)

PAULOWNIA IMPERIALIS EXTRACT

Definition: Paulownia Imperialis Extract is
an extract of the leaves, bark and flowers
of *Paulownia imperialis*. See *"Regulatory
and Ingredient Use Information," regarding
the labeling names for botanical ingredients
in Volume 1, Introduction, Part A.*

Chemical Class: Biological Products

Function: Not Reported

Technical/Other Name:
Extract of Paulownia Imperialis

Trade Name Mixture:
Campo I Tung Extract (Campo)

PAULOWNIA TOMENTOSA LEAF EXTRACT

Definition: Paulownia Tomentosa Leaf
Extract is an extract of the leaves of
Paulownia tomentosa. See *"Regulatory and
Ingredient Use Information," regarding the
labeling names for botanical ingredients in
Volume 1, Introduction, Part A.*

Chemical Class: Biological Products

Function: Skin-Conditioning Agent -
Humectant

Technical/Other Name:
Extract of Paulownia Tomentosa Leaf

Trade Name Mixture:
Kiri-you Extract (Fuji Sangyo)

PAVONIA ODORATA ROOT POWDER

Definition: Pavonia Odorata Root Powder
is the powder derived from the roots of
Pavonia odorata. See *"Regulatory and
Ingredient Use Information," regarding the
labeling names for botanical ingredients in
Volume 1, Introduction, Part A.*

Chemical Class: Biological Products

Function: Not Reported

PCA

CAS Nos.	EINECS Nos.
98-79-3	202-700-3
149-87-1 (dl-alpha)	205-748-3

JPN Translation:
PCA

Empirical Formula:
$C_5H_7NO_3$

Definition: PCA is the cyclic organic
compound that conforms to the formula:

Information Sources: CIR: [SQ] IJT-18
(SUPPL. 2)1999, INN, JCLS, JSCI, MI-13
(8091), TSCA

Chemical Classes: Amides; Heterocyclic
Compounds

Function: Skin-Conditioning Agent -
Humectant

Reported Product Categories: Hair Conditioners; Moisturizing Preparations

Technical/Other Names:
Glutimic Acid
Glutiminic Acid
5-Oxo-L-Proline
L-Proline, 5-Oxo-
L-Pyroglutamic Acid
2-Pyrrolidone-5-Carboxylic Acid
DL-Pyrrolidonecarboxylic Acid

Trade Names:
Ajidew A-100 (Ajinomoto)
Pidolidone (UCIB (Solabia))

Trade Name Mixtures:
Ajidew SP-100 (Ajinomoto)
Fluxhydran (Laboratoires Serobiologiques)
Hydro-Diffuser Microreservoir (Sederma)
Hydroveg VV (Variati)
Osmhydran (Laboratoires Serobiologiques)
Osmhydran LS 8453 (Laboratoires Sero-
biologiques)
P.A. Antifroid LS 9224B (Laboratoires
Serobiologiques)

PCA DIMETHICONE

CAS No.: 179005-03-9

Definition: PCA Dimethicone is the siloxane
polymer that conforms to the formula:

Chemical Classes: Amides; Heterocyclic
Compounds; Siloxanes and Silanes;
Synthetic Polymers

Functions: Hair Conditioning Agent; Skin-
Conditioning Agent - Miscellaneous

Technical/Other Name:
Siloxanes, and Silicones, 3-(4-Carboxy-2-
Oxo-1-Pyrrolidinyl)Propyl Me, Di-Me

Trade Name:
Monasil PCA (Uniqema)

PCA ETHYL COCOYL ARGINATE

JPN Translation:
ココイルアルギニンエチルPCA

Definition: PCA Ethyl Cocoyl Arginate is a
salt of PCA and ethyl cocoyl arginate that
conforms generally to the formula:

where RCO- represents the fatty acids
derived from coconut oil.

Information Sources: JCIC, JCLS, JSQI

Chemical Classes: Amides; Amino Acids;
Heterocyclic Compounds; Organic Salts

Functions: Hair Conditioning Agent; Skin-
Conditioning Agent - Miscellaneous

Technical/Other Name:
DL-Pyrrolidonecarboxylic Acid Salt of L-
Cocoyl Arginine Ethyl Ester

Trade Name:
CAE (Ajinomoto)

PCA GLYCERYL OLEATE

JPN Translation:
PCA オレイン酸グリセリル

Empirical Formula:
$C_{26}H_{45}NO_6$

Definition: PCA Glyceryl Oleate is the
organic compound that conforms to the
formula:

Information Sources: JCIC, JCLS, JSQI

Chemical Classes: Esters; Heterocyclic
Compounds

Function: Skin-Conditioning Agent - Miscellaneous

Technical/Other Name:
Glyceryl Monopyroglutamate Monooleate

Trade Name:
Amifat P-30 (Ajinomoto)

PEANUT ACID

CAS No.   **EINECS No.**
91051-35-3   293-087-1

Definition: Peanut Acid is a mixture of fatty acids derived from Arachis Hypogaea (Peanut) Oil (q.v.).

Information Source: CIR: [S] IJT-20 (SUPPL. 2) 2001

Chemical Class: Fatty Acids

Function: Surfactant - Cleansing Agent

Technical/Other Names:
Acids, Peanut
Arachis Hypogaea (Peanut) Acid

Trade Name:
PRIFAC 7911 (Uniqema Europe)

PEANUTAMIDE MEA

Definition: Peanutamide MEA is a mixture of ethanolamides of the fatty acids derived from Arachis Hypogaea (Peanut) Oil (q.v.). It conforms generally to the formula:

$$RC \overset{O}{\overset{\|}{-}} NHCH_2CH_2OH$$

where RCO- represents the fatty acids derived from peanut oil.

Information Source: CTFA D

Chemical Class: Alkanolamides

Functions: Surfactant - Foam Booster; Viscosity Increasing Agent - Aqueous

Technical/Other Names:
N-(2 Hydroxyethyl) Peanut Acid Amide
Monoethanolamine Peanut Acid Amide
Peanut Fatty Acid Amide, N-(2-Hydroxyethyl)-
Peanut Fatty Acid Monoethanolamide

PEANUTAMIDE MIPA

Definition: Peanutamide MIPA is a mixture of isopropanolamides of the fatty acids derived from Arachis Hypogaea (Peanut) Oil (q.v.). It conforms generally to the formula:

$$RC \overset{O}{\overset{\|}{-}} NHCH_2\underset{CH_3}{CHOH}$$

where RCO- represents the fatty acids derived from peanut oil.

Chemical Class: Alkanolamides

Functions: Surfactant - Foam Booster; Viscosity Increasing Agent - Aqueous

Technical/Other Names:
N-(2-Hydroxypropyl)Peanut Acid Amide
Peanut Fatty Acid Amide, N-(2-Hydroxypropyl)-

PEANUT GLYCERIDES

CAS No.   **EINECS No.**
91744-77-3   294-643-6

Definition: Peanut Glycerides is a mixture of mono-, di- and triglycerides derived from Arachis Hypogaea (Peanut) Oil (q.v.).

Information Source: CIR: [S] IJT-20 (SUPPL. 2)2001

Chemical Class: Glyceryl Esters and Derivatives

Function: Skin-Conditioning Agent - Occlusive

Technical/Other Name:
Glycerides, Peanut Oil, Mono-, Di- and Tri-

Trade Name:
Olicine (Gattefosse s.a.)

PEANUT OIL PEG-6 ESTERS

JPN Translation:
ピーナッツ油 PEG - 6
Definition: Peanut Oil PEG-6 Esters is a complex mixture obtained from the transesterification of Arachis Hypogaea (Peanut) Oil (q.v.) and PEG-6 (q.v.).

Chemical Class: Glyceryl Esters and Derivatives

Functions: Skin-Conditioning Agent - Emollient; Surfactant - Emulsifying Agent

PEA PALMITATE

Definition: Pea Palmitate is the product obtained by the reaction of crushed peas with palmitic acid chloride.

Chemical Classes: Amides; Carbohydrates; Esters; Protein Derivatives

Function: Skin-Conditioning Agent - Miscellaneous

PEARL POWDER

Definition: Pearl Powder is the dried powder obtained from freshwater pearls.

Chemical Class: Biological Products

Function: Not Reported

Trade Names:
AEC Pearl Powder (A & E Connock)
OriStar PP (Orient Stars)
Pearl Plus SL (Sino Lion)
PPP-100 (Sino Lion)

Trade Name Mixture:
Pearl Powder Extract (Carrubba)

PEAT

Definition: Peat is a highly organic material found in marshy or damp regions, composed of partially decayed vegetable matter. *See Reported Ingredient Functions-The Cosmetic Drug Distinction, in Regulatory and Ingredient Use Information, Volume I, Part A.*

Chemical Class: Biological Products

Functions: Antiacne Agent; Skin-Conditioning Agent - Miscellaneous

Trade Names:
Golden Moor (Golden Moor)
Neydhartinger Heilmoor (HEILMOORBAD)
PELAVIE PEAT (Creations Couleurs)
PELAVIE PEAT FINE (Creations Couleurs)

PEAT EXTRACT

Definition: Peat Extract is an extract of Peat (q.v.).

Chemical Class: Biological Products

Function: Skin-Conditioning Agent - Miscellaneous

Technical/Other Name:
Extract of Peat

Trade Name:
White Peat Extract (CO2 extract) (Aromtech)

PEAT WATER

Definition: Peat Water is an aqueous suspension of Peat (q.v.). *See Reported Ingredient Functions-The Cosmetic Drug Distinction, in Regulatory and Ingredient Use Information, Volume I, Part A.*

Chemical Class: Biological Products

Functions: Antiacne Agent; Skin-Conditioning Agent - Miscellaneous

Trade Names:
Bonaparte Peat extract (Charter-Pacific)

Neydhartinger Heilmoor-Schwarzwasser (HEILMOORBAD)

Trade Name Mixture:
Herbasol Extract Peat (Cosmetochem) (Cosmetochem International Ltd.)

PECTIN

CAS No.	EINECS No.
9000-69-5	232-553-0

JPN Translation:
ペクチン

Definition: Pectin is a purified carbo-hydrate product obtained from the dilute acid extract of the inner portion of the rind of citrus fruits or from apple pomace. It consists chiefly of partially methoxylated polygalacturonic acids. In the United States, Pectin may be used as an active ingredient in OTC drug products. When used as an active drug ingredient, the established name is *Pectin. See "Regulatory and Ingredient Use Information," regarding the labeling names for U.S. OTC Drug Ingredients in Volume 1, Introduction, Part A.*

Information Sources: AUS, BRA, 21CFR135.140, 21CFR145, 21CFR150, 21CFR150.110, 21CFR150.140, 21CFR150.141, 21CFR150.160, 21CFR150.161, 21CFR173.385, 21CFR184.1588, 21CFR310.545, 27CFR21.141, FCC, JCIC, JCLS, JSQI, MAR, MI-13(7135), OTC-I-OH, TSCA, USAN, USD, USP XXIV

Chemical Class: Gums, Hydrophilic Colloids and Derivatives

Functions: Binder; Emulsion Stabilizer; Oral Health Care Drug; Viscosity Increasing Agent - Aqueous

Reported Product Categories: Shampoos (Non-coloring); Hair Conditioners; Permanent Waves; Hair Preparations (Non-coloring), Misc.; Tonics, Dressings, and Other Hair Grooming Aids

Technical/Other Name:
Citrus Pectin

Trade Names:
Genu (Hercules)
Genu-2 (CP Kelco U.S.)
Pectins (Herbstreith & Fox)

Trade Name Mixture:
Kollosin - RS (Kramer)

PEG-4

CAS Nos.	EINECS No.
112-60-7	203-989-9
25322-68-3 (Generic)	

JPN Translation:
PEG - 4

Empirical Formula:
$C_8H_{18}O_5$

Definition: PEG-4 is the polymer of ethylene oxide that conforms generally to the formula:

$$H(OCH_2CH_2)_nOH$$

where n has an average value of 4.

Information Sources: BAN, 21CFR73.1, 21CFR73.2180, 21CFR172.210, 21CFR172.770, 21CFR172.820, 21CFR173.310, 21CFR173.340, 21CFR175.105, 21CFR175.300, 21CFR178.3520, 21CFR178.3750, CIR: [S], CTFA S, FCC, JAN, JCLS, JSCI, MAR, MI-13(7651), NF XIX, ROM, TSCA, USAN

Chemical Classes: Alkoxylated Alcohols; Polymeric Ethers

Functions: Humectant; Solvent

Reported Product Categories: Bath Preparations, Misc.; Body and Hand Preparations (Excluding Shaving Preparations); Aftershave Lotions; Baby Shampoos; Moisturizing Preparations; Hair Conditioners; Skin Care Preparations, Misc.; Deodorants (Underarm); Shampoos (Non-coloring)

Technical/Other Names:
Ethanol, 2,2'-[Oxybis(2,1-Ethanediyloxy)-Bis-2,2'-[Oxybis(2,1-Ethanediyloxy)]Bisethanol
Polyethylene Glycol 200
Polyoxyethylene (4)
Tetraethylene Glycol

Trade Names:
CARBOWAX PEG 200 (Dow Chemical)
DePEG 200 (DeForest)
Hetoxide PEG-200 (Heterene)
Lipo Polyglycol 200 (Lipo)
Lumulse PEG 200 (Lambent)
Macrogol 200 (NOF)
Pluracol E 200 (BASF)
Polyglycol E-200 (Dow Chemical)
Polyglykol 200 USP (Clariant)
Polyglykol 200 USP (Clariant GmbH, Personal Care)
Sabopeg 200 (Sabo)
TOHO PEG#200 (Toho)
Unipeg-200 X (Universal Preserv-A-Chem)
Upiwax 200 (Universal Preserv-A-Chem)

Trade Name Mixtures:
Chemyde Preservative (Chemeq Limited)
Dowicil QK-20 Antimicrobial (The Dow Chemical Co.)
Eyebright Extract HS 2727 G (Grau)

PEG-6

CAS Nos.	EINECS No.
2615-15-8	220-045-1
25322-68-3 (Generic)	

JPN Translation:
PEG - 6

Empirical Formula:
$C_{12}H_{26}O_7$

Definition: PEG-6 is the polymer of ethylene oxide that conforms generally to the formula:

$$H(OCH_2CH_2)_nOH$$

where n has an average value of 6.

Information Sources: BAN, BP, BPC, 21CFR172.210, 21CFR172.770, 21CFR172.820, 21CFR173.310, 21CFR173.340, 21CFR175.105, 21CFR175.300, 21CFR178.3570, 21CFR178.3910, CIR: [SQ] JACT-12(5)-1993, CTFA S, CZE, FCC, JAN, JCLS, JSCI, MAR, MI-13(7651), NF XVIII, OTC-I-OP, ROM, TSCA, USAN, USD

Chemical Classes: Alkoxylated Alcohols; Polymeric Ethers

Functions: Humectant; Solvent

Reported Product Categories: Bath Oils, Tablets, and Salts; Moisturizing Preparations; Bath Capsules; Cleansing Products (Cold Creams, Cleansing Lotions, Liquids and Pads); Bath Soaps and Detergents; Face and Neck Preparations (Excluding Shaving Preparations); Eyebrow Pencils; Skin Care Preparations, Misc.; Suntan Gels, Creams, and Liquids; Bath Preparations, Misc.; Body and Hand Preparations (Excluding Shaving Preparations); Eye Makeup Preparations, Misc.; Paste Masks (Mud Packs)

Technical/Other Names:
Hexaethylene Glycol
3,6,9,12,15-Pentaoxaheptadecane-1,17-Diol
Polyethylene Glycol 300
Polyoxyethylene (6)

Trade Names:
CARBOWAX PEG 300 (Dow Chemical)
DePEG 300 (DeForest)
Hetoxide PEG-300 (Heterene)
Lipo Polyglycol 300 (Lipo)
Lipoxol 300 MED (Sasol GmbH - Marl)
Lumulse PEG 300 (Lambent)
Macrogol 300 (NOF)
Pluracare E 300 (BASF)
Pluracol E 300 (BASF)
Polyglykol 300 (Clariant)
Polyglykol 300 (Clariant GmbH, Personal Care)
Sabopeg 300 (Sabo)
TOHO PEG#300 (Toho)
Upiwax 300 (Universal Preserv-A-Chem)

Trade Name Mixtures:
AZG-7190 PEG-6 Solution (Summit Research Labs)
CARBOWAX PEG 540 Blend (Dow Chemical)
Cellulinol (Prod'Hyg)

CREALBA AQUASOL (Creations Couleurs)
Filladyn (Laboratoires Serobiologiques)
Lanobase S.E. (Lanaetex)
Lanogen 1500 (Clariant)
Lanogen 1500 (Clariant GmbH, Personal Care)
Swertianin P (Ichimaru Pharcos)
Unipeg-1500 X (Universal Preserv-A-Chem)
Uniwax 1450 (Universal Preserv-A-Chem)
Vegeles SR (Laboratoires Serobiologiques)

PEG-7

CAS No.: 25322-68-3 (Generic)

Definition: PEG-7 is the polymer of ethylene oxide that conforms generally to the formula:

$$H(OCH_2CH_2)_nOH$$

where n has an average value of 7.

Chemical Classes: Alkoxylated Alcohols; Polymeric Ethers

Functions: Humectant; Solvent

Technical/Other Names:
Polyethylene Glycol (7)
Polyoxyethylene (7)

Trade Name:
Jeechem 300 (Jeen)

PEG-8

CAS Nos.   **EINECS No.**
5117-19-1   225-856-4
25322-68-3 (Generic)

JPN Translation:
PEG - 8

Empirical Formula:
$C_{16}H_{34}O_9$

Definition: PEG-8 is the polymer of ethylene oxide that conforms generally to the formula:

$$H(OCH_2CH_2)_nOH$$

where n has an average value of 8.

Information Sources: BAN, BRA, 21CFR172.210, 21CFR172.770, 21CFR172.820, 21CFR173.310, 21CFR173.340, 21CFR175.105, 21CFR175.300, 21CFR178.3750, 21CFR178.3910, 21CFR181.22, 21CFR181.30, CIR: [SQ] JACT-12(5)1993, CTFA S, FCC, HUN, JAN, JCLS, JSCI, MAR, MI-13(7651), NF XVIII, NFJ, OTC-I-OP, PN, POL, ROM, TSCA, USAN, USD

Chemical Classes: Alkoxylated Alcohols; Polymeric Ethers

Functions: Humectant; Solvent

Reported Product Categories: Bath Oils, Tablets, and Salts; Cleansing Products (Cold Creams, Cleansing Lotions, Liquids and Pads); Deodorants (Underarm); Foundations; Bath Preparations, Misc.; Body and Hand Preparations (Excluding Shaving Preparations); Moisturizing Preparations; Skin Care Preparations, Misc.; Makeup Bases; Bath Soaps and Detergents; Paste Masks (Mud Packs); Bath Capsules; Hair Conditioners; Bubble Baths; Eye Makeup Removers; Face and Neck Preparations (Excluding Shaving Preparations); Night Skin Care Preparations; Skin Fresheners; Lipsticks; Shaving Cream (Aerosol, Brushless and Lather)

Technical/Other Names:
3,6,9,12,15,18,21-Heptaoxatricosane-1,23-diol
Octaethylene Glycol
Polyethylene Glycol 400
Polyoxyethylene (8)

Trade Names:
CARBOWAX PEG 400 (Dow Chemical)
DePEG 400 (DeForest)
Jeechem 400 (Jeen)
Lipo Polyglycol 400 (Lipo)
Lipoxol 400 MED (Sasol GmbH - Marl)
Lumulse PEG 400 (Lambent)
Macrogol 400 (NOF)
Pluracare E 400 (BASF)
Pluracol E 400 (BASF)
Polyglykol 400 (Clariant)
Polyglykol 400 (Clariant GmbH, Personal Care)
Prochem 400 (Protameen)
Renex PEG 400 (Uniqema Americas)
RPF Complex (Greentech)
Sabopeg 400 (Sabo)
Sympatens-PEG/400 (Kolb)
TOHO PEG#400 (Toho)
Unipeg-400 X (Universal Preserv-A-Chem)
Upiwax 400 (Universal Preserv-A-Chem)

Trade Name Mixtures:
Activated Botanicals Estroherb Complex AB 106 (Norjin)
Activated Botanicals P8-MIN AB 102 (Norjin)
Afron 22 (Vevy)
Afron-A (Vevy)
Afron-LS (Vevy)
Afron-N (Vevy)
Bio-Bustyl (Sederma)
Biopeptide-EL (Sederma)
Carbossalina (Vevy)
Ceramide A2 (Sederma)
Ceramide 2 Sol 2% (Sederma)
Dermocide L (Fabriquimica)
Hair Complex Aquosum (CLR)
Hostapon SCI-40 L (Clariant GmbH, Personal Care)
JM ActiCare Plus (Microbial Systems)
Kalixide Idrata (Vevy)
Kava Kava (Sederma)

Melibion (Vevy)
Osmohair (Sederma)
Oxynex K Liquid (Merck KGaA/EMD Chemicals Inc.)
Polysol GL (Polygon)
Seromarine (Sederma)
Tensioplastidina Avena (Vevy)
Vegewhite (LCW)
Vitaderm (Fabriquimica)

PEG-9

CAS Nos.   **EINECS No.**
3386-18-3   222-206-1
25322-68-3 (Generic)

Empirical Formula:
$C_{18}H_{38}O_{10}$

Definition: PEG-9 is the polymer of ethylene oxide that conforms generally to the formula:

$$H(OCH_2CH_2)_nOH$$

where n has an average value of 9.

Information Sources: BAN, 21CFR172.210, 21CFR172.770, 21CFR172.820, 21CFR173.310, 21CFR173.340, 21CFR175.105, 21CFR175.300, 21CFR178.3750, 21CFR178.3910, JAN, MI-13(7651), NF XVIII, TSCA, USAN

Chemical Classes: Alkoxylated Alcohols; Polymeric Ethers

Functions: Humectant; Solvent

Technical/Other Names:
Nonaethylene Glycol
3,6,9,12,15,18,21,24-Octaoxahexacosane-1,26-diol
Polyoxyethylene (9)

Trade Name:
Sipol PEG 400 (Specialty Industrial)

Trade Name Mixture:
MACKERNIUM SFES (McIntyre)

PEG-10

CAS Nos.   **EINECS No.**
5579-66-8   226-962-3
25322-68-3 (Generic)

Empirical Formula:
$C_{20}H_{42}O_{11}$

Definition: PEG-10 is the polymer of ethylene oxide that conforms generally to the formula:

$$H(OCH_2CH_2)_nOH$$

where n has an average value of 10.

Information Sources: BAN, 21CFR172.210, 21CFR172.770,

21CFR172.820, 21CFR173.310,
21CFR173.340, 21CFR175.105,
21CFR175.300, 21CFR178.3750,
21CFR178.3910, JAN, MI-13(7651), NF
XVIII, TSCA, USAN

Chemical Classes: Alkoxylated Alcohols;
Polymeric Ethers

Functions: Humectant; Solvent

Technical/Other Names:
Decaethylene Glycol
3,6,9,12,15,18,21,24,27-
Nonaoxanonacosane-1,29-diol
Polyethylene Glycol 500
Polyoxyethylene (10)

PEG-12

CAS Nos.	EINECS No.
6790-09-6	229-859-1
25322-68-3 (Generic)	

JPN Translation:
PEG - 12

Empirical Formula:
$C_{24}H_{50}O_{13}$

Definition: PEG-12 is the polymer of
ethylene oxide that conforms generally to the
formula:

$$H(OCH_2CH_2)_nOH$$

where n has an average value of 12.

Information Sources: BAN,
21CFR172.210, 21CFR172.770,
21CFR172.820, 21CFR173.310,
21CFR173.340, 21CFR175.105,
21CFR175.300, 21CFR178.3750,
21CFR178.3910, CTFA S, FCC, JAN,
JCLS, JSCI, MI-13(7651), NF XIX, ROM,
TSCA, USAN

Chemical Classes: Alkoxylated Alcohols;
Polymeric Ethers

Functions: Humectant; Solvent

Reported Product Categories: Bath Soaps
and Detergents; Moisturizing Preparations

Technical/Other Names:
Dodecaethylene Glycol
Polyethylene Glycol 600
Polyoxyethylene (12)
3,6,9,12,15,18,21,24,27,30,33-
Undecaoxapentatriacontane-1,35-Diol
3,6,9,12,15,18,21,24,27,30,33-
Undecaoxapentatricontane-1,35-diol

Trade Names:
CARBOWAX PEG 600 (Dow Chemical)
DePEG 600 (DeForest)
Jeechem 600 (Jeen)
Lipo Polyglycol 600 (Lipo)
Lipoxol 600 MED (Sasol GmbH - Marl)

Macrogol 600 (NOF)
Norfox E-600 (Norman, Fox & Co.)
Pluracare E 600 (BASF)
Pluracol E 600 (BASF)
Polyglykol 600 (Clariant)
Polyglykol 600 (Clariant GmbH, Personal
Care)
Renex PEG 600 (Uniqema Americas)
Sabopeg 600 (Sabo)
Sipol PEG-600 (Specialty Industrial)
TOHO PEG#600 (Toho)
Unipeg-600 (Universal Preserv-A-Chem)
Upiwax 600 (Universal Preserv-A-Chem)

PEG-14

CAS No.: 25322-68-3 (Generic)

Definition: PEG-14 is the polymer of
ethylene oxide that conforms generally to the
formula:

$$H(OCH_2CH_2)_nOH$$

where n has an average value of 14.

Information Sources: BAN,
21CFR172.210, 21CFR172.770,
21CFR172.820, 21CFR173.310,
21CFR173.340, 21CFR175.105,
21CFR175.300, 21CFR178.3750,
21CFR178.3910, JAN, MI-13(7651), NF
XVIII, TSCA, USAN

Chemical Classes: Alkoxylated Alcohols;
Polymeric Ethers

Functions: Humectant; Solvent

Reported Product Category: Foot Powders
and Sprays

Technical/Other Names:
Polyethylene Glycol (14)
Polyoxyethylene (14)

Trade Name Mixtures:
Ultrasil Copolyol-1 Silicone (Noveon)
Ultrasil Copolyol-7 Silicone (Noveon)

PEG-16

CAS No.: 25322-68-3 (Generic)

Definition: PEG-16 is the polymer of
ethylene oxide that conforms generally to the
formula:

$$H(OCH_2CH_2)_nOH$$

where n has an average value of 16.

Information Sources: BAN,
21CFR172.210, 21CFR172.770,
21CFR172.820, 21CFR173.310,
21CFR173.340, 21CFR175.105,
21CFR175.300, 21CFR178.3750,
21CFR178.3910, JAN, MI-13(7651), NF
XVIII, TSCA, USAN

Chemical Classes: Alkoxylated Alcohols;
Polymeric Ethers

Functions: Humectant; Solvent

Reported Product Categories: Bath Preparations, Misc.; Body and Hand Preparations
(Excluding Shaving Preparations)

Technical/Other Names:
Polyethylene Glycol (16)
Polyoxyethylene (16)

Trade Names:
Polyglykol 800 (Clariant)
Polyglykol 800 (Clariant GmbH, Personal
Care)

PEG-18

CAS No.: 25322-68-3 (Generic)

Definition: PEG-18 is the polymer of
ethylene oxide that conforms generally to the
formula:

$$H(OCH_2CH_2)_nOH$$

where n has an average value of 18.

Information Sources: BAN,
21CFR172.210, 21CFR172.770,
21CFR172.820, 21CFR173.310,
21CFR173.340, 21CFR175.105,
21CFR175.300, 21CFR178.3750,
21CFR178.3910, JAN, MI-13(7651), NF
XVIII, TSCA, USAN

Chemical Classes: Alkoxylated Alcohols;
Polymeric Ethers

Functions: Humectant; Solvent

Technical/Other Names:
Polyethylene Glycol (18)
Polyoxyethylene (18)

Trade Name:
CARBOWAX PEG 900 (Dow Chemical)

PEG-20

CAS No.: 25322-68-3 (Generic)

JPN Translation:
PEG - 20

Definition: PEG-20 is the polymer of
ethylene oxide that conforms generally to the
formula:

$$H(OCH_2CH_2)_nOH$$

where n has an average value of 20.

Information Sources: BAN,
21CFR73.1001, 21CFR172.210,
21CFR172.770, 21CFR172.820,
21CFR173.310, 21CFR173.340,
21CFR175.105, 21CFR175.300,
21CFR178.3750, 21CFR178.3910, CTFA S,

FCC, JAN, JCLS, JSCI, MI-13(7651), NF XVIII, ROM, TSCA, USAN

Chemical Classes: Alkoxylated Alcohols; Polymeric Ethers

Functions: Humectant; Solvent

Reported Product Categories: Moisturizing Preparations; Foundations; Bath Capsules; Bath Oils, Tablets, and Salts; Cleansing Products (Cold Creams, Cleansing Lotions, Liquids and Pads); Hair Wave Sets; Personal Cleanliness Products, Misc.

Technical/Other Names:
Polyethylene Glycol 1000
Polyoxyethylene (20)

Trade Names:
CARBOWAX PEG 1000 (Dow Chemical)
Lipo Polyglycol 1000 (Lipo)
Lipoxol 1000 MED (Sasol GmbH - Marl)
Macrogol 1000 (NOF)
Pluracol E 1000 (BASF)
Polyglykol 1000 (Clariant)
Polyglykol 1000 (Clariant GmbH, Personal Care)
Renex PEG 1000 (Uniqema Americas)
Sabopeg 1000 (Sabo)
Sipol PEG 1000 (Specialty Industrial)
TOHO PEG#1000 (Toho)
Unipeg-1000 X (Universal Preserv-A-Chem)
Upiwax 1000 (Universal Preserv-A-Chem)

Trade Name Mixtures:
Suncaps 664 (Particle Sciences)
Suncaps 903 (Particle Sciences)

PEG-32

CAS No.: 25322-68-3 (Generic)

JPN Translation:
PEG - 32

Definition: PEG-32 is the polymer of ethylene oxide that conforms generally to the formula:

$$H(OCH_2CH_2)_nOH$$

where n has an average value of 32.

Information Sources: BAN, BP, BPC, 21CFR172.210, 21CFR172.770, 21CFR172.820, 21CFR173.310, 21CFR173.340, 21CFR175.105, 21CFR175.300, 21CFR178.3750, 21CFR178.3910, CIR: [SQ] JACT-12(5)-1993, CTFA S, CZE, FCC, HUN, JAN, JCIC, JCLS, JSQI, MAR, MI-13(7651), NF XVIII, TSCA, USAN, USD

Chemical Classes: Alkoxylated Alcohols; Polymeric Ethers

Functions: Binder; Humectant; Solvent

Reported Product Categories: Bath Oils, Tablets, and Salts; Moisturizing Preparations; Cleansing Products (Cold Creams, Cleansing Lotions, Liquids and Pads); Bath Capsules; Skin Care Preparations, Misc.; Dentifrices (Aerosol, Liquid, Pastes and Powders); Bath Preparations, Misc.; Body and Hand Preparations (Excluding Shaving Preparations); Face and Neck Preparations (Excluding Shaving Preparations); Paste Masks (Mud Packs); Mascara

Technical/Other Names:
Polyethylene Glycol 1540
Polyoxyethylene (32)

Trade Names:
CARBOWAX PEG 1450 (Dow Chemical)
Jeechem 1450 NF (Jeen)
LipoPolyglycol 1500 (Lipo)
Lipo Polyglycol 3350 (Lipo)
Lipoxol 1500 MED (Sasol GmbH - Marl)
Lumulse PEG 1450 (Lambent)
Macrogol 1500 (NOF)
Macrogol 1540 (NOF)
Pluracare E 1500 Flakes (BASF)
Pluracol E 1450 (BASF)
Polyglycol E1450 (Dow Chemical)
Polyglykol 1500 (Clariant)
Polyglykol 1500 (Clariant GmbH, Personal Care)
Protachem 1450 NF (Protameen)
Renex PEG 1500 (Uniqema Americas)
Sabopeg 1500 (Sabo)
Sympatens-PEG/1500 G (Kolb)
TOHO PEG#1540 (Toho)
Unipeg-1540 X (Universal Preserv-A-Chem)

Trade Name Mixtures:
CARBOWAX PEG 540 Blend (Dow Chemical)
Lanobase S.E. (Lanaetex)
Lanogen 1500 (Clariant)
Lanogen 1500 (Clariant GmbH, Personal Care)
Swertianin P (Ichimaru Pharcos)
Unipeg-1500 X (Universal Preserv-A-Chem)
Uniwax 1450 (Universal Preserv-A-Chem)

PEG-33

Definition: PEG-33 is the polymer of ethylene oxide that conforms generally to the formula:

$$H(OCH_2CH_2)_nOH$$

where n has an average value of 33.

Chemical Classes: Alkoxylated Alcohols; Polymeric Ethers

Functions: Binder; Humectant; Solvent

Technical/Other Names:
Polyethylene Glycol (33)
Polyoxyethylene (33)

Trade Name Mixtures:
Ultrasil Copolyol-1 Silicone (Noveon)
Ultrasil Copolyol-7 Silicone (Noveon)

PEG-40

CAS No.: 25322-68-3 (Generic)

JPN Translation:
PEG - 40

Definition: PEG-40 is the polymer of ethylene oxide that conforms generally to the formula:

$$H(OCH_2CH_2)_nOH$$

where n has an average value of 40.

Information Sources: BAN, 21CFR172.210, 21CFR172.770, 21CFR172.820, 21CFR173.310, 21CFR173.340, 21CFR175.105, 21CFR175.300, 21CFR176.200, 21CFR178.3750, 21CFR178.3910, JAN, JCIC, JCLS, MI-13(7651), NF XVIII, ROM, TSCA, USAN

Chemical Classes: Alkoxylated Alcohols; Polymeric Ethers

Functions: Binder; Humectant; Solvent

Technical/Other Names:
Polyethylene Glycol (2000)
Polyoxyethylene (40)

Trade Names:
Pluracol E 2000 (BASF)
Polyglykol 2000 (Clariant)
Polyglykol 2000 (Clariant GmbH, Personal Care)

PEG-45

CAS No.: 25322-68-3 (Generic)

Definition: PEG-45 is the polymer of ethylene oxide that conforms generally to the formula:

$$H(OCH_2CH_2)_nOH$$

where n has an average value of 45.

Chemical Classes: Alkoxylated Alcohols; Polymeric Ethers

Functions: Binder; Humectant; Solvent

Technical/Other Names:
Polyethylene Glycol (45)
Polyoxyethylene (45)

Trade Name:
Toho PEG#2000 (Toho)

PEG-55

CAS No.: 25322-68-3 (Generic)

Definition: PEG-55 is the polymer of ethylene oxide that conforms generally to the formula:

$$H(OCH_2CH_2)_nOH$$

where n has an average value of 55.

Information Sources: BAN, JAN, NF XVIII, USAN

Chemical Classes: Alkoxylated Alcohols; Polymeric Ethers

Functions: Binder; Humectant; Solvent

Technical/Other Names:
Polyethylene Glycol (55)
Polyoxyethylene (55)

Trade Names:
Jeechem 3350 NF (Jeen)
Renex PEG 3350 (Uniqema Americas)

PEG-60

CAS No.: 25322-68-3 (Generic)

Definition: PEG-60 is the polymer of ethylene oxide that conforms generally to the formula:

$$H(OCH_2CH_2)_nOH$$

where n has an average value of 60.

Information Sources: BAN, JAN, MI-13 (7651), NF XVIII, USAN

Chemical Classes: Alkoxylated Alcohols; Polymeric Ethers

Functions: Binder; Humectant; Solvent

Technical/Other Names:
Polyethylene Glycol 3000
Polyoxyethylene (60)

Trade Names:
Polyglykol 3000 (Clariant)
Polyglykol 3000 (Clariant GmbH, Personal Care)

PEG-75

CAS No.: 25322-68-3 (Generic)

JPN Translation:
PEG - 75

Definition: PEG-75 is the polymer of ethylene oxide that conforms generally to the formula:

$$H(OCH_2CH_2)_nOH$$

where n has an average value of 75.

Information Sources: BAN, BP, BPC, BRA, 21CFR172.210, 21CFR172.770, 21CFR172.820, 21CFR173.310, 21CFR173.340, 21CFR175.105, 21CFR175.300, 21CFR178.3750, 21CFR178.3910, CIR: [SQ] JACT-12(5)-1993, CTFA S, FCC, HUN, JAN, JCLS, JSCI, MAR, MI-13(7651), NF XVIII, NFJ, PN, POL, ROM, TSCA, USAN, USD

Chemical Classes: Alkoxylated Alcohols; Polymeric Ethers

Functions: Binder; Humectant; Solvent

Reported Product Categories: Skin Care Preparations, Misc.; Paste Masks (Mud Packs); Bath Oils, Tablets, and Salts; Cleansing Products (Cold Creams, Cleansing Lotions, Liquids and Pads); Moisturizing Preparations

Technical/Other Names:
Polyethylene Glycol 4000
Polyoxyethylene (75)

Trade Names:
CARBOWAX PEG 3350 (Dow Chemical)
Lipoxol 3350 MED (Sasol GmbH - Marl)
Lumulse PEG 3350 (Lambent)
Pluracare E 3400 Flakes (BASF)
Pluracol E 4000 (BASF)
Polyglykol 3350 (Clariant)
Polyglykol 3350 (Clariant GmbH, Personal Care)
Protachem 75 (Protameen)
Renex PEG 4000 (Uniqema Americas)
Sabopeg 4000 (Sabo)
Sympatens-PEG/4000 G (Kolb)
Upiwax 3350 (Universal Preserv-A-Chem)

Trade Name Mixture:
Suncaps C (Particle Sciences)

PEG-80

CAS No.: 25322-68-3 (Generic)

Definition: PEG-80 is the polymer of ethylene oxide that conforms generally to the formula:

$$H(OCH_2CH_2)_nOH$$

where n has an average value of 80.

Chemical Classes: Alkoxylated Alcohols; Polymeric Ethers

Functions: Binder; Humectant; Solvent

Technical/Other Names:
Polyethylene Glycol (80)
Polyethylene Glycol 4000
Polyoxyethylene (80)

Trade Name:
Protachem 400 (Protameen)

PEG-90

CAS No.: 25322-68-3

Definition: PEG-90 is the polymer of ethylene oxide that conforms to the formula:

$$H(OCH_2CH_2)_nOH$$

where n has an average value of 90.

Information Sources: BAN, INN, JAN, NF XVIII, USAN

Chemical Classes: Alkoxylated Alcohols; Polymeric Ethers

Functions: Binder; Humectant; Solvent

Technical/Other Names:
Polyethylene Glycol (90)
Polyoxyethylene (90)

Trade Names:
Lipoxol 4000 MED (Sasol GmbH - Marl)
Macrogol 4000 (NOF)
Pluracare E 4000 Flakes (BASF)
Polyglycol E-4000 (Dow Chemical)
Polyglykol 4000 (Clariant)
Polyglykol 4000 (Clariant GmbH, Personal Care)
TOHO PEG #4000 (Toho)
Unipeg-4000 X (Universal Preserv-A-Chem)

PEG-100

CAS No.: 25322-68-3 (Generic)

Definition: PEG-100 is the polymer of ethylene oxide that conforms generally to the formula:

$$H(OCH_2CH_2)_nOH$$

where n has an average value of 100.

Information Sources: BAN, 21CFR172.210, 21CFR172.770, 21CFR172.820, 21CFR173.310, 21CFR173.340, 21CFR175.105, 21CFR175.300, 21CFR178.3750, 21CFR178.3910, JAN, MI-13(7651), NF XVIII, USAN

Chemical Classes: Alkoxylated Alcohols; Polymeric Ethers

Functions: Binder; Humectant; Solvent

Technical/Other Names:
Polyethylene Glycol (100)
Polyoxyethylene (100)

Trade Names:
CARBOWAX PEG 4600 (Dow Chemical)
Polyglycol E-4500 (Dow Chemical)

PEG-135

CAS No.: 25322-68-3 (Generic)

Definition: PEG-135 is the polymer of ethylene oxide that conforms to the formula:

$$H(OCH_2CH_2)_nOH$$

where n has an average value of 135.

Information Sources: BAN, 21CFR172.210, 21CFR172.770, 21CFR172.820, 21CFR173.310, 21CFR173.340, 21CFR175.105, 21CFR175.300, 21CFR178.3750, 21CFR178.3910, JAN, MI-13(7651), NF XVIII, USAN

Chemical Classes: Alkoxylated Alcohols; Polymeric Ethers

Functions: Binder; Humectant; Solvent

Technical/Other Names:
Polyethylene Glycol (135)
Polyoxyethylene (135)

Trade Names:
Lipoxol 6000 MED (Sasol GmbH - Marl)
Macrogol 6000 (NOF)

PEG-150

CAS No.: 25322-68-3 (Generic)

JPN Translation:
PEG - 150

Definition: PEG-150 is the polymer of ethylene oxide that conforms generally to the formula:

$$H(OCH_2CH_2)_nOH$$

where n has an average value of 150.

Information Sources: BAN, 21CFR172.210, 21CFR172.770, 21CFR172.820, 21CFR173.310, 21CFR173.340, 21CFR175.300, 21CFR177.2420, 21CFR178.3750, 21CFR178.3910, CIR: [SQ] JACT-12(5)-1993, CTFA S, FCC, JAN, JCLS, JSCI, MAR, MI-13(7651), NF XVIII, OTC-I-OP, PN, ROM, TSCA, USAN

Chemical Classes: Alkoxylated Alcohols; Polymeric Ethers

Functions: Binder; Humectant; Solvent

Reported Product Category: Bath Oils, Tablets, and Salts

Technical/Other Names:
Polyethylene Glycol 6000
Polyoxyethylene (150)

Trade Names:
Pluracare E 6000 Flakes (BASF)
Pluracol E 8000 (BASF)
Polyglykol 6000 (Clariant)
Polyglykol 6000 (Clariant GmbH, Personal Care)
Renex PEG 6000 (Uniqema Americas)
Sabopeg 6000 (Sabo)
TOHO PEG#6000 (Toho)

Unipeg-6000 X (Universal Preserv-A-Chem)

PEG-180

CAS No.: 25322-68-3 (Generic)

Definition: PEG-180 is the polymer of ethylene oxide that conforms generally to the formula:

$$H(OCH_2CH_2)_nOH$$

where n has an average value of 180.

Information Sources: BAN, 21CFR172.210, 21CFR172.770, 21CFR172.820, 21CFR173.310, 21CFR173.340, 21CFR175.300, 21CFR178.3750, 21CFR178.3910, JAN, MI-13(7651), NF XVIII, USAN

Chemical Classes: Alkoxylated Alcohols; Polymeric Ethers

Functions: Binder; Humectant; Solvent

Technical/Other Names:
Polyethylene Glycol (180)
Polyoxyethylene (180)

Trade Names:
CARBOWAX PEG 8000 (Dow Chemical)
Lipo Polyglycol 8000 (Lipo)
Lumulse PEG 8000 (Lambent)
Pluracare E8000 Flakes (BASF)
Polyglykol 8000 (Clariant)
Polyglykol 8000 (Clariant GmbH, Personal Care)
Renex PEG 8000 (Uniqema Americas)
Upiwax 8000 (Universal Preserv-A-Chem)

Trade Name Mixtures:
Aqua-Thik (Guardian)
Suncaps C (Particle Sciences)

PEG-200

CAS No.: 25322-68-3 (Generic)

Definition: PEG-200 is the polymer of ethylene oxide that conforms generally to the formula:

$$H(OCH_2CH_2)_nOH$$

where n has an average value of 200.

Information Sources: BAN, 21CFR172.210, 21CFR172.770, 21CFR172.820, 21CFR173.310, 21CFR173.340, 21CFR175.300, 21CFR178.3750, 21CFR178.3910, CTFA D, FCC, JAN, MI-13(7651), NF XVIII, TSCA, USAN

Chemical Classes: Alkoxylated Alcohols; Polymeric Ethers

Functions: Binder; Humectant; Solvent

Technical/Other Names:
Polyethylene Glycol 9000
Polyoxyethylene (200)

Trade Name Mixtures:
Germinol (Dr. Gerhard Steidl)
Harmonic ASP (Dr. Gerhard Steidl)
Hexatrate (Vevy)
Hexatrate Al-Free (Vevy)
Jonat AS (Dr. Gerhard Steidl)

PEG-220

CAS No.: 25322-68-3 (Generic)

Definition: PEG-220 is the polymer of ethylene oxide that conforms generally to the formula:

$$H(OCH_2CH_2)_nOH$$

where n has an average value of 220.

Chemical Classes: Alkoxylated Alcohols; Polymeric Ethers

Functions: Binder; Humectant; Solvent

Technical/Other Names:
Polyethylene Glycol (220)
Polyoxyethylene (220)

Trade Names:
Polyglykol 10000 (Clariant)
Polyglykol 10000 (Clariant GmbH, Personal Care)

PEG-240

CAS No.: 25322-68-3 (Generic)

JPN Translation:
PEG - 240

Definition: PEG-240 is the polymer of ethylene oxide that conforms generally to the formula:

$$H(OCH_2CH_2)_nOH$$

where n has an average value of 240.

Information Sources: 21CFR172.770, 21CFR175.300, 21CFR178.3910, JCIC, JCLS, MI-13(7651)

Chemical Classes: Alkoxylated Alcohols; Polymeric Ethers

Functions: Binder; Humectant; Solvent

Technical/Other Names:
Polyethylene Glycol (240)
Polyethylene Glycol 11000
Polyoxyethylene (240)

Trade Names:
Polyglykol 12000 (Clariant)

Polyglykol 12000 (Clariant GmbH, Personal Care)
TOHO PEG#11000 (Toho)

PEG-350

CAS No.: 25322-68-3 (Generic)

Definition: PEG-350 is the polymer of ethylene oxide that conforms generally to the formula:

$$H(OCH_2CH_2)_nOH$$

where n has an average value of 350.

Information Sources: 21CFR172.770, 21CFR173.310, 21CFR175.300, 21CFR178.3910, JCLS, JSCI, MI-13(7651), TSCA

Chemical Classes: Alkoxylated Alcohols; Polymeric Ethers

Functions: Binder; Emulsion Stabilizer; Solvent

Technical/Other Names:
Polyethylene Glycol 20000
Polyoxyethylene (350)

Trade Names:
Polyglykol 20000 (Clariant)
Polyglykol 20000 (Clariant GmbH, Personal Care)
Upiwax 20000 (Universal Preserv-A-Chem)

PEG-400

CAS No.: 25322-68-3 (Generic)

JPN Translation:
PEG - 400

Definition: PEG-400 is the polymer of ethylene oxide that conforms generally to the formula:

$$H(OCH_2CH_2)_nOH$$

where n has an average value of 400.

Chemical Classes: Alkoxylated Alcohols; Polymeric Ethers

Functions: Binder; Emulsion Stabilizer; Solvent

Technical/Other Names:
Polyethylene Glycol 400
Polyoxyethylene (400)

Trade Name Mixture:
Lacteclat Gel (I.D. bio)

PEG-450

Definition: PEG-450 is the polymer of ethylene oxide that conforms generally to the formula:

$$H(OCH_2CH_2)_nOH$$

where n has an average value of 450.

Chemical Classes: Alkoxylated Alcohols; Polymeric Ethers

Functions: Binder; Emulsion Stabilizer; Solvent

Technical/Other Names:
Polyethylene Glycol 20000
Polyoxyethylene (450)

Trade Names:
Polyglykol 20000 P (Clariant GmbH Functional Chemicals)
Polyglykol 20000 S (Clariant GmbH Functional Chemicals)

PEG-500

CAS No.: 25322-68-3 (Generic)

Definition: PEG-500 is the polymer of ethylene oxide that conforms generally to the formula:

$$H(OCH_2CH_2)_nOH$$

where n has an average value of 500.

Chemical Classes: Alkoxylated Alcohols; Polymeric Ethers

Functions: Binder; Emulsion Stabilizer; Solvent

Technical/Other Names:
Polyethylene Glycol 500
Polyoxyethylene (500)

Trade Name:
Toho PEG#20000 (Toho)

PEG-800

CAS No.: 25322-68-3 (Generic)

Definition: PEG-800 is the polymer of ethylene oxide that conforms generally to the formula:

$$H(OCH_2CH_2)_nOH$$

where n has an average value of 800.

Chemical Classes: Alkoxylated Alcohols; Polymeric Ethers

Functions: Anticaking Agent; Binder; Humectant; Plasticizer; Viscosity Increasing Agent - Aqueous

Technical/Other Names:
Polyethylene Glycol (800)
Polyoxyethylene (800)

Trade Names:
Polyglykol 35000 (Clariant)
Polyglykol 35000 (Clariant GmbH, Personal Care)
Polyglykol 35000 S (Clariant)
Polyglykol 35000 S (Clariant GmbH, Personal Care)

PEG-6 ALMOND GLYCERIDES

Definition: PEG-6 Almond Glycerides is a polyethylene glycol derivative of the mono- and diglycerides from almond oil with an average of 6 moles of ethylene oxide.

Chemical Classes: Alkoxylated Alcohols; Glyceryl Esters and Derivatives

Functions: Skin-Conditioning Agent - Emollient; Surfactant - Emulsifying Agent

Technical/Other Names:
Polyethylene Glycol (6) Almond Glycerides
Polyoxyethylene (6) Almond Glycerides

Trade Name:
ESTOL 3657 (Uniqema Europe)

PEG-20 ALMOND GLYCERIDES

CAS No.: 124046-50-0

Definition: PEG-20 Almond Glycerides is a polyethylene glycol derivative of the mono- and diglycerides from almond oil with an average of 20 moles of ethylene oxide.

Chemical Classes: Alkoxylated Alcohols; Glyceryl Esters and Derivatives

Functions: Skin-Conditioning Agent - Emollient; Surfactant - Emulsifying Agent

Reported Product Categories: Bath Oils, Tablets, and Salts; Cleansing Products (Cold Creams, Cleansing Lotions, Liquids and Pads)

Technical/Other Names:
Polyethylene Glycol 1000 Almond Glycerides
Polyoxyethylene (20) Almond Glycerides

Trade Names:
Crovol A40 (Croda Chemicals)
Crovol A-40 (Croda, Inc.)

PEG-35 ALMOND GLYCERIDES

CAS No.: 124046-50-0

Definition: PEG-35 Almond Glycerides is a polyethylene glycol derivative of the mono- and diglycerides derived from almond oil with an average of 35 moles of ethylene oxide.

Chemical Class: Glyceryl Esters and Derivatives

Functions: Skin-Conditioning Agent - Emollient; Surfactant - Emulsifying Agent

Technical/Other Names:
Polyethylene Glycol (35) Almond Glycerides
Polyoxyethylene (35) Almond Glycerides

Trade Name:
Sympatens-TAL/350 (Kolb)

PEG-60 ALMOND GLYCERIDES

CAS No.: 124046-50-0

JPN Translation:
PEG - 60 アーモンド脂肪酸グリセリル

Definition: PEG-60 Almond Glycerides is a polyethylene glycol derivative of the mono- and diglycerides from almond oil with an average of 60 moles of ethylene oxide.

Chemical Classes: Alkoxylated Alcohols; Glyceryl Esters and Derivatives

Functions: Skin-Conditioning Agent - Emollient; Surfactant - Emulsifying Agent

Reported Product Categories: Tonics, Dressings, and Other Hair Grooming Aids; Shampoos (Non-coloring); Skin Fresheners

Technical/Other Names:
Polyethylene Glycol 3000 Almond Glycerides
Polyoxyethylene (60) Almond Glycerides

Trade Names:
Crovol A70 (Croda Chemicals)
Crovol A-70 (Croda, Inc.)

PEG-7 AMODIMETHICONE

CAS No.: 133779-14-3

Definition: PEG-7 Amodimethicone is the polyethylene glycol derivative of Amodimethicone (q.v.) containing an average of 7 moles of ethylene oxide.

Chemical Class: Siloxanes and Silanes

Function: Skin-Conditioning Agent - Emollient

Technical/Other Name:
Siloxanes and Silicones, Dimethyl, 3-Hydroxypropyl Methyl, Ethers with Polyethylene Glycol Mono[3-[Bis(2-Carboxyethyl)Amino]Propyl] Ether

Trade Names:
Ultrasil A-100 (Noveon)
Ultrasil A-300 (Noveon)
Ultrasil A-21 Silicone (Noveon)
Ultrasil A-23 Silicone (Noveon)

PEG-192 APRICOT KERNEL GLYCERIDES

Definition: PEG-192 Apricot Kernel Glycerides is a polyethylene glycol derivative of the mono- and diglycerides from apricot kernel oil with an average of 192 moles of ethylene oxide.

Chemical Classes: Alkoxylated Alcohols; Glyceryl Esters and Derivatives

Functions: Skin-Conditioning Agent - Emollient; Surfactant - Emulsifying Agent

Technical/Other Names:
Polyethylene Glycol (192) Apricot Kernel Glycerides
Polyoxyethylene (192) Apricot Kernel Glycerides

Trade Name:
Crovol AK90 (Croda Chemicals)

PEG-9 AVOCADOATE

Definition: PEG-9 Avocadoate is the polyethylene glycol ester of the fatty acids derived from Persea Gratissima (Avocado) Oil that conforms generally to the formula:

$$RC\overset{\displaystyle O}{\overset{\displaystyle \|}{-}}(OCH_2CH_2)_nOH$$

where RCO- represents the avocado oil fatty acids and n has an average value of 9.

Chemical Class: Alkoxylated Carboxylic Acids

Function: Surfactant - Emulsifying Agent

Technical/Other Names:
Polyethylene Glycol (9) Monoavocadoate
Polyoxyethylene (9) Monoavocadoate

Trade Name Mixture:
Silwax DMC-AV (Siltech LLC)

PEG-11 AVOCADO GLYCERIDES

CAS No.: 103819-44-9 (Generic)

Definition: PEG-11 Avocado Glycerides is a polyethylene glycol derivative of mono- and diglycerides from avocado oil with an average of 11 moles of ethylene oxide.

Chemical Classes: Alkoxylated Alcohols; Glyceryl Esters and Derivatives

Functions: Skin-Conditioning Agent - Emollient; Surfactant - Emulsifying Agent

Technical/Other Names:
Polyethylene Glycol (11) Avocado Glycerides
Polyoxyethylene (11) Avocado Glycerides

Trade Names:
AEC PEG-11 Avocado Glycerides (A & E Connock)

Avocado Oil W (Cosmetochem)
Oxypon 365 (Zschimmer & Schwarz)

PEG-14 AVOCADO GLYCERIDES

CAS No.: 103819-44-9 (Generic)

Definition: PEG-14 Avocado Glycerides is a polyethylene glycol derivative of the mono- and diglycerides derived from avocado oil with an average of 14 moles of ethylene oxide.

Chemical Classes: Alkoxylated Alcohols; Glyceryl Esters and Derivatives

Functions: Skin-Conditioning Agent - Emollient; Surfactant - Emulsifying Agent

Technical/Other Names:
Polyethylene Glycol (14) Avocado Glycerides
Polyoxyethylene (14) Avocado Glycerides

PEG-11 BABASSU GLYCERIDES

Definition: PEG-11 Babassu Glycerides is a polyethylene glycol derivative of the mono- and diglycerides derived from babassu oil with an average of 11 moles of ethylene oxide.

Chemical Classes: Alkoxylated Alcohols; Glyceryl Esters and Derivatives

Functions: Skin-Conditioning Agent - Emollient; Surfactant - Emulsifying Agent

Technical/Other Names:
Polyethylene Glycol (11) Babassu Glycerides
Polyoxyethylene (11) Babassu Glycerides

PEG-42 BABASSU GLYCERIDES

Definition: PEG-42 Babassu Glycerides is a polyethylene glycol derivative of the mono- and diglycerides derived from babassu oil with an average of 42 moles of ethylene oxide.

Chemical Classes: Alkoxylated Alcohols; Glyceryl Esters and Derivatives

Functions: Skin-Conditioning Agent - Emollient; Surfactant - Emulsifying Agent

Technical/Other Names:
Polyethylene Glycol (42) Babassu Glycerides
Polyoxyethylene (42) Babassu Glycerides

Trade Name:
Crovol BA70G (Croda Chemicals)

PEG-6 BEESWAX

Definition: PEG-6 Beeswax is a polyethylene glycol derivative of Beeswax (q.v.) with an average of 6 moles of ethylene oxide.

Chemical Classes: Alkoxylated Alcohols; Waxes

Function: Surfactant - Emulsifying Agent

Technical/Other Names:
Polyethylene Glycol 300 Beeswax
Polyoxyethylene (6) Beeswax

Trade Name:
ESTOL 3751 (Uniqema Europe)

PEG-8 BEESWAX

JPN Translation:
PEG - 8 ミツロウ

Definition: PEG-8 Beeswax is a poly-ethylene glycol derivative of Beeswax (q.v.) with an average of 8 moles of ethylene oxide.

Information Sources: JCIC, JCLS

Chemical Classes: Alkoxylated Alcohols; Waxes

Function: Surfactant - Emulsifying Agent

Technical/Other Names:
Polyethylene Glycol 400 Beeswax
Polyoxyethylene (8) Beeswax
Polyoxyethylene Beeswax (8E.O.)

Trade Names:
Abesin AE (Fabriquimica)
Apifil (Gattefosse s.a.)
ESTOL 3752 (Uniqema Europe)

PEG-12 BEESWAX

Definition: PEG-12 Beeswax is a poly-ethylene glycol derivative of Beeswax (q.v.) with an average of 12 moles of ethylene oxide.

Chemical Classes: Alkoxylated Alcohols; Waxes

Function: Surfactant - Emulsifying Agent

Technical/Other Names:
Polyethylene Glycol 600 Beeswax
Polyoxyethylene (12) Beeswax

Trade Name:
ESTOL 3753 (Uniqema Europe)

PEG-20 BEESWAX

Definition: PEG-20 Beeswax is a poly-ethylene glycol derivative of Beeswax (q.v.) with an average of 20 moles of ethylene oxide.

Chemical Classes: Alkoxylated Alcohols; Waxes

Function: Surfactant - Emulsifying Agent

Technical/Other Names:
Polyethylene Glycol 1000 Beeswax
Polyoxyethylene (20) Beeswax

PEG-8 BEHENATE

Empirical Formula:
$C_{38}H_{76}O_{10}$

Definition: PEG-8 Behenate is polyethylene glycol ester of behenic acid that conforms to the formula:

$$CH_3(CH_2)_{20}\overset{\overset{\textstyle O}{\|}}{C}-(OCH_2CH_2)_nOH$$

where n has an average value of 8.

Information Source: MI-13(7660)

Chemical Class: Alkoxylated Carboxylic Acids

Function: Surfactant - Emulsifying Agent

Technical/Other Names:
Polyethylene Glycol 400 Behenate
Polyoxyethylene (8) Behenate

Trade Name Mixture:
Compritol HD5 ATO (Gattefosse s.a.)

PEG-105 BEHENYL PROPYLENEDIAMINE

Definition: PEG-105 Behenyl Propylene-diamine is the ethoxylated amine that conforms generally to the formula:

$$CH_3(CH_2)_{21}\overset{\overset{\textstyle (CH_2CH_2O)_xH}{|}}{N}(CH_2)_3\overset{\overset{\textstyle |}{N}(CH_2CH_2O)_yH}{\underset{\textstyle (CH_2CH_2O)_zH}{|}}$$

where x+y+z has an average value of 105.

Chemical Class: Alkoxylated Amines

Function: Hair Conditioning Agent

Technical/Other Names:
Polyethylene Glycol (105) Behenyl Propyl-enediamine
Polyoxyethylene (105) Behenyl Propylene-diamine

Trade Names:
Sandogen NH (Clariant)
Sandogen NH (Clariant GmbH, Personal Care)

PEG-2 BENZYL ETHER

Definition: PEG-2 Benzyl Ether is the organic compound that conforms generally to the formula:

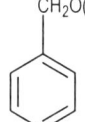

where n has an average value of 2.

Chemical Class: Alkoxylated Alcohols

Function: Solvent

Technical/Other Names:
Polyethylene Glycol 100 Benzyl Ether
Polyoxypropylene (2) Benzyl Ether

Trade Name:
Nikkol BZ-2 (Nikko)

PEG-9 BORAGEATE

Definition: PEG-9 Borageate is the poly-ethylene glycol ester of the fatty acids derived from Borago Officinalis Seed Oil (q.v.) that conforms generally to the formula:

$$RC\overset{\overset{\textstyle O}{\|}}{}-(OCH_2CH_2)_nOH$$

where RCO- represents the fatty acids derived from borage seed oil and n has an average value of 9.

Chemical Class: Alkoxylated Carboxylic Acids

Function: Surfactant - Emulsifying Agent

Technical/Other Names:
Polyethylene Glycol (9) Monoborageate
Polyoxyethylene (9) Monoborageate

Trade Name Mixture:
Silwax DMC-BOR (Siltech LLC)

PEG-15 BUTANEDIOL

Definition: PEG-15 Butanediol is the ethoxylated diol that conforms generally to the formula:

$$CH_3CH\overset{\overset{\textstyle (OCH_2CH_2)_yOH}{|}}{(CH_2)_2}(OCH_2CH_2)_xOH$$

where x+y has an average value of 15.

Chemical Class: Alkoxylated Alcohols

Functions: Humectant; Solvent

Technical/Other Names:
Polyethylene Glycol (15) Butanediol
Polyoxyethylene (15) Butanediol

PEG-3 BUTYLENE GLYCOL LAURATE

Definition: PEG-3 Butylene Glycol Laurate is the organic compound that conforms generally to the formula:

$$CH_3(CH_2)_{10}\overset{\displaystyle O}{\overset{\|}{C}}\text{---}(OCH_2CH_2)_nOCH_2CH_2CHCH_3$$
$$\underset{\displaystyle OH}{|}$$

where n has an average value of 3.

Chemical Classes: Alkoxylated Alcohols; Esters

Functions: Skin-Conditioning Agent - Emollient; Surfactant - Emulsifying Agent

Trade Name:
ANIMUS BL-207 (NOF America Corp.)

PEG-6 BUTYLENE GLYCOL LAURATE

Definition: PEG-6 Butylene Glycol Laurate is the organic compound that conforms generally to the formula:

$$CH_3(CH_2)_{10}\overset{\displaystyle O}{\overset{\|}{C}}\text{---}(OCH_2CH_2)_nOCH_2CH_2CHCH_3$$
$$\underset{\displaystyle OH}{|}$$

where n has an average value fo 6.

Chemical Classes: Alkoxylated Alcohols; Esters

Functions: Skin-Conditioning Agent - Emollient; Surfactant - Emulsifying Agent

Trade Name:
ANIMUS BL-210 (NOF America Corp.)

PEG-9 BUTYLENE GLYCOL LAURATE

Definition: PEG-9 Butylene Glycol Laurate is the organic compound that conforms generally to the formula:

$$CH_3(CH_2)_{10}\overset{\displaystyle O}{\overset{\|}{C}}\text{---}(OCH_2CH_2)_nOCH_2CH_2CHCH_3$$
$$\underset{\displaystyle OH}{|}$$

where n has an average value of 9.

Chemical Classes: Alkoxylated Alcohols; Esters

Trade Name:
ANIMUS BL-212 (NOF America Corp.)

PEG-9 BUTYLOCTANOATE

Definition: PEG-9 Butyloctanoate is the polyethylene glycol ester of Butyloctanoic Acid (q.v.) that conforms generally to the formula:

$$CH_3(CH_2)_5\overset{\displaystyle O}{\overset{\|}{C}}HC\text{---}(OCH_2CH_2)_nOH$$
$$\underset{\displaystyle (CH_2)_3CH_3}{|}$$

where n has an average value of 9.

Chemical Class: Alkoxylated Carboxylic Acids

Function: Surfactant - Emulsifying Agent

Technical/Other Name:
Polyoxyethylene (9) Butyloctanoate

Trade Name:
Silube G12-9 (Siltech LLC)

PEG-40 BUTYLOCTANOL WHEAT GERM ESTERS

Definition: PEG-40 Butyloctanol Wheat Germ Esters is a the product obtained by the reaction of a mixture of Triticum Vulgare (Wheat) Germ Oil (q.v.) and Butyloctanol (q.v.) with PEG-40.

Chemical Class: Esters

Function: Skin-Conditioning Agent - Emollient

Trade Name:
Peg-40 Butyloctanol Wheat Germ Glycerides (Symrise)

Trade Name Mixture:
Dragocare W 2/032700 (Symrise)

PEG-8 CAPRATE

Empirical Formula:
$C_{26}H_{54}O_{10}$

Definition: PEG-8 Caprate is the polyethylene glycol ester of capric acid that conforms to the formula:

$$CH_3(CH_2)_8\overset{\displaystyle O}{\overset{\|}{C}}\text{---}(OCH_2CH_2)_nOH$$

where n has an average value of 8.

Information Sources: 21CFR175.105, 21CFR177.2260, MI-13(7660)

Chemical Class: Alkoxylated Carboxylic Acids

Function: Surfactant - Emulsifying Agent

Technical/Other Names:
Polyethylene Glycol 400 Monocaprate
Polyoxyethylene (8) Monocaprate

Trade Name:
AEC PEG-8 Caprate (A & E Connock)

PEG-8 CAPRYLATE

Empirical Formula:
$C_{24}H_{48}O_{10}$

Definition: PEG-8 Caprylate is the polyethylene glycol ester of caprylic acid that conforms to the formula:

$$CH_3(CH_2)_6\overset{\displaystyle O}{\overset{\|}{C}}\text{---}(OCH_2CH_2)_nOH$$

where n has an average value of 8.

Information Sources: 21CFR175.105, MI-13(7660)

Chemical Class: Alkoxylated Carboxylic Acids

Function: Surfactant - Emulsifying Agent

Technical/Other Names:
Polyethylene Glycol 400 Monocaprylate
Polyoxyethylene (8) Monocaprylate

Trade Name:
AEC PEG-8 Caprylate (A & E Connock)

PEG-8 CAPRYLATE/CAPRATE

Definition: PEG-8 Caprylate/Caprate is the polyethylene glycol ester of a mixture of caprylic and capric acids. It conforms generally to the formula:

$$R\overset{\displaystyle O}{\overset{\|}{C}}\text{---}(OCH_2CH_2)_nOH$$

where RCO- represents a mixture of caprylic and capric acid radicals and n has an average value of 8.

Information Sources: 21CFR175.105, MI-13(7660)

Chemical Class: Alkoxylated Carboxylic Acids

Function: Surfactant - Emulsifying Agent

Technical/Other Names:
Polyethylene Glycol 400 Caprylate/Caprate
Polyoxyethylene (8) Caprylate/Caprate

Trade Name:
AEC PEG-8 Caprylate/Caprate (A & E Connock)

PEG-4 CAPRYLIC/CAPRIC GLYCERIDES

JPN Translation:
PEG - 4（カプリル酸 / カプリン酸）グリセリズ

Definition: PEG-4 Caprylic/Capric Glycerides is a polyethylene glycol derivative of a mixture of mono-, di-, and triglycerides of caprylic and capric acids with an average of 4 moles of ethylene oxide.

Chemical Class: Glyceryl Esters and Derivatives

The inclusion of any compound in the Dictionary and Handbook does not indicate that use of that substance as a cosmetic ingredient complies with the laws and regulations governing such use in the United States or any other country.

Functions: Skin-Conditioning Agent - Emollient; Surfactant - Emulsifying Agent

PEG-6 CAPRYLIC/CAPRIC GLYCERIDES

JPN Translation:
PEG - 6（カプリル酸 / カプリン酸）グリセリズ

Definition: PEG-6 Caprylic/Capric Glycerides is a polyethylene glycol derivative of a mixture of mono-, di-, and triglycerides of caprylic and capric acids with an average of 6 moles of ethylene oxide.

Information Source: JSQI

Chemical Class: Glyceryl Esters and Derivatives

Functions: Skin-Conditioning Agent - Emollient; Surfactant - Emulsifying Agent

Reported Product Categories: Moisturizing Preparations; Bath Soaps and Detergents; Bath Preparations, Misc.; Skin Care Preparations, Misc.

Technical/Other Names:
Polyethylene Glycol 300 Caprylic/Capric Glycerides
Polyoxyethylene (6) Caprylic/Capric Glycerides

Trade Names:
Acconon CC-6 (Abitec Corporation)
ESTOL 3684 (Uniqema Europe)
Glycerox 767 (Croda Chemicals)
Glycerox 767 (Croda, Inc.)
Saboderm CC (Sabo)
Softigen 767 (Sasol GmbH - Witten)
Sterol CC 595 (Cesalpinia)
Tegosoft GMC 6 (Degussa Care Specialties)

Trade Name Mixtures:
Crothix Liquid (Croda, Inc.)
Dioscorea Extract (Sederma)
Orchid Complex WS (Guardian)

PEG-7 CAPRYLIC/CAPRIC GLYCERIDES

Definition: PEG-7 Caprylic/Capric Glycerides is a polyethylene glycol derivative of a mixture of mono-, di- and triglycerides of caprylic and capric acids with an average of 7 moles of ethylene oxide.

Chemical Class: Glyceryl Esters and Derivatives

Functions: Skin-Conditioning Agent - Emollient; Surfactant - Emulsifying Agent

Trade Name:
Cetiol HE 810 (Cognis Care Chemicals/PA)

PEG-8 CAPRYLIC/CAPRIC GLYCERIDES

JPN Translation:
PEG - 8（カプリル酸 / カプリン酸）グリセリズ

Definition: PEG-8 Caprylic/Capric Glycerides is a polyethylene glycol derivative of a mixture of mono-, di- and triglycerides of caprylic and capric acids with an average of 8 moles of ethylene oxide.

Information Sources: JCIC, JCLS

Chemical Class: Glyceryl Esters and Derivatives

Functions: Skin-Conditioning Agent - Emollient; Surfactant - Emulsifying Agent

Technical/Other Names:
Polyethylene Glycol 400 Caprylate/Caprate Glycerides
Polyoxyethylene (Caprylate/Caprate) Glycerides
Polyoxyethylene (8) Caprylate/Caprate Glycerides

Trade Names:
Labrasol (Gattefosse s.a.)
L.A.S. (Gattefosse s.a.)
Trivent BW (Trivent)

PEG-12 CARNAUBA

Definition: PEG-12 Carnauba is the polyethylene glycol derivative of carnauba with an average of 12 moles of ethylene oxide.

Chemical Classes: Alkoxylated Alcohols; Waxes

Function: Emulsion Stabilizer

Technical/Other Names:
Polyethylene Glycol (12) Carnauba
Polyoxyethylene (12) Carnauba

Trade Name:
PEG-Carnauba (Koster Keunen)

PEG-2 CASTOR OIL

CAS No.: 61791-12-6 (Generic)

JPN Translation:
PEG - 2 ヒマシ油

Definition: PEG-2 Castor Oil is a polyethylene glycol derivative of Ricinus Communis (Castor) Oil (q.v.) with an average of 2 moles of ethylene oxide.

Information Sources: 21CFR175.300, 21CFR176.210, 21CFR177.2800, JCLS

Chemical Classes: Alkoxylated Alcohols; Glyceryl Esters and Derivatives

Functions: Skin-Conditioning Agent - Emollient; Surfactant - Emulsifying Agent

Technical/Other Names:
Polyethylene Glycol (100) Castor Oil
Polyoxyethylene (2) Castor Oil

Trade Name:
Hetoxide C-2 (Heterene)

PEG-3 CASTOR OIL

CAS No.: 61791-12-6 (Generic)

JPN Translation:
PEG - 3 ヒマシ油

Definition: PEG-3 Castor Oil is a polyethylene glycol derivative of Ricinus Communis (Castor) Oil (q.v.) with an average of 3 moles of ethylene oxide.

Information Sources: 21CFR175.300, JCLS, TSCA

Chemical Classes: Alkoxylated Alcohols; Glyceryl Esters and Derivatives

Functions: Skin-Conditioning Agent - Emollient; Surfactant - Emulsifying Agent

Technical/Other Names:
Polyethylene Glycol (3) Castor Oil
Polyoxyethylene (3) Castor Oil

Trade Name:
Nikkol CO-3 (Nikko)

PEG-4 CASTOR OIL

CAS No.: 61791-12-6 (Generic)

JPN Translation:
PEG - 4 ヒマシ油

Definition: PEG-4 Castor Oil is a polyethylene glycol derivative of Ricinus Communis (Castor) Oil (q.v.) with an average of 4 moles of ethylene oxide.

Information Sources: 21CFR175.105, 21CFR175.300, 21CFR176.210, 21CFR177.2800, JCLS, TSCA

Chemical Classes: Alkoxylated Alcohols; Glyceryl Esters and Derivatives

Functions: Skin-Conditioning Agent - Emollient; Surfactant - Emulsifying Agent

Technical/Other Names:
Polyethylene Glycol 200 Castor Oil
Polyoxyethylene (4) Castor Oil

PEG-5 CASTOR OIL

CAS No.: 61791-12-6 (Generic)

JPN Translation:
PEG - 5 ヒマシ油

Definition: PEG-5 Castor Oil is a polyethylene glycol derivative of Ricinus Communis (Castor) Oil (q.v.) with an average of 5 moles of ethylene oxide.

Information Sources: 21CFR175.105, 21CFR175.300, JCLS, TSCA

Chemical Classes: Alkoxylated Alcohols; Glyceryl Esters and Derivatives

Functions: Skin-Conditioning Agent - Emollient; Surfactant - Emulsifying Agent

Technical/Other Names:
Polyethylene Glycol (5) Castor Oil
Polyoxyethylene (5) Castor Oil

Trade Names:
Ethox CO-5 (Ethox)
Etocas 5 (Croda Chemicals)
Jeechem CA-5 (Jeen)
Surfactol 318 (CasChem)

PEG-8 CASTOR OIL

CAS No.: 61791-12-6 (Generic)

JPN Translation:
PEG - 8 ヒマシ油

Definition: PEG-8 Castor Oil is a poly-ethylene glycol derivative of Ricinus Communis (Castor) Oil (q.v.) with an average of 8 moles of ethylene oxide.

Information Sources: 21CFR175.105, 21CFR175.300, 21CFR176.210, 21CFR177.2800, JCLS, TSCA

Chemical Classes: Alkoxylated Alcohols; Glyceryl Esters and Derivatives

Functions: Skin-Conditioning Agent - Emollient; Surfactant - Emulsifying Agent

Technical/Other Names:
Polyethylene Glycol 400 Castor Oil
Polyoxyethylene (8) Castor Oil

Trade Name:
Unipeg-CO-8 (Universal Preserv-A-Chem)

PEG-9 CASTOR OIL

CAS No.: 61791-12-6 (Generic)

JPN Translation:
PEG - 9 ヒマシ油

Definition: PEG-9 Castor Oil is a poly-ethylene glycol derivative of Ricinus Communis (Castor) Oil (q.v.) with an average of 9 moles of ethylene oxide.

Information Sources: 21CFR175.105, 21CFR175.300, 21CFR177.2800, JCLS, TSCA

Chemical Classes: Alkoxylated Alcohols; Glyceryl Esters and Derivatives

Functions: Skin-Conditioning Agent - Emollient; Surfactant - Emulsifying Agent

Technical/Other Names:
Polyethylene Glycol 450 Castor Oil
Polyoxyethylene (9) Castor Oil

Trade Names:
Acconon CA-9 (Abitec Corporation)
Hetoxide C-9 (Heterene)
Jeechem CA-9 (Jeen)
Protachem CO 9 (Protameen)

PEG-10 CASTOR OIL

CAS No.: 61791-12-6 (Generic)

JPN Translation:
PEG - 10 ヒマシ油

Definition: PEG-10 Castor Oil is a poly-ethylene glycol derivative of Ricinus Communis (Castor) Oil (q.v.) with an average of 10 moles of ethylene oxide.

Information Sources: 21CFR175.105, 21CFR175.300, 21CFR177.2800, JCLS, TSCA

Chemical Classes: Alkoxylated Alcohols; Glyceryl Esters and Derivatives

Functions: Skin-Conditioning Agent - Emollient; Surfactant - Cleansing Agent; Surfactant - Emulsifying Agent

Technical/Other Names:
Polyethylene Glycol 500 Castor Oil
Polyoxyethylene (10) Castor Oil

Trade Names:
Etocas 10 (Croda Chemicals)
Nikkol CO-10 (Nikko)

PEG-11 CASTOR OIL

CAS No.: 61791-12-6 (Generic)

JPN Translation:
PEG - 11 ヒマシ油

Definition: PEG-11 Castor Oil is a poly-ethylene glycol derivative of Ricinus Communis (Castor) Oil (q.v.) with an average of 11 moles of ethylene oxide.

Information Sources: 21CFR175.105, 21CFR175.300, 21CFR177.2800

Chemical Classes: Alkoxylated Alcohols; Glyceryl Esters and Derivatives

Functions: Skin-Conditioning Agent - Emollient; Surfactant - Emulsifying Agent

Technical/Other Names:
Polyethylene Glycol (11) Castor Oil
Polyoxyethylene (11) Castor Oil

PEG-15 CASTOR OIL

CAS No.: 61791-12-6 (Generic)

JPN Translation:
PEG - 15 ヒマシ油

Definition: PEG-15 Castor Oil is a poly-ethylene glycol derivative of Ricinus Communis (Castor) Oil (q.v.) with an average of 15 moles of ethylene oxide.

Information Sources: 21CFR175.105, 21CFR175.300, 21CFR176.210, 21CFR177.2800, JCLS

Chemical Classes: Alkoxylated Alcohols; Glyceryl Esters and Derivatives

Functions: Skin-Conditioning Agent - Emollient; Surfactant - Emulsifying Agent

Technical/Other Names:
Polyethylene Glycol (15) Castor Oil
Polyoxyethylene (15) Castor Oil

Trade Names:
Etocas 15 (Croda Chemicals)
Protachem CO 15 (Protameen)
Triglicoleum (Vevy)

PEG-16 CASTOR OIL

CAS No.: 61791-12-6 (Generic)

JPN Translation:
PEG - 16 ヒマシ油

Definition: PEG-16 Castor Oil is the poly-ethylene glycol derivative of Ricinus Communis (Castor) Oil (q.v.) with an average of 16 moles of ethylene oxide.

Information Source: JCLS

Chemical Classes: Alkoxylated Alcohols; Glyceryl Esters and Derivatives

Functions: Skin-Conditioning Agent - Emollient; Surfactant - Emulsifying Agent

Technical/Other Names:
Polyethylene Glycol (16) Castor Oil
Polyoxyethylene (16) Castor Oil

Trade Names:
Ethox CO-16 (Ethox)
Jeechem CA-16 (Jeen)

PEG-20 CASTOR OIL

CAS No.: 61791-12-6 (Generic)

JPN Translation:
PEG - 20 ヒマシ油

Definition: PEG-20 Castor Oil is a poly-ethylene glycol derivative of Ricinus Communis (Castor) Oil (q.v.) with an average of 20 moles of ethylene oxide.

Information Sources: 21CFR175.105, 21CFR175.300, 21CFR176.210, 21CFR177.2800, JCLS, TSCA

Chemical Classes: Alkoxylated Alcohols; Glyceryl Esters and Derivatives

Function: Surfactant - Emulsifying Agent

Technical/Other Names:
Polyethylene Glycol 1000 Castor Oil
Polyoxyethylene (20) Castor Oil

Trade Names:
Atlas G-1281 (Uniqema Americas)
DePEG 20-CO (DeForest)
Etocas 20 (Croda Chemicals)
Nikkol CO-20TX (Nikko)

Trade Name Mixture:
Complex 5 - Vitaminic (Provital/
Centerchem)

PEG-25 CASTOR OIL

CAS No.: 61791-12-6 (Generic)

JPN Translation:
PEG - 25 ヒマシ油

Definition: PEG-25 Castor Oil is a polyethylene glycol derivative of Ricinus Communis (Castor) Oil (q.v.) with an average of 25 moles of ethylene oxide.

Information Sources: 21CFR175.105, 21CFR175.300, 21CFR177.2800, JCLS, TSCA

Chemical Classes: Alkoxylated Alcohols; Glyceryl Esters and Derivatives

Function: Surfactant - Emulsifying Agent

Technical/Other Names:
Polyethylene Glycol (25) Castor Oil
Polyoxyethylene (25) Castor Oil

Trade Names:
Ethox CO-25 (Ethox)
Hetoxide C-25 (Heterene)
Jeechem CA-25 (Jeen)
Lumulse GR-25 (Lambent)
Protachem CO 25 (Protameen)
Ricino Viscoil (Vevy)
Unipeg-CO-25 (Universal Preserv-A-Chem)

Trade Name Mixture:
Synthro-Pon A 1820 N 75 (Synthron)

PEG-26 CASTOR OIL

CAS No.: 61791-12-6

JPN Translation:
PEG - 26 ヒマシ油

Definition: PEG-26 Castor Oil is a polyethylene glycol derivative of Ricinus Communis (Castor) Oil (q.v.) with an average of 26 moles of ethylene oxide.

Information Source: JCLS

Chemical Classes: Alkoxylated Alcohols; Glyceryl Esters and Derivatives

Functions: Surfactant - Emulsifying Agent; Surfactant - Solubilizing Agent

Technical/Other Names:
Polyethylene Glycol (26) Castor Oil
Polyoxyethylene (26) Castor Oil

PEG-29 CASTOR OIL

CAS No.: 61791-12-6 (Generic)

JPN Translation:
PEG - 29 ヒマシ油

Definition: PEG-29 Castor Oil is a polyethylene glycol derivative of Ricinus Communis (Castor) Oil (q.v.) with an average of 29 moles of ethylene oxide.

Information Sources: JCLS, TSCA

Chemical Classes: Alkoxylated Alcohols; Glyceryl Esters and Derivatives

Functions: Surfactant - Emulsifying Agent; Surfactant - Solubilizing Agent

Technical/Other Names:
Polyethylene Glycol (29) Castor Oil
Polyoxyethylene (29) Castor Oil

Trade Name:
Etocas 29 (Croda Chemicals)

PEG-30 CASTOR OIL

CAS No.: 61791-12-6 (Generic)

JPN Translation:
PEG - 30 ヒマシ油

Definition: PEG-30 Castor Oil is a polyethylene glycol derivative of Ricinus Communis (Castor) Oil (q.v.) with an average of 30 moles of ethylene oxide.

Information Sources: 21CFR175.105, 21CFR175.300, 21CFR177.2800, CIR: [SQ] IJT-16(3)1997, JCLS, TSCA

Chemical Classes: Alkoxylated Alcohols; Glyceryl Esters and Derivatives

Function: Surfactant - Emulsifying Agent

Reported Product Categories: Hair Dyes and Colors (All Types Requiring Caution Statements and Patch Tests); Hair Tints

Technical/Other Names:
Polyethylene Glycol (30) Castor Oil
Polyoxyethylene (30) Castor Oil

Trade Names:
Alkamuls B (Rhodia)
Alkamuls EL-620 (Rhodia)
Chemax CO-30 (Chemax)

DePEG 30-CO (DeForest)
Ethox CO-30 (Ethox)
Fancol CO-30 (Fanning)
Hetoxide C-30 (Heterene)
Incrocas-30 (Croda, Inc.)
Jeechem CA-30 (Jeen)
Protachem CO 30 (Protameen)
Sabowax EL 30 (Sabo)

PEG-33 CASTOR OIL

CAS No.: 61791-12-6 (Generic)

JPN Translation:
PEG - 33 ヒマシ油

Definition: PEG-33 Castor Oil is a polyethylene glycol derivative of Ricinus Communis (Castor) Oil (q.v.) with an average of 33 moles of ethylene oxide.

Information Sources: 21CFR175.105, 21CFR175.300, 21CFR176.210, 21CFR177.2800, CIR: [SQ] IJT-16(3)1997, JCLS

Chemical Classes: Alkoxylated Alcohols; Glyceryl Esters and Derivatives

Function: Surfactant - Emulsifying Agent

Reported Product Category: Skin Care Preparations, Misc.

Technical/Other Names:
Polyethylene Glycol (33) Castor Oil
Polyoxyethylene (33) Castor Oil

Trade Names:
Mergital EL 33 (Cognis France)
Ricinion (Gattefosse s.a.)

PEG-35 CASTOR OIL

CAS No.: 61791-12-6 (Generic)

JPN Translation:
PEG - 35 ヒマシ油

Definition: PEG-35 Castor Oil is the polyethylene glycol derivative of Ricinus Communis (Castor) Oil (q.v.) with an average of 35 moles of ethylene oxide.

Information Sources: 21CFR175.105, 21CFR175.300, 21CFR177.2800, CIR: [SQ] IJT-16(3)1997, JCLS, NF XIX, USAN

Chemical Classes: Alkoxylated Alcohols; Glyceryl Esters and Derivatives

Functions: Surfactant - Emulsifying Agent; Surfactant - Solubilizing Agent

Technical/Other Names:
Polyethylene Glycol (35) Castor Oil
Polyoxyethylene (35) Castor Oil

Trade Names:
Cremophor EL (BASF)

Etocas 35 (Croda Chemicals)
Eumulgin RO 35 (Cognis Deutschland)
Mergital EL 35 (Cognis France)

Trade Name Mixtures:
Guaiazulene 25% WS 2/013000 (Symrise)
Phenolines The Vert (Coletica SA)
Soluvit Richter (CLR)
Sulfoconcentrol (2/380011) (Symrise)

PEG-36 CASTOR OIL

CAS No.: 61791-12-6 (Generic)

JPN Translation:
PEG - 36 ヒマシ油

Definition: PEG-36 Castor Oil is a polyethylene glycol derivative of Ricinus Communis (Castor) Oil (q.v.) with an average of 36 moles of ethylene oxide.

Information Sources: 21CFR175.105, 21CFR175.300, 21CFR176.210, 21CFR177.2800, CIR: [SQ] IJT-16(3)1997, CTFA D, JCLS, TSCA

Chemical Classes: Alkoxylated Alcohols; Glyceryl Esters and Derivatives

Functions: Surfactant - Emulsifying Agent; Surfactant - Solubilizing Agent

Technical/Other Names:
Polyethylene Glycol 1800 Castor Oil
Polyoxyethylene (36) Castor Oil

Trade Names:
Alpicare CO 36 (Cesalpinia)
Arlatone 650 (Uniqema Americas)
Chemax CO-36 (Chemax)
Emulsogen EL (Clariant)
Emulsogen EL (Clariant GmbH, Personal Care)
Ethox CO-36 (Ethox)
Unipeg-CO-36 (Universal Preserv-A-Chem)

Trade Name Mixtures:
Lipoderma PF-WS (Lipo)
Safester A-75 watersoluble (Induchem)
Unibiovit B-332 watersoluble (Induchem)
Unibiovit B-332 WS (Lipo)
Uniprosolv U-20 (Induchem)
Unitrienol T-272 watersoluble (Induchem)
Unitrienol T-272 WS (Lipo)
Vitasol, Vitamin-Horsechestnut-Complex (Crodarom)

PEG-40 CASTOR OIL

CAS No.: 61791-12-6 (Generic)

JPN Translation:
PEG - 40 ヒマシ油

Definition: PEG-40 Castor Oil is a polyethylene glycol derivative of Ricinus Communis (Castor) Oil (q.v.) with an average of 40 moles of ethylene oxide.

Information Sources: 21CFR175.105, 21CFR175.300, 21CFR176.170, 21CFR176.180, 21CFR176.210, 21CFR177.2800, CIR: [SQ] IJT-16(3)1997, CTFA S, JCLS, TSCA

Chemical Classes: Alkoxylated Alcohols; Glyceryl Esters and Derivatives

Functions: Surfactant - Cleansing Agent; Surfactant - Solubilizing Agent

Reported Product Categories: Skin Care Preparations, Misc.; Bath Oils, Tablets, and Salts; Cleansing Products (Cold Creams, Cleansing Lotions, Liquids and Pads); Bath Preparations, Misc.; Body and Hand Preparations (Excluding Shaving Preparations); Permanent Waves; Moisturizing Preparations; Hair Conditioners; Paste Masks (Mud Packs); Tonics, Dressings, and Other Hair Grooming Aids; Hair Preparations (Non-coloring), Misc.; Face and Neck Preparations (Excluding Shaving Preparations); Hair Wave Sets

Technical/Other Names:
Polyethylene Glycol 2000 Castor Oil
Polyoxyethylene (40) Castor Oil

Trade Names:
Alkamuls EL-719 (Rhodia)
Alpicare CO 40 (Cesalpinia)
Atlas G-1284 (Uniqema Americas)
Chemax CO-40 (Chemax)
DePEG 40-CO (DeForest)
T-DET C-40 (Harcros)
Ethox CO-40 (Ethox)
Etocas 40 (Croda Chemicals)
Eumulgin RO 40 (Cognis Deutschland)
Hetoxide C-40 (Heterene)
Jeechem CA-40 (Jeen)
Lanaetex CO-40 (Lanaetex)
Lumulse GR-40 (Lambent)
Marlowet R 40 (Sasol GmbH - Marl)
Nikkol CO-40TX (Nikko)
Protachem CO 40 (Protameen)
Sabowax EL 40 (Sabo)
Simulsol OL 50 (SEPPIC)
Surfactol 365 (CasChem)
Sympatens-TR/400 (Kolb)
Unipeg-CO-40 (Universal Preserv-A-Chem)

Trade Name Mixtures:
Andiroba Water Soluble Oil CPH (Centroflora)
Andiroba Water Soluble Oil SPH (Centroflora)
Brazil Nut Water Soluble Oil CPH (Centroflora)
Brazil Nut Water Soluble Oil SPH (Centroflora)
Ceral EFN (Fabriquimica)
CoChem SCS (Costec)
Copaiba Water Soluble Oil CPH (Centroflora)
Copaiba Water Soluble Oil SPH (Centroflora)
Emulgade F (Cognis Care Chemicals/NJ)
Emulgade F (Cognis Care Chemicals/PA)
Emulgade F (Cognis Deutschland)
Emulgade F Special (Cognis Deutschland)
Incroquat CR Concentrate (Croda, Inc.)
Lecithin water-dispersible CLR (CLR)
Maquat SC 1632 (Mason)
Proceramide L2 (Sederma)
Quatrex CRC (Chemron)
Sabosol HO (Sabo)
Sabosolv EEM (Sabo)
Simulgel A (SEPPIC)
Tioxolone Water Soluble 5% (Provital/Centerchem)
Unimulgade-F (Universal Preserv-A-Chem)
Unimulgade-F SPECIAL (Universal Preserv-A-Chem)

PEG-44 CASTOR OIL

CAS No.: 61791-12-6 (Generic)

JPN Translation:
PEG - 44 ヒマシ油

Definition: PEG-44 Castor Oil is a polyethylene glycol derivative of Ricinus Communis (Castor) Oil (q.v.) with an average of 44 moles of ethylene oxide.

Information Sources: JCLS, TSCA

Chemical Classes: Alkoxylated Alcohols; Glyceryl Esters and Derivatives

Functions: Surfactant - Cleansing Agent; Surfactant - Solubilizing Agent

Technical/Other Names:
Polyethylene Glycol (44) Castor Oil
Polyoxyethylene (44) Castor Oil

Trade Name:
Eumulgin RO 44 (Cognis France)

PEG-50 CASTOR OIL

CAS No.: 61791-12-6 (Generic)

JPN Translation:
PEG - 50 ヒマシ油

Definition: PEG-50 Castor Oil is a polyethylene glycol derivative of Ricinus Communis (Castor) Oil (q.v.) with an average of 50 moles of ethylene oxide.

Information Sources: 21CFR175.105, 21CFR175.300, 21CFR177.2800, CTFA D, JCLS, TSCA

Chemical Classes: Alkoxylated Alcohols; Glyceryl Esters and Derivatives

Functions: Surfactant - Cleansing Agent; Surfactant - Solubilizing Agent

Technical/Other Names:
Polyethylene Glycol (50) Castor Oil
Polyoxyethylene (50) Castor Oil

Trade Names:
Ethox CO-50 (Ethox)
Nikkol CO-50TX (Nikko)

PEG-54 CASTOR OIL

CAS No.: 61791-12-6 (Generic)

JPN Translation:
PEG - 54 ヒマシ油

Definition: PEG-54 Castor Oil is a polyethylene glycol derivative of Ricinus Communis (Castor) Oil (q.v.) with an average of 54 moles of ethylene oxide.

Information Sources: 21CFR175.105, 21CFR175.300, 21CFR177.2800, JCLS

Chemical Classes: Alkoxylated Alcohols; Glyceryl Esters and Derivatives

Functions: Surfactant - Cleansing Agent; Surfactant - Solubilizing Agent

Technical/Other Names:
Polyethylene Glycol (54) Castor Oil
Polyoxyethylene (54) Castor Oil

PEG-55 CASTOR OIL

CAS No.: 61791-12-6 (Generic)

JPN Translation:
PEG - 55 ヒマシ油

Definition: PEG-55 Castor Oil is a polyethylene glycol derivative of Ricinus Communis (Castor) Oil (q.v.) with an average of 55 moles of ethylene oxide.

Information Sources: 21CFR175.105, 21CFR175.300, 21CFR177.2800

Chemical Classes: Alkoxylated Alcohols; Glyceryl Esters and Derivatives

Functions: Surfactant - Cleansing Agent; Surfactant - Solubilizing Agent

Technical/Other Names:
Polyethylene Glycol (55) Castor Oil
Polyoxyethylene (55) Castor Oil

PEG-60 CASTOR OIL

CAS No.: 61791-12-6 (Generic)

JPN Translation:
PEG - 60 ヒマシ油

Definition: PEG-60 Castor Oil is a polyethylene glycol derivative of Ricinus Communis (Castor) Oil (q.v.) with an average of 60 moles of ethylene oxide.

Information Sources: 21CFR175.105, 21CFR175.300, 21CFR177.2800, CTFA D, TSCA

Chemical Classes: Alkoxylated Alcohols; Glyceryl Esters and Derivatives

Functions: Surfactant - Cleansing Agent; Surfactant - Solubilizing Agent

Technical/Other Names:
Polyethylene Glycol 3000 Castor Oil
Polyoxyethylene (60) Castor Oil

Trade Names:
Jeechem CA-60 (Jeen)
Nikkol CO-60TX (Nikko)
Protachem CO 60 (Protameen)
Simulsol 1285 (Solubilizers) (SEPPIC)

PEG-75 CASTOR OIL

CAS No.: 61791-12-6 (Generic)

JPN Translation:
PEG - 75 ヒマシ油

Definition: PEG-75 Castor Oil is a polyethylene glycol derivative of Ricinus Communis (Castor) Oil (q.v.) with an average of 75 moles of ethylene oxide.

Information Sources: JCLS, TSCA

Chemical Classes: Alkoxylated Alcohols; Glyceryl Esters and Derivatives

Function: Surfactant - Solubilizing Agent

Technical/Other Names:
Polyethylene Glycol 4000 Castor Oil
Polyoxyethylene (75) Castor Oil

PEG-80 CASTOR OIL

CAS No.: 61791-12-6 (Generic)

Definition: PEG-80 Castor Oil is a polyethylene glycol derivative of Ricinus Communis (Castor) Oil (q.v.), with an average of 80 moles of ethylene oxide.

Chemical Classes: Alkoxylated Alcohols; Glyceryl Esters and Derivatives

Functions: Surfactant - Cleansing Agent; Surfactant - Solubilizing Agent

Technical/Other Names:
Polyethylene Glycol (80) Castor Oil
Polyoxyethylene (80) Castor Oil

Trade Name:
Ethox CO-81 (Ethox)

PEG-100 CASTOR OIL

CAS No.: 61791-12-6 (Generic)

JPN Translation:
PEG - 100 ヒマシ油

Definition: PEG-100 Castor Oil is a polyethylene glycol derivative of Ricinus Communis (Castor) Oil (q.v.) with an average of 100 moles of ethylene oxide.

Information Sources: 21CFR175.300, 21CFR176.210, JCLS, TSCA

Chemical Classes: Alkoxylated Alcohols; Glyceryl Esters and Derivatives

Functions: Surfactant - Cleansing Agent; Surfactant - Solubilizing Agent

Technical/Other Names:
Polyethylene Glycol (100) Castor Oil
Polyoxyethylene (100) Castor Oil

Trade Names:
Jeechem CA-100 (Jeen)
Protachem CO 100 (Protameen)

PEG-200 CASTOR OIL

CAS No.: 61791-12-6 (Generic)

JPN Translation:
PEG - 200 ヒマシ油

Definition: PEG-200 Castor Oil is a polyethylene glycol derivative of Ricinus Communis (Castor) Oil (q.v.) with an average of 200 moles of ethylene oxide.

Information Sources: 21CFR175.300, CTFA D, JCLS, TSCA

Chemical Classes: Alkoxylated Alcohols; Glyceryl Esters and Derivatives

Functions: Surfactant - Cleansing Agent; Surfactant - Solubilizing Agent

Technical/Other Names:
Polyethylene Glycol 200 Castor Oil
Polyoxyethylene (200) Castor Oil

Trade Names:
Atlas G-1300 (Uniqema Americas)
Ethox CO-200 (Ethox)
Hetoxide C-200 (Heterene)
Jeechem CA-200 (Jeen)
Protachem CO 200 (Protameen)
Unipeg-CO-200 (Universal Preserv-A-Chem)

PEG-18 CASTOR OIL DIOLEATE

Definition: PEG-18 Castor Oil Dioleate is the oleic acid diester of ethoxylated castor oil in which the average ethoxylation value is 18.

Chemical Classes: Alkoxylated Alcohols; Esters; Glyceryl Esters and Derivatives

Functions: Suspending Agent - Nonsurfactant; Viscosity Increasing Agent - Nonaqueous

Technical/Other Names:
Polyethylene Glycol (18) Castor Oil Dioleate
Polyoxyethylene (18) Castor Oil Dioleate

Trade Name:
Marlowet LVS (Sasol GmbH - Marl)

PEG-60 CASTOR OIL ISOSTEARATE

Definition: PEG-60 Castor Oil Isostearate is the ester of Isostearic Acid (q.v.) and PEG-60 Castor Oil (q.v.).

Chemical Classes: Alkoxylated Alcohols; Glyceryl Esters and Derivatives

Function: Not Reported

Technical/Other Names:
Polyethylene Glycol (60) Castor Oil Isostearate
Polyoxyethylene (60) Castor Oil Isostearate

PEG-8 C12-18 ESTER

Definition: PEG-8 C12-18 Ester is the polyethylene glycol ester of a synthetic mixture of saturated acids containing 12 to 18 carbons in the alkyl chain, with an average of 8 moles of ethylene oxide.

Chemical Class: Alkoxylated Carboxylic Acids

Function: Surfactant - Emulsifying Agent

Reported Product Categories: Moisturizing Preparations; Bath Preparations, Misc.; Body and Hand Preparations (Excluding Shaving Preparations); Night Skin Care Preparations; Bath Oils, Tablets, and Salts; Cleansing Products (Cold Creams, Cleansing Lotions, Liquids and Pads); Skin Care Preparations, Misc.; Eye Makeup Preparations, Misc.

Technical/Other Names:
C12-18 Acid PEG-8 Ester
PEG-8 C12-18 Alkyl Ester

PEG-8 CETYL DIMETHICONE

Definition: PEG-8 Cetyl Dimethicone is the polydimethylsiloxane that conforms to the formula:

$$(CH_3)_3SiO{-}\left[\underset{\overset{|}{(CH_2)_{15}}}{\overset{\overset{CH_3}{|}}{Si}O}\right]_x\left[\underset{\overset{|}{O}}{\overset{\overset{CH_3}{|}}{Si}O}\right] {-} Si(CH_3)_3$$

where n has an average value of 8.

Chemical Class: Siloxanes and Silanes

Function: Skin-Conditioning Agent - Miscellaneous

Trade Name Mixtures:
Suncaps C (Particle Sciences)
Zenibee D (Zenitech)

PEG-2 COCAMIDE

Definition: PEG-2 Cocamide is the polyethylene glycol amide of coconut acid that conforms generally to the formula:

$$\underset{RC}{\overset{\overset{O}{\|}}{}} {-} NH(CH_2CH_2O)_nH$$

where RCO- represents the fatty acids derived from coconut oil and n has an average value of 2.

Information Source: JCLS

Chemical Class: Alkoxylated Amides

Functions: Surfactant - Emulsifying Agent; Surfactant - Foam Booster

Technical/Other Names:
Polyethylene Glycol 100 Coconut Acid Amide
Polyoxyethylene (2) Coconut Acid Amide

Trade Name:
Norfox DGCA (Norman, Fox & Co.)

PEG-3 COCAMIDE

CAS No.: 61791-08-0 (Generic)

JPN Translation:
PEG - 3 コカミド

Definition: PEG-3 Cocamide is the polyethylene glycol amide of Cocos Nucifera (Coconut) Acid (q.v.) that conforms generally to the formula:

$$\underset{RC}{\overset{\overset{O}{\|}}{}} {-} NH(CH_2CH_2O)_nH$$

where RCO- represents the fatty acids derived from coconut oil and n has an average value of 3.

Information Sources: JCLS, JSCI, JSQI, TSCA

Chemical Class: Alkoxylated Amides

Functions: Surfactant - Emulsifying Agent; Surfactant - Foam Booster

Technical/Other Names:
Polyethylene Glycol (3) Coconut Amide
Polyoxyethylene (3) Coconut Amide
Polyoxyethylene (2) Coconut Fatty Acid Monoethanolamide

Trade Names:
Amadol CMA-2 (Akzo Nobel Surface AB)
Hetoxamide CD-4 (Heterene)
NINOL C-2 (Stepan)
Upamide C-2 (Universal Preserv-A-Chem)

PEG-4 COCAMIDE

Definition: PEG-4 Cocamide is the polyethylene glycol amide of Cocos Nucifera (Coconut) Acid (q.v.) that conforms generally to the formula:

$$\underset{RC}{\overset{\overset{O}{\|}}{}} {-} NH(CH_2CH_2O)_nH$$

where RCO- represents the fatty acids derived from coconut oil and n has an average value of 4.

Chemical Class: Alkoxylated Amides

Function: Surfactant - Emulsifying Agent

Technical/Other Names:
Polyethylene Glycol 200 Coconut Amide
Polyoxyethylene (4) Coconut Amide

Trade Name:
Eumulgin C4 (Cognis Deutschland)

PEG-5 COCAMIDE

CAS No.: 61791-08-0 (Generic)

JPN Translation:
PEG - 5 コカミド

Definition: PEG-5 Cocamide is the polyethylene glycol amide of Cocos Nucifera (Coconut) Acid (q.v.) that conforms generally to the formula:

$$\underset{RC}{\overset{\overset{O}{\|}}{}} {-} NH(CH_2CH_2O)_nH$$

where RCO- represents the fatty acids derived from coconut oil and n has an average value of 5.

Information Sources: JCIC, JCLS, JSQI, TSCA

Chemical Class: Alkoxylated Amides

Function: Surfactant - Emulsifying Agent

Technical/Other Names:
Polyethylene Glycol (5) Coconut Amide
Polyoxyethylene (5) Coconut Amide
Polyoxyethylene (5) Coconut Fatty Acid Amide

Trade Names:
Empilan MAA (Albright & Wilson UK)
Genagen CA-050 (Clariant)
Genagen CA-050 (Clariant GmbH, Personal Care)

PEG-6 COCAMIDE

CAS No.: 61791-08-0 (Generic)

JPN Translation:
PEG - 6 コカミド

Definition: PEG-6 Cocamide is the polyethylene glycol amide of Cocos Nucifera (Coconut) Acid (q.v.) that conforms generally to the formula:

$$RC \overset{\overset{\displaystyle O}{\|}}{-} NH(CH_2CH_2O)_nH$$

where RCO- represents the fatty acids derived from coconut oil and n has an average value of 6.

Information Sources: JCLS, JSCI, TSCA

Chemical Class: Alkoxylated Amides

Function: Surfactant - Emulsifying Agent

Reported Product Category: Bubble Baths

Technical/Other Names:
Polyethylene Glycol 300 Coconut Amide
Polyoxyethylene (6) Coconut Amide
Polyoxyethylene (5) Coconut Fatty Acid
   Monoethanolamide

Trade Names:
Hetoxamide CD-6 (Heterene)
NINOL 1301 (Stepan)
NINOL C-5 (Stepan)
Upamide C-5 (Universal Preserv-A-Chem)

PEG-7 COCAMIDE

CAS No.: 61791-08-0 (Generic)

Definition: PEG-7 Cocamide is the polyethylene glycol amide of Cocos Nucifera (Coconut) Acid (q.v.) that conforms generally to the formula:

$$RC \overset{\overset{\displaystyle O}{\|}}{-} NH(CH_2CH_2O)_nH$$

where RCO- represents the fatty acids derived from coconut oil and n has an average value of 7.

Information Sources: JCLS, TSCA

Chemical Class: Alkoxylated Amides

Function: Surfactant - Emulsifying Agent

Technical/Other Names:
Polyethylene Glycol (7) Coconut Amide
Polyoxyethylene (7) Coconut Amide

PEG-11 COCAMIDE

CAS No.: 61791-08-0 (Generic)

JPN Translation:
PEG - 11 コカミド

Definition: PEG-11 Cocamide is the polyethylene glycol amide of Cocos Nucifera (Coconut) Acid (q.v.) that conforms generally to the formula:

$$RC \overset{\overset{\displaystyle O}{\|}}{-} NH(CH_2CH_2O)_nH$$

where RCO- represents the fatty acids derived from coconut oil and n has an average value of 11.

Information Sources: JCLS, JSCI, TSCA

Chemical Class: Alkoxylated Amides

Functions: Surfactant - Cleansing Agent; Surfactant - Emulsifying Agent

Technical/Other Names:
Polyethylene Glycol (11) Coconut Amide
Polyoxyethylene (11) Coconut Amide
Polyoxyethylene (10) Coconut Fatty Acid
   Monoethanolamide

PEG-20 COCAMIDE

CAS No.: 61791-08-0 (Generic)

Definition: PEG-20 Cocamide is the polyethylene glycol amide of Cocos Nucifera (Coconut) Acid (q.v.) that conforms generally to the formula:

$$RC \overset{\overset{\displaystyle O}{\|}}{-} NH(CH_2CH_2O)_nH$$

where RCO- represents the fatty acids derived from coconut oil and n has an average value of 20.

Information Sources: JCLS, JSCI

Chemical Class: Alkoxylated Amides

Function: Surfactant - Emulsifying Agent

Technical/Other Names:
Polyethylene Glycol 1000 Coconut Amide
Polyoxyethylene (20) Coconut Amide
Polyoxyethylene (20) Coconut Fatty Acid
   Monoethanolamide

PEG-3 COCAMIDE DEA

Definition: PEG-3 Cocamide DEA is the polyethylene glycol derivative of Cocamide DEA (q.v.) with an average of 3 moles of ethylene oxide.

Information Source: JCLS

Chemical Class: Alkoxylated Amines

Function: Surfactant - Emulsifying Agent

PEG-20 COCAMIDE MEA

JPN Translation:
PEG - 20 ヤシ脂肪酸アミド MEA

Definition: PEG-20 Cocamide MEA is the monoethanolamine salt of PEG-20 Cocamide (q.v.).

Information Source: JCIC

Chemical Class: Alkanolamides

Function: Surfactant - Emulsifying Agent

Technical/Other Name:
Polyoxyethylene (20) Coconut Fatty Acid
   Monoethanolamide

PEG-6 COCAMIDE PHOSPHATE

JPN Translation:
PEG - 5 ヤシ脂肪酸アミド MEA リン酸

Definition: PEG-6 Cocamide Phosphate is the phosphate ester of the polyethylene glycol amide of Coconut Acid (q.v.) with an average 6 moles of ethylene oxide.

Information Source: JCLS

Chemical Classes: Alkoxylated Amides; Phosphorus Compounds

Function: Surfactant - Cleansing Agent

PEG-2 COCAMINE

CAS No.: 61791-14-8 (Generic)

JPN Translation:
PEG - 2 コカミン

Definition: PEG-2 Cocamine is the polyethylene glycol derivative of Cocamine (q.v.) that conforms generally to the formula:

$$R - N \overset{\displaystyle (CH_2CH_2O)_xH}{\underset{\displaystyle (CH_2CH_2O)_yH}{}}$$

where R represents the alkyl groups derived from coconut oil and x + y has an average value of 2.

Information Sources: CIR: [I] IJT-18 (SUPPL. 1)1999, JCLS, JSQI, TSCA

Chemical Class: Alkoxylated Amines

Function: Surfactant - Emulsifying Agent

Reported Product Categories: Hair Tints; Hair Dyes and Colors (All Types Requiring Caution Statements and Patch Tests)

Technical/Other Names:
Polyethylene Glycol 100 Coconut Amine
Polyoxyethylene (2) Coconut Amine

Trade Names:
Chemeen C-2 (Chemax)
Ethomeen C/12 (Akzo Nobel Surface AB)
Ethox CAM-2 (Ethox)
Hetoxamine C-2 (Heterene)
Jeetox C-2 (Jeen)

Protox C-2 (Protameen)
Sabopal NC 2 (Sabo)
Unizeen C-2 (Universal Preserv-A-Chem)

PEG-3 COCAMINE

CAS No.: 61791-14-8 (Generic)

JPN Translation:
PEG - 3 コカミン

Definition: PEG-3 Cocamine is the polyethylene glycol derivative of Cocamine (q.v.)
that conforms generally to the formula:

$$R-N \begin{array}{c} (CH_2CH_2O)_xH \\ \\ (CH_2CH_2O)_yH \end{array}$$

where R represents the alkyl groups derived
from coconut oil and x + y has an average
value of 3.

Information Sources: CIR: [I] IJT-18
(SUPPL. 1)1999, JCLS, TSCA

Chemical Class: Alkoxylated Amines

Function: Surfactant - Emulsifying Agent

Reported Product Categories: Hair Dyes
and Colors (All Types Requiring Caution
Statements and Patch Tests); Hair Tints

Technical/Other Names:
Polyethylene Glycol (3) Coconut Amine
Polyoxyethylene (3) Coconut Amine

Trade Name:
Lowenol C-243 (Lowenstein)

PEG-5 COCAMINE

CAS No.: 61791-14-8 (Generic)

JPN Translation:
PEG - 5 コカミン

Definition: PEG-5 Cocamine is the polyethylene glycol derivative of Cocamine (q.v.)
that conforms generally to the formula:

$$R-N \begin{array}{c} (CH_2CH_2O)_xH \\ \\ (CH_2CH_2O)_yH \end{array}$$

where R represents the alkyl groups derived
from coconut oil and x + y has an average
value of 5.

Information Sources: CIR: [I] IJT-18
(SUPPL. 1)1999, JCLS, TSCA

Chemical Class: Alkoxylated Amines

Function: Surfactant - Emulsifying Agent

Reported Product Categories: Hair Dyes
and Colors (All Types Requiring Caution
Statements and Patch Tests); Hair Tints

Technical/Other Names:
Polyethylene Glycol (5) Coconut Amine
Polyoxyethylene (5) Coconut Amine

Trade Names:
Chemeen C-5 (Chemax)
DeTHOX AMINE C-5 (DeForest)
Ethomeen C/15 (Akzo Nobel)
Ethox CAM-5 (Ethox)
Hetoxamine C-5 (Heterene)
Jeetox C-5 (Jeen)
Protox C-5 (Protameen)
Unizeen C-5 (Universal Preserv-A-Chem)

PEG-10 COCAMINE

CAS No.: 61791-14-8 (Generic)

JPN Translation:
PEG - 10 コカミン

Definition: PEG-10 Cocamine is the polyethylene glycol derivative of Cocamine (q.v.)
that conforms generally to the formula:

$$R-N \begin{array}{c} (CH_2CH_2O)_xH \\ \\ (CH_2CH_2O)_yH \end{array}$$

where R represents the alkyl groups derived
from coconut oil and x + y has an average
value of 10 .

Information Sources: CIR: [I] IJT-18
(SUPPL. 1)1999, JCLS, TSCA

Chemical Class: Alkoxylated Amines

Function: Surfactant - Emulsifying Agent

Reported Product Categories: Hair Tints;
Hair Dyes and Colors (All Types Requiring
Caution Statements and Patch Tests)

Technical/Other Names:
Polyethylene Glycol 500 Coconut Amine
Polyoxyethylene (10) Coconut Amine

Trade Names:
Chemeen C-10 (Chemax)
Jeetox C-10 (Jeen)
Protox C-10 (Protameen)
Unizeen C-10 (Universal Preserv-A-Chem)

PEG-15 COCAMINE

CAS Nos.: 8051-52-3 (Generic); 61791-14-8
(Generic)

JPN Translation:
PEG - 15 コカミン

Definition: PEG-15 Cocamine is the polyethylene glycol derivative of Cocamine (q.v.)
that conforms generally to the formula:

$$R-N \begin{array}{c} (CH_2CH_2O)_xH \\ \\ (CH_2CH_2O)_yH \end{array}$$

where R represents the alkyl groups derived
from coconut oil and x + y has an average
value of 15 .

Information Sources: CIR: [I] IJT-18
(SUPPL. 1)1999, CTFA D, JCLS, TSCA

Chemical Class: Alkoxylated Amines

Function: Surfactant - Emulsifying Agent

Reported Product Categories: Hair Tints;
Hair Dyes and Colors (All Types Requiring
Caution Statements and Patch Tests);
Powders (Dusting and Talcum, Excluding
Aftershave Talcs); Moisturizing Preparations;
Tonics, Dressings, and Other Hair Grooming
Aids; Bath Oils, Tablets, and Salts; Cleansing
Products (Cold Creams, Cleansing Lotions,
Liquids and Pads); Nail Polish and Enamel
Removers; Skin Fresheners

Technical/Other Names:
Polyethylene Glycol (15) Coconut Amine
Polyoxyethylene (15) Coconut Amine

Trade Names:
AEC PEG-15 Cocamine (A & E Connock)
Chemeen C-15 (Chemax)
DeTHOX AMINE C-15 (DeForest)
Ethomeen C/25 (Akzo Nobel)
Ethox CAM-15 (Ethox)
Hetoxamine C-15 (Heterene)
Jeetox C-15 (Jeen)
Protox C-15 (Protameen)
Sabopal NC 15 (Sabo)

Trade Name Mixture:
Lubrasil DS (Guardian)

PEG-20 COCAMINE

CAS No.: 61791-14-8 (Generic)

Definition: PEG-20 Cocamine is the polyethylene glycol derivative of Cocamine (q.v.)
that conforms generally to the formula:

$$R-N \begin{array}{c} (CH_2CH_2O)_xH \\ \\ (CH_2CH_2O)_yH \end{array}$$

where R represents the alkyl groups derived
from coconut oil and x + y has an average
value of 20.

Information Sources: CIR: [I] IJT-18
(SUPPL. 1)1999, JCLS

Chemical Class: Alkoxylated Amines

Functions: Surfactant - Emulsifying Agent;
Surfactant - Solubilizing Agent

Reported Product Categories: Hair Dyes
and Colors (All Types Requiring Caution
Statements and Patch Tests); Hair Tints

Technical/Other Names:
Polyethylene Glycol 1000 Cocamine

Polyoxyethylene (20) Cocamine
Polyoxyethylene (20) Coconut Amine

PEG-15 COCAMINE OLEATE/PHOSPHATE

Definition: PEG-15 Cocamine Oleate/ Phosphate is a mixture of oleic acid and phosphoric acid esters of PEG-15 Cocamine (q.v.).

Chemical Classes: Alkoxylated Amines; Phosphorus Compounds

Function: Surfactant - Emulsifying Agent

Technical/Other Names:
Polyethylene Glycol (15) Cocamine Oleate/ Phosphate
Polyoxyethylene (15) Cocamine Oleate/ Phosphate

Trade Name:
Lanaetex CPS (Lanaetex)

PEG-11 COCOA BUTTER GLYCERIDES

Definition: PEG-11 Cocoa Butter Glycerides is a polyethylene glycol derivative of the mono- and diglycerides derived from Theobroma Cacao (Cocoa) Butter (q.v.) with an average of 11 moles of ethylene oxide.

Chemical Classes: Alkoxylated Alcohols; Glyceryl Esters and Derivatives

Functions: Skin-Conditioning Agent - Emollient; Surfactant - Emulsifying Agent

Technical/Other Names:
Polyethylene Glycol (11) Cocoa Butter Glycerides
Polyoxyethylene (11) Cocoa Butter Glycerides

Trade Name:
Oxypon 315 (Zschimmer & Schwarz)

PEG-75 COCOA BUTTER GLYCERIDES

Definition: PEG-75 Cocoa Butter Glycerides is a polyethylene glycol derivative of the glycerides derived from Theobroma Cacao (Cocoa) Butter (q.v.) with an average of 75 moles of ethylene oxide.

Chemical Classes: Alkoxylated Alcohols; Glyceryl Esters and Derivatives

Functions: Skin-Conditioning Agent - Emollient; Surfactant - Emulsifying Agent

Technical/Other Names:
Polyethylene Glycol 4000 Cocoa Butter Glycerides
Polyoxyethylene (75) Cocoa Butter Glycerides

PEG-5 COCOATE

CAS No.: 61791-29-5 (Generic)

Definition: PEG-5 Cocoate is the polyethylene glycol ester of Coconut Acid (q.v.) that conforms generally to the formula:

$$\overset{\displaystyle O}{\overset{\displaystyle \|}{RC}} - (OCH_2CH_2)_nOH$$

where RCO- represents the fatty acids derived from coconut oil and n has an average value of 5.

Information Sources: 21CFR175.105, 21CFR175.300, 21CFR177.2260, MI-13 (7660), TSCA

Chemical Class: Alkoxylated Carboxylic Acids

Function: Surfactant - Emulsifying Agent

Technical/Other Names:
Polyethylene Glycol (5) Monococoate
Polyoxyethylene (5) Monococoate

Trade Name:
Ethofat C/15 (Akzo Nobel)

PEG-8 COCOATE

CAS No.: 61791-29-5 (Generic)

Definition: PEG-8 Cocoate is the polyethylene glycol ester of Coconut Acid (q.v.) that conforms generally to the formula:

$$\overset{\displaystyle O}{\overset{\displaystyle \|}{RC}} - (OCH_2CH_2)_nOH$$

where RCO- represents the fatty acids derived from coconut oil and n has an average value of 8.

Information Sources: 21CFR175.105, 21CFR175.300, 21CFR176.170, 21CFR176.200, 21CFR176.210, 21CFR177.1210, 21CFR177.2260, 21CFR177.2800, CTFA D, MI-13(7660), TSCA

Chemical Class: Alkoxylated Carboxylic Acids

Function: Surfactant - Emulsifying Agent

Technical/Other Names:
Polyethylene Glycol 400 Monococoate
Polyoxyethylene (8) Monococoate

Trade Names:
Dermol HC (Fabriquimica)
ROL L 40 (Fabriquimica)
Waglinol 488 (Industrial Quimica)

PEG-9 COCOATE

CAS No.: 61791-29-5 (Generic)

Definition: PEG-9 Cocoate is the polyethylene glycol ester of Coconut Acid (q.v.) that conforms generally to the formula:

$$\overset{\displaystyle O}{\overset{\displaystyle \|}{RC}} - (OCH_2CH_2)_nOH$$

where RCO- represents the fatty acids derived from coconut oil and n has an average value of 9.

Information Source: TSCA

Chemical Class: Alkoxylated Carboxylic Acids

Function: Surfactant - Emulsifying Agent

Technical/Other Names:
Polyethylene Glycol 450 Monococoate
Polyoxyethylene (9) Monococoate

Trade Name:
Sabowax CE 9 (Sabo)

Trade Name Mixture:
Silwax DMC-C (Siltech LLC)

PEG-10 COCOATE

CAS No.: 61791-29-5 (Generic)

Definition: PEG-10 Cocoate is the polyethylene glycol ester of Coconut Acid (q.v.) that conforms generally to the formula:

$$\overset{\displaystyle O}{\overset{\displaystyle \|}{RC}} - (OCH_2CH_2)_nOH$$

where RCO- represents the fatty acids derived from coconut oil and n has an average value of 10.

Information Source: TSCA

Chemical Class: Alkoxylated Carboxylic Acids

Function: Surfactant - Emulsifying Agent

Technical/Other Names:
Polyethylene Glycol 500 Monococoate
Polyoxyethylene (10) Monococoate

Trade Name:
Sympatens-BCO/100 (Kolb)

PEG-15 COCOATE

CAS No.: 61791-29-5 (Generic)

Definition: PEG-15 Cocoate is the polyethylene glycol ester of Coconut Acid (q.v.) that conforms generally to the formula:

$$\overset{\displaystyle O}{\overset{\displaystyle \|}{RC}} - (OCH_2CH_2)_nOH$$

where RCO- represents fatty acids derived from the coconut oil and n has an average value of 15.

Information Sources: 21CFR175.300, 21CFR176.210, 21CFR177.2260, 21CFR177.2800, MI-13(7660), TSCA

Chemical Class: Alkoxylated Carboxylic Acids

Function: Surfactant - Emulsifying Agent

Technical/Other Names:
Polyethylene Glycol (15) Monococoate
Polyoxyethylene (15) Monococoate

PEG-2 COCO-BENZONIUM CHLORIDE

Definition: PEG-2 Coco-Benzonium Chloride is the quaternary ammonium salt that conforms generally to the formula:

$$\left[R - N \begin{array}{c} (CH_2CH_2O)_xH \\ (CH_2CH_2O)_yH \\ CH_2 \\ \end{array} \right]^+ \quad Cl^-$$

where R represents the alkyl groups derived from coconut oil and x+y has an average value of 2.

Chemical Classes: Alkoxylated Amines; Quaternary Ammonium Compounds

Functions: Antistatic Agent; Cosmetic Biocide

Technical/Other Names:
Polyethylene Glycol 100 Coco-Benzonium Chloride
Polyoxyethylene (2) Coco-Benzonium Chloride

Trade Name Mixture:
Ethoquad C/12B (Akzo Nobel)

PEG-10 COCO-BENZONIUM CHLORIDE

Definition: PEG-10 Coco-Benzonium Chloride is the quaternary ammonium salt that conforms generally to the formula:

$$\left[R - N \begin{array}{c} (CH_2CH_2O)_xH \\ (CH_2CH_2O)_yH \\ CH_2 \\ \end{array} \right]^+ \quad Cl^-$$

where R represents the alkyl groups derived coconut oil and x+y has an average value of 10.

Chemical Classes: Alkoxylated Amines; Quaternary Ammonium Compounds

Functions: Antistatic Agent; Cosmetic Biocide

Technical/Other Names:
Polyethylene Glycol 500 Coco-Benzonium Chloride
Polyoxyethylene (10) Coco-Benzonium Chloride

PEG-9 COCOGLYCERIDES

CAS No.: 67762-35-0 (Generic)

Definition: PEG-9 Cocoglycerides is the polyethylene glycol derivative of the mono- and diglycerides of coconut oil with and average ethoxylation value of 9.

Chemical Classes: Alkoxylated Alcohols; Glyceryl Esters and Derivatives

Functions: Skin-Conditioning Agent - Emollient; Surfactant - Emulsifying Agent

Technical/Other Names:
Polyethylene Glycol (9) Coconut Glycerides
Polyoxyethylene (9) Coconut Glycerides

Trade Names:
Lexol EC (Inolex)
Oxypon 401 (Zschimmer & Schwarz)

PEG-2 COCOMONIUM CHLORIDE

CAS No.: 61791-10-4 (Generic)

Definition: PEG-2 Cocomonium Chloride is the quaternary ammonium salt that conforms generally to the formula:

$$\left[R - N \begin{array}{c} (CH_2CH_2O)_xH \\ (CH_2CH_2O)_yH \\ CH_3 \\ \end{array} \right]^+ \quad Cl^-$$

where R represents the alkyl groups derived from coconut oil and x + y has an average value of 2.

Information Source: TSCA

Chemical Classes: Alkoxylated Amines; Quaternary Ammonium Compounds

Function: Antistatic Agent

Technical/Other Names:
PEG-2 Cocoyl Quaternium-4
Polyethylene Glycol 100 Cocomonium Chloride
Polyoxyethylene (2) Cocomonium Chloride

Trade Name Mixture:
Ethoquad C/12 (Akzo Nobel)

PEG-15 COCOMONIUM CHLORIDE

CAS No.: 61791-10-4 (Generic)

Definition: PEG-15 Cocomonium Chloride is the quaternary ammonium salt that conforms generally to the formula:

$$\left[R - N \begin{array}{c} (CH_2CH_2O)_xH \\ (CH_2CH_2O)_yH \\ CH_3 \\ \end{array} \right]^+ \quad Cl^-$$

where R represents the alkyl groups derived from coconut oil and x + y has an average value of 15.

Information Source: TSCA

Chemical Classes: Alkoxylated Amines; Quaternary Ammonium Compounds

Function: Antistatic Agent

Reported Product Categories: Hair Sprays (Aerosol Fixatives); Hair Preparations (Non-coloring), Misc.

Technical/Other Names:
PEG-15 Cocoyl Quaternium-4
Polyethylene Glycol (15) Cocomonium Chloride
Polyoxyethylene (15) Cocomonium Chloride

Trade Names:
AEC PEG-15 Cocomonium Chloride (A & E Connock)
Ethoquad C/25 (Akzo Nobel)

Trade Name Mixture:
SM2112 (GE Silicones)

PEG-5 COCOMONIUM METHOSULFATE

CAS No.: 68989-03-7

Definition: PEG-5 Cocomonium Methosulfate is the quaternary ammonium salt that conforms generally to the formula:

$$\left[R - N \begin{array}{c} (CH_2CH_2O)_xH \\ (CH_2CH_2O)_yH \\ CH_3 \\ \end{array} \right]^+ \quad CH_3OSO_3^-$$

where R represents the alkyl groups derived from coconut oil and x + y has an average value of 5.

Chemical Classes: Alkoxylated Amines; Quaternary Ammonium Compounds

Function: Antistatic Agent

Technical/Other Names:
Polyethylene Glycol (5) Cocomonium Methosulfate
Polyoxyethylene (5) Cocomonium Methosulfate

Trade Name:
Varisoft CPEM (Degussa Care Specialties)

PEG-15 COCOMONIUM METHOSULFATE

Definition: PEG-15 Cocomonium Methosulfate is the quaternary ammonium salt that conforms generally to the formula:

$$\left[R - \overset{\overset{\displaystyle (CH_2CH_2O)_xH}{|}}{\underset{\underset{\displaystyle CH_3}{|}}{N}} - (CH_2CH_2O)_yH \right]^+ \quad CH_3OSO_3^-$$

where R represents the alkyl groups derived from coconut oil and x + y has an average value of 15.

Chemical Classes: Alkoxylated Amines; Quaternary Ammonium Compounds

Function: Antistatic Agent

Trade Name:
Servamine KW 100 (Sasol Servo)

PEG-15 COCOPOLYAMINE

JPN Translation:
PEG - 15 ココポリアミン

Definition: PEG-15 Cocopolyamine is the polyethylene glycol polyamine that conforms generally to the formula:

$$RNH - \underset{\displaystyle C_3H_5OH}{|} $$
O(CH₂CH₂O)ₙC₃H₅OH
NH(CH₂)ₘNH
C₃H₅OH
O(CH₂CH₂O)ₙ
C₃H₅OH
RNH

where R represents the alkyl groups derived from coconut oil, n has a total value between 10 and 20, m has a value between 2 and 6 and x has a value between 2 and 4.

Information Sources: JCIC, JCLS, JSQI

Chemical Class: Alkoxylated Amines

Functions: Antistatic Agent; Surfactant - Emulsifying Agent

Reported Product Categories: Hair Dyes and Colors (All Types Requiring Caution Statements and Patch Tests); Shampoos (Non-coloring); Hair Shampoos (Coloring)

Technical/Other Names:
Polyethylene Glycol (15) Coconut Polyamine
Polyoxyethylene (15) Coconut Polyamine

Trade Names:
Polyquart H (Cognis Care Chemicals/NJ)
Polyquart H (Cognis Care Chemicals/PA)

Trade Name Mixture:
Uniquart H (Universal Preserv-A-Chem)

PEG-20 CORN GLYCERIDES

Definition: PEG-20 Corn Glycerides is a polyethylene glycol derivative of Corn Glycerides (q.v.) with an average of 20 moles of ethylene oxide.

Information Sources: 21CFR175.300, 21CFR176.210

Chemical Classes: Alkoxylated Alcohols; Glyceryl Esters and Derivatives

Functions: Skin-Conditioning Agent - Emollient; Surfactant - Emulsifying Agent

Technical/Other Names:
Polyethylene Glycol 1000 Corn Glycerides
Polyoxyethylene (20) Corn Glycerides

PEG-60 CORN GLYCERIDES

Definition: PEG-60 Corn Glycerides is a polyethylene glycol derivative of Corn Glycerides (q.v.) with an average of 60 moles of ethylene oxide.

Information Sources: 21CFR175.300, 21CFR176.210

Chemical Classes: Alkoxylated Alcohols; Glyceryl Esters and Derivatives

Functions: Skin-Conditioning Agent - Emollient; Surfactant - Emulsifying Agent; Surfactant - Solubilizing Agent

Technical/Other Names:
Polyethylene Glycol 3000 Corn Glycerides
Polyoxyethylene (60) Corn Glycerides

Trade Names:
Crovol M70 (Croda Chemicals)
Crovol M-70 (Croda, Inc.)

PEG-CROSSPOLYMER

Definition: PEG-Crosspolymer is a cross-linked polymer of polyethylene glycol.

Chemical Class: Synthetic Polymers

Function: Viscosity Increasing Agent - Aqueous

Trade Name:
Cool-Jel (Nepera)

PEG-150/DECYL ALCOHOL/SMDI COPOLYMER

Definition: PEG-150/Decyl Alcohol/SMDI Copolymer is a copolymer of PEG-150, decyl alcohol, and saturated methylene diphenyldiisocyanate monomers.

Chemical Class: Synthetic Polymers

Functions: Film Former; Suspending Agent - Nonsurfactant; Viscosity Increasing Agent - Aqueous

Trade Name:
Aculyn 44 Polymer (Rohm and Haas)

PEG-5 DEDM HYDANTOIN

Empirical Formula:
$C_{19}H_{35}N_2O_9$

Definition: PEG-5 DEDM Hydantoin is the polyethylene glycol ether of DEDM Hydantoin (q.v.) that has an average of 5 moles of ethylene oxide.

Chemical Classes: Alkoxylated Alcohols; Alkoxylated Amides; Heterocyclic Compounds

Function: Preservative

Technical/Other Names:
Polyethylene Glycol (5) DEDM Hydantoin
Polyoxyethylene (5) DEDM Hydantoin

PEG-15 DEDM HYDANTOIN

CAS No.: 68130-12-1

Definition: PEG-15 DEDM Hydantoin is the polyethylene glycol ether of DEDM Hydantoin (q.v.) that has an average of 15 moles of ethylene oxide.

Chemical Classes: Alkoxylated Alcohols; Alkoxylated Amides; Heterocyclic Compounds

Function: Preservative

Technical/Other Names:
Polyethylene Glycol (15) DEDM Hydantoin
Polyoxyethylene (15) DEDM Hydantoin

PEG-5 DEDM HYDANTOIN OLEATE

Empirical Formula:
$C_{37}H_{67}N_2O_{10}$

Definition: PEG-5 DEDM Hydantoin Oleate is the ester of PEG-5 DEDM Hydantoin (q.v.) and Oleic Acid (q.v.).

Chemical Classes: Alkoxylated Alcohols; Alkoxylated Amides; Esters; Heterocyclic Compounds

Function: Preservative

Technical/Other Names:
Polyethylene Glycol (5) DEDM Hydantoin Oleate
Polyoxyethylene (5) DEDM Hydantoin Oleate

Trade Name:
Dantosperse DHE(5)MO (Lonza Inc./Lonza Ltd.)

PEG-15 DEDM HYDANTOIN STEARATE

Definition: PEG-15 DEDM Hydantoin Stearate is the ester of PEG-15 DEDM Hydantoin (q.v.) and Stearic Acid (q.v.).

Chemical Classes: Alkoxylated Alcohols; Alkoxylated Amides; Esters; Heterocyclic Compounds

Function: Preservative

Trade Names:
Dantosperse DHE(15)DS (Lonza Inc./Lonza Ltd.)
Polyethylene Glycol (15) DEDM Hydantoin Stearate (Lonza Inc./Lonza Ltd.)
Polyoxyethylene (15) DEDM Hydantoin Stearate (Lonza Inc./Lonza Ltd.)

PEG-150 DIBEHENATE

Definition: PEG-150 Dibehenate is the polyethylene glycol diester of behenic acid that conforms generally to the formula:

$$CH_3(CH_2)_{20}\overset{\text{O}}{\overset{\|}{C}} - (OCH_2CH_2)_nO - \overset{\text{O}}{\overset{\|}{C}}(CH_2)_{20}CH_3$$

where n has an average value of 150.

Chemical Class: Alkoxylated Carboxylic Acids

Functions: Surfactant - Cleansing Agent; Surfactant - Solubilizing Agent

Technical/Other Names:
Polyethylene Glycol 6000 Dibehenate
Polyoxyethylene (150) Dibehenate

Trade Name:
Ethox P-6000 DB (Ethox)

PEG-8 DICOCOATE

Definition: PEG-8 Dicocoate is the polyethylene glycol diester of Coconut Acid (q.v.) that conforms to the formula:

$$RC\overset{\text{O}}{\overset{\|}{}} - (OCH_2CH_2)_nO - \overset{\text{O}}{\overset{\|}{C}}R$$

where n has an average value of 8 and RCO- represents the fatty acids derived from coconut oil.

Information Sources: 21CFR175.105, 21CFR175.300, 21CFR176.210, 21CFR177.1210, 21CFR177.2260, 21CFR177.2800, MI-13(7660)

Chemical Class: Alkoxylated Carboxylic Acids

Function: Surfactant - Emulsifying Agent

Technical/Other Names:
Polyethylene Glycol 400 Dicocoate
Polyoxyethylene (8) Dicocoate

Trade Name:
ROL DL 40 (Fabriquimica)

PEG-3 DIETHYLENETRIAMINE DIPALMAMIDE

Definition: PEG-3 Diethylenetriamine Dipalmamide is the organic compound that conforms generally to the formula:

$$RC\overset{\text{O}}{\overset{\|}{}} - NH(CH_2)_2N(CH_2)_2NH - \overset{\text{O}}{\overset{\|}{C}}R$$
$$|$$
$$(CH_2CH_2O)_xH$$

where RCO- represents the fatty acids derived from palm oil and x has an average value of 3.

Chemical Classes: Alkoxylated Amines; Amides

Function: Hair Conditioning Agent

PEG-2 DIETHYLHEXANOATE

Empirical Formula:
$C_{20}H_{38}O_5$

Definition: PEG-2 Diethylhexanoate is the polyethylene glycol diester of 2-ethylhexanoic acid that conforms to the formula:

$$CH_3(CH_2)_3\overset{}{C}H\overset{\text{O}}{\overset{\|}{C}} - (OCH_2CH_2)_nO - \overset{\text{O}}{\overset{\|}{C}}CH(CH_2)_3CH_3$$
$$|\qquad\qquad\qquad\qquad\qquad |$$
$$CH_2CH_3\qquad\qquad\qquad\qquad CH_2CH_3$$

where n has an average value of 2.

Information Source: MI-13(7660)

Chemical Class: Alkoxylated Carboxylic Acids

Function: Surfactant - Emulsifying Agent

Technical/Other Names:
PEG-2 Dioctanoate
Polyoxyethylene (2) Dioctanoate

Trade Name:
Dermol 488 (Alzo)

Trade Name Mixtures:
Dermol 334 (Alzo)
Jeechem IDO (Jeen)

PEG-9 DIETHYLMONIUM CHLORIDE

Empirical Formula:
$C_{23}H_{50}NO_9 \cdot Cl$

Definition: PEG-9 Diethylmonium Chloride is a quaternary ammonium salt that conforms to the formula:

$$\left[\begin{array}{c} CH_2CH_3 \\ | \\ CH_3-N-(CH_2CH_2O)_nH \\ | \\ CH_2CH_3 \end{array}\right]^+ Cl^-$$

where n has an average value of 9.

Chemical Classes: Alkoxylated Amines; Quaternary Ammonium Compounds

Function: Antistatic Agent

Technical/Other Names:
Polyethylene Glycol 450 Diethylmonium Chloride
Polyoxyethylene (9) Diethylmonium Chloride

PEG-25 DIETHYLMONIUM CHLORIDE

Definition: PEG-25 Diethylmonium Chloride is a quaternary ammonium salt that conforms to the formula:

$$\left[\begin{array}{c} CH_2CH_3 \\ | \\ CH_3-N-(CH_2CH_2O)_nH \\ | \\ CH_2CH_3 \end{array}\right]^+ Cl^-$$

where n has an average value of 25.

Chemical Classes: Alkoxylated Amines; Quaternary Ammonium Compounds

Function: Antistatic Agent

Technical/Other Names:
Polyethylene Glycol (25) Diethylmonium Chloride

Polyoxyethylene (25) Diethylmonium Chloride

PEG-4 DIHEPTANOATE

CAS No.
70729-68-9

EINECS No.
274-829-3

Empirical Formula:
$C_{22}H_{42}O_7$

Definition: PEG-4 Diheptanoate is the polyethylene glycol diester of heptanoic acid that conforms to the formula:

$$CH_3(CH_2)_5\overset{O}{\overset{||}{C}}-(OCH_2CH_2O)_4-\overset{O}{\overset{||}{C}}(CH_2)_5CH_3$$

Chemical Class: Alkoxylated Carboxylic Acids

Functions: Skin-Conditioning Agent - Emollient; Surfactant - Emulsifying Agent

Reported Product Category: Lipsticks

Technical/Other Names:
Heptanoic Acid, Oxybis(2,1-Ethanediyloxy-2,1-Ethanediyl)Ester
Oxybis(2,1-Ethanediyloxy-2,1-Ethanediyl)Heptanoate
Polyethylene Glycol 200 Diheptanoate
Polyoxyethylene (4) Diheptanoate
Tetraethylene Glycol Diheptanoate

Trade Names:
ESTOL 3777 (Uniqema Europe)
Hatcol 5174 (Hatco)
Liponate 2-DH (Lipo)

PEG-2 DIISONONANOATE

Empirical Formula:
$C_{22}H_{20}O_5$

Definition: PEG-2 Diisononanoate is the polyethylene glycol diester of isononanoic acid that conforms to the formula:

$$C_8H_{17}\overset{O}{\overset{||}{C}}-(OCH_2CH_2)_nO-\overset{O}{\overset{||}{C}}C_8H_{17}$$

where n has an average value of 2.

Chemical Class: Alkoxylated Carboxylic Acids

Function: Surfactant - Emulsifying Agent

Technical/Other Names:
Polyethylene Glycol 100 Diisononanoate
Polyoxyethylene (2) Diisononanoate

Trade Name Mixtures:
Dermol 334 (Alzo)
Jeechem IDO (Jeen)

PEG-2 DIISOSTEARATE

Definition: PEG-2 Diisostearate is the polyethylene glycol diester of isostearic acid that conforms generally to the formula:

$$C_{17}H_{35}\overset{O}{\overset{||}{C}}-(OCH_2CH_2)_nO-\overset{O}{\overset{||}{C}}C_{17}H_{35}$$

where n has an average value of 2.

Chemical Class: Alkoxylated Carboxylic Acids

Function: Surfactant - Emulsifying Agent

Technical/Other Names:
Polyethylene Glycol 100 Diisostearate
Polyoxyethylene (2) Diisostearate

Trade Name: Emalex DEG-di-IS (Nihon Emulsion)

PEG-3 DIISOSTEARATE

Definition: PEG-3 Diisostearate is the polyethylene glycol diester of isostearic acid that conforms generally to the formula:

$$C_{17}H_{35}\overset{O}{\overset{||}{C}}-(OCH_2CH_2)_nO-\overset{O}{\overset{||}{C}}C_{17}H_{35}$$

where n has an average value of 3.

Chemical Class: Alkoxylated Carboxylic Acids

Function: Surfactant - Emulsifying Agent

Technical/Other Names:
Polyethylene Glycol (3) Diisostearate
Polyoxyethylene (3) Diisostearate

Trade Name: Emalex TEG-di-IS (Nihon Emulsion)

PEG-4 DIISOSTEARATE

Definition: PEG-4 Diisostearate is the polyethylene glycol diester of isostearic acid that conforms generally to the formula:

$$C_{17}H_{35}\overset{O}{\overset{||}{C}}-(OCH_2CH_2)_nO-\overset{O}{\overset{||}{C}}C_{17}H_{35}$$

where n has an average value of 4.

Chemical Classes: Alkoxylated Carboxylic Acids; Esters

Function: Surfactant - Emulsifying Agent

Technical/Other Names:
Polyethylene Glycol 200 Diisostearate
Polyoxyethylene (4) Diisostearate

Trade Name: Emalex 200di - IS (Nihon Emulsion)

PEG-6 DIISOSTEARATE

Definition: PEG-6 Diisostearate is the polyethylene glycol diester of isostearic acid that conforms generally to the formula:

$$C_{17}H_{35}\overset{O}{\overset{||}{C}}-(OCH_2CH_2)_nO-\overset{O}{\overset{||}{C}}C_{17}H_{35}$$

where n has an average value of 6.

Chemical Classes: Alkoxylated Carboxylic Acids; Esters

Function: Surfactant - Emulsifying Agent

Technical/Other Names:
Polyethylene Glycol (6) Diisostearate
Polyoxyethylene (6) Diisostearate

Trade Name: Emalex 300di - IS (Nihon Emulsion)

PEG-8 DIISOSTEARATE

JPN Translation:
ジイソステアリン酸 PEG - 8

Empirical Formula:
$C_{52}H_{102}O_{11}$

Definition: PEG-8 Diisostearate is the polyethylene glycol diester of isostearic acid that conforms to the formula:

$$C_{17}H_{35}\overset{O}{\overset{||}{C}}-(OCH_2CH_2)_nO-\overset{O}{\overset{||}{C}}C_{17}H_{35}$$

where n has an average value of 8.

Information Sources: JCIC, JCLS

Chemical Class: Alkoxylated Carboxylic Acids

Function: Surfactant - Emulsifying Agent

Reported Product Categories: Bath Oils, Tablets, and Salts; Cleansing Products (Cold Creams, Cleansing Lotions, Liquids and Pads)

Technical/Other Names:
Polyethylene Glycol Diisostearate
Polyethylene Glycol 400 Diisostearate
Polyoxyethylene (8) Diisostearate

Trade Name: Nikkol CDIS-400 (Nikko)

PEG-12 DIISOSTEARATE

JPN Translation:
ジイソステアリン酸 PEG - 12

Definition: PEG-12 Diisostearate is the polyethylene glycol diester of isostearic acid that conforms generally to the formula:

$$C_{17}H_{35}\overset{O}{\overset{||}{C}}-(OCH_2CH_2)_nO-\overset{O}{\overset{||}{C}}C_{17}H_{35}$$

where n has an average value of 12.

Chemical Class: Alkoxylated Carboxylic Acids

Function: Surfactant - Emulsifying Agent

Technical/Other Names:
Polyethylene Glycol 600 Diisostearate
Polyoxyethylene (12) Diisostearate

Trade Name:
Emalex 600di-ISEX (Nihon Emulsion)

PEG-90 DIISOSTEARATE

Definition: PEG-90 Diisostearate is the polyethylene glycol diester of isostearic acid that conforms generally to the formula:

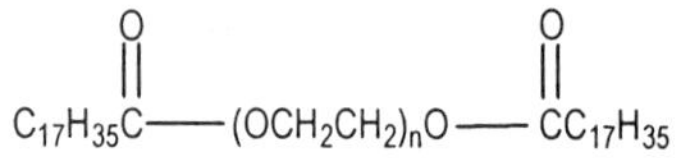

where n has an average value of 90.

Chemical Class: Alkoxylated Carboxylic Acids

Function: Surfactant - Cleansing Agent

Trade Name:
Hydramol PGDS (Scher)

PEG-175 DIISOSTEARATE

Definition: PEG-175 Diisostearate is the polyethylene glycol diester of isostearic acid that conforms generally to the formula:

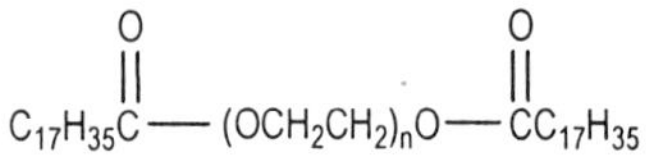

where n has an average value of 175.

Chemical Class: Alkoxylated Carboxylic Acids

Functions: Surfactant - Emulsifying Agent; Viscosity Increasing Agent - Aqueous

Technical/Other Names:
Polyethylene Glycol (175) Diisostearate
Polyoxyethylene (175) Diisostearate

Trade Name:
Ethox HVB (Ethox)

PEG-2 DILAURATE

CAS Nos.	**EINECS No.**
6281-04-5	228-486-1
9005-02-1 (Generic)	

JPN Translation:
ジラウリン酸 PEG - 2

Empirical Formula:
$C_{28}H_{54}O_5$

Definition: PEG-2 Dilaurate is the polyethylene glycol diester of lauric acid that conforms to the formula:

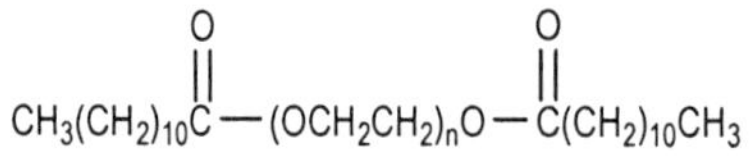

where n has an average value of 2.

Information Sources: 21CFR175.300, 21CFR176.210, 21CFR177.2800, CIR: [SQ]-IJT-19(SUPPL.2)2000, JCIC, JCLS, MI-13 (7660)

Chemical Class: Alkoxylated Carboxylic Acids

Function: Surfactant - Emulsifying Agent

Technical/Other Names:
Diethylene Glycol Dilaurate
Dodecanoic Acid, Oxydi-2,1-Ethanediyl Ester
Oxydi-2,1-Ethanediyl Dodecanoate
Polyethylene Glycol (2) Dilaurate
Polyethylene Glycol 100 Dilaurate
Polyoxyethylene (2) Dilaurate

Trade Name:
Jeemate 200-DL (Jeen)

PEG-4 DILAURATE

CAS No.: 9005-02-1 (Generic)

JPN Translation:
ジラウリン酸 PEG - 4

Empirical Formula:
$C_{32}H_{62}O_7$

Definition: PEG-4 Dilaurate is the polyethylene glycol diester of lauric acid that conforms to the formula:

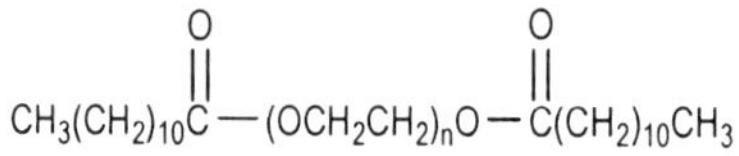

where n has an average value of 4.

Information Sources: 21CFR175.105, 21CFR175.300, 21CFR176.170, 21CFR176.180, 21CFR176.200, 21CFR176.210, CIR: [SQ] IJT-19(SUPPL.2)-2000, CTFA S, JCLS, MI-13(7660), TSCA

Chemical Class: Alkoxylated Carboxylic Acids

Function: Surfactant - Emulsifying Agent

Reported Product Categories: Bath Oils, Tablets, and Salts; Perfumes; Bath Preparations, Misc.; Body and Hand Preparations (Excluding Shaving Preparations)

Technical/Other Names:
Polyethylene Glycol 200 Dilaurate
Polyoxyethylene (4) Dilaurate

Trade Names:
AEC PEG-4 Dilaurate (A & E Connock)
Chemax PEG-200-DL (Chemax)
Emerest 2704 (Cognis Care Chemicals/NJ)
Emerest 2704 (Cognis Care Chemicals/PA)
Ethox DL-5 (Ethox)
Hetoxamate 200 DL (Heterene)
Jeemate 400-DL (Jeen)
LIPOPEG 2-DL (Lipo)
Pegosperse 200 DL (Lonza Inc./Lonza Ltd.)
Protamate 200 DL (Protameen)
Sipoest PEG 200 DL (Specialty Industrial)
STEPAN PEG 200 DL (Stepan)

PEG-6 DILAURATE

CAS No.: 9005-02-1 (Generic)

JPN Translation:
ジラウリン酸 PEG - 6

Empirical Formula:
$C_{36}H_{70}O_9$

Definition: PEG-6 Dilaurate is the polyethylene glycol diester of lauric acid that conforms to the formula:

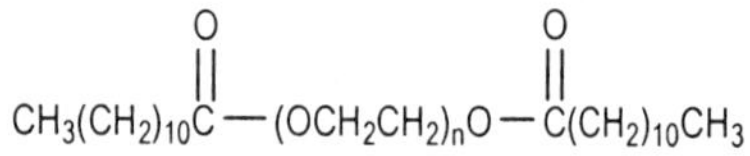

where n has an average value of 6.

Information Sources: 21CFR175.105, 21CFR175.300, 21CFR176.210, CIR: [SQ] IJT-19(SUPPL. 2)2000, JCLS, JSQI, MI-13 (7660), TSCA

Chemical Class: Alkoxylated Carboxylic Acids

Function: Surfactant - Emulsifying Agent

Technical/Other Names:
Polyethylene Glycol 300 Dilaurate
Polyoxyethylene (6) Dilaurate

Trade Names:
AEC PEG-6 Dilaurate (A & E Connock)
Sipoest PEG 300 DL (Specialty Industrial)
STEPAN PEG 300 DL (Stepan)

PEG-8 DILAURATE

CAS No.: 9005-02-1 (Generic)

JPN Translation:
ジラウリン酸 PEG - 8

Empirical Formula:
$C_{40}H_{78}O_{11}$

Definition: PEG-8 Dilaurate is the polyethylene glycol diester of lauric acid that conforms to the formula:

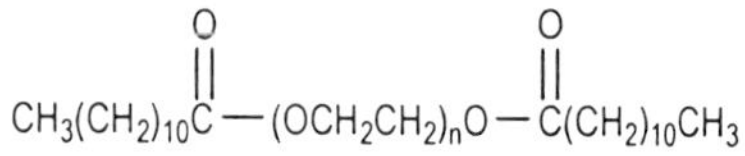

where n has an average value of 8.

Information Sources: 21CFR175.105, 21CFR175.300, 21CFR176.210, 21CFR177.1210, 21CFR177.2260, 21CFR177.2800, 21CFR178.3520, CIR: [SQ] IJT-19(SUPPL.2)2000, CTFA D, JCLS, JSQI, MI-13(7660), TSCA

Chemical Class: Alkoxylated Carboxylic Acids

Function: Surfactant - Emulsifying Agent

Reported Product Categories: Tonics, Dressings, and Other Hair Grooming Aids; Bath Oils, Tablets, and Salts; Cleansing Products (Cold Creams, Cleansing Lotions, Liquids and Pads); Hair Conditioners

Technical/Other Names:
Polyethylene Glycol 400 Dilaurate
Polyoxyethylene (8) Dilaurate

Trade Names:
AEC PEG-8 Dilaurate (A & E Connock)
Cithrol 4DL (Croda Chemicals)
Ethox DL-9 (Ethox)
Hetoxamate 400 DL (Heterene)
LIPOPEG 4-DL (Lipo)
Lumulse 42-L (Lambent)
Pegosperse 400 DL (Lonza Inc./Lonza Ltd.)
Protamate 400-DL (Protameen)
Sipoest PEG 400 DL (Specialty Industrial)
STEPAN PEG 400 DL (Stepan)
Unipeg-400 DL (Universal Preserv-A-Chem)

PEG-12 DILAURATE

CAS No.: 9005-02-1 (Generic)

JPN Translation:
ジラウリン酸 PEG - 12

Definition: PEG-12 Dilaurate is the polyethylene glycol diester of lauric acid that conforms to the formula:

$$CH_3(CH_2)_{10}C\overset{O}{\overset{\|}{}} - (OCH_2CH_2)_nO - \overset{O}{\overset{\|}{}}C(CH_2)_{10}CH_3$$

where n has an average value of 12.

Information Sources: 21CFR175.105, 21CFR175.300, 21CFR176.210, 21CFR177.2260, 21CFR177.2800, CIR: [SQ] IJT-19(SUPPL.2)2000, JCLS, JSQI, MI-13(7660), TSCA

Chemical Class: Alkoxylated Carboxylic Acids

Function: Surfactant - Emulsifying Agent

Technical/Other Names:
Polyethylene Glycol 600 Dilaurate
Polyoxyethylene (12) Dilaurate

Trade Names:
AEC PEG-12 Dilaurate (A & E Connock)

Ethox DL-14 (Ethox)
Ethox PEG-600 DL (Ethox)
Jeemate 600-DL (Jeen)
Protamate 600 DL (Protameen)
STEPAN PEG 600 DL (Stepan)

PEG-16 DILAURATE

CAS No.: 9005-02-1 (Generic)

Definition: PEG-16 Dilaurate is the polyethylene glycol diester of lauric acid that conforms generally to the formula:

$$CH_3(CH_2)_{10}\overset{O}{\overset{\|}{C}} - (OCH_2CH_2)_nO - \overset{O}{\overset{\|}{C}}(CH_2)_{10}CH_3$$

where n has an average value of 16.

Chemical Class: Alkoxylated Carboxylic Acids

Function: Surfactant - Emulsifying Agent

Technical/Other Names:
Polyethylene Glycol (16) Dilaurate
Polyoxyethylene(16) Dilaurate

Trade Name:
Emalex 800di - L (Nihon Emulsion)

PEG-20 DILAURATE

CAS No.: 9005-02-1 (Generic)

JPN Translation:
ジラウリン酸 PEG - 20

Definition: PEG-20 Dilaurate is the polyethylene glycol diester of lauric acid that conforms to the formula:

$$CH_3(CH_2)_{10}\overset{O}{\overset{\|}{C}} - (OCH_2CH_2)_nO - \overset{O}{\overset{\|}{C}}(CH_2)_{10}CH_3$$

where n has an average value of 20.

Information Sources: 21CFR175.300, 21CFR176.210, 21CFR177.2260, 21CFR177.2800, CIR: [SQ]IJT-19 (SUPPL.2)2000, JCLS, JSQI, MI-13(7660), TSCA

Chemical Class: Alkoxylated Carboxylic Acids

Function: Surfactant - Emulsifying Agent

Technical/Other Names:
Polyethylene Glycol 1000 Dilaurate
Polyoxyethylene (20) Dilaurate

Trade Names:
AEC PEG-20 Dilaurate (A & E Connock)
STEPAN PEG 1000 DL (Stepan)

PEG-32 DILAURATE

CAS No.: 9005-02-1 (Generic)

JPN Translation:
ジラウリン酸 PEG - 32

Definition: PEG-32 Dilaurate is the polyethylene glycol diester of lauric acid that conforms to the formula:

$$CH_3(CH_2)_{10}C\overset{O}{\overset{\|}{}} - (OCH_2CH_2)_nO - \overset{O}{\overset{\|}{}}C(CH_2)_{10}CH_3$$

where n has an average value of 32.

Information Sources: 21CFR175.300, 21CFR176.210, 21CFR177.2260, 21CFR177.2800, CIR: [SQ]IJT-19 (SUPPL.2)2000, JCLS, JSQI, MI-13(7660), TSCA

Chemical Class: Alkoxylated Carboxylic Acids

Function: Surfactant - Emulsifying Agent

Technical/Other Names:
Polyethylene Glycol 1540 Dilaurate
Polyoxyethylene (32) Dilaurate

Trade Names:
AEC PEG-32 Dilaurate (A & E Connock)
STEPAN PEG 1540 DL (Stepan)

PEG-75 DILAURATE

CAS No.: 9005-02-1 (Generic)

JPN Translation:
ジラウリン酸 PEG - 75

Definition: PEG-75 Dilaurate is the polyethylene glycol diester of lauric acid that conforms to the formula:

$$CH_3(CH_2)_{10}\overset{O}{\overset{\|}{C}} - (OCH_2CH_2)_nO - \overset{O}{\overset{\|}{C}}(CH_2)_{10}CH_3$$

where n has an average value of 75.

Information Sources: 21CFR175.300, CIR: [SQ]IJT-19(SUPPL.2)2000, JCLS, JSQI, MI-13(7660), TSCA

Chemical Class: Alkoxylated Carboxylic Acids

Functions: Surfactant - Cleansing Agent; Surfactant - Solubilizing Agent

Technical/Other Names:
Polyethylene Glycol 4000 Dilaurate
Polyoxyethylene (75) Dilaurate

Trade Names:
AEC PEG-75 Dilaurate (A & E Connock)
STEPAN PEG 4000 DL (Stepan)

PEG-150 DILAURATE

CAS No.: 9005-02-1 (Generic)

ジラウリン酸 PEG - 150

JPN Translation:

Definition: PEG-150 Dilaurate is the polyethylene glycol diester of lauric acid that conforms to the formula:

$$CH_3(CH_2)_{10}\overset{O}{\overset{\|}{C}}-(OCH_2CH_2)_nO-\overset{O}{\overset{\|}{C}}(CH_2)_{10}CH_3$$

where n has an average value of 150.

Information Sources: 21CFR175.300, CIR: [SQ]IJT-19(SUPPL.2)2000, CTFA D, JCLS, JSQI, MI-13(7660), TSCA

Chemical Class: Alkoxylated Carboxylic Acids

Functions: Surfactant - Cleansing Agent; Surfactant - Solubilizing Agent

Technical/Other Names:
Polyethylene Glycol 6000 Dilaurate
Polyoxyethylene (150) Dilaurate

Trade Names:
AEC PEG-150 Dilaurate (A & E Connock)
STEPAN PEG 6000 DL (Stepan)

PEG-2 DIMEADOWFOAMAMIDOETHYL-MONIUM METHOSULFATE

Definition: PEG-2 Dimeadowfoamamido-ethylmonium Methosulfate is the quaternary ammonium salt that conforms generally to the formula:

$$\left[\overset{O}{\overset{\|}{RCNH(CH_2)_2}}-\underset{\underset{CH_3}{|}}{\overset{(CH_2CH_2O)_2H}{N}}-(CH_2)_2NH\overset{O}{\overset{\|}{CR}}\right]^{+}CH_3OSO_3^{-}$$

where RCO- represents the fatty acids derived from meadowfoam seed oil.

Chemical Classes: Amides; Quaternary Ammonium Compounds

Functions: Antistatic Agent; Hair Conditioning Agent

Trade Name:
Meadowquat HG (Fanning)

PEG-4 DIMETHACRYLATE

CAS No.	EINECS No.
109-17-1	203-653-1

Definition: PEG-4 Dimethacrylate is the organic compound that conforms generally to the formula:

$$CH_2=\underset{\underset{CH_3}{|}}{\overset{O}{\overset{\|}{CC}}}-O(CH_2CH_2O)_n-\underset{\underset{CH_3}{|}}{\overset{O}{\overset{\|}{CC}}}=CH_2$$

where n has an average value of 4.

Information Source: CIR: [SQ]

Chemical Class: Alkoxylated Carboxylic Acids

Function: Not Reported

Technical/Other Names:
Polyethylene Glycol (4) Dimethacrylate
Polyoxyethylene (4) Dimethacrylate
2-Propenoic Acid, 2-Methyl-, Oxybis(2,1-Ethanediyloxy-2,1-Ethanediyl) Ester

Trade Name:
Tetraethylene Glycol Dimethacrylate (Wilde Cosmetics)

PEG-9 DIMETHACRYLATE

Definition: PEG-9 Dimethacrylate is the organic compound that conforms generally to the formula:

$$CH_2=\underset{\underset{CH_3}{|}}{\overset{O}{\overset{\|}{CC}}}-O(CH_2CH_2O)_n-\underset{\underset{CH_3}{|}}{\overset{O}{\overset{\|}{CC}}}=CH_2$$

where n has an average value of 9.

Chemical Class: Alkoxylated Carboxylic Acids

Function: Not Reported

Technical/Other Names:
Polyethylene Glycol (9) Dimethacrylate
Polyoxyethylene (9) Dimethacrylate

Trade Name:
Polyethylene Glycol 400 Dimethacrylate (Wilde Cosmetics)

PEG-3 DIMETHICONE

Definition: PEG-3 Dimethicone is the siloxane polymer that conforms generally to the formula:

$$(CH_3)_3SiO-\left[\underset{\underset{CH_3}{|}}{\overset{\overset{CH_3}{|}}{SiO}}\right]_x\left[\underset{\underset{(CH_2CH_2O)_nH}{|}}{\overset{\overset{CH_3}{|}}{SiO}}\right]_y-Si(CH_3)_3$$

where n has an average value of 3.

Chemical Classes: Siloxanes and Silanes; Synthetic Polymers

Functions: Hair Conditioning Agent; Skin-Conditioning Agent - Miscellaneous

Trade Names:
KF-6015 (Shin Etsu)
KF-945A (Shin Etsu)

PEG-7 DIMETHICONE

Definition: PEG-7 Dimethicone is the polyethylene glycol derivative of Dimethicone (q.v.) containing an average of 7 moles of ethylene oxide.

Chemical Class: Siloxanes and Silanes

Function: Film Former

Reported Product Category: Eye Makeup Removers

Technical/Other Names:
Polyethylene Glycol (7) Dimethicone
Polyoxyethylene (7) Dimethicone

Trade Name Mixtures:
Biosil Basics Cocosil (Biosil Technologies, Inc.)
Silsoft ME-5 Silicone Microemulsion (Crompton Corporation)

PEG-8 DIMETHICONE

CAS No.: 68937-54-2

Definition: PEG-8 Dimethicone is the polyethylene glycol derivative of Dimethicone (q.v.) containing an average of 8 moles of ethylene oxide.

Chemical Class: Siloxanes and Silanes

Functions: Hair Conditioning Agent; Skin-Conditioning Agent - Miscellaneous

Reported Product Categories: Hair Sprays (Aerosol Fixatives); Moisturizing Preparations; Skin Fresheners

Technical/Other Names:
Polyethylene Glycol (8) Dimethicone
Polyoxyethylene (8) Dimethicone

Trade Names:
Biowax 754 (Biosil Technologies, Inc.)
Biowax Liquid 754 (Biosil Technologies, Inc.)
Biowax Liquid D (Biosil Technologies, Inc.)
FZ-2404 (Nippon Unicar)
Silsoft 805 dimethicone copolyol (OSi Specialties)
Silsoft 810 dimethicone copolyol (OSi Specialties)
Silsoft 840 dimethicone copolyol (OSi Specialties)
SS-2805 (Nippon Unicar)
Ultrasil Copolyol-1 (Noveon)
Zenicone DMC-1 (Zenitech)

Trade Name Mixtures:
Biosil Basics Cetylsil J (Biosil Technologies, Inc.)
C9833 Hydrophylic Black Iron Oxide (LCW)
C9828 Hydrophylic Titanium Dioxide (LCW)
C9804 Hydrophylic Ultramarine Blue (LCW)
C9855 Hydrophylic Yellow Iron Oxide (LCW)

The inclusion of any compound in the *Dictionary and Handbook* does not indicate that use of that substance as a cosmetic ingredient complies with the laws and regulations governing such use in the United States or any other country.

Silsoft MSC (OSi Specialties)
Ultrasil Copolyol-1 Silicone (Noveon)
Ultrasil Copolyol-7 Silicone (Noveon)
Zenicone IX (Zenitech)
Zenicone IX (Zenitech)
Zenicone X (Zenitech)
Zenicone XQ (Zenitech)
Zenicone XX (Zenitech)

PEG-9 DIMETHICONE

CAS No.: 68937-54-2

Definition: PEG-9 Dimethicone is the siloxane polymer that conforms generally to the formula:

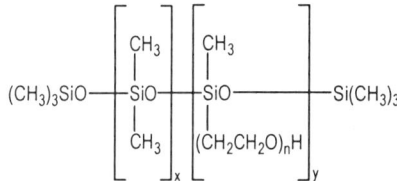

where n has an average value of 9.

Chemical Classes: Siloxanes and Silanes; Synthetic Polymers

Functions: Hair Conditioning Agent; Skin-Conditioning Agent - Miscellaneous

Trade Names:
KF-6005 (Shin-Etsu Chemical Co.)
KF-6009 (Shin-Etsu Chemical Co.)
KF-6013 (Shin Etsu)
KF-6019 (Shin-Etsu Chemical Co.)
KF-6029 (Shin-Etsu Chemical Co.)

Trade Name Mixtures:
Silwax DMC-AV (Siltech LLC)
Silwax DMC-BOR (Siltech LLC)
Silwax DMC-C (Siltech LLC)
Silwax DMC-IS (Siltech LLC)
Silwax DMC-O (Siltech LLC)
Silwax DMC-SOY (Siltech LLC)

PEG-10 DIMETHICONE

Definition: PEG-10 Dimethicone is the polyethylene glycol derivative of Dimethicone (q.v.) containing an average of 10 moles of ethylene oxide.

Chemical Class: Siloxanes and Silanes

Functions: Hair Conditioning Agent; Skin-Conditioning Agent - Miscellaneous

Reported Product Category: Suntan Gels, Creams, and Liquids

Trade Names:
KF-6017 (Shin Etsu)
KF353A (Shin Etsu)
KF354A (Shin Etsu)
KF-355A (Shin Etsu)

Trade Name Mixture:
Nikkol Nikkomulese WO (Nikko)

PEG-12 DIMETHICONE

Definition: PEG-12 Dimethicone is the polyethylene glycol derivative of Dimethicone (q.v.) containing an average of 12 moles of ethylene oxide.

Chemical Class: Siloxanes and Silanes

Functions: Hair Conditioning Agent; Skin-Conditioning Agent - Miscellaneous

Reported Product Categories: Aftershave Lotions; Bath Preparations, Misc.; Bath Soaps and Detergents; Body and Hand Preparations (Excluding Shaving Preparations); Cleansing Products (Cold Creams, Cleansing Lotions, Liquids and Pads); Colognes and Toilet Waters; Foundations; Fragrance Preparations, Misc.; Hair Straighteners; Makeup Preparations (Not eye), Misc.; Moisturizing Preparations; Personal Cleanliness Products, Misc.; Shampoos (Non-coloring); Shaving Preparations, Misc.; Skin Fresheners; Tonics, Dressings, and Other Hair Grooming Aids

Trade Names:
AEC PEG-12 Dimethicone (A & E Connock)
Botanisil S-19 (Botanigenics)
Dow Corning 5324 Fluid (Dow Corning)
Dow Corning 5329 Performance Modifier (Dow Corning)
Dow Corning 193 Surfactant (Dow Corning)
FZ-2411 (Nippon Unicar)
FZ-2412 (Nippon Unicar)
SF1288 (GE Silicones)
SH3746 (DCTS)
SH3771C (DCTS)
SH3772C (DCTS)
SH3773C (DCTS)
SH3775C (DCTS)
SH3771M (DCTS)
SH3772M (DCTS)
SH3773M (DCTS)
SH3775M (DCTS)
Silsoft 870 dimethicone copolyol (OSi Specialties)
Silsoft 880 dimethicone copolyol (OSi Specialties)
Silsurf D-212-CG (Siltech LLC)
SS-2804 (Nippon Unicar)
TAYLOR T-Wet 643 (Taylor Chemical Company)

Trade Name Mixtures:
A31003.90 Tudor Aspen (Kingfisher Colours)
Activated Botanicals AB 109 (Norjin)
Activated Botanicals AB 110 (Norjin)
Activated Botanicals Awapuhi AB 108 (Norjin)
Activated Botanicals Si-LIP AB 100 (Norjin)

Activated Botanicals Si-MIN AB 101 (Norjin)
Hydro Trichosyl (Laboratoires Sero-biologiques)
Liposomes Trichogen Veg (Laboratoires Serobiologiques)
Trichogen Veg (Laboratoires Sero-biologiques)
Yodosol PUD (National Starch)

PEG-14 DIMETHICONE

Definition: PEG-14 Dimethicone is the polyethylene glycol derivative of Dimethicone (q.v.) containing an average of 14 moles of ethylene oxide.

Chemical Class: Siloxanes and Silanes

Functions: Hair Conditioning Agent; Skin-Conditioning Agent - Miscellaneous

Trade Name:
Abil B 8843 (Degussa Care Specialties)

PEG-17 DIMETHICONE

Definition: PEG-17 Dimethicone is the polyethylene glycol derivative of Dimethicone (q.v.) containing an average of 17 moles of ethylene oxide.

Chemical Class: Siloxanes and Silanes

Functions: Hair Conditioning Agent; Skin-Conditioning Agent - Miscellaneous

Trade Name:
Silsoft 895 dimethicone copolyol (Crompton Corporation)

PEG-10 DIMETHICONE CROSSPOLYMER

Definition: PEG-10 Dimethicone Crosspolymer is a crosslinked copolymer formed from PEG-10 and Dimethicone (q.v.).

Chemical Class: Siloxanes and Silanes

Function: Viscosity Increasing Agent - Nonaqueous

Trade Name:
KSG-20 (Shin Etsu)

PEG-12 DIMETHICONE CROSSPOLYMER

Definition: PEG-12 Dimethicone Crosspolymer is a copolymer of PEG-12 dimethicone crosslinked with a C3-20 diene.

Chemical Classes: Siloxanes and Silanes; Synthetic Polymers

Functions: Emulsion Stabilizer; Surfactant - Emulsifying Agent; Suspending Agent - Nonsurfactant; Viscosity Increasing Agent - Nonaqueous

Trade Name:
Dow Corning 9010 Silicone Elastomer Blend (Dow Corning)

Trade Name Mixtures:
Dow Corning 9010 Silicone Elastomer Blend (Dow Corning)
Dow Corning 9011 Silicone Elastomer Blend (Dow Corning)

PEG-2 DIOLEATE

Definition: PEG-2 Dioleate is the polyethylene glycol diester of oleic acid that conforms generally to the formula:

$$CH(CH_2)_7C(=O)-(OCH_2CH_2)_nO-C(=O)(CH_2)_7CH$$
$$\overset{\|}{CH}(CH_2)_7CH_3 \qquad CH_3(CH_2)_7\overset{\|}{CH}$$

where n has an average value of 2.

Chemical Class: Alkoxylated Carboxylic Acids

Function: Surfactant - Emulsifying Agent

Technical/Other Names:
Polyethylene Glycol 100 Dioleate
Polyoxyethylene (2) Dioleate

Trade Name:
Emalex DEG-di-O (Nihon Emulsion)

PEG-3 DIOLEATE

Definition: PEG-3 Dioleate is the polyethylene glycol diester of oleic acid that conforms generally to the formula:

$$CH(CH_2)_7C(=O)-(OCH_2CH_2)_nO-C(=O)(CH_2)_7CH$$
$$\overset{\|}{CH}(CH_2)_7CH_3 \qquad CH_3(CH_2)_7\overset{\|}{CH}$$

where n has an average value of 3.

Chemical Class: Alkoxylated Carboxylic Acids

Function: Surfactant - Emulsifying Agent

Technical/Other Names:
Polyethylene Glycol (3) Dioleate
Polyoxyethylene (3) Dioleate

Trade Name:
Emalex TEG-di-O (Nihon Emulsion)

PEG-4 DIOLEATE

CAS Nos.: 9005-07-6 (Generic); 52668-97-0 (Generic); 134141-38-1

JPN Translation:
ジオレイン酸 PEG - 4

Empirical Formula:
$C_{44}H_{82}O_7$

Definition: PEG-4 Dioleate is the polyethylene glycol diester of oleic acid that conforms to the formula:

$$CH(CH_2)_7C(=O)-(OCH_2CH_2)_nO-C(=O)(CH_2)_7CH$$
$$\overset{\|}{CH}(CH_2)_7CH_3 \qquad CH_3(CH_2)_7\overset{\|}{CH}$$

where n has an average value of 4.

Information Sources: 21CFR173.340, 21CFR175.105, 21CFR175.300, 21CFR176.210, JCLS, JSQI, MI-13(7660), TSCA

Chemical Class: Alkoxylated Carboxylic Acids

Function: Surfactant - Emulsifying Agent

Technical/Other Names:
9-Octadecenoic Acid, Oxybis(2,1-Ethanediyloxy-2,1-Ethanediyl)Ester
Polyethylene Glycol 200 Dioleate
Polyoxyethylene (4) Dioleate

Trade Names:
AEC PEG-4 Dioleate (A & E Connock)
Jeemate 400-DO (Jeen)
STEPAN PEG 200 DO (Stepan)

PEG-6 DIOLEATE

CAS Nos.: 9005-07-6 (Generic); 52668-97-0 (Generic)

JPN Translation:
ジオレイン酸 PEG - 6

Empirical Formula:
$C_{48}H_{90}O_9$

Definition: PEG-6 Dioleate is the polyethylene glycol diester of oleic acid that conforms to the formula:

$$CH(CH_2)_7C(=O)-(OCH_2CH_2)_nO-C(=O)(CH_2)_7CH$$
$$\overset{\|}{CH}(CH_2)_7CH_3 \qquad CH_3(CH_2)_7\overset{\|}{CH}$$

where n has an average value of 6.

Information Sources: 21CFR175.105, 21CFR175.300, 21CFR176.210, JCLS, JSQI, MI-13(7660), TSCA

Chemical Class: Alkoxylated Carboxylic Acids

Function: Surfactant - Emulsifying Agent

Technical/Other Names:
Polyethylene Glycol 300 Dioleate
Polyoxyethylene (6) Dioleate

Trade Names:
AEC PEG-6 Dioleate (A & E Connock)
STEPAN PEG 300 DO (Stepan)

PEG-8 DIOLEATE

CAS Nos.: 9005-07-6 (Generic); 52668-97-0 (Generic)

JPN Translation:
ジオレイン酸 PEG - 8

Empirical Formula:
$C_{52}H_{98}O_{11}$

Definition: PEG-8 Dioleate is the polyethylene glycol diester of oleic acid that conforms to the formula:

$$CH(CH_2)_7C(=O)-(OCH_2CH_2)_nO-C(=O)(CH_2)_7CH$$
$$\overset{\|}{CH}(CH_2)_7CH_3 \qquad CH_3(CH_2)_7\overset{\|}{CH}$$

where n has an average value of 8.

Information Sources: 21CFR173.340, 21CFR175.105, 21CFR175.300, 21CFR176.170, 21CFR176.200, 21CFR176.210, 21CFR177.1210, 21CFR177.2260, 21CFR177.2800, CTFA D, JCLS, JSQI, MI-13(7660), TSCA

Chemical Class: Alkoxylated Carboxylic Acids

Function: Surfactant - Emulsifying Agent

Reported Product Categories: Mascara; Bath Oils, Tablets, and Salts

Technical/Other Names:
Polyethylene Glycol 400 Dioleate
Polyoxyethylene (8) Dioleate

Trade Names:
AEC PEG-8 Dioleate (A & E Connock)
Alkamuls 400 DO (Rhodia)
Cithrol 4DO (Croda Chemicals)
Ethox DO-9 (Ethox)
LIPOPEG 4-DO (Lipo)
Pegnol 24-O (Toho)
Pegosperse 400 DO (Lonza Inc./Lonza Ltd.)
Protamate 400 DO (Protameen)
ROL DO 40 (Fabriquimica)
STEPAN PEG 400 DO (Stepan)
Unipeg-400 DO (Universal Preserv-A-Chem)

PEG-10 DIOLEATE

CAS Nos.: 9005-07-6 (Generic); 52668-97-0 (Generic)

JPN Translation:
ジオレイン酸 PEG - 10

Empirical Formula:
$C_{56}H_{106}O_{13}$

Definition: PEG-10 Dioleate is the polyethylene glycol diester of oleic acid that conforms to the formula:

$$CH(CH_2)_7C \overset{O}{\overset{||}{\ }} \!\!-\!\!-\!\!(OCH_2CH_2)_nO \!-\!\!-\!\! \overset{O}{\overset{||}{C}}(CH_2)_7CH$$
$$\overset{||}{CH(CH_2)_7CH_3} \qquad CH_3(CH_2)_7\overset{||}{CH}$$

where n has an average value of 10.

Information Sources: 21CFR175.105, 21CFR175.300, 21CFR177.2260, JCLS, JSQI, MI-13(7660), TSCA

Chemical Class: Alkoxylated Carboxylic Acids

Function: Surfactant - Emulsifying Agent

Technical/Other Names:
Polyethylene Glycol 500 Dioleate
Polyoxyethylene (10) Dioleate

Trade Name:
AEC PEG-10 Dioleate (A & E Connock)

PEG-12 DIOLEATE

CAS Nos.: 9005-07-6 (Generic); 52668-97-0 (Generic)

JPN Translation:
ジオレイン酸 PEG - 12

Definition: PEG-12 Dioleate is the polyethylene glycol diester of oleic acid that conforms to the formula:

$$CH(CH_2)_7C \overset{O}{\overset{||}{\ }} \!\!-\!\!-\!\!(OCH_2CH_2)_nO \!-\!\!-\!\! \overset{O}{\overset{||}{C}}(CH_2)_7CH$$
$$\overset{||}{CH(CH_2)_7CH_3} \qquad CH_3(CH_2)_7\overset{||}{CH}$$

where n has an average value of 12.

Information Sources: 21CFR173.340, 21CFR175.105, 21CFR175.300, 21CFR176.200, 21CFR176.210, 21CFR177.2260, 21CFR177.2800, JCLS, JSQI, MI-13(7660), TSCA

Chemical Class: Alkoxylated Carboxylic Acids

Function: Surfactant - Emulsifying Agent

Technical/Other Names:
Polyethylene Glycol 600 Dioleate
Polyoxyethylene (12) Dioleate

Trade Names:
AEC PEG-12 Dioleate (A & E Connock)
Alkamuls 600 DO (Rhodia)
Ethox DO-14 (Ethox)
Lipopeg 6-DO (Lipo)
Marlosol FS (Sasol GmbH - Marl)

Protamate 600 DO (Protameen)
STEPAN PEG 600 DO (Stepan)
Sympatens-EDO/120 (Kolb)
Unipeg-600 DO (Universal Preserv-A-Chem)

PEG-20 DIOLEATE

CAS Nos.: 9005-07-6 (Generic); 52668-97-0 (Generic)

JPN Translation:
ジオレイン酸 PEG - 20

Definition: PEG-20 Dioleate is the polyethylene glycol diester of oleic acid that conforms to the formula:

$$CH(CH_2)_7C \overset{O}{\overset{||}{\ }} \!\!-\!\!-\!\!(OCH_2CH_2)_nO \!-\!\!-\!\! \overset{O}{\overset{||}{C}}(CH_2)_7CH$$
$$\overset{||}{CH(CH_2)_7CH_3} \qquad CH_3(CH_2)_7\overset{||}{CH}$$

where n has an average value of 20.

Information Sources: 21CFR175.300, 21CFR176.210, 21CFR177.2260, 21CFR177.2800, JCLS, JSQI, MI-13(7660), TSCA

Chemical Class: Alkoxylated Carboxylic Acids

Function: Surfactant - Emulsifying Agent

Technical/Other Names:
Polyethylene Glycol 1000 Dioleate
Polyoxyethylene (20) Dioleate

Trade Names:
AEC PEG-20 Dioleate (A & E Connock)
STEPAN PEG 1000 DO (Stepan)

PEG-32 DIOLEATE

CAS Nos.: 9005-07-6 (Generic); 52668-97-0 (Generic)

JPN Translation:
ジオレイン酸 PEG - 32

Definition: PEG-32 Dioleate is the polyethylene glycol diester of oleic acid that conforms to the formula:

$$CH(CH_2)_7C \overset{O}{\overset{||}{\ }} \!\!-\!\!-\!\!(OCH_2CH_2)_nO \!-\!\!-\!\! \overset{O}{\overset{||}{C}}(CH_2)_7CH$$
$$\overset{||}{CH(CH_2)_7CH_3} \qquad CH_3(CH_2)_7\overset{||}{CH}$$

where n has an average value of 32.

Information Sources: 21CFR175.300, 21CFR176.210, 21CFR177.2260, 21CFR177.2800, JCLS, JSQI, MI-13(7660), TSCA

Chemical Class: Alkoxylated Carboxylic Acids

Function: Surfactant - Emulsifying Agent

Technical/Other Names:
Polyethylene Glycol 1540 Dioleate
Polyoxyethylene (32) Dioleate

Trade Names:
AEC PEG-32 Dioleate (A & E Connock)
STEPAN PEG 1540 DO (Stepan)

PEG-75 DIOLEATE

CAS Nos.: 9005-07-6 (Generic); 52668-97-0 (Generic)

JPN Translation:
ジオレイン酸 PEG - 75

Definition: PEG-75 Dioleate is the polyethylene glycol diester of oleic acid that conforms to the formula:

$$CH(CH_2)_7C \overset{O}{\overset{||}{\ }} \!\!-\!\!-\!\!(OCH_2CH_2)_nO \!-\!\!-\!\! \overset{O}{\overset{||}{C}}(CH_2)_7CH$$
$$\overset{||}{CH(CH_2)_7CH_3} \qquad CH_3(CH_2)_7\overset{||}{CH}$$

where n has an average value of 75.

Information Sources: 21CFR175.300, JCLS, JSQI, MI-13(7660), TSCA

Chemical Class: Alkoxylated Carboxylic Acids

Functions: Surfactant - Cleansing Agent; Surfactant - Solubilizing Agent

Technical/Other Names:
Polyethylene Glycol 4000 Dioleate
Polyoxyethylene (75) Dioleate

Trade Names:
AEC PEG-75 Dioleate (A & E Connock)
STEPAN PEG 4000 DO (Stepan)

PEG-150 DIOLEATE

CAS Nos.: 9005-07-6 (Generic); 52668-97-0 (Generic)

JPN Translation:
ジオレイン酸 PEG - 150

Definition: PEG-150 Dioleate is the polyethylene glycol diester of oleic acid that conforms to the formula:

$$CH(CH_2)_7C \overset{O}{\overset{||}{\ }} \!\!-\!\!-\!\!(OCH_2CH_2)_nO \!-\!\!-\!\! \overset{O}{\overset{||}{C}}(CH_2)_7CH$$
$$\overset{||}{CH(CH_2)_7CH_3} \qquad CH_3(CH_2)_7\overset{||}{CH}$$

where n has an average value of 150.

Information Sources: 21CFR175.300, JCLS, JSQI, MI-13(7660), TSCA

Chemical Class: Alkoxylated Carboxylic Acids

Function: Surfactant - Cleansing Agent

Technical/Other Names:
Polyethylene Glycol 6000 Dioleate
Polyoxyethylene (150) Dioleate

Trade Names:
AEC PEG-150 Dioleate (A & E Connock)
STEPAN PEG 6000 DO (Stepan)

PEG-3 DIOLEOYLAMIDOETHYLMONIUM METHOSULFATE

Definition: PEG-3 Dioleoylamidoethyl-monium Methosulfate is the quaternary ammonium salt that conforms to the formula:

$$\left[\begin{array}{c} O \\ \parallel \\ RCNH(CH_2)_2 \end{array} \begin{array}{c} (CH_2CH_2O)_nH \quad O \\ | \quad\quad \parallel \\ -N-(CH_2)_2NHCR \\ | \\ CH_3 \end{array} \right]^{+} CH_3OSO_3^{-}$$

where RCO- represents the oleoyl moiety and n has an average value of 3.

Chemical Class: Quaternary Ammonium Compounds

Functions: Antistatic Agent; Hair Conditioning Agent

Reported Product Category: Hair Dyes and Colors (All Types Requiring Caution Statements and Patch Tests)

Trade Name:
Incroquat HO-80PG (Croda, Inc.)

PEG-3 DIPALMITATE

CAS No.: 32628-06-1 (Generic)

JPN Translation:
ジパルミチン酸 PEG - 3

Definition: PEG-3 Dipalmitate is the polyethylene glycol diester of palmitic acid that conforms generally to the formula:

$$CH_3(CH_2)_{14}C \begin{array}{c} O \\ \parallel \end{array} -(OCH_2CH_2)_nO- \begin{array}{c} O \\ \parallel \end{array} C(CH_2)_{14}CH_3$$

where n has an average value of 3.

Information Sources: 21CFR175.300, JCIC, JCLS, JSQI, MI-13(7660), TSCA

Chemical Class: Alkoxylated Carboxylic Acids

Function: Surfactant - Emulsifying Agent

Technical/Other Names:
Polyethylene Glycol (3) Dipalmitate
Polyethylene Glycol 150 Dipalmitate
Polyoxyethylene (3) Dipalmitate

Trade Name:
AEC PEG-3 Dipalmitate (A & E Connock)

PEG-3 2,2'-DI-p-PHENYLENEDIAMINE

Empirical Formula:
$C_{18}H_{26}N_4O_4$

Definition: PEG-3 2,2'-Di-p-Phenylenediamine is the organic compound that conforms generally to the formula:

$$NH_2 \quad\quad\quad\quad NH_2$$
$$O(CH_2CH_2O)_3$$
$$NH_2 \quad\quad\quad\quad NH_2$$

See "Regulatory and Ingredient Use Information," for Colorants in Volume 1, Introduction, Part A.

Chemical Classes: Amines; Color Additives - Hair; Ethers

Function: Hair Colorant

PEG-13 DIPHENYLOL PROPANE

CAS No.: 9014-86-2

Definition: PEG-13 Diphenylol Propane is the organic compound that conforms generally to the formula:

$$(OCH_2CH_2)_xOH$$
$$CH_3CCH_3$$
$$(OCH_2CH_2)_yOH$$

where x+y has an average value 13.

Chemical Class: Alkoxylated Alcohols

Function: Surfactant - Emulsifying Agent

Technical/Other Names:
Polyethylene Glycol (13) Diphenylol Propane
Polyoxyethylene (13) Diphenylol Propane

PEG-30 DIPOLYHYDROXYSTEARATE

Definition: PEG-30 Dipolyhydroxystearate is the polyethylene glycol diester of Polyhydroxystearic Acid (q.v.) that conforms generally to the formula:

$$RC \begin{array}{c} O \\ \parallel \end{array} -(OCH_2CH_2)_nO- \begin{array}{c} O \\ \parallel \end{array} CR$$

where RCO- represents the alkyl groups derived from Polyhydroxystearic Acid (q.v.) and n has an average value of 30.

Chemical Classes: Alkoxylated Carboxylic Acids; Esters

Function: Surfactant - Emulsifying Agent

Technical/Other Names:
Polyethylene Glycol (30) Dipolyhydroxy-stearate
Polyoxyethylene (30) Dipolyhydroxy-stearate

Trade Name:
Arlacel P135 (Uniqema Americas)

PEG-20 DIRICINOLEATE

JPN Translation:
ジリシノレイン酸 PEG-20

Definition: PEG-20 Diricinoleate is the diester of Ricinoleic Acid (q.v.) and PEG-20 (q.v.).

Information Source: JCLS、

Chemical Class: Esters

Function: Skin-Conditioning Agent - Miscellaneous

PEG-2 DIROSINATE

Definition: PEG-2 Dirosinate is the polyethylene glycol diester of the acids derived from Rosin (q.v.). It conforms generally to the formula:

$$RC \begin{array}{c} O \\ \parallel \end{array} -(OCH_2CH_2)_nO- \begin{array}{c} O \\ \parallel \end{array} CR$$

where RCO- represents the acids derived from Rosin (q.v.) and has an average value of 2.

Chemical Class: Alkoxylated Carboxylic Acids

Functions: Skin-Conditioning Agent - Occlusive; Viscosity Increasing Agent - Nonaqueous

Technical/Other Names:
Polyethylene Glycol 100 Dirosinate
Polyoxyethylene (2) Dirosinate

Trade Name Mixture:
Recol T 3 (Granel Derivados)

PEG-3 DIROSINATE

CAS No.	EINECS No.
8050-25-7	232-478-3

Definition: PEG-3 Dirosinate is the poly-ethylene glycol diester of the acids derived from Rosin (q.v.). It conforms generally to the formula:

$$RC \overset{\overset{O}{\|}}{-} (OCH_2CH_2)_nO \overset{\overset{O}{\|}}{-} CR$$

where n has an average value of 3 and RCO-represents the acids derived from Rosin (q.v.).

Chemical Class: Alkoxylated Carboxylic Acids

Functions: Skin-Conditioning Agent - Occlusive; Viscosity Increasing Agent - Nonaqueous

Technical/Other Names:
Polyethylene Glycol (3) Dirosinate
Polyoxyethylene (3) Dirosinate
Resin Acids and Rosin Acids, Esters with Triethylene Glycol
Rosin, Triethylene Glycol Ester

Trade Names:
Ennesin DP.829 HV (Lawter)
Tergum 45 (Cray Valley Iberica)

Trade Name Mixture:
Recol T 3 (Granel Derivados)

PEG-2 DISTEARATE

CAS Nos.   **EINECS No.**
109-30-8   203-663-6
9005-08-7
52668-97-0

JPN Translation:
ジステアリン酸 PEG - 2

Empirical Formula:
$C_{40}H_{78}O_5$

Definition: PEG-2 Distearate is the poly-ethylene glycol diester of stearic acid that conforms to the formula:

$$CH_3(CH_2)_{16}\overset{\overset{O}{\|}}{C} - (OCH_2CH_2)_nO - \overset{\overset{O}{\|}}{C}(CH_2)_{16}CH_3$$

where n has an average value of 2.

Information Sources: 21CFR73.1, 21CFR175.300, 21CFR176.210, 21CFR177.2800, CIR: [S] IJT-18(SUPPL. 1)-1999, JCIC, JCLS, JSQI, MI-13(7660), TSCA

Chemical Class: Alkoxylated Carboxylic Acids

Function: Surfactant - Emulsifying Agent

Technical/Other Names:
Diethylene Glycol Distearate
Octadecanoic Acid, Oxydi-2,1-Ethanediyl Ester
Polyethylene Glycol 100 Distearate
Polyoxyethylene (2) Distearate

Trade Names:
AEC PEG-2 Distearate (A & E Connock)
STEPAN DGDS (Stepan)

PEG-3 DISTEARATE

CAS No.: 9005-08-7 (Generic)

JPN Translation:
ジステアリン酸 PEG - 3

Empirical Formula:
$C_{42}H_{82}O_6$

Definition: PEG-3 Distearate is the poly-ethylene glycol diester of stearic acid that conforms to the formula:

$$CH_3(CH_2)_{16}\overset{\overset{O}{\|}}{C} - (OCH_2CH_2)_nO - \overset{\overset{O}{\|}}{C}(CH_2)_{16}CH_3$$

where n has an average value of 3.

Information Sources: 21CFR175.300, CIR: [S] IJT-18(SUPPL. 1)1999, JCLS, JSQI, MI-13(7660), TSCA

Chemical Class: Alkoxylated Carboxylic Acids

Function: Surfactant - Emulsifying Agent

Reported Product Categories: Shampoos (Non-coloring); Bath Oils, Tablets, and Salts; Cleansing Products (Cold Creams, Cleansing Lotions, Liquids and Pads)

Technical/Other Names:
Polyethylene Glycol (3) Distearate
Polyoxyethylene (3) Distearate
Triglycol Distearate

Trade Names:
AEC PEG-3 Distearate (A & E Connock)
Cutina TS (Cognis Deutschland)
Genapol TS Powder (Clariant)
Genapol TS Powder (Clariant GmbH, Personal Care)
Nikkol Estepearl 30 (Nikko)
Phoenate 3 DSA (Phoenix)
Tegin D 1102 (Degussa Care Specialties)

Trade Name Mixtures:
Base Nacrante 6030CP (SEPPIC)
Euperlan PK-900 BENZ-W (Cognis Care Chemicals/NJ)
Euperlan PK-900 BENZ-W (Cognis Care Chemicals/PA)
Euperlan PK-900 BENZ-W (Cognis Deutschland)
Genapol TSM (Clariant)
Genapol TSM (Clariant GmbH, Personal Care)

PEG-4 DISTEARATE

CAS Nos.: 142-20-1; 9005-08-7 (Generic)

JPN Translation:
ジステアリン酸 PEG - 4

Empirical Formula:
$C_{44}H_{86}O_7$

Definition: PEG-4 Distearate is the poly-ethylene glycol diester of stearic acid that conforms to the formula:

$$CH_3(CH_2)_{16}\overset{\overset{O}{\|}}{C} - (OCH_2CH_2)_nO - \overset{\overset{O}{\|}}{C}(CH_2)_{16}CH_3$$

where n has an average value of 4.

Information Sources: 21CFR175.105, 21CFR175.300, 21CFR176.210, CIR: [S] IJT-18(SUPPL. 1)1999, JCLS, JSQI, MI-13 (7660), TSCA

Chemical Class: Alkoxylated Carboxylic Acids

Function: Surfactant - Emulsifying Agent

Reported Product Category: Hair Conditioners

Technical/Other Names:
Octadecanoic Acid, Oxybis(2,1-Ethane-diyloxy-2,1-Ethanediyl) Ester
Polyethylene Glycol 200 Distearate
Polyoxyethylene (4) Distearate

Trade Names:
Protamate 200 DS (Protameen)
STEPAN PEG 200 DS (Stepan)

PEG-6 DISTEARATE

CAS No.: 9005-08-7 (Generic)

JPN Translation:
ジステアリン酸 PEG - 6

Empirical Formula:
$C_{48}H_{94}O_9$

Definition: PEG-6 Distearate is the poly-ethylene glycol diester of stearic acid that conforms to the formula:

$$CH_3(CH_2)_{16}\overset{\overset{O}{\|}}{C} - (OCH_2CH_2)_nO - \overset{\overset{O}{\|}}{C}(CH_2)_{16}CH_3$$

where n has an average value of 6.

Information Sources: 21CFR175.105, 21CFR175.300, 21CFR176.210, CIR: [S] IJT-18(SUPPL. 1)1999, JCLS, JSQI, MI-13 (7660), TSCA

Chemical Class: Alkoxylated Carboxylic Acids

Function: Surfactant - Emulsifying Agent

Technical/Other Names:
Acetylbutyrylcellulose
Polyethylene Glycol 300 Distearate
Polyoxyethylene (6) Distearate

Trade Names:
AEC PEG-6 Distearate (A & E Connock)
Sipoest PEG 300 DS (Specialty Industrial)
STEPAN PEG 300 DS (Stepan)

PEG-8 DISTEARATE

CAS No.: 9005-08-7 (Generic)

JPN Translation:
ジステアリン酸 PEG - 8

Empirical Formula:
$C_{52}H_{102}O_{11}$

Definition: PEG-8 Distearate is the poly-ethylene glycol diester of stearic acid that conforms to the formula:

$$CH_3(CH_2)_{16}\overset{O}{\overset{||}{C}}\!-\!(OCH_2CH_2)_nO\!-\!\overset{O}{\overset{||}{C}}(CH_2)_{16}CH_3$$

where n has an average value of 8.

Information Sources: 21CFR175.105, 21CFR175.300, 21CFR176.210, 21CFR177.1210, 21CFR177.2260, 21CFR177.2800, CIR: [S] IJT-18(SUPPL. 1)-1999, CTFA S, JCLS, JSQI, MI-13(7660), TSCA

Chemical Class: Alkoxylated Carboxylic Acids

Function: Surfactant - Emulsifying Agent

Reported Product Categories: Personal Cleanliness Products, Misc.; Deodorants (Underarm); Aftershave Lotions; Baby Shampoos; Bath Preparations, Misc.; Body and Hand Preparations (Excluding Shaving Preparations); Hair Conditioners; Shaving Preparations, Misc.

Technical/Other Names:
Polyethylene Glycol 400 Distearate
Polyoxyethylene (8) Distearate

Trade Names:
AEC PEG-8 Distearate (A & E Connock)
Cithrol 4DS (Croda Chemicals)
Emerest 2712 (Cognis Care Chemicals/NJ)
Emerest 2712 (Cognis Care Chemicals/PA)
ESTOL 3724 (Uniqema Europe)
Ethox DS-9 (Ethox)
Hetoxamate 400 DS (Heterene)
Jeemate 400-DS (Jeen)
Lipopeg 4-DS (Lipo)
Nikkol CDS-400 (Nikko)
Pegosperse 400 DS (Lonza Inc./Lonza Ltd.)
Protamate 400 DS (Protameen)
Radiasurf 7453 (Oleon NV)
Sipoest PEG 400 DS (Specialty Industrial)
STEPAN PEG 400 DS (Stepan)
Unipeg-400 DS (Universal Preserv-A-Chem)

PEG-9 DISTEARATE

CAS Nos.: 109-34-2; 9005-08-7 (Generic)

JPN Translation:
ジステアリン酸 PEG - 9

Empirical Formula:
$C_{54}H_{106}O_{12}$

Definition: PEG-9 Distearate is the poly-ethylene glycol diester of stearic acid that conforms to the formula:

$$CH_3(CH_2)_{16}\overset{O}{\overset{||}{C}}\!-\!(OCH_2CH_2)_nO\!-\!\overset{O}{\overset{||}{C}}(CH_2)_{16}CH_3$$

where n has an average value of 9.

Information Sources: 21CFR175.105, 21CFR175.300, 21CFR177.2260, 21CFR177.2800, CIR: [S] IJT-18(SUPPL. 1)-1999, JCLS, JSQI, MI-13(7660), TSCA

Chemical Class: Alkoxylated Carboxylic Acids

Function: Surfactant - Emulsifying Agent

Technical/Other Names:
Nonaethylene Glycol Stearate
Octadecanoic Acid, 3,6,9,12,15,18,21,24-Octaoxahexacosane-1,26-Diyl Ester
3,6,9,12,15,18,21,24-Octaoxahexacosane-1,26-Diyl Octadecanoate
Polyethylene Glycol 450 Distearate
Polyoxyethylene (9) Distearate

Trade Name:
AEC PEG-9 Distearate (A & E Connock)

PEG-12 DISTEARATE

CAS No.: 9005-08-7 (Generic)

JPN Translation:
ジステアリン酸 PEG - 12

Definition: PEG-12 Distearate is the poly-ethylene glycol diester of stearic acid that conforms to the formula:

$$CH_3(CH_2)_{16}\overset{O}{\overset{||}{C}}\!-\!(OCH_2CH_2)_nO\!-\!\overset{O}{\overset{||}{C}}(CH_2)_{16}CH_3$$

where n has an average value of 12.

Information Sources: 21CFR175.105, 21CFR175.300, 21CFR176.210, 21CFR177.2260, 21CFR177.2800, CIR: [S] IJT-18(SUPPL. 1)1999, CTFA S, JCLS, JSQI, MI-13(7660), TSCA

Chemical Class: Alkoxylated Carboxylic Acids

Function: Surfactant - Emulsifying Agent

Reported Product Categories: Hair Conditioners; Hair Rinses (Non-coloring)

Technical/Other Names:
Polyethylene Glycol 600 Distearate
Polyoxyethylene (12) Distearate

Trade Names:
AEC PEG-12 Distearate (A & E Connock)
Ethox DS-14 (Ethox)
Jeemate 600-DS (Jeen)
Protamate 600 DS (Protameen)
Radiasurf 7454 (Oleon NV)
STEPAN PEG 600 DS (Stepan)
Unipeg-600 DS (Universal Preserv-A-Chem)

PEG-20 DISTEARATE

CAS No.: 9005-08-7 (Generic)

JPN Translation:
ジステアリン酸 PEG - 20

Definition: PEG-20 Distearate is the poly-ethylene glycol diester of stearic acid that conforms to the formula:

$$CH_3(CH_2)_{16}\overset{O}{\overset{||}{C}}\!-\!(OCH_2CH_2)_nO\!-\!\overset{O}{\overset{||}{C}}(CH_2)_{16}CH_3$$

where n has an average value of 20.

Information Sources: 21CFR175.300, 21CFR176.210, 21CFR177.2260, 21CFR177.2800, CIR: [S] IJT-18(SUPPL. 1)-1999, JCLS, JSQI, MI-13(7660), TSCA

Chemical Class: Alkoxylated Carboxylic Acids

Function: Surfactant - Emulsifying Agent

Technical/Other Names:
Polyethylene Glycol 1000 Distearate
Polyoxyethylene (20) Distearate

Trade Names:
AEC PEG-20 Distearate (A & E Connock)
Ethox DS-20 (Ethox)
STEPAN PEG 1000 DS (Stepan)

PEG-32 DISTEARATE

CAS No.: 9005-08-7 (Generic)

JPN Translation:
ジステアリン酸 PEG - 32

Definition: PEG-32 Distearate is the poly-ethylene glycol diester of stearic acid that conforms to the formula:

$$CH_3(CH_2)_{16}\overset{O}{\overset{||}{C}}\!-\!(OCH_2CH_2)_nO\!-\!\overset{O}{\overset{||}{C}}(CH_2)_{16}CH_3$$

where n has an average value of 32.

Information Sources: 21CFR175.300, 21CFR176.210, 21CFR177.2260, 21CFR177.2800, CIR: [S] IJT-18(SUPPL. 1)-1999, CTFA S, JCLS, JSQI, MI-13(7660), TSCA

Chemical Class: Alkoxylated Carboxylic Acids

Function: Surfactant - Emulsifying Agent

Technical/Other Names:
Polyethylene Glycol 1540 Distearate
Polyoxyethylene (32) Distearate

Trade Name:
STEPAN PEG 1540 DS (Stepan)

PEG-75 DISTEARATE

CAS No.: 9005-08-7 (Generic)

JPN Translation:
ジステアリン酸 PEG - 75

Definition: PEG-75 Distearate is the polyethylene glycol diester of stearic acid that conforms to the formula:

$$CH_3(CH_2)_{16}\overset{\displaystyle O}{\overset{\displaystyle \|}{C}} - (OCH_2CH_2)_nO - \overset{\displaystyle O}{\overset{\displaystyle \|}{C}}(CH_2)_{16}CH_3$$

where n has an average value of 75.

Information Sources: 21CFR175.300, CIR: [S] IJT-18(SUPPL. 1)1999, JCLS, JSQI, MI-13(7660), TSCA

Chemical Class: Alkoxylated Carboxylic Acids

Functions: Surfactant - Cleansing Agent; Surfactant - Solubilizing Agent

Technical/Other Names:
Polyethylene Glycol 4000 Distearate
Polyoxyethylene (75) Distearate

Trade Names:
AEC PEG-75 Distearate (A & E Connock)
STEPAN PEG 4000 DS (Stepan)
Unipeg-4000 DS (Universal Preserv-A-Chem)

PEG-120 DISTEARATE

CAS No.: 9005-08-7 (Generic)

JPN Translation:
ジステアリン酸 PEG - 120

Definition: PEG-120 Distearate is the polyethylene glycol diester of stearic acid that conforms to the formula:

$$CH_3(CH_2)_{16}\overset{\displaystyle O}{\overset{\displaystyle \|}{C}} - (OCH_2CH_2)_nO - \overset{\displaystyle O}{\overset{\displaystyle \|}{C}}(CH_2)_{16}CH_3$$

where n has an average value of 120.

Information Sources: CIR: [S] IJT-18 (SUPPL. 1)1999, JCLS, TSCA

Chemical Class: Alkoxylated Carboxylic Acids

Functions: Surfactant - Cleansing Agent; Surfactant - Solubilizing Agent

Technical/Other Names:
Polyethylene Glycol (120) Distearate
Polyoxyethylene (120) Distearate

PEG-150 DISTEARATE

CAS No.: 9005-08-7 (Generic)

JPN Translation:
ジステアリン酸 PEG - 150

Definition: PEG-150 Distearate is the polyethylene glycol diester of stearic acid that conforms to the formula:

$$CH_3(CH_2)_{16}\overset{\displaystyle O}{\overset{\displaystyle \|}{C}} - (OCH_2CH_2)_nO - \overset{\displaystyle O}{\overset{\displaystyle \|}{C}}(CH_2)_{16}CH_3$$

where n has an average value of 150.

Information Sources: 21CFR175.300, CIR: [S] IJT-18(SUPPL. 1)1999, CTFA S, JCLS, JSQI, MI-13(7660), TSCA

Chemical Class: Alkoxylated Carboxylic Acids

Functions: Surfactant - Cleansing Agent; Surfactant - Solubilizing Agent

Reported Product Categories: Shampoos (Non-coloring); Baby Shampoos; Blushers (All types); Bath Oils, Tablets, and Salts; Bath Soaps and Detergents; Cleansing Products (Cold Creams, Cleansing Lotions, Liquids and Pads); Shaving Cream (Aerosol, Brushless and Lather); Baby Products, Misc.; Shaving Preparations, Misc.; Moisturizing Preparations; Hair Conditioners; Bath Preparations, Misc.; Bubble Baths; Foundations; Hair Preparations (Non-coloring), Misc.; Paste Masks (Mud Packs); Skin Care Preparations, Misc.

Technical/Other Names:
Polyethylene Glycol 6000 Distearate
Polyoxyethylene (150) Distearate

Trade Names:
AEC PEG-150 Distearate (A & E Connock)
Atlas G-1821 (Uniqema Americas)
Customulse 6000 DS (PEG-150 Distearate) (Custom Ingredients)
ESTOL 3734 (Uniqema Europe)
Ethox P-6000 DS (Ethox)
Ethox P-6000 DS Special (Ethox)
Jeemate 6000-DS (Jeen)
LIPOPEG 6000-DS (Lipo)
Lumulse 602-S (Lambent)
Nikkol CDS-6000P (Nikko)
Norfox 6000 DS (Norman, Fox & Co.)
Pegnol PDS-60A (Toho)
Pegnol 6000S (Toho)
Protamate 6000 DS (Protameen)
Rewopal PEG 6000 DS (Degussa Care Specialties)
Ritapeg 150DS (RITA)
Rol D 600 (Fabriquimica)
Sipoest PEG 6000 DS (Specialty Industrial)
STEPAN PEG 6000 DS (Stepan)
Unipeg-6000 DS (Universal Preserv-A-Chem)

Trade Name Mixtures:
Atlas G-1823 (Uniqema Americas)
Custoblend AEG (ALS Blend) (Custom Ingredients)
Custoblend BAC (Anti-Bac Blend) (Custom Ingredients)
Custoblend BAT (Anti-Bac Blend) (Custom Ingredients)
Custoblend BSC-50 (Baby Shampoo) (Custom Ingredients)
Custoblend UB (Universal Blend) (Custom Ingredients)
DeCONC BSC-50 (DeForest)
DeCONC SCE-40 (DeForest)
Lowenol 4667 (Lowenstein)
Miracare BC-10 (Rhodia)
Miracare BC-20 (Rhodia)
Miracare BC-27 (Rhodia)
Miracare MS-2 (Rhodia)
Miracare MS-4 (Rhodia)
Viscolene (Vevy)

PEG-175 DISTEARATE

CAS No.: 9005-08-7 (Generic)

JPN Translation:
ジステアリン酸 PEG - 175

Definition: PEG-175 Distearate is the polyethylene glycol diester of stearic acid that conforms to the formula:

$$CH_3(CH_2)_{16}\overset{\displaystyle O}{\overset{\displaystyle \|}{C}} - (OCH_2CH_2)_nO - \overset{\displaystyle O}{\overset{\displaystyle \|}{C}}(CH_2)_{16}CH_3$$

where n has an average value of 175.

Information Sources: 21CFR175.300, CIR: [S] IJT-18(SUPPL. 1)1999, JCLS, JSQI, MI-13(7660), TSCA

Chemical Class: Alkoxylated Carboxylic Acids

Functions: Surfactant - Cleansing Agent; Surfactant - Solubilizing Agent

Technical/Other Names:
Polyethylene Glycol (175) Distearate
Polyoxyethylene (175) Distearate

Trade Name:
AEC PEG-175 Distearate (A & E Connock)

PEG-190 DISTEARATE

CAS No.: 9005-08-7 (Generic)

Definition: PEG-190 Distearate is the polyethylene glycol diester of stearic acid that conforms generally to the formula:

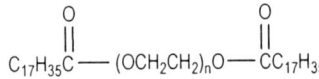

where n has an average value of 190.

Chemical Class: Alkoxylated Carboxylic Acids

Functions: Surfactant - Cleansing Agent; Surfactant - Emulsifying Agent; Surfactant - Solubilizing Agent; Viscosity Increasing Agent - Aqueous

Technical/Other Names:
Polyethylene Glycol (190) Distearate
Polyoxyethylene (190) Distearate

Trade Name:
Emulmin 862 (Sanyo Chemical)

PEG-250 DISTEARATE

CAS No.: 9005-08-7 (Generic)

JPN Translation:
ジステアリン酸 PEG - 250

Definition: PEG-250 Distearate is the polyethylene glycol diester of stearic acid that conforms generally to the formula:

$$CH_3(CH_2)_{16}C\overset{O}{\overset{\|}{-}}(OCH_2CH_2)_nO-\overset{O}{\overset{\|}{C}}(CH_2)_{16}CH_3$$

where n has an average value of 250.

Information Source: JCLS

Chemical Class: Alkoxylated Carboxylic Acids

Functions: Surfactant - Cleansing Agent; Surfactant - Solubilizing Agent

Technical/Other Names:
Polyethylene Glycol (250) Distearate
Polyoxyethylene (250) Distearate

Trade Name:
Emanon 3299R (Kao Corp.)

PEG-3 DISTEAROYLAMIDOETHYL-MONIUM METHOSULFATE

Definition: PEG-3 Distearoylamidoethylmonium Methosulfate is the quaternary ammonium salt that conforms generally to the formula:

$$\left[RCNH(CH_2)_2\overset{O}{\overset{\|}{-}}\overset{(CH_2CH_2O)_nH}{\underset{CH_3}{\overset{|}{N}}}-(CH_2)_2NHCR\overset{O}{\overset{\|}{}}\right]^+ CH_3OSO_3^-$$

where n has an average value of 3.

Chemical Class: Quaternary Ammonium Compounds

Functions: Antistatic Agent; Hair Conditioning Agent

Trade Name Mixture:
Incroquat OSC (Croda, Inc.)

PEG-4 DISTEARYL ETHER

Definition: PEG-4 Distearyl Ether is the polyethylene glycol ether of stearyl alcohol that conforms generally to the formula:

$$CH_3(CH_2)_{17}O(CH_2CH_2O)_n(CH_2)_{17}CH_3$$

where n has an average value of 4.

Chemical Class: Ethers

Functions: Hair Conditioning Agent; Opacifying Agent; Plasticizer; Skin-Conditioning Agent - Miscellaneous; Slip Modifier

Trade Name:
Cutina DSE 4 (Cognis Deutschland)

PEG-4 DISTEARYLETHONIUM ETHOSULFATE

Empirical Formula:
$C_{46}H_{95}NO_4 \cdot C_2H_5O_4S$

Definition: PEG-4 Distearylethonium Ethosulfate is the quaternary ammonium salt that conforms generally to the formula:

$$\left[CH_3(CH_2)_{17}\overset{(CH_2CH_2O)_nH}{\underset{CH_2CH_3}{\overset{|}{N}}}-(CH_2)_{17}CH_3\right]^+ CH_3CH_2OSO_3^-$$

where n has an average value of 4.

Chemical Class: Esters

Functions: Antistatic Agent; Hair Conditioning Agent

Trade Name:
Mackernium CK-31 (McIntyre)

PEG-8 DITALLATE

CAS No.: 61791-01-3 (Generic)

Definition: PEG-8 Ditallate is the polyethylene glycol diester of Tall Oil Acid (q.v.) that conforms generally to the formula:

$$RC\overset{O}{\overset{\|}{-}}(OCH_2CH_2)_nO-\overset{O}{\overset{\|}{C}}R$$

where RCO- represents the tall oil fatty radicals and n has an average value of 8.

Information Sources: 21CFR175.105, 21CFR175.300, 21CFR176.210, 21CFR177.1210, 21CFR177.2800, MI-13 (7660), TSCA

Chemical Class: Alkoxylated Carboxylic Acids

Function: Surfactant - Emulsifying Agent

Technical/Other Names:
Polyethylene Glycol 400 Ditallate
Polyoxyethylene (8) Ditallate

Trade Names:
AEC PEG-8 Ditallate (A & E Connock)
Chemax PEG-400-DT (Chemax)
Pegosperse 400 DOT (Lonza Inc./Lonza Ltd.)

PEG-12 DITALLATE

CAS No.: 61791-01-3 (Generic)

Definition: PEG-12 Ditallate is the polyethylene glycol diester of Tall Oil Acid (q.v.) that conforms generally to the formula:

$$RC\overset{O}{\overset{\|}{-}}(OCH_2CH_2)_nO-\overset{O}{\overset{\|}{C}}R$$

where RCO- represents the tall oil fatty radicals and n has an average value of 12.

Information Sources: 21CFR175.105, 21CFR175.300, 21CFR176.210, 21CFR177.2800, MI-13(7660), TSCA

Chemical Class: Alkoxylated Carboxylic Acids

Function: Surfactant - Emulsifying Agent

Technical/Other Names:
Polyethylene Glycol 600 Ditallate
Polyoxyethylene (12) Ditallate

Trade Names:
Chemax PEG-600-DT (Chemax)
Lipopeg 6-DT (Lipo)
Pegosperse 600 DOT (Lonza Inc./Lonza Ltd.)

PEG-4 DITALLOW ETHER

Definition: PEG-4 Ditallow Ether is the polyethylene glycol ether of Tallow Alcohol (q.v.). It conforms generally to the formula:

$$RO(CH_2CH_2O)_nR$$

where R represents the alkyl groups derived from tallow and n has an average value of 4.

Chemical Class: Ethers

Function: Skin-Conditioning Agent - Occlusive

PEG-5 DITRIDECYLMONIUM CHLORIDE

Empirical Formula:
$C_{37}H_{78}NO_5 \cdot Cl$

Definition: PEG-5 Ditridecylmonium Chloride is the ethoxylated quaternary ammonium compound that conforms generally to the formula:

$$\left[CH_3(CH_2)_{12} - \overset{\overset{\displaystyle (CH_2CH_2O)_nH}{|}}{\underset{\underset{\displaystyle CH_3}{|}}{N}} - (CH_2)_{12}CH_3 \right]^{+} \quad Cl^{-}$$

where n has an average value of 5.

Chemical Class: Quaternary Ammonium Compounds

Function: Hair Conditioning Agent

Technical/Other Names:
Polyethylene Glycol (5) Ditridecylmonium Chloride
Polyoxyethylene (5) Ditridecylmonium Chloride

PEG-8 DI/TRIRICINOLEATE

Definition: PEG-8 Di/Triricinoleate is a polyethylene glycol ester of a mixture of dimer and trimer acids derived from ricinoleic acid.

Information Sources: 21CFR175.300, 21CFR176.210, 21CFR177.1210, 21CFR177.2800, MI-13(7660)

Chemical Class: Alkoxylated Carboxylic Acids

Functions: Skin-Conditioning Agent - Emollient; Surfactant - Emulsifying Agent

Technical/Other Names:
Polyethylene Glycol 400 Di-Tri-Ricinoleate
Polyoxyethylene (8) Di-Tri-Ricinoleate

PEG-22/DODECYL GLYCOL COPOLYMER

CAS No.: 78336-31-9

Definition: PEG-22/Dodecyl Glycol Copolymer is the polyoxyethylene, poly-dodecyl glycol block polymer that conforms generally to the formula:

$$H(OCHCH_2)_xO(CH_2CH_2O)_yCH_2CHO)_zH$$
$$\underset{C_{10}H_{21}}{|} \qquad \underset{C_{10}H_{21}}{|}$$

in which the average value of x, y, and z are 4.5, 22 and 4.5 respectively.

Chemical Classes: Alkoxylated Alcohols; Synthetic Polymers

Functions: Emulsion Stabilizer; Skin-Conditioning Agent - Emollient

Reported Product Category: Night Skin Care Preparations

Technical/Other Name:
Poly(Oxy-1,2-Ethanediyl), α-(12-Hydroxydodecyl)-θ-[(12-Hydroxydodecyl)Oxy]-

Trade Names:
AEC PEG-22/Dodecyl Glycol Copolymer (A & E Connock)
Elfacos ST-37 (Akzo Nobel)
Elfacos ST37 (Akzo Nobel Surface AB)

PEG-45/DODECYL GLYCOL COPOLYMER

CAS No.: 78336-31-9

Definition: PEG-45/Dodecyl Glycol Copolymer is the polyoxyethylene, poly-dodecyl glycol block polymer that conforms generally to the formula:

$$H(OCHCH_2)_xO(CH_2CH_2O)_yCH_2CHO)_zH$$
$$\underset{C_{10}H_{21}}{|} \qquad \underset{C_{10}H_{21}}{|}$$

in which the average value of x, y, and z are 11, 45 and 11 respectively.

Chemical Classes: Alkoxylated Alcohols; Synthetic Polymers

Functions: Emulsion Stabilizer; Skin-Conditioning Agent - Emollient

Reported Product Category: Night Skin Care Preparations

Trade Names:
AEC PEG-45/Dodecyl Glycol Copolymer (A & E Connock)
Elfacos ST-9 (Akzo Nobel)
Elfacos ST9 (Akzo Nobel Surface AB)

PEG-4 ETHYLHEXANOATE

JPN Translation:
オクタン酸 PEG - 4

Empirical Formula:
$C_{16}H_{32}O_5$

Definition: PEG-4 Ethylhexanoate is the polyethylene glycol ester of 2-ethylhexanoic acid that conforms generally to the formula:

$$CH_3(CH_2)_3\overset{\overset{\displaystyle O}{\|}}{\underset{\underset{\displaystyle CH_2CH_3}{|}}{CHC}} - (OCH_2CH_2)_nH$$

where n has an average value of 4.

Information Sources: 21CFR175.105, 21CFR175.300, 21CFR176.210, MI-13 (7660)

Chemical Class: Alkoxylated Carboxylic Acids

Function: Surfactant - Emulsifying Agent

Reported Product Categories: Skin Care Preparations, Misc.; Bath Oils, Tablets, and Salts; Bath Soaps and Detergents; Cleansing Products (Cold Creams, Cleansing Lotions, Liquids and Pads); Hair Conditioners

Technical/Other Names:
PEG-4 Octanoate
Polyethylene Glycol 200 Monooctanoate
Polyoxyethylene (4) Monooctanoate

Trade Name:
AEC PEG-4 Ethylhexanoate (A & E Connock)

PEG-5 ETHYLHEXANOATE

Empirical Formula:
$C_{18}H_{36}O_6$

Definition: PEG-5 Ethylhexanoate is poly-ethylene glycol ester of 2-ethyl hexanoic acid that conforms to the formula:

$$CH_3(CH_2)_3\overset{\overset{\displaystyle O}{\|}}{\underset{\underset{\displaystyle CH_2CH_3}{|}}{CHC}} - (OCH_2CH_2)_nH$$

where n has an average value of 5.

Information Source: MI-13(7660)

Chemical Class: Alkoxylated Carboxylic Acids

Function: Surfactant - Emulsifying Agent

Reported Product Categories: Bath Soaps and Detergents; Aftershave Lotions; Body and Hand Preparations (Excluding Shaving Preparations); Colognes and Toilet Waters; Deodorants (Underarm); Fragrance Preparations, Misc.; Night Skin Care Preparations; Paste Masks (Mud Packs); Perfumes

Technical/Other Names:
PEG-5 Octanoate
Polyethylene Glycol (5) Octanoate
Polyoxyethylene (5) Octanoate

Trade Name Mixtures:
Neo PCL Prime 2/789230 (Symrise)
Neo-PCL Watersoluble N 2/966211 (Symrise)

PEG-13 ETHYLHEXANOATE

Definition: PEG-13 Ethylhexanoate is the polyethylene glycol ester of 2-ethylhexanoic acid that conforms generally to the formula:

$$CH_3(CH_2)_3\overset{\overset{\displaystyle O}{\|}}{\underset{\underset{\displaystyle CH_2CH_3}{|}}{CHC}} - (OCH_2CH_2)_nH$$

where n has an average value of 13.

Information Sources: 21CFR175.105, 21CFR175.300, 21CFR177.2800, MI-13 (7660)

Chemical Class: Alkoxylated Carboxylic Acids

Function: Surfactant - Emulsifying Agent

Technical/Other Names:
PEG-13 Octanoate
Polyethylene Glycol (13) Monooctanoate
Polyoxyethylene (13) Monooctanoate

Trade Name:
PCL-Liquid Watersoluble 2/966213 (Symrise)

PEG-20 EVENING PRIMROSE GLYCERIDES

Definition: PEG-20 Evening Primrose Glycerides is a polyethylene glycol derivative of the mono and diglycerides from evening primrose oil with an average of 20 moles of ethylene oxide.

Chemical Classes: Alkoxylated Alcohols; Glyceryl Esters and Derivatives

Function: Skin-Conditioning Agent - Emollient

Technical/Other Names:
Polyethylene Glycol 1000 Evening Primrose Glycerides
Polyoxyethylene (20) Evening Primrose Glycerides

PEG-60 EVENING PRIMROSE GLYCERIDES

Definition: PEG-60 Evening Primrose Glycerides is a polyethylene glycol derivative of the mono and diglycerides from evening primrose oil with an average of 60 moles of ethylene oxide.

Chemical Classes: Alkoxylated Alcohols; Glyceryl Esters and Derivatives

Functions: Surfactant - Emulsifying Agent; Surfactant - Solubilizing Agent

Technical/Other Names:
Polyethylene Glycol 3000 Evening Primrose Glycerides
Polyoxyethylene (60) Evening Primrose Glycerides

Trade Name:
Crovol EP70 (Croda Chemicals)

PEG-6 GLYCERYL CAPRATE

JPN Translation:
PEG - 6 カプリン酸グリセリル

Definition: PEG-6 Glyceryl Caprate is the polyethylene glycol ether of Glyceryl Caprate (q.v.) that conforms generally to the formula:

$$CH_3(CH_2)_8C(=O)-OCH_2CHCH_2(OCH_2CH_2)_nOH$$
$$|$$
$$OH$$

where n has an average value of 6.

Information Source: JCLS

Chemical Classes: Alkoxylated Alcohols; Glyceryl Esters and Derivatives

Functions: Skin-Conditioning Agent - Emollient; Surfactant - Emulsifying Agent

Technical/Other Names:
Polyethylene Glycol (6) Glyceryl Caprate
Polyoxyethylene (6) Glyceryl Caprate

PEG-3 GLYCERYL COCOATE

Definition: PEG-3 Glyceryl Cocoate is the polyethylene glycol ether of Glyceryl Cocoate (q.v.) that conforms generally to the formula:

$$RC(=O)-OCH_2CHCH_2(OCH_2CH_2)_nOH$$
$$|$$
$$OH$$

where RCO- represents the fatty acids derived from coconut oil and n has an average value of 3.

Chemical Classes: Alkoxylated Alcohols; Glyceryl Esters and Derivatives

Functions: Skin-Conditioning Agent - Emollient; Surfactant - Emulsifying Agent

Technical/Other Names:
Polyethylene Glycol (3) Glyceryl Cocoate
Polyoxyethylene (3) Glyceryl Cocoate

Trade Name:
Chemonic LI-3 (Chemron)

PEG-7 GLYCERYL COCOATE

CAS Nos.: 66105-29-1; 68201-46-7 (Generic)

JPN Translation:
PEG - 7 グリセリルココエート

Definition: PEG-7 Glyceryl Cocoate is the polyethylene glycol ether of Glyceryl Cocoate (q.v.) that conforms generally to the formula:

$$RC(=O)-OCH_2CHCH_2(OCH_2CH_2)_nOH$$
$$|$$
$$OH$$

where RCO- represents the fatty acids derived from coconut oil and n has an average value of 7.

Information Sources: 21CFR175.300, CIR: [SQ] IJT-18(SUPPL. 1)1999, CTFA D, JCLS, JSQI

Chemical Classes: Alkoxylated Alcohols; Glyceryl Esters and Derivatives

Functions: Skin-Conditioning Agent - Emollient; Surfactant - Emulsifying Agent

Reported Product Categories: Hair Dyes and Colors (All Types Requiring Caution Statements and Patch Tests); Shampoos (Non-coloring); Bath Oils, Tablets, and Salts; Personal Cleanliness Products, Misc.; Cleansing Products (Cold Creams, Cleansing Lotions, Liquids and Pads); Skin Care Preparations, Misc.; Bubble Baths; Bath Soaps and Detergents; Hair Shampoos (Coloring); Bath Preparations, Misc.; Body and Hand Preparations (Excluding Shaving Preparations); Baby Shampoos; Bath Capsules; Face and Neck Preparations (Excluding Shaving Preparations); Tonics, Dressings, and Other Hair Grooming Aids; Aftershave Lotions; Hair Conditioners; Paste Masks (Mud Packs)

Technical/Other Names:
Polyethylene Glycol (7) Glyceryl Monococoate
Polyoxyethylene (7) Glyceryl Monococoate

Trade Names:
Acconon CO-7 (Abitec Corporation)
AEC PEG-7 Glyceryl Cocoate (A & E Connock)
Cetiol HE (Cognis Care Chemicals/NJ)
Cetiol HE (Cognis Care Chemicals/PA)
Cetiol HE (Cognis Deutschland)
Chemonic LI-7 (Chemron)
ESTOL 3606 (Uniqema Europe)
Ethox GC-7 (Ethox)
Glycerox HE (Croda Chemicals)
Glycerox HE (Croda, Inc.)
Lumulse POE (7) GML (Lambent)
Protachem GC-7 (Protameen)
Radiasurf 7318 (Oleon NV)
Saboderm HE (Sabo)
Sterol LG 492 (Cesalpinia)
Tegosoft GC (Degussa Care Specialties)
Unimul-HE (Universal Preserv-A-Chem)
Unitolate HE (Universal Preserv-A-Chem)

Trade Name Mixtures:
Antil 200 (Degussa Care Specialties)
Centaurium (Sederma)
Dermoprotectine (Sederma)
Plantaren LSC (Cognis Care Chemicals/NJ)
Plantaren LSC (Cognis Care Chemicals/PA)
Polysol GL (Polygon)
Protaderm HA (Protameen)

Rewoderm LI S 80 (Degussa Care Specialties)
Saboderm SHO (Sabo)
Sabosol HO (Sabo)
Sabosol SDP (Sabo)
Solvariane (LCW)

PEG-30 GLYCERYL COCOATE

CAS No.: 68201-46-7 (Generic)

JPN Translation:
PEG - 30 グリセリルココエート

Definition: PEG-30 Glyceryl Cocoate is the polyethylene glycol ether of Glyceryl Cocoate (q.v.) that conforms generally to the formula:

$$RC \overset{O}{\underset{\|}{}} OCH_2CHCH_2(OCH_2CH_2)_nOH$$
$$\underset{OH}{|}$$

where RCO- represents the fatty acids derived from coconut oil and n has an average value of 30.

Information Sources: 21CFR175.300, 21CFR177.2800, CIR: [SQ] IJT-18(SUPPL. 1)1999, JCLS, JSQI

Chemical Classes: Alkoxylated Alcohols; Glyceryl Esters and Derivatives

Functions: Surfactant - Cleansing Agent; Surfactant - Solubilizing Agent

Reported Product Categories: Shampoos (Non-coloring); Permanent Waves; Hair Straighteners

Technical/Other Names:
Polyethylene Glycol (30) Glyceryl Monococoate
Polyoxyethylene (30) Glyceryl Monococoate

Trade Names:
Acconon C-30 (Abitec Corporation)
AEC PEG-30 Glyceryl Cocoate (A & E Connock)
Chemonic LI-63 (Chemron)
Ethox GC-30 (Ethox)
Jeechem GC-30 (Jeen)
Protachem GC-30 (Protameen)
Rewoderm LI 63 (Degussa Care Specialties)

PEG-40 GLYCERYL COCOATE

CAS No.: 68201-46-7 (Generic)

JPN Translation:
PEG - 40 グリセリルココエート

Definition: PEG-40 Glyceryl Cocoate is the polyethylene glycol ether of Glyceryl Cocoate (q.v.) that conforms generally to the formula:

$$RC \overset{O}{\underset{\|}{}} OCH_2CHCH_2(OCH_2CH_2)_nOH$$
$$\underset{OH}{|}$$

where RCO- represents the fatty acids derived from coconut oil and n has an average value of 40.

Information Sources: 21CFR175.300, 21CFR176.210, 21CFR177.2800, CIR: [SQ] IJT-18(SUPPL. 1)1999, JCLS

Chemical Classes: Alkoxylated Alcohols; Glyceryl Esters and Derivatives

Functions: Surfactant - Cleansing Agent; Surfactant - Solubilizing Agent

Reported Product Category: Eye Makeup Removers

Technical/Other Names:
Polyethylene Glycol 2000 Glyceryl Cocoate
Polyoxyethylene (40) Glyceryl Cocoate

Trade Name Mixtures:
Base SP 100 (LCW)
Oronal LCG (SEPPIC)

PEG-78 GLYCERYL COCOATE

CAS No.: 68201-46-7 (Generic)

JPN Translation:
PEG - 78 グリセリルココエート

Definition: PEG-78 Glyceryl Cocoate is the polyethylene glycol ether of Glyceryl Cocoate (q.v.) that conforms generally to the formula:

$$RC \overset{O}{\underset{\|}{}} OCH_2CHCH_2(OCH_2CH_2)_nOH$$
$$\underset{OH}{|}$$

where RCO- represents the fatty acids derived from coconut oil and n has an average value of 78.

Information Sources: 21CFR175.300, CIR: [SQ] IJT-18(SUPPL. 1)1999, JCLS

Chemical Classes: Alkoxylated Alcohols; Glyceryl Esters and Derivatives

Functions: Surfactant - Cleansing Agent; Surfactant - Solubilizing Agent

Technical/Other Names:
Polyethylene Glycol (78) Glyceryl Cocoate
Polyoxyethylene (78) Glyceryl Cocoate

Trade Name:
Simulsol CG (SEPPIC)

PEG-80 GLYCERYL COCOATE

CAS No.: 68201-46-7 (Generic)

Definition: PEG-80 Glyceryl Cocoate is the polyethylene glycol ether of Glyceryl Cocoate (q.v.) that conforms generally to the formula:

$$RC \overset{O}{\underset{\|}{}} OCH_2CHCH_2(OCH_2CH_2)_nOH$$
$$\underset{OH}{|}$$

where R represents the coconut fatty radical and n has an average value of 80.

Information Sources: 21CFR175.300, CIR: [SQ] IJT-18(SUPPL. 1)1999

Chemical Classes: Alkoxylated Alcohols; Glyceryl Esters and Derivatives

Functions: Surfactant - Cleansing Agent; Surfactant - Solubilizing Agent

Reported Product Category: Shampoos (Non-coloring)

Technical/Other Names:
Polyethylene Glycol (80) Glyceryl Monococoate
Polyoxyethylene (80) Glyceryl Monococoate

Trade Names:
Acconon C-80 (Abitec Corporation)
Chemonic LI-6875 (Chemron)
Ethox GC-80 (Ethox)
Rewoderm LI 67-75 (Degussa Care Specialties)

PEG-12 GLYCERYL DIOLEATE

Definition: PEG-12 Glyceryl Dioleate is the polyethylene glycol ether of Glyceryl Dioleate (q.v.) that conforms generally to the formula:

$$CH_2O \overset{O}{\underset{\|}{}} C(CH_2)_7CH = CH(CH_2)_7CH_3$$
$$|$$
$$HC-O \overset{O}{\underset{\|}{}} C(CH_2)_7CH = CH(CH_2)_7CH_3$$
$$|$$
$$CH_2(OCH_2CH_2)_nOH$$

where n has an average value of 12.

Information Sources: 21CFR175.300, 21CFR176.210, 21CFR177.2800

Chemical Classes: Alkoxylated Alcohols; Glyceryl Esters and Derivatives

Functions: Skin-Conditioning Agent - Emollient; Surfactant - Emulsifying Agent

Technical/Other Names:
Polyethylene Glycol 600 Glyceryl Dioleate
Polyoxyethylene (12) Glyceryl Dioleate

Trade Name:
Inflacin (BioZone Laboratories)

PEG-4 GLYCERYL DISTEARATE

Definition: PEG-4 Glyceryl Distearate is the polyethylene glycol ether of Glyceryl Distearate (q.v.) that conforms generally to the formula:

$$CH_2O-\overset{\overset{O}{\|}}{C}(CH_2)_{16}CH_3$$
$$CHO-\overset{\overset{O}{\|}}{C}(CH_2)_{16}CH_3$$
$$CH_2(OCH_2CH_2)_nOH$$

where n has an average value or 4.

Chemical Classes: Alkoxylated Alcohols; Glyceryl Esters and Derivatives

Function: Skin-Conditioning Agent - Emollient

Technical/Other Names:
Polyethylene Glycol (4) Glyceryl Distearate
Polyoxyethylene (4) Glyceryl Distearate

Trade Name:
Emalex GWS-204 (Nihon Emulsion)

PEG-12 GLYCERYL DISTEARATE

Definition: PEG-12 Glyceryl Distearate is the polyethylene glycol ether of Glyceryl Distearate (q.v.) that conforms generally to the formula:

$$CH_3(CH_2)_{16}\overset{\overset{O}{\|}}{C}-OCH_2CHCH_2(OCH_2CH_2)_nOH$$
$$O-\overset{\overset{O}{\|}}{C}(CH_2)_{16}CH_3$$

where n has an average value of 12.

Chemical Classes: Alkoxylated Alcohols; Glyceryl Esters and Derivatives

Function: Skin-Conditioning Agent - Emollient

Technical/Other Names:
Polyethylene Glycol (12) Glyceryl Distearate
Polyoxyethylene (12) Glyceryl Distearate

Trade Names:
Alkoxylated Diester (BioZone Laboratories)
Evacin (BioZone Laboratories)
Qusome (BioZone Laboratories)

PEG-3 GLYCERYL ISOSTEARATE

Definition: PEG-3 Glyceryl Isostearate is the polyethylene glycol ether of Glyceryl Isostearate (q.v.) that conforms generally to the formula:

$$C_{17}H_{35}\overset{\overset{O}{\|}}{C}-OCH_2CHCH_2(OCH_2CH_2)_nOH$$
$$OH$$

where n has an average value of 3.

Chemical Classes: Alkoxylated Alcohols; Glyceryl Esters and Derivatives

Function: Surfactant - Emulsifying Agent

Technical/Other Names:
Polyethylene Glycol (3) Glyceryl Isostearate
Polyoxyethylene (3) Glyceryl Isostearate

Trade Name:
Emalex GWIS-103 (Nihon Emulsion)

PEG-5 GLYCERYL ISOSTEARATE

Definition: PEG-5 Glyceryl Isostearate is the polyethylene glycol ether of Glyceryl Isostearate (q.v.) that conforms generally to the formula:

$$C_{17}H_{35}\overset{\overset{O}{\|}}{C}-OCH_2CHCH_2(OCH_2CH_2)_nOH$$
$$OH$$

where n has an average value of 5.

Chemical Classes: Alkoxylated Alcohols; Glyceryl Esters and Derivatives

Function: Surfactant - Emulsifying Agent

Technical/Other Names:
Polyethylene Glycol (5) Glyceryl Isostearate
Polyoxyethylene (5) Glyceryl Isostearate

Trade Name:
Emalex GWIS-105 (Nihon Emulsion)

PEG-6 GLYCERYL ISOSTEARATE

Definition: PEG-6 Glyceryl Isostearate is the polyethylene glycol ether of Glyceryl Isostearate (q.v.) that conforms generally to the formula:

$$C_{17}H_{35}\overset{\overset{O}{\|}}{C}-OCH_2CHCH_2(OCH_2CH_2)_nOH$$
$$OH$$

where n has an average value of 6.

Chemical Classes: Alkoxylated Alcohols; Glyceryl Esters and Derivatives

Function: Surfactant - Emulsifying Agent

Technical/Other Names:
Polyethylene Glycol (6) Glyceryl Isostearate

Polyethylene Glycol (6) Glyceryl Stearate
Polyoxyethylene (6) Glyceryl Isostearate

Trade Name:
Emalex GWIS-106 (Nihon Emulsion)

PEG-8 GLYCERYL ISOSTEARATE

Definition: PEG-8 Glyceryl Isostearate is the polyethylene glycol ether of Glyceryl Isostearate (q.v.) that conforms generally to the formula:

$$C_{17}H_{35}\overset{\overset{O}{\|}}{C}-OCH_2CHCH_2(OCH_2CH_2)_nOH$$
$$OH$$

where n has an average value of 8.

Chemical Classes: Alkoxylated Alcohols; Glyceryl Esters and Derivatives

Function: Surfactant - Emulsifying Agent

Technical/Other Names:
Polyethylene Glycol (8) Glyceryl Monoisostearate
Polyoxyethylene (8) Glyceryl Monoisostearate

Trade Name:
Emalex GWIS-108 (Nihon Emulsion)

PEG-9 GLYCERYL ISOSTEARATE

Definition: PEG-9 Glyceryl Isostearate is the polyethylene glycol ether of Glyceryl Isostearate (q.v.) that conforms generally to the formula:

$$C_{17}H_{35}\overset{\overset{O}{\|}}{C}-OCH_2CHCH_2(OCH_2CH_2)_nOH$$
$$OH$$

where n has an average value of 9.

Chemical Classes: Alkoxylated Alcohols; Glyceryl Esters and Derivatives

Function: Surfactant - Emulsifying Agent

Technical/Other Names:
Polyethylene Glycol (9) Glyceryl Isostearate
Polyoxyethylene (9) Glyceryl Isostearate

Trade Name:
Emalex GWIS-109 (Nihon Emulsion)

PEG-10 GLYCERYL ISOSTEARATE

Definition: PEG-10 Glyceryl Isostearate is the polyethylene glycol ether of Glyceryl

Isostearate (q.v.) that conforms generally to the formula:

$$C_{17}H_{35}C(O)-OCH_2CHCH_2(OCH_2CH_2)_nOH$$

with OH on the central carbon

where n has an average value of 10.

Chemical Classes: Alkoxylated Alcohols; Glyceryl Esters and Derivatives

Function: Surfactant - Emulsifying Agent

Technical/Other Names:
Polyethylene Glycol 500 Glyceryl Isostearate
Polyoxyethylene (10) Glyceryl Isostearate

Trade Name:
Emalex GWIS-110 (Nihon Emulsion)

PEG-15 GLYCERYL ISOSTEARATE

CAS No.: 68958-58-7 (Generic)

JPN Translation:
イソステアリン酸 PEG - 15 グリセリル

Definition: PEG-15 Glyceryl Isostearate is the polyethylene glycol ether of Glyceryl Isostearate (q.v.) that conforms generally to the formula:

$$C_{17}H_{35}C(O)-OCH_2CHCH_2(OCH_2CH_2)_nOH$$

with OH on the central carbon

where n has an average value of 15.

Information Sources: JCLS, JSQI, TSCA

Chemical Classes: Alkoxylated Alcohols; Glyceryl Esters and Derivatives

Function: Surfactant - Emulsifying Agent

Technical/Other Names:
Polyethylene Glycol (15) Glyceryl Isostearate
Polyoxyethylene (15) Glyceryl Isostearate

Trade Name:
Oxypon 2145 (Zschimmer & Schwarz)

PEG-20 GLYCERYL ISOSTEARATE

CAS No.: 68958-58-7 (Generic)

JPN Translation:
イソステアリン酸 PEG - 20 グリセリル

Definition: PEG-20 Glyceryl Isostearate is the polyethylene glycol ether of Glyceryl Isostearate (q.v.) that conforms to the formula:

$$C_{17}H_{35}C(O)-OCH_2CHCH_2(OCH_2CH_2)_nOH$$

with OH on the central carbon

where n has an average value of 20.

Information Sources: JCLS, JSQI, TSCA

Chemical Classes: Alkoxylated Alcohols; Glyceryl Esters and Derivatives

Functions: Surfactant - Emulsifying Agent; Surfactant - Solubilizing Agent

Technical/Other Names:
Polyethylene Glycol 1000 Glyceryl Isostearate
Polyoxyethylene (20) Glyceryl Isostearate

Trade Name:
Sympatens-GMIS/200 (Kolb)

PEG-25 GLYCERYL ISOSTEARATE

Definition: PEG-25 Glyceryl Isostearate is the polyethylene glycol ether of Glyceryl Isostearate (q.v.) that conforms generally to the formula:

$$C_{17}H_{35}C(O)-OCH_2CHCH_2(OCH_2CH_2)_nOH$$

with OH on the central carbon

where n has an average value of 25.

Chemical Classes: Alkoxylated Alcohols; Glyceryl Esters and Derivatives

Function: Surfactant - Emulsifying Agent

Technical/Other Names:
Polyethylene Glycol 910 Glyceryl Isostearate
Polyoxyethylene (25) Glyceryl Isostearate

Trade Name:
Emalex GWIS-125 (Nihon Emulsion)

PEG-30 GLYCERYL ISOSTEARATE

CAS No.: 689-58-7 (Generic)

JPN Translation:
イソステアリン酸 PEG - 30 グリセリル

Definition: PEG-30 Glyceryl Isostearate is the polyethylene glycol ether of Glyceryl Isostearate (q.v.) that conforms generally to the formula:

$$C_{17}H_{35}C(O)-OCH_2CHCH_2(OCH_2CH_2)_nOH$$

with OH on the central carbon

where n has an average value of 30.

Information Sources: JCLS, JSQI, TSCA

Chemical Classes: Alkoxylated Alcohols; Glyceryl Esters and Derivatives

Functions: Surfactant - Cleansing Agent; Surfactant - Solubilizing Agent

Technical/Other Names:
Polyethylene Glycol (30) Glyceryl Isostearate
Polyoxyethylene (30) Glyceryl Isostearate

Trade Name:
Emalex Gwis-130 (Ikeda)

PEG-40 GLYCERYL ISOSTEARATE

Definition: PEG-40 Glyceryl Isostearate is the polyethylene glycol ether of Glyceryl Isostearate (q.v.) that conforms generally to the formula:

$$C_{17}H_{35}C(O)-OCH_2CHCH_2(OCH_2CH_2)_nOH$$

with OH on the central carbon

where n has an average value of 40.

Chemical Classes: Alkoxylated Alcohols; Glyceryl Esters and Derivatives

Functions: Surfactant - Cleansing Agent; Surfactant - Emulsifying Agent

Technical/Other Names:
Polyethylene Glycol (40) Glyceryl Monoiso-stearate
Polyoxyethylene (40) Glyceryl Monoiso-stearate

Trade Name:
Emalex GWIS-140 (Nihon Emulsion)

PEG-50 GLYCERYL ISOSTEARATE

Definition: PEG-50 Glyceryl Isostearate is the polyethylene glycol ether of Glyceryl Isostearate (q.v.) that conforms generally to the formula:

$$C_{17}H_{35}C(O)-OCH_2CHCH_2(OCH_2CH_2)_nOH$$

with OH on the central carbon

where n has an average value of 50.

Chemical Classes: Alkoxylated Alcohols; Glyceryl Esters and Derivatives

Functions: Surfactant - Cleansing Agent; Surfactant - Emulsifying Agent

Technical/Other Names:
Polyethylene Glycol (50) Glyceryl Monoiso-stearate
Polyoxyethylene (50) Glyceryl Monoiso-stearate

Trade Name:
Emalex GWIS-150 (Nihon Emulsion)

PEG-60 GLYCERYL ISOSTEARATE

CAS No.: 68958-58-7 (Generic)

JPN Translation:
イソステアリン酸 PEG - 60 グリセリル

Definition: PEG-60 Glyceryl Isostearate is the polyethylene glycol ether of Glyceryl Isostearate (q.v.) that conforms generally to the formula:

$$C_{17}H_{35}C-OCH_2CHCH_2(OCH_2CH_2)_nOH$$
$$|$$
$$OH$$

where n has an average value of 60.

Information Sources: JCLS, JSQI, TSCA

Chemical Classes: Alkoxylated Alcohols; Glyceryl Esters and Derivatives

Function: Surfactant - Cleansing Agent

Reported Product Categories: Bath Oils, Tablets, and Salts; Cleansing Products (Cold Creams, Cleansing Lotions, Liquids and Pads); Moisturizing Preparations; Bath Capsules; Bath Preparations, Misc.; Body and Hand Preparations (Excluding Shaving Preparations); Fragrance Preparations, Misc.

Technical/Other Names:
Polyethylene Glycol 3000 Glyceryl Isostearate
Polyoxyethylene (60) Glyceryl Isostearate

Trade Name:
Emalex GWIS-160N (Nihon Emulsion)

PEG-90 GLYCERYL ISOSTEARATE

Definition: PEG-90 Glyceryl Isostearate is the polyethylene glycol ether of Glyceryl Isostearate (q.v.) that conforms generally to the formula:

$$C_{17}H_{35}C-OCH_2CHCH_2(OCH_2CH_2)_nOH$$
$$|$$
$$OH$$

where n has an average value of 90.

Information Source: JCLS

Chemical Classes: Alkoxylated Alcohols; Glyceryl Esters and Derivatives

Function: Surfactant - Cleansing Agent

Technical/Other Names:
Polyethylene Glycol (90) Glyceryl Isostearate
Polyoxyethylene (90) Glyceryl Isostearate

Trade Name:
Emalex GWIS-190 (Nihon Emulsion)

Trade Name Mixture:
Oxetal VD 92 (Zschimmer & Schwarz)

PEG-8 GLYCERYL LAURATE

CAS No.: 57107-95-6 (Generic)

JPN Translation:
ラウリン酸 PEG - 8 グリセリル

Empirical Formula:
$C_{31}H_{62}O_{12}$

Definition: PEG-8 Glyceryl Laurate is the polyethylene glycol ether of Glyceryl Laurate that conforms generally to the formula:

$$CH_3(CH_2)_{10}C-OCH_2CHCH_2(OCH_2CH_2)_nOH$$
$$|$$
$$OH$$

where n has an average value of 8.

Information Sources: JCLS, TSCA

Chemical Classes: Alkoxylated Alcohols; Glyceryl Esters and Derivatives

Functions: Skin-Conditioning Agent - Emollient; Surfactant - Emulsifying Agent

Technical/Other Names:
Polyethylene Glycol 400 Glyceryl Laurate
Polyoxyethylene (8) Glyceryl Laurate

Trade Name:
Glycerox L8 (Croda Chemicals)

PEG-12 GLYCERYL LAURATE

CAS Nos.: 51248-32-9; 59070-56-3 (Generic)

JPN Translation:
ラウリン酸 PEG - 12 グリセリル

Definition: PEG-12 Glyceryl Laurate is the polyethylene glycol ether of Glyceryl Laurate that conforms generally to the formula:

$$CH_3(CH_2)_{10}C-OCH_2CHCH_2(OCH_2CH_2)_nOH$$
$$|$$
$$OH$$

where n has an average value of 12.

Information Sources: 21CFR175.300, 21CFR176.210, 21CFR177.2800, JCLS, JSQI, TSCA

Chemical Classes: Alkoxylated Alcohols; Glyceryl Esters and Derivatives

Function: Surfactant - Emulsifying Agent

Technical/Other Names:
Polyethylene Glycol 600 Glyceryl Laurate
Polyoxyethylene (12) Glyceryl Laurate

Trade Name:
AEC PEG-12 Glyceryl Laurate (A & E Connock)

Trade Name Mixtures:
Lipoderma PF-WS (Lipo)
Safester A-75 watersoluble (Induchem)
Unibiovit B-332 watersoluble (Induchem)
Unibiovit B-332 WS (Lipo)
Uniprosolv U-20 (Induchem)
Unitrienol T-272 watersoluble (Induchem)
Unitrienol T-272 WS (Lipo)

PEG-15 GLYCERYL LAURATE

CAS No.: 57107-95-6 (Generic)

JPN Translation:
ラウリン酸 PEG - 15 グリセリル

Definition: PEG-15 Glyceryl Laurate is the polyethylene glycol ether of Glyceryl Laurate (q.v.) that conforms generally to the formula:

$$CH_3(CH_2)_{10}C-OCH_2CHCH_2(OCH_2CH_2)_nOH$$
$$|$$
$$OH$$

where n has an average value of 15.

Information Sources: JCLS, TSCA

Chemical Classes: Alkoxylated Alcohols; Glyceryl Esters and Derivatives

Functions: Skin-Conditioning Agent - Emollient; Surfactant - Emulsifying Agent

Technical/Other Names:
Polyethylene Glycol (15) Glyceryl Laurate
Polyoxyethylene (15) Glyceryl Laurate

Trade Name:
Glycerox L15 (Croda Chemicals)

PEG-20 GLYCERYL LAURATE

CAS Nos.: 51248-32-9; 59070-56-3 (Generic)

JPN Translation:
ラウリン酸 PEG - 20 グリセリル

Definition: PEG-20 Glyceryl Laurate is the polyethylene glycol ether of Glyceryl Laurate (q.v.) that conforms generally to the formula:

$$CH_3(CH_2)_{10}C-OCH_2CHCH_2(OCH_2CH_2)_nOH$$
$$|$$
$$OH$$

where n has an average value of 20.

Information Sources: 21CFR175.300, 21CFR176.210, 21CFR177.2800, JCLS, JSQI, TSCA

Chemical Classes: Alkoxylated Alcohols; Glyceryl Esters and Derivatives

Functions: Surfactant - Emulsifying Agent; Surfactant - Solubilizing Agent

Technical/Other Names:
Polyethylene Glycol 1000 Glyceryl Monolaurate
Polyoxyethylene (20) Glyceryl Monolaurate

Trade Names:
AEC PEG-20 Glyceryl Laurate (A & E Connock)
Lamacit GML 20 (Cognis Deutschland)
Tagat L 2 (Degussa Care Specialties)

PEG-23 GLYCERYL LAURATE

CAS Nos.: 51248-32-9; 59070-56-3 (Generic)

JPN Translation:
ラウリン酸 PEG - 23 グリセリル

Definition: PEG-23 Glyceryl Laurate is the polyethylene glycol ether of Glyceryl Laurate (q.v.) that conforms generally to the formula:

$$CH_3(CH_2)_{10}C - OCH_2CHCH_2(OCH_2CH_2)_nOH$$

where n has an average value of 23.

Information Sources: 21CFR175.300, 21CFR177.2800, JCLS

Chemical Classes: Alkoxylated Alcohols; Glyceryl Esters and Derivatives

Functions: Surfactant - Cleansing Agent; Surfactant - Solubilizing Agent

Technical/Other Names:
Polyethylene Glycol (23) Glyceryl Laurate
Polyoxyethylene (23) Glyceryl Laurate

Trade Name:
Aldosperse ML-23 (Lonza Inc./Lonza Ltd.)

PEG-30 GLYCERYL LAURATE

CAS Nos.: 51248-32-9; 59070-56-3 (Generic)

JPN Translation:
ラウリン酸 PEG - 30 グリセリル

Definition: PEG-30 Glyceryl Laurate is the polyethylene glycol ether of Glyceryl Laurate (q.v.) that conforms generally to the formula:

$$CH_3(CH_2)_{10}C - OCH_2CHCH_2(OCH_2CH_2)_nOH$$

where n has an average value of 30.

Information Sources: 21CFR175.300, 21CFR177.2800, JCLS, JSQI, TSCA

Chemical Classes: Alkoxylated Alcohols; Glyceryl Esters and Derivatives

Functions: Surfactant - Cleansing Agent; Surfactant - Solubilizing Agent

Technical/Other Names:
Polyethylene Glycol (30) Glyceryl Mono-laurate
Polyoxyethylene (30) Glyceryl Monolaurate

PEG-5 GLYCERYL OLEATE

Definition: PEG-5 Glyceryl Oleate is the polyethylene glycol ether of Glyceryl Oleate (q.v.) that conforms generally to the formula:

$$CH(CH_2)_7C - OCH_2CHCH_2(OCH_2CH_2)_nOH$$

where n has an average value of 5.

Chemical Classes: Alkoxylated Alcohols; Glyceryl Esters and Derivatives

Function: Surfactant - Emulsifying Agent

Technical/Other Names:
Polyethylene Glycol (5) Glyceryl Oleate
Polyoxyethylene (5) Glyceryl Oleate

Trade Name:
Nikkol TMGO-5 (Nikko)

PEG-10 GLYCERYL OLEATE

CAS No.: 68889-49-6 (Generic)

JPN Translation:
オレイン酸 PEG - 10 グリセリル

Empirical Formula:
$C_{41}H_{80}O_{14}$

Definition: PEG-10 Glyceryl Oleate is the polyethylene glycol ether of Glyceryl Oleate (q.v.) that conforms generally to the formula:

$$CH = CH(CH_2)_7C - OCH_2CHCH_2(OCH_2CH_2)_nOH$$

where n has an average value of 10.

Information Sources: 21CFR175.300, 21CFR177.2800, JCLS

Chemical Classes: Alkoxylated Alcohols; Glyceryl Esters and Derivatives

Function: Surfactant - Emulsifying Agent

Technical/Other Names:
Polyethylene Glycol 500 Glyceryl Monooleate
Polyoxyethylene (10) Glyceryl Monooleate

Trade Name:
Nikkol TMGO-10 (Nikko)

PEG-15 GLYCERYL OLEATE

CAS No.: 68889-49-6 (Generic)

JPN Translation:
オレイン酸 PEG - 15 グリセリル

Definition: PEG-15 Glyceryl Oleate is the polyethylene glycol ether of Glyceryl Oleate (q.v.) that conforms generally to the formula:

$$CH = CH(CH_2)_7C - OCH_2CHCH_2(OCH_2CH_2)_nOH$$

where n has an average value of 15.

Information Sources: 21CFR175.300, 21CFR176.210, 21CFR177.2800, JCLS

Chemical Classes: Alkoxylated Alcohols; Glyceryl Esters and Derivatives

Function: Surfactant - Emulsifying Agent

Technical/Other Names:
Polyethylene Glycol (15) Glyceryl Monooleate
Polyoxyethylene (15) Glyceryl Monooleate

Trade Name:
Nikkol TMGO-15 (Nikko)

PEG-20 GLYCERYL OLEATE

CAS No.: 68889-49-6 (Generic)

JPN Translation:
オレイン酸 PEG - 20 グリセリル

Definition: PEG-20 Glyceryl Oleate is the polyethylene glycol ether of Glyceryl Oleate (q.v.) that conforms generally to the formula:

$$CH = CH(CH_2)_7C - OCH_2CHCH_2(OCH_2CH_2)_nOH$$

where n has an average value of 20.

Information Sources: 21CFR175.300, 21CFR176.210, 21CFR177.2800, JCLS, JSQI, TSCA

Chemical Classes: Alkoxylated Alcohols; Glyceryl Esters and Derivatives

Functions: Surfactant - Emulsifying Agent; Surfactant - Solubilizing Agent

Technical/Other Names:
Polyethylene Glycol 1000 Glyceryl Monooleate
Polyoxyethylene (20) Glyceryl Monooleate

Trade Names:
AEC PEG-20 Glyceryl Oleate (A & E Connock)
Tagat O2 V (Degussa Care Specialties)

PEG-25 GLYCERYL OLEATE

CAS No.: 68889-49-6 (Generic)

JPN Translation:
オレイン酸 PEG - 25 グリセリル

Definition: PEG-25 Glyceryl Oleate is the polyethylene glycol ether of Glyceryl Oleate (q.v.) that conforms generally to the formula:

$$CH_3(CH_2)_7\quad O$$
$$CH=CH(CH_2)_7C - OCH_2CHCH_2(OCH_2CH_2)_nOH$$
$$OH$$

where n has an average value of 25.

Information Sources: 21CFR175.300, 21CFR177.2800, JCLS, JSQI, TSCA

Chemical Classes: Alkoxylated Alcohols; Glyceryl Esters and Derivatives

Functions: Surfactant - Cleansing Agent; Surfactant - Solubilizing Agent

Technical/Other Names:
Polyethylene Glycol (25) Glyceryl Monooleate
Polyoxyethylene (25) Glyceryl Monooleate

Trade Name:
AEC PEG-25 Glyceryl Oleate (A & E Connock)

PEG-30 GLYCERYL OLEATE

CAS No.: 68889-49-6 (Generic)

JPN Translation:
オレイン酸 PEG - 30 グリセリル

Definition: PEG-30 Glyceryl Oleate is the polyethylene glycol ether of Glyceryl Oleate (q.v.) that conforms generally to the formula:

$$CH_3(CH_2)_7\quad O$$
$$CH=CH(CH_2)_7C - OCH_2CHCH_2(OCH_2CH_2)_nOH$$
$$OH$$

where n has an average value of 30.

Information Sources: 21CFR175.300, 21CFR177.2800, JCLS, JSQI, TSCA

Chemical Classes: Alkoxylated Alcohols; Glyceryl Esters and Derivatives

Functions: Surfactant - Cleansing Agent; Surfactant - Solubilizing Agent

Technical/Other Names:
Polyethylene Glycol (30) Glyceryl Monooleate
Polyoxyethylene (30) Glyceryl Monooleate

PEG-18 GLYCERYL OLEATE/COCOATE

Definition: PEG-18 Glyceryl Oleate/Cocoate is the ethoxylated glyceryl ester that conforms generally to the formula:

$$(CH_2)_7CH_3\quad O$$
$$CH=CH(CH_2)_7C - OCH_2CHCH_2(OCH_2CH_2)_nOH$$
$$O-CR$$
$$O$$

where RCO- represents the fatty acids derived from coconut oil and n has an average value of 18.

Chemical Classes: Alkoxylated Alcohols; Glyceryl Esters and Derivatives

Functions: Skin-Conditioning Agent - Emollient; Surfactant - Emulsifying Agent

Reported Product Category: Shampoos (Non-coloring)

Trade Names:
AEC PEG-18 Glyceryl Oleate/Cocoate (A & E Connock)
Antil 171 (Degussa Care Specialties)

Trade Name Mixture:
Lowenol 4667 (Lowenstein)

PEG-15 GLYCERYL RICINOLEATE

CAS No.: 51142-51-9 (Generic)

Definition: PEG-15 Glyceryl Ricinoleate is the polyethylene glycol ether of Glyceryl Ricinoleate (q.v.) that conforms generally to the formula:

$$OH$$
$$CH_2CH(CH_2)_5CH_3$$
$$CH\quad O$$
$$CH(CH_2)_7C - OCH_2CHCH_2(OCH_2CH_2)_nOH$$
$$OH$$

where n has an average value of 15.

Information Sources: 21CFR175.300, 21CFR176.210, 21CFR177.2800, TSCA

Chemical Classes: Alkoxylated Alcohols; Glyceryl Esters and Derivatives

Function: Surfactant - Emulsifying Agent

Technical/Other Names:
Polyethylene Glycol (15) Glyceryl Monoricinoleate
Polyoxyethylene (15) Glyceryl Monoricinoleate

Trade Name:
AEC PEG-15 Glyceryl Ricinoleate (A & E Connock)

PEG-20 GLYCERYL RICINOLEATE

CAS No.: 51142-51-9 (Generic)

Definition: PEG-20 Glyceryl Ricinoleate is the polyethylene glycol ether of Glyceryl Ricinoleate (q.v.) that conforms generally to the formula:

$$OH$$
$$CH_2CH(CH_2)_5CH_3$$
$$CH\quad O$$
$$CH(CH_2)_7C - OCH_2CHCH_2(OCH_2CH_2)_nOH$$
$$OH$$

where n has an average value of 20.

Information Sources: 21CFR175.300, 21CFR176.210, 21CFR177.2800, TSCA

Chemical Classes: Alkoxylated Alcohols; Glyceryl Esters and Derivatives

Functions: Surfactant - Emulsifying Agent; Surfactant - Solubilizing Agent

Technical/Other Names:
Polyethylene Glycol 1000 Glyceryl Monoricinoleate
Polyoxyethylene (20) Glyceryl Monoricinoleate

Trade Name:
AEC PEG-20 Glyceryl Ricinoleate (A & E Connock)

PEG-5 GLYCERYL SESQUIOLEATE

Definition: PEG-5 Glyceryl Sesquioleate is the polyethylene glycol ether of Glyceryl Sesquioleate (q.v.) with an average value of 5 moles of ethylene oxide.

Information Source: 21CFR175.300

Chemical Classes: Alkoxylated Alcohols; Glyceryl Esters and Derivatives

Function: Surfactant - Emulsifying Agent

Technical/Other Names:
Polyethylene Glycol (5) Glyceryl Sesquioleate
Polyoxyethylene (5) Glyceryl Sesquioleate

PEG-7 GLYCERYL SOYATE

Definition: PEG-7 Glyceryl Soyate is the polyethylene glycol ether of glyceryl soyate that conforms generally to the formula:

$$O$$
$$RC - OCH_2CHCH_2(OCH_2CH_2)_nOH$$
$$OH$$

where R represents the soy fatty radical and and n has an average value of 7.

Chemical Classes: Alkoxylated Alcohols; Glyceryl Esters and Derivatives

Functions: Skin-Conditioning Agent - Emollient; Surfactant - Emulsifying Agent

Technical/Other Names:
Polyethylene Glycol (7) Glyceryl Monosoyate
Polyoxyethylene (7) Glyceryl Monosoyate

Trade Name:
Chemonic SI-7 (Chemron)

PEG-5 GLYCERYL STEARATE

JPN Translation:
ステアリン酸 PEG - 5 グリセリル

Empirical Formula:
$C_{31}H_{62}O_9$

Definition: PEG-5 Glyceryl Stearate is the polyethylene glycol ether of Glyceryl Stearate (q.v.) that conforms generally to the formula:

$$CH_3(CH_2)_{16}C(=O) - OCH_2CHCH_2(OCH_2CH_2)_nOH$$
$$|$$
$$OH$$

where n has an average value of 5.

Information Sources: 21CFR175.300, JCLS

Chemical Classes: Alkoxylated Alcohols; Glyceryl Esters and Derivatives

Function: Surfactant - Emulsifying Agent

Reported Product Categories: Moisturizing Preparations; Makeup Bases; Bath Capsules; Bath Oils, Tablets, and Salts; Cleansing Products (Cold Creams, Cleansing Lotions, Liquids and Pads); Bath Preparations, Misc.; Body and Hand Preparations (Excluding Shaving Preparations); Fragrance Preparations, Misc.

Technical/Other Names:
Polyethylene Glycol (5) Glyceryl Monostearate
Polyoxyethylene (5) Glyceryl Monostearate

Trade Names:
AEC PEG-5 Glyceryl Stearate (A & E Connock)
Nikkol TMGS-5 (Nikko)
POEM-S-105 (Riken Vitamin Oil)
Sabowax FL 5 (Sabo)

PEG-10 GLYCERYL STEARATE

JPN Translation:
ステアリン酸 PEG - 10 グリセリル

Empirical Formula:
$C_{41}H_{82}O_{14}$

Definition: PEG-10 Glyceryl Stearate is the polyethylene glycol ether of Glyceryl Stearate (q.v.) that conforms generally to the formula:

$$CH_3(CH_2)_{16}C(=O) - OCH_2CHCH_2(OCH_2CH_2)_nOH$$
$$|$$
$$OH$$

where n has an average value of 10.

Information Sources: 21CFR175.300, 21CFR177.2800, JCLS

Chemical Classes: Alkoxylated Alcohols; Glyceryl Esters and Derivatives

Function: Surfactant - Emulsifying Agent

Technical/Other Names:
Polyethylene Glycol 500 Glyceryl Monostearate
Polyoxyethylene (10) Glyceryl Monostearate

Trade Name:
AEC PEG-10 Glyceryl Stearate (A & E Connock)

PEG-15 GLYCERYL STEARATE

Definition: PEG-15 Glyceryl Stearate is the polyethylene glycol ether of Glyceryl Stearate (q.v.) that conforms generally to the formula:

$$CH_3(CH_2)_{16}C(=O) - OCH_2CHCH_2(OCH_2CH_2)_nOH$$
$$|$$
$$OH$$

where n has an average value of 15.

Chemical Classes: Alkoxylated Alcohols; Glyceryl Esters and Derivatives

Function: Surfactant - Emulsifying Agent

Technical/Other Names:
Polyethylene Glycol (15) Glyceryl Stearate
Polyoxyethylene (15) Glyceryl Stearate

Trade Names:
Emalex GM-15 (Nihon Emulsion)
Nikkol TMGS-15 (Nikko)

PEG-20 GLYCERYL STEARATE

JPN Translation:
ステアリン酸 PEG - 20 グリセリル

Definition: PEG-20 Glyceryl Stearate is the polyethylene glycol ether of Glyceryl Stearate (q.v.) that conforms generally to the formula:

$$CH_3(CH_2)_{16}C(=O) - OCH_2CHCH_2(OCH_2CH_2)_nOH$$
$$|$$
$$OH$$

where n has an average value of 20.

Information Source: JCLS

Chemical Classes: Alkoxylated Alcohols; Glyceryl Esters and Derivatives

Functions: Skin-Conditioning Agent - Emollient; Surfactant - Emulsifying Agent

Technical/Other Names:
Polyethylene Glycol 1000 Glyceryl Stearate
Polyoxyethylene (20) Glyceryl Stearate

Trade Names:
AEC PEG-20 Glyceryl Stearate (A & E Connock)
Aldosperse MS-20 (Lonza Inc./Lonza Ltd.)
Capmul EMG (Abitec Corporation)
Cutina E 24 (Cognis Deutschland)
Jeechem GMS-20 (Jeen)
Sympatens-GMS/200 (Kolb)
Tagat S 2 (Degussa Care Specialties)
Unitina E-24 (Universal Preserv-A-Chem)
Unitolate 80MG (Universal Preserv-A-Chem)

Trade Name Mixtures:
Aldosperse 40/60 (Lonza Inc./Lonza Ltd.)
Emulgade CBN (Cognis Deutschland)

PEG-25 GLYCERYL STEARATE

JPN Translation:
ステアリン酸 PEG - 25 グリセリル

Definition: PEG-25 Glyceryl Stearate is the polyethylene glycol ether of Glyceryl Stearate (q.v.) that conforms generally to the formula:

$$CH_3(CH_2)_{16}C(=O) - OCH_2CHCH_2(OCH_2CH_2)_nOH$$
$$|$$
$$OH$$

where n has an average value of 25.

Information Sources: 21CFR175.300, 21CFR177.2800, JCLS

Chemical Classes: Alkoxylated Alcohols; Glyceryl Esters and Derivatives

Functions: Surfactant - Cleansing Agent; Surfactant - Solubilizing Agent

Technical/Other Names:
Polyethylene Glycol (25) Glyceryl Monostearate
Polyoxyethylene (25) Glyceryl Monostearate

PEG-30 GLYCERYL STEARATE

JPN Translation:
ステアリン酸 PEG - 30 グリセリル

Definition: PEG-30 Glyceryl Stearate is the polyethylene glycol ether of Glyceryl Stearate (q.v.) that conforms generally to the formula:

$$CH_3(CH_2)_{16}C - OCH_2CHCH_2(OCH_2CH_2)_nOH$$
$$| $$
$$OH$$

where n has an average value of 30.

Information Sources: 21CFR175.300, 21CFR177.2800, JCLS

Chemical Classes: Alkoxylated Alcohols; Glyceryl Esters and Derivatives

Functions: Surfactant - Cleansing Agent; Surfactant - Solubilizing Agent

Reported Product Categories: Bath Preparations, Misc.; Body and Hand Preparations (Excluding Shaving Preparations); Moisturizing Preparations

Technical/Other Names:
  Polyethylene Glycol (30) Glyceryl Mono-stearate
  Polyoxyethylene (30) Glyceryl Monostearate

Trade Name:
  Tagat S (Degussa Care Specialties)

Trade Name Mixture:
  Polymoist Marine (Cognis Deutschland)

PEG-40 GLYCERYL STEARATE

Definition: PEG-40 Glyceryl Stearate is the polyethylene glycol ether of Glyceryl Stearate (q.v.) that conforms generally to the formula:

$$CH_3(CH_2)_{16}C - OCH_2CHCH_2(OCH_2CH_2)_nOH$$
$$|$$
$$OH$$

where n has an average value of 40.

Chemical Classes: Alkoxylated Alcohols; Glyceryl Esters and Derivatives

Functions: Surfactant - Cleansing Agent; Surfactant - Solubilizing Agent

Technical/Other Names:
  Polyethylene Glycol (40) Glyceryl Stearate
  Polyoxyethylene (40) Glyceryl Stearate

Trade Name:
  Emalex GM-40 (Nihon Emulsion)

PEG-120 GLYCERYL STEARATE

JPN Translation:
  ステアリン酸 PEG - 120 グリセリル

Definition: PEG-120 Glyceryl Stearate is the polyethylene glycol ether of Glyceryl Stearate (q.v.) that conforms generally to the formula:

$$CH_3(CH_2)_{16}C - OCH_2CHCH_2(OCH_2CH_2)_nOH$$
$$|$$
$$OH$$

where n has an average value of 120.

Information Sources: 21CFR175.300, JCLS

Chemical Classes: Alkoxylated Alcohols; Glyceryl Esters and Derivatives

Functions: Surfactant - Cleansing Agent; Surfactant - Solubilizing Agent

Technical/Other Names:
  Polyethylene Glycol (120) Glyceryl Mono-stearate
  Polyoxyethylene (120) Glyceryl Mono-stearate

Trade Name:
  AEC PEG-120 Glyceryl Stearate (A & E Connock)

PEG-200 GLYCERYL STEARATE

JPN Translation:
  ステアリン酸 PEG - 200 グリセリル

Definition: PEG-200 Glyceryl Stearate is the polyethylene glycol ether of Glyceryl Stearate (q.v.) that conforms generally to the formula:

$$CH_3(CH_2)_{16}C - OCH_2CHCH_2(OCH_2CH_2)_nOH$$
$$|$$
$$OH$$

where n has an average value of 200.

Information Sources: 21CFR175.300, JCLS

Chemical Classes: Alkoxylated Alcohols; Glyceryl Esters and Derivatives

Functions: Surfactant - Cleansing Agent; Surfactant - Solubilizing Agent

Technical/Other Names:
  Polyethylene Glycol (200) Glyceryl Mono-stearate
  Polyoxyethylene (200) Glyceryl Mono-stearate

Trade Name:
  Simusol 220 TM (SEPPIC)

PEG-28 GLYCERYL TALLOWATE

Definition: PEG-28 Glyceryl Tallowate is the polyethylene glycol ether of Tallow Glyceride (q.v.) that conforms generally to the formula:

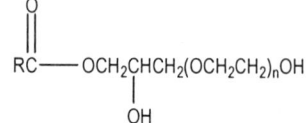

$$RC - OCH_2CHCH_2(OCH_2CH_2)_nOH$$
$$|$$
$$OH$$

where RCO- represents the fatty acids derived from tallow and n has an average value of 28.

Information Sources: 21CFR175.300, 21CFR177.2800

Chemical Classes: Alkoxylated Alcohols; Glyceryl Esters and Derivatives

Functions: Surfactant - Cleansing Agent; Surfactant - Solubilizing Agent

Technical/Other Names:
  Polyethylene Glycol (28) Glyceryl Mono-tallowate
  Polyoxyethylene (28) Glyceryl Monotallow-ate

Trade Name:
  AEC PEG-28 Glyceryl Tallowate (A & E Connock)

PEG-80 GLYCERYL TALLOWATE

Definition: PEG-80 Glyceryl Tallowate is the polyethylene glycol ether of Tallow Glyceride (q.v.) that conforms generally to the formula:

$$RC - OCH_2CHCH_2(OCH_2CH_2)_nOH$$
$$|$$
$$OH$$

where RCO- represents the fatty acids derived from tallow and n has an average value of 80.

Information Sources: 21CFR175.300, JCLS, JSQI

Chemical Classes: Alkoxylated Alcohols; Glyceryl Esters and Derivatives

Functions: Surfactant - Cleansing Agent; Surfactant - Solubilizing Agent

Technical/Other Names:
  Polyethylene Glycol (80) Glyceryl Mono-tallowate
  Polyoxyethylene (80) Glyceryl Monotallow-ate

Trade Name:
  Rewoderm LI 48 (Degussa Care Specialties)

PEG-82 GLYCERYL TALLOWATE

JPN Translation:
  PEG - 82 牛脂脂肪酸グリセリル

Definition: PEG-82 Glyceryl Tallowate is the polyethylene glycol ether of Tallow Glyceride (q.v.) that conforms generally to the formula:

$$RC\overset{O}{\overset{||}{-}}OCH_2CHCH_2(OCH_2CH_2)_nOH$$
$$|$$
$$OH$$

where RCO- represents the fatty acids derived from tallow and n has an average value of 82.

Information Source: JCLS

Chemical Classes: Alkoxylated Alcohols; Glyceryl Esters and Derivatives

Functions: Surfactant - Cleansing Agent; Surfactant - Solubilizing Agent

Technical/Other Names:
Polyethylene Glycol (82) Glyceryl Tallowate
Polyoxyethylene (82) Glyceryl Tallowate

PEG-130 GLYCERYL TALLOWATE

JPN Translation:
ベニバナ赤

Definition: PEG-130 Glyceryl Tallowate is the polyethylene glycol ether of Tallow Glycerides (q.v.) that conforms generally to the formula:

$$RC\overset{O}{\overset{||}{-}}OCH_2CHCH_2(OCH_2CH_2)_nOH$$
$$|$$
$$OH$$

where RCO- represents the fatty acids derived from tallow and n has an average value of 130.

Information Source: JCLS

Chemical Classes: Alkoxylated Alcohols; Glyceryl Esters and Derivatives

Functions: Surfactant - Cleansing Agent; Surfactant - Solubilizing Agent

Technical/Other Names:
Polyethylene Glycol (130) Glyceryl Tallowate
Polyoxyethylene (130) Glyceryl Tallowate

PEG-200 GLYCERYL TALLOWATE

Definition: PEG-200 Glyceryl Tallowate is the polyethylene glycol ether of Tallow Glyceride (q.v.) that conforms generally to the formula:

$$RC\overset{O}{\overset{||}{-}}OCH_2CHCH_2(OCH_2CH_2)_nOH$$
$$|$$
$$OH$$

where RCO- represents the fatty acids derived from tallow and n has an average value of 200.

Information Sources: 21CFR175.300, JCLS

Chemical Classes: Alkoxylated Alcohols; Glyceryl Esters and Derivatives

Functions: Surfactant - Cleansing Agent; Surfactant - Solubilizing Agent

Reported Product Category: Shampoos (Non-coloring)

Technical/Other Names:
Polyethylene Glycol (200) Glyceryl Mono-tallowate
Polyoxyethylene (200) Glyceryl Monotallowate

PEG-3 GLYCERYL TRIISOSTEARATE

Definition: PEG-3 Glyceryl Triisostearate is the triester of Isostearic Acid (q.v.) and a polyethylene glycol ether of glycerin. It conforms generally to the formula:

$$O(CH_2CH_2O)_x\overset{O}{\overset{||}{-}}CC_{17}H_{35}$$
$$|$$
$$CH_2CHCH_2O(CH_2CH_2O)_y\overset{O}{\overset{||}{-}}CC_{17}H_{35}$$
$$|$$
$$O(CH_2CH_2O)_z\overset{}{-}CC_{17}H_{35}$$
$$\overset{||}{O}$$

where x+y+z has an average value of 3.

Chemical Class: Glyceryl Esters and Derivatives

Functions: Skin-Conditioning Agent - Emollient; Surfactant - Emulsifying Agent

Technical/Other Names:
Polyethylene Glycol (3) Glyceryl Triisostearate
Polyoxyethylene (3) Glyceryl Triisostearate

Trade Name:
Emalex GWIS-303 (Nihon Emulsion)

PEG-5 GLYCERYL TRIISOSTEARATE

CAS No.: 86846-21-1

JPN Translation:
トリイソステアリン酸 PEG - 5 グリセリル

Empirical Formula:
$C_{67}H_{130}O_{11}$

Definition: PEG-5 Glyceryl Triisostearate is the triester of Isostearic Acid (q.v.) and a polyethylene glycol ether of glycerin. It conforms generally to the formula:

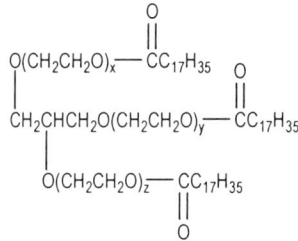

where x + y + z has an average value of 5.

Information Sources: JCIC, JCLS, JSQI

Chemical Class: Glyceryl Esters and Derivatives

Functions: Skin-Conditioning Agent - Emollient; Surfactant - Emulsifying Agent

Technical/Other Names:
Polyethylene Glycol (5) Glyceryl Triisostearate
Polyoxyethylene Glyceryl Triisostearate
Polyoxyethylene (5) Glyceryl Triisostearate

Trade Name:
Emalex GWIS-305 (Nihon Emulsion)

PEG-10 GLYCERYL TRIISOSTEARATE

JPN Translation:
トリイソステアリン酸 PEG - 10 グリセリル

Definition: PEG-10 Glyceryl Triisostearate is the triester of Isostearic Acid (q.v.) and a polyethylene glycol ether of Glycerin (q.v.). It conforms generally to the formula:

$$O(CH_2CH_2O)_x\overset{O}{\overset{||}{-}}CC_{17}H_{35}$$
$$|$$
$$CH_2CHCH_2O(CH_2CH_2O)_y\overset{O}{\overset{||}{-}}CC_{17}H_{35}$$
$$|$$
$$O(CH_2CH_2O)_z\overset{}{-}CC_{17}H_{35}$$
$$\overset{||}{O}$$

where x + y+ z has an average value of 10.

Chemical Class: Glyceryl Esters and Derivatives

Functions: Skin-Conditioning Agent - Emollient; Surfactant - Emulsifying Agent

Technical/Other Names:
Polyethylene Glycol (10) Glyceryl Triisostearate
Polyoxyethylene (10) Glyceryl Triisostearate

Trade Name:
Emalex GWIS-310 (Nihon Emulsion)

PEG-15 GLYCERYL TRIISOSTEARATE

Definition: PEG-15 Glyceryl Triisostearate is the triester of Isostearic Acid (q.v.) and a polyethylene glycol ether of glycerin. It conforms generally to the formula:

$$O(CH_2CH_2O)_x\text{---}\overset{\overset{\displaystyle O}{\|}}{C}C_{17}H_{35}$$

$$CH_2CHCH_2O(CH_2CH_2O)_y\text{---}\overset{\overset{\displaystyle O}{\|}}{C}C_{17}H_{35}$$

$$O(CH_2CH_2O)_{\overline{z}}\text{---}\underset{\underset{\displaystyle O}{\|}}{C}C_{17}H_{35}$$

where x+y+z has an average value of 15.

Chemical Class: Glyceryl Esters and Derivatives

Functions: Skin-Conditioning Agent - Emollient; Surfactant - Emulsifying Agent

Trade Name:
Emalex GWIS-315 (Nihon Emulsion)

PEG-20 GLYCERYL TRIISOSTEARATE

JPN Translation:
トリイソステアリン酸 PEG - 20 グリセリル

Definition: PEG-20 Glyceryl Triisostearate is the triester of Isostearic Acid (q.v.) and a polyethylene glycol ether of glycerin. It conforms generally to the formula:

$$O(CH_2CH_2O)_x\text{---}\overset{\overset{\displaystyle O}{\|}}{C}C_{17}H_{35}$$

$$CH_2CHCH_2O(CH_2CH_2O)_y\text{---}\overset{\overset{\displaystyle O}{\|}}{C}C_{17}H_{35}$$

$$O(CH_2CH_2O)_{\overline{z}}\text{---}\underset{\underset{\displaystyle O}{\|}}{C}C_{17}H_{35}$$

where x+y+z has an average value of 20.

Chemical Class: Glyceryl Esters and Derivatives

Functions: Skin-Conditioning Agent - Emollient; Surfactant - Emulsifying Agent

Technical/Other Names:
Polyethylene Glycol 1000 Glyceryl Triisostearate
Polyoxyethylene (20) Glyceryl Triisostearate

Trade Name:
Emalex GWIS-320 (Nihon Emulsion)

PEG-30 GLYCERYL TRIISOSTEARATE

JPN Translation:
トリイソステアリン酸 PEG - 30 グリセリル

Definition: PEG-30 Glyceryl Triisostearate is the triester of isostearic acid and a polyethylene glycol ether of glycerin. It conforms generally to the formula:

$$O(CH_2CH_2O)_x\text{---}\overset{\overset{\displaystyle O}{\|}}{C}C_{17}H_{35}$$

$$CH_2CHCH_2O(CH_2CH_2O)_y\text{---}\overset{\overset{\displaystyle O}{\|}}{C}C_{17}H_{35}$$

$$O(CH_2CH_2O)_{\overline{z}}\text{---}\underset{\underset{\displaystyle O}{\|}}{C}C_{17}H_{35}$$

where x + y + z has an average value of 30.

Chemical Class: Glyceryl Esters and Derivatives

Functions: Skin-Conditioning Agent - Emollient; Surfactant - Emulsifying Agent

Technical/Other Names:
Polyethylene Glycol (30) Glyceryl Triisostearate
Polyoxyethylene (30) Glyceryl Triisostearate

Trade Name:
Emalex GWIS-330 (Nihon Emulsion)

PEG-40 GLYCERYL TRIISOSTEARATE

JPN Translation:
トリイソステアリン酸 PEG - 40 グリセリル

Definition: PEG-40 Glyceryl Triisostearate is the triester of isostearic acid and a polyethylene glycol ether of glycerin. It conforms generally to the formula:

$$O(CH_2CH_2O)_x\text{---}\overset{\overset{\displaystyle O}{\|}}{C}C_{17}H_{35}$$

$$CH_2CHCH_2O(CH_2CH_2O)_y\text{---}\overset{\overset{\displaystyle O}{\|}}{C}C_{17}H_{35}$$

$$O(CH_2CH_2O)_{\overline{z}}\text{---}\underset{\underset{\displaystyle O}{\|}}{C}C_{17}H_{35}$$

where x + y+ z has an average value of 40.

Chemical Class: Glyceryl Esters and Derivatives

Functions: Skin-Conditioning Agent - Emollient; Surfactant - Emulsifying Agent

Trade Name:
Emalex GWIS-340 (Nihon Emulsion)

PEG-50 GLYCERYL TRIISOSTEARATE

JPN Translation:
トリイソステアリン酸 PEG - 50 グリセリル

Definition: PEG-50 Glyceryl Triisostearate is the triester of isostearic acid and a polyethylene glycol ether of glycerin. It conforms generally to the formula:

$$O(CH_2CH_2O)_x\text{---}\overset{\overset{\displaystyle O}{\|}}{C}C_{17}H_{35}$$

$$CH_2CHCH_2O(CH_2CH_2O)_y\text{---}\overset{\overset{\displaystyle O}{\|}}{C}C_{17}H_{35}$$

$$O(CH_2CH_2O)_{\overline{z}}\text{---}\underset{\underset{\displaystyle O}{\|}}{C}C_{17}H_{35}$$

where x + y + z has an average value of 50.

Chemical Class: Glyceryl Esters and Derivatives

Functions: Skin-Conditioning Agent - Emollient; Surfactant - Emulsifying Agent

Technical/Other Names:
Polyethylene Glycol (50) Glyceryl Triisostearate
Polyoxyethylene (50) Glyceryl Triisostearate

Trade Name:
Emalex GWIS-350 (Nihon Emulsion)

PEG-60 GLYCERYL TRIISOSTEARATE

JPN Translation:
トリイソステアリン酸 PEG - 60 グリセリル

Definition: PEG-60 Glyceryl Triisostearate is the triester of isostearic acid and a polyethylene glycol ether of glycerin. It conforms generally to the formula:

$$O(CH_2CH_2O)_x\text{---}\overset{\overset{\displaystyle O}{\|}}{C}C_{17}H_{35}$$

$$CH_2CHCH_2O(CH_2CH_2O)_y\text{---}\overset{\overset{\displaystyle O}{\|}}{C}C_{17}H_{35}$$

$$O(CH_2CH_2O)_{\overline{z}}\text{---}\underset{\underset{\displaystyle O}{\|}}{C}C_{17}H_{35}$$

where x + y + z has an average value of 60.

Chemical Class: Glyceryl Esters and Derivatives

Functions: Skin-Conditioning Agent - Emollient; Surfactant - Emulsifying Agent

Technical/Other Names:
Polyethylene Glycol (60) Glyceryl Triisostearate
Polyoxyethylene (60) Glyceryl Triisostearate

Trade Name:
Emalex GWIS-360 (Nihon Emulsion)

PEG-15 GLYCERYL TRIOLEATE

Definition: PEG-15 Glyceryl Trioleate is the triester of oleic acid and a polyethylene glycol ether of glycerin. It conforms generally to the formula:

where x+y+z has an average value of 15.

Chemical Class: Glyceryl Esters and Derivatives

Functions: Skin-Conditioning Agent - Emollient; Surfactant - Emulsifying Agent

Technical/Other Names:
Polyethylene Glycol (15) Glyceryl Trioleate
Polyoxyethylene (15) Glyceryl Trioleate

PEG-25 GLYCERYL TRIOLEATE

Definition: PEG-25 Glyceryl Trioleate is the triester of oleic acid and a polyethylene glycol ether of glycerin. It conforms generally to the formula:

where x + y + z has an average value of 25.

Information Sources: 21CFR175.300, 21CFR177.2800

Chemical Class: Glyceryl Esters and Derivatives

Functions: Skin-Conditioning Agent - Emollient; Surfactant - Emulsifying Agent

Technical/Other Names:
Polyethylene Glycol (25) Glyceryl Trioleate
Polyoxyethylene (25) Glyceryl Trioleate

Trade Name:
AEC PEG-25 Glyceryl Trioleate (A & E Connock)

PEG-3 GLYCERYL TRISTEARATE

Definition: PEG-3 Glyceryl Tristearate is the triester of stearic acid and a polyethylene

glycol ether of glycerin. It conforms generally to the formula:

where x + y + z has an average value of 3.

Chemical Class: Glyceryl Esters and Derivatives

Functions: Skin-Conditioning Agent - Emollient; Surfactant - Emulsifying Agent

Technical/Other Names:
Polyethylene Glycol (3) Glyceryl Tristearate
Polyoxyethylene (3) Glyceryl Tristearate

Trade Name:
Emalex GWS-303 (Nihon Emulsion)

PEG-4 GLYCERYL TRISTEARATE

Definition: PEG-4 Glyceryl Tristearate is the triester of stearic acid and a polyethylene glycol ether of glycerin. It conforms generally to the formula:

where x + y + z has an average value of 4.

Chemical Class: Glyceryl Esters and Derivatives

Functions: Skin-Conditioning Agent - Emollient; Surfactant - Emulsifying Agent

Technical/Other Names:
Polyethylene Glycol (4) Glyceryl Tristearate
Polyoxyethylene (4) Glyceryl Tristearate

Trade Name:
Emalex GWS-304 (Nihon Emulsion)

PEG-5 GLYCERYL TRISTEARATE

Definition: PEG-5 Glyceryl Tristearate is the triester of stearic acid and a polyethylene glycol ether of glycerin. It conforms generally to the formula:

where x+y+z has an average value of 5.

Chemical Class: Glyceryl Esters and Derivatives

Functions: Skin-Conditioning Agent - Emollient; Surfactant - Emulsifying Agent

Technical/Other Names:
Polyethylene Glycol (5) Glyceryl Tristearate
Polyoxyethylene (5) Glyceryl Tristearate

Trade Name:
Emalex GWS-305 (Nihon Emulsion)

PEG-6 GLYCERYL TRISTEARATE

Definition: PEG-6 Glyceryl Tristearate is the triester of stearic acid and a polyethylene glycol ether of glycerin. It conforms generally to the formula:

where x+y+z has an average value of 6.

Chemical Class: Glyceryl Esters and Derivatives

Functions: Skin-Conditioning Agent - Emollient; Surfactant - Emulsifying Agent

Technical/Other Names:
Polyethylene Glycol (6) Glyceryl Tristearate
Polyoxyethylene (6) Glyceryl Tristearate

Trade Name:
Emalex GWS-306 (Nihon Emulsion)

PEG-10 GLYCERYL TRISTEARATE

Definition: PEG-10 Glyceryl Tristearate is the triester of stearic acid and a polyethylene glycol ether of glycerin. It conforms generally to the formula:

$$CH_2O(CH_2CH_2O)_x \!-\! \overset{O}{\overset{\|}{C}}(CH_2)_{16}CH_3$$
$$CHO(CH_2CH_2O)_y \!-\! \overset{O}{\overset{\|}{C}}(CH_2)_{16}CH_3$$
$$CH_2O(CH_2CH_2O)_z \!-\! \overset{O}{\overset{\|}{C}}(CH_2)_{16}CH_3$$

where x+y+z has an average value of 10.

Chemical Class: Glyceryl Esters and Derivatives

Functions: Skin-Conditioning Agent - Emollient; Surfactant - Emulsifying Agent

Technical/Other Names:
Polyethylene Glycol (10) Glyceryl Tristearate
Polyoxyethylene (10) Glyceryl Tristearate

Trade Name:
Emalex GWS-310 (Nihon Emulsion)

PEG-15 GLYCERYL TRISTEARATE

Definition: PEG-15 Glyceryl Tristearate is the triester of stearic acid and a polyethylene glycol ether of glycerin. It conforms generally to the formula:

where x+y+z has an average value of 15.

Chemical Class: Glyceryl Esters and Derivatives

Functions: Surfactant - Emulsifying Agent; Viscosity Increasing Agent - Aqueous

Technical/Other Names:
Polyethylene Glycol (15) Glyceryl Tristearate
Polyoxyethylene (15) Glyceryl Tristearate

Trade Name:
Emalex GWS-315 (Nihon Emulsion)

PEG-20 GLYCERYL TRISTEARATE

Definition: PEG-20 Glyceryl Tristearate is the triester of stearic acid and a polyethylene glycol ether of glycerin. It conforms generally to the formula:

where x+y+z has an average value of 20.

Chemical Class: Glyceryl Esters and Derivatives

Functions: Skin-Conditioning Agent - Emollient; Surfactant - Emulsifying Agent

Technical/Other Names:
Polyethylene Glycol (20) Glyceryl Tristearate
Polyoxyethylene(20) Glyceryl Tristearate

Trade Name:
Emalex GWS-320 (Nihon Emulsion)

PEG-140 GLYCERYL TRISTEARATE

CAS No.: 41080-66-4

JPN Translation:
トリステアリン酸 PEG - 140 グリセリル

Definition: PEG-140 Glyceryl Tristearate is the triester of stearic acid an a polyethylene glycol ether of glycerin. It conforms generally to the formula:

where x+y+z has an average value of 140.

Information Sources: JCIC, JCLS

Chemical Class: Glyceryl Esters and Derivatives

Functions: Surfactant - Emulsifying Agent; Viscosity Increasing Agent - Aqueous

Technical/Other Names:
Polyethylene Glycol (140) Glyceryl Tristearate
Polyoxyethylene Glyceryl Tristearate
Polyoxyethylene (140) Glyceryl Tristearate

Trade Name:
Eumulgin EO-35 (Pulcra)

PEG-240/HDI COPOLYMER BIS-DECYLTETRADECETH-20 ETHER

Definition: PEG-240/HDI Copolymer Bis-Decyltetradeceth-20 Ether is a copolymer of PEG-240, decyltetradeceth-20 and hexamethylene diisocyanate monomers.

Chemical Class: Synthetic Polymers

Function: Viscosity Increasing Agent - Aqueous

Trade Name:
Adekanol GT-700 (Asahi Denka Kogyo)

PEG-20 HEXADECENYLSUCCINATE

CAS No.: 178254-04-1

Empirical Formula:
$C_{20}H_{36}O_4(C_2H_4O)_n$

Definition: PEG-20 Hexadecenylsuccinate is the reaction product of hexadecenylsuccinic anhydride and PEG-20 (q.v.). It conforms generally to the formula:

where n has an average value of 20.

Chemical Class: Alkoxylated Carboxylic Acids

Function: Surfactant - Emulsifying Agent

Technical/Other Name:
Poly(Oxy-1,2-Ethanediyl), α-Hydro-θ-Hydroxy-, Monoester with 2-Hexadecenylbutanedioic Acid

PEG-23 HEXADECYLEICOSANOATE

Definition: PEG-23 Hexadecyleicosanoate is the polyethylene glycol ester of hexadecyleicosanoic acid that conforms generally to the formula:

where n has an average value of 23.

Chemical Class: Alkoxylated Carboxylic Acids

Function: Surfactant - Emulsifying Agent

Technical/Other Names:
Polyethylene Glycol (23) Hexadecyleicosanoate
Polyoxyethylene (23) Hexadecyleicosanoate

Trade Name:
Silube G36-23 (Siltech LLC)

PEG-2 HYDROGENATED CASTOR OIL

CAS No.: 61788-85-0 (Generic)

JPN Translation:
PEG - 2 水添ヒマシ油

Definition: PEG-2 Hydrogenated Castor Oil is a polyethylene glycol derivative of Hydrogenated Castor Oil (q.v.) with an average of 2 moles of ethylene oxide.

Information Sources: JCLS, RIFM, TSCA

Chemical Classes: Alkoxylated Alcohols; Glyceryl Esters and Derivatives

Functions: Fragrance Ingredient; Skin-Conditioning Agent - Emollient; Surfactant - Emulsifying Agent

Technical/Other Names:
Castor oil, hydrogenated, ethoxylated (RIFM)
Polyethylene Glycol (100) Hydrogenated Castor Oil
Polyoxyethylene (2) Hydrogenated Castor Oil

Trade Names:
Atlas G-4831 (Uniqema Americas)
Sympatens-TRH/020 (Kolb)

PEG-5 HYDROGENATED CASTOR OIL

CAS No.: 61788-85-0 (Generic)

JPN Translation:
PEG - 5 水添ヒマシ油

Definition: PEG-5 Hydrogenated Castor Oil is a polyethylene glycol derivative of Hydrogenated Castor Oil (q.v.) with an average of 5 moles of ethylene oxide.

Information Sources: JCLS, RIFM, TSCA

Chemical Classes: Alkoxylated Alcohols; Glyceryl Esters and Derivatives

Functions: Fragrance Ingredient; Skin-Conditioning Agent - Emollient; Surfactant - Emulsifying Agent

Technical/Other Names:
Castor oil, hydrogenated, ethoxylated (RIFM)
Polyethylene Glycol (5) Hydrogenated Castor Oil
Polyoxyethylene (5) Hydrogenated Castor Oil

Trade Names:
Ethox HCO-5 (Ethox)
Jeechem CAH-5 (Jeen)
Nikkol HCO-5 (Nikko)

PEG-6 HYDROGENATED CASTOR OIL

CAS No.: 61788-85-0 (Generic)

JPN Translation:
PEG - 6 水添ヒマシ油

Definition: PEG-6 Hydrogenated Castor Oil is the polyethylene glycol derivative of Hydrogenated Castor Oil (q.v.) with an average of 6 moles of ethylene oxide.

Information Sources: JCLS, RIFM, TSCA

Chemical Classes: Alkoxylated Alcohols; Glyceryl Esters and Derivatives

Functions: Fragrance Ingredient; Skin-Conditioning Agent - Emollient; Surfactant - Emulsifying Agent

Technical/Other Names:
Castor oil, hydrogenated, ethoxylated (RIFM)
Polyethylene Glycol 300 Hydrogenated Castor Oil
Polyoxyethylene (6) Hydrogenated Castor Oil

Trade Name:
Sabowax ELH 6 (Sabo)

PEG-7 HYDROGENATED CASTOR OIL

CAS No.: 61788-85-0 (Generic)

JPN Translation:
PEG - 7 水添ヒマシ油

Definition: PEG-7 Hydrogenated Castor Oil is a polyethylene glycol derivative of Hydrogenated Castor Oil (q.v.) with an average of 7 moles of ethylene oxide.

Information Sources: JCLS, RIFM, TSCA

Chemical Classes: Alkoxylated Alcohols; Glyceryl Esters and Derivatives

Functions: Fragrance Ingredient; Skin-Conditioning Agent - Emollient; Surfactant - Emulsifying Agent

Reported Product Categories: Eyebrow Pencils; Makeup Preparations (Not eye), Misc.; Suntan Gels, Creams, and Liquids; Moisturizing Preparations; Night Skin Care Preparations

Technical/Other Names:
Castor oil, hydrogenated, ethoxylated (RIFM)
Polyethylene Glycol (7) Hydrogenated Castor Oil
Polyoxyethylene (7) Hydrogenated Castor Oil

Trade Names:
Arlacel 989 (Uniqema Americas)
Cremophor WO-7 (BASF)
Croduret 7 Special (Croda Chemicals)
Dehymuls HRE 7 (Cognis Care Chemicals/NJ)
Dehymuls HRE 7 (Cognis Care Chemicals/PA)
Dehymuls HRE 7 (Cognis Deutschland)
Simulsol 989 (SEPPIC)

Sympatens-TRH/070 (Kolb)
Tagat R 7 (Degussa Care Specialties)

Trade Name Mixture:
Eusolex T-Olio P (Merck KGaA)

PEG-8 HYDROGENATED CASTOR OIL

Definition: PEG-8 Hydrogenated Castor Oil is a polyethylene glycol derivative of Hydrogenated Castor Oil (q.v.) with an average of 8 moles of ethylene oxide.

Chemical Classes: Alkoxylated Alcohols; Glyceryl Esters and Derivatives

Functions: Skin-Conditioning Agent - Emollient; Surfactant - Emulsifying Agent

Technical/Other Names:
Polyethylene Glycol (8) Hydrogenated Castor Oil
Polyoxyethylene (8) Hydrogenated Castor Oil

Trade Name:
Emalex HC-7.5 (Nihon Emulsion)

PEG-10 HYDROGENATED CASTOR OIL

CAS No.: 61788-85-0 (Generic)

JPN Translation:
PEG - 10 水添ヒマシ油

Definition: PEG-10 Hydrogenated Castor Oil is a polyethylene glycol derivative of Hydrogenated Castor Oil (q.v.) with an average of 10 moles of ethylene oxide.

Information Sources: JCLS, RIFM

Chemical Classes: Alkoxylated Alcohols; Glyceryl Esters and Derivatives

Functions: Fragrance Ingredient; Skin-Conditioning Agent - Emollient; Surfactant - Emulsifying Agent

Technical/Other Names:
Castor oil, hydrogenated, ethoxylated (RIFM)
Polyoxyethylene (10) Hydrogenated Castor Oil

Trade Names:
Emalex HC-10 (Nihon Emulsion)
Nikkol HCO-10 (Nikko)
Pegnol HC-10 (Toho)

Trade Name Mixture:
Hydrocos "P" (Cosmetochem) (Cosmetochem International Ltd.)

PEG-16 HYDROGENATED CASTOR OIL

CAS No.: 61788-85-0 (Generic)

JPN Translation:
PEG - 16 水添ヒマシ油

Definition: PEG-16 Hydrogenated Castor Oil is a polyethylene glycol derivative of Hydrogenated Castor Oil (q.v.) with an average of 16 moles of ethylene oxide.

Information Sources: 21CFR177.2800, JCLS, RIFM, TSCA

Chemical Classes: Alkoxylated Alcohols; Glyceryl Esters and Derivatives

Functions: Fragrance Ingredient; Skin-Conditioning Agent - Emollient; Surfactant - Emulsifying Agent

Reported Product Category: Shampoos (Non-coloring)

Technical/Other Names:
Castor oil, hydrogenated, ethoxylated (RIFM)
Polyethylene Glycol (16) Hydrogenated Castor Oil
Polyoxyethylene (16) Hydrogenated Castor Oil

Trade Names:
Chemax HCO-16 (Chemax)
Ethox HCO-16 (Ethox)
Hetoxide HC-16 (Heterene)
Jeechem CAH-16 (Jeen)
Protachem HCO 16 (Protameen)

PEG-20 HYDROGENATED CASTOR OIL

CAS No.: 61788-85-0 (Generic)

JPN Translation:
PEG - 20 水添ヒマシ油

Definition: PEG-20 Hydrogenated Castor Oil is a polyethylene glycol derivative of Hydrogenated Castor Oil (q.v.) with an average value of 20 moles of ethylene oxide.

Information Sources: 21CFR177.2800, JCLS, RIFM, TSCA

Chemical Classes: Alkoxylated Alcohols; Glyceryl Esters and Derivatives

Functions: Fragrance Ingredient; Surfactant - Emulsifying Agent

Technical/Other Name:
Castor oil, hydrogenated, ethoxylated (RIFM)

Trade Names:
Nikkol HCO-20 (Nikko)
Pegnol HC-20 (Toho)

PEG-25 HYDROGENATED CASTOR OIL

CAS No.: 61788-85-0 (Generic)

JPN Translation:
PEG - 25 水添ヒマシ油

Definition: PEG-25 Hydrogenated Castor Oil is a polyethylene glycol derivative of Hydrogenated Castor Oil (q.v.) with an average of 25 moles of ethylene oxide.

Information Sources: 21CFR177.2800, JCLS, RIFM, TSCA

Chemical Classes: Alkoxylated Alcohols; Glyceryl Esters and Derivatives

Functions: Fragrance Ingredient; Surfactant - Emulsifying Agent

Reported Product Category: Tonics, Dressings, and Other Hair Grooming Aids

Technical/Other Names:
Castor oil, hydrogenated, ethoxylated (RIFM)
Polyethylene Glycol (25) Hydrogenated Castor Oil
Polyoxyethylene (25) Hydrogenated Castor Oil

Trade Names:
AEC PEG-25 Hydrogenated Castor Oil (A & E Connock)
Arlatone G (Uniqema Americas)
Chemax HCO-25 (Chemax)
Ethox HCO-25 (Ethox)
Jeechem CAH-25 (Jeen)
Lanaetex CO-25 (Lanaetex)
Lumulse GRH-25 (Lambent)
Merpoxen HRO 250 (Wall)
Pegnol HC-25 (Toho)
Protachem HCO 25 (Protameen)
Simulsol 1292 (Solubilizers) (SEPPIC)
Sympatens-TRH/250 (Kolb)
Unipeg-CO-25H (Universal Preserv-A-Chem)

PEG-30 HYDROGENATED CASTOR OIL

CAS No.: 61788-85-0 (Generic)

JPN Translation:
PEG - 30 水添ヒマシ油

Definition: PEG-30 Hydrogenated Castor Oil is a polyethylene glycol derivative of Hydrogenated Castor Oil (q.v.) with an average of 30 moles of ethylene oxide.

Information Sources: 21CFR177.2800, CIR: [SQ] IJT-16(3)1997, JCLS, RIFM, TSCA

Chemical Classes: Alkoxylated Alcohols; Glyceryl Esters and Derivatives

Functions: Fragrance Ingredient; Surfactant - Emulsifying Agent

Technical/Other Names:
Castor oil, hydrogenated, ethoxylated (RIFM)

Polyethylene Glycol (30) Hydrogenated Castor Oil
Polyoxyethylene (30) Hydrogenated Castor Oil

Trade Names:
Emalex HC-30 (Nihon Emulsion)
Jeechem CAH-30 (Jeen)
Nikkol HCO-30 (Nikko)

Trade Name Mixture:
Pronalen Silymarin HSC (Provital/Centerchem)

PEG-35 HYDROGENATED CASTOR OIL

CAS No.: 61788-85-0 (Generic)

JPN Translation:
PEG - 35 水添ヒマシ油

Definition: PEG-35 Hydrogenated Castor Oil is the polyethylene glycol derivative of Hydrogenated Castor Oil (q.v.) with an average of 35 moles of ethylene oxide.

Information Sources: 21CFR177.2800, JCLS, RIFM

Chemical Classes: Alkoxylated Alcohols; Glyceryl Esters and Derivatives

Functions: Fragrance Ingredient; Surfactant - Emulsifying Agent; Surfactant - Solubilizing Agent

Technical/Other Names:
Castor oil, hydrogenated, ethoxylated (RIFM)
Polyethylene Glycol (35) Hydrogenated Castor Oil
Polyoxyethylene (35) Hydrogenated Castor Oil

PEG-40 HYDROGENATED CASTOR OIL

CAS No.: 61788-85-0 (Generic)

JPN Translation:
PEG - 40 水添ヒマシ油

Definition: PEG-40 Hydrogenated Castor Oil is a polyethylene glycol derivative of Hydrogenated Castor Oil (q.v.) with an average of 40 moles of ethylene oxide.

Information Sources: 21CFR177.2800, CIR: [SQ] IJT-16(3)1997, CTFA S, NF XIX, RIFM, TSCA, USAN

Chemical Classes: Alkoxylated Alcohols; Glyceryl Esters and Derivatives

Functions: Fragrance Ingredient; Surfactant - Emulsifying Agent; Surfactant - Solubilizing Agent

Reported Product Categories: Tonics, Dressings, and Other Hair Grooming Aids;

Bath Oils, Tablets, and Salts; Baby Shampoos; Aftershave Lotions; Bath Preparations, Misc.; Body and Hand Preparations (Excluding Shaving Preparations); Cleansing Products (Cold Creams, Cleansing Lotions, Liquids and Pads); Skin Care Preparations, Misc.; Moisturizing Preparations; Hair Preparations (Non-coloring), Misc.; Colognes and Toilet Waters; Skin Fresheners; Fragrance Preparations, Misc.; Shampoos (Non-coloring); Bath Soaps and Detergents; Mascara; Bath Capsules; Face and Neck Preparations (Excluding Shaving Preparations); Hair Conditioners; Hair Wave Sets; Personal Cleanliness Products, Misc.; Paste Masks (Mud Packs); Shaving Preparations, Misc.; Deodorants (Underarm); Bubble Baths; Eye Makeup Preparations, Misc.; Eye Shadows; Eyeliners; Face Powders; Foundations; Hair Sprays (Aerosol Fixatives); Indoor Tanning Preparations

Technical/Other Names:
Castor oil, hydrogenated, ethoxylated (RIFM)
Polyethylene Glycol 2000 Hydrogenated Castor Oil
Polyoxyethylene (40) Hydrogenated Castor Oil

Trade Names:
Akyporox CO 400 (Kao GmbH)
Alkamuls CRH-40 (Rhodia)
Alpicare COH 40 (Cesalpinia)
Alpicare COH 410 (Cesalpinia)
Cremophor CO 40 (BASF)
Cremophor CO 410 (BASF)
Cremophor RH 40 (BASF)
Croduret 40 (Croda Chemicals)
Emulsogen HCO 40 (Clariant)
Emulsogen HCO 40 (Clariant GmbH, Personal Care)
Eumulgin HRE 40 (Cognis Deutschland)
Fancol HCO-40 (Fanning)
Jeechem CAH-40 (Jeen)
Lipocol HCO-40 (Lipo)
Lipocol LAV HCO-40 (Lipo)
Lumulse GRH-40 (Lambent)
Merpoxen HRO 400 (Wall)
Nikkol HCO-40 (Nikko)
Protachem HCO 40 (Protameen)
Sabowax ELH 40 (Sabo)
Sabowax ELH 40/L (Sabo)
Simulsol 1293 (Sol) (SEPPIC)
Sipotrig HCO-40 (Specialty Industrial)
Surfactol 365-H (CasChem)
Sympatens-TRH/400 (Kolb)
Tagat CH 40 (Degussa Care Specialties)
Unipeg-CO-40H (Universal Preserv-A-Chem)

Trade Name Mixtures:
AEC Menthyl Lactate (Water Soluble) (A & E Connock)
ASCORBIC ACID MICROEMULSION (Ennagram)
Biobranil Watersoluble 2/012600 (Symrise)

Bio-Sulphur Liquid (Crodarom)
BISABOLOL MICROEMULSION (Ennagram)
Chamazulene - HCE (Vincience)
Chamomile CL 2/033026 (Symrise)
Clary Sage CL Forte 2/033010 (Symrise)
CoAXEL Q (Sederma)
COSMETIC ESSENTIAL VITAMINS MICROEMULSION (Ennagram)
CoSolv (RTD Hall Star)
Covabsorb EW (LCW)
Covafresh (LCW)
Covafresh II (LCW)
Creasoluble No.1 (Creations Couleurs)
Cremogen Camomile MEW Special New (739027) (Haarmann & Reimer GmbH)
Cremophor CO 455 (BASF)
C8 Soie Hydro (Phytocos)
DHEA MICROEMULSION (Ennagram)
Ederline-H (Vincience)
Emulsogen HCP 049 (Clariant)
Emulsogen HCP 049 (Clariant GmbH, Personal Care)
Eumulgin HPS (Cognis Deutschland)
Eumulgin HRE 455 (Cognis Deutschland)
Extrapone Acacia 2/033650 (Symrise)
Extrapone Acacia Milk 2/032300 (Symrise)
Extrapone Acacia Milk 2/033805 (Symrise)
Extrapone Almond Milk 2/033132 (Symrise)
Extrapone Almond Milk 2/033895 (Symrise)
Extrapone Aloe Vera Milk 2/033890 (Symrise)
Extrapone Caramel Milk 2/033834 (Symrise)
Extrapone Cedarwood 2/033660 (Symrise)
Extrapone Cereals GW 2/031300 (Symrise)
Extrapone Chamomille Romaine GW 2/033585 (Symrise)
Extrapone Copaiba 2/B16991 (Symrise)
Extrapone Corn GW 2/031303 (Symrise)
Extrapone Fig-Mile 2/033875 (Symrise)
Extrapone Honey/Rice Blend GW 2/031306 (Symrise)
Extrapone Honey Rice Milk P 2/030830 (Symrise)
Extrapone Jojoba/Rose Blend GW 2/031326 (Symrise)
Extrapone Jojoba Special (2/032990) (Symrise)
Extrapone Lemongrass 2/0333086 (Symrise)
Extrapone Linseed GW 2/031305 (Symrise)
Extrapone Macadamia Nut 2/032160 (Symrise)
Extrapone Macadamia Nut 2/032160 (Symrise)
Extrapone Macadamia Nut Milk 2/033848 (Symrise)
Extrapone #4 Herbs 2/032495 (Symrise)
Extrapone Olive GW 2/033655 (Symrise)
Extrapone Orris 2/033460 (Symrise)
Extrapone Patchouli GW 2/033657 (Symrise)
Extrapone Rice GW 2/031228 (Symrise)

Extrapone Rosemary/Aloe Vera Blend 2/033197 (Symrise)
Extrapone Sandalwood 2/032161 (Symrise)
Extrapone Sesame GW 2/031304 (Symrise)
Extrapone Soy Milk 2/033858 (Symrise)
Extrapone Vanilla 2/032162 (Symrise)
Extrapone Wheat GWP 2/032680 (Symrise)
Ferulan Proactiv (GfN)
Fruitapone Blackthorn B 2/036700 (Symrise)
Fruitapone Cherry B 2/036440 (Symrise)
Fruitapone Cranberry B 2/036600 (Symrise)
Fruitapone Elder B 2/036220 (Symrise)
Fruitapone Gooseberry B 2/036710 (Symrise)
Fruitapone Grapefruit B 2/036150 (Symrise)
Fruitapone Grape Red B 2/036900 (Symrise)
Fruitapone Grape White B 2/036950 (Symrise)
Fruitapone Guarana B 2/036200 (Symrise)
Fruitapone Kiwi B 2/036250 (Symrise)
Fruitapone Lime B 2/036300 (Symrise)
Fruitapone Mandarine B 2/036340 (Symrise)
Fruitapone Mandarin Orange B 2/036340 (Symrise)
Fruitapone Mango B 2/036350 (Symrise)
Fruitapone Orange B 2/036400 (Symrise)
Fruitapone Papaya B 2/036450 (Symrise)
Fruitapone Passionfruit B 2/036500 (Symrise)
Fruitapone Peach B 2/036550 (Symrise)
Fruitapone Pineapple B 2/036000 (Symrise)
Fruitapone Quince B 2/036630 (Symrise)
Fruitapone Strawberry B 2/036120 (Symrise)
Germinol (Dr. Gerhard Steidl)
GSP-T (Mibelle AG)
Harmonic ASP (Dr. Gerhard Steidl)
Herbasol Extract Apricot (Cosmetochem) (Cosmetochem International Ltd.)
Herbasol Extract Myrrh (Cosmetochem) (Cosmetochem International Ltd.)
Herbasol Extract Orange Pulp (Cosmetochem) (Cosmetochem International Ltd.)
Hydrocos (Cosmetochem)
Ivy CL 2/032763 (Symrise)
Jasmine CL 2/033081 (Symrise)
Jonat AS (Dr. Gerhard Steidl)
Lime Blossom CL 2/033093 (Symrise)
Melaclear (Sederma)
Melissa Cl 2/033128 (Symrise)
Microemulsion A (Ennagram)
Microemulsion E (Ennagram)
Microfolia (LCW)
Peppermint CL 2/033173 (Symrise)
Peppermint Cl Forte 2/J33172 (Symrise)

Phenolines The Vert (Coletica SA)
Protaquat ASP (Protameen)
Rose CL 2/033395 (Symrise)
Rose CL Forte 2/J33295 (Symrise)
Rosemary CL 2/033253 (Symrise)
Saboperl 450 (Sabo)
Sabosolv EEM (Sabo)
Sage Cl 2/033294 (Symrise)
Solubilisant Gamma 2428 (Gattefosse s.a.)
Solubilisant LRI (LCW)
Solubilizer 2/014160 (Symrise)
Solubilizer HP 2/014180 (Symrise)
Spicypone Laurel 2/035500 (Symrise)
Spicypone Sweet Marjoram 2/035600 (Symrise)
Standamul Conc. 1002 (Cognis Care Chemicals/NJ)
Standamul Conc. 1002 (Cognis Care Chemicals/PA)
TOCOPHERYL ACETATE MICROEMULSION (Ennagram)
UBIQUINONE MICROEMULSION (Ennagram)
Unimul-1002 Conc. (Universal Preserv-A-Chem)
Vamasol IT (Ultra Chemical)
VITAMIN A MICROEMULSION (Ennagram)
Water-Soluble Tea Tree Oil (Southern Cross Botanicals)
Witchhazel CL (2/033900) (Symrise)
Witch Hazel CL 2/033900 (Symrise)

PEG-45 HYDROGENATED CASTOR OIL

CAS No.: 61788-85-0 (Generic)

JPN Translation:
　PEG - 45 水添ヒマシ油

Definition: PEG-45 Hydrogenated Castor Oil is a polyethylene glycol derivative of Hydrogenated Castor Oil (q.v.) with an average of 45 moles of ethylene oxide.

Information Sources: 21CFR177.2800, JCLS, RIFM

Chemical Classes: Alkoxylated Alcohols; Glyceryl Esters and Derivatives

Functions: Fragrance Ingredient; Surfactant - Cleansing Agent; Surfactant - Solubilizing Agent

Technical/Other Names:
　Castor oil, hydrogenated, ethoxylated (RIFM)
　Polyethylene Glycol (45) Hydrogenated Castor Oil
　Polyoxyethylene (45) Hydrogenated Castor Oil

PEG-50 HYDROGENATED CASTOR OIL

CAS No.: 61788-85-0 (Generic)

JPN Translation:
　PEG - 50 水添ヒマシ油

Definition: PEG-50 Hydrogenated Castor Oil is a polyethylene glycol derivative of Hydrogenated Castor Oil (q.v.) with an average of 50 moles of ethylene oxide.

Information Sources: 21CFR177.2800, JCLS, RIFM, TSCA

Chemical Classes: Alkoxylated Alcohols; Glyceryl Esters and Derivatives

Functions: Fragrance Ingredient; Surfactant - Cleansing Agent; Surfactant - Solubilizing Agent

Technical/Other Names:
　Castor oil, hydrogenated, ethoxylated (RIFM)
　Polyethylene Glycol (50) Hydrogenated Castor Oil
　Polyoxyethylene (50) Hydrogenated Castor Oil

Trade Names:
　Croduret 50 Special (Croda Chemicals)
　Nikkol HCO-50 (Nikko)
　Protachem HCO 50 (Protameen)

PEG-54 HYDROGENATED CASTOR OIL

CAS No.: 61788-85-0 (Generic)

JPN Translation:
　PEG - 54 水添ヒマシ油

Definition: PEG-54 Hydrogenated Castor Oil is a polyethylene glycol derivative of Hydrogenated Castor Oil (q.v.) with an average of 54 moles of ethylene oxide.

Information Sources: 21CFR177.2800, JCLS, RIFM

Chemical Classes: Alkoxylated Alcohols; Glyceryl Esters and Derivatives

Functions: Fragrance Ingredient; Surfactant - Cleansing Agent; Surfactant - Solubilizing Agent

Technical/Other Names:
　Castor oil, hydrogenated, ethoxylated (RIFM)
　Polyethylene Glycol (54) Hydrogenated Castor Oil
　Polyoxyethylene (54) Hydrogenated Castor Oil

Trade Name:
　Arlatone 289 (Uniqema Americas)

PEG-55 HYDROGENATED CASTOR OIL

CAS No.: 61788-85-0 (Generic)

JPN Translation:
　PEG - 55 水添ヒマシ油

Definition: PEG-55 Hydrogenated Castor Oil is a polyethylene glycol derivative of Hydrogenated Castor Oil (q.v.) with an average of 55 moles of ethylene oxide.

Information Sources: 21CFR177.2800, JCLS, RIFM

Chemical Classes: Alkoxylated Alcohols; Glyceryl Esters and Derivatives

Functions: Fragrance Ingredient; Surfactant - Cleansing Agent; Surfactant - Solubilizing Agent

Technical/Other Names:
　Castor oil, hydrogenated, ethoxylated (RIFM)
　Polyethylene Glycol (55) Hydrogenated Castor Oil
　Polyoxyethylene (55) Hydrogenated Castor Oil

PEG-60 HYDROGENATED CASTOR OIL

CAS No.: 61788-85-0 (Generic)

JPN Translation:
　PEG - 60 水添ヒマシ油

Definition: PEG-60 Hydrogenated Castor Oil is a polyethylene glycol derivative of Hydrogenated Castor Oil (q.v.) with an average of 60 moles of ethylene oxide.

Information Sources: 21CFR177.2800, JCLS, RIFM, TSCA

Chemical Classes: Alkoxylated Alcohols; Glyceryl Esters and Derivatives

Functions: Fragrance Ingredient; Surfactant - Cleansing Agent; Surfactant - Solubilizing Agent

Reported Product Categories: Bath Oils, Tablets, and Salts; Cleansing Products (Cold Creams, Cleansing Lotions, Liquids and Pads); Foundations; Moisturizing Preparations; Skin Care Preparations, Misc.; Skin Fresheners; Tonics, Dressings, and Other Hair Grooming Aids; Aftershave Lotions; Baby Shampoos; Bath Preparations, Misc.; Deodorants (Underarm); Hair Conditioners; Personal Cleanliness Products, Misc.; Shampoos (Non-coloring)

Technical/Other Names:
　Castor oil, hydrogenated, ethoxylated (RIFM)
　Polyethylene Glycol 3000 Hydrogenated Castor Oil
　Polyoxyethylene (60) Hydrogenated Castor Oil

Trade Names:
　Cremophor CO-60 (BASF)
　Croduret 60 (Croda Chemicals)
　Customulse HCO-60 (PEG-60 hydrogenated castor oil) (Custom Ingredients)

Emalex RWIS-160 (Nihon Emulsion)
Emulsogen HCO 60 (Clariant)
Emulsogen HCO 60 (Clariant GmbH, Personal Care)
Eumulgin HRE 60 (Cognis Deutschland)
Hetoxide HC-60 (Heterene)
Jeechem CAH-60 (Jeen)
Lipocol HCO-60 (Lipo)
Nikkol HCO-60 (Nikko)
Protachem HCO 60 (Protameen)
Sabowax ELH 60 (Sabo)
Simulsol 1294 (SEPPIC)
Sipotrig HCO-60 (Specialty Industrial)
Tagat CH 60 (Degussa Care Specialties)

Trade Name Mixtures:
Antiac - Ig (CNTEC Inc.)
Ceramide 2 Sol 2% (Sederma)
Complex 1 - Antipollution (Provital/Centerchem)
Complex 5 - Vitaminic (Provital/Centerchem)
Extrapone Coco-Nut Special 2/033055 (Symrise)
Neo Extrapone Chamomile Liquid 2/070350 (Symrise)
Neo Extrapone Chamomille 2/060350 (Symrise)
NIKKOL Aquasome AE (Nikko)
NIKKOL Aquasome BH (Nikko)
NIKKOL Aquasome VA (Nikko)
Nikkol Aquasome VE (Nikko)
Pronalen Capsicum HSC (Provital/Centerchem)

PEG-65 HYDROGENATED CASTOR OIL

Definition: PEG-65 Hydrogenated Castor Oil is the polyethylene glycol derivative of Hydrogenated Castor Oil (q.v.) with average of 65 moles of ethylene oxide.

Chemical Classes: Alkoxylated Alcohols; Glyceryl Esters and Derivatives

Functions: Skin-Conditioning Agent - Emollient; Surfactant - Emulsifying Agent

Technical/Other Names:
Polyethylene Glycol (65) Hydrogenated Castor Oil
Polyoxyethylene (65) Hydrogenated Castor Oil

Trade Name:
Emalex HC-65 (Nihon Emulsion)

PEG-80 HYDROGENATED CASTOR OIL

CAS No.: 61788-85-0 (Generic)

JPN Translation:
PEG - 80 水添ヒマシ油

Definition: PEG-80 Hydrogenated Castor Oil is a polyethylene glycol derivative of Hydrogenated Castor Oil (q.v.) with an average of 80 moles of ethylene oxide.

Information Sources: JCLS, RIFM, TSCA

Chemical Classes: Alkoxylated Alcohols; Glyceryl Esters and Derivatives

Functions: Fragrance Ingredient; Surfactant - Cleansing Agent; Surfactant - Solubilizing Agent

Technical/Other Names:
Castor oil, hydrogenated, ethoxylated (RIFM)
Polyethylene Glycol (80) Hydrogenated Castor Oil
Polyoxyethylene (80) Hydrogenated Castor Oil

Trade Names:
Nikkol HCO-80 (Nikko)
Protachem CO 80 (Protameen)

PEG-100 HYDROGENATED CASTOR OIL

CAS No.: 61788-85-0 (Generic)

JPN Translation:
PEG - 100 水添ヒマシ油

Definition: PEG-100 Hydrogenated Castor Oil is a polyethylene glycol derivative of Hydrogenated Castor Oil (q.v.) with an average of 100 moles of ethylene oxide.

Information Sources: JCLS, RIFM, TSCA

Chemical Classes: Alkoxylated Alcohols; Glyceryl Esters and Derivatives

Functions: Fragrance Ingredient; Surfactant - Cleansing Agent; Surfactant - Solubilizing Agent

Technical/Other Names:
Castor oil, hydrogenated, ethoxylated (RIFM)
Jeechem CAH-100
Polyethylene Glycol (100) Hydrogenated Castor Oil
Polyoxyethylene (100) Hydrogenated Castor Oil

Trade Names:
Nikkol HCO-100 (Nikko)
Protachem HCO 100 (Protameen)

Trade Name Mixtures:
Nikkol Aquasome EC-5 (Nikko)
NIKKOL Aquasome EC-30 (Nikko)

PEG-200 HYDROGENATED CASTOR OIL

CAS No.: 61788-85-0 (Generic)

JPN Translation:
PEG - 200 水添ヒマシ油

Definition: PEG-200 Hydrogenated Castor Oil is a polyethylene glycol derivative of Hydrogenated Castor Oil (q.v.) with an average of 200 moles of ethylene oxide.

Information Sources: JCLS, RIFM, TSCA

Chemical Classes: Alkoxylated Alcohols; Glyceryl Esters and Derivatives

Functions: Fragrance Ingredient; Surfactant - Cleansing Agent; Surfactant - Solubilizing Agent

Technical/Other Names:
Castor oil, hydrogenated, ethoxylated (RIFM)
Polyethylene Glycol (200) Hydrogenated Castor Oil
Polyoxyethylene (200) Hydrogenated Castor Oil

Trade Names:
Chemax HCO-200/50 (Chemax)
Ethox HCO-200 (Ethox)
Jeechem CAH-200 (Jeen)
Protachem HCO 200 (Protameen)

PEG-5 HYDROGENATED CASTOR OIL ISOSTEARATE

Definition: PEG-5 Hydrogenated Castor Oil Isostearate is a polyethylene glycol derivative of the isostearic acid ester of Hydrogenated Castor Oil (q.v.) with an average ethoxylation value of 5.

Chemical Classes: Alkoxylated Alcohols; Glyceryl Esters and Derivatives

Functions: Surfactant - Emulsifying Agent; Viscosity Increasing Agent - Nonaqueous

Technical/Other Names:
Polyethylene Glycol (5) Hydrogenated Castor Oil Isostearate
Polyoxyethylene (5) Hydrogenated Castor Oil Isostearate

Trade Name:
Emalex RWIS - 105 (Nihon Emulsion)

PEG-10 HYDROGENATED CASTOR OIL ISOSTEARATE

Definition: PEG-10 Hydrogenated Castor Oil Isostearate is a polyethylene glycol derivative of the isostearic acid ester of Hydrogenated Castor Oil (q.v.) with an average ethoxylation value of 10.

Chemical Classes: Alkoxylated Alcohols; Glyceryl Esters and Derivatives

Functions: Surfactant - Emulsifying Agent; Viscosity Increasing Agent - Nonaqueous

Technical/Other Names:
Polyethylene Glycol (10) Hydrogenated Castor Oil Isostearate
Polyoxyethylene (10) Hydrogenated Castor Oil Isostearate

Trade Name:
Emalex RWIS-110 (Nihon Emulsion)

PEG-15 HYDROGENATED CASTOR OIL ISOSTEARATE

Definition: PEG-15 Hydrogenated Castor Oil Isostearate is a polyethylene glycol derivative of the isostearic acid ester of Hydrogenated Castor Oil (q.v.) with an average of 15 moles of ethylene oxide.

Chemical Classes: Alkoxylated Alcohols; Glyceryl Esters and Derivatives

Function: Surfactant - Emulsifying Agent

Technical/Other Names:
Polyethylene Glycol (15) Hydrogenated Castor Oil Isostearate
Polyoxyethylene (15) Hydrogenated Castor Oil Isostearate

Trade Name:
Emalex RWIS-115 (Nihon Emulsion)

PEG-20 HYDROGENATED CASTOR OIL ISOSTEARATE

JPN Translation:
イソステアリン酸 PEG - 20 水添ヒマシ油

Definition: PEG-20 Hydrogenated Castor Oil Isostearate is a polyethylene glycol derivative of the isostearic acid ester of Hydrogenated Castor Oil (q.v.) with an average ethoxylation value of 20.

Information Sources: JCIC, JCLS

Chemical Classes: Alkoxylated Alcohols; Esters; Glyceryl Esters and Derivatives

Functions: Surfactant - Emulsifying Agent; Viscosity Increasing Agent - Nonaqueous

Technical/Other Names:
Polyethylene Glycol (20) Hydrogenated Castor Oil Isostearate
Polyoxyethylene (20) Hydrogenated Castor Oil Isostearate
Polyoxyethylene Hydrogenated Castor Oil Monoisostearate (20E.O.)

PEG-30 HYDROGENATED CASTOR OIL ISOSTEARATE

JPN Translation:
トリイソステアリン酸 PEG - 30 水添ヒマシ油

Definition: PEG-30 Hydrogenated Castor Oil Isostearate is a polyethylene glycol derivative of the isostearic acid ester of Hydrogenated Castor Oil (q.v.) with an average ethoxylation value of 30.

Information Source: JCLS

Chemical Classes: Alkoxylated Alcohols; Glyceryl Esters and Derivatives

Functions: Surfactant - Emulsifying Agent; Viscosity Increasing Agent - Nonaqueous

Technical/Other Names:
Polyethylene Glycol (30) Hydrogenated Castor Oil Isostearate
Polyoxyethylene (30) Hydrogenated Castor Oil Isostearate

Trade Name:
Emalex RWIS - 130 (Nihon Emulsion)

PEG-40 HYDROGENATED CASTOR OIL ISOSTEARATE

Definition: PEG-40 Hydrogenated Castor Oil Isostearate is a polyethylene glycol derivative of the isostearic acid ester of Hydrogenated Castor Oil (q.v.) with an average ethoxylation value of 40.

Chemical Classes: Alkoxylated Alcohols; Glyceryl Esters and Derivatives

Functions: Surfactant - Emulsifying Agent; Viscosity Increasing Agent - Nonaqueous

Technical/Other Names:
Polyethylene Glycol (40) Hydrogenated Castor Oil Isostearate
Polyoxyethylene (40) Hydrogenated Castor Oil Isostearate

Trade Name:
Emalex RWIS - 140 (Nihon Emulsion)

PEG-50 HYDROGENATED CASTOR OIL ISOSTEARATE

JPN Translation:
イソステアリン酸 PEG - 50 水添ヒマシ油

Definition: PEG-50 Hydrogenated Castor Oil Isostearate is a polyethylene glycol derivative of the isostearic acid ester of Hydrogenated Castor Oil (q.v.) with an average ethoxylation value of 50.

Information Sources: JCIC, JCLS

Chemical Classes: Alkoxylated Alcohols; Esters; Glyceryl Esters and Derivatives

Functions: Surfactant - Emulsifying Agent; Viscosity Increasing Agent - Nonaqueous

Technical/Other Names:
Polyethylene Glycol (50) Hydrogenated Castor Oil Isostearate
Polyoxyethylene (50) Hydrogenated Castor Oil Isostearate
Polyoxyethylene Hydrogenated Castor Oil Monoisostearate (50E.O.)

Trade Name:
Emalex RWIS-150 (Nihon Emulsion)

PEG-58 HYDROGENATED CASTOR OIL ISOSTEARATE

JPN Translation:
イソステアリン酸 PEG - 58 水添ヒマシ油

Definition: PEG-58 Hydrogenated Castor Oil Isostearate is a polyethylene glycol derivative of the isostearic acid ester of Hydrogenated Castor Oil (q.v.) with an average ethoxylation value of 58.

Information Source: JCLS

Chemical Classes: Alkoxylated Alcohols; Glyceryl Esters and Derivatives

Functions: Surfactant - Emulsifying Agent; Viscosity Increasing Agent - Nonaqueous

Technical/Other Names:
Polyethylene Glycol (58) Hydrogenated Castor Oil Isostearate
Polyoxyethylene (58) Hydrogenated Castor Oil Isostearate

Trade Name:
Emalex RWIS - 158 (Nihon Emulsion)

PEG-20 HYDROGENATED CASTOR OIL PCA ISOSTEARATE

Definition: PEG-20 Hydrogenated Castor Oil PCA Isostearate is the diester of PEG-20 Hydrogenated Castor Oil (q.v.) and a mixture of PCA (q.v.) and Isostearic Acid (q.v.).

Chemical Classes: Alkoxylated Alcohols; Glyceryl Esters and Derivatives; Heterocyclic Compounds

Function: Surfactant - Emulsifying Agent

Trade Name:
Pyroter CPI-20 (Nihon Emulsion)

PEG-30 HYDROGENATED CASTOR OIL PCA ISOSTEARATE

Definition: PEG-30 Hydrogenated Castor Oil PCA Isostearate is a diester of PEG-30 Hydrogenated Castor Oil (q.v.) and a mixture of PCA (q.v.) and Isostearic Acid (q.v.).

Chemical Classes: Alkoxylated Alcohols; Glyceryl Esters and Derivatives; Heterocyclic Compounds

Function: Skin-Conditioning Agent - Miscellaneous

Trade Name:
Pyroter CPI-30 (Nihon Emulsion)

PEG-40 HYDROGENATED CASTOR OIL PCA ISOSTEARATE

JPN Translation:
PCA イソステアリン酸 PEG - 40 水添ヒマシ油

Definition: PEG-40 Hydrogenated Castor Oil PCA Isostearate is a diester of PEG-40 Hydrogenated Castor Oil (q.v.) and a mixture of PCA (q.v.) and Isostearic Acid (q.v.).

Information Sources: JCIC, JCLS, JSQI

Chemical Classes: Alkoxylated Alcohols; Glyceryl Esters and Derivatives; Heterocyclic Compounds

Function: Skin-Conditioning Agent - Miscellaneous

Technical/Other Names:
Polyethylene Glycol 2000 Hydrogenated Castor Oil PCA Isostearate
Polyoxyethylene Hydrogenated Castor Oil Monopyroglutamate Monoisostearate
Polyoxyethylene (40) Hydrogenated Castor Oil PCA Isostearate

Trade Name:
Pyroter CPI-40 (Ajinomoto)

PEG-60 HYDROGENATED CASTOR OIL PCA ISOSTEARATE

Definition: PEG-60 Hydrogenated Castor Oil PCA Isostearate is a diester of PEG-60 Hydrogenated Castor Oil and a mixture of PCA (q.v.) and Isostearic Acid (q.v.).

Chemical Classes: Alkoxylated Alcohols; Glyceryl Esters and Derivatives; Heterocyclic Compounds

Function: Skin-Conditioning Agent - Miscellaneous

Trade Name:
Pyroter CPI-60 (Nihon Emulsion)

PEG-50 HYDROGENATED CASTOR OIL SUCCINATE

JPN Translation:
コハク酸 PEG - 50 水添ヒマシ油

Definition: PEG-50 Hydrogenated Castor Oil Succinate is a polyethylene glycol derivtive of the succinic acid ester of Hydrogenated Castor Oil (q.v.) with an average ethoxylation value of 50.

Information Sources: JCIC, JCLS

Chemical Classes: Alkoxylated Alcohols; Glyceryl Esters and Derivatives

Functions: Surfactant - Emulsifying Agent; Viscosity Increasing Agent - Nonaqueous

Technical/Other Names:
Polyethylene Glycol (50) Hydrogenated Castor Oil Succinate
Polyoxyethylene (50) Hydrogenated Castor Oil Succinate
Polyoxyethylene Hydrogenated Castor Oil Succinate (50E.O.)

PEG-5 HYDROGENATED CASTOR OIL TRIISOSTEARATE

Definition: PEG-5 Hydrogenated Castor Oil Triisostearate is the triester of isostearic acid and Hydrogenated Castor Oil (q.v.) with an average of 5 moles of ethylene oxide.

Chemical Classes: Alkoxylated Alcohols; Glyceryl Esters and Derivatives

Functions: Surfactant - Emulsifying Agent; Viscosity Increasing Agent - Nonaqueous

Technical/Other Names:
Polyethylene Glycol (5) Hydrogenated Castor Oil Triisostearate
Polyoxyethylene (5) Hydrogenated Castor Oil Triisostearate

Trade Name:
Emalex RWIS-305 (Nihon Emulsion)

PEG-10 HYDROGENATED CASTOR OIL TRIISOSTEARATE

Definition: PEG-10 Hydrogenated Castor Oil Triisostearate is the triester of isostearic acid and Hydrogenated Castor Oil (q.v.) with an average of 10 moles of ethylene oxide.

Chemical Classes: Alkoxylated Alcohols; Glyceryl Esters and Derivatives

Functions: Surfactant - Emulsifying Agent; Viscosity Increasing Agent - Nonaqueous

Technical/Other Names:
Polyethylene Glycol (10) Hydrogenated Castor Oil Triisostearate
Polyoxyethylene (10) Hydrogenated Castor Oil Triisostearate

Trade Name:
Emalex RWIS-310 (Nihon Emulsion)

PEG-15 HYDROGENATED CASTOR OIL TRIISOSTEARATE

JPN Translation:
トリイソステアリン酸 PEG - 15 水添ヒマシ油

Definition: PEG-15 Hydrogenated Castor Oil Triisostearate is the triester of isostearic acid and Hydrogenated Castor Oil (q.v.) with an average of 15 moles of ethylene oxide.

Chemical Classes: Alkoxylated Alcohols; Glyceryl Esters and Derivatives

Functions: Surfactant - Emulsifying Agent; Viscosity Increasing Agent - Nonaqueous

Technical/Other Names:
Polyethylene Glycol (15) Hydrogenated Castor Oil Triisostearate
Polyoxyethylene (15) Hydrogenated Castor Oil Triisostearate

Trade Name:
Emalex RWIS-315 (Nihon Emulsion)

PEG-20 HYDROGENATED CASTOR OIL TRIISOSTEARATE

JPN Translation:
トリイソステアリン酸 PEG - 20 水添ヒマシ油

Definition: PEG-20 Hydrogenated Castor Oil Triisostearate is the isostearic acid triester of Hydrogenated Castor Oil (q.v.) with an average ethoxylation value of 20.

Information Source: JCLS

Chemical Classes: Alkoxylated Alcohols; Esters; Glyceryl Esters and Derivatives

Functions: Surfactant - Emulsifying Agent; Viscosity Increasing Agent - Nonaqueous

Technical/Other Names:
Polyethylene Glycol 1000 Hydrogenated Castor Oil Triisostearate
Polyoxyethylene (20) Hydrogenated Castor Oil Triisostearate

Trade Names:
Emalex RWIS-320 (Ikeda)
Emalex RWIS-320 (Nihon Emulsion)

PEG-30 HYDROGENATED CASTOR OIL TRIISOSTEARATE

JPN Translation:
トリイソステアリン酸 PEG - 30 水添ヒマシ油

Definition: PEG-30 Hydrogenated Castor Oil Triisostearate is the triester of isostearic acid and Hydrogenated Castor Oil (q.v.) with an average of 30 moles of ethylene oxide.

Chemical Classes: Alkoxylated Alcohols; Glyceryl Esters and Derivatives

Functions: Surfactant - Emulsifying Agent; Viscosity Increasing Agent - Nonaqueous

Technical/Other Names:
Polyethylene Glycol (30) Hydrogenated Castor Oil Triisostearate

Polyoxyethylene (30) Hydrogenated Castor Oil Triisostearate

Trade Name:
Emalex RWIS-330 (Nihon Emulsion)

PEG-40 HYDROGENATED CASTOR OIL TRIISOSTEARATE

JPN Translation:
トリイソステアリン酸 PEG - 40 水添ヒマシ油

Definition: PEG-40 Hydrogenated Castor Oil Triisostearate is the triester of isostearic acid and Hydrogenated Castor Oil (q.v.) with an average of 40 moles of ethylene oxide.

Information Source: JCLS

Chemical Classes: Alkoxylated Alcohols; Glyceryl Esters and Derivatives

Functions: Surfactant - Emulsifying Agent; Viscosity Increasing Agent - Nonaqueous

Technical/Other Names:
Polyethylene Glycol (40) Hydrogenated Castor Oil Triisostearate
Polyoxyethylene (40) Hydrogenated Castor Oil Triisostearate

Trade Name:
Emalex RWIS-340 (Nihon Emulsion)

PEG-50 HYDROGENATED CASTOR OIL TRIISOSTEARATE

JPN Translation:
トリイソステアリン酸 PEG - 50 水添ヒマシ油

Definition: PEG-50 Hydrogenated Castor Oil Triisostearate is the isostearic acid triester of Hydrogenated Castor Oil (q.v.) with an average of 50 moles of ethylene oxide.

Information Source: JCLS

Chemical Classes: Alkoxylated Alcohols; Glyceryl Esters and Derivatives

Function: Surfactant - Emulsifying Agent

Technical/Other Names:
Polyethylene Glycol (50) Hydrogenated Castor Oil Triisostearate
Polyoxyethylene (50) Hydrogenated Castor Oil Triisostearate

Trade Name:
Emalex RWIS-350 (Nihon Emulsion)

PEG-60 HYDROGENATED CASTOR OIL TRIISOSTEARATE

JPN Translation:
トリイソステアリン酸 PEG - 60 水添ヒマシ油

Definition: PEG-60 Hydrogenated Castor Oil Triisostearate is the isostearic acid triester of Hydrogenated Castor Oil (q.v.) with an average of 60 moles of ethylene oxide.

Information Source: JCLS

Chemical Classes: Alkoxylated Alcohols; Glyceryl Esters and Derivatives

Functions: Surfactant - Emulsifying Agent; Viscosity Increasing Agent - Nonaqueous

Technical/Other Names:
Polyethylene Glycol (60) Hydrogenated Castor Triisostearate
Polyoxyethylene (60) Hydrogenated Castor Oil Triisostearate

Trade Name:
Emalex RWIS-360 (Nihon Emulsion)

PEG-5 HYDROGENATED CORN GLYCERIDES

Definition: PEG-5 Hydrogenated Corn Glycerides is the polyethylene glycol derivative of mixed glycerides derived from hydrogenated corn oil. It has an average of 5 moles of ethylene oxide.

Chemical Classes: Alkoxylated Alcohols; Glyceryl Esters and Derivatives

Functions: Skin-Conditioning Agent - Emollient; Surfactant - Emulsifying Agent

Technical/Other Names:
Polyethylene Glycol (5) Hydrogenated Corn Glycerides
Polyoxyethylene (5) Hydrogenated Corn Glycerides

PEG-8 HYDROGENATED FISH GLYCERIDES

Definition: PEG-8 Hydrogenated Fish Glycerides is a polyethylene glycol derivative of Hydrogenated Fish Oil (q.v.) with an average of 8 moles of ethylene oxide.

Chemical Classes: Alkoxylated Alcohols; Glyceryl Esters and Derivatives

Functions: Skin-Conditioning Agent - Emollient; Surfactant - Emulsifying Agent

Technical/Other Names:
Polyethylene Glycol 400 Hydrogenated Fish Glycerides
Polyoxyethylene (8) Hydrogenated Fish Glycerides

PEG-80 HYDROGENATED GLYCERYL PALMATE

Definition: PEG-80 Hydrogenated Glyceryl Palmate is the polyethylene glycol derivative of Hydrogenated Glyceryl Palmate (q.v.). It has an average of 80 moles of ethylene oxide.

Chemical Class: Glyceryl Esters and Derivatives

Function: Surfactant - Solubilizing Agent

Technical/Other Names:
Polyethylene Glycol (80) Hydrogenated Glyceryl Palmate
Polyoxyethylene (80) Hydrogenated Glyceryl Palmate

PEG-200 HYDROGENATED GLYCERYL PALMATE

Definition: PEG-200 Hydrogenated Glyceryl Palmate is the polyethylene glycol derivative of Hydrogenated Palm Glyceride (q.v.). It has an average of 200 moles of ethylene oxide.

Chemical Classes: Alkoxylated Alcohols; Glyceryl Esters and Derivatives

Function: Surfactant - Solubilizing Agent

Reported Product Categories: Shampoos (Non-coloring); Bath Oils, Tablets, and Salts

Trade Name:
Rewoderm LI 520-70 (Degussa Care Specialties)

Trade Name Mixtures:
Antil 200 (Degussa Care Specialties)
Rewoderm LI S 80 (Degussa Care Specialties)
Saboderm SHO (Sabo)

PEG-5 HYDROGENATED LANOLIN

CAS No.: 68648-27-1 (Generic)

JPN Translation:
PEG - 5 水添ラノリン

Definition: PEG-5 Hydrogenated Lanolin is a polyethylene glycol derivative of Hydrogenated Lanolin (q.v.) with an average of 5 moles of ethylene oxide.

Information Sources: CIR: [S] IJT-18 (SUPPL. 1)1999, JCLS, JSQI, TSCA

Chemical Class: Lanolin and Lanolin Derivatives

Functions: Hair Conditioning Agent; Skin-Conditioning Agent - Emollient; Surfactant - Emulsifying Agent

Technical/Other Names:
Polyethylene Glycol (5) Hydrogenated Lanolin
Polyoxyethylene (5) Hydrogenated Lanolin

Trade Name:
Polychol WH-50 (Croda Japan)

PEG-10 HYDROGENATED LANOLIN

CAS No.: 68648-27-1 (Generic)

JPN Translation:
PEG - 10 水添ラノリン

Definition: PEG-10 Hydrogenated Lanolin is a polyethylene glycol derivative of Hydrogenated Lanolin (q.v.) with an average of 10 moles of ethylene oxide.

Information Sources: CIR: [S] IJT-18 (SUPPL. 1)1999, JCLS, JSQI, TSCA

Chemical Class: Lanolin and Lanolin Derivatives

Functions: Hair Conditioning Agent; Skin-Conditioning Agent - Emollient; Surfactant - Emulsifying Agent

Technical/Other Names:
Polyethylene Glycol 500 Hydrogenated Lanolin
Polyoxyethylene (10) Hydrogenated Lanolin

PEG-15 HYDROGENATED LANOLIN

Definition: PEG-15 Hydrogenated Lanolin is a polyethylene glycol derivative of Hydrogenated Lanolin (q.v.) with an average of 15 moles of ethylene oxide.

Chemical Class: Lanolin and Lanolin Derivatives

Functions: Hair Conditioning Agent; Surfactant - Emulsifying Agent

Technical/Other Names:
Polyethylene Glycol (15) Hydrogenated Lanolin
Polyoxyethylene (15) Hydrogenated Lanolin

Trade Name:
Polychol WH-150 (Croda Japan)

PEG-20 HYDROGENATED LANOLIN

CAS No.: 68648-27-1 (Generic)

JPN Translation:
PEG - 20 水添ラノリン

Definition: PEG-20 Hydrogenated Lanolin is a polyethylene glycol derivative of Hydrogenated Lanolin (q.v.) with an average of 20 moles of ethylene oxide.

Information Sources: CIR: [S] IJT-18 (SUPPL. 1)1999, JCLS, JSQI, TSCA

Chemical Class: Lanolin and Lanolin Derivatives

Functions: Hair Conditioning Agent; Surfactant - Emulsifying Agent

Reported Product Category: Hair Straighteners

Technical/Other Names:
Polyethylene Glycol 1000 Hydrogenated Lanolin
Polyoxyethylene (20) Hydrogenated Lanolin

Trade Names:
Fancol HL-20 (Fanning)
Ivarlan 3450 (Arch Personal Care Products)
Ivarlan HL-20 (Arch Personal Care Products)
Lanidrol (Esperis)
Lipolan 31-20 (Lipo)
Supersat AWS-4 (RITA)

Trade Name Mixture:
Ritachol 5000 (RITA)

PEG-24 HYDROGENATED LANOLIN

CAS No.: 68648-27-1 (Generic)

JPN Translation:
PEG - 24 水添ラノリン

Definition: PEG-24 Hydrogenated Lanolin is a polyethylene glycol derivative of Hydrogenated Lanolin (q.v.) with an average of 24 moles of ethylene oxide.

Information Sources: CIR: [S] IJT-18 (SUPPL. 1)1999, JCLS, JSQI, TSCA

Chemical Class: Lanolin and Lanolin Derivatives

Functions: Hair Conditioning Agent; Surfactant - Emulsifying Agent

Technical/Other Names:
Polyethylene Glycol (24) Hydrogenated Lanolin
Polyoxyethylene (24) Hydrogenated Lanolin

Trade Names:
Fancol HL-24 (Fanning)
Ivarlan 3452 (Arch Personal Care Products)
Lipolan 31 (Lipo)
Supersat AWS-24 (RITA)

PEG-30 HYDROGENATED LANOLIN

CAS No.: 68648-27-1 (Generic)

JPN Translation:
PEG - 30 水添ラノリン

Definition: PEG-30 Hydrogenated Lanolin is a polyethylene glycol derivative of Hydrogenated Lanolin (q.v.) with an average of 30 moles of ethylene oxide.

Information Sources: CIR: [S] IJT-18 (SUPPL. 1)1999, JCLS, JSQI, TSCA

Chemical Class: Lanolin and Lanolin Derivatives

Functions: Hair Conditioning Agent; Surfactant - Cleansing Agent; Surfactant - Solubilizing Agent

Technical/Other Names:
Polyethylene Glycol (30) Hydrogenated Lanolin
Polyoxyethylene (30) Hydrogenated Lanolin

PEG-40 HYDROGENATED LANOLIN

Definition: PEG-40 Hydrogenated Lanolin is a polyethylene glycol derivative of Hydrogenated Lanolin (q.v.) with an average of 40 moles of ethylene oxide.

Chemical Class: Lanolin and Lanolin Derivatives

Functions: Hair Conditioning Agent; Surfactant - Emulsifying Agent

Technical/Other Names:
Polyethylene Glycol (40) Hydrogenated Lanolin
Polyoxyethylene (40) Hydrogenated Lanolin

Trade Name:
Polychol WH-400 (Croda Japan)

PEG-70 HYDROGENATED LANOLIN

CAS No.: 68648-27-1 (Generic)

JPN Translation:
PEG - 70 水添ラノリン

Definition: PEG-70 Hydrogenated Lanolin is a polyethylene glycol derivative of Hydrogenated Lanolin (q.v.) with an average of 70 moles of ethylene oxide.

Information Sources: CIR: [S] IJT-18 (SUPPL. 1)1999, JCLS, TSCA

Chemical Class: Lanolin and Lanolin Derivatives

Functions: Hair Conditioning Agent; Surfactant - Cleansing Agent; Surfactant - Solubilizing Agent

Technical/Other Names:
Polyethylene Glycol (70) Hydrogenated Lanolin
Polyoxyethylene (70) Hydrogenated Lanolin

PEG-6 HYDROGENATED PALMAMIDE

Definition: PEG-6 Hydrogenated Palmamide is the polyethylene glycol amide of Hydrogenated Palm Oil (q.v.) that conforms generally to the formula:

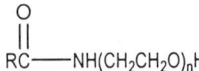

where RCO- represents the fatty acids derived from hydrogenated palm oil and n has an average value of 6.

Chemical Class: Alkoxylated Amides

Functions: Emulsion Stabilizer; Surfactant - Emulsifying Agent

Technical/Other Names:
Polyethylene Glycol (6) Hydrogenated Palmamide
Polyoxyethylene (6) Hydrogenated Palmamide

Trade Name:
Ethomid HP/15 (Akzo Nobel Surface AB)

PEG-50 HYDROGENATED PALMAMIDE

Definition: PEG-50 Hydrogenated Palmamide is the polyethylene glycol amide of hydrogenated palm oil that conforms generally to the formula:

$$\underset{RC}{\overset{O}{\|}}-NH(CH_2CH_2O)_nH$$

where RCO- represents the fatty acids from hydrogenated palm oil and n has an average value of 50.

Chemical Class: Alkoxylated Amides

Functions: Surfactant - Cleansing Agent; Surfactant - Solubilizing Agent

Technical/Other Names:
Polyethylene Glycol (50) Hydrogenated Palm Amide
Polyoxyethylene (50) Hydrogenated Palm Amide

Trade Name:
Ethomid HP-60 (Akzo Nobel)

PEG-20 HYDROGENATED PALM GLYCERIDES

Definition: PEG-20 Hydrogenated Palm Glycerides is a polyethylene glycol ether of Hydrogenated Palm Glycerides (q.v.) with an average of 20 moles of ethylene oxide.

Chemical Classes: Alkoxylated Alcohols; Glyceryl Esters and Derivatives

Functions: Skin-Conditioning Agent - Emollient; Surfactant - Emulsifying Agent

Technical/Other Names:
PEG-20 Hydrogenated Palm Oil Glycerides

Polyethylene Glycol 1000 Hydrogenated Palm Oil Glycerides
Polyoxyethylene (20) Hydrogenated Palm Oil Glycerides

PEG-6 HYDROGENATED PALM/PALM KERNEL GLYCERIDE

JPN Translation:
PEG‑6 水添（パーム / パーム核）脂肪酸グリセリル

Definition: PEG-6 Hydrogenated Palm/Palm Kernel Glyceride is a polyethylene glycol derivative of a mixture of Hydrogenated Palm Glyceride (q.v.) and Hydrogenated Palm Kernel Glyceride (q.v.) containing an average of 6 moles of ethylene oxide.

Information Source: JCIC

Chemical Classes: Alkoxylated Alcohols; Glyceryl Esters and Derivatives

Functions: Skin-Conditioning Agent - Emollient; Surfactant - Emulsifying Agent

Technical/Other Names:
Polyethylene Glycol (6) Hydrogenated Palm/Palm Kernel Glyceride
Polyoxyethylene Hydrogenated (Palm Oil/Palm Kernel Oil) Fatty Acid Mono-glycerides

PEG-13 HYDROGENATED TALLOW AMIDE

CAS No.: 68783-22-2 (Generic)

Definition: PEG-13 Hydrogenated Tallow Amide is the polyethylene glycol amide of Hydrogenated Tallow Amide (q.v.) that conforms generally to the formula:

$$\underset{RC}{\overset{O}{\|}}-NH(CH_2CH_2O)_nH$$

where RCO- represents the fatty acids derived from hydrogenated tallow and n has an average value of 13.

Information Source: TSCA

Chemical Class: Alkoxylated Amides

Function: Surfactant - Emulsifying Agent

Technical/Other Names:
Amides, Tallow, Hydrogenated, N,N-Bis (Hydroxyethyl), Ethoxylated
Polyethylene Glycol (13) Hydrogenated Tallow Amide
Polyoxyethylene (13) Hydrogenated Tallow Amide

Trade Name:
Ethomid HT/23 (Akzo Nobel)

PEG-2 HYDROGENATED TALLOW AMINE

CAS No.: 61791-26-2 (Generic)

Definition: PEG-2 Hydrogenated Tallow Amine is the polyethylene glycol amine of Hydrogenated Tallow (q.v.) that conforms generally to the formula:

$$R-N\underset{(CH_2CH_2O)_yH}{\overset{(CH_2CH_2O)_xH}{<}}$$

where R represents the alkyl groups derived from hydrogenated tallow and x + y has an average value of 2.

Information Source: TSCA

Chemical Class: Alkoxylated Amines

Functions: Antistatic Agent; Surfactant - Foam Booster

Reported Product Category: Hair Dyes and Colors (All Types Requiring Caution Statements and Patch Tests)

Technical/Other Names:
PEG-2 Tallow Amine
Polyethylene Glycol 100 Hydrogenated Tallow Amine
Polyoxyethylene (2) Hydrogenated Tallow Amine

Trade Names:
Chemeen T-2 (Chemax)
DeTHOX AMINE T-2 (DeForest)
Ethomeen T/12 (Akzo Nobel)
Hetoxamine T-2 (Heterene)
Jeetox T-2 (Jeen)
Protox T-2 (Protameen)
Unizeen T-2 (Universal Preserv-A-Chem)

PEG-5 HYDROGENATED TALLOW AMINE

CAS No.: 61791-26-2 (Generic)

Definition: PEG-5 Hydrogenated Tallow Amine is the polyethylene glycol amine of Hydrogenated Tallow (q.v.) that conforms generally to the formula:

$$R-N\underset{(CH_2CH_2O)_yH}{\overset{(CH_2CH_2O)_xH}{<}}$$

where R represents the alkyl groups derived from hydrogenated tallow and x + y has an average value of 5.

Information Sources: 21CFR178.3910, TSCA

Chemical Class: Alkoxylated Amines

Function: Antistatic Agent

Reported Product Category: Hair Tints

Technical/Other Names:
PEG-5 Tallow Amine

Polyethylene Glycol (5) Hydrogenated Tallow Amine

Polyoxyethylene (5) Hydrogenated Tallow Amine

Trade Names:
Chemeen T-5 (Chemax)
DeTHOX AMINE T-5 (DeForest)
Ethomeen T/15 (Akzo Nobel)
Ethox HTAM-5 (Ethox)
Hetoxamine T-5 (Heterene)
Protox T-5 (Protameen)
Unizeen T-5 (Universal Preserv-A-Chem)

PEG-8 HYDROGENATED TALLOW AMINE

CAS No.: 61791-26-2 (Generic)

Definition: PEG-8 Hydrogenated Tallow Amine is the polyethylene glycol amine of Hydrogenated Tallow (q.v.) that conforms generally to the formula:

$$R-N \begin{matrix} (CH_2CH_2O)_xH \\ (CH_2CH_2O)_yH \end{matrix}$$

where R represents the alkyl groups derived from hydrogenated tallow and x + y has an average value of 8.

Information Source: TSCA

Chemical Class: Alkoxylated Amines

Functions: Antistatic Agent; Surfactant - Emulsifying Agent

Reported Product Category: Hair Dyes and Colors (All Types Requiring Caution Statements and Patch Tests)

Technical/Other Names:
PEG-8 Tallow Amine
Polyethylene Glycol 400 Hydrogenated Tallow Amine
Polyoxyethylene (8) Hydrogenated Tallow Amine

Trade Name:
Lowenol 1985 (Lowenstein)

PEG-10 HYDROGENATED TALLOW AMINE

CAS No.: 61791-26-2 (Generic)

Definition: PEG-10 Hydrogenated Tallow Amine is the polyethylene glycol amine of Hydrogenated Tallow (q.v.) that conforms generally to the formula:

$$R-N \begin{matrix} (CH_2CH_2O)_xH \\ (CH_2CH_2O)_yH \end{matrix}$$

where R represents the fatty alkyl groups derived from hydrogenated tallow and x + y has an average value of 10.

Chemical Class: Alkoxylated Amines

Functions: Antistatic Agent; Surfactant - Emulsifying Agent

Technical/Other Names:
Polyethylene Glycol 500 Hydrogenated Tallow Amine
Polyoxyethylene (10) Hydrogenated Tallow Amine

Trade Names:
Ethox HTAM-10 (Ethox)
Jeetox T-10 (Jeen)

PEG-15 HYDROGENATED TALLOW AMINE

CAS No.: 61791-26-2 (Generic)

Definition: PEG-15 Hydrogenated Tallow Amine is the polyethylene glycol amine of Hydrogenated Tallow (q.v.) that conforms generally to the formula:

$$R-N \begin{matrix} (CH_2CH_2O)_xH \\ (CH_2CH_2O)_yH \end{matrix}$$

where R represents the alkyl groups derived from hydrogenated tallow and x + y has an average value of 15.

Information Source: TSCA

Chemical Class: Alkoxylated Amines

Functions: Antistatic Agent; Surfactant - Emulsifying Agent

Technical/Other Names:
PEG-15 Tallow Amine
Polyethylene Glycol (15) Hydrogenated Tallow Amine
Polyoxyethylene (15) Hydrogenated Tallow Amine

Trade Names:
Chemeen T-15 (Chemax)
DeTHOX AMINE T-15 (DeForest)
Ethomeen T/25 (Akzo Nobel)
Ethox HTAM-15 (Ethox)
Hetoxamine T-15 (Heterene)
Jeetox T-15 (Jeen)
Protox T-15 (Protameen)
Rhodameen T-15 (Rhodia)
Unizeen T-15 (Universal Preserv-A-Chem)

PEG-20 HYDROGENATED TALLOW AMINE

CAS No.: 61791-26-2 (Generic)

Definition: PEG-20 Hydrogenated Tallow Amine is the polyethylene glycol amine of Hydrogenated Tallow (q.v.) that conforms generally to the formula:

$$R-N \begin{matrix} (CH_2CH_2O)_xH \\ (CH_2CH_2O)_yH \end{matrix}$$

where R represents the alkyl groups derived from hydrogenated tallow and x + y has an average value of 20.

Information Source: TSCA

Chemical Class: Alkoxylated Amines

Functions: Antistatic Agent; Surfactant - Emulsifying Agent; Surfactant - Solubilizing Agent

Technical/Other Names:
PEG-20 Tallow Amine
Polyethylene Glycol 1000 Hydrogenated Tallow Amine
Polyoxyethylene (20) Hydrogenated Tallow Amine

Trade Names:
Ethox HTAM-20 (Ethox)
Hetoxamine T-20 (Heterene)
Jeetox T-20 (Jeen)
Protox T-20 (Protameen)

PEG-30 HYDROGENATED TALLOW AMINE

CAS No.: 61791-26-2 (Generic)

Definition: PEG-30 Hydrogenated Tallow Amine is the polyethylene glycol amine of Hydrogenated Tallow (q.v) that conforms generally to the formula:

$$R-N \begin{matrix} (CH_2CH_2O)_xH \\ (CH_2CH_2O)_yH \end{matrix}$$

where R represents the alkyl groups derived from hydrogenated tallow and x + y has an average value of 30.

Chemical Class: Alkoxylated Amines

Functions: Antistatic Agent; Surfactant - Solubilizing Agent

Technical/Other Names:
PEG-30 Tallow Amine
Polyethylene Glycol (30) Hydrogenated Tallow Amine
Polyoxyethylene (30) Hydrogenated Tallow Amine

Trade Name:
Hetoxamine T-30 (Heterene)

PEG-40 HYDROGENATED TALLOW AMINE

CAS No.: 61791-26-2 (Generic)

Definition: PEG-40 Hydrogenated Tallow Amine is the polyethylene glycol amine of Hydrogenated Tallow (q.v.) that conforms generally to the formula:

$$R-N \begin{array}{c} (CH_2CH_2O)_xH \\ (CH_2CH_2O)_yH \end{array}$$

where R represents the alkyl groups derived from hydrogenated tallow and x + y has an average value of 40.

Information Source: TSCA

Chemical Class: Alkoxylated Amines

Functions: Antistatic Agent; Surfactant - Solubilizing Agent

Technical/Other Names:
PEG-40 Tallow Amine
Polyethylene Glycol 2000 Hydrogenated Tallow Amine
Polyoxyethylene (40) Hydrogenated Tallow Amine

Trade Names:
Jeetox T-40 (Jeen)
Protox T-40 (Protameen)

PEG-50 HYDROGENATED TALLOW AMINE

CAS No.: 61791-26-2 (Generic)

Definition: PEG-50 Hydrogenated Tallow Amine is the polyethylene glycol amine of Hydrogenated Tallow (q.v.) that conforms generally to the formula:

$$R-N \begin{array}{c} (CH_2CH_2O)_xH \\ (CH_2CH_2O)_yH \end{array}$$

where R represents the alkyl groups derived from hydrogenated tallow and x + y has an average value of 50.

Information Sources: CTFA D, TSCA

Chemical Class: Alkoxylated Amines

Functions: Antistatic Agent; Surfactant - Solubilizing Agent

Technical/Other Names:
PEG-50 Tallow Amine
Polyethylene Glycol (50) Hydrogenated Tallow Amine
Polyoxyethylene (50) Hydrogenated Tallow Amine

Trade Names:
Jeetox T-50 (Jeen)
Protox T-50 (Protameen)

PEG-15 HYDROGENATED TALLOWMONIUM CHLORIDE

CAS No.: 68187-69-9

Definition: PEG-15 Hydrogenated Tallowmonium Chloride is the quaternary ammonium salt that conforms generally to the formula:

$$\left[R-N \begin{array}{c} (CH_2CH_2O)_x \\ | \\ CH_3 \\ (CH_2CH_2O)_y \end{array} \right]^+ Cl^-$$

where R represents the alkyl groups derived from Hydrogenated Tallow (q.v.) and x + y has an average value of 15.

Information Source: TSCA

Chemical Class: Quaternary Ammonium Compounds

Functions: Antistatic Agent; Hair Conditioning Agent

PEG-15 HYDROXYSTEARATE

Definition: PEG-15 Hydroxystearate is a polyethylene glycol ester of hydroxystearic acid that conforms to the formula:

$$CH_3(CH_2)_5CH(CH_2)_{10}\overset{\overset{\displaystyle O}{\|}}{C}-(OCH_2CH_2)_nOH$$
$$|$$
$$OH$$

where n has an average value of 15.

Chemical Class: Alkoxylated Carboxylic Acids

Function: Surfactant - Emulsifying Agent

Technical/Other Names:
Polyethylene Glycol (15) Hydroxystearate
Polyoxyethylene (15) Hydroxystearate

Trade Name:
Atlas G-4972 (Uniqema Americas)

PEG-100/IPDI COPOLYMER

Definition: PEG-100/IPDI Copolymer is a copolymer of isophorone diisocyanate and PEG-100 (q.v.) monomers.

Chemical Classes: Polymeric Ethers; Synthetic Polymers

Functions: Binder; Emulsion Stabilizer; Viscosity Increasing Agent - Aqueous

PEG-5 ISODECYLOXYPROPYLAMINE

Definition: PEG-5 Isodecyloxypropylamine is the amine that conforms generally to the formula:

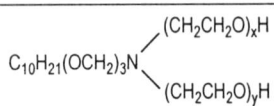

Chemical Class: Alkoxylated Amines

Function: Antistatic Agent

Trade Name:
E-14-5 (Tomah)

PEG-6 ISOLAURYL THIOETHER

Empirical Formula:
$C_{24}H_{50}O_6S$

Definition: PEG-6 Isolauryl Thioether is the polyethylene glycol ether of a branched chain dodecyl mercaptan. It conforms generally to the formula:

$$C_{12}H_{26}S(CH_2CH_2O)_nH$$

where n has an average value of 6.

Chemical Classes: Alkoxylated Alcohols; Thio Compounds

Function: Surfactant - Emulsifying Agent

Technical/Other Names:
Polyethylene Glycol 300 Isolauryl Thioether
Polyoxyethylene (6) Isolauryl Thioether

Trade Names:
Sebum Control COS-218/2-A (Cosmetochem)
Sulfocos 2B (Cosmetochem)

PEG-8 ISOLAURYL THIOETHER

Empirical Formula:
$C_{28}H_{58}O_8S$

Definition: PEG-8 Isolauryl Thioether is the polyethylene glycol ether of a branched chain dodecyl mercaptan. It conforms generally to the formula:

$$C_{12}H_{26}S(CH_2CH_2O)_nH$$

where n has an average value of 8.

Chemical Classes: Alkoxylated Alcohols; Thio Compounds

Function: Surfactant - Emulsifying Agent

Technical/Other Names:
Polyethylene Glycol 400 Isolauryl Thioether
Polyoxyethylene (8) Isolauryl Thioether

PEG-10 ISOLAURYL THIOETHER

Empirical Formula:
$C_{32}H_{66}O_{10}S$

Definition: PEG-10 Isolauryl Thioether is the polyethylene glycol ether of a branched chain dodecyl mercaptan. It conforms generally to the formula:

$$C_{12}H_{26}S(CH_2CH_2O)_nH$$

where n has an average value of 10.

Chemical Classes: Alkoxylated Alcohols; Thio Compounds

Function: Surfactant - Emulsifying Agent

Technical/Other Names:
Polyethylene Glycol 500 Isolauryl Thioether
Polyoxyethylene (10) Isolauryl Thioether

PEG-6 ISOPALMITATE

Empirical Formula:
$C_{28}H_{56}O_8$

Definition: PEG-6 Isopalmitate is the polyethylene glycol ester of isopalmitic acid that conforms generally to the formula:

$$C_{15}H_{31}C \overset{O}{\overset{\|}{}} - (OCH_2CH_2)_nOH$$

where n has an average value of 6.

Information Source: CTFA D

Chemical Class: Alkoxylated Carboxylic Acids

Function: Surfactant - Emulsifying Agent

Technical/Other Names:
Polyethylene Glycol 300 Monoisopalmitate
Polyoxyethylene (6) Monoisopalmitate

PEG-2 ISOSTEARATE

JPN Translation:
ラウレス-１０カルボン酸

Definition: PEG-2 Isostearate is the polyethyleneglycol ester of Isostearic Acid (q.v.) that conforms generally to the formula:

$$C_{17}H_{20}C \overset{O}{\overset{\|}{}} - (OCH_2CH_2)_nOH$$

where n has an average value of 2.

Information Source: JCLS

Chemical Class: Alkoxylated Carboxylic Acids

Functions: Surfactant - Cleansing Agent; Surfactant - Emulsifying Agent

PEG-3 ISOSTEARATE

JPN Translation:
イソステアリン酸 PEG - 3

Definition: PEG-3 Isostearate is the polyethylene glycol ester of Isostearic Acid (q.v.) that conforms generally to the formula:

$$C_{17}H_{20}C \overset{O}{\overset{\|}{}} - (OCH_2CH_2)_nOH$$

where n has an average value of 3.

Chemical Class: Alkoxylated Carboxylic Acids

Functions: Surfactant - Cleansing Agent; Surfactant - Emulsifying Agent

Technical/Other Names:
Polyethylene Glycol (3) Monoisostearate
Polyoxyethylene (3) Monoisostearate

Trade Name:
Emalex PEIS-3 (Nihon Emulsion)

PEG-4 ISOSTEARATE

CAS No.: 56002-14-3 (Generic)

JPN Translation:
イソステアリン酸 PEG - 4

Empirical Formula:
$C_{26}H_{52}O_6$

Definition: PEG-4 Isostearate is the polyethylene glycol ester of Isostearic Acid that conforms generally to the formula:

$$C_{17}H_{20}C \overset{O}{\overset{\|}{}} - (OCH_2CH_2)_nOH$$

where n has an average value of 4.

Information Source: JCLS

Chemical Class: Alkoxylated Carboxylic Acids

Functions: Surfactant - Cleansing Agent; Surfactant - Emulsifying Agent

Technical/Other Names:
Polyethylene Glycol 200 Isostearate
Polyoxyethylene (4) Isostearate

PEG-6 ISOSTEARATE

CAS No.: 56002-14-3 (Generic)

JPN Translation:
イソステアリン酸 PEG - 6

Empirical Formula:
$C_{30}H_{60}O_8$

Definition: PEG-6 Isostearate is the polyethylene glycol ester of Isostearic Acid (q.v.) that conforms generally to the formula:

$$C_{17}H_{20}C \overset{O}{\overset{\|}{}} - (OCH_2CH_2)_nOH$$

where n has an average value of 6.

Information Sources: JCLS, TSCA

Chemical Class: Alkoxylated Carboxylic Acids

Functions: Surfactant - Cleansing Agent; Surfactant - Emulsifying Agent

Technical/Other Names:
Polyethylene Glycol 300 Monoisostearate
Polyoxyethylene (6) Monoisostearate

Trade Names:
AEC PEG-6 Isostearate (A & E Connock)
Olepal Isostearique (Gattefosse s.a.)

Trade Name Mixtures:
Flavagrum PEG (Coletica SA)
Solubilisant WL 3215 (Gattefosse s.a.)

PEG-8 ISOSTEARATE

CAS No.: 56002-14-3 (Generic)

JPN Translation:
イソステアリン酸 PEG - 8

Empirical Formula:
$C_{34}H_{68}O_{10}$

Definition: PEG-8 Isostearate is the polyethylene glycol ester of Isostearic Acid (q.v.) that conforms generally to the formula:

$$C_{17}H_{20}C \overset{O}{\overset{\|}{}} - (OCH_2CH_2)_nOH$$

where n has an average value of 8.

Information Source: JCLS

Chemical Class: Alkoxylated Carboxylic Acids

Functions: Surfactant - Cleansing Agent; Surfactant - Emulsifying Agent

Technical/Other Names:
Polyethylene Glycol 400 Isostearate
Polyoxyethylene (8) Isostearate

Trade Name:
Ethox MI-9 (Ethox)

PEG-9 ISOSTEARATE

Definition: PEG-9 Isostearate is the polyethylene glycol ester of isostearic acid that conforms generally to the formula:

$$C_{17}H_{35}C \overset{O}{\overset{\|}{}} - (OCH_2CH_2)_nOH$$

where n has an average value of 9.

Chemical Class: Alkoxylated Carboxylic Acids

Function: Surfactant - Emulsifying Agent

Technical/Other Names:
Polyethylene Glycol (9) Isostearate
Polyoxyethylene (9) Isostearate

Trade Name Mixture:
Silwax DMC-IS (Siltech LLC)

PEG-10 ISOSTEARATE

CAS No.: 56002-14-3 (Generic)

JPN Translation:
イソステアリン酸 PEG - 10

Empirical Formula:
$C_{38}H_{76}O_{12}$

Definition: PEG-10 Isostearate is the poly-ethylene glycol ester of Isostearic Acid (q.v.) that conforms generally to the formula:

$$C_{17}H_{20}C \overset{O}{\overset{\|}{—}} (OCH_2CH_2)_nOH$$

where n has an average value of 10.

Information Source: JCLS

Chemical Class: Alkoxylated Carboxylic Acids

Functions: Surfactant - Cleansing Agent; Surfactant - Emulsifying Agent

Technical/Other Names:
Polyethylene Glycol 500 Isostearate
Polyoxyethylene (10) Isostearate

PEG-12 ISOSTEARATE

CAS No.: 56002-14-3 (Generic)

JPN Translation:
イソステアリン酸 PEG - 12

Definition: PEG-12 Isostearate is the poly-ethylene glycol ester of Isostearic Acid (q.v.) that conforms generally to the formula:

$$C_{17}H_{20}C \overset{O}{\overset{\|}{—}} (OCH_2CH_2)_nOH$$

where n has an average value of 12.

Information Sources: JCLS, TSCA

Chemical Class: Alkoxylated Carboxylic Acids

Functions: Surfactant - Cleansing Agent; Surfactant - Emulsifying Agent

Technical/Other Names:
Polyethylene Glycol 600 Monoisostearate
Polyoxyethylene (12) Monoisostearate

Trade Names:
AEC PEG-12 Isostearate (A & E Connock)
Ethox MI-14 (Ethox)

PEG-20 ISOSTEARATE

JPN Translation:
イソステアリン酸 PEG - 20

Definition: PEG-20 Isostearate is the poly-ethylene glycol ester of Isostearic Acid (q.v.) that conforms generally to the formula:

$$C_{17}H_{20}C \overset{O}{\overset{\|}{—}} (OCH_2CH_2)_nOH$$

where n has an average value of 20.

Chemical Class: Alkoxylated Carboxylic Acids

Functions: Surfactant - Cleansing Agent; Surfactant - Emulsifying Agent

Technical/Other Names:
Polyethylene Glycol 1000 Monoisostearate
Polyoxyethylene (20) Monoisostearate

Trade Name:
Emalex PEIS-20 (Nihon Emulsion)

PEG-30 ISOSTEARATE

JPN Translation:
イソステアリン酸 PEG - 30

Definition: PEG-30 Isostearate is the poly-ethylene glycol ester of Isosearic Acid (q.v.) that conforms generally to the formula:

$$C_{17}H_{20}C \overset{O}{\overset{\|}{—}} (OCH_2CH_2)_nOH$$

Where n has an average value of 30.

Information Source: JCLS

Chemical Class: Alkoxylated Carboxylic Acids

Functions: Surfactant - Cleansing Agent; Surfactant - Emulsifying Agent

PEG-40 ISOSTEARATE

JPN Translation:
イソステアリン酸 PEG - 40

Definition: PEG-40 Isostearate is the poly-ethylene glycol ester of Isostearic Acid (q.v.) that conforms generally to the formula:

$$C_{17}H_{20}C \overset{O}{\overset{\|}{—}} (OCH_2CH_2)_nOH$$

where n has an average value of 40.

Information Source: JCLS

Chemical Class: Alkoxylated Carboxylic Acids

Functions: Surfactant - Cleansing Agent; Surfactant - Emulsifying Agent

PEG-15 JOJOBA ACID

Definition: PEG-15 Jojoba Acid is the poly-ethylene glycol derivative of the acids obtained from the saponification of jojoba oil with an average ethoxylation value of 15.

Chemical Class: Alkoxylated Carboxylic Acids

Function: Surfactant - Emulsifying Agent

Technical/Other Names:
Polyethylene Glycol (15) Jojoba Acid
Polyoxyethylene (15) Jojoba Acid

PEG-26 JOJOBA ACID

Definition: PEG-26 Jojoba Acid is the poly-ethylene glycol derivative of the acids obtained from the saponification of jojoba oil with an average ethoxylation value of 26.

Chemical Class: Alkoxylated Carboxylic Acids

Functions: Surfactant - Cleansing Agent; Surfactant - Emulsifying Agent; Surfactant - Solubilizing Agent

Technical/Other Names:
Polyethylene Glycol (26) Jojoba Acid
Polyoxyethylene (26) Jojoba Acid

Trade Name Mixtures:
Extrapone Jojoba Special (2/032990) (Symrise)
Oxypon 328 (Zschimmer & Schwarz)

PEG-40 JOJOBA ACID

Definition: PEG-40 Jojoba Acid is the poly-ethylene glycol derivative of the acids obtained from the saponification of jojoba oil with an average ethoxylation value of 40.

Chemical Class: Alkoxylated Carboxylic Acids

Functions: Surfactant - Cleansing Agent; Surfactant - Emulsifying Agent; Surfactant - Solubilizing Agent

Technical/Other Names:
Polyethylene Glycol 2000 Jojoba Acid
Polyoxyethylene (40) Jojoba Acid

PEG-15 JOJOBA ALCOHOL

Definition: PEG-15 Jojoba Alcohol is the polyethylene glycol derivative of jojoba alcohol with an average ethoxylation value of 15.

Chemical Class: Alkoxylated Alcohols

Functions: Surfactant - Cleansing Agent; Surfactant - Emulsifying Agent

Technical/Other Names:
Polyethylene Glycol (15) Jojoba Alcohol
Polyoxyethylene (15) Jojoba Alcohol

PEG-26 JOJOBA ALCOHOL

Definition: PEG-26 Jojoba Alcohol is the polyethylene glycol derivative of jojoba alcohol with an average ethoxylation value of 26.

Chemical Class: Alkoxylated Alcohols

Functions: Surfactant - Cleansing Agent; Surfactant - Emulsifying Agent; Surfactant - Solubilizing Agent

Technical/Other Names:
Polyethylene Glycol (26) Jojoba Alcohol
Polyoxyethylene (26) Jojoba Alcohol

Trade Name Mixtures:
Extrapone Jojoba Special (2/032990) (Symrise)
Oxypon 328 (Zschimmer & Schwarz)

PEG-40 JOJOBA ALCOHOL

Definition: PEG-40 Jojoba Alcohol is the polyethylene glycol derivative of jojoba alcohol with an average ethoxylation value of 40.

Chemical Class: Alkoxylated Alcohols

Functions: Surfactant - Cleansing Agent; Surfactant - Solubilizing Agent

Technical/Other Names:
Polyethylene Glycol 2000 Jojoba Alcohol
Polyoxyethylene (40) Jojoba Alcohol

PEG-3 LANOLATE

CAS No.: 68459-50-7 (Generic)

Definition: PEG-3 Lanolate is the polyethylene glycol ester of Lanolin Acid (q.v.) that conforms generally to the formula:

$$RC \overset{O}{\overset{\|}{-}} (OCH_2CH_2)_nOH$$

where RCO- represents the fatty acids derived from lanolin and n has an average value of 3.

Chemical Classes: Alkoxylated Carboxylic Acids; Lanolin and Lanolin Derivatives

Functions: Skin-Conditioning Agent - Emollient; Surfactant - Emulsifying Agent

Technical/Other Names:
Polyethylene Glycol (3) Lanolate
Polyoxyethylene (3) Lanolate

PEG-4 LANOLATE

CAS No.: 68459-50-7 (Generic)

JPN Translation:
ラノリン脂肪酸 PEG - 4

Definition: PEG-4 Lanolate is the polyethylene glycol ester of Lanolin Acid (q.v.) that conforms generally to the formula:

$$RC \overset{O}{\overset{\|}{-}} (OCH_2CH_2)_nOH$$

where RCO- represents the fatty acids derived from lanolin and n has an average value of 4.

Information Sources: JCLS, JSCI

Chemical Classes: Alkoxylated Carboxylic Acids; Lanolin and Lanolin Derivatives

Functions: Skin-Conditioning Agent - Emollient; Surfactant - Emulsifying Agent

Technical/Other Names:
Polyethylene Glycol 200 Lanolate
Polyethylene Glycol 200 Lanolin Fatty Acid
Polyoxyethylene (4) Lanolate

PEG-5 LANOLATE

CAS No.: 68459-50-7 (Generic)

Definition: PEG-5 Lanolate is the polyethylene glycol ester of Lanolin Acid (q.v.) that conforms generally to the formula:

$$RC \overset{O}{\overset{\|}{-}} (OCH_2CH_2)_nOH$$

where RCO- represents the fatty acids derived from lanolin and n has an average value of 5.

Information Source: TSCA

Chemical Classes: Alkoxylated Carboxylic Acids; Lanolin and Lanolin Derivatives

Function: Surfactant - Emulsifying Agent

Technical/Other Names:
Polyethylene Glycol (5) Lanolate
Polyoxyethylene (5) Lanolate

Trade Name:
Ritalafa 5 (RITA)

PEG-6 LANOLATE

CAS No.: 68459-50-7 (Generic)

JPN Translation:
ラノリン脂肪酸 PEG - 6

Definition: PEG-6 Lanolate is the polyethylene glycol ester of Lanolin Acid (q.v.). that conforms generally to the formula:

$$RC \overset{O}{\overset{\|}{-}} (OCH_2CH_2)_nOH$$

where RCO- represents the fatty acids derived from lanolin and n has an average value of 6.

Information Sources: JCLS, JSCI

Chemical Classes: Alkoxylated Carboxylic Acids; Lanolin and Lanolin Derivatives

Function: Surfactant - Emulsifying Agent

Technical/Other Names:
Polyethylene Glycol (6) Lanolate
Polyethylene Glycol 300 Lanolin Fatty Acid
Polyoxyethylene (6) Lanolate

PEG-7 LANOLATE

CAS No.: 68459-50-7 (Generic)

Definition: PEG-7 Lanolate is the polyethylene glycol ester of Lanolin Acid (q.v.) that conforms generally to the formula:

$$RC \overset{O}{\overset{\|}{-}} (OCH_2CH_2)_nOH$$

where RCO- represents the fatty acids derived from lanolin and n has an average value of 7.

Chemical Classes: Alkoxylated Carboxylic Acids; Lanolin and Lanolin Derivatives

Function: Surfactant - Emulsifying Agent

Technical/Other Names:
Polyethylene Glycol (7) Lanolate
Polyoxyethylene (7) Lanolate

PEG-8 LANOLATE

CAS No.: 68459-50-7 (Generic)

JPN Translation:
ラノリン脂肪酸 PEG - 8

Definition: PEG-8 Lanolate is the polyethylene glycol ester of Lanolin Acid (q.v.) that conforms generally to the formula:

$$RC \overset{O}{\overset{\|}{-}} (OCH_2CH_2)_nOH$$

where RCO- represents the fatty acids derived from lanolin and n has an average value of 8.

Information Sources: JCLS, JSCI

Chemical Classes: Alkoxylated Carboxylic Acids; Lanolin and Lanolin Derivatives

Function: Surfactant - Emulsifying Agent

Technical/Other Names:
   Polyethylene Glycol (8) Lanolate
   Polyethylene Glycol 400 Lanolin Fatty Acid
   Polyoxyethylene (8) Lanolate

PEG-10 LANOLATE

CAS No.: 68459-50-7 (Generic)

Definition: PEG-10 Lanolate is the polyethylene glycol ester of Lanolin Acid (q.v.) that conforms generally to the formula:

$$\underset{RC}{\overset{O}{\underset{\|}{}}} - (OCH_2CH_2)_nOH$$

where RCO- represents the fatty acids derived from lanolin and n has an average value of 10.

Information Source: TSCA

Chemical Classes: Alkoxylated Carboxylic Acids; Lanolin and Lanolin Derivatives

Function: Surfactant - Emulsifying Agent

Technical/Other Names:
   Polyethylene Glycol 500 Lanolate
   Polyoxyethylene (10) Lanolate

PEG-12 LANOLATE

CAS No.: 68459-50-7 (Generic)

JPN Translation:
   ラノリン脂肪酸 PEG - 12

Definition: PEG-12 Lanolate is the polyethylene glycol ester of Lanolin Acid (q.v.) that conforms generally to the formula:

$$\underset{RC}{\overset{O}{\underset{\|}{}}} - (OCH_2CH_2)_nOH$$

where RCO- represents the fatty acids derived from lanolin and n has as an average value of 12.

Information Sources: JCLS, JSCI

Chemical Classes: Alkoxylated Carboxylic Acids; Lanolin and Lanolin Derivatives

Function: Surfactant - Emulsifying Agent

Technical/Other Names:
   Polyethylene Glycol (12) Lanolate
   Polyethylene Glycol 600 Lanolin Fatty Acid
   Polyoxyethylene (12) Lanolate

PEG-15 LANOLATE

CAS No.: 68459-50-7 (Generic)

Definition: PEG-15 Lanolate is the polyethylene glycol ester of Lanolin Acid (q.v.). that conforms generally to the formula:

$$\underset{RC}{\overset{O}{\underset{\|}{}}} - (OCH_2CH_2)_nOH$$

where RCO- represents the fatty acids derived from lanolin and n has an average value of 15.

Chemical Classes: Alkoxylated Carboxylic Acids; Lanolin and Lanolin Derivatives

Function: Surfactant - Emulsifying Agent

Technical/Other Names:
   Polyethylene Glycol (15) Lanolate
   Polyoxyethylene (15) Lanolate

PEG-20 LANOLATE

CAS No.: 68459-50-7 (Generic)

JPN Translation:
   ラノリン脂肪酸 PEG - 20

Definition: PEG-20 Lanolate is the polyethylene glycol ester of Lanolin Acid (q.v.) that conforms generally to the formula:

$$\underset{RC}{\overset{O}{\underset{\|}{}}} - (OCH_2CH_2)_nOH$$

where RCO- represents the fatty acids derived from lanolin and n has an average value of 20.

Information Sources: JCLS, JSCI, TSCA

Chemical Classes: Alkoxylated Carboxylic Acids; Lanolin and Lanolin Derivatives

Functions: Surfactant - Emulsifying Agent; Surfactant - Solubilizing Agent

Technical/Other Names:
   Polyethylene Glycol 1000 Lanolate
   Polyethylene Glycol 1000 Lanolin Fatty Acid
   Polyoxyethylene (20) Lanolate

PEG-5 LANOLIN

CAS No.: 61790-81-6 (Generic)

JPN Translation:
   PEG - 5 ラノリン

Definition: PEG-5 Lanolin is a polyethylene glycol derivative of Lanolin (q.v.) with an average of 5 moles of ethylene oxide.

Information Sources: CIR: [S] IJT-18 (SUPPL. 1)1999, JCLS, TSCA

Chemical Classes: Alkoxylated Alcohols; Lanolin and Lanolin Derivatives

Function: Surfactant - Emulsifying Agent

Reported Product Categories: Shampoos (Non-coloring); Hair Conditioners

Technical/Other Names:
   Polyethylene Glycol (5) Lanolin
   Polyoxyethylene (5) Lanolin

PEG-10 LANOLIN

CAS No.: 61790-81-6 (Generic)

JPN Translation:
   PEG - 10 ラノリン

Definition: PEG-10 Lanolin is a polyethylene glycol derivative of Lanolin (q.v.) with an average of 10 moles of ethylene oxide.

Information Sources: CIR: [S] IJT-18 (SUPPL. 1)1999, JCLS

Chemical Classes: Alkoxylated Alcohols; Lanolin and Lanolin Derivatives

Function: Surfactant - Emulsifying Agent

Technical/Other Names:
   Polyethylene Glycol 500 Lanolin
   Polyoxyethylene (10) Lanolin

Trade Name:
   Nikkol TW-10 (Nikko)

PEG-20 LANOLIN

CAS No.: 61790-81-6 (Generic)

JPN Translation:
   PEG - 20 ラノリン

Definition: PEG-20 Lanolin is a polyethylene glycol derivative of Lanolin (q.v.) with an average of 20 moles of ethylene oxide.

Information Sources: 21CFR175.105, CIR: [S] JACT-1(4)1982, TSCA

Chemical Classes: Alkoxylated Alcohols; Lanolin and Lanolin Derivatives

Function: Surfactant - Emulsifying Agent

Technical/Other Names:
   Polyethylene Glycol 1000 Lanolin
   Polyoxyethylene (20) Lanolin

Trade Name:
   Nikkol TW-20 (Nikko)

PEG-24 LANOLIN

CAS No.: 61790-81-6 (Generic)

JPN Translation:
   PEG - 24 ラノリン

Definition: PEG-24 Lanolin is a poly-ethylene glycol derivative of Lanolin (q.v.) with an average of 24 moles of ethylene oxide.

Information Sources: CIR: [S] IJT-18 (SUPPL. 1)1999, JCLS, TSCA

Chemical Classes: Alkoxylated Alcohols; Lanolin and Lanolin Derivatives

Function: Surfactant - Emulsifying Agent

Technical/Other Names:
Polyethylene Glycol (24) Lanolin
Polyoxyethylene (24) Lanolin

PEG-27 LANOLIN

CAS Nos.: 8051-81-8; 61790-81-6 (Generic)

JPN Translation:
PEG - 27 ラノリン

Definition: PEG-27 Lanolin is a poly-ethylene glycol derivative of Lanolin (q.v.) with an average of 27 moles of ethylene oxide.

Information Sources: CIR: [S] JACT-1(4)-1982, JCLS, TSCA

Chemical Classes: Alkoxylated Alcohols; Lanolin and Lanolin Derivatives

Functions: Surfactant - Emulsifying Agent; Surfactant - Solubilizing Agent

Reported Product Category: Eye Shadows

Technical/Other Names:
Polyethylene Glycol (27) Lanolin
Polyoxyethylene (27) Lanolin

Trade Name:
Lanogel 21 (Amerchol)

PEG-30 LANOLIN

CAS No.: 61790-81-6 (Generic)

JPN Translation:
PEG - 30 ラノリン

Definition: PEG-30 Lanolin is a poly-ethylene glycol derivative of Lanolin (q.v.) with an average of 30 moles of ethylene oxide.

Information Sources: CIR: [S] JACT-1(4)-1982, JCLS, TSCA

Chemical Classes: Alkoxylated Alcohols; Lanolin and Lanolin Derivatives

Functions: Surfactant - Emulsifying Agent; Surfactant - Solubilizing Agent

Technical/Other Names:
Polyethylene Glycol (30) Lanolin
Polyoxyethylene (30) Lanolin

Trade Names:
Bellpol L-30 (Nippon Chemical)
Ivarlan 3405-L30 (Arch Personal Care Products)
Jeelan L-30 (Jeen)
Nikkol TW-30 (Nikko)

Trade Name Mixtures:
C-Base (Maybrook)
Ivarbase 3230 (Arch Personal Care Products)
Sebase (Croda Chemicals)

PEG-35 LANOLIN

CAS No.: 61790-81-6 (Generic)

JPN Translation:
PEG - 35 ラノリン

Definition: PEG-35 Lanolin is the poly-ethylene glycol derivative of Lanolin (q.v.) with an average of 35 moles of ethylene oxide.

Information Sources: CIR: [S] IJT-18 (SUPPL. 1)1999, JCLS

Chemical Classes: Alkoxylated Alcohols; Lanolin and Lanolin Derivatives

Functions: Surfactant - Cleansing Agent; Surfactant - Solubilizing Agent

Technical/Other Names:
Polyethylene Glycol (35) Lanolin
Polyoxyethylene (35) Lanolin

PEG-40 LANOLIN

CAS Nos.: 8051-82-9; 61790-81-6

JPN Translation:
PEG - 40 ラノリン

Definition: PEG-40 Lanolin is a poly-ethylene glycol derivative of Lanolin (q.v.) with an average of 40 moles of ethylene oxide.

Information Sources: CIR: [S] JACT-1(4)-1982, JCLS, TSCA

Chemical Classes: Alkoxylated Alcohols; Lanolin and Lanolin Derivatives

Functions: Surfactant - Cleansing Agent; Surfactant - Emulsifying Agent; Surfactant - Solubilizing Agent

Technical/Other Names:
Polyethylene Glycol 2000 Lanolin
Polyoxyethylene (40) Lanolin

Trade Name:
Laneto 40 (RITA)

PEG-50 LANOLIN

CAS No.: 61790-81-6 (Generic)

JPN Translation:
PEG - 50 ラノリン

Definition: PEG-50 Lanolin is a poly-ethylene glycol derivative of Lanolin (q.v.) with an average of 50 moles of ethylene oxide.

Information Sources: CIR: [S] JACT-1(4)-1982, JCLS, TSCA

Chemical Classes: Alkoxylated Alcohols; Lanolin and Lanolin Derivatives

Functions: Surfactant - Cleansing Agent; Surfactant - Solubilizing Agent

Reported Product Category: Hair Straighteners

Technical/Other Names:
Polyethylene Glycol (50) Lanolin
Polyoxyethylene (50) Lanolin

Trade Name:
Bellpol L-50 (Nippon Chemical)

PEG-55 LANOLIN

CAS No.: 61790-81-6 (Generic)

JPN Translation:
PEG - 55 ラノリン

Definition: PEG-55 Lanolin is the poly-ethylene glycol derivative of Lanolin (q.v.) with an average of 55 moles of ethylene oxide.

Information Sources: CIR: [S] IJT-18 (SUPPL. 1)1999, JCLS

Chemical Classes: Alkoxylated Alcohols; Lanolin and Lanolin Derivatives

Functions: Surfactant - Cleansing Agent; Surfactant - Solubilizing Agent

Technical/Other Names:
Polyethylene Glycol (55) Lanolin
Polyoxyethylene (55) Lanolin

PEG-60 LANOLIN

CAS No.: 61790-81-6 (Generic)

JPN Translation:
PEG - 60 ラノリン

Definition: PEG-60 Lanolin is a poly-ethylene glycol derivative of Lanolin (q.v.) with an average of 60 moles of ethylene oxide.

Information Sources: CIR: [S] JACT-1(4)-1982, CTFA D, JCLS, MI-13(8776), TSCA

Chemical Classes: Alkoxylated Alcohols; Lanolin and Lanolin Derivatives

Function: Surfactant - Cleansing Agent

Reported Product Categories: Hair Straighteners; Baby Products, Misc.; Personal Cleanliness Products, Misc.

Technical/Other Names:
Polyethylene Glycol 3000 Lanolin
Polyoxyethylene (60) Lanolin

Trade Names:
Ivarlan 3406 (Arch Personal Care Products)
Ivarlan 3407 (Arch Personal Care Products)
Ivarlan 3409-60 (Arch Personal Care Products)
Jeelan L-60 (Jeen)
Laneto 60 (RITA)
Solan (Croda, Inc.)
Solan 50 (Croda, Inc.)
Sympatens-LAN/600 (Kolb)

Trade Name Mixture:
Relaxer Concentrate #2 (Arch Personal Care Products)

PEG-75 LANOLIN

CAS Nos.: 8039-09-6; 61790-81-6 (Generic)

JPN Translation:
PEG - 75 ラノリン

Definition: PEG-75 Lanolin is a polyethylene glycol derivative of Lanolin (q.v.) with an average of 75 moles of ethylene oxide.

Information Sources: CIR: [S] JACT-1(4)-1982, CTFA S, JCLS, TSCA

Chemical Classes: Alkoxylated Alcohols; Lanolin and Lanolin Derivatives

Function: Surfactant - Cleansing Agent

Reported Product Categories: Tonics, Dressings, and Other Hair Grooming Aids; Hair Straighteners; Shampoos (Noncoloring); Baby Products, Misc.; Personal Cleanliness Products, Misc.; Hair Bleaches; Hair Conditioners; Hair Coloring Preparations, Misc.; Bath Oils, Tablets, and Salts; Cleansing Products (Cold Creams, Cleansing Lotions, Liquids and Pads); Skin Fresheners; Bath Preparations, Misc.; Body and Hand Preparations (Excluding Shaving Preparations); Hair Preparations (Noncoloring), Misc.; Permanent Waves

Technical/Other Names:
Polyethylene Glycol 4000 Lanolin
Polyoxyethylene (75) Lanolin

Trade Names:
AC PEG-75 Lanolin (Active Concepts)
Brooksgel 41 (Arch Personal Care Products)
Colonial LANCO-75 (Colonial Chemical Inc)

Ethoxylan 1685 (Cognis Care Chemicals/NJ)
Ethoxylan 1685 (Cognis Care Chemicals/PA)
Ethoxylan 1686 (Cognis Care Chemicals/NJ)
Ethoxylan 1686 (Cognis Care Chemicals/PA)
Ivarlan 3400 (Arch Personal Care Products)
Ivarlan 3401 (Arch Personal Care Products)
Ivarlan 3407-E (Arch Personal Care Products)
Ivarlan L575 (Arch Personal Care Products)
Ivarlan 3408W (Arch Personal Care Products)
Jeelan L-75 (Jeen)
Jeelan L75/50 (Jeen)
Lan-Aqua-Sol 75:50 (Fanning)
Lan-Aqua-Sol 75:100 (Fanning)
Laneto 50 (RITA)
Laneto 100 (RITA)
Lanogel 41 (Amerchol)
Lanopol-500 (Nippon Chemical)
Lanor OE 75 (Lanolines de la Tossee)
Lanoxal 75 (SEPPIC)
Lantox 55 (Lanaetex)
Lantox 110 (Lanaetex)
Oxypon 440 (Zschimmer & Schwarz)
PEG-75 Lanolin/50 (ChemMark)
Protalan L-75 (Protameen)
Solan E (Croda Chemicals)
Solulan 75 (Amerchol)
Solulan L-575 (Amerchol)
Super Solan (Croda, Inc.)
Sympatens-LAN/750 (Kolb)

Trade Name Mixtures:
Lanatein-25 (Lanaetex)
Lanobase S.E. (Lanaetex)
Lipowax R-2 (Lipo)
Proto-Lan 20 (Maybrook)
Relaxer Concentrate No. 1 (Arch Personal Care Products)

PEG-85 LANOLIN

CAS No.: 61790-81-6 (Generic)

JPN Translation:
PEG - 85 ラノリン

Definition: PEG-85 Lanolin is a polyethylene glycol derivative of Lanolin (q.v.) with an average of 85 moles of ethylene oxide.

Information Sources: CIR: [S] JACT-1(4)-1982, JCLS, TSCA

Chemical Classes: Alkoxylated Alcohols; Lanolin and Lanolin Derivatives

Function: Surfactant - Cleansing Agent

Technical/Other Names:
Polyethylene Glycol (85) Lanolin
Polyoxyethylene (85) Lanolin

Trade Names:
Brooksgel 61 (Arch Personal Care Products)
Ivarlan 3410 (Arch Personal Care Products)
Jeelan 85 (Jeen)

PEG-100 LANOLIN

CAS No.: 61790-81-6 (Generic)

JPN Translation:
PEG - 100 ラノリン

Definition: PEG-100 Lanolin is a polyethylene glycol derivative of Lanolin (q.v.) with an average of 100 moles of ethylene oxide.

Information Sources: CIR: [S] IJT-18 (SUPPL. 1)1999, JCLS, TSCA

Chemical Classes: Alkoxylated Alcohols; Lanolin and Lanolin Derivatives

Function: Surfactant - Cleansing Agent

Technical/Other Names:
Polyethylene Glycol (100) Lanolin
Polyoxyethylene (100) Lanolin

PEG-150 LANOLIN

CAS No.: 61790-81-6 (Generic)

JPN Translation:
PEG - 150 ラノリン

Definition: PEG-150 Lanolin is a polyethylene glycol derivative of Lanolin (q.v.) with an average of 150 moles of ethylene oxide.

Information Sources: CIR: [S] IJT-18 (SUPPL. 1)1999, JCLS

Chemical Classes: Alkoxylated Alcohols; Lanolin and Lanolin Derivatives

Function: Surfactant - Cleansing Agent

Technical/Other Names:
Polyethylene Glycol (6000) Lanolin
Polyoxyethylene (150) Lanolin

Trade Name:
Solan X (Croda Chemicals)

PEG-5 LANOLINAMIDE

Definition: PEG-5 Lanolinamide is the polyethylene glycol amide of Lanolin Acid (q.v.) with an average of 5 moles of ethylene oxide.

Chemical Classes: Alkoxylated Amides; Lanolin and Lanolin Derivatives

Functions: Hair Conditioning Agent; Viscosity Increasing Agent - Nonaqueous

Technical/Other Name:
 Polyethylene Glycol (5) Lanolinamide

PEG-75 LANOLIN OIL

CAS No.: 68648-38-4 (Generic)

JPN Translation:
 PEG - 75 液状ラノリン

Definition: PEG-75 Lanolin Oil is a polyethylene glycol derivative of Lanolin Oil (q.v.) with an average of 75 moles of ethylene oxide.

Information Sources: CIR: [S] IJT-18 (SUPPL. 1)1999, JCIC, JCLS, JSQI, TSCA

Chemical Classes: Alkoxylated Alcohols; Lanolin and Lanolin Derivatives

Functions: Surfactant - Cleansing Agent; Surfactant - Solubilizing Agent

Reported Product Categories: Tonics, Dressings, and Other Hair Grooming Aids; Hair Preparations (Non-coloring), Misc.; Hair Sprays (Aerosol Fixatives)

Technical/Other Names:
 Polyethylene Glycol 4000 Lanolin Oil
 Polyoxyethylene (75) Lanolin Oil
 Polyoxyethylene Liquid Lanolin (75E.O.)

Trade Name:
 Megalan E50 (Megachem)

PEG-75 LANOLIN WAX

Definition: PEG-75 Lanolin Wax is a polyethylene glycol derivative of Lanolin Wax (q.v.) with an average of 75 moles of ethylene oxide.

Information Source: CIR: [S] IJT-18 (SUPPL. 1)1999

Chemical Classes: Alkoxylated Alcohols; Lanolin and Lanolin Derivatives

Functions: Surfactant - Emulsifying Agent; Surfactant - Solubilizing Agent

Technical/Other Names:
 Polyethylene Glycol 4000 Lanolin Wax
 Polyoxyethylene (75) Lanolin Wax

PEG-2 LAURAMIDE

JPN Translation:
 PEG - 1 ラウリン酸アミド MEA

Definition: PEG-2 Lauramide is the polyethylene glycol amide of Lauric Acid (q.v.) that conforms to the formula:

$$CH_3(CH_2)_{10}C(O)-NH(CH_2CH_2O)_nH$$

where n have an average value of 2.

Information Source: JCLS

Chemical Class: Alkoxylated Carboxylic Acids

Functions: Surfactant - Cleansing Agent; Surfactant - Emulsifying Agent

PEG-3 LAURAMIDE

CAS No.: 26635-75-6 (Generic)

JPN Translation:
 PEG - 3 ラウラミド

Empirical Formula:
 $C_{18}H_{37}NO_4$

Definition: PEG-3 Lauramide is the polyethylene glycol amide of lauric acid that conforms to the formula:

$$CH_3(CH_2)_{10}C(O)-NH(CH_2CH_2O)_nH$$

where n has an average value of 3.

Information Sources: CTFA D, TSCA

Chemical Class: Alkoxylated Amides

Functions: Surfactant - Emulsifying Agent; Surfactant - Foam Booster

Technical/Other Names:
 Polyethylene Glycol (3) Lauryl Amide
 Polyoxyethylene (3) Lauryl Amide

Trade Name:
 Upamide L-2 (Universal Preserv-A-Chem)

PEG-5 LAURAMIDE

CAS No.: 26635-75-6 (Generic)

JPN Translation:
 PEG - 5 ラウラミド

Empirical Formula:
 $C_{22}H_{45}NO_6$

Definition: PEG-5 Lauramide is the polyethylene glycol amide of lauric acid that conforms generally to the formula:

$$CH_3(CH_2)_{10}C(O)-NH(CH_2CH_2O)_nH$$

where n has an average value of 5.

Information Source: TSCA

Chemical Class: Alkoxylated Amides

Function: Surfactant - Emulsifying Agent

Technical/Other Names:
 Polyethylene Glycol (5) Lauryl Amide
 Polyoxyethylene (5) Lauryl Amide

PEG-6 LAURAMIDE

CAS No.: 26635-75-6 (Generic)

JPN Translation:
 PEG - 6 ラウラミド

Empirical Formula:
 $C_{24}H_{49}NO_7$

Definition: PEG-6 Lauramide is the polyethylene glycol amide of Lauric Acid (q.v.) that conforms to the formula:

$$CH_3(CH_2)_{10}C(O)-NH(CH_2CH_2O)_nH$$

where n has an average value of 6.

Information Source: TSCA

Chemical Class: Alkoxylated Amides

Function: Surfactant - Emulsifying Agent

Technical/Other Names:
 Polyethylene Glycol 300 Lauryl Amide
 Polyoxyethylene (6) Lauryl Amide

Trade Names:
 NINOL L-5 (Stepan)
 Upamide L-5 (Universal Preserv-A-Chem)

PEG-11 LAURAMIDE

Definition: PEG-11 Lauramide is the polyethylene glycol amide of Lauric Acid (q.v.) that conforms to the formula:

$$CH_3(CH_2)_{10}C(O)-NH(CH_2CH_2O)_nH$$

where n has an average value of 11.

Information Source: JCLS

Chemical Class: Alkoxylated Amides

Function: Surfactant - Emulsifying Agent

PEG-2 LAURAMINE

Definition: PEG-2 Lauramine is the polyethylene glycol derivative of lauryl amine that conforms to the formula:

$$CH_3(CH_2)_{11}-N{\overset{\displaystyle (CH_2CH_2O)_xH}{\underset{\displaystyle (CH_2CH_2O)_yH}{}}}$$

where x + y has an average value of 2.

Information Sources: JCIC, JCLS

Chemical Class: Alkoxylated Amines

Functions: Antistatic Agent; Surfactant - Foam Booster

Technical/Other Names:
bis(2-Hydroxyethyl)dodecylamine
N,N-bis(2-Hydroxyethyl)lauramine
Ethanol, 2,2'-(Dodecylimino)bis-
N-Lauryl Diethanolamine
Polyethylene Glycol 100 Lauryl Amine
Polyoxyethylene (2) Lauryl Amine

PEG-3 LAURAMINE OXIDE

JPN Translation:
PEG - 3 ラウラミンオキシド

Empirical Formula:
$C_{18}H_{37}NO_4$

Definition: PEG-3 Lauramine Oxide is the tertiary amine oxide that conforms generally to the formula:

$$CH_3(CH_2)_{11}-N \begin{matrix} (CH_2CH_2O)_xH \\ | \\ \rightarrow O \\ | \\ (CH_2CH_2O)_yH \end{matrix}$$

where x + y has an average value of 3.

Information Source: JSQI

Chemical Class: Amine Oxides

Functions: Hair Conditioning Agent; Surfactant - Cleansing Agent; Surfactant - Foam Booster; Surfactant - Hydrotrope

PEG-2 LAURATE

CAS No. 141-20-8   **EINECS No.** 205-468-1

JPN Translation:
ラウリン酸 PEG - 2

Empirical Formula:
$C_{16}H_{32}O_4$

Definition: PEG-2 Laurate is the polyethylene glycol ester of lauric acid that conforms to the formula:

$$CH_3(CH_2)_{10}C \overset{O}{\overset{||}{-}} (OCH_2CH_2)_nOH$$

where n has an average value of 2.

Information Sources: 21CFR175.105, 21CFR175.300, 21CFR176.200, 21CFR176.210, 21CFR177.2800, 21CFR178.3910, CIR: [SQ] IJT-19(SUPPL.

2)2000, CTFA D, JCIC, JCLS, MI-13(3147), TSCA

Chemical Class: Alkoxylated Carboxylic Acids

Function: Surfactant - Emulsifying Agent

Technical/Other Names:
Diethylene Glycol Monolaurate
Diglycol Laurate
Diglycol Monolaurate
Dodecanoic Acid, 2-(2-Hydroxyethoxy)Ethyl Ester
Polyethylene Glycol 100 Monolaurate
Polyoxyethylene (2) Monolaurate

Trade Name:
Unipeg-DGL (Universal Preserv-A-Chem)

Trade Name Mixture:
Pegosperse 100 L (Lonza Inc./Lonza Ltd.)

PEG-4 LAURATE

CAS Nos.: 9004-81-3 (Generic); 10108-24-4

JPN Translation:
ラウリン酸 PEG - 4

Empirical Formula:
$C_{20}H_{40}O_6$

Definition: PEG-4 Laurate is the polyethylene glycol ester of lauric acid that conforms to the formula:

$$CH_3(CH_2)_{10}C \overset{O}{\overset{||}{-}} (OCH_2CH_2)_nOH$$

where n has an average value of 4.

Information Sources: 21CFR175.105, 21CFR175.300, 21CFR176.210, 21CFR178.3910, CIR: [SQ] IJT-19(SUPPL. 2)2000, CTFA D, JCLS, MI-13(7660), TSCA

Chemical Class: Alkoxylated Carboxylic Acids

Function: Surfactant - Emulsifying Agent

Reported Product Category: Cleansing Products (Cold Creams, Cleansing Lotions, Liquids and Pads)

Technical/Other Names:
Dodecanoic Acid, 2-[2-[2-(2-Hydroxyethoxy)Ethoxy]Ethoxy]Ethyl Ester
2-[2-[2-(2-Hydroxyethoxy)Ethoxy]Ethoxy]-Ethyl Dodecanoate
Polyethylene Glycol 200 Monolaurate
Polyoxyethylene (4) Monolaurate
Tetraethylene Glycol Laurate

Trade Names:
Chemax E-200-ML (Chemax)
Hetoxamate LA-4 (Heterene)
Jeemate 200-ML (Jeen)
LIPOPEG 2-L (Lipo)
Pegosperse 200 ML (Lonza Inc./Lonza Ltd.)

Protamate 200 ML (Protameen)
Sipoic ML-5 (Specialty Industrial)
STEPAN PEG 200 ML (Stepan)
Unipeg-200 ML (Universal Preserv-A-Chem)

Trade Name Mixtures:
Biodocarb L 1045 (G+G)
Nipaguard IPF (Clariant)
Nipaguard IPF (Clariant GmbH, Personal Care)

PEG-6 LAURATE

CAS Nos. 2370-64-1 9004-81-3 (Generic)   **EINECS No.** 219-136-9

JPN Translation:
ラウリン酸 PEG - 6

Empirical Formula:
$C_{24}H_{48}O_8$

Definition: PEG-6 Laurate is the polyethylene glycol ester of lauric acid that conforms to the formula:

$$CH_3(CH_2)_{10}C \overset{O}{\overset{||}{-}} (OCH_2CH_2)_nOH$$

where n has an average value of 6.

Information Sources: 21CFR175.105, 21CFR175.300, 21CFR176.210, 21CFR178.3910, CIR: [SQ] IJT-19(SUPPL. 2)2000, JCLS, MI-13(7660), TSCA

Chemical Class: Alkoxylated Carboxylic Acids

Function: Surfactant - Emulsifying Agent

Technical/Other Names:
Dodecanoic Acid, 17-Hydroxy-3,6,9,12,15-Pentaoxaheptadec-1-yl Ester
Hexaethylene Glycol Monolaurate
17-Hydroxy-3,6,9,12,15-Pentaoxaheptadec-1-yl Dodecanoate
Polyethylene Glycol 300 Monolaurate
Polyoxyethylene (6) Monolaurate

Trade Names:
STEPAN PEG 300 ML (Stepan)
Waglinol 312 (Industrial Quimica)

Trade Name Mixture:
Amisol HS-6 (Lucas Meyer GmbH)

PEG-8 LAURATE

CAS Nos.: 9004-81-3 (Generic); 35179-86-3

JPN Translation:
ラウリン酸 PEG - 8

Empirical Formula:
$C_{28}H_{56}O_{10}$

Definition: PEG-8 Laurate is the poly-ethylene glycol ester of lauric acid that conforms to the formula:

$$CH_3(CH_2)_{10}\overset{\displaystyle O}{\overset{\|}{C}}\!\!-\!\!(OCH_2CH_2)_nOH$$

where n has an average value of 8.

Information Sources: 21CFR175.105, 21CFR175.300, 21CFR176.170, 21CFR176.210, 21CFR177.1200, 21CFR177.1210, 21CFR177.2260, 21CFR177.2800, 21CFR178.3520, 21CFR178.3760, 21CFR178.3910, CIR: [SQ] IJT-19(SUPPL. 2)2000, CTFA D, JCLS, MI-13(7660), TSCA

Chemical Class: Alkoxylated Carboxylic Acids

Function: Surfactant - Emulsifying Agent

Technical/Other Names:
Dodecanoic Acid, 23-Hydroxy-3,6,9,12,15, 18,21-Heptaoxatricos-1-yl Ester
23-Hydroxy-3,6,9,12,15,18,21-Heptaoxa-tricos-1-yl Dodecanoate
Octaethylene Glycol Laurate
Polyethylene Glycol 400 Monolaurate
Polyoxyethylene (8) Monolaurate

Trade Names:
AEC PEG-8 Laurate (A & E Connock)
Chemax E-400-ML (Chemax)
Cithrol 4ML (Croda Chemicals)
Ethox ML-9 (Ethox)
Jeemate 400-ML (Jeen)
Laurate De PEG 400 (Prod'Hyg)
LIPOPEG 4-L (Lipo)
Lumulse 40-L (Lambent)
Pegosperse 400 ML (Lonza Inc./Lonza Ltd.)
Protamate 400 ML (Protameen)
Sipoic ML-9 (Specialty Industrial)
STEPAN PEG 400 ML (Stepan)
Unipeg-400 ML (Universal Preserv-A-Chem)

PEG-9 LAURATE

CAS Nos.   **EINECS No.**
106-08-1   203-359-3
9004-81-3 (Generic)

JPN Translation:
ラウリン酸 PEG - 9

Empirical Formula:
$C_{30}H_{60}O_{11}$

Definition: PEG-9 Laurate is the poly-ethylene glycol ester of lauric acid that conforms to the formula:

$$CH_3(CH_2)_{10}\overset{\displaystyle O}{\overset{\|}{C}}\!\!-\!\!(OCH_2CH_2)_nOH$$

where n has an average value of 9.

Information Sources: 21CFR175.105, 21CFR175.300, 21CFR177.2260, 21CFR177.2800, 21CFR178.3910, CIR: [SQ] IJT-19(SUPPL. 2)2000, JCLS, MI-13 (7660), TSCA

Chemical Class: Alkoxylated Carboxylic Acids

Function: Surfactant - Emulsifying Agent

Technical/Other Names:
Polyethylene Glycol 450 Monolaurate
Polyoxyethylene (9) Monolaurate

Trade Names:
AEC PEG-9 Laurate (A & E Connock)
DeTHOX ACID L-9 (DeForest)
Hetoxamate LA-9 (Heterene)

PEG-10 LAURATE

CAS No.: 9004-81-3 (Generic)

JPN Translation:
ラウリン酸 PEG - 10

Empirical Formula:
$C_{32}H_{64}O_{12}$

Definition: PEG-10 Laurate is the poly-ethylene glycol ester of lauric acid that conforms to the formula:

$$CH_3(CH_2)_{10}\overset{\displaystyle O}{\overset{\|}{C}}\!\!-\!\!(OCH_2CH_2)_nOH$$

where n has an average value of 10.

Information Sources: 21CFR175.105, 21CFR175.300, 21CFR177.2260, 21CFR177.2800, 21CFR178.3910, CIR: [SQ] IJT-19(SUPPL. 2)2000, JCLS, MI-13 (7660), TSCA

Chemical Class: Alkoxylated Carboxylic Acids

Function: Surfactant - Emulsifying Agent

Technical/Other Names:
Polyethylene Glycol 500 Monolaurate
Polyoxyethylene (10) Monolaurate

Trade Names:
AEC PEG-10 Laurate (A & E Connock)
Nikkol MYL-10 (Nikko)

PEG-12 LAURATE

CAS No.: 9004-81-3 (Generic)

JPN Translation:
ラウリン酸 PEG - 12

Definition: PEG-12 Laurate is the poly-ethylene glycol ester of lauric acid that conforms to the formula:

$$CH_3(CH_2)_{10}\overset{\displaystyle O}{\overset{\|}{C}}\!\!-\!\!(OCH_2CH_2)_nOH$$

where n has an average value of 12.

Information Sources: 21CFR175.105, 21CFR175.300, 21CFR176.170, 21CFR176.210, 21CFR177.1200, 21CFR177.2260, 21CFR177.2800, 21CFR178.3910, CIR: [SQ] IJT-19(SUPPL. 2)2000, CTFA D, JCLS, MI-13(7660), TSCA

Chemical Class: Alkoxylated Carboxylic Acids

Function: Surfactant - Emulsifying Agent

Technical/Other Names:
Polyethylene Glycol 600 Monolaurate
Polyoxyethylene (12) Monolaurate

Trade Names:
AEC PEG-12 Laurate (A & E Connock)
Ethox ML-14 (Ethox)
Pegosperse 600 ML (Lonza Inc./Lonza Ltd.)
Protamate 600 ML (Protameen)
STEPAN PEG 600 ML (Stepan)
Unipeg-600 ML (Universal Preserv-A-Chem)

PEG-14 LAURATE

CAS No.: 9004-81-3 (Generic)

JPN Translation:
ラウリン酸 PEG - 14

Definition: PEG-14 Laurate is a poly-ethylene glycol ester of lauric acid that conforms to the formula:

$$CH_3(CH_2)_{10}\overset{\displaystyle O}{\overset{\|}{C}}\!\!-\!\!(OCH_2CH_2)_nOH$$

where n has an average value of 14.

Information Sources: 21CFR175.105, 21CFR175.300, 21CFR177.2260, 21CFR177.2800, 21CFR178.3910, CIR: [SQ] IJT-19(SUPPL. 2)2000, JCLS, MI-13 (7660), TSCA

Chemical Class: Alkoxylated Carboxylic Acids

Function: Surfactant - Emulsifying Agent

Technical/Other Names:
Polyethylene Glycol (14) Monolaurate
Polyoxyethylene (14) Monolaurate

Trade Names:
AEC PEG-14 Laurate (A & E Connock)
Jeemate 600-ML (Jeen)

PEG-20 LAURATE

CAS No.: 9004-81-3 (Generic)

JPN Translation:
ラウリン酸 PEG - 20

Definition: PEG-20 Laurate is the poly-ethylene glycol ester of lauric acid that conforms to the formula:

$$CH_3(CH_2)_{10}C(=O){-}(OCH_2CH_2)_nOH$$

where n has an average value of 20.

Information Sources: 21CFR175.300, 21CFR176.210, 21CFR177.2260, 21CFR177.2800, 21CFR178.3910, CIR: [SQ] IJT-19(SUPPL. 2)2000, JCLS, MI-13 (7660), TSCA

Chemical Class: Alkoxylated Carboxylic Acids

Functions: Surfactant - Cleansing Agent; Surfactant - Solubilizing Agent

Technical/Other Names:
Polyethylene Glycol 1000 Monolaurate
Polyoxyethylene (20) Monolaurate

Trade Names:
AEC PEG-20 Laurate (A & E Connock)
Chemax E-1000-ML (Chemax)
Jeemate 1000-ML (Jeen)
Protamate 1000 ML (Protameen)
STEPAN PEG 1000 ML (Stepan)

PEG-32 LAURATE

CAS No.: 9004-81-3 (Generic)

JPN Translation:
ラウリン酸 PEG - 32

Definition: PEG-32 Laurate is the poly-ethylene glycol ester of lauric acid that conforms to the formula:

$$CH_3(CH_2)_{10}C(=O){-}(OCH_2CH_2)_nOH$$

where n has an average value of 32.

Information Sources: 21CFR175.300, 21CFR176.210, 21CFR177.2260, 21CFR177.2800, 21CFR178.3910, CIR: [SQ] IJT-19(SUPPL. 2)2000, JCLS, MI-13 (7660), TSCA

Chemical Class: Alkoxylated Carboxylic Acids

Functions: Surfactant - Cleansing Agent; Surfactant - Solubilizing Agent

Technical/Other Names:
Polyethylene Glycol 1540 Monolaurate
Polyoxyethylene (32) Monolaurate

Trade Names:
AEC PEG-32 Laurate (A & E Connock)
STEPAN PEG 1540 ML (Stepan)

PEG-75 LAURATE

CAS No.: 9004-81-3 (Generic)

JPN Translation:
ラウリン酸 PEG - 75

Definition: PEG-75 Laurate is the poly-ethylene glycol ester of lauric acid that conforms to the formula:

$$CH_3(CH_2)_{10}C(=O){-}(OCH_2CH_2)_nOH$$

where n has an average value of 75.

Information Sources: 21CFR175.300, 21CFR178.3910, CIR: [SQ] IJT-19(SUPPL. 2)2000, JCLS, MI-13(7660), TSCA

Chemical Class: Alkoxylated Carboxylic Acids

Function: Surfactant - Cleansing Agent

Technical/Other Names:
Polyethylene Glycol 4000 Monolaurate
Polyoxyethylene (75) Monolaurate

Trade Names:
AEC PEG-75 Laurate (A & E Connock)
STEPAN PEG 4000 ML (Stepan)

PEG-150 LAURATE

CAS No.: 9004-81-3 (Generic)

JPN Translation:
ラウリン酸 PEG - 150

Definition: PEG-150 Laurate is the poly-ethylene glycol ester of lauric acid that conforms to the formula:

$$CH_3(CH_2)_{10}C(=O){-}(OCH_2CH_2)_nOH$$

where n has an average value of 150.

Information Sources: 21CFR175.300, 21CFR178.3910, CIR: [SQ] IJT-19(SUPPL. 2)2000, JCLS, MI-13(7660), TSCA

Chemical Class: Alkoxylated Carboxylic Acids

Function: Surfactant - Cleansing Agent

Technical/Other Names:
Polyethylene Glycol 6000 Monolaurate
Polyoxyethylene (150) Monolaurate

Trade Names:
AEC PEG-150 Laurate (A & E Connock)
STEPAN PEG 6000 ML (Stepan)
Unipeg-6000 ML (Universal Preserv-A-Chem)

PEG-2 LAURATE SE

Definition: PEG-2 Laurate SE is a self-emulsifying grade of PEG-2 Laurate (q.v.)

that contains some sodium and/or potassium laurate.

Information Sources: CIR: [SQ] IJT-19 (SUPPL. 2)2000, JCLS

Chemical Class: Alkoxylated Carboxylic Acids

Function: Surfactant - Emulsifying Agent

Technical/Other Names:
Diethylene Glycol Monolaurate Self-Emulsifying
Polyethylene Glycol 100 Monolaurate Self-Emulsifying
Polyoxyethylene (2) Monolaurate Self-Emulsifying

Trade Name:
Lipo DGLS (Lipo)

Trade Name Mixture:
Pegosperse 100 L (Lonza Inc./Lonza Ltd.)

PEG-6 LAURATE/TARTRATE

Definition: PEG-6 Laurate/Tartrate is the mixed ester of PEG-6 and lauric and tartaric acids that conforms generally to the formula:

$$CH_3(CH_2)_{10}C(=O){-}(OCH_2CH_2)_nO{-}CCH(OH)CH(OH)COOH$$

where n has an average value of 6.

Chemical Class: Alkoxylated Carboxylic Acids

Function: Surfactant - Emulsifying Agent

Technical/Other Name:
PEG-6 Laurate/Tartarate

Trade Name:
Hydrophore 312 (Prod'Hyg)

PEG-180/LAURETH-50/TMMG COPOLYMER

Definition: PEG-180/Laureth-50/TMMG Copolymer is a copolymer of PEG-180 (q.v.), a polyethylene glycol ether of lauryl alcohol with an average ethoxylation value of 50, and tetramethoxymethylglycouril monomers.

Chemical Class: Synthetic Polymers

Function: Viscosity Increasing Agent - Aqueous

Trade Names:
Pure Thix 1450 (Sud-Chemie, United Catalysts)
Pure-Thix M (Sud-Chemie, United Catalysts)

PEG-10/LAURYL DIMETHICONE CROSS-POLYMER

Definition: PEG-10/Lauryl Dimethicone Crosspolymer is a copolymer of Lauryl Dimethicone (q.v.) crosslinked wtih diallyl PEG-10.

Chemical Classes: Siloxanes and Silanes; Synthetic Polymers

Functions: Surfactant - Suspending Agent; Viscosity Increasing Agent - Aqueous

Trade Name Mixtures:
KSG-34 (Shin-Etsu Chemical Co.)
KSG-340 (Shin-Etsu Chemical Co.)

PEG-15/LAURYL DIMETHICONE CROSS-POLYMER

Definition: PEG-15/Lauryl Dimethicone Crosspolymer is a copolymer of Lauryl Dimethicone (q.v.) crosslinked with diallyl PEG-15.

Chemical Classes: Siloxanes and Silanes; Synthetic Polymers

Function: Viscosity Increasing Agent - Aqueous

Trade Name Mixtures:
KSG-31 (Shin-Etsu Chemical Co.)
KSG-32 (Shin-Etsu Chemical Co.)
KSG-33 (Shin-Etsu Chemical Co.)
KSG-34 (Shin-Etsu Chemical Co.)
KSG-310 (Shin-Etsu Chemical Co.)
KSG-320 (Shin-Etsu Chemical Co.)
KSG-330 (Shin-Etsu Chemical Co.)
KSG-340 (Shin-Etsu Chemical Co.)

PEG-8 LINOLEATE

Empirical Formula:
$C_{34}H_{64}O_{10}$

Definition: PEG-8 Linoleate is the polyethylene glycol ester of linoleic acid that conforms to the formula:

$$CH_3(CH_2)_4CH$$
$$\|$$
$$CHCH_2CH \quad O$$
$$\| \quad \|$$
$$CH(CH_2)_7C-(OCH_2CH_2)_nOH$$

where n has an average value of 8.

Information Source: MI-13(7660)

Chemical Class: Alkoxylated Carboxylic Acids

Function: Surfactant - Emulsifying Agent

Technical/Other Names:
Polyethylene Glycol 400 Linoleate
Polyoxyethylene (8) Linoleate

Trade Name Mixture:
Efevit S (Fabriquimica)

PEG-8 LINOLENATE

Empirical Formula:
$C_{34}H_{62}O_{10}$

Definition: PEG-8 Linolenate is polyethylene glycol ester of linolenic acid that conforms to the formula:

$$CHCH_2CH_3$$
$$\|$$
$$CHCH_2CH$$
$$\|$$
$$CHCH_2CH \quad O$$
$$\| \quad \|$$
$$CH(CH_2)_7C-(OCH_2CH_2)_nOH$$

where n has an average value of 8.

Information Source: MI-13(7660)

Chemical Class: Alkoxylated Carboxylic Acids

Function: Surfactant - Emulsifying Agent

Technical/Other Names:
Polyethylene Glycol 400 Linolenate
Polyoxyethylene (8) Linolenate

Trade Name Mixture:
Efevit S (Fabriquimica)

PEG-2M

CAS No.: 25322-68-3 (Generic)

JPN Translation:
PEG - 2M

Definition: PEG-2M is the polymer of ethylene oxide that conforms generally to the formula:

$$H(OCH_2CH_2)_nCOOH$$

where n has an average value of 2000.

Information Sources: 21CFR172.770, 21CFR173.310, 21CFR175.300, 21CFR178.3910, JSQI, MI-13(7651), NF XVIII, TSCA, USAN

Chemical Classes: Alkoxylated Alcohols; Polymeric Ethers

Functions: Binder; Emulsion Stabilizer; Viscosity Increasing Agent - Aqueous

Reported Product Category: Hair Conditioners

Technical/Other Names:
PEG-2000
Polyethylene Glycol (2000)
Polyoxyethylene (2000)

Trade Name:
Polyox WSR N-10 (Amerchol)

Trade Name Mixture:
Spectraveil AQ (Uniqema, Belgium)

PEG-5M

CAS No.: 25322-68-3 (Generic)

JPN Translation:
PEG - 5M

Definition: PEG-5M is the polymer of ethylene oxide that conforms generally to the formula:

$$H(OCH_2CH_2)_nCOOH$$

where n has an average value of 5000.

Information Sources: 21CFR172.770, 21CFR173.310, 21CFR175.300, 21CFR178.3910, JSQI, MI-13(7651), NF XVIII, TSCA, USAN

Chemical Classes: Alkoxylated Alcohols; Polymeric Ethers

Functions: Binder; Emulsion Stabilizer; Viscosity Increasing Agent - Aqueous

Reported Product Categories: Shampoos (Non-coloring); Hair Conditioners

Technical/Other Names:
PEG-5000
Polyethylene Glycol (5000)
Polyoxyethylene (5000)

Trade Names:
Polyox WSR N-80 (Amerchol)
RITA PEO-1 (RITA)

PEG-7M

CAS No.: 25322-68-3 (Generic)

JPN Translation:
PEG - 7M

Definition: PEG-7M is the polymer of ethylene oxide that conforms generally to the formula:

$$H(OCH_2CH_2)_nCOOH$$

where n has an average value of 7000.

Information Sources: 21CFR172.770, 21CFR173.310, 21CFR175.300, 21CFR178.3910, JSQI, MI-13(7651), NF XVIII, TSCA, USAN

Chemical Classes: Alkoxylated Alcohols; Polymeric Ethers

Functions: Binder; Emulsion Stabilizer; Viscosity Increasing Agent - Aqueous

Reported Product Category: Shampoos (Non-coloring)

Technical/Other Names:
PEG-7000

Polyethylene Glycol (7000)
Polyoxyethylene (7000)

Trade Name:
Polyox WSR N-750 (Amerchol)

PEG-9M

CAS No.: 25322-68-3 (Generic)

JPN Translation:
PEG - 9M

Definition: PEG-9M is the polymer of ethylene oxide that conforms generally to the formula:

$$H(OCH_2CH_2)_nCOOH$$

where n has an average value of 9000.

Information Sources: 21CFR172.770, 21CFR173.310, 21CFR175.300, 21CFR178.3910, JSQI, MI-13(7651), NF XVIII, USAN

Chemical Classes: Alkoxylated Alcohols; Polymeric Ethers

Functions: Binder; Emulsion Stabilizer; Viscosity Increasing Agent - Aqueous

Technical/Other Names:
PEG-9000
Polyethylene Glycol 9000
Polyoxyethylene (9000)

Trade Names:
Alkox E-30G (Meisei)
RITA PEO-2 (RITA)

PEG-14M

CAS No.: 25322-68-3 (Generic)

JPN Translation:
PEG - 14M

Definition: PEG-14M is the polymer of ethylene oxide that conforms generally to the formula:

$$H(OCH_2CH_2)_nCOOH$$

where n has an average value of 14000.

Information Sources: 21CFR172.770, 21CFR173.310, 21CFR175.300, 21CFR178.3910, CIR: [SQ] JACT-12(5)-1993, JSQI, MI-13(7651), NF XVIII, TSCA, USAN

Chemical Classes: Alkoxylated Alcohols; Polymeric Ethers

Functions: Binder; Emulsion Stabilizer; Viscosity Increasing Agent - Aqueous

Reported Product Categories: Shampoos (Non-coloring); Shaving Preparations, Misc.; Shaving Cream (Aerosol, Brushless and Lather); Bath Oils, Tablets, and Salts; Bath

Soaps and Detergents; Cleansing Products (Cold Creams, Cleansing Lotions, Liquids and Pads)

Technical/Other Names:
PEG-14000
Polyethylene Glycol (14000)
Polyoxyethylene (14000)

Trade Names:
Polyox WSR-205 (Amerchol)
Polyox WSR N-3000 (Amerchol)

PEG-20M

CAS No.: 25322-68-3 (Generic)

JPN Translation:
PEG - 20M

Definition: PEG-20M is the polymer of ethylene oxide that conforms generally to the formula:

$$H(OCH_2CH_2)_nCOOH$$

where n has an average value of 20000.

Information Sources: 21CFR172.770, 21CFR173.310, 21CFR175.300, 21CFR178.3910, CIR: [SQ] JACT-12(5)-1993, JSQI, MI-13(7651), NF XIX, TSCA, USAN

Chemical Classes: Alkoxylated Alcohols; Polymeric Ethers

Functions: Binder; Emulsion Stabilizer; Viscosity Increasing Agent - Aqueous

Technical/Other Names:
PEG-20000
Polyethylene Glycol 20000
Polyoxyethylene (20000)

Trade Name Mixture:
Vegeles SR (Laboratoires Serobiologiques)

PEG-23M

CAS No.: 25322-68-3 (Generic)

JPN Translation:
PEG - 23M

Definition: PEG-23M is the polymer of ethylene oxide that conforms generally to the formula:

$$H(OCH_2CH_2)_nCOOH$$

where n has an average value of 23000.

Information Sources: 21CFR172.770, 21CFR173.310, 21CFR175.300, 21CFR178.3910, JSQI, MI-13(7651), NF XVIII, USAN

Chemical Classes: Alkoxylated Alcohols; Polymeric Ethers

Functions: Binder; Emulsion Stabilizer; Viscosity Increasing Agent - Aqueous

Technical/Other Names:
PEG-23000
Polyethylene Glycol (23000)
Polyoxyethylene (23000)

Trade Names:
Polyox WSR N-12K (Amerchol)
RITA PEO-3 (RITA)

PEG-25M

CAS No.: 25322-68-3 (Generic)

JPN Translation:
PEG - 25M

Definition: PEG-25M is the polymer of ethylene oxide that conforms generally to the formula:

$$H(OCH_2CH_2)_nCOOH$$

where n has a value of 25000.

Information Source: JSQI

Chemical Classes: Alkoxylated Alcohols; Polymeric Ethers

Functions: Binder; Emulsion Stabilizer; Viscosity Increasing Agent - Aqueous

Technical/Other Names:
PEG-25000
Polyethylene Glycol (25000)
Polyoxyethylene (25000)

PEG-45M

CAS No.: 25322-68-3 (Generic)

JPN Translation:
PEG - 45M

Definition: PEG-45M is the polymer of ethylene oxide that conforms generally to the formula:

$$H(OCH_2CH_2)_nCOOH$$

where n has an average value of 45000.

Information Sources: 21CFR172.770, 21CFR173.310, 21CFR175.300, 21CFR178.3910, JSQI, MI-13(7651), NF XVIII, USAN

Chemical Classes: Alkoxylated Alcohols; Polymeric Ethers

Functions: Binder; Emulsion Stabilizer; Viscosity Increasing Agent - Aqueous

Reported Product Category: Shampoos (Non-coloring)

Technical/Other Names:
PEG-45000

Polyethylene Glycol (45000)
Polyoxyethylene (45000)

Trade Names:
Polyox WSR N-60K (Amerchol)
RITA PEO-8 (RITA)

PEG-65M

CAS No.: 25322-68-3 (Generic)

Definition: PEG-65M is the polymer of ethylene oxide that conforms generally to the formula:

$$H(OCH_2CH_2)_nCOOH$$

where n has an average value of 65000.

Chemical Classes: Alkoxylated Alcohols; Polymeric Ethers

Functions: Binder; Emulsion Stabilizer; Viscosity Increasing Agent - Aqueous

Technical/Other Names:
Polyethylene Glycol (65000)
Polyoxyethylene (65000)

Trade Name:
Alkox E-100 (Meisei)

PEG-90M

CAS No.: 25322-68-3 (Generic)

JPN Translation:
PEG - 90M

Definition: PEG-90M is the polymer of ethylene oxide that conforms generally to the formula:

$$H(OCH_2CH_2)_nCOOH$$

where n has an average value of 90000.

Information Sources: 21CFR172.770, 21CFR173.310, 21CFR175.300, 21CFR178.3910, JSQI, MI-13(7651), NF XVIII, TSCA, USAN

Chemical Classes: Alkoxylated Alcohols; Polymeric Ethers

Functions: Binder; Emulsion Stabilizer; Viscosity Increasing Agent - Aqueous

Technical/Other Names:
PEG-90000
Polyethylene Glycol (90000)
Polyoxyethylene (90000)

Trade Names:
Polyox WSR-301 (Amerchol)
RITA PEO-18 (RITA)

PEG-115M

CAS No.: 25322-68-3 (Generic)

JPN Translation:
PEG - 115M

Definition: PEG-115M is the polymer of ethylene oxide that conforms generally to the formula:

$$H(OCH_2CH_2)_nCOOH$$

where n has an average value of 115000.

Information Sources: 21CFR172.770, 21CFR173.310, 21CFR175.300, 21CFR178.3910, JSQI, MI-13(7651)

Chemical Classes: Alkoxylated Alcohols; Polymeric Ethers

Functions: Binder; Emulsion Stabilizer; Viscosity Increasing Agent - Aqueous

Technical/Other Names:
PEG-115000
Polyethylene Glycol (115000)
Polyoxyethylene (115000)

Trade Name:
Alkox E-240 (Meisei)

PEG-160M

CAS No.: 25322-68-3 (Generic)

Definition: PEG-160M is a polymer of ethylene oxide that conforms generally to the formula:

$$H(OCH_2CH_2)_nCOOH$$

where n has an average value of 160000.

Chemical Classes: Alkoxylated Alcohols; Polymeric Ethers

Functions: Binder; Emulsion Stabilizer; Viscosity Increasing Agent - Aqueous

Technical/Other Names:
Polyethylene Glycol (160000)
Polyoxyethylene (160000)

Trade Name:
RITA PEO-27 (RITA)

PEG-16 MACADAMIA GLYCERIDES

Definition: PEG-16 Macadamia Glycerides is the polyethylene glycol derivative of the mono- and diglycerides derived from macadamia nut oil with an average of 16 moles of ethylene oxide.

Chemical Classes: Alkoxylated Alcohols; Glyceryl Esters and Derivatives

Functions: Skin-Conditioning Agent - Emollient; Surfactant - Emulsifying Agent

Technical/Other Names:
Polyethylene Glycol (16) Macadamia Glycerides
Polyoxyethylene (16) Macadamia Glycerides

Trade Name:
Florasolvs PEG-16 Macadamia (Floratech)

PEG-70 MANGO GLYCERIDES

Definition: PEG-70 Mango Glycerides is a polyethylene glycol derivative of the mono- and diglycerides from mango seed oil containing an average of 70 moles of ethylene oxide.

Chemical Classes: Alkoxylated Alcohols; Glyceryl Esters and Derivatives

Functions: Skin-Conditioning Agent - Emollient; Surfactant - Cleansing Agent; Surfactant - Solubilizing Agent

Technical/Other Names:
Polyethylene Glycol (70) Mango Glycerides
Polyoxyethylene (70) Mango Glycerides

Trade Name:
Lipex 203 E-70 (Karlshamns AB)

PEG-20 MANNITAN LAURATE

Definition: PEG-20 Mannitan Laurate is an ethoxylated mannitan ester of lauric acid with an average of 20 moles of ethylene oxide. It conforms generally to the formula:

where x+y+z has an average value of 20.

Chemical Classes: Alkoxylated Alcohols; Esters

Functions: Surfactant - Cleansing Agent; Surfactant - Emulsifying Agent

Technical/Other Names:
Polyethylene Glycol 1000 Mannitan Laurate
Polyoxyethylene (20) Mannitan Laurate

PEG-75 MEADOWFOAM OIL

Definition: PEG-75 Meadowfoam Oil is a polyethylene glycol derivative of Limnanthes Alba (Meadowfoam) Seed Oil (q.v.) with an average of 75 moles of ethylene oxide.

Chemical Classes: Alkoxylated Alcohols; Glyceryl Esters and Derivatives

Functions: Skin-Conditioning Agent - Emollient; Surfactant - Emulsifying Agent; Surfactant - Solubilizing Agent

Technical/Other Names:
Polyethylene Glycol 4000 Meadowfoam Oil
Polyoxyethylene (75) Meadowfoam Oil

Trade Names:
Meadow-Sol 75:50 (Fanning)
Meadowsol 75:75 (Fanning)
Meadow-Sol 75:100 (Fanning)

PEG-8 METHICONE

Definition: PEG-8 Methicone is the polyethylene glycol derivative of Methicone (q.v.) containing an average of 8 moles of ethylene oxide.

Chemical Class: Siloxanes and Silanes

Functions: Hair Conditioning Agent; Skin-Conditioning Agent - Emollient

Trade Name:
Masil SF 19 CG (BASF)

PEG-6 METHICONE ACETATE

Definition: PEG-6 Methicone Acetate is a partial ester of acetic acid and a polyethylene glycol derivative of Methicone (q.v.) containing an average of 6 moles of ethylene oxide.

Chemical Classes: Esters; Siloxanes and Silanes

Functions: Hair Conditioning Agent; Skin-Conditioning Agent - Miscellaneous; Surfactant - Emulsifying Agent

Trade Name:
Wacker-Belsil DMC 6035 (Wacker-Chemie)

PEG-1182 METHYL ESTER SERICIN

Definition: PEG-1182 Methyl Ester Sericin is a polyethylene glycol derivative of the methyl ester of Sericin (q.v.) containing an average of 1182 moles of ethylene oxide.

Chemical Classes: Alkoxylated Alcohols; Esters

Function: Skin-Conditioning Agent - Miscellaneous

Trade Name:
PEG-Sericin Nanoparticle (Bioland)

PEG-3 METHYL ETHER

CAS No.	EINECS No.
112-35-6	203-962-1

Definition: PEG-3 Methyl Ether is the polyethylene glycol ether of Methyl Alcohol (q.v.) that conforms generally to the formula:

$$CH_3(OCH_2CH_2)_nOH$$

where n has an average value of 3.

Information Source: TSCA

Chemical Class: Alkoxylated Alcohols

Functions: Humectant; Solvent

Technical/Other Names:
Polyethylene Glycol (3) Methyl Ether
Polyoxyethylene (3) Methyl Ether
Triethylene Glycol Methyl Ether

Trade Name:
Hymol TM (Toho)

PEG-4 METHYL ETHER

Definition: PEG-4 Methyl Ether is the polyethylene glycol ether of Methyl Alcohol (q.v.) that conforms generally to the formula:

$$CH_3(OCH_2CH_2)_nOH$$

where n has an average value of 4.

Chemical Class: Alkoxylated Alcohols

Functions: Humectant; Solvent

Technical/Other Names:
Polyethylene Glycol 200 Monomethyl Ether
Polyoxyethylene (4) Monomethyl Ether

Trade Names:
Hymol PM (Toho)
Hymol TPM (Toho)

PEG-6 METHYL ETHER

CAS No.: 9004-74-4 (Generic)

Empirical Formula:
$C_{13}H_{28}O_4$

Definition: PEG-6 Methyl Ether is the polyethylene glycol ether of Methyl Alcohol (q.v.) that conforms to the formula:

$$CH_3(OCH_2CH_2)_nOH$$

where n has an average value of 6.

Information Sources: NF XVIII, TSCA, USAN

Chemical Class: Alkoxylated Alcohols

Function: Solvent

Reported Product Category: Bath Soaps and Detergents

Technical/Other Names:
Polyethylene Glycol 300 Methyl Ether
Polyoxyethylene (6) Methyl Ether

Trade Names:
CARBOWAX MPEG 350 (Dow Chemical)
Upiwax 350 (Universal Preserv-A-Chem)

PEG-7 METHYL ETHER

Definition: PEG-7 Methyl Ether is the polyethylene glycol ether of Methyl Alcohol (q.v.) that conforms generally to the formula:

$$CH_3(OCH_2CH_2)_nOH$$

where n has an average value of 7.

Chemical Class: Alkoxylated Alcohols

Function: Solvent

Technical/Other Names:
Polyethylene Glycol (7) Methyl Ether
Polyxoyethylene (7) Methyl Ether

Trade Name:
Marlipal 1/7 (Sasol GmbH - Marl)

PEG-6 METHYL ETHER DIMETHICONE

Definition: PEG-6 Methyl Ether Dimethicone is the methyl ether of a polyethylene glycol derivative of Dimethicone (q.v.) containing an average of 6 moles of ethylene oxide.

Chemical Classes: Ethers; Siloxanes and Silanes

Functions: Hair Conditioning Agent; Skin-Conditioning Agent - Miscellaneous; Surfactant - Emulsifying Agent

Trade Name:
KF-618 (Shin-Etsu Chemical Co.)

Trade Name Mixture:
KM-906 (Shin-Etsu Chemical Co.)

PEG-7 METHYL ETHER DIMETHICONE

Definition: PEG-7 Methyl Ether Dimethicone is the methyl ether of a derivative of Dimethicone (q.v.) containing an average of 7 moles of ethylene oxide.

Chemical Classes: Ethers; Siloxanes and Silanes

Functions: Hair Conditioning Agent; Skin-Conditioning Agent - Humectant; Surfactant - Emulsifying Agent

Trade Names:
FZ-2405 (Nippon Unicar)
SS-2803 (Nippon Unicar)

PEG-8 METHYL ETHER DIMETHICONE

Definition: PEG-8 Methyl Ether Dimethicone is the methyl ether of a derivative of

Dimethicone (q.v.) containing an average of 8 moles of ethylene oxide.

Chemical Class: Siloxanes and Silanes

Functions: Hair Conditioning Agent; Humectant; Surfactant - Cleansing Agent; Surfactant - Suspending Agent

Technical/Other Name:
Polyethylene Glycol (8) Methyl Ether Dimethicone

Trade Names:
FZ-2408 (Nippon Unicar)
FZ-2423 (Nippon Unicar)

PEG-9 METHYL ETHER DIMETHICONE

Definition: PEG-9 Methyl Ether Dimethicone is a siloxane polymer that conforms generally to the formula:

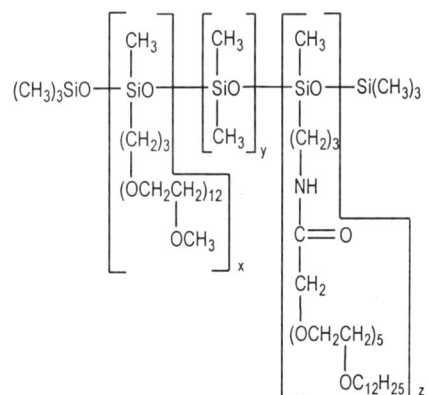

where n has an average value of 9.

Chemical Classes: Ethers; Siloxanes and Silanes

Functions: Hair Conditioning Agent; Skin-Conditioning Agent - Miscellaneous; Surfactant - Emulsifying Agent

Trade Name:
KF-6016 (Shin Etsu)

PEG-10 METHYL ETHER DIMETHICONE

Definition: PEG-10 Methyl Ether Dimethicone is the methyl ether of a derivative of Dimethicone (q.v.) containing an average of 10 moles of ethylene oxide.

Chemical Class: Siloxanes and Silanes

Functions: Hair Conditioning Agent; Skin-Conditioning Agent - Miscellaneous; Surfactant - Emulsifying Agent

Trade Names:
KF-351AS (Shin Etsu)
SS-2801 (Nippon Unicar)
SS-2802 (Nippon Unicar)

PEG-11 METHYL ETHER DIMETHICONE

Definition: PEG-11 Methyl Ether Dimethicone is the methyl ether of a derivative of

Dimethicone (q.v.) containing an average of 11 moles of ethylene oxide.

Chemical Class: Siloxanes and Silanes

Functions: Hair Conditioning Agent; Skin-Conditioning Agent - Miscellaneous; Surfactant - Emulsifying Agent

Trade Names:
KF-6011 (Shin Etsu)
KF-6018 (Shin-Etsu Chemical Co.)
KF-351A (Shin Etsu)

PEG-32 METHYL ETHER DIMETHICONE

Definition: PEG-32 Methyl Ether Dimethicone is the methyl ether of a derivative of Dimethicone (q.v.) containing an average of 32 moles of ethylene oxide.

Chemical Class: Siloxanes and Silanes

Functions: Hair Conditioning Agent; Skin-Conditioning Agent - Emollient; Surfactant - Emulsifying Agent

Trade Name:
KF-6004 (Shin-Etsu Chemical Co.)

PEG-12 METHYL ETHER LAUROXY PEG-5 AMIDOPROPYL DIMETHICONE

Definition: PEG-12 Methyl Ether Lauroxy PEG-5 Amidopropyl Dimethicone is the silicone polymer that conforms generally to the formula:

Chemical Class: Synthetic Polymers

Function: Hair Conditioning Agent

Trade Name:
BY16-906 (DCTS)

PEG-114 METHYLETHER POLYEPSILON CAPRALACTONE

Definition: PEG-114 Methylether Polyepsilon Capralactone is the block

copolymer that conforms generally to the formula:

where x has an average value of 114.

Chemical Class: Synthetic Polymers

Functions: Buffering Agent; Film Former; Hair Conditioning Agent; Hair Fixative; Hair-Waving/Straightening Agent; pH Adjuster; Suspending Agent - Nonsurfactant

Trade Name:
PCG-101 (Pacific)

PEG-120 METHYL GLUCOSE DIOLEATE

CAS No.: 86893-19-8

JPN Translation:
ジオレイン酸 PEG - 120 メチルグルコース

Definition: PEG-120 Methyl Glucose Dioleate is the polyethylene glycol ether of the diester of methyl glucose and oleic acid with an average of 120 moles of ethylene oxide.

Information Sources: JCIC, JCLS, JSQI

Chemical Classes: Alkoxylated Alcohols; Carbohydrates

Function: Surfactant - Cleansing Agent

Reported Product Categories: Bath Soaps and Detergents; Bath Oils, Tablets, and Salts; Bath Preparations, Misc.; Shampoos (Non-coloring); Cleansing Products (Cold Creams, Cleansing Lotions, Liquids and Pads); Hair Shampoos (Coloring); Personal Cleanliness Products, Misc.; Bubble Baths; Fragrance Preparations, Misc.

Technical/Other Names:
Polyethylene Glycol (120) Methyl Glucose Dioleate
Polyoxyethylene (120) Methyl Glucose Dioleate
Polyoxyethylene Methylglucoside Dioleate (120E.O.)

Trade Names:
AEC PEG-120 Methyl Glucose Dioleate (A & E Connock)
Antil 120 (Degussa Care Specialties)
Antil 127 (Degussa Care Specialties)
Glucamate DOE-120 (Amerchol)
Glutamate DOE-120 Syrup (Amerchol)

PEG-20 METHYL GLUCOSE DISTEARATE

JPN Translation:
ジステアリン酸 PEG - 20 メチルグルコ - ス

Definition: PEG-20 Methyl Glucose Distearate is the polyethylene glycol ether of the diester of methyl glucose and stearic acid with an average of 20 moles of ethylene oxide.

Information Sources: JCIC, JCLS

Chemical Classes: Carbohydrates; Esters

Functions: Skin-Conditioning Agent - Emollient; Surfactant - Emulsifying Agent

Technical/Other Name:
Polyoxyethylene Methylglucoside Distearate

Trade Name:
Glucam E-20 Distearate (Amerchol)

PEG-80 METHYL GLUCOSE LAURATE

Definition: PEG-80 Methyl Glucose Laurate is the polyethylene glycol ether of the ester of methyl glucose and lauric acid with an average of 80 moles of ethylene oxide.

Chemical Classes: Carbohydrates; Esters

Functions: Skin-Conditioning Agent - Emollient; Surfactant - Cleansing Agent; Surfactant - Solubilizing Agent

PEG-20 METHYL GLUCOSE SESQUI-CAPRYLATE/SESQUICAPRATE

Definition: PEG-20 Methyl Glucose Sesquicaprylate/Sesquicaprate is the polyethylene glycol ether of the mono and diesters of methyl glucose and caprylic and capric acids with an average of 20 moles of ethylene oxide.

Chemical Classes: Carbohydrates; Esters

Functions: Skin-Conditioning Agent - Emollient; Surfactant - Emulsifying Agent

PEG-20 METHYL GLUCOSE SESQUI-LAURATE

Definition: PEG-20 Methyl Glucose Sesquilaurate is the polyethylene glycol ether of the mono and diesters of methyl glucose and lauric acid with an average of 20 moles of ethylene oxide.

Chemical Classes: Carbohydrates; Esters

Functions: Skin-Conditioning Agent - Emollient; Surfactant - Emulsifying Agent

PEG-20 METHYL GLUCOSE SESQUISTEARATE

JPN Translation:
セスキステアリン酸 PEG - 20 メチルグルコース

Definition: PEG-20 Methyl Glucose Sesquistearate is the polyethylene glycol ether of the mono and diesters of methyl glucose and stearic acid with an average of 20 moles of ethylene oxide.

Information Sources: JCIC, JCLS, JSQI

Chemical Classes: Alkoxylated Alcohols; Carbohydrates

Functions: Skin-Conditioning Agent - Emollient; Surfactant - Emulsifying Agent

Reported Product Categories: Bath Oils, Tablets, and Salts; Cleansing Products (Cold Creams, Cleansing Lotions, Liquids and Pads); Foundations; Moisturizing Preparations; Bath Capsules; Face and Neck Preparations (Excluding Shaving Preparations); Eye Makeup Preparations, Misc.; Eyeliners; Indoor Tanning Preparations; Tonics, Dressings, and Other Hair Grooming Aids

Technical/Other Names:
PEG-20 Methyl Glucoside Sesquistearate
Polyethylene Glycol 1000 Methyl Glucose Sesquistearate
Polyoxyethylene (20) Methyl Glucose Sesquistearate

Trade Name:
Glucamate SSE-20 (Amerchol)

PEG-120 METHYL GLUCOSE TRIOLEATE

Definition: PEG-120 Methyl Glucose Trioleate is the polyethylene glycol ether of the triester of methyl glucose and oleic acid with an average of 120 moles of ethylene oxide.

Chemical Classes: Alkoxylated Alcohols; Carbohydrates

Functions: Skin-Conditioning Agent - Emollient; Surfactant - Cleansing Agent; Viscosity Increasing Agent - Aqueous

Technical/Other Names:
Polyethylene Glycol (120) Methyl Glucose Trioleate
Polyoxyethylene (120) Methyl Glucose Trioleate

Trade Name:
Glucamate LT (Amerchol)

PEG-2 MILK SOLIDS

Definition: PEG-2 Milk Solids is a polyethylene glycol derivative of milk solids.

Chemical Class: Protein Derivatives

Functions: Hair Conditioning Agent; Skin-Conditioning Agent - Miscellaneous

Reported Product Category: Paste Masks (Mud Packs)

Trade Name:
Galactene (Vevy)

PEG-13 MINK GLYCERIDES

CAS No.: 103819-45-0

Definition: PEG-13 Mink Glycerides is a polyethylene glycol derivative of mono and diglycerides derived from Mink Oil (q.v.) with an average of 13 moles of ethylene oxide.

Chemical Classes: Alkoxylated Alcohols; Glyceryl Esters and Derivatives

Functions: Skin-Conditioning Agent - Emollient; Surfactant - Emulsifying Agent

Technical/Other Name:
Polyethylene Glycol (13) Mink Glycerides

Trade Name:
Mink Oil W (Cosmetochem)

PEG-4 MONTANATE

CAS No.: 68476-04-0

Definition: PEG-4 Montanate is the polyethylene glycol ester of Montan Acid Wax (q.v.) with an average of 4 moles of ethylene oxide.

Chemical Class: Alkoxylated Carboxylic Acids

Function: Surfactant - Cleansing Agent

Technical/Other Names:
Fatty Acids, Montan-Wax, Ethoxylated
Montan-Wax Fatty Acids, Ethoxylated
PEG-200 Montanate
Polyethylene Glycol 200 Montanate
Polyoxyethylene (4) Montanate

Trade Name:
Hoechst Wax KST (Path Silicones)

PEG-25 MORINGA GLYCERIDES

Definition: PEG-25 Moringa Glycerides is a polyethylene glycol derivative of the fatty oil obtained from the seeds of *Moringa pterygosperma* containing an average of 25 moles of ethylene oxide.

Chemical Classes: Alkoxylated Alcohols; Glyceryl Esters and Derivatives

Functions: Skin-Conditioning Agent - Emollient; Surfactant - Emulsifying Agent

Technical/Other Names:
Polyethylene Glycol (25) Moringa Glycerides
Polyoxyethylene (25) Moringa Glycerides

Trade Name:
Florasolvs PEG-25 Moringa (Floratech)

PEG-42 MUSHROOM GLYCERIDES

Definition: PEG-42 Mushroom Glycerides is the polyethylene glycol derivative of the mono- and diglycerides of mushroom oil with an average of 42 moles of ethylene oxide.

Chemical Classes: Alkoxylated Alcohols; Glyceryl Esters and Derivatives

Functions: Skin-Conditioning Agent - Emollient; Surfactant - Emulsifying Agent

Technical/Other Names:
Polyethylene Glycol (42) Mushroom Glycerides
Polyoxyethylene (42) Mushroom Glyceride

Trade Name:
Campo Eburiko Oil Soluble (Campo)

PEG-8 MYRISTATE

JPN Translation:
ミリスチン酸 PEG - 8

Definition: PEG-8 Myristate is the polyethylene glycol ester of myristic acid that conforms generally to the formula:

$$CH_3(CH_2)_{12}C \overset{O}{\overset{\|}{}} - (OCH_2CH_2)_nOH$$

where n has an average value of 8.

Information Sources: 21CFR175.105, 21CFR175.300, 21CFR176.210, 21CFR177.1210, 21CFR177.2260, 21CFR177.2800, JCLS, MI-13(7660)

Chemical Class: Alkoxylated Carboxylic Acids

Function: Surfactant - Emulsifying Agent

Technical/Other Names:
Polyethylene Glycol 400 Myristate
Polyoxyethylene (8) Myristate

PEG-20 MYRISTATE

JPN Translation:
ミリスチン酸 PEG - 20

Definition: PEG-20 Myristate is the polyethylene glycol ester of myristic acid that conforms generally to the formula:

$$CH_3(CH_2)_{12}C \overset{O}{\overset{\|}{}} - (OCH_2CH_2)_nOH$$

where n has an average value of 20.

Information Sources: 21CFR175.300, 21CFR176.210, 21CFR177.2260, 21CFR177.2800, JCLS, MI-13(7660)

Chemical Class: Alkoxylated Carboxylic Acids

Functions: Surfactant - Emulsifying Agent; Surfactant - Solubilizing Agent

Technical/Other Names:
Polyethylene Glycol 1000 Myristate
Polyoxyethylene (20) Myristate

Trade Name Mixture:
Xalidrene (Vevy)

PEG-7 BETA-NAPHTHOL

Empirical Formula:
$C_{24}H_{36}O_8$

Definition: PEG-7 Beta-Naphthol is the polyethylene glycol ether of 2-Naphthol (q.v.) that conforms generally to the formula:

$$\text{(OCH}_2\text{CH}_2)_7\text{OH}$$

Chemical Class: Alkoxylated Alcohols

Function: Not Reported

Reported Product Category: Hair Dyes and Colors (All Types Requiring Caution Statements and Patch Tests)

Technical/Other Names:
PEG-7-2-Naphthol
Polyethylene Glycol (7) Betanaphthol
Polyoxyethylene (7) Betanaphthol

PEG-10 NONAFLUOROHEXYL DIMETHICONE COPOLYMER

Definition: PEG-10 Nonafluorohexyl Dimethicone Copolymer is the siloxane polymer that conforms generally to the formula:

$$(CH_3)_3SiO - \begin{bmatrix} CH_3 \\ | \\ SiO \\ | \\ CH_3 \end{bmatrix}_x \begin{bmatrix} CH_3 \\ | \\ SiO \\ | \\ (CH_2)_2 \\ | \\ C_4F_9 \end{bmatrix}_y \begin{bmatrix} CH_3 \\ | \\ SiO \\ | \\ (CH_2)_3 \\ | \\ O \\ | \\ (CH_2CH_2O)_nH \end{bmatrix}_z - Si(CH_3)_3$$

where n has an average value of 10.

Chemical Classes: Halogen Compounds; Siloxanes and Silanes; Synthetic Polymers

Functions: Hair Conditioning Agent; Skin-Conditioning Agent - Miscellaneous

Trade Name:
FPD-4668 (Shin Etsu)

PEG-180/OCTOXYNOL-40/TMMG COPOLYMER

Definition: PEG-180/Octoxynol-40/TMMG Copolymer is a copolymer of PEG-180 (q.v.), Octoxynol-40 (q.v.) and tetramethoxymethylglycouril monomers.

Chemical Class: Synthetic Polymers

Function: Viscosity Increasing Agent - Aqueous

Trade Name:
Pure-Thix L (Sud-Chemie, United Catalysts)

PEG-9 OCTYLDODECANOATE

Definition: PEG-9 Octyldodecanoate is the polyethylene glycol ester of octyldodecanoic acid that conforms generally to the formula:

$$CH_3(CH_2)_9\overset{O}{\overset{\|}{C}}HC - (OCH_2CH_2)_nOH$$
$$| \atop (CH_2)_7CH_3$$

where n has an average value of 9.

Chemical Class: Alkoxylated Carboxylic Acids

Function: Surfactant - Emulsifying Agent

Technical/Other Names:
Polyethylene Glycol (9) Octyldodecanoate
Polyoxyethylene (9) Octyldodecanoate

Trade Name:
Silube G20-9 (Siltech LLC)

PEG-23 OCTYLDODECANOATE

Definition: PEG-23 Octyldodecanoate is the polyethylene glycol ester of octyldodecanoic acid that conforms generally to the formula:

$$CH_3(CH_2)_9\overset{O}{\overset{\|}{C}}HC - (OCH_2CH_2)_nOH$$
$$| \atop (CH_2)_7CH_3$$

where n has an average value of 23.

Chemical Class: Alkoxylated Carboxylic Acids

Function: Surfactant - Emulsifying Agent

Technical/Other Names:
Polyethylene Glycol (23) Octyldodecanoate
Polyoxyethylene (23) Octyldodecanoate

Trade Name:
  Silube G20-23 (Siltech LLC)

PEG-3 OLEAMIDE

Empirical Formula:
  $C_{24}H_{47}NO_4$

Definition: PEG-3 Oleamide is the poly-ethylene glycol amide of oleic acid that conforms to the formula:

$$CH_3(CH_2)_7CH = CH(CH_2)_7\overset{\overset{\displaystyle O}{\|}}{C} - NH(CH_2CH_2O)_nH$$

where n has an average value of 3.

Chemical Class: Alkoxylated Amides

Functions: Surfactant - Emulsifying Agent; Surfactant - Foam Booster

Technical/Other Names:
  Polyethylene Glycol (3) Oleamide
  Polyoxyethylene (3) Oleamide

PEG-4 OLEAMIDE

Empirical Formula:
  $C_{28}H_{55}NO_6$

Definition: PEG-4 Oleamide is the poly-ethylene glycol amide of oleic acid that conforms to the formula:

$$CH_3(CH_2)_7CH = CH(CH_2)_7\overset{\overset{\displaystyle O}{\|}}{C} - NH(CH_2CH_2O)_nH$$

where n has an average value of 4.

Chemical Class: Alkoxylated Amides

Function: Surfactant - Emulsifying Agent

Technical/Other Names:
  Polyethylene Glycol 200 Oleyl Amide
  Polyoxyethylene (4) Oleyl Amide

Trade Name:
  Serdox NXC 3 K (Sasol Servo)

PEG-5 OLEAMIDE

Empirical Formula:
  $C_{64}H_{109}NO_8$

Definition: PEG-5 Oleamide is the poly-ethylene glycol amide of oleic acid that conforms to the formula:

$$CH_3(CH_2)_7CH = CH(CH_2)_7\overset{\overset{\displaystyle O}{\|}}{C} - NH(CH_2CH_2O)_nH$$

where n has an average value of 5.

Chemical Class: Alkoxylated Amides

Function: Surfactant - Emulsifying Agent

Technical/Other Names:
  Polyethylene Glycol (5) Oleyl Amide
  Polyoxyethylene (5) Oleyl Amide

Trade Name:
  Nikkol Tamdo-5 (Nikko)

PEG-6 OLEAMIDE

Empirical Formula:
  $C_{30}H_{59}NO_7$

Definition: PEG-6 Oleamide is the poly-ethylene glycol amide of oleic acid that conforms to the formula:

$$CH_3(CH_2)_7CH = CH(CH_2)_7\overset{\overset{\displaystyle O}{\|}}{C} - NH(CH_2CH_2O)_nH$$

where n has an average value of 6.

Chemical Class: Alkoxylated Amides

Function: Surfactant - Emulsifying Agent

Technical/Other Names:
  Polyethylene Glycol 300 Oleyl Amide
  Polyoxyethylene (6) Oleyl Amide

PEG-7 OLEAMIDE

Empirical Formula:
  $C_{32}H_{63}NO_8$

Definition: PEG-7 Oleamide is the poly-ethylene glycol amide of oleic acid that conforms to the formula:

$$CH_3(CH_2)_7CH = CH(CH_2)_7\overset{\overset{\displaystyle O}{\|}}{C} - NH(CH_2CH_2O)_nH$$

where n has an average value of 7.

Chemical Class: Alkoxylated Amides

Function: Surfactant - Emulsifying Agent

Technical/Other Names:
  Polyethylene Glycol (7) Oleyl Amide
  Polyoxyethylene (7) Oleyl Amide

Trade Name:
  Ethomid O/17 (Akzo Nobel)

PEG-9 OLEAMIDE

Empirical Formula:
  $C_{36}H_{71}NO_{10}$

Definition: PEG-9 Oleamide is the poly-ethylene glycol amide of oleic acid that conforms generally to the formula:

$$CH_3(CH_2)_7CH = CH(CH_2)_7\overset{\overset{\displaystyle O}{\|}}{C} - NH(CH_2CH_2O)_nH$$

where n has an average value of 9.

Chemical Class: Alkoxylated Amides

Function: Surfactant - Emulsifying Agent

Technical/Other Names:
  Polyethylene Glycol 450 Oleyl Amide
  Polyoxyethylene (9) Oleyl Amide

PEG-5 OLEAMIDE DIOLEATE

Definition: PEG-5 Oleamide Dioleate is the ethoxylated amide that conforms generally to the formula:

$$\begin{array}{c}(CH_2)_7CH_3\\ |\\ CH \quad\quad O \quad\quad\quad O \quad\quad CH_3\\ \|\quad\quad \|\quad\quad\quad \|\quad\quad |\\ CH(CH_2)_7C - N(CH_2CH_2O)_xC(CH_2)_7CH = CH(CH_2)_7\\ |\\ (CH_2CH_2O)_yC(CH_2)_7CH = CH(CH_2)_7\\ \|\quad\quad\quad\quad |\\ O \quad\quad\quad\quad CH_3\end{array}$$

where x + y has an average value of 5.

Chemical Classes: Alkoxylated Amides; Esters

Functions: Hair Conditioning Agent; Viscosity Increasing Agent - Nonaqueous

Technical/Other Names:
  Polyethylene Glycol (5) Oleamide Dioleate
  Polyoxyethylene (5) Oleamide Dioleate

Trade Name:
  Lowenol OT-216 (Lowenstein)

PEG-2 OLEAMINE

CAS No.: 26635-93-8 (Generic)

Empirical Formula:
  $C_{22}H_{55}NO_2$

Definition: PEG-2 Oleamine is the poly-ethylene glycol amine of oleic acid that conforms to the formula:

$$CH_3(CH_2)_7CH == CH(CH_2)_8N(CH_2CH_2O)_xH\\ |\\ (CH_2CH_2O)_yH$$

where x + y has an average value of 2.

Information Source: TSCA

Chemical Class: Alkoxylated Amines

Functions: Surfactant - Emulsifying Agent; Surfactant - Foam Booster

Technical/Other Names:
  Polyethylene Glycol 100 Oleyl Amine
  Polyoxyethylene (2) Oleyl Amine

Trade Names:
  Ethomeen O/12 (Akzo Nobel)

Ethox OAM-2 (Ethox)
Hetoxamine O-2 (Heterene)
Jeetox O-2 (Jeen)
Protox O-2 (Protameen)
Rhodameen O2 V (Rhodia)
Sabopal NO 2 (Sabo)

Trade Name Mixture:
Sabowax GT (Sabo)

PEG-5 OLEAMINE

CAS No.: 26635-93-8 (Generic)

JPN Translation:
PEG - 5 オレアミン

Empirical Formula:
$C_{28}H_{57}NO_5$

Definition: PEG-5 Oleamine is the polyethylene glycol amine of oleic acid that conforms to the formula:

$$CH_3(CH_2)_7CH = CH(CH_2)_8N(CH_2CH_2O)_xH$$
$$(CH_2CH_2O)_yH$$

where x + y has an average value of 5.

Information Sources: JCIC, JCLS, TSCA

Chemical Class: Alkoxylated Amines

Functions: Antistatic Agent; Surfactant - Emulsifying Agent

Technical/Other Names:
Polyethylene Glycol (5) Oleyl Amine
Polyoxyethylene Oleylamine
Polyoxyethylene (5) Oleyl Amine

Trade Names:
Ethomeen O/15 (Akzo Nobel)
Hetoxamine O-5 (Heterene)
Jeetox O-5 (Jeen)
Nikkol Tamno-5 (Nikko)
Protox O-5 (Protameen)

PEG-6 OLEAMINE

CAS No.: 26635-93-8 (Generic)

Definition: PEG-6 Oleamine is the polyethylene glycol derivative of Oleamine (q.v.) that conforms generally to the formula:

$$CH_3(CH_2)_7CH = CH(CH_2)_8N(CH_2CH_2O)_xH$$
$$(CH_2CH_2O)_yH$$

where x+y has an average value of 6.

Chemical Class: Alkoxylated Amines

Functions: Surfactant - Emulsifying Agent; Surfactant - Foam Booster

Trade Name:
Sabopal NO 6 (Sabo)

PEG-10 OLEAMINE

CAS No.: 26635-93-8 (Generic)

Definition: PEG-10 Oleamine is the polyethylene glycol derivative of Oleamine (q.v.) that conforms generally to the formula:

$$CH_3(CH_2)_7CH = CH(CH_2)_8N(CH_2CH_2O)_xH$$
$$(CH_2CH_2O)_yH$$

where x+y has an average value of 10.

Chemical Class: Alkoxylated Amines

Function: Not Reported

Technical/Other Names:
Polyethylene Glycol (10) Oleyl Amine
Polyoxyethylene (10) Oleyl Amine

Trade Name:
Sabopal NO 10 (Sabo)

PEG-15 OLEAMINE

CAS No.: 26635-93-8 (Generic)

Definition: PEG-15 Oleamine is the polyethylene glycol amine of oleic acid that conforms to the formula:

$$CH_3(CH_2)_7CH = CH(CH_2)_8N(CH_2CH_2O)_xH$$
$$(CH_2CH_2O)_yH$$

where x + y has an average value of 15.

Information Sources: JSQI, TSCA

Chemical Class: Alkoxylated Amines

Functions: Antistatic Agent; Surfactant - Emulsifying Agent

Technical/Other Names:
Polyethylene Glycol (15) Oleyl Amine
Polyoxyethylene (15) Oleyl Amine

Trade Names:
Ethomeen O/25 (Akzo Nobel)
Hetoxamine O-15 (Heterene)
Jeetox O-15 (Jeen)
Nikkol Tamno-15 (Nikko)
Protox O-15 (Protameen)

PEG-20 OLEAMINE

CAS No.: 26635-93-8 (Generic)

Definition: PEG-20 Oleamine is the polyethylene glycol derivative of Oleamine (q.v.) that conforms generally to the formula:

$$CH_3(CH_2)_7CH = CH(CH_2)_8N(CH_2CH_2O)_xH$$
$$(CH_2CH_2O)_yH$$

where x+y has an average value of 20.

Chemical Class: Alkoxylated Amines

Function: Not Reported

Technical/Other Names:
Polyethylene Glycol (20) Oleyl Amine
Polyoxyethylene Glycol (20) Oleyl Amine

Trade Name:
Sabopal NO 20 (Sabo)

PEG-25 OLEAMINE

CAS No.: 26635-93-8 (Generic)

Definition: PEG-25 Oleamine is the polyethylene glycol derivative of Oleamine (q.v.) that conforms generally to the formula:

$$CH_3(CH_2)_7CH = CH(CH_2)_8N(CH_2CH_2O)_xH$$
$$(CH_2CH_2O)_yH$$

where x+y has an average value of 25.

Chemical Class: Alkoxylated Amines

Function: Not Reported

Technical/Other Names:
Polyethylene Glycol (25) Oleyl Amine
Polyoxyethylene Glycol (25) Oleyl Amine

Trade Name:
Sabopal NO 25 (Sabo)

PEG-30 OLEAMINE

CAS No.: 26635-93-8 (Generic)

Definition: PEG-30 Oleamine is the polyethylene glycol amine of oleic acid that conforms to the formula:

$$CH_3(CH_2)_7CH = CH(CH_2)_8N(CH_2CH_2O)_xH$$
$$(CH_2CH_2O)_yH$$

where x + y has an average value of 30.

Information Source: TSCA

Chemical Class: Alkoxylated Amines

Functions: Antistatic Agent; Surfactant - Cleansing Agent; Surfactant - Solubilizing Agent

Technical/Other Names:
Polyethylene Glycol (30) Oleyl Amine
Polyoxyethylene (30) Oleyl Amine

Trade Names:
Ethox OAM-308 (Ethox)
Sabopal NO 30 (Sabo)

PEG-2 OLEAMINE HYDROFLUORIDE

CAS No.: 207916-33-4

Definition: PEG-2 Oleamine Hydrofluoride is the hydrofluoric acid salt of PEG-2 Oleamine (q.v.)

Chemical Class: Alkoxylated Amines

Function: Oral Care Agent

PEG-2 OLEAMMONIUM CHLORIDE

JPN Translation:
PEG - 2 オレアンモニウムクロリド

Empirical Formula:
$C_{23}H_{48}NO_2 \cdot Cl$

Definition: PEG-2 Oleammonium Chloride is the quaternary ammonium salt that conforms generally to the formula:

$$\left[\begin{array}{c} CH_3(CH_2)_7CH \qquad (CH_2CH_2O)_x \\ \| \qquad\qquad | \\ CH(CH_2)_7CH_2{-}N{-}CH_3 \\ | \\ (CH_2CH_2O)_y \end{array} \right]^+ Cl^-$$

where x + y has an average value of 2.

Information Sources: JCIC, JCLS, JSQI

Chemical Classes: Alkoxylated Amines; Quaternary Ammonium Compounds

Functions: Antistatic Agent; Hair Conditioning Agent

Reported Product Categories: Hair Preparations (Non-coloring), Misc.; Hair Conditioners; Hair Sprays (Aerosol Fixatives); Permanent Waves

Technical/Other Names:
Di(Polyoxyethylene) Oleyl Methyl Ammonium Chloride (2E.O.)
PEG-2 Oleyl Quaternium-4
Polyethylene Glycol 100 Oleammonium Chloride
Polyoxyethylene (2) Oleammonium Chloride

Trade Name:
AEC PEG-2 Oleammonium Chloride (A & E Connock)

Trade Name Mixture:
Ethoquad 0/12 PG (Akzo Nobel)

PEG-15 OLEAMMONIUM CHLORIDE

Definition: PEG-15 Oleammonium Chloride is the quaternary ammonium salt that conforms generally to the formula:

$$\left[\begin{array}{c} CH_3(CH_2)_7CH \qquad (CH_2CH_2O)_x \\ \| \qquad\qquad | \\ CH(CH_2)_7CH_2{-}N{-}CH_3 \\ | \\ (CH_2CH_2O)_y \end{array} \right]^+ Cl^-$$

where x + y has an average value of 15.

Chemical Class: Quaternary Ammonium Compounds

Function: Antistatic Agent

Technical/Other Names:
PEG-15 Oleamonium Chloride
PEG-15 Oleyl Quaternium-4
PEG-15 Stearyl Quaternium-4
Polyethylene Glycol (15) Oleamonium Chloride
Polyoxyethylene (15) Oleamonium Chloride

Trade Name:
Ethoquad O/25 (Akzo Nobel)

PEG-2 OLEATE

CAS No.	EINECS No.
106-12-7	203-364-0

JPN Translation:
オレイン酸 PEG - 2

Empirical Formula:
$C_{22}H_{42}O_4$

Definition: PEG-2 Oleate is the polyethylene glycol ester of oleic acid that conforms to the formula:

$$CH_3(CH_2)_7CH = CH(CH_2)_7\overset{\displaystyle O}{\overset{\|}{C}}{-}(OCH_2CH_2)_nOH$$

where n has an average value of 2.

Information Sources: 21CFR175.105, 21CFR175.300, 21CFR176.210, 21CFR177.2800, JCLS, MI-13(7660), TSCA

Chemical Class: Alkoxylated Carboxylic Acids

Function: Surfactant - Emulsifying Agent

Technical/Other Names:
Diethylene Glycol Monooleate
Diglycol Oleate
9-Octadecenoic Acid, 2-(2-Hydroxyethoxy)Ethyl Ester
Polyethylene Glycol 100 Monooleate
Polyoxyethylene (2) Monooleate

Trade Names:
AEC PEG-2 Oleate (A & E Connock)
Emalex 200 (Nihon Emulsion)
Hetoxamate MO-2 (Heterene)
Nikkol MYO-2 (Nikko)
Pegosperse 100 O (Lonza Inc./Lonza Ltd.)

PEG-3 OLEATE

CAS Nos.	EINECS No.
9004-96-0 (Generic)	
10233-14-4	233-561-7

JPN Translation:
オレイン酸 PEG - 3

Empirical Formula:
$C_{24}H_{46}O_5$

Definition: PEG-3 Oleate is the polyethylene glycol ester of oleic acid that conforms to the formula:

$$CH_3(CH_2)_7CH = CH(CH_2)_7\overset{\displaystyle O}{\overset{\|}{C}}{-}(OCH_2CH_2)_nOH$$

where n has an average value of 3.

Information Sources: 21CFR175.300, JCLS, MI-13(7660), TSCA

Chemical Class: Alkoxylated Carboxylic Acids

Function: Surfactant - Emulsifying Agent

Technical/Other Names:
9-Octadecenoic Acid, 2-[2-(2-Hydroxyethoxy)Ethoxy]Ethyl Ester
Polyethylene Glycol (3) Monooleate
Polyoxyethylene (3) Monooleate
Triethylene Glycol Oleate

Trade Name:
AEC PEG-3 Oleate (A & E Connock)

PEG-4 OLEATE

CAS Nos.	EINECS No.
9004-96-0 (Generic)	
10108-25-5	233-293-0

JPN Translation:
オレイン酸 PEG - 4

Empirical Formula:
$C_{26}H_{50}O_6$

Definition: PEG-4 Oleate is the polyethylene glycol ester of oleic acid that conforms to the formula:

$$CH_3(CH_2)_7CH = CH(CH_2)_7\overset{\displaystyle O}{\overset{\|}{C}}{-}(OCH_2CH_2)_nOH$$

where n has an average value of 4.

Information Sources: 21CFR175.105, 21CFR175.300, 21CFR176.210, JCLS, MI-13(7660), TSCA

Chemical Class: Alkoxylated Carboxylic Acids

Function: Surfactant - Emulsifying Agent

Technical/Other Names:
9-Octadecenoic Acid, 2-[2-[2-(2-Hydroxyethoxy)Ethoxy]Ethoxy]Ethyl Ester
Polyethylene Glycol 200 Monooleate
Polyoxyethylene (4) Monooleate
Tetraethylene Glycol Oleate

Trade Names:
AEC PEG-4 Oleate (A & E Connock)

Chemax E-200-MO (Chemax)
Empilan BQ4.5 (Albright & Wilson UK)
Ethox MO-5 (Ethox)
Jeemate 200-OC (Jeen)
Protamate 200 OC (Protameen)
Radiasurf 7402 (Oleon NV)
STEPAN PEG 200 MO (Stepan)
Unipeg-200 MO (Universal Preserv-A-
Chem)

PEG-5 OLEATE

CAS Nos.: 9004-96-0 (Generic); 23336-36-9

JPN Translation:
オレイン酸 PEG - 5

Empirical Formula:
$C_{28}H_{54}O_7$

Definition: PEG-5 Oleate is the poly-ethylene glycol ester of oleic acid that conforms to the formula:

$$CH_3(CH_2)_7CH = CH(CH_2)_7C\overset{\displaystyle O}{\overset{\|}{}}-(OCH_2CH_2)_nOH$$

where n has an average value of 5.

Information Sources: 21CFR175.105, 21CFR175.300, JCLS, MI-13(7660), TSCA

Chemical Class: Alkoxylated Carboxylic Acids

Function: Surfactant - Emulsifying Agent

Technical/Other Names:
14-Hydroxy-3,6,9,12-Tetraoxatetradec-1-yl-9-Octadecenoic Acid
9-Octadecenoic Acid, 14-Hydroxy-3,6,9,12-Tetraoxatetradec-1-yl-
Pentaethylene Glycol Oleate
Polyethylene Glycol (5) Monooleate
Polyoxyethylene (5) Monooleate

Trade Names:
AEC PEG-5 Oleate (A & E Connock)
Hetoxamate MO-5 (Heterene)
Sympatens-BO/050 (Kolb)

PEG-6 OLEATE

CAS Nos.: 9004-96-0 (Generic); 60344-26-5

JPN Translation:
オレイン酸 PEG - 6

Empirical Formula:
$C_{30}H_{58}O_8$

Definition: PEG-6 Oleate is the poly-ethylene glycol ester of oleic acid that conforms to the formula:

$$CH_3(CH_2)_7CH = CH(CH_2)_7C\overset{\displaystyle O}{\overset{\|}{}}-(OCH_2CH_2)_nOH$$

where n has an average value of 6.

Information Sources: 21CFR175.105, 21CFR175.300, 21CFR176.210, JCLS, MI-13(7660), TSCA

Chemical Class: Alkoxylated Carboxylic Acids

Function: Surfactant - Emulsifying Agent

Reported Product Categories: Bath Preparations, Misc.; Body and Hand Preparations (Excluding Shaving Preparations)

Technical/Other Names:
Hexaethylene Glycol Oleate
9-Octadecenoic Acid, 17-Hydroxy-3,6,9,12,15-Pentaoxaheptadec-1-yl Ester
Polyethylene Glycol 300 Monooleate
Polyoxyethylene (6) Monooleate

Trade Names:
AEC PEG-6 Oleate (A & E Connock)
Empilan BQ6 (Albright & Wilson UK)
Jeemate 300-OC (Jeen)
Nikkol MYO-6 (Nikko)
Pegosperse 300 MO (Lonza Inc./Lonza Ltd.)
Prodhyphore B (Prod'Hyg)
Protamate 300 OC (Protameen)
STEPAN PEG 300 MO (Stepan)

PEG-7 OLEATE

CAS No.: 9004-96-0 (Generic)

JPN Translation:
オレイン酸 PEG - 7

Empirical Formula:
$C_{32}H_{62}O_9$

Definition: PEG-7 Oleate is the poly-ethylene glycol ester of oleic acid that conforms to the formula:

$$CH_3(CH_2)_7CH = CH(CH_2)_7C\overset{\displaystyle O}{\overset{\|}{}}-(OCH_2CH_2)_nOH$$

where n has an average value of 7.

Information Sources: 21CFR175.300, JCLS, MI-13(7660), TSCA

Chemical Class: Alkoxylated Carboxylic Acids

Function: Surfactant - Emulsifying Agent

Technical/Other Names:
Polyethylene Glycol (7) Monooleate
Polyoxyethylene (7) Monooleate

Trade Names:
AEC PEG-7 Oleate (A & E Connock)
Atlas G-5507 (Uniqema Americas)
Marlosol OL 7 (Sasol GmbH - Marl)

PEG-8 OLEATE

CAS No.: 9004-96-0 (Generic)

JPN Translation:
オレイン酸 PEG - 8

Empirical Formula:
$C_{34}H_{66}O_{10}$

Definition: PEG-8 Oleate is the poly-ethylene glycol ester of oleic acid that conforms to the formula:

$$CH_3(CH_2)_7CH = CH(CH_2)_7C\overset{\displaystyle O}{\overset{\|}{}}-(OCH_2CH_2)_nOH$$

where n has an average value of 8.

Information Sources: 21CFR175.105, 21CFR175.300, 21CFR176.170, 21CFR176.200, 21CFR177.1200, 21CFR177.1210, 21CFR177.2260, 21CFR177.2800, JCLS, MI-13(7660), TSCA

Chemical Class: Alkoxylated Carboxylic Acids

Function: Surfactant - Emulsifying Agent

Technical/Other Names:
Polyethylene Glycol 400 Monooleate
Polyoxyethylene (8) Monooleate

Trade Names:
Acconon 400 MO (Abitec Corporation)
AEC PEG-8 Oleate (A & E Connock)
Chemax E-400-MO (Chemax)
Ethox MO-9 (Ethox)
Jeemate 400-OC (Jeen)
LIPOPEG 4-O (Lipo)
Lumulse 40-O (Lambent)
Pegosperse 400 MO (Lonza Inc./Lonza Ltd.)
Protamate 400 OC (Protameen)
ROL AG (Fabriquimica)
Sipoic MO-9 (Specialty Industrial)
STEPAN PEG 400 MO (Stepan)
Unipeg-400 MO (Universal Preserv-A-Chem)

PEG-9 OLEATE

CAS No.: 9004-96-0 (Generic)

JPN Translation:
オレイン酸 PEG - 9

Empirical Formula:
$C_{36}H_{70}O_{11}$

Definition: PEG-9 Oleate is the poly-ethylene glycol ester of oleic acid that conforms to the formula:

$$CH_3(CH_2)_7CH = CH(CH_2)_7C\overset{\displaystyle O}{\overset{\|}{}}-(OCH_2CH_2)_nOH$$

where n has an average value of 9.

Information Sources: 21CFR175.105, 21CFR175.300, 21CFR176.200,

21CFR177.2260, 21CFR177.2800, JCLS, MI-13(7660), TSCA

Chemical Class: Alkoxylated Carboxylic Acids

Function: Surfactant - Emulsifying Agent

Technical/Other Names:
Polyethylene Glycol 450 Monooleate
Polyoxyethylene (9) Monooleate

Trade Names:
AEC PEG-9 Oleate (A & E Connock)
Alkamuls 400 MO (Rhodia)
DeTHOX ACID O-9 (DeForest)
Empilan BQ9 (Albright & Wilson UK)
Hetoxamate MO-9 (Heterene)

PEG-10 OLEATE

CAS No.: 9004-96-0 (Generic)

JPN Translation:
オレイン酸 PEG - 10

Empirical Formula:
$C_{38}H_{74}O_{12}$

Definition: PEG-10 Oleate is the polyethylene glycol ester of oleic acid that conforms to the formula:

$$CH_3(CH_2)_7CH = CH(CH_2)_7C \overset{O}{\overset{\|}{-}} (OCH_2CH_2)_nOH$$

where n has an average value of 10.

Information Sources: 21CFR175.105, 21CFR175.300, 21CFR176.200, 21CFR177.2260, 21CFR177.2800, JCLS, MI-13(7660), TSCA

Chemical Class: Alkoxylated Carboxylic Acids

Function: Surfactant - Emulsifying Agent

Technical/Other Names:
Polyethylene Glycol 500 Monooleate
Polyoxyethylene (10) Monooleate

Trade Names:
AEC PEG-10 Oleate (A & E Connock)
Nikkol MYO-10 (Nikko)

PEG-11 OLEATE

CAS No.: 9004-96-0 (Generic)

JPN Translation:
オレイン酸 PEG - 11

Definition: PEG-11 Oleate is polyethylene glycol ester of oleic acid that conforms generally to the formula:

$$CH_3(CH_2)_7CH = CH(CH_2)_7C \overset{O}{\overset{\|}{-}} (OCH_2CH_2)_nOH$$

where n has an average value of 11.

Information Sources: JCLS, MI-13(7660)

Chemical Class: Alkoxylated Carboxylic Acids

Function: Surfactant - Emulsifying Agent

Technical/Other Names:
Polyethylene Glycol (11) Oleate
Polyoxyethylene (11) Oleate

PEG-12 OLEATE

CAS No.: 9004-96-0 (Generic)

JPN Translation:
オレイン酸 PEG - 12

Definition: PEG-12 Oleate is the polyethylene glycol ester of oleic acid that conforms to the formula:

$$CH_3(CH_2)_7CH = CH(CH_2)_7C \overset{O}{\overset{\|}{-}} (OCH_2CH_2)_nOH$$

where n has an average value of 12.

Information Sources: 21CFR175.105, 21CFR175.300, 21CFR176.170, 21CFR176.200, 21CFR177.1200, 21CFR177.2260, 21CFR177.2800, JCLS, MI-13(7660), TSCA

Chemical Class: Alkoxylated Carboxylic Acids

Function: Surfactant - Emulsifying Agent

Reported Product Category: Shampoos (Non-coloring)

Technical/Other Names:
Polyethylene Glycol 600 Monooleate
Polyoxyethylene (12) Monooleate

Trade Names:
AEC PEG-12 Oleate (A & E Connock)
Alkamuls 600 MO (Rhodia)
Chemax E-600-MO (Chemax)
Ethox MO-14 (Ethox)
Jeemate 600-OC (Jeen)
Pegosperse 600 MO (Lonza Inc./Lonza Ltd.)
Protamate 600 OC (Protameen)
Unipeg-600 MO (Universal Preserv-A-Chem)

Trade Name Mixtures:
Amisol HS-2 (Lucas Meyer GmbH)
Amisol HS-3 (Lucas Meyer GmbH)
Amisol HS3US (Lucas Meyer)

PEG-14 OLEATE

CAS No.: 9004-96-0 (Generic)

JPN Translation:
オレイン酸 PEG - 14

Definition: PEG-14 Oleate is the polyethylene glycol ester of oleic acid that conforms to the formula:

$$CH_3(CH_2)_7CH = CH(CH_2)_7C \overset{O}{\overset{\|}{-}} (OCH_2CH_2)_nOH$$

where n has an average value of 14.

Information Sources: 21CFR175.105, 21CFR175.300, 21CFR176.200, 21CFR177.2260, 21CFR177.2800, JCLS, MI-13(7660), TSCA

Chemical Class: Alkoxylated Carboxylic Acids

Function: Surfactant - Emulsifying Agent

Technical/Other Names:
Polyethylene Glycol (14) Monooleate
Polyoxyethylene (14) Monooleate

Trade Names:
AEC PEG-14 Oleate (A & E Connock)
DeTHOX ACID O-14 (DeForest)
Hetoxamate MO-14 (Heterene)
Jeemate 600-DO (Jeen)

PEG-15 OLEATE

CAS No.: 9004-96-0 (Generic)

JPN Translation:
オレイン酸 PEG - 15

Definition: PEG-15 Oleate is a polyethylene glycol ester of oleic acid that conforms generally to the formula:

$$CH_3(CH_2)_7CH = CH(CH_2)_7C \overset{O}{\overset{\|}{-}} (OCH_2CH_2)_nOH$$

where n has an average value of 15.

Information Sources: 21CFR175.300, 21CFR176.200, 21CFR176.210, 21CFR177.2260, 21CFR177.2800, JCLS, MI-13(7660)

Chemical Class: Alkoxylated Carboxylic Acids

Function: Surfactant - Emulsifying Agent

Technical/Other Names:
Polyethylene Glycol (15) Oleate
Polyoxyethylene (15) Oleate

PEG-16 OLEATE

CAS No.: 9004-96-0 (Generic)

JPN Translation:
オレイン酸 PEG - 16

Definition: PEG-16 Oleate is the polyethylene glycol ester of oleic acid that conforms to the formula:

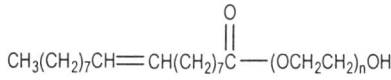

where n has an average value of 16.

Information Source: JCLS

Chemical Class: Alkoxylated Carboxylic Acids

Function: Surfactant - Emulsifying Agent

Technical/Other Names:
Polyethylene Glycol (16) Monooleate
Polyoxyethylene (16) Monooleate

Trade Name:
DeTHOX ACID O-16 (DeForest)

PEG-20 OLEATE

CAS No.: 9004-96-0 (Generic)

JPN Translation:
オレイン酸 PEG - 20

Definition: PEG-20 Oleate is the polyethylene glycol ester of oleic acid that conforms to the formula:

$$CH_3(CH_2)_7CH=CH(CH_2)_7C(=O)-(OCH_2CH_2)_nOH$$

where n has an average value of 20.

Information Sources: 21CFR175.300, 21CFR176.200, 21CFR177.2260, 21CFR177.2800, JCLS, MI-13(7660), TSCA

Chemical Class: Alkoxylated Carboxylic Acids

Functions: Surfactant - Cleansing Agent; Surfactant - Emulsifying Agent; Surfactant - Solubilizing Agent

Technical/Other Names:
Polyethylene Glycol 1000 Monooleate
Polyoxyethylene (20) Monooleate

Trade Names:
AEC PEG-20 Oleate (A & E Connock)
Chemax E-1000-MO (Chemax)
Jeemate 1000-OC (Jeen)
Protamate 1000 OC (Protameen)

PEG-23 OLEATE

CAS No.: 9004-96-0 (Generic)

JPN Translation:
オレイン酸 PEG - 23

Definition: PEG-23 Oleate is the polyethylene glycol ester of oleic acid that conforms generally to the formula:

$$CH_3(CH_2)_7CH=CH(CH_2)_7C(=O)-(OCH_2CH_2)_nOH$$

where n has an average value of 23.

Information Source: JCLS

Chemical Class: Alkoxylated Carboxylic Acids

Functions: Surfactant - Cleansing Agent; Surfactant - Solubilizing Agent

Technical/Other Names:
Polyethylene Glycol (23) Oleate
Polyoxyethylene (23) Oleate

PEG-32 OLEATE

CAS No.: 9004-96-0 (Generic)

JPN Translation:
オレイン酸 PEG - 32

Definition: PEG-32 Oleate is the polyethylene glycol ester of oleic acid that conforms to the formula:

$$CH_3(CH_2)_7CH=CH(CH_2)_7C(=O)-(OCH_2CH_2)_nOH$$

where n has an average value of 32.

Information Sources: 21CFR175.300, 21CFR176.200, 21CFR177.2800, JCLS, MI-13(7660), TSCA

Chemical Class: Alkoxylated Carboxylic Acids

Functions: Surfactant - Cleansing Agent; Surfactant - Solubilizing Agent

Technical/Other Names:
Polyethylene Glycol 1540 Monooleate
Polyoxyethylene (32) Monooleate

Trade Name:
AEC PEG-32 Oleate (A & E Connock)

PEG-36 OLEATE

CAS No.: 9004-96-0 (Generic)

JPN Translation:
オレイン酸 PEG - 36

Definition: PEG-36 Oleate is the polyethylene glycol ester of oleic acid that conforms to the formula:

$$CH_3(CH_2)_7CH=CH(CH_2)_7C(=O)-(OCH_2CH_2)_nOH$$

where n has an average value of 36.

Information Sources: 21CFR175.300, 21CFR176.200, 21CFR177.2260, 21CFR177.2800, JCLS, MI-13(7660), TSCA

Chemical Class: Alkoxylated Carboxylic Acids

Functions: Surfactant - Cleansing Agent; Surfactant - Solubilizing Agent

Technical/Other Names:
Polyethylene Glycol 1800 Monooleate
Polyoxyethylene (36) Monooleate

Trade Name:
AEC PEG-36 Oleate (A & E Connock)

PEG-75 OLEATE

CAS No.: 9004-96-0 (Generic)

JPN Translation:
オレイン酸 PEG - 75

Definition: PEG-75 Oleate is the polyethylene glycol ester of oleic acid that conforms to the formula:

$$CH_3(CH_2)_7CH=CH(CH_2)_7C(=O)-(OCH_2CH_2)_nOH$$

where n has an average value of 75.

Information Sources: 21CFR175.300, 21CFR176.200, JCLS, MI-13(7660), TSCA

Chemical Class: Alkoxylated Carboxylic Acids

Function: Surfactant - Cleansing Agent

Technical/Other Names:
Polyethylene Glycol 4000 Monooleate
Polyoxyethylene (75) Monooleate

Trade Name:
AEC PEG-75 Oleate (A & E Connock)

PEG-150 OLEATE

CAS No.: 9004-96-0 (Generic)

JPN Translation:
オレイン酸 PEG - 150

Definition: PEG-150 Oleate is the polyethylene glycol ester of oleic acid that conforms to the formula:

$$CH_3(CH_2)_7CH=CH(CH_2)_7C(=O)-(OCH_2CH_2)_nOH$$

where n has an average value of 150.

Information Sources: 21CFR175.300, 21CFR176.200, JCLS, MI-13(7660), TSCA

Chemical Class: Alkoxylated Carboxylic Acids

Function: Surfactant - Cleansing Agent

Technical/Other Names:
Polyethylene Glycol 6000 Monooleate
Polyoxyethylene (150) Monooleate

Trade Name:
AEC PEG-150 Oleate (A & E Connock)

PEG-2 OLEATE SE

Definition: PEG-2 Oleate SE is a self-emulsifying grade of PEG-2 Oleate (q.v.) that contains some sodium and/or potassium oleate.

Information Source: JCLS

Chemical Class: Alkoxylated Carboxylic Acids

Function: Surfactant - Emulsifying Agent

Technical/Other Names:
  Diethylene Glycol Monooleate Self-Emulsifying
  Polyethylene Glycol 100 Monooleate Self-Emulsifying
  Polyoxyethylene (2) Monooleate Self-Emulsifying

PEG-4 OLIVATE

Definition: PEG-4 Olivate is the polyethylene glycol ester of the fatty acids derived from olive oil that conforms generally to the formula:

$$RC \overset{O}{\underset{||}{}} - (OCH_2CH_2)_nOH$$

where RCO- represents the fatty acids derived from Olea Europaea (Olive) Oil (q.v.) and n has an average value of 4.

Chemical Class: Alkoxylated Carboxylic Acids

Function: Surfactant - Emulsifying Agent

Technical/Other Names:
  Polyethylene Glycol (4) Monoolivate
  Polyoxyethylene (4) Monoolivate

Trade Name:
  Olivem 700 (B & T)

PEG-7 OLIVATE

Definition: PEG-7 Olivate is the polyethylene glycol ester of the fatty acids derived from olive oil that conforms generally to the formula:

$$RC \overset{O}{\underset{||}{}} - (OCH_2CH_2)_nOH$$

where RCO- represents the fatty acids derived from Olea Europaea (Olive) Oil (q.v.) and n has an average value of 7.

Chemical Class: Alkoxylated Carboxylic Acids

Function: Surfactant - Emulsifying Agent

Trade Name:
  Olivem 300 N-Sol (B & T)

PEG-23 OLIVATE

Definition: PEG-23 Olivate is the polyethylene glycol ester of Olive Acid (q.v.) that conforms generally to the formula:

$$RC \overset{O}{\underset{||}{}} - (OCH_2CH_2)_nOH$$

where RCO- represents the olive acid moiety and n has an average value of 23.

Chemical Class: Alkoxylated Carboxylic Acids

Functions: Skin-Conditioning Agent - Emollient; Surfactant - Emulsifying Agent; Surfactant - Solubilizing Agent

Technical/Other Names:
  Polyethylene Glycol (23) Olivate
  Polyoxyethylene (23) Olivate

Trade Name:
  Polygreen Emulive (Polytechno Ind.)

PEG-9 OLIVEATE

Definition: PEG-9 Oliveate is the polyethylene glycol ester of the fatty acids derived from Olea Europaea (Olive) Oil (q.v.) that conforms generally to the formula:

$$RC \overset{O}{\underset{||}{}} - (OCH_2CH_2)_nOH$$

where RCO- represents the olive oil fatty acids and n has an average value of 9.

Chemical Class: Alkoxylated Carboxylic Acids

Function: Surfactant - Emulsifying Agent

Technical/Other Names:
  Polyethylene Glycol (9) Monoliveate
  Polyoxyethylene (9) Monoliveate

Trade Name Mixture:
  Silwax DMC-O (Siltech LLC)

PEG-2 OLIVE GLYCERIDES

CAS No.: 103819-46-1

Definition: PEG-2 Olive Glycerides is a polyethylene glycol derivative of mono- and diglycerides derived from olive oil with an average of 2 moles of ethylene oxide.

Chemical Classes: Alkoxylated Alcohols; Glyceryl Esters and Derivatives

Functions: Skin-Conditioning Agent - Emollient; Surfactant - Emulsifying Agent

Technical/Other Names:
  Polyethylene Glycol 100 Olive Glycerides
  Polyoxyethylene (2) Olive Glycerides

PEG-6 OLIVE GLYCERIDES

CAS No.: 103819-46-1

Definition: PEG-6 Olive Glycerides is a polyethylene glycol derivative of mono- and diglycerides derived from olive oil with an average of 6 moles of ethylene oxide.

Chemical Classes: Alkoxylated Alcohols; Glyceryl Esters and Derivatives

Functions: Skin-Conditioning Agent - Emollient; Surfactant - Emulsifying Agent

Technical/Other Names:
  Polyethylene Glycol 300 Olive Glycerides
  Polyoxyethylene (6) Olive Glycerides

PEG-7 OLIVE GLYCERIDES

Definition: PEG-7 Olive Glycerides is a polyethylene glycol derivative of the mono- and diglycerides derived from olive oil with an average of 7 moles of ethylene oxide.

Chemical Classes: Alkoxylated Alcohols; Glyceryl Esters and Derivatives

Functions: Skin-Conditioning Agent - Emollient; Surfactant - Emulsifying Agent

Technical/Other Names:
  Polyethylene Glycol (7) Olive Glycerides
  Polyoxyethylene (7) Olive Glycerides

PEG-10 OLIVE GLYCERIDES

Definition: PEG-10 Olive Glycerides is a polyethylene glycol derivative of mono and diglycerides of olive oil with an average of 10 moles of ethylene oxide.

Chemical Classes: Alkoxylated Alcohols; Glyceryl Esters and Derivatives

Functions: Skin-Conditioning Agent - Emollient; Surfactant - Emulsifying Agent

Technical/Other Names:
  Polyethylene Glycol 500 Olive Glycerides
  Polyoxyethylene (10) Olive Glycerides

Trade Names:
  AEC PEG-10 Olive Glycerides (A & E Connock)
  Lexol EO (Inolex)
  Olive Oil W (Cosmetochem)
  Olive Oil "W" Watersoluble (Cosmetochem) (Cosmetochem International Ltd.)
  Oxypon 288 (Zschimmer & Schwarz)

Trade Name Mixtures:
  Almond Oil "W" Water Soluble (Cosmetochem) (Cosmetochem International Ltd.)
  Apricot Kernel Oil "W" Watersoluble (Cosmetochem) (Cosmetochem International Ltd.)
  Carrot Oil "W" Watersoluble (Cosmetochem) (Cosmetochem International Ltd.)

The inclusion of any compound in the *Dictionary and Handbook* does not indicate that use of that substance as a cosmetic ingredient complies with the laws and regulations governing such use in the United States or any other country.

Evening Primrose Oil "W" Watersoluble (Cosmetochem) (Cosmetochem International Ltd.)
Herbasol Extract Apricot Kernel Oil (Cosmetochem) (Cosmetochem International Ltd.)
Herbasol Extract Betelnut/Pepper Betle (Cosmetochem) (Cosmetochem International Ltd.)
Herbasol Extract Cotton Seeds (Cosmetochem) (Cosmetochem International Ltd.)
Herbasol Extract Grape Seed (Cosmetochem) (Cosmetochem International Ltd.)
Jojobaoil "W" Watersoluble (Cosmetochem) (Cosmetochem International Ltd.)

PEG-40 OLIVE GLYCERIDES

Definition: PEG-40 Olive Glycerides is a polyethylene glycol derivative of the mono- and diglycerides derived from olive oil with an average of 40 moles of ethylene oxide.

Chemical Classes: Alkoxylated Alcohols; Glyceryl Esters and Derivatives

Functions: Skin-Conditioning Agent - Emollient; Surfactant - Emulsifying Agent

Technical/Other Names:
Polyethylene Glycol 2000 Olive Glycerides
Polyoxyethylene (40) Olive Glycerides

PEG-25 PABA

CAS Nos.: 113010-52-9; 116242-27-4

Definition: PEG-25 PABA is the polyethylene glycol derivative of PABA (q.v.). It conforms generally to the formula:

$$CH_3(CH_2)_{15}... $$

where x+y+z has an average value of 25.

Information Source: EEC(VII/1-13)

Chemical Class: PABA Derivatives

Function: Ultraviolet Light Absorber

Reported Product Category: Tonics, Dressings, and Other Hair Grooming Aids

---

Technical/Other Names:
Ethoxylated Ethyl-4-Aminobenzoate
Polyethylene Glycol (25) PABA
Polyoxyethylene (25) PABA

Trade Names:
Liposcreen PABA-25 (Lipo)
Unipabol U-17 (Induchem)
Uvinul P 25 (BASF)

PEG-18 PALM GLYCERIDES

Definition: PEG-18 Palm Glycerides is a polyethylene glycol derivative of Palm Glycerides (q.v.) with an average of 18 moles of ethylene oxide.

Chemical Classes: Alkoxylated Alcohols; Glyceryl Esters and Derivatives

Functions: Skin-Conditioning Agent - Emollient; Surfactant - Emulsifying Agent

Trade Name:
Resassol BPE (Res Pharma)

PEG-12 PALMITAMINE

CAS No.: 68155-33-9 (Generic)

Definition: PEG-12 Palmitamine is the polyethylene glycol derivative of Palmitamine (q.v.) that conforms generally to the formula:

$$(CH_2CH_2O)_xH$$
$$|$$
$$CH_3(CH_2)_{15}N(CH_2CH_2O)_yH$$

where x + y has an average value of 12.

Information Source: TSCA

Chemical Class: Alkoxylated Amines

Functions: Antistatic Agent; Surfactant - Emulsifying Agent

Technical/Other Names:
Polyethylene Glycol (12) Palmityl Amine
Polyoxyethyene (12) Palmityl Amine

Trade Name:
Pegnol HA-120 (Toho)

PEG-6 PALMITATE

CAS No.: 9004-94-8 (Generic)

JPN Translation:
パルミチン酸 PEG - 6

Empirical Formula:
$C_{28}H_{56}O_8$

Definition: PEG-6 Palmitate is the polyethylene glycol ester of palmitic acid that conforms to the formula:

$$CH_3(CH_2)_{14}C \overset{O}{\overset{||}{-}} (OCH_2CH_2)_nOH$$

where n has an average value of 6.

Information Sources: 21CFR175.105, 21CFR175.300, 21CFR176.210, CTFA D, JCLS, JSQI, MI-13(7660), TSCA

Chemical Class: Alkoxylated Carboxylic Acids

Function: Surfactant - Emulsifying Agent

Technical/Other Names:
Polyethylene Glycol 300 Monopalmitate
Polyoxyethylene (6) Monopalmitate

Trade Name:
AEC PEG-6 Palmitate (A & E Connock)

PEG-18 PALMITATE

CAS No.: 9004-94-8 (Generic)

JPN Translation:
パルミチン酸 PEG - 18

Definition: PEG-18 Palmitate is the polyethylene glycol ester of palmitic acid that conforms to the formula:

$$CH_3(CH_2)_{14}C \overset{O}{\overset{||}{-}} (OCH_2CH_2)_nOH$$

where n has an average value of 18.

Information Sources: 21CFR175.300, 21CFR177.2260, 21CFR177.2800, JCLS, JSQI, MI-13(7660), TSCA

Chemical Class: Alkoxylated Carboxylic Acids

Function: Surfactant - Emulsifying Agent

Technical/Other Names:
Polyethylene Glycol (18) Monopalmitate
Polyoxyethylene (18) Monopalmitate

Trade Name:
AEC PEG-18 Palmitate (A & E Connock)

PEG-20 PALMITATE

CAS No.: 9004-94-8 (Generic)

JPN Translation:
パルミチン酸 PEG - 20

Definition: PEG-20 Palmitate is the polyethylene glycol ester of palmitic acid that conforms to the formula:

$$CH_3(CH_2)_{14}C \overset{O}{\overset{||}{-}} (OCH_2CH_2)_nOH$$

where n has an average value of 20.

Information Sources: 21CFR175.300, 21CFR176.210, 21CFR177.2260, 21CFR177.2800, JCLS, MI-13(7660), TSCA

Chemical Class: Alkoxylated Carboxylic Acids

Functions: Surfactant - Cleansing Agent; Surfactant - Emulsifying Agent; Surfactant - Solubilizing Agent

Technical/Other Names:
Polyethylene Glycol 1000 Monopalmitate
Polyoxyethylene (20) Monopalmitate

Trade Name:
AEC PEG-20 Palmitate (A & E Connock)

Trade Name Mixture:
Xalidrene (Vevy)

PEG-8 PALMITOYL METHYL DIETHONIUM METHOSULFATE

Empirical Formula:
$C_{37}H_{72}NO_9 \cdot CH_3O_4S$

Definition: PEG-8 Palmitoyl Methyl Diethonium Methosulfate is the quaternary ammonium salt that conforms generally to the formula:

Chemical Class: Quaternary Ammonium Compounds

Function: Antistatic Agent

PEG-12 PALM KERNEL GLYCERIDES

CAS No.: 124046-52-2

Definition: PEG-12 Palm Kernel Glycerides is a polyethylene glycol derivative of the mono and diglycerides of palm kernel oil with an average of 12 moles of ethylene oxide.

Chemical Classes: Alkoxylated Alcohols; Glyceryl Esters and Derivatives

Functions: Skin-Conditioning Agent - Emollient; Surfactant - Emulsifying Agent

Technical/Other Names:
Polyethylene Glycol 600 Palm Kernel Glycerides
Polyoxyethylene (12) Palm Kernel Glycerides

PEG-45 PALM KERNEL GLYCERIDES

CAS No.: 124046-52-2

Definition: PEG-45 Palm Kernel Glycerides is a polyethylene glycol derivative of the mono and diglycerides of palm kernel oil with an average of 45 moles of ethylene oxide.

Chemical Classes: Alkoxylated Alcohols; Glyceryl Esters and Derivatives

Functions: Skin-Conditioning Agent - Emollient; Surfactant - Emulsifying Agent

Reported Product Categories: Bath Preparations, Misc.; Shampoos (Non-coloring)

Technical/Other Names:
Polyethylene Glycol (45) Palm Kernel Glycerides
Polyoxyethylene (45) Palm Kernel Glycerides

Trade Names:
Crovol PK70 (Croda Chemicals)
Crovol PK-70 (Croda, Inc.)

PEG-5 PENTAERYTHRITYL ETHER

Definition: PEG-5 Pentaerythrityl Ether is the polyethylene glycol ether of pentaerythritol that has an average of 5 moles of ethylene oxide.

Chemical Class: Alkoxylated Alcohols

Function: Humectant

Technical/Other Names:
PEG-5 Pentaerythritol Ether
Polyethylene Glycol (5) Pentaerythritol Ether
Polyoxyethylene (5) Pentaerythritol Ether

Trade Name Mixtures:
Lanolide (Vevy)
Seboside (Vevy)

PEG-150 PENTAERYTHRITYL TETRASTEARATE

JPN Translation:
テトラステアリン酸 PEG - 150 ペンタエリスリチル

Definition: PEG-150 Pentaerythrityl Tetrastearate is the tetraester of stearic acid and a polyethylene glycol ether of pentaerythritol with an average of 150 moles of ethylene oxide.

Chemical Classes: Esters; Ethers

Function: Viscosity Increasing Agent - Aqueous

Reported Product Categories: Bath Preparations, Misc.; Bath Oils, Tablets, and Salts; Hair Shampoos (Coloring); Cleansing Products (Cold Creams, Cleansing Lotions, Liquids and Pads); Shampoos (Non-coloring); Bath Soaps and Detergents

Technical/Other Names:
Pigment White 31
Polyethylene Glycol 6000 Pentaerythrityl Tetrastearate
Polyoxyethylene (150) Pentaerythrityl Tetrastearate

Trade Name:
Crothix (Croda, Inc.)

Trade Name Mixture:
Crothix Liquid (Croda, Inc.)

PEG-5 PHYTOSTEROL

JPN Translation:
PEG - 5 フィトステロール

Definition: PEG-5 Phytosterol is the polyethylene glycol ether of Phytosterol (q.v.) with an average ethoxylation value of 5.

Information Source: JCLS

Chemical Classes: Alkoxylated Alcohols; Sterols

Function: Surfactant - Emulsifying Agent

PEG-10 PHYTOSTEROL

JPN Translation:
PEG - 10 フィトステロール

Definition: PEG-10 Phytosterol is the polethylene glycol ether of Phytosterols (q.v.) with an average ethoxyloation value of 10.

Information Source: JCLS

Chemical Classes: Alkoxylated Alcohols; Sterols

Function: Surfactant - Emulsifying Agent

PEG-20 PHYTOSTEROL

JPN Translation:
ペンタオレイン酸 PEG - 40 ソルビット

Definition: PEG-20 Phytosterol is the polyethylene glycol ether of Phytosterols (q.v.) with an average ethoxylation value of 20.

Information Source: JCLS

Chemical Classes: Alkoxylated Alcohols; Sterols

Function: Not Reported

PEG-25 PHYTOSTEROL

JPN Translation:
PEG - 25 フィトステロール

Definition: PEG-25 Phytosterol is the poly-ethylene glycol ether of phytosterol with an average ethoxylation value of 25.

Information Source: JCLS

Chemical Classes: Alkoxylated Alcohols; Sterols

Function: Surfactant - Emulsifying Agent

Trade Name:
Nikkol BPSH-25 (Nikko)

PEG-30 PHYTOSTEROL

JPN Translation:
PEG - 30 フィトステロール

Definition: PEG-30 Phytosterol is the poly-ethylene glycol ether of Phytosterols (q.v.) with an average ethoxylation value of 30.

Information Source: JCLS

Chemical Classes: Alkoxylated Alcohols; Sterols

Function: Surfactant - Emulsifying Agent

PEG-9 POLYDIMETHYLSILOXYETHYL DIMETHICONE

Definition: PEG-9 Polydimethylsiloxyethyl Dimethicone is the siloxane polymer that conforms generally to the formula:

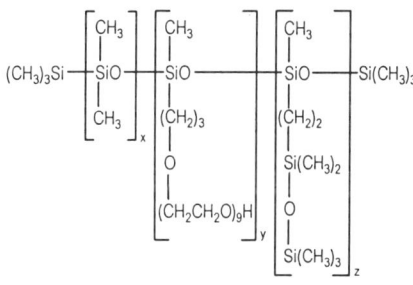

Chemical Classes: Siloxanes and Silanes; Synthetic Polymers

Functions: Skin-Conditioning Agent - Miscellaneous; Surfactant - Emulsifying Agent

Trade Name:
KF-6028 (Shin Etsu)

PEG-4 POLYGLYCERYL-2 DISTEARATE

CAS No.: 72828-11-6

Definition: PEG-4 Polyglyceryl-2 Distearate is the polyethylene glycol ether of Poly-glyceryl-2 Distearate (q.v.) with an average ethoxylation value of 4.

Information Source: TSCA

Chemical Classes: Alkoxylated Alcohols; Glyceryl Esters and Derivatives

Function: Surfactant - Emulsifying Agent

Technical/Other Names:
Polyethylene Glycol 200 Polyglyceryl-2 Distearate
Polyoxyethylene (4) Polyglyceryl-2 Distearate

PEG-10 POLYGLYCERYL-2 LAURATE

Definition: PEG-10 Polyglyceryl-2 Laurate is a polyethylene glycol ether of a dimer of glycerin esterified with lauric acid containing an average ethoxylation value of 10.

Chemical Classes: Alkoxylated Alcohols; Glyceryl Esters and Derivatives

Function: Surfactant - Emulsifying Agent

Technical/Other Names:
Polyethylene Glycol 500 Polyglyceryl-2 Laurate
Polyoxyethylene (10) Polyglyceryl-2 Laurate

Trade Names:
Hostacerin DGL (Clariant)
Hostacerin DGL (Clariant GmbH, Personal Care)

PEG-4 POLYGLYCERYL-2 STEARATE

Definition: PEG-4 Polyglyceryl-2 Stearate is a polyethylene glycol ether of a dimer of glycerin esterified with stearic acid. It has an average of 4 moles of ethylene oxide.

Chemical Classes: Alkoxylated Alcohols; Glyceryl Esters and Derivatives

Function: Surfactant - Emulsifying Agent

Technical/Other Names:
Polyethylene Glycol 200 Polyglyceryl-2 Stearate
Polyoxyethylene (4) Polyglyceryl-2 Stearate

Trade Names:
Hostacerin DGSB (Clariant)
Hostacerin DGSB (Clariant GmbH, Personal Care)

PEG-800/POLYVINYL ALCOHOL COPOLYMER

Definition: PEG-800/Polyvinyl Alcohol Copolymer is a copolymer of PEG-800 and vinyl alcohol monomers.

Chemical Class: Synthetic Polymers

Function: Hair Fixative

PEG/PPG-28/21 ACETATE DIMETHICONE

CAS No.: 68037-64-9

Definition: PEG/PPG-28/21 Acetate Dimethicone is the silicone polymer that conforms generally to the formula:

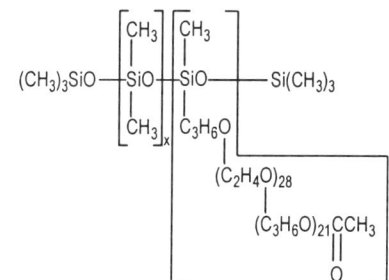

Chemical Class: Synthetic Polymers

Functions: Hair Conditioning Agent; Humectant; Surfactant - Cleansing Agent; Surfactant - Emulsifying Agent; Surfactant - Foam Booster; Surfactant - Suspending Agent

Technical/Other Name:
Siloxanes and Silicones, di-Me, Me Hydrogen, Reaction Products with Polyethylene-Polypropylene Glycol Monoacetate Allyl Ether

Trade Name:
FZ-2409 (Nippon Unicar)

PEG/PPG/BUTYLENE/DIMETHICONE COPOLYMER

JPN Translation:
ハマメリス水

Definition: *See "Regulatory and Ingredient Use Information," regarding use of Japan Trivial names in Volume 1, Introduction, Part A.*

Information Source: JCLS

Chemical Class: Synthetic Polymers

Function: Not Reported

PEG/PPG-20/22 BUTYL ETHER DIMETHICONE

CAS No.: 67762-87-2

Definition: PEG/PPG-20/22 Butyl Ether Dimethicone is the butyl ether of a derivative of Dimethicone (q.v.) containing an average of 20 moles of ethylene oxide and 22 moles of propylene oxide.

Chemical Classes: Alkoxylated Alcohols; Siloxanes and Silanes

Functions: Hair Conditioning Agent; Skin-Conditioning Agent - Miscellaneous

Technical/Other Name:
Siloxanes and Silicones, Di-Me, 3-Hydroxypropyl Me, Ethers wtih Polyethylene-Polypropylene Glycol Mono-Butyl Ether

Trade Name:
KF-6012 (Shin Etsu)

PEG/PPG-22/22 BUTYL ETHER DIMETHICONE

Definition: PEG/PPG-22/22 Butyl Ether Dimethicone is the butyl ether of a derivative of Dimethicone (q.v.) containing an average of 22 moles of ethylene oxide and 22 moles of propylene oxide.

Chemical Class: Siloxanes and Silanes

Functions: Hair Conditioning Agent; Skin-Conditioning Agent - Miscellaneous; Surfactant - Emulsifying Agent

Trade Name:
KF352A (Shin Etsu)

PEG/PPG-23/23 BUTYL ETHER DIMETHICONE

Definition: PEG/PPG-23/23 Butyl Ether Dimethicone is the butyl ether of a derivative of Dimethicone (q.v.) containing an average of 23 moles of ethylene oxide and 23 moles of propylene oxide.

Chemical Classes: Alkoxylated Alcohols; Siloxanes and Silanes

Functions: Hair Conditioning Agent; Humectant; Skin-Conditioning Agent - Miscellaneous; Surfactant - Emulsifying Agent; Surfactant - Suspending Agent

Technical/Other Name:
Polyethylene Glycol (23) Polypropylene Glycol (23) Butyl Ether Dimethicone

Trade Name:
FZ-2420 (Nippon Unicar)

PEG/PPG-24/18 BUTYL ETHER DIMETHICONE

CAS No.: 67762-87-2

Definition: PEG/PPG-24/18 Butyl Ether Dimethicone is the butyl ether of a derivative of Dimethicone (q.v.) containing an average of 24 moles of ethylene oxide and 18 moles of propylene oxide.

Chemical Class: Siloxanes and Silanes

Functions: Hair Conditioning Agent; Humectant; Surfactant - Cleansing Agent; Surfactant - Emulsifying Agent; Surfactant - Suspending Agent

Technical/Other Names:
Polyethylene Glycol (24) Polypropylene Glycol (18) Butyl Ether Dimethicone
Siloxanes and Silicones, di-Me, 3-Hydroxypropyl Me, Ethers with Polyethylene-Polypropylene Glycol mono-Bu Ether

Trade Name:
FZ-2402 (Nippon Unicar)

PEG/PPG-27/9 BUTYL ETHER DIMETHICONE

Definition: PEG/PPG-27/9 Butyl Ether Dimethicone is the butyl ether of a derivative of Dimethicone (q.v.) containing an average of 27 moles of ethylene oxide and 9 moles of propylene oxide.

Chemical Class: Siloxanes and Silanes

Functions: Hair Conditioning Agent; Skin-Conditioning Agent - Miscellaneous; Surfactant - Emulsifying Agent

Trade Name:
KF-615A (Shin Etsu)

PEG-4-PPG-7 C13/C15 ALCOHOL

Definition: PEG-4-PPG-7 C13/C15 Alcohol is the polyoxypropylene, polyoxyethylene ether of a mixture of synthetic C13/C15 alcohols with an average propoxylation value of 7 and an average ethoxylation value of 4.

Chemical Class: Alkoxylated Alcohols

Function: Surfactant - Emulsifying Agent

Technical/Other Names:
Polyoxyethylene (4) Polyoxypropylene (7) C13/15 Alcohol
Polyoxypropylene (7) Polyoxyethylene (4) C13/15 Alcohol

PEG/PPG-1/2 COPOLYMER

CAS No.: 9003-11-6 (Generic)

JPN Translation:
PEG / PPG - 1 / 2 コポリマー

Definition: PEG/PPG-1/2 Copolymer is the random copolymer produced by the interaction of an average of 1 mole of ethylene oxide with 2 moles of propylene oxide.

Information Source: JCIC

Chemical Class: Polymeric Ethers

Function: Solvent

Technical/Other Name:
Polyoxyethylene (1) Polyoxypropylene (2)

PEG/PPG-4/2 COPOLYMER

CAS No.: 9003-11-6 (Generic)

JPN Translation:
PEG / PPG - 4 / 2 コポリマー

Definition: PEG/PPG-4/2 Copolymer is the random copolymer produced by the interaction of an average of 4 moles of ethylene oxide with 2 moles of propylene oxide.

Information Source: JCIC

Chemical Class: Polymeric Ethers

Function: Solvent

Technical/Other Name:
Polyoxyethylene (4) Polyoxypropylene (2)

PEG/PPG-5/30 COPOLYMER

CAS No.: 9003-11-6

JPN Translation:
PEG / PPG - 5 / 30 コポリマー

Definition: PEG/PPG-5/30 Copolymer is the polyoxyethylene, polyoxypropylene block polymer that conforms generally to the formula:

$$HO(CH_2CH_2O)_x(CHCH_2O)_y(CH_2CH_2O)_zH$$
$$CH_3$$

where x+z has an average value of 5 and y has an average value of 30.

Chemical Class: Polymeric Ethers

Function: Solvent

Trade Name:
Pepol B-181 (Toho)

PEG/PPG-6/2 COPOLYMER

CAS No.: 9003-11-6 (Generic)

JPN Translation:
PEG / PPG - 6 / 2 コポリマー

Definition: PEG/PPG-6/2 Copolymer is the random copolymer produced by the interaction of an average of 6 moles of ethylene oxide with 2 moles of propylene oxide.

Information Source: JCIC

Chemical Class: Polymeric Ethers

Function: Solvent

Technical/Other Name:
Polyoxyethylene (6) Polyoxypropylene (2)

PEG/PPG-7/50 COPOLYMER

CAS No.: 9003-11-6 (Generic)

JPN Translation:
PEG / PPG - 7 / 50 コポリマー

Definition: PEG/PPG-7/50 Copolymer is the random copolymer produced by the interaction of an average of 7 moles of ethylene oxide with 50 moles of propylene oxide.

Information Source: JCIC

Chemical Class: Polymeric Ethers

Function: Solvent

Technical/Other Name:
Polyoxyethylene (7) Polyoxypropylene (50)

PEG/PPG-8/17 COPOLYMER

CAS No.: 9003-11-6 (Generic)

JPN Translation:
PEG / PPG - 8 / 17 コポリマー

Definition: PEG/PPG-8/17 Copolymer is the random copolymer produced by the interaction of an average of 4 moles of ethylene oxide with 2 moles of propylene oxide.

Information Source: JCIC

Chemical Class: Polymeric Ethers

Function: Solvent

Technical/Other Name:
Polyoxyethylene (8) Polyoxypropylene (17)

PEG/PPG-10/2 COPOLYMER

CAS No.: 9003-11-6 (Generic)

JPN Translation:
PEG / PPG - 10 / 2 コポリマー

Definition: PEG/PPG-10/2 Copolymer is the polyoxyethylene, polyoxypropylene block copolymer that conforms generally to the formula:

$$HO(CH_2CH_2O)_x(CHCH_2O)_y(CH_2CH_2O)_zH$$
$$|$$
$$CH_3$$

where x+z has an average value of 10 and y has an average value of 2.

Chemical Class: Polymeric Ethers

Function: Surfactant - Emulsifying Agent

Trade Name:
Zenimer 210 (Zenitech)

PEG/PPG-10/70 COPOLYMER

CAS No.: 9003-11-6 (Generic)

JPN Translation:
PEG / PPG - 10 / 70 コポリマー

Definition: PEG/PPG-10/70 Copolymer is the random copolymer produced by the interaction of an average of 10 moles of ethylene oxide with 70 moles of propylene oxide.

Information Source: JCIC

Chemical Class: Polymeric Ethers

Function: Solvent

Technical/Other Name:
Polyoxyethylene (10) Polyoxypropylene (70)

PEG/PPG-17/6 COPOLYMER

CAS No.: 9003-11-6 (Generic)

Definition: PEG/PPG-17/6 Copolymer is the random copolymer produced by the interaction of an average of 17 moles of ethylene oxide with 6 moles of propylene oxide.

Information Sources: CTFA D, TSCA

Chemical Class: Polymeric Ethers

Function: Solvent

Trade Name:
UCON Fluid 75-H-450 (Amerchol)

PEG/PPG-18/4 COPOLYMER

CAS No.: 9003-11-6 (Generic)

Definition: PEG/PPG-18/4 Copolymer is the random copolymer produced by the interaction of an average of 18 moles of ethylene oxide with 4 moles of propylene oxide.

Information Source: TSCA

Chemical Class: Polymeric Ethers

Function: Solvent

Trade Name:
UCON 75-H-450 Lubricant (Dow Chemical)

PEG/PPG-19/21 COPOLYMER

JPN Translation:
ヘチマエキス

Definition: PEG/PPG-19/21 Copolymer is the random copolymer produced by the interaction fo 19 moles of ethylene oxide and 21 moles of propylene oxide.

Information Source: JCLS

Chemical Class: Polymeric Ethers

Function: Solvent

PEG/PPG-23/17 COPOLYMER

CAS No.: 9003-11-6 (Generic)

JPN Translation:
PEG / PPG - 23 / 17 コポリマー

Definition: PEG/PPG-23/17 Copolymer is the random copolymer produced the interaction of an average of 23 moles of ethylene oxide with 17 moles of propylene oxide.

Information Source: JCIC

Chemical Class: Polymeric Ethers

Function: Solvent

Technical/Other Name:
Polyoxyethylene (23) Polyoxypropylene (17)

PEG/PPG-23/50 COPOLYMER

CAS No.: 9003-11-6 (Generic)

Definition: PEG/PPG-23/50 Copolymer is the random copolymer produced by the interaction of an average of 23 moles of ethylene oxide with 50 moles of propylene oxide.

Information Source: TSCA

Chemical Class: Polymeric Ethers

Function: Solvent

PEG/PPG-25/30 COPOLYMER

CAS No.: 9003-11-6

JPN Translation:
PEG / PPG - 25 / 30 コポリマー

Definition: PEG/PPG-25/30 Copolymer is the polyoxyethylene, polyoxypropylene block copolymer that conforms generally to the formula:

$$HO(CH_2CH_2O)_x(CHCH_2O)_y(CH_2CH_2O)_zH$$
$$|$$
$$CH_3$$

where x+z has an average value of 25 and y has an average value of 30.

Chemical Class: Polymeric Ethers

Function: Solvent

Trade Name:
Pepol B-184 (Toho)

PEG/PPG-26/31 COPOLYMER

CAS No.: 9003-11-6 (Generic)

JPN Translation:
PEG / PPG - 26 / 31 コポリマー

Definition: PEG/PPG-26/31 Copolymer is the random copolymer produced by the interaction of an average of 26 moles of ethylene oxide with 31 moles of propylene oxide.

Information Source: JCIC

Chemical Class: Polymeric Ethers

Function: Solvent

Technical/Other Name:
Polyoxyethylene (26) Polyoxypropylene (31)

PEG/PPG-30/33 COPOLYMER

CAS No.: 9003-11-6 (Generic)

JPN Translation:
PEG / PPG - 30 / 33 コポリマー

Definition: PEG/PPG-30/33 Copolymer is the random copolymer produced by the interaction of an average of 30 moles of ethylene oxide with 33 moles of propylene oxide.

Information Source: JCIC

Chemical Class: Polymeric Ethers

Function: Solvent

Technical/Other Name:
Polyoxyethylene (30) Polyoxypropylene (33)

PEG/PPG-30/160 COPOLYMER

CAS No.: 9003-11-6 (Generic)

Definition: PEG/PPG-30/160 Copolymer is the random copolymer produced by the interaction of an average of 30 moles of ethylene oxide with 160 moles of propylene oxide.

Chemical Class: Polymeric Ethers

Functions: Surfactant - Emulsifying Agent; Surfactant - Solubilizing Agent; Surfactant - Suspending Agent

Trade Names:
Pepol B-182 (Toho)
Pepol B-188 (Toho)

PEG/PPG-32/3 COPOLYMER

CAS No.: 9003-11-6 (Generic)

Definition: PEG/PPG-32/3 Copolymer is the polyoxyethylene, polyoxypropylene block copolymer that conforms generally to the formula:

$$HO(CH_2CH_2O)_x(CHCH_2O)_y(CH_2CH_2O)_zH$$
$$|$$
$$CH_3$$

where x+z has an average value of 32 and y has an average value of 3.

Chemical Class: Polymeric Ethers

Function: Surfactant - Hydrotrope

Trade Name:
Zenimer 332 (Zenitech)

PEG/PPG-35/9 COPOLYMER

CAS No.: 9003-11-6 (Generic)

Definition: PEG/PPG-35/9 Copolymer is the random copolymer produced by the interaction of an average of 35 moles of ethylene oxide with 9 moles of propylene oxide.

Information Source: TSCA

Chemical Class: Polymeric Ethers

Function: Solvent

Trade Name:
UCON 75-H-1400 Lubricant (Dow Chemical)

PEG/PPG-38/8 COPOLYMER

CAS No.: 9003-11-6 (Generic)

Definition: PEG/PPG-38/8 Copolymer is a random copolymer produced by the interaction of an average of 38 moles of ethylene oxide with 8 moles of propylene oxide.

Chemical Class: Polymeric Ethers

Function: Solvent

Trade Name:
Pluraflo L4370 (BASF)

PEG/PPG-116/66 COPOLYMER

CAS No.: 9003-11-6 (Generic)

Definition: PEG/PPG-116/66 Copolymer is a random copolymer produced by the interaction of an average of 116 moles of ethylene oxide with 66 moles of propylene oxide.

Chemical Class: Polymeric Ethers

Function: Solvent

Trade Name:
Pluraflo L1220 (BASF)

PEG/PPG-125/30 COPOLYMER

CAS No.: 9003-11-6 (Generic)

Definition: PEG/PPG-125/30 Copolymer is the random copolymer produced by the interaction of an average of 125 moles of ethylene oxide with 30 moles of propylene oxide.

Information Source: TSCA

Chemical Class: Polymeric Ethers

Function: Solvent

Trade Name:
UCON 75-H-90,000 Lubricant (Dow Chemical)

PEG/PPG-150/30 COPOLYMER

CAS No.: 9003-11-6 (Generic)

JPN Translation:
PEG / PPG - 150 / 30 コポリマー

Definition: PEG/PPG-150/30 Copolymer is the random copolymer produced by the interaction of an average of 150 moles of ethylene oxide with 30 moles of propylene oxide.

Information Sources: JCLS, TSCA

Chemical Class: Polymeric Ethers

Functions: Surfactant - Emulsifying Agent; Surfactant - Solubilizing Agent

Trade Names:
Genapol PF 80 Powder (Clariant)
Genapol PF 80 Powder (Clariant GmbH, Personal Care)

PEG/PPG-160/31 COPOLYMER

CAS No.: 9003-11-6 (Generic)

JPN Translation:
PEG / PPG - 160 / 31 コポリマー

Definition: PEG/PPG-160/31 Copolymer is the random copolymer produced by the interaction of an average of 160 moles of ethylene oxide with 31 moles of propylene oxide.

Information Source: JCIC

Chemical Class: Polymeric Ethers

Function: Solvent

Technical/Other Name:
Polyoxyethylene (160) Polyoxypropylene (31)

PEG/PPG-200/70 COPOLYMER

CAS No.: 9003-11-6 (Generic)

JPN Translation:
PEG / PPG - 200 / 70 コポリマー

Definition: PEG/PPG-200/70 Copolymer is the random copolymer produced by the interaction of an average of 200 moles of ethylene oxide with 70 moles of propylene oxide.

Information Source: JCIC

Chemical Class: Polymeric Ethers

Function: Solvent

Technical/Other Name:
Polyoxyethylene (200) Polyoxypropylene (70)

PEG/PPG-240/60 COPOLYMER

CAS No.: 9003-11-6 (Generic)

JPN Translation:
PEG / PPG - 240 / 60 コポリマー

Definition: PEG/PPG-240/60 Copolymer is a random copolymer produced by the interaction of an average of 240 moles of ethylene oxide and 60 moles of propylene oxide.

Chemical Class: Polymeric Ethers

Function: Solvent

Trade Name:
Unilube 75DE-2620R (NOF)

PEG/PPG-300/55 COPOLYMER

CAS No.: 9003-11-6

JPN Translation:
PEG / PPG - 300 / 55 コポリマー

Definition: PEG/PPG-300/55 Copolymer is the polyoxyethylene, polyoxypropylene block polymer that conforms generally to the formula:

$$HO(CH_2CH_2O)_x(CHCH_2O)_y(CH_2CH_2O)_zH$$
$$|$$
$$CH_3$$

where x+z has an average value of 300 and y has an average value of 55.

Information Source: JCLS

Chemical Class: Polymeric Ethers

Function: Surfactant - Cleansing Agent

Trade Name:
Empilan P7108 (Albright & Wilson UK)

PEG/PPG-1/25 DIETHYLMONIUM CHLORIDE

JPN Translation:
PEG - 1 - PPG - 25 ジエチルメチルアンモ
ニウムクロリド

Definition: PEG/PPG-1/25 Diethylmonium Chloride is the quaternary ammonium salt that conforms generally to the formula:

$$\left[CH_3CH_2-\overset{\overset{\displaystyle CH_3}{|}}{\underset{\underset{\displaystyle CH_2CH_3}{|}}{N}}-O(CH_2CH_2O)_x(CH_2CHO)_yH \right]^+ \quad Cl^-$$
$$\underset{\displaystyle CH_3}{}$$

where x has an average value of 1 and y has an average value of 25.

Information Source: JCIC

Chemical Class: Quaternary Ammonium Compounds

Function: Antistatic Agent

Technical/Other Names:
Polyethylene Glycol (1) Polypropylene Glycol (25) Diethylmonium Chloride
Polyoxyethylene (1) Polyoxypropylene (25) Diethyl Methyl Ammonium Chloride
Polyoxyethylene (1) Polyoxypropylene (25) Diethylmonium Chloride

PEG/PPG-8/3 DIISOSTEARATE

Definition: PEG/PPG-8/3 Diisostearate is the polyethylene glycol ether of the propoxylated diester of isostearic acid containing an average ethoxylation value of 8 and propoxylation value of 3.

Chemical Class: Alkoxylated Alcohols

Function: Surfactant - Emulsifying Agent

Trade Name:
Hydramol PGPD (Scher)

PEG/PPG-3/10 DIMETHICONE

Definition: PEG/PPG-3/10 Dimethicone is the alkoxylated derivative of Dimethicone (q.v.) containing an average of 3 moles of ethylene oxide and 12 moles of propylene oxide.

Chemical Class: Siloxanes and Silanes

Function: Surfactant - Emulsifying Agent

Trade Name:
TAYLOR T-Wet 820 (Taylor Chemical Company)

PEG/PPG-4/12 DIMETHICONE

Definition: PEG/PPG-4/12 Dimethicone is the alkoxylated derivative of Dimethicone (q.v.) containing an average of 4 moles of ethylene oxide and 12 moles of propylene oxide.

Chemical Class: Siloxanes and Silanes

Function: Surfactant - Emulsifying Agent

Reported Product Categories: Aftershave Lotions; Bath Soaps and Detergents; Body and Hand Preparations (Excluding Shaving Preparations); Cleansing Products (Cold Creams, Cleansing Lotions, Liquids and Pads); Colognes and Toilet Waters; Deodorants (Underarm); Mascara; Skin Care Preparations, Misc.

Trade Name:
Abil B 8852 (Degussa Care Specialties)

PEG/PPG-6/11 DIMETHICONE

Definition: PEG/PPG-6/11 Dimethicone is the alkoxylated derivative of Dimethicone (q.v.) containing an average of 6 moles of ethylene oxide and 11 moles of propylene oxide.

Chemical Classes: Siloxanes and Silanes; Synthetic Polymers

Function: Surfactant - Emulsifying Agent

PEG/PPG-8/14 DIMETHICONE

Definition: PEG/PPG-8/14 Dimethicone is the alkoxylated derivative of Dimethicone (q.v.) containing an average of 8 moles of ethylene oxide and 14 moles of propylene oxide.

Chemical Class: Siloxanes and Silanes

Function: Surfactant - Emulsifying Agent

PEG/PPG-12/16 DIMETHICONE

Definition: PEG/PPG-12/16 Dimethicone is the alkoxylated derivative of Dimethicone (q.v.) containing an average of 12 moles of ethylene oxide and 16 moles of propylene oxide.

Chemical Class: Siloxanes and Silanes

Functions: Antifoaming Agent; Skin-Conditioning Agent - Miscellaneous; Slip Modifier; Surfactant - Emulsifying Agent

Trade Name Mixture:
MagiSil 451S (Diow)

PEG/PPG-12/18 DIMETHICONE

Definition: PEG/PPG-12/18 Dimethicone is the alkoxylated derivative of Dimethicone (q.v.) containing an average of 12 moles of ethylene oxide and 18 moles of propylene oxide.

Chemical Class: Siloxanes and Silanes

Functions: Antifoaming Agent; Skin-Conditioning Agent - Miscellaneous; Slip Modifier; Surfactant - Emulsifying Agent

Trade Name Mixture:
MagiSil 451SS (Diow)

PEG/PPG-14/4 DIMETHICONE

Definition: PEG/PPG-14/4 Dimethicone is the alkoxylated derivative of Dimethicone (q.v.) containing an average of 14 moles of ethylene oxide and 4 moles of propylene oxide.

Chemical Class: Siloxanes and Silanes

Function: Surfactant - Emulsifying Agent

Reported Product Categories: Bath Soaps and Detergents; Body and Hand Preparations (Excluding Shaving Preparations); Cleansing Products (Cold Creams, Cleansing Lotions, Liquids and Pads); Eye Lotions; Hair Sprays (Aerosol Fixatives); Hair Straighteners; Moisturizing Preparations; Shampoos (Non-coloring); Skin Fresheners; Tonics, Dressings, and Other Hair Grooming Aids

Trade Name:
Abil B 8851 (Degussa Care Specialties)

PEG/PPG-15/15 DIMETHICONE

Definition: PEG/PPG-15/15 Dimethicone is the alkoxylated derivative of Dimethicone (q.v.) containing an average of 15 moles of ethylene oxide and 15 moles of propylene oxide.

Chemical Class: Siloxanes and Silanes

Functions: Anticaking Agent; Surfactant - Emulsifying Agent

Trade Name:
Dow Corning 5330 Fluid (Dow Corning)

PEG/PPG-16/2 DIMETHICONE

Definition: PEG/PPG-16/2 Dimethicone is the alkoxylated derivative of Dimethicone (q.v.) containing an average of 16 moles of ethylene oxide and 2 moles of propylene oxide.

Chemical Class: Siloxanes and Silanes

Function: Surfactant - Emulsifying Agent

Trade Name:
TAYLOR T-Wet 910 (Taylor Chemical Company)

PEG/PPG-16/8 DIMETHICONE

Definition: PEG/PPG-16/8 Dimethicone is the alkoxylated derivative of Dimethicone (q.v.) containing an average of 16 moles of ethylene oxide and 8 moles of propylene oxide.

Chemical Class: Siloxanes and Silanes

Functions: Antifoaming Agent; Skin-Conditioning Agent - Miscellaneous; Slip Modifier; Surfactant - Emulsifying Agent

Trade Name Mixture:
MagiSil 451SW (Diow)

PEG/PPG-17/18 DIMETHICONE

Definition: PEG/PPG-17/18 Dimethicone is the alkoxylated derivative of Dimethicone (q.v.) containing an average of 17 moles of ethylene oxide and 18 moles of propylene oxide.

Chemical Class: Siloxanes and Silanes

Function: Surfactant - Emulsifying Agent

Reported Product Categories: Hair Conditioners; Hair Sprays (Aerosol Fixatives); Shampoos (Non-coloring); Tonics, Dressings, and Other Hair Grooming Aids

Trade Name:
Dow Corning 2-5220 Resin Modifier (Dow Corning)

PEG/PPG-18/18 DIMETHICONE

Definition: PEG/PPG-18/18 Dimethicone is the alkoxylated derivative of Dimethicone (q.v.) containing an average of 18 moles of ethylene oxide and 18 moles of propylene oxide.

Chemical Class: Siloxanes and Silanes

Function: Surfactant - Emulsifying Agent

Reported Product Categories: Aftershave Lotions; Bath Preparations, Misc.; Cleansing Products (Cold Creams, Cleansing Lotions, Liquids and Pads); Eye Shadows; Face Powders; Foundations; Fragrance Preparations, Misc.; Indoor Tanning Preparations; Lipsticks; Makeup Preparations (Not eye), Misc.; Personal Cleanliness Products, Misc.; Shaving Preparations, Misc.; Skin Care Preparations, Misc.

Trade Names:
Dow Corning 190 Surfactant (Dow Corning)
Silsurf J-1013-V-CG (Siltech LLC)

Trade Name Mixtures:
DCH45TS (Kobo)

DC65ZC1 (Kobo)
Dow Corning 3225C Formulation Aid (Dow Corning)
Dow Corning 5225C Formulation Aid (Dow Corning)
Lipofirm LCW (LCW)
Vitacap (LCW)

PEG/PPG-19/19 DIMETHICONE

Definition: PEG/PPG-19/19 Dimethicone is the alkoxylated derivative of Dimethicone (q.v.) containing an average of 19 moles of ethylene oxide and 19 moles of propylene oxide.

Chemical Class: Siloxanes and Silanes

Function: Surfactant - Emulsifying Agent

Trade Name Mixtures:
BY22-008M (DCTS)
Dow Corning-Toray Silicone BY 11-030 (DCTS)

PEG/PPG-20/6 DIMETHICONE

Definition: PEG/PPG-20/6 Dimethicone is the alkoxylated derivative of Dimethicone (q.v.) containing an average of 20 moles of ethylene oxide and 6 moles of propylene oxide.

Chemical Class: Siloxanes and Silanes

Function: Surfactant - Emulsifying Agent

Reported Product Categories: Bath Preparations, Misc.; Bath Soaps and Detergents; Cleansing Products (Cold Creams, Cleansing Lotions, Liquids and Pads)

Trade Names:
Abil B 88183 (Degussa Care Specialties)
Abil B 88184 (Degussa Care Specialties)

PEG/PPG-20/15 DIMETHICONE

Definition: PEG/PPG-20/15 Dimethicone is the alkoxylated derivative of Dimethicone (q.v.) containing an average of 20 moles of ethylene oxide and 15 moles of propylene oxide.

Chemical Class: Siloxanes and Silanes

Function: Surfactant - Emulsifying Agent

Reported Product Categories: Hair Sprays (Aerosol Fixatives); Tonics, Dressings, and Other Hair Grooming Aids

Trade Names:
AEC PEG/PPG-20/15 Dimethicone (A & E Connock)

Botanisil S-18 (Botanigenics)
SF1188A (GE Silicones)

Trade Name Mixtures:
AEC Cyclomethicone (&) PEG/PPG-20/15 Dimethicone (A & E Connock)
Botanisil CD-80 (Botanigenics)
Botanisil CD-90 (Botanigenics)
SF1328 (GE Silicones)
SF1528 (GE Silicones)
SF1540 (GE Silicones)

PEG/PPG-20/20 DIMETHICONE

Definition: PEG/PPG-20/20 Dimethicone is the alkoxylated derivative of Dimethicone (q.v.) containing an average of 20 moles of ethylene oxide and 20 moles of propylene oxide.

Chemical Class: Siloxanes and Silanes

Function: Surfactant - Emulsifying Agent

Trade Names:
Abil B 8863 (Degussa Care Specialties)
SH 3749 (DCTS)

Trade Name Mixtures:
BY22-008 (DCTS)
BY22-012 (DCTS)

PEG/PPG-20/23 DIMETHICONE

Definition: PEG/PPG-20/23 Dimethicone is the alkoxylated derivative of Dimethicone (q.v.) containing an average of 20 moles of ethylene oxide and 23 moles of propylene oxide.

Chemical Class: Siloxanes and Silanes

Functions: Emulsion Stabilizer; Hair Conditioning Agent; Skin-Conditioning Agent - Miscellaneous; Slip Modifier; Surface Modifier; Surfactant - Solubilizing Agent

Trade Names:
Silsoft 430 dimethicone copolyol (Crompton Corporation)
Silsoft 440 dimethicone copolyol (Crompton Corporation)

PEG/PPG-20/29 DIMETHICONE

Definition: PEG/PPG-20/29 Dimethicone is the alkoxylated derivative of Dimethicone (q.v.) containing an average of 20 moles of ethylene oxide and 29 moles of propylene oxide.

Chemical Classes: Siloxanes and Silanes; Synthetic Polymers

Function: Surfactant - Emulsifying Agent

PEG/PPG-22/23 DIMETHICONE

Definition: PEG/PPG-22/23 Dimethicone is the alkoxylated derivative of Dimethicone (q.v.) containing an average of 22 moles of ethylene oxide and 23 moles of propylene oxide.

Chemical Class: Siloxanes and Silanes

Function: Surfactant - Emulsifying Agent

Reported Product Categories: Skin Care Preparations, Misc.; Skin Fresheners

Trade Name:
TAYLOR T-Wet 640 (Taylor Chemical Company)

PEG/PPG-22/24 DIMETHICONE

Definition: PEG/PPG-22/24 Dimethicone is the alkoxylated derivative of Dimethicone (q.v.) containing an average of 22 moles of ethylene oxide and 24 moles of propylene oxide.

Chemical Class: Siloxanes and Silanes

Function: Surfactant - Emulsifying Agent

Trade Name:
Mirasil DMCO (Rhodia)

PEG/PPG-23/6 DIMETHICONE

Definition: PEG/PPG-23/6 Dimethicone is the alkoxylated derivative of Dimethicone (q.v.) containing an average of 23 moles of ethylene oxide and 6 moles of propylene oxide.

Chemical Class: Siloxanes and Silanes

Functions: Emulsion Stabilizer; Slip Modifier; Surface Modifier; Surfactant - Solubilizing Agent

Trade Name:
Silsoft 475 dimethicone copolyol (Crompton Corporation)

PEG/PPG-25/25 DIMETHICONE

Definition: PEG/PPG-25/25 Dimethicone is the alkoxylated derivative of Dimethicone (q.v.) containing an average of 25 moles of ethylene oxide and 25 moles of propylene oxide.

Chemical Class: Siloxanes and Silanes

Function: Surfactant - Emulsifying Agent

Trade Name:
Wacker Belsil DMC 6031 (Wacker-Chemie)

PEG/PPG-27/27 DIMETHICONE

Definition: PEG/PPG-27/27 Dimethicone is the alkoxylated derivative of Dimethicone (q.v.) containing an average of 27 moles of ethylene oxide and 27 moles of propylene oxide.

Chemical Class: Siloxanes and Silanes

Function: Surfactant - Emulsifying Agent

PEG/PPG-25/25 DIMETHICONE/ ACRYLATES COPOLYMER

Definition: PEG/PPG-25/25 Dimethicone/ Acrylates Copolymer is a copolymer of PEG-25, PPG-25, dimethicone, and one or more monomers consisting of acrylic acid, methacrylic acid or one of their simple esters.

Chemical Classes: Siloxanes and Silanes; Synthetic Polymers

Functions: Film Former; Hair Fixative

Trade Name:
Luviflex Silk (BASF)

PEG/PPG-3/6 DIMETHYL ETHER

CAS No.: 61419-46-3

Empirical Formula:
$C_{26}H_{54}O_{10}$

Definition: PEG/PPG-3/6 Dimethyl Ether is the copolymer produced by the interaction of 3 moles of ethylene oxide with 6 moles of propylene oxide end-blocked with methyl ether.

Chemical Class: Alkoxylated Alcohols

Function: Skin-Conditioning Agent - Miscellaneous

Trade Name:
MACBIOBRIDE E-530 (NOF America Corp.)

PEG/PPG-9/2 DIMETHYL ETHER

CAS No.: 61419-46-3

Empirical Formula:
$C_{26}H_{54}O_{12}$

Definition: PEG/PPG-9/2 Dimethyl Ether is the copolymer produced by the interaction of 9 moles of ethylene oxide with 2 moles of propylene oxide end-blocked with dimethyl ether.

Chemical Class: Alkoxylated Alcohols

Function: Skin-Conditioning Agent - Miscellaneous

Trade Name:
 MACBIOBRIDE E-575 (NOF America Corp.)

PEG/PPG-14/7 DIMETHYL ETHER

CAS No.: 61419-46-3

Empirical Formula:
 $C_{51}H_{104}O_{22}$

Definition: PEG/PPG-14/7 Dimethyl Ether is the copolymer produced by the interaction of 14 moles of ethylene oxide with 7 moles of propylene oxide end-blocked with dimethyl ether.

Chemical Class: Alkoxylated Alcohols

Function: Skin-Conditioning Agent - Miscellaneous

Trade Name:
 MACBIOBRIDE E-1060 (NOF America Corp.)

PEG/PPG-27/14 DIMETHYL ETHER

Definition: PEG/PPG-27/14 Dimethyl Ether is a copolymer produced by the interaction of 27 moles of ethylene oxide with 14 moles of propylene oxide end-blocked with methyl ether.

Chemical Class: Alkoxylated Alcohols

Function: Skin-Conditioning Agent - Miscellaneous

Trade Name:
 Macbiobride E-2060 (NOF.)

PEG/PPG-36/41 DIMETHYL ETHER

Definition: PEG/PPG-36/41 Dimethyl Ether is the copolymer produced by the interaction of 36 moles of ethylene oxide and 41 moles of propylene oxide end-blocked with methyl ether.

Chemical Class: Alkoxylated Alcohols

Function: Skin-Conditioning Agent - Miscellaneous

Trade Name:
 Macbiobride E-4040 (NOF)

PEG/PPG-55/28 DIMETHYL ETHER

Definition: PEG/PPG-55/28 Dimethyl Ether is a copolymer produced by the interaction of 55 moles of ethylene oxide with 28 moles of propylene oxide end-blocked with methyl ether.

Chemical Class: Alkoxylated Alcohols

Function: Skin-Conditioning Agent - Miscellaneous

Trade Name:
 Macbiobride E-4060 (NOF.)

PEG/PPG-10/2 DIRICINOLEATE

Definition: PEG/PPG-10/2 Diricinoleate is the diester obtained by the interaction of 10 moles of ethylene oxide and 2 moles of propylene oxide with Ricinoleic Acid (q.v.).

Chemical Class: Alkoxylated Carboxylic Acids

Function: Surfactant - Emulsifying Agent

Trade Names:
 Zenimer R-210-D (Zenitech)
 Zenimer R-210-M (Zenitech)

PEG/PPG-32/3 DIRICINOLEATE

Definition: PEG/PPG-32/3 Diricinoleate is the diester obtained by the interaction of 32 moles of ethylene oxide and 3 moles of propylene oxide with Ricinoleic Acid (q.v.).

Chemical Class: Alkoxylated Carboxylic Acids

Function: Surfactant - Emulsifying Agent

Trade Name:
 Zenimer R-332-D (Zenitech)

PEG-3/PPG-2 GLYCERYL/SORBITOL HYDROXYSTEARATE/ISOSTEARATE

Definition: PEG-3/PPG-2 Glyceryl/Sorbitol Hydroxystearate/Isostearate is the polyethylene glycol, polypropylene glycol derivative of a mixture of glycerol and sorbitol esters of hydroxystearic and isostearic acids.

Chemical Class: Esters

Function: Surfactant - Emulsifying Agent

Technical/Other Names:
 Polyoxyethylene (3) Polyoxypropylene (2) Glyceryl/Sorbitol Hydroxystearate/ Isostearate
 Polyoxypropylene (2) Polyoxyethylene (3) Glyceryl/Sorbitol Hydroxystearate/ Isostearate

PEG-20-PPG-10 GLYCERYL STEARATE

Definition: PEG-20-PPG-10 Glyceryl Stearate is the polyoxypropylene, polyoxyethylene ether of Glyceryl Stearate (q.v.) with an average propoxylation value of 10 and an average ethoxylation value of 20.

Chemical Classes: Alkoxylated Alcohols; Glyceryl Esters and Derivatives

Function: Surfactant - Emulsifying Agent

Trade Name:
 Acconon TGH (Abitec Corporation)

PEG/PPG-8/3 LAURATE

Definition: PEG/PPG-8/3 Laurate is the product formed by the reaction of lauric acid with an average of 8 moles of ethylene oxide and 3 moles of propylene oxide.

Chemical Class: Alkoxylated Carboxylic Acids

Functions: Emulsion Stabilizer; Hair Conditioning Agent; Skin-Conditioning Agent - Humectant; Surfactant - Emulsifying Agent; Surfactant - Solubilizing Agent

Technical/Other Names:
 Polyoxyethylene (8) Polyoxypropylene (3) Monolaurate
 Polyoxypropylene (3) Polyoxyethylene (8) Monolaurate

Trade Name:
 Hydramol PGPL (Scher)

PEG/PPG-20/22 METHYL ETHER DIMETHICONE

CAS No.: 125857-75-2

Definition: PEG/PPG-20/22 Methyl Ether Dimethicone is the methyl ether of a derivative of Dimethicone (q.v.) containing an average of 20 moles of ethylene oxide and 22 moles of propylene oxide.

Chemical Class: Siloxanes and Silanes

Functions: Hair Conditioning Agent; Surfactant - Cleansing Agent; Surfactant - Emulsifying Agent; Surfactant - Suspending Agent

Technical/Other Names:
 Polyethylene Glycol (20) Polypropylene Glycol (22) Methyl Ether Dimethicone
 Siloxanes and Silicones, di-Me, 3-Hydroxypropyl Me, Ethers with Polyethylene-Polypropylene Glycol mono-Me Ether

Trade Name:
 FZ-2401 (Nippon Unicar)

Trade Name Mixture:
 FZ-2406 (Nippon Unicar)

The inclusion of any compound in the *Dictionary and Handbook* does not indicate that use of that substance as a cosmetic ingredient complies with the laws and regulations governing such use in the United States or any other country.

PEG/PPG-24/24 METHYL ETHER GLYCIDOXY DIMETHICONE

Definition: PEG/PPG-24/24 Methyl Ether Glycidoxy Dimethicone is the siloxane polymer that conforms to the formula:

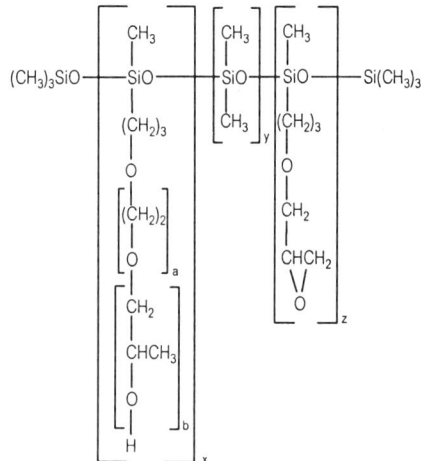

where a has an average value of 24 and b has an average value of 24.

Chemical Class: Siloxanes and Silanes

Function: Hair Conditioning Agent

Trade Name Mixture:
Dow Corning 8600 Hydrophilic Softener (Dow Corning)

PEG/PPG-10/3 OLEYL ETHER DIMETHICONE

Definition: PEG/PPG-10/3 Oleyl Ether Dimethicone is the oleyl ether of a derivative of Dimethicone (q.v.) containing an average of 10 moles of ethylene oxide and 3 moles of propylene oxide.

Chemical Class: Siloxanes and Silanes

Functions: Emulsion Stabilizer; Hair Conditioning Agent; Skin-Conditioning Agent - Miscellaneous; Surfactant - Emulsifying Agent; Surfactant - Solubilizing Agent

Technical/Other Name:
Polydimethylsiloxane, compound with Polyoxypropylene (3) Polyoxyethylene (10) Oleyl Ether

Trade Name:
KF-6026 (Shin Etsu)

PEG/PPG-3/1 OLIVE OIL ESTERS

JPN Translation:
オレイン酸 (トリエチレングリコール / PG)

Definition: PEG/PPG-3/1 Olive Oil Esters is a mixture of esters obtained by the interaction of olive oil with 3 moles of ethylene oxide and 1 mole of propylene oxide.

Information Source: JCIC

Chemical Class: Glyceryl Esters and Derivatives

Functions: Skin-Conditioning Agent - Emollient; Surfactant - Emulsifying Agent

Technical/Other Name:
(Triethyleneglycol.Propyleneglycol) Oleate

PEG-26-PPG-30 PHOSPHATE

Definition: PEG-26-PPG-30 Phosphate is a copolymer of polyoxyethylene and polyoxypropylene terminated with polyphosphoric acid and containing an average of 26 moles of ethylene oxide and 30 moles of propylene oxide.

Chemical Class: Phosphorus Compounds

Functions: Surfactant - Cleansing Agent; Surfactant - Hydrotrope

Trade Name:
Solsperse 41090 (Avecia)

Trade Name Mixtures:
Colourspheres FD&C Green No. 3 HL 25% (Creations Couleurs)
Colourspheres Alizuridine HL 25% (Creations Couleurs)
Colourspheres Black HB 25% (Creations Couleurs)
Colourspheres Black HB 70% (Creations Couleurs)
Colourspheres Black HL 25% (Creations Couleurs)
Colourspheres Black HL 70% (Creations Couleurs)
Colourspheres Blue HB 70% (Creations Couleurs)
Colourspheres Blue HL 25% (Creations Couleurs)
Colourspheres Blue HL 70% (Creations Couleurs)
Colourspheres Blue Sky Dye HL 25% (Creations Couleurs)
Colourspheres Bright Yellow HL 25% (Creations Couleurs)
Colourspheres Bright Yellow Lake HL 25% (Creations Couleurs)
Colourspheres Coral Dye HL 25% (Creations Couleurs)
Colourspheres Electric Pink Lake HL 25% (Creations Couleurs)
Colourspheres Eosin Dye HL 25% (Creations Couleurs)
Colourspheres Geranium HB 25% (Creations Couleurs)
Colourspheres Gold Grass Dye HL 25% (Creations Couleurs)
Colourspheres Green HL 25% (Creations Couleurs)
Colourspheres Harvest Dye HL 25% (Creations Couleurs)
Colourspheres Helindone Lake HL 25% (Creations Couleurs)
Colourspheres Magenta Dye HL 25% (Creations Couleurs)
Colourspheres Magenta HB 25% (Creations Couleurs)
Colourspheres Magenta Lake HL 25% (Creations Couleurs)
Colourspheres Mandarin Dye HL 25% (Creations Couleurs)
Colourspheres Melon Dye HL 25% (Creations Couleurs)
Colourspheres Melon Lake HL 25% (Creations Couleurs)
Colourspheres Ocean Blue Dye HL 25% (Creations Couleurs)
Colourspheres Ocean Blue HL 25% (Creations Couleurs)
Colourspheres Orange Dye HL 25% (Creations Couleurs)
Colourspheres Orange Lake HL 25% (Creations Couleurs)
Colourspheres Pinky Dye HL 25% (Creations Couleurs)
Colourspheres Poppy Lake HL 25% (Creations Couleurs)
Colourspheres Prussian Blue HL 25% (Creations Couleurs)
Colourspheres Red HB 25% (Creations Couleurs)
Colourspheres Red HB 70% (Creations Couleurs)
Colourspheres Red HL 25% (Creations Couleurs)
Colourspheres Red HL 70% (Creations Couleurs)
Colourspheres Safran Dye HL 25% (Creations Couleurs)
Colourspheres Safran HB 25% (Creations Couleurs)
Colourspheres Safran HL 25% (Creations Couleurs)
Colourspheres White R HB 25% (Creations Couleurs)
Colourspheres White R HB 70% (Creations Couleurs)
Colourspheres White R HL 25% (Creations Couleurs)
Colourspheres White R HL 70% (Creations Couleurs)
Colourspheres Wood Dye HL 25% (Creations Couleurs)
Colourspheres Yellow Grass Dye HL 25% (Creations Couleurs)
Colourspheres Yellow HB 25% (Creations Couleurs)
Colourspheres Yellow HB 70% (Creations Couleurs)
Colourspheres Yellow HL 25% (Creations Couleurs)

Coloursphere Yellow HL 70% (Creations Couleurs)
Eospoly UV Cristal HB 50% (Creations Couleurs)
Eospoly UV Cristal HL 50% (Creations Couleurs)
Eospoly UV Shadow HL 12,5% (Creations Couleurs)
Eospoly UV Zinc HB 50% (Creations Couleurs)
Eospoly UV Zinc HL 50% (Creations Couleurs)

PEG/PPG-10/2 RICINOLEATE

Definition: PEG/PPG-10/2 Ricinoleate is the monoester obtained by the interaction of 10 moles of ethylene oxide and 2 moles of propylene oxide with Ricinoleic Acid (q.v.).

Chemical Class: Alkoxylated Carboxylic Acids

Function: Surfactant - Cleansing Agent

Trade Name Mixture:
Zenicone IX (Zenitech)

PEG/PPG-32/3 RICINOLEATE

Definition: PEG/PPG-32/3 Ricinoleate is the monoester obtained by the interaction of 32 moles of ethylene oxide and 3 moles of propylene oxide with Ricinoleic Acid (q.v.).

Chemical Class: Alkoxylated Carboxylic Acids

Function: Surfactant - Cleansing Agent

Trade Name:
Zenimer R-332-M (Zenitech)

Trade Name Mixture:
Zenicone X (Zenitech)

PEG-3 PPG-20 SUCCINATE

JPN Translation:
オリゴコハク酸 PEG - 3 - PPG - 20

Definition: PEG-3 PPG-20 Succinate is the alkoxylated derivative of succinic acid that conforms generally to the formula:

$$(OCH_2CH_2)_xO-\overset{\overset{O}{\|}}{C}(CH_2)_2\overset{\overset{O}{\|}}{C}-O(CH_2CH_2O)_x$$
$$\underset{CH_3}{\overset{|}{(CH_2CHO)_y}} \qquad \underset{CH_3}{\overset{|}{H(OCHCH_2)_y}}$$

where x has an average value of 1.5 and y has an average value of 10.

Information Source: JCLS

Chemical Class: Alkoxylated Carboxylic Acids

Function: Skin-Conditioning Agent - Emollient

Technical/Other Names:
Polyethylene Glycol (3) Polypropylene Glycol (4) Succinate
Polyoxyethylene (3) Polyoxypropylene (20) Succinate

Trade Name:
Estemol 50 (Nisshin OilliO)

PEG/PPG-2/5 TOCOPHERYL ETHER

Definition: PEG/PPG-2/5 Tocopheryl Ether is the product formed by the reaction of Tocopherol (q.v.) with an average of 2 moles of ethylene oxide and 5 moles of propylene oxide.

Chemical Classes: Alkoxylated Alcohols; Heterocyclic Compounds

Functions: Anticaking Agent; Antioxidant; Binder; Emulsion Stabilizer; Plasticizer; Skin-Conditioning Agent - Miscellaneous; Surface Modifier; Surfactant - Emulsifying Agent; Surfactant - Suspending Agent; Ultraviolet Light Absorber

Technical/Other Name:
Polyoxyethylene (2) Polyoxypropylene (5) Tocopheryl Ether

Trade Name:
Vitanicsr-5/2 (Vitacos Corporation)

PEG/PPG-5/10 TOCOPHERYL ETHER

Definition: PEG/PPG-5/10 Tocopheryl Ether is the product formed by the reaction of Tocopherol (q.v.) with an average of 5 moles of ethylene oxide and 10 moles of propylene oxide.

Chemical Classes: Alkoxylated Alcohols; Heterocyclic Compounds

Functions: Anticaking Agent; Antioxidant; Binder; Emulsion Stabilizer; Plasticizer; Skin-Conditioning Agent - Miscellaneous; Surface Modifier; Surfactant - Emulsifying Agent; Surfactant - Suspending Agent; Ultraviolet Light Absorber

Technical/Other Name:
Polyoxyethylene (5) Polyoxypropylene (10) Tocopheryl Ether

Trade Name:
Vitanicsr-10/5 (Vitacos Corporation)

PEG/PPG-5/20 TOCOPHERYL ETHER

Definition: PEG/PPG-5/20 Tocopheryl Ether is the product formed by the reaction of

Tocopherol (q.v.) with an average of 5 moles of ethylene oxide and 20 moles of propylene oxide.

Chemical Classes: Alkoxylated Alcohols; Heterocyclic Compounds

Functions: Anticaking Agent; Antioxidant; Binder; Emulsion Stabilizer; Plasticizer; Skin-Conditioning Agent - Miscellaneous; Surface Modifier; Surfactant - Suspending Agent; Ultraviolet Light Absorber

Technical/Other Name:
Polyoxyethylene (5) Polyoxypropylene (20) Tocopheryl Ether

Trade Name:
Vitanicsr-20/5 (Vitacos Corporation)

PEG/PPG-5/30 TOCOPHERYL ETHER

Definition: PEG/PPG-5/30 Tocopheryl Ether is the product formed by the reaction of Tocopherol (q.v.) with an average of 5 moles of ethylene oxide and 30 moles of propylene oxide.

Chemical Classes: Alkoxylated Alcohols; Heterocyclic Compounds

Functions: Anticaking Agent; Antioxidant; Binder; Emulsion Stabilizer; Plasticizer; Skin-Conditioning Agent - Miscellaneous; Surface Modifier; Surfactant - Emulsifying Agent; Surfactant - Suspending Agent; Ultraviolet Light Absorber

Technical/Other Name:
Polyoxyethylene (5) Polyoxypropylene (30) Tocopheryl Ether

Trade Name:
Vitanicsr-30/5 (Vitacos Corporation)

PEG/PPG-30/10 TOCOPHERYL ETHER

Definition: PEG/PPG-30/10 Tocopheryl Ether is the product formed by the reaction of Tocopherol (q.v.) with an average of 30 moles of ethylene oxide and 10 moles of propylene oxide.

Chemical Classes: Alkoxylated Alcohols; Heterocyclic Compounds

Functions: Anticaking Agent; Antioxidant; Binder; Emulsion Stabilizer; Plasticizer; Skin-Conditioning Agent - Miscellaneous; Surface Modifier; Surfactant - Emulsifying Agent; Surfactant - Suspending Agent; Ultraviolet Light Absorber

Technical/Other Name:
Polyoxyethylene (30) Polyoxypropylene (10) Tocopheryl Ether

Trade Name:
Vitanicsr-10/30 (Vitacos Corporation)

PEG/PPG-50/20 TOCOPHERYL ETHER

Definition: PEG/PPG-50/20 Tocopheryl Ether is the product formed by the reaction of Tocopherol (q.v.) with an average of 50 moles of ethylene oxide and 20 moles of propylene oxide.

Chemical Classes: Alkoxylated Alcohols; Heterocyclic Compounds

Functions: Anticaking Agent; Antioxidant; Binder; Emulsion Stabilizer; Plasticizer; Skin-Conditioning Agent - Miscellaneous; Surface Modifier; Surfactant - Emulsifying Agent; Surfactant - Suspending Agent; Ultraviolet Light Absorber

Trade Name:
Vitanicsr-20/50 (Vitacos Corporation)

PEG/PPG-70/30 TOCOPHERYL ETHER

Definition: PEG/PPG-70/30 Tocopheryl Ether is the product formed by the reaction of Tocopherol (q.v.) with an average of 70 moles of ethylene oxide and 30 moles of propylene oxide.

Chemical Classes: Alkoxylated Alcohols; Heterocyclic Compounds

Functions: Anticaking Agent; Antioxidant; Binder; Emulsion Stabilizer; Plasticizer; Skin-Conditioning Agent - Miscellaneous; Surface Modifier; Surfactant - Suspending Agent; Ultraviolet Light Absorber

Technical/Other Name:
Polyoxyethylene (70) Polyoxypropylene (30) Tocopheryl Ether

Trade Name:
Vitanicsr-30/70 (Vitacos Corporation)

PEG/PPG-100/70 TOCOPHERYL ETHER

Definition: PEG/PPG-100/70 Tocopheryl Ether is the product formed by the reaction of Tocopherol (q.v.) with 100 moles of ethylene oxide and 70 moles of propylene oxide.

Chemical Classes: Alkoxylated Alcohols; Heterocyclic Compounds

Functions: Anticaking Agent; Antioxidant; Binder; Emulsion Stabilizer; Plasticizer; Skin-Conditioning Agent - Miscellaneous; Surface Modifier; Surfactant - Emulsifying Agent; Surfactant - Suspending Agent; Ultraviolet Light Absorber

Technical/Other Name:
Polyoxyethylene (100) Polyoxypropylene (70) Tocopheryl Ether

Trade Name:
Vitanicsr-70/100 (Vitacos Corporation)

PEG/PPG-5/3 TRISILOXANE

Definition: PEG/PPG-5/3 Trisiloxane is the alkoxylated derivative of Trisiloxane (q.v.) containing an average of 5 moles of ethylene oxide and 3 moles of propylene oxide.

Chemical Class: Siloxanes and Silanes

Function: Surfactant - Emulsifying Agent

Trade Name:
Silsoft 305 dimethicone copolyol (Crompton Corporation)

PEG-4 PROLINE LINOLEATE

Empirical Formula:
$C_{31}H_{55}NO_7$

Definition: PEG-4 Proline Linoleate is the organic compound that conforms generally to the formula:

where R represents the linoleic group and n has an average value of 4.

Chemical Classes: Amino Acids; Esters; Heterocyclic Compounds

Functions: Skin-Conditioning Agent - Miscellaneous; Surfactant - Emulsifying Agent

Technical/Other Names:
Polyethylene Glycol 200 Proline Linoleate
Polyoxyethylene (4) Proline Linoleate

Trade Name Mixture:
Aminoefaderma (Vevy)

PEG-4 PROLINE LINOLENATE

Empirical Formula:
$C_{31}H_{53}NO_7$

Definition: PEG-4 Proline Linolenate is the organic compound that conforms generally to the formula:

where R represents the linolenic group and n has an average value of 4.

Chemical Classes: Amino Acids; Esters; Heterocyclic Compounds

Functions: Skin-Conditioning Agent - Miscellaneous; Surfactant - Emulsifying Agent

Technical/Other Names:
Polyethylene Glycol 200 Proline Linolenate
Polyoxyethylene (4) Proline Linolenate

Trade Name Mixture:
Aminoefaderma (Vevy)

PEG-10 PROPYLENE GLYCOL

Empirical Formula:
$C_{23}H_{48}O_{12}$

Definition: PEG-10 Propylene Glycol is the polyethylene glycol ether of propylene glycol that conforms generally to the formula:

$$HO(CH_2CH_2O)_yCHCH_2(OCH_2CH_2)_xOH$$
$$|$$
$$CH_3$$

where x + y has an average value of 10.

Information Source: CIR: [S] IJT-20 (SUPPL. 4)2001

Chemical Class: Alkoxylated Alcohols

Functions: Skin-Conditioning Agent - Humectant; Solvent

Technical/Other Names:
Polyethylene Glycol 500 Propylene Glycol
Polyoxyethylene (10) Propylene Glycol

Trade Name Mixture:
Acconon CON (Abitec Corporation)

PEG-6 PROPYLENE GLYCOL CAPRYLATE/CAPRATE

Definition: PEG-6 Propylene Glycol Caprylate/Caprate is the organic compound that conforms generally to the formula:

where RCO- represents the capryloyl/caproyl moiety and n has an average value of 6.

Chemical Classes: Alkoxylated Alcohols; Esters

Functions: Skin-Conditioning Agent - Emollient; Surfactant - Emulsifying Agent

Trade Name:
Acconon 200 E-6 (Abitec Corporation)

PEG-8 PROPYLENE GLYCOL COCOATE

CAS No.: 126645-98-5

Definition: PEG-8 Propylene Glycol Cocoate is the polyethylene glycol ether of propylene glycol cocoate that conforms generally to the formula:

$$RC \overset{O}{\underset{\|}{}} - OCH_2CH(OCH_2CH_2)_nOH$$
$$\underset{CH_3}{|}$$

where RCO- represents the coconut fatty radical and n has an average value of 8.

Information Sources: CIR: [S] IJT-20 (SUPPL. 4)2001, CTFA D

Chemical Classes: Alkoxylated Alcohols; Esters

Functions: Skin-Conditioning Agent - Emollient; Surfactant - Emulsifying Agent

Reported Product Categories: Eye Makeup Preparations, Misc.; Eye Shadows; Face Powders

Technical/Other Names:
Polyethylene Glycol 400 Propylene Glycol Cocoate
Polyoxyethylene (8) Propylene Glycol Cocoate

Trade Name Mixtures:
AEC Polyglyceryl-4 Oleate (&) PEG-8 Propylene Glycol Cocoate (A & E Connock)
Emulsynt 1055 (International Specialty Products)

PEG-55 PROPYLENE GLYCOL OLEATE

Definition: PEG-55 Propylene Glycol Oleate is the polyethylene glycol ether of propylene glycol oleate. It conforms generally to the formula:

$$\underset{}{(CH_2)_7CH_3} \quad \overset{O}{\underset{\|}{}}$$
$$CH = CH(CH_2)_7C - OCH_2CH(OCH_2CH_2)_nOH$$
$$\underset{CH_3}{|}$$

where n has an average value of 55.

Information Source: CIR: [S] IJT-20 (SUPPL. 4)2001

Chemical Classes: Alkoxylated Alcohols; Esters

Functions: Surfactant - Cleansing Agent; Surfactant - Solubilizing Agent

Reported Product Categories: Shampoos (Non-coloring); Bath Soaps and Detergents; Fragrance Preparations, Misc.

Technical/Other Names:
Polyethylene Glycol (55) Propylene Glycol Oleate
Polyoxyethylene (55) Propylene Glycol Oleate

Trade Name Mixture:
Antil 141 Liquid (Degussa Care Specialties)

PEG-25 PROPYLENE GLYCOL STEARATE

Definition: PEG-25 Propylene Glycol Stearate is the polyethylene glycol ether of Propylene Glycol Stearate (q.v.) that conforms generally to the formula:

$$CH_3(CH_2)_{16}C \overset{O}{\underset{\|}{}} - OCH_2CH(OCH_2CH_2)_nOH$$
$$\underset{CH_3}{|}$$

where n has an average value of 25.

Information Sources: CIR: [S] IJT-20 (SUPPL. 4)2001, CTFA D

Chemical Classes: Alkoxylated Alcohols; Esters

Functions: Surfactant - Cleansing Agent; Surfactant - Solubilizing Agent

Reported Product Categories: Personal Cleanliness Products, Misc.; Deodorants (Underarm); Shampoos (Non-coloring)

Technical/Other Names:
Polyethylene Glycol (25) Propylene Glycol Monostearate
Polyoxyethylene (25) Propylene Glycol Monostearate

Trade Names:
AEC PEG-25 Propylene Glycol Stearate (A & E Connock)
Atlas G-2162 (Uniqema Americas)

PEG-75 PROPYLENE GLYCOL STEARATE

Definition: PEG-75 Propylene Glycol Stearate is the polyethylene glycol ester of Propylene Glycol Stearate (q.v.) that conforms to the formula:

$$CH_3(CH_2)_{16}C \overset{O}{\underset{\|}{}} - OCH_2CH(OCH_2CH_2)_nOH$$
$$\underset{CH_3}{|}$$

where n has an average value of 75.

Information Source: CIR: [S] IJT-20 (SUPPL. 4)2001

Chemical Classes: Alkoxylated Alcohols; Esters

Functions: Surfactant - Cleansing Agent; Surfactant - Solubilizing Agent

Technical/Other Names:
Polyethylene Glycol 4000 Propylene Glycol Monostearate
Polyoxyethylene (75) Propylene Glycol Monostearate

Trade Name:
AEC PEG-75 Propylene Glycol Stearate (A & E Connock)

PEG-120 PROPYLENE GLYCOL STEARATE

Definition: PEG-120 Propylene Glycol Stearate is the polyethylene glycol ether of Propylene Glycol Stearate (q.v.) that conforms generally to the formula:

$$CH_3(CH_2)_{16}C \overset{O}{\underset{\|}{}} - OCH_2CH(OCH_2CH_2)_nOH$$
$$\underset{CH_3}{|}$$

where n has an average value of 120.

Information Source: CIR: [S] IJT-20 (SUPPL. 4)2001

Chemical Classes: Alkoxylated Alcohols; Esters

Functions: Surfactant - Cleansing Agent; Surfactant - Solubilizing Agent

Technical/Other Names:
Polyethylene Glycol (120) Propylene Glycol Monostearate
Polyoxyethylene (120) Propylene Glycol Monostearate

Trade Name:
AEC PEG-120 Propylene Glycol Stearate (A & E Connock)

PEG-4 RAPESEEDAMIDE

Definition: PEG-4 Rapeseedamide is the polyethylene glycol amide of the fatty acids derived from rapeseed oil with an average of 4 moles of ethylene oxide.

Chemical Class: Alkoxylated Amides

Functions: Surfactant - Emulsifying Agent; Viscosity Increasing Agent - Aqueous

Reported Product Categories: Hair Dyes and Colors (All Types Requiring Caution Statements and Patch Tests); Shampoos (Non-coloring)

Technical/Other Names:
Polyethylene Glycol 200 Rapeseedamide
Polyoxyethylene (4) Rapeseedamide

Trade Name:
  Aminol N (Kao GmbH)

PEG-2 RAPESEEDAMINE

Definition: PEG-2 Rapeseedamine is the polyethylene glycol derivative of rapeseedamine that conforms generally to the formula:

$$R-N \begin{cases} (CH_2CH_2O)_xH \\ (CH_2CH_2O)_yH \end{cases}$$

where R represents the alkyl groups derived from rapeseed oil and x+y has an average value of 2.

Chemical Class: Alkoxylated Amines

Function: Antistatic Agent

Reported Product Category: Hair Dyes and Colors (All Types Requiring Caution Statements and Patch Tests)

Technical/Other Names:
  Polyethylene Glycol 100 Rapeseed Amine
  Polyoxyethylene (2) Rapeseed Amine

Trade Name:
  Ethomeen OV/12 (Akzo Nobel)

PEG-3 RAPESEED AMINOPROPYLAMINE

Definition: PEG-3 Rapeseed Aminopropylamine is the polyethylene glycol amine that conforms generally to the formula:

$$R-N(CH_2)_3N \begin{cases} (CH_2CH_2O)_xH \\ (CH_2CH_2O)_yH \end{cases}$$
$$(CH_2CH_2O)_zH$$

where R represents the alkyl groups derived from rapeseed oil and x+y+z has an average value of 3.

Chemical Class: Alkoxylated Amines

Functions: Antistatic Agent; Surfactant - Emulsifying Agent; Surfactant - Suspending Agent

Technical/Other Names:
  Polyethylene Glycol (3) Rapeseed Amino-
    propylamine
  Polyethylene Glycol (3) Rapeseed
    Aminoproylamine
  Polyoxyethylene (3) Rapeseed Amino-
    propylamine
  Polyoxyethylene (3) Rapessed Amino-
    propylamine

Trade Name:
  Ethoduomeen OV/13 (Akzo Nobel Surface
    AB)

PEG-5 RAPESEED STEROL

Definition: PEG-5 Rapeseed Sterol is a polyethylene glycol ether of Rapeseed Sterols (q.v.) containing an average of 5 moles of ethylene oxide.

Chemical Classes: Alkoxylated Alcohols; Sterols

Functions: Surfactant - Cleansing Agent; Surfactant - Solubilizing Agent

Trade Name:
  Generol R E5 (Cognis Deutschland)

PEG-10 RAPESEED STEROL

Definition: PEG-10 Rapeseed Sterol is a polyethylene glycol ether of Rapeseed Sterol (q.v.) containing an average of 10 moles of ethylene oxide.

Chemical Classes: Alkoxylated Alcohols; Sterols

Functions: Surfactant - Cleansing Agent; Surfactant - Solubilizing Agent

Trade Name:
  Generol R E10 (Cognis Deutschland)

PEG-7 RICINOLEAMIDE

JPN Translation:
  PEG - 7 リシノレイン酸アミド

Definition: PEG-7 Ricinoleamide is the polyethylene glycol amide of ricinoleic acid that conforms generally to the formula:

$$(CH_2)_5CHCH_2CH=CH(CH_2)_7C-NH(CH_2CH_2O)_nH$$
with CH_3 and OH substituents and carbonyl O.

where n has an average value of 7.

Information Source: JCIC

Chemical Class: Alkoxylated Amides

Function: Surfactant - Emulsifying Agent

Technical/Other Names:
  Polyethylene Glycol (7) Ricinoleamide
  Polyoxyethylene (7) Ricinoleamide
  Polyoxyethylene Ricinoleate Amide

PEG-40 RICINOLEAMIDE

Definition: PEG-40 Ricinoleamide is the polyethylene glycol amide of ricinoleic acid that conforms generally to the formula:

$$(CH_2)_5CHCH_2CH=CH(CH_2)_7C-NH(CH_2CH_2O)_nH$$
with CH_3 and OH substituents and carbonyl O.

where n has an average value of 40.

Information Source: JCLS

Chemical Class: Alkoxylated Amides

Functions: Surfactant - Cleansing Agent; Surfactant - Solubilizing Agent

Technical/Other Names:
  Polyethylene Glycol 2000 Ricinoleamide
  Polyoxyethylene (40) Ricinoleamide

PEG-2 RICINOLEATE

CAS Nos.	EINECS No.
5401-17-2	226-448-9
42426-59-5 (Generic)	

Empirical Formula:
  $C_{22}H_{42}O_5$

Definition: PEG-2 Ricinoleate is the polyethylene glycol ester of ricinoleic acid that conforms to the formula:

$$(CH_2)_5CHCH_2CH=CH(CH_2)_7C-(OCH_2CH_2O)_nH$$
with CH_3 and OH substituents and carbonyl O.

where n has an average value of 2.

Information Sources: 21CFR175.300, MI-13(7660), TSCA

Chemical Class: Alkoxylated Carboxylic Acids

Function: Surfactant - Emulsifying Agent

Technical/Other Names:
  Diethylene Glycol Monoricinoleate
  9-Octadecanoic Acid, 12-Hydroxy-, 2-(2-
    Hydroxyethoxy)ethyl Ester
  Polyethylene Glycol 100 Monoricinoleate
  Polyoxyethylene (2) Monoricinoleate

Trade Name:
  AEC PEG-2 Ricinoleate (A & E Connock)

PEG-7 RICINOLEATE

CAS Nos.: 9004-97-1 (Generic); 42426-59-5 (Generic)

Empirical Formula:
  $C_{32}H_{62}O_{10}$

Definition: PEG-7 Ricinoleate is the polyethylene glycol ester of ricinoleic acid that conforms to the formula:

$$(CH_2)_5CHCH_2CH=CH(CH_2)_7C-(OCH_2CH_2O)_nH$$
with CH_3 and OH substituents and carbonyl O.

The inclusion of any compound in the *Dictionary and Handbook* does not indicate that use of that substance as a cosmetic ingredient complies with the laws and regulations governing such use in the United States or any other country.

---

where n has an average value of 7.

Information Sources: 21CFR173.340, 21CFR175.300, MI-13(7660), TSCA

Chemical Class: Alkoxylated Carboxylic Acids

Function: Surfactant - Emulsifying Agent

Trade Name:
AEC PEG-7 Ricinoleate (A & E Connock)

PEG-8 RICINOLEATE

CAS Nos.: 9004-97-1 (Generic); 42426-59-5 (Generic)

Empirical Formula:
$C_{34}H_{66}O_{11}$

Definition: PEG-8 Ricinoleate is a polyethylene glycol ester of ricinoleic acid that conforms generally to the formula:

$$CH_3$$
$$|$$
$$(CH_2)_5CHCH_2CH=CH(CH_2)_7C-(OCH_2CH_2O)_nH$$
$$|$$
$$OH$$

where n has an average value of 8.

Information Sources: 21CFR175.300, 21CFR176.210, 21CFR177.1210, 21CFR177.2800, MI-13(7660)

Chemical Class: Alkoxylated Carboxylic Acids

Function: Surfactant - Emulsifying Agent

Technical/Other Names:
Polyethylene Glycol 400 Ricinoleate
Polyoxyethylene (8) Ricinoleate

Trade Name Mixtures:
Zenicone IX (Zenitech)
Zenicone XQ (Zenitech)
Zenicone XX (Zenitech)
Zenigloss Q-SE (Zenitech)
Zenigloss SE (Zenitech)
Zeniquat Base (Zenitech)

PEG-9 RICINOLEATE

CAS Nos.: 9004-97-1 (Generic); 42426-59-5 (Generic)

Empirical Formula:
$C_{36}H_{70}O_{12}$

Definition: PEG-9 Ricinoleate is a polyethylene glycol ester of ricinoleic acid that conforms generally to the formula:

$$CH_3$$
$$|$$
$$(CH_2)_5CHCH_2CH=CH(CH_2)_7C-(OCH_2CH_2O)_nH$$
$$|$$
$$OH$$

where n has an average value of 9.

Information Sources: 21CFR175.300, 21CFR177.2800, MI-13(7660)

Chemical Class: Alkoxylated Carboxylic Acids

Function: Surfactant - Emulsifying Agent

Technical/Other Names:
Polyethylene Glycol 450 Ricinoleate
Polyoxyethylene (9) Ricinoleate

Trade Name:
Atlas G-4929 (Uniqema Americas)

PEG-45 SAFFLOWER GLYCERIDES

Definition: PEG-45 Safflower Glycerides is a polyethylene glycol derivative of the mono- and diglycerides derived from safflower oil with an average of 45 moles of ethylene oxide.

Chemical Classes: Alkoxylated Alcohols; Glyceryl Esters and Derivatives

Functions: Skin-Conditioning Agent - Emollient; Surfactant - Emulsifying Agent

Technical/Other Names:
Polyethylene Glycol (45) Safflower Glycerides
Polyoxyethylene (45) Safflower Glycerides

PEG-8 SESQUILAURATE

Definition: PEG-8 Sesquilaurate is a mixture of the polyethylene glycol mono and diesters of lauric acid with an average of 8 moles of ethylene oxide.

Information Sources: 21CFR175.105, 21CFR175.300, 21CFR176.210, 21CFR177.1210, 21CFR177.2260, 21CFR177.2800, MI-13(7660)

Chemical Class: Alkoxylated Carboxylic Acids

Function: Surfactant - Emulsifying Agent

Technical/Other Names:
Polyethylene Glycol 400 Sesquilaurate
Polyoxyethylene (8) Sesquilaurate

Trade Name:
AEC PEG-8 Sesquilaurate (A & E Connock)

PEG-8 SESQUIOLEATE

Definition: PEG-8 Sesquioleate is a mixture of the polyethylene glycol mono and diesters of oleic acid with an average of 8 moles of ethylene oxide.

Information Sources: 21CFR175.105, 21CFR175.300, 21CFR176.210, 21CFR177.1210, 21CFR177.2260, 21CFR177.2800, MI-13(7660)

Chemical Class: Alkoxylated Carboxylic Acids

Function: Surfactant - Emulsifying Agent

Technical/Other Names:
Polyethylene Glycol 400 Sesquioleate
Polyoxyethylene (8) Sesquioleate

Trade Names:
AEC PEG-8 Sesquioleate (A & E Connock)
Ethox SO-9 (Ethox)
Sesquiol 20 (Fabriquimica)
Unipeg-400 SO (Universal Preserv-A-Chem)

PEG-50 SHEA BUTTER

Definition: PEG-50 Shea Butter is the polyethylene glycol derivative of Butyrospermum Parkii (Shea Butter) (q.v.) with an average of 50 moles of ethylene oxide.

Chemical Classes: Alkoxylated Alcohols; Glyceryl Esters and Derivatives

Function: Surfactant - Emulsifying Agent

Reported Product Category: Paste Masks (Mud Packs)

Technical/Other Names:
Polyethylene Glycol 4000 Shea Butter
Polyoxyethylene (50) Shea Butter

Trade Name:
Shebu WS (RITA)

PEG-60 SHEA BUTTER GLYCERIDES

Definition: PEG-60 Shea Butter Glycerides is a polyethylene glycol derivative of the glycerides derived from Butyrospermum Parkii (Shea Butter) (q.v.) with an average ethoxylation value of 60.

Chemical Classes: Alkoxylated Alcohols; Glyceryl Esters and Derivatives

Functions: Skin-Conditioning Agent - Emollient; Surfactant - Emulsifying Agent; Surfactant - Solubilizing Agent

Technical/Other Names:
Polyethylene Glycol (60) Shea Butter Glycerides
Polyoxyethylene (60) Shea Butter Glycerides

Trade Name:
Crovol SB70 (Croda Chemicals)

PEG-75 SHEA BUTTER GLYCERIDES

Definition: PEG-75 Shea Butter Glycerides is a polyethylene glycol derivative of the glycerides derived from Butyrospermum Parkii (Shea Butter) (q.v.) with an average of 75 moles of ethylene oxide.

Chemical Classes: Alkoxylated Alcohols; Glyceryl Esters and Derivatives

Functions: Skin-Conditioning Agent - Emollient; Surfactant - Cleansing Agent; Surfactant - Solubilizing Agent

Technical/Other Names:
Polyethylene Glycol 4000 Shea Butter Glycerides
Polyoxyethylene (75) Shea Butter Glycerides

Trade Name:
Lipex 102 E-75 (Karlshamns AB)

Trade Name Mixture:
Shea Butter Milk (Cosmetochem) (Cosmetochem International Ltd.)

PEG-75 SHOREA BUTTER GLYCERIDES

Definition: PEG-75 Shorea Butter Glycerides is a polyethylene glycol derivative of the glycerides derived from Butyrospermum Parkii (Shorea Butter) (q.v.) with an average of 75 moles of ethylene oxide.

Chemical Classes: Alkoxylated Alcohols; Glyceryl Esters and Derivatives

Functions: Skin-Conditioning Agent - Emollient; Surfactant - Cleansing Agent; Surfactant - Solubilizing Agent

Technical/Other Names:
Polyethylene Glycol (75) Shorea Butter Glycerides
Polyoxyethylene (75) Shorea Butter Glycerides

PEG-75 BETA-SITOSTEROL

Definition: PEG-75 Beta-Sitosterol is a polyethylene glycol derivative of Beta-Sitosterol (q.v.) containing an average of 75 moles of ethylene oxide.

Chemical Classes: Alkoxylated Alcohols; Sterols

Function: Surfactant - Emulsifying Agent

Technical/Other Names:
Polyethylene Glycol 4000 Beta-Sitosterol
Polyoxyethylene (75) Beta-Sitosterol

Trade Name:
Plant Sterol L-75 (Arch Personal Care Products)

PEG-8/SMDI COPOLYMER

CAS No.: 39444-87-6 (Generic)

Empirical Formula:
$[C_{15}H_{22}N_2O_2 \bullet (C_2H_4O)_nH_2O]_x$

Definition: PEG-8/SMDI Copolymer is a copolymer of PEG-8 and saturated methylene diphenyldiisocyanate (dicyclohexylmethane diisocyanate) monomer.

Information Source: TSCA

Chemical Class: Synthetic Polymers

Functions: Hair Conditioning Agent; Hair Fixative; Plasticizer; Skin-Conditioning Agent - Emollient; Skin-Conditioning Agent - Miscellaneous

Technical/Other Names:
Methylene Dicyclohexylene Diisocyanate, Polyethylene Glycol Polymer
Polyethylene Glycol 400 SMDI Copolymer
Poly(Oxy-1,2-Ethanediyl)-α-Hydro-ω-Hydroxy, Polymer with 1, 1'Methylenebis-(4-Isocyanatocyclo-hexane)
Polyoxyethylene (8) SMDI Copolymer

Trade Name:
Polyolprepolymer-15 (Penederm)

Trade Name Mixtures:
Ceramide A2 (Sederma)
Kava Kava (Sederma)

PEG-20 SORBITAN COCOATE

JPN Translation:
PEG - 20 ソルビタンココエート

Definition: PEG-20 Sorbitan Cocoate is an ethoxylated sorbitan ester of coconut acid with an average of 20 moles of ethylene oxide.

Information Sources: CIR: [S] IJT-19 (SUPPL.2)2000, JCIC, JCLS, JSQI

Chemical Class: Sorbitan Derivatives

Functions: Surfactant - Cleansing Agent; Surfactant - Solubilizing Agent

Technical/Other Names:
Polyethylene Glycol 1000 Sorbitan Cocoate
Polyoxyethylene (20) Sorbitan Cocoate
Polyoxyethylene Sorbitan Monococoate (20E.O.)

PEG-40 SORBITAN DIISOSTEARATE

Definition: PEG-40 Sorbitan Diisostearate is an ethoxylated sorbitan diester of Isostearic Acid (q.v.) with an average of 40 moles of ethylene oxide.

Information Source: CIR: [S] IJT-19 (SUPPL.2)2000

Chemical Class: Sorbitan Derivatives

Functions: Surfactant - Emulsifying Agent; Surfactant - Solubilizing Agent

Trade Names:
Emsorb 2726 (Cognis Care Chemicals/NJ)
Emsorb 2726 (Cognis Care Chemicals/PA)

Trade Name Mixture:
Lubrasil DS (Guardian)

PEG-2 SORBITAN ISOSTEARATE

CAS No.: 66794-58-9

Empirical Formula:
$C_{28}H_{54}O_8$

Definition: PEG-2 Sorbitan Isostearate is an ethoxylated sorbitan monoester of Isostearic Acid (q.v.). with an average of 2 moles of ethylene oxide.

Information Source: CIR: [S] IJT-19 (SUPPL.2)2000

Chemical Class: Sorbitan Derivatives

Function: Surfactant - Emulsifying Agent

Technical/Other Names:
Polyethylene Glycol 100 Sorbitan Isostearate
Polyoxyethylene (2) Sorbitan Isostearate

Trade Names:
Atlas G-4832 (Uniqema Americas)
Kotilen-I/1/020 (Kolb)

PEG-5 SORBITAN ISOSTEARATE

CAS No.: 66794-58-9 (Generic)

Definition: PEG-5 Sorbitan Isostearate is an ethoxylated sorbitan monoester of Isostearic Acid (q.v.) with an average of 5 moles of ethylene oxide.

Information Source: CIR: [S] IJT-19 (SUPPL.2)2000

Chemical Class: Sorbitan Derivatives

Function: Surfactant - Emulsifying Agent

Technical/Other Names:
Polyethylene Glycol (5) Sorbitan Isostearate
Polyoxyethylene (5) Sorbitan Isostearate

Trade Name:
AEC PEG-5 Sorbitan Isostearate (A & E Connock)

PEG-20 SORBITAN ISOSTEARATE

CAS No.: 66794-58-9 (Generic)

JPN Translation:
イソステアリン酸 PEG - 20 ソルビタン

Definition: PEG-20 Sorbitan Isostearate is an ethoxylated sorbitan monoester of Isostearic Acid (q.v.) with an average of 20 moles of ethylene oxide.

Information Sources: CIR: [S] IJT-19 (SUPPL.2)2000, JCIC, JCLS, JSQI

Chemical Class: Sorbitan Derivatives

Functions: Surfactant - Cleansing Agent; Surfactant - Emulsifying Agent; Surfactant - Solubilizing Agent

Technical/Other Names:
Polyethylene Glycol 1000 Sorbitan Mono-isostearate
Polyoxyethylene Sorbitan Isostearate (20E.O.)
Polyoxyethylene (20) Sorbitan Monoiso-stearate

Trade Names:
AEC PEG-20 Sorbitan Isostearate (A & E Connock)
Crillet 6 (Croda Chemicals)
Crillet 6 (Croda, Inc.)
Isoixol 6 (Vevy)
Nikkol TI-10 (Nikko)

PEG-40 SORBITAN LANOLATE

CAS No.: 8036-77-9

JPN Translation:
PEG - 40 ソルビットラノリン

Definition: PEG-40 Sorbitan Lanolate is an ethoxylated sorbitan derivative of Lanolin Acid (q.v.) with an average of 40 moles of ethylene oxide.

Information Sources: CIR: [S] IJT-19 (SUPPL.2)2000, CTFA D, JCIC, JCLS, JSQI

Chemical Classes: Lanolin and Lanolin Derivatives; Sorbitan Derivatives

Functions: Surfactant - Cleansing Agent; Surfactant - Solubilizing Agent

Reported Product Category: Hair Preparations (Non-coloring), Misc.

Technical/Other Names:
Polyethylene Glycol 2000 Sorbitan Lanolate
Polyoxyethylene (40) Sorbitol Lanolate
Polyoxyethylene Sorbitol Lanolate (40E.O.)

Trade Names:
AEC PEG-40 Sorbitan Lanolate (A & E Connock)
Atlas G-1441 (Uniqema Americas)

PEG-75 SORBITAN LANOLATE

CAS No.: 8051-13-6

Definition: PEG-75 Sorbitan Lanolate is an ethoxylated sorbitan derivative of Lanolin Acid (q.v.) with an average of 75 moles of ethylene oxide.

Information Source: CIR: [S] IJT-19 (SUPPL.2)2000

Chemical Classes: Lanolin and Lanolin Derivatives; Sorbitan Derivatives

Functions: Surfactant - Cleansing Agent; Surfactant - Solubilizing Agent

Technical/Other Names:
Polyethylene Glycol 4000 Sorbitan Lanolate
Polyoxyethylene (75) Sorbitol Lanolate

Trade Name:
AEC PEG-75 Sorbitan Lanolate (A & E Connock)

PEG-10 SORBITAN LAURATE

CAS No.: 9005-64-5 (Generic)

JPN Translation:
ラウリン酸 PEG - 10 ソルビタン

Empirical Formula:
$C_{38}H_{74}O_{16}$

Definition: PEG-10 Sorbitan Laurate is an ethoxylated sorbitan ester of lauric acid with an average of 10 moles of ethylene oxide.

Information Sources: 21CFR172.515, 21CFR175.300, CIR: [S] IJT-19(SUPPL.2)-2000, CTFA S, JCLS, RIFM, TSCA

Chemical Class: Sorbitan Derivatives

Functions: Fragrance Ingredient; Surfactant - Cleansing Agent; Surfactant - Solubilizing Agent

Reported Product Categories: Tonics, Dressings, and Other Hair Grooming Aids; Bath Oils, Tablets, and Salts; Cleansing Products (Cold Creams, Cleansing Lotions, Liquids and Pads); Powders (Dusting and Talcum, Excluding Aftershave Talcs); Shampoos (Non-coloring); Skin Care Preparations, Misc.; Hair Conditioners; Hair Dyes and Colors (All Types Requiring Caution Statements and Patch Tests); Bath Preparations, Misc.; Body and Hand Preparations (Excluding Shaving Preparations); Moisturizing Preparations; Skin Fresheners; Makeup Bases; Foundations; Hair Preparations (Non-coloring), Misc.; Bath Soaps and Detergents; Bubble Baths; Hair Sprays (Aerosol Fixatives); Bath Capsules; Eye Makeup Preparations, Misc.; Eye Makeup Removers; Face and Neck Preparations (Excluding Shaving Preparations); Paste Masks (Mud Packs); Personal Cleanliness Products, Misc.; Perfumes; Permanent Waves; After-shave Lotions; Baby Shampoos; Hair Wave Sets; Mouthwashes and Breath Fresheners (Liquids and Sprays); Baby Products, Misc.; Eyebrow Pencils; Shaving Cream (Aerosol, Brushless and Lather); Suntan Gels, Creams, and Liquids; Colognes and Toilet Waters; Eye Lotions; Eyeliners; Indoor Tanning Preparations; Mascara; Night Skin Care Preparations; Blushers (All types); Eye Shadows; Fragrance Preparations, Misc.; Hair Coloring Preparations, Misc.; Makeup Preparations (Not eye), Misc.; Suntan Preparations, Misc.; Dentifrices (Aerosol, Liquid, Pastes and Powders); Deodorants (Underarm); Makeup Fixatives; Shaving Preparations, Misc.

Technical/Other Names:
Polyethylene Glycol 500 Sorbitan Mono-laurate
Polyoxyethylene (10) Sorbitan Monolaurate
Polysorbate 20 (RIFM)

Trade Names:
AEC PEG-10 Sorbitan Laurate (A & E Connock)
Hetsorb L-10 (Heterene)
Liposorb L-10 (Lipo)
Polisorbac 25 (Specialty Industrial)

PEG-40 SORBITAN LAURATE

CAS No.: 9005-64-5 (Generic)

JPN Translation:
ラウリン酸 PEG - 40 ソルビタン

Definition: PEG-40 Sorbitan Laurate is an ethoxylated sorbitan ester of lauric acid with an average of 40 moles of ethylene oxide.

Information Sources: 21CFR175.300, 21CFR176.210, CIR: [S] IJT-19(SUPPL.2)-2000, JCLS, RIFM

Chemical Class: Sorbitan Derivatives

Functions: Fragrance Ingredient; Surfactant - Cleansing Agent; Surfactant - Solubilizing Agent

Reported Product Categories: Tonics, Dressings, and Other Hair Grooming Aids; Bath Oils, Tablets, and Salts; Cleansing Products (Cold Creams, Cleansing Lotions, Liquids and Pads); Powders (Dusting and Talcum, Excluding Aftershave Talcs); Shampoos (Non-coloring); Skin Care Preparations, Misc.; Hair Conditioners; Hair Dyes and Colors (All Types Requiring Caution Statements and Patch Tests); Bath Preparations, Misc.; Body and Hand Preparations (Excluding Shaving Preparations); Moisturizing Preparations; Skin Fresheners; Makeup Bases; Foundations; Hair Preparations (Non-coloring), Misc.; Bath Soaps and Detergents; Bubble Baths; Hair Sprays (Aerosol Fixatives); Bath Capsules; Eye Makeup Preparations, Misc.; Eye Makeup Removers;

Face and Neck Preparations (Excluding Shaving Preparations); Paste Masks (Mud Packs); Personal Cleanliness Products, Misc.; Perfumes; Permanent Waves; After-shave Lotions; Baby Shampoos; Hair Wave Sets; Mouthwashes and Breath Fresheners (Liquids and Sprays); Baby Products, Misc.; Eyebrow Pencils; Shaving Cream (Aerosol, Brushless and Lather); Suntan Gels, Creams, and Liquids; Colognes and Toilet Waters; Eye Lotions; Eyeliners; Indoor Tanning Preparations; Mascara; Night Skin Care Preparations; Blushers (All types); Eye Shadows; Fragrance Preparations, Misc.; Hair Coloring Preparations, Misc.; Makeup Preparations (Not eye), Misc.; Suntan Preparations, Misc.; Dentifrices (Aerosol, Liquid, Pastes and Powders); Deodorants (Underarm); Makeup Fixatives; Shaving Preparations, Misc.

Technical/Other Names:
Polyethylene Glycol 2000 Sorbitan Laurate
Polyoxyethylene (40) Sorbitan Laurate
Polysorbate 20 (RIFM)

PEG-44 SORBITAN LAURATE

CAS No.: 9005-64-5 (Generic)

JPN Translation:
ラウリン酸 PEG - 44 ソルビタン

Definition: PEG-44 Sorbitan Laurate is an ethoxylated sorbitan ester of lauric acid with an average of 44 moles of ethylene oxide.

Information Sources: 21CFR175.300, CIR: [S] IJT-19(SUPPL.2)2000, JCLS, RIFM, TSCA

Chemical Class: Sorbitan Derivatives

Functions: Fragrance Ingredient; Surfactant - Cleansing Agent; Surfactant - Solubilizing Agent

Reported Product Categories: Tonics, Dressings, and Other Hair Grooming Aids; Bath Oils, Tablets, and Salts; Cleansing Products (Cold Creams, Cleansing Lotions, Liquids and Pads); Powders (Dusting and Talcum, Excluding Aftershave Talcs); Shampoos (Non-coloring); Skin Care Preparations, Misc.; Hair Conditioners; Hair Dyes and Colors (All Types Requiring Caution Statements and Patch Tests); Bath Preparations, Misc.; Body and Hand Preparations (Excluding Shaving Preparations); Moisturizing Preparations; Skin Fresheners; Makeup Bases; Foundations; Hair Preparations (Non-coloring), Misc.; Bath Soaps and Detergents; Bubble Baths; Hair Sprays (Aerosol Fixatives); Bath Capsules; Eye Makeup Preparations, Misc.; Eye Makeup Removers; Face and Neck Preparations (Excluding Shaving Preparations); Paste Masks (Mud

Packs); Personal Cleanliness Products, Misc.; Perfumes; Permanent Waves; After-shave Lotions; Baby Shampoos; Hair Wave Sets; Mouthwashes and Breath Fresheners (Liquids and Sprays); Baby Products, Misc.; Eyebrow Pencils; Shaving Cream (Aerosol, Brushless and Lather); Suntan Gels, Creams, and Liquids; Colognes and Toilet Waters; Eye Lotions; Eyeliners; Indoor Tanning Preparations; Mascara; Night Skin Care Preparations; Blushers (All types); Eye Shadows; Fragrance Preparations, Misc.; Hair Coloring Preparations, Misc.; Makeup Preparations (Not eye), Misc.; Suntan Preparations, Misc.; Dentifrices (Aerosol, Liquid, Pastes and Powders); Deodorants (Underarm); Makeup Fixatives; Shaving Preparations, Misc.

Technical/Other Names:
Polyethylene Glycol (44) Sorbitan Monolaurate
Polyoxyethylene (44) Monolaurate
Polysorbate 20 (RIFM)

Trade Names:
AEC PEG-44 Sorbitan Laurate (A & E Connock)
Hetsorb L-44 (Heterene)

PEG-75 SORBITAN LAURATE

CAS No.: 9005-64-5 (Generic)

JPN Translation:
ラウリン酸 PEG - 75 ソルビタン

Definition: PEG-75 Sorbitan Laurate is an ethoxylated sorbitan monoester of Lauric Acid (q.v.) with an average of 75 moles of ethylene oxide.

Information Sources: 21CFR175.300, CIR: [S] IJT-19(SUPPL.2)2000, JCLS, RIFM, TSCA

Chemical Class: Sorbitan Derivatives

Functions: Fragrance Ingredient; Surfactant - Cleansing Agent; Surfactant - Solubilizing Agent

Reported Product Categories: Tonics, Dressings, and Other Hair Grooming Aids; Bath Oils, Tablets, and Salts; Cleansing Products (Cold Creams, Cleansing Lotions, Liquids and Pads); Powders (Dusting and Talcum, Excluding Aftershave Talcs); Shampoos (Non-coloring); Skin Care Preparations, Misc.; Hair Conditioners; Hair Dyes and Colors (All Types Requiring Caution Statements and Patch Tests); Bath Preparations, Misc.; Body and Hand Preparations (Excluding Shaving Preparations); Moisturizing Preparations; Skin Fresheners; Makeup Bases; Foundations; Hair Preparations (Non-coloring), Misc.; Bath Soaps and Detergents; Bubble Baths; Hair Sprays (Aerosol

Fixatives); Bath Capsules; Eye Makeup Preparations, Misc.; Eye Makeup Removers; Face and Neck Preparations (Excluding Shaving Preparations); Paste Masks (Mud Packs); Personal Cleanliness Products, Misc.; Perfumes; Permanent Waves; After-shave Lotions; Baby Shampoos; Hair Wave Sets; Mouthwashes and Breath Fresheners (Liquids and Sprays); Baby Products, Misc.; Eyebrow Pencils; Shaving Cream (Aerosol, Brushless and Lather); Suntan Gels, Creams, and Liquids; Colognes and Toilet Waters; Eye Lotions; Eyeliners; Indoor Tanning Preparations; Mascara; Night Skin Care Preparations; Blushers (All types); Eye Shadows; Fragrance Preparations, Misc.; Hair Coloring Preparations, Misc.; Makeup Preparations (Not eye), Misc.; Suntan Preparations, Misc.; Dentifrices (Aerosol, Liquid, Pastes and Powders); Deodorants (Underarm); Makeup Fixatives; Shaving Preparations, Misc.

Technical/Other Names:
Polyethylene Glycol 4000 Sorbitan Monolaurate
Polyoxyethylene (75) Sorbitan Monolaurate
Polysorbate 20 (RIFM)

Trade Name:
AEC PEG-75 Sorbitan Laurate (A & E Connock)

PEG-80 SORBITAN LAURATE

CAS No.: 9005-64-5 (Generic)

JPN Translation:
ラウリン酸 PEG - 80 ソルビタン

Definition: PEG-80 Sorbitan Laurate is an ethoxylated sorbitan monoester of Lauric Acid (q.v.) with an average of 80 moles of ethylene oxide.

Information Sources: 21CFR175.300, CIR: [S] IJT-19(SUPPL.2)2000, JCLS, RIFM, TSCA

Chemical Class: Sorbitan Derivatives

Functions: Fragrance Ingredient; Surfactant - Cleansing Agent; Surfactant - Solubilizing Agent

Reported Product Categories: Tonics, Dressings, and Other Hair Grooming Aids; Bath Oils, Tablets, and Salts; Cleansing Products (Cold Creams, Cleansing Lotions, Liquids and Pads); Powders (Dusting and Talcum, Excluding Aftershave Talcs); Shampoos (Non-coloring); Skin Care Preparations, Misc.; Hair Conditioners; Hair Dyes and Colors (All Types Requiring Caution Statements and Patch Tests); Bath Preparations, Misc.; Body and Hand Preparations (Excluding Shaving Preparations); Moisturizing Preparations; Skin Fresheners; Makeup Bases; Foundations; Hair Preparations (Non-coloring), Misc.; Bath Soaps and Detergents; Bubble Baths; Hair Sprays (Aerosol Fixatives); Bath Capsules; Eye Makeup Preparations, Misc.; Eye Makeup Removers; Face and Neck Preparations (Excluding Shaving Preparations); Paste Masks (Mud Packs); Personal Cleanliness Products, Misc.; Perfumes; Permanent Waves; After-shave Lotions; Baby Shampoos; Hair Wave Sets; Mouthwashes and Breath Fresheners (Liquids and Sprays); Baby Products, Misc.; Eyebrow Pencils; Shaving Cream (Aerosol, Brushless and Lather); Suntan Gels, Creams, and Liquids; Colognes and Toilet Waters; Eye Lotions; Eyeliners; Indoor Tanning Preparations; Mascara; Night Skin Care Preparations; Blushers (All types); Eye Shadows; Fragrance Preparations, Misc.; Hair Coloring Preparations, Misc.; Makeup Preparations (Not eye), Misc.; Suntan Preparations, Misc.; Dentifrices (Aerosol, Liquid, Pastes and Powders); Deodorants (Underarm); Makeup Fixatives; Shaving Preparations, Misc.

Technical/Other Names:
Polyethylene Glycol (80) Sorbitan Mono-laurate
Polyoxyethylene (80) Sorbitan Monolaurate
Polysorbate 20 (RIFM)

Trade Names:
AEC PEG-80 Sorbitan Laurate (A & E Connock)
Alkamuls PSML-80/72% (Rhodia)
Atlas G-4280 (Uniqema Americas)
Ethsorbox PSML-80 (Ethox)
Hetsorb L-80 (Heterene)
Laxan-S (Lanaetex)
Liposorb L-80 (Lipo)
Lumisorb PSML-80 (Lambent)

Trade Name Mixtures:
Custoblend BSC-50 (Baby Shampoo) (Custom Ingredients)
Miracare BC-10 (Rhodia)
Miracare BC-20 (Rhodia)
Miracare BC-27 (Rhodia)
Miracare MS-2 (Rhodia)
Miracare MS-4 (Rhodia)

PEG-3 SORBITAN OLEATE

CAS No.: 9005-65-6 (Generic)

Empirical Formula:
$C_{30}H_{56}O_8$

Definition: PEG-3 Sorbitan Oleate is an ethoxylated sorbitan ester of oleic acid with an average of 3 moles of ethylene oxide.

Information Sources: 21CFR73.1001, 21CFR172.515, 21CFR172.623, 21CFR173.340, 21CFR175.105, 21CFR175.300, 21CFR176.180, 21CFR178.3400, 21CFR573.860, CIR: [S] IJT-19(SUPPL.2)2000, MI-13(7664), RIFM, TSCA

Chemical Class: Sorbitan Derivatives

Functions: Fragrance Ingredient; Surfactant - Emulsifying Agent

Reported Product Categories: Hair Conditioners; Moisturizing Preparations; Hair Sprays (Aerosol Fixatives); Tonics, Dressings, and Other Hair Grooming Aids; Bath Preparations, Misc.; Body and Hand Preparations (Excluding Shaving Preparations); Eyeliners; Mouthwashes and Breath Fresheners (Liquids and Sprays); Hair Dyes and Colors (All Types Requiring Caution Statements and Patch Tests); Hair Preparations (Non-coloring), Misc.; Shampoos (Non-coloring); Skin Fresheners; Powders (Dusting and Talcum, Excluding Aftershave Talcs); Bubble Baths; Colognes and Toilet Waters; Paste Masks (Mud Packs); Bath Capsules; Bath Oils, Tablets, and Salts; Cleansing Products (Cold Creams, Cleansing Lotions, Liquids and Pads); Makeup Preparations (Not eye), Misc.; Night Skin Care Preparations; Aftershave Lotions; Baby Lotions, Oils, Powders and Creams; Baby Shampoos; Eye Makeup Preparations, Misc.; Face and Neck Preparations (Excluding Shaving Preparations); Fragrance Preparations, Misc.; Shaving Cream (Aerosol, Brushless and Lather); Skin Care Preparations, Misc.

Technical/Other Names:
Polyethylene Glycol (3) Sorbitan Monooleate
Polyoxyethylene (3) Sorbitan Monooleate
Polysorbate 80 (RIFM)

Trade Name:
AEC PEG-3 Sorbitan Oleate (A & E Connock)

PEG-6 SORBITAN OLEATE

CAS No.: 9005-65-6 (Generic)

JPN Translation:
オレイン酸 PEG - 6 ソルビタン

Empirical Formula:
$C_{36}H_{68}O_{11}$

Definition: PEG-6 Sorbitan Oleate is an ethoxylated sorbitan ester of oleic acid with an average of 6 moles of ethylene oxide.

Information Sources: 21CFR175.300, 21CFR176.210, CIR: [S] IJT-19(SUPPL.2)-2000, JCLS, JSCI, RIFM, TSCA

Chemical Class: Sorbitan Derivatives

Functions: Fragrance Ingredient; Surfactant - Emulsifying Agent

Reported Product Categories: Hair Conditioners; Moisturizing Preparations; Hair Sprays (Aerosol Fixatives); Tonics, Dressings, and Other Hair Grooming Aids; Bath Preparations, Misc.; Body and Hand Preparations (Excluding Shaving Preparations); Eyeliners; Mouthwashes and Breath Fresheners (Liquids and Sprays); Hair Dyes and Colors (All Types Requiring Caution Statements and Patch Tests); Hair Preparations (Non-coloring), Misc.; Shampoos (Non-coloring); Skin Fresheners; Powders (Dusting and Talcum, Excluding Aftershave Talcs); Bubble Baths; Colognes and Toilet Waters; Paste Masks (Mud Packs); Bath Capsules; Bath Oils, Tablets, and Salts; Cleansing Products (Cold Creams, Cleansing Lotions, Liquids and Pads); Makeup Preparations (Not eye), Misc.; Night Skin Care Preparations; Aftershave Lotions; Baby Lotions, Oils, Powders and Creams; Baby Shampoos; Eye Makeup Preparations, Misc.; Face and Neck Preparations (Excluding Shaving Preparations); Fragrance Preparations, Misc.; Shaving Cream (Aerosol, Brushless and Lather); Skin Care Preparations, Misc.

Technical/Other Names:
Polyethylene Glycol 300 Sorbitan Monooleate
Polyoxyethylene Sorbitan Monooleate (6E.O.)
Polysorbate 80 (RIFM)

Trade Names:
AEC PEG-6 Sorbitan Oleate (A & E Connock)
Nikkol TO-106 (Nikko)

PEG-20 SORBITAN OLEATE

Definition: PEG-20 Sorbitan Oleate is an ethoxylated sorbitan monoester of Oleic Acid (q.v.) with an average of 20 moles of ethylene oxide.

Information Source: JCIC

Chemical Class: Sorbitan Derivatives

Functions: Surfactant - Cleansing Agent; Surfactant - Emulsifying Agent; Surfactant - Solubilizing Agent

PEG-40 SORBITAN OLEATE

Definition: PEG-40 Sorbitan Oleate is an ethoxylated sorbitan monoester of Oleic Acid (q.v.) containing an average of 40 moles of ethylene oxide.

Chemical Class: Sorbitan Derivatives

Function: Surfactant - Emulsifying Agent

Technical/Other Names:
Polyethylene Glycol 2000 Sorbitan Monooleate
Polyoxyethylene (40) Sorbitan Monooleate

Trade Name:
Emalex ET-8040 (Nihon Emulsion)

PEG-80 SORBITAN PALMITATE

CAS No.: 9005-66-7 (Generic)

Definition: PEG-80 Sorbitan Palmitate is an ethoxylated sorbitan monoester of palmitic acid with an average of 80 moles of ethylene oxide.

Information Sources: 21CFR175.300, CIR: [S] IJT-19(SUPPL.2)2000

Chemical Class: Sorbitan Derivatives

Functions: Surfactant - Cleansing Agent; Surfactant - Solubilizing Agent

Reported Product Categories: Bath Oils, Tablets, and Salts; Cleansing Products (Cold Creams, Cleansing Lotions, Liquids and Pads); Moisturizing Preparations; Skin Care Preparations, Misc.; Fragrance Preparations, Misc.; Bath Preparations, Misc.; Body and Hand Preparations (Excluding Shaving Preparations)

Technical/Other Names:
Polyethylene Glycol (80) Sorbitan Monopalmitate
Polyoxyethylene (80) Sorbitan Monopalmitate

Trade Names:
AEC PEG-80 Sorbitan Palmitate (A & E Connock)
Atlas G-4252 (Uniqema Americas)

PEG-40 SORBITAN PERISOSTEARATE

Definition: PEG-40 Sorbitan Perisostearate is a mixture of isostearic acid esters of sorbitol condensed with an average of 40 moles of ethylene oxide.

Information Source: CIR: [S] IJT-19 (SUPPL.2)2000

Chemical Class: Sorbitan Derivatives

Function: Surfactant - Emulsifying Agent

Technical/Other Names:
Polyethylene Glycol 2000 Sorbitan Perisostearate
Polyoxyethylene (40) Sorbitan Perisostearate

Trade Names:
Atlas G-1049 (Uniqema Americas)
Sympatens-SIS/400 (Kolb)

PEG-40 SORBITAN PEROLEATE

JPN Translation:
オレイン酸 PEG - 40 ソルビット

Definition: PEG-40 Sorbitan Peroleate is a mixture of oleic acid esters of sorbitol condensed with an average of 40 moles of ethylene oxide.

Information Sources: 21CFR175.300, 21CFR176.210, CIR: [S] IJT-19(SUPPL.2)-2000, CTFA D

Chemical Class: Sorbitan Derivatives

Functions: Surfactant - Emulsifying Agent; Surfactant - Solubilizing Agent

Reported Product Categories: Bath Oils, Tablets, and Salts; Moisturizing Preparations

Technical/Other Names:
Polyethylene Glycol 2000 Sorbitan Peroleate
Polyoxyethylene (40) Sorbitan Peroleate

Trade Names:
AEC PEG-40 Sorbitan Peroleate (A & E Connock)
Arlatone T (Uniqema Americas)
Sympatens-SPO/400 (Kolb)

Trade Name Mixture:
Polymoist Mask (Cognis Deutschland)

PEG-3 SORBITAN STEARATE

CAS No.: 9005-67-8 (Generic)

Definition: PEG-3 Sorbitan Stearate is an ethoxylated sorbitan monoester of stearic acid with an average of 3 moles of ethylene oxide.

Information Sources: 21CFR73.1001, 21CFR163.123, 21CFR163.130, 21CFR163.135, 21CFR163.140, 21CFR163.145, 21CFR163.150, 21CFR163.153, 21CFR163.155, 21CFR172.515, 21CFR172.836, 21CFR173.340, 21CFR175.300, 21CFR573.840, CIR: [S] IJT-19(SUPPL.2)-2000, RIFM, TSCA

Chemical Class: Sorbitan Derivatives

Functions: Fragrance Ingredient; Surfactant - Emulsifying Agent

Reported Product Categories: Moisturizing Preparations; Bath Preparations, Misc.; Body and Hand Preparations (Excluding Shaving Preparations); Bath Oils, Tablets, and Salts; Cleansing Products (Cold Creams, Cleansing Lotions, Liquids and Pads); Foundations; Skin Care Preparations, Misc.; Bath Capsules; Mascara; Face and Neck Preparations (Excluding Shaving Preparations); Hair Conditioners; Shaving Cream (Aerosol, Brushless and Lather); Eyebrow Pencils; Paste Masks (Mud Packs); Night Skin Care Preparations; Eyeliners; Indoor Tanning Preparations; Tonics, Dressings, and Other Hair Grooming Aids; Eye Makeup Preparations, Misc.; Fragrance Preparations, Misc.; Makeup Preparations (Not eye), Misc.; Baby Lotions, Oils, Powders and Creams; Cuticle Softeners; Foot Powders and Sprays; Hair Straighteners

Technical/Other Names:
Polyethylene Glycol (3) Sorbitan Monostearate
Polyoxyethylene (3) Sorbitan Monostearate
Polysorbate 60 (RIFM)

Trade Name:
AEC PEG-3 Sorbitan Stearate (A & E Connock)

PEG-4 SORBITAN STEARATE

CAS No.: 9005-67-8 (Generic)

Definition: PEG-4 Sorbitan Stearate is an ethoxylated sorbitan monoester of stearic acid with an average of 4 moles of ethylene oxide.

Information Source: RIFM

Chemical Class: Sorbitan Derivatives

Functions: Fragrance Ingredient; Surfactant - Emulsifying Agent

Reported Product Categories: Bath Preparations, Misc.; Bath Oils, Tablets, and Salts; Foundations; Bath Capsules; Mascara; Hair Conditioners; Eyebrow Pencils; Eyeliners; Tonics, Dressings, and Other Hair Grooming Aids; Eye Makeup Preparations, Misc.; Fragrance Preparations, Misc.; Makeup Preparations (Not eye), Misc.; Baby Lotions, Oils, Powders and Creams; Cuticle Softeners; Hair Straighteners

Technical/Other Names:
Polyethylene Glycol 200 Sorbitan Monostearate
Polyoxyethylene (4) Sorbitan Monostearate
Polysorbate 60 (RIFM)

Trade Name:
Kotilen-S/1/040 (Kolb)

PEG-6 SORBITAN STEARATE

CAS No.: 9005-67-8 (Generic)

JPN Translation:
ステアリン酸 PEG - 6 ソルビタン

Empirical Formula:
$C_{36}H_{70}O_{12}$

Definition: PEG-6 Sorbitan Stearate is an ethoxylated sorbitan ester of Stearic Acid (q.v.) with an average of 6 moles of ethylene oxide.

Information Sources: 21CFR175.300, 21CFR176.210, CIR: [S] IJT-19(SUPPL.2)-2000, JCLS, JSCI, RIFM

Chemical Class: Sorbitan Derivatives

Functions: Fragrance Ingredient; Surfactant - Emulsifying Agent

Reported Product Categories: Moisturizing Preparations; Bath Preparations, Misc.; Body and Hand Preparations (Excluding Shaving Preparations); Bath Oils, Tablets, and Salts; Cleansing Products (Cold Creams, Cleansing Lotions, Liquids and Pads); Foundations; Skin Care Preparations, Misc.; Bath Capsules; Mascara; Face and Neck Preparations (Excluding Shaving Preparations); Hair Conditioners; Shaving Cream (Aerosol, Brushless and Lather); Eyebrow Pencils; Paste Masks (Mud Packs); Night Skin Care Preparations; Eyeliners; Indoor Tanning Preparations; Tonics, Dressings, and Other Hair Grooming Aids; Eye Makeup Preparations, Misc.; Fragrance Preparations, Misc.; Makeup Preparations (Not eye), Misc.; Baby Lotions, Oils, Powders and Creams; Cuticle Softeners; Foot Powders and Sprays; Hair Straighteners

Technical/Other Names:
Polyethylene Glycol 300 Sorbitan Monostearate
Polyoxyethylene (6) Sorbitan Monostearate
Polyoxyethylene Sorbitan Monostearate (6E.O.)
Polysorbate 60 (RIFM)

Trade Name:
Nikkol TS-106 (Nikko)

PEG-40 SORBITAN STEARATE

CAS No.: 9005-67-8 (Generic)

Definition: PEG-40 Sorbitan Stearate is an ethoxylated sorbitan ester of stearic acid with an average of 40 moles of ethylene oxide.

Information Sources: 21CFR175.300, 21CFR176.210, CIR: [S] IJT-19(SUPPL.2)-2000, RIFM, TSCA

Chemical Class: Sorbitan Derivatives

Functions: Fragrance Ingredient; Surfactant - Cleansing Agent; Surfactant - Solubilizing Agent

Reported Product Categories: Moisturizing Preparations; Bath Preparations, Misc.; Body and Hand Preparations (Excluding Shaving Preparations); Bath Oils, Tablets, and Salts; Cleansing Products (Cold Creams, Cleansing Lotions, Liquids and Pads); Foundations; Skin Care Preparations, Misc.; Bath Capsules; Mascara; Face and Neck Preparations (Excluding Shaving Preparations); Hair Conditioners; Shaving Cream (Aerosol, Brushless and Lather); Eyebrow Pencils; Paste Masks (Mud Packs); Night Skin Care Preparations; Eyeliners; Indoor Tanning Preparations; Tonics, Dressings, and Other Hair Grooming Aids; Eye Makeup Preparations, Misc.; Fragrance Preparations, Misc.; Makeup Preparations (Not eye), Misc.; Baby Lotions, Oils, Powders and Creams; Cuticle Softeners; Foot Powders and Sprays; Hair Straighteners

Technical/Other Names:
Polyethylene Glycol 2000 Sorbitan Stearate
Polyoxyethylene (40) Sorbitan Stearate
Polysorbate 60 (RIFM)

Trade Name:
AEC PEG-40 Sorbitan Stearate (A & E Connock)

PEG-60 SORBITAN STEARATE

CAS No.: 9005-67-8 (Generic)

Definition: PEG-60 Sorbitan Stearate is an ethoxylated sorbitan ester of Stearic Acid (q.v.) with an average of 60 moles of ethylene oxide.

Information Sources: 21CFR175.300, CIR: [S] IJT-19(SUPPL.2)2000, RIFM

Chemical Class: Sorbitan Derivatives

Functions: Fragrance Ingredient; Surfactant - Cleansing Agent; Surfactant - Solubilizing Agent

Reported Product Categories: Moisturizing Preparations; Bath Preparations, Misc.; Body and Hand Preparations (Excluding Shaving Preparations); Bath Oils, Tablets, and Salts; Cleansing Products (Cold Creams, Cleansing Lotions, Liquids and Pads); Foundations; Skin Care Preparations, Misc.; Bath Capsules; Mascara; Face and Neck Preparations (Excluding Shaving Preparations); Hair Conditioners; Shaving Cream (Aerosol, Brushless and Lather); Eyebrow Pencils; Paste Masks (Mud Packs); Night Skin Care Preparations; Eyeliners; Indoor Tanning Preparations; Tonics, Dressings, and Other Hair Grooming Aids; Eye Makeup Preparations, Misc.; Fragrance Preparations, Misc.; Makeup Preparations (Not eye), Misc.; Baby Lotions, Oils, Powders and Creams; Cuticle Softeners; Foot Powders and Sprays; Hair Straighteners

Technical/Other Names:
Polyethylene Glycol 3000 Sorbitan Monostearate
Polyoxyethylene (60) Sorbitan Monostearate
Polysorbate 60 (RIFM)

Trade Name:
Tego SMS 60 (Degussa Care Specialties)

PEG-30 SORBITAN TETRAOLEATE

Empirical Formula:
$C_{17}H_{44}O_8$

Definition: PEG-30 Sorbitan Tetraoleate is the tetraester of oleic acid and a polyethylene glycol ether of sorbitol, with an average of 30 moles of ethylene oxide.

Information Sources: 21CFR175.300, CIR: [S] IJT-19(SUPPL.2)2000

Chemical Class: Sorbitan Derivatives

Function: Surfactant - Emulsifying Agent

Technical/Other Names:
Polyethylene Glycol (30) Sorbitan Tetraoleate
Polyoxyethylene (30) Sorbitan Tetraoleate

Trade Name:
AEC PEG-30 Sorbitan Tetraoleate (A & E Connock)

PEG-40 SORBITAN TETRAOLEATE

Definition: PEG-40 Sorbitan Tetraoleate is the tetraester of oleic acid and a polyethylene glycol ether of sorbitol, with an average of 40 moles of ethylene oxide.

Information Sources: 21CFR175.300, 21CFR176.210, CIR: [S] IJT-19(SUPPL.2)-2000

Chemical Class: Sorbitan Derivatives

Function: Surfactant - Emulsifying Agent

Technical/Other Names:
Polyethylene Glycol 2000 Sorbitan Tetraoleate
Polyoxyethylene (40) Sorbitan Tetraoleate

Trade Names:
AEC PEG-40 Sorbitan Tetraoleate (A & E Connock)
Rheodol 440 (Kao Corp.)

PEG-60 SORBITAN TETRAOLEATE

Definition: PEG-60 Sorbitan Tetraoleate is the tetraester of oleic acid and a polyethylene glycol ether of sorbitol, with an average of 60 moles of ethylene oxide.

Information Sources: 21CFR175.300, CIR: [S] IJT-19(SUPPL.2)2000

Chemical Class: Sorbitan Derivatives

Function: Surfactant - Emulsifying Agent

Technical/Other Names:
Polyethylene Glycol 3000 Sorbitan Tetraoleate
Polyoxyethylene (60) Sorbitan Tetraoleate

Trade Names:
AEC PEG-60 Sorbitan Tetraoleate (A & E Connock)
Rheodol 460 (Kao Corp.)

PEG-60 SORBITAN TETRASTEARATE

Definition: PEG-60 Sorbitan Tetrastearate is the tetraester of stearic acid and a polyethylene glycol ether of sorbitol, with an average of 60 moles of ethylene oxide.

Information Sources: 21CFR175.300, CIR: [S] IJT-19(SUPPL.2)2000

Chemical Class: Sorbitan Derivatives

Function: Surfactant - Emulsifying Agent

Technical/Other Names:
Polyethylene Glycol (60) Sorbitan Tetrastearate
Polyoxyethylene (60) Sorbitan Tetrastearate

Trade Name:
AEC PEG-60 Sorbitan Tetrastearate (A & E Connock)

PEG-4 SORBITAN TRIISOSTEARATE

Definition: PEG-4 Sorbitan Triisostearate is the triester of isostearic acid and a polyethylene glycol ether of sorbitol with an average of 4 moles of ethylene oxide.

Chemical Class: Sorbitan Derivatives

Function: Surfactant - Emulsifying Agent

Technical/Other Names:
Polyethylene Glycol 200 Sorbitan Triisostearate
Polyoxyethylene (4) Sorbitan Triisostearate

Trade Name:
Emalex EG-2854-IS (Nihon Emulsion)

PEG-20 SORBITAN TRIISOSTEARATE

Definition: PEG-20 Sorbitan Triisostearate is the triester of isostearic acid and a polyethylene glycol ether of sorbitol with an average of 20 moles of ethylene oxide.

Information Sources: CIR: [S] IJT-19 (SUPPL.2)2000, JCLS

Chemical Class: Sorbitan Derivatives

Function: Surfactant - Emulsifying Agent

Technical/Other Names:
Polyethylene Glycol 1000 Sorbitan Triisostearate
Polyoxyethylene (20) Sorbitan Triisostearate

Trade Name:
Crillet 65 (Croda Chemicals)

PEG-160 SORBITAN TRIISOSTEARATE

JPN Translation:
トリイソステアリン酸 PEG - 160 ソルビタン

Definition: PEG-160 Sorbitan Triisostearate is the triester of isostearic acid and a polyethylene glycol ether of sorbitol with an average of 160 moles of ethylene oxide.

Information Sources: CIR: [S] IJT-19 (SUPPL.2)2000, JCLS

Chemical Class: Sorbitan Derivatives

Functions: Surfactant - Cleansing Agent; Surfactant - Solubilizing Agent

Reported Product Category: Shampoos (Non-coloring)

Trade Name:
Rheodol TW-IS399C (Kao Corp.)

PEG-2 SORBITAN TRIOLEATE

CAS No.: 9005-70-3

Definition: PEG-2 Sorbitan Trioleate is a triester of oleic acid and a polyethylene glycol ether of sorbitol with an average of 2 moles of ethylene oxide.

Information Source: TSCA

Chemical Class: Sorbitan Derivatives

Function: Surfactant - Emulsifying Agent

Reported Product Categories: Eyeliners; Indoor Tanning Preparations; Bath Oils, Tablets, and Salts; Cleansing Products (Cold Creams, Cleansing Lotions, Liquids and Pads); Bath Preparations, Misc.; Body and Hand Preparations (Excluding Shaving Preparations); Makeup Preparations (Not eye), Misc.; Moisturizing Preparations

Trade Name:
Kotilen-O/3/020 (Kolb)

PEG-3 SORBITAN TRISTEARATE

JPN Translation:
トリステアリン酸 PEG - 3 ソルビット

Definition: PEG-3 Sorbitan Tristearate is the triester of stearic acid and a polyethylene glycol ether of sorbitol with an average of 3 moles of ethylene oxide.

Information Source: JCIC

Chemical Class: Sorbitan Derivatives

Function: Skin-Conditioning Agent - Emollient

Technical/Other Names:
Polyethylene Glycol (3) Sorbitan Tristearate
Polyoxyethylene (3) Sorbitan Tristearate

PEG-2 SOYAMINE

CAS No.: 61791-24-0 (Generic)

Definition: PEG-2 Soyamine is the polyethylene glycol amine of Soy Acid (q.v.) that conforms generally to the formula:

$$R-N \underset{(CH_2CH_2O)_yH}{\overset{(CH_2CH_2O)_xH}{}}$$

where R represents the alkyl groups derived from soy and x + y has an average value of 2.

Information Source: TSCA

Chemical Class: Alkoxylated Amines

Functions: Antistatic Agent; Surfactant - Foam Booster

Technical/Other Names:
Polyethylene Glycol 100 Soy Amine
Polyoxyethylene (2) Soy Amine

Trade Names:
AEC PEG-2 Soyamine (A & E Connock)
Ethomeen S/12 (Akzo Nobel)
Hetoxamine S-2 (Heterene)
Protox S-2 (Protameen)

PEG-5 SOYAMINE

CAS No.: 61791-24-0 (Generic)

Definition: PEG-5 Soyamine is the polyethylene glycol amine of Soy Acid (q.v.) that conforms generally to the formula:

$$R-N \underset{(CH_2CH_2O)_yH}{\overset{(CH_2CH_2O)_xH}{}}$$

where R represents the alkyl groups derived from soy and x + y has an average value of 5.

Information Sources: JSQI, TSCA

Chemical Class: Alkoxylated Amines

Functions: Antistatic Agent; Surfactant - Emulsifying Agent

Reported Product Category: Hair Bleaches

Technical/Other Names:
Polyethylene Glycol (5) Soy Amine
Polyoxyethylene (5) Soy Amine

Trade Names:
Ethomeen S/15 (Akzo Nobel)
Hetoxamine S-5 (Heterene)
Protox S-5 (Protameen)

PEG-8 SOYAMINE

CAS No.: 61791-24-0 (Generic)

Definition: PEG-8 Soyamine is the polyethylene glycol amine of Soy Acid (q.v.) that conforms generally to the formula:

$$R-N \Big\langle \begin{array}{l} (CH_2CH_2O)_xH \\ (CH_2CH_2O)_yH \end{array}$$

where R represents the alkyl groups derived from soy and x + y has an average value of 8.

Information Source: TSCA

Chemical Class: Alkoxylated Amines

Functions: Antistatic Agent; Surfactant - Emulsifying Agent

Technical/Other Names:
Polyethylene Glycol 400 Soy Amine
Polyoxyethylene (8) Soy Amine

PEG-10 SOYAMINE

CAS No.: 61791-24-0 (Generic)

Definition: PEG-10 Soyamine is the polyethylene glycol amine of Soy Acid (q.v.) that conforms generally to the formula:

$$R-N \Big\langle \begin{array}{l} (CH_2CH_2O)_xH \\ (CH_2CH_2O)_yH \end{array}$$

where R represents the alkyl groups derived from soy and x + y has an average value of 10.

Information Source: TSCA

Chemical Class: Alkoxylated Amines

Functions: Antistatic Agent; Surfactant - Emulsifying Agent

Technical/Other Names:
Polyethylene Glycol 500 Soy Amine
Polyoxyethylene (10) Soy Amine

Trade Names:
Protox S-10 (Protameen)
Unizeen S-10 (Universal Preserv-A-Chem)

PEG-15 SOYAMINE

CAS No.: 61791-24-0 (Generic)

Definition: PEG-15 Soyamine is the polyethylene glycol amine of Soy Acid (q.v.) that conforms generally to the formula:

$$R-N \Big\langle \begin{array}{l} (CH_2CH_2O)_xH \\ (CH_2CH_2O)_yH \end{array}$$

where R represents the alkyl groups derived from soy and x + y has an average value of 15.

Information Source: TSCA

Chemical Class: Alkoxylated Amines

Functions: Antistatic Agent; Surfactant - Emulsifying Agent

Technical/Other Names:
Polyethylene Glycol (15) Soy Amine
Polyoxyethylene (15) Soy Amine

Trade Names:
Ethomeen S/25 (Akzo Nobel)
Hetoxamine S-15 (Heterene)
Protox S-15 (Protameen)
Unizeen S-15 (Universal Preserv-A-Chem)

PEG-9 SOYATE

Definition: PEG-9 Soyate is the polyethylene glycol ester of the fatty acids derived from Glycine Soja (Soy) Oil (q.v.) that conforms generally to the formula:

$$RC \overset{\displaystyle O}{\underset{\displaystyle \|}{}} - (OCH_2CH_2)_nOH$$

where R represents the fatty acids derived from Glycine Soja (Soy) Oil (q.v.) and n has an average value of 9.

Chemical Class: Alkoxylated Carboxylic Acids

Function: Surfactant - Emulsifying Agent

Technical/Other Names:
Polyethylene Glycol (9) Soyate
Polyoxyethylene (9) Soyate

Trade Name Mixture:
Silwax DMC-SOY (Siltech LLC)

PEG-35 SOY GLYCERIDES

Definition: PEG-35 Soy Glycerides is a polyethylene glycol derivative of the mono- and diglycerides derived from Glycine Soja (Soybean) Oil (q.v.) containing an average of 35 moles of ethylene oxide.

Chemical Classes: Alkoxylated Alcohols; Glyceryl Esters and Derivatives

Functions: Skin-Conditioning Agent - Miscellaneous; Surfactant - Emulsifying Agent; Surfactant - Solubilizing Agent

Trade Name:
Acconon S-35 (Abitec Corporation)

PEG-75 SOY GLYCERIDES

Definition: PEG-75 Soy Glycerides is a polyethylene glycol derivative of the mono- and diglycerides from Glycine Soja (Soybean) Oil (q.v.) with an average of 75 moles of ethylene oxide.

Chemical Classes: Alkoxylated Alcohols; Glyceryl Esters and Derivatives

Functions: Skin-Conditioning Agent - Emollient; Surfactant - Emulsifying Agent

Technical/Other Names:
Polyethylene Glycol 4000 Soy Glycerides
Polyoxyethylene (75) Soy Glycerides

Trade Name:
Acconon S-75 (Abitec Corporation)

PEG-5 SOY STEROL

Definition: PEG-5 Soy Sterol is a polyethylene glycol derivative of sterols found in Glycine Soja (Soybean) Oil (q.v.) with an average of 5 moles of ethylene oxide.

Information Sources: CIR: [I] IJT-19 (SUPPL.1)2000, CIR: [S], CTFA D, JSQI

Chemical Classes: Alkoxylated Alcohols; Sterols

Functions: Skin-Conditioning Agent - Miscellaneous; Surfactant - Emulsifying Agent

Reported Product Categories: Hair Conditioners; Skin Care Preparations, Misc.; Moisturizing Preparations; Bath Capsules; Face and Neck Preparations (Excluding Shaving Preparations); Bath Preparations, Misc.; Body and Hand Preparations (Excluding Shaving Preparations); Mascara

Technical/Other Names:
PEG-5 Soya Sterol
Polyethylene Glycol (5) Soy Sterol
Polyoxyethylene (5) Soy Sterol

Trade Names:
Genurol-122E5 (Universal Preserv-A-Chem)
Nikkol BPS-5 (Nikko)
Sitostene 5-OE (Vevy)

PEG-10 SOY STEROL

Definition: PEG-10 Soy Sterol is a polyethylene glycol derivative of sterols found in Glycine Soja (Soybean) Oil (q.v.) with an average of 10 moles of ethylene oxide.

Information Sources: CIR: [I] IJT-19 (SUPPL.1)2000, CIR: [S], CTFA D, JSQI

Chemical Classes: Alkoxylated Alcohols; Sterols

Functions: Skin-Conditioning Agent - Miscellaneous; Surfactant - Emulsifying Agent

Reported Product Categories: Bath Oils, Tablets, and Salts; Cleansing Products (Cold Creams, Cleansing Lotions, Liquids and Pads); Moisturizing Preparations; Body and Hand Preparations (Excluding Shaving Preparations); Eyebrow Pencils; Suntan Gels, Creams, and Liquids

Technical/Other Names:
PEG-10 Soya Sterol
Polyethylene Glycol 500 Soy Sterol
Polyoxyethylene (10) Soy Sterol

Trade Names:
Genurol-122E10 (Universal Preserv-A-Chem)
Nikkol BPS-10 (Nikko)

PEG-16 SOY STEROL

Definition: PEG-16 Soy Sterol is a polyethylene glycol derivative of sterols found in Glycine Soja (Soybean) Oil (q.v.) with an average of 16 moles of ethylene oxide.

Information Sources: CIR: [I] IJT-19 (SUPPL.1)2000, CIR: [S], JSQI

Chemical Classes: Alkoxylated Alcohols; Sterols

Function: Surfactant - Emulsifying Agent

Reported Product Categories: Bath Oils, Tablets, and Salts; Cleansing Products (Cold Creams, Cleansing Lotions, Liquids and Pads); Eye Makeup Preparations, Misc.

Technical/Other Names:
PEG-16 Soya Sterol
Polyethylene Glycol (16) Soy Sterol
Polyoxyethylene (16) Soy Sterol

Trade Names:
Generol 122 N E16 (Cognis Care Chemicals/NJ)
Generol 122 N E16 (Cognis Care Chemicals/PA)
Genurol-122E16 (Universal Preserv-A-Chem)

PEG-20 SOY STEROL

Definition: PEG-20 Soy Sterol is a polyethylene glycol derivative of the sterols obtained from Glycine Soja (Soybean) Oil (q.v.) with an average of 20 moles of ethylene oxide.

Chemical Classes: Alkoxylated Alcohols; Sterols

Function: Surfactant - Emulsifying Agent

Technical/Other Names:
Polyethylene Glycol (20) Soy Sterol
Polyoxyethylene (20) Soy Sterol

Trade Name:
Nikkol BPS-20 (Nikko)

PEG-25 SOY STEROL

Definition: PEG-25 Soy Sterol is a polyethylene glycol derivative of sterols found in Glycine Soja (Soybean) Oil (q.v.) with an average of 25 moles of ethylene oxide.

Information Sources: CIR: [I] IJT-19 (SUPPL.1)2000, CIR: [S], JSQI

Chemical Classes: Alkoxylated Alcohols; Sterols

Function: Surfactant - Emulsifying Agent

Reported Product Category: Paste Masks (Mud Packs)

Technical/Other Names:
PEG-25 Soya Sterol
Polyethylene Glycol (25) Soy Sterol
Polyoxyethylene (25) Soy Sterol

Trade Names:
Generol 122 N E25 (Cognis Care Chemicals/NJ)
Generol 122 N E25 (Cognis Care Chemicals/PA)
Genurol-122E25 (Universal Preserv-A-Chem)

Trade Name Mixture:
Liposomes Anti-Age Veg (Laboratoires Serobiologiques)

PEG-30 SOY STEROL

Definition: PEG-30 Soy Sterol is a polyethylene glycol derivative of sterols obtained from Glycine Soja (Soybean) Oil (q.v.) with an average of 30 moles of ethylene oxide.

Information Sources: CIR: [I] IJT-19 (SUPPL.1)2000, CIR: [S]

Chemical Classes: Alkoxylated Alcohols; Sterols

Function: Surfactant - Emulsifying Agent

Technical/Other Names:
PEG-30 Soya Sterol
Polyethylene Glycol (30) Soy Sterol
Polyoxyethylene (30) Soy Sterol

Trade Name:
Nikkol BPS-30 (Nikko)

Trade Name Mixture:
Ivarbase 3231 (Arch Personal Care Products)

PEG-40 SOY STEROL

Definition: PEG-40 Soy Sterol is a polyethylene glycol derivative of sterols found in Glycine Soja (Soybean) Oil (q.v.) with an average of 40 moles of ethylene oxide.

Information Sources: CIR: [I] IJT-19 (SUPPL.1)2000, CIR: [S], JSQI

Chemical Classes: Alkoxylated Alcohols; Sterols

Functions: Surfactant - Cleansing Agent; Surfactant - Solubilizing Agent

Technical/Other Names:
PEG-40 Soya Sterol
Polyethylene Glycol 2000 Soy Sterol
Polyoxyethylene (40) Soy Sterol

PEG-4 STEARAMIDE

JPN Translation:
PEG - 4 ステアラミド

Empirical Formula:
$C_{26}H_{53}NO_5$

Definition: PEG-4 Stearamide is the polyethylene glycol amide of stearic acid that conforms generally to the formula:

$$CH_3(CH_2)_{16}C\overset{\displaystyle O}{\overset{\|}{}}-NH(CH_2CH_2O)_nH$$

where n has an average value of 4.

Information Sources: JCLS, JSCI

Chemical Class: Alkoxylated Amides

Function: Surfactant - Emulsifying Agent

Technical/Other Names:
Polyethylene Glycol 200 Stearamide
Polyethylene Glycol 200 Stearyl Amide
Polyoxyethylene (4) Stearamide
Polyoxyethylene Stearoylamide
Polyoxyethylene (4) Stearyl Amide

Trade Name:
Nikkol TAMDS-4 (Nikko)

PEG-10 STEARAMIDE

Definition: PEG-10 Stearamide is the polyethylene glycol amide of stearic acid that conforms generally to the formula:

$$CH_3(CH_2)_{16}C\overset{\displaystyle O}{\overset{\|}{}}-NH(CH_2CH_2O)_nH$$

where n has an average value of 10.

Chemical Class: Alkoxylated Amides

Function: Surfactant - Emulsifying Agent

Technical/Other Names:
Polyethylene Glycol (15) Stearyl Amide
Polyoxyethylene (15) Stearyl Amide

Trade Name:
Nikkol Tamds-10 (Nikko)

PEG-15 STEARAMIDE

Definition: PEG-15 Stearamide is the polyethylene glycol amide of stearic acid that conforms generally to the formula:

$$CH_3(CH_2)_{16}C-NH(CH_2CH_2O)_nH$$
$$O$$

where n ;has an average value of 15.

Chemical Class: Alkoxylated Amides

Function: Surfactant - Emulsifying Agent

Technical/Other Names:
Polyethylene Glycol (15) Stearamide
Polyoxyethylene (15) Stearamide

Trade Name:
, Nikkol TAMDS-15 (Nikko)

PEG-2 STEARAMIDE CARBOXYLIC ACID

CAS No.: 90453-59-1

Empirical Formula:
$C_{22}H_{43}NO_4$

Definition: PEG-2 Stearamide Carboxylic Acid is the organic acid that conforms generally to the formula:

$$CH_3(CH_2)_{16}C-NH(CH_2)_2OCH_2COOH$$
$$O$$

Chemical Classes: Alkoxylated Amides; Carboxylic Acids

Function: Surfactant - Cleansing Agent

Technical/Other Names:
Polyethylene Glycol 100 Stearamide
Carboxylic Acid
Polyoxyethylene (2) Stearamide Carboxylic
Acid

PEG-9 STEARAMIDE CARBOXYLIC ACID

CAS No.: 90453-59-1

Empirical Formula:
$C_{36}H_{71}NO_{11}$

Definition: PEG-9 Stearamide Carboxylic Acid is the organic acid that conforms generally to the formula:

$$CH_3(CH_2)_{16}C-NH(CH_2)_nOCH_2COOH$$
$$O$$

where n has an average value of 8.

Chemical Classes: Alkoxylated Amides; Carboxylic Acids

Function: Surfactant - Cleansing Agent

Technical/Other Names:
Polyethylene Glycol 450 Stearamide
Carboxylic Acid
Polyoxyethylene (9) Stearamide Carboxylic
Acid

Trade Name:
Akypo Muls 400 (Kao GmbH)

Trade Name Mixture:
Hydromyristenol N (2/014090) (Symrise)

PEG-2 STEARAMINE

CAS Nos.	**EINECS No.**
9003-93-4 (Generic)	
10213-78-2	233-520-3

JPN Translation:
PEG - 2 ステアラミン

Empirical Formula:
$C_{22}H_{47}NO_2$

Definition: PEG-2 Stearamine is the polyethylene glycol derivative of stearyl amine that conforms to the formula:

$$CH_3(CH_2)_{17}-N \begin{array}{c} (CH_2CH_2O)_xH \\ (CH_2CH_2O)_yH \end{array}$$

where x + y has an average value of 2.

Information Sources: JCIC, JCLS

Chemical Class: Alkoxylated Amines

Functions: Antistatic Agent; Surfactant - Foam Booster

Technical/Other Names:
N,N-Bis(2-Hydroxyethyl)-N-Octadecylamine
Ethanol, 2,2'-(Octadecylimino)Bis-
2,2'-(Octadecylimino)Bisethanol
Polyethylene Glycol 100 Stearyl Amine
Polyoxyethylene Stearylamine
Polyoxyethylene (2) Stearyl Amine
N-Stearyldiethanolamine

Trade Names:
Chemeen 18-2 (Chemax)
Ethomeen 18/12 (Akzo Nobel)
Ethox SAM-2 (Ethox)
Hetoxamine ST-2 (Heterene)
Jeetox HTA-2 (Jeen)
Protox HTA-2 (Protameen)

PEG-5 STEARAMINE

CAS No.: 9003-93-4 (Generic)

Empirical Formula:
$C_{28}H_{59}NO_5$

Definition: PEG-5 Stearamine is the polyethylene glycol derivative of stearyl amine that conforms to the formula:

$$CH_3(CH_2)_{17}-N \begin{array}{c} (CH_2CH_2O)_xH \\ (CH_2CH_2O)_yH \end{array}$$

where x + y has an average value of 5.

Chemical Class: Alkoxylated Amines

Functions: Antistatic Agent; Surfactant - Emulsifying Agent

Reported Product Categories: Hair Conditioners; Hair Rinses (Non-coloring)

Technical/Other Names:
Polyethylene Glycol (5) Stearyl Amine
Polyoxyethylene (5) Stearyl Amine

Trade Names:
Chemeen 18-5 (Chemax)
Ethomeen 18/15 (Akzo Nobel)
Hetoxamine ST-5 (Heterene)
Jeetox HTA-5 (Jeen)
Nikkol Tamns-5 (Nikko)
Protox HTA-5 (Protameen)

PEG-10 STEARAMINE

CAS No.: 9003-93-4 (Generic)

Empirical Formula:
$C_{38}H_{79}NO_{10}$

Definition: PEG-10 Stearamine is the polyethylene glycol derivative of stearyl amine that conforms to the formula:

$$CH_3(CH_2)_{17}-N \begin{array}{c} (CH_2CH_2O)_xH \\ (CH_2CH_2O)_yH \end{array}$$

where x + y has an average value of 10.

Chemical Class: Alkoxylated Amines

Functions: Antistatic Agent; Surfactant - Emulsifying Agent

Technical/Other Names:
Polyethylene Glycol 500 Stearyl Amine
Polyoxyethylene (10) Stearyl Amine

Trade Names:
Ethox SAM-10 (Ethox)
Jeetox HTA-10 (Jeen)
Nikkol Tamns-10 (Nikko)
Protox HTA-10 (Protameen)

PEG-15 STEARAMINE

CAS No.: 9003-93-4 (Generic)

Definition: PEG-15 Stearamine is the poly-ethylene glycol derivative of stearyl amine that conforms to the formula:

$$CH_3(CH_2)_{17}-N\begin{matrix} (CH_2CH_2O)_xH \\ (CH_2CH_2O)_yH \end{matrix}$$

where x + y has an average value of 15.

Chemical Class: Alkoxylated Amines

Functions: Antistatic Agent; Surfactant - Emulsifying Agent

Technical/Other Names:
Polyethylene Glycol (15) Stearyl Amine
Polyoxyethylene (15) Stearyl Amine

Trade Names:
Ethomeen 18/25 (Akzo Nobel)
Hetoxamine ST-15 (Heterene)
Jeetox HTA-15 (Jeen)
Nikkol Tamns-15 (Nikko)
Protox HTA-15 (Protameen)

PEG-50 STEARAMINE

CAS No.: 9003-93-4 (Generic)

Definition: PEG-50 Stearamine is the poly-ethylene glycol derivative of stearyl amine that conforms to the formula:

$$CH_3(CH_2)_{17}-N\begin{matrix} (CH_2CH_2O)_xH \\ (CH_2CH_2O)_yH \end{matrix}$$

where x + y has an average value of 50.

Information Source: CTFA D

Chemical Class: Alkoxylated Amines

Functions: Antistatic Agent; Surfactant - Solubilizing Agent

Technical/Other Names:
Polyethylene Glycol (50) Stearyl Amine
Polyoxyethylene (50) Stearyl Amine

Trade Names:
Chemeen 18-50 (Chemax)
Ethomeen 18/60 (Akzo Nobel)
Ethox SAM-50 (Ethox)
Hetoxamine ST-50 (Heterene)
Jeetox HTA-50 (Jeen)
Protox HTA-50 (Protameen)

PEG-2 STEARATE

CAS Nos.
106-11-6
9004-99-3 (Generic)

EINECS No.
203-363-5

JPN Translation:
ステアリン酸 PEG - 2

Empirical Formula:
$C_{22}H_{44}O_4$

Definition: PEG-2 Stearate is the poly-ethylene glycol ester of stearic acid that conforms to the formula:

$$CH_3(CH_2)_{16}\overset{O}{\overset{||}{C}}-(OCH_2CH_2)_nOH$$

where n has an average value of 2.

Information Sources: 21CFR175.300, 21CFR176.200, 21CFR176.210, 21CFR177.2800, CIR: [S] JACT-2(7)1983, CTFA D, JCIC, JCLS, JSQI, MI-13(7660), TSCA

Chemical Class: Alkoxylated Carboxylic Acids

Function: Surfactant - Emulsifying Agent

Reported Product Categories: Moisturizing Preparations; Night Skin Care Preparations; Skin Care Preparations, Misc.; Bath Preparations, Misc.; Bath Oils, Tablets, and Salts; Cleansing Products (Cold Creams, Cleansing Lotions, Liquids and Pads); Body and Hand Preparations (Excluding Shaving Preparations); Foundations; Paste Masks (Mud Packs); Aftershave Lotions; Baby Shampoos

Technical/Other Names:
Diethylene Glycol Monostearate
Diglycol Stearate
Octadecanoic Acid, 2-(2-Hydroxyethoxy)-Ethyl Ester
Polyethylene Glycol 100 Monostearate
Polyoxyethylene (2) Monostearate

Trade Names:
AEC PEG-2 Stearate (A & E Connock)
Cithrol DGMS N/E (Croda Chemicals)
DUB SDEG (Stearinerie Dubois Fils)
ESTOL 3710 (Uniqema Europe)
Glycosterine (Vevy)
Hydrine (Gattefosse s.a.)
Jeechem DGS (Jeen)
Lipo DGS (Lipo)
Nikkol DEGS (Nikko)
Nikkol MYS-2 (Nikko)
Prodhybase A (Prod'Hyg)
Protachem DGS (Protameen)
Radiasurf 7410 (Oleon NV)
ROL DGE (Fabriquimica)
STEPAN DGMS (Stepan)
STEPAN DGS NEUTRAL (Stepan)

Trade Name Mixtures:
Base RW 135 (LCW)
Base RW 136 (LCW)
Lipopeg 2-DEGS (Lipo)
Monamilk (Argeville)
Pegosperse 100 S (Lonza Inc./Lonza Ltd.)
Sedefos 75 (Gattefosse s.a.)

PEG-3 STEARATE

CAS Nos.
9004-99-3 (Generic)
10233-24-6

EINECS No.
233-562-2

JPN Translation:
ステアリン酸 PEG - 3

Empirical Formula:
$C_{24}H_{48}O_5$

Definition: PEG-3 Stearate is the poly-ethylene glycol ester of stearic acid that conforms generally to the formula:

$$CH_3(CH_2)_{16}\overset{O}{\overset{||}{C}}-(OCH_2CH_2)_nOH$$

where n has an average value of 3.

Information Sources: 21CFR175.300, JCLS, MI-13(7660)

Chemical Class: Alkoxylated Carboxylic Acids

Function: Surfactant - Emulsifying Agent

Technical/Other Names:
2-[2-(2-Hydroxyethoxy)Ethoxy]Ethyl Octadecanoate
Octadecanoic Acid, 2-[2-(2-Hydroxyethoxy)Ethoxy]Ethyl Ester
Polyethylene Glycol (3) Stearate
Polyoxyethylene (3) Stearate
Triethylene Glycol Monooctadecanoate
Triethylene Glycol Stearate

Trade Name:
Serdox NSG 100 (Sasol Servo)

PEG-4 STEARATE

CAS Nos.
106-07-0
9004-99-3 (Generic)

EINECS No.
203-358-8

JPN Translation:
ステアリン酸 PEG - 4

Empirical Formula:
$C_{26}H_{52}O_6$

Definition: PEG-4 Stearate is the poly-ethylene glycol ester of stearic acid that conforms to the formula:

$$CH_3(CH_2)_{16}\overset{O}{\overset{||}{C}}-(OCH_2CH_2)_nOH$$

where n has an average value of 4.

Information Sources: 21CFR175.105, 21CFR175.300, 21CFR176.210, JCLS, MI-13(7660), TSCA

Chemical Class: Alkoxylated Carboxylic Acids

Function: Surfactant - Emulsifying Agent

Reported Product Category: Hair Coloring Preparations, Misc.

Technical/Other Names:
2-[2-[2-[2-(2-Hydroxyethoxy)Ethoxy]Ethoxy]-Ethyl Octadecanoate

Octadecanoic Acid, 2-[2-[2-(2-Hydroxyeth-oxy)Ethoxy]Ethoxy]Ethyl Ester
Polyethylene Glycol 200 Monostearate
Polyoxyethylene (4) Monostearate
Tetraethylene Glycol, Monostearate

Trade Names:
Acconon 200 MS (Abitec Corporation)
AEC PEG-4 Stearate (A & E Connock)
Chemax E-200-MS (Chemax)
Jeemate 200-DPS (Jeen)
Nikkol MYS-4 (Nikko)
Pegosperse 200 MS (Lonza Inc./Lonza Ltd.)
Protamate 200 DPS (Protameen)
Sabowax SE 4 (Sabo)
Sipoic MS-5 (Specialty Industrial)
STEPAN PEG 200 MS (Stepan)
Unipeg-200 MS (Universal Preserv-A-Chem)

Trade Name Mixture:
Lipopeg 2-DEGS (Lipo)

PEG-5 STEARATE

CAS No.: 9004-99-3 (Generic)

JPN Translation:
ステアリン酸 PEG - 5

Empirical Formula:
$C_{28}H_{56}O_7$

Definition: PEG-5 Stearate is the poly-ethylene glycol ester of stearic acid that conforms to the formula:

$$CH_3(CH_2)_{16}\overset{\displaystyle O}{\overset{\|}{C}}\!\!-\!\!(OCH_2CH_2)_nOH$$

where n has an average value of 5.

Information Sources: 21CFR173.340, 21CFR175.105, 21CFR175.300, JCLS, MI-13(7660), TSCA

Chemical Class: Alkoxylated Carboxylic Acids

Function: Surfactant - Emulsifying Agent

Reported Product Categories: Bath Preparations, Misc.; Body and Hand Preparations (Excluding Shaving Preparations)

Technical/Other Names:
Polyethylene Glycol (5) Monostearate
Polyoxyethylene (5) Monostearate

Trade Names:
AEC PEG-5 Stearate (A & E Connock)
Hetoxamate SA-5 (Heterene)

PEG-6 STEARATE

CAS Nos.: 9004-99-3 (Generic); 10108-28-8

JPN Translation:
ステアリン酸 PEG - 6

Empirical Formula:
$C_{30}H_{60}O_8$

Definition: PEG-6 Stearate is the poly-ethylene glycol ester of stearic acid that conforms to the formula:

$$CH_3(CH_2)_{16}\overset{\displaystyle O}{\overset{\|}{C}}\!\!-\!\!(OCH_2CH_2)_nOH$$

where n has an average value of 6.

Information Sources: 21CFR175.105, 21CFR175.300, 21CFR176.210, CIR: [S] JACT-2(7)1983, JCLS, MI-13(7660), TSCA

Chemical Class: Alkoxylated Carboxylic Acids

Function: Surfactant - Emulsifying Agent

Reported Product Categories: Skin Care Preparations, Misc.; Cleansing Products (Cold Creams, Cleansing Lotions, Liquids and Pads); Moisturizing Preparations; Night Skin Care Preparations

Technical/Other Names:
Hexaethylene Glycol, Monostearate
17-Hydroxy-3,6,9,12,15-Pentaoxaheptadec-1-yl Octadecanoate
Octadecanoic Acid, 17-Hydroxy-3,6,9,12, 15-Pentaoxaheptadec-1-yl Ester
Polyethylene Glycol 300 Monostearate
Polyoxyethylene (6) Monostearate

Trade Names:
AEC PEG-6 Stearate (A & E Connock)
Hetoxamate SA-7 (Heterene)
Jeemate 300-DPS (Jeen)
Polystate B (Gattefosse s.a.)
Polystate C (Gattefosse s.a.)
Prodhybase Pga (Prod'Hyg)
Protamate 300 DPS (Protameen)
Sabowax SE 6 (Sabo)
Serdox NSG 300 (Sasol Servo)
STEPAN PEG 300 MS (Stepan)
Superpolystate (Gattefosse s.a.)

Trade Name Mixtures:
LIPOPEG 15-S (Lipo)
Pegosperse 1500 MS, B (Lonza Inc./Lonza Ltd.)
Pegosperse 1500 MS Grade B (Lonza Inc./Lonza Ltd.)
Prodhy 206 (Prod'Hyg)
Prodhybase 1500 (Prod'Hyg)
Sabowax FL 84 (Sabo)
Tefose 1500 (Gattefosse s.a.)
Tefose 2000 (Gattefosse s.a.)
Tefose 2561 (Gattefosse s.a.)

PEG-7 STEARATE

CAS No.: 9004-99-3 (Generic)

JPN Translation:
ステアリン酸 PEG - 7

Empirical Formula:
$C_{32}H_{64}O_9$

Definition: PEG-7 Stearate is the poly-ethylene glycol ester of stearic acid that conforms to the formula:

$$CH_3(CH_2)_{16}\overset{\displaystyle O}{\overset{\|}{C}}\!\!-\!\!(OCH_2CH_2)_nOH$$

where n has an average value of 7.

Information Sources: 21CFR175.105, 21CFR175.300, JCLS, MI-13(7660), TSCA, USAN

Chemical Class: Alkoxylated Carboxylic Acids

Function: Surfactant - Emulsifying Agent

Technical/Other Names:
Polyethylene Glycol (7) Monostearate
Polyoxyethylene (7) Monostearate

Trade Name:
AEC PEG-7 Stearate (A & E Connock)

PEG-8 STEARATE

CAS Nos.: 9004-99-3 (Generic); 70802-40-3

JPN Translation:
ステアリン酸 PEG - 8

Empirical Formula:
$C_{34}H_{68}O_{10}$

Definition: PEG-8 Stearate is the poly-ethylene glycol ester of stearic acid that conforms to the formula:

$$CH_3(CH_2)_{16}\overset{\displaystyle O}{\overset{\|}{C}}\!\!-\!\!(OCH_2CH_2)_nOH$$

where n has an average value of 8.

Information Sources: AUS, BAN, 21CFR175.105, 21CFR175.300, 21CFR176.170, 21CFR176.200, 21CFR176.210, 21CFR177.1200, 21CFR177.1210, 21CFR177.2260, 21CFR177.2800, 21CFR178.3910, CIR: [S] JACT-2(7)1983, CTFA S, DA, HUN, INN, JCLS, MAR, MI-13(7660), TSCA, USAN, USP XXIV

Chemical Class: Alkoxylated Carboxylic Acids

Function: Surfactant - Emulsifying Agent

Reported Product Categories: Moisturizing Preparations; Body and Hand Preparations (Excluding Shaving Preparations); Hair Conditioners

Technical/Other Names:
23-Hydroxy-3,6,9,12,15,18,21-Heptaoxa-tricos-1-yl Octadecanoate

Macrogol Stearate 400
Octadecanoic Acid, 23-Hydroxy-3,6,9,12,
15,18,21-Heptaoxatricos-1-yl Ester
Octaethylene Glycol Stearate
Polyethylene Glycol 400 Monostearate
Polyoxyethylene (8) Monostearate
Polyoxyl 8 Stearate

Trade Names:
Acconon 400 MS (Abitec Corporation)
AEC PEG-8 Stearate (A & E Connock)
Chemax E-400-MS (Chemax)
Cithrol 4MS (Croda Chemicals)
DeTHOX ACID S-8 (DeForest)
ESTOL 3723 (Uniqema Europe)
Ethox MS-8 (Ethox)
Hetoxamate SA-9 (Heterene)
Jeemate 400-DPS (Jeen)
Lasemul 400 E (Industrial Quimica)
LIPOPEG 4-S (Lipo)
Lumulse 40-S (Lambent)
Myrj 45 (Uniqema Americas)
Pegosperse 400 MS (Lonza Inc./Lonza
Ltd.)
Prodhybase 400 (Prod'Hyg)
Protamate 400 DPS (Protameen)
Radiasurf 7473 (Oleon NV)
ROL 400 (Fabriquimica)
Rol E 40 (Fabriquimica)
Sabowax SE 8 (Sabo)
Simulsol M 45 (SEPPIC)
Sipoic MS-9 (Specialty Industrial)
STEPAN PEG 400 MS (Stepan)
Sympatens-BS/080 (Kolb)
Unipeg-400 MS (Universal Preserv-A-
Chem)

Trade Name Mixture:
Lanola 90 (Lanaetex)

PEG-9 STEARATE

CAS Nos.   **EINECS No.**
5349-52-0   226-312-9
9004-99-3 (Generic)

JPN Translation:
ステアリン酸 PEG - 9

Empirical Formula:
$C_{36}H_{72}O_{11}$

Definition: PEG-9 Stearate is the polyethylene glycol ester of stearic acid that
conforms to the formula:

$$CH_3(CH_2)_{16}\overset{\displaystyle O}{\overset{\|}{C}}\!\!-\!\!(OCH_2CH_2)_nOH$$

where n has an average value of 9.

Information Sources: 21CFR175.105,
21CFR175.300, 21CFR177.2260,
21CFR177.2800, JCLS, MI-13(7660), TSCA

Chemical Class: Alkoxylated Carboxylic
Acids

Function: Surfactant - Emulsifying Agent

Technical/Other Names:
Nonaethylene Glycol Monostearate
Polyethylene Glycol 450 Monostearate
Polyoxyethylene (9) Monostearate

PEG-10 STEARATE

CAS No.: 9004-99-3 (Generic)

JPN Translation:
ステアリン酸 PEG - 10

Empirical Formula:
$C_{38}H_{76}O_{12}$

Definition: PEG-10 Stearate is the polyethylene glycol ester of stearic acid that
conforms to the formula:

$$CH_3(CH_2)_{16}\overset{\displaystyle O}{\overset{\|}{C}}\!\!-\!\!(OCH_2CH_2)_nOH$$

where n has an average value of 10.

Information Sources: 21CFR175.105,
21CFR175.300, 21CFR177.2260,
21CFR177.2800, JCLS, MI-13(7660), TSCA

Chemical Class: Alkoxylated Carboxylic
Acids

Function: Surfactant - Emulsifying Agent

Technical/Other Names:
Polyethylene Glycol 500 Monostearate
Polyoxyethylene (10) Monostearate

Trade Names:
AEC PEG-10 Stearate (A & E Connock)
Nikkol MYS-10 (Nikko)

Trade Name Mixture:
Nikkol MGS-DEX (Nikko)

PEG-12 STEARATE

CAS No.: 9004-99-3 (Generic)

JPN Translation:
ステアリン酸 PEG - 12

Definition: PEG-12 Stearate is the polyethylene glycol ester of stearic acid that
conforms to the formula:

$$CH_3(CH_2)_{16}\overset{\displaystyle O}{\overset{\|}{C}}\!\!-\!\!(OCH_2CH_2)_nOH$$

where n has an average value of 12.

Information Sources: 21CFR175.105,
21CFR175.300, 21CFR176.170,
21CFR176.210, 21CFR177.1200,
21CFR177.2260, 21CFR177.2800, CIR: [S]
JACT-2(7)1983, CTFA S, JCLS, MI-13
(7660), TSCA

Chemical Class: Alkoxylated Carboxylic
Acids

Function: Surfactant - Emulsifying Agent

Reported Product Categories: Shaving
Cream (Aerosol, Brushless and Lather); Bath
Preparations, Misc.; Body and Hand Prepa-
rations (Excluding Shaving Preparations)

Technical/Other Names:
Polyethylene Glycol 600 Monostearate
Polyoxyethylene (12) Monostearate

Trade Names:
AEC PEG-12 Stearate (A & E Connock)
Chemax E-600-MS (Chemax)
Cithrol 6MS (Croda Chemicals)
Ethox MS-14 (Ethox)
Hetoxamate SA-13 (Heterene)
Jeemate 600-DPS (Jeen)
Pegosperse 600 MS (Lonza Inc./Lonza
Ltd.)
Prodhybase 600 (Prod'Hyg)
Protamate 600 DPS (Protameen)
Radiasurf 7414 (Oleon NV)
STEPAN PEG 600 MS (Stepan)
Unipeg-600 MS (Universal Preserv-A-
Chem)

Trade Name Mixture:
Upiwax 163 (Universal Preserv-A-Chem)

PEG-14 STEARATE

CAS Nos.   **EINECS No.**
9004-99-3 (Generic)
10289-94-8   233-641-1

JPN Translation:
ステアリン酸 PEG - 14

Empirical Formula:
$C_{46}H_{92}O_{16}$

Definition: PEG-14 Stearate is the polyethylene glycol ester of stearic acid that
conforms to the formula:

$$CH_3(CH_2)_{16}\overset{\displaystyle O}{\overset{\|}{C}}\!\!-\!\!(OCH_2CH_2)_nOH$$

where n has an average value of 14.

Information Sources: 21CFR175.105,
21CFR175.300, 21CFR177.2260,
21CFR177.2800, JCLS, MI-13(7660), TSCA

Chemical Class: Alkoxylated Carboxylic
Acids

Function: Surfactant - Emulsifying Agent

Technical/Other Names:
41-Hydroxy-3,6,9,12,15,18,21,24,-27,30,
33,36,39-Tridecaoxahentetracont-1-yl
Octadecanoate
Octadecanoic Acid, 41-Hydroxy-3,6,9,12,
15,18,21,24,27,-30,33,36,39-
Tridecaoxahentetracont-1-yl Ester
Polyethylene Glycol (14) Monostearate

Polyoxyethylene (14) Monostearate
Tetradecaethylene Glycol Monostearate

Trade Name:
AEC PEG-14 Stearate (A & E Connock)

PEG-15 STEARATE

CAS No.: 9004-99-3 (Generic)

Definition: PEG-15 Stearate is the polyethylene glycol ester of stearic acid that conforms generally to the formula:

$$CH_3(CH_2)_{16}\overset{\displaystyle O}{\overset{\displaystyle \|}{C}}\!-\!(OCH_2CH_2)_nOH$$

where n has an average value of 15.

Chemical Class: Alkoxylated Carboxylic Acids

Function: Surfactant - Emulsifying Agent

Technical/Other Name:
Polyoxyethylene (15) Monostearate

Trade Name:
Emalex 815 (Nihon Emulsion)

PEG-18 STEARATE

CAS No.: 9004-99-3 (Generic)

JPN Translation:
ステアリン酸 PEG - 18

Definition: PEG-18 Stearate is the polyethylene glycol ester of stearic acid that conforms to the formula:

$$CH_3(CH_2)_{16}\overset{\displaystyle O}{\overset{\displaystyle \|}{C}}\!-\!(OCH_2CH_2)_nOH$$

where n has an average value of 18.

Information Sources: 21CFR175.300, 21CFR177.2260, 21CFR177.2800, JCLS, MI-13(7660), TSCA

Chemical Class: Alkoxylated Carboxylic Acids

Function: Surfactant - Emulsifying Agent

Technical/Other Names:
Polyethylene Glycol (18) Monostearate
Polyoxyethylene (18) Monostearate

Trade Name:
AEC PEG-18 Stearate (A & E Connock)

PEG-20 STEARATE

CAS No.: 9004-99-3 (Generic)

JPN Translation:
ステアリン酸 PEG - 20

Definition: PEG-20 Stearate is the polyethylene glycol ester of stearic acid that conforms to the formula:

$$CH_3(CH_2)_{16}\overset{\displaystyle O}{\overset{\displaystyle \|}{C}}\!-\!(OCH_2CH_2)_nOH$$

where n has an average value of 20.

Information Sources: 21CFR175.300, 21CFR176.210, 21CFR177.2260, 21CFR177.2800, CIR: [S] JACT-2(7)1983, CTFA S, JCLS, MI-13(7660), TSCA

Chemical Class: Alkoxylated Carboxylic Acids

Functions: Surfactant - Cleansing Agent; Surfactant - Emulsifying Agent; Surfactant - Solubilizing Agent

Reported Product Categories: Moisturizing Preparations; Hair Conditioners; Bath Oils, Tablets, and Salts; Bath Preparations, Misc.; Body and Hand Preparations (Excluding Shaving Preparations); Cleansing Products (Cold Creams, Cleansing Lotions, Liquids and Pads); Paste Masks (Mud Packs); Personal Cleanliness Products, Misc.; Skin Care Preparations, Misc.

Technical/Other Names:
Polyethylene Glycol 1000 Monostearate
Polyoxyethylene (20) Monostearate

Trade Names:
AEC PEG-20 Stearate (A & E Connock)
Cerasynt 840 (International Specialty Products)
Chemax E-1000-MS (Chemax)
Cithrol 10MS (Croda Chemicals)
ESTOL 3727 (Uniqema Europe)
Hetoxamate SA-23 (Heterene)
Jeemate 1000-DPS (Jeen)
LIPOPEG 10-S (Lipo)
Lumulse 100-S (Lambent)
Myrj 49 (Uniqema Americas)
Protamate 1000 DPS (Protameen)
Sabowax SE 20 (Sabo)
Simulsol M 49 (SEPPIC)
STEPAN PEG 1000 MS (Stepan)
Sympatens-BS/200 (Kolb)
Unipeg-1000 MS (Universal Preserv-A-Chem)

Trade Name Mixtures:
AEC PEG-20 Stearate (&) Cetearyl Alcohol (A & E Connock)
Conditioner Base (Croda Chemicals)
Megasperse 1402 (Megachem)
Sabowax FL 20 (Sabo)

PEG-23 STEARATE

CAS No.: 9004-99-3 (Generic)

JPN Translation:
ステアリン酸 PEG - 23

Definition: PEG-23 Stearate is the polyethylene glycol ester of stearic acid that conforms to the formula:

$$CH_3(CH_2)_{16}\overset{\displaystyle O}{\overset{\displaystyle \|}{C}}\!-\!(OCH_2CH_2)_nOH$$

where n has an average value of 23.

Information Sources: JCLS, TSCA

Chemical Class: Alkoxylated Carboxylic Acids

Functions: Surfactant - Cleansing Agent; Surfactant - Solubilizing Agent

Technical/Other Names:
Polyethylene Glycol (23) Monostearate
Polyoxyethylene (23) Stearate

Trade Name:
Ethox MS-23 (Ethox)

PEG-25 STEARATE

CAS No.: 9004-99-3 (Generic)

JPN Translation:
ステアリン酸 PEG - 25

Definition: PEG-25 Stearate is the polyethylene glycol ester of stearic acid that conforms to the formula:

$$CH_3(CH_2)_{16}\overset{\displaystyle O}{\overset{\displaystyle \|}{C}}\!-\!(OCH_2CH_2)_nOH$$

where n has an average value of 25.

Information Sources: 21CFR175.300, 21CFR177.2260, 21CFR177.2800, JCLS, MI-13(7660), TSCA

Chemical Class: Alkoxylated Carboxylic Acids

Functions: Surfactant - Cleansing Agent; Surfactant - Solubilizing Agent

Technical/Other Names:
Polyethylene Glycol (25) Monostearate
Polyoxyethylene (25) Monostearate

Trade Names:
AEC PEG-25 Stearate (A & E Connock)
Nikkol MYS-25 (Nikko)

PEG-30 STEARATE

CAS No.: 9004-99-3 (Generic)

JPN Translation:
ステアリン酸 PEG - 30

Definition: PEG-30 Stearate is the polyethylene glycol ester of stearic acid that conforms to the formula:

$$CH_3(CH_2)_{16}\overset{\displaystyle O}{\overset{\displaystyle \|}{C}}\!-\!(OCH_2CH_2)_nOH$$

where n has an average value of 30.

Information Sources: 21CFR175.300, 21CFR177.2260, 21CFR177.2800, CTFA D, JCLS, MI-13(7660), TSCA

Chemical Class: Alkoxylated Carboxylic Acids

Functions: Surfactant - Cleansing Agent; Surfactant - Solubilizing Agent

Technical/Other Names:
Polyethylene Glycol (30) Monostearate
Polyoxyethylene (30) Monostearate

Trade Names:
AEC PEG-30 Stearate (A & E Connock)
Myrj 51 (Uniqema Americas)
Pegosperse 1500 MS (Lonza Inc./Lonza Ltd.)
Radiasurf 7417 (Oleon NV)
Sympatens-BS/300 (Kolb)

Trade Name Mixtures:
Cutina E 5 (Cognis Care Chemicals/NJ)
Cutina E 5 (Cognis Deutschland)

PEG-32 STEARATE

CAS No.: 9004-99-3 (Generic)

JPN Translation:
ステアリン酸 PEG - 32

Definition: PEG-32 Stearate is the polyethylene glycol ester of stearic acid that conforms to the formula:

$$CH_3(CH_2)_{16}C\overset{O}{\overset{\|}{}}{-}(OCH_2CH_2)_nOH$$

where n has an average value of 32.

Information Sources: 21CFR175.300, 21CFR176.210, 21CFR177.2260, 21CFR177.2800, CIR: [S] JACT-2(7)1983, JCLS, MI-13(7660), TSCA

Chemical Class: Alkoxylated Carboxylic Acids

Functions: Surfactant - Cleansing Agent; Surfactant - Solubilizing Agent

Reported Product Categories: Moisturizing Preparations; Cleansing Products (Cold Creams, Cleansing Lotions, Liquids and Pads); Face and Neck Preparations (Excluding Shaving Preparations)

Technical/Other Names:
Polyethylene Glycol 1540 Monostearate
Polyoxyethylene (32) Monostearate

Trade Names:
AEC PEG-32 Stearate (A & E Connock)
Jeemate 1540-DPS (Jeen)
Prodhybase 1540 (Prod'Hyg)
STEPAN PEG 1540 MS (Stepan)
Unipeg-1540 MS (Universal Preserv-A-Chem)

Trade Name Mixtures:
LIPOPEG 15-S (Lipo)
Pegosperse 1500 MS, B (Lonza Inc./Lonza Ltd.)
Pegosperse 1500 MS Grade B (Lonza Inc./Lonza Ltd.)
Prodhybase 1500 (Prod'Hyg)
Tefose 1500 (Gattefosse s.a.)

PEG-35 STEARATE

CAS No.: 9004-99-3 (Generic)

JPN Translation:
ステアリン酸 PEG - 35

Definition: PEG-35 Stearate is the polyethylene glycol ester of stearic acid that conforms to the formula:

$$CH_3(CH_2)_{16}C\overset{O}{\overset{\|}{}}{-}(OCH_2CH_2)_nOH$$

where n has an average value of 35.

Information Sources: 21CFR175.300, 21CFR177.2260, 21CFR177.2800, JCLS, MI-13(7660), TSCA

Chemical Class: Alkoxylated Carboxylic Acids

Functions: Surfactant - Cleansing Agent; Surfactant - Solubilizing Agent

Technical/Other Names:
Polyethylene Glycol (35) Monostearate
Polyoxyethylene (35) Monostearate

Trade Names:
AEC PEG-35 Stearate (A & E Connock)
Hetoxamate SA-35 (Heterene)

PEG-36 STEARATE

CAS No.: 9004-99-3 (Generic)

JPN Translation:
ステアリン酸 PEG - 36

Definition: PEG-36 Stearate is the polyethylene glycol ester of stearic acid that conforms to the formula:

$$CH_3(CH_2)_{16}C\overset{O}{\overset{\|}{}}{-}(OCH_2CH_2)_nOH$$

where n has an average value of 36.

Information Sources: 21CFR175.300, 21CFR177.2260, 21CFR177.2800, JCLS, MI-13(7660), TSCA

Chemical Class: Alkoxylated Carboxylic Acids

Functions: Surfactant - Cleansing Agent; Surfactant - Solubilizing Agent

Technical/Other Names:
Polyethylene Glycol 1800 Monostearate
Polyoxyethylene (36) Monostearate

Trade Name:
AEC PEG-36 Stearate (A & E Connock)

PEG-40 STEARATE

CAS No.: 9004-99-3 (Generic)

JPN Translation:
ステアリン酸 PEG - 40

Definition: PEG-40 Stearate is the polyethylene glycol ester of stearic acid that conforms to the formula:

$$CH_3(CH_2)_{16}C\overset{O}{\overset{\|}{}}{-}(OCH_2CH_2)_nOH$$

where n has an average value of 40.

Information Sources: BAN, BRA, 21CFR173.340, 21CFR175.105, 21CFR175.300, 21CFR176.200, 21CFR176.210, 21CFR177.2260, 21CFR177.2800, CIR: [S] JACT-2(7)1983, CTFA S, JAN, JCLS, JP, MAR, NF XIX, TSCA, USAN, USD

Chemical Class: Alkoxylated Carboxylic Acids

Functions: Surfactant - Cleansing Agent; Surfactant - Solubilizing Agent

Reported Product Categories: Moisturizing Preparations; Bath Preparations, Misc.; Body and Hand Preparations (Excluding Shaving Preparations); Aftershave Lotions; Baby Shampoos; Skin Care Preparations, Misc.; Bath Oils, Tablets, and Salts; Cleansing Products (Cold Creams, Cleansing Lotions, Liquids and Pads); Bath Capsules; Face and Neck Preparations (Excluding Shaving Preparations); Perfumes; Paste Masks (Mud Packs); Foundations; Hair Conditioners; Night Skin Care Preparations; Fragrance Preparations, Misc.; Eyeliners; Indoor Tanning Preparations; Mascara; Personal Cleanliness Products, Misc.

Technical/Other Names:
·Macrogol Stearate 2000
Polyethylene Glycol 2000 Monostearate
Polyoxyethylene (40) Monostearate
Polyoxyl 40 Stearate
Stearethate 40

Trade Names:
AEC PEG-40 Stearate (A & E Connock)
Crodet S40 (Croda Chemicals)
Emerest 2715 (Cognis Care Chemicals/NJ)
Emerest 2715 (Cognis Care Chemicals/PA)
Ethox MS-40 (Ethox)
Hetoxamate SA-40 (Heterene)
Jeemate 2000-DPS (Jeen)

The inclusion of any compound in the *Dictionary and Handbook* does not indicate that use of that substance as a cosmetic ingredient complies with the laws and regulations governing such use in the United States or any other country.

Lanoxide-52 (Lanaetex)
Lipopeg 39-S (Lipo)
Lumulse POE (40) MS (Lambent)
Myrj 52 (Uniqema Americas)
Myrj 52S (Uniqema Americas)
Nikkol MYS-40 (Nikko)
Pegosperse 1750 MS (Lonza Inc./Lonza Ltd.)
Protamate 1540 DPS (Protameen)
Protamate 2000 DPS (Protameen)
Ritox 52 (RITA)
ROL 52 (Fabriquimica)
Sabowax SE 40 (Sabo)
Simulsol M 52 (SEPPIC)
Sipoic MS-40 (Specialty Industrial)
Sympatens-BS/400 (Kolb)
Tego Acid S 40 P (Degussa Care Specialties)
Unipeg-S-40 (Universal Preserv-A-Chem)

Trade Name Mixtures:
AF75 (GE Silicones)
Ceral ML (Fabriquimica)
Ritachol 5000 (RITA)

PEG-45 STEARATE

CAS No.: 9004-99-3 (Generic)

JPN Translation:
ステアリン酸 PEG - 45

Definition: PEG-45 Stearate is the polyethylene glycol ester of stearic acid that conforms to the formula:

$$CH_3(CH_2)_{16}C \overset{O}{\overset{\|}{\quad}} (OCH_2CH_2)_nOH$$

where n has an average value of 45.

Information Sources: 21CFR175.300, 21CFR177.2260, 21CFR177.2800, JCLS, MI-13(7660), TSCA

Chemical Class: Alkoxylated Carboxylic Acids

Functions: Surfactant - Cleansing Agent; Surfactant - Solubilizing Agent

Technical/Other Names:
Polyethylene Glycol (45) Monostearate
Polyoxyethylene (45) Monostearate

Trade Names:
AEC PEG-45 Stearate (A & E Connock)
Nikkol MYS-45 (Nikko)

PEG-50 STEARATE

CAS No.: 9004-99-3 (Generic)

JPN Translation:
ステアリン酸 PEG - 50

Definition: PEG-50 Stearate is the polyethylene glycol ester of stearic acid that conforms to the formula:

$$CH_3(CH_2)_{16}C \overset{O}{\overset{\|}{\quad}} (OCH_2CH_2)_nOH$$

where n has an average value of 50.

Information Sources: 21CFR175.300, 21CFR177.2260, 21CFR177.2800, CIR: [S] JACT-2(7)1983, CTFA D, JCLS, MI-13 (7660), NF XVIII, TSCA, USAN

Chemical Class: Alkoxylated Carboxylic Acids

Function: Surfactant - Cleansing Agent

Reported Product Categories: Moisturizing Preparations; Body and Hand Preparations (Excluding Shaving Preparations)

Technical/Other Names:
Polyethylene Glycol (50) Monostearate
Polyoxyethylene (50) Monostearate

Trade Names:
AEC PEG-50 Stearate (A & E Connock)
Lanoxide-53 (Lanaetex)
Myrj 53 (Uniqema Americas)
Ritox 53 (RITA)
ROL 53 (Fabriquimica)
Sympatens-BS/500 (Kolb)

PEG-55 STEARATE

CAS No.: 9004-99-3 (Generic)

Definition: PEG-55 Stearate is the polyethylene glycol ester of stearic acid that conforms generally to the formula:

$$CH_3(CH_2)_{16}C \overset{O}{\overset{\|}{\quad}} (OCH_2CH_2)_nOH$$

where n has an average value of 55.

Information Source: JCLS

Chemical Class: Alkoxylated Carboxylic Acids

Function: Surfactant - Cleansing Agent

Reported Product Category: Eyeliners

Technical/Other Names:
Polyethylene Glycol (55) Monostearate
Polyoxyethylene (55) Monostearate

Trade Name:
Nikkol MYS-55 (Nikko)

PEG-75 STEARATE

CAS No.: 9004-99-3 (Generic)

JPN Translation:
ステアリン酸 PEG - 75

Definition: PEG-75 Stearate is the polyethylene glycol ester of stearic acid that conforms to the formula:

$$CH_3(CH_2)_{16}C \overset{O}{\overset{\|}{\quad}} (OCH_2CH_2)_nOH$$

where n has an average value of 75.

Information Sources: 21CFR175.300, JCLS, MI-13(7660), TSCA

Chemical Class: Alkoxylated Carboxylic Acids

Function: Surfactant - Cleansing Agent

Technical/Other Names:
Polyethylene Glycol 4000 Monostearate
Polyoxyethylene (75) Monostearate

Trade Names:
AEC PEG-75 Stearate (A & E Connock)
Jeemate 4000-DPS (Jeen)
Prodhybase 4000 (Prod'Hyg)
Protamate 4000 DPS (Protameen)
STEPAN PEG 4000 MS (Stepan)
Unipeg-4000 MS (Universal Preserv-A-Chem)

Trade Name Mixtures:
Emulium Delta (Gattefosse s.a.)
Gelot 64 (Gattefosse s.a.)

PEG-90 STEARATE

CAS No.: 9004-99-3 (Generic)

JPN Translation:
ステアリン酸 PEG - 90

Definition: PEG-90 Stearate is the polyethylene glycol ester of stearic acid that conforms to the formula:

$$CH_3(CH_2)_{16}C \overset{O}{\overset{\|}{\quad}} (OCH_2CH_2)_nOH$$

where n has an average value of 90.

Information Sources: 21CFR175.300, JCLS, MI-13(7660), TSCA

Chemical Class: Alkoxylated Carboxylic Acids

Function: Surfactant - Cleansing Agent

Technical/Other Names:
Polyethylene Glycol (90) Monostearate
Polyoxyethylene (90) Monostearate

Trade Names:
AEC PEG-90 Stearate (A & E Connock)
Hetoxamate SA-90 (Heterene)

PEG-100 STEARATE

CAS No.: 9004-99-3 (Generic)

JPN Translation:
ステアリン酸 PEG - 100

Definition: PEG-100 Stearate is the polyethylene glycol ester of stearic acid that conforms to the formula:

$$CH_3(CH_2)_{16}\overset{\displaystyle O}{\overset{\displaystyle \|}{C}}\!\!-\!\!(OCH_2CH_2)_nOH$$

where n has an average value of 100.

Information Sources: 21CFR175.300, 21CFR176.210, CIR: [S] JACT-2(7)1983, CTFA D, JCLS, MI-13(7660), TSCA

Chemical Class: Alkoxylated Carboxylic Acids

Function: Surfactant - Cleansing Agent

Reported Product Categories: Moisturizing Preparations; Bath Oils, Tablets, and Salts; Cleansing Products (Cold Creams, Cleansing Lotions, Liquids and Pads); Skin Care Preparations, Misc.; Bath Preparations, Misc.; Body and Hand Preparations (Excluding Shaving Preparations); Hair Conditioners; Bath Capsules; Face and Neck Preparations (Excluding Shaving Preparations); Personal Cleanliness Products, Misc.; Paste Masks (Mud Packs); Night Skin Care Preparations; Eyeliners; Indoor Tanning Preparations; Eyebrow Pencils; Fragrance Preparations, Misc.; Suntan Gels, Creams, and Liquids; Mascara; Eye Shadows; Suntan Preparations, Misc.; Eye Makeup Preparations, Misc.; Foundations; Baby Lotions, Oils, Powders and Creams; Hair Preparations (Non-coloring), Misc.; Aftershave Lotions; Deodorants (Underarm); Eye Lotions

Technical/Other Names:
Polyethylene Glycol 100 Monostearate
Polyoxyethylene (100) Monostearate

Trade Names:
AEC PEG-100 Stearate (A & E Connock)
Crodet S100 (Croda Chemicals)
Emerest 2717 (Cognis Care Chemicals/PA)
Jeemate 4400-DPS (Jeen)
Lanoxide-59 (Lanaetex)
LIPOPEG-100-S (Lipo)
Myrj 59 (Uniqema Americas)
Protamate 4400 DPS (Protameen)
Ritox 59 (RITA)
ROL 59 (Fabriquimica)
Sabowax SE 100 (Sabo)
Simulsol M 59 (SEPPIC)
Sipoic MS-100 (Specialty Industrial)
Sympatens-BS/1000 (Kolb)
Tego Acid S 100 P (Degussa Care Specialties)

Trade Name Mixtures:
AEC Glyceryl Stearate (&) PEG-100 Stearate (A & E Connock)
Arlacel 165 (Uniqema Americas)
Base TC 100 (LCW)
Ceral 165 (Fabriquimica)
Cithrol GMS A/S (Croda Chemicals)
Customulse 165 (glyceryl stearate and PEG-100 Stearate) (Custom Ingredients)
Dracorin 100 SE P 2/008480 (Symrise)
Extan-GMS (Lanaetex)
Jeechem GMS-165 (Jeen)
Lexemul 561 (Inolex)
Lipomulse 165 (Lipo)
Lonzest MSA (Lonza Inc./Lonza Ltd.)
Megasperse Z (Megachem)
Nano-emulsion Concentrate (Active Concepts)
Nano-emulsion Concentrate Sun (Active Concepts)
Protachem GMS-165 (Protameen)
Ritapro 165 (RITA)
Sabowax FL 65/K (Sabo)
Simulsol 165 (SEPPIC)
STEPAN GMS SE/AS (Stepan)
Suncaps 664 (Particle Sciences)
Suncaps 903 (Particle Sciences)
Tewax TC 60 (Cesalpinia)
Tewax TC 65 (Cesalpinia)
Unitolate 165-C (Universal Preserv-A-Chem)

PEG-120 STEARATE

CAS No.: 9004-99-3 (Generic)

JPN Translation:
ステアリン酸 PEG - 120

Definition: PEG-120 Stearate is the polyethylene glycol ester of stearic acid that conforms to the formula:

$$CH_3(CH_2)_{16}\overset{\displaystyle O}{\overset{\displaystyle \|}{C}}\!\!-\!\!(OCH_2CH_2)_nOH$$

where n has an average value of 120.

Information Sources: 21CFR175.300, JCLS, MI-13(7660), TSCA

Chemical Class: Alkoxylated Carboxylic Acids

Function: Surfactant - Cleansing Agent

Technical/Other Names:
Polyethylene Glycol (120) Monostearate
Polyoxyethylene (120) Monostearate

Trade Name:
AEC PEG-120 Stearate (A & E Connock)

PEG-150 STEARATE

CAS No.: 9004-99-3 (Generic)

JPN Translation:
ステアリン酸 PEG - 150

Definition: PEG-150 Stearate is the polyethylene glycol ester of stearic acid that conforms to the formula:

$$CH_3(CH_2)_{16}\overset{\displaystyle O}{\overset{\displaystyle \|}{C}}\!\!-\!\!(OCH_2CH_2)_nOH$$

where n has an average value of 150.

Information Sources: 21CFR175.300, CIR: [S] JACT-2(7)1983, JCLS, MI-13 (7660), TSCA

Chemical Class: Alkoxylated Carboxylic Acids

Function: Surfactant - Cleansing Agent

Reported Product Categories: Moisturizing Preparations; Bath Preparations, Misc.; Body and Hand Preparations (Excluding Shaving Preparations); Hair Conditioners; Hair Straighteners; Shampoos (Non-coloring)

Technical/Other Names:
Polyethylene Glycol 6000 Monostearate
Polyoxyethylene (150) Monostearate

Trade Names:
AEC PEG-150 Stearate (A & E Connock)
Jeemate 6000 DPS (Jeen)
Lumulse 600-S (Lambent)
Prodhybase 6000 (Prod'Hyg)
STEPAN PEG 6000 MS (Stepan)
Unipeg-6000 MS (Universal Preserv-A-Chem)

Trade Name Mixtures:
Brookswax R (Arch Personal Care Products)
Jeecol P (Jeen)
Lipowax PR (Lipo)
Lipowax R-2 (Lipo)
Procol P (Protameen)
Relaxer Concentrate #2 (Arch Personal Care Products)
Relaxer Concentrate #3 (Arch Personal Care Products)
Ritachol 1000 (RITA)
Ritachol 5000 (RITA)
Teinowax (Lanaetex)
Unicol CPS (Universal Preserv-A-Chem)
Viscolene (Vevy)

PEG-45 STEARATE PHOSPHATE

Definition: PEG-45 Stearate Phosphate is the complex mixture of esters of phosphoric acid and PEG-45 Stearate (q.v.).

Chemical Classes: Alkoxylated Carboxylic Acids; Phosphorus Compounds

Function: Surfactant - Cleansing Agent

Technical/Other Names:
Polyethylene Glycol (45) Stearate Phosphate
Polyoxyethylene (45) Stearate Phosphate

Trade Names:
Emulsifier DMR (Clariant)

Emulsifier DMR (Clariant GmbH, Personal Care)

PEG-2 STEARATE SE

JPN Translation:
ステアリン酸 PEG - 2（SE）

Definition: PEG-2 Stearate SE is a self-emulsifying grade of PEG-2 Stearate (q.v.) that contains some sodium and/or potassium stearate.

Information Sources: CTFA D, JCIC, JCLS

Chemical Class: Alkoxylated Carboxylic Acids

Function: Surfactant - Emulsifying Agent

Reported Product Categories: Skin Care Preparations, Misc.; Cleansing Products (Cold Creams, Cleansing Lotions, Liquids and Pads); Bath Preparations, Misc.; Body and Hand Preparations (Excluding Shaving Preparations)

Technical/Other Names:
Diethylene Glycol Monostearate Self-Emulsifying
Polyethylene Glycol Monostearate (2E.O.), Self-emulsifying
Polyethylene Glycol 100 Monostearate Self-Emulsifying
Polyoxyethylene (2) Monostearate Self-Emulsifying

Trade Names:
AEC PEG-2 Stearate SE (A & E Connock)
ESTOL 3712 (Uniqema Europe)
Prodhybase Na (Prod'Hyg)

Trade Name Mixtures:
Base O/W 097 (LCW)
Pegosperse 100 S (Lonza Inc./Lonza Ltd.)
Prodhycreme (Prod'Hyg)
Prodhycreme 013 (Prod'Hyg)
STEPAN DGS SE and Stearic Acid (Stepan)

PEG-2 STEARMONIUM CHLORIDE

CAS No.: 60687-87-8

Empirical Formula:
$C_{23}H_{50}NO_2$

Definition: PEG-2 Stearmonium Chloride is the quaternary ammonium salt that conforms generally to the formula:

$$\left[CH_3(CH_2)_{17}-\underset{\underset{(CH_2CH_2O)_yH}{|}}{\overset{\overset{CH_3}{|}}{N}}-(CH_2CH_2O)_xH \right]^+ \quad Cl^-$$

where x + y has an average value of 2.

Chemical Classes: Alkoxylated Amines; Quaternary Ammonium Compounds

Function: Antistatic Agent

Technical/Other Names:
N,N-Bis(2-Hydroxyethyl)-N-Methylocta-decanaminium Chloride
Octadecanaminium, N,N-Bis(2-Hydroxy-ethyl)-N-Methyl-, Chloride
PEG-2 Stearyl Quaternium-4
Polyethylene Glycol 100 Stearmonium Chloride
Polyoxyethylene (2) Stearmonium Chloride

Trade Name Mixture:
Ethoquad 18/12 (Akzo Nobel)

PEG-15 STEARMONIUM CHLORIDE

Definition: PEG-15 Stearmonium Chloride is the quaternary ammonium salt that conforms generally to the formula:

$$\left[CH_3(CH_2)_{17}-\underset{\underset{(CH_2CH_2O)_yH}{|}}{\overset{\overset{CH_3}{|}}{N}}-(CH_2CH_2O)_xH \right]^+ \quad Cl^-$$

where x + y has an average value of 15.

Chemical Class: Quaternary Ammonium Compounds

Functions: Antistatic Agent; Hair Conditioning Agent

Reported Product Category: Hair Dyes and Colors (All Types Requiring Caution Statements and Patch Tests)

Technical/Other Names:
Polyethylene Glycol (15) Stearmonium Chloride
Polyoxyethylene (15) Stearmonium Chloride

Trade Name:
Ethoquad 18/25 (Akzo Nobel)

PEG-150/STEARYL ALCOHOL/SMDI COPOLYMER

Definition: PEG-150/Stearyl Alcohol/SMDI Copolymer is a copolymer of PEG-150, saturated methylene diphenyldiisocyanate, and stearyl alcohol monomers.

Chemical Class: Synthetic Polymers

Functions: Film Former; Suspending Agent - Nonsurfactant; Viscosity Increasing Agent - Aqueous

Trade Names:
Aculyn 46 (Rohm and Haas)
ACULYN 48 Polymer (Rohm and Haas)

PEG-5 STEARYL AMMONIUM CHLORIDE

CAS Nos.: 80238-02-4; 80462-94-8

JPN Translations:
PEG - 5 ステアリルアンモニウムクロリド
PEG - 5 ステアリルメチルアンモニウムク
ロリド

Empirical Formula:
$C_{2860}NO_5 \cdot Cl$

Definition: PEG-5 Stearyl Ammonium Chloride is the quaternary ammonium salt that conforms generally to the formula:

$$\left[CH_3(CH_2)_{17}-\underset{\underset{(CH_2CH_2O)_zH}{|}}{\overset{\overset{(CH_2CH_2O)_xH}{|}}{N}}-(CH_2CH_2O)_yH \right]^+ \quad Cl^-$$

where x+y+z has an average value of 5.

Information Sources: JCIC, JCLS, JSQI

Chemical Class: Quaternary Ammonium Compounds

Function: Antistatic Agent

Technical/Other Names:
N,N-Bis[2-(2-Hydroxyethoxy)Ethyl]-N-(2-Hydroxyethyl)Octadecanaminium Chloride
Octadecanaminium, N,N-Bis[2-(2-Hydroxy-ethoxy)Ethyl]-N-(2-Hydroxyethyl)-, Chloride
Pentaethoxystearylammonium Chloride
Quaternium-36
Tri (Polyoxyethylene) Stearyl Ammonium Chloride (5E.O.)

PEG-5 STEARYL AMMONIUM LACTATE

Empirical Formula:
$C_{28}H_{60}NO_5 \cdot C_3H_6O_3$

Definition: PEG-5 Stearyl Ammonium Lactate is the quaternary ammonium salt that conforms generally to the formula:

$$\left[CH_3(CH_2)_{17}-\underset{\underset{(CH_2CH_2O)_zH}{|}}{\overset{\overset{(CH_2CH_2O)_xH}{|}}{N}}-(CH_2CH_2O)_yH \right]^+ \quad CH_3\underset{\underset{OH}{|}}{CH}COO^-$$

where x+y+z has an average value of 5.

Chemical Class: Quaternary Ammonium Compounds

Function: Antistatic Agent

Technical/Other Names:
Polyethylene Glycol (5) Stearyl Ammonium Lactate
Polyoxyethylene (5) Stearyl Ammonium Lactate

Trade Names:
Genamin KSL (Clariant)
Genamin KSL (Clariant GmbH, Personal Care)

PEG-10 STEARYL BENZONIUM CHLORIDE

Empirical Formula:
$C_{45}H_{86}NO_{10} \cdot Cl$

Definition: PEG-10 Stearyl Benzonium Chloride is the quaternary ammonium salt that conforms generally to the formula:

$$\left[CH_3(CH_2)_{17}-\overset{\displaystyle (CH_2CH_2O)_xH}{\underset{\displaystyle (CH_2CH_2O)_yH}{N}}-CH_2-\bigcirc \right]^+ \quad Cl^-$$

where x+y has an average value of 10.

Chemical Classes: Alkoxylated Amines; Quaternary Ammonium Compounds

Function: Antistatic Agent

Technical/Other Names:
Polyethylene Glycol 500 Stearyl Benzonium Chloride
Polyoxyethylene (10) Stearyl Benzonium Chloride

PEG-6 STEARYLGUANIDINE

Definition: PEG-6 Stearylguanidine is the polyethylene glycol derivative of stearyl guanidine with an average of 6 moles of ethylene oxide.

Chemical Class: Alkoxylated Amines

Functions: Antistatic Agent; Surfactant - Emulsifying Agent

Trade Name:
Aerosol C-61 Surfactant (Cytec Industries)

PEG-2 SUNFLOWER GLYCERIDES

CAS Nos.: 180254-52-8 (Generic); 186511-05-7 (Generic)

Definition: PEG-2 Sunflower Glycerides is a polyethylene glycol derivative of the mono- and diglycerides derived from sunflower seed oil with an average of 2 moles of ethylene oxide.

Chemical Classes: Alkoxylated Alcohols; Glyceryl Esters and Derivatives

Functions: Skin-Conditioning Agent - Emollient; Surfactant - Emulsifying Agent

Technical/Other Names:
Polyethylene Glycol 100 Sunflower Glycerides
Polyoxyethylene (2) Sunflower Glycerides

PEG-7 SUNFLOWER GLYCERIDES

Definition: PEG-7 Sunflower Glycerides is the polyethylene glycol derivative of the mono- and diglycerides derived from sunflower seed oil with an average of 7 moles of ethylene oxide.

Chemical Classes: Alkoxylated Alcohols; Glyceryl Esters and Derivatives

Functions: Skin-Conditioning Agent - Emollient; Surfactant - Emulsifying Agent

Technical/Other Names:
Polyethylene Glycol (7) Sunflower Glycerides
Polyoxyethylene (7) Sunflower Glycerides

PEG-10 SUNFLOWER GLYCERIDES

CAS Nos.: 180254-52-8 (Generic); 186511-05-7 (Generic)

Definition: PEG-10 Sunflower Glycerides is the polyethylene glycol derivative of the mono- and diglycerides derived from sunflower seed oil with an average of 10 moles of ethylene oxide.

Chemical Classes: Alkoxylated Alcohols; Glyceryl Esters and Derivatives

Functions: Skin-Conditioning Agent - Emollient; Surfactant - Emulsifying Agent

Technical/Other Names:
Polyethylene Glycol (10) Sunflower Glycerides
Polyoxyethylene (10) Sunflower Glycerides

Trade Name:
Florasolvs PEG-10 Sunflower (Floratech)

PEG-13 SUNFLOWER GLYCERIDES

CAS Nos.: 70377-91-2 (Generic); 186511-05-7 (Generic)

Definition: PEG-13 Sunflower Glycerides is the polyethylene glycol derivative of the mono- and diglycerides of sunflower seed oil with an average ethoxylation value of 13.

Chemical Classes: Alkoxylated Alcohols; Glyceryl Esters and Derivatives

Functions: Skin-Conditioning Agent - Emollient; Surfactant - Emulsifying Agent

Technical/Other Names:
Polyethylene Glycol (13) Sunflower Glycerides
Polyoxyethylene (13) Sunflower Glycerides

Trade Name:
Lexol ES (Inolex)

PEG-4 TALLATE

CAS No.: 61791-00-2 (Generic)

Definition: PEG-4 Tallate is the poly-ethylene glycol ester of Tall Oil Acid (q.v.) that conforms generally to the formula:

$$\overset{\displaystyle O}{\underset{\displaystyle \|}{RC}}-(OCH_2CH_2)_nOH$$

where RCO- represents the fatty acids derived from tall oil and n has an average value of 4.

Information Sources: 21CFR175.105, 21CFR175.300, 21CFR176.210, MI-13 (7660), TSCA

Chemical Class: Alkoxylated Carboxylic Acids

Function: Surfactant - Emulsifying Agent

Technical/Other Names:
Polyethylene Glycol 200 Monotallate
Polyoxyethylene (4) Monotallate

Trade Names:
AEC PEG-4 Tallate (A & E Connock)
Hetoxamate FA 2-5 (Heterene)
Jeemate 200-T (Jeen)

PEG-5 TALLATE

CAS No.: 61791-00-2 (Generic)

Definition: PEG-5 Tallate is the poly-ethylene glycol ester of Tall Oil Acid (q.v.) that conforms generally to the formula:

$$\overset{\displaystyle O}{\underset{\displaystyle \|}{RC}}-(OCH_2CH_2)_nOH$$

where RCO- represents the fatty acids derived from tall oil and n has an average value of 5.

Information Sources: 21CFR175.105, 21CFR175.300, MI-13(7660)

Chemical Class: Alkoxylated Carboxylic Acids

Function: Surfactant - Emulsifying Agent

Technical/Other Names:
Polyethylene Glycol (5) Monotallate
Polyoxyethylene (5) Monotallate

PEG-8 TALLATE

CAS No.: 61791-00-2 (Generic)

Definition: PEG-8 Tallate is the polyethylene glycol ester of Tall Oil Acid (q.v.) that conforms generally to the formula:

$$RC \overset{O}{\overset{\|}{{}}} \!\!-\!\! (OCH_2CH_2)_nOH$$

where RCO- represents the fatty acids derived from tall oil and n has an average value of 8.

Information Sources: 21CFR175.105, 21CFR175.300, 21CFR176.210, 21CFR177.1210, 21CFR177.2800, MI-13 (7660), TSCA

Chemical Class: Alkoxylated Carboxylic Acids

Function: Surfactant - Emulsifying Agent

Technical/Other Names:
Polyethylene Glycol 400 Monotallate
Polyoxyethylene (8) Monotallate

Trade Names:
AEC PEG-8 Tallate (A & E Connock)
Chemax E-400-MT (Chemax)
DeTHOX ACID TO-8.5 (DeForest)
Ethox TO-8 (Ethox)
Jeemate 400-T (Jeen)
Lipo 142 (Lipo)
Pegosperse 400 MOT (Lonza Inc./Lonza Ltd.)
Unipeg-400 MOT (Universal Preserv-A-Chem)

PEG-10 TALLATE

CAS No.: 61791-00-2 (Generic)

Definition: PEG-10 Tallate is the polyethylene glycol ester of Tall Oil Acid (q.v.) that conforms generally to the formula:

$$RC \overset{O}{\overset{\|}{{}}} \!\!-\!\! (OCH_2CH_2)_nOH$$

where RCO- represents the fatty acids derived from tall oil and n has an average value of 10.

Information Sources: 21CFR175.105, 21CFR175.300, 21CFR177.2800, MI-13 (7660)

Chemical Class: Alkoxylated Carboxylic Acids

Function: Surfactant - Emulsifying Agent

Technical/Other Names:
Polyethylene Glycol 500 Monotallate
Polyoxyethylene (10) Monotallate

PEG-12 TALLATE

CAS No.: 61791-00-2 (Generic)

Definition: PEG-12 Tallate is the polyethylene glycol ester of Tall Oil Acid (q.v.) that conforms generally to the formula:

$$RC \overset{O}{\overset{\|}{{}}} \!\!-\!\! (OCH_2CH_2)_nOH$$

where RCO- represents the fatty acids derived from tall oil and n has an average value of 12.

Information Sources: 21CFR175.105, 21CFR175.300, 21CFR176.170, 21CFR176.180, 21CFR176.210, 21CFR177.2800, MI-13(7660), TSCA

Chemical Class: Alkoxylated Carboxylic Acids

Function: Surfactant - Emulsifying Agent

Technical/Other Names:
Polyethylene Glycol 600 Monotallate
Polyoxyethylene (12) Monotallate

Trade Names:
AEC PEG-12 Tallate (A & E Connock)
Ethox TO-14 (Ethox)
Jeemate 600-T (Jeen)
Lipopeg 6-T (Lipo)
Norfox 600 MOT (Norman, Fox & Co.)

PEG-14 TALLATE

CAS No.: 61791-00-2 (Generic)

Definition: PEG-14 Tallate is the polyethylene glycol ester of Tall Oil Acid (q.v.) that conforms generally to the formula:

$$RC \overset{O}{\overset{\|}{{}}} \!\!-\!\! (OCH_2CH_2)_nOH$$

where RCO- represents the fatty acids derived from tall oil and n has an average value of 14.

Chemical Class: Alkoxylated Carboxylic Acids

Function: Surfactant - Emulsifying Agent

Technical/Other Names:
Polyethylene Glycol (14) Monotallate
Polyoxyethylene (14) Monotallate

Trade Name:
DeTHOX ACID TO-14 (DeForest)

PEG-15 TALLATE

CAS No.: 61791-00-2 (Generic)

Definition: PEG-15 Tallate is the polyethylene glycol ester of Tall Oil Acid (q.v.) that conforms generally to the formula:

$$RC \overset{O}{\overset{\|}{{}}} \!\!-\!\! (OCH_2CH_2)_nOH$$

where RCO- represents the fatty acids derived from tall oil and n has an average value of 15.

Chemical Class: Alkoxylated Carboxylic Acids

Function: Surfactant - Emulsifying Agent

Technical/Other Names:
Polyethylene Glycol (15) Tallate
Polyoxyethylene (15) Tallate

Trade Name:
Ethofat 242/25 (Akzo Nobel)

PEG-16 TALLATE

CAS No.: 61791-00-2 (Generic)

Definition: PEG-16 Tallate is the polyethylene glycol ester of Tall Oil Acid (q.v.) that conforms generally to the formula:

$$RC \overset{O}{\overset{\|}{{}}} \!\!-\!\! (OCH_2CH_2)_nOH$$

where RCO- represents fatty acids derived from the tall oil and n has an average value of 16.

Information Sources: 21CFR175.300, 21CFR176.180, 21CFR176.210, 21CFR177.2800, CTFA D, MI-13(7660), TSCA

Chemical Class: Alkoxylated Carboxylic Acids

Function: Surfactant - Emulsifying Agent

Technical/Other Names:
Polyethylene Glycol (16) Monotallate
Polyoxyethylene (16) Monotallate

Trade Names:
AEC PEG-16 Tallate (A & E Connock)
DeTHOX ACID TO-16.5 (DeForest)
Ethox TO-16 (Ethox)

PEG-20 TALLATE

CAS No.: 61791-00-2 (Generic)

Definition: PEG-20 Tallate is the poly-ethylene glycol ester of Tall Oil Acid (q.v.) that conforms generally to the formula:

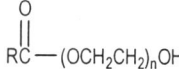

$$RC\overset{O}{\overset{||}{-}}(OCH_2CH_2)_nOH$$

where RCO- represents the fatty acids derived from tall oil and n has an average value of 20.

Information Sources: 21CFR175.300, 21CFR176.180, 21CFR176.210, 21CFR177.2800, MI-13(7660), TSCA

Chemical Class: Alkoxylated Carboxylic Acids

Functions: Surfactant - Cleansing Agent; Surfactant - Emulsifying Agent; Surfactant - Solubilizing Agent

Technical/Other Names:
Polyethylene Glycol 1000 Monotallate
Polyoxyethylene (20) Monotallate

Trade Names:
AEC PEG-20 Tallate (A & E Connock)
Hetoxamate FA-20 (Heterene)
Jeemate 1000-T (Jeen)

PEG-5 TALL OIL STEROL

Definition: PEG-5 Tall Oil Sterol is the poly-ethylene glycol ether of Tall Oil Sterol (q.v.) containing an average of 5 moles of ethylene oxide.

Chemical Classes: Alkoxylated Alcohols; Sterols

Function: Skin-Conditioning Agent - Miscellaneous

Technical/Other Names:
PEG-5 Tall Oil Sterol Ether
Polyethylene Glycol (5) Tall Oil Sterol
Polyoxyethylene (5) Tall Oil Sterol

PEG-5 TALLOW AMIDE

CAS No.: 8051-61-4

Definition: PEG-5 Tallow Amide is the poly-ethylene glycol amide of tallow acid that conforms generally to the formula:

$$RC\overset{O}{\overset{||}{-}}NH(CH_2CH_2O)_nH$$

where RCO- represents the fatty acids derived from tallow and n has an average value of 5.

Chemical Class: Alkoxylated Amides

Functions: Antistatic Agent; Surfactant - Emulsifying Agent

Technical/Other Names:
Polyethylene Glycol (5) Hydrogenated Tallow Amide
Polyoxyethylene (5) Hydrogenated Tallow Amide

PEG-8 TALLOW AMIDE

Definition: PEG-8 Tallow Amide is the poly-ethylene glycol amide of tallow acid that conforms generally to the formula:

$$RC\overset{O}{\overset{||}{-}}NH(CH_2CH_2O)_nH$$

where RCO- represents the fatty acids derived from tallow and n has an average value of 8.

Chemical Class: Alkoxylated Amides

Function: Surfactant - Emulsifying Agent

Technical/Other Names:
Polyethylene Glycol 400 Tallow Amide
Polyoxyethylene (8) Tallow Amide

Trade Name Mixture:
Lowenol C-77 (Lowenstein)

PEG-50 TALLOW AMIDE

CAS No.: 8051-63-6

Definition: PEG-50 Tallow Amide is the polyethylene glycol amide of tallow acid that conforms generally to the formula:

$$RC\overset{O}{\overset{||}{-}}NH(CH_2CH_2O)_nH$$

where RCO- represents the fatty acids derived from tallow and n has an average value of 50.

Information Source: CTFA D

Chemical Class: Alkoxylated Amides

Functions: Surfactant - Cleansing Agent; Surfactant - Solubilizing Agent

Reported Product Category: Hair Dyes and Colors (All Types Requiring Caution Statements and Patch Tests)

Technical/Other Names:
Polyethylene Glycol (50) Hydrogenated Tallow Amide
Polyoxyethylene (50) Hydrogenated Tallow Amide

Trade Names:
Ethomid HT/60 (Akzo Nobel)
Schercomid HT-60 (Scher)

Trade Name Mixtures:
Lowenol T-163A (Lowenstein)

Lowenol Copolymer 1985-A (Lowenstein)
Lowenol Copolymer 1985-B (Lowenstein)
Lowenol T-163 (Lowenstein)

PEG-2 TALLOWAMIDE DEA

JPN Translation:
PEG - 2 牛脂アルキル DEA

Definition: PEG-2 Tallowamide DEA is the polyethylene glycol amine derived from Tallow Acid (q.v.) that conforms generally to the formula:

$$RC\overset{O}{\overset{||}{-}}N(CH_2CH_2OCH_2CH_2OH)_2$$

where RCO represents tallowoyl moiety.

Information Source: JCLS

Chemical Class: Alkanolamines

Function: Surfactant - Cleansing Agent

PEG-2 TALLOW AMINE

CAS No.: 61791-26-2 (Generic)

Definition: PEG-2 Tallow Amine is the poly-ethylene glycol amine of Tallow (q.v.) that conforms generally to the formula:

$$R-N\overset{\displaystyle (CH_2CH_2O)_xH}{\underset{\displaystyle (CH_2CH_2O)_yH}{}}$$

where R represents the alkyl groups derived from tallow and x+y has an average value of 2.

Chemical Class: Alkoxylated Amines

Function: Antistatic Agent

Reported Product Category: Hair Dyes and Colors (All Types Requiring Caution Statements and Patch Tests)

Technical/Other Names:
Polyethylene Glycol (2) Tallow Amine
Polyoxyethylene (2) Tallow Amine

PEG-7 TALLOW AMINE

CAS No.: 61791-26-2 (Generic)

Definition: PEG-7 Tallow Amine is the poly-ethylene glycol amine of Tallow (q.v.) that conforms generally to the formula:

$$R-N\overset{\displaystyle (CH_2CH_2O)_xH}{\underset{\displaystyle (CH_2CH_2O)_yH}{}}$$

where R represents the alkyl groups derived from tallow and x + y has an average value of 7.

Chemical Class: Alkoxylated Amines

Function: Antistatic Agent

Technical/Other Names:
Polyethylene Glycol (7) Tallow Amine
Polyoxyethylene (7) Tallow Amine

PEG-11 TALLOW AMINE

CAS No.: 61791-26-2 (Generic)

Definition: PEG-11 Tallow Amine is the polyethylene glycol amine of Tallow (q.v.) that conforms generally to the formula:

$$R-N \begin{cases} (CH_2CH_2O)_xH \\ (CH_2CH_2O)_yH \end{cases}$$

where R represents the alkyl groups derived from tallow and x + y has an average value of 11.

Chemical Class: Alkoxylated Amines

Function: Antistatic Agent

Technical/Other Names:
Polyethylene Glycol (11) Tallow Amine
Polyoxyethylene (11) Tallow Amine

Trade Name:
Empilan AMT 11 (Albright & Wilson UK)

PEG-15 TALLOW AMINE

CAS No.: 61791-26-2 (Generic)

Definition: PEG-15 Tallow Amine is the polyethylene glycol amine of Tallow (q.v.) that conforms generally to the formula:

$$R-N \begin{cases} (CH_2CH_2O)_xH \\ (CH_2CH_2O)_yH \end{cases}$$

where R represents the alkyl groups derived from tallow and x+y has an average value of 15.

Chemical Class: Alkoxylated Amines

Function: Antistatic Agent

Trade Names:
Empilan AMT 15 (Albright & Wilson UK)
Ethox TAM-15 (Ethox)

PEG-20 TALLOW AMINE

CAS No.: 61791-26-2 (Generic)

Definition: PEG-20 Tallow Amine is the polyethylene glycol amine that conforms generally to the formula:

$$R-N \begin{cases} (CH_2CH_2O)_xH \\ (CH_2CH_2O)_yH \end{cases}$$

where R represents the alkyl groups derived from Tallow (q.v.) and x + y has an average value of 20.

Chemical Class: Alkoxylated Amines

Function: Antistatic Agent

Technical/Other Names:
Polyethylene Glycol 1000 Tallow Amine
Polyoxyethylene (20) Tallow Amine

Trade Name:
Ethox TAM-20 (Ethox)

PEG-22 TALLOW AMINE

CAS No.: 61791-26-2 (Generic)

Definition: PEG-22 Tallow Amine is the polyethylene glycol amine of Tallow (q.v.) that conforms generally to the formula:

$$R-N \begin{cases} (CH_2CH_2O)_xH \\ (CH_2CH_2O)_yH \end{cases}$$

where R represents the alkyl groups derived from Tallow (q.v.) and x+y has an average value of 22.

Chemical Class: Alkoxylated Amines

Function: Hair Conditioning Agent

Technical/Other Names:
Polyethylene Glycol (22) Tallow Amine
Polyoxyethylene Glycol (22) Tallow Amine

PEG-25 TALLOW AMINE

CAS No.: 61791-26-2 (Generic)

Definition: PEG-25 Tallow Amine is the polyethylene glycol amine that conforms generally to the formula:

$$R-N \begin{cases} (CH_2CH_2O)_xH \\ (CH_2CH_2O)_yH \end{cases}$$

where R represents the alkyl groups derived from Tallow (q.v.) and x + y has an average value of 25.

Chemical Class: Alkoxylated Amines

Function: Antistatic Agent

Technical/Other Names:
Polyethylene Glycol (25) Tallow Amine
Polyoxyethylene (25) Tallow Amine

Trade Name:
Ethox TAM-25 (Ethox)

PEG-30 TALLOW AMINE

CAS No.: 61791-26-2 (Generic)

Definition: PEG-30 Tallow Amine is the polyethylene glycol amine of Tallow (q.v.) that conforms generally to the formula:

$$R-N \begin{cases} (CH_2CH_2O)_xH \\ (CH_2CH_2O)_yH \end{cases}$$

where R represents the alkyl groups derived from tallow and x+y has an average value of 30.

Chemical Class: Alkoxylated Amines

Function: Hair Conditioning Agent

Technical/Other Names:
Polyethylene Glycol (30) Tallow Amine
Polyoxyethylene (30) Tallow Amine

PEG-3 TALLOW AMINOPROPYLAMINE

CAS No. 90367-27-4

EINECS No. 291-275-8

Definition: PEG-3 Tallow Aminopropylamine is the polyethylene glycol amine that conforms generally to the formula:

$$R-N(CH_2)_3N \begin{cases} (CH_2CH_2O)_x \\ (CH_2CH_2O)_y \end{cases}$$
$$(CH_2CH_2O)_z$$

where R represents the alkyl groups derived from tallow and x + y + z has an average value of 3.

Chemical Class: Alkoxylated Amines

Function: Antistatic Agent

Technical/Other Names:
Ethanol, 2,2'-[[3-[(2-Hydroxyethyl)Amino]-Propyl]Imino]Bis-, N-Tallow Alkyl Derivs.
Polyethylene Glycol (3) Tallow Amino-propylamine
Polyoxyethylene (3) Tallow Aminopropyl-amine
N-Tallow Alkyl 2,2'-[[3-[(2-Hydroxyethyl)-Amino]Propyl]Imino]Bisethanol

Trade Name:
Ethoduomeen T/13 (Akzo Nobel)

PEG-10 TALLOW AMINOPROPYLAMINE

Definition: PEG-10 Tallow Aminopropyl-amine is the polyethylene glycol amine that conforms generally to the formula:

$$R-N(CH_2)_3N \begin{cases} (CH_2CH_2O)_x \\ (CH_2CH_2O)_y \end{cases}$$
$$(CH_2CH_2O)_z$$

where R represents the alkyl groups derived from tallow and x + y + z has an average value of 10.

Chemical Class: Alkoxylated Amines

Function: Antistatic Agent

Technical/Other Names:
Polyethylene Glycol 500 Tallow Amino-
propylamine
Polyoxyethylene (10) Tallow Aminopropyl-
amine

Trade Name:
Ethoduomeen T/20 (Akzo Nobel)

PEG-15 TALLOW AMINOPROPYLAMINE

Definition: PEG-15 Tallow Aminopropyl-
amine is the polyethylene glycol amine that
conforms generally to the formula:

$$R-N(CH_2)_3N \underset{(CH_2CH_2O)_z}{\overset{(CH_2CH_2O)_x}{\langle}} (CH_2CH_2O)_y$$

where R represents the alkyl groups derived
from tallow and x + y + z has an average
value of 15.

Chemical Class: Alkoxylated Amines

Function: Antistatic Agent

Technical/Other Names:
Polyethylene Glycol (15) Tallow Amino-
propylamine
Polyoxyethylene (15) Tallow Aminopropyl-
amine

Trade Names:
Chemeen DT-15 (Chemax)
Ethoduomeen T/25 (Akzo Nobel)

PEG-20 TALLOW AMMONIUM ETHOSULFATE

Definition: PEG-20 Tallow Ammonium
Ethosulfate is the quaternary ammonium salt
that conforms generally to the formula:

$$\left[R-N \underset{(CH_2CH_2O)_zH}{\overset{(CH_2CH_2O)_xH}{\langle}} (CH_2CH_2O)_yH \right]^+ CH_3CH_2OSO_3^-$$

where R represents the alkyl groups derived
from tallow and x+y+z has an average value
of 20.

Chemical Classes: Alkoxylated Amines;
Quaternary Ammonium Compounds

Functions: Antistatic Agent; Hair Conditioning Agent

Technical/Other Names:
Polyethylene Glycol 1000 Tallow
Ammonium Ethosulfate
Polyoxyethylene (20) Tallow Ammonium
Ethosulfate

Trade Names:
Atlas G-265 (Uniqema Americas)
Ethox TAM-20 DQ (Ethox)

PEG-20 TALLOWATE

CAS No.: 68153-64-0 (Generic)

Definition: PEG-20 Tallowate is the polyethylene glycol ester of Tallow Acid (q.v.) that
conforms generally to the formula:

$$RC \overset{O}{\overset{\|}{-}} (OCH_2CH_2)_nOH$$

where RCO- represents the alkyl groups
derived from tallow and n has an average
value of 20.

Information Sources: 21CFR176.210,
21CFR177.2260, 21CFR177.2800, TSCA

Chemical Class: Alkoxylated Carboxylic
Acids

Functions: Surfactant - Cleansing Agent;
Surfactant - Emulsifying Agent; Surfactant -
Solubilizing Agent

Technical/Other Names:
Polyethylene Glycol 1000 Monotallowate
Polyoxyethylene (20) Monotallowate

PEG-5 TALLOW BENZONIUM CHLORIDE

Definition: PEG-5 Tallow Benzonium
Chloride is the quaternary ammonium salt
that conforms generally to the formula:

$$\left[R-\underset{(CH_2CH_2O)_yH}{\overset{(CH_2CH_2O)_xH}{N}}-CH_2-\bigcirc \right]^+ Cl^-$$

where R represents the alkyl groups derived
from tallow and x + y has an average value of
5.

Chemical Classes: Alkoxylated Amines;
Quaternary Ammonium Compounds

Function: Antistatic Agent

Technical/Other Names:
Polyethylene Glycol (5) Tallow Benzonium
Chloride
Polyoxyethylene (5) Tallow Benzonium
Chloride

PEG-15 TALLOW POLYAMINE

JPN Translation:
PEG - 15 タロウポリアミン

Definition: PEG-15 Tallow Polyamine is
polyethylene glycol polyamine that conforms
generally to the formula:

$$\left[RNH \overset{}{\underset{}{-}} \begin{array}{l} C_3H_5OH \\ O(CH_2CH_2O)_nC_3H_5OH \\ NH(CH_2)_mNH \\ C_3H_5OH \quad OH \\ O(CH_2CH_2O)C_3H_5 \\ NHR \end{array} \right]_x$$

where R represents the alkyl groups derived
from tallow, n has a total value between 10
and 20, m has a value between 2 and 6 and
x has a value between 2 and 4.

Information Sources: CTFA D, JCIC,
JCLS, JSQI

Chemical Class: Alkoxylated Amines

Functions: Antistatic Agent; Hair Conditioning Agent

Reported Product Categories: Shampoos
(Non-coloring); Hair Conditioners

Technical/Other Names:
Polyethylene Glycol•Epichlorohydrine•
Tallow Alkyl Amine•Dipropylene Triamine
Condensation Product
Polyethylene Glycol (15) Tallow Polyamine
Polyoxyethylene (15) Tallow Polyamine

PEG-3 TALLOW PROPYLENEDIMONIUM DIMETHOSULFATE

CAS No.	EINECS No.
93572-63-5	297-495-0

Definition: PEG-3 Tallow Propylene-
dimonium Dimethosulfate is the quaternary
ammonium salt that conforms to the formula:

$$\left[R-\underset{CH_3}{\overset{CH_2CH_2OH}{N}}-(CH_2)_3-\underset{CH_2CH_2OH}{\overset{CH_2CH_2OH}{N}}-CH_3 \right]^{+2} 2CH_3OSO_3^-$$

where R represents the alkyl groups derived
from tallow.

Chemical Classes: Alkoxylated Amines;
Quaternary Ammonium Compounds

Functions: Antistatic Agent; Hair Conditioning Agent

Technical/Other Names:
Polyethylene Glycol (3) Tallow Propylene-
dimonium Dimethosulfate

Polyoxyethylene (3) Tallow Propylene-
dimonium Dimethosulfate
N-Tallowalkyl-N,N'-Dimethyl-N,N'-Poly-
ethyleneglycol-propylenebis-ammonium-
bismethosulfate

PEG-5 TRICAPRYL CITRATE

Definition: PEG-5 Tricapryl Citrate is the
organic compound that conforms generally to
the formula:

$$CH_2 - \overset{\overset{\displaystyle O}{\|}}{C} - O(CH_2CH_2O)_x(CH_2)_9CH_3$$
$$HOC - \overset{\overset{\displaystyle O}{\|}}{C} - O(CH_2CH_2O)_y(CH_2)_9CH_3$$
$$CH_2 - \overset{\overset{\displaystyle O}{\|}}{C} - O(CH_2CH_2O)_z(CH_2)_9CH_3$$

where x+y+z has an average value of 5.

Chemical Class: Esters

Functions: Skin-Conditioning Agent -
Emollient; Surfactant - Emulsifying Agent

Technical/Other Names:
PEG-5 Tridecyl Citrate
Polyethylene Glycol (5) Capryl Tricitrate
Polyoxyethylene (5) Capryl Tricitrate

PEG-5 TRICETYL CITRATE

Definition: PEG-5 Tricetyl Citrate is the
organic compound that conforms generally to
the formula:

$$CH_2 - \overset{\overset{\displaystyle O}{\|}}{C} - O(CH_2CH_2O)_x(CH_2)_{15}CH_3$$
$$HOC - \overset{\overset{\displaystyle O}{\|}}{C} - O(CH_2CH_2O)_y(CH_2)_{15}CH_3$$
$$CH_2 - \overset{\overset{\displaystyle O}{\|}}{C} - O(CH_2CH_2O)_z(CH_2)_{15}CH_3$$

where x+y+z has an average value of 5.

Chemical Class: Esters

Functions: Skin-Conditioning Agent -
Emollient; Surfactant - Emulsifying Agent

Technical/Other Names:
Polyethylene Glycol (5) Cetyl Tricitrate
Polyoxyethylene (5) Cetyl Tricitrate

Trade Name Mixture:
Trioxene S (Vevy)

PEG-4 TRIFLUOROPROPYL DIMETHI-
CONE COPOLYMER

Definition: PEG-4 Trifluoropropyl Dimethicone Copolymer is the siloxane polymer that
conforms generally to the formula:

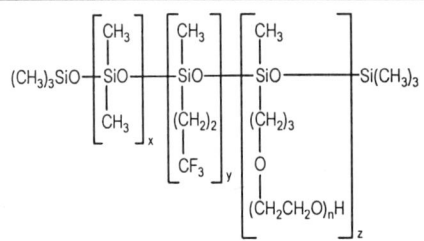

where n has an average value of 4.

Chemical Classes: Halogen Compounds;
Siloxanes and Silanes; Synthetic Polymers

Functions: Hair Conditioning Agent; Skin-
Conditioning Agent - Miscellaneous

Trade Name:
FPD-4694 (Shin Etsu)

PEG-8 TRIFLUOROPROPYL DIMETHI-
CONE COPOLYMER

Definition: PEG-8 Trifluoropropyl Dimethicone Copolymer is the siloxane polymer that
conforms to the formula:

$$(CH_3)_3SiO - \begin{bmatrix} CH_3 \\ | \\ SiO \\ | \\ CH_3 \end{bmatrix}_x \begin{bmatrix} CH_3 \\ | \\ SiO \\ | \\ (CH_2)_2 \\ | \\ CF_3 \end{bmatrix}_y \begin{bmatrix} CH_3 \\ | \\ SiO \\ | \\ (CH_2)_3 \\ | \\ O \\ | \\ (CH_2CH_2O)_nH \end{bmatrix}_z - Si(CH_3)_3$$

where n has an average value of 8.

Chemical Classes: Halogen Compounds;
Siloxanes and Silanes; Synthetic Polymers

Functions: Hair Conditioning Agent; Skin-
Conditioning Agent - Miscellaneous

Trade Name:
FPD-6131 (Shin Etsu)

PEG-10 TRIFLUOROPROPYL DIMETHI-
CONE COPOLYMER

Definition: PEG-10 Trifluoropropyl
Dimethicone Copolymer is the siloxane
polymer that conforms generally to the
formula:

$$(CH_3)_3SiO - \begin{bmatrix} CH_3 \\ | \\ SiO \\ | \\ CH_3 \end{bmatrix}_x \begin{bmatrix} CH_3 \\ | \\ SiO \\ | \\ (CH_2)_2 \\ | \\ CF_3 \end{bmatrix}_y \begin{bmatrix} CH_3 \\ | \\ SiO \\ | \\ (CH_2)_3 \\ | \\ O \\ | \\ (CH_2CH_2O)_nH \end{bmatrix}_z - Si(CH_3)_3$$

where n has an average value of 10.

Chemical Classes: Halogen Compounds;
Siloxanes and Silanes; Synthetic Polymers

Functions: Hair Conditioning Agent; Skin-
Conditioning Agent - Miscellaneous

Trade Name:
FPD-4970 (Shin Etsu)

PEG-66 TRIHYDROXYSTEARIN

Definition: PEG-66 Trihydroxystearin is a
polyethylene glycol derivative of Trihydroxystearin (q.v.) with an average of 66 moles of
ethylene oxide.

Chemical Classes: Alkoxylated Alcohols;
Glyceryl Esters and Derivatives

Functions: Surfactant - Cleansing Agent;
Surfactant - Solubilizing Agent

Technical/Other Names:
Polyethylene Glycol (66) Trihydroxystearin
Polyoxyethylene (66) Trihydroxystearin

PEG-200 TRIHYDROXYSTEARIN

Definition: PEG-200 Trihydroxystearin is a
polyethylene glycol derivative of Trihydroxystearin (q.v.) with an average of 200 moles of
ethylene oxide.

Chemical Classes: Alkoxylated Alcohols;
Glyceryl Esters and Derivatives

Functions: Surfactant - Cleansing Agent;
Surfactant - Solubilizing Agent

Technical/Other Names:
Polyethylene Glycol (200) Trihydroxystearin
Polyoxyethylene (200) Trihydroxystearin

PEG-5 TRILAURYL CITRATE

Empirical Formula:
$C_{52}H_{100}O_{12}$

Definition: PEG-5 Trilauryl Citrate is the
organic compound that conforms generally to
the formula:

$$CH_2 - \overset{\overset{\displaystyle O}{\|}}{C} - O(CH_2CH_2O)_x(CH_2)_{11}CH_3$$
$$HOC - \overset{\overset{\displaystyle O}{\|}}{C} - O(CH_2CH_2O)_y(CH_2)_{11}CH_3$$
$$CH_2 - \overset{\overset{\displaystyle O}{\|}}{C} - O(CH_2CH_2O)_z(CH_2)_{11}CH_3$$

wwhere x+y+z has an average value of 5.

Chemical Classes: Alkoxylated Alcohols; Esters

Functions: Skin-Conditioning Agent - Emollient; Surfactant - Emulsifying Agent

Technical/Other Names:
Polyethylene Glycol (5) Lauryl Tricitrate
Polyoxyethylene (5) Lauryl Tricitrate

PEG-4 TRIMETHYLOLPROPANE DISTEARATE

JPN Translation:
ジステアリン酸 PEG - 4 トリメチロールプ
ロパン

Definition: PEG-4 Trimethylolpropane Distearate is the organic compound that conforms generally to the formula:

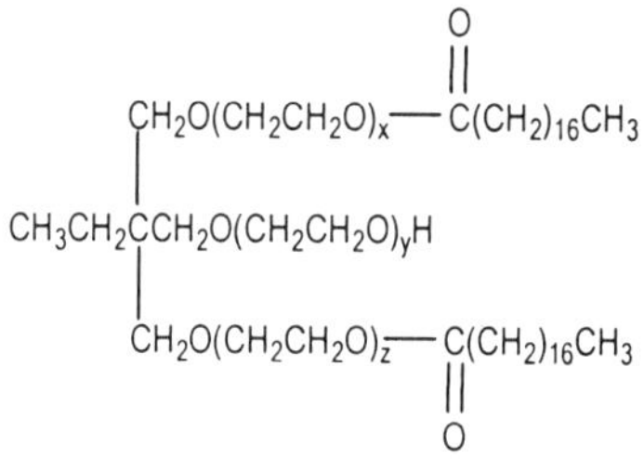

where x+y+z has an average value of 4.

Information Source: JCLS

Chemical Class: Esters

Function: Skin-Conditioning Agent - Emollient

Technical/Other Names:
Polyethylene Glycol (4) Trimethylolpropane Distearate
Polyoxyethylene (4) Trimethylolpropane Distearate

Trade Name:
Emalex TPS-204 (Nihon Emulsion)

PEG-3 TRIMETHYLOLPROPANE TRIISOSTEARATE

JPN Translation:
トリ（イソステアリン酸 PEG - 3）トリメ
チロールプロパン

Definition: PEG-3 Trimethylolpropane Triisostearate is the triester of Isostearic Acid (q.v.) and a polyethylene glycol ether of trimethylolpropan with an average of 3 moles of ethylene oxide.

Information Source: jcls

Chemical Class: Esters

Function: Skin-Conditioning Agent - Miscellaneous

PEG-5 TRIMETHYLOLPROPANE TRI-MYRISTATE

JPN Translation:
トリミリスチン酸 PEG - 5 トリメチロール
プロパン

Empirical Formula:
$C_{58}H_{112}O_{11}$

Definition: PEG-5 Trimethylolpropane Trimyristate is the ethoxylated triester that conforms generally to the formula:

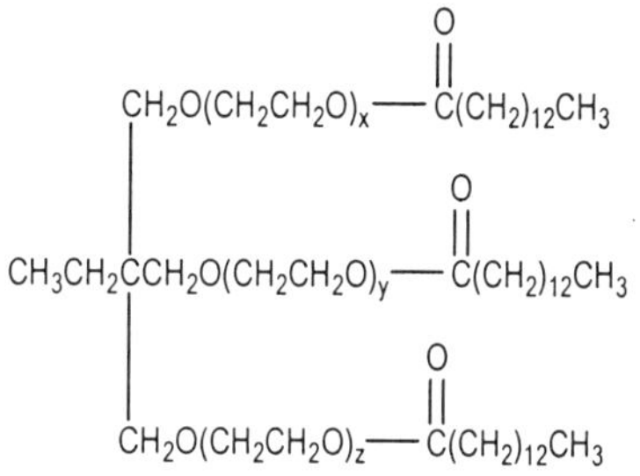

where x+y+z has an average value of 5.

Information Sources: JCIC, JCLS

Chemical Class: Esters

Function: Skin-Conditioning Agent - Emollient

Technical/Other Names:
Polyethylene Glycol (5) Trimethylolpropane Trimyristate
Polyoxyethylene (5) Trimethylolpropane Trimyristate
Polyoxyethylene 1,1,1-Trimethylolpropane Trimyristate (5E.O.)

PEG-3 TRIMETHYLOLPROPANE TRISTEARATE

Definition: PEG-3 Trimethylolpropane Tristearate is the organic compound that conforms generally to the formula:

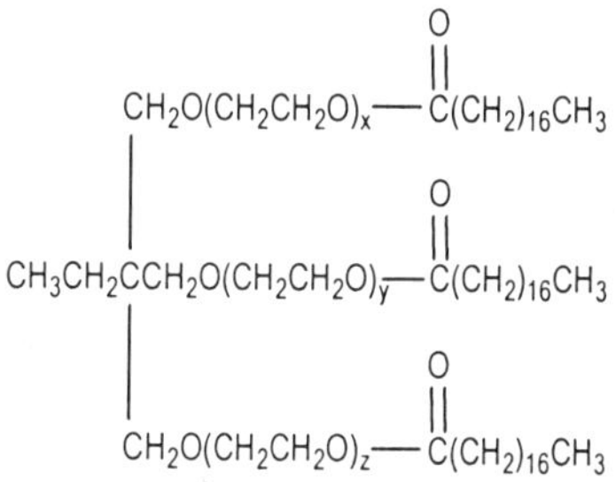

where x+y+z has an average value of 3.

Chemical Class: Esters

Function: Skin-Conditioning Agent - Emollient

Technical/Other Names:
Polyethylene Glycol (3) Trimethylolpropane Tristearate

Polyoxyethylene (3) Trimethylolpropane Tristearate

PEG-5 TRIMYRISTYL CITRATE

Empirical Formula:
$C_{58}H_{112}O_{12}$

Definition: PEG-5 Trimyristyl Citrate is the organic compound that conforms generally to the formula:

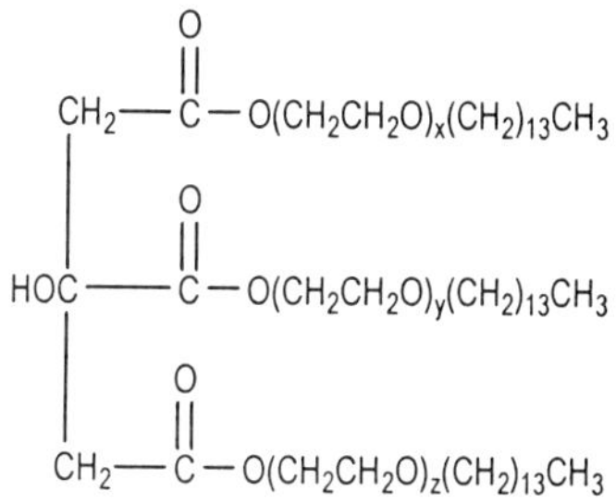

where x+y+z has an average value of 5.

Chemical Classes: Alkoxylated Alcohols; Esters

Functions: Skin-Conditioning Agent - Emollient; Surfactant - Emulsifying Agent

Technical/Other Names:
Polyethylene Glycol (5) Myristyl Tricitrate
Polyoxyethylene (5) Myristyl Tricitrate

Trade Name Mixture:
Trioxene S (Vevy)

PEG-5 TRISTEARYL CITRATE

Empirical Formula:
$C_{70}H_{136}O_{12}$

Definition: PEG-5 Tristearyl Citrate is the organic compound that conforms generally to the formula:

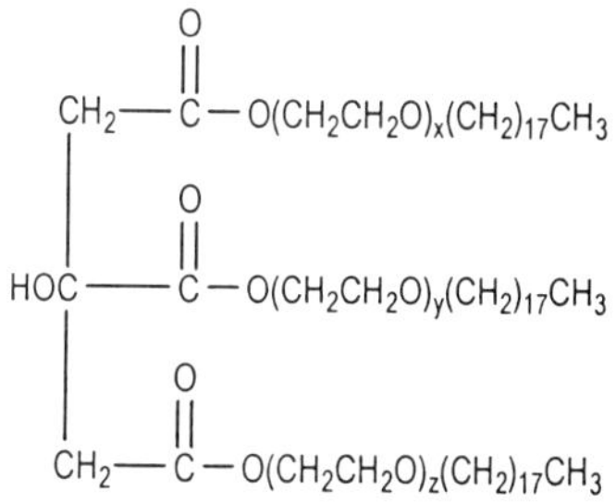

where x+y+z has an average value of 5.

Chemical Classes: Alkoxylated Alcohols; Esters

Functions: Skin-Conditioning Agent - Emollient; Surfactant - Emulsifying Agent

Technical/Other Names:
Polyethylene Glycol (5) Stearyl Tricitrate
Polyoxyethylene (5) Stearyl Tricitrate

Trade Name Mixture:
  Trioxene S (Vevy)

PEG-5 TSUBAKIATE GLYCERIDES

Definition: PEG-5 Tsubakiate Glycerides is the polyethylene glycol derivative of the mono- and diglycerides derived from Camellia Japonica Seed Oil (q.v.) containing an average of 5 moles of ethylene oxide.

Chemical Class: Alkoxylated Carboxylic Acids

Function: Surfactant - Emulsifying Agent

Trade Name:
  PEG-5 Tsubaki Oil (Oshima)

PEG-10 TSUBAKIATE GLYCERIDES

Definition: PEG-10 Tsubakiate Glycerides is the polyethylene glycol derivative of the mono- and diglycerides derived from Camellia Japonica Seed Oil (q.v.) containing an average of 10 moles of ethylene oxide.

Chemical Class: Alkoxylated Carboxylic Acids

Function: Surfactant - Emulsifying Agent

Trade Name:
  PEG-10 Tsubaki Oil (Oshima)

PEG-20 TSUBAKIATE GLYCERIDES

Definition: PEG-20 Tsubakiate Glycerides is the polyethylene glycol derivative of the mono- and diglycerides derived from Camellia Japonica Seed Oil (q.v.) containing an average of 20 moles of ethylene oxide.

Chemical Class: Alkoxylated Carboxylic Acids

Function: Surfactant - Emulsifying Agent

Trade Name:
  PEG-20 Tsubaki Oil (Oshima)

PEG-60 TSUBAKIATE GLYCERIDES

Definition: PEG-60 Tsubakiate Glycerides is the polyethylene glycol derivative of the mono- and diglycerides derived from Camellia Japonica Seed Oil (q.v.) containing an average of 60 moles of ethylene oxide.

Chemical Class: Alkoxylated Carboxylic Acids

Function: Surfactant - Cleansing Agent

Trade Name:
  PEG-60 Tsubaki Oil (Oshima)

PEG-6 UNDECYLENATE

Definition: PEG-6 Undecylenate is the polyethylene glycol ester of undecylenic acid that conforms generally to the formula:

$$CH_2{=}CH(CH_2)_8\overset{\displaystyle O}{\overset{\displaystyle \|}{C}}{-}(OCH_2CH_2)_nOH$$

where n has an average value of 6.

Chemical Class: Esters

Functions: Cosmetic Biocide; Skin-Conditioning Agent - Miscellaneous; Surfactant - Emulsifying Agent

Trade Names:
  Undelene (Vevy)
  Undexal (Vevy)

PEG-8 UNDECYLENATE

Empirical Formula:
  $C_{27}H_{52}O_{10}$

Definition: PEG-8 Undecylenate is the polyethylene glycol ester of undecylenic acid that conforms generally to the formula:

$$CH_2{=}CH(CH_2)_8\overset{\displaystyle O}{\overset{\displaystyle \|}{C}}{-}(OCH_2CH_2)_nOH$$

where n has an average value of 8.

Chemical Class: Alkoxylated Carboxylic Acids

Functions: Cosmetic Biocide; Surfactant - Emulsifying Agent

Technical/Other Names:
  Polyethylene Glycol 400 Monoundecylenate
  Polyoxyethylene (8) Monoundecylenate

Trade Name:
  Dermocide EU (Fabriquimica)

PEI-7

CAS No.: 9002-98-6 (Generic)

JPN Translation:
  PEI - 7

Empirical Formula:
  $C_{14}H_{35}N_7$

Definition: PEI-7 is the polymer of ethylenimine that conforms generally to the formula:

$$(CH_2CH_2NH)_n$$

where n has an average value of 7.

Information Sources: 21CFR175.105, 21CFR175.320, 21CFR176.170, 21CFR177.1200, TSCA

Chemical Classes: Amines; Synthetic Polymers

Function: Suspending Agent - Nonsurfactant

Technical/Other Name:
  Polyethylenimine 7

PEI-10

CAS No.: 9002-98-6 (Generic)

JPN Translation:
  PEI - 10

Empirical Formula:
  $C_{20}H_{50}N_{10}$

Definition: PEI-10 is the polymer of ethylenimine that conforms generally to the formula:

$$(CH_2CH_2NH)_n$$

where n has an average value of 10.

Chemical Classes: Amines; Synthetic Polymers

Function: Suspending Agent - Nonsurfactant

Technical/Other Name:
  Polyethylenimine 10

Trade Name:
  Lupasol FG (BASF)

PEI-15

CAS No.: 9002-98-6 (Generic)

JPN Translation:
  PEI - 15

Definition: PEI-15 is the polymer of ethylenimine that conforms generally to the formula:

$$(CH_2CH_2NH)_n$$

where n has an average value of 15.

Information Sources: 21CFR175.105, 21CFR175.320, 21CFR176.170, 21CFR177.1200, TSCA

Chemical Classes: Amines; Synthetic Polymers

Function: Suspending Agent - Nonsurfactant

Technical/Other Name:
  Polyethylenimine 15

Trade Name:
  Epomin SP-006 (Aceto)

PEI-30

CAS No.: 9002-98-6 (Generic)

JPN Translation:
PEI - 30

Definition: PEI-30 is the polymer of ethylenimine that conforms generally to the formula:

$$(CH_2CH_2NH)_n$$

where n has an average value of 30.

Information Sources: 21CFR175.105, 21CFR175.320, 21CFR176.170, 21CFR177.1200, TSCA

Chemical Classes: Amines; Synthetic Polymers

Function: Suspending Agent - Nonsurfactant

Technical/Other Name:
Polyethylenimine 30

Trade Name:
Epomin SP-012 (Aceto)

PEI-35

CAS No.: 9002-98-6 (Generic)

JPN Translation:
PEI - 35

Definition: PEI-35 is the polymer of ethylenimine that conforms generally to the formula:

$$(CH_2CH_2NH)_n$$

where n has an average value of 35.

Chemical Classes: Amines; Synthetic Polymers

Function: Suspending Agent - Nonsurfactant

Technical/Other Name:
Polyethylenimine 35

Trade Name:
Lupasol G-35 (BASF)

PEI-45

CAS No.: 9002-98-6 (Generic)

JPN Translation:
PEI - 45

Definition: PEI-45 is the polymer of ethylenimine that conforms generally to the formula:

$$(CH_2CH_2NH)_n$$

where n has an average value of 45.

Information Sources: 21CFR175.105, 21CFR175.320, 21CFR176.170, 21CFR177.1200, TSCA

Chemical Classes: Amines; Synthetic Polymers

Function: Suspending Agent - Nonsurfactant

Technical/Other Name:
Polyethylenimine 45

Trade Name:
Epomin SP-018 (Aceto)

PEI-250

CAS No.: 9002-98-6 (Generic)

JPN Translation:
PEI - 250

Definition: PEI-250 is the polymer of ethylenimine that conforms generally to the formula:

$$(CH_2CH_2NH)_n$$

where n has an average value of 250.

Chemical Classes: Amines; Synthetic Polymers

Function: Suspending Agent - Nonsurfactant

Technical/Other Name:
Polyethylenimine 250

Trade Name:
Lupasol Waterfree (BASF)

PEI-275

CAS No.: 9002-98-6 (Generic)

JPN Translation:
PEI - 275

Definition: PEI-275 is the polymer of ethylenimine that conforms generally to the formula:

$$(CH_2CH_2NH)_n$$

where n has an average value of 275.

Information Sources: 21CFR175.105, 21CFR175.320, 21CFR176.170, 21CFR177.1200

Chemical Classes: Amines; Synthetic Polymers

Function: Suspending Agent - Nonsurfactant

Technical/Other Name:
Polyethylenimine 275

PEI-700

CAS No.: 9002-98-6 (Generic)

JPN Translation:
PEI - 700

Definition: PEI-700 is the polymer of ethylenimine that conforms generally to the formula:

$$(CH_2CH_2NH)_n$$

where n has an average value of 700.

Information Sources: 21CFR175.105, 21CFR175.320, 21CFR176.170, 21CFR177.1200

Chemical Classes: Amines; Synthetic Polymers

Function: Suspending Agent - Nonsurfactant

Technical/Other Name:
Polyethylenimine 700

PEI-1000

CAS No.: 9002-98-6 (Generic)

JPN Translation:
PEI - 1000

Definition: PEI-1000 is the polymer of ethylenimine that conforms generally to the formula:

$$(CH_2CH_2NH)_n$$

where n has an average value of 1000.

Information Sources: 21CFR175.105, 21CFR175.320, 21CFR176.170, 21CFR177.1200, TSCA

Chemical Classes: Amines; Synthetic Polymers

Function: Suspending Agent - Nonsurfactant

Technical/Other Names:
Aziridine, Homopolymer
Polyethylenimine 1000

PEI-1400

CAS No.: 9002-98-6 (Generic)

JPN Translation:
PEI - 1400

Definition: PEI-1400 is the polymer of ethylenimine that conforms generally to the formula:

$$(CH_2CH_2NH)_n$$

where n has an average value of 1400.

Information Sources: 21CFR175.105, 21CFR175.320, 21CFR176.170, 21CFR177.1200

Chemical Classes: Amines; Synthetic Polymers

Function: Suspending Agent - Nonsurfactant

Technical/Other Name:
Polyethylenimine 1400

PEI-1500

CAS No.: 9002-98-6 (Generic)

JPN Translation:
PEI - 1500

Definition: PEI-1500 is the polymer of ethylenimine that conforms generally to the formula:

$$(CH_2CH_2NH)_n$$

where n has an average value of 1500.

Information Sources: 21CFR175.105, 21CFR175.320, 21CFR176.170, 21CFR177.1200, TSCA

Chemical Classes: Amines; Synthetic Polymers

Function: Suspending Agent - Nonsurfactant

Technical/Other Name:
Polyethylenimine 1500

Trade Names:
Lupasol P (BASF)
Lupasol PS (BASF)

PEI-1750

CAS No.: 9002-98-6 (Generic)

JPN Translation:
PEI - 1750

Definition: PEI-1750 is the polymer of ethylenimine that conforms generally to the formula:

$$(CH_2CH_2NH)_n$$

where n has an average value of 1750.

Information Sources: 21CFR175.105, 21CFR175.320, 21CFR176.170, 21CFR177.1200

Chemical Classes: Amines; Synthetic Polymers

Function: Suspending Agent - Nonsurfactant

Reported Product Category: Hair Conditioners

Technical/Other Name:
Polyethylenimine 1750

Trade Name:
Epomin P-1000 (Aceto)

PEI-2500

CAS No.: 9002-98-6 (Generic)

JPN Translation:
PEI - 2500

Definition: PEI-2500 is the polymer of ethylenimine that conforms generally to the formula:

$$(CH_2CH_2NH)_n$$

where n has an average value of 2500.

Information Sources: 21CFR175.105, 21CFR175.320, 21CFR176.170, 21CFR177.1200, TSCA

Chemical Classes: Amines; Synthetic Polymers

Function: Suspending Agent - Nonsurfactant

Technical/Other Name:
Polyethylenimine 2500

PEI-14M

CAS No.: 9002-98-6

JPN Translation:
PEI - 14M

Definition: PEI-14M is the polymer of ethylenimine that conforms generally to the formula:

$$(CH_2CH_2NH)_n$$

where n has an average value of 14000.

Chemical Classes: Amines; Synthetic Polymers

Function: Suspending Agent - Nonsurfactant

Technical/Other Name:
Polyethylenimine 14000

PELARGONIC ACID

CAS No.   **EINECS No.**
112-05-0   203-931-2

Empirical Formula:
$C_9H_{18}O_2$

Definition: Pelargonic Acid is an acid that conforms to the formula:

$$CH_3(CH_2)_7COOH$$

Information Sources: 21CFR172.515, 21CFR173.315, 21CFR178.1010, MI-13 (7141), RIFM, TSCA

Chemical Class: Fatty Acids

Functions: Fragrance Ingredient; Surfactant - Cleansing Agent; Surfactant - Emulsifying Agent

Surfactant-Cleansing Agent is included as a function for the soap form of Pelargonic Acid.

Technical/Other Names:
Nonanoic acid (RIFM)
Nonanoic Acid
Nonoic Acid
Nonylic Acid
1-Octanecarboxylic Acid
Pelargic Acid
Pergonic Acid

PELARGONIUM CAPITATUM LEAF EXTRACT

Definition: Pelargonium Capitatum Leaf Extract is an extract of the leaves of the rose geranium, *Pelargonium capitatum*. See "Regulatory and Ingredient Use Information," regarding the labeling names for botanical ingredients in Volume 1, Introduction, Part A.

Chemical Class: Biological Products

Function: Not Reported

Technical/Other Names:
Extract of Rose Geranium
Rose Geranium Leaf Extract

Trade Name Mixtures:
Actiphyte of Rose Geranium BG50 (Active Organics)
Actiphyte of Rose Geranium GL50 (Active Organics)
Actiphyte of Rose Geranium Lipo S (Active Organics)
Actiphyte of Rose Geranium PG50 (Active Organics)

PELARGONIUM GRAVEOLENS EXTRACT

CAS No.   **EINECS No.**
90082-51-2   290-140-0

Definition: Pelargonium Graveolens Extract is an extract of of *Pelargonium graveolens*. See "Regulatory and Ingredient Use Information," regarding the labeling names for botanical ingredients in Volume 1, Introduction, Part A.

Chemical Class: Biological Products

Functions: Fragrance Ingredient; Skin-Conditioning Agent - Emollient

Technical/Other Names:
Extract of Pelargonium Graveolens
Rose Geranium Extract

PELARGONIUM GRAVEOLENS FLOWER OIL

CAS No.: 8000-46-2

Definition: Pelargonium Graveolens Flower Oil is the volatile oil obtained from the flowers of *Pelargonium graveolens.* See *"Regulatory and Ingredient Use Information,"* regarding the labeling names for botanical ingredients in Volume 1, Introduction, Part A.

Information Source: RIFM

Chemical Class: Essential Oils

Function: Fragrance Ingredient

Reported Product Categories: Skin Care Preparations, Misc.; Bath Preparations, Misc.; Body and Hand Preparations (Excluding Shaving Preparations); Bath Capsules; Face and Neck Preparations (Excluding Shaving Preparations); Moisturizing Preparations; Bath Soaps and Detergents; Bubble Baths

Technical/Other Names:
Geranium absolute (Pelaragonium graveolens) (RIFM)
Geranium oil, African (RIFM)
Geranium oil, bourbon (RIFM)
Geranium oil, Chinese (RIFM)
Oils, Pelargonium Graveolens
Rose Geranium Flower Oil

Trade Name Mixtures:
Essentiaderm N.14 (Universal Flavors)
Essentiaderm N.19 (Universal Flavors)
Germinol (Dr. Gerhard Steidl)

PELARGONIUM GRAVEOLENS WATER

Definition: Pelargonium Graveolens Water is an aqueous solution of the steam distillate obtained from *Pelargonium graveolens.* See *"Regulatory and Ingredient Use Information,"* regarding the labeling names for botanical ingredients in Volume 1, Introduction, Part A.

Chemical Class: Biological Products

Functions: Cosmetic Astringent; Deodorant Agent; Fragrance Ingredient; Hair Conditioning Agent

Trade Name:
Rose Geranium Hydroflorate (Bayliss Ranch)

PELARGONIUM GRAVEOLENS WAX

Definition: Pelargonium Graveolens Wax is a wax obtained from the whole Geranium plant *Pelargonium graveolens.* See *"Regulatory and Ingredient Use Information,"* regarding the labeling names for botanical ingredients in Volume 1, Introduction, Part A.

Chemical Class: Essential Oils

Function: Fragrance Ingredient

Technical/Other Names:
Rose Geranium Wax
Waxes, Pelargonium Graveolens

Trade Names:
AEC Cire Essentielle De Geranium Rosat (A & E Connock)
Cire Essentielle de Geranium Rosat (Bertin)

PELARGONIUM PELTATUM EXTRACT

Definition: Pelargonium Peltatum Extract is an extract of the pelargonium, *Pelargonium peltatum.* See *"Regulatory and Ingredient Use Information,"* regarding the labeling names for botanical ingredients in Volume 1, Introduction, Part A.

Chemical Class: Biological Products

Function: Skin-Conditioning Agent - Occlusive

Technical/Other Names:
Extract of Pelargonium
Extract of Pelargonium Peltatum
Pelargonium Extract

Trade Name:
Hydroessential Pelargonium (Vevy)

PELVETIA CANALICULATA EXTRACT

Definition: Pelvetia Canaliculata Extract is an extract of the alga, *Pelvetia canaliculata.* See *"Regulatory and Ingredient Use Information,"* regarding the labeling names for botanical ingredients in Volume 1, Introduction, Part A.

Chemical Class: Biological Products

Function: Not Reported

Technical/Other Name:
Extract of Pelvetia Canaliculata

Trade Name Mixtures:
Bio-Energizer (SECMA)
Bioenergizer BG (SECMA)
Efficiensea (GELYMA)
Pelvetiane (SECMA)
Phycol PC (SECMA)
Phycol PC BG (SECMA)

PENTADECALACTONE

CAS No. 106-02-5   **EINECS No.** 203-354-6

JPN Translation:
ペンタデカラクトン

Function: Fragrance Ingredient

Empirical Formula:
$C_{15}H_{28}O_2$

Definition: Pentadecalactone is the lactone of 15-hydroxypentadecanoic acid that conforms to the formula:

$$CH_2(CH_2)_{13}CO$$
$$\llcorner\text{—}O\text{—}\lrcorner$$

Information Sources: 21CFR172.515, JCIC, JCLS, MI-13(3942), RIFM, TSCA

Chemical Classes: Esters; Heterocyclic Compounds

Function: Fragrance Ingredient

Reported Product Categories: Personal Cleanliness Products, Misc.; Moisturizing Preparations

Technical/Other Names:
Oxacyclohexadecan-2-one
omega-Pentadecalactone (RIFM)

Trade Names:
Exaltex (Firmenich)
Exaltolide (Firmenich)

PENTADECYL ALCOHOL

CAS No. 629-76-5   **EINECS No.** 211-107-9

Empirical Formula:
$C_{15}H_{32}O$

Definition: Pentadecyl Alcohol is the aliphatic alcohol that conforms to the formula:
$$CH_3(CH_2)_{14}OH$$

Information Source: RIFM

Chemical Class: Alcohols

Functions: Emulsion Stabilizer; Fragrance Ingredient; Skin-Conditioning Agent - Emollient

Technical/Other Names:
Pentadecanol
1-Pentadecanol (RIFM)

PENTADESMA BUTYRACEA

Definition: See *"Regulatory and Ingredient Use Information,"* regarding EU labeling names for botanical ingredients in Volume 1, Introduction, Part A.

Chemical Class: Biological Products

Technical/Other Name:
Pentadesma Butyracea Seed Butter (U.S.)

PENTADESMA BUTYRACEA SEED BUTTER

Definition: Pentadesma Butyracea Seed Butter is the oily fat extracted from the nut of *Pentadesma butyracea.* See "Regulatory and Ingredient Use Information," regarding the labeling names for botanical ingredients in Volume 1, Introduction, Part A.

Chemical Class: Biological Products

Function: Skin-Conditioning Agent - Occlusive

Technical/Other Names:
Butter, Pentadesma Butyracea
Kanya Butter
Pentadesma Butyracea (EU)

PENTADIPLANDRA BRAZZEANA ROOT EXTRACT

Definition: Pentadiplandra Brazzeana Root Extract is an extract of the roots of *Pentadiplandra brazzeana.* See "Regulatory and Ingredient Use Information," regarding the labeling names for botanical ingredients in Volume 1, Introduction, Part A.

Chemical Class: Biological Products

Function: Skin-Conditioning Agent - Miscellaneous

Technical/Other Name:
Extract of Pentadiplandra Brazzeana Root

Trade Name Mixture:
Pentadiplandra Brazzeana Root Extract (Plantes et Industrie)

PENTADOXYNOL-200

CAS No.: 40160-92-7 (Generic)

Empirical Formula:
$(C_2H_4O)_nC_{21}H_{36}O$

Definition: Pentadoxynol-200 is the ethoxylated alkyl phenol that conforms generally to the formula:

$$C_{15}H_{31}C_6H_4(OCH_2CH_2)_nOH$$

where n has an average value of 200.

Chemical Class: Alkoxylated Alcohols

Functions: Surfactant - Cleansing Agent; Surfactant - Solubilizing Agent

Reported Product Category: Deodorants (Underarm)

Technical/Other Names:
PEG-200 Pentadecyl Phenyl Ether
Polyethylene Glycol (200) Pentadecyl Phenyl Ether
Polyoxyethylene (200) Pentadecyl Phenyl Ether

Trade Names:
Clarit PDP-200 (Application)
Clarit PDP-200 (RTD Chemicals)

PENTAERYTHRITOL/TEREPHTHALIC ACID COPOLYMER

Definition: Pentaerythritol/Terephthalic Acid Copolymer is a copolymer of pentaerythritol and terephthalic acid monomers.

Chemical Class: Synthetic Polymers

Function: Film Former

PENTAERYTHRITYL ADIPATE/CAPRATE/CAPRYLATE/HEPTANOATE

CAS No.	EINECS No.
68130-55-2	268-597-2

Definition: Pentaerythrityl Adipate/Caprate/Caprylate/Heptanoate is the mixed ester of pentaerythritol and adipic, capric, caprylic and heptanoic acids.

Information Source: TSCA

Chemical Class: Esters

Functions: Skin-Conditioning Agent - Occlusive; Viscosity Increasing Agent - Nonaqueous

Trade Names:
PureSyn ME100 (ExxonMobil/Synthetics)
PureSyn ME450 (ExxonMobil/Synthetics)
PureSyn ME2500 (ExxonMobil/Synthetics)

PENTAERYTHRITYL COCOATE

Definition: Pentaerythrityl Cocoate is the ester of Coconut Acid (q.v.) and pentaerythritol.

Information Source: JCLS

Chemical Class: Esters

Function: Skin-Conditioning Agent - Miscellaneous

PENTAERYTHRITYL CYCLOHEXANE DI-CARBOXYLATE

Definition: Pentaerythrityl Cyclohexane Dicarboxylate is the product formed from the esterification of Pentaerythritol (q.v.) and cyclohexane-1,2-dicarbonic acid.

Information Source: JCIC

Chemical Class: Esters

Function: Film Former

Technical/Other Name:
Cyclohexane Alkyd Resin

PENTAERYTHRITYL DIOLEATE

CAS No.	EINECS No.
25151-96-6	246-665-2

Empirical Formula:
$C_{41}H_{76}O_6$

Definition: Pentaerythrityl Dioleate is diester of pentaerythritol and oleic acid that conforms to the formula:

$$RC{-}OCH_2{-}\overset{\displaystyle CH_2OH}{\underset{\displaystyle CH_2OH}{\overset{O}{\|}C}}{-}CH_2O{-}\overset{O}{\overset{\|}{C}}R$$

where RCO- represents the oleic acid radical.

Information Source: 21CFR176.210

Chemical Class: Esters

Functions: Skin-Conditioning Agent - Emollient; Viscosity Increasing Agent - Nonaqueous

Technical/Other Name:
9-Octadecenoic Acid, 2,2-Bis(Hydroxymethyl)-1,3-PropanediylEster

Trade Names:
AEC Pentaerythrityl Dioleate (A & E Connock)
Radiasurf 7156 (Oleon NV)

PENTAERYTHRITYL DISTEARATE

CAS No.	EINECS No.
13081-97-5	235-991-0

Empirical Formula:
$C_{41}H_{80}O_6$

Definition: Pentaerythrityl Distearate is the diester of pentaerythritol and stearic acid. It conforms to the formula:

$$RC{-}OCH_2{-}\overset{\displaystyle CH_2OH}{\underset{\displaystyle CH_2OH}{\overset{O}{\|}C}}{-}CH_2O{-}\overset{O}{\overset{\|}{C}}R$$

where RCO- represents the stearic acid radical.

Information Sources: 21CFR176.210, 21CFR178.2010

Chemical Class: Esters

Functions: Skin-Conditioning Agent - Emollient; Viscosity Increasing Agent - Nonaqueous

Technical/Other Names:
Octadecanoic Acid, 2,2-Bis (Hydroxymethyl)-1,3-Propanediyl Ester
Stearic Acid, 2,2-bis(Hydroxymethyl)-trimethylene Ester

Trade Name:
Radiasurf 7175 (Oleon NV)

PENTAERYTHRITYL HYDROGENATED ROSINATE

CAS No.	EINECS No.
64365-17-9	264-848-5

Definition: Pentaerythrityl Hydrogenated Rosinate is the ester of pentaerythritol and hydrogenated acids derived from Rosin (q.v.).

Information Sources: 21CFR176.210, 21CFR178.3120, 21CFR178.3800, 21CFR178.3870

Chemical Class: Esters

Functions: Skin-Conditioning Agent - Occlusive; Viscosity Increasing Agent - Nonaqueous

Reported Product Category: Mascara

Trade Names:
Foral 105 (Hercules)
Foral 105-E (Eastman Chemical)
Foralyn 110 (Eastman Chemical)
Pentalyn H-E (Eastman Chemical)
Pentalyn H (Hercules)

PENTAERYTHRITYL ISOSTEARATE/ CAPRATE/CAPRYLATE/ADIPATE

Definition: Pentaerythrityl Isostearate/ Caprate/Caprylate/Adipate is the mixed ester of pentaerythritol and isostearic, capric, caprylic and adipic acids.

Chemical Class: Esters

Functions: Skin-Conditioning Agent - Occlusive; Viscosity Increasing Agent - Nonaqueous

Trade Name:
Supermol L (Croda Chemicals)

PENTAERYTHRITYL ROSINATE

CAS No.	EINECS No.
8050-26-8	232-479-9

JPN Translation:
ロジン酸ペンタエリスリチル

Definition: Pentaerythrityl Rosinate is the ester of rosin acids derived from Rosin (q.v.), with the polyol, pentaerythritol.

Information Sources: 21CFR172.615, 21CFR175.105, 21CFR175.300, 21CFR176.170, 21CFR176.210, 21CFR178.3120, 21CFR178.3800, 21CFR178.3870, CIR: [I] IJT-17(Suppl. 4)-1998, CIR: [I] JACT-13(5)1994, FCC, JCIC, JCLS, JSQI

Chemical Class: Esters

Functions: Skin-Conditioning Agent - Emollient; Viscosity Increasing Agent - Nonaqueous

Reported Product Category: Mascara

Technical/Other Name:
Pentaerythritol Rosinate

Trade Names:
Pentalyn 344 (Eastman Chemical)
Pentalyn A (Hercules)
Resiester N 35 S (La Union Resinera)
Zonester 100 (Arizona)

Trade Name Mixture:
Rosin Esters (Arakawa)

PENTAERYTHRITYL STEARATE

Empirical Formula:
$C_{23}H_{46}O_5$

Definition: Pentaerythrityl Stearate is the ester of pentaerythritol and stearic acid that conforms to the formula:

$$HOCH_2 - \overset{\displaystyle CH_2OH}{\underset{\displaystyle CH_2OH}{C}} - CH_2O - \overset{\displaystyle O}{\overset{\|}{C}}(CH_2)_{16}CH_3$$

Chemical Class: Esters

Functions: Skin-Conditioning Agent - Emollient; Surfactant - Emulsifying Agent

PENTAERYTHRITYL STEARATE/ CAPRATE/CAPRYLATE/ADIPATE

Definition: Pentaerythrityl Stearate/Caprate/ Caprylate/Adipate is the mixed ester of pentaerythritol and stearic, capric, caprylic and adipic acids.

Chemical Class: Esters

Functions: Skin-Conditioning Agent - Occlusive; Viscosity Increasing Agent - Nonaqueous

Trade Name:
Supermol S (Croda Chemicals)

PENTAERYTHRITYL STEARATE/ ISOSTEARATE/ADIPATE/ HYDROXYSTEARATE

Definition: Pentaerythrityl Stearate/ Isostearate/Adipate/Hydroxystearate is the mixed ester of Pentaerythritol (q.v.) with Stearic Acid (q.v.), Isostearic Acid (q.v.), Adipic Acid (q.v.), and Hydroxystearic Acid (q.v.).

Chemical Class: Esters

Functions: Skin-Conditioning Agent - Occlusive; Viscosity Increasing Agent - Nonaqueous

PENTAERYTHRITYL TETRAABIETATE

CAS No.	EINECS No.
127-23-1	204-830-6

Empirical Formula:
$C_{85}H_{124}O_8$

Definition: Pentaerythrityl Tetraabietate is the tetraester of pentaerythritol and abietic acid. It conforms to the formula:

$$RC - OCH_2 - \overset{\displaystyle CH_2O - CR}{\underset{\displaystyle CH_2O - CR}{C}} - CH_2O - CR$$

where RCO- represents the abietic acid radical.

Information Source: TSCA

Chemical Class: Esters

Functions: Binder; Skin-Conditioning Agent - Occlusive; Viscosity Increasing Agent - Nonaqueous

Reported Product Category: Mascara

Technical/Other Names:
Pentaerythritol Tetraabietate
Podocarpa-8c12-dien-15-oic Acid, 13-Isopropyl-, Neopentanetetrayl Ester

Trade Name:
Pentalyn H (Hercules)

PENTAERYTHRITYL TETRAACETATE

CAS No.	EINECS No.
597-71-7	209-907-8

Empirical Formula:
$C_{13}H_{20}O_8$

Definition: Pentaerythrityl Tetraacetate is the tetraester of pentaerythritol and acetic acid. It conforms to the formula:

$$CH_3C - OCH_2 - \overset{\displaystyle CH_2O - CCH_3}{\underset{\displaystyle CH_2O - CCH_3}{C}} - CH_2O - CCH_3$$

Chemical Class: Esters

Function: Skin-Conditioning Agent - Miscellaneous

Technical/Other Names:
2,2-Bis[(Acetyloxy)Methyl]-1,3-Propanediyl Diacetate

1,3-Propanediyl, 2,2-Bis[(Acetyloxy)-Methyl]-, Diacetate

Trade Name:
Pelemol PTA (Phoenix)

PENTAERYTHRITYL TETRABEHENATE

CAS No.   **EINECS No.**
61682-73-3   262-895-6

Empirical Formula:
$C_{93}H_{180}O_8$

Definition: Pentaerythrityl Tetrabehenate is the tetraester of pentaerythritol and Behenic Acid (q.v.). It conforms to the formula:

where RCO- represents the behenic acid radical.

Chemical Class: Esters

Functions: Binder; Skin-Conditioning Agent - Occlusive; Viscosity Increasing Agent - Nonaqueous

Technical/Other Names:
2,2-Bis[[(1-Oxodocosyl)Oxy]Methyl]-1,3-Propanediyl Docosanoate
Docosanoic Acid, 2,2-Bis[[(1-Oxodocosyl)oxy]Methyl]-1,3-Propanediyl Ester
Pentaerythritol Tetradocosanoate

Trade Names:
Ethox PB-4 (Ethox)
Liponate PB-4 (Lipo)

PENTAERYTHRITYL TETRABENZOATE

Empirical Formula:
$C_{33}H_{28}O_8$

Definition: Pentaerythrityl Tetrabenzoate is the tetraester of pentaerythritol and benzoic acid. It conforms to the formula:

where RCO- represents the benzoic acid radical.

Chemical Class: Esters

Functions: Binder; Skin-Conditioning Agent - Emollient; Viscosity Increasing Agent - Nonaqueous

Trade Name:
Uniplex 552 (Unitex)

PENTAERYTHRITYL TETRA C5-9 ACID ESTERS

CAS No.   **EINECS No.**
67762-53-2   267-022-2

Definition: Pentaerythrityl Tetra C5-9 Acid Esters is the tetraester of branched and linear C5-9 acids with Pentaerythritol (q.v.).

Chemical Class: Esters

Functions: Hair Conditioning Agent; Plasticizer; Skin-Conditioning Agent - Miscellaneous

Technical/Other Name:
Carboxylic Acids, C5-9, Tetraesters wtih Pentaerythritol

Trade Name Mixture:
LexFilm 59 (Inolex)

PENTAERYTHRITYL TETRACAPRYLATE/ TETRACAPRATE

CAS No.   **EINECS No.**
68441-68-9   270-474-3

Definition: Pentaerythrityl Tetracaprylate/Tetracaprate is the tetraester of pentaerythritol and a blend of caprylic and capric acids.

Chemical Class: Esters

Functions: Skin-Conditioning Agent - Occlusive; Viscosity Increasing Agent - Nonaqueous

Technical/Other Name:
Decanoic Acid, Mixed Esters with Octanoic Acid and Pentaerythritol

Trade Names:
Crodamol PTC (Croda Chemicals)
Crodamol PTC (Croda, Inc.)
Hatcol 5190 (Hatco)
Liponate PE-810 (Lipo)
PureSyn 4E30 (ExxonMobil/Synthetics)
Radia 7178 (Oleon NV)
STEPAN PTC (Stepan)

PENTAERYTHRITYL TETRACOCOATE

Definition: Pentaerythrityl Tetracocoate is the tetraester of pentaerythritol and coconut fatty acid. It conforms generally to the formula:

where RCO- represents the coconut acid radical.

Information Sources: 21CFR175.300, 21CFR176.210

Chemical Class: Esters

Functions: Binder; Skin-Conditioning Agent - Occlusive; Viscosity Increasing Agent - Nonaqueous

PENTAERYTHRITYL TETRA-DI-t-BUTYL HYDROXYHYDROCINNAMATE

CAS No.: 6683-19-8

Empirical Formula:
$C_{73}H_{108}O_{12}$

Definition: Pentaerythrityl Tetra-di-t-butyl Hydroxyhydrocinnamate is the organic compound that conforms to the formula:

Chemical Classes: Esters; Phenols

Function: Antioxidant

Technical/Other Name:
Pentaerythritol Tetrakis(4-Hydroxy-3,5-Di-Tert-Butyl)Hydrocinnamate

Trade Names:
ANOX 20 (Coloplast Consumer)
Irganox 1010 (Wilde Cosmetics)

PENTAERYTHRITYL TETRAETHYLHEXANOATE

CAS No.   **EINECS No.**
7299-99-2   230-743-8

JPN Translation:
テトラオクタン酸ペンタエリスリチル

Empirical Formula:
$C_{37}H_{68}O_8$

The inclusion of any compound in the *Dictionary and Handbook* does not indicate that use of that substance as a cosmetic ingredient complies with the laws and regulations governing such use in the United States or any other country.

Definition: Pentaerythrityl Tetraethylhexanoate is the tetraester of pentaerythritol and 2-ethylhexanoic acid. It conforms generally to the formula:

where RCO- represents the 2-ethylhexanoic acid radical.

Information Sources: JCIC, JCLS, JSQI, TSCA

Chemical Class: Esters

Functions: Binder; Skin-Conditioning Agent - Occlusive; Viscosity Increasing Agent - Nonaqueous

Reported Product Categories: Bath Capsules; Moisturizing Preparations; Skin Care Preparations, Misc.; Bath Oils, Tablets, and Salts; Cleansing Products (Cold Creams, Cleansing Lotions, Liquids and Pads); Baby Lotions, Oils, Powders and Creams; Blushers (All types); Body and Hand Preparations (Excluding Shaving Preparations); Colognes and Toilet Waters; Eye Shadows; Face and Neck Preparations (Excluding Shaving Preparations); Face Powders; Foundations; Hair Coloring Preparations, Misc.; Hair Conditioners; Hair Rinses (Non-coloring); Lipsticks; Makeup Bases; Makeup Preparations (Not eye), Misc.; Men's Talcum; Night Skin Care Preparations; Paste Masks (Mud Packs); Perfumes; Powders (Dusting and Talcum, Excluding Aftershave Talcs); Shaving Cream (Aerosol, Brushless and Lather); Suntan Preparations, Misc.; Tonics, Dressings, and Other Hair Grooming Aids

Technical/Other Names:
2,2-Bis[[(1-Oxo-2-Ethylhexyl)Oxy]Methyl]-1,3-Propanediyl 2-Ethylhexanoate
2-Ethylhexanoic Acid, 2,2-Bis[[(1-Oxo-2-Ethylhexyl)Oxy]Methyl]-1,3-Propanediyl Ester
Hexanoic Acid, 2-Ethyl-, 2,2-Bis[[(2-Ethyl-1-Oxohexyl)Oxy]Methyl]-1,3-Propandiyl Ester
Pentaerythritol Tetra-2-ethylhexanoate
Pentaerythritol Tetraoctanoate
Pentaerythrityl Tetraoctanoate

Trade Names:
AEC Pentaerythrityl Tetraethylhexanoate (A & E Connock)
DUB PTO (Stearinerie Dubois Fils)
Hatcol 5140 (Hatco)
KAK PTO (Kokyu Alcohol)
NIKKOL Pentalan-408 (Nikko)

Nikkol Pentarate 408 (Nikko)
NS-408 (Nippon Chemical)
Salacos 5408 (Nisshin OilliO)
Trivent PE-48 (Trivent)

PENTAERYTHRITYL TETRAETHYLHEXANOATE/ TETRAMETHOXYCINNAMATE

Definition: Pentaerythrityl Tetraethylhexanoate/Tetramethoxycinnamate is the tetraester of pentaerythritol and a blend of ethylhexanoic and methoxycinnamic acids.

Chemical Class: Esters

Function: Skin-Conditioning Agent - Emollient

Trade Name:
Nomcort 58K (Nisshin OilliO)

PENTAERYTHRITYL TETRAISONONANOATE

CAS No. 93803-89-5　　**EINECS No.** 298-364-0

Empirical Formula:
$C_{41}H_{76}O_8$

Definition: Pentaerythrityl Tetraisononanoate is the tetraester of pentaerythritol and a branched chain nonanoic acid. It conforms generally to the formula:

where RCO- represents the isononanoic acid radical.

Chemical Class: Esters

Functions: Skin-Conditioning Agent - Occlusive; Viscosity Increasing Agent - Nonaqueous

Technical/Other Names:
2,2-Bis[[(1-Oxoisononyl)Oxy]Methyl]-1,3-Propanediyl Isononanoate
Isononanoic Acid, 2,2-Bis[[(1-Oxoisononyl)Oxy]Methyl]-1,3-Propanediyl Ester

Trade Name:
Pelemol P-49 (Phoenix)

PENTAERYTHRITYL TETRA-ISOSTEARATE

CAS No.: 62125-22-8

Empirical Formula:
$C_{77}H_{148}O_8$

Definition: Pentaerythrityl Tetraisostearate is the tetraester of isostearic acid and pentaerythritol that conforms to the formula:

where RCO- represents the isostearic acid radical.

Information Sources: JCIC, JCLS

Chemical Class: Esters

Functions: Binder; Skin-Conditioning Agent - Occlusive; Viscosity Increasing Agent - Nonaqueous

Reported Product Categories: Fragrance Preparations, Misc.; Eye Shadows; Makeup Preparations (Not eye), Misc.; Face Powders

Technical/Other Names:
2,2-Bis[[(1-Isooctadecyl)Oxy]Methyl]-1,3-Propanediyl Isooctadecanoate
Isooctadecanoic Acid, 2,2-Bis[[(1-Oxoisooctadecyl)Oxy]Methyl]-1,3-Propanediyl Ester
Pentaerythritol Tetraisooctanoate
Pentaerythritol Tetraisostearate

Trade Names:
AEC Pentaerythrityl Tetraisostearate (A & E Connock)
Crodamol PTIS (Croda Chemicals)
Crodamol PTIS (Croda, Inc.)
DUB PTIS (Stearinerie Dubois Fils)
KAK PTI (Kokyu Alcohol)
Kessco PTIS (Akzo Nobel Surface AB)
PRISORINE 3631 (Uniqema Europe)

Trade Name Mixtures:
Color Marine Filling Spheres (Coletica SA)
Color Marine Vitamine CPMG Spheres (Coletica SA)
Color Vegetal Filling Spheres (Coletica SA)
Complexe Hydroxy-Salicylique Color (Coletica SA)
Covaclear (LCW)

PENTAERYTHRITYL TETRALAURATE

CAS No. 13057-50-6　　**EINECS No.** 235-946-5

Empirical Formula:
$C_{53}H_{100}O_8$

The inclusion of any compound in the *Dictionary and Handbook* does not indicate that use of that substance as a cosmetic ingredient complies with the laws and regulations governing such use in the United States or any other country.

Definition: Pentaerythrityl Tetralaurate is the tetraester of pentaerythritol and lauric acid. It conforms to the formula:

where RCO- represents the lauric acid radical.

Information Source: 21CFR176.210

Chemical Class: Esters

Functions: Binder; Skin-Conditioning Agent - Occlusive; Viscosity Increasing Agent - Nonaqueous

Reported Product Category: Lipsticks

Technical/Other Names:
2,2-Bis[[(1-Oxododecyl)Oxy]Methyl]-1,3-Propanediyl Dodecanoate
Dodecanoic Acid, 2,2-Bis[[(1-Oxododecyl)Oxy]Methyl]-1,3-Propanediyl Ester

Trade Name:
Pelemol PTL (Phoenix)

PENTAERYTHRITYL TETRAMYRISTATE

CAS No.: 18641-59-3

JPN Translation:
テトラミリスチン酸ペンタエリスリチル

Empirical Formula:
$C_{61}H_{116}O_8$

Definition: Pentaerythrityl Tetramyristate is the tetraester of pentaerythritol and myristic acid. It conforms to the formula:

where RCO- represents the myristic acid radical.

Information Sources: JCIC, JCLS

Chemical Class: Esters

Functions: Skin-Conditioning Agent - Occlusive; Viscosity Increasing Agent - Nonaqueous

Technical/Other Names:
2,2-Bis[[(1-Oxotetradecyl)Oxy]Methyl]-1,3-Propanediyl Tetradecanoate

Pentaerythritol Tetramyristate
Tetradecanoic Acid, 2,2-Bis[[(1-Oxotetradecyl)Oxy]Methyl]-1,3-Propanediyl Ester

Trade Name:
Technol PTM (Technonet)

PENTAERYTHRITYL TETRAOLEATE

CAS No.	EINECS No.
19321-40-5	242-960-5

Empirical Formula:
$C_{77}H_{140}O_8$

Definition: Pentaerythrityl Tetraoleate is the tetraester of pentaerythritol and oleic acid. It conforms to the formula:

where RCO- represents the oleic acid radical.

Information Sources: 21CFR176.210, TSCA

Chemical Class: Esters

Functions: Binder; Skin-Conditioning Agent - Occlusive; Viscosity Increasing Agent - Nonaqueous

Technical/Other Names:
2,2-Bis[[(1-Oxo-9-Octadecenyl)Oxy]Methyl]-1,3-Propanediyl 9-Octadecenoate
9-Octadecenoic Acid, 2,2-Bis[[(1-Oxo-9-Octadecenyl)Oxy]Methyl]-1,3-Propanediyl Ester
Pentaerythritol Tetraoleate

Trade Names:
Liponate PO-4 (Lipo)
PureSyn 4E68 (ExxonMobil/Synthetics)

PENTAERYTHRITYL TETRA-PELARGONATE

CAS No.	EINECS No.
14450-05-6	238-430-8

Empirical Formula:
$C_{41}H_{76}O_8$

Definition: Pentaerythrityl Tetrapelargonate is the tetraester of pentaerythritol and pelargonic acid. It conforms to the formula:

where RCO- represents the pelargonic acid radical.

Chemical Class: Esters

Functions: Binder; Skin-Conditioning Agent - Occlusive; Viscosity Increasing Agent - Nonaqueous

Technical/Other Names:
2,2-Bis[[(1-Oxononyl)Oxy]Methyl]-1,3-Propanediyl Nonanoate
Nonanoic Acid, 2,2-Bis[[(1-Oxononyl)Oxy]Methyl]-1,3-Propanediyl Ester

Trade Name:
Pelemol PTP (Phoenix)

PENTAERYTHRITYL TETRASTEARATE

CAS No.	EINECS No.
115-83-3	204-110-1

Empirical Formula:
$C_{77}H_{148}O_8$

Definition: Pentaerythrityl Tetrastearate is the tetraester of pentaerythritol and stearic acid. It conforms to the formula:

where RCO- represents the stearic acid radical.

Information Sources: 21CFR175.105, 21CFR176.170, 21CFR176.210, 21CFR177.1200, 21CFR177.1580, 21CFR178.2010, TSCA

Chemical Class: Esters

Functions: Binder; Skin-Conditioning Agent - Occlusive; Viscosity Increasing Agent - Nonaqueous

Reported Product Category: Mascara

Technical/Other Names:
2,2-Bis[[(1-Oxooctadecyl)Oxy]Methyl]-1,3-Propanediyl Octadecanoate
Octadecanoic Acid, 2,2-Bis[[(1-Oxooctadecyl)Oxy]Methyl]-1,3-Propanediyl Ester
Pentaerythritol Tetrastearate

Trade Names:
AEC Pentaerythrityl Tetrastearate (A & E Connock)
Alkamuls PETS (Rhodia)
DUB PTS (Stearinerie Dubois Fils)
Liponate PS-4 (Lipo)
Radiasurf 7176 (Oleon NV)

PENTAERYTHRITYL TRIOLEATE

CAS No. 39874-62-9 **EINECS No.** 254-664-3

Empirical Formula:
$C_{59}H_{108}O_7$

Definition: Pentaerythrityl Trioleate is the triester of pentaerythritol and oleic acid. It conforms to the formula:

where RCO- represents the oleic acid radical.

Information Source: 21CFR176.210

Chemical Class: Esters

Functions: Binder; Skin-Conditioning Agent - Emollient; Viscosity Increasing Agent - Nonaqueous

Technical/Other Name:
9-Octadecenoic Acid, 2-(Hydroxymethyl)-2-[[(1-Oxo-9-Octadecenyl)Oxy]Methyl]-1, 3-Propanediyl Ester

PENTAHYDROSQUALENE

CAS No. 68629-07-2 **EINECS No.** 271-913-1

JPN Translation:
部分水添スクワレン

Empirical Formula:
$C_{30}H_{60}$

Definition: Pentahydrosqualene is the end product of the controlled hydrogenation of Squalene (q.v.).

Information Sources: JCIC, JCLS, JSQI, TSCA

Chemical Class: Hydrocarbons

Function: Skin-Conditioning Agent - Occlusive

Reported Product Categories: Blushers (All types); Moisturizing Preparations; Cleansing Products (Cold Creams, Cleansing Lotions, Liquids and Pads); Face Powders; Foundations

Technical/Other Names:
2,6,10,15,19,23-Hexamethyltetracosene Partially Hydrogenated Squalene Tetracosene, 2,6,10,15,19,23-Hexamethyl-

Trade Name:
AEC Pentahydrosqualene (A & E Connock)

PENTAMETHYLHEPTENONE

CAS Nos. 81786-73-4 **EINECS Nos.** 279-822-9
81786-74-5 279-823-4

Empirical Formula:
$C_{12}H_{22}O$

Definition: Pentamethylheptenone is the organic compound that conforms to the formula:

Chemical Class: Ketones

Function: Fragrance Ingredient

Trade Name:
Koavone (International Flavors)

PENTANE

CAS No. 109-66-0 **EINECS No.** 203-692-4

JPN Translation:
ペンタン

Empirical Formula:
C_5H_{12}

Definition: Pentane is the aliphatic hydrocarbon that conforms to the formula:

$$CH_3(CH_2)_3CH_3$$

Information Sources: 21CFR172.515, 21CFR175.105, 21CFR178.3010, JCIC, JCLS, MI-13(7193), TSCA

Chemical Class: Hydrocarbons

Functions: Propellant; Solvent; Viscosity Decreasing Agent

Reported Product Categories: Hair Sprays (Aerosol Fixatives); Shaving Cream (Aerosol, Brushless and Lather)

Technical/Other Name:
n-Pentane

Trade Name:
n-Pentane Cosmetic (Haltermann)

1,5-PENTANEDIOL

CAS No. 111-29-5 **EINECS No.** 203-854-4

Empirical Formula:
$C_5H_{12}O_2$

Definition: 1,5-Pentanediol is the organic compound that conforms to the formula:

$$HOCH_2CH_2CH_2CH_2CH_2OH$$

Information Source: MI-13(7194)

Chemical Class: Alcohols

Function: Solvent

Trade Name:
1,5-pentanediol (Natumin)

PENTAPEPTIDE-1

Definition: Pentapeptide-1 is the synthetic peptide containing these five amino acid residues: arginine, aspartic acid, lysine, tyrosine, and valine.

Chemical Class: Protein Derivatives

Function: Skin-Conditioning Agent - Miscellaneous

PENTAPOTASSIUM TRIPHOSPHATE

CAS No. 13845-36-8 **EINECS No.** 237-574-9

Empirical Formula:
$H_5O_{10}P_3 \cdot 5K$

Definition: Pentapotassium Triphosphate is the inorganic salt that conforms to the formula:

$$K_5P_3O_{10}$$

Information Sources: 21CFR173.310, 21CFR175.105, FCC, TSCA

Chemical Classes: Inorganic Salts; Phosphorus Compounds

Functions: Chelating Agent; pH Adjuster

Technical/Other Names:
Potassium Tripolyphosphate
Triphosphoric Acid, Pentapotassium Salt

PENTASODIUM AMINOTRIMETHYLENE PHOSPHONATE

CAS No. 2235-43-0 **EINECS No.** 218-791-8

Empirical Formula:
$C_3H_{12}NO_9P_3 \cdot 5Na$

Definition: Pentasodium Aminotrimethylene Phosphonate is the organic compound that conforms generally to the formula:

Chemical Classes: Amines; Phosphorus Compounds

Function: Chelating Agent

The inclusion of any compound in the *Dictionary and Handbook* does not indicate that use of that substance as a cosmetic ingredient complies with the laws and regulations governing such use in the United States or any other country.

Technical/Other Names:
Aminotri(Methylenephosphonic Acid)
Pentasodium Salt
Pentasodium [Nitrilotris(Methylene)]Tris-
Phosphonate
Phosphonic Acid, [Nitrilotris(Methylene)]-
Tris-, Pentasodium Salt

Trade Name:
Dequest 2006 (Solutia)

PENTASODIUM ETHYLENEDIAMINE TETRAMETHYLENE PHOSPHONATE

CAS No.
7651-99-2

EINECS No.
231-615-4

Empirical Formula:
$C_6H_{15}N_2O_{12}P_4Na_5$

Definition: Pentasodium Ethylenediamine Tetramethylene Phosphonate is the organic compound that conforms to the formula:

$$NaHO_3PCH_2 \diagdown \qquad \diagup CH_2PO_3HNa$$
$$NCH_2CH_2N$$
$$NaHO_3PCH_2 \diagup \qquad \diagdown CH_2PO_3Na_2$$

Chemical Classes: Amines; Organic Salts; Phosphorus Compounds

Function: Chelating Agent

Technical/Other Name:
Phosphonic Acid, (1,2-Ethanediylbis (nitrilobis(methylene)))tetrakis-, Penta-sodium Salt

Trade Name:
Dequest 2046 (Solutia)

PENTASODIUM PENTETATE

CAS No.
140-01-2

EINECS No.
205-391-3

JPN Translation:
ペンテト酸 5Na

Empirical Formula:
$C_{14}H_{23}N_3O_{10} \cdot 5Na$

Definition: Pentasodium Pentetate is the pentasodium salt of diethylenetriamine-pentaacetic acid. It conforms to the formula:

$$NaOOCCH_2 \diagdown \qquad \qquad \diagup CH_2COONa$$
$$NCH_2CH_2NCH_2CH_2N$$
$$NaOOCCH_2 \diagup \qquad | \qquad \diagdown CH_2COONa$$
$$CH_2COONa$$

Information Sources: 21CFR175.105, 21CFR176.150, CTFA D, JCIC, JCLS, JSCI, JSQI, TSCA

Chemical Classes: Amines; Organic Salts

Function: Chelating Agent

Reported Product Categories: Hair Dyes and Colors (All Types Requiring Caution Statements and Patch Tests); Bath Soaps and Detergents; Permanent Waves; Hair Color Sprays (Aerosol); Bath Oils, Tablets, and Salts; Colognes and Toilet Waters; Aftershave Lotions; Baby Shampoos; Hair Straighteners; Hair Wave Sets; Tonics, Dressings, and Other Hair Grooming Aids; Body and Hand Preparations (Excluding Shaving Preparations); Cleansing Products (Cold Creams, Cleansing Lotions, Liquids and Pads); Hair Coloring Preparations, Misc.; Hair Bleaches; Hair Preparations (Non-coloring), Misc.

Technical/Other Names:
N,N-Bis[2-[Bis(Carboxymethyl)Amino]-
Ethyl]Glycine, Pentasodium Salt
Glycine, N,N-Bis[2-[Bis(Carboxymethyl)-
Amino]Ethyl]-, Pentasodium Salt
Pentasodium Diethylenetriamine-
pentaacetate
Pentasodium Diethylenetriamine Penta-
acetate Solution

Trade Names:
AEC Pentasodium Pentetate (A & E
Connock)
Dissolvine D-40 (Akzo Nobel)
Mayoquest 300 (Vulcan Performance)
VERSENEX 80 (Dow Chemical)

Trade Name Mixture:
Mayoquest 1545M (Vulcan Performance)

PENTASODIUM TRIPHOSPHATE

CAS No.
7758-29-4

EINECS No.
231-838-7

JPN Translation:
三リン酸 5Na

Empirical Formula:
$H_5O_{10}P_3 \cdot 5Na$

Definition: Pentasodium Triphosphate is the inorganic salt that conforms to the formula:

$$Na_5P_3O_{10}$$

Information Sources: 21CFR172.892, 21CFR173.310, 21CFR182.70, 21CFR182.90, 21CFR182.1810, 21CFR182.6810, CTFA D, FCC, MI-13 (8771), TSCA

Chemical Classes: Inorganic Salts; Phosphorus Compounds

Functions: Chelating Agent; pH Adjuster

Reported Product Categories: Bath Oils, Tablets, and Salts; Bath Preparations, Misc.

Technical/Other Names:
Sodium Tripolyphosphate
Triphosphoric Acid, Pentasodium Salt

PENTETIC ACID

CAS No.
67-43-6

EINECS No.
200-652-8

JPN Translation:
ペンテト酸

Empirical Formula:
$C_{14}H_{23}N_3O_{10}$

Definition: Pentetic Acid is the substituted amine that conforms to the formula:

$$HOOCCH_2 \diagdown \qquad \qquad \diagup CH_2COOH$$
$$NCH_2CH_2NCH_2CH_2N$$
$$HOOCCH_2 \diagup \qquad \qquad \diagdown CH_2COOH$$

Information Sources: BAN, CTFA D, INN, JCIC, JCLS, JSQI, MI-13(7202), USAN, USP XXIV

Chemical Classes: Alkyl-Substituted Amino Acids; Amines

Function: Chelating Agent

Reported Product Categories: Hair Dyes and Colors (All Types Requiring Caution Statements and Patch Tests); Permanent Waves; Shampoos (Non-coloring)

Technical/Other Names:
N,N-Bis[2-[Bis(Carboxymethyl)Amino]-
Ethyl]Glycine
Diethylenetriaminepentaacetic Acid
Glycine, N,N-Bis[2-[Bis(Carboxymethyl)-
Amino]Ethyl]-
3,6,9-Triazaundecanedioic Acid, 3,6,9-tris
(Carboxymethyl)-

PENTYL DIMETHYL PABA

CAS No.
14779-78-3

EINECS No.
238-849-6

JPN Translation:
ジメチル PABA アミル

Empirical Formula:
$C_{14}H_{21}NO_2$

Definition: Pentyl Dimethyl PABA is the ester of pentyl alcohol and dimethyl p-amino-benzoic acid. It conforms to the formula:

$$\begin{array}{c} O \\ \| \\ C-OC_5H_{11} \end{array}$$

$$N(CH_3)_2$$

Information Sources: BAN, EEC(II-381), INN, JCIC, JCLS, MHLW-331/4, USAN

Chemical Class: PABA Derivatives

Function: Ultraviolet Light Absorber

Reported Product Category: Suntan Gels, Creams, and Liquids

Technical/Other Names:
Amyl 4-Dimethylaminobenzoate, Mixed Isomers
Amyl p-Dimethylamino Benzoate
Amyl Dimethyl PABA
Benzoic Acid, 4-(Dimethylamino)-, Pentyl Ester
Padimate A
Pentyl p-(Dimethylamino)Benzoate

PENTYLENE GLYCOL

CAS No. 5343-92-0   **EINECS No.** 226-285-3

JPN Translation:
ペンチレングリコール

Empirical Formula:
$C_5H_{12}O_2$

Definition: Pentylene Glycol is the organic compound that conforms to the formula:

$$HOCH_2CH(CH_2)_2CH_3$$
$$|$$
$$OH$$

Information Source: JCLS

Chemical Class: Alcohols

Function: Solvent

Technical/Other Names:
1,2-Dihydroxypentane
1,2-Pentanediol

Trade Name:
Hydrolite-5 (Item No. 2/016020) (Symrise)

Trade Name Mixtures:
AC 305 (Active Concepts)
Deolite 2/027095 (Symrise)
Dermosoft MCA 2 (Straetmans)
Extrapone Acacia Milk 2/033805 (Symrise)
Extrapone Almond Milk 2/033895 (Symrise)
Extrapone Aloe Vera Milk 2/033890 (Symrise)
Extrapone Butcher's Broom GW P 2/030700 (Symrise)
Extrapone Caramel Milk 2/033834 (Symrise)
Extrapone Fig-Mile 2/033875 (Symrise)
Extrapone Honey Rice Milk P 2/030830 (Symrise)
Extrapone Macadamia Nut Milk 2/033848 (Symrise)
Extrapone Mallow 2/030112 (Symrise)
Extrapone Soy Milk 2/033858 (Symrise)
Extrpone Avacado Milk 2/033820 (Symrise)
Hydroviton 24 (2/059351) (Symrise)
Peru Liana (Coletica SA)
Rovisome Retinol Moist (Rovi)
Solubilizer HP 2/014180 (Symrise)
Unitamuron H-22 (Induchem)

PENTYLPHENYL METHOXYBENZOATE

CAS No. 38444-13-2   **EINECS No.** 253-932-7

Empirical Formula:
$C_{19}H_{22}O_3$

Definition: Pentylphenyl Methoxybenzoate is the organic compound that conforms to the formula:

Chemical Class: Esters

Functions: Skin-Conditioning Agent - Miscellaneous; Skin-Conditioning Agent - Occlusive

Technical/Other Name:
Benzoic Acid, 4-Methoxy-, 4-Pentylphenyl Ester

Trade Name Mixtures:
Liquid Crystal BN 600 (Hallcrest Limited)
Liquid Crystal BN 823 (Hallcrest Limited)
Liquid Crystal BN 825 (Hallcrest Limited)
Liquid Crystal BN 826 (Hallcrest Limited)
Liquid Crystal BN 1001 (Hallcrest Limited)

PENTYLPHENYL OCTYLOXYBENZOATE

CAS No. 50649-56-4   **EINECS No.** 256-682-7

Empirical Formula:
$C_{26}H_{36}O_3$

Definition: Pentylphenyl Octyloxybenzoate is the organic compound that conforms to the formula:

Chemical Class: Esters

Function: Skin-Conditioning Agent - Miscellaneous

Technical/Other Name:
Benzoic acid, 4-(Octyloxy)-, 4-Pentylphenyl Ester

Trade Name Mixture:
Liquid Crystal BN 1001 (Hallcrest Limited)

PEPSIN

CAS No. 9001-75-6   **EINECS No.** 232-629-3

JPN Translation:
ペプシン

Definition: Pepsin is a digestive enzyme found in gastric juice.

Information Sources: 21CFR184.1595, 21CFR310.545, MI-13(7225), TSCA

Chemical Class: Proteins

Functions: Hair Conditioning Agent; Skin-Conditioning Agent - Miscellaneous

PERFLUOROALKYLSILYL MICA

JPN Translation:
牛乳糖タンパク

Definition: *See "Regulatory and Ingredient Use Information," regarding use of Japan Trivial names in Volume 1, Introduction, Part A.*

Information Source: JCLS

Chemical Class: Inorganics

Function: Not Reported

PERFLUOROCAPRYLYL BROMIDE

CAS No. 423-55-2   **EINECS No.** 207-028-4

Empirical Formula:
C_8BrF_{17}

Definition: Perfluorocaprylyl Bromide is the organic compound that conforms to the formula:

$$CF_3(CF_2)_6CF_2Br$$

Information Source: MI-13(7235)

Chemical Class: Halogen Compounds

Function: Solvent

Technical/Other Name:
1-Bromoheptadecafluorooctane

PERFLUOROCAPRYLYL TRIETHOXYSILYLETHYL METHICONE

Definition: Perfluorocaprylyl Triethoxysilylethyl Methicone is the siloxane polymer that conforms to the formula:

The inclusion of any compound in the *Dictionary and Handbook* does not indicate that use of that substance as a cosmetic ingredient complies with the laws and regulations governing such use in the United States or any other country.

Chemical Classes: Halogen Compounds; Siloxanes and Silanes

Functions: Binder; Skin-Conditioning Agent - Emollient

Trade Name:
FLS-630 (Asahi Glass)

PERFLUOROCYCLOHEXYLMETHANOL

CAS No.: 28788-68-3

Empirical Formula:
$C_7H_3F_{11}O$

Definition: Perfluorocyclohexylmethanol is the organic compound that conforms to the formula:

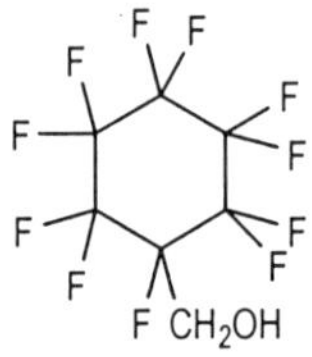

Chemical Classes: Alcohols; Halogen Compounds

Functions: Emulsion Stabilizer; Skin-Conditioning Agent - Miscellaneous; Slip Modifier; Surface Modifier

Technical/Other Name:
Cyclohexanemethanol, 1,2,2,3,3,4,4,5,5,6,6-Undecafluoro-

Trade Name:
Fiflow PB 145 (C.I.T.)

PERFLUORODECALIN

CAS No.   **EINECS No.**
306-94-5   206-192-4

Empirical Formula:
$C_{10}F_{18}$

Definition: Perfluorodecalin is the organic compound that conforms to the formula:

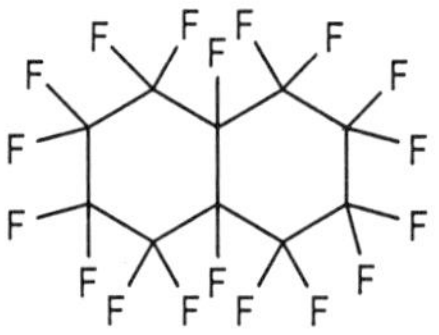

Information Sources: BAN, INN

Chemical Class: Halogen Compounds

Functions: Skin-Conditioning Agent - Miscellaneous; Solvent

Technical/Other Names:
Decalin Perfluoride

Naphthalene, Octadecafluorodecahydro-Perfluorodecahydronaphthalene

Trade Names:
Fiflow PB 140 (C.I.T.)
Flutec PC 6 (F2 Chemicals)

Trade Name Mixtures:
CREAGEL EZ PFC (Creations Couleurs)
Micropearl PA (C.I.T.)
Micropearl PN (C.I.T.)
Micropearl PO (C.I.T.)
Nanosource PA (C.I.T.)
Nanosource PN (C.I.T.)
Nanosource PO (C.I.T.)
SUPER A (C.I.T.)
SUPER ARBUTIN (C.I.T.)
SUPER C (C.I.T.)

PERFLUORODIMETHYLCYCLOHEXANE

CAS Nos.   **EINECS Nos.**
335-27-3   206-386-9
26637-68-3   247-863-1

Empirical Formula:
C_8F_{16}

Definition: Perfluorodimethylcyclohexane is the organic compound that conforms to the formula:

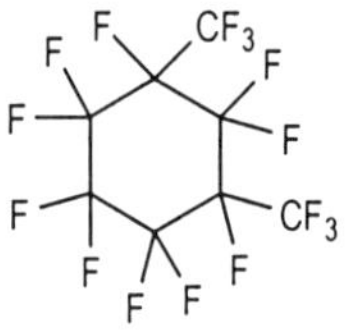

Chemical Class: Halogen Compounds

Function: Solvent

Technical/Other Names:
Cyclohexane, Decafluorobis (trifluoromethyl)-
Cyclohexane, 1,1,2,2,3,3,4,5,5,6-Decafluoro-4,6-Bis(Trifluoromethyl)-
1,1,2,2,3,3,4,5,5,6-Decafluor-4,6-Bis (Trifluormethyl)Cyclohexane
Perfluoro(1,3-Dimethylcyclohexane)

Trade Names:
Fiflow PB 100 (C.I.T.)
Flutec PC-3 (F2 Chemicals)

PERFLUOROHEPTANE

CAS No.   **EINECS No.**
335-57-9   206-392-1

Empirical Formula:
C_7F_{16}

Definition: Perfluoroheptane is the organic compound that conforms to the formula:

$$CF_3(CF_2)_5CF_3$$

Chemical Class: Halogen Compounds

Functions: Absorbent; Anticaking Agent; Emulsion Stabilizer; Skin-Conditioning Agent - Miscellaneous; Slip Modifier; Surface Modifier

Technical/Other Name:
Heptane, Hexadecafluoro-

Trade Name:
Fiflow PB 78 (C.I.T.)

PERFLUOROHEXANE

CAS No.   **EINECS No.**
355-42-0   206-585-0

Empirical Formula:
C_6F_{14}

Definition: Perfluorohexane is the organic compound that conforms to the formula:

$$CF_3(CF_2)_4CF_3$$

Information Sources: 21CFR173.342, TSCA, USAN

Chemical Class: Halogen Compounds

Function: Solvent

Technical/Other Names:
Hexane, Tetradecafluoro-
Perfluoro-n-hexane
Tetradecafluorohexane

Trade Names:
Fiflow PB 60 (C.I.T.)
Flutec PC-1 (F2 Chemicals)

PERFLUOROHEXYLETHYL DIMETHYL-BUTYL ETHER

CAS No.: 210896-25-6

Empirical Formula:
$C_{14}H_{17}F_{13}O$

Definition: Perfluorohexylethyl Dimethylbutyl Ether is the organic compound that conforms to the formula:

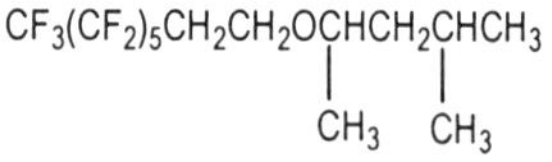

Chemical Classes: Ethers; Halogen Compounds

Function: Skin-Conditioning Agent - Miscellaneous

Technical/Other Name:
Octane, 8-(1,3-Dimethylbutoxy)-1,1,1,2,2,3,3,4,4,5,5,6,6,-Tridecafluoro-

PERFLUOROMETHYLCYCLOHEXANE

CAS No.   **EINECS No.**
355-02-2   206-573-5

Empirical Formula:
C_7F_{14}

Definition: Perfluoromethylcyclohexane is the organic compound that conforms to the formula:

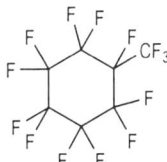

Chemical Class: Halogen Compounds

Functions: Absorbent; Anticaking Agent; Binder; Emulsion Stabilizer; Skin-Conditioning Agent - Miscellaneous; Slip Modifier; Surface Modifier

Technical/Other Name:
Cyclohexane, Undecafluoro (Trifluoromethyl)-

Trade Names:
Fiflow PB 75 (C.I.T.)
Fiflow PB 80 (C.I.T.)

PERFLUOROMETHYLCYCLOPENTANE

CAS No.	EINECS No.
1805-22-7	217-298-5

Empirical Formula:
C_6F_{12}

Definition: Perfluoromethylcyclopentane is the organic compound that conforms to the formula:

Chemical Class: Halogen Compounds

Functions: Skin-Conditioning Agent - Miscellaneous; Solvent

Technical/Other Names:
Cyclopentane, Nonafluoro(Trifluoromethyl)-
Nonofluoro(Trifluoromethyl)Cyclopentane

Trade Names:
Fiflow PB 50 (C.I.T.)
Flutec PC-1C (F2 Chemicals)

PERFLUOROMETHYLDECALIN

CAS No.: 51294-16-7

Empirical Formula:
$C_{11}F_{20}$

Definition: Perfluoromethyldecalin is the organic compound that conforms to the formula:

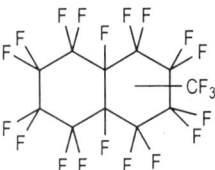

Chemical Class: Halogen Compounds

Functions: Absorbent; Anticaking Agent; Binder; Emulsion Stabilizer; Skin-Conditioning Agent - Miscellaneous; Slip Modifier; Surface Modifier

Technical/Other Name:
Naphthalene, Heptadecafluorodecahydro (Trifluoromethyl)-

Trade Names:
Fiflow PB 160 (C.I.T.)
Flutec PC9 (F2 Chemicals)

PERFLUORONONYL DIMETHICONE

Definition: Perfluorononyl Dimethicone is the fluorinated siloxane polymer that conforms generally to the formula:

Chemical Classes: Halogen Compounds; Siloxanes and Silanes; Synthetic Polymers

Functions: Skin-Conditioning Agent - Miscellaneous; Skin-Conditioning Agent - Occlusive; Slip Modifier

Trade Names:
Biosil Basics Fluorosil 14 (Biosil Technologies, Inc.)
Biosil Basics Fluorosil 35 (Biosil Technologies, Inc.)
Biosil Basics Fluorosil LF (Biosil Technologies, Inc.)
Fluorosil D-2 (Siltech)
Fluorosil H-4 (Siltech)
Fluorosil J-15 (Siltech)
Pecosil FSH-150 (Phoenix)
Pecosil FSH-300 (Phoenix)
Pecosil FSL-150 (Phoenix)
Pecosil FSL-300 (Phoenix)
Pecosil FSU-150 (Phoenix)
Pecosil FSU-300 (Phoenix)

PERFLUORONONYLETHYL CARBOXY-DECYL PEG-8 DIMETHICONE

Definition: Perfluorononylethyl Carboxy-decyl PEG-8 Dimethicone is the polysiloxane that conforms to the formula:

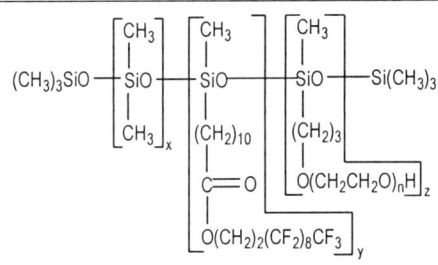

where n has an average value of 8.

Chemical Classes: Halogen Compounds; Siloxanes and Silanes

Function: Skin-Conditioning Agent - Emollient

Trade Name:
Silwax C-F-1 (Siltech)

PERFLUORONONYLETHYL CARBOXY-DECYL PEG-10 DIMETHICONE

Definition: Perfluorononylethyl Carboxy-decyl PEG-10 Dimethicone is the polysiloxane that conforms to the formula:

where n has an average value of 10.

Chemical Classes: Halogen Compounds; Siloxanes and Silanes

Functions: Skin-Conditioning Agent - Emollient; Surface Modifier; Surfactant - Emulsifying Agent

Trade Names:
Pecosil FDM-5 (Phoenix)
Pecosil FDM-15 (Phoenix)
Pecosil FDM-30 (Phoenix)

PERFLUORONONYLETHYL PEG-8 DIMETHICONE

Definition: Perfluorononylethyl PEG-8 Dimethicone is the fluorinated siloxane polymer that conforms generally to the formula:

where n has an average value of 8.

Chemical Classes: Halogen Compounds; Siloxanes and Silanes

Function: Not Reported

PERFLUORONONYLETHYL STEARYL DIMETHICONE

Definition: Perfluorononylethyl Stearyl Dimethicone is the siloxane polymer that conforms generally to the formula:

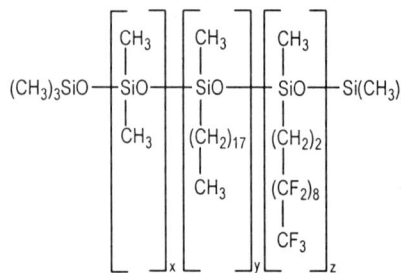

Chemical Classes: Halogen Compounds; Siloxanes and Silanes; Synthetic Polymers

Functions: Film Former; Skin-Conditioning Agent - Occlusive

Trade Name:
Silwax FS-1615 (Siltech)

PERFLUORONONONYL OCTYLDODECYL GLYCOL MEADOWFOAMATE

Definition: Perfluorononyl Octyldodecyl Glycol Meadowfoamate is the ester that conforms generally to the formula:

$$RC - OCHCH_2OCH_2CH(CH_2)_9CH_3$$

with O double bonded to RC, (CH_2)_7CH_3 branch, and (CF_2)_8CF_3 branch

where RCO- represents the fatty acids derived from Limnanthes Alba (Meadowfoam) Seed Oil (q.v.).

Chemical Classes: Esters; Ethers; Halogen Compounds

Functions: Hair Conditioning Agent; Skin-Conditioning Agent - Miscellaneous; Suspending Agent - Nonsurfactant

Trade Names:
Fancor Meadowester GME-F (Fanning)
Silube GME-F (Siltech)

PERFLUOROOCTYLETHYL/DIPHENYL DIMETHICONE COPOLYMER

Definition: Perfluorooctylethyl/Diphenyl Dimethicone Copolymer is a copolymer of perfluorooctylethylsiloxane and diphenyl-dimethylsiloxane monomers.

Chemical Classes: Halogen Compounds; Siloxanes and Silanes; Synthetic Polymers

Function: Skin-Conditioning Agent - Emollient

Reported Product Category: Lipsticks

Trade Name:
FMPS (Shiseido Company)

PERFLUOROOCTYLETHYL TRIETHOXYSILANE

CAS No.: 101947-16-4

Empirical Formula:
$C_{16}H_{19}F_{17}O_3Si$

Definition: Perfluorooctylethyl Triethoxysilane is the organic compound that conforms to the formula:

$$CF_3(CF_2)_7CH_2CH_2SiOCH_2CH_3$$

with OCH_2CH_3 above and OCH_2CH_3 below the Si

Chemical Classes: Halogen Compounds; Siloxanes and Silanes

Function: Anticaking Agent

Technical/Other Name:
Silane, Triethoxy(3,3,4,4,5,5,6,6,7,7,8,8,9,9,10,10-Heptadecafluorodecyl)-

Trade Name:
Dow Corning-Toray Silicone AY 43-158 E (DCTS)

PERFLUOROOCTYLETHYL TRIMETHOXYSILANE

CAS No.: 83048-65-1

Empirical Formula:
$C_{13}H_{13}F_{17}O_3Si$

Definition: Perfluorooctylethyl Trimethoxysilane is the organic compound that conforms to the formula:

$$CF_3(CF_2)_7CH_2CH_2SiOCH_3$$

with OCH_3 above and OCH_3 below the Si

Chemical Classes: Halogen Compounds; Siloxanes and Silanes

Function: Bulking Agent

Technical/Other Name:
Silane, (3,3,4,4,5,5,6,6,7,7,8,8,9,9,10,10,10-Heptadecafluorodecyl)Trimethoxy-

Trade Name Mixtures:
NFI-MICA (US Cosmetics)
NFI-Sericite (US Cosmetics)
NFI-TALC (US Cosmetics)
PFI-Powder La Vie (US Cosmetics)

PERFLUOROOCTYLETHYL TRISILOXANE

CAS No.: 163921-85-5

Empirical Formula:
$C_{17}H_{26}F_{17}O_2Si_3$

Definition: Perfluorooctylethyl Trisiloxane is the siloxane compound that conforms to the formula:

$$(CH_3)_3SiO - SiO - Si(CH_3)_3$$

with CH_3 above and CH_2CH_3 and (CF_2)_7CF_3 below the central Si

Chemical Classes: Halogen Compounds; Siloxanes and Silanes

Function: Binder

Technical/Other Name:
Trisiloxane, 2-(3,3,4,4,5,5,6,6,7,7,8,8,9,9,10,10,10-Heptadecafluorodecyl)-1,1,1,3,5,5,5-Heptamethyl-

Trade Name:
FLS 614 (Asahi Glass)

PERFLUOROOCTYL TRIETHOXYSILANE

CAS No.	EINECS No.
51851-37-7	257-473-3

Empirical Formula:
$C_{14}H_{19}F_{13}O_3Si$

Definition: Perfluorooctyl Triethoxysilane is the organic compound that conforms to the formula:

$$CH_3CH_2O - Si - (CH_2)_2(CF_2)_5CF_3$$

with OCH_2CH_3 above and OCH_2CH_3 below the Si

Chemical Classes: Ethers; Siloxanes and Silanes

Function: Binder

Trade Name Mixtures:
Cardre Black Iron Oxide FS (Cardre)
Cardre Red Iron Oxide FS (Cardre)
Cardre Talc FS (Cardre)
Cardre Titanium Dioxide FS (Cardre)
Cardre Ultra Blue FS (Cardre)
Cardre Yellow Iron Oxide FS (Cardre)

PERFLUOROPERHYDROBENZYL TETRALIN

CAS No.: 116265-66-8

Empirical Formula:
$C_{17}F_{30}$

Definition: Perfluoroperhydrobenzyl Tetralin is the organic compound that conforms to the formula:

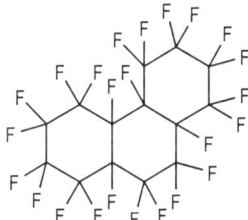

Chemical Class: Halogen Compounds

Functions: Skin-Conditioning Agent - Miscellaneous; Solvent

Trade Names:
Fiflow PB 260 (C.I.T.)
Flutec PC-25 (F2 Chemicals)

PERFLUOROPER-HYDROPHENANTHRENE

CAS No.: 306-91-2

Empirical Formula:
$C_{14}F_{24}$

Definition: Perfluoroperhydrophenanthrene is the organic compound that conforms to the formula:

Chemical Class: Halogen Compounds

Functions: Skin-Conditioning Agent - Miscellaneous; Solvent

Technical/Other Names:
Phenanthrene, Tetracosafluoro-tetradecahydro-
Tetracosafluorotetra-decahydrophenanthrene

Trade Names:
Fiflow PB 220 (C.I.T.)
Flutec PC-11 (F2 Chemicals)

Trade Name Mixtures:
CREABN PFC (Creations Couleurs)
MICAPOLY WL 12 PFC (Creations Couleurs)
NYLONPOLY WL 5 PFC (Creations Couleurs)

TALCPPOLY WL 8 PFC (Creations Couleurs)
Tefpoly Begonia (C.I.T.)
Tefpoly Black (C.I.T.)
Tefpoly Blue (C.I.T.)
Tefpoly Blue Green (C.I.T.)
Tefpoly Blue Sky Dye (C.I.T.)
Tefpoly Bright Yellow Dye (C.I.T.)
Tefpoly Bright Yellow Lake (C.I.T.)
Tefpoly Brown (C.I.T.)
Tefpoly Coral Dye (C.I.T.)
Tefpoly Electric Pink Lake (C.I.T.)
TefPoly Eosin Dye (C.I.T.)
Tefpoly Geranium Lake (C.I.T.)
Tefpoly Gold Grass Dye (C.I.T.)
Tefpoly Harvest Dye (C.I.T.)
Tefpoly Helindone Lake (C.I.T.)
Tefpoly Magenta Dye (C.I.T.)
Tefpoly Magenta Lake (C.I.T.)
Tefpoly Mandarin Dye (C.I.T.)
Tefpoly Melon Dye (C.I.T.)
Tefpoly Melon Lake (C.I.T.)
Tefpoly Ocean Blue Dye (C.I.T.)
Tefpoly Ocean Blue Lake (C.I.T.)
Tefpoly Orange Dye (C.I.T.)
Tefpoly Orange Lake (C.I.T.)
TEFPOLY PFC 5 (Creations Couleurs)
Tefpoly Pinky Dye (C.I.T.)
Tefpoly Poppy Lake (C.I.T.)
Tefpoly Prussian Blue (C.I.T.)
Tefpoly Red (C.I.T.)
Tefpoly Red Wood Dye (C.I.T.)
Tefpoly Safran Lake (C.I.T.)
Tefpoly Tomato Dye (C.I.T.)
Tefpoly Violet (C.I.T.)
Tefpoly White (C.I.T.)
Tefpoly Yellow (C.I.T.)
Tefpoly Yellow Grass Dye (C.I.T.)

PERFLUOROTETRALIN

Empirical Formula:
$C_{10}F_{12}$

Definition: Perfluorotetralin is the organic compound that conforms to the formula:

Chemical Class: Halogen Compounds

Functions: Skin-Conditioning Agent - Miscellaneous; Solvent

Technical/Other Names:
Dodecafluorotetrahydronaphthalene
Naphthalene, Dodecafluorotetrahydro-

PERILLALDEHYDE

CAS No. 2111-75-3
EINECS No. 218-302-8

Empirical Formula:
$C_{10}H_{14}O$

Definition: Perillaldehyde is the organic compound that conforms to the formula:

Information Sources: MI-13(7244), RIFM

Chemical Class: Aldehydes

Functions: Fragrance Ingredient; Skin-Conditioning Agent - Miscellaneous

Technical/Other Names:
1-Cyclohexene-1-Carboxaldehyde, 4-(1-Methylethenyl)-
Dihydrocuminyl Aldehyde
p-Mentha-1,8-dien-7-al (RIFM)

PERILLA OCYMOIDES LEAF EXTRACT

CAS No. 90082-61-4
EINECS No. 290-151-0

JPN Translation:
シソ葉エキス

Definition: Perilla Ocymoides Leaf Extract is an extract of the leaves of the perilla, *Perilla ocymoides*. See "Regulatory and Ingredient Use Information," regarding the labeling names for botanical ingredients in Volume 1, Introduction, Part A.

Information Sources: JCIC, JCLS

Chemical Class: Biological Products

Function: Not Reported

Technical/Other Names:
Extract of Perilla
Perilla Extract
Perilla Extract (2)
Perilla Frutescens Extract
Soyou Ekisu (JPN)

Trade Name Mixtures:
Bathgranue Shiso (Ichimaru Pharcos)
Beefsteak Plant Extract (Maruzen Pharmaceuticals Co., Ltd.)
Beefsteak Plant Extract BG (Maruzen Pharmaceuticals Co., Ltd.)
Beefsteak Plant Extract BG-J (Maruzen Pharmaceuticals Co., Ltd.)
Beefsteak Plant Extract LA (Maruzen Pharmaceuticals Co., Ltd.)
Beefsteak Plant Extract Powder-S (Maruzen Pharmaceuticals Co., Ltd.)
Beefsteak Plant Extract SQ (Maruzen Pharmaceuticals Co., Ltd.)

Beefsteak Plant Extract W-A (Maruzen
Pharmaceuticals Co., Ltd.)
Extrait Hydroglycolique de Perilla
Fructescens (Greentech)
Micropearl P (C.I.T.)
Nanosource P (C.I.T.)
Shiso Liquid (Ichimaru Pharcos)
Shiso Liquid B (Ichimaru Pharcos)
Shiso Liquid WA (Ichimaru Pharcos)

PERILLA OCYMOIDES LEAF POWDER

Definition: Perilla Ocymoides Leaf
Powder is the powder obtained from the
dehydrated, crushed leaves of *Perilla
ocymoides*. See "*Regulatory and Ingredient
Use Information,*" regarding the labeling
names for botanical ingredients in Volume
1, Introduction, Part A.

Chemical Class: Biological Products

Function: Skin-Conditioning Agent - Miscellaneous

Trade Name:
Shiso Leaf Powder (Caring Japan)

PERILLA OCYMOIDES SEED EXTRACT

CAS No.	EINECS No.
90082-61-4	290-151-0

Definition: Perilla Ocymoides Seed
Extract is the extract of the seeds of *Perilla
ocymoides*. See "*Regulatory and Ingredient
Use Information,*" regarding the labeling
names for botanical ingredients in Volume
1, Introduction, Part A.

Chemical Class: Biological Products

Function: Antioxidant

Technical/Other Names:
Extract of Perilla Ocymoides Seed
Perilla Frutescens Seed Extract

Trade Name:
Perilla Seed Extract - P (YZ Connection)

Trade Name Mixtures:
Perilla Seed Extract - L (YZ Connection)
Perilla Seed Extract-LC (Nikko)
Perilla Seed Extract-LC (YZ Connection)
Tri-K Perilla Seed Extract (Tri-K)

PERILLA OCYMOIDES SEED OIL

JPN Translation:
エゴマ油

Definition: Perilla Ocymoides Seed Oil is
the fixed oil obtained from the seeds of
Perilla ocymoides. See "*Regulatory and
Ingredient Use Information,*" regarding the

labeling names for botanical ingredients in
Volume 1, Introduction, Part A.

Chemical Class: Fats and Oils

Function: Skin-Conditioning Agent - Miscellaneous

Technical/Other Name:
Oil, Perilla Ocymoides

Trade Names:
AEC Perilla Seed Oil (A & E Connock)
Maruta Shisoyu (Ohta Oil)
Perilla Oil (Maruzen Pharmaceuticals Co.,
Ltd.)

Trade Name Mixture:
Laminactine (Advanced Beauty)

PERLITE

CAS Nos.	EINECS No.
93763-70-3	
130885-09-5	310-127-6

Definition: Perlite is a chemically inert
siliceous mineral consisting chiefly of:

$$SiO_2 \cdot Al_2O_3 \cdot Na_2O \cdot K_2O$$

Chemical Class: Inorganics

Functions: Abrasive; Absorbent; Bulking
Agent; Suspending Agent - Nonsurfactant

Trade Names:
Europerl-50 (Europerlite)
Perlite MBK (MBK)

PERMETHRIN

CAS No.	EINECS No.
52645-53-1	258-067-9

Empirical Formula:
$C_{22}H_{20}Cl_2NO_3$.

Definition: Permethrin is the organic
comound that conforms to the formula:

Information Sources: BAN,
40CFR180.378, INN, MI-13(7257), USAN

Chemical Classes: Esters; Ethers; Halogen
Compounds

Function: Pesticide

Technical/Other Names:
Cyclopropanecarboxylic Acid, 3-(2,2-
Dichloroethenyl)-2,2-Dimethyl-, (3-
Phenoxyphenyl)Methyl Ester

m-Phenoxybenzyl 3-(2,2-Dichlorovinyl)-2,2-
Dimethylcyclopropanecarboxylate

Trade Name:
Permethrin Technical (Sumitomo
Chemical)

PERSEA GRATISSIMA (AVOCADO) BUTTER

Definition: Persea Gratissima (Avocado)
Butter is the fat obtained from the seed of
the avocado, *Persea gratissimia*. See
"*Regulatory and Ingredient Use
Information,*" regarding the labeling names
for botanical ingredients in Volume 1, Introduction, Part A.

Chemical Class: Fats and Oils

Function: Skin-Conditioning Agent -
Occlusive

Trade Name:
AEC Avocado Butter (A & E Connock)

PERSEA GRATISSIMA (AVOCADO) FRUIT EXTRACT

JPN Translation:
アボカドエキス

Definition: Persea Gratissima (Avocado)
Fruit Extract is an extract of the fruit of the
avocado, *Persea gratissima*. See
"*Regulatory and Ingredient Use
Information,*" regarding the labeling names
for botanical ingredients in Volume 1, Introduction, Part A.

Information Sources: JCIC, JCLS, JSQI

Chemical Class: Biological Products

Function: Not Reported

Technical/Other Names:
Avocado Extract
Avocado Fruit Extract
Extract of Avocado
Persea Americana Extract
Tolune-2,4-Diamine,5,5'-((4-Methyl-m-
Phenylene)Bis(Azo))Bis-

Trade Name Mixtures:
Actiphyte of Avocado BG50 (Active
Organics)
Actiphyte of Avocado GL50 (Active
Organics)
Actiphyte of Avocado Lipo S (Active
Organics)
Actiphyte of Avocado PG50 (Active
Organics)
Avocado Extract HG (Provital/Centerchem)
Avocado Extract HS 2384 G (Grau)
Avocado Fruit LS (Alban Muller)

Avocado Liquid B (Ichimaru Pharcos)
Avocado Liquid E (Ichimaru Pharcos)
Avocado Milk (CEP (Solabia))
Cremogen Avocado (PN 773 521)
   (Haarmann & Reimer GmbH)
DRY HAIR (Greentech S.A)
Extrait D'Avocat MPE100 (Yves Rocher)
Extrapone Avocado Special 2/034599
   (Symrise)
Glycolysat of Avocado (CEP (Solabia))
Noster A (Ennagram)
Noster MX (Ennagram)
Oleat of Avocado (CEP (Solabia))
Phytelene of Avocado EG 460 Liquid
   (Indena SA)
Phytogreen 55 of Avocado EXH 681 Liquid
   (Phytochim)
Prodhy Extract Avocat (Prod'Hyg)
VT-074 Extract of Avocado (Vege-Tech)

PERSEA GRATISSIMA (AVOCADO) FRUIT POWDER

Definition: Persea Gratissima (Avocado) Fruit Powder is the dried plant material obtained from the fruit of *Persea gratissima. See "Regulatory and Ingredient Use Information," regarding the labeling names for botanical ingredients in Volume 1, Introduction, Part A.*

Chemical Class: Biological Products

Function: Abrasive

Technical/Other Names:
Avocado Fruit Powder
Avocado Powder

Trade Name:
Avocado Powder (Expanscience)

PERSEA GRATISSIMA (AVOCADO) FRUIT WATER

Definition: Persea Gratissima (Avocado) Fruit Water is an aqueous solution of the steam distillate obtained from the fruit of *Persea gratissima. See "Regulatory and Ingredient Use Information," regarding the labeling names for botanical ingredients in Volume 1, Introduction, Part A.*

Chemical Class: Biological Products

Function: Fragrance Ingredient

Technical/Other Names:
Avocado Fruit Water
Avocado Water
Persea Gratissima Water
Water, Avocado
Water, Persea Gratissima

Trade Name:
Vegebios of Avocado (CEP (Solabia))

PERSEA GRATISSIMA (AVOCADO) LEAF EXTRACT

Definition: Persea Gratissima (Avocado) Leaf Extract is an extract of the leaves of the avocado, *Persea gratissima. See "Regulatory and Ingredient Use Information," regarding the labeling names for botanical ingredients in Volume 1, Introduction, Part A.*

Chemical Class: Biological Products

Function: Not Reported

Technical/Other Names:
Avocado Leaf Extract
Extract of Avocado Leaves

Trade Name Mixture:
Herbasol Extract Avocado (Cosmetochem)
   (Cosmetochem International Ltd.)

PERSEA GRATISSIMA (AVOCADO) OIL

CAS No. 8024-32-6   **EINECS No.** 232-428-0

JPN Translation:
アボカド油

Definition: Persea Gratissima (Avocado) Oil is the fixed oil obtained by pressing the dehydrated sliced flesh of the avocado pear *Persea americana*. It consists principally of the glycerides of fatty acids. *See "Regulatory and Ingredient Use Information," regarding the labeling names for botanical ingredients in Volume 1, Introduction, Part A.*

Information Sources: CIR: [S] JEPT-4(4)-1980, CTFA S, JCLS, JSCI

Chemical Class: Fats and Oils

Function: Skin-Conditioning Agent - Occlusive

Reported Product Categories: Lipsticks; Bath Preparations, Misc.; Body and Hand Preparations (Excluding Shaving Preparations); Moisturizing Preparations; Skin Care Preparations, Misc.; Bath Capsules; Bath Oils, Tablets, and Salts; Cleansing Products (Cold Creams, Cleansing Lotions, Liquids and Pads); Face and Neck Preparations (Excluding Shaving Preparations); Hair Dyes and Colors (All Types Requiring Caution Statements and Patch Tests); Hair Conditioners; Eyebrow Pencils; Night Skin Care Preparations; Suntan Gels, Creams, and Liquids; Foundations; Makeup Bases; Paste Masks (Mud Packs); Shampoos (Noncoloring); Blushers (All types); Eye Lotions; Eye Shadows; Face Powders; Suntan Preparations, Misc.; Tonics, Dressings, and Other Hair Grooming Aids

Technical/Other Names:
Alligator Pear Oil
Avocado Oil
Oils, Avocado

Trade Names:
AEC Avocado Oil (A & E Connock)
Avocado Butter - ASAB001 (Aloestar)
Avocado Oil (Dekker)
Avocado Oil (Desert Whale)
Avocado Oil (Protameen)
Avocado Oil (Provital/Centerchem)
Avocado Oil 2/012400 (Symrise)
Avocado Oil CLR (CLR)
Avocado Oil "COS" (Cosmetochem)
   (Cosmetochem International Ltd.)
Certified Organic Avocado Oil (Formula
   One Sciences)
Cropure Avocado (Croda Chemicals)
Cropure Avocado (Croda, Inc.)
Huile D'Avocat Vierge (Bertin)
Jeen Avocado Oil (Jeen)
Lipovol A (Lipo)
Nikkol Avocado Oil (Nikko)
Persea Gratissima (Avocado) Oil ies (IES
   LABO)
Phytol AVO (avocado oil) (Custom Ingredients)
Tri-OL AVO (Tri-K)

Trade Name Mixtures:
Avamid 150 (Uniqema)
Avocado Milk (Cosmetochem)
   (Cosmetochem International Ltd.)
Avocado Oil Cosmetic Blend (Henry
   Lamotte)
Avocado Softcream (CEP (Solabia))
Capispheres (Barnet)
Chelonine (JUVEX)
Colamid AVCO (Colonial Chemical Inc)
Dragobotania 2/H00005 (Symrise)
Extrpone Avacado Milk 2/033820 (Symrise)
Gelhyperm (Avocado oil) (Novoselect)
Mackamide I-141 (McIntyre)
Novarom Sea Buckthorn Oil forte
   (Crodarom)
Phytogreen 55 of Red Vine EXH 642 Liquid
   (Phytochim)

PERSEA GRATISSIMA (AVOCADO) OIL UNSAPONIFIABLES

CAS No. 91770-40-0   **EINECS No.** 294-825-5

Definition: Persea Gratissima (Avocado) Oil Unsaponifiables is the fraction of Persea Gratissima (Avocado) Oil (q.v.) which is not saponified in the refining recovery of avocado oil fatty acids. *See "Regulatory and Ingredient Use Information," regarding the labeling names for botanical ingredients in Volume 1, Introduction, Part A.*

Chemical Class: Unsaponifiables

Functions: Hair Conditioning Agent; Skin-Conditioning Agent - Miscellaneous

Technical/Other Names:
Avocado Oil Unsaponifiables
Persea Gratissima Unsaponifiables
Unsaponifiables, Avocado Oil
Unsaponifiables, Persea Gratissima

Trade Names:
Avocadin (Crodarom)
Avocadol (CosmoCare)
Avocadol-B (CosmoCare)
Crodarom Avocadin (Croda, Inc.)
Unsaponifiable Avocado Oil
(Expanscience)

Trade Name Mixtures:
ASU Complex (Expanscience)
Avocadol - SLO (CosmoCare)
Avocadol - VLO (CosmoCare)

PERSEA GRATISSIMA (AVOCADO) OIL UNSAPONIFIABLES WATER

Definition: Persea Gratissima (Avocado) Oil Unsaponifiables Water is the aqueous solution containing the volatile oils obtained by the distillation of Persea Gratissima (Avocado) Oil Unsaponifiables (q.v.). *See "Regulatory and Ingredient Use Information," regarding the labeling names for botanical ingredients in Volume 1, Introduction, Part A.*

Chemical Class: Biological Products

Function: Skin-Conditioning Agent - Miscellaneous

PERSEA GRATISSIMA (AVOCADO) STEROLS

Definition: Persea Gratissima (Avocado) Sterols is a mixture of sterols obtained from Persea Gratissima (Avocado) Oil (q.v.). *See "Regulatory and Ingredient Use Information," regarding the labeling names for botanical ingredients in Volume 1, Introduction, Part A.*

Chemical Class: Sterols

Function: Skin-Conditioning Agent - Emollient

Trade Name:
Crodarom Avocadin HS-80 (Croda, Inc.)

PERSEA GRATISSIMA (AVOCADO) WAX

Definition: Persea Gratissima (Avocado) Wax is the semi-solid fraction of Persea Gratissima (Avocado) Oil (q.v.). *See "Regulatory and Ingredient Use Information," regarding the labeling names for botanical ingredients in Volume 1, Introduction, Part A.*

Chemical Class: Waxes

Function: Skin-Conditioning Agent - Emollient

Technical/Other Name:
Avocado Wax

Trade Name:
Avocatine (Expanscience)

PERSEA THUNBERGII EXTRACT

Definition: Persea Thunbergii Extract is an extract of the roots, stems, leaves and branches of *Persea thunbergii*. *See "Regulatory and Ingredient Use Information," regarding the labeling names for botanical ingredients in Volume 1, Introduction, Part A.*

Chemical Class: Biological Products

Function: Humectant

Technical/Other Name:
Extract of Persea Thunbergii

Trade Name Mixture:
Kondokikusui (Estate Chemical)

PETASITES HYBRIDUS ROOT EXTRACT

Definition: Petasites Hybridus Root Extract is an extract of the rhizomes of *Petasites hybridus*. *See "Regulatory and Ingredient Use Information," regarding the labeling names for botanical ingredients in Volume 1, Introduction, Part A.*

Chemical Class: Biological Products

Function: Not Reported

Technical/Other Name:
Extract of Petasites Hybridus

Trade Name Mixtures:
Butterbur Extract (Cosmetic Developments)
Cosflor Butterbur HGS (A & E Connock)

PETASITES VULGARIS LEAF EXTRACT

Definition: Petasites Vulgaris Leaf Extract is an extract of the leaves of *Petasites vulgaris*. *See "Regulatory and Ingredient Use Information," regarding the labeling names for botanical ingredients in Volume 1, Introduction, Part A.*

Chemical Class: Biological Products

Functions: Cosmetic Astringent; Skin-Conditioning Agent - Miscellaneous

Trade Name Mixture:
Actiphyte of Butter Bur Leaves (Active Organics)

PETROLATUM

CAS No.	EINECS No.
8009-03-8 (NF)	232-373-2

JPN Translation:
ワセリン

Definition: Petrolatum is a semisolid mixture of hydrocarbons obtained from petroleum. In the United States, Petrolatum may be used as an active ingredient in OTC drug products. When used as an active drug ingredient, the established name is *Petrolatum*. *See "Regulatory and Ingredient Use Information," regarding the labeling names for U.S. OTC Drug Ingredients in Volume 1, Introduction, Part A.*

Information Sources: ARG, AUS, BEL, BP, BPC, 21CFR172.615, 21CFR172.880, 21CFR173.340, 21CFR175.105, 21CFR175.125, 21CFR175.250, 21CFR175.300, 21CFR176.170, 21CFR176.180, 21CFR176.200, 21CFR176.210, 21CFR177.1200, 21CFR177.2600, 21CFR177.2800, 21CFR178.3570, 21CFR178.3700, 21CFR178.3910, 21CFR346.14, 21CFR349.14, 21CFR573.720, CTFA S, CZE, DA, DDR, EGY, FCC, FI, FIN, HUN, ITA, JAN, JCLS, JP, JSCI, MAR, MEX, MI-13(7262), NFJ, OTC-I-AR, OTC-I-OP, OTC-I-SK, PF, PN, POR, ROM, SNPF, TSCA, USAN, USD, USP XXIV, YUG

Chemical Class: Hydrocarbons

Functions: Hair Conditioning Agent; Skin-Conditioning Agent - Occlusive; Skin Protectant

Reported Product Categories: Moisturizing Preparations; Bath Preparations, Misc.; Body and Hand Preparations (Excluding Shaving Preparations); Lipsticks; Eye Shadows; Skin Care Preparations, Misc.; Bath Oils, Tablets, and Salts; Cleansing Products (Cold Creams, Cleansing Lotions, Liquids and Pads); Hair Conditioners; Tonics, Dressings, and Other Hair Grooming Aids; Night Skin Care Preparations; Bath Capsules; Foundations; Perfumes; Face and Neck Preparations (Excluding Shaving Preparations); Makeup Preparations (Not eye), Misc.; Hair Straighteners; Eyebrow Pencils; Eyeliners; Suntan Gels, Creams, and Liquids; Makeup Bases; Face Powders; Hair Coloring Preparations,

Misc.; Shaving Cream (Aerosol, Brushless and Lather); Eye Makeup Preparations, Misc.; Paste Masks (Mud Packs); Fragrance Preparations, Misc.; Permanent Waves; Bath Soaps and Detergents; Eye Makeup Removers; Hair Preparations (Non-coloring), Misc.; Blushers (All types); Eye Lotions; Baby Lotions, Oils, Powders and Creams; Deodorants (Underarm); Nail Creams and Lotions

Technical/Other Names:
Mineral Jelly
Petrolatum Amber
Petrolatum White
Petroleum Jelly
White Petrolatum
Yellow Petrolatum

Trade Names:
Chevron Petrolatum Amber (Chevron Lubricants)
Chevron Petrolatum Snow White (Chevron Lubricants)
Fonoline (Crompton Corporation)
Merkur 115 (MERKUR Vaseline)
Merkur 500 (MERKUR Vaseline)
Merkur 525 (MERKUR Vaseline)
Merkur 527 (MERKUR Vaseline)
Merkur 546 (MERKUR Vaseline)
Merkur 555 (MERKUR Vaseline)
Merkur 560 (MERKUR Vaseline)
Merkur 600 (MERKUR Vaseline)
Merkur 620 (MERKUR Vaseline)
Merkur 638 (MERKUR Vaseline)
Merkur 639 (MERKUR Vaseline)
Merkur 640 (MERKUR Vaseline)
Merkur 674 (MERKUR Vaseline)
Merkur 746 (MERKUR Vaseline)
Merkur 770 (MERKUR Vaseline)
Merkur 771 (MERKUR Vaseline)
Merkur 775 (MERKUR Vaseline)
Merkur 776 (MERKUR Vaseline)
Merkur 831 (MERKUR Vaseline)
MERKUR; VARA (MERKUR Vaseline)
Mineral Jelly (Crompton Corporation)
Mineral Jelly No. 5 (Penreco)
Mineral Jelly No. 10 (Penreco)
Mineral Jelly No. 15 (Penreco)
Mineral Jelly No. 20 (Penreco)
Nomcort W (Nisshin OilliO)
Penreco Amber (Penreco)
Penreco Blond (Penreco)
Penreco Cream (Penreco)
Penreco Lily (Penreco)
Penreco Regent (Penreco)
Penreco Royal (Penreco)
Penreco Snow (Penreco)
Penreco Super (Penreco)
Penreco Ultima (Penreco)
Perfecta (Crompton Corporation)
Petrolan USP (RITA)
PIONIER Vaseline/Petrolatum (Hansen & Rosenthal)
Protopet (Crompton Corporation)
Snow White Petrolatum (Stevenson-Cooper)

Sonojell (Crompton Corporation)
Sun White P-150 (Nikko Rica)
Sun White P-200 (Nikko Rica)
Ultrapure L Amber Petrolatum USP (Ultra Chemical)
Ultrapure ES Liquid Petrolatum USP (Ultra Chemical)
Ultrapure K (Ultra Chemical)
Ultrapure Liquid Petrolatum USP (Ultra Chemical)
Ultrapure SC White Petrolatum USP (Ultra Chemical)
Ultrapure L White Petrolatum USP (Ultra Chemical)
WeiBes Vaselin DAB 10 (Retterspitz)

Trade Name Mixtures:
Aloe Extract #102 (Florida Food Products)
Amerchol C (Amerchol)
Amerchol CAB (Amerchol)
Amphocerin K (Cognis Deutschland)
Argobase EU (Croda Chemicals)
Argobase L2 (Croda Chemicals)
Crosterol SFA (Croda Chemicals)
Dehymuls K (Cognis Deutschland)
Destressine 2000 (Sederma)
Emery 1740 (Cognis Care Chemicals/NJ)
Emery 1740 (Cognis Care Chemicals/PA)
Emulgator Apicerol 2/014081 (Symrise)
ESP Dry Wax hi vis (Earth Supplied Products)
ESP Dry Wax low vis (Earth Supplied Products)
ESP Dry Wax med vis (Earth Supplied Products)
Fancol C (Fanning)
Fancol CAB (Fanning)
Forlan 200 (RITA)
Forlan 300 (RITA)
Forlan 500 (RITA)
Forlan L (RITA)
Forlan LM (RITA)
Gel Base BSM-PT (Arch Personal Care Products)
Homulgator 1330 G (Grau)
Isocreme CB 0279 (Croda Chemicals)
Ivarbase T (Arch Personal Care Products)
Lanaetex CLC (Lanaetex)
Lanaetex-H (Lanaetex)
Lanaetex L-15 (Lanaetex)
Lexate PX (Inolex)
Metasomes-1-STD (Floratech)
Metaspheres-1-STD (Floratech)
Pionier MAA (Hansen & Rosenthal)
Polytrap 6500 Dimethicone/Petrolatum Powder (EDT, Inc.)
Protegin V (Degussa Care Specialties)
Protegin W (Degussa Care Specialties)
Protegin WX (Degussa Care Specialties)
Protegin XV (Degussa Care Specialties)
Ritaderm (RITA)
SanSurf Petrolatum-25 (Collaborative Labs)
San Surf Petrolatum-50 (Collaborative Labs)

Ultrapure HMP (Ultra Chemical)
Ultrapure HMP-S (Ultra Chemical)
Unieucerin (Chemyunion)
Unigel (Chemyunion)

PETROLEUM DISTILLATES

CAS No.	EINECS No.
8002-05-9	232-298-5

JPN Translation:
LPG

Definition: Petroleum Distillates is a mixture of volatile hydrocarbons obtained from petroleum.

Information Sources: CIR: [S] JACT-5(3)-1986, CTFA D, JCIC, JCLS, JSQI, KOR, MAR

Chemical Class: Hydrocarbons

Functions: Antifoaming Agent; Solvent

Reported Product Categories: Mascara; Eye Shadows

Technical/Other Names:
Light Liquid Paraffin (1)
Petroleum Distillate

Trade Names:
Drakesol 165 (Penreco)
Drakesol 165 AT (Penreco)
Halpanal 23/31 (Haltermann)
PD-23 (Crompton Corporation)
PD-25 (Crompton Corporation)
PD-28 (Crompton Corporation)
Shell Sol 71 (Shell)
Solvesso 100 (Gramos Surface)

Trade Name Mixtures:
Bentone Gel SS-71 (ELE)
Bentone Gel 10ST (ELE)
Tixogel OMS 1562 (Sud-Chemie, United Catalysts)

PEUCEDANUM GRAVEOLENS (DILL) EXTRACT

CAS No.	EINECS No.
90028-03-8	289-790-8

Definition: Peucedanum Graveolens (Dill) Extract is an extract of the dill, *Peucedanum graveolens.* See "Regulatory and Ingredient Use Information," regarding the labeling names for botanical ingredients in Volume 1, Introduction, Part A.

Chemical Class: Biological Products

Function: Not Reported

Technical/Other Names:
Anethum Graveolens Extract
Dill Extract
Extract of Dill

The inclusion of any compound in the *Dictionary and Handbook* does not indicate that use of that substance as a cosmetic ingredient complies with the laws and regulations governing such use in the United States or any other country.

Trade Name Mixtures:
Actiphyte of Dill BG50 (Active Organics)
Actiphyte of Dill GL50 (Active Organics)
Actiphyte of Dill Lipo S (Active Organics)
Actiphyte of Dill PG50 (Active Organics)

PEUMUS BOLDUS LEAF EXTRACT

Definition: Peumus Boldus Leaf Extract is an extract of the leaves of *Peumus boldus. See "Regulatory and Ingredient Use Information," regarding the labeling names for botanical ingredients in Volume 1, Introduction, Part A.*

Chemical Class: Biological Products

Function: Not Reported

Technical/Other Name:
Extract of Peumus Boldus Leaf

Trade Name Mixtures:
Defensine3 (Coletica SA)
Sculpturine (Silab)
Slimactive (Silab)

PFAFFIA PANICULATA EXTRACT

Definition: Pfaffia Paniculata Extract is an extract of the roots of *Pfaffia paniculata. See "Regulatory and Ingredient Use Information," regarding the labeling names for botanical ingredients in Volume 1, Introduction, Part A.*

Chemical Class: Biological Products

Function: Not Reported

Technical/Other Name:
Extract of Pfaffia Paniculata

Trade Name:
Amazon Extract Suma (E.U.K)

Trade Name Mixtures:
Actiphyte of Suma BG50 (Active Organics)
Actiphyte of Suma GL50 (Active Organics)
Actiphyte of Suma Lipo S (Active Organics)
Actiphyte of Suma PG50 (Active Organics)
VT-200 Extract of Suma (Vege-Tech)

PG-HYDROXYETHYLCELLULOSE COCO-DIMONIUM CHLORIDE

CAS No.: 130353-64-9

Definition: PG-Hydroxyethylcellulose Coco-dimonium Chloride is the quaternary ammonium salt that conforms generally to the formula:

$$\left[R'-\underset{\underset{CH_3}{|}}{\overset{\overset{CH_3}{|}}{N}}-CH_2CHCH_2OR \atop \underset{OH}{|} \right]^+ \quad Cl^-$$

where R represents the hydroxyethylcellulose moiety and R' represents the alkyl groups derived from coconut oil.

Chemical Class: Quaternary Ammonium Compounds

Functions: Antistatic Agent; Hair Conditioning Agent

Technical/Other Name:
Quaternary Ammonium Compounds, Coco Alkyl (2,3-Dihydroxypropyl)Dimethyl, 3-Ethers with Cellulose 2-Hydroxyethyl Ether, Chlorides

Trade Names:
Crodacel QM (Croda Chemicals)
Crodacel QM (Croda, Inc.)

PG-HYDROXYETHYLCELLULOSE LAURYLDIMONIUM CHLORIDE

CAS No.: 137802-19-8

Definition: PG-Hydroxyethylcellulose Lauryldimonium Chloride is the quaternary ammonium salt that conforms generally to the formula:

$$\left[(CH_3(CH_2)_{11}-\underset{\underset{CH_3}{|}}{\overset{\overset{CH_3}{|}}{N}}-CH_2CHCH_2OR \atop \underset{OH}{|} \right]^+ \quad Cl^-$$

where R represents the hydroxyethylcellulose moiety.

Chemical Class: Quaternary Ammonium Compounds

Functions: Antistatic Agent; Hair Conditioning Agent

Technical/Other Name:
Crodacel QL

Trade Name:
Crodacel QL (Croda Chemicals)

PG-HYDROXYETHYLCELLULOSE STEARYLDIMONIUM CHLORIDE

Definition: PG-Hydroxyethylcellulose Stearyldimonium Chloride is the quaternary ammonium salt that conforms generally to the formula:

$$\left[(CH_3(CH_2)_{17}-\underset{\underset{CH_3}{|}}{\overset{\overset{CH_3}{|}}{N}}-CH_2CHCH_2OR \atop \underset{OH}{|} \right]^+ \quad Cl^-$$

where R represents the hydroxyethylcellulose moiety.

Chemical Class: Quaternary Ammonium Compounds

Functions: Antistatic Agent; Hair Conditioning Agent

Trade Names:
Crodacel QS (Croda Chemicals)
Crodacel QS (Croda, Inc.)

PHAEODACTYLUM TRICORNUTUM EXTRACT

Definition: Phaeodactylum Tricornutum Extract is an extract of the alga, *Phaeodactylum tricornutum. See "Regulatory and Ingredient Use Information," regarding the labeling names for botanical ingredients in Volume 1, Introduction, Part A.*

Chemical Class: Biological Products

Function: Not Reported

Technical/Other Name:
Extract of Phaeodactylum Tricornutum

Trade Name Mixture:
Phaeodactylum HPG Titrated (Alban Muller)

PHALAENOPSIS AMABILIS EXTRACT

Definition: Phalaenopsis Amabilis Extract is an extract of the whole plant of *Phalaenopsis amabilis. See "Regulatory and Ingredient Use Information," regarding the labeling names for botanical ingredients in Volume 1, Introduction, Part A.*

Chemical Class: Biological Products

Function: Skin-Conditioning Agent - Humectant

Trade Name Mixtures:
Extrait Hydroglycolique d' Orchidee blanche (Greentech)
Phalaenopsis Extract (Nonogawa)

PHALAENOSPIS LOBBI EXTRACT

Definition: Phalaenospis Lobbi Extract is an extract of the whole plant, *Phalaenospis lobbi. See "Regulatory and Ingredient Use Information," regarding the labeling names for botanical ingredients in Volume 1, Introduction, Part A. See Reported Ingredient Functions-The Cosmetic Drug Distinction, in Regulatory and Ingredient Use Information, Volume I, Part A.*

Chemical Class: Biological Products

Function: Skin Bleaching Agent

Technical/Other Name:
Extract of Phalaenospis Lobbi

Trade Name:
Orchideline (Greentech)

PHASEOLUS ACONITIFOLIUS SEED EXTRACT

Definition: Phaseolus Aconitifolius Seed Extract is an extract of the seeds of *Phaseolus aconitifolius. See "Regulatory and Ingredient Use Information," regarding the labeling names for botanical ingredients in Volume 1, Introduction, Part A.*

Chemical Class: Biological Products

Function: Skin-Conditioning Agent - Miscellaneous

Trade Name Mixture:
Vitoptine LS (Laboratoires Serobiologiques)

PHASEOLUS ANGULARIS SEED POWDER

JPN Translation:
アズキ

Definition: Phaseolus Angularis Seed Powder is the powder obtained from the seed of *Phaseolus angularis. See "Regulatory and Ingredient Use Information," regarding the labeling names for botanical ingredients in Volume 1, Introduction, Part A.*

Information Sources: JCIC, JCLS

Chemical Class: Biological Products

Function: Viscosity Increasing Agent - Aqueous

Technical/Other Names:
Adzuki Bean Powder
Phaseolus Angularis Bean Powder

Trade Name:
AEC Adzuki Beans Coarse Milled (A & E Connock)

PHASEOLUS ANGULARIS SEED STARCH

JPN Translation:
アズキデンプン

Definition: Phaseolus Angularis Seed Starch is a starch obtained from the bean, *Phaseolus angularis. See "Regulatory and Ingredient Use Information," regarding the*

labeling names for botanical ingredients in *Volume 1, Introduction, Part A.*

Information Sources: JCIC, JCLS

Chemical Class: Carbohydrates

Function: Absorbent

Technical/Other Names:
Adzuki Bean Starch
Phaseolus Angularis Bean Starch

PHASEOLUS LUNATUS (GREEN BEAN) SEED EXTRACT

CAS No. 85085-22-9
EINECS No. 285-354-6

Definition: Phaseolus Lunatus (Green Bean) Seed Extract is an extract of the unripe beans of the green bean, *Phaseolus lunatus. See "Regulatory and Ingredient Use Information," regarding the labeling names for botanical ingredients in Volume 1, Introduction, Part A.*

Chemical Class: Biological Products

Function: Not Reported

Technical/Other Names:
Extract of Green Beans
Green Bean Extract
Green Bean Seed Extract

Trade Name Mixture:
Lipoplastidine Phaseolus (Vevy)

PHASEOLUS RADIATUS EXTRACT

JPN Translation:
モヤシエキス

Definition: Phaseolus Radiatus Extract is an extract of the bean sprout, *Phaseolus radiatus. See "Regulatory and Ingredient Use Information," regarding the labeling names for botanical ingredients in Volume 1, Introduction, Part A.*

Information Source: JCIC

Chemical Class: Biological Products

Function: Not Reported

Technical/Other Names:
Bean Sprouts Extract
Extract of Phaseolus Radiatus

Trade Name:
SFE Green Gram Extract (Green Tek 21)

PHASEOLUS RADIATUS SEED STARCH

Definition: Phaseolus Radiatus Seed Starch is the starch obtained from the

seeds of the bean, *Phaseolus radiatus. See "Regulatory and Ingredient Use Information," regarding the labeling names for botanical ingredients in Volume 1, Introduction, Part A.*

Information Source: JCIC

Chemical Class: Carbohydrates

Functions: Abrasive; Bulking Agent

Technical/Other Name:
Adzuki Bean Starch

PHASEOLUS TRILOBUS SEED EXTRACT

Definition: Phaseolus Trilobus Seed Extract is an extract of the seeds of *Phaseolus trilobus. See "Regulatory and Ingredient Use Information," regarding the labeling names for botanical ingredients in Volume 1, Introduction, Part A.*

Chemical Class: Biological Products

Functions: Skin-Conditioning Agent - Humectant; Skin-Conditioning Agent - Miscellaneous

Technical/Other Name:
Extract of Phaseolus Trilobus Seed

Trade Name Mixture:
Vitoptine LS (Laboratoires Serobiologiques)

PHASEOLUS VULGARIS (KIDNEY BEAN) SEED EXTRACT

CAS No. 85085-22-9
EINECS No. 285-354-6

Definition: Phaseolus Vulgaris (Kidney Bean) Seed Extract is an extract of the seeds of the kidney bean, *Phaseolus vulgaris. See "Regulatory and Ingredient Use Information," regarding the labeling names for botanical ingredients in Volume 1, Introduction, Part A.*

Chemical Class: Biological Products

Function: Skin-Conditioning Agent - Miscellaneous

Technical/Other Names:
Extract of Kidney Beans
Extract of Phaseolus Vulgaris
Kidney Bean Extract
Kidney Bean Seed Extract
Phaseolus Vulgaris Extract

Trade Name Mixtures:
287 Demaquillant HS (Alban Muller)
687 Demaquillant LS (Alban Muller)
238 Emollient HS (Alban Muller)

Herbasol Extract Red Bean (Cosmetochem) (Cosmetochem International Ltd.)
PW 2000(Phyto-Whitening) (Proimex GmbH)

PHELLINUS LINTEUS EXTRACT

Definition: Phellinus Linteus Extract is an extract of the mushroom, *Phellinus linteus*. See "*Regulatory and Ingredient Use Information*," regarding the labeling names for botanical ingredients in Volume 1, Introduction, Part A.

Chemical Class: Biological Products

Function: Skin-Conditioning Agent - Miscellaneous

Trade Name Mixtures:
DB-PL-compound-1 (Dong Bang Future)
Natural Phellinus Linteus Extract (Dong Bang Future)

PHELLODENDRON AMURENSE BARK

Definition: Phellodendron Amurense Bark is the bark of *Phellodendron amurense*. See "*Regulatory and Ingredient Use Information*," regarding the labeling names for botanical ingredients in Volume 1, Introduction, Part A.

Chemical Class: Biological Products

Function: Not Reported

Technical/Other Name:
Oubaku (JPN)

PHELLODENDRON AMURENSE BARK EXTRACT

CAS No.: 164288-52-2

JPN Translation:
キハダ樹皮エキス

Definition: Phellodendron Amurense Bark Extract is an extract of the powdered bark of the phellodendron, *Phellodendron amurense*. See "*Regulatory and Ingredient Use Information*," regarding the labeling names for botanical ingredients in Volume 1, Introduction, Part A.

Information Sources: JCIC, JCLS, JSQI

Chemical Class: Biological Products

Function: Not Reported

Reported Product Category: Cleansing Products (Cold Creams, Cleansing Lotions, Liquids and Pads)

Technical/Other Names:
Extract of Phellodendron
Oubaku Ekisu (JPN)
Phellodendron Amureuse Extract
Phellodendron Bark Extract
Phellodendron Extract

Trade Name:
Phellodendron Extract Powder (Maruzen Pharmaceuticals Co., Ltd.)

Trade Name Mixtures:
Actiphyte Phellodendron Amurense (Active Organics)
Bathgranue Oubaku (Ichimaru Pharcos)
Bois II (Barnet)
Boumdan (Bioland)
Granule AA (Ichimaru Pharcos)
Ohbaku (Koei Perfumery)
Oubaku Liquid B (Ichimaru Pharcos)
Oubaku Liquid E (Ichimaru Pharcos)
Phellodendron Extract (Maruzen Pharmaceuticals Co., Ltd.)
Phellodendron Extract BG (Maruzen Pharmaceuticals Co., Ltd.)
Phellodendron Extract BG-J (Maruzen Pharmaceuticals Co., Ltd.)
Phellodendron Extract-J (Maruzen Pharmaceuticals Co., Ltd.)
Phellodendron Extract LA (Maruzen Pharmaceuticals Co., Ltd.)
Phellodendron Extract LA-J (Maruzen Pharmaceuticals Co., Ltd.)
Phellodendron Extract Powder-S (Maruzen Pharmaceuticals Co., Ltd.)
Phellodendron Extract SQ (Maruzen Pharmaceuticals Co., Ltd.)
Platycodon Extract BG (Maruzen Pharmaceuticals Co., Ltd.)

PHENACETIN

CAS No.
62-44-2

EINECS No.
200-533-0

Empirical Formula:
$C_{10}H_{13}NO_2$

Definition: Phenacetin is the aromatic compound that conforms to the formula:

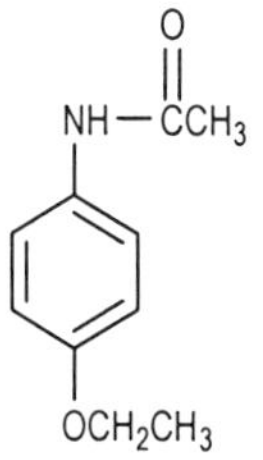

Information Sources: ARG, AUS, BP, BPC, BRA, 21CFR201.309, 21CFR310.545, CZE, DA, DDR, EGY, EP, FIN, HUN, IND, INN, ITA, JAN, MAR, MEX, MI-13(7281), PF, PN, ROM, TSCA, USD, USP XXIV, WHO, YUG

Chemical Class: Amides

Function: Not Reported

Reported Product Categories: Hair Dyes and Colors (All Types Requiring Caution Statements and Patch Tests); Permanent Waves

Technical/Other Names:
Acetamide, N-(4-Ethoxyphenyl)-
Acetophenetidin
N-Acetyl-p-Ethoxyaniline
N-(4-Ethoxyphenyl)Acetamide

PHENETH-6 PHOSPHATE

Definition: Pheneth-6 Phosphate is a complex mixture of esters of phosphoric acid and a polyethylene glycol derivative of phenol containing an average of 6 moles of ethylene oxide.

Chemical Class: Phosphorus Compounds

Functions: Surfactant - Cleansing Agent; Surfactant - Solubilizing Agent; Surfactant - Suspending Agent

Trade Name:
Phosphanol LP-700 (Toho)

PHENETHYL ACETATE

CAS No.
103-45-7

EINECS No.
203-113-5

JPN Translation:
酢酸フェネチル

Empirical Formula:
$C_{10}H_{12}O_2$

Definition: Phenethyl Acetate is the ester of Phenethyl Alcohol (q.v.) and acetic acid.

Information Sources: JCLS, JSCI, RIFM, TSCA

Chemical Class: Esters

Function: Fragrance Ingredient

Technical/Other Names:
Acetic Acid, 2-Phenylethyl Ester
Phenethyl acetate (RIFM)
β-Phenylethyl Acetate
2-Phenylethyl Acetate
Phenylethyl Ethanoate

PHENETHYL ALCOHOL

CAS No.
60-12-8

EINECS No.
200-456-2

JPN Translation:
フェネチルアルコール

Empirical Formula:
$C_8H_{10}O$

Definition: Phenethyl Alcohol is the aromatic alcohol that conforms to the formula:

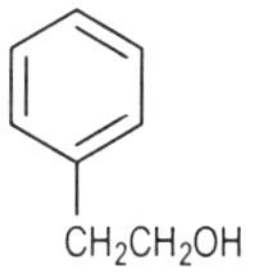

Information Sources: BAN, BP 1963, 21CFR172.515, CIR: [SQ] JACT-9(2)1990, FCC, JCLS, JSCI, MAR, MI-13(7304), PN, RIFM, TSCA, USAN, USD, USP XXIV

Chemical Class: Alcohols

Functions: Fragrance Ingredient; Preservative

Reported Product Categories: Mascara; Makeup Preparations (Not eye), Misc.; Eye Makeup Removers; Moisturizing Preparations; Skin Care Preparations, Misc.

Technical/Other Names:
Benzeneethanol
Benzyl Carbinol
(2-Hydroxyethyl)benzene
Phenethyl alcohol (RIFM)
2-Phenylethanol
Phenyl Ethyl Alcohol

Trade Name:
Etaphen (Vevy)

Trade Name Mixtures:
Farnesol KSN 2/060060 (Symrise)
Fenilight (Sinerga)
Hexatrate (Vevy)
Hexatrate Al-Free (Vevy)
Polysol FS (Polygon)
Polysol GL (Polygon)

PHENETHYL DIMETHICONE

Definition: Phenethyl Dimethicone is the siloxane polymer that conforms generally to the formula:

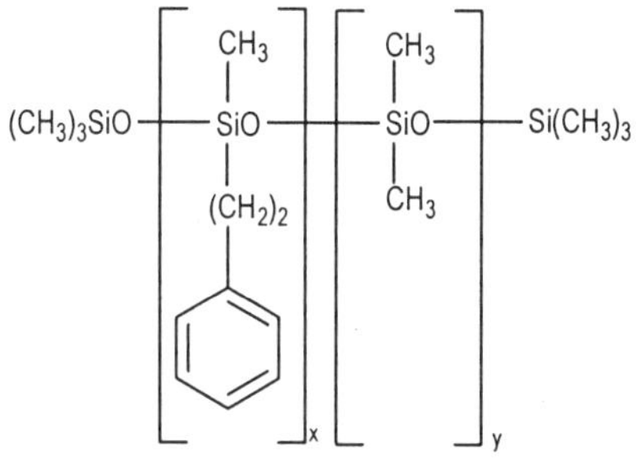

Chemical Classes: Siloxanes and Silanes; Synthetic Polymers

Function: Skin-Conditioning Agent - Emollient

Trade Name:
Silsoft PEDM (OSi Specialties)

PHENETHYL DISILOXANE

Definition: Phenethyl Disiloxane is the organic compound that conforms to the formula:

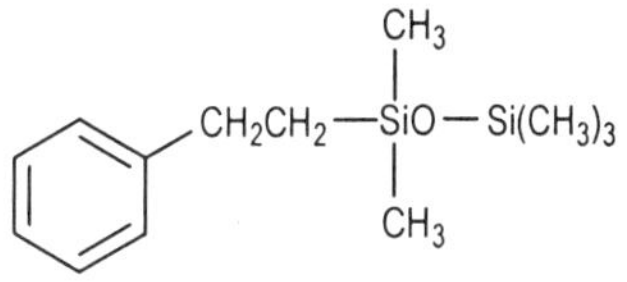

Chemical Class: Siloxanes and Silanes

Functions: Antifoaming Agent; Skin-Conditioning Agent - Occlusive

Technical/Other Name:
Phenyl Ethyl Disiloxane

p-PHENETIDINE

CAS Nos.	EINECS No.
156-43-4	205-855-5
659-34-7	

Empirical Formula:
$C_8H_{11}NO$

Definition: p-Phenetidine is the organic compound that conforms to the formula:

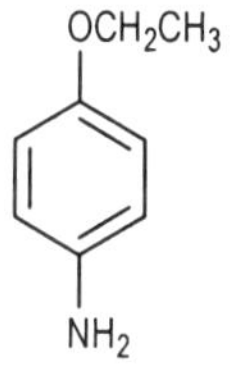

See "Regulatory and Ingredient Use Information," for Colorants in Volume 1, Introduction, Part A.

Information Source: MI-13(7307)

Chemical Classes: Amines; Color Additives - Hair; Ethers

Function: Hair Colorant

Technical/Other Names:
p-Aminophenetole
4-Ethoxyaniline
4-Ethoxybenzenamine
4-Ethoxyphenylamine
Ethyl p-Aminophenol

PHENOL

CAS No.	EINECS No.
108-95-2	203-632-7

JPN Translation:
フェノール

Empirical Formula:
C_6H_6O

Definition: Phenol is the aromatic compound that conforms to the formula:

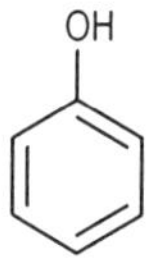

In the United States, Phenol may be used as an active ingredient in OTC drug products. When used as an active drug ingredient, the established name is *Phenol*. *See "Regulatory and Ingredient Use Information," regarding the labeling names for U.S. OTC Drug Ingredients in Volume 1, Introduction, Part A.*

Information Sources: ARG, AUS, BP, BPC, BRA, 21CFR133.5, 21CFR165.110, 21CFR172.105, 21CFR175.105, 21CFR175.300, 21CFR175.380, 21CFR175.390, 21CFR176.170, 21CFR177.1210, 21CFR177.1580, 21CFR177.2260, 21CFR177.2400, 21CFR177.2410, 21CFR177.2600, 21CFR178.3870, 21CFR310.545, 21CFR369.20, 21CFR436.515, 21CFR452.75, 27CFR21.65, 27CFR21.151, CZE, DA, DDR, EEC(III/1-19), EGY, FIN, HP, HUN, IND, ITA, JAN, JCLS, JSCI, MAR, MEX, MHLW-331/3, MI-13(7323), OTC-I-AM, OTC-I-EA, OTC-I-OD, OTC-I-OH, PF, PN, POR, RIFM, ROM, TSCA, USAN, USD, USP XXIV, WHO, YUG

Chemical Class: Phenols

Functions: Antimicrobial Agent; Cosmetic Biocide; Denaturant; Deodorant Agent; Exfoliant; External Analgesic; Fragrance Ingredient; Oral Health Care Drug; Preservative

Reported Product Categories: Bath Oils, Tablets, and Salts; Cleansing Products (Cold Creams, Cleansing Lotions, Liquids and Pads); Skin Care Preparations, Misc.

Technical/Other Names:
Benzenol
Carbolic Acid
Hydroxybenzene
Liquid Phenol
Oxybenzene
Phenol (RIFM)
Phenyl Alcohol

PHENOLPHTHALEIN

CAS Nos.	EINECS No.
77-09-8	201-004-7
5768-87-6	

The inclusion of any compound in the *Dictionary and Handbook* does not indicate that use of that substance as a cosmetic ingredient complies with the laws and regulations governing such use in the United States or any other country.

Empirical Formula:
$C_{20}H_{14}O_4$

Definition: Phenolphthalein is the organic compound that conforms to the formula:

Information Sources: BAN, EEC(II-417), INN, MI-13(7325), OTC-I-LX, TSCA, USAN, USP XXIV

Chemical Classes: Esters; Phenols

Function: Not Reported

Technical/Other Names:
Benzoic Acid, 2-((4-Hydroxyphenyl)(4-Oxo-2,5-cyclohexadien-1-ylidene)methyl)-
3,3-Bis(4-Hydroxyphenyl)-1(3H)-Isobenzofuranone
3,3-Bis(4-hydroxyphenyl) Phthalide
1(3H)-Isobenzofuranone, 3,3-Bis(4-Hydroxyphenyl)-

PHENOXYETHANOL

CAS No.	EINECS No.
122-99-6	204-589-7

JPN Translation:
フェノキシエタノール

Empirical Formula:
$C_8H_{10}O_2$

Definition: Phenoxyethanol is the aromatic ether alcohol that conforms to the formula:

OCH_2CH_2OH

Information Sources: BPC, 21CFR175.105, CIR: [S] JACT-9(2)1990, CTFA D, EEC(VI/1-29), JCLS, JSCI, JSQI, MAR, MHLW-331/3, MI-13(7341), RIFM, SNPF, TSCA

Chemical Classes: Alcohols; Ethers

Functions: Fragrance Ingredient; Preservative

Reported Product Categories: Bath Oils, Tablets, and Salts; Moisturizing Preparations; Skin Care Preparations, Misc.; Bath Preparations, Misc.; Cleansing Products (Cold Creams, Cleansing Lotions, Liquids and Pads); Body and Hand Preparations (Excluding Shaving Preparations); Shampoos (Non-coloring); Bath Capsules; Hair Conditioners; Face and Neck Preparations (Excluding Shaving Preparations); Mascara; Hair Dyes and Colors (All Types Requiring Caution Statements and Patch Tests); Paste Masks (Mud Packs); Night Skin Care Preparations; Tonics, Dressings, and Other Hair Grooming Aids; Foundations; Face Powders; Eyeliners; Eye Makeup Removers; Indoor Tanning Preparations; Hair Preparations (Non-coloring), Misc.; Fragrance Preparations, Misc.; Eye Makeup Preparations, Misc.; Bubble Baths; Eyebrow Pencils; Personal Cleanliness Products, Misc.; Suntan Gels, Creams, and Liquids; Makeup Preparations (Not eye), Misc.; Blushers (All types); Baby Lotions, Oils, Powders and Creams; Aftershave Lotions; Baby Shampoos; Eye Shadows; Makeup Bases; Skin Fresheners; Suntan Preparations, Misc.; Deodorants (Underarm); Foot Powders and Sprays; Makeup Fixatives; Powders (Dusting and Talcum, Excluding Aftershave Talcs); Bath Soaps and Detergents; Lipsticks; Eye Lotions; Douches; Shaving Cream (Aerosol, Brushless and Lather); Shaving Preparations, Misc.; Hair Sprays (Aerosol Fixatives); Rouges; Colognes and Toilet Waters; Cuticle Softeners; Hair Rinses (Non-coloring); Manicuring Preparations, Misc.

Technical/Other Names:
Ethanol, 2-Phenoxy-
Ethylene Glycol Monophenyl Ether
2-Hydroxyethyl Phenyl Ether
2-Phenoxyethanol (RIFM)
2-Phenoxyethanol
2-Phenoxyethyl Alcohol
Phenoxytol

Trade Names:
CoSept PHE (Costec)
Emeressence 1160 (Cognis Care Chemicals/NJ)
Emeressence 1160 (Cognis Care Chemicals/PA)
Hisolve EPH (Toho)
Igepal OD-410 (Rhodia)
Phenoxen (Vevy)
Phenoxetol (Clariant)
Phenoxetol (Clariant GmbH, Personal Care)
Polioxol F-01 (Cognis Iberia/Centerchem)
Protacide P-OH (Protameen)
Sepicide LD (SEPPIC)
S&M Phenoxyethanol (Schulke & Mayr)
Tri-K Phenoxyethanol (Tri-K)

Trade Name Mixtures:
Aceromine (Greentech)
AEC Papaya Extract (Papain 0.25%) (A & E Connock)
AEC Pineapple Extract (Bromelain 0.25%) (A & E Connock)
Bactecar 125S (Phytocos)
Bactiphen 2506 G (Grau)
Chemynol (Chemyunion)
Collagen S (PF) (Nitta Gelatin)
Collagen S-03 (PF) (Nitta Gelatin)
Collagen S-06 (PF) (Nitta Gelatin)
Collagen SP-03 (PF) (Nitta Gelatin)
Conservateur GD500 (Phytocos)
Conservateur GD700 (Phytocos)
CoSept PEP (RTD Hall Star)
Cosmocil AF (Zeneca)
Custoblend BAC (Anti-Bac Blend) (Custom Ingredients)
Dekaben (Dekker)
Dekaben GN (Dekker)
Dekaben IGN (Dekker)
Dekaben LMP (Dekker)
Dekaben LMP-5 (Dekker)
Dekaben P (Dekker)
Dekacydol (Dekker)
Elestab 388 (Laboratoires Serobiologiques)
Emericide 1199 (Cognis Care Chemicals/NJ)
Emericide 1199 (Cognis Care Chemicals/PA)
Eusolex T-Aqua (Merck KGaA/EMD Chemicals Inc.)
Euxyl K 300 (Schulke & Mayr)
Euxyl K400 (Schulke & Mayr)
Euxyl K 700 (Schulke & Mayr)
Euxyl K 702 (Schulke & Mayr)
Euxyl K 727 (Schulke & Mayr)
2/027030 Farnesol plus (Symrise)
2/027020 Farnesol plus 50 % in Dipropylene glycol (Symrise)
Fenossiparaben (Sinerga)
Geogard 361 Preservative (Lonza Inc./Lonza Ltd.)
Germazide MPB (Collaborative Labs)
Germazide WS (Collaborative Labs)
Greenosome Ace (Greentech)
Hexatrate (Vevy)
Hexatrate Al-Free (Vevy)
Killitol II (Collaborative Labs)
Liposerve PP (Lipo)
Liquapar PE (Sutton)
Lowenol 4558 (Lowenstein)
LOWENOL T-1106-A (Lowenstein)
Magnalys (Greentech)
Merguard 1200 (ONDEO Nalco)
Merguard 1200 (ONDEO Nalco Europe)
Microcare MG (Acti-Chem)
Microcare MGI (Acti-Chem)
Microcare PM (Acti-Chem)
Microcare PM5 (Acti-Chem)
Midpol PHN (Microbial Systems)
Midtect TFP (Microbial Systems)
Nipaguard BPX (Clariant)
Nipaguard BPX (Clariant GmbH, Personal Care)
Nipaguard DCB (Clariant)
Nipaguard DCB (Clariant GmbH, Personal Care)
Nipaguard IPP2 (Clariant)

Nipaguard IPP2 (Clariant GmbH, Personal Care)
Paragon III (McIntyre)
Paragon MEPB (McIntyre)
Phenagon IPBC (McIntyre)
Phenagon PDI (McIntyre)
Phenonip (Clariant)
Phenonip (Clariant GmbH, Personal Care)
Phenosept 25P (Clariant)
Phenosept 25P (Clariant GmbH, Personal Care)
Phenova (Crodarom)
Phytoconcentrol Seaweed Oil Soluble 2/012300 (Symrise)
Retimine II (Greentech)
Retimine III (Greentech)
RonaCare ASC III (Merck KGaA)
RonaCare VTA (Merck KGaA)
Rosamine (Greentech)
Saccaluronate CC (LCW)
Saccaluronate LC (LCW)
Self Tanning Complex (Greentech)
Sepicide HB (SEPPIC)
Sepicide HB2 (SEPPIC)
Sepicide WP1 (SEPPIC)
Talcoseptic C (Vevy)
Undebenzofene C (Vevy)
Uniphen P-23 (Induchem)
Unisuprol S-25 (Induchem)

PHENOXYETHYLPARABEN

Empirical Formula:
$C_{15}H_{14}O_3$

Definition: Phenoxyethylparaben is the organic ester of Phenoxyethanol (q.v.) and p-hydroxybenzoic acid. It conforms to the formula:

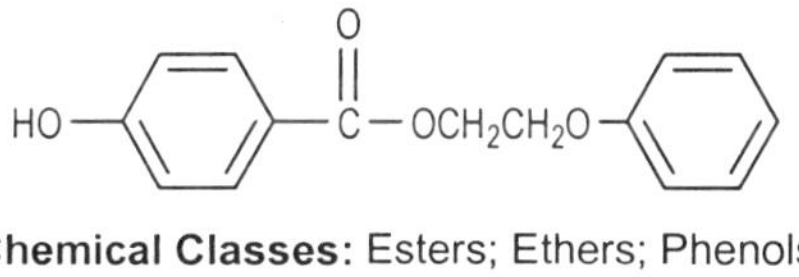

Chemical Classes: Esters; Ethers; Phenols

Function: Preservative

Technical/Other Name:
p-Hydroxybenzoic Acid, Phenoxyethyl Ester

Trade Name Mixture:
Undebenzofene (Vevy)

PHENOXYISOPROPANOL

CAS No.   **EINECS No.**
770-35-4   212-222-7

JPN Translation:
フェノキシイソプロパノール

Empirical Formula:
$C_9H_{12}O_2$

Definition: Phenoxyisopropanol is the aromatic ether alcohol that conforms to the formula:

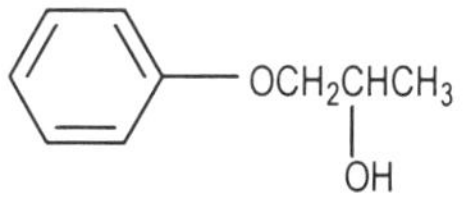

Information Sources: EEC(III/1-54), EEC(VI/1-43), JCIC, JCLS, TSCA

Chemical Classes: Alcohols; Ethers

Functions: Preservative; Solvent

Reported Product Categories: Bath Preparations, Misc.; Body and Hand Preparations (Excluding Shaving Preparations)

Technical/Other Names:
2-Phenoxy-1-Methylethanol
1-Phenoxy-2-Propanol

Trade Names:
DOWANOL PPh (Dow Chemical)
Propylene Phenoxetol (Clariant)
Propylene Phenoxetol (Clariant GmbH, Personal Care)

Trade Name Mixtures:
Phenosept PG (Clariant)
Phenosept PG (Clariant GmbH, Personal Care)

PHENYLALANINE

CAS Nos.   **EINECS Nos.**
63-91-2 (L-Form)   200-568-1
150-30-1 (dl-alpha)   205-756-7
62056-68-2 (L-Form)
167088-01-9 (dl-alpha)

JPN Translation:
フェニルアラニン

Empirical Formula:
$C_9H_{11}NO_2$

Definition: Phenylalanine is the amino acid that conforms to the formula:

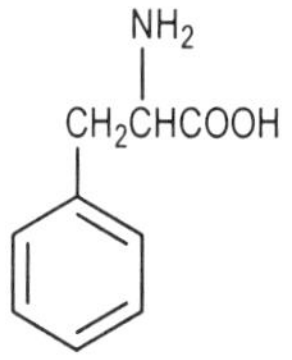

Information Sources: 21CFR172.320, 21CFR201.21, 21CFR310.545, 21CFR582.5590, INN, JAN, JCLS, JP, MI-13(7355), RIFM, TSCA, USAN, USP XXIV

Chemical Class: Amino Acids

Functions: Fragrance Ingredient; Hair Conditioning Agent; Skin-Conditioning Agent - Miscellaneous

Reported Product Categories: Hair Conditioners; Permanent Waves; Suntan Preparations, Misc.; Indoor Tanning Preparations

Technical/Other Names:
Alanine, Phenyl Ester
α-Aminobenzenepropanoic Acid
α-Aminohydrocinnamic Acid
2-Amino-3-Phenylpropionic Acid
D,L-Phenylalanine (RIFM)
L-Phenylalanine (RIFM)

Trade Name:
AEC Phenylalanine (A & E Connock)

Trade Name Mixtures:
Omega-CHS-Activator (GfN)
Dermawhite HS (Laboratoires Serobiologiques)
Photonyl (Laboratoires Serobiologiques)
Sel-Smooth (Seltzer)

PHENYLBENZIMIDAZOLE SULFONIC ACID

CAS No.   **EINECS No.**
27503-81-7   248-502-0

Empirical Formula:
$C_{13}H_{10}N_2O_3S$

Definition: Phenylbenzimidazole Sulfonic Acid is the aromatic organic compound that conforms to the formula:

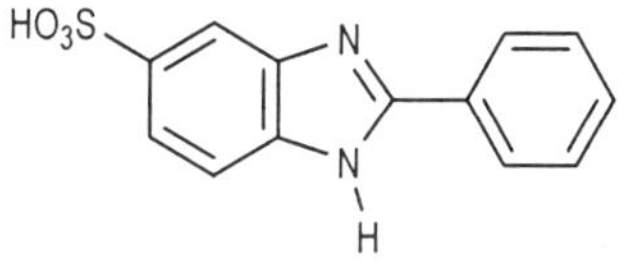

In the United States, Phenylbenzimidazole Sulfonic Acid may be used as an active ingredient in OTC drug products. When used as an active drug ingredient, the established name is *Ensulizole*. See *"Regulatory and Ingredient Use Information,"* regarding the labeling names for U.S. OTC Drug Ingredients in Volume 1, Introduction, Part A.

Information Sources: EEC(VII/1-6), MHLW-331/4, OTC-I-SU, USAN, USP XXIV

Chemical Classes: Heterocyclic Compounds; Sulfonic Acids

Functions: Sunscreen Agent; Ultraviolet Light Absorber

Reported Product Categories: Eyebrow Pencils; Moisturizing Preparations; Suntan Gels, Creams, and Liquids; Body and Hand Preparations (Excluding Shaving Preparations); Cleansing Products (Cold Creams, Cleansing Lotions, Liquids and Pads)

Technical/Other Names:
Ensulizole

2-Phenylbenzimidazole-5-Sulfonic Acid
2-Phenyl-5-Sulfobenzimidazole

Trade Names:
Eusolex 232 (Merck KGaA/EMD Chemicals Inc.)
Neo Heliopan, Type Hydro (Haarmann & Reimer GmbH)
Parsol HS (Roche)

PHENYL BENZOATE

CAS No.
93-99-2

EINECS No.
202-293-2

Empirical Formula:
$C_{13}H_{10}O_2$

Definition: Phenyl Benzoate is the ester of phenol and Benzoic Acid (q.v.). It conforms to the formula:

Information Sources: EEC(VI/1-1), MI-13 (7359), RIFM, TSCA

Chemical Class: Esters

Functions: Fragrance Ingredient; Preservative

Technical/Other Names:
Benzoic Acid, Phenyl Ester
Phenyl benzoate (RIFM)

PHENYL DIMETHICONE

CAS No.: 9005-12-3

JPN Translation:
フェニルジメチコン

Definition: Phenyl Dimethicone is the siloxane polymer that conforms generally to the formula:

Chemical Class: Siloxanes and Silanes

Functions: Antifoaming Agent; Skin-Conditioning Agent - Occlusive

Reported Product Categories: Eye Shadows; Moisturizing Preparations; Face Powders; Hair Sprays (Aerosol Fixatives); Night Skin Care Preparations; Lipsticks;

Tonics, Dressings, and Other Hair Grooming Aids; Foundations; Bath Preparations, Misc.; Blushers (All types); Body and Hand Preparations (Excluding Shaving Preparations); Eyebrow Pencils; Suntan Gels, Creams, and Liquids; Hair Conditioners; Skin Care Preparations, Misc.

p-PHENYLENEDIAMINE HCl

CAS No.
624-18-0

EINECS No.
210-834-9

Empirical Formula:
$C_6H_8N_2$ • 2ClH

Definition: p-Phenylenediamine HCl is the aromatic amine salt that conforms to the formula:

See "Regulatory and Ingredient Use Information," for Colorants in Volume 1, Introduction, Part A.

Information Sources: CI 76061, EEC(III/1-8), MI-13(7369), TSCA

Chemical Classes: Amines; Color Additives - Hair

Function: Hair Colorant

Technical/Other Names:
1,4-Benzenediamine Dihydrochloride
CI 76061
1,4-Diaminobenzene Dihydrochloride
Oxidation Base 10A
p-Phenylenediamine Dihydrochloride
1,4-Phenylenediamine Hydrochloride

Trade Name:
Rodol DC (Lowenstein)

m-PHENYLENEDIAMINE

CAS No.
108-45-2

EINECS No.
203-584-7

Empirical Formula:
$C_6H_8N_2$

Definition: m-Phenylenediamine is the aromatic amine that conforms to the formula:

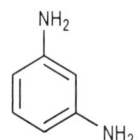

See "Regulatory and Ingredient Use Information," for Colorants in Volume 1, Introduction, Part A.

Information Sources: 21CFR177.2280, CI 76025, CIR: [SQ] IJT-16(Suppl. 1)1997, EEC(III/1-8), MI-13(7367), TSCA

Chemical Classes: Amines; Color Additives - Hair

Function: Hair Colorant

Reported Product Category: Hair Dyes and Colors (All Types Requiring Caution Statements and Patch Tests)

Technical/Other Names:
m-Aminoaniline
1,3-Benzenediamine
CI 76025
Developer 11
1,3-Diaminobenzene
1,3-Phenylenediamine

Trade Names:
Colorex MPD (Chemical Compounds, Inc.)
Covastyle MPD (LCW)
Rodol MPD (Lowenstein)

p-PHENYLENEDIAMINE

CAS No.
106-50-3

EINECS No.
203-404-7

Empirical Formula:
$C_6H_8N_2$

Definition: p-Phenylenediamine is the aromatic amine that conforms to the formula:

See "Regulatory and Ingredient Use Information," for Colorants in Volume 1, Introduction, Part A.

Information Sources: CI 76060, CIR: [S] JACT-4(3)1985, EEC(III/1-8), MAR, MI-13 (7369), TSCA

Chemical Classes: Amines; Color Additives - Hair

Function: Hair Colorant

Reported Product Categories: Hair Dyes and Colors (All Types Requiring Caution Statements and Patch Tests); Hair Color Sprays (Aerosol); Hair Tints

Technical/Other Names:
p-Aminoaniline
1,4-Benzenediamine
CI 76060

p-Diaminobenzene
Oxidation Base 10
1,4-Phenylenediamine

Trade Names:
Colorex PPD-CG (Chemical Compounds, Inc.)
Covastyle PPD (LCW)
Jarocol PPD (Robinson)
Rodol D (Lowenstein)
RODOL D-99 (Lowenstein)

Trade Name Mixtures:
Blonde 90 (Fusion) (Lowenstein)
Blonde R-50 (Fusion) (Lowenstein)
Brown GE (Fusion) (Lowenstein)
Brown R-36 (Fusion) (Lowenstein)
Rodol Black 2/0A (Lowenstein)
Rodol Brown DB (3/A) (Lowenstein)
Rodol Chestnut Brown 5/41 (Lowenstein)
Rodol Light Brown 5/01 (Lowenstein)
Rodol Medium Brown 4/0-PW (Lowenstein)

m-PHENYLENEDIAMINE SULFATE

CAS Nos.	EINECS No.
541-70-8	208-791-6
25723-55-1	

Empirical Formula:
$C_6H_8N_2 \cdot H_2O_4S$

Definition: m-Phenylenediamine Sulfate is the aromatic amine salt that conforms to the formula:

See "Regulatory and Ingredient Use Information," for Colorants in Volume 1, Introduction, Part A.

Information Sources: CIR: [SQ] IJT-16 (Suppl. 1)1997, EEC(III/1-8), TSCA

Chemical Classes: Amines; Color Additives - Hair

Function: Hair Colorant

Reported Product Categories: Hair Dyes and Colors (All Types Requiring Caution Statements and Patch Tests); Hair Tints

Technical/Other Names:
1,3-Benzenediamine, Sulfate
1,3-Phenylenediamine Sulfate
m-Xylenediamine Sulfate

Trade Names:
Colorex MPDS (Chemical Compounds, Inc.)
Covastyle MPDS (LCW)
Rodol MPDS (Lowenstein)

Trade Name Mixture:
Rodol Burgundy 4/5 (Lowenstein)

p-PHENYLENEDIAMINE SULFATE

CAS Nos.	EINECS No.
16245-77-5	240-357-1
50994-40-6	

Empirical Formula:
$C_6H_8N_2 \cdot H_2O_4S$

Definition: p-Phenylenediamine Sulfate is the aromatic amine salt that conforms to the formula:

See "Regulatory and Ingredient Use Information," for Colorants in Volume 1, Introduction, Part A.

Information Sources: EEC(III/1-8), TSCA

Chemical Classes: Amines; Color Additives - Hair

Function: Hair Colorant

Technical/Other Names:
1,4-Benzenediamine, Sulfate
1,4-Benzenediamine Sulfate (1:1)

Trade Names:
Colorex PPDS (Chemical Compounds, Inc.)
Covastyle PPDS (LCW)
Jarocol PPDS (Robinson)
Rodol DS (Lowenstein)

Trade Name Mixture:
Rodol Burgundy 4/5 (Lowenstein)

PHENYLISOHEXANOL

CAS No.	EINECS No.
55066-48-3	259-461-3

Empirical Formula:
$C_{12}H_{18}O$

Definition: Phenylisohexanol is the organic compound that conforms to the formula:

Information Source: RIFM

Chemical Class: Alcohols

Function: Fragrance Ingredient

Technical/Other Names:
Benzenepentanol, γ-Methyl-
γ-Methylbenzenepentanol
3-Methyl-5-phenylpentanol (RIFM)
3-Methyl-5-Phenyl-n-Pentanol

Trade Name:
Phenylhexanol (Firmenich)

PHENYL MERCURIC ACETATE

CAS No.	EINECS No.
62-38-4	200-532-5

Empirical Formula:
$C_8H_8HgO_2$

Definition: Phenyl Mercuric Acetate is the metallo-organic compound that conforms to the formula:

Information Sources: AUS, BPC, DDR, EEC(VI/1-17), MAR, MHLW-331/1, MI-13 (7383), NF XIX, NFJ, POR, TSCA, USAN

Chemical Class: Organic Salts

Function: Preservative

Reported Product Categories: Mascara; Eyeliners

Technical/Other Names:
(Acetato-o)Phenylmercury
Acetoxyphenylmercury

PHENYL MERCURIC BENZOATE

CAS No.	EINECS No.
94-43-9	202-331-8

Empirical Formula:
$C_{13}H_{10}HgO_2$

Definition: Phenyl Mercuric Benzoate is the metallo-organic compound that conforms to the formula:

The inclusion of any compound in the *Dictionary and Handbook* does not indicate that use of that substance as a cosmetic ingredient complies with the laws and regulations governing such use in the United States or any other country.

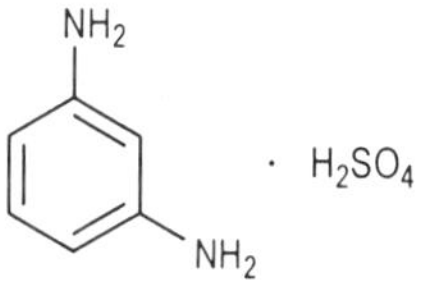

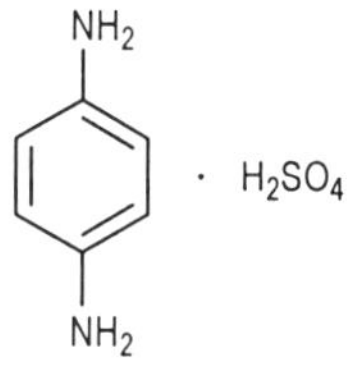

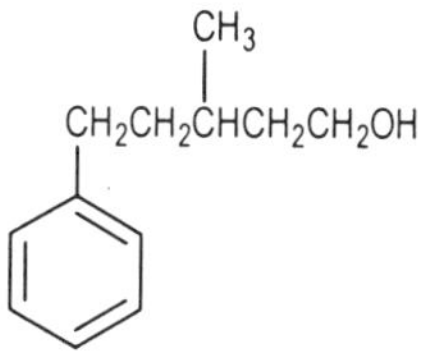

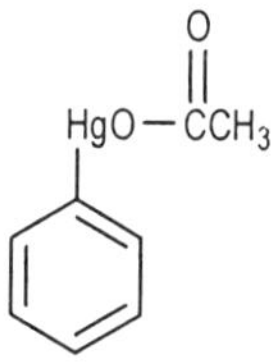

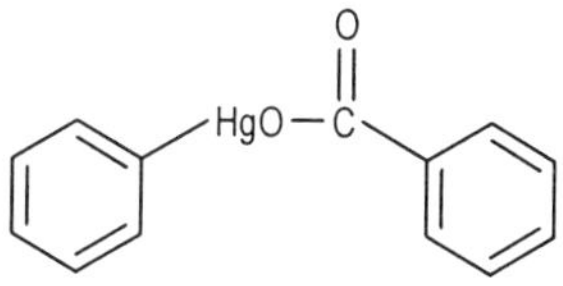

Information Sources: 27CFR21.120, MHLW-331/1

Chemical Class: Organic Salts

Function: Preservative

Technical/Other Names:
(Benzoato-o)Phenylmercury
Mercury, (Benzoato-o)Phenyl-

PHENYL MERCURIC BORATE

CAS Nos.	EINECS Nos.
102-98-7	203-068-1
6273-99-0	228-465-7

Empirical Formula:
$C_6H_7BHgO_3$

Definition: Phenyl Mercuric Borate is the metallo-organic compound that conforms to the formula:

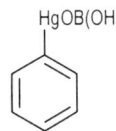

$HgOB(OH)_2$

Information Sources: BRA, CZE, DDR, EEC(VI/1-17), FIN, INN, MAR, MHLW-331/1, MI-13(7386)

Chemical Class: Organic Salts

Function: Preservative

Technical/Other Names:
Mercurate(2-), [Orthoboroato(3-)-O]Phenyl, Dihydrogen
[Orthoborato(3-)-O]Phenylmercurate(2-), Dihydrogen

PHENYL MERCURIC BROMIDE

CAS No.	EINECS No.
1192-89-8	214-760-8

Empirical Formula:
C_6H_5BrHg

Definition: Phenyl Mercuric Bromide is the metallo-organic compound that conforms to the formula:

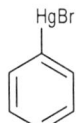

HgBr

Information Sources: EEC(VI/1-17), MHLW-331/1

Chemical Class: Organic Salts

Function: Preservative

Technical/Other Name:
Bromophenylmercury

PHENYL MERCURIC CHLORIDE

CAS No.	EINECS No.
100-56-1	202-865-1

Empirical Formula:
C_6H_5ClHg

Definition: Phenyl Mercuric Chloride is the metallo-organic compound that conforms to the formula:

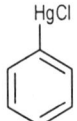

HgCl

Information Sources: EEC(VI/1-17), MHLW-331/1, MI-13(7384)

Chemical Class: Organic Salts

Function: Preservative

Technical/Other Name:
Chlorophenylmercury

PHENYL METHICONE

CAS Nos.: 31230-04-3; 63148-58-3

JPN Translation:
フェニルメチコン

Empirical Formula:
$(C_7H_8OSi)_xC_4H_{12}Si$

Definition: Phenyl Methicone is the siloxane polymer that conforms generally to the formula:

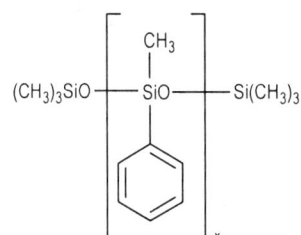

$(CH_3)_3SiO$ — SiO — $Si(CH_3)_3$

Information Sources: JCLS, JSCI

Chemical Class: Siloxanes and Silanes

Function: Skin-Conditioning Agent - Emollient

Technical/Other Names:
Methyl Phenyl Polysiloxane
Phenyl Methyl Polysiloxane
Polymethylphenyl Siloxane
Polyphenylmethyl Siloxane

Trade Name Mixtures:
MagiSil 451 (Diow)
MagiSil 509 (Diow)
MagiSil 451S (Diow)
MagiSil 451SS (Diow)
MagiSil 451SW (Diow)

StrataGel (Diow)
StrataGel 171 (Diow)
StrataGel 500 (Diow)

PHENYL METHICONOL

CAS No.: 80801-30-5

Definition: Phenyl Methiconol is the organic compound that conforms to the formula:

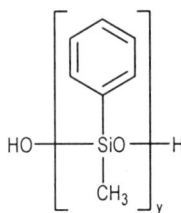

Chemical Class: Siloxanes and Silanes

Functions: Hair Conditioning Agent; Skin-Conditioning Agent - Miscellaneous

Technical/Other Name:
Siloxanes and Silicones, Me Ph, Hydroxy-Terminated

Trade Name Mixture:
Dow Corning 1-2294 Fluid (Dow Corning)

PHENYLMETHYLPENTANAL

CAS No.: 55066-49-4

Empirical Formula:
$C_{12}H_{16}O$

Definition: Phenylmethylpentanal is the aldehyde that conforms to the formula:

$CH_2CH_2CHCH_2CH$

Information Sources: RIFM, TSCA

Chemical Class: Aldehydes

Function: Fragrance Ingredient

Technical/Other Names:
Benzenepentanal, Beta, -Methyl-
Benzenepentanal, .beta.-methyl- (RIFM)
3-Methyl-5-Phenyl-Pentanal

PHENYL METHYL PYRAZOLONE

CAS No.	EINECS No.
89-25-8	201-891-0

Empirical Formula:
$C_{10}H_{10}N_2O$

Definition: Phenyl Methyl Pyrazolone is the heterocyclic amine that conforms to the formula:

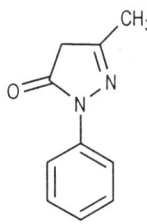

See "Regulatory and Ingredient Use Information," for Colorants in Volume 1, Introduction, Part A.

Information Sources: CIR: [S] JACT-11(4)-1992, EEC(III/2-20), MI-13(6746), TSCA

Chemical Classes: Color Additives - Hair; Heterocyclic Compounds

Function: Hair Colorant

Reported Product Categories: Hair Dyes and Colors (All Types Requiring Caution Statements and Patch Tests); Hair Color Sprays (Aerosol)

Technical/Other Names:
Developer 1
2,4-Dihydro-5-Methyl-2-Phenyl-3H-Pyrazol-3-one
3H-Pyrazol-3-one, 2,4-Dihydro-5-Methyl-2-Phenyl-
Norphenazone

Trade Names:
Colorex PMP (Chemical Compounds, Inc.)
Jarocol PMP (James Robinson)
Rodol PMP (Lowenstein)

Trade Name Mixtures:
Lowenol Stabilizer L-536 (Lowenstein)
Lowenol Stabilizer L-552 (Lowenstein)
Rodol Blue 2/X (Lowenstein)
Rodol Chestnut Brown 5/41 (Lowenstein)
Rodol Light Brown 5/01 (Lowenstein)

PHENYLPARABEN

CAS No.   **EINECS No.**
17696-62-7   241-698-9

Empirical Formula:
$C_{13}H_{10}O_3$

Definition: Phenylparaben is the ester of phenol and 4-Hydroxybenzoic Acid (q.v.). It conforms to the formula:

Information Sources: EEC(VI/1-12), MHLW-331/3

Chemical Classes: Esters; Phenols

Function: Preservative

Technical/Other Names:
Benzoic Acid, 4-Hydroxy-, Phenyl Ester
4-Hydroxybenzoic Acid, Phenyl Ester
p-Hydroxybenzoic Acid, Phenyl Ester
Phenyl p-Hydroxybenzoate

o-PHENYLPHENOL

CAS No.   **EINECS No.**
90-43-7   201-993-5

JPN Translation:
フェニルフェノール

Empirical Formula:
$C_{12}H_{10}O$

Definition: o-Phenylphenol is the substituted aromatic compound that conforms to the formula:

Information Sources: 21CFR175.105, 21CFR176.210, 21CFR177.2600, CTFA D, EEC(VI/1-7), JCLS, JSCI, MAR, MHLW-331/3, MI-13(7388), RIFM, TSCA

Chemical Class: Phenols

Functions: Cosmetic Biocide; Fragrance Ingredient; Preservative

Reported Product Category: Hair Coloring Preparations, Misc.

Technical/Other Names:
(1,1'-Biphenyl)-2-ol
o-Diphenylol
Orthophenylphenol
2-Phenylphenol (RIFM)
2-Phenyl Phenol

Trade Names:
DOWICIDE 1 (Dow Chemical)
Preventol O Extra (Bayer AG)

N-PHENYL-p-PHENYLENEDIAMINE HCl

CAS Nos.   **EINECS Nos.**
2198-59-6   218-599-4
56426-15-4   260-173-5

Empirical Formula:
$C_{12}H_{12}N_2 \cdot ClH$

Definition: N-Phenyl-p-Phenylenediamine HCl is the aromatic amine salt that conforms to the formula:

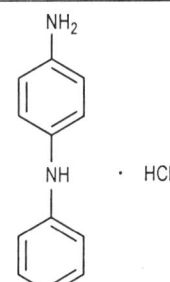

See "Regulatory and Ingredient Use Information," for Colorants in Volume 1, Introduction, Part A.

Information Sources: CI 76086, CIR: [SQ] JACT-13(5)1994, EEC(III/1-8), TSCA

Chemical Classes: Amines; Color Additives - Hair

Function: Hair Colorant

Reported Product Category: Hair Dyes and Colors (All Types Requiring Caution Statements and Patch Tests)

Technical/Other Names:
p-Aminodiphenylamine HCl
1,4-Benzenediamine, N-Phenyl-, Hydrochloride
CI 76086
N-Phenyl-1,4-Benzenediamine Hydrochloride

Trade Name:
Rodol Gray BC (Lowenstein)

N-PHENYL-p-PHENYLENEDIAMINE

CAS No.   **EINECS No.**
101-54-2   202-951-9

Empirical Formula:
$C_{12}H_{12}N_2$

Definition: N-Phenyl-p-Phenylenediamine is the aromatic amine salt that conforms to the formula:

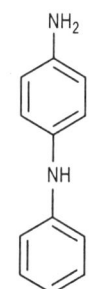

See "Regulatory and Ingredient Use Information," for Colorants in Volume 1, Introduction, Part A.

Information Sources: CI 76085, CIR: [SQ] JACT-13(5)1994, EEC(III/1-8), TSCA

Chemical Classes: Amines; Color Additives - Hair

Function: Hair Colorant

Reported Product Category: Hair Dyes and Colors (All Types Requiring Caution Statements and Patch Tests)

Technical/Other Names:
4-Aminodiphenylamine
p-Aminodiphenylamine
1,4-Benzenediamine, N-Phenyl-
N,4'-Bianiline
CI 76085
Oxidation Base 2
4-(Phenylamino)aniline
N-Phenyl-1,4-Benzenediamine

Trade Names:
Colorex 4AD (Chemical Compounds, Inc.)
Rodol Gray B Base (Lowenstein)

N-PHENYL-p-PHENYLENEDIAMINE SULFATE

CAS No.   **EINECS No.**
4698-29-7   225-173-1

Empirical Formula:
$C_{12}H_{12}N_2 \cdot H_2O_4S$

Definition: N-Phenyl-p-Phenylenediamine Sulfate is the aromatic amine salt that conforms to the formula:

See "Regulatory and Ingredient Use Information," for Colorants in Volume 1, Introduction, Part A.

Information Sources: CIR: [SQ] JACT-13 (5)1994, EEC(III/1-8), TSCA

Chemical Classes: Amines; Color Additives - Hair

Function: Hair Colorant

Reported Product Category: Hair Dyes and Colors (All Types Requiring Caution Statements and Patch Tests)

Technical/Other Names:
p-Aminodiphenylamine Sulfate
1,4-Benzenediamine, N-Phenyl-, Sulfate
1,4-Benzenediamine, N-Phenyl-, Sulfate (2:1)
N-Phenyl-1,4-Benzenediamine Sulfate

Trade Names:
Colorex 4ADS (Chemical Compounds, Inc.)
Rodol Gray BS (Lowenstein)

PHENYLPROPANOL

CAS No.   **EINECS No.**
1335-12-2   215-621-4

Empirical Formula:
$C_9H_{12}O$

Definition: Phenylpropanol is the organic compound that conforms to the formula:

$CH_2CH_2CH_2OH$

Information Sources: 21CFR1310.02, 21CFR1310.04, JAN, RIFM

Chemical Class: Alcohols

Functions: Fragrance Ingredient; Solvent

Technical/Other Name:
Phenylpropanol (RIFM)

Trade Name Mixtures:
Dermosoft 690 (Straetmans)
Dermosoft MCA (Straetmans)
Lexgard 690 (Inolex)
Lexgard MCA (Inolex)

PHENYLPROPYL-DIMETHYLSILOXYSILICATE

Definition: Phenylpropyl-dimethylsiloxysilicate is the siloxane polymer that conforms to the formula:

Chemical Classes: Siloxanes and Silanes; Synthetic Polymers

Functions: Film Former; Hair Conditioning Agent; Skin-Conditioning Agent - Emollient; Skin-Conditioning Agent - Miscellaneous

Trade Name:
Baysilone CF-1301 (GE Silicones)

PHENYLPROPYL ETHYL METHICONE

CAS No.: 68037-77-4

Definition: Phenylpropyl Ethyl Methicone is the organic compound that conforms to the formula:

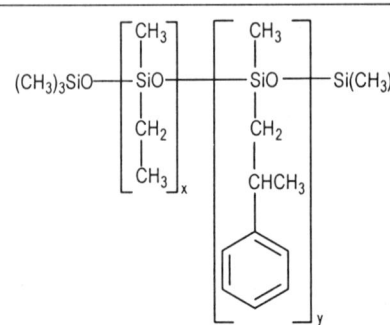

Chemical Class: Siloxanes and Silanes

Functions: Hair Conditioning Agent; Skin-Conditioning Agent - Miscellaneous

Technical/Other Name:
Siloxanes and Silicones, Et Me, Me 2-Phenylpropyl

Trade Name:
Dow Corning 230 Fluid (Dow Corning)

PHENYL PROPYL TRIMETHICONE

Definition: Phenyl Propyl Trimethicone is the siloxane polymer that conforms generally to the formula:

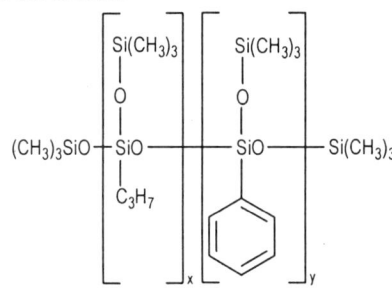

Chemical Classes: Siloxanes and Silanes; Synthetic Polymers

Functions: Hair Conditioning Agent; Skin-Conditioning Agent - Miscellaneous

Trade Name:
Dow Corning 560 Cosmetic Grade Fluid (Dow Corning)

PHENYL PROPYL TRIMETHICONE/DIPHENYLMETHICONE

Definition: Phenyl Propyl Trimethicone/Diphenylmethicone is mixture of siloxane polymers that conform generally to the formula:

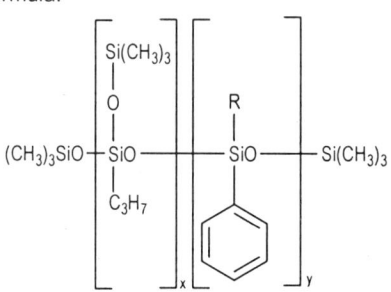

where R represents either (CH₃)₃SiO-or (CH₃)(C₆H₅)₂SiO-.

Chemical Class: Siloxanes and Silanes

Functions: Anticaking Agent; Skin-Conditioning Agent - Emollient

Trade Name:
  Dow Corning 2-2043 Cosmetic Grade Fluid (Dow Corning)

PHENYL SALICYLATE

CAS No.	**EINECS No.**
118-55-8	204-259-2

JPN Translation:
サリチル酸フェニル

Empirical Formula:
$C_{13}H_{10}O_3$

Definition: Phenyl Salicylate is the organic compound that conforms to the formula:

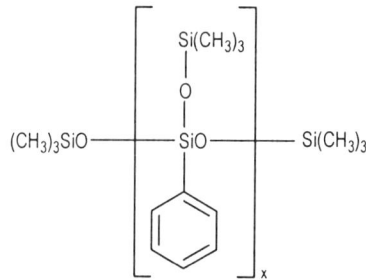

Information Sources: 21CFR177.1010, 27CFR21.65, 27CFR21.151, JCLS, JSCI, MHLW-331/2, MI-13(7394), RIFM

Chemical Classes: Esters; Phenols

Functions: Cosmetic Biocide; Denaturant; Fragrance Ingredient

Technical/Other Names:
  Benzoic Acid, 2-Hydroxy-, Phenyl Ester
  2-Hydroxybenzoic Acid, Phenyl Ester
  2-Phenoxycarbonylphenol
  Phenyl-2-Hydroxybenzoate
  Phenyl salicylate (RIFM)

PHENYLTHIOGLYCOLIC ACID

CAS No.	**EINECS No.**
103-04-8	203-073-9

Empirical Formula:
$C_8H_8O_2S$

Definition: Phenylthioglycolic Acid is the aromatic acid that conforms to the formula:

SHCHCOOH

Information Source: CTFA D

Chemical Classes: Carboxylic Acids; Thio Compounds

Function: Antioxidant

Technical/Other Names:
  Acetic Acid, (Phenylthio)-
  α-Carboxythioanisole
  Phenyl Mercaptoacetic Acid
  Phenyl Thioacetic Acid

PHENYL TRIMETHICONE

CAS Nos.	**EINECS No.**
2116-84-9	218-320-6
18758-91-3	
18876-34-1	
70131-69-0	
73559-47-4	
195868-36-1	

JPN Translation:
フェニルトリメチコン

Definition: Phenyl Trimethicone is the siloxane polymer that conforms generally to the formula:

where n is predominantly 1 to 3.

Information Sources: 21CFR175.105, 21CFR175.300, CIR: [S] JACT-5(5)1986, CTFA S, TSCA

Chemical Class: Siloxanes and Silanes

Functions: Antifoaming Agent; Hair Conditioning Agent; Skin-Conditioning Agent - Occlusive

Reported Product Categories: Eye Shadows; Lipsticks; Tonics, Dressings, and Other Hair Grooming Aids; Foundations; Hair Sprays (Aerosol Fixatives); Makeup Preparations (Not eye), Misc.; Eye Makeup Preparations, Misc.; Moisturizing Preparations; Face Powders; Blushers (All types); Eyeliners; Hair Conditioners; Hair Preparations (Noncoloring), Misc.; Makeup Bases; Indoor Tanning Preparations; Rouges; Bath Oils, Tablets, and Salts; Bath Preparations, Misc.; Body and Hand Preparations (Excluding Shaving Preparations); Deodorants (Underarm); Skin Care Preparations, Misc.; Bath Capsules; Cleansing Products (Cold Creams, Cleansing Lotions, Liquids and Pads); Face and Neck Preparations (Excluding Shaving Preparations)

Trade Names:
  AEC Phenyl Trimethicone (A & E Connock)
  Baysilone PK5 (GE Bayer Silicones)
  Baysilone PK20 (GE Bayer Silicones)
  Botanisil PT-100 (Botanigenics)
  Dow Corning 556 Cosmetic Grade Fluid (Dow Corning)
  Dow Corning 558 Fluid (Dow Corning)
  KF56 (Shin Etsu)
  KF58 (Shin Etsu)
  Mirasil PTM (Rhodia)
  Organza (Diow)
  Organza CFT (Diow)
  Organza XT (Diow)
  SF1550 (GE Silicones)
  SilCare Silicone 15M30 Phenyl Trimethicone (Clariant)
  SilCare Silicone 15M30 Phenyl Trimethicone (Clariant GmbH, Personal Care)
  SilCare Silicone 15M40 Phenyl Trimethicone (Clariant)
  SilCare Silicone 15M40 Phenyl Trimethicone (Clariant GmbH, Personal Care)
  SilCare Silicone 15M50 Phenyl Trimethicone (Clariant)
  SilCare Silicone 15M50 Phenyl Trimethicone (Clariant GmbH, Personal Care)
  SilCare Silicone 15M60 Phenyl Trimethicone (Clariant)
  SilCare Silicone 15M60 Phenyl Trimethicone (Clariant GmbH, Personal Care)
  Ultrapure KF 54 (Ultra Chemical)
  Ultrapure KF 56 (Ultra Chemical)
  Wacker-Belsil PDM 20 (Wacker-Chemie)
  Wacker-Belsil PDM 200 (Wacker-Chemie)
  Wacker-Belsil PDM 1000 (Wacker-Chemie)
Trade Name Mixtures:
  Actiprime 100 (Active Organics)
  Catiosphere (Coletica SA)
  Gel Base BSM 5 (Arch Personal Care Products)
  Gransil PM-GEL (Grant)
  KSG-18 (Shin Etsu)
  NanoGard 556 (Nanophase)
  NanoGard FE45B 556 (Nanophase)
  NanoGard FE45BL 556 (Nanophase)
  NanoGard FE45R 556 (Nanophase)
  Satin Finish (Collaborative Labs)
  Silatex 17 (Cosmetic Ingredient Resources/ Centerchem)
  Silatex-D30 (Cosmetic Ingredient Resources/Centerchem)
  TAYLOR T-Sil CD-5 (Taylor Chemical Company)

PHILADELPHUS CORONARIUS FLOWER EXTRACT

CAS No.	**EINECS No.**
94279-93-3	304-839-6

Definition: Philadelphus Coronarius Flower Extract is an extract of the flowers of *Philadelphus coronarius*. See *"Regulatory and Ingredient Use Information,"* regarding the labeling names for botanical ingredients in Volume 1, Introduction, Part A.

Chemical Class: Biological Products

Function: Not Reported

Technical/Other Name:
Extract of Philadelphus Coronarius

Trade Name Mixture:
Extrait De Seringa (Yves Rocher)

PHLOROGLUCINOL

CAS No.	EINECS No.
108-73-6	203-611-2

Empirical Formula:
$C_6H_6O_3$

Definition: Phloroglucinol is the aromatic alcohol that conforms to the formula:

See *"Regulatory and Ingredient Use Information,"* for Colorants in Volume 1, Introduction, Part A.

Information Sources: CIR: [I] JACT-14(6)-1995, MI-13(7413), TSCA

Chemical Classes: Color Additives - Hair; Phenols

Functions: Antioxidant; Hair Colorant

Reported Product Category: Hair Dyes and Colors (All Types Requiring Caution Statements and Patch Tests)

Technical/Other Names:
1,3,5-Benzenetriol
1,3,5-Trihydroxybenzene

Trade Name:
Rodol PHG (Lowenstein)

PHLOROGLUCINOL TRIMETHYL ETHER

CAS No.	EINECS No.
621-23-8	210-673-4

Empirical Formula:
$C_9H_{12}O_3$

Definition: Phloroglucinol Trimethyl Ether is the organic compound that conforms to the formula:

Chemical Class: Ethers

Function: Skin-Conditioning Agent - Miscellaneous

Technical/Other Name:
Benzene, 1,3,5-Trimethoxy-

PHOENIX DACTYLIFERA (DATE) FRUIT EXTRACT

CAS No.	EINECS No.
90027-90-0	289-776-1

Definition: Phoenix Dactylifera (Date) Fruit Extract is an extract of the fruit of the date, *Phoenix dactylifera*. See *"Regulatory and Ingredient Use Information,"* regarding the labeling names for botanical ingredients in Volume 1, Introduction, Part A.

Chemical Class: Biological Products

Function: Not Reported

Technical/Other Names:
Date Extract
Date Fruit Extract
Extract of Dates
Phoenix Dactylifera Extract

Trade Name Mixtures:
Actiphyte of Date BG50 (Active Organics)
Actiphyte of Date GL50 (Active Organics)
Actiphyte of Date Lipo S (Active Organics)
Actiphyte of Date PG50 (Active Organics)
Enna 216 Breast Development Extract (Ennagram)
Glycolysat of Date (CEP (Solabia))

PHOENIX DACTYLIFERA (DATE) SEED

Definition: Phoenix Dactylifera (Date) Seed is the dried ground seeds of the date, *Phoenix dactylifera*. See *"Regulatory and Ingredient Use Information,"* regarding the labeling names for botanical ingredients in Volume 1, Introduction, Part A.

Chemical Class: Biological Products

Function: Skin-Conditioning Agent - Miscellaneous

Technical/Other Name:
Date Seed Powder

Trade Name:
Ground Date Seeds (Macoma)

PHONOLITE

Definition: Phonolite is a mineral consisting chiefly of SiO2, Al2O3, Fe2O3, CaO, Na2O, and K2O.

Chemical Class: Inorganics

Function: Absorbent

Trade Name:
Vulkanit (Hans G. Hauri)

PHOSPHATIDYLCHOLINE

Definition: Phosphatidylcholine is a purified grade of Lecithin (q.v.) containing no less than 95% of the phospholipid that conforms to the formula:

where RCO- represents various naturally occurring fatty acids.

Chemical Classes: Glyceryl Esters and Derivatives; Phosphorus Compounds

Functions: Skin-Conditioning Agent - Miscellaneous; Surfactant - Emulsifying Agent

Trade Names:
Emulmetik 930 (Lucas Meyer GmbH)
Lipoid S 100 (Lipoid)
Phospholipon 90G (Phospholipid)

Trade Name Mixtures:
Kryosome 1708 (Lipoid)
Membranol PC-30 (Formula One Sciences)
Membranol PC-35 (Formula One Sciences)
Ultraspheres 8028 (Lipoid)

PHOSPHOLIPIDS

CAS No.: 123465-35-0

Definition: Phospholipids are complex lipids in which one of the primary hydroxyl groups of glycerin is esterified with phosphoric acid which carries an additional ester grouping. The two remaining hydroxyl groups are esterified with long chain, saturated or unsaturated fatty acids.

Chemical Classes: Fats and Oils; Phosphorus Compounds

Function: Skin-Conditioning Agent - Miscellaneous

Reported Product Categories: Moisturizing Preparations; Bath Capsules; Face and Neck Preparations (Excluding Shaving Preparations); Bath Oils, Tablets, and Salts; Cleansing Products (Cold Creams, Cleansing Lotions, Liquids and Pads); Eye Makeup Preparations, Misc.; Night Skin Care Preparations

Trade Names:
Alpha Lipid 100 (Bay Milk)
Ceramax (Quest International)
Phospholipon 80 (Phospholipid)
Sojaphos (Vevy)

Trade Name Mixtures:
AC Allantoin Liposome (Active Concepts)
AC Aloe Liposome (Active Concepts)
AC Caffeine Liposome (Active Concepts)
AC CC Liposome (Active Concepts)
AC EFA Liposome (Active Concepts)
AC Folic Liposome (Active Concepts)
AC Liposome DRF (Active Concepts)
AC Liposome GSL (Active Concepts)
AC Melanin Liposome (Active Concepts)
AC Panthenol Liposome (Active Concepts)
AC Retinol Liposome (Active Concepts)
AC Saw Palmetto Liposome (Active Concepts)
AC SOD Liposome (Active Concepts)
AC Unloaded Liposome (Active Concepts)
AC Ursolic Acid Liposome (Active Concepts)
AC Vitamin ABCDE Liposome (Active Concepts)
AC Vitamin ACE-BC Liposome (Active Concepts)
AC Vitamin ACE Liposome (Active Concepts)
AC Vitamin AE Liposome (Active Concepts)
AC Vitamin A Liposome (Active Concepts)
AC Vitamin B5 Liposome (Active Concepts)
AC Vitamin C Liposome (Active Concepts)
AC Vitamin E Liposome (Active Concepts)
Alpha Hydroxy Acid Liposomes (Collaborative Labs)
Alpha Lipid 100 (Lucas Meyer)
Anti-Irritant Liposomes (Collaborative Labs)
Botaniceutical BR-1 (Botanigenics)
Botaniceutical BR-2 (Botanigenics)
Botaniceutical BR-T (Botanigenics)
Brookosome A (Arch Personal Care Products)
Brookosome ACE (Arch Personal Care Products)
Brookosome ACEBC (Arch Personal Care Products)
Brookosome ACEBC Plus (Arch Personal Care Products)
Brookosome BC (Arch Personal Care Products)
Brookosome Biophos (Arch Personal Care Products)
Brookosome DHA (Arch Personal Care Products)
Brookosome E (Arch Personal Care Products)
Brookosome EFA (Arch Personal Care Products)
Brookosome ELL (Arch Personal Care Products)
Brookosome EPO (Arch Personal Care Products)
Brookosome FIH (Arch Personal Care Products)
Brookosome Fucus (Arch Personal Care Products)
Brookosome GSL (Arch Personal Care Products)
Brookosome H (Arch Personal Care Products)
Brookosome Herbal 1 (Arch Personal Care Products)
Brookosome MPS (Arch Personal Care Products)
Brookosome MT (Arch Personal Care Products)
Brookosome P (Arch Personal Care Products)
Brookosome RP (Arch Personal Care Products)
Brookosome S (Arch Personal Care Products)
Brookosome SC (Arch Personal Care Products)
Brookosome SE (Arch Personal Care Products)
Brookosome Serum (Arch Personal Care Products)
Brookosome TE (Arch Personal Care Products)
Brookosome TRF (Arch Personal Care Products)
Brookosome TYE (Arch Personal Care Products)
Brookosome U (Arch Personal Care Products)
Brookosome V (Arch Personal Care Products)
Brookosome Willow Bark (Arch Personal Care Products)
Catezomes SA-20 Japan (Collaborative Labs)
Centella Phytosome (Indena SpA)
Ceramide Complex CLR (P) (CLR)
Ceramides LS (Laboratoires Sero-biologiques)
Escin/B-Sitosterol Phytosome (Indena SpA)
Firming Liposomes (Collaborative Labs)
Ginkgo Biloba Dimeric Flavonoids Phytosome (Indena SpA)
Ginkgo Biloba Phytosome (Indena SpA)
Ginseng Phytosome (Indena SpA)
Glycoderm P (CLR)
Glycosome (Pentapharm/Centerchem)
Glycyrrhetinic Acid Phytosome (Indena SpA)
Heliogel (Advanced Beauty)
Leucocyanidins Phytosome (Indena SpA)
Matipure (Advanced Beauty)
Melarrest L (Collaborative Labs)
Moisturizing Liposomes (Collaborative Labs)
Molecularsource LPC (C.I.T.)
MPC-Liposomes (CLR)
Nano-emulsion Concentrate (Active Concepts)
Natipide II (Phospholipid)
Palmisomes (Alban Muller)
Proceramide L2 (Sederma)
Pro-Lipo H (Lucas Meyer GmbH)
Pro-Lipo S (Lucas Meyer GmbH)
Questamix H (Quest International)
Salicylic Acid Liposome (Active Concepts)
Sansurf 260 (Collaborative Labs)
SanSurf Petrolatum-25 (Collaborative Labs)
San Surf Petrolatum-50 (Collaborative Labs)
Sansurf PFMP (Collaborative Labs)
Satin Finish (Collaborative Labs)
Sericoside Phytosome (Indena SpA)
Silatex 17 (Cosmetic Ingredient Resources/Centerchem)
Silatex-D30 (Cosmetic Ingredient Resources/Centerchem)
Silymarin Phytosome (Indena SpA)
Solarease II (Collaborative Labs)
Solarease OMC (Collaborative Labs)
Thiosome P (CLR)
Visnadex (Indena SpA)
Vitamin C & E Liposomes (Collaborative Labs)

PHOSPHONOBUTANETRICARBOXYLIC ACID

CAS No. 37971-36-1   **EINECS No.** 253-733-5

Empirical Formula:
$C_7H_{11}O_9P$

Definition: Phosphonobutanetricarboxylic Acid is the organic compound that conforms to the formula:

$$\begin{array}{c} COOH \\ | \\ HOOCCH_2CH_2CCH_2COOH \\ | \\ O{=}P{-}OH \\ | \\ OH \end{array}$$

Chemical Classes: Carboxylic Acids; Phosphorus Compounds

Functions: Buffering Agent; Chelating Agent; Corrosion Inhibitor; Suspending Agent - Nonsurfactant

Technical/Other Name:
1,2,4-Butanetricarboxylic Acid, 2-Phosphono-

Trade Name:
Bayhibit AM (Bayer AG)

PHOSPHORIC ACID

CAS No. 7664-38-2   **EINECS No.** 231-633-2

JPN Translation:
リン酸

Empirical Formula:
H_3O_4P

Definition: Phosphoric Acid is the inorganic acid that conforms to the formula:

H_3PO_4

Information Sources: ARG, AUS, BP, BPC, BRA, 21CFR133, 21CFR175.300, 21CFR177.2260, 21CFR178.1010, 21CFR178.3520, 21CFR182.1073, 21CFR558.311, DA, DDR, EGY, EP, FCC, HP, HUN, IND, ITA, JCLS, JSCI, MAR, MEX, MI-13(7430), NF XIX, NFJ, PF, POR, RIFM, ROM, TSCA, USAN, USD, YUG

Chemical Classes: Inorganic Acids; Phosphorus Compounds

Functions: Fragrance Ingredient; pH Adjuster

Reported Product Categories: Shampoos (Non-coloring); Hair Bleaches; Bubble Baths; Permanent Waves; Hair Conditioners; Hair Coloring Preparations, Misc.; Hair Dyes and Colors (All Types Requiring Caution Statements and Patch Tests); Hair Preparations (Non-coloring), Misc.; Bath Oils, Tablets, and Salts; Cleansing Products (Cold Creams, Cleansing Lotions, Liquids and Pads); Eyeliners; Indoor Tanning Preparations; Tonics, Dressings, and Other Hair Grooming Aids; Aftershave Lotions; Baby Shampoos; Bath Preparations, Misc.; Hair Wave Sets; Moisturizing Preparations; Nail Polish and Enamels; Body and Hand Preparations (Excluding Shaving Preparations); Personal Cleanliness Products, Misc.; Skin Fresheners; Hair Lighteners with Color; Hair Straighteners; Night Skin Care Preparations; Skin Care Preparations, Misc.

Technical/Other Name:
Phosphoric acid (RIFM)

Trade Name Mixtures:
Extrapone Cucumber Special 2/032878 (Symrise)
Hydrocos (Cosmetochem)
pHL 100 (pH Solutions)

PHOSPHORUS

CAS No.	EINECS No.
7723-14-0	231-768-7

Definition: Phosphorus is a naturally occurring element.

Information Sources: EEC(II-279), MI-13 (7433)

Chemical Class: Elements

Function: Not Reported

PHOSPHORUS PENTOXIDE

CAS No.: 1314-56-3

Empirical Formula:
P_2O_5

Definition: Phosphorus Pentoxide is the inorganic oxide that conforms to the formula:

P_2O_5

Information Source: MI-13(7441)

Chemical Class: Inorganics

Function: Not Reported

Trade Name Mixture:
SilPower Plus (LG Cosmetic)

PHRAGMITES COMMUNIS EXTRACT

CAS No.	EINECS No.
84604-02-4	283-278-8

Definition: Phragmites Communis Extract is an extract of the aerial parts of *Phragmites communis. See "Regulatory and Ingredient Use Information," regarding the labeling names for botanical ingredients in Volume 1, Introduction, Part A.*

Chemical Class: Biological Products

Function: Not Reported

Technical/Other Name:
Extract of Phragmites Communis

Trade Name Mixtures:
Extract Hydroglycolique De Roseau (Greentech)
Extrait aqueux de Roseau (Greentech)
Extrait Hydroglycolique de Roseau (Greentech)
IAA 50 (Greentech)

PHTHALIC ACID DENATURED WITH EPOXY RESIN ALKYD RESIN

Definition: *See "Regulatory and Ingredient Use Information," regarding use of Japan Trivial names in Volume 1, Introduction, Part A.*

Information Source: JCIC

Chemical Class: Synthetic Polymers

Function: Film Former

Technical/Other Name:
Isophthalic Acid Alkyd Resin

PHTHALIC ANHYDRIDE/ADIPIC ACID/CASTOR OIL/NEOPENTYL GLYCOL/PEG-3/TRIMETHYLOLPROPANE COPOLYMER

Definition: Phthalic Anhydride/Adipic Acid/Castor Oil/Neopentyl Glycol/PEG-3/

Trimethylolpropane Copolymer is a copolymer formed from phthalic anhydride, adipic acid, castor oil, neopentyl glycol, PEG-3, and trimethylolpropane monomers.

Chemical Class: Synthetic Polymers

Function: Film Former

Trade Name:
Rokraphen 365 (Kraemer)

PHTHALIC ANHYDRIDE/BENZOIC ACID/GLYCERIN COPOLYMER

Definition: Phthalic Anhydride/Benzoic Acid/Glycerin Copolymer is a copolymer of phthalic anhydride, benzoic acid and glycerin monomers.

Chemical Class: Synthetic Polymers

Functions: Binder; Film Former

Trade Name Mixture:
Beckosol DB-174 (Ina Trading)

PHTHALIC ANHYDRIDE/BENZOIC ACID/TRIMETHYLOLPROPANE COPOLYMER

Definition: Phthalic Anhydride/Benzoic Acid/Trimethylolpropane Copolymer is a copolymer formed from phthalic anhydride, benzoic acid, amd trimethylolpropane monomers.

Chemical Class: Synthetic Polymers

Function: Film Former

Trade Name:
Erkarex 1105 (Kraemer)

PHTHALIC ANHYDRIDE/BUTYL BENZOIC ACID/PROPYLENE GLYCOL COPOLYMER

Definition: Phthalic Anhydride/Butyl Benzoic Acid/Propylene Glycol Copolymer is a polymer of phthalic anhydride, butyl benzoic acid and propylene glycol monomers.

Chemical Class: Synthetic Polymers

Functions: Film Former; Viscosity Increasing Agent - Nonaqueous

PHTHALIC ANHYDRIDE/GLYCERIN/GLYCIDYL DECANOATE COPOLYMER

Definition: Phthalic Anhydride/Glycerin/Glycidyl Decanoate Copolymer is a copolymer of phthalic anhydride, glycerin and glycidyl decanoate monomers.

Chemical Class: Synthetic Polymers

Functions: Film Former; Viscosity Increasing Agent - Nonaqueous

Reported Product Category: Nail Polish and Enamels

PHTHALIC ANHYDRIDE/TRIMELLITIC ANHYDRIDE/GLYCOLS COPOLYMER

Definition: Phthalic Anhydride/Trimellitic Anhydride/Glycols Copolymer is a copolymer of phthalic anhydride, trimellitic anhydride, ethylene glycol, and neopentyl glycol monomers.

Chemical Class: Synthetic Polymers

Functions: Film Former; Viscosity Increasing Agent - Nonaqueous

Reported Product Category: Nail Polish and Enamels

Technical/Other Names:
Phathlic Anhydride/Trimellitic Anhydride/ Glycol/Neopentyl Glycol Copolymer
Phthalic/Trimelitic/Glycols Copolymer

Trade Name:
Polynex Resin (Estron)

Trade Name Mixture:
7809 Polyester Resin (Degen)

PHTHALIMIDOPEROXYCAPROIC ACID

CAS No.	EINECS No.
128275-31-0	410-850-8

Empirical Formula:
$C_{14}H_{15}NO_5$

Definition: Phthalimidoperoxycaproic Acid is the organic compound that conforms to the formula:

Chemical Classes: Amides; Carboxylic Acids; Heterocyclic Compounds

Function: Oxidizing Agent

Technical/Other Names:
2H-Isoindole-2-Hexaneperoxoic Acid, 1,3-Dihydro-1,3-Dioxo
6-Phthalimidohexaneperoxoic Acid

Trade Name:
Eureco HC (Solvay Solexis SpA)

PHYLLANTHUS EMBLICA EXTRACT

Definition: Phyllanthus Emblica Extract is an extract of the herb, *Phyllanthus emblica.* See "Regulatory and Ingredient Use Information," regarding the labeling names for botanical ingredients in Volume 1, Introduction, Part A.

Chemical Class: Biological Products

Function: Not Reported

Technical/Other Name:
Extract of Phyllanthus Emblica

Trade Name Mixtures:
Amla/Amalaki (Carlisle)
Amla-Extrakt (Glycolic extract of Amla fruits) (Laboratoires Cosmedic)
Amla-Extrakt (Oily Extract of Amla Fruits) (Laboratoires Cosmedic)
Amla-Tinktur (Laboratoires Cosmedic)

PHYLLANTHUS EMBLICA FRUIT EXTRACT

CAS No.	EINECS No.
90028-28-7	289-817-3

Definition: Phyllanthus Emblica Fruit Extract is an extract of the fruit of *Phyllanthus emblica.* See "Regulatory and Ingredient Use Information," regarding the labeling names for botanical ingredients in Volume 1, Introduction, Part A.

Chemical Class: Biological Products

Function: Not Reported

Technical/Other Name:
Extract of Phyllanthus Emblica Fruit

Trade Name:
Rona Care Emblica (Merck KGaA/EMD Chemicals Inc.)

Trade Name Mixtures:
Actiphyte Amla Fruit (Active Organics)
Campo Shripala (Campo)

PHYLLANTHUS EMBLICA FRUIT JUICE

Definition: Phyllanthus Emblica Fruit Juice is the juice expressed from the fruit of *Phyllanthus emblica.* See "Regulatory and Ingredient Use Information," regarding the labeling names for botanical ingredients in Volume 1, Introduction, Part A.

Chemical Class: Biological Products

Function: Proprietary

Technical/Other Name:
Juice, Phyllanthus Emblica

Trade Name:
Amla-Fruchtsaft (Amla fruit juice/pulp) (Laboratoires Cosmedic)

Trade Name Mixtures:
Amla Extract HJK-3061 (Inabata Koryo)
Amla Extract HJK-3062 (Inabata Koryo)
Amla Extract HJK-3063 (Inabata Koryo)
Amla Extract HJK-3064 (Inabata Koryo)

PHYLLANTHUS NIRURI EXTRACT

Definition: Phyllanthus Niruri Extract is an extract of the whole plant, *Phyllanthus niruri.* See "Regulatory and Ingredient Use Information," regarding the labeling names for botanical ingredients in Volume 1, Introduction, Part A. See Reported Ingredient Functions-The Cosmetic Drug Distinction, in Regulatory and Ingredient Use Information, Volume I, Part A.

Chemical Class: Biological Products

Functions: Antimicrobial Agent; Skin Protectant

Technical/Other Name:
Extract of Phyllanthus Niruri

Trade Name Mixture:
Bhuiamla (Heritage Bio-Natural)

PHYLLOSTACHIS BAMBUSOIDES EXTRACT

Definition: Phyllostachis Bambusoides Extract is an extract of the rhizomes of *Phyllostachis bambusoides.* See "Regulatory and Ingredient Use Information," regarding the labeling names for botanical ingredients in Volume 1, Introduction, Part A.

Chemical Class: Biological Products

Function: Skin-Conditioning Agent - Miscellaneous

Technical/Other Name:
Extract of Phyllostachis Bambusoides

Trade Name:
Takechikakei Extract Powder (Takara Belmont)

Trade Name Mixtures:
Bambusoides Water (Bioland)
Takechikakei Extract (Takara Belmont)
Takechikakei Extract BG (Takara Belmont)

PHYLLOSTACHIS BAMBUSOIDES JUICE

Definition: Phyllostachis Bambusoides Juice is the sap of the bamboo tree, *Phyllostachis bambusoides.* See "Regulatory and Ingredient Use Information," regarding the labeling names for botanical ingredients in Volume 1, Introduction, Part A.

Chemical Class: Biological Products

Function: Skin-Conditioning Agent - Miscellaneous

Trade Name:
Bamboo Water (Bioland)

PHYSALIS ALKEKENGI CALYX EXTRACT

CAS No.	EINECS No.
90082-67-0	290-157-3

Definition: Physalis Alkekengi Calyx Extract is an extract of the calyxes of *Physalis alkekengi*. See *"Regulatory and Ingredient Use Information," regarding the labeling names for botanical ingredients in Volume 1, Introduction, Part A.*

Chemical Class: Biological Products

Function: Not Reported

Technical/Other Name:
Extract of Physalis Alkekengi Calyx

Trade Name Mixture:
Extrait De Calices De Physalis (Silab)

PHYSALIS ALKEKENGI FRUIT EXTRACT

Definition: Physalis Alkekengi Fruit Extract is an extract of the fruit of *Physalis alkekengi*. See *"Regulatory and Ingredient Use Information," regarding the labeling names for botanical ingredients in Volume 1, Introduction, Part A.*

Chemical Class: Biological Products

Function: Not Reported

Technical/Other Name:
Extract of Physalis Alkekengi

Trade Name Mixtures:
Herbasol Extract Alkekengi (Winter Cherry) (Cosmetochem) (Cosmetochem International Ltd.)
Winter Cherries Extract HS 2951 G (Grau)

PHYSALIS PUBESCENS FRUIT JUICE

Definition: Physalis Pubescens Fruit Juice is the juice obtained from the fruit of *Physalis pubescens*. See *"Regulatory and Ingredient Use Information," regarding the labeling names for botanical ingredients in Volume 1, Introduction, Part A.*

Chemical Class: Biological Products

Functions: Hair Conditioning Agent; Skin-Conditioning Agent - Miscellaneous

Technical/Other Name:
Physalis Peruviana Fruit Juice

Trade Name:
Uchuba Juice (Provital/Centerchem)

PHYTANTRIOL

CAS No.	EINECS No.
74563-64-7	277-932-2

JPN Translation:
フィタントリオール

Empirical Formula:
$C_{20}H_{42}O_3$

Definition: Phytantriol is the aliphatic alcohol that conforms to the formula:

$$CH_3CH(CH_2)_3CH(CH_2)_3CH(CH_2)_3C-CHCH_2OH$$

with CH₃ groups and OH OH substituents as shown

Information Sources: JCIC, JCLS

Chemical Class: Polyols

Functions: Anticaking Agent; Hair Conditioning Agent; Skin-Conditioning Agent - Miscellaneous

Reported Product Categories: Hair Conditioners; Shampoos (Non-coloring); Tonics, Dressings, and Other Hair Grooming Aids; Hair Preparations (Non-coloring), Misc.; Lipsticks; Hair Sprays (Aerosol Fixatives); Manicuring Preparations, Misc.; Aftershave Lotions; Bath Preparations, Misc.; Deodorants (Underarm); Face and Neck Preparations (Excluding Shaving Preparations); Personal Cleanliness Products, Misc.; Rouges

Technical/Other Names:
3,7,11,15,-Tetramethyl-1,2,3-Hexadecanetriol
Tetramethyl Trihydroxyhexadecane

Trade Names:
AEC Phytantriol (A & E Connock)
Phytantriol (Roche)
PTL (Kuraray)

Trade Name Mixtures:
Curasan (CLR)
Haircare Complex CLR (CLR)

PHYTIC ACID

CAS No.	EINECS No.
83-86-3	201-506-6

JPN Translation:
フィチン酸

Empirical Formula:
$C_6H_{18}O_{24}P_6$

Definition: Phytic Acid is the hexaphosphoric acid ester of Inositol (q.v.). It conforms to the formula:

[Structural formula showing cyclohexane ring with OPO₃H₂ and H₂O₃PO substituents]

Information Sources: INN, JCIC, JCLS, JSQI, MI-13(7471), TSCA

Chemical Class: Phosphorus Compounds

Functions: Chelating Agent; Oral Care Agent

Technical/Other Names:
Inositol Hexaphosphate
myo-Inositol, Hexakis(Dihydrogen Phosphate)

Trade Name:
Dermofeel PA (Straetmans)

Trade Name Mixtures:
Biosil Basics SMC (Biosil Technologies, Inc.)
Phytic Acid Complex (Biosil Technologies, Inc.)

PHYTOL

CAS No.: 7541-49-3

Empirical Formula:
$C_{20}H_{40}O$

Definition: Phytol is the organic compound that conforms to the formula:

$$CH_3CH(CH_2)_3CH(CH_2)_3CH(CH_2)_3C=CHCH_2OH$$

with CH₃ substituents as shown

Information Sources: MI-13(7474), RIFM

Chemical Class: Alcohols

Functions: Fragrance Ingredient; Skin-Conditioning Agent - Emollient

Technical/Other Names:
2-Hexadecen-1-ol, 3,7,11,15-tetramethyl- (RIFM)
Phytol (RIFM)

PHYTOLACCA DECANDRA EXTRACT

Definition: Phytolacca Decandra Extract is an extract of the roots of the poke root, *Phytolacca decandra*. See *"Regulatory and Ingredient Use Information," regarding the labeling names for botanical ingredients in Volume 1, Introduction, Part A.*

Information Source: EEC(II-374)

Chemical Class: Biological Products

Function: Not Reported

Technical/Other Names:
Extract of Poke Root
Phytolacca Decandra Root Extract
Poke Root Extract
Poke Root (Phytolacca Decandra) Extract

PHYTONADIONE

CAS Nos.   **EINECS Nos.**
84-80-0   201-564-2
11104-38-4   234-330-3
81818-54-4   279-833-9

Empirical Formula:
$C_{31}H_{46}O_2$

Definition: Phytonadione is the organic compound that conforms to the formula:

Information Sources: BAN, INN, JAN, MI-13(7465), USAN, USP XXIV

Chemical Class: Ketones

Function: Skin-Conditioning Agent - Miscellaneous

Technical/Other Names:
2-Methyl-3-(3,7,11,15-Tetramethyl-2-Hexadecenyl)-1,4-Naphthalenedione
1,4-Naphthalenedione, 2-Methyl-3-(3,7,11, 15-Tetramethyl-2-Hexadecenyl)-
Phylloquinone
Vitamin K1

Trade Names:
OriStar VK1 (Orient Stars)
Vitamin K1 (Roche)

PHYTOSPHINGOSINE

CAS No.: 13552-11-9

Definition: Phytosphingosine is a synthetic sphingoid that conforms generally to the formula:

where n has a value ranging from 10 to 20.

Chemical Classes: Alcohols; Amines

Functions: Hair Conditioning Agent; Skin-Conditioning Agent - Miscellaneous

Technical/Other Names:
4-Hydroxysphinganine
1,3,4-Octadecanetriol, 2-Amino-,

Trade Names:
DS-Phytosphingosine (Doosan)
DS-Phytosphingosine.HCl (Doosan)
Phytosphingosine (Degussa Care Specialties)

Trade Name Mixtures:
DS-CERIX (Doosan)
Phytosphingosine ETP (Degussa Goldschmidt AG)
Rece!derm - 503 (Bioland)
SK - Influx (Degussa Care Specialties)

PHYTOSPHINGOSINE GLYCOLATE

Definition: Phytosphingosine Glycolate is the salt that conforms to the formula:

Chemical Classes: Amines; Organic Salts

Function: Skin-Conditioning Agent - Miscellaneous

Trade Name:
Phytosphingosine Glycolate (Degussa Care Specialties)

PHYTOSPHINGOSINE HCl

Definition: Phytosphingosine HCl is the salt that conforms to the formula:

Chemical Classes: Alcohols; Amines; Organic Salts

Function: Skin-Conditioning Agent - Miscellaneous

Trade Name:
Phytosphingosine-Hydrochloride (Degussa Care Specialties)

PHYTOSPHINGOSINE LACTATE

CAS No.   **EINECS No.**
100403-19-8   309-560-3

Definition: Phytosphingosine Lactate is the salt that conforms to the formula:

Chemical Classes: Amines; Organic Salts

Function: Skin-Conditioning Agent - Miscellaneous

Reported Product Categories: Moisturizing Preparations; Face and Neck Preparations (Excluding Shaving Preparations); Cleansing Products (Cold Creams, Cleansing Lotions, Liquids and Pads); Eye Makeup Preparations, Misc.; Hair Conditioners; Hair Preparations (Non-coloring), Misc.

Trade Name:
Phytosphingosine-(L)-Lactate (Degussa Care Specialties)

PHYTOSPHINGOSINE PCA

CAS No.   **EINECS No.**
100403-19-8   309-560-3

Definition: Phytosphingosine PCA is the salt that conforms to the formula:

Chemical Classes: Amines; Heterocyclic Compounds

Function: Skin-Conditioning Agent - Miscellaneous

Reported Product Categories: Moisturizing Preparations; Face and Neck Preparations (Excluding Shaving Preparations); Cleansing Products (Cold Creams, Cleansing Lotions, Liquids and Pads); Eye Makeup Preparations, Misc.; Hair Conditioners; Hair Preparations (Non-coloring), Misc.

Trade Name:
Phytosphingosine PCA (Degussa Care Specialties)

PHYTOSTEROLS

Definition: Phytosterols is a mixture of sterols obtained from higher plants.

Chemical Class: Sterols

Function: Skin-Conditioning Agent - Miscellaneous

Trade Name:
Tamasterol (Tama Biochemical)

Trade Name Mixtures:
Jojobasomes (Desert Whale)
Nutriene Tocotrienols (Eastrnan Chemical)
Steromax 80% (Carotech)
Steromax 90% (Carotech)
Tocomin 30% (Carotech)
Tocomin 50% (Carotech)

PHYTOSTERYL/BEHENYL/OCTYLDODECYL LAUROYL GLUTAMATE

Definition: Phytosteryl/Behenyl/Octyldodecyl Lauroyl Glutamate is the mixed ester of phytosterol, Behenyl Alcohol (q.v.), and Octyldodecanol (q.v.) with Lauroyl Glutamic Acid (q.v.).

Chemical Classes: Amino Acids; Esters

Function: Skin-Conditioning Agent - Occlusive

Trade Name:
ELDEW PS-304 (Ajinomoto)

PHYTOSTERYL CANOLA GLYCERIDES

Definition: Phytosteryl Canola Glycerides is the reaction product of Canola Oil (q.v.) with Phytosterols (q.v.).

Chemical Class: Glyceryl Esters and Derivatives

Functions: Hair Conditioning Agent; Skin-Conditioning Agent - Miscellaneous

Trade Name:
Lipex cellect (Karlshamns AB)

PHYTOSTERYL GLUCOSIDE

Definition: Phytosteryl Glucoside is the product obtained by the condensation of Phytosterols (q.v.) with glucose.

Chemical Classes: Carbohydrates; Ethers; Sterols

Function: Skin-Conditioning Agent - Miscellaneous

Trade Name:
Phytosteside (Okayasu Shotenn)

PHYTOSTERYL ISOSTEARATE

JPN Translation:
イソステアリン酸フィトステリル

Definition: Phytosteryl Isostearate is the ester of phytosterol and Isostearic Acid (q.v.).

Chemical Classes: Esters; Sterols

Functions: Hair Conditioning Agent; Skin-Conditioning Agent - Occlusive

Technical/Other Name:
Isostearic Acid, Phytosteryl Ester

Trade Name:
Nikkol Phytosteryl Isostearate (Nikko)

PHYTOSTERYL/ISOSTEARYL/CETYL/STEARYL/BEHENYL DIMER DILINOLEATE

Definition: Phytosteryl/Isostearyl/Cetyl/Stearyl/Behenyl Dimer Dilinoleate is the ester of Dilinoleic Acid (q.v.) with a mixture of Phytosterols (q.v.), Isostearyl Alcohol (q.v.), Cetyl Alcohol (q.v.), Stearyl Alcohol (q.v.) and Behenyl Alcohol (q.v.)

Chemical Class: Esters

Functions: Hair Conditioning Agent; Skin-Conditioning Agent - Occlusive; Viscosity Increasing Agent - Nonaqueous

Trade Names:
Plandool-H (Nippon Chemical)
Plandool-S (Nippon Chemical)

PHYTOSTERYL ISOSTEARYL DIMER DILINOLEATE

Definition: Phytosteryl Isostearyl Dimer Dilinoleate is the diester of Dilinoleic Acid (q.v.). with phytosterol and isostearyl alcohol.

Chemical Class: Esters

Functions: Binder; Hair Conditioning Agent; Skin-Conditioning Agent - Emollient; Skin-Conditioning Agent - Occlusive; Viscosity Increasing Agent - Nonaqueous

Trade Names:
Lusplan PI-DA (Nippon Chemical)
PHY/IS-DA (Nippon Chemical)

PHYTOSTERYL MACADAMIATE

JPN Translation:
マカデミアナッツ脂肪酸フィトステリル

Definition: Phytosteryl Macadamiate is the ester of phytosterol and the fatty acids derived macadamia seed oil.

Information Sources: JCIC, JCLS

Chemical Classes: Esters; Sterols

Functions: Hair Conditioning Agent; Skin-Conditioning Agent - Miscellaneous

Reported Product Categories: Lipsticks; Makeup Preparations (Not eye), Misc.

Technical/Other Name:
Phytosterol Macadamia Nut Fatty Acid Ester

Trade Names:
Technol PSM (Technonet)
Yofco MAS (Nippon Chemical)

Trade Name Mixture:
Complex 3C (Vincience)

PHYTOSTERYL/OCTYLDODECYL LAUROYL GLUTAMATE

JPN Translation:
ラウロイルグルタミン酸ジ（フィトステリル／オクチルドデシル）

Definition: Phytosteryl/Octyldodecyl Lauroyl Glutamate is the mixed ester of phytosterol and Octyldodecanol (q.v.) with Lauroyl Glutamic Acid (q.v.).

Information Source: JCLS

Chemical Classes: Amino Acids; Esters

Function: Skin-Conditioning Agent - Occlusive

Reported Product Categories: Lipsticks; Eye Shadows; Blushers (All types); Eyeliners; Mascara; Face Powders; Rouges; Makeup Fixatives; Eyebrow Pencils; Eye Makeup Preparations, Misc.; Makeup Preparations (Not eye), Misc.

Trade Name:
Eldew PS-203 (Ajinomoto)

PHYTOSTERYL OLEATE

JPN Translation:
オレイン酸フィトステリル

Definition: Phytosteryl Oleate is the ester of phytosterol and oleic acid.

Information Source: JCLS

Chemical Classes: Esters; Sterols

Functions: Hair Conditioning Agent; Skin-Conditioning Agent - Miscellaneous

Trade Names:
AEC Phytosteryl Oleate (A & E Connock)
Emacol ZR-02076 (San-Ei Kagaku)
Emacol ZR-02077 (San-Ei Kagaku)
Salacos PO (Nisshin OilliO)
Salacos PO (T) (Nisshin OilliO)
Tamasterol W (Tama Biochemical)

Trade Name Mixture:
Emacol ZR-02075 (San-Ei Kagaku)

PHYTOSTERYL RICINOLEATE

Definition: Phytosteryl Ricinoleate is the ester of phytosterol and ricinoleic acid.

Chemical Classes: Esters; Sterols

Functions: Hair Conditioning Agent; Skin-Conditioning Agent - Miscellaneous

Trade Name:
Salacos PR (Nisshin OilliO)

PICEA EXCELSA EXTRACT

CAS No.	EINECS No.
91770-69-3	294-855-9

Definition: Picea Excelsa Extract is an extract of the buds of the Norway spruce,

Picea excelsa. See "Regulatory and Ingredient Use Information," regarding the labeling names for botanical ingredients in Volume 1, Introduction, Part A.

Chemical Class: Biological Products

Function: Not Reported

Technical/Other Names:
Extract of Norway Spruce
Extract of Picea Abies
Norway Spruce Extract
Norway Spruce (Picea Excelsa) Extract
Picea Abies Extract

Trade Name Mixtures:
Actiphyte of Norway Spruce Cone BG50 (Active Organics)
Actiphyte of Norway Spruce Cone GL50 (Active Organics)
Actiphyte of Norway Spruce Cone Lipo S (Active Organics)
Actiphyte of Norway Spruce Cone PG50 (Active Organics)
Actiphyte of Spruce Cone BG50 (Active Organics)
Actiphyte of Spruce Cone GL50 (Active Organics)
Actiphyte of Spruce Cone Lipo S (Active Organics)
Actiphyte of Spruce Cone PG50 (Active Organics)
Epicea Extract (Greentech S.A)
Extrapone Alpine Herbs GW 2/031560 (Symrise)
Extrapone Spruce Needle-Special (2/034831) (Symrise)
Pine Extract PG (Rahn)

PICEA EXCELSA LEAF EXTRACT

Definition: Picea Excelsa Leaf Extract is an extract of the needles of the Norway spruce, *Picea excelsa.* See "Regulatory and Ingredient Use Information," regarding the labeling names for botanical ingredients in Volume 1, Introduction, Part A.

Chemical Class: Biological Products

Function: Not Reported

Reported Product Category: Shampoos (Non-coloring)

Technical/Other Names:
Extract of Norway Spruce Needle
Extract of Picea Excelsa Needles
Norway Spruce (Picea Excelsa) Needle Extract
Picea Excelsa Needle Extract

Trade Name Mixtures:
Herbasol Extract Fir Needles (Cosmetochem) (Cosmetochem International Ltd.)

Herbasol Extract Spruce Buds (Cosmetochem) (Cosmetochem International Ltd.)
Herbasol Extract Spruce Needles (Cosmetochem) (Cosmetochem International Ltd.)
Pine Needles Extract HS 1779 AT (Grau)

PICEA EXCELSA LEAF OIL

Definition: Picea Excelsa Leaf Oil is the volatile oil expressed from the needles of *Picea excelsa.* See "Regulatory and Ingredient Use Information," regarding the labeling names for botanical ingredients in Volume 1, Introduction, Part A.

Chemical Class: Essential Oils

Function: Fragrance Ingredient

Technical/Other Names:
Norway Spruce Oil
Norway Spruce (Picea Excelsa) Oil
Oil of Norway Spruce
Oil of Picea Abies

Trade Name:
Epicea Oil (Robertet S.A.)

PICOLINAMIDE

CAS No.
1452-77-3

EINECS No.
215-921-5

Empirical Formula:
$C_6H_6N_2O$

Definition: Picolinamide is the heterocyclic compound that conforms to the formula:

Chemical Classes: Amides; Heterocyclic Compounds

Function: Skin-Conditioning Agent - Emollient

Technical/Other Name:
2-Pyridinecarboxamide

Trade Name:
Picolinamide (PIAS)

PICRAMIC ACID

CAS No.
96-91-3

EINECS No.
202-544-6

Empirical Formula:
$C_6H_5N_3O_5$

Definition: Picramic Acid is the substituted phenolic compound that conforms to the formula:

See "Regulatory and Ingredient Use Information," for Colorants in Volume 1, Introduction, Part A.

Information Sources: CI 76540, MI-13 (7491)

Chemical Classes: Color Additives - Hair; Phenols

Function: Hair Colorant

Technical/Other Names:
1-Amino-3,5-Dinitro-2-Hydroxybenzene
2-Amino-4,6-Dinitrophenol
CI 76540
2,4-Dinitro-6-Aminophenol
Oxidation Base 21
Phenol, 2-Amino-4,6-Dinitro-

PICRASMA EXCELSA WOOD EXTRACT

Definition: Picrasma Excelsa Wood Extract is an extract of the wood of *Picrasma excelsa.* See "Regulatory and Ingredient Use Information," regarding the labeling names for botanical ingredients in Volume 1, Introduction, Part A.

Chemical Class: Biological Products

Function: Not Reported

Technical/Other Name:
Extract of Picrasma Excelsa

Trade Name Mixture:
VT-036, Extract of Quassia (Vege-Tech)

PIGMENT BLUE 15

CAS No.
147-14-8

EINECS No.
205-685-1

Empirical Formula:
$C_{32}H_{16}CuN_8$

Definition: Pigment Blue 15 is classed chemically as a phthalocyanine (copper complex) color. It conforms generally to the formula:

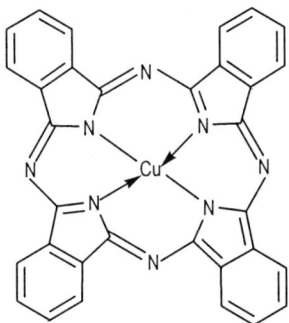

See "Regulatory and Ingredient Use Information," for Colorants in Volume 1, Introduction, Part A. To identify the colorant allowed for use in the European Union (EU), the INCI Name CI 74160 must be used, except for hair dye products. To identify the colorant allowed for use in Japan, the INCI name Ao404 must be used.

Information Sources: 21CFR74.3045, 21CFR175.300, 21CFR177.1680, 21CFR177.2260, 21CFR177.2600, CI 74160, M3, MI-13(2546), TSCA

Chemical Class: Color Additives - Miscellaneous

Function: Hair Colorant

Reported Product Category: Hair Dyes and Colors (All Types Requiring Caution Statements and Patch Tests)

Technical/Other Names:
Blue No. 404
CI 74160
Copper Phthalocyanine
Heliogen Blue B

Trade Names:
Monastral Fast Blue B (DuPont de Nemours)
Monastral Fast Blue G (DuPont de Nemours)
Monastral Fast Blue R (DuPont de Nemours)
Phthalocyanine Blue (Noveon Hilton Davis)
Sicomet Blue P 74160 (BASF)
Sunfast Blue (Sun Pigments)

Trade Name Mixtures:
210 Blue 60 (Sterling)
Creasparkles Hologram Blue Sky (Creations Couleurs)
FTX Comet Blue 60 (Swada)
FTX Violet 45 (Swada)
LPF Comet Blue 60 (Swada)
RTS-60 Comet Blue (Swada)

PIGMENT BLUE 15:1

CAS No.
147-14-8

EINECS No.
205-685-1

Empirical Formula:
$C_{32}H_{16}CuN_8$

Definition: Pigment Blue 15:1 is classed chemically as a phthalocyanine color. It conforms to the formula:

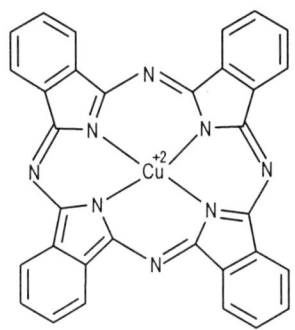

See "Regulatory and Ingredient Use Information," for Colorants in Volume 1, Introduction, Part A.

Information Source: CI 74160:1

Chemical Class: Color Additives - Hair

Function: Hair Colorant

Reported Product Category: Hair Dyes and Colors (All Types Requiring Caution Statements and Patch Tests)

Technical/Other Names:
Copper, [29H, 31H-Phthalocyaninato(2-)-κN29,κN30,κN31,κN32]-, (SP-4-1)-Tetrabenzo-5,10,15,20-Diazaporphyrine-phthalocyanine

Trade Name Mixtures:
A-60 Comet Blue (Swada)
610 Blue 60 (Sterling)
650 Blue 60 (Sterling)
810 Blue 60 (Sterling)
911 Blue 60 (Sterling)
T-60 Comet Blue (Swada)
E-45 Violet (Swada)
FEX-60 Comet Blue (Swada)
HMP-60 Comet Blue (Swada)
710 Violet 45 (Sterling)

PIGMENT BLUE 15:2

Empirical Formula:
$C_{32}H_{16}CuN_8$

Definition: Pigment Blue 15:2 is a solvent stable form of Pigment Blue 15 (q.v.) prepared by the introduction of a small amount of chlorine into the molecule. *See "Regulatory and Ingredient Use Information," for Colorants in Volume 1, Introduction, Part A. To identify the colorant allowed for use in the European Union (EU), the INCI Name CI 74160 must be used, except for hair dye products. To identify the colorant allowed for use in Japan, the INCI name Ao404 must be used.*

Information Source: CI 74160:2

Chemical Class: Color Additives - Miscellaneous

Function: Colorant

Technical/Other Name:
CI 74160:2

PIGMENT GREEN 7

CAS No.
1328-53-6

EINECS No.
215-524-7

Empirical Formula:
$C_{32}H_{16}CuN_8$

Definition: Pigment Green 7 is classed chemically as a phthalocyanine color. It conforms generally to the formula:

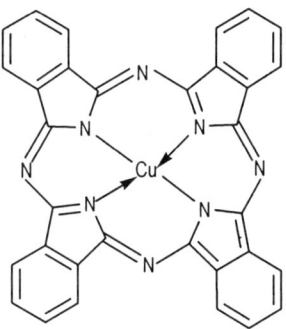

See "Regulatory and Ingredient Use Information," for Colorants in Volume 1, Introduction, Part A. To identify the colorant allowed for use in the European Union (EU), the INCI Name CI 74260 must be used, except for hair dye products.

Information Sources: 21CFR73.3124, 21CFR701.3, CI 74260, M3, TSCA

Chemical Class: Color Additives - Miscellaneous

Function: Colorant

Technical/Other Names:
CI 74260
Heliogen Green G

Trade Names:
Monastral Fast Green B (DuPont de Nemours)
Monastral Fast Green G (DuPont de Nemours)
Monastral Fast Green Y (DuPont de Nemours)
Phthalocyanine Green (Noveon Hilton Davis)
Sicomet Green P 74260 (BASF Corporation)
Sunfast Green (Sun Pigments)

Trade Name Mixtures:
A-8 Stellar Green (Swada)
E-8 Stellar Green (Swada)
FEX-8 Stellar Green (Swada)
Fluorescent Pigments (Day Light) - Green
   (Aron Universal)
210 Green 8 (Sterling)
610 Green 8 (Sterling)
650 Green 8 (Sterling)
710 Green 8 (Sterling)
810 Green 8 (Sterling)
911 Green 8 (Sterling)
915 Green 8 (Sterling)
HMP-8 Stellar Green (Swada)
LMP-8 Stellar Green (Swada)
RTS-8 Stellar Green (Swada)
T-8 Stellar Green (Swada)

PIGMENT ORANGE 5

CAS No. 3468-63-1   **EINECS No.** 222-429-4

Empirical Formula:
$C_{16}H_{10}N_4O_5$

Definition: Pigment Orange 5 is classed chemically as a monoazo color. It conforms to the formula:

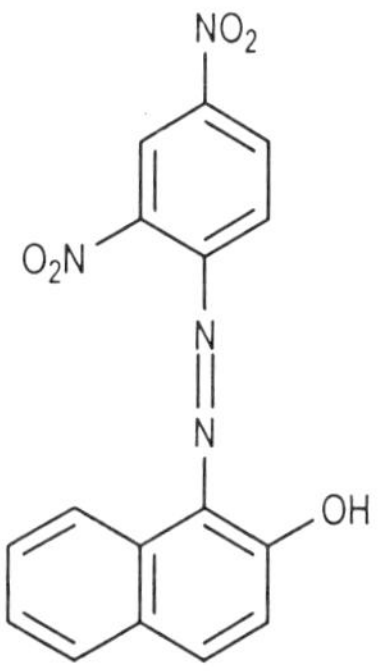

See "Regulatory and Ingredient Use Information," for Colorants in Volume 1, Introduction, Part A. To identify the colorant allowed for use in Japan, the INCI name Daidai203 must be used.

Information Sources: CI 12075, EEC(II-397), TSCA

Chemical Class: Color Additives - Miscellaneous

Function: Colorant

Technical/Other Names:
CI 12075
Dinitroaniline Orange
1-[(2,4-Dinitrophenyl)Azo]-2-Naphthalenol
2-Naphthalenol, 1-[(2,4-Dinitrophenyl)Azo]-
Orange No. 203

PIGMENT RED 4

CAS No. 2814-77-9   **EINECS No.** 220-562-2

Empirical Formula:
$C_{16}H_{10}ClN_3O_3$

Definition: Pigment Red 4 is classed chemically as a monoazo color. It conforms to the formula:

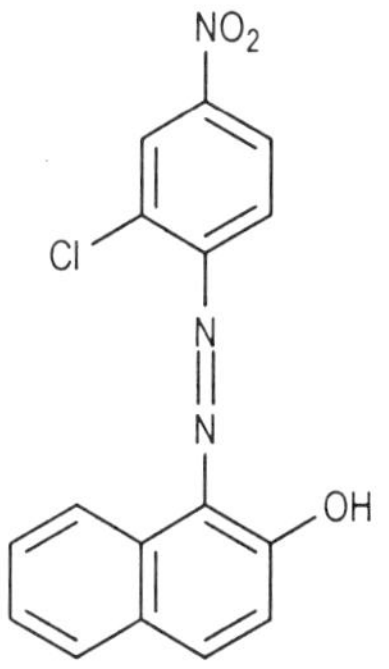

See "Regulatory and Ingredient Use Information," for Colorants in Volume 1, Introduction, Part A. To identify the certified colorant for labeling purposes in the US, the INCI Name Red 36 must be used. To identify the colorant allowed for use in the European Union (EU), the INCI Name CI 12085 must be used, except for hair dye products. The INCI Name for batches of this colorant that have not been certified is Pigment Red 4. To identify the colorant allowed for use in Japan, the INCI name Aka228 must be used.

Information Sources: CI 12085, M3, TSCA

Chemical Classes: Color Additives - Miscellaneous; Halogen Compounds

Function: Colorant

Reported Product Categories: Lipsticks; Blushers (All types); Makeup Preparations (Not eye), Misc.; Face Powders

Technical/Other Names:
1-[(2-Chloro-4-Nitrophenyl)Azo]-2-Naphthalenol
Chloroparanitraniline Red
CI 12085
2-Naphthalenol, 1-[(2-Chloro-4-Nitrophenyl)Azo]-
Permanent Red R
Red No. 228

PIGMENT RED 5

CAS No. 6410-41-9   **EINECS No.** 229-107-2

Empirical Formula:
$C_{30}H_{31}ClN_4O_7S$

Definition: Pigment Red 5 is classed chemically as a monoazo color. It conforms to the formula:

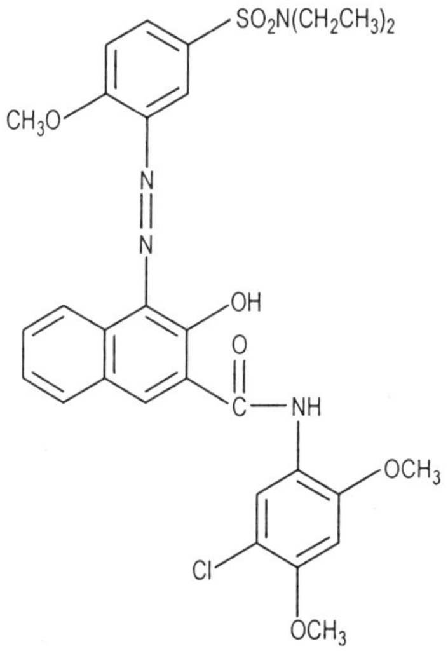

See "Regulatory and Ingredient Use Information," for Colorants in Volume 1, Introduction, Part A. To identify the colorant allowed for use in the European Union (EU), the INCI Name CI 12490 must be used, except for hair dye products.

Information Sources: CI 12490, M3, TSCA

Chemical Classes: Color Additives - Miscellaneous; Halogen Compounds

Function: Colorant

Technical/Other Names:
N-(5-Chloro-2,4-Dimethoxyphenyl)-4-[[5-[(Diethylamino)Sulfonyl]-2-Methoxyphenyl]Azo]-3-Hydroxy-2-Naphthalenecarboxamide
CI 12490
2-Naphthalenecarboxamide, N-(5-Chloro-2,4-Dimethoxyphenyl)-4-[[5-[(Diethylamino)Sulfonyl]-2-Methoxyphenyl]Azo]-3-Hydroxy-
Permanent Carmine FB Extra

PIGMENT RED 48

CAS Nos. 3564-21-4   **EINECS Nos.** 222-642-2
5280-66-0   226-102-7

Empirical Formula:
$C_{18}H_{13}ClN_2O_6S \cdot 2Na$

Definition: Pigment Red 48 is classed chemically as a monoazo color. It conforms to the formula:

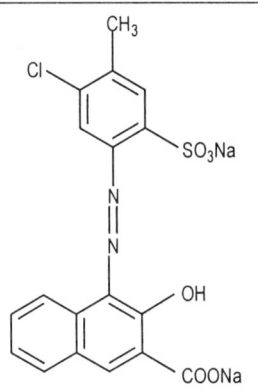

See "Regulatory and Ingredient Use Information," for Colorants in Volume 1, Introduction, Part A. To identify the colorant allowed for use in the European Union (EU), the INCI Name CI 15865 must be used, except for hair dye products. To identify the colorant allowed for use in Japan, the INCI name Aka405 must be used.

Information Sources: CI 15865, M3, MHLW Ord. No. 30, TSCA

Chemical Classes: Color Additives - Miscellaneous; Halogen Compounds

Function: Colorant

Technical/Other Names:
4-[(5-Chloro-4-Methyl-2-Sulfophenyl)Azo]-3-Hydroxy-2-Naphthalenecarboxylic Acid, Disodium Salt
CI 15865
2-Naphthalenecarboxylic Acid, 4-[(5-Chloro-4-Methyl-2-Sulfophenyl)Azo]-3-Hydroxy-, Disodium Salt

PIGMENT RED 53

CAS No.	EINECS No.
2092-56-0	218-248-5

Empirical Formula:
$C_{17}H_{13}ClN_2O_4S \cdot Na$

Definition: Pigment Red 53 is classed chemically as a monoazo color. It conforms to the formula:

The barium salt of this color is Pigment Red 53:1. *See "Regulatory and Ingredient Use Information," for Colorants in Volume 1, Introduction, Part A. To identify the colorant allowed for use in Japan, the INCI name Aka203 must be used.*

Information Sources: CI 15585, EEC(II-401), M3, TSCA

Chemical Classes: Color Additives - Miscellaneous; Halogen Compounds

Function: Colorant

Technical/Other Names:
Benzenesulfonic Acid, 5-Chloro-2-[(2-Hydroxy-1-Naphthalenyl)Azo]-4-Methyl-, Monosodium Salt
Betanine
5-Chloro-2-[(2-Hydroxy-1-Naphthalenyl)-Azo]-4-Methylbenzenesulfonic Acid, Monosodium Salt
CI 15585
Red No. 203

PIGMENT RED 53:1

CAS No.	EINECS No.
5160-02-1	225-935-3

Empirical Formula:
$C_{17}H_{13}ClN_2O_4S \cdot \frac{1}{2}Ba$

Definition: Pigment Red 53:1 is classed chemically as a monoazo color. It conforms to the formula:

This color is the barium salt of Pigment Red 53. *See "Regulatory and Ingredient Use Information," for Colorants in Volume 1, Introduction, Part A. To identify the colorant allowed for use in Japan, the INCI name Aka204 must be used.*

Information Sources: CI 15585:1, M3, TSCA

Chemical Classes: Color Additives - Miscellaneous; Halogen Compounds

Function: Colorant

Technical/Other Names:
Benzenesulfonic Acid, 5-Chloro-2-[(2-Hydroxy-1-Naphthalenyl)Azo]-4-Methyl-, Barium Salt(2:1)
5-Chloro-2-[(2-Hydroxy-1-Naphthalenyl)-Azo]-4-Methylbenzenesulfonic Acid, Barium Salt(2:1)
CI 15585:1
Red No. 204

Trade Name:
Rouge Covanor W 3600 (LCW)

PIGMENT RED 57

CAS No.	EINECS No.
5858-81-1	227-497-9

Empirical Formula:
$C_{18}H_{14}N_2O_6S \cdot 2Na$

Definition: Pigment Red 57 is classed chemically as a monoazo color. It conforms to the formula:

See "Regulatory and Ingredient Use Information," for Colorants in Volume 1, Introduction, Part A. To identify the certified colorant for labeling purposes in the US, the INCI Name Red 6 must be used. To identify the certified colorant (calcium salt) for labeling purposes in the US, the INCI Name Red 7 must be used. To identify the colorant allowed for use in the European Union (EU), the INCI Name CI 15850 must be used, except for hair dye products. The INCI Name for batches of this colorant that have not been certified is Pigment Red 57.

Information Sources: CI 15850, M3, MHLW Ord. No. 30, TSCA

Chemical Class: Color Additives - Miscellaneous

Function: Colorant

Reported Product Categories: Lipsticks; Nail Polish and Enamels; Blushers (All types); Makeup Preparations (Not eye), Misc.; Face Powders

Technical/Other Names:
CI 15850
3-Hydroxy-4-[(4-Methyl-2-Sulfophenyl)Azo]-2-Naphthalenecarboxylic Acid, Disodium Salt
Lithol Rubin B
2-Naphthalenecarboxylic Acid, 3-Hydroxy-4-[(4-Methyl-2-Sulfophenyl)Azo]-, Disodium Salt

PIGMENT RED 57:1

CAS Nos.	EINECS No.
5281-04-9	226-109-5
29092-56-6	

Empirical Formula:
$C_{18}H_{14}N_2O_6S \cdot Ca$

Definition: Pigment Red 57:1 is classed chemically as a monoazo color. It conforms to the formula:

See "Regulatory and Ingredient Use Information," for Colorants in Volume 1, Introduction, Part A. To identify the certified colorant for labeling purposes in the US, the INCI Name Red 7 must be used. To identify the certified colorant (sodium salt) for labeling purposes in the US, the INCI Name Red 6 must be used. To identify the colorant allowed for use in the European Union (EU), the INCI Name CI 15850 must be used, except for hair dye products. The INCI Name for batches of this colorant that have not been certified is Pigment Red 57:1. To identify the colorant allowed for use in Japan, the INCI name Aka202 must be used.

Information Sources: CI 15850:1, M3, TSCA

Chemical Class: Color Additives - Miscellaneous

Function: Colorant

Reported Product Categories: Nail Polish and Enamels; Lipsticks; Blushers (All types); Face Powders; Makeup Preparations (Not eye), Misc.

Technical/Other Names:
CI 15850:1
3-Hydroxy-4-[(4-Methyl-2-Sulfophenyl)Azo]-2-Naphthalenecarboxylic Acid, Calcium Salt
Lithol Rubin B Ca
2-Naphthalenecarboxylic Acid, 3-Hydroxy-4-[(4-Methyl-2-Sulfophenyl)Azo]-, Calcium Salt
Pigment Red 57, Calcium Salt
Red No. 202

Trade Name:
Sicomet Red P 15850 Ca (BASF)

PIGMENT RED 63:1

CAS No.
6417-83-0

EINECS No.
229-142-3

Empirical Formula:
$C_{21}H_{14}N_2O_6S \cdot Ca$

Definition: Pigment Red 63:1 is classed chemically as a monoazo color. It conforms to the formula:

See "Regulatory and Ingredient Use Information," for Colorants in Volume 1, Introduction, Part A. To identify the certified colorant for labeling purposes in the US, the INCI Name Red 34 must be used. To identify the colorant allowed for use in the European Union (EU), the INCI Name CI 15880 must be used, except for hair dye products. The INCI Name for batches of this colorant that have not been certified is Pigment Red 63:1. To identify the colorant allowed for use in Japan, the INCI name Aka220 must be used.

Information Sources: CI 15880:1, M3, TSCA

Chemical Class: Color Additives - Miscellaneous

Function: Colorant

Reported Product Category: Nail Polish and Enamels

Technical/Other Names:
CI 15880:1
3-Hydroxy-4-[(1-Sulfo-2-Naphthalenyl)Azo]-2-Naphthalenecarboxylic Acid, Calcium Salt
Lithol Bordeaux Toner R
2-Naphthalenecarboxylic Acid, 3-Hydroxy-4-[(1-Sulfo-2-Naphthalenyl)Azo]-, Calcium Salt
Pigment Red 63, Calcium Salt (1:1)
Red No. 220

PIGMENT RED 64:1

CAS No.
6371-76-2

EINECS No.
228-899-7

Empirical Formula:
$C_{17}H_{12}N_2O_3 \cdot \frac{1}{2}Ca$

Definition: Pigment Red 64:1 is classed chemically as a monoazo color. It conforms to the formula:

See "Regulatory and Ingredient Use Information," for Colorants in Volume 1, Introduction, Part A. To identify the certified colorant for labeling purposes in the US, the INCI Name Red 31 must be used. To identify the colorant allowed for use in the European Union (EU), the INCI Name CI 15800 must be used, except for hair dye products. The INCI Name for batches of this colorant that have not been certified is Pigment Red 64:1. To identify the colorant allowed for use in Japan, the INCI name Aka219 must be used.

Information Sources: CI 15800:1, M3, TSCA

Chemical Class: Color Additives - Miscellaneous

Function: Colorant

Technical/Other Names:
CI 15800:1
3-Hydroxy-4-(Phenylazo)-2-Naphthalenecarboxylic Acid, Calcium Salt(2:1)
2-Naphthalenecarboxylic Acid, 3-Hydroxy-4-(Phenylazo)-, Calcium Salt(2:1)
Pigment Red 64, Calcium Salt (2:1)
Red No. 219

PIGMENT RED 68

CAS No.
5850-80-6

EINECS No.
227-456-5

Empirical Formula:
$C_{17}H_{11}ClN_2O_6S \cdot \frac{1}{2}Ca \cdot Na$

Definition: Pigment Red 68 is classed chemically as a monoazo color. It conforms to the formula:

See "Regulatory and Ingredient Use Information," for Colorants in Volume 1, Introduction, Part A. To identify the colorant allowed for use in the European Union (EU), the INCI Name CI 15525 must be used, except for hair dye products.

Information Sources: CI 15525, M3

Chemical Classes: Color Additives - Miscellaneous; Halogen Compounds

---

Function: Colorant

Technical/Other Names:
Benzoic Acid, 2-Chloro-5-[(2-Hydroxy-1-Naphthalenyl)Azo]-4-Sulfo-, Calcium Sodium Salt (2:1:2)
2-Chloro-5-[(2-Hydroxy-1-Naphthalenyl)-Azo]-4-Sulfobenzoic Acid, Calcium Sodium Salt (2:1:2)
CI 15525
Permanent Red Toner NCR
Pigment Red 68, Calcium Sodium Salt (2:1:2)

PIGMENT RED 83

CAS Nos.	EINECS No.
72-48-0	200-782-5
104074-25-1	

Empirical Formula:
$C_{14}H_8O_4$

Definition: Pigment Red 83 is classed chemically as an anthraquinone color. It conforms to the formula:

See "Regulatory and Ingredient Use Information," for Colorants in Volume 1, Introduction, Part A. To identify the colorant allowed for use in the European Union (EU), the INCI Name CI 58000 must be used, except for hair dye products.

Information Sources: CI 58000:1, M3, MI-13(246), TSCA

Chemical Class: Color Additives - Miscellaneous

Function: Colorant

Technical/Other Names:
Alizarin
9,10-Anthracenedione, 1,2-Dihydroxy-
1,2-Anthraquinonediol
CI 58000:1
1,2-Dihydroxy-9,10-Anthracenedione
1,2-Dihydroxyanthraquinone
Mordant Red 11

PIGMENT RED 88

CAS No.	EINECS No.
14295-43-3	238-222-7

Empirical Formula:
$C_{16}H_4Cl_4O_2S_2$

Definition: Pigment Red 88 is classed chemically as a thioindigoid (Pigment Red 88) color. It conforms to the formula:

See "Regulatory and Ingredient Use Information," for Colorants in Volume 1, Introduction, Part A.

Information Source: CI 73312

Chemical Classes: Color Additives - Miscellaneous; Halogen Compounds

Function: Colorant

Technical/Other Names:
Benzo(b)thiophen-3(2H)-one, 4,7-dichloro-2(4,7-dichloro-3-oxobenzo(b)thien-2(3H)-ylidene)-
CI 73312
4,7-Dichloro-2-(4,7-Dichloro-3-Oxobenzo[b]Thien-2(3H)-ylidene)Benzo[b]Thiophene-3(2H)-one

Trade Name Mixtures:
Siliglit - Aluminium-Glitter epoxy (Sigmund Lindner GmbH)
Siliglit - Polyester-Glitter (Sigmund Lindner GmbH)

PIGMENT RED 90:1 ALUMINUM LAKE

CAS No.	EINECS No.
16508-80-8	240-569-4

Empirical Formula:
$C_{20}H_8Br_4O_5 \cdot {}^1/_3Al$

Definition: Pigment Red 90:1 Aluminum Lake is an insoluble pigment composed of the aluminum salt of Solvent Red 43 (q.v.) extended on an appropriate substrate. It conforms to the formula:

See "Regulatory and Ingredient Use Information," for Colorants in Volume 1, Introduction, Part A. To identify the certified colorant for labeling purposes in the US, the INCI Name Red 21 Lake must be used. To identify the colorant allowed for use in

the European Union (EU), the INCI Name CI 45380 must be used, except for hair dye products.

Information Sources: CI 45380:3, M3

Chemical Classes: Color Additives - Miscellaneous; Halogen Compounds

Function: Colorant

Technical/Other Names:
CI 45380:3
Eosine, Aluminum Salt
Red 230 Aluminum Lake
2',4',5',7'-Tetrabromofluorescein, Aluminum Salt

PIGMENT RED 112

CAS No.	EINECS No.
6535-46-2	229-440-3

Empirical Formula:
$C_{24}H_{16}Cl_3N_3O_2$

Definition: Pigment Red 112 is classed chemically as a monoazo color. It conforms to the formula:

See "Regulatory and Ingredient Use Information," for Colorants in Volume 1, Introduction, Part A. To identify the colorant allowed for use in the European Union (EU), the INCI Name CI 12370 must be used, except for hair dye products.

Information Sources: CI 12370, M3, TSCA

Chemical Classes: Color Additives - Miscellaneous; Halogen Compounds

Function: Colorant

Technical/Other Names:
CI 12370
3-Hydroxy-N-(2-Methylphenyl)-4-[(2,4,5-Trichlorophenyl)Azo]-2-Naphthalene-carboxamide
2-Naphthalenecarboxamide, 3-Hydroxy-N-(2-Methylphenyl)-4-[(2,4,5-Trichloro-phenyl)Azo]-
Permanent Red FGR

PIGMENT RED 172 ALUMINUM LAKE

CAS No.   **EINECS No.**
12227-78-0   235-440-4

Empirical Formula:
$C_{20}H_8I_4O_5 \cdot xAl$

Definition: Pigment Red 172 Aluminum Lake is an insoluble pigment composed of the aluminum salt of Acid Red 51 (q.v.) extended on an appropriate substrate. It conforms to the formula:

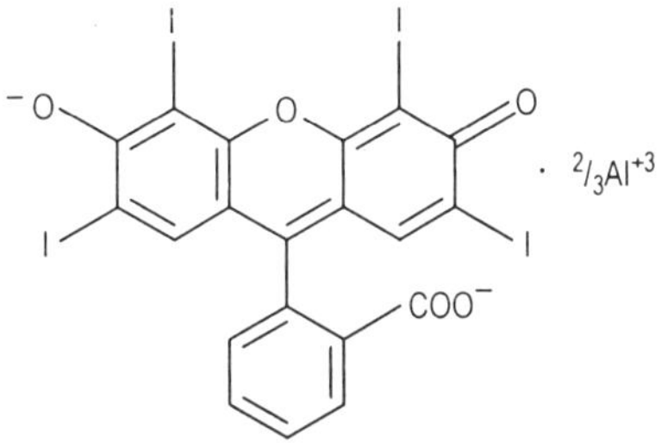

See "Regulatory and Ingredient Use Information," for Colorants in Volume 1, Introduction, Part A. To identify the colorant allowed for use in the European Union (EU), the INCI Name CI 45430 must be used, except for hair dye products. To identify the colorant allowed for use in Japan, the INCI name Aka3 must be used.

Information Sources: CI 45430:1, M3

Chemical Classes: Color Additives - Miscellaneous; Halogen Compounds

Function: Colorant

Technical/Other Names:
CI 45430:1
3',6'-Dihydroxy-2',4',5',7'-Tetraiodospiro [Isobenzofuran-1(3H),9'-[9H]Xanthen-3-one, Aluminum Salt
Red No. 3
Spiro[Isobenzofuran-1(3H),9'-[9H]Xanthen-3-one, 3',6'-Dihydroxy-2',4',5',7'-Tetraiodo-, Aluminum Salt
2',4',5',7'-Tetraiodofluorescein, Aluminum Salt

PIGMENT RED 173 ALUMINUM LAKE

Empirical Formula:
$C_{28}H_{31}ClN_2O_3 \cdot \frac{1}{3}Al$

Definition: Pigment Red 173 Aluminum Lake is an insoluble pigment composed of the aluminum salt of Basic Violet 10 (q.v.) extended on an appropriate substrate. It conforms to the formula:

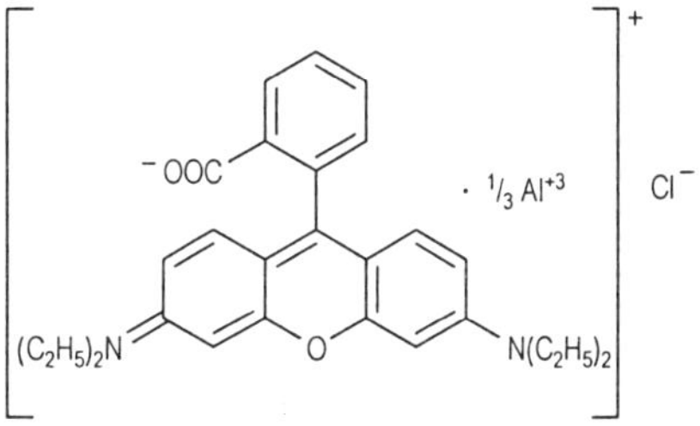

See "Regulatory and Ingredient Use Information," for Colorants in Volume 1, Introduction, Part A.

Information Source: CI 45170:3

Chemical Class: Color Additives - Miscellaneous

Function: Colorant

Technical/Other Name:
CI 45170:3

Trade Name:
Rose Covalac W 4507 (LCW)

PIGMENT RED 190

CAS No.   **EINECS No.**
6424-77-7   229-187-9

Empirical Formula:
$C_{38}H_{22}N_2O_6$

Definition: Pigment Red 190 is classed chemically as a perylene (Pigment Red 190) color. It conforms to the formula:

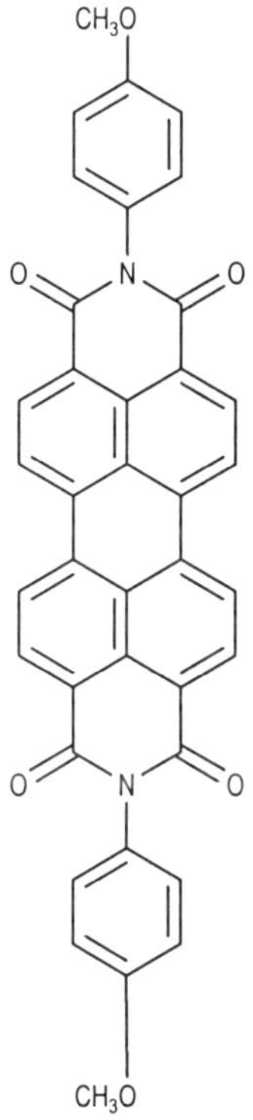

Information Source: CI 71140

Chemical Class: Color Additives - Miscellaneous

Function: Colorant

Technical/Other Names:
N,N'-bis(p-Methoxyphenyl)perylene'3,4,9, 10-Tetracarboxylic Diimide
CI 71140
17,18-Dihydrodinaphtho[1',2',3':3,4;3",2", 1":9,10]Perylo[1,12-*efg*][1,4]Dioxocin-5, 10-Dione
4,4'7,7'-Tetrachlorothioindigo

Trade Name Mixtures:
Siliglit - Aluminium-Glitter epoxy (Sigmund Lindner GmbH)
Siliglit - Polyester-Glitter (Sigmund Lindner GmbH)

PIGMENT VIOLET 19

CAS No.   **EINECS No.**
1047-16-1   213-879-2

Empirical Formula:
$C_{20}H_{12}N_2O_2$

Definition: Pigment Violet 19 is classed as a quinacridone color. It conforms to the formula:

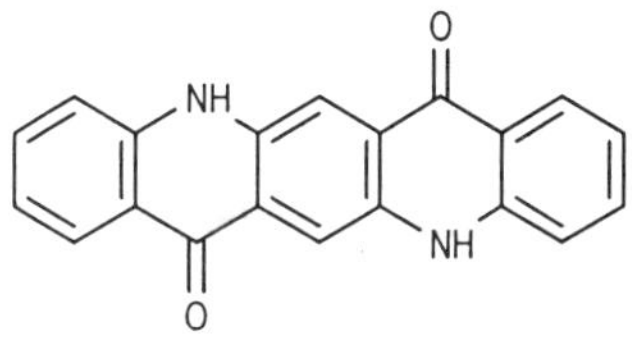

See "Regulatory and Ingredient Use Information," for Colorants in Volume 1, Introduction, Part A. To identify the colorant allowed for use in the European Union (EU), the INCI Name CI 73900 must be used, except for hair dye products.

Information Sources: CI 73900, TSCA

Chemical Class: Color Additives - Miscellaneous

Function: Colorant

Technical/Other Names:
CI 46500
5,12-Dihydroquino[2,3-b]Acridine-7,14-dione
Quinacridone Violet 19
Quino[2,3-b]Acridine-7,14-dione, 5,12-Dihydro-

Trade Names:
Sunfast Quinacridone Red Zeta (Sun Pigments)
Sunfast Quinacridone Violet (Sun Pigments)

PIGMENT VIOLET 23

CAS No.   **EINECS No.**
6358-30-1   228-767-9

Empirical Formula:
$C_{34}H_{22}Cl_2N_4O_2$

Definition: Pigment Violet 23 is classed chemically as an dioxazine color. It conforms to the formula:

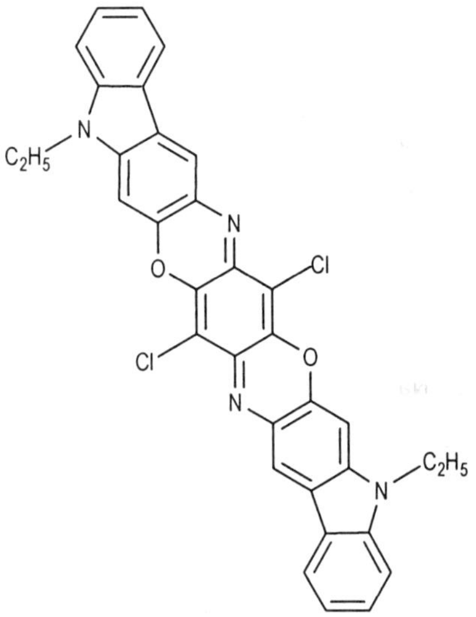

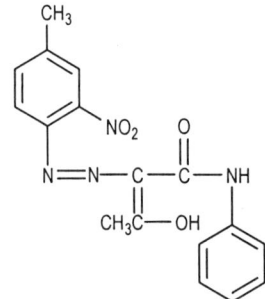

See "Regulatory and Ingredient Use Information," for Colorants in Volume 1, Introduction, Part A. To identify the colorant allowed for use in the European Union (EU), the INCI Name CI 51319 must be used, except for hair dye products.

Information Sources: 21CFR73.3107, CI 51319, M3, TSCA

Chemical Class: Color Additives - Hair

Function: Hair Colorant

Reported Product Category: Hair Dyes and Colors (All Types Requiring Caution Statements and Patch Tests)

Technical/Other Names:
Carbazole Violet
CI 51319
8,18-Dichloro-5,15-Diethyl-5,15-
   Dihydrodiindolo[3,2-b:3',2'-m]-
   Triphenodioxazine
Diindolo[3,2-b:3',2'-m-]Triphenodioxazine,
   8,18-Dichloro-5,15-Diethyl-5,15-Dihydro-
Dioxazine Purple
Helio Fast Violet EB

PIGMENT YELLOW 1

CAS No.
2512-29-0

EINECS No.
219-730-8

Empirical Formula:
$C_{17}H_{16}N_4O_4$

Definition: Pigment Yellow 1 is classed chemically as a monoazo color. It conforms to the formula:

See "Regulatory and Ingredient Use Information," for Colorants in Volume 1, Introduction, Part A. To identify the colorant allowed for use in the European Union (EU), the INCI Name CI 11680 must be used, except for hair dye products. To identify the colorant allowed for use in Japan, the INCI name Ki401 must be used.

Information Sources: 21CFR178.3297, CI 11680, M3, TSCA

Chemical Class: Color Additives - Miscellaneous

Function: Colorant

Technical/Other Names:
Butanamide, 2-[(4-Methyl-2-Nitrophenyl)-
   Azo]-3-Oxo-N-Phenyl-
CI 11680
Hansa Yellow
Hansa Yellow G
2-[(4-Methyl-2-Nitrophenyl)Azo]-3-Oxo-N-
   Phenylbutanamide
Yellow No. 401

Trade Names:
Hancock Yellow G (Sun Pigments)
Sicomet Yellow P 11680 (BASF)

PIGMENT YELLOW 3

CAS No.
6486-23-3

EINECS No.
229-355-1

Empirical Formula:
$C_{16}H_{12}Cl_2N_4O_4$

Definition: Pigment Yellow 3 is classed chemically as a monoazo color. It conforms to the formula:

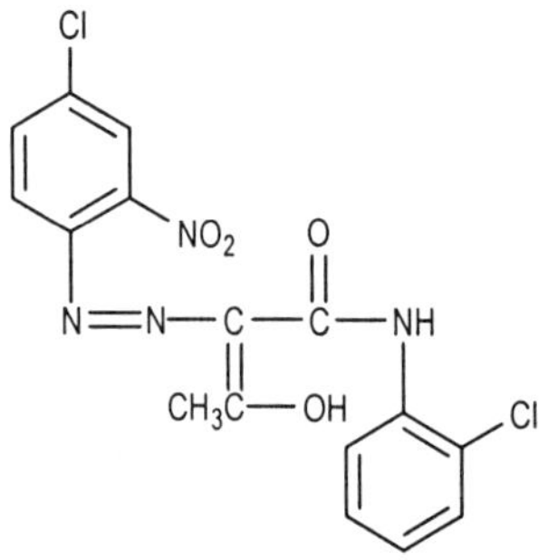

See "Regulatory and Ingredient Use Information," for Colorants in Volume 1, Introduction, Part A. To identify the colorant allowed for use in the European Union (EU), the INCI Name CI 11710 must be used, except for hair dye products.

Information Sources: CI 11710, M3, TSCA

Chemical Classes: Color Additives - Miscellaneous; Halogen Compounds

Function: Colorant

Technical/Other Names:
Acetoacetanilide, 2'-Chloro-2-(4-chloro-2-
   nitrophenylazo)-
Butanamide, 2-[(4-Chloro-2-Nitrophenyl)-
   Azo]-N-(2-Chlorophenyl)-3-Oxo-
2-[(4-Chloro-2-Nitrophenyl)Azo]-N-(2-
   Chlorophenyl)-3-Oxobutanamide
CI 11710

PIGMENT YELLOW 12

CAS No.
6358-85-6

EINECS No.
228-787-8

Empirical Formula:
$C_{32}H_{26}Cl_2N_6O_4$

Definition: Pigment Yellow 12 is classed chemically as a disazo color. It conforms to the formula:

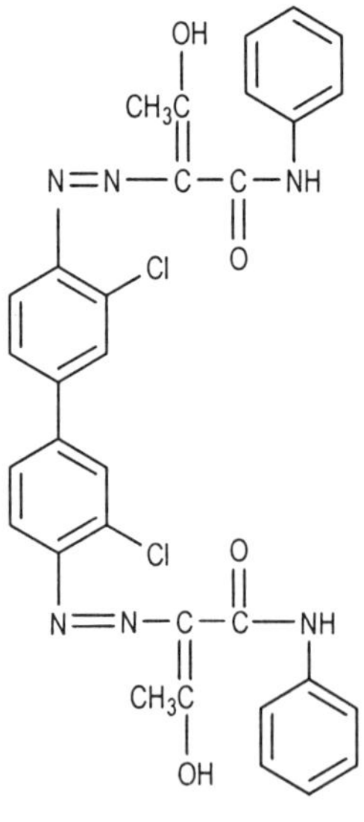

See "Regulatory and Ingredient Use Information," for Colorants in Volume 1, Introduction, Part A. To identify the colorant allowed for use in Japan, the INCI name Ki205 must be used.

Information Sources: CI 21090, TSCA

Chemical Classes: Color Additives - Miscellaneous; Halogen Compounds

---

Function: Colorant

Technical/Other Names:
Acetoacetanilide, 2,2"-(3,3'-dichloro-4,4'-biphenylylenebisazo)bis-
Butanamide, 2,2'-[(3,3'-Dichloro[1,1'-Biphenyl]4,4'-diyl)Bis(Azo)]Bis[3-Oxo-N-Phenyl-
CI 21090
2,2'-[(3,3'-Dichloro[1,1'-Biphenyl]-4,4'-diyl)Bis(Azo)]Bis[3-Oxo-N-Phenylbutanamide]
Yellow No. 205

PIGMENT YELLOW 13

CAS No. 5102-83-0   **EINECS No.** 225-822-9

Empirical Formula:
$C_{36}H_{34}Cl_2N_6O_4$

Definition: Pigment Yellow 13 is classed chemically as a disazo color. It conforms to the formula:

See "Regulatory and Ingredient Use Information," for Colorants in Volume 1, Introduction, Part A. To identify the colorant allowed for use in the European Union (EU), the INCI Name CI 21100 must be used, except for hair dye products.

Information Sources: CI 21100, TSCA

Chemical Classes: Color Additives - Hair; Color Additives - Miscellaneous

Function: Hair Colorant

Technical/Other Names:
Butanamide, 2,2'-[(3,3'-Dichloro[1,1'-Biphenyl]-4,4'-diyl)Bis(Azo)]Bis[N-(2,4-Dimethylphenyl)-3-Oxo-
CI 21100
2,2'-[(3,3'-Dichloro[1,1'-Biphenyl]-4,4'-diyl)Bis(Azo)]Bis[N-(2,4-Dimethylphenyl)-3-Oxobutanamide

PIGMENT YELLOW 73

CAS No. 13515-40-7   **EINECS No.** 236-852-7

Empirical Formula:
$C_{17}H_{15}ClN_4O_5$

Definition: Pigment Yellow 73 is classed as a monoazo color. It conforms to the formula:

See "Regulatory and Ingredient Use Information," for Colorants in Volume 1, Introduction, Part A.

Information Sources: CI 11738, TSCA

Chemical Classes: Color Additives - Miscellaneous; Halogen Compounds

Function: Colorant

Technical/Other Names:
Butanamide, 2-[(4-Chloro-2-Nitrophenyl)-Azo]-N-(2-Methoxyphenyl)-3-Oxo-
2-[(4-Chloro-2-Nitrophenyl)Azo]-N-(2-Methoxyphenyl)-3-Oxobutanamide
CI 11738

Trade Name:
Hancock Yellow 73 (Sun Pigments)

PIGSKIN EXTRACT

CAS No. 91081-63-9   **EINECS No.** 293-670-0

Definition: Pigskin Extract is an extract of the skin of young pigs. *See "Regulatory and Ingredient Use Information," regarding use of EU Trivial names in Volume 1, Introduction, Part A.*

Chemical Class: Biological Products

Function: Not Reported

Reported Product Categories: Body and Hand Preparations (Excluding Shaving Preparations); Skin Care Preparations, Misc.; Hair Conditioners

Technical/Other Name:
Extract of Pigskin

PIKEA ROBUSTA EXTRACT

Definition: Pikea Robusta Extract is an extract of the alga, *Pikea robusta*. *See "Regulatory and Ingredient Use Information," regarding the labeling names*
for botanical ingredients in Volume 1, Introduction, Part A. See Reported Ingredient Functions-The Cosmetic Drug Distinction, in Regulatory and Ingredient Use Information, Volume I, Part A.*

Chemical Class: Biological Products

Functions: Antioxidant; Skin-Conditioning Agent - Miscellaneous; Skin Protectant

Technical/Other Name:
Extract of Pikea Robusta

Trade Name Mixture:
NAB Pikea Robusta Extract (Arch Personal Care Products)

PILOCARPUS PENNATIFOLIUS LEAF EXTRACT

CAS No. 84696-42-4   **EINECS No.** 283-649-4

Definition: Pilocarpus Pennatifolius Leaf Extract is an extract of the leaves of *Pilocarpus pennatifolius*. *See "Regulatory and Ingredient Use Information," regarding the labeling names for botanical ingredients in Volume 1, Introduction, Part A.*

Information Source: EEC(II-311)

Chemical Class: Biological Products

Function: Not Reported

Reported Product Categories: Shampoos (Non-coloring); Tonics, Dressings, and Other Hair Grooming Aids

Technical/Other Names:
Extract of Jaborandi
Jaborandi Extract
Jaborandi (Pilocarpus Pennatifolius) Extract

Trade Name Mixtures:
Actiphyte of Jaborandi BG50 (Active Organics)
Actiphyte of Jaborandi GL50 (Active Organics)
Actiphyte of Jaborandi Lipo S (Active Organics)
Actiphyte of Jaborandi PG50 (Active Organics)
Herbasol Extract Jaborandi (Cosmetochem) (Cosmetochem International Ltd.)
Pilocarpus Pennatifolius Leaf Extract ies (IES LABO)
VT-172 Extract of Jaborandi (Vege-Tech)

PIMENTA ACRIS

Definition: *See "Regulatory and Ingredient Use Information," regarding EU labeling*

The inclusion of any compound in the Dictionary and Handbook does not indicate that use of that substance as a cosmetic ingredient complies with the laws and regulations governing such use in the United States or any other country.

names for botanical ingredients in Volume 1, Introduction, Part A.

Chemical Class: Biological Products

Technical/Other Name:
Pimenta Acris (Bay) Leaf Oil (U.S.)

PIMENTA ACRIS (BAY) LEAF OIL

CAS No.: 8006-78-8

Definition: Pimenta Acris (Bay) Leaf Oil is the volatile oil distilled from the leaves of *Pimenta acris*. See "Regulatory and Ingredient Use Information," regarding the labeling names for botanical ingredients in Volume 1, Introduction, Part A.

Information Sources: 21CFR182.20, 27CFR21.65, 27CFR21.151, FCC, MAR, MI-13(6809), RIFM, TSCA

Chemical Class: Essential Oils

Function: Fragrance Ingredient

Technical/Other Names:
Bay Leaf Oil
Bay Oil
Bay oil (Pimenta racemosa) (RIFM)
Bay oil terpenes (RIFM)
Myrica Oil
Oil, Essential, Bay
Oil of Bay, Terpeneless
Oils, Bay
Pimenta Acris (EU)

Trade Name:
AEC Bay Leaf Oil (A & E Connock)

Trade Name Mixtures:
Bay Leaf Water (Carrubba)
Nail Regenerative Complex (Alban Muller)
Premier Caribbean Denatured Rum (Premier Specialties)

PIMENTA DIOICA

Definition: See "Regulatory and Ingredient Use Information," regarding EU labeling names for botanical ingredients in Volume 1, Introduction, Part A.

Chemical Class: Biological Products

Technical/Other Names:
Pimenta Dioica Fruit Extract (U.S.)
Pimenta Dioica Leaf Extract (U.S.)

PIMENTA DIOICA FRUIT EXTRACT

Definition: Pimenta Dioica Fruit Extract is an extract of the unripe fruit of *Pimenta dioica*. See "Regulatory and Ingredient Use Information," regarding the labeling names

for botanical ingredients in Volume 1, Introduction, Part A.

Chemical Class: Biological Products

Function: Not Reported

Technical/Other Name:
Pimenta Dioica (EU)

PIMENTA DIOICA LEAF EXTRACT

Definition: Pimenta Dioica Leaf Extract is an extract of the leaves of *Pimento dioica*. See "Regulatory and Ingredient Use Information," regarding the labeling names for botanical ingredients in Volume 1, Introduction, Part A.

Chemical Class: Biological Products

Function: Not Reported

Technical/Other Names:
Extract of Pimento Dioica
Pimenta Dioica (EU)

Trade Name Mixture:
Cronatural Allspice O (Croda, Inc.)

PIMENTA OFFICINALIS

Definition: See "Regulatory and Ingredient Use Information," regarding EU labeling names for botanical ingredients in Volume 1, Introduction, Part A.

Chemical Class: Biological Products

Technical/Other Name:
Pimenta Officinalis (Pimento) Fruit Extract (U.S.)

PIMENTA OFFICINALIS (PIMENTO) FRUIT EXTRACT

CAS No. 84929-57-7
EINECS No. 284-540-4

Definition: Pimenta Officinalis (Pimento) Fruit Extract is an extract of the fruit of the pimento, *Pimenta officinalis*. See "Regulatory and Ingredient Use Information," regarding the labeling names for botanical ingredients in Volume 1, Introduction, Part A.

Information Source: 21CFR182.20

Chemical Class: Biological Products

Function: Not Reported

Technical/Other Names:
Extract of Pimento
Pimenta Officinalis (EU)
Pimenta Officinalis Extract

Pimento Extract
Pimento Fruit Extract

Trade Name Mixtures:
Herbasol-Extract Jamaica Pepper (Cosmetochem)
Herbasol-Extract Pimento (Jamuican Pepper) (Cosmetochem)
Phytotal RS (Phybiotex/Sederma)
Pimento Extract HS 2388 G (Grau)

PIMPINELLA ANISUM (ANISE) FRUIT EXTRACT

CAS No. 84775-42-8
EINECS No. 283-872-7

Definition: Pimpinella Anisum (Anise) Fruit Extract is an extract of the dried ripe fruit of the anise, *Pimpinella anisum*. See "Regulatory and Ingredient Use Information," regarding the labeling names for botanical ingredients in Volume 1, Introduction, Part A.

Information Sources: BRA, 21CFR182.20, ITA, RIFM

Chemical Class: Biological Products

Functions: Fragrance Ingredient; Skin-Conditioning Agent - Miscellaneous

Technical/Other Names:
Anies Extract
Aniseed Extract
Anise Extract
Anise Fruit Extract
Anise oil (Pimpinella anisum L.) (RIFM)
Anise (Pimpinella anisum L.) (RIFM)
Extract of Anise
Pimpinella Anisum Extract

Trade Name Mixtures:
Anise HS (Alban Muller)
Anise LS (Alban Muller)
Anise Seed Extract HS 2712 G (Grau)
Anise Softcream (CEP (Solabia))
Herbasol-Extract Aniseed (Anise) (Cosmetochem)
Vegebios of Green Anise (CEP (Solabia))
VT-094 Extract of Anise (Vege-Tech)

PIMPINELLA ANISUM (ANISE) SEED EXTRACT

Definition: Pimpinella Anisum (Anise) Seed Extract is an extract of the seeds of *Pimpinella anisum*. See "Regulatory and Ingredient Use Information," regarding the labeling names for botanical ingredients in Volume 1, Introduction, Part A.

Chemical Class: Biological Products

Function: Cosmetic Astringent

Technical/Other Name:
Extract of Pimpinella Anisum Seed

Trade Name Mixtures:
Actiphyte Anise AQ (Active Organics)
Actiphyte Anise Lipo M (Active Organics)
Actiphyte of Anise (Active Organics)
Actiphyte of Anise BG50 (Active Organics)
Actiphyte of Anise GL50 (Active Organics)
Actiphyte of Anise Lipo S (Active Organics)
Vegetol Anise ME 158 Hydro (Gattefosse s.a.)
Vegetol Anise ME 160 Oily (Gattefosse s.a.)

PIMPINELLA SAXIFRAGA EXTRACT

Definition: Pimpinella Saxifraga Extract is the extract obtained from the aerial parts of *Pimpinella saxifraga*. See "Regulatory and Ingredient Use Information," regarding the labeling names for botanical ingredients in Volume 1, Introduction, Part A.

Chemical Class: Biological Products

Function: Not Reported

Technical/Other Name:
Extract of Pimpinella Saxifraga

Trade Name Mixture:
Cosflor Burnet Saxifrage HGS (A & E Connock)

PINELLIA TERNATA EXTRACT

Definition: Pinellia Ternata Extract is an extract of the tuber of *Pinellia ternata*. See "Regulatory and Ingredient Use Information," regarding the labeling names for botanical ingredients in Volume 1, Introduction, Part A.

Chemical Class: Biological Products

Function: Skin-Conditioning Agent - Miscellaneous

Technical/Other Name:
Extract of Pinellia Ternata

Trade Name Mixture:
Pinellia Tuber Extract (LG Cosmetic)

PINELLIA TERNATA ROOT EXTRACT

Definition: Pinellia Ternata Root Extract is an extract of the tubers of *Pinellia ternata*. See "Regulatory and Ingredient Use Information," regarding the labeling names for botanical ingredients in Volume 1, Introduction, Part A.

Chemical Class: Biological Products

Function: Not Reported

Technical/Other Name:
Extract of Pinellia Ternata

Trade Name Mixture:
Hange Liquid E (Ichimaru Pharcos)

PINUS HAEDA BARK EXTRACT

Definition: Pinus Haeda Bark Extract is an extract of the bark of the pine, *Pinus haeda*. See "Regulatory and Ingredient Use Information," regarding the labeling names for botanical ingredients in Volume 1, Introduction, Part A.

Chemical Class: Biological Products

Function: Not Reported

Technical/Other Names:
Extract of Pinus Haeda
Pine (Pinus Haeda) Bark Extract

Trade Name Mixture:
VT-255 Extract of Pine Bark (Vege-Tech)

PINUS KORAIENSIS SEED EXTRACT

Definition: Pinus Koraiensis Seed Extract is an extract of the seeds of the conifer, *Pinus koraiensis*. See "Regulatory and Ingredient Use Information," regarding the labeling names for botanical ingredients in Volume 1, Introduction, Part A.

Chemical Class: Biological Products

Function: Not Reported

Technical/Other Name:
Pine (Pinus Koraiensis) Extract

Trade Name Mixture:
Huile de Pinus Koraiensis (Bertin)

PINUS KORAIENSIS SEED OIL

Definition: Pinus Koraiensis Seed Oil is the fixed oil obtained from the seeds of *Pinus koraiensis*. See "Regulatory and Ingredient Use Information," regarding the labeling names for botanical ingredients in Volume 1, Introduction, Part A.

Chemical Class: Fats and Oils

Function: Skin-Conditioning Agent - Occlusive

Technical/Other Name:
Oils, Pinus Koraiensis Seed

Trade Name:
Dermaprotectoil (Katakura)

PINUS LAMBERTIANA WOOD EXTRACT

Definition: Pinus Lambertiana Wood Extract is an extract of the wood of *Pinus lambertiana*. See "Regulatory and Ingredient Use Information," regarding the labeling names for botanical ingredients in Volume 1, Introduction, Part A.

Chemical Class: Biological Products

Function: Skin-Conditioning Agent - Miscellaneous

Technical/Other Names:
Extract of Pinus Lambertiana Wood
Giant Sugar Pine Extract

Trade Name Mixture:
Pinus Lambertiana Extract (Sederma)

PINUS PALUSTRIS LEAF EXTRACT

Definition: Pinus Palustris Leaf Extract is an extract of the needles of the pine, *Pinus palustris*. See "Regulatory and Ingredient Use Information," regarding the labeling names for botanical ingredients in Volume 1, Introduction, Part A.

Information Source: 21CFR172.510

Chemical Class: Biological Products

Function: Not Reported

Reported Product Category: Cleansing Products (Cold Creams, Cleansing Lotions, Liquids and Pads)

Technical/Other Names:
Extract of Pine Needles
Extract of Pinus Palustris Needles
Pine Needle Extract
Pine (Pinus Palustris) Needle Extract
Pinus Palustris Needle Extract

Trade Name Mixtures:
Extrapone Alpine Herbs Special 2/032561 (Symrise)
VT-142 Extract of Pine Needle (Vege-Tech)

PINUS PALUSTRIS OIL

CAS No.: 8002-09-3

Definition: Pinus Palustris Oil is the volatile oil obtained from distillation of the species *Pinus palustris*. See "Regulatory and Ingredient Use Information," regarding the labeling names for botanical ingredients in Volume 1, Introduction, Part A.

Information Sources: ARG, AUS, BEL, BPC, BRA, 21CFR172.510, 21CFR175.105, 21CFR176.180, 21CFR176.200, 21CFR176.210, 21CFR177.2800,

27CFR21.65, 27CFR21.151, CZE, DA, DDR, EGY, FI, FIN, HUN, ITA, KOR, MAR, MEX, MI-13(7529), NF XII, PF, PN, POL, POR, RIFM, ROM, TSCA, YUG

Chemical Class: Essential Oils

Function: Fragrance Ingredient

Reported Product Category: Bath Oils, Tablets, and Salts

Technical/Other Names:
Oils, Pine
Pine oil (RIFM)
Pine (Pinus Palustris) Oil
Yarmor pine oil (RIFM)

Trade Names:
AEC Pine Oil (A & E Connock)
Herco Pine Oil (Hercules)
Phytol PINE (pine oil) (Custom Ingredients)

Trade Name Mixtures:
Essentiaderm Capillare N.18 (Universal Flavors)
Essentiaderm n.2 (Universal Flavors)
Essentiaderm n.7 (Universal Flavors)
Essentiaderm n.9 (Universal Flavors)
Essentiaderm N.20 (Universal Flavors)
Essentiaderm N.21 (Universal Flavors)

PINUS PALUSTRIS TAR OIL

Definition: Pinus Palustris Tar Oil is the volatile oil from steam distillation of Pinus Palustris Tar (q.v.). *See "Regulatory and Ingredient Use Information," regarding the labeling names for botanical ingredients in Volume 1, Introduction, Part A.*

Information Sources: 21CFR172.515, MI-13(9154)

Chemical Class: Essential Oils

Function: Fragrance Ingredient

Technical/Other Names:
Oils, Pine Tar
Pine (Pinus Palustris) Tar Oil
Pine Tar Oil
Tar Oil Rectified

PINUS PALUSTRIS WOOD TAR

CAS No.	EINECS No.
8011-48-1	232-374-8

Definition: Pinus Palustris Wood Tar is the product obtained by the destructive distillation of the wood of *Pinus palustris*. *See "Regulatory and Ingredient Use Information," regarding the labeling names for botanical ingredients in Volume 1, Introduction, Part A.*

Information Sources: BP, BPC, 21CFR177.2600, EGY, IND, JAN, MAR, MI-13(7530), NF XV, USD

Chemical Class: Biological Products

Function: Denaturant

Technical/Other Names:
Pine (Pinus Palustris) Tar
Pine Tar
Tar, Pinus Palustris

Trade Name:
AEC Pine Tar (A & E Connock)

PINUS PENTAPHYLLA SEED OIL

Definition: Pinus Pentaphylla Seed Oil is the fixed oil expressed from the seeds of *Pinus pentaphylla*. *See "Regulatory and Ingredient Use Information," regarding the labeling names for botanical ingredients in Volume 1, Introduction, Part A.*

Chemical Class: Fats and Oils

Function: Skin-Conditioning Agent - Emollient

Technical/Other Name:
Oils, Pinus Pentaphylla Seed

Trade Name:
Finepinoleoil (Katakura)

PINUS PINASTER BARK EXTRACT

Definition: Pinus Pinaster Bark Extract is an extract of the bark and pine buds of the maritime pine, *Pinus pinaster. See "Regulatory and Ingredient Use Information," regarding the labeling names for botanical ingredients in Volume 1, Introduction, Part A. See Reported Ingredient Functions-The Cosmetic Drug Distinction, in Regulatory and Ingredient Use Information, Volume I, Part A.*

Chemical Class: Biological Products

Functions: Anticaries Agent; Antidandruff Agent; Antifungal Agent; Antimicrobial Agent; Antioxidant; Oral Care Agent; Oral Health Care Drug; Skin-Conditioning Agent - Miscellaneous; Sunscreen Agent; Ultraviolet Light Absorber

Technical/Other Names:
Extract of Maritime Pine
Maritime Pine Extract
Maritime Pine (Pinus Pinaster) Extract
Pinus Maritima Extract

Trade Names:
OPC ecorce de Pin (Berkem)
Pine Bark Extract (Euromed)

Trade Name Mixtures:
Maritime Pine HS/Sea Pine HS (Alban Muller)
Noster MX (Ennagram)
Sea Pine LS/Maritime Pine LS (Alban Muller)

PINUS PINEA KERNEL OIL

Definition: Pinus Pinea Kernel Oil is the fixed oil obtained from the kernels of the pine, *Pinus pinea. See "Regulatory and Ingredient Use Information," regarding the labeling names for botanical ingredients in Volume 1, Introduction, Part A.*

Chemical Class: Essential Oils

Function: Skin-Conditioning Agent - Occlusive

Technical/Other Names:
Oils, Pine
Pine (Pinus Pinea) Kernel Oil

Trade Names:
AEC Pine Nut Oil (A & E Connock)
Huile Vierge de Pignon de Pin Parasol (Bertin)

PINUS PUMILIO BARK EXTRACT

Definition: Pinus Pumilio Bark Extract is an extract of the bark of the Pine, *Pinus pumilio. See "Regulatory and Ingredient Use Information," regarding the labeling names for botanical ingredients in Volume 1, Introduction, Part A.*

Chemical Class: Biological Products

Function: Not Reported

Technical/Other Names:
Extract of Pine Bark
Pine (Pinus Pumilio) Bark Extract

PINUS PUMILIO LEAF EXTRACT

Definition: Pinus Pumilio Leaf Extract is an extract of the needles of the pine, *Pinus pumilio. See "Regulatory and Ingredient Use Information," regarding the labeling names for botanical ingredients in Volume 1, Introduction, Part A.*

Information Source: 21CFR172.510

Chemical Class: Biological Products

Function: Not Reported

Reported Product Category: Cleansing Products (Cold Creams, Cleansing Lotions, Liquids and Pads)

Technical/Other Names:
Extract of Pine Needles
Extract of Pinus Pumilio Needles
Pine Needle Extract
Pine (Pinus Pumilio) Needle Extract
Pinus Pumilio Needle Extract

Trade Name Mixture:
Cosflor Dwarf Pine HGS (A & E Connock)

PINUS PUMILIO OIL

Definition: Pinus Pumilio Oil is the volatile oil distilled from the needles and branches of *Pinus pumilio. See "Regulatory and Ingredient Use Information," regarding the labeling names for botanical ingredients in Volume 1, Introduction, Part A.*

Chemical Class: Essential Oils

Function: Fragrance Ingredient

Technical/Other Names:
Oils, Pinus Pumilio
Pine (Pinus Pumilio) Oil

Trade Names:
Latschen Kiefer Oil (Kneipp)
Latschenkiefer-Ol (Oleum Pini montanae/
Pini pumiliones) Dwarf Mountain Pine Oil
(B. Mueller KG)
Oil Pinus Pumilionis (Erste Tiroler
Latschenolbrennerei)

PINUS RADIATA BARK EXTRACT

Definition: Pinus Radiata Bark Extract is an extract of the bark of *Pinus radiata. See "Regulatory and Ingredient Use Information," regarding the labeling names for botanical ingredients in Volume 1, Introduction, Part A.*

Chemical Class: Biological Products

Function: Antioxidant

Technical/Other Name:
Extract of Pinus Radiata Bark

Trade Name:
Enzogenol (Nippon Bio)

PINUS STROBUS BARK EXTRACT

CAS No.   **EINECS No.**
90082-77-2   290-168-3

Definition: Pinus Strobus Bark Extract is an extract of the bark of the eastern pine, *Pinus strobus. See "Regulatory and Ingredient Use Information," regarding the labeling names for botanical ingredients in Volume 1, Introduction, Part A.*

Chemical Class: Biological Products

Function: Not Reported

Technical/Other Names:
Eastern Pine Extract
Eastern Pine (Pinus Strobus) Extract
Extract of Eastern Pine

Trade Name Mixtures:
Actiphyte of Pine Bark BG50 (Active
Organics)
Actiphyte of Pine Bark GL50 (Active
Organics)
Actiphyte of Pine Bark Lipo S (Active
Organics)
Actiphyte of Pine Bark PG50 (Active
Organics)

PINUS STROBUS CONE EXTRACT

CAS No.   **EINECS No.**
94266-48-5   304-455-9

Definition: Pinus Strobus Cone Extract is an extract of the cone of the eastern pine, *Pinus strobus. See "Regulatory and Ingredient Use Information," regarding the labeling names for botanical ingredients in Volume 1, Introduction, Part A.*

Information Sources: JSQI, RIFM

Chemical Class: Biological Products

Functions: Fragrance Ingredient; Not Reported

Technical/Other Names:
Eastern Pine (Pinus Strobus) Cone Extract
Pine Cone Extract
Pine, ext. (RIFM)

Trade Name Mixtures:
Actiphyte of Pine Cone BG50 (Active
Organics)
Actiphyte of Pine Cone GL50 (Active
Organics)
Actiphyte of Pine Cone Lipo S (Active
Organics)
Actiphyte of Pine Cone PG50 (Active
Organics)

PINUS SYLVESTRIS BARK EXTRACT

Definition: Pinus Sylvestris Bark Extract is an extract of the bark of *Pinus sylvestris. See "Regulatory and Ingredient Use Information," regarding the labeling names for botanical ingredients in Volume 1, Introduction, Part A.*

Chemical Class: Biological Products

Function: Not Reported

Technical/Other Name:
Extract of Pinus Sylvestris Bark

Trade Name Mixture:
EPICA (Greentech S.A)

PINUS SYLVESTRIS BUD EXTRACT

Definition: Pinus Sylvestris Bud Extract is an extract of the buds of the Scotch pine, *Pinus sylvestris. See "Regulatory and Ingredient Use Information," regarding the labeling names for botanical ingredients in Volume 1, Introduction, Part A.*

Chemical Class: Biological Products

Function: Not Reported

Technical/Other Name:
Extract of Pinus Sylvestris Bud

Trade Name Mixtures:
Extrait Hydroglycolique de Bourgeons de
Pin (Greentech)
Pine Extract HG (Provital/Centerchem)
Polyplant Anti-Wrinkles (Provital/
Centerchem)
Polyplant Hair (Provital/Centerchem)
Prodhy Extract PIN (Prod'Hyg)
Scotch Pine Extract HS 2643 G (Grau)
Vegetol Pine GR 300 Hydro (Gattefosse
s.a.)

PINUS SYLVESTRIS CONE EXTRACT

CAS No.   **EINECS No.**
94266-48-5   304-455-9

JPN Translation:
マツエキス

Definition: Pinus Sylvestris Cone Extract is an extract of the cones of the Scotch pine, *Pinus sylvestris. See "Regulatory and Ingredient Use Information," regarding the labeling names for botanical ingredients in Volume 1, Introduction, Part A.*

Information Sources: JCIC, JCLS, JSQI, RIFM

Chemical Class: Biological Products

Functions: Fragrance Ingredient; Not Reported

Technical/Other Names:
Pine Cone Extract
Pine, ext. (RIFM)
Pine Extract
Pine (Pinus Sylvestris) Cone Extract

Trade Name Mixtures:
Antiwrinkles Phytogreen Complex GXH
263 (Phytochim)
Aromaphyte of Pine Cone (Active
Organics)
Hair Treatment Phytogreen Complex GXH
260 (Phytochim)

Microcirculation Factor No. 5 (Indena SA)
Pharcolex BX32 (Ichimaru Pharcos)
Pharcolex BX46 (Ichimaru Pharcos)
Phytelene Complex EGX 232 (Indena SA)
Phytelene Complex EGX 246 (Indena SA)
Phytelene Complex EGX 254 (Indena SA)
Phytelene of Pine Tree EG 028 Liquid
  (Indena SA)
Phytogreen 55 of Pine Tree EXH 616
  Liquid (Phytochim)
Pine Cone Liquid B (Ichimaru Pharcos)
Transomes Anti-Age EGX 246-TR (Indena
  SA)
Transomes of Factor of Micro-circulation
  No 5 (Indena SA)
VT-155 Extract of Pinecone (Vege-Tech)

PINUS SYLVESTRIS CONE OIL

Definition: Pinus Sylvestris Cone Oil is
the fixed oil obtained from the cone of the
pine, *Pinus sylvestris. See "Regulatory and
Ingredient Use Information," regarding the
labeling names for botanical ingredients in
Volume 1, Introduction, Part A.*

Chemical Class: Essential Oils

Function: Fragrance Ingredient

Technical/Other Names:
  Oil of Pine Cone
  Oil of Pinus Sylvestris
  Oils, Pine Cone
  Oils, Pinus Sylvestris
  Pine (Pinus Sylvestris) Cone Oil

Trade Name Mixture:
  Aromaphyte of Pine Cone (Active
    Organics)

PINUS SYLVESTRIS LEAF EXTRACT

CAS No.	EINECS No.
84012-35-1	281-679-2

Definition: Pinus Sylvestris Leaf Extract is
an extract of the needles of the Scotch
pine, *Pinus sylvestris. See "Regulatory and
Ingredient Use Information," regarding the
labeling names for botanical ingredients in
Volume 1, Introduction, Part A.*

Information Source: 21CFR172.510

Chemical Class: Biological Products

Function: Not Reported

Reported Product Category: Cleansing
Products (Cold Creams, Cleansing Lotions,
Liquids and Pads)

Technical/Other Names:
  Extract of Pine Needles
  Extract of Pinus Sylvestris Needles
  Pine Needle Extract

Pine (Pinus Sylvestris) Needle Extract
Pinus Sylvestris Needle Extract

Trade Name Mixtures:
  Anti Wrinkles Complex 263 (Ennagram)
  Cremogen Pine Needle (PN 739 007)
    (Haarmann & Reimer GmbH)
  658 Demaquillant LS (Alban Muller)
  Hair Treatment Complex 260 (Ennagram)
  Heavy Legs Complex 269 (Ennagram)
  Heavy Legs Phytogreen Complex GXH 269
    (Phytochim)
  Herbaliquid Alpine Herbs Special
    (Crodarom)
  Herbaliquid Pine Needle Special
    (Crodarom)
  Herbasol Extract Oil Soluble Pine Needle
    (Cosmetochem) (Cosmetochem
    International Ltd.)
  Herbasol Extract Pine Needles
    (Cosmetochem) (Cosmetochem
    International Ltd.)
  Novaplant Pine Needle Extract (Crodarom)

PINUS SYLVESTRIS LEAF OIL

CAS No.: 8023-99-2

Definition: Pinus Sylvestris Leaf Oil is the
volatile oil obtained from the needles of
*Pinus sylvestris. See "Regulatory and
Ingredient Use Information," regarding the
labeling names for botanical ingredients in
Volume 1, Introduction, Part A.*

Information Sources: NF XVIII, RIFM

Chemical Class: Essential Oils

Function: Fragrance Ingredient

Technical/Other Names:
  Pine Needle Oil
  Pine (Pinus Sylvestris) Oil
  Pine scotch oil (Pinus sylvestris L.) (RIFM)
  Pinus Sylvestris Oil

Trade Names:
  Norway Pine Essential Oil (Alban Muller)
  Scotch Pine Essential Oil (Alban Muller)

PINUS TABULAEFORMIS BARK EXTRACT

Definition: Pinus Tabulaeformis Bark
Extract is an extract of the bark of *Pinus
tabulaeformis. See "Regulatory and Ingredient Use Information," regarding the
labeling names for botanical ingredients in
Volume 1, Introduction, Part A.*

Chemical Class: Biological Products

Function: Not Reported

Technical/Other Name:
  Extract of Pinus Tabulaeformis Bark

Trade Name Mixture:
  Draco Pine Bark Full Spectrum
    Standardized Extract (Draco)

PIPER BETLE LEAF OIL

Definition: Piper Betle Leaf Oil is the
volatile oil obtained by the steam distillation
of the leaves of *Piper betle. See
"Regulatory and Ingredient Use
Information," regarding the labeling names
for botanical ingredients in Volume 1, Introduction, Part A.*

Chemical Class: Essential Oils

Functions: Deodorant Agent; Fragrance
Ingredient; Oral Care Agent; Skin-Conditioning Agent - Miscellaneous

Technical/Other Name:
  Oils, Piper Betle Leaf

Trade Name:
  Betel Pepper (Haldin Pacific)

PIPER LONGUM FRUIT

Definition: Piper Longum Fruit is the fruit
of *Piper longum. See "Regulatory and
Ingredient Use Information," regarding the
labeling names for botanical ingredients in
Volume 1, Introduction, Part A.*

Chemical Class: Biological Products

Function: Not Reported

Trade Name:
  Pipli (Long Pepper) (Heritage Bio-Natural)

PIPER METHYSTICUM EXTRACT

CAS No.	EINECS No.
84696-40-2	283-648-9

Definition: Piper Methysticum Extract is
an extract of the leaves, roots and stems of
the kawa, *Piper methysticum. See
"Regulatory and Ingredient Use
Information," regarding the labeling names
for botanical ingredients in Volume 1, Introduction, Part A.*

Chemical Class: Biological Products

Function: Skin-Conditioning Agent - Miscellaneous

Technical/Other Names:
  Extract of Kawa
  Kawa Extract
  Kawa (Piper Methisticum) Extract

Trade Name Mixtures:
  Actiphyte Kava Kava AL (Active Organics)

Actiphyte of Kava Kava BG50 (Active
Organics)
Actiphyte of Kava Kava GL50 (Active
Organics)
Actiphyte of Kava Kava Lipo S (Active
Organics)
Actiphyte of Kava Kava PG50 (Active
Organics)
Far East Extract Kava Kava (E.U.K)
Herbasol Extract Kava Kava
(Cosmetochem) (Cosmetochem
International Ltd.)
Kava Extract BG9010 (Concentrated Aloe
Corp. (CAC))
Kava Extract HG776PB (Ennagram)
Kava Extract PG9010 (Concentrated Aloe
Corp. (CAC))
Kava Kava (Sederma)
Kawa-Kawa LS (Alban Muller)
Piper Extract HG 776 PB (Ennagram)
Piper Methysticum Extract (IES LABO)
Vegebios of Kawa Kawa (CEP (Solabia))
VT-132 Extract of Kava-Kava (Vege-Tech)

PIPER METHYSTICUM ROOT EXTRACT

Definition: Piper Methysticum Root
Extract is an extract of the roots of *Piper
methysticum. See "Regulatory and Ingredient Use Information," regarding the
labeling names for botanical ingredients in
Volume 1, Introduction, Part A.*

Chemical Class: Biological Products

Function: Skin-Conditioning Agent - Miscellaneous

Technical/Other Names:
Extract of Kava Root
Extract of Piper Methysticum Root

Trade Names:
Ground Kava Root (Concentrated Aloe
Corp. (CAC))
Micronized Kava Root-MC (Concentrated
Aloe Corp. (CAC))

Trade Name Mixture:
Kawa Kawa Milk (CEP (Solabia))

PIPER NIGRUM (BLACK PEPPER) FRUIT

Definition: Piper Nigrum (Black Pepper)
Fruit is the fruit of *Piper nigrum. See
"Regulatory and Ingredient Use
Information," regarding the labeling names
for botanical ingredients in Volume 1, Intro-
duction, Part A.*

Chemical Class: Biological Products

Function: Not Reported

Trade Name:
Kalimirich (Black Pepper) (Heritage Bio-
Natural)

PIPER NIGRUM (BLACK PEPPER) FRUIT OIL

Definition: Piper Nigrum (Black Pepper)
Fruit Oil is the volatile oil distilled from the
dried ripe fruit of *Piper nigrum.*

Chemical Class: Essential Oils

Function: Fragrance Ingredient

Technical/Other Name:
Oil, Piper Nigrum Fruit

Trade Name:
Oil of Black Pepper (Quest International)

PIPER NIGRUM (BLACK PEPPER) SEED

Definition: Piper Nigrum (Black Pepper)
Seed is the seed obtained from *Piper
nigrum. See "Regulatory and Ingredient
Use Information," regarding the labeling
names for botanical ingredients in Volume
1, Introduction, Part A. See Reported
Ingredient Functions-The Cosmetic Drug
Distinction, in Regulatory and Ingredient
Use Information, Volume I, Part A.*

Chemical Class: Biological Products

Functions: Antioxidant; External Analgesic

PIPER NIGRUM (BLACK PEPPER) SEED EXTRACT

Definition: Piper Nigrum (Black Pepper)
Seed Extract is an extract of the seeds of
the black pepper, *Piper nigrum. See
"Regulatory and Ingredient Use
Information," regarding the labeling names
for botanical ingredients in Volume 1, Introduction, Part A.*

Chemical Class: Biological Products

Function: Not Reported

Technical/Other Names:
Black Pepper Extract
Black Pepper Seed Extract
Extract of Black Pepper
Piper Nigrum Extract

Trade Name Mixtures:
Actiphyte of Black Pepper BG50 (Active
Organics)
Actiphyte of Black Pepper GL50 (Active
Organics)
Actiphyte of Black Pepper Lipo S (Active
Organics)
Actiphyte of Black Pepper PG50 (Active
Organics)
Caospice (CEP (Solabia))

PIPERONYL BUTOXIDE

CAS No.	EINECS No.
51-03-6	200-076-7

Empirical Formula:
$C_{19}H_{30}O_5$

Definition: Piperonyl Butoxide is the organic
compound that conforms to the formula:

$$CH_3(CH_2)_3O(CH_2)_2O(CH_2)_2OCH_2$$

$$CH_3(CH_2)_2$$

Information Sources: BAN,
21CFR178.3730, 21CFR358.610,
21CFR524.2140, 40CFR180.127,
40CFR185.4900, 40CFR186.4900, MI-13
(7557)

Chemical Class: Ethers

Function: Not Reported

Technical/Other Names:
1,3-Benzodioxole, 5-[[2-(2-Butoxyethoxy)-
Ethoxy]Methyl]-6-Propyl-
2,(2-Butoxyethoxy)Ethyl 6-Propylpiperonyl
Ether
Butyl Carbitol 6-Propylpiperonyl Ether
6-Propylpiperonyl Butyl Diethylene Glycol
Ether

Trade Name:
Piperonyl Butoxide Technical (Sumitomo
Chemical)

PIPER RETROFRACTUM FRUIT POWDER

Definition: Piper Retrofractum Fruit
Powder is the powder obtained from the
crushed, dried fruit of *Piper retrofractum.
See "Regulatory and Ingredient Use Information," regarding the labeling names for
botanical ingredients in Volume 1, Introduction, Part A.*

Chemical Class: Biological Products

Function: Skin-Conditioning Agent - Miscellaneous

Trade Name:
Javanese Long Pepper (Haldin Pacific)

PIPERYLENE/BUTENE/PENTENE COPOLYMER

CAS No.: 152698-66-3

Definition: Piperylene/Butene/Pentene
Copolymer is a copolymer of piperylene,
butene and pentene monomers.

Chemical Class: Synthetic Polymers

Functions: Binder; Film Former; Viscosity
Increasing Agent - Nonaqueous

Technical/Other Name:
Distillates (Petroleum), C3-6, Piperylene-
Rich, Polymers with Isobutylene

Trade Names:
Piccotac 1094-E Hydrocarbon Resin (Eastman Chemical)
Piccotac 1095-N Hydrocarbon Resin (Eastman Chemical)

PIPERYLENE/BUTENE/PENTENE/ PENTADIENE COPOLYMER

Definition: Piperylene/Butene/Pentene/ Pentadiene Copolymer is a copolymer of piperylene, butene, pentene and pentadiene monomers.

Chemical Class: Synthetic Polymers

Function: Film Former

Trade Name:
Piccotac 1020-E Hydrocarbon Resin (Eastman Chemical)

PIROCTONE OLAMINE

CAS No.
68890-66-4

EINECS No.
272-574-2

Empirical Formula:
$C_{14}H_{23}NO_2 \cdot C_2H_7NO$

Definition: Piroctone Olamine is the amine salt that conforms to the formula:

$$CH_3CHCH_2CH_2CHCH_2 \quad N \quad O \cdot NH_3CH_2CH_2OH$$

Information Sources: EEC(VI/1-35), MI-13 (7585), USAN

Chemical Classes: Heterocyclic Compounds; Organic Salts

Function: Cosmetic Biocide

Reported Product Categories: Shampoos (Non-coloring); Tonics, Dressings, and Other Hair Grooming Aids

Technical/Other Name:
1-Hydroxy-4-Methyl-6-(2,4,4-Trimethyl-pentyl)-2-(1H)Pyridinone, 2-Aminoethanol Salt

Trade Names:
Octopirox (Clariant)
Octopirox (Clariant GmbH, Personal Care)

Trade Name Mixture:
Antidandruff Agent NOVA (Crodarom)

PISTACIA LENTISCUS LEAF WAX

Definition: Pistacia Lentiscus Leaf Wax is a wax obtained from the leaf of *Pistacia*

lentiscus. See "Regulatory and Ingredient Use Information," regarding the labeling names for botanical ingredients in Volume 1, Introduction, Part A.

Chemical Class: Waxes

Function: Not Reported

Technical/Other Names:
Pistachio (Pistacia Lentiscus) Wax
Wax, Pistachio (Pistacia Lentiscus)

Trade Names:
AEC Cire Essentielle De Feuilles De Lentisque (A & E Connock)
Cire Essentielle de feuilles de Lentisque (Bertin)

PISTACIA LENTISCUS (MASTIC) GUM

CAS No.
61789-92-2

EINECS No.
263-098-6

Definition: Pistacia Lentiscus (Mastic) Gum is a gum obtained from the tree, *Pistacia lentiscus. See "Regulatory and Ingredient Use Information," regarding the labeling names for botanical ingredients in Volume 1, Introduction, Part A.*

Information Source: RIFM

Chemical Class: Biological Polymers and their Derivatives

Functions: Adhesive; Film Former; Fragrance Ingredient

Technical/Other Names:
Mastic absolute (RIFM)
Mastic oil (RIFM)
Mastic (Resin)

Trade Name:
Gum Mastic PLT (P.L. Thomas)

PISTACIA VERA SEED OIL

CAS No.: 129871-01-8

Definition: Pistacia Vera Seed Oil is the oil obtained from the nuts of *Pistacia vera. See "Regulatory and Ingredient Use Information," regarding the labeling names for botanical ingredients in Volume 1, Introduction, Part A.*

Chemical Class: Fats and Oils

Function: Skin-Conditioning Agent - Occlusive

Technical/Other Names:
Oils, Pistachio Nut
Pistachio Nut Oil
Pistachio (Pistacia Vera) Nut Oil

Trade Names:
AEC Pistachio Nut Oil (A & E Connock)

Huile de Pistache Vierge (Bertin)
Phytol Pistachio (pistachio nut oil) (Custom Ingredients)
Pistachio Nut Oil (Desert Whale)

Trade Name Mixture:
Pistachio Softcream (CEP (Solabia))

PISUM SATIVUM (PEA) EXTRACT

CAS No.
90082-41-0

EINECS No.
290-130-6

JPN Translation:
エンドウエキス

Definition: Pisum Sativum (Pea) Extract is an extract of the pea, *Pisum sativum. See "Regulatory and Ingredient Use Information," regarding the labeling names for botanical ingredients in Volume 1, Introduction, Part A. See Reported Ingredient Functions-The Cosmetic Drug Distinction, in Regulatory and Ingredient Use Information, Volume I, Part A.*

Chemical Class: Biological Products

Functions: Skin-Conditioning Agent - Miscellaneous; Skin Protectant

Technical/Other Names:
Extract of Pea
Pea Extract
Pisum Sativum Extract

Trade Name:
Etival (Coletica SA)

Trade Name Mixtures:
Actiphyte of Pea (Active Organics)
Blond Pea Phytolait (Alban Muller)
Connexan LS (Laboratoires Serobiologiques)
Connexan LS (Laboratoires Serobiologiques)
Cryocytol (Laboratoires Serobiologiques)
Etival 5L (Coletica SA)
Etival 15L (Coletica SA)
Etival 15LBG (Coletica SA)
Extraliss (Coletica SA)
Extraliss NAT (Coletica SA)
Golden Pea Milk (CEP (Solabia))
Hydroplastidine Pisum (Vevy)
Lipoplastidine Pisum (Vevy)
Proteasyl LS 9055 (Laboratoires Serobiologiques)
Proteasyl PW (Laboratoires Serobiologiques)
Sunactyl LS 9610 (Laboratoires Serobiologiques)
Vegetensor (Alban Muller)

PISUM SATIVUM SYMBIOSOME EXTRACT

Definition: Pisum Sativum Symbiosome Extract is an extract of the symbiosome (root nodule) of *Pisum sativum. See "Regulatory and Ingredient Use Information," regarding the labeling names for botanical ingredients in Volume 1, Introduction, Part A.*

Chemical Class: Biological Products

Functions: Antioxidant; Skin-Conditioning Agent - Miscellaneous

Technical/Other Name:
Extract of Pisum Sativum Symtiosome

Trade Name Mixture:
Pea Zymbiozome Fermentum (Arch Personal Care Products)

PLACENTAL ENZYMES

Definition: Placental Enzymes is a mixture of enzymes obtained from an aqueous extraction of animal placentas.

Information Source: CIR: [I] IJT-21 (SUPPL. 1)2002

Chemical Class: Biological Products

Functions: Hair Conditioning Agent; Skin-Conditioning Agent - Miscellaneous

Reported Product Category: Face and Neck Preparations (Excluding Shaving Preparations)

Technical/Other Name:
Enzymes, Placental

Trade Name:
Placentol (Pentapharm/Centerchem)

PLACENTAL LIPIDS

Definition: Placental Lipids is a mixture of lipids from animal placentas.

Information Source: CIR: [I] IJT-21 (SUPPL. 1)2002

Chemical Class: Fats and Oils

Functions: Hair Conditioning Agent; Skin-Conditioning Agent - Miscellaneous

Reported Product Category: Face and Neck Preparations (Excluding Shaving Preparations)

Technical/Other Name:
Lipids, Placental

Trade Name Mixtures:
Pharconix OV (Ichimaru Pharcos)
Placenta Extract, Oil- Soluble (Bottger)

PLACENTAL PROTEIN

JPN Translation:
プラセンタエキス

Definition: Placental Protein is a mixture of proteins derived from animal placentas.

Information Source: CIR: [I] IJT-21 (SUPPL. 1)2002

Chemical Class: Proteins

Functions: Hair Conditioning Agent; Skin-Conditioning Agent - Miscellaneous

Reported Product Categories: Face and Neck Preparations (Excluding Shaving Preparations); Moisturizing Preparations; Night Skin Care Preparations; Paste Masks (Mud Packs)

Technical/Other Name:
Proteins, Placental

Trade Names:
Bio-BPL (Snowden)
Bio-BPL-X (Snowden)
Bio-BPS (Snowden)
Bio-BPS-X (Snowden)
Biopharco PQ-1 (PFE) (Ichimaru Pharcos)
Lyophilized Bovine Placenta Extract (Marcor)
NICHIREI WATER-SOLUBLE PLACENTA EXTRACT A (Nichirei)
NICHIREI WATER-SOLUBLE PLACENTA EXTRACT B (Nichirei)
Pharconix BPS PF (Ichimaru Pharcos)
Pharconix CPS (Ichimaru Pharcos)
Placenta-Extract, Water Soluble with 200 KAE 7383 (Crodarom)
Placentaliquid water-soluble (CLR)
Placentormon Liquite (Labopharma)
Snowden Placenta Extract (Snowden)
Snowden Placenta Extract BS-3 (Snowden)
Snowden Placenta Extract BS-11 (Snowden)

Trade Name Mixtures:
Biopharco BP (Ichimaru Pharcos)
Biopharco BP 20 E (Ichimaru Pharcos)
Biopharco CP-12 (Ichimaru Pharcos)
Biopharco CP-12 20E (Ichimaru Pharcos)
Biopharco PQ-1 (Ichimaru Pharcos)
Biopharco PQ-1 (PF) (Ichimaru Pharcos)
Biopharco PQ-1 (PFE) (Ichimaru Pharcos)
Hair Complex 20/70 n (CLR)
Hair Complex NOVA (Crodarom)
Pharconix BPS (Ichimaru Pharcos)
Pharconix BPS 20E (Ichimaru Pharcos)
Pharconix CPS 20E (Ichimaru Pharcos)
Pharconix CPS PF (Ichimaru Pharcos)
Pharconix PC-1 (Ichimaru Pharcos)
Pharconix PC-1 (PF) (Ichimaru Pharcos)
Pharconix PC-1 (PFE) (Ichimaru Pharcos)
PRE Complex (Atrium)

PLANKTON EXTRACT

CAS No.	EINECS No.
91079-57-1	293-445-7

Definition: Plankton Extract is an extract of marine biomass which includes one or more of the following organisms: Thalassoplankton, green micro-algae, diatoms, greenish-blue and nitrogen-fixing seaweed.

Chemical Class: Biological Products

Function: Skin-Conditioning Agent - Miscellaneous

Reported Product Categories: Bath Oils, Tablets, and Salts; Cleansing Products (Cold Creams, Cleansing Lotions, Liquids and Pads); Skin Care Preparations, Misc.; Body and Hand Preparations (Excluding Shaving Preparations); Moisturizing Preparations; Face and Neck Preparations (Excluding Shaving Preparations); Suntan Gels, Creams, and Liquids

Trade Name:
Bioplasma (SECMA)

Trade Name Mixtures:
Actiphyte Plankton (Active Organics)
Actiphyte Plankton BG50P (Active Organics)
Actiphyte Plankton Lipo S (Active Organics)
Complexe Anti-Pollution (Codif)
Nutri-Plancton (Somaig)
Oleocomplex 1 (Somaig)
Omegaplancton (SECMA)
Phormidium C. extract (Codif)
Photosomes (Barnet)
Phycol Omega Plancton (SECMA)
Phycol Omega Plancton BG (SECMA)
Phytoplankton HPG Titrated (Alban Muller)
Plancton Extract HG (Provital/Centerchem)
Thalaton (Vevy)

PLANTAGO ASIATICA EXTRACT

Definition: Plantago Asiatica Extract is an extract of the whole plant, *Plantago asiatica. See "Regulatory and Ingredient Use Information," regarding the labeling names for botanical ingredients in Volume 1, Introduction, Part A.*

Chemical Class: Biological Products

Function: Skin-Conditioning Agent - Miscellaneous

Technical/Other Name:
Extract of Plantago Asiatica

Trade Name Mixture:
Shazensou Liquid B (Ichimaru Pharcos)

PLANTAGO ASIATICA SEED EXTRACT

Definition: Plantago Asiatica Seed Extract is an extract of the seeds of *Plantago*

asiatica. See "Regulatory and Ingredient Use Information," regarding the labeling names for botanical ingredients in Volume 1, Introduction, Part A.

Chemical Class: Biological Products

Function: Skin-Conditioning Agent - Miscellaneous

Technical/Other Name:
Extract of Plantago Asiatica Seed

Trade Name Mixture:
Shazenshi Liquid B (Ichimaru Pharcos)

PLANTAGO LANCEOLATA LEAF EXTRACT

CAS No. 85085-64-9   **EINECS No.** 285-388-1

Definition: Plantago Lanceolata Leaf Extract is an extract of the leaves of the plantain, *Plantago lanceolata. See "Regulatory and Ingredient Use Information," regarding the labeling names for botanical ingredients in Volume 1, Introduction, Part A.*

Chemical Class: Biological Products

Function: Not Reported

Reported Product Categories: Bath Oils, Tablets, and Salts; Cleansing Products (Cold Creams, Cleansing Lotions, Liquids and Pads); Skin Care Preparations, Misc.; Baby Lotions, Oils, Powders and Creams

Technical/Other Names:
Extract of Plantago Lanceolata Leaves
Extract of Plantain Leaves
Plantain Extract
Plantain Leaf Extract
Plantain (Plantago Lanceolata) Extract

Trade Name Mixtures:
Glycolysat of Plantain (CEP (Solabia))
Herbasol Extract Ribwort (Plantain) (Cosmetochem) (Cosmetochem International Ltd.)
Natupure Plantain (E.U.K)
Plantain Extract PG (Rahn)
Plantain Tincture (Rahn)
Ribwort Extract HS 2441 G (Grau)
Seboclear (Rahn)
Sederma Plantain (Sederma)

PLANTAGO MAJOR LEAF EXTRACT

CAS No. 84929-43-1   **EINECS No.** 284-526-8

Definition: Plantago Major Leaf Extract is an extract of the leaves of the plantain,

Plantago major. See "Regulatory and Ingredient Use Information," regarding the labeling names for botanical ingredients in Volume 1, Introduction, Part A.

Chemical Class: Biological Products

Function: Skin-Conditioning Agent - Miscellaneous

Reported Product Categories: Bath Oils, Tablets, and Salts; Cleansing Products (Cold Creams, Cleansing Lotions, Liquids and Pads); Skin Care Preparations, Misc.; Baby Lotions, Oils, Powders and Creams

Technical/Other Names:
Extract of Plantago Major Leaves
Extract of Plantain Leaves
Plantain Extract
Plantain Leaf Extract
Plantain (Plantago Major) Extract

Trade Name Mixtures:
Actiphyte of Plantain BG50 (Active Organics)
Actiphyte of Plantain GL50 (Active Organics)
Actiphyte of Plantain Lipo S (Active Organics)
Actiphyte of Plantain PG50 (Active Organics)
Activated Botanicals AB 110 (Norjin)
250 Blend for Greasy Skin HS (Alban Muller)
650 Blend for Greasy Skin LS (Alban Muller)
Phytogreen 55 of Plantain EXH 670 Liquid (Phytochim)
Plantain HS (Alban Muller)
Plantain LS (Alban Muller)
VT-035 Extract of Plantain (Vege-Tech)

PLANTAGO OVATA LEAF EXTRACT

Definition: Plantago Ovata Leaf Extract is an extract of the leaves of the plantain, *Plantago ovata. See "Regulatory and Ingredient Use Information," regarding the labeling names for botanical ingredients in Volume 1, Introduction, Part A.*

Chemical Class: Biological Products

Function: Not Reported

Reported Product Categories: Cleansing Products (Cold Creams, Cleansing Lotions, Liquids and Pads); Skin Care Preparations, Misc.

Technical/Other Names:
Extract of Plantago Ovata Leaves
Extract of Plantain Leaves
Plantain Leaf Extract
Plantain (Plantago Ovata) Extract

PLANTAGO OVATA SEED EXTRACT

Definition: Plantago Ovata Seed Extract is an extract of the seeds of the plantain,

Plantago ovata. See "Regulatory and Ingredient Use Information," regarding the labeling names for botanical ingredients in Volume 1, Introduction, Part A.

Chemical Class: Biological Products

Function: Skin-Conditioning Agent - Miscellaneous

Reported Product Categories: Bath Oils, Tablets, and Salts; Cleansing Products (Cold Creams, Cleansing Lotions, Liquids and Pads); Skin Care Preparations, Misc.; Baby Lotions, Oils, Powders and Creams

Technical/Other Names:
Extract of Plantago Ovata Seeds
Extract of Plantain Seeds
Plantain Extract
Plantain (Plantago Ovata) Seed Extract
Plantain Seed Extract

Trade Name Mixtures:
Bio-Chelated Derma-Plex I (Bio-Botanica)
Extrait De Plantain MP PG 40 (Yves Rocher)
Indian Plantain HPG Titrated (Alban Muller)
Indian Plantain HS (Alban Muller)
Moisturizing Phytoamine Biocomplex (Alban Muller)
Phytelene of Plantain EG 417 liquid (Indena SA)
Plantago Ovata HS 2669 G (Grau)
2500 Vege-Plex Skin Complex (Vege-Tech)

PLANTAGO PSYLLIUM SEED EXTRACT

Definition: Plantago Psyllium Seed Extract is an extract of the seed of *Plantago psyllium. See "Regulatory and Ingredient Use Information," regarding the labeling names for botanical ingredients in Volume 1, Introduction, Part A.*

Chemical Class: Biological Products

Function: Skin-Conditioning Agent - Miscellaneous

Technical/Other Name:
Extract of Plantago Psyllium

Trade Name Mixture:
Flea Wort HS (Alban Muller)

PLATINUM POWDER

CAS No. 7440-06-4   **EINECS No.** 231-116-1

Definition: Platinum Powder is a metallic element.

Information Source: MI-13(7612)

The inclusion of any compound in the *Dictionary and Handbook* does not indicate that use of that substance as a cosmetic ingredient complies with the laws and regulations governing such use in the United States or any other country.

Chemical Class: Inorganics

Functions: Abrasive; Skin-Conditioning Agent - Miscellaneous

Technical/Other Name:
CI 77795

Trade Name:
Platina L (Bisyo Chemical)

Trade Name Mixture:
HAKKIN - GENSUI (Iwase Cosfa)

PLATONIN

CAS No. 3571-88-8

EINECS No. 222-681-5

JPN Translation:
感光素 101 号

Empirical Formula:
$C_{38}H_{61}N_3S_3 \cdot 2I$

Definition: Platonin is the quaternary heterocyclic compound that conforms to the formula:

Information Sources: JAN, MHLW-331/3, MI-13(7613)

Chemical Classes: Heterocyclic Compounds; Quaternary Ammonium Compounds; Thio Compounds

Function: Skin-Conditioning Agent - Miscellaneous

Technical/Other Names:
2,2'-[3-[(3-Heptyl-4-Methyl-3H-Thiazol-2-ylidene)Ethylidene]Propylene] Bis[3-Heptyl-4-Methylthiazolium]Diiodide
2,2'-Pentamethinethiazolocyanine 3,3"-Diiodide, 4,4',4"-Trimethyl-3,3',3"-Triheptyl-8-(2"-Thiazolyl)-Platonin Diiodide

Trade Name:
Platonin-IK (Ikeda)

PLATYCODON GRANDIFLORUM ROOT EXTRACT

Definition: Platycodon Grandiflorum Root Extract is an extract of the roots of *Platycodon grandiflorum. See "Regulatory and Ingredient Use Information," regarding the labeling names for botanical ingredients in Volume 1, Introduction, Part A.*

Chemical Class: Biological Products

Function: Antioxidant

Technical/Other Name:
Extract of Platycodon Grandiflorum Root

Trade Name:
Premier Platycodon 100% Extract (Premier Specialties)

Trade Name Mixtures:
Draco Platycodon Grandiflorum Full Spectrum Standardized Extract (Draco)
Platycodon Extract BG (Maruzen Pharmaceuticals Co., Ltd.)

PLECTRANTHUS BARBATUS ROOT EXTRACT

Definition: Plectranthus Barbatus Root Extract is an extract of the roots of *Plectranthus barbatus. See "Regulatory and Ingredient Use Information," regarding the labeling names for botanical ingredients in Volume 1, Introduction, Part A.*

Chemical Class: Biological Products

Function: Proprietary

Technical/Other Name:
Extract of Plectranthus Barbatus

Trade Name:
Phytoselect Plectranthus (Indena SpA)

PLEUROCHRYSIS CARTERAE EXTRACT

Definition: Pleurochrysis Carterae Extract is an extract of the alga, *Pleurochrysis carterae. See "Regulatory and Ingredient Use Information," regarding the labeling names for botanical ingredients in Volume 1, Introduction, Part A.*

Chemical Class: Biological Products

Function: Not Reported

Technical/Other Name:
Extract of Pleurochrysis Carterae

Trade Name Mixture:
Dunaliella and Pleurochrysis Extract Powder (MicroAlgae)

PLEUROTUS CYSTIDIOSUS FERMENT FILTRATE

Definition: Pleurotus Cystidiosus Ferment Filtrate is a filtrate of the fermentation of *Pleurotus cystidiosus.*

Chemical Class: Biological Products

Function: Skin-Conditioning Agent - Humectant

Trade Name Mixture:
Ohhiratake Culture Solution (Sansho Seiyaku)

PLEUROTUS SAJOR-CAJU FERMENT FILTRATE

Definition: Pleurotus Sajor-Caju Ferment Filtrate is a filtrate of the fermentation of *Pleurotus sajor-caju.*

Chemical Class: Biological Products

Function: Skin-Conditioning Agent - Humectant

Trade Name Mixture:
Himarayahiratake Culture Solution (Sansho Seiyaku)

PLUCHEA INDICA LEAF EXTRACT

Definition: Pluchea Indica Leaf Extract is an extract of the leaves of *Pluchea indica. See "Regulatory and Ingredient Use Information," regarding the labeling names for botanical ingredients in Volume 1, Introduction, Part A.*

Chemical Class: Biological Products

Function: Skin-Conditioning Agent - Miscellaneous

Technical/Other Name:
Extract of Pluchea Indica Leaf

Trade Name Mixture:
Pluchea Indica Extract (Sansho Seiyaku)

PLUMERIA ACUTIFOLIA FLOWER EXTRACT

Definition: Plumeria Acutifolia Flower Extract is an extract of the flowers of *Plumeria acutifolia. See "Regulatory and Ingredient Use Information," regarding the labeling names for botanical ingredients in Volume 1, Introduction, Part A.*

Chemical Class: Biological Products

Function: Not Reported

Technical/Other Name:
Extract of Plumeria Acutifolia Flowers

Trade Name Mixtures:
Cosflor Frangipani HGS (A & E Connock)
VT-130 Extract of Plumeria Flower (Vege-Tech)

PLUMERIA ALBA FLOWER EXTRACT

Definition: Plumeria Alba Flower Extract is an extract of the flowers of *Plumeria alba*. See *"Regulatory and Ingredient Use Information,"* regarding the labeling names for botanical ingredients in Volume 1, Introduction, Part A.

Chemical Class: Biological Products

Function: Skin-Conditioning Agent - Miscellaneous

Trade Name Mixtures:
Frangipani Flowers Milk (CEP (Solabia))
Glycolysat of Frangipani Flowers (CEP (Solabia))
Phytelene of Frangipana BG 761 (Indena SA)
Phytelene of Frangipana EG 761 (Indena SA)
Vegebios of Frangipani Flowers (CEP (Solabia))

PLUMERIA RUBRA FLOWER EXTRACT

Definition: Plumeria Rubra Flower Extract is an extract of the flowers of *Plumeria rubra*. See *"Regulatory and Ingredient Use Information,"* regarding the labeling names for botanical ingredients in Volume 1, Introduction, Part A.

Chemical Class: Biological Products

Function: Not Reported

Trade Name Mixtures:
Actiphyte of Plumeria (Active Organics)
Extrait De Frangipanier (Yves Rocher)
Frangipany Phytolait (Alban Muller)
Herbal Extract Frangipani (Jarvis)

POA ANNUA EXTRACT

Definition: Poa Annua Extract is an extract of the herb, *Poa annua*. See *"Regulatory and Ingredient Use Information,"* regarding the labeling names for botanical ingredients in Volume 1, Introduction, Part A.

Chemical Class: Biological Products

Function: Not Reported

Technical/Other Name:
Extract of Poa Annua

Trade Name Mixture:
Cosflor Meadowgrass HGS (A & E Connock)

PODOPHYLLUM PELTATUM

Definition: See *"Regulatory and Ingredient Use Information,"* regarding EU labeling names for botanical ingredients in Volume 1, Introduction, Part A.

Chemical Class: Biological Products

Technical/Other Name:
Podophyllum Peltatum Extract (U.S.)

PODOPHYLLUM PELTATUM EXTRACT

Definition: Podophyllum Peltatum Extract is an extract of the rhizomes and roots of the podophyllum, *Podophyllum peltatum*. See *"Regulatory and Ingredient Use Information,"* regarding the labeling names for botanical ingredients in Volume 1, Introduction, Part A.

Information Source: HP

Chemical Class: Biological Products

Function: Not Reported

Technical/Other Names:
Extract of Podophyllum
Podophyllum Extract
Podophyllum Peltatum (EU)

POGOSTEMON CABLIN LEAF EXTRACT

Definition: Pogostemon Cablin Leaf Extract is an extract of the leaves of the patchouli, *Pogostemon cablin*. See *"Regulatory and Ingredient Use Information,"* regarding the labeling names for botanical ingredients in Volume 1, Introduction, Part A.

Chemical Class: Biological Products

Function: Not Reported

Technical/Other Names:
Extract of Patchouli
Patchouli Extract
Patchouli (Pogostemon Cablin) Extract
Pogostemon Patchouli Extract

Trade Name Mixtures:
Actiphyte of Patchouli BG50 (Active Organics)
Actiphyte of Patchouli GL50 (Active Organics)
Actiphyte of Patchouli Lipo S (Active Organics)
Actiphyte of Patchouli PG50 (Active Organics)
Aromaphyte of Patchouli (Active Organics)
Herbasol Extract Patchouli (Cosmetochem) (Cosmetochem International Ltd.)
Patchouli HS (Alban Muller)
VT-262 Extract of Patchouly (Vege-Tech)

POGOSTEMON CABLIN OIL

CAS No.: 8014-09-3

Definition: Pogostemon Cablin Oil is the volatile oil obtained from *Pogostemon patchouli*. See *"Regulatory and Ingredient Use Information,"* regarding the labeling names for botanical ingredients in Volume 1, Introduction, Part A.

Information Sources: MI-13(6862), RIFM

Chemical Class: Essential Oils

Function: Fragrance Ingredient

Technical/Other Names:
Oil of Patchouli
Patchouli Oil
Patchouli (Pogostemon Cablin) Oil
Patchouly oil (RIFM)
Pogostemon Patchouli Oil

Trade Name:
AEC Patchouli Oil (A & E Connock)

Trade Name Mixtures:
Aromaphyte of Patchouli (Active Organics)
Covazen Relax (LCW)
Essentiaderm Capillare N.18 (Universal Flavors)
Essentiaderm N.19 (Universal Flavors)
Extrapone Patchouli GW 2/033657 (Symrise)

POLIANTHES TUBEROSA EXTRACT

CAS No.	EINECS No.
94334-35-7	305-108-4

Definition: Polianthes Tuberosa Extract is an extract of the tuberose, *Polianthes tuberosa*. See *"Regulatory and Ingredient Use Information,"* regarding the labeling names for botanical ingredients in Volume 1, Introduction, Part A. See Reported Ingredient Functions-The Cosmetic Drug Distinction, in Regulatory and Ingredient Use Information, Volume I, Part A.

Information Source: RIFM

Chemical Class: Biological Products

Functions: Film Former; Fragrance Ingredient; Skin-Conditioning Agent - Miscellaneous; Skin-Conditioning Agent - Occlusive; Skin Protectant; Viscosity Controlling Agent; Viscosity Increasing Agent - Aqueous

Technical/Other Names:
Extract of Tuberose
Tuberose absolute (RIFM)
Tuberose concrete (RIFM)
Tuberose Extract
Tuberose (Polianthes Tubarosa) Extract

Trade Name:
Hydroessential Polyanthes (Vevy)

Trade Name Mixture:
Actiphyte of Tuberose (Active Organics)

POLIANTHES TUBEROSA FLOWER WAX

Definition: Polianthes Tuberosa Flower Wax is a wax obtained from the flower of *Polianthes tuberosa*. See "Regulatory and Ingredient Use Information," regarding the labeling names for botanical ingredients in Volume 1, Introduction, Part A.

Chemical Class: Waxes

Function: Not Reported

Technical/Other Names:
Tuberose (Polianthes Tuberosa) Wax
Tuberose Wax
Waxes, Tuberose (Polianthes Tuberosa)

Trade Names:
AEC Cire Essentielle De Fleurs De Tubereuse (A & E Connock)
AEC Tuberose Wax (A & E Connock)
Cire Essentielle de fleurs de Tubereuse (Bertin)
Tuberose Wax (Koster Keunen Holland)

POLIANTHES TUBEROSA POLYSACCHARIDE

Definition: Polianthes Tuberosa Polysaccharide is the polysaccharide fraction produced by the cultured cells of *Polianthes tuberosa*. See "Regulatory and Ingredient Use Information," regarding the labeling names for botanical ingredients in Volume 1, Introduction, Part A.

Chemical Class: Biological Products

Function: Skin-Conditioning Agent - Miscellaneous

Trade Name Mixture:
TPS-1 (Kao Corp.)

POLLEN

Definition: Pollen is the fertilizing element of flowering plants consisting of fine, powdery, yellowish grains or spores.

Chemical Class: Biological Products

Function: Skin-Conditioning Agent - Miscellaneous

Trade Names:
AEC Bees Pollen (A & E Connock)
Ennapollen (Ennagram)

POLLEN EXTRACT

Definition: Pollen Extract is an extract of Pollen (q.v.).

Chemical Class: Biological Products

Function: Not Reported

Technical/Other Name:
Extract of Pollen

Trade Name:
Pollen Extract (Uniom)

Trade Name Mixtures:
Actiphyte of Bee Pollen (Active Organics)
Actiphyte of Bee Pollen BG50 (Active Organics)
Actiphyte of Bee Pollen GL50 (Active Organics)
Actiphyte of Bee Pollen Lipo S (Active Organics)
Actiphyte of Bee Pollen PG50 (Active Organics)
Aquabios of Pollen (CEP (Solabia))
Biofloreol hydrosoluble (Esperis)
Biofloreol liposoluble (Esperis)
Biofloreol liquid liposoluble (Esperis)
Filagrinol (Vevy)
Filagrinol-Ext (Vevy)
Glycolysat of Pollen (CEP (Solabia))
Herbasol Extract Bee Pollen (Cosmetochem) (Cosmetochem International Ltd.)
Hydroplastidine Pollen (Vevy)
Lipoplastidine Pollen (Vevy)
Oleat of Pollen (CEP (Solabia))
Phytelene of Pollen EG 406 liquid (Indena SA)
Phytoderm P/25 Hydroalcoholic (Universal Flavors)
Phytogreen 55 of Pollen EXH 667 Liquid (Phytochim)
Pollen Extract HG (Provital/Centerchem)
Pollen Extract HS 2586 G (Grau)
Pollen Extract HS 3383 G (Grau)
Pollen Extract ies (IES LABO)
Pronalen Moisturizing HSC (Provital/Centerchem)
Vegebois of Pollen (CEP (Solabia))
VT-087 Extract of Bee Pollen (Vege-Tech)

POLOXAMER 101

CAS No.: 9003-11-6 (Generic)

Definition: Poloxamer 101 is the polyoxyethylene, polyoxypropylene block polymer that conforms generally to the formula:

$$HO(CH_2CH_2O)_x(CHCH_2O)_y(CH_2CH_2O)_zH$$
$$|$$
$$CH_3$$

in which the average value of x, y and z are respectively 2, 16 and 2.

Information Sources: 21CFR176.210, MI-13(7645), TSCA

Chemical Class: Polymeric Ethers

Function: Surfactant - Emulsifying Agent

Technical/Other Name:
Poly(oxyethylene)m-block-poly (oxypropylene)n-block-poly (oxyethylene)p

Trade Name:
Pluronic L-31 (BASF)

POLOXAMER 105

CAS No.: 9003-11-6 (Generic)

JPN Translation:
ポロキサマー105

Definition: Poloxamer 105 is the polyoxyethylene, polyoxypropylene block polymer that conforms generally to the formula:

$$HO(CH_2CH_2O)_x(CHCH_2O)_y(CH_2CH_2O)_zH$$
$$|$$
$$CH_3$$

in which the average values of x, y and z are respectively 11, 16 and 11.

Information Sources: 21CFR172.808, 21CFR173.340, 21CFR175.105, 21CFR176.180, 21CFR176.200, 21CFR176.210, 21CFR177.1200, JCIC, JCLS, MI-13(7645), TSCA

Chemical Class: Polymeric Ethers

Functions: Surfactant - Cleansing Agent; Surfactant - Emulsifying Agent; Surfactant - Solubilizing Agent

Reported Product Categories: Hair Conditioners; Shaving Cream (Aerosol, Brushless and Lather)

Technical/Other Name:
Polyoxyethylene Polyoxypropylene Glycol (22E.O.) (16P.O.)

Trade Names:
Pluronic L-35 (BASF)
Synperonic PE/L35 (Uniqema Americas)

POLOXAMER 108

CAS No.: 9003-11-6 (Generic)

Definition: Poloxamer 108 is the polyoxyethylene, polyoxypropylene block polymer that conforms generally to the formula:

$$HO(CH_2CH_2O)_x(CHCH_2O)_y(CH_2CH_2O)_zH$$
$$|$$
$$CH_3$$

in which the average values of x, y and z are respectively 46, 16 and 46.

Information Sources: 21CFR172.808, 21CFR173.340, 21CFR175.105, 21CFR176.180, 21CFR176.200, 21CFR176.210, 21CFR177.1200, 21CFR177.1210, MI-13(7645), TSCA

Chemical Class: Polymeric Ethers

Function: Surfactant - Cleansing Agent

Trade Names:
Pluronic F-38 (BASF)
Synperonic PE/F38 (Uniqema Americas)

POLOXAMER 122

CAS No.: 9003-11-6 (Generic)

Definition: Poloxamer 122 is the polyoxyethylene, polyoxypropylene block polymer that conforms generally to the formula:

$$HO(CH_2CH_2O)_x(CHCH_2O)_y(CH_2CH_2O)_zH$$
$$|$$
$$CH_3$$

in which the average values of x, y and z are respectively 5, 21 and 5.

Information Sources: 21CFR172.808, 21CFR173.340, 21CFR175.105, 21CFR176.180, 21CFR176.200, 21CFR176.210, 21CFR177.1200, MI-13 (7645), TSCA

Chemical Class: Polymeric Ethers

Functions: Surfactant - Emulsifying Agent; Surfactant - Solubilizing Agent

Trade Name:
Synperonic PE/L42 (Uniqema Americas)

POLOXAMER 123

CAS No.: 9003-11-6 (Generic)

Definition: Poloxamer 123 is the polyoxyethylene, polyoxypropylene block polymer that conforms generally to the formula:

$$HO(CH_2CH_2O)_x(CHCH_2O)_y(CH_2CH_2O)_zH$$
$$|$$
$$CH_3$$

in which the average values of x, y and z are respectively 7, 21 and 7.

Information Sources: 21CFR172.808, 21CFR173.340, 21CFR175.105, 21CFR176.180, 21CFR176.200, 21CFR176.210, 21CFR177.1200, MI-13 (7645), TSCA

Chemical Class: Polymeric Ethers

Functions: Surfactant - Emulsifying Agent; Surfactant - Solubilizing Agent

Trade Name:
Pluronic L-43 (BASF)

POLOXAMER 124

CAS No.: 9003-11-6 (Generic)

JPN Translation:
ボロキサマー124

Definition: Poloxamer 124 is the polyoxyethylene, polyoxypropylene block polymer that conforms generally to the formula:

$$HO(CH_2CH_2O)_x(CHCH_2O)_y(CH_2CH_2O)_zH$$
$$|$$
$$CH_3$$

in which the average values of x, y and z are respectively 11, 21 and 11.

Information Sources: BAN, 21CFR172.808, 21CFR173.340, 21CFR175.105, 21CFR176.180, 21CFR176.200, 21CFR176.210, 21CFR177.1200, INN, MI-13(7469), NF XVIII, TSCA, USAN

Chemical Class: Polymeric Ethers

Functions: Surfactant - Emulsifying Agent; Surfactant - Solubilizing Agent

Trade Names:
Pluronic L-44 (BASF)
Synperonic PE/L44 (Uniqema Americas)

POLOXAMER 181

CAS No.: 9003-11-6 (Generic)

JPN Translation:
ボロキサマー181

Definition: Poloxamer 181 is the polyoxyethylene, polyoxypropylene block polymer that conforms generally to the formula:

$$HO(CH_2CH_2O)_x(CHCH_2O)_y(CH_2CH_2O)_zH$$
$$|$$
$$CH_3$$

in which the average values of x, y and z are respectively 3, 30 and 3.

Information Sources: 21CFR172.808, 21CFR173.340, 21CFR175.105, 21CFR176.180, 21CFR176.200, 21CFR176.210, 21CFR177.1200, MI-13 (7469), TSCA

Chemical Class: Polymeric Ethers

Function: Surfactant - Emulsifying Agent

Reported Product Categories: Moisturizing Preparations; Skin Care Preparations, Misc.

Trade Names:
Chemal BP-261 (Chemax)
DelONIC EO-PO-1 (DeForest)
Ethox L-61 (Ethox)
Pluronic L-61 (BASF)
Synperonic PE/L61 (Uniqema Americas)

POLOXAMER 182

CAS No.: 9003-11-6 (Generic)

JPN Translation:
ボロキサマー182

Definition: Poloxamer 182 is the polyoxyethylene, polyoxypropylene block polymer that conforms generally to the formula:

$$HO(CH_2CH_2O)_x(CHCH_2O)_y(CH_2CH_2O)_zH$$
$$|$$
$$CH_3$$

in which the average values of x, y and z are respectively 8, 30 and 8.

Information Sources: 21CFR172.808, 21CFR173.340, 21CFR175.105, 21CFR176.180, 21CFR176.200, 21CFR176.210, 21CFR177.1200, CTFA D, MI-13(7469), TSCA

Chemical Class: Polymeric Ethers

Functions: Surfactant - Cleansing Agent; Surfactant - Solubilizing Agent

Reported Product Categories: Paste Masks (Mud Packs); Skin Care Preparations, Misc.; Suntan Gels, Creams, and Liquids

Trade Names:
Chemal BP-262 (Chemax)
DelONIC EO-PO-2 (DeForest)
Ethox L-62 (Ethox)
Pluronic L-62 (BASF)
Synperonic PE/L62 (Uniqema Americas)

POLOXAMER 183

CAS No.: 9003-11-6 (Generic)

Definition: Poloxamer 183 is the polyoxyethylene, polyoxypropylene block polymer that conforms generally to the formula:

$$HO(CH_2CH_2O)_x(CHCH_2O)_y(CH_2CH_2O)_zH$$
$$|$$
$$CH_3$$

in which the average values of x, y and z are respectively 10, 30 and 10.

Information Sources: 21CFR172.808, 21CFR173.340, 21CFR175.105, 21CFR176.180, 21CFR176.200, 21CFR176.210, 21CFR177.1200, CTFA D, MI-13(7645), TSCA

Chemical Class: Polymeric Ethers

Functions: Surfactant - Cleansing Agent; Surfactant - Solubilizing Agent

POLOXAMER 184

CAS No.: 9003-11-6 (Generic)

JPN Translation:
ボロキサマー184

Definition: Poloxamer 184 is the polyoxyethylene, polyoxypropylene block polymer that conforms generally to the formula:

$$HO(CH_2CH_2O)_x(CHCH_2O)_y(CH_2CH_2O)_zH$$
$$|$$
$$CH_3$$

in which the average values of x, y and z are respectively 13, 30 and 13.

Information Sources: 21CFR172.808, 21CFR173.340, 21CFR175.105, 21CFR176.180, 21CFR176.200, 21CFR176.210, 21CFR177.1200, 21CFR177.1210, CTFA D, JCIC, JCLS, JSQI, MI-13(7645), TSCA

Chemical Class: Polymeric Ethers

Functions: Surfactant - Cleansing Agent; Surfactant - Solubilizing Agent

Reported Product Categories: Bath Oils, Tablets, and Salts; Cleansing Products (Cold Creams, Cleansing Lotions, Liquids and Pads); Bath Preparations, Misc.; Eye Makeup Removers; Hair Conditioners; Shampoos (Non-coloring); Body and Hand Preparations (Excluding Shaving Preparations); Makeup Fixatives; Makeup Preparations (Not eye), Misc.; Moisturizing Preparations; Paste Masks (Mud Packs); Skin Care Preparations, Misc.; Suntan Gels, Creams, and Liquids

Technical/Other Name:
Polyoxyethylene Polyoxypropylene Glycol (26E.O.) (30P.O.)

Trade Names:
Pluracare L-64 (BASF)
Synperonic PE/L64 (Uniqema Americas)

Trade Name Mixtures:
Lubrajel WA (Guardian)
Solimate E (Guardian)

POLOXAMER 185

CAS No.: 9003-11-6 (Generic)

JPN Translation:
ボロキサマー185

Definition: Poloxamer 185 is the polyoxyethylene, polyoxypropylene block polymer that conforms generally to the formula:

$$HO(CH_2CH_2O)_x(CHCH_2O)_y(CH_2CH_2O)_zH$$
$$|$$
$$CH_3$$

in which the average values of x, y and z are respectively 19, 30 and 19.

Information Sources: 21CFR172.808, 21CFR173.340, 21CFR175.105, 21CFR176.180, 21CFR176.200, 21CFR176.210, 21CFR177.1200, 21CFR177.1210, MI-13(7645), TSCA

Chemical Class: Polymeric Ethers

Functions: Surfactant - Emulsifying Agent; Surfactant - Solubilizing Agent

Reported Product Categories: Paste Masks (Mud Packs); Eye Lotions; Eye Makeup Removers; Skin Care Preparations, Misc.

Trade Name:
Pluracare P-65 (BASF)

POLOXAMER 188

CAS No.: 9003-11-6 (Generic)

JPN Translation:
ボロキサマー188

Definition: Poloxamer 188 is the polyoxyethylene, polyoxypropylene block polymer that conforms generally to the formula:

$$HO(CH_2CH_2O)_x(CHCH_2O)_y(CH_2CH_2O)_zH$$
$$|$$
$$CH_3$$

in which the average values of x, y and z are respectively 75, 30 and 75. In the United States, Poloxamer 188 may be used as an active ingredient in OTC drug products. When used as an active drug ingredient, the established name is *Poloxamer 188. See "Regulatory and Ingredient Use Information," regarding the labeling names for U.S. OTC Drug Ingredients in Volume 1, Introduction, Part A.*

Information Sources: BAN, 21CFR172.808, 21CFR173.340, 21CFR175.105, 21CFR176.180, 21CFR176.200, 21CFR176.210, 21CFR177.1200, 21CFR177.1210, 21CFR310.545, CTFA D, INN, JAN, MI-13(7645), NF XVIII, OTC-I-AM, TSCA, USAN, USD

Chemical Class: Polymeric Ethers

Functions: Antimicrobial Agent; Surfactant - Cleansing Agent

Reported Product Categories: Permanent Waves; Bath Oils, Tablets, and Salts; Cleansing Products (Cold Creams, Cleansing Lotions, Liquids and Pads); Indoor Tanning Preparations; Skin Care Preparations, Misc.

Technical/Other Name:
Poloxalkol

Trade Names:
Pluronic F-68 (BASF)
Synperonic PE/F68 (Uniqema Americas)

POLOXAMER 212

CAS No.: 9003-11-6 (Generic)

Definition: Poloxamer 212 is the polyoxyethylene, polyoxypropylene block polymer that conforms generally to the formula:

$$HO(CH_2CH_2O)_x(CHCH_2O)_y(CH_2CH_2O)_zH$$
$$|$$
$$CH_3$$

in which the average values of x, y and z are respectively 8, 35 and 8.

Information Sources: 21CFR172.808, 21CFR173.340, 21CFR175.105, 21CFR176.180, 21CFR176.200, 21CFR176.210, 21CFR177.1200, 21CFR177.1210, MI-13(7645), TSCA

Chemical Class: Polymeric Ethers

Functions: Surfactant - Emulsifying Agent; Surfactant - Solubilizing Agent

Reported Product Category: Skin Care Preparations, Misc.

POLOXAMER 215

CAS No.: 9003-11-6 (Generic)

JPN Translation:
ボロキサマー215

Definition: Poloxamer 215 is the polyoxyethylene, polyoxypropylene block polymer that conforms generally to the formula:

$$HO(CH_2CH_2O)_x(CHCH_2O)_y(CH_2CH_2O)_zH$$
$$|$$
$$CH_3$$

in which the average values of x, y and z are respectively 24, 35 and 24.

Information Sources: 21CFR172.808, 21CFR173.340, 21CFR175.105, 21CFR176.180, 21CFR176.200, 21CFR176.210, 21CFR177.1200, 21CFR177.1210, MI-13(7645), TSCA

Chemical Class: Polymeric Ethers

Functions: Surfactant - Emulsifying Agent; Surfactant - Solubilizing Agent

Trade Name:
Synperonic PE/P75 (Uniqema Americas)

POLOXAMER 217

CAS No.: 9003-11-6 (Generic)

Definition: Poloxamer 217 is the polyoxyethylene, polyoxypropylene block polymer that conforms generally to the formula:

$$HO(CH_2CH_2O)_x(CHCH_2O)_y(CH_2CH_2O)_zH$$
$$|$$
$$CH_3$$

in which the average values of x, y and z are respectively 52, 35 and 52.

Information Sources: 21CFR172.808, 21CFR173.340, 21CFR175.105, 21CFR176.180, 21CFR176.200, 21CFR176.210, 21CFR177.1200, 21CFR177.1210, MI-13(7645), TSCA

Chemical Class: Polymeric Ethers

Functions: Surfactant - Cleansing Agent; Surfactant - Solubilizing Agent

Trade Name:
Pluronic F-77 (BASF)

POLOXAMER 231

CAS No.: 9003-11-6 (Generic)

Definition: Poloxamer 231 is the polyoxyethylene, polyoxypropylene block polymer that conforms generally to the formula:

$$HO(CH_2CH_2O)_x(CHCH_2O)_y(CH_2CH_2O)_zH$$
$$|$$
$$CH_3$$

in which the average values of x, y and z are respectively 6, 39 and 6.

Information Sources: 21CFR172.808, 21CFR173.340, 21CFR175.105, 21CFR176.180, 21CFR176.200, 21CFR176.210, 21CFR177.1200, 21CFR177.1210, MI-13(7645), TSCA

Chemical Class: Polymeric Ethers

Function: Surfactant - Emulsifying Agent

Trade Names:
Pluronic L-81 (BASF)
Synperonic PE/L81 (Uniqema Americas)

POLOXAMER 234

CAS No.: 9003-11-6 (Generic)

Definition: Poloxamer 234 is the polyoxyethylene, polyoxypropylene block polymer that conforms generally to the formula:

$$HO(CH_2CH_2O)_x(CHCH_2O)_y(CH_2CH_2O)_zH$$
$$|$$
$$CH_3$$

in which the average values of x, y and z are respectively 22, 39 and 22.

Information Sources: 21CFR172.808, 21CFR173.340, 21CFR175.105, 21CFR176.180, 21CFR176.200, 21CFR176.210, 21CFR177.1200, 21CFR177.1210, MI-13(7645), TSCA

Chemical Class: Polymeric Ethers

Functions: Surfactant - Cleansing Agent; Surfactant - Solubilizing Agent

Reported Product Categories: Body and Hand Preparations (Excluding Shaving Preparations); Cleansing Products (Cold Creams, Cleansing Lotions, Liquids and Pads); Face and Neck Preparations (Excluding Shaving Preparations); Moisturizing Preparations; Shampoos (Non-coloring); Skin Care Preparations, Misc.; Suntan Preparations, Misc.; Tonics, Dressings, and Other Hair Grooming Aids

Trade Names:
Pluronic P-84 (BASF)
Synperonic PE/P84 (Uniqema Americas)

POLOXAMER 235

CAS No.: 9003-11-6 (Generic)

Definition: Poloxamer 235 is the polyoxyethylene, polyoxypropylene block polymer that conforms generally to the formula:

$$HO(CH_2CH_2O)_x(CHCH_2O)_y(CH_2CH_2O)_zH$$
$$|$$
$$CH_3$$

in which the average values of x, y and z are respectively 27, 39 and 27.

Information Sources: 21CFR172.808, 21CFR173.340, 21CFR175.105, 21CFR176.180, 21CFR176.200, 21CFR176.210, 21CFR177.1200, 21CFR177.1210, MI-13(7645), TSCA

Chemical Class: Polymeric Ethers

Functions: Surfactant - Cleansing Agent; Surfactant - Solubilizing Agent

Trade Names:
Pluronic P-85 (BASF)
Synperonic PE/P85 (Uniqema Americas)

POLOXAMER 237

CAS No.: 9003-11-6 (Generic)

Definition: Poloxamer 237 is the polyoxyethylene, polyoxypropylene block polymer that conforms generally to the formula:

$$HO(CH_2CH_2O)_x(CHCH_2O)_y(CH_2CH_2O)_zH$$
$$|$$
$$CH_3$$

in which the average values of x, y and z are respectively 62, 39 and 62.

Information Sources: BAN, 21CFR172.808, 21CFR173.340, 21CFR175.105, 21CFR176.180, 21CFR176.200, 21CFR176.210, 21CFR177.1200, 21CFR177.1210, INN, MI-13(7645), NF XVIII, TSCA, USAN

Chemical Class: Polymeric Ethers

Functions: Surfactant - Cleansing Agent; Surfactant - Solubilizing Agent

Trade Names:
Pluronic F-87 (BASF)
Synperonic PE/F87 (Uniqema Americas)

POLOXAMER 238

CAS No.: 9003-11-6 (Generic)

Definition: Poloxamer 238 is the polyoxyethylene, polyoxypropylene block polymer that conforms generally to the formula:

$$HO(CH_2CH_2O)_x(CHCH_2O)_y(CH_2CH_2O)_zH$$
$$|$$
$$CH_3$$

in which the average values of x, y and z are respectively 97, 39 and 97.

Information Sources: 21CFR172.808, 21CFR173.340, 21CFR175.105, 21CFR176.180, 21CFR176.200, 21CFR176.210, 21CFR177.1200, 21CFR177.1210, MI-13(7645), TSCA

Chemical Class: Polymeric Ethers

Function: Surfactant - Cleansing Agent

Trade Names:
Pluronic F-88 (BASF)
Synperonic PE/F88 (Uniqema Americas)

POLOXAMER 282

CAS No.: 9003-11-6 (Generic)

Definition: Poloxamer 282 is the polyoxyethylene, polyoxypropylene block polymer that conforms generally to the formula:

$$HO(CH_2CH_2O)_x(CHCH_2O)_y(CH_2CH_2O)_zH$$
$$|$$
$$CH_3$$

in which the average values of x, y and z are respectively 10, 47 and 10.

Information Sources: 21CFR172.808, 21CFR173.340, 21CFR175.105, 21CFR176.180, 21CFR176.200, 21CFR176.210, 21CFR177.1200, 21CFR177.1210, MI-13(7645), TSCA

Chemical Class: Polymeric Ethers

Function: Surfactant - Emulsifying Agent

Trade Names:
Pluronic L-92 (BASF)
Synperonic PE/L92 (Uniqema Americas)

POLOXAMER 284

CAS No.: 9003-11-6 (Generic)

Definition: Poloxamer 284 is the polyoxyethylene, polyoxypropylene block polymer that conforms generally to the formula:

$$HO(CH_2CH_2O)_x(CHCH_2O)_y(CH_2CH_2O)_zH$$
$$|$$
$$CH_3$$

in which the average values of x, y and z are respectively 21, 47 and 21.

Information Sources: 21CFR172.808, 21CFR173.340, 21CFR175.105, 21CFR176.180, 21CFR176.200, 21CFR176.210, 21CFR177.1200, 21CFR177.1210, MI-13(7645), TSCA

Chemical Class: Polymeric Ethers

Functions: Surfactant - Emulsifying Agent; Surfactant - Solubilizing Agent

POLOXAMER 288

CAS No.: 9003-11-6 (Generic)

Definition: Poloxamer 288 is the polyoxyethylene, polyoxypropylene block polymer that conforms generally to the formula:

$$HO(CH_2CH_2O)_x(CHCH_2O)_y(CH_2CH_2O)_zH$$
$$|$$
$$CH_3$$

in which the average values of x, y and z are respectively 122, 47 and 122.

Information Sources: 21CFR172.808, 21CFR173.340, 21CFR175.105, 21CFR176.180, 21CFR176.200, 21CFR176.210, 21CFR177.1200, 21CFR177.1210, MI-13(7645), TSCA

Chemical Class: Polymeric Ethers

Function: Surfactant - Cleansing Agent

Trade Name:
Pluronic F-98 (BASF)

POLOXAMER 331

CAS No.: 9003-11-6 (Generic)

Definition: Poloxamer 331 is the polyoxyethylene, polyoxypropylene block polymer that conforms generally to the formula:

$$HO(CH_2CH_2O)_x(CHCH_2O)_y(CH_2CH_2O)_zH$$
$$|$$
$$CH_3$$

in which the average values of x, y and z are respectively 7, 54 and 7.

Information Sources: 21CFR172.808, 21CFR173.340, 21CFR175.105, 21CFR176.180, 21CFR176.200, 21CFR176.210, 21CFR177.1200, 21CFR177.1210, FCC, MI-13(5875), TSCA

Chemical Class: Polymeric Ethers

Function: Surfactant - Emulsifying Agent

Trade Names:
Pluronic L-101 (BASF)
Synperonic PE/L101 (Uniqema Americas)

Trade Name Mixtures:
Syntran 5900 (Interpolymer)
Syntran 5902 (Interpolymer)

POLOXAMER 333

CAS No.: 9003-11-6 (Generic)

JPN Translation:
ポロキサマー333

Definition: Poloxamer 333 is the polyoxyethylene, polyoxypropylene block polymer that conforms generally to the formula:

$$HO(CH_2CH_2O)_x(CHCH_2O)_y(CH_2CH_2O)_zH$$
$$|$$
$$CH_3$$

in which the average values of x, y and z are respectively 20, 54 and 20.

Information Sources: 21CFR172.808, 21CFR173.340, 21CFR175.105, 21CFR176.180, 21CFR176.200, 21CFR176.210, 21CFR177.1200, 21CFR177.1210, MI-13(7645), TSCA

Chemical Class: Polymeric Ethers

Function: Surfactant - Emulsifying Agent

Reported Product Categories: Moisturizing Preparations; Body and Hand Preparations (Excluding Shaving Preparations)

Trade Names:
Pluronic P-103 (BASF)
Synperonic PE/P103 (Uniqema Americas)

POLOXAMER 334

CAS No.: 9003-11-6 (Generic)

Definition: Poloxamer 334 is the polyoxyethylene, polyoxypropylene block polymer that conforms generally to the formula:

$$HO(CH_2CH_2O)_x(CHCH_2O)_y(CH_2CH_2O)_zH$$
$$|$$
$$CH_3$$

in which the average values of x, y and z are respectively 31, 54 and 31.

Information Sources: 21CFR172.808, 21CFR173.340, 21CFR175.105, 21CFR176.180, 21CFR176.200, 21CFR176.210, 21CFR177.1200, 21CFR177.1210, MI-13(7645), TSCA

Chemical Class: Polymeric Ethers

Functions: Surfactant - Cleansing Agent; Surfactant - Solubilizing Agent

Reported Product Category: Tonics, Dressings, and Other Hair Grooming Aids

Trade Name:
Pluronic P-104 (BASF)

POLOXAMER 335

CAS No.: 9003-11-6 (Generic)

Definition: Poloxamer 335 is the polyoxyethylene, polyoxypropylene block polymer that conforms generally to the formula:

$$HO(CH_2CH_2O)_x(CHCH_2O)_y(CH_2CH_2O)_zH$$
$$|$$
$$CH_3$$

in which the average values of x, y and z are respectively 38, 54 and 38.

Information Sources: 21CFR172.808, 21CFR173.340, 21CFR175.105, 21CFR176.180, 21CFR176.200, 21CFR176.210, 21CFR177.1200, 21CFR177.1210, MI-13(7645), TSCA

Chemical Class: Polymeric Ethers

Functions: Surfactant - Cleansing Agent; Surfactant - Solubilizing Agent

Trade Name:
Pluronic P-105 (BASF)

POLOXAMER 338

CAS No.: 9003-11-6 (Generic)

Definition: Poloxamer 338 is the polyoxyethylene, polyoxypropylene block polymer that conforms generally to the formula:

$$HO(CH_2CH_2O)_x(CHCH_2O)_y(CH_2CH_2O)_zH$$
$$|$$
$$CH_3$$

in which the average values of x, y and z are respectively 128, 54 and 128.

Information Sources: BAN, 21CFR172.808, 21CFR173.340, 21CFR175.105, 21CFR176.180, 21CFR176.200, 21CFR176.210, 21CFR177.1200, 21CFR177.1210, CTFA D, INN, MI-13(7645), NF XVIII, TSCA, USAN

Chemical Class: Polymeric Ethers

Function: Surfactant - Cleansing Agent

Trade Names:
Pluronic F-108 (BASF)
Synperonic PE/F108 (Uniqema Americas)

Trade Name Mixture:
HVM 4852 Emulsion (OSi Specialties)

POLOXAMER 401

CAS No.: 9003-11-6 (Generic)

Definition: Poloxamer 401 is the polyoxyethylene, polyoxypropylene block polymer that conforms generally to the formula:

$$HO(CH_2CH_2O)_x(CHCH_2O)_y(CH_2CH_2O)_zH$$
$$|$$
$$CH_3$$

in which the average values of x, y and z are respectively 6, 67 and 6.

Information Sources: 21CFR172.808, 21CFR173.340, 21CFR175.105, 21CFR176.180, 21CFR176.200, 21CFR176.210, 21CFR177.1200, 21CFR177.1210, CTFA D, MI-13(7645), TSCA

Chemical Class: Polymeric Ethers

Function: Surfactant - Emulsifying Agent

Trade Names:
Ethox L-121 (Ethox)
Pluronic L-121 (BASF)
Synperonic PE/L121 (Uniqema Americas)

POLOXAMER 402

CAS No.: 9003-11-6 (Generic)

Definition: Poloxamer 402 is the polyoxyethylene, polyoxypropylene block polymer that conforms generally to the formula:

$$HO(CH_2CH_2O)_x(CHCH_2O)_y(CH_2CH_2O)_zH$$
$$|$$
$$CH_3$$

in which the average values of x, y and z are respectively 13, 67 and 13.

Information Sources: 21CFR172.808, 21CFR173.340, 21CFR175.105, 21CFR176.180, 21CFR176.200, 21CFR176.210, 21CFR177.1200, 21CFR177.1210, TSCA

Chemical Class: Polymeric Ethers

Function: Surfactant - Emulsifying Agent

Trade Name:
Ethox L-122 (Ethox)

POLOXAMER 403

CAS No.: 9003-11-6 (Generic)

Definition: Poloxamer 403 is the polyoxyethylene, polyoxypropylene block polymer that conforms generally to the formula:

$$HO(CH_2CH_2O)_x(CHCH_2O)_y(CH_2CH_2O)_zH$$
$$|$$
$$CH_3$$

in which the average values of x, y and z are respectively 21, 67 and 21.

Information Sources: 21CFR172.808, 21CFR173.340, 21CFR175.105, 21CFR176.180, 21CFR176.200, 21CFR176.210, 21CFR177.1200, 21CFR177.1210, MI-13(7645), TSCA

Chemical Class: Polymeric Ethers

Function: Surfactant - Emulsifying Agent

Trade Name:
Pluronic P-123 (BASF)

POLOXAMER 407

CAS No.: 9003-11-6 (Generic)

Definition: Poloxamer 407 is the polyoxyethylene, polyoxypropylene block polymer that conforms generally to the formula:

$$HO(CH_2CH_2O)_x(CHCH_2O)_y(CH_2CH_2O)_zH$$
$$|$$
$$CH_3$$

in which the average values of x, y and z are respectively 98, 67 and 98.

Information Sources: BAN, 21CFR172.808, 21CFR173.340, 21CFR175.105, 21CFR176.180, 21CFR176.200, 21CFR176.210, 21CFR177.1200, 21CFR177.1210, CTFA D, FCC, INN, MI-13(7645), NF XVIII, TSCA, USAN

Chemical Class: Polymeric Ethers

Functions: Surfactant - Emulsifying Agent; Surfactant - Solubilizing Agent

Reported Product Categories: Mouthwashes and Breath Fresheners (Liquids and Sprays); Deodorants (Underarm); Skin Care Preparations, Misc.; Cleansing Products (Cold Creams, Cleansing Lotions, Liquids and Pads); Dentifrices (Aerosol, Liquid, Pastes and Powders); Eye Makeup Removers; Manicuring Preparations, Misc.; Moisturizing Preparations

Trade Names:
Pluracare F-127 (BASF)
Synperonic PE/F127 (Uniqema Americas)

POLOXAMER 105 BENZOATE

Definition: Poloxamer 105 Benzoate is the ester of Poloxamer 105 (q.v.) and benzoic acid.

Chemical Classes: Esters; Polymeric Ethers

Functions: Surfactant - Emulsifying Agent; Surfactant - Solubilizing Agent

Trade Name:
Finsolv PL-355 (Finetex)

POLOXAMER 182 DIBENZOATE

Definition: Poloxamer 182 Dibenzoate is the diester of Poloxamer 182 (q.v.) and benzoic acid.

Chemical Classes: Esters; Polymeric Ethers

Function: Skin-Conditioning Agent - Emollient

Trade Name:
Finsolv PL-62 (Finetex)

POLOXAMINE 304

CAS No.: 11111-34-5 (Generic)

JPN Translation:
ポロキサミン 304

Definition: Poloxamine 304 is the polyoxyethylene, polyoxypropylene block polymer of ethylene diamine that conforms to the formula:

in which the values of x and y are respectively 4 and 3.

Information Sources: 21CFR178.1010, TSCA

Chemical Class: Alkoxylated Amines

Function: Surfactant - Emulsifying Agent

Trade Names:
Synperonic T/304 (Uniqema Americas)
Tetronic 304 (BASF)

POLOXAMINE 504

CAS No.: 11111-34-5 (Generic)

JPN Translation:
ポロキサミン 504

Definition: Poloxamine 504 is the polyoxyethylene, polyoxypropylene block polymer of ethylene diamine that conforms to the formula:

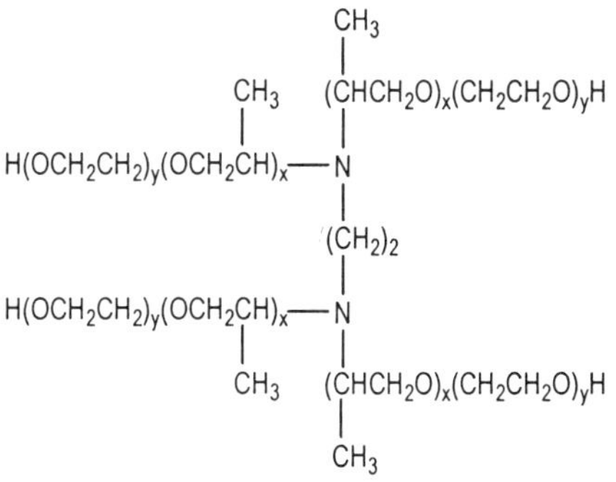

in which the values of x and y are respectively 8 and 7.

Information Source: TSCA

Chemical Class: Alkoxylated Amines

Function: Surfactant - Emulsifying Agent

POLOXAMINE 604

CAS No.: 11111-34-5 (Generic)

Definition: Poloxamine 604 is the polyoxyethylene, polyoxypropylene block polymer of ethylene diamine that conforms generally to the formula:

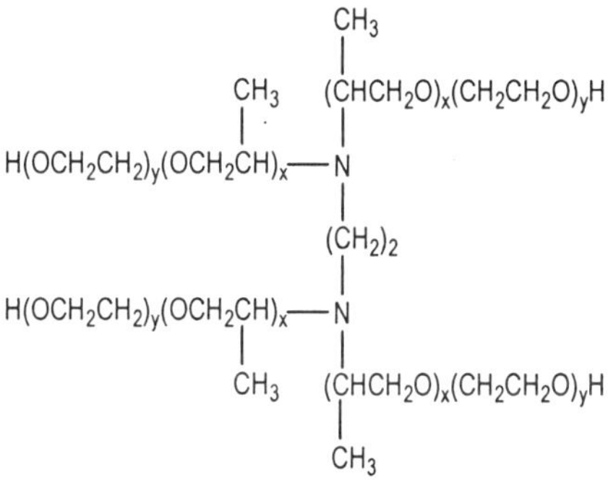

in which the values of x and y are 16 and 51, respectively.

Chemical Class: Alkoxylated Amines

Function: Surfactant - Emulsifying Agent

Trade Names:
Pepol D-301A (Toho)
Pepol D-304 (Toho)

POLOXAMINE 701

CAS No.: 11111-34-5 (Generic)

JPN Translation:
ボロキサミン 701

Definition: Poloxamine 701 is the polyoxyethylene, polyoxypropylene block polymer of

ethylene diamine that conforms to the formula:

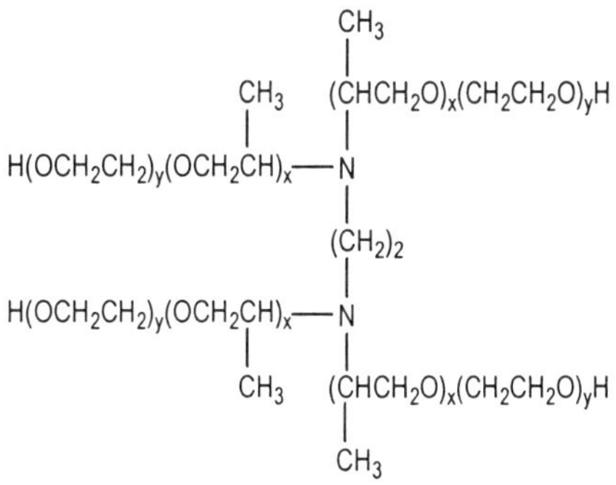

in which the values of x and y are respectively 12 and 2.

Information Source: TSCA

Chemical Class: Alkoxylated Amines

Function: Surfactant - Emulsifying Agent

Trade Names:
Synperonic T/701 (Uniqema Americas)
Tetronic 701 (BASF)

POLOXAMINE 702

CAS No.: 11111-34-5 (Generic)

JPN Translation:
ボロキサミン 702

Definition: Poloxamine 702 is the polyoxyethylene, polyoxypropylene block polymer of ethylene diamine that conforms to the formula:

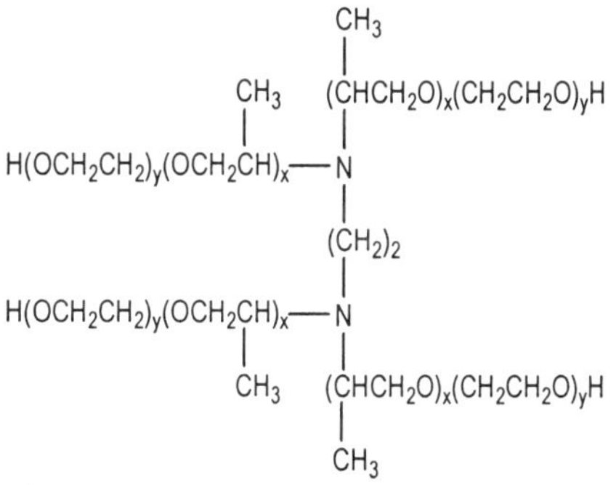

in which the values of x and y are respectively 13 and 4.

Information Source: TSCA

Chemical Class: Alkoxylated Amines

Function: Surfactant - Emulsifying Agent

POLOXAMINE 704

CAS No.: 11111-34-5 (Generic)

JPN Translation:
ボロキサミン 704

Definition: Poloxamine 704 is the polyoxyethylene, polyoxypropylene block polymer of ethylene diamine that conforms to the formula:

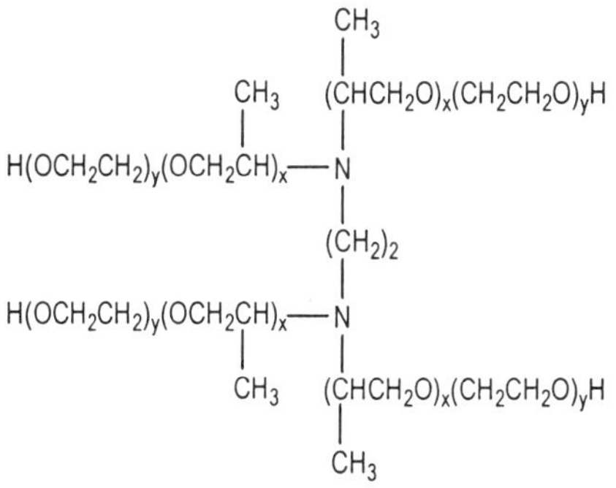

in which the values of x and y are respectively 14 and 12.

Information Source: TSCA

Chemical Class: Alkoxylated Amines

Function: Surfactant - Emulsifying Agent

POLOXAMINE 707

CAS No.: 11111-34-5 (Generic)

JPN Translation:
ボロキサミン 707

Definition: Poloxamine 707 is the polyoxyethylene, polyoxypropylene block polymer of ethylene diamine that conforms to the formula:

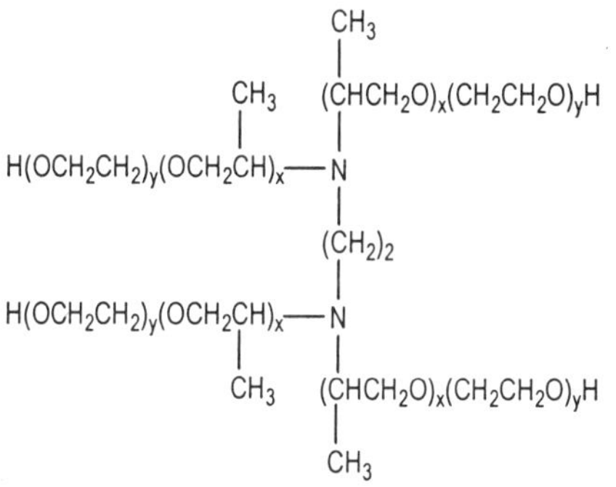

in which the values of x and y are respectively 19 and 47.

Information Source: TSCA

Chemical Class: Alkoxylated Amines

Functions: Surfactant - Emulsifying Agent; Surfactant - Solubilizing Agent

Trade Name:
Synperonic T/707 (Uniqema Americas)

POLOXAMINE 901

CAS No.: 11111-34-5 (Generic)

JPN Translation:
ボロキサミン 901

Definition: Poloxamine 901 is the polyoxy-ethylene, polyoxypropylene block polymer of ethylene diamine that conforms to the formula:

$$\begin{array}{c} CH_3 \\ | \\ CH_3 \quad (CHCH_2O)_x(CH_2CH_2O)_yH \\ | \quad | \\ H(OCH_2CH_2)_y(OCH_2CH)_x\!-\!N \\ | \\ (CH_2)_2 \\ | \\ H(OCH_2CH_2)_y(OCH_2CH)_x\!-\!N \\ | \quad | \\ CH_3 \quad (CHCH_2O)_x(CH_2CH_2O)_yH \\ | \\ CH_3 \end{array}$$

in which the values of x and y are respectively 18 and 2.

Information Source: TSCA

Chemical Class: Alkoxylated Amines

Function: Surfactant - Emulsifying Agent

Trade Name:
Tetronic 901 (BASF)

POLOXAMINE 904

CAS No.: 11111-34-5 (Generic)

JPN Translation:
ボロキサミン 904

Definition: Poloxamine 904 is the polyoxy-ethylene, polyoxypropylene block polymer of ethylene diamine that conforms to the formula:

$$\begin{array}{c} CH_3 \\ | \\ CH_3 \quad (CHCH_2O)_x(CH_2CH_2O)_yH \\ | \quad | \\ H(OCH_2CH_2)_y(OCH_2CH)_x\!-\!N \\ | \\ (CH_2)_2 \\ | \\ H(OCH_2CH_2)_y(OCH_2CH)_x\!-\!N \\ | \quad | \\ CH_3 \quad (CHCH_2O)_x(CH_2CH_2O)_yH \\ | \\ CH_3 \end{array}$$

in which the values of x and y are respectively 19 and 16.

Information Source: TSCA

Chemical Class: Alkoxylated Amines

Function: Surfactant - Emulsifying Agent

Trade Names:
Synperonic T/904 (Uniqema Americas)
Tetronic 904 (BASF)

POLOXAMINE 908

CAS No.: 11111-34-5 (Generic)

JPN Translation:
ボロキサミン 908

Definition: Poloxamine 908 is the polyoxy-ethylene, polyoxypropylene block polymer of ethylene diamine that conforms to the formula:

$$\begin{array}{c} CH_3 \\ | \\ CH_3 \quad (CHCH_2O)_x(CH_2CH_2O)_yH \\ | \quad | \\ H(OCH_2CH_2)_y(OCH_2CH)_x\!-\!N \\ | \\ (CH_2)_2 \\ | \\ H(OCH_2CH_2)_y(OCH_2CH)_x\!-\!N \\ | \quad | \\ CH_3 \quad (CHCH_2O)_x(CH_2CH_2O)_yH \\ | \\ CH_3 \end{array}$$

in which the values of x and y are respectively 22 and 122.

Information Source: TSCA

Chemical Class: Alkoxylated Amines

Functions: Surfactant - Cleansing Agent; Surfactant - Solubilizing Agent

Trade Names:
Synperonic T/908 (Uniqema Americas)
Tetronic 908 (BASF)

POLOXAMINE 1101

CAS No.: 11111-34-5 (Generic)

JPN Translation:
ボロキサミン 1101

Definition: Poloxamine 1101 is the polyoxy-ethylene, polyoxypropylene block polymer of ethylene diamine that conforms to the formula:

$$\begin{array}{c} CH_3 \\ | \\ CH_3 \quad (CHCH_2O)_x(CH_2CH_2O)_yH \\ | \quad | \\ H(OCH_2CH_2)_y(OCH_2CH)_x\!-\!N \\ | \\ (CH_2)_2 \\ | \\ H(OCH_2CH_2)_y(OCH_2CH)_x\!-\!N \\ | \quad | \\ CH_3 \quad (CHCH_2O)_x(CH_2CH_2O)_yH \\ | \\ CH_3 \end{array}$$

in which the values of x and y are respectively 21 and 3.

Information Source: TSCA

Chemical Class: Alkoxylated Amines

Function: Surfactant - Emulsifying Agent

POLOXAMINE 1102

CAS No.: 11111-34-5 (Generic)

JPN Translation:
ボロキサミン 1102

Definition: Poloxamine 1102 is the polyoxy-ethylene, polyoxypropylene block polymer of ethylene diamine that conforms to the formula:

$$\begin{array}{c} CH_3 \\ | \\ CH_3 \quad (CHCH_2O)_x(CH_2CH_2O)_yH \\ | \quad | \\ H(OCH_2CH_2)_y(OCH_2CH)_x\!-\!N \\ | \\ (CH_2)_2 \\ | \\ H(OCH_2CH_2)_y(OCH_2CH)_x\!-\!N \\ | \quad | \\ CH_3 \quad (CHCH_2O)_x(CH_2CH_2O)_yH \\ | \\ CH_3 \end{array}$$

in which the values of x and y are respectively 21 and 7.

Information Source: TSCA

Chemical Class: Alkoxylated Amines

Function: Surfactant - Emulsifying Agent

POLOXAMINE 1104

CAS No.: 11111-34-5 (Generic)

JPN Translation:
ボロキサミン 1104

Definition: Poloxamine 1104 is the polyoxy-ethylene, polyoxypropylene block polymer of ethylene diamine that conforms to the formula:

$$\begin{array}{c} CH_3 \\ | \\ CH_3 \quad (CHCH_2O)_x(CH_2CH_2O)_yH \\ | \quad | \\ H(OCH_2CH_2)_y(OCH_2CH)_x\!-\!N \\ | \\ (CH_2)_2 \\ | \\ H(OCH_2CH_2)_y(OCH_2CH)_x\!-\!N \\ | \quad | \\ CH_3 \quad (CHCH_2O)_x(CH_2CH_2O)_yH \\ | \\ CH_3 \end{array}$$

in which the values of x and y are respectively 21 and 19.

Information Source: TSCA

Chemical Class: Alkoxylated Amines

Function: Surfactant - Emulsifying Agent

POLOXAMINE 1301

CAS No.: 11111-34-5 (Generic)

JPN Translation:
ボロキサミン 1301

Definition: Poloxamine 1301 is the polyoxyethylene, polyoxypropylene block polymer of ethylene diamine that conforms to the formula:

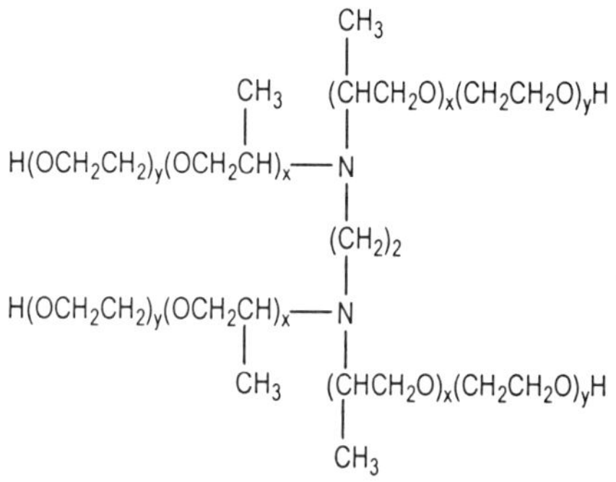

$$H(OCH_2CH_2)_y(OCH_2CH)_x-N \begin{matrix} CH_3 \\ (CHCH_2O)_x(CH_2CH_2O)_yH \end{matrix}$$

with CH_3, $(CH_2)_2$ bridge to second nitrogen bearing $(CHCH_2O)_x(CH_2CH_2O)_yH$ and CH_3 groups

in which the values of x and y are respectively 25 and 3.

Information Source: TSCA

Chemical Class: Alkoxylated Amines

Function: Surfactant - Emulsifying Agent

Trade Name:
Synperonic T/1301 (Uniqema Americas)

POLOXAMINE 1302

CAS No.: 11111-34-5 (Generic)

JPN Translation:
ボロキサミン 1302

Definition: Poloxamine 1302 is the polyoxyethylene, polyoxypropylene block polymer of ethylene diamine that conforms to the formula:

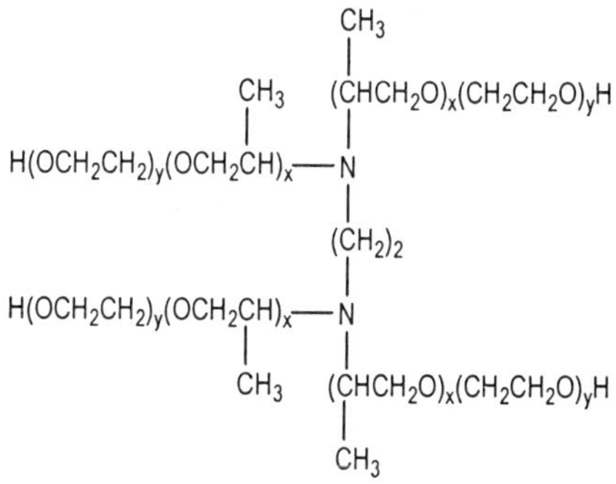

$$H(OCH_2CH_2)_y(OCH_2CH)_x-N \begin{matrix} CH_3 \\ (CHCH_2O)_x(CH_2CH_2O)_yH \end{matrix}$$

with CH_3, $(CH_2)_2$ bridge to second nitrogen

in which the values of x and y are respectively 26 and 8.

Information Source: TSCA

POLOXAMINE 1304

CAS No.: 11111-34-5 (Generic)

JPN Translation:
ボロキサミン 1304

Definition: Poloxamine 1304 is the polyoxyethylene, polyoxypropylene block polymer of ethylene diamine that conforms to the formula:

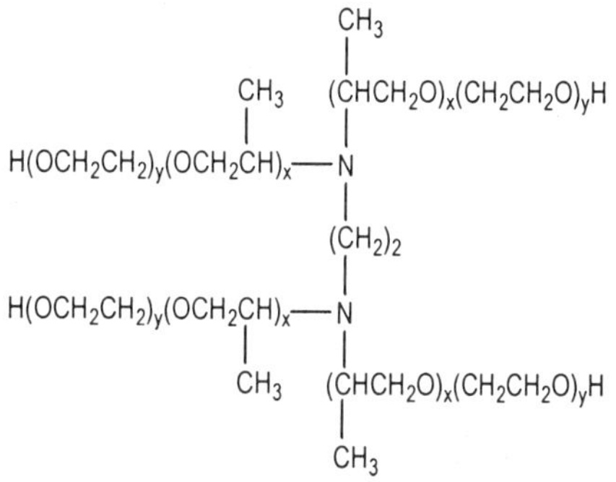

$$H(OCH_2CH_2)_y(OCH_2CH)_x-N \begin{matrix} CH_3 \\ (CHCH_2O)_x(CH_2CH_2O)_yH \end{matrix}$$

with CH_3, $(CH_2)_2$ bridge to second nitrogen

in which the values of x and y are respectively 26 and 24.

Information Source: TSCA

Chemical Class: Alkoxylated Amines

Function: Surfactant - Emulsifying Agent

POLOXAMINE 1307

CAS No.: 11111-34-5 (Generic)

JPN Translation:
ボロキサミン 1307

Definition: Poloxamine 1307 is the polyoxyethylene, polyoxypropylene block polymer of ethylene diamine that conforms to the formula:

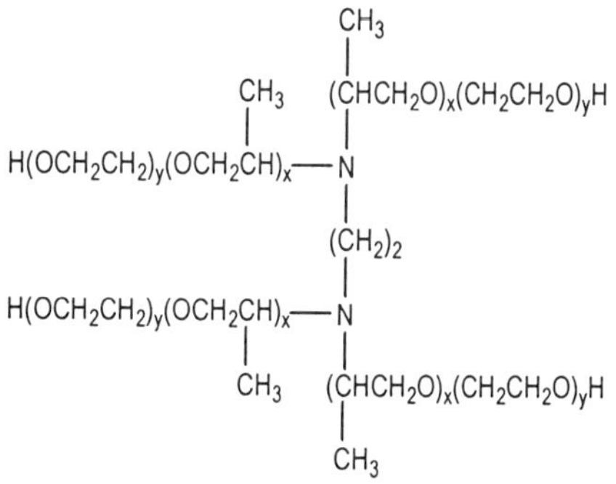

$$H(OCH_2CH_2)_y(OCH_2CH)_x-N \begin{matrix} CH_3 \\ (CHCH_2O)_x(CH_2CH_2O)_yH \end{matrix}$$

with CH_3, $(CH_2)_2$ bridge to second nitrogen

in which the values of x and y are respectively 23 and 74.

Information Source: TSCA

Chemical Class: Alkoxylated Amines

Functions: Surfactant - Emulsifying Agent; Surfactant - Solubilizing Agent

Reported Product Category: Deodorants (Underarm)

Trade Name:
Tetronic 1307 (BASF)

POLOXAMINE 1501

CAS No.: 11111-34-5 (Generic)

JPN Translation:
ボロキサミン 1501

Definition: Poloxamine 1501 is the polyoxyethylene, polyoxypropylene block polymer of ethylene diamine that conforms to the formula:

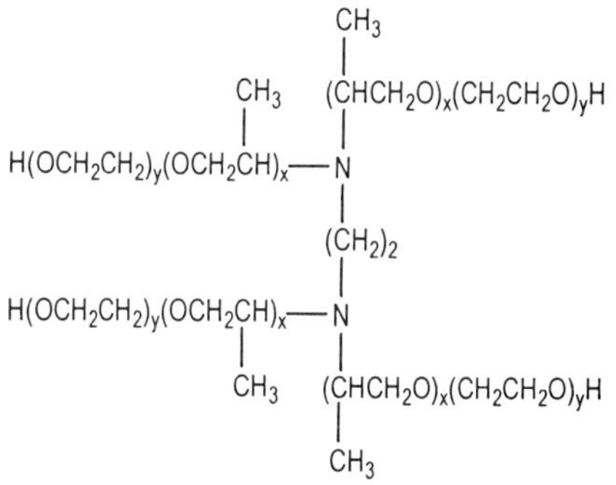

$$H(OCH_2CH_2)_y(OCH_2CH)_x-N \begin{matrix} CH_3 \\ (CHCH_2O)_x(CH_2CH_2O)_yH \end{matrix}$$

with CH_3, $(CH_2)_2$ bridge to second nitrogen

in which the values of x and y are respectively 30 and 4.

Information Source: TSCA

Chemical Class: Alkoxylated Amines

Function: Surfactant - Emulsifying Agent

POLOXAMINE 1502

CAS No.: 11111-34-5 (Generic)

JPN Translation:
ボロキサミン 1502

Definition: Poloxamine 1502 is the polyoxyethylene, polyoxypropylene block polymer of ethylene diamine that conforms to the formula:

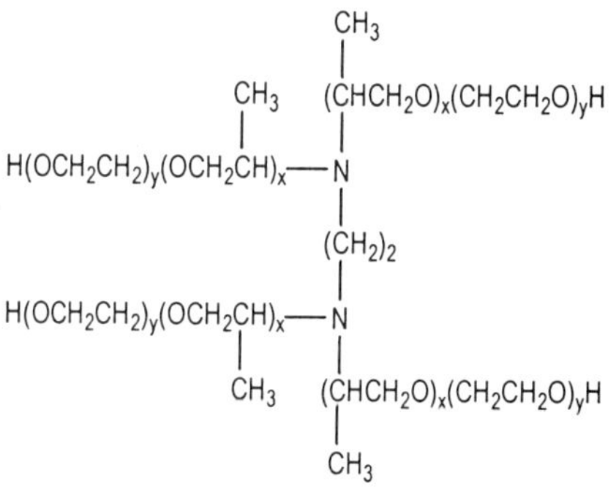

$$H(OCH_2CH_2)_y(OCH_2CH)_x-N \begin{matrix} CH_3 \\ (CHCH_2O)_x(CH_2CH_2O)_yH \end{matrix}$$

with CH_3, $(CH_2)_2$ bridge to second nitrogen

in which the values of x and y are respectively 30 and 10.

Information Source: TSCA

Chemical Class: Alkoxylated Amines

Function: Surfactant - Emulsifying Agent

POLOXAMINE 1504

CAS No.: 11111-34-5 (Generic)

JPN Translation:
ポロキサミン 1504

Definition: Poloxamine 1504 is the polyoxyethylene, polyoxypropylene block polymer of ethylene diamine that conforms to the formula:

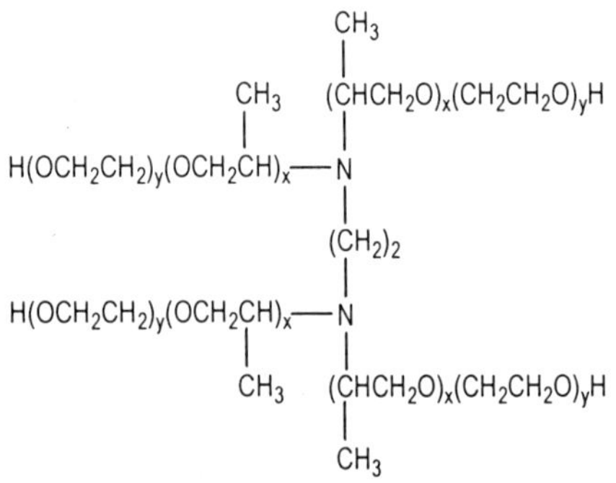

in which the values of x and y are respectively 32 and 28.

Information Source: TSCA

Chemical Class: Alkoxylated Amines

Function: Surfactant - Emulsifying Agent

POLOXAMINE 1508

CAS No.: 11111-34-5 (Generic)

JPN Translation:
ポロキサミン 1508

Definition: Poloxamine 1508 is the polyoxyethylene, polyoxypropylene block polymer of ethylene diamine that conforms to the formula:

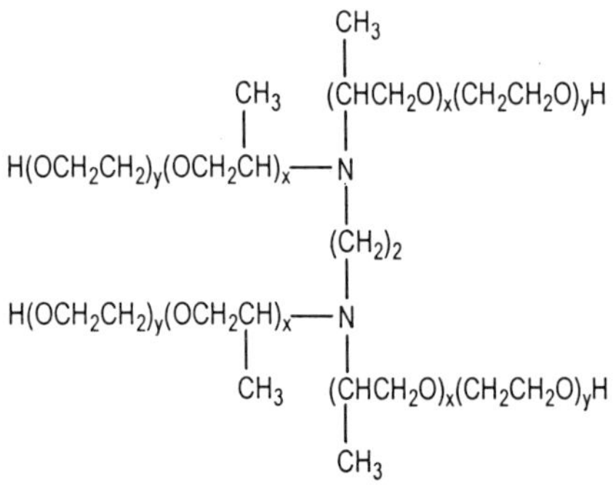

in which the values of x and y are respectively 22 and 122.

Information Source: TSCA

Chemical Class: Alkoxylated Amines

Functions: Surfactant - Cleansing Agent; Surfactant - Solubilizing Agent

POLYACRYLAMIDE

CAS No.: 9003-05-8

JPN Translation:
ポリアクリルアミド

Empirical Formula:
$(C_3H_5NO)_x$

Definition: Polyacrylamide is the polymer of acrylamide monomers that conforms generally to the formula:

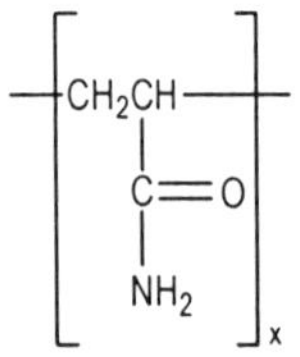

Information Sources: 21CFR172.255, 21CFR173.10, 21CFR173.315, 21CFR175.105, 21CFR176.180, CIR: [SQ], CIR: [SQ] JACT-10(1)1991, EEC(III/1-66), JCIC, JCLS, JSQI, TSCA

Chemical Classes: Amides; Synthetic Polymers

Functions: Binder; Film Former; Hair Fixative

Reported Product Categories: Moisturizing Preparations; Hair Dyes and Colors (All Types Requiring Caution Statements and Patch Tests); Bath Preparations, Misc.; Body and Hand Preparations (Excluding Shaving Preparations); Hair Rinses (Coloring); Bath Capsules; Face and Neck Preparations (Excluding Shaving Preparations); Eyeliners; Indoor Tanning Preparations; Night Skin Care Preparations; Paste Masks (Mud Packs); Skin Care Preparations, Misc.; Bath Oils, Tablets, and Salts; Cleansing Products (Cold Creams, Cleansing Lotions, Liquids and Pads); Foundations; Tonics, Dressings, and Other Hair Grooming Aids; Bath Soaps and Detergents; Eye Makeup Preparations, Misc.; Hair Conditioners; Suntan Preparations, Misc.

Technical/Other Names:
Acrylamide Homopolymer
2-Propenamide, Homopolymer

Trade Names:
Flocare FL 920 3% (SNF)

Flocare T 920 GC (SNF)
Octacare F100 (Associated Octel)

Trade Name Mixtures:
Akypomine P 191 (Kao GmbH)
Anise Softcream (CEP (Solabia))
Apricot Softcream (CEP (Solabia))
Avocado Softcream (CEP (Solabia))
Bitter Orange Softcream (CEP (Solabia))
Cardamom Softcream (CEP (Solabia))
Cinnamon Softcream (CEP (Solabia))
CREAGEL C 13-14 (Creations Couleurs)
Creagel EZ 7 (C.I.T.)
CREAGEL EZ PFC (Creations Couleurs)
Erase (Degussa Care Specialties)
Eucalyptus Softcream (CEP (Solabia))
Ginger Softcream (CEP (Solabia))
Grapefruit Softcream (CEP (Solabia))
Grape Seed Softcream (CEP (Solabia))
Lavender Softcream (CEP (Solabia))
Lime Softcream (CEP (Solabia))
Peppermint Softcream (CEP (Solabia))
Pistachio Softcream (CEP (Solabia))
RheoThik 110 (International Additive)
Rice Bran Softcream (CEP (Solabia))
Rosemary Softcream (CEP (Solabia))
Sepigel 305 (SEPPIC)
Sepigel 501 (SEPPIC)
Sepigel 502 (SEPPIC)
Sweet Almond Softcream (CEP (Solabia))
White Tea Softcream (CEP (Solabia))

POLYACRYLAMIDOMETHYL BENZYLIDENE CAMPHOR

CAS Nos.: 113783-61-2; 147897-12-9

Empirical Formula:
$(C_{21}H_{25}NO_2)_x$

Definition: Polyacrylamidomethyl Benzylidene Camphor is the polymer that conforms generally to the formula:

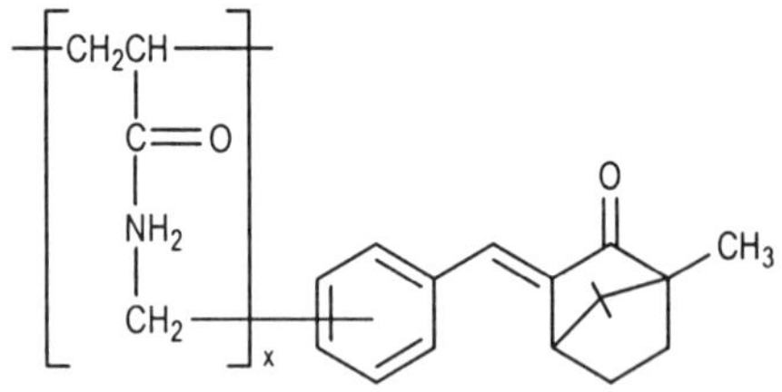

Information Source: EEC(VII/1-11)

Chemical Classes: Amides; Synthetic Polymers

Function: Ultraviolet Light Absorber

Technical/Other Names:
Polymer of N-{(2 and 4)-[2-Oxoborn-3-ylidene)Methyl]Benzyl}Acrylamide
2-Propenamide, N-[[4-[(4,7,7-Trimethyl-3-Oxobicyclo[2.2.1]Hept-2-ylidene)-Methyl]Phenyl]Methyl]-, Homopolymer

Trade Name:
Mexoryl SW (Chimex)

POLYACRYLAMIDOMETHYLPROPANE SULFONIC ACID

CAS No.: 27119-07-9

Empirical Formula:
$(C_7H_{13}NO_4S)_x$

Definition: Polyacrylamidomethylpropane Sulfonic Acid is a homopolymer of acrylamidomethylpropane sulfonic acid that conforms generally to the formula:

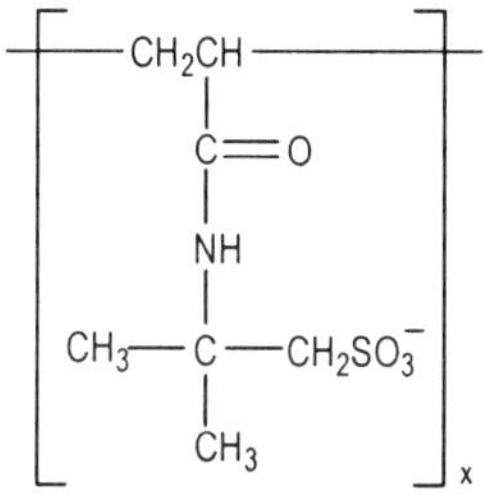

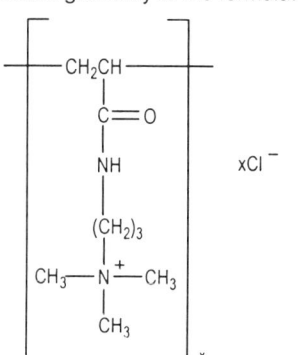

Chemical Classes: Sulfonic Acids; Synthetic Polymers

Functions: Film Former; Suspending Agent - Nonsurfactant

Reported Product Categories: Bath Soaps and Detergents; Bath Preparations, Misc.; Body and Hand Preparations (Excluding Shaving Preparations)

Trade Names:
Cosmedia Polymer HSP-1180 (Cognis Care Chemicals/NJ)
Cosmedia Polymer HSP-1180 (Cognis Care Chemicals/PA)

Trade Name Mixture:
Plexajel ASC (Guardian)

POLYACRYLAMIDOPROPYLTRIMONIUM CHLORIDE

Definition: Polyacrylamidopropyltrimonium Chloride is the quaternary ammonium salt that conforms generally to the formula:

Chemical Class: Quaternary Ammonium Compounds

Function: Hair Conditioning Agent

Trade Name:
Fancor Polyquat LC (Fanning)

POLYACRYLATE-1

Definition: Polyacrylate-1 is a copolymer of vinyl pyrrolidone, dimethylaminoethyl methacrylate, stearyl acrylate and PPG-3 diacrylate monomers.

Chemical Class: Synthetic Polymers

Functions: Film Former; Hair Conditioning Agent

Trade Name:
Cosquat GA467 (Osaka Organic)

POLYACRYLATE-2

CAS No.: 31759-42-9

Definition: Polyacrylate-2 is a copolymer of styrene, acrylamide, octyl acrylate and methyl methacrylate monomers.

Chemical Class: Synthetic Polymers

Function: Film Former

Technical/Other Name:
2-Propenoic Acid, 2-Methyl-, Polymer with Ethenylbenzene, 2-Ethylhexyl 2-Propenoate, Methyl 2-Methyl-2-Propenoate and 2-Propenamide

Trade Name:
Yodosol GH 265 (National Starch)

POLYACRYLATE-3

Definition: Polyacrylate-3 is a copolymer of methacrylic acid, methyl methacrylate, methylstyrene isopropylisocyanate and PEG-40 behenate monomers.

Chemical Class: Synthetic Polymers

Function: Viscosity Increasing Agent - Aqueous

Trade Name:
Viscophobe DB-1000 (Amerchol)

POLYACRYLATE-4

CAS No.: 228863-31-8

Empirical Formula:
$(C_{24}H_{34}O_{10} \cdot C_{24}H_{22}O_9 \cdot C_{24}H_{22}O_9)_x$

Definition: Polyacrylate-4 is a polymer of the acrylic monomers represented by the following structures:

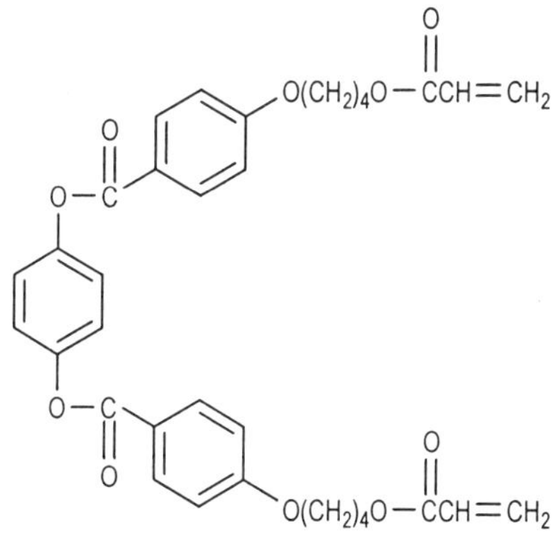

(and)

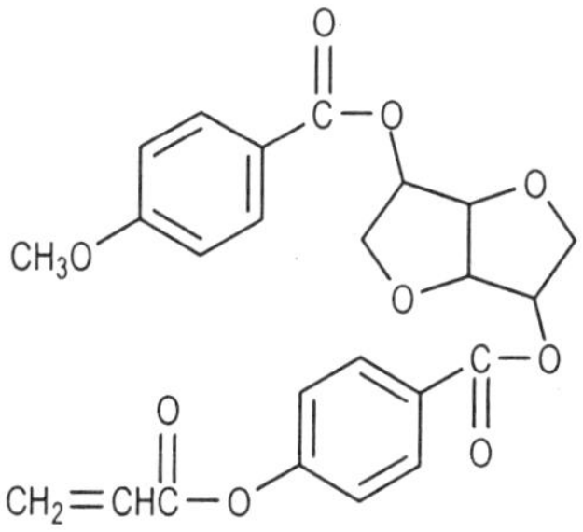

Chemical Class: Synthetic Polymers

Function: Opacifying Agent

Technical/Other Name:
D-Glucitol, 1,4:3,6-Dianhydro-,2-(4-Methoxybenzoate)5-[4-[(1-Oxo-2-Propenyl)Oxy]Benzoate], Polymer with 1,4:3,6-Dianhydro-D-Glucitol 5-(4-Methoxybenzoate) 2-[4-[(1-Oxo-2-Propenyl)Oxy]Benzoate] & 1,4-Phenylene Bis[4-[4-[(1-Oxo-2-Propenyl)OxyButoxy]Benzoate

Trade Name:
Helicone HC (Wacker-Chemie)

POLYACRYLATE-5

Definition: Polyacrylate-5 is a copolymer of styrene, ethylhexyl acrylate, hydroxyethyl acrylate, and one or more monomers of acrylic acid, methacrylic acid, or one of their simple esters.

Chemical Class: Synthetic Polymers

Function: Film Former

Technical/Other Name:
Acrylates/Styrene/Ethylhexyl Acrylate/Hydroxyethyl Acrylate Copolymer

Trade Names:
Avalure EX-612 (Noveon)
Helicone HCP (Wacker-Chemie)

POLYACRYLATE-6

Definition: Polyacrylate-6 is a copolymer of methylmethacrylate, triethoxysilyl-propylmethacrylate, tris(trimethylsiloxy)-silylpropylmethacrylate and acryloyloxyethyl (trimethyl)ammonium chloride monomers.

Chemical Classes: Siloxanes and Silanes; Synthetic Polymers

Functions: Film Former; Hair Fixative

Trade Name:
MK Polymer (Shin Etsu)

POLYACRYLATE-7

CAS No.: 243140-33-2

Definition: Polyacrylate-7 is th copolymer formed by polymerization of 2-acryloylethyl trimethylammonium chloride, acrylic acid, acrylamide, and 2-acrylamido-2-methylprop-ane sulfonic acid.

Chemical Class: Synthetic Polymers

Functions: Hair Conditioning Agent; Skin-Conditioning Agent - Miscellaneous

Trade Name:
OF-420 (WSP Chemicals & Technology)

POLYACRYLATE-9

Definition: Polyacrylate-9 is a copolymer of octylpropenamide, butylaminoethyl-methacrylate, hydroxypropylmethacrylate, and one or more momoners of acrylic acid, methacrylic acid or one of thier simple esters.

Chemical Class: Synthetic Polymers

Functions: Film Former; Hair Fixative

Trade Name:
Amphomer 28-4920 (National Starch)

POLYACRYLIC ACID

CAS No.: 9003-01-4

JPN Translation:
ポリアクリル酸

Empirical Formula:
$(C_3H_4O_2)_x$

Definition: Polyacrylic Acid is the polymer of acrylic acid that conforms generally to the formula:

$$\left[\begin{array}{c} CH_2CH \\ | \\ COOH \end{array} \right]_x$$

Information Sources: 21CFR175.105, 21CFR175.300, 21CFR175.320, 21CFR176.180, CIR: [SQ] IJT 21(SUPPL. 3) 2002, CTFA D, JCIC, JCLS, TSCA

Chemical Classes: Carboxylic Acids; Synthetic Polymers

Functions: Binder; Emulsion Stabilizer; Film Former; Viscosity Increasing Agent - Aqueous

Reported Product Categories: Hair Dyes and Colors (All Types Requiring Caution Statements and Patch Tests); Bath Oils, Tablets, and Salts; Bath Preparations, Misc.; Body and Hand Preparations (Excluding Shaving Preparations); Cleansing Products (Cold Creams, Cleansing Lotions, Liquids and Pads); Skin Care Preparations, Misc.

Technical/Other Names:
Acrylic Acid Homopolymer
Acrylic Acid Resin
Carbopol 907
Carboxypolymethylene
2-Propenoic Acid, Homopolymer

Trade Names:
CREAGEL TN 500 (C.I.T.)
Lowenol P-1030 (Lowenstein)
Lowenol P-1095 (Lowenstein)
Modarez V 200 PX (Synthron)
Modarez V 600 PX (Synthron)
Modarez V 1250 PX (Synthron)
Modarez V 1300 PX (Synthron)
Modarez V 1400 PX (Synthron)
Modarez V 2000 PX (Synthron)
Polycarbophil (Acid) (Block Drug)

POLYAMIDE-1

Definition: Polyamide-1 is a polymer formed by the reaction of isophoronediamine, cyclohexylamine, trimethylolpropane, adipic acid and isophthalic acid.

Chemical Classes: Amides; Synthetic Polymers

Function: Film Former

Trade Name Mixtures:
HMP-1 Astral Pink (Swada)
HMP-5 Blaze (Swada)
HMP-60 Comet Blue (Swada)
HMP-4 Flame Orange (Swada)
HMP-3 Laser Red (Swada)
HMP-27 Lunar Yellow (Swada)
HMP-10 Magenta (Swada)
HMP-2 Nova Red (Swada)
HMP-8 Stellar Green (Swada)
920 Strong Magenta 21 (Sterling)
920 Strong Red 23 (Sterling)
920 Strong Yellow 29 (Sterling)
XSP-21 Strong Magenta (Swada)
XSP-23 Strong Red (Swada)
XSP-29 Strong Yellow (Swada)

POLYAMINOPROPYL BIGUANIDE

CAS No.: 133029-32-0

Definition: Polyaminopropyl Biguanide is the organic compound that conforms to the formula:

$$\left[\begin{array}{c} (CH_2)_3 \\ | \\ NH_2 \\ \cdot \\ HCl \end{array} (CH_2)_3NHCNHCNH(CH_2)_3 \begin{array}{c} NH\ NH \\ || \ \ || \\ \cdot HCl \end{array} \right]_x \begin{array}{c} (CH_2)_3 \\ | \\ NH \\ | \\ NH=C \\ | \\ CN \cdot NH \end{array}$$

Information Source: EEC(VI/1-28)

Chemical Class: Synthetic Polymers

Function: Preservative

Reported Product Category: Baby Products, Misc.

Technical/Other Name:
Imidodicarbonimidic Diamide, N-(3-Amino-propyl)- Homopolymer

Trade Names:
AC Polyaminopropyl Biguanide (Active Concepts)
Cosmocil CQ (Zeneca)
Mikrokill (Arch Personal Care Products)
Mikrokill 20 (Arch Personal Care Products)

Trade Name Mixtures:
Cosmocil AF (Zeneca)
Euxyl K 600 (Schulke & Mayr)
Mikrokill 2 (Arch Personal Care Products)
Mikrokill 300 (Arch Personal Care Products)

POLYAMINOPROPYL BIGUANIDE STEARATE

Definition: Polyaminopropyl Biguanide Stearate is the product obtained by the reaction of Polyaminopropyl Biguanide (q.v.) with stearic acid.

Chemical Class: Synthetic Polymers

Function: Preservative

Trade Name:
Cosmocil Stearate (Zeneca)

POLYAMINO SUGAR CONDENSATE

CAS No.: 120022-92-6

JPN Translation:
糖 / アミノ酸エステル - 1

Definition: Polyamino Sugar Condensate is a condensation product of several of the following sugars: fructose, galactose, glucose, lactose, maltose, mannose, rhamnose, ribose, or xylose, with a mixture consisting of several of the following amino acids: alanine, arginine, aspartic acid, glutamic acid, glycine, histidine, hydroxy-proline, isoleucine, leucine, lysine, methionine, phenylalanine, proline, pyro-glutamic acid, serine, threonine, tyrosine, or valine.

Information Sources: 21CFR701.3, CIR: [S] JACT-1(4)1982, JCIC, JCLS, JSQI

Chemical Class: Synthetic Polymers

Function: Skin-Conditioning Agent - Humectant

Reported Product Categories: Moisturizing Preparations; Bath Preparations, Misc.; Body and Hand Preparations (Excluding Shaving Preparations); Eye Makeup Preparations, Misc.; Skin Care Preparations, Misc.

Technical/Other Name:
Aminoic Acid•Sugar Mixture

Trade Name Mixture:
Aqualizer EJ (Kolmar)

POLYBETA-ALANINE

Empirical Formula:
$(C_3H_5NO)_x$

Definition: Polybeta-Alanine is the polyamide that conforms to the formula:

$$\left[\begin{array}{c} O \\ \| \\ -NHCH_2CH_2C- \end{array}\right]_x$$

Chemical Classes: Amides; Amino Acids; Synthetic Polymers

Functions: Film Former; Hair Conditioning Agent

Technical/Other Name:
Poly β-alanine

Trade Names:
Mexomere PAC (Chimex)
Mexomere PZ (Chimex)

POLYBETA-ALANINE/GLUTARIC ACID CROSSPOLYMER

Definition: Polybeta-Alanine/Glutaric Acid Crosspolymer is the polymer of beta-alanine crosslinked with glutaric acid.

Chemical Class: Synthetic Polymers

Functions: Film Former; Hair Fixative

Trade Name:
Mexomere PAG (Chimex)

POLYBUTENE

CAS Nos.: 9003-28-5; 9003-29-6

JPN Translation:
ポリブテン

Empirical Formula:
$(C_4H_8)_x$

Definition: Polybutene is the polymer formed by the polymerization of a mixture of iso- and normal butenes.

Information Sources: 21CFR175.105, 21CFR175.125, 21CFR175.300, 21CFR176.170, 21CFR176.180, 21CFR176.210, 21CFR177.1570, 21CFR177.2260, 21CFR177.2600, 21CFR177.2800, 21CFR178.3570, 21CFR178.3740, 21CFR178.3750, 21CFR178.3860, 21CFR178.3910, CIR: [S] JACT-1(4)1982, CTFA D, JCLS, TSCA

Chemical Class: Synthetic Polymers

Functions: Binder; Epilating Agent; Viscosity Increasing Agent - Nonaqueous

Reported Product Categories: Lipsticks; Mascara; Makeup Preparations (Not eye), Misc.; Eye Shadows; Eye Makeup Preparations, Misc.; Foundations; Eyebrow Pencils; Eyeliners; Moisturizing Preparations

Technical/Other Name:
1-Butene, Homopolymer

Trade Names:
Indopol H-100 (Amoco Chemical)
Indopol H-100 (Lipo)
Indopol H-300 (Lipo)
Indopol H-1500 (Lipo)
Indopol H-1900 (Lipo)

Trade Name Mixtures:
Covaclear (LCW)
Covagloss (LCW)
Lipshine (LCW)
Tixogel PIB - 1461 (Sud-Chemie, United Catalysts)

POLYBUTYL ACRYLATE

CAS No.: 9003-49-0

JPN Translation:
ポリアクリル酸ブチル

Empirical Formula:
$(C_7H_{12}O_2)_x$

Definition: Polybutyl Acrylate is a polymer of n-butyl acrylate that conforms generally to the formula:

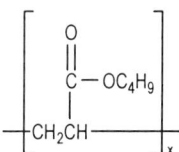

Information Source: JCLS

Chemical Class: Synthetic Polymers

Functions: Binder; Film Former

Technical/Other Names:
Acrylic Acid, Butyl Ester, Polymer
Butyl Acrylate Polymer
2-Propenoic Acid, Butyl Ester, Homo-polymer

Trade Name:
Acronal 4 F (BASF AG)

POLYBUTYLENE GLYCOL-10

Definition: Polybutylene Glycol-10 is a polymer of butylene glycol that conforms generally to the formula:

$$HO(C_4H_8O)_nH$$

where n has an average value of 10.

Chemical Classes: Polyols; Synthetic Polymers

Function: Skin-Conditioning Agent - Miscellaneous

Trade Name:
Uniol B-700 (NOF.)

POLYBUTYLENE GLYCOL/PPG-9/1 COPOLYMER

Definition: Polybutylene Glycol/PPG-9/1 Copolymer is a copolymer produced by the interaction of an average of 9 moles of butylene glycol and 1 mole of propylene glycol.

Chemical Class: Synthetic Polymers

Function: Skin-Conditioning Agent - Miscellaneous

Trade Name:
Uniol PB-700 (NOF)

POLYBUTYLENE TEREPHTHALATE

CAS Nos.: 24968-12-5; 26062-94-2

Empirical Formula:
$(C_8H_6O_4 \cdot C_4H_{10}O_2)_x$

Definition: Polybutylene Terephthalate is the polymer that conforms generally to the formula:

Chemical Class: Synthetic Polymers

Functions: Film Former; Hair Fixative; Viscosity Increasing Agent - Nonaqueous

Technical/Other Names:
1,4-Benzenedicarboxylic Acid, Polymer with 1,4-Butanediol
1,4-Butanediol-Terephthalic Acid Copolymer
Butylene Glycol-Terephthalic Acid Copolymer
Butylene Glycol-Terephthalic Acid Polymer

Trade Name Mixtures:
AEC Glitter Opal (A & E Connock)
SiLiGlit Iris Glitter (Series 400) (Sigmund Lindner GmbH)

POLYBUTYL METHACRYLATE

Definition: Polybutyl Methacrylate is the homopolymer of butyl methacrylate.

Information Source: JCLS

Chemical Class: Synthetic Polymers

Function: Film Former

POLY C10-30 ALKYL ACRYLATE

Definition: Poly C10-30 Alkyl Acrylate is a polymer of the ester of acrylic acid and C10-30 alcohol.

Chemical Class: Synthetic Polymers

Functions: Binder; Emulsion Stabilizer; Viscosity Increasing Agent - Nonaqueous

Trade Names:
Doresco IPA 13-1 (Landec)
Doresco IPA 13-6 (Landec)
Intelimer IPA 13-1 (Landec)
Intelimer IPA 13-6 (Landec)

POLYCAPROLACTONE

CAS Nos.: 24980-41-4; 25248-42-4

Definition: Polycaprolactone is a polymer of caprolactone. It conforms generally to the formula:

Chemical Classes: Esters; Synthetic Polymers

Function: Suspending Agent - Nonsurfactant

Technical/Other Names:
Caprolactone Homopolymer
6-Hydroxycaproic Acid Homopolymer
2-Oxepanone Homopolymer
Poly(oxy(1-oxo-1,6-Hexanediyl

POLYCHLOROTRIFLUOROETHYLENE

CAS No.: 9002-83-9

Empirical Formula:
$(C_2ClF_3)_x$

Definition: Polychlorotrifluoroethylene is the polymer of chlorotrifluoroethylene that conforms generally to the formula:

Information Sources: 21CFR177.1380, TSCA

Chemical Classes: Halogen Compounds; Synthetic Polymers

Functions: Film Former; Skin-Conditioning Agent - Occlusive

Technical/Other Names:
Chlorotrifluoroethene, Homopolymer
Chlorotrifluoroethylene Polymer
Ethene, Chlorotrifluoro-, Homopolymer
Poly(Ethylene Trifluoride Chloride)

POLYCYCLOPENTADIENE

CAS Nos.: 25568-84-7; 68132-00-3

Definition: Polycyclopentadiene is a homopolymer of cyclopentadiene.

Chemical Class: Synthetic Polymers

Functions: Epilating Agent; Viscosity Increasing Agent - Aqueous; Viscosity Increasing Agent - Nonaqueous

Technical/Other Name:
1,3-Cyclopentadiene Homopolymer

POLYDECENE

CAS Nos.: 25189-70-2; 37309-58-3

Empirical Formula:
$(C_{10}H_{20})_x$

Definition: Polydecene is the polymer formed by the polymerization of decene. It conforms to the formula:

Chemical Class: Synthetic Polymers

Function: Skin-Conditioning Agent - Occlusive

Reported Product Categories: Lipsticks; Body and Hand Preparations (Excluding Shaving Preparations); Face Powders

Technical/Other Name:
Decene, Homopolymer

Trade Names:
Pionier 0030 SYN-FG (Hansen & Rosenthal)
PureSyn 1000 (ExxonMobil/Synthetics)
PureSyn 3000 (ExxonMobil/Synthetics)
Synton PAO-100 (Uniroyal)

Trade Name Mixtures:
IBR-AAC Anti-Aging Carotenoids (IBR)
IBR-CLC ColorLess Carotenoids (IBR)
PIONIER 17106 Mineral Oil Free Vaseline (Hansen & Rosenthal)
PIONIER PLW Polydecene Oleo Gel (Hansen & Rosenthal)

POLYDEXTROSE

CAS No.: 68424-04-4

Definition: Polydextrose is a random polymer formed from the condensation of D-glucose.

Information Sources: 21CFR172.841, FCC, MI-13(7648), USAN

Chemical Class: Carbohydrates

Function: Bulking Agent

Reported Product Category: Skin Care Preparations, Misc.

POLYDIETHYLENEGLYCOL ADIPATE/IPDI COPOLYMER

CAS No.: 55636-50-5

Definition: Polydiethyleneglycol Adipate/IPDI Copolymer is a copolymer of polydiethylene glycol adipate and isophorone diisocyanate monomers.

Information Source: TSCA

Chemical Class: Synthetic Polymers

Function: Film Former

Technical/Other Name:
Hexanedioic Acid, Polymer with 5-Isocyanato-1-(Isocyanatomethyl)-1,3,3-Triemthylcyclohexane and 2,2'Oxybis [Ethanol]

Trade Name:
Polyderm PPI-PE (Alzo)

POLYDIHYDROXYINDOLE

CAS No.	**EINECS No.**
8049-97-6	232-473-6

Definition: Polydihydroxyindole is a polymer of dihydroxyindole that conforms to the formula:

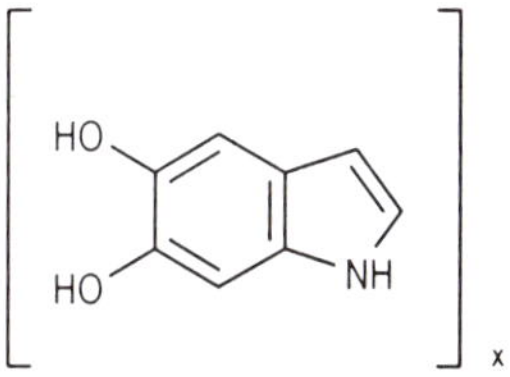

Chemical Classes: Heterocyclic Compounds; Phenols; Synthetic Polymers

Function: Skin-Conditioning Agent - Miscellaneous

Reported Product Categories: Foundations; Eyebrow Pencils; Suntan Gels, Creams, and Liquids; Mascara

Trade Name:
Mexanyl GW (Chimex)

POLYDIMETHYLAMINOETHYL METHACRYLATE

Empirical Formula:
$(C_8H_{15}NO_2)_x$

Definition: Polydimethylaminoethyl Methacrylate is a polymer of Dimethylaminoethyl Methacrylate (q.v.).

Information Source: TSCA

Chemical Classes: Esters; Synthetic Polymers

Function: Film Former

Technical/Other Names:
2-Methyl-2-Propenoic Acid, 2-(Dimethylamino)Ethyl Ester, Homopolymer
2-Propenoic Acid, 2-Methyl-, 2-(Dimethylamino)Ethyl Ester, Homopolymer

POLYDIMETHYLSILOXYETHYL DIMETHICONE/METHICONE COPOLYMER

Definition: Polydimethylsiloxyethyl Dimethicone/Methicone Copolymer is a copolymer of polydimethylsiloxyethyl dimethicone and Methicone (q.v.) .

Chemical Classes: Siloxanes and Silanes; Synthetic Polymers

Function: Skin-Conditioning Agent - Occlusive

Trade Name:
KF - 9906 (Shin-Etsu Chemical Co.)

POLYDIMETHYLSILOXY PEG/PPG-24/19 BUTYL ETHER SILSESQUIOXANE

CAS No.: 68554-65-4

Definition: Polydimethylsiloxy PEG/PPG-24/19 Butyl Ether Silsesquioxane is the silicone polymer that conforms generally to the formula:

Chemical Classes: Siloxanes and Silanes; Synthetic Polymers

Functions: Skin-Conditioning Agent - Humectant; Surfactant - Cleansing Agent; Surfactant - Emulsifying Agent; Surfactant - Suspending Agent

Technical/Other Name:
Siloxanes and Silicones, di-Me, Polymers with Me Silsesquioxanes and Polyethylene-Polypropylene Glycol Mono-Bu Ether

Trade Name:
FZ-2403 (Nippon Unicar)

POLYDIMETHYLSILOXY PPG-13 BUTYL ETHER SILSESQUIOXANE

Definition: Polydimethylsiloxy PPG-13 Butyl Ether Silsesquioxane is the silicone polymer that conforms generally to the formula:

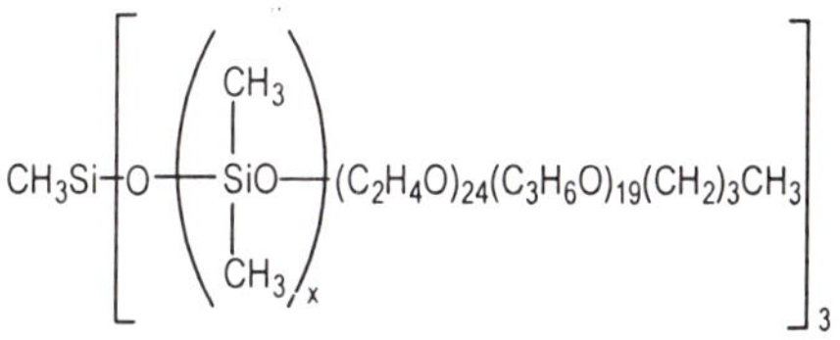

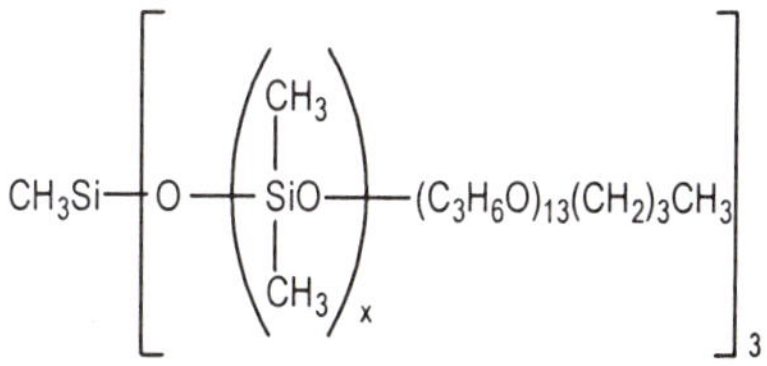

Chemical Class: Siloxanes and Silanes

Functions: Hair Conditioning Agent; Humectant; Surfactant - Cleansing Agent; Surfactant - Emulsifying Agent; Surfactant - Suspending Agent

Technical/Other Name:
Siloxanes and Silicones, di-Me, Polymers with Me Silsesquioxanes and Polypropylene Glycol Mono-Bu Ether

Trade Name:
FZ-2407 (Nippon Unicar)

POLYDIPENTENE

CAS No.: 9003-73-0

Definition: Polydipentene is the product formed by the polymerization of terpene hydrocarbons.

Chemical Class: Synthetic Polymers

Functions: Binder; Viscosity Increasing Agent - Nonaqueous

Technical/Other Names:
Cyclohexene, 1-Methyl-4-(1-Methylethenyl)-, Homopolymer
Polylimonene
Poly-1,8-Paramenthadiene

Trade Names:
Piccolyte C115 (Hercules)
Zonarez 7085 (Arizona)

POLYDODECANAMIDEAMINIUM TRIAZADIPHENYLETHENESULFONATE

CAS No.: 367952-88-3

Definition: Polydodecanamideaminium Triazadiphenylethenesulfonate is the reaction product of Nylon-12 (q.v.) and the compound that conforms to the formula:

The inclusion of any compound in the *Dictionary and Handbook* does not indicate that use of that substance as a cosmetic ingredient complies with the laws and regulations governing such use in the United States or any other country.

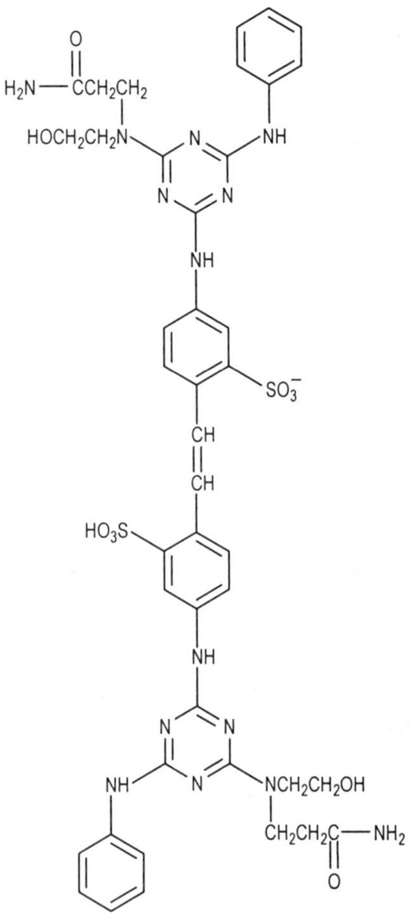

Chemical Classes: Heterocyclic Compounds; Synthetic Polymers

Function: Colorant

Trade Name:
Lipolight OAP (Lipo)

Trade Name Mixtures:
LipoLight OAP/C (Lipo)
LipoLight OAP/PVA (Lipo)

POLYEPSILON-LYSINE

CAS No.: 28211-04-3

Definition: Polyepsilon-Lysine is the organic compound that conforms generally to the formula:

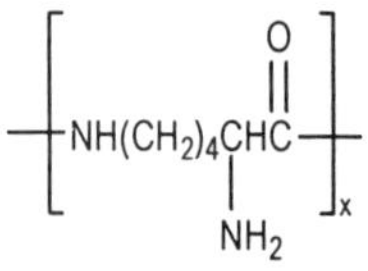

Chemical Classes: Amides; Synthetic Polymers

Functions: Hair Conditioning Agent; Skin-Conditioning Agent - Miscellaneous; Surfactant - Emulsifying Agent

Technical/Other Name:
Poly[Imino[(2S)-2-Amino-1-Oxo-1,6-Hexanediyl)]]

Trade Name:
e-polylysine (Chisso)

POLYESTER-1

Definition: Polyester-1 is a copolymer of t-butylacrylamide, cyclohexane dimethanol, diethylene glycol, isophthalic acid, sodium isophthalic acid sulfonate and one or more monomers of acrylic acid, methacrylic acid or one of their simple esters.

Chemical Class: Synthetic Polymers

Functions: Film Former; Hair Fixative

POLYESTER-2

Definition: Polyester-2 is a copolymer of cyclohexanedicarboxylic acid, Adipic Acid (q.v.), Trimethylolpropane (q.v.) and Hexanediol (q.v.).

Chemical Class: Synthetic Polymers

Function: Film Former

Trade Name Mixture:
RL-4051-1 (Akzo Nobel Coatings)

POLYESTER-3

Definition: Polyester-3 is a polymer formed by the reaction of ethylene glycol, terephthalic acid, isophthalic acid, cyclohex-anedimethanol and norboranediamine.

Chemical Class: Synthetic Polymers

Function: Film Former

Trade Name Mixtures:
Dermaglo Red (Day-Glo)
Dermaglo Red 222 (Day-Glo)
Dermaglo Red 228 (Day-Glo)
Dermaglo Red 422 (Day-Glo)
Dermaglo Red 428 (Day-Glo)
Dermaglo Violet 402 (Day-Glo)
Dermaglo Yellow 311 (Day-Glo)

POLYETHER-1

Definition: Polyether-1 is a copolymer of PEG-180 (q.v.), Dodoxynol-5 (q.v.), PEG-25 tristyrylphenol and tetramethoxy-methylglycouril monomers.

Chemical Class: Synthetic Polymers

Function: Viscosity Increasing Agent - Aqueous

Trade Names:
Pure Thix 1442 (Sud-Chemie, United Catalysts)
Pure-Thix HH (Sud-Chemie, United Catalysts)

POLYETHYLACRYLATE

CAS No.: 9003-32-1

JPN Translation:
ポリアクリル酸エチル

Empirical Formula:
$(C_5H_8O_2)_x$

Definition: Polyethylacrylate is the polymer of ethyl acrylate that conforms generally to the formula:

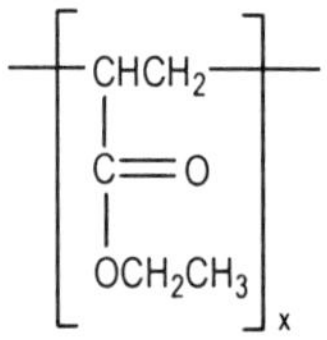

Information Sources: 21CFR175.105, 21CFR175.300, 21CFR175.320, 21CFR176.180, 21CFR177.1010, 21CFR177.2260, 21CFR178.3790, JCLS, JSQI, TSCA

Chemical Classes: Esters; Synthetic Polymers

Functions: Binder; Film Former; Hair Fixative; Suspending Agent - Nonsurfactant

Reported Product Categories: Mascara; Face Powders

Technical/Other Names:
Ethyl Acrylate Homopolymer
2-Propenoic Acid, Ethyl Ester, Homo-polymer

POLYETHYLENE

CAS No.: 9002-88-4

JPN Translation:
ポリエチレン

Empirical Formula:
$(C_2H_4)_x$

Definition: Polyethylene is a polymer of ethylene monomers that conforms generally to the formula:

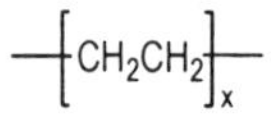

Information Sources: 21CFR73.1, 21CFR73.2180, 21CFR172.615, 21CFR173.20, 21CFR175.105, 21CFR175.300, 21CFR176.180,

21CFR176.200, 21CFR176.210, 21CFR177.1200, 21CFR177.1520, 21CFR177.2600, 21CFR178.3570, 21CFR178.3850, 21CFR310.545, 21CFR349.12, 21CFR573.780, 21CFR573.800, FCC, JCIC, JCLS, JSCI, JSQI, MI-13(7650), TSCA

Chemical Class: Synthetic Polymers

Functions: Abrasive; Adhesive; Binder; Bulking Agent; Emulsion Stabilizer; Film Former; Oral Care Agent; Viscosity Increasing Agent - Nonaqueous

Reported Product Categories: Lipsticks; Bath Oils, Tablets, and Salts; Cleansing Products (Cold Creams, Cleansing Lotions, Liquids and Pads); Face Powders; Eyeliners; Skin Care Preparations, Misc.; Eye Makeup Preparations, Misc.; Eyebrow Pencils; Bath Preparations, Misc.; Body and Hand Preparations (Excluding Shaving Preparations); Personal Cleanliness Products, Misc.; Moisturizing Preparations; Paste Masks (Mud Packs); Blushers (All types); Eye Shadows; Foundations; Mascara; Powders (Dusting and Talcum, Excluding Aftershave Talcs); Bath Capsules; Eye Makeup Removers; Makeup Bases; Rouges; Face and Neck Preparations (Excluding Shaving Preparations); Makeup Fixatives; Makeup Preparations (Not eye), Misc.; Tonics, Dressings, and Other Hair Grooming Aids; Night Skin Care Preparations; Bath Soaps and Detergents; Deodorants (Underarm); Foot Powders and Sprays; Hair Coloring Preparations, Misc.; Skin Fresheners; Suntan Gels, Creams, and Liquids

Technical/Other Names:
Ethene, Homopolymer
High Melting Point Polyethylene Powder
Polyethylene Powder
Polyethylene Wax
Synthetic Wax

Trade Names:
Abifor 1070 (Billeter)
Abifor 1200 (Billeter)
A-C Polyethylene 6 (Honeywell)
A-C Polyethylene 7 (Honeywell)
A-C Polyethylene 8 (Honeywell)
A-C Polyethylene 9 (Honeywell)
A-C Polyethylene 617 (Honeywell)
A-C Polyethylene 6A (Honeywell)
A-C Polyethylene 7A (Honeywell)
A-C Polyethylene 8A (Honeywell)
A-C Polyethylene 9A (Honeywell)
A-C Polyethylene 617A (Honeywell)
ACumist B-6 (Honeywell)
ACumist B-12 (Honeywell)
ACumist B-18 (Honeywell)
ACumist C-5 (Honeywell)
ACumist C-12 (Honeywell)
ACumist C-18 (Honeywell)
AEC Polyethylene (A & E Connock)

AEC Polyethylene Spheres (A & E Connock)
CL-2080 (Kobo)
CREABASE 70 (Creations Couleurs)
Epolene Polyethylene Wax (Eastman Chemical)
ESP Polyethylene C-30 (Earth Supplied Products)
Fabras 30 (Fabriquimica)
Hoechst Wax PE130 (Clariant)
Hoechst Wax PE130 (Clariant GmbH, Personal Care)
Hoechst Wax PE 190 (Clariant)
Hoechst Wax PE 190 (Clariant GmbH, Personal Care)
Hoechst Wax PE 520 (Clariant)
Hoechst Wax PE 520 (Clariant GmbH, Personal Care)
Inducos 13-14 Series (Induchem)
LipoSatin PE35 (Lipo)
Liposcrub HDW-200 (Lipo)
Liposcrub HDW-400 (Lipo)
Lupolen 1800 SP, Low Density Polyethylene (Kjemi)
Metapearls-1-STD (Floratech)
Micropoly 200 (Micro Powders)
Micropoly 200 (Presperse)
Micropoly 220 (Micro Powders)
Micropoly 220 (Presperse)
Micropoly 220L (Presperse)
Micropoly 250S (Micro Powders)
Microscrub 20PC (Micro Powders)
Microscrub 50PC (Micro Powders)
Microscrub 80PC (Micro Powders)
Microscrub 100PC (Micro Powders)
Microscrub H-50PCS (Micro Powders)
Polyscrub (Chrystal)
Polywax 1000 Polyethylene (Baker Petrolite)
Siltek GR Polymer (Baker Petrolite)
Siltek PL Polymer (Baker Petrolite)
Siltek M Polymer (Baker Petrolite)
Toshiki PFL-White (Nikko)
Toshiki PFM-White (Nikko)
Toshiki PFS-White (Nikko)
Ultrapure LDP (Ultra Chemical)

Trade Name Mixtures:
AEC Polyethylene HPC Granules (A & E Connock)
Cera Albalate 105 (Koster Keunen)
Covagloss (LCW)
CREABASE NTL 80 (Creations Couleurs)
Crodabase SQ (Croda Brasil)
Ethylene MC Green (Ichimaru Pharcos)
Etylene MC Blue (Ichimaru Pharcos)
Etylene MC Orange (Ichimaru Pharcos)
Etylene MC Pink (Ichimaru Pharcos)
Etylene MC Yellow (Ichimaru Pharcos)
Iridescent Glitter IF4101 (Mitsubishi Petrochemical)
Lipcare Wax 7782 (Kahl)
Liposcrub LDB-315 (Lipo)
Metasomes-1-STD (Floratech)
Metaspheres-1-STD (Floratech)

Microcare 300 (Micro Powders)
Microcare 310 (Micro Powders)
Microsilk 418 (Micro Powders)
Microsilk 418 (Presperse)
Microsilk 419 (Micro Powders)
MPG Granule Blue (Ichimaru Pharcos)
MPG Granule Green (Ichimaru Pharcos)
MPG Granule Pink (Ichimaru Pharcos)
MPG Granule Red (Ichimaru Pharcos)
MPG Granule White (Ichimaru Pharcos)
MPG Granule Yellow (Ichimaru Pharcos)
PEC-1414 (Sud-Chemie, United Catalysts)
Pionier PLW (Hansen & Rosenthal)
PIONIER PLW Polydecene Oleo Gel (Hansen & Rosenthal)
PM Wax 82 (Nikko Rica)
Tixogel IDP - 1388 (Sud-Chemie, United Catalysts)
Tixogel PIB - 1461 (Sud-Chemie, United Catalysts)
Toshiki Mica HP-18K (Nikko)
Toshiki Nylon AP-18K (Nikko)
Toshiki PFL-Green-7002 (Nikko)
Toshiki PFL-Red-7044 (Nikko)
Toshiki PFM-Blue-6010 (Nikko)
Toshiki PFM-Green-6018 (Nikko)
Toshiki PFM-Pink 6055 (Nikko)
Toshiki PFM-Red-6011 (Nikko)
Toshiki PFM-Yellow-6017 (Nikko)
Toshiki PFS-Green-7008 (Nikko)
Toshiki PFS-Red-7001 (Nikko)
Ultracolor Soft Focus (Ultra Chemical)
Unicid 425 Acid (Baker Petrolite)
Unicid 700 Acid (Baker Petrolite)
Unigel (Chemyunion)

POLYETHYLENE/ISOPROPYL MALEATE/ MA COPOLYOL

Definition: Polyethylene/Isopropyl Maleate/ MA Copolyol is a copolymer formed by the polymerization of ethylene and maleic acid, partially esterified with isopropanol and further ethoxylated and/or propoxylated.

Chemical Class: Synthetic Polymers

Functions: Emulsion Stabilizer; Viscosity Increasing Agent - Aqueous

Trade Name:
Ceramer 67 Polymer (Baker Petrolite)

POLYETHYLENE NAPHTHALATE

Definition: Polyethylene Naphthalate is the polymer that conforms generally to the formula:

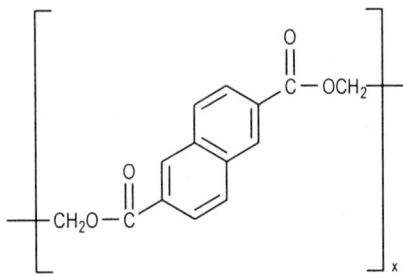

Chemical Class: Synthetic Polymers

Function: Film Former

Trade Name Mixture:
3M Highly Reflective Visible Mirror Film, 3M Colored Mirror (3M)

POLYETHYLENE/POLYETHYLENE TERE-PHTHALATE LAMINATED POWDER

Definition: Polyethylene/Polyethylene Terephthalate Laminated Powder is the powder obtained from the laminated film of Polyethylene (q.v.) and Polyethylene Terephthalate (q.v.). *See "Regulatory and Ingredient Use Information," regarding use of Japan Trivial names in Volume 1, Introduction, Part A.*

Information Source: JCLS

Chemical Class: Synthetic Polymers

Function: Not Reported

POLYETHYLENE/ POLYPENTAERYTHRITYL TERE-PHTHALATE LAMINATED POWDER

Definition: Polyethylene/ Polypentaerythrityl Terephthalate Laminated Powder is the powder obtained from the laminated film of Polyethylene (q.v) and Polypentaerythrityl Terephthalate. *See "Regulatory and Ingredient Use Information," regarding use of Japan Trivial names in Volume 1, Introduction, Part A.*

Information Source: JCLS

Chemical Class: Synthetic Polymers

Function: Not Reported

POLYETHYLENE TEREPHTHALATE

CAS No.: 25038-59-9

Empirical Formula:
$(C_{10}H_8O_4)_n$

Definition: Polyethylene Terephthalate is the organic compound that conforms to the formula:

Information Sources: 21CFR177.1630, 21CFR177.2260, 21CFR177.2800, MI-13 (7652)

Chemical Class: Synthetic Polymers

Functions: Adhesive; Film Former; Hair Fixative; Viscosity Increasing Agent - Nonaqueous

Reported Product Categories: Foundations; Hair Coloring Preparations, Misc.; Indoor Tanning Preparations; Leg and Body Paints; Makeup Preparations (Not eye), Misc.; Nail Polish and Enamels

Technical/Other Names:
Ethylene Glycol-Terephthalic Acid Polymer
Poly(Oxy-1,2-Ethanediyloxycarbonyl-1,4-Phenylenecarbonyl)

Trade Names:
SiLiGlit-Hologram Glitter Gold (Sigmund Lindner GmbH)
SiLiGlit-Hologram Glitter Silver (Sigmund Lindner GmbH)
SiLiGlit-Polyester Glitter (#3) (Sigmund Lindner GmbH)
SiLiGlit-Polyester Glitter (Red 122) (Sigmund Lindner GmbH)

Trade Name Mixtures:
AEC Glitter Silver (A & E Connock)
Creasparkles Coloures Green (Creations Couleurs)
Creasparkles Colours Blue Aqua (Creations Couleurs)
Creasparkles Colours Coral Pink (Creations Couleurs)
Creasparkles Colours Lilac Blue (Creations Couleurs)
Creasparkles Colours Magenta (C.I.T.)
Creasparkles Colours Mint (Creations Couleurs)
Creasparkles Colours Raspberry (Creations Couleurs)
Creasparkles Colours Rubis (Creations Couleurs)
Creasparkles Hologram Blue Sky (Creations Couleurs)
Creasparkles Hologram Silver (Creations Couleurs)
Creasparkles Metallic Gold 600 (Creations Couleurs)
Creasparkles Metallic Reddish Gold 590 (Creations Couleurs)
Creasparkles Metallic Silver 700 (Creations Couleurs)
Creasparkles Metallic Silver 700 THIN (Creations Couleurs)
Creasparlkes Colours Magenta (Creations Couleurs)
Crystal Color (Daiya Kogyo)

Daiya Hologram EP Type (Daiya Kogyo)
Daiya Hologram HG-S Type (Daiya Kogyo)
DC Glitter (Mitsubishi Petrochemical)
Diamond Piece Co-EP Type, Silver (Daiya Kogyo)
Diamond Piece CO Type, Blue (Daiya Kogyo)
Diamond Piece CO Type, DG Gold (Daiya Kogyo)
Diamond Piece CO Type, Green (Daiya Kogyo)
Diamond Piece CO Type, LG Gold (Daiya Kogyo)
Diamond Piece CO Type, Pink (Daiya Kogyo)
Diamond Piece CO Type, Red (Daiya Kogyo)
Diamond Piece CO Type, Silver (Daiya Kogyo)
Diamond Piece CO Type, Violet (Daiya Kogyo)
Epocrystal Blue (Nikko)
Epocrystal Brown (Nikko)
Epocrystal Purple (Nikko)
Iridescent Glitter IF4101 (Mitsubishi Petrochemical)
Iridescent Glitter IF8101 (Mitsubishi Petrochemical)
3M brand Colored Mirror Film (3M)
Morphotex (Kishimoto)
Morphotex Blue (Kishimoto)
Morphotex Red (Kishimoto)
Morphotex Yellow (Kishimoto)
Pearl Color (Nikko)
Rainbow Flake, Crystal (Daiya Kogyo)
Siliglit - Iris-Glitter (Sigmund Lindner GmbH)
Siliglit - Polyester-Glitter (Sigmund Lindner GmbH)
SiLiGlit-Trend Glitter (Coloured Iris-Glitter) (Sigmund Lindner GmbH)
Spectra F/X (Spectratek)
Spectra F/X Copper (Spectratek)
Spectra F/X Gold (Spectratek)

POLYETHYLGLUTAMATE

JPN Translation:
ポリグルタミン酸エチル

Empirical Formula:
$(C_7H_{13}NO_4)_x$

Definition: Polyethylglutamate is the polymer that conforms to the formula:

Information Sources: JCIC, JCLS

Chemical Classes: Esters; Synthetic Polymers

Function: Film Former

Technical/Other Name:
Poly-γ-Ethyl Glutamate

POLYETHYLHEXYL ACRYLATE

Definition: Polyethylhexyl Acrylate is the homopolymer of ethylhexyl acrylate.

Chemical Class: Synthetic Polymers

Function: Film Former

POLYETHYLHEXYL METHACRYLATE

Definition: Polyethylhexyl Methacrylate is the homopolymer of 2-ethylhexyl methacrylate.

Information Source: JCLS

Chemical Class: Synthetic Polymers

Function: Film Former

POLYETHYLMETHACRYLATE

CAS Nos.: 9003-42-3; 197098-43-4

JPN Translation:
ポリメタクリル酸エチル

Empirical Formula:
$(C_6H_{10}O_2)_x$

Definition: Polyethylmethacrylate is the polymer of ethyl methacrylate that conforms to the formula:

$$\left[-CH_2\underset{\underset{COOCH_2CH_3}{|}}{\overset{\overset{CH_3}{|}}{C}}- \right]_x$$

Chemical Classes: Esters; Synthetic Polymers

Functions: Artificial Nail Builder; Film Former

Technical/Other Names:
Ethyl Methacrylate Homopolymer
2-Propenoic Acid, 2-Methyl, Homopolymer, Ethyl Ester

POLYGALA SENEGA ROOT EXTRACT

CAS Nos.	EINECS No.
8057-58-7	
90082-95-4	290-188-2

Definition: Polygala Senega Root Extract is an extract of the roots of the senega, *Polygala senega. See "Regulatory and Ingredient Use Information," regarding the labeling names for botanical ingredients in Volume 1, Introduction, Part A.*

Chemical Class: Biological Products

Function: Not Reported

Technical/Other Names:
Extract of Polygala Senega Roots
Extract of Senega
Senega Extract
Senega (Polygala Senega) Extract

Trade Name Mixtures:
Actiphyte of Senega BG50 (Active Organics)
Actiphyte of Senega GL50 (Active Organics)
Actiphyte of Senega Lipo S (Active Organics)
Actiphyte of Senega PG50 (Active Organics)
Actiphyte of Snakeroot BG50 (Active Organics)
Actiphyte of Snakeroot GL50 (Active Organics)
Actiphyte of Snakeroot Lipo S (Active Organics)
Actiphyte of Snakeroot PG50 (Active Organics)

POLYGALA TENUIFOLIA ROOT EXTRACT

Definition: Polygala Tenuifolia Root Extract is an extract of the roots of *Polygala tenuifolia. See "Regulatory and Ingredient Use Information," regarding the labeling names for botanical ingredients in Volume 1, Introduction, Part A.*

Chemical Class: Biological Products

Function: Not Reported

Technical/Other Name:
Extract of Polygala Tenuifolia

Trade Name Mixture:
Onji Liquid E (Ichimaru Pharcos)

POLYGLUCURONIC ACID

CAS No.: 36655-86-4 (D-Form)

Definition: Polyglucuronic Acid is a polymer of Glucuronic Acid (q.v.).

Chemical Classes: Carbohydrates; Carboxylic Acids; Polyols

Functions: Film Former; Skin-Conditioning Agent - Humectant

Technical/Other Names:
D-Glucuronic Acid, Homopolymer
Mucronic Acid

POLYGLUTAMIC ACID

CAS No.: 25513-46-6

Definition: Polyglutamic Acid is the polymer of glutamic acid that conforms to the formula:

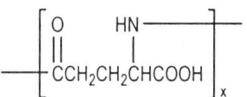

Chemical Class: Synthetic Polymers

Function: Skin-Conditioning Agent - Miscellaneous

Trade Names:
Bio PGA (Ichimaru Pharcos)
Polyglutamic Acid MZ (Maruzen Pharmaceuticals Co., Ltd.)

POLYGLYCERIN-3

CAS Nos.	EINECS No.
25618-55-7 (Generic)	
56090-54-1	259-986-8

JPN Translation:
ポリグリセリン - 3

Definition: Polyglycerin-3 is a glycerin polymer containing 3 glycerin units.

Information Sources: JCLS, TSCA

Chemical Classes: Ethers; Polyols

Function: Skin-Conditioning Agent - Humectant

Reported Product Category: Shampoos (Non-coloring)

Technical/Other Names:
Triglycerin
Triglycerol

Trade Name:
Triglycerine (Solvay GmbH)

POLYGLYCERIN-4

CAS No.: 56491-53-3

JPN Translation:
ポリグリセリン - 4

Definition: Polyglycerin-4 is a glycerin polymer containing 4 glycerin units.

Information Source: JCLS

Chemical Classes: Ethers; Polyols

Function: Skin-Conditioning Agent - Humectant

Technical/Other Names:
Tetraglycerin
Tetraglycerol

Trade Name:
Polyglycerin #310 (Sakamoto Yakuhin Kogyo)

POLYGLYCERIN-6

CAS No.	EINECS No.
36675-34-0	253-154-8

JPN Translation:
ポリグリセリン - 6

Definition: Polyglycerin-6 is a glycerin polymer containing 6 glycerin units.

Information Source: JCLS

Chemical Classes: Ethers; Polyols

Function: Skin-Conditioning Agent - Humectant

Technical/Other Names:
Hexaglycerin
Hexaglycerol

Trade Name:
Polyglycerin #500 (Sakamoto Yakuhin Kogyo)

POLYGLYCERIN-10

CAS No.	EINECS No.
9041-07-0	232-921-0

JPN Translation:
ポリグリセリン - 10

Definition: Polyglycerin-10 is a glycerin polymer containing 10 glycerin units.

Information Source: JCLS

Chemical Classes: Ethers; Polyols

Function: Skin-Conditioning Agent - Humectant

Technical/Other Names:
Decaglycerin
Decaglycerol

Trade Name:
Polyglycerin #750 (Sakamoto Yakuhin Kogyo)

Trade Name Mixture:
Nikkol Decaglyn 1-50VS (Nikko)

POLYGLYCERYL-4 ALMONDATE/SHEA BUTTERATE

Definition: Polyglyceryl-4 Almondate/Shea Butterate is an ester of a mixture of fatty acids derived from almond oil and Butyrospermum Parkii (Shea Butter) (q.v.) with Polyglycerin-4 (q.v.).

Chemical Class: Glyceryl Esters and Derivatives

Function: Surfactant - Emulsifying Agent

Trade Name:
Polyphytol AK-40-DS (Blendit)

POLYGLYCERYL-3 BEESWAX

CAS No.: 136097-93-3

Definition: Polyglyceryl-3 Beeswax is an ester of beeswax fatty acids and Polyglyceryl-3 (q.v).

Chemical Class: Glyceryl Esters and Derivatives

Function: Surfactant - Emulsifying Agent

Reported Product Category: Mascara

Trade Names:
Cera Bellina (Dekker)
Cera Bellina (Koster Keunen)

POLYGLYCERYL-6 BEHENATE

Definition: Polyglyceryl-6 Behenate is the monoester of Behenic Acid (q.v.) and Polyglycerin-6 (q.v.).

Chemical Class: Glyceryl Esters and Derivatives

Functions: Emulsion Stabilizer; Slip Modifier; Surface Modifier

Technical/Other Names:
Behenic Acid, Monoester with Hexaglycerol
Hexaglyceryl Monobehenate

Trade Name:
Pelemol 6G22 (Phoenix)

POLYGLYCERYL-10 BEHENATE/ EICOSADIOATE

Definition: Polyglyceryl-10 Behenate/ Eicosadioate is the monoester of Polyglycerin-10 (q.v.) and a blend of behenic and eicosadioic acids.

Chemical Class: Glyceryl Esters and Derivatives

Functions: Skin-Conditioning Agent - Miscellaneous; Surfactant - Emulsifying Agent

Technical/Other Name:
Decaglyceryl Behenate/Eicosadioate

Trade Names:
AEC Polyglyceryl-10 Behenate/ Eicosadioate (A & E Connock)

Nomcort HK-P (Nisshin OilliO)
Technol DG20 (Technonet)

POLYGLYCERYL-2 CAPRATE

CAS No.: 156153-06-9

Empirical Formula:
$C_{16}H_{32}O_6$

Definition: Polyglyceryl-2 Caprate is the ester of capric acid and Diglycerin (q.v.).

Chemical Class: Glyceryl Esters and Derivatives

Functions: Skin-Conditioning Agent - Emollient; Surfactant - Emulsifying Agent

Technical/Other Names:
Capric Acid, Monoester with Diglycerol
Diglyceryl Monocaprate

Trade Names:
Dermosoft DGMC (Straetmans)
Sunsoft Q-10D (Taiyo Kagaku)

POLYGLYCERYL-3 CAPRATE

CAS Nos.: 51033-30-8; 133654-02-1

Empirical Formula:
$C_{19}H_{38}O_8$

Definition: Polyglyceryl-3 Caprate is an ester of capric acid and Polyglycerin-3 (q.v.).

Chemical Class: Glyceryl Esters and Derivatives

Functions: Skin-Conditioning Agent - Emollient; Surfactant - Emulsifying Agent

Technical/Other Names:
Decanoic Acid, Monoester with Triglycerol
Triglyceryl Monocaprate

Trade Name:
Tegosoft PC 31 (Degussa Care Specialties)

POLYGLYCERYL-4 CAPRATE

CAS No.: 160391-93-5

Empirical Formula:
$C_{22}H_{44}O_{10}$

Definition: Polyglyceryl-4 Caprate is the ester of Capric Acid (q.v.) and Polyglycerin-4 (q.v.).

Chemical Class: Glyceryl Esters and Derivatives

Functions: Skin-Conditioning Agent - Emollient; Surfactant - Emulsifying Agent

Technical/Other Names:
Decanoic Acid, Monoester with Tetraglycerol
Tetraglyceryl Monocaprate

Trade Name:
Tegosoft PC 41 (Degussa Care Specialties)

POLYGLYCERYL-5 CAPRATE

Definition: Polyglyceryl-5 Caprate is the monoester of capric acid and Polyglycerin-5 (q.v.).

Chemical Class: Glyceryl Esters and Derivatives

Functions: Skin-Conditioning Agent - Emollient; Surfactant - Emulsifying Agent

Technical/Other Names:
Decanoic Acid, Monoester with Pentaglycerol
Pentaglyceryl Monocaprate

Trade Name:
Sunsoft A-10E (Taiyo Kagaku)

POLYGLYCERYL-10 CAPRATE

Definition: Polyglyceryl-10 Caprate is the ester of capric acid and Polyglyceryl-10 (q.v.).

Chemical Class: Glyceryl Esters and Derivatives

Functions: Skin-Conditioning Agent - Emollient; Surfactant - Emulsifying Agent

Technical/Other Names:
Capric Acid, Monoester with Decaglycerol
Decaglyceryl Monocaprate

Trade Names:
Sunsoft Q-10S (Taiyo Kagaku)
SY-Glyster MD-750 (Sakamoto Yakuhin Kogyo) (Celless)

POLYGLYCERYL-2 CAPRYLATE

Empirical Formula:
$C_{14}H_{28}O_6$

Definition: Polyglyceryl-2 Caprylate is the ester of Caprylic Acid (q.v.) and Diglycerin (q.v.).

Chemical Class: Glyceryl Esters and Derivatives

Functions: Skin-Conditioning Agent - Emollient; Surfactant - Emulsifying Agent

Technical/Other Name:
Diglyceryl Caprylate

POLYGLYCERYL-3 CAPRYLATE

Definition: Polyglyceryl-3 Caprylate is the ester of caprylic acid and Polyglycerin-3 (q.v.).

Chemical Class: Glyceryl Esters and Derivatives

Function: Surfactant - Emulsifying Agent

Technical/Other Name:
Triglyceryl Monocaprylate

Trade Name:
Cosmocair P 813 (Degussa Care Specialties)

POLYGLYCERYL-6 CAPRYLATE

Definition: Polyglyceryl-6 Caprylate is the monoester of caprylic acid and Polyglycerin-6 (q.v.).

Chemical Class: Glyceryl Esters and Derivatives

Functions: Skin-Conditioning Agent - Emollient; Surfactant - Emulsifying Agent

Technical/Other Names:
Caprylic Acid, Monoester with Hexaglycerol
Hexaglycerol Caprylate

Trade Names:
Dermofeel G 6 CY (Straetmans)
Sunsoft Q-81F (Taiyo Kagaku)

POLYGLYCERYL-10 CAPRYLATE

CAS No.: 51033-41-1

Definition: Polyglyceryl-10 Caprylate is the monoester of caprylic acid and Polyglycerin-10 (q.v.).

Chemical Class: Glyceryl Esters and Derivatives

Function: Surfactant - Emulsifying Agent

Technical/Other Names:
Decaglyceryl Monocaprylate
Octanoic Acid, Monoester, with Decaglycerol

Trade Name:
SY-GLYSTER MCA-750 (Sakamoto Yakuhin Kogyo)

POLYGLYCERYL-3 CETYL ETHER

CAS No.: 128895-87-4

Empirical Formula:
$C_{25}H_{52}O_7$

Definition: Polyglyceryl-3 Cetyl Ether is an ether of cetyl alcohol and Polyglycerin-3 (q.v.).

Chemical Class: Ethers

Function: Surfactant - Emulsifying Agent

Reported Product Category: Indoor Tanning Preparations

Technical/Other Names:
Triglycerol, Monohexadecyl Ether
Triglyceryl Cetyl Ether

Trade Name:
Chimexane NL (Chimex)

POLYGLYCERYL-3 COCOATE

Definition: Polyglyceryl-3 Cocoate is the ester of Coconut Acid (q.v.). and Polyglycerin-3 (q.v.).

Chemical Class: Glyceryl Esters and Derivatives

Function: Surfactant - Emulsifying Agent

Technical/Other Names:
Coconut Fatty Acids, Monoester with Triglycerol
Triglyceryl Monococoate

POLYGLYCERYL-4 COCOATE

Definition: Polyglyceryl-4 Cocoate is an ester of Coconut Acid (q.v.) and Polyglycerin-4 (q.v.).

Chemical Class: Glyceryl Esters and Derivatives

Function: Surfactant - Emulsifying Agent

Technical/Other Names:
Coconut Fatty Acids, Monoester with Tetraglycerol
Tetraglyceryl Monococoate

Trade Name:
AEC Polyglyceryl-4 Cocoate (A & E Connock)

POLYGLYCERYL-8 DECAERUCATE/ DECAISOSTEARATE/DECARICINOLEATE

Definition: Polyglyceryl-8 Decaerucate/ Decaisostearate/Decaricinoleate is the decaester of Polyglycerin-8 (q.v.) with a mixture of Erucic Acid (q.v.), Isostearic Acid (q.v.) and Ricinoleic Acid (q.v.).

Chemical Class: Glyceryl Esters and Derivatives

Function: Skin-Conditioning Agent - Emollient

Trade Name:
S-Face MX-10 (Sakamoto Yakuhin Kogyo)

POLYGLYCERYL-10 DECAHYDROXY-STEARATE

Definition: Polyglyceryl-10 Decahydroxy-stearate is the decaester of Hydroxystearic Acid (q.v.) and Polyglycerin-10 (q.v.).

Chemical Class: Glyceryl Esters and Derivatives

Functions: Skin-Conditioning Agent - Emollient; Surfactant - Emulsifying Agent

Technical/Other Names:
Decaglyceryl Decahydroxystearate
Hydroxystearic Acid, Decaester with Deca-glycerol

Trade Name:
Nikkol Decaglyn 10-HS (Nikko)

POLYGLYCERYL-10 DECAISOSTEARATE

JPN Translation:
デカイソステアリン酸ポリグリセリル - 10

Definition: Polyglyceryl-10 Decaisostearate is the ester of Polyglycerin-10 (q.v.) and Isostearic Acid (q.v.).

Information Source: JCLS

Chemical Class: Glyceryl Esters and Derivatives

Function: Skin-Conditioning Agent - Emollient

Trade Name:
Nikkol Decaglyn 10-IS (Nikko)

POLYGLYCERYL-10 DECALINOLEATE

CAS No.	EINECS No.
68900-96-9	272-645-8

Empirical Formula:
$C_{210}H_{362}O_{31}$

Definition: Polyglyceryl-10 Decalinoleate is a decaester of linoleic acid and Polyglycerin-10 (q.v.).

Chemical Class: Glyceryl Esters and Derivatives

Functions: Skin-Conditioning Agent - Emollient; Surfactant - Emulsifying Agent

Technical/Other Names:
Decaglyceryl Decalinoleate
9,12-Octadecadienoic Acid, Decaester with Decaglycerol
1,2,3-Propanetriol, Decamer, Deca-9,12-Octadecadienoate

Trade Name:
AEC Polyglyceryl-10 Decalinoleate (A & E Connock)

POLYGLYCERYL-10 DECAMACADAMIATE

Definition: Polyglyceryl-10 Decamacadamiate is a decaester of Poly-glycerin-10 (q.v.) and the fatty acids derived from macadamia nut oil.

Chemical Class: Glyceryl Esters and Derivatives

Functions: Skin-Conditioning Agent - Emollient; Surfactant - Emulsifying Agent

Trade Name:
Nikkol Decaglyn 10-MAC (Nikko)

POLYGLYCERYL-10 DECAOLEATE

CAS No.	EINECS No.
11094-60-3	234-316-7

JPN Translation:
デカオレイン酸ポリグリセリル - 10

Empirical Formula:
$C_{210}H_{382}O_{31}$

Definition: Polyglyceryl-10 Decaoleate is a decaester of oleic acid and Polyglycerin-10 (q.v.).

Information Sources: 21CFR172.854, CTFA D, JCIC, JCLS, JSQI

Chemical Class: Glyceryl Esters and Derivatives

Functions: Skin-Conditioning Agent - Emollient; Surfactant - Emulsifying Agent

Technical/Other Names:
Decaglyceryl Decaoleate
9-Octadecenoic Acid, Decaester with Decaglycerol
9-Octadecenoic Acid, Decaester with 4,8,12,16,20,24,28,32,36-Nonaoxa-nonatriacontane-1,2,6,10,14,18,22,26,30,34,38,39-Dodecol
Oleic Acid, Decaester with Decaglycerol
Polyglyceryl Decaoleate

Trade Names:
AEC Polyglyceryl-10 Decaoleate (A & E Connock)
Caprol 10G100 (Abitec Corporation)
Nikkol Decaglyn 10-O (Nikko)
Polyaldo DGDO (10-10-0) (Lonza Inc./Lonza Ltd.)
Sunsoft Q-1710S (Taiyo Kagaku)
Unitolate PGO-1010 (Universal Preserv-A-Chem)

POLYGLYCERYL-10 DECASTEARATE

CAS No.	EINECS No.
39529-26-5	254-495-5

JPN Translation:
デカステアリン酸ポリグリセリル - 10

Empirical Formula:
$C_{210}H_{402}O_{31}$

Definition: Polyglyceryl-10 Decastearate is a decaester of stearic acid and Polyglycerin-10 (q.v.).

Information Sources: 21CFR172.854, JCIC, JCLS

Chemical Class: Glyceryl Esters and Derivatives

Functions: Skin-Conditioning Agent - Emollient; Surfactant - Emulsifying Agent

Technical/Other Names:
Decaglyceryl Decastearate
Octadecanoic Acid, Decaester with Deca-glycerol
Octadecanoic Acid, Decaester with 4,8,12,16,20,24,28,32,36-Nonaoxa-nonatriacontane-1,2,6,10,14,18,22,26,30,34,38,39-Dodecol

Trade Names:
AEC Polyglyceryl-10 Decastearate (A & E Connock)
Nikkol Decaglyn 10-S (Nikko)
Sunsoft Q-1810S (Taiyo Kagaku)

POLYGLYCERYL-3 DECYLTETRADECYL ETHER

Definition: Polyglyceryl-3 Decyltetradecyl Ether is the ether of decyltetradecanol and Polyglycerin-3 (q.v.).

Chemical Class: Ethers

Functions: Skin-Conditioning Agent - Emollient; Surfactant - Emulsifying Agent

Technical/Other Names:
Polyglyceryl-3 Decyltetradecanol
Triglyceryl Decyltetradecyl Ether

Trade Name:
Chimexane NR (Chimex)

POLYGLYCERYL-3 DICAPRATE

Empirical Formula:
$C_{29}H_{56}O_9$

Definition: Polyglyceryl-3 Dicaprate is the diester of capric acid and Polyglycerin-3 (q.v.).

Chemical Class: Glyceryl Esters and Derivatives

Functions: Skin-Conditioning Agent - Emollient; Surfactant - Emulsifying Agent

Technical/Other Name:
Triglyceryl Dicaprate

POLYGLYCERYL-6 DICAPRATE

Definition: Polyglyceryl-6 Dicaprate is the diester of Capric Acid (q.v.) and Polyglycerin-6 (q.v.).

Chemical Class: Glyceryl Esters and Derivatives

Functions: Skin-Conditioning Agent - Emollient; Surfactant - Emulsifying Agent

Technical/Other Names:
Capric Acid, Diester with Hexaglycerol
Hexaglycerol Dicaprate

Trade Name:
Dermofeel 81F (Straetmans)

POLYGLYCERYL-3 DICOCOATE

Definition: Polyglyceryl-3 Dicocoate is the diester of Coconut Acid (q.v.) and Polyglycerin-3 (q.v.).

Chemical Class: Glyceryl Esters and Derivatives

Functions: Skin-Conditioning Agent - Emollient; Surfactant - Emulsifying Agent

Technical/Other Names:
Coconut Fatty Acids, Diester with Triglycerol
Triglyceryl Dicocoate

POLYGLYCERYL-10 DIDECANOATE

CAS No.: 182015-59-4

Definition: Polyglyceryl-10 Didecanoate is the diester of decanoic acid and Polyglycerin-10 (q.v.).

Chemical Class: Glyceryl Esters and Derivatives

Functions: Skin-Conditioning Agent - Emollient; Surfactant - Emulsifying Agent

Technical/Other Names:
Decaglycerol Didecanoate
Decanoic Acid, Diester with Decaglycerol

Trade Name:
Nikkol Decaglyn-2D (Nikko)

POLYGLYCERYL-2 DIISOSTEARATE

CAS Nos.   **EINECS No.**
63705-03-3 (Generic)
67938-21-0   267-821-6

JPN Translation:
ジイソステアリン酸ポリグリセリル - 2

Empirical Formula:
$C_{42}H_{82}O_7$

Definition: Polyglyceryl-2 Diisostearate is the diester of Isostearic Acid (q.v.) and Diglycerin (q.v.).

Information Sources: JCLS, JSQI, TSCA

Chemical Class: Glyceryl Esters and Derivatives

Functions: Skin-Conditioning Agent - Emollient; Surfactant - Emulsifying Agent

Reported Product Categories: Mascara; Moisturizing Preparations; Skin Care Preparations, Misc.

Technical/Other Names:
Diglycerin Diisostearate
Diglyceryl Diisostearate
Isooctadecanoic Acid, Diester with Diglycerol
Isooctadecanoic Acid, Diester with Oxybis (Propanediol)
Isooctadecanoic Acid, Diester with Oxybis (Propenediol)

Trade Names:
AEC Polyglyceryl-2 Diisostearate (A & E Connock)
Cosmol 42 (Nisshin OilliO)
Dermol DGDIS (Alzo)
Matsunate DI (Matsumoto Trading)
Matsunate DI-N (Matsumoto Trading)
PRISORINE 3792 (Uniqema Europe)
Salacos 42 (Nisshin OilliO)
Technol F2 (Technonet)
Wogel 18D (Matsumoto Trading)
Wogel 18DV (Matsumoto Trading)

POLYGLYCERYL-3 DIISOSTEARATE

CAS Nos.: 63705-03-3 (Generic); 66082-42-6

JPN Translation:
ジイソステアリン酸ポリグリセリル - 3

Empirical Formula:
$C_{45}H_{88}O_9$

Definition: Polyglyceryl-3 Diisostearate is a diester of Isostearic Acid (q.v.) and Polyglycerin-3 (q.v.).

Information Sources: CTFA D, JCLS, TSCA

Chemical Class: Glyceryl Esters and Derivatives

Functions: Skin-Conditioning Agent - Emollient; Surfactant - Emulsifying Agent

Reported Product Categories: Lipsticks; Makeup Preparations (Not eye), Misc.; Eye Shadows; Eye Makeup Preparations, Misc.; Foundations; Face Powders; Moisturizing Preparations; Blushers (All types)

Technical/Other Names:
Isooctadecanoic Acid, Diester with 1,2,3-Propanetriol Trimer
Isooctadecanoic Acid, Diester with Triglycerol
Triglycerin Diisostearate
Triglyceryl Diisostearate

Trade Names:
AEC Polyglyceryl-3 Diisostearate (A & E Connock)
DUB ISO G3 (Stearinerie Dubois Fils)
Emerest 2452 (Cognis Care Chemicals/NJ)
Emerest 2452 (Cognis Care Chemicals/PA)
Hostacerin TGI (Clariant)
Hostacerin TGI (Clariant GmbH, Personal Care)
Lameform TGI (Cognis Care Chemicals/PA)
Lameform TGI (Cognis Deutschland)
Plurol Diisostearique (Gattefosse s.a.)
PRISORINE 3700 (Uniqema Europe)

Trade Name Mixtures:
Anise Softcream (CEP (Solabia))
Apricot Softcream (CEP (Solabia))
Avocado Softcream (CEP (Solabia))
Bitter Orange Softcream (CEP (Solabia))
Cardamom Softcream (CEP (Solabia))
Cinnamon Softcream (CEP (Solabia))
Eucalyptus Softcream (CEP (Solabia))
Grapefruit Softcream (CEP (Solabia))
Grape Seed Softcream (CEP (Solabia))
Lavender Softcream (CEP (Solabia))
Lime Softcream (CEP (Solabia))
Peppermint Softcream (CEP (Solabia))
Pistachio Softcream (CEP (Solabia))
Rice Bran Softcream (CEP (Solabia))
Rosemary Softcream (CEP (Solabia))
Sweet Almond Softcream (CEP (Solabia))

POLYGLYCERYL-6 DIISOSTEARATE

Definition: Polyglyceryl-6 Diisostearate is the diester of Isostearic Acid (q.v.) and Polyglycerin-6 (q.v.).

Chemical Class: Glyceryl Esters and Derivatives

Functions: Skin-Conditioning Agent - Emollient; Surfactant - Emulsifying Agent

Technical/Other Names:
Hexaglyceryl Diisostearate
Isostearic Acid, Diester, with Hexaglycerol

Trade Name:
Emalex DISG-6 (Nihon Emulsion)

POLYGLYCERYL-10 DIISOSTEARATE

CAS Nos.: 63705-03-3 (Generic); 102033-55-6

JPN Translation:
ジイソステアリン酸ポリグリセリル - 10

Empirical Formula:
$C_{66}H_{130}O_{23}$

Definition: Polyglyceryl-10 Diisostearate is a diester of Isostearic Acid (q.v.) and Polyglycerin-10 (q.v.).

Information Sources: JCLS, JSQI, TSCA

Chemical Class: Glyceryl Esters and Derivatives

Functions: Skin-Conditioning Agent - Emollient; Surfactant - Emulsifying Agent

Technical/Other Names:
Decaglycerin Diisostearate
Decaglyceryl Diisostearate
Isooctadecanoic Acid, Diester with Deca-glycerol

Trade Names:
Emalex DISG-10 (Nihon Emulsion)
Matsunate MI-102 (Matsumoto Trading)
Nikkol Decaglyn 2-IS (Nikko)

POLYGLYCERYL-2 DIISOSTEARATE/IPDI COPOLYMER

Definition: Polyglyceryl-2 Diisostearate/IPDI Copolymer is a copolymer of isophorone diisocyanate and Polyglyceryl-2 Diisostearate (q.v.).

Chemical Class: Synthetic Polymers

Functions: Binder; Emulsion Stabilizer; Film Former

Trade Name:
Polyderm PPI-DGDIS (Alzo)

POLYGLYCERYL-4 DILAURATE

Definition: Polyglyceryl-4 Dilaurate is the diester of Lauric Acid (q.v.) and Polyglycerin-4 (q.v.).

Chemical Class: Glyceryl Esters and Derivatives

Functions: Skin-Conditioning Agent - Emollient; Surfactant - Emulsifying Agent

Technical/Other Names:
Lauric Acid, Diester with Tetraglycerol
Tetraglyceryl Dilaurate

POLYGLYCERYL DIMER SOYATE

Definition: Polyglyceryl Dimer Soyate is a complex mixture of esters of polyglycerin and the fatty acids derived from dimerized soybean oil.

Chemical Class: Glyceryl Esters and Derivatives

Function: Surfactant - Emulsifying Agent

Trade Names:
Akoline PG 3 (Karlshamns AB)
Rylo PG 11 (Danisco)

POLYGLYCERYL-2 DIOLEATE

CAS Nos.: 60219-68-3; 67965-56-4

Empirical Formula:
$C_{42}H_{78}O_7$

Definition: Polyglyceryl-2 Dioleate is a diester of oleic acid and Diglycerin (q.v.).

Chemical Class: Glyceryl Esters and Derivatives

Functions: Skin-Conditioning Agent - Emollient; Surfactant - Emulsifying Agent

Technical/Other Names:
Diglycerin Dioleate
Diglyceryl Dioleate
9-Octadecenoic Acid, Diester with Oxybis (Propanediol)
9-Octadecenoic Acid, Diester with 3,3'-Oxybis[1,2-Propanediol]

Trade Name:
Nikkol DGDO (Nikko)

POLYGLYCERYL-3 DIOLEATE

CAS No.: 79665-94-4

Empirical Formula:
$C_{45}H_{84}O_9$

Definition: Polyglyceryl-3 Dioleate is a diester of oleic acid and Polyglycerin-3 (q.v.).

Information Source: 21CFR172.854

Chemical Class: Glyceryl Esters and Derivatives

Functions: Skin-Conditioning Agent - Emollient; Surfactant - Emulsifying Agent

Technical/Other Names:
9-Octadecenoic Acid, Diester with Triglycerol
Triglyceryl Dioleate

POLYGLYCERYL-6 DIOLEATE

CAS No. 76009-37-5 **EINECS No.** 278-358-4

JPN Translation:
ジオレイン酸ポリグリセリル - 6

Empirical Formula:
$C_{54}H_{102}O_{15}$

Definition: Polyglyceryl-6 Dioleate is a diester of oleic acid and Polyglycerin-6 (q.v.).

Information Sources: 21CFR172.854, JCIC, JCLS

Chemical Class: Glyceryl Esters and Derivatives

Functions: Skin-Conditioning Agent - Emollient; Surfactant - Emulsifying Agent

Technical/Other Names:
Hexaglycerin Dioleate
Hexaglyceryl Dioleate
9-Octadecenoic Acid, Diester with Hexaglycerol
Polyglyceryl Dioleate

Trade Names:
AEC Polyglyceryl-6 Dioleate (A & E Connock)
Caprol MPGO (Abitec Corporation)
Plurol Oleique (Gattefosse s.a.)
Polyaldo 6-2-0 (Lonza Inc./Lonza Ltd.)

Trade Name Mixtures:
Aquafix Marin (Coletica SA)
Aquafix Vegetal (Coletica SA)

POLYGLYCERYL-10 DIOLEATE

CAS No. 33940-99-7 **EINECS No.** 251-750-2

Empirical Formula:
$C_{66}H_{126}O_{23}$

Definition: Polyglyceryl-10 Dioleate is a diester of oleic acid and Polyglycerin-10 (q.v.).

Chemical Class: Glyceryl Esters and Derivatives

Functions: Skin-Conditioning Agent - Emollient; Surfactant - Emulsifying Agent

Technical/Other Names:
Decaglyceryl Dioleate
9-Octadecenoic Acid, Diester with Deca-glycerol
9-Octadecenoic Acid, Diester with 4,8,12,16,20,24,28,32,36-Nonaoxa-nonatriacontane-1,2,6,10,14,18,22,26,30,34,38,39-Dodecol

Trade Name:
Nikkol Decaglyn 2-O (Nikko)

POLYGLYCERYL-6 DIPALMITATE

Definition: Polyglyceryl-6 Dipalmitate is the diester of Palmitic Acid (q.v.) and Poly-glycerin-6 (q.v.).

Chemical Class: Glyceryl Esters and Derivatives

Functions: Skin-Conditioning Agent - Emollient; Surfactant - Emulsifying Agent

Technical/Other Names:
Hexaglycerol Dipalmitate
Palmitic Acid, Diester with Hexaglycerol

Trade Name:
Polyaldo 6-2-P (Lonza Inc./Lonza Ltd.)

POLYGLYCERYL-10 DIPALMITATE

Definition: Polyglyceryl-10 Dipalmitate is the diester of Palmitic Acid (q.v.) and Poly-glycerin-10 (q.v.).

Chemical Class: Glyceryl Esters and Derivatives

Functions: Skin-Conditioning Agent - Emollient; Surfactant - Emulsifying Agent

Technical/Other Names:
Decaglycerol Dipalmitate
Palmitic Acid, Diester with Decaglycerol

POLYGLYCERYL-2 DIPOLYHYDROXY-STEARATE

Definition: Polyglyceryl-2 Dipolyhydroxy-stearate is the diester of Polyhydroxystearic Acid (q.v.) and Polyglyceryl-2 (q.v.).

Chemical Class: Glyceryl Esters and Derivatives

Function: Skin-Conditioning Agent - Occlusive

Trade Names:
AEC Polyglyceryl-2 Dipolyhydroxystearate (A & E Connock)
Dehymuls PGPH (Cognis Care Chemicals/PA)
Dehymuls PGPH (Cognis Deutschland)

Trade Name Mixtures:
Cutina LM Conc (Cognis Deutschland)
Dehymuls SBL (Cognis Care Chemicals/NJ)
Eumulgin VL 75 (Cognis Deutschland)

POLYGLYCERYL-3 DISILOXANE DIMETHICONE

Definition: Polyglyceryl-3 Disiloxane Dimethicone is the siloxane polymer that conforms generally to the formula:

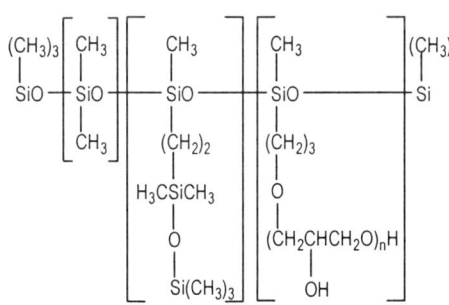

Chemical Classes: Glyceryl Esters and Derivatives; Siloxanes and Silanes

Functions: Hair Conditioning Agent; Skin-Conditioning Agent - Miscellaneous; Surfactant - Emulsifying Agent; Surfactant - Solubilizing Agent; Viscosity Increasing Agent - Aqueous

Trade Name:
KF-6100 (Shin-Etsu Chemical Co.)

POLYGLYCERYL-2 DISTEARATE

CAS No.: 9009-32-9

Empirical Formula:
$C_{42}H_{82}O_7$

Definition: Polyglyceryl-2 Distearate is the diester of stearic acid and Diglycerin (q.v.).

Information Source: JCLS

Chemical Class: Glyceryl Esters and Derivatives

Function: Surfactant - Emulsifying Agent

Technical/Other Names:
Diglyceryl Distearate
Stearic Acid, Diester with Diglycerol

Trade Names:
Emalex DSG-2 (Ikeda)
Emalex PGSA (Ikeda)
Sunsoft Q-192B (Taiyo Kagaku)

POLYGLYCERYL-3 DISTEARATE

CAS Nos.: 9009-32-9 (Generic); 94423-19-5

Empirical Formula:
$C_{45}H_{88}O_9$

Definition: Polyglyceryl-3 Distearate is the diester of stearic acid and Polyglycerin-3 (q.v.).

Information Source: JCLS

Chemical Class: Glyceryl Esters and Derivatives

Functions: Skin-Conditioning Agent - Emollient; Surfactant - Emulsifying Agent

Technical/Other Names:
Octadecanoic Acid, Diester with Triglycerol
Triglycerin Distearate
Triglyceryl Distearate

Trade Name:
Cremophor GS 32 (BASF)

Trade Name Mixture:
Capispheres (Barnet)

POLYGLYCERYL-6 DISTEARATE

CAS Nos.	**EINECS No.**
9009-32-9 (Generic)	
34424-97-0	252-010-1

JPN Translation:
ジステアリン酸ポリグリセリル - 6

Empirical Formula:
$C_{54}H_{106}O_{15}$

Definition: Polyglyceryl-6 Distearate is a diester of stearic acid and Polyglycerin-6 (q.v.).

Information Sources: 21CFR172.854, JCLS

Chemical Class: Glyceryl Esters and Derivatives

Functions: Skin-Conditioning Agent - Emollient; Surfactant - Emulsifying Agent

Technical/Other Names:
Hexaglyceryl Distearate
Octadecanoic Acid, Diester with Hexaglycerol
Stearic Acid, Diester with Hexaglycerol

Trade Names:
Caprol 6G2S (Abitec Corporation)
Plurol Stearique WL 1009 (Gattefosse s.a.)
Polyaldo HGDS (6-2-S) (Lonza Inc./Lonza Ltd.)

POLYGLYCERYL-10 DISTEARATE

CAS Nos.: 9009-32-9 (Generic); 12764-60-2

JPN Translation:
ジステアリン酸ポリグリセリル - 10

Empirical Formula:
$C_{66}H_{130}O_{23}$

Definition: Polyglyceryl-10 Distearate is the diester of stearic acid and Polyglycerin-10 (q.v.).

Information Sources: 21CFR172.854, JCLS, JSQI

Chemical Class: Glyceryl Esters and Derivatives

Functions: Skin-Conditioning Agent - Emollient; Surfactant - Emulsifying Agent

Technical/Other Names:
Decaglycerin Distearate
Decaglyceryl Distearate
Octadecanoic Acid, Diester with Decaglycerol
9-Octadecanoic Acid, Diester with 4,8,12,16,20,24,28,32,36-Nonaoxa-nonatriacontane-1,2,6,10,14,18,22,26,30,34,38,39-Dodecol

Trade Names:
Emalex DSG-10 (Nihon Emulsion)
Nikkol Decaglyn 2-S (Nikko)
Salacos PGDSV (Nisshin OilliO)
Sunsoft Q-182S (Taiyo Kagaku)

POLYGLYCERYL-10 DODECABEHENATE

Definition: Polyglyceryl-10 Dodecabehenate is the dodecaester of behenic acid and Polyglycerin-10 (q.v.).

Chemical Class: Glyceryl Esters and Derivatives

Function: Surfactant - Emulsifying Agent

Trade Name:
SY-GLYSTER DDB-750 (Sakamoto Yakuhin Kogyo)

POLYGLYCERYL-4 HAZELSEEDATE

Definition: Polyglyceryl-4 Hazelseedate is an ester of the fatty acids derived from hazel seed oil with Polyglycerin-4 q.v.).

Chemical Class: Glyceryl Esters and Derivatives

Function: Surfactant - Emulsifying Agent

Technical/Other Name:
Polyglyceryl-4 Hazelnutate

Trade Name:
Polyphytol N-40-D (Blendit)

POLYGLYCERYL-10 HEPTAHYDROXY-STEARATE

Definition: Polyglyceryl-10 Heptahydroxy-stearate is a heptaester of Hydroxystearic Acid (q.v.). and Polyglycerin-10 (q.v.).

Chemical Class: Glyceryl Esters and Derivatives

Functions: Skin-Conditioning Agent - Humectant; Surfactant - Emulsifying Agent

Technical/Other Names:
Decaglyceryl Heptahydroxystearate
Hydroxystearic Acid, Heptaester with Decaglycerol

Trade Name Mixture:
Nikkol Nikkowax W-40 (Nikko)

POLYGLYCERYL-10 HEPTAOLEATE

CAS No.: 103175-09-3

Empirical Formula:
$C_{156}H_{286}O_{28}$

Definition: Polyglyceryl-10 Heptaoleate is a heptaester of oleic acid and Polyglycerin-10 (q.v.).

Chemical Class: Glyceryl Esters and Derivatives

Functions: Skin-Conditioning Agent - Emollient; Surfactant - Emulsifying Agent

Technical/Other Names:
Decaglyceryl Heptaoleate
9-Octadecenoic Acid, Heptaester with Decaglycerol

Trade Names:
Nikkol Decaglyn 7-O (Nikko)
Sunsoft Q-177S (Taiyo Kagaku)

POLYGLYCERYL-10 HEPTASTEARATE

CAS Nos.: 9009-32-9 (Generic); 99126-54-2

JPN Translation:
ヘプタステアリン酸ポリグリセリル - 10

Empirical Formula:
$C_{156}H_{300}O_{28}$

Definition: Polyglyceryl-10 Heptastearate is the heptaester of stearic acid and Polyglycerin-10 (q.v.).

Information Sources: JCIC, JCLS

Chemical Class: Glyceryl Esters and Derivatives

Functions: Skin-Conditioning Agent - Emollient; Surfactant - Emulsifying Agent

Technical/Other Names:
Decaglyceryl Heptastearate
Octadecanoic Acid, Heptaester with Decaglycerol
Polyglyceryl Heptastearate

Trade Name:
Nikkol Decaglyn 7-S (Nikko)

POLYGLYCERYL-10 HEXAERUCATE

Definition: Polyglyceryl-10 Hexaerucate is the hexaester of Polyglycerin-10 (q.v.) and Erucic Acid (q.v.).

Chemical Class: Glyceryl Esters and Derivatives

Functions: Surfactant - Emulsifying Agent; Surfactant - Suspending Agent

Technical/Other Names:
Decaglyceryl Hexaerucate
Erucic Acid, Hexaester with Decaglycerol

Trade Name:
Ryoto Polyglycerol Ester ER-60D (Mitsubishi-Kagaku)

POLYGLYCERYL-6 HEXAOLEATE

CAS No.: 95482-05-6

Empirical Formula:
$C_{126}H_{230}O_{19}$

Definition: Polyglyceryl-6 Hexaoleate is a hexaester of oleic acid and Polyglycerin-6 (q.v.).

Information Source: 21CFR172.854

Chemical Class: Glyceryl Esters and Derivatives

Functions: Skin-Conditioning Agent - Emollient; Surfactant - Emulsifying Agent

Technical/Other Names:
Hexaglyceryl Hexaoleate

9-Octadecenoic Acid, Hexaester with Hexaglycerol

POLYGLYCERYL-10 HEXAOLEATE

CAS No.	EINECS No.
65573-03-7	265-820-5

Definition: Polyglyceryl-10 Hexaoleate is the hexaester of Oleic Acid (q.v.) and Polyglycerin-10 (q.v.).

Chemical Class: Glyceryl Esters and Derivatives

Functions: Skin-Conditioning Agent - Emollient; Surfactant - Emulsifying Agent

Technical/Other Names:
Decaglycerol Hexaoleate
9-Octadecenoic Acid, Hexaester with Decaglycerol
Oleic Acid, Hexaester with Decaglycerol

POLYGLYCERYL-5 HEXASTEARATE

Definition: Polyglyceryl-5 Hexastearate i s the hexaester of stearic acid and Polyglycerin-5 (q.v.).

Chemical Class: Glyceryl Esters and Derivatives

Functions: Skin-Conditioning Agent - Emollient; Surfactant - Emulsifying Agent

Technical/Other Names:
Pentaglycerol Hexastearate
Stearic Acid, Hexaester with Pentaglycerol

Trade Name:
Sunsoft A-186E (Taiyo Kagaku)

POLYGLYCERYL-6 HEXASTEARATE

Definition: Polyglyceryl-6 Hexastearate is the hexaester of stearic acid and Polyglyceryl-6 (q.v.).

Chemical Class: Glyceryl Esters and Derivatives

Functions: Skin-Conditioning Agent - Emollient; Surfactant - Emulsifying Agent

Technical/Other Names:
Hexaglycerol Hexastearic Acid Ester
Stearic Acid, Hexaester with Hexaglycerol

Trade Name:
Sunfat PS-66 (Taiyo Kagaku)

POLYGLYCERYL-3 HYDROXYLAURYL ETHER

Empirical Formula:
$C_{21}H_{44}O_7$

Definition: Polyglyceryl-3 Hydroxylauryl Ether is an ether of hydroxylauryl alcohol and Polyglycerin-3 (q.v.).

Chemical Class: Ethers

Functions: Skin-Conditioning Agent - Emollient; Surfactant - Emulsifying Agent

Technical/Other Names:
Triglyceryl Hydroxylauryl Ether
Triglyceryl Monohydroxylauryl Ether

Trade Name:
Chimexane NF (Chimex)

POLYGLYCERYL-2 ISOPALMITATE

JPN Translation:
イソパルミチン酸ポリグリセリル - 2

Empirical Formula:
$C_{22}H_{44}O_6$

Definition: Polyglyceryl-2 Isopalmitate is an ester of isopalmitic acid and Diglycerin (q.v.).

Information Sources: JCIC, JCLS

Chemical Class: Glyceryl Esters and Derivatives

Functions: Skin-Conditioning Agent - Emollient; Surfactant - Emulsifying Agent

Technical/Other Names:
Diglyceryl Isopalmitate
Diglyceryl Monoisopalmitate

POLYGLYCERYL-2 ISOPALMITATE/ SEBACATE

Definition: Polyglyceryl-2 Isopalmitate/ Sebacate is the mixed ester of Isopalmitic Acid (q.v.), Sebacic Acid (q.v.) and Diglycerin (q.v.).

Chemical Class: Glyceryl Esters and Derivatives

Function: Surfactant - Emulsifying Agent

Technical/Other Name:
Diglyceryl Isopalmitate/Sebacate

Trade Name:
Salacos DGS16 (Nisshin OilliO)

POLYGLYCERYL-2 ISOSTEARATE

CAS Nos.	EINECS No.
73296-86-3	277-361-8
81752-33-2	

JPN Translation:
イソステアリン酸ポリグリセリル - 2

Empirical Formula:
$C_{24}H_{48}O_6$

Definition: Polyglyceryl-2 Isostearate is the ester of isostearic acid and Diglycerin (q.v.).

Information Sources: JCLS, JSQI

Chemical Class: Glyceryl Esters and Derivatives

Functions: Skin-Conditioning Agent - Emollient; Surfactant - Emulsifying Agent

Technical/Other Names:
Diglyceryl Isostearate
Isooctadecanoic Acid, Ester with Oxybis (Propanediol)
Isooctanoic Acid, Monoester with Diglycerol

Trade Names:
AEC Polyglyceryl-2 Isostearate (A & E Connock)
Cosmol 41 (Nisshin OilliO)
Matsunate DI-1 (Matsumoto Trading)
Nikkol DGMIS (Nikko)
PRISORINE 3791 (Uniqema Europe)
Salacos 41 (Nisshin OilliO)

POLYGLYCERYL-3 ISOSTEARATE

CAS No.: 127512-63-4

JPN Translation:
イソステアリン酸ポリグリセリル - 3

Empirical Formula:
$C_{27}H_{54}O_8$

Definition: Polyglyceryl-3 Isostearate is the ester of isostearic acid and Polyglycerin-3 (q.v.).

Information Source: JCLS

Chemical Class: Glyceryl Esters and Derivatives

Functions: Skin-Conditioning Agent - Emollient; Surfactant - Emulsifying Agent

Reported Product Category: Moisturizing Preparations

Technical/Other Names:
Isooctadecanoic Acid, Monoester with Triglycerol
Triglyceryl Monoisostearate

POLYGLYCERYL-4 ISOSTEARATE

CAS No.: 91824-88-3

JPN Translation:
イソステアリン酸ポリグリセリル - 4

Empirical Formula:
$C_{30}H_{60}O_{10}$

Definition: Polyglyceryl-4 Isostearate is an ester of Isostearic Acid (q.v.) and Poly-glycerin-4 (q.v.).

Information Source: JCLS

Chemical Class: Glyceryl Esters and Derivatives

Functions: Skin-Conditioning Agent - Emollient; Surfactant - Emulsifying Agent

Reported Product Categories: Foundations; Makeup Bases; Bath Capsules; Face and Neck Preparations (Excluding Shaving Preparations)

Technical/Other Names:
Isooctanoic Acid, Monoester with Tetragly-cerol
Tetraglyceryl Monoisostearate

Trade Names:
AEC Polyglyceryl-4 Isostearate (A & E Connock)
Isolan GI 34 (Degussa Care Specialties)

Trade Name Mixture:
Abil WE 09 (Degussa Care Specialties)

POLYGLYCERYL-5 ISOSTEARATE

JPN Translation:
イソステアリン酸ポリグリセリル - 5

Empirical Formula:
$C_{33}H_{66}O_{12}$

Definition: Polyglyceryl-5 Isostearate is the ester of isostearic acid and a glycerin polymer containing an average of 5 glycerin units.

Information Source: JCLS

Chemical Class: Glyceryl Esters and Derivatives

Functions: Skin-Conditioning Agent - Emollient; Surfactant - Emulsifying Agent

Technical/Other Names:
Isooctadecanoic Acid, Monoester with Pentaglycerol
Pentaglyceryl Monoisostearate

Trade Name:
Sunsoft A-19E (Taiyo Kagaku)

POLYGLYCERYL-6 ISOSTEARATE

CAS No.: 126928-07-2

JPN Translation:
イソステアリン酸ポリグリセリル - 6

Empirical Formula:
$C_{36}H_{72}O_{14}$

Definition: Polyglyceryl-6 Isostearate is the ester of isostearic acid and Polyglycerin-6 (q.v.).

Information Source: JCLS

Chemical Class: Glyceryl Esters and Derivatives

Functions: Skin-Conditioning Agent - Emollient; Surfactant - Emulsifying Agent

Technical/Other Names:
Hexaglyceryl Isostearate
Isooctadecanoic Acid, Monoester with Hexaglycerol

Trade Names:
Matsunate MI-610 (Matsumoto Trading)
Plurol Isostearique (Gattefosse s.a.)

POLYGLYCERYL-10 ISOSTEARATE

CAS No.: 133738-23-5

JPN Translation:
イソステアリン酸ポリグリセリル - 10

Empirical Formula:
$C_{48}H_{96}O_{22}$

Definition: Polyglyceryl-10 Isostearate is the ester of isostearic acid and Polyglyceirn-10 (q.v.).

Information Source: JCLS

Chemical Class: Glyceryl Esters and Derivatives

Functions: Skin-Conditioning Agent - Emollient; Surfactant - Emulsifying Agent

Technical/Other Names:
Decaglyceryl Monoisostearate
Isooctadecanoic Acid, Monoester with Decaglycerol

Trade Name:
Nikkol Decaglyn 1-IS (Nikko)

POLYGLYCERYL-2 LANOLIN ALCOHOL ETHER

Definition: Polyglyceryl-2 Lanolin Alcohol Ether is an ether of Lanolin Alcohol (q.v.) and Diglycerin (q.v.).

Chemical Classes: Ethers; Lanolin and Lanolin Derivatives

Functions: Skin-Conditioning Agent - Emollient; Surfactant - Emulsifying Agent

Technical/Other Names:
Diglyceryl Monolanolin Ether
Polyglyceryl-2 Lanolin Ether

POLYGLYCERYL-2 LAURATE

CAS No.: 96499-68-2

Empirical Formula:
$C_{18}H_{38}O_6$

Definition: Polyglyceryl-2 Laurate is the ester of lauric acid and Diglycerin (q.v.).

Information Source: JCLS

Chemical Class: Glyceryl Esters and Derivatives

Functions: Skin-Conditioning Agent - Emollient; Surfactant - Emulsifying Agent

Technical/Other Names:
Decanoic Acid, monoester with Oxybis (Propanediol)
Diglyceryl Monolaurate
Lauric Acid, Monoester with Diglycerol

Trade Names:
Dermofeel G 2 L (Straetmans)
Sunsoft Q-12D (Taiyo Kagaku)

Trade Name Mixture:
Nikkol Nikkoguard DL (Nikko)

POLYGLYCERYL-3 LAURATE

CAS No.: 51033-31-9

Empirical Formula:
$C_{21}H_{42}O_8$

Definition: Polyglyceryl-3 Laurate is the ester of lauric acid and Polyglycerin-3 (q.v.).

Information Source: JCLS

Chemical Class: Glyceryl Esters and Derivatives

Functions: Skin-Conditioning Agent - Emollient; Surfactant - Emulsifying Agent

Technical/Other Names:
Dodecanoic Acid, Monoester with 3,3'-[(2-Hydroxy-1,3-Propanediyl)Bis(Oxy)]Bis[1,2-Propanediol]
Dodecanoic Acid, Monoester with Triglycerol
Triglycerol Laurate
Triglyceryl Monolaurate

Trade Names:
Dermofeel G 3 L (Straetmans)
Hydramol TGL (Scher)
Sunsoft A-12C (Taiyo Kagaku)
Sunsoft A-121C (Taiyo Kagaku)

POLYGLYCERYL-4 LAURATE

CAS No.: 75798-42-4

JPN Translation:
ラウリン酸ポリグリセリル - 4

Empirical Formula:
$C_{24}H_{48}O_{10}$

Definition: Polyglyceryl-4 Laurate is the ester of lauric acid and Polyglycerin-4 (q.v.).

Information Sources: JCLS, JSQI

Chemical Class: Glyceryl Esters and Derivatives

Functions: Skin-Conditioning Agent - Emollient; Surfactant - Emulsifying Agent

Technical/Other Names:
Dodecanoic Acid, Monoester with Tetra-glycerol
Tetraglyceryl Monolaurate

POLYGLYCERYL-5 LAURATE

CAS No.: 128738-83-0

JPN Translation:
ラウリン酸ポリグリセリル - 5

Empirical Formula:
$C_{27}H_{54}O_{12}$

Definition: Polyglyceryl-5 Laurate is the ester of lauric acid and a glycerin polymer containing an average of 5 glycerin units.

Information Source: JCLS

Chemical Class: Glyceryl Esters and Derivatives

Functions: Skin-Conditioning Agent - Emollient; Surfactant - Emulsifying Agent

Technical/Other Names:
Dodecanoic Acid, Monoester with Pentaglycerol
Pentaglyceryl Monolaurate

Trade Names:
Dermofeel G 5 L (Straetmans)
Sunsoft A-12E (Taiyo Kagaku)
Sunsoft A-121E (Taiyo Kagaku)

POLYGLYCERYL-6 LAURATE

CAS No.: 51033-38-6

JPN Translation:
ラウリン酸ポリグリセリル - 6

Empirical Formula:
$C_{30}H_{60}O_{14}$

Definition: Polyglyceryl-6 Laurate is the ester of lauric acid and Polyglycerin-6 (q.v.).

Information Source: JCLS

Chemical Class: Glyceryl Esters and Derivatives

Functions: Skin-Conditioning Agent - Emollient; Surfactant - Emulsifying Agent

Technical/Other Names:
Dodecanoic Acid, Monoester with Hexaglycerol
Dodecanoic Acid, Monoester with 4,8,12,16,20-Pentaoxatridecane-1,2,6,10,14,18,22,23-Octol
Hexaglyceryl Monolaurate

Trade Names:
Nikkol Hexaglyn 1-L (Nikko)
Sunsoft Q-12F (Taiyo Kagaku)

POLYGLYCERYL-10 LAURATE

CAS No.: 34406-66-1

JPN Translation:
ラウリン酸ポリグリセリル - 10

Empirical Formula:
$C_{42}H_{84}O_{22}$

Definition: Polyglyceryl-10 Laurate is an ester of lauric acid and Polyglycerin-10 (q.v.).

Information Sources: 21CFR172.854, JCLS, JSQI

Chemical Class: Glyceryl Esters and Derivatives

Functions: Skin-Conditioning Agent - Miscellaneous; Surfactant - Emulsifying Agent

Technical/Other Names:
Decaglycerin Monolaurate
Dodecanoic Acid, Monoester with Decaglycerol
Dodecanoic Acid, Monoester with 4,8,12,16,20,24,28,32,36-Nanoxanonatriacontane-1,2,6,10,14,18,22,26,30,34,38,39-Dodecol

Trade Names:
Dermofeel G 10 L (Straetmans)
Nikkol Decaglyn 1-L (Nikko)
Sunsoft M-12J (Taiyo Kagaku)
Sunsoft Q-12S (Taiyo Kagaku)

Trade Name Mixtures:
Dermofeel G 10 LW (Straetmans)
Nikkol Nikkoguard DL (Nikko)

POLYGLYCERYL-4 LAURYL ETHER

CAS Nos.: 9022-75-7 (Generic); 95012-79-6

Empirical Formula:
$C_{24}H_{50}O_9$

Definition: Polyglyceryl-4 Lauryl Ether is an ether of lauryl alcohol and Polyglycerin-4 (q.v.).

Chemical Class: Ethers

Functions: Skin-Conditioning Agent - Emollient; Surfactant - Emulsifying Agent

Technical/Other Names:
Tetraglycerol, Monododecyl Ether
Tetraglycerol, Monoether with 1-Dodecanol
Tetraglyceryl Monolauryl Ether

Trade Name:
Chimexane NA (Chimex)

POLYGLYCERYL-10 LINOLEATE

Definition: Polyglyceryl-10 Linoleate is the monoester of linoleic acid and Polyglycerin-10 (q.v.).

Chemical Class: Glyceryl Esters and Derivatives

Functions: Skin-Conditioning Agent - Miscellaneous; Surfactant - Emulsifying Agent

Technical/Other Names:
Decaglyceryl Linoleate
Decaglyceryl 9,12-Octadecadienoate

Trade Name:
NIKKOL Decaglyn 1-LN (Nikko)

POLYGLYCERYL-3 METHYLGLUCOSE DISTEARATE

Definition: Polyglyceryl-3 Methylglucose Distearate is the diester of stearic acid and the condensation product of methylglucose and Polyglycerin-3 (q.v.).

Chemical Class: Glyceryl Esters and Derivatives

Functions: Skin-Conditioning Agent - Emollient; Surfactant - Emulsifying Agent

Trade Name:
Tego Care 450 (Degussa Care Specialties)

POLYGLYCERYL-10 MONO/DIOLEATE

Definition: Polyglyceryl-10 Mono/Dioleate is a mixture of mono- and diesters of oleic acid and Polyglycerin-10 (q.v.).

Chemical Class: Glyceryl Esters and Derivatives

Functions: Skin-Conditioning Agent - Emollient; Surfactant - Emulsifying Agent

Trade Name:
Caprol PGE 860 (Abitec Corporation)

POLYGLYCERYL-2 MYRISTATE

Definition: Polyglyceryl-2 Myristate is the monoester of myristic acid and Diglycerol.

Chemical Class: Glyceryl Esters and Derivatives

Functions: Skin-Conditioning Agent - Emollient; Surfactant - Emulsifying Agent

Technical/Other Names:
Diglyceryl Monomyristate
Myristic Acid, Monoester, Monoester, with Diglycerol

Trade Name:
Sunsoft Q-14D (Taiyo Kagaku)

POLYGLYCERYL-3 MYRISTATE

Empirical Formula:
$C_{23}H_{46}O_8$

Definition: Polyglyceryl-3 Myristate is the ester of myristic acid and Polyglycerin-3 (q.v.).

Chemical Class: Glyceryl Esters and Derivatives

Functions: Skin-Conditioning Agent - Emollient; Surfactant - Emulsifying Agent

Technical/Other Names:
Myristic Acid, Monoester with Triglycerol
Triglyceryl Monomyristate

Trade Names:
Sunsoft A-14C (Taiyo Kagaku)
Sunsoft A-141C (Taiyo Kagaku)

POLYGLYCERYL-5 MYRISTATE

Definition: Polyglyceryl-5 Myristate is the monoester of myristic acid and Polyglycerin-5 (q.v.).

Chemical Class: Glyceryl Esters and Derivatives

Functions: Skin-Conditioning Agent - Emollient; Surfactant - Emulsifying Agent

Technical/Other Names:
Myristic Acid, Monoester with Pentaglycerol
Pentaglyceryl Monomyristate

Trade Names:
Sunsoft A-14E (Taiyo Kagaku)
Sunsoft A-141E (Taiyo Kagaku)

POLYGLYCERYL-6 MYRISTATE

Definition: Polyglyceryl-6 Myristate is the monoester of myristic acid and Polyglycerin-6 (q.v.).

Chemical Class: Glyceryl Esters and Derivatives

Functions: Skin-Conditioning Agent - Emollient; Surfactant - Emulsifying Agent

Technical/Other Names:
Hexaglycerol Monomyristate
Myristic Acid Monoester with Hexaglycerol

Trade Names:
NIKKOL Hexaglyn 1-M (Nikko)
Sunsoft Q-14F (Taiyo Kagaku)

POLYGLYCERYL-10 MYRISTATE

CAS No.: 87390-32-7

JPN Translation:
ミリスチン酸ポリグリセリル - 10

Empirical Formula:
$C_{44}H_{88}O_{22}$

Definition: Polyglyceryl-10 Myristate is an ester of myristic acid and Polyglycerin-10 (q.v.).

Information Sources: 21CFR172.854, JCIC, JCLS, JSQI

Chemical Class: Glyceryl Esters and Derivatives

Functions: Skin-Conditioning Agent - Miscellaneous; Surfactant - Emulsifying Agent

Technical/Other Names:
Decaglycerin Monomyristate
Polyglyceryl Monomyristate
Tetradecanoic Acid, Monoester with Decaglycerol

Trade Names:
Nikkol Decaglyn 1-M (Nikko)
Sunsoft Q-14S (Taiyo Kagaku)

POLYGLYCERYL-10 NONAERUCATE

CAS No.: 155808-79-0

Definition: Polyglyceryl-10 Nonaerucate is the nonaester of Erucic Acid (q.v.) and Polyglycerin-10 (q.v.).

Chemical Class: Glyceryl Esters and Derivatives

Functions: Skin-Conditioning Agent - Emollient; Surfactant - Emulsifying Agent

Technical/Other Name:
1,2,3-Propanetriol, Homopolymer, (13Z)-13-Docosenoate

Trade Name:
SY-GLYSTER NE-750 (Sakamoto Yakuhin Kogyo)

POLYGLYCERYL-6 OCTASTEARATE

Definition: Polyglyceryl-6 Octastearate is the octaester of stearic acid and Polyglycerin-6 (q.v.).

Chemical Class: Glyceryl Esters and Derivatives

Functions: Skin-Conditioning Agent - Emollient; Surfactant - Emulsifying Agent

Technical/Other Names:
Hexaglycerol Octastearate
Stearic Acid, Octaester with Hexaglycerol

Trade Names:
Caprol ET (Abitec Corporation)
Pelemol 6G818 (Phoenix)
Sunfat PS-68 (Taiyo Kagaku)

POLYGLYCERYL-2 OLEATE

CAS Nos.: 9007-48-1 (Generic); 49553-76-6

JPN Translation:
オレイン酸ポリグリセリル - 2

Empirical Formula:
$C_{24}H_{46}O_6$

Definition: Polyglyceryl-2 Oleate is an ester of oleic acid and Diglycerin (q.v.).

Information Sources: 21CFR172.854, JCLS, JSQI

Chemical Class: Glyceryl Esters and Derivatives

Functions: Skin-Conditioning Agent - Emollient; Surfactant - Emulsifying Agent

Reported Product Category: Foundations

Technical/Other Names:
Diglyceryl Monooleate
9-Octadecenoic Acid, Ester with 1,2,3-Propanetriol (1:2)
9-Octadecenoic Acid, monoester with Oxybis(Propanediol)

Trade Names:
Matsunate DO-1 (Matsumoto Trading)
Nikkol DGMO-90 (Nikko)
Nikkol DGMO-C (Nikko)
RYLO PG 29 Polyglycerol Ester (Danisco)
Sunsoft Q-17D (Taiyo Kagaku)

POLYGLYCERYL-3 OLEATE

CAS Nos.: 9007-48-1 (Generic); 33940-98-6

JPN Translation:
オレイン酸ポリグリセリル - 3

Empirical Formula:
$C_{27}H_{52}O_8$

Definition: Polyglyceryl-3 Oleate is an ester of oleic acid and Polyglycerin-3 (q.v.).

Information Sources: 21CFR172.854, JCLS, TSCA

Chemical Class: Glyceryl Esters and Derivatives

Functions: Skin-Conditioning Agent - Emollient; Surfactant - Emulsifying Agent

Technical/Other Names:
9-Octadecenoic Acid, Monoester with 3,3'-[(2-Hydroxy-1,3-Propanediyl)Bis (Oxy)Bis[1,2-Propanediol]
9-Octadecenoic Acid, Monoester with Triglycerol
Sunsoft A-17C
Triglyceryl Monooleate

Trade Names:
AEC Polyglyceryl-3 Oleate (A & E Connock)
Caprol 3GO (Abitec Corporation)
Cremophor GO 31 (BASF)
Isolan GO 33 (Degussa Care Specialties)
Rylo PG 13 (Danisco)
Sunsoft A-17C (Taiyo Kagaku)
Sunsoft A-171C (Taiyo Kagaku)

Trade Name Mixtures:
Protegin W (Degussa Care Specialties)
Protegin WX (Degussa Care Specialties)

POLYGLYCERYL-4 OLEATE

CAS Nos.: 9007-48-1 (Generic); 71012-10-7

JPN Translation:
オレイン酸ポリグリセリル - 4

Empirical Formula:
$C_{30}H_{58}O_{10}$

Definition: Polyglyceryl-4 Oleate is an ester of oleic acid and Polyglycerin-4 (q.v.).

Information Sources: 21CFR172.854, CTFA D, JCLS, TSCA

Chemical Class: Glyceryl Esters and Derivatives

Functions: Skin-Conditioning Agent - Emollient; Surfactant - Emulsifying Agent

Reported Product Categories: Hair Sprays (Aerosol Fixatives); Tonics, Dressings, and Other Hair Grooming Aids

Technical/Other Names:
9-Octadecenoic Acid, Monoester with Tetraglycerol
Tetraglyceryl Monooleate

Trade Names:
AEC Polyglyceryl-4 Oleate (A & E Connock)
Jeechem 100 (Jeen)
Nikkol Tetraglyn 1-O (Nikko)
Protachem 100 (Protameen)

Trade Name Mixtures:
AEC Polyglyceryl-4 Oleate (&) PEG-8 Propylene Glycol Cocoate (A & E Connock)
Emulsynt 1055 (International Specialty Products)

POLYGLYCERYL-5 OLEATE

CAS Nos.: 9007-48-1 (Generic); 86529-98-8

JPN Translation:
オレイン酸ポリグリセリル - 5

Empirical Formula:
$C_{33}H_{64}O_{12}$

Definition: Polyglyceryl-5 Oleate is the ester of oleic acid and a glycerin polymer containing an average of 5 glycerin units.

Information Sources: JCLS, TSCA

Chemical Class: Glyceryl Esters and Derivatives

Functions: Skin-Conditioning Agent - Emollient; Surfactant - Emulsifying Agent

Technical/Other Names:
9-Octadecenoic Acid, Monoester with Pentaglyceryl
Pentaglyceryl Monooleate

Trade Names:
Sunsoft A-17E (Taiyo Kagaku)
Sunsoft A-171E (Taiyo Kagaku)

POLYGLYCERYL-6 OLEATE

CAS Nos.: 9007-48-1 (Generic); 79665-92-2

JPN Translation:
オレイン酸ポリグリセリル - 6

Empirical Formula:
$C_{36}H_{70}O_{14}$

Definition: Polyglyceryl-6 Oleate is the ester of oleic acid and Polyglycerin-6 (q.v.).

Information Sources: 21CFR172.854, JCLS, JSQI

Chemical Class: Glyceryl Esters and Derivatives

Functions: Skin-Conditioning Agent - Emollient; Surfactant - Emulsifying Agent

Technical/Other Names:
Hexaglyceryl Oleate
9-Octadecenoic Acid, Monoester with Hexaglycerol

Trade Names:
Nikkol Hexaglyn 1-O (Nikko)
Sunsoft Q-17F (Taiyo Kagaku)

POLYGLYCERYL-8 OLEATE

CAS Nos.: 9007-48-1 (Generic); 75719-56-1

JPN Translation:
オレイン酸ポリグリセリル - 8

Empirical Formula:
$C_{42}H_{82}O_{18}$

Definition: Polyglyceryl-8 Oleate is an ester of oleic acid and a glycerin polymer containing an average of 8 glycerin units.

Information Sources: 21CFR172.854, JCLS, TSCA

Chemical Class: Glyceryl Esters and Derivatives

Functions: Skin-Conditioning Agent - Miscellaneous; Surfactant - Emulsifying Agent

Technical/Other Names:
9-Octadecenoic Acid, Monoester with Octaglycerol
Octaglycerol Oleate
Octaglyceryl Oleate

POLYGLYCERYL-10 OLEATE

CAS Nos.	EINECS No.
9007-48-1 (Generic)	
79665-93-3	279-230-0

JPN Translation:
オレイン酸ポリグリセリル - 10

Empirical Formula:
$C_{48}H_{94}O_{22}$

Definition: Polyglyceryl-10 Oleate is an ester of oleic acid and Polyglycerin-10 (q.v.)

Information Sources: 21CFR172.854, JCLS

Chemical Class: Glyceryl Esters and Derivatives

Functions: Skin-Conditioning Agent - Miscellaneous; Surfactant - Emulsifying Agent

Technical/Other Names:
Decaglyceryl Monooleate
9-Octadecenoic Acid, Monoester with Decaglycerol

Trade Names:
Nikkol Decaglyn 1-O (Nikko)
Nikkol Decaglyn 1-VO (Nikko)
Polyaldo 10-1-0 (Lonza Inc./Lonza Ltd.)
Sunsoft Q-17S (Taiyo Kagaku)
Sunsoft Q-171S (Taiyo Kagaku)

Trade Name Mixtures:
Lyc-O-Mato 2-4%SG NG(LycoRed USA)
Oryza Ceramide-LC (Ikeda)
Oryza Gamma Milky (Ichimaru Pharcos)
Oryza Tocotrienol-L (Oryza Oil)

POLYGLYCERYL-2 OLEYL ETHER

CAS Nos.: 9022-76-8 (Generic); 71032-90-1

JPN Translation:
ポリグリセリル - 2 オレイル

Empirical Formula:
$C_{24}H_{48}O_5$

Definition: Polyglyceryl-2 Oleyl Ether is the ether of oleyl alcohol and Diglycerin (q.v.).

Information Sources: JCIC, JCLS, JSQI

Chemical Class: Ethers

Functions: Hair Conditioning Agent; Skin-Conditioning Agent - Emollient; Surfactant - Emulsifying Agent

Reported Product Category: Hair Dyes and Colors (All Types Requiring Caution Statements and Patch Tests)

Technical/Other Names:
Diglycerin Monooleylether
Diglycerol, Monoether with 9-Octadecen-1-ol
Diglyceryl Monooleyl Ether

1,2-Propanediol, 3-[3-Hydroxy-2-(9-Octadecenyloxy)Propoxy]-

Trade Name:
Chimexane NB (Chimex)

POLYGLYCERYL-4 OLEYL ETHER

CAS Nos.: 9022-76-8 (Generic); 112708-25-5

Empirical Formula:
$C_{30}H_{60}O_9$

Definition: Polyglyceryl-4 Oleyl Ether is an ether of oleyl alcohol and Polyglycerin-4 (q.v.).

Chemical Class: Ethers

Functions: Hair Conditioning Agent; Skin-Conditioning Agent - Emollient; Surfactant - Emulsifying Agent

Reported Product Category: Hair Dyes and Colors (All Types Requiring Caution Statements and Patch Tests)

Technical/Other Names:
Tetraglycerol, Monoether with 9-Octadecen-1-ol
Tetraglyceryl Monooleyl Ether
Tetraglyceryl Oleyl Ether

Trade Name:
Chimexane NC (Chimex)

POLYGLYCERYL-10 PALMATE

Definition: Polyglyceryl-10 Palmate is the ester of Palm Acid (q.v.) and Polyglyceryl-10 (q.v.).

Chemical Class: Glyceryl Esters and Derivatives

Functions: Skin-Conditioning Agent - Miscellaneous; Surfactant - Emulsifying Agent

Technical/Other Names:
Decaglyceryl Palmate
Palm Fatty Acids, Monoester with Decaglycerol

Trade Name Mixture:
Polydermanol ME-11-DA (Creaderm)

POLYGLYCERYL-2 PALMITATE

Definition: Polyglyceryl-2 Palmitate is the monoester of palmitic acid and Diglycerol.

Chemical Class: Glyceryl Esters and Derivatives

Functions: Skin-Conditioning Agent - Emollient; Surfactant - Emulsifying Agent

Technical/Other Names:
Diglyceryl Monopalmitate
Palmitic Acid, Monoester with Diglycerol

Trade Name:
Sunsoft Q-16D (Taiyo Kagaku)

POLYGLYCERYL-3 PALMITATE

Definition: Polyglyceryl-3 Palmitate is an ester of Palmitic Acid (q.v.) and Polyglycerin-3 (q.v.)

Chemical Class: Glyceryl Esters and Derivatives

Functions: Skin-Conditioning Agent - Emollient; Surfactant - Emulsifying Agent

Trade Name:
Dermofeel PP (Straetmans)

POLYGLYCERYL-6 PALMITATE

Definition: Polyglyceryl-6 Palmitate is the ester of Palmitic Acid (q.v.) and Polyglycerin-6 (q.v.).

Chemical Class: Glyceryl Esters and Derivatives

Functions: Skin-Conditioning Agent - Emollient; Surfactant - Emulsifying Agent

Technical/Other Names:
Hexaglycerol Monopalmitate
Palmitic Acid, Monoester, with Hexaglycerol

POLYGLYCERYL-4-PEG-2 COCAMIDE

Definition: Polyglyceryl-4-PEG-2 Cocamide is an ether of PEG-2 cocamide and Polyglycerin-4 (q.v.).

Chemical Class: Alkoxylated Amides

Function: Surfactant - Emulsifying Agent

Technical/Other Names:
Polyglyceryl-3-PEG-2 Cocamide
Tetraglyceryl PEG-2 Cocamide

Trade Name:
Chimexane NJ (Chimex)

POLYGLYCERYL-2-PEG-4 STEARATE

Empirical Formula:
$C_{32}H_{64}O_{10}$

Definition: Polyglyceryl-2-PEG-4 Stearate is an ether of PEG-4 Stearate (q.v.) and Diglycerin (q.v.).

Chemical Classes: Alkoxylated Carboxylic Acids; Glyceryl Esters and Derivatives

Function: Surfactant - Emulsifying Agent

Technical/Other Name:
Diglyceryl PEG-4 Stearate

Trade Names:
Hostacerin DGS (Clariant)
Hostacerin DGS (Clariant GmbH, Personal Care)

POLYGLYCERYL-10 PENTAHYDROXY-STEARATE

Definition: Polyglyceryl-10 Pentahydroxy-stearate is the pentaester of hydroxystearic acid and Polyglycerin-10 (q.v.).

Chemical Class: Glyceryl Esters and Derivatives

Functions: Skin-Conditioning Agent - Emollient; Surfactant - Emulsifying Agent

Technical/Other Names:
Decaglyceryl Pentahydroxystearate
Hydroxystearic Acid, Pentaester with Decaglycerol

Trade Name:
Nikkol Decaglyn-5HS (Nikko)

POLYGLYCERYL-10 PENTAISOSTEARATE

JPN Translation:
ペンタイソステアリン酸ポリグリセリル-10

Definition: Polyglyceryl-10 Pentaisostearate is the pentaester of isostearic acid and Polyglycerin-10 (q.v.).

Information Source: JCLS

Chemical Class: Glyceryl Esters and Derivatives

Functions: Skin-Conditioning Agent - Emollient; Surfactant - Emulsifying Agent

Technical/Other Names:
Decaglyceryl Pentaisostearate
Isostearic Acid, Pentaester with Decaglycerol

Trade Name:
Nikkol Decaglyn 5-IS (Nikko)

POLYGLYCERYL-10 PENTALAURATE

Definition: Polyglyceryl-10 Pentalaurate is the pentaester of Lauric Acid (q.v.) and Polyglycerin-10.

Chemical Class: Glyceryl Esters and Derivatives

Functions: Skin-Conditioning Agent - Emollient; Surfactant - Emulsifying Agent

Technical/Other Names:
Decaglyceryl Pentalaurate
Lauric Acid, Pentaester with Decaglycerol

Trade Name:
Nikkol Decaglyn 5-L (Nikko)

POLYGLYCERYL-10 PENTALINOLEATE

Definition: Polyglyceryl-10 Pentalinoleate is the pentaester of Linoleic Acid (q.v.) and Polyglycerin-10 (q.v.).

Chemical Class: Glyceryl Esters and Derivatives

Functions: Skin-Conditioning Agent - Emollient; Surfactant - Emulsifying Agent

Technical/Other Names:
Decaglyceryl Pentalinoleate
9,12-Octadecadienoic Acid, Pentaester with Decaglycerol

Trade Name:
Nikkol Decaglyn 5-LN (Nikko)

POLYGLYCERYL-5 PENTAMYRISTATE

Definition: Polyglyceryl-5 Pentamyristate is the pentaester of myristic acid and Polyglycerin-5 (q.v.).

Chemical Class: Glyceryl Esters and Derivatives

Functions: Skin-Conditioning Agent - Emollient; Surfactant - Emulsifying Agent

Technical/Other Names:
Myristic Acid, Pentaester with Pentaglycerol
Pentaglyceryl Pentamyristate

Trade Name:
Sunsoft A-145E (Taiyo Kagaku)

POLYGLYCERYL-4 PENTAOLEATE

CAS No.: 103230-29-1

Empirical Formula:
$C_{102}H_{186}O_{14}$

Definition: Polyglyceryl-4 Pentaoleate is the pentaester of Oleic Acid (q.v.) and Polyglyceryl-4 (q.v.).

Information Source: JCLS

Chemical Class: Glyceryl Esters and Derivatives

Functions: Skin-Conditioning Agent -
Emollient; Surfactant - Emulsifying Agent

Technical/Other Names:
9-Octadecanoic Acid (Z)-, Pentaester with
Tetraglycerol
Oleic Acid, Pentaester with Tetraglycerol
Tetraglyceryl Pentaoleate

Trade Names:
Nikkol Tetraglyn 5-O (Nikko)
SY-GLYSTER PO-310 (Sakamoto Yakuhin
Kogyo)

POLYGLYCERYL-6 PENTAOLEATE

CAS No.: 104934-17-0

JPN Translation:
ペンタオレイン酸ポリグリセリル - 6

Empirical Formula:
$C_{108}H_{198}O_{18}$

Definition: Polyglyceryl-6 Pentaoleate is the
pentaester of oleic acid and Polyglycerin-6
(q.v.).

Information Source: JCLS

Chemical Class: Glyceryl Esters and
Derivatives

Functions: Skin-Conditioning Agent -
Emollient; Surfactant - Emulsifying Agent

Technical/Other Names:
Hexaglyceryl Pentaoleate
9-Octadecenoic Acid, Pentaester with
Hexaglycerol

Trade Name:
Nikkol Hexaglyn 5-O (Nikko)

POLYGLYCERYL-10 PENTAOLEATE

CAS No.: 86637-84-5

JPN Translation:
ペンタオレイン酸ポリグリセリル - 10

Empirical Formula:
$C_{120}H_{222}O_{26}$

Definition: Polyglyceryl-10 Pentaoleate is
the pentaester of oleic acid and Polyglycerin-
10 (q.v.).

Information Sources: 21CFR172.854,
JCLS

Chemical Class: Glyceryl Esters and
Derivatives

Functions: Skin-Conditioning Agent -
Emollient; Surfactant - Emulsifying Agent

Technical/Other Names:
Decaglyceryl Pentaoleate
9-Octadecenoic Acid, Pentaester with
Decaglycerol

Trade Names:
Nikkol Decaglyn 5-O (Nikko)
Nikkol Decaglyn 5-O-R (Nikko)
Sunsoft Q-175S (Taiyo Kagaku)

POLYGLYCERYL-3 PENTARICINOLEATE

Definition: Polyglyceryl-3 Pentaricinoleate
is the pentaester of Ricinoleic Acid (q.v.) and
Polyglycerin-3 (q.v.).

Chemical Class: Glyceryl Esters and
Derivatives

Functions: Skin-Conditioning Agent -
Emollient; Surfactant - Emulsifying Agent

Technical/Other Name:
Triglyceryl Pentaricinoleate

Trade Name:
Admul Wol 1403 (Quest International)

POLYGLYCERYL-6 PENTARICINOLEATE

Definition: Polyglyceryl-6 Pentaricinoleate
is the pentaester of Ricinoleic Acid (q.v.) and
Polyglycerin-6 (q.v.).

Information Sources: JCIC, JCLS

Chemical Class: Glyceryl Esters and
Derivatives

Functions: Skin-Conditioning Agent -
Emollient; Surfactant - Emulsifying Agent

Technical/Other Names:
Concentrated Polyglyceryl Pentaricinoleate
Hexaglyceryl Pentaricinoleate
12-Hydroxy-9-Octadecenoic Acid,
Pentaester with Hexaglycerol

POLYGLYCERYL-10 PENTARICINOLEATE

Definition: Polyglyceryl-10 Pentaricinoleate
is the pentaester of Ricinoleic Acid (q.v.) and
Polyglycerin-10 (q.v.).

Chemical Class: Glyceryl Esters and
Derivatives

Functions: Skin-Conditioning Agent -
Emollient; Surfactant - Emulsifying Agent

Technical/Other Names:
Decaglycerol Pentaricinoleate
Ricinoleic Acid, Pentaester with Decagly-
cerol

Trade Name:
Nikkol Decaglyn 5-RN (Nikko)

POLYGLYCERYL-4 PENTASTEARATE

CAS No.: 99570-00-0

Empirical Formula:
$C_{102}H_{196}O_{14}$

Definition: Polyglyceryl-4 Pentastearate is
the pentaester of Stearic Acid (q.v.) and
Polyglycerin-4 (q.v.).

Chemical Class: Glyceryl Esters and
Derivatives

Functions: Skin-Conditioning Agent -
Emollient; Surfactant - Emulsifying Agent

Technical/Other Names:
Octadecanoic Acid, Pentaester with Tetra-
glycerol
Stearic Acid, Pentaester with Tetraglycerol
Tetraglyceryl Pentastearate

Trade Names:
Nikkol Tetraglyn 5-S (Nikko)
SY-GLYSTER PS-310 (Sakamoto Yakuhin
Kogyo)

POLYGLYCERYL-6 PENTASTEARATE

CAS Nos.: 9009-32-9 (Generic); 99734-30-2

Empirical Formula:
$C_{108}H_{208}O_{18}$

Definition: Polyglyceryl-6 Pentastearate is
the pentaester of stearic acid and Poly-
glycerin-6 (q.v.).

Chemical Class: Glyceryl Esters and
Derivatives

Functions: Skin-Conditioning Agent -
Emollient; Surfactant - Emulsifying Agent

Technical/Other Names:
Hexaglyceryl Pentastearate
Octadecanoic Acid, Pentaester with
Hexaglycerol

Trade Names:
Nikkol Hexaglyn 5-S (Nikko)
SY-GLYSTER PS-500 (Sakamoto Yakuhin
Kogyo)

POLYGLYCERYL-10 PENTASTEARATE

CAS Nos.: 9009-32-9 (Generic); 95461-64-6

JPN Translation:
ペンタステアリン酸ポリグリセリル - 10

Empirical Formula:
$C_{120}H_{232}O_{26}$

Definition: Polyglyceryl-10 Pentastearate is
a pentaester of stearic acid and Polyglycerin-
10 (q.v.).

Information Sources: JCIC, JCLS

Chemical Class: Glyceryl Esters and
Derivatives

Functions: Skin-Conditioning Agent - Emollient; Surfactant - Emulsifying Agent

Technical/Other Names:
Decaglyceryl Pentastearate
Octadecanoic Acid, Pentaester with Decaglycerol

Trade Names:
Nikkol Decaglyn 5-S (Nikko)
Sunsoft Q-185S (Taiyo Kagaku)

Trade Name Mixtures:
Nikkol Nikkomulese 41 (Nikko)
Nikkol Nikkomulese 61H (Nikko)

POLYGLYCERYL-3 POLYDIMETHYLSILOXYETHYL DIMETHICONE

Definition: Polyglyceryl-3 Polydimethylsiloxyethyl Dimethicone is the organic compound that conforms generally to the formula:

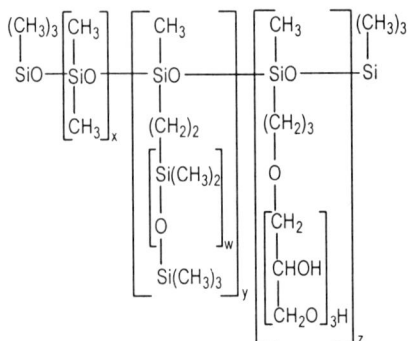

Chemical Classes: Siloxanes and Silanes; Synthetic Polymers

Functions: Hair Conditioning Agent; Skin-Conditioning Agent - Miscellaneous; Surfactant - Cleansing Agent; Surfactant - Emulsifying Agent; Surfactant - Solubilizing Agent; Viscosity Increasing Agent - Aqueous

Trade Names:
KF-6104 (Shin-Etsu Chemical Co.)
KF-6106 (Shin-Etsu Chemical Co.)

Trade Name Mixtures:
KF-6108 (Shin-Etsu Chemical Co.)
SPD-T5 (Shin-Etsu Chemical Co.)
SPD-Z5 (Shin-Etsu Chemical Co.)

POLYGLYCERYL-10 POLYHYDROXYSTEARATE

Definition: Polyglyceryl-10 Polyhydroxystearate is the complex ester formed by the reaction of Polyglycerin-10 and Polyhydroxystearic Acid (q.v.).

Chemical Class: Glyceryl Esters and Derivatives

Function: Surfactant - Emulsifying Agent

Technical/Other Name:
Decaglyceryl Polyhydroxstearate

Trade Name:
Polycare S7 (Biogenico World Wide)

POLYGLYCERYL-3 POLYRICINOLEATE

CAS No.: 29894-35-7 (Generic)

Definition: Polyglyceryl-3 Polyricinoleate is a complex ester of Polyglycerin-3 (q.v.) and a polymer of Ricinoleic Acid (q.v.).

Chemical Classes: Glyceryl Esters and Derivatives; Synthetic Polymers

Functions: Surfactant - Emulsifying Agent; Viscosity Increasing Agent - Nonaqueous

Trade Names:
Cithrol PR (Croda Chemicals)
Dermofeel PR (Straetmans)
Imwitor 600 (Sasol GmbH - Witten)
Radiamuls 2253 (Oleon NV)
Rylo PG 19 (Danisco)

POLYGLYCERYL-4 POLYRICINOLEATE

CAS No.: 29894-35-7 (Generic)

Definition: Polyglyceryl-4 Polyricinoleate is a complex ester of Polyglycerin-4 (q.v.) and a polymer of Ricinoleic Acid (q.v.).

Chemical Classes: Glyceryl Esters and Derivatives; Synthetic Polymers

Functions: Surfactant - Emulsifying Agent; Viscosity Increasing Agent - Nonaqueous

Trade Name:
Sunsoft 818DG (Taiyo Kagaku)

POLYGLYCERYL-5 POLYRICINOLEATE

CAS No.: 29894-35-7 (Generic)

Definition: Polyglyceryl-5 Polyricinoleate is a complex ester of Polyglycerin-5 (q.v.) and a polymer of Ricinoleic Acid (q.v.).

Chemical Classes: Glyceryl Esters and Derivatives; Synthetic Polymers

Functions: Surfactant - Emulsifying Agent; Viscosity Increasing Agent - Nonaqueous

Trade Name:
Sunsoft AZ-18E (Taiyo Kagaku)

POLYGLYCERYL-6 POLYRICINOLEATE

CAS Nos.: 29894-35-7 (Generic); 107615-51-0; 114355-43-0

JPN Translation:
ポリリシノレイン酸ポリグリセリル

Definition: Polyglyceryl-6 Polyricinoleate is a complex ester of a polymer of Ricinoleic Acid (q.v.) and Polyglycerin-6 (q.v.).

Information Sources: JCIC, JCLS

Chemical Class: Glyceryl Esters and Derivatives

Functions: Skin-Conditioning Agent - Emollient; Surfactant - Emulsifying Agent

Technical/Other Names:
Hexaglyceryl Polyricinoleate
9-Octadecenoic Acid, 12-Hydroxy-, polyester with Hexaglycerol

Trade Names:
Nikkol Hexaglyn PR-15 (Nikko)
PELEMOL 6GPR (Phoenix)
Polycare P 5 (Biogenico World Wide)
Sunsof 818H (Taiyo Kagaku)
Sunsoft 818SK (Taiyo Kagaku)
Sunsoft 818TY (Taiyo Kagaku)

Trade Name Mixtures:
DCH45TS (Kobo)
INH73MZ (Kobo)
Sunkatol No.1 (Taiyo Kagaku)

POLYGLYCERYL-10 POLYRICINOLEATE

CAS No.: 29894-35-7 (Generic)

Definition: Polyglyceryl-10 Polyricinoleate is a complex ester of Polyglycerin-10 (q.v.) and a polymer of Ricinoleic Acid (q.v.).

Chemical Class: Glyceryl Esters and Derivatives

Functions: Surfactant - Emulsifying Agent; Viscosity Increasing Agent - Nonaqueous

Trade Name:
Nikkol Decaglyn PR-20 (Nikko)

Trade Name Mixture:
Phytoglyn AO (Maruzen Pharmaceuticals Co., Ltd.)

POLYGLYCERYL-3 RICINOLEATE

CAS No.: 29894-35-7 (Generic)

Empirical Formula:
$C_{27}H_{52}O_9$

Definition: Polyglyceryl-3 Ricinoleate is an ester of ricinoleic acid and Polyglycerin-3 (q.v.).

Chemical Class: Glyceryl Esters and Derivatives

Functions: Skin-Conditioning Agent - Emollient; Surfactant - Emulsifying Agent

Technical/Other Names:
9-Octadecenoic Acid, 12-Hydroxy-, Monoester with Triglycerol
Triglyceryl Monoricinoleate

POLYGLYCERYL-6 RICINOLEATE

JPN Translation:
リシノレイン酸ポリグリセリル - 6

Definition: Polyglyceryl-6 Ricinoleate is the ester of Polyglycerin-6 (q.v.) and Ricinoleic Acid (q.v.).

Chemical Class: Glyceryl Esters and Derivatives

Functions: Skin-Conditioning Agent - Miscellaneous; Surfactant - Emulsifying Agent

Technical/Other Name:
Hexaglyceryl Monoricinoleate

Trade Name Mixture:
Eusolex T-45D (Merck KGaA/EMD Chemicals Inc.)

POLYGLYCERYL-2 SESQUIISOSTEARATE

Definition: Polyglyceryl-2 Sesquiisostearate is a mixture of mono and diesters of Isostearic Acid (q.v.) and Diglycerin (q.v.).

Information Source: CTFA D

Chemical Class: Glyceryl Esters and Derivatives

Functions: Skin-Conditioning Agent - Emollient; Surfactant - Emulsifying Agent

Technical/Other Names:
Diglyceryl Sesquiisostearate
Isooctadecanoic Acid, Sesquiester with Diglycerol

Trade Names:
AEC Polyglyceryl-2 Sesquiisostearate (A & E Connock)
Hostacerin DGI (Clariant)
Hostacerin DGI (Clariant GmbH, Personal Care)

Trade Name Mixtures:
Hostacerin WO (Clariant)
Hostacerin WO (Clariant GmbH, Personal Care)
Hostaphat KML (Clariant)
Hostaphat KML (Clariant GmbH, Personal Care)

POLYGLYCERYL-2 SESQUIOLEATE

JPN Translation:
セスキオレイン酸ポリグリセリル - 2

Definition: Polyglyceryl-2 Sesquioleate is a mixture of mono and diesters of oleic acid and Diglycerin (q.v.).

Information Sources: 21CFR172.854, JCIC, JCLS

Chemical Class: Glyceryl Esters and Derivatives

Functions: Skin-Conditioning Agent - Emollient; Surfactant - Emulsifying Agent

Technical/Other Names:
Diglyceryl Sesquioleate
9-Octadecenoic Acid, Sesquiester with Diglycerol

Trade Names:
AEC Polyglyceryl-2 Sesquioleate (A & E Connock)
Emulsogen OG (Clariant)
Emulsogen OG (Clariant GmbH, Personal Care)
Sunsoft Q-17B (Taiyo Kagaku)

Trade Name Mixture:
Emulgator Apicerol 2/014081 (Symrise)

POLYGLYCERYL-2 SESQUISTEARATE

CAS No.: 9009-32-9 (Generic)

Definition: Polyglyceryl-2 Sesquistearate is a mixture of mono- and diesters of stearic acid and Diglycerin (q.v.).

Chemical Class: Glyceryl Esters and Derivatives

Functions: Skin-Conditioning Agent - Emollient; Surfactant - Emulsifying Agent

Technical/Other Name:
Diglyceryl Monosesquistearate

Trade Name:
Sunsoft Q-18B (Taiyo Kagaku)

POLYGLYCERYL-6 SESQUISTEARATE

CAS No.: 112939-69-2

Definition: Polyglyceryl-6 Sesquistearate is a mixture of mono- and diesters of stearic acid and Polyglycerin-6 (q.v.).

Chemical Class: Glyceryl Esters and Derivatives

Function: Surfactant - Emulsifying Agent

Trade Name:
SY-GLYSTER SS-500 (Sakamoto Yakuhin Kogyo)

POLYGLYCERYL-2 SORBITAN TETRAETHYLHEXANOATE

JPN Translation:
テトラオクタン酸ジグリセロールソルビタン

Definition: Polyglyceryl-2 Sorbitan Tetraethylhexanoate is the tetraester of ethylhexanoic acid and a polyglyceryl ether of sorbitol with an average of 2 moles of glycerin.

Information Source: JCIC

Chemical Class: Sorbitan Derivatives

Function: Surfactant - Emulsifying Agent

Technical/Other Name:
Diglyceryl Sorbitan Tetraoctanoate

POLYGLYCERYL SORBITOL

Definition: Polyglyceryl Sorbitol is a condensation product of glycerin and sorbitol.

Chemical Class: Ethers

Functions: Humectant; Viscosity Decreasing Agent

POLYGLYCERYL-2 STEARATE

CAS Nos.	EINECS No.
9009-32-9 (Generic)	
12694-22-3	235-777-7

JPN Translation:
ステアリン酸ポリグリセリル - 2

Empirical Formula:
$C_{24}H_{48}O_6$

Definition: Polyglyceryl-2 Stearate is the ester of stearic acid and Diglycerin (q.v.).

Information Sources: JCLS, JSQI

Chemical Class: Glyceryl Esters and Derivatives

Functions: Skin-Conditioning Agent - Emollient; Surfactant - Emulsifying Agent

Reported Product Categories: Moisturizing Preparations; Paste Masks (Mud Packs)

Technical/Other Names:
Diglyceryl Stearate
Octadecanoic Acid, Monoester with Diglycerol
9-Octadecanoic Acid, Monoester with Oxybis[Propanediol]

Trade Names:
Hostacerin DGMS (Clariant)
Hostacerin DGMS (Clariant GmbH, Personal Care)
Nikkol DGMS (Nikko)
Sunsoft Q-18D (Taiyo Kagaku)

POLYGLYCERYL-3 STEARATE

CAS Nos.	EINECS No.
26855-43-6	

27321-72-8   248-403-2
37349-34-1 (Generic)

JPN Translation:
ステアリン酸ポリグリセリル - 3

Empirical Formula:
$C_{27}H_{54}O_8$

Definition: Polyglyceryl-3 Stearate is an ester of stearic acid and Polyglycerin-3 (q.v.).

Information Sources: 21CFR172.854, JCLS, TSCA

Chemical Class: Glyceryl Esters and Derivatives

Functions: Skin-Conditioning Agent - Emollient; Surfactant - Emulsifying Agent

Technical/Other Names:
Octadecanoic Acid, Monoester with 3, 3'-[(2-Hydroxy-1,3-Propanediyl)Bis (Oxy)]Bis[1,2-Propanediol-
Octadecanoic Acid, Monoester with Triglycerol
Octadecanoic Acid, Monoester with Tri-1,2, 3-Propanetriol Diether
Triglyceryl Monostearate

Trade Names:
AEC Polyglyceryl-3 Stearate (A & E Connock)
Akoline PG 7 (Karlshamns AB)
Caprol 3GS (Abitec Corporation)
Dermofeel PS (Straetmans)
Polyaldo 3-1-S (TGMS) (Lonza Inc./Lonza Ltd.)
Radiasurf 7248 (Oleon NV)
Rylo PG 18 (Danisco)
Sunsoft A-18C (Taiyo Kagaku)
Sunsoft A-181C (Taiyo Kagaku)

POLYGLYCERYL-4 STEARATE

CAS Nos.: 26855-44-7; 37349-34-1 (Generic); 68004-11-5

JPN Translation:
ステアリン酸ポリグリセリル - 4

Empirical Formula:
$C_{30}H_{60}O_{10}$

Definition: Polyglyceryl-4 Stearate is an ester of stearic acid and Polyglycerin-4 (q.v.).

Information Sources: 21CFR172.854, JCLS, TSCA

Chemical Class: Glyceryl Esters and Derivatives

Function: Surfactant - Emulsifying Agent

Technical/Other Names:
Octadecanoic Acid, Monoester with Tetra-glycerol
Tetraglyceryl Monostearate
Tetraglyceryl Stearate

Trade Names:
AEC Polyglyceryl-4 Stearate (A & E Connock)
Nikkol Tetraglyn 1-S (Nikko)

Trade Name Mixture:
Nikkol Nikkomulese GT (Nikko)

POLYGLYCERYL-5 STEARATE

CAS No.: 37349-34-1 (Generic)

Definition: Polyglyceryl-5 Stearate is the monoester of stearic acid and Polyglycerin-5 (q.v.).

Chemical Class: Glyceryl Esters and Derivatives

Function: Surfactant - Emulsifying Agent

Technical/Other Names:
Pentaglyceryl Monostearate
Stearic Acid, Monoester with Pentaglycerol

Trade Names:
Sunsoft A-18E (Taiyo Kagaku)
Sunsoft A-181E (Taiyo Kagaku)

POLYGLYCERYL-6 STEARATE

CAS No.: 95461-65-7

JPN Translation:
ステアリン酸ポリグリセリル - 6

Definition: Polyglyceryl-6 Stearate is the ester of Stearic Acid (q.v.) and Polyglycerin-6 (q.v.).

Information Source: JCLS

Chemical Class: Glyceryl Esters and Derivatives

Functions: Skin-Conditioning Agent - Emollient; Surfactant - Emulsifying Agent

Technical/Other Names:
Hexaglyceryl Monooctadecanoate
Hexaglyceryl Monostearate
Octadecanoic Acid, Monoester with Hexaglycerol

Trade Names:
NIKKOL Hexaglyn 1-S (Nikko)
Polydermanol PS-104-DA (Blendit)
Sunsoft Q-18F (Taiyo Kagaku)

POLYGLYCERYL-8 STEARATE

CAS Nos.: 37349-34-1 (Generic); 75719-57-2

JPN Translation:
ステアリン酸ポリグリセリル - 8

Empirical Formula:
$C_{42}H_{84}O_{18}$

Definition: Polyglyceryl-8 Stearate is an ester of stearic acid and a glycerin polymer containing an average of 8 glycerin units.

Information Sources: 21CFR172.854, JCLS, TSCA

Chemical Class: Glyceryl Esters and Derivatives

Functions: Skin-Conditioning Agent - Miscellaneous; Surfactant - Emulsifying Agent

Technical/Other Names:
Octadecanoic Acid, Monoester with Octa-glycerol
Octaglycerin Monostearate
Octaglyceryl Stearate

Trade Name:
AEC Polyglyceryl-8 Stearate (A & E Connock)

POLYGLYCERYL-10 STEARATE

CAS Nos.: 9009-32-9 (Generic); 79777-30-3

JPN Translation:
ステアリン酸ポリグリセリル - 10

Empirical Formula:
$C_{48}H_{96}O_{22}$

Definition: Polyglyceryl-10 Stearate is an ester of stearic acid and Polyglycerin-10 (q.v.).

Information Sources: 21CFR172.854, JCLS, JSQI

Chemical Class: Glyceryl Esters and Derivatives

Functions: Skin-Conditioning Agent - Miscellaneous; Surfactant - Emulsifying Agent

Technical/Other Names:
Decaglyceryl Monostearate
Octadecanoic Acid, Monoester with Deca-glycerol

Trade Names:
Nikkol Decaglyn 1-S (Nikko)
Nikkol Decaglyn 1-VS (Nikko)
Polyaldo 10-1-S (Lonza Inc./Lonza Ltd.)
Salacos PGMSV (Nisshin OilliO)
Sunsoft Q-18S (Taiyo Kagaku)

Trade Name Mixtures:
Essential Vital Elements - S (Dipta)
Heliogel (Advanced Beauty)
Nikkol Decaglyn 1-50VS (Nikko)

POLYGLYCERYL-3 STEARATE SE

Definition: Polyglyceryl-3 Stearate SE is a self-emulsifying grade of Polyglyceryl-3 Stearate (q.v.) that contains some sodium and/or potassium stearate.

Information Source: JCLS

Chemical Class: Glyceryl Esters and Derivatives

Function: Surfactant - Emulsifying Agent

Technical/Other Name:
Triglyceryl Monostearate SE

Trade Name:
AEC Polyglyceryl-3 Stearate SE (A & E Connock)

POLYGLYCERYL-4 SWEET ALMONDATE

Definition: Polyglyceryl-4 Sweet Almondate is an ester of the fatty acids derived from sweet almond and Polyglycerin-4 (q.v.).

Chemical Class: Glyceryl Esters and Derivatives

Functions: Skin-Conditioning Agent - Miscellaneous; Surfactant - Emulsifying Agent

Technical/Other Names:
Sweet Almond Fatty Acids, Monoester with Tetraglycerin
Tetraglyceryl Sweet Almondate

Trade Name Mixture:
Polydermanol ME-11-DA (Creaderm)

POLYGLYCERYL-6 TETRABEHENATE

Definition: Polyglyceryl-6 Tetrabehenate is the tetraester of Behenic Acid (q.v.) and Polyglycerin-6 (q.v.).

Chemical Class: Glyceryl Esters and Derivatives

Functions: Skin-Conditioning Agent - Emollient; Surfactant - Emulsifying Agent

Trade Name:
Nikkol Hexaglyn 4-B (Nikko)

POLYGLYCERYL-2 TETRAISOSTEARATE

CAS No.: 121440-30-0

JPN Translation:
テトライソステアリン酸ポリグリセリル -2

Empirical Formula:
$C_{78}H_{150}O_9$

Definition: Polyglyceryl-2 Tetraisostearate is the tetraester of isostearic acid and a dimer of glycerin.

Information Sources: JCIC, JCLS

Chemical Class: Glyceryl Esters and Derivatives

Functions: Skin-Conditioning Agent - Emollient; Surfactant - Emulsifying Agent

Technical/Other Names:
Diglyceryl Tetraisostearate
Isooctadecanoic Acid, Tetraester with Diglycerol
Isooctadecanoic Acid, tetraester with Oxybis(Propanediol)
Polyglyceryl Tetraisostearate

Trade Names:
AEC Polyglyceryl-2 Tetraisostearate (A & E Connock)
Cosmol 44 (Nisshin OilliO)
PRISORINE 3794 (Uniqema Europe)
Salacos 44 (Nisshin OilliO)

POLYGLYCERYL-2 TETRAOLEATE

Definition: Polyglyceryl-2 Tetraoleate is the tetraester of oleic acid and Diglycerin (q.v.).

Chemical Class: Glyceryl Esters and Derivatives

Functions: Skin-Conditioning Agent - Miscellaneous; Surfactant - Emulsifying Agent

Trade Name:
SY-DGTO (Sakamoto Yakuhin Kogyo)

POLYGLYCERYL-6 TETRAOLEATE

CAS No.: 128774-95-8

Definition: Polyglyceryl-6 Tetraoleate is the tetraester of Oleic Acid (q.v.) and Polyglycerin-6 (q.v.).

Chemical Class: Glyceryl Esters and Derivatives

Functions: Skin-Conditioning Agent - Emollient; Surfactant - Emulsifying Agent

Technical/Other Names:
Hexaglycerol Tetraoleate
9-Octadecenoic Acid, Tetraester with Hexaglycerol
Oleic Acid, Tetraester with Hexaglycerol

POLYGLYCERYL-10 TETRAOLEATE

CAS No.	EINECS No.
34424-98-1	252-011-7

Empirical Formula:
$C_{102}H_{190}O_{25}$

Definition: Polyglyceryl-10 Tetraoleate is a tetraester of oleic acid and Polyglycerin-10 (q.v.).

Information Sources: 21CFR172.854, CTFA D, TSCA

Chemical Class: Glyceryl Esters and Derivatives

Functions: Skin-Conditioning Agent - Emollient; Surfactant - Emulsifying Agent

Technical/Other Names:
Decaglyceryl Tetraoleate
9-Octadecenoic Acid, Tetraester with Decaglycerol
9-Octadecenoic Acid, Tetraester with 4,8,12,16,20,24,28,32,36-Nonaoxa-nonatriacontane-1,2,6,10,14,18,22,26,30,34,38,39-Dodecol
Oleic Acid, Tetraester with Decaglycerol

Trade Names:
AEC Polyglyceryl-10 Tetraoleate (A & E Connock)
Caprol 10G40 (Abitec Corporation)

POLYGLYCERYL-2 TETRASTEARATE

CAS Nos.: 9009-32-9 (Generic); 72347-89-8

Empirical Formula:
$C_{78}H_{150}O_9$

Definition: Polyglyceryl-2 Tetrastearate is the tetraester of stearic acid and Diglycerin (q.v.).

Information Source: 21CFR172.854

Chemical Class: Glyceryl Esters and Derivatives

Functions: Skin-Conditioning Agent - Emollient; Surfactant - Emulsifying Agent

Technical/Other Names:
Diglycerin Tetrastearate
Diglyceryl Tetrastearate
Octadecanoic Acid, Tetraester with Diglycerol

Trade Name:
AEC Polyglyceryl-2 Tetrastearate (A & E Connock)

POLYGLYCERYL-10 TRIDECANOATE

CAS No.: 217782-56-4

Definition: Polyglyceryl-10 Tridecanoate is the triester of decanoic acid and Polyglyceryl-10 (q.v.).

Chemical Class: Glyceryl Esters and Derivatives

Functions: Skin-Conditioning Agent - Emollient; Surfactant - Emulsifying Agent

Technical/Other Names:
Decaglyceryl Tridecanoate
Decanoic Acid, Triester with Decaglycerol

Trade Name:
Nikkol Decaglyn-3D (Nikko)

POLYGLYCERYL-10 TRIERUCATE

Definition: Polyglyceryl-10 Trierucate is the triester of Polyglycerin-10 (q.v.) and Erucic Acid (q.v.).

Chemical Class: Glyceryl Esters and Derivatives

Functions: Surfactant - Emulsifying Agent; Surfactant - Suspending Agent

Technical/Other Names:
Decaglyceryl Trierucate
Erucic Acid, Triester with Decaglycerol

Trade Name:
Ryoto Polyglycerol Ester ER-30D (Mitsubishi-Kagaku)

POLYGLYCERYL-2 TRIISOSTEARATE

CAS No.: 120486-24-0

JPN Translation:
トリイソステアリン酸ポリグリセリル - 2

Empirical Formula:
$C_{60}H_{116}O_8$

Definition: Polyglyceryl-2 Triisostearate is the triester of isostearic acid and Diglycerin (q.v.).

Information Sources: JCIC, JCLS, JSQI

Chemical Class: Glyceryl Esters and Derivatives

Functions: Skin-Conditioning Agent - Emollient; Surfactant - Emulsifying Agent

Reported Product Categories: Lipsticks; Rouges; Makeup Preparations (Not eye), Misc.

Technical/Other Names:
Diglyceryl Triisostearate
Isooctadecanoic Acid, Triester with Diglycerol
Isooctadecanoic Acid, Triester with Oxybis (Propanediol)

Trade Names:
AEC Polyglyceryl-2 Triisostearate (A & E Connock)
Cosmol 43 (Nisshin OilliO)
PRISORINE 3793 (Uniqema Europe)
Salacos 43 (Nisshin OilliO)
Technol F3 (Technonet)

POLYGLYCERYL-3 TRIISOSTEARATE

CAS No.: 66082-43-7

Definition: Polyglyceryl-3 Triisostearate is the triester of Isostearic Acid (q.v.) and Poly-glycerin-3 (q.v.).

Information Source: JCLS

Chemical Class: Glyceryl Esters and Derivatives

Functions: Skin-Conditioning Agent - Emollient; Surfactant - Emulsifying Agent

Technical/Other Names:
Isooctadecanoic Acid, Triester with Triglycerol
Triglyceryl Triisostearate

Trade Name:
PRISORINE 3701 (Uniqema Europe)

POLYGLYCERYL-10 TRIISOSTEARATE

Definition: Polyglyceryl-10 Triisostearate is the triester of Polyglycerin-10 (q.v.) and Isostearic Acid (q.v.).

Chemical Class: Glyceryl Esters and Derivatives

Function: Surfactant - Emulsifying Agent

Technical/Other Names:
Decaglyceryl Triisostearate
Isostearic Acid, Triester with Decaglycerol

Trade Name:
Emalex TISG-10 (Nihon Emulsion)

POLYGLYCERYL-5 TRIMYRISTATE

Definition: Polyglyceryl-5 Trimyristate is the triester of myristic acid and Polyglycerin-5 (q.v.).

Chemical Class: Glyceryl Esters and Derivatives

Functions: Skin-Conditioning Agent - Emollient; Surfactant - Emulsifying Agent

Technical/Other Names:
Myristic Acid, Triester with Pentaglycerol
Pentaglyceryl Trimyristate

Trade Name:
Sunsoft A-143E (Taiyo Kagaku)

POLYGLYCERYL-5 TRIOLEATE

Definition: Polyglyceryl-5 Trioleate is the triester of oleic acid and Polyglycerin-5 (q.v.).

Chemical Class: Glyceryl Esters and Derivatives

Functions: Skin-Conditioning Agent - Emollient; Surfactant - Emulsifying Agent

Technical/Other Names:
Oleic Acid, Triester with Pentaglycerol
Pentaglyceryl Trioleate

Trade Name:
Sunsoft A-173E (Taiyo Kagaku)

POLYGLYCERYL-10 TRIOLEATE

CAS No.: 102051-00-3

JPN Translation:
トリオレイン酸ポリグリセリル - 10

Empirical Formula:
$C_{84}H_{158}O_{24}$

Definition: Polyglyceryl-10 Trioleate is the triester of oleic acid and Polyglycerin-10 (q.v.).

Information Sources: JCIC, JCLS

Chemical Class: Glyceryl Esters and Derivatives

Function: Surfactant - Emulsifying Agent

Technical/Other Names:
Decaglyceryl Trioleate
9-Octadecenoic Acid, Triester with Deca-glycerol
Polyglyceryl Trioleate

Trade Name:
Nikkol Decaglyn 3-O (Nikko)

POLYGLYCERYL-4 TRISTEARATE

CAS No.: 99734-29-9

Empirical Formula:
$C_{66}H_{128}O_{12}$

Definition: Polyglyceryl-4 Tristearate is the triester of Stearic Acid (q.v.) and Polyglycerin-4 (q.v.).

Chemical Class: Glyceryl Esters and Derivatives

Functions: Skin-Conditioning Agent - Emollient; Surfactant - Emulsifying Agent

Technical/Other Names:
Octadecanoic Acid, Triester with Tetragly-cerol
Stearic Acid, Triester with Tetraglycerol
Tetraglyceryl Tristearate

Trade Names:
Nikkol Tetraglyn 3-S (Nikko)
SY-GLYSTER TS-310 (Sakamoto Yakuhin Kogyo)

POLYGLYCERYL-5 TRISTEARATE

CAS No.: 9009-32-9 (Generic)

Definition: Polyglyceryl-5 Tristearate is the triester of stearic acid and Polyglycerin-5 (q.v.).

Chemical Class: Glyceryl Esters and Derivatives

Function: Surfactant - Emulsifying Agent

International Cosmetic Ingredient Dictionary and Handbook

Technical/Other Names:
Octadecanoic Acid, Triester with Pentaglycerol
Pentaglyceryl Tristearate

Trade Name:
Sunsoft A-183E (Taiyo Kagaku)

POLYGLYCERYL-6 TRISTEARATE

CAS Nos.: 9009-32-9 (Generic); 71185-87-0

Empirical Formula:
$C_{72}H_{140}O_{16}$

Definition: Polyglyceryl-6 Tristearate is the triester of stearic acid and Polyglycerin-6 (q.v.).

Chemical Class: Glyceryl Esters and Derivatives

Function: Surfactant - Emulsifying Agent

Technical/Other Names:
Hexaglyceryl Tristearate
Octadecanoic Acid, Triester with Hexaglycerol

Trade Names:
Nikkol Hexaglyn 3-S (Nikko)
SY-GLYSTER TS-500 (Sakamoto Yakuhin Kogyo)

POLYGLYCERYL-10 TRISTEARATE

CAS Nos.: 9009-32-9 (Generic); 12709-64-7

JPN Translation:
トリステアリン酸ポリグリセリル - 10

Empirical Formula:
$C_{84}H_{164}O_{24}$

Definition: Polyglyceryl-10 Tristearate is the triester of stearic acid and Polyglycerin-10 (q.v.).

Information Sources: JCIC, JCLS

Chemical Class: Glyceryl Esters and Derivatives

Functions: Skin-Conditioning Agent - Emollient; Surfactant - Emulsifying Agent

Technical/Other Names:
Decaglyceryl Tristearate
Octadecanoic Acid, Triester with Decaglycerol
Octadecanoic Acid, Triester with 4,8,12,16,20,24,28,32,36-Nonaoxanonatriacontane-1,2,6,10,14,18,22,26,30,34,38,39-Dodecol
Polyglyceryl Tristearate

Trade Name:
Nikkol Decaglyn 3-S (Nikko)

POLYGLYCERYL-6 UNDECYLENATE

Definition: Polyglyceryl-6 Undecylenate is the an ester of Undecylenic Acid (q.v.) and Polyglycerin-6 (q.v.).

Chemical Class: Glyceryl Esters and Derivatives

Function: Surfactant - Emulsifying Agent

Technical/Other Names:
Hexaglyceryl Monoundecylenate
Undecylenic Acid, Monoester, with Hexaglycerol

Trade Name:
Polyglycerine Mono Undecylenate-6 (Hokoku)

POLYGLYCERYL-10 UNDECYLENATE

Definition: Polyglyceryl-10 Undecylenate is an ester of Undecylenic Acid (q.v.) and Polyglycerin-10 (q.v.).

Chemical Class: Glyceryl Esters and Derivatives

Function: Surfactant - Emulsifying Agent

Technical/Other Names:
Decaglyceryl Monoundecylenate
Undecylenic Acid, Monoester, with Decaglycerol

Trade Name:
Polyglycerine Mono Undecylenate-10 (Hokoku)

POLYGONATUM MULTIFLORUM EXTRACT

Definition: Polygonatum Multiflorum Extract is an extract of the rhizomes and roots of the solomon's seal, *Polygonatum multiflorum. See "Regulatory and Ingredient Use Information," regarding the labeling names for botanical ingredients in Volume 1, Introduction, Part A.*

Chemical Class: Biological Products

Function: Not Reported

Technical/Other Names:
Extract of Polygonatum Multiflorum
Extract of Solomon's Seal
Solomon's Seal Extract
Solomon's Seal (Polygonatum Multiflorum) Extract

Trade Name Mixtures:
Actiphyte of Solomon's Seal BG50 (Active Organics)
Actiphyte of Solomon's Seal GL50 (Active Organics)
Actiphyte of Solomon's Seal Lipo S (Active Organics)
Actiphyte of Solomon's Seal PG50 (Active Organics)
Phytotonine (Greentech S.A)

POLYGONATUM OFFICINALE EXTRACT

CAS No.: 90082-98-7

Definition: Polygonatum Officinale Extract is an extract of the rhizomes and roots of the solomon's seal, *Polygonatum officinale. See "Regulatory and Ingredient Use Information," regarding the labeling names for botanical ingredients in Volume 1, Introduction, Part A.*

Chemical Class: Biological Products

Function: Skin-Conditioning Agent - Miscellaneous

Technical/Other Names:
Extract of Polygonatum Officinale
Extract of Solomon's Seal
Solomon's Seal Extract
Solomon's Seal (Polygonatum Officinale) Extract

Trade Name Mixtures:
Heavy Legs Complex MU 3422 (Greentech S.A)
ROSACEA (Greentech S.A)
Solomon's Seal HG (Alban Muller)
Solomon's Seal HS (Alban Muller)

POLYGONUM AVICULARE EXTRACT

CAS No.	EINECS No.
84604-04-6	283-280-9

Definition: Polygonum Aviculare Extract is an extract of the knotweed, *Polygonum aviculare. See "Regulatory and Ingredient Use Information," regarding the labeling names for botanical ingredients in Volume 1, Introduction, Part A.*

Chemical Class: Biological Products

Function: Not Reported

Technical/Other Names:
Extract of Knotweed
Extract of Polygonum Aviculare
Knotweed Extract
Knotweed (Polygonum Aviculare) Extract

Trade Name Mixtures:
Bird's Knotgrass HS (Alban Muller)
Solomons Seal Extract HS 2432 G (Grau)

POLYGONUM BISTORTA ROOT EXTRACT

CAS No.	EINECS No.
84012-36-2	281-680-8

Definition: Polygonum Bistorta Root Extract is an extract of the roots of the bistort, *Polygonum bistorta. See "Regulatory and Ingredient Use Information," regarding the labeling names for botanical ingredients in Volume 1, Introduction, Part A.*

Chemical Class: Biological Products

Function: Skin-Conditioning Agent - Miscellaneous

Technical/Other Names:
Bistort Extract
Bistort (Polygonum Bistorta) Extract
Extract of Bistort
Extract of Polygonum Bistorta

Trade Name:
Ibukitoranoo Extract (Koei Perfumery)

Trade Name Mixtures:
Actiphyte of Bistort BG50 (Active Organics)
Actiphyte of Bistort GL50 (Active Organics)
Actiphyte of Bistort Lipo S (Active Organics)
Actiphyte of Bistort PG50 (Active Organics)
Bistort HS (Alban Muller)
Phytelene of Bistort Root EG 485 liquid (Indena SA)
Phytogreen 55 of Bistort Root EXH 694 Liquid (Phytochim)
Polygonum Bistorta Root Extract ies (IES LABO)
VT-122 Extract of Bistort (Vege-Tech)
VT-125 Extract of Snakeroot (Vege-Tech)

POLYGONUM CUSPIDATUM EXTRACT

Definition: Polygonum Cuspidatum Extract is an extract of the whole plant, *Polygonum cuspidatum. See "Regulatory and Ingredient Use Information," regarding the labeling names for botanical ingredients in Volume 1, Introduction, Part A.*

Chemical Class: Biological Products

Functions: Cosmetic Astringent; Skin-Conditioning Agent - Miscellaneous

Trade Name Mixture:
Actiphyte of Japanese Knotweed (Active Organics)

POLYGONUM CUSPIDATUM ROOT EXTRACT

JPN Translation:
イタドリエキス

Definition: Polygonum Cuspidatum Root Extract is an extract of the roots of *Polygonum cuspidatum. See "Regulatory and Ingredient Use Information," regarding the labeling names for botanical ingredients in Volume 1, Introduction, Part A.*

Chemical Class: Biological Products

Function: Antioxidant

Trade Names:
Extract PC-80 (Sino Lion)
Resveratrol 98-OSST (Hunan Osst)

POLYGONUM FAGOPYRUM (BUCKWHEAT) LEAF EXTRACT

CAS No. 89958-09-8
EINECS No. 289-631-2

Definition: Polygonum Fagopyrum (Buckwheat) Leaf Extract is an extract of the leaves and shoots of the buckwheat, *Polygonum fagopyrum. See "Regulatory and Ingredient Use Information," regarding the labeling names for botanical ingredients in Volume 1, Introduction, Part A.*

Chemical Class: Biological Products

Function: Skin-Conditioning Agent - Miscellaneous

Technical/Other Names:
Buckwheat Extract
Buckwheat Leaf Extract
Extract of Buckwheat
Extract of Polygonum Fagopyrum
Fagopyrum Sagitatum Extract

Trade Name Mixtures:
Buckwheat Extract HS 2861 G (Grau)
Buckwheat Extract, Watersoluble (Crodarom)
Buckwheat HS (Alban Muller)
Buckwheat LS (Alban Muller)
Extrait De Sarrasin PPE PG40 (Yves Rocher)
Glycolysat of 7 Cereals (CEP (Solabia))

POLYGONUM FAGOPYRUM SEED EXTRACT

Definition: Polygonum Fagopyrum Seed Extract is an extract of the seeds of *Polygonum fagopyrum. See "Regulatory and Ingredient Use Information," regarding the labeling names for botanical ingredients in Volume 1, Introduction, Part A.*

Chemical Class: Biological Products

Function: Not Reported

Trade Name:
Polygonum Fagopyrum Seed Extract (Codif)

Trade Name Mixtures:
Buckwheat Extract HS 28606 (Grau)
Buckwheat Extract HS 2860 G (Grau)
Cire De Ble Noir/Buckwheat Wax (Codif)

POLYGONUM MULTIFLORUM EXTRACT

Definition: Polygonum Multiflorum Extract is an extract of the herb, *Polygonum multiflorum. See "Regulatory and Ingredient Use Information," regarding the labeling names for botanical ingredients in Volume 1, Introduction, Part A.*

Chemical Class: Biological Products

Function: Not Reported

Technical/Other Name:
Extract of Polygonum Multiflorum

Trade Name Mixtures:
Campo He Shou Wu Extract (Campo)
Polygonum Phytexcell (Crodarom)
VT-238 Extract of Fo-Ti (Vege-Tech)

POLYGONUM MULTIFLORUM ROOT EXTRACT

Definition: Polygonum Multiflorum Root Extract is an extract of the roots of the polygonum, *Polygonum multiflorum. See "Regulatory and Ingredient Use Information," regarding the labeling names for botanical ingredients in Volume 1, Introduction, Part A.*

Chemical Class: Biological Products

Function: Not Reported

Technical/Other Names:
Extract of Polygonum
Extract of Polygonum Multiflorum
Polygonum Extract
Polygonum Multiflorum Extract

Trade Name:
Viapure Polygonum (Actives International)

Trade Name Mixtures:
Actiphyte Fo-Ti Root BG50P (Active Organics)
Actiphyte Fo-Ti Root GL (Active Organics)
Actiphyte Fo-Ti Root Lipo S (Active Organics)
Actiphyte Fo-Ti Root Root (Active Organics)
China Extract Ho Shou Wu (E.U.K)
Chine Extract Polygonum Multiflorum (Ennagram)
Polygoni Multiflori Extract (Maruzen Pharmaceuticals Co., Ltd.)
POLYGONI MULTIFLORI EXTRACT BG (Maruzen Pharmaceuticals Co., Ltd.)

POLYGONUM PERSICARIA EXTRACT

CAS No. 90082-99-8
EINECS No. 290-192-4

Definition: Polygonum Persicaria Extract is an extract of the plant, *Polygonum persicaria. See "Regulatory and Ingredient Use Information," regarding the labeling names for botanical ingredients in Volume 1, Introduction, Part A.*

Chemical Class: Biological Products

Function: Skin-Conditioning Agent - Miscellaneous

Technical/Other Name:
Extract of Polygonum Persicaria

Trade Name Mixture:
Persicary HS (Alban Muller)

POLY HEMA GLUCOSIDE

Definition: Poly HEMA Glucoside is a polymer of glucosyl hydroxyethyl methacrylate.

Chemical Classes: Carbohydrates; Synthetic Polymers

Functions: Hair Conditioning Agent; Skin-Conditioning Agent - Emollient

Trade Name:
P-GEMA-S (Nippon Chemical)

POLYHYDROXYSTEARIC ACID

CAS Nos.: 27924-99-8; 58128-22-6

JPN Translation:
ポリヒドロキシステアリン酸

Definition: Polyhydroxystearic Acid is a polymer of Hydroxystearic Acid (q.v.)

Information Source: TSCA

Chemical Class: Synthetic Polymers

Function: Suspending Agent - Nonsurfactant

Technical/Other Name:
Polyhydroxyoctadecanoic Acid

Trade Names:
Arlacel P100 (Uniqema Americas)
Solsperse 21000 (Zeneca)

Trade Name Mixtures:
Eusolex T-Olio F (Merck KGaA)
Solaveil CT-100 (Uniqema Europe)
Spectraveil FIN (Uniqema, Belgium)
Spectraveil IPM (Uniqema, Belgium)
Spectraveil MOTG (Uniqema, Belgium)
Spectraveil OP (Uniqema, Belgium)
Spectraveil TG (Uniqema, Belgium)
Tioveil EUT (Uniqema, Belgium)
Tioveil FCM (Uniqema, Belgium)
Tioveil 50 FCM (Uniqema, Belgium)
Tioveil FIN (Uniqema, Belgium)
Tioveil 50 FIN (Uniqema, Belgium)
Tioveil GCM (Uniqema, Belgium)
Tioveil 50 GCM (Uniqema, Belgium)
Tioveil IPM (Uniqema, Belgium)
Tioveil 50 IPM (Uniqema, Belgium)
Tioveil MOTG (Uniqema, Belgium)
Tioveil 50 MOTG (Uniqema, Belgium)
Tioveil OP (Uniqema, Belgium)
Tioveil 50 OP (Uniqema, Belgium)
Tioveil TG (Uniqema, Belgium)
Tioveil 50 TG (Uniqema, Belgium)
Tioveil TGOP (Uniqema, Belgium)
Tioveil 50 TGOP (Uniqema, Belgium)

POLYIMIDE-1

CAS No.: 497926-97-3

Definition: Polyimide-1 is a terpolymer that is made by reacting poly(isobutylene-alt-maleic anhydride) with dimethylamino-propylamine and methoxy-PEG/PPG-31/9-2-propylamine in a mixture of ethanol and Water (q.v.). The resulting polymer contains both imide, ester, and acid functionality.

Chemical Class: Synthetic Polymers

Functions: Film Former; Hair Fixative

Technical/Other Name:
2,5-Furandione, Polymer with 2-Methyl-1-Propene, Ethyl Ester, Reaction Products with N,N-Dimethyl-1,3-Propanediamine and Polyethylene-Polypropylene Glycol 2-Aminopropyl Me Ether

Trade Name:
Aquaflex XL-30 (International Specialty Products)

POLYISOBUTENE

CAS No.: 9003-27-4

JPN Translation:
ポリイソブテン

Empirical Formula:
$(C_4H_8)_x$

Definition: Polyisobutene is the homopolymer of isobutylene that conforms generally to the formula:

$$\left[\begin{array}{c} CH_2CHCH_2 \\ | \\ CH_3 \end{array}\right]_x$$

Information Sources: 21CFR172.615, 21CFR175.105, 21CFR175.125, 21CFR175.300, 21CFR176.180, 21CFR177.1200, 21CFR177.1210, 21CFR177.1420, 21CFR178.3570, 21CFR178.3740, 21CFR178.3910, FCC, TSCA

Chemical Class: Synthetic Polymers

Functions: Binder; Film Former; Viscosity Increasing Agent - Nonaqueous

Reported Product Categories: Lipsticks; Mascara

Technical/Other Names:
Isobutene Homopolymer
Isobutylene Homopolymer
2-Methyl-1-Propene, Homopolymer
Permethyl 108A
Polyisobutylene
1-Propene,-2-Methyl-, Homopolymer

Trade Names:
AEC Polyisobutene (A & E Connock)
Creasil IC (C.I.T.)
Creasil IP (C.I.T.)
ESP P IB 731 (Earth Supplied Products)
ESP PIB 0611 (Earth Supplied Products)
ESP PIB 731 (Earth Supplied Products)
ESP PIB 5011 (Earth Supplied Products)
Permethyl 104A (Presperse)
Permethyl 106A (Presperse)
Permethyl 108A (Presperse)
Rewopal PIB 1000 (Degussa Care Specialties)

Trade Name Mixtures:
Fancorsil P (Fanning)
Simulgel EPG (SEPPIC)

POLYISOPRENE

CAS No.: 9003-31-0

JPN Translation:
ポリイソプレン

Empirical Formula:
$(C_5H_8)_x$

Definition: Polyisoprene is the polymer of isoprene that conforms generally to the formula:

$$\left[\begin{array}{c} CH_2C = CHCH_2 \\ | \\ CH_3 \end{array}\right]_x$$

Information Sources: 21CFR175.105, 21CFR175.125, 21CFR176.180, 21CFR177.2600, JCIC, JCLS, TSCA

Chemical Class: Synthetic Polymers

Function: Viscosity Increasing Agent - Nonaqueous

Technical/Other Names:
1,3-Butadiene, 2-Methyl-, Homopolymer
Isoprene Homopolymer
Liquid Polyisoprene
2-Methyl-1,3-Butadiene, Homopolymer
Poly(isopentadiene)

Trade Name:
Syntesqual (Vevy)

Trade Name Mixtures:
Lipogelag (Vevy)
Sebopessina (Vevy)

POLYLACTATE/RICINOLEATE

Definition: Polylactate/Ricinoleate is a polymer of the ester formed by the reaction of Lactic Acid (q.v.). with Ricinoleic Acid (q.v.).

Chemical Class: Synthetic Polymers

Function: Skin-Conditioning Agent - Miscellaneous

Trade Name:
Ethox PLPR (Ethox)

POLY LACTOBIONAMIDOMETHYL-STYRENE

Definition: Poly Lactobionamidomethyl-styrene is the polymer that conforms to the formula:

Chemical Class: Synthetic Polymers

Function: Skin-Conditioning Agent - Humectant

Trade Name Mixture:
Cyto-effector LA (Pola Chemical.)

POLYLYSINE

CAS No.: 25104-18-1 (L-Form)

Empirical Formula:
$(C_6H_{12}N_2O)_x$

Definition: Polylysine is the polyamide that conforms to the formula:

Information Source: MI-13(7654)

Chemical Classes: Amides; Synthetic Polymers

Functions: Film Former; Hair Conditioning Agent

Technical/Other Names:
Lysine Homopolymer
Poly(Imino(-1-(4-Aminobutyl)-2-Oxo-1,2-Ethanediyl

Trade Names:
CPC Peptide (Solabia)
Oligolysine (Chisso)
Polylysine (Ajinomoto)

POLYLYSINE HBR

Definition: Polylysine HBr is the amine salt that conforms to the formula:

Chemical Classes: Amines; Amino Acids
Function: Not Reported

POLYMETHACRYLAMIDE

JPN Translation:
ポリメタクリルアミド
Definition: Polymethacrylamide is the polymer that conforms to the formula:

Information Source: JCIC
Chemical Classes: Amides; Synthetic Polymers
Function: Film Former

POLYMETHACRYLAMIDOPROPYL-TRIMONIUM CHLORIDE

CAS No.: 68039-13-4
Empirical Formula:
$(C_{10}H_{21}N_2O \cdot Cl)_x$
Definition: Polymethacrylamidopropyl-trimonium Chloride is the polymeric quaternary ammonium salt that conforms to the formula:

Chemical Classes: Quaternary Ammonium Compounds; Synthetic Polymers

Functions: Antistatic Agent; Hair Conditioning Agent

Reported Product Categories: Hair Dyes and Colors (All Types Requiring Caution Statements and Patch Tests); Shampoos (Non-coloring); Hair Conditioners; Hair Preparations (Non-coloring), Misc.; Permanent Waves

Technical/Other Names:
Polydimethylaminopropyl Methacrylamide Methylchloride Quaternium
1-Propanaminium, N,N,N-Trimethyl-3-[(2-Methyl-1-Oxo-2-Propenyl)Amino]-, Chloride, Homopolymer

Trade Names:
Polycare 133 (Rhodia)
Polycare S (Rhodia)

POLYMETHACRYLAMIDOPROPYL-TRIMONIUM METHOSULFATE

CAS No.: 181788-04-5

Definition: Polymethacrylamidopropyl-trimonium Methosulfate is the polymeric quaternary ammonium salt that conforms to the formula:

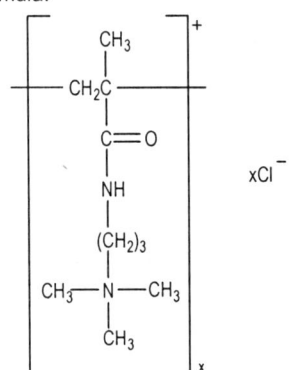

Chemical Classes: Quaternary Ammonium Compounds; Synthetic Polymers

Functions: Antistatic Agent; Film Former; Hair Conditioning Agent

Technical/Other Name:
Sulfuric Acid, Methyl Ester, Compd. with N-[3-(Dimethylamino)Propyl]-2-Methyl-2-Propenamide Homopolymer

POLYMETHACRYLIC ACID

CAS No.: 25087-26-7

JPN Translation:
ポリメタクリル酸

Empirical Formula:
$(C_4H_6O_2)_x$

Definition: Polymethacrylic Acid is the polymer that conforms to the formula:

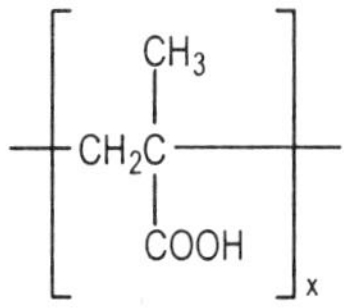

Information Sources: JCIC, JCLS

Chemical Class: Synthetic Polymers

Functions: Film Former; Viscosity Increasing Agent - Aqueous

Reported Product Categories: Bath Oils, Tablets, and Salts; Bath Preparations, Misc.; Body and Hand Preparations (Excluding Shaving Preparations); Cleansing Products (Cold Creams, Cleansing Lotions, Liquids and Pads); Skin Care Preparations, Misc.

Technical/Other Name:
Poly(acrylic Acid)

POLYMETHACRYLOYL ETHYL BETAINE

JPN Translation:
ポリメタクリロイルエチルベタイン

Definition: Polymethacryloyl Ethyl Betaine is a homopolymer of methacryloyl ethyl betaine.

Chemical Class: Synthetic Polymers

Functions: Anticaking Agent; Hair Fixative

Trade Name:
Yukaformer 530 (Mitsubishi Chemical)

POLYMETHACRYLOYL LYSINE

Definition: Polymethacryloyl Lysine is the polymer that conforms generally to the formula:

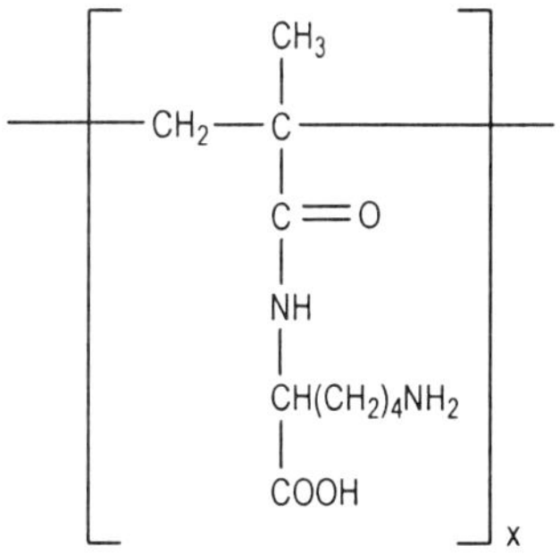

Chemical Classes: Amides; Synthetic Polymers

Function: Skin-Conditioning Agent - Humectant

Trade Name:
PM-Lysine (Gifu Shellac)

POLYMETHOXY BICYCLIC OXAZOLIDINE

CAS No.: 56709-13-8

Empirical Formula:
$(CH_2O)_n \cdot C_6H_{11}NO_3$

Definition: Polymethoxy Bicyclic Oxazolidine is the organic compound that conforms to the formula:

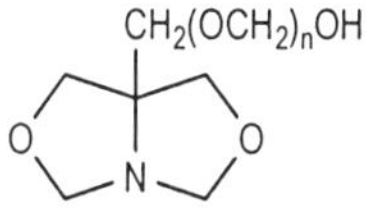

where n has a value of 0 through 5.

Chemical Class: Heterocyclic Compounds

Function: Preservative

Reported Product Categories: Shampoos (Non-coloring); Cleansing Products (Cold Creams, Cleansing Lotions, Liquids and Pads)

Technical/Other Name:
5-Hydroxypoly[Methyleneoxy]Methyl-1-Aza-3,7-Dioxabicyclo-3,3-Octane

Trade Name:
Nuosept C Preservative (Creanova)

POLYMETHYL ACRYLATE

CAS No.: 9003-21-8

JPN Translation:
ポリアクリル酸メチル

Empirical Formula:
$(C_4H_6O_2)_x$

Definition: Polymethyl Acrylate is the polymer that conforms to the formula:

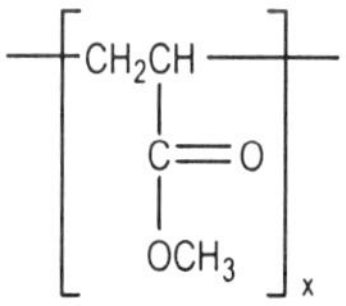

Information Sources: JCLS, TSCA

Chemical Classes: Esters; Synthetic Polymers

Function: Film Former

Technical/Other Names:
Methyl Acrylate Homopolymer
Methyl 2-Propenoate Homopolymer
2-Propenoic Acid, Methyl Ester, Homopolymer

POLYMETHYLGLUTAMATE

CAS No.: 25086-16-2

Empirical Formula:
$(C_6H_{11}NO_4)_x$

Definition: Polymethylglutamate is the polymer that conforms to the formula:

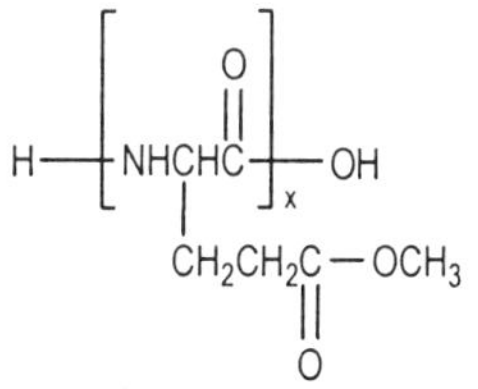

Information Source: TSCA

Chemical Classes: Amides; Esters; Synthetic Polymers

Functions: Film Former; Hair Conditioning Agent

Technical/Other Names:
Glutamic Acid, 5-Methyl Ester, Homopolymer
Poly-γ-Methyl Glutamate

POLYMETHYL METHACRYLATE

CAS No.: 9011-14-7

JPN Translation:
ポリメタクリル酸メチル

Empirical Formula:
$(C_5H_8O_2)_x$

Definition: Polymethyl Methacrylate is the polymer of methyl methacrylate that conforms to the formula:

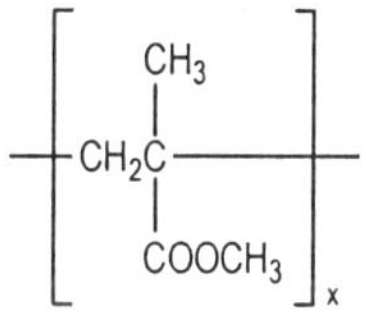

Information Sources: 21CFR175.105, 21CFR175.300, 21CFR176.180, 21CFR177.1010, 21CFR177.2465, 21CFR178.3790, JSQI

Chemical Class: Synthetic Polymers

Function: Film Former

Reported Product Categories: Eye Shadows; Face Powders; Blushers (All types); Foundations; Eye Makeup Preparations, Misc.; Eyeliners; Lipsticks; Makeup Preparations (Not eye), Misc.; Bath Capsules; Face and Neck Preparations (Excluding Shaving Preparations); Makeup Fixatives; Eyebrow Pencils; Makeup Bases; Rouges; Skin Care Preparations, Misc.; Suntan Gels, Creams, and Liquids; Night Skin Care Preparations; Paste Masks (Mud Packs); Powders (Dusting and Talcum, Excluding Aftershave Talcs)

Technical/Other Name:
2-Propenoic Acid, 2-Methyl-, Methyl Ester, Homopolymer

Trade Names:
BPA-500 (Kobo)
Covabead PMMA (LCW)
Ganzpearl GM-0600 (Presperse)
Jurymer MB-1 (Nihon Junyaku)
Microsphere M (Tomen America)
Microsphere M-100 (Matsumoto Yushi-Seiyaku)
Polymethylmethacrylate (Tagra)
TECHPOLYMER MB-C series (Sekisui)

Trade Name Mixtures:
Crystal Color (Daiya Kogyo)
Iridescent Glitter IF8101 (Mitsubishi Petrochemical)
Jurymer MB-1 (Ti) (Nihon Junyaku)
Jurymer MB-1 (UAV) (Nihon Junyaku)
Jurymer MB-1 (UPA) (Nihon Junyaku)
MCP-45 (Presperse)
Pearl Color (Nikko)
PW Covasil S (LCW)
Rainbow Flake, Crystal (Daiya Kogyo)
Tagravit A1 (Tagra)
Tagravit A2 (Tagra)
Tagravit E1 (Tagra)
Tagravit F1 (Tagra)
Tagrol B1 (Tagra)
Tagrol EPO1 (Tagra)
Tagrol H1 (Tagra)

POLYMETHYL METHACRYLATE/ POLYPENTAERYTHRITYL TEREPHTHALATE/STEARATE/PALMITATE LAMIATED POWDER

Definition: Polymethyl Methacrylate/Polypentaerythrityl Terephthalate/Stearate/Palmitate Lamiated Powder is the powder of the laminated film of Polymethyl Methacrylate and Polypentaerythrityl. *See "Regulatory and Ingredient Use Information," regarding use of Japan Trivial names in Volume 1, Introduction, Part A.*

Information Source: JCLS

Chemical Class: Synthetic Polymers

Function: Not Reported

POLYMETHYLSILSESQUIOXANE

CAS No.: 68554-70-1

JPN Translation:
ポリメチルシルセスキオキサン

Definition: Polymethylsilsesquioxane is a polymer formed by the hydrolysis and condensation of methyltrimethoxysilane.

Chemical Class: Siloxanes and Silanes

Function: Opacifying Agent

Reported Product Categories: Foundations; Blushers (All types)

Trade Names:
AEC Silicone Resin Spheres (A & E Connock)
KMP-590 (Shin Etsu)
KMP-599 (Shin Etsu)
MSP-K050 (Nikko Rica)
Tospearl 2000 (Toshiba)
Tospearl 120A (Toshiba)
Tospearl 130A (Toshiba)
Tospearl 145A (Toshiba)
Wacker - Belsil PMS MK (Wacker-Chemie)

POLYOXYETHYLENE CETYL STEARYL DIETHER

Definition: *See "Regulatory and Ingredient Use Information," regarding use of Japan Trivial names in Volume 1, Introduction, Part A.*

Information Source: JCLS

Chemical Class: Ethers

Function: Not Reported

POLYOXYISOBUTYLENE/METHYLENE UREA COPOLYMER

Definition: Polyoxyisobutylene/Methylene Urea Copolymer is the polymer formed from the condensation of urea, isobutyraldehyde and formaldehyde.

Chemical Class: Synthetic Polymers

Function: Film Former

POLYOXYMETHYLENE MELAMINE

CAS No.: 9003-08-1

Empirical Formula:
$(C_3H_6N_6 \cdot CH_2O)_x$

Definition: Polyoxymethylene Melamine is a reaction product of melamine and formaldehyde.

Information Sources: 21CFR175.105, 21CFR175.300, 21CFR175.320, 21CFR177.1200, 21CFR177.1460, 21CFR177.2260, 21CFR177.2470, 21CFR181.22, 21CFR181.30, CIR: [I] JACT-14(5)1995, TSCA

Chemical Class: Synthetic Polymers

Function: Film Former

Technical/Other Names:
Melamine/Formaldehyde Resin

1,3,5-Triazine-2,4,6-Triamine, Polymer with Formaldehyde

POLYOXYMETHYLENE MELAMINE UREA

CAS No.: 25036-13-9

Empirical Formula:
$(C_3H_6N_6 \cdot CH_4N_2O \cdot CH_2O)$

Definition: Polyoxymethylene Melamine Urea is a reaction product of urea, Melamine (q.v.) and formaldehyde.

Information Source: TSCA

Chemical Class: Synthetic Polymers

Function: Bulking Agent

Technical/Other Names:
Carbamide-Formaldehyde-Melamine Copolymer
Urea/Melamine/Formaldehyde Resin
Urea, Polymer with Formaldehyde and 1,3,5-Triazine-2,4,6-Triamine

Trade Name:
3M Brand PMMU Capsules (3M)

POLYOXYMETHYLENE UREA

CAS Nos.	EINECS No.
9011-05-6	
68611-64-3	271-898-1

Empirical Formula:
$(CH_4N_2O \cdot CH_2O)_x$

Definition: Polyoxymethylene Urea is a reaction product of urea and formaldehyde.

Information Sources: BAN, 21CFR175.105, 21CFR175.300, 21CFR177.1200, 21CFR177.1650, 21CFR177.1900, 21CFR181.30, CIR: [SQ] JACT-14(3)1995, INN, MI-13(7657), MI-13 (9936), TSCA

Chemical Class: Synthetic Polymers

Function: Bulking Agent

Reported Product Categories: Blushers (All types); Face Powders; Eye Shadows; Rouges; Lipsticks

Technical/Other Names:
Carbamide-Formaldehyde Copolymer
Polynoxylin
Urea-Formaldehyde Resin
Urea, Polymer with Formaldehyde

Trade Name:
3M Brand PMU Capsules (3M)

Trade Name Mixtures:
AEC Jojoba Oil Microencapsulated (A & E Connock)

LipoLight OAP/C (Lipo)
3M PMU Eicosane Cooling Microcapsules, 9850Q (3M)

POLYOXYPROPYLENE GLYCERYL ETHER PHOSPHATE

JPN Translation:
ポリオキシプロピレングリセリルエーテルリン酸

Definition: *See "Regulatory and Ingredient Use Information," regarding use of Japan Trivial names in Volume 1, Introduction, Part A.*

Information Source: JCLS

Chemical Classes: Alkoxylated Alcohols; Phosphorus Compounds

Function: Not Reported

POLYOXYPROPYLENE SORBITOL RICINOLEATE

JPN Translation:
リシノレイン酸ポリオキシプロピレンソルビット

Definition: *See "Regulatory and Ingredient Use Information," regarding use of Japan Trivial names in Volume 1, Introduction, Part A.*

Information Source: JCLS

Chemical Class: Esters

Function: Not Reported

POLYPENTAERYTHRITYL TEREPHTHALATE

Definition: Polypentaerythrityl Terephthalate is the polyester of Pentaerythritol (q.v.) and Terephthalic Acid (q.v.).

Information Source: JCIC

Chemical Class: Synthetic Polymers

Functions: Film Former; Hair Fixative

POLYPENTENE

CAS No.: 9078-70-0

Empirical Formula:
$(C_5H_{10})_x$

Definition: Polypentene is the polymer formed by the polymerization of pentene. It conforms to the formula:

$$\left[-CH-CH- \right]_x$$ with CH_3, CH_2CH_3 substituents

Chemical Class: Synthetic Polymers

Functions: Film Former; Viscosity Increasing Agent - Nonaqueous

Technical/Other Name:
Pentene, Homopolymer

Trade Name:
Escorez 1271U (ExxonMobil Japan)

POLYPERFLUOROETHOXYMETHOXY DIFLUOROETHYL PEG ETHER

CAS Nos.: 88645-29-8; 162492-15-1

Definition: Polyperfluoroethoxymethoxy Difluoroethyl PEG Ether is the polymer that conforms generally to the formula:

$$(CH_2CH_2O)_nOH \qquad (CH_2CH_2O)_nOH$$
$$CH_2CF_2O(CF_2CF_2O_p(CF_2O)_8CF_2CH_2$$

where n has an average value of 1/2 and p/q has an average value of 1.

Chemical Classes: Halogen Compounds; Synthetic Polymers

Functions: Hair Conditioning Agent; Skin-Conditioning Agent - Miscellaneous; Skin-Conditioning Agent - Occlusive

Technical/Other Name:
Ethene, Tetrafluoro-, Oxidized, Polymd., Reduced, Me Esters, Reduced Ethoxylated

Trade Names:
Fomblin HC/OL-1000 TX (Solvay Solexis SpA)
Fomblin HC/OL-2000 TX (Solvay Solexis SpA)
Fomblin HC/OL-4000 TX (Solvay Solexis SpA)

POLYPERFLUOROETHOXYMETHOXY DIFLUOROETHYL PEG PHOSPHATE

JPN Translation:
パーフルオロアルキル PEG リン酸

Definition: Polyperfluoroethoxymethoxy Difluoroethyl PEG Phosphate is a complex mixture of esters of phosphoric acid and Polyperfluoroethoxymethoxy Difluoroethyl PEG Ether (q.v.).

Chemical Classes: Halogen Compounds; Phosphorus Compounds; Synthetic Polymers

Functions: Hair Conditioning Agent; Skin-Conditioning Agent - Miscellaneous

Trade Names:
Fomblin HC/P2-1000 (Solvay Solexis SpA)
Fomblin HC/P2-2000 (Solvay Solexis SpA)
Fomblin HC/P2-4000 (Solvay Solexis SpA)

POLYPERFLUOROETHOXYMETHOXY DIFLUOROHYDROXYETHYL ETHER

Definition: Polyperfluoroethoxymethoxy Difluorohydroxyethyl Ether is the polymer that conforms generally to the formula:

$$HOCH_2CF_2O(CF_2CF_2O)_p(CF_2O)_qCF_2CH_2OH$$

where p/q has an average value of 1.

Chemical Classes: Halogen Compounds; Synthetic Polymers

Functions: Hair Conditioning Agent; Skin-Conditioning Agent - Miscellaneous

Trade Names:
Fomblin HC/OH-1000 (Solvay Solexis SpA)
Fomblin HC/OH 2000 (Solvay Solexis SpA)
Fomblin HC/OH 4000 (Solvay Solexis SpA)

POLYPERFLUOROETHOXYMETHOXY DIFLUOROMETHYL DISTEARAMIDE

CAS No.: 220207-10-3

Definition: Polyperfluoroethoxymethoxy Difluoromethyl Distearamide is the polymer that conforms generally to the formula:

$$RC\overset{O}{\underset{||}{}}\!-\!NHCF_2O(CF_2CF_2O)_x(CF_2O)_yCF_2NH\!-\!\overset{O}{\underset{||}{}}\!CR$$

where RCO represents the stearoyl moiety.

Chemical Classes: Halogen Compounds; Synthetic Polymers

Functions: Skin-Conditioning Agent - Miscellaneous; Viscosity Increasing Agent - Nonaqueous

Technical/Other Name:
1-Octadecanamine, Reaction Products with Et Esters of reduced Polymd. Oxidized Tetrafluoroethylene

Trade Name:
Fomblin HC/SA-18 (Solvay Solexis SpA)

POLYPERFLUOROETHOXYMETHOXY DIFLUOROMETHYL ETHER

CAS No.: 161075-02-1

Definition: Polyperfluoroethoxymethoxy Difluoromethyl Ether is the polymer that conforms generally to the formula:

$$HF_2C(OCF_2CF_2)_x(OCF_2)_yOCF_2H$$

Chemical Classes: Halogen Compounds; Synthetic Polymers

Function: Solvent

Trade Names:
Fomblin HC/H-50 (Solvay Solexis SpA)

Fomblin HC/H-100 (Solvay Solexis SpA)
Fomblin HC/H-130 (Solvay Solexis SpA)
Fomblin HC/H-200 (Solvay Solexis SpA)

POLYPERFLUOROETHOXYMETHOXY PEG-2 PHOSPHATE

CAS No.: 162567-74-0

Definition: Polyperfluoroethoxymethoxy PEG-2 Phosphate is a complex mixture of esters of phosphoric acid and perfluoropolymethylisopropyl ether.

Chemical Classes: Halogen Compounds; Phosphorus Compounds; Polymeric Ethers

Function: Suspending Agent - Nonsurfactant

Trade Name Mixtures:
Cardre Black Iron Oxide RFHC (Cardre)
Cardre Carmine RFHC (Cardre)
Cardre Manganese Violet RFHC (Cardre)
Cardre Mica 8 RFHC (Cardre)
Cardre Red Iron Oxide RFHC (Cardre)
Cardre Russet RFHC (Cardre)
Cardre Sericite RFHC (Cardre)
Cardre Titanium Dioxide RFHC (Cardre)
Cardre Ultramarine Blue RFHC (Cardre)
Cardre Yellow Iron Oxide RFHC (Cardre)

POLYPERFLUOROISOPROPYL ETHER

Empirical Formula:
$(C_3F_6O)_xC_2F_6$

Definition: Polyperfluoroisopropyl Ether is a fluorine-based polymer that conforms to the formula:

Information Source: JCLS

Chemical Classes: Halogen Compounds; Polymeric Ethers

Functions: Skin-Conditioning Agent - Miscellaneous; Skin-Conditioning Agent - Occlusive

Trade Names:
Koflube (Kobo)
Krytox GPL-101 (DuPont de Nemours)
Krytox GPL-102 (DuPont de Nemours)
Krytox GPL-103 (DuPont de Nemours)
Krytox GPL-104 (DuPont de Nemours)
Krytox GPL-105 (DuPont de Nemours)
Krytox GPL-106 (DuPont de Nemours)
Krytox GPL-107 (DuPont de Nemours)
Krytox GPL-108 (DuPont de Nemours)

POLYPERFLUOROMETHYLISOPROPYL ETHER

CAS No.: 69991-67-9

JPN Translation:
パーフルオロポリメチルイソプロピル

Definition: Polyperfluoromethylisopropyl Ether is the fluorine-based polymer that conforms to the formula:

Information Sources: JCLS, TSCA

Chemical Classes: Halogen Compounds; Polymeric Ethers

Functions: Skin-Conditioning Agent - Miscellaneous; Skin-Conditioning Agent - Occlusive

Reported Product Categories: Makeup Preparations (Not eye), Misc.; Foundations; Eye Shadows; Face Powders; Moisturizing Preparations; Lipsticks

Technical/Other Names:
1,1,2,3,3,3-Hexafluoro-1-Propene, Oxidized, Polymd.
1-Propene, 1,1,2,3,3,3-Hexafluoro-, Oxidized, Polymd.

Trade Names:
Fomblin HC/01 (Solvay Solexis SpA)
Fomblin HC/02 (Solvay Solexis SpA)
Fomblin HC/03 (Solvay Solexis SpA)
Fomblin HC/04 (Solvay Solexis SpA)
Fomblin HC/04 (Uniqema Americas)
Fomblin HC/25 (Solvay Solexis SpA)
Fomblin HC/25 (Uniqema Americas)
Fomblin HC/R (Solvay Solexis SpA)
Fomblin HC/R (Uniqema Americas)

Trade Name Mixtures:
Black Iron Oxide FHC (Cardre)
Chronosphere FHC/HA Blend (Arch Personal Care Products)
Fomblin HC/P-R (Solvay Solexis SpA)
Kelifluo 1500 (Kelisema Italy)
Kelifluo 3200 (Kelisema Italy)
Kelifluo 6250 (Kelisema Italy)
Mica FHC (Cardre)
Red Iron Oxide FHC (Cardre)
Sansurf PFMP (Collaborative Labs)
Sericite FHC (Cardre)
Talc FHC (Cardre)
Titanium Dioxide FHC (Cardre)
Toshiki Sericite JF-25-3 (Nikko)
Toshiki Sericite JSF-25-3 (Nikko)
Toshiki Talc LF-25-3 (Nikko)
Toshiki Talc LSF-25-3 (Nikko)
Toshiki TiO2 AF-25-3 (Nikko)
Toshiki TiO2 ASF-25-3 (Nikko)
Ultrafine Titanium Dioxide FHC (Cardre)
Yellow Iron Oxide FHC (Cardre)

POLYPERFLUOROPERHYDRO-PHENANTHRENE

Definition: Polyperfluoroperhydrophenanthrene is a polymer of Perfluoroperhydrophenanthrene (q.v.).

Chemical Classes: Halogen Compounds; Synthetic Polymers

Functions: Film Former; Hair Fixative; Skin-Conditioning Agent - Miscellaneous

POLY-p-PHENYLENE TEREPHTHALAMIDE

CAS Nos.: 24938-64-5; 25035-37-4

Empirical Formula:
$(C_{14}H_{10}N_2O_2)_x$

Definition: Poly-p-Phenylene Terephthalamide is the polyamide that conforms generally to the formula:

Chemical Classes: Amides; Synthetic Polymers

Function: Film Former

Technical/Other Names:
1,4-Benzenediamine-Terephthalic Acid Copolymer
1,4-Diaminobenzene-Terephthalic Acid Copolymer
1,4-Phenylenediamine-Terephthalic Acid Copolymer
Poly(Imino-1,4-Phenyleneiminocarbonyl-1, 4-Phenylenecarbonyl)

Trade Name:
Kevlar brand pulp 1F543 (DuPont de Nemours)

POLYPHOSPHORYLCHOLINE GLYCOL ACRYLATE

CAS No.: 67881-99-6

Definition: Polyphosphorylcholine Glycol Acrylate is the polymer that conforms generally to the formula:

Chemical Classes: Phosphorus Compounds; Synthetic Polymers

Function: Film Former

Technical/Other Name:
3,5,8-Trioxa-4-Phosphaundec-10-en-1-aminium, 4-Hydroxy-N,N,N,10-Tetramethyl-9-Oxo, Inner Salt, 4-Oxide, Homopolymer

Trade Name Mixture:
Lipidure-HM (NOF)

POLYPODIUM LEUCOTOMOS LEAF EXTRACT

Definition: Polypodium Leucotomos Leaf Extract is an extract of the leaves of *Polypodium leucotomos. See "Regulatory and Ingredient Use Information," regarding the labeling names for botanical ingredients in Volume 1, Introduction, Part A.*

Chemical Class: Biological Products

Function: Not Reported

Technical/Other Name:
Extract of Polypodium Leucotomos Leaf

Trade Name:
Polypodium Leucotomos Extract (Industrial Farmaceutica)

POLYPORUS UMBELLATUS (MUSHROOM) EXTRACT

JPN Translation:
チョレイエキス

Definition: Polyporus Umbellatus (Mushroom) Extract is an extract of the mushroom, *Polyporus umbellatus. See "Regulatory and Ingredient Use Information," regarding the labeling names for botanical ingredients in Volume 1, Introduction, Part A.*

Chemical Class: Biological Products

Function: Not Reported

Technical/Other Names:
Extract of Mushroom
Extract of Polyporus Umbellatus
Mushroom Extract
Polyporus Umbellatus Extract

Trade Name Mixture:
Polyporus Extract (Koshiro)

POLYPROPYLENE

CAS No.: 9003-07-0

JPN Translation:
ポリプロピレン

Empirical Formula:
$(C_3H_6)_x$

Definition: Polypropylene is a polymer of propylene monomers that conforms generally to the formula:

$$\left[CH_2CH \atop CH_3 \right]_x$$

Information Sources: 21CFR173.340, 21CFR175.105, 21CFR175.300, 21CFR176.170, 21CFR177.1200, 21CFR177.1520, 21CFR177.1680, 21CFR178.3740, 21CFR179.45, JCIC, JCLS, JSQI, MI-13(7663), USAN

Chemical Class: Synthetic Polymers

Functions: Bulking Agent; Viscosity Increasing Agent - Nonaqueous

Reported Product Categories: Eye Shadows; Lipsticks

Technical/Other Names:
Polypropylene Powder
1-Propene, Homopolymer

Trade Names:
Amoco Polypropene 9011 (Amoco Chemical)
Amoco Polypropene 9012 (Amoco Chemical)
Amoco Polypropene 9013 (Amoco Chemical)
Epolene Polypropylene Wax (Eastman Chemical)
Hoechst Wax PP 230 (Clariant)
Hoechst Wax PP 230 (Clariant GmbH, Personal Care)
Hoechst Wax PP 690 (Clariant)
Hoechst Wax PP 690 (Clariant GmbH, Personal Care)

Trade Name Mixtures:
CREASPARKLES IRIDESCENT BLUE 560 (Creations Couleurs)
CREASPARKLES IRIDESCENT GREEN 510 (Creations Couleurs)
CREASPARKLES IRIDESCENT RAINBOW 530 (Creations Couleurs)
CREASPARKLES IRIDESCENT VIOLET 610 (Creations Couleurs)
CREASPARKLES IRIDESCENT YELLOW 460 (Creations Couleurs)

POLYPROPYLENE TEREPHTHALATE

Definition: Polypropylene Terephthalate is the homopolymer that conforms generally to the formula:

$$\left[\overset{O}{\underset{}{C}} - C_6H_4 - \overset{O}{\underset{}{C}} - OCHCH_2O \atop CH_3 \right]_x$$

Chemical Class: Synthetic Polymers

Functions: Emulsion Stabilizer; Skin-Conditioning Agent - Miscellaneous

Trade Names:
Aristoflex PEA (Clariant)
Aristoflex PEA (Clariant GmbH, Personal Care)
Aristoflex PEA 70 (Clariant GmbH, Fine Chemicals)

POLYPROPYL METHACRYLATE

Definition: Polypropyl Methacrylate is the homopolymer of propyl acrylate.

Information Source: JCLS

Chemical Class: Synthetic Polymers

Function: Film Former

POLYPROPYLSILSESQUIOXANE

CAS No.: 36088-62-7

Definition: Polypropylsilsesquioxane is a polymer formed by the hydrolysis and condensation of propyltrichlorosilane.

Chemical Classes: Siloxanes and Silanes; Synthetic Polymers

Functions: Binder; Film Former

Technical/Other Name:
Silanetriol, Propyl-, Homopolymer

Trade Name Mixture:
Dow Corning 670 Fluid (Dow Corning)

POLYQUATERNIUM-1

CAS Nos.: 68518-54-7; 75345-27-6

Definition: Polyquaternium-1 is the polymeric quaternary ammonium salt that conforms generally to the formula:

$$\left[\begin{array}{c} CH_3 \\ CH_3-N^+ \\ CH_2 \\ CH \\ \| \\ CH \\ CH_2 \end{array} - (HOCH_2CH_2)_3N^+ - CH_2CH=CHCH_2 \quad \overset{+}{N}(CH_2CH_2OH)_3 \right]_x \quad (x+2)\ Cl^-$$

Chemical Classes: Alkoxylated Amines; Quaternary Ammonium Compounds; Synthetic Polymers

---

Functions: Antistatic Agent; Film Former; Hair Fixative

POLYQUATERNIUM-2

CAS No.: 63451-27-4

Definition: Polyquaternium-2 is the polymeric quaternary ammonium salt that conforms generally to the formula:

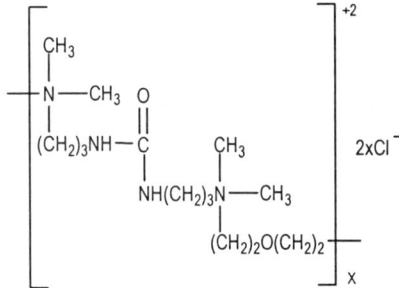

Chemical Classes: Quaternary Ammonium Compounds; Synthetic Polymers

Functions: Antistatic Agent; Film Former; Hair Fixative

Reported Product Categories: Hair Dyes and Colors (All Types Requiring Caution Statements and Patch Tests); Bath Preparations, Misc.; Hair Conditioners; Hair Preparations (Non-coloring), Misc.

Trade Names:
Ethpol PQ-2 (Ethox)
Mirapol A-15 (Rhodia)

POLYQUATERNIUM-4

CAS No.: 92183-41-0

JPN Translation:
ポリクオタニウム - 4

Definition: Polyquaternium-4 is a copolymer of hydroxyethylcellulose and diallyldimethyl ammonium chloride.

Information Sources: CTFA D, JCLS, JSCI, JSQI

Chemical Classes: Quaternary Ammonium Compounds; Synthetic Polymers

Functions: Antistatic Agent; Film Former; Hair Fixative

Reported Product Categories: Tonics, Dressings, and Other Hair Grooming Aids; Hair Preparations (Non-coloring), Misc.; Hair Sprays (Aerosol Fixatives); Hair Conditioners; Shampoos (Non-coloring); Hair Wave Sets; Permanent Waves

Technical/Other Names:
Cellulose, 2-Hydroxyethyl Ether, Polymer with N,N-Dimethyl-N-2-Propenyl-2-Propen-1-aminium Chloride

Diallyldimonium Chloride/Hydroxyethyl-cellulose Copolymer
Hydroxyethylcellulose Dimethyldiallyl-ammonium Chloride Copolymer

Trade Names:
Celquat H-100 (National Starch)
Celquat L-200 (National Starch)

Trade Name Mixtures:
Plexajel ASC (Guardian)
Wave Control (Chemyunion)

POLYQUATERNIUM-5

CAS No.: 26006-22-4

Definition: Polyquaternium-5 is the copolymer of acrylamide and beta-methacrylyloxyethyl trimethyl ammonium methosulfate.

Information Source: TSCA

Chemical Classes: Quaternary Ammonium Compounds; Synthetic Polymers

Functions: Antistatic Agent; Film Former; Hair Fixative

Technical/Other Names:
Chloine, Methyl Sulfate, Methacrylate, Polymer with Acrylamide
Ethanaminium, N,N,N-Trimethyl-2-[(2-Methyl-1-Oxo-2-Propenyl)Oxy]-, Methyl Sulfate, Polymer with 2-Propenamide
Quaternium-39

Trade Names:
Merquat 5 (ONDEO Nalco)
Merquat 5 (ONDEO Nalco Europe)

POLYQUATERNIUM-6

CAS No.: 26062-79-3

JPN Translation:
ポリクオタニウム - 6

Empirical Formula:
$(C_8H_{16}N \cdot Cl)_x$

Definition: Polyquaternium-6 is a polymer of dimethyl diallyl ammonium chloride.

Information Sources: CTFA D, JCIC, JCLS, JSQI, TSCA

Chemical Classes: Quaternary Ammonium Compounds; Synthetic Polymers

Functions: Antistatic Agent; Film Former; Hair Fixative

Reported Product Categories: Hair Conditioners; Permanent Waves; Tonics, Dressings, and Other Hair Grooming Aids; Hair Preparations (Non-coloring), Misc.

Technical/Other Names:
N,N-Dimethyl-N-2-Propenyl-2-Propen-1-aminium Chloride, Homopolymer

Poly(Dimethyl Diallyl Ammonium Chloride)
Poly N,N'-Dimethyl-3,5-methylene-piper-idinium Chloride Solution
Poly(DMDAAC)
2-Propen-1-aminium, N,N-Dimethyl-N-2-Propenyl-, Chloride, Homopolymer
Quaternium-40

Trade Names:
AEC Polyquaternium-6 (A & E Connock)
Agequat 400 (CPS)
Alcofix 131 (Ciba Specialty Chemicals)
Conditioner P6 (3V Inc.)
Flocare C 106 (SNF)
Genamin PDAC (Clariant)
Genamin PDAC (Clariant GmbH, Personal Care)
Mackernium 006 (McIntyre)
Merquat 100 (ONDEO Nalco)
Merquat 100 (ONDEO Nalco Europe)
Mirapol 100 (Rhodia)
Octacare PQ6 (Associated Octel)
Rheocare CC6 (Cosmetic Rheologies)
Rheocare CC6P (Cosmetic Rheologies)
Salcare SC30 (Ciba Specialty Chemicals)

POLYQUATERNIUM-7

CAS No.: 26590-05-6

JPN Translation:
ポリクオタニウム - 7

Empirical Formula:
$(C_8H_{16}N \cdot C_3H_5NO \cdot Cl)_x$

Definition: Polyquaternium-7 is the polymeric quaternary ammonium salt consisting of acrylamide and dimethyl diallyl ammonium chloride monomers.

Information Sources: 21CFR176.170, CIR: [S] JACT-14(6)1995, CTFA S, JCIC, JCLS, JSQI, TSCA

Chemical Classes: Quaternary Ammonium Compounds; Synthetic Polymers

Functions: Antistatic Agent; Film Former; Hair Fixative

Reported Product Categories: Shampoos (Non-coloring); Bath Oils, Tablets, and Salts; Permanent Waves; Tonics, Dressings, and Other Hair Grooming Aids; Bath Soaps and Detergents; Cleansing Products (Cold Creams, Cleansing Lotions, Liquids and Pads); Hair Sprays (Aerosol Fixatives); Hair Conditioners; Bath Preparations, Misc.; Bubble Baths; Personal Cleanliness Products, Misc.; Baby Shampoos; Hair Preparations (Non-coloring), Misc.; Hair Shampoos (Coloring); Shaving Cream (Aerosol, Brushless and Lather)

Technical/Other Names:
Dimethyldiallyl Ammonium Chloride·Acrylamide Copolymer

Dimethyldiallyl Ammonium Chloride•
Acrylamide Copolymer Solution
N,N-Dimethyl-N-2-Propenyl-2-Propen-1-
aminium Chloride, Polymer with 2-Prop-
enamide
2-Propen-1-aminium, N,N-Dimethyl-N-2-
Propenyl-, Chloride, Polymer with 2-
Propenamide
Quaternium-41

Trade Names:
AEC Polyquaternium-7 (A & E Connock)
Agequat-5008 (CPS)
Agequat C-505 (CPS)
Conditioner P7 (3V Inc.)
Flocare C 107 (SNF)
Mackernium 007 (McIntyre)
Mackernium 007S (McIntyre)
ME Polymer 09W (Toho)
Merquat 550 (ONDEO Nalco)
Merquat 550 (ONDEO Nalco Europe)
Merquat 2200 (ONDEO Nalco)
Merquat 2200 (ONDEO Nalco Europe)
Merquat 550L (ONDEO Nalco)
Merquat S (ONDEO Nalco)
Merquat S (ONDEO Nalco Europe)
Mirapol 550 (Rhodia)
Mirapol 550E (Rhodia)
Octacare PQ7 (Associated Octel)
Rheocare CC7 (Cosmetic Rheologies)
Rheocare CC7P (Cosmetic Rheologies)
Salcare SC10 (Ciba Specialty Chemicals)
Salcare SC11 (Ciba Specialty Chemicals)
Salcare Super 7 (Ciba Specialty
Chemicals)

Trade Name Mixtures:
BGW60CQ (Kobo)
Integrahair RE (Chemyunion)
Integrahair Sphere (Chemyunion)
Protequat 21 (Variati)
Wave Control (Chemyunion)

POLYQUATERNIUM-8

Definition: Polyquaternium-8 is the
polymeric quaternary ammonium salt of
methyl and stearyl dimethylaminoethyl
methacrylate quaternized with dimethyl
sulfate.

Chemical Classes: Quaternary Ammonium
Compounds; Synthetic Polymers

Functions: Antistatic Agent; Film Former;
Hair Fixative

Technical/Other Names:
Methyl and Stearyl Dimethylaminoethyl
Methacrylate Quaternized with Dimethyl
Sulfate
2-Propenoic Acid, 2-Methyl-, 2-(Dimethyl-
amino)Ethyl Ester, Polymer with Methyl
2-Methyl-2-Propenoate, compd. with
Dimethyl Sulfate
Quaternium-42

POLYQUATERNIUM-9

Definition: Polyquaternium-9 is the
polymeric quaternary ammonium salt of
polydimethylaminoethyl methacrylate
quaternized with methyl bromide.

Chemical Classes: Quaternary Ammonium
Compounds; Synthetic Polymers

Functions: Antistatic Agent; Film Former;
Hair Fixative

Technical/Other Names:
Polydimethylaminoethyl Methacrylate
Quaternized with Methyl Bromide
2-Propenoic Acid, 2-Methyl-, 2-(Dimethyl-
amino)Ethyl Ester, Homopolymer,
Compd. with Bromomethane
Quaternium-49

POLYQUATERNIUM-10

CAS Nos.: 53568-66-4; 54351-50-7; 55353-
19-0; 68610-92-4; 81859-24-7

JPN Translation:
ポリクオタニウム - 10

Definition: Polyquaternium-10 is a
polymeric quaternary ammonium salt of
hydroxyethyl cellulose reacted with a
trimethyl ammonium substituted epoxide.

Information Sources: CIR: [S] JACT-7(3)-
1988, CTFA D, JCIC, JCLS, JSQI

Chemical Classes: Quaternary Ammonium
Compounds; Synthetic Polymers

Functions: Antistatic Agent; Film Former;
Hair Fixative

Reported Product Categories: Shampoos
(Non-coloring); Hair Conditioners; Hair Tints;
Bath Soaps and Detergents; Tonics,
Dressings, and Other Hair Grooming Aids;
Bath Oils, Tablets, and Salts; Cleansing
Products (Cold Creams, Cleansing Lotions,
Liquids and Pads); Hair Preparations (Non-
coloring), Misc.; Personal Cleanliness
Products, Misc.; Baby Shampoos; Hair
Shampoos (Coloring); Mascara; Hair Wave
Sets; Permanent Waves; Bath Preparations,
Misc.; Body and Hand Preparations
(Excluding Shaving Preparations); Face and
Neck Preparations (Excluding Shaving Prep-
arations); Fragrance Preparations, Misc.;
Hair Dyes and Colors (All Types Requiring
Caution Statements and Patch Tests); Hair
Sprays (Aerosol Fixatives); Hair Straighten-
ers; Shaving Preparations, Misc.; Skin Care
Preparations, Misc.

Technical/Other Names:
Cellulose, 2-Hydroxyethyl 2-(2-Hydroxy-
3-(Trimethylammonio)Propoxy)Ethyl 2-
Hydroxy-3-(Trimethylammonio)Propyl
Ether, Chloride

Cellulose, 2-Hydroxyethyl 2-Hydroxy-
3-(Trimethylammonio)Propyl Ether,
Chloride
Cellulose, 2-Hydroxyethyl 23-Hydroxy-
3-(Trimethylammonio)Propyl Ether,
Chloride
Cellulose, 2-[2-Hydroxy-3-(Trimethylam-
monio)Propoxy]Ethyl Ether, Chloride
Cellulose, 2-[2-Hydroxy-3-
Trimethylammono)propoxy] Ethyl ether,
chloride
O-[2-Hydroxy-3-(trimethylammonio)propyl]
Hydroxyethylcellulose Chloride
Quaternium-19

Trade Names:
AEC Polyquaternium-10 (A & E Connock)
Catinal C-100 (Toho)
Catinal HC-35 (Toho)
Catinal HC-100 (Toho)
Catinal HC-200 (Toho)
Catinal LC-100 (Toho)
Catinal LC-200 (Toho)
Celquat SC-240C (National Starch)
Celquat SC-230M (National Starch)
Dekaquat 400 (Dekker)
Dekaquat 3000 (Dekker)
Leogard GP (Akzo Nobel Surface AB)
RITA Polyquta 400 (RITA)
RITA Polyquta 3000 (RITA)
UCARE Polymer JR-125 (Amerchol)
UCARE Polymer JR-400 (Amerchol)
UCARE Polymer JR-30M (Amerchol)
UCARE Polymer LK (Amerchol)
UCARE Polymer LR 400 (Amerchol)
UCARE Polymer LR 30M (Amerchol)

Trade Name Mixtures:
AC Colorplex (Active Concepts)
BioCare BHA-10 (Amerchol)
BioCare Polymer BHA-10 (Amerchol)
Covathick 421 (LCW)

POLYQUATERNIUM-11

CAS No.: 53633-54-8

JPN Translation:
ポリクオタニウム - 11

Empirical Formula:
$(C_8H_{15}NO_2 \cdot C_6H_9NO)_x \cdot xC_4H_{10}O_4S$

Definition: Polyquaternium-11 is a
quaternary ammonium polymer formed by
the reaction of diethyl sulfate and a
copolymer of vinyl pyrrolidone and dimethyl
aminoethylmethacrylate.

Information Sources: CIR: [S] JACT-2(5)-
1983, CTFA D, JCIC, JCLS, JSQI

Chemical Classes: Quaternary Ammonium
Compounds; Synthetic Polymers

Functions: Antistatic Agent; Film Former;
Hair Fixative

Reported Product Categories: Tonics, Dressings, and Other Hair Grooming Aids; Hair Conditioners; Hair Preparations (Non-coloring), Misc.; Shampoos (Non-coloring); Permanent Waves; Hair Sprays (Aerosol Fixatives); Hair Wave Sets; Hair Tints; Hair Bleaches; Personal Cleanliness Products, Misc.; Shaving Cream (Aerosol, Brushless and Lather)

Technical/Other Names:
Polyvinylpyrrolidone N,N-Dimethyl Aminoethyl Methacrylic Acid Copolymer Diethyl Sulfate Solution
2-Propenol Acid, 2-Methyl-2-(Dimethlamino) Ethyl Ester, Polymer and 1-Ethenyl-2-Pyrrolidinone, Compound with Diethyl Sulfate
2-Pyrrolidinone, 1-Ethenyl- Polymer and 2-(Dimethylamino) Ethyl 2-Methyl-2-Propenoate, Compound and Diethyl Sulfate
2-Pyrrolidinone, 1-Ethenyl-, Polymer and 2-(Dimethylamino) Ethyl 2-Methyl-2-Propenoate, compound with Diethyl Sulfate
Quaternium-23

Trade Names:
AEC Polyquaternium-11 (A & E Connock)
Gafquat 440 (International Specialty Products)
Gafquat 734 (International Specialty Products)
Gafquat 755 (International Specialty Products)
Gafquat 755N (International Specialty Products)
Luviquat PQ 11 PN (BASF)
Polyquat-11 SL (Sino Lion)

Trade Name Mixtures:
Aminoresin K (Cosmetochem) (Cosmetochem International Ltd.)
Condipon (Cosmetochem)
Condisoap (Cosmetochem) (Cosmetochem International Ltd.)
Liposomes Trichogen Veg (Laboratoires Serobiologiques)
Trichogen Veg (Laboratoires Sero-biologiques)

POLYQUATERNIUM-12

CAS No.: 68877-50-9

Definition: Polyquaternium-12 is a polymeric quaternary ammonium salt prepared by the reaction of ethyl methacrylate/abietyl methacrylate/diethyl-aminoethyl methacrylate copolymer with dimethyl sulfate.

Information Source: TSCA

Chemical Classes: Quaternary Ammonium Compounds; Synthetic Polymers

Functions: Antistatic Agent; Film Former; Hair Fixative

Technical/Other Names:
Ethyl Methacrylate/Abietyl Methacrylate/Diethylaminoethyl Methacrylate-Quaternized with Dimethyl Sulfate
2-Propenoic Acid, 2-Methyl-, Decahydro-1,4-Dimethyl-7-(1-Methylethyl)-1-Phenanthrenyl)Methyl Ester, Polymer with 2-(Diethylamino)Ethyl 2-Methyl-2-Propenoate and Ethyl 2-Methyl-2-Propenoate, compd. with Dimethyl Sulfate
Quaternium-37

POLYQUATERNIUM-13

CAS No.: 68877-47-4

Definition: Polyquaternium-13 is a polymeric quaternary ammonium salt prepared by the reaction of ethyl methacrylate/oleyl methacrylate/diethyl-aminoethyl methacrylate copolymer with dimethyl sulfate.

Information Source: TSCA

Chemical Classes: Quaternary Ammonium Compounds; Synthetic Polymers

Functions: Antistatic Agent; Film Former; Hair Fixative

Technical/Other Names:
Ethyl Methacrylate/Oleyl Methacrylate/Diethylaminoethyl Methacrylate-Quaternized with Dimethyl Sulfate
2-Propenoic Acid, 2-Methyl-, 2-(Diethyl-amino)Ethyl Ester, Polymer with Ethyl 2-Methyl-2-Propenoate and 9-Octadecenyl 2-Methyl-2-Propenoate, compd. with Dimethyl Sulfate
Quaternium-38

POLYQUATERNIUM-14

CAS No.: 27103-90-8

Definition: Polyquaternium-14 is the polymeric quaternary ammonium salt that conforms generally to the formula:

Chemical Classes: Quaternary Ammonium Compounds; Synthetic Polymers

Functions: Antistatic Agent; Film Former; Hair Fixative

Technical/Other Names:
Choline, Methyl Sulfate, Methacrylate, Polymer
Ethanaminium, N,N,N-Trimethyl-2-[(2-Methyl-1-Oxo-2-Propenyl)Oxy]-, Methyl Sulfate, Homopolymer

POLYQUATERNIUM-15

CAS Nos.: 35429-19-7; 67504-24-9

Definition: Polyquaternium-15 is the copolymer of methacrylamide and betameth-acrylyloxyethyl trimethyl ammonium chloride.

Chemical Classes: Quaternary Ammonium Compounds; Synthetic Polymers

Functions: Antistatic Agent; Film Former; Hair Fixative

Technical/Other Names:
Acrylamide-Dimethylaminoethyl Methacrylate Methyl Chloride Copolymer
Ethanaminium, N,N,N-Trimethyl-2-[(2-Methyl-1-Oxo-2-Propenyl)Oxy]-, Chloride, Polymer with 2-Propenamide

Trade Name:
Rohagit KF 720F (Rohm GmbH)

POLYQUATERNIUM-16

CAS No.: 95144-24-4

Definition: Polyquaternium-16 is a polymeric quaternary ammonium salt formed from methylvinylimidazolium chloride and vinylpyrrolidone.

Information Source: CTFA D

Chemical Classes: Quaternary Ammonium Compounds; Synthetic Polymers

Functions: Antistatic Agent; Film Former; Hair Fixative

Reported Product Categories: Tonics, Dressings, and Other Hair Grooming Aids; Hair Preparations (Non-coloring), Misc.; Permanent Waves; Hair Conditioners; Shampoos (Non-coloring)

Technical/Other Names:
1H-Imidazolium, 1-Ethenyl-3-Methyl-, Chloride, Polymer with 1-Ethenyl-2-Pyrrolidinone
3-Methyl-1-Vinylimidazolium Chloride-1-Vinyl-2-Pyrrolidinone Copolymer

Trade Names:
Luviquat Excellence (BASF)

Luviquat FC 370 (BASF)
Luviquat FC 550 (BASF)
Luviquat Style (BASF)

POLYQUATERNIUM-17

CAS No.: 148506-50-7

Definition: Polyquaternium-17 is a polymeric quaternary salt prepared by the reaction of adipic acid and dimethylamino-propylamine, reacted with dichloroethyl ether. It conforms generally to the formula:

Chemical Classes: Quaternary Ammonium Compounds; Synthetic Polymers

Functions: Antistatic Agent; Film Former; Hair Fixative

Technical/Other Name:
Poly(Oxy-1,2-Ethanediyl(Dimethyliminio)-1, 3-Propanediylimino(1,6-Dioxo-1,6-Hexanediyl)Imino-1,3-Propanediyl (Dimethyliminio)-1,2-Ethanediyl Dichloride

POLYQUATERNIUM-18

Definition: Polyquaternium-18 is a polymeric quaternary salt prepared by the reaction of azelaic acid and dimethylamino-propylamine reacted with dichloroethyl ether. It conforms generally to the formula:

Chemical Classes: Quaternary Ammonium Compounds; Synthetic Polymers

Functions: Antistatic Agent; Film Former; Hair Fixative

POLYQUATERNIUM-19

CAS No.: 110736-85-1

Definition: Polyquaternium-19 is the polymeric quaternary ammonium salt prepared by the reaction of polyvinyl alcohol with 2,3- epoxypropylamine.

Chemical Classes: Quaternary Ammonium Compounds; Synthetic Polymers

Functions: Antistatic Agent; Film Former; Hair Fixative

POLYQUATERNIUM-20

CAS No.: 110736-86-2

Definition: Polyquaternium-20 is the polymeric quaternary ammonium salt prepared by the reaction of polyvinyl octadecyl ether with 2,3-epoxypropylamine.

Chemical Classes: Quaternary Ammonium Compounds; Synthetic Polymers

Functions: Antistatic Agent; Film Former; Hair Fixative

POLYQUATERNIUM-22

CAS No.: 53694-17-0

JPN Translation:
ポリクオタニウム - 22

Empirical Formula:
$(C_8H_{16}NCl)(C_3H_4O_2)$

Definition: Polyquaternium-22 is a copolymer of dimethyldiallyl ammonium chloride and acrylic acid. It conforms generally to the formula:

Chemical Classes: Quaternary Ammonium Compounds; Synthetic Polymers

Functions: Antistatic Agent; Film Former; Hair Fixative

Reported Product Categories: Hair Dyes and Colors (All Types Requiring Caution Statements and Patch Tests); Permanent Waves; Hair Conditioners; Shampoos (Non-coloring)

Technical/Other Name:
Acrylic Acid-Diallyldimethylammonium Chloride Polymer

Trade Names:
Merquat 280 (ONDEO Nalco)
Merquat 280 (ONDEO Nalco Europe)
Merquat 281 (ONDEO Nalco)
Merquat 295 (ONDEO Nalco)
Merquat 295 (ONDEO Nalco Europe)

Trade Name Mixture:
Polysol HQ (Polygon)

POLYQUATERNIUM-24

JPN Translation:
ポリクオタニウム - 24

Definition: Polyquaternium-24 is a polymeric quaternary ammonium salt of hydroxyethyl cellulose reacted with a lauryl dimethyl ammonium substituted epoxide.

Chemical Classes: Quaternary Ammonium Compounds; Synthetic Polymers

Functions: Antistatic Agent; Film Former; Hair Fixative

Reported Product Categories: Personal Cleanliness Products, Misc.; Moisturizing Preparations

Technical/Other Name:
Cellulose, 2-[2-Hydroxy-3-(Trimethylam-monio)Propoxy]Ethyl Ether, Chloride

Trade Name:
Quatrisoft Polymer LM-200 (Amerchol)

Trade Name Mixtures:
BioCare HA-24 Bio (Amerchol)
BioCare Polymer HA-24 (Amerchol)

POLYQUATERNIUM-27

CAS No.: 132977-85-6

Definition: Polyquaternium-27 is the block copolymer formed by the reaction of Poly-quaternium-2 (q.v.) with Polyquaternium-17 (q.v.).

Chemical Classes: Quaternary Ammonium Compounds; Synthetic Polymers

Functions: Antistatic Agent; Film Former; Hair Fixative

Technical/Other Name:
Hexanediamide, N,N'-bis (3-(Dimethylamino)Propyl)-, Polymer with N,N'-bis(3-Dimethylamino)Propyl Urea and 1,1'-Oxybis(2-Chloroethane), Block

POLYQUATERNIUM-28

CAS No.: 131954-48-8

Definition: Polyquaternium-28 is a polymeric quaternary ammonium salt consisting of vinylpyrrolidone and dimethyl-aminopropyl methacrylamide monomers. It conforms generally to the formula:

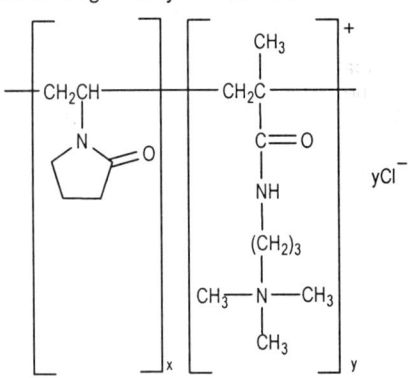

Chemical Classes: Quaternary Ammonium Compounds; Synthetic Polymers

Functions: Antistatic Agent; Film Former; Hair Fixative

Reported Product Categories: Permanent Waves; Shampoos (Non-coloring)

Technical/Other Names:
1-Propanaminium, N,N,N-Trimethyl-3-((2-Methyl-1-Oxo-2-Propenyl)Amino)-, Chloride, Polymer with 1-Ethenyl-2-Pyr-rolidinone
Vinylpyrrolidone/Methacrylamidopropyl-trimethylammonium Chloride Copolymer

Trade Names:
AEC Polyquaternium-28 (A & E Connock)
Gafquat HS-100 (International Specialty Products)

Trade Name Mixture:
Gafquat HSi (International Specialty Products)

POLYQUATERNIUM-29

CAS Nos.: 92091-36-6; 148880-30-2

Definition: Polyquaternium-29 is Chitosan (q.v.) that has been reacted with propylene oxide and quaternized with epichlorohydrin.

Chemical Classes: Quaternary Ammonium Compounds; Synthetic Polymers

Functions: Antistatic Agent; Film Former; Hair Fixative

Reported Product Categories: Body and Hand Preparations (Excluding Shaving Preparations); Face and Neck Preparations (Excluding Shaving Preparations); Foundations; Indoor Tanning Preparations; Lipsticks; Moisturizing Preparations; Night Skin Care Preparations; Paste Masks (Mud Packs); Suntan Gels, Creams, and Liquids

Technical/Other Names:
Chitosan, 2,3-Dihydroxypropyl-2-Hydroxy-3-(Trimethylammonio)Propyl Ether, Chloride

Dihydroxypropyl Chitosan Trimonium Chloride
Quaternized Chitosan

POLYQUATERNIUM-30

CAS No.: 147398-77-4

Definition: Polyquaternium-30 is the polymeric quaternary ammonium salt that conforms generally to the formula:

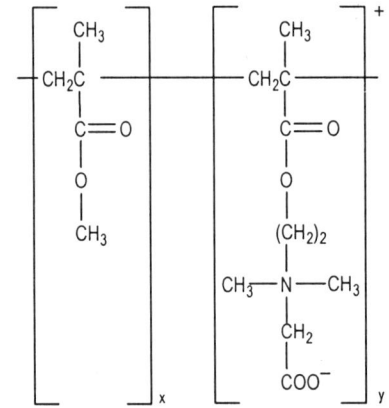

Chemical Classes: Quaternary Ammonium Compounds; Synthetic Polymers

Functions: Antistatic Agent; Film Former; Hair Fixative

Reported Product Categories: Shampoos (Non-coloring); Hair Conditioners

Technical/Other Name:
Ethanaminium, N-Carboxymethyl)-N,N-Dimethyl-2-((2-Methyl-1-Oxo-2-Propenyl)Oxy)-, Inner Salt, Polymer with Methyl 2-Methyl-2-Propenoate

Trade Name:
Mexomere PX (Chimex)

POLYQUATERNIUM-31

CAS Nos.: 136505-02-7; 189767-67-7

Definition: Polyquaternium-31 is a polymeric quaternary ammonium salt prepared by the reaction of DMAPA Acrylates/Acrylic Acid/Acrylonitrogens Copolymer (q.v.) and diethyl sulfate.

Chemical Classes: Quaternary Ammonium Compounds; Synthetic Polymers

Functions: Antistatic Agent; Film Former; Hair Fixative

Technical/Other Name:
2-Propenenitrile, Homopolymer, Hydrolyzed, Block, Reaction Products with N,N-Dimethyl-1,3-Propanediamine, Di-Et Sulfate-Quaternized

Trade Name:
Hypan QT100 (Lipo)

POLYQUATERNIUM-32

CAS No.: 35429-19-7

Empirical Formula:
$(C_9H_{18}NO_2 \cdot C_3H_5NO)_x \cdot xCl$

Definition: Polyquaternium-32 is the polymeric quaternary ammonium salt that conforms generally to the formula:

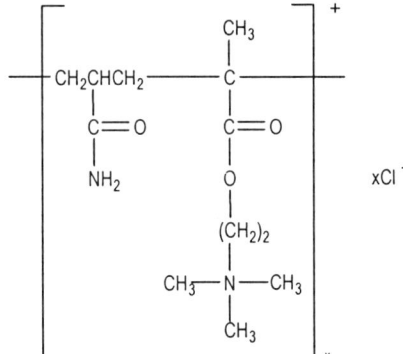

Information Source: TSCA

Chemical Classes: Quaternary Ammonium Compounds; Synthetic Polymers

Functions: Antistatic Agent; Film Former; Hair Fixative

Reported Product Category: Hair Conditioners

Technical/Other Names:
Acrylamide-Dimethylaminoethyl Methacrylate Methyl Chloride Copolymer
Ethanaminium, N,N,N-Trimethyl-2-[(2-Methyl-1-Oxo-2-Propenyl)Oxy]-, Chloride, Polymer with 2-Propenamide

Trade Name Mixtures:
Rheocare CTC (Cosmetic Rheologies)
Salcare SC92 (Ciba Specialty Chemicals)

POLYQUATERNIUM-33

CAS No.: 69418-26-4

Empirical Formula:
$(C_8H_{16}NO_2 \cdot C_3H_5NO)_x \cdot xCl$

Definition: Polyquaternium-33 is the polymeric quaternary ammonium salt that conforms generally to the formula:

Chemical Classes: Quaternary Ammonium Compounds; Synthetic Polymers

Functions: Antistatic Agent; Film Former; Hair Fixative

Technical/Other Names:
Acrylamide-Dimethylaminoethyl Ethanaminium, N,N,N-Trimethyl-2-[1-Oxo-2-Propenyl)Oxy]-, Chloride, Polymer with 2-Propenamide

POLYQUATERNIUM-34

Definition: Polyquaternium-34 is the polymeric quaternary ammonium salt that conforms generally to the formula:

$$\left[\begin{array}{c} CH_2CH_3 \qquad CH_3 \\ | \qquad\qquad | \\ -N-(CH_2)_3-N-(CH_2)_3- \\ | \qquad\qquad | \\ CH_2CH_3 \qquad CH_3 \end{array} \right]_x^{2+} 2x Br^-$$

Chemical Classes: Quaternary Ammonium Compounds; Synthetic Polymers

Functions: Antistatic Agent; Film Former; Hair Fixative

Trade Name:
Mexomere PAK (Chimex)

POLYQUATERNIUM-35

Definition: Polyquaternium-35 is the polymeric quaternary ammonium salt that conforms generally to the formula:

(x+y) $CH_3OSO_3^-$

Chemical Classes: Quaternary Ammonium Compounds; Synthetic Polymers

Functions: Antistatic Agent; Film Former; Hair Fixative

Trade Name:
Plex 3074 L (Rohm GmbH)

POLYQUATERNIUM-36

Definition: Polyquaternium-36 is the polymeric quaternary ammonium salt that conforms generally to the formula:

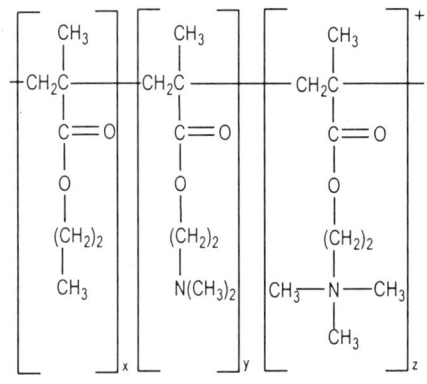

$zCH_3OSO_3^-$

Chemical Classes: Quaternary Ammonium Compounds; Synthetic Polymers

Functions: Antistatic Agent; Film Former; Hair Fixative

Technical/Other Name:
2-Propenoic Acid, 2-Methyl-, 2-(Dimethyl-amino)Ethyl Ester, Polymer with Methyl 2-Methyl-2-Propenoate, compd. with Dimethyl Sulfate

Trade Name:
Plex 4739L (Rohm GmbH)

POLYQUATERNIUM-37

CAS No.: 26161-33-1

Empirical Formula:
$(C_9H_{18}NO_2 \cdot Cl)_x$

Definition: Polyquaternium-37 is the polymeric quaternary ammonium salt that conforms generally to the formula:

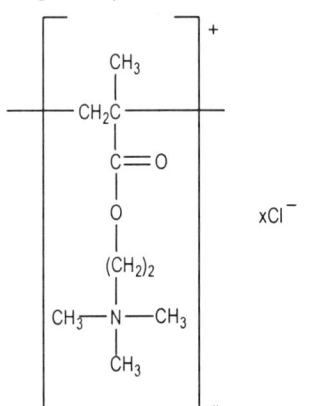

Information Source: TSCA

Chemical Classes: Quaternary Ammonium Compounds; Synthetic Polymers

Functions: Antistatic Agent; Film Former; Hair Fixative

Reported Product Category: Hair Conditioners

Technical/Other Names:
Choline, Chloride, Methacrylate, Polymer
Ethanaminium, N,N,N-Trimethyl-2-[(Methyl-1-Oxo-2-Propenyl)Oxy]-, Chloride, Homopolymer
Trimethylaminoethyl Methacrylate Chloride Polymer
N,N,N-Trimethyl-2-[(Methyl-1-Oxo-2-Propenyl)Oxy]Ethanaminium Chloride, Homopolymer

Trade Names:
Synthalen CN (3V Sigma S.P.A.)
Synthalen CR (3V Sigma S.P.A.)
Synthalen CU (3V Sigma S.P.A.)
Ultragel 300 (Cosmetic Rheologies)

Trade Name Mixtures:
Rheocare CTH(E) (Cosmetic Rheologies)
Salcare SC95 (Ciba Specialty Chemicals)
Salcare SC96 (Ciba Specialty Chemicals)

POLYQUATERNIUM-39

CAS No.: 25136-75-8

JPN Translation:
ポリクオタニウム - 39

Definition: Polyquaternium-39 is a polymeric quaternary ammonium salt of acrylic acid, diallyl dimethyl ammonium chloride and acrylamide.

Chemical Classes: Quaternary Ammonium Compounds; Synthetic Polymers

Functions: Antistatic Agent; Film Former; Hair Fixative

Reported Product Category: Shampoos (Non-coloring)

Technical/Other Names:
Acrylic Acid, Polymer with Acrylamide and Diallyldimethylammonium Chloride
2-Propen-1-aminium, N,N-Dimethyl-N-2-Propenyl-, Chloride, Polymer with 2-Propenamide and 2-Propenoic Acid

Trade Names:
Merquat 3333 (ONDEO Nalco)
Merquat Plus 3330 (ONDEO Nalco)
Merquat Plus 3330 (ONDEO Nalco Europe)
Merquat Plus 3331 (ONDEO Nalco)
Merquat Plus 3331 (ONDEO Nalco Europe)

POLYQUATERNIUM-42

Empirical Formula:
$(C_{10}H_{24}N_2O)_x \cdot (2Cl)_x$

Definition: Polyquaternium-42 is the quaternary ammonium compound that conforms to the formula:

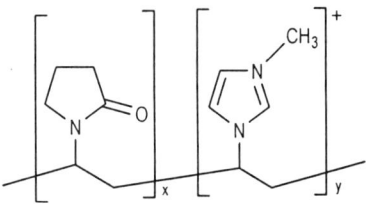

Chemical Classes: Quaternary Ammonium Compounds; Synthetic Polymers

Function: Preservative

Trade Name:
Busan 1507 (Buckman Labs)

POLYQUATERNIUM-43

Definition: Polyquaternium-43 is the copolymer of acrylamide, acrylamidopropyl-trimonium chloride, 2-amidopropylacrylamide sulfonate, and DMAPA monomers.

Chemical Classes: Quaternary Ammonium Compounds; Synthetic Polymers

Functions: Antistatic Agent; Film Former; Hair Conditioning Agent

Trade Name:
Bozequat 4000 (Clariant S.A.)

POLYQUATERNIUM-44

Definition: Polyquaternium-44 is the polymeric quaternary ammonium salt consisting of vinylpyrrolidone and quaternized imidazoline monomers. It conforms generally to the formula:

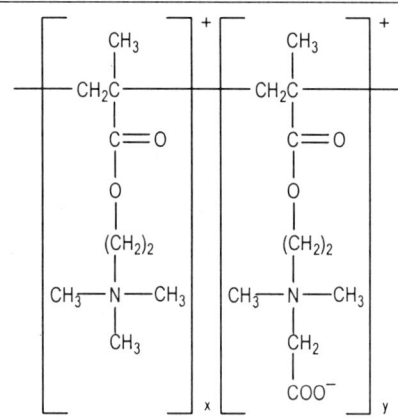

$$yCH_3OSO_3^-$$

Chemical Classes: Quaternary Ammonium Compounds; Synthetic Polymers

Functions: Antistatic Agent; Film Former; Hair Conditioning Agent

Trade Names:
Luviquat Care (BASF)
Luviquat UltraCare (BASF Corporation)

POLYQUATERNIUM-45

Definition: Polyquaternium-45 is the polymeric quaternary ammonium salt that conforms generally to the formula:

$$(x+y)\ CH_3OSO_3^-$$

Chemical Classes: Quaternary Ammonium Compounds; Synthetic Polymers

Functions: Antistatic Agent; Film Former; Hair Fixative

Trade Name:
Plex 3073L (Rohm GmbH)

POLYQUATERNIUM-46

CAS No.: 174761-16-1

Definition: Polyquaternium-46 is a polymeric quaternary ammonium salt prepared by the reaction of vinylcaprolactam and vinylpyrrolidone with methylvinylimidazolium methosulfate.

Chemical Classes: Quaternary Ammonium Compounds; Synthetic Polymers

Functions: Antistatic Agent; Film Former; Hair Fixative

Technical/Other Name:
1H-Imidazolium, 1-Ethenyl-3-Methyl-, Methyl Sulfate, Polymer with 1-Ethenylhexahydro-2H-Azepin-2-one and 1-Ethenyl-2-Pyrrolildinone

Trade Name:
Luviquat Hold (BASF)

POLYQUATERNIUM-47

Definition: Polyquaternium-47 is a polymeric quaternary ammonium chloride formed by the polymerization of acrylic acid, methyl acrylate and methacrylamidopropyl-trimonium chloride.

Chemical Classes: Quaternary Ammonium Compounds; Synthetic Polymers

Functions: Film Former; Hair Fixative; Skin-Conditioning Agent - Miscellaneous

Technical/Other Name:
1-Propanaminium, N,N,N-Trimethyl-3-((2-Methyl-1-Oxo-2-Propenyl)Amino)-, Chloride, Polymer with Methyl 2-Propenoate and 2-Propenoic Acid

Trade Names:
Merquat 2001 (ONDEO Nalco)
Merquat 2001 (ONDEO Nalco Europe)
Merquat 2001N (ONDEO Nalco)
Merquat 2001N (ONDEO Nalco Europe)

POLYQUATERNIUM-48

JPN Translation:
ポリクオタニウム - 48
Definition: Polyquaternium-48 is a copolymer of methacryloyl ethyl betaine, 2-hydroxyethyl methacrylate and methacryloyl ethyl trimethyl ammonium chloride.

Chemical Classes: Quaternary Ammonium Compounds; Synthetic Polymers

Functions: Antistatic Agent; Film Former; Hair Fixative

Trade Name:
Plascize L-450 (Goo Chemical)

POLYQUATERNIUM-49

JPN Translation:
ポリクオタニウム - 49
Definition: Polyquaternium-49 is a copolymer of methacryloyl ethyl betaine, PEG-9 methacrylate and methacryloyl ethyl trimethyl ammonium chloride.

Chemical Classes: Quaternary Ammonium Compounds; Synthetic Polymers

Functions: Antistatic Agent; Film Former; Hair Fixative

Trade Name:
Plascize L-440 (Goo Chemical)

POLYQUATERNIUM-50

Definition: Polyquaternium-50 is the polymeric quaternary ammonium salt that conforms generally to the formula:

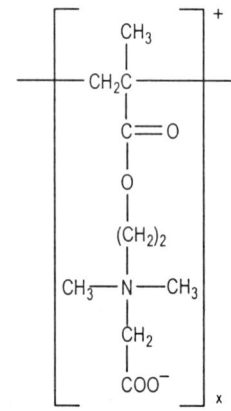

Chemical Classes: Quaternary Ammonium Compounds; Synthetic Polymers

Functions: Antistatic Agent; Film Former; Hair Fixative

Trade Name:
Plascize L-401 (Goo Chemical)

POLYQUATERNIUM-51

CAS No.: 125275-25-4

Definition: Polyquaternium-51 is the polymeric quaternary ammonium salt that conforms generally to the formula:

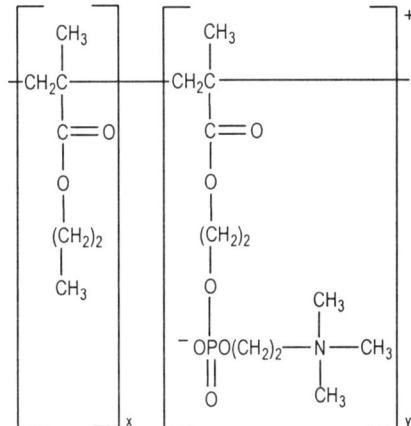

Chemical Classes: Phosphorus Compounds; Quaternary Ammonium Compounds; Synthetic Polymers

Functions: Film Former; Skin-Conditioning Agent - Humectant

Technical/Other Name:
3,5,8-Triox-4-Phosphaundec-10-en-1-aminium, 4-Hydroxy-N,N,N,10-Tetra-methyl-9-Oxo, Inner Salt, 4-Oxide, Polymer with Butyl 2-Methyl-2-Propenoate

Trade Name:
Lipidure-PMB (NOF)

Trade Name Mixture:
Advanced Moisture Complex (Collaborative Labs)

POLYQUATERNIUM-52

Definition: Polyquaternium-52 is the polymeric quaternary ammonium salt that conforms generally to the formula:

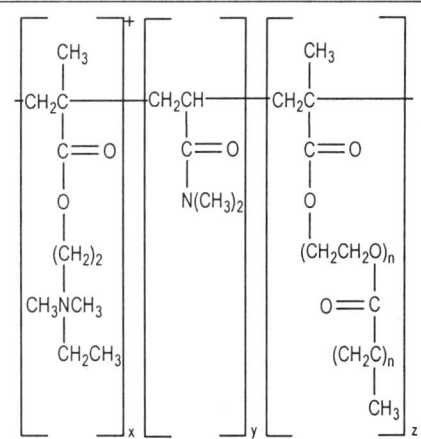

$$xCH_3CH_2OSO_3{}^-$$

where n has an average value between 1 and 30.

Chemical Classes: Quaternary Ammonium Compounds; Synthetic Polymers

Functions: Hair Conditioning Agent; Viscosity Increasing Agent - Aqueous

POLYQUATERNIUM-53

CAS No.: 84647-38-1

Definition: Polyquaternium-53 is a copolymer of acrylic acid, acrylamide and methacrylamidopropyltrimonium chloride monomers.

Chemical Classes: Quaternary Ammonium Compounds; Synthetic Polymers

Function: Hair Conditioning Agent

Technical/Other Name:
Acrylic Acid/Acrylamide/Methacrylamido-propyltrimonium Chloride Copolymer

Trade Names:
Merquat 2003 (ONDEO Nalco)
Merquat 2003 (ONDEO Nalco Europe)

POLYQUATERNIUM-54

Definition: Polyquaternium-54 is the polymeric quaternary ammonium salt prepared by the reaction of aspartic acid and C6-18 alkylamine with dimethylaminopropyl-amine and sodium chloroacetate.

Chemical Class: Quaternary Ammonium Compounds

Functions: Antistatic Agent; Hair Conditioning Agent

Trade Name:
Quilty-Hy (Mitsui)

POLYQUATERNIUM-55

CAS No.: 306769-73-3

Definition: Polyquaternium-55 is the polymeric quaternary ammonium chloride formed by the reaction of vinylpyrrolidone, dimethylaminopropyl methacrylamide and methacryloylaminopropyl lauryldimonium chloride.

Chemical Classes: Quaternary Ammonium Compounds; Synthetic Polymers

Function: Hair Fixative

Technical/Other Name:
1-Dodecanaminium, N,N-Dimethyl-N-[3-[(2-Methyl-1-Oxo-2-Propenyl)AminoPropyl]-, Chloride, Polymer with N-[3-(Dimethyl-amino)Propyl]-2-Methyl-2-Propenamide and 1-Ethenyl-2-Pyrrolidinone

Trade Names:
Styleze W-10 (International Specialty Products)
Styleze W-20 (International Specialty Products)

POLYQUATERNIUM-56

Definition: Polyquaternium-56 is a polymeric quaternary ammonium salt consisting of isophorone diisocyanate, butylene glycol and dihydroxyethyldimonium methosulfate monomers.

Chemical Classes: Quaternary Ammonium Compounds; Synthetic Polymers

Functions: Film Former; Hair Conditioning Agent; Hair Fixative

Trade Name:
Hairrol UC-4 (Sanyo Chemical)

POLYQUATERNIUM-57

Definition: Polyquaternium-57 is the polymeric quaternary ammonium salt consisting of Castor Isostearate Succinate (q.v.) and Ricinoleamidopropyltrimonium Chloride (q.v.) monomers.

Chemical Classes: Quaternary Ammonium Compounds; Synthetic Polymers

Functions: Film Former; Hair Conditioning Agent

Trade Name:
Zenigloss Q (Zenitech)

Trade Name Mixtures:
Zenibee Cream - Q (Zenitech)
Zenicone XQ (Zenitech)
Zenigloss Q-SE (Zenitech)

Zenigloss-SQ (Zenitech)
Zeniquat Base (Zenitech)

POLYQUATERNIUM-58

CAS No.: 88230-37-9

Definition: Polyquaternium-58 is the polymeric quaternary ammonium salt prepared by the reaction of methyl 2-propenoate, 2,2-bis[(2-propenyloxy)methyl]-1-butanol, diethenylbenzene, and N,N-dimethyl-1,3-propanediamine quaternized with chloromethane.

Chemical Class: Quaternary Ammonium Compounds

Function: Hair Conditioning Agent

Technical/Other Name:
2-Propenoic Acid, Methyl Ester, Polymer with 2,2-Bis[(2-Propenyloxy)Methyl]-1-Butanol and Diethenylbenzene, Reaction Products with N,N-Dimethyl-1,3-Propanediamine, Chloromethane-Quaternized

Trade Name Mixture:
Lowenol Conditioner PWW (Lowenstein)

POLYQUATERNIUM-59

Definition: Polyquaternium-59 is the polymeric quaternary ammonium salt that conforms to the formula:

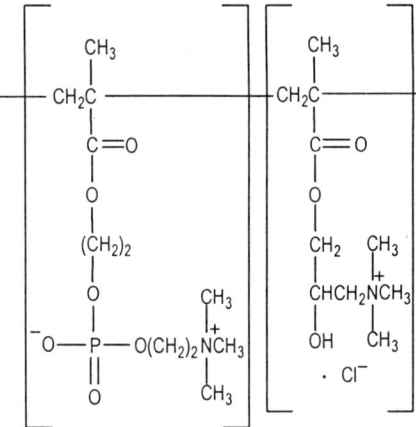

Chemical Class: Quaternary Ammonium Compounds

Function: Ultraviolet Light Absorber

Trade Name Mixture:
Crodasorb UV-HPP (Croda, Inc.)

POLYQUATERNIUM-60

CAS No.: 351425-04-2

Definition: Polyquaternium-60 is the polymeric quaternary ammonium compound prepared by the reaction of TEA-Diricinoleate/IPDI Copolymer (q.v.) with diethyl sulfate.

Chemical Classes: Quaternary Ammonium Compounds; Synthetic Polymers

Functions: Antistatic Agent; Hair Conditioning Agent

Technical/Other Name:
9-Octadecenoic Acid, 12-Hydroxy-, [(2-Hydroxyethyl)Imino]Di-2,1-Ethanediyl Ester, Polymer with 5-Isocyanato-1-(Isocyanatomethyl)-1,3,3-Trimethyl-cyclohexane, Compd. with Diethyl Sulfate

Trade Name Mixture:
Polylipid PPI-RC (Alzo)

POLYQUATERNIUM-61

Definition: Polyquaternium-61 is the polymeric quaternary ammonium salt that conforms generally to the formula:

Chemical Classes: Phosphorus Compounds; Quaternary Ammonium Compounds; Synthetic Polymers

Functions: Film Former; Skin-Conditioning Agent - Humectant

Trade Name:
Lipidure-S (NOF America Corp.)

POLYQUATERNIUM-62

Definition: Polyquaternium-62 is the polymeric quaternary ammonium salt

prepared by the reaction of butyl methacrylate, polyethylene glycol methyl ether methacrylate, ethylene glycol dimeth-acrylate and 2-methacryloyethyl trimonium chloride with 2,2'-azobis(2-methyl propionamidine) dihydrochloride.

Chemical Classes: Quaternary Ammonium Compounds; Synthetic Polymers

Functions: Film Former; Hair Fixative; Hair-Waving/Straightening Agent; Opacifying Agent

Trade Name:
Nanoaquasome (Pacific)

POLYQUATERNIUM-63

Definition: Polyquaternium-63 is a copolymer of acrylamide, acrylic acid and ethyltrimonium chloride acrylate.

Chemical Class: Synthetic Polymers

Functions: Hair Conditioning Agent; Skin-Conditioning Agent - Miscellaneous

Trade Name:
OF-308 (WSP Chemicals & Technology)

POLYQUATERNIUM-64

Definition: Polyquaternium-64 is the polymeric quaternary ammonium salt that conforms to the formula:

Chemical Class: Quaternary Ammonium Compounds

Functions: Hair Conditioning Agent; Humectant

Trade Name:
Lipidure-C (NOF)

POLYQUATERNIUM-65

Definition: Polyquaternium-65 is the polymeric quaternary ammonium salt

---

consisting of 2-methacryloyloxyethyl-phosphorylcholine, butyl methacrylate and sodium methacrylate monomers.

Chemical Class: Quaternary Ammonium Compounds

Functions: Emulsion Stabilizer; Hair Conditioning Agent; Humectant

Trade Name:
Lipidure-A (NOF)

POLYQUATERNIUM-4/HYDROXYPROPYL STARCH COPOLYMER

Definition: Polyquaternium-4/Hydroxypropyl Starch Copolymer is the polymer formed by the reaction of Polyquaternium-4 (q.v.) and Hydroxypropyl Starch (q.v.)

Chemical Classes: Carbohydrates; Quaternary Ammonium Compounds

Functions: Film Former; Hair Conditioning Agent; Hair Fixative

POLYSILICONE-1

Definition: Polysilicone-1 is a propylammonium dimethicone amine salt of the succinic acid ester of a polyoxypropylene, polyoxyethylene ether of glycerin.

Chemical Class: Siloxanes and Silanes

Functions: Antifoaming Agent; Antistatic Agent; Hair Conditioning Agent

POLYSILICONE-2

Definition: Polysilicone-2 is the polymer formed by the reaction of tetradecene with polymerized tetramethylcyclotetrasiloxane.

Chemical Class: Siloxanes and Silanes

Functions: Antifoaming Agent; Hair Conditioning Agent

Reported Product Categories: Lipsticks; Eye Shadows; Eyeliners; Foundations; Face Powders; Eye Makeup Preparations, Misc.; Makeup Preparations (Not eye), Misc.; Rouges; Blushers (All types); Eyebrow Pencils

Trade Name:
EP 1 (Shiseido Company)

POLYSILICONE-3

Definition: Polysilicone-3 is prepared from an alkylthiosulfate substituted N-acetyl-

methionyl silanol by hydrolysis in the presence of Dextran (q.v.). It conforms to the formula:

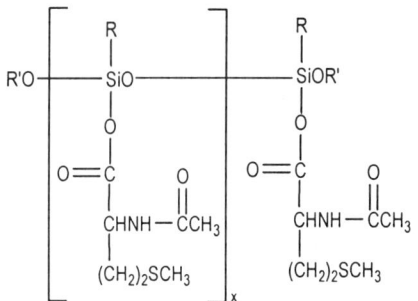

where R is $(CH_2)_3OCH_2CHOHCH_2S_2O_3Na$ and R' is Dextran (q.v.).

Chemical Class: Siloxanes and Silanes

Functions: Hair Conditioning Agent; Skin-Conditioning Agent - Emollient

Trade Name:
Methiosil C+ (Exsymol)

POLYSILICONE-4

Definition: Polysilicone-4 is the reaction product of the polymerization of tetramethylcyclotetrasiloxane.

Chemical Class: Siloxanes and Silanes

Functions: Hair Conditioning Agent; Viscosity Increasing Agent - Nonaqueous

Trade Name:
EP 0 (Shiseido Company)

POLYSILICONE-5

Definition: Polysilicone-5 is the reaction product of Polysilicone-4 (q.v.) and glyceryl monoallyl ether.

Chemical Class: Siloxanes and Silanes

Functions: Hair Conditioning Agent; Viscosity Increasing Agent - Nonaqueous

Reported Product Categories: Foundations; Face Powders; Rouges; Makeup Preparations (Not eye), Misc.

Trade Name:
EP 2 (Shiseido Company)

POLYSILICONE-6

CAS No.: 146632-09-9

Definition: Polysilicone-6 is a copolymer of dimethylsiloxane and methyl 3-

mercaptopropylsiloxane reacted with isobutyl methacrylate.

Chemical Class: Siloxanes and Silanes

Functions: Film Former; Hair Conditioning Agent

POLYSILICONE-7

CAS No.: 146632-08-8

Definition: Polysilicone-7 is a copolymer of isobutyl methacrylate, and n-butyl end blocked dimethylsiloxane propyl methacrylate.

Chemical Class: Siloxanes and Silanes

Functions: Antifoaming Agent; Hair Conditioning Agent

Trade Names:
AEC Polysilicone-7 (A & E Connock)
3M Silicones "Plus" SA70 Polymer (3M)

POLYSILICONE-8

Definition: Polysilicone-8 is the siloxane polymer that conforms to the formula:

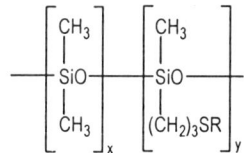

where R represents the Acrylates Copolymer (q.v.) radical.

Chemical Classes: Siloxanes and Silanes; Synthetic Polymers; Thio Compounds

Functions: Antifoaming Agent; Film Former; Hair Conditioning Agent

Trade Name:
3M Brand Silicones "Plus" Polymer VS 80 (3M)

POLYSILICONE-9

Definition: Polysilicone-9 is the siloxane polymer that conforms generally to the formula:

Chemical Classes: Siloxanes and Silanes; Synthetic Polymers

Function: Hair Fixative

Trade Name:
Elastomer OS (Kao Corp.)

POLYSILICONE-10

Definition: Polysilicone-10 is the siloxane polymer that conforms generally to the formula:

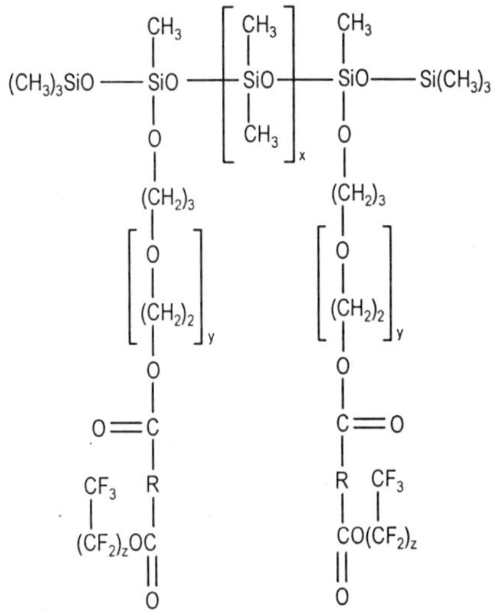

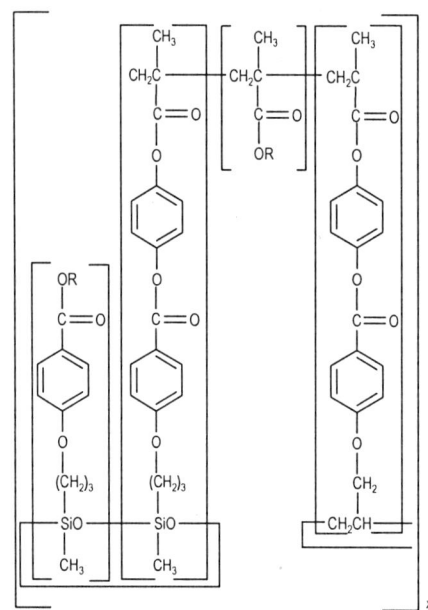

where -OCRCO- repesents the dilinoleoyl moiety.

Chemical Classes: Halogen Compounds; Siloxanes and Silanes; Synthetic Polymers

Functions: Antifoaming Agent; Hair Conditioning Agent

POLYSILICONE-11

Definition: Polysilicone-11 is a crosslinked siloxane rubber formed by the reaction of vinyl-terminated siloxane and methylhydroxydimethyl siloxane in the presence of cyclomethicone.

Chemical Classes: Siloxanes and Silanes; Synthetic Polymers

Function: Film Former

Trade Name Mixtures:
Actiprime 100 (Active Organics)
Gransil DMG-6 (Grant)
Gransil GCM (Grant)
Gransil PM-GEL (Grant)

POLYSILICONE-12

Definition: Polysilicone-12 is a complex copolymer consisting of esters of methacrylic acid, including cholesteryl methacrylate and Cyclomethicone (q.v.). It conforms generally to the formula:

where R represents the cholesteryl moiety.

Chemical Classes: Siloxanes and Silanes; Synthetic Polymers

Function: Opacifying Agent

POLYSILICONE-13

CAS No.: 158451-77-5

Definition: Polysilicone-13 is the siloxane polymer that conforms generally to the formula:

$$-[(CH_2CHO)_x(CH_2CH_2O)_y CHCH_2(SiO)_z Si-]_x$$

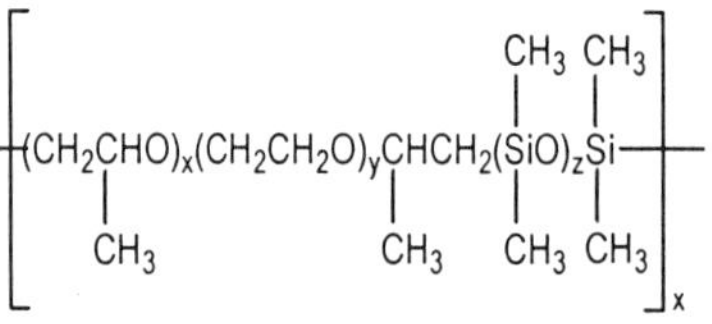

Chemical Class: Siloxanes and Silanes

Function: Hair Conditioning Agent

Technical/Other Name:
Siloxanes and Silicones, Di-Me, Hydrogen-Terminated, Polymers with Polyethylene-Polypropylene Glycol Bis(2-Methyl-2-Propenyl)Ether

Trade Names:
FZ-2203 (Nippon Unicar)
FZ-2207 (Nippon Unicar)
FZ-2222 (Nippon Unicar)
FZ-2231 (Nippon Unicar)
FZ-2232 (Nippon Unicar)
FZ-2233 (Nippon Unicar)
FZ-2234 (Nippon Unicar)
FZ-2235 (Nippon Unicar)
FZ-2236 (Nippon Unicar)
FZ-2237 (Nippon Unicar)
FZ-2238 (Nippon Unicar)
FZ-2239 (Nippon Unicar)
Silwet 236-L (Nippon Unicar)

Trade Name Mixture:
Silsoft 477 (OSi Specialties)

POLYSILICONE-14

Definition: Polysilicone-14 is the siloxane polymer that conforms generally to the formula:

$$CH_2CHCH_2(NHCHCO)_xOH$$

$$(CH_2)_3(SiO_{3/2})_a(SiO_{3/2})_b(SiO_{3/2})_c(SiO_{4/2/})_d(SiO_{1/2})_e$$

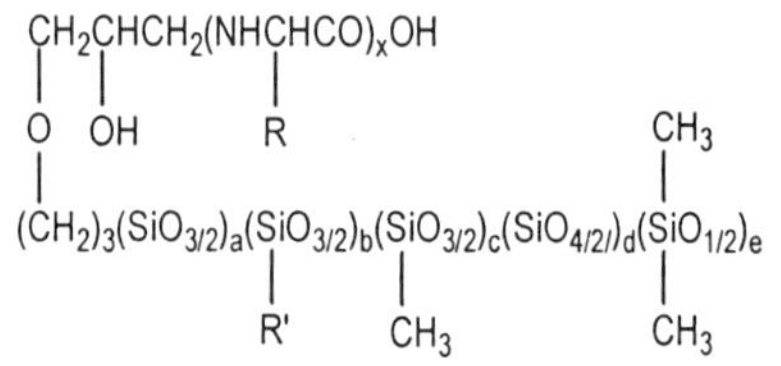

where R represents a hydrolyzed silk moiety and R' represents a C6-10 alkyl chain.

Chemical Classes: Siloxanes and Silanes; Synthetic Polymers

Functions: Film Former; Hair Conditioning Agent; Skin-Conditioning Agent - Emollient

Trade Name:
Promois SLA-SE (Seiwa Kasei)

POLYSILICONE-15

CAS No.: 207574-74-1

Definition: Polysilicone-15 is the siloxane polymer that conforms generally to the formula:

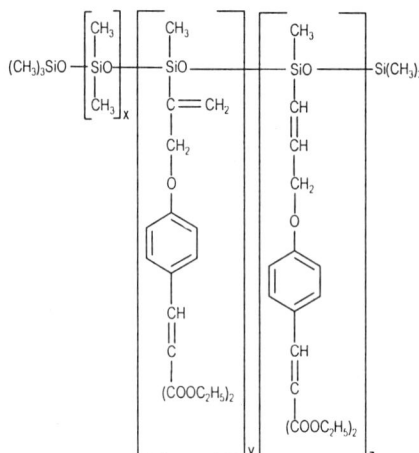

Information Source: EEC(VII/1-26)

Chemical Classes: Siloxanes and Silanes; Synthetic Polymers

Function: Ultraviolet Light Absorber

Technical/Other Names:
Diethylbenzylidene Malonate Dimethicone
Diethylmalonylbenzylidene Oxypropene
  Dimethicone
Dimethicodiethylbenzalmalonate

Trade Names:
Dimethicodiethylbenzalmalonate (Roche
  Vitamins)
Parsol-SLX (Hoffmann-La Roche)

Trade Name Mixture:
Parsol-SLX (Roche)

POLYSILICONE-16

Definition: Polysilicone-16 is the poly-siloxane that conforms to the formula:

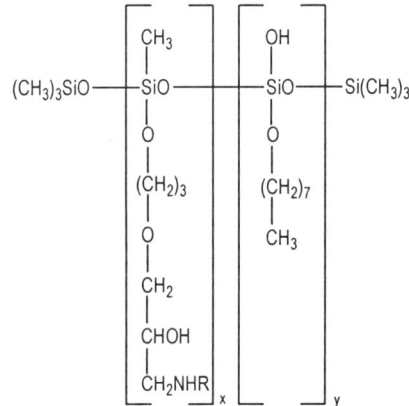

where R represents the hydrolyzed collagen moiety.

Chemical Class: Siloxanes and Silanes

Functions: Bulking Agent; Emulsion Stabilizer; Skin-Conditioning Agent - Occlusive; Viscosity Controlling Agent

Technical/Other Name:
Hydrolyzed Collagen PG-Propyl Caprylyl
  Dimethicone Silsesquioxane

Trade Name:
Protesil C (Seiwa Kasei)

POLYSILICONE-17

Definition: Polysilicone-17 is a crosslinked methylphenyldiphenylpolysiloxane obtained from trichloromethylsilane, trichlorophenyl-silane, and dichlorophenylsilane.

Information Source: JCIC

Chemical Class: Siloxanes and Silanes

Functions: Skin-Conditioning Agent - Miscellaneous; Viscosity Increasing Agent - Nonaqueous

Technical/Other Name:
  Crosslinked Methylphenylpolysiloxane

POLYSIPHONIA LANOSA EXTRACT

Definition: Polysiphonia Lanosa Extract is an extract of the alga, *Polysiphonia lanosa. See "Regulatory and Ingredient Use Information," regarding the labeling names for botanical ingredients in Volume 1, Introduction, Part A.*

Chemical Class: Biological Products

Function: Skin-Conditioning Agent - Miscellaneous

Technical/Other Name:
  Extract of Polysiphonia Lanosa Extract

Trade Name Mixture:
  Extract of Polysiphonia lanosa Sun'Ytol
    (GELYMA)

POLYSORBATE 20

CAS No.: 9005-64-5 (Generic)

JPN Translation:
  ポリソルベート 20

Definition: Polysorbate 20 is a mixture of laurate esters of sorbitol and sorbitol anhydrides, consisting predominantly of the monoester, condensed with approximately 20 moles of ethylene oxide. It conforms generally to the formula:

$$(OCH_2CH_2)_wOH$$
$$(OCH_2CH_2)_xOH$$
$$CH(OCH_2CH_2)_yOH$$
$$CH_2(OCH_2CH_2)_zO-C(CH_2)_{10}CH_3$$
$$O$$

where w + x + y + z has an average value of 20.

Information Sources: BAN, BPC, 21CFR172.515, 21CFR175.105, 21CFR175.300, 21CFR178.3400, CIR: [S] JACT-3(5)1984, CTFA S, FCC, INN, ITA, JCLS, JSCI, MAR, NF XIX, RIFM, TSCA, USAN

Chemical Class: Sorbitan Derivatives

Functions: Fragrance Ingredient; Surfactant - Emulsifying Agent; Surfactant - Solubilizing Agent

Reported Product Categories: Tonics, Dressings, and Other Hair Grooming Aids; Bath Oils, Tablets, and Salts; Cleansing Products (Cold Creams, Cleansing Lotions, Liquids and Pads); Powders (Dusting and Talcum, Excluding Aftershave Talcs); Shampoos (Non-coloring); Skin Care Preparations, Misc.; Hair Conditioners; Hair Dyes and Colors (All Types Requiring Caution Statements and Patch Tests); Bath Preparations, Misc.; Body and Hand Preparations (Excluding Shaving Preparations); Moisturizing Preparations; Skin Fresheners; Makeup Bases; Foundations; Hair Preparations (Non-coloring), Misc.; Bath Soaps and Detergents; Bubble Baths; Hair Sprays (Aerosol Fixatives); Bath Capsules; Eye Makeup Preparations, Misc.; Eye Makeup Removers; Face and Neck Preparations (Excluding Shaving Preparations); Paste Masks (Mud Packs); Personal Cleanliness Products, Misc.; Perfumes; Permanent Waves; Aftershave Lotions; Baby Shampoos; Hair Wave Sets; Mouthwashes and Breath Fresheners (Liquids and Sprays); Baby Products, Misc.; Eyebrow Pencils; Shaving Cream (Aerosol, Brushless and Lather); Suntan Gels, Creams, and Liquids; Colognes and Toilet Waters; Eye Lotions; Eyeliners; Indoor Tanning Preparations; Mascara; Night Skin Care Preparations; Blushers (All types); Douches; Eye Shadows; Fragrance Preparations, Misc.; Hair Coloring Preparations, Misc.; Hair Rinses (Non-coloring); Makeup Preparations (Not eye), Misc.; Suntan Preparations, Misc.; Dentifrices (Aerosol, Liquid, Pastes and Powders); Deodorants (Underarm); Makeup Fixatives; Manicuring Preparations, Misc.; Shaving Preparations, Misc.

Technical/Other Names:
Polyoxyethylene (20) Sorbitan Monolaurate
Polyoxyethylene Sorbitan Monolaurate
  (20E.O.)
Polysorbate 20 (RIFM)
Sorbimacrogol Laurate 300
Sorbitan, Monododecanoate, Poly(Oxy-1,2-
  Ethanediyl) Derivs.

Trade Names:
AEC Polysorbate 20 (A & E Connock)
Alkamuls PSML-20 (Rhodia)
Alkamuls T20 (Rhodia)
Cremophor PS 20 (BASF)
Crillet 1 (Croda Chemicals)
Customulse L-20 (Polysorbate 20) (Custom
  Ingredients)
DeMULS PSML-20 (DeForest)
Emsorb 2720 (Cognis Care Chemicals/PA)
Ethsorbox L-20 (Ethox)
Glycosperse L-20K (Lonza Inc./Lonza Ltd.)
Hetsorb L-20 (Heterene)
Ixol 2 (Vevy)
Jeesorb L-20 (Jeen)
Kotilen-L/1 (Kolb)
Laxan ESL (Lanaetex)
Liposorb L-20 (Lipo)
Lonzest SML-20 (Lonza Inc./Lonza Ltd.)
Lumisorb PSML-20 (Lambent)
Merpoxen SML 200 (Wall)

Montanox 20 (SEPPIC)
Mulsifan RT 141 (Zschimmer & Schwarz)
Nikkol TL-10 (Nikko)
Nikkol TL-10EX (Nikko)
Polisorbac 20 (Specialty Industrial)
Protasorb L-20 (Protameen)
Radiasurf 7137 (Oleon NV)
Ritabate 20 (RITA)
Sabosorb MLE (Sabo)
Sorbax PML-20 (Chemax)
Sorbilene L (Cesalpinia)
Sorbon T-20 (Toho)
Tego SML 20 (Degussa Care Specialties)
Tween 20 (Uniqema Americas)

Trade Name Mixtures:
Ami Nail Bioregenerator (Alban Muller)
Amisol 4135 (Lucas Meyer GmbH)
Biron NLD (Merck KGaA/EMD Chemicals Inc.)
Blue Green Klamath Extract (CEP (Solabia))
Complex Capsico (Fabriquimica)
Concentrate for Dandruff (Crodarom)
Concentrate for Greasy Hair (Crodarom)
DeCONC BSC-50 (DeForest)
Deperoxidium Marin (Coletica SA)
Deperoxidium Vegetal (Coletica SA)
Eucalyptus-Extract, Water Soluble 4786 (Crodarom)
Extrapone Copaiba 2/B16991 (Symrise)
Herbaliquid Lavender Special (Crodarom)
Herbaliquid Watercress Special (Crodarom)
Integryssime (Sederma)
Lowenol 4558 (Lowenstein)
Lubrasil (Guardian)
Novaplant Balm Mint Extract (Crodarom)
Novaplant-Birch Leaves Extract, Water Soluble (Crodarom)
Novaplant-Fennel Extract Fluid (Crodarom)
Novaplant Juniper Extract (Crodarom)
Novaplant Rosemary Extract (Crodarom)
Novaplant Wild Thyme Extract Water Soluble (Crodarom)
Phenolines The Vert (Coletica SA)
Post-Depilatory (Greentech S.A)
Retinol 50 C (BASF)
Rovisome Retinol Moist (Rovi)
Sabosol HO (Sabo)
Sabosolv EEM (Sabo)
SilDerm Formulating Base (Active Concepts)
SilDerm Formulating Base IF (Active Concepts)
Solubilisant Gamma 2420 (Gattefosse s.a.)
Solubilisant Gamma 2428 (Gattefosse s.a.)
Soluvit Richter (CLR)
SUPER A (C.I.T.)
VitAine (AGI Dermatics)
Vitamin Extract AEFH' Water Soluble (Crodarom)
Vitaminextract VC, Water Soluble (Crodarom)
Vitamin F, Water Soluble (Crodarom)
Vitamin F water-soluble CLR (CLR)
Vitamin F Watersoluble (Cosmetochem) (Cosmetochem International Ltd.)
Water-Soluble Tea Tree Oil (Southern Cross Botanicals)
Ylang Extract (Carrubba)
Ylang Ylang Extract Water Soluble 2/ 033610 (Symrise)

POLYSORBATE 21

CAS No.: 9005-64-5 (Generic)

Definition: Polysorbate 21 is a mixture of laurate esters of sorbitol and sorbitol anhydrides, consisting predominantly of the monoester, condensed with approximately 4 moles of ethylene oxide. It conforms generally to the formula:

$$(OCH_2CH_2)_wOH$$
$$(OCH_2CH_2)_xOH$$
$$CH(OCH_2CH_2)_yOH$$
$$CH_2(OCH_2CH_2)_zO\text{---}\underset{O}{\overset{\|}{C}}(CH_2)_{10}CH_3$$

where w + x + y + z has an average value of 4.

Information Sources: 21CFR175.300, CIR: [S] JACT-3(5)1984, CTFA D, RIFM, TSCA

Chemical Class: Sorbitan Derivatives

Functions: Fragrance Ingredient; Surfactant - Emulsifying Agent

Reported Product Categories: Tonics, Dressings, and Other Hair Grooming Aids; Bath Oils, Tablets, and Salts; Cleansing Products (Cold Creams, Cleansing Lotions, Liquids and Pads); Powders (Dusting and Talcum, Excluding Aftershave Talcs); Shampoos (Non-coloring); Skin Care Preparations, Misc.; Hair Conditioners; Hair Dyes and Colors (All Types Requiring Caution Statements and Patch Tests); Bath Preparations, Misc.; Body and Hand Preparations (Excluding Shaving Preparations); Moisturizing Preparations; Skin Fresheners; Makeup Bases; Foundations; Hair Preparations (Non-coloring), Misc.; Bath Soaps and Detergents; Bubble Baths; Hair Sprays (Aerosol Fixatives); Bath Capsules; Eye Makeup Preparations, Misc.; Eye Makeup Removers; Face and Neck Preparations (Excluding Shaving Preparations); Paste Masks (Mud Packs); Personal Cleanliness Products, Misc.; Perfumes; Permanent Waves; Aftershave Lotions; Baby Shampoos; Hair Wave Sets; Mouthwashes and Breath Fresheners (Liquids and Sprays); Baby Products, Misc.; Eyebrow Pencils; Shaving Cream (Aerosol, Brushless and Lather); Suntan Gels, Creams, and Liquids; Colognes and Toilet Waters; Eye Lotions; Eyeliners; Indoor Tanning Preparations; Mascara; Night Skin Care Preparations; Blushers (All types); Eye Shadows; Fragrance Preparations, Misc.; Hair Coloring Preparations, Misc.; Makeup Preparations (Not eye), Misc.; Suntan Preparations, Misc.; Dentifrices (Aerosol, Liquid, Pastes and Powders); Deodorants (Underarm); Makeup Fixatives; Shaving Preparations, Misc.

Technical/Other Names:
Polyoxyethylene (4) Sorbitan Monolaurate
Polysorbate 20 (RIFM)

Trade Names:
Crillet 11 (Croda Chemicals)
Hetsorb L-4 (Heterene)
Jeesorb L-5 (Jeen)
Liposorb L-4 (Lipo)
Polisorbac 21 (Specialty Industrial)
Tween 21 (Uniqema Americas)

Trade Name Mixtures:
Amisol MS-10 (Lucas Meyer GmbH)
Atlas G-1823 (Uniqema Americas)

POLYSORBATE 40

CAS No.: 9005-66-7

JPN Translation:
ポリソルベート40

Definition: Polysorbate 40 is a mixture of palmitate esters of sorbitol and sorbitol anhydrides, consisting predominantly of the monoester, condensed with approximately 20 moles of ethylene oxide. It conforms generally to the formula:

$$(OCH_2CH_2)_wOH$$
$$(OCH_2CH_2)_xOH$$
$$CH(OCH_2CH_2)_yOH$$
$$CH_2(OCH_2CH_2)_zO\text{---}\underset{O}{\overset{\|}{C}}(CH_2)_{14}CH_3$$

where w + x + y + z has an average value of 20.

Information Sources: BAN, 21CFR175.105, 21CFR175.300, 21CFR178.3400, CIR: [S] JACT-3(5)1984, CTFA S, INN, JCLS, JSCI, MAR, NF XIX, TSCA, USAN

Chemical Class: Sorbitan Derivatives

Functions: Surfactant - Emulsifying Agent; Surfactant - Solubilizing Agent

Reported Product Categories: Bath Oils, Tablets, and Salts; Cleansing Products (Cold

Creams, Cleansing Lotions, Liquids and Pads); Moisturizing Preparations; Skin Care Preparations, Misc.; Fragrance Preparations, Misc.; Indoor Tanning Preparations; Bath Preparations, Misc.; Body and Hand Preparations (Excluding Shaving Preparations)

Technical/Other Names:
Polyoxyethylene (20) Sorbitan Monopalmitate
Polyoxyethylene Sorbitan Monopalmitate (20E.O.)
Sorbimacrogol Palmitate 300
Sorbitan, Monohexadecanoate, Poly(Oxy-1,2-Ethanediyl) Derivs.

Trade Names:
Crillet 2 (Croda Chemicals)
Dehymuls SMP 20 (Cognis France)
Hetsorb P-20 (Heterene)
Ixol 4 (Vevy)
Jeesorb P-20 (Jeen)
Kotilen-P/1 (Kolb)
Laxan ESP (Lanaetex)
Liposorb P-20 (Lipo)
Lonzest SMP-20 (Lonza Inc./Lonza Ltd.)
Montanox 40 (SEPPIC)
Nikkol TP-10 (Nikko)
Polisorbac 40 (Specialty Industrial)
Protasorb P-20 (Protameen)
Ritabate 40 (RITA)
Sabosorb MPE (Sabo)
Sorbax PMP-20 (Chemax)
Sorbilene P (Cesalpinia)
Sorbon T-40 (Toho)
Tween 40 (Uniqema Americas)

POLYSORBATE 60

CAS No.: 9005-67-8 (Generic)

JPN Translation:
ポリソルベート 60

Definition: Polysorbate 60 is a mixture of stearate esters of sorbitol and sorbitol anhydrides, consisting predominantly of the monoester, condensed with approximately 20 moles of ethylene oxide. It conforms generally to the formula:

where w + x + y + z has an average value of 20.

Information Sources: BAN, BPC, 21CFR73.1001, 21CFR172.515, 21CFR172.836, 21CFR172.838, 21CFR172.840, 21CFR172.842, 21CFR173.340, 21CFR175.105, 21CFR175.300, 21CFR178.3400, 21CFR573.840, CIR: [S] JACT-3(5)1984, CTFA S, FCC, INN, ITA, JCLS, JSCI, MAR, NF XIX, RIFM, TSCA, USAN

Chemical Class: Sorbitan Derivatives

Functions: Fragrance Ingredient; Surfactant - Emulsifying Agent; Surfactant - Solubilizing Agent

Reported Product Categories: Moisturizing Preparations; Bath Preparations, Misc.; Body and Hand Preparations (Excluding Shaving Preparations); Bath Oils, Tablets, and Salts; Cleansing Products (Cold Creams, Cleansing Lotions, Liquids and Pads); Foundations; Skin Care Preparations, Misc.; Bath Capsules; Mascara; Face and Neck Preparations (Excluding Shaving Preparations); Hair Conditioners; Shaving Cream (Aerosol, Brushless and Lather); Eyebrow Pencils; Paste Masks (Mud Packs); Night Skin Care Preparations; Eyeliners; Indoor Tanning Preparations; Suntan Gels, Creams, and Liquids; Tonics, Dressings, and Other Hair Grooming Aids; Eye Makeup Preparations, Misc.; Makeup Bases; Eye Shadows; Fragrance Preparations, Misc.; Makeup Preparations (Not eye), Misc.; Baby Lotions, Oils, Powders and Creams; Blushers (All types); Cuticle Softeners; Face Powders; Foot Powders and Sprays; Hair Straighteners; Manicuring Preparations, Misc.; Suntan Preparations, Misc.

Technical/Other Names:
Polyoxyethylene (20) Sorbitan Monostearate
Polyoxyethylene Sorbitan Monostearate (20E.O.)
Polysorbate 60 (RIFM)
Sorbimacrogol Stearate 300
Sorbitan, Monooctadecanoate, Poly(Oxy-1,2-Ethanediyl) Derivs.

Trade Names:
AEC Polysorbate 60 (A & E Connock)
Capmul POE-S (Abitec Corporation)
Crillet 3 (Croda Chemicals)
Crillet 3 (Croda, Inc.)
DeMULS PSMS-60 (DeForest)
Emsorb 2728 (Cognis Care Chemicals/PA)
Ethsorbox S-20 (Ethox)
Hetsorb S-20 (Heterene)
Ixol 6 (Vevy)
Jeesorb S-20 (Jeen)
Kotilen-S/1 (Kolb)
Laxan ESS (Lanaetex)
Liposorb S-20 (Lipo)
Lonzest SMS-20 (Lonza Inc./Lonza Ltd.)
Lumisorb PSMS-20 (Lambent)
Merpoxen SMS 200 (Wall)
Montanox 60 (SEPPIC)
Nikkol TS-10 (Nikko)
Polisorbac 60 (Specialty Industrial)

Protasorb S-20 (Protameen)
Radiasurf 7147 (Oleon NV)
Ritabate 60 (RITA)
Sabosorb MSE (Sabo)
Sorbax PMS-20 (Chemax)
Sorbilene S (Cesalpinia)
Sorbon T-60 (Toho)
Tween 60 (Uniqema Americas)

Trade Name Mixtures:
Brookswax P (Arch Personal Care Products)
Brookswax R (Arch Personal Care Products)
Capispheres (Barnet)
Cosmowax S (Croda, Inc.)
Covawax 501 (LCW)
Dispersen-S (Lanaetex)
Incroquat OSC (Croda, Inc.)
Jeecol P (Jeen)
Lipowax P-B Pastilles (Lipo)
Lipowax ES (Lipo)
Lipowax P (Lipo)
Lipowax P-31 (Lipo)
Lipowax PA Pastilles (Lipo)
Lipowax PR (Lipo)
Lipowax R-2 (Lipo)
Marine Phospholipids Deodorized (MMP)
Procol P (Protameen)
Relaxer Concentrate #2 (Arch Personal Care Products)
Relaxer Concentrate #3 (Arch Personal Care Products)
Relaxer Concentrate #4 (Arch Personal Care Products)
Ritachol 1000 (RITA)
Ritachol 2000 (RITA)
Ritachol 5000 (RITA)
Sepigel 502 (SEPPIC)
Simulgel NS (SEPPIC)
Sipowax-P (Specialty Industrial)
Teinowax (Lanaetex)
Upiwax 163 R (Universal Preserv-A-Chem)

POLYSORBATE 61

CAS No.: 9005-67-8 (Generic)

Definition: Polysorbate 61 is a mixture of stearate esters of sorbitol and sorbitol anhydrides, consisting predominantly of the monoester, condensed with approximately 4 moles of ethylene oxide. It conforms generally to the formula:

where w + x + y + z has an average value of 4.

Information Sources: 21CFR175.300, CIR: [S] JACT-3(5)1984, CTFA S, RIFM, TSCA

Chemical Class: Sorbitan Derivatives

Functions: Fragrance Ingredient; Surfactant - Emulsifying Agent

Reported Product Categories: Moisturizing Preparations; Bath Oils, Tablets, and Salts; Cleansing Products (Cold Creams, Cleansing Lotions, Liquids and Pads); Foundations; Skin Care Preparations, Misc.; Bath Capsules; Mascara; Face and Neck Preparations (Excluding Shaving Preparations); Hair Conditioners; Shaving Cream (Aerosol, Brushless and Lather); Eyebrow Pencils; Paste Masks (Mud Packs); Night Skin Care Preparations; Eyeliners; Indoor Tanning Preparations; Tonics, Dressings, and Other Hair Grooming Aids; Eye Makeup Preparations, Misc.; Baby Lotions, Oils, Powders and Creams; Bath Preparations, Misc.; Body and Hand Preparations (Excluding Shaving Preparations); Fragrance Preparations, Misc.; Makeup Preparations (Not eye), Misc.; Cuticle Softeners; Foot Powders and Sprays; Hair Straighteners

Technical/Other Names:
Polyoxyethylene (4) Sorbitan Monostearate
Polysorbate 60 (RIFM)

Trade Names:
Crillet 31 (Croda Chemicals)
Jeesorb S-4 (Jeen)
Laxan-ESE (Lanaetex)
Liposorb S-4 (Lipo)
Polisorbac 61 (Specialty Industrial)
Sabosorb MSE/4 (Sabo)
Tween 61 (Uniqema Americas)

POLYSORBATE 65

CAS No.: 9005-71-4

JPN Translation:
ポリソルベート 65

Definition: Polysorbate 65 is a mixture of stearate esters of sorbitol and sorbitol anhydrides, consisting predominantly of the triester, condensed with approximately 20 moles of ethylene oxide. It conforms generally to the formula:

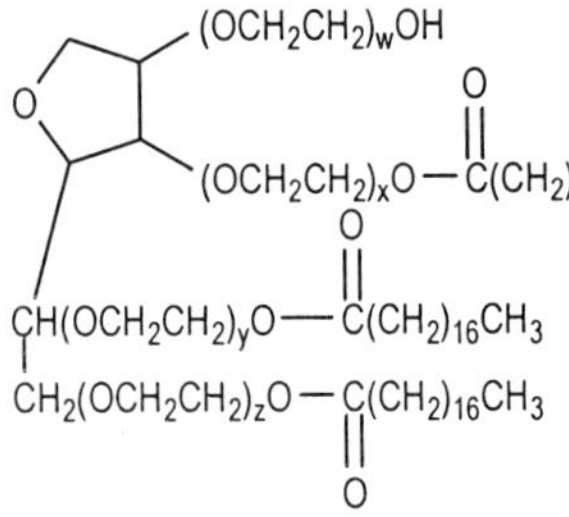

$(OCH_2CH_2)_wOH$
$(OCH_2CH_2)_xO-C(CH_2)_{16}CH_3$
$CH(OCH_2CH_2)_yO-C(CH_2)_{16}CH_3$
$CH_2(OCH_2CH_2)_zO-C(CH_2)_{16}CH_3$

where w + x + y + z has an average value of 20.

Information Sources: BAN, 21CFR73.1001, 21CFR172.836, 21CFR172.838, 21CFR172.840, 21CFR172.842, 21CFR173.340, 21CFR175.300, 21CFR178.3400, CIR: [S] JACT-3(5)1984, CTFA S, FCC, INN, JCIC, JCLS, JSCI, MAR, TSCA, USAN

Chemical Class: Sorbitan Derivatives

Function: Surfactant - Emulsifying Agent

Technical/Other Names:
Polyoxyethylene Sorbitan Tristearate
Polyoxyethylene (20) Sorbitan Tristearate
Polyoxyethylene Sorbitan Tristearate (20E.O.)
Sorbimacrogol Tristearate 300
Sorbitan, Trioctadecanoate, Poly(Oxy-1,2-Ethanediyl) Derivs.

Trade Names:
Crillet 35 (Croda Chemicals)
DeMULS PSTS-65 (DeForest)
Hetsorb TS-20 (Heterene)
Jeesorb STS-20 (Jeen)
Kotilen-S/3 (Kolb)
Laxan-ESR (Lanaetex)
Liposorb TS-20 (Lipo)
Lonzest STS-20 (Lonza Inc./Lonza Ltd.)
Lumisorb PSTS-20 (Lambent)
Montanox 65 (SEPPIC)
Nikkol TS-30 (Nikko)
Protasorb STS-20 (Protameen)
Sabosorb TSE (Sabo)
Sorbax PTS-20 (Chemax)
Tween 65 (Uniqema Americas)

Trade Name Mixtures:
Aldosperse TS-20 (Lonza Inc./Lonza Ltd.)
Aldosperse TS-40 (Lonza Inc./Lonza Ltd.)

POLYSORBATE 80

CAS No.: 9005-65-6 (Generic)

JPN Translation:
ポリソルベート 80

Definition: Polysorbate 80 is a mixture of oleate esters of sorbitol and sorbitol anhydrides, consisting predominantly of the monoester, condensed with approximately 20 moles of ethylene oxide. It conforms generally to the formula:

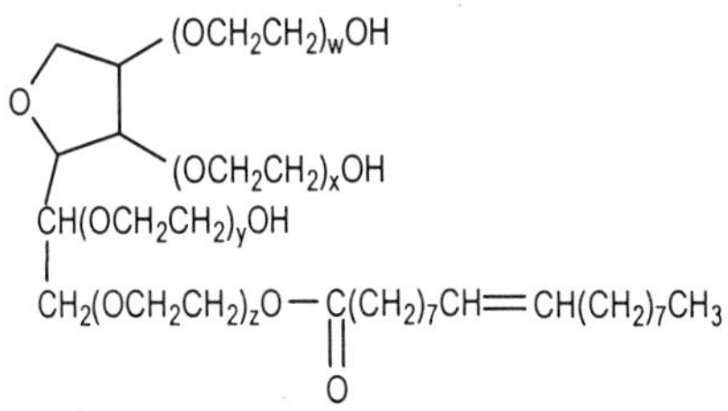

$(OCH_2CH_2)_wOH$
$(OCH_2CH_2)_xOH$
$CH(OCH_2CH_2)_yOH$
$CH_2(OCH_2CH_2)_zO-C(CH_2)_7CH=CH(CH_2)_7CH_3$

where w + x + y + z has an average value of 20.

Information Sources: AUS, BAN, BPC, BRA, 21CFR73.1, 21CFR73.1001, 21CFR172.515, 21CFR172.623, 21CFR172.836, 21CFR172.838, 21CFR172.840, 21CFR172.842, 21CFR173.340, 21CFR175.105, 21CFR175.300, 21CFR178.3400, 21CFR349.12, 21CFR436.102, 21CFR460.6, 21CFR573.860, 27CFR21.68, 27CFR21.151, CIR: [S] JACT-3(5)1984, CTFA S, CZE, DA, FCC, INN, ITA, JCLS, JP, JSCI, MAR, MI-13(7664), NF XIX, NFJ, OTC-I-OP, PN, RIFM, ROM, TSCA, USAN, USD

Chemical Class: Sorbitan Derivatives

Functions: Denaturant; Fragrance Ingredient; Surfactant - Emulsifying Agent; Surfactant - Solubilizing Agent

Reported Product Categories: Hair Conditioners; Moisturizing Preparations; Hair Sprays (Aerosol Fixatives); Tonics, Dressings, and Other Hair Grooming Aids; Bath Preparations, Misc.; Body and Hand Preparations (Excluding Shaving Preparations); Eyeliners; Mouthwashes and Breath Fresheners (Liquids and Sprays); Hair Dyes and Colors (All Types Requiring Caution Statements and Patch Tests); Hair Preparations (Non-coloring), Misc.; Shampoos (Non-coloring); Skin Fresheners; Permanent Waves; Powders (Dusting and Talcum, Excluding Aftershave Talcs); Bubble Baths; Colognes and Toilet Waters; Paste Masks (Mud Packs); Bath Capsules; Bath Oils, Tablets, and Salts; Cleansing Products (Cold Creams, Cleansing Lotions, Liquids and Pads); Makeup Preparations (Not eye), Misc.; Night Skin Care Preparations; Personal Cleanliness Products, Misc.; Shaving Preparations, Misc.; Aftershave Lotions; Baby Lotions, Oils, Powders and Creams; Baby Shampoos; Eye Makeup Preparations, Misc.; Face and Neck Preparations (Excluding Shaving Preparations); Fragrance Preparations, Misc.; Hair Wave Sets; Makeup Bases; Shaving Cream (Aerosol, Brushless and Lather); Skin Care Preparations, Misc.

Technical/Other Names:
Polyoxyethylene (20) Sorbitan Monooleate
Polyoxyethylene Sorbitan Monooleate (20E.O.)
Polysorbate 80 (RIFM)
Sorbimacrogol Oleate 300
Sorbitan, Mono-9-Octadecenoate, Poly(Oxy-1,2-Ethanediyl) Derivs.

Trade Names:
AEC Polysorbate 80 (A & E Connock)
Alkamuls PSMO-20 (Rhodia)
Alkamuls T80 (Rhodia)

Capmul POE-O (Abitec Corporation)
Crillet 4 (Croda Chemicals)
Customulse O-20 (Polysorbate 80)
(Custom Ingredients)
DeMULS PSMO-80 (DeForest)
Emsorb 2722 (Cognis Care Chemicals/PA)
Ethsorbox O-20 (Ethox)
Glycosperse O-20K (Lonza Inc./Lonza Ltd.)
Hetsorb O-20 (Heterene)
Ixol 8 (Vevy)
Jeesorb O-20 (Jeen)
Kotilen-O/1 (Kolb)
Laxan ESO (Lanaetex)
Liposorb O-20 (Lipo)
Lonzest SMO-20 (Lonza Inc./Lonza Ltd.)
Lumisorb PSMO-20 (Lambent)
Merpoxen SMO 200 (Wall)
Montanox 80 (SEPPIC)
Nikkol TO-10 (Nikko)
Nikkol TO-10M (Nikko)
Polisorbac 80 (Specialty Industrial)
Polisorbac 80K (Specialty Industrial)
Protasorb O-20 (Protameen)
Radiasurf 7157 (Oleon NV)
Radiasurf 7757 (Oleon NV)
Ritabate 80 (RITA)
Sabosorb MOE (Sabo)
Sorbax PMO-20 (Chemax)
Sorbilene O (Cesalpinia)
Sorbilene O/CR (Cesalpinia)
Sorbon T-80 (Toho)
Tego SMO 80 V (Degussa Care
Specialties)
Tween 80 (Uniqema Americas)

Trade Name Mixtures:
AEC Sodium Acrylate/Acryloyldimethyl
Taurate Copolymer (&) Isohexadecane
(&) Polysorbate 80 (A & E Connock)
Aldosperse O-20 (Lonza Inc./Lonza Ltd.)
Aroma Fennel E (Ichimaru Pharcos)
Aroma Lavender B (Ichimaru Pharcos)
Aroma Orange B (Ichimaru Pharcos)
Aroma Rosemary B (Ichimaru Pharcos)
Azufre Soluble (Fabriquimica)
Base W/O 126 (LCW)
Biosulphur Fluid (CLR)
Complex Zanahoria (Fabriquimica)
Crodalan AWS (Croda Chemicals)
Crodalan AWS (Croda, Inc.)
Dermol Jojoba S (Fabriquimica)
Emulmetik 910 (Lucas Meyer GmbH)
Esterlan SN (Fabriquimica)
Extrapone Cedarwood 2/B04080 (Symrise)
Extrapone Cotton 2/033327 (Symrise)
Extrapone Sandalcomplex 2/B04081
(Symrise)
Fancol ALA-10 (Fanning)
Fruit Vinegar (Provital/Centerchem)
Guardian O9 (Earth Supplied Products)
HAKKIN - GENSUI (Iwase Cosfa)
Ivarbase 98 (Arch Personal Care Products)
Jeelan 98 (Jeen)
Lanalene 98 (Maybrook)

Lecsoy S (Fabriquimica)
Liant TW 406 (LCW)
Liant TW 729 (LCW)
Lipobelle Soyaglycone (Mibelle AG)
Lipolan 98 (Lipo)
Mascawax 012 (LCW)
Mearlmaid OL (Engelhard Corp.)
Natuchrom Super Green (Quest
International)
Protalan 98 (Protameen)
Quassia Vinegar (Provital/Centerchem)
Ritawax AEO (RITA)
SilDerm Formulating Base (Active
Concepts)
SilDerm Formulating Base IF (Active
Concepts)
Simulgel 600 (SEPPIC)
Simulgel EG (SEPPIC)
Solubilisant WL 3215 (Gattefosse s.a.)
Solulan 98 (Amerchol)
Vitaphyle ACE (LCW)
Ylang-Ylang DG 2/B04068 (Symrise)

POLYSORBATE 81

CAS No.: 9005-65-6 (Generic)

Definition: Polysorbate 81 is a mixture of oleate esters of sorbitol and sorbitol anhydrides, consisting predominantly of the monoester, condensed with approximately 5 moles of ethylene oxide. It conforms generally to the formula:

where w + x + y + z has an average value of 5.

Information Sources: 21CFR175.300, CIR: [S] JACT-3(5)1984, CTFA S, RIFM, TSCA

Chemical Class: Sorbitan Derivatives

Functions: Fragrance Ingredient; Surfactant - Emulsifying Agent

Reported Product Categories: Hair Conditioners; Moisturizing Preparations; Hair Sprays (Aerosol Fixatives); Tonics, Dressings, and Other Hair Grooming Aids; Bath Preparations, Misc.; Body and Hand Preparations (Excluding Shaving Preparations); Eyeliners; Mouthwashes and Breath Fresheners (Liquids and Sprays); Hair Dyes and Colors (All Types Requiring Caution Statements and Patch Tests); Hair Prepara-

tions (Non-coloring), Misc.; Shampoos (Non-coloring); Skin Fresheners; Powders (Dusting and Talcum, Excluding Aftershave Talcs); Bubble Baths; Colognes and Toilet Waters; Paste Masks (Mud Packs); Bath Capsules; Bath Oils, Tablets, and Salts; Cleansing Products (Cold Creams, Cleansing Lotions, Liquids and Pads); Makeup Preparations (Not eye), Misc.; Night Skin Care Preparations; Aftershave Lotions; Baby Lotions, Oils, Powders and Creams; Baby Shampoos; Eye Makeup Preparations, Misc.; Face and Neck Preparations (Excluding Shaving Preparations); Fragrance Preparations, Misc.; Shaving Cream (Aerosol, Brushless and Lather); Skin Care Preparations, Misc.

Technical/Other Names:
Polyoxyethylene (5) Sorbitan Monooleate
Polysorbate 80 (RIFM)

Trade Names:
Crillet 41 (Croda Chemicals)
Ethsorbox O-5 (Ethox)
Hetsorb O-5 (Heterene)
Jeesorb O-5 (Jeen)
Kotilen-O/1050 (Kolb)
Laxan-ESM (Lanaetex)
Liposorb O-5 (Lipo)
Lonzest SMO-5 (Lonza Inc./Lonza Ltd.)
Lumisorb PSMO-5 (Lambent)
Sorbax PMO-5 (Chemax)
Tego SMO 81 (Degussa Care Specialties)
Tween 81 (Uniqema Americas)

Trade Name Mixtures:
Lubrasil DS (Guardian)
Novaplant-Yarrow Extract BG 91286
(Crodarom)

POLYSORBATE 85

CAS No.: 9005-70-3

JPN Translation:
ポリソルベート 85

Definition: Polysorbate 85 is a mixture of oleate esters of sorbitol and sorbitol anhydrides, consisting predominantly of the triester, condensed with approximately 20 moles of ethylene oxide. It conforms generally to the formula:

where w + x + y + z has an average value of 20.

Information Sources: BAN, 21CFR175.300, 21CFR178.3400, CIR: [S] JACT-3(5)1984, CTFA S, INN, JCLS, JSCI, MAR, TSCA, USAN

Chemical Class: Sorbitan Derivatives

Functions: Surfactant - Emulsifying Agent; Surfactant - Suspending Agent

Reported Product Categories: Eyeliners; Indoor Tanning Preparations; Bath Oils, Tablets, and Salts; Cleansing Products (Cold Creams, Cleansing Lotions, Liquids and Pads); Bath Preparations, Misc.; Body and Hand Preparations (Excluding Shaving Preparations); Makeup Fixatives; Makeup Preparations (Not eye), Misc.; Moisturizing Preparations; Tonics, Dressings, and Other Hair Grooming Aids

Technical/Other Names:
Polyoxyethylene (20) Sorbitan Trioleate
Polyoxyethylene Sorbitan Trioleate (20E.O.)
Sorbimacrogol Trioleate 300
Sorbitan, Tri-9-Octadecenoate, Poly(Oxy-1, 2-Ethanediyl) Derivs.

Trade Names:
Alkamuls PSTO-20 (Rhodia)
Crillet 45 (Croda Chemicals)
DeMULS PSTO-85 (DeForest)
Ethsorbox TO-20 (Ethox)
Hetsorb TO-20 (Heterene)
Jeesorb TO-20 (Jeen)
Kotilen-O/3 (Kolb)
Laxan EST (Lanaetex)
Liposorb TO-20 (Lipo)
Lonzest STO-20 (Lonza Inc./Lonza Ltd.)
Montanox 85 (SEPPIC)
Nikkol TO-30 (Nikko)
Protasorb TO-20 (Protameen)
Sabosorb TOE (Sabo)
Sorbax PTO-20 (Chemax)
Sorbon T-85 (Toho)
Tween 85 (Uniqema Americas)

Trade Name Mixtures:
Flamenco Pearl SEC (Engelhard Corp.)
Polymer EX-617 (Noveon)
Ritaderm (RITA)
Sepigel 501 (SEPPIC)

POLYSORBATE 80 ACETATE

Definition: Polysorbate 80 Acetate is the acetyl ester of Polysorbate 80 (q.v.).

Chemical Classes: Esters; Sorbitan Derivatives

Function: Surfactant - Emulsifying Agent

Reported Product Category: Hair Sprays (Aerosol Fixatives)

Technical/Other Names:
Polyethylene 100 Sorbitan Monooleate Acetate

Polyoxyethylene (20) Sorbitan Monooleate Acetate

POLYSTYRENE

CAS No.: 9003-53-6

JPN Translation:
ポリスチレン

Empirical Formula:
$(C_8H_8)_x$

Definition: Polystyrene is the polymer that conforms to the formula:

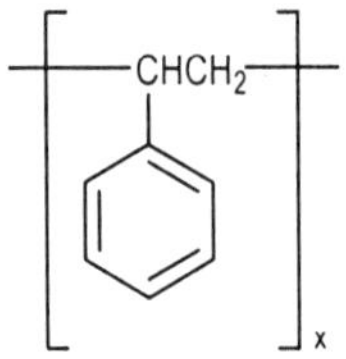

Information Sources: 21CFR175.105, 21CFR175.125, 21CFR175.300, 21CFR175.320, 21CFR176.180, 21CFR177.1200, 21CFR177.1640, 21CFR177.1990, 21CFR177.2600, 21CFR179.45, JCIC, JCLS, JSQI, MI-13 (8944), TSCA

Chemical Class: Synthetic Polymers

Functions: Film Former; Viscosity Increasing Agent - Nonaqueous

Reported Product Categories: Moisturizing Preparations; Night Skin Care Preparations; Skin Care Preparations, Misc.; Eye Shadows; Face Powders; Paste Masks (Mud Packs)

Technical/Other Names:
Benzene, Ethenyl-, Homopolymer
Ethenylbenzene, Homopolymer
Polystyrene Emulsion

Trade Names:
AEC Styrene Granules (A & E Connock)
Modarez OS 197 (Synthron)

Trade Name Mixtures:
CREASPARKLES IRIDESCENT BLUE 560 (Creations Couleurs)
CREASPARKLES IRIDESCENT GREEN 510 (Creations Couleurs)
CREASPARKLES IRIDESCENT RAINBOW 530 (Creations Couleurs)
CREASPARKLES IRIDESCENT VIOLET 610 (Creations Couleurs)
CREASPARKLES IRIDESCENT YELLOW 460 (Creations Couleurs)
Nanospheres 100 Hydrophilic (Exsymol)
Styrene MC Green (Ichimaru Pharcos)
Syntran 5900 (Interpolymer)
Syntran 5902 (Interpolymer)

POLYSTYRENE/HYDROGENATED POLYISOPENTENE COPOLYMER

Definition: Polystyrene/Hydrogenated Polyisopentene Copolymer is a copolymer of

polystyrene and hydrogenated polyisopentene.

Chemical Class: Synthetic Polymers

Function: Not Reported

Trade Name Mixtures:
Flexiplas Matrix (Arch Personal Care Products)
Flexismooth (Arch Personal Care Products)
Gel Base I (Arch Personal Care Products)

POLYTETRAFLUOROETHYLENE ACETOXYPROPYL BETAINE

Definition: Polytetrafluoroethylene Acetoxypropyl Betaine is the polymeric betaine that conforms to the formula:

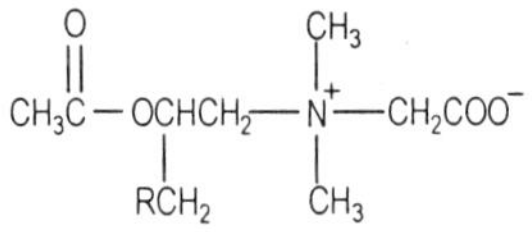

where R is polytetrafluoroethylene.

Chemical Classes: Betaines; Halogen Compounds; Polymeric Ethers

Function: Not Reported

Trade Name:
Zonyl FSK (DuPont de Nemours)

POLYURETHANE-1

Definition: Polyurethane-1 is a copolymer consisting of isophthalic acid, adipic acid, hexylene glycol, neopentyl glycol, dimethylolpropanoic acid, and isophorone diisocyanate monomers.

Chemical Class: Synthetic Polymers

Functions: Binder; Film Former; Hair Fixative

Trade Name:
Luviset P.U.R. (BASF)

POLYURETHANE-2

Definition: Polyurethane-2 is a copolymer of hexylene glycol, neopentyl glycol, adipic acid, saturated methylene diphenyldiisocyanate and dimethylolpropionic acid monomers.

Chemical Class: Synthetic Polymers

Function: Film Former

Technical/Other Name:
Adipic Acid/DMPA/Hexylene Glycol/ Neopentyl Glycol/SMDI Copolymer

Trade Names:
Avalure UR 405 (Noveon)

Avalure UR 410 (Noveon)
Avalure UR 425 (Noveon)
Avalure UR 430 (Noveon)

POLYURETHANE-4

Definition: Polyurethane-4 is a copolymer of PPG-17, PPG-34, isophorone diisocyanate and dimethylolpropionic acid monomers.

Chemical Class: Synthetic Polymers

Function: Film Former

Technical/Other Name:
DMPA/IPDI/PPG-17/PPG-34 Copolymer

Trade Name:
Avalure UR 445 (Noveon)

POLYURETHANE-5

Definition: Polyurethane-5 is a copolymer of Hexylene Glycol (q.v.), Neopentyl Glycol (q.v.), Adipic Acid (q.v.) and isophorone diisocyanate monomers.

Chemical Class: Synthetic Polymers

Function: Film Former

POLYURETHANE-6

Definition: Polyurethane-6 is a copolymer consisting of isophthalic acid, adipic acid, hexylene glycol, neopentyl glycol, dimethoylolpropanoic acid, isophorone diisocyanate and bis-ethylaminoisobutyl-dimethicone monomers.

Chemical Class: Synthetic Polymers

Functions: Binder; Film Former; Hair Fixative

POLYURETHANE-7

Definition: Polyurethane-7 is a copolymer of Hexylene Glycol (q.v.), Neopentyl Glycol (q.v.), Adipic Acid (q.v.), isophorone diisocyanate and dimethylol propionic acid monomers.

Chemical Class: Synthetic Polymers

Function: Film Former

Trade Name:
Avalure EX-608 (Noveon)

POLYURETHANE-8

Definition: Polyurethane-8 is a copolymer of polyethylene-poly(tetramethylene)glycol,

propanoic anhydride, dibutyl tindilaurate, isophorone diisocyanate, and isophorone diamine.

Chemical Class: Synthetic Polymers

Functions: Binder; Film Former; Plasticizer

Trade Name Mixture:
KFILM 2071 (Kane)

POLYURETHANE-9

CAS No.: 69011-31-0

Definition: Polyurethane-9 is the copolymer formed from adipic acid, toluene diisocyanate, propylene glycol, ethylene glycol and hydroxyethyl acrylate monomers.

Chemical Class: Synthetic Polymers

Function: Artificial Nail Builder

Trade Name:
Actilane 170 (Wilde Cosmetics)

POLYURETHANE-10

Definition: Polyurethane-10 is a copolymer of isophorone diisocyanate, cyclohexane-dimethanol, dimethylol butanoic acid, polyalkylene glycol and N-methyl diethanol-amine monomers.

Chemical Classes: Siloxanes and Silanes; Synthetic Polymers

Functions: Film Former; Hair Fixative

Trade Names:
Cydrothane CG 5500 polyurethane dispersion (Cytec Industries)
Cydrothane CG 6500 polyurethane dispersion (Cytec Industries)
Cydrothane CG 5000 polyurethane resin (Cytec Industries)

Trade Name Mixture:
Yodosol PUD (National Starch)

POLYURETHANE-11

Definition: Polyurethane-11 is a copolymer of Adipic Acid (q.v.), 1,4-Butanediol (q.v.), isophthalic acid, methylene bis-(4-cyclohexylisocyanate), Neopentyl Glycol (q.v.) and trimethylolpropane monomers.

Chemical Class: Synthetic Polymers

Function: Film Former

Trade Name:
WSR Coating (Glitterex Corporation)

POLYURETHANE-12

Definition: Polyurethane-12 is a copolymer of trimethylolpropane, neopentyl glycol,

dimethylol propionic acid, polytetramethylene ether glycol and isocyanato methylethyl-benzene monomers.

Chemical Class: Synthetic Polymers

Functions: Binder; Film Former

Trade Names:
Cydrothane CG 4300 polyurethane dispersion (Cytec Industries)
Cydrothane CG 4000 polyurethane resin (Cytec Industries)

POLYURETHANE-13

Definition: Polyurethane-13 is a copolymer of trimethylolpropane, dimethylol propionic acid, hexanediol, adipic acid, polyester diol, and isocyanato methylethylbenzene monomers.

Chemical Class: Synthetic Polymers

Functions: Binder; Film Former

Trade Name:
Cydrothane CG 1500 polyurethane dispersion (Cytec Industries)

POLYURETHANE-14

Definition: Polyurethane-14 is a copolymer formed from isophorone diisocyanate, dimethylol propionic acid, and 4,4'-isopropylidenediphenol reacted with propylene oxide, ethylene oxide and PEG/PPG-17/3.

Chemical Class: Synthetic Polymers

Functions: Film Former; Hair Conditioning Agent

Trade Name Mixture:
DynamX (National Starch)

POLYURETHANE-15

Definition: Polyurethane-15 is a copolymer of isophorone diisocyanate, adipic acid, tri-ethylene glycol, and dimethylolpropionic acid.

Chemical Class: Synthetic Polymers

Function: Film Former

Trade Name:
Soft Touch WS (Catalysts & Chemicals)

Trade Name Mixture:
MPU-WS2 (Catalysts & Chemicals)

POLYVINYLACETAL DIETHYLAMINO-ACETATE

JPN Translation:
ポリビニルアセタールジエチルアミノアセテート

Definition: Polyvinylacetal Diethylamino-acetate is the synthetic polymer that conforms to the formula:

$$\left[CH_2CHCH_2CH - CH_2CH \right]_x \left[CH_2CH \quad C_2H_5 \right]_y$$
$$O-CH-O \qquad OH \qquad O-CCH_2N$$
$$CH_3 \qquad\qquad O \quad C_2H_5$$

Information Source: JCIC

Chemical Class: Synthetic Polymers

Function: Film Former

POLYVINYL ACETATE

CAS No.: 9003-20-7

JPN Translation:
ポリ酢酸ビニル

Empirical Formula:
$(C_4H_6O_2)_x$

Definition: Polyvinyl Acetate is the homopolymer of Vinyl Acetate (q.v.) that conforms generally to the formula:

$$\left[CH_2CH \right]_x$$
$$O$$
$$C=O$$
$$CH_3$$

Information Sources: 21CFR73.1, 21CFR172.615, 21CFR175.105, 21CFR175.300, 21CFR175.320, 21CFR176.170, 21CFR176.180, 21CFR177.1200, 21CFR177.2800, 21CFR181.22, 21CFR181.30, CIR: [S] JACT-11(4)1992, CIR: [S] JACT-15(2)1996, FCC, JCIC, JCLS, JSQI, TSCA

Chemical Classes: Esters; Synthetic Polymers

Functions: Binder; Emulsion Stabilizer; Film Former; Hair Fixative

Reported Product Categories: Mascara; Eyeliners

Technical/Other Names:
Acetic Acid, Ethenyl Ester, Homopolymer
Acetylated Polyvinyl Alcohol
Ethenyl Acetate, Homopolymer
Polyvinyl Acetate Emulsion
Polyvinyl Acetate Solution

Trade Names:
UCAR 130 Latex Resin (Dow Chemical)
Vinac (Air Products)

POLYVINYL ALCOHOL

CAS No.: 9002-89-5

JPN Translation:
ポリビニルアルコール

Empirical Formula:
$(C_2H_4O)_x$

Definition: Polyvinyl Alcohol is the polymer conforming generally to the formula:

$$\left[CH_2CH \right]_x$$
$$OH$$

It is generally produced by the controlled hydrolysis of Polyvinyl Acetate (q.v.) and normally contains unhydrolyzed acetate groups.

Information Sources: 21CFR73.1, 21CFR175.105, 21CFR175.300, 21CFR175.320, 21CFR176.170, 21CFR176.180, 21CFR177.1200, 21CFR177.1670, 21CFR177.2260, 21CFR177.2800, 21CFR178.3910, 21CFR181.22, 21CFR181.30, CIR: [S] IJT-17(Suppl. 5)1998, CTFA D, DDR, JCLS, JSCI, MAR, MI-13(7667), OTC-I-OP, TSCA, USAN, USP XXIV

Chemical Classes: Alcohols; Synthetic Polymers

Functions: Binder; Film Former; Viscosity Increasing Agent - Aqueous

Reported Product Categories: Mascara; Paste Masks (Mud Packs); Nail Polish and Enamels; Moisturizing Preparations; Skin Care Preparations, Misc.; Bath Oils, Tablets, and Salts; Makeup Preparations (Not eye), Misc.

Technical/Other Name:
Ethenol, Homopolymer

Trade Names:
Airvol 523 (Air Products)
Airvol 540 (Air Products)
Elvanol (DuPont de Nemours)

Trade Name Mixture:
Vinex 2019 (Air Products)

POLYVINYLALCOHOL CROSSPOLYMER

CAS No.: 93409-71-3

Definition: Polyvinylalcohol Crosspolymer is a polymer of vinyl alcohol crosslinked with glyoxal.

Chemical Class: Synthetic Polymers

Function: Bulking Agent

Technical/Other Names:
Ethanedial, Polymer with Ethenol
Ethenol-Glyoxal Copolymer

Trade Name:
Lipocapsule PVA Crosspolymer (Lipo)

Trade Name Mixture:
LipoLight OAP/PVA (Lipo)

POLYVINYL BUTYRAL

CAS No.: 63148-65-2

JPN Translation:
ポリビニルブチラール

Definition: Polyvinyl Butyral is a polymer produced by the condensation of Polyvinyl Alcohol (q.v.) and butyraldehyde.

Information Sources: 21CFR175.105, 21CFR175.300, 21CFR176.170, JCIC, JCLS, JSQI, TSCA

Chemical Class: Synthetic Polymers

Functions: Binder; Film Former; Hair Fixative; Viscosity Increasing Agent - Nonaqueous

Reported Product Categories: Basecoats and Undercoats; Nail Polish and Enamels; Manicuring Preparations, Misc.

Technical/Other Name:
Vinyl Acetal Polymers, Butyrals

POLYVINYLCAPROLACTAM

CAS No.: 25189-83-7

Definition: Polyvinylcaprolactam is a polymer of vinylcaprolactam that conforms generally to the formula:

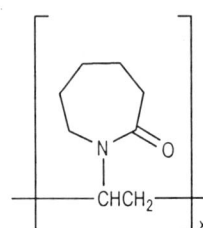

Chemical Classes: Amides; Heterocyclic Compounds; Synthetic Polymers

Functions: Film Former; Hair Fixative

Technical/Other Names:
Chloroethylene Polymer
2H-Azepin-2-one, 1-Ethenylhexahydro-, Homopolymer
N-Vinylcaprolactam Homopolymer
Vinyl Chloride Homopolymer

Trade Name:
Luviskol Plus (BASF)

POLYVINYL CHLORIDE

CAS No.: 9002-86-2

The inclusion of any compound in the *Dictionary and Handbook* does not indicate that use of that substance as a cosmetic ingredient complies with the laws and regulations governing such use in the United States or any other country.

JPN Translation:
ポリ塩化ビニル

Definition: Polyvinyl Chloride is a homo-polymer of vinyl chloride that conforms to the formula:

$$\left[-CH_2CH- \atop \quad\;\; Cl \right]_x$$

Information Sources: JCIC, JCLS, MI-13 (7668)

Chemical Classes: Halogen Compounds; Synthetic Polymers

Function: Film Former

Technical/Other Name:
Polyvinyl Chloride Resin

Trade Names:
Pionier P (Hansen & Rosenthal)
Solvic 271 GB (Solvay France)

Trade Name Mixture:
AEC Stars Silver Spectrum (A & E Connock)

POLYVINYLFORMAMIDE

Definition: Polyvinylformamide is the homo-polymer that conforms to the formula:

$$\left[\begin{array}{c} O \\ \| \\ NH-CH \\ | \\ CH_2CH \end{array} \right]_x$$

Chemical Classes: Amides; Synthetic Polymers

Function: Hair Fixative

Technical/Other Name:
Formamide, N-Ethenyl-, Homopolymer

POLYVINYL IMIDAZOLINIUM ACETATE

Definition: Polyvinyl Imidazolinium Acetate is the polymer of vinyl imidazolinium acetate that conforms generally to the formula:

$$\left[CHCH_2 \atop \cdots \right]_x \quad CH_3COO^-$$

Information Source: TSCA

Chemical Class: Synthetic Polymers

Functions: Antistatic Agent; Film Former; Hair Fixative; Suspending Agent - Nonsurfactant

POLYVINYL ISOBUTYL ETHER

JPN Translation:
ポリビニルイソブチルエーテル

Definition: Polyvinyl Isobutyl Ether is the polymer produced by the polymerization of isobutyl alcohol with vinyl isobutyl ether.

Information Source: JCIC

Chemical Class: Synthetic Polymers

Function: Film Former

POLYVINYL LAURATE

Definition: Polyvinyl Laurate is the polyester of Polyvinyl Alcohol (q.v.) and lauric acid.

Chemical Class: Synthetic Polymers

Functions: Binder; Film Former; Viscosity Increasing Agent - Nonaqueous

Technical/Other Name:
Dodecanoic Acid, Ethenyl Ester, Homopolymer

Trade Name:
Mexomere PP (Chimex)

POLYVINYL METHYL ETHER

CAS No.: 9003-09-2

Empirical Formula:
$(C_3H_6O)_x$

Definition: Polyvinyl Methyl Ether is the polymer of vinyl methyl ether that conforms generally to the formula:

$$\left[-CH_2CH- \atop \quad\;\; OCH_3 \right]_x$$

Information Sources: 21CFR175.105, 21CFR177.1680, TSCA

Chemical Classes: Ethers; Synthetic Polymers

Functions: Binder; Film Former; Hair Fixative; Suspending Agent - Nonsurfactant

Technical/Other Names:
Ethene, Methoxy-, Homopolymer
Methoxyethene, Homopolymer

Trade Name:
Gantrez M-154 (International Specialty Products)

POLYVINYL STEARYL ETHER

CAS No.: 9003-96-7

Definition: Polyvinyl Stearyl Ether is the homopolymer of vinyl stearyl ether that conforms generally to the formula:

$$\left[-CH_2CH- \atop \quad\;\; O(CH_2)_{17}CH_3 \right]_x$$

Chemical Classes: Ethers; Synthetic Polymers

Functions: Film Former; Skin-Conditioning Agent - Miscellaneous

Technical/Other Names:
Octadecane, 1-(Ethenyloxy)-() Homopolymer
Polyvinyl Octadecyl Ether

Trade Name:
GIOVAREZ 1800 (Phoenix)

PONCEAU SX

CAS No.	EINECS No.
4548-53-2	224-909-9

Empirical Formula:
$C_{18}H_{16}N_2O_7S_2 \cdot 2Na$

Definition: Ponceau SX is classed chemically as a monoazo color. It conforms to the formula:

See "Regulatory and Ingredient Use Information," for Colorants in Volume 1, Introduction, Part A. To identify the certified colorant for labeling purposes in the US, the INCI Name Red 4 must be used. To identify the colorant allowed for use in the European Union (EU), the INCI Name CI 14700 must be used, except for hair dye products. The INCI Name for batches of this colorant that have not been certified is Ponceau SX. To identify the colorant allowed for use in Japan, the INCI name Aka504 must be used.

Information Sources: CI 14700, EEC(III/2-59), M3, MI-13(7672), TSCA

Chemical Class: Color Additives - Miscellaneous

Function: Colorant

Reported Product Categories: Colognes and Toilet Waters; Bath Preparations, Misc.;

Body and Hand Preparations (Excluding Shaving Preparations); Shampoos (Non-coloring); Perfumes; Moisturizing Preparations; Hair Conditioners; Skin Care Preparations, Misc.; Bath Oils, Tablets, and Salts; Bath Soaps and Detergents; Aftershave Lotions; Baby Shampoos; Cleansing Products (Cold Creams, Cleansing Lotions, Liquids and Pads); Bubble Baths; Bath Capsules; Face and Neck Preparations (Excluding Shaving Preparations); Fragrance Preparations, Misc.; Paste Masks (Mud Packs); Deodorants (Underarm); Skin Fresheners; Personal Cleanliness Products, Misc.; Blushers (All types); Night Skin Care Preparations; Hair Rinses (Coloring); Tonics, Dressings, and Other Hair Grooming Aids; Baby Products, Misc.; Hair Preparations (Non-coloring), Misc.; Nail Creams and Lotions; Nail Polish and Enamel Removers

Technical/Other Names:
CI 14700
3-[(2,4-Dimethyl-5-Sulfophenyl)Azo]-4-Hydroxy-1-Naphthalenesulfonic Acid, Disodium Salt
Food Red 1
1-Naphthalenesulfonic Acid, 3-[(2,4-Dimethyl-5-Sulfophenyl)Azo]-4-Hydroxy-, Disodium Salt
Red No. 504

PONGAMIA GLABRA SEED OIL

Definition: Pongamia Glabra Seed Oil is the fixed oil expressed from seeds of *Pongamia glabra. See "Regulatory and Ingredient Use Information," regarding the labeling names for botanical ingredients in Volume 1, Introduction, Part A.*

Chemical Class: Fats and Oils

Functions: Hair Conditioning Agent; Skin-Conditioning Agent - Miscellaneous

Technical/Other Name:
Oils, Pongamia Glabra Seed

Trade Names:
Karanja Oil (Aldivia)
Karanja Oil (Quest International)

PONGAMIA PINNATA EXTRACT

Definition: Pongamia Pinnata Extract is an extract of the fruit of *Pongamia pinnata. See "Regulatory and Ingredient Use Information," regarding the labeling names for botanical ingredients in Volume 1, Introduction, Part A.*

Chemical Class: Biological Products

Function: Hair Conditioning Agent

Technical/Other Name:
Extract of Pongamia Pinnata

Trade Name Mixtures:
Pongamia Complex (Greentech)
RPF Complex II (Greentech S.A)

PONGAMIA PINNATA SEED EXTRACT

Definition: Pongamia Pinnata Seed Extract is an extract of the seeds of *Pongamia pinnata. See "Regulatory and Ingredient Use Information," regarding the labeling names for botanical ingredients in Volume 1, Introduction, Part A.*

Chemical Class: Biological Products

Function: Skin-Conditioning Agent - Miscellaneous

Technical/Other Name:
Extract of Pongamia Pinnata Seed

Trade Name:
RPF Complex (Greentech)

PONGAMOL

CAS No.: 484-33-3

Empirical Formula:
$C_{18}H_{14}O_4$

Definition: Pongamol is an organic compound obtained from the seeds of various species of *Pongamia*. It conforms to the formula:

Chemical Classes: Heterocyclic Compounds; Ketones

Function: Fragrance Ingredient

Technical/Other Names:
1-(4-Methoxy-5-Benzofuranyl)-3-Phenyl-1,3-Propanedione
1,3-Propanedione, 1-(4-Methoxy-5-Benzofuranyl)-3-Phenyl-

Trade Name:
Pongamia Extract (Quest International)

POPULUS NIGRA EXTRACT

CAS No.	EINECS No.
84650-39-5	283-509-2

Definition: Populus Nigra Extract is an extract of the leaves, twigs, bark, and buds of the black poplar, *Populus nigra. See "Regulatory and Ingredient Use Information," regarding the labeling names for botanical ingredients in Volume 1, Introduction, Part A.*

Chemical Class: Biological Products

Function: Skin-Conditioning Agent - Miscellaneous

Technical/Other Names:
Black Poplar Extract
Black Poplar (Populus Nigra) Extract
Extract of Black Poplar
Extract of Populus Nigra
Poplar Extract

Trade Name Mixtures:
Black Poplar HPG Titrated (Alban Muller)
Black Poplar HS (Alban Muller)
Extrait De Peuplier MPE100 (Yves Rocher)
Herbasol-Extract Poplar (Cosmetochem)
Poplar Buds Extract HS 2641 G (Grau)
Populus Nigra Extract ies (IES LABO)
VT-213 Extract of Poplar Buds (Vege-Tech)

POPULUS NIGRA FLOWER EXTRACT

Definition: Populus Nigra Flower Extract is an extract of the flowers of *Populus nigra. See "Regulatory and Ingredient Use Information," regarding the labeling names for botanical ingredients in Volume 1, Introduction, Part A.*

Chemical Class: Biological Products

Function: Skin-Conditioning Agent - Miscellaneous

Technical/Other Name:
Extract of Populus Nigra Flower

POPULUS TREMULOIDES BARK EXTRACT

Definition: Populus Tremuloides Bark Extract is an extract of the bark of the aspen, *Populus tremuloides. See "Regulatory and Ingredient Use Information," regarding the labeling names for botanical ingredients in Volume 1, Introduction, Part A.*

Chemical Class: Biological Products

Function: Not Reported

Technical/Other Names:
Aspen Bark Extract
Aspen (Populus Tremuloides) Extract
Extract of Aspen Bark
Extract of Populus Tremuloides Bark

PORIA COCOS ROOT EXTRACT

Definition: Poria Cocos Root Extract is an extract of the roots of *Poria cocos. See*

"Regulatory and Ingredient Use Information," regarding the labeling names for botanical ingredients in Volume 1, Introduction, Part A.

Chemical Class: Biological Products

Functions: Anticaking Agent; Skin-Conditioning Agent - Emollient; Skin-Conditioning Agent - Miscellaneous; Surfactant - Cleansing Agent

Technical/Other Name:
Extract of Poria Cocos

Trade Name Mixtures:
Bukuryou Liquid B (Ichimaru Pharcos)
Bukuryou Liquid E (Ichimaru Pharcos)
Chinese White Tuckahoe Extract (Alban Muller)
Fu Ling Extract (Alban Muller)
Hoelen Extract (Maruzen Pharmaceuticals Co., Ltd.)
Hoelen Extract BG (Maruzen Pharmaceuticals Co., Ltd.)
Hoelen Extract BG-J (Maruzen Pharmaceuticals Co., Ltd.)
Hoelen Extract LA (Maruzen Pharmaceuticals Co., Ltd.)
Hoelen Extract SQ (Maruzen Pharmaceuticals Co., Ltd.)
Hoelen Extract W (Maruzen Pharmaceuticals Co., Ltd.)
YSK Magic 6 (Phyto-Technologies)

PORIA COCOS SCLEROTIUM EXTRACT

JPN Translation:
ブクリョウエキス

Definition: Poria Cocos Sclerotium Extract is an extract of the sclerotium of the fungus, *Poria cocos.*

Information Source: JCLS

Chemical Class: Biological Products

Function: Cosmetic Astringent

Technical/Other Name:
Extract of Poria Cocos Sclerotium

Trade Name Mixture:
Phytostan (EUROCOSTECH)

PORPHYRA UMBILICALIS EXTRACT

Definition: Porphyra Umbilicalis Extract is an extract of the alga, *Porphyra umbilicalis.* See *"Regulatory and Ingredient Use Information,"* regarding the labeling names for botanical ingredients in Volume 1, Introduction, Part A.

Chemical Class: Biological Products

Function: Skin-Conditioning Agent - Miscellaneous

Trade Name Mixtures:
Helionori (GELYMA)
Porphyra HS (Alban Muller)
Porphyra MBE BG 40 (Yves Rocher)
Rhodofiltrat Porphyra HG (Codif)
Sea Extract Nori (E.U.K)
Vege Plex VP-1297.050WB Sea Plex in Butylene Glycol (Vege-Tech)
Vege-Tech VT-351 (Vege-Tech)

PORPHYRIDIUM CRUENTUM EXTRACT

Definition: Porphyridium Cruentum Extract is an extract of *Porphyridium cruentum.* See *"Regulatory and Ingredient Use Information,"* regarding the labeling names for botanical ingredients in Volume 1, Introduction, Part A.

Chemical Class: Biological Products

Function: Skin-Conditioning Agent - Miscellaneous

Technical/Other Name:
Extract of Porphyridium Cruentum

Trade Name:
Porphyridium SOD (I.D. bio)

Trade Name Mixtures:
Hydrocomplex 2 (Somaig)
Hydrocomplex 3 (Somaig)
Porphyridium HS (Alban Muller)
Porphyridium LS (Alban Muller)

PORPHYRIDIUM PURPUREUM EXTRACT

Definition: Porphyridium Purpureum Extract is the extract of the alga, *Porphyridium purpureum.* See *"Regulatory and Ingredient Use Information,"* regarding the labeling names for botanical ingredients in Volume 1, Introduction, Part A.

Chemical Class: Biological Products

Function: Skin-Conditioning Agent - Miscellaneous

Technical/Other Name:
Extract of Porphyridium Purpureum

Trade Name:
Porphyridium Extract Powder (MicroAlgae)

Trade Name Mixture:
Porphyridium Extract (MicroAlgae)

PORPHYRIDIUM/ZINC FERMENT

Definition: Porphyridium/Zinc Ferment is an extract of the fermentation product of porphyridium in the presence of zinc ions.

Chemical Class: Biological Products

Function: Not Reported

Trade Name:
Algualane Zinc (Vincience)

PORTULACA GRANDIFLORA EXTRACT

Definition: Portulaca Grandiflora Extract is an extract of the whole plant, *Portulaca grandiflora.* See *"Regulatory and Ingredient Use Information,"* regarding the labeling names for botanical ingredients in Volume 1, Introduction, Part A.

Chemical Class: Biological Products

Function: Skin-Conditioning Agent - Humectant

Technical/Other Name:
Extract of Portulaca Grandiflora

Trade Name Mixture:
Grandiflora Extract (Bioland)

PORTULACA OLERACEA EXTRACT

CAS No.	EINECS No.
90083-07-1	290-201-1

Definition: Portulaca Oleracea Extract is an extract of the whole plant *Portulaca oleracea.* See *"Regulatory and Ingredient Use Information,"* regarding the labeling names for botanical ingredients in Volume 1, Introduction, Part A.

Chemical Class: Biological Products

Function: Skin-Conditioning Agent - Humectant

Technical/Other Name:
Extract of Portulaca Oleracea

Trade Name Mixture:
Portulaca Extract (Bioland)

POTASSIUM ABIETOYL HYDROLYZED COLLAGEN

Definition: Potassium Abietoyl Hydrolyzed Collagen is the potassium salt of the condensation product of abietic acid chloride and Hydrolyzed Collagen (q.v.).

Chemical Class: Protein Derivatives

Functions: Hair Conditioning Agent; Skin-Conditioning Agent - Miscellaneous; Surfactant - Cleansing Agent

Technical/Other Name:
Proteins, Hydrolysates, Reaction Products with Abietoyl Chloride, Compounds with Potassium

POTASSIUM ABIETOYL HYDROLYZED SOY PROTEIN

Definition: Potassium Abietoyl Hydrolyzed Soy Protein is the potassium salt of the condensation product of abietic acid chloride and Hydrolyzed Soy Protein (q.v.).

Chemical Class: Protein Derivatives

Function: Skin-Conditioning Agent - Miscellaneous

Technical/Other Name:
Proteins, Hydrolysates, Reaction Products with Abietic Acid Chloride

Trade Name:
Abietoilpolipeptide Di Soja (Sinerga)

POTASSIUM ACESULFAME

CAS No.	EINECS No.
55589-62-3	259-715-3

Empirical Formula:
$C_4H_5NO_4S \cdot K$

Definition: Potassium Acesulfame is the organic salt that conforms to the formula:

Information Sources: 21CFR172.800, MI-13(38)

Chemical Classes: Heterocyclic Compounds; Organic Salts; Sulfonic Acids

Function: Flavoring Agent

Technical/Other Names:
Acesulfame-K
6-Methyl-1,2,2-Oxathiazin-4(3H)-one 2,2'-Dioxide, Potassium Salt
1,2,3-Oxathiazin-4(3H)-one, 6-Methyl-, 2,2'Dioxide, Potassium Salt
Potassium 6-Methyl-1,2,2-Oxathiazin-4(3H)-one 2,2'-Dioxide

POTASSIUM ACETATE

CAS No.	EINECS No.
127-08-2	204-822-2

Empirical Formula:
$C_2H_4O_2 \cdot K$

Definition: Potassium Acetate is the potassium salt of acetic acid that conforms to the formula:

$$CH_3COOK$$

Information Sources: JAN, MI-13(7687), RIFM, TSCA, USAN, USP XXIV

Chemical Class: Organic Salts

Functions: Buffering Agent; Fragrance Ingredient

Technical/Other Names:
Acetic Acid, Potassium Salt
Potassium acetate (RIFM)
Potassium Ethanoate

POTASSIUM ACRYLATES/ACRYLAMIDE COPOLYMER

JPN Translation:
(アクリル酸アルキル / アクリルアミド) コポリマーK

Definition: Potassium Acrylates/Acrylamide Copolymer is the potassium salt of Acrylates/Acrylamide Copolymer (q.v.).

Information Source: JCIC

Chemical Class: Synthetic Polymers

Function: Film Former

Technical/Other Name:
Acrylic Acid.Acrylamide.Ethyl Acrylate Copolymer Potassium Salt Solution

POTASSIUM ACRYLATES/C10-30 ALKYL ACRYLATE CROSSPOLYMER

Definition: Potassium Acrylates/C10-30 Alkyl Acrylate Crosspolymer is the potassium salt of Acrylates/C10-30 Alkyl Acrylate Crosspolymer (q.v.).

Chemical Class: Synthetic Polymers

Function: Film Former

POTASSIUM ACRYLATES/ETHYLHEXYL ACRYLATE COPOLYMER

Definition: Potassium Acrylates/Ethylhexyl Acrylate Copolymer is the potassium salt of Acrylates/Ethylhexyl Acrylate Copolymer (q.v.).

Chemical Class: Synthetic Polymers

Function: Film Former

POTASSIUM ALGINATE

CAS No.: 9005-36-1

JPN Translation:
アルギン酸K

Empirical Formula:
$H_2SO_4 \cdot Al \cdot K$

Definition: Potassium Alginate is the potassium salt of Alginic Acid (q.v.).

Information Sources: 21CFR184.1610, FCC, JCIC, JCLS, TSCA

Chemical Class: Gums, Hydrophilic Colloids and Derivatives

Functions: Binder; Emulsion Stabilizer; Viscosity Increasing Agent - Aqueous

Technical/Other Name:
Alginic Acid, Potassium Salt

Trade Names:
COS-Kelp PA-5020 - ISP (International Specialty Products)
Protanal KF 200 RBS (Pronova Biopolymer Inc.)
Protanal KF 200 S (Pronova Biopolymer Inc.)

POTASSIUM ALUM

CAS No.: 7784-24-9

JPN Translation:
硫酸 (Al / K)

Definition: Potassium Alum is the inorganic salt that conforms to the formula:

$$KAl(SO_4)_2 \cdot 12H_2O$$

In the United States, Potassium Alum may be used as an active ingredient in OTC drug products. When used as an active drug ingredient, the established name for Potassium Alum is *Alum, Potassium. See "Regulatory and Ingredient Use Information," regarding the labeling names for U.S. OTC Drug Ingredients in Volume 1, Introduction, Part A.*

Information Sources: ARG, AUS, BP, BPC, BRA, 21CFR133.102, 21CFR133.106, 21CFR133.111, 21CFR133.141, 21CFR133.165, 21CFR133.181, 21CFR133.183, 21CFR133.195, 21CFR137.105, 21CFR137.155, 21CFR137.160, 21CFR137.165, 21CFR137.170, 21CFR137.175, 21CFR137.180, 21CFR137.185, 21CFR178.3120, 21CFR182.90, 21CFR182.1129, CZE, DDR, EGY, EP, FCC, FIN, HP, HUN, IND, ITA, JAN, JCLS, JSCI, MAR, MEX, MI-13(359), OTC-I-OH, PF, PN, POR, ROM, TSCA, USAN, USD, USP XXIV, USSR, YUG

Chemical Class: Inorganic Salts

Functions: Cosmetic Astringent; Drug Astringent - Oral Health Care Drugs; Oral Health Care Drug

Technical/Other Names:
Alum
Aluminum Potassium Sulfate
Alum, Potassium
Exsiccated Alum
Potassium Aluminum Sulfate
Sulfuric Acid, Aluminum Potassium Salt
(2:1:1), Dodecahydrate

POTASSIUM ALUMINUM POLYACRYLATE

Definition: Potassium Aluminum Polyacrylate is a mixture of the potassium and aluminum salts of Polyacrylic Acid (q.v.).

Information Source: CIR: [SQ] IJT 21 (SUPPL. 3) 2002

Chemical Class: Synthetic Polymers

Functions: Absorbent; Binder; Viscosity Increasing Agent - Aqueous

POTASSIUM ASCORBYLBORATE

Definition: Potassium Ascorbylborate is the potassium salt of the reaction product of Boric Acid (q.v.) and Ascorbic Acid (q.v.).

Chemical Class: Organic Salts

Function: Skin-Conditioning Agent - Humectant

Trade Name:
Collagain (VDF FutureCeuticals)

POTASSIUM ASCORBYL TOCOPHERYL PHOSPHATE

Empirical Formula:
$C_{34}H_{57}O_{10}P \cdot 2K$

Definition: Potassium Ascorbyl Tocopheryl Phosphate is the organic compound that conforms to the formula:

Information Source: CIR: [S] IJT 21 (SUPPL. 3) 2002

Chemical Classes: Heterocyclic Compounds; Organic Salts; Phosphorus Compounds

---

Function: Antioxidant

Reported Product Categories: Moisturizing Preparations; Skin Care Preparations, Misc.

Trade Names:
EPC-K (Senju)
Sepivital (SEPPIC)

POTASSIUM ASPARTATE

CAS Nos.	EINECS Nos.
923-09-1 (dl-alpha)	213-088-2
1115-63-5	214-226-4
14007-45-5	237-814-2
14434-35-6 (dl-alpha)	238-407-2

JPN Translation:
アスパラギン酸K

Empirical Formula:
$C_4H_7NO_4 \cdot K$

Definition: Potassium Aspartate is the potassium salt of aspartic acid that conforms to the formula:

$$KOOCCH_2\underset{\underset{NH_2}{|}}{C}HCOOH$$

Information Sources: 21CFR172.320, JAN, JCIC, JCLS, TSCA

Chemical Class: Amino Acids

Function: Skin-Conditioning Agent - Miscellaneous

Reported Product Category: Body and Hand Preparations (Excluding Shaving Preparations)

Technical/Other Names:
DL-Aspartic Acid, Monopotassium Salt
L-Aspartic Acid, Monopotassium Salt
Aspartic Acid, Potassium Salt
Potassium L-Aspartate

Trade Names:
Givobio AKDL (SEPPIC)
Givobio AKL (SEPPIC)

Trade Name Mixtures:
Givobio AMgKL (SEPPIC)
Moisturizing Factor Hygro-Complex ARO 5272 (Crodarom)
Sepicalm S (SEPPIC)

POTASSIUM AZELOYL DIGLYCINATE

Empirical Formula:
$C_{13}H_{21}N_2O_6K$

Definition: Potassium Azeloyl Diglycinate is the organic compound that conforms to the formula:

Chemical Classes: Amides; Organic Salts

Function: Skin-Conditioning Agent - Miscellaneous

Trade Name:
Azeloglicina (Sinerga)

POTASSIUM BABASSUATE

Definition: Potassium Babassuate is the potassium salt of the fatty acids derived from Orbignya oleifera (babassu) oil.

Chemical Class: Soaps

Functions: Surfactant - Cleansing Agent; Surfactant - Emulsifying Agent

POTASSIUM BEHENATE

Definition: Potassium Behenate is the potassium salt of Behenic Acid (q.v.).

Chemical Class: Soaps

Function: Surfactant - Cleansing Agent

Technical/Other Name:
Behenic Acid, Potassium Salt

POTASSIUM BENZOATE

CAS No.	EINECS No.
582-25-2	209-431-3

Empirical Formula:
$C_7H_6O_2 \cdot K$

Definition: Potassium Benzoate is the potassium salt of Benzoic Acid (q.v.). It conforms to the formula:

Information Sources: EEC(VI/1-1), MHLW-331/3, NF XIX, TSCA, USAN

Chemical Class: Organic Salts

Function: Preservative

Technical/Other Name:
Benzoic Acid, Potassium Salt

POTASSIUM BICARBONATE

CAS No.	EINECS No.
298-14-6	206-059-0

Empirical Formula:
$CH_2O_3 \cdot K$

Definition: Potassium Bicarbonate is the inorganic salt that conforms to the formula:

$$KHCO_3$$

Information Sources: ARG, 21CFR163.110, 21CFR184.1613, DA, DDR, EP, FIN, IND, MEX, MI-13(7691), NED, POR, ROM, TSCA, USAN, USP XXIV

Chemical Class: Inorganic Salts

Functions: Buffering Agent; pH Adjuster

Technical/Other Names:
Carbonic Acid, Monopotassium Salt
Potassium Hydrogen Carbonate

POTASSIUM BIPHTHALATE

CAS No.	EINECS No.
877-24-7	212-889-4

Empirical Formula:
$C_8H_6O_4 \cdot K$

Definition: Potassium Biphthalate is the salt that conforms to the formula:

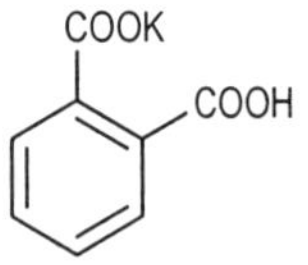

Information Sources: MI-13(7694), TSCA

Chemical Class: Organic Salts

Functions: Buffering Agent; pH Adjuster

Technical/Other Names:
1,2-Benzenedicarboxylic Acid, Monopotassium Salt
Potassium Acid Phthalate

POTASSIUM BORATE

CAS No.	EINECS No.
1332-77-0	215-575-5

Empirical Formula:
$B_4H_2O_7 \cdot 2K$

Definition: Potassium Borate is the inorganic salt that conforms to the formula:

$$K_2B_4O_7 \quad \cdot \quad 5H_2O$$

Information Sources: EEC(III/1-1), MI-13 (7765), TSCA

Chemical Class: Inorganic Salts

Function: pH Adjuster

Technical/Other Names:
Boric Acid, Dipotassium Salt
Potassium Tetraborate

POTASSIUM BROMATE

CAS No.	EINECS No.
7758-01-2	231-829-8

Empirical Formula:
$BrHO_3 \cdot K$

Definition: Potassium Bromate is the inorganic salt that conforms to the formula:

$$KBrO_3$$

Information Sources: 21CFR136.110, 21CFR136.115, 21CFR136.130, 21CFR136.160, 21CFR136.180, 21CFR137.155, 21CFR137.205, 21CFR172.730, CIR: [SQ] JACT-13(5)1994, FCC, MAR, MI-13(7700), TSCA

Chemical Class: Inorganic Salts

Function: Oxidizing Agent

Technical/Other Name:
Bromic Acid, Potassium Salt

POTASSIUM BROMIDE

CAS No.	EINECS No.
7758-02-3	231-830-3

Empirical Formula:
KBr

Definition: Potassium Bromide is the inorganic salt that conforms to the formula:

$$KBr$$

Information Sources: 21CFR178.1010, JAN, MI-13(7701), TSCA

Chemical Class: Inorganic Salts

Function: Proprietary

POTASSIUM BUTYL ESTER OF PVM/MA COPOLYMER

Definition: Potassium Butyl Ester of PVM/MA Copolymer is the potassium salt of Butyl Ester of PVM/MA Copolymer (q.v.).

Chemical Class: Synthetic Polymers

Functions: Binder; Film Former; Hair Fixative

POTASSIUM BUTYLPARABEN

Empirical Formula:
$C_{11}H_{14}O_3 \cdot K$

Definition: Potassium Butylparaben is the potassium salt of Butylparaben (q.v.) that conforms to the formula:

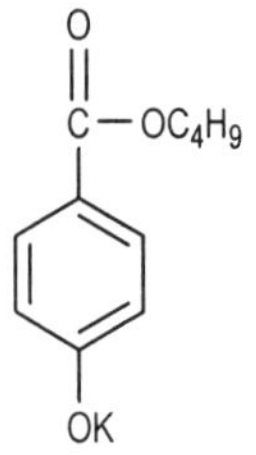

Information Source: EEC(VI/1-12)

Chemical Classes: Esters; Organic Salts; Phenols

Function: Preservative

Technical/Other Name:
Butylparaben, Potassium Salt

POTASSIUM C9-15 ALKYL PHOSPHATE

Definition: Potassium C9-15 Alkyl Phosphate is the potassium salt of a complex mixture of esters of synthetic C9-15 alcohols with phosphoric acid.

Information Source: JCLS

Chemical Class: Phosphorus Compounds

Function: Surfactant - Cleansing Agent

Trade Names:
Arlatone MAP (Uniqema Americas)
MAP 115K (Kao Corp.)

POTASSIUM C11-15 ALKYL PHOSPHATE

JPN Translation:
メチルヘスペリジン

Definition: Potassium C11-15 Alkyl Phosphate is the potassium salt of the phosphoric ester of C11-15 alcohol.

Information Source: JCLS

Chemical Class: Phosphorus Compounds

Functions: Surfactant - Cleansing Agent; Surfactant - Emulsifying Agent

POTASSIUM C12-13 ALKYL PHOSPHATE

Definition: Potassium C12-13 Alkyl Phosphate is the potassium salt of a complex mixture of esters of phosphoric acid and C12-13 Alcohols (q.v.).

Chemical Classes: Organic Salts; Phosphorus Compounds

Function: Surfactant - Cleansing Agent

Trade Names:
Arlatone MAP 230K (Uniqema Americas)
Arlatone MAP 230K-40 (Uniqema Americas)

POTASSIUM CAPRATE

CAS No.	EINECS No.
13040-18-1	235-910-9

Definition: Potassium Caprate is the potassium salt of Capric Acid (q.v.).

Chemical Class: Soaps

Function: Surfactant - Cleansing Agent

Technical/Other Names:
Capric Acid, Potassium Salt
n-Decanoic Acid, Potassium Salt
Potassium Decanoate

POTASSIUM CAPROYL TYROSINE

Empirical Formula:
$C_{19}H_{28}NO_4K$

Definition: Potassium Caproyl Tyrosine is the organic compound that conforms to the formula:

Chemical Classes: Amides; Phenols

Function: Skin-Conditioning Agent - Miscellaneous

Trade Name:
Tyrostan (Sinerga)

POTASSIUM CAPRYLOYL GLUTAMATE

Definition: Potassium Capryloyl Glutamate is the substituted amino acid that conforms to the formula:

Chemical Class: Amino Acids

Functions: Deodorant Agent; Surfactant - Cleansing Agent

Trade Name:
Protelan AG 80/K (Zschimmer & Schwarz Italiana)

POTASSIUM CAPRYLOYL HYDROLYZED RICE PROTEIN

Definition: Potassium Capryloyl Hydrolyzed Rice Protein is the potassium salt of the condensation product of caprylic acid chloride and Hydrolyzed Rice Protein (q.v.).

Chemical Class: Protein Derivatives

Functions: Hair Conditioning Agent; Skin-Conditioning Agent - Miscellaneous; Surfactant - Cleansing Agent

Trade Name:
Protelan R 8 (Zschimmer & Schwarz Italiana)

POTASSIUM CARBOMER

Definition: Potassium Carbomer is the potassium salt of Carbomer (q.v.).

Chemical Classes: Organic Salts; Synthetic Polymers

Functions: Emulsion Stabilizer; Film Former; Viscosity Increasing Agent - Aqueous

POTASSIUM CARBONATE

CAS No. 584-08-7

EINECS No. 209-529-3

JPN Translation:
炭酸 K

Empirical Formula:
$CH_2O_3 \cdot 2K$

Definition: Potassium Carbonate is the inorganic salt that conforms to the formula:

$$K_2CO_3$$

Information Sources: ARG, AUS, BRA, 21CFR163.110, 21CFR172.560, 21CFR173.310, 21CFR184.1619, DDR, EGY, FCC, FIN, HP, HUN, JCLS, JP, KOR, MAR, MEX, MI-13(7702), PF, PN, POR, ROM, TSCA, USAN, USP XXIV, YUG

Chemical Class: Inorganic Salts

Function: pH Adjuster

Technical/Other Names:
Carbonic Acid, Dipotassium Salt
Dipotassium Carbonate
Potash

Trade Name:
Unichem POCARB (Universal Preserv-A-Chem)

POTASSIUM CAROATE

CAS No. 70693-62-8

EINECS No. 274-778-7

Definition: Potassium Caroate is a mixture of potassium monopersulfate ($KHSO_5$), potassium sulfate (K_2SO_4) and potassium bisulfate ($KHSO_4$).

Chemical Class: Inorganic Salts

Function: Oxidizing Agent

Technical/Other Names:
Caro's Acid, Potasssium Salt
Curox
Pentapotassium Bis(Peroxymonosulphate) Bis(Sulphate)
Potassium Hydrogen Peroxymonosulfate
Potassium Peroxymonosulfate Sulfate

POTASSIUM CARRAGEENAN

CAS No.: 64366-24-1

Definition: Potassium Carrageenan is the potassium salt of Carrageenan (q.v.).

Information Sources: 21CFR172.623, 21CFR172.626, 21CFR176.170, TSCA

Chemical Class: Gums, Hydrophilic Colloids and Derivatives

Functions: Binder; Emulsion Stabilizer; Film Former; Viscosity Increasing Agent - Aqueous

Technical/Other Name:
Carrageenan, Potassium Salts

POTASSIUM CASEINATE

CAS No.: 68131-54-4

Definition: Potassium Caseinate is the potassium salt of Casein (q.v.).

Information Sources: 21CFR135.110, 21CFR135.140, CTFA D, TSCA

Chemical Class: Protein Derivatives

Functions: Hair Conditioning Agent; Skin-Conditioning Agent - Miscellaneous

Technical/Other Name:
Casein, Potassium Salt

Trade Name:
Alanate 351 (New Zealand)

POTASSIUM CASTORATE

CAS No. 8013-05-6

EINECS No. 232-388-4

Definition: Potassium Castorate is the potassium salt of the fatty acids derived from Ricinus Communis (Castor) Oil (q.v.).

Information Sources: 21CFR175.105, 21CFR176.170, 21CFR176.200, 21CFR176.210, 21CFR177.1200, 21CFR177.2600, 21CFR177.2800, 21CFR178.3910, TSCA

Chemical Class: Soaps

Functions: Surfactant - Cleansing Agent; Surfactant - Emulsifying Agent

Technical/Other Name:
 Castor Oil, Potassium Salt

POTASSIUM CELLULOSE SUCCINATE

Definition: Potassium Cellulose Succinate is the potassium salt of Cellulose Succinate (q.v.).

Chemical Classes: Carbohydrates; Organic Salts

Functions: Opacifying Agent; Skin-Conditioning Agent - Humectant

Trade Name:
 Moiscell Suc-K (Liba)

POTASSIUM CETYL PHOSPHATE

CAS Nos.	EINECS No.
17026-85-6	
19035-79-1	242-768-1

JPN Translation:
 セチルリン酸 K

Empirical Formula:
 $C_{16}H_{35}O_4PK$

Definition: Potassium Cetyl Phosphate is the potassium salt of a complex mixture of esters of phosphoric acid and cetyl alcohol.

Information Sources: JCIC, JCLS

Chemical Class: Phosphorus Compounds

Function: Surfactant - Emulsifying Agent

Reported Product Categories: Skin Care Preparations, Misc.; Bath Capsules; Face and Neck Preparations (Excluding Shaving Preparations); Bath Preparations, Misc.; Moisturizing Preparations; Night Skin Care Preparations; Paste Masks (Mud Packs); Body and Hand Preparations (Excluding Shaving Preparations); Eyebrow Pencils; Foundations; Suntan Gels, Creams, and Liquids

Technical/Other Names:
 1-Hexadecanol, Dihydrogen Phosphate, Monopotassium Salt
 Phosphoric Acid, Cetyl Ester, Potassium Salt

Trade Names:
 Aamphisol K (LaRoche)
 Amphisol K (Roche)

POTASSIUM CHLORATE

CAS No.	EINECS No.
3811-04-9	223-289-7

Empirical Formula:
 $HClO_3 \cdot K$

Definition: Potassium Chlorate is the inorganic salt that conforms to the formula:

$$KClO_3$$

Information Sources: BPC, BRA, CIR: [I] JACT-14(3)1995, EEC(III/1-6), EGY, HP, ITA, MAR, MI-13(7703), PF, POR, TSCA

Chemical Class: Inorganic Salts

Function: Oxidizing Agent

Technical/Other Names:
 Berthollet Salt
 Chloric Acid, Potassium Salt
 Potassium Oxymuriate

POTASSIUM CHLORIDE

CAS No.	EINECS No.
7447-40-7	231-211-8

JPN Translation:
 塩化 K

Empirical Formula:
 ClK

Definition: Potassium Chloride is the inorganic salt that conforms to the formula:

$$KCl$$

Information Sources: ARG, AUS, BP, BPC, BRA, 21CFR150.141, 21CFR150.161, 21CFR166.110, 21CFR184.1622, 21CFR558.311, CZE, DA, DDR, EGY, EP, FCC, FIN, HP, HUN, IND, ITA, JCLS, JSCI, MAR, MEX, MI-13(7704), NED, PN, POR, ROM, TSCA, USAN, USD, USP XXIV, WHO

Chemical Class: Inorganic Salts

Function: Viscosity Increasing Agent - Aqueous

Reported Product Categories: Hair Conditioners; Bath Oils, Tablets, and Salts; Cleansing Products (Cold Creams, Cleansing Lotions, Liquids and Pads); Eye Makeup Removers; Skin Fresheners

Trade Name:
 Unichem POCHLOR (Universal Preserv-A-Chem)

Trade Name Mixtures:
 Dacriosalt (Vevy)
 Essential Vital Elements (Dipta)
 Essential Vital Elements - S (Dipta)
 Saltworks brine Bad Rothenfelde (Ultra-Pharm)
 Schercotaine CAB-K (Scher)
 Schercotaine SCAB-KG (Scher)

POTASSIUM CITRATE

CAS No.	EINECS No.
866-84-2	212-755-5

JPN Translation:
 クエン酸 K

Empirical Formula:
 $C_6H_8O_7 \cdot 3K$

Definition: Potassium Citrate is the potassium salt of citric acid that conforms to the formula:

$$\begin{array}{c} COOK \\ | \\ KOOCCH_2CCH_2COOK \\ | \\ OH \end{array}$$

Information Sources: MI-13(7706), USAN, USP XXIV

Chemical Class: Organic Salts

Functions: Buffering Agent; Chelating Agent; pH Adjuster

Technical/Other Names:
 Citric Acid, Tripotassium Salt
 2-Hydroxy-1,2,3-Propanetricarboxylic Acid, Tripotassium Salt
 Potassium Tribasic Citrate
 1,2,3-Propanetricarboxylic Acid, 2-Hydroxy-, Tripotassium Salt
 Tripotassium Citrate

Trade Names:
 ADM Potassium Citrate (Archer Daniels Midland)
 Tripotassium Citrate (Jungbunzlauer)
 Tripotassium Citrate Monohydrate (Jungbunzlauer)

Trade Name Mixtures:
 Organo Silanetriol (Carilene)
 Organo Silanetriol (Synthesa)

POTASSIUM COCOATE

CAS No.	EINECS No.
61789-30-8	263-049-9

JPN Translations:
 ヤシカリ石ケン
 ヤシ脂肪酸 K

Definition: Potassium Cocoate is the potassium salt of Coconut Acid (q.v.).

Information Sources: 21CFR175.105, 21CFR176.170, 21CFR176.200, 21CFR176.210, 21CFR177.1200, 21CFR177.2600, 21CFR177.2800, 21CFR178.3910, JCIC, JCLS, JSQI, TSCA

Chemical Class: Soaps

Functions: Surfactant - Cleansing Agent; Surfactant - Emulsifying Agent

Reported Product Categories: Bath Soaps and Detergents; Shaving Cream (Aerosol, Brushless and Lather)

Technical/Other Names:
 Fatty Acids, Coconut Oil, Potassium Salts
 Potassium Cocoate Solution

Trade Names:
AEC Potassium Cocoate (A & E Connock)
Custoblend 40K (Cocoate soap) (Custom Ingredients)
Jeechem KC-40 (Jeen)
Mackadet 40K (McIntyre)
Nansa PC38 (Albright & Wilson UK)
Norfox 1101 (Norman, Fox & Co.)

Trade Name Mixtures:
Akypogene ZA 97 SP (Kao GmbH)
Collagen-CCK-Complex (Kelisema Italy)
Emulgade CL Special (Cognis Deutschland)
Gluplex AC (Kelisema Italy)
Nikkol MNK-40 (Nikko)
Sabosol RIS (Sabo)
Vegetable Soapbase OC (Weleda)

POTASSIUM COCOYL GLUTAMATE

JPN Translation:
ココイルグルタミン酸 K

Definition: Potassium Cocoyl Glutamate is the mixed potassium salts of the coconut acid amide of glutamic acid. It conforms generally to the formula:

$$RC \overset{\overset{\displaystyle O}{\|}}{—} NHCH(CH_2)_2COOH$$
$$| $$
$$COOK$$

where RCO- represents the fatty acids derived from coconut oil.

Information Sources: JCIC, JCLS

Chemical Classes: Amides; Amino Acids; Carboxylic Acids; Organic Salts

Functions: Hair Conditioning Agent; Surfactant - Cleansing Agent

Technical/Other Name:
Potassium N-Cocoacyl-L-glutamate

Trade Names:
Amisoft CK-11 (Ajinomoto)
Amisoft CK-22 (Ajinomoto)
Eversoft UCK (Sino Lion)

Trade Name Mixture:
Aminosurfact ACDP-L (Asahi Denka Kogyo)

POTASSIUM COCOYL GLYCINATE

JPN Translation:
ココイルグリシン K

Definition: Potassium Cocoyl Glycinate is the organic compound that conforms to the formula:

$$RC \overset{\overset{\displaystyle O}{\|}}{—} NHCH_2COOK$$

where RCO- represents the cocoyl moiety.

Chemical Classes: Amides; Amino Acids; Organic Salts

Functions: Hair Conditioning Agent; Surfactant - Cleansing Agent

Trade Names:
Amilite Gck-12 (Ajinomoto)
Amilite GCK-IIF (Ajinomoto)
Eversoft YCK (Sino Lion)

POTASSIUM COCOYL HYDROLYZED CASEIN

JPN Translation:
ココイル加水分解カゼイン K

Definition: Potassium Cocoyl Hydrolyzed Casein is the potassium salt of the condensation product of coconut acid chloride and Hydrolyzed Casein (q.v.).

Information Sources: JCIC, JCLS

Chemical Class: Protein Derivatives

Functions: Hair Conditioning Agent; Skin-Conditioning Agent - Miscellaneous; Surfactant - Cleansing Agent

Technical/Other Name:
Potassium N-Cocoyl-hydrolyzed Casein Solution

Trade Name:
Promois EMCP (Seiwa Kasei)

POTASSIUM COCOYL HYDROLYZED COLLAGEN

CAS No.: 68920-65-0

JPN Translation:
ココイル加水分解コラーゲン K

Definition: Potassium Cocoyl Hydrolyzed Collagen is the potassium salt of the condensation product of coconut acid chloride and Hydrolyzed Collagen (q.v.).

Information Sources: CIR: [S] JACT-2(7)-1983, CTFA D, JCIC, JCLS, JSQI, TSCA

Chemical Class: Protein Derivatives

Functions: Hair Conditioning Agent; Skin-Conditioning Agent - Miscellaneous; Surfactant - Cleansing Agent

Reported Product Categories: Hair Dyes and Colors (All Types Requiring Caution Statements and Patch Tests); Permanent Waves; Hair Tints; Hair Straighteners; Shampoos (Non-coloring); Bath Oils, Tablets, and Salts; Cleansing Products (Cold Creams, Cleansing Lotions, Liquids and Pads); Hair Conditioners; Tonics, Dressings, and Other Hair Grooming Aids

Technical/Other Names:
Acid Chlorides, Coco, Reaction Products with Protein Hydrolyzates, Potassium Salts
Potassium Coco-Hydrolyzed Animal Protein
Potassium Coco-hydrolyzed Animal Protein Solution
Potassium Coco Hydrolyzed Protein
Potassium Cocoyl Hydrolyzed Animal Protein

Trade Names:
AC Foaming Collagen (Active Concepts)
AEC Potassium Cocoyl Hydrolyzed Collagen (A & E Connock)
Coccocollagene K 1500 (Sinerga)
Foam-Coll 4C (Arch Personal Care Products)
Liprot CK (Fabriquimica)
Maypon 4C (Inolex)
May-Tein C (Maybrook)
Potassium Coco-Hydrolyzed Animal Protein (ChemMark)
Promois ECP (Seiwa Kasei)
Promois EC-P (Seiwa Kasei)
Promois ECP-C (Seiwa Kasei)
Promois ECP-CF (Seiwa Kasei)
Promois ECP-K (Seiwa Kasei)
Promois ECP-P (Seiwa Kasei)
Promois EUCP (Seiwa Kasei)
Promois EUCP-C (Seiwa Kasei)
Promois EUCP-CF (Seiwa Kasei)
Promois EUCP-P (Seiwa Kasei)
Promois Marine Collagen EUCP (Seiwa Kasei)
Texatein C (Lanaetex)

Trade Name Mixture:
Lamesoft LMG (Cognis Deutschland)

POTASSIUM COCOYL HYDROLYZED CORN PROTEIN

JPN Translation:
ココイル加水分解トウモロコシタンパク K

Definition: Potassium Cocoyl Hydrolyzed Corn Protein is the potassium salt of the condensation product of coconut acid chloride and Hydrolyzed Corn Protein (q.v.).

Information Sources: JCIC, JCLS

Chemical Class: Protein Derivatives

Functions: Hair Conditioning Agent; Skin-Conditioning Agent - Miscellaneous; Surfactant - Cleansing Agent

Technical/Other Name:
Potassium N-Cocoyl Hydrolyzed Corn Protein

Trade Name:
Promois EZCP-P (Seiwa Kasei)

POTASSIUM COCOYL HYDROLYZED KERATIN

JPN Translation:
ココイル加水分解ケラチン K

Definition: Potassium Cocoyl Hydrolyzed Keratin is the potassium salt of the condensation product of coconut acid chloride and Hydrolyzed Keratin (q.v.).

Information Source: JCLS

Chemical Class: Protein Derivatives

Functions: Hair Conditioning Agent; Skin-Conditioning Agent - Miscellaneous; Surfactant - Cleansing Agent

Trade Name:
May-Tein KK (Maybrook)

POTASSIUM COCOYL HYDROLYZED OAT PROTEIN

Definition: Potassium Cocoyl Hydrolyzed Oat Protein is the potassium salt of the condensation product of coconut acid chloride and Hydrolyzed Oat Protein (q.v.).

Chemical Class: Protein Derivatives

Functions: Hair Conditioning Agent; Skin-Conditioning Agent - Miscellaneous; Surfactant - Cleansing Agent

Trade Name Mixture:
Avenopac (Sinerga)

POTASSIUM COCOYL HYDROLYZED POTATO PROTEIN

JPN Translations:
ココイル加水分解ジャガイモタンパク K
ココイル加水分解バレイショタンパク K

Definition: Potassium Cocoyl Hydrolyzed Potato Protein is the potassium salt of the condensation product of coconut acid chloride and Hydrolyzed Potato Protein (q.v.).

Information Sources: JCIC, JCLS

Chemical Class: Protein Derivatives

Functions: Hair Conditioning Agent; Skin-Conditioning Agent - Miscellaneous; Surfactant - Cleansing Agent

Technical/Other Name:
Potassium N-Cocoyl Hydrolyzed Potato Protein

Trade Name:
Promois EPCP-P (Seiwa Kasei)

POTASSIUM COCOYL HYDROLYZED RICE BRAN PROTEIN

Definition: Potassium Cocoyl Hydrolyzed Rice Bran Protein is the potassium salt of the condensation product of coconut acid chloride and Hydrolyzed Rice Bran Protein (q.v.).

Chemical Class: Protein Derivatives

Functions: Hair Conditioning Agent; Skin-Conditioning Agent - Miscellaneous; Surfactant - Cleansing Agent

Trade Name:
Promois ERCP (Seiwa Kasei)

POTASSIUM COCOYL HYDROLYZED RICE PROTEIN

Definition: Potassium Cocoyl Hydrolyzed Rice Protein is the potassium salt of the condensation product of coconut acid chloride and Hydrolyzed Rice Protein (q.v.).

Chemical Class: Protein Derivatives

Functions: Hair Conditioning Agent; Skin-Conditioning Agent - Miscellaneous; Surfactant - Cleansing Agent

POTASSIUM COCOYL HYDROLYZED SILK

Definition: Potassium Cocoyl Hydrolyzed Silk is the potassium salt of the condensation product of coconut acid chloride and Hydrolyzed Silk (q.v.).

Chemical Class: Protein Derivatives

Functions: Hair Conditioning Agent; Skin-Conditioning Agent - Miscellaneous; Surfactant - Cleansing Agent

Trade Name:
Promois EFCP (Seiwa Kasei)

POTASSIUM COCOYL HYDROLYZED SOY PROTEIN

JPN Translation:
ココイル加水分解ダイズタンパク K

Definition: Potassium Cocoyl Hydrolyzed Soy Protein is the potassium salt of the condensation product of coconut acid chloride and Hydrolyzed Soy Protein (q.v.).

Information Sources: JCIC, JCLS

Chemical Class: Protein Derivatives

Functions: Hair Conditioning Agent; Skin-Conditioning Agent - Miscellaneous; Surfactant - Cleansing Agent

Technical/Other Name:
Potassium N-Cocoyl-hydrolyzed Soybean Protein Solution

Trade Names:
Coccopolipeptide di Soja 30% (Sinerga)
Promois ESCP (Seiwa Kasei)

Trade Name Mixture:
Stearilpolipeptide Di Soja (Sinerga)

POTASSIUM COCOYL HYDROLYZED WHEAT PROTEIN

JPN Translation:
ココイル加水分解コムギタンパク K

Definition: Potassium Cocoyl Hydrolyzed Wheat Protein is the potassium salt of the condensation product of coconut acid chloride and Hydrolyzed Wheat Protein (q.v.).

Chemical Class: Protein Derivatives

Functions: Hair Conditioning Agent; Skin-Conditioning Agent - Miscellaneous; Surfactant - Cleansing Agent

Trade Name:
Promois EGCP (Seiwa Kasei)

POTASSIUM COCOYL HYDROLYZED YEAST PROTEIN

JPN Translations:
ココイル加水分解酵母 K
ココイル加水分解酵母タンパク K

Definition: Potassium Cocoyl Hydrolyzed Yeast Protein is the potassium salt of the condensation product of coconut acid chloride and Hydrolyzed Yeast Protein (q.v.).

Information Source: JCLS

Chemical Class: Protein Derivatives

Functions: Hair Conditioning Agent; Skin-Conditioning Agent - Miscellaneous; Surfactant - Cleansing Agent

Trade Name:
Promois EYCP-P (Seiwa Kasei)

POTASSIUM COCOYL PCA

Definition: Potassium Cocoyl PCA is the organic compound that conforms to the formula:

where RCO represents the cocoyl radical.

Chemical Classes: Amides; Heterocyclic Compounds; Organic Salts

Functions: Skin-Conditioning Agent - Emollient; Skin-Conditioning Agent - Humectant; Surfactant - Cleansing Agent; Surfactant - Emulsifying Agent

Technical/Other Name:
N-Cocoyl L-Pyroglutamic Acid, Potassium Salt

Trade Name:
Protelan NMA/C (Zschimmer & Schwarz Italiana)

Trade Name Mixture:
Emulvama AGW (Vama Farmacosmetica)

POTASSIUM COCOYL SARCOSINATE

Definition: Potassium Cocoyl Sarcosinate is the potassium salt of Cocoyl Sarcosine (q.v.).

Chemical Class: Sarcosinates and Sarcosine Derivatives

Functions: Hair Conditioning Agent; Surfactant - Cleansing Agent

Technical/Other Names:
Amides, Coconut Oil, with Sarcosine, Potassium Salt
Potassium N-Cocoyl Sarcosinate

POTASSIUM COCOYL TAURATE

Definition: Potassium Cocoyl Taurate is the organic salt that conforms to the formula:

$$RC-NHCH_2CH_2SO_3K$$
$$\overset{O}{\overset{\|}{}}$$

where RCO- represents the coconut acid radical.

Chemical Classes: Amides; Sulfonic Acids

Function: Surfactant - Cleansing Agent

Trade Name:
Neoscoap CTP (Toho)

POTASSIUM CORNATE

CAS No.	EINECS No.
61789-23-9	263-044-1

Definition: Potassium Cornate is the potassium salt of Corn Acid (q.v.).

Information Sources: 21CFR175.105, 21CFR176.200, 21CFR176.210, 21CFR177.1200, 21CFR177.2260, 21CFR178.3910, TSCA

Chemical Class: Soaps

Functions: Surfactant - Cleansing Agent; Surfactant - Emulsifying Agent

Technical/Other Names:
Fatty Acids, Corn Oil, Potassium Salts
Potassium Corn Acid Soap

POTASSIUM CUMENESULFONATE

CAS No.	EINECS No.
28085-69-0	248-827-8

Definition: Potassium Cumenesulfonate is the substituted aromatic compound that conforms generally to the formula:

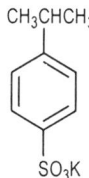

Chemical Class: Alkyl Aryl Sulfonates

Function: Surfactant - Hydrotrope

Technical/Other Name:
Benzenesulfonic Acid, (1-Methylethyl)-, Potassium Salt

Trade Name Mixture:
KNa-Cumosulfonat (Sasol GmbH - Marl)

POTASSIUM CYANATE

CAS No.	EINECS No.
590-28-3	209-677-9

Empirical Formula:
CHNO · K

Definition: Potassium Cyanate is the organic compound that conforms to the formula:

KCNO

Information Sources: MI-13(7708), TSCA

Chemical Class: Organic Salts

Function: Not Reported

Technical/Other Name:
Cyanic Acid, Potassium Salt

Trade Name Mixture:
Capigen CG (Sederma)

POTASSIUM CYCLOCARBOXYPROPYL-OLEATE

CAS No.	EINECS No.
68127-33-3	268-571-0

Empirical Formula:
$C_{21}H_{38}O_4$ · K

Definition: Potassium Cyclocarboxypropyl-oleate is the potassium salt of Cyclocarboxy-propyloleic Acid (q.v.). It conforms generally to the formula:

$(CH_2)_7COOK$

$(CH_2)_5CH_3$

COOH

Chemical Class: Soaps

Function: Surfactant - Cleansing Agent

Reported Product Category: Bath Soaps and Detergents

Technical/Other Names:
Potassium Acrylinoleate
Potassium C21-Dicarboxylate

POTASSIUM DECETH-4 PHOSPHATE

Definition: Potassium Deceth-4 Phosphate is the potassium salt of Deceth-4 Phosphate (q.v.).

Chemical Classes: Esters; Phosphorus Compounds

Function: Surfactant - Emulsifying Agent

Technical/Other Names:
Potassium Polyethylene Glycol 200 Decyl Ether Phosphate
Potassium Polyoxyethylene (4) Decyl Ether Phosphate

Trade Name:
Phosfetal 201 K (Zschimmer & Schwarz)

POTASSIUM DEXTRIN OCTENYL-SUCCINATE

Definition: Potassium Dextrin Octenyl-succinate is the potassium salt of the reaction product of octenylsuccinic anhydride with Dextrin (q.v.).

Chemical Classes: Biological Polymers and their Derivatives; Carbohydrates

Functions: Emulsion Stabilizer; Hair Conditioning Agent; Humectant; Skin-Conditioning Agent - Emollient; Surfactant - Emulsifying Agent

Trade Name:
Naturalnisk K (Nippon Starch)

POTASSIUM DIHYDROXYETHYL COCAMINE OXIDE PHOSPHATE

Definition: Potassium Dihydroxyethyl Cocamine Oxide Phosphate is the organic compound that conforms to the formula:

$$RNCH_2CH_2O-\overset{O}{\overset{\|}{P}}-O^- \cdot 2K^+$$
$$\overset{|}{CH_2CH_2OH} \quad \overset{|}{O^-}$$

where R represents the coconut radical.

Chemical Classes: Amine Oxides; Phosphorus Compounds

Functions: Hair Conditioning Agent; Surfactant - Cleansing Agent; Surfactant - Foam Booster; Surfactant - Hydrotrope

POTASSIUM DIMETHICONE PEG-7 PANTHENYL PHOSPHATE

Definition: Potassium Dimethicone PEG-7 Panthenyl Phosphate is the potassium salt of a complex mixture of phosphate esters of PEG-7 Dimethicone (q.v.) and Panthenol (q.v.).

Chemical Classes: Siloxanes and Silanes; Synthetic Polymers

Functions: Hair Conditioning Agent; Skin-Conditioning Agent - Miscellaneous

Trade Names:
Pecosil PAN-400 (Phoenix)
Pecosil SPP-50 (Phoenix)

Trade Name Mixture:
Sepicap MP (SEPPIC)

POTASSIUM DIMETHICONE PEG-7 PHOS-PHATE

Definition: Potassium Dimethicone PEG-7 Phosphate is the potassium salt of Dimethicone PEG-7 Phosphate (q.v.).

Chemical Classes: Phosphorus Compounds; Siloxanes and Silanes

Functions: Surfactant - Cleansing Agent; Surfactant - Emulsifying Agent

Trade Name:
Pecosil PS-100K (Phoenix)

POTASSIUM DNA

JPN Translation:
DNA - K

Definition: Potassium DNA is the potassium salt of DNA (q.v.).

Information Sources: JCIC, JCLS, JSQI

Chemical Classes: Biological Polymers and their Derivatives; Organic Salts

Function: Skin-Conditioning Agent - Miscellaneous

Technical/Other Names:
DNA, Potassium Salt
Potassium Deoxyribonucleic Acid

POTASSIUM DODECYLBENZENE-SULFONATE

CAS No.	EINECS No.
27177-77-1	248-296-2

Empirical Formula:
$C_{18}H_{30}O_3S \cdot K$

Definition: Potassium Dodecylbenzene-sulfonate is the substituted aromatic compound that conforms generally to the formula:

$$SO_3K - C_6H_4 - (CH_2)_{11}CH_3$$

Information Sources: 21CFR178.3400, TSCA

Chemical Class: Alkyl Aryl Sulfonates

Function: Surfactant - Cleansing Agent

Technical/Other Names:
Benzenesulfonic Acid, Dodecyl-, Potassium Salt
Dodecylbenzenesulfonic Acid, Potassium Salt

POTASSIUM EDTMP

CAS No.	EINECS No.
34274-30-1	251-910-1

Empirical Formula:
$C_6H_{20}N_2O_{12}P_4 \cdot xK$

Definition: Potassium EDTMP is the substituted diamine that conforms to the formula:

$$KHO_3PCH_2, \quad CH_2PO_3H_2$$
$$NCH_2CH_2N$$
$$H_2O_3PCH_2 \quad CH_2PO_3H_2$$

Information Source: TSCA

Chemical Classes: Amines; Phosphorus Compounds

Function: Chelating Agent

Technical/Other Names:
[Ethylenebis[Nitrilobis(Methylene)]]-Tetrakisphosphonic Acid, Potassium Salt
Phosphonic Acid, [1,2-Ethanediylbis [Nitrilobis(Methylene)]]Tetrakis-, Potassium Salt
Potassium Ethylenediamine Tetramethylene Phosphonate

Trade Name:
Sequion P 30 (Bozzetto)

POTASSIUM ETHYL ESTER OF PVM/MA COPOLYMER

Definition: Potassium Ethyl Ester of PVM/MA Copolymer is the potassium salt of Ethyl Ester of PVM/MA Copolymer (q.v.).

Chemical Class: Synthetic Polymers

Functions: Binder; Film Former; Hair Fixative

POTASSIUM ETHYLPARABEN

Empirical Formula:
$C_9H_{10}O_3 \cdot K$

Definition: Potassium Ethylparaben is the potassium salt of Ethylparaben (q.v.) that conforms to the formula:

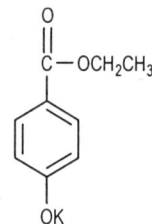

Information Source: EEC(VI/1-12)

Chemical Classes: Esters; Organic Salts; Phenols

Function: Preservative

Technical/Other Name:
Ethylparaben, Potassium Salt

POTASSIUM FLUORIDE

CAS No.	EINECS No.
7789-23-3	232-151-5

Empirical Formula:
FK

Definition: Potassium Fluoride is the inorganic salt that conforms to the formula:

$$KF$$

Information Sources: EEC(III/1-32), MI-13 (7714), TSCA

Chemical Class: Inorganic Salts

Function: Oral Care Agent

Technical/Other Name:
Potassium Monofluoride

POTASSIUM FLUOROSILICATE

CAS No.	EINECS No.
16871-90-2	240-896-2

Empirical Formula:
$F_6H_2Si \cdot 2K$

Definition: Potassium Fluorosilicate is the inorganic salt that conforms to the formula:

$$K_2F_6Si$$

Information Sources: EEC(III/1-41), MI-13 (7722), TSCA

Chemical Class: Inorganic Salts

Function: Oral Care Agent

Technical/Other Names:
Potassium Hexafluorosilicate
Potassium Silicofluoride
Silicate, Hexafluoro-, Dipotassium

POTASSIUM FRUCTOBORATE

Definition: Potassium Fructoborate is the potassium salt of the reaction product of Boric Acid (q.v.) and Fructose (q.v.).

Chemical Classes: Carbohydrates; Organic Salts

Function: Skin-Conditioning Agent - Humectant

Trade Name:
Straw-BX (VDF FutureCeuticals)

POTASSIUM GLUCOHEPTONATE

CAS No. 23167-96-6   **EINECS No.** 245-473-6

Empirical Formula:
$C_7H_{13}O_8 \cdot K$

Definition: Potassium Glucoheptonate is a mixture of potassium salts of α- and β-glucoheptonic acid. It conforms generally to the formula:

$$\left[CH_2OH(CHOH)_5COO^- \right] K^+$$

Chemical Classes: Carboxylic Acids; Organic Salts; Polyols

Function: Skin-Conditioning Agent - Miscellaneous

Technical/Other Name:
Glucoheptonic Acid, Potassium Salt

Trade Name:
Givobio GHK (SEPPIC)

POTASSIUM GLUCONATE

CAS No. 299-27-4   **EINECS No.** 206-074-2

Empirical Formula:
$C_6H_{12}O_7 \cdot K$

Definition: Potassium Gluconate is the potassium salt of Gluconic Acid (q.v.). *See Reported Ingredient Functions-The Cosmetic Drug Distinction, in Regulatory and Ingredient Use Information, Volume I, Part A.*

Information Sources: MI-13(7716), TSCA, USAN, USP XXIV

Chemical Classes: Carbohydrates; Organic Salts

Functions: Chelating Agent; Skin Protectant

Technical/Other Name:
D-Gluconic Acid, Monopotassium Salt

Trade Names:
Givobio GK (SEPPIC)
Gluconal K (Glucona America)
Jungbunzlauer Potassium Gluconate (Jungbunzlauer)

Trade Name Mixture:
Givobio GMgK (SEPPIC)

POTASSIUM GLYCEROPHOSPHATE

CAS No.: 1319-69-3

Definition: Potassium Glycerophosphate is the organic compound that conforms generally to the formula:

$$\left[HOCH_2CHCH_2OPO_3H^- \atop \qquad\quad OH \right] K^+$$

Information Source: MI-13(7717)

Chemical Classes: Glyceryl Esters and Derivatives; Organic Salts; Phosphorus Compounds

Function: Oral Care Agent

Technical/Other Name:
1,2,3-Propanetriol, Mono(Dihydrogen Phosphate), Dipotassium Salt

Trade Name:
Givobio GPK (SEPPIC)

POTASSIUM GLYCOL SULFATE

CAS No.: 59599-54-1

Empirical Formula:
$C_2H_6O_5S \cdot K$

Definition: Potassium Glycol Sulfate is the salt that conforms to the formula:

$$HOCH_2CH_2OSO_3K$$

Chemical Classes: Organic Salts; Sulfuric Acid Esters

Function: Not Reported

Technical/Other Name:
1,2-Ethanediol, Mono(Hydrogen Sulfate), Monopotassium Salt

POTASSIUM GLYCYRRHETINATE

CAS No.: 85985-61-1

Definition: Potassium Glycyrrhetinate is the potassium salt of Glycyrrhetinic Acid (q.v.).

Chemical Class: Organic Salts

Functions: Flavoring Agent; Skin-Conditioning Agent - Miscellaneous

Technical/Other Name:
Olean-12-en-29-oic Acid, 3-Hydroxy-1,1-Oxo-, Monopotassium Salt

Trade Names:
Glycyrrhetic Acid Potassium Salt Vinyals (Vinyals)
Plantactiv GLAP 18 (Cognis Deutschland)

POTASSIUM GLYCYRRHIZINATE

CAS No. 68039-19-0   **EINECS No.** 268-237-4

Definition: Potassium Glycyrrhizinate is an enzymatically produced mixture of potassium salts of Glycyrrhizic Acid (q.v) and glycyrrhetinic acid monoglucoside.

Chemical Classes: Ketones; Organic Salts

Function: Skin-Conditioning Agent - Miscellaneous

Technical/Other Name:
Glycopyranosiduronic Acid, 20-Carboxy-11-Oxo-30-Norolean-12-en-3-yl 2Glucopyranuronosyl-, Potassium Salt

Trade Name:
Monopotassium Glycyrrhizinate (Maruzen Pharmaceuticals Co., Ltd.)

POTASSIUM HEMPSEEDATE

CAS No. 68424-21-5   **EINECS No.** 270-281-4

Definition: Potassium Hempseedate is the potassium salt of the fatty acids derived from the seeds of *Cannabis sativa.*

Chemical Class: Soaps

Functions: Surfactant - Cleansing Agent; Surfactant - Emulsifying Agent

Trade Name Mixture:
Potassium Hempate (Chemron)

POTASSIUM HYALURONATE

CAS No.: 31799-91-4

Definition: Potassium Hyaluronate is the potassium salt of Hyaluronic Acid (q.v.).

Chemical Class: Biological Polymers and their Derivatives

Function: Skin-Conditioning Agent - Miscellaneous

Reported Product Categories: Moisturizing Preparations; Bath Capsules

Technical/Other Name:
Hyaluronic Acid, Potassium Salt

POTASSIUM HYDROGENATED COCOATE

Definition: Potassium Hydrogenated Cocoate is the potassium salt of Hydrogenated Coconut Acid (q.v.).

Chemical Class: Soaps

Function: Surfactant - Cleansing Agent

POTASSIUM HYDROGENATED PALMATE

Definition: Potassium Hydrogenated Palmate is the potassium salt of Hydrogenated Palm Acid (q.v.).

Chemical Class: Soaps

Function: Surfactant - Cleansing Agent

POTASSIUM HYDROGENATED TALLOWATE

Definition: Potassium Hydrogenated Tallowate is the potassium salt of Hydrogenated Tallow Acid (q.v.).

Chemical Class: Soaps

Function: Surfactant - Cleansing Agent

POTASSIUM HYDROXIDE

CAS No. 1310-58-3   **EINECS No.** 215-181-3

JPN Translation:
水酸化 K

Empirical Formula:
HKO

Definition: Potassium Hydroxide is the inorganic base that conforms to the formula:

$$KOH$$

Information Sources: ARG, AUS, BP, BPC, 21CFR114.90, 21CFR163.110, 21CFR175.210, 21CFR184.1631, DDR, EEC(III/1-15a), EGY, FCC, FIN, HP, HUN, IND, JAN, JCLS, JSCI, MAR, MEX, MI-13 (7724), NF XIX, PF, PN, POR, TSCA, USAN, USD, WHO, YUG

Chemical Class: Inorganic Bases

Function: pH Adjuster

Reported Product Categories: Bath Oils, Tablets, and Salts; Lipsticks; Moisturizing Preparations; Cleansing Products (Cold Creams, Cleansing Lotions, Liquids and Pads); Skin Care Preparations, Misc.; Bath Capsules; Shaving Cream (Aerosol, Brushless and Lather); Bath Preparations, Misc.; Body and Hand Preparations (Excluding Shaving Preparations); Makeup Bases; Hair Sprays (Aerosol Fixatives); Mascara; Shampoos (Non-coloring); Tonics, Dressings, and Other Hair Grooming Aids; Hair Dyes and Colors (All Types Requiring Caution Statements and Patch Tests); Bath Soaps and Detergents; Eyeliners; Indoor Tanning Preparations; Makeup Preparations (Not eye), Misc.; Paste Masks (Mud Packs); Personal Cleanliness Products, Misc.; Eye Lotions; Fragrance Preparations, Misc.; Shaving Preparations, Misc.

Technical/Other Name:
Caustic Potash

Trade Name:
Unichem POHYD (Universal Preserv-A-Chem)

Trade Name Mixture:
SC-1000 (Gemtek)

POTASSIUM HYDROXYCITRATE

Empirical Formula:
$C_6H_5O_8K_3$

Definition: Potassium Hydroxycitrate is the organic compound that conforms to the formula:

$$KOOCCH_2C-CHCOOK$$

with COOK above and OH OH below

Chemical Class: Organic Salts

Function: Skin-Conditioning Agent - Miscellaneous

Technical/Other Names:
2,3-Dihydroxypropane Tricarboxylic Acid, Tripotssium Salt
Tripotassium Hydroxycitrate

Trade Name:
Citrin (Sabinsa)

POTASSIUM HYDROXYSTEARATE

Definition: Potassium Hydroxystearate is the potassium salt of Hydroxystearic Acid (q.v.).

Chemical Class: Soaps

Function: Surfactant - Cleansing Agent

Technical/Other Names:
12-Hydroxystearic Acid, Potassium Salt
Potassium 12-Hydroxystearate

POTASSIUM IODIDE

CAS No. 7681-11-0   **EINECS No.** 231-659-4

Empirical Formula:
IK

Definition: Potassium Iodide is the inorganic salt that conforms to the formula:

$$KI$$

Information Sources: ARG, AUS, BP, BPC, BRA, 21CFR172.375, 21CFR178.1010, 21CFR184.1634, 21CFR582.80, CZE, DA, DDR, EGY, EP, FCC, FIN, HP, HUN, IND, ITA, JAN, MAR, MEX, MI-13(7727), PF, PN, POR, ROM, TSCA, USAN, USD, USP XXIV, USSR, WHO, YUG

Chemical Class: Inorganic Salts

Function: Not Reported

Trade Name:
Unichem KI (Universal Preserv-A-Chem)

POTASSIUM ISOSTEARATE

CAS No. 68413-46-7   **EINECS No.** 270-218-0

Definition: Potassium Isostearate is the potassium salt of Isostearic Acid.

Chemical Class: Soaps

Function: Surfactant - Cleansing Agent

Technical/Other Name:
Isostearic Acid, Potassium Salt

POTASSIUM ISOSTEARETH-2 PHOSPHATE

Definition: Potassium Isosteareth-2 Phosphate is the potassium salt of Isosteareath-2 Phosphate (q.v.).

Chemical Class: Phosphorus Compounds

Function: Surfactant - Emulsifying Agent

POTASSIUM LACTATE

CAS Nos.	EINECS Nos.
996-31-6	213-631-3
85895-78-9	288-752-8

Empirical Formula:
$C_3H_6O_3 \cdot K$

Definition: Potassium Lactate is the potassium salt of lactic acid that conforms to the formula:

$$CH_3CHCOOK$$
$$|$$
$$OH$$

Information Sources: 21CFR184.1639, CIR: [SQ] IJT-17(Suppl. 1)1998

Chemical Class: Organic Salts

Functions: Buffering Agent; Exfoliant; Skin-Conditioning Agent - Humectant

Technical/Other Names:
Potassium α-Hydroxypropionate
Propanoic Acid, 2-Hydroxy-, Monopotassium Salt

Trade Name:
PURASAL P (Purac)

Trade Name Mixture:
PURAC BF/P (Purac)

POTASSIUM LANOLATE

Definition: Potassium Lanolate is the potassium salt of Lanolin Acid (q.v.).

Chemical Classes: Organic Salts; Soaps

Function: Surfactant - Cleansing Agent

Technical/Other Name:
Lanolin Acid, Potassium Salt

POTASSIUM LAURATE

CAS No. 10124-65-9
EINECS No. 233-344-7

JPN Translation:
ラウリン酸K

Empirical Formula:
$C_{12}H_{24}O_2 \cdot K$

Definition: Potassium Laurate is the potassium salt of lauric acid. It conforms generally to the formula:

$$CH_3(CH_2)_{10}COOK$$

Information Sources: 21CFR172.863, 21CFR175.105, 21CFR176.170, 21CFR176.200, 21CFR176.210, 21CFR177.1200, 21CFR177.2260, 21CFR177.2600, 21CFR177.2800, 21CFR178.3910, JCIC, JCLS, JSQI, TSCA

Chemical Class: Soaps

Functions: Surfactant - Cleansing Agent; Surfactant - Emulsifying Agent

Reported Product Categories: Bath Oils, Tablets, and Salts; Cleansing Products (Cold Creams, Cleansing Lotions, Liquids and Pads); Moisturizing Preparations

Technical/Other Name:
Dodecanoic Acid, Potassium Salt

Trade Name Mixture:
Rosemary Extract Powder-S (Maruzen Pharmaceuticals Co., Ltd.)

POTASSIUM LAURETH-3 CARBOXYLATE

Definition: Potassium Laureth-3 Carboxylate is the potassium salt of Laureth-3 Carboxylic Acid (q.v.).

Chemical Class: Organic Salts

Function: Surfactant - Cleansing Agent

Technical/Other Names:
Potassium Polyethylene Glycol (3) Lauryl Ether Carboxylate
Potassium Polyoxyethylene (3) Lauryl Ether Carboxylate

POTASSIUM LAURETH-4 CARBOXYLATE

Definition: Potassium Laureth-4 Carboxylate is the potassium salt of Laureth-4 Carboxylic Acid (q.v.).

Chemical Class: Organic Salts

Function: Surfactant - Cleansing Agent

Technical/Other Names:
Potassium Polyethylene Glycol 200 Lauryl Ether Carboxylate
Potassium Polyoxyethylene (4) Lauryl Ether Carboxylate

POTASSIUM LAURETH-5 CARBOXYLATE

Definition: Potassium Laureth-5 Carboxylate is the potassium salt of Laureth-5 Carboxylic Acid (q.v.).

Chemical Class: Organic Salts

Function: Surfactant - Cleansing Agent

Technical/Other Names:
Potassium Polyethylene Glycol (5) Lauryl Ether Carboxylate
Potassium Polyoxyethylene (5) Lauryl Ether Carboxylate

POTASSIUM LAURETH-6 CARBOXYLATE

Definition: Potassium Laureth-6 Carboxylate is the potassium salt of Laureth-6 Carboxylic Acid (q.v.).

Chemical Class: Organic Salts

Function: Surfactant - Cleansing Agent

Technical/Other Names:
Potassium Polyethylene Glycol 300 Lauryl Ether Carboxylate
Potassium Polyoxyethylene (6) Lauryl Ether Carboxylate

POTASSIUM LAURETH-10 CARBOXYLATE

Definition: Potassium Laureth-10 Carboxylate is the potassium salt of Laureth-10 Carboxylic Acid (q.v.).

Chemical Class: Organic Salts

Function: Surfactant - Cleansing Agent

Technical/Other Names:
Potassium Polyethylene Glycol 500 Lauryl Ether Carboxylate
Potassium Polyoxyethylene (10) Lauryl Ether Carboxylate

POTASSIUM LAURETH PHOSPHATE

CAS No.: 68954-87-0

Definition: Potassium Laureth Phosphate is the potassium salt of a mixture of phosphate esters of ethoxylated lauryl alcohol with an average ethoxylation value between 1 and 3.

Chemical Classes: Alkoxylated Alcohols; Phosphorus Compounds

Functions: Surfactant - Cleansing Agent; Surfactant - Emulsifying Agent; Surfactant - Foam Booster; Surfactant - Solubilizing Agent

Technical/Other Names:
Potassium Polyethylene Glycol Lauryl Ether Phosphate
Potassium Polyoxyethylene Lauryl Ether Phosphate

Trade Name:
Dermalcare MAP L-213/K (Rhodia)

POTASSIUM LAUROYL COLLAGEN AMINO ACIDS

Definition: Potassium Lauroyl Collagen Amino Acids is the potassium salt of the condensation product of lauric acid chloride and Collagen Amino Acids (q.v.).

Chemical Class: Amino Acids

Functions: Hair Conditioning Agent; Skin-Conditioning Agent - Miscellaneous; Surfactant - Cleansing Agent

POTASSIUM LAUROYL GLUTAMATE

CAS No.: 89187-78-0 (L-Form)

JPN Translation:
ラウロイルグルタミン酸 K

Empirical Formula:
$C_{17}H_{30}NO_5 \cdot K$

Definition: Potassium Lauroyl Glutamate is the substituted amino acid that conforms to the formula:

$$CH_3(CH_2)_{10}\overset{\displaystyle O}{\overset{\|}{C}}-NHCHCOOK$$
$$|$$
$$CH_2CH_2COOH$$

Information Sources: JCIC, JCLS

Chemical Class: Amino Acids

Functions: Hair Conditioning Agent; Surfactant - Cleansing Agent

Technical/Other Names:
Glutamic Acid, N-(1-Oxododecyl)-, Potassium Salt
Potassium N-Lauroyl-L-glutamate

Trade Name:
Amisoft LK-11 (Ajinomoto)

Trade Name Mixture:
Sel-Smooth (Seltzer)

POTASSIUM LAUROYL HYDROLYZED COLLAGEN

JPN Translation:
ラウロイル加水分解コラーゲン K

Definition: Potassium Lauroyl Hydrolyzed Collagen is the potassium salt of the condensation product of lauric acid chloride and Hydrolyzed Collagen (q.v.).

Information Sources: JCIC, JCLS

Chemical Class: Protein Derivatives

Functions: Hair Conditioning Agent; Skin-Conditioning Agent - Miscellaneous; Surfactant - Cleansing Agent

Technical/Other Names:
Potassium Lauroyl Hydrolyzed Animal Protein
Proteins, Hydrolysates, Reaction Products with Lauroyl Chloride, Compds. with Potassium

Trade Names:
Promois ELP (Seiwa Kasei)
Promois ELP-C (Seiwa Kasei)
Promois ELP-P (Seiwa Kasei)
Promois EULP (Seiwa Kasei)

POTASSIUM LAUROYL HYDROLYZED SOY PROTEIN

Definition: Potassium Lauroyl Hydrolyzed Soy Protein is the potassium salt of the

reaction product of lauric acid chloride with Hydrolyzed Soy Protein (q.v.).

Chemical Class: Protein Derivatives

Functions: Hair Conditioning Agent; Skin-Conditioning Agent - Miscellaneous; Surfactant - Cleansing Agent

POTASSIUM LAUROYL METHYL BETA-ALANINE

Definition: Potassium Lauroyl Methyl Beta-Alanine is the potassium salt of the lauric acid amide of N-methyl beta-alanine. It conforms to the formula:

$$CH_3(CH_2)_{10}\overset{\displaystyle O}{\overset{\|}{C}}-NCH_2CH_2\overset{\displaystyle O}{\overset{\|}{C}}-OK$$
$$|$$
$$CH_3$$

Chemical Classes: Amino Acids; Organic Salts

Function: Skin-Conditioning Agent - Miscellaneous

Technical/Other Name:
Potassium Lauroyl Methylaminopropionate

POTASSIUM LAUROYL PCA

Empirical Formula:
$C_{17}H_{29}NO_4 \cdot K$

Definition: Potassium Lauroyl PCA is the organic compound that conforms to the formula:

Chemical Classes: Heterocyclic Compounds; Organic Salts

Functions: Skin-Conditioning Agent - Emollient; Skin-Conditioning Agent - Humectant; Surfactant - Cleansing Agent; Surfactant - Emulsifying Agent

Technical/Other Name:
N-Lauroyl L-Pyroglutamic Acid, Potassium Salt

Trade Name:
Protelan NMA (Zschimmer & Schwarz Italiana)

POTASSIUM LAUROYL SARCOSINATE

Empirical Formula:
$C_{15}H_{29}NO_3K$

Definition: Potassium Lauroyl Sarcosinate is the potassium salt of Lauroyl Sarcosine (q.v.). It conforms to the formula:

$$CH_3(CH_2)_{10}\overset{\displaystyle O}{\overset{\|}{C}}-NCH_2COOK$$
$$|$$
$$CH_3$$

Chemical Class: Sarcosinates and Sarcosine Derivatives

Functions: Hair Conditioning Agent; Surfactant - Cleansing Agent

Trade Name:
Nikkol Sarcosinate LK-30 (Nikko)

POTASSIUM LAUROYL WHEAT AMINO ACIDS

Definition: Potassium Lauroyl Wheat Amino Acids is the potassium salt of the condensation product of lauric acid chloride and Wheat Amino Acids (q.v.).

Chemical Class: Amino Acids

Functions: Hair Conditioning Agent; Skin-Conditioning Agent - Miscellaneous; Surfactant - Cleansing Agent

Reported Product Category: Cleansing Products (Cold Creams, Cleansing Lotions, Liquids and Pads)

Trade Names:
Aminofoam W (Croda Chemicals)
Aminofoam W (Croda, Inc.)

POTASSIUM LAURYL HYDROXYPROPYL SULFONATE

Empirical Formula:
$C_{15}H_{32}O_5S \cdot K$

Definition: Potassium Lauryl Hydroxypropyl Sulfonate is the organic compound that conforms to the formula:

$$CH_3(CH_2)_{11}OCH_2CHCH_2SO_3K$$
$$|$$
$$OH$$

Chemical Classes: Alcohols; Ethers; Sulfonic Acids

Function: Surfactant - Emulsifying Agent

Trade Name:
Ages 2006K (Solvay GmbH)

POTASSIUM LAURYL PHOSPHATE

Definition: Potassium Lauryl Phosphate is the potassium salt of Lauryl Phosphate (q.v.).

The inclusion of any compound in the *Dictionary and Handbook* does not indicate that use of that substance as a cosmetic ingredient complies with the laws and regulations governing such use in the United States or any other country.

Chemical Classes: Organic Salts; Phosphorus Compounds

Function: Surfactant - Cleansing Agent

POTASSIUM LAURYL SULFATE

CAS No.	EINECS No.
4706-78-9	225-190-4

JPN Translation:
ラウリル硫酸 K

Empirical Formula:
$C_{12}H_{26}O_4S \cdot K$

Definition: Potassium Lauryl Sulfate is the potassium salt of lauryl sulfate that conforms to the formula:

$$CH_3(CH_2)_{11}OSO_3K$$

Information Sources: 21CFR175.105, 21CFR176.170, 21CFR177.1200, 21CFR177.2800, JCIC, JCLS, JSQI, TSCA

Chemical Class: Alkyl Sulfates

Functions: Surfactant - Cleansing Agent; Surfactant - Emulsifying Agent

Reported Product Categories: Bath Oils, Tablets, and Salts; Cleansing Products (Cold Creams, Cleansing Lotions, Liquids and Pads)

Technical/Other Names:
Potassium Dodecyl Sulfate
Sulfuric Acid, Monododecyl Ester, Potassium Salt

Trade Names:
Alscoap CPSK-90 (Toho)
Nikkol KLS (Nikko)

POTASSIUM LINOLEATE

CAS Nc.	EINECS No.
3414-89-9	222-308-6

Definition: Potassium Linoleate is the potassium salt of Linoleic Acid (q.v.).

Chemical Class: Soaps

Functions: Surfactant - Cleansing Agent; Surfactant - Emulsifying Agent; Viscosity Increasing Agent - Nonaqueous

Technical/Other Name:
Linoleic Acid, Potassium Salt

Trade Name:
PRISAVON KL (Uniqema Europe)

POTASSIUM MAGNESIUM ASPARTATE

CAS No.: 67528-13-6

Empirical Formula:
$C_4H_5NO_4 \cdot K \cdot \frac{1}{2}Mg$

Definition: Potassium Magnesium Aspartate is the organic compound that conforms to the formula:

$$^-OOCCH_2CHCOO^- \cdot K^+ \cdot \tfrac{1}{2}Mg^{+2}$$
$$|$$
$$NH_2$$

See Reported Ingredient Functions-The Cosmetic Drug Distinction, in Regulatory and Ingredient Use Information, Volume I, Part A.

Chemical Class: Amines

Functions: pH Adjuster; Skin Protectant

Technical/Other Name:
L-Aspartic Acid, Magnesium Salt (2:1), mixt. with L-Aspartic Acid Monopotassium Salt

Trade Name:
Potassium Magnesium Aspartate (Sabinsa)

POTASSIUM METABISULFITE

CAS Nos.	EINECS No.
4429-42-9	
16731-55-8	240-795-3

Empirical Formula:
$H_2O_5S_2 \cdot 2K$

Definition: Potassium Metabisulfite is the inorganic salt that conforms to the formula:

$$K_2S_2O_5$$

Information Sources: 21CFR182.3637, CIR: [S], MAR, MI-13(7729), NF XIX, TSCA, USAN

Chemical Class: Inorganic Salts

Functions: Hair-Waving/Straightening Agent; Reducing Agent

Technical/Other Names:
Dipotassium Disulfite
Dipotassium Pyrosulfite
Disulfurous Acid, Dipotassium Salt
Potassium Pyrosulfite

Trade Name:
Uantox POMEBIS (Universal Preserv-A-Chem)

POTASSIUM METHOXYCINNAMATE

Empirical Formula:
$C_{10}H_{10}O_3 \cdot K$

Definition: Potassium Methoxycinnamate is the organic compound that conforms to the formula:

Chemical Classes: Ethers; Organic Salts

Function: Ultraviolet Light Absorber

POTASSIUM METHYL COCOYL TAURATE

JPN Translation:
ココイルメチルタウリン K

Definition: Potassium Methyl Cocoyl Taurate is the potassium salt of the coconut acid amide of N-methyl taurine. It conforms to the formula:

where RCO- represents the fatty acids derived from coconut oil.

Information Sources: JCIC, JCLS, JSQI

Chemical Classes: Amides; Sulfonic Acids

Function: Surfactant - Cleansing Agent

Technical/Other Name:
Potassium N-Cocoyl-N-methyl Taurate

POTASSIUM METHYLPARABEN

CAS No.	EINECS No.
26112-07-2	247-464-2

Empirical Formula:
$C_8H_8O_3 \cdot K$

Definition: Potassium Methylparaben is the potassium salt of Methylparaben (q.v.) that conforms to the formula:

Information Source: EEC(VI/1-12)

Chemical Classes: Esters; Organic Salts; Phenols

Function: Preservative

Reported Product Category: Skin Care Preparations, Misc.

The inclusion of any compound in the *Dictionary and Handbook* does not indicate that use of that substance as a cosmetic ingredient complies with the laws and regulations governing such use in the United States or any other country.

Technical/Other Names:
Benzoic Acid, 4-Hydroxy-, Methyl Ester,
Potassium Salt
4-Hydroxybenzoic Acid, Methyl Ester,
Potassium Salt
Methyl p-Hydroxybenzoate Potassium Salt
Methylparaben, Potassium Salt

POTASSIUM MONOFLUOROPHOSPHATE

CAS Nos.
14104-28-0
14306-73-1
20859-37-4

EINECS No.
237-957-0

Empirical Formula:
$FH_2O_3P \cdot 2K$

Definition: Potassium Monofluorophosphate
is the inorganic salt that conforms to the
formula:

$$K_2PO_3F$$

Information Source: EEC(III/1-28)

Chemical Classes: Inorganic Salts;
Phosphorus Compounds

Function: Oral Care Agent

Technical/Other Names:
Phosphorofluoridic Acid, Potassium Salt
Potassium Phosphorofluoridate

POTASSIUM MYRISTATE

CAS No.
13429-27-1

EINECS No.
236-550-5

JPN Translation:
ミリスチン酸 K

Empirical Formula:
$C_{14}H_{28}O_2 \cdot K$

Definition: Potassium Myristate is the
potassium salt of myristic acid. It conforms
generally to the formula:

$$CH_3(CH_2)_{12}COOK$$

Information Sources: 21CFR172.863,
21CFR175.105, 21CFR176.170,
21CFR176.200, 21CFR176.210,
21CFR177.1200, 21CFR177.2260,
21CFR177.2600, 21CFR177.2800,
21CFR178.3910, JCIC, JCLS, JSQI, TSCA

Chemical Class: Soaps

Functions: Surfactant - Cleansing Agent;
Surfactant - Emulsifying Agent

Reported Product Categories: Bath Oils,
Tablets, and Salts; Cleansing Products (Cold
Creams, Cleansing Lotions, Liquids and
Pads); Moisturizing Preparations

Technical/Other Names:
Potassium Tetradecanoate
Tetradecanoic Acid, Potassium Salt

Trade Name Mixture:
Nikkol MNK-40 (Nikko)

POTASSIUM MYRISTOYL GLUTAMATE

JPN Translation:
ミリストイルグルタミン酸 K

Empirical Formula:
$C_{19}H_{34}NO_5 \cdot K$

Definition: Potassium Myristoyl Glutamate
is the potassium salt of the myristic acid
amide of glutamic acid. It conforms to the
formula:

$$CH_3(CH_2)_{12}\overset{\displaystyle O}{\overset{\displaystyle \|}{C}}-NHCHCOOK \\ | \\ CH_2CH_2COOH$$

Chemical Class: Amino Acids

Functions: Hair Conditioning Agent;
Surfactant - Cleansing Agent

Trade Names:
Acylglutamate MK-11 (Ajinomoto)
Aminosurfact AMMP (Asahi Denka Kogyo)

POTASSIUM MYRISTOYL HYDROLYZED COLLAGEN

JPN Translation:
ミリストイル加水分解コラーゲン K

Definition: Potassium Myristoyl Hydrolyzed
Collagen is the potassium salt of the
condensation product of myristic acid
chloride and Hydrolyzed Collagen (q.v.).

Information Sources: JCIC, JCLS

Chemical Class: Protein Derivatives

Functions: Hair Conditioning Agent; Skin-
Conditioning Agent - Miscellaneous;
Surfactant - Cleansing Agent

Technical/Other Names:
Potassium Myristoyl Hydrolyzed Animal
Protein
Potassium Myristoyl Hydrolyzed Collagen
Solution
Proteins, Hydrolysates, Reaction Products
with Myristoyl Chloride, Compounds with
Potassium

Trade Names:
Promois EMP (Seiwa Kasei)
Promois EUMP (Seiwa Kasei)

POTASSIUM NITRATE

CAS No.
7757-79-1

EINECS No.
231-818-8

JPN Translation:
硝酸 K

Empirical Formula:
KNO_3

Definition: Potassium Nitrate is the
potassium salt of nitric acid that conforms to
the formula:

$$KNO_3$$

Information Source: MI-13(7733)

Chemical Class: Inorganic Salts

Function: Not Reported

Technical/Other Name:
Nitric Acid, Potassium Salt

POTASSIUM OCTOXYNOL-12 PHOS-PHATE

CAS No.: 68891-73-6

Definition: Potassium Octoxynol-12 Phosphate is the potassium salt of a complex
mixture of esters of phosphoric acid and
Octoxynol-12 (q.v.).

Information Sources: 21CFR175.105,
CIR: [S]

Chemical Class: Phosphorus Compounds

Functions: Surfactant - Cleansing Agent;
Surfactant - Emulsifying Agent; Surfactant -
Hydrotrope

Reported Product Categories: Mascara;
Eyeliners; Eye Shadows; Eyebrow Pencils;
Suntan Gels, Creams, and Liquids

POTASSIUM OLEATE

CAS Nos.
143-18-0
23282-35-1

EINECS No.
205-590-5

JPN Translation:
オレイン酸 K

Empirical Formula:
$C_{18}H_{34}O_2 \cdot K$

Definition: Potassium Oleate is the
potassium salt of oleic acid. It conforms
generally to the formula:

$$CH_3(CH_2)_7CH=CH(CH_2)_7COOK$$

Information Sources: 21CFR172.863,
21CFR175.105, 21CFR175.300,
21CFR176.170, 21CFR176.200,
21CFR176.210, 21CFR177.1200,
21CFR177.2260, 21CFR177.2600,
21CFR177.2800, 21CFR178.3910,
21CFR181.22, 21CFR181.29, JCIC, JCLS,
JSQI, MI-13(7735), TSCA

Chemical Class: Soaps

Functions: Surfactant - Cleansing Agent; Surfactant - Emulsifying Agent

Reported Product Categories: Hair Dyes and Colors (All Types Requiring Caution Statements and Patch Tests); Permanent Waves

Technical/Other Names:
9-Octadecenoic Acid, Potassium Salt
Potassium 9-Octadecenoate

Trade Names:
AEC Potassium Oleate (A & E Connock)
Norfox KO (Norman, Fox & Co.)

POTASSIUM OLEOYL HYDROLYZED COLLAGEN

Definition: Potassium Oleoyl Hydrolyzed Collagen is the potassium salt of the condensation product of oleic acid chloride and Hydrolyzed Collagen (q.v.).

Chemical Class: Protein Derivatives

Functions: Hair Conditioning Agent; Skin-Conditioning Agent - Miscellaneous; Surfactant - Cleansing Agent

Technical/Other Names:
Potassium Oleoyl Hydrolyzed Animal Collagen
Proteins, Hydrolysates, Reaction Products with Oleoyl Chloride, Compounds with Potassium

Trade Name:
Promois EOP (Seiwa Kasei)

POTASSIUM OLIVATE

Definition: Potassium Olivate is the potassium salt of the fatty acids derived from Olea Europaea (Olive) Oil (q.v.).

Chemical Class: Soaps

Functions: Surfactant - Cleansing Agent; Surfactant - Emulsifying Agent

Technical/Other Name:
Fatty Acids, Olive Oil, Potassium Salts

Trade Name:
PRISAVON KOL (Uniqema Europe)

Trade Name Mixtures:
Vegetable Soapbase OC (Weleda)
Vegetable Soapbase OP (Weleda)

POTASSIUM OXIDIZED MICROCRYSTALLINE WAX

Definition: Potassium Oxidized Microcrystalline Wax is the potassium salt of Oxidized Microcrystalline Wax (q.v.).

Chemical Classes: Organic Salts; Waxes

Function: Viscosity Increasing Agent - Nonaqueous

Trade Name:
Polymekon Polymer (Baker Petrolite)

POTASSIUM PALMATE

Definition: Potassium Palmate is the potassium salt of Palm Acid (q.v.).

Chemical Class: Soaps

Functions: Surfactant - Cleansing Agent; Surfactant - Emulsifying Agent; Viscosity Increasing Agent - Aqueous

Trade Name:
PRISAVON KP (Uniqema Europe)

POTASSIUM PALMITATE

CAS No.	EINECS No.
2624-31-9	220-088-6

JPN Translation:
パルミチン酸K

Empirical Formula:
$C_{16}H_{32}O_2 \cdot K$

Definition: Potassium Palmitate is the potassium salt of palmitic acid. It conforms generally to the formula:

$$CH_3(CH_2)_{14}COOK$$

Information Sources: 21CFR172.863, 21CFR175.105, 21CFR176.170, 21CFR176.200, 21CFR176.210, 21CFR177.1200, 21CFR177.2260, 21CFR177.2600, 21CFR177.2800, 21CFR178.3910, JCIC, JCLS, JSQI, TSCA

Chemical Class: Soaps

Functions: Surfactant - Cleansing Agent; Surfactant - Emulsifying Agent

Reported Product Categories: Bath Oils, Tablets, and Salts; Cleansing Products (Cold Creams, Cleansing Lotions, Liquids and Pads); Moisturizing Preparations

Technical/Other Names:
Hexadecanoic Acid, Potassium Salt
Potassium Hexadecanoate

POTASSIUM PALMITOYL HYDROLYZED CORN PROTEIN

Definition: Potassium Palmitoyl Hydrolyzed Corn Protein is the potassium salt of the condensation product of palmitic acid chloride and Hydrolyzed Corn Protein (q.v.).

Chemical Class: Protein Derivatives

Functions: Hair Conditioning Agent; Skin-Conditioning Agent - Miscellaneous; Surfactant - Cleansing Agent

Trade Name Mixture:
Zealat (Sinerga)

POTASSIUM PALMITOYL HYDROLYZED OAT PROTEIN

Definition: Potassium Palmitoyl Hydrolyzed Oat Protein is the potassium salt of the condensation product of palmitic acid chloride and Hydrolyzed Oat Protein (q.v.).

Chemical Class: Protein Derivatives

Functions: Hair Conditioning Agent; Skin-Conditioning Agent - Miscellaneous; Surfactant - Cleansing Agent

Trade Name Mixture:
Aveno-LAT (Sinerga)

POTASSIUM PALMITOYL HYDROLYZED WHEAT PROTEIN

Definition: Potassium Palmitoyl Hydrolyzed Wheat Protein is the potassium salt of the condensation product of palmitic acid chloride and Hydrolyzed Wheat Protein (q.v.).

Chemical Class: Protein Derivatives

Functions: Hair Conditioning Agent; Skin-Conditioning Agent - Miscellaneous; Surfactant - Cleansing Agent

Trade Name Mixtures:
Granolat (Sinerga)
Mandor-Lat (Sinerga)
Orzolat (Sinerga)
Pescolat (Sinerga)
Phytocream (Sinerga)

POTASSIUM PALM KERNELATE

CAS No.	EINECS No.
70969-43-6	275-067-4

Definition: Potassium Palm Kernelate is the potassium salt of the fatty acids derived from Elaeis Guineensis (Palm) Kernel Oil (q.v.).

Chemical Class: Soaps

Functions: Surfactant - Cleansing Agent; Viscosity Increasing Agent - Nonaqueous

Technical/Other Name:
Fatty Acids, Palm Kernel Oil, Potassium Salt

Trade Name:
PRISAVON 9275 (Uniqema Europe)

POTASSIUM PARABEN

CAS No.	EINECS No.
16782-08-4	240-830-2

Empirical Formula:
$C_7H_6O_3 \cdot K$

Definition: Potassium Paraben is the organic salt that conforms to the formula:

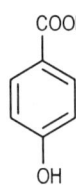

Information Source: EEC(VI/1-12)

Chemical Classes: Organic Salts; Phenols

Function: Preservative

Technical/Other Names:
Benzoic Acid, 4-Hydroxy-, Potassium Salt
4-Hydroxybenzoic Acid, Potassium Salt
p-Hydroxybenzoic Acid, Potassium Salt
Potassium p-Hydroxybenzoate

POTASSIUM PCA

CAS No.	EINECS No.
4810-50-8	225-373-9

JPN Translation:
PCA - K

Empirical Formula:
$C_5H_7NO_3 \cdot K$

Definition: Potassium PCA is the potassium salt of PCA (q.v.).

Information Sources: JCIC, JCLS

Chemical Classes: Amides; Heterocyclic Compounds; Organic Salts

Functions: Humectant; Skin-Conditioning Agent - Humectant

Technical/Other Names:
Potassium L-Pyroglutamate
Potassium 2-Pyrrolidone-5-Carboxylate
Potassium DL-Pyrrolidonecarboxylate
L-Proline, 5-Oxo-, Monopotassium Salt

Trade Name:
Kalidone (UCIB (Solabia))

POTASSIUM PEANUTATE

Definition: Potassium Peanutate is the potassium salt of the fatty acids derived from Arachis Hypogaea (Peanut) Oil (q.v.).

Chemical Class: Soaps

Functions: Surfactant - Cleansing Agent; Surfactant - Emulsifying Agent

Trade Name Mixture:
Vegetable Soapbase OP (Weleda)

POTASSIUM PEG-50 HYDROGENATED CASTOR OIL SUCCINATE

Definition: Potassium PEG-50 Hydrogenated Castor Oil Succinate is the potassium salt of PEG-50 Hydrogenated Castor Oil Succinate (q.v.).

Chemical Classes: Alkoxylated Alcohols; Organic Salts

Functions: Surfactant - Emulsifying Agent; Viscosity Increasing Agent - Nonaqueous

POTASSIUM PERSULFATE

CAS No.	EINECS No.
7727-21-1	231-781-8

Empirical Formula:
$H_2O_8S_2 \cdot 2K$

Definition: Potassium Persulfate is the inorganic salt that conforms to the formula:

$$K_2S_2O_8$$

Information Sources: 21CFR172.210, 21CFR175.105, 21CFR175.210, 21CFR176.170, 21CFR177.1210, 21CFR177.2600, CIR: [SQ] IJT-20(SUPPL. 3)2001, MI-13(7741), TSCA

Chemical Class: Inorganic Salts

Function: Oxidizing Agent

Reported Product Categories: Hair Bleaches; Hair Coloring Preparations, Misc.; Hair Lighteners with Color

Technical/Other Name:
Peroxydisulfuric Acid, Dipotassium Salt

POTASSIUM PHENOXIDE

CAS No.	EINECS No.
100-67-4	202-877-7

Empirical Formula:
$C_6H_6O \cdot K$

Definition: Potassium Phenoxide is the aromatic compound that conforms to the formula:

Information Sources: EEC(III/1-19), MI-13 (7742), TSCA

Chemical Classes: Organic Salts; Phenols

Functions: Cosmetic Biocide; Preservative

Technical/Other Names:
Phenol, Potassium Salt
Potassium Phenate
Potassium Phenylate

POTASSIUM PHENYLBENZIMIDAZOLE SULFONATE

Empirical Formula:
$C_{13}H_{10}N_2O_3S \cdot K$

Definition: Potassium Phenylbenzimidazole Sulfonate is the organic salt that conforms to the formula:

Information Source: EEC(VII/1-6)

Chemical Classes: Heterocyclic Compounds; Organic Salts; Sulfonic Acids

Function: Ultraviolet Light Absorber

Technical/Other Name:
2-Phenylbenzimidazole-5-Sulfonic Acid, Potassium Salt

Trade Name:
Phenylbenzimidazole Sulfonic Acid Potassium Salt (Merck KGaA)

POTASSIUM o-PHENYLPHENATE

CAS No.	EINECS No.
13707-65-8	237-243-9

Empirical Formula:
$C_{12}H_{10}O \cdot K$

Definition: Potassium o-Phenylphenate is the potassium salt of o-phenylphenol that conforms to the formula:

Information Sources: EEC(VI/1-7), TSCA

Chemical Classes: Organic Salts; Phenols

Functions: Cosmetic Biocide; Preservative

Technical/Other Names:
(1,1'-Biphenyl)-2-ol, Potassium Salt
o-Phenylphenol, Potassium Salt
Potassium o-Phenylphenol
Potassium 2-Phenylphenoxide

POTASSIUM PHOSPHATE

CAS Nos.	EINECS Nos.
7778-77-0	231-913-4
16068-46-5	240-213-8

JPN Translation:
リン酸 K

Empirical Formula:
$H_3O_4P \cdot K$

Definition: Potassium Phosphate is the inorganic salt that conforms generally to the formula:

$$KH_2PO_4$$

Information Sources: 21CFR160.110, 21CFR175.105, 21CFR610.12, CTFA D, FCC, JCLS, JSCI, MAR, MI-13(7744), NF XIX, TSCA, USAN

Chemical Classes: Inorganic Salts; Phosphorus Compounds

Function: pH Adjuster

Reported Product Categories: Eye Makeup Removers; Bath Oils, Tablets, and Salts; Cleansing Products (Cold Creams, Cleansing Lotions, Liquids and Pads); Moisturizing Preparations; Skin Care Preparations, Misc.

Technical/Other Names:
Monobasic Potassium Phosphate
Phosphoric Acid, Monopotassium Salt
Potassium Phosphate, Monobasic

Trade Name Mixtures:
Afron 22 (Vevy)
Afron-A (Vevy)
Afron-LS (Vevy)
Afron-N (Vevy)
Tensioplastidina Avena (Vevy)
Vegetal Tensor (Inocosm)

POTASSIUM POLYACRYLATE

CAS No.: 25608-12-2

JPN Translation:
ポリアクリル酸 K

Empirical Formula:
$(C_3H_4O_2)_x \cdot xK$

Definition: Potassium Polyacrylate is the potassium salt of Polyacrylic Acid (q.v.).

Information Sources: CIR: [SQ] IJT 21 (SUPPL. 3) 2002, JCLS

Chemical Class: Synthetic Polymers

Functions: Emulsion Stabilizer; Viscosity Increasing Agent - Aqueous

Technical/Other Names:
Polyacrylic Acid, Potassium Salt
2-Propenoic Acid, Homopolymer, Potassium Salt

POTASSIUM POLYPHOSPHATE

CAS No.	EINECS No.
68956-75-2	273-317-7

Definition: Potassium Polyphosphate is the potassium salt of polyphosphoric acid.

Information Source: TSCA

Chemical Classes: Inorganic Salts; Phosphorus Compounds

Function: Chelating Agent

Technical/Other Name:
Polyphosphoric Acids, Potassium Salts

Trade Name:
Kalipol 18 (Albright & Wilson UK)

Trade Name Mixtures:
Lipothix 100-B (Lipo)
Lipothix 200-S (Lipo)

POTASSIUM PROPIONATE

CAS No.	EINECS No.
327-62-8	206-323-5

Empirical Formula:
$C_3H_6O_2 \cdot K$

Definition: Potassium Propionate is the potassium salt of propionic acid that conforms to the formula:

$$CH_3CH_2COOK$$

Information Source: EEC(VI/1-2)

Chemical Class: Organic Salts

Function: Preservative

Technical/Other Names:
Potassium Propanoate
Propanoic Acid, Potassium Salt

POTASSIUM PROPYLPARABEN

Empirical Formula:
$C_{10}H_{12}O_3 \cdot K$

Definition: Potassium Propylparaben is the potassium salt of Propylparaben (q.v.) that conforms to the formula:

Information Source: EEC(VI/1-12)

Chemical Classes: Esters; Organic Salts; Phenols

Function: Preservative

Technical/Other Name:
Propylparaben, Potassium Salt

POTASSIUM RAPESEEDATE

Definition: Potassium Rapeseedate is the potassium salt of the fatty acids derived from Brassica Campestris (Rapeseed) Oil (q.v.).

Chemical Class: Soaps

Functions: Surfactant - Cleansing Agent; Surfactant - Emulsifying Agent; Viscosity Increasing Agent - Nonaqueous

Technical/Other Name:
Fatty Acids, Rapeseed Oil, Potassium Salt

Trade Name:
PRISAVON KRP (Uniqema Europe)

POTASSIUM RICINOLEATE

CAS No.	EINECS No.
7492-30-0	231-314-8

Empirical Formula:
$C_{18}H_{34}O_3 \cdot K$

Definition: Potassium Ricinoleate is the potassium salt of ricinoleic acid. It conforms generally to the formula:

$$CH_3(CH_2)_5CHCH_2CH = CH(CH_2)_7COOK$$
$$| $$
$$OH$$

Information Sources: 21CFR175.300, TSCA

Chemical Class: Soaps

Functions: Surfactant - Cleansing Agent; Surfactant - Emulsifying Agent

Reported Product Category: Hair Dyes and Colors (All Types Requiring Caution Statements and Patch Tests)

Technical/Other Names:
12-Hydroxy-9-Octadecenoic Acid, Mono-potassium Salt
9-Octadecenoic Acid, 12-Hydroxy-, Mono-potassium Salt

POTASSIUM SAFFLOWERATE

Definition: Potassium Safflowerate is the potassium salt of the fatty acids derived from Carthamus Tinctorius (Safflower) Seed Oil (q.v.).

Chemical Class: Soaps

Function: Surfactant - Cleansing Agent

POTASSIUM SALICYLATE

CAS No. 578-36-9 **EINECS No.** 209-421-6

Empirical Formula:
$C_7H_6O_3 \cdot K$

Definition: Potassium Salicylate is the potassium salt of Salicylic Acid (q.v.) that conforms to the formula:

Information Sources: CIR: [SQ], EEC(VI/1-3), MHLW-331/3, MI-13(7750)

Chemical Classes: Organic Salts; Phenols

Functions: Cosmetic Biocide; Preservative

Technical/Other Names:
Benzoic Acid, 2-Hydroxy-, Potassium Salt
Potassium 2-Hydroxybenzoate
Salicylic Acid, Potassium Salt

POTASSIUM SILICATE

CAS No. 1312-76-1 **EINECS No.** 215-199-1

Definition: Potassium Silicate is a potassium salt of silicic acid.

Information Sources: CIR: [SQ], MI-13(7753), TSCA

Chemical Class: Inorganic Salts

Function: Corrosion Inhibitor

Technical/Other Name:
Silicic Acid, Potassium Salt

POTASSIUM SODIUM TARTRATE

CAS Nos. 304-59-6, 6381-59-5 **EINECS No.** 206-156-8

Empirical Formula:
$C_4H_6O_6 \cdot KNa$

Definition: Potassium Sodium Tartrate is the organic salt that conforms to the formula:

$$KOOCCHCHCOONa \cdot 4H_2O$$ (with OH on each CH)

Information Sources: AUS, BPC, 21CFR133.169, 21CFR133.173, 21CFR133.179, 21CFR150.141, 21CFR150.161, 21CFR184.1804, CZE, EGY, FCC, HUN, IND, ITA, KOR, MAR, MEX, MI-13(7755), OTC-I-AA, PF, TSCA, USAN, USD, USP XXIV

Chemical Class: Organic Salts

Function: Not Reported

Technical/Other Names:
Butanedioic Acid, 2,3-Dihydroxy-, Monopotassium Monosodium Salt
2,3-Dihydroxybutanedioic Acid, Monopotassium Monosodium Salt
Rochelle Salt

POTASSIUM SORBATE

CAS Nos. 590-00-1, 24634-61-5 **EINECS No.** 246-376-1

JPN Translation:
ソルビン酸K

Empirical Formula:
$C_6H_8O_2 \cdot K$

Definition: Potassium Sorbate is the organic salt that conforms generally to the formula:

$$CH_3CH=CHCH=CHCOOK$$

Information Sources: BPC, 21CFR133, 21CFR150.141, 21CFR150.161, 21CFR166.110, 21CFR182.90, 21CFR182.3640, CIR: [S] JACT-7(6)1988, CTFA S, EEC(VI/1-4), FCC, ITA, JCLS, JSCI, MAR, MHLW-331/3, MI-13(7756), NF XIX, RIFM, TSCA, USAN

Chemical Class: Organic Salts

Functions: Fragrance Ingredient; Preservative

Reported Product Categories: Hair Dyes and Colors (All Types Requiring Caution Statements and Patch Tests); Bath Oils, Tablets, and Salts; Moisturizing Preparations; Bath Capsules; Face and Neck Preparations (Excluding Shaving Preparations); Aftershave Lotions; Baby Shampoos; Cleansing Products (Cold Creams, Cleansing Lotions, Liquids and Pads); Skin Care Preparations, Misc.; Powders (Dusting and Talcum, Excluding Aftershave Talcs); Paste Masks (Mud Packs); Bath Preparations, Misc.; Body and Hand Preparations (Excluding Shaving Preparations); Hair Conditioners; Night Skin Care Preparations; Face Powders; Fragrance Preparations, Misc.; Tonics, Dressings, and Other Hair Grooming Aids; Eye Makeup Removers; Foundations; Baby Products, Misc.; Shampoos (Non-coloring)

Technical/Other Names:
2,4-Hexadienoic Acid, Potassium Salt
Potassium sorbate (RIFM)

Trade Names:
AEC Potassium Sorbate (A & E Connock)
Jeen Potassium Sorbate USP (Jeen)
Nutrinova Potassium Sorbate (Nutrinova)
Potassium sorbate granular material DAB (Merck KGaA)
Potassium Sorbate PC (Protameen)
Tristat K (Tri-K)
Unistat K (Universal Preserv-A-Chem)

Trade Name Mixture:
Euxyl K 700 (Schulke & Mayr)

POTASSIUM SOYATE

Definition: Potassium Soyate is the potassium salt of the fatty acids derived from Glycine Soja (Soybean) Oil (q.v.).

Chemical Class: Soaps

Functions: Surfactant - Cleansing Agent; Surfactant - Emulsifying Agent; Viscosity Increasing Agent - Nonaqueous

POTASSIUM STEARATE

CAS No. 593-29-3 **EINECS No.** 209-786-1

JPN Translation:
ステアリン酸K

Empirical Formula:
$C_{18}H_{36}O_2 \cdot K$

Definition: Potassium Stearate is the potassium salt of stearic acid. It conforms generally to the formula:

$$CH_3(CH_2)_{16}COOK$$

Information Sources: 21CFR172.615, 21CFR172.863, 21CFR173.340, 21CFR175.105, 21CFR175.300, 21CFR176.170, 21CFR176.200, 21CFR176.210, 21CFR177.1200, 21CFR177.2260, 21CFR177.2600, 21CFR177.2800, 21CFR178.3910, 21CFR179.45, 21CFR181.22, 21CFR181.29, CIR: [S] JACT-1(2)1982, JCIC, JCLS, MI-13(7758), TSCA

Chemical Class: Soaps

Functions: Surfactant - Cleansing Agent; Surfactant - Emulsifying Agent

Reported Product Categories: Bath Oils, Tablets, and Salts; Cleansing Products (Cold Creams, Cleansing Lotions, Liquids and Pads); Bath Preparations, Misc.; Body and Hand Preparations (Excluding Shaving Preparations); Moisturizing Preparations; Shaving Cream (Aerosol, Brushless and Lather); Skin Care Preparations, Misc.; Mascara

Technical/Other Names:
Octadecanoic Acid, Potassium Salt
Potassium Octadecanoate

Trade Name:
AEC Potassium Stearate (A & E Connock)

Trade Name Mixtures:
Ceral TK (Fabriquimica)
Cutina GMS SE C (Cognis Deutschland)
Emulgade CL Special (Cognis Deutschland)

POTASSIUM STEAROYL GLUTAMATE

JPN Translation:
ステアロイルグルタミン酸K

Definition: Potassium Stearoyl Glutamate is the potassium salt of Stearoyl Glutamic Acid (q.v.).

Information Source: JCLS

Chemical Class: Amino Acids

Functions: Hair Conditioning Agent; Skin-Conditioning Agent - Miscellaneous

POTASSIUM STEAROYL HYDROLYZED COLLAGEN

JPN Translation:
ステアロイル加水分解コラーゲンK

Definition: Potassium Stearoyl Hydrolyzed Collagen is the potassium salt of the condensation product of stearic acid chloride and Hydrolyzed Collagen (q.v.).

Information Sources: JCIC, JCLS

Chemical Class: Protein Derivatives

Functions: Hair Conditioning Agent; Skin-Conditioning Agent - Miscellaneous; Surfactant - Cleansing Agent

Technical/Other Names:
Potassium Stearoyl Hydrolyzed Collagen Solution
Proteins, Hydrolysates, Reaction Products with Stearoyl Chloride, Compounds with Potassium

Trade Name:
Promois ESP (Seiwa Kasei)

POTASSIUM SULFATE

CAS No.
7778-80-5

EINECS No.
231-915-5

JPN Translation:
硫酸K

Empirical Formula:
$H_2O_4S \cdot 2K$

Definition: Potassium Sulfate is the inorganic salt that conforms to the formula:

$$K_2SO_4$$

Information Sources: ARG, AUS, 21CFR184.1643, 21CFR558.311, DDR, EGY, FCC, FIN, HP, HUN, JAN, JCLS, JP, MAR, MI-13(7759), PF, PN, POR, TSCA, YUG

Chemical Class: Inorganic Salts

Function: Viscosity Increasing Agent - Aqueous

Technical/Other Name:
Sulfuric Acid, Dipotassium Salt

POTASSIUM SULFIDE

CAS Nos.
1312-73-8
37248-34-3

EINECS No.
215-197-0

Empirical Formula:
K_2S

Definition: Potassium Sulfide is the inorganic salt that conforms to the formula:

$$K_2S$$

Information Sources: EEC(III/1-23), MI-13 (7760), TSCA

Chemical Class: Inorganic Salts

Function: Depilating Agent

POTASSIUM SULFITE

CAS Nos.
10117-38-1
23873-77-0

EINECS No.
233-321-1

Empirical Formula:
$H_2O_3S \cdot 2K$

Definition: Potassium Sulfite is the inorganic salt that conforms to the formula:

$$K_2SO_3$$

Information Sources: CIR: [S], EEC(VI/1-9), FCC, MI-13(7761), POR, TSCA

Chemical Class: Inorganic Salts

Functions: Antioxidant; Hair-Waving/Straightening Agent; Reducing Agent

Technical/Other Name:
Sulfurous Acid, Potassium Salt

POTASSIUM TALLATE

CAS No.: 61790-44-1

Definition: Potassium Tallate is the potassium salt of Tall Oil Acid (q.v.).

Information Sources: 21CFR176.170, 21CFR177.1200, 21CFR177.2600, 21CFR177.2800, 21CFR178.3910

Chemical Class: Soaps

Functions: Surfactant - Cleansing Agent; Surfactant - Emulsifying Agent

Technical/Other Name:
Tall Oil Acid, Potassium Salt

Trade Name Mixture:
Akypogene ZA 97 SP (Kao GmbH)

POTASSIUM TALLOWATE

CAS No.
61790-32-7

EINECS No.
263-124-6

Definition: Potassium Tallowate is the potassium salt of Tallow Acid (q.v.).

Information Sources: 21CFR175.105, 21CFR176.170, 21CFR176.200, 21CFR177.1200, 21CFR177.2260, 21CFR177.2600, 21CFR177.2800, 21CFR178.3910, TSCA

Chemical Class: Soaps

Functions: Surfactant - Cleansing Agent; Surfactant - Emulsifying Agent

Reported Product Category: Bath Soaps and Detergents

Technical/Other Names:
Fatty Acids, Tallow, Potassium Salts
Tallow Fatty Acids, Potassium Salts

POTASSIUM TARTRATE

CAS No.
921-53-9

EINECS No.
213-067-8

JPN Translation:
酒石酸K

Definition: Potassium Tartrate is the potassium salt of Tartaric Acid (q.v.). It conforms to the formula:

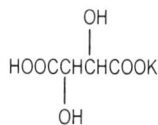

Information Sources: JCLS, MI-13(7762)

Chemical Class: Organic Salts

Function: pH Adjuster

POTASSIUM TAURATE

Empirical Formula:
$C_2H_7NSO_3K$

Definition: Potassium Taurate is the organic salt that conforms to the formula:

$NH_2CH_2CH_2SO_3K$

Chemical Classes: Organic Salts; Sulfonic Acids

Functions: Surfactant - Cleansing Agent; Surfactant - Foam Booster

Technical/Other Name:
2-Aminoethanesulfonic Acid, Potassium Salt

POTASSIUM TAURINE LAURATE

Definition: Potassium Taurine Laurate is the organic salt that conforms to the formula:

$$CH_3(CH_2)_{10}COOH \cdot NH_2CH_2CH_2SO_3K$$

Chemical Classes: Organic Salts; Sulfonic Acids

Functions: Surfactant - Cleansing Agent; Surfactant - Emulsifying Agent

Technical/Other Name:
Dodecanoic Acid, 2-Aminoethanesulfonic, Potassium Salt

Trade Name:
L-TK (NOF)

POTASSIUM TETRATHIONATE

CAS No. 13932-13-3
EINECS No. 237-702-3

Empirical Formula:
$K_2O_6S_4$

Definition: Potassium Tetrathionate is the inorganic salt that conforms to the formula:

$$K_2O_6S_4$$

Chemical Class: Inorganic Salts

Function: Proprietary

Technical/Other Names:
Dipotassium Tetrathionate
Tetrathionic Acid, Dipotassium Salt

POTASSIUM THIOCYANATE

CAS No. 333-20-0
EINECS No. 206-370-1

Empirical Formula:
$CHNS \cdot K$

Definition: Potassium Thiocyanate is the inorganic salt that conforms to the formula:

$$KSCN$$

Information Source: MI-13(7776)

Chemical Class: Inorganic Salts

Function: Not Reported

Reported Product Category: Moisturizing Preparations

Technical/Other Name:
Thiocyanic Acid, Potassium Salt

Trade Name Mixture:
SB-12 (Sederma)

POTASSIUM THIOGLYCOLATE

CAS No. 34452-51-2
EINECS No. 252-038-4

Empirical Formula:
$C_2H_4O_2S \cdot K$

Definition: Potassium Thioglycolate is the organic salt that conforms to the formula:

$$HSCH_2COOK$$

Information Source: EEC(III/1-2a)

Chemical Classes: Organic Salts; Thio Compounds

Functions: Depilating Agent; Hair-Waving/Straightening Agent; Reducing Agent

Reported Product Categories: Bubble Baths; Depilatories

Technical/Other Names:
Acetic Acid, Mercapto-, Monopotassium Salt
Mercaptoacetic Acid, Monopotassium Salt
Potassium Mercaptoacetate
Potassium Thioglycollate

POTASSIUM TOLUENESULFONATE

CAS No. 16106-44-8
EINECS No. 240-273-5

Empirical Formula:
$C_7H_8O_3S \cdot K$

Definition: Potassium Toluenesulfonate is the substituted aromatic compound that conforms generally to the formula:

Information Source: TSCA

Chemical Class: Alkyl Aryl Sulfonates

Function: Surfactant - Hydrotrope

Technical/Other Names:
Benzenesulfonic Acid, 4-Methyl-, Potassium Salt
4-Methylbenzenesulfonic Acid, Potassium Salt
Potassium Tosylate

Trade Name:
Eltesol PT40 (Albright & Wilson UK)

POTASSIUM TRIDECETH-3 CARBOXYLATE

Definition: Potassium Trideceth-3 Carboxylate is the potassium salt of Trideceth-3 Carboxylic Acid (q.v.).

Chemical Class: Organic Salts

Function: Surfactant - Cleansing Agent

POTASSIUM TRIDECETH-4 CARBOXYLATE

Definition: Potassium Trideceth-4 Carboxylate is the potassium salt of Trideceth-4 Carboxylic Acid (q.v.).

Chemical Class: Organic Salts

Function: Surfactant - Cleansing Agent

POTASSIUM TRIDECETH-7 CARBOXYLATE

Definition: Potassium Trideceth-7 Carboxylate is the potassium salt of Trideceth-7 Carboxylic Acid (q.v.).

Chemical Class: Organic Salts

Function: Surfactant - Cleansing Agent

POTASSIUM TRIDECETH-15 CARBOXYLATE

Definition: Potassium Trideceth-15 Carboxylate is the potassium salt of Trideceth-15 Carboxylic Acid (q.v.).

Chemical Class: Organic Salts

Function: Surfactant - Cleansing Agent

POTASSIUM TRIDECETH-19 CARBOXYLATE

Definition: Potassium Trideceth-19 Carboxylate is the potassium salt of Trideceth-19 Carboxylic Acid (q.v.).

Chemical Class: Organic Salts

Function: Surfactant - Cleansing Agent

POTASSIUM TRIDECETH-6 PHOSPHATE

Definition: Potassium Trideceth-6 Phosphate is the potassium salt of Trideceth-6 Phosphate (q.v.).

Chemical Classes: Alkoxylated Alcohols; Esters; Phosphorus Compounds

Function: Surfactant - Cleansing Agent

Technical/Other Name:
Potassium Polyoxyethylene (6) Tridecyl Ether Phosphate

Trade Name:
Ethfac 353 (Ethox)

POTASSIUM TRIDECETH-7 PHOSPHATE

Definition: Potassium Trideceth-7 Phosphate is the potassium salt of a complex mixture of esters of Trideceth-7 (q.v.) and phosphoric acid.

Chemical Classes: Alkoxylated Alcohols; Phosphorus Compounds

Function: Surfactant - Cleansing Agent

Trade Name:
AquaPhos TD (Arch Personal Care Products)

POTASSIUM TRISPHOSPHONO-METHYLAMINE OXIDE

CAS No.
15834-10-3

EINECS No.
239-940-3

Empirical Formula:
$C_3H_{12}NO_{10}P_3 \cdot 3K$

Definition: Potassium Trisphosphono-methylamine Oxide is the amine oxide that conforms to the formula:

$$3K \cdot \quad HO_3PCH_2N \longrightarrow O$$

Chemical Classes: Amine Oxides; Phosphorus Compounds

Function: Chelating Agent

Technical/Other Name:
[Nitrilotris(Methylene)Trisphosphonic Acid N-Oxide

Trade Name:
Sequion CLR (Bozzetto)

POTASSIUM TROCLOSENE

CAS No.
2244-21-5

EINECS No.
218-828-8

Empirical Formula:
$C_3HCl_2N_3O_3 \cdot K$

Definition: Potassium Troclosene is the heterocyclic compound that conforms to the formula:

Information Sources: INN, MI-13(9836), TSCA, USAN

Chemical Classes: Heterocyclic Compounds; Organic Salts

Function: Cosmetic Biocide

Technical/Other Names:
1,3-Dichloro-1,3,5-Triazine-2,4,6(1H,3H, 5H)-Trione, Potassium Salt
Potassium Dichloroisocyanurate
1,3,5-Triazine-2,4,6(1H,3H,5H)-Trione, 1,3-Dichloro-, Potassium Salt
Troclosene Potassium

POTASSIUM TSUBAKIATE

Definition: Potassium Tsubakiate is the potassium salt of the fatty acids derived from tsubaki oil.

Chemical Class: Soaps

Functions: Surfactant - Cleansing Agent; Surfactant - Emulsifying Agent

Trade Name:
Tsubaki Soap (Oshima)

POTASSIUM UNDECYLENATE

CAS No.: 6159-41-7

Definition: Potassium Undecylenate is the potassium salt of Undecylenic Acid (q.v.).

Information Source: EEC(VI/1-18)

Chemical Class: Soaps

Functions: Surfactant - Cleansing Agent; Surfactant - Emulsifying Agent

Technical/Other Name:
Undecylenic acid, Potassium Salt

POTASSIUM UNDECYLENOYL ALGINATE

Definition: Potassium Undecylenoyl Alginate is the potassium salt of the condensation product of undecylenic acid chloride and Alginic Acid (q.v.).

Chemical Classes: Gums, Hydrophilic Colloids and Derivatives; Organic Salts

Functions: Emulsion Stabilizer; Hair Conditioning Agent; Skin-Conditioning Agent - Miscellaneous

Trade Name:
Lifidrem XLUK (Coletica SA)

POTASSIUM UNDECYLENOYL CARRA-GEENAN

Definition: Potassium Undecylenoyl Carrageenan is the potassium salt of the condensation product of undecylenic acid chloride and Carrageenan (q.v.).

Chemical Class: Gums, Hydrophilic Colloids and Derivatives

Functions: Emulsion Stabilizer; Hair Conditioning Agent; Skin-Conditioning Agent - Miscellaneous

Trade Name:
Lifidrem ARUK (Coletica SA)

POTASSIUM UNDECYLENOYL GLUTAMATE

Definition: Potassium Undecylenoyl Glutamate is the substituted amino acid that conforms to the formula:

$$CH_2 = CH(CH_2)_8C - NHCH(CH_2)_2COOK$$
$$\underset{COOH}{|}$$

Chemical Class: Amino Acids

Functions: Abrasive; Hair Conditioning Agent

POTASSIUM UNDECYLENOYL HYDROLYZED COLLAGEN

CAS No.: 68951-92-8

JPN Translation:
ウンデシレノイル加水分解コラーゲンK

Definition: Potassium Undecylenoyl Hydrolyzed Collagen is the potassium salt of the condensation product of undecylenic acid chloride and Hydrolyzed Collagen (q.v.).

Information Sources: JCIC, JCLS, JSQI, TSCA

Chemical Class: Protein Derivatives

Functions: Hair Conditioning Agent; Skin-Conditioning Agent - Miscellaneous; Surfactant - Cleansing Agent

Reported Product Category: Shampoos (Non-coloring)

Technical/Other Names:
Potassium Undecylenoyl Hydrolyzed
Animal Protein
Proteins, Hydrolysates, Reaction Products
with 10-Undecenoyl Chloride, Potassium
Salts

Trade Names:
Liprot UK (Fabriquimica)
Promois EUP (Seiwa Kasei)
Promois EUUP (Seiwa Kasei)

POTASSIUM UNDECYLENOYL HYDROLYZED CORN PROTEIN

Definition: Potassium Undecylenoyl
Hydrolyzed Corn Protein is the potassium
salt of the condensation product of undecylenic chloride and Hydrolyzed Corn Protein
(q.v.).

Chemical Class: Protein Derivatives

Functions: Emulsion Stabilizer; Hair
Conditioning Agent; Skin-Conditioning Agent
- Miscellaneous

Trade Name Mixture:
Lifidrem PVUK (Coletica SA)

POTASSIUM UNDECYLENOYL HYDROLYZED RICE PROTEIN

Definition: Potassium Undecylenoyl
Hydrolyzed Rice Protein is the potassium salt
of the condensation product of undecylenic
acid chloride and Hydrolyzed Rice Protein
(q.v.).

Chemical Class: Protein Derivatives

Functions: Hair Conditioning Agent; Skin-
Conditioning Agent - Miscellaneous;
Surfactant - Cleansing Agent

Trade Name:
Protelan AG 11/R (Zschimmer & Schwarz
Italiana)

POTASSIUM UNDECYLENOYL HYDROLYZED SOY PROTEIN

Definition: Potassium Undecylenoyl
Hydrolyzed Soy Protein is the potassium salt
of the condensation product of undecylenic
acid chloride and Hydrolyzed Soy Protein
(q.v.).

Chemical Class: Protein Derivatives

Functions: Emulsion Stabilizer; Hair
Conditioning Agent; Skin-Conditioning Agent
- Miscellaneous

Trade Name:
Undecilenoilpolipeptide Di Soja (Sinerga)

Trade Name Mixture:
Lifidrem PVUK (Coletica SA)

POTASSIUM UNDECYLENOYL HYDROLYZED WHEAT PROTEIN

Definition: Potassium Undecylenoyl
Hydrolyzed Wheat Protein is the potassium
salt of the condensation product of undecylenic acid chloride and Hydrolyzed Wheat
Protein (q.v.).

Chemical Class: Protein Derivatives

Functions: Emulsion Stabilizer; Hair
Conditioning Agent; Skin-Conditioning Agent
- Miscellaneous

Trade Names:
Lifidrem BLUK (Coletica SA)
Protelan AG 11 (Zschimmer & Schwarz
Italiana)

POTASSIUM XYLENE SULFONATE

CAS No.   **EINECS No.**
30346-73-7   250-140-3

Empirical Formula:
$C_8H_{10}O_3S \cdot K$

Definition: Potassium Xylene Sulfonate is
the potassium salt of xylene sulfonic acid that
conforms to the formula:

Chemical Class: Sulfonic Acids

Function: Surfactant - Hydrotrope

Reported Product Category: Bath Soaps
and Detergents

Technical/Other Names:
Benzenesulfonic Acid, Dimethyl-,
Potassium Salt
Xylene Sulfonic Acid, Potassium Salt

Trade Names:
Eltesol PX40 (Albright & Wilson UK)
Eltesol PX93 (Albright & Wilson UK)

POTATO STARCH MODIFIED

Definition: Potato Starch Modified is the
ether formed from the reaction of
haloethylaminodipropionic acid and potato
starch in which the degree of substitution per
glucose unit is less than 0.1.

Information Sources: 21CFR172.892,
21CFR178.3520

Chemical Classes: Carbohydrates; Ethers

Function: Viscosity Increasing Agent -
Aqueous

Trade Name:
Structure Solanace (National Starch)

POTENTILLA ANSERINA EXTRACT

CAS No.   **EINECS No.**
85085-65-0   285-389-7

Definition: Potentilla Anserina Extract is
an extract of the herb of the wild agrimony,
Potentilla anserina. See *"Regulatory and
Ingredient Use Information,"* regarding the
labeling names for botanical ingredients in
Volume 1, Introduction, Part A.

Chemical Class: Biological Products

Function: Skin-Conditioning Agent - Miscellaneous

Technical/Other Names:
Extract of Potentilla Anserina
Extract of Wild Pansy
Wild Agrimony Extract
Wild Agrimony (Potentilla Anserina) Extract

Trade Name Mixtures:
Cinquefoil Extract HS 2436 G (Grau)
Herbasol Extract Oil Soluble Pontentilla
(Cosmetochem) (Cosmetochem
International Ltd.)
Potentilla Anserina Extract ies (IES LABO)
Silver Weed HS (Alban Muller)

POTENTILLA ERECTA EXTRACT

CAS No.   **EINECS No.**
90083-09-3   290-203-2

Definition: Potentilla Erecta Extract is an
extract of the whole plant, *Potentilla erecta*.
See *"Regulatory and Ingredient Use Information,"* regarding the labeling names for
botanical ingredients in Volume 1, Introduction, Part A.

Chemical Class: Biological Products

Function: Cosmetic Astringent

Reported Product Categories: Shampoos
(Non-coloring); Bath Preparations, Misc.; Hair
Conditioners; Tonics, Dressings, and Other
Hair Grooming Aids

Technical/Other Name:
Extract of Potentilla Erecta

Trade Name Mixture:
Actiphyte Tormentil (Active Organics)

POTENTILLA ERECTA ROOT EXTRACT

CAS No.	EINECS No.
90083-09-3	290-203-2

JPN Translation:
トルメンチラエキス

Definition: Potentilla Erecta Root Extract is an extract of the roots of the tormentil, *Potentilla erecta*. See "Regulatory and Ingredient Use Information," regarding the labeling names for botanical ingredients in Volume 1, Introduction, Part A.

Information Sources: JCIC, JCLS

Chemical Class: Biological Products

Function: Skin-Conditioning Agent - Miscellaneous

Reported Product Categories: Shampoos (Non-coloring); Bath Preparations, Misc.; Body and Hand Preparations (Excluding Shaving Preparations); Hair Conditioners; Tonics, Dressings, and Other Hair Grooming Aids

Technical/Other Names:
Extract of Potentilla Erecta
Extract of Tormentil
Tormentil Extract
Tormentilla Extract
Tormentil (Potentilla Erecta) Extract

Trade Name:
OPC de potentille (Berkem)

Trade Name Mixtures:
Connective Tissue Complex 268 (Ennagram)
Connective Tissue Phytogreen Complex GXH 268 (Phytochim)
Cremogen MZ (PN 739 032) (Haarmann & Reimer GmbH)
Cremogen MZ/N (PN 755 321) (Haarmann & Reimer GmbH)
Extrapure 2 Special 2/789480 (same as 2/032471) (Symrise)
Glycolysat of Tormentil (CEP (Solabia))
Herbalcomplex 2 special (Crodarom)
Herbaliquid Tormentil Root Special (Crodarom)
Herbasol-Extract Potentilla (Cosmetochem)
Herbasol-Extract Tormentill (Cosmetochem)
Hydroplastidine Potentilla (Vevy)
Nat Tormentil Extract (Natiris)
Oral Mucous Protection Complex MU 3776 (Greentech S.A)
Phytelene Complex EGX 253 (Indena SA)
Phytelene of Tormentil EG 470 liquid (Indena SA)
Phytogreen 55 of Tormentil EXH 684 Liquid (Phytochim)
Potentilla Erecta Root Extract ies (IES LABO)
Prodhy Extract Tormentille (Prod'Hyg)
285 Stimulant HS (Alban Muller)

Toothpaste Complex MU 3319 (Greentech S.A)
Tormentil Extract HG (Provital/Centerchem)
Tormentil HS (Alban Muller)
Tormentil Liquid B (Ichimaru Pharcos)
Tormentil Liquid E (Ichimaru Pharcos)
Tormentill Root Extract HS 2775 G (Grau)
Vegebios of Tormentil (CEP (Solabia))
2310 Vege-Plex Body Complex (Vege-Tech)
2325 Vege-Plex Body Complex (Vege-Tech)
2215 Vege-Plex Hair Complex (Vege-Tech)
2240 Vege-Plex Hair Complex Conditioner (Vege-Tech)
2400 Vege-Plex Skin Complex (Vege-Tech)
2500 Vege-Plex Skin Complex (Vege-Tech)
VT-044 Extract of Tormentil (Vege-Tech)

POTERIUM OFFICINALE

Definition: See "Regulatory and Ingredient Use Information," regarding EU labeling names for botanical ingredients in Volume 1, Introduction, Part A.

Chemical Class: Biological Products

Technical/Other Names:
Poterium Officinale Leaf Extract (U.S.)
Poterium Officinale Root Extract (U.S.)

POTERIUM OFFICINALE LEAF EXTRACT

Definition: Poterium Officinale Leaf Extract is an extract of the leaves of *Poterium officinale*. See "Regulatory and Ingredient Use Information," regarding the labeling names for botanical ingredients in Volume 1, Introduction, Part A.

Chemical Class: Biological Products

Function: Not Reported

Technical/Other Names:
Extract of Great Burnet Leaf
Extract of Poterium Officinale Leaf
Great Burnet (Poterium Officinale) Leaf Extract
Poterium Officinale (EU)

Trade Name Mixture:
Glycolysat BG of Garden Burnet (CEP (Solabia))

POTERIUM OFFICINALE ROOT EXTRACT

Definition: Poterium Officinale Root Extract is an extract of the roots of the

great burnet, *Poterium officinale*. See "Regulatory and Ingredient Use Information," regarding the labeling names for botanical ingredients in Volume 1, Introduction, Part A.

Information Sources: JCIC, JCLS, JSQI

Chemical Class: Biological Products

Function: Skin-Conditioning Agent - Miscellaneous

Technical/Other Names:
Burnet Extract
Extract of Great Burnet
Extract of Poterium Officinale
Great Burnet Extract
Great Burnet (Poterium Officinale) Extract
Poterium Officinale (EU)

Trade Name Mixtures:
Great Burnet HS (Alban Muller)
Sebustop (CEP (Solabia))

PPG-3

CAS No.: 25322-69-4 (Generic)

Definition: PPG-3 is the polymer of propylene oxide that conforms generally to the formula:

$$H(OCH_2CH)_nOH$$
$$|$$
$$CH_3$$

where n has an average value of 3.

Chemical Classes: Alkoxylated Alcohols; Polymeric Ethers

Function: Solvent

Reported Product Category: Hair Dyes and Colors (All Types Requiring Caution Statements and Patch Tests)

Technical/Other Names:
Polyoxypropylene (3)
Polypropylene Glycol (3)

Trade Name:
Newpol PP-200 (Sanyo Chemical)

PPG-7

CAS No.: 25322-69-4 (Generic)

JPN Translation:
PPG - 7

Definition: PPG-7 is the polymer of propylene oxide that conforms generally to the formula:

$$H(OCH_2CH)_nOH$$
$$|$$
$$CH_3$$

where n has an average value of 7.

Information Source: JCLS

Chemical Classes: Alkoxylated Alcohols; Polymeric Ethers

Functions: Skin-Conditioning Agent - Miscellaneous; Solvent

Reported Product Category: Hair Dyes and Colors (All Types Requiring Caution Statements and Patch Tests)

Technical/Other Names:
Polyoxypropylene (7)
Polypropylene Glycol (7)

PPG-9

CAS No.: 25322-69-4 (Generic)

JPN Translation:
PPG - 9

Empirical Formula:
$C_{27}H_{47}O_{10}$

Definition: PPG-9 is the polymer of propylene oxide that conforms generally to the formula:

$$H(OCH_2CH)_nOH$$
$$|$$
$$CH_3$$

where n has an average value of 9.

Information Sources: 21CFR173.310, 21CFR175.105, 21CFR175.300, 21CFR176.200, 21CFR176.210, CIR: [SQ] JACT-13(6)1994, JCLS, JSQI, TSCA

Chemical Classes: Alkoxylated Alcohols; Polymeric Ethers

Function: Skin-Conditioning Agent - Miscellaneous

Reported Product Categories: Shampoos (Non-coloring); Hair Dyes and Colors (All Types Requiring Caution Statements and Patch Tests); Deodorants (Underarm); Bath Preparations, Misc.

Technical/Other Names:
Polyoxypropylene (9)
Polypropylene Glycol (9)

Trade Names:
Polyglycol P-425 (Dow Chemical)
Unicol P-400 (Universal Preserv-A-Chem)

PPG-12

CAS No.: 25322-69-4 (Generic)

JPN Translation:
PPG - 12

Definition: PPG-12 is the polymer of propylene oxide that conforms generally to the formula:

$$H(OCH_2CH)_nOH$$
$$|$$
$$CH_3$$

where n has an average value of 12.

Information Sources: 21CFR173.310, 21CFR175.105, 21CFR176.200, 21CFR176.210, CIR: [SQ] JACT-13(6)1994, JCLS, JSQI, TSCA

Chemical Classes: Alkoxylated Alcohols; Polymeric Ethers

Function: Skin-Conditioning Agent - Miscellaneous

Reported Product Category: Hair Dyes and Colors (All Types Requiring Caution Statements and Patch Tests)

Technical/Other Names:
Polyoxypropylene (12)
Polypropylene Glycol (12)

PPG-13

CAS No.: 25322-69-4 (Generic)

JPN Translation:
PPG - 13

Definition: PPG-13 is the polymer of propylene oxide that conforms generally to the formula:

$$H(OCH_2CH)_nOH$$
$$|$$
$$CH_3$$

where n has an average value of 13.

Information Source: JCLS

Chemical Classes: Alkoxylated Alcohols; Polymeric Ethers

Function: Skin-Conditioning Agent - Miscellaneous

Reported Product Category: Hair Dyes and Colors (All Types Requiring Caution Statements and Patch Tests)

Technical/Other Names:
Polyoxypropylene (13)
Polypropylene Glycol (13)

PPG-15

CAS No.: 25322-69-4 (Generic)

JPN Translation:
PPG - 15

Definition: PPG-15 is the polymer of propylene oxide that conforms generally to the formula:

$$H(OCH_2CH)_nOH$$
$$|$$
$$CH_3$$

where n has an average value of 15.

Information Sources: 21CFR173.310, 21CFR175.105, 21CFR176.170, 21CFR176.200, 21CFR176.210, CIR: [SQ] JACT-13(6)1994, JCLS, JSQI, TSCA

Chemical Classes: Alkoxylated Alcohols; Polymeric Ethers

Function: Skin-Conditioning Agent - Miscellaneous

Reported Product Category: Hair Dyes and Colors (All Types Requiring Caution Statements and Patch Tests)

Technical/Other Names:
Polyoxypropylene (15)
Polypropylene Glycol (15)

PPG-16

CAS No.: 25322-69-4 (Generic)

Definition: PPG-16 is the polymer of propylene oxide that conforms generally to the formula:

$$H(OCH_2CH)_nOH$$
$$|$$
$$CH_3$$

where n has an average value of 16.

Chemical Classes: Alkoxylated Alcohols; Polymeric Ethers

Function: Skin-Conditioning Agent - Miscellaneous

Reported Product Category: Hair Dyes and Colors (All Types Requiring Caution Statements and Patch Tests)

Technical/Other Names:
Polyoxypropylene (16)
Polypropylene Glycol (16)

Trade Name:
Newpol PP-950 (Sanyo Chemical)

PPG-17

CAS No.: 25322-69-4 (Generic)

JPN Translation:
PPG - 17

Definition: PPG-17 is the polymer of propylene oxide that conforms generally to the formula:

$$H(OCH_2CH)_nOH$$
$$|$$
$$CH_3$$

where n has an average value of 17.

Information Sources: 21CFR173.310, 21CFR175.105, 21CFR176.170, 21CFR176.200, 21CFR176.210, CIR: [SQ] JACT-13(6)1994, JCLS, JSQI, TSCA

Chemical Classes: Alkoxylated Alcohols; Polymeric Ethers

Function: Skin-Conditioning Agent - Miscellaneous

Reported Product Category: Hair Dyes and Colors (All Types Requiring Caution Statements and Patch Tests)

Technical/Other Names:
Polyoxypropylene (17)
Polypropylene Glycol (17)

PPG-20

CAS No.: 25322-69-4 (Generic)

JPN Translation:
PPG - 20

Definition: PPG-20 is the polymer of propylene oxide that conforms generally to the formula:

$$H(OCH_2CH)_nOH$$
$$|$$
$$CH_3$$

where n has an average value of 20.

Information Sources: 21CFR173.310, 21CFR173.340, 21CFR175.105, 21CFR176.170, 21CFR176.200, 21CFR176.210, 21CFR178.3740, CIR: [SQ] JACT-13(6)1994, JCLS, JSQI, TSCA

Chemical Classes: Alkoxylated Alcohols; Polymeric Ethers

Function: Skin-Conditioning Agent - Miscellaneous

Reported Product Category: Hair Dyes and Colors (All Types Requiring Caution Statements and Patch Tests)

Technical/Other Names:
Polyoxypropylene (20)
Polypropylene Glycol (20)

Trade Names:
Polyglycol P-1200 (Dow Chemical)
Unicol 1200 (Universal Preserv-A-Chem)

PPG-26

CAS No.: 25322-69-4 (Generic)

JPN Translation:
PPG - 26

Definition: PPG-26 is the polymer of propylene oxide that conforms generally to the formula:

$$H(OCH_2CH)_nOH$$
$$|$$
$$CH_3$$

where n has an average value of 26.

Information Sources: 21CFR173.310, 21CFR173.340, 21CFR175.105, 21CFR176.170, 21CFR176.200, 21CFR176.210, 21CFR178.3740, CIR: [SQ] JACT-13(6)1994, CTFA D, JCLS, JSQI, TSCA

Chemical Classes: Alkoxylated Alcohols; Polymeric Ethers

Function: Skin-Conditioning Agent - Miscellaneous

Reported Product Category: Hair Dyes and Colors (All Types Requiring Caution Statements and Patch Tests)

Technical/Other Names:
Polyoxypropylene (26)
Polypropylene Glycol (26)

Trade Names:
JEFFOX PPG-2000 (Texaco)
Polyglycol P-2000 (Dow Chemical)
Unicol P-2000 (Universal Preserv-A-Chem)

PPG-30

CAS No.: 25322-69-4 (Generic)

JPN Translation:
PPG - 30

Definition: PPG-30 is the polymer of propylene oxide that conforms generally to the formula:

$$H(OCH_2CH)_nOH$$
$$|$$
$$CH_3$$

where n has an average value of 30.

Information Sources: 21CFR173.310, 21CFR173.340, 21CFR175.105, 21CFR176.170, 21CFR176.200, 21CFR178.3740, CIR: [SQ] JACT-13(6)-1994, JCLS, JSQI, TSCA

Chemical Classes: Alkoxylated Alcohols; Polymeric Ethers

Function: Skin-Conditioning Agent - Miscellaneous

Reported Product Category: Hair Dyes and Colors (All Types Requiring Caution Statements and Patch Tests)

Technical/Other Names:
Polyoxypropylene (30)
Polypropylene Glycol (30)

Trade Names:
Polyglycol P-4000 (Dow Chemical)
Unicol P-4000 (Universal Preserv-A-Chem)

PPG-33

CAS No.: 25322-69-4 (Generic)

JPN Translation:
PPG - 33

Definition: PPG-33 is a polymer of propylene oxide that conforms generally to the formula:

$$H(OCH_2CH)_nOH$$
$$|$$
$$CH_3$$

where n has an average value of 33.

Information Source: JCLS

Chemical Classes: Alkoxylated Alcohols; Polymeric Ethers

Functions: Skin-Conditioning Agent - Emollient; Skin-Conditioning Agent - Miscellaneous

Reported Product Category: Hair Dyes and Colors (All Types Requiring Caution Statements and Patch Tests)

Technical/Other Names:
Polyoxypropylene (33)
Polypropylene Glycol (33)

PPG-34

CAS No.: 25322-69-4 (Generic)

JPN Translation:
PPG - 34

Definition: PPG-34 is the polymer of propylene oxide that conforms generally to the formula:

$$H(OCH_2CH)_nOH$$
$$|$$
$$CH_3$$

where n has an average value of 34.

Information Sources: 21CFR173.310, 21CFR173.340, 21CFR175.105, 21CFR176.170, 21CFR176.200, 21CFR178.3740, CIR: [SQ] JACT-13(6)-1994, JCLS, JSQI, TSCA

Chemical Classes: Alkoxylated Alcohols; Polymeric Ethers

Function: Skin-Conditioning Agent - Miscellaneous

Reported Product Category: Hair Dyes and Colors (All Types Requiring Caution Statements and Patch Tests)

Technical/Other Names:
Polyoxypropylene (34)
Polypropylene Glycol (34)

PPG-51

CAS No.: 25322-69-4 (Generic)

JPN Translation:
PPG - 51

Definition: PPG-51 is a polymer of propylene oxide that conforms to the formula:

$$H(OCH_2CH)_nOH$$
$$|$$
$$CH_3$$

where n has an average value of 51.

Chemical Classes: Alkoxylated Alcohols; Polymeric Ethers

Functions: Skin-Conditioning Agent - Emollient; Skin-Conditioning Agent - Miscellaneous

Reported Product Category: Hair Dyes and Colors (All Types Requiring Caution Statements and Patch Tests)

Technical/Other Names:
Polyoxypropylene (51)
Polypropylene Glycol (51)

PPG-69

CAS No.: 25322-69-4 (Generic)

JPN Translation:
PPG - 69

Definition: PPG-69 is a polymer of propylene oxide that conforms generally to the formula:

$$H(OCH_2CH)_nOH$$
$$|$$
$$CH_3$$

where n has an average value of 69.

Information Source: JCLS

Chemical Classes: Alkoxylated Alcohols; Polymeric Ethers

Functions: Skin-Conditioning Agent - Emollient; Skin-Conditioning Agent - Miscellaneous

Reported Product Category: Hair Dyes and Colors (All Types Requiring Caution Statements and Patch Tests)

Technical/Other Name:
Polypropylene Glycol (69)

PPG-1 BEHENETH-15

Definition: PPG-1 Beheneth-15 is the polyoxypropylene, polyoxyethylene ether of behenyl alcohol that conforms generally to the formula:

$$CH_3(CH_2)_{21}(OCHCH_2)_x(OCH_2CH_2)_yOH$$
$$|$$
$$CH_3$$

where x has an average value of 1 and y has an average value of 15.

Chemical Class: Alkoxylated Alcohols

Functions: Skin-Conditioning Agent - Emollient; Surfactant - Emulsifying Agent

Technical/Other Names:
PEG-15 PPG-1 Behenyl Ether
Polyoxyethylene (15) Polyoxypropylene (1) Behenyl Ether
Polyoxypropylene (1) Polyoxyethylene (15) Behenyl Ether

Trade Name:
Pepol BEP-0115 (Toho)

PPG-10 BUTANEDIOL

Empirical Formula:
$C_{34}H_{70}O_{12}$

Definition: PPG-10 Butanediol is the polyoxypropylene ether of butanediol that conforms generally to the formula:

$$HO(CH_2CHO)_x(CH_2)_4(OCHCH_2)_yOH$$
$$|\qquad\qquad\qquad|$$
$$CH_3\qquad\qquad CH_3$$

where x+y has an average value of 10.

Chemical Class: Alkoxylated Alcohols

Function: Solvent

Reported Product Categories: Bath Oils, Tablets, and Salts; Deodorants (Underarm)

Trade Names:
Luvitol BD 10 P (BASF)
Probutyl DB-10 (Croda, Inc.)

PPG-2-BUTETH-1

Definition: PPG-2-Buteth-1 is the polyoxypropylene, polyoxyethylene ether of butyl alcohol that conforms generally to the formula:

$$C_4H_9(OCHCH_2)_x(OCH_2CH_2)_yOH$$
$$|$$
$$CH_3$$

where x has an average value of 2 and y has an average value of 1.

Chemical Class: Alkoxylated Alcohols

Functions: Hair Conditioning Agent; Skin-Conditioning Agent - Miscellaneous

Technical/Other Names:
Polyoxyethylene (1) Polyoxypropylene (2) Monobutyl Ether
Polypropylene (2) Polyoxyethylene (1) Monobutyl Ether

Trade Name:
Adeka Carpol MH-4 (Asahi Denka Kogyo)

PPG-2-BUTETH-2

CAS Nos.: 9038-95-3 (Generic); 9065-63-8 (Generic)

JPN Translation:
PPG - 2 ブテス - 2

Empirical Formula:
$C_{14}H_{30}O_5$

Definition: PPG-2-Buteth-2 is the polyoxypropylene, polyoxyethylene ether of butyl alcohol that conforms generally to the formula:

$$C_4H_9(OCHCH_2)_x(OCH_2CH_2)_yOH$$
$$|$$
$$CH_3$$

where x has an average value of 2 and y has an average value of 2.

Information Sources: 21CFR173.310, 21CFR173.340, 21CFR175.105, 21CFR176.210, 21CFR178.3570, JCLS, JSCI, RIFM, TSCA

Chemical Class: Alkoxylated Alcohols

Functions: Fragrance Ingredient; Hair Conditioning Agent; Skin-Conditioning Agent - Miscellaneous; Surfactant - Emulsifying Agent

Technical/Other Names:
Oxirane, methyl-, polymer with oxirane, monobutyl ether (RIFM)
Polyoxyethylene (2) Polyoxypropylene (2) Butyl Ether
Polyoxyethylene (2) Polyoxypropylene (2) Monobutyl Ether
Polyoxypropylene (2) Polyoxyethylene (2) Monobutyl Ether

PPG-2-BUTETH-3

CAS Nos.: 9038-95-3 (Generic); 9065-63-8 (Generic)

JPN Translation:
PPG - 2 ブテス - 3

Empirical Formula:
$C_{16}H_{34}O_6$

Definition: PPG-2-Buteth-3 is the polyoxypropylene, polyoxyethylene ether of butyl alcohol that conforms generally to the formula:

$$C_4H_9(OCHCH_2)_x(OCH_2CH_2)_yOH$$
$$|$$
$$CH_3$$

where x has an average value of 2 and y has an average value of 3.

Information Sources: 21CFR173.310, 21CFR173.340, 21CFR175.105, 21CFR176.210, 21CFR178.3570, CTFA D, JCLS, JSQI, RIFM, TSCA

Chemical Class: Alkoxylated Alcohols

Functions: Fragrance Ingredient; Hair Conditioning Agent; Skin-Conditioning Agent - Miscellaneous; Solvent

Technical/Other Names:
Oxirane,Methyl,Polymer and Oxibane, Butyl Ether
Oxirane, methyl-, polymer with oxirane, monobutyl ether (RIFM)
Polyoxyethylene (3) Polyoxypropylene (2) Monobutyl Ether
Polyoxypropylene (2) Polyoxyethylene (3) Monobutyl Ether
PPG Buteth-55

Trade Name:
UCON 50-HB-55 Lubricant Inh. (Dow Chemical)

PPG-3-BUTETH-5

CAS Nos.: 9038-95-3 (Generic); 9065-63-8 (Generic)

JPN Translation:
PPG - 3 ブテス - 5

Empirical Formula:
$C_{23}H_{48}O_9$

Definition: PPG-3-Buteth-5 is the polyoxypropylene, polyoxyethylene ether of butyl alcohol that conforms generally to the formula:

$$C_4H_9(OCHCH_2)_x(OCH_2CH_2)_yOH$$
$$|$$
$$CH_3$$

where x has an average value of 3 and y has as average value of 5.

Information Sources: 21CFR173.310, 21CFR175.105, 21CFR176.210, 21CFR178.3570, JCLS, JSQI, RIFM, TSCA

Chemical Class: Alkoxylated Alcohols

Functions: Fragrance Ingredient; Hair Conditioning Agent; Skin-Conditioning Agent - Miscellaneous; Solvent

Technical/Other Names:
Oxirane, methyl-, polymer with oxirane, monobutyl ether (RIFM)
Polyoxyethylene (5) Polyoxypropylene (3) Monobutyl Ether
Polyoxypropylene (3) Polyoxyethylene (5) Monobutyl Ether

Trade Name:
UCON 50-HB-100 Lubricant (Dow Chemical)

PPG-4-BUTETH-4

CAS Nos.: 9038-95-3 (Generic); 9065-63-8 (Generic)

JPN Translation:
PPG - 4 ブテス - 4

Empirical Formula:
$C_{24}H_{50}O_9$

Definition: PPG-4-Buteth-4 is the polyoxypropylene, polyoxyethylene ether of butyl alcohol that conforms generally to the formula:

$$C_4H_9(OCHCH_2)_x(OCH_2CH_2)_yOH$$
$$|$$
$$CH_3$$

where x has an average value of 4 and y has an average value of 4.

Information Sources: JCLS, JSCI, RIFM

Chemical Class: Alkoxylated Alcohols

Functions: Fragrance Ingredient; Hair Conditioning Agent; Skin-Conditioning Agent - Miscellaneous; Surfactant - Emulsifying Agent

Technical/Other Names:
Oxirane, methyl-, polymer with oxirane, monobutyl ether (RIFM)
Polyoxyethylene (4) Polyoxypropylene (4) Butyl Ether
Polyoxyethylene (4) Polyoxypropylene (4) Monobutyl Ether
Polyoxypropylene (4) Polyoxyethylene (4) Monobutyl Ether

PPG-5-BUTETH-5

CAS Nos.: 9038-95-3 (Generic); 9065-63-8 (Generic)

JPN Translation:
PPG - 5 ブテス - 5

Empirical Formula:
$C_{29}H_{60}O_{11}$

Definition: PPG-5-Buteth-5 is the polyoxypropylene, polyoxyethylene ether of butyl alcohol that conforms generally to the formula:

$$C_4H_9(OCHCH_2)_x(OCH_2CH_2)_yOH$$
$$|$$
$$CH_3$$

where x has an average value of 5 and y has an average value of 5.

Information Sources: JCLS, JSCI, RIFM

Chemical Class: Alkoxylated Alcohols

Functions: Fragrance Ingredient; Hair Conditioning Agent; Skin-Conditioning Agent - Miscellaneous; Surfactant - Emulsifying Agent

Technical/Other Names:
Oxirane, methyl-, polymer with oxirane, monobutyl ether (RIFM)

Polyoxyethylene (5) Polyoxypropylene (5) Butyl Ether
Polyoxyethylene (5) Polyoxypropylene (5) Monobutyl Ether
Polyoxypropylene (5) Polyoxyethylene (5) Monobutyl Ether

PPG-5-BUTETH-7

CAS Nos.: 9038-95-3 (Generic); 9065-63-8 (Generic)

JPN Translation:
PPG - 5 ブテス - 7

Empirical Formula:
$C_{33}H_{68}O_{13}$

Definition: PPG-5-Buteth-7 is the polyoxypropylene, polyoxyethylene ether of butyl alcohol that conforms generally to the formula:

$$C_4H_9(OCHCH_2)_x(OCH_2CH_2)_yOH$$
$$|$$
$$CH_3$$

where x has an average value of 5 and y has an average value of 7.

Information Sources: 21CFR173.310, 21CFR175.105, 21CFR176.210, 21CFR178.3570, JCLS, JSQI, RIFM, TSCA

Chemical Class: Alkoxylated Alcohols

Functions: Fragrance Ingredient; Hair Conditioning Agent; Skin-Conditioning Agent - Miscellaneous; Solvent

Technical/Other Names:
Oxirane, methyl-, polymer with oxirane, monobutyl ether (RIFM)
Polyoxyethylene (7) Polyoxypropylene (5) Monobutyl Ether
Polyoxypropylene (5) Polyoxyethylene (7) Monobutyl Ether

Trade Name:
UCON 50-HB-170 Lubricant (Dow Chemical)

PPG-7-BUTETH-4

CAS Nos.: 9038-95-3 (Generic); 9065-63-8 (Generic)

Definition: PPG-7-Buteth-4 is the polyoxypropylene, polyoxyethylene ether of butyl alcohol that conforms generally to the formula:

$$C_4H_9(OCHCH_2)_x(OCH_2CH_2)_yOH$$
$$|$$
$$CH_3$$

where x has an average value of 7 and y has an average value of 4.

Information Source: RIFM

Chemical Class: Alkoxylated Alcohols

Functions: Fragrance Ingredient; Hair Conditioning Agent; Skin-Conditioning Agent - Miscellaneous; Solvent; Surfactant - Emulsifying Agent

Technical/Other Names:
Oxirane, methyl-, polymer with oxirane, monobutyl ether (RIFM)
Polyoxyethylene (4) Polyoxypropylene (7) Butyl Ether
Polyoxypropylene (7) Polyoxyethylene (4) Butyl Ether

PPG-7-BUTETH-10

CAS Nos.: 9038-95-3 (Generic); 9065-63-8 (Generic)

JPN Translation:
PPG - 7 ブテス - 10

Empirical Formula:
$C_{45}H_{92}O_{18}$

Definition: PPG-7-Buteth-10 is the polyoxypropylene, polyoxyethylene ether of butyl alcohol that conforms generally to the formula:

$$C_4H_9(OCHCH_2)_x(OCH_2CH_2)_yOH$$
$$|$$
$$CH_3$$

where x has an average value of 7 and y has an average value of 10.

Information Sources: 21CFR173.310, 21CFR175.105, 21CFR176.210, 21CFR178.3570, JCLS, JSCI, RIFM, TSCA

Chemical Class: Alkoxylated Alcohols

Functions: Fragrance Ingredient; Hair Conditioning Agent; Skin-Conditioning Agent - Miscellaneous; Solvent; Surfactant - Emulsifying Agent

Technical/Other Names:
Oxirane, methyl-, polymer with oxirane, monobutyl ether (RIFM)
Polyoxyethylene (10) Polyoxypropylene (7) Butyl Ether
Polyoxyethylene (10) Polyoxypropylene (7) Monobutyl Ether
Polyoxypropylene (7) Polyoxyethylene (10) Monobutyl Ether
PPG Buteth-260

Trade Name:
UCON 50-HB-260 Lubricant (Dow Chemical)

PPG-9-BUTETH-12

CAS Nos.: 9038-95-3 (Generic); 9065-63-8 (Generic)

JPN Translation:
PPG - 9 ブテス - 12

Definition: PPG-9-Buteth-12 is the polyoxypropylene, polyoxyethylene ether of butyl alcohol that conforms generally to the formula:

$$C_4H_9(OCHCH_2)_x(OCH_2CH_2)_yOH$$
$$|$$
$$CH_3$$

where x has an average value of 9 and y has an average value of 12.

Information Sources: 21CFR173.310, 21CFR175.105, 21CFR176.210, 21CFR178.3570, CIR: [I] IJT-19(SUPPL.1)-2000, CIR: [S], JCLS, JSQI, RIFM, TSCA

Chemical Class: Alkoxylated Alcohols

Functions: Fragrance Ingredient; Hair Conditioning Agent; Skin-Conditioning Agent - Miscellaneous; Surfactant - Emulsifying Agent

Technical/Other Names:
Oxirane, methyl-, polymer with oxirane, monobutyl ether (RIFM)
Polyoxyethylene (12) Polyoxypropylene (9) Monobutyl Ether
Polyoxypropylene (9) Polyoxyethylene (12) Monobutyl Ether

PPG-10-BUTETH-9

CAS Nos.: 9038-95-3 (Generic); 9065-63-8 (Generic)

JPN Translation:
PPG - 10 ブテス - 9

Empirical Formula:
$C_{52}H_{106}O_{20}$

Definition: PPG-10-Buteth-9 is the polyoxypropylene, polyoxyethylene ether of butyl alcohol that conforms generally to the formula:

$$C_4H_9(OCHCH_2)_x(OCH_2CH_2)_yOH$$
$$|$$
$$CH_3$$

where x has an average value of 10 and y has an average value of 9.

Information Sources: JCLS, JSCI, RIFM

Chemical Class: Alkoxylated Alcohols

Functions: Fragrance Ingredient; Hair Conditioning Agent; Skin-Conditioning Agent - Miscellaneous; Surfactant - Emulsifying Agent

Technical/Other Names:
Oxirane, methyl-, polymer with oxirane, monobutyl ether (RIFM)
Polyoxyethylene (9) Polyoxypropylene (10) Butyl Ether
Polyoxyethylene (9) Polyoxypropylene (10) Monobutyl Ether

Polyoxypropylene (10) Polyoxyethylene (9) Monobutyl Ether

PPG-12-BUTETH-12

CAS Nos.: 9038-95-3 (Generic); 9065-63-8 (Generic)

JPN Translation:
PPG - 12 ブテス - 12

Definition: PPG-12-Buteth-12 is the polyoxypropylene, polyoxyethylene ether of butyl alcohol that conforms generally to the formula:

$$C_4H_9(OCHCH_2)_x(OCH_2CH_2)_yOH$$
$$|$$
$$CH_3$$

where x has an average value of 12 and y has an average value of 12.

Information Sources: JCLS, JSCI, RIFM

Chemical Class: Alkoxylated Alcohols

Functions: Fragrance Ingredient; Hair Conditioning Agent; Skin-Conditioning Agent - Miscellaneous; Surfactant - Emulsifying Agent

Technical/Other Names:
Oxirane, methyl-, polymer with oxirane, monobutyl ether (RIFM)
Polyoxyethylene (12) Polyoxypropylene (12) Butyl Ether
Polyoxyethylene (12) Polypropylene (12) Monobutyl Ether
Polyoxypropylene (12) Polyoxyethylene (12) Monobutyl Ether

PPG-12-BUTETH-16

CAS Nos.: 9038-95-3 (Generic); 9065-63-8 (Generic)

JPN Translation:
PPG - 12 ブテス - 16

Definition: PPG-12-Buteth-16 is the polyoxypropylene, polyoxyethylene ether of butyl alcohol that conforms generally to the formula:

$$C_4H_9(OCHCH_2)_x(OCH_2CH_2)_yOH$$
$$|$$
$$CH_3$$

where x has an average value of 12 and y has an average value of 16.

Information Sources: 21CFR173.310, 21CFR175.105, 21CFR176.210, 21CFR178.3570, CIR: [I] IJT-19(SUPPL.1)-2000, CIR: [S], CTFA D, JCLS, JSQI, RIFM, TSCA

Chemical Class: Alkoxylated Alcohols

Functions: Fragrance Ingredient; Hair Conditioning Agent; Skin-Conditioning Agent - Miscellaneous; Solvent; Surfactant - Emulsifying Agent

Reported Product Categories: Bath Preparations, Misc.; Bubble Baths; Shampoos (Non-coloring); Tonics, Dressings, and Other Hair Grooming Aids; Bath Soaps and Detergents; Hair Conditioners; Eyeliners; Indoor Tanning Preparations

Technical/Other Names:
Oxirane, methyl-, polymer with oxirane, monobutyl ether (RIFM)
Polyoxyethylene (16) Polyoxypropylene (12) Monobutyl Ether
Polyoxypropylene (12) Polyoxyethylene (16) Monobutyl Ether
PPG Buteth-600

Trade Name:
UCON Fluid 50-HB-660 (Amerchol)

PPG-15-BUTETH-20

CAS Nos.: 9038-95-3 (Generic); 9065-63-8 (Generic)

JPN Translation:
PPG - 15 ブテス - 20

Definition: PPG-15-Buteth-20 is the polyoxypropylene, polyoxyethylene ether of butyl alcohol that conforms generally to the formula:

$$C_4H_9(OCHCH_2)_x(OCH_2CH_2)_yOH$$
$$|$$
$$CH_3$$

where x has an average value of 15 and y has an average value of 20.

Information Sources: 21CFR173.310, 21CFR175.105, 21CFR176.210, 21CFR178.3570, JCLS, JSCI, RIFM, TSCA

Chemical Class: Alkoxylated Alcohols

Functions: Fragrance Ingredient; Hair Conditioning Agent; Skin-Conditioning Agent - Miscellaneous; Solvent; Surfactant - Emulsifying Agent

Technical/Other Names:
Oxirane, methyl-, polymer with oxirane, monobutyl ether (RIFM)
Polyoxyethylene (20) Polyoxypropylene (15) Butyl Ether
Polyoxyethylene (20) Polyoxypropylene (15) Monobutyl Ether
Polyoxypropylene (15) Polyoxyethylene (20) Monobutyl Ether
PPG Buteth-1000

PPG-17-BUTETH-17

CAS Nos.: 9038-95-3 (Generic); 9065-63-8 (Generic)

JPN Translation:
PPG - 17 ブテス - 17

Definition: PPG-17-Buteth-17 is the polyoxypropylene, polyoxyethylene ether of butyl alcohol that conforms generally to the formula:

$$C_4H_9(OCHCH_2)_x(OCH_2CH_2)_yOH$$
$$|$$
$$CH_3$$

where x has an average value of 17 and y has an average value of 17.

Information Sources: JCLS, JSCI, RIFM

Chemical Class: Alkoxylated Alcohols

Functions: Fragrance Ingredient; Hair Conditioning Agent; Skin-Conditioning Agent - Miscellaneous; Surfactant - Emulsifying Agent

Technical/Other Names:
Oxirane, methyl-, polymer with oxirane, monobutyl ether (RIFM)
Polyoxyethylene (17) Polyoxypropylene (17) Butyl Ether
Polyoxyethylene (17) Polyoxypropylene (17) Monobutyl Ether
Polyoxypropylene (17) Polyoxyethylene (17) Monobutyl Ether

PPG-19-BUTETH-19

CAS Nos.: 9038-95-3 (Generic); 9065-63-8 (Generic)

Definition: PPG-19-Buteth-19 is the polyoxypropylene, polyoxyethylene ether of butyl alcohol that conforms generally to the formula:

$$C_4H_9(OCHCH_2)_x(OCH_2CH_2)_yOH$$
$$|$$
$$CH_3$$

where x has an average value of 19 and y has an average value of 19.

Information Sources: JCLS, RIFM

Chemical Class: Alkoxylated Alcohols

Functions: Fragrance Ingredient; Hair Conditioning Agent; Skin-Conditioning Agent - Miscellaneous; Surfactant - Emulsifying Agent

Technical/Other Names:
Oxirane, methyl-, polymer with oxirane, monobutyl ether (RIFM)
PEG-19 PPG-19 Butyl Ether
Polyoxyethylene (19) Polyoxypropylene (19) Butyl Ether
Polyoxypropylene (19) Polyoxyethylene (19) Butyl Ether

Trade Name:
Oleotex SL-19 (Cesalpinia)

PPG-20-BUTETH-30

CAS Nos.: 9038-95-3 (Generic); 9065-63-8 (Generic)

JPN Translation:
PPG - 20 ブテス - 30

Definition: PPG-20-Buteth-30 is the polyoxypropylene, polyoxyethylene ether of butyl alcohol that conforms generally to the formula:

$$C_4H_9(OCHCH_2)_x(OCH_2CH_2)_yOH$$
$$|$$
$$CH_3$$

where x has an average value of 20 and y has an average value of 30.

Information Sources: 21CFR173.310, 21CFR175.105, 21CFR176.210, 21CFR178.3570, JCLS, JSQI, RIFM, TSCA

Chemical Class: Alkoxylated Alcohols

Functions: Fragrance Ingredient; Hair Conditioning Agent; Skin-Conditioning Agent - Miscellaneous; Solvent; Surfactant - Emulsifying Agent

Technical/Other Names:
Oxirane, methyl-, polymer with oxirane, monobutyl ether (RIFM)
Polyoxyethylene (30) Polyoxypropylene (20) Monobutyl Ether
Polyoxypropylene (20) Polyoxyethylene (30) Monobutyl Ether
PPG Buteth-2000

PPG-24-BUTETH-27

CAS Nos.: 9038-95-3 (Generic); 9065-63-8 (Generic)

JPN Translation:
PPG - 24 ブテス - 27

Definition: PPG-24-Buteth-27 is the polyoxypropylene, polyoxyethylene ether of butyl alcohol that conforms generally to the formula:

$$C_4H_9(OCHCH_2)_x(OCH_2CH_2)_yOH$$
$$|$$
$$CH_3$$

where x has an average value of 24 and y has an average value of 27.

Information Sources: 21CFR173.310, 21CFR175.105, 21CFR176.210, 21CFR178.3570, JCLS, JSQI, RIFM, TSCA

Chemical Class: Alkoxylated Alcohols

Functions: Fragrance Ingredient; Hair Conditioning Agent; Skin-Conditioning Agent - Miscellaneous; Surfactant - Emulsifying Agent

Technical/Other Names:
Oxirane, methyl-, polymer with oxirane, monobutyl ether (RIFM)

Polyoxyethylene (27) Polyoxypropylene (24) Monobutyl Ether
Polyoxypropylene (24) Polyoxyethylene (27) Monobutyl Ether

Trade Name:
TERGITOL XD Surfactant (Dow Chemical)

PPG-26-BUTETH-26

CAS Nos.: 9038-95-3 (Generic); 9065-63-8 (Generic)

JPN Translation:
PPG - 26 ブテス - 26

Definition: PPG-26-Buteth-26 is the polyoxypropylene, polyoxyethylene ether of butyl alcohol that conforms generally to the formula:

$$C_4H_9(OCHCH_2)_x(OCH_2CH_2)_yOH$$
$$|$$
$$CH_3$$

where x has an average value of 26 and y has an average value of 26.

Information Sources: 21CFR173.310, 21CFR175.105, 21CFR176.210, CIR: [S], CIR: [S] IJT-19(SUPPL. 1)2000, JCLS, RIFM

Chemical Class: Alkoxylated Alcohols

Functions: Fragrance Ingredient; Hair Conditioning Agent; Skin-Conditioning Agent - Miscellaneous; Surfactant - Emulsifying Agent

Reported Product Categories: Bath Oils, Tablets, and Salts; Cleansing Products (Cold Creams, Cleansing Lotions, Liquids and Pads); Skin Fresheners

Technical/Other Names:
Oxirane, methyl-, polymer with oxirane, monobutyl ether (RIFM)
Polyoxyethylene (26) Polyoxypropylene (26) Monobutyl Ether
Polyoxypropylene (26) Polyoxyethylene (26) Monobutyl Ether

Trade Name Mixtures:
AEC Menthyl Lactate (Water Soluble) (A & E Connock)
Covabsorb EW (LCW)
Covafresh (LCW)
Covafresh II (LCW)
Creasoluble No.1 (Creations Couleurs)
Microfolia (LCW)
Solubilisant LRI (LCW)
Vamasol IT (Ultra Chemical)

PPG-28-BUTETH-35

CAS Nos.: 9038-95-3 (Generic); 9065-63-8 (Generic)

JPN Translation:
PPG - 28 ブテス - 35

Definition: PPG-28-Buteth-35 is the polyoxypropylene, polyoxyethylene ether of butyl alcohol that conforms generally to the formula:

$$C_4H_9(OCHCH_2)_x(OCH_2CH_2)_yOH$$
$$|$$
$$CH_3$$

where x has an average value of 28 and y has an average value of 35.

Information Sources: 21CFR173.310, 21CFR175.105, 21CFR176.210, 21CFR178.3570, CIR: [S], CIR: [S] IJT-19 (SUPPL. 1)2000, JCLS, JSCI, RIFM, TSCA

Chemical Class: Alkoxylated Alcohols

Functions: Fragrance Ingredient; Hair Conditioning Agent; Skin-Conditioning Agent - Miscellaneous; Surfactant - Emulsifying Agent

Reported Product Categories: Shampoos (Non-coloring); Hair Rinses (Non-coloring)

Technical/Other Names:
Oxirane, methyl-, polymer with oxirane, monobutyl ether (RIFM)
Polyoxyethylene (35) Polyoxypropylene (28) Butyl Ether
Polyoxyethylene (35) Polyoxypropylene (28) Monobutyl Ether
Polyoxypropylene (28) Polyoxyethylene (35) Monobutyl Ether
PPG Buteth-3520

Trade Name:
UCON Fluid 50-HB-3520 (Amerchol)

PPG-30-BUTETH-30

CAS Nos.: 9038-95-3 (Generic); 9065-63-8 (Generic)

JPN Translation:
PPG - 30 ブテス - 30

Definition: PPG-30-Buteth-30 is the polyoxypropylene, polyoxyethylene ether of butyl alcohol that conforms generally to the formula:

$$C_4H_9(OCHCH_2)_x(OCH_2CH_2)_yOH$$
$$|$$
$$CH_3$$

where x has an average value of 30 and y has an average value of 30.

Information Sources: JCLS, JSCI, RIFM

Chemical Class: Alkoxylated Alcohols

Functions: Fragrance Ingredient; Hair Conditioning Agent; Skin-Conditioning Agent - Miscellaneous; Surfactant - Cleansing Agent; Surfactant - Solubilizing Agent

Technical/Other Names:
Oxirane, methyl-, polymer with oxirane, monobutyl ether (RIFM)
Polyoxyethylene (30) Polyoxypropylene (30) Butyl Ether
Polyoxyethylene (30) Polyoxypropylene (30) Monobutyl Ether
Polyoxypropylene (30) Polyoxyethylene (30) Monobutyl Ether

PPG-33-BUTETH-45

CAS Nos.: 9038-95-3 (Generic); 9065-63-8 (Generic)

JPN Translation:
PPG - 33 ブテス - 45

Definition: PPG-33-Buteth-45 is the polyoxypropylene, polyoxyethylene ether of butyl alcohol that conforms generally to the formula:

$$C_4H_9(OCHCH_2)_x(OCH_2CH_2)_yOH$$
$$|$$
$$CH_3$$

where x has an average value of 33 and y has an average value of 45.

Information Sources: 21CFR173.310, 21CFR173.340, 21CFR175.105, 21CFR176.210, 21CFR178.3570, JCLS, JSCI, RIFM, TSCA

Chemical Class: Alkoxylated Alcohols

Functions: Fragrance Ingredient; Hair Conditioning Agent; Skin-Conditioning Agent - Miscellaneous

Technical/Other Names:
Oxirane, methyl-, polymer with oxirane, monobutyl ether (RIFM)
Polyoxyethylene (45) Polyoxypropylene (33) Butyl Ether
Polyoxyethylene (45) Polyoxypropylene (33) Monobutyl Ether
Polyoxypropylene (33) Polyoxyethylene (45) Monobutyl Ether
PPG Buteth-5100

Trade Names:
Pluracol W-5100N (BASF)
UCON Fluid 50-HB-5100 (Amerchol)

PPG-36-BUTETH-36

CAS Nos.: 9038-95-3 (Generic); 9065-63-8 (Generic)

JPN Translation:
PPG - 36 ブテス - 36

Definition: PPG-36-Buteth-36 is the polyoxypropylene, polyoxyethylene ether of butyl alcohol that conforms generally to the formula:

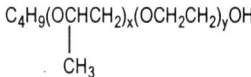

$$C_4H_9(OCHCH_2)_x(OCH_2CH_2)_yOH$$
$$|$$
$$CH_3$$

where x has an average value of 36 and y has an average value of 36.

Information Sources: JCLS, JSCI, RIFM

Chemical Class: Alkoxylated Alcohols

Functions: Fragrance Ingredient; Hair Conditioning Agent; Skin-Conditioning Agent - Miscellaneous; Surfactant - Cleansing Agent; Surfactant - Solubilizing Agent

Technical/Other Names:
Oxirane, methyl-, polymer with oxirane, monobutyl ether (RIFM)
Polyoxyethylene (36) Polyoxypropylene (36) Butyl Ether
Polyoxyethylene (36) Polyoxypropylene (36) Monobutyl Ether
Polyoxypropylene (36) Polyoxyethylene (36) Monobutyl Ether

PPG-38-BUTETH-37

CAS Nos.: 9038-95-3 (Generic); 9065-63-8 (Generic)

JPN Translation:
PPG - 38 ブテス - 37

Definition: PPG-38-Buteth-37 is the polyoxypropylene, polyoxyethylene ether of butyl alcohol that conforms generally to the formula:

$$C_4H_9(OCHCH_2)_x(OCH_2CH_2)_yOH$$
$$|$$
$$CH_3$$

where x has an average value of 38 and y has an average value of 37.

Information Sources: JCLS, JSCI, RIFM

Chemical Class: Alkoxylated Alcohols

Functions: Fragrance Ingredient; Hair Conditioning Agent; Skin-Conditioning Agent - Miscellaneous; Surfactant - Cleansing Agent; Surfactant - Solubilizing Agent

Technical/Other Names:
Oxirane, methyl-, polymer with oxirane, monobutyl ether (RIFM)
Polyoxyethylene (37) Polyoxypropylene (38) Butyl Ether
Polyoxyethylene (37) Polyoxypropylene (38) Monobutyl Ether
Polyoxypropylene (38) Polyoxyethylene (37) Monobutyl Ether

PPG-2 BUTYL ETHER

CAS No.: 9003-13-8 (Generic)

JPN Translation:
PPG - 2 ブチル

Empirical Formula:
$C_{10}H_{22}O_3$

Definition: PPG-2 Butyl Ether is the polypropylene glycol ether of butyl alcohol that conforms generally to the formula:

$$C_4H_9(OCHCH_2)_nOH$$
$$|$$
$$CH_3$$

where n has an average value of 2.

Information Sources: CIR: [SQ] IJT-20 (SUPPL. 4)2001, JCLS, JSCI

Chemical Class: Alkoxylated Alcohols

Functions: Hair Conditioning Agent; Skin-Conditioning Agent - Miscellaneous; Solvent

Technical/Other Names:
Polyoxypropylene (2) Butyl Ether
Polypropylene Glycol (2) Butyl Ether

Trade Names:
Arcosolv DPNB (Lyondell Chemical)
DOWANOL DPnB (Dow Chemical)

PPG-3 BUTYL ETHER

CAS No.	EINECS No.
55934-93-5	259-910-3

Definition: PPG-3 Butyl Ether is the polypropylene glycol ether of butyl alcohol that conforms to the formula:

$$C_4H_9(OCHCH_2)_nOH$$
$$|$$
$$CH_3$$

where n has an averagae value of 3.

Information Source: JCLS

Chemical Class: Alkoxylated Alcohols

Functions: Hair Conditioning Agent; Skin-Conditioning Agent - Miscellaneous; Solvent

Technical/Other Names:
Polyoxypropylene (3) Butyl Ether
Polypropylene Glycol (3) Butyl Ether
Propanol, [2-(2-Butoxymethylethoxy)-Methylethoxy]-

Trade Name:
DOWANOL TPnB (Dow Chemical)

PPG-4 BUTYL ETHER

CAS No.: 9003-13-8 (Generic)

JPN Translation:
PPG - 4 ブチル

Empirical Formula:
$C_{16}H_{34}O_5$

Definition: PPG-4 Butyl Ether is the polypropylene glycol ether of butyl alcohol that conforms generally to the formula:

$$C_4H_9(OCHCH_2)_nOH$$
$$|$$
$$CH_3$$

where n has an average value of 4.

Information Sources: CIR: [SQ] IJT-20 (SUPPL. 4)2001, JCLS, JSCI, TSCA

Chemical Class: Alkoxylated Alcohols

Functions: Hair Conditioning Agent; Skin-Conditioning Agent - Miscellaneous

Technical/Other Names:
Polyoxypropylene (4) Butyl Ether
Polypropylene Glycol (4) Butyl Ether

PPG-5 BUTYL ETHER

CAS No.: 9003-13-8 (Generic)

JPN Translation:
PPG - 5 ブチル

Empirical Formula:
$C_{19}H_{40}O_6$

Definition: PPG-5 Butyl Ether is the polypropylene glycol ether of butyl alcohol that conforms generally to the formula:

$$C_4H_9(OCHCH_2)_nOH$$
$$|$$
$$CH_3$$

where n has an average value of 5.

Information Sources: CIR: [SQ] IJT-20 (SUPPL. 4)2001, JCLS, JSQI, TSCA

Chemical Class: Alkoxylated Alcohols

Functions: Hair Conditioning Agent; Skin-Conditioning Agent - Miscellaneous

Technical/Other Names:
Polyoxypropylene (5) Butyl Ether
Polypropylene Glycol (5) Butyl Ether

Trade Name:
UCON LB-65 Lubricant (Dow Chemical)

PPG-9 BUTYL ETHER

CAS No.: 9003-13-8 (Generic)

JPN Translation:
PPG - 9 ブチル

Empirical Formula:
$C_{31}H_{64}O_{10}$

Definition: PPG-9 Butyl Ether is the polypropylene glycol ether of butyl alcohol that conforms generally to the formula:

$$C_4H_9(OCHCH_2)_nOH$$
$$|$$
$$CH_3$$

where n has an average value of 9.

Information Sources: CIR: [SQ] IJT-20 (SUPPL. 4)2001, JCLS, JSQI, TSCA

Chemical Class: Alkoxylated Alcohols

Functions: Hair Conditioning Agent; Skin-Conditioning Agent - Miscellaneous

Technical/Other Names:
Polyoxypropylene (9) Butyl Ether
Polypropylene Glycol (9) Butyl Ether

Trade Name:
UCON LB-135 Lubricant (Dow Chemical)

PPG-12 BUTYL ETHER

CAS No.: 9003-13-8 (Generic)

JPN Translation:
PPG - 12 ブチル

Definition: PPG-12 Butyl Ether is the polypropylene glycol ether of butyl alcohol that conforms generally to the formula:

$$C_4H_9(OCHCH_2)_nOH$$
$$|$$
$$CH_3$$

where n has an average value of 12.

Information Sources: CIR: [SQ] IJT-20 (SUPPL. 4)2001, JCLS, JSCI

Chemical Class: Alkoxylated Alcohols

Functions: Hair Conditioning Agent; Skin-Conditioning Agent - Miscellaneous

Technical/Other Names:
Polyoxypropylene (12) Butyl Ether
Polypropylene Glycol (12) Butyl Ether

PPG-14 BUTYL ETHER

CAS No.: 9003-13-8 (Generic)

JPN Translation:
PPG - 14 ブチル

Definition: PPG-14 Butyl Ether is the polypropylene glycol ether of butyl alcohol that conforms generally to the formula:

$$C_4H_9(OCHCH_2)_nOH$$
$$|$$
$$CH_3$$

where n has an average value of 14.

Information Sources: CIR: [SQ] IJT-20 (SUPPL. 4)2001, CTFA D, JCLS, JSQI, TSCA

Chemical Class: Alkoxylated Alcohols

Functions: Hair Conditioning Agent; Skin-Conditioning Agent - Miscellaneous

Reported Product Categories: Personal Cleanliness Products, Misc.; Colognes and Toilet Waters; Deodorants (Underarm)

Technical/Other Names:
Polyoxypropylene (14) Butyl Ether
Polypropylene Glycol (14) Butyl Ether
PPG Butyl Ether-200

Trade Names:
AEC PPG-14 Butyl Ether (A & E Connock)
Probutyl 14 (Croda, Inc.)

PPG-15 BUTYL ETHER

CAS No.: 9003-13-8 (Generic)

JPN Translation:
PPG - 15 ブチル

Definition: PPG-15 Butyl Ether is the polypropylene glycol ether of butyl alcohol that conforms generally to the formula:

$$C_4H_9(OCHCH_2)_nOH$$
$$|$$
$$CH_3$$

where n has an average value of 15.

Information Sources: CIR: [SQ] IJT-20 (SUPPL. 4)2001, JCLS, JSCI, TSCA

Chemical Class: Alkoxylated Alcohols

Functions: Hair Conditioning Agent; Skin-Conditioning Agent - Miscellaneous

Technical/Other Names:
Polyoxypropylene (15) Butyl Ether
Polypropylene Glycol (15) Butyl Ether

Trade Name:
UCON LB-285 Lubricant (Dow Chemical)

PPG-16 BUTYL ETHER

CAS No.: 9003-13-8 (Generic)

JPN Translation:
PPG - 16 ブチル

Definition: PPG-16 Butyl Ether is the polypropylene glycol ether of butyl alcohol that conforms generally to the formula:

$$C_4H_9(OCHCH_2)_nOH$$
$$|$$
$$CH_3$$

where n has an average value of 16.

Information Sources: CIR: [SQ] IJT-20 (SUPPL. 4)2001, JCLS, JSQI, TSCA

Chemical Class: Alkoxylated Alcohols

Functions: Hair Conditioning Agent; Skin-Conditioning Agent - Miscellaneous

Technical/Other Names:
Polyoxypropylene (16) Butyl Ether
Polypropylene Glycol (16) Butyl Ether
PPG Butyl Ether-300

PPG-17 BUTYL ETHER

CAS No.: 9003-13-8 (Generic)

JPN Translation:
PPG - 17 ブチル

Definition: PPG-17 Butyl Ether is the polypropylene glycol ether of butyl alcohol that conforms generally to the formula:

$$C_4H_9(OCHCH_2)_nOH$$
$$|$$
$$CH_3$$

where n has an average value of 17.

Information Sources: CIR: [SQ] IJT-20 (SUPPL. 4)2001, JCLS, JSCI

Chemical Class: Alkoxylated Alcohols

Functions: Hair Conditioning Agent; Skin-Conditioning Agent - Miscellaneous

Technical/Other Names:
Polyoxypropylene (17) Butyl Ether
Polypropylene Glycol (17) Butyl Ether

Trade Name:
Tegosoft PBE (Degussa Care Specialties)

PPG-18 BUTYL ETHER

CAS No.: 9003-13-8 (Generic)

JPN Translation:
PPG - 18 ブチル

Definition: PPG-18 Butyl Ether is the polypropylene glycol ether of butyl alcohol that conforms generally to the formula:

$$C_4H_9(OCHCH_2)_nOH$$
$$|$$
$$CH_3$$

where n has an average value of 18.

Information Sources: 21CFR175.105, 21CFR176.210, CIR: [SQ] IJT-20(SUPPL. 4)2001, JCLS, JSQI, TSCA

Chemical Class: Alkoxylated Alcohols

Functions: Hair Conditioning Agent; Skin-Conditioning Agent - Miscellaneous

Technical/Other Names:
Polyoxypropylene (18) Butyl Ether
Polypropylene Glycol (18) Butyl Ether
PPG Butyl Ether 385

Trade Name:
UCON LB-385 Lubricant (Dow Chemical)

PPG-20 BUTYL ETHER

CAS No.: 9003-13-8 (Generic)

JPN Translation:
PPG - 20 ブチル

Definition: PPG-20 Butyl Ether is the polypropylene glycol ether of butyl alcohol that conforms generally to the formula:

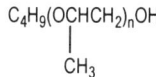

where n has an average value of 20.

Information Sources: CIR: [SQ] IJT-20 (SUPPL. 4)2001, JCLS, JSCI

Chemical Class: Alkoxylated Alcohols

Functions: Hair Conditioning Agent; Skin-Conditioning Agent - Miscellaneous

Technical/Other Names:
Polyoxypropylene (20) Butyl Ether
Polypropylene Glycol (20) Butyl Ether

PPG-22 BUTYL ETHER

CAS No.: 9003-13-8 (Generic)

JPN Translation:
PPG - 22 ブチル

Definition: PPG-22 Butyl Ether is the polypropylene glycol ether of butyl alcohol that conforms generally to the formula:

$$C_4H_9(OCHCH_2)_nOH$$
$$|$$
$$CH_3$$

where n has an average value of 22.

Information Sources: 21CFR175.105, 21CFR176.200, 21CFR176.210, CIR: [SQ] IJT-20(SUPPL. 4)2001, JCLS, JSQI, TSCA

Chemical Class: Alkoxylated Alcohols

Functions: Hair Conditioning Agent; Skin-Conditioning Agent - Miscellaneous

Technical/Other Names:
Polyoxypropylene (22) Butyl Ether
Polypropylene Glycol (22) Butyl Ether

Trade Name:
UCON LB-525 Lubricant (Dow Chemical)

PPG-24 BUTYL ETHER

CAS No.: 9003-13-8 (Generic)

JPN Translation:
PPG - 24 ブチル

Definition: PPG-24 Butyl Ether is the polypropylene glycol ether of butyl alcohol that conforms generally to the formula:

$$C_4H_9(OCHCH_2)_nOH$$
$$|$$
$$CH_3$$

where n has an average value of 24.

Information Sources: 21CFR175.105, 21CFR176.210, 21CFR178.3570, CIR: [SQ] IJT-20(SUPPL. 4)2001, JCLS, JSCI, TSCA

Chemical Class: Alkoxylated Alcohols

Functions: Hair Conditioning Agent; Skin-Conditioning Agent - Miscellaneous

Technical/Other Names:
Polyoxypropylene (24) Butyl Ether
Polypropylene Glycol (24) Butyl Ether

Trade Name:
UCON LB-625 Lubricant (Dow Chemical)

PPG-26 BUTYL ETHER

CAS No.: 9003-13-8 (Generic)

JPN Translation:
PPG - 26 ブチル

Definition: PPG-26 Butyl Ether is the polypropylene glycol ether of butyl alcohol that conforms generally to the formula:

$$C_4H_9(OCHCH_2)_nOH$$
$$|$$
$$CH_3$$

where n has an average value of 26.

Information Sources: CIR: [SQ] IJT-20 (SUPPL. 4)2001, JCLS, JSCI

Chemical Class: Alkoxylated Alcohols

Functions: Hair Conditioning Agent; Skin-Conditioning Agent - Miscellaneous

Technical/Other Names:
Polyoxypropylene (26) Butyl Ether
Polypropylene Glycol (26) Butyl Ether

PPG-30 BUTYL ETHER

CAS No.: 9003-13-8 (Generic)

JPN Translation:
PPG - 30 ブチル

Definition: PPG-30 Butyl Ether is the polypropylene glycol ether of butyl alcohol that conforms generally to the formula:

$$C_4H_9(OCHCH_2)_nOH$$
$$|$$
$$CH_3$$

where n has an average value of 30.

Information Sources: 21CFR175.105, 21CFR176.210, 21CFR178.3570, CIR: [SQ] IJT-20(SUPPL. 4)2001, JCLS, JSCI, TSCA

Chemical Class: Alkoxylated Alcohols

Functions: Hair Conditioning Agent; Skin-Conditioning Agent - Miscellaneous

Technical/Other Names:
Polyoxypropylene (30) Butyl Ether
Polypropylene Glycol (30) Butyl Ether

Trade Name:
Nissan Unilube MB-38 (NOF)

PPG-33 BUTYL ETHER

CAS No.: 9003-13-8 (Generic)

JPN Translation:
PPG - 33 ブチル

Definition: PPG-33 Butyl Ether is the polypropylene glycol ether of butyl alcohol that conforms generally to the formula:

$$C_4H_9(OCHCH_2)_nOH$$
$$|$$
$$CH_3$$

where n has an average value of 33.

Information Sources: 21CFR166.110, 21CFR172.615, 21CFR175.105, 21CFR175.125, 21CFR175.300, 21CFR175.380, 21CFR175.390, 21CFR176.170, 21CFR176.210, 21CFR177.1010, 21CFR177.1210, 21CFR177.1350, 21CFR178.3570, 21CFR184.1660, CIR: [SQ] IJT-20(SUPPL. 4)2001, JCLS, JSCI, TSCA

Chemical Class: Alkoxylated Alcohols

Functions: Hair Conditioning Agent; Skin-Conditioning Agent - Miscellaneous

Reported Product Category: Colognes and Toilet Waters

Technical/Other Names:
Polyoxypropylene (33) Butyl Ether
Polypropylene Glycol (33) Butyl Ether
PPG Butyl Ether-1145

Trade Name:
UCON Fluid LB-1145 (Amerchol)

PPG-40 BUTYL ETHER

CAS No.: 9003-13-8 (Generic)

JPN Translation:
PPG - 40 ブチル

Definition: PPG-40 Butyl Ether is the polypropylene glycol ether of butyl alcohol that conforms generally to the formula:

$$C_4H_9(OCHCH_2)_nOH$$
$$|$$
$$CH_3$$

where n has an average value of 40.

Information Sources: 21CFR175.105, 21CFR176.200, 21CFR176.210, 21CFR178.3570, CIR: [SQ] IJT-20(SUPPL. 4)2001, CIR: [SQ] JACT-12(3)1993, JCLS, JSCI, TSCA

Chemical Class: Alkoxylated Alcohols

Functions: Hair Conditioning Agent; Skin-Conditioning Agent - Miscellaneous

Reported Product Categories: Hair Dyes and Colors (All Types Requiring Caution

Statements and Patch Tests); Tonics, Dressings, and Other Hair Grooming Aids

Technical/Other Names:
Polyoxypropylene (40) Butyl Ether
Polypropylene Glycol (40) Butyl Ether
PPG Butyl Ether 1715

Trade Names:
UCON Fluid LB-1715 (Amerchol)
Unilube MB-370 (NOF)

PPG-52 BUTYL ETHER

CAS No.: 9003-13-8 (Generic)

JPN Translation:
PPG - 52 ブチル

Definition: PPG-52 Butyl Ether is the polypropylene glycol ether of butyl alcohol that conforms generally to the formula:

$$C_4H_9(OCHCH_2)_nOH$$
$$|$$
$$CH_3$$

where n has an average value of 52.

Information Sources: CIR: [SQ] IJT-20 (SUPPL. 4)2001, JCLS, JSCI

Chemical Class: Alkoxylated Alcohols

Functions: Hair Conditioning Agent; Skin-Conditioning Agent - Miscellaneous

Technical/Other Names:
Polyoxypropylene (52) Butyl Ether
Polypropylene Glycol (52) Butyl Ether

PPG-53 BUTYL ETHER

CAS No.: 9003-13-8 (Generic)

JPN Translation:
PPG - 53 ブチル

Definition: PPG-53 Butyl Ether is the polypropylene glycol ether of butyl alcohol that conforms generally to the formula:

$$C_4H_9(OCHCH_2)_nOH$$
$$|$$
$$CH_3$$

where n has an average value of 53.

Information Sources: 21CFR178.3570, CIR: [SQ] IJT-20(SUPPL. 4)2001, JCLS, JSQI, TSCA

Chemical Class: Alkoxylated Alcohols

Functions: Hair Conditioning Agent; Skin-Conditioning Agent - Miscellaneous

Technical/Other Names:
Polyoxypropylene (53) Butyl Ether
Polypropylene Glycol (53) Butyl Ether

Trade Name:
UCON LB-3000 Lubricant (Dow Chemical)

PPG-12 BUTYL ETHER DIMETHICONE

Definition: PPG-12 Butyl Ether Dimethicone is a butyl ether of a derivative of Dimethicone (q.v.) containing an average of 12 moles of propylene oxide.

Chemical Class: Siloxanes and Silanes

Functions: Hair Conditioning Agent; Humectant; Surfactant - Emulsifying Agent; Surfactant - Suspending Agent

Technical/Other Name:
Polypropylene Glycol (12) Butyl Ether Dimethicone

Trade Name:
FZ-2410 (Nippon Unicar)

PPG-21 BUTYL ETHER PHOSPHATE

Definition: PPG-21 Butyl Ether Phosphate is a complex mixture of esters of phosphoric acid and PPG-21 butyl ether.

Chemical Classes: Alkoxylated Alcohols; Phosphorus Compounds

Function: Surfactant - Emulsifying Agent

Technical/Other Name:
Polyoxypropylene (21) Butyl Ether Phosphate

Trade Name:
Parlon EP-70 (Goo Chemical)

PPG-25 BUTYL ETHER PHOSPHATE

JPN Translation:
PPG - 25 ブチルリン酸

Definition: PPG-25 Butyl Ether Phosphate is a complex mixture of esters of phosphoric acid and PPG-25 butyl ether.

Information Sources: JCIC, JCLS, JSQI

Chemical Classes: Alkoxylated Alcohols; Phosphorus Compounds

Function: Surfactant - Emulsifying Agent

Technical/Other Name:
Polyoxypropylene Butyl Ether Phosphate

PPG-35 BUTYL ETHER PHOSPHATE

Definition: PPG-35 Butyl Ether Phosphate is a complex mixture of esters of phosphoric acid and PPG-35 butyl ether.

Chemical Classes: Alkoxylated Alcohols; Phosphorus Compounds

Function: Surfactant - Emulsifying Agent

Technical/Other Name:
Polyoxypropylene (35) Butyl Ether Phosphate

Trade Name:
Parlon EP-71 (Goo Chemical)

PPG-12 CAPRYLETH-18

Definition: PPG-12 Capryleth-18 is the polyoxypropylene, polyoxyethylene ether of Caprylic Alcohol (q.v.) that conforms generally to the formula:

$$CH_3(CH_2)_7(OCHCH_2)_x(OCH_2CH_2)_yOH$$
$$|$$
$$CH_3$$

where x has an average value of 12 and y has an average value of 18.

Chemical Class: Alkoxylated Alcohols

Function: Surfactant - Emulsifying Agent

Technical/Other Names:
PEG-18 PPG-12 Caprylic Ether
Polyoxyethylene (18) Polyoxypropylene (12) Caprylic Ether
Polyoxypropylene (12) Polyoxyethylene (18) Caprylic Ether

Trade Name:
Pepol A-0858 (Toho)

PPG-30-CAPRYLETH-4 PHOSPHATE

Definition: PPG-30-Capryleth-4 Phosphate is a complex mixture of esters of phosphoric acid and a polyoxyethylene, polyoxypropylene derivative of caprylyl alcohol containing an average of 30 moles of propylene oxide and 4 moles of ethylene oxide.

Information Source: JCIC

Chemical Classes: Alkoxylated Alcohols; Phosphorus Compounds

Function: Surfactant - Cleansing Agent

Technical/Other Names:
Polyoxyethylene (4) Polyoxypropylene (30) Caprylyl Ether Phosphate
Polyoxyethylene Polyoxypropylene Octyl Ether Phosphate (4 E.O.) (30 P.O.)
Polyoxypropylene (30) Polyoxyethylene (4) Caprylyl Ether Phosphate

PPG-6 CASTORATE

JPN Translation:
ヒマシ脂肪酸 PPG - 5.5

Definition: PPG-6 Castorate is the polypropylene glycol ester of the fatty acids derived from Ricinus Communis (Castor) Oil (q.v.). It conforms to the formula:

$$\begin{array}{c} O \\ \| \\ RC—(OCHCH_2)_nOH \\ | \\ CH_3 \end{array}$$

---

where RCO- represents the castor oil fatty acids and n has an average value of 6.

Information Source: JCLS

Chemical Class: Esters

Function: Skin-Conditioning Agent - Miscellaneous

PPG-2-CETEARETH-9

Definition: PPG-2-Ceteareth-9 is the polyoxypropylene, polyoxyethylene ether of Cetearyl Alcohol (q.v.) that conforms generally to the formula:

$$R(OCHCH_2)_x(OCH_2CH_2)_yOH$$
$$|$$
$$CH_3$$

where R represents a blend of cetyl and stearyl radicals, x has an average value of 2 and y has an average value of 9.

Chemical Class: Alkoxylated Alcohols

Function: Surfactant - Emulsifying Agent

Reported Product Category: Face and Neck Preparations (Excluding Shaving Preparations)

Technical/Other Names:
Polyoxyethylene (9) Polyoxypropylene (2) Cetyl/Stearyl Ether
Polyoxypropylene (2) Polyoxyethylene (9) Cetyl/Stearyl Ether

Trade Name Mixtures:
Chamazulene - HCE (Vincience)
Dermoprotectine (Sederma)
Ederline-H (Vincience)

PPG-4-CETEARETH-12

Definition: PPG-4-Ceteareth-12 is the polyoxypropylene, polyoxyethylene ether of Cetearyl Alcohol (q.v.) that conforms generally to the formula:

$$R(OCHCH_2)_x(OCH_2CH_2)_yOH$$
$$|$$
$$CH_3$$

where R represents a blend of cetyl and stearyl radicals, x has an average value of 4 and y has an average value of 12.

Information Source: TSCA

Chemical Class: Alkoxylated Alcohols

Function: Surfactant - Emulsifying Agent

Technical/Other Names:
Polyoxyethylene (12) Polyoxypropylene (4) Cetyl/Stearyl Ether
Polyoxypropylene (4) Polyoxyethylene (12) Cetyl/Stearyl Ether

PPG-10-CETEARETH-20

Definition: PPG-10-Ceteareth-20 is the polyoxypropylene, polyoxyethylene ether of Cetearyl Alcohol (q.v.) that conforms generally to the formula:

$$R(OCHCH_2)_x(OCH_2CH_2)_yOH$$
$$|$$
$$CH_3$$

where R represents a blend of cetyl and stearyl radicals, x has an average value of 10 and y has an average value of 20.

Chemical Class: Alkoxylated Alcohols

Function: Surfactant - Emulsifying Agent

Technical/Other Names:
Polyoxyethylene (20) Polyoxypropylene (10) Cetyl/Stearyl Ether
Polyoxypropylene (10) Polyoxyethylene (20) Cetyl/Stearyl Ether

PPG-1-CETETH-1

CAS Nos.: 9087-53-0 (Generic); 37311-01-6 (Generic)

JPN Translation:
PPG - 1 セテス - 1

Empirical Formula:
$C_{21}H_{44}O_3$

Definition: PPG-1-Ceteth-1 is the polyoxypropylene, polyoxyethylene ether of cetyl alcohol that conforms generally to the formula:

$$CH_3(CH_2)_{15}(OCHCH_2)_x(OCH_2CH_2)_yOH$$
$$|$$
$$CH_3$$

where x has an average value of 1 and y has an average value of 1.

Information Sources: JCLS, JSCI

Chemical Class: Alkoxylated Alcohols

Functions: Skin-Conditioning Agent - Emollient; Surfactant - Emulsifying Agent

Technical/Other Names:
Polyoxyethylene (1) Polyoxypropylene (1) Cetyl Ether
Polyoxypropylene (1) Polyoxyethylene (1) Cetyl Ether

PPG-1-CETETH-5

CAS Nos.: 9087-53-0 (Generic); 37311-01-6 (Generic)

JPN Translation:
PPG - 1 セテス - 5

Empirical Formula:
$C_{21}H_{60}O_4$

Definition: PPG-1-Ceteth-5 is the polyoxypropylene, polyoxyethylene ether of cetyl alcohol that conforms generally to the formula:

$$CH_3(CH_2)_{15}(OCHCH_2)_x(OCH_2CH_2)_yOH$$
$$|$$
$$CH_3$$

where x has an average value of 1 and y has an average value of 5.

Information Sources: JCLS, JSCI

Chemical Class: Alkoxylated Alcohols

Functions: Skin-Conditioning Agent - Emollient; Surfactant - Emulsifying Agent

Technical/Other Names:
Polyoxyethylene (5) Polyoxypropylene (1) Cetyl Ether
Polyoxypropylene (1) Polyoxyethylene (5) Cetyl Ether

PPG-1-CETETH-10

CAS Nos.: 9087-53-0 (Generic); 37311-01-6 (Generic)

JPN Translation:
PPG - 1 セテス - 10

Empirical Formula:
$C_{34}H_{80}O_{12}$

Definition: PPG-1-Ceteth-10 is the polyoxypropylene, polyoxyethylene ether of cetyl alcohol that conforms generally to the formula:

$$CH_3(CH_2)_{15}(OCHCH_2)_x(OCH_2CH_2)_yOH$$
$$|$$
$$CH_3$$

where x has an average value of 1 and y has an average value of 10.

Information Sources: JCLS, JSCI

Chemical Class: Alkoxylated Alcohols

Functions: Skin-Conditioning Agent - Emollient; Surfactant - Emulsifying Agent

Technical/Other Names:
Polyoxyethylene (10) Polyoxypropylene (1) Cetyl Ether
Polyoxypropylene (1) Polyoxyethylene (10) Cetyl Ether

PPG-1-CETETH-20

CAS Nos.: 9087-53-0 (Generic); 37311-01-6 (Generic)

JPN Translation:
PPG - 1 セテス - 20

Definition: PPG-1-Ceteth-20 is the polyoxypropylene, polyoxyethylene ether of cetyl alcohol that conforms generally to the formula:

$$CH_3(CH_2)_{15}(OCHCH_2)_x(OCH_2CH_2)_yOH$$
$$|$$
$$CH_3$$

where x has an average value of 1 and y has an average value of 20.

Information Sources: JCLS, JSCI

Chemical Class: Alkoxylated Alcohols

Functions: Skin-Conditioning Agent - Emollient; Surfactant - Emulsifying Agent; Surfactant - Solubilizing Agent

Technical/Other Names:
Polyoxyethylene (20) Polyoxypropylene (1) Cetyl Ether
Polyoxypropylene (1) Polyoxyethylene (20) Cetyl Ether

PPG-2-CETETH-1

CAS Nos.: 9087-53-0 (Generic); 37311-01-6 (Generic)

JPN Translation:
PPG - 2 セテス - 1

Empirical Formula:
$C_{24}H_{44}O_4$

Definition: PPG-2-Ceteth-1 is the polyoxy-propylene, polyoxyethylene ether of cetyl alcohol that conforms generally to the formula:

$$CH_3(CH_2)_{15}(OCHCH_2)_x(OCH_2CH_2)_yOH$$
$$|$$
$$CH_3$$

where x has an average value of 2 and y has an average value 1.

Information Sources: JCLS, JSCI

Chemical Class: Alkoxylated Alcohols

Functions: Skin-Conditioning Agent - Emollient; Surfactant - Emulsifying Agent

Technical/Other Names:
Polyoxyethylene (1) Polyoxypropylene (2) Cetyl Ether
Polyoxypropylene (2) Polyoxyethylene (1) Cetyl Ether

PPG-2-CETETH-5

CAS Nos.: 9087-53-0 (Generic); 37311-01-6 (Generic)

JPN Translation:
PPG - 2 セテス - 5

Empirical Formula:
$C_{32}H_{60}O_8$

Definition: PPG-2-Ceteth-5 is the polyoxy-propylene, polyoxyethylene ether of cetyl alcohol that conforms generally to the formula:

$$CH_3(CH_2)_{15}(OCHCH_2)_x(OCH_2CH_2)_yOH$$
$$|$$
$$CH_3$$

where x has an average value of 2 and y has an average value of 5.

Information Sources: JCLS, JSCI

Chemical Class: Alkoxylated Alcohols

Functions: Skin-Conditioning Agent - Emollient; Surfactant - Emulsifying Agent

Technical/Other Names:
Polyoxyethylene (5) Polyoxypropylene (2) Cetyl Ether
Polyoxypropylene (2) Polyoxyethylene (5) Cetyl Ether

PPG-2-CETETH-10

CAS Nos.: 9087-53-0 (Generic); 37311-01-6 (Generic)

JPN Translation:
PPG - 2 セテス - 10

Empirical Formula:
$C_{42}H_{80}O_{13}$

Definition: PPG-2-Ceteth-10 is the polyoxy-propylene, polyoxyethylene ether of cetyl alcohol that conforms generally to the formula:

$$CH_3(CH_2)_{15}(OCHCH_2)_x(OCH_2CH_2)_yOH$$
$$|$$
$$CH_3$$

where x has an average value of 2 and y has an average value of 10.

Information Sources: JCLS, JSCI

Chemical Class: Alkoxylated Alcohols

Functions: Skin-Conditioning Agent - Emollient; Surfactant - Emulsifying Agent

Technical/Other Names:
Polyoxyethylene (10) Polyoxypropylene (2) Cetyl Ether
Polyoxypropylene (2) Polyoxyethylene (10) Cetyl Ether

PPG-2-CETETH-20

CAS Nos.: 9087-53-0 (Generic); 37311-01-6 (Generic)

JPN Translation:
PPG - 2 セテス - 20

Definition: PPG-2-Ceteth-20 is the polyoxy-propylene, polyoxyethylene ether of cetyl

alcohol that conforms generally to the formula:

$$CH_3(CH_2)_{15}(OCHCH_2)_x(OCH_2CH_2)_yOH$$
$$|$$
$$CH_3$$

where x has an average value of 2 and y has an average value of 20.

Information Sources: JCLS, JSCI

Chemical Class: Alkoxylated Alcohols

Functions: Skin-Conditioning Agent - Emollient; Surfactant - Emulsifying Agent; Surfactant - Solubilizing Agent

Technical/Other Names:
Polyoxyethylene (20) Polyoxypropylene (2) Cetyl Ether
Polyoxypropylene (2) Polyoxyethylene (20) Cetyl Ether

PPG-4-CETETH-1

CAS Nos.: 9087-53-0 (Generic); 37311-01-6 (Generic)

JPN Translation:
PPG - 4 セテス - 1

Definition: PPG-4-Ceteth-1 is the polyoxy-propylene polyoxyethylene ether of cetyl alcohol that conforms generally to the formula:

$$CH_3(CH_2)_{15}(OCHCH_2)_x(OCH_2CH_2)_yOH$$
$$|$$
$$CH_3$$

where x has an average value of 4 and y has an average value of 1.

Information Sources: JCLS, JSCI, JSQI, TSCA

Chemical Class: Alkoxylated Alcohols

Functions: Skin-Conditioning Agent - Emollient; Surfactant - Emulsifying Agent

Technical/Other Names:
Polyoxyethylene (1) Polyoxypropylene (4) Cetyl Ether
Polyoxypropylene (4) Polyoxyethylene (1) Cetyl Ether

Trade Name:
Nikkol PBC-31 (Nikko)

PPG-4-CETETH-5

CAS Nos.: 9087-53-0 (Generic); 37311-01-6 (Generic)

JPN Translation:
PPG - 4 セテス - 5

Empirical Formula:
$C_{38}H_{78}O_{10}$

Definition: PPG-4-Ceteth-5 is the polyoxypropylene, polyoxyethylene ether of cetyl alcohol that conforms generally to the formula:

$$CH_3(CH_2)_{15}(OCHCH_2)_x(OCH_2CH_2)_yOH$$
$$|$$
$$CH_3$$

where x has an average value of 4 and y has an average value of 5.

Information Sources: JCLS, JSCI, JSQI, TSCA

Chemical Class: Alkoxylated Alcohols

Functions: Skin-Conditioning Agent - Emollient; Surfactant - Emulsifying Agent

Technical/Other Names:
Pclyoxyethylene (5) Polyoxypropylene (4) Cetyl Ether
Polyoxypropylene (4) Polyoxyethylene (5) Cetyl Ether

PPG-4-CETETH-10

CAS Nos.: 9087-53-0 (Generic); 37311-01-6 (Generic)

JPN Translation:
PPG - 4 セテス - 10

Empirical Formula:
$C_{48}H_{98}O_{15}$

Definition: PPG-4-Ceteth-10 is the polyoxypropylene, polyoxyethylene ether of cetyl alcohol that conforms generally to the formula:

$$CH_3(CH_2)_{15}(OCHCH_2)_x(OCH_2CH_2)_yOH$$
$$|$$
$$CH_3$$

where x has an average value of 4 and y has an average value of 10.

Information Sources: JCLS, JSCI, JSQI, TSCA

Chemical Class: Alkoxylated Alcohols

Function: Surfactant - Emulsifying Agent

Technical/Other Names:
Polyoxyethylene (10) Polyoxypropylene (4) Cetyl Ether
Polyoxypropylene (4) Polyoxyethylene (10) Cetyl Ether

Trade Name:
Nikkol PBC-33 (Nikko)

PPG-4-CETETH-20

CAS Nos.: 9087-53-0 (Generic); 37311-01-6 (Generic)

JPN Translation:
PPG - 4 セテス - 20

Definition: PPG-4-Ceteth-20 is the polyoxypropylene, polyoxyethylene ether of cetyl alcohol that conforms generally to the formula:

$$CH_3(CH_2)_{15}(OCHCH_2)_x(OCH_2CH_2)_yOH$$
$$|$$
$$CH_3$$

where x has an average value of 4 and y has an average value of 20.

Information Sources: JCLS, JSCI

Chemical Class: Alkoxylated Alcohols

Function: Surfactant - Emulsifying Agent

Technical/Other Names:
Polyoxyethylene (20) Polyoxypropylene (4) Cetyl Ether
Polyoxypropylene (4) Polyoxyethylene (20) Cetyl Ether

Trade Name:
Nikkol PBC-34 (Nikko)

PPG-5-CETETH-20

CAS Nos.: 9087-53-0 (Generic); 37311-01-6 (Generic)

JPN Translation:
PPG - 5 セテス - 20

Definition: PPG-5-Ceteth-20 is the polyoxypropylene, polyoxyethylene ether of cetyl alcohol that conforms generally to the formula:

$$CH_3(CH_2)_{15}(OCHCH_2)_x(OCH_2CH_2)_yOH$$
$$|$$
$$CH_3$$

where x has an average value of 5 and y has an average value of 20.

Information Sources: CTFA D, JCLS, JSQI, TSCA

Chemical Class: Alkoxylated Alcohols

Function: Surfactant - Emulsifying Agent

Reported Product Categories: Bath Oils, Tablets, and Salts; Tonics, Dressings, and Other Hair Grooming Aids; Hair Conditioners; Bath Soaps and Detergents; Bubble Baths; Colognes and Toilet Waters; Skin Care Preparations, Misc.; Skin Fresheners; Bath Preparations, Misc.; Body and Hand Preparations (Excluding Shaving Preparations); Deodorants (Underarm); Hair Preparations (Non-coloring), Misc.; Moisturizing Preparations

Technical/Other Names:
Polyoxyethylene (20) Polyoxypropylene (5) Cetyl Ether
Polyoxypropylene (5) Polyoxyethylene (20) Cetyl Ether

Trade Names:
AEC PPG-5-Ceteth-20 (A & E Connock)
Procetyl AWS (Croda Chemicals)
Procetyl AWS (Croda, Inc.)
Protachem AWS-100 (Protameen)

Trade Name Mixture:
Crodapearl N.I. Liquid (Croda, Inc.)

PPG-8-CETETH-1

CAS Nos.: 9087-53-0 (Generic); 37311-01-6 (Generic)

JPN Translation:
PPG - 8 セテス - 1

Empirical Formula:
$C_{42}H_{86}O_{10}$

Definition: PPG-8-Ceteth-1 is the polyoxypropylene, polyoxyethylene ether of cetyl alcohol that conforms generally to the formula:

$$CH_3(CH_2)_{15}(OCHCH_2)_x(OCH_2CH_2)_yOH$$
$$|$$
$$CH_3$$

where x has an average value of 8 and y has an average value of 1.

Information Sources: JCLS, JSCI, JSQI, TSCA

Chemical Class: Alkoxylated Alcohols

Functions: Skin-Conditioning Agent - Emollient; Surfactant - Emulsifying Agent

Technical/Other Names:
Polyoxyethylene (1) Polyoxypropylene (8) Cetyl Ether
Polyoxypropylene (8) Polyoxyethylene (1) Cetyl Ether

Trade Name:
Nikkol PBC-41 (Nikko)

PPG-8-CETETH-2

CAS Nos.: 9087-53-0 (Generic); 37311-01-6 (Generic)

JPN Translation:
PPG - 8 セテス - 2

Empirical Formula:
$C_{44}H_{90}O_{11}$

Definition: PPG-8-Ceteth-2 is the polyoxypropylene, polyoxyethylene ether of cetyl alcohol that conforms generally to the formula:

$$CH_3(CH_2)_{15}(OCHCH_2)_x(OCH_2CH_2)_yOH$$
$$|$$
$$CH_3$$

where x has an average value of 8 and y has an average value of 2.

Information Sources: JCLS, JSQI, TSCA

Chemical Class: Alkoxylated Alcohols

Functions: Skin-Conditioning Agent - Emollient; Surfactant - Emulsifying Agent

Technical/Other Names:
Polyoxyethylene (2) Polyoxypropylene (8) Cetyl Ether
Polyoxypropylene (8) Polyoxyethylene (2) Cetyl Ether

PPG-8-CETETH-5

CAS Nos.: 9087-53-0 (Generic); 37311-01-6 (Generic)

JPN Translation:
PPG - 8 セテス - 5

Empirical Formula:
$C_{50}H_{102}O_{14}$

Definition: PPG-8-Ceteth-5 is the polyoxypropylene, polyoxyethylene ether of cetyl alcohol that conforms generally to the formula:

$$CH_3(CH_2)_{15}(OCHCH_2)_x(OCH_2CH_2)_yOH$$
$$|$$
$$CH_3$$

where x has an average value of 8 and y has an average value of 5.

Information Sources: JCLS, JSCI, JSQI, TSCA

Chemical Class: Alkoxylated Alcohols

Functions: Skin-Conditioning Agent - Emollient; Surfactant - Emulsifying Agent

Technical/Other Names:
Polyoxyethylene (5) Polyoxypropylene (8) Cetyl Ether
Polyoxypropylene (8) Polyoxyethylene (5) Cetyl Ether

PPG-8-CETETH-10

CAS Nos.: 9087-53-0 (Generic); 37311-01-6 (Generic)

JPN Translation:
PPG - 8 セテス - 10

Empirical Formula:
$C_{60}H_{122}O_{19}$

Definition: PPG-8-Ceteth-10 is the polyoxypropylene, polyoxyethylene ether of cetyl alcohol that conforms generally to the formula:

$$CH_3(CH_2)_{15}(OCHCH_2)_x(OCH_2CH_2)_yOH$$
$$|$$
$$CH_3$$

where x has an average value of 8 and y has an average value of 10.

Information Sources: JCLS, JSCI, JSQI, TSCA

Chemical Class: Alkoxylated Alcohols

Functions: Skin-Conditioning Agent - Emollient; Surfactant - Emulsifying Agent

Technical/Other Names:
Polyoxyethylene (10) Polyoxypropylene (8) Cetyl Ether
Polyoxypropylene (8) Polyoxyethylene (10) Cetyl Ether

PPG-8-CETETH-20

CAS Nos.: 9087-53-0 (Generic); 37311-01-6 (Generic)

JPN Translation:
PPG - 8 セテス - 20

Definition: PPG-8-Ceteth-20 is the polyoxypropylene, polyoxyethylene ether of cetyl alcohol that conforms generally to the formula:

$$CH_3(CH_2)_{15}(OCHCH_2)_x(OCH_2CH_2)_yOH$$
$$|$$
$$CH_3$$

where x has an average value of 8 and y has an average value of 20.

Information Sources: JCLS, JSCI, JSQI, TSCA

Chemical Class: Alkoxylated Alcohols

Function: Surfactant - Emulsifying Agent

Technical/Other Names:
Polyoxyethylene (20) Polyoxypropylene (8) Cetyl Ether
Polyoxypropylene (8) Polyoxyethylene (20) Cetyl Ether

Trade Name:
Nikkol PBC-44 (Nikko)

PPG-5-CETETH-10 PHOSPHATE

JPN Translation:
PPG - 5 セテス - 10 リン酸

Definition: PPG-5-Ceteth-10 Phosphate is a complex mixture of esters of phosphoric acid and the polyoxypropylene, polyoxyethylene ether of cetyl alcohol.

Information Sources: JCIC, JCLS, JSQI

Chemical Class: Phosphorus Compounds

Function: Surfactant - Emulsifying Agent

Reported Product Categories: Shampoos (Non-coloring); Hair Dyes and Colors (All Types Requiring Caution Statements and Patch Tests); Permanent Waves; Eyeliners; Indoor Tanning Preparations; Bath Oils, Tablets, and Salts; Cleansing Products (Cold Creams, Cleansing Lotions, Liquids and Pads); Hair Conditioners; Hair Straighteners; Paste Masks (Mud Packs)

Technical/Other Names:
Polyoxyethylene Polyoxypropylene Cetyl Ether Phosphate
Polyoxyethylene (10) Polyoxypropylene (5) Cetyl Ether Phosphate
Polyoxypropylene (5) Polyoxyethylene (10) Cetyl Ether Phosphate

Trade Names:
Brophos 5C10 (Arch Personal Care Products)
Crodafos SG (Croda Chemicals)
Crodafos SG (Croda, Inc.)

PPG-10 CETYL ETHER

CAS No.: 9035-85-2 (Generic)

JPN Translation:
PPG - 10 セチル

Empirical Formula:
$C_{46}H_{94}O_{11}$

Definition: PPG-10 Cetyl Ether is the polypropylene glycol ether of cetyl alcohol that conforms to the formula:

$$CH_3(CH_2)_{15}(OCHCH_2)_nOH$$
$$|$$
$$CH_3$$

where n has an average value of 10.

Information Sources: CTFA D, JCIC, JCLS, JSQI, TSCA

Chemical Class: Alkoxylated Alcohols

Function: Skin-Conditioning Agent - Emollient

Reported Product Category: Deodorants (Underarm)

Technical/Other Names:
Polyoxypropylene (10) Cetyl Ether
Polyoxypropylene Cetyl Ether (10P.O.)
Polypropylene Glycol (10) Cetyl Ether

Trade Names:
Acconon CA 10 (Abitec Corporation)
Jeecol PCA-10 (Jeen)
Procol PCA-10 (Protameen)

PPG-20 CETYL ETHER

CAS No.: 9035-85-2 (Generic)

Definition: PPG-20 Cetyl Ether is the polypropylene glycol ether of cetyl alcohol that conforms to the formula:

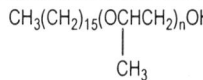

where n has an average value of 20.

Chemical Class: Alkoxylated Alcohols

Function: Skin-Conditioning Agent - Emollient

Technical/Other Names:
Polyoxypropylene (20) Cetyl Ether
Polypropylene Glycol (20) Cetyl Ether

Trade Name:
Procetyl 20 (Croda Chemicals)

PPG-28 CETYL ETHER

CAS No.: 9035-85-2 (Generic)

Definition: PPG-28 Cetyl Ether is the polypropylene glycol ether of cetyl alcohol that conforms to the formula:

$$CH_3(CH_2)_{15}(OCHCH_2)_nOH$$
$$|$$
$$CH_3$$

where n has an average value of 28.

Information Source: TSCA

Chemical Class: Alkoxylated Alcohols

Function: Skin-Conditioning Agent - Emollient

Technical/Other Names:
Polyoxypropylene (28) Cetyl Ether
Polypropylene Glycol (28) Cetyl Ether

PPG-30 CETYL ETHER

CAS No.: 9035-85-2 (Generic)

Definition: PPG-30 Cetyl Ether is the polypropylene glycol ether of cetyl alcohol that conforms to the formula:

$$CH_3(CH_2)_{15}(OCHCH_2)_nOH$$
$$|$$
$$CH_3$$

where n has an average value of 30.

Information Source: TSCA

Chemical Class: Alkoxylated Alcohols

Function: Skin-Conditioning Agent - Emollient

Reported Product Categories: Hair Preparations (Non-coloring), Misc.; Hair Conditioners; Shampoos (Non-coloring); Bath Soaps and Detergents

Technical/Other Names:
Polyoxypropylene (30) Cetyl Ether
Polypropylene Glycol (30) Cetyl Ether

PPG-50 CETYL ETHER

CAS No.: 9035-85-2 (Generic)

Definition: PPG-50 Cetyl Ether is the polypropylene glycol ether of cetyl alcohol that conforms to the formula:

$$CH_3(CH_2)_{15}(OCHCH_2)_nOH$$
$$|$$
$$CH_3$$

where n has an average value of 50.

Information Source: TSCA

Chemical Class: Alkoxylated Alcohols

Function: Skin-Conditioning Agent - Emollient

Technical/Other Names:
Polyoxypropylene (50) Cetyl Ether
Polypropylene Glycol (50) Cetyl Ether

PPG-10 CETYL ETHER PHOSPHATE

Definition: PPG-10 Cetyl Ether Phosphate is a complex mixture of esters of phosphoric acid and PPG-10 Cetyl Ether (q.v.)

Information Source: CTFA D

Chemical Class: Phosphorus Compounds

Functions: Surfactant - Cleansing Agent; Surfactant - Emulsifying Agent

Technical/Other Names:
Polyoxypropylene (10) Cetyl Ether Phosphate
Polypropylene Glycol (10) Cetyl Ether Phosphate

Trade Name Mixtures:
Covascreen TI (LCW)
Covascreen TIYO (LCW)
Covascreen ZN (LCW)

PPG-2 COCAMIDE

Definition: PPG-2 Cocamide is the dipropylene glycol amide of Coconut Acid (q.v.) that conforms generally to the formula:

$$
\begin{array}{c}
O \\
\| \\
RC-NH(OCHCH_2)_nOH \\
| \\
CH_3
\end{array}
$$

where n has an average value of 2 and RCO- represents the cocoyl moiety.

Chemical Class: Alkoxylated Amides

Functions: Surfactant - Foam Booster; Viscosity Increasing Agent - Aqueous

Trade Name:
Amizett 1PC (Kawaken)

PPG-2 COCAMINE

Definition: PPG-2 Cocamine is the dipropylene glycol amine of Cocos Nucifera (Coconut) Alcohol (q.v.) that conforms generally to the formula:

$$
\begin{array}{c}
CH_3 \\
| \\
(OCHCH_2)_xH \\
/ \\
R-N \\
\backslash \\
(OCHCH_2)_yH \\
| \\
CH_3
\end{array}
$$

where R represents the alkyl groups derived from coconut oil and x+y has an average value of 2.

Chemical Class: Alkoxylated Amines

Function: Antistatic Agent

Trade Name:
Propomeen C/12 (Akzo Nobel)

PPG-2 C12-13 PARETH-8

Definition: PPG-2 C12-13 Pareth-8 is the polypropylene glycol ether a mixture of synthetic C12-13 ethoxylated fatty alcohols containing an average of 8 moles of ethylene oxide and 2 moles of propylene oxide.

Chemical Class: Alkoxylated Alcohols

Function: Surfactant - Emulsifying Agent

Trade Name:
Sunmorl DB-80 (Nicca Chemical)

PPG-2 C12-15 PARETH-6

Definition: PPG-2 C12-15 Pareth-6 is a polyoxyethylene, polyoxypropylene ether of a mixture of synthetic alcohols that conforms generally to the formula:

$$R(OCHCH_2)_x(OCH_2CH_2)_yOH$$
$$|$$
$$CH_3$$

where R represents a C12-15 alcohol, x has an average value of 2 and y has an average value of 6.

Chemical Class: Alkoxylated Alcohols

Function: Surfactant - Emulsifying Agent

Technical/Other Name:
Polyoxypropylene (2) Polyoxyethylene (6) C12-15 Alcohol Ether

Trade Name:
Lorodac/L6-S-50 (Sasol Italy)

PPG-4 C13-15 PARETH-15

Definition: PPG-4 C13-15 Pareth-15 is the polyoxyethylene, polyoxypropylene ether of a mixture of synthetic alcohols that conforms generally to the formula:

$$R(OCHCH_2)_x(OCH_2CH_2)_yOH$$
$$|$$
$$CH_3$$

where R represents an alcohol stem with 13-15 carbons in the alkyl chain, x has an average value of 4, and y has an average value of 15.

Chemical Class: Alkoxylated Alcohols

Function: Surfactant - Emulsifying Agent

Trade Name:
Synperonic LF/RA 320 (Uniqema Americas)

PPG-5 C9-15 PARETH-6

Definition: PPG-5 C9-15 Pareth-6 is the polyoxyethylene, polyoxypropylene ether of a mixture of synthetic alcohols that conforms generally to the formula:

$$R(OCHCH_2)_x(OCH_2CH_2)_yOH$$
$$|$$
$$CH_3$$

where R represents a fatty alcohol group with 9 to 15 carbons in the alkyl chain, x has an average value of 5 and y has an average value of 6.

Chemical Class: Alkoxylated Alcohols

Function: Surfactant - Emulsifying Agent

Trade Name:
Propetal 281 (Zschimmer & Schwarz)

PPG-6 C9-11 PARETH-5

CAS No.: 154518-36-2 (Generic)

Definition: PPG-6 C9-11 Pareth-5 is the polyoxypropylene, polyoxyethylene ether of a mixture of synthetic alcohols that conforms generally to the formula:

$$R(OCHCH_2)_x(OCH_2CH_2)_yOH$$
$$|$$
$$CH_3$$

where R represents the C9-11 fatty alcohol group, x has an average value of 6 and y has an average value of 5.

Information Source: TSCA

Chemical Class: Alkoxylated Alcohols

Function: Surfactant - Emulsifying Agent

Trade Name:
Pepol AS-053X (Toho)

PPG-6 C12-15 PARETH-12

Definition: PPG-6 C12-15 Pareth-12 is the polyoxyethylene, polyoxypropylene ether of

a mixture of synthetic alcohols that conforms generally to the formula:

$$R(OCHCH_2)_x(OCH_2CH_2)_yOH$$
$$|$$
$$CH_3$$

where R represents an alkyl stem with 12-15 carbons, x has an average value of 6 and y has an average value of 12.

Chemical Class: Alkoxylated Alcohols

Function: Surfactant - Emulsifying Agent

PPG-6 C12-18 PARETH-11

Definition: PPG-6 C12-18 Pareth-11 is the polyoxyethylene, polyoxypropylene ether of a mixture of synthetic alcohols that conforms generally to the formula:

$$R(OCHCH_2)_x(OCH_2CH_2)_yOH$$
$$|$$
$$CH_3$$

where R represents an alcohol stem with 12 to 18 carbons in the alkyl chain, x has an average value of 6 and y has an average value of 11.

Chemical Class: Alkoxylated Alcohols

Function: Surfactant - Emulsifying Agent

Technical/Other Name:
PPG-6-Pareth-28-11

Trade Name:
Plurafac D-25 (BASF)

PPG-3 C12-14 SEC-PARETH-7

Definition: PPG-3 C12-14 Sec-Pareth-7 is the polyoxyethylene, polyoxypropylene ether of a mixture of synthetic secondary C12-14 alcohols with an average ethoxylation value of 7 and an average propoxylation value of 3.

Chemical Class: Alkoxylated Alcohols

Functions: Emulsion Stabilizer; Surfactant - Emulsifying Agent

Trade Name:
SOFTANOL EP7025 (Nippon Shokubai)

PPG-4 C12-14 SEC-PARETH-5

Definition: PPG-4 C12-14 Sec-Pareth-5 is the polyoxyethylene, polyoxypropylene ether of a mixture of synthetic secondary C12-14 alcohols with an average ethoxylation value of 5 and an average propoxylation value of 4.

Chemical Class: Alkoxylated Alcohols

Functions: Emulsion Stabilizer; Surfactant - Emulsifying Agent

Trade Name:
SOFTANOL EP5035 (Nippon Shokubai)

PPG-5 C12-14 SEC-PARETH-7

Definition: PPG-5 C12-14 Sec-Pareth-7 is the polyoxyethylene, polyoxypropylene ether of a mixture of synthetic secondary C12-14 alcohols with an average ethoxylation value of 7 and an average propoxylation value of 5.

Chemical Class: Alkoxylated Alcohols

Functions: Emulsion Stabilizer; Surfactant - Emulsifying Agent

Trade Name:
SOFTANOL EP7045 (Nippon Shokubai)

PPG-5 C12-14 SEC-PARETH-9

Definition: PPG-5 C12-14 Sec-Pareth-9 is the polyoxyethylene, polyoxypropylene ether of a mixture of synthetic secondary C12-14 alcohols with an average ethoxylation value of 9 and an average propoxylation value of 5.

Chemical Class: Alkoxylated Alcohols

Functions: Emulsion Stabilizer; Surfactant - Emulsifying Agent

Trade Name:
SOFTANOL EP9050 (Nippon Shokubai)

PPG-1-DECETH-6

Definition: PPG-1-Deceth-6 is the polyoxyethylene, polyoxypropylene ether of decyl alcohol that conforms generally to the formula:

$$CH_3(CH_2)_9(OCHCH_2)_x(OCH_2CH_2)_yOH$$
$$|$$
$$CH_3$$

where x has an average value of 1 and y has an average value of 6.

Chemical Class: Alkoxylated Alcohols

Function: Surfactant - Emulsifying Agent

Technical/Other Names:
Polyoxyethylene (6) Polyoxypropylene (1) Decyl Ether
Polyoxypropylene (1) Polyoxyethylene (6) Decyl Ether

Trade Name:
TRITON XL-80N Surfactact (Dow Chemical)

Trade Name Mixture:
Dow Corning 9090 Silicone Elastomer Emulsion (Dow Corning)

PPG-2-DECETH-3

Definition: PPG-2-Deceth-3 is the polyoxypropylene, polyoxyethylene ether of decyl alcohol that conforms generally to the formula:

$$CH_3(CH_2)_8CH_2(OCHCH_2)_x(OCH_2CH_2)_yOH$$
$$|$$
$$CH_3$$

where x has an average value of 2 and y has an average value of 3.

Chemical Class: Alkoxylated Alcohols

Function: Surfactant - Emulsifying Agent

Technical/Other Name:
Polyoxypropylene (2) Polyoxyethylene (3) Decyl Ether

Trade Names:
Emalex DAPE-0203 (Nihon Emulsion)
Emalex DAPE-0205 (Nihon Emulsion)

PPG-2-DECETH-7

Definition: PPG-2-Deceth-7 is the polyoxypropylene, polyoxyethylene ether of decyl alcohol that conforms generally to the formula:

$$CH_3(CH_2)_8CH_2(OCHCH_2)_x(OCH_2CH_2)_yOH$$
$$|$$
$$CH_3$$

where x has an average value of 2 and y has an average value of 7.

Chemical Class: Alkoxylated Alcohols

Function: Surfactant - Emulsifying Agent

Technical/Other Name:
Polyoxypropylene (2) Polyoxyethylene (7) Decyl Ether

Trade Name:
Emalex DAPE-0207 (Nihon Emulsion)

PPG-2-DECETH-10

Empirical Formula:
$C_{36}H_{74}O_{13}$

Definition: PPG-2-Deceth-10 is the polyoxypropylene, polyoxyethylene ether of decyl alcohol that conforms generally to the formula:

$$CH_3(CH_2)_9(OCHCH_2)_x(OCH_2CH_2)_yOH$$
$$|$$
$$CH_3$$

where x has an average value of 2 and y has an average value of 10.

Chemical Class: Alkoxylated Alcohols

Function: Surfactant - Emulsifying Agent

Technical/Other Names:
Polyoxyethylene (10) Polyoxypropylene (2) Decyl Ether
Polyoxypropylene (2) Polyoxyethylene (10) Decyl Ether

Trade Name:
Emalex DAPE-0210 (Nihon Emulsion)

PPG-2-DECETH-12

Definition: PPG-2-Deceth-12 is the polyoxypropylene, polyoxyethylene ether of decyl alcohol that conforms generally to the formula:

$$CH_3(CH_2)_8CH_2(OCHCH_2)_x(OCH_2CH_2)_yOH$$
$$|$$
$$CH_3$$

where x has an average value of 2 and y has an average value of 12.

Chemical Class: Alkoxylated Alcohols

Function: Surfactant - Emulsifying Agent

Technical/Other Name:
Polyoxypropylene (2) Polyoxyethylene (12) Decyl Ether

Trade Name:
Emalex DAPE-0212 (Nihon Emulsion)

PPG-2-DECETH-15

Definition: PPG-2-Deceth-15 is the polyoxypropylene, polyoxyethylene ether of decyl alcohol that conforms generally to the formula:

$$CH_3(CH_2)_8CH_2(OCHCH_2)_x(OCH_2CH_2)_yOH$$
$$|$$
$$CH_3$$

where x has an average value of 2 and y has an average value of 15.

Chemical Class: Alkoxylated Alcohols

Function: Surfactant - Emulsifying Agent

Technical/Other Name:
Polyoxypropylene (2) Polyoxyethylene (15) Decyl Ether

Trade Name:
Emalex DAPE-0215 (Nihon Emulsion)

PPG-2-DECETH-20

Definition: PPG-2-Deceth-20 is the polyoxypropylene, polyoxyethylene ether of decyl alcohol that conforms generally to the formula:

$$CH_3(CH_2)_8CH_2(OCHCH_2)_x(OCH_2CH_2)_yOH$$
$$|$$
$$CH_3$$

where x has an average value of 2 and y has an average value of 20.

Chemical Class: Alkoxylated Alcohols

Function: Surfactant - Emulsifying Agent

Technical/Other Name:
Polyoxypropylene (2) Polyoxyethylene (20) Decyl Ether

Trade Name:
Emalex DAPE-0220 (Nihon Emulsion)

PPG-2-DECETH-30

Definition: PPG-2-Deceth-30 is the polyoxyethylene, polyoxypropylene ether of decyl alcohol that conforms generally to the formula:

$$CH_3(CH_2)_8CH_2(OCHCH_2)_x(OCH_2CH_2)_yOH$$
$$|$$
$$CH_3$$

where x has an average value of 2 and y has an average value of 30.

Chemical Class: Alkoxylated Alcohols

Function: Surfactant - Emulsifying Agent

Technical/Other Name:
Polyoxypropylene (2) Polyoxyethylene (30) Decyl Ether

Trade Name:
Emalex DAPE-0230 (Nihon Emulsion)

PPG-4-DECETH-4

Empirical Formula:
$C_{30}H_{62}O_9$

Definition: PPG-4-Deceth-4 is the polyoxypropylene, polyoxyethylene ether of decyl alcohol that conforms generally to the formula:

$$CH_3(CH_2)_9(OCHCH_2)_x(OCH_2CH_2)_yOH$$
$$|$$
$$CH_3$$

where x has an average value of 4 and y has an average value of 4.

Chemical Class: Alkoxylated Alcohols

Function: Surfactant - Emulsifying Agent

Technical/Other Name:
Polyoxyethylene (4) Polyoxypropylene (4) Decyl Ether

Trade Name Mixtures:
Genapol PGL (Clariant)
Genapol PGL (Clariant GmbH, Personal Care)

PPG-4 DECETH-6

Definition: PPG-4 Deceth-6 is the polyoxypropylene, polyoxyethylene ether of decyl alcohol that conforms generally to the formula:

$$CH_3(CH_2)_9(OCHCH_2)_x(OCH_2CH_2)_yOH$$
$$|$$
$$CH_3$$

where x has an average value of 4 and y has an average value of 6.

Chemical Class: Alkoxylated Alcohols

Function: Surfactant - Emulsifying Agent

Technical/Other Name:
Polyoxypropylene (4) Polyoxyethylene (6) Decyl Ether

Trade Name:
Biodac OP 1 (Sasol Italy)

PPG-6-DECETH-4

Empirical Formula:
$C_{36}H_{79}O_{11}$

Definition: PPG-6-Deceth-4 is the polyoxypropylene, polyoxyethylene ether of decyl alcohol that conforms generally to the formula:

$$CH_3(CH_2)_9(OCHCH_2)_x(OCH_2CH_2)_yOH$$
$$|$$
$$CH_3$$

where x has an average value of 6 and y has an average value of 4.

Chemical Class: Alkoxylated Alcohols

Function: Surfactant - Emulsifying Agent

Technical/Other Name:
Polyoxyethylene (4) Polyoxypropylene (6) Decyl Ether

Trade Name:
Marlox FK 64 (Sasol GmbH - Marl)

PPG-6-DECETH-9

Empirical Formula:
$C_{46}H_{94}O_{16}$

Definition: PPG-6-Deceth-9 is the polyoxypropylene, polyoxyethylene ether of decyl alcohol that conforms generally to the formula:

$$CH_3(CH_2)_9(OCHCH_2)_x(OCH_2CH_2)_yOH$$
$$|$$
$$CH_3$$

where x has an average value of 6 and y has an average value of 9.

Chemical Class: Alkoxylated Alcohols

Function: Surfactant - Emulsifying Agent

Technical/Other Name:
Polyoxyethylene (9) Polyoxypropylene (6) Decyl Ether

Trade Name:
Marlox FK 69 (Sasol GmbH - Marl)

PPG-8 DECETH-6

Empirical Formula:
$C_{46}H_{94}O_{15}$

Definition: PPG-8 Deceth-6 is the polyoxypropylene, polyoxyethylene ether of decyl alcohol that conforms generally to the formula:

$$CH_3(CH_2)_9(OCHCH_2)_x(OCH_2CH_2)_yOH$$
$$|$$
$$CH_3$$

where x has an average value of 8 and y has an average value of 6.

Chemical Class: Alkoxylated Alcohols

Function: Surfactant - Emulsifying Agent

Technical/Other Names:
Polyoxyethylene (6) Polyoxypropylene (8) Decyl Ether
Polyoxypropylene (8) Polyoxyethylene (6) Decyl Ether

Trade Name:
Marlox FK 86 (Sasol GmbH - Marl)

PPG-14 DECETH-6

Definition: PPG-14 Deceth-6 is the polyoxypropylene, polyoxyethylene ether of decyl alcohol that conforms generally to the formula:

$$CH_3(CH_2)_9(OCHCH_2)_x(OCH_2CH_2)_yOH$$
$$|$$
$$CH_3$$

where x has an average value of 14 and y has an average value of 6.

Chemical Class: Alkoxylated Alcohols

Function: Surfactant - Emulsifying Agent

Technical/Other Name:
Polyoxypropylene (14) Polyoxyethylene (6) Decyl Ether

PPG-3-DECETH-2 CARBOXYLIC ACID

Empirical Formula:
$C_{23}H_{46}O_7$

Definition: PPG-3-Deceth-2 Carboxylic Acid is the carboxylic acid of a propoxylated, ethoxylated decyl alcohol that conforms generally to the formula:

$$CH_3(CH_2)_9(OCHCH_2)_x(OCH_2CH_2)_yOCH_2COOH$$
$$|$$
$$CH_3$$

where x has an average value of 3 and y has an average value of 1.

Chemical Class: Carboxylic Acids

Function: Surfactant - Cleansing Agent

Technical/Other Names:
PEG-2-PPG-3 Decyl Ether Carboxylic Acid
Polyoxyethylene (2) Polyoxypropylene (3) Decyl Ether Carboxylic Acid
Polyoxypropylene (3) Polyoxyethylene (2) Decyl Ether Carboxylic Acid

PPG-6-DECYLTETRADECETH-12

JPN Translation:
PPG - 6 デシルテトラデセス - 12

Definition: PPG-6-Decyltetradeceth-12 is the polyoxypropylene, polyoxyethylene ether of Decyltetradecanol (q.v.) that conforms generally to the formula:

$$(CH_2)_9CH_3$$
$$|$$
$$CH_3(CH_2)_{11}CHCH_2O(CH_2CHO)_x(CH_2CH_2O)_yH$$
$$|$$
$$CH_3$$

where x has an average value of 6 and y has an average value of 12.

Information Source: JCLS

Chemical Class: Alkoxylated Alcohols

Function: Surfactant - Emulsifying Agent

Technical/Other Names:
Polyoxyethylene (12) Polyoxypropylene (6) Tetradecyl Ether
Polyoxypropylene (6) Polyoxyethylene (12) Tetradecyl Ether

Trade Name:
Nikkol PEN-4612 (Nikko)

PPG-6-DECYLTETRADECETH-20

JPN Translation:
PPG - 6 デシルテトラデセス - 20

Definition: PPG-6-Decyltetradeceth-20 is the polyoxypropylene, polyoxyethylene ether of Decyltetradecanol (q.v.) that conforms generally to the formula:

$$(CH_2)_9CH_3$$
$$|$$
$$CH_3(CH_2)_{11}CHCH_2O(CH_2CHO)_x(CH_2CH_2O)_yH$$
$$|$$
$$CH_3$$

where x has an average value of 6 and y has an average value of 20.

Information Source: JCLS

Chemical Class: Alkoxylated Alcohols

Function: Surfactant - Emulsifying Agent

Technical/Other Names:
Polyoxyethylene (20) Polyoxypropylene (6) Tetradecyl Ether
Polyoxypropylene (6) Polyoxyethylene (20) Tetradecyl Ether

Trade Name:
Nikkol PEN-4620 (Nikko)

PPG-6-DECYLTETRADECETH-30

JPN Translation:
PPG - 6 デシルテトラデセス - 30

Definition: PPG-6-Decyltetradeceth-30 is the polyoxypropylene, polyoxyethylene ether of Decyltetradecanol (q.v.) that conforms generally to the formula:

$$CH_3(CH_2)_{11}CHCH_2O(CH_2CHO)_x(CH_2CH_2O)_yH$$

with $(CH_2)_9CH_3$ and CH_3 side groups

where x has an average value of 6 and y has an average value of 30.

Information Source: JCLS

Chemical Class: Alkoxylated Alcohols

Function: Surfactant - Emulsifying Agent

Technical/Other Names:
Polyoxyethlene (30) Polyoxypropylene (6) Tetradecyl Ether
Polyoxypropylene (6) Polyoxyethylene (30) Tetradecyl Ether

Trade Name:
Nikkol PEN-4630 (Nikko)

PPG-13-DECYLTETRADECETH-24

JPN Translation:
PPG - 13 デシルテトラデセス - 24

Definition: PPG-13-Decyltetradeceth-24 is the polyoxypropylene, polyoxyethylene ether of Decyltetradecanol (q.v.) that conforms generally to the formula:

$$CH_3(CH_2)_{11}CHCH_2O(CH_2CHO)_x(CH_2CH_2O)_yH$$

with $(CH_2)_9CH_3$ and CH_3 side groups

where x has an average value of 13 and y has an average value 24.

Information Source: JCLS

Chemical Class: Alkoxylated Alcohols

Function: Surfactant - Emulsifying Agent

Reported Product Categories: Moisturizing Preparations; Bath Capsules; Skin Care Preparations, Misc.

Technical/Other Names:
Polyoxyethylene (24) Polyoxypropylene (13) Decyltetradecyl Ether
Polyoxypropylene (13) Polyoxyethylene (24) Decyltetradecyl Ether

Trade Name:
S-Safe 1324D (NOF)

PPG-20-DECYLTETRADECETH-10

JPN Translation:
PPG - 20 デシルテトラデセス - 10

Definition: PPG-20-Decyltetradeceth-10 is the polyoxypropylene, polyoxyethylene ether of Decyltetradecanol (q.v.) that conforms generally to the formula:

$$CH_3(CH_2)_{11}CHCH_2O(CH_2CHO)_x(CH_2CH_2O)_yH$$

with $(CH_2)_9CH_3$ and CH_3 side groups

where x has an average value of 20 and y has an average value of 10.

Information Source: JCLS

Chemical Class: Alkoxylated Alcohols

Function: Surfactant - Emulsifying Agent

Trade Name:
S-Safe 2010 (NOF)

PPG-9 DIETHYLMONIUM CHLORIDE

CAS No.: 9042-76-6

JPN Translation:
PPG - 9 ジエチルモニウムクロリド

Empirical Formula:
$C_{32}H_{68}NO_9 \cdot Cl$

Definition: PPG-9 Diethylmonium Chloride is the quaternary ammonium salt that conforms generally to the formula:

$$\left[CH_3CH_2 - \overset{\overset{\displaystyle CH_2CH_3}{|}}{\underset{\underset{\displaystyle CH_3}{|}}{N}} - (CH_2CHO)_nH \right]^+ Cl^-$$

with CH_3 side group

where n has an average value of 9.

Information Sources: CIR: [I] IJT-18 (SUPPL. 3)1999, JCLS, JSQI

Chemical Class: Quaternary Ammonium Compounds

Functions: Antistatic Agent; Hair Conditioning Agent

Reported Product Categories: Permanent Waves; Hair Sprays (Aerosol Fixatives); Hair Preparations (Non-coloring), Misc.; Shampoos (Non-coloring); Tonics, Dressings, and Other Hair Grooming Aids

Technical/Other Names:
Polyoxypropylene (9) Methyl Diethyl Ammonium Chloride
Quaternium-6

PPG-25 DIETHYLMONIUM CHLORIDE

JPN Translation:
PPG - 25 ジエチルモニウムクロリド

Definition: PPG-25 Diethylmonium Chloride is the quaternary ammonium salt that conforms generally to the formula:

$$\left[CH_3CH_2 - \overset{\overset{\displaystyle CH_2CH_3}{|}}{\underset{\underset{\displaystyle CH_3}{|}}{N}} - (CH_2CHO)_nH \right]^+ Cl^-$$

with CH_3 side group

where n has an average value of 25.

Information Sources: CIR: [I] IJT-18 (SUPPL. 3)1999, CTFA D, JCLS, JSQI

Chemical Class: Quaternary Ammonium Compounds

Function: Antistatic Agent

Technical/Other Names:
Polyoxypropylene (25) Methyl Diethyl Ammonium Chloride
Quaternium-20

PPG-40 DIETHYLMONIUM CHLORIDE

CAS Nos.: 9042-76-6; 9076-43-1

JPN Translation:
PPG - 40 ジエチルモニウムクロリド

Definition: PPG-40 Diethylmonium Chloride is the quaternary ammonium salt that conforms generally to the formula:

$$\left[CH_3CH_2 - \overset{\overset{\displaystyle CH_2CH_3}{|}}{\underset{\underset{\displaystyle CH_3}{|}}{N}} - (CH_2CHO)_nH \right]^+ Cl^-$$

with CH_3 side group

where n has an average value of 40.

Information Sources: CIR: [I] IJT-18 (SUPPL. 3)1999, JCLS, JSQI

Chemical Class: Quaternary Ammonium Compounds

Function: Antistatic Agent

Reported Product Categories: Hair Sprays (Aerosol Fixatives); Tonics, Dressings, and Other Hair Grooming Aids

Technical/Other Names:
Polyoxypropylene (40) Methyl Diethyl Ammonium Chloride
Quaternium-21

PPG-12 DIGLUCOSYL C14-18 ACIDATE

JPN Translation:
ポリオキシプロピレンカルボキシアルキル
（C14 - 18）ジグルコシド

Definition: PPG-12 Diglucosyl C14-18 Acidate is the propoxylated ester of diglucose and a C14-18 acid with an average propoxylation value of 12.

Information Source: JCIC

Chemical Classes: Carbohydrates; Esters

Function: Surfactant - Emulsifying Agent

Technical/Other Name:
Polyoxypropylene Carboxyalkyl (14-18) Diglucoside

PPG-9 DIGLYCERYL ETHER

CAS No.: 61710-63-2

JPN Translation:
PPG - 9 ジグリセリル

Empirical Formula:
$C_{33}H_{68}O_{14}$

Definition: PPG-9 Diglyceryl Ether is the polypropylene glycol ether of a dimer of glycerin containing an average of 9 moles of propylene oxide.

Information Sources: JCIC, JCLS, JSQI

Chemical Classes: Alkoxylated Alcohols; Ethers

Function: Skin-Conditioning Agent - Emollient

Technical/Other Names:
Polyoxypropylene Diglyceryl Ether
Polyoxypropylene (9) Diglyceryl Ether
Polypropylene Glycol (9) Diglyceryl Ether

PPG-14 DIGLYCERYL ETHER

JPN Translation:
PPG - 14 ジグリセリル

Definition: PPG-14 Diglyceryl Ether is the polypropylene glycol ether of a dimer of glycerin containing an average of 14 moles of propylene oxide.

Information Source: JCLS

Chemical Classes: Alkoxylated Alcohols; Ethers

Function: Surfactant - Emulsifying Agent

PPG-12 DIMETHICONE

Definition: PPG-12 Dimethicone is the polypropylene glycol derivative of Dimethicone (q.v.) containing an average of 12 moles of propylene oxide.

Chemical Class: Siloxanes and Silanes

Functions: Hair Conditioning Agent; Skin-Conditioning Agent - Emollient

Trade Names:
Silsoft 900 (OSi Specialties)
Silsoft 900 dimethicone copolyol (OSi Specialties)
Silsoft 910 dimethicone copolyol (OSi Specialties)

PPG-27 DIMETHICONE

Definition: PPG-27 Dimethicone is the polypropylene glycol derivative of Dimethicone (q.v.) containing an average of 27 moles of propylene oxide.

Chemical Class: Siloxanes and Silanes

Functions: Hair Conditioning Agent; Skin-Conditioning Agent - Miscellaneous

Trade Name:
KF-625A (Shin Etsu)

PPG-17 DIOLEATE

CAS No.: 26571-49-3 (Generic)

Definition: PPG-17 Dioleate is the polypropylene glycol diester of oleic acid that conforms generally to the formula:

$$CH(CH_2)_7C \overset{\overset{O}{\|}}{\underset{\underset{CH(CH_2)_7CH_3}{\|}}{}} — (OCHCH_2)_nO \overset{CH_3}{\underset{}{|}} — \overset{\overset{O}{\|}}{C}(CH_2)_7CH \underset{\underset{CH_3(CH_2)_7CH}{}}{}$$

where n has an average value of 17.

Information Sources: 21CFR175.300, 21CFR176.210, TSCA

Chemical Class: Alkoxylated Carboxylic Acids

Function: Skin-Conditioning Agent - Emollient

Technical/Other Names:
Polyoxypropylene (17) Dioleate
Polypropylene Glycol (17) Dioleate

PPG-3 DIPIVALATE

Empirical Formula:
$C_{19}H_{36}O_6$

Definition: PPG-3 Dipivalate is the organic compound that conforms to the formula:

$$CH_3C \overset{\overset{CH_3}{|}}{\underset{\underset{CH_3}{|}}{}} — \overset{\overset{O}{\|}}{C} — O(CHCH_2O)_3 \overset{CH_3}{\underset{}{|}} — \overset{\overset{O}{\|}}{C} — CCH_3 \overset{\overset{CH_3}{|}}{\underset{}{}}$$

Chemical Class: Esters

Function: Skin-Conditioning Agent - Emollient

Trade Name:
Salacos TPG-521 (Nisshin OilliO)

PPG-9-ETHYLHEXETH-5

CAS No.: 64366-70-7 (Generic)

Definition: PPG-9-Ethylhexeth-5 is the polyoxypropylene, polyoxyethylene ether of octyl alcohol that conforms generally to the formula:

$$CH_3(CH_2)_3CHCH_2(OCHCH_2)_x(OCH_2CH_2)_yOH \\ \underset{CH_2CH_3 \quad CH_3}{|\qquad\quad|}$$

where x has an average value of 9 and y has an average value of 5.

Information Source: TSCA

Chemical Class: Alkoxylated Alcohols

Function: Surfactant - Emulsifying Agent

Technical/Other Names:
PEG-5 PPG-9 Octyl Ether
Polyoxyethylene (5) Polyoxypropylene (9) Octyl Ether
Polyoxypropylene (9) Polyoxyethylene (5) Octyl Ether
PPG-9-Octeth-5

Trade Name:
Pepol A-0638 (Toho)

PPG-30 ETHYLHEXETH-4 PHOSPHATE

JPN Translation:
PEG - 4 - PPG - 30 オクチルリン酸

Definition: PPG-30 Ethylhexeth-4 Phosphate is a complex mixture of esters of phosphoric acid and PPG-30 ethylhexeth-4.

Information Source: JCLS

Chemical Class: Phosphorus Compounds

Functions: Surfactant - Cleansing Agent; Surfactant - Emulsifying Agent

PPG-20-GLYCERETH-30

CAS Nos.: 9082-00-2 (Generic); 51258-15-2 (Generic)

Definition: PPG-20-Glycereth-30 is the polyoxypropylene, polyoxyethylene ether of glycerin that conforms generally to the formula:

$$HOCH_2CHCH_2(OCHCH_2)_x(OCH_2CH_2)_yOH$$
$$\quad\quad\; |\quad\quad\quad |$$
$$\quad\quad OH\quad\quad CH_3$$

where x has an average value of 20 and y has an average value of 30.

Information Sources: 21CFR177.1680, TSCA

Chemical Class: Alkoxylated Alcohols

Function: Surfactant - Cleansing Agent

Technical/Other Names:
Polyoxyethylene (30) Polyoxypropylene (20) Glyceryl Ether
Polyoxypropylene (20) Polyoxyethylene (30) Glyceryl Ether

PPG-24-GLYCERETH-24

CAS Nos.: 9082-00-2 (Generic); 51258-15-2 (Generic)

JPN Translation:
PPG - 24 グリセレス - 24

Definition: PPG-24-Glycereth-24 is the polyoxypropylene, polyoxyethylene ether of glycerin that conforms generally to the formula:

$$HOCH_2CHCH_2(OCHCH_2)_x(OCH_2CH_2)_yOH$$
$$\quad\quad\; |\quad\quad\quad |$$
$$\quad\quad OH\quad\quad CH_3$$

where x has an average value of 24 and y has an average value of 24.

Information Sources: CTFA D, JCIC, JCLS, JSQI, TSCA

Chemical Class: Alkoxylated Alcohols

Functions: Solvent; Surfactant - Emulsifying Agent

Technical/Other Names:
Polyoxyethylene (24) Polyoxypropylene (24) Glyceryl Ether
Polyoxyethylene Polyoxypropylene Glyceryl Ether (24E.O.)(24P.O.)
Polyoxypropylene (24) Polyoxyethylene (24) Glyceryl Ether

Trade Names:
Adeka GH-200 (Asahi Denka Kogyo)
PEG/PPG-24/24 Glycerine (Kyowa Hakko Kogyo)
Polyglycol 15-200 (Dow Chemical)

PPG-25-GLYCERETH-22

Definition: PPG-25-Glycereth-22 is the polyoxypropylene, polyoxyethylene ether of glycerin that conforms generally to the formula:

$$HOCH_2CHCH_2(OCHCH_2)_x(OCH_2CH_2)_yOH$$
$$\quad\quad\; |\quad\quad\quad |$$
$$\quad\quad OH\quad\quad CH_3$$

where x has an average value of 25 and y has an average value of 22.

Chemical Class: Alkoxylated Alcohols

Functions: Solvent; Surfactant - Emulsifying Agent

Technical/Other Name:
Polyoxyethylene (22) Polyoxypropylene (25) Glyceryl Ether

Trade Name Mixture:
FZ-2406 (Nippon Unicar)

PPG-66-GLYCERETH-12

CAS No.: 51258-15-2 (Generic)

Definition: PPG-66-Glycereth-12 is the polyoxypropylene, polyoxyethylene ether of glycerin that conforms generally to the formula:

$$HOCH_2CHCH_2(OCHCH_2)_x(OCH_2CH_2)_yOH$$
$$\quad\quad\; |\quad\quad\quad |$$
$$\quad\quad OH\quad\quad CH_3$$

where x has an average value of 66 and y has an average value of 12.

Information Source: TSCA

Chemical Class: Alkoxylated Alcohols

Function: Surfactant - Emulsifying Agent

Technical/Other Names:
Polyoxyethylene (12) Polyoxypropylene (66) Glyceryl Ether
Polyoxypropylene (66) Polyoxyethylene (12) Glyceryl Ether

PPG-3 GLYCERYL ETHER

CAS No.: 25791-96-2 (Generic)

Definition: PPG-3 Glyceryl Ether is the polypropylene glycol ether of glycerin that conforms generally to the formula:

$$C_3H_7O_2(OCHCH_2)_n OH$$
$$\quad\quad\quad\quad |$$
$$\quad\quad\quad CH_3$$

where n has an average value of 3.

Chemical Class: Glyceryl Esters and Derivatives

Function: Surfactant - Emulsifying Agent

Technical/Other Name:
Polyoxypropylene (3) Glyceryl Ether

Trade Name:
Newpol GP-250 (Sanyo Chemical)

PPG-6 GLYCERYL ETHER

CAS No.: 25791-96-2 (Generic)

Definition: PPG-6 Glyceryl Ether is the polypropylene glycol ether of glycerin that conforms generally to the formula:

$$C_3H_7O_2(OCHCH_2)_n OH$$
$$\quad\quad\quad\quad |$$
$$\quad\quad\quad CH_3$$

where n has an average value of 6.

Chemical Class: Glyceryl Esters and Derivatives

Function: Surfactant - Emulsifying Agent

Technical/Other Name:
Polyoxypropylene (6) Glyceryl Ether

Trade Name:
Newpol GP-400 (Sanyo Chemical)

PPG-9 GLYCERYL ETHER

CAS No.: 25791-96-2 (Generic)

Definition: PPG-9 Glyceryl Ether is the polypropylene glycol ether of glycerin that conforms generally to the formula:

$$C_3H_7O_2(OCHCH_2)_nOH$$
$$\quad\quad\quad\quad |$$
$$\quad\quad\quad CH_3$$

where n has an average value of 9.

Information Source: JCLS

Chemical Class: Glyceryl Esters and Derivatives

Function: Surfactant - Emulsifying Agent

Trade Names:
Mushul C-060 (Toho)
Newpol GP-600 (Sanyo Chemical)
Toho #200T (Toho)

PPG-10 GLYCERYL ETHER

CAS No.: 25791-96-2 (Generic)

JPN Translation:
PPG - 10 グリセリル

Empirical Formula:
$C_{33}H_{68}O_{13}$

Definition: PPG-10 Glyceryl Ether is the polypropylene glycol ether of glycerin that conforms generally to the formula:

$$C_3H_7O_2(OCHCH_2)_nOH$$
$$\quad\quad\quad\quad |$$
$$\quad\quad\quad CH_3$$

where n has an average value of 10.

Information Sources: JCLS, JSQI

Chemical Class: Alkoxylated Alcohols

Functions: Solvent; Surfactant - Emulsifying Agent

Technical/Other Names:
Polyoxypropylene (10) Glyceryl Ether
Polypropylene Glycol (10) Glyceryl Ether

PPG-16 GLYCERYL ETHER

CAS No.: 25791-96-2 (Generic)

Definition: PPG-16 Glyceryl Ether is the polypropylene glycol ether of glycerin that conforms generally to the formula:

$$C_3H_7O_2(OCHCH_2)_n OH$$
$$|$$
$$CH_3$$

where n has an average value of 16.

Chemical Class: Glyceryl Esters and Derivatives

Function: Surfactant - Emulsifying Agent

Technical/Other Name:
Polyoxypropylene (16) Glyceryl Ether

Trade Names:
Newpol GP-1000 (Sanyo Chemical)
Uniol TG-1000 (NOF.)

PPG-27 GLYCERYL ETHER

CAS No.: 25791-96-2 (Generic)

JPN Translation:
PPG - 27 グリセリル

Definition: PPG-27 Glyceryl Ether is a polypropylene glycol ether of glycerin that conforms generally to the formula:

$$C_3H_7O_2(OCHCH_2)_n OH$$
$$|$$
$$CH_3$$

where n has an average value of 27.

Information Sources: 21CFR175.105, JCLS, JSQI, TSCA

Chemical Class: Alkoxylated Alcohols

Function: Surfactant - Emulsifying Agent

Reported Product Category: Shaving Cream (Aerosol, Brushless and Lather)

Technical/Other Names:
Polyoxypropylene (27) Glyceryl Ether
Polypropylene Glycol (27) Glyceryl Ether

PPG-50 GLYCERYL ETHER

CAS No.: 25791-96-2 (Generic)

Definition: PPG-50 Glyceryl Ether is the polypropylene glycol ether of glycerin that conforms generally to the formula:

$$C_3H_7O_2(OCHCH_2)_n OH$$
$$|$$
$$CH_3$$

where n has an average value of 50.

Chemical Class: Glyceryl Esters and Derivatives

Function: Surfactant - Emulsifying Agent

Technical/Other Name:
Polyoxypropylene (50) Glyceryl Ether

Trade Name:
Newpol GP-3000 (Sanyo Chemical)

PPG-55 GLYCERYL ETHER

CAS No.: 25791-96-2 (Generic)

JPN Translation:
PPG - 55 グリセリル

Definition: PPG-55 Glyceryl Ether is the polypropylene glycol ether of glycerin that conforms generally to the formula:

$$C_3H_7O_2(OCHCH_2)_n OH$$
$$|$$
$$CH_3$$

where n has an average value of 55.

Information Sources: JCLS, JSQI, TSCA

Chemical Class: Alkoxylated Alcohols

Function: Solvent

Technical/Other Names:
Polyoxypropylene (55) Glyceryl Ether
Polypropylene Glycol (9) Glyceryl Ether
Polypropylene Glycol (55) Glyceryl Ether

Trade Name:
Toho #200 (Toho)

PPG-67 GLYCERYL ETHER

CAS No.: 25791-96-2 (Generic)

Definition: PPG-67 Glyceryl Ether is the polypropylene glycol ether of glycerin that conforms generally to the formula:

$$C_3H_7O_2(OCHCH_2)_n OH$$
$$|$$
$$CH_3$$

where n has an average value of 67.

Information Source: JCLS

Chemical Class: Glyceryl Esters and Derivatives

Functions: Skin-Conditioning Agent - Emollient; Solvent

Technical/Other Names:
Polyoxypropylene (67) Glyceryl Ether
Polypropylene Glycol (67) Glyceryl Ether

Trade Names:
Mushul C-400 (Toho)
Newpol GP-4000 (Sanyo Chemical)

PPG-70 GLYCERYL ETHER

CAS No.: 25791-96-2 (Generic)

Definition: PPG-70 Glyceryl Ether is the polypropylene glycol ether that conforms generally to the formula:

$$C_3H_7O_2(OCHCH_2)_n OH$$
$$|$$
$$CH_3$$

where n has an average value of 70.

Chemical Class: Glyceryl Esters and Derivatives

Functions: Skin-Conditioning Agent - Emollient; Solvent

Technical/Other Names:
Polyoxypropylene (70) Glyceryl Ether
Polypropylene Glycol (70) Glyceryl Ether

Trade Name:
Uniol TG-4000R (NOF)

PPG-26/HDI COPOLYMER

Definition: PPG-26/HDI Copolymer is a copolymer of hexamethylene diisocyanate and PPG-26 (q.v.) monomers.

Chemical Class: Synthetic Polymers

Functions: Film Former; Plasticizer

PPG-3 HYDROGENATED CASTOR OIL

Definition: PPG-3 Hydrogenated Castor Oil is a polypropylene glycol derivative of Hydrogenated Castor Oil (q.v.) with an average propoxylation value of 3.

Information Source: 21CFR175.300

Chemical Classes: Alkoxylated Alcohols; Glyceryl Esters and Derivatives

Function: Skin-Conditioning Agent - Emollient

Reported Product Category: Lipsticks

Technical/Other Names:
Polyoxypropylene (3) Hydrogenated Castor Oil
Polypropylene Glycol (3) Hydrogenated Castor Oil

Trade Names:
   Hetester HCP (Bernel)
   Procas H3 (Croda Chemicals)

PPG-20 HYDROGENATED CASTOR OIL LAURATE

Definition: PPG-20 Hydrogenated Castor Oil Laurate is a polypropylene glycol derivative of the ester of Lauric Acid (q.v.) and Hydrogenated Castor Oil (q.v.), with an average ethoxylation value of 20.

Information Source: JCLS

Chemical Class: Esters

Function: Skin-Conditioning Agent - Miscellaneous

PPG-30 HYDROGENATED CASTOR OIL LAURATE

Definition: PPG-30 Hydrogenated Castor Oil Laurate is the polypropylene glycol derivative of the lauric acid ester of Hydrogenated Castor Oil (q.v.), with an average ethoxylation value of 30.

Information Source: JCLS

Chemical Class: Esters

Function: Skin-Conditioning Agent - Miscellaneous

PPG-40 HYDROGENATED CASTOR OIL LAURATE

Definition: PPG-40 Hydrogenated Castor Oil Laurate is a polypropylene glycol derivative of the Lauric Acid (q.v.) ester of Hydrogenated Castor Oil (q.v.), with an average ethoxylation of 40.

Information Source: jcls

Chemical Class: Esters

Function: Skin-Conditioning Agent - Miscellaneous

PPG-50 HYDROGENATED CASTOR OIL LAURATE

Definition: PPG-50 Hydrogenated Castor Oil Laurate is a polypropylene glycol derivative of the Lauric Acid(q.v.). ester of Hydrogenated Castor Oil (q.v.), with an average ethoxylation value of 50.

Information Source: JCLS

Chemical Class: Esters

Function: Skin-Conditioning Agent - Miscellaneous

PPG-60 HYDROGENATED CASTOR OIL LAURATE

Definition: PPG-60 Hydrogenated Castor Oil Laurate is the polypropylene glycol derivative of the Lauric Acid (q.v.) ester of Hydrogenated Castor Oil (q.v.), with an average ethoxylation value of 60.

Information Source: JCLS

Chemical Class: Esters

Function: Skin-Conditioning Agent - Miscellaneous

PPG-2 HYDROGENATED TALLOWAMINE

Definition: PPG-2 Hydrogenated Tallowamine is the dipropylene glycol amine of Hydrogenated Tallow (q.v.) that conforms generally to the formula:

$$HO(CH_2CHO)_yN(OCHCH_2)_xOH$$

with a CH$_3$ group on each propylene unit and an R group on the nitrogen.

where R represents the alkyl groups derived from hydrogenated tallow and x+y has an average value of 2.

Chemical Class: Alkoxylated Amines

Function: Antistatic Agent

Trade Name:
   Propomeen HT/12 (Akzo Nobel)

PPG-1 HYDROXYETHYL CAPRYLAMIDE

Definition: PPG-1 Hydroxyethyl Caprylamide is the organic compound that conforms generally to the formula:

$$CH_3(CH_2)_6C - NH(CH_2)_2O(CH_2CHO)_nH$$

with a C=O group and a CH$_3$ group.

where n has an average value of 1.

Chemical Class: Alkoxylated Amides

Functions: Surfactant - Emulsifying Agent; Surfactant - Foam Booster; Viscosity Increasing Agent - Aqueous

Trade Name:
   Promidium CC (Uniqema)

PPG-2 HYDROXYETHYL COCAMIDE

Definition: PPG-2 Hydroxyethyl Cocamide is the organic compound that conforms generally to the formula:

$$RC - NH(CH_2)_2O(CH_2CHO)_nH$$

with a C=O group and a CH$_3$ group.

where RCO- represents the fatty acids derived from coconut oil and n has an average value of 2.

Chemical Class: Alkoxylated Amides

Functions: Surfactant - Emulsifying Agent; Surfactant - Foam Booster; Viscosity Increasing Agent - Aqueous

Trade Name:
   Promidium CO (Uniqema)

PPG-2 HYDROXYETHYL COCO/ISO-STEARAMIDE

Definition: PPG-2 Hydroxyethyl Coco/Isostearamide is the organic compound that conforms generally to the formula:

$$RC - NH(CH_2)_2O(CH_2CHO)_nH$$

with a C=O group and a CH$_3$ group.

where RCO- represents a mixture of isostearic acid and coconut acid and n has an average value of 2.

Chemical Class: Alkoxylated Amides

Functions: Surfactant - Cleansing Agent; Surfactant - Foam Booster; Surfactant - Solubilizing Agent; Viscosity Increasing Agent - Aqueous

Trade Name:
   Promidium 2 (Uniqema DE)

PPG-3 HYDROXYETHYL SOYAMIDE

Definition: PPG-3 Hydroxyethyl Soyamide is the organic compound that conforms to the formula:

$$RC - NHCH_2CH_2O(CH_2CHO)_nH$$

with a C=O group and a CH$_3$ group.

where RCO- represents the fatty acids derived from soybean oil and n has an average value of 3.

Chemical Class: Alkoxylated Amides

Functions: Surfactant - Emulsifying Agent; Surfactant - Foam Booster; Viscosity Increasing Agent - Aqueous

Trade Name:
   Promidium SY (Uniqema)

PPG-17/IPDI/DMPA COPOLYMER

Definition: PPG-17/IPDI/DMPA Copolymer is a copolymer of PPG-17, isophorone diisocyanate and dimethylol propionic acid monomers.

Chemical Class: Synthetic Polymers

Function: Film Former

Trade Name:
   Avalure UR 450 (Noveon)

PPG-2 ISOCETETH-20 ACETATE

JPN Translation:
   酢酸 PPG - 2 イソセテス - 20

Definition: PPG-2 Isoceteth-20 Acetate is the acetic acid ester of the polyoxypropylene, polyoxyethylene ether of Isocetyl Alcohol (q.v.) that conforms generally to the formula:

$$CH_3C - O(CH_2CH_2O)_y(CH_2CHO)_xC_{16}H_{33}$$
$$\overset{O}{\overset{\|}{}} \qquad \overset{|}{CH_3}$$

where x has an average value of 2 and y has an average value of 20.

Information Source: JCLS

Chemical Classes: Alkoxylated Alcohols; Esters

Functions: Skin-Conditioning Agent - Emollient; Surfactant - Emulsifying Agent

Trade Name:
   CUPL PIC (Bernel)

PPG-30 ISOCETYL ETHER

Definition: PPG-30 Isocetyl Ether is the polypropylene glycol ether of Isocetyl Alcohol (q.v.). It conforms generally to the formula:

$$C_{16}H_{33}(OCHCH_2)_nOH$$
$$\overset{|}{CH_3}$$

where n has an average value of 30.

Chemical Class: Alkoxylated Alcohols

Function: Skin-Conditioning Agent - Emollient

Technical/Other Names:
   Polyoxypropylene (30) Isocetyl Ether
   Polypropylene Glycol (30) Isocetyl Ether

PPG-2-ISODECETH-4

Empirical Formula:
   $C_{24}H_{50}O_7$

Definition: PPG-2-Isodeceth-4 is the polyoxypropylene, polyoxyethylene glycol ether of isodecyl alcohol that conforms generally to the formula:

$$C_{10}H_{21}(OCHCH_2)_x(OCH_2CH_2)_yOH$$
$$\overset{|}{CH_3}$$

where x has an average value of 2 and y has an average value of 4.

Chemical Class: Alkoxylated Alcohols

Function: Surfactant - Emulsifying Agent

Technical/Other Names:
   Polyoxyethylene (4) Polyoxypropylene (2) Isodecyl Ether
   Polyoxypropylene (2) Polyoxyethylene (4) Isodecyl Ether

Trade Names:
   Sandoxylate SX-408 (Clariant)
   Sandoxylate SX-408 (Clariant GmbH, Personal Care)

PPG-2-ISODECETH-6

Empirical Formula:
   $C_{28}H_{58}O_9$

Definition: PPG-2-Isodeceth-6 is the polyoxyethylene, polyoxypropylene ether of isodecyl alcohol that conforms generally to the formula:

$$C_{10}H_{21}(OCHCH_2)_x(OCH_2CH_2)_yOH$$
$$\overset{|}{CH_3}$$

where x has an average value of 2 and y has an average value of 6.

Chemical Class: Alkoxylated Alcohols

Function: Surfactant - Emulsifying Agent

Technical/Other Names:
   Polyoxyethylene (6) Polyoxypropylene (2) Isodecyl Ether
   Polyoxypropylene (2) Polyoxyethylene (6) Isodecyl Ether

Trade Names:
   Sandoxylate SX-412 (Clariant)
   Sandoxylate SX-412 (Clariant GmbH, Personal Care)

PPG-2-ISODECETH-9

Empirical Formula:
   $C_{34}H_{70}O_{12}$

Definition: PPG-2-Isodeceth-9 is the polyoxypropylene, polyoxyethylene ether of isodecyl alcohol that conforms generally to the formula:

$$C_{10}H_{21}(OCHCH_2)_x(OCH_2CH_2)_yOH$$
$$\overset{|}{CH_3}$$

where x has an average value of 2 and y has an average value of 9.

Chemical Class: Alkoxylated Alcohols

Function: Surfactant - Emulsifying Agent

Technical/Other Names:
   Polyoxyethylene (9) Polyoxypropylene (2) Isodecyl Ether
   Polyoxypropylene (2) Polyoxyethylene (9) Isodecyl Ether

Trade Names:
   Sandoxylate SX-418 (Clariant)
   Sandoxylate SX-418 (Clariant GmbH, Personal Care)

PPG-2-ISODECETH-12

CAS No.: 155683-77-5

Definition: PPG-2-Isodeceth-12 is the polyoxypropylene, polyoxyethylene ether of isodecyl alcohol that conforms generally to the formula:

$$C_{10}H_{21}(OCHCH_2)_x(OCH_2CH_2)_yOH$$
$$\overset{|}{CH_3}$$

where x has an average value of 2 and y has an average value of 12.

Chemical Class: Alkoxylated Alcohols

Function: Surfactant - Emulsifying Agent

Technical/Other Names:
   Polyoxyethylene (12) Polyoxypropylene (2) Isodecyl Ether
   Polyoxypropylene (2) Polyoxyethylene (12) Isodecyl Ether

Trade Names:
   Sandoxylate SX-424 (Clariant)
   Sandoxylate SX-424 (Clariant GmbH, Personal Care)

PPG-3-ISODECETH-1

Empirical Formula:
   $C_{21}H_{43}O_5$

Definition: PPG-3-Isodeceth-1 is the polyoxypropylene, polyoxyethylene ether of isodecyl alcohol that conforms generally to the formula:

$$C_{10}H_{21}(OCHCH_2)_x(OCH_2CH_2)_yOH$$
$$\overset{|}{CH_3}$$

where x has an average value of 3 and y has an average value of 1.

Chemical Class: Alkoxylated Alcohols

Function: Skin-Conditioning Agent - Emollient

Technical/Other Names:
Polyoxyethylene (3) Poloxypropylene (1) Isodecyl Ether
Polyoxypropylene (1) Poloxyethylene (3) Isodecyl Ether

PPG-2 ISOSTEARATE

Empirical Formula:
$C_{24}H_{48}O_4$

Definition: PPG-2 Isostearate is the dipropylene glycol ester of Isostearic Acid (q.v.) that conforms generally to the formula:

$$C_{17}H_{35}\overset{\displaystyle O}{\overset{\displaystyle \|}{C}}-(OCHCH_2)_nOH$$
$$\underset{\displaystyle CH_3}{|}$$

where n has an average value of 2.

Information Source: JCLS

Chemical Class: Alkoxylated Carboxylic Acids

Function: Skin-Conditioning Agent - Emollient

PPG-15 ISOSTEARATE

Definition: PPG-15 Isostearate is the polypropylene glycol ester of isostearic acid that conforms generally to the formula:

$$C_{17}H_{35}\overset{\displaystyle O}{\overset{\displaystyle \|}{C}}-(OCHCH_2)_nOH$$
$$\underset{\displaystyle CH_3}{|}$$

where n has an average value of 15.

Information Source: JCLS

Chemical Class: Alkoxylated Carboxylic Acids

Function: Skin-Conditioning Agent - Emollient

Technical/Other Names:
Polyethylene Glycol (15) Isostearate
Polyoxyethylene (15) Isostearate

PPG-3-ISOSTEARETH-9

Empirical Formula:
$C_{45}H_{92}O_{13}$

Definition: PPG-3-Isosteareth-9 is the polyoxypropylene, polyoxyethylene ether of

Isostearyl Alcohol (q.v.) that conforms generally to the formula:

$$C_{17}H_{35}\overset{\displaystyle O}{\overset{\displaystyle \|}{C}}-(OCHCH_2)_x(OCH_2CH_2)_yOH$$
$$\underset{\displaystyle CH_3}{|}$$

where x has an average value of 3 and y has an average value of 9.

Chemical Class: Alkoxylated Alcohols

Function: Surfactant - Emulsifying Agent

Technical/Other Names:
Polyoxyethylene (9) Polyoxypropylene (3) Isostearyl Ether
Polyoxypropylene (3) Polyoxyethylene (9) Isostearyl Ether

PPG-4 JOJOBA ACID

Definition: PPG-4 Jojoba Acid is the polypropylene glycol derivative of the acids obtained from the saponification of Simmondsia Chinensis (Jojoba) Oil (q.v.) with an average propoxylation value of 4.

Chemical Class: Alkoxylated Carboxylic Acids

Functions: Skin-Conditioning Agent - Emollient; Surfactant - Emulsifying Agent

Technical/Other Names:
Polyoxypropylene (4) Jojoba Acid
Polypropylene Glycol (4) Jojoba Acid

PPG-10 JOJOBA ACID

Definition: PPG-10 Jojoba Acid is the polypropylene glycol derivative of the acids obtained from the saponification of Simmondsia Chinensis (Jojoba) Oil (q.v.) with an average propoxylation value of 10.

Chemical Class: Alkoxylated Carboxylic Acids

Function: Skin-Conditioning Agent - Emollient

Technical/Other Names:
Polyoxypropylene (10) Jojoba Acid
Polypropylene Glycol (10) Jojoba Acid

PPG-4 JOJOBA ALCOHOL

Definition: PPG-4 Jojoba Alcohol is the polypropylene glycol ether of Jojoba Alcohol (q.v.) with an average propoxylation value of 4.

Chemical Class: Alkoxylated Alcohols

Functions: Skin-Conditioning Agent - Emollient; Surfactant - Emulsifying Agent

Technical/Other Names:
Polyoxypropylene (4) Jojoba Alcohol
Polypropylene Glycol (4) Jojoba Alcohol

PPG-10 JOJOBA ALCOHOL

Definition: PPG-10 Jojoba Alcohol is the polypropylene glycol ether of Jojoba Alcohol (q.v.) with an average propoxylation value of 10.

Chemical Class: Alkoxylated Alcohols

Function: Surfactant - Emulsifying Agent

Technical/Other Names:
Polyoxypropylene (10) Jojoba Alcohol
Polypropylene Glycol (10) Jojoba Alcohol

PPG-12-LANETH-50

Definition: PPG-12-Laneth-50 is the polyoxypropylene, polyoxyethylene ether of Lanolin Alcohol (q.v.) that conforms generally to the formula:

$$R(OCHCH_2)_x(OCH_2CH_2)_yOH$$
$$\underset{\displaystyle CH_3}{|}$$

where R represents the lanolin alcohol radical, x has an average value of 12 and y has an average value of 50.

Chemical Classes: Alkoxylated Alcohols; Lanolin and Lanolin Derivatives

Function: Surfactant - Emulsifying Agent

Technical/Other Names:
Polyoxyethylene (50) Polyoxypropylene (12) Lanolin Ether
Polyoxypropylene (12) Polyoxyethylene (50) Lanolin Ether

PPG-5 LANOLATE

JPN Translation:
PPG - 5 ラノリン

Definition: PPG-5 Lanolate is the polypropylene glycol ester of Lanolin Acid (q.v.) with an average propoxylation value of 5.

Information Sources: JCIC, JCLS

Chemical Classes: Alkoxylated Carboxylic Acids; Lanolin and Lanolin Derivatives

Function: Skin-Conditioning Agent - Emollient

Technical/Other Names:
Polyoxypropylene Lanolin (5P.O.)

Polyoxypropylene (5) Monolanolate
Polypropylene Glycol (5) Monolanolate

PPG-2 LANOLIN ALCOHOL ETHER

CAS No.: 68439-53-2 (Generic)

JPN Translation:
PPG - 2 ラノリル

Definition: PPG-2 Lanolin Alcohol Ether is the polypropylene glycol ether of Lanolin Alcohol (q.v.) with an average propoxylation value of 2.

Information Sources: JCLS, JSQI, TSCA

Chemical Classes: Alkoxylated Alcohols; Lanolin and Lanolin Derivatives

Functions: Hair Conditioning Agent; Skin-Conditioning Agent - Emollient

Reported Product Category: Hair Conditioners

Technical/Other Names:
Polyoxypropylene (2) Lanolin Ether
Polypropylene Glycol (2) Lanolin Ether
PPG-2 Lanolin Ether

PPG-5 LANOLIN ALCOHOL ETHER

CAS No.: 68439-53-2 (Generic)

JPN Translation:
PPG - 5 ラノリル

Definition: PPG-5 Lanolin Alcohol Ether is the polypropylene glycol ether of Lanolin Alcohol (q.v.) with an average propoxylation value of 5.

Information Sources: JCLS, JSQI, TSCA

Chemical Classes: Alkoxylated Alcohols; Lanolin and Lanolin Derivatives

Functions: Hair Conditioning Agent; Skin-Conditioning Agent - Emollient

Technical/Other Names:
Polyoxypropylene (5) Lanolin Ether
Polypropylene Glycol (5) Lanolin Ether
PPG-5 Lanolin Ether

PPG-10 LANOLIN ALCOHOL ETHER

CAS No.: 68439-53-2 (Generic)

JPN Translation:
PPG - 10 ラノリル

Definition: PPG-10 Lanolin Alcohol Ether is the polypropylene glycol ether of Lanolin Alcohol (q.v.) with an average propoxylation value of 10.

Information Sources: JCLS, JSQI, TSCA

Chemical Classes: Alkoxylated Alcohols; Lanolin and Lanolin Derivatives

Functions: Hair Conditioning Agent; Skin-Conditioning Agent - Emollient

Reported Product Category: Hair Sprays (Aerosol Fixatives)

Technical/Other Names:
Polyoxypropylene (10) Lanolin Ether
Polypropylene Glycol (10) Lanolin Ether
PPG-10 Lanolin Ether

PPG-20 LANOLIN ALCOHOL ETHER

CAS No.: 68439-53-2 (Generic)

JPN Translation:
PPG - 20 ラノリル

Definition: PPG-20 Lanolin Alcohol Ether is the polypropylene glycol ether of Lanolin Alcohol (q.v.) with an average propoxylation value of 20.

Information Sources: JCLS, JSQI, TSCA

Chemical Classes: Alkoxylated Alcohols; Lanolin and Lanolin Derivatives

Functions: Hair Conditioning Agent; Skin-Conditioning Agent - Emollient

Reported Product Category: Bath Oils, Tablets, and Salts

Technical/Other Names:
Polyoxypropylene (20) Lanolin Ether
Polypropylene Glycol (20) Lanolin Ether
PPG-20 Lanolin Ether

PPG-30 LANOLIN ALCOHOL ETHER

CAS No.: 68439-53-2 (Generic)

JPN Translation:
PPG - 30 ラノリル

Definition: PPG-30 Lanolin Alcohol Ether is the polypropylene glycol ether of Lanolin Alcohol (q.v.) with an average propoxylation value of 30.

Information Sources: CTFA D, JCLS, JSQI, TSCA

Chemical Classes: Alkoxylated Alcohols; Lanolin and Lanolin Derivatives

Functions: Hair Conditioning Agent; Skin-Conditioning Agent - Emollient

Technical/Other Names:
Polyoxypropylene (30) Lanolin Ether
Polypropylene Glycol (30) Lanolin Ether
PPG-30 Lanolin Ether

PPG-5 LANOLIN WAX

JPN Translation:
PPG - 5 ラノリンロウ

Definition: PPG-5 Lanolin Wax is a polypropylene glycol derivative of Lanolin Wax (q.v.) with an average propoxylation value of 5.

Information Sources: CIR: [S] IJT-16(3)-1997, JCIC, JCLS

Chemical Classes: Alkoxylated Alcohols; Lanolin and Lanolin Derivatives

Function: Skin-Conditioning Agent - Emollient

Reported Product Category: Lipsticks

Technical/Other Names:
Polyoxypropylene Lanolin Wax
Polyoxypropylene (5) Lanolin Wax
Polypropylene Glycol (5) Lanolin Wax

Trade Names:
Propoxyol 1695 (Cognis Care Chemicals/NJ)
Propoxyol 1695 (Cognis Care Chemicals/PA)

PPG-5 LANOLIN WAX GLYCERIDE

Definition: PPG-5 Lanolin Wax Glyceride is the polypropylene glycol ether of the condensation product of Lanolin Wax (q.v.) and Glycerin (q.v.) with an average propoxylation value of 5.

Information Source: CIR: [S] IJT-16(3)-1997

Chemical Classes: Alkoxylated Alcohols; Glyceryl Esters and Derivatives; Lanolin and Lanolin Derivatives

Function: Skin-Conditioning Agent - Emollient

Reported Product Category: Lipsticks

Technical/Other Names:
Polyoxypropylene (5) Lanolin Wax Glyceride
Polypropylene Glycol (5) Lanolin Wax Glyceride

PPG-9 LAURATE

CAS No.: 9035-84-1

Empirical Formula:
$C_{39}H_{78}O_{11}$

Definition: PPG-9 Laurate is the polypropylene glycol ester of lauric acid that conforms to the formula:

$$CH_3(CH_2)_{10}\overset{\displaystyle O}{\overset{\|}{C}}-(OCHCH_2)_nOH$$
$$\underset{CH_3}{|}$$

where n has an average value of 9.

Information Sources: 21CFR175.300, 21CFR176.210, JCLS

Chemical Class: Alkoxylated Carboxylic Acids

Function: Skin-Conditioning Agent - Emollient

Technical/Other Names:
Polyoxypropylene (9) Monolaurate
Polypropylene Glycol (9) Monolaurate

Trade Name:
AEC PPG-9 Laurate (A & E Connock)

PPG-2-LAURETH-5

Definition: PPG-2-Laureth-5 is the polyoxypropylene, polyoxyethylene ether of Lauryl Alcohol (q.v.) that conforms generally to the formula:

$$CH_3(CH_2)_{10}CH_2(OCHCH_2)_x(OCH_2CH_2)_yOH$$
$$|$$
$$CH_3$$

where x has an average value of 2 and y has an average value of 5.

Chemical Class: Alkoxylated Alcohols

Functions: Skin-Conditioning Agent - Emollient; Surfactant - Emulsifying Agent

Trade Name:
Noigen LP-70 (Dai-Ichi Kogyo)

PPG-2-LAURETH-8

Definition: PPG-2-Laureth-8 is the polyoxyethylene, polyoxypropylene ether of Lauryl Alcohol (q.v.) that conforms generally to the formula:

$$CH_3(CH_2)_{10}CH_2(OCHCH_2)_x(OCH_2CH_2)_yOH$$
$$|$$
$$CH_3$$

where x has an average value of 2 and y has an average value of 8.

Chemical Class: Alkoxylated Alcohols

Functions: Skin-Conditioning Agent - Emollient; Surfactant - Emulsifying Agent

Trade Names:
Noigen LP-100 (Dai-Ichi Kogyo)
Wondersurf 80 (Aoki Oil Industrial)

PPG-3-LAURETH-9

Empirical Formula:
$C_{39}H_{80}O_{13}$

Definition: PPG-3-Laureth-9 is the polyoxypropylene, polyoxyethylene ether of Lauryl Alcohol (q.v.) that conforms generally to the formula:

$$CH_3(CH_2)_{11}(OCHCH_2)_x(OCH_2CH_2)_yOH$$
$$|$$
$$CH_3$$

where x has an average value of 3 and y has an average value of 9.

Information Source: JCLS

Chemical Class: Alkoxylated Alcohols

Function: Surfactant - Emulsifying Agent

Technical/Other Names:
Polyoxyethylene (9) Polyoxypropylene (3) Lauryl Ether
Polyoxypropylene (3) Polyoxyethylene (9) Lauryl Ether

Trade Name:
Acconon 1300 (Abitec Corporation)

PPG-4 LAURETH-2

Empirical Formula:
$C_{28}H_{58}O_7$

Definition: PPG-4 Laureth-2 is the polyoxypropylene, polyoxyethylene ether of Lauryl Alcohol (q.v.) that conforms to the formula:

$$CH_3(CH_2)_{11}(OCHCH_2)_x(OCH_2CH_2)_yOH$$
$$|$$
$$CH_3$$

where x has an average value of 4 and y has an average of value of 2.

Information Source: JCLS

Chemical Class: Alkoxylated Alcohols

Functions: Skin-Conditioning Agent - Emollient; Surfactant - Emulsifying Agent

Technical/Other Names:
Polyoxyethylene (2) Polyoxypropylene (4) Lauryl Ether
Polyoxypropylene (4) Polyoxyethylene (2) Lauryl Ether

Trade Name:
Marlox MO 124 (Sasol GmbH - Marl)

PPG-4 LAURETH-5

Empirical Formula:
$C_{34}H_{70}O_{10}$

Definition: PPG-4 Laureth-5 is the polyoxypropylene, polyoxyethylene ether of Lauryl Alcohol (q.v.) that conforms generally to the formula:

$$CH_3(CH_2)_{11}(OCHCH_2)_x(OCH_2CH_2)_yOH$$
$$|$$
$$CH_3$$

where x has an average value of 4 and y has an average value of 5.

Information Source: JCLS

Chemical Class: Alkoxylated Alcohols

Functions: Skin-Conditioning Agent - Emollient; Surfactant - Emulsifying Agent

Technical/Other Name:
Polyoxyethylene (5) Polyoxypropylene (4) Lauryl Ether

Trade Name:
Marlox MO 154 (Sasol GmbH - Marl)

PPG-4 LAURETH-7

Empirical Formula:
$C_{38}H_{78}O_{12}$

Definition: PPG-4 Laureth-7 is the polyoxypropylene, polyoxyethylene ether of Lauryl Alcohol (q.v.) that conforms generally to the formula:

$$CH_3(CH_2)_{11}(OCHCH_2)_x(OCH_2CH_2)_yOH$$
$$|$$
$$CH_3$$

where x has an average value of 4 and y has an average value of 7.

Information Source: JCLS

Chemical Class: Alkoxylated Alcohols

Functions: Skin-Conditioning Agent - Emollient; Surfactant - Emulsifying Agent

Technical/Other Name:
Polyoxypropylene (4) Polyoxyethylene (7) Lauryl Ether

PPG-4-LAURETH-15

Definition: PPG-4-Laureth-15 is the polyoxyethylene, polyoxypropylene ether of Lauryl Alcohol (q.v.) that conforms generally to the formula:

$$CH_3(CH_2)_{10}CH_2(OCHCH_2)_x(OCH_2CH_2)_yOH$$
$$|$$
$$CH_3$$

where x has an average value of 4 and y has an average value of 15.

Chemical Class: Alkoxylated Alcohols

Functions: Skin-Conditioning Agent - Emollient; Surfactant - Emulsifying Agent

Trade Name:
Noigen LP-180 (Dai-Ichi Kogyo)

PPG-5-LAURETH-5

JPN Translation:
PPG - 5 ラウレス - 5

The inclusion of any compound in the *Dictionary and Handbook* does not indicate that use of that substance as a cosmetic ingredient complies with the laws and regulations governing such use in the United States or any other country.

Empirical Formula:
$C_{37}H_{76}O_{11}$

Definition: PPG-5-Laureth-5 is the polyoxyethylene, polyoxypropylene ether of lauryl alcohol that conforms generally to the formula:

$$CH_3(CH_2)_{11}(OCHCH_2)_x(OCH_2CH_2)_yOH$$
$$|$$
$$CH_3$$

where x has an average value of 5 and y has an average value of 5.

Information Source: JCLS

Chemical Class: Alkoxylated Alcohols

Functions: Skin-Conditioning Agent - Emollient; Surfactant - Emulsifying Agent

Technical/Other Names:
Polyoxyethylene (5) Polyoxypropylene (5) Lauryl Ether
Polyoxypropylene (5) Polyoxyethylene (5) Lauryl Ether

Trade Names:
Aethoxal B (Cognis Care Chemicals/NJ)
Aethoxal B (Cognis Care Chemicals/PA)
Aethoxal B (Cognis Deutschland)

Trade Name Mixture:
Cosmedia SPL (Cognis France)

PPG-6-LAURETH-3

JPN Translation:
PPG - 6 ラウレス - 3

Empirical Formula:
$C_{36}H_{74}O_{10}$

Definition: PPG-6-Laureth-3 is the polyoxypropylene, polyoxyethylene ether of lauryl alcohol that conforms generally to the formula:

$$CH_3(CH_2)_{11}(OCHCH_2)_x(OCH_2CH_2)_yOH$$
$$|$$
$$CH_3$$

where x has an average value of 6 and y has an average value of 3.

Information Source: JCLS

Chemical Class: Alkoxylated Alcohols

Functions: Skin-Conditioning Agent - Emollient; Surfactant - Emulsifying Agent

Technical/Other Names:
Polyoxyethylene (3) Polyoxypropylene (6) Lauryl Ether
Polyoxypropylene (6) Polyoxyethylene (3) Lauryl Ether

PPG-25-LAURETH-25

CAS No.: 37311-00-5

JPN Translation:
PPG - 25 ラウレス - 25

Definition: PPG-25-Laureth-25 is the polyoxypropylene, polyoxyethylene ether of Lauryl Alcohol (q.v.) that conforms generally to the formula:

$$CH_3(CH_2)_{11}(OCHCH_2)_x(OCH_2CH_2)_yOH$$
$$|$$
$$CH_3$$

where x has an average value of 25 and y has an average value of 25.

Information Source: JCLS

Chemical Class: Alkoxylated Alcohols

Function: Surfactant - Emulsifying Agent

Reported Product Category: Bath Preparations, Misc.

Technical/Other Names:
Polyoxyethylene (25) Polyoxypropylene (25) Lauryl Ether
Polyoxypropylene (25) Polyoxyethylene (25) Lauryl Ether

Trade Name:
ADF Oleile (Vevy)

Trade Name Mixture:
Cosmacol PSE (Sasol Italy)

PPG-14 LAURETH-60 HEXYL DICARBAMATE

Definition: PPG-14 Laureth-60 Hexyl Dicarbamate is the carbamic acid diester of the polyoxypropylene, polyoxyethylene ether of lauryl alcohol. It conforms generally to the formula:

$$\left[R(OCHCH_2)_y(OCH_2CH_2)_xO - CNH(CH_2)_6 \right]_2$$

where x has an average value of 60, y has an average value of 14, and R represents the lauryl alkyl moiety.

Chemical Classes: Alkoxylated Alcohols; Esters

Function: Viscosity Increasing Agent - Aqueous

PPG-14 LAURETH-60 ISOPHORYL DICARBAMATE

Definition: PPG-14 Laureth-60 Isophoryl Dicarbamate is the diester of isophorone diisocyanate and the polyoxyethylene, polyoxypropylene derivative of Lauryl Alcohol (q.v.), in which the average ethoxylation value is 60, and the average propoxylation value is 14.

Chemical Class: Alkoxylated Amides

Function: Viscosity Increasing Agent - Aqueous

PPG-4 LAURYL ETHER

CAS No.: 9064-14-6 (Generic)

Empirical Formula:
$C_{24}H_{50}O_5$

Definition: PPG-4 Lauryl Ether is the polypropylene glycol ether of lauryl alcohol that conforms to the formula:

$$CH_3(CH_2)_{11}(OCHCH_2)_nOH$$
$$|$$
$$CH_3$$

where n has an average value of 4.

Information Source: TSCA

Chemical Class: Alkoxylated Alcohols

Function: Skin-Conditioning Agent - Emollient

Technical/Other Names:
Polyoxypropylene (4) Lauryl Ether
Polypropylene Glycol (4) Lauryl Ether

PPG-7 LAURYL ETHER

CAS No.: 9064-14-6 (Generic)

Empirical Formula:
$C_{33}H_{68}O_8$

Definition: PPG-7 Lauryl Ether is the polypropylene glycol ether of lauryl alcohol that conforms generally to the formula:

$$CH_3(CH_2)_{11}(OCHCH_2)_nOH$$
$$|$$
$$CH_3$$

where n has an average value of 7.

Chemical Class: Alkoxylated Alcohols

Functions: Skin-Conditioning Agent - Emollient; Surfactant - Emulsifying Agent

Technical/Other Names:
Polyoxypropylene (7) Lauryl Ether
Polypropylene Glycol (7) Lauryl Ether

PPG-2 METHYL ETHER

CAS Nos.	EINECS Nos.
13429-07-7	236-547-9
34590-94-8	252-104-2

Empirical Formula:
$C_7H_{16}O_3$

Definition: PPG-2 Methyl Ether is the poly-propylene glycol ether of methyl alcohol that conforms to the formula:

$$CH_3(OCHCH_2)_nOH$$
$$|$$
$$CH_3$$

where n has an average value of 2.

Information Sources: 21CFR175.105, 21CFR181.22, 21CFR181.30, CTFA D, MI-13(3378), RIFM, TSCA

Chemical Class: Alkoxylated Alcohols

Functions: Fragrance Ingredient; Solvent

Reported Product Categories: Hair Condi-tioners; Hair Dyes and Colors (All Types Requiring Caution Statements and Patch Tests)

Technical/Other Names:
Dipropylene glycol monomethyl ether (RIFM)
Dipropylene Glycol Monomethyl Ether
Methoxy Dipropylene Glycol
(2-Methoxymethylethoxy)Propanol
1-(2-Methoxypropoxy)-2-Propanol
Polyoxypropylene (2) Methyl Ether
Polypropylene Glycol (2) Methyl Ether
Propanol, (2-Methoxymethylethoxy)-
2-Propanol, 1-(2-Methoxypropoxy)-

Trade Names:
DOWANOL DPM (Dow Chemical)
Hisolve DPM (Toho)

Trade Name Mixtures:
Synthro-Pon A 1820 N 75 (Synthron)
Wheat Protein "COS" Liquid
(Cosmetochem) (Cosmetochem International Ltd.)

PPG-3 METHYL ETHER

CAS Nos.   **EINECS No.**
25498-49-1   247-045-4
37286-64-9 (Generic)

Empirical Formula:
$C_{10}H_{22}O_4$

Definition: PPG-3 Methyl Ether is the poly-propylene glycol ether of methyl alcohol that conforms to the formula:

$$CH_3(OCHCH_2)_nOH$$
$$|$$
$$CH_3$$

where n has an average value of 3.

Information Sources: 21CFR181.22, 21CFR181.30, TSCA

Chemical Class: Alkoxylated Alcohols

Function: Solvent

Technical/Other Names:
Polyoxypropylene (3) Methyl Ether
Polypropylene Glycol (3) Methyl Ether
Tripropylene Glycol Monomethyl Ether

Trade Name:
DOWANOL TPM (Dow Chemical)

PPG-2 METHYL ETHER ACETATE

CAS No.: 88917-22-0

Empirical Formula:
$C_9H_{18}O_4$

Definition: PPG-2 Methyl Ether Acetate is the ester of PPG-2 Methyl Ether (q.v.) and Acetic Acid (q.v.).

Chemical Classes: Alkoxylated Alcohols; Esters

Function: Solvent

Technical/Other Name:
Dipropylene Glycol Monomethyl Ether Acetate

Trade Name:
DOWANOL DPMA (Dow Chemical)

PPG-10 METHYL GLUCOSE ETHER

JPN Translation:
PPG - 10 メチルグルコース

Empirical Formula:
$C_{37}H_{74}O_{16}$

Definition: PPG-10 Methyl Glucose Ether is the polypropylene glycol ether of methyl glucose that conforms generally to the formula:

$$CH_3(C_6H_{10}O_5)(OCHCH_2)_nOH$$
$$|$$
$$CH_3$$

where n has an average value of 10.

Information Source: JSQI

Chemical Classes: Alkoxylated Alcohols; Carbohydrates

Functions: Hair Conditioning Agent; Skin-Conditioning Agent - Miscellaneous

Reported Product Categories: Bath Oils, Tablets, and Salts; Hair Sprays (Aerosol Fixatives); Tonics, Dressings, and Other Hair Grooming Aids

Technical/Other Names:
Polyoxypropylene (10) Methyl Glucose Ether
Polypropylene Glycol (10) Methyl Glucose Ether

Trade Name:
Glucam P-10 (Amerchol)

PPG-20 METHYL GLUCOSE ETHER

JPN Translation:
PPG - 20 メチルグルコース

Definition: PPG-20 Methyl Glucose Ether is the polypropylene glycol ether of methyl glucose that conforms generally to the formula:

$$CH_3(C_6H_{10}O_5)(OCHCH_2)_nOH$$
$$|$$
$$CH_3$$

where n has an average value of 20.

Information Source: JSQI

Chemical Classes: Alkoxylated Alcohols; Carbohydrates

Functions: Hair Conditioning Agent; Skin-Conditioning Agent - Miscellaneous

Reported Product Categories: Bath Oils, Tablets, and Salts; Hair Preparations (Non-coloring), Misc.; Hair Sprays (Aerosol Fixatives); Tonics, Dressings, and Other Hair Grooming Aids

Technical/Other Names:
Polyoxypropylene (20) Methyl Glucose Ether
Polypropylene Glycol (20) Methyl Glucose Ether

Trade Names:
AEC PPG-20 Methyl Glucose Ether (A & E Connock)
Glucam P-20 (Amerchol)

PPG-25 METHYL GLUCOSE ETHER

JPN Translation:
PPG - 25 メチルグルコース

Definition: PPG-25 Methyl Glucose Ether is the polypropylene glycol ether of methyl glucose that conforms generally to the formula:

$$CH_3(C_6H_{10}O_5)(OCHCH_2)_nOH$$
$$|$$
$$CH_3$$

where n has na average value of 25.

Information Source: JCLS

Chemical Classes: Alkoxylated Alcohols; Carbohydrates

Functions: Hair Conditioning Agent; Skin-Conditioning Agent - Miscellaneous

Technical/Other Name:
Polyoxypropylene (25) Methyl Glycose Ether

PPG-20 METHYL GLUCOSE ETHER ACETATE

Definition: PPG-20 Methyl Glucose Ether Acetate is the ester of PPG-20 Methyl Glucose Ether (q.v.) and acetic acid.

Chemical Classes: Alkoxylated Alcohols; Carbohydrates; Esters

Function: Skin-Conditioning Agent - Miscellaneous

PPG-20 METHYL GLUCOSE ETHER DISTEARATE

JPN Translation:
ジステアリン酸 PPG - 20 メチルグルコ - ス

Definition: PPG-20 Methyl Glucose Ether Distearate is the diester of PPG-20 Methyl Glucose Ether (q.v.) and stearic acid.

Information Sources: JCIC, JCLS

Chemical Classes: Alkoxylated Alcohols; Carbohydrates; Esters

Function: Skin-Conditioning Agent - Emollient

Reported Product Categories: Moisturizing Preparations; Eyeliners; Indoor Tanning Preparations

Technical/Other Name:
Polyoxypropylene Methyl Glucoside Distearate

Trade Name:
Glucam P-20 Distearate (Amerchol)

PPG-3-MYRETH-3

CAS No.: 37311-04-9 (Generic)

Empirical Formula:
$C_{29}H_{60}O_7$

Definition: PPG-3-Myreth-3 is the polyoxypropylene, polyoxyethylene ether of myristyl alcohol that conforms generally to the formula:

$$CH_3(CH_2)_{13}(OCHCH_2)_x(OCH_2CH_2)_yOH$$
$$|$$
$$CH_3$$

where x has an average value of 3 and y has an average value of 3.

Information Source: TSCA

Chemical Class: Alkoxylated Alcohols

Functions: Skin-Conditioning Agent - Emollient; Surfactant - Emulsifying Agent

Technical/Other Names:
Polyoxyethylene (3) Polyoxypropylene (3) Myristyl Ether

Polyoxypropylene (3) Polyoxyethylene (3) Myristyl Ether

PPG-3-MYRETH-11

CAS No.: 37311-04-9 (Generic)

Definition: PPG-3-Myreth-11 is the polyoxypropylene, polyoxyethylene ether of myristyl alcohol that conforms generally to the formula:

$$CH_3(CH_2)_{13}(OCHCH_2)_x(OCH_2CH_2)_yOH$$
$$|$$
$$CH_3$$

where x has an average value of 3 and y has an average value of 11.

Information Source: TSCA

Chemical Class: Alkoxylated Alcohols

Function: Surfactant - Emulsifying Agent

Technical/Other Names:
Polyoxyethylene (11) Polyoxypropylene (3) Myristyl Ether
Polyoxypropylene (3) Polyoxyethylene (11) Myristyl Ether

PPG-3 MYRISTYL ETHER

CAS Nos.: 63793-60-2 (Generic); 74790-85-5

JPN Translation:
PPG - 3 ミリスチル

Empirical Formula:
$C_{23}H_{48}O_4$

Definition: PPG-3 Myristyl Ether is the polypropylene glycol ether of myristyl alcohol that conforms to the formula:

$$CH_3(CH_2)_{13}(OCHCH_2)_nOH$$
$$|$$
$$CH_3$$

where n has an average value of 3.

Information Sources: JCIC, JCLS, TSCA

Chemical Class: Alkoxylated Alcohols

Function: Skin-Conditioning Agent - Emollient

Reported Product Categories: Deodorants (Underarm); Colognes and Toilet Waters; Personal Cleanliness Products, Misc.

Technical/Other Names:
Polyoxypropylene (3) Myristyl Ether
Polyoxypropylene Myristylether (3P.O.)
Polypropylene Glycol (3) Myristyl Ether
Tripropylene Glycol Myristyl Ether

Trade Names:
Acconon MA 3 (Abitec Corporation)

Jeecol PMA-3 (Jeen)
Procol PMA-3 (Protameen)
Promyristyl PM3 (Croda Chemicals)
Promyristyl PM-3 (Croda, Inc.)
Varionic APM (Degussa Care Specialties)

Trade Name Mixtures:
Dow Corning 2-8178 Gellant (Dow Corning)
Varisoft Clear (Degussa Care Specialties)

PPG-4 MYRISTYL ETHER

CAS No.: 63793-60-2 (Generic)

Empirical Formula:
$C_{26}H_{54}O_5$

Definition: PPG-4 Myristyl Ether is the polypropylene glycol ether of myristyl alcohol that conforms to the formula:

$$CH_3(CH_2)_{13}(OCHCH_2)_nOH$$
$$|$$
$$CH_3$$

where n has an average value of 4.

Information Source: TSCA

Chemical Class: Alkoxylated Alcohols

Function: Skin-Conditioning Agent - Emollient

Technical/Other Names:
Polyoxypropylene (4) Myristyl Ether
Polypropylene Glycol (4) Myristyl Ether

PPG-3 MYRISTYL ETHER NEO-HEPTANOATE

CAS No.: 325726-83-8

Definition: PPG-3 Myristyl Ether Neoheptanoate is the ester of PPG-3 Myristyl Ether (q.v.) and neoheptanoic acid. It conforms generally to the formula:

$$CH_3(CH_2)_{12}CH_2(OCHCH_2)_nO-\overset{\overset{O}{\|}}{C}C_6H_{13}$$
$$|$$
$$CH_3$$

where n has an average value of 3.

Chemical Classes: Alkoxylated Alcohols; Esters

Functions: Skin-Conditioning Agent - Emollient; Suspending Agent - Nonsurfactant

Technical/Other Name:
Polypropylene Glycol (3) Myristyl Ether Neoheptanoate

Trade Name:
Trivasperse NH (Trivent)

PPG-2 MYRISTYL ETHER PROPIONATE

CAS No.: 74775-06-7

JPN Translation:
プロピオン酸 PPG - 2 ミリスチル

Empirical Formula:
$C_{23}H_{46}O_4$

Definition: PPG-2 Myristyl Ether Propionate is the ester of propionic acid and the polypropylene glycol ether of myristyl alcohol that conforms to the formula:

$$CH_3(CH_2)_{13}(OCHCH_2)_2O-CCH_2CH_3$$

(with O double-bonded to the carbonyl C, and CH_3 branch on the $OCHCH_2$ group)

Information Sources: JCIC, JCLS

Chemical Classes: Alkoxylated Alcohols; Esters

Function: Skin-Conditioning Agent - Miscellaneous

Reported Product Categories: Eyebrow Pencils; Lipsticks; Eyeliners; Foundations; Night Skin Care Preparations; Bath Oils, Tablets, and Salts; Cleansing Products (Cold Creams, Cleansing Lotions, Liquids and Pads); Eye Shadows; Makeup Preparations (Not eye), Misc.; Moisturizing Preparations

Technical/Other Names:
Dipropylene Glycol, Myristyl Ether, Propionate
Polyoxypropylene (2) Myristyl Ether Propionate
Polyoxypropylene Myristyl Ether Propionate (2P.O.)
Polypropylene Glycol (2) Myristyl Ether Propionate

Trade Names:
Crodamol PMP (Croda Chemicals)
Crodamol PMP (Croda, Inc.)

PPG-26 OLEATE

CAS No.: 31394-71-5 (Generic)

JPN Translation:
オレイン酸 PPG - 26

Definition: PPG-26 Oleate is the polypropylene glycol ester of oleic acid that conforms to the formula:

$$CH_3(CH_2)_7CH=CH(CH_2)_7C-(OCHCH_2)_nOH$$

(with O double-bonded to the carbonyl C, and CH_3 branch on the $OCHCH_2$ group)

where n has an average value of 26.

Information Sources: 21CFR175.300, 21CFR176.210, JCIC, JCLS, TSCA

Chemical Class: Alkoxylated Carboxylic Acids

Function: Skin-Conditioning Agent - Emollient

Reported Product Categories: Bath Preparations, Misc.; Body and Hand Preparations (Excluding Shaving Preparations); Fragrance Preparations, Misc.; Moisturizing Preparations; Hair Sprays (Aerosol Fixatives); Shaving Cream (Aerosol, Brushless and Lather); Hair Preparations (Non-coloring), Misc.; Paste Masks (Mud Packs)

Technical/Other Names:
Polyoxypropylene (26) Monooleate
Polyoxypropylene Monooleate (26P.O.)
Polypropylene Glycol (26) Monooleate

Trade Names:
AEC PPG-26 Oleate (A & E Connock)
Lumulse CSA-80 (Lambent)
Lutrol OP 2000 (BASF)

PPG-36 OLEATE

CAS No.: 31394-71-5 (Generic)

JPN Translation:
オレイン酸 PPG - 36

Definition: PPG-36 Oleate is the polypropylene glycol ester of oleic acid that conforms to the formula:

$$CH_3(CH_2)_7CH=CH(CH_2)_7C-(OCHCH_2)_nOH$$

(with O double-bonded to the carbonyl C, and CH_3 branch on the $OCHCH_2$ group)

where n has an average value of 36.

Information Sources: 21CFR175.300, 21CFR176.210, CTFA D, JCIC, JCLS, JSQI, TSCA

Chemical Class: Alkoxylated Carboxylic Acids

Function: Skin-Conditioning Agent - Emollient

Technical/Other Names:
Polyoxypropylene (36) Monooleate
Polyoxypropylene Monooleate (36P.O.)
Polypropylene Glycol (36) Monooleate

Trade Name:
AEC PPG-36 Oleate (A & E Connock)

PPG-10 OLEYL ETHER

CAS No.: 52581-71-2 (Generic)

Empirical Formula:
$C_{48}H_{96}O_{11}$

Definition: PPG-10 Oleyl Ether is the polypropylene glycol ether of oleyl alcohol that conforms to the formula:

$$CH_3(CH_2)_7CH=CH(CH_2)_8(OCHCH_2)_nOH$$

(with CH_3 branch on the $OCHCH_2$ group)

where n has an average value of 10.

Information Sources: CTFA D, TSCA

Chemical Class: Alkoxylated Alcohols

Function: Skin-Conditioning Agent - Emollient

Technical/Other Names:
Polyoxypropylene (10) Oleyl Ether
Polypropylene Glycol (10) Oleyl Ether

PPG-20 OLEYL ETHER

CAS No.: 52581-71-2 (Generic)

Definition: PPG-20 Oleyl Ether is the polypropylene glycol ether of oleyl alcohol that conforms generally to the formula:

$$CH_3(CH_2)_7CH=CH(CH_2)_8(OCHCH_2)_nOH$$

(with CH_3 branch on the $OCHCH_2$ group)

where n has an average value of 20.

Information Source: TSCA

Chemical Class: Alkoxylated Alcohols

Function: Skin-Conditioning Agent - Emollient

Technical/Other Names:
Polyoxypropylene (20) Oleyl Ether
Polypropylene Glycol (20) Oleyl Ether

PPG-23 OLEYL ETHER

CAS No.: 52581-71-2 (Generic)

Definition: PPG-23 Oleyl Ether is the polypropylene glycol ether of oleyl alcohol that conforms generally to the formula:

$$CH_3(CH_2)_7CH=CH(CH_2)_8(OCHCH_2)_nOH$$

(with CH_3 branch on the $OCHCH_2$ group)

where n has an average value of 23.

Information Source: TSCA

Chemical Class: Alkoxylated Alcohols

Function: Skin-Conditioning Agent - Emollient

Technical/Other Names:
Polyoxypropylene (23) Oleyl Ether
Polypropylene Glycol (23) Oleyl Ether

PPG-30 OLEYL ETHER

CAS No.: 52581-71-2 (Generic)

Definition: PPG-30 Oleyl Ether is the polypropylene glycol ether of oleyl alcohol that conforms generally to the formula:

$$CH_3(CH_2)_7CH=CH(CH_2)_8(OCHCH_2)_nOH$$
$$|$$
$$CH_3$$

where n has an average value of 30.

Information Source: TSCA

Chemical Class: Alkoxylated Alcohols

Function: Skin-Conditioning Agent - Emollient

Technical/Other Names:
Polyoxypropylene (30) Oleyl Ether
Polypropylene Glycol (30) Oleyl Ether

PPG-37 OLEYL ETHER

CAS No.: 52581-71-2 (Generic)

Definition: PPG-37 Oleyl Ether is the polypropylene glycol ether of oleyl alcohol that conforms generally to the formula:

$$CH_3(CH_2)_7CH=CH(CH_2)_8(OCHCH_2)_nOH$$
$$|$$
$$CH_3$$

where n has an average value of 37.

Information Source: TSCA

Chemical Class: Alkoxylated Alcohols

Function: Skin-Conditioning Agent - Emollient

Technical/Other Names:
Polyoxypropylene (37) Oleyl Ether
Polypropylene Glycol (37) Oleyl Ether

PPG-50 OLEYL ETHER

CAS No.: 52581-71-2 (Generic)

Definition: PPG-50 Oleyl Ether is the polypropylene glycol ether of oleyl alcohol that conforms generally to the formula:

$$CH_3(CH_2)_7CH=CH(CH_2)_8(OCHCH_2)_nOH$$
$$|$$
$$CH_3$$

where n has an average value of 50.

Information Source: TSCA

Chemical Class: Alkoxylated Alcohols

Function: Skin-Conditioning Agent - Emollient

Technical/Other Names:
Polyoxypropylene (50) Oleyl Ether
Polypropylene Glycol (50) Oleyl Ether

PPG-14 PALMETH-60 HEXYL DICARBAMATE

CAS No.: 205599-29-7

Definition: PPG-14 Palmeth-60 Hexyl Dicarbamate is the carbamic acid diester of the polyoxypropylene, polyoxyethylene ether of the fatty alcohols derived from Elaeis Guineensis (Palm) Kernel Oil (q.v.). It conforms generally to the formula:

$$\left[\begin{array}{cc} CH_3 & O \\ | & \| \\ R(OCHCH_2)_y(OCH_2CH_2)_xO - CNH(CH_2)_6 \end{array} \right]_2$$

where x has an average value of 60, y has an average value of 14, and R represents the fatty alcohols derived from palm kernel oil.

Chemical Classes: Alkoxylated Alcohols; Esters

Function: Viscosity Increasing Agent - Aqueous

Trade Name:
Elfacos T212 (Akzo Nobel Surface AB)

PPG-2-PEG-6 COCONUT OIL ESTERS

Definition: PPG-2-PEG-6 Coconut Oil Esters is the end product of the controlled ethoxylation, propoxylation of Cocos Nucifera (Coconut) Oil (q.v.).

Chemical Class: Glyceryl Esters and Derivatives

Functions: Skin-Conditioning Agent - Emollient; Surfactant - Emulsifying Agent

Technical/Other Names:
Polyoxyethylene (6) Polyoxypropylene (2) Coconut Oil
Polyoxypropylene (2) Polyoxyethylene (6) Coconut Oil
PPG-2-PEG-6 Coconut Oil Complex

Trade Name:
Colan 32 (Lanaetex)

PPG/PEG-10/2 GLYCERYL COCOATE

Definition: PPG/PEG-10/2 Glyceryl Cocoate is the polyoxypropylene, polyoxyethylene derivative of Glyceryl Cocoate (q.v.) containing an average of 2 moles of propylene oxide and 10 moles of ethylene oxide.

Chemical Classes: Alkoxylated Alcohols; Glyceryl Esters and Derivatives

Function: Surfactant - Emulsifying Agent

Technical/Other Name:
Polyoxypropylene (2) Polyoxyethylene (10) Glyceryl Cocoate

Trade Name:
Unigly 10MK-2B (NOF)

PPG-75-PEG-300 HEXYLENE GLYCOL

JPN Translation:
PPG - 75 - PEG - 300 ヘキシレングリコール

Definition: PPG-75-PEG-300 Hexylene Glycol is the polyoxypropylene, polyoxyethylene ether of hexylene glycol that has an average of 75 moles of propylene oxide and 300 moles of ethylene oxide.

Information Sources: JCIC, JCLS, JSQI

Chemical Class: Alkoxylated Alcohols

Function: Surfactant - Solubilizing Agent

Technical/Other Names:
Polyoxyethylene (300) Polyoxypropylene (75) Hexylene Glycol
Polyoxyethylene Polyoxypropylene Hexyleneglycol Ether (300E.O.) (75P.O.)
Polyoxypropylene (75) Polyoxyethylene (300) Hexylene Glycol

PPG-20-PEG-20 HYDROGENATED LANOLIN

JPN Translation:
PPG - 20 - PEG - 20 水添ラノリン

Definition: PPG-20-PEG-20 Hydrogenated Lanolin is the polyoxypropylene, polyoxyethylene derivative of hydrogenated lanolin that conforms generally to the formula:

$$R(OCHCH_2)_x(OCH_2CH_2)_yOH$$
$$|$$
$$CH_3$$

where R represents the hydrogenated lanolin radical, x has an average value of 20 and y has an average value of 20.

Information Sources: JCIC, JCLS, JSQI

Chemical Class: Lanolin and Lanolin Derivatives

Functions: Hair Conditioning Agent; Skin-Conditioning Agent - Emollient; Surfactant - Emulsifying Agent

Technical/Other Names:
Polyoxyethylene (20) Polyoxypropylene (20) Hydrogenated Lanolin
Polyoxypropylene (20) Polyoxyethylene (20) Hydrogenated Lanolin

PPG-2-PEG-11 HYDROGENATED LAURYL ALCOHOL ETHER

Definition: PPG-2-PEG-11 Hydrogenated Lauryl Alcohol Ether is a polyoxypropylene, polyoxyethylene ether of Hydrogenated Lauryl Alcohol (q.v.) that conforms generally to the formula:

$$R(OCHCH_2)_x(OCH_2CH_2)_yOH$$
$$|$$
$$CH_3$$

where x has an average value of 2, y has an average value of 11, and R represents the alkyl groups derived from hydrogenated lauryl alcohol.

Chemical Class: Alkoxylated Alcohols

Function: Surfactant - Emulsifying Agent

Technical/Other Names:
Polyoxyethylene (11) Polyoxypropylene (2) Hydrogenated Lauryl Alcohol Ether
Polyoxypropylene (2) Polyoxyethylene (11) Hydrogenated Lauryl Alcohol Ether

PPG-12-PEG-50 LANOLIN

CAS No.: 68458-88-8

JPN Translation:
PPG - 12 - PEG - 50 ラノリン

Definition: PPG-12-PEG-50 Lanolin is the polyoxypropylene, polyoxyethylene derivative of Lanolin (q.v.) that conforms generally to the formula:

$$R(OCHCH_2)_x(OCH_2CH_2)_yOH$$
$$|$$
$$CH_3$$

where R represents the lanolin radical, x has an average value of 12 and y has an average value of 50.

Information Sources: JCIC, JCLS, JSQI, TSCA

Chemical Class: Lanolin and Lanolin Derivatives

Functions: Hair Conditioning Agent; Surfactant - Emulsifying Agent

Reported Product Categories: Hair Sprays (Aerosol Fixatives); Permanent Waves; Tonics, Dressings, and Other Hair Grooming Aids; Hair Conditioners; Shampoos (Non-coloring); Hair Straighteners

Technical/Other Names:
Polyoxyethylene (50) Polyoxypropylene (12) Lanolin
Polyoxypropylene (12) Polyoxyethylene (50) Lanolin

Trade Names:
AC Lanolin AWS (Active Concepts)
Ivarlan 3420 (Arch Personal Care Products)
Jeelan AWS (Jeen)
Laneto AWS (RITA)
Lanexol (Croda Chemicals)
Lanexol AWS (Croda, Inc.)
Lanoil AWS (Lanaetex)
Lanolox AWS (Maybrook)
Protalan AWS (Protameen)

PPG-12-PEG-65 LANOLIN OIL

JPN Translation:
PPG - 12 - PEG - 65 液状ラノリン

Definition: PPG-12-PEG-65 Lanolin Oil is the polyoxypropylene, polyoxyethylene derivative of lanolin oil that conforms generally to the formula:

$$R(OCHCH_2)_x(OCH_2CH_2)_yOH$$
$$|$$
$$CH_3$$

where R represents the lanolin radical and x has an average value of 12 and y has an average value of 65.

Information Sources: JCLS, JSQI

Chemical Class: Lanolin and Lanolin Derivatives

Functions: Hair Conditioning Agent; Surfactant - Emulsifying Agent

Reported Product Categories: Hair Preparations (Non-coloring), Misc.; Bath Oils, Tablets, and Salts; Bath Preparations, Misc.; Body and Hand Preparations (Excluding Shaving Preparations); Hair Conditioners; Tonics, Dressings, and Other Hair Grooming Aids

Technical/Other Names:
Polyoxyethylene (65) Polyoxypropylene (12) Lanolin Oil
Polyoxypropylene (12) Polyoxyethylene (65) Lanolin Oil

Trade Names:
Fluilan AWS (Croda, Inc.)
Ivarlan AWS (Arch Personal Care Products)
Jeelan AWS-LO (Jeen)
Lanalene Liquid Super AWS (Maybrook)
Lanor Crystal AWS (Lanolines de la Tossee)
Lantrol AWS 1692 (Cognis Care Chemicals/NJ)
Lantrol AWS 1692 (Cognis Care Chemicals/PA)
Ritalan AWS (RITA)
Vigilan AWS (Fanning)

Trade Name Mixtures:
Collamoist WS (Arch Personal Care Products)
Proto-Lan 30 (Maybrook)

PPG-40-PEG-60 LANOLIN OIL

JPN Translation:
PPG - 40 - PEG - 60 液状ラノリン

Definition: PPG-40-PEG-60 Lanolin Oil is the polyoxypropylene, polyoxyethylene derivative of Lanolin Oil (q.v.) that conforms generally to the formula:

$$R(OCHCH_2)_x(OCH_2CH_2)_yOH$$
$$|$$
$$CH_3$$

where R represents the lanolin oil radical, x has an average value of 40 and y has an average value of 60.

Information Sources: JCLS, JSQI

Chemical Class: Lanolin and Lanolin Derivatives

Functions: Hair Conditioning Agent; Skin-Conditioning Agent - Emollient; Surfactant - Emulsifying Agent

Technical/Other Names:
Polyoxyethylene (60) Polyoxypropylene (40) Lanolin Oil
Polyoxypropylene (40) Polyoxyethylene (60) Lanolin Oil

Trade Name:
Aqualose LL 100 (Croda Chemicals)

PPG-1-PEG-9 LAURYL GLYCOL ETHER

Definition: PPG-1-PEG-9 Lauryl Glycol Ether is the ethoxylated, propoxylated ether of a lauryl epoxide and ethylene glycol reaction product.

Chemical Classes: Alkoxylated Alcohols; Ethers

Function: Surfactant - Emulsifying Agent

Technical/Other Names:
Polyoxyethylene (9) Polyoxypropylene (1) Lauryl Glycol Ether
Polyoxypropylene (1) Polyoxyethylene (9) Lauryl Glycol Ether

Trade Names:
Eumulgin L (Cognis Care Chemicals/NJ)
Eumulgin L (Cognis Care Chemicals/PA)
Eumulgin L (Cognis Deutschland)

Trade Name Mixture:
Eumulgin HPS (Cognis Deutschland)

PPG-3-PEG-6 OLEYL ETHER

Definition: PPG-3-PEG-6 Oleyl Ether is the polyoxypropylene, polyoxyethylene derivative of oleyl alcohol that conforms to the formula:

$$CH_3(CH_2)_7CH = CH(CH_2)_8(OCHCH_2)_x(OCH_2CH_2)_yH$$
$$|$$
$$CH_3$$

where x has an average value of 3 and y has an average value of 6.

Chemical Class: Alkoxylated Alcohols

Function: Surfactant - Emulsifying Agent

Technical/Other Names:
Polyoxyethylene (6) Polyoxypropylene (3) Oleyl Ether

Polyoxypropylene (3) Polyoxyethylene (6) Oleyl Ether

PPG-65-PEG-5 PENTAERYTHRITYL ETHER

JPN Translation:
PEG - 5 - PPG - 65 ペンタエリスリチル

Definition: PPG-65-PEG-5 Pentaerythrityl Ether is the polyoxypropylene, polyoxyethylene derivative of pentaerythritol that has an average of 65 moles of propylene oxide and 5 moles of ethylene oxide.

Information Source: JCLS

Chemical Class: Alkoxylated Alcohols

Functions: Anticaking Agent; Hair Conditioning Agent; Skin-Conditioning Agent - Emollient; Surfactant - Solubilizing Agent; Surfactant - Suspending Agent

Technical/Other Names:
Polyoxyethylene (5) Polyoxypropylene (65) Pentaerythritol Ether
Polyoxypropylene (65) Polyethylene (5) Pentaerythritol Ether

Trade Names:
Beltamol P-700 (NOF)
Mushul T-401 (Toho)

PPG-24-PEG-21 TALLOWAMINOPROPYL-AMINE

Definition: PPG-24-PEG-21 Tallowaminopropylamine is the ethoxylated, propoxylated amine that conforms generally to the formula:

$$CH_3$$
$$(OCHCH_2)_x(OCH_2CH_2)_mOH$$
$$RN(CH_2)_3N(OCHCH_2)_y(OCH_2CH_2)_nOH$$
$$CH_3$$
$$(OCHCH_2)_z(OCH_2CH_2)_oOH$$
$$CH_3$$

where $x+y+z$ has an average value of 24 and $m+n+o$ has an average value of 21.

Chemical Class: Alkoxylated Amines

Function: Antistatic Agent

PPG-23-PEG-4 TRIMETHYLOLPROPANE

Definition: PPG-23-PEG-4 Trimethylolpropane is the polyoxypropylene, polyoxyethylene derivative of trimethylolpropane that conforms generally to the formula:

$$CH_2O(CH_2CHO)_x(CH_2CH_2O)_lH$$
$$CH_3$$
$$CH_3CH_2CCH_2O(CH_2CHO)_y(CH_2CH_2O)_mH$$
$$CH_3$$
$$CH_2O(CH_2CHO)_z(CH_2CH_2O)_nH$$
$$CH_3$$

where $x+y+z$ has an average value of 23 and $l+m+n$ has an average value of 4.

Information Source: JSQI

Chemical Class: Alkoxylated Alcohols

Functions: Skin-Conditioning Agent - Emollient; Surfactant - Emulsifying Agent

Technical/Other Names:
Polyoxyethylene (4) Polyoxypropylene (23) Trimethylolpropane
Polyoxypropylene (23) Polyoxyethylene (4) Trimethylolpropane

PPG-25-PEG-25 TRIMETHYLOLPROPANE

JPN Translation:
ポリオキシプロピレングリセリルエーテルリン酸

Definition: PPG-25-PEG-25 Trimethylolpropane is the polypropylene, polyoxyethylene derivative of trimethylolpropane that conforms generally to the formula:

$$CH_2O(CH_2CHO)_x(CH_2CH_2O)_lH$$
$$CH_3$$
$$CH_3CH_2CCH_2O(CH_2CHO)_y(CH_2CH_2O)_mH$$
$$CH_3$$
$$CH_2O(CH_2CHO)_z(CH_2CH_2O)_nH$$
$$CH_3$$

where $x+y+z$ has an average value of 25 and $l+m+n$ has an average value of 25.

Information Source: JCLS

Chemical Class: Alkoxylated Alcohols

Functions: Skin-Conditioning Agent - Emollient; Surfactant - Emulsifying Agent

PPG-68-PEG-10 TRIMETHYLOLPROPANE

JPN Translation:
PPG - 68 - PEG - 10 トリメチロールプロパン

Definition: PPG-68-PEG-10 Trimethylolpropane is the polyoxypropylene, polyoxyethylene derivative of trimethylolpropane that conforms generally to the formula:

$$CH_2O(CH_2CHO)_x(CH_2CH_2O)_lH$$
$$CH_3$$
$$CH_3CH_2CCH_2O(CH_2CHO)_y(CH_2CH_2O)_mH$$
$$CH_3$$
$$CH_2O(CH_2CHO)_z(CH_2CH_2O)_nH$$
$$CH_3$$

where $x+y+z$ has an average value of 68 and $l+m+n$ has an average value of 10.

Information Sources: JCIC, JCLS, JSQI

Chemical Class: Alkoxylated Alcohols

Functions: Skin-Conditioning Agent - Emollient; Surfactant - Emulsifying Agent

Technical/Other Names:
Polyoxyethylene Polyoxypropylene
· Trimethylolpropane
Polyoxyethylene (10) Polyoxypropylene (68) Trimethylolpropane
Polyoxypropylene (68) Polyoxyethylene (10) Trimethylolpropane

PPG-5 PENTAERYTHRITYL ETHER

Empirical Formula:
$C_{20}H_{42}O_9$

Definition: PPG-5 Pentaerythrityl Ether is the 5 mole propylene glycol ether of pentaerythritol.

Chemical Class: Alkoxylated Alcohols

Function: Skin-Conditioning Agent - Emollient

Technical/Other Names:
Polyoxypropylene (5) Pentaerythritol Ether
Polyoxypropylene (5) Pentaerythrityl Ether
Polypropylene Glycol (5) Pentaerythritol Ether
Polypropylene Glycol (5) Pentaerythrityl Ether

Trade Name Mixtures:
Lanolide (Vevy)
Seboside (Vevy)

PPG-8 POLYGLYCERYL-2 ETHER

Empirical Formula:
$C_{30}H_{62}O_{13}$

Definition: PPG-8 Polyglyceryl-2 Ether is the polypropylene glycol ether of a dimer of Glycerin (q.v.) with an average propoxylation value of 8.

Chemical Class: Alkoxylated Alcohols

Function: Surfactant - Emulsifying Agent

Reported Product Category: Nail Polish and Enamels

Technical/Other Names:
Polyoxypropylene (8) Diglycerin
Polypropylene Glycol (8) Diglycerin

Trade Name:
Savondol GP-9 (NOF)

PPG-14 POLYGLYCERYL-2 ETHER

Definition: PPG-14 Polyglyceryl-2 Ether is the polypropylene glycol ether of a dimer of Glycerin (q.v.) with an average propoxylation value of 14.

Chemical Class: Alkoxylated Alcohols

Function: Surfactant - Emulsifying Agent

Technical/Other Names:
Polyoxypropylene (14) Diglycerin
Polypropylene Glycol (14) Diglycerin

Trade Name:
SY-DP14 (Sakamoto Yakuhin Kogyo)

Trade Name Mixture:
SY-DP14T (US Cosmetics)

PPG-70 POLYGLYCERYL-10 ETHER

Definition: PPG-70 Polyglyceryl-10 Ether is the propoxylated ether of Polyglycerin-10 (q.v.) containing an average of 70 moles of propylene oxide.

Information Source: JCLS

Chemical Class: Alkoxylated Alcohols

Function: Hair Fixative

Technical/Other Name:
Polyoxypropylene (70) Decaglyceryl Ether

Trade Name:
Beltamol DC 25 (NOF)

PPG-35/PPG-51 GLYCERYL ETHER/IPDI CROSSPOLYMER

CAS No.: 66101-64-2

Definition: PPG-35/PPG-51 Glyceryl Ether/IPDI Crosspolymer is a crosslinked copolymer of isophorone diisocyanate, polyoxypropylene and PPG-51 glyceryl ether monomers.

Chemical Class: Synthetic Polymers

Functions: Plasticizer; Skin-Conditioning Agent - Miscellaneous

PPG-2 PROPYL ETHER

CAS Nos.	EINECS No.
29911-27-1	249-949-4
127303-87-1	

Empirical Formula:
$C_9H_{20}O_3$

Definition: PPG-2 Propyl Ether is the polypropylene glycol ether of n-propanol that conforms to the formula:

$$CH_3(CH_2)_2(OCHCH_2)_nOH$$
$$|$$
$$CH_3$$

where n has an average value of 2.

Chemical Class: Alkoxylated Alcohols

Function: Solvent

Technical/Other Names:
Dipropylene Glycol Monopropyl Ether
1-(1-Methyl-2-Propoxyethoxy)Propan-2-ol
Polyoxypropylene (2) Propyl Ether
Polypropylene Glycol (2) Propyl Ether
Propanol, (Methyl-2-Propoxyethoxy)-
2-Propanol, 1-(1-Methyl-2-Propoxyethoxy)-
Propoxy Dipropylene Glycol

Trade Names:
Arcosolv DPNP (Lyondell Chemical)
DOWANOL DPnP (Dow Chemical)

PPG-12/SMDI COPOLYMER

CAS No.: 9042-82-4 (Generic)

JPN Translation:
(PPG - 12 / SMDI) コポリマー

Definition: PPG-12/SMDI Copolymer is a copolymer of saturated methylene diphenyldiisocyanate and PPG-12 monomers.

Information Source: JCLS

Chemical Class: Synthetic Polymers

Functions: Film Former; Hair Conditioning Agent; Hair Fixative; Plasticizer; Skin-Conditioning Agent - Emollient; Skin-Conditioning Agent - Miscellaneous

Reported Product Categories: Eye Shadows; Foundations; Makeup Preparations (Not eye), Misc.; Moisturizing Preparations

Trade Names:
AEC PPG-12/SMDI Copolymer (A & E Connock)
Polyolprepolymer-2 (Penederm)

Trade Name Mixture:
Protaderm HA (Protameen)

PPG-51/SMDI COPOLYMER

CAS No.: 9042-82-4 (Generic)

JPN Translation:
(PPG - 51 / SMDI) コポリマー

Definition: PPG-51/SMDI Copolymer is a copolymer of PPG-51 and saturated methylene diphenyldiisocyanate (dicyclohexylmethane diisocyanate) monomer.

Information Source: JCLS

Chemical Class: Synthetic Polymers

Functions: Hair Conditioning Agent; Hair Fixative; Skin-Conditioning Agent - Emollient; Skin-Conditioning Agent - Miscellaneous

Reported Product Category: Lipsticks

Technical/Other Names:
Polyoxypropylene (51) SMDI Copolymer
Polypropylene Glycol (51) SMDI Copolymer

Trade Name:
Polyolprepolymer-14 (Penederm)

PPG-6-SORBETH-245

CAS No.: 56449-05-9 (Generic)

Definition: PPG-6-Sorbeth-245 is the polyoxypropylene, polyoxyethylene ether of Sorbitol (q.v.) that conforms generally to the formula:

$$C_6H_{13}O_6(CH_2CH_2O)_x(CH_2CHO)_yH$$
$$|$$
$$CH_3$$

where x has an average value of 245 and y has an average value of 6.

Chemical Class: Alkoxylated Alcohols

Functions: Binder; Humectant

Technical/Other Names:
Polyoxyethylene (245) Polyoxypropylene (6) Sorbitol
Polyoxypropylene (6) Polyoxyethylene (245) Sorbitol

PPG-6-SORBETH-500

CAS No.: 56449-05-9 (Generic)

Definition: PPG-6-Sorbeth-500 is the polyoxypropylene, polyoxyethylene ether of Sorbitol (q.v.) that conforms generally to the formula:

$$C_6H_{13}O_6(CH_2CH_2O)_x(CH_2CHO)_yH$$
$$|$$
$$CH_3$$

where x has an average value of 500 and y has an average value of 6.

Chemical Class: Alkoxylated Alcohols

Functions: Binder; Humectant

Technical/Other Names:
Polyoxyethylene (500) Polyoxypropylene (6) Sorbitol
Polyoxypropylene (6) Polyoxyethylene (500) Sorbitol

PPG-10 SORBITOL

Empirical Formula:
$C_{36}H_{74}O_{16}$

Definition: PPG-10 Sorbitol is the ether of Sorbitol (q.v.) and polypropylene glycol containing an average of 10 moles of propylene oxide.

Chemical Class: Alkoxylated Alcohols

Function: Hair Fixative

Technical/Other Name:
Polypropylene Glycol (10) Sorbitol

Trade Name:
SMACK SP-10P (Kao Corp.)

PPG-33 SORBITOL

Definition: PPG-33 Sorbitol is the ether of Sorbitol (q.v.) and polypropylene glycol containing an average of 33 moles of propylene oxide.

Information Source: JCLS

Chemical Class: Alkoxylated Alcohols

Function: Hair Fixative

PPG-41 SORBITOL

Definition: PPG-41 Sorbitol is the ether of Sorbitol (q.v.) and polypropylene glycol containing an average of 41 moles of propylene oxide.

Information Source: JCLS

Chemical Class: Alkoxylated Alcohols

Function: Hair Fixative

PPG-8 SORBITOL CASTOR OIL

JPN Translation:
PPG - 8 ソルビットヒマシ油

Definition: PPG-8 Sorbitol Castor Oil is the polypropylene glycol ether of the reaction product of Sorbitol (q.v.) and Ricinus Communis (Castor) Oil (q.v.) containing an average of 8 moles of propylene oxide.

Information Source: JCLS

Chemical Classes: Alkoxylated Alcohols; Sorbitan Derivatives

Function: Skin-Conditioning Agent - Miscellaneous

PPG-15 STEARATE

Definition: PPG-15 Stearate is the polypropylene glycol ester of stearic acid that conforms to the formula:

$$CH_3(CH_2)_{16}\overset{\displaystyle O}{\overset{\displaystyle \|}{C}} - (OCHCH_2)_n OH$$
$$\underset{CH_3}{|}$$

where n has an average value of 15.

Information Source: JCLS

Chemical Class: Alkoxylated Carboxylic Acids

Function: Skin-Conditioning Agent - Emollient

Technical/Other Names:
Polyoxypropylene (15) Monostearate
Polypropylene Glycol (15) Stearate

PPG-9-STEARETH-3

CAS No.: 9038-43-1 (Generic)

JPN Translation:
PPG - 9 ステアレス - 3

Empirical Formula:
$C_{51}H_{104}O_{13}$

Definition: PPG-9-Steareth-3 is the polyoxypropylene, polyoxyethylene ether of stearyl alcohol that conforms generally to the formula:

$$CH_3(CH_2)_{17}(OCHCH_2)_x(OCH_2CH_2)_y OH$$
$$\underset{CH_3}{|}$$

where x has an average value of 9 and y has an average value of 3.

Information Source: JCLS

Chemical Class: Alkoxylated Alcohols

Function: Skin-Conditioning Agent - Emollient

Technical/Other Names:
Polyoxyethylene (3) Polyoxypropylene (9) Stearyl Ether
Polyoxypropylene (9) Polyoxyethylene (3) Stearyl Ether

PPG-23-STEARETH-34

CAS No.: 9038-43-1

JPN Translation:
PPG - 23 ステアレス - 34

Definition: PPG-23-Steareth-34 is the polyoxypropylene, polyoxyethylene ether of stearyl alcohol that conforms generally to the formula:

$$CH_3(CH_2)_{17}(OCHCH_2)_x(OCH_2CH_2)_y OH$$
$$\underset{CH_3}{|}$$

where x has an average value of 23 and y has an average value of 34.

Information Sources: JCIC, JCLS, JSQI

Chemical Class: Alkoxylated Alcohols

Functions: Skin-Conditioning Agent - Emollient; Surfactant - Emulsifying Agent

Technical/Other Names:
Polyoxyethylene Polyoxypropylene Stearyl Ether (34E.O.) (23P.O.)
Polyoxypropylene (23) Polyoxyethylene (34) Stearyl Ether
Polyoxyethylene (34) Polyoxypropylene (23) Stearyl Ether

Trade Name:
Unisafe 34S-23 (Pola)

PPG-30 STEARETH-4

JPN Translation:
PPG - 30 ステアレス - 4

Definition: PPG-30 Steareth-4 is the polyoxyethylene, polyoxypropylene ether of stearyl alcohol that conforms generally to the formula:

$$CH_3(CH_2)_{17}(OCHCH_2)_x(OCH_2CH_2)_y OH$$
$$\underset{CH_3}{|}$$

where x has an average value of 30 and y has an average value of 4.

Information Source: JCLS

Chemical Class: Alkoxylated Alcohols

Functions: Skin-Conditioning Agent - Emollient; Surfactant - Emulsifying Agent

Technical/Other Name:
Polyoxyethylene (4) Polyoxypropylene (30) Stearyl Ether

Trade Name:
Kao Polyether SP-2010 (Kao Corp.)

PPG-34-STEARETH-3

JPN Translation:
PPG - 34 ステアレス - 3

Definition: PPG-34-Steareth-3 is the polyoxypropylene, polyoxyethylene ether of stearyl alcohol that conforms generally to the formula:

$$CH_3(CH_2)_{17}(OCHCH_2)_x(OCH_2CH_2)_yOH$$
$$| \atop CH_3$$

where x has an average value of 34 and y has an average value of 3.

Information Source: JCLS

Chemical Class: Alkoxylated Alcohols

Functions: Skin-Conditioning Agent - Emollient; Surfactant - Emulsifying Agent

Technical/Other Names:
Polyoxyethylene (3) Polyoxypropylene (34) Stearyl Ether
Polyoxypropylene (34) Polyoxyethylene (9) Stearyl Ether

Trade Name:
Unilube 10MS-250KB (NOF)

PPG-38 STEARETH-6

JPN Translation:
PPG - 38 ステアレス - 6

Definition: PPG-38 Steareth-6 is the polyoxyethylene, polyoxypropylene ether of stearyl alcohol that conforms to the formula:

$$CH_3(CH_2)_{17}(OCHCH_2)_x(OCH_2CH_2)_yOH$$
$$| \atop CH_3$$

where x has an average value of 38 and y has an average value of 6.

Information Source: JCLS

Chemical Class: Alkoxylated Alcohols

Functions: Skin-Conditioning Agent - Emollient; Surfactant - Emulsifying Agent

PPG-11 STEARYL ETHER

CAS No.: 25231-21-4 (Generic)

JPN Translation:
PPG - 11 ステアリル

Definition: PPG-11 Stearyl Ether is the polypropylene glycol ether of stearyl alcohol that conforms to the formula:

$$CH_3(CH_2)_{17}(OCHCH_2)_nOH$$
$$| \atop CH_3$$

where n has an average value of 11.

Information Sources: CIR: [S] IJT-20 (SUPPL. 4)2001, CTFA S, JCLS, JSQI, TSCA

Chemical Class: Alkoxylated Alcohols

Function: Skin-Conditioning Agent - Emollient

Reported Product Category: Personal Cleanliness Products, Misc.

Technical/Other Names:
Polyoxypropylene (11) Stearyl Ether
Polypropylene Glycol (11) Stearyl Ether

Trade Names:
Arlamol F (Uniqema Americas)
Jeecol PSA-11 (Jeen)
Procol PSA-11 (Protameen)
Varionic APS (Degussa Care Specialties)

PPG-15 STEARYL ETHER

CAS No.: 25231-21-4 (Generic)

JPN Translation:
PPG - 15 ステアリル

Definition: PPG-15 Stearyl Ether is the polypropylene glycol ether of stearyl alcohol that conforms to the formula:

$$CH_3(CH_2)_{17}(OCHCH_2)_nOH$$
$$| \atop CH_3$$

where n has an average value of 15.

Information Sources: CIR: [S] IJT-20 (SUPPL. 4)2001, CTFA S, JCLS, JSQI, TSCA, USAN

Chemical Class: Alkoxylated Alcohols

Function: Skin-Conditioning Agent - Emollient

Reported Product Categories: Moisturizing Preparations; Bath Oils, Tablets, and Salts; Cleansing Products (Cold Creams, Cleansing Lotions, Liquids and Pads); Face and Neck Preparations (Excluding Shaving Preparations); Bath Capsules; Personal Cleanliness Products, Misc.; Bath Preparations, Misc.; Body and Hand Preparations (Excluding Shaving Preparations); Skin Care Preparations, Misc.

Technical/Other Names:
Polyoxypropylene (15) Stearyl Ether
Polypropylene Glycol (15) Stearyl Ether

Trade Names:
Acconon E (Abitec Corporation)
Arlamol E (Uniqema Americas)
Fancol SA-15 (Fanning)
Hetoxol SP-15 (Heterene)
Jeecol PSA-15 (Jeen)
Lipocol P-15 (Lipo)
Procol PSA-15 (Protameen)
Prostearyl 15 (Croda Chemicals)
Sympatens-ASP/150 (Kolb)

Trade Name Mixtures:
Arlamol S7 (Uniqema Americas)
Polymoist Mask (Cognis Deutschland)

PPG-15 STEARYL ETHER BENZOATE

Definition: PPG-15 Stearyl Ether Benzoate is the ester of PPG-15 Stearyl Ether (q.v.)

and benzoic acid. It conforms generally to the formula:

$$\begin{array}{c} O \\ \| \\ C_6H_5\!-\!C-O(CHCH_2O)_n(CH_2)_{17}CH_3 \\ | \\ CH_3 \end{array}$$

where n has an average value of 15.

Chemical Classes: Alkoxylated Alcohols; Esters

Function: Skin-Conditioning Agent - Emollient

Reported Product Category: Tonics, Dressings, and Other Hair Grooming Aids

Technical/Other Names:
Polyoxypropylene (15) Stearyl Ether Benzoate
Polypropylene Glycol (15) Stearyl Ether Benzoate

Trade Name:
Finsolv P (Finetex)

Trade Name Mixtures:
Finsolv TPP (Finetex)
Natrlfine 137-T (Finetex)
Natrlfine TP-T (Finetex)
Natrlfine TP-Z (Finetex)

PPG-7/SUCCINIC ACID COPOLYMER

JPN Translation:
(PPG - 7 / コハク酸) コポリマー

Definition: PPG-7/Succinic Acid Copolymer is a copolymer of PPG-7 and Succinic Acid (q.v.).

Information Sources: JCIC, JCLS, JSQI

Chemical Class: Synthetic Polymers

Functions: Film Former; Hair Conditioning Agent

Technical/Other Name:
Polypropyleneglycol Oligosuccinate (35P.O.)

Trade Names:
AEC PPG-7/Succinic Acid Copolymer (A & E Connock)
Cosmol 102 (Nisshin OilliO)
Salacos 102 (Nisshin OilliO)

PPG-2 TALLOWAMINE

Definition: PPG-2 Tallowamine is the derivative of Tallow Amine (q.v.) that conforms generally to the formula:

$$R\!-\!N \begin{array}{c} CH_3 \\ | \\ (OCHCH_2)_xH \\ \\ (OCHCH_2)_yH \\ | \\ CH_3 \end{array}$$

where R represents the alkyl groups derived from tallow and x+y has an average value of 2.

Chemical Class: Alkoxylated Amines

Function: Antistatic Agent

Trade Name:
Propomeen T/12 (Akzo Nobel)

PPG-3 TALLOW AMINOPROPYLAMINE

Definition: PPG-3 Tallow Aminopropylamine is the polypropylene glycol amine that conforms generally to the formula:

where R represents the tallow radical and x+y+z has an average value of 3.

Chemical Class: Alkoxylated Amines

Function: Antistatic Agent

Technical/Other Names:
Polyoxypropylene (3) Tallow Aminopropyl-amine
Polypropylene Glycol (3) Tallow Amino-propylamine

Trade Name:
Propoduomeen T/13 (Akzo Nobel)

PPG-26/TDI COPOLYMER

Definition: PPG-26/TDI Copolymer is a copolymer of PPG-26 (q.v.) and toluene diisocyanate.

Chemical Class: Synthetic Polymers

Functions: Film Former; Plasticizer

PPG-2 TOCOPHERETH-5

Definition: PPG-2 Tocophereth-5 is the alkoxylated derivative of Tocopherol (q.v.) containing an average of 2 moles of propylene oxide and 5 moles of ethylene oxide.

Chemical Classes: Alkoxylated Alcohols; Heterocyclic Compounds

Functions: Anticaking Agent; Antioxidant; Binder; Emulsion Stabilizer; Humectant; Plasticizer; Skin-Conditioning Agent - Humectant; Surfactant - Emulsifying Agent; Ultraviolet Light Absorber

Technical/Other Name:
Polyoxypropylene (2) Polyoxyethylene (5) Tocopherol Ether

Trade Name:
Vitanics- 5/2 (Vitacos Corporation)

PPG-5 TOCOPHERETH-2

Definition: PPG-5 Tocophereth-2 is the alkoxylated derivative of Tocopherol (q.v.) containing an average of 5 moles of propylene oxide and 2 moles of ethylene oxide.

Chemical Classes: Alkoxylated Alcohols; Heterocyclic Compounds

Functions: Anticaking Agent; Antioxidant; Binder; Emulsion Stabilizer; Skin-Conditioning Agent - Humectant; Surfactant - Emulsifying Agent; Ultraviolet Light Absorber

Technical/Other Name:
Polyoxypropylene (5) Polyoxyethylene (2) Tocopherol Ether

Trade Name:
Vitanics 2/5 (Vitacos Corporation)

PPG-10 TOCOPHERETH-30

Definition: PPG-10 Tocophereth-30 is the alkoxylated derivative of Tocopherol (q.v.) containing an average of 10 moles of propylene oxide and 30 moles of ethylene oxide.

Chemical Classes: Alkoxylated Alcohols; Heterocyclic Compounds

Functions: Antioxidant; Binder; Emulsion Stabilizer; Film Former; Skin-Conditioning Agent - Humectant; Surfactant - Emulsifying Agent; Ultraviolet Light Absorber

Technical/Other Name:
Polyoxypropylene (10) Polyoxyethylene (30) Tocopherol Ether

Trade Name:
Vitanics 30/10 (Vitacos Corporation)

PPG-20 TOCOPHERETH-50

Definition: PPG-20 Tocophereth-50 is the alkoxylated derivative of Tocopherol (q.v.) containing an average of 20 moles of propylene oxide and 50 moles of ethylene oxide.

Chemical Classes: Alkoxylated Alcohols; Heterocyclic Compounds

Functions: Antioxidant; Binder; Emulsion Stabilizer; Film Former; Skin-Conditioning

Agent - Humectant; Surfactant - Emulsifying Agent; Ultraviolet Light Absorber

Technical/Other Name:
Polyoxypropylene (20) Polyoxyethylene (50) Tocopherol Ether

Trade Name:
Vitanics 50/20 (Vitacos Corporation)

PPG-30 TOCOPHERETH-70

Definition: PPG-30 Tocophereth-70 is the alkoxylated derivative of Tocopherol (q.v.) containing an average of 30 moles of propylene oxide and 70 moles of ethylene oxide.

Chemical Classes: Alkoxylated Alcohols; Heterocyclic Compounds

Functions: Antioxidant; Binder; Plasticizer; Skin-Conditioning Agent - Humectant; Surfactant - Solubilizing Agent; Ultraviolet Light Absorber

Technical/Other Name:
Polyoxypropylene (30) Polyoxyethylene (70) Tocopherol Ether

Trade Name:
Vitanics 70/30 (Vitacos Corporation)

PPG-70 TOCOPHERETH-100

Definition: PPG-70 Tocophereth-100 is the alkoxylated derivative of Tocopherol (q.v.) containing an average of 70 moles of propylene oxide and 100 moles of ethylene oxide.

Chemical Classes: Alkoxylated Alcohols; Heterocyclic Compounds

Functions: Anticaking Agent; Antioxidant; Binder; Skin-Conditioning Agent - Humectant; Ultraviolet Light Absorber

Technical/Other Name:
Polyoxypropylene (70) Polyoxyethylene (100) Tocopherol Ether

Trade Name:
Vitanics 100/70 (Vitacos Corporation)

PPG-5 TOCOPHERYL ETHER

Definition: PPG-5 Tocopheryl Ether is the polypropylene glycol ether of Tocopherol (q.v.) that conforms to the formula:

where n has an average value of 5.

Chemical Classes: Alkoxylated Alcohols; Heterocyclic Compounds

Functions: Anticaking Agent; Antioxidant; Binder; Plasticizer; Skin-Conditioning Agent - Humectant; Ultraviolet Light Absorber

Technical/Other Name:
Polypropylene Glycol (5) Tocopherol Ether

Trade Name:
Vitanics 0/5 (Vitacos Corporation)

PPG-1 TRIDECETH-6

Empirical Formula:
$C_{28}H_{58}O_8$

Definition: PPG-1 Trideceth-6 is the polyoxypropylene, polyoxyethylene ether of tridecyl alcohol that conforms generally to the formula:

$$CH_3(CH_2)_{12}(OCHCH_2)_x(OCH_2CH_2)_yOH$$
$$|$$
$$CH_3$$

where x has an average value of 1 and y has an average value of 6.

Chemical Class: Alkoxylated Alcohols

Functions: Skin-Conditioning Agent - Emollient; Surfactant - Emulsifying Agent

Reported Product Categories: Hair Conditioners; Hair Dyes and Colors (All Types Requiring Caution Statements and Patch Tests)

Technical/Other Names:
Polyoxyethyene (6) Polyoxypropylene (1) Tridecyl Ether
Polyoxypropylene (1) Polyoxyethylene (6) Tridecyl Ether

Trade Name Mixtures:
DP 705-9339 (Ciba Specialty Chemicals)
Fancorgel A (Fanning)
Lowenol Copolymer 725 (Lowenstein)
Lowenol Copolymer 1097 (Lowenstein)
Rheocare CTH(E) (Cosmetic Rheologies)
Salcare AST (Ciba Specialty Chemicals)
Salcare SC91 (Ciba Specialty Chemicals)
Salcare SC92 (Ciba Specialty Chemicals)
Salcare SC95 (Ciba Specialty Chemicals)
Salcare SC96 (Ciba Specialty Chemicals)

PPG-4 TRIDECETH-6

Definition: PPG-4 Trideceth-6 is the polyoxyethylene, polyoxypropylene ether of tridecyl alcohol that conforms generally to the formula:

$$CH_3(CH_2)_{12}(OCHCH_2)_x(OCH_2CH_2)_yOH$$
$$|$$
$$CH_3$$

where x has an average value of 4 and y has an average value of 6.

Chemical Class: Alkoxylated Alcohols

Functions: Skin-Conditioning Agent - Emollient; Surfactant - Emulsifying Agent

Technical/Other Names:
Polyoxyethylene (6) Polyoxypropylene (4) Tridedcyl Ether
Polyoxypropylene (4) Polyoxyethylene (6) Tridecyl Ether

PPG-6 TRIDECETH-8

Definition: PPG-6 Trideceth-8 is the polyoxyethylene, polyoxypropylene ether of tridecyl alcohol that conforms generally to the formula:

$$CH_3(CH_2)_{12}(OCHCH_2)_x(OCH_2CH_2)_yOH$$
$$|$$
$$CH_3$$

where x has an average value of 6 and y has an average value of 8.

Chemical Class: Alkoxylated Alcohols

Functions: Skin-Conditioning Agent - Emollient; Surfactant - Emulsifying Agent

Technical/Other Names:
PEG-8 PPG-6 Tridecyl Ether
Polyoxyethylene (8) Polyoxypropylene (6) Tridecyl Ether
Polyoxypropylene (6) Polyoxyethylene (8) Tridecyl Ether

Trade Name:
Pepol AS-054C (Toho)

PPG-77 TRIMETHYLOLPROPANE ETHER

Definition: PPG-77 Trimethylolpropane Ether is a polyoxypropylene derivative of trimethylolpropane that conforms generally to the formula:

$$CH_2O(CH_2CHO)_xH$$
$$|$$
$$CH_3$$
$$CH_3CH_2CCH_2O(CH_2CHO)_yH$$
$$|$$
$$CH_3$$
$$CH_2O(CH_2CHO)_zH$$
$$|$$
$$CH_3$$

where x+y+z has an average value of 77.

Chemical Class: Alkoxylated Alcohols

Function: Skin-Conditioning Agent - Miscellaneous

Technical/Other Names:
Polyoxypropylene (77) Trimethylolpropane Ether
Polypropylene Glycol (77) Trimethylolpropane Ether

Trade Name:
Nikkol TMP-EP (Nikko)

PRAMOXINE HCl

CAS No. 637-58-1   **EINECS No.** 211-293-1

Empirical Formula:
$C_{17}H_{27}NO_3 \cdot HCl$

Definition: Pramoxine HCl is the heterocyclic organic compound that conforms to the formula:

See "Regulatory and Ingredient Use Information," regarding the labeling names for U.S. OTC Drug Ingredients in Volume 1, Introduction, Part A. See Reported Ingredient Functions-The Cosmetic Drug Distinction, in Regulatory and Ingredient Use Information, Volume I, Part A.

Information Source: MI-13(7794)

Chemical Classes: Ethers; Heterocyclic Compounds

Function: External Analgesic

Technical/Other Names:
Morpholine, 4-[3-(4-Butoxyphenoxy)-Propyl]-, Hydrochloride
Pramocaine Hydrochloride
Tronothane Hydrochloride

Trade Name:
Pramoxine HCl (Abbott Lab)

PRASTERONE

CAS No. 53-43-0   **EINECS No.** 200-175-5

Empirical Formula:
$C_{19}H_{28}O_2$

Definition: Prasterone the sterol that conforms to the formula:

Information Sources: INN, MI-13(7798)

Chemical Class: Sterols

Function: Skin-Conditioning Agent - Miscellaneous

Technical/Other Names:
Androst-5-en-17-one, 3-Hydroxy-, (3/B)-
Dehydroepiandrosterone
Dehydroisoandrosterone
DHEA

Trade Name:
DHEA SL (Sino Lion)

Trade Name Mixture:
DHEA MICROEMULSION (Ennagram)

PREGNENOLONE ACETATE

CAS No. 1778-02-5

EINECS No. 217-212-6

Empirical Formula:
$C_{23}H_{34}O_3$

Definition: Pregnenolone Acetate is the acetate ester of pregnenolone that conforms to the formula:

Information Source: MHLW-331/1

Chemical Classes: Esters; Sterols

Function: Skin-Conditioning Agent - Miscellaneous

PRIMULA SIKKIMENSIS FLOWER EXTRACT

Definition: Primula Sikkimensis Flower Extract is an extract of the flowers of *Primula sikkimensis*. See *"Regulatory and Ingredient Use Information,"* regarding the labeling names for botanical ingredients in Volume 1, Introduction, Part A.

Chemical Class: Biological Products

Function: Skin-Conditioning Agent - Humectant

Technical/Other Name:
Extract of Primula Sikkimensis Flower

Trade Name Mixture:
Oukahoushun Ekisu (Noevir)

PRIMULA VERIS EXTRACT

CAS No. 84787-68-8

EINECS No. 284-109-0

Definition: Primula Veris Extract is an extract of the primula, *Primula veris*. See *"Regulatory and Ingredient Use Information,"* regarding the labeling names for botanical ingredients in Volume 1, Introduction, Part A.

Chemical Class: Biological Products

Function: Skin-Conditioning Agent - Miscellaneous

Technical/Other Names:
Extract of Primula Veris
Primula Extract

Trade Name Mixtures:
Actiphyte of Cowslip (Active Organics)
Actiphyte of Primrose (Active Organics)
Cowslip Extract HS 2435 G (Grau)
Gigawhite (Cosmetic Ingredient Resources/ Centerchem)
Herbasol-Extract Primose (Cosmetochem)
Phytelene of Primula EG 509 liquid (Indena SA)
Phytoderm Primula Glycolic (Universal Flavors)
Phytogreen 55 of Primula EXH 709 Liquid (Phytochim)
Primrose Extract HG (Provital/Centerchem)
Primrose Flower HS (Alban Muller)
Primrose Milk (CEP (Solabia))

PRIMULA VULGARIS EXTRACT

CAS No. 84604-06-8

EINECS No. 283-283-5

Definition: Primula Vulgaris Extract is an extract of the primula, *Primula vulgaris*. See *"Regulatory and Ingredient Use Information,"* regarding the labeling names for botanical ingredients in Volume 1, Introduction, Part A.

Chemical Class: Biological Products

Function: Not Reported

Technical/Other Names:
Extract of Primula Vulgaris
Primula Extract

Trade Name Mixture:
Primula Vulgaris Extract ies (IES LABO)

PRISTANE

CAS No. 1921-70-6

EINECS No. 217-650-8

JPN Translation:
プリスタン

Empirical Formula:
$C_{19}H_{40}$

Definition: Pristane is a saturated branched chain hydrocarbon found in shark liver oil and other natural oils. It conforms to the formula:

$$CH_3CH(CH_2)_3CH(CH_2)_3CH(CH_2)_3CHCH_3$$

Information Sources: JCIC, JCLS, JSQI, MI-13(7841), TSCA

Chemical Class: Hydrocarbons

Function: Skin-Conditioning Agent - Occlusive

Technical/Other Names:
Norphytane
Pentadecane, 2,6,10,14-Tetramethyl-
2,6,10,14-Tetramethylpentadecane

PROCOLLAGEN

Definition: Procollagen is the precursor of Collagen (q.v.). It contains N-terminal and C-terminal telopeptides which provide the template for the construction of the triple-helix during synthesis.

Chemical Class: Protein Derivatives

Functions: Hair Conditioning Agent; Skin-Conditioning Agent - Miscellaneous

Reported Product Categories: Body and Hand Preparations (Excluding Shaving Preparations); Foundations

Trade Names:
Solu-Coll P (Arch Personal Care Products)
Solu-Mar P (Arch Personal Care Products)

PROGESTERONE

CAS No. 57-83-0

EINECS No. 200-350-6

Empirical Formula:
$C_{21}H_{30}O_2$

Definition: Progesterone is the steroid that conforms to the formula:

Information Sources: EEC(II-194), MI-13 (7864)

Chemical Class: Sterols

Function: Skin-Conditioning Agent - Miscellaneous

Technical/Other Name:
Pregn-4-ene3,20-dione

Trade Names:
Progesterone (Norjin)
Scalp Bioactivator SI 1007 (Norjin)

PROLINAMIDOETHYL IMIDAZOLE

Empirical Formula:
$C_{10}H_{16}N_4O$

Definition: Prolinamidoethyl Imidazole is the heterocyclic compound that conforms to the formula:

See Reported Ingredient Functions-The Cosmetic Drug Distinction, in Regulatory and Ingredient Use Information, Volume I, Part A.

Chemical Classes: Amides; Heterocyclic Compounds

Function: Skin Protectant

Trade Name:
Prolyglyoxaline AE (Exsymol)

PROLINE

CAS Nos.	EINECS Nos.
147-85-3 (L-Form)	205-702-2
609-36-9 (dl-alpha)	210-189-3

JPN Translation:
プロリン

Empirical Formula:
$C_5H_9NO_2$

Definition: Proline is the amino acid that conforms to the formula:

Information Sources: 21CFR172.320, 21CFR582.5650, INN, JCIC, JCLS, JSQI, MI-13(7871), RIFM, TSCA, USAN, USP XXIV

Chemical Classes: Amino Acids; Heterocyclic Compounds

Functions: Fragrance Ingredient; Hair Conditioning Agent; Skin-Conditioning Agent - Miscellaneous

Reported Product Categories: Moisturizing Preparations; Hair Conditioners; Skin Care Preparations, Misc.; Night Skin Care Preparations; Permanent Waves; Bath Capsules; Bath Oils, Tablets, and Salts; Body and Hand Preparations (Excluding Shaving Preparations); Cleansing Products (Cold Creams, Cleansing Lotions, Liquids and Pads); Face and Neck Preparations (Excluding Shaving Preparations); Paste Masks (Mud Packs)

Technical/Other Name:
L-Proline (RIFM)

Trade Name:
AEC Proline (A & E Connock)

Trade Name Mixtures:
Cryocytol (Laboratoires Serobiologiques)
Hydro-Diffuser Microreservoir (Sederma)
Moisturizing Factor Hygro-Complex ARO 5272 (Crodarom)
Moisturizing Liposomes (Collaborative Labs)
Moisturizing Phytoamine Biocomplex (Alban Muller)
P.A. Antifroid LS 9224B (Laboratoires Serobiologiques)
Prodew 100 (Ajinomoto)
Prodew 200 (Ajinomoto)
Prodew 300 (Ajinomoto)
Prodew 400 (Ajinomoto)
Rovisome AA (Rovi)
Sel-Smooth (Seltzer)
Slimming Phytoamine Biocomplex (Alban Muller)

PROLYL HISTAMINE HCl

Empirical Formula:
$C_{10}H_{18}N_4OCl_2$

Definition: Prolyl Histamine HCl is the heterocyclic compound that conforms to the formula:

Chemical Class: Heterocyclic Compounds

Function: Skin-Conditioning Agent - Miscellaneous

Trade Name:
Prolylhistamine Chlorhydrate (Exsymol)

PROPAGERMANIUM

CAS No.: 126595-07-1

Empirical Formula:
$C_6H_{10}Ge_2O_6$

Definition: Propagermanium is the organic compound that conforms to the formula:

Information Sources: INN, MI-13(7887)

Chemical Class: Carboxylic Acids

Function: Skin-Conditioning Agent - Miscellaneous

Technical/Other Names:
3,3'-(1,3-Dioxo-1,3-Digermoxanediyl)-Dipropionic Acid
Propanoic Acid, 3,3'-(1,3-Dioxo-1,3-Digermoxanediyl)bis-, Homopolymer

Trade Names:
Arlamol GEO (Uniqema Americas)
Organic Germanium (Tokai)

PROPANE

CAS No.	EINECS No.
74-98-6	200-827-9

JPN Translation:
プロパン

Empirical Formula:
C_3H_8

Definition: Propane is the hydrocarbon that conforms to the formula:

$$CH_3CH_2CH_3$$

Information Sources: 21CFR173.350, 21CFR175.105, 21CFR177.1680, 21CFR178.2010, 21CFR184.1655, 21CFR582.1655, CIR: [S] JACT-1(4)1982, JCIC, JCLS, MI-13(7891), NF XIX, TSCA, USAN

Chemical Class: Hydrocarbons

Function: Propellant

Reported Product Categories: Hair Sprays (Aerosol Fixatives); Shaving Cream (Aerosol, Brushless and Lather); Tonics, Dressings, and Other Hair Grooming Aids; Deodorants (Underarm); Colognes and Toilet Waters; Hair Preparations (Non-coloring), Misc.; Personal Cleanliness Products, Misc.; Perfumes; Hair Conditioners; Hair Wave Sets; Fragrance Preparations, Misc.; Sachets; Hair Tints

Trade Name Mixture:
Drivosol (Creanova)

PROPANEDIOL

CAS Nos.	EINECS No.
504-63-2	207-997-3
26264-14-2	

The inclusion of any compound in the *Dictionary and Handbook* does not indicate that use of that substance as a cosmetic ingredient complies with the laws and regulations governing such use in the United States or any other country.

Empirical Formula:
$C_3H_8O_2$

Definition: Propanediol is the organic compound that conforms to the formula:

$HOCH_2CH_2CH_2OH$

Information Sources: MI-13(9786), TSCA

Chemical Class: Alcohols

Functions: Solvent; Viscosity Decreasing Agent

Technical/Other Names:
1,3-Dihydroxypropane
1,3-Propylene Glycol

PROPANE TRICARBOXYLIC ACID

CAS Nos.	EINECS No.
99-14-9	202-733-3
51750-56-2	

Empirical Formula:
$C_6H_8O_6$

Definition: Propane Tricarboxylic Acid is the organic compound that conforms to the formula:

$HOOCCH_2CH(COOH)CH_2COOH$

Information Source: MI-13(9695)

Chemical Class: Carboxylic Acids

Function: Not Reported

Technical/Other Names:
β-Carboxyglutaric Acid
1,2,3-Propanetricarboxylic Acid
Tricarballylic Acid
1,2,3-Tricarboxypropane

PROPIONIC ACID

CAS No.	EINECS No.
79-09-4	201-176-3

JPN Translation:
プロピオン酸

Empirical Formula:
$C_3H_6O_2$

Definition: Propionic Acid is the organic acid that conforms to the formula:

CH_3CH_2COOH

Information Sources: 21CFR184.1081, EEC(VI/1-2), JCLS, MI-13(7917), NF XIX, RIFM, TSCA, USAN

Chemical Class: Carboxylic Acids

Functions: Cosmetic Biocide; Fragrance Ingredient; pH Adjuster; Preservative

Technical/Other Names:
Carboxyethane
Methylacetic Acid
Propanoic Acid
Propionic acid (RIFM)

PROPIONYL COLLAGEN AMINO ACIDS

Definition: Propionyl Collagen Amino Acids is the condensation product of propionic acid chloride with Collagen Amino Acids (q.v.).

Chemical Class: Amino Acids

Function: Skin-Conditioning Agent - Occlusive

PROPOLIS EXTRACT

Definition: Propolis Extract is an extract of Propolis Wax (q.v.).

Chemical Class: Biological Products

Function: Skin-Conditioning Agent - Miscellaneous

Reported Product Categories: Bath Preparations, Misc.; Body and Hand Preparations (Excluding Shaving Preparations); Cleansing Products (Cold Creams, Cleansing Lotions, Liquids and Pads); Paste Masks (Mud Packs); Face and Neck Preparations (Excluding Shaving Preparations); Moisturizing Preparations; Night Skin Care Preparations; Skin Fresheners

Trade Name Mixtures:
ABS Propolis Extract (Active Concepts)
Propolis Extract (Burgundy)
Propolis Extract "COS" (Cosmetochem) (Cosmetochem International Ltd.)
Propolis Extract ies (IES LABO)
Propolis LS (Alban Muller)
VT-270 Extract of Propolis (Vege-Tech)

PROPOLIS WAX

Definition: Propolis Wax is the material obtained from the extraction of propolis, a resinous substance found in beehives. *See "Regulatory and Ingredient Use Information," regarding use of EU Trivial names in Volume 1, Introduction, Part A.*

Information Source: MI-13(7928)

Chemical Class: Biological Products

Function: Proprietary

Technical/Other Names:
Extract of Propolis Wax
Propolis Resin

Trade Name Mixtures:
AEC Propolis Extract (A & E Connock)
Glycolysat of Propolis (CEP (Solabia))
Phytoderm Propolis Glycolic (Universal Flavors)
Propoli Extract G 5 (Esperis)
Propoli Extract H 5 (Esperis)
Propoli Extract L 10 (Esperis)
Propolis Extract HS 2602 G (Grau)
Propolis Extract Powder Pan 80 (Paninkret)
Propolis Extract Powder Pan 90 (Paninkret)

PROPOXYTETRAMETHYL PIPERIDINYL DIMETHICONE

CAS No.: 171543-65-0

Definition: Propoxytetramethyl Piperidinyl Dimethicone is the siloxane polymer that conforms generally to the formula:

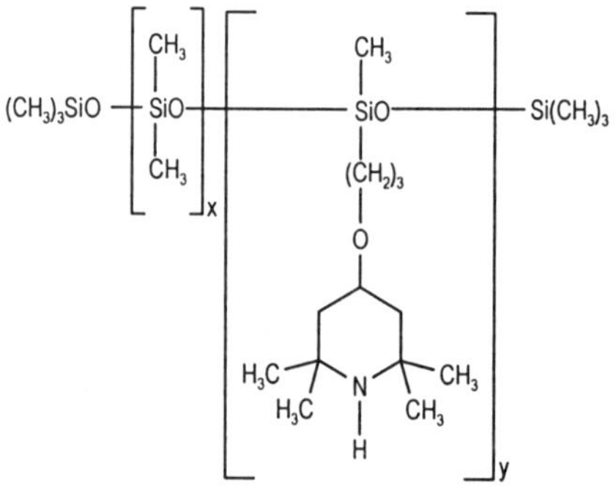

Information Source: TSCA

Chemical Classes: Siloxanes and Silanes; Synthetic Polymers

Function: Hair Conditioning Agent

Technical/Other Name:
Siloxanes and Silicones, Di-Me, Me 3-[(2,2,6,6-Tetramethyl-4-Piperidinyl)Oxy]Propyl

Trade Name:
Jeesilc UAF (Jeen)

Trade Name Mixtures:
TME-2028 (Taylor)
TME-4052 (Taylor)

PROPYL ACETATE

CAS No.	EINECS No.
109-60-4	203-686-1

Empirical Formula:
$C_5H_{10}O_2$

Definition: Propyl Acetate is the ester of propyl alcohol and acetic acid that conforms to the formula:

$CH_3C(=O)-OCH_2CH_2CH_3$

The inclusion of any compound in the *Dictionary and Handbook* does not indicate that use of that substance as a cosmetic ingredient complies with the laws and regulations governing such use in the United States or any other country.

Information Sources: 21CFR172.515, 21CFR177.1200, MI-13(7933), RIFM, TSCA

Chemical Class: Esters

Functions: Fragrance Ingredient; Solvent

Reported Product Categories: Nail Polish and Enamels; Basecoats and Undercoats; Manicuring Preparations, Misc.; Nail Polish and Enamel Removers

Technical/Other Names:
Acetic Acid, Propyl Ester
Propyl acetate (RIFM)
n-Propyl Acetate
Propyl Ethanoate

Trade Name:
Eastman Propyl Acetate (Eastman Chemical)

Trade Name Mixture:
KFilm 2001 (Kane)

PROPYL ALCOHOL

CAS No.
71-23-8

EINECS No.
200-746-9

JPN Translation:
プロパノール

Empirical Formula:
C_3H_8O

Definition: Propyl Alcohol is the aliphatic alcohol that conforms to the formula:

$$CH_3CH_2CH_2OH$$

Information Sources: 21CFR172.515, 21CFR175.105, 21CFR177.1200, 21CFR573.880, DDR, JCIC, JCLS, JSQI, MI-13(7934), PN, RIFM, TSCA

Chemical Class: Alcohols

Functions: Antifoaming Agent; Fragrance Ingredient; Solvent; Viscosity Decreasing Agent

Reported Product Category: Skin Care Preparations, Misc.

Technical/Other Names:
1-Hydroxypropane
1-Propanol
Propyl alcohol (RIFM)
n-Propyl Alcohol

Trade Names:
Eastman n-Propyl Alcohol (Eastman Chemical)
n-Propanol Purest (extra pure) (BASF)

PROPYL BENZOATE

CAS No.
2315-68-6

EINECS No.
219-020-8

Empirical Formula:
$C_{10}H_{12}O_2$

Definition: Propyl Benzoate is the ester of n-propyl alcohol and Benzoic Acid (q.v.). It conforms to the formula:

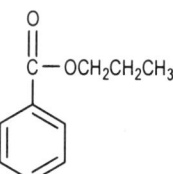

Information Sources: EEC(VI/1-1), RIFM, TSCA

Chemical Class: Esters

Functions: Fragrance Ingredient; Preservative

Technical/Other Names:
Benzoic Acid, n-Propyl Ester
Propyl benzoate (RIFM)

PROPYL C12-15 PARETH-8 CARBOXYLATE

Definition: Propyl C12-15 Pareth-8 Carboxylate is the propyl ester of C12-15 Pareth-8 Carboxylic Acid (q.v.) that conforms generally to the formula:

$$R(OCH_2CH_2)_nOCH_2C—OCH_2CH_2CH_3$$

where R represents the C12-15 alkyl group and n has an average value of 7.

Information Source: JCIC

Chemical Class: Esters

Function: Skin-Conditioning Agent - Emollient

Technical/Other Name:
Propyl Polyoxyethylene Alkyl (12-15) Ether Acetate

PROPYLENE CARBONATE

CAS No.
108-32-7

EINECS No.
203-572-1

JPN Translation:
炭酸プロピレン

Empirical Formula:
$C_4H_6O_3$

Definition: Propylene Carbonate is the organic compound that conforms to the formula:

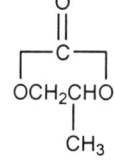

Information Sources: 21CFR175.105, CIR: [S] JACT-6(1)1987, JCIC, JCLS, JSQI, NF XIX, TSCA, USAN

Chemical Class: Esters

Functions: Solvent; Viscosity Decreasing Agent

Reported Product Categories: Lipsticks; Personal Cleanliness Products, Misc.; Mascara; Makeup Preparations (Not eye), Misc.; Eyeliners; Eye Makeup Preparations, Misc.; Eye Shadows; Eyebrow Pencils; Nail Polish and Enamel Removers; Makeup Bases; Moisturizing Preparations; Eye Makeup Removers; Foundations

Technical/Other Names:
Carbonic Acid Cyclic Methylethylene Ester
1,3-Dioxolan-2-one, 4-Methyl-
4-Methyl-1,3-Dioxolan-2-one

Trade Names:
Jeffsol Propylene Carbonate (Huntsman)
4-Methyl-1,3-Dioxolan-2-one (Huntsman)

Trade Name Mixtures:
Bentone Gel CAO (ELE)
Bentone Gel DOA (ELE)
Bentone Gel EUG (ELE)
Bentone Gel IPM (ELE)
Bentone Gel ISD (ELE)
Bentone Gel LOI (ELE)
Bentone Gel M-20 (ELE)
Bentone Gel MIO (ELE)
Bentone Gel SS-71 (ELE)
Bentone Gel 10ST (ELE)
Bentone Gel TN (ELE)
Bentone Gel VS-5 PC (ELE)
Covalip LL 48 (LCW)
Miglyol Gel B (Sasol GmbH - Witten)
Miglyol 840 Gel B (Sasol GmbH - Witten)
Miglyol Gel T (Sasol GmbH - Witten)
Simagel C (Biophil)
Simagel IM (Biophil)
Simagel NO (Biophil)
Softisan Gel (Sasol GmbH - Witten)
Tixogel CCT 6030 (Sud-Chemie, United Catalysts)
Tixogel COG 1540 (Sud-Chemie, United Catalysts)
Tixogel FTN (Sud-Chemie, United Catalysts)
Tixogel FTN 1564 (Sud-Chemie, United Catalysts)
Tixogel HXL-1209 (Sud-Chemie, United Catalysts)
Tixogel IDD - 1168 (Sud-Chemie, United Catalysts)
Tixogel IDD 1538 (Sud-Chemie, United Catalysts)
Tixogel IHD-1235 (Sud-Chemie, United Catalysts)
Tixogel IIN 1578 (Sud-Chemie, United Catalysts)
Tixogel IPM (Sud-Chemie, United Catalysts)
Tixogel JOG 1583 (Sud-Chemie, United Catalysts)

Tixogel LAN (Sud-Chemie, United Catalysts)
Tixogel MIO (Sud-Chemie, United Catalysts)
Tixogel MIO 1584 (Sud-Chemie, United Catalysts)
Tixogel ODD-1520 (Sud-Chemie, United Catalysts)
Tixogel OMS (Sud-Chemie, United Catalysts)
Tixogel OMS 1562 (Sud-Chemie, United Catalysts)
Tixogel VSP (Sud-Chemie, United Catalysts)
Tixogel VSP-1438-V (Sud-Chemie, United Catalysts)

PROPYLENE GLYCOL

CAS No.   **EINECS No.**
57-55-6   200-338-0

JPN Translation:
PG

Empirical Formula:
$C_3H_8O_2$

Definition: Propylene Glycol is the aliphatic alcohol that conforms generally to the formula:

$$CH_3CHCH_2OH$$
$$|$$
$$OH$$

Information Sources: ARG, AUS, BP, BPC, BRA, 21CFR169.175, 21CFR169.176, 21CFR169.177, 21CFR169.178, 21CFR169.180, 21CFR169.181, 21CFR175.300, 21CFR177.2600, 21CFR178.3300, 21CFR184.1666, 21CFR582.4666, CIR: [SQ] JACT-13(6)-1994, CTFA S, CZE, DDR, FCC, HUN, ITA, JAN, JCLS, JSCI, MAR, MI-13(7947), NFJ, OTC-I-OP, PF, PN, POR, RIFM, ROM, SNPF, TSCA, USAN, USD, USP XXIV, WHO

Chemical Class: Alcohols

Functions: Fragrance Ingredient; Humectant; Skin-Conditioning Agent - Humectant; Skin-Conditioning Agent - Miscellaneous; Solvent; Viscosity Decreasing Agent

Reported Product Categories: Hair Dyes and Colors (All Types Requiring Caution Statements and Patch Tests); Bath Preparations, Misc.; Body and Hand Preparations (Excluding Shaving Preparations); Moisturizing Preparations; Skin Care Preparations, Misc.; Bath Oils, Tablets, and Salts; Cleansing Products (Cold Creams, Cleansing Lotions, Liquids and Pads); Hair Conditioners; Deodorants (Underarm); Tonics, Dressings, and Other Hair Grooming Aids; Foundations; Bath Capsules; Paste Masks (Mud Packs); Face and Neck Preparations (Excluding Shaving Preparations); Bath Soaps and Detergents; Baby Shampoos; Aftershave Lotions; Night Skin Care Preparations; Skin Fresheners; Colognes and Toilet Waters; Makeup Bases; Mascara; Lipsticks; Hair Preparations (Non-coloring), Misc.; Eye Makeup Preparations, Misc.; Eyebrow Pencils; Suntan Gels, Creams, and Liquids; Fragrance Preparations, Misc.; Makeup Preparations (Not eye), Misc.; Bubble Baths; Permanent Waves; Perfumes; Hair Straighteners; Shaving Cream (Aerosol, Brushless and Lather); Eye Makeup Removers; Eyeliners; Hair Sprays (Aerosol Fixatives); Indoor Tanning Preparations; Shaving Preparations, Misc.; Baby Lotions, Oils, Powders and Creams; Blushers (All types); Hair Shampoos (Coloring); Eye Shadows; Suntan Preparations, Misc.; Hair Tints; Eye Lotions; Hair Wave Sets; Baby Products, Misc.; Hair Coloring Preparations, Misc.; Hair Rinses (Non-coloring); Cuticle Softeners; Hair Bleaches; Sachets; Foot Powders and Sprays; Manicuring Preparations, Misc.; Nail Polish and Enamel Removers; Hair Rinses (Coloring); Mouthwashes and Breath Fresheners (Liquids and Sprays); Nail Polish and Enamels; Douches; Shampoos (Non-coloring); Depilatories; Makeup Fixatives; Nail Creams and Lotions; Basecoats and Undercoats; Face Powders; Personal Cleanliness Products, Misc.; Powders (Dusting and Talcum, Excluding Aftershave Talcs); Rouges

Technical/Other Names:
1,2-Dihydroxypropane
2-Hydroxypropanol
Methylethyl Glycol
1,2-Propanediol
Propylene glycol (RIFM)

Trade Names:
1,2-Propylene Glycol Care (BASF)
Propylene Glycol USP (Jeen)
Propylene Glycol USP (Protameen)
Propylene Glycol USP (RITA)

Trade Name Mixtures:
ABS Green Tea Extract (Active Concepts)
ABS Propolis Extract (Active Concepts)
ABS Sea Rocket Extract (Active Concepts)
ABS Sea Rocket Extract PG (Active Concepts)
ABS White Tea Extract (Active Concepts)
Acacia Collagen 97 (Cosmetochem) (Cosmetochem International Ltd.)
Acacia Flower Extract HS 2744 G (Grau)
Acacia Phytolait (Alban Muller)
AC Dermal Respiratory Factor 5% (Active Concepts)
Acerola Extract HS 3629 G (Grau)
Acerola HS (Alban Muller)
Acerola HSA (Alban Muller)
Acifructol Complex P 63 (Gattefosse s.a.)
Acifructol Complex P 67 (Gattefosse s.a.)
Acifructol Tomato P 62 (Gattefosse s.a.)
Acnacidol PG (Vincience)
Acneous Phytogreen Complex GXH 267 (Phytochim)
AC RNA/DNA Complex (Active Concepts)
Actifirm TSJ (Active Organics)
Actiphpyte Damiana (Active Organics)
Actiphyte Algae (Active Organics)
Actiphyte Amla Fruit (Active Organics)
Actiphyte Artichoke (Active Organics)
Actiphyte Asafoetida (Active Organics)
Actiphyte Ascophyllum (Active Organics)
Actiphyte Autolyzed Yeast (Active Organics)
Actiphyte Barberry (Active Organics)
Actiphyte Bayberry Bark (Active Organics)
Actiphyte Bearberry (Active Organics)
Actiphyte Beet Root (Active Organics)
Actiphyte Berberis (Active Organics)
Actiphyte Bilberry Leaves (Active Organics)
Actiphyte Bitter Orange (Active Organics)
Actiphyte Bitter Orange Peel PG50P (Active Organics)
Actiphyte Bittersweet (Active Organics)
Actiphyte Black Cohosh (Active Organics)
Actiphyte Black Currant (Active Organics)
Actiphyte Blue Vervain (Active Organics)
Actiphyte Brazil Nut (Active Organics)
Actiphyte Broccoli (Active Organics)
Actiphyte Brussel Sprouts PG50NP (Active Organics)
Actiphyte Buchu (Active Organics)
Actiphyte Cabbage (Active Organics)
Actiphyte Cats Claw (Active Organics)
Actiphyte Chaste Tree Berry (Active Organics)
Actiphyte Chinchona (Active Organics)
Actiphyte Cotton Seed (Active Organics)
Actiphyte Cranesbill (Active Organics)
Actiphyte Cypress (Active Organics)
Actiphyte Daisy (Active Organics)
Actiphyte Dulse (Active Organics)
Actiphyte European Centaury (Active Organics)
Actiphyte Fo-Ti Root Root (Active Organics)
Actiphyte Freesia (Active Organics)
Actiphyte Goldenrod (Active Organics)
Actiphyte Grape Seed (Active Organics)
Actiphyte Honey (Active Organics)
Actiphyte Hypericum (Active Organics)
Actiphyte Indian Cress (Active Organics)
Actiphyte Indian Hemp Root (Active Organics)
Actiphyte Jojoba Meal (Active Organics)
Actiphyte Kukui (Active Organics)
Actiphyte Lemon Myrtle (Active Organics)
Actiphyte Lilac Blossom (Active Organics)
Actiphyte Lily of the Valley (Active Organics)
Actiphyte Linden Tree (Active Organics)
Actiphyte Maidenhair (Active Organics)

Actiphyte Melaleuca (Active Organics)
Actiphyte Milk Thistle (Active Organics)
Actiphyte Molasses (Active Organics)
Actiphyte Mugwort (Active Organics)
Actiphyte Mulberry Leaf (Active Organics)
Actiphyte Myrtle (Active Organics)
Actiphyte Neem (Active Organics)
Actiphyte Noni (Active Organics)
Actiphyte of Acacia PG50 (Active Organics)
Actiphyte of Acerola (Active Organics)
Actiphyte of Agrimony PG50 (Active Organics)
Actiphyte of Alfalfa PG50 (Active Organics)
Actiphyte of Alkanet Root PG50 (Active Organics)
Actiphyte of Almond PG50 (Active Organics)
Actiphyte of Aloe Vera 10-Fold (Active Organics)
Actiphyte of Aloe Vera PG50 (Active Organics)
Actiphyte of Andrographis Paniculata (Active Organics)
Actiphyte of Angelica PG50 (Active Organics)
Actiphyte of Anise (Active Organics)
Actiphyte of Apple Fruit (Active Organics)
Actiphyte of Apple Green (Active Organics)
Actiphyte of Apple Leaves (Active Organics)
Actiphyte of Apple Leaves PG50 (Active Organics)
Actiphyte of Apple Mint (Active Organics)
Actiphyte of Apple PG50 (Active Organics)
Actiphyte of Apricot Kernel PG50 (Active Organics)
Actiphyte of Apricot Leaves PG50 (Active Organics)
Actiphyte of Apricot PG50 (Active Organics)
Actiphyte of Arbutus PG50 (Active Organics)
Actiphyte of Arnica PG50 (Active Organics)
Actiphyte of Arrowroot PG50 (Active Organics)
Actiphyte of Ashwanganda (Active Organics)
Actiphyte of Asparagus PG50 (Active Organics)
Actiphyte of Astragalus PG50 (Active Organics)
Actiphyte of Avocado PG50 (Active Organics)
Actiphyte of Balm Mint PG50 (Active Organics)
Actiphyte of Balm of Gilead PG50 (Active Organics)
Actiphyte of Balsam Canada PG50 (Active Organics)
Actiphyte of Bamboo PG50 (Active Organics)
Actiphyte of Banana Leaves PG50 (Active Organics)

Actiphyte of Banana PG50 (Active Organics)
Actiphyte of Barley Malt PG50 (Active Organics)
Actiphyte of Basil PG50 (Active Organics)
Actiphyte of Bayberry PG50 (Active Organics)
Actiphyte of Bay Laurel PG50 (Active Organics)
Actiphyte of Bee Balm PG50 (Active Organics)
Actiphyte of Beechnut (Active Organics)
Actiphyte of Bee Pollen (Active Organics)
Actiphyte of Bee Pollen PG50 (Active Organics)
Actiphyte of Benzoin PG50 (Active Organics)
Actiphyte of Bergamot PG50 (Active Organics)
Actiphyte of Bilberry PG50 (Active Organics)
Actiphyte of Birch Bark PG50 (Active Organics)
Actiphyte of Birch Leaves PG50 (Active Organics)
Actiphyte of Bistort PG50 (Active Organics)
Actiphyte of Blackberry Leaves PG50 (Active Organics)
Actiphyte of Blackberry PG50 (Active Organics)
Actiphyte of Black Henna PG50 (Active Organics)
Actiphyte of Black Indian Hemp PG50 (Active Organics)
Actiphyte of Black Pepper PG50 (Active Organics)
Actiphyte of Black Raspberry (Active Organics)
Actiphyte of Black Snakeroot PG50 (Active Organics)
Actiphyte of Black Walnut Hull PG50 (Active Organics)
Actiphyte of Black Walnut Leaves PG50 (Active Organics)
Actiphyte of Bladderwrack PG50 (Active Organics)
Actiphyte of Bloodroot PG50 (Active Organics)
Actiphyte of Blueberry PG50 (Active Organics)
Actiphyte of Blue Corn PG50 (Active Organics)
Actiphyte of Blue Flag PG50 (Active Organics)
Actiphyte of Blue Malva PG50 (Active Organics)
Actiphyte of Borage PG50 (Active Organics)
Actiphyte of Boswellia Serrata (Active Organics)
Actiphyte of Bougainvillea PG50 (Active Organics)
Actiphyte of Brazil Nut (Active Organics)

Actiphyte of Brown Mustard Seed PG50 (Active Organics)
Actiphyte of Buckthorn PG50 (Active Organics)
Actiphyte of Burdock PG50 (Active Organics)
Actiphyte of Burnet Root (Active Organics)
Actiphyte of Butcher's Broom PG50 (Active Organics)
Actiphyte of Butter Bur Leaves (Active Organics)
Actiphyte of Cactus PG50 (Active Organics)
Actiphyte of Calendula PG50 (Active Organics)
Actiphyte of Camphor AL (Active Organics)
Actiphyte of Cantaloupe PL (Active Organics)
Actiphyte of Capsicum PG50 (Active Organics)
Actiphyte of Caraway Seed PG50 (Active Organics)
Actiphyte of Cardamom Seed (Active Organics)
Actiphyte of Carrot PG50 (Active Organics)
Actiphyte of Carrot Seed PG50 (Active Organics)
Actiphyte of Cascara Sagrada (Active Organics)
Actiphyte of Cassia PG50 (Active Organics)
Actiphyte of Catnip PG50 (Active Organics)
Actiphyte of Cedar Wood PG50 (Active Organics)
Actiphyte of Celandine PG50 (Active Organics)
Actiphyte of Celery PG50 (Active Organics)
Actiphyte of Centaury (Active Organics)
Actiphyte of Centaury PG50 (Active Organics)
Actiphyte of Chamomile (Active Organics)
Actiphyte of Chamomile PG50 (Active Organics)
Actiphyte of Chaparral PG50 (Active Organics)
Actiphyte of Chardonnay (Active Organics)
Actiphyte of Cherimoya PG50 (Active Organics)
Actiphyte of Chesnut (Active Organics)
Actiphyte of Chickweed PG50 (Active Organics)
Actiphyte of Chrysanthemum PG50 (Active Organics)
Actiphyte of Cinnamon PG50 (Active Organics)
Actiphyte of Clary Sage PG50 (Active Organics)
Actiphyte of Cleavers PG50 (Active Organics)
Actiphyte of Clemantine PG50 (Active Organics)
Actiphyte of Clove PG50 (Active Organics)
Actiphyte of Coconut PG50 (Active Organics)
Actiphyte of Coffee PG50 (Active Organics)

Actiphyte of Comfrey Leaf PG50 (Active Organics)
Actiphyte of Comfrey Root PG50 (Active Organics)
Actiphyte of Condurango PG50 (Active Organics)
Actiphyte of Coneflower PG50 (Active Organics)
Actiphyte of Coriander PG50 (Active Organics)
Actiphyte of Corn (Active Organics)
Actiphyte of Cornflower PG50 (Active Organics)
Actiphyte of Cornsilk (Active Organics)
Actiphyte of Cowslip (Active Organics)
Actiphyte of Cranberry PG50 (Active Organics)
Actiphyte of Cucumber PG50 (Active Organics)
Actiphyte of Cumin Seed (Active Organics)
Actiphyte of Cypress Cone PG50 (Active Organics)
Actiphyte of Damiana (Active Organics)
Actiphyte of Dandelion PG50 (Active Organics)
Actiphyte of Date PG50 (Active Organics)
Actiphyte of Devil's Claw (Active Organics)
Actiphyte of Dill PG50 (Active Organics)
Actiphyte of Dong Quai PG50 (Active Organics)
Actiphyte of Eggplant PG50 (Active Organics)
Actiphyte of Elderberry PG50 (Active Organics)
Actiphyte of Elderflower PG50 (Active Organics)
Actiphyte of Elecampane PG50 (Active Organics)
Actiphyte of English Cucumber PG50 (Active Organics)
Actiphyte of Ephedra PG50 (Active Organics)
Actiphyte of Eucalyptus PG50 (Active Organics)
Actiphyte of Evening Primrose PG50 (Active Organics)
Actiphyte of Eyebright PG50 (Active Organics)
Actiphyte of Fennel PG50 (Active Organics)
Actiphyte of Fenugreek PG50 (Active Organics)
Actiphyte of Feverfew PG50 (Active Organics)
Actiphyte of Fig PG50 (Active Organics)
Actiphyte of Fir Needle PG50 (Active Organics)
Actiphyte of Flaxseed PG50 (Active Organics)
Actiphyte of Fo-Ti-Tieng PG50 (Active Organics)
Actiphyte of Frankincense PG50 (Active Organics)

Actiphyte of Fumitory PG50 (Active Organics)
Actiphyte of Gardenia PG50 (Active Organics)
Actiphyte of Garlic PG50 (Active Organics)
Actiphyte of Gentian PG50 (Active Organics)
Actiphyte of Geranium PG50 (Active Organics)
Actiphyte of Ginger PG50 (Active Organics)
Actiphyte of Ginkgo Biloba Concentrate YR (Active Organics)
Actiphyte of Ginkgo PG50 (Active Organics)
Actiphyte of Ginseng PG50 (Active Organics)
Actiphyte of Goldenseal PG50 (Active Organics)
Actiphyte of Gotu Kola PG50 (Active Organics)
Actiphyte of Grapefruit Peel PG50 (Active Organics)
Actiphyte of Grapefruit PG50 (Active Organics)
Actiphyte of Grapefruit Seed PG50 (Active Organics)
Actiphyte of Grape Leaf (Active Organics)
Actiphyte of Grape Skin PG (Active Organics)
Actiphyte of Grapes PG50 (Active Organics)
Actiphyte of Grindelia PG50 (Active Organics)
Actiphyte of Ground Ivy PG50 (Active Organics)
Actiphyte of Guarana PG50 (Active Organics)
Actiphyte of Guava PG50 (Active Organics)
Actiphyte of Haritaki (Active Organics)
Actiphyte of Hawaiian Ti (Active Organics)
Actiphyte of Hawaiian White Ginger PG50 (Active Organics)
Actiphyte of Hawthorn Berry PG50 (Active Organics)
Actiphyte of Hazel Nut PG50 (Active Organics)
Actiphyte of Heather PG50 (Active Organics)
Actiphyte of Hibiscus PG50 (Active Organics)
Actiphyte of Honeydew PG50 (Active Organics)
Actiphyte of Honeysuckle PG50 (Active Organics)
Actiphyte of Hops PG50 (Active Organics)
Actiphyte of Horehound PG50 (Active Organics)
Actiphyte of Horse Chestnut PG50 (Active Organics)
Actiphyte of Horseradish PG50 (Active Organics)
Actiphyte of Horsetail PG50 (Active Organics)
Actiphyte of Huckleberry (Active Organics)

Actiphyte of Hyacinth (Active Organics)
Actiphyte of Hyssop PG50 (Active Organics)
Actiphyte of Iceland Moss PG50 (Active Organics)
Actiphyte of Impatiens PG50 (Active Organics)
Actiphyte of Irish Moss PG50 (Active Organics)
Actiphyte of Italian Olive PG50 (Active Organics)
Actiphyte of Ivy PG50 (Active Organics)
Actiphyte of Jaborandi PG50 (Active Organics)
Actiphyte of Japanese Green Tea PG50 (Active Organics)
Actiphyte of Japanese Knotweed (Active Organics)
Actiphyte of Jasmine PG50 (Active Organics)
Actiphyte of Jewel Weed BG (Active Organics)
Actiphyte of Jujube Fruit (Active Organics)
Actiphyte of Juniper Berry PG50 (Active Organics)
Actiphyte of Kaffir Lime Leaves (Active Organics)
Actiphyte of Kangaroo Paw PG50 (Active Organics)
Actiphyte of Kava Kava PG50 (Active Organics)
Actiphyte of Kiwi PG50 (Active Organics)
Actiphyte of Kola Nut PG50 (Active Organics)
Actiphyte of Lady's Mantle PG50 (Active Organics)
Actiphyte of Lady's Slipper PG50 (Active Organics)
Actiphyte of Laminaria (Active Organics)
Actiphyte of Lavender PG50 (Active Organics)
Actiphyte of Lemon Bioflavonoids PG50 (Active Organics)
Actiphyte of Lemongrass PG50 (Active Organics)
Actiphyte of Lemon Peel PG50 (Active Organics)
Actiphyte of Lemon PG50 (Active Organics)
Actiphyte of Lemon Verbena PG50 (Active Organics)
Actiphyte of Lettuce PG50 (Active Organics)
Actiphyte of Lichen PG50 (Active Organics)
Actiphyte of Licorice PG50 (Active Organics)
Actiphyte of Ligusticum PG50 (Active Organics)
Actiphyte of Lime Flower (Active Organics)
Actiphyte of Lime Peel PG50 (Active Organics)
Actiphyte of Lime PG50 (Active Organics)
Actiphyte of Linden PG50 (Active Organics)
Actiphyte of Locust Bean (Active Organics)
Actiphyte of Lotus Blossom PG50 (Active Organics)

Actiphyte of Lovage PG50 (Active Organics)
Actiphyte of Lungwort PG50 (Active Organics)
Actiphyte of Lupine PG50 (Active Organics)
Actiphyte of Macadamia Nut PG50 (Active Organics)
Actiphyte of Magnolia Bark PG50 (Active Organics)
Actiphyte of Mandarin Orange PG50 (Active Organics)
Actiphyte of Mango PG50 (Active Organics)
Actiphyte of Marjoram PG50 (Active Organics)
Actiphyte of Marsh Mallow PG50 (Active Organics)
Actiphyte of Meadowsweet PG50 (Active Organics)
Actiphyte of Mexican Tea PG50 (Active Organics)
Actiphyte of Milk Vetch PG50 (Active Organics)
Actiphyte of Millet PG50 (Active Organics)
Actiphyte of Mimosa Leaf PG50 (Active Organics)
Actiphyte of Mistletoe PG50 (Active Organics)
Actiphyte of Moonflower PG50 (Active Organics)
Actiphyte of Mullein PG50 (Active Organics)
Actiphyte of Mushroom PG50 (Active Organics)
Actiphyte of Myrrh PG50 (Active Organics)
Actiphyte of Neroli (Active Organics)
Actiphyte of Nettle PG50 (Active Organics)
Actiphyte of Neutral Henna PG50 (Active Organics)
Actiphyte of Norway Spruce Cone PG50 (Active Organics)
Actiphyte of Nutmeg PG50 (Active Organics)
Actiphyte of Oak Bark PG50 (Active Organics)
Actiphyte of Oat Flour PG50 (Active Organics)
Actiphyte of Oatmeal PG50 (Active Organics)
Actiphyte of Oat PG50 (Active Organics)
Actiphyte of Olive Leaf PG50 (Active Organics)
Actiphyte of Orange Bioflavonoids PG50 (Active Organics)
Actiphyte of Orange Blossom PG50 (Active Organics)
Actiphyte of Orange Peel PG50 (Active Organics)
Actiphyte of Orange PG50 (Active Organics)
Actiphyte of Orris Root PG50 (Active Organics)
Actiphyte of Pansy (Active Organics)
Actiphyte of Pansy PG50 (Active Organics)

Actiphyte of Papaya Fruit (Active Organics)
Actiphyte of Papaya Leaves (Active Organics)
Actiphyte of Papaya Leaves PG50 (Active Organics)
Actiphyte of Papaya PG50 (Active Organics)
Actiphyte of Paraguay Tea (Active Organics)
Actiphyte of Paraguay Tea Concentrate PG50 (Active Organics)
Actiphyte of Parsley (Active Organics)
Actiphyte of Parsley PG50 (Active Organics)
Actiphyte of Passionflower PG50 (Active Organics)
Actiphyte of Passion Fruit Hull PG50 (Active Organics)
Actiphyte of Passion Fruit PG50 (Active Organics)
Actiphyte of Patchouli PG50 (Active Organics)
Actiphyte of Pea (Active Organics)
Actiphyte of Peach Leaves PG50 (Active Organics)
Actiphyte of Peach PG50 (Active Organics)
Actiphyte of Pellitory PG50 (Active Organics)
Actiphyte of Pennyroyal PG50 (Active Organics)
Actiphyte of Peony Flower PG50 (Active Organics)
Actiphyte of Peony Root PG50 (Active Organics)
Actiphyte of Peppermint PG50 (Active Organics)
Actiphyte of Persimmon Leaf PG50 (Active Organics)
Actiphyte of Perwinkle PG50 (Active Organics)
Actiphyte of Pichi (Active Organics)
Actiphyte of Pichi PG50 (Active Organics)
Actiphyte of Pineapple (Active Organics)
Actiphyte of Pineapple PG50 (Active Organics)
Actiphyte of Pine Bark PG50 (Active Organics)
Actiphyte of Pine Cone PG50 (Active Organics)
Actiphyte of Pink Peppercorn (Active Organics)
Actiphyte of Plantain PG50 (Active Organics)
Actiphyte of Plumeria (Active Organics)
Actiphyte of Pomegranate (Active Organics)
Actiphyte of Pomegranate PG50 (Active Organics)
Actiphyte of Poppy PG50 (Active Organics)
Actiphyte of Potato (Active Organics)
Actiphyte of Prickly Ash PG50 (Active Organics)
Actiphyte of Primrose (Active Organics)
Actiphyte of Prune (Active Organics)

Actiphyte of Prune PG50 (Active Organics)
Actiphyte of Psoralea (Active Organics)
Actiphyte of Psoralea PG50 (Active Organics)
Actiphyte of Pueraria PG (Active Organics)
Actiphyte of Pyrethrum (Active Organics)
Actiphyte of Pyrethrum PG50 (Active Organics)
Actiphyte of Queen of the Meadow PG50 (Active Organics)
Actiphyte of Quince Seed PG50 (Active Organics)
Actiphyte of Quinoa PG50 (Active Organics)
Actiphyte of Raspberry Leaves PG50 (Active Organics)
Actiphyte of Raspberry PG50 (Active Organics)
Actiphyte of Red Beet PG50 (Active Organics)
Actiphyte of Red Clover Blossom PG50 (Active Organics)
Actiphyte of Red Henna PG50 (Active Organics)
Actiphyte of Red Root PG50 (Active Organics)
Actiphyte of Red Sandalwood PG50 (Active Organics)
Actiphyte of Red Vine PG50P (Active Organics)
Actiphyte of Rehmannia PG50 (Active Organics)
Actiphyte of Rest Harrow PG50 (Active Organics)
Actiphyte of Rhatany PG50 (Active Organics)
Actiphyte of Rhododendron PG50 (Active Organics)
Actiphyte of Rhubarb PG50 (Active Organics)
Actiphyte of Rhubarb Stalks PG50 (Active Organics)
Actiphyte of Rice Bran PG50 (Active Organics)
Actiphyte of Roman Chamomile PG50 (Active Organics)
Actiphyte of Rose Geranium PG50 (Active Organics)
Actiphyte of Rosehips PG50 (Active Organics)
Actiphyte of Rosemary PG50 (Active Organics)
Actiphyte of Rose PG50 (Active Organics)
Actiphyte of Roucouyer Seed (Active Organics)
Actiphyte of Saffron PG50 (Active Organics)
Actiphyte of Sage PG50 (Active Organics)
Actiphyte of Sandalwood PG50 (Active Organics)
Actiphyte of Sanguinaria PG50 (Active Organics)

Actiphyte of Sarsaparilla PG50 (Active Organics)

Actiphyte of Sassafras PG50 (Active Organics)

Actiphyte of Schisandra PG50 (Active Organics)

Actiphyte of Sea Kelp PG50 (Active Organics)

Actiphyte of Senega PG50 (Active Organics)

Actiphyte of Sesame PG50 (Active Organics)

Actiphyte of Sheep's Sorrel PG50 (Active Organics)

Actiphyte of Shiitake Mushroom (Active Organics)

Actiphyte of Siberian Ginseng PG50 (Active Organics)

Actiphyte of Skullcap PG50 (Active Organics)

Actiphyte of Slippery Elm PG50 (Active Organics)

Actiphyte of Snakeroot PG50 (Active Organics)

Actiphyte of Soap Bark PG50 (Active Organics)

Actiphyte of Soapwort PG50 (Active Organics)

Actiphyte of Solomon's Seal PG50 (Active Organics)

Actiphyte of Southernwood PG50 (Active Organics)

Actiphyte of Spanish Moss PG50 (Active Organics)

Actiphyte of Spearmint PG50 (Active Organics)

Actiphyte of Spinach PG50 (Active Organics)

Actiphyte of Spirulina PG50 (Active Organics)

Actiphyte of Spruce Cone PG50 (Active Organics)

Actiphyte of Star Gazer (Active Organics)

Actiphyte of Star of Bethlehem PG50 (Active Organics)

Actiphyte of St. John's Wort PG50 (Active Organics)

Actiphyte of Strawberry Leaves PG50 (Active Organics)

Actiphyte of Strawberry PG50 (Active Organics)

Actiphyte of Sugar Beet PG50 (Active Organics)

Actiphyte of Sugar Maple (Active Organics)

Actiphyte of Sumach PG50 (Active Organics)

Actiphyte of Suma PG50 (Active Organics)

Actiphyte of Sunflower Petal PG50 (Active Organics)

Actiphyte of Sunflower Seed PG50 (Active Organics)

Actiphyte of Surinam Cherry GL (Active Organics)

Actiphyte of Sweet Flag PG50 (Active Organics)

Actiphyte of Tamarind (Active Organics)

Actiphyte of Tangerine Peel PG50 (Active Organics)

Actiphyte of Tangerine PG50 (Active Organics)

Actiphyte of Tansy PG50 (Active Organics)

Actiphyte of Taro Root PG50 (Active Organics)

Actiphyte of Tea, Black PG50 (Active Organics)

Actiphyte of Tea, Green PG50 (Active Organics)

Actiphyte of Tea, Oolong PG50 (Active Organics)

Actiphyte of Tea PG50 (Active Organics)

Actiphyte of Thai Tea (Active Organics)

Actiphyte of Thistle PG50 (Active Organics)

Actiphyte of Thyme PG50 (Active Organics)

Actiphyte of Tobacco PG50 (Active Organics)

Actiphyte of Tomato PG50 (Active Organics)

Actiphyte of Tuberose (Active Organics)

Actiphyte of Valerian PG50 (Active Organics)

Actiphyte of Vanilla (Active Organics)

Actiphyte of Vanilla Cactus PG50 (Active Organics)

Actiphyte of Viburnum (Active Organics)

Actiphyte of Violet (Active Organics)

Actiphyte of Violet PG50 (Active Organics)

Actiphyte of Walnut Seed (Active Organics)

Actiphyte of Wasabi (Active Organics)

Actiphyte of Water Chestnut PG50 (Active Organics)

Actiphyte of Watercress PG50 (Active Organics)

Actiphyte of Water Lily (Active Organics)

Actiphyte of Watermelon PG50 (Active Organics)

Actiphyte of Wheat (Active Organics)

Actiphyte of Wheat Bran (Active Organics)

Actiphyte of Wheat Germ (Active Organics)

Actiphyte of Wheat Germ PG50 (Active Organics)

Actiphyte of Wheat Grass (Active Organics)

Actiphyte of Wheat PG50 (Active Organics)

Actiphyte of White Balsam (Active Organics)

Actiphyte of White Balsam PG50 (Active Organics)

Actiphyte of White Lily PG50 (Active Organics)

Actiphyte of White Nettle (Active Organics)

Actiphyte of White Nettle PG50 (Active Organics)

Actiphyte of White Oak (Active Organics)

Actiphyte of White Pond Lily PG50 (Active Organics)

Actiphyte of White Pond Lily Root (Active Organics)

Actiphyte of White Tea (Active Organics)

Actiphyte of Wild Cherry Bark PG50 (Active Organics)

Actiphyte of Wild Cherry PG50 (Active Organics)

Actiphyte of Wild Mint (Active Organics)

Actiphyte of Wild Thyme (Active Organics)

Actiphyte of Wild Yam PG50 (Active Organics)

Actiphyte of Willow Bark (Active Organics)

Actiphyte of Willow Bark PG50 (Active Organics)

Actiphyte of Wintergreen PG50 (Active Organics)

Actiphyte of Witch Hazel PG50 (Active Organics)

Actiphyte of Yarrow PG50 (Active Organics)

Actiphyte of Yellow Dock PG50 (Active Organics)

Actiphyte of Yerba Santa PG50 (Active Organics)

Actiphyte of Yucca PG50 (Active Organics)

Actiphyte Orchid (Active Organics)

Actiphyte Oregon Grape (Active Organics)

Actiphyte Peach Kernel (Active Organics)

Actiphyte Pear Fruit (Active Organics)

Actiphyte Phellodendron Amurense (Active Organics)

Actiphyte Plankton (Active Organics)

Actiphyte Plum (Active Organics)

Actiphyte Poppy Seed (Active Organics)

Actiphyte Prickly Pear (Active Organics)

Actiphyte Prickly Pear Cactus (Active Organics)

Actiphyte Pumpkin (Active Organics)

Actiphyte Red Sage (Active Organics)

Actiphyte Rhubarb Root (Active Organics)

Actiphyte Royal Jelly (Active Organics)

Actiphyte Sacred Lotus (Active Organics)

Actiphyte Sacred Lotus Seed (Active Organics)

Actiphyte Safflower (Active Organics)

Actiphyte Saw Palmetto (Active Organics)

Actiphyte Sea Buckthorn (Active Organics)

Actiphyte Sea Fennel (Active Organics)

Actiphyte Shave Grass (Active Organics)

Actiphyte Siberian Ginseng (Active Organics)

Actiphyte Soybean (Active Organics)

Actiphyte Star Fruit (Active Organics)

Actiphyte Stone Root (Active Organics)

Actiphyte Sugar Cane (Active Organics)

Actiphyte Szechuan Lovage Root (Active Organics)

Actiphyte Tormentil (Active Organics)

Actiphyte Uva Ursi (Active Organics)

AEC Papaya Extract (Papain 0.25%) (A & E Connock)

AEC Pineapple Extract (Bromelain 0.25%) (A & E Connock)

292 Aftershave HS (Alban Muller)

Agave Extract HS 2817 G (Grau)

Agrimony Leaves Extract HS 2891 G (Grau)

A.H.A. 40 (Ennagram)

A.H.A. Extracts (Ennagram)
Alchemilla Extract HG (Provital/ Centerchem)
Alder Buckthorn HS (Alban Muller)
ALE-75 (OSi Specialties)
Alfalfa Herb Extract HS 2967 G (Grau)
Almond Botanical Milk (Cosmetochem) (Cosmetochem International Ltd.)
Almond Bran Extract HS 2496 G (Grau)
Almond Phytolait (Alban Muller)
Aloe Dew AD2WOO1 (Aloestar)
Aloe Extract HS 2386 G (Grau)
Aloe Ferox/Cape Aloe HPG Titrated (Alban Muller)
Aloe Ferox HS/Cape Aloe HS (Alban Muller)
Aloe Flower Extract HS (Tri-K)
Aloe-Moist (Terry)
Amazon Extract Acerola (E.U.K)
Amazon Extract Caju (E.U.K)
Amazon Extract Carqueja (E.U.K)
Amazon Extract Gomphrena (E.U.K)
Amazon Extract Guarana (E.U.K)
Amazon Extract Jatoba (E.U.K)
Ambrose Extract HS 3719 G (Grau)
Amiderm Phytoamine Biocomplex (Alban Muller)
Amidroxy 4 Flowers (Alban Muller)
Amidroxy 4 Flowers (Alban Muller)
Aminat (Vedeqsa)
Aminoefaderma (Vevy)
Amisol 641-A (Lucas Meyer GmbH)
Amisol HS-2 (Lucas Meyer GmbH)
Amisol HS-3 (Lucas Meyer GmbH)
Amisol HS-6 (Lucas Meyer GmbH)
Amisol HS3US (Lucas Meyer)
Amisol 406-N (Lucas Meyer GmbH)
Amla/Amalaki (Carlisle)
Amla-Extrakt (Glycolic extract of Amla fruits) (Laboratoires Cosmedic)
Angelica Archangelica Root Extract ies (IES LABO)
Angelica Extract HG (Provital/Centerchem)
Angelica HS (Alban Muller)
Angelica Root Extract HS 3797 G (Grau)
Anise HS (Alban Muller)
Anise Seed Extract HS 2712 G (Grau)
Annatto HS/Rocou HS (Alban Muller)
Anthemis Nobilis Flower Extract ies (IES LABO)
Anthriscus Cerefolium Extract (IES LABO)
Antiblotchiness Phytogreen Complex GXH 261 (Phytochim)
ANTI-DANDRUFF (Greentech S.A)
Anti-Inflammatory Phytoamine Biocomplex (Alban Muller)
Antil 141 Liquid (Degussa Care Specialties)
ANTI-PERSPIRANT (Greentech S.A)
Antiphlogistic "aro" (Crodarom)
Anti-Reddening Complex 261 (Ennagram)
Anti-Seborrhoeic Phytoamine Biocomplex (Alban Muller)
Anti-Stress Phytoamine Biocomplex (Alban Muller)

ANTI-STRETCH MARKS (Greentech S.A)
338 Antistretchmarks HS (Alban Muller)
ANTI-WRINKLE (Greentech S.A)
Anti Wrinkles Complex 263 (Ennagram)
Antiwrinkles Phytogreen Complex GXH 263 (Phytochim)
Apium Graveolens (Celery) Extract ies (IES LABO)
Apple Extract HG (Provital/Centerchem)
Apple Extract HS 1806 AT (Grau)
Apple HS (Alban Muller)
Applemint Extract (Cosmetic Developments)
Apricot Extract G (Provital/Centerchem)
Apricot Extract HS 2509 G (Grau)
Apricot HG (Alban Muller)
Apricot HPG Titrated (Alban Muller)
Apricot HS (Alban Muller)
Apricot Kernel Extract HS 2813 G (Grau)
A.P.S. LS 8425 (Laboratoires Sero-biologiques)
Aqua-Tein C (Maybrook)
Aqua-Tein S (Maybrook)
Aragoline (Siera)
Arbute Tree HS (Alban Muller)
Arctium Majus Root Extract ies (IES LABO)
Aritha Extract (Carlisle)
Arlacel 186 (Uniqema Americas)
Arnica Chamissonis Flower Extract ies (IES LABO)
Arnica Chamissonis Glycolyc Extract (Bernett)
Arnica Chamissonis HS (Alban Muller)
Arnica Extract HG (Provital/Centerchem)
Arnica Extract HS 2397 G (Grau)
Arnica Montana Flower Extract ies (IES LABO)
Arnica Montana HS (Alban Muller)
Aromaphyte of Almond (Active Organics)
Aromaphyte of Basil (Active Organics)
Aromaphyte of Bay Laurel (Active Organics)
Aromaphyte of Bergamot (Active Organics)
Aromaphyte of Birch (Active Organics)
Aromaphyte of Chamomile (Active Organics)
Aromaphyte of Cinnamon (Active Organics)
Aromaphyte of Cypress (Active Organics)
Aromaphyte of Eucalyptus (Active Organics)
Aromaphyte of Fennel (Active Organics)
Aromaphyte of Fir (Active Organics)
Aromaphyte of Gardenia (Active Organics)
Aromaphyte of Geranium (Active Organics)
Aromaphyte of Ginger (Active Organics)
Aromaphyte of Grapefruit (Active Organics)
Aromaphyte of Grapefruit Peel (Active Organics)
Aromaphyte of Jasmine (Active Organics)
Aromaphyte of Juniper (Active Organics)
Aromaphyte of Lavender (Active Organics)
Aromaphyte of Lemon (Active Organics)
Aromaphyte of Lemongrass (Active Organics)

Aromaphyte of Neroli (Active Organics)
Aromaphyte of Nutmeg (Active Organics)
Aromaphyte of Orange (Active Organics)
Aromaphyte of Orange Peel (Active Organics)
Aromaphyte of Patchouli (Active Organics)
Aromaphyte of Pennyroyal (Active Organics)
Aromaphyte of Peppermint (Active Organics)
Aromaphyte of Pine Cone (Active Organics)
Aromaphyte of Rose (Active Organics)
Aromaphyte of Rosemary (Active Organics)
Aromaphyte of Sage (Active Organics)
Aromaphyte of Sandalwood (Active Organics)
Aromaphyte of Spearmint (Active Organics)
Aromaphyte of Tangerine (Active Organics)
Aromaphyte of Thyme (Active Organics)
Aromaphyte of Violet (Active Organics)
Artemisia Vulgaris Extract ies (IES LABO)
Artichoke Extract HS 2564 G (Grau)
Artichoke HS (Alban Muller)
Ascophyllum HS (Alban Muller)
Asebiol LS 2539 BT2 (Laboratoires Sero-biologiques)
Asiatic Centella Extract HG (Provital/ Centerchem)
Asparagus Extract HS 2777 G (Grau)
Asparagus HS (Alban Muller)
Asparagus Stem AMW (Pentapharm/ Centerchem)
Asperula Odora Extract ies (IES LABO)
Astragalus Extract (Carrubba)
ASTRINGENT EMOLLIENT COMPLEX (Greentech S.A)
ATP Nucleotides (Croda, Inc.)
Aubergine Extract HS 2477 G (Grau)
Avena Sativa (Oak) Kernel Extract ies (IES LABO)
Avocado Extract HG (Provital/Centerchem)
Avocado Extract HS 2384 G (Grau)
Awapuhi Extract (Bell Flavors)
265 Babyderme HS (Alban Muller)
342 Babyderme HS (Alban Muller)
Bactecar 125S (Phytocos)
Balm Mint Extract HG (Provital/ Centerchem)
Balm Mint Extract PG (Rahn)
Balm Mint HS (Alban Muller)
Bamboo HS (Alban Muller)
Bamboo Manna Extract (CEP (Solabia))
Bamboo Milk (Cosmetochem) (Cosmetochem International Ltd.)
Bambusa Vulgaris Extract ies (IES LABO)
Banaba Glycolic Extract (Vinyals)
Banana Extract HG (Provital/Centerchem)
Banana Extract HS 3381 G (Grau)
Banana HS (Alban Muller)
Barley HS (Alban Muller)
Barm Extract HS 2963 G (Grau)
Basil HS (Alban Muller)
Bearberry Leaves Extract HS 2513 G (Grau)

Beet HS/Red Beet HS (Alban Muller)
Beet LS Titrated/Red Beet LS Titrated (Alban Muller)
Benzoin HS (Alban Muller)
Beta Hydroxy Acid (BHA) Willow Bark (Cosmetochem) (Cosmetochem International Ltd.)
Betula Alba Extract ies (IES LABO)
Bilberry HS/Blueberry HS (Alban Muller)
Bilberry Leaf HG (Alban Muller)
Biobranil Watersoluble 2/012600 (Symrise)
Bio-Bustyl (Sederma)
Biocrystal (Siera)
Biodermine (Sederma)
Bio-Energizer (SECMA)
Bio-Keratin Haircomplex 230389 (Crodarom)
BIOMEDULINE (Libiol)
Biomin Cu/P/C Liquid (Arch Personal Care Products)
Biopeptide-CL (Sederma)
Bio-Plex RNA (Arch Personal Care Products)
Birch Bark Extract HS 2780 G (Grau)
Birch Bark Extract PG (Maruzen Pharmaceuticals Co., Ltd.)
Birch Extract PG (Rahn)
Birch Leaves Extract HS 2429 G (Grau)
Bird's Knotgrass HS (Alban Muller)
Bistort HS (Alban Muller)
Bitter Almond HS (Alban Muller)
Bitter Melon (Greentech)
Bitter Orange Extract HG (Provital/ Centerchem)
Bitter Orange Peel HS (Alban Muller)
Bittersweet HS (Alban Muller)
Bixa Extract (Kelisema Italy)
Black Currant Extract HS 2955 G (Grau)
Black Currant Fruit HPG Titrated (Alban Muller)
Black Currant Leaves Extract HS 2475 G (Grau)
Black Mulberry Leaf HS (Alban Muller)
Black Mulberry Root HS (Alban Muller)
Black Mustard HS (Alban Muller)
Black Orchid Flower HG (Alban Muller)
Black Poplar HPG Titrated (Alban Muller)
Black Poplar HS (Alban Muller)
Black Raspberry Extract (Carrubba)
Black Sampson Extract HS 2502G (Grau)
Black Sesame Seed Extract (Carrubba)
Black Snakeroot HS (Alban Muller)
Black Tea Extract HS 2516 G (Grau)
Black Tea HPG Titrated (Alban Muller)
Black Tea HS (Alban Muller)
Blackberry Extract HG (Provital/ Centerchem)
Blackcurrant HS (Alban Muller)
Blackcurrant Leaf HPG Titrated (Alban Muller)
Blackcurrant Leaf HS (Alban Muller)
Bladderwrack Extract HS 2836 G (Grau)
Bladderwrack HS (Alban Muller)
Blanc Covasop W 9775 (LCW)

201 Blend for Bust Cares HS (Alban Muller)
361 Blend for Chapped Skin HS (Alban Muller)
215 Blend for Delicate Skin HS (Alban Muller)
275 Blend for Deodorant HS (Alban Muller)
296 Blend for Elderly Skin HS (Alban Muller)
312 Blend for Greasy Hair HS (Alban Muller)
250 Blend for Greasy Skin HS (Alban Muller)
315 Blend for Greasy Skin Imperfections HS (Alban Muller)
255 Blend for Slenderizing Products HS (Alban Muller)
316 Blend for Slimming Products HS (Alban Muller)
Blend 3250 HS (Alban Muller)
Blessed Thistle HS (Alban Muller)
Bleu Covasop W 6775 (LCW)
Bleu Covasop W 6776 (LCW)
Blue Poppy Seed HS (Alban Muller)
Borage Extract HS 2562 G (Grau)
Borage HS (Alban Muller)
Boswellia Carterii Extract ies (IES LABO)
Box Tree Leaves HS 2838 G (Grau)
Brazil Nut Extract (Bell Flavors)
Brazil Nut Extract (Carrubba)
Brocose Q (Arch Personal Care Products)
Bronidox L (Cognis Care Chemicals/NJ)
Bronidox L (Cognis Care Chemicals/PA)
Bronidox L (Cognis Deutschland)
Brooklime Extract (Cosmetic Developments)
Broom HS (Alban Muller)
Broom Tops Extract HS 2645 G (Grau)
BROWN HAIR (Greentech S.A)
Brumble Concentrate HS 3670 G (Grau)
Brumble Leaves Extract HS 2527 G (Grau)
Brun Covasop W 8770 (LCW)
Buchu HPG Titrated (Alban Muller)
Buchu HS (Alban Muller)
Buckbean Extract HG (Provital/ Centerchem)
Buckthorn Extract HS 2520 G (Grau)
Buckwheat Extract HS 28606 (Grau)
Buckwheat Extract HS 2860 G (Grau)
Buckwheat Extract HS 2861 G (Grau)
Buckwheat Extract, Watersoluble (Crodarom)
Buckwheat HS (Alban Muller)
Bugloss HS (Alban Muller)
Burdock Extract PG (Rahn)
Burdock Root Extract HS 2443 G (Grau)
Burnet Rose Extract (Cosmetic Developments)
Butcherbroom Extract PG (Rahn)
Butcher's Broom Extract HG (Provital/ Centerchem)
Butcher's Broom HS (Alban Muller)
Butea Superba Extract (Tropical Herbal)
Butterbur Extract (Cosmetic Developments)

Buxus Sempervirens Leaf Extract ies (IES LABO)
Cabbage HS (Alban Muller)
Cabbage Rose HS (Alban Muller)
Cactus Extract HS 2692 G (Grau)
Cactus Milk (Cosmetochem) (Cosmetochem International Ltd.)
Cade HS (Alban Muller)
Cajuput Extract (Cosmetic Developments)
Calamint Extract (Cosmetic Developments)
Calamus Root Extract HS 3044 G (Grau)
Calendula Extract HG (Provital/ Centerchem)
Calendula Extract HS 2380 G (Grau)
Calendula Officinalis Flower Extract ies (IES LABO)
Calluna Vulgaris Extract ies (IES LABO)
Camellia Japonica Leaf Extract ies (IES LABO)
Camellia Sinensis Leaf Extract ies (IES LABO)
Camomile Extract PG (Rahn)
Campo Ai Yen Extract (Campo)
Campo An Mo Le Extract (Campo)
Campo Australian Neem Tree (Campo)
Campo Bai Qi Extract (Campo)
Campo Balada Turagogandha (Campo)
Campo Biao Beng Li Extract (Campo)
Campo Bijaka (Campo)
Campo Bimbi (Campo)
Campo Black Tea Tree (Campo)
Campo Bo Hai Cai Extract (Campo)
Campo Brazil Nut (Campo)
Campo Broad Leafed Tea Tree (Campo)
Campo Broom Brush Ti-Tri (Campo)
Campo Cao Hua Extract (Campo)
Campo Citisu (Campo)
Campo Dong Hua Zei Extract (Campo)
Campo Engosaku (Campo)
Campo Gao Ben Hua Extract (Campo)
Campo Gorga (Campo)
Campo He Shou Wu Extract (Campo)
Campo Hua Gua Extract (Campo)
Campo Hua Jiao Extract (Campo)
Campo Hui Xian Extract (Campo)
Campo Inga (Campo)
Campo I Tung Extract (Campo)
Campo Jambu (Campo)
Campo Jatoba (Campo)
Campo Ju Hua Extract (Campo)
Campo Kovil Tulsi (Campo)
Campo Ling Ling Xiang Extract (Campo)
Campo Liniment Ti-Tri (Campo)
Campo Long Xu Cai Extract (Campo)
Campo Mahakannii (Campo)
Campo Mahanimba (Campo)
Campo Maka (Campo)
Campo Malkagni (Campo)
Campo Mao Xiang Extract (Campo)
Campo Muruity-Muruity (Campo)
Campo Nu Chen Extract (Campo)
Campo Pei Lan Extract (Campo)
Campo Po Ku Cao Extract (Campo)
Campo Po Zhulin Hua Extract (Campo)

Campo Ryutan (Campo)
Campo Sala Siddha (Campo)
Campo Shako Nakai (Campo)
Campo Shan Cha Yao Extract (Campo)
Campo Shao Yao Extract (Campo)
Campo Shripala (Campo)
Campo Sunisannaka (Campo)
Campo Tan Shen Extract (Campo)
Campo Tien Shi Li Extract (Campo)
Campo Tohasaku (Campo)
Campo Tung Kua Extract (Campo)
Campo Udumbara (Campo)
Campo Vaipillai (Campo)
Campo Vasa Kovil Tulsi (Campo)
Campo Wu Qing Extract (Campo)
Campo Wu Tung Extract (Campo)
Campo Wu Wei Zi Extract (Campo)
Campo Zao Jiao Extract (Campo)
Cane Sugar Extract HS 3711 G (Grau)
Caomint (CEP (Solabia))
Caophenol (CEP (Solabia))
Capigen CG (Sederma)
Capigen CS (Sederma)
220 Capilotonique HS (Alban Muller)
226 Capilotonique HS (Alban Muller)
245 Capilotonique HS (Alban Muller)
Capsicum Frutescens Fruit Extract ies (IES LABO)
Capsicum HPG Titrated (Alban Muller)
Capucine Extract HG (Provital/ Centerchem)
Carambola Extract (Cosmetic Developments)
Carica Papaya Fruit Extract ies (IES LABO)
Carnation HS (Alban Muller)
Carob HS (Alban Muller)
Carrot Extract HG (Provital/Centerchem)
Carrot Extract PG90 (Maruzen Pharmaceuticals Co., Ltd.)
Carrot Liquid (Ichimaru Pharcos)
Carrot Seed HS (Alban Muller)
Carrots Extract HS 2597 G (Grau)
Cassia HS (Alban Muller)
Castanea Sativa (Chestnut) Extract ies (IES LABO)
Catnip Herb Extract HS 2805 G (Grau)
Cat's Claw HS (Alban Muller)
Caviar HS (Alban Muller)
Cedarwood Extract (Cosmetic Developments)
Celeriac Extract HS 3369 G (Grau)
Cellulinol (Prod'Hyg)
Centaurea Cyanus Flower Extract ies (IES LABO)
Centaury Extract HG (Provital/Centerchem)
Centella Asiatica Extract HS 2772G (Grau)
Centella Asiatica Extract ies (IES LABO)
Centella Asiatica Glycolic Extract Vinyals (Vinyals)
Centella Asiatica HS (Alban Muller)
Centella Extract PG (Maruzen Pharmaceuticals Co., Ltd.)
Ceramiane (SECMA)

Ceraphyl 65 (International Specialty Products)
Ceraphyl 70 (International Specialty Products)
Ceraphyl 85 (International Specialty Products)
Ceraphyl RBO (International Specialty Products)
4-Cereals Milk (Cosmetochem) (Cosmetochem International Ltd.)
Chamomile Extract 21006 (Fragrance Oils Int. Ltd.)
Chamomile Extract HG (Provital/ Centerchem)
Chamomile Extract HS 2382 G (Grau)
Chamomile Extract HS 2779 G (Grau)
Chelactyl (Silab)
Cherry Seed Extract (Bell)
Chestnut HS (Alban Muller)
Chestnut Leaves Extract HS 2729 G (Grau)
Chicory HPG Titrated (Alban Muller)
Chicory HS (Alban Muller)
China Bark Extract HS 2410 G (Grau)
China Extract Amonum Tsao Ko (E.U.K)
China Extract Baio Beng Li (E.U.K)
China Extract Bai QI (E.U.K)
China Extract Ho Shou Wu (E.U.K)
China Extract Hua Jiao (E.U.K)
China Extract Hui Xian (E.U.K)
China Extract Lung Xu Cai (E.U.K)
China Extract Nu Chen (E.U.K)
China Extract Po Zhu Lin Hua (E.U.K)
China Extract Tan Shen (E.U.K)
China Extract Xin Yi (E.U.K)
Chine Extract Amomum Aromaticum (Ennagram)
Chine Extract Bletilla Striata (Ennagram)
Chine Extract Eriobotrya Japonica (Ennagram)
Chine Extract Foeniculum Vulgare (Ennagram)
Chine Extract Laminaria Japonica (Ennagram)
Chine Extract Lugustrum Lucidum (Ennagram)
Chine Extract Magnolia Liliflora (Ennagram)
Chine Extract Nelumbium Nelumbo (Ennagram)
Chine Extract Polygonum Multiflorum (Ennagram)
Chine Extract Rubus Chingli (Ennagram)
Chine Extract Salva Miltiorhiza (Ennagram)
Chine Extract Xanthoxylum Alatum (Ennagram)
Chinese Magnolia Flower Extract HS (Tri-K)
Chinese Peony Extract (Alban Muller)
Chinese White Tuckahoe Extract (Alban Muller)
Chlorella HPG Titrated (Alban Muller)
Chlorophyll Extract, Watersoluble (Crodarom)

Chrysanthemum Flower Extract (IES LABO)
Chrysanthemum Flower Extract HS (Tri-K)
Chrysanthemum Parthenium (Feverfew) (IES LABO)
Chrysocalmine (Greentech)
Cinchona Extract HG (Provital/ Centerchem)
Cinchona Succirubra Bark Extract ies (IES LABO)
Cinnamomum Zeylanicum Bark Extract ies (IES LABO)
Cinquefoil Extract HS 2436 G (Grau)
318 Circulatory Blend HS (Alban Muller)
Citrus Aurantium Amara (Bitter Orange) Fruit Extract ies (IES LABO)
Citrus Aurantium Dulcis (Orange) Flower (IES LABO)
Citrus Grandis (Grapefruit) Fruit Extract ies (IES LABO)
Citrus Medica Extract (Plantes et Industrie)
Citrus Medica Limonum (Lemon) Extract ies (IES LABO)
Citrus Nobilis (Mandarin Orage) Fruit Extract ies (IES LABO)
Clariline (CEP (Solabia))
ClariTea (CEP (Solabia))
Clary Sage CL Forte 2/033010 (Symrise)
Clary Sage Extract (Cosmetic Developments)
Clercicyne L (Greentech)
Clover Flower Extract HS 2525 G (Grau)
Cloves Extract HS 3316 G (Grau)
Cocoa HPG Titrated (Alban Muller)
Cocoa Milk (Cosmetochem) (Cosmetochem International Ltd.)
Cocoa Phytolait (Alban Muller)
Cocoa Shell Extract HS 2968 G (Grau)
Coconut Extract HS 3117 G (Grau)
Coffea Arabica (Coffee) Seed Extract ies (IES LABO)
Coffee Extract (Kelisema Italy)
Cola Extract (Kelisema Italy)
Cola Nuts Extract HS 3154 G (Grau)
Collamino Complex ESC (Arch Personal Care Products)
Collamino Complex S (Arch Personal Care Products)
Collamino Complex SS (Arch Personal Care Products)
Collamoist WS (Arch Personal Care Products)
Coltsfoot Extract HG (Provital/Centerchem)
Colt's Foot Extract HS 2378 G (Grau)
Coltsfoot HS (Alban Muller)
Combretum HS (Alban Muller)
Comfrey HS (Alban Muller)
Comfrey Root Extract HS 2446 G (Grau)
Comfrey Root Extract Liquid (Crodarom)
Commiphora Myrrha Extract ies (IES LABO)
Common Ash Extract HS 2950 G (Grau)

Propylene Glycol (Cont.)

Common Avens Herb Extract HS 2901 G
 (Grau)
Common Basil Herb Extract HS 2580 G
 (Grau)
Common Bean HS (Alban Muller)
Common Blue Berries Extract HS 2774 G
 (Grau)
Common Fig Wort HS 2581 G (Grau)
Common Fumitory HS (Alban Muller)
Common Hazel HS (Alban Muller)
Common Heath Extract HS 2629 G (Grau)
Common Ladies Mantle HS 2900 G (Grau)
Common Millet Extract HS 3460 G (Grau)
Common Nettle Extract HG (Provital/
 Centerchem)
Common Soapwort Extract HG (Provital/
 Centerchem)
Common Speedwell HS (Alban Muller)
Common Tansy Extract HS 2389 G (Grau)
Common Verbena HS (Alban Muller)
Common Wormwood Extract HS 2850 G
 (Grau)
Complex 5 (Fabriquimica)
Complex Aloe (Fabriquimica)
Complex 3 - Anti-Aging (Provital/
 Centerchem)
Complex 1 - Antipollution (Provital/
 Centerchem)
Complex Arnica (Fabriquimica)
Complex Calendula (Fabriquimica)
Complex Centella (Fabriquimica)
Complexe DM 50 (Gattefosse s.a.)
Complex Ginseng (Fabriquimica)
Complex GT (Fabriquimica)
Complex Henna (Fabriquimica)
Complex Hiedra (Fabriquimica)
Complex Manzanilla (Fabriquimica)
Complex 2 - Moisturizing (Provital/
 Centerchem)
Complex Ortiga (Fabriquimica)
Complex 4 - Oxygenant (Provital/
 Centerchem)
Complex Pepino (Fabriquimica)
Complex Quassia (Fabriquimica)
Complex Romero (Fabriquimica)
Complex Salvia (Fabriquimica)
Complex 6 - Sensitive Skin (Provital/
 Centerchem)
Complex 5 - Vitaminic (Provital/
 Centerchem)
Complex Zanahoria (Fabriquimica)
Compositum (Vevy)
Concentrate for Greasy Hair (Crodarom)
Condurango Bark Extract HS 2663 G
 (Grau)
Condurango HPG Titrated (Alban Muller)
Condurango HS (Alban Muller)
Coneflower/Echinacea HS (Alban Muller)
Coneflower Extract HG (Provital/
 Centerchem)
Coneflower Extract PG (Rahn)
ConeflowerHPG Titrated/Echinacea HPG
 Titrated (Alban Muller)

Coneflower Liquid (Crodarom)
Connective Tissue Complex 268
 (Ennagram)
Connective Tissue Phytogreen Complex
 GXH 268 (Phytochim)
Conservateur GD500 (Phytocos)
Conservateur GD700 (Phytocos)
Coobato Angelica (Universal Flavors)
Coobato Camomilla (Universal Flavors)
Coobato Fiordaliso (Universal Flavors)
Coobato Fiori d'Arancio (Universal Flavors)
Coobato Hamamelis (Universal Flavors)
Coobato Meliloto (Universal Flavors)
Coobato Melissa (Universal Flavors)
Coobato Rosa (Universal Flavors)
Coobato Sambuco (Universal Flavors)
Coobato Speciale (Universal Flavors)
Coobato Tiglio (Universal Flavors)
Cordical (Cosmecal Sarl)
Corn Extract HS 3006 G (Grau)
Cornflower Extract HS 2369 G (Grau)
Cornflower HPG Titrated (Alban Muller)
Cornflower HS (Alban Muller)
Corn Poppy HPG Titrated (Alban Muller)
Corn Poppy HS (Alban Muller)
Corn Silk Extract HS 2969 G (Grau)
Corsican Moss HS (Alban Muller)
Cosflor Brooklime HGS (A & E Connock)
Cosflor Burnet Rose HGS (A & E Connock)
Cosflor Burnet Saxifrage HGS (A & E
 Connock)
Cosflor Butterbur HGS (A & E Connock)
Cosflor Cajuput HGS (A & E Connock)
Cosflor Calamint HGS (A & E Connock)
Cosflor Chumba HGS (A & E Connock)
Cosflor Creosote Bush HGS (A & E
 Connock)
Cosflor Crest Marine HGS (A & E
 Connock)
Cosflor Damask Rose HGS (A & E
 Connock)
Cosflor Damiana HGS (A & E Connock)
Cosflor Dwarf Pine HGS (A & E Connock)
Cosflor Forget-Me-Not HGS (A & E
 Connock)
Cosflor Frangipani HGS (A & E Connock)
Cosflor Freesia HGS (A & E Connock)
Cosflor Green Ginger HGS (A & E
 Connock)
Cosflor Hollyhock HGS (A & E Connock)
Cosflor Huckleberry HGS (A & E Connock)
Cosflor Japonica HGS (A & E Connock)
Cosflor Kumquat HGS (A & E Connock)
Cosflor Lily Root HGS (A & E Connock)
Cosflor Loganberry HGS (A & E Connock)
Cosflor Lychee HGS (A & E Connock)
Cosflor May HGS (A & E Connock)
Cosflor Meadowgrass HGS (A & E
 Connock)
Cosflor Mignonette HGS (A & E Connock)
Cosflor Muira Puama HGS (A & E
 Connock)
Cosflor Musk Mallow HGS (A & E
 Connock)
Cosflor Neroli HGS (A & E Connock)

Cosflor Orange Barley HGS (A & E
 Connock)
Cosflor Orange Blossom HGS (A & E
 Connock)
Cosflor Pennywort HGS (A & E Connock)
Cosflor Pepino HGS (A & E Connock)
Cosflor Pomelo HGS (A & E Connock)
Cosflor Potency Wood HGS (A & E
 Connock)
Cosflor Rose Mallow HGS (A & E
 Connock)
Cosflor Rose Petal HGS (A & E Connock)
Cosflor Sea Lavender HGS (A & E
 Connock)
Cosflor Sea Minerals HGS (A & E
 Connock)
Cosflor Sea Pink HGS (A & E Connock)
Cosflor Shave Grass HGS (A & E
 Connock)
Cosflor Spike Lavender HGS (A & E
 Connock)
Cosflor Starfruit HGS (A & E Connock)
Cosflor Sweet Briar HGS (A & E Connock)
Cosflor Sweet Chestnut HGS (A & E
 Connock)
Cosflor Sweet Fennel HGS (A & E
 Connock)
Cosflor Sweet Gale HGS (A & E Connock)
Cosflor Thrift HGS (A & E Connock)
Cosflor Tiger Lily HGS (A & E Connock)
Cosflor Tree Balsam HGS (A & E Connock)
Cosflor Verveine HGS (A & E Connock)
Cosflor White Almond HGS (A & E
 Connock)
Cosflor White Pond Lily HGS (A & E
 Connock)
Cosflor Willow Herb HGS (A & E Connock)
Cosflor Wisteria HGS (A & E Connock)
Cosflor Woodsage HGS (A & E Connock)
Cosflor Yellow Water Lily HG-1 (A & E
 Connock)
Cosflor Yerba Buena HGS (A & E
 Connock)
Cosflor Yuzu HGS (A & E Connock)
Couch Grass HS (Alban Muller)
Covafix 123 (LCW)
Cowslip Extract HS 2435 G (Grau)
Cranesbill Extract HS 2830 G (Grau)
Crape Myrtle Extract (Carrubba)
Crataegus Monogyna Flower Extract ies
 (IES LABO)
CREAGEL DV (Creations Couleurs)
CREAGEL MA (C.I.T.)
Creagel MS (C.I.T.)
Creagel RT PA/ID (C.I.T.)
Creagel RT PA/ISO (C.I.T.)
Creagel RT PA/SIL (C.I.T.)
Cremogen AF (PN 736567) (Haarmann &
 Reimer GmbH)
Cremogen Aloe Vera (PN 734 514)
 (Haarmann & Reimer GmbH)
Cremogen Burdock Root (PN 702 228)
 (Haarmann & Reimer GmbH)
Cremogen Camomile (739012) (Haarmann
 & Reimer GmbH)

Cremogen Camomile Dist. PF (PN 317631)
(Haarmann & Reimer GmbH)
Cremogen Camomile forte (PN 728 790)
(Haarmann & Reimer GmbH)
Cremogen Camomile MEW Special New
(739027) (Haarmann & Reimer GmbH)
Cremogen Gentian Root (PN 739 006)
(Haarmann & Reimer GmbH)
Cremogen Ginseng (PN 739 020)
(Haarmann & Reimer GmbH)
Cremogen Hops (PN 739 010) (Haarmann
& Reimer GmbH)
Cremogen Horse Chestnut (PN 739 015)
(Haarmann & Reimer GmbH)
Cremogen M-I (PN 739 001) (Haarmann
& Reimer GmbH)
Cremogen Melissa Balm Mint (PN 739 013)
(Haarmann & Reimer GmbH)
Cremogen MZ (PN 739 032) (Haarmann &
Reimer GmbH)
Cremogen MZ/N (PN 755 321) (Haarmann
& Reimer GmbH)
Cremogen Pine Needle (PN 739 007)
(Haarmann & Reimer GmbH)
Cremogen M-2 (PN 739 029) (Haarmann
& Reimer GmbH)
Cremogen M-5 (PN 739 035) (Haarmann
& Reimer GmbH)
Cremogen M-82 (PN 730 337) (Haarmann
& Reimer GmbH)
Cremogen M-2/N (PN 755326) (Haarmann
& Reimer GmbH)
Cremogen Rose Hip (PN 774 200)
(Haarmann & Reimer GmbH)
Cremogen Rosemary (PN 739 014)
(Haarmann & Reimer GmbH)
Cremogen Sage (PN 739 016) (Haarmann
& Reimer GmbH)
Cremogen Stinging Nettle (PN 739 005)
(Haarmann & Reimer GmbH)
Cremophor CO 455 (BASF)
Crocus Sativus Flower Extract ies (IES
LABO)
Crodarom Chamomile A (Croda, Inc.)
Crodarom Nut A (Croda, Inc.)
Cryo-Bourgeon De Houblon (Greentech)
Cryo-Bourgeon De Myrtilles (Greentech)
Cryo-Bourgeon De Sequoia (Greentech)
Cryo-Bourgeon D'Olivier (Greentech)
Cryolidone (UCIB (Solabia))
Cucumber Extract G (Provital/Centerchem)
Cucumber Extract HS 1436 AT (Grau)
Cucumber Extract HS 3351 G (Grau)
Cucumber Extract PG (Rahn)
Cucumber HS (Alban Muller)
Cucumis Sativus (Cucumber) Fruit Extract
ies (IES LABO)
Cumin HS (Alban Muller)
Curcuma Root Extract HS 2533 G (Grau)
Curled Dock HS (Alban Muller)
Currant Berries Extract HS 3298 G (Grau)
Custoblend BAT (Anti-Bac Blend) (Custom
Ingredients)

Cymbopogon Schoenanthus Extract ies
(IES LABO)
Cynara Scolymus (Artichoke) Leaf Extract
ies (IES LABO)
Cypress Extract HG (Provital/Centerchem)
Cypress HS (Alban Muller)
CYTOBIOL BURDOCK (Libiol)
Cytobiol Iris (Libiol)
CYTOBIOL ULMAIRE (Libiol)
Cytoflavin-C (Chemyunion)
Cytoflavin-CR (Chemyunion)
Daffodil Extract HS 3633 G (Grau)
Daisy Flower Extract PG (Rahn)
Daisy HS (Alban Muller)
Dalmatian Pyrethrum HS (Alban Muller)
Damask Rose Extract (Cosmetic Developments)
Damiana Extract (Cosmetic Developments)
Damiana HS (Alban Muller)
Damson Extract (Cosmetic Developments)
Dandelion HS (Alban Muller)
Daucus Carota Sativa (Carrot) Root Extract
(IES LABO)
Dead Nettle Extract HS 2430 G (Grau)
Dehymuls F (Cognis Care Chemicals/NJ)
Dehyquart L 80 (Cognis Care Chemicals/
NJ)
Dehyquart L 80 (Cognis Deutschland)
Dekaben B-30 (Dekker)
Dekaben DMM (Dekker)
Dekaben P (Dekker)
Dekacydol (Dekker)
Dekasol 5 (Dekker)
Dekasol 10 (Dekker)
287 Demaquillant HS (Alban Muller)
Deo-Usnate (Cosmetochem)
Depil Enzyme (I.R.A.)
Dermaperline (Siera)
Dermascreen (Indena SA)
Dermocenta (Libiol)
Dermo Fruit Lime (Esperis)
Dermonectin (Vevy)
Dermotenseur (Libiol)
Devil's Apron HS (Alban Muller)
Devil's Claw HS (Alban Muller)
Dewberry Extract (Cosmetic
Developments)
Dicopamine DP (Phoenix)
Dioscorea Villosa (Wild Yam) Root Extract
ies (IES LABO)
Diospyros Kaki Fruit Extract ies (IES
LABO)
Dog Rose HS / Rose Hips HS (Alban
Muller)
Dow Corning 7224 Conditioning Agent
(Dow Corning)
Dragon Extract (Greentech)
DRY HAIR (Greentech S.A)
Dryopteris Filix-mas Root Extract ies (IES
LABO)
"Dulse Extract" (Bio-Botanica)
East Indian Balmony Extract HS 2748 G
(Grau)
Ecofloral Acerola (Centroflora)

Ecofloral Anti-Cellulite Complex
(Centroflora)
Ecofloral Cashew Fruit (Laboratorio
Centroflora Ltda) (Centroflora)
Ecofloral Guarana (Centroflora)
Ecofloral Guassatonga (Centroflora)
Ecofloral Mulateiro (Centroflora)
Ecofloral Raspa De Jua (Centroflora)
Edelweiss HPG Titrated (Alban Muller)
Edelweiss HS (Alban Muller)
Eggplant HS (Alban Muller)
EHG Ananas / Bromelain (I.D. bio)
EHG Papaye / Papain (I.D. bio)
Elastin PG 2000 (GfN)
Elderberry Extract (Bell Flavors)
Elder Flower Extract HS 2440 G (Grau)
Elder HS (Alban Muller)
Elder Tree Extract HG (Provital/
Centerchem)
Elecampane Extract (Kelisema Italy)
Elestab 388 (Laboratoires Serobiologiques)
Eleutherococcus Extract HG (Provital/
Centerchem)
Eleutherococcus Extract HS 2933 G (Grau)
Eleutherococcus HS (Alban Muller)
Elm HS (Alban Muller)
235 Emollient HS (Alban Muller)
238 Emollient HS (Alban Muller)
Emulmetik 310 (Lucas Meyer GmbH)
Emulsogen HCP 049 (Clariant)
Emulsogen HCP 049 (Clariant GmbH,
Personal Care)
Enna 201 Slimming Vegetal Extract
(Ennagram)
Ephedra Herb Extract HS 2971 G (Grau)
Epicea Extract (Greentech S.A)
Epilami (Alban Muller)
Epilobium Angustifolium Extract ies (IES
LABO)
Eriobotrya Japonica Extract ies (IES LABO)
Eschscholtzia Californica Extract ies (IES
LABO)
Eschscholtzia Extract HS 3605 G (Grau)
Eschscholtzia HS (Alban Muller)
Ethoquad 0/12 PG (Akzo Nobel)
Etival 5L (Coletica SA)
Etival 15L (Coletica SA)
Eucalyptus Extract HG (Provital/
Centerchem)
Eucalyptus Globulus Leaf Extract ies (IES
LABO)
Eucalyptus HS (Alban Muller)
Eucalyptus Leaves Extract HS 2646 G
(Grau)
Eugenia Caryophyllus (Clove) Flower (IES
LABO)
Eumulgin HRE 455 (Cognis Deutschland)
Euperlan PL-1000 (Cognis Care
Chemicals/PA)
Euphrabiol (Libiol)
Euphrasy HS (Alban Muller)
Eurotiale PG20 (Yves Rocher)
Euxyl K 600 (Schulke & Mayr)
Evergreen Extract HS 2709 G (Grau)
Everlasting HS (Alban Muller)

Evosina NA2GP (Variati)
Evosina SBS (Variati)
Extan-GO (Lanaetex)
Extract Hydroglycolique De Roseau (Greentech)
Extract of Bambusa Vulgaris (Arda Natura)
Extract of Camelia Sinesis (Arda Natura)
Extract of Jerico's Rose (Arda Natura)
Extract of Sage/Indian Cress (CEP (Solabia))
Extract of Sweet Almond Fruit (Arda Natura)
Extrait D'Abricot MPEPG10 (Yves Rocher)
Extrait D'Althea PPPG40 (Yves Rocher)
Extrait D'Amandes Deshuilees MP PG 10 (Yves Rocher)
Extrait D'Amarante PPE PG 10 (Yves Rocher)
Extrait D'Armoise MP PG 40 (Yves Rocher)
Extrait d'Artemia MPE PG5 (Yves Rocher)
Extrait d' ARUM - GT10-P (Greentech)
Extrait De Bourrache PPE PG 40 (Yves Rocher)
Extrait De Camomille Romaine PPE PG 40 (Yves Rocher)
Extrait de Cardamone MPE PG 40 (Yves Rocher)
Extrait De Cedre PPG PG 10 (Yves Rocher)
Extrait de Centella MP PG 40 (Yves Rocher)
Extrait De Chatons De Saule MP PG 40 (Yves Rocher)
Extrait de Chene PPE PG20 (Yves Rocher)
Extrait de Chevrefeuille PPE PG 20 (Yves Rocher)
Extrait de Coleus (Greentech)
Extrait De Concombre (Yves Rocher)
Extrait de Coquelicot MP PG 40 (Yves Rocher)
Extrait De Coralline PPE PG 40 (Yves Rocher)
Extrait De Coton MP PG 40 (Yves Rocher)
Extrait de Cypres MPE PG 10 (Yves Rocher)
Extrait de Cypres PPE PG 20 (Yves Rocher)
Extrait De Diginee MPE PG 40 (Yves Rocher)
Extrait De Fenouil Des Sables PPE PG 40 (Yves Rocher)
Extrait de Ficaire MPE PG 40 (Yves Rocher)
Extrait de Fleur d'Oranger PPE PG10 (Yves Rocher)
Extrait de Frene PPE GP 20 (Yves Rocher)
Extrait De Fucus MP PG 20 (Yves Rocher)
Extrait de Genet PPE PG 40 (Yves Rocher)
Extrait De Genevrier PPE PG 40 (Yves Rocher)
Extrait De Germe De BLE MPE PG40 (Yves Rocher)

Extrait de Ginseng MPE PG 40 (Yves Rocher)
Extrait de Houblon MPE PG40 (Yves Rocher)
Extrait de Kiwi MPE PG 40 (Yves Rocher)
Extrait De Lavande PP PG40 (Yves Rocher)
Extrait De Lierre Grimpant PPE PG40 (Yves Rocher)
Extrait de Marjolaine PPE PG 100 (Yves Rocher)
Extrait de Marronnier D'Inde MPE PG 20 (Yves Rocher)
Extrait De Mauve PP PG40 (Yves Rocher)
Extrait De Melilot PPE PG40 (Yves Rocher)
Extrait de Myrte MPE PG 40 (Yves Rocher)
Extrait de Nigella Sativa (Greentech)
Extrait De Pamplemousse MP PG 40 (Yves Rocher)
Extrait de Parietaire PPE PG 40 (Yves Rocher)
Extrait De Peche MP PG SS (Yves Rocher)
Extrait De Pensee Sauvage PP PG40 (Yves Rocher)
Extrait De Pervenche PPE PG40 (Yves Rocher)
Extrait de Petit Houx MPE PG 40 (Yves Rocher)
Extrait De Plantain MP PG 40 (Yves Rocher)
Extrait De Pomme MPE PG10 (Yves Rocher)
Extrait de Pulpe de Baobab (Greentech)
Extrait de Ratanhia MPE PG 40 (Yves Rocher)
Extrait De Romarin PPE PG10 (Yves Rocher)
Extrait De Sabline PPE PG 40 (Yves Rocher)
Extrait de Salicorne MPE PG10 (Yves Rocher)
Extrait De Santoline MP PG 40 (Yves Rocher)
Extrait De Sarrasin PPE PG40 (Yves Rocher)
Extrait De Sauge PP PG40 (Yves Rocher)
Extrait De Sureau PPE PG40 (Yves Rocher)
Extrait De Terminalia Chebula (Greentech)
Extrait de Tilleul PP PG 40 (Yves Rocher)
Extrait de Tournesol MPE PG 30 (Yves Rocher)
Extrait de Verveine MP PG 40 (Yves Rocher)
Extrait D'Eglantier MPE PG 40 (Yves Rocher)
Extrait D'Erable PPG PG 20 (Yves Rocher)
Extrait D'Eucalyptus MPE PG10 (Yves Rocher)
Extrait Fluide De Morinda Citrifolia (Plantes et Industrie)
Extrait D'Hamamelis PPE PG40 (Yves Rocher)
Extrait D'Hibiscus MP PG 40 (Yves Rocher)

Extrait Hydrogelycolique de Neflier du Japon (Greentech)
Extrait Hydroglycolique d'Agrumes (Greentech)
Extrait Hydroglycolique d'Algue Bleue (Greentech)
Extrait Hydroglycolique d' Asperule (Greentech)
Extrait Hydroglycolique de Bletilla (Greentech)
Extrait Hydroglycolique de Bourgeons de Noisetier (Greentech)
Extrait Hydroglycolique de Bourgeons de Pecher (Greentech)
Extrait Hydroglycolique de Bourgeons de Pin (Greentech)
Extrait Hydroglycolique de Cactus (Greentech)
Extrait Hydroglycolique de Cedrat (Greentech)
Extrait Hydroglycolique de Coton (Greentech)
Extrait Hydroglycolique de Fraises (Greentech)
Extrait Hydroglycolique De Grenade (Greentech)
Extrait Hydroglycolique de Jeunes Pousses d'Airelles (Greentech)
Extrait Hydroglycolique de Lavande (Greentech)
Extrait Hydroglycolique de Litchis-GT10 (Greentech)
Extrait Hydroglycolique De Lotus-GT 10 (Greentech)
Extrait Hydroglycolique de Lys d'eau - GT10 (Greentech)
Extrait Hydroglycolique De Magnolia (Greentech)
Extrait Hydroglycolique de menthe de la baie saint John's-GT110 (Greentech)
Extrait Hydroglycolique de Mucilage de Figue (Greentech)
Extrait Hydroglycolique de Paprika (Greentech)
Extrait Hydroglycolique de Perilla Fructescens (Greentech)
Extrait Hydroglycolique de Roseau (Greentech)
Extrait Hydroglycolique de Rue Officinale (Greentech)
Extrait Hydroglycolique d'Edelweiss (Greentech)
Extrait Hydroglycolique D'Hevea (Greentech)
Extrait Hydroglycolique d' Orchidee blanche (Greentech)
Extrait hydroglycolique d'Orchidee (Greentech)
Extrait D'Hysope MP PG 40 (Yves Rocher)
Extrait D'Immortelle MP PG 40 (Yves Rocher)
Extrait d'Orange Sanguine - GT10 (Greentech)
Extrait Proteique de Cacao (Greentech)
Extranatura Baobab (E.U.K)

Extranatura Jackal (E.U.K)
Extranatura Kigelia (E.U.K)
Extranatura Maytenus (E.U.K)
Extranatura Tamarind (E.U.K)
Extraot De Prele PPE PG40 (Yves Rocher)
Extrapone Acacia Honey 2/391310
(Symrise)
Extrapone Aloe Vera 2/B06500 (Symrise)
Extrapone Alpine Herbs Special 2/032561
(Symrise)
Extrapone Apple 2/033317 (Symrise)
Extrapone Apricot 2/03365 (Symrise)
Extrapone Arnica Special 2/034591
(Symrise)
Extrapone Avocado Special 2/034599
(Symrise)
Extrapone Babosa 2/386380 (Symrise)
Extrapone Balm Mint 2/033121 (Symrise)
Extrapone Birch Special 2/032681
(Symrise)
Extrapone Burdock Root Special 2/033041
(Symrise)
Extrapone Calendula 2/B04042 (Symrise)
Extrapone Calendula Special 2/033231
(Symrise)
Extrapone Cedarwood 2/033660 (Symrise)
Extrapone Chamomile P 2/033023
(Symrise)
Extrapone Chamomile Special 2/033021
(Symrise)
Extrapone Chamomile UK 2/033025
(Symrise)
Extrapone Cinchona Bark Special (2/
032751) (Symrise)
Extrapone Coco-Nut Special 2/033055
(Symrise)
Extrapone Coltsfoot 2/032981 (Symrise)
Extrapone Cooling Complex 2/B16500
(Symrise)
Extrapone Cooling Complex N 2/B16501
(Symrise)
Extrapone Copaiba 2/B16991 (Symrise)
Extrapone Cotton 2/033327 (Symrise)
Extrapone Cucumber Special 2/032878
(Symrise)
Extrapone Fruit Mixture 2/033257
(Symrise)
Extrapone Gentian Special 2/032801
(Symrise)
Extrapone Ginkgo Biloba 2/032851
(Symrise)
Extrapone Ginseng Special 2/032861
(Symrise)
Extrapone Grapefruit P 2/030072 (Symrise)
Extrapone Hawthorn Special 2/033481
(Symrise)
Extrapone Hayflower Special 2/032941
(Symrise)
Extrapone Henna Special 2/032930
(Symrise)
Extrapone Hibiscus-Special (2/033115)
(Symrise)
Extrapone Hops Special 2/032971
(Symrise)

Extrapone Horse Chestnut Special 2/
033261 (Symrise)
Extrapone Horsetail Special 2/033321
(Symrise)
Extrapone Ivy-Special (2/032760)
(Symrise)
Extrapone Jatoba 2/B04152 (Symrise)
Extrapone Jojoba Special (2/032990)
(Symrise)
Extrapone Juniper-Special (2/033461)
(Symrise)
Extrapone Kiwi 2/033035 (Symrise)
Extrapone Lavender 2/033080 (Symrise)
Extrapone Lemon 2/032130 (Symrise)
Extrapone Lemongrass 2/0333086
(Symrise)
Extrapone Licorice Root 2/033925
(Symrise)
Extrapone Lime 2/032303 (Symrise)
Extrapone Linden Blossom Special 2/
034091 (Symrise)
Extrapone Linseed GW 2/031305
(Symrise)
Extrapone Mallow 2/030112 (Symrise)
Extrapone Mallow-Special (2/033111)
(Symrise)
Extrapone Mango 2/033123 (Symrise)
Extrapone Marshmallow Special 2/032781
(Symrise)
Extrapone Marygold-Special (2/033231)
(Symrise)
Extrapone Mimose 2/0333577 (Symrise)
Extrapone Mistletoe Special 2/033141
(Symrise)
Extrapone #4 Herbs 2/032495 (Symrise)
Extrapone #1 Special 2/032451 (Symrise)
Extrapone #3 Special 2/034481 (Symrise)
Extrapone #3 Special 2/789490 (Symrise)
Extrapone #5 Special 2/032501 (Symrise)
Extrapone Orange Flower 2/033200
(Symrise)
Extrapone Orris 2/033460 (Symrise)
Extrapone Palmfruit Milk 2/033056
(Symrise)
Extrapone Palmfruit Milk 2/033056
(Symrise)
Extrapone Pansy 2/033440 (Symrise)
Extrapone Pansy (2/033445) (Symrise)
Extrapone Passionflower 2/033165
(Symrise)
Extrapone Peach 2/033319 (Symrise)
Extrapone Peppermint-Special (2/033171)
(Symrise)
Extrapone Raspberry 2/033318 (Symrise)
Extrapone Rose Hip 2/032121 (Symrise)
Extrapone Rosemary 2/783630 (Symrise)
Extrapone Rosemary/Aloe Vera Blend 2/
033197 (Symrise)
Extrapone Rosemary Special 2/033251
(Symrise)
Extrapone Sage Special 2/033291
(Symrise)
Extrapone Sandalwood 2/032161
(Symrise)

Extrapone Seven Herbs Special 2/032527
(Symrise)
Extrapone 1 Special 2/789470 (Symrise)
Extrapone 3 Special 2/789490 (Symrise)
Extrapone 4 Special 2/788400 (Symrise)
Extrapone 5 Special 2/789500 (Symrise)
Extrapone 3 Special New 2/034484
(Symrise)
Extrapone 2 Special 2/789480 (same as 2/
032471) (Symrise)
Extrapone Spruce Needle-Special (2/
034831) (Symrise)
Extrapone Stinging Nettle Special 2/
032721 (Symrise)
Extrapone St. John's Wort P/0230985
(Symrise)
Extrapone St. John's Wort Watersoluble (2/
032985) (Symrise)
Extrapone Thyme Special 2/033401
(Symrise)
Extrapone Walnut 2/033451 (Symrise)
Extrapone Watercress 2/032122 (Symrise)
Extrapone White Nettle 2/034017 (Symrise)
Extrapone Witch Hazel Dist. Colorless
Special 2/032891 (Symrise)
Extrapone Witch Hazel Special 2/032901
(Symrise)
Extrapone Yarrow Special (2/033341)
(Symrise)
Eyebright Extract HS 2810 G (Grau)
Eyelid Tone Complex 271 (Ennagram)
FAIR HAIR (Greentech S.A)
Far East Extract Andawali (E.U.K)
Far East Extract Ginkgo Biloba (E.U.K)
Far East Extract Green Tea (E.U.K)
Far East Extract Indian Pennywort (E.U.K)
Far East Extract Kaffir (E.U.K)
Far East Extract Lime (E.U.K)
Far East Extract Reishi (E.U.K)
Far East Extract Shii-Take (E.U.K)
Far East Extract Tohasaku (E.U.K)
Fennel Extract HG (Provital/Centerchem)
Fennel Extract HS 2657 G (Grau)
Fennel HS (Alban Muller)
Fenugreek Extract HG (Provital/
Centerchem)
Fenugreek Extract HS 2447 G (Grau)
Fenugreek HPG Titrated (Alban Muller)
Fenugreek HS (Alban Muller)
Fern Extract HS 2630 G (Grau)
Fieldpoppy Extract HG (Provital/
Centerchem)
Fig Milk (Cosmetochem) (Cosmetochem
International Ltd.)
Fig Tree HS (Alban Muller)
Flaxseed HS (Alban Muller)
Flea Wort HS (Alban Muller)
Foeniculum Vulgare (Fennel) Fruit Extract
ies (IES LABO)
Fondix G Bis (Gattefosse s.a.)
Forget-Me-Not Extract (Cosmetic Develop-
ments)
Frangipany Phytolait (Alban Muller)
Free Radical Scavenger Phytoamine
Biocomplex (Alban Muller)

Freesia Extract (Cosmetic Developments)
French Rose HS (Alban Muller)
Fresh Plant Cells (Libiol)
Fructus Hippophae Rhamnoides Extract HS 2666 G (Grau)
Fruitapone Acerola GT 2/037010 (Symrise)
Fruitapone Blackberry GT 2/037070 (Symrise)
Fruitapone Black Currant GT 2/037100 (Symrise)
Fruitapone Blackthorn B 2/036700 (Symrise)
Fruitapone Blackthorn GT 2/037700 (Symrise)
Fruitapone Cherry B 2/036440 (Symrise)
Fruitapone Cherry GT 2/037440 (Symrise)
Fruitapone Cranberry B 2/036600 (Symrise)
Fruitapone Cranberry GT 2/037600 (Symrise)
Fruitapone Elder B 2/036220 (Symrise)
Fruitapone Elder GT 2/037220 (Symrise)
Fruitapone Gooseberry B 2/036710 (Symrise)
Fruitapone Gooseberry G 2/036710 (Symrise)
Fruitapone Gooseberry GT 2/037710 (Symrise)
Fruitapone Grapefruit B 2/036150 (Symrise)
Fruitapone Grapefruit GT 2/037150 (Symrise)
Fruitapone Grape Red B 2/036900 (Symrise)
Fruitapone Grape Red GT 2/037900 (Symrise)
Fruitapone Grape White B 2/036950 (Symrise)
Fruitapone Grape White GT 2/037950 (Symrise)
Fruitapone Guarana B 2/036200 (Symrise)
Fruitapone Guarana GT 2/037200 (Symrise)
Fruitapone Kiwi B 2/036250 (Symrise)
Fruitapone Kiwi GT 2/037250 (Symrise)
Fruitapone Lime B 2/036300 (Symrise)
Fruitapone Lime GT 2/037300 (Symrise)
Fruitapone Mandarine B 2/036340 (Symrise)
Fruitapone Mandarin Orange B 2/036340 (Symrise)
Fruitapone Mandarin orange GT 2/037340 (Symrise)
Fruitapone Mango B 2/036350 (Symrise)
Fruitapone Mango GT 2/037350 (Symrise)
Fruitapone Orange B 2/036400 (Symrise)
Fruitapone OrangeGT 2/037400 (Symrise)
Fruitapone Papaya B 2/036450 (Symrise)
Fruitapone Papaya GT 2/037450 (Symrise)
Fruitapone Passionfruit B 2/036500 (Symrise)
Fruitapone Passionfruit GT 2/037500 (Symrise)
Fruitapone Peach B 2/036550 (Symrise)

Fruitapone Peach GT 2/037550 (Symrise)
Fruitapone Pineapple B 2/036000 (Symrise)
Fruitapone Pineapple GT 2/037000 (Symrise)
Fruitapone Quince B 2/036630 (Symrise)
Fruitapone Quince GT 2/037630 (Symrise)
Fruitapone Raspberry GT 2/037215 (Symrise)
Fruitapone Strawberry B 2/036120 (Symrise)
Fruitapone Strawberry GT 2/037120 (Symrise)
Fruitliquid Cherimoya (Crodarom)
Fu Ling Extract (Alban Muller)
Gallnut HPG Titrated (Alban Muller)
Gallnut HS (Alban Muller)
Garden Daisy Extract HS 3019 G (Grau)
Gardenia HPG Titrated (Alban Muller)
Gardenia HS (Alban Muller)
Garlic Extract G (Provital/Centerchem)
Garlic Extract HS 2710 G (Grau)
Gemtex PA-70P (Finetex)
Gemtex PA-85P (Finetex)
Gentian Extract HG (Provital/Centerchem)
Gentian Extract HS 2414 G (Grau)
Gentian Extract PG (Rahn)
Gentian HS (Alban Muller)
Gentiana Lutea Extract ies (IES LABO)
Geranium Robertianum Extract ies (IES LABO)
Germaben II (Sutton)
Germaben II-E (Sutton)
German Chamomile HPG Titrated (Alban Muller)
German Chamomile HPG Titrated Bisabolol (Alban Muller)
German Chamomile HS (Alban Muller)
Germander Extract HS 2970 G (Grau)
Germazide WS (Collaborative Labs)
Geum Urbanum Extract ies (IES LABO)
Geum Urbanum Extract (Root) ies (IES LABO)
Ginger Root Extract HS 2514 G (Grau)
Ginkgo Biloba Extract HG (Provital/Centerchem)
Ginkgo Extract HS 2464 G (Grau)
Ginkgo Extract PG (Rahn)
Ginseng Extract HS 2457 G (Grau)
Ginseng HPG Titrated (Alban Muller)
Ginseng HS (Alban Muller)
Gleditschia (Greentech)
GLEDITSCHIA (Greentech S.A)
Glucose Tyrosinate AMI (Alban Muller)
Glycine Soja (Soybean) Germ Extract ies (IES LABO)
Glycolic Extract R.U.P. (CEP (Solabia))
Glycolysat Complex 85-5-10 (CEP (Solabia))
Glycolysat of Acacia (CEP (Solabia))
Glycolysat of Aloe (CEP (Solabia))
Glycolysat Of Apple (CEP (Solabia))
Glycolysat of Apricot (CEP (Solabia))
Glycolysat of Arnica (CEP (Solabia))

Glycolysat of Artichoke (CEP (Solabia))
Glycolysat of Avocado (CEP (Solabia))
Glycolysat of Bamboo Leaves (CEP (Solabia))
Glycolysat of Bearberry (CEP (Solabia))
Glycolysat of Beechwood (CEP (Solabia))
Glycolysat of Birch (CEP (Solabia))
Glycolysat of Black Currant (CEP (Solabia))
Glycolysat of Black Currant Leaves (CEP (Solabia))
Glycolysat of Bladderwrack (CEP (Solabia))
Glycolysat of Broom (CEP (Solabia))
Glycolysat of Burdock (CEP (Solabia))
Glycolysat of Butcher's Broom (CEP (Solabia))
Glycolysat of Cabbage Rose (CEP (Solabia))
Glycolysat of Cactus (CEP (Solabia))
Glycolysat of Camphor Tree (CEP (Solabia))
Glycolysat of Carline Thistle (CEP (Solabia))
Glycolysat of Carrot (CEP (Solabia))
Glycolysat of Caviar (CEP (Solabia))
Glycolysat of Celandine (CEP (Solabia))
Glycolysat of 7 Cereals (CEP (Solabia))
Glycolysat of Chamomile (CEP (Solabia))
Glycolysat of Cherry (CEP (Solabia))
Glycolysat of Cherry Stalks (CEP (Solabia))
Glycolysat of Chinese Tea (CEP (Solabia))
Glycolysat of Clementine (CEP (Solabia))
Glycolysat of Coltsfoot (CEP (Solabia))
Glycolysat of Condurango (CEP (Solabia))
Glycolysat of Corn (CEP (Solabia))
Glycolysat of Cornflower (CEP (Solabia))
Glycolysat of Cucumber (CEP (Solabia))
Glycolysat of Cypress (CEP (Solabia))
Glycolysat of Dalmatian Pyrethrum (CEP (Solabia))
Glycolysat of Date (CEP (Solabia))
Glycolysat of Elder (FL/BE) (CEP (Solabia))
Glycolysat of Eyebright (CEP (Solabia))
Glycolysat of Fenugreek (CEP (Solabia))
Glycolysat of Fig (CEP (Solabia))
Glycolysat of Florentine Orris (CEP (Solabia))
Glycolysat of Frangipani Flowers (CEP (Solabia))
Glycolysat of Galanga (CEP (Solabia))
Glycolysat of Gardenia Jasminoides (CEP (Solabia))
Glycolysat of Gardenia Tahitensis (CEP (Solabia))
Glycolysat of Gentian (CEP (Solabia))
Glycolysat of Ginger (CEP (Solabia))
Glycolysat of Gingko Biloba (CEP (Solabia))
Glycolysat of Ginseng (CEP (Solabia))
Glycolysat of Goat Milk (CEP (Solabia))
Glycolysat of Grape (CEP (Solabia))
Glycolysat of Grapefruit (CEP (Solabia))
Glycolysat of Grape Leaf (CEP (Solabia))
Glycolysat of Green Coffee (CEP (Solabia))
Glycolysat of Guarana (CEP (Solabia))

Glycolysat Of Hibiscus Flowers (CEP (Solabia))
Glycolysat of Holly (CEP (Solabia))
Glycolysat of Honey (CEP (Solabia))
Glycolysat of Hops (CEP (Solabia))
Glycolysat of Horse Chestnut (CEP (Solabia))
Glycolysat of Hydrocotyl (CEP (Solabia))
Glycolysat of indigo (CEP (Solabia))
Glycolysat of Laminaria (CEP (Solabia))
Glycolysat of Lemon (CEP (Solabia))
Glycolysat of Lemon Verbena (CEP (Solabia))
Glycolysat of Lettuce (CEP (Solabia))
Glycolysat of Licorice (CEP (Solabia))
Glycolysat of Lime (CEP (Solabia))
Glycolysat of Linden (CEP (Solabia))
Glycolysat of Mallow (CEP (Solabia))
Glycolysat of Mango (CEP (Solabia))
Glycolysat of Marigold (CEP (Solabia))
Glycolysat of Marjoram (CEP (Solabia))
Glycolysat of Marshmallow (CEP (Solabia))
Glycolysat of Matricaria (CEP (Solabia))
Glycolysat of Melegueta (CEP (Solabia))
Glycolysat of Mint (CEP (Solabia))
Glycolysat of Natural Henna (CEP (Solabia))
Glycolysat of Neutral Henna (CEP (Solabia))
Glycolysat of Nutmeg (CEP (Solabia))
Glycolysat of Orange (CEP (Solabia))
Glycolysat of Oysters (CEP (Solabia))
Glycolysat of Papaya (CEP (Solabia))
Glycolysat of Passionflower (CEP (Solabia))
Glycolysat of Passion Fruit (CEP (Solabia))
Glycolysat of Peach (CEP (Solabia))
Glycolysat of (Perles de Caviar) (CEP (Solabia))
Glycolysat of Pineapple (CEP (Solabia))
Glycolysat of Plantain (CEP (Solabia))
Glycolysat of Pollen (CEP (Solabia))
Glycolysat of Pomegranate (CEP (Solabia))
Glycolysat of Propolis (CEP (Solabia))
Glycolysat of Raspberry (CEP (Solabia))
Glycolysat of Red Currant (CEP (Solabia))
Glycolysat of Red Sandalwood (CEP (Solabia))
Glycolysat of Rhatany (CEP (Solabia))
Glycolysat of Rice (CEP (Solabia))
Glycolysat of Rosemary (CEP (Solabia))
Glycolysat of Royal Gelly (CEP (Solabia))
Glycolysat of Sage (CEP (Solabia))
Glycolysat of Saponaria (CEP (Solabia))
Glycolysat of Sea Fennel (CEP (Solabia))
Glycolysat of Sea Rocket (CEP (Solabia))
Glycolysat of Sea Urchins (CEP (Solabia))
Glycolysat of Shrimp (CEP (Solabia))
Glycolysat of Soothing Plants (CEP (Solabia))
Glycolysat of Spinach (CEP (Solabia))
Glycolysat of St. John's Wort (CEP (Solabia))
Glycolysat of Sweet Almond (CEP (Solabia))

Glycolysat of Sweet Clover (CEP (Solabia))
Glycolysat of Tamarind (CEP (Solabia))
Glycolysat of Tepescohuite (CEP (Solabia))
Glycolysat of Thuya (CEP (Solabia))
Glycolysat of Tormentil (CEP (Solabia))
Glycolysat of Valerian (CEP (Solabia))
Glycolysat of Walnut Husk (CEP (Solabia))
Glycolysat of Walnut Tree (CEP (Solabia))
Glycolysat of Wheat Germs (CEP (Solabia))
Glycolysat of White Lily Bulb (CEP (Solabia))
Glycolysat of White Nettle (CEP (Solabia))
Glycolysat of Wild Pansy (CEP (Solabia))
Glycolysat Of Willow (CEP (Solabia))
Glycolysat of Witch Hazel (CEP (Solabia))
Glycolysat of Yarrow (CEP (Solabia))
Glycophytolo-BHE (Vevy)
Glycyrrhiza Glabra (Licorice) Extract ies (IES LABO)
Goat's Rue HS (Alban Muller)
Golden Rod HPG Titrated (Alban Muller)
Golden Rod HS (Alban Muller)
Goose Grass HS (Alban Muller)
Gramben II (Sinerga)
Granoliquid (Wheat Germ Extract, Water Soluble) (Crodarom)
Grapefruit Extract HG (Provital/ Centerchem)
Grapefruit Extract HS 1807 AT (Grau)
Grapefruit HS (Alban Muller)
Grapefruit Pulp Extract (Alban Muller)
Grapes Extract HS 2803 G (Grau)
Grass Polly HS (Alban Muller)
GREASY HAIR (Greentech S.A)
Great Burdock Extract HG (Provital/ Centerchem)
Great Burdock HS (Alban Muller)
Great Burnet HS (Alban Muller)
Green Coffee Extract HS 3416 G (Grau)
Green Coffee HPG Titrated (Alban Muller)
Green Coffee HS (Alban Muller)
Green Tea (Arch Personal Care Products)
Green Tea Extract 101266 (Fragrance Oils Int. Ltd.)
Green Tea Extract HG (Provital/ Centerchem)
Green Tea Extract HS 2614 G (Grau)
Green Tea Glycolic Extract (Euromed)
Green Tea HPG Titrated (Alban Muller)
Green Tea Phytolait (Alban Muller)
Gromwell HS (Alban Muller)
Guaiac HS (Alban Muller)
Guana Extract (Bio-Botanica)
Guarana Glycolic Extract NTR2 (Centroflora Group)
Guarana HPG Titrated (Alban Muller)
Guardian GP (Earth Supplied Products)
Guardian O9 (Earth Supplied Products)
Guavacal (Cosmecal Sarl)
Guava HS (Alban Muller)
Gulfweed Extract 101873 (Fragrance Oils Int. Ltd.)
Gypsophila HS (Alban Muller)

Hair Care Blend (Alban Muller)
Hair Care Phytoamine Biocomplex SP Lotion (Alban Muller)
Haircomplex AKS (Crodarom)
Hair Complex NOVA (Crodarom)
Hair Treatment Complex 260 (Ennagram)
Hair Treatment Phytogreen Complex GXH 260 (Phytochim)
HAIR VOLUME (Greentech S.A)
Hamamelis Extract HG (Provital/ Centerchem)
Hamamelis Extract HS 2456 G (Grau)
Harpagophytum Extract (CFEB Sisley)
Harpagophytum Extract HG (Provital/ Centerchem)
Harpagophytum Extract HS 2664 G (Grau)
Harungana Extract (Exsymol)
Hawaian Ginger PG (Alban Muller)
Hawthorn Extract HG (Provital/ Centerchem)
Hawthorn Extract HS 2398 G (Grau)
Hawthorn HS (Alban Muller)
Hay Extract HG (Provital/Centerchem)
Hay Flower Extract HS 2451 G (Grau)
Hazel Leaves Extract HS 2563 G (Grau)
Hazel Nut HS (Alban Muller)
Heather Extract (Kelisema Italy)
Heather HS (Alban Muller)
Heavy Legs Complex 269 (Ennagram)
Heavy Legs Complex MU 3422 (Greentech S.A)
Heavy Legs Phytogreen Complex GXH 269 (Phytochim)
Helianthus Annuus (Sunflower) Seed Extract (IES LABO)
Helychrisum Extract HG (Provital/ Centerchem)
Hemp Agrimony HS (Alban Muller)
Henna Extract HG (Provital/Centerchem)
Henna Extract HS 2448 G Red Coloring (Grau)
Henna Extract Not Coloring HS 2522 G (Grau)
Henna Extract, Water Soluble (Crodarom)
Henna HS (Alban Muller)
Herba Hieracii Extract HS 3337 G (Grau)
Herbalcomplex 1 Special (Crodarom)
Herbalcomplex 2 special (Crodarom)
Herbalcomplex 3 Special (Crodarom)
Herbal-Complex 4 special (Crodarom)
Herbalcomplex 5 Special (Crodarom)
Herbal Extract for Normal Hair (Crodarom)
Herbal Extract Frangipani (Jarvis)
Herbal Extract Glycolic - Article 251172 (Plantextrakt)
Herbaliquid Alant Root Special (Crodarom)
Herbaliquid Aloe Special (Crodarom)
Herbaliquid Alpine Herbs Special (Crodarom)
Herbaliquid Arnica Special (Crodarom)
Herbaliquid Balm Mint Special (Crodarom)
Herbaliquid Birch Special (Crodarom)
Herbaliquid Burdock Root Special (Crodarom)

Herbaliquid Calendula Special (Crodarom)
Herbaliquid Camomile Special (Crodarom)
Herbaliquid China Bark Special (Crodarom)
Herbaliquid Coltsfoot Special (Crodarom)
Herbaliquid Dandelion Special (Crodarom)
Herbaliquid Eyebright Special (Crodarom)
Herbaliquid Ginseng Special (Crodarom)
Herbaliquid Hawthorn Special (Crodarom)
Herbaliquid Hayflower Special (Crodarom)
Herbaliquid Hops Special (Crodarom)
Herbaliquid Horsechestnut Special
   (Crodarom)
Herbaliquid Horsetail Special (Crodarom)
Herbaliquid Ivy Special (Crodarom)
Herbaliquid Juniper Special (Crodarom)
Herbaliquid Lavender Special (Crodarom)
Herbaliquid Limetree Blossom Special
   (Crodarom)
Herbaliquid Marsh Mallow Special
   (Crodarom)
Herbaliquid Melilot Special (Crodarom)
Herbaliquid Mountain Gentian Special
   (Crodarom)
Herbaliquid Oak Bark Special (Crodarom)
Herbaliquid Pansy Special (Crodarom)
Herbaliquid Peppermint Special
   (Crodarom)
Herbaliquid Pine Needle Special
   (Crodarom)
Herbaliquid Quillaja Special (Crodarom)
Herbaliquid Rhatany Root (Crodarom)
Herbaliquid Rosemary Special (Crodarom)
Herbaliquid Sage Special (Crodarom)
Herbaliquid Sambucus Special (Crodarom)
Herbaliquid Soap Wort special (Crodarom)
Herbaliquid Stinging Nettle Special
   (Crodarom)
Herbaliquid St. John's Wort Special
   (Crodarom)
Herbaliquid Tormentil Root Special
   (Crodarom)
Herbaliquid Valerian Special (Crodarom)
Herbaliquid Watercress Special
   (Crodarom)
Herbaliquid Wild Thyme Special
   (Crodarom)
Herbaliquid Willow Special (Crodarom)
Herbaliquid Witch Hazel Special
   (Crodarom)
Herbaliquid Yarrow Special (Crodarom)
Herbasol Arnica (Mexican) (Cosmetochem
   International Ltd.)
Herbasol Complex GU-61-A
   (Cosmetochem)
Herbasol Complex GU-61-Standard
   (Cosmetochem)
Herbasol Complex "Herbes de Provence"
   (Cosmetochem) (Cosmetochem
   International Ltd.)
Herbasol Complex 7-Herbs
   (Cosmetochem) (Cosmetochem
   International Ltd.)
Herbasol Complex "Sedative/Relaxing"
   (Cosmetochem) (Cosmetochem
   International Ltd.)

Herbasol Dead Nettle Extract (Leaf)
   (Cosmetochem) (Cosmetochem
   International Ltd.)
Herbasol Evening Primrose (Leaf)
   (Cosmetochem) (Cosmetochem
   International Ltd.)
Herbasol Extract Acacia Catechu
   (Cosmetochem) (Cosmetochem
   International Ltd.)
Herbasol Extract Acacia (Cosmetochem)
   (Cosmetochem International Ltd.)
Herbasol-Extract Agave (Cosmetochem)
Herbasol Extract Agrimony (Cosmetochem)
   (Cosmetochem International Ltd.)
Herbasol Extract Alfalfa (Cosmetochem)
   (Cosmetochem International Ltd.)
Herbasol-Extract Algae (Cosmetochem)
Herbasol Extract Alkanna (Alkanet)
   (Cosmetochem) (Cosmetochem
   International Ltd.)
Herbasol Extract Alkekengi (Winter Cherry)
   (Cosmetochem) (Cosmetochem
   International Ltd.)
Herbasol-Extract Almond (Cosmetochem)
Herbasol-Extract Aloe (Cosmetochem)
Herbasol Extract Amaranth
   (Cosmetochem) (Cosmetochem
   International Ltd.)
Herbasol Extract Ammi Visnaga
   (Cosmetochem International Ltd.)
Herbasol Extract Angelica (Cosmetochem)
   International Ltd.)
Herbasol Extract Angelica Leaf
   (Cosmetochem) (Cosmetochem
   International Ltd.)
Herbasol-Extract Angelica (Root)
   (Cosmetochem)
Herbasol-Extract Aniseed (Anise)
   (Cosmetochem)
Herbasol Extract Apple (Cosmetochem)
   (Cosmetochem International Ltd.)
Herbasol Extract Apricot (Cosmetochem)
   (Cosmetochem International Ltd.)
Herbasol Extract Apricot Kernel Oil
   (Cosmetochem) (Cosmetochem
   International Ltd.)
Herbasol-Extract Arnica (Cosmetochem)
Herbasol Extract Artichoke (Leaf)
   (Cosmetochem) (Cosmetochem
   International Ltd.)
Herbasol Extract Ash (Cosmetochem)
   (Cosmetochem International Ltd.)
Herbasol-Extract Asparagus (Root)
   (Cosmetochem)
Herbasol-Extract Avens (Cosmetochem)
Herbasol Extract Avens (Cosmetochem)
   (Cosmetochem International Ltd.)
Herbasol Extract Avocado (Cosmetochem)
   (Cosmetochem International Ltd.)
Herbasol-Extract Balm Mint
   (Cosmetochem)
Herbasol Extract Bamboo (Cosmetochem)
   (Cosmetochem International Ltd.)

Herbasol-Extract Banana (Cosmetochem)
Herbasol Extract Barberry Fruit
   (Cosmetochem International Ltd.)
Herbasol Extract Barley (Cosmetochem)
   (Cosmetochem International Ltd.)
Herbasol Extract Basil (Cosmetochem)
   (Cosmetochem International Ltd.)
Herbasol Extract Bearberry
   (Cosmetochem) (Cosmetochem
   International Ltd.)
Herbasol Extract Bear's Ears (Hawkweed)
   (Cosmetochem) (Cosmetochem
   International Ltd.)
Herbasol Extract Bee Pollen
   (Cosmetochem) (Cosmetochem
   International Ltd.)
Herbasol Extract Bergamot
   (Cosmetochem) (Cosmetochem
   International Ltd.)
Herbasol Extract Betelnut/Pepper Betle
   (Cosmetochem) (Cosmetochem
   International Ltd.)
Herbasol Extract Birch Bark
   (Cosmetochem) (Cosmetochem
   International Ltd.)
Herbasol-Extract Birch (Leaf)
   (Cosmetochem)
Herbasol-Extract Blackberry/Bramble
   (Cosmetochem)
Herbasol Extract Black Currant Fruit
   (Cosmetochem) (Cosmetochem
   International Ltd.)
Herbasol Extract Black Currant Leaf
   (Cosmetochem) (Cosmetochem
   International Ltd.)
Herbasol Extract Black Tea
   (Cosmetochem) (Cosmetochem
   International Ltd.)
Herbasol Extract Bladder Wrack (Sea
   Weed) (Cosmetochem) (Cosmetochem
   International Ltd.)
Herbasol Extract Blueberry
   (Cosmetochem) (Cosmetochem
   International Ltd.)
Herbasol Extract Borage (Cosmetochem)
   (Cosmetochem International Ltd.)
Herbasol Extract Box Tree (Cosmetochem)
   (Cosmetochem International Ltd.)
Herbasol Extract Broom (Cosmetochem)
   (Cosmetochem International Ltd.)
Herbasol-Extract Buckbean
   (Cosmetochem)
Herbasol-Extract Buckthorn
   (Cosmetochem)
Herbasol-Extract Burcher's Broom
   (Ruscus) (Cosmetochem)
Herbasol-Extract Burdock (Root)
   (Cosmetochem)
Herbasol Extract Cabbage (Cosmetochem)
   (Cosmetochem International Ltd.)
Herbasol Extract Cactus Flower
   (Cosmetochem) (Cosmetochem
   International Ltd.)

Herbasol-Extract Calamus (Root) (Cosmetochem)

Herbasol-Extract Calendula/Marigold (Cosmetochem)

Herbasol Extract Capsicum (Red Pepper) (Cosmetochem) (Cosmetochem International Ltd.)

Herbasol Extract Caraway (Cosmetochem) (Cosmetochem International Ltd.)

Herbasol Extract Cardamom (Cosmetochem) (Cosmetochem International Ltd.)

Herbasol Extract Carob (Cosmetochem International Ltd.)

Herbasol Extract Carrot (Cosmetochem) (Cosmetochem International Ltd.)

Herbasol Extract Cat's Foot (Everlasting) (Cosmetochem) (Cosmetochem International Ltd.)

Herbasol Extract Cedar (Thuja Leaf) (Cosmetochem) (Cosmetochem International Ltd.)

Herbasol Extract Celandine/Tetterwort (Cosmetochem) (Cosmetochem International Ltd.)

Herbasol Extract Celery (Root) (Cosmetochem) (Cosmetochem International Ltd.)

Herbasol Extract Centaury (Cosmetochem) (Cosmetochem International Ltd.)

Herbasol Extract Centella Asiatica (Gotukola) (Cosmetochem) (Cosmetochem International Ltd.)

Herbasol Extract Chamomile (Cosmetochem) (Cosmetochem International Ltd.)

Herbasol Extract Chaste Tree (Monk's Pepper) (Cosmetochem) (Cosmetochem International Ltd.)

Herbasol Extract Cherry (Cosmetochem) (Cosmetochem International Ltd.)

Herbasol Extract Cherry Stalks (Cosmetochem) (Cosmetochem International Ltd.)

Herbasol Extract Chervil (Cosmetochem) (Cosmetochem International Ltd.)

Herbasol Extract Chickweed (Cosmetochem) (Cosmetochem International Ltd.)

Herbasol-Extract China Bark (Cosmetochem)

Herbasol-Extract Cinchona Bark (Cosmetochem)

Herbasol Extract Cinnamon (Cosmetochem) (Cosmetochem International Ltd.)

Herbasol Extract Clary Sage (Cosmetochem) (Cosmetochem International Ltd.)

Herbasol Extract Clematis (Traveller's Joy) Leaf (Cosmetochem) (Cosmetochem International Ltd.)

Herbasol Extract Cloves (Cosmetochem) (Cosmetochem International Ltd.)

Herbasol Extract Cocoa (Cosmetochem) (Cosmetochem International Ltd.)

Herbasol Extract Cocos (Coconut Milk) (Cosmetochem) (Cosmetochem International Ltd.)

Herbasol Extract Coffee (Cosmetochem) (Cosmetochem International Ltd.)

Herbasol Extract Cola Nuts (Cosmetochem) (Cosmetochem International Ltd.)

Herbasol-Extract Coltsfoot (Cosmetochem)

Herbasol Extract Combrete (Cosmetochem) (Cosmetochem International Ltd.)

Herbasol-Extract Comfrey (Cosmetochem)

Herbasol-Extract Condurango (Cosmetochem)

Herbasol-Extract Coriander (Cosmetochem)

Herbasol-Extract Cornflower (Cosmetochem)

Herbasol Extract Cotton Seeds (Cosmetochem) (Cosmetochem International Ltd.)

Herbasol-Extract Couch Grass (Cosmetochem)

Herbasol-Extract Curcuma (Cosmetochem)

Herbasol Extract Cypress-Cedar (Cosmetochem) (Cosmetochem International Ltd.)

Herbasol-Extract Daisy (Cosmetochem)

Herbasol-Extract Dandelion (Cosmetochem)

Herbasol Extract Devil's Claw (Cosmetochem) (Cosmetochem International Ltd.)

Herbasol-Extract Dwarf Pine (Cosmetochem)

Herbasol-Extract Echinacea (Cosmetochem)

Herbasol Extract Echinacea Purpurea Radix (Cosmetochem) (Cosmetochem International Ltd.)

Herbasol Extract Edelweiss (Cosmetochem) (Cosmetochem International Ltd.)

Herbasol-Extract Elder (Cosmetochem)

Herbasol Extract Elecampane Root (Alant) (Cosmetochem) (Cosmetochem International Ltd.)

Herbasol-Extract Elm (Cosmetochem)

Herbasol-Extract Eucalyptus (Cosmetochem)

Herbasol-Extract Eyebright (Cosmetochem)

Herbasol-Extract Fennel (Cosmetochem)

Herbasol-Extract Fenugreek (Cosmetochem)

Herbasol Extract Fig (Fruit) (Cosmetochem) (Cosmetochem International Ltd.)

Herbasol Extract Fig (Leaves) (Cosmetochem) (Cosmetochem International Ltd.)

Herbasol Extract Fig Wort (Cosmetochem) (Cosmetochem International Ltd.)

Herbasol Extract Fir Needles (Cosmetochem) (Cosmetochem International Ltd.)

Herbasol Extract Fumitory (Cosmetochem) (Cosmetochem International Ltd.)

Herbasol Extract Galangal (Cosmetochem) (Cosmetochem International Ltd.)

Herbasol Extract Garlic (Cosmetochem) (Cosmetochem International Ltd.)

Herbasol-Extract Gentian (Cosmetochem)

Herbasol-Extract Geranium (Cosmetochem)

Herbasol Extract Ginger (Cosmetochem) (Cosmetochem International Ltd.)

Herbasol-Extract Ginseng (Cosmetochem)

Herbasol-Extract Goatweed (Cosmetochem)

Herbasol Extract Golden Seal (Cosmetochem) (Cosmetochem International Ltd.)

Herbasol Extract Golden Seal Root Colouring (Cosmetochem) (Cosmetochem International Ltd.)

Herbasol Extract Grapefruit (Cosmetochem) (Cosmetochem International Ltd.)

Herbasol Extract Grape (Fruit) (Cosmetochem) (Cosmetochem International Ltd.)

Herbasol Extract Grape Seed (Cosmetochem) (Cosmetochem International Ltd.)

Herbasol Extract Green Tea (Cosmetochem) (Cosmetochem International Ltd.)

Herbasol Extract Groundsel (Cosmetochem) (Cosmetochem International Ltd.)

Herbasol Extract Guajara (Cosmetochem) (Cosmetochem International Ltd.)

Herbasol Extract Guarana (Cosmetochem) (Cosmetochem International Ltd.)

Herbasol Extract Harpagophytum (Cosmetochem) (Cosmetochem International Ltd.)

Herbasol Extract Hawkweed (Cosmetochem) (Cosmetochem International Ltd.)

Herbasol-Extract Hawthorn (Cosmetochem)

Herbasol Extract Hawthorn (Cosmetochem International Ltd.)

Herbasol Extract Hawthorn (Leaf) (Cosmetochem) (Cosmetochem International Ltd.)

Herbasol-Extract Hay Flowers (Cosmetochem)

Herbasol-Extract Hazel (Cosmetochem)
Herbasol Extract Heather (Common Health) (Cosmetochem) (Cosmetochem International Ltd.)
Herbasol-Extract Henna (Cosmetochem)
Herbasol Extract Hibiscus (Sudanese Tea) (Cosmetochem) (Cosmetochem International Ltd.)
Herbasol Extract Holly (Cosmetochem) (Cosmetochem International Ltd.)
Herbasol-Extract Hops (Cosmetochem)
Herbasol-Extract Horse Chestnut (Cosmetochem)
Herbasol-Extract Horse Radish (Cosmetochem)
Herbasol-Extract Horse Tail (Cosmetochem)
Herbasol Extract Hyssop (Cosmetochem) (Cosmetochem International Ltd.)
Herbasol-Extract Iceland Moss (Cosmetochem)
Herbasol-Extract Indian Cress (Cosmetochem)
Herbasol Extract Irish Moss (Cosmetochem) (Cosmetochem International Ltd.)
Herbasol-Extract Iris (Orris) (Cosmetochem)
Herbasol-Extract Ivy (Cosmetochem)
Herbasol Extract Jaborandi (Cosmetochem) (Cosmetochem International Ltd.)
Herbasol-Extract Jamaica Pepper (Cosmetochem)
Herbasol Extract Japanese Pagoda (Cosmetochem) (Cosmetochem International Ltd.)
Herbasol Extract Jasmine Flowers (Cosmetochem) (Cosmetochem International Ltd.)
Herbasol Extract Java Tea (Cosmetochem) (Cosmetochem International Ltd.)
Herbasol Extract Jojoba (Cosmetochem) (Cosmetochem International Ltd.)
Herbasol-Extract Juniper (Cosmetochem)
Herbasol Extract Kava Kava (Cosmetochem) (Cosmetochem International Ltd.)
Herbasol Extract Kiwi (Cosmetochem) (Cosmetochem International Ltd.)
Herbasol-Extract Laborador/Marsu Tea (Cosmetochem)
Herbasol-Extract Lady's Mantle (Cosmetochem)
Herbasol Extract Laurel (Cosmetochem) (Cosmetochem International Ltd.)
Herbasol-Extract Lavender (Cosmetochem)
Herbasol Extract Leek (Cosmetochem International Ltd.)
Herbasol Extract Lemon Grass (Cosmetochem) (Cosmetochem International Ltd.)

Herbasol Extract Lemon (Peel) (Cosmetochem) (Cosmetochem International Ltd.)
Herbasol Extract Lichen (Alpine) (Cosmetochem) (Cosmetochem International Ltd.)
Herbasol Extract Lilac (Cosmetochem) (Cosmetochem International Ltd.)
Herbasol-Extract Lime Blossom (Cosmetochem)
Herbasol Extract Lime (Cosmetochem) (Cosmetochem International Ltd.)
Herbasol Extract Linseed (Cosmetochem) (Cosmetochem International Ltd.)
Herbasol-Extract Liquorice (Cosmetochem)
Herbasol Extract Logwood (Cosmetochem) (Cosmetochem International Ltd.)
Herbasol Extract Lotus (Cosmetochem) (Cosmetochem International Ltd.)
Herbasol Extract Lovage (Leaf) (Cosmetochem International Ltd.)
Herbasol Extract Madder (Root) (Cosmetochem) (Cosmetochem International Ltd.)
Herbasol-Extract Maiden Hair Fern (Cosmetochem)
Herbasol Extract Maize (Cosmetochem) (Cosmetochem International Ltd.)
Herbasol-Extract Mallow Flowers (Cosmetochem)
Herbasol Extract Mango (Cosmetochem) (Cosmetochem International Ltd.)
Herbasol Extract Mansarin/Tangerine (Peel) (Cosmetochem) (Cosmetochem International Ltd.)
Herbasol-Extract Marigold (Calendula) (Cosmetochem)
Herbasol-Extract Marjoram (Cosmetochem)
Herbasol-Extract Marrow (Cosmetochem)
Herbasol-Extract Marsh Mallow (Cosmetochem)
Herbasol Extract Mate(Cosmetochem) (Cosmetochem International Ltd.)
Herbasol-Extract Meadowsweet (Cosmetochem)
Herbasol Extract Melon (Honey Dew) (Cosmetochem) (Cosmetochem International Ltd.)
Herbasol-Extract Mercury (Cosmetochem)
Herbasol Extract Milk Thistle (St Mary's Thistle) (Cosmetochem) (Cosmetochem International Ltd.)
Herbasol-Extract Millet (Cosmetochem)
Herbasol Extract Mimose Tenuiflora (Cosmetochem) (Cosmetochem International Ltd.)
Herbasol-Extract Mistletoe (Cosmetochem)
Herbasol-Extract Mountain Ash (Cosmetochem)
Herbasol Extract Mugwort (Cosmetochem) (Cosmetochem International Ltd.)

Herbasol Extract Muira-Puama (Cosmetochem International Ltd.)
Herbasol Extract Mulberry (Leaf) (Cosmetochem) (Cosmetochem International Ltd.)
Herbasol Extract Mullein, Great (Cosmetochem) (Cosmetochem International Ltd.)
Herbasol Extract Mushroom (Cosmetochem) (Cosmetochem International Ltd.)
Herbasol-Extract Mustard (Cosmetochem)
Herbasol Extract Nutmeg (Cosmetochem) (Cosmetochem International Ltd.)
Herbasol-Extract Oak Bark (Cosmetochem)
Herbasol Extract Oats (Cosmetochem) (Cosmetochem International Ltd.)
Herbasol-Extract Olive (Leaf) (Cosmetochem)
Herbasol Extract Olive (Leaf) (Cosmetochem) (Cosmetochem International Ltd.)
Herbasol Extract Olive Wood (Cosmetochem International Ltd.)
Herbasol Extract Onion (Cosmetochem) (Cosmetochem International Ltd.)
Herbasol Extract Orange Flower (Cosmetochem) (Cosmetochem International Ltd.)
Herbasol Extract Orange Peel Bitter (Cosmetochem) (Cosmetochem International Ltd.)
Herbasol Extract Orange Peel Sweet (Cosmetochem) (Cosmetochem International Ltd.)
Herbasol Extract Orange Pulp (Cosmetochem) (Cosmetochem International Ltd.)
Herbasol Extract Paeony (Root) (Cosmetochem International Ltd.)
Herbasol-Extract Panama Bark (Cosmetochem)
Herbasol-Extract Pansy (Cosmetochem)
Herbasol Extract Papaya (Cosmetochem) (Cosmetochem International Ltd.)
Herbasol-Extract Parsley (Cosmetochem)
Herbasol Extract Parsley (Cosmetochem) (Cosmetochem International Ltd.)
Herbasol-Extract Passion Flower (Cosmetochem)
Herbasol Extract Passion Fruit (Cosmetochem) (Cosmetochem International Ltd.)
Herbasol Extract Patchouli (Cosmetochem) (Cosmetochem International Ltd.)
Herbasol Extract Peach (Cosmetochem) (Cosmetochem International Ltd.)
Herbasol Extract Pear (Cosmetochem) (Cosmetochem International Ltd.)
Herbasol Extract Peat (Cosmetochem) (Cosmetochem International Ltd.)

Herbasol Extract Pennyroyal Mint (Cosmetochem) (Cosmetochem International Ltd.)

Herbasol-Extract Peppermint (Cosmetochem)

Herbasol Extract Periwinkle (Evergreen) (Cosmetochem) (Cosmetochem International Ltd.)

Herbasol-Extract Pimento (Jamuican Pepper) (Cosmetochem)

Herbasol Extract Pineapple (Cosmetochem) (Cosmetochem International Ltd.)

Herbasol Extract Pine Needles (Cosmetochem) (Cosmetochem International Ltd.)

Herbasol Extract Plum (Cosmetochem) (Cosmetochem International Ltd.)

Herbasol Extract Pomegranate (Cosmetochem) (Cosmetochem International Ltd.)

Herbasol Extract Pond Lily (Cosmetochem) (Cosmetochem International Ltd.)

Herbasol-Extract Poplar (Cosmetochem)

Herbasol Extract Poppy (Cosmetochem) (Cosmetochem International Ltd.)

Herbasol Extract Potato (Cosmetochem) (Cosmetochem International Ltd.)

Herbasol-Extract Potentilla (Cosmetochem)

Herbasol-Extract Primose (Cosmetochem)

Herbasol Extract Pumpkin (Cosmetochem) (Cosmetochem International Ltd.)

Herbasol Extract Quillaja (Cosmetochem)

Herbasol Extract Quince (Seed) (Cosmetochem) (Cosmetochem International Ltd.)

Herbasol Extract Radish (Cosmetochem) (Cosmetochem International Ltd.)

Herbasol-Extract Raspberry (Fruit) (Cosmetochem)

Herbasol Extract Raspberry Leaf (Cosmetochem) (Cosmetochem International Ltd.)

Herbasol Extract Red Bean (Cosmetochem) (Cosmetochem International Ltd.)

Herbasol-Extract Red Clover (Cosmetochem)

Herbasol-Extract Restharrow (Cosmetochem)

Herbasol Extract Rhatany (Cosmetochem) (Cosmetochem International Ltd.)

Herbasol Extract Rhubarb (Cosmetochem) (Cosmetochem International Ltd.)

Herbasol Extract Ribwort (Plantain) (Cosmetochem) (Cosmetochem International Ltd.)

Herbasol Extract Rice (Cosmetochem) (Cosmetochem International Ltd.)

Herbasol Extract Rooibos (Cosmetochem) (Cosmetochem International Ltd.)

Herbasol-Extract Rose Flowers (Cosmetochem)

Herbasol-Extract Rose Hips (Cosmetochem)

Herbasol-Extract Rosemary (Cosmetochem)

Herbasol Extract Rue (Cosmetochem) (Cosmetochem International Ltd.)

Herbasol Extract Russian Ginseng (Cosmetochem) (Cosmetochem International Ltd.)

Herbasol Extract Rye (Cosmetochem) (Cosmetochem International Ltd.)

Herbasol Extract Sabal (Saw Palmetto) (Cosmetochem) (Cosmetochem International Ltd.)

Herbasol-Extract Sage (Cosmetochem)

Herbasol Extract Sandalwood Red (Cosmetochem) (Cosmetochem International Ltd.)

Herbasol-Extract Sarsaparilla (Cosmetochem)

Herbasol Extract Savory (Wort) (Cosmetochem) (Cosmetochem International Ltd.)

Herbasol-Extract Scurvy Grass (Cosmetochem)

Herbasol Extract Sea Buckthorn (Sallowthorn) (Cosmetochem) (Cosmetochem International Ltd.)

Herbasol-Extract Sea Weed (Cosmetochem)

Herbasol Extract Senna (Cosmetochem) (Cosmetochem International Ltd.)

Herbasol Extract Sesame Seed (Cosmetochem) (Cosmetochem International Ltd.)

Herbasol-Extract Shepherd's Purse (Cosmetochem)

Herbasol Extract Soap Root (Cosmetochem) (Cosmetochem International Ltd.)

Herbasol Extract Soap Wort (Cosmetochem) (Cosmetochem International Ltd.)

Herbasol Extract Soybean (Cosmetochem) (Cosmetochem International Ltd.)

Herbasol-Extract Speedwell (Cosmetochem)

Herbasol Extract Spruce Buds (Cosmetochem) (Cosmetochem International Ltd.)

Herbasol Extract Spruce Needles (Cosmetochem) (Cosmetochem International Ltd.)

Herbasol-Extract Stinging Nettle (Cosmetochem)

Herbasol-Extract St. John's Wort (Cosmetochem)

Herbasol-Extract Strawberry (Fruit) (Cosmetochem)

Herbasol Extract Strawberry Leaves (Cosmetochem) (Cosmetochem International Ltd.)

Herbasol Extract Sunflower (Cosmetochem) (Cosmetochem International Ltd.)

Herbasol Extract Sweet Chestnut (Cosmetochem) (Cosmetochem International Ltd.)

Herbasol Extract Sweet Orange Leaves (Cosmetochem International Ltd.)

Herbasol Extract Sweet Violet (Cosmetochem) (Cosmetochem International Ltd.)

Herbasol Extract Tea Tree Oil (Cosmetochem) (Cosmetochem International Ltd.)

Herbasol-Extract Thyme (Cosmetochem)

Herbasol Extract Tomato (Cosmetochem) (Cosmetochem International Ltd.)

Herbasol-Extract Tormentill (Cosmetochem)

Herbasol Extract Traveller's Joy (Cosmetochem) (Cosmetochem International Ltd.)

Herbasol-Extract Valerian (Cosmetochem)

Herbasol Extract Vanilla (Orchid) (Cosmetochem) (Cosmetochem International Ltd.)

Herbasol-Extract Veronica (Cosmetochem)

Herbasol-Extract Vervain (Cosmetochem)

Herbasol-Extract Vine Leaves (Cosmetochem)

Herbasol Extract Walnut Leaves (Cosmetochem) (Cosmetochem International Ltd.)

Herbasol-Extract Walnut Shells (Cosmetochem)

Herbasol-Extract Watercress (Cosmetochem)

Herbasol-Extract watercress (Nasturtium) (Cosmetochem)

Herbasol Extract Water Melon (Cosmetochem) (Cosmetochem International Ltd.)

Herbasol Extract Water Mint (Cosmetochem) (Cosmetochem International Ltd.)

Herbasol Extract Wheat Bran (Cosmetochem) (Cosmetochem International Ltd.)

Herbasol-Extract Wheat Germ (Cosmetochem)

Herbasol-Extract White/Dead Nettle (Cosmetochem)

Herbasol Extract White Tea (Cosmetochem) (Cosmetochem International Ltd.)

Herbasol-Extract Wild Mint (Cosmetochem)

Herbasol Extract Willow Bark (Cosmetochem) (Cosmetochem International Ltd.)

Herbasol-Extract Willow Leaf (Cosmetochem)

Herbasol Extract Wintergreen (Cosmetochem) (Cosmetochem International Ltd.)

Herbasol-Extract Wirch Itazel (Hamamelis) (Cosmetochem)

Herbasol-Extract Witch Hazel (Cosmetochem)
Herbasol-Extract Woodruff (Cosmetochem)
Herbasol Extract Yam Root (Cosmetochem) (Cosmetochem International Ltd.)
Herbasol-Extract Yarrow (Cosmetochem)
Herbasol Extract Ylang Ylang (Cosmetochem) (Cosmetochem International Ltd.)
Herbasol Extract Yucca (Cosmetochem) (Cosmetochem International Ltd.)
Herbassol Black Cohosh (Cosmetochem) (Cosmetochem International Ltd.)
Herb Robert HS (Alban Muller)
Hibiscus Extract HG (Provital/Centerchem)
Hibiscus Extract HS 2361 G (Grau)
Hibiscus HPG Titrated (Alban Muller)
Hibiscus HS (Alban Muller)
Hibiscus Sabdariffa Flower Extract ies (IES LABO)
Hierochloe Herb Extract HS 2628 G (Grau)
Hip Extract HS 2576 G (Grau)
Holly HS (Alban Muller)
Holly Leaves Extract HS 2656 G (Grau)
Honey Extract "COS" (Cosmetochem) (Cosmetochem International Ltd.)
Honey Extract HG (Provital/Centerchem)
Honey Extract HS 2660 G (Grau)
Honey Extract 2/S00004 (Symrise)
Honey HG (Alban Muller)
Honey HS (Alban Muller)
Honey Melon Extract HS 3301 G (Grau)
Honey Suckle Extract HS 2551 G (Grau)
Hop CL 2/032978 (Symrise)
Hops Extract HG (Provital/Centerchem)
Hops Extract HS 2367 G (Grau)
Hops Extract PG (Rahn)
Hops HPG Titrated (Alban Muller)
Hops HS (Alban Muller)
Hops Malt Extract HS 2518 G (Grau)
Hordeum Vulgare Seed Extract ies (IES LABO)
Horse Chestnut Extract HG (Provital/Centerchem)
Horse Chestnut Extract HS 2704 G spir spiss. (Grau)
Horse Chestnut Extract PG (Rahn)
Horse Chestnut HPG Titrated (Alban Muller)
Horse Chestnut HS (Alban Muller)
Horse Chestnut Nut HPG Titrated (Alban Muller)
Horse Radish Extract HS 2490 G (Grau)
Horsetail Extract HG (Provital/Centerchem)
Horse Tail Extract HS 2374 G (Grau)
Horsetail Extract PG (Rahn)
Horsetail HPG Titrated (Alban Muller)
Horsetail HS (Alban Muller)
House-Leek HS (Alban Muller)
Hyacinth HS (Alban Muller)
Hydralphatine Asiatique (Lanatech)
Hydralphatine 3P (Lanatech)
Hydrangea Extract (Bell)

Hydratherm CGI Glycolic (Universal Flavors)
Hydrocos "P" (Cosmetochem) (Cosmetochem International Ltd.)
Hydrocotyl HPG Titrated/Centella HPG Titrated (Alban Muller)
Hydrocotyl HS/Centella HS (Alban Muller)
Hydroglycolic extract of Ylang (Greentech)
Hydromarine (Sederma)
Hydroxan (Lanatech)
Hydroxan CH (Lanatech)
Hydrumine Arnica (Exsymol)
Hydrumine Calendula (Exsymol)
Hydrumine Centella Asiatica (Exsymol)
Hydrumine Hazelnut Tree (Exsymol)
Hydrumine Hops (Exsymol)
Hydrumine Horsetail (Exsymol)
Hydrumine Liquorice (Exsymol)
Hydrumine Sage (Exsymol)
Hydrumine Witchazel (Exsymol)
Hyssop Extract HS 2848 G (Grau)
Hyssop HS (Alban Muller)
Hyssopus Offinalis Extract ies (IES LABO)
Iceland Moss Extract HS 2818 G (Grau)
Ilex Paraguariensis Leaf Extract ies (IES LABO)
Immune System (Sederma)
Incense HS (Alban Muller)
Incroquat Behenyl BDQ/P (Croda Chemicals)
Incroquat Behenyl TMC/P (Croda Chemicals)
Indian Cress Extract HS 2469 G (Grau)
Indian Cress Extract PG (Rahn)
Indian Cress HS (Alban Muller)
Indian Plantain HPG Titrated (Alban Muller)
Indian Plantain HS (Alban Muller)
Induxin (Laboratoires Serobiologiques)
Irgasan PG 60 (Ciba Specialty Chemicals)
Irish Moss Extract HS 3639 G (Grau)
Irish Moss HS (Alban Muller)
Italian Orange Extract (Kelisema Italy)
Ivy Extract HG (Provital/Centerchem)
Ivy Extract HS 2387 G (Grau)
Ivy Extract PG (Rahn)
Ivy HS (Alban Muller)
Jambul Bark Extract HS 2670 G (Grau)
Jasmine HS (Alban Muller)
Jasmine Phytolait (Alban Muller)
Jasmin Extract HS 2862 G (Grau)
Jaune Covasop W 1770 (LCW)
Jaune Covasop W 1771 (LCW)
Jaune Covasop W 1775 (LCW)
Java Tea Extract HS 3739 G (Grau)
Jeechem 186 (Jeen)
JM ActiCare (Microbial Systems)
Jojoba Extract HS 3496 G (Grau)
Juglans Regia (Walnut) Leaf Extract ies (IES LABO)
Juniper Extract HG (Provital/Centerchem)
Juniper Extract HS 2444 G (Grau)
Juniper HS (Alban Muller)
263 Kalokiros HS (Alban Muller)
Kapillarine (Greentech)
Kapur Kachri (Carlisle)

Kava Extract PG9010 (Concentrated Aloe Corp. (CAC))
Keratoline (Sederma)
Keratoline CL (Sederma)
Keshastim (Bio-Botanica)
Khus (Carlisle)
Kigelia Africana Fruit Extract (Libiol)
Kigeline (Greentech)
Kinkeliba HS (Alban Muller)
Kiwi Extract (Bell Flavors)
Kiwi Extract, Glycolic (Plantextrakt)
Kiwi Extract HG (Provital/Centerchem)
Kiwi Extract HS 2519 G (Grau)
Kiwi HPG Titrated (Alban Muller)
Kiwi HS (Alban Muller)
Kiwi Pulp Extract (Alban Muller)
Kola (Nut) Extract HG (Provital/Centerchem)
Kola Nut HPG Titrated (Alban Muller)
Kola Nut HS (Alban Muller)
Kumquat Extract (Bell Flavors & Fragrances Europe)
Lacteclat (I.D. bio)
Lacteclat Gel (I.D. bio)
Lactuca Scariola Sativa (Lettuce) Extract ies (IES LABO)
Lady's Mantle HS (Alban Muller)
Laminarine (Vevy)
Lanachrys (Lanatech)
Langherine (Sederma)
Lanoquat 1751A (Cognis Care Chemicals/ NJ)
Lanoquat 1751A (Cognis Care Chemicals/ PA)
Lapacho Bark Extract HS 3793 G (Grau)
Lapacho Extract, Glycolic (Plantextrakt)
Lapacho HPG Titrated (Alban Muller)
Laurel Extract HS 2385 G (Grau)
Laurel HS (Alban Muller)
Laurus Nobilis Leaf Extract ies (IES LABO)
Lavender Extract HG (Provital/Centerchem)
Lavender Flowers Extract HS 2565 G (Grau)
Lavender HS (Alban Muller)
Lemon Extract HS 2364 G (Grau)
Lemon Extract PG (Maruzen Pharmaceuticals Co., Ltd.)
Lemon (Fruit) Extract HG (Provital/Centerchem)
Lemongrass Extract HS 2523 G (Grau)
Lemongrass HS (Alban Muller)
Lemon HS (Alban Muller)
Lemon (Peel) Extract HG (Provital/Centerchem)
Lemon Peel HS (Alban Muller)
Lemon Thyme Extract (Carrubba)
Lemon Verbena HS (Alban Muller)
Lentil HS (Alban Muller)
Lettuce Extract HG (Provital/Centerchem)
Lettuce Extract HS 2575 G (Grau)
Lettuce HS (Alban Muller)
Licorice Extract HS 2511 G (Grau)
Licorice HS (Alban Muller)
Lilac Extract (Carrubba)

Lilium Candidum Flower Extract ies (IES LABO)
Lime HS (Alban Muller)
Lime Pulp Extract (Alban Muller)
Lime Tree Extract HG (Provital/Centerchem)
Lime Tree Extract HS 2381 G (Grau)
Linden Extract PG (Rahn)
Linden HS (Alban Muller)
Lipobrite PG (Lipo)
Lipoderma - Shield PG (Lipo)
Lipoid Liposome 0040 (Lipoid)
Lipokel 12 G (Bozzetto)
Lipo PE Base GP-55 (Lipo)
Liposerve DUP (Lipo)
Liposerve MB (Lipo)
Liposome Concentrate (Cosmetochem) (Cosmetochem International Ltd.)
Liposome/Oxylho3 (Sederma)
Liquid Guaranine (Libiol)
Litchee HS (Alban Muller)
Litchi Chinensis Fruit Extract ies (IES LABO)
Litsea Glutinosa Extract (Libiol)
Logwood Extract HS 2711 G (Grau)
Lonicera Caprifolium (Honey-Suckle) Flower Extract ies (IES LABO)
Lotus Corniculatus Extract HS 3130 G (Grau)
Lotus Extract (IES LABO)
Lotus HPG Titrated (Alban Muller)
Lotus HS (Alban Muller)
Lotus Phytolait (Alban Muller)
Lowenol 4558 (Lowenstein)
Lowenol Conditioner 288 (Lowenstein)
Lubrajel CG (Guardian)
Lubrajel DV (Guardian)
Lubrajel MS (Guardian)
Lubrajel Oil (Guardian)
Lubrajel TW (Guardian)
Lubrajel WA (Guardian)
Lubrasil (Guardian)
Luffa Extract 3516 G (Grau)
Lupine Extract HS 3007 G (Grau)
Lupine HS (Alban Muller)
Lychee Extract (Cosmetic Developments)
Lysofat OL (CEP (Solabia))
Lythrum Salicaria Extract ies (IES LABO)
Mackernium CG32 (McIntyre)
Mackernium CG34 (McIntyre)
Madder Extract (Kelisema Italy)
Maerl MP PG 40 (Yves Rocher)
Magilyne (Greentech)
Magnalight 200 (MAFCO)
Magnalys (Greentech)
Magnolia HG (Alban Muller)
Maka/Bhringraj (Carlisle)
Male Fern HG (Alban Muller)
Male Speedwell Wort Extract HS 2728 G (Grau)
Mallow Extract HG (Provital/Centerchem)
Mallow Extract HS 2387G (Grau)
Mallow Extract PG (Rahn)

Mallow HS (Alban Muller)
Mandarin Extract HG (Provital/Centerchem)
Mandarin HS (Alban Muller)
Mandarin HSA (Alban Muller)
Mandragora Extract HS 3131 G (Grau)
Mangifera Indica (Mango) Fruit Extract ies (IES LABO)
Mango Extract (Bio-Botanica)
Mango Extract HG (Provital/Centerchem)
Mango Extract HS 2756 G (Grau)
Mango HPG Titrated (Alban Muller)
Mango HS (Alban Muller)
Mango Milk (Cosmetochem) (Cosmetochem International Ltd.)
Maple Extract (Cosmetic Developments)
Maracuja Extract HS 3349 G (Grau)
Marigold Extract PG (Rahn)
Maritime Pine HS/Sea Pine HS (Alban Muller)
Marjoram Extract HS 2474 G (Grau)
Marjoram HS (Alban Muller)
Marlinat 242/90M (Sasol GmbH - Marl)
Marlinat 242/90T (Sasol Servo)
Marshmallow Extract HG (Provital/Centerchem)
Marsh Mallow HS (Alban Muller)
Marshmallow Root Extract HS 2787 G (Grau)
Mate HS (Alban Muller)
Maypop Extract HS 2828 G (Grau)
Meadowsweet Extract HG (Provital/Centerchem)
Meadow Sweet Extract HS 2473 G (Grau)
Meadow Sweet HPG Titrated (Alban Muller)
Meadow Sweet HS (Alban Muller)
Medicago Sativa (Alfalfa) Extract ies (IES LABO)
Melaleuca Alternifolia (Tea Tree) Extract ies (IES LABO)
Melange Epices HG (Alban Muller)
Melilot Extract HS 2376 G (Grau)
Melilot HS (Alban Muller)
Melilotus Officinalis Extract ies (IES LABO)
Melissa Extract HS 2453 G (Grau)
Melon Extract HG (Provital/Centerchem)
Melon HPG Titrated (Alban Muller)
Melon HS (Alban Muller)
Melon Pulp Extract (Alban Muller)
Melscreen Black (Chemyunion)
Melscreen Gold (Chemyunion)
Melscreen Red (Chemyunion)
Mentha Piperita (Peppermint) Leaf Extract IES (IES LABO)
Meristematic Extract HG (Provital/Centerchem)
Microadsorbed Cucumber HS (Alban Muller)
Microcirculation Factor No. 3 (Indena SA)
Microcirculation Factor No. 5 (Indena SA)
Milfoil HS (Alban Muller)
Milk Thistle HPG Titrated (Alban Muller)
Milk Thistle HS (Alban Muller)
Millet Extract HS 3460 G (Grau)

Mimosa Extract (Plantextrakt)
Mimosa Extract, Glycolic (Plantextrakt)
Mimosa Tenuiflora Extract HS 3386 G (Grau)
Mimosa Tenuiflora Extract-NOVA (Crodarom)
Mimosa Tenuiflora 5% Solution (Libiol)
Mimosoie (Alban Muller)
Mint Extract HG (Provital/Centerchem)
Miracare 2MCAS (Rhodia)
Miracare UM-140 (Rhodia)
Mistletoe Extract HS 2399 G (Grau)
Mistletoe Extract PG (Rahn)
Mistletoe Extract, Water Soluble (Crodarom)
Moisturizing Complex 266 (Ennagram)
Moisturizing Factor Hygro-Complex ARO 5272 (Crodarom)
Moisturizing Phytoamine Biocomplex (Alban Muller)
Moisturizing Phytogreen Complex GXH 266 (Phytochim)
Moor Extract HS 3665 G (Grau)
Morello Cherry Extract HS 2851 G (Grau)
Morning Glory Extract (Bell)
Morus Alba Leaf Extract ies (IES LABO)
Morus Nigra Root Extract ies (IES LABO)
Mother of Thyme Extract HS 3171 G (Grau)
Mountain Ash Fruit Extract HS 2655 G (Grau)
Mugwort HS (Alban Muller)
Muira Puama Extract (Cosmetic Developments)
Muira Puama HS (Alban Muller)
Mulberry Concentrate (CEP (Solabia))
Mullein HS (Alban Muller)
Murumuru Extract 136286 (Fragrance Oils Int. Ltd.)
Murva Extract (Carlisle)
Musa Sapientum (Banana) Fruit Extract ies (IES LABO)
Musk Mallow Extract (Cosmetic Developments)
Mustard Seed Extract HS 2371 G (Grau)
Myrrh Extract 16780 (Crodarom)
Myrrh Extract HA (Provital/Centerchem)
Myrrh Extract HS 2598 G (Grau)
Myrrh HS (Alban Muller)
Myrtle Extract HG (Provital/Centerchem)
Myrtle HS (Alban Muller)
Myrtus Communis Extract ies (IES LABO)
Nagarmotha Extract (Carlisle)
Nail Regenerative Complex (Alban Muller)
Nano-emulsion Concentrate (Active Concepts)
Nasturtium Officinale Extract ies (IES LABO)
Nat Aloe Vera Extract (Natiris)
Nat Arnica Extract (Natiris)
Nat Ginseng Extract (Natiris)
Nat Gotu Kola Extract (Natiris)
Nat Ground Ivy Extract (Natiris)
Nat Horse Chestnut Extract (Natiris)
Nat Horsetail Extract (Natiris)
Nat Matricaria Chamomila Extract (Natiris)

Nat Ruscus Extract (Natiris)
Nat Seaweed Extract (Natiris)
Nat Stinging Nettle Extract (Natiris)
Nat Tormentil Extract (Natiris)
Natuchrom Super Green (Quest International)
Natuchrom Turmeric Yellow (Quest International)
Natural Custard Apple Extract (Manheimer)
Nat Witch Hazel Extract (Natiris)
Nectarin Extract HG (Provital/Centerchem)
Neem Tree Leaf HS (Alban Muller)
Neo Extrapone Chamomile Liquid 2/ 070350 (Symrise)
Neo Extrapone Lemon Liquid 2/070130 (Symrise)
Neo Placenta LS 8407 (Laboratoires Sero-biologiques)
Nettle Wort Extract HS 2455 G (Grau)
Neutral Henna HS (Alban Muller)
Nicotiana Tabacum (Tobacco) Leaf Extract ies (IES LABO)
Nipaguard CMB (Clariant)
Nipaguard CMB (Clariant GmbH, Personal Care)
Nipaguard MPS (Clariant)
Nipaguard MPS (Clariant GmbH, Personal Care)
Nipaguard PDU (Clariant)
Nipaguard PDU (Clariant GmbH, Personal Care)
Nipanox S1 (Clariant)
Nipanox S1 (Clariant GmbH, Personal Care)
NMF/ZI (Variati)
Noir Covasop W 9774 (LCW)
Novamed-Camomile Extract Water Soluble (Crodarom)
Novaplant Arnica Extract (Crodarom)
Novaplant Balm Mint Extract (Crodarom)
Novaplant-Birch Leaves Extract, Water Soluble (Crodarom)
Novaplant-Cornflower-Extract (Crodarom)
Novaplant-Fennel Extract Fluid (Crodarom)
Novaplant Ginkgo Biloba Leaves Extract (Crodarom)
Novaplant Ginseng Extract (Crodarom)
Novaplant Horsechestnut Extract (Crodarom)
Novaplant Juniper Extract (Crodarom)
Novaplant Mallow Extract (Crodarom)
Novaplant Parsley Extract (Crodarom)
Novaplant Pine Needle Extract (Crodarom)
Novaplant Rosemary Extract (Crodarom)
Novaplant Sambucus Extract (Crodarom)
Novaplant Wild Thyme Extract Water Soluble (Crodarom)
Novapur Centella Asiatica PG (Crodarom)
NOVAPUR Da Zao (Crodarom)
Novapur Echinacea-PG (Crodarom)
NOVAPUR Gan Jiang (Crodarom)
NOVAPUR Jixuecao (Crodarom)
NOVAPUR Ren Shen (Crodarom)

NOVAPUR Zi Cao water soluble (Crodarom)
Novarom-Haircomplex Special (Crodarom)
Novozym 809 (Novozymes)
Nut-Extract, Water Soluble (Crodarom)
Nutmeg HS (Alban Muller)
210 Nutriderme HS (Alban Muller)
240 Nutriderme HS (Alban Muller)
243 Nutriderme HS (Alban Muller)
313 Nutriderme HS (Alban Muller)
Nutriplant (Bio-Botanica)
Nymphaeae-Alba-Root Extract HS 3136 G (Grau)
Nympheline (CEP (Solabia))
Oak Bark Extract HS 2716 G (Grau)
Oakmoss Extract HS 2935 G (Grau)
Oat Beta Glucan 1% (Arch Personal Care Products)
Oat Extract HG (Provital/Centerchem)
Oat Extract HS 2675 G (Grau)
Oat HS (Alban Muller)
Oat Milk (Cosmetochem) (Cosmetochem International Ltd.)
Oat Phytolait (Alban Muller)
Oat Protein "COS" (Cosmetochem) (Cosmetochem International Ltd.)
Oat Protein Extract (Bell Flavors)
Oenotherol (Somaig)
OILY SKIN (Greentech S.A)
Oily Skins Complex 264 (Ennagram)
Oily Skins Phytogreen Complex GXH 264 (Phytochim)
Okoume HG (Gaboon (Mahogany) (Alban Muller)
Olea Europaea Olive (Leaf) Extract ies (IES LABO)
Olibanum Extract HS 3718 G (Grau)
Oligoceane (Sederma)
Olive Leaf Extract HG (Provital/Centerchem)
Olive Tree HS (Alban Muller)
Olive Tree Leaf HPG Titrated (Alban Muller)
Onion Extract HS 3332 G (Grau)
Onion HS (Alban Muller)
Optigel BEN - 1255 (Sud-Chemie, United Catalysts)
Optivegetol Cannelle P110 Hydro (Gattefosse s.a.)
Optivegetol Guarana P107 Hydro (Gattefosse s.a.)
Optivegetol The Vert P108 Hydro (Gattefosse s.a.)
Opuntia Extract HG (Provital/Centerchem)
Oral Mucous Protection Complex MU 3776 (Greentech S.A)
Orange Extract HG (Provital/Centerchem)
Orange Extract HS 3331 G (Grau)
Orange Extract PG100 (Maruzen Pharmaceuticals Co., Ltd.)
Orange Flowers Extract HS 2746 G (Grau)
Orange HS (Alban Muller)
Orange Liquid (Ichimaru Pharcos)
Orchid Flower HG (Alban Muller)

Orchid Flower HPG Titrated (Alban Muller)
Orchid Flower HS (Alban Muller)
Orchid Flower HSB (Alban Muller)
Orchid Phalaenopsis Extract HS 3634 G (Grau)
Oregano HS (Alban Muller)
Origan-Extract HS 2465 G (Grau)
Origanum Vulgare Flower Extract ies (IES LABO)
Ormagel SHE (Assessa-Industria)
Ormagel XPX (Assessa-Industria)
Orris Root Extract HS 2479 G (Grau)
Oryza Sativa (Rice) Extract ies (IES LABO)
Oxylastil (Sederma)
Oxynex 2004 (Merck KGaA/EMD Chemicals Inc.)
Pacific Sea Kelp Extract (Bell Flavors)
Paeonia Albiflora Flower Extract ies (IES LABO)
Palmaria Palmata HS (Alban Muller)
Palm Phytolait (Alban Muller)
Palm Sap HPG Titrated (Alban Muller)
Panama Bark Extract (Crodarom)
Panama Bark Extract HS 2383 G (Grau)
Panama Wood HS (Alban Muller)
Panicum Miliaceum (Millet) Seed Extract ies (IES LABO)
Pansy Extract HS 2366 G (Grau)
Pansy HS (Alban Muller)
D-Panthenol 50P (BASF)
Papaver Rhoeas Extract ies (IES LABO)
Papaya Extract (Bio-Botanica)
Papaya Extract HG (Provital/Centerchem)
Papaya Extract HS 2625 G (Grau)
Papaya HS (Alban Muller)
Para Cress Extract HS 3738 G (Grau)
Paragon (McIntyre)
Paragon II (McIntyre)
Paraoxiben (Vevy)
Parietaria Officinalis Extract ies (IES LABO)
Parsley Extract HS 2476 G (Grau)
Parsley HS (Alban Muller)
Passion Flower HS (Alban Muller)
Passion Fruit HG (Alban Muller)
Patchouli HS (Alban Muller)
Pau d'Arco Extract (Carrubba)
Peach Extract HG (Provital/Centerchem)
Peach Extract HS 2791 G (Grau)
Peach Flower HS (Alban Muller)
Pear HS (Alban Muller)
Pearl Powder Extract (Carrubba)
Pellitory of the Wall HS (Alban Muller)
Pellitory Root Extract HS 2601 G (Grau)
Pennyroyal Extract HS 2390 G (Grau)
Pennyroyal HS (Alban Muller)
Peony Root HPG Titrated (Alban Muller)
Peony Root HS (Alban Muller)
Peppermint Extract HS 2365 G (Grau)
Peppermint HS (Alban Muller)
Perlglanzmittel GM 4175 (Zschimmer & Schwarz)
Persicary HS (Alban Muller)
Phaeodactylum HPG Titrated (Alban Muller)
Phlorogine (SECMA)

Phosal 50PG (Phospholipid)
Phycol CC (SECMA)
Phycol CM (SECMA)
Phycol Durvillea Antartica (SECMA)
Phycol EC (SECMA)
Phycol FV (SECMA)
Phycol LD (SECMA)
Phycol Omega Plancton (SECMA)
Phycol PC (SECMA)
Phycol Phytoplancton (SECMA)
Phycol UL (SECMA)
Phycol UP (SECMA)
Phytelene Complex EGX 232 (Indena SA)
Phytelene Complex EGX 243 (Indena SA)
Phytelene Complex EGX 244 (Indena SA)
Phytelene Complex EGX 246 (Indena SA)
Phytelene Complex EGX 247 (Indena SA)
Phytelene Complex EGX 250 (Indena SA)
Phytelene Complex EGX 251 (Indena SA)
Phytelene Complex EGX 252 (Indena SA)
Phytelene Complex EGX 253 (Indena SA)
Phytelene Complex EGX 254 (Indena SA)
Phytelene Complex EGX 255 (Indena SA)
Phytelene of Agrimony EG 416 liquid (Indena SA)
Phytelene of Ail EG 225 Liquid (Indena SA)
Phytelene of Alchemilla EG 208 (Indena SA)
Phytelene of Aloes Colourless EG 543 (Indena SA)
Phytelene of Aloes EG 273 liquid (Indena SA)
Phytelene of Angelica EG 510 Liquid (Indena SA)
Phytelene of Apricot EG 473 liquid (Indena SA)
Phytelene of Arbutus EG 479 liquid (Indena SA)
Phytelene of Arnica EG 001 Liquid (Indena SA)
Phytelene of Asiatic Hydrocotyl EG 356 Liquid (Indena SA)
Phytelene of Averrhoa EG 736 (Indena SA)
Phytelene of Avocado EG 460 Liquid (Indena SA)
Phytelene of Balm Mint EG 456 liquid (Indena SA)
Phytelene of Bardane EG 195 Liquid (Indena SA)
Phytelene of Bearberry EG 563 Liquid (Indena SA)
Phytelene of Bilberry EG 348 liquid (Indena SA)
Phytelene of Birch-Leaves EG 496 liquid (Indena SA)
Phytelene of Bistort Root EG 485 liquid (Indena SA)
Phytelene of Biting Clematis EG 241 liquid (Indena SA)
Phytelene of Bitter Orange EG 090 Liquid (Indena SA)
Phytelene of Black-Currant EG 513 liquid (Indena SA)

Phytelene of Bloodroot EG 531 Liquid (Indena SA)
Phytelene of Bluet EG 002 liquid (Indena SA)
Phytelene of Bramble, Blackberry-Bush EG 482 liquid (Indena SA)
Phytelene of Burdock EG 195 liquid (Indena SA)
Phytelene of Butcher's Broom EG 440 liquid (Indena SA)
Phytelene of Calendula EG 003 liquid (Indena SA)
Phytelene of Capsicum EG 487 liquid (Indena SA)
Phytelene of Capucine EG 005 liquid (Indena SA)
Phytelene of Carrot EG 428 liquid (Indena SA)
Phytelene of Catleya Orchid EG 568 Liquid (Indena SA)
Phytelene of Christophine EG 739 (Indena SA)
Phytelene of Citron EG 240 liquid (Indena SA)
Phytelene of Coltsfoot EG 512 liquid (Indena SA)
Phytelene of Comfrey EG 032 liquid (Indena SA)
Phytelene of Concombre EG 108 liquid (Indena SA)
Phytelene of Corn-Germs EG 492 liquid (Indena SA)
Phytelene of Cresson EG 224 liquid (Indena SA)
Phytelene of Cucumber EG 108 liquid (Indena SA)
Phytelene of Cymbidium Orchid EG 590 Liquid (Indena SA)
Phytelene of Cypress EG 338 liquid (Indena SA)
Phytelene of Devil's Claw EG 522 Liquid (Indena SA)
Phytelene of Elder Tree EG 050 liquid (Indena SA)
Phytelene of Eucalyptus EG 524 liquid (Indena SA)
Phytelene of Evening Primrose EG 564 Liquid (Indena SA)
Phytelene of Fennel EG 471 liquid (Indena SA)
Phytelene of Fenugreek EG 234 liquid (Indena SA)
Phytelene of Florentine Orris EG 089 liquid (Indena SA)
Phytelene of Frangipana EG 761 (Indena SA)
Phytelene of Fucus Vesiculosus EG 006 liquid (Indena SA)
Phytelene of Garlic EG 225 liquid (Indena SA)
Phytelene of Gentian EG 156 liquid (Indena SA)
Phytelene of Geranium EG 480 liquid (Indena SA)

Phytelene of Ginkgo Biloba EG 489 liquid (Indena SA)
Phytelene of Ginseng EG 209 liquid (Indena SA)
Phytelene of Golden Rod EG 474 liquid (Indena SA)
Phytelene of Gourd EG 504 liquid (Indena SA)
Phytelene of Guava EG 717 (Indena SA)
Phytelene of Gum Benzoin EG 500 liquid (Indena SA)
Phytelene of Hamamelis EG 138 liquid (Indena SA)
Phytelene of Hawthorn EG 449 Liquid (Indena SA)
Phytelene of Henna Leaf EG 405 liquid (Indena SA)
Phytelene of Hops EG 136 liquid (Indena SA)
Phytelene of Horse Chestnut EG 042 liquid (Indena SA)
Phytelene of Horsetail EG 199 liquid (Indena SA)
Phytelene of Houblon EG 136 liquid (Indena SA)
Phytelene of Indian Cress EG 005 liquid (Indena SA)
Phytelene of Ipomea Sweet Purple EG 738 (Indena SA)
Phytelene of Ivy EG 008 liquid (Indena SA)
Phytelene of Jojoba EG 541 Liquid (Indena SA)
Phytelene of Juniper EG 388 liquid (Indena SA)
Phytelene of Junquil EG 433 liquid (Indena SA)
Phytelene of Kola EG 576 Liquid (Indena SA)
Phytelene of Laminaria EG 483 Liquid (Indena SA)
Phytelene of Lavender EG 503 Liquid (Indena SA)
Phytelene of Lemon EG 240 liquid (Indena SA)
Phytelene of Lierre Grimpant EG 008 liquid (Indena SA)
Phytelene of Lily EG 162 Liquid (Indena SA)
Phytelene of Limetree EG 023 liquid (Indena SA)
Phytelene of Liquorice EG 527 liquid (Indena SA)
Phytelene of Maize-Stigmas EG 212 liquid (Indena SA)
Phytelene of Mango EG 740 (Indena SA)
Phytelene of Marigold EG 003 liquid (Indena SA)
Phytelene of Marjoram EG 478 Liquid (Indena SA)
Phytelene of Marronnier EG 042 liquid (Indena SA)
Phytelene of Marshmallow EG 039 liquid (Indena SA)
Phytelene of Matricaire EG 004 liquid (Indena SA)

Phytelene of Mauve EG 216 liquid (Indena SA)

Phytelene of Melilot EG 451 liquid (Indena SA)

Phytelene of Milfoil EG 472 liquid (Indena SA)

Phytelene of Millepertuis EG 217 liquid (Indena SA)

Phytelene of Mimosa Tenuiflora EG 363 Liquid (Indena SA)

Phytelene of Mouse-Ear EG 403 Liquid (Indena SA)

Phytelene of Myrrh EG 499 liquid (Indena SA)

Phytelene of Myrtle EG 519 liquid (Indena SA)

Phytelene of Neutral Henna EG 525 Liquid (Indena SA)

Phytelene of Oak EG 014 Liquid (Indena SA)

Phytelene of Orthosiphon EG 211 liquid (Indena SA)

Phytelene of Ortie Blanche EG 157 liquid (Indena SA)

Phytelene of Panama EG 476 liquid (Indena SA)

Phytelene of Papaw EG 402 liquid (Indena SA)

Phytelene of Paraguay Tea EG 390 liquid (Indena SA)

Phytelene of Parietary EG 233 liquid (Indena SA)

Phytelene of Passion Flower EG 389 liquid (Indena SA)

Phytelene of Peach EG 491 liquid (Indena SA)

Phytelene of Pellitory of Spain EG 481 liquid (Indena SA)

Phytelene of Pensee Sauvage EG 230 liquid (Indena SA)

Phytelene of Pepper Mint EG 339 liquid (Indena SA)

Phytelene of Periwinkle EG 045 liquid (Indena SA)

Phytelene of Pine Apple EG 401 liquid (Indena SA)

Phytelene of Pine Tree EG 028 Liquid (Indena SA)

Phytelene of Plantain EG 417 liquid (Indena SA)

Phytelene of Pollen EG 406 liquid (Indena SA)

Phytelene of Prele EG 199 Liquid (Indena SA)

Phytelene of Primula EG 509 liquid (Indena SA)

Phytelene of Queen Meadow EG 213 liquid (Indena SA)

Phytelene of Quince EG 511 liquid (Indena SA)

Phytelene of Redbark EG 331 liquid (Indena SA)

Phytelene of Red Poppy EG 517 Liquid (Indena SA)

Phytelene of Red Vine EG 218 liquid (Indena SA)

Phytelene of Rhatany EG 488 Liquid (Indena SA)

Phytelene of Roman Camomile EG 004 Liquid (Indena SA)

Phytelene of Romarin EG 009 liquid (Indena SA)

Phytelene of Rose EG 518 Liquid (Indena SA)

Phytelene of Rosemary EG 009 liquid (Indena SA)

Phytelene of Sage EG 010 liquid (Indena SA)

Phytelene of Sage EN 101 powder (Indena SA)

Phytelene of Saponaire EG 048 liquid (Indena SA)

Phytelene of Sauge EG 010 Liquid (Indena SA)

Phytelene of Savory EG 502 liquid (Indena SA)

Phytelene of Sea Ware EG 006 liquid (Indena SA)

Phytelene of Soapwort EG 048 liquid (Indena SA)

Phytelene of Soja EG 275 liquid (Indena SA)

Phytelene of Soya Germ EG 275 liquid (Indena SA)

Phytelene of Spirulina EG 718 Liquid (Indena SA)

Phytelene of Stinging Nettle EG 443 liquid (Indena SA)

Phytelene of St.John's Wort EG 217 liquid (Indena SA)

Phytelene of Sureau EG 050 liquid (Indena SA)

Phytelene of Sweet Lime EG 750 (Indena SA)

Phytelene of Tea EG 226 liquid (Indena SA)

Phytelene of Thyme EG 484 liquid (Indena SA)

Phytelene of Tilleul EG 023 liquid (Indena SA)

Phytelene of Tormentil EG 470 liquid (Indena SA)

Phytelene of Ulmaire EG 213 liquid (Indena SA)

Phytelene of Vervain EG 521 liquid (Indena SA)

Phytelene of Walnut EG 462 Liquid (Indena SA)

Phytelene of Watercress EG 224 liquid (Indena SA)

Phytelene of White Nettle EG 157 liquid (Indena SA)

Phytelene of Wild Chamomile EG 004 liquid (Indena SA)

Phytelene of Wild Pansy EG 230 liquid (Indena SA)

Phytelene of Wild Rose EG 494 Liquid (Indena SA)

Phytelene of Wild Thyme EG 469 liquid (Indena SA)

Phytelene of Witch Hazel EG 138 liquid (Indena SA)

Phytelene of Wood Mallow EG 216 liquid (Indena SA)

Phytelene of Yeast EG 407 liquid (Indena SA)

Phytelene (R) of Lily (Indena SA)

Phytelenes of Irish Moss EG 701 Liquid (Indena SA)

Phytelenes of Laminaria EG 749 Liquid (Indena SA)

Phytelenes of Sea Fennel EG 756 Liquid (Indena SA)

Phytelenes of Spirulina EG 718 Liquid (Indena SA)

Phytelenes of Ulva EG 746 Liquid (Indena SA)

Phytobel-Complex G (Crodarom)

Phytoconcentrol Aloe Water Soluble 2/ 070450 (Symrise)

Phytoderm Avena Glycolic (Universal Flavors)

Phytoderm Camomilla Glycolic (Universal Flavors)

Phytoderm Centella Glicolico (Universal Flavors)

Phytoderm-Complex "G" (Cosmetochem)

Phytoderm Crusca Glycolic (Universal Flavors)

Phytoderm Edera Glicolico (Universal Flavors)

Phytoderm Finocchio Glycolic (Universal Flavors)

Phytoderm Ginkgo Biloba Glycolic (Universal Flavors)

Phytoderm Ginseng Glycolic (Universal Flavors)

Phytoderm Glicolico Elicriso (Universal Flavors)

Phytoderm Harpago Phyton Glycolic (Universal Flavors)

Phytoderm Liquirizia Glycolic (Universal Flavors)

Phytoderm Maggiorana Glycolic (Universal Flavors)

Phytoderm Mandorla dolce Glycolic (Universal Flavors)

Phytoderm Mate (Agipal)

Phytoderm Mirra Glicolico (Universal Flavors)

Phytoderm Primula Glycolic (Universal Flavors)

Phytoderm Propolis Glycolic (Universal Flavors)

Phytoderm Tan Complex Glycolic (Universal Flavors)

Phytoderm The Glycolic (Universal Flavors)

Phytoderm UV Complex Glycolic (Universal Flavors)

Phytogreen 55 of Agrimony EXH 669 Liquid (Phytochim)

Phytogreen 55 of Alchemilla EXH 635 Liquid (Phytochim)

Phytogreen 55 of Aloes Colourless EXH 731 Liquid (Phytochim)

Phytogreen 55 of Aloes EXH 652 Liquid (Phytochim)

Phytogreen 55 of Angelica EXH 710 Liquid (Phytochim)

Phytogreen 55 of Apricot EXH 687 Liquid (Phytochim)

Phytogreen 55 of Arnica EXH 605 Liquid (Phytochim)

Phytogreen of Asiatic Hydrocotyl EXH 659 Liquid (Phytochim)

Phytogreen 55 of Avocado EXH 681 Liquid (Phytochim)

Phytogreen 55 of Balm EXH 680 Liquid (Phytochim)

Phytogreen 55 of Bilberry EXH 658 Liquid (Phytochim)

Phytogreen 55 of Birch-Leaves EXH 701 Liquid (Phytochim)

Phytogreen 55 of Bistort Root EXH 694 Liquid (Phytochim)

Phytogreen 55 of Biting Clematis EXH 651 Liquid (Phytochim)

Phytogreen of Bitter Orange EXH 626 Liquid (Phytochim)

Phytogreen 55 of Black-Currant EXH 713 Liquid (Phytochim)

Phytogreen 55 of Bluet EXH 606 Liquid (Phytochim)

Phytogreen 55 of Burdock EXH 633 Liquid (Phytochim)

Phytogreen 55 of Butcher's Broom EXH 676 Liquid (Phytochim)

Phytogreen 55 of Calendula EXH 607 Liquid (Phytochim)

Phytogreen 55 of Capsicum EXH 695 Liquid (Phytochim)

Phytogreen 55 of Carrot EXH 673 Liquid (Phytochim)

Phytogreen 55 of Catleya Orchid EXH 742 Liquid (Phytochim)

Phytogreen 55 of Coltsfoot EXH 712 Liquid (Phytochim)

Phytogreen 55 of Comfrey EXH 617 Liquid (Phytochim)

Phytogreen 55 of Corn-Germs EXH 699 Liquid (Phytochim)

Phytogreen 55 of Cucumber EXH 627 Liquid (Phytochim)

Phytogreen 55 of Cypress EXH 655 Liquid (Phytochim)

Phytogreen 55 of Devil's Claw EXH 721 Liquid (Phytochim)

Phytogreen 55 of Elder Tree EXH 623 Liquid (Phytochim)

Phytogreen 55 of Eucalyptus EXH 722 Liquid (Phytochim)

Phytogreen 55 of Fennel EXH 685 Liquid (Phytochim)

Phytogreen 55 of Fenugreek EXH 649 Liquid (Phytochim)

Phytogreen 55 of Florentine Orris EXH 625 Liquid (Phytochim)

Phytogreen 55 of Fucus Vesiculosus EXH 610 Liquid (Phytochim)

Phytogreen 55 of Garlic EXH 644 Liquid (Phytochim)

Phytogreen 55 of Gentian EXH 630 Liquid (Phytochim)

Phytogreen 55 of Geranium EXH 691 Liquid (Phytochim)

Phytogreen 55 of Ginkgo Biloba EXH 697 Liquid (Phytochim)

Phytogreen 55 of Ginseng EXH 636 Liquid (Phytochim)

Phytogreen 55 of Golden Rod EXH 688 Liquid (Phytochim)

Phytogreen 55 of Gourd EXH 707 Liquid (Phytochim)

Phytogreen 55 of Gum Benzoin EXH 704 Liquid (Phytochim)

Phytogreen 55 of Hawthorn EXH 678 Liquid (Phytochim)

Phytogreen 55 of Henna Leaf EXH 666 Liquid (Phytochim)

Phytogreen 55 of Hops EXH 628 Liquid (Phytochim)

Phytogreen 55 of Horse Chestnut EXH 620 Liquid (Phytochim)

Phytogreen 55 of Horsetail EXH 634 Liquid (Phytochim)

Phytogreen 55 of Indian Cress EXH 609 Liquid (Phytochim)

Phytogreen 55 of Ivy EXH 650 Liquid (Phytochim)

Phytogreen 55 of Jojoba EXH Liquid (Phytochim)

Phytogreen 55 of Juniper EXH 660 Liquid (Phytochim)

Phytogreen 55 of Junquil EXH 674 Liquid (Phytochim)

Phytogreen 55 of Kola EXH 745 Liquid (Phytochim)

Phytogreen 55 of Laminaria EXH 692 Liquid (Phytochim)

Phytogreen 55 of Lavender EXH 706 Liquid (Phytochim)

Phytogreen 55 of Lemon EXH 650 Liquid (Phytochim)

Phytogreen 55 of Lily EXH 632 Liquid (Phytochim)

Phytogreen 55 of Limetree EXH 615 Liquid (Phytochim)

Phytogreen 55 of Linden EXH 615 Liquid (Phytochim)

Phytogreen 55 of Liquorice EXH 725 Liquid (Phytochim)

Phytogreen 55 of Maize-Stigmas EXH 638 Liquid (Phytochim)

Phytogreen 55 of Mallow EXH 640 Liquid (Phytochim)

Phytogreen 55 of Marigold EXH 607 Liquid (Phytochim)

Phytogreen 55 of Marjoram EXH 690 Liquid (Phytochim)

Phytogreen 55 of Marshmallow EXH 619 Liquid (Phytochim)

Phytogreen 55 of Matricaria EXH 608 Liquid (Phytochim)

Phytogreen 55 of Melilot EXH 679 Liquid (Phytochim)

Phytogreen 55 of Molfoil EXH 686 Liquid (Phytochim)

Phytogreen 55 of Mouse-Ear EXH 665 Liquid (Phytochim)

Phytogreen 55 of Myrrh EXH 703 Liquid (Phytochim)

Phytogreen 55 of Myrtle EXH 718 Liquid (Phytochim)

Phytogreen 55 of Neutral Henna EXH 723 Liquid (Phytochim)

Phytogreen 55 of Oak EXH 614 Liquid (Phytochim)

Phytogreen 55 of Orthosiphon EXH 637 Liquid (Phytochim)

Phytogreen 55 of Panama EXH 689 Liquid (Phytochim)

Phytogreen 55 of Papaw EXH 664 Liquid (Phytochim)

Phytogreen 55 of Paraguay EXH 662 Liquid (Phytochim)

Phytogreen 55 of Parietary EXH 648 Liquid (Phytochim)

Phytogreen 55 of Passion Flower EXH 661 Liquid (Phytochim)

Phytogreen 55 of Peach EXH 698 Liquid (Phytochim)

Phytogreen 55 of Pepper Mint EXH 656 Liquid (Phytochim)

Phytogreen 55 of Periwinkle EXH 621 Liquid (Phytochim)

Phytogreen 55 of Pine Apple EXH 663 Liquid (Phytochim)

Phytogreen 55 of Pine Tree EXH 616 Liquid (Phytochim)

Phytogreen 55 of Plantain EXH 670 Liquid (Phytochim)

Phytogreen 55 of Pollen EXH 667 Liquid (Phytochim)

Phytogreen 55 of Primula EXH 709 Liquid (Phytochim)

Phytogreen 55 of Queen Meadow EXH 639 Liquid (Phytochim)

Phytogreen 55 of Quince EXH 711 Liquid (Phytochim)

Phytogreen 55 of Redbark EXH 654 Liquid (Phytochim)

Phytogreen 55 of Red Poppy EXH 716 Liquid (Phytochim)

Phytogreen 55 of Rhatany EXH 696 Liquid (Phytochim)

Phytogreen 55 of Roman Camomile EXH 624 Liquid (Phytochim)

Phytogreen 55 of Rose EXH 717 Liquid (Phytochim)

Phytogreen 55 of Rosemary EXH 612 Liquid (Phytochim)

Phytogreen 55 of Sage EXH 613 Liquid (Phytochim)

Phytogreen 55 of Saponaire EXH 622 Liquid (Phytochim)

Phytogreen 55 of Savory EXH 705 Liquid (Phytochim)

Phytogreen 55 of Sea Ware EXH 610 Liquid (Phytochim)

Phytogreen 55 of Smaller Nettle EXH 677 Liquid (Phytochim)

Phytogreen 55 of Soapwort EXH 622 Liquid (Phytochim)

Phytogreen 55 of Soja EXH 653 Liquid (Phytochim)

Phytogreen 55 of Soya Germ EXH 275 Liquid (Phytochim)

Phytogreen 55 of Spirulina EXH 747 Liquid (Phytochim)

Phytogreen 55 of St. John's Wort EXH 641 Liquid (Phytochim)

Phytogreen 55 of Sureau EXH 623 Liquid (Phytochim)

Phytogreen 55 of Tea EXH 645 Liquid (Phytochim)

Phytogreen 55 of Thyme EXH 693 Liquid (Phytochim)

Phytogreen 55 of Tormentil EXH 684 Liquid (Phytochim)

Phytogreen 55 of Ulmarie EXH 639 Liquid (Phytochim)

Phytogreen 55 of Vervain EXH 720 Liquid (Phytochim)

Phytogreen 55 of Vigne Rouge EXH 642 Liquid (Phytochim)

Phytogreen 55 of Walnut EXH 682 Liquid (Phytochim)

Phytogreen 55 of Watercress EXH 643 Liquid (Phytochim)

Phytogreen 55 of White Nettle EXH 631 Liquid (Phytochim)

Phytogreen 55 of Wild Chamomile EXH 624 Liquid (Phytochim)

Phytogreen 55 of Wild Pansy EXH 647 Liquid (Phytochim)

Phytogreen 55 of Witch Hazel EXH 629 Liquid (Phytochim)

Phytogreen 55 of Wood Mallow EXH 640 Liquid (Phytochim)

Phytogreen 55 of Yeast EXH 668 Liquid (Phytochim)

Phytolight (Coletica SA)

Phytoselect Aloe (Indena SpA)

Phytotonine (Greentech S.A)

Pilinhib Veg (Laboratoires Serobiologiques)

Pilocarpus Pennatifolius Leaf Extract ies (IES LABO)

Pimento Extract HS 2388 G (Grau)

Pineapple Extract HG (Provital/Centerchem)

Pine Apple Extract HS 1809 AT (Grau)

Pine Extract HG (Provital/Centerchem)

Pine Extract PG (Rahn)

Pine Needles Extract HS 1779 AT (Grau)

Piper Methysticum Extract (IES LABO)

Pitanga folha extrato glicolico NTR (Centroflora Group)

Plancton Extract HG (Provital/Centerchem)

Plantago Ovata HS 2669 G (Grau)

Plantain Extract PG (Rahn)

Plantain HS (Alban Muller)

Plum Tree HS (Alban Muller)

Pod Pepper HS (Alban Muller)

Pollen Extract HG (Provital/Centerchem)

Pollen Extract HS 2586 G (Grau)

Pollen Extract HS 3383 G (Grau)

Pollen Extract ies (IES LABO)

Poly-Extract for Hair NOVA (Crodarom)

Polygonum Bistorta Root Extract ies (IES LABO)

Polylipid PPI-RC (Alzo)

Polymox PPI-SA-15 (Alzo)

Polyplant Anti-Acne (Provital/Centerchem)

Polyplant Anti-Cellulite (Provital/Centerchem)

Polyplant Anti-Couperosis (Provital/Centerchem)

Polyplant Anti-Inflammation (Provital/Centerchem)

Polyplant Anti-Seborrhoea (Provital/Centerchem)

Polyplant Anti-Wrinkles (Provital/Centerchem)

Polyplant Astringent (Provital/Centerchem)

Polyplant Base (Provital/Centerchem)

Polyplant Base S.E. (Provital/Centerchem)

Polyplant Descongestant (Provital/Centerchem)

Polyplant Desensitizer (Provital/Centerchem)

Polyplant Emollient (Provital/Centerchem)

Polyplant Epithelizing (Provital/Centerchem)

Polyplant Hair (Provital/Centerchem)

Polyplant Intimate Hygiene (Provital/Centerchem)

Polyplant Moisturizing (Provital/Centerchem)

Polyplant Oily Skin (Provital/Centerchem)

Polyplant Refirming (Provital/Centerchem)

Polyplant Sedative (Provital/Centerchem)

Polyplant Skin Purifying (Provital/Centerchem)

Polyplant Slimming (Provital/Centerchem)

Polyplant 5 Special (Provital/Centerchem)

Polyplant Stimulant (Provital/Centerchem)

Polyplant Suntan (Provital/Centerchem)

Polyplant Vasoconstrictor (Provital/Centerchem)

Polyquat PPI-SA-15 (Alzo)

Pomegranate HS (Alban Muller)

Poplar Buds Extract HS 2641 G (Grau)

Populus Nigra Extract ies (IES LABO)

Porphyra HS (Alban Muller)

Porphyridium HS (Alban Muller)

Post-Depilatory (Greentech S.A)

Potato Extract HS 3327 G (Grau)

Potato HS (Alban Muller)

Potentilla Anserina Extract ies (IES LABO)

Potentilla Erecta Root Extract ies (IES LABO)

Potimarron HS (Alban Muller)

Pot Marigold HS (Alban Muller)

Primrose Extract HG (Provital/Centerchem)

Primrose Flower HS (Alban Muller)

Primula Vulgaris Extract ies (IES LABO)

Prodhy Extract Alchemille (Prod'Hyg)

Prodhy Extract Aloes (Prod'Hyg)

Prodhy Extract Arnica (Prod'Hyg)

Prodhy Extract Aubepine (Prod'Hyg)

Prodhy Extract Avocat (Prod'Hyg)

Prodhy Extract Bletilla (Prod'Hyg)

Prodhy Extract Bleuet (Prod'Hyg)

Prodhy Extract Bouleau (Prod'Hyg)

Prodhy Extract Bruyere (Prod'Hyg)

Prodhy Extract Busserole (Prod'Hyg)

Prodhy Extract Calendula (Prod'Hyg)

Prodhy Extract Camomille (Prod'Hyg)

Prodhy Extract Capucine (Prod'Hyg)

Prodhy Extract Citron (Prod'Hyg)

Prodhy Extract Clematite (Prod'Hyg)

Prodhy Extract Concombre (Prod'Hyg)

Prodhy Extract Consoude (Prod'Hyg)

Prodhy Extract Coquelicot (Prod'Hyg)

Prodhy Extract Cresson (Prod'Hyg)

Prodhy Extract Cypres (Prod'Hyg)

Prodhy Extract Eglantier (Prod'Hyg)

Prodhy Extract Eucalyptus (Prod'Hyg)

Prodhy Extract Fenouil (Prod'Hyg)

Prodhy Extract Fenugrec (Prod'Hyg)

Prodhy Extract Fucus (Prod'Hyg)

Prodhy Extract Genievre (Prod'Hyg)

Prodhy Extract Gentiane (Prod'Hyg)

Prodhy Extract Geranium (Prod'Hyg)

Prodhy Extract Ginkgo Biloba (Prod'Hyg)

Prodhy Extract Ginseng (Prod'Hyg)

Prodhy Extract Gui (Prod'Hyg)

Prodhy Extract Guimauve (Prod'Hyg)

Prodhy Extract Hamamelis (Prod'Hyg)

Prodhy Extract Harpagophytum (Prod'Hyg)

Prodhy Extract Hibiscus (Prod'Hyg)

Prodhy Extract Houblon (Prod'Hyg)

Prodhy Extract Hydrocotyle Asiatique (Prod'Hyg)

Prodhy Extract Iris (Prod'Hyg)

Prodhy Extract Jojoba (Prod'Hyg)

Prodhy Extract Laminaire (Prod'Hyg)

Prodhy Extract Lavande (Prod'Hyg)

Prodhy Extract Lierre (Prod'Hyg)

Prodhy Extract LYS (Prod'Hyg)

Prodhy Extract Marron D'Inde (Prod'Hyg)

Prodhy Extract Matricaire (Prod'Hyg)

Prodhy Extract Mauve (Prod'Hyg)

Prodhy Extract Melilot (Prod'Hyg)

Prodhy Extract Melisse (Prod'Hyg)

Prodhy Extract Menthe (Prod'Hyg)

Prodhy Extract Millefeuille (Prod'Hyg)

Prodhy Extract Millepertuis (Prod'Hyg)

Prodhy Extract Murier Blanc (Prod'Hyg)

Prodhy Extract Ortie Blanche (Prod'Hyg)

Prodhy Extract Ortie Piquante (Prod'Hyg)

Prodhy Extract Pamplemousse (Prod'Hyg)

Prodhy Extract Panama (Prod'Hyg)

Prodhy Extract Papaye (Prod'Hyg)

Prodhy Extract Parietaire (Prod'Hyg)

Prodhy Extract Passiflore (Prod'Hyg)
Prodhy Extract Pensee Sauvage (Prod'Hyg)
Prodhy Extract Petit Houx (Prod'Hyg)
Prodhy Extract PIN (Prod'Hyg)
Prodhy Extract Prele (Prod'Hyg)
Prodhy Extract Ratanhia (Prod'Hyg)
Prodhy Extract Reglisse (Prod'Hyg)
Prodhy Extract Romarin (Prod'Hyg)
Prodhy Extract Rose (Prod'Hyg)
Prodhy Extract Saponaire (Prod'Hyg)
Prodhy Extract Suage (Prod'Hyg)
Prodhy Extract Sureau (Prod'Hyg)
Prodhy Extract The (Prod'Hyg)
Prodhy Extract Thym (Prod'Hyg)
Prodhy Extract Tilleul (Prod'Hyg)
Prodhy Extract Tormentille (Prod'Hyg)
Prodhy Extract Ulmaire (Prod'Hyg)
Prodhy Extract Verveine (Prod'Hyg)
Prodhy Extract Vigne Rouge (Prod'Hyg)
Promarine (Sederma)
Pronalen A/C HSC (Provital/Centerchem)
Pronalen Aesculus HSC (Provital/Centerchem)
Pronalen Anti-Cellulite HSC (Provital/Centerchem)
Pronalen Anti-Fatigue HSC (Provital/Centerchem)
Pronalen Bilberry HSC (Provital/Centerchem)
Pronalen Ginseng HSC (Provital/Centerchem)
Pronalen Guarana (Provital/Centerchem)
Pronalen Licorice HSC (Provital/Centerchem)
Pronalen Moisturizing HSC (Provital/Centerchem)
PRONALEN MOISTURIZING-II (Provital/Centerchem)
Pronalen Refirming HSC (Provital/Centerchem)
Propoli Extract G 5 (Esperis)
Propolis Extract ies (IES LABO)
Protaquat 70 (Protameen)
PROTECTIVE MOISTURIZER (Greentech S.A)
Protelin L (Greentech)
Proto-Lan 20 (Maybrook)
Proto-Lan 30 (Maybrook)
Prunus Armeniaca (Apricot) Fruit Extract ies (IES LABO)
Prunus Domestica Fruit Extract ies (IES LABO)
Psidium Guajava Fruit Extract ies (IES LABO)
Pueraria Mirifica Extract (Tropical Herbal)
Pumpkin Carota Extract HS 3627 G (Grau)
Punica Granatum Extract ies (IES LABO)
Purisoft LS 9272 B (Laboratoires Sero-biologiques)
PW Covasop (LCW)
Pyrus Malus (Apple) Fruit Extract ies (IES LABO)
Quercus Robur Bark Extract ies (IES LABO)

Quillaja Extract (Libiol)
Quillaja Extract HG (Provital/Centerchem)
Quillaja Saponaria Bark Extract ies (IES LABO)
Quince HS (Alban Muller)
Quinoa Extract HG (Provital/Centerchem)
Quitchroot Extract HS 2512 G (Grau)
Radicaptol (Solabia)
Radish HS (Alban Muller)
Radix Rusci Extract HS 3027 G (Grau)
Raphanus Sativus (Radish) Root Extract ies (IES LABO)
Raspberry Extract HG (Provital/Centerchem)
Raspberry Extract HS 2574 G (Grau)
Raspberry Fruit HPG Titrated (Alban Muller)
Raspberry HS (Alban Muller)
Raspberry Leaf HPG Titrated (Alban Muller)
Raspberry Leaf HS (Alban Muller)
Rauwolfia Extract HS 2400 G (Grau)
RAYOLYS (Greentech S.A)
Reasun (CEP (Solabia))
Red Bark HS (Alban Muller)
Red Beet Extract HS 3405 G (Grau)
RED-BROWN HAIR (Greentech S.A)
Red Clover HS (Alban Muller)
Red Currant HS (Alban Muller)
Red Pepper Extract HS 2738 G (Grau)
Red Pepper HS (Alban Muller)
Red Poppy Extract HS 2472 G (Grau)
Red Rose Petals Extract HS 2531 G (Grau)
Red Sandal Wood Extract HS 2510 G (Grau)
Red Vine Leaf Extract PG (Rahn)
Red Vine Leaf HPG Titrated (Alban Muller)
Red Vine Leaves Extract HS 2578 G (Grau)
236 Regederme HS (Alban Muller)
330 Regederme HS (Alban Muller)
Regenerative Phytoamine Biocomplex (Alban Muller)
278 Relaxant HS (Alban Muller)
Relax Complex RPJ (CEP (Solabia))
Relaxing Extract (CEP (Solabia))
Remoduline (Silab)
Rhatany Extract (Kelisema Italy)
Rhatany Extract HG (Provital/Centerchem)
Rhatany Root Extract HS 2613 G (Grau)
Rheumatism Complex 270 (Ennagram)
Rheutmatism Phytogreen Complex GXH 270 (Phytochim)
Rhizodermin (Laboratoires Serobiologiques)
Rhubarb Root Extract HS 2804 G (Grau)
328 Rhumacalm HS (Alban Muller)
Ribes Nigrum (Black Curant) Fruit Extract (IES LABO)
Ribwort Extract HS 2441 G (Grau)
Rice Extract "COS" (Cosmetochem) (Cosmetochem International Ltd.)
Rice HS (Alban Muller)
Rice Phytolait (Alban Muller)

Robinia Pseudacacia Flower Extract ies (IES LABO)
Rocket HS (Alban Muller)
Rocou HPG Titrated (Alban Muller)
Roman Chamomile HPG Titrated (Alban Muller)
Roman Chamomile HS (Alban Muller)
RonaCare Olaflur (Merck KGaA)
ROSACEA (Greentech S.A)
Rosebay Leaves Extract HS 2715 G (Grau)
Rose Extract HG (Provital/Centerchem)
Rose Mallow Extract (Cosmetic Developments)
Rose-Mallow HPG Titrated (Alban Muller)
Rose-Mallow HS (Alban Muller)
Rosemary Extract HG (Provital/Centerchem)
Rosemary Extract HS 2379 G (Grau)
Rosemary Extract PG (Rahn)
Rosemary HPG Titrated (Alban Muller)
Rosemary HS (Alban Muller)
Rosewood Extract (Cosmetic Developments)
Rosewood HS (Alban Muller)
Rouge Covasop W 3774 (LCW)
Rubus Idaeus (Raspberry) Fruit Extract ies (IES LABO)
Rupture Wort HS (Alban Muller)
Rye-Extract HS 3461 G (Grau)
Sabowax GT (Sabo)
Saccharum Officinarum (Sugar Canne) (IES LABO)
Sage Extract HG (Provital/Centerchem)
Sage Extract HS 2428 G (Grau)
Sage Extract PG (Rahn)
Sage HPG Titrated (Alban Muller)
Sage HS (Alban Muller)
Saint John's Wort Extract PG (Rahn)
Salix Alba (Willow) Bark Extract ies (IES LABO)
Sambucus Nigra Fruit Extract ies (IES LABO)
Samphire Extract (Cosmetic Developments)
Sandalwood Extract (Kelisema Italy)
Sanguinaria Root Extract HS 2778 G (Grau)
Sanicle-Wort-Extract HS 3139 G (Grau)
Sarsaparilla Extract HS 3334 G (Grau)
Sarsaparilla HS (Alban Muller)
Satureia Hortensis Extract ies (IES LABO)
Savory HS (Alban Muller)
Saw Palmetto Berries Extract HS 2792 G (Grau)
Saw Palmetto HPG Titrated (Alban Muller)
Saw Palmetto HS (Alban Muller)
Scabiosa Herb Extract HS 2680 G (Grau)
Scotch Pine Extract HS 2643 G (Grau)
Scrophularia Nodosa Extract ies (IES LABO)
Scurvy Grass Extract HS 2952 G (Grau)
Sea Cabbage HS (Alban Muller)
Sea Extract AO-Nori (E.U.K)
Sea Extract Ascophyllum (E.U.K)
Sea Extract Chrondus (E.U.K)

Sea Extract Dulse (E.U.K)
Sea Extract Fucus (E.U.K)
Sea Extract Himanthalia (E.U.K)
Sea Extract Laminaria (E.U.K)
Sea Extract Lithothamnium (E.U.K)
Sea Extract Nori (E.U.K)
Sea Extract Samphire (E.U.K)
Sea Extract Ulva (E.U.K)
Sea Extract Wakame (E.U.K)
Sea Fennel Extract (CFEB Sisley)
Sea Holly Extract (Cosmetic Developments)
Sea Lavender Extract (Cosmetic Developments)
Sea Lettuce HS (Alban Muller)
Sea Rocket Extract (CFEB Sisley)
Seaweedex (Assessa-Industria)
Sea-Weed Extract HS 2449 G (Grau)
Sea-Weed Extract HS 2796 G (Grau)
Seaweed Extract NOVA (Crodarom)
Sebaryl FL (Laboratoires Serobiologiques)
Sederma Balm Mint (Sederma)
Sederma Birch Leaf (Sederma)
Sederma Burdock Root (Sederma)
Sederma Butcherbroom (Sederma)
Sederma Chamomile (Sederma)
Sederma Comfrey (Sederma)
Sederma Cucumber (Sederma)
Sederma Gentian (Sederma)
Sederma Hops (Sederma)
Sederma Horse Chestnut (Sederma)
Sederma Horsetail Extract (Sederma)
Sederma Indian Cress (Sederma)
Sederma Ivy (Sederma)
Sederma Mallow (Sederma)
Sederma Marigold (Sederma)
Sederma Plantain (Sederma)
Sederma Rosemary (Sederma)
Sederma Sage (Sederma)
Sederma Stinging Nettle Extract (Sederma)
Sederma St. John's Wort Extract (Sederma)
Sederma Thyme (Sederma)
Sederma Walnut (Sederma)
Sederma Watercress (Sederma)
Sederma Wild Pansy Extract (Sederma)
Sederma Witch Hazel (Sederma)
Sederma Yarrow (Sederma)
Self Heal Extract (Cosmetic Developments)
Senna HPG Titrated (Alban Muller)
Senna HS (Alban Muller)
Senna Leaves Extract HS 2515 G (Grau)
Senn Extract HG (Provital/Centerchem)
Sepicide WP1 (SEPPIC)
Seromarine (Sederma)
SEVAFOTSY (Christian Dior)
Shepherd's Burse Herb Extract HS 3354 G (Grau)
Shepherd's Purse HS (Alban Muller)
Shiitake HG (Alban Muller)
Shikakai Extract (Carlisle)
Silk Tree Extract (Carrubba)
Silver Birch Bark HPG Titrated (Alban Muller)
Silver Birch Bark HS (Alban Muller)

Silver Birch Leaf HS (Alban Muller)
Silver Weed HS (Alban Muller)
Sinominceur (I.D. bio)
Sinopurete (I.D. bio)
Six Fruit Concentrate 3519 G (Grau)
Skin Acne Complex MU 3690 (Greentech S.A)
Skin Tree Extract PG (Rahn)
Slimfit LS 9509 (Laboratoires Serobiologiques)
Slimmigen (Laboratoires Serobiologiques)
SLIMMING (Greentech S.A)
Slimming Complex 265 (Ennagram)
Slimming Factor "T" (Cosmetochem) (Cosmetochem International Ltd.)
Slimming Factor Karkade (Cosmetochem) (Cosmetochem International Ltd.)
Slimming Phytogreen Complex GXH 265 (Phytochim)
Snake Weed Extract HS 2890 G (Grau)
Soap Nut Extract HS 2754 G (Grau)
Soap Wort Extract HS 2577 G (Grau)
Soapwort HS (Alban Muller)
Softening Complex 262 (Ennagram)
Softening Complex MU 3658 (Greentech S.A)
Softening Phytogreen Complex GXH 262 (Phytochim)
269 Solarium HS (Alban Muller)
270 Solarium HS (Alban Muller)
Solomons Seal Extract HS 2432 G (Grau)
Solomon's Seal HG (Alban Muller)
Solomon's Seal HS (Alban Muller)
Soothing Complex MU 3682 (Greentech S.A)
Sophora Japonica Extract HS 2665 G (Grau)
Sophora Japonica Flower Extract ies (IES LABO)
Sorbus Aucuparia Fruit Extract ies (IES LABO)
Sorrel Extract HS 2582 G (Grau)
Soy Extract HS 2478 G (Grau)
Soy Milk Protein Extract (Carrubba)
SPD (SECMA)
4-Spices Milk (Cosmetochem) (Cosmetochem International Ltd.)
Spike Lavender Extract (Cosmetic Developments)
Spinach Extract HS 3538 G (Grau)
Spinach HS (Alban Muller)
Spirulina Extract (CEP (Solabia))
Spirulina Extract HS 3511 G (Grau)
Spirulina HPG Titrated (Alban Muller)
Spirulina Maxima Extract ies (IES LABO)
Stabolec C (Lucas Meyer GmbH)
Standamid LD 80/20 (Cognis Care Chemicals/PA)
Standapol Pearl Conc. 7130 (Cognis Care Chemicals/NJ)
Standapol Pearl Conc. 7130 (Cognis Care Chemicals/PA)
280 Stimulant HS (Alban Muller)
285 Stimulant HS (Alban Muller)

Stinging Nettle Extract PG (Rahn)
Stinging Nettle HS (Alban Muller)
St.John's Wort Extract HG (Provital/Centerchem)
St. John's Wort Extract HS 2304 G (Grau)
St. John's Wort HPG Titrated (Alban Muller)
St. John's Wort HS (Alban Muller)
St. Mary's Thistle Extract HS 2693 G (Grau)
St. Mary's Thistle Extract HS 3687 G (Grau)
Strawberry Extract HG (Provital/Centerchem)
Strawberry Extract HS 2573 G (Grau)
Strawberry HS (Alban Muller)
Strawflower HS (Alban Muller)
Styrax Benzoin Resin Extract ies (IES LABO)
Succory Leaves Extract HS 2912 G (Grau)
Succory Leaves Extract HS 2953 G (Grau)
Sugar Cane HS (Alban Muller)
Sulfoconcentrol (2/380011) (Symrise)
Sundew Herb Extract HS 3688 G (Grau)
Sunflower Extract HG (Provital/Centerchem)
Sunflower Extract HS 2362 G (Grau)
Sunflower HS (Alban Muller)
SUN PROTECTION (Greentech S.A)
Suntan Bioactivator AMI (Alban Muller)
Super Extrapone White Nettle 2/500016 (Symrise)
Surfactol Q1 (CasChem)
Sweet Almond HS (Alban Muller)
Sweet Almonds Extract HS 2740 G (Grau)
Sweet Briar Extract (Cosmetic Developments)
Sweet Cherry Extract HS 3330 G (Grau)
Sweet Cicely Extract (Cosmetic Developments)
Sweet Cicely Extract HS 2431 G (Grau)
Sweet Dumpling HS / Patidou HS (Alban Muller)
Sweet Gale Extract (Cosmetic Developments)
Sweet Marjoram Extract HG (Provital/Centerchem)
Sweet Orange Peel Extract HS 2524 G (Grau)
Sweet Orange Peel HPG Titrated (Alban Muller)
Sweet Orange Peel HS (Alban Muller)
Sweet Orange Pulp Extract (Alban Muller)
Symphytum Officinale Extract ies (IES LABO)
Syntran 5190 (Interpolymer)
Syntran KL-219C (Interpolymer)
Tagetes HS (Alban Muller)
Taraxacum Officinale (Dandelion) Extract ies (IES LABO)
Taraxacum Root Extract HS 3020 G (Grau)
Taro HS (Alban Muller)
Tarragon HS (Alban Muller)
TEA Extract HG (Provital/Centerchem)
Tea Tree Extract (Carrubba)

Technobion BL (Laboratoires Sero-
biologiques)
Tech-0 #6-040L (Beacon)
Tech-0 #6-080L Oat Extract (Beacon)
Teen Age Skin Disorders Complex 267
(Ennagram)
Tegodeo CW 90 (Degussa Care Special-
ties)
Tegodeo LYS (Degussa Care Specialties)
Tenox 2 (Eastman Chemical)
Tenox 6 (Eastman Chemical)
Tenox 7 (Eastman Chemical)
Tenox 20 (Eastman Chemical)
Tenox 21 (Eastman Chemical)
Tenox 22 (Eastman Chemical)
Tenox 25 (Eastman Chemical)
Tenox A (Eastman Chemical)
Tenox 20A (Eastman Chemical)
Tenox R (Eastman Chemical)
Tenox S-1 (Eastman Chemical)
Tepescohuite Extract HG (Provital/
Centerchem)
Tepescohuite HPG Titrated (Alban Muller)
Tetterwort Extract HS 3619 G (Grau)
THIN AND DULL HAIR (Greentech S.A)
Thistle Extract HS 2481 G (Grau)
Thrift Extract (Cosmetic Developments)
Thuja Leaves Extract HS 3384 G (Grau)
Thyme Extract HG (Provital/Centerchem)
Thyme Extract HS 2442 G (Grau)
Thyme Extract PG (Rahn)
Thyme HG (Alban Muller)
Thyme HS (Alban Muller)
Thyme Liquid P (Ichimaru Pharcos)
Thymus Serpillum Extract ies (IES LABO)
Tiger Lily Extract (Cosmetic Developments)
Tilicine (Greentech)
TIRED LEGS (Greentech S.A)
Tomato HS (Alban Muller)
Tonic Antiseptic Complex MU 3660
(Greentech S.A)
290 Tonique HS (Alban Muller)
Toothpaste Complex MU 3319 (Greentech
S.A)
Tormentil Extract HG (Provital/
Centerchem)
Tormentil HS (Alban Muller)
Tormentill Root Extract HS 2775 G (Grau)
Toshiki BINS-2 (Nikko)
Toshiki BINS-3 (Nikko)
Transomes Anti-Age EGX 246-TR (Indena
SA)
Transomes of Factor of MIcro-circulation
No 3 (Indena SA)
Transomes of Factor of Micro-circulation
No 5 (Indena SA)
Traveller's Joy HS (Alban Muller)
Triticum Vulgare (wheat) Germ Extract ies
(IES LABO)
Tropaeolum Majus Extract ies (IES LABO)
Tropical Resins HS (Alban Muller)
Tulsi Extract (Carlisle)
Tussilago Farfara (Coltsfoot) Flower
Extract ies (IES LABO)

Uantox 3 (Universal Preserv-A-Chem)
Uantox 20 (Universal Preserv-A-Chem)
Uantox W-1 (Universal Preserv-A-Chem)
Uikyou Liquid PO (Ichimaru Pharcos)
Ultraspheres 8028 (Lipoid)
Ulva Lactuca ME PG 40 (Yves Rocher)
Umordant P (Cosmetochem)
(Cosmetochem International Ltd.)
Uncaria HS (Alban Muller)
Unicontrozon C-49 (Induchem)
Uninontan U-34 (Induchem)
Urchin HS (Alban Muller)
Usnidin - N (Crodarom)
Vaccinium Myrtillus Extract ies (IES LABO)
Valerian Extract HG (Provital/Centerchem)
Valerian Extract HS 2396 G (Grau)
Valerian HS (Alban Muller)
Vanilla Beans Extract HS 3141 G (Grau)
Vanilla HS (Alban Muller)
Vanirea (CEP (Solabia))
Vegebios of Cucumber (CEP (Solabia))
Vegebios of Kiwi (CEP (Solabia))
Vegebios of Lemon (CEP (Solabia))
Vegebios of Melon (CEP (Solabia))
Vegebios of Orange (CEP (Solabia))
Vegebios of Pineapple (CEP (Solabia))
Vegecel 13 (Variati)
Vegeles Phytofiltre AD LS (Laboratoires
Serobiologiques)
Vegeles Phytofiltre AD LS (Laboratoires
Serobiologiques)
Vegeles SR (Laboratoires Serobiologiques)
Vegeles WP (Laboratoires
Serobiologiques)
Vegetol Agrimony GR 437 Hydro
(Gattefosse s.a.)
Vegetol Aloe GR 198 Hydro (Gattefosse
s.a.)
Vegetol Alpine Veronica MCF 792 Hydro
(Gattefosse s.a.)
Vegetol Anise ME 158 Hydro (Gattefosse
s.a.)
Vegetol Ap GR 048 Hydro (Gattefosse s.a.)
Vegetol Apple GR 294 Hydro (Gattefosse
s.a.)
Vegetol Arnica MCF 1157 Hydro
(Gattefosse s.a.)
Vegetol Bearberry GR 457 Hydro
(Gattefosse s.a.)
Vegetol Bilberry MCF 1838 Hydro
(Gattefosse s.a.)
Vegetol Birch Tree MCF 787 Hydro
(Gattefosse s.a.)
Vegetol Burdock MCF 777 Hydro
(Gattefosse s.a.)
Vegetol Butcher's Broom GR 089 Hydro
(Gattefosse s.a.)
Vegetol Calendula MCF 774 Hydro
(Gattefosse s.a.)
Vegetol Coltsfoot MCF 1391 Hydro
(Gattefosse s.a.)
Vegetol Common Nettle MCF 801 Hydro
(Gattefosse s.a.)
Vegetol Cornflower MCF 783 Hydro
(Gattefosse s.a.)

Vegetol Cp GR 049 Hydro (Gattefosse s.a.)
Vegetol Cress LC 422 Hydro (Gattefosse
s.a.)
Vegetol Elder MCF 1238 Hydro
(Gattefosse s.a.)
Vegetol Eucalyptus GR 452 Hydro
(Gattefosse s.a.)
Vegetol Flowers MCF 1161 Hydro
(Gattefosse s.a.)
Vegetol Gentian MCF 817 Hydro
(Gattefosse s.a.)
Vegetol Ginkgo Biloba LC 416 Hydro
(Gattefosse s.a.)
Vegetol Ginseng GR 471 Hydro
(Gattefosse s.a.)
Vegetol Grapefruit GR 403 Hydro
(Gattefosse s.a.)
Vegetol Harpagophytum GR 472 Hydro
(Gattefosse s.a.)
Vegetol Hawthorn MCF 859 Hydro
(Gattefosse s.a.)
Vegetol Henna MCF 1232 Hydro
(Gattefosse s.a.)
Vegetol Hibiscus GR 235 Hydro
(Gattefosse s.a.)
Vegetol Honeysuckle GR 115 Hydro
(Gattefosse s.a.)
Vegetol Hop MCF 778 Hydro (Gattefosse
s.a.)
Vegetol Horse Chestnut MCF 1972 Hydro
(Gattefosse s.a.)
Vegetol Horseradish MCF 1191 Hydro
(Gattefosse s.a.)
Vegetol Horsetail MCF 860 Hydro
(Gattefosse s.a.)
Vegetol Hp GR 050 Hydro (Gattefossé s.a.)
Vegetol Hydrocotyl GR 040 Hydro
(Gattefosse s.a.)
Vegetol Ivy MCF 775 Hydro (Gattefosse
s.a.)
Vegetol Jasmine GR 151 Hydro
(Gattefosse s.a.)
Vegetol Juniper GR 482 Hydro (Gattefosse
s.a.)
Vegetol Lavender MCF 1484 Hydro
(Gattefosse s.a.)
Vegetol Lemon GR 020 Hydro (Gattefosse
s.a.)
Vegetol Licorice GR 456 Hydro (Gattefosse
s.a.)
Vegetol Lily MCF 1968 Hydro (Gattefosse
s.a.)
Vegetol Linden MCF 762 Hydro
(Gattefosse s.a.)
Vegetol Lotus ME 169 Hydro (Gattefosse
s.a.)
Vegetol Lp GR 223 Hydro (Gattefosse s.a.)
Vegetol Mallow MCF 784 Hydro
(Gattefosse s.a.)
Vegetol Marsh Mallow MCF 770 Hydro
(Gattefosse s.a.)
Vegetol Matricaria MCF 793 Hydro
(Gattefosse s.a.)
Vegetol Matricaria ME 106 Hydro
(Gattefosse s.a.)

Vegetol Milfoil MCF 1327 Hydro (Gattefosse s.a.)

Vegetol Myrtle GR 288 Hydro (Gattefosse s.a.)

Vegetol Nasturtium MCF 1270 Hydro (Gattefosse s.a.)

Vegetol Passionflower MCF 773 Hydro (Gattefosse s.a.)

Vegetol Peach Tree LC 339 Hydro (Gattefosse s.a.)

Vegetol Peony ME 170 Hydro (Gattefosse s.a.)

Vegetol Peppermint GR 105 Hydro (Gattefosse s.a.)

Vegetol Pine GR 300 Hydro (Gattefosse s.a.)

Vegetol Quillai GR 038 Hydro (Gattefosse s.a.)

Vegetol Quinquina MCF 1330 Hydro (Gattefosse s.a.)

Vegetol Red Grapevine MCF 1159 Hydro (Gattefosse s.a.)

Vegetol Red Poppy GR 106 Hydro (Gattefosse s.a.)

Vegetol Red Quinquina GR 013 Hydro (Gattefosse s.a.)

Vegetol Roman Chamomile LC 376 Hydro (Gattefosse s.a.)

Vegetol Rosemary MCF 772 Hydro (Gattefosse s.a.)

Vegetol Rose MCF 789 Hydro (Gattefosse s.a.)

Vegetol Sage MCF 776 Hydro (Gattefosse s.a.)

Vegetol Saponaria LC 386 Hydro (Gattefosse s.a.)

Vegetol Seaweed CB 4136 Hydro (Gattefosse s.a.)

Vegetol Sp GR 051 Hydro (Gattefosse s.a.)

Vegetol St. John's Wort MCF 786 Hydro (Gattefosse s.a.)

Vegetol Sunflower ME 177 Hydro (Gattefosse s.a.)

Vegetol Sweet Clover GR 436 Hydro (Gattefosse s.a.)

Vegetol Tea LC 412 Hydro (Gattefosse s.a.)

Vegetol Thyme GR 208 Hydro (Gattefosse s.a.)

Vegetol Tomato GR 207 Hydro (Gattefosse s.a.)

Vegetol Tp GR 052 Hydro (Gattefosse s.a.)

Vegetol Ulmary LC 383 Hydro (Gattefosse s.a.)

Vegetol Walnut Tree MCF 861 Hydro (Gattefosse s.a.)

Vegetol White Nettle MCF 796 Hydro (Gattefosse s.a.)

Vegetol Wild Pansy MCF 790 Hydro (Gattefosse s.a.)

Vegetol Wild Roseberry MCF 1833 Hydro (Gattefosse s.a.)

Vegetol Witch Hazel MCF 2032 Hydro (Gattefosse s.a.)

Vegewhite (LCW)

Vert Covasop W 7770 (LCW)

Vert Covasop W 7775 (LCW)

Vert Covasop W 7776 (LCW)

Vervain Extract HG (Provital/Centerchem)

Vervain Extract HS 2433 G (Grau)

Vexel (Sederma)

Vinca Minor Extract ies (IES LABO)

Vine Extract HG (Provital/Centerchem)

Viola Odorata Extract ies (IES LABO)

Violet Extract HS 2482 G (Grau)

Violet HS (Alban Muller)

Vitis Vinifera (Grape) Seed Extract ies (IES LABO)

V-Protein Liquid (Cosmetochem) (Cosmetochem International Ltd.)

VTA0043.050WP (Vege-Tech)

VTA0143.050WP (Vege-Tech)

Wall Pellitory Extract HG (Provital/Centerchem)

Walnut Extract HG (Provital/Centerchem)

Walnut Extract HS 2454 G (Grau)

Walnut Extract PG (Rahn)

Walnut HS (Alban Muller)

Walnut Husk HS (Alban Muller)

Walnut Leaves Extract HS 2849 G (Grau)

Watercress Extract G (Provital/Centerchem)

Water Cress Extract HS 2345 G (Grau)

Watercress Extract PG (Rahn)

Watercress HS (Alban Muller)

Watermelon HS (Alban Muller)

Water Melon Seed Extract HS 2795 G (Grau)

Watermint Extract (Cosmetic Developments)

Watermint HS (Alban Muller)

Wheat Bran Extract HS 2493 G/A (Grau)

Wheat Germ Extract HG (Provital/Centerchem)

Wheat Germ Extract HS 3499 G (Grau)

Wheat HS (Alban Muller)

Wheat Placenta "COS" (Cosmetochem) (Cosmetochem International Ltd.)

Wheat/Sweet Almond Bran Extract HS 2816 G (Grau)

White Dead Nettle HS (Alban Muller)

White Grape Extract HS 3302 G (Grau)

White Hore Hound Wort Extract HS 3350 G (Grau)

White Lily Bulb HPG Titrated (Alban Muller)

White Lily Bulb HS (Alban Muller)

White Lily Flower HS (Alban Muller)

White Mulberry HS (Alban Muller)

White Nettle Extract HG (Provital/Centerchem)

White Truffle Extract (Carrubba)

White Water Lily Rhizome HS (Alban Muller)

Wichai I and II Butea Superba (American Phyto)

Wichai III Cultivar Pueraria (American Phyto)

Wild Pansy Extract PG (Rahn)

Wild Thyme HS (Alban Muller)

Wild Yam (Greentech)

Wild Yam HPG Titrated (Alban Muller)

Wild Yam HS (Alban Muller)

Willow Extract HG (Provital/Centerchem)

Willow Extract HS 2377 G (Grau)

Willowherb Extract (Cosmetic Developments)

Winter Cherries Extract HS 2951 G (Grau)

Wintergreen HPG Titrated (Alban Muller)

Wistaria Extract (Cosmetic Developments)

Witch Hazel Extract PG (Maruzen Pharmaceuticals Co., Ltd.)

Witch Hazel Extract PG (Rahn)

Witch Hazel HS (Alban Muller)

Woodruff Extract HS 2373 G (Grau)

Yarrow Extract HG (Provital/Centerchem)

Yarrow Extract HS 2375 G (Grau)

Yarrow Extract PG (Rahn)

Yeast Extract "COS" (Cosmetochem) (Cosmetochem International Ltd.)

Yellow Chaste Weed Extract HS 3140 G (Grau)

Yellow Tarweed Extract HS 2731 G (Grau)

Yerba Santa Extract HS 3717 G (Grau)

Ylang-Ylang DG 2/B04068 (Symrise)

Ylang Ylang Extract HS 3673 G (Grau)

Ylang Ylang Extract Water Soluble 2/033610 (Symrise)

Ylang Ylang HG (Alban Muller)

Ylang Ylang HS (Alban Muller)

Yucca Extract HS 3647 G (Grau)

Yucca HPG Titrated (Alban Muller)

Yucca HS (Alban Muller)

Zea Mays (Corn) Kernel Extract ies (IES LABO)

Zeinquat (Variati)

Zetesol TP 200 (Zschimmer & Schwarz)

Zingiber Officinale (Ginger) Root Extract ies (IES LABO)

PROPYLENE GLYCOL ALGINATE

CAS No.: 9005-37-2

JPN Translation:
アルギン酸PG

Definition: Propylene Glycol Alginate is a mixture of the propylene glycol esters of alginic acid.

Information Sources: 21CFR133.133, 21CFR133.134, 21CFR133.162, 21CFR133.178, 21CFR133.179, 21CFR172.210, 21CFR172.858, 21CFR173.340, 21CFR176.170, FCC, JCLS, JSCI, NF XIX, RIFM, TSCA, USAN

Chemical Classes: Esters; Gums, Hydrophilic Colloids and Derivatives

Functions: Binder; Fragrance Ingredient; Viscosity Increasing Agent - Aqueous

Technical/Other Names:
Alginic Acid, Ester with 1,2-Propanediol
Propylene glycol alginate (RIFM)

Trade Names:
COS-Kelp PGA 5021- ISP (International Specialty Products)
COS-Kelp PGA-5022 - ISP (International Specialty Products)
COS-Kelp PGA-5023 - ISP (International Specialty Products)

Trade Name Mixture:
Retinol Cylasphere (Coletica SA)

PROPYLENE GLYCOL BEHENATE

CAS Nos.: 27923-61-1; 100214-87-7

Empirical Formula:
$C_{25}H_{50}O_3$

Definition: Propylene Glycol Behenate is the ester of propylene glycol and behenic acid. It conforms to the formula:

$$CH_3(CH_2)_{20}\overset{\overset{\displaystyle O}{\|}}{C}-OCH_2\underset{\underset{\displaystyle OH}{|}}{C}HCH_3$$

Chemical Class: Esters

Functions: Skin-Conditioning Agent - Emollient; Surfactant - Emulsifying Agent

Technical/Other Names:
Docosanoic Acid, Monoester with 1,2-Propanediol
Propylene Glycol Monodocosanoate

Trade Name:
Sunsoft 25BD (Taiyo Kagaku)

PROPYLENE GLYCOL BUTYL ETHER

CAS Nos.	EINECS Nos.
5131-66-8	225-878-4
15821-83-7	
29387-86-8	249-598-7

Empirical Formula:
$C_7H_{16}O_2$

Definition: Propylene Glycol Butyl Ether is the propylene glycol ether of n-butyl alcohol that conforms to the formula:

$$C_4H_9OCH_2\underset{\underset{\displaystyle OH}{|}}{C}HCH_3$$

Information Source: RIFM

Chemical Classes: Alkoxylated Alcohols; Ethers

Functions: Fragrance Ingredient; Solvent

Technical/Other Names:
Butoxyisopropanol
3-Butoxypropan-2-ol (RIFM)
2-Propanol, 1-Butoxy-
Propylene Glycol Monobutyl Ether

Trade Names:
Arcosolv PNB (Lyondell Chemical)
DOWANOL PnB (Dow Chemical)

PROPYLENE GLYCOL CAPRETH-4

Definition: Propylene Glycol Capreth-4 is the propylene glycol ether of a polyethylene glycol derivative of capryl alcohol that conforms generally to the formula:

$$CH_3(CH_2)_9OCH_2\underset{\underset{\displaystyle CH_3}{|}}{C}H(OCH_2CH_2)_nOH$$

where n has an average value of 4.

Chemical Class: Alkoxylated Alcohols

Function: Surfactant - Emulsifying Agent

Technical/Other Name:
Propylene Glycol Polyethylene Glycol (4) Capryl Ether

PROPYLENE GLYCOL CAPRYLATE

CAS Nos.: 31565-12-5; 68332-79-6

Empirical Formula:
$C_{11}H_{22}O_3$

Definition: Propylene Glycol Caprylate is the ester of caprylic acid and propylene glycol that conforms to the formula:

$$CH_3(CH_2)_6\overset{\overset{\displaystyle O}{\|}}{C}-OCH_2\underset{\underset{\displaystyle OH}{|}}{C}HCH_3$$

Chemical Class: Esters

Functions: Skin-Conditioning Agent - Emollient; Surfactant - Emulsifying Agent

Technical/Other Names:
Caprylic Acid, Monoester with 1,2-Propanediol
Octanoic Acid, Monoester with 1,2-Propanediol

Trade Names:
Capmul 908P (Abitec Corporation)
Imwitor 408 (Sasol GmbH - Witten)
Nikkol Sefsol-218 (Nikko)

PROPYLENE GLYCOL CETETH-3 ACETATE

CAS No.: 93385-03-6

JPN Translation:
酢酸 PG セテス - 3

Empirical Formula:
$C_{27}H_{54}O_6$

Definition: Propylene Glycol Ceteth-3 Acetate is the acetic acid ester of the ether of a propylene glycol, polyethylene glycol derivative of cetyl alcohol. It conforms generally to the formula:

$$CH_3(CH_2)_{15}OCH_2\underset{\underset{\displaystyle CH_3}{|}}{C}H(OCH_2CH_2)_nO-\overset{\overset{\displaystyle O}{\|}}{C}CH_3$$

where n has an average value of 3.

Chemical Class: Esters

Function: Skin-Conditioning Agent - Emollient

Reported Product Category: Cleansing Products (Cold Creams, Cleansing Lotions, Liquids and Pads)

Technical/Other Name:
Propylene Glycol Polyethylene Glycol (3) Cetyl Ether Acetate

Trade Name:
Hetester PCA (Bernel)

PROPYLENE GLYCOL CETETH-3 PROPIONATE

Empirical Formula:
$C_{28}H_{56}O_6$

Definition: Propylene Glycol Ceteth-3 Propionate is the propionic acid ester of an ether of a propylene glycol, polyethylene glycol derivative of cetyl alcohol. It conforms generally to the formula:

$$CH_3(CH_2)_{15}OCH_2\underset{\underset{\displaystyle CH_3}{|}}{C}H(OCH_2CH_2)_nO-\overset{\overset{\displaystyle O}{\|}}{C}CH_2CH_3$$

where n has an average value of 3.

Chemical Class: Esters

Function: Skin-Conditioning Agent - Emollient

Technical/Other Name:
Propylene Glycol Polyethylene Glycol (3) Cetyl Ether Propionate

Trade Name:
Hetester PCP (Bernel)

PROPYLENE GLYCOL CITRATE

Empirical Formula:
$C_9H_{14}O_8$

Definition: Propylene Glycol Citrate is the ester of citric acid and propylene glycol that conforms to the formula:

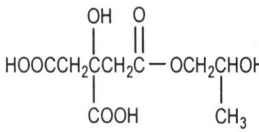

Chemical Class: Esters

Function: Skin-Conditioning Agent - Humectant

Technical/Other Name:
Citric Acid, Monoester with 1,2-Propanediol

Trade Name Mixtures:
Arrectosina (Vevy)
Zincotrat (Vevy)

PROPYLENE GLYCOL COCOATE

Definition: Propylene Glycol Cocoate is the ester of propylene glycol and Coconut Acid (q.v.).

Chemical Class: Esters

Functions: Skin-Conditioning Agent - Emollient; Surfactant - Emulsifying Agent

Technical/Other Name:
Coconut Fatty Acid, Monoester with 1,2-Propanediol

Trade Name:
ROL LP (Fabriquimica)

PROPYLENE GLYCOL DICAPRATE

CAS Nos.
53824-77-4
56519-72-3

EINECS No.
258-814-9

JPN Translation:
ジカプリン酸 PG

Empirical Formula:
$C_{23}H_{44}O_4$

Definition: Propylene Glycol Dicaprate is the diester of propylene glycol and capric acid. It conforms to the formula:

$$CH_3(CH_2)_8C-OCH_2CHO-C(CH_2)_8CH_3$$
$$|$$
$$CH_3$$

Information Sources: CIR: [S] IJT-18 (SUPPL. 2)1999, JCIC, JCLS, TSCA

Chemical Class: Esters

Functions: Skin-Conditioning Agent - Occlusive; Viscosity Increasing Agent - Nonaqueous

Technical/Other Names:
Decanoic Acid, 1-Methyl-1,2-Ethanediyl Ester

Decanoic Acid, 1-Methylo-1,2-Ethanediyl Ester
n-Decanoic Acid, 1,2-Propanediyl Ester
1,2-Propanediol Dicaprate
1,2-Propanediol Didecanoate
Propylene Glycol Didecanoate

Trade Names:
Captex 100 (Abitec Corporation)
Nikkol PDD (Nikko)

PROPYLENE GLYCOL DICAPROATE

CAS No.: 50343-36-7

Empirical Formula:
$C_{15}H_{28}O_4$

Definition: Propylene Glycol Dicaproate is the diester of propylene glycol and caproic acid. It conforms to the formula:

$$CH_3(CH_2)_4C-OCH_2CHO-C(CH_2)_4CH_3$$
$$|$$
$$CH_3$$

Chemical Class: Esters

Functions: Skin-Conditioning Agent - Occlusive; Viscosity Increasing Agent - Nonaqueous

Technical/Other Names:
Hexanoic Acid, 1-Methyl-1,2-Ethanediyl Ester
n-Hexanoic Acid, 1,2-Propanediyl Ester
1,2-Propanediol Dihexanoate

PROPYLENE GLYCOL DICAPRYLATE

CAS No.
7384-98-7

EINECS No.
230-962-9

JPN Translation:
ジカプリル酸 PG

Empirical Formula:
$C_{19}H_{36}O_4$

Definition: Propylene Glycol Dicaprylate is the diester of propylene glycol and caprylic acid that conforms generally to the formula:

$$CH_3(CH_2)_6C-OCH_2CHO-C(CH_2)_6CH_3$$
$$|$$
$$CH_3$$

Information Sources: 21CFR172.856, 21CFR173.340, 21CFR175.300, 21CFR176.170, 21CFR176.210, 21CFR177.2800, CIR: [S] IJT-18(SUPPL. 2)-1999, JCIC, JCLS, JSQI, TSCA

Chemical Class: Esters

Functions: Skin-Conditioning Agent - Occlusive; Viscosity Increasing Agent - Nonaqueous

Technical/Other Name:
Octanoic Acid, 1-Methyl-1,2-Ethanediyl Ester

Trade Names:
AEC Propylene Glycol Dicaprylate (A & E Connock)
Nikkol Sefsol-228 (Nikko)

PROPYLENE GLYCOL DICAPRYLATE/ DICAPRATE

CAS Nos.
58748-27-9
68583-51-7
68988-72-7

EINECS No.
271-516-3

JPN Translation:
ジ (カプリル / カプリン酸) PG

Definition: Propylene Glycol Dicaprylate/ Dicaprate is a mixture of the propylene glycol diesters of caprylic and capric acids.

Information Sources: 21CFR172.856, 21CFR173.340, 21CFR175.300, 21CFR176.170, 21CFR176.210, 21CFR177.2800, CIR: [S] IJT-18(SUPPL. 2)-1999, CTFA D, JCIC, JCLS, JSQI, TSCA

Chemical Class: Esters

Function: Skin-Conditioning Agent - Occlusive

Reported Product Categories: Moisturizing Preparations; Lipsticks; Foundations; Bath Preparations, Misc.; Body and Hand Preparations (Excluding Shaving Preparations); Bath Oils, Tablets, and Salts; Cleansing Products (Cold Creams, Cleansing Lotions, Liquids and Pads); Fragrance Preparations, Misc.; Hair Conditioners; Suntan Gels, Creams, and Liquids; Bath Capsules; Blushers (All types); Eye Makeup Removers; Eyeliners; Face and Neck Preparations (Excluding Shaving Preparations); Indoor Tanning Preparations; Skin Care Preparations, Misc.; Eye Shadows; Makeup Preparations (Not eye), Misc.; Eye Makeup Preparations, Misc.; Foot Powders and Sprays; Night Skin Care Preparations

Technical/Other Names:
Decanoic Acid, Mixed Diesters with Octanoic Acid and Propylene Glycol
Octanoic Acid, Mixed Diesters with Decanoic Acid and Propylene Glycol

Trade Names:
Acconon 200 (Abitec Corporation)
AEC Propylene Glycol Dicaprylate/ Dicaprate (A & E Connock)
Crodamol PC (Croda Chemicals)

Dermol F 3 (Fabriquimica)
DUB 810 PG (Stearinerie Dubois Fils)
ESTOL 1526 (Uniqema Europe)
Ethox PC (Ethox)
Hatcol 5128 (Hatco)
Jeechem PGDD (Jeen)
Lexol PG 865 (Inolex)
Liponate PC (Lipo)
Miglyol 840 (Sasol GmbH - Witten)
Myritol PC (Cognis Care Chemicals/NJ)
Myritol PC (Cognis Care Chemicals/PA)
Myritol PC (Cognis Deutschland)
NEOBEE M-20 (Stepan)
Pelemol PDD (Phoenix)
Procol PGDD (Protameen)
Radia 7208 (Oleon NV)
Unitolate 380 (Universal Preserv-A-Chem)
Waglinol 2/7680 (Industrial Quimica)

Trade Name Mixtures:
Acacia Flowers Milk (CEP (Solabia))
Anise Softcream (CEP (Solabia))
Apricot Milk (CEP (Solabia))
Avocado Milk (CEP (Solabia))
Bamboo Manna Milk (CEP (Solabia))
Banana (Tree) Flowers Milk (CEP (Solabia))
Bentone Gel M-20 (ELE)
Bitter Orange Softcream (CEP (Solabia))
Cactus Milk (CEP (Solabia))
Cardamom Softcream (CEP (Solabia))
Chestnut Milk (CEP (Solabia))
Chinese Tea Milk (CEP (Solabia))
Cinnamon Softcream (CEP (Solabia))
Clementine Milk (CEP (Solabia))
Cocoa Milk (CEP (Solabia))
Condurango Milk (CEP (Solabia))
Eucalyptus Softcream (CEP (Solabia))
Fig Milk (CEP (Solabia))
Florentine Orris Milk (CEP (Solabia))
Frangipani Flowers Milk (CEP (Solabia))
Fruits Milk (CEP (Solabia))
Gentle Cleansing Milk (CEP (Solabia))
Ginger Softcream (CEP (Solabia))
Golden Pea Milk (CEP (Solabia))
Grapefruit Softcream (CEP (Solabia))
Hibiscus Flowers Milk (CEP (Solabia))
Honey Milk (CEP (Solabia))
Incense Milk (CEP (Solabia))
Jasmine Flowers Milk (CEP (Solabia))
Jujube Milk (CEP (Solabia))
Kawa Kawa Milk (CEP (Solabia))
Kiwi Milk (CEP (Solabia))
Lavender Softcream (CEP (Solabia))
Lime Softcream (CEP (Solabia))
Lotus Flowers Milk (CEP (Solabia))
Mandarin Milk (CEP (Solabia))
Marshmallow Milk (CEP (Solabia))
Miglyol 840 Gel B (Sasol GmbH - Witten)
Mushroom Milk (CEP (Solabia))
Myrrh Milk (CEP (Solabia))
Myrrh Milk (CEP (Solabia))
Oleat J of Cotton (CEP (Solabia))
Oleat of Acacia (CEP (Solabia))

Oleat of Almond (CEP (Solabia))
Oleat of Aloe (CEP (Solabia))
Oleat of Apricot (CEP (Solabia))
Oleat of Arnica (CEP (Solabia))
Oleat of Avocado (CEP (Solabia))
Oleat of Balm Mint (CEP (Solabia))
Oleat of Bamboo Leaves (CEP (Solabia))
Oleat of Bladderwrack (CEP (Solabia))
Oleat of Broom (CEP (Solabia))
Oleat of Carrot (CEP (Solabia))
Oleat of Caviar (CEP (Solabia))
Oleat of Chamomile (CEP (Solabia))
Oleat of Chinese Tea (CEP (Solabia))
Oleat of Cornflower (CEP (Solabia))
Oleat of Cotton (CEP (Solabia))
Oleat of Ginkgo Biloba (CEP (Solabia))
Oleat of Ginseng (CEP (Solabia))
Oleat of Grape Leaf (CEP (Solabia))
Oleat of Green Coffee (CEP (Solabia))
Oleat of Hops (CEP (Solabia))
Oleat of Horse Chestnut (CEP (Solabia))
Oleat of Horsetail (CEP (Solabia))
Oleat of Hydrocotyl (CEP (Solabia))
Oleat of Incense (CEP (Solabia))
Oleat of Ivy (CEP (Solabia))
Oleat of Klamath (CEP (Solabia))
Oleat of Laminaria (CEP (Solabia))
Oleat of Linden (CEP (Solabia))
Oleat of Lotus Flowers (CEP (Solabia))
Oleat of Marigold (CEP (Solabia))
Oleat of Myrrh (CEP (Solabia))
Oleat of Orange (CEP (Solabia))
Oleat of Peach (CEP (Solabia))
Oleat of Pollen (CEP (Solabia))
Oleat of Sage (CEP (Solabia))
Oleat of Saint John's Wort (CEP (Solabia))
Oleat of Tahiti Vanilla (CEP (Solabia))
Oleat of Vanilla (CEP (Solabia))
Oleat of Water Lily (CEP (Solabia))
Oleat of Wheat (Bran) (CEP (Solabia))
Oleat of White Lily Bulb (CEP (Solabia))
Oleat of White Tea (CEP (Solabia))
Oleat of Witch Hazel (CEP (Solabia))
Oleat of Yarrow (CEP (Solabia))
Oleoplex Aloe (Fabriquimica)
Oleoplex Arnica (Fabriquimica)
Oleoplex Calendula (Fabriquimica)
Oleoplex Centella (Fabriquimica)
Oleoplex Manzanilla (Fabriquimica)
Oleoplex Romero (Fabriquimica)
Papaya Milk (CEP (Solabia))
Passion Flowers Milk (CEP (Solabia))
Peach Milk (CEP (Solabia))
Peach Tree Milk (CEP (Solabia))
Peppermint Softcream (CEP (Solabia))
Pomegranate Milk (CEP (Solabia))
Primrose Milk (CEP (Solabia))
Rheocare CTH(E) (Cosmetic Rheologies)
Rosemary Softcream (CEP (Solabia))
Salcare SC96 (Ciba Specialty Chemicals)
Silk Milk (CEP (Solabia))
Simagel NO (Biophil)
Softisan Gel (Sasol GmbH - Witten)
Soothing Milk (CEP (Solabia))

Tiare Flowers Milk (CEP (Solabia))
Vanilla Milk (CEP (Solabia))
Water Lily Flowers Milk (CEP (Solabia))
White Tea Softcream (CEP (Solabia))
Ylang-Ylang Milk (CEP (Solabia))

PROPYLENE GLYCOL DICOCOATE

CAS No.	EINECS No.
68953-19-5	273-191-3

Definition: Propylene Glycol Dicocoate is the diester of propylene glycol and Coconut Acid (q.v.). It conforms to the formula:

$$RC(=O)-OCH_2CHO-CR(=O), \ CH_3$$

where RCO- represents the fatty acids derived from coconut oil.

Information Sources: 21CFR172.856, 21CFR175.105, 21CFR175.300, 21CFR176.170, 21CFR176.210, 21CFR177.2800, CIR: [S] IJT-18(SUPPL. 2)-1999

Chemical Class: Esters

Functions: Skin-Conditioning Agent - Occlusive; Viscosity Increasing Agent - Nonaqueous

Technical/Other Names:
Coconut Fatty Acids, 1-Methyl-1,2-Ethanediyl Ester
Propylene Glycol Dicoconate
Propylene Glycol Diester Coconut Acids

Trade Name:
AEC Propylene Glycol Dicocoate (A & E Connock)

PROPYLENE GLYCOL DIETHYLHEXANOATE

CAS No.	EINECS No.
93981-97-6	301-185-3

Empirical Formula:
$C_{19}H_{36}O_4$

Definition: Propylene Glycol Diethylhexanoate is the diester of Propylene Glycol (q.v.) and 2-ethylhexanoic acid. It conforms generally to the formula:

$$CH_3(CH_2)_3CHC(=O)-OCH_2CHO-C(=O)CH(CH_2)_3CH_3$$

Chemical Class: Esters

Function: Skin-Conditioning Agent - Occlusive

Reported Product Categories: Fragrance Preparations, Misc.; Moisturizing Preparations; Eyebrow Pencils; Suntan Gels, Creams, and Liquids; Bath Preparations, Misc.; Body and Hand Preparations (Excluding Shaving Preparations); Night Skin Care Preparations; Paste Masks (Mud Packs); Skin Care Preparations, Misc.

Technical/Other Names:
2-Ethylhexanoic Acid, 1-Methyl-1,2-Ethandiyl Ester
Propylene Glycol Dioctanoate

Trade Names:
AEC Propylene Glycol Diethylhexanoate (A & E Connock)
Captex 800 (Abitec Corporation)
Dekamol 8 (Dekker)
DUB DOPG (Stearinerie Dubois Fils)
Lexol PG 800 (Inolex)

Trade Name Mixture:
Rhodysterol (SECMA)

PROPYLENE GLYCOL DIISONONANOATE

CAS No.: 125804-17-3

Empirical Formula:
$C_{21}H_{40}O_4$

Definition: Propylene Glycol Diisononanoate is the diester of propylene glycol and branched chain nonanoic acids. It conforms generally to the formula:

$$C_8H_{17}\overset{\overset{O}{\|}}{C}-OCH_2\underset{\underset{CH_3}{|}}{C}HO-\overset{\overset{O}{\|}}{C}C_8H_{17}$$

Chemical Class: Esters

Functions: Skin-Conditioning Agent - Occlusive; Viscosity Increasing Agent - Nonaqueous

Technical/Other Name:
Isononanoic Acid, 1-Methyl-1,2-Ethanediyl Ester

Trade Names:
AEC Propylene Glycol Diisononanoate (A & E Connock)
Dermol PGDI (Alzo)

PROPYLENE GLYCOL DIISOSTEARATE

CAS No. 68958-54-3
EINECS No. 273-373-2

JPN Translation:
ジイソステアリン酸 PG

Empirical Formula:
$C_{39}H_{76}O_4$

Definition: Propylene Glycol Diisostearate is the diester of propylene glycol and isostearic acid. It conforms generally to the formula:

$$C_{17}H_{35}\overset{\overset{O}{\|}}{C}-OCH_2\underset{\underset{CH_3}{|}}{C}HO-\overset{\overset{O}{\|}}{C}C_{17}H_{35}$$

Information Sources: CIR: [S] IJT-18 (SUPPL. 2)1999, JCIC, JCLS

Chemical Class: Esters

Functions: Skin-Conditioning Agent - Occlusive; Viscosity Increasing Agent - Nonaqueous

Technical/Other Name:
Isooctadecanoic Acid, 1,2-Propanediyl Ester

Trade Name:
PRISORINE 2035 (Uniqema Europe)

PROPYLENE GLYCOL DILAURATE

CAS No. 22788-19-8
EINECS No. 245-217-3

Empirical Formula:
$C_{27}H_{52}O_4$

Definition: Propylene Glycol Dilaurate is the diester of propylene glycol and lauric acid that conforms generally to the formula:

$$CH_3(CH_2)_{10}\overset{\overset{O}{\|}}{C}-OCH_2\underset{\underset{O-\overset{}{C}(CH_2)_{10}CH_3}{|}}{C}HCH_3$$

Information Sources: 21CFR172.856, 21CFR173.340, 21CFR175.300, 21CFR176.170, 21CFR176.210, 21CFR177.2800, CIR: [S] IJT-18(SUPPL. 2)-1999

Chemical Class: Esters

Functions: Skin-Conditioning Agent - Occlusive; Viscosity Increasing Agent - Nonaqueous

Technical/Other Name:
Dodecanoic Acid, 1-Methyl-1,2-Ethanediyl Ester

Trade Names:
AEC Propylene Glycol Dilaurate (A & E Connock)
Emalex PG-di-L (Nihon Emulsion)

PROPYLENE GLYCOL DIOLEATE

CAS No. 105-62-4
EINECS No. 203-315-3

JPN Translation:
ジオレイン酸 PG

Empirical Formula:
$C_{39}H_{72}O_4$

Definition: Propylene Glycol Dioleate is the diester of propylene glycol and oleic acid. It conforms to the formula:

$$CH(CH_2)_7\overset{\overset{O}{\|}}{C}-OCH_2\underset{\underset{CH(CH_2)_7CH_3}{\|}}{C}H\overset{\overset{CH_3}{|}}{O}-\overset{\overset{O}{\|}}{C}(CH_2)_7CH$$

Information Sources: CIR: [S] IJT-18 (SUPPL. 2)1999, JCLS, JSCI

Chemical Class: Esters

Functions: Skin-Conditioning Agent - Occlusive; Viscosity Increasing Agent - Nonaqueous

Technical/Other Names:
1,2-Dioleoyloxypropane
1-Methyl-1,2-Ethanediyl 9-Octadecenoate
9-Octadecenoic Acid, 1-Methyl-1,2-Ethanediyl Ester
9-Octadecenoic Acid, 1,3-Propanediyl Ester

Trade Name:
Adiplas CD (Fabriquimica)

PROPYLENE GLYCOL DIPELARGONATE

CAS No. 41395-83-9
EINECS No. 255-350-9

JPN Translation:
ジノナン酸 PG

Empirical Formula:
$C_{21}H_{40}O_4$

Definition: Propylene Glycol Dipelargonate is the diester of propylene glycol and Pelargonic Acid (q.v.) that conforms generally to the formula:

$$CH_3(CH_2)_7\overset{\overset{O}{\|}}{C}-OCH_2\underset{\underset{CH_3}{|}}{C}HO-\overset{\overset{O}{\|}}{C}(CH_2)_7CH_3$$

Information Sources: CIR: [S] IJT-18 (SUPPL. 2)1999, CTFA D, JCIC, JCLS, JSQI, TSCA

Chemical Class: Esters

Functions: Skin-Conditioning Agent - Occlusive; Viscosity Increasing Agent - Nonaqueous

Reported Product Categories: Foundations; Lipsticks; Skin Care Preparations, Misc.; Bath Oils, Tablets, and Salts; Cleansing Products (Cold Creams, Cleansing Lotions, Liquids and Pads); Fragrance Preparations, Misc.; Moisturizing Preparations; Bath Preparations, Misc.; Blushers (All types); Body and Hand Preparations (Excluding Shaving Preparations)

Technical/Other Names:
Nonanoic Acid, 1-Methyl-1,2-Ethanediyl Ester
1,2-Propanediol Dinonanoate

Trade Names:
AEC Propylene Glycol Dipelargonate (A & E Connock)
Dermol PGDP (Alzo)
D.P.P.G. (Gattefosse s.a.)
Emerest 2388 (Cognis Care Chemicals/NJ)
Emerest 2388 (Cognis Care Chemicals/PA)
Pelemol PGDP (Phoenix)
Schercemol PGDP (Scher)

PROPYLENE GLYCOL DIRICINOLEATE/IPDI COPOLYMER

CAS No.: 351425-00-8

Definition: Propylene Glycol Diricinoleate/IPDI Copolymer is a copolymer of propylene glycol diricinoleate and isophorone diisocyanate (IPDI) monomers.

Chemical Class: Synthetic Polymers

Functions: Binder; Film Former; Skin-Conditioning Agent - Emollient

Technical/Other Name:
9-Octadecenoic Acid, 12-Hydroxy-, 1-Methyl-1,2-Ethanediyl Ester, Polymer with 5-Isocyanato-1-(Isocyanatomethyl)-1,3,3-Trimethylcyclohexane

Trade Name:
Polyderm PPI-PGR (Alzo)

PROPYLENE GLYCOL DISTEARATE

CAS No.	EINECS No.
6182-11-2	228-229-3

JPN Translation:
ジステアリン酸 PG

Empirical Formula:
$C_{39}H_{76}O_4$

Definition: Propylene Glycol Distearate is the diester of propylene glycol and stearic acid. It conforms to the formula:

$$CH_3(CH_2)_{16}\overset{\displaystyle O}{\overset{\|}{C}}-OCH_2\overset{\displaystyle CH_3}{\underset{|}{C}}HO-\overset{\displaystyle O}{\overset{\|}{C}}(CH_2)_{16}CH_3$$

Information Sources: 21CFR172.856, 21CFR173.340, 21CFR175.300, 21CFR176.170, 21CFR176.210, 21CFR177.2800, JCIC, JCLS, TSCA

Chemical Class: Esters

Functions: Opacifying Agent; Skin-Conditioning Agent - Occlusive; Viscosity Increasing Agent - Nonaqueous

Technical/Other Name:
Octadecanoic Acid, 1-Methyl-1,2-Ethanediyl Ester

Trade Names:
AEC Propylene Glycol Distearate (A & E Connock)
Emalex PG-di-S (Nihon Emulsion)

PROPYLENE GLYCOL DIUNDECANOATE

CAS No.	EINECS No.
68227-47-4	269-359-0

Empirical Formula:
$C_{25}H_{48}O_4$

Definition: Propylene Glycol Diundecanoate is the diester of propylene glycol and Undecanoic Acid (q.v.). It conforms to the formula:

$$CH_3(CH_2)_9\overset{\displaystyle O}{\overset{\|}{C}}-OCH_2\overset{\displaystyle CH_3}{\underset{|}{C}}HO-\overset{\displaystyle O}{\overset{\|}{C}}(CH_2)_9CH_3$$

Information Source: TSCA

Chemical Class: Esters

Functions: Skin-Conditioning Agent - Occlusive; Viscosity Increasing Agent - Nonaqueous

Technical/Other Name:
Undecanoic Acid, 1-Methyl-1,2-Ethanediyl Ester

PROPYLENE GLYCOL FATTY ACIDATE

JPN Translation:
脂肪酸 PG

Definition: *See "Regulatory and Ingredient Use Information," regarding use of Japan Trivial names in Volume 1, Introduction, Part A.*

Information Source: JCLS

Chemical Class: Esters

Function: Not Reported

PROPYLENE GLYCOL HEPTANOATE

Empirical Formula:
$C_{10}H_{20}O_3$

Definition: Propylene Glycol Heptanoate is the ester of propylene glycol and heptanoic acid that conforms to the formula:

$$CH_3(CH_2)_5\overset{\displaystyle O}{\overset{\|}{C}}-OCH_2\overset{\displaystyle}{\underset{\underset{\textstyle OH}{|}}{C}}HCH_3$$

Chemical Class: Esters

Functions: Skin-Conditioning Agent - Emollient; Surfactant - Emulsifying Agent

Technical/Other Name:
Heptanoic Acid, Monoester with 1,2-Propanediol

Trade Name:
Capmul 907P (Abitec Corporation)

PROPYLENE GLYCOL HYDROXYSTEARATE

CAS Nos.	EINECS Nos.
33907-47-0	251-734-5
38621-51-1	254-045-8

Empirical Formula:
$C_{21}H_{42}O_4$

Definition: Propylene Glycol Hydroxystearate is the ester of propylene glycol and Hydroxystearic Acid (q.v.). It conforms generally to the formula:

$$CH_3(CH_2)_5\underset{\underset{\textstyle OH}{|}}{C}H(CH_2)_{10}\overset{\displaystyle O}{\overset{\|}{C}}-OCH_2\underset{\underset{\textstyle OH}{|}}{C}HCH_3$$

Information Source: CTFA D

Chemical Class: Esters

Functions: Skin-Conditioning Agent - Emollient; Surfactant - Emulsifying Agent

Technical/Other Name:
Octadecanoic Acid, 12-Hydroxy-, Monoester with 1,2-Propanediol

Trade Names:
AEC Propylene Glycol Hydroxystearate (A & E Connock)
Naturechem PGHS (CasChem)

PROPYLENE GLYCOL ISOCETETH-3 ACETATE

CAS Nos.: 93385-13-8; 178900-23-7

JPN Translation:
酢酸 PG イソセテス - 3

Empirical Formula:
$C_{27}H_{54}O_6$

Definition: Propylene Glycol Isoceteth-3 Acetate is the acetic acid ester of an ether of a propylene glycol, polyethylene glycol derivative of isocetyl alcohol. It conforms generally to the formula:

$$C_{16}H_{33}(OCHCH_2)_x(OCH_2CH_2)_yO-CCH_3$$
$$| \qquad\qquad\qquad\qquad\qquad\; \| $$
$$CH_3 \qquad\qquad\qquad\qquad\qquad O$$

where x has an average value of 1 and y has an average value of 3.

Information Source: JCLS

Chemical Class: Esters

Function: Skin-Conditioning Agent - Emollient

Reported Product Category: Foundations

Technical/Other Name:
Propylene Glycol Polyethylene Glycol (3) Isocetyl Ether Acetate

Trade Name:
Hetester PHA (Bernel)

PROPYLENE GLYCOL ISODECETH-4

Empirical Formula:
$C_{21}H_{44}O_6$

Definition: Propylene Glycol Isodeceth-4 is the propylene glycol ether of ethoxylated isodecyl alcohol that conforms generally to the formula:

$$C_{10}H_{21}(OCHCH_2)_x(OCH_2CH_2)_yOH$$
$$|$$
$$CH_3$$

where x has an average value of 1 and y has an average value of 4.

Chemical Class: Alkoxylated Alcohols

Function: Surfactant - Emulsifying Agent

Technical/Other Names:
Polyoxyethylene (4) Polyoxypropylene (1) Isodecyl Ether
Polyoxypropylene (1) Polyoxyethylene (4) Isodecyl Ether

PROPYLENE GLYCOL ISODECETH-12

Definition: Propylene Glycol Isodeceth-12 is the propylene glycol ether of ethoxylated isodecyl alcohol that conforms generally to the formula:

$$C_{10}H_{21}(OCHCH_2)_x(OCH_2CH_2)_yOH$$
$$|$$
$$CH_3$$

where x has an average value of 1 and y has an average value of 12.

Chemical Class: Alkoxylated Alcohols

Function: Surfactant - Emulsifying Agent

Technical/Other Names:
Polyoxyethylene (12) Polyoxypropylene (1) Isodecyl Ether
Polyoxypropylene (1) Polyoxyethylene (12) Isodecyl Ether

PROPYLENE GLYCOL ISOSTEARATE

CAS Nos.	EINECS No.
63799-53-1	
68171-38-0	269-027-5

JPN Translation:
イソステアリン酸 PG

Empirical Formula:
$C_{21}H_{42}O_3$

Definition: Propylene Glycol Isostearate is the ester of propylene glycol and Isostearic Acid (q.v.).

Information Sources: CIR: [S] IJT-18 (SUPPL. 2)1999, JCIC, JCLS, TSCA

Chemical Class: Esters

Functions: Skin-Conditioning Agent - Emollient; Surfactant - Emulsifying Agent

Reported Product Categories: Shaving Preparations, Misc.; Shaving Cream (Aerosol, Brushless and Lather)

Technical/Other Names:
Isooctadecanoic Acid, Monoester with 1,2-Propanediol
Propylene Glycol Monoisostearate

Trade Names:
AEC Propylene Glycol Isostearate (A & E Connock)
Emerest 2384 (Cognis Care Chemicals/NJ)
Emerest 2384 (Cognis Care Chemicals/PA)
Hydrophilol Isostearique (Gattefosse s.a.)
PRISORINE 2034 (Uniqema Europe)

Trade Name Mixtures:
Emulfree P (Gattefosse s.a.)
Hydrolactol 70 (Gattefosse s.a.)

PROPYLENE GLYCOL LAURATE

CAS Nos.	EINECS Nos.
142-55-2	205-542-3
27194-74-7	248-315-4
37321-62-3	
199282-83-2	

JPN Translation:
ラウリン酸 PG

Empirical Formula:
$C_{15}H_{30}O_3$

Definition: Propylene Glycol Laurate is the ester of propylene glycol and lauric acid that conforms generally to the formula:

$$CH_3(CH_2)_{10}C-OCH_2CHOH$$
$$\| \qquad\qquad\qquad\qquad |$$
$$O \qquad\qquad\qquad\qquad CH_3$$

Information Sources: 21CFR172.856, 21CFR173.340, 21CFR175.105, 21CFR175.300, 21CFR176.170, 21CFR176.210, 21CFR177.2800, CIR: [S] IJT-18(SUPPL. 2)1999, CTFA D, JCIC, JCLS, JSQI, TSCA

Chemical Class: Esters

Functions: Skin-Conditioning Agent - Emollient; Surfactant - Emulsifying Agent

Reported Product Categories: Bath Preparations, Misc.; Body and Hand Preparations (Excluding Shaving Preparations); Colognes and Toilet Waters; Makeup Bases; Bath Oils, Tablets, and Salts; Cleansing Products (Cold Creams, Cleansing Lotions, Liquids and Pads); Foundations; Lipsticks; Mascara; Moisturizing Preparations; Blushers (All types); Makeup Fixatives; Makeup Preparations (Not eye), Misc.; Perfumes; Skin Care Preparations, Misc.; Suntan Gels, Creams, and Liquids; Tonics, Dressings, and Other Hair Grooming Aids

Technical/Other Names:
Dodecanoic Acid, 2-Hydroxypropyl Ester
Dodecanoic Acid, Monoester with 1,2-Propanediol
2-Hydroxypropyl Dodecanoate
Lauroglycol
Propylene Glycol Monododecanoate
Propylene Glycol Monolaurate

Trade Names:
AEC Propylene Glycol Laurate (A & E Connock)
DUB LPG (Stearinerie Dubois Fils)
Imwitor 412 (Sasol GmbH - Witten)
Jeechem PGML (Jeen)
Laurate De Propylene Glycol (Prod'Hyg)
Protachem PGML (Protameen)
Schercemol PGML (Scher)
Sipoest PGML (Specialty Industrial)
Unipeg-PGML (Universal Preserv-A-Chem)

Trade Name Mixtures:
Amisol 4135 (Lucas Meyer GmbH)
Aquafix Marin (Coletica SA)
Aquafix Vegetal (Coletica SA)
Emulfree P (Gattefosse s.a.)

PROPYLENE GLYCOL LAURETH-6

Definition: Propylene Glycol Laureth-6 is the propylene glycol ether of Laureth-6 (q.v.) that conforms generally to the formula:

$$CH_3(CH_2)_{11}(OCHCH_2)_x(OCH_2CH_2)_yOH$$
$$|$$
$$CH_3$$

where x has an average value of 1 and y has an average value of 6.

Chemical Class: Alkoxylated Alcohols

Function: Surfactant - Emulsifying Agent

Technical/Other Name:
Propylene Glycol Polyethylene Glycol (6) Lauryl Ether

Trade Name:
Dehydol 980 (Cognis Deutschland)

PROPYLENE GLYCOL LINOLEATE

Empirical Formula:
$C_{21}H_{38}O_3$

Definition: Propylene Glycol Linoleate is the ester of propylene glycol and linoleic acid that conforms to the formula:

$$CH_3(CH_2)_4CH$$
$$||$$
$$CHCH_2CH \quad\quad O$$
$$|| \quad\quad ||$$
$$CH(CH_2)_7C - OCH_2CHCH_3$$
$$|$$
$$OH$$

Chemical Class: Esters

Functions: Skin-Conditioning Agent - Emollient; Surfactant - Emulsifying Agent

Technical/Other Name:
Linoleic Acid, Monoester with 1,2-Propanediol

Trade Name Mixture:
Efevit E (Fabriquimica)

PROPYLENE GLYCOL LINOLENATE

Empirical Formula:
$C_{21}H_{36}O_3$

Definition: Propylene Glycol Linolenate is the ester of propylene glycol and linolenic acid that conforms to the formula:

$$CHCH_2CH_3$$
$$||$$
$$CHCH_2CH$$
$$||$$
$$CHCH_2CH \quad\quad O$$
$$|| \quad\quad ||$$
$$CH(CH_2)_7C - OCH_2CHCH_3$$
$$|$$
$$OH$$

Chemical Class: Esters

Functions: Skin-Conditioning Agent - Emollient; Surfactant - Emulsifying Agent

Technical/Other Names:
Linolenic Acid, Monoester with 1,2-Propanediol
9,12-Octadecadienoic Acid, Oxiranylmethyl Ester

Trade Name Mixture:
Efevit E (Fabriquimica)

PROPYLENE GLYCOL MYRISTATE

CAS No.	EINECS No.
29059-24-3	249-395-3

Empirical Formula:
$C_{17}H_{34}O_3$

Definition: Propylene Glycol Myristate is the ester of propylene glycol and myristic acid that conforms generally to the formula:

$$O$$
$$||$$
$$CH_3(CH_2)_{12}C - OCH_2CHCH_3$$
$$|$$
$$OH$$

Information Sources: 21CFR172.856, 21CFR173.340, 21CFR175.300, 21CFR176.170, 21CFR176.210, 21CFR177.2800, CIR: [S] IJT-18(SUPPL. 2)-1999, TSCA

Chemical Class: Esters

Functions: Skin-Conditioning Agent - Emollient; Surfactant - Emulsifying Agent

Reported Product Categories: Skin Care Preparations, Misc.; Bath Preparations, Misc.; Body and Hand Preparations (Excluding Shaving Preparations)

Technical/Other Names:
Propylene Glycol Monomyristate
Propylene Glycol Monotetradecanoate
Tetradecanoic Acid, Monoester with 1,2-Propanediol

Trade Names:
AEC Propylene Glycol Myristate (A & E Connock)
DUB MPG 19 (Stearinerie Dubois Fils)
Megester 1324 (Megachem)
Prodhysolve (Prod'Hyg)

PROPYLENE GLYCOL MYRISTYL ETHER

CAS No.: 63793-60-2 (Generic)

Empirical Formula:
$C_{17}H_{36}O_2$

Definition: Propylene Glycol Myristyl Ether is the propylene glycol ether of myristyl alcohol that conforms to the formula:

$$CH_3(CH_2)_{13}OCHCH_2OH$$
$$|$$
$$CH_3$$

Information Source: TSCA

Chemical Class: Ethers

Function: Skin-Conditioning Agent - Emollient

Technical/Other Name:
PPG-1 Myristyl Ether

PROPYLENE GLYCOL MYRISTYL ETHER ACETATE

CAS No.: 135326-54-4

JPN Translation:
酢酸 PG ミリスチル

Definition: Propylene Glycol Myristyl Ether Acetate is the ester of Propylene Glycol Myristyl Ether (q.v.) and acetic acid.

Chemical Class: Esters

Function: Skin-Conditioning Agent - Emollient

Reported Product Categories: Fragrance Preparations, Misc.; Lipsticks; Eye Shadows; Suntan Preparations, Misc.

Technical/Other Names:
PPG-1 Myristyl Ether Acetate
Propanol, (Tetradecyloxy)-Acetate

Trade Name:
Hetester PMA (Bernel)

PROPYLENE GLYCOL OLEATE

CAS No.	EINECS No.
1330-80-9	215-549-3

JPN Translation:
オレイン酸 PG

Empirical Formula:
$C_{21}H_{40}O_3$

Definition: Propylene Glycol Oleate is the ester of propylene glycol and oleic acid. It conforms to the formula:

$$O$$
$$||$$
$$CH_3(CH_2)_7CH = CH(CH_2)_7C - OCH_2CHOH$$
$$|$$
$$CH_3$$

Information Sources: 21CFR172.856, 21CFR173.340, 21CFR175.300, 21CFR176.210, 21CFR177.2800, CIR: [S] IJT-18(SUPPL. 2)1999, JCIC, JCLS, TSCA

Chemical Class: Esters

Functions: Skin-Conditioning Agent - Emollient; Surfactant - Emulsifying Agent

Technical/Other Names:
9-Octadecenoic Acid, Monoester with 1,2-Propanediol
Propylene Glycol Monooleate

Trade Names:
  AEC Propylene Glycol Oleate (A & E Connock)
  Sunsoft 25OD (Taiyo Kagaku)

Trade Name Mixture:
  Sunkatol No.1 (Taiyo Kagaku)

PROPYLENE GLYCOL OLEATE SE

Definition: Propylene Glycol Oleate SE is a self-emulsifying grade of Propylene Glycol Oleate (q.v.) that contains some sodium and/or potassium oleate.

Information Source: CIR: [S] IJT-18 (SUPPL. 2)1999

Chemical Class: Esters

Function: Surfactant - Emulsifying Agent

PROPYLENE GLYCOL OLETH-5

Empirical Formula:
  $C_{31}H_{62}O_7$

Definition: Propylene Glycol Oleth-5 is the propylene glycol ether of Oleth-5 (q.v.) that conforms generally to the formula:

$$CH(CH_2)_7CH_3$$
$$CH(CH_2)_8(OCHCH_2)_x(OCH_2CH_2)_yOH$$
$$CH_3$$

where x has an average value of 1 and y has an average value of 5.

Chemical Class: Alkoxylated Alcohols

Function: Surfactant - Emulsifying Agent

Technical/Other Name:
  Propylene Glycol Polyethylene Glycol (5) Oleyl Ether

Trade Name:
  Marlowet 5001 (Sasol GmbH - Marl)

PROPYLENE GLYCOL PROPYL ETHER

CAS Nos.	EINECS Nos.
1569-01-3	216-372-4
30136-13-1	250-069-8

Empirical Formula:
  $C_6H_{14}O_2$

Definition: Propylene Glycol Propyl Ether is the propylene glycol ether of n-propanol that conforms to the formula:

$$CH_3(CH_2)_2OCH_2CHCH_3$$
$$OH$$

Chemical Classes: Alkoxylated Alcohols; Ethers

Function: Solvent

Technical/Other Names:
  2-Propanol, 1-Propoxy-
  Propoxypropanol
  1-Propoxypropan-2-ol
  Propylene Glycol Monopropyl Ether
  1,2-Propylene Glycol 1-Propyl Ether

Trade Names:
  Arcosolv PNP (Lyondell Chemical)
  DOWANOL PnP (Dow Chemical)

Trade Name Mixture:
  ChromaFlair Light Interference Pigment (Flex Products)

PROPYLENE GLYCOL RICINOLEATE

CAS No.	EINECS No.
26402-31-3	247-669-7

JPN Translation:
  リシノレイン酸PG

Empirical Formula:
  $C_{21}H_{40}O_4$

Definition: Propylene Glycol Ricinoleate is the ester of propylene glycol and ricinoleic acid that conforms generally to the formula:

$$CH_3(CH_2)_5CHCH_2CH \qquad O$$
$$OH \quad CH(CH_2)_7C-OCH_2CHOH$$
$$CH_3$$

Information Sources: JCIC, JCLS, JSQI, TSCA

Chemical Class: Esters

Functions: Skin-Conditioning Agent - Emollient; Surfactant - Emulsifying Agent; Suspending Agent - Nonsurfactant

Reported Product Category: Lipsticks

Technical/Other Names:
  12-Hydroxy-9-Octadecenoic Acid, Monoester with 1,2-Propanediol
  9-Octadecenoic Acid, 12-Hydroxy-, Monoester with 1,2-Propanediol
  Propylene Glycol Monoricinoleate

Trade Names:
  AEC Propylene Glycol Ricinoleate (A & E Connock)
  DUB RPG (Stearinerie Dubois Fils)
  Jeechem PGR (Jeen)
  Naturechem PGR (CasChem)
  Protachem PGR (Protameen)
  ROL RP (Fabriquimica)
  Schercemol PGRI (Scher)

PROPYLENE GLYCOL SOYATE

CAS No.	EINECS No.
67784-79-6	267-054-7

Definition: Propylene Glycol Soyate is the ester of propylene glycol and Soy Acid (q.v.).

Information Sources: 21CFR172.856, 21CFR173.340, 21CFR175.105, 21CFR175.300, 21CFR176.170, 21CFR176.200, 21CFR176.210, 21CFR177.2800

Chemical Class: Esters

Functions: Skin-Conditioning Agent - Emollient; Surfactant - Emulsifying Agent

Technical/Other Name:
  Soy Fatty Acid, Monoester with 1,2-Propanediol

Trade Name:
  AEC Propylene Glycol Soyate (A & E Connock)

PROPYLENE GLYCOL STEARATE

CAS Nos.	EINECS Nos.
142-75-6	205-557-5
1323-39-3	215-354-3

JPN Translation:
  ステアリン酸PG

Empirical Formula:
  $C_{21}H_{42}O_3$

Definition: Propylene Glycol Stearate is the ester of propylene glycol and stearic acid that conforms generally to the formula:

$$O$$
$$CH_3(CH_2)_{16}C-OCH_2CHCH_3$$
$$OH$$

Information Sources: 21CFR172.856, 21CFR173.340, 21CFR175.105, 21CFR175.300, 21CFR176.170, 21CFR176.210, 21CFR177.2800, CIR: [S] JACT-2(5)1983, CTFA S, FCC, JCLS, JSCI, MAR, NF XIX, RIFM, TSCA, USAN

Chemical Class: Esters

Functions: Fragrance Ingredient; Skin-Conditioning Agent - Emollient; Surfactant - Emulsifying Agent

Reported Product Categories: Makeup Bases; Moisturizing Preparations; Foundations; Bath Preparations, Misc.; Body and Hand Preparations (Excluding Shaving Preparations); Fragrance Preparations, Misc.; Mascara; Perfumes; Shaving Cream (Aerosol, Brushless and Lather); Skin Care Preparations, Misc.; Makeup Preparations (Not eye), Misc.; Cleansing Products (Cold Creams, Cleansing Lotions, Liquids and Pads); Hair Conditioners; Bath Oils, Tablets, and Salts; Night Skin Care Preparations;

Sachets; Bath Capsules; Eye Lotions; Eye Makeup Preparations, Misc.; Face and Neck Preparations (Excluding Shaving Preparations); Lipsticks; Makeup Fixatives; Shampoos (Non-coloring)

Technical/Other Names:
Octadecanoic Acid, Monoester with 1,2-Propanediol
Propylene Glycol Monooctadecanoate
Propylene Glycol Monostearate
Propylene glycol stearate (RIFM)

Trade Names:
AEC Propylene Glycol Stearate (A & E Connock)
Capmul PGMS (Abitec Corporation)
Ceral P (Fabriquimica)
Cerasynt PA (International Specialty Products)
Emerest 2380 (Cognis Care Chemicals/NJ)
Emerest 2380 (Cognis Care Chemicals/PA)
ESTOL 3737 (Uniqema Europe)
Jeechem PGMS (Jeen)
Lipo PGMS (Lipo)
Monosteol (Gattefosse s.a.)
Nikkol PMS-1C (Nikko)
Nikkol PMS-FR (Nikko)
Prodhybase Pra (Prod'Hyg)
Protachem PGMS (Protameen)
Radiasurf 7201 (Oleon NV)
Rylo PR 18 (Danisco)
Schercemol PGMS (Scher)
Sipoest PGMS (Specialty Industrial)
STEPAN PGMS PURE (Stepan)
Sunsoft 25CD (Taiyo Kagaku)
Unipeg-PGMS (Universal Preserv-A-Chem)

Trade Name Mixtures:
Amisol 4135 (Lucas Meyer GmbH)
Cetasal (Gattefosse s.a.)
Hydrolactol 70 (Gattefosse s.a.)

PROPYLENE GLYCOL STEARATE SE

JPN Translation:
ステアリン酸 PG（SE）

Definition: Propylene Glycol Stearate SE is a self-emulsifying grade of Propylene Glycol Stearate (q.v.) that contains some sodium and/or potassium stearate.

Information Sources: CIR: [S] JACT-2(5)-1983, JCIC, JCLS, JSQI

Chemical Class: Esters

Function: Surfactant - Emulsifying Agent

Reported Product Categories: Foundations; Moisturizing Preparations; Face and Neck Preparations (Excluding Shaving Preparations); Night Skin Care Preparations; Makeup Fixatives; Mascara

Trade Names:
AEC Propylene Glycol Stearate SE (A & E Connock)

Ceral PA (Fabriquimica)
Ceral TP (Fabriquimica)
Lexemul P (Inolex)
Nikkol PMS-1CSE (Nikko)
Prodhybase Prn (Prod'Hyg)
Tesal (Gattefosse s.a.)

PROPYL GALLATE

CAS No.	EINECS No.
121-79-9	204-498-2

JPN Translation:
没食子酸プロピル

Empirical Formula:
$C_{10}H_{12}O_5$

Definition: Propyl Gallate is the aromatic ester of propyl alcohol and Gallic Acid (q.v.). It conforms generally to the formula:

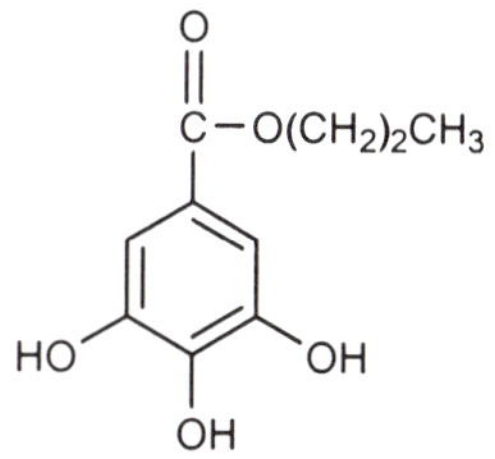

Information Sources: AUS, BP, BPC, 21CFR172.615, 21CFR175.125, 21CFR175.300, 21CFR181.22, 21CFR181.24, 21CFR184.1660, CIR: [SQ] JACT-4(3)1985, CTFA S, CZE, FCC, JCLS, JSCI, MAR, MI-13(7951), NF XIX, PN, RIFM, TSCA, USAN

Chemical Classes: Esters; Phenols

Functions: Antioxidant; Fragrance Ingredient

Reported Product Categories: Lipsticks; Bath Preparations, Misc.; Body and Hand Preparations (Excluding Shaving Preparations); Bath Capsules; Moisturizing Preparations; Skin Care Preparations, Misc.; Makeup Preparations (Not eye), Misc.; Eye Makeup Preparations, Misc.; Face and Neck Preparations (Excluding Shaving Preparations); Bath Oils, Tablets, and Salts; Cleansing Products (Cold Creams, Cleansing Lotions, Liquids and Pads); Eyeliners; Night Skin Care Preparations; Blushers (All types); Eye Shadows; Eyebrow Pencils; Face Powders; Foundations; Indoor Tanning Preparations; Mascara; Suntan Gels, Creams, and Liquids

Technical/Other Names:
Benzoic Acid, 3,4,5-Trihydroxy-, Propyl Ester
Propyl gallate (RIFM)
3,4,5-Trihydroxybenzoic Acid, Propyl Ester

Trade Names:
Progallin P (Clariant)

Progallin P (Clariant GmbH, Personal Care)
Tenox PG (Eastman Chemical)
Uantox PG (Universal Preserv-A-Chem)

Trade Name Mixtures:
Nipanox S1 (Clariant)
Nipanox S1 (Clariant GmbH, Personal Care)
Tenox 2 (Eastman Chemical)
Tenox 6 (Eastman Chemical)
Tenox 7 (Eastman Chemical)
Tenox S-1 (Eastman Chemical)
Uantox 3 (Universal Preserv-A-Chem)

PROPYL METHACRYLATE

CAS No.	EINECS No.
2210-28-8	218-639-0

Definition: Propyl Methacrylate is the homopolymer of propyl methacrylate.

Information Source: JCLS

Chemical Class: Synthetic Polymers

Function: Film Former

PROPYLPARABEN

CAS No.	EINECS No.
94-13-3	202-307-7

JPN Translation:
プロピルパラベン

Empirical Formula:
$C_{10}H_{12}O_3$

Definition: Propylparaben is the ester of n-propyl alcohol and p-hydroxybenzoic acid. It conforms to the formula:

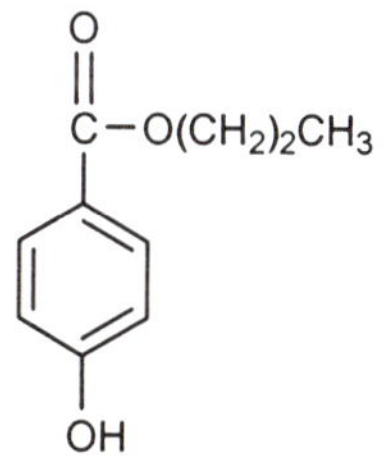

Information Sources: AUS, BP, BPC, BRA, 21CFR150.141, 21CFR150.161, 21CFR172.515, 21CFR181.22, 21CFR181.23, 21CFR184.1670, 21CFR310.545, 21CFR556.550, 21CFR582.3670, CIR: [S] JACT-3(5)1984, CTFA S, CZE, DA, DDR, EEC(VI/1-12), FCC, FIN, HUN, IND, JAN, JCLS, JSCI, MAR, MEX, MHLW-331/3, MI-13(7958), NF XIX, NFJ, PN, RIFM, ROM, TSCA, USAN, USD, YUG

Chemical Classes: Esters; Phenols

Functions: Fragrance Ingredient; Preservative

Reported Product Categories: Moisturizing Preparations; Bath Preparations, Misc.; Lipsticks; Body and Hand Preparations (Excluding Shaving Preparations); Bath Oils, Tablets, and Salts; Cleansing Products (Cold Creams, Cleansing Lotions, Liquids and Pads); Skin Care Preparations, Misc.; Foundations; Blushers (All types); Hair Conditioners; Face Powders; Shampoos (Non-coloring); Bath Capsules; Face and Neck Preparations (Excluding Shaving Preparations); Mascara; Paste Masks (Mud Packs); Makeup Preparations (Not eye), Misc.; Night Skin Care Preparations; Eye Makeup Preparations, Misc.; Tonics, Dressings, and Other Hair Grooming Aids; Makeup Bases; Hair Dyes and Colors (All Types Requiring Caution Statements and Patch Tests); Powders (Dusting and Talcum, Excluding Aftershave Talcs); Bath Soaps and Detergents; Eyebrow Pencils; Suntan Gels, Creams, and Liquids; Personal Cleanliness Products, Misc.; Fragrance Preparations, Misc.; Eyeliners; Eye Makeup Removers; Shaving Cream (Aerosol, Brushless and Lather); Indoor Tanning Preparations; Skin Fresheners; Deodorants (Underarm); Aftershave Lotions; Baby Lotions, Oils, Powders and Creams; Baby Shampoos; Bubble Baths; Hair Preparations (Non-coloring), Misc.; Eye Shadows; Suntan Preparations, Misc.; Shaving Preparations, Misc.; Hair Coloring Preparations, Misc.; Colognes and Toilet Waters; Hair Rinses (Non-coloring); Eye Lotions; Cuticle Softeners; Foot Powders and Sprays; Nail Creams and Lotions; Sachets; Baby Products, Misc.; Makeup Fixatives; Rouges; Dentifrices (Aerosol, Liquid, Pastes and Powders); Hair Sprays (Aerosol Fixatives); Hair Straighteners; Manicuring Preparations, Misc.; Perfumes; Douches; Hair Wave Sets; Nail Polish and Enamels; Permanent Waves; Depilatories; Hair Bleaches; Hair Tints; Leg and Body Paints; Mouthwashes and Breath Fresheners (Liquids and Sprays); Preshave Lotions (All types)

Technical/Other Names:
Benzoic Acid, 4-Hydroxy-, Propyl Ester
4-Hydroxybenzoic Acid, Propyl Ester
Parahydroxybenzoate Ester
Propyl p-hydroxybenzoate (RIFM)
Propyl p-Hydroxybenzoate
Propyl Parahydroxybenzoate

Trade Names:
Botanistat PP (Botanigenics)
CoSept P (Costec)
Jeen Propyl Paraben NF (Jeen)
Lexgard P (Inolex)
Nipasol M (Clariant)

Nipasol M (Clariant GmbH, Personal Care)
Paridol P (Dekker)
Propyl Aseptoform (Greeff)
Propyl-4-hydroxybenzoate (Merck KGaA)
Propylparaben NF (RITA)
Propylparaben NF-PC (Protameen)
Propyl Parasept (Tenneco)
Solbrol P (Bayer AG)
S&M Propylparaben (Schulke & Mayr)
Unisept P (Universal Preserv-A-Chem)

Trade Name Mixtures:
Bactecar 125S (Phytocos)
Bactiphen 2506 G (Grau)
Bentone Gel LOI (ELE)
Chemynol (Chemyunion)
Compositum (Vevy)
Conservateur GD500 (Phytocos)
Conservateur GD700 (Phytocos)
CoSept PEP (RTD Hall Star)
Cosmocil AF (Zeneca)
Covalip LL 48 (LCW)
Dekaben (Dekker)
Dekaben P (Dekker)
Dekacydol (Dekker)
Dermocide L (Fabriquimica)
Dragocid Forte 2/027045 (Symrise)
Elastase Inhibitor-3 (Arval)
Elestab 305 (Laboratoires Serobiologiques)
Elestab 4112 (Laboratoires Serobiologiques)
Erase (Degussa Care Specialties)
Euxyl K 300 (Schulke & Mayr)
Fenilight (Sinerga)
Fenossiparaben (Sinerga)
Germaben II (Sutton)
Germaben II-E (Sutton)
Gramben II (Sinerga)
Greenosome Ace (Greentech)
Killitol II (Collaborative Labs)
Liposerve DUP (Lipo)
Liposerve PP (Lipo)
Microcare DMP (Acti-Chem)
Microcare IMP (Acti-Chem)
Microcare PM (Acti-Chem)
Microcare PM5 (Acti-Chem)
Mikrokill 300 (Arch Personal Care Products)
Neo Dragocide Powder 2/060100 (Symrise)
Neo Dragocid Liquid 2/060110 (Symrise)
Nipaguard BPX (Clariant)
Nipaguard BPX (Clariant GmbH, Personal Care)
Nipaguard MPA (Clariant)
Nipaguard MPA (Clariant GmbH, Personal Care)
Nipaguard MPS (Clariant)
Nipaguard MPS (Clariant GmbH, Personal Care)
Nipaguard PDU (Clariant)
Nipaguard PDU (Clariant GmbH, Personal Care)
Nipasept (Clariant)

Nipasept (Clariant GmbH, Personal Care)
Nipastat (Clariant)
Nipastat (Clariant GmbH, Personal Care)
Paragon II (McIntyre)
Paragon III (McIntyre)
Paragon MEPB (McIntyre)
Paraoxiben (Vevy)
Phenonip (Clariant)
Phenonip (Clariant GmbH, Personal Care)
Phenova (Crodarom)
Pongamia Complex (Greentech)
RonaCare ASC III (Merck KGaA)
RonaCare VTA (Merck KGaA)
Saccaluronate CC (LCW)
Saccaluronate LC (LCW)
Sepicide HB (SEPPIC)
Sepicide HB2 (SEPPIC)
Sepicide WP1 (SEPPIC)
Talcoseptic C (Vevy)
Undebenzofene C (Vevy)
Uniphen P-23 (Induchem)

PROPYLTRIMONIUM HYDROLYZED COLLAGEN

Definition: Propyltrimonium Hydrolyzed Collagen is the quaternary ammonium chloride that conforms generally to the formula:

$$\left[R(CH_2)_3 - \overset{\overset{\displaystyle CH_3}{|}}{\underset{\underset{\displaystyle CH_3}{|}}{N}} - CH_3 \right]^{+} \quad Cl^{-}$$

where R represents the hydrolyzed collagen moiety.

Chemical Classes: Protein Derivatives; Quaternary Ammonium Compounds

Functions: Antistatic Agent; Hair Conditioning Agent; Skin-Conditioning Agent - Miscellaneous; Surfactant - Cleansing Agent

Trade Name:
Colla-Quat PT (Maybrook)

PROPYLTRIMONIUM HYDROLYZED SOY PROTEIN

Definition: Propyltrimonium Hydrolyzed Soy Protein is the quaternary ammonium chloride that conforms generally to the formula:

$$\left[R(CH_2)_3 - \overset{\overset{\displaystyle CH_3}{|}}{\underset{\underset{\displaystyle CH_3}{|}}{N}} - CH_3 \right]^{+} \quad Cl^{-}$$

where R represents the hydrolyzed soy protein moiety.

Chemical Classes: Protein Derivatives; Quaternary Ammonium Compounds

Functions: Antistatic Agent; Hair Conditioning Agent

PROPYLTRIMONIUM HYDROLYZED WHEAT PROTEIN

Definition: Propyltrimonium Hydrolyzed Wheat Protein is the quaternary ammonium chloride that conforms generally to the formula:

$$\left[R(CH_2)_3 - \overset{\overset{\textstyle CH_3}{|}}{\underset{\underset{\textstyle CH_3}{|}}{N}} - CH_3 \right]^+ \quad Cl^-$$

where R represents the hydrolyzed wheat protein moiety.

Chemical Classes: Protein Derivatives; Quaternary Ammonium Compounds

Functions: Antistatic Agent; Hair Conditioning Agent

PROSTANTHERA INCISA LEAF EXTRACT

Definition: Prostanthera Incisa Leaf Extract is an extract of the leaves of *Prostanthera incisa.* See *"Regulatory and Ingredient Use Information," regarding the labeling names for botanical ingredients in Volume 1, Introduction, Part A.*

Chemical Class: Biological Products

Function: Skin-Conditioning Agent - Humectant

Technical/Other Name:
Extract of Prostanthera Incisa Leaf

Trade Name Mixture:
Australian Bush Mint Extract (Bronson and Jacobs)

PROTAMINE SULFATE

CAS No.: 9009-65-8

Definition: Protamine Sulfate is the salt of protamine with sulfuric acid.

Information Source: USP XXIV

Chemical Class: Sulfuric Acid Esters

Function: Skin-Conditioning Agent - Miscellaneous

PROTEASE

CAS No.	EINECS No.
9001-92-7	232-642-4

JPN Translation:
プロテアーゼ

Definition: Protease is an enzyme capable of hydrolyzing a range of nonspecific proteins.

Information Sources: 21CFR310.545, JCLS

Chemical Class: Proteins

Functions: Lytic Agent; Skin-Conditioning Agent - Miscellaneous

Trade Name:
Arazyme (Insect Biotech)

Trade Name Mixtures:
Bellsilk EZ-M (Ichimaru Pharcos)
Bellsilk EZ-P (Ichimaru Pharcos)
Corneosine (Solabia)
E3M-MP (Active Organics)
Kerasoft (Bioland)
Prozymex HBT (Laboratoires Serobiologiques)

PROTEASE DEXTRAN

Definition: Protease Dextran is an adduct of Protease (q.v.) and Dextran (q.v.).

Chemical Classes: Carbohydrates; Proteins

Function: Skin-Conditioning Agent - Miscellaneous

Trade Name:
Aquafit Protease (Takasago)

PRUNELLA VULGARIS LEAF EXTRACT

Definition: Prunella Vulgaris Leaf Extract is an extract of the leaves of *Prunella vulgaris.* See *"Regulatory and Ingredient Use Information," regarding the labeling names for botanical ingredients in Volume 1, Introduction, Part A.*

Chemical Class: Biological Products

Function: Not Reported

Technical/Other Name:
Extract of Prunella Vulgaris

Trade Name Mixture:
Self Heal Extract (Cosmetic Developments)

PRUNUS AFRICANA BARK EXTRACT

CAS No.	EINECS No.
94279-95-5	304-841-7

Definition: Prunus Africana Bark Extract is an extract of the bark of *Prunus africana.* See *"Regulatory and Ingredient Use Information," regarding the labeling names for botanical ingredients in Volume 1, Introduction, Part A.*

Chemical Class: Biological Products

Function: Skin-Conditioning Agent - Miscellaneous

Trade Names:
Phytoselect Prunus (Indena SpA)
Pygeum Africanum Lipid Sterolic Extract (Euromed)

Trade Name Mixtures:
Activated Botanicals Estroherb Complex AB 106 (Norjin)
Phytosterol Complex Concentrate SI 1006 (Norjin)

PRUNUS AMYGDALUS AMARA (BITTER ALMOND) KERNEL OIL

CAS Nos.: 8013-76-1; 8015-75-6

Definition: Prunus Amygdalus Amara (Bitter Almond) Kernel Oil is a volatile oil obtained from kernels of *Prunus amygdalus amara.* See *"Regulatory and Ingredient Use Information," regarding the labeling names for botanical ingredients in Volume 1, Introduction, Part A.*

Information Sources: 21CFR182.20, 27CFR21.65, 27CFR21.151, FCC, HP, MAR, MI-13(6810), RIFM, TSCA, YUG

Chemical Class: Fats and Oils

Function: Fragrance Ingredient

Reported Product Categories: Bath Preparations, Misc.; Body and Hand Preparations (Excluding Shaving Preparations)

Technical/Other Names:
Almond Oil, Bitter
Almond oil, bitter (FFPA) (Prunus spp.) (RIFM)
Bitter Almond Kernel Oil
Bitter Almond Oil
Oil, Essential, Bitter Almond
Oil of Bitter Almond

PRUNUS AMYGDALUS AMARA (BITTER ALMOND) SEED EXTRACT

Definition: Prunus Amygdalus Amara (Bitter Almond) Seed Extract is an extract of the seeds of the bitter almond, *Prunus amygdalus amara.* See *"Regulatory and Ingredient Use Information," regarding the labeling names for botanical ingredients in Volume 1, Introduction, Part A.*

Information Source: 21CFR182.20

Chemical Class: Biological Products

Function: Skin-Conditioning Agent - Miscellaneous

Technical/Other Names:
Bitter Almond Extract
Bitter Almond Seed Extract
Extract of Bitter Almond
Extract of Prunus Amygdalus Amara
Prunus Amygdalus Amara Extract

Trade Name Mixtures:
Bitter Almond HS (Alban Muller)
Bitter Almond LS (Alban Muller)
VT-272 Extract of Bitter Almond (Vege-Tech)

PRUNUS AMYGDALUS DULCIS (SWEET ALMOND) BARK POWDER

Definition: Prunus Amygdalus Dulcis (Sweet Almond) Bark Powder is the powder obtained from the finely, ground bark of *Prunus amygdalus dulcis. See "Regulatory and Ingredient Use Information," regarding the labeling names for botanical ingredients in Volume 1, Introduction, Part A.*

Chemical Class: Biological Products

Function: Skin-Conditioning Agent - Miscellaneous

Trade Name:
Almond Bark Powder (Aveda)

PRUNUS AMYGDALUS DULCIS (SWEET ALMOND) FRUIT EXTRACT

CAS No.	EINECS No.
90320-37-9	291-063-5

Definition: Prunus Amygdalus Dulcis (Sweet Almond) Fruit Extract is an extract of the fruit of the sweet almond, *Prunus amygdalus dulcis. See "Regulatory and Ingredient Use Information," regarding the labeling names for botanical ingredients in Volume 1, Introduction, Part A.*

Chemical Class: Biological Products

Functions: Hair Conditioning Agent; Skin-Conditioning Agent - Miscellaneous

Reported Product Categories: Moisturizing Preparations; Bath Oils, Tablets, and Salts; Cleansing Products (Cold Creams, Cleansing Lotions, Liquids and Pads); Fragrance Preparations, Misc.; Hair Conditioners

Technical/Other Names:
Extract of Prunus Amygdalus Dulcis
Extract of Sweet Almonds
Prunus Amygdalus Dulcis Extract
Sweet Almond Extract
Sweet Almond Fruit Extract

Trade Name Mixtures:
Actiphyte of Almond BG50 (Active Organics)
Actiphyte of Almond GL50 (Active Organics)
Actiphyte of Almond Lipo S (Active Organics)
Actiphyte of Almond PG50 (Active Organics)
Almond Bran Extract HS 2496 G (Grau)
Almondermin LS (Laboratoires Sero-biologiques)
Almond Milk (Weleda)
Almond Phytolait (Alban Muller)
Amanduline SG (Silab)
Aromaphyte of Almond (Active Organics)
Cosflor White Almond HGS (A & E Connock)
Eau D'Anande Mee (Yves Rocher)
Elespher Almondermin (Laboratoires Sero-biologiques)
Extract of Sweet Almond Fruit (Arda Natura)
Extrait D'Amandes Deshuilees MP PG 10 (Yves Rocher)
Extrapone Almond Milk 2/033132 (Symrise)
Extrapone Almond Milk 2/033895 (Symrise)
Glycolysat of Sweet Almond (CEP (Solabia))
Herbasol Distillate Almond (Cosmetochem) (Cosmetochem International Ltd.)
Herbasol-Extract Almond (Cosmetochem)
Oleat of Almond (CEP (Solabia))
Phytoderm Mandorla dolce Glycolic (Universal Flavors)
Sweet Almond HS (Alban Muller)
Sweet Almond LS (Alban Muller)
Sweet Almond Milk (CEP (Solabia))
Sweet Almonds Extract HS 2740 G (Grau)
VT-114 Extract of Almond (Vege-Tech)
Wheat/Sweet Almond Bran Extract HS 2816 G (Grau)

PRUNUS AMYGDALUS DULCIS (SWEET ALMOND) OIL

CAS No.: 8007-69-0

JPN Translation:
アーモンド油

Definition: Prunus Amygdalus Dulcis (Sweet Almond) Oil is the fixed oil obtained from the ripe seed kernel of *Prunus amygdalus dulcis. See "Regulatory and Ingredient Use Information," regarding the labeling names for botanical ingredients in Volume 1, Introduction, Part A.*

Information Sources: ARG, BP, BPC, CIR: [S] JACT-2(5)1983, CTFA S, EGY, FIN, ITA, JCLS, JSCI, MAR, MEX, MI-13 (6878), NF XV, NF XVIII, POR, RIFM, SNPF, TSCA, USAN, USD

Chemical Class: Fats and Oils

Functions: Fragrance Ingredient; Skin-Conditioning Agent - Occlusive

Reported Product Categories: Bath Preparations, Misc.; Body and Hand Preparations (Excluding Shaving Preparations); Moisturizing Preparations; Skin Care Preparations, Misc.; Bath Oils, Tablets, and Salts; Cleansing Products (Cold Creams, Cleansing Lotions, Liquids and Pads); Bath Capsules; Face and Neck Preparations (Excluding Shaving Preparations); Hair Conditioners; Eyebrow Pencils; Suntan Gels, Creams, and Liquids; Night Skin Care Preparations; Tonics, Dressings, and Other Hair Grooming Aids; Bath Soaps and Detergents; Paste Masks (Mud Packs); Indoor Tanning Preparations; Powders (Dusting and Talcum, Excluding Aftershave Talcs); Shampoos (Non-coloring); Baby Lotions, Oils, Powders and Creams; Fragrance Preparations, Misc.; Eye Makeup Preparations, Misc.; Hair Sprays (Aerosol Fixatives); Lipsticks; Personal Cleanliness Products, Misc.; Suntan Preparations, Misc.

Technical/Other Names:
Almond Oil
Almond Oil, Sweet
Almond oil, sweet (Prunis species) (RIFM)
Oil of Sweet Almond
Oils, Almond, Sweet
Sweet Almond Oil

Trade Names:
AEC Sweet Almond Oil (A & E Connock)
Almond Oil (Cosmetochem) (Cosmetochem International Ltd.)
Cropure Almond (Croda Chemicals)
Cropure Almond (Croda, Inc.)
Huile d'Amandes Douces Extra Vierge (Sictia)
Huile Vierge D'Amandes Douces (Bertin)
Jeen Sweet Almond Oil (Jeen)
Lait d'oleosomes (Yves Rocher)
Lipovol ALM (Lipo)
Nikkol Sweet Almond Oil (Nikko)
Phytol ALS (sweet almond oil) (Custom Ingredients)
Prunus Amygdalus Dulcis Oil ies (IES LABO)
Sweet Almond Oil (Dekker)
Sweet Almond Oil (Desert Whale)
Sweet Almond Oil Cosmetic Blend (Henry Lamotte)
Sweet Almond Oil, Premier (Premier Specialties)
Tri-OL ALM (Tri-K)
Unioil Almond (Chemyunion)

Trade Name Mixtures:
Actiphyte Lemon Verbena Lipo AK (Active Organics)

Almond Milk (Cosmetochem)
(Cosmetochem International Ltd.)
Almond Milk J (Alban Muller)
Almond Oil "W" Water Soluble
(Cosmetochem) (Cosmetochem
International Ltd.)
Almond Phytolait (Alban Muller)
Aloe Vera Milk (Cosmetochem)
(Cosmetochem International Ltd.)
Aromaphyte of Almond (Active Organics)
Bamboo Milk (Cosmetochem)
(Cosmetochem International Ltd.)
Cactus Milk (Cosmetochem)
(Cosmetochem International Ltd.)
4-Cereals Milk (Cosmetochem)
(Cosmetochem International Ltd.)
Cocoa Milk (Cosmetochem)
(Cosmetochem International Ltd.)
Cocos Milk (Cosmetochem)
(Cosmetochem International Ltd.)
Eglantineol (Esperis)
Extrapone Almond Milk 2/033132 (Symrise)
Extrapone Almond Milk 2/033895 (Symrise)
Fig Milk (Cosmetochem) (Cosmetochem
International Ltd.)
Germamande (Bertin)
Gilugel ALM (Giulini/Giulini Chemie)
Macadamia Milk (Cosmetochem)
(Cosmetochem International Ltd.)
Mandor-Lat (Sinerga)
Mango Milk (Cosmetochem)
(Cosmetochem International Ltd.)
Oat Milk (Cosmetochem) (Cosmetochem
International Ltd.)
Olio Di Mandorleda Frutti (Vama Farma-
cosmetica)
Ouo Di Mandorue Da Frutti (Va Ma Farma-
cosmetica S.R.L)
4-Spices Milk (Cosmetochem)
(Cosmetochem International Ltd.)
Sweet Almond Softcream (CEP (Solabia))

PRUNUS AMYGDALUS DULCIS (SWEET ALMOND) OIL UNSAPONIFIABLES

Definition: Prunus Amygdalus Dulcis
(Sweet Almond) Oil Unsaponifiables is the
fraction of Prunus Amygdalus Dulcis
(Sweet Almond) Oil (q.v.) which is not
saponified in the refining recovery of sweet
almond oil fatty acids. *See "Regulatory and
Ingredient Use Information," regarding the
labeling names for botanical ingredients in
Volume 1, Introduction, Part A.*

Chemical Class: Unsaponifiables

Functions: Emulsion Stabilizer; Viscosity
Increasing Agent - Nonaqueous

Trade Name Mixture:
Mandorwax (Sinerga)

PRUNUS AMYGDALUS DULCIS (SWEET ALMOND) PROTEIN

Definition: Prunus Amygdalus Dulcis
(Sweet Almond) Protein is a protein

obtained from the sweet almond *Prunus
amygdalus dulcis*. *See "Regulatory and
Ingredient Use Information," regarding the
labeling names for botanical ingredients in
Volume 1, Introduction, Part A.*

Chemical Class: Proteins

Functions: Hair Conditioning Agent; Skin-
Conditioning Agent - Miscellaneous

Technical/Other Names:
Proteins, Prunus Amygdalus Dulcis
Protiens, Sweet Almond
Prunus Amygdalus Dulcis Protein
Sweet Almond Protein

Trade Name Mixtures:
Cerealmilk Premium (Chemyunion)
Dulcemin (Laboratoires Serobiologiques)
P.A. Antifroid LS 9224B (Laboratoires
Serobiologiques)

PRUNUS AMYGDALUS DULCIS (SWEET ALMOND) SEED EXTRACT

JPN Translation:
アーモンドエキス

Definition: Prunus Amygdalus Dulcis
(Sweet Almond) Seed Extract is an extract
of the dried ripe seeds of the sweet
almond, *Prunus amygdalus dulcis*. *See
"Regulatory and Ingredient Use
Information," regarding the labeling names
for botanical ingredients in Volume 1, Introduction, Part A.*

Chemical Class: Biological Products

Function: Not Reported

Technical/Other Names:
Extract of Prunus Amygdalus Dulcis Seed
Extract of Sweet Almond Seed
Prunus Amygdalis Dulcis Extract
Sweet Almond Seed Extract

Trade Name Mixtures:
Actiphyte Almond AQ (Active Organics)
Actiphyte Almond BG100P (Active
Organics)
Actiphyte Almond Lipo M (Active Organics)
Almond Liquid B (Ichimaru Pharcos)
Almond Liquid E (Ichimaru Pharcos)
Almond Milk Protein (Warner Graham)
Eau d'Amande Mee (Yves Rocher)
Kahelift (Silab)
Polylift (Silab)

PRUNUS AMYGDALUS DULCIS (SWEET ALMOND) SEED MEAL

Definition: Prunus Amygdalus Dulcis
(Sweet Almond) Seed Meal is the residue
from the expression of oil from the dried
ripe seeds of sweet almonds, *Prunus

amygdalus dulcis*. *See "Regulatory and
Ingredient Use Information," regarding the
labeling names for botanical ingredients in
Volume 1, Introduction, Part A.*

Information Sources: CIR: [S] JACT-2(5)-
1983, CTFA D, JSQI

Chemical Class: Biological Products

Functions: Abrasive; Bulking Agent

Reported Product Category: Paste Masks
(Mud Packs)

Technical/Other Names:
Almond Meal
Sweet Almond Meal
Sweet Almond Seed Meal

Trade Names:
AEC Sweet Almond Meal (A & E Connock)
Almond Meal (Cosmetochem)
(Cosmetochem International Ltd.)
ESP Almond Meal-Fine Grind (Earth
Supplied Products)
Lipo AMS (Lipo)

Trade Name Mixture:
AEC Cream of Almonds (A & E Connock)

PRUNUS AMYGDALUS DULCIS (SWEET ALMOND) SHELL POWDER

Definition: Prunus Amygdalus Dulcis
(Sweet Almond) Shell Powder is a powder
of the shell of *Prunus amygdalus dulcis*.
*See "Regulatory and Ingredient Use Infor-
mation," regarding the labeling names for
botanical ingredients in Volume 1, Introduction, Part A.*

Information Source: JCLS

Chemical Class: Biological Products

Function: Abrasive

Technical/Other Names:
Powdered Sweet Almond (Prunus
Amygdalus Dulcis)
Prunus Amygdalus Dulcis Powder
Sweet Almond Powder
Sweet Almond Shell Powder

Trade Names:
Actiscrub Sweet Almond Shell (Active
Organics)
AEC Almond Shell Granules (A & E
Connock)
Almond Shell Granules (Brightstern)
Almond Shell Powder (Alban Muller)
Almond Shell Powder 50/200 (Alban
Muller)
Almond Shell Powder 200/300 (Alban
Muller)
Almond Shell Powder 300/400 (Alban
Muller)

PRUNUS ARMENIACA (APRICOT) FRUIT

Definition: Prunus Armeniaca (Apricot) Fruit is the fruit of the apricot, *Prunus armeniaca. See "Regulatory and Ingredient Use Information," regarding the labeling names for botanical ingredients in Volume 1, Introduction, Part A.*

Chemical Class: Biological Products

Functions: Cosmetic Astringent; Skin-Conditioning Agent - Miscellaneous

Technical/Other Names:
 Apricot
 Apricot Fruit

Trade Name:
 AEC Apricot Puree (A & E Connock)

PRUNUS ARMENIACA (APRICOT) FRUIT EXTRACT

CAS No.	EINECS No.
68650-44-2	272-046-1

Definition: Prunus Armeniaca (Apricot) Fruit Extract is an extract of the fruit of the apricot, *Prunus armeniaca. See "Regulatory and Ingredient Use Information," regarding the labeling names for botanical ingredients in Volume 1, Introduction, Part A.*

Information Sources: JSQI, TSCA

Chemical Class: Biological Products

Function: Skin-Conditioning Agent - Miscellaneous

Reported Product Categories: Bath Preparations, Misc.; Hair Conditioners; Skin Care Preparations, Misc.; Tonics, Dressings, and Other Hair Grooming Aids

Technical/Other Names:
 Apricot Extract
 Apricot Fruit Extract
 Extract of Apricot
 Extract of Apricot Fruit
 Extract of Prunus Armeniaca
 Prunus Armeniaca Extract

Trade Name Mixtures:
 ABS Apricot Extract (Active Concepts)
 Actiphyte of Apricot BG50 (Active Organics)
 Actiphyte of Apricot GL50 (Active Organics)
 Actiphyte of Apricot Lipo S (Active Organics)
 Actiphyte of Apricot PG50 (Active Organics)
 A.H.A. Extracts (Ennagram)
 Apricot Extract (Quest International)
 Apricot Extract G (Provital/Centerchem)
 Apricot Extract HS 2509 G (Grau)
 Apricot HG (Alban Muller)
 Apricot HPG Titrated (Alban Muller)
 Apricot HS (Alban Muller)
 Apricot Milk (CEP (Solabia))
 Extrait D'Abricot MPEPG10 (Yves Rocher)
 Extrapone Apricot 2/03365 (Symrise)
 Fruits Milk (CEP (Solabia))
 Fruit Vinegar (Provital/Centerchem)
 Glycolysat of Apricot (CEP (Solabia))
 Herbasol Distillate Apricot (Cosmetochem) (Cosmetochem International Ltd.)
 Herbasol Extract Apricot (Cosmetochem) (Cosmetochem International Ltd.)
 Hormo Fruit Apricot (Esperis)
 Oleat of Apricot (CEP (Solabia))
 Phytelene of Apricot EG 473 liquid (Indena SA)
 Phytogreen 55 of Apricot EXH 687 Liquid (Phytochim)
 Prunus Armeniaca (Apricot) Fruit Extract ies (IES LABO)
 Quassia Vinegar (Provital/Centerchem)
 Vegeles SR (Laboratoires Serobiologiques)
 VT-064 Extract of Apricot (Vege-Tech)

PRUNUS ARMENIACA (APRICOT) FRUIT WATER

Definition: Prunus Armeniaca (Apricot) Fruit Water is an aqueous solution of the steam distillate obtained from the fruit of *Prunus armeniaca. See "Regulatory and Ingredient Use Information," regarding the labeling names for botanical ingredients in Volume 1, Introduction, Part A.*

Chemical Class: Biological Products

Function: Fragrance Ingredient

Technical/Other Names:
 Apricot Fruit Water
 Apricot Water
 Prunus Armeniaca Water
 Water, Apricot
 Water, Prunus Armeniaca

Trade Name:
 Vegebios of Apricot (CEP (Solabia))

PRUNUS ARMENIACA (APRICOT) JUICE

JPN Translation:
 アンズ果汁

Definition: Prunus Armeniaca (Apricot) Juice is the liquid expressed from the fresh pulp of *Prunus armeniaca. See "Regulatory and Ingredient Use Information," regarding the labeling names for botanical ingredients in Volume 1, Introduction, Part A.*

Information Sources: JCIC, JCLS

Chemical Class: Biological Products

Function: Proprietary

Technical/Other Names:
 Apricot Juice
 Juice, Apricot
 Juice, Prunus Armeniaca
 Prunus Armeniaca Juice

Trade Names:
 AEC Apricot Conc. (A & E Connock)
 Authenticals of Apricot (CEP (Solabia))

PRUNUS ARMENIACA (APRICOT) KERNEL EXTRACT

CAS No.	EINECS No.
68650-44-2	272-046-1

JPN Translation:
 アンズ種子エキス

Definition: Prunus Armeniaca (Apricot) Kernel Extract is an extract of the kernels of the apricot, *Prunus armeniaca. See "Regulatory and Ingredient Use Information," regarding the labeling names for botanical ingredients in Volume 1, Introduction, Part A.*

Information Sources: JCIC, JCLS, JSQI, TSCA

Chemical Class: Biological Products

Function: Not Reported

Reported Product Categories: Bath Preparations, Misc.; Skin Care Preparations, Misc.

Technical/Other Names:
 Apricot Kernel Extract
 Extract of Apricot Kernels
 Extract of Prunus Armeniaca Kernel
 Kyounin Ekisu (JPN)
 Prunus Armeniaca Kernel Extract

Trade Name Mixtures:
 Actiphyte of Apricot Kernel BG50 (Active Organics)
 Actiphyte of Apricot Kernel GL50 (Active Organics)
 Actiphyte of Apricot Kernel Lipo S (Active Organics)
 Actiphyte of Apricot Kernel PG50 (Active Organics)
 Apricot Kernel Extract (Maruzen Pharmaceuticals Co., Ltd.)
 Apricot Kernel Extract BG (Maruzen Pharmaceuticals Co., Ltd.)
 Apricot Kernel Extract HS 2813 G (Grau)
 Apricot Kernel Extract LA (Maruzen Pharmaceuticals Co., Ltd.)
 Apricot Kernel Extract Powder-S (Maruzen Pharmaceuticals Co., Ltd.)

PRUNUS ARMENIACA (APRICOT) KERNEL OIL

CAS No.: 72869-69-3

JPN Translation:
パーシック油

Definition: Prunus Armeniaca (Apricot) Kernel Oil is the fixed oil expressed from the kernels of varieties of *Prunus armeniaca*. See *"Regulatory and Ingredient Use Information,"* regarding the labeling names for botanical ingredients in Volume 1, Introduction, Part A.

Information Sources: 21CFR182.40, CTFA S, JCIC, JCLS, JSCI, JSQI, MAR, RIFM, TSCA, USD

Chemical Class: Fats and Oils

Functions: Fragrance Ingredient; Skin-Conditioning Agent - Occlusive

Reported Product Categories: Lipsticks; Bath Preparations, Misc.; Body and Hand Preparations (Excluding Shaving Preparations); Moisturizing Preparations; Bath Capsules; Face and Neck Preparations (Excluding Shaving Preparations); Tonics, Dressings, and Other Hair Grooming Aids; Hair Conditioners; Skin Care Preparations, Misc.; Bath Oils, Tablets, and Salts; Manicuring Preparations, Misc.; Nail Polish and Enamels; Basecoats and Undercoats; Cleansing Products (Cold Creams, Cleansing Lotions, Liquids and Pads); Night Skin Care Preparations; Foundations; Paste Masks (Mud Packs); Suntan Gels, Creams, and Liquids

Technical/Other Names:
Apricot Kernel Oil
Apricot kernel oil (Prunus armeniaca L.) (RIFM)
Kyounin Yu (JPN)
Oil of Apricot Kernels
Persic Oil
Prunus Armeniaca Oil
Vegetol Calendula Me 209 Oily

Trade Names:
AEC Apricot Kernel Oil (A & E Connock)
Apricot Kernel Oil (Dekker)
Apricot Kernel Oil (Desert Whale)
Apricot Kernel Oil (Nestle World Trade)
Apricot Kernel Oil (Cosmetochem) (Cosmetochem International Ltd.)
Apricot oil, Premier (Premier Specialties)
Cold-Pressed Apricot Kernel Oil (Naturex)
Cropure Apricot Kernel (Croda Chemicals)
Cropure Apricot Kernel (Croda, Inc.)
Huile de Noyaux d'Abricot Extra Vierge (Sictia)
Huile de Noyaux D'Abricot Vierge (Bertin)
Lipovol P (Lipo)
Nikkol Apricot Kernel Oil (Nikko)
Phytol APR (apricot seed oil) (Custom Ingredients)
Prunus Armeniaca Kernel Oil ies (IES LABO)
Refined Apricot Kernel Oil (Naturex)

Trade Name Mixtures:
Albiwax (Sinerga)
Apricot Kernel Oil "W" Watersoluble (Cosmetochem) (Cosmetochem International Ltd.)
Apricot Milk (Cosmetochem) (Cosmetochem International Ltd.)
Apricot Milkl (Cosmetochem) (Cosmetochem International Ltd.)
Apricot Softcream (CEP (Solabia))
Carotene Huileux 10000 (Gattefosse s.a.)
Cosmefas 40 (Bertin)
Herbasol Extract Apricot Kernel Oil (Cosmetochem) (Cosmetochem International Ltd.)
Lyco-Sol (Libiol)
Melange Huile de Noyau D'Abricot / Huile de Pepins de Cassis (Bertin)
Olio Di Mandorleda Frutti (Vama Farmacosmetica)
Ouo Di Mandorue Da Frutti (Va Ma Farmacosmetica S.R.L)
Pacific Sea Kelp Oil Extract (Bell Flavors)
Tea Tree Extract (Bell Flavors)
Vegetol Arnica MCF 1397 Oily (Gattefosse s.a.)
Vegetol Burdock 4148 Oily (Gattefosse s.a.)
Vegetol Calendula WL 1072 Oily (Gattefosse s.a.)
Vegetol Capsicum LC 481 Oily (Gattefosse s.a.)
Vegetol Elder 4144 Oily (Gattefosse s.a.)
Vegetol Hawthorn MCF 066 Oily (Gattefosse s.a.)
Vegetol Linden 4141 Oily (Gattefosse s.a.)
Vegetol Mallow 4142 Oily (Gattefosse s.a.)
Vegetol Matricaria 4140 Oily (Gattefosse s.a.)
Vegetol Rosemary 4145 Oily (Gattefosse s.a.)
Vegetol Sage 4138 Oily (Gattefosse s.a.)
Vegetol St. John's Wort MCF 883 Oily (Gattefosse s.a.)
Vegetol White Nettle MCF 1233 Oily (Gattefosse s.a.)
Vegetol White Willow 4151 Oily (Gattefosse s.a.)
Vegetol Wild Roseberry MCF 1837 Oily (Gattefosse s.a.)

PRUNUS ARMENIACA (APRICOT) KERNEL OIL UNSAPONIFIABLES

Definion: Prunus Armeniaca (Apricot) Kernel Oil Unsaponifiables iis the fraction of apricot kernel oil which is not saponified in the refining recovery of apricot kernel oil fatty acids. See *"Regulatory and Ingredient Use Information,"* regarding the labeling names for botanical ingredients in Volume 1, Introduction, Part A.

Chemical Class: Unsaponifiables

Functions: Hair Conditioning Agent; Skin-Conditioning Agent - Miscellaneous

Trade Name Mixture:
Albiwax (Sinerga)

PRUNUS ARMENIACA (APRICOT) LEAF EXTRACT

CAS No.	EINECS No.
68650-44-2	272-046-1

Definition: Prunus Armeniaca (Apricot) Leaf Extract is an extract of the leaves of the apricot, *Prunus armeniaca*. See *"Regulatory and Ingredient Use Information,"* regarding the labeling names for botanical ingredients in Volume 1, Introduction, Part A.

Information Source: TSCA

Chemical Class: Biological Products

Function: Not Reported

Reported Product Categories: Bath Preparations, Misc.; Skin Care Preparations, Misc.

Technical/Other Names:
Apricot Leaf Extract
Extract of Apricot Leaves
Extract of Prunus Armeniaca Leaf
Prunus Armeniaca Leaf Extract

Trade Name Mixtures:
Actiphyte of Apricot Leaves BG50 (Active Organics)
Actiphyte of Apricot Leaves GL50 (Active Organics)
Actiphyte of Apricot Leaves Lipo S (Active Organics)
Actiphyte of Apricot Leaves PG50 (Active Organics)

PRUNUS ARMENIACA (APRICOT) SEED POWDER

JPN Translation:
アンズ核

Definition: Prunus Armeniaca (Apricot) Seed Powder is the powder ground from the seeds of the apricot, *Prunus armeniaca*. See *"Regulatory and Ingredient Use Information,"* regarding the labeling names for botanical ingredients in Volume 1, Introduction, Part A.

Information Sources: JCIC, JCLS, JSQI

Chemical Class: Biological Products

Function: Abrasive

Technical/Other Names:
Apricot Core Grain

Apricot Seed Powder
Powdered Apricot Seed
Prunus Armeniaca Seed

Trade Names:
AEC Apricot Stone Granules (A & E Connock)
AP Grit (Ichimaru Pharcos)
Apricot Abrasives "GBU" (Coaarse, medium-fine & Standard) (Cosmetochem) (Cosmetochem International Ltd.)
Apricot Kernel Powder (Alban Muller)
Apricot Meal (Desert Whale)
Apricot Seed 40-60 Mesh Size (Ikeda)
Apricot Seed 60-80 Mesh Size (Ikeda)
Aprispheres (Naturex)
Lipo APS 40/60 (Lipo)

PRUNUS AVIUM (SWEET CHERRY) FRUIT EXTRACT

Definition: Prunus Avium (Sweet Cherry) Fruit Extract is an extract of the fruit of the sweet cherry, *Prunus avium. See "Regulatory and Ingredient Use Information," regarding the labeling names for botanical ingredients in Volume 1, Introduction, Part A.*

Chemical Class: Biological Products

Function: Not Reported

Technical/Other Names:
Extract of Prunus Avium
Extract of Sweet Cherries
Prunus Avium Extract
Sweet Cherry Extract
Sweet Cherry Fruit Extract

Trade Name Mixtures:
Herbasol Extract Cherry (Cosmetochem) (Cosmetochem International Ltd.)
Sweet Cherry Extract HS 3330 G (Grau)

PRUNUS AVIUM (SWEET CHERRY) FRUIT JUICE

Definition: Prunus Avium (Sweet Cherry) Fruit Juice is the liquid expressed from the fruit of the sweet cherry, *Prunus avium. See "Regulatory and Ingredient Use Information," regarding the labeling names for botanical ingredients in Volume 1, Introduction, Part A.*

Chemical Class: Biological Products

Functions: Cosmetic Astringent; Skin-Conditioning Agent - Miscellaneous

Technical/Other Names:
Juice, Prunus Avium
Juice, Sweet Cherry
Prunus Avium Juice

Sweet Cherry Fruit Juice
Sweet Cherry Juice

Trade Name:
AEC Sweet Cherry Conc (A & E Connock)

PRUNUS AVIUM (SWEET CHERRY) SEED EXTRACT

Definition: Prunus Avium (Sweet Cherry) Seed Extract is an extract of the seeds of *Prunus avium. See "Regulatory and Ingredient Use Information," regarding the labeling names for botanical ingredients in Volume 1, Introduction, Part A.*

Chemical Class: Biological Products

Function: Skin-Conditioning Agent - Miscellaneous

Technical/Other Name:
Extract of Prunus Avium (Sweet Cherry) Seed

Trade Name Mixture:
Cherry Seed Extract (Bell)

PRUNUS AVIUM (SWEET CHERRY) SEED OIL

Definition: Prunus Avium (Sweet Cherry) Seed Oil is the fixed oil obtained from the kernels of cherries. *See "Regulatory and Ingredient Use Information," regarding the labeling names for botanical ingredients in Volume 1, Introduction, Part A.*

Information Source: 21CFR172.510

Chemical Class: Fats and Oils

Function: Skin-Conditioning Agent - Occlusive

Technical/Other Names:
Cherry Pit Oil
Oils, Cherry, Sweet
Sweet Cherry (Prunus Avium) Pit Oil
Sweet Cherry Seed Oil

Trade Names:
AEC Sweet Cherry Kernel Oil (A & E Connock)
Cherry Kernel Oil (Desert Whale)

PRUNUS CERASUS (BITTER CHERRY) EXTRACT

CAS No.	EINECS No.
89997-53-5	289-688-3

Definition: Prunus Cerasus (Bitter Cherry) Extract is an extract of the bitter cherry, *Prunus cerasus. See "Regulatory and Ingredient Use Information," regarding the*

labeling names for botanical ingredients in Volume 1, Introduction, Part A.

Chemical Class: Biological Products

Functions: Antioxidant; Proprietary

Technical/Other Names:
Bitter Cherry Extract
Extract of Bitter Cherry
Extract of Prunus Cerasus
Prunus Cerasus Extract

Trade Name:
Michigan Tart Cherry Extract (Prunus Cerasus) (Formula One Sciences)

Trade Name Mixtures:
Fruitapone Cherry B 2/036440 (Symrise)
Fruitapone Cherry GT 2/037440 (Symrise)
Glycolysat of Cherry (CEP (Solabia))
Herbasol Extract Cherry Stalks (Cosmetochem) (Cosmetochem International Ltd.)
Morello Cherry Extract HS 2851 G (Grau)
Rosamine (Greentech)

PRUNUS CERASUS (BITTER CHERRY) FRUIT

Definition: Prunus Cerasus (Bitter Cherry) Fruit is the fruit of the sour cherry, *Prunus Cerasus. See "Regulatory and Ingredient Use Information," regarding the labeling names for botanical ingredients in Volume 1, Introduction, Part A.*

Chemical Class: Biological Products

Functions: Cosmetic Astringent; Skin-Conditioning Agent - Miscellaneous

Technical/Other Names:
Bitter Cherry
Bitter Cherry Fruit

Trade Name:
AEC Sour Cherry Puree (A & E Connock)

PRUNUS CERASUS (BITTER CHERRY) JUICE

Definition: Prunus Cerasus (Bitter Cherry) Juice is the juice expressed from the fruit of *Prunus cerasus, See "Regulatory and Ingredient Use Information," regarding the labeling names for botanical ingredients in Volume 1, Introduction, Part A.*

Chemical Class: Biological Products

Function: Skin-Conditioning Agent - Miscellaneous

Trade Names:
AEC Red Cherry Conc (A & E Connock)
Authenticals of Cherry (CEP (Solabia))

PRUNUS CERASUS (BITTER CHERRY) SEED OIL

Definition: Prunus Cerasus (Bitter Cherry) Seed Oil is the fixed oil obtained from the pit of *Prunus cerasus*. See "Regulatory and Ingredient Use Information," regarding the labeling names for botanical ingredients in Volume 1, Introduction, Part A.

Chemical Class: Fats and Oils

Function: Fragrance Ingredient

Technical/Other Names:
Bitter Cherry Oil
Bitter Cherry Seed Oil
Oils, Bitter Cherry
Prunus Cerasus Oil

Trade Names:
AEC Bitter Cherry Kernel Oil (A & E Connock)
Huile de Noyaux de Cerise (Bertin)

PRUNUS CERASUS (BITTER CHERRY) STALK EXTRACT

Definition: Prunus Cerasus (Bitter Cherry) Stalk Extract is an extract of the stalks of the bitter cherry, *Prunus cerasus*. See "Regulatory and Ingredient Use Information," regarding the labeling names for botanical ingredients in Volume 1, Introduction, Part A.

Chemical Class: Biological Products

Function: Skin-Conditioning Agent - Miscellaneous

Technical/Other Names:
Bitter Cherry Stalk Extract
Extract of Prunus Cerasus (Bitter Cherry) Stalk

Trade Name Mixture:
Glycolysat of Cherry Stalks (CEP (Solabia))

PRUNUS DOMESTICA FRUIT EXTRACT

CAS No.	EINECS No.
90082-87-4	290-179-3

JPN Translation:
プルーンエキス

Definition: Prunus Domestica Fruit Extract is an extract of the fruit of the plum, *Prunus domestica*. See "Regulatory and Ingredient Use Information," regarding the labeling names for botanical ingredients in Volume 1, Introduction, Part A.

Information Sources: JCIC, JCLS, JSQI

Chemical Class: Biological Products

Functions: Proprietary; Skin-Conditioning Agent - Miscellaneous

Technical/Other Names:
Extract of Plum
Extract of Prunus Domestica
Plum Extract
Plum (Prunus Domestica) Extract
Prune Extract

Trade Names:
AEC Yellow Plum Conc 68 Brix Pasteurized, Unpreserved (A & E Connock)
Dried Plum Extract (Formula One Sciences)
Plum Liquid E (Ichimaru Pharcos)

Trade Name Mixtures:
Actiphyte of Prune (Active Organics)
Actiphyte of Prune BG50 (Active Organics)
Actiphyte of Prune GL50 (Active Organics)
Actiphyte of Prune Lipo S (Active Organics)
Actiphyte of Prune PG50 (Active Organics)
Actiphyte Plum (Active Organics)
Actiphyte Plum BG50P (Active Organics)
Damson Extract (Cosmetic Developments)
Herbasol Extract Plum (Cosmetochem) (Cosmetochem International Ltd.)
Plum Liquid B (Ichimaru Pharcos)
Plum Tree HS (Alban Muller)
Prunus Domestica Fruit Extract ies (IES LABO)
VT-301 Extract of Plum (Vege-Tech)

PRUNUS DOMESTICA FRUIT JUICE

Definition: Prunus Domestica Fruit Juice is the juice expressed from the fruit of *Prunus domestica*. See "Regulatory and Ingredient Use Information," regarding the labeling names for botanical ingredients in Volume 1, Introduction, Part A.

Chemical Class: Biological Products

Function: Skin-Conditioning Agent - Miscellaneous

Trade Names:
AEC Prune Conc (A & E Connock)
AEC Red Plum Conc (A & E Connock)

PRUNUS DOMESTICA SEED EXTRACT

Definition: Prunus Domestica Seed Extract is an extract of the seeds of *Prunus domestica*. See "Regulatory and Ingredient Use Information," regarding the labeling names for botanical ingredients in Volume 1, Introduction, Part A.

Chemical Class: Biological Products

Function: Skin-Conditioning Agent - Emollient

Technical/Other Name:
Extract of Prunus Domestica Seed

Trade Name:
Virgin Prune Oil (Huile Viergeale Prune) (Expanscience)

PRUNUS DOMESTICA SEED OIL

Definition: Prunus Domestica Seed Oil is the fixed oil expressed from the seeds of *Prunus domestica*. See "Regulatory and Ingredient Use Information," regarding the labeling names for botanical ingredients in Volume 1, Introduction, Part A.

Chemical Class: Fats and Oils

Function: Skin-Conditioning Agent - Occlusive

Technical/Other Name:
Oils, Prunus Domestica Seed

Trade Names:
Lipofructyl DP LS (Laboratoires Sero-biologiques)
Prune Oil (IES LABO)

PRUNUS INSITITIA JUICE

Definition: Prunus Insititia Juice is the juice expressed from the pulp of the plum, *Prunus insititia*. See "Regulatory and Ingredient Use Information," regarding the labeling names for botanical ingredients in Volume 1, Introduction, Part A.

Chemical Class: Biological Products

Function: Skin-Conditioning Agent - Miscellaneous

Trade Name:
Authenticals of Mirabelle Plum (CEP (Solabia))

PRUNUS INSITITIA SEED OIL

CAS No.: 216446-21-8

Definition: Prunus Insititia Seed Oil is the fixed oil expressed from the seeds of *Prunus insititia*. See "Regulatory and Ingredient Use Information," regarding the labeling names for botanical ingredients in Volume 1, Introduction, Part A.

Chemical Class: Fats and Oils

Functions: Skin-Conditioning Agent - Emollient; Skin-Conditioning Agent - Occlusive

Technical/Other Name:
Oils, Prunus Insititia

Trade Name:
Lipofructyl PI (Laboratoires Sero-biologiques)

PRUNUS MUME FRUIT

JPN Translation:
ウメ

Definition: Prunus Mume Fruit is the dried powder obtained from the fruit of *Prunus mume. See "Regulatory and Ingredient Use Information," regarding the labeling names for botanical ingredients in Volume 1, Introduction, Part A.*

Information Source: JCIC

Chemical Class: Biological Products

Function: Abrasive

PRUNUS MUME FRUIT EXTRACT

Definition: Prunus Mume Fruit Extract is an extract of the fruit of *Prunus mume. See "Regulatory and Ingredient Use Information," regarding the labeling names for botanical ingredients in Volume 1, Introduction, Part A.*

Chemical Class: Biological Products

Function: Humectant

Technical/Other Name:
Extract of Prunus Mume Fruit

Trade Name Mixture:
Mume Extract (Wamiles)

PRUNUS MUME FRUIT WATER

Definition: Prunus Mume Fruit Water is an aqueous solution of the steam distillate obtained from the fruit of *Prunus mume. See "Regulatory and Ingredient Use Information," regarding the labeling names for botanical ingredients in Volume 1, Introduction, Part A.*

Chemical Class: Biological Products

Function: Skin-Conditioning Agent - Humectant

Trade Names:
Maesil Water (Korea Kolmar)
Ume Fruit Water F (Maruzen Pharmaceuticals Co., Ltd.)
Umesui (Iwase Cosfa)

PRUNUS MUME SEED EXTRACT

Definition: Prunus Mume Seed Extract is an extract of the seeds of *Prunus mume. See "Regulatory and Ingredient Use Information," regarding the labeling names for botanical ingredients in Volume 1, Introduction, Part A.*

Chemical Class: Biological Products

Function: Skin-Conditioning Agent - Miscellaneous

Technical/Other Name:
Extract of Prunus Mume Seed

Trade Name:
Ume Seed Extract (Naris)

PRUNUS PERSICA NECTARINA EXTRACT

Definition: Prunus Persica Nectarina Extract is an extract of the nectarine, *Prunus persica nectarina. See "Regulatory and Ingredient Use Information," regarding the labeling names for botanical ingredients in Volume 1, Introduction, Part A.*

Chemical Class: Biological Products

Function: Not Reported

Technical/Other Name:
Nectarine (Prunus Persica Nectarina) Extract

Trade Name Mixture:
Nectarin Extract HG (Provital/Centerchem)

PRUNUS PERSICA (PEACH) BUD EXTRACT

Definition: Prunus Persica (Peach) Bud Extract is an extract of the buds of *Prunus persica. See "Regulatory and Ingredient Use Information," regarding the labeling names for botanical ingredients in Volume 1, Introduction, Part A.*

Chemical Class: Biological Products

Function: Skin-Conditioning Agent - Humectant

Technical/Other Name:
Extract of Prunus Persica (Peach) Bud

Trade Name Mixture:
Extrait Hydroglycolique de Bourgeons de Pecher (Greentech)

PRUNUS PERSICA (PEACH) FLOWER EXTRACT

Definition: Prunus Persica (Peach) Flower Extract is an extract of the flowers of the peach, *Prunus persica. See "Regulatory and Ingredient Use Information," regarding the labeling names for botanical ingredients in Volume 1, Introduction, Part A.*

Chemical Class: Biological Products

Function: Skin-Conditioning Agent - Miscellaneous

Technical/Other Name:
Peach Flower Extract

Trade Name Mixture:
Peach Flower HS (Alban Muller)

PRUNUS PERSICA (PEACH) FRUIT

Definition: Prunus Persica (Peach) Fruit is the fruit of the peach, *Prunus persica. See "Regulatory and Ingredient Use Information," regarding the labeling names for botanical ingredients in Volume 1, Introduction, Part A.*

Chemical Class: Biological Products

Functions: Cosmetic Astringent; Skin-Conditioning Agent - Miscellaneous

Technical/Other Names:
Peach
Peach Fruit

Trade Name:
AEC Peach Puree (A & E Connock)

PRUNUS PERSICA (PEACH) FRUIT EXTRACT

CAS No.	EINECS No.
84012-34-0	281-678-7

Definition: Prunus Persica (Peach) Fruit Extract is an extract of the fruit of the peach, *Prunus persica. See "Regulatory and Ingredient Use Information," regarding the labeling names for botanical ingredients in Volume 1, Introduction, Part A.*

Information Sources: 21CFR172.510, JSQI

Chemical Class: Biological Products

Function: Not Reported

Reported Product Category: Paste Masks (Mud Packs)

Technical/Other Names:
Extract of Peach
Extract of Prunus Persica
Peach Extract
Peach Fruit Extract
Prunus Persica Extract

Trade Name Mixtures:
Actiphyte of Peach BG50 (Active Organics)
Actiphyte of Peach GL50 (Active Organics)
Actiphyte of Peach Lipo S (Active Organics)
Actiphyte of Peach PG50 (Active Organics)
A.H.A. Extracts (Ennagram)
A.H.A. Extracts (Phytochim)
Extrait De Peche MP PG SS (Yves Rocher)
Fruitapone Peach B 2/036550 (Symrise)
Fruitapone Peach GT 2/037550 (Symrise)

Fruits Milk (CEP (Solabia))
Fruit Vinegar (Provital/Centerchem)
Glycolysat of Peach (CEP (Solabia))
Herbasol Extract Peach (Cosmetochem)
(Cosmetochem International Ltd.)
Hydralphatine 3P (Lanatech)
Oleat of Peach (CEP (Solabia))
Peach Extract HG (Provital/Centerchem)
Peach Extract HS 2791 G (Grau)
Peach Fruit Extract (Maruzen
Pharmaceuticals Co., Ltd.)
Peach Fruit Extract BG (Maruzen
Pharmaceuticals Co., Ltd.)
Peach Milk (CEP (Solabia))
Peach Tree Milk (CEP (Solabia))
Pescolat (Sinerga)
Phytelene of Peach EG 491 liquid (Indena
SA)
Phytogreen 55 of Peach EXH 698 Liquid
(Phytochim)
Quassia Vinegar (Provital/Centerchem)
Six Fruit Concentrate 3519 G (Grau)
Vegebios of Peach (CEP (Solabia))
VT-105 Extract of Peach (Vege-Tech)

PRUNUS PERSICA (PEACH) JUICE

JPN Translation:
モモ果汁

Definition: Prunus Persica (Peach) Juice
is the liquid expressed from the fresh pulp
of the peach, *Prunus persica. See
"Regulatory and Ingredient Use
Information," regarding the labeling names
for botanical ingredients in Volume 1, Introduction, Part A.*

Information Sources: JCIC, JCLS

Chemical Class: Biological Products

Function: Skin-Conditioning Agent - Miscellaneous

Technical/Other Names:
Juice, Peach
Juice, Prunus Persica
Peach Juice
Prunus Persica Juice

Trade Names:
AEC Peach Juice (A & E Connock)
Peach Juice Concentrate (Symrise)

Trade Name Mixtures:
Complex 1 - Antipollution (Provital/
Centerchem)
Extrapone Peach 2/033319 (Symrise)
Hormo Fruit Peach (Esperis)
Pronalen Bio-Protect (Provital/Centerchem)

PRUNUS PERSICA (PEACH) KERNEL EXTRACT

JPN Translation:
モモ種子エキス

Definition: Prunus Persica (Peach) Kernel
Extract is an extract of the kernels of the
peach, *Prunus persica. See "Regulatory
and Ingredient Use Information," regarding
the labeling names for botanical ingredients
in Volume 1, Introduction, Part A.*

Chemical Class: Biological Products

Function: Not Reported

Technical/Other Names:
Extract of Peach Kernels
Extract of Prunus Persica Kernels
Peach Kernel Extract
Prunus Persica Kernel Extract
Tounin Ekisu (JPN)

Trade Name Mixtures:
Actiphyte Peach Kernel (Active Organics)
Actiphyte Peach Kernel AL (Active
Organics)
Actiphyte Peach Kernel BG50P (Active
Organics)
Actiphyte Peach Kernel Lipo S (Active
Organics)
Peach Seed Extract (Maruzen
Pharmaceuticals Co., Ltd.)
Peach Seed Extract BG (Maruzen
Pharmaceuticals Co., Ltd.)
Peach Seed Extract-J (Maruzen
Pharmaceuticals Co., Ltd.)
Peach Seed Extract-JC (Maruzen
Pharmaceuticals Co., Ltd.)
Tounin Liquid B (Ichimaru Pharcos)
Tounin Liquid E (Ichimaru Pharcos)

PRUNUS PERSICA (PEACH) KERNEL OIL

CAS Nos.: 8002-78-6; 8023-98-1

Definition: Prunus Persica (Peach) Kernel
Oil is the oil expressed from the kernels of
*Prunus persica. See "Regulatory and
Ingredient Use Information," regarding the
labeling names for botanical ingredients in
Volume 1, Introduction, Part A.*

Information Sources: BEL, 21CFR182.40,
JCLS, JSCI, MAR, MI-13(7130), NF XV, PF,
USD

Chemical Class: Fats and Oils

Function: Skin-Conditioning Agent -
Occlusive

Reported Product Categories: Moisturizing
Preparations; Lipsticks

Technical/Other Names:
Oils, Peach Kernel
Peach Oil, Expressed
Persic Oil

Trade Names:
AEC Peach Kernel Oil (A & E Connock)
Peach Kernel Oil (Desert Whale)

Trade Name Mixtures:
Olio Di Mandorleda Frutti (Vama Farma-
cosmetica)
Ouo Di Mandorue Da Frutti (Va Ma Farma-
cosmetica S.R.L)
Peach Milk (Cosmetochem)
(Cosmetochem International Ltd.)
Pescolat (Sinerga)
Protachem 35A (Protameen)

PRUNUS PERSICA (PEACH) LEAF EXTRACT

CAS No.	EINECS No.
84012-34-0	281-678-7

JPN Translation:
モモ葉エキス

Definition: Prunus Persica (Peach) Leaf
Extract is an extract of the leaves of the
peach, *Prunus persica. See "Regulatory
and Ingredient Use Information," regarding
the labeling names for botanical ingredients
in Volume 1, Introduction, Part A.*

Information Sources: JCIC, JCLS, JSQI

Chemical Class: Biological Products

Function: Not Reported

Technical/Other Names:
Extract of Peach Leaves
Extract of Prunus Persica Leaf
Peach Leaf Extract
Prunus Persica Leaf Extract

Trade Name Mixtures:
Actiphyte of Peach Leaves BG50 (Active
Organics)
Actiphyte of Peach Leaves GL50 (Active
Organics)
Actiphyte of Peach Leaves Lipo S (Active
Organics)
Actiphyte of Peach Leaves PG50 (Active
Organics)
Anti-wrinkles Blend (Alban Muller)
Bathgranue Peach Leaf (Ichimaru Pharcos)
Flat Peach Extract (Nonogawa)
Peach Leaf Extract (Maruzen
Pharmaceuticals Co., Ltd.)
Peach Leaf Extract BG (Maruzen
Pharmaceuticals Co., Ltd.)
Peach Leaf Extract BG100 (Maruzen
Pharmaceuticals Co., Ltd.)
Peach Leaf Extract ET100 (Maruzen
Pharmaceuticals Co., Ltd.)
Peach Leaf Extract LA (Maruzen
Pharmaceuticals Co., Ltd.)
Peach Leaf Extract Powder-S (Maruzen
Pharmaceuticals Co., Ltd.)
Peach Leaf Extract SQ (Maruzen
Pharmaceuticals Co., Ltd.)
Peach Leaf Liquid-B (Ichimaru Pharcos)
Peach Tree Milk (CEP (Solabia))
RAYOLYS (Greentech S.A)

Vegebios Peach Tree (CEP (Solabia))
Vege Plex VP#1341 (Vege-Tech)
Vegetol Peach Tree LC 339 Hydro
(Gattefosse s.a.)
VT-210 Extract of Peach Leaves (Vege-Tech)

PRUNUS PERSICA (PEACH) SEED POWDER

JPN Translation:
モモ核

Definition: Prunus Persica (Peach) Seed Powder is a powder of the ground pits of the peach, *Prunus persica*. See *"Regulatory and Ingredient Use Information,"* regarding the labeling names for botanical ingredients in Volume 1, Introduction, Part A.

Information Sources: JCIC, JCLS, JSQI

Chemical Class: Biological Products

Functions: Abrasive; Bulking Agent

Technical/Other Names:
Peach Core Grain
Peach Pit Powder
Peach (Prunus Persica) Pit Powder
Peach Seed Powder
Powdered Peach Pits

Trade Names:
AEC Peach Stone Granules (A & E Connock)
Peach Seed 40-60 Mesh Size (Ikeda)
Peach Seed 60-80 Mesh Size (Ikeda)
PS Grit (Ichimaru Pharcos)

PRUNUS SEROTINA (WILD CHERRY)

Definition: Prunus Serotina (Wild Cherry) is a plant material derived from the dried stem bark of the wild cherry, *Prunus serotina*. See *"Regulatory and Ingredient Use Information,"* regarding the labeling names for botanical ingredients in Volume 1, Introduction, Part A.

Information Sources: BPC, HP, MAR, MI-13(10104), USD

Chemical Class: Biological Products

Function: Not Reported

Technical/Other Name:
Wild Cherry

PRUNUS SEROTINA (WILD CHERRY) BARK EXTRACT

CAS Nos.	EINECS No.
8000-44-0	
84604-07-9	283-284-0

Definition: Prunus Serotina (Wild Cherry) Bark Extract is an extract of the bark of the wild cherry, *Prunus serotina*. See *"Regulatory and Ingredient Use Information,"* regarding the labeling names for botanical ingredients in Volume 1, Introduction, Part A.

Information Sources: 21CFR182.20, HP, RIFM, TSCA

Chemical Class: Biological Products

Functions: Fragrance Ingredient; Not Reported

Technical/Other Names:
Cherry bark, wild, extract (Prunus serotina Ehrh.) (RIFM)
Cherry laurel oil (FFPA) (Prunus laurocerasus L.) (RIFM)
Extract of Prunus Serotina Bark
Extract of Wild Cherry
Extract of Wild Cherry Bark
Prunus Serotina Extract
Wild Cherry Bark Extract

Trade Name Mixtures:
Actiphyte of Wild Cherry Bark BG50 (Active Organics)
Actiphyte of Wild Cherry Bark GL50 (Active Organics)
Actiphyte of Wild Cherry Bark Lipo S (Active Organics)
Actiphyte of Wild Cherry Bark PG50 (Active Organics)
Activated Botanicals P8-MIN AB 102 (Norjin)
VT-012 Cherry Bark (Vege-Tech)
VT-012, Extract Of Cherry Bark (Vege-Tech)
VT-300 Extract of Wild Cherry Bark (Vege-Tech)

PRUNUS SEROTINA (WILD CHERRY) FRUIT EXTRACT

CAS No.	EINECS No.
84604-07-9	283-284-0

Definition: Prunus Serotina (Wild Cherry) Fruit Extract is an extract of the dried fruit of the wild cherry, *Prunus serotina*. See *"Regulatory and Ingredient Use Information,"* regarding the labeling names for botanical ingredients in Volume 1, Introduction, Part A.

Information Source: RIFM

Chemical Class: Biological Products

Functions: Fragrance Ingredient; Not Reported

Technical/Other Names:
Cherry bark, wild, extract (Prunus serotina Ehrh.) (RIFM)
Extract of Prunus Serotina Fruit
Extract of Wild Cherry
Extract of Wild Cherry Fruit
Prunus Serotina Extract
Wild Cherry Extract
Wild Cherry Fruit Extract

Trade Name Mixtures:
Actiphyte of Wild Cherry BG50 (Active Organics)
Actiphyte of Wild Cherry GL50 (Active Organics)
Actiphyte of Wild Cherry Lipo S (Active Organics)
Actiphyte of Wild Cherry PG50 (Active Organics)

PRUNUS SERRULATA BARK EXTRACT

Definition: Prunus Serrulata Bark Extract is an extract of the bark of *Prunus serrulata*. See *"Regulatory and Ingredient Use Information,"* regarding the labeling names for botanical ingredients in Volume 1, Introduction, Part A.

Chemical Class: Biological Products

Function: Skin-Conditioning Agent - Miscellaneous

Trade Name:
Sakura Bark Extract (Naris)

PRUNUS SERRULATA FLOWER EXTRACT

Definition: Prunus Serrulata Flower Extract is an extract of the flowers of *Prunus serrulata*. See *"Regulatory and Ingredient Use Information,"* regarding the labeling names for botanical ingredients in Volume 1, Introduction, Part A.

Chemical Class: Biological Products

Function: Skin-Conditioning Agent - Miscellaneous

Trade Name:
Sakura Flower Extract (Naris)

PRUNUS SPECIOSA LEAF EXTRACT

Definition: Prunus Speciosa Leaf Extract is an extract of the leaves of *Prunus speciosa*. See *"Regulatory and Ingredient Use Information,"* regarding the labeling names for botanical ingredients in Volume 1, Introduction, Part A.

Chemical Class: Biological Products

Function: Not Reported

Technical/Other Names:
Extract of Prunus Speciosa
Oshima Cherry Leaf Extract

Trade Name Mixtures:
Sakura Leaf Liquid B (Ichimaru Pharcos)
Sakura Leaf Liquid E (Ichimaru Pharcos)

PRUNUS SPINOSA FLOWER WATER

Definition: Prunus Spinosa Flower Water is an aqueous solution containing volatile oils obtained by the distillation of the flowers of *Prunus spinosa. See "Regulatory and Ingredient Use Information," regarding the labeling names for botanical ingredients in Volume 1, Introduction, Part A.*

Chemical Class: Biological Products

Function: Not Reported

Trade Name Mixture:
Blackthorn Blossom Distillate (Weleda)

PRUNUS SPINOSA FRUIT JUICE

Definition: Prunus Spinosa Fruit Juice is the liquid expressed from the fruit of the blackthorn, *Prunus spinosa. See "Regulatory and Ingredient Use Information," regarding the labeling names for botanical ingredients in Volume 1, Introduction, Part A.*

Chemical Class: Biological Products

Functions: Cosmetic Astringent; Flavoring Agent

Technical/Other Names:
Blackthorn Juice
Juice, Blackthorn
Juice, Prunus Spinosa
Juice, Prunus Spinoza

Trade Names:
Blackthorn Juice (Weleda)
Sloe Juice Concentrate (Bayernwald)

Trade Name Mixtures:
Fruitapone Blackthorn B 2/036700 (Symrise)
Fruitapone Blackthorn GT 2/037700 (Symrise)

PSALLIOTA CAMPESTRIS (MUSHROOM) EXTRACT

Definition: Psalliota Campestris (Mushroom) Extract is an extract of the mushroom, *Psalliota campestris. See "Regulatory and Ingredient Use Information," regarding the labeling names for botanical ingredients in Volume 1, Introduction, Part A.*

Chemical Class: Biological Products

Function: Not Reported

Technical/Other Names:
Extract of Mushroom
Extract of Psalliota Campestris
Mushroom Extract
Psalliota Campestris Extract

Trade Name Mixtures:
Extrait N30 (CEP (Solabia))
Mushroom Milk (CEP (Solabia))

PSEUDANABAENA GALEATA EXTRACT

Definition: Pseudanabaena Galeata Extract is an extract of the whole plant, *Pseudanabaena galeata. See "Regulatory and Ingredient Use Information," regarding the labeling names for botanical ingredients in Volume 1, Introduction, Part A.*

Chemical Class: Biological Products

Function: Not Reported

Technical/Other Name:
Extract of Pseudanabaena Galeata

Trade Name Mixture:
Pseudanabaena Cafeine 400 EtOH (Eclosarium)

PSEUDOALTEROMONAS FERMENT EXTRACT

Definition: Pseudoalteromonas Ferment Extract is an extract of the fermentation product of *Pseudoalteromonas.*

Chemical Class: Biological Products

Functions: Humectant; Skin-Conditioning Agent - Humectant

Trade Name:
Antarcticine (Lipotec/Centerchem)

PSEUDOTSUGA MENZIESII

Definition: *See "Regulatory and Ingredient Use Information," regarding EU labeling names for botanical ingredients in Volume 1, Introduction, Part A.*

Chemical Class: Biological Products

Technical/Other Name:
Pseudotsuga Menziesii (Balsam Oregon) Resin (U.S.)

PSEUDOTSUGA MENZIESII (BALSAM OREGON) RESIN

CAS No.: 8050-89-3

Definition: Pseudotsuga Menziesii (Balsam Oregon) Resin is an oleoresin obtained from *Pseudotsuga menziesi. See "Regulatory and Ingredient Use Information," regarding the labeling names for botanical ingredients in Volume 1, Introduction, Part A.*

Chemical Class: Biological Products

Function: Film Former

Technical/Other Names:
Balsam Fir Oregon
Balsam Oregon
Balsam Oregon Resin
Balsams, Pseudotsuga Menziesi
Douglas Fir Oil
Oregon Balsam
Pseudotsuga Menziesi Balsam
Pseudotsuga Menziesii (EU)
Pseudotsuga Menziesi Resin

PSIDIUM GUAJAVA FRUIT EXTRACT

CAS No. 90045-46-8 **EINECS No.** 289-907-2

Definition: Psidium Guajava Fruit Extract is an extract of the fruit of the guava, *Psidium guajava. See "Regulatory and Ingredient Use Information," regarding the labeling names for botanical ingredients in Volume 1, Introduction, Part A.*

Chemical Class: Biological Products

Function: Skin-Conditioning Agent - Miscellaneous

Reported Product Categories: Bath Oils, Tablets, and Salts; Eyeliners; Indoor Tanning Preparations

Technical/Other Names:
Extract of Guava
Extract of Psidium Guajava
Guava Extract
Guava (Psidium Guajava) Extract

Trade Name Mixtures:
Actiphyte of Guava BG50 (Active Organics)
Actiphyte of Guava GL50 (Active Organics)
Actiphyte of Guava Lipo S (Active Organics)
Actiphyte of Guava PG50 (Active Organics)
A.H.A. Extracts (Ennagram)
A.H.A. Extracts (Phytochim)
Extrait de Goyave ME 100 (Yves Rocher)
Guana Extract (Bio-Botanica)
Guava HS (Alban Muller)
Herbasol Extract Guajara (Cosmetochem) (Cosmetochem International Ltd.)
Phytelene of Guava BG 717 (Indena SA)
Phytelene of Guava EG 717 (Indena SA)
Psidium Guajava Fruit Extract ies (IES LABO)
VT-066 Extract of Guava (Vege-Tech)

PSIDIUM GUAJAVA FRUIT JUICE

Definition: Psidium Guajava Fruit Juice is the juice expressed from the fruit of the guava, *Psidium guajava. See "Regulatory and Ingredient Use Information," regarding the labeling names for botanical ingredients in Volume 1, Introduction, Part A.*

Chemical Class: Biological Products

Functions: Cosmetic Astringent; Skin-Conditioning Agent - Miscellaneous

Technical/Other Names:
Guava (Psidium Guajava) Juice
Juice, Guava
Juice, Psidium Guajava
Psidium Guajava Juice

Trade Name:
AEC Guava Conc (A & E Connock)

PSIDIUM GUAJAVA LEAF EXTRACT

CAS No.	EINECS No.
90045-46-8	289-907-2

Definition: Psidium Guajava Leaf Extract is an extract of the leaves of the guava, *Psidium guajava. See "Regulatory and Ingredient Use Information," regarding the labeling names for botanical ingredients in Volume 1, Introduction, Part A.*

Chemical Class: Biological Products

Function: Not Reported

Reported Product Categories: Bath Oils, Tablets, and Salts; Eyeliners; Indoor Tanning Preparations

Technical/Other Names:
Extract of Guava Leaf
Extract of Psidium Guajava Leaf
Guava (Psidium Guajava) Leaf Extract

Trade Name Mixtures:
Guavacal (Cosmecal Sarl)
Guava Extract (Yamakawa)
PSIDIUM GUAJAVA EXTRACT LA
(Maruzen Pharmaceuticals Co., Ltd.)

PSORALEA CORYLIFOLIA

Definition: *See "Regulatory and Ingredient Use Information," regarding EU labeling names for botanical ingredients in Volume 1, Introduction, Part A.*

Chemical Class: Biological Products

Technical/Other Name:
Psoralea Corylifolia Extract (U.S.)

PSORALEA CORYLIFOLIA EXTRACT

Definition: Psoralea Corylifolia Extract is an extract of the fruit and seeds of the psoralea, *Psoralea corylifolia. See "Regulatory and Ingredient Use Information," regarding the labeling names for botanical ingredients in Volume 1, Introduction, Part A.*

Chemical Class: Biological Products

Function: Not Reported

Technical/Other Names:
Extract of Psoralea
Extract of Psoralea Corylifolia
Psoralea Corylifolia (EU)
Psoralea Extract

Trade Name Mixtures:
Actiphyte of Psoralea (Active Organics)
Actiphyte of Psoralea BG50 (Active Organics)
Actiphyte of Psoralea GL50 (Active Organics)
Actiphyte of Psoralea Lipo S (Active Organics)
Actiphyte of Psoralea PG50 (Active Organics)

PTEROCARPUS MARSUPIUM BARK EXTRACT

CAS No.	EINECS No.
84604-08-0	283-285-6

Definition: Pterocarpus Marsupium Bark Extract is an extract of the bark of *Pterocarpus marsupium. See "Regulatory and Ingredient Use Information," regarding the labeling names for botanical ingredients in Volume 1, Introduction, Part A.*

Chemical Class: Biological Products

Function: Hair Conditioning Agent

Technical/Other Name:
Extract of Pterocarpus Marsupium Bark

Trade Name Mixtures:
Indian Kino Tree Extract (Nonogawa)
Trichodyn LS (Laboratoires Sero-biologiques)

PTEROCARPUS MARSUPIUM LEAF EXTRACT

Definition: Pterocarpus Marsupium Leaf Extract is an extract of the leaves of *Pterocarpus marsupium. See "Regulatory and Ingredient Use Information," regarding the labeling names for botanical ingredients in Volume 1, Introduction, Part A.*

Chemical Class: Biological Products

Function: Not Reported

Technical/Other Name:
Extract of Pterocarpus Marsupium

Trade Name Mixture:
Campo Bijaka (Campo)

PTEROCARPUS SANTALINUS WOOD EXTRACT

CAS No.	EINECS No.
84650-41-9	283-511-3

Definition: Pterocarpus Santalinus Wood Extract is an extract of the wood of the red sandalwood, *Pterocarpus santalinus. See "Regulatory and Ingredient Use Information," regarding the labeling names for botanical ingredients in Volume 1, Introduction, Part A.*

Chemical Class: Biological Products

Function: Not Reported

Technical/Other Names:
Extract of Pterocarpus Santalinus
Extract of Red Sandalwood
Red Sandalwood Extract
Red Sandalwood (Pterocarpus Santalinus) Extract

Trade Name Mixtures:
Actiphyte of Red Sandalwood BG50 (Active Organics)
Actiphyte of Red Sandalwood GL50 (Active Organics)
Actiphyte of Red Sandalwood Lipo S (Active Organics)
Actiphyte of Red Sandalwood PG50 (Active Organics)
Glycolysat of Red Sandalwood (CEP (Solabia))
Red Sandal Wood Extract HS 2510 G (Grau)
Sandalwood Extract (Kelisema Italy)
VT-128 Extract of Sandalwood (Vege-Tech)

PTFE

CAS No.: 9002-84-0

JPN Translation:
ポリテトラフルオロエチレン

Empirical Formula:
$(C_2F_4)_x$

Definition: PTFE is the polymer of tetra-fluoroethylene that conforms to the formula:

$$\left[C_2F_4 \right]_x$$

Information Sources: 21CFR175.300, 21CFR177.1520, 21CFR556.760, INN, JCIC, JCLS, MI-13(7665), USAN

Chemical Classes: Halogen Compounds; Synthetic Polymers

Functions: Bulking Agent; Slip Modifier

Reported Product Categories: Eye Shadows; Face Powders; Blushers (All types); Mascara; Makeup Bases

Technical/Other Names:
Polytetrafluoroethylene
Teflon
Tetrafluoroethene Homopolymer

Trade Names:
Algoflon HC (Solvay Solexis SpA)
Ceridust 9202F (Path Silicones)
Ceridust 9205F (Path Silicones)
Microslip 519 (Micro Powders)
Microslip 519 (Presperse)
Microslip 519L (Micro Powders)
TEFPOLY PUR WL 3 (Creations Couleurs)

Trade Name Mixtures:
Fomblin HC/P-R (Solvay Solexis SpA)
Lubraslide (Guardian)
Microsilk 418 (Micro Powders)
Microsilk 418 (Presperse)
Microsilk 419 (Micro Powders)
Tefpoly Begonia (C.I.T.)
Tefpoly Black (C.I.T.)
Tefpoly Blue (C.I.T.)
Tefpoly Blue Green (C.I.T.)
Tefpoly Blue Sky Dye (C.I.T.)
Tefpoly Bright Yellow Dye (C.I.T.)
Tefpoly Bright Yellow Lake (C.I.T.)
Tefpoly Brown (C.I.T.)
Tefpoly Coral Dye (C.I.T.)
Tefpoly Electric Pink Lake (C.I.T.)
TefPoly Eosin Dye (C.I.T.)
Tefpoly FLUO Pink 65% (Creations Couleurs)
Tefpoly Geranium Lake (C.I.T.)
Tefpoly Gold Grass Dye (C.I.T.)
Tefpoly Harvest Dye (C.I.T.)
Tefpoly Helindone Lake (C.I.T.)
Tefpoly Magenta Dye (C.I.T.)
Tefpoly Magenta Lake (C.I.T.)
Tefpoly Mandarin Dye (C.I.T.)
Tefpoly Melon Dye (C.I.T.)
Tefpoly Melon Lake (C.I.T.)
Tefpoly Ocean Blue Dye (C.I.T.)
Tefpoly Ocean Blue Lake (C.I.T.)
Tefpoly Orange Dye (C.I.T.)
Tefpoly Orange Lake (C.I.T.)
TEFPOLY PFC 5 (Creations Couleurs)
Tefpoly Pinky Dye (C.I.T.)
Tefpoly Poppy Lake (C.I.T.)
Tefpoly Prussian Blue (C.I.T.)
Tefpoly Red (C.I.T.)
Tefpoly Red Wood Dye (C.I.T.)
Tefpoly Safran Lake (C.I.T.)
Tefpoly Tomato Dye (C.I.T.)
Tefpoly Violet (C.I.T.)
Tefpoly White (C.I.T.)
Tefpoly Yellow (C.I.T.)
Tefpoly Yellow Grass Dye (C.I.T.)

PTYCHOPETALUM OLACOIDES EXTRACT

Definition: Ptychopetalum Olacoides Extract is an extract of the bark and root of *Ptychopetalum olacoides*. See "Regulatory and Ingredient Use Information," regarding the labeling names for botanical ingredients in Volume 1, Introduction, Part A.

Chemical Class: Biological Products

Function: Not Reported

Technical/Other Name:
Extract of Ptychopetalum Olacoides

Trade Name Mixtures:
Cosflor Muira Puama HGS (A & E Connock)
Cosflor Potency Wood HGS (A & E Connock)
Huile de Muira Puama (Greentech)
Muira Puama Extract (Bell Flavors)
Muira Puama Extract (Cosmetic Developments)

PTYCHOPETALUM OLACOIDES LEAF EXTRACT

Definition: Ptychopetalum Olacoides Leaf Extract is an extract of the leaves of *Ptychopetalum olacoides*. See "Regulatory and Ingredient Use Information," regarding the labeling names for botanical ingredients in Volume 1, Introduction, Part A.

Chemical Class: Biological Products

Function: Skin-Conditioning Agent - Miscellaneous

Technical/Other Name:
Extract of Ptychopetalum Olacoides Leaf

Trade Name Mixture:
Muira Puama HS (Alban Muller)

PUERARIA LOBATA ROOT EXTRACT

JPN Translation:
カッコンエキス

Definition: Pueraria Lobata Root Extract is an extract of the roots of *Pueraria lobata*. See "Regulatory and Ingredient Use Information," regarding the labeling names for botanical ingredients in Volume 1, Introduction, Part A.

Chemical Class: Biological Products

Function: Skin-Conditioning Agent - Humectant

Technical/Other Name:
Extract of Pueraria Lobata

Trade Name:
Kakkon Extract Powder (Ichimaru Pharcos)

Trade Name Mixtures:
Acticlear P (Active Organics)
Actiphyte of Pueraria PG (Active Organics)
Bathgranue Kakkon (Ichimaru Pharcos)
Bio antiage WE (Ichimaru Pharcos)
Bio antige B (Ichimaru Pharcos)
Harmowhite (Ichimaru Pharcos)
Harmowhite (Ichimaru Pharcos)
Inhipase (Coletica SA)
Kakkon Liquid B (Ichimaru Pharcos)
Kakkon Liquid E (Ichimaru Pharcos)
Pharcolex PSP (Ichimaru Pharcos)
Phytosterone (Ichimaru Pharcos)

PUERARIA MIRIFICA ROOT EXTRACT

Definition: Pueraria Mirifica Root Extract is an extract of the roots of *Pueraria Mirifica*. See "Regulatory and Ingredient Use Information," regarding the labeling names for botanical ingredients in Volume 1, Introduction, Part A.

Chemical Class: Biological Products

Function: Skin-Conditioning Agent - Miscellaneous

Technical/Other Name:
Extract of Pueraria Mirifica

Trade Name:
Pueraria Mirifica Tuber Powder (Tropical Herbal)

Trade Name Mixtures:
Pueraria Mirifica Extract (Shiratori Pharmaceutical)
Pueraria Mirifica Extract (Tropical Herbal)
Wichai III Cultivar Pueraria (American Phyto)

PUERARIA THUNBERGIANA EXTRACT

Definition: Pueraria Thunbergiana Extract is an extract of the aerial parts of *Pueraria thunbergiana*. See "Regulatory and Ingredient Use Information," regarding the labeling names for botanical ingredients in Volume 1, Introduction, Part A.

Chemical Class: Biological Products

Function: Skin-Conditioning Agent - Miscellaneous

Technical/Other Name:
Extract of Pueraria Thunbergiana

Trade Name Mixture:
Activated Botanicals Estroherb Complex AB 106 (Norjin)

PUERARIA THUNBERGIANA ROOT EXTRACT

Definition: Pueraria Thunbergiana Root Extract is an extract of the roots of *Pueraria*

thunbergiana. *See "Regulatory and Ingredient Use Information," regarding the labeling names for botanical ingredients in Volume 1, Introduction, Part A.*

Chemical Class: Biological Products

Function: Skin-Conditioning Agent - Miscellaneous

Technical/Other Name:
Extract of Pueraria Thunbergiana Root

Trade Name Mixture:
Phytosterol Complex Concentrate SI 1006 (Norjin)

PULLULAN

CAS No.   **EINECS No.**
9057-02-7   232-945-1

Definition: Pullulan is a polysaccharide produced from starch by cultivating the yeast, *Aureobasidium pullulans.*

Information Source: TSCA

Chemical Class: Carbohydrates

Functions: Binder; Film Former

Trade Names:
Pullulan PF-20 (Hayashibara)
Pullulan PI-20 (Hayashibara)

PULMONARIA OFFICINALIS EXTRACT

CAS No.   **EINECS No.**
84604-09-1   283-286-1

Definition: Pulmonaria Officinalis Extract is an extract of the lungwort, *Pulmonaria officinalis. See "Regulatory and Ingredient Use Information," regarding the labeling names for botanical ingredients in Volume 1, Introduction, Part A.*

Chemical Class: Biological Products

Function: Not Reported

Technical/Other Names:
Extract of Lungwort
Extract of Pulmonaria Officinalis
Lungwort Extract
Lungwort (Pulmonaria Officinalis) Extract
Virgin Mary's Honeysuckle Extract

Trade Name Mixtures:
Actiphyte of Lungwort BG50 (Active Organics)
Actiphyte of Lungwort GL50 (Active Organics)
Actiphyte of Lungwort Lipo S (Active Organics)
Actiphyte of Lungwort PG50 (Active Organics)

PUMICE

CAS No.: 1332-09-8

JPN Translation:
軽石

Definition: Pumice is a substance of volcanic origin consisting chiefly of complex silicates of aluminum and alkali metals.

Information Sources: JCIC, JCLS, JSQI, MAR, MI-13(8031), USAN, USP XXIV

Chemical Class: Inorganics

Functions: Abrasive; Bulking Agent

Reported Product Categories: Manicuring Preparations, Misc.; Personal Cleanliness Products, Misc.

Technical/Other Names:
Pumice 8849 1/2
Pumice Powder

Trade Names:
AEC Pumice Granules (A & E Connock)
AEC Pumice Powder (A & E Connock)
AEC Pumice Sand (A & E Connock)
Navajo Brand 0 (CR Minerals)
Navajo Brand 0-1/2 (CR Minerals)
Navajo Brand 0-3/4 (CR Minerals)
Navajo Brand 1 (CR Minerals)
Navajo Brand 1-1/2 (CR Minerals)
Navajo Brand 1/2 (CR Minerals)
Navajo Brand 3 (CR Minerals)
Navajo Brand 4 (CR Minerals)
Navajo Brand 5 (CR Minerals)
Navajo Brand FF (CR Minerals)
Navajo Brand FFF (CR Minerals)
Navajo Brand FFFF (CR Minerals)
Navajo Brand NB005 (CR Minerals)
Navajo Brand NB012 (CR Minerals)

PUMPKIN SEED OIL PEG-8 ESTERS

Definition: Pumpkin Seed Oil PEG-8 Esters is the product obtained by the transesterification of Cucurbita Pepo (Pumpkin) Seed Oil (q.v.) and PEG-8 (q.v.).

Chemical Class: Glyceryl Esters and Derivatives

Functions: Skin-Conditioning Agent - Emollient; Surfactant - Emulsifying Agent

Trade Name:
Viatenza Pumpkin PE8 (Aldivia)

PUNICA GRANATUM

Definition: *See "Regulatory and Ingredient Use Information," regarding EU labeling names for botanical ingredients in Volume 1, Introduction, Part A.*

Chemical Class: Biological Products

Technical/Other Names:
Punica Granatum Bark Extract (U.S.)
Punica Granatum Extract (U.S.)
Punica Granatum Fruit Juice (U.S.)
Punica Granatum Seed Powder (U.S.)

PUNICA GRANATUM BARK EXTRACT

CAS No.   **EINECS No.**
84961-57-9   284-646-0

Definition: Punica Granatum Bark Extract is an extract of the bark of *Punica granatum. See "Regulatory and Ingredient Use Information," regarding the labeling names for botanical ingredients in Volume 1, Introduction, Part A.*

Information Source: RIFM

Chemical Class: Biological Products

Functions: Fragrance Ingredient; Skin-Conditioning Agent - Miscellaneous

Technical/Other Names:
Extract of Punica Granatum Bark
Pomegranate bark extract (Punica granatum L.) (RIFM)
Punica Granatum (EU)

Trade Name Mixtures:
Glycolysat of Pomegranate (CEP (Solabia))
Vegebios of Pomegranate (CEP (Solabia))

PUNICA GRANATUM EXTRACT

CAS No.   **EINECS No.**
84961-57-9   284-646-0

Definition: Punica Granatum Extract is an extract of the pomegranate, *Punica granatum. See "Regulatory and Ingredient Use Information," regarding the labeling names for botanical ingredients in Volume 1, Introduction, Part A.*

Information Source: RIFM

Chemical Class: Biological Products

Functions: Fragrance Ingredient; Not Reported

Technical/Other Names:
Extract of Pomegranate
Extract of Punica Granatum
Pomegranate bark extract (Punica granatum L.) (RIFM)
Pomegranate Extract
Pomegranate (Punica Granatum) Extract
Punica Granatum (EU)

Trade Name:
California Pomegranate Extract (Formula One Sciences)

Trade Name Mixtures:
   Actiphyte of Pomegranate (Active
      Organics)
   Actiphyte of Pomegranate BG50 (Active
      Organics)
   Actiphyte of Pomegranate GL50 (Active
      Organics)
   Actiphyte of Pomegranate Lipo S (Active
      Organics)
   Actiphyte of Pomegranate PG50 (Active
      Organics)
   Extrait Hydroglycolique De Grenade
      (Greentech)
   Extrapone Cactus Flower/Pomegranate
      Blend GW 2/031337 (Symrise)
   Herbasol Extract Pomegranate
      (Cosmetochem) (Cosmetochem
      International Ltd.)
   POMEGRANATE FRUIT EXTRACT (Libiol)
   Pomegranate HS (Alban Muller)
   Pomegranate Milk (CEP (Solabia))
   Punica Granatum Extract ies (IES LABO)
   VT-188 Extract of Pomegranate (Vege-
      Tech)

PUNICA GRANATUM FRUIT JUICE

Definition: Punica Granatum Fruit Juice is
the juice expressed from the fruit of the
pomegranate, *Punica granatum. See
"Regulatory and Ingredient Use
Information," regarding the labeling names
for botanical ingredients in Volume 1, Introduction, Part A.*

Chemical Class: Biological Products

Functions: Flavoring Agent; Skin-Conditioning Agent - Miscellaneous

Technical/Other Name:
   Punica Granatum (EU)

Trade Names:
   Pomegranate Juice Concentrate (Madera
      Enterprises)
   Pomegranate Juice Concentrate 4/160016
      (Symrise)

PUNICA GRANATUM SEED POWDER

Definition: Punica Granatum Seed
Powder is the powder obtained from the
crushed seeds of *Punica granatum. See
"Regulatory and Ingredient Use
Information," regarding the labeling names
for botanical ingredients in Volume 1, Introduction, Part A.*

Chemical Class: Biological Products

Function: Abrasive

Technical/Other Name:
   Punica Granatum (EU)

Trade Name:
   Pomegranate Scrub (American Natural
      Products)

PVM/MA COPOLYMER

CAS No.: 9011-16-9

JPN Translation:
   (メチルビニルエーテル / マレイン酸) コ
   ポリマー

Empirical Formula:
   $(C_4H_2O_3 \cdot C_3H_6O)_x$

Definition: PVM/MA Copolymer is a
copolymer of methyl vinyl ether and maleic
anhydride.

Information Sources: CTFA D, MAR,
TSCA

Chemical Class: Synthetic Polymers

Functions: Binder; Emulsion Stabilizer; Film
Former; Hair Fixative; Suspending Agent -
Nonsurfactant

Technical/Other Names:
   2,5-Furandione, Polymer with Methoxyeth-
      ylene
   Methyl Vinyl Ether/Maleic Anhydride
      Copolymer

Trade Names:
   Gantrez AN-119 (International Specialty
      Products)
   Gantrez AN-139 (International Specialty
      Products)
   Gantrez AN-149 (International Specialty
      Products)
   Gantrez AN-169 (International Specialty
      Products)
   Gantrez S-95 (International Specialty
      Products)
   Gantrez S-97 (International Specialty
      Products)

Trade Name Mixtures:
   Lubrajel Oil (Guardian)
   Zilgel Oil (Zile)

PVM/MA DECADIENE CROSSPOLYMER

Definition: PVM/MA Decadiene Crosspolymer is a polymer of maleic anhydride and
methyl vinyl ether crosslinked with 1,9-
decadiene.

Chemical Class: Synthetic Polymers

Functions: Film Former; Viscosity
Increasing Agent - Nonaqueous

Reported Product Category: Shampoos
(Non-coloring)

Trade Names:
   Stabileze 06 (International Specialty
      Products)

   Stabileze QM (International Specialty
      Products)
   Stabileze XL-80W (International Specialty
      Products)

PVP

CAS No.: 9003-39-8

JPN Translation:
   PVP

Empirical Formula:
   $(C_6H_9NO)_x$

Definition: PVP is the linear polymer that
consists of 1-vinyl-2-pyrrolidone monomers
conforming generally to the formula:

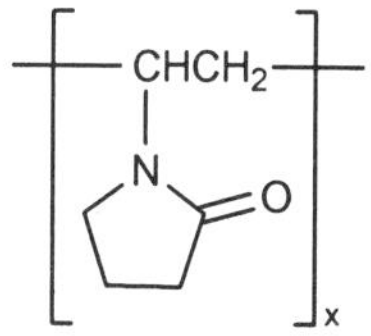

Information Sources: BAN, BPC,
21CFR73.1, 21CFR73.1001,
21CFR172.210, 21CFR173.50,
21CFR173.55, 21CFR175.105,
21CFR175.300, 21CFR176.170,
21CFR176.180, 21CFR176.210,
21CFR310.545, 21CFR333.210,
21CFR349.12, CIR: [S] IJT-17(Suppl. 4)-
1998, CTFA S, FCC, INN, ITA, JAN, JCLS,
JSCI, MAR, MI-13(7783), NF XIX, OTC-I-
OP, TSCA, USAN, USD, USP XXIV

Chemical Class: Synthetic Polymers

Functions: Binder; Emulsion Stabilizer; Film
Former; Hair Fixative; Suspending Agent -
Nonsurfactant

Reported Product Categories: Tonics,
Dressings, and Other Hair Grooming Aids;
Mascara; Hair Conditioners; Hair Prepara-
tions (Non-coloring), Misc.; Hair Sprays
(Aerosol Fixatives); Eye Makeup Prepara-
tions, Misc.; Eyeliners; Shampoos (Non-
coloring); Hair Wave Sets; Paste Masks (Mud
Packs); Cleansing Products (Cold Creams,
Cleansing Lotions, Liquids and Pads); Hair
Rinses (Coloring); Moisturizing Preparations;
Bath Capsules; Bath Preparations, Misc.;
Body and Hand Preparations (Excluding
Shaving Preparations); Face and Neck Prep-
arations (Excluding Shaving Preparations);
Skin Care Preparations, Misc.; Bath Soaps
and Detergents; Eyebrow Pencils; Founda-
tions; Hair Rinses (Non-coloring); Shaving
Cream (Aerosol, Brushless and Lather)

Technical/Other Names:
   1-Ethenyl-2-Pyrrolidinone, Homopolymer
   Polyvinylpyrrolidone
   Povidone

2-Pyrrolidinone, 1-Ethenyl-, Homopolymer
N-Vinylbutyrolactam Polymer

Trade Names:
Luviskol K30, 30% (BASF)
Luviskol K60, 45% (BASF)
Luviskol K 85,20% (BASF)
Luviskol K90, 20% (BASF)
Luviskol K17, Powder (BASF)
Luviskol K30 Powder (BASF)
Luviskol K80, Powder (BASF)
Luviskol K90 Powder (BASF)
Plasdone K26-28 (International Specialty Products)
Plasdone K29-32 (International Specialty Products)
PVP K-15 (International Specialty Products)
PVP K-30 (International Specialty Products)
PVP K-60 (International Specialty Products)
PVP K-90 (International Specialty Products)

Trade Name Mixtures:
AMPEA 80401 (Fine Oligomers)
Aquatrix II Hydrogel (Hydromer)
Eusolex UV-Pearls OMC (Merck KGaA/ EMD Chemicals Inc.)
Firmiderm LS 9120 (Laboratoires Sero-biologiques)
Kerafix 620 (Variati)
PME-1 (Vevy)
Pronalen Flash-Tense Plus (Provital/ Centerchem)
PVP Si-10 (International Specialty Products)
Slimmigen (Laboratoires Serobiologiques)

PVP/DECENE COPOLYMER

Empirical Formula:
$(C_6H_9NO \cdot C_{10}H_{20})_x$

Definition: PVP/Decene Copolymer is a polymer of vinylpyrrolidone and decene monomers. It conforms generally to the formula:

Chemical Class: Synthetic Polymers

Functions: Binder; Emulsion Stabilizer; Viscosity Increasing Agent - Aqueous; Viscosity Increasing Agent - Nonaqueous

PVP-HYDROGEN PEROXIDE

CAS No.: 135927-36-5

Empirical Formula:
$(C_6H_9NO)_x \cdot \frac{1}{2}H_2O_2$

Definition: PVP-Hydrogen Peroxide is a complex of polyvinylpyrrolidone and hydrogen peroxide.

Information Source: EEC(III/1-12)

Chemical Class: Synthetic Polymers

Functions: Cosmetic Astringent; Oxidizing Agent

Technical/Other Name:
2-Pyrrolidinone, 1-Ethenyl-, Homopolymer, Compd. with Hydrogen Peroxide (H2O2) (2:1)

PVP-IODINE

CAS No.: 25655-41-8

Empirical Formula:
$(C_6H_9NO)_x \cdot xI_2$

Definition: PVP-Iodine is a complex of polyvinylpyrrolidone and iodine. In the United States, PVP-Iodine may be used as an active ingredient in OTC drug products. When used as an active drug ingredient, the established name for PVP-Iodine is *Povidone-Iodine. See "Regulatory and Ingredient Use Information," regarding the labeling names for U.S. OTC Drug Ingredients in Volume 1, Introduction, Part A.*

Information Sources: BAN, 21CFR333.210, JAN, MAR, MI-13(7784), OTC-I-AF, OTC-I-AM, TSCA, USAN, USD, USP XXIV

Chemical Class: Synthetic Polymers

Functions: Antifungal Agent; Antimicrobial Agent; Cosmetic Biocide

Technical/Other Names:
Polyvinylpyrrolidone-Iodine Complex
2-Pyrrolidinone, 1-Ethenyl-, Homopolymer, Compd. with Iodine

Trade Names:
Povidone Iodine (Huntington)
PVP-Iodine 17/12 (BASF)
PVP-Iodine 30/06 (BASF)

PVP MONTMORILLONITE

Definition: PVP Montmorillonite is Montmorillonite (q.v.) that has been surface-treated with PVP (q.v.).

Chemical Classes: Inorganics; Synthetic Polymers

Functions: Emulsion Stabilizer; Film Former; Viscosity Increasing Agent - Aqueous

Trade Name:
Polargel IVP (Amcol)

PVP/VA/ITACONIC ACID COPOLYMER

CAS No.: 68928-72-3

Empirical Formula:
$(C_6H_9NO \cdot C_4H_6O_3 \cdot C_5H_6O_4)_x$

Definition: PVP/VA/Itaconic Acid Copolymer is a polymer formed from vinylpyrrolidone, vinyl acetate and itaconic acid monomers.

Chemical Class: Synthetic Polymers

Functions: Binder; Film Former; Hair Fixative; Suspending Agent - Nonsurfactant

Technical/Other Names:
Butanedioic Acid, Methylene-, Polymer with Ethenyl Acetate and 1-Ethenyl-2-Pyrroli-dinone
Methylenebutanedioic Acid, Polymer with Ethenyl Acetate and 1-Ethenyl-2-Pyrroli-dinone
PVP/Vinyl Acetate/Itaconic Acid Copolymer

PVP/VA/VINYL PROPIONATE COPOLYMER

Definition: PVP/VA/Vinyl Propionate Copolymer is a polymer of vinylpyrrolidone, vinyl acetate and vinyl propionate monomers.

Chemical Class: Synthetic Polymers

Functions: Film Former; Hair Fixative

PYRACANTHA FORTUNEANA FRUIT EXTRACT

Definition: Pyracantha Fortuneana Fruit Extract is an extract of the fruit of *Pyracantha fortuneana. See "Regulatory and Ingredient Use Information," regarding the labeling names for botanical ingredients in Volume 1, Introduction, Part A.*

Chemical Class: Biological Products

Function: Skin-Conditioning Agent - Miscellaneous

Technical/Other Name:
Extract of Pyrocantha Fortuneana

Trade Name Mixture:
Pyracantha Extract (Maruzen Pharmaceuticals Co., Ltd.)

PYRETHRINS

CAS No.	EINECS No.
8003-34-7	232-319-8

Definition: Pyrethrins are esters isolated from pyrethrum flowers consisting chiefly of the following components:

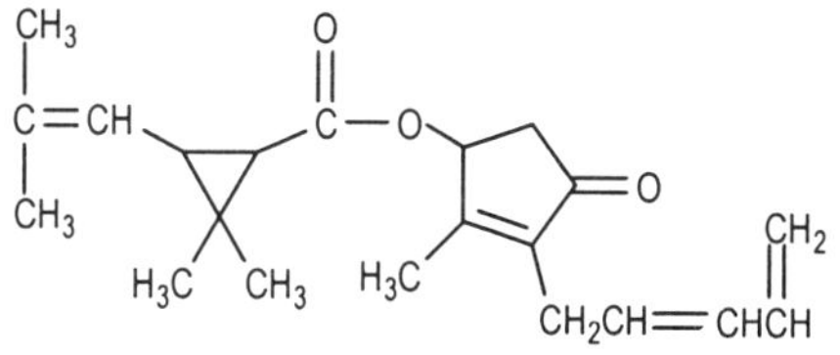

Pyrethrin I

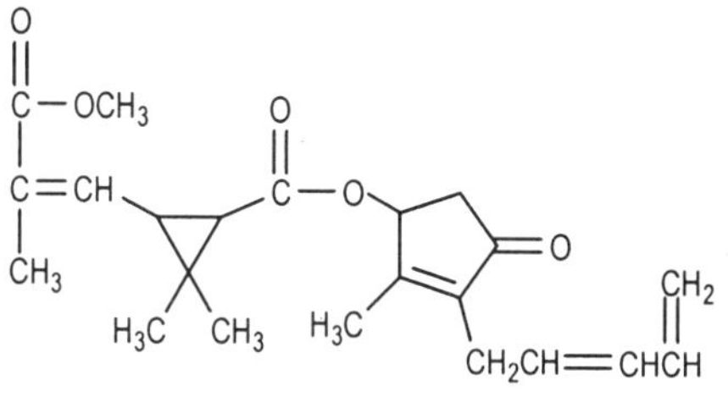

Pyrethrin II

Information Sources: 21CFR178.3730, 21CFR358.610, 21CFR524.2140, 40CFR180.128, 40CFR185.5200, 40CFR186.5200, MI-13(8054)

Chemical Classes: Esters; Ketones

Function: Pesticide

Trade Name:
Pyrethrum Extract 25% Pale (Sumitomo Chemical)

PYRICARBATE

CAS No.	EINECS No.
1882-26-4	217-538-9

Empirical Formula:
$C_{11}H_{15}N_3O_4$

Definition: Pyricarbate is the heterocyclic compound that conforms to the formula:

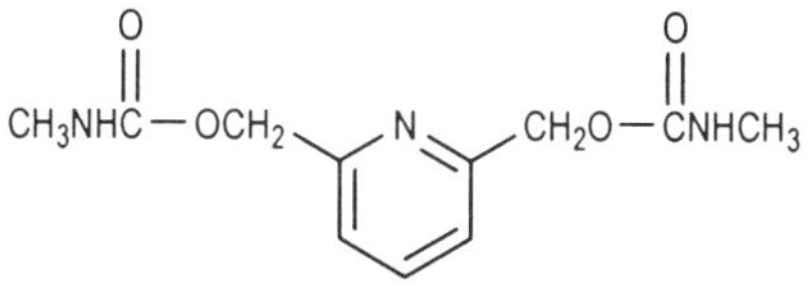

Information Sources: INN, MI-13(8065), USAN

Chemical Classes: Amides; Esters; Heterocyclic Compounds

Function: Skin-Conditioning Agent - Miscellaneous

Technical/Other Names:
2,6-Bis(Hydroxymethyl)Pyridine Bis(N-Methylcarbamate)
Carbamic Acid, Methyl-, 2,6-Pyridinediyl-dimethylene Ester
2,6-Pyridinedimethanol, Bis (Methylcarbamate) (Ester)

Trade Name:
Soothing Cream (Prodotti Gianni)

PYRIDINEDICARBOXYLIC ACID

CAS Nos.	EINECS No.
100-26-5	202-834-2
28605-84-7	

Empirical Formula:
$C_7H_5NO_4$

Definition: Pyridinedicarboxylic Acid is the organic compound that conforms to the formula:

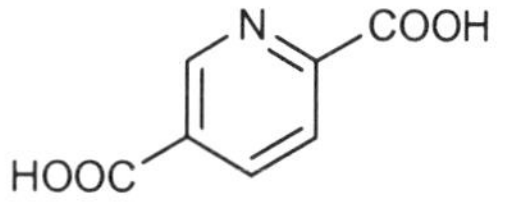

Information Sources: MI-13(5177), TSCA

Chemical Classes: Carboxylic Acids; Heterocyclic Compounds

Function: Proprietary

Technical/Other Names:
Isocinchomeronic Acid
2,5-Pyridinedicarboxylic Acid

Trade Name:
2,5-Pyridinedicarboxylic Acid (Merck KGaA)

PYRIDOXAL 5-PHOSPHATE

CAS No.	EINECS No.
54-47-7	200-208-3

JPN Translation:
リン酸ピリドキサール

Empirical Formula:
$C_8H_{10}NO_6P$

Definition: Pyridoxal 5-Phosphate is the organic compound that conforms to the formula:

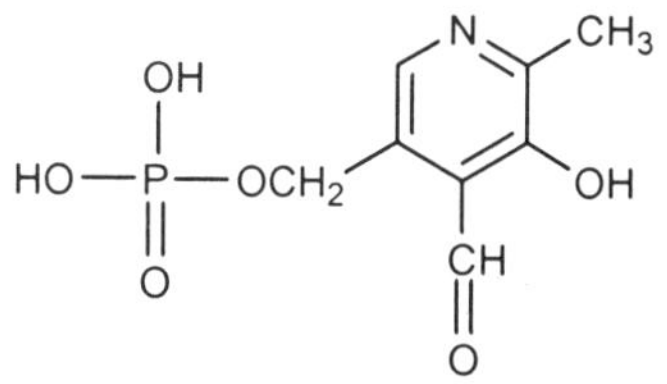

Information Sources: JAN, JCIC, JCLS, MI-13(8069), TSCA

Chemical Classes: Aldehydes; Heterocyclic Compounds; Phosphorus Compounds

Function: Skin-Conditioning Agent - Miscellaneous

Technical/Other Names:
3-Hydroxy-2-Methyl-5-[(Phosphonooxy)-Methyl]-4-Pyridinecarboxaldehyde

Phosphopyridoxal
4-Pyridinecarboxaldehyde, 3-Hydroxy-2-Methyl-5-[(Phosphonooxy)Methyl]-
Pyridoxal Phosphate
Vitamin B6 Phosphate

PYRIDOXINE

CAS Nos.	EINECS Nos.
65-23-6	200-603-0
8059-24-3	232-503-8

JPN Translation:
ピリドキシン

Empirical Formula:
$C_8H_{11}NO_3$

Definition: Pyridoxine is the substituted aromatic compound that conforms to the formula:

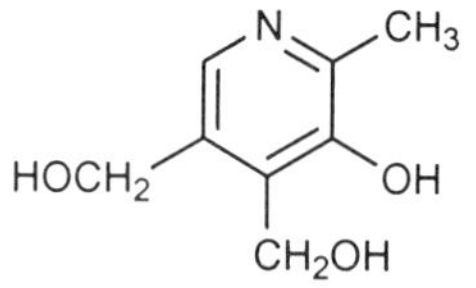

Information Sources: BAN, 21CFR184.1555, 21CFR184.1676, 21CFR310.545, 21CFR582.5676, INN, JCIC, JCLS, JSQI, TSCA

Chemical Class: Heterocyclic Compounds

Functions: Hair Conditioning Agent; Skin-Conditioning Agent - Miscellaneous

Technical/Other Names:
5-Hydroxy-6-Methyl-3,4-Pyridinedimethanol
3,4-Pyridinedimethanol, 5-Hydroxy-6-Methyl-
Pyridoxol
Vitamin B6

Trade Name Mixtures:
Hair Care Blend (Alban Muller)
Hair Care Phytoamine Biocomplex (Alban Muller)
Hair Care Phytoamine Biocomplex SP Lotion (Alban Muller)

PYRIDOXINE DICAPRYLATE

CAS No.: 106483-04-9

JPN Translation:
ジカプリル酸ピリドキシン

Empirical Formula:
$C_{24}H_{37}NO_5$

Definition: Pyridoxine Dicaprylate is the substituted aromatic compound that conforms generally to the formula:

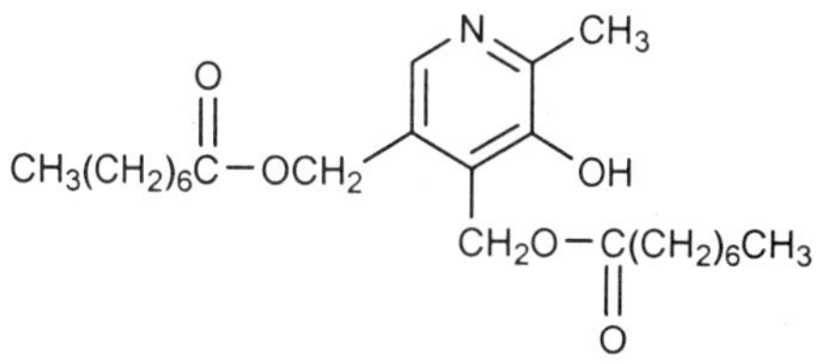

Information Sources: JCLS, JSCI

Chemical Classes: Esters; Heterocyclic Compounds

Functions: Hair Conditioning Agent; Skin-Conditioning Agent - Miscellaneous

Technical/Other Names:
Octanoic Acid, Diester with 5-Hydroxy-6-Methyl-3,4-Pyridinedimethanol
Pyridoxine Dioctanoate

Trade Name:
Nikkol DK (Nikko)

PYRIDOXINE DILAURATE

Empirical Formula:
$C_{32}H_{55}NO_5$

Definition: Pyridoxine Dilaurate is the substituted aromatic compound that conforms generally to the formula:

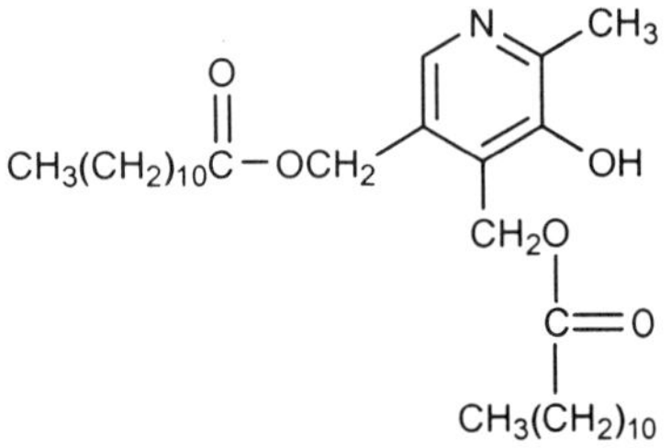

Chemical Classes: Esters; Heterocyclic Compounds

Functions: Hair Conditioning Agent; Skin-Conditioning Agent - Miscellaneous

Technical/Other Names:
Dodecanoic Acid, Diester with 5-Hydroxy-6-Methyl-pyridinedimethanol
Dodecanoic Acid, Diester with Pyridoxol
Vitamin B6 Dilaurate

Trade Name:
Nikkol DL (Nikko)

PYRIDOXINE DIOCTENOATE

CAS No.: 59599-61-0

Empirical Formula:
$C_{24}H_{35}NO_5$

Definition: Pyridoxine Dioctenoate is the substituted aromatic compound that conforms generally to the formula:

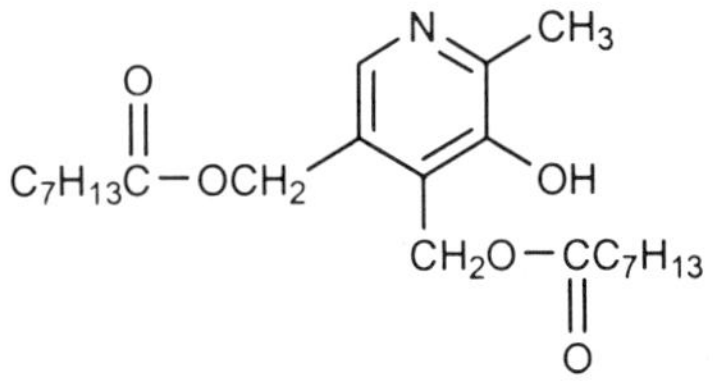

Chemical Classes: Esters; Heterocyclic Compounds

Functions: Hair Conditioning Agent; Skin-Conditioning Agent - Miscellaneous

Technical/Other Names:
Octenoic Acid, (5-Hydroxy-6-Methyl-3,4-Pyridinediyl)bis(Methylene) Ester
Vitamin B6 Dioctenoate

PYRIDOXINE DIPALMITATE

CAS Nos.	EINECS No.
635-38-1	
31229-74-0	250-520-9
39379-66-3	

JPN Translation:
ジパルミチン酸ピリドキシン

Empirical Formula:
$C_{40}H_{71}NO_5$

Definition: Pyridoxine Dipalmitate is the substituted aromatic compound that conforms generally to the formula:

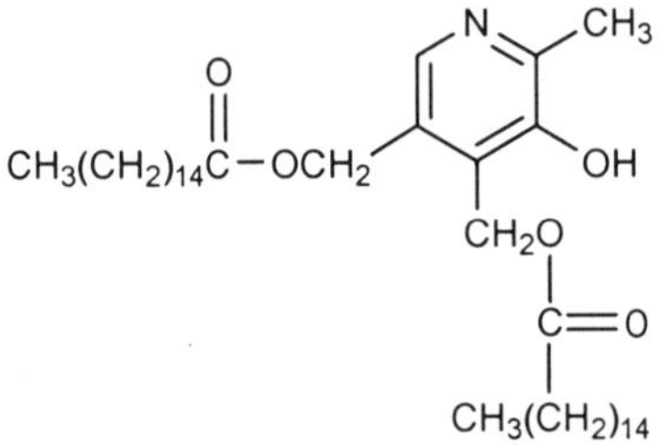

Information Sources: JCLS, JSCI

Chemical Classes: Esters; Heterocyclic Compounds

Functions: Hair Conditioning Agent; Skin-Conditioning Agent - Miscellaneous

Reported Product Category: Moisturizing Preparations

Technical/Other Names:
Hexadecanoic Acid, (5-Hydroxy-6-Methyl-3, 4-Pyridinediyl)bis(Methylene) Ester
Palmitic Acid, Diester with Pyridoxol
Vitamin B6 Dipalmitate

Trade Name:
Nikkol DP (Nikko)

PYRIDOXINE GLYCYRRHETINATE

JPN Translation:
グリチルレチン酸ピリドキシン

Definition: Pyridoxine Glycyrrhetinate is the ester of Pyridoxine (q.v.) and Glycyrrhetinic Acid (q.v.).

Information Sources: JCIC, JCLS

Chemical Classes: Alcohols; Carboxylic Acids; Esters; Heterocyclic Compounds; Ketones

Function: Skin-Conditioning Agent - Miscellaneous

PYRIDOXINE HCl

CAS Nos.	EINECS No.
58-56-0	200-386-2
12001-77-3	

JPN Translation:
ピリドキシン HCl

Empirical Formula:
$C_8H_{11}NO_3 \cdot ClH$

Definition: Pyridoxine HCl is the substituted aromatic compound that conforms to the formula:

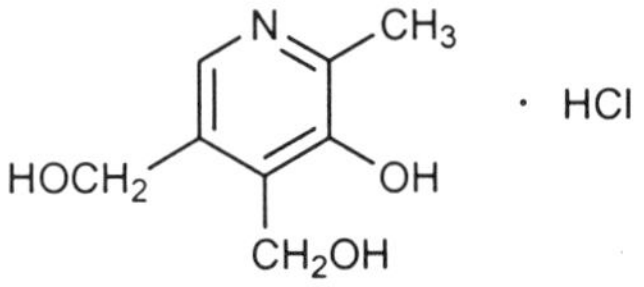

Information Sources: AUS, BP, BPC, BRA, 21CFR184.1676, CZE, DA, DDR, EP, FCC, HUN, IND, ITA, JAN, JCLS, JSCI, MAR, MI-13(8072), NFJ, PN, POR, ROM, TSCA, USAN, USD, USP XXIV, WHO

Chemical Classes: Heterocyclic Compounds; Organic Salts

Functions: Hair Conditioning Agent; Skin-Conditioning Agent - Miscellaneous

Reported Product Categories: Tonics, Dressings, and Other Hair Grooming Aids; Bath Oils, Tablets, and Salts; Bath Capsules; Body and Hand Preparations (Excluding Shaving Preparations); Cleansing Products (Cold Creams, Cleansing Lotions, Liquids and Pads); Hair Conditioners; Hair Preparations (Non-coloring), Misc.; Paste Masks (Mud Packs); Shampoos (Non-coloring)

Technical/Other Names:
Aderomine Hydrochloride
Aderoxine
5-Hydroxy-6-Methyl-3,4-Pyridinedimethanol Hydrochloride
3,4-Pyridinedimethanol, 5-Hydroxy-6-Methyl-, Hydrochloride
Pyridoxine Hydrochloride
Vitamin B6 Hydrochloride

Trade Name:
Vitamin B6 Hydrochloride (Roche)

Trade Name Mixtures:
Amaryl Hydro (Laboratoires Sero-biologiques)
Asebiol LS 2539 BT2 (Laboratoires Sero-biologiques)
Elespher Vitaplex Hydro (Laboratoires Serobiologiques)

Fortified Yeast T-6361 (Universal Foods)
HAIR LOSS PREVENTION (Greentech S.A)
Photonyl (Laboratoires Serobiologiques)
Sunactyl LS 9610 (Laboratoires Sero-biologiques)

PYRIDOXINE SERINATE

CAS No.: 14942-12-2

Empirical Formula:
$C_{11}H_{16}N_2O_5$

Definition: Pyridoxine Serinate is the organic compound that conforms to the formula:

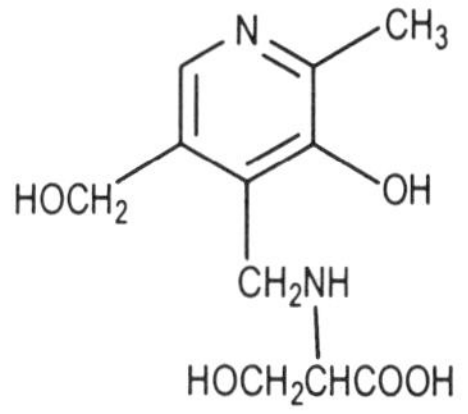

Chemical Classes: Amino Acids; Heterocyclic Compounds

Function: Antioxidant

Technical/Other Name:
L-Serine, N-[[3-Hydroxy-5-(Hydroxymethyl)-2-Methyl-4-Pyridinyl]Methyl]-

Trade Name:
Vitamin B6 Serine (Ajinomoto)

PYRIDOXINE TRIPALMITATE

CAS No.	EINECS No.
4372-46-7	224-470-3

JPN Translation:
トリパルミチン酸ピリドキシン

Empirical Formula:
$C_{56}H_{101}NO_6$

Definition: Pyridoxine Tripalmitate is the substituted aromatic compound that conforms generally to the formula:

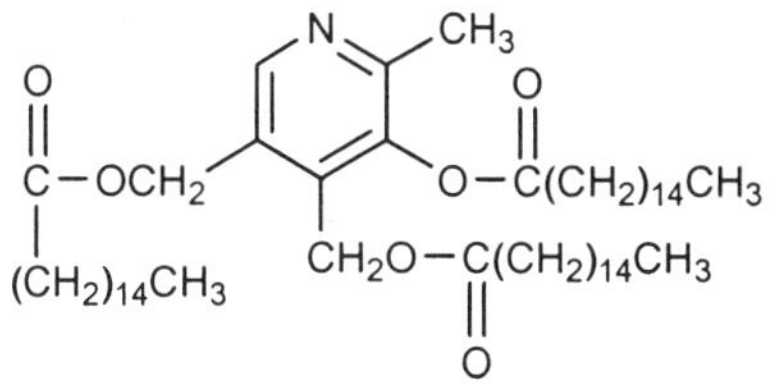

Information Sources: JCIC, JCLS

Chemical Classes: Esters; Heterocyclic Compounds

Functions: Hair Conditioning Agent; Skin-Conditioning Agent - Miscellaneous

Reported Product Category: Moisturizing Preparations

Technical/Other Names:
Hexadecanoic Acid, (6-Methyl-5-((1-Oxo-hexadecyl)Oxy)-3,4-Pyridinediyl)bis(Methylene) Ester
Palmitic Acid, Triester with Pyridoxol
Vitamin B6 Tripalmitate

PYROCATECHOL

CAS No.	EINECS No.
120-80-9	204-427-5

Empirical Formula:
$C_6H_6O_2$

Definition: Pyrocatechol is the phenol that conforms to the formula:

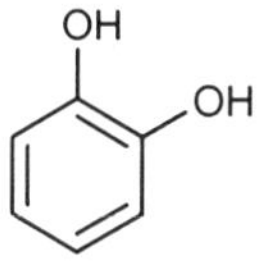

See "Regulatory and Ingredient Use Information," for Colorants in Volume 1, Introduction, Part A.

Information Sources: CI 76500, CIR: [U] IJT-16(Suppl. 1)1997, CTFA D, EEC(II-408), MI-13(8089), RIFM, TSCA

Chemical Classes: Color Additives - Hair; Phenols

Functions: Fragrance Ingredient; Hair Colorant

Reported Product Category: Hair Dyes and Colors (All Types Requiring Caution Statements and Patch Tests)

Technical/Other Names:
Benzene-1,2-Diol
1,2-Benzenediol
Catechol (RIFM)
CI 76500
1,2-Dihydroxybenzene
o-Hydroxyphenol
Phthalic Alcohol
Pyrocatechin

Trade Name:
Rodol C (Lowenstein)

PYROGALLOL

CAS No.	EINECS No.
87-66-1	201-762-9

Empirical Formula:
$C_6H_6O_3$

Definition: Pyrogallol is the phenol that conforms to the formula:

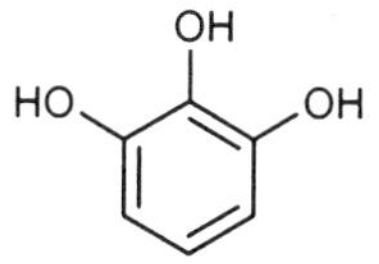

See "Regulatory and Ingredient Use Information," for Colorants in Volume 1, Introduction, Part A.

Information Sources: AUS, 21CFR73.1375, CI 76515, CIR: [S] JACT-10(1)1991, DDR, EEC(II-409), EGY, MAR, MHLW-331/1, MI-13(8090), POR, RIFM, TSCA

Chemical Classes: Color Additives - Hair; Phenols

Functions: Fragrance Ingredient; Hair Colorant

Reported Product Category: Hair Dyes and Colors (All Types Requiring Caution Statements and Patch Tests)

Technical/Other Names:
Benzene-1,2,3-Triol
1,2,3-Benzenetriol
CI 76515
2,3-Dihydroxyphenol
Oxidation Base 32
Pyrogallic Acid
Pyrogallol (RIFM)
1,2,3-Trihydroxybenzene

Trade Names:
Colorex PYRGL (Chemical Compounds, Inc.)
Rodol PG (Lowenstein)

PYROLA INCARNATA EXTRACT

Definition: Pyrola Incarnata Extract is an extract of the herb, *Pyrola incarnata*. See *"Regulatory and Ingredient Use Information,"* regarding the labeling names for botanical ingredients in Volume 1, Introduction, Part A.

Chemical Class: Biological Products

Function: Skin-Conditioning Agent - Miscellaneous

Technical/Other Names:
Extract of Pyrola Incarnata
Pyrola Japonica Extract

Trade Name Mixture:
Ichiyakusou Extract (Shiseido Company)

PYROPHYLLITE

CAS Nos.: 12269-78-2; 113349-10-3; 113349-11-4; 113349-12-5; 141040-73-5; 141040-74-6

JPN Translation:
パイロフェライト

Empirical Formula:
$Al_2O_3 \cdot 4SiO_2 \cdot H_2O$

Definition: Pyrophyllite is a naturally occurring mineral substance consisting predominantly of a hydrous aluminum silicate represented as:

$$Al_2O_3 \quad \cdot \quad 4SiO_2 \quad \cdot \quad H_2O$$

See "Regulatory and Ingredient Use Information," for Colorants in Volume 1, Introduction, Part A. To identify the colorant meeting the requirements for labeling purposes in the US, the INCI Name Pyrophyllite must be used.

Information Sources: 21CFR73.1400, 21CFR73.2400, 21CFR573.900, CIR: [S] IJT-22(SUPPL. 1)2003, JCIC, JCLS, MI-13 (361)

Chemical Classes: Color Additives - Exempt from Batch Certification by the U.S. Food and Drug Administration; Inorganics

Functions: Absorbent; Colorant; Opacifying Agent

Technical/Other Name:
Pyrophyllite Clay

Trade Name:
Brilliant White Sericite (Chrystal)

PYRROLIDINYL DIAMINOPYRIMIDINE OXIDE

CAS No.: 55921-65-8

Empirical Formula:
$C_8H_{13}N_5O$

Definition: Pyrrolidinyl Diaminopyrimidine Oxide is the heterocyclic compound that conforms to the formula:

Chemical Class: Heterocyclic Compounds

Function: Hair-Waving/Straightening Agent

Technical/Other Name:
2,4-Pyrimidinediamine, 6-(1-Pyrrolidinyl)-, 3-Oxide

Trade Name:
Triaminodil (Proderma Srl)

PYRUS COMMUNIS (PEAR) FRUIT

Definition: Pyrus Communis (Pear) Fruit is the fruit of the pear, *Pyrus communis.*

See "Regulatory and Ingredient Use Information," regarding the labeling names for botanical ingredients in Volume 1, Introduction, Part A.

Chemical Class: Biological Products

Functions: Cosmetic Astringent; Skin-Conditioning Agent - Miscellaneous

Technical/Other Names:
Pear
Pear Fruit

Trade Name:
AEC Pear Puree (A & E Connock)

PYRUS COMMUNIS (PEAR) FRUIT EXTRACT

CAS No.	EINECS No.
90082-43-2	290-131-1

Definition: Pyrus Communis (Pear) Fruit Extract is an extract of the fruit of the pear, *Pyrus communis. See "Regulatory and Ingredient Use Information," regarding the labeling names for botanical ingredients in Volume 1, Introduction, Part A.*

Chemical Class: Biological Products

Function: Skin-Conditioning Agent - Miscellaneous

Technical/Other Names:
Extract of Pear
Extract of Pyrus Communis
Pear Extract
Pear Fruit Extract
Pyrus Communis Extract

Trade Name Mixtures:
ABS Pear Fruit Extract (Active Concepts)
Actiphyte Pear Fruit (Active Organics)
Actiphyte Pear Fruit AQ (Active Organics)
Actiphyte Pear Fruit BG50P (Active Organics)
Complex 2 - Moisturizing (Provital/ Centerchem)
Herbasol Extract Pear (Cosmetochem) (Cosmetochem International Ltd.)
Pear HS (Alban Muller)
PRONALEN MOISTURIZING-II (Provital/ Centerchem)
VT-226 Extract of Pear (Vege-Tech)

PYRUS COMMUNIS (PEAR) FRUIT JUICE

Definition: Pyrus Communis (Pear) Fruit Juice is the liquid expressed from the fruit of the pear, *Pyrus communis. See "Regulatory and Ingredient Use Information," regarding the labeling names for botanical ingredients in Volume 1, Introduction, Part A.*

Chemical Class: Biological Products

Functions: Cosmetic Astringent; Skin-Conditioning Agent - Miscellaneous

Technical/Other Names:
Juice, Pear
Juice, Pyrus Communis
Pear Fruit Juice
Pear Juice
Pyrus Communis Juice

Trade Names:
AEC Pear Conc (A & E Connock)
Authenticals of Pear (CEP (Solabia))

PYRUS COMMUNIS (PEAR) WATER

Definition: Pyrus Communis (Pear) Water is an aqueous solution of the steam distillate obtained from *Pyrus communis. See "Regulatory and Ingredient Use Information," regarding the labeling names for botanical ingredients in Volume 1, Introduction, Part A.*

Chemical Class: Biological Products

Function: Skin-Conditioning Agent - Miscellaneous

Trade Name:
Essential Pear Nectar (Libiol)

PYRUS CYDONIA FRUIT JUICE

Definition: Pyrus Cydonia Fruit Juice is the juice expressed from the fruit of *Pyrus cydonia. See "Regulatory and Ingredient Use Information," regarding the labeling names for botanical ingredients in Volume 1, Introduction, Part A.*

Chemical Class: Biological Products

Function: Proprietary

Technical/Other Names:
Juice, Quince (Pyrus Cydonia)
Quince Juice
Quince (Pyrus Cydonia) Juice

Trade Names:
AEC Quince Conc (A & E Connock)
Quince Juice Clarified Concentrate (Tradimpex)

Trade Name Mixtures:
Fruitapone Quince B 2/036630 (Symrise)
Fruitapone Quince GT 2/037630 (Symrise)

PYRUS CYDONIA SEED

JPN Translation:
クインスシード

Definition: Pyrus Cydonia Seed is the dried seed of the quince, *Pyrus cydonia*. See "Regulatory and Ingredient Use Information," regarding the labeling names for botanical ingredients in Volume 1, Introduction, Part A.

Information Sources: 21CFR182.40, JCIC, JCLS, JSQI, MAR, MI-13(8143)

Chemical Class: Biological Products

Functions: Skin-Conditioning Agent - Miscellaneous; Viscosity Increasing Agent - Aqueous

Technical/Other Names:
Cydonia Oblonga Seed
Quince (Pyrus Cydonia) Seed
Quince Seed
Seeds, Pyrus Cydonia
Seeds, Quince

PYRUS CYDONIA SEED EXTRACT

CAS No. 85117-13-1
EINECS No. 285-564-8

JPN Translation:
クインスシードエキス

Definition: Pyrus Cydonia Seed Extract is an extract of the seeds of the quince, *Pyrus cydonia*. See "Regulatory and Ingredient Use Information," regarding the labeling names for botanical ingredients in Volume 1, Introduction, Part A.

Information Sources: JCIC, JCLS, JSQI, RIFM

Chemical Class: Biological Products

Functions: Fragrance Ingredient; Skin-Conditioning Agent - Miscellaneous

Technical/Other Names:
Cydonia Oblonga Extract
Extract of Pyrus Cydonia
Extract of Quince
Extract of Quince (Pyrus Cydonia) Seed
Quince Extract
Quince (Pyrus Cydonia) Seed Extract
Quince Seed Extract
Quince seed extract (Cydonia spp.) (RIFM)

Trade Name Mixtures:
Cosflor Japonica HGS (A & E Connock)
Herbasol Extract Quince (Seed) (Cosmetochem) (Cosmetochem International Ltd.)
Phytelene of Quince EG 511 liquid (Indena SA)
Phytogreen 55 of Quince EXH 711 Liquid (Phytochim)
Quince Extract GfN (GfN)
Quince HS (Alban Muller)
Quince Seed Extract LA (Maruzen Pharmaceuticals Co., Ltd.)

VT-054 Extract of Quince Seed (Vege-Tech)

PYRUS GERMANICA EXTRACT

Definition: Pyrus Germanica Extract is an extract of the medlar, *Pyrus germanica*. See "Regulatory and Ingredient Use Information," regarding the labeling names for botanical ingredients in Volume 1, Introduction, Part A.

Chemical Class: Biological Products

Function: Not Reported

Trade Name Mixture:
Phytotal FM (Phybiotex/Sederma)

PYRUS MALUS (APPLE) FIBER

Definition: Pyrus Malus (Apple) Fiber is the finely ground fiber obtained from the dried fruit of *Pyrus malus*. See "Regulatory and Ingredient Use Information," regarding the labeling names for botanical ingredients in Volume 1, Introduction, Part A.

Chemical Class: Biological Products

Functions: Binder; Emulsion Stabilizer; Viscosity Controlling Agent; Viscosity Increasing Agent - Aqueous

Technical/Other Name:
Apple Fiber

Trade Names:
GT Pomme F 145 (Greentech)
Vitacel AF 12 (Rettenmaier)
Vitacel AF 400 (Rettenmaier)
Vitacel AF 400/30 (Rettenmaier)
Vitacel AFE 400 (Rettenmaier)
Vitacel AFG 750 (Rettenmaier)

PYRUS MALUS (APPLE) FRUIT

Definition: Pyrus Malus (Apple) Fruit is the fruit of the apple, *Pyrus malus*. See "Regulatory and Ingredient Use Information," regarding the labeling names for botanical ingredients in Volume 1, Introduction, Part A.

Chemical Class: Biological Products

Function: Cosmetic Astringent

Technical/Other Names:
Apple
Apple Fruit

Trade Names:
AEC Apple Puree (A & E Connock)
Edible Apple (Link Brand Solutions)

PYRUS MALUS (APPLE) FRUIT EXTRACT

CAS No. 85251-63-4
EINECS No. 286-475-7

JPN Translation:
リンゴエキス

Definition: Pyrus Malus (Apple) Fruit Extract is an extract of the fruit of the apple, *Pyrus malus*. See "Regulatory and Ingredient Use Information," regarding the labeling names for botanical ingredients in Volume 1, Introduction, Part A.

Information Sources: JCIC, JCLS, JSQI

Chemical Class: Biological Products

Functions: Proprietary; Skin-Conditioning Agent - Miscellaneous

Reported Product Categories: Skin Care Preparations, Misc.; Shampoos (Non-coloring)

Technical/Other Names:
Apple Extract
Apple Fruit Extract
Extract of Apple
Extract of Apple Fruit
Extract of Pyrus Malus
Pyrus Malus Extract

Trade Names:
AEC Apple Juice Conc (A & E Connock)
Applephenon C-100 (Maruzen Pharmaceuticals Co., Ltd.)
Apple Secrets (Gattefosse s.a.)

Trade Name Mixtures:
Actiphyte of Apple BG50 (Active Organics)
Actiphyte of Apple Fruit (Active Organics)
Actiphyte of Apple GL50 (Active Organics)
Actiphyte of Apple Green (Active Organics)
Actiphyte of Apple Lipo S (Active Organics)
Actiphyte of Apple PG50 (Active Organics)
Actiplex 1071 Fruit Acid Complex (Active Organics)
A.H.A. 40 (Ennagram)
Apple Concentrate (Quest International)
Apple Extract BG (Maruzen Pharmaceuticals Co., Ltd.)
Apple Extract BG-J (Maruzen Pharmaceuticals Co., Ltd.)
Apple Extract HG (Provital/Centerchem)
Apple Extract HS 1806 AT (Grau)
Apple HS (Alban Muller)
Bioprotectyl (Silab)
Ederline-H (Vincience)
Ederline-L (Vincience)
Elastin Factor (Silab)
Extrait De Pomme MPE PG10 (Yves Rocher)
Firming Liposomes (Collaborative Labs)
Fruit Vinegar (Provital/Centerchem)
Fruity Sun (Silab)
Glycolysat Of Apple (CEP (Solabia))
Herbasol Extract Apple (Cosmetochem) (Cosmetochem International Ltd.)
Hydralphatine 3P (Lanatech)

Micromerol (Collaborative Labs)
Multifruit MFA (Arch Personal Care
  Products)
Phytolight (Coletica SA)
Phytolight BG (Coletica SA)
Pyrus Malus (Apple) Fruit Extract ies (IES
  LABO)
Quassia Vinegar (Provital/Centerchem)
RAYOLYS (Greentech S.A)
Ringo Liquid B (Ichimaru Pharcos)
Six Fruit Concentrate 3519 G (Grau)
Vegetol Apple GR 294 Hydro (Gattefosse
  s.a.)
VT-079 Extract of Apple (Vege-Tech)

PYRUS MALUS (APPLE) FRUIT WATER

JPN Translation:
リンゴ水

Definition: Pyrus Malus (Apple) Fruit
Water is an aqueous solution of the steam
distillate obtained from the fruit of *Pyrus
malus*. See "Regulatory and Ingredient Use
Information," regarding the labeling names
for botanical ingredients in Volume 1, Introduction, Part A.

Chemical Class: Biological Products

Function: Fragrance Ingredient

Technical/Other Names:
Apple Fruit Water
Apple Water
Water, Apple

Trade Names:
Extrait Originel Pomme (Gattefosse s.a.)
Extrait Originel Pomme HF (Gattefosse
  s.a.)
Nomcort APW (Nisshin OilliO)

Trade Name Mixture:
Nomcort APW-DPG (Nisshin OilliO)

PYRUS MALUS (APPLE) JUICE

JPN Translation:
リンゴ果汁

Definition: Pyrus Malus (Apple) Juice is
the liquid expressed from the fresh pulp of
the apple, *Pyrus malus*. See "Regulatory
and Ingredient Use Information," regarding
the labeling names for botanical ingredients
in Volume 1, Introduction, Part A.

Chemical Class: Biological Products

Function: Skin-Conditioning Agent - Miscellaneous

Technical/Other Names:
Apple Juice
Juice, Apple
Juice, Pyrus Malus
Pyrus Malus Juice

Trade Name:
Apple Juice Concentrate (Symrise)

Trade Name Mixtures:
Complex 1 - Antipollution (Provital/
  Centerchem)
Extrapone Apple 2/033317 (Symrise)
Extrapone Fruit Mixture 2/033257
  (Symrise)
Green Apple Complex (Greentech)
Pronalen Bio-Protect (Provital/Centerchem)
Red Apple Complex (Greentech)

PYRUS MALUS (APPLE) LEAF EXTRACT

CAS No. 85251-63-4
EINECS No. 286-475-7

Definition: Pyrus Malus (Apple) Leaf
Extract is an extract of the leaves of the
apple, *Pyrus malus*. See "Regulatory and
Ingredient Use Information," regarding the
labeling names for botanical ingredients in
Volume 1, Introduction, Part A.

Chemical Class: Biological Products

Function: Not Reported

Reported Product Category: Skin Care
Preparations, Misc.

Technical/Other Names:
Apple Leaf Extract
Extract of Apple Leaves
Extract of Pyrus Malus Leaf
Pyrus Malus Leaf Extract

Trade Name Mixtures:
Actiphyte of Apple Leaves (Active
  Organics)
Actiphyte of Apple Leaves BG50 (Active
  Organics)
Actiphyte of Apple Leaves GL50 (Active
  Organics)
Actiphyte of Apple Leaves Lipo S (Active
  Organics)
Actiphyte of Apple Leaves PG50 (Active
  Organics)

PYRUS MALUS (APPLE) OIL

Definition: Pyrus Malus (Apple) Oil is the
oil expressed from the apple, *Pyrus malus*.
See "Regulatory and Ingredient Use Information," regarding the labeling names for
botanical ingredients in Volume 1, Introduction, Part A.

Chemical Class: Fats and Oils

Function: Skin-Conditioning Agent - Miscellaneous

Technical/Other Names:
Apple Oil
Oils, Apple (Pyrus Malus)

Trade Name:
Lipofructyl PM (Laboratoires Sero-
  biologiques)

PYRUS MALUS (APPLE) PECTIN EXTRACT

Definition: Pyrus Malus (Apple) Pectin
Extract is an extract of the pectin of the
apple, *Pyrus malus*. See "Regulatory and
Ingredient Use Information," regarding the
labeling names for botanical ingredients in
Volume 1, Introduction, Part A.

Chemical Class: Biological Products

Function: Not Reported

Technical/Other Names:
Apple Pectin Extract
Extract of Apple Pectin
Extract of Pyrus Malus Pectin
Pectin, Apple, Extract
Pyrus Malus Pectin Extract

PYRUS MALUS (APPLE) PEEL WAX

Definition: Pyrus Malus (Apple) Peel Wax
is a wax obtained from the peel of the
apple, *Pyrus malus*. See "Regulatory and
Ingredient Use Information," regarding the
labeling names for botanical ingredients in
Volume 1, Introduction, Part A.

Chemical Class: Waxes

Function: Not Reported

Technical/Other Names:
Apfelwachs
Apple Peel Wax
Pyrus Malus Peel Wax
Wax, Apple Peel
Waxes, Apple Peel
Wax, Pyrus Malus Peel

PYRUS MALUS (APPLE) ROOT EXTRACT

Definition: Pyrus Malus (Apple) Root
Extract is an extract of the roots of *Pyrus
malus*. See "Regulatory and Ingredient Use
Information," regarding the labeling names
for botanical ingredients in Volume 1, Introduction, Part A.

Chemical Class: Biological Products

Function: Skin-Conditioning Agent - Miscellaneous

Technical/Other Names:
Apple Root Extract
Extract of Pyrus Malus (Apple) Root

Trade Name:
ViaPure Apple (Actives International)

PYRUS MALUS (APPLE) SEED EXTRACT

Definition: Pyrus Malus (Apple) Seed Extract is an extract of the seeds of *Pyrus malus*. See "*Regulatory and Ingredient Use Information,*" *regarding the labeling names for botanical ingredients in Volume 1, Introduction, Part A.*

Chemical Class: Biological Products

Function: Skin-Conditioning Agent - Miscellaneous

Technical/Other Names:
Apple Seed Extract
Extract of Pyrus Malus (Apple) Seed

Trade Name Mixture:
Seedex-Senior (Vincience)

PYRUS MALUS (APPLE) STEM EXTRACT

Definition: Pyrus Malus (Apple) Stem Extract is an extract of the stems of *Pyrus malus*. See "*Regulatory and Ingredient Use Information,*" *regarding the labeling names for botanical ingredients in Volume 1, Introduction, Part A.*

Chemical Class: Biological Products

Function: Skin-Conditioning Agent - Miscellaneous

Technical/Other Names:
Apple Stem Extract
Extract of Pyrus Malus (Apple) Stem

Trade Name Mixtures:
Extrait Alcolique De Pommier (Plantes et Industrie)
Extrait Alcoolique De Pommier (Plantes et Industrie)

PYRUS SORBUS EXTRACT

CAS No.   **EINECS No.**
90131-20-7   290-330-3

Definition: Pyrus Sorbus Extract is an extract of the sorbus, *Pyrus sorbus*. See "*Regulatory and Ingredient Use Information,*" *regarding the labeling names for botanical ingredients in Volume 1, Introduction, Part A.*

Chemical Class: Biological Products

Function: Not Reported

Technical/Other Names:
Extract of Pyrus Sorbus
Extract of Sorbus
Service Tree Extract
Sorbus Domestica Extract
Sorbus Extract
Sorbus (Pyrus Sorbus) Extract

PYRUVIC ACID

CAS No.   **EINECS No.**
127-17-3   204-824-3

Empirical Formula:
$C_3H_4O_3$

Definition: Pyruvic Acid is the organic compound that conforms to the formula:

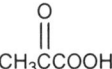

$$CH_3CCOOH$$

Information Sources: MI-13(8110), RIFM, TSCA

Chemical Class: Carboxylic Acids

Functions: Fragrance Ingredient; Not Reported

Technical/Other Names:
Propanoic Acid, 2-Oxo-
Pyruvic acid (RIFM)

Trade Name:
Pyruvic Acid (General Topics)

The inclusion of any compound in the *Dictionary and Handbook* does not indicate that use of that substance as a cosmetic ingredient complies with the laws and regulations governing such use in the United States or any other country.

- Q -

QUARK CHEESE

Definition: Quark Cheese is a sharp, acidic skim milk cheese.

Chemical Class: Biological Products

Function: Skin-Conditioning Agent - Miscellaneous

Trade Name:
Miferm Quark (Heidelberger)

QUARTZ

CAS No.
14808-60-7

EINECS No.
238-878-4

Definition: Quartz is a mineral consisting chiefly of silicon dioxide.

Information Source: MI-13(8567)

Chemical Class: Inorganics

Function: Abrasive

QUASSIA AMARA WOOD EXTRACT

Definition: Quassia Amara Wood Extract is an extract of the wood of *Quassia amara. See "Regulatory and Ingredient Use Information," regarding the labeling names for botanical ingredients in Volume 1, Introduction, Part A.*

Chemical Class: Biological Products

Function: Not Reported

Technical/Other Name:
Extract of Quassia Amara Wood

Trade Name Mixtures:
Complex Quassia (Fabriquimica)
Quassia Vinegar (Provital/Centerchem)

QUASSIN

CAS Nos.
76-78-8
75991-65-0 (dl-alpha)

EINECS No.
200-985-9

Empirical Formula:
$C_{22}H_{28}O_6$

Definition: Quassin is a bitter alkaloid obtained from the wood of *Quassia amara.* It is chiefly used as a denaturant for ethyl alcohol.

Information Sources: ARG, 27CFR21.74, 27CFR21.123, 27CFR21.151, MI-13(8115), POR, TSCA

Chemical Classes: Ethers; Heterocyclic Compounds; Ketones

Function: Denaturant

Technical/Other Names:
2,12-Dimethoxypicrasa-2,12-diene-1,11,16-trione
Picrasa-2,12-diene-1,11,16-trione, 2,12-Dimethoxy-

Trade Name Mixture:
Bio-Dandra Plex (Bio-Botanica)

QUATERNIUM-8

Definition: Quaternium-8 is the quaternary ammonium salt that conforms generally to the formula:

$$\left[\begin{array}{c} CH_3 \\ | \\ R-N-CH_2C_6H_4CH_2CH_3 \\ | \\ CH_3 \end{array} \right]^+ \begin{array}{c} SO_3^- \\ | \\ NH \\ | \\ C_6H_{11} \end{array}$$

where R represents a mixture of c12-18 fatty alkyl radicals.

Information Sources: 21CFR175.105, CTFA D

Chemical Class: Quaternary Ammonium Compounds

Functions: Antistatic Agent; Cosmetic Biocide; Hair Conditioning Agent; Preservative

Technical/Other Name:
Alkyl Dimethyl Ethylbenzyl Ammonium Cyclohexyl Sulfamate

QUATERNIUM-14

CAS No.
27479-28-3

EINECS No.
248-486-5

JPN Translation:
クオタニウム - 14

Empirical Formula:
$C_{23}H_{42}N \cdot Cl$

Definition: Quaternium-14 is the quaternary ammonium salt that conforms generally to the formula:

$$\left[CH_3CH_2-\bigcirc-CH_2\overset{\displaystyle (CH_2)_{11}CH_3}{\underset{\displaystyle CH_3}{N}CH_3} \right]^+ Cl^-$$

Information Sources: 21CFR172.165, 21CFR173.320, JCIC, JCLS, TSCA

Chemical Class: Quaternary Ammonium Compounds

Functions: Antistatic Agent; Cosmetic Biocide; Hair Conditioning Agent; Preservative

Technical/Other Names:
Benzenemethanaminium, N-Dodecyl-*ar*-Ethyl-N,N-Dimethyl-, Chloride
N-Dodecyl-*ar*-Ethyl-N,N-Dimethyl-benzenemethanaminium Chloride
Dodecyl Dimethyl Ethylbenzyl Ammonium Chloride
Lauryl Dimethyl Ethylbenzyl Ammonium Chloride Solution

Trade Names:
Maquat MQ2525M-50% (Mason)
Maquat MQ2525M-80% (Mason)

Trade Name Mixtures:
BTC 2125 (Stepan)
BTC 2125 M (Stepan)
FMB 3328-5 Quat (Huntington)
FMB 3328-8 Quat (Huntington)

QUATERNIUM-15

CAS Nos.
4080-31-3
51229-78-8

EINECS No.
223-805-0

Empirical Formula:
$C_9H_{16}ClN_4 \cdot Cl$

Definition: Quaternium-15 is the quaternary ammonium salt that conforms to the formula:

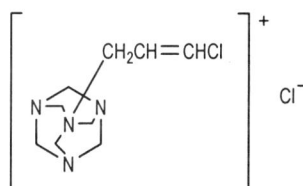

Information Sources: 21CFR175.105, 21CFR176.170, CIR: [S] JACT-5(3)1986, CTFA D, EEC(VI/1-31), MI-13(2135), TSCA

Chemical Classes: Halogen Compounds; Heterocyclic Compounds; Quaternary Ammonium Compounds

Functions: Antistatic Agent; Preservative

Reported Product Categories: Shampoos (Non-coloring); Body and Hand Preparations (Excluding Shaving Preparations); Mascara; Moisturizing Preparations; Bath Preparations, Misc.; Tonics, Dressings, and Other Hair Grooming Aids; Hair Conditioners; Bath Oils, Tablets, and Salts; Cleansing Products (Cold Creams, Cleansing Lotions, Liquids and Pads); Baby Shampoos; Hair Dyes and Colors (All Types Requiring Caution Statements and Patch Tests); Hair Sprays (Aerosol Fixatives); Bath Capsules; Face and

Neck Preparations (Excluding Shaving Preparations); Bath Soaps and Detergents; Face Powders; Skin Care Preparations, Misc.; Foundations; Paste Masks (Mud Packs); Bubble Baths; Eyeliners; Eye Makeup Preparations, Misc.; Eyebrow Pencils; Night Skin Care Preparations; Permanent Waves; Suntan Gels, Creams, and Liquids; Baby Products, Misc.; Hair Preparations (Non-coloring), Misc.; Blushers (All types); Eye Shadows; Hair Bleaches; Hair Wave Sets; Powders (Dusting and Talcum, Excluding Aftershave Talcs); Shaving Cream (Aerosol, Brushless and Lather); Eye Makeup Removers; Makeup Bases; Manicuring Preparations, Misc.; Personal Cleanliness Products, Misc.; Skin Fresheners; Suntan Preparations, Misc.

Technical/Other Names:
N-(3-Chloroallyl)Hexaminium Chloride
Chloroallyl Methenamine Chloride
1-(3-Chloroallyl)-3,5,7-Triaza-1-Azoniaadamantane Chloride
1-(3-Chloro-2-Propenyl)-3,5,7-Triaza-1-Azoniatricyclo[3.3.1.1]Decane Chloride
Methenamine 3-Chloroallylchloride
3,5,7-Triaza-1-Azoniatricyclo[3.3.1.1]-Decane, 1-(3-Chloro-2-Propenyl)-

Trade Names:
CoSept 200 (Costec)
DOWICIL 200 (Dow Chemical)

Trade Name Mixture:
Miracare BC-27 (Rhodia)

QUATERNIUM-16

CAS No.: 35239-12-4

Definition: Quaternium-16 is the quaternary ammonium salt that conforms to the formula:

$$\left[\begin{array}{c} CH_2CH_2OH \\ | \\ R\text{---}N\text{---}CH_2CH_2OH \\ | \\ CH_2CH_2OH \end{array} \right]^+ \quad Cl^-$$

where R represents tallow fatty radicals.

Chemical Class: Quaternary Ammonium Compounds

Functions: Antistatic Agent; Hair Conditioning Agent

Technical/Other Name:
Octadecanaminium, N,N,N-tris(2-Hydroxyethyl)-, Chloride

QUATERNIUM-18

CAS No. 61789-80-8 **EINECS No.** 263-090-2

JPN Translation:
クオタニウム - 18

Definition: Quaternium-18 is the quaternary ammonium salt that conforms generally to the formula:

$$\left[\begin{array}{c} CH_3 \\ | \\ R\text{---}N\text{---}R \\ | \\ CH_3 \end{array} \right]^+ \quad Cl^-$$

where R represents hydrogenated tallow fatty radicals.

Information Sources: CIR: [S] JACT-1(2)-1982, TSCA

Chemical Class: Quaternary Ammonium Compounds

Function: Antistatic Agent

Reported Product Categories: Hair Conditioners; Foundations; Hair Rinses (Non-coloring); Hair Preparations (Non-coloring), Misc.; Makeup Fixatives; Tonics, Dressings, and Other Hair Grooming Aids; Permanent Waves

Technical/Other Names:
Dimethyl Di(Hydrogenated Tallow) Ammonium Chloride
Quaternary Ammonium Compounds, Bis (Hydrogenated Tallow Alkyl)Dimethyl, Chlorides

Trade Names:
AEC Quaternium-18 (A & E Connock)
Varisoft DHT (Degussa Care Specialties)
Varisoft 442 100P (Degussa Care Specialties)

Trade Name Mixtures:
AEC Quaternium-18 & IPA (A & E Connock)
Arquad 2HT-75 (Akzo Nobel)

QUATERNIUM-18 BENTONITE

CAS Nos. 1340-69-8
68953-58-2   **EINECS No.** 273-219-4

JPN Translation:
クオタニウム - 18 ベントナイト

Definition: Quaternium-18 Bentonite is a reaction product of Bentonite (q.v.) and Quaternium-18 (q.v.).

Information Sources: 21CFR175.300, 21CFR178.3570, CIR: [S] JACT-1(2)1982, CTFA S, JCIC, JCLS, JSQI, MI-13(1054), TSCA

Chemical Classes: Inorganics; Quaternary Ammonium Compounds

Function: Suspending Agent - Nonsurfactant

Reported Product Categories: Lipsticks; Eyeliners; Eye Shadows; Eyebrow Pencils; Makeup Preparations (Not eye), Misc.; Mascara; Personal Cleanliness Products, Misc.

Technical/Other Names:
Dimethyldistearyl Ammonium Bentonite
Quaternary Ammonium Compounds, Bis (Hydrogenated Tallow Alkyl)Dimethyl, Chlorides, Reaction Products with Bentonite

Trade Names:
Bentone 34 (ELE)
Claytone 40 (Southern Clay)
Claytone SO (Southern Clay)
Tixogel MP (Sud-Chemie, United Catalysts)
Tixogel MP - 100 (Sud-Chemie, United Catalysts)
Tixogel TE (Sud-Chemie, United Catalysts)
Tixogel VP (Sud-Chemie, United Catalysts)

Trade Name Mixtures:
Tixogel CYM (Sud-Chemie, United Catalysts)
Tixogel CYM - 1374 (Sud-Chemie, United Catalysts)
Tixogel HXL-1209 (Sud-Chemie, United Catalysts)
Tixogel IDD - 1168 (Sud-Chemie, United Catalysts)
Tixogel MIO (Sud-Chemie, United Catalysts)
Tixogel OMS (Sud-Chemie, United Catalysts)
Tixogel VSP (Sud-Chemie, United Catalysts)

QUATERNIUM-18/BENZALKONIUM BENTONITE

Definition: Quaternium-18/Benzalkonium Bentonite is a reaction product of Bentonite (q.v.) and Quaternium-18 (q.v.) and Benzalkonium Chloride (q.v.).

Chemical Classes: Inorganics; Quaternary Ammonium Compounds

Function: Suspending Agent - Nonsurfactant

Trade Name:
Claytone HT (Southern Clay)

QUATERNIUM-18 HECTORITE

CAS Nos. 12001-31-9
71011-27-3   **EINECS No.** 234-406-6

The inclusion of any compound in the *Dictionary and Handbook* does not indicate that use of that substance as a cosmetic ingredient complies with the laws and regulations governing such use in the United States or any other country.

JPN Translation:
クオタニウム - 18 ヘクトライト

Definition: Quaternium-18 Hectorite is a reaction product of Hectorite (q.v.) and Quaternium-18 (q.v.).

Information Sources: CIR: [S] JACT-1(2)-1982, CTFA D, JCIC, JCLS, JSQI

Chemical Classes: Inorganics; Quaternary Ammonium Compounds

Function: Suspending Agent - Nonsurfactant

Reported Product Categories: Mascara; Personal Cleanliness Products, Misc.; Lipsticks; Eyeliners; Makeup Preparations (Not eye), Misc.; Eye Shadows; Makeup Bases; Foundations; Eye Makeup Preparations, Misc.; Bath Capsules; Deodorants (Underarm); Night Skin Care Preparations; Skin Care Preparations, Misc.; Cleansing Products (Cold Creams, Cleansing Lotions, Liquids and Pads); Eye Makeup Removers; Eyebrow Pencils; Moisturizing Preparations; Suntan Gels, Creams, and Liquids

Technical/Other Names:
Dimethyldistearyl Ammonium Hectorite
Quaternary Ammonium Compounds, Bis (Hydrogenated Tallow Alkyl)Dimethyl, Chlorides, Reaction Products with Hectorite

Trade Names:
Bentone 38 (ELE)
LUCENTITE SAN (Co-Op/Kobo)

Trade Name Mixtures:
Bentone Gel DOA (ELE)
Bentone Gel EUG (ELE)
Bentone Gel ISD (ELE)
Bentone Gel MIO (ELE)
Bentone Gel MIO-A40 (ELE)
Bentone Gel SS-71 (ELE)
Bentone Gel 10ST (ELE)
Bentone Gel VS-5 (ELE)
Bentone Gel VS-5 PC (ELE)
Covalip LL 48 (LCW)
DUB GEL SI 1400 (Stearinerie Dubois Fils)
Simagel M (Biophil)
Simagel SI 345 (Biophil)

QUATERNIUM-18 METHOSULFATE

CAS No.	EINECS No.
61789-81-9	263-091-8

Definition: Quaternium-18 Methosulfate is the quaternary ammonium salt that conforms to the formula:

where R represents the hydrogenated tallow radical.

Information Source: TSCA

Chemical Class: Quaternary Ammonium Compounds

Function: Antistatic Agent

Technical/Other Name:
Quaternary Ammonium Compounds, Bis (Hydrogenated Tallow Alkyl)Dimethyl, Methyl Sulfates

QUATERNIUM-22

CAS Nos.	EINECS No.
51812-80-7	257-440-3
82970-95-4	

JPN Translation:
クオタニウム - 22

Empirical Formula:
$C_{13}H_{29}N_2O_7 \cdot Cl$

Definition: Quaternium-22 is the quaternary ammonium salt reported to conform generally to the formula:

Information Sources: CIR: [S] JACT-14(6)-1995, JCIC, JCLS, JSQI, TSCA

Chemical Classes: Carbohydrates; Quaternary Ammonium Compounds

Functions: Antistatic Agent; Film Former; Hair Conditioning Agent

Reported Product Categories: Mascara; Hair Conditioners; Moisturizing Preparations; Bath Soaps and Detergents; Shampoos (Non-coloring); Bath Preparations, Misc.

Technical/Other Names:
γ-Gluconamidopropyl Dimethyl 2-Hydroxyethyl Ammonium Chloride
N-γ-Gluconamidopropyl-N,N-dimethyl-N-hydroxyethyl Ammonium Chloride
3-(D-Gluconoylamino)-N-(2-Hydroxyethyl)-N,N-Dimethyl-1-Propanaminium Chloride
1-Propanaminium, 3-(D-Gluconoylamino)-N-(2-Hydroxyethyl)-N,N-Dimethyl-, Chloride

Trade Name:
Ceraphyl 60 (International Specialty Products)

QUATERNIUM-24

CAS No.	EINECS No.
32426-11-2	251-035-5

Empirical Formula:
$C_{20}H_{44}N \cdot Cl$

Definition: Quaternium-24 is the quaternary ammonium salt that conforms generally to the formula:

Chemical Class: Quaternary Ammonium Compounds

Functions: Antistatic Agent; Cosmetic Biocide; Surfactant - Cleansing Agent

Technical/Other Names:
Ammonium, Decyldimethyloctyl, Chloride
Decyl Dimethyl Octyl Ammonium Chloride

Trade Names:
AEC Quaternium-24 (A & E Connock)
Bardac 2050 (Lonza Inc./Lonza Ltd.)
Maquat 4050 (Mason)

QUATERNIUM-26

CAS Nos.	EINECS No.
64156-20-3	
68953-64-0	273-222-0

Definition: Quaternium-26 is the quaternary ammonium salt that conforms generally to the formula:

where RCO- represents the fatty acid groups derived from mink oil.

Information Sources: CIR: [SQ] IJT-19 (SUPPL.1)2000, TSCA

Chemical Class: Quaternary Ammonium Compounds

The inclusion of any compound in the *Dictionary and Handbook* does not indicate that use of that substance as a cosmetic ingredient complies with the laws and regulations governing such use in the United States or any other country.

Functions: Antistatic Agent; Hair Conditioning Agent

Reported Product Categories: Tonics, Dressings, and Other Hair Grooming Aids; Hair Conditioners; Shampoos (Non-coloring); Bath Oils, Tablets, and Salts; Cleansing Products (Cold Creams, Cleansing Lotions, Liquids and Pads); Hair Preparations (Non-coloring), Misc.; Permanent Waves

Technical/Other Names:
Minkamidopropyl Dimethyl 2-Hydroxyethyl Ammonium Chloride
Quaternary Ammonium Compounds, (Hydroxyethyl)Dimethyl(3-Mink Oil Amidopropyl), Chlorides

Trade Name:
Incroquat 26 (Croda, Inc.)

Trade Name Mixture:
Ceraphyl 65 (International Specialty Products)

QUATERNIUM-27

CAS Nos.	EINECS Nos.
68122-86-1	268-531-2
86088-85-9	289-151-3

Definition: Quaternium-27 is the quaternary ammonium salt that conforms generally to the formula:

where RCO- and R are derived from the tallow.

Chemical Classes: Heterocyclic Compounds; Quaternary Ammonium Compounds

Functions: Antistatic Agent; Hair Conditioning Agent

Reported Product Category: Hair Conditioners

Technical/Other Names:
4,5-Dihydro-1-Methyl-2-Nortallow Alkyl-1-(2-Tallow Amidoethyl)Imidazolium Compounds, Methyl Sulfates
Imidazolium Compounds, 4,5-Dihydro-1-Methyl-2-Nortallow Alkyl-1-(2-Tallow Amidoethyl), Methyl Sulfates

Trade Names:
AEC Quaternium-27 (A & E Connock)

Incrosoft S-90 (Croda, Inc.)
Unsoft 475 (Universal Preserv-A-Chem)

Trade Name Mixture:
Condiquat (Cosmetochem) (Cosmetochem International Ltd.)

QUATERNIUM-30

CAS Nos.: 50978-31-9; 58748-40-6

Empirical Formula:
$C_{25}H_{46}NO_3 \cdot Cl$

Definition: Quaternium-30 is the quaternary ammonium salt that conforms generally to the formula:

Chemical Classes: Alkoxylated Amines; Quaternary Ammonium Compounds

Function: Antistatic Agent

Technical/Other Names:
Benzenemethanaminium, N,N,N-Tris(2-Hydroxyethyl)-4-Isododecyl-, Chloride
Isododecylbenzyl Triethanolammonium Chloride
N,N,N-Tris(2-Hydroxyethyl)-4-Isododecyl-benzenemethanaminium Chloride

QUATERNIUM-33

CAS No.: 86221-07-0

JPN Translation:
クオタニウム - 33

Definition: Quaternium-33 is the quaternary ammonium salt that conforms to the formula:

where RCO- represents the lanolin acid radical.

Information Sources: JCIC, JCLS, JSQI

Chemical Classes: Lanolin and Lanolin Derivatives; Quaternary Ammonium Compounds

Functions: Antistatic Agent; Hair Conditioning Agent

Reported Product Category: Hair Straighteners

Technical/Other Names:
N (N'-Lanolin Fatty Acid Amide Propyl) N-Ethyl-N,N-Dimethyl Ammonium Ethyl Sulfate (1)
N (N'-Lanolin Fatty Acid Amide Propyl) N-Ethyl-N,N-Dimethyl Ammonium Ethyl Sulfate (2)

Trade Names:
Lanostar DP-50 (Nippon Chemical)
Nikkol Lanoquat DES-50 (Nikko)

Trade Name Mixture:
Incroquat Behenyl 18-MEA (Croda, Inc.)

QUATERNIUM-43

Definition: Quaternium-43 is the quaternary ammonium salt that conforms to the formula:

where RCO- represents the cocoyl radical.

Chemical Class: Quaternary Ammonium Compounds

Function: Antistatic Agent

Technical/Other Name:
Cocamidopropyl Dimethyl Acetamido Ammonium Chloride

Trade Name:
Chimexane CK (Chimex)

QUATERNIUM-45

CAS No.	EINECS No.
21034-17-3	244-158-0

JPN Translation:
クオタニウム - 45

Empirical Formula:
$C_{13}H_{15}N_2O \cdot I$

Definition: Quaternium-45 is the quaternary ammonium salt that conforms to the formula:

Information Source: MHLW-331/3

Chemical Classes: Heterocyclic Compounds; Quaternary Ammonium Compounds

Function: Antistatic Agent

Trade Name:
Luminex (Ikeda)

QUATERNIUM-51

CAS No.	EINECS No.
1463-95-2	215-976-5

JPN Translation:
クオタニウム - 51

Empirical Formula:
$C_{15}H_{17}BrN_3 \cdot I$

Definition: Quaternium-51 is the quaternary ammonium salt that conforms to the formula:

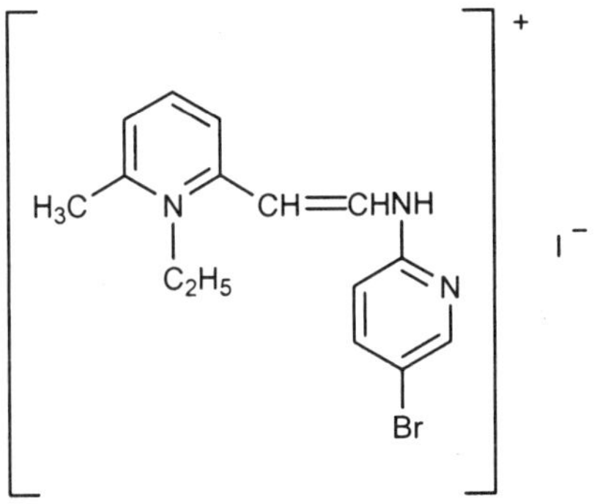

Information Source: MHLW-331/3

Chemical Class: Quaternary Ammonium Compounds

Function: Antistatic Agent

Reported Product Category: Tonics, Dressings, and Other Hair Grooming Aids

QUATERNIUM-52

CAS No.: 58069-11-7

Definition: Quaternium-52 is the quaternary ammonium salt that conforms generally to the formula:

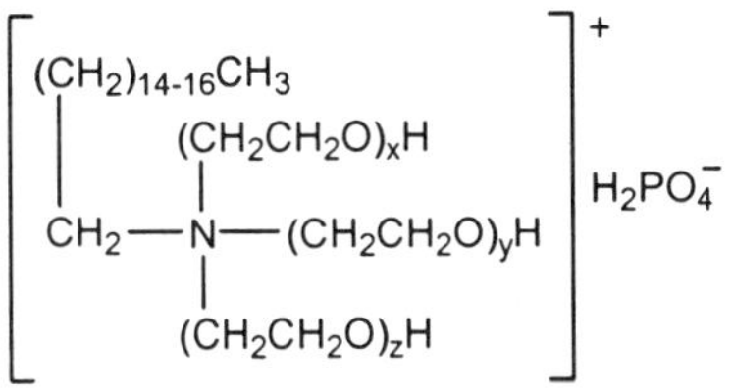

where x+y+z has an average value of 10.

Chemical Class: Quaternary Ammonium Compounds

Functions: Antistatic Agent; Hair Conditioning Agent; Surfactant - Cleansing Agent

Reported Product Categories: Permanent Waves; Tonics, Dressings, and Other Hair Grooming Aids; Hair Preparations (Non-coloring), Misc.; Shampoos (Non-coloring)

Technical/Other Name:
Poly(Oxy-1,2-Ethanediyl), ((Octadecylnitrilo)tri-2,1-Ethanediyl)tris (Hydroxy)- Phosphate (1:1) (Salt)

Trade Names:
Dehyquart SP (Cognis Care Chemicals/NJ)
Dehyquart SP (Cognis Care Chemicals/PA)
Dehyquart SP (Cognis Deutschland)

Trade Name Mixture:
Condipon (Cosmetochem)

QUATERNIUM-53

CAS No.: 68410-69-5

Definition: Quaternium-53 is the quaternary ammonium salt that conforms generally to the formula:

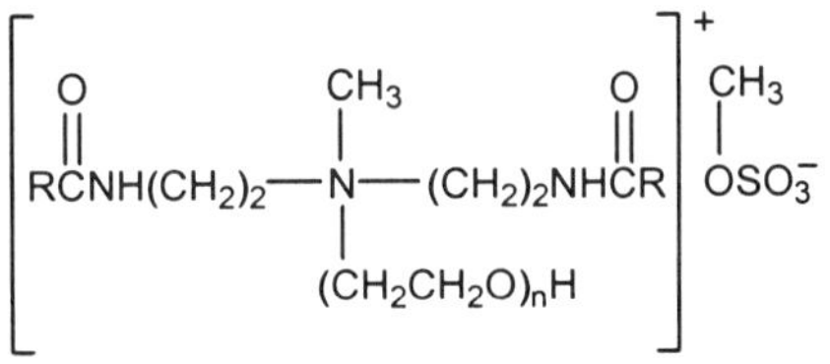

where RCO- represents the tallow acid radical and n has an average value of 3.

Chemical Class: Quaternary Ammonium Compounds

Functions: Antistatic Agent; Hair Conditioning Agent

Technical/Other Name:
Poly(Oxy-1,2-Ethanediyl),α-[2-[Bis(2-Aminoethyl)Methylammonio]Ethyl]-ω-Hydroxy, N,N'-Ditallow Acyl Derivs., Methyl Sulfates

Trade Name:
Incrosoft T-90 (Croda, Inc.)

QUATERNIUM-56

Definition: Quaternium-56 is the quaternary ammonium salt that conforms generally to the formula:

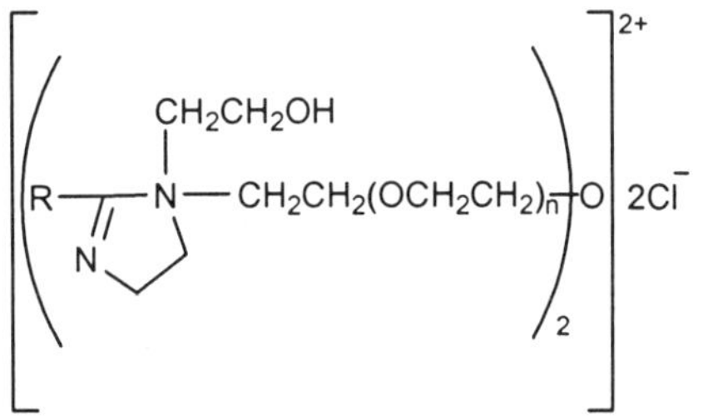

where R is derived from Oleic Acid (q.v.) and n has an average value of 12.

Chemical Classes: Imidazoline Compounds; Quaternary Ammonium Compounds

Functions: Antistatic Agent; Hair Conditioning Agent

Technical/Other Name:
Bis(N-Hydroxyethyl-2-Oleyl Imidazolinium Chloride) Polyethylene Glycol 600

QUATERNIUM-60

Definition: Quaternium-60 is the quaternary ammonium salt that conforms generally to the formula:

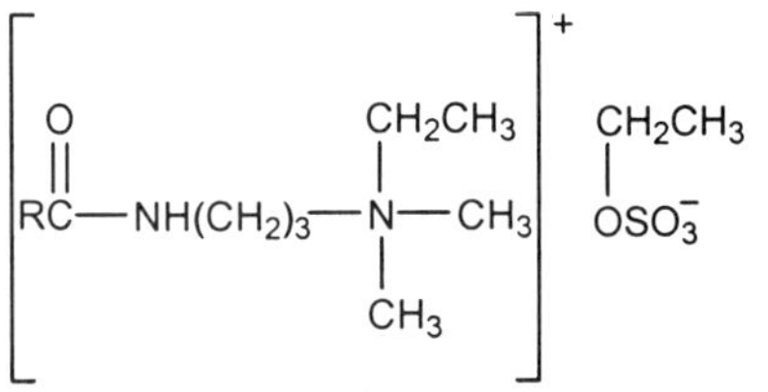

where RCO- represents a mixture of lanolin and isostearic fatty acid radicals.

Chemical Classes: Lanolin and Lanolin Derivatives; Quaternary Ammonium Compounds

Functions: Antistatic Agent; Hair Conditioning Agent

Reported Product Category: Shampoos (Non-coloring)

Technical/Other Name:
Lanolin/Isostearamidopropyl Ethyl Dimethyl Ammonium Ethosulfate

Trade Name Mixtures:
Lanoquat 1751A (Cognis Care Chemicals/NJ)
Lanoquat 1751A (Cognis Care Chemicals/PA)

QUATERNIUM-61

CAS No.: 111905-55-6

Definition: Quaternium-61 is the quaternary ammonium salt that conforms generally to the formula:

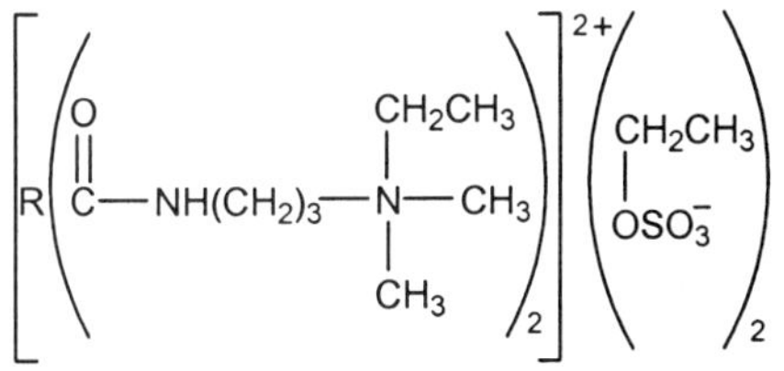

where R is the alkyl portion of Dimer Acid (q.v.).

Chemical Class: Quaternary Ammonium Compounds

Functions: Antistatic Agent; Hair Conditioning Agent

Reported Product Category: Hair Conditioners

Technical/Other Name:
Dimer Acid, Bis[Amidopropyl-N,N-Dimethyl-N-Ethyl Ammonium Ethosulfate]

Trade Name:
Schercoquat DAS (Scher)

QUATERNIUM-63

Empirical Formula:
$C_{62}H_{102}N_4O_4 \cdot Cl$

Definition: Quaternium-63 is the quaternary ammonium salt that conforms to the formula:

where R is derived from Dilinoleic Acid (q.v.).

Chemical Class: Quaternary Ammonium Compounds

Functions: Antistatic Agent; Hair Conditioning Agent

QUATERNIUM-70

CAS No.	**EINECS No.**
68921-83-5	272-964-2

Empirical Formula:
$C_{39}H_{79}N_2O_3 \cdot Cl$

Definition: Quaternium-70 is the quaternary ammonium salt that conforms generally to the formula:

where RCO- represents the stearic acid radical.

Chemical Class: Quaternary Ammonium Compounds

Functions: Antistatic Agent; Hair Conditioning Agent

Technical/Other Names:
N,N-Dimethyl-3-[(1-Oxooctadecyl)Amino]-N-[2-Oxo-2-Tetradecenyloxy)Ethyl]-1-Propanaminium Chloride
1-Propanaminium, N,N-Dimethyl-3-[(1-Oxooctadecyl)Amino]-N-[2-Oxo-2-(Tetradecenyloxy)Ethyl]-, Chloride
Stearamidopropyl Dimethyl (Myristyl Acetate) Ammonium Chloride

Trade Name Mixtures:
Ceraphyl 70 (International Specialty Products)
Protaquat 70 (Protameen)

QUATERNIUM-71

Empirical Formula:
$C_{39}H_{75}N_2O_3 \cdot Cl$

Definition: Quaternium-71 is the quaternary ammonium salt that conforms generally to the formula:

Chemical Class: Quaternary Ammonium Compounds

Function: Antistatic Agent

QUATERNIUM-72

CAS No.: 92201-88-2

Empirical Formula:
$C_{41}H_{78}N_3O \cdot CH_3O_4S$

Definition: Quaternium-72 is the quaternary ammonium compound that conforms to the formula:

where R represents the oleyl group.

Chemical Classes: Imidazoline Compounds; Quaternary Ammonium Compounds

Functions: Antistatic Agent; Hair Conditioning Agent

Trade Name:
Incrosoft CFI 75 (Croda, Inc.)

QUATERNIUM-73

CAS No.	**EINECS No.**
15763-48-1	239-852-5

JPN Translation:
クオタニウム - 73

Empirical Formula:
$C_{23}H_{39}N_2S_2 \cdot I$

Definition: Quaternium-73 is the quaternary ammonium salt that conforms generally to the formula:

Information Source: MHLW-331/3

Chemical Classes: Heterocyclic Compounds; Quaternary Ammonium Compounds

Function: Antistatic Agent

Technical/Other Names:
4,4'-Dimethyl-3,3'-Diheptyl-2, 2'Thiazolocyanine Iodide
2-[2-(3-Heptyl-4-Methyl-2-Thiazolin-2-ylidene)Methene]-3-Heptyl- 4-Methyl Thiazolinium Iodide

Trade Name:
Kankoh SO 201 (Hippon)

QUATERNIUM-75

Definition: Quaternium-75 is the quaternary ammonium compound that conforms generally to the formula:

where R represents a mixture of C14 and C16 alkyl groups.

Chemical Class: Quaternary Ammonium Compounds

Functions: Antistatic Agent; Hair Conditioning Agent

Reported Product Categories: Permanent Waves; Shampoos (Non-coloring)

Trade Names:
Finquat CT (Finetex)
Finquat LM-730 (Finetex)
Finquat LMP-532 (Finetex)

Trade Name Mixtures:
Lowenol Conditioner C-33 (Lowenstein)
Lowenol Conditioner C-727 (Lowenstein)

QUATERNIUM-76 HYDROLYZED COLLAGEN

Definition: Quaternium-76 Hydrolyzed Collagen is the quaternary ammonium salt that conforms generally to the formula:

$$\left[R-\overset{\overset{\displaystyle CH_3}{|}}{\underset{\underset{\displaystyle CH_3}{|}}{N}}-CH_2\overset{\overset{\displaystyle}{}}{CH}\underset{\underset{\displaystyle OH}{|}}{}CH_2R' \right]^+ X^-$$

where R represents the alkyl groups derived from coconut oil and X represents the hydrolyzed collagen moiety.

Chemical Classes: Protein Derivatives; Quaternary Ammonium Compounds

Functions: Antistatic Agent; Hair Conditioning Agent; Skin-Conditioning Agent - Miscellaneous

Reported Product Categories: Permanent Waves; Hair Conditioners

Trade Name:
Lexein QX-3000 (Inolex)

QUATERNIUM-77

Definition: Quaternium-77 is the quaternary ammonium salt that conforms generally to the formula:

$$\left[R\overset{\overset{\displaystyle O}{||}}{C}NH(CH_2)_2-\overset{\overset{\displaystyle CH_2CH_3}{|}}{\underset{\underset{\displaystyle CH_2CHCH_3}{|}}{N}}-(CH_2)_2NH\overset{\overset{\displaystyle O}{||}}{C}R \right]^+ \begin{array}{c} CH_3CH_2 \\ | \\ OSO_3^- \end{array}$$

where RCO- represents a mixture of stearoyl and palmitoyl acid radicals.

Chemical Class: Quaternary Ammonium Compounds

Functions: Antistatic Agent; Hair Conditioning Agent

Trade Name Mixture:
Finsoft HCM-100 (Finetex)

QUATERNIUM-78

Definition: Quaternium-78 is the quaternary ammonium salt that conforms generally to the formula:

$$\left[RCNH(CH_2)_2N(CH_2)_2-\overset{\overset{\displaystyle CH_2CH_3}{|}}{\underset{\underset{\displaystyle}{}}{N}}-CH_2CHCH_3 \right]^+ \begin{array}{c} CH_3 \\ | \\ CH_2 \\ | \\ OSO_3^- \end{array}$$

where RCO- represents a mixture of stearoyl and palmitoyl acid radicals.

Chemical Classes: Alkoxylated Amines; Quaternary Ammonium Compounds

Functions: Antistatic Agent; Hair Conditioning Agent

Trade Name Mixture:
Finsoft HCM-100 (Finetex)

QUATERNIUM-79 HYDROLYZED COLLAGEN

Definition: Quaternium-79 Hydrolyzed Collagen is the quaternary ammonium chloride that conforms generally to the formula:

$$\left[R-\overset{\overset{\displaystyle CH_3}{|}}{\underset{\underset{\displaystyle CH_3}{|}}{N}}-CH_2CHCH_2R' \right]^+ Cl^-$$

where R represents an oleyl/palmityl/palmitoleyl radical and R' represents the hydrolyzed collagen moiety.

Chemical Classes: Protein Derivatives; Quaternary Ammonium Compounds

Functions: Antistatic Agent; Hair Conditioning Agent; Skin-Conditioning Agent - Miscellaneous

Trade Name:
Mackpro NLP (McIntyre)

QUATERNIUM-79 HYDROLYZED KERATIN

Definition: Quaternium-79 Hydrolyzed Keratin the quaternary ammonium chloride/lactate that conforms to the formula:

$$\left[R-\overset{\overset{\displaystyle CH_3}{|}}{\underset{\underset{\displaystyle CH_3}{|}}{N}}-CH_2CHCH_2R' \right]^+ X^-$$

where R represents an oleyl/palmityl/palmitoleyl radical, R' represents the hydrolyzed keratin moiety and X is either a chloride or lactate ion.

Chemical Classes: Protein Derivatives; Quaternary Ammonium Compounds

Functions: Antistatic Agent; Hair Conditioning Agent; Skin-Conditioning Agent - Miscellaneous

Reported Product Categories: Hair Dyes and Colors (All Types Requiring Caution Statements and Patch Tests); Hair Conditioners

Trade Name:
Mackpro KLP (McIntyre)

QUATERNIUM-79 HYDROLYZED MILK PROTEIN

Definition: Quaternium-79 Hydrolyzed Milk Protein is the quaternary ammonium chloride that conforms generally to the formula:

$$\left[R\overset{\overset{\displaystyle O}{||}}{C}-NH(CH_2)_3-\overset{\overset{\displaystyle CH_3}{|}}{\underset{\underset{\displaystyle CH_3}{|}}{N}}-CH_2CHCH_2R' \right]^+ Cl^-$$

where RCO- represents an oleoyl/palmitoyl/palmitoleoyl acid radical and R' represents the hydrolyzed milk protein moiety.

Chemical Classes: Protein Derivatives; Quaternary Ammonium Compounds

Functions: Antistatic Agent; Hair Conditioning Agent; Skin-Conditioning Agent - Miscellaneous

Trade Name:
Mackpro MLP (McIntyre)

QUATERNIUM-79 HYDROLYZED SILK

Definition: Quaternium-79 Hydrolyzed Silk is the quaternary ammonium chloride that conforms generally to the formula:

$$\left[R\overset{\overset{\displaystyle O}{||}}{C}-NH(CH_2)_3-\overset{\overset{\displaystyle CH_3}{|}}{\underset{\underset{\displaystyle CH_3}{|}}{N}}-CH_2CHCH_2R' \right]^+ Cl^-$$

where RCO- represents an oleoyl/palmitoyl/ palmitoleoyl radical and R' represents the hydrolyzed silk protein moiety.

Chemical Classes: Protein Derivatives; Quaternary Ammonium Compounds

Functions: Antistatic Agent; Hair Conditioning Agent; Skin-Conditioning Agent - Miscellaneous

Reported Product Category: Tonics, Dressings, and Other Hair Grooming Aids

Trade Name:
  Mackpro NSP (McIntyre)

QUATERNIUM-79 HYDROLYZED SOY PROTEIN

Definition: Quaternium-79 Hydrolyzed Soy Protein is the quaternary ammonium chloride conforms generally to the formula:

$$\left[RC \underset{O}{\overset{O}{\parallel}} NH(CH_2)_3 \underset{CH_3}{\overset{CH_3}{\underset{|}{\overset{|}{N}}}} CH_2CHCH_2R' \right]^+ Cl^-$$

where RCO- represents an oleoyl/palmitoyl/ palmitoleoyl radical and R' represents the hydrolyzed soy protein moiety.

Chemical Classes: Protein Derivatives; Quaternary Ammonium Compounds

Functions: Antistatic Agent; Hair Conditioning Agent; Skin-Conditioning Agent - Miscellaneous

Trade Name:
  Mackpro SLP (McIntyre)

QUATERNIUM-79 HYDROLYZED WHEAT PROTEIN

Definition: Quaternium-79 Hydrolyzed Wheat Protein is the quaternary ammonium chloride that conforms generally to the formula:

$$\left[RC \underset{O}{\overset{O}{\parallel}} NH(CH_2)_3 \underset{CH_3}{\overset{CH_3}{\underset{|}{\overset{|}{N}}}} CH_2CHCH_2R' \right]^+ Cl^-$$

where RCO- represents the oleoyl/palmitoyl/ palmitoleoyl acid radical and R' represents the wheat protein moiety.

Chemical Classes: Protein Derivatives; Quaternary Ammonium Compounds

Functions: Antistatic Agent; Hair Conditioning Agent; Skin-Conditioning Agent - Miscellaneous

Trade Name:
  Mackpro NLW (McIntyre)

QUATERNIUM-80

CAS No.: 134737-05-6

Definition: Quaternium-80 is the quaternary ammonium salt that conforms to the formula:

where R represents the alkyl groups derived from coconut oil.

Chemical Classes: Amides; Quaternary Ammonium Compounds; Siloxanes and Silanes

Functions: Antistatic Agent; Hair Conditioning Agent

Reported Product Categories: Hair Conditioners; Hair Shampoos (Coloring); Shampoos (Non-coloring); Hair Tints

Technical/Other Name:
  Siloxanes and Silicones, di-Me, 3-(3-((3-Coco Amidopropyl)Dimethylammonio)-2-Hydroxypropoxy)Propyl Group-terminated, Acetates (Salts)

Trade Names:
  Abil Quat 3272 (Degussa Care Specialties)
  Abil Quat 3474 (Degussa Care Specialties)
  AEC Quaternium-80 (A & E Connock)

Trade Name Mixtures:
  Covafix 123 (LCW)
  Solvariane (LCW)

QUATERNIUM-81

Definition: Quaternium-81 is the quaternary ammonium salt prepared by reacting oleic acid with diethylenetriamine followed by quaternization with dimethyl sulfate.

Chemical Class: Quaternary Ammonium Compounds

Functions: Antistatic Agent; Hair Conditioning Agent

QUATERNIUM-82

CAS No.: 173833-36-8

Empirical Formula:
  $C_{51}H_{99}N_2O_6 \cdot CH_3O_4S$

Definition: Quaternium-82 is the quaternary ammonium salt that conforms generally to the formula:

where RCO- represents an oleic acid radical.

Chemical Classes: Alkoxylated Amines; Quaternary Ammonium Compounds

Functions: Antistatic Agent; Hair Conditioning Agent

Reported Product Category: Makeup Preparations (Not eye), Misc.

Trade Names:
  Amonyl DM (SEPPIC)
  STEPANQUAT ML (Stepan)

Trade Name Mixture:
  Stepanquat DC1 (Stepan)

QUATERNIUM-83

CAS No.	EINECS No.
91723-55-6	294-563-1

Definition: Quaternium-83 is the quaternary ammonium salt that conforms to the formula:

where R represents the alkyl groups derived from hydrogenated tallow.

Chemical Classes: Heterocyclic Compounds; Quaternary Ammonium Compounds

Functions: Antistatic Agent; Hair Conditioning Agent

Technical/Other Name:
Imidazolinium Compounds, 4,5-Dihydro-2-(Hydrogenated Nortallowalkyl)-1-(2-(Hydrogenated Tallow Amido)Ethyl)-1-Methyl Methosulfates

QUATERNIUM-84

Definition: Quaternium-84 is the quaternary ammonium salt that conforms generally to the formula:

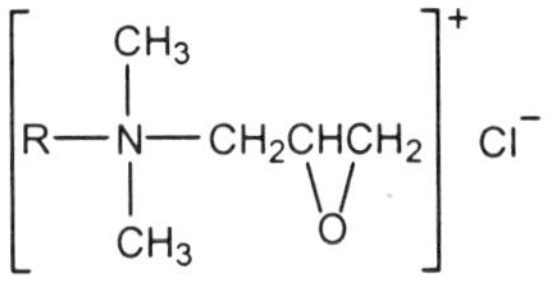

where R represents an oleyl/palmityl/palmitoleyl alkyl radical.

Chemical Class: Quaternary Ammonium Compounds

Function: Hair Conditioning Agent

Trade Name:
Mackernium NLE (McIntyre)

QUATERNIUM-85

CAS No.: 141890-30-4

Empirical Formula:
$C_{24}H_{52}N_3O_3 \cdot Cl$

Definition: Quaternium-85 is the quaternary ammonium compound that conforms to the formula:

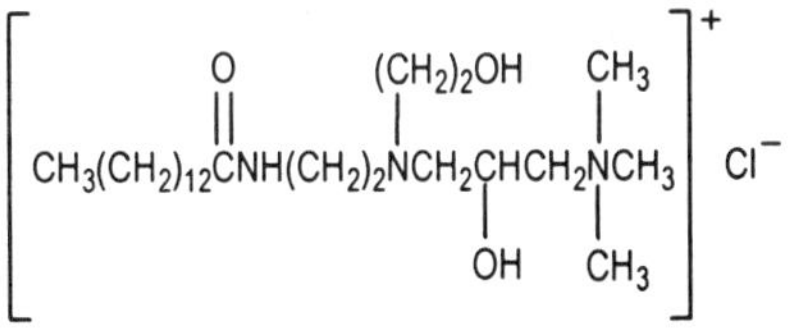

Chemical Classes: Alkylamido Alkylamines; Quaternary Ammonium Compounds

Function: Hair Conditioning Agent

Technical/Other Name:
1-Propanaminium, 2-Hydroxy-3-((2-Hydroxyethyl)(2-((1-Oxotetradecyl (Amino)Ethyl)Amino)-N,N,N-Trimethyl-, Chloride

Trade Name:
Aminocation S-36 (Kao Corp.)

QUATERNIUM-86

Definition: Quaternium-86 is the quaternary ammonium salt formed by the reaction of dimethicone copolyol acetyl chloride and hydrolyzed wheat protein, quaternized with chlorohydrin.

Chemical Class: Quaternary Ammonium Compounds

Functions: Antistatic Agent; Hair Conditioning Agent

Trade Name:
Pecosil SWQ-40 (Phoenix)

QUATERNIUM-87

CAS No.: 92201-88-2

Definition: Quaternium-87 is the organic compound that conforms to the formula:

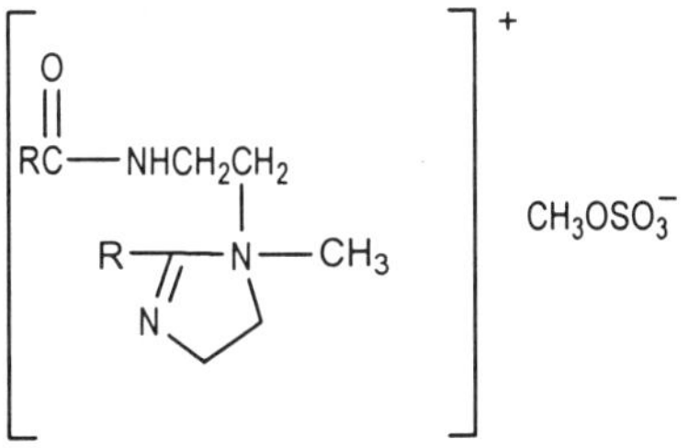

where RCO- and R are derived from palm oil.

Chemical Class: Quaternary Ammonium Compounds

Functions: Hair Conditioning Agent; Surfactant - Cleansing Agent

Technical/Other Name:
Imidazolium Compounds, 2-(C9-19 and C9-19-unsatd. Alkyl)-1-[(C10-20 and C10-20-unsatd. Amido)Ethyl]-4,5-Dihydro-1-Me, Me Sulfates

Trade Name:
Varisoft W 575 PG (Degussa Care Specialties)

QUATERNIUM-88

Definition: Quaternium-88 is the quaternary ammonium salt that conforms generally to the formula:

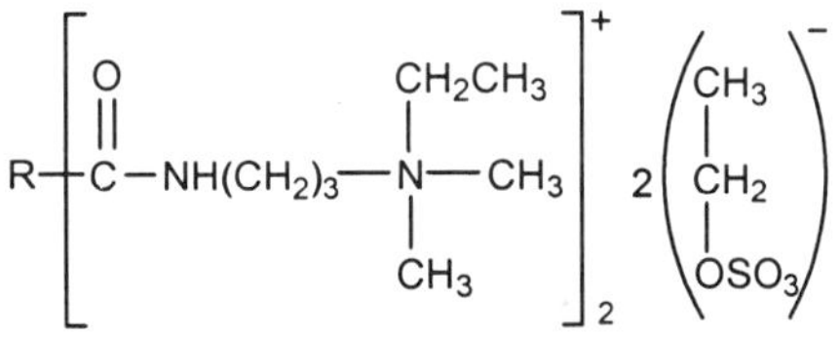

where RCO- represents Dilinoleic Acid (q.v.).

Chemical Class: Quaternary Ammonium Compounds

Functions: Antistatic Agent; Hair Conditioning Agent

Trade Name:
Nequat DAS-D (Alzo)

QUATERNIUM-89

Definition: Quaternium-89 is the quaternary ammonium salt that conforms generally to the formula:

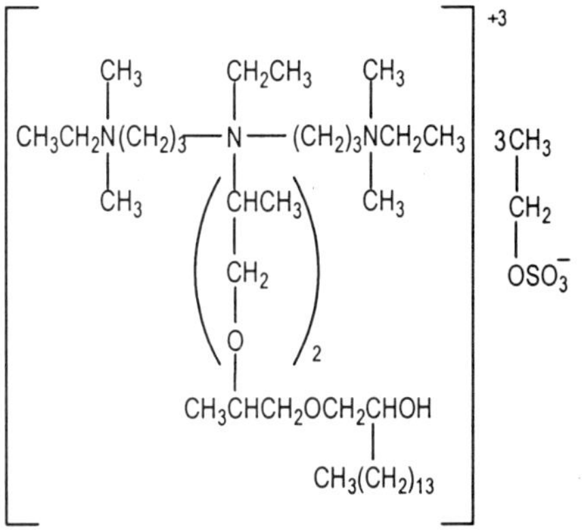

Chemical Class: Quaternary Ammonium Compounds

Functions: Antistatic Agent; Hair Conditioning Agent

Trade Name:
Finquat CT-P (Finetex)

QUATERNIUM-90

Definition: Quaternium-90 is the quaternary ammonium salt that conforms generally to the formula:

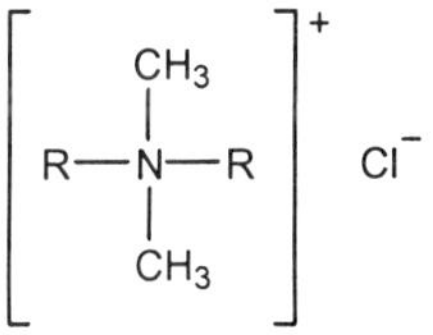

where R represents the alkyl groups derived from Elaesis Guineensis (Palm) Oil (q.v.).

Chemical Class: Quaternary Ammonium Compounds

Function: Antistatic Agent

QUATERNIUM-90 BENTONITE

CAS No.: 226226-22-8

Definition: Quaternium-90 Bentonite is a reaction product of Bentonite (q.v.) and Quaternium-90 (q.v.).

Chemical Classes: Inorganics; Quaternary Ammonium Compounds

Function: Suspending Agent - Nonsurfactant

Trade Name:
Tixogel VP-V (Sud-Chemie, United Catalysts)

Trade Name Mixtures:
Tixogel COG 1540 (Sud-Chemie, United Catalysts)
Tixogel IDD 1538 (Sud-Chemie, United Catalysts)
Tixogel IHD-1235 (Sud-Chemie, United Catalysts)
Tixogel IIN 1578 (Sud-Chemie, United Catalysts)
Tixogel JOG 1583 (Sud-Chemie, United Catalysts)
Tixogel MIO 1584 (Sud-Chemie, United Catalysts)
Tixogel ODD-1520 (Sud-Chemie, United Catalysts)
Tixogel OMS 1562 (Sud-Chemie, United Catalysts)
Tixogel VSP-1438-V (Sud-Chemie, United Catalysts)

QUATERNIUM-91

Definition: Quaternium-91 is the quaternary ammonium salt that conforms to the formula:

Chemical Class: Quaternary Ammonium Compounds

Function: Hair Conditioning Agent

Trade Name Mixture:
Crodazosoft DBQ (Croda, Inc.)

QUATERNIUM-92

Definition: Quaternium-92 is the quaternary ammonium salt that conforms generally to the formula:

where RCO- represents the lanolin acid moiety.

Chemical Class: Quaternary Ammonium Compounds

Functions: Antistatic Agent; Hair Conditioning Agent; Skin-Conditioning Agent - Miscellaneous; Surfactant - Emulsifying Agent; Surfactant - Foam Booster

QUERCETIN

CAS No.	EINECS No.
117-39-5	204-187-1

Empirical Formula:
$C_{15}H_{10}O_7$

Definition: Quercetin is the organic compound that conforms to the formula:

Information Sources: 21CFR330.12, MI-13(8122)

Chemical Classes: Ethers; Heterocyclic Compounds; Ketones; Phenols

Functions: Antioxidant; Skin-Conditioning Agent - Miscellaneous

Technical/Other Names:
4H-1-Benzopyran-4-one, 2-(3,4-Dihydroxy-phenyl)-3,5,7-Trihydroxy
3,3',4',5,7-Pentahydroxyflavone
Quercetol

Trade Name:
Antrancine Q (Jan Dekker)

QUERCETIN CAPRYLATE

Definition: Quercetin Caprylate is a mixture of esters of Quercetin (q.v.) and Caprylic Acid (q.v.).

Chemical Classes: Esters; Heterocyclic Compounds; Ketones

Functions: Antioxidant; Skin-Conditioning Agent - Emollient

Trade Name Mixture:
Flavenger (Coletica SA)

QUERCUS ACUTISSIMA EXTRACT

Definition: Quercus Acutissima Extract is an extract of the fruit of *Quercus accutissima*. See *"Regulatory and Ingredient Use Information,"* regarding the labeling names for botanical ingredients in Volume 1, Introduction, Part A.

Chemical Class: Biological Products

Function: Cosmetic Astringent

Technical/Other Name:
Extract of Quercus Acutissima

Trade Name Mixture:
Quercus Extract (Bioland)

QUERCUS ALBA BARK

Definition: Quercus Alba Bark is the bark of *Quercus alba*. See *"Regulatory and Ingredient Use Information,"* regarding the labeling names for botanical ingredients in Volume 1, Introduction, Part A.

Chemical Class: Biological Products

Function: Skin-Conditioning Agent - Miscellaneous

Trade Name:
White Oak Bark (Aveda)

QUERCUS ALBA BARK EXTRACT

Definition: Quercus Alba Bark Extract is the extract of the bark of the oak tree, *Quercus alba*. See *"Regulatory and Ingredient Use Information,"* regarding the labeling names for botanical ingredients in Volume 1, Introduction, Part A.

Information Source: 21CFR172.510

Chemical Class: Biological Products

Function: Not Reported

Technical/Other Names:
Extract of Oak Bark
Extract of Quercus Alba Bark
Extract of White Oak Bark
Oak Bark Extract
White Oak Bark Extract
White Oak (Quercus Alba) Bark Extract

Trade Name Mixtures:
Actiphyte of Oak Bark BG50 (Active Organics)
Actiphyte of Oak Bark GL50 (Active Organics)
Actiphyte of Oak Bark Lipo S (Active Organics)
Actiphyte of Oak Bark PG50 (Active Organics)
Actiphyte of White Oak (Active Organics)
Extrait de Chene PPE PG20 (Yves Rocher)
Herbaliquid Oak Bark Special (Crodarom)
Herbasol-Extract Oak Bark (Cosmetochem)
VT-157 Extract of Oak Bark (Vege-Tech)

QUERCUS GLAUCA SEED EXTRACT

Definition: Quercus Glauca Seed Extract is an extract of the seeds of *Quercus glauca*. See *"Regulatory and Ingredient Use Information,"* regarding the labeling names for botanical ingredients in Volume 1, Introduction, Part A.

Chemical Class: Biological Products

Function: Skin-Conditioning Agent - Miscellaneous

Technical/Other Name:
Extract of Quercus Glauca Seed

Trade Name:
Quercus Extract G (Kotobuki Chemical)

QUERCUS INFECTORIA (OAK) GALL EXTRACT

Definition: Quercus Infectoria (Oak) Gall Extract is an extract of the gall of the oak, *Quercus infectoria. See "Regulatory and Ingredient Use Information,"* regarding the labeling names for botanical ingredients in Volume 1, Introduction, Part A.

Chemical Class: Biological Products

Functions: Cosmetic Astringent; Skin-Conditioning Agent - Miscellaneous

Technical/Other Names:
Extract of Oak Gall
Extract of Quercus Infectoria Gall
Oak Gall Extract
Quercus Lusitanica Gall Extract

Trade Name:
Oak Galls Supextrat (Indena SA)

Trade Name Mixtures:
Amiderm Phytoamine Biocomplex (Alban Muller)
Epilami (Alban Muller)
Gallnut HPG Titrated (Alban Muller)
Gallnut HS (Alban Muller)

QUERCUS PETRAEA BARK EXTRACT

Definition: Quercus Petraea Bark Extract is an extract of the bark of the oak, *Quercus petraea. See "Regulatory and Ingredient Use Information,"* regarding the labeling names for botanical ingredients in Volume 1, Introduction, Part A.

Chemical Class: Biological Products

Function: Not Reported

QUERCUS ROBUR BARK EXTRACT

Definition: Quercus Robur Bark Extract is an extract of the bark of the english oak,

Quercus robur. See "Regulatory and Ingredient Use Information," regarding the labeling names for botanical ingredients in Volume 1, Introduction, Part A.

Information Source: 21CFR172.510

Chemical Class: Biological Products

Function: Not Reported

Technical/Other Names:
English Oak Extract
English Oak (Quercus Robur) Extract
Extract of English Oak
Extract of Quercus Robur

Trade Name:
Questex Oak Bark Concentrate (Quest International)

Trade Name Mixtures:
Oak Bark Extract HS 2716 G (Grau)
Phytelene of Oak EG 014 Liquid (Indena SA)
Phytogreen 55 of Oak EXH 614 Liquid (Phytochim)
Quercus Robur Bark Extract ies (IES LABO)

QUERCUS SERRATA EXTRACT

Definition: Quercus Serrata Extract is an extract of the oak tree, *Quercus serrata. See "Regulatory and Ingredient Use Information,"* regarding the labeling names for botanical ingredients in Volume 1, Introduction, Part A.

Chemical Class: Biological Products

Function: Skin-Conditioning Agent - Miscellaneous

Technical/Other Name:
Extract of Quercus Serrata

Trade Name Mixture:
Nati Oak Tree Extract (New Aqua Technical Institute)

QUERCUS SERRATA SEED EXTRACT

Definition: Quercus Serrata Seed Extract is an extract of the seeds of *Quercus serrata. See "Regulatory and Ingredient Use Information,"* regarding the labeling names for botanical ingredients in Volume 1, Introduction, Part A.

Chemical Class: Biological Products

Function: Skin-Conditioning Agent - Miscellaneous

Technical/Other Name:
Extract of Quercus Serrata Seed

Trade Name:
Quercus Extract S (Kotobuki Chemical)

QUILLAJA SAPONARIA BARK

Definition: Quillaja Saponaria Bark is a plant material derived from the dried bark of the quillaja, *Quillaja saponaria. See "Regulatory and Ingredient Use Information,"* regarding the labeling names for botanical ingredients in Volume 1, Introduction, Part A.

Information Sources: ARG, BP, BPC, 21CFR172.510, EGY, HP, IND, MAR, MI-13 (8128), NED, PF, POR, YUG

Chemical Class: Biological Products

Function: Not Reported

Technical/Other Names:
Quillaja
Quillaja Bark

QUILLAJA SAPONARIA BARK EXTRACT

CAS No.	EINECS No.
68990-67-0	273-620-4

Definition: Quillaja Saponaria Bark Extract is an extract of the bark of the quillaja, *Quillaja saponaria. See "Regulatory and Ingredient Use Information,"* regarding the labeling names for botanical ingredients in Volume 1, Introduction, Part A.

Information Sources: 21CFR172.510, RIFM

Chemical Class: Biological Products

Functions: Fragrance Ingredient; Skin-Conditioning Agent - Miscellaneous

Reported Product Categories: Shampoos (Non-coloring); Personal Cleanliness Products, Misc.

Technical/Other Names:
Extract of Quillaja
Extract of Quillaja Saponaria
Panama Wood Extract
Quillaia (Quillaja saponaria Molina) (RIFM)
Quillaja Extract BG

Trade Names:
AEC Quillaia Extract (A & E Connock)
Phytelene of Panama EN 259 powder (Indena SA)
Phytogreen of Panama EP 502 Powder (Phytochim)

Trade Name Mixtures:
Actiphyte of Soap Bark BG50 (Active Organics)

Actiphyte of Soap Bark GL50 (Active Organics)
Actiphyte of Soap Bark Lipo S (Active Organics)
Actiphyte of Soap Bark PG50 (Active Organics)
Dermotenseur (Libiol)
Extrapone Soap Bark GW 2/031850 (Symrise)
Herbaliquid Quillaja Special (Crodarom)
Herbasol-Extract Panama Bark (Cosmetochem)
Herbasol Extract Quillaja (Cosmetochem)
Panama Bark Extract (Crodarom)
Panama Bark Extract HS 2383 G (Grau)
Panama Wood HS (Alban Muller)
Phytelene of Panama EG 476 liquid (Indena SA)
Phytogreen 55 of Panama EXH 689 Liquid (Phytochim)
Prodhy Extract Panama (Prod'Hyg)
Quillaja Extract (Libiol)
Quillaja Saponaria Bark Extract ies (IES LABO)
Quillayanin C-100 (Maruzen Pharmaceuticals Co., Ltd.)
Vegetol Quillai GR 038 Hydro (Gattefosse s.a.)
VT-042 Extract of Soap Bark (Vege-Tech)

QUILLAJA SAPONARIA ROOT EXTRACT

CAS No.   **EINECS No.**
68990-67-0   273-620-4

Definition: Quillaja Saponaria Root Extract is an extract of the roots of *Quillaja saponaria. See "Regulatory and Ingredient Use Information," regarding the labeling names for botanical ingredients in Volume 1, Introduction, Part A.*

Information Source: RIFM

Chemical Class: Biological Products

Functions: Fragrance Ingredient; Skin-Conditioning Agent - Miscellaneous

Reported Product Categories: Shampoos (Non-coloring); Personal Cleanliness Products, Misc.

Technical/Other Names:
Extract of Quillaja Saponaria Root
Quillaia (Quillaja saponaria Molina) (RIFM)

Trade Name Mixture:
Quillaja Extract HG (Provital/Centerchem)

QUINIC ACID

CAS Nos.   **EINECS Nos.**
77-95-2   201-072-8

562-73-2   209-233-4
36413-60-2

Empirical Formula:
$C_7H_{12}O_6$

Definition: Quinic Acid is the organic compound that conforms to the formula:

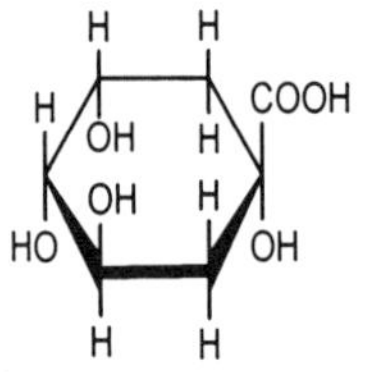

Information Source: MI-13(8149)

Chemical Class: Carboxylic Acids

Function: pH Adjuster

Technical/Other Names:
Cyclohexanecarboxylic Acid, 1,3,4,5-Tetrahydroxy-, [1R-(1α, 3α, 4α, 5β)]-
1,3,4,5-Tetrahydroxycyclohexanecarboxylic Acid

QUININE

CAS No.   **EINECS No.**
130-95-0   205-003-2

Empirical Formula:
$C_{20}H_{24}N_2O_2$

Definition: Quinine is an alkaloid from the bark of *Cinchona officinalis.* It conforms to the formula:

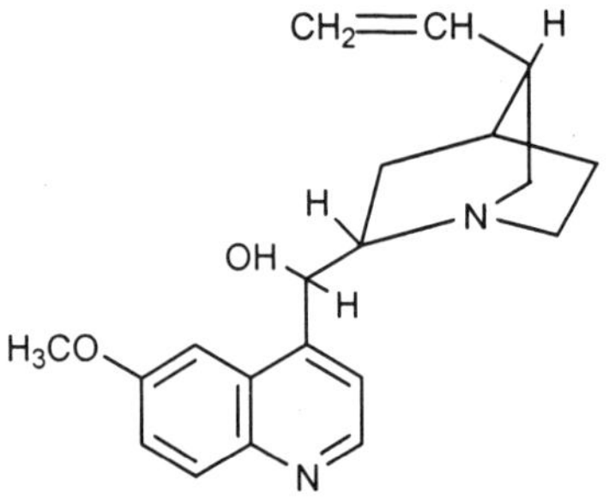

Information Sources: ARG, AUS, BAN, BEL, BRA, 21CFR172.575, 21CFR310.545, 21CFR310.546, 21CFR310.547, 27CFR21.70, 27CFR21.151, DA, DDR, EEC(III/1-21), EP, FIN, HUN, IND, ITA, KOR, MEX, MI-13(8151), NFJ, POR, RIFM, TSCA

Chemical Classes: Amines; Heterocyclic Compounds

Functions: Denaturant; Fragrance Ingredient; Hair Conditioning Agent

Technical/Other Names:
Cinchonan-9-ol, 6'-Methoxy-
6'-Methoxycinchonan-9-ol
Quinine (RIFM)

- R -

RABBIT FAT

Definition: Rabbit Fat is the fat obtained from rabbits.

Chemical Class: Fats and Oils

Function: Skin-Conditioning Agent - Occlusive

Trade Name:
Rablu (Croda Japan)

RAFFINOSE

CAS No.
512-69-6

EINECS No.
208-146-9

Empirical Formula:
$C_{18}H_{32}O_{16}$

Definition: Raffinose is the trisaccharide formed from D-galactose, D-fructose, and D-glucose. It conforms to the formula:

Information Source: MI-13(8188)

Chemical Classes: Carbohydrates; Polyols

Function: Skin-Conditioning Agent - Emollient

Technical/Other Names:
α-D-Glucopyranoside, β-fructofuranosyl O-α-D-Galactopyranosyl-
Melitriose
D-Raffinose

Trade Name:
Oligo GGF (Asahi Kasei)

RAFFINOSE MYRISTATE

CAS No.: 91433-10-2

JPN Translation:
ミリスチン酸ラフィノーズ

Definition: Raffinose Myristate is the ester of raffinose and myristic acid.

Information Sources: JCIC, JCLS

Chemical Classes: Carbohydrates; Esters

Functions: Skin-Conditioning Agent - Emollient; Surfactant - Emulsifying Agent

Technical/Other Name:
D-Glucopyranoside, D-fructofurannosyl D-Galactopyranosyl Tetradecanoate

RAFFINOSE OLEATE

CAS No.: 96352-58-8

Definition: Raffinose Oleate is the ester of raffinose and oleic acid.

Chemical Classes: Carbohydrates; Esters

Functions: Skin-Conditioning Agent - Emollient; Surfactant - Emulsifying Agent

Technical/Other Name:
Glucopyranoside, D-Fructofuranosyl D-Galactopyranosyl 9-Octadecenoate

RAHNELLA/SOY PROTEIN FERMENT

Definition: Rahnella/Soy Protein Ferment is the product derived by the fermentation of soy protein by the organism, *Rahnella*.

Chemical Class: Biological Products

Function: Skin-Conditioning Agent - Miscellaneous

Trade Name Mixture:
Bio-Bustyl (Sederma)

RANSOU EKISU

JPN Translation:
ラン藻エキス

Definition: Ransou Ekisu is an extract of the whole blue-green alga, *Oyanophyceae*. *See "Regulatory and Ingredient Use Information," regarding use of Japan Trivial names in Volume 1, Introduction, Part A.*

Chemical Class: Biological Products

Function: Skin-Conditioning Agent - Miscellaneous

Reported Product Categories: Hair Dyes and Colors (All Types Requiring Caution Statements and Patch Tests); Bath Capsules; Bath Preparations, Misc.; Shampoos (Non-coloring); Bath Oils, Tablets, and Salts; Hair Conditioners; Mascara; Tonics, Dressings, and Other Hair Grooming Aids; Eyebrow Pencils; Baby Shampoos; Bubble Baths; Fragrance Preparations, Misc.; Hair Rinses (Non-coloring)

Technical/Other Name:
Algae Extract (U.S.)

RANUNCULUS FICARIA EXTRACT

CAS No.
84929-74-8

EINECS No.
284-553-5

Definition: Ranunculus Ficaria Extract is an extract of the pilewort, *Ranunculus ficuria*. *See "Regulatory and Ingredient Use Information," regarding the labeling names for botanical ingredients in Volume 1, Introduction, Part A.*

Chemical Class: Biological Products

Function: Not Reported

Technical/Other Names:
Extract of Pilewort
Extract of Ranunculus Ficuria
Pilewort Extract
Pilewort (Ranunculis Ficuria) Extract
Ranunculus Ficuria Extract

Trade Name:
Gatuline A (Gattefosse s.a.)

Trade Name Mixture:
Extrait de Ficaire MPE PG 40 (Yves Rocher)

RAPESEED ACID

Definition: Rapeseed Acid is a mixture of fatty acids derived from Brassica Campestris (Rapeseed) Oil (q.v.).

Chemical Class: Fatty Acids

Function: Surfactant - Cleansing Agent

Technical/Other Names:
Acids, Rapeseed
Brassica Campestris (Rapeseed) Acid

Trade Name:
PRIFAC 8944 (Uniqema Europe)

RAPESEEDAMIDOPROPYL BENZYL-DIMONIUM CHLORIDE

Definition: Rapeseedamidopropyl Benzyl-dimonium Chloride is the quaternary ammonium salt that conforms generally to the formula:

where RCO- represents the rapeseed acid radical.

Chemical Class: Quaternary Ammonium Compounds

Function: Antistatic Agent

RAPESEEDAMIDOPROPYL EPOXY-PROPYL DIMONIUM CHLORIDE

CAS No.: 112324-11-5

Definition: Rapeseedamidopropyl Epoxy-propyl Dimonium Chloride is the quaternary ammonium salt that conforms generally to the formula:

$$\left[\begin{array}{c} O \\ \parallel \\ RCNH(CH_2)_3 \end{array} - \overset{\displaystyle CH_3}{\underset{\displaystyle CH_3}{N}} - CH_2CHCH_2 \right]^{+} \quad Cl^{-}$$

where RCO- represents the fatty acids derived from rapeseed oil.

Chemical Class: Quaternary Ammonium Compounds

Function: Antistatic Agent

Technical/Other Names:
N-(3-Aminopropyl)-N,N-Dimethyloxirane-methanaminium, N-Rape-Oil Acyl Derivs., Chloride
Oxiranemethanaminium, N-(3-Aminopropyl)-N,N-Dimethyl-, N-Rape Oil Acyl Derivs., Chlorides

Trade Name:
Schercoquat ROEP (Scher)

RAPESEEDAMIDOPROPYL ETHYLDIMONIUM ETHOSULFATE

CAS No.	EINECS No.
94552-41-7	305-488-1

Definition: Rapeseedamidopropyl Ethyldimonium Ethosulfate is the quaternary ammonium salt that conforms generally to the formula:

$$\left[\begin{array}{c} O \\ \parallel \\ RC \end{array} - NH(CH_2)_3 - \overset{\displaystyle CH_3}{\underset{\displaystyle CH_3}{N}} - CH_2CH_3 \right]^{+} \quad \underset{\displaystyle OSO_3^{-}}{CH_3CH_2}$$

where RCO- represents the rapeseed acid radical.

Chemical Class: Quaternary Ammonium Compounds

Function: Antistatic Agent

Technical/Other Names:
3-Amino-N-Ethyl-N,N-Dimethyl-1-Propana-minium, N-Rape-Oil Acyl Derivs., Ethyl Sulfate
1-Propanaminium, 3-Amino-N-Ethyl-N,N-Dimethyl-, N-Rape-Oil Acyl Derivs., Ethyl Sulfate
1-Propanaminium, 3-Amino-N-Ethyl-N,N-Dimethyl-, N-Rape Oil Acyl Derivs., Ethyl Sulfates
3-(Rape Oil Acyl)-Amino-N-Ethyl-N,N-Dimethyl-1-Propaneaminium Ethosulfate

Trade Name:
Schercoquat ROAS (Scher)

RAPESEED GLYCERIDE

Definition: Rapeseed Glyceride is the monoglyceride derived from Brassica Campestris (Rapeseed) Oil (q.v.).

Chemical Class: Glyceryl Esters and Derivatives

Functions: Skin-Conditioning Agent - Emollient; Surfactant - Emulsifying Agent

Technical/Other Names:
Glycerides, Rapeseed Mono-
Rapeseed Monoglyceride

RAPESEED GLYCERIDES

Definition: Rapeseed Glycerides is a mixture of mono-, di- and triglycerides derived from Brassica Campestris (Rapeseed) Oil (q.v.).

Chemical Class: Glyceryl Esters and Derivatives

Function: Skin-Conditioning Agent - Emollient

Technical/Other Name:
Glycerides, Rapeseed Mono-, Di- and Tri-

RAPESEED OIL SORBITOL ESTERS

Definition: Rapeseed Oil Sorbitol Esters are the mono- and diesters obtained by the transesterification of Sorbitol (q.v.) with Brassica Campestris (Rapeseed) Oil (q.v.).

Chemical Class: Esters

Functions: Skin-Conditioning Agent - Emollient; Surfactant - Emulsifying Agent

Trade Names:
Emulsogen SRO (Clariant)
Emulsogen SRO (Clariant GmbH, Personal Care)

RAPE SHUSHI YU

JPN Translation:
ナタネ油

Definition: Rape Shushi Yu is the fixed oil obtained from the seeds of *Brassica campestris* or of *Brassica napus*. See *"Regulatory and Ingredient Use Information,"* regarding use of Japan Trivial names in Volume 1, Introduction, Part A.

Chemical Class: Fats and Oils

Function: Skin-Conditioning Agent - Occlusive

Technical/Other Names:
Brassica Campestris (Rapeseed) Seed Oil (U.S.)
Rape Shusi Yu

RAPHANUS SATIVUS (RADISH) ROOT EXTRACT

CAS No.	EINECS No.
84775-94-0	283-918-6

Definition: Raphanus Sativus (Radish) Root Extract is an extract of the roots of the radish, *Raphanus sativus*. See *"Regulatory and Ingredient Use Information,"* regarding the labeling names for botanical ingredients in Volume 1, Introduction, Part A.

Chemical Class: Biological Products

Function: Skin-Conditioning Agent - Miscellaneous

Technical/Other Names:
Extract of Radish
Extract of Raphanus Sativus
Radish Extract
Radish Root Extract
Raphanus Sativus Extract

Trade Name Mixtures:
Herbasol Extract Radish (Cosmetochem) (Cosmetochem International Ltd.)
Radish HS (Alban Muller)
Raphanus Sativus (Radish) Root Extract ies (IES LABO)

RAPHANUS SATIVUS (RADISH) SEED EXTRACT

Definition: Raphanus Sativus (Radish) Seed Extract is an extract of the seeds of the radish, *Raphanus sativus*. See *"Regulatory and Ingredient Use Information,"* regarding the labeling names for botanical ingredients in Volume 1, Introduction, Part A.

Chemical Class: Biological Products

Function: Not Reported

Technical/Other Name:
Radish Seed Extract

Trade Name Mixtures:
Pronalen Refirming BG (Provital/
Centerchem)
Pronalen Refirming HSC (Provital/
Centerchem)

RASPBERRY KETONE

CAS No. 5471-51-2   **EINECS No.** 226-806-4

Empirical Formula:
$C_{10}H_{12}O_2$

Definition: Raspberry Ketone is the organic compound that conforms to the formula:

Information Source: RIFM

Chemical Classes: Ketones; Phenols

Functions: Fragrance Ingredient; Skin-Conditioning Agent - Miscellaneous

Technical/Other Names:
4-(4-Hydroxyphenyl)-2-Butanone
4-(p-Hydroxyphenyl)-2-butanone (RIFM)

Trade Name:
Raspberry Ketone TIC (Takasago)

RASPBERRYKETONE GLUCOSIDE

CAS No.: 38963-94-9

Empirical Formula:
$C_{16}H_{22}O_7$

Definition: Raspberryketone Glucoside is the organic compound that conforms to the formula:

Chemical Classes: Carbohydrates; Ketones

Functions: Fragrance Ingredient; Skin-Conditioning Agent - Humectant

Technical/Other Name:
2-Butanone, 4-[4-β-D-Glucopyranosyloxy)Phenyl]-

Trade Name:
Raspberryketone Glucoside TH
(T.HASEGAWA)

RASPBERRY SEED OIL PEG-8 ESTERS

Definition: Raspberry Seed Oil PEG-8 Esters is the product obtained by the transesterification of Rubus Idaeus (Raspberry) Seed Oil (q.v.) and PEG-8 (q.v.).

Chemical Class: Glyceryl Esters and Derivatives

Functions: Skin-Conditioning Agent - Emollient; Surfactant - Emulsifying Agent

Trade Name:
Viatenza Raspberry PE8 (Aldivia)

RAUWOLFIA SERPENTINA

Definition: *See "Regulatory and Ingredient Use Information," regarding EU labeling names for botanical ingredients in Volume 1, Introduction, Part A.*

Information Source: MI-13(8206)

Chemical Class: Biological Products

Technical/Other Name:
Rauwolfia Serpentina Root Extract (U.S.)

RAUWOLFIA SERPENTINA ROOT EXTRACT

CAS No. 90106-13-1   **EINECS No.** 290-234-1

Definition: Rauwolfia Serpentina Root Extract is an extract of the roots of the rauwolfia, *Rauwolfia serpentina.* See *"Regulatory and Ingredient Use Information," regarding the labeling names for botanical ingredients in Volume 1, Introduction, Part A.*

Chemical Class: Biological Products

Function: Not Reported

Technical/Other Names:
Extract of Rauwolfia Serpentina
Rauwolfia Extract
Rauwolfia Serpentina (EU)

Trade Name Mixture:
Rauwolfia Extract HS 2400 G (Grau)

RAYON

CAS Nos.: 9006-02-4; 61788-77-0

JPN Translation:
レーヨン

Definition: Rayon is a composition of synthetic fibers and filaments composed of regenerated cellulose.

Information Sources: 21CFR176.170, 21CFR177.2260, 21CFR177.2800, DDR, JCIC, JCLS, MI-13(8207), USAN, USP XXIV

Chemical Class: Synthetic Polymers

Functions: Adhesive; Bulking Agent

Reported Product Category: Mascara

Technical/Other Names:
Cellulose Regenerated
Rayon Flock

RED 4

CAS No.: 4548-53-2

Empirical Formula:
$C_{18}H_{16}N_2O_7S_2 \cdot 2Na$

Definition: Red 4 is classed chemically as a monoazo color. It conforms to the formula:

See "Regulatory and Ingredient Use Information," for Colorants in Volume 1, Introduction, Part A. To identify the certified colorant for labeling purposes in the US, the INCI Name Red 4 must be used. The INCI Name for batches of this colorant that have not been certified is Ponceaux SX. To identify the colorant allowed for use in the European Union (EU), the INCI Name CI 14700 must be used, except for hair dye products. To identify the colorant allowed for use in Japan, the INCI name Aka504 must be used.

Information Sources: 21CFR74.340, 21CFR74.1304, 21CFR74.1340, 21CFR74.2304, 21CFR74.2340, 21CFR81.10, 21CFR81.30, 21CFR82.304, CI 14700, M3, MI-13(7672), TSCA

Chemical Class: Color Additives - Batch Certified by the U.S. Food and Drug Administration

Function: Colorant

Reported Product Categories: Colognes and Toilet Waters; Bath Preparations, Misc.; Body and Hand Preparations (Excluding Shaving Preparations); Shampoos (Non-coloring); Perfumes; Moisturizing Preparations; Hair Conditioners; Skin Care Preparations, Misc.; Bath Oils, Tablets, and Salts; Bath Soaps and Detergents; Aftershave Lotions; Baby Shampoos; Cleansing Products (Cold Creams, Cleansing Lotions, Liquids and Pads); Bubble Baths; Bath Capsules; Face and Neck Preparations (Excluding Shaving Preparations); Fragrance Preparations, Misc.; Paste Masks (Mud Packs); Deodorants (Underarm); Skin Fresheners; Personal Cleanliness Products, Misc.; Blushers (All types); Night Skin Care Preparations; Hair Rinses (Coloring); Tonics, Dressings, and Other Hair Grooming Aids; Makeup Preparations (Not eye), Misc.; Suntan Gels, Creams, and Liquids; Suntan Preparations, Misc.; Baby Products, Misc.; Hair Preparations (Non-coloring), Misc.; Hair Straighteners; Nail Creams and Lotions; Nail Polish and Enamel Removers; Shaving Cream (Aerosol, Brushless and Lather); Shaving Preparations, Misc.

Technical/Other Name:
FD&C Red No. 4

Trade Name:
FD & C Red 4 W 084 (LCW)

Trade Name Mixture:
Tefpoly Tomato Dye (C.I.T.)

RED 4 LAKE

Definition: Red 4 Lake is the salt of Red 4 extended on an appropriate substrate in compliance with 21CFR82.1051. *See "Regulatory and Ingredient Use Information," for Colorants in Volume 1, Introduction, Part A. To identify the certified colorant for labeling purposes in the US, the INCI Name Red 4 Lake must be used. To identify the colorant allowed for use in the European Union (EU), the INCI Name CI 14700 must be used, except for hair dye products. To identify the colorant allowed for use in Japan, the INCI name Aka504 must be used.*

Information Sources: 21CFR74.1304, 21CFR74.2304, 21CFR81.1, 21CFR82.1051

Chemical Class: Color Additives Lakes - Batch Certified by the U.S. Food and Drug Administration

Function: Colorant

Technical/Other Names:
D&C Red No. 4 Aluminum Lake
FD&C Red No. 4 Aluminum Lake

RED 6

CAS No.: 5858-81-1

Empirical Formula:
$C_{18}H_{14}N_2O_6S \cdot 2Na$

Definition: Red 6 is classed chemically as a monoazo color. It conforms to the formula:

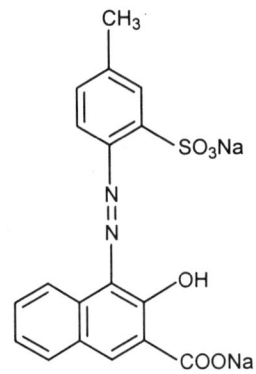

See "Regulatory and Ingredient Use Information," for Colorants in Volume 1, Introduction, Part A. The INCI Name for batches of this colorant that have not been certified is Pigment Red 57. The INCI Name for batches of this colorant (calcium salt) that have not been certified is Pigment Red 57:1. To identify the colorant allowed for use in the European Union (EU), the INCI Name CI 15850 must be used, except for hair dye products. To identify the certified colorant for labeling purposes in the US, the INCI Name Red 6 must be used. To identify the certified colorant (calcium salt) for labeling purposes in the US, the INCI Name Red 7 must be used. To identify the colorant allowed for use in Japan, the INCI name Aka201 must be used.

Information Sources: 21CFR74.1306, 21CFR74.1307, 21CFR74.2306, 21CFR82.1306, 21CFR82.1307, CI 15850, MHLW Ord. No. 30, TSCA

Chemical Class: Color Additives - Batch Certified by the U.S. Food and Drug Administration

Function: Colorant

Reported Product Categories: Lipsticks; Nail Polish and Enamels; Blushers (All types); Makeup Preparations (Not eye), Misc.; Face Powders

Technical/Other Name:
D&C Red No. 6

Trade Names:
A506.90 Tudor Orchid (Kingfisher Colours)
C19-6619 Original Orange (Sun Pigments)
D&C Red 6 W 003 (LCW)
D & C Red #6 K7034 (LCW)
Unipure Red LC 303 (LCW)

RED 6 LAKE

Definition: Red 6 Lake is the salt of Red 6 extended on an appropriate substrate in

compliance with 21CFR82.1051. *See "Regulatory and Ingredient Use Information," for Colorants in Volume 1, Introduction, Part A. To identify the certified colorant for labeling purposes in the US, the INCI Name Red 6 Lake must be used. To identify the colorant allowed for use in the European Union (EU), the INCI Name CI 15850 must be used, except for hair dye products.*

Information Sources: 21CFR81.1, 21CFR82.1051

Chemical Class: Color Additives Lakes - Batch Certified by the U.S. Food and Drug Administration

Function: Colorant

Reported Product Categories: Blushers (All types); Nail Polish and Enamels; Face Powders; Makeup Preparations (Not eye), Misc.; Lipsticks; Basecoats and Undercoats; Manicuring Preparations, Misc.; Powders (Dusting and Talcum, Excluding Aftershave Talcs); Personal Cleanliness Products, Misc.; Rouges

Technical/Other Names:
D&C Red No. 6 Aluminum Lake
D&C Red No. 6 Barium Lake
D&C Red No. 6 Barium/Strontium Lake
D&C Red No. 6 Potassium Lake
D&C Red No. 6 Strontium Lake

Trade Names:
A506.12 Tudor Orchid (Kingfisher Colours)
A506 Tudor Orchid (Kingfisher Colours)
C 6506 D&C Red 6 Barium Lake (LCW)
C19-012 Light Rubine Lake (Sun Pigments)
C19-022 Light Rubine Lake (Sun Pigments)
D+C RED 6 Ba LAKE 10-31-DA-3006 (Noveon Hilton Davis)
D & C Red 6 Barium Lake W004 (LCW)
D & C Red #6 Barium Lake K7096 (LCW)
D & C Red No. 6 Potassium Lake C6406 (LCW)
Unipure Red LC 304 (LCW)

Trade Name Mixtures:
Colourspheres Poppy Lake HL 25% (Creations Couleurs)
C6506 Red 6BA-I2 (Kobo)
Tefpoly Poppy Lake (C.I.T.)

RED 7

CAS Nos.: 5281-04-9; 29092-56-6

Empirical Formula:
$C_{18}H_{14}N_2O_6S \cdot Ca$

Definition: Red 7 is classed chemically as a monoazo color. It conforms to the formula:

See "Regulatory and Ingredient Use Information," for Colorants in Volume 1, Introduction, Part A. To identify the certified colorant for labeling purposes in the US, the INCI Name Red 7 must be used. To identify the certified colorant (sodium salt) for labeling purposes in the US, the INCI Name Red 6 must be used. The INCI Name for batches of this colorant that have not been certified is Pigment Red 57:1. The INCI Name for batches of this colorant (sodium salt) that have not been certified is Pigment Red 57. To identify the colorant allowed for use in the European Union (EU), the INCI Name CI 15850 must be used, except for hair dye products. To identify the colorant allowed for use in Japan, the INCI name Aka202 must be used.

Information Sources: 21CFR74.1306, 21CFR74.1307, 21CFR74.2307, 21CFR82.1306, 21CFR82.1307, 21CFR178.3297, CI 15850:1, TSCA

Chemical Class: Color Additives - Batch Certified by the U.S. Food and Drug Administration

Function: Colorant

Reported Product Categories: Nail Polish and Enamels; Lipsticks; Blushers (All types); Face Powders; Makeup Preparations (Not eye), Misc.

Technical/Other Name:
D&C Red No. 7

Trade Names:
D & C Red 7 Calcium Lake W 023 (LCW)
Unipure Red LC 3079 (LCW)

Trade Name Mixtures:
Creasparkles Colours Magenta (C.I.T.)
Creasparkles Colours Rubis (Creations Couleurs)
Creasparkles Metallic Reddish Gold 590 (Creations Couleurs)
Creasparlkes Colours Magenta (Creations Couleurs)
Diamond Piece CO Type, DG Gold (Daiya Kogyo)
Diamond Piece CO Type, Red (Daiya Kogyo)
LipoCrystals LIP Peach (Lipo)
LipoCrystals LIP Violet (Lipo)

RED 7 LAKE

Definition: Red 7 Lake is the salt of Red 7 extended on an appropriate substrate in compliance with 21CFR82.1051. See "Regulatory and Ingredient Use Information," for Colorants in Volume 1, Introduction, Part A. To identify the certified colorant for labeling purposes in the US, the INCI Name Red 7 Lake must be used. To identify the colorant allowed for use in the European Union (EU), the INCI Name CI 15850 must be used, except for hair dye products.

Information Source: TSCA

Chemical Class: Color Additives Lakes - Batch Certified by the U.S. Food and Drug Administration

Function: Colorant

Reported Product Categories: Lipsticks; Makeup Preparations (Not eye), Misc.; Blushers (All types)

Technical/Other Names:
D&C Red No. 7 Aluminum Lake
D&C Red No. 7 Barium Lake
D&C Red No. 7 Calcium Lake
D&C Red No. 7 Calcium/Strontium Lake
D&C Red No. 7 Zirconium Lake

Trade Names:
A502.03 Tudor Paeony (Kingfisher Colours)
A502.04 Tudor Paeony (Kingfisher Colours)
A502.11 Tudor Paeony (Kingfisher Colours)
A502.25 Tudor Paeony (Kingfisher Colours)
C 6507 D&C Red No.7 Calcium Lake (LCW)
C6607 D&C Red No. 7 Calcium Lake (LCW)
D+C RED 7 Ca LAKE 10-31-DA-3007 (Noveon Hilton Davis)
C19-003 Rubine Lake (Sun Pigments)
C19-011 Rubine Lake (Sun Pigments)
C19-021 Rubine Lake (Sun Pigments)
C19-025 Rubine Lake (Sun Pigments)
D&C Red 7 Calcium Lake 10-31-DA-3507 (Noveon Hilton Davis)
D&C Red 7 Calcium Lake W 005 (LCW)
D&C Red 7 Calcium Lake W 048 (LCW)
D & C Red No. 7, Ca Lake C6507 (LCW)
D & C Red #7 Calcium Lake K7044 (LCW)
D & C Red #7 Calcium Lake K7121 (LCW)
D & C Red #7 Calcium Lake K7183 (LCW)
Unipure Red LC 3071 (LCW)
Unipure Red LC 3075 (LCW)

Trade Name Mixtures:
Cellini Red (Engelhard Corp.)
Colourspheres Geranium HB 25% (Creations Couleurs)
C6507 Red 7CA-I2 (Kobo)
Micapoly WOE Geranium (Creations Couleurs)
Tefpoly Begonia (C.I.T.)
Tefpoly Geranium Lake (C.I.T.)

RED 17

CAS No.: 85-86-9

Empirical Formula:
$C_{22}H_{16}N_4O$

Definition: Red 17 is classed chemically as a disazo color. It conforms to the formula:

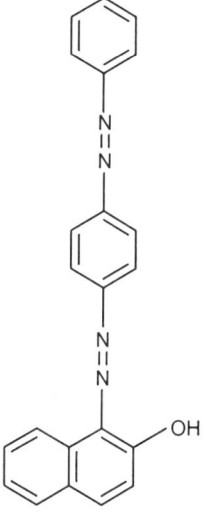

See "Regulatory and Ingredient Use Information," for Colorants in Volume 1, Introduction, Part A. To identify the certified colorant for labeling purposes in the US, the INCI Name Red 17 must be used. The INCI Name for batches of this colorant that have not been certified is Solvent Red 23. To identify the colorant allowed for use in the European Union (EU), the INCI Name CI 26100 must be used, except for hair dye products. To identify the colorant allowed for use in Japan, the INCI name Aka225 must be used.

Information Sources: 21CFR74.1317, 21CFR74.2317, 21CFR74.3230, 21CFR81.30, 21CFR82.1317, CI 26100, MI-13(8969), TSCA

Chemical Class: Color Additives - Batch Certified by the U.S. Food and Drug Administration

Function: Colorant

Reported Product Categories: Tonics, Dressings, and Other Hair Grooming Aids; Colognes and Toilet Waters; Nail Polish and Enamel Removers; Bath Oils, Tablets, and Salts; Perfumes; Bath Soaps and Detergents; Fragrance Preparations, Misc.; Cuticle Softeners; Eyebrow Pencils; Hair Conditioners; Hair Preparations (Non-coloring), Misc.; Moisturizing Preparations; Nail Polish and Enamels; Night Skin Care Preparations; Skin Care Preparations, Misc.; Suntan Gels, Creams, and Liquids

Technical/Other Name:
D&C Red No. 17

Trade Names:
D & C Red 17 W 085 (LCW)
D & C Red #17 K7007 (LCW)

Trade Name Mixtures:
Colourspheres Wood Dye HL 25% (Creations Couleurs)
Dermaglo Red (Day-Glo)
LipoCrystals EU Orange (Lipo)
LipoCrystals EU Pink (Lipo)
Tefpoly Red Wood Dye (C.I.T.)

RED 21

CAS No.: 15086-94-9

Empirical Formula:
$C_{20}H_8Br_4O_5$

Definition: Red 21 is classed chemically as a fluoran color. It conforms to the formula:

See "Regulatory and Ingredient Use Information," for Colorants in Volume 1, Introduction, Part A. To identify the certified colorant for labeling purposes in the US, the INCI Name Red 21 must be used. To identify the certified colorant (sodium salt) for labeling purposes in the US, the INCI Name Red 22 must be used. The INCI Name for batches of this colorant that have not been certified is Solvent Red 43. The INCI Name for batches of this colorant (sodium salt) that have not been certified is Acid Red 87. To identify the colorant allowed for use in the European Union (EU), the INCI Name CI 45380 must be used, except for hair dye products. To

identify the colorant allowed for use in Japan, the INCI name Aka223 must be used.

Information Sources: 21CFR74.1255, 21CFR74.1321, 21CFR74.1322, 21CFR74.2321, 21CFR82.1321, CI 45380:2, TSCA

Chemical Classes: Color Additives - Batch Certified by the U.S. Food and Drug Administration; Halogen Compounds

Function: Colorant

Reported Product Categories: Lipsticks; Makeup Preparations (Not eye), Misc.; Rouges

Technical/Other Name:
D&C Red No. 21

Trade Names:
A515 Tudor Cherry (Kingfisher Colours)
C14-032 Bromo (Sun Pigments)
D & C Red 21 W 031 (LCW)
D & C Red #21 K7061 (LCW)

Trade Name Mixtures:
Colourspheres Eosin Dye HL 25% (Creations Couleurs)
TefPoly Eosin Dye (C.I.T.)

RED 21 LAKE

Definition: Red 21 Lake is the salt of Red 21 extended on an appropriate substrate in compliance with 21CFR82.1051. See "Regulatory and Ingredient Use Information," for Colorants in Volume 1, Introduction, Part A. To identify the certified colorant for labeling purposes in the US, the INCI Name Red 21 Lake must be used. To identify the colorant allowed for use in the European Union (EU), the INCI Name CI 45380 must be used, except for hair dye products.

Information Sources: 21CFR81.1, 21CFR82.1051

Chemical Classes: Color Additives Lakes - Batch Certified by the U.S. Food and Drug Administration; Halogen Compounds

Function: Colorant

Reported Product Categories: Lipsticks; Blushers (All types); Makeup Preparations (Not eye), Misc.

Technical/Other Names:
D&C Red No. 21 Aluminum Lake
D&C Red No. 21 Zirconium Lake

Trade Names:
A504.01 Tudor Rose (Kingfisher Colours)
C 6521 D&C Red No. 21 Aluminum Lake (LCW)
C 6821 D&C Red No. 21 Aluminum Lake (LCW)

C14-034 Dell Red (Sun Pigments)
D & C Red 21 Aluminium Lake W 011 (LCW)
Unipure Red LC 321 (LCW)

Trade Name Mixture:
Tefpoly FLUO Pink 65% (Creations Couleurs)

RED 22

CAS Nos.: 548-26-5; 17372-87-1

Empirical Formula:
$C_{20}H_8Br_4O_5 \cdot 2Na$

Definition: Red 22 is classed chemically as a xanthene color. It conforms to the formula:

See "Regulatory and Ingredient Use Information," for Colorants in Volume 1, Introduction, Part A. The INCI Name for batches of this colorant that have not been certified is Acid Red 87. The INCI Name for batches of this colorant (acid form) that have not been certified is Solvent Red 43. To identify the certified colorant for labeling purposes in the US, the INCI Name Red 22 must be used. To identify the certified colorant (acid form) for labeling purposes in the US, the INCI Name Red 21 must be used. To identify the colorant allowed for use in the European Union (EU), the INCI Name CI 45380 must be used, except for hair dye products. To identify the colorant allowed for use in Japan, the INCI name Aka230(1) must be used.

Information Sources: 21CFR74.1322, 21CFR74.2322, 21CFR82.1322, CI 45380, MAR, MI-13(3634), TSCA

Chemical Classes: Color Additives - Batch Certified by the U.S. Food and Drug Administration; Halogen Compounds

Function: Colorant

Reported Product Categories: Lipsticks; Makeup Preparations (Not eye), Misc.

Technical/Other Name:
D&C Red No. 22

Trade Names:
D+C RED 22 10-25-DA-0500 (Noveon Hilton Davis)
D & C Red #22 K7008 (LCW)

Trade Name Mixtures:
 Colourspheres Pinky Dye HL 25%
  (Creations Couleurs)
 Dermaglo Red 222 (Day-Glo)
 Dermaglo Red 422 (Day-Glo)
 Tefpoly Pinky Dye (C.I.T.)

RED 22 LAKE

Definition: Red 22 Lake is the salt of Red 22 extended on an appropriate substrate in compliance with 21CFR82.1051. *See "Regulatory and Ingredient Use Information," for Colorants in Volume 1, Introduction, Part A. To identify the certified colorant for labeling purposes in the US, the INCI Name Red 22 Lake must be used. To identify the colorant allowed for use in the European Union (EU), the INCI Name CI 45380 must be used, except for hair dye products. To identify the colorant allowed for use in Japan, the INCI name Aka230(1) must be used.*

Chemical Class: Color Additives Lakes - Batch Certified by the U.S. Food and Drug Administration

Function: Colorant

Reported Product Category: Lipsticks

Technical/Other Name:
 D&C Red No. 22 Aluminum Lake

Trade Name:
 A504.02 Tudor Rose (Kingfisher Colours)

RED 27

CAS Nos.: 2134-15-8; 13473-26-2

Empirical Formula:
 $C_{20}H_4Br_4Cl_4O_5$

Definition: Red 27 is classed chemically as a fluoran color. It conforms to the formula:

See "Regulatory and Ingredient Use Information," for Colorants in Volume 1, Introduction, Part A. To identify the certified colorant for labeling purposes in the US, the INCI Name Red 27 must be used. To identify the certified colorant (sodium salt) for labeling purposes in the US, the INCI Name Red 28 must be used. The INCI Name for batches of this colorant that have not been certified is Solvent Red 48. The INCI Name for batches of this colorant (sodium salt) that have not been certified is Acid Red 92. To identify the colorant allowed for use in the European Union (EU), the INCI Name CI 45410 must be used, except for hair dye products. To identify the colorant allowed for use in Japan, the INCI name Aka218 must be used.

Information Sources: 21CFR74.1327, 21CFR74.2327, 21CFR82.1327, CI 45410:1, TSCA

Chemical Classes: Color Additives - Batch Certified by the U.S. Food and Drug Administration; Halogen Compounds

Function: Colorant

Reported Product Categories: Lipsticks; Permanent Waves

Technical/Other Name:
 D&C Red No. 27

Trade Names:
 A516 Tudor Plum (Kingfisher Colours)
 D+C RED 27 10-25-DA 3427 (Noveon Hilton Davis)
 D & C Red 27 W 039 (LCW)
 D & C Red #27 K7053 (LCW)

RED 27 LAKE

Definition: Red 27 Lake is the salt of Red 27 extended on an appropriate substrate in compliance with 21CFR82.1051. *See "Regulatory and Ingredient Use Information," for Colorants in Volume 1, Introduction, Part A. To identify the certified colorant for labeling purposes in the US, the INCI Name Red 27 Lake must be used. To identify the colorant allowed for use in the European Union (EU), the INCI Name CI 45410 must be used, except for hair dye products.*

Information Sources: 21CFR81.1, 21CFR82.1051

Chemical Classes: Color Additives Lakes - Batch Certified by the U.S. Food and Drug Administration; Halogen Compounds

Function: Colorant

Reported Product Categories: Lipsticks; Blushers (All types); Makeup Preparations (Not eye), Misc.; Hair Dyes and Colors (All Types Requiring Caution Statements and Patch Tests); Rouges; Face Powders; Bath Oils, Tablets, and Salts

Technical/Other Names:
 D&C Red No. 27 Aluminum Lake
 D&C Red No. 27 Aluminum/Titanium/Zirconium Lake
 D&C Red No. 27 Barium Lake
 D&C Red No. 27 Calcium Lake
 D&C Red No. 27 Zirconium Lake

Trade Names:
 A511.01 Tudor Foxglove (Kingfisher Colours)
 C6527 D&C Red No. 27 Aluminum Lake (LCW)
 C6627 D&C Red No. 27 Aluminum Lake (LCW)
 C14-023 Mystic Red (Sun Pigments)
 D+C RED 27 Al LAKE 10-31-DA-3127 (Noveon Hilton Davis)
 D&C Red 27 Al. Lake 10-30-DA-3117 (Noveon Hilton Davis)
 D&C Red 27 Al/Zr/Ti Lake 10-31-DA-7047 (Noveon Hilton Davis)
 D & C Red 27 Aluminium Lake W 030 (LCW)
 Unipure Red LC 327 (LCW)

Trade Name Mixtures:
 Colourspheres Electric Pink Lake HL 25% (Creations Couleurs)
 Tefpoly Electric Pink Lake (C.I.T.)

RED 28

CAS No.: 18472-87-2

Empirical Formula:
 $C_{20}H_4Br_4Cl_4O_5 \cdot 2Na$

Definition: Red 28 is classed chemically as a xanthene color. It conforms to the formula:

See "Regulatory and Ingredient Use Information," for Colorants in Volume 1, Introduction, Part A. The INCI Name for batches of this colorant that have not been certified is Acid Red 92. The INCI Name for batches of this colorant (acid form) that have not been certified is Solvent Red 48. To identify the certified colorant for labeling purposes in the US, the INCI Name Red 28 must be used. To identify the certified colorant (acid form) for labeling purposes in the US, the INCI Name Red 27 must be used. To

---

identify the colorant allowed for use in the European Union (EU), the INCI Name CI 45410 must be used, except for hair dye products. To identify the colorant allowed for use in Japan, the INCI name Aka104(1) must be used.

Information Sources: 21CFR74.1328, 21CFR74.2328, 21CFR82.1328, CI 45410, TSCA

Chemical Classes: Color Additives - Batch Certified by the U.S. Food and Drug Administration; Halogen Compounds

Function: Colorant

Reported Product Categories: Lipsticks; Hair Dyes and Colors (All Types Requiring Caution Statements and Patch Tests); Permanent Waves; Bath Soaps and Detergents; Bubble Baths; Hair Bleaches; Hair Rinses (Coloring); Paste Masks (Mud Packs)

Technical/Other Name:
D&C Red No. 28

Trade Names:
D+C RED 28 10-25-DA-0906 (Noveon Hilton Davis)
D & C Red #28 K7054 (LCW)

Trade Name Mixtures:
Dermaglo Red (Day-Glo)
Dermaglo Red 228 (Day-Glo)
Dermaglo Red 428 (Day-Glo)

RED 28 LAKE

Definition: Red 28 Lake is the salt of Red 28 extended on an appropriate substrate in compliance with 21CFR82.1051. See "Regulatory and Ingredient Use Information," for Colorants in Volume 1, Introduction, Part A. To identify the certified colorant for labeling purposes in the US, the INCI Name Red 28 Lake must be used. To identify the colorant allowed for use in the European Union (EU), the INCI Name CI 45410 must be used, except for hair dye products. To identify the colorant allowed for use in Japan, the INCI name Aka104(1) must be used.

Information Sources: 21CFR81.1, 21CFR82.1051

Chemical Class: Color Additives Lakes - Batch Certified by the U.S. Food and Drug Administration

Function: Colorant

Reported Product Categories: Lipsticks; Blushers (All types); Makeup Preparations (Not eye), Misc.

Technical/Other Name:
D&C Red No. 28 Aluminum Lake

Trade Names:
A511.02 Tudor Foxglove (Kingfisher Colours)
D&C Red 28 Aluminum Lake 10-31-DA-3128 (Noveon Hilton Davis)
Unipure Red LC 328 (LCW)

RED 30

CAS No.: 2379-74-0

Empirical Formula:
$C_{18}H_{10}Cl_2O_2S_2$

Definition: Red 30 is classed chemically as an indigoid color. It conforms to the formula:

See "Regulatory and Ingredient Use Information," for Colorants in Volume 1, Introduction, Part A. To identify the certified colorant for labeling purposes in the US, the INCI Name Red 30 must be used. The INCI Name for batches of this colorant that have not been certified is Vat Red 1. To identify the colorant allowed for use in the European Union (EU), the INCI Name CI 73360 must be used, except for hair dye products. To identify the colorant allowed for use in Japan, the INCI name Aka226 must be used.

Information Sources: 21CFR74.1330, 21CFR74.2330, 21CFR82.1330, CI 73360, M3, TSCA

Chemical Class: Color Additives - Batch Certified by the U.S. Food and Drug Administration

Function: Colorant

Reported Product Categories: Blushers (All types); Lipsticks; Face Powders; Foundations; Nail Polish and Enamels; Powders (Dusting and Talcum, Excluding Aftershave Talcs); Bath Oils, Tablets, and Salts; Cleansing Products (Cold Creams, Cleansing Lotions, Liquids and Pads); Bath Preparations, Misc.; Bath Soaps and Detergents

Technical/Other Name:
D&C Red No. 30

Trade Names:
A505.10 Tudor Geranium (Kingfisher Colours)
Ultrapearl Red 30 Pink Luster (Ultra Chemical)

Trade Name Mixtures:
Colorona Imperial Red (Merck KGaA/EMD Chemicals Inc.)

Hot Dots - Red (Charm Girl)
LipoCrystals LIP Coral (Lipo)
LipoCrystals LIP Pink (Lipo)
LipoSphere Glitter Pink (Lipo)
Toshiki PFL-Red-7044 (Nikko)
Toshiki PFM-Pink 6055 (Nikko)
Toshiki PFM-Red-6011 (Nikko)
Toshiki PFS-Red-7001 (Nikko)
Toshiki SP-565-41 Red (Nikko)

RED 30 LAKE

Definition: Red 30 Lake is Red 30 extended on an appropriate substrate in compliance with 21CFR82.1051. See "Regulatory and Ingredient Use Information," for Colorants in Volume 1, Introduction, Part A. To identify the certified colorant for labeling purposes in the US, the INCI Name Red 30 Lake must be used. To identify the colorant allowed for use in the European Union (EU), the INCI Name CI 73360 must be used, except for hair dye products.

Information Sources: 21CFR81.1, 21CFR82.1051

Chemical Class: Color Additives Lakes - Batch Certified by the U.S. Food and Drug Administration

Function: Colorant

Reported Product Categories: Lipsticks; Blushers (All types); Face Powders; Bath Soaps and Detergents; Hair Bleaches; Hair Dyes and Colors (All Types Requiring Caution Statements and Patch Tests); Makeup Bases; Rouges; Makeup Preparations (Not eye), Misc.; Nail Polish and Enamels; Powders (Dusting and Talcum, Excluding Aftershave Talcs)

Technical/Other Name:
D&C Red No. 30 Lake

Trade Names:
A505.01 Tudor Geranium (Kingfisher Colours)
A505 Tudor Geranium (Kingfisher Colours)
C6530 D&C Red No. 30 Aluminum Lake (LCW)
C37-038 Permanent Pink (Sun Pigments)
C37-5290 Permanent Pink (Sun Pigments)
D+C RED 30TALC LAKE 10-31-DA-3130 (Noveon Hilton Davis)
D & C Red #30 Alum Lake K7156 (LCW)
D & C Red #30 Talc Lake K7094 (LCW)
Unipure Red LC 300 (LCW)

Trade Name Mixtures:
Colourspheres Helindone Lake HL 25% (Creations Couleurs)
Tefpoly Helindone Lake (C.I.T.)

RED 31

CAS No.: 6371-76-2

Empirical Formula:
$C_{17}H_{12}N_2O_3 \cdot \frac{1}{2}Ca$

Definition: Red 31 is classed chemically as a monoazo color. It conforms to the formula:

See "Regulatory and Ingredient Use Information," for Colorants in Volume 1, Introduction, Part A. To identify the certified colorant for labeling purposes in the US, the INCI Name Red 31 must be used. The INCI Name for batches of this colorant that have not been certified is Pigment Red 64:1. To identify the colorant allowed for use in the European Union (EU), the INCI Name CI 15800 must be used, except for hair dye products. To identify the colorant allowed for use in Japan, the INCI name Aka219 must be used.

Information Sources: 21CFR74.1331, 21CFR74.2331, 21CFR81.30, 21CFR82.1331, CI 15800, TSCA

Chemical Class: Color Additives - Batch Certified by the U.S. Food and Drug Administration

Function: Colorant

Technical/Other Name:
D&C Red No. 31

RED 31 LAKE

Definition: Red 31 Lake is the salt of Red 31 extended on an appropriate substrate in compliance with 21CFR 82.1051. *See "Regulatory and Ingredient Use Information," for Colorants in Volume 1, Introduction, Part A. To identify the certified colorant for labeling purposes in the US, the INCI Name Red 31 Lake must be used. To identify the colorant allowed for use in the European Union (EU), the INCI Name CI 15800 must be used, except for hair dye products.*

Information Sources: 21CFR81.1, 21CFR82.1051

Chemical Class: Color Additives Lakes - Batch Certified by the U.S. Food and Drug Administration

Function: Colorant

Technical/Other Name:
D&C Red No. 31 Calcium Lake

RED 33

CAS No.: 3567-66-6

Empirical Formula:
$C_{16}H_{13}N_3O_7S_2 \cdot 2Na$

Definition: Red 33 is classed chemically as a monoazo color. It conforms to the formula:

See "Regulatory and Ingredient Use Information," for Colorants in Volume 1, Introduction, Part A. To identify the certified colorant for labeling purposes in the US, the INCI Name Red 33 must be used. The INCI Name for batches of this colorant that have not been certified is Acid Red 33. To identify the colorant allowed for use in the European Union (EU), the INCI Name CI 17200 must be used, except for hair dye products. To identify the colorant allowed for use in Japan, the INCI name Aka227 must be used.

Information Sources: 21CFR74.1333, 21CFR74.2333, 21CFR82.1333, CI 17200, TSCA

Chemical Class: Color Additives - Batch Certified by the U.S. Food and Drug Administration

Function: Colorant

Reported Product Categories: Shampoos (Non-coloring); Colognes and Toilet Waters; Bath Soaps and Detergents; Moisturizing Preparations; Bath Oils, Tablets, and Salts; Bubble Baths; Cleansing Products (Cold Creams, Cleansing Lotions, Liquids and Pads); Hair Conditioners; Bath Preparations, Misc.; Body and Hand Preparations (Excluding Shaving Preparations); Skin Care Preparations, Misc.; Personal Cleanliness Products, Misc.; Skin Fresheners; Tonics, Dressings, and Other Hair Grooming Aids; Bath Capsules; Face and Neck Preparations (Excluding Shaving Preparations); Hair Preparations (Non-coloring), Misc.; Paste Masks (Mud Packs); Fragrance Preparations, Misc.; Deodorants (Underarm); Perfumes; Nail Polish and Enamel Removers; Nail Polish and Enamels; Aftershave Lotions; Baby Shampoos; Blushers (All types); Hair Wave Sets; Baby Lotions, Oils, Powders and Creams; Hair Rinses (Coloring); Makeup Preparations (Not eye), Misc.; Shaving Preparations, Misc.; Hair Dyes and Colors (All Types Requiring Caution Statements and Patch Tests); Manicuring Preparations, Misc.; Mouthwashes and Breath Fresheners (Liquids and Sprays); Baby Products, Misc.; Cuticle Softeners; Hair Bleaches; Hair Rinses (Non-coloring); Hair Straighteners; Lipsticks; Night Skin Care Preparations; Permanent Waves; Powders (Dusting and Talcum, Excluding Aftershave Talcs); Rouges; Shaving Cream (Aerosol, Brushless and Lather); Suntan Gels, Creams, and Liquids

Technical/Other Name:
D&C Red No. 33

Trade Names:
D&C Red 33 10-25-DA-3733 (Noveon Hilton Davis)
D & C Red 33 W 083 (LCW)
D & C Red #33 K7057 (LCW)

Trade Name Mixtures:
Colourspheres Magenta Dye HL 25% (Creations Couleurs)
Tefpoly Magenta Dye (C.I.T.)

RED 33 LAKE

Definition: Red 33 Lake is the salt of Red 33 extended on an appropriate substrate in compliance with 21CFR 82.1051. *See "Regulatory and Ingredient Use Information," for Colorants in Volume 1, Introduction, Part A. To identify the certified colorant for labeling purposes in the US, the INCI Name Red 33 Lake must be used. To identify the colorant allowed for use in the European Union (EU), the INCI Name CI 17200 must be used, except for hair dye products. To identify the colorant allowed for use in Japan, the INCI name Aka227 must be used.*

Information Sources: 21CFR74.1333, 21CFR74.2333

Chemical Class: Color Additives Lakes - Batch Certified by the U.S. Food and Drug Administration

Function: Colorant

Reported Product Categories: Lipsticks; Blushers (All types); Bath Oils, Tablets, and

---

Salts; Face Powders; Makeup Preparations (Not eye), Misc.

Technical/Other Name:
D&C Red No. 33 Aluminum Lake

Trade Names:
A529 Tudor Betony (Kingfisher Colours)
D & C Red #33 Aluminum Lake K7192 (LCW)
Unipure Red LC 323 (LCW)

Trade Name Mixtures:
Colourspheres Magenta HB 25% (Creations Couleurs)
Colourspheres Magenta Lake HL 25% (Creations Couleurs)
Tefpoly Magenta Lake (C.I.T.)

RED 34

CAS No.: 6417-83-0

Empirical Formula:
$C_{21}H_{14}N_2O_6S \cdot Ca$

Definition: Red 34 is classed chemically as a monoazo color. It conforms to the formula:

See "Regulatory and Ingredient Use Information," for Colorants in Volume 1, Introduction, Part A. To identify the certified colorant for labeling purposes in the US, the INCI Name Red 34 must be used. The INCI Name for batches of this colorant that have not been certified is Pigment Red 63:1. To identify the colorant allowed for use in the European Union (EU), the INCI Name CI 15880 must be used, except for hair dye products. To identify the colorant allowed for use in Japan, the INCI name Aka220 must be used.

Information Sources: 21CFR74.1334, 21CFR74.2334, 21CFR81.30, 21CFR82.1334, CI 15880, M3, TSCA

Chemical Class: Color Additives - Batch Certified by the U.S. Food and Drug Administration

Function: Colorant

Reported Product Category: Nail Polish and Enamels

Technical/Other Name:
D&C Red No. 34

Trade Name:
D & C Red 34 Calcium Lake W 014 (LCW)

Trade Name Mixtures:
Creasparkles Colours Coral Pink (Creations Couleurs)
Creasparkles Colours Lilac Blue (Creations Couleurs)
Creasparkles Colours Magenta (C.I.T.)
Creasparkles Colours Raspberry (Creations Couleurs)
Creasparkles Colours Rubis (Creations Couleurs)
Creasparlkes Colours Magenta (Creations Couleurs)
Diamond Piece CO Type, Pink (Daiya Kogyo)
Diamond Piece CO Type, Red (Daiya Kogyo)
Diamond Piece CO Type, Violet (Daiya Kogyo)

RED 34 LAKE

Definition: Red 34 Lake is the salt of Red 34 extended on an appropriate substrate in compliance with 21CFR 82.1051. See "Regulatory and Ingredient Use Information," for Colorants in Volume 1, Introduction, Part A. To identify the certified colorant for labeling purposes in the US, the INCI Name Red 34 Lake must be used. To identify the colorant allowed for use in the European Union (EU), the INCI Name CI 15880 must be used, except for hair dye products.

Information Sources: 21CFR81.1, 21CFR82.1051

Chemical Class: Color Additives Lakes - Batch Certified by the U.S. Food and Drug Administration

Function: Colorant

Reported Product Categories: Nail Polish and Enamels; Blushers (All types); Face Powders

Technical/Other Name:
D&C Red No. 34 Calcium Lake

Trade Names:
A518 Tudor Iris (Kingfisher Colours)
C24-012 Rangoon Maroon (Sun Pigments)
D & C Red #34 Calcium Lake K7122 (LCW)

RED 36

CAS No.: 2814-77-9

Empirical Formula:
$C_{16}H_{10}ClN_3O_3$

Definition: Red 36 is classed chemically as a monoazo color. It conforms to the formula:

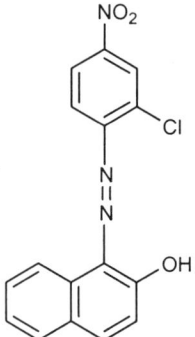

See "Regulatory and Ingredient Use Information," for Colorants in Volume 1, Introduction, Part A. To identify the certified colorant for labeling purposes in the US, the INCI Name Red 36 must be used. The INCI Name for batches of this colorant that have not been certified is Pigment Red 4. To identify the colorant allowed for use in the European Union (EU), the INCI Name CI 12085 must be used, except for hair dye products. To identify the colorant allowed for use in Japan, the INCI name Aka228 must be used.

Information Sources: 21CFR74.1336, 21CFR74.2336, 21CFR82.1336, CI 12085, M3, TSCA

Chemical Classes: Color Additives - Batch Certified by the U.S. Food and Drug Administration; Halogen Compounds

Function: Colorant

Reported Product Categories: Lipsticks; Blushers (All types); Makeup Preparations (Not eye), Misc.; Face Powders

Technical/Other Name:
D&C Red No. 36

Trade Names:
A513 Tudor Aster (Kingfisher Colours)
C6636 D&C Red No. 36 (LCW)
C23-009 Flaming Red (Sun Pigments)
D & C Red 36 W 008 (LCW)
D & C Red #36 C6636 (LCW)

RED 36 LAKE

Definition: Red 36 Lake is an insoluble pigment composed of Red 36 extended on an appropriate substrate in compliance with 21CFR82.1051. See "Regulatory and Ingredient Use Information," for Colorants in Volume 1, Introduction, Part A. To identify the certified colorant for labeling purposes in the US, the INCI Name Red 36

Lake must be used. To identify the colorant allowed for use in the European Union (EU), the INCI Name CI 12085 must be used, except for hair dye products.

Information Sources: 21CFR81.1, 21CFR82.1051

Chemical Classes: Color Additives Lakes - Batch Certified by the U.S. Food and Drug Administration; Halogen Compounds

Function: Colorant

Reported Product Categories: Lipsticks; Blushers (All types)

Technical/Other Name:
D&C Red No. 36 Lake

Trade Name:
Unipure Red LC 306 (LCW)

RED 40

CAS No.: 25956-17-6

Empirical Formula:
$C_{18}H_{16}N_2O_8S_2 \cdot 2Na$

Definition: Red 40 is classed chemically as a monoazo color. It conforms to the formula:

See "Regulatory and Ingredient Use Information," for Colorants in Volume 1, Introduction, Part A. To identify the certified colorant for labeling purposes in the US, the INCI Name Red 40 must be used. The INCI Name for batches of this colorant that have not been certified is Curry Red. To identify the colorant allowed for use in the European Union (EU), the INCI Name CI 16035 must be used, except for hair dye products.

Information Sources: 21CFR74.340, 21CFR74.1340, 21CFR74.2340, CI 16035, MI-13(281), TSCA

Chemical Class: Color Additives - Batch Certified by the U.S. Food and Drug Administration

Function: Colorant

Reported Product Categories: Shampoos (Non-coloring); Hair Conditioners; Moisturizing Preparations; Bath Soaps and Detergents; Colognes and Toilet Waters; Bath Oils, Tablets, and Salts; Cleansing Products (Cold Creams, Cleansing Lotions, Liquids and Pads); Bath Preparations, Misc.; Body and Hand Preparations (Excluding Shaving Preparations); Bubble Baths; Personal Cleanliness Products, Misc.; Hair Sprays (Aerosol Fixatives); Skin Care Preparations, Misc.; Aftershave Lotions; Baby Shampoos; Mouthwashes and Breath Fresheners (Liquids and Sprays); Skin Fresheners; Bath Capsules; Face and Neck Preparations (Excluding Shaving Preparations); Hair Preparations (Non-coloring), Misc.; Dentifrices (Aerosol, Liquid, Pastes and Powders); Tonics, Dressings, and Other Hair Grooming Aids; Fragrance Preparations, Misc.; Lipsticks; Nail Polish and Enamel Removers; Shaving Preparations, Misc.

Technical/Other Name:
FD&C Red No. 40

Trade Names:
FD&C Red 40 10-21-DA-6056 (Noveon Hilton Davis)
FD & C Red 40 W 093 (LCW)
Ultrapearl Red 40 Luster (Ultra Chemical)

Trade Name Mixtures:
Colourspheres Coral Dye HL 25% (Creations Couleurs)
Tefpoly Coral Dye (C.I.T.)

RED 40 LAKE

Definition: Red 40 Lake is the salt of Red 40 extended on an appropriate substrate in compliance with 21CFR82.1051. *See "Regulatory and Ingredient Use Information," for Colorants in Volume 1, Introduction, Part A. To identify the certified colorant for labeling purposes in the US, the INCI Name Red 40 Lake must be used. To identify the colorant allowed for use in the European Union (EU), the INCI Name CI 16035 must be used, except for hair dye products.*

Information Sources: 21CFR74.2340, 21CFR81.1

Chemical Class: Color Additives Lakes - Batch Certified by the U.S. Food and Drug Administration

Function: Colorant

Reported Product Category: Lipsticks

Technical/Other Names:
D&C Red No. 40 Aluminum Lake
FD&C Red No. 40 Aluminum Lake

Trade Names:
A530 Tudor Geum (Kingfisher Colours)
09310 D&C Red No. 40 Aluminum Lake (LCW)
FD&C Red 40 Aluminum Lake 10-21-DB-6807 (Noveon Hilton Davis)
FD&C Red 40 Aluminum Lake 10-21-DB-6808 (Noveon Hilton Davis)
Unipure Red LC 324 (LCW)

Trade Name Mixtures:
Cellini Coral (Engelhard Corp.)
Spectra F/X Copper (Spectratek)
Spectra F/X Gold (Spectratek)
Ultracolor Red 40 (Ultra Chemical)

RED PETROLATUM

Definition: Red Petrolatum is a minimally refined variety of Petrolatum (q.v.). *See Reported Ingredient Functions-The Cosmetic Drug Distinction, in Regulatory and Ingredient Use Information, Volume I, Part A.*

Information Source: OTC-I-SU

Chemical Class: Hydrocarbons

Functions: Sunscreen Agent; Ultraviolet Light Absorber

REHMANNIA CHINENSIS ROOT EXTRACT

JPN Translation:
アカヤジオウ根エキス

Definition: Rehmannia Chinensis Root Extract is an extract of the roots of *Rehmannia chinensis*. *See "Regulatory and Ingredient Use Information," regarding the labeling names for botanical ingredients in Volume 1, Introduction, Part A.*

Information Sources: JCIC, JCLS, JSQI

Chemical Class: Biological Products

Function: Not Reported

Technical/Other Names:
Extract of Rehmannia Chinensis
Extract of Rehmannia Glutinosa
Jiou Ekisu (JPN)
Rehmannia Glutinosa Extract
Rehmannia Root Extract

Trade Name Mixtures:
JIOH Extract BG (Ikeda)
Jiou Liquid (Ichimaru Pharcos)
Jiou Liquid B (Ichimaru Pharcos)

REHMANNIA ELATA ROOT EXTRACT

Definition: Rehmannia Elata Root Extract is an extract of the whole plant without the roots of *Rehmannia elata*. See *"Regulatory and Ingredient Use Information," regarding the labeling names for botanical ingredients in Volume 1, Introduction, Part A.*

Chemical Class: Biological Products

Function: Not Reported

Technical/Other Name:
Extract of Rehmannia Elata

Trade Name Mixtures:
Actiphyte of Rehmannia BG50 (Active Organics)
Actiphyte of Rehmannia GL50 (Active Organics)
Actiphyte of Rehmannia Lipo S (Active Organics)
Actiphyte of Rehmannia PG50 (Active Organics)

REHMANNIA GLUTINOSA ROOT EXTRACT

Definition: Rehmannia Glutinosa Root Extract is an extract of the roots of *Rehmannia glutinosa*. See *"Regulatory and Ingredient Use Information," regarding the labeling names for botanical ingredients in Volume 1, Introduction, Part A.*

Chemical Class: Biological Products

Function: Skin-Conditioning Agent - Miscellaneous

Technical/Other Name:
Extract of Rehmannia Glutinosa

Trade Name Mixtures:
Boumdan (Bioland)
Rehmannia Extract (Maruzen Pharmaceuticals Co., Ltd.)
Rehmannia Extract BG (Maruzen Pharmaceuticals Co., Ltd.)
Rehmannia Extract BG-J (Maruzen Pharmaceuticals Co., Ltd.)
Rehmannia Extract-J (Maruzen Pharmaceuticals Co., Ltd.)
Rehmannia Extract Powder-S (Maruzen Pharmaceuticals Co., Ltd.)
Rehmannia Extract W (Maruzen Pharmaceuticals Co., Ltd.)

REPAGERMANIUM

CAS No.: 12758-40-6

Definition: Repagermanium is the polymer that conforms to the formula:

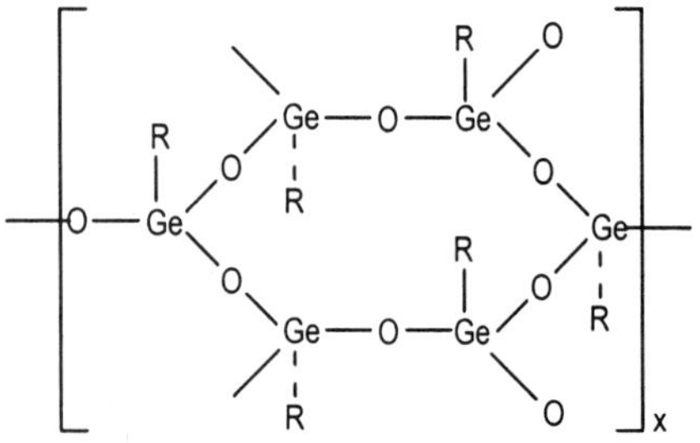

where R represents a carboxyethyl moiety.

Information Sources: INN, MI-13(7887)

Chemical Classes: Carboxylic Acids; Synthetic Polymers

Function: Skin-Conditioning Agent - Miscellaneous

Technical/Other Name:
Poly-trans-[(2-Carboxyethyl) Germasesquioxane]

Trade Name:
Asai Germanium (Asai Germanium Research Institute)

RESEDA LUTEOLA

Definition: *See "Regulatory and Ingredient Use Information," regarding EU labeling names for botanical ingredients in Volume 1, Introduction, Part A.*

Chemical Class: Biological Products

Technical/Other Name:
Reseda Luteola Extract (U.S.)

RESEDA LUTEOLA EXTRACT

CAS No. 90106-16-4
EINECS No. 290-237-8

Definition: Reseda Luteola Extract is an extract of the whole plant of the dyer's rocket, *Reseda luteola*. See *"Regulatory and Ingredient Use Information," regarding the labeling names for botanical ingredients in Volume 1, Introduction, Part A.*

Chemical Class: Biological Products

Function: Not Reported

Technical/Other Names:
Extract of Reseda Luteola
Reseda Luteola (EU)

Trade Name:
ViaPure Reseda (Actives International)

Trade Name Mixtures:
Clariline (CEP (Solabia))
Glycolic Extract R.U.P. (CEP (Solabia))

RESEDA ODORATA EXTRACT

Definition: Reseda Odorata Extract is the extract obtained from the aerial parts of *Reseda odorata*. See *"Regulatory and Ingredient Use Information," regarding the labeling names for botanical ingredients in Volume 1, Introduction, Part A.*

Chemical Class: Biological Products

Function: Not Reported

Technical/Other Name:
Extract of Reseda Odorata

Trade Name Mixture:
Cosflor Mignonette HGS (A & E Connock)

RESMETHRIN

CAS No. 10453-86-8
EINECS No. 233-940-7

Empirical Formula:
$C_{22}H_{26}O_3$

Definition: Resmethrin is the organic compound that conforms to the formula:

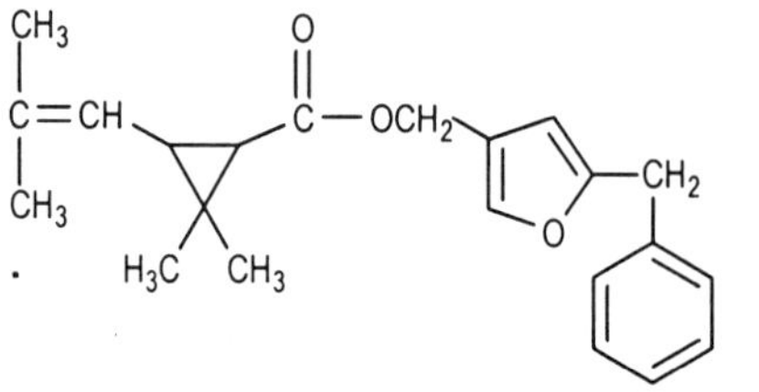

Information Source: 40CFR180.525

Chemical Classes: Esters; Heterocyclic Compounds; Ketones

Function: Pesticide

Technical/Other Names:
5-Benzylfurfuryl Chrysanthemate
Cyclopropanecarboxylic Acid, 2,2-Dimethyl-3-(2-Methyl-1-Propenyl)-, [5-(Phenylmethyl)-3-Furanyl]Methyl Ester

Trade Name:
Resmethrin Technical (Sumitomo Chemical)

RESORCINOL

CAS No. 108-46-3
EINECS No. 203-585-2

JPN Translation:
レゾルシン

Empirical Formula:
$C_6H_6O_2$

Definition: Resorcinol is the phenol that conforms to the formula:

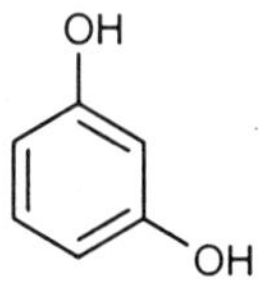

In the United States, Resorcinol may be used as an active ingredient in OTC drug products. When used as an active drug ingredient, the established drug name is

---

The inclusion of any compound in the *Dictionary and Handbook* does not indicate that use of that substance as a cosmetic ingredient complies with the laws and regulations governing such use in the United States or any other country.

Resorcinol. See "Regulatory and Ingredient Use Information," regarding the labeling names for U.S. OTC Drug Ingredients in Volume 1, Introduction, Part A. See "Regulatory and Ingredient Use Information," for Colorants in Volume 1, Introduction, Part A.

Information Sources: ARG, AUS, BP, BPC, BRA, 21CFR74.1707, 21CFR74.1708, 21CFR74.2151, 21CFR177.1210, 21CFR310.545, 21CFR333.310, 21CFR333.320, 21CFR346.20, 21CFR533.310, 27CFR21.151, CI 76505, CIR: [S] JACT-5(3)1986, CTFA S, CZE, DA, DDR, EEC(III/1-22), EGY, FIN, HP, HUN, IND, ITA, JAN, JCLS, JSCI, MAR, MEX, MHLW-331/3, MI-13(8240), OTC-I-AR, OTC-I-EA, PF, PN, POR, RIFM, ROM, TSCA, USAN, USD, USP XXIV, USSR, YUG

Chemical Classes: Color Additives - Hair; Phenols

Functions: Antiacne Agent; Denaturant; External Analgesic; Fragrance Ingredient; Hair Colorant

Reported Product Categories: Hair Dyes and Colors (All Types Requiring Caution Statements and Patch Tests); Hair Color Sprays (Aerosol); Hair Tints; Cleansing Products (Cold Creams, Cleansing Lotions, Liquids and Pads)

Technical/Other Names:
1,3-Benzenediol
CI Developer 4
m-Dihydroxybenzene
m-Hydroquinone
3-Hydroxyphenol
Oxidation Base 31
m-Phenylenediol
Resorcin
Resorcinol (RIFM)

Trade Names:
Colorex RES-CG (Chemical Compounds, Inc.)
Covastyle RCN (LCW)
Jarocol RL (James Robinson)
Rodol RS (Lowenstein)
Rodol RS TECH (Lowenstein)
Rodol RS TECH SP (Lowenstein)
Rodol RS USP-C (Lowenstein)
Rodol RS USP-F (Lowenstein)
Unichem RSC (Universal Preserv-A-Chem)

Trade Name Mixtures:
Blonde 90 (Fusion) (Lowenstein)
Blonde R-50 (Fusion) (Lowenstein)
Brown R-36 (Fusion) (Lowenstein)
Rodol Black 2/0A (Lowenstein)
Rodol Brown DB (3/A) (Lowenstein)
Rodol Chestnut Brown 5/42A (Lowenstein)
Rodol Light Brown 5/02A (Lowenstein)
Rodol Medium Brown 4/0-PW (Lowenstein)

RESORCINOL ACETATE

CAS No. 102-29-4   **EINECS No.** 203-022-0

Empirical Formula: $C_8H_8O_3$

Definition: Resorcinol Acetate is the monoester of resorcinol and acetic acid. It conforms to the formula:

See "Regulatory and Ingredient Use Information," regarding the labeling names for U.S. OTC Drug Ingredients in Volume 1, Introduction, Part A.

Information Sources: MAR, MI-13(8240), OTC-I-AK, TSCA, USAN, USD, USP XXIV

Chemical Class: Esters

Functions: Antiacne Agent; Hair Conditioning Agent; Skin-Conditioning Agent - Miscellaneous

Technical/Other Names:
3-Acetoxyphenol
1,3-Benzenediol, Monoacetate
m-Hydroxyphenyl Acetate
Resorcinol Monoacetate
Resorcitate

RESVERATROL

CAS No.: 501-36-0

Definition: Resveratrol is the organic compound that conforms to the formula:

See Reported Ingredient Functions-The Cosmetic Drug Distinction, in Regulatory and Ingredient Use Information, Volume I, Part A.

Information Source: MI-13(8243)

Chemical Class: Phenols

Functions: Antioxidant; Skin Protectant

Technical/Other Names:
1,3-Benzenediol, 5-[(1E)-2-(4-Hydroxyphenyl)ethenyl]-
3,5,4'-Trihydroxystilbene

Trade Name:
Resveratrol 98% (Lalilab, Inc.)

RETINAL

CAS No. 116-31-4   **EINECS No.** 204-135-8

Empirical Formula: $C_{20}H_{28}O$

Definition: Retinal is the organic compound that conforms to the formula:

Information Sources: MI-13(8249), TSCA

Chemical Class: Aldehydes

Function: Skin-Conditioning Agent - Miscellaneous

Technical/Other Names:
2,4,6,8-Nonatetraenal, 3,7-Dimethyl-9-(2,6,6-Trimethyl-1-Cyclohexen-1-yl)-, (all-E)-
Retinaldehyde
trans-Retinal
Vitamin A Aldehyde

Trade Name Mixture:
Retinal Microcapsule (Coletica SA)

RETINOIC ACID

CAS No. 302-79-4   **EINECS No.** 206-129-0

Empirical Formula: $C_{19}H_{26}O_2$

Definition: Retinoic Acid is the organic compound that conforms to the formula:

See Reported Ingredient Functions-The Cosmetic Drug Distinction, in Regulatory and Ingredient Use Information, Volume I, Part A.

Information Sources: EEC(II-375), MI-13(8251)

Chemical Class: Carboxylic Acids

Function: Antiacne Agent

Technical/Other Names:
3,7-Dimethyl-9-(2,6,6-Trimethyl-1-Cyclohexen-1-yl)-2,4,6,8-Nonatetraenoic Acid
Retin A
Tretinoin
Vitamin A Acid

Trade Name Mixture:
Glycosan Retinoico (Chemyunion)

RETINOL

CAS Nos.	EINECS Nos.
68-26-8	200-683-7
11103-57-4	234-328-2

Empirical Formula:
$C_{20}H_{30}O$

Definition: Retinol is the organic compound that conforms to the formula:

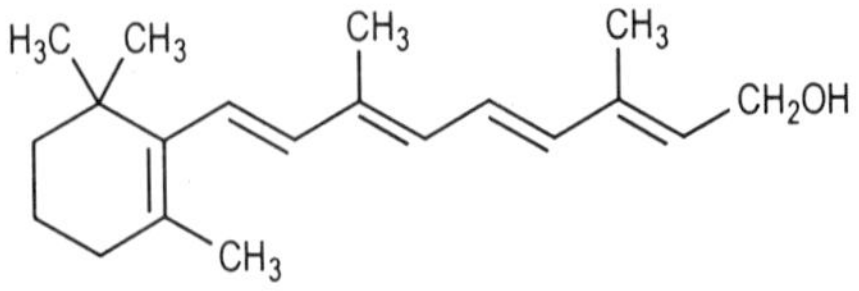

Information Sources: ARG, BAN, BRA, 21CFR101.9, 21CFR101.36, 21CFR104.20, 21CFR104.47, 21CFR107.10, 21CFR107.100, 21CFR131, 21CFR131.127, 21CFR133, 21CFR135.130, 21CFR166.40, 21CFR166.110, 21CFR184.1245, 21CFR184.1930, 21CFR310.545, 21CFR582.5930, 21CFR582.5933, 21CFR582.5936, CIR: [S] JACT-6(3)1987, EGY, FCC, INN, JAN, MAR, MI-13(10073), PF, POL, POR, TSCA, USAN, USD, USP XXIV

Chemical Class: Alcohols

Function: Skin-Conditioning Agent - Miscellaneous

Reported Product Categories: Skin Care Preparations, Misc.; Moisturizing Preparations; Bath Preparations, Misc.; Body and Hand Preparations (Excluding Shaving Preparations); Bath Oils, Tablets, and Salts; Bath Soaps and Detergents; Night Skin Care Preparations

Technical/Other Names:
3,7-Dimethyl-9-(2,6,6-Trimethyl-1-Cyclohexen-1-yl)-2,4,6,8-Nonatetraen-1-ol
Dry Formed Vitamin A
Vitamin A

Trade Name Mixtures:
AC Retinol Liposome (Active Concepts)
Color Retinol Thalasphere (Coletica SA)
Complex 5 - Vitaminic (Provital/Centerchem)
Retinol 50 C (BASF)
Retinol Cylasphere (Coletica SA)
Retinol 15 D (BASF)
Retinol Microcapsule (Coletica SA)
RetiSTAR (BASF Corporation)
Rovisome Retinol Moist (Rovi)

SUPER A (C.I.T.)
VitAine (AGI Dermatics)
Vitazyme ABC (Arch Personal Care Products)

RETINOXYTRIMETHYLSILANE

CAS No.: 16729-19-4

Empirical Formula:
$C_{23}H_{38}OSi$

Definition: Retinoxytrimethylsilane is the organic compound that conforms to the formula:

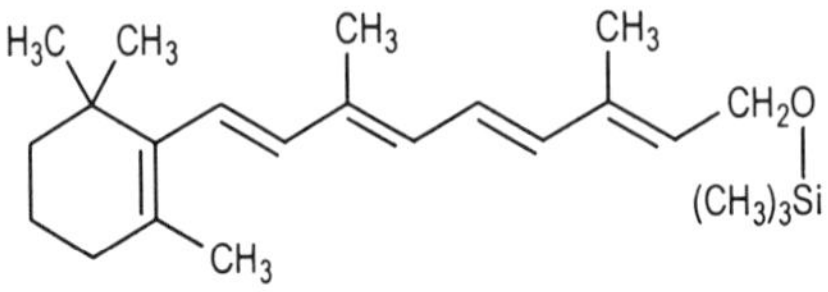

Chemical Class: Siloxanes and Silanes

Function: Skin-Conditioning Agent - Miscellaneous

Technical/Other Name:
Retinol, trans-7, trans-9, cis-11, cis-13, trimethylsilyl-

Trade Name Mixtures:
SilCare Silicone 1M75 Retinoxytrimethylsilane in Soybean Oil (Clariant)
SilCare Silicone 1M75 Retinoxytrimethylsilane in Soybean Oil (Clariant GmbH, Personal Care)

RETINYL ACETATE

CAS No.	EINECS No.
127-47-9	204-844-2

JPN Translation:
酢酸レチノール

Empirical Formula:
$C_{22}H_{32}O_2$

Definition: Retinyl Acetate is the ester of Retinol (q.v.) and acetic acid.

Information Sources: 21CFR184.1930, JAN, JCLS, JSCI

Chemical Class: Esters

Function: Skin-Conditioning Agent - Miscellaneous

Reported Product Categories: Bath Capsules; Face and Neck Preparations (Excluding Shaving Preparations)

Technical/Other Names:
Acetic Acid, Retinyl Ester
Retinol Acetate
Vitamin A Acetate

Trade Name Mixtures:
AEC Retinyl Acetate (&) Peanut Oil (A & E Connock)
Dry Vitamin A Acetate + D3 500/50 (Roche)
Nanocos A-73 (Induchem)
Vitamin A Acetate 500/BASF (BASF Corporation)

RETINYL LINOLEATE

CAS No.: 631-89-0

Empirical Formula:
$C_{38}H_{60}O_2$

Definition: Retinyl Linoleate is the ester of Retinol (q.v.) and Linoleic Acid (q.v.).

Chemical Class: Esters

Function: Skin-Conditioning Agent - Miscellaneous

Technical/Other Names:
Linoleic Acid, Ester with Retinol
Retinol, 9,12-Octadecadienoate, (Z,Z)-

Trade Name:
Retinyl-18 (National Starch)

RETINYL PALMITATE

CAS No.	EINECS No.
79-81-2	201-228-5

JPN Translation:
パルミチン酸レチノール

Empirical Formula:
$C_{36}H_{60}O_2$

Definition: Retinyl Palmitate is the ester of Retinol (q.v.) and palmitic acid.

Information Sources: 21CFR184.1930, CIR: [S] JACT-6(3)1987, CZE, JAN, JCIC, JCLS, MI-13(10073), POL, TSCA

Chemical Class: Esters

Function: Skin-Conditioning Agent - Miscellaneous

Reported Product Categories: Moisturizing Preparations; Bath Preparations, Misc.; Body and Hand Preparations (Excluding Shaving Preparations); Skin Care Preparations, Misc.; Foundations; Lipsticks; Bath Capsules; Face and Neck Preparations (Excluding Shaving Preparations); Night Skin Care Preparations; Hair Conditioners; Bath Oils, Tablets, and Salts; Paste Masks (Mud Packs); Shampoos (Non-coloring); Cleansing Products (Cold Creams, Cleansing Lotions, Liquids and Pads); Face Powders; Makeup Preparations (Not eye), Misc.; Bath Soaps and Detergents; Eye Makeup Preparations, Misc.; Makeup

Bases; Tonics, Dressings, and Other Hair Grooming Aids; Blushers (All types); Eye Lotions; Eye Shadows; Suntan Preparations, Misc.; Powders (Dusting and Talcum, Excluding Aftershave Talcs); Hair Preparations (Non-coloring), Misc.; Eyebrow Pencils; Mascara; Suntan Gels, Creams, and Liquids; Aftershave Lotions; Baby Shampoos; Fragrance Preparations, Misc.; Hair Sprays (Aerosol Fixatives); Nail Creams and Lotions; Baby Lotions, Oils, Powders and Creams; Cuticle Softeners; Manicuring Preparations, Misc.; Nail Polish and Enamels; Skin Fresheners

Technical/Other Names:
  Axerophthol Palmitate
  Retinol, Hexadecanoate
  Retinol Palmitate
  Vitamin A Palmitate

Trade Names:
  AEC Retinyl Palmitate (A & E Connock)
  TINODERM A (Ciba Specialty Chemicals)
  Vitamin A Palmitate 1.7 Mio I.U./g/BASF (BASF)
  Vitamin A Palmitate Oily Concentrate (Merck KGaA)
  Vitamin A Palmitate Type P 1.7/E (Roche)
  Vitamin A Palmitate, Type P1.7 (Roche)
  Vitamine A Palmitate (Roche)

Trade Name Mixtures:
  ACB Bio-Chelate VA (Active Concepts)
  AC Vitamin ABCDE Liposome (Active Concepts)
  AC Vitamin ACE-BC Liposome (Active Concepts)
  AC Vitamin ACE Liposome (Active Concepts)
  AC Vitamin AE Liposome (Active Concepts)
  AC Vitamin A Liposome (Active Concepts)
  AEC Retinyl Palmitate & Peanut Oil (A & E Connock)
  Brookosome ACE (Arch Personal Care Products)
  Brookosome ACEBC (Arch Personal Care Products)
  Brookosome ACEBC Concentrate (Arch Personal Care Products)
  Brookosome ACEBC Plus (Arch Personal Care Products)
  Brookosome ELL (Arch Personal Care Products)
  Brookosome RP (Arch Personal Care Products)
  Ceraspheres-G 9507 (Lipoid)
  Chronosphere RP (Arch Personal Care Products)
  COSMETIC ESSENTIAL VITAMINS MICROEMULSION (Ennagram)
  Crodasome A/E (Croda, Inc.)
  Cutavit Forte (CLR)
  Cutavit Richter (CLR)

Deperoxidium Marin (Coletica SA)
Deperoxidium Vegetal (Coletica SA)
Derma-Vitamincomplex, Oil Soluble (Crodarom)
Dry Vitamin A Palmitate 500 (Roche)
Dry Vitamin A Palmitate, Type 250-SD (Roche)
Elespher Vitaplex Lipo (Laboratoires Sero-biologiques)
Facteur Arl (LCW)
Germinol (Dr. Gerhard Steidl)
Glycosan VIT A, E, F-12 (Chemyunion)
Greenosome Ace (Greentech)
Hot Dots - Blue (Charm Girl)
Hot Dots - Green (Charm Girl)
Hot Dots - Red (Charm Girl)
Integrahair Sphere (Chemyunion)
Lipobead Blue-AE (Lipo)
Lipobead Blue-AEL (Lipo)
Lipobead Pink-AEC (Lipo)
Lipodermol Veg (Laboratoires Sero-biologiques)
Lipofacteur Vitentiel (LCW)
Microderm AE (Zehentmayer)
Microemulsion A (Ennagram)
NIKKOL Aquasome AE (Nikko)
Nikkol Aquasome AE Conc (Nikko)
Retimine III (Greentech)
Rovisome ACE (Rovi)
Soluvit Richter (CLR)
Tagravit A1 (Tagra)
Tagravit A2 (Tagra)
Ultraspheres-5012 (Lipoid)
Ultraspheres-8009 (Lipoid)
Ultraspheres 8028 (Lipoid)
Vitamin A and D3 Blend (Roche)
VITAMIN A MICROEMULSION (Ennagram)
Vitamin A Palmitate Lipomicron (Sederma)
Vitamin A Palmitate 1.0 Mio I.U./g/BASF (BASF)
Vitamin A Palmitate 1.5 Mio I.U./g/BASF AG (BASF)
Vitamin A Palmitate, with Vitamin D3 Blend 10:1 (Roche)
Vitamin A Palmitate w/Vitamin D3 (BASF)
Vitamin A Plamitate 1.0 Mio I.U./g/BASF AG (BASF)
Vitamin Concentrate "O" (Cosmetochem) (Cosmetochem International Ltd.)
Vitamin Extract AEFH' Water Soluble (Crodarom)
Vitamin Extract AEF, Oil Soluble (Crodarom)
Vitaminextract VC, Oil Soluble (Crodarom)
Vitaminextract VC, Water Soluble (Crodarom)
Vitaphyle ACE (LCW)
Vitasol, Vitamin-Horsechestnut-Complex (Crodarom)
Vitazyme A-Plus (Arch Personal Care Products)

RETINYL PROPIONATE

CAS No.	EINECS No.
7069-42-3	230-363-2

Empirical Formula:
  $C_{23}H_{34}O_2$

Definition: Retinyl Propionate is the ester of Retinol (q.v.) and propionic acid.

Chemical Class: Esters

Function: Skin-Conditioning Agent - Miscellaneous

Technical/Other Names:
  Propionic Acid, Retinol Ester
  Retinol, Propanoate
  Vitamin A Propionate

RHAMNOSE

CAS Nos.	EINECS No.
3615-41-6 (L-Form)	222-793-4
10030-85-0	

Empirical Formula:
  $C_6H_{12}O_5$

Definition: Rhamnose is the organic compound that conforms to the formula:

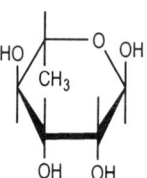

Information Sources: MI-13(8256), RIFM

Chemical Class: Carbohydrates

Functions: Flavoring Agent; Fragrance Ingredient

Technical/Other Names:
  Isodulcitol
  Locaose
  Mannomethylose
  L-Mannose, 6-Deoxy-, Monohydrate
  L-Rhamnose (RIFM)
  L-Rhamnose Monohydrate

Trade Name:
  L-Rhamnose (PVP Sociedade Anonima)

RHAMNUS PURSHIANA BARK EXTRACT

CAS No.	EINECS No.
84650-55-5	283-515-5

Definition: Rhamnus Purshiana Bark Extract is an extract of the dried bark of the cascara, *Rhamnus purshiana*. See *"Regulatory and Ingredient Use Information,"* regarding the labeling names for botanical ingredients in Volume 1, Introduction, Part A.

Information Sources: 21CFR172.510, OTC-I-LX

Chemical Class: Biological Products

Function: Not Reported

Technical/Other Names:
Cascara Extract
Cascara (Rhamnus Purshiana) Extract
Extract of Cascara
Extract of Cascara Bark
Extract of Rhamnus Purshiana

Trade Name Mixtures:
Actiphyte of Cascara Sagrada (Active Organics)
Phytoderm UV Complex Glycolic (Universal Flavors)

RHEUM PALMATUM EXTRACT

CAS Nos.	EINECS No.
8016-55-5	
90106-27-7	290-249-3

Definition: Rheum Palmatum Extract is an extract of the stalks or roots of rhubarb, *Rheum palmatum. See "Regulatory and Ingredient Use Information," regarding the labeling names for botanical ingredients in Volume 1, Introduction, Part A.*

Information Source: 21CFR172.510

Chemical Class: Biological Products

Function: Not Reported

Technical/Other Names:
Extract of Rheum Palmatum
Extract of Rhubarb
Rhubarb Extract
Rhubarb (Rheum Palmatum) Extract

Trade Name Mixtures:
Actiphyte of Rhubarb BG50 (Active Organics)
Actiphyte of Rhubarb GL50 (Active Organics)
Actiphyte of Rhubarb Lipo S (Active Organics)
Actiphyte of Rhubarb PG50 (Active Organics)
Actiphyte of Rhubarb Stalks BG50 (Active Organics)
Actiphyte of Rhubarb Stalks GL50 (Active Organics)
Actiphyte of Rhubarb Stalks Lipo S (Active Organics)
Actiphyte of Rhubarb Stalks PG50 (Active Organics)
Rhubarb Extract BG (Maruzen Pharmaceuticals Co., Ltd.)
Rhubarb Root Extract HS 2804 G (Grau)

RHEUM PALMATUM ROOT

Definition: Rheum Palmatum Root is the dried, crushed roots of *Rheum palmatum.*

See "Regulatory and Ingredient Use Information," regarding the labeling names for botanical ingredients in Volume 1, Introduction, Part A.

Information Source: MI-13(8280)

Chemical Class: Biological Products

Function: Not Reported

Technical/Other Names:
Rhubarb
Rhubarb (Rheum Palmatum)

Trade Name:
Logoplant Rhabarber Pulver (Logona)

RHEUM PALMATUM ROOT EXTRACT

Definition: Rheum Palmatum Root Extract is an extract of the root of the rhubarb, *Rheum palmatum. See "Regulatory and Ingredient Use Information," regarding the labeling names for botanical ingredients in Volume 1, Introduction, Part A.*

Chemical Class: Biological Products

Function: Not Reported

Technical/Other Names:
Extract of Rheum Palmatum Root
Extract of Rhubarb
Rhubarb (Rheum Palmatum) Root Extract
Rhubarb Root Extract

Trade Name Mixtures:
Actiphyte Rhubarb Root (Active Organics)
Herbasol Extract Rhubarb (Cosmetochem) (Cosmetochem International Ltd.)
Rhubarb Extract (Maruzen Pharmaceuticals Co., Ltd.)

RHEUM UNDULATUM EXTRACT

CAS No.: 8016-55-5

Definition: Rheum Undulatum Extract is an extract of the stalks or roots of the rhubarb, *Rheum undulatum. See "Regulatory and Ingredient Use Information," regarding the labeling names for botanical ingredients in Volume 1, Introduction, Part A.*

Information Source: 21CFR172.510

Chemical Class: Biological Products

Function: Not Reported

Technical/Other Names:
Extract of Rheum Undulatum
Extract of Rhubarb
Rhubarb Extract
Rhubarb (Rheum Undulatum) Extract

RHINACANTHUS COMMUNIS EXTRACT

Definition: Rhinacanthus Communis Extract is an extract of the plant, *Rhinacanthus communis. See "Regulatory and Ingredient Use Information," regarding the labeling names for botanical ingredients in Volume 1, Introduction, Part A.*

Chemical Class: Biological Products

Function: Not Reported

Technical/Other Name:
Extract of Rhinacanthus Communis

Trade Name Mixtures:
Rhinacanthus Nastus Extract (Maruzen Pharmaceuticals Co., Ltd.)
Rhinacanthus Nasutus Extract (Maruzen Pharmaceuticals Co., Ltd.)

RHIZOBIAN GUM

Definition: Rhizobian Gum is the polysaccharide gum produced by the fermentation by *Rhizobium bacterium.*

Chemical Classes: Biological Polymers and their Derivatives; Gums, Hydrophilic Colloids and Derivatives

Functions: Film Former; Hair Fixative; Plasticizer; Suspending Agent - Nonsurfactant; Viscosity Increasing Agent - Aqueous

Trade Name:
Soligel (Soliance)

RHODIOLA CRENULATA ROOT EXTRACT

Definition: Rhodiola Crenulata Root Extract is an extract of the roots of *Rhodoila crenulata. See "Regulatory and Ingredient Use Information," regarding the labeling names for botanical ingredients in Volume 1, Introduction, Part A.*

Chemical Class: Biological Products

Function: Skin-Conditioning Agent - Emollient

Trade Name Mixture:
Rhodiola Crenulata Extract (Coletica SA)

RHODIOLA SACRA ROOT EXTRACT

Definition: Rhodiola Sacra Root Extract is an extract of the roots of *Rhodiola sacra. See "Regulatory and Ingredient Use Information," regarding the labeling names for botanical ingredients in Volume 1, Introduction, Part A.*

Chemical Class: Biological Products

Function: Skin-Conditioning Agent - Emollient

Technical/Other Name:
Extract of Rhodiola Sacra Root

Trade Name Mixtures:
Rhodiola Sacra Extract BG30 (Maruzen Pharmaceuticals Co., Ltd.)
Rhodiola Sacra Extract W-BG30 (Maruzen Pharmaceuticals Co., Ltd.)

RHODOCHROSITE

CAS Nos.: 598-62-9; 14476-12-1

Definition: Rhodochrosite is the mineral that consists chiefly of manganese carbonate.

Information Source: MI-13(5749)

Chemical Class: Inorganics

Function: Skin-Conditioning Agent - Miscellaneous

Trade Name:
Rhodochrosite (Aurora Mineral)

RHODOCHROSITE EXTRACT

CAS Nos.: 598-62-9; 14476-12-1

Definition: Rhodochrosite Extract is an extract of Rhodochrosite (q.v.). *See Reported Ingredient Functions-The Cosmetic Drug Distinction, in Regulatory and Ingredient Use Information, Volume I, Part A.*

Chemical Class: Inorganics

Functions: Antioxidant; Skin-Conditioning Agent - Miscellaneous; Skin Protectant

Technical/Other Name:
Extract of Rhodochrosite

Trade Name Mixture:
Rhodo'Lite (Libiol)

RHODODENDRON CHRYSANTHUM

Definition: *See "Regulatory and Ingredient Use Information," regarding EU labeling names for botanical ingredients in Volume 1, Introduction, Part A.*

Chemical Class: Biological Products

Technical/Other Name:
Rhododendron Chrysanthum Leaf Extract (U.S.)

RHODODENDRON CHRYSANTHUM LEAF EXTRACT

CAS No.	EINECS No.
90106-20-0	290-242-5

Definition: Rhododendron Chrysanthum Leaf Extract is an extract of the leaves of the rhododendron, *Rhododendron chrysanthum*. *See "Regulatory and Ingredient Use Information," regarding the labeling names for botanical ingredients in Volume 1, Introduction, Part A.*

Information Source: RIFM

Chemical Class: Biological Products

Functions: Fragrance Ingredient; Not Reported

Technical/Other Names:
Extract of Rhododendron
Extract of Rhododendron Chrysanthum
Rhododendron Chrysanthum (EU)
Rhododendron chrysanthum, ext. (RIFM)
Rhododendron Extract

Trade Name Mixtures:
Actiphyte of Rhododendron BG50 (Active Organics)
Actiphyte of Rhododendron GL50 (Active Organics)
Actiphyte of Rhododendron Lipo S (Active Organics)
Actiphyte of Rhododendron PG50 (Active Organics)
Hydroplastidine Rhododendron (Vevy)

RHODODENDRON FERRUGINEUM

Definition: *See "Regulatory and Ingredient Use Information," regarding EU labeling names for botanical ingredients in Volume 1, Introduction, Part A.*

Chemical Class: Biological Products

Technical/Other Name:
Rhododendron Ferrugineum Extract (U.S.)

RHODODENDRON FERRUGINEUM EXTRACT

CAS No.	EINECS No.
90106-21-1	290-243-0

Definition: Rhododendron Ferrugineum Extract is an extract of the rhododendron, *Rhododendron ferrugineum*. *See "Regulatory and Ingredient Use Information," regarding the labeling names for botanical ingredients in Volume 1, Introduction, Part A.*

Information Source: RIFM

Chemical Class: Biological Products

Functions: Fragrance Ingredient; Not Reported

Technical/Other Names:
Rhododendron Ferrugineum (EU)
Rhododendron ferrugineum, ext. (RIFM)

RHUS GLABRA

Definition: *See "Regulatory and Ingredient Use Information," regarding EU labeling names for botanical ingredients in Volume 1, Introduction, Part A.*

Chemical Class: Biological Products

Technical/Other Name:
Rhus Glabra (Sumac) Extract (U.S.)

RHUS GLABRA (SUMAC) EXTRACT

CAS No.	EINECS No.
90106-33-5	290-256-1

Definition: Rhus Glabra (Sumac) Extract is an extract of the bark, roots and berries of the sumac, *Rhus glabra*. *See "Regulatory and Ingredient Use Information," regarding the labeling names for botanical ingredients in Volume 1, Introduction, Part A.*

Chemical Class: Biological Products

Function: Not Reported

Technical/Other Names:
Extract of Rhus Glabra
Extract of Sumac
Rhus Glabra (EU)
Rhus Glabra Extract
Sumac Extract

Trade Name Mixtures:
Actiphyte of Sumach BG50 (Active Organics)
Actiphyte of Sumach GL50 (Active Organics)
Actiphyte of Sumach Lipo S (Active Organics)
Actiphyte of Sumach PG50 (Active Organics)
VT-082 Extract of Sumac (Vege-Tech)

RHUS SEMIALATA EXTRACT

Definition: Rhus Semialata Extract is an extract of the plant, *Rhus semialata*. *See "Regulatory and Ingredient Use Information," regarding the labeling names for botanical ingredients in Volume 1, Introduction, Part A.*

Chemical Class: Biological Products

Function: Skin-Conditioning Agent - Miscellaneous

Technical/Other Name:
Extract of Rhus Semialata

Trade Name Mixture:
Astrin-Ag (Bioland)

RHUS SUCCEDANEA FRUIT WAX

CAS No.: 8001-39-6

JPN Translation:
モクロウ

Definition: Rhus Succedanea Fruit Wax is a wax obtained from the mesocarp of the fruit of *Rhus succedanea. See "Regulatory and Ingredient Use Information," regarding the labeling names for botanical ingredients in Volume 1, Introduction, Part A.*

Information Sources: 21CFR73.1, 21CFR175.105, 21CFR175.350, 21CFR176.170, 21CFR182.70, 21CFR186.1555, CIR: [S] JACT-3(3)1984, CTFA D, JCLS, JSCI, MAR, MI-13(5275), TSCA

Chemical Class: Biological Products

Functions: Binder; Viscosity Increasing Agent - Nonaqueous

Reported Product Categories: Eyeliners; Lipsticks; Eyebrow Pencils; Eye Shadows; Makeup Preparations (Not eye), Misc.; Rouges; Eye Makeup Preparations, Misc.; Makeup Fixatives; Face and Neck Preparations (Excluding Shaving Preparations)

Technical/Other Names:
Japan (Rhus Succedanea) Wax
Japan Wax
Waxes, Japan
Waxes, Rhus Succedanea

Trade Names:
AEC Japan Wax (A & E Connock)
Hansonwax JH-7 (Hansotech)
Japan Wax (Megachem)
Japan Wax (Ross)
Japan Wax Genuine (Ross)
Japan Wax NJ-9002 (Ikeda)
Japan Wax - STRALPITZ (Strahl & Pitsch)

Trade Name Mixture:
Japan Wax Y (Yokozeki)

RHUS VERNICIFLUA PEEL WAX

Definition: Rhus Verniciflua Peel Wax is a wax obtained from the peel of the fruit of *Rhus verniciflua. See "Regulatory and Ingredient Use Information," regarding the labeling names for botanical ingredients in Volume 1, Introduction, Part A.*

Chemical Class: Biological Products

Function: Not Reported

Trade Names:
Berry Wax 6290 (KAHL & CO) (Kahl)
Botanol OT (Botanigenics)
NC-1220 (Nikko)

Trade Name Mixtures:
AEC Kukui Nut Wax (A & E Connock)
AEC Macadamia Nut Wax (A & E Connock)
AEC Mac-Kui Nut Wax (A & E Connock)
Japan Wax Y (Yokozeki)

RIBES GROSSULARIA FRUIT EXTRACT

Definition: Ribes Grossularia Fruit Extract is an extract of the fruit of *Ribes grossularia. See "Regulatory and Ingredient Use Information," regarding the labeling names for botanical ingredients in Volume 1, Introduction, Part A.*

Chemical Class: Biological Products

Function: Proprietary

Technical/Other Name:
Extract of Ribes Grossularia

Trade Name:
Gooseberry Concentrate HS 3611 G (Grau)

Trade Name Mixture:
Fruitapone Gooseberry GT 2/037710 (Symrise)

RIBES GROSSULARIA FRUIT JUICE

Definition: Ribes Grossularia Fruit Juice is the juice expressed from the fruit of *Ribes grossularia. See "Regulatory and Ingredient Use Information," regarding the labeling names for botanical ingredients in Volume 1, Introduction, Part A.*

Chemical Class: Biological Products

Function: Skin-Conditioning Agent - Miscellaneous

Trade Name:
AEC Gooseberry Conc 68 Brix Pasteurized, Unpreserved (A & E Connock)

Trade Name Mixture:
Fruitapone Gooseberry B 2/036710 (Symrise)

RIBES NIGRUM (BLACK CURRANT) BUD WAX

Definition: Ribes Nigrum (Black Currant) Bud Wax is a wax obtained from the bud of *Ribes nigrum. See "Regulatory and Ingredient Use Information," regarding the labeling names for botanical ingredients in Volume 1, Introduction, Part A.*

Chemical Class: Waxes

Function: Not Reported

Technical/Other Names:
Black Currant Wax
Ribes Nigrum (Black Currant) Wax
Ribes Nigrum Wax
Wax, Black Currant (Ribes Nigrum)

Trade Names:
AEC Cire Essentielle De Bourgeons De Cassis (A & E Connock)

Cire Essentielle de bourgeons de Cassis (Bertin)

RIBES NIGRUM (BLACK CURRANT) FRUIT

Definition: Ribes Nigrum (Black Currant) Fruit is the fruit of the currant, *Ribes nigrum. See "Regulatory and Ingredient Use Information," regarding the labeling names for botanical ingredients in Volume 1, Introduction, Part A.*

Chemical Class: Biological Products

Functions: Cosmetic Astringent; Skin-Conditioning Agent - Miscellaneous

Technical/Other Names:
Black Currant
Black Currant Fruit

Trade Name:
AEC Blackcurrant Puree (A & E Connock)

RIBES NIGRUM (BLACK CURRANT) FRUIT EXTRACT

Definition: Ribes Nigrum (Black Currant) Fruit Extract is an extract of the fruit of the black currant, *Ribes nigrum. See "Regulatory and Ingredient Use Information," regarding the labeling names for botanical ingredients in Volume 1, Introduction, Part A.*

Information Sources: 21CFR172.510, 21CFR182.20

Chemical Class: Biological Products

Functions: Skin-Conditioning Agent - Miscellaneous; Skin-Conditioning Agent - Occlusive

Reported Product Category: Moisturizing Preparations

Technical/Other Names:
Black Currant Extract
Black Currant Fruit Extract
Extract of Black Current
Extract of Ribes Nigrum

Trade Name:
Frabes Oil (Vevy)

Trade Name Mixtures:
Actiphyte Black Currant (Active Organics)
Actiphyte Black Currant AQ (Active Organics)
Actiphyte Black Currant BG50P (Active Organics)
Actiphyte Black Currant Lipo S (Active Organics)
BC Cassis Extract (Ichimaru Pharcos)

Black Currant Extract HS 2955 G (Grau)
Black Currant Extract W-BG (Maruzen
Pharmaceuticals Co., Ltd.)
Black Currant Fruit HPG Titrated (Alban
Muller)
Blackcurrant HS (Alban Muller)
Blackcurrant LS (Alban Muller)
Extrait huileux de Cassis (Greentech)
Glycolysat of Black Currant (CEP (Solabia))
Herbasol Extract Black Currant Fruit
(Cosmetochem) (Cosmetochem
International Ltd.)
Phytelene of Black-Currant EG 513 liquid
(Indena SA)
Phytogreen 55 of Black-Currant EXH 713
Liquid (Phytochim)
Planell Oil - Black Currant (Arch Personal
Care Products)
Radicaptol (Solabia)

RIBES NIGRUM (BLACK CURRANT) JUICE

Definition: Ribes Nigrum (Black Currant) Juice is the liquid expressed from the black currant, *Ribes nigrum. See "Regulatory and Ingredient Use Information," regarding the labeling names for botanical ingredients in Volume 1, Introduction, Part A.*

Chemical Class: Biological Products

Function: Skin-Conditioning Agent - Miscellaneous

Technical/Other Names:
Black Currant Juice
Juice, Black Currant
Juice, Ribes Nigrum
Ribes Nigrum Juice

Trade Names:
AEC Blackcurrant Conc. (A & E Connock)
Black Currant Concentrate (Symrise)

Trade Name Mixture:
Fruitapone Black Currant GT 2/037100
(Symrise)

RIBES NIGRUM (BLACK CURRANT) LEAF EXTRACT

Definition: Ribes Nigrum (Black Currant) Leaf Extract is an extract of the leaf of *Ribes nigrum. See "Regulatory and Ingredient Use Information," regarding the labeling names for botanical ingredients in Volume 1, Introduction, Part A.*

Chemical Class: Biological Products

Function: Skin-Conditioning Agent - Miscellaneous

Technical/Other Names:
Black Currant Leaf Extract
Ribes Nigrum Leaf Extract

Trade Name Mixtures:
Black Currant Leaves Extract HS 2475 G
(Grau)
Blackcurrant Leaf HPG Titrated (Alban
Muller)
Blackcurrant Leaf HS (Alban Muller)
EPICA (Greentech S.A)
Glycolysat of Black Currant Leaves (CEP
(Solabia))
Herbasol Extract Black Currant Leaf
(Cosmetochem) (Cosmetochem
International Ltd.)
Natupure Black Currant (E.U.K)
Ribes Nigrum (Black Curant) Fruit Extract
(IES LABO)

RIBES NIGRUM (BLACK CURRANT) SEED EXTRACT

Definition: Ribes Nigrum (Black Currant) Seed Extract is an extract of the seeds of *Ribes nigrum. See "Regulatory and Ingredient Use Information," regarding the labeling names for botanical ingredients in Volume 1, Introduction, Part A.*

Chemical Class: Biological Products

Functions: Skin-Conditioning Agent - Emollient; Skin-Conditioning Agent - Humectant

Reported Product Category: Moisturizing Preparations

Technical/Other Names:
Black Currant Extract
Extract of Black Currant Seed
Ribes Nigrum Seed Extract

RIBES NIGRUM (BLACK CURRANT) SEED OIL

CAS No.: 97676-19-2

Definition: Ribes Nigrum (Black Currant) Seed Oil is the fixed oil obtained from the seeds of *Ribes nigrum. See "Regulatory and Ingredient Use Information," regarding the labeling names for botanical ingredients in Volume 1, Introduction, Part A.*

Chemical Class: Fats and Oils

Function: Skin-Conditioning Agent - Emollient

Technical/Other Names:
Black Currant Oil
Black Currant Seed Oil
Oils, Black Currant
Ribes Nigrum Oil

Trade Names:
AEC Blackcurrant Seed Oil (A & E
Connock)

Black Currant Oil (Desert Whale)
Blackcurrant Seed Oil (Nestle World Trade)
Efaduo Blackcurrant Seed Oil (CO2
extract) (Aromtech)
Huile de Pepins de Cassis (Bertin)

Trade Name Mixture:
Melange Huile de Noyau D'Abricot / Huile
de Pepins de Cassis (Bertin)

RIBES RUBRUM (CURRANT) FRUIT

Definition: Ribes Rubrum (Currant) Fruit is the fruit of the currant, *Ribes rubrum. See "Regulatory and Ingredient Use Information," regarding the labeling names for botanical ingredients in Volume 1, Introduction, Part A.*

Chemical Class: Biological Products

Function: Cosmetic Astringent

Technical/Other Names:
Currant
Currant Fruit

Trade Name:
AEC Redcurrant Puree (A & E Connock)

RIBES RUBRUM (CURRANT) FRUIT EXTRACT

Definition: Ribes Rubrum (Currant) Fruit Extract is an extract of the berries of the red currant, *Ribes rubrum. See "Regulatory and Ingredient Use Information," regarding the labeling names for botanical ingredients in Volume 1, Introduction, Part A.*

Chemical Class: Biological Products

Function: Proprietary

Technical/Other Names:
Currant Extract
Currant Fruit Extract
Extract of Ribes Rubrum
Ribes Rubrum Extract
Ribes Sylvestre Extract

Trade Name Mixtures:
Currant Berries Extract HS 3298 G (Grau)
Glycolysat of Red Currant (CEP (Solabia))
Red Currant HS (Alban Muller)

RIBES RUBRUM (CURRANT) JUICE

Definition: Ribes Rubrum (Currant) Juice is the juice expressed from the berries of *Ribes rubrum. See "Regulatory and Ingredient Use Information," regarding the labeling names for botanical ingredients in Volume 1, Introduction, Part A.*

Chemical Class: Biological Products

Function: Skin-Conditioning Agent - Miscellaneous

Trade Name:
Authenticals of Currant (CEP (Solabia))

RIBOFLAVIN

CAS No.
83-88-5

EINECS No.
201-507-1

JPN Translation:
リボフラビン

Empirical Formula:
$C_{17}H_{20}N_4O_6$

Definition: Riboflavin is the organic compound that conforms to the formula:

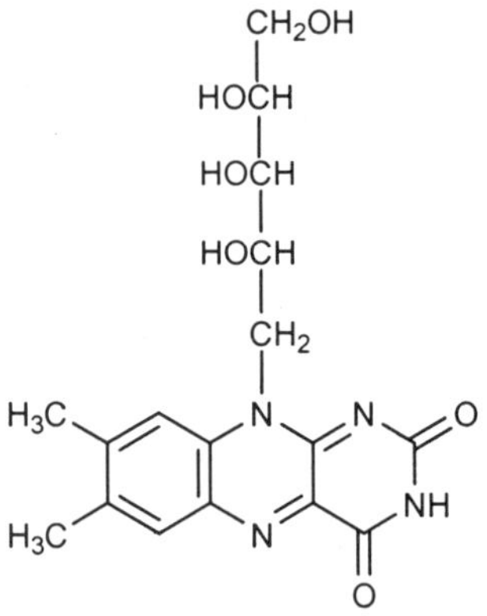

See "Regulatory and Ingredient Use Information," for Colorants in Volume 1, Introduction, Part A.

Information Sources: ARG, AUS, BAN, BP, BPC, BRA, 21CFR73.450, 21CFR101.9, 21CFR104.20, 21CFR104.47, 21CFR107.10, 21CFR107.100, 21CFR136.115, 21CFR137, 21CFR137.165, 21CFR137.185, 21CFR139, 21CFR184.1695, 21CFR184.1697, 21CFR310.545, 21CFR582.5695, 21CFR582.5697, CZE, DA, DDR, EGY, EP, FCC, FIN, HUN, IND, INN, ITA, JAN, JCLS, JP, MAR, MEX, MI-13 (8284), PN, POR, ROM, TSCA, USAN, USD, USP XXIV, WHO, YUG

Chemical Classes: Amides; Heterocyclic Compounds; Polyols

Function: Skin-Conditioning Agent - Miscellaneous

Technical/Other Names:
Beflavin
6,7-Dimethyl-9-ribitylisoalloxazine
Food Yellow 15
Lactoflavin
Vitamin B2
Vitamin G

Trade Name:
Riboflavin, USP, FCC (Roche)

Trade Name Mixtures:
Fortified Yeast T-6361 (Universal Foods)
Lipobronze AB (Lipo)
Lipobronze B (Lipo)
Lipobronze Veg-AB (Lipo)
Lipobronze Veg-B (Lipo)
Unipertan P-24 (Induchem)
Unipertan P-2002 (Induchem)
Unipertan VEG-24 (Induchem)
Unipertan VEG-2002 (Induchem)

RIBOFLAVIN TETRAACETATE

CAS No.
752-13-6

EINECS No.
212-032-4

Empirical Formula:
$C_{25}H_{28}N_4O_{10}$

Definition: Riboflavin Tetraacetate is the tetraester of acetic acid and Riboflavin (q.v.) that conforms to the formula:

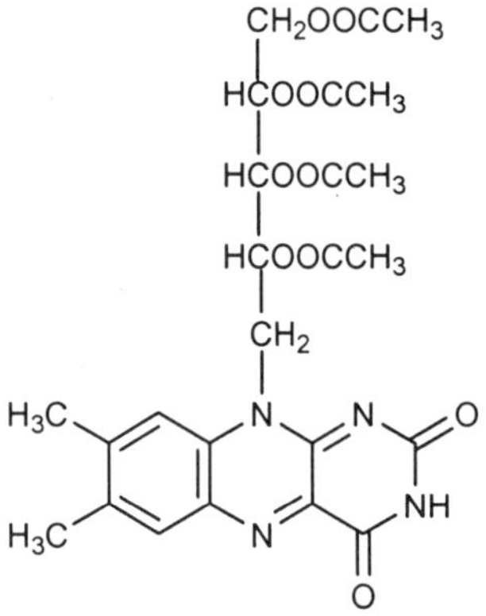

Chemical Classes: Amides; Esters; Heterocyclic Compounds

Function: Skin-Conditioning Agent - Miscellaneous

Technical/Other Names:
Acetic Acid, Riboflavin Tetraester
Riboflavine, 2',3',4',5'-Tetraacetate
Vitamin B2 Tetraacetate

RIBONIC ACID

CAS No.: 17812-24-7

Empirical Formula:
$C_5H_{10}O_6$

Definition: Ribonic Acid is the organic compound that conforms to the formula:

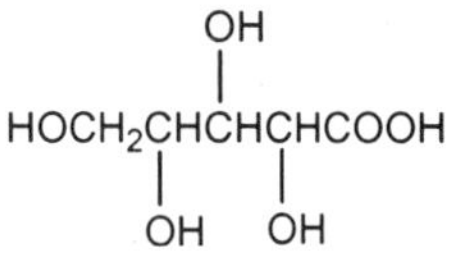

Chemical Classes: Carboxylic Acids; Polyols

Functions: Chelating Agent; pH Adjuster

Technical/Other Name:
2,3,4,5-Tetrahydroxypentanoic Acid

RIBONOLACTONE

CAS No.
5336-08-3

EINECS No.
226-256-5

Empirical Formula:
$C_5H_8O_5$

Definition: Ribonolactone is the organic compound that conforms to the formula:

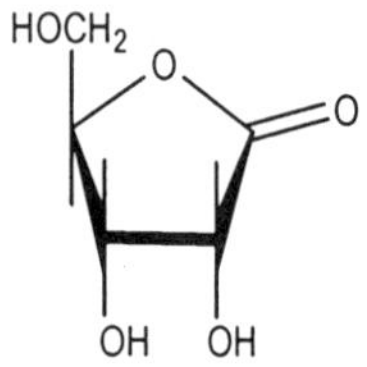

Chemical Classes: Alcohols; Esters

Function: Not Reported

Technical/Other Names:
D-Ribonic Acid, γ-Lactone
D-Ribono-1,4-Lactone

RICE AMINO ACIDS

Definition: Rice Amino Acids is the mixture of amino acids derived from the complete hydrolysis of rice protein.

Chemical Class: Amino Acids

Functions: Hair Conditioning Agent; Skin-Conditioning Agent - Miscellaneous

Trade Names:
Aqua Pro II RAA (MGP)
Hydroryza AA (Croda, Inc.)
Rice-Tein AA (Maybrook)

RICE BRAN ACID

CAS No.
93165-33-4

EINECS No.
296-918-6

Definition: Rice Bran Acid is a mixture of fatty acids derived from Oryza Sativa (Rice) Bran Oil (q.v.).

Information Source: CIR: [S]

Chemical Class: Fatty Acids

Function: Surfactant - Cleansing Agent

Technical/Other Names:
Acids, Rice Bran
Oryza Sativa (Rice) Bran Acid

RICEBRANAMIDE DEA

Definition: Ricebranamide DEA is a mixture of ethanolamides of Rice Bran Acid (q.v.). It conforms generally to the formula:

$$RC \overset{\overset{O}{\|}}{}-N(CH_2CH_2OH)_2$$

where RCO represents the fatty acids derived from rice bran oil.

Information Source: JCIC

Chemical Class: Alkanolamides

Functions: Surfactant - Foam Booster; Viscosity Increasing Agent - Aqueous

RICEBRANAMIDOPROPYL HYDROXY-ETHYL DIMONIUM CHLORIDE

Definition: Ricebranamidopropyl Hydroxy-ethyl Dimonium Chloride is the quaternary ammonium salt that conforms generally to the formula:

$$\left[RC\overset{\overset{O}{\|}}{}-NH(CH_2)_3-\overset{\overset{CH_3}{|}}{\underset{\underset{CH_3}{|}}{N}}-CH_2CH_2OH \right]^+ \quad Cl^-$$

where RCO- represents the fatty acids derived from rice bran oil.

Chemical Class: Quaternary Ammonium Compounds

Functions: Antistatic Agent; Hair Conditioning Agent

Trade Name Mixture:
Ceraphyl RBO (International Specialty Products)

RICE FERMENT FILTRATE (SAKE)

Definition: Rice Ferment Filtrate (Sake) is a filtrate of the fermentation of *Oryza sativa*.

Chemical Class: Biological Products

Function: Skin-Conditioning Agent - Miscellaneous

Trade Name:
Ozeki Sake (Ozeki Sake)

RICINOLEAMIDE DEA

CAS No. 40716-42-5   **EINECS No.** 255-051-3

Empirical Formula:
$C_{22}H_{43}NO_4$

Definition: Ricinoleamide DEA is a mixture of ethanolamides of ricinoleic acid. It conforms generally to the formula:

$$CH_2CH(CH_2)_5CH_3 \quad (OH)$$
$$CH(CH_2)_7\overset{\overset{O}{\|}}{C}-N(CH_2CH_2OH)_2$$

Information Sources: EEC(III/1-60), TSCA

Chemical Class: Alkanolamides

Functions: Surfactant - Foam Booster; Viscosity Increasing Agent - Aqueous

Technical/Other Names:
N,N-Bis(2-Hydroxyethyl)Ricinoleamide
Diethanolamine Ricinoleic Acid Amide
12-Hydroxy-N,N-Bis(2-Hydroxyethyl)-9-Octadecenamide
9-Octadecenamide, 12-Hydroxy-N,N-Bis(2-Hydroxyethyl)-,
Ricinoleoyl Diethanolamide

Trade Names:
Aminol CA-2 (Finetex)
Mackamide R (McIntyre)
Olamida RD (Fabriquimica)

RICINOLEAMIDE MEA

CAS Nos.   **EINECS No.**
106-16-1   203-368-2
75033-33-9

Empirical Formula:
$C_{20}H_{39}NO_3$

Definition: Ricinoleamide MEA is a mixture of ethanolamides of ricinoleic acid. It conforms generally to the formula:

$$CH_2CH(CH_2)_5CH_3 \quad (OH)$$
$$CH(CH_2)_7\overset{\overset{O}{\|}}{C}-NHCH_2CH_2OH$$

Information Source: TSCA

Chemical Class: Alkanolamides

Functions: Surfactant - Foam Booster; Viscosity Increasing Agent - Aqueous

Technical/Other Names:
N-(2-Hydroxyethyl)-12-Hydroxy-9-Octadecenamide
N-(2-Hydroxyethyl)Ricinoleamide
12-Hydroxy-9-Octadecenamide, N-(2-Hydroxyethyl)-
Monoethanolamine Ricinoleic Acid Amide
9-Octadecenamide, 12-Hydroxy-N-(2-Hydroxyethyl)-,

Ricinoleic Ethylolamide
Ricinoleoyl Monoethanolamide

RICINOLEAMIDE MIPA

CAS No.: 40986-29-6

Empirical Formula:
$C_{21}H_{41}NO_3$

Definition: Ricinoleamide MIPA is a mixture of isopropanolamides of Ricinoleic Acid (q.v.). It conforms generally to the formula:

$$CH_2CH(CH_2)_5CH_3 \quad (OH)$$
$$CH(CH_2)_7\overset{\overset{O}{\|}}{C}-NHCH_2\underset{\underset{CH_3}{|}}{C}HOH$$

Chemical Class: Alkanolamides

Functions: Surfactant - Foam Booster; Viscosity Increasing Agent - Aqueous

Technical/Other Names:
12-Hydroxy-N-(2-Hydroxy-1-Methylethyl)-9-Octadecenamide
Monoisopropanolamine Ricinoleic Acid Amide
9-Octadecenamide, 12-Hydroxy-N-(2-Hydroxy-1-Methylethyl)-
Ricinoleoyl Monoisopropanolamide

RICINOLEAMIDOPROPYL BETAINE

CAS No.: 71850-81-2

Empirical Formula:
$C_{25}H_{48}N_2O_4$

Definition: Ricinoleamidopropyl Betaine is the zwitterion (inner salt) that conforms generally to the formula:

$$CH_2CH(CH_2)_5CH_3 \quad (OH)$$
$$CH(CH_2)_7\overset{\overset{O}{\|}}{C}-NH(CH_2)_3-\overset{\overset{CH_3}{|}}{\underset{\underset{CH_3}{|}}{\overset{+}{N}}}-CH_2COO^-$$

Information Sources: JCIC, JCLS

Chemical Class: Betaines

Functions: Antistatic Agent; Hair Conditioning Agent; Skin-Conditioning Agent - Miscellaneous; Surfactant - Cleansing Agent; Surfactant - Foam Booster; Viscosity Increasing Agent - Aqueous

---

Technical/Other Names:
N-(Carboxymethyl)-N,N-Dimethyl-3-[(1-Oxoricinoleyl)Amino]-1-Propanaminium Hydroxide, Inner Salt
1-Propanaminium, N-(Carboxymethyl)-N,N-Dimethyl-3-[(1-Oxoricinoleyl)Amino]-, Hydroxide, Inner Salt
Propyl Betaine Ricinoleate Amide Solution
Ricinoleamidopropyl Dimethyl Glycine

Trade Names:
Mackam RA (McIntyre)
Rewoteric AM R 40 (Degussa Care Specialties)

RICINOLEAMIDOPROPYL DIMETHYLAMINE

CAS No.	EINECS No.
20457-75-4	243-835-8

Empirical Formula:
$C_{23}H_{46}N_2O_2$

Definition: Ricinoleamidopropyl Dimethylamine is the amidoamine that conforms generally to the formula:

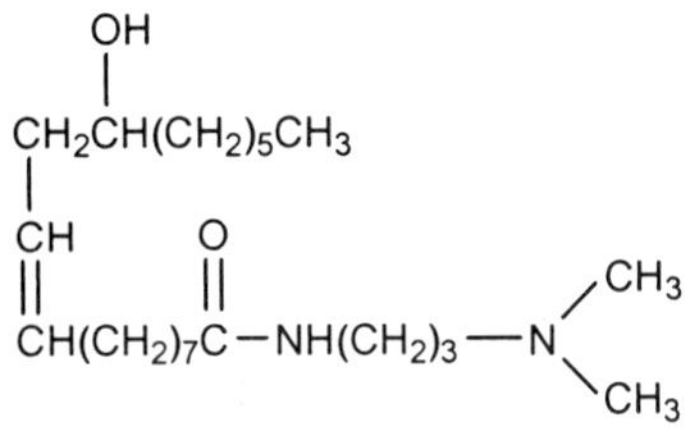

Chemical Class: Amines

Function: Antistatic Agent

Technical/Other Names:
N-[3-(Dimethylamino)Propyl]Ricinoleamide
9-Octadecenamide, N-(3-(Dimethylamino)Propyl)-12-Hydroxy-
Ricinoleamide, N-[3-(Dimethylamino)-Propyl]-

Trade Name:
Mackine 201 (McIntyre)

RICINOLEAMIDOPROPYL DIMETHYLAMINE LACTATE

Empirical Formula:
$C_{23}H_{46}N_2O_2 \cdot C_3H_6O_3$

Definition: Ricinoleamidopropyl Dimethylamine Lactate is the lactic acid salt of Ricinoleamidopropyl Dimethylamine (q.v.).

Chemical Class: Amines

Function: Antistatic Agent

Technical/Other Names:
N-[3-(Dimethylamino)Propyl]Ricinoleamide Lactate

Ricinoleamide, N-[3-(Dimethylamino)-Propyl]-, Lactate

Trade Name:
Mackalene 216 (McIntyre)

RICINOLEAMIDOPROPYL ETHYLDIMONIUM ETHOSULFATE

CAS No.: 112324-16-0

Empirical Formula:
$C_{25}H_{51}N_2O_2 \cdot C_2H_5O_4S$

Definition: Ricinoleamidopropyl Ethyldimonium Ethosulfate is the quaternary ammonium salt that conforms generally to the formula:

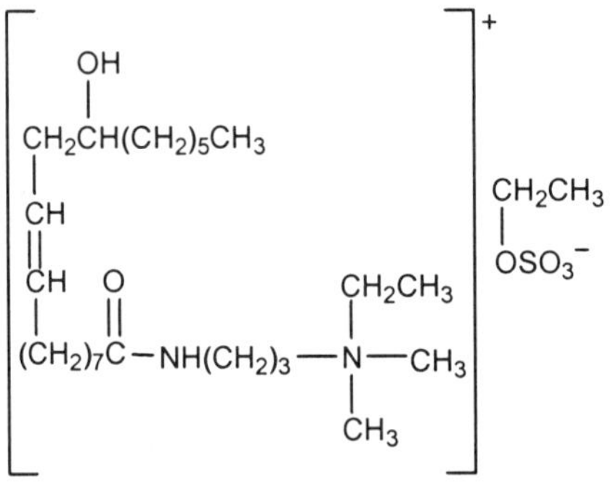

Chemical Class: Quaternary Ammonium Compounds

Function: Antistatic Agent

Reported Product Category: Shampoos (Non-coloring)

Technical/Other Names:
N-Ethyl-N,N-Dimethyl-3-[(1-Oxoricinoleyl)Amino]-1-Propanaminium Ethosulfate
N-Ethyl-3-[(12-Hydroxy-1-Oxo-9-Octadecenyl)Amino-N,N-Dimethyl-1-Propanaminium Ethyl Sulfate (Salt)
[(12-Hydroxy-1-Oxo-9-Octadecenyl)Amino]-N,N-Dimethyl, Ethyl Sulfate (Salt)
1-Propanaminium, N-Ethyl-N,N-Dimethyl-3-[(1-Oxoricinoleyl)Amino]-, Ethosulfate

Trade Names:
Lipoquat R (Lipo)
Mackernium DC-159 (McIntyre)
Surfactol Q4 (CasChem)

RICINOLEAMIDOPROPYLTRIMONIUM CHLORIDE

CAS No.: 127311-98-2

Empirical Formula:
$C_{24}H_{49}N_2O_2 \cdot Cl$

Definition: Ricinoleamidopropyltrimonium Chloride is the quaternary ammonium salt that conforms to the formula:

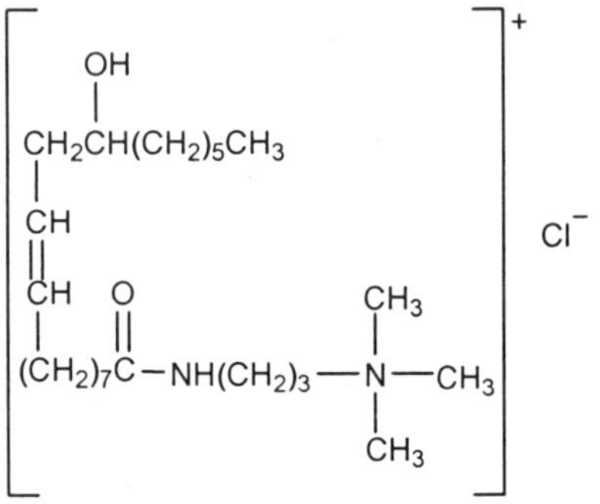

Chemical Class: Quaternary Ammonium Compounds

Functions: Antistatic Agent; Hair Conditioning Agent

Technical/Other Names:
3-[(12-Hydroxy-1-Oxo-9-Octadecenyl)-Amino]-N,N,N-Trimethyl-2-Propanaminium Chloride
1-Propanaminium, 3-[(12-Hydroxy-1-Oxo-9-Octadecenyl)Amino]-N,N,N-Trimethyl-, Chloride
Ricinoleamidopropyl Trimethylammonium Chloride

Trade Name Mixture:
Surfactol Q1 (CasChem)

RICINOLEAMIDOPROPYLTRIMONIUM METHOSULFATE

CAS No.	EINECS No.
85508-38-9	287-462-9

Empirical Formula:
$C_{24}H_{49}N_2O_2 \cdot CH_3O_4S$

Definition: Ricinoleamidopropyltrimonium Methosulfate is the quaternary ammonium salt that conforms to the formula:

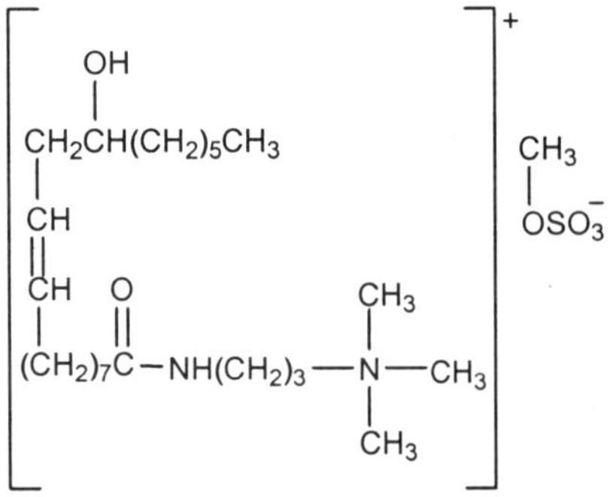

Chemical Class: Quaternary Ammonium Compounds

Functions: Antistatic Agent; Hair Conditioning Agent

Technical/Other Names:
3-[(12-Hydroxy-1-Oxo-9-Octadecenyl)-Amino]-N,N,N-Trimethyl-1-Propanaminium Methyl Sulfate

1-Propanaminium, 3-[(12-Hydroxy-1-Oxo-9-Octadecenyl)Amino]-N,N,N-Trimethyl-, Methyl Sulfate

Ricinoleamidopropyl Trimethylammonium Methyl Sulfate

Trade Name:
Varisoft RTM 50 (Degussa Care Specialties)

RICINOLEIC ACID

CAS Nos.   **EINECS No.**
141-22-0   205-470-2
7431-95-0

Empirical Formula:
$C_{18}H_{34}O_3$

Definition: Ricinoleic Acid is the unsaturated fatty acid that conforms generally to the formula:

$$CH_3(CH_2)_5CHCH_2CH=CH(CH_2)_7COOH$$
$$|$$
$$OH$$

Information Sources: CTFA D, MAR, MI-13(8295), TSCA

Chemical Class: Fatty Acids

Function: Surfactant - Cleansing Agent

Surfactant-Cleansing Agent is included as a function for the soap form of Ricinoleic Acid.

Reported Product Category: Fragrance Preparations, Misc.

Technical/Other Names:
12-Hydroxy-9-Octadecenoic Acid
9-Octadecenoic Acid, 12-Hydroxy-
Ricinic Acid
Ricinolic Acid

RICINOLEIC ACID/ADIPIC ACID/AEEA COPOLYMER

Definition: Ricinoleic Acid/Adipic Acid/AEEA Copolymer is a copolymer of ricinoleic acid, adipic acid and aminoethylethanolamine monomers.

Chemical Class: Synthetic Polymers

Function: Viscosity Increasing Agent - Aqueous

Trade Name:
Saboamid acid (Sabo)

Trade Name Mixture:
Saboamid RIT (Sabo)

RICINOLEIC/CAPROIC/CAPRYLIC/CAPRIC TRIGLYCERIDE

JPN Translation:
トリ (リシノレイン / カプロン / カプリル / カプリン酸) グリセリル

Definition: Ricinoleic/Caproic/Caprylic/Capric Triglyceride is the mixed triester of glycerin with ricinoleic, caproic, caprylic and capric acids.

Information Source: JCIC

Chemical Class: Fats and Oils

Function: Skin-Conditioning Agent - Occlusive

RICINOLEOYL EPOXY RESIN

JPN Translation:
エポキシエステル - 5

Definition: Ricinoleoyl Epoxy Resin is the ricinoleic acid ester of 4,4'-Isopropylidenediphenol/Epichlorohydrin Copolymer (q.v.), also known as epoxy resin.

Information Source: JCLS

Chemical Class: Synthetic Polymers

Function: Film Former

RICINOLETH-40

Definition: Ricinoleth-40 is the polyethylene glycol ether of ricinoleyl alcohol that conforms to the formula:

$$OH$$
$$|$$
$$CH_2CH(CH_2)_5CH_3$$
$$|$$
$$CH=CH(CH_2)_7CH_2(OCH_2CH_2)_nOH$$

where n has an average value of 40.

Information Source: 21CFR177.2800

Chemical Class: Alkoxylated Alcohols

Functions: Surfactant - Cleansing Agent; Surfactant - Solubilizing Agent

Reported Product Category: Bubble Baths

Technical/Other Names:
PEG-40 Ricinoleyl Ether
Polyethylene Glycol 2000 Ricinoleyl Ether
Polyoxyethylene (40) Ricinoleyl Ether

Trade Name:
Poliglicoleum (Vevy)

RICINUS COMMUNIS (CASTOR) SEED OIL

CAS No.   **EINECS No.**
8001-79-4   232-293-8

JPN Translations:
酵母処理ヒマシ油
ヒマシ油

Definition: Ricinus Communis (Castor) Seed Oil is the fixed oil obtained from the seeds of *Ricinus communis. See "Regulatory and Ingredient Use Information," regarding the labeling names for botanical ingredients in Volume 1, Introduction, Part A.*

Information Sources: ARG, AUS, BEL, BP, BPC, BRA, 21CFR73.1, 21CFR172.510, 21CFR172.876, 21CFR175.300, 21CFR176.170, 21CFR176.210, 21CFR177.2600, 21CFR177.2800, 21CFR178.3120, 21CFR178.3570, 21CFR178.3910, 21CFR181.22, 21CFR181.28, CTFA S, CZE, DA, DDR, EGY, FCC, FI, FIN, HP, HUN, IND, ITA, JAN, JCLS, JSCI, MAR, MEX, MI-13(1908), OTC-I-LX, PF, PN, POL, POR, RIFM, ROM, SNPF, TSCA, USAN, USD, USP XXIV, WHO, YUG

Chemical Class: Fats and Oils

Functions: Fragrance Ingredient; Skin-Conditioning Agent - Occlusive

Reported Product Categories: Lipsticks; Tonics, Dressings, and Other Hair Grooming Aids; Makeup Preparations (Not eye), Misc.; Hair Dyes and Colors (All Types Requiring Caution Statements and Patch Tests); Hair Conditioners; Bath Soaps and Detergents; Mascara; Eyeliners; Eye Makeup Preparations, Misc.; Blushers (All types); Eyebrow Pencils; Eye Shadows; Moisturizing Preparations; Nail Polish and Enamel Removers; Skin Care Preparations, Misc.; Foundations; Hair Preparations (Non-coloring), Misc.; Shampoos (Non-coloring); Bath Oils, Tablets, and Salts; Cleansing Products (Cold Creams, Cleansing Lotions, Liquids and Pads); Makeup Bases; Suntan Preparations, Misc.; Bath Preparations, Misc.; Body and Hand Preparations (Excluding Shaving Preparations); Manicuring Preparations, Misc.; Aftershave Lotions; Deodorants (Underarm); Face and Neck Preparations (Excluding Shaving Preparations); Oral Hygiene Products, Misc.; Perfumes; Powders (Dusting and Talcum, Excluding Aftershave Talcs); Suntan Gels, Creams, and Liquids

Technical/Other Names:
Castor Oil
Castor oil (Ricinus communis L.) (RIFM)
Castor Seed Oil
Ricinus Communis Oil

Trade Names:
AEC Castor Oil (A & E Connock)
Botanol CO (Botanigenics)
Castor Oil USP (Greeff)
Castor Oil USP (Sud-Chemie, United Catalysts)
Crystal Crown (CasChem)
Crystal LC (CasChem)
Crystal O (CasChem)

Hansonol JH-PPCO (Hansotech)
Huile de Ricin (Bertin)
Huile de Ricin (Sictia)
Lanaetex CO (Lanaetex)
Lipovol CO (Lipo)
Ricinus Communis Seed Oil ies (IES LABO)
Unicast-CO (Universal Preserv-A-Chem)

Trade Name Mixtures:
Activera 107 Lipo C (Active Organics)
Aloe Oil Extract, Castor (Concentrated Aloe Corp. (CAC))
Base 323 MS (LCW)
Base 323 MTC (LCW)
Base RAL W 323 T (LCW)
Base Rouge A Levres 323 TAL (LCW)
Base RW 101 (LCW)
Bentone Gel CAO (ELE)
Biron Silver CO (Merck KGaA/EMD Chemicals Inc.)
Blanc Covapate W 9765 (LCW)
Blue Covapate W 6763 (LCW)
Botanivera 107 (Botanigenics)
Brun Covapate W 8760 (LCW)
Casto Cetyle (LCW)
Castorcet (Lanaetex)
Castorlatum (CasChem)
Cloisonne Gold CC (Engelhard Corp.)
Cloisonne Red CC (Engelhard Corp.)
Covalip 22 (LCW)
Covalip 99 (LCW)
Flamenco Gold CC (Engelhard Corp.)
Flamenco Orange CC (Engelhard Corp.)
Flamenco Pearl CC (Engelhard Corp.)
Flamenco Superpearl CC (Engelhard Corp.)
Flamenco Super Red CC (Engelhard Corp.)
Flamenco Twilight Red CC (Engelhard Corp.)
Gemtone Sunstone CC (Engelhard Corp.)
Gemtone Tan Opal CC (Engelhard Corp.)
Gilugel CAO (Giulini/Giulini Chemie)
Gransil CS Fluid (Grant)
Liant TW 729 (LCW)
Mearlite GEH (Engelhard Corp.)
Mearlite LEM (Engelhard Corp.)
Natunola Castor 1023 (Natunola)
Natunola Maple 2701 (Natunola)
Natural Wax Jelly SP 511 (Strahl & Pitsch)
Natural Wax Jelly SP 512 (Strahl & Pitsch)
Nikkol Nikkosome VCLI (Nikko)
Noir Covapate W 9764 (LCW)
Orange Covapate W 2762 (LCW)
Prodhyrouge 2000 (Prod'Hyg)
Ricino - Cetyle (Prod'Hyg)
Rubis Covapate W 4760 (LCW)
Rubis Covapate W 4765 (LCW)
Simagel C (Biophil)
Tixogel COG 1540 (Sud-Chemie, United Catalysts)
Unitina LM (Universal Preserv-A-Chem)

RNA

CAS No.: 63231-63-0

JPN Translation:
RNA

Definition: RNA is the polynucleotide involved in protein synthesis. It is found in both the nucleus and cytoplasm of cells. The four primary nucleosides are adenosine, guanosine, cytidine and uridine.

Information Sources: 21CFR369.20, 21CFR369.21, JCIC, JCLS, JSQI, MI-13 (8287), TSCA

Chemical Class: Biological Polymers and their Derivatives

Function: Skin-Conditioning Agent - Miscellaneous

Reported Product Categories: Hair Sprays (Aerosol Fixatives); Bath Capsules; Moisturizing Preparations; Face and Neck Preparations (Excluding Shaving Preparations); Night Skin Care Preparations; Tonics, Dressings, and Other Hair Grooming Aids; Permanent Waves; Bath Oils, Tablets, and Salts; Cleansing Products (Cold Creams, Cleansing Lotions, Liquids and Pads); Skin Care Preparations, Misc.; Shampoos (Non-coloring); Hair Conditioners; Paste Masks (Mud Packs)

Technical/Other Names:
Ribonucleic Acid
Ribonucleic Acid (1)
Ribonucleic Acid (2)

Trade Name Mixture:
Photonyl (Laboratoires Serobiologiques)

ROBINIA PSEUDACACIA FLOWER EXTRACT

CAS No.	EINECS No.
89957-93-7	289-615-5

Definition: Robinia Pseudacacia Flower Extract is an extract of the flowers of the black locust, *Robinia pseudacacia. See "Regulatory and Ingredient Use Information," regarding the labeling names for botanical ingredients in Volume 1, Introduction, Part A.*

Chemical Class: Biological Products

Function: Not Reported

Technical/Other Names:
Black Locust Extract
Black Locust (Robinia Pseudacacia) Extract
Extract of Black Locust
Extract of Robinia Pseudoacacia
Robinia Pseudoacacia Extract

Trade Name Mixtures:
Acacia Flowers Milk (CEP (Solabia))
Acacia Phytolait (Alban Muller)
Glycolysat of Acacia (CEP (Solabia))
Herbasol Extract Acacia (Cosmetochem) (Cosmetochem International Ltd.)
Oleat of Acacia (CEP (Solabia))
Robinia Pseudacacia Flower Extract ies (IES LABO)
Vegebios of Black Locust (CEP (Solabia))

ROE EXTRACT

Definition: Roe Extract is an extract of fish eggs.

Chemical Class: Biological Products

Function: Not Reported

Technical/Other Name:
Extract of Roe

ROMNEYA COULTERI

Definition: *See "Regulatory and Ingredient Use Information," regarding EU labeling names for botanical ingredients in Volume 1, Introduction, Part A.*

Chemical Class: Biological Products

Technical/Other Name:
Romneya Coulteri Flower Extract (U.S.)

ROMNEYA COULTERI FLOWER EXTRACT

Definition: Romneya Coulteri Flower Extract is an extract of the flowers of the matilija poppy, *Romneya coulteri. See "Regulatory and Ingredient Use Information," regarding the labeling names for botanical ingredients in Volume 1, Introduction, Part A.*

Chemical Class: Biological Products

Function: Not Reported

Technical/Other Names:
Extract of Matilija Poppy
Extract of Romneya Coulteri
Matilija Poppy Extract
Matilija Poppy (Romneya Coulteri) Extract
Romneya Coulteri (EU)

Trade Name Mixtures:
Actiphyte of Poppy BG50 (Active Organics)
Actiphyte of Poppy GL50 (Active Organics)
Actiphyte of Poppy Lipo S (Active Organics)
Actiphyte of Poppy PG50 (Active Organics)

ROSA CANINA FLOWER

Definition: Rosa Canina Flower is the petals of the flower of *Rosa canina. See*

"Regulatory and Ingredient Use Information," regarding the labeling names for botanical ingredients in Volume 1, Introduction, Part A.

Chemical Class: Biological Products

Function: Fragrance Ingredient

ROSA CANINA FLOWER EXTRACT

Definition: Rosa Canina Flower Extract is an extract of the flowers of *Rosa canina. See "Regulatory and Ingredient Use Information," regarding the labeling names for botanical ingredients in Volume 1, Introduction, Part A.*

Chemical Class: Biological Products

Function: Cosmetic Astringent

Technical/Other Name:
Extract of Rosa Canina Flower

Trade Name Mixtures:
Actiphyte of Rose BG50 (Active Organics)
Actiphyte of Rose GL50 (Active Organics)
Actiphyte of Rosehips Lipo S (Active Organics)
Actiphyte of Rose PG50 (Active Organics)
Actiphyte Rose Lipo M (Active Organics)
Extrait Huileux de Rose (Greentech)

ROSA CANINA FLOWER OIL

Definition: Rosa Canina Flower Oil is the volatile oil obtained from the flowers of *Rosa canina. See "Regulatory and Ingredient Use Information," regarding the labeling names for botanical ingredients in Volume 1, Introduction, Part A.*

Chemical Class: Essential Oils

Functions: Fragrance Ingredient; Skin-Conditioning Agent - Emollient

Trade Name Mixture:
Unioil Roses (Chemyunion)

ROSA CANINA FRUIT

Definition: Rosa Canina Fruit is the fleshy fruit of *Rosa canina. See "Regulatory and Ingredient Use Information," regarding the labeling names for botanical ingredients in Volume 1, Introduction, Part A.*

Chemical Class: Biological Products

Function: Cosmetic Astringent

Technical/Other Name:
Dog Rose (Rosa Canina) Hips

Trade Names:
AEC Rose Hips Puree (A & E Connock)
AEC Rose Hips Whole (A & E Connock)

ROSA CANINA FRUIT EXTRACT

CAS No. 84696-47-9
EINECS No. 283-652-0

JPN Translation:
ノバラエキス

Definition: Rosa Canina Fruit Extract is an extract of dog rose hips, *Rosa canina. See "Regulatory and Ingredient Use Information," regarding the labeling names for botanical ingredients in Volume 1, Introduction, Part A.*

Information Sources: JCIC, JCLS, JSQI

Chemical Class: Biological Products

Function: Skin-Conditioning Agent - Miscellaneous

Reported Product Category: Bath Capsules

Technical/Other Names:
Dog Rose Extract
Dog Rose (Rosa Canina) Hips Extract
Extract of Dog Rose
Extract of Rosa Canina
Extract of Rosa Canina Hips
Extract of Rose Hips
Rose Hips Extract
Wild Rose Extract

Trade Name:
Flavex Rosehip CO2-to extract, Type 046.001 (Flavex)

Trade Name Mixtures:
Actiphyte of Rosehips BG50 (Active Organics)
Actiphyte of Rosehips GL50 (Active Organics)
Actiphyte of Rosehips PG50 (Active Organics)
Actiphyte of Rose Lipo S (Active Organics)
Aromaphyte of Rose (Active Organics)
Bio-Chelated Neutral Henna Plus II (Bio-Botanica)
215 Blend for Delicate Skin HS (Alban Muller)
Cremogen Rose Hip (PN 774 200) (Haarmann & Reimer GmbH)
Cytoflavin-CR (Chemyunion)
Dog Rose HS / Rose Hips HS (Alban Muller)
Eglantineol (Esperis)
Extrapone Rose Hip 2/032121 (Symrise)
Herbasol Distillate Rose Hips (Cosmetochem) (Cosmetochem International Ltd.)
Herbasol Extract Oil Soluble Rose Hips (Cosmetochem) (Cosmetochem International Ltd.)
Herbasol-Extract Rose Hips (Cosmetochem)

Hip Extract HS 2576 G (Grau)
Phytelene of Wild Rose EG 494 Liquid (Indena SA)
Prodhy Extract Eglantier (Prod'Hyg)
Rose Hips Liquid B (Ichimaru Pharcos)
Rose Hips Liquid E (Ichimaru Pharcos)
Vegeles SR (Laboratoires Serobiologiques)
2110 Vege-Plex Hair Complex Shampoo (Vege-Tech)
Vegetol Wild Roseberry MCF 1833 Hydro (Gattefosse s.a.)
Vegetol Wild Roseberry MCF 1837 Oily (Gattefosse s.a.)
VT-038 Extract of Rose Hips (Vege-Tech)
Wild Rose Extract (Maruzen Pharmaceuticals Co., Ltd.)
Wild Rose Extract BG (Maruzen Pharmaceuticals Co., Ltd.)
Wild Rose Extract BG100 (Maruzen Pharmaceuticals Co., Ltd.)
Wild Rose Extract Powder-S (Maruzen Pharmaceuticals Co., Ltd.)

ROSA CANINA FRUIT JUICE

Definition: Rosa Canina Fruit Juice is the liquid expressed from the hips of *Rosa canina. See "Regulatory and Ingredient Use Information," regarding the labeling names for botanical ingredients in Volume 1, Introduction, Part A.*

Chemical Class: Biological Products

Function: Cosmetic Astringent

Technical/Other Names:
Dog Rose (Rosa Canina) Hips Juice
Juice, Dog Rose Hips (Rosa Canina)
Juice, Rosa Canina
Rosa Canina Juice

Trade Name:
AEC Rose Hips Conc (A & E Connock)

ROSA CANINA FRUIT OIL

JPN Translations:
ノバラ油
ローズヒップ油

Definition: Rosa Canina Fruit Oil is an oil derived from rose hips, *Rosa canina. See "Regulatory and Ingredient Use Information," regarding the labeling names for botanical ingredients in Volume 1, Introduction, Part A.*

Information Sources: JCIC, JCLS, JSQI

Chemical Class: Fats and Oils

Function: Skin-Conditioning Agent - Miscellaneous

Technical/Other Names:
Dog Rose (Rosa Canina) Hips Oil

Oils, Rose Hips
Rose Hips Oil

Trade Names:
AEC Rose Hips Oil (A & E Connock)
Huile de Rosa Mosqueta (LCW)
Lipovol RHO (Lipo)
Nikkol Rose Hips Oil (Nikko)
Phytol RHO (rose hip oil) (Custom Ingredients)
Rosa Canina Fruit Oil ies (IES LABO)
Rose Hip oil, Premier (Premier Specialties)
Rose Hips Oil (Freeman)
Tri-K Rose Hip Seed Oil (Tri-K)

Trade Name Mixtures:
Aromaphyte of Rose (Active Organics)
Lipofacteur Vitentiel (LCW)
Macarose (LCW)
Ultraspheres-6100 (Lipoid)

ROSA CANINA LEAF EXTRACT

Definition: Rosa Canina Leaf Extract is an extract of the leaves of the dog rose, *Rosa canina*. See *"Regulatory and Ingredient Use Information,"* regarding the labeling names for botanical ingredients in Volume 1, Introduction, Part A.

Chemical Class: Biological Products

Function: Not Reported

Technical/Other Names:
Dog Rose Leaf Extract
Dog Rose (Rosa Canina) Leaf Extract
Extract of Dog Rose Leaf
Extract of Rosa Canina
Rosa Canina Extract

Trade Name Mixture:
Extrait D'Eglantier MPE PG 40 (Yves Rocher)

ROSA CANINA SEED EXTRACT

CAS No.	EINECS No.
84696-47-9	283-652-0

Definition: Rosa Canina Seed Extract is an extract of the seeds of the dog rose, *Rosa canina*. See *"Regulatory and Ingredient Use Information,"* regarding the labeling names for botanical ingredients in Volume 1, Introduction, Part A.

Chemical Class: Biological Products

Functions: Humectant; Skin-Conditioning Agent - Emollient

Reported Product Categories: Bath Capsules; Bath Oils, Tablets, and Salts; Bath Preparations, Misc.

Technical/Other Names:
Dog Rose (Rosa Canina) Seed Extract

Extract of Rosa Canina Seed
Extract of Rose Seed
Rose Seed Extract

Trade Names:
Flavex Rosehip Seed CO2-to extract, Type 046.002 (Flavex)
Rose Hip Oil (Desert Whale)

Trade Name Mixture:
Rose Hip Oil (Desert Whale)

ROSA CANINA SEED POWDER

Definition: Rosa Canina Seed Powder is the powder obtained from the ground seeds of *Rosa canina*. See *"Regulatory and Ingredient Use Information,"* regarding the labeling names for botanical ingredients in Volume 1, Introduction, Part A.

Chemical Class: Biological Products

Functions: Abrasive; Exfoliant

Trade Name:
Lipo RHS 60/100 (Lipo)

ROSA CENTIFOLIA FLOWER EXTRACT

CAS No.	EINECS No.
84604-12-6	283-289-8

JPN Translation:
バラエキス

Definition: Rosa Centifolia Flower Extract is an extract of the flowers of the cabbage rose, *Rosa centifolia*. See *"Regulatory and Ingredient Use Information,"* regarding the labeling names for botanical ingredients in Volume 1, Introduction, Part A.

Information Sources: JCIC, JCLS, RIFM

Chemical Class: Biological Products

Functions: Fragrance Ingredient; Skin-Conditioning Agent - Miscellaneous

Reported Product Categories: Bath Oils, Tablets, and Salts; Cleansing Products (Cold Creams, Cleansing Lotions, Liquids and Pads); Bath Preparations, Misc.; Skin Care Preparations, Misc.; Bath Capsules; Face and Neck Preparations (Excluding Shaving Preparations); Eye Makeup Preparations, Misc.; Eye Makeup Removers; Hair Conditioners; Moisturizing Preparations

Technical/Other Names:
Cabbage Rose Extract
Cabbage Rose (Rosa Centifolia) Extract
Extract of Cabbage Rose
Extract of Rosa Centifolia
Rose Extract
Rose water, stronger (Rosa centifolia L.) (RIFM)

Trade Name:
Hydroessential Flores Rosae (Vevy)

Trade Name Mixtures:
AEC Cosflor Blend 017 Moisture Factor WSS (A & E Connock)
AEC Moisture Factor HV (A & E Connock)
Cabbage Rose HS (Alban Muller)
Crodarom Hygroderm (Croda, Inc.)
Hydroessential Balsamic-1 (Vevy)
Hydroplastidine Flores Rosae (Vevy)
Moisturising Factor Hydrogerm (Crodarom)
Phytelene of Rose EG 518 Liquid (Indena SA)
Phytogreen 55 of Rose EXH 717 Liquid (Phytochim)
Prodhy Extract Rose (Prod'Hyg)
Red Rose Petals Extract HS 2531 G (Grau)
Rose Flower Red Rose Pigment (Naris)
Vegebios of Cabbage rose (CEP (Solabia))
Vegetol Rose MCF 789 Hydro (Gattefosse s.a.)

ROSA CENTIFOLIA FLOWER JUICE

Definition: Rosa Centifolia Flower Juice is the juice expressed from the flower of *Rosa centifolia*. See *"Regulatory and Ingredient Use Information,"* regarding the labeling names for botanical ingredients in Volume 1, Introduction, Part A.

Chemical Class: Biological Products

Function: Skin-Conditioning Agent - Miscellaneous

Trade Name:
Authenticals of Cabbage Rose (CEP (Solabia))

ROSA CENTIFOLIA FLOWER OIL

Definition: Rosa Centifolia Flower Oil is the volatile oil obtained from the flowers of *Rosa centifolia*. See *"Regulatory and Ingredient Use Information,"* regarding the labeling names for botanical ingredients in Volume 1, Introduction, Part A.

Chemical Class: Essential Oils

Function: Fragrance Ingredient

Technical/Other Names:
Cabbage Rose Oil
Cabbage Rose (Rosa Centifolia) Oil
Oils, Cabbage Rose

Trade Name Mixture:
Essentiaderm N.14 (Universal Flavors)

ROSA CENTIFOLIA FLOWER WATER

JPN Translation:
ローズ水

Definition: Rosa Centifolia Flower Water is an aqueous solution of the steam distillate obtained from the flowers of the rose, *Rosa centifolia*. See *"Regulatory and Ingredient Use Information,"* regarding the labeling names for botanical ingredients in Volume 1, Introduction, Part A.

Information Sources: ARG, 21CFR182.20, FI, JCIC, JCLS, JSQI, NF XIX, USAN, USD

Chemical Class: Biological Products

Function: Skin-Conditioning Agent - Miscellaneous

Reported Product Categories: Skin Fresheners; Bath Oils, Tablets, and Salts; Cleansing Products (Cold Creams, Cleansing Lotions, Liquids and Pads); Bath Capsules; Face and Neck Preparations (Excluding Shaving Preparations); Eye Makeup Removers; Skin Care Preparations, Misc.; Bath Preparations, Misc.; Body and Hand Preparations (Excluding Shaving Preparations); Moisturizing Preparations; Paste Masks (Mud Packs)

Technical/Other Names:
Cabbage Rose (Rosa Centifolia) Water
Rose Water
Water, Rose

Trade Names:
AEC Rose Water (A & E Connock)
Rose Water (Alban Muller)
Rose Water D (Maruzen Pharmaceuticals Co., Ltd.)

Trade Name Mixtures:
Elder and Rose Water (Indena SA)
Rose Water (Ichimaru Pharcos)

ROSA CENTIFOLIA FLOWER WAX

Definition: Rosa Centifolia Flower Wax is a wax obtained from the flower of *Rosa centifolia*. See *"Regulatory and Ingredient Use Information,"* regarding the labeling names for botanical ingredients in Volume 1, Introduction, Part A.

Chemical Class: Waxes

Function: Not Reported

Technical/Other Names:
Cabbage Rose (Rosa Centifolia) Wax
Cabbage Rose Wax
Wax, Cabbage Rose (Rosa Centifolia)

Trade Name Mixtures:
AEC Cire Essentielle De Fleurs De Rose (A & E Connock)
AEC Cire Essentielle De NCR (A & E Connock)
Cire Essentielle de fleurs de Rose (Bertin)
Cire Essentielle NCR (Bertin)

ROSA DAMASCENA EXTRACT

Definition: Rosa Damascena Extract is an extract of the rose, *Rosa damascena*. See *"Regulatory and Ingredient Use Information,"* regarding the labeling names for botanical ingredients in Volume 1, Introduction, Part A.

Chemical Class: Biological Products

Function: Fragrance Ingredient

Reported Product Categories: Cleansing Products (Cold Creams, Cleansing Lotions, Liquids and Pads); Skin Care Preparations, Misc.; Face and Neck Preparations (Excluding Shaving Preparations); Bath Preparations, Misc.; Eye Makeup Preparations, Misc.; Hair Conditioners; Moisturizing Preparations; Eye Makeup Removers

Technical/Other Names:
Extract of Rosa Damascena
Rose (Rosa Damascena) Extract

Trade Names:
Rose Absolute (Chauvet)
Rose Absolute 4/311208 (Symrise)
Rose Ecoconcentrate (Robertet, Inc.)

Trade Name Mixtures:
Cosflor Damask Rose HGS (A & E Connock)
Cosflor Rose Petal HGS (A & E Connock)
Damask Rose Extract (Cosmetic Developments)

ROSA DAMASCENA FLOWER OIL

CAS Nos. **EINECS No.**
8007-01-0
90106-38-0   290-260-3

Definition: Rosa Damascena Flower Oil is the volatile oil obtained from the flowers of *Rosa damascena*. See *"Regulatory and Ingredient Use Information,"* regarding the labeling names for botanical ingredients in Volume 1, Introduction, Part A.

Information Sources: NF XVIII, RIFM, TSCA

Chemical Class: Essential Oils

Function: Fragrance Ingredient

Reported Product Categories: Skin Care Preparations, Misc.; Moisturizing Preparations; Bath Oils, Tablets, and Salts; Cleansing Products (Cold Creams, Cleansing Lotions, Liquids and Pads); Baby Lotions, Oils, Powders and Creams; Paste Masks (Mud Packs)

Technical/Other Names:
Oils, Rosa Damascena
Rose absolute (Rosa spp.) (RIFM)
Rose concrete (RIFM)
Rose hips extract (Rosa spp.) (RIFM)
Rose oil (Rosa damascena Mill.) (RIFM)
Rose (Rosa Damascena) Oil

Trade Name:
Rose Oil (Chauvet)

Trade Name Mixtures:
Extrapone Jojoba/Rose Blend GW 2/031326 (Symrise)
Rose CL 2/033395 (Symrise)
Rose CL Forte 2/J33295 (Symrise)

ROSA DAMASCENA FLOWER WATER

JPN Translation:
ローズ水

Definition: Rosa Damascena Flower Water is an aqueous solution of the steam distillate obtained from the flowers of *Rosa damascena*. See *"Regulatory and Ingredient Use Information,"* regarding the labeling names for botanical ingredients in Volume 1, Introduction, Part A.

Chemical Class: Biological Products

Function: Fragrance Ingredient

Reported Product Categories: Skin Fresheners; Bath Oils, Tablets, and Salts; Cleansing Products (Cold Creams, Cleansing Lotions, Liquids and Pads); Bath Capsules; Face and Neck Preparations (Excluding Shaving Preparations); Eye Makeup Removers; Skin Care Preparations, Misc.; Bath Preparations, Misc.; Body and Hand Preparations (Excluding Shaving Preparations); Moisturizing Preparations

Technical/Other Name:
Water, Rosa Damascena Flower

Trade Names:
Essential Rose Nectar (Libiol)
Rose Ecoconcentrate Natural (Robertet S.A.)
Rose Oil and Rose Water (Floressence)
Rose Water Concentrate (Biolandes Parfumerie)

Trade Name Mixture:
Rose Water (Sensory)

ROSA DAMASCENA FLOWER WAX

Definition: Rosa Damascena Flower Wax is a wax obtained from the flower of *Rosa damascena*. See *"Regulatory and Ingredient Use Information,"* regarding the labeling names for botanical ingredients in Volume 1, Introduction, Part A.

Chemical Class: Essential Oils

Function: Fragrance Ingredient

Reported Product Category: Moisturizing Preparations

Technical/Other Name:
Rose (Rosa Damascena) Wax

Trade Name Mixtures:
AEC Cire Essentielle De Fleurs De Rose (A & E Connock)
AEC Cire Essentielle De NCR (A & E Connock)
AEC Rose Cire Essentielle (A & E Connock)
Cire Essentielle de fleurs de Rose (Bertin)
Cire Essentielle NCR (Bertin)

ROSA EGLENTARIA EXTRACT

Definition: Rosa Eglentaria Extract is an extract of the leaves, flowers, and stem of *Rosa eglanteria. See "Regulatory and Ingredient Use Information," regarding the labeling names for botanical ingredients in Volume 1, Introduction, Part A.*

Chemical Class: Biological Products

Function: Fragrance Ingredient

Technical/Other Name:
Sweet Briar Rose Extract

Trade Name Mixtures:
Cosflor Sweet Briar HGS (A & E Connock)
Sweet Briar Extract (Cosmetic Developments)

ROSA EGLENTARIA SEED OIL

Definition: Rosa Eglentaria Seed Oil is the oil expressed from the seeds of *Rosa eglantaria. See "Regulatory and Ingredient Use Information," regarding the labeling names for botanical ingredients in Volume 1, Introduction, Part A.*

Chemical Class: Fats and Oils

Function: Skin-Conditioning Agent - Occlusive

Trade Name:
Rosehip Seed Oil (Siber Hegner)

ROSA GALLICA FLOWER EXTRACT

CAS No.	EINECS No.
84604-13-7	283-290-3

Definition: Rosa Gallica Flower Extract is an extract of the flowers of the French rose, *Rosa gallica. See "Regulatory and Ingredient Use Information," regarding the labeling names for botanical ingredients in Volume 1, Introduction, Part A.*

Chemical Class: Biological Products

Function: Skin-Conditioning Agent - Miscellaneous

Reported Product Category: Bath Oils, Tablets, and Salts

Technical/Other Names:
Extract of French Rose
Extract of Rosa Gallica
French Rose Extract
French Rose (Rosa Gallica) Extract

Trade Name Mixtures:
French Rose HS (Alban Muller)
Herbasec Rose Flower (Cosmetochem) (Cosmetochem International Ltd.)
Herbasol Distillate Rose Flower (Cosmetochem) (Cosmetochem International Ltd.)
Herbasol-Extract Rose Flowers (Cosmetochem)
Rose Petal Extract (Toyo Hakko)

ROSA GALLICA FLOWER OIL

Definition: Rosa Gallica Flower Oil is the volatile oil obtained from the flowers of *Rosa gallica. See "Regulatory and Ingredient Use Information," regarding the labeling names for botanical ingredients in Volume 1, Introduction, Part A.*

Chemical Class: Essential Oils

Functions: Fragrance Ingredient; Skin-Conditioning Agent - Emollient

Trade Name Mixture:
Unioil Roses (Chemyunion)

ROSA MOSCHATA LEAF EXTRACT

Definition: Rosa Moschata Leaf Extract is an extract of the leaves of *Rosa moschata. See "Regulatory and Ingredient Use Information," regarding the labeling names for botanical ingredients in Volume 1, Introduction, Part A.*

Chemical Class: Biological Products

Function: Cosmetic Astringent

Technical/Other Name:
Extract of Rosa Moschata Leaf

Trade Name Mixture:
Musk Rose (Rosa Moschata) Leaf Extract (Weleda)

ROSA MOSCHATA OIL

Definition: Rosa Moschata Oil is the oil obtained from the musk rose, *Rosa moschata. See "Regulatory and Ingredient Use Information," regarding the labeling names for botanical ingredients in Volume 1, Introduction, Part A.*

Chemical Class: Essential Oils

Function: Fragrance Ingredient

Technical/Other Names:
Musk Rose Oil
Musk Rose (Rosa Moschata) Oil

Trade Name:
Delta Rosa Moschata (Vevy)

ROSA MOSCHATA SEED OIL

Definition: Rosa Moschata Seed Oil is the oil expressed from the seeds of *Rosa moschata* or *Rosa eglanteria. See "Regulatory and Ingredient Use Information," regarding the labeling names for botanical ingredients in Volume 1, Introduction, Part A.*

Chemical Class: Fats and Oils

Function: Skin-Conditioning Agent - Emollient

Technical/Other Names:
Musk Rose (Rosa Moschata) Seed Oil
Rose Seed Oil

Trade Name:
Rosehip Seed Oil (Nestle World Trade)

ROSA MULTIFLORA FLOWER WAX

Definition: Rosa Multiflora Flower Wax is the wax obtained from the flowers of *Rosa multiflora. See "Regulatory and Ingredient Use Information," regarding the labeling names for botanical ingredients in Volume 1, Introduction, Part A.*

Chemical Class: Waxes

Functions: Skin-Conditioning Agent - Emollient; Skin-Conditioning Agent - Occlusive; Viscosity Controlling Agent; Viscosity Increasing Agent - Aqueous; Viscosity Increasing Agent - Nonaqueous

Trade Name:
Rose Wax (Koster Keunen Holland)

ROSA MULTIFLORA FRUIT

Definition: Rosa Multiflora Fruit is the fruit of *Rosa multiflora. See "Regulatory and Ingredient Use Information," regarding the labeling names for botanical ingredients in Volume 1, Introduction, Part A.*

Chemical Class: Biological Products

Function: Not Reported

Technical/Other Name:
Eijitsu (JPN)

ROSA MULTIFLORA FRUIT EXTRACT

JPN Translation:
ノイバラ果実エキス

Definition: Rosa Multiflora Fruit Extract is an extract of the fruit of *Rosa multiflora*. See "Regulatory and Ingredient Use Information," regarding the labeling names for botanical ingredients in Volume 1, Introduction, Part A.

Information Source: JCLS

Chemical Class: Biological Products

Function: Fragrance Ingredient

Reported Product Categories: Cleansing Products (Cold Creams, Cleansing Lotions, Liquids and Pads); Skin Care Preparations, Misc.; Face and Neck Preparations (Excluding Shaving Preparations); Bath Preparations, Misc.; Eye Makeup Preparations, Misc.; Hair Conditioners; Eye Makeup Removers

Technical/Other Names:
Eijitsu Ekisu (JPN)
Extract of Rosa Multiflora
Rose (Rosa Multiflora) Extract

Trade Name Mixtures:
DB-PL-compound-1 (Dong Bang Future)
Eijitsu Liquid B (Ichimaru Pharcos)
Eijitsu Liquid E (Ichimaru Pharcos)
Rose Fruit Extract (Maruzen Pharmaceuticals Co., Ltd.)
Rose Fruit Extract BG (Maruzen Pharmaceuticals Co., Ltd.)
Rose Fruit Extract LA (Maruzen Pharmaceuticals Co., Ltd.)

ROSA ROXBURGHII FRUIT EXTRACT

JPN Translation:
イザヨイバラエキス

Definition: Rosa Roxburghii Fruit Extract is an extract of the fruit of the chestnut rose, *Rosa roxburghii*. See "Regulatory and Ingredient Use Information," regarding the labeling names for botanical ingredients in Volume 1, Introduction, Part A.

Information Source: JCLS

Chemical Class: Biological Products

Function: Not Reported

Technical/Other Names:
Chestnut Rose Fruit Extract
Extract of Rosa Roxburgii

Trade Name:
Chestnut Rose Fruit Extract (Maruzen Pharmaceuticals Co., Ltd.)

Trade Name Mixtures:
Cili Extract (Maruzen Pharmaceuticals Co., Ltd.)
Cili Extract BG (Maruzen Pharmaceuticals Co., Ltd.)

ROSA RUBIGINOSA SEED OIL

Definition: Rosa Rubiginosa Seed Oil is the oil expressed from the seeds of *Rosa rubiginosa*. See "Regulatory and Ingredient Use Information," regarding the labeling names for botanical ingredients in Volume 1, Introduction, Part A.

Chemical Class: Fats and Oils

Function: Skin-Conditioning Agent - Emollient

Reported Product Categories: Bath Capsules; Bath Oils, Tablets, and Salts; Bath Preparations, Misc.

Technical/Other Name:
Oils, Rosa Rubiginosa Seed

Trade Name:
Rose Hip Oil (Aldivia)

ROSA RUBIGINOSA SEED OIL PEG-8 ESTERS

Definition: Rosa Rubiginosa Seed Oil PEG-8 Esters is the complex mixture obtained from the transesterification of Rosa Rubiginosa Seed Oil (q.v.) and PEG-8 (q.v.).

Chemical Class: Glyceryl Esters and Derivatives

Functions: Skin-Conditioning Agent - Emollient; Surfactant - Emulsifying Agent

Trade Name:
Viatenza Rose PE8 (Aldivia)

ROSA RUGOSA FLOWER EXTRACT

JPN Translation:
ハマナスエキス

Definition: Rosa Rugosa Flower Extract is an extract of the flowers of *Rosa rugosa*. See "Regulatory and Ingredient Use Information," regarding the labeling names for botanical ingredients in Volume 1, Introduction, Part A.

Information Source: JCLS

Chemical Class: Biological Products

Function: Skin-Conditioning Agent - Miscellaneous

Technical/Other Name:
Extract of Rosa Rugosa Flower

Trade Name Mixture:
Rosa Rugosa Extract LA (Maruzen Pharmaceuticals Co., Ltd.)

ROSA RUGOSA LEAF EXTRACT

Definition: Rosa Rugosa Leaf Extract is an extract of the leaves of *Rosa rugosa*. See "Regulatory and Ingredient Use Information," regarding the labeling names for botanical ingredients in Volume 1, Introduction, Part A.

Chemical Class: Biological Products

Function: Skin-Conditioning Agent - Humectant

Technical/Other Name:
Extract of Rosa Rugosa Leaves

Trade Name Mixture:
Hamanasu Extract (Nonogawa)

ROSA SPINOSISSIMA FRUIT EXTRACT

Definition: Rosa Spinosissima Fruit Extract is an extract of the fruit of *Rosa spinosissima*. See "Regulatory and Ingredient Use Information," regarding the labeling names for botanical ingredients in Volume 1, Introduction, Part A.

Chemical Class: Biological Products

Function: Not Reported

Technical/Other Name:
Burnet Rose Extract

Trade Name Mixtures:
Burnet Rose Extract (Cosmetic Developments)
Cosflor Burnet Rose HGS (A & E Connock)

ROSE EXTRACT

CAS No. 84696-47-9
EINECS No. 283-652-0

Definition: Rose Extract is an extract of various species of rose, *Rosa spp*. See "Regulatory and Ingredient Use Information," regarding the labeling names for botanical ingredients in Volume 1, Introduction, Part A.

Information Sources: 21CFR182.20, JCLS

Chemical Class: Biological Products

Function: Skin-Conditioning Agent - Occlusive

Reported Product Categories: Bath Oils, Tablets, and Salts; Cleansing Products (Cold Creams, Cleansing Lotions, Liquids and Pads); Bath Preparations, Misc.; Skin Care Preparations, Misc.; Bath Capsules; Face and Neck Preparations (Excluding Shaving Preparations); Body and Hand Preparations (Excluding Shaving Preparations); Eye Makeup Preparations, Misc.; Eye Makeup Removers; Hair Conditioners; Skin Fresheners; Moisturizing Preparations

Technical/Other Names:
Extract of Rose
Rosa Canina Extract

Trade Name Mixtures:
BBC Moisture Trol (Bio-Botanica)
BBC Relaxing Complex (Bio-Botanica)
Coobato Rosa (Universal Flavors)
Coobato Speciale (Universal Flavors)
Glycolysat of Cabbage Rose (CEP (Solabia))
Rose Extract HG (Provital/Centerchem)
Rose Wax (Weleda)
VT-090 Extract of Red Rose (Vege-Tech)

ROSE FLOWER OIL

CAS No.: 8007-01-0

Definition: Rose Flower Oil is the volatile oil obtained from the flowers of *Rosa spp.* See *"Regulatory and Ingredient Use Information,"* regarding the labeling names for botanical ingredients in Volume 1, Introduction, Part A.

Information Sources: BRA, 21CFR182.20, EP, FI, MI-13(6870), NF XIX, RIFM, TSCA, USAN, USSR

Chemical Class: Essential Oils

Function: Fragrance Ingredient

Reported Product Categories: Skin Care Preparations, Misc.; Moisturizing Preparations; Bath Oils, Tablets, and Salts; Cleansing Products (Cold Creams, Cleansing Lotions, Liquids and Pads); Hair Conditioners; Baby Lotions, Oils, Powders and Creams; Paste Masks (Mud Packs)

Technical/Other Names:
Oils, Rose
Rose absolute (Rosa spp.) (RIFM)
Rose concrete (RIFM)
Rose hips extract (Rosa spp.) (RIFM)
Rose oil (Rosa damascena Mill.) (RIFM)

ROSIN

CAS No.	EINECS No.
8050-09-7	232-475-7

JPN Translation:
ロジン

Definition: Rosin is the residue left after distilling off the volatile oil from the oleoresin obtained from *Pinus palustris* and other species of *Pinaceae*. See *"Regulatory and Ingredient Use Information,"* regarding use of EU Trivial names in Volume 1, Introduction, Part A.

Information Sources: AUS, 21CFR73.1, 21CFR172.210, 21CFR172.510, 21CFR172.615, 21CFR172.862, 21CFR175.105, 21CFR175.125, 21CFR175.300, 21CFR176.170, 21CFR176.200, 21CFR176.210, 21CFR177.1200, 21CFR177.1210, 21CFR177.2600, 21CFR178.2010, 21CFR178.3120, 21CFR178.3800, 21CFR178.3850, 21CFR178.3870, 27CFR21.126, 27CFR21.141, CTFA D, EGY, KOR, MAR, MI-13(8347), PF, PN, TSCA, USP XX, YUG

Chemical Class: Biological Products

Functions: Binder; Epilating Agent; Film Former; Viscosity Increasing Agent - Nonaqueous

Reported Product Categories: Mascara; Bubble Baths; Depilatories

Technical/Other Names:
Colophony
Gum Rosin
Rosin Gum
WW Wood Rosin

Trade Name:
AEC Gum Rosin (A & E Connock)

ROSIN ACRYLATE

CAS No.	EINECS No.
83137-13-7	280-192-2

Definition: Rosin Acrylate is the ester of rosin and acrylic acid.

Information Source: TSCA

Chemical Class: Esters

Function: Hair Fixative

Technical/Other Names:
Acrylic Acid, Rosin Ester
Rosin, Reaction Products with Acrylic Acid
Rosin, Reaction Product with Acrylic Acid

ROSIN/FORMALDEHYDE COPOLYMER

CAS No.: 65997-07-1

Definition: Rosin/Formaldehyde Copolymer is a copolymer of Rosin (q.v.) and Formaldehdye (q.v.) monomers.

Chemical Class: Synthetic Polymers

Function: Depilating Agent

Technical/Other Name:
Rosin, Polymer with Formaldehyde

Trade Name Mixture:
Dermulsene RA 405 (Les Derives Resiniques)

ROSIN HYDROLYZED COLLAGEN

JPN Translation:
ロジン加水分解コラーゲン

Definition: Rosin Hydrolyzed Collagen is the condensation product of rosin acid chloride and Hydrolyzed Collagen (q.v.).

Chemical Class: Protein Derivatives

Functions: Hair Conditioning Agent; Skin-Conditioning Agent - Miscellaneous

Trade Names:
Promois Resin A (Seiwa Kasei)
Promois Resin AC (Seiwa Kasei)

ROSMARINIC ACID

CAS No.: 20283-92-5

Empirical Formula:
$C_{18}H_{16}O_8$

Definition: Rosmarinic Acid is the acid obtained from *Melissa officinalis* or *Rosmarinis officinalis*. It conforms to the formula:

Chemical Class: Carboxylic Acids

Function: Antioxidant

Technical/Other Names:
Benzenepropanoic Acid, Alpha-[[(3-,4- Dihydroxyphenyl)-1-Oxo-2-Propenyl]oxy]-3, 4-Dihydroxy-
α-[[(3,4-Dihydroxyphenyl)-1-Oxo-2-Propenyl]Oxy]-3,4-Dihydroxy-benzenepropanoic Acid
Rosemary Acid

ROSMARINUS OFFICINALIS

Definition: See *"Regulatory and Ingredient Use Information,"* regarding EU labeling

names for botanical ingredients in Volume 1, Introduction, Part A.

Chemical Class: Biological Products

Technical/Other Names:
Rosmarinus Officinalis (Rosemary) Flower Extract (U.S.)
Rosmarinus Officinalis (Rosemary) Flower Wax (U.S.)
Rosmarinus Officinalis (Rosemary) Leaf Extract (U.S.)
Rosmarinus Officinalis (Rosemary) Leaf Oil (U.S.)
Rosmarinus Officinalis (Rosemary) Leaf Powder (U.S.)
Rosmarinus Officinalis (Rosemary) Leaf Water (U.S.)
Rosmarinus Officinalis (Rosemary) Water (U.S.)

Trade Name Mixture:
Activated Botanicals AB 109 (Norjin)

ROSMARINUS OFFICINALIS (ROSEMARY) FLOWER EXTRACT

Definition: Rosmarinus Officinalis (Rosemary) Flower Extract is an extract of the flowers of *Rosmarinus officinalis*. See "Regulatory and Ingredient Use Information," regarding the labeling names for botanical ingredients in Volume 1, Introduction, Part A.

Chemical Class: Biological Products

Functions: Antioxidant; Deodorant Agent; Skin-Conditioning Agent - Miscellaneous

Technical/Other Names:
Extract of Rosemary Flower
Extract of Rosmarinus Officinalis Flower
Rosemary Flower Extract
Rosmarinus Officinalis (EU)

Trade Name Mixtures:
Guardian 12 (Earth Supplied Products)
Guardian GP (Earth Supplied Products)
Guardian O9 (Earth Supplied Products)

ROSMARINUS OFFICINALIS (ROSEMARY) FLOWER WAX

Definition: Rosmarinus Officinalis (Rosemary) Flower Wax is a wax obtained from the flower of *Rosmarinus officinalis*. See "Regulatory and Ingredient Use Information," regarding the labeling names for botanical ingredients in Volume 1, Introduction, Part A.

Chemical Class: Essential Oils

Function: Fragrance Ingredient

Technical/Other Names:
Rosemary Flower Wax
Rosemary Wax
Rosmarinus Officinalis (EU)
Rosmarinus Officinalis Wax
Waxes, Rosemary (Rosmarinus Officinalis)

Trade Names:
AEC Cire Essentielle De Fleurs De Romarin (A & E Connock)
Cire Essentielle de fleurs de Romarin (Bertin)

ROSMARINUS OFFICINALIS (ROSEMARY) LEAF EXTRACT

CAS No.	EINECS No.
84604-14-8	283-291-9

JPN Translation:
ローズマリーエキス

Definition: Rosmarinus Officinalis (Rosemary) Leaf Extract is an extract of the leaves of the rosemary, *Rosmarinus officinalis*. See "Regulatory and Ingredient Use Information," regarding the labeling names for botanical ingredients in Volume 1, Introduction, Part A. See Reported Ingredient Functions-The Cosmetic Drug Distinction, in Regulatory and Ingredient Use Information, Volume I, Part A.

Information Sources: 21CFR182.20, HP, JCIC, JCLS, JSQI, RIFM

Chemical Class: Biological Products

Functions: Antimicrobial Agent; Antioxidant; Fragrance Ingredient; Skin-Conditioning Agent - Miscellaneous; Skin-Conditioning Agent - Occlusive

Reported Product Categories: Skin Care Preparations, Misc.; Hair Dyes and Colors (All Types Requiring Caution Statements and Patch Tests); Bath Preparations, Misc.; Body and Hand Preparations (Excluding Shaving Preparations); Bath Soaps and Detergents; Shampoos (Non-coloring); Personal Cleanliness Products, Misc.; Skin Fresheners; Bath Capsules; Face and Neck Preparations (Excluding Shaving Preparations); Night Skin Care Preparations; Paste Masks (Mud Packs); Bath Oils, Tablets, and Salts; Bubble Baths; Cleansing Products (Cold Creams, Cleansing Lotions, Liquids and Pads); Hair Preparations (Non-coloring), Misc.; Hair Rinses (Non-coloring); Tonics, Dressings, and Other Hair Grooming Aids; Hair Conditioners; Hair Sprays (Aerosol Fixatives); Moisturizing Preparations; Permanent Waves; Lipsticks; Suntan Preparations, Misc.

Technical/Other Names:
Extract of Rosemary
Extract of Rosmarinus Officinalis
Rosemary concrete (RIFM)
Rosemary Extract
Rosemary Leaf Extract
Rosmarinus Officinalis (EU)
Rosmarinus Officinalis Extract

Trade Names:
AEC Rosemary Extract Powder (A & E Connock)
Flavex Rosemary Antioxidant Extract, oil soluble, 14%Ditperpene Phenols, Type 027.002 (Flavex)
Flavex Rosemary CO2-se extract, Cineole Type, Type 027.005 (Flavex)
Hydroessential Rosmarinus (Vevy)
Oxy'Less .R (Naturex)
Phytelene of Romarin EN 194 powder (Indena SA)
Phytelene of Rosemary EN 194 powder (Indena SA)
Phytogreen of Rosemary EP 493 Powder (Phytochim)
Pronalen Rosemary SPE (Provital/Centerchem)
Rosemary Ecoconcentrate (Robertet, Inc.)
Rosemary Ecoconcentrate Natural (Robertet S.A.)
Rosemary Extract CG (Sabinsa)
Rosemary leaf (Kansai Koso)

Trade Name Mixtures:
Actifirm Ultra (Active Organics)
Actiphyte of Rosemary BG50 (Active Organics)
Actiphyte of Rosemary GL50 (Active Organics)
Actiphyte of Rosemary Lipo S (Active Organics)
Actiphyte of Rosemary PG50 (Active Organics)
Amiox (Alban Muller)
ANTI-WRINKLE (Greentech S.A)
Anti Wrinkles Complex 263 (Ennagram)
Antiwrinkles Phytogreen Complex GXH 263 (Phytochim)
Aqueous Spray - Dried G Mix (Indena SA)
Aromaphyte of Rosemary (Active Organics)
Aroma Rosemary B (Ichimaru Pharcos)
Bio-Chelated Neutral Henna Plus I (Bio-Botanica)
Bio-Chelated Nutra Plant Complex I (Bio-Botanica)
Bio-Chelated Nutra Plant Complex II (Bio-Botanica)
Bio-Dandra Plex (Bio-Botanica)
Biopein (Bio-Botanica)
Camomille Ecoconcentrate (Robertet, Inc.)
226 Capilotonique HS (Alban Muller)
245 Capilotonique HS (Alban Muller)
Carnosol/Rosmanol Extract (Robertet, Inc.)
Chronosphere Rosemary Extract (Arch Personal Care Products)
Complex 5 (Fabriquimica)
Complex Romero (Fabriquimica)
C-Protect (Gattefosse s.a.)

Cremogen AF (PN 736567) (Haarmann & Reimer GmbH)

Cremogen MZ (PN 739 032) (Haarmann & Reimer GmbH)

Cremogen MZ/N (PN 755 321) (Haarmann & Reimer GmbH)

Cremogen M-2 (PN 739 029) (Haarmann & Reimer GmbH)

Cremogen M-82 (PN 730 337) (Haarmann & Reimer GmbH)

Cremogen M-88(PN 458 469) (Haarmann & Reimer GmbH)

Cremogen M-2/N (PN 755326) (Haarmann & Reimer GmbH)

Cremogen Rosemary (PN 739 014) (Haarmann & Reimer GmbH)

Crodarom Rosemary Oil forte (Croda, Inc.)

Extrait De Romarin PPE PG10 (Yves Rocher)

Extrapone Cooling Complex 2/B16500 (Symrise)

Extrapone Cooling Complex N 2/B16501 (Symrise)

Extrapone #4 GW 2/031040 (Symrise)

Extrapone #7 Herbs 2/032535 (Symrise)

Extrapone #3 Special 2/034481 (Symrise)

Extrapone #3 Special 2/789490 (Symrise)

Extrapone #5 Special 2/032501 (Symrise)

Extrapone Rosemary 2/783630 (Symrise)

Extrapone Rosemary/Aloe Vera Blend 2/ 033197 (Symrise)

Extrapone Rosemary GW 2/031740 (Symrise)

Extrapone Rosemary Special 2/033251 (Symrise)

Extrapone Seven Herbs Special 2/032527 (Symrise)

Extrapone 3 Special 2/789490 (Symrise)

Extrapone 4 Special 2/788400 (Symrise)

Extrapone 5 Special 2/789500 (Symrise)

Extrapone 2 Special J (2/032473) (Symrise)

Extrapone 3 Special New 2/034484 (Symrise)

Extrapone 2 Special 2/789480 (same as 2/ 032471) (Symrise)

Glycolysat of Rosemary (CEP (Solabia))

Hair Treatment Complex 260 (Ennagram)

Hair Treatment Phytogreen Complex GXH 260 (Phytochim)

Herbalcomplex 5 Special (Crodarom)

Herbal Extract for Normal Hair (Crodarom)

Herbal Extract Glycolic - Article 251172 (Plantextrakt)

Herbaliquid Rosemary Special (Crodarom)

Herbal Vinegar (Provital/Centerchem)

Herbasol Complex "Herbes de Provence" (Cosmetochem) (Cosmetochem International Ltd.)

Herbasol Complex 7-Herbs (Cosmetochem) (Cosmetochem International Ltd.)

Herbasol Extract Oil Soluble Rosemary (Cosmetochem) (Cosmetochem International Ltd.)

Herbasol-Extract Rosemary (Cosmetochem)

Hydroplastidine Rosmarinus (Vevy)

Macerat Huileux de Feuilles de Romarin (Bertin)

Merospheres (Barnet)

Microcirculation Factor No. 3 (Indena SA)

Microcirculation Factor No. 5 (Indena SA)

Natuscreen (Pronectar)

Novaplant Rosemary Extract (Crodarom)

Novarom Rosemary Oil forte (Crodarom)

Nutriplant (Bio-Botanica)

Oleoplex Romero (Fabriquimica)

Oxy'Less .Clear (Naturex)

Oxy'Less .RD (Naturex)

Oxy'Less .RW (Naturex)

Pharcolex BX32 (Ichimaru Pharcos)

Pharcolex BX46 (Ichimaru Pharcos)

Phytelene Complex EGX 232 (Indena SA)

Phytelene Complex EGX 246 (Indena SA)

Phytelene of Romarin EG 009 liquid (Indena SA)

Phytelene of Rosemary EG 009 liquid (Indena SA)

Phytoderm P/25 Hydroalcoholic (Universal Flavors)

Phytogreen 55 of Rosemary EXH 612 Liquid (Phytochim)

Polyplant Anti-Cellulite (Provital/ Centerchem)

Polyplant Anti-Wrinkles (Provital/ Centerchem)

Polyplant Astringent (Provital/Centerchem)

Polyplant Epithelizing (Provital/ Centerchem)

Polyplant Hair (Provital/Centerchem)

Polyplant Skin Purifying (Provital/ Centerchem)

Polyplant 5 Special (Provital/Centerchem)

Polyplant Stimulant (Provital/Centerchem)

Prodhy Extract Romarin (Prod'Hyg)

Rosemary Antioxidant 20% (Robertet, Inc.)

Rosemary CL 2/033253 (Symrise)

Rosemary Extract (Maruzen Pharmaceuticals Co., Ltd.)

Rosemary Extract BG (Maruzen Pharmaceuticals Co., Ltd.)

Rosemary Extract BG 211087 (Crodarom)

Rosemary Extract BG-J (Maruzen Pharmaceuticals Co., Ltd.)

Rosemary Extract HG (Provital/ Centerchem)

Rosemary Extract HS 2379 G (Grau)

Rosemary Extract-J (Maruzen Pharmaceuticals Co., Ltd.)

Rosemary Extract LA (Maruzen Pharmaceuticals Co., Ltd.)

Rosemary Extract PG (Rahn)

Rosemary Extract Powder-S (Maruzen Pharmaceuticals Co., Ltd.)

Rosemary Extract SQ (Maruzen Pharmaceuticals Co., Ltd.)

Rosemary Extract W (Maruzen Pharmaceuticals Co., Ltd.)

Rosemary HPG Titrated (Alban Muller)

Rosemary HS (Alban Muller)

Rosemary Liquid B (Ichimaru Pharcos)

Rosemary Liquid E (Ichimaru Pharcos)

Rosemary LS (Alban Muller)

Rosemary Oil (Provital/Centerchem)

Rosemary Phytexcell (Crodarom)

Rosemary Tincture (Rahn)

Sederma Rosemary (Sederma)

280 Stimulant HS (Alban Muller)

680 Stimulant LS (Alban Muller)

Toothpaste Complex MU 3319 (Greentech S.A)

Transomes Anti-Age EGX 246-TR (Indena SA)

Transomes of Factor of MIcro-circulation No 3 (Indena SA)

Transomes of Factor of Micro-circulation No 5 (Indena SA)

Vegebios of Rosemary (CEP (Solabia))

2305 Vege-Plex Body Complex (Vege-Tech)

2350 Vege-Plex Body Complex (Vege-Tech)

2210 Vege-Plex Hair Complex (Vege-Tech)

2270 Vege-Plex Hair Complex (Vege-Tech)

2230 Vege-Plex Hair Complex Conditioner (Vege-Tech)

2240 Vege-Plex Hair Complex Conditioner (Vege-Tech)

2110 Vege-Plex Hair Complex Shampoo (Vege-Tech)

2120 Vege-Plex Hair Complex Shampoo (Vege-Tech)

2600 Vege-Plex Skin Complex (Vege-Tech)

2610 Vege-Plex Skin Complex (Vege-Tech)

Vegetol Rosemary GR 376 Hydro (Gattefosse s.a.)

Vegetol Rosemary MCF 772 Hydro (Gattefosse s.a.)

Vegetol Rosemary 4145 Oily (Gattefosse s.a.)

Vivaderm (Bio-Botanica)

VT-039 Extract of Rosemary (Vege-Tech)

ROSMARINUS OFFICINALIS (ROSEMARY) LEAF OIL

CAS No.: 8000-25-7

JPN Translation:
ローズマリー油

Definition: Rosmarinus Officinalis (Rosemary) Leaf Oil is the essential oil obtained from the flowering tops and leaves of *Rosmarinus officinalis. See*

"Regulatory and Ingredient Use Information," regarding the labeling names for botanical ingredients in Volume 1, Introduction, Part A.

Information Sources: AUS, BRA, 21CFR182.20, 27CFR21.65, 27CFR21.151, EP, FCC, JCIC, JCLS, JSQI, MEX, MI-13 (6871), RIFM, ROM, TSCA

Chemical Class: Essential Oils

Functions: Fragrance Ingredient; Skin-Conditioning Agent - Miscellaneous

Reported Product Categories: Skin Care Preparations, Misc.; Bath Preparations, Misc.; Body and Hand Preparations (Excluding Shaving Preparations); Shampoos (Non-coloring); Bath Capsules; Face and Neck Preparations (Excluding Shaving Preparations); Bath Oils, Tablets, and Salts; Cleansing Products (Cold Creams, Cleansing Lotions, Liquids and Pads); Tonics, Dressings, and Other Hair Grooming Aids; Hair Conditioners; Moisturizing Preparations; Paste Masks (Mud Packs); Aftershave Lotions

Technical/Other Names:
Oils, Rosemary
Rosemarinus Officinalis Oil
Rosemary absolute (RIFM)
Rosemary Leaf Oil
Rosemary Oil
Rosemary oil (Rosemarinus officinalis L.) (RIFM)
Rosemary (Rosemarinus officinalis L.) (RIFM)
Rosmarinus Officinalis (EU)

Trade Names:
AEC Rosemary Oil (A & E Connock)
Custosense Rosemary (Rosemary Oil) (Custom Ingredients)
Rosemary Essential Oil (Alban Muller)

Trade Name Mixtures:
Anti-Rheumatic Phytospa (Alban Muller)
Anti-Rheumatic Phytospa NaCl (Alban Muller)
Aromaphyte of Rosemary (Active Organics)
Covazen Detox (LCW)
C-Protect (Gattefosse s.a.)
Essentiaderm n.2 (Universal Flavors)
Essentiaderm n.7 (Universal Flavors)
Essentiaderm n.9 (Universal Flavors)
Essentiaderm N.20 (Universal Flavors)
Essentiaderm N.21 (Universal Flavors)
Extrapone Rosemary 2/783630 (Symrise)
Extrapone Rosemary/Aloe Vera Blend 2/ 033197 (Symrise)
Rosemary Softcream (CEP (Solabia))
Slimming/Tonic Phytospa (Alban Muller)
Slimming/Tonic Phytospa NaCl (Alban Muller)
Unioil Herbs (Chemyunion)

ROSMARINUS OFFICINALIS (ROSEMARY) LEAF POWDER

JPN Translation:
ローズマリー

Definition: Rosmarinus Officinalis (Rosemary) Leaf Powder is the powder derived from the leaf of *Rosmarinus officinalis*. *See "Regulatory and Ingredient Use Information," regarding the labeling names for botanical ingredients in Volume 1, Introduction, Part A.*

Information Sources: JCIC, JCLS

Chemical Class: Biological Products

Function: Flavoring Agent

Technical/Other Names:
Rosemary Leaf Powder
Rosemary Powder
Rosmarinus Officinalis (EU)

ROSMARINUS OFFICINALIS (ROSEMARY) LEAF WATER

JPN Translation:
ローズマリー水

Definition: Rosmarinus Officinalis (Rosemary) Leaf Water is an aqueous solution of the steam distillate obtained from the leaves of *Rosmarinus officinalis*. *See "Regulatory and Ingredient Use Information," regarding the labeling names for botanical ingredients in Volume 1, Introduction, Part A.*

Chemical Class: Biological Products

Function: Fragrance Ingredient

Technical/Other Names:
Rosemary Leaf Water
Rosinarinus Officinalis (EU)
Water, Rosemary Leaf
Water, Rosmarinus Officinalis Leaf

Trade Names:
Extrait Originel Romarin (Gattefosse s.a.)
Rosemary Water (Maruzen Pharmaceuticals Co., Ltd.)
Rosemary Water K (Koei Perfumery)

Trade Name Mixture:
Aroma Rosemary B (Ichimaru Pharcos)

ROSMARINUS OFFICINALIS (ROSEMARY) WATER

Definition: Rosmarinus Officinalis (Rosemary) Water is an aqueous solution of the steam distillate obtained from *Rosmarinus officinalis*. *See "Regulatory and Ingredient Use Information," regarding the labeling names for botanical ingredients in Volume 1, Introduction, Part A.*

Chemical Class: Biological Products

Function: Fragrance Ingredient

Technical/Other Name:
Rosmarinus Officinalis (EU)

Trade Name:
Rosemary Hydroflorate (Bayliss Ranch)

ROYAL JELLY

CAS No.: 8031-67-2

JPN Translation:
ローヤルゼリー

Definition: Royal Jelly is the pharyngeal secretion of worker bees. *See "Regulatory and Ingredient Use Information," regarding use of EU Trivial names in Volume 1, Introduction, Part A.*

Information Sources: JCIC, JCLS, JSQI, MAR, MI-13(8357)

Chemical Class: Biological Products

Function: Skin-Conditioning Agent - Miscellaneous

Reported Product Categories: Skin Care Preparations, Misc.; Moisturizing Preparations; Night Skin Care Preparations; Bath Capsules; Bath Oils, Tablets, and Salts; Bath Preparations, Misc.; Body and Hand Preparations (Excluding Shaving Preparations); Cleansing Products (Cold Creams, Cleansing Lotions, Liquids and Pads); Face and Neck Preparations (Excluding Shaving Preparations)

Technical/Other Name:
Queen Bee Jelly

Trade Name:
AEC Royal Jelly Fresh (A & E Connock)

Trade Name Mixture:
Campo Citisu (Campo)

ROYAL JELLY EXTRACT

CAS No.	EINECS No.
91081-56-0	293-662-7

JPN Translation:
ローヤルゼリーエキス

Definition: Royal Jelly Extract is an extract of Royal Jelly (q.v.). *See "Regulatory and Ingredient Use Information," regarding use of EU Trivial names in Volume 1, Introduction, Part A.*

Information Sources: JCIC, JCLS, JSQI

Chemical Class: Biological Products

Function: Not Reported

Reported Product Categories: Skin Care Preparations, Misc.; Moisturizing Preparations; Night Skin Care Preparations; Body and Hand Preparations (Excluding Shaving Preparations); Cleansing Products (Cold Creams, Cleansing Lotions, Liquids and Pads); Face and Neck Preparations (Excluding Shaving Preparations)

Technical/Other Name:
Extract of Royal Jelly

Trade Names:
Royal Jelly Extract Powder F (Maruzen Pharmaceuticals Co., Ltd.)
Royal Jelly Liquid (Ichimaru Pharcos)

Trade Name Mixtures:
Actiphyte Royal Jelly (Active Organics)
Actiphyte Royal Jelly BG50P (Active Organics)
Actiphyte Royal Jelly GL (Active Organics)
Actiphyte Royal Jelly Lipo S (Active Organics)
Glycolysat of Royal Gelly (CEP (Solabia))
Lipoplastidine Pappa Regalis (Vevy)
Pharconix Royal Jelly (Ichimaru Pharcos)
Royal Jelly Extract BG (Maruzen Pharmaceuticals Co., Ltd.)
Royal Jelly Extract "COS (Cosmetochem) (Cosmetochem International Ltd.)
Royal Jelly Extract Powder-S (Maruzen Pharmaceuticals Co., Ltd.)
Royal Jelly Extract SQ (Maruzen Pharmaceuticals Co., Ltd.)
Royal Jelly Extract W (Maruzen Pharmaceuticals Co., Ltd.)
VT-268 Extract of Royal Jelly (Vege-Tech)

ROYAL JELLY POWDER

CAS No.: 8031-67-2

Definition: Royal Jelly Powder is the the powder obtained by lyophilizing Royal Jelly (q.v.). *See "Regulatory and Ingredient Use Information," regarding use of EU Trivial names in Volume 1, Introduction, Part A.*

Chemical Class: Biological Products

Function: Proprietary

Reported Product Categories: Skin Care Preparations, Misc.; Moisturizing Preparations; Night Skin Care Preparations; Bath Capsules; Bath Oils, Tablets, and Salts; Bath Preparations, Misc.; Body and Hand Preparations (Excluding Shaving Preparations); Cleansing Products (Cold Creams, Cleansing Lotions, Liquids and Pads); Face and Neck Preparations (Excluding Shaving Preparations)

Trade Name:
AEC Royal Jelly Powder (A & E Connock)

RUBBER LATEX

CAS No.	EINECS No.
9006-04-6	232-689-0

JPN Translation:
ゴムラテックス

Definition: Rubber Latex is the milky juice exuded from various species of the rubber tree. It has been processed to increase its solids content and stability.

Information Sources: 21CFR175.105, 21CFR175.125, JCLS, JSCI, MI-13(2064)

Chemical Class: Biological Polymers and their Derivatives

Functions: Film Former; Opacifying Agent

Technical/Other Names:
Natural Rubber Latex
Rubber, Natural

Trade Names:
Chicle (Avantgarde)
Sorva/Sorvinha (Avantgarde)

RUBIA CORDIFOLIA ROOT EXTRACT

Definition: Rubia Cordifolia Root Extract is an extract of the roots of *Rubia cordifolia. See "Regulatory and Ingredient Use Information," regarding the labeling names for botanical ingredients in Volume 1, Introduction, Part A. See Reported Ingredient Functions-The Cosmetic Drug Distinction, in Regulatory and Ingredient Use Information, Volume I, Part A.*

Chemical Class: Biological Products

Functions: Colorant; Skin Protectant

Technical/Other Name:
Extract of Rubia Cordifolia Root

Trade Name Mixtures:
Manjishta (Heritage Bio-Natural)
VT- 0687 Extract of Indian Madder Root (Vege)

RUBIA TINCTORUM ROOT

Definition: Rubia Tinctorum Root is the dried, crushed roots of madder, *Rubia tinctorum. See "Regulatory and Ingredient Use Information," regarding the labeling names for botanical ingredients in Volume 1, Introduction, Part A.*

Chemical Class: Biological Products

Function: Not Reported

RUBIA TINCTORUM ROOT EXTRACT

Definition: Rubia Tinctorum Root Extract is an extract of the roots of the madder, *Rubia tinctorum. See "Regulatory and Ingredient Use Information," regarding the labeling names for botanical ingredients in Volume 1, Introduction, Part A.*

Chemical Class: Biological Products

Function: Not Reported

Technical/Other Name:
Extract of Rubia Tinctorum

Trade Name Mixtures:
Herbasol Extract Madder (Root) (Cosmetochem) (Cosmetochem International Ltd.)
Madder Extract (Kelisema Italy)

RUBUS CHAMAEMORUS SEED EXTRACT

Definition: Rubus Chamaemorus Seed Extract is an extract of the seeds of *Rubus chamaemorus. See "Regulatory and Ingredient Use Information," regarding the labeling names for botanical ingredients in Volume 1, Introduction, Part A.*

Chemical Class: Biological Products

Function: Skin-Conditioning Agent - Miscellaneous

Technical/Other Name:
Extract of Rubus Chamaemorus Seed

Trade Name Mixture:
Actiphyte of Arctic Cloudberry Lipo O (Active Organics)

RUBUS CHAMAEMORUS SEED OIL

Definition: Rubus Chamaemorus Seed Oil is the fixed oil expressed from the seeds of *Rubus chamaemorus. See "Regulatory and Ingredient Use Information," regarding the labeling names for botanical ingredients in Volume 1, Introduction, Part A.*

Chemical Class: Fats and Oils

Function: Skin-Conditioning Agent - Miscellaneous

Technical/Other Name:
Oils, Rubus Chamaemorus Seed

Trade Name:
Cloudberry Seed Oil (CO2 Extract) (Aromtech)

RUBUS CHINGII FRUIT EXTRACT

Definition: Rubus Chingii Fruit Extract is an extract of the berries of *Rubus chingii. See "Regulatory and Ingredient Use Information," regarding the labeling names for*

botanical ingredients in Volume 1, Introduction, Part A.

Chemical Class: Biological Products

Function: Not Reported

Technical/Other Name:
Extract of Rubus Chingii

Trade Name Mixtures:
China Extract Baio Beng Li (E.U.K)
Chine Extract Rubus Chingli (Ennagram)
Fukubonshi Liquid E (Ichimaru Pharcos)

RUBUS DELICIOSUS (BOYSENBERRY) EXTRACT

Definition: Rubus Deliciosus (Boysenberry) Extract is an extract of the boysenberry, *Rubus deliciosus. See "Regulatory and Ingredient Use Information," regarding the labeling names for botanical ingredients in Volume 1, Introduction, Part A.*

Chemical Class: Biological Products

Function: Not Reported

Technical/Other Names:
Boysenberry Extract
Extract of Boysenberry
Extract of Rubus Deliciosus
Rubus Deliciosus Extract

Trade Name Mixtures:
Cosflor Loganberry HGS (A & E Connock)
VT-251 Extract of Boysenberry (Vege-Tech)

RUBUS DELICIOSUS (BOYSENBERRY) FRUIT JUICE

Definition: Rubus Deliciosus (Boysenberry) Fruit Juice is the liquid expressed from the fruit of *Rubus deliciosus. See "Regulatory and Ingredient Use Information," regarding the labeling names for botanical ingredients in Volume 1, Introduction, Part A.*

Chemical Class: Biological Products

Functions: Cosmetic Astringent; Skin-Conditioning Agent - Miscellaneous

Technical/Other Names:
Boysenberry Fruit Juice
Boysenberry Juice
Juice, Boysenberry
Juice, Rubus Deliciosus
Rubus Deliciosus Juice

Trade Name:
AEC Boysenberry Conc (A & E Connock)

RUBUS FRUTICOSUS (BLACKBERRY) FRUIT EXTRACT

CAS No.	EINECS No.
84787-69-9	284-110-6

Definition: Rubus Fruticosus (Blackberry) Fruit Extract is an extract of the fruit of the blackberry, *Rubus fruticosus. See "Regulatory and Ingredient Use Information," regarding the labeling names for botanical ingredients in Volume 1, Introduction, Part A.*

Information Sources: 21CFR172.510, RIFM

Chemical Class: Biological Products

Functions: Fragrance Ingredient; Not Reported

Technical/Other Names:
Blackberry bark extract (Rubus, spp. of section Eubatus) (RIFM)
Blackberry Extract
Blackberry Fruit Extract
Extract of Blackberry
Extract of Rubus Fruticosus
Rubus Fruticosus Extract

Trade Name Mixtures:
Blackberry Extract HG (Provital/Centerchem)
Brumble Concentrate HS 3670 G (Grau)
Dewberry Extract (Cosmetic Developments)
Herbasol-Extract Blackberry/Bramble (Cosmetochem)
Phytelene of Bramble, Blackberry-Bush EG 482 liquid (Indena SA)

RUBUS FRUTICOSUS (BLACKBERRY) JUICE

Definition: Rubus Fruticosus (Blackberry) Juice is the liquid expressed from the juice of the blackberry, *Rubus fruticosus. See "Regulatory and Ingredient Use Information," regarding the labeling names for botanical ingredients in Volume 1, Introduction, Part A.*

Chemical Class: Biological Products

Function: Cosmetic Astringent

Technical/Other Names:
Blackberry Juice
Juice, Blackberry
Juice, Rubus Fruticosus
Rubus Fruticosus Juice

Trade Name:
AEC Blackberry Conc (A & E Connock)

Trade Name Mixture:
Fruitapone Blackberry GT 2/037070 (Symrise)

RUBUS FRUTICOSUS (BLACKBERRY) LEAF EXTRACT

CAS No.	EINECS No.
84787-69-9	284-110-6

Definition: Rubus Fruticosus (Blackberry) Leaf Extract is an extract of the leaves of the blackberry, *Rubus fruticosus. See "Regulatory and Ingredient Use Information," regarding the labeling names for botanical ingredients in Volume 1, Introduction, Part A.*

Information Source: RIFM

Chemical Class: Biological Products

Functions: Fragrance Ingredient; Not Reported

Technical/Other Names:
Blackberry bark extract (Rubus, spp. of section Eubatus) (RIFM)
Blackberry Leaf Extract
Bramble Leaf Extract
Extract of Blackberry Leaves
Extract of Bramble Leaf
Extract of Rubus Fruticosus Leaves
Rubus Fruticosus Leaf Extract

Trade Name Mixtures:
Brumble Leaves Extract HS 2527 G (Grau)
Dewberry Extract (Cosmetic Developments)
Vegebios of Bramble (CEP (Solabia))

RUBUS FRUTICOSUS/IDAEUS EXTRACT

Definition: Rubus Fruticosus/Idaeus Extract is an extract of a hybrid of Rubus fruticosus (Blackberry) and Rubus Idaeus (Raspberry). *See "Regulatory and Ingredient Use Information," regarding the labeling names for botanical ingredients in Volume 1, Introduction, Part A.*

Chemical Class: Biological Products

Function: Proprietary

Technical/Other Name:
Extract of Rubus Fruticosus/Idaeus

Trade Name:
Cosflor Tayberry HGS (A & E Connock)

RUBUS IDAEUS (RASPBERRY) FRUIT

Definition: Rubus Idaeus (Raspberry) Fruit is the fruit of the raspberry, *Rubus idaeus. See "Regulatory and Ingredient Use Information," regarding the labeling names for botanical ingredients in Volume 1, Introduction, Part A.*

Information Source: MI-13(8204)

Chemical Class: Biological Products

Functions: Cosmetic Astringent; Skin-Conditioning Agent - Miscellaneous

Technical/Other Names:
Raspberry
Raspberry Fruit

Trade Name:
 AEC Raspberry Puree (A & E Connock)

RUBUS IDAEUS (RASPBERRY) FRUIT EXTRACT

CAS No.	EINECS No.
84929-76-0	284-554-0

JPN Translation:
キイチゴエキス

Definition: Rubus Idaeus (Raspberry) Fruit Extract is an extract of the fruit of the red raspberry, *Rubus idaeus. See "Regulatory and Ingredient Use Information," regarding the labeling names for botanical ingredients in Volume 1, Introduction, Part A.*

Information Sources: JCIC, JCLS, JSQI

Chemical Class: Biological Products

Function: Skin-Conditioning Agent - Miscellaneous

Reported Product Categories: Moisturizing Preparations; Bath Preparations, Misc.; Hair Conditioners; Skin Care Preparations, Misc.

Technical/Other Names:
 Extract of Red Raspberry
 Extract of Rubus Idaeus Fruit
 Raspberry Extract
 Raspberry Fruit Extract
 Red Raspberry Extract
 Red Raspberry Fruit Extract
 Rubus Idaeus Extract

Trade Name Mixtures:
 Actiphyte of Raspberry BG50 (Active Organics)
 Actiphyte of Raspberry GL50 (Active Organics)
 Actiphyte of Raspberry Lipo S (Active Organics)
 Actiphyte of Raspberry PG50 (Active Organics)
 A.H.A. Extracts (Ennagram)
 A.H.A. Extracts (Phytochim)
 Extrait De Framboise MPE 100 (Yves Rocher)
 Glycolysat of Raspberry (CEP (Solabia))
 Herbasol-Extract Raspberry (Fruit) (Cosmetochem)
 Natural Raspberry Aqueous Extract (Flores)
 Raspberry BG (Alban Muller)
 Raspberry Extract BG (Maruzen Pharmaceuticals Co., Ltd.)
 Raspberry Extract BG100 (Maruzen Pharmaceuticals Co., Ltd.)
 Raspberry Extract HG (Provital/ Centerchem)
 Raspberry Extract HS 2574 G (Grau)

 Raspberry Fruit HPG Titrated (Alban Muller)
 Raspberry HS (Alban Muller)
 Raspberry Liquid B (Ichimaru Pharcos)
 RAYOLYS (Greentech S.A)
 Rubus Idaeus (Raspberry) Fruit Extract ies (IES LABO)
 VT-086 Extract of Raspberry Fruit (Vege-Tech)

RUBUS IDAEUS (RASPBERRY) FRUIT WATER

Definition: Rubus Idaeus (Raspberry) Fruit Water is an aqueous solution of the steam distillate obtained from the fruit of *Rubus idaeus. See "Regulatory and Ingredient Use Information," regarding the labeling names for botanical ingredients in Volume 1, Introduction, Part A.*

Chemical Class: Biological Products

Functions: Fragrance Ingredient; Skin-Conditioning Agent - Miscellaneous

Technical/Other Names:
 Raspberry Water
 Water, Raspberry
 Water, Rubus Idaeus Fruit

Trade Name:
 Extrait Originel Framboise (Gattefosse s.a.)

RUBUS IDAEUS (RASPBERRY) JUICE

CAS No.: 8027-46-1

JPN Translation:
キイチゴ果汁

Definition: Rubus Idaeus (Raspberry) Juice is the liquid expressed from the fresh pulp of the red raspberry, *Rubus idaeus. See "Regulatory and Ingredient Use Information," regarding the labeling names for botanical ingredients in Volume 1, Introduction, Part A.*

Information Sources: BPC, JCIC, JCLS, MAR, USD, USP XVIII

Chemical Class: Biological Products

Function: Skin-Conditioning Agent - Miscellaneous

Technical/Other Names:
 Juice, Raspberry
 Juice, Rubus Idaeus
 Raspberry Juice
 Rubus Idaeus Juice

Trade Name:
 AEC Raspberry Conc (A & E Connock)

Trade Name Mixtures:
 Extrapone Raspberry 2/033318 (Symrise)

 Fruitapone Raspberry GT 2/037215 (Symrise)

RUBUS IDAEUS (RASPBERRY) LEAF EXTRACT

CAS No.	EINECS No.
84929-76-0	284-554-0

Definition: Rubus Idaeus (Raspberry) Leaf Extract is an extract of the leaves of the red raspberry, *Rubus idaeus. See "Regulatory and Ingredient Use Information," regarding the labeling names for botanical ingredients in Volume 1, Introduction, Part A.*

Chemical Class: Biological Products

Function: Not Reported

Technical/Other Names:
 Extract of Red Raspberry Leaf
 Extract of Rubus Idaeus Leaves
 Raspberry Leaf Extract
 Red Raspberry Leaf Extract

Trade Name Mixtures:
 Actiphyte of Raspberry Leaves BG50 (Active Organics)
 Actiphyte of Raspberry Leaves GL50 (Active Organics)
 Actiphyte of Raspberry Leaves Lipo S (Active Organics)
 Actiphyte of Raspberry Leaves PG50 (Active Organics)
 BBC Mineral Complex (Bio-Botanica)
 Herbasol Extract Raspberry Leaf (Cosmetochem) (Cosmetochem International Ltd.)
 Raspberry Leaf HPG Titrated (Alban Muller)
 Raspberry Leaf HS (Alban Muller)
 VT-085 Extract of Raspberry Leaves (Vege-Tech)

RUBUS IDAEUS (RASPBERRY) LEAF POWDER

Definition: Rubus Idaeus (Raspberry) Leaf Powder is the powder derived from the ground leaves of *Rubus idaeus. See "Regulatory and Ingredient Use Information," regarding the labeling names for botanical ingredients in Volume 1, Introduction, Part A.*

Chemical Class: Biological Products

Function: Skin-Conditioning Agent - Miscellaneous

Trade Name:
 Raspberry Leaf Powder (Aveda)

RUBUS IDAEUS (RASPBERRY) LEAF WAX

Definition: Rubus Idaeus (Raspberry) Leaf Wax is a wax obtained from the leaf of

---

Rubus idaeus. See "Regulatory and Ingredient Use Information," regarding the labeling names for botanical ingredients in Volume 1, Introduction, Part A.

Chemical Class: Essential Oils

Function: Fragrance Ingredient

Technical/Other Names:
Raspberry Leaf Wax
Raspberry Wax
Rubus Idaeus Wax
Waxes, Raspberry (Rubus Idaeus)

Trade Names:
AEC Cire Essentielle De Feuilles De Framboisier (A & E Connock)
Cire Essentielle de feuilles de Framboisier (Bertin)

RUBUS IDAEUS (RASPBERRY) SEED

Definition: Rubus Idaeus (Raspberry) Seed is the dried seeds of the raspberry, *Rubus idaeus. See "Regulatory and Ingredient Use Information," regarding the labeling names for botanical ingredients in Volume 1, Introduction, Part A.*

Chemical Class: Biological Products

Function: Abrasive

Technical/Other Names:
Raspberry Seed
Rubus Idaeus Seed
Seeds, Raspberry
Seeds, Rubus Idaeus

Trade Names:
Actiscrub R (Active Organics)
AEC Raspberry Seed (A & E Connock)

RUBUS IDAEUS (RASPBERRY) SEED OIL

Definition: Rubus Idaeus (Raspberry) Seed Oil is the fixed oil obtained from the seeds of *Rubus idaeus. See "Regulatory and Ingredient Use Information," regarding the labeling names for botanical ingredients in Volume 1, Introduction, Part A.*

Chemical Class: Fats and Oils

Function: Skin-Conditioning Agent - Emollient

Technical/Other Names:
Oils, Raspberry Seed
Oils, Rubus Idaeus Seed
Raspberry Seed Oil

Trade Names:
AEC Raspberry Seed Oil (A & E Connock)
Lipovol RASP (Lipo)
Raspberry Oil (IES LABO)

Raspberry Seed Oil (Aldivia)
Red Gamma Raspberry Seed Oil (CO2 extract) (Aromtech)
Sympholip FR (Symphonia)

Trade Name Mixture:
Raspberry Butter (Zenitech)

RUBUS IDAEUS SACHALINENSIS FRUIT JUICE

Definition: Rubus Idaeus Sachalinensis Fruit Juice is the juice expressed from the fruit of *Rubus idaeus sachalinensis. See "Regulatory and Ingredient Use Information," regarding the labeling names for botanical ingredients in Volume 1, Introduction, Part A.*

Chemical Class: Biological Products

Functions: Flavoring Agent; Skin-Conditioning Agent - Miscellaneous

Trade Name Mixture:
Boysenberry Fruit Water F (Maruzen Pharmaceuticals Co., Ltd.)

RUBUS OCCIDENTALIS FRUIT EXTRACT

Definition: Rubus Occidentalis Fruit Extract is an extract of the fruit of *Rubus occidentalis. See "Regulatory and Ingredient Use Information," regarding the labeling names for botanical ingredients in Volume 1, Introduction, Part A.*

Chemical Class: Biological Products

Functions: Cosmetic Astringent; Skin-Conditioning Agent - Miscellaneous

Technical/Other Name:
Extract of Rubus Occidentalis Fruit

Trade Name Mixtures:
Actiphyte of Black Raspberry (Active Organics)
Black Raspberry Extract (Carrubba)

RUBUS PARVIFOLIUS FRUIT EXTRACT

Definition: Rubus Parvifolius Fruit Extract is an extract of the berries of *Rubus parvifolius. See "Regulatory and Ingredient Use Information," regarding the labeling names for botanical ingredients in Volume 1, Introduction, Part A.*

Chemical Class: Biological Products

Function: Not Reported

Technical/Other Name:
Extract of Rubus Parvifolius

Trade Name Mixture:
Campo Biao Beng Li Extract (Campo)

RUBUS STRIGOSUS (RASPBERRY) SEED

Definition: Rubus Strigosus (Raspberry) Seed is the crushed seed obtained from the fruit of *Rubus strigosus. See "Regulatory and Ingredient Use Information," regarding the labeling names for botanical ingredients in Volume 1, Introduction, Part A.*

Chemical Class: Biological Products

Function: Skin-Conditioning Agent - Miscellaneous

Trade Name:
Red Raspberry Seeds, Premier (Premier Specialties)

RUBUS SUAVISSIMUS (RASPBERRY) LEAF EXTRACT

JPN Translation:
テンチャエキス

Definition: Rubus Suavissimus (Raspberry) Leaf Extract is an extract of the leaves of *Rubus suavissimus. See "Regulatory and Ingredient Use Information," regarding the labeling names for botanical ingredients in Volume 1, Introduction, Part A.*

Chemical Class: Biological Products

Function: Not Reported

Reported Product Categories: Moisturizing Preparations; Bath Preparations, Misc.; Hair Conditioners; Skin Care Preparations, Misc.

Technical/Other Names:
Extract of Rubus Sauvissimus
Raspberry Extract
Raspberry Leaf Extract
Rubus Suavissimus Extract

Trade Name Mixtures:
Tiencha Extract BG (Maruzen Pharmaceuticals Co., Ltd.)
Tiencha Extract BGW (Maruzen Pharmaceuticals Co., Ltd.)

RUBUS VILLOSUS (BLACKBERRY) FRUIT EXTRACT

Definition: Rubus Villosus (Blackberry) Fruit Extract is an extract of the fruit of the blackberry, *Rubus villosus. See "Regulatory and Ingredient Use Information," regarding the labeling names for botanical ingredients in Volume 1, Introduction, Part A.*

Information Source: 21CFR172.510

Chemical Class: Biological Products

Function: Not Reported

Technical/Other Names:
Blackberry Extract
Blackberry Fruit Extract
Extract of Blackberry
Extract of Rubus Villosus
Rubus Villosus Extract

Trade Name Mixtures:
Actiphyte of Blackberry BG50 (Active
   Organics)
Actiphyte of Blackberry GL50 (Active
   Organics)
Actiphyte of Blackberry Lipo S (Active
   Organics)
Actiphyte of Blackberry PG50 (Active
   Organics)
VT-136 Extract of Blackberry (Vege-Tech)

RUBUS VILLOSUS (BLACKBERRY) LEAF EXTRACT

Definition: Rubus Villosus (Blackberry) Leaf Extract is an extract of the leaves of the blackberry, *Rubus villosus*. See *"Regulatory and Ingredient Use Information," regarding the labeling names for botanical ingredients in Volume 1, Introduction, Part A.*

Chemical Class: Biological Products

Function: Not Reported

Technical/Other Names:
Blackberry Leaf Extract
Bramble Leaf Extract
Extract of Blackberry Leaves
Extract of Bramble Leaf
Extract of Rubus Villosus Leaf
Rubus Villosus Leaf Extract

Trade Name Mixtures:
Actiphyte of Blackberry Leaves BG50
   (Active Organics)
Actiphyte of Blackberry Leaves GL50
   (Active Organics)
Actiphyte of Blackberry Leaves Lipo S
   (Active Organics)
Actiphyte of Blackberry Leaves PG50
   (Active Organics)

RUBUS VILLOSUS (BLACKBERRY) ROOT EXTRACT

Definition: Rubus Villosus (Blackberry) Root Extract is an extract of the roots of the blackberry, *Rubus villosus. See "Regulatory and Ingredient Use Information," regarding the labeling names for botanical ingredients in Volume 1, Introduction, Part A.*

Chemical Class: Biological Products

Function: Not Reported

Technical/Other Names:
Blackberry Root Extract
Extract of Blackberry Roots
Extract of Rubus Villosus Root
Rubus Villosus Root Extract

RUBY POWDER

CAS No.: 12174-49-1

Definition: Ruby Powder is the powder obtained from crushed rubies and consists chiefly of aluminum oxide.

Chemical Class: Inorganics

Function: Skin-Conditioning Agent - Miscellaneous

Trade Name:
Ruby Av-1 (Aveda)

RUMEX ACETOSELLA EXTRACT

Definition: Rumex Acetosella Extract is an extract of the sorrel, *Rumex acetosella. See "Regulatory and Ingredient Use Information," regarding the labeling names for botanical ingredients in Volume 1, Introduction, Part A.*

Chemical Class: Biological Products

Function: Not Reported

Technical/Other Names:
Extract of Rumex Acetosella
Extract of Sorrel
Rumex Extract
Sorrel Extract
Sorrel (Rumex Acetosella) Extract

Trade Name Mixtures:
Actiphyte of Sheep's Sorrel BG50 (Active
   Organics)
Actiphyte of Sheep's Sorrel GL50 (Active
   Organics)
Actiphyte of Sheep's Sorrel Lipo S (Active
   Organics)
Actiphyte of Sheep's Sorrel PG50 (Active
   Organics)
Sorrel Extract HS 2582 G (Grau)
VT-174 Extract of Sheep Sorrel (Vege-
   Tech)

RUMEX CRISPUS ROOT EXTRACT

CAS No.	EINECS No.
90106-41-5	290-264-5

Definition: Rumex Crispus Root Extract is an extract of the roots of the curled dock, *Rumex crispus. See "Regulatory and Ingredient Use Information," regarding the labeling names for botanical ingredients in Volume 1, Introduction, Part A.*

Chemical Class: Biological Products

Function: Skin-Conditioning Agent - Miscellaneous

Technical/Other Names:
Curled Dock Extract
Curled Dock (Rumex Crispus) Extract
Extract of Curled Dock
Extract of Rumex Crispus

Trade Name Mixtures:
Actiphyte of Yellow Dock BG50 (Active
   Organics)
Actiphyte of Yellow Dock GL50 (Active
   Organics)
Actiphyte of Yellow Dock Lipo S (Active
   Organics)
Actiphyte of Yellow Dock PG50 (Active
   Organics)
Curled Dock HS (Alban Muller)
Dragostat-9 (2/050900) (Symrise)
Dragostat-11 (2/050950) (Symrise)
VT-242 Extract of Yellow Dock (Vege-
   Tech)

RUMEX OCCIDENTALIS EXTRACT

Definition: Rumex Occidentalis Extract is an extract of the plant *Rumex occidentalis. See "Regulatory and Ingredient Use Information," regarding the labeling names for botanical ingredients in Volume 1, Introduction, Part A.*

Chemical Class: Biological Products

Function: Not Reported

Technical/Other Name:
Extract of Rumex Occidentalis

Trade Name Mixtures:
Tyrostat - 08 (Fytokem)
Tyrostat - 09 (Fytokem)
Tyrostat-10 (Fytokem)
Tyrostat-11 (Fytokem)
Tyrostat - 20 (Fytokem)
Tyrostat - 21 (Fytokem)

RUSCOGENIN

CAS No.	EINECS No.
472-11-7	207-447-2

Empirical Formula:
$C_{27}H_{42}O_4$

Definition: Ruscogenin is the sterol that conforms to the formula:

Chemical Class: Sterols

Function: Skin-Conditioning Agent - Miscellaneous

Technical/Other Names:
(25R)-Spirost-5-ene-1β,3β-diol
Spirost-5-ene-1,3-diol, (1β-,3β,25R)-

Trade Name Mixtures:
Plantactiv Ruscus (Cognis Deutschland)
Ruscogenines Vinyals (Vinyals)
Ruscogenins (Indena SpA)

RUSCUS ACULEATUS ROOT EXTRACT

CAS No.	EINECS No.
84012-38-4	281-682-9

JPN Translation:
ブッチャーブルームエキス

Definition: Ruscus Aculeatus Root Extract is an extract of the rhizomes of the butcherbroom, *Ruscus aculeatus. See "Regulatory and Ingredient Use Information," regarding the labeling names for botanical ingredients in Volume 1, Introduction, Part A.*

Information Sources: JCIC, JCLS

Chemical Class: Biological Products

Function: Skin-Conditioning Agent - Miscellaneous

Reported Product Categories: Bath Preparations, Misc.; Body and Hand Preparations (Excluding Shaving Preparations); Skin Care Preparations, Misc.; Bath Capsules; Paste Masks (Mud Packs); Face and Neck Preparations (Excluding Shaving Preparations); Moisturizing Preparations

Technical/Other Names:
Butcherbroom Extract
Butcherbroom (Ruscus Aculeatus) Extract
Extract of Butcherbroom
Extract of Ruscus Aculeatus
Rusco Extract

Trade Name:
Herbalia Butcher's Broom (Cognis Deutschland)

Trade Name Mixtures:
Actiphyte of Butcher's Broom BG50 (Active Organics)
Actiphyte of Butcher's Broom GL50 (Active Organics)
Actiphyte of Butcher's Broom Lipo S (Active Organics)
Actiphyte of Butcher's Broom PG50 (Active Organics)
Butcher Broom Extract (Maruzen Pharmaceuticals Co., Ltd.)
Butcherbroom Extract PG (Rahn)
Butcherbroom Liquid E (Ichimaru Pharcos)
Butcherbroom Tincture (Rahn)
Butcher's Broom Extract HG (Provital/Centerchem)
Butcher's Broom HS (Alban Muller)
Butcher's Broom Phytexcell (Crodarom)
318 Circulatory Blend HS (Alban Muller)
Extrait de Petit Houx MPE PG 40 (Yves Rocher)
Extrapone Butcher's Broom GW P 2/030700 (Symrise)
Glycolysat of Butcher's Broom (CEP (Solabia))
Heavy Legs Complex 269 (Ennagram)
Heavy Legs Phytogreen Complex GXH 269 (Phytochim)
Herbasol-Extract Burcher's Broom (Ruscus) (Cosmetochem)
Nat Ruscus Extract (Natiris)
Natupure Butcher's Broom (E.U.K)
Perfeline (Rahn)
Phytelene Complex EGX 254 (Indena SA)
Phytelene of Butcher's Broom EG 440 liquid (Indena SA)
Phytogreen 55 of Butcher's Broom EXH 676 Liquid (Phytochim)
Prodhy Extract Petit Houx (Prod'Hyg)
Pronalen A/C HSC (Provital/Centerchem)
Pronalen Anti-Cellulite HSC (Provital/Centerchem)
Pronalen Anti-Fatigue HSC (Provital/Centerchem)
Pronalen Ruscus HSC (Provital/Centerchem)
Pronalen Ruscus SPE (Provital/Centerchem)
Pronalen Slimming (Provital/Centerchem)
Radix Rusci Extract HS 3027 G (Grau)
Sederma Butcherbroom (Sederma)
TIRED LEGS (Greentech S.A)
Vegebois of Butcher's Broom (CEP (Solabia))
Vegetol Butcher's Broom GR 089 Hydro (Gattefosse s.a.)
VT-062 Extract of Butcherbroom (Vege-Tech)

RUTA GRAVEOLENS

Definition: *See "Regulatory and Ingredient Use Information," regarding EU labeling names for botanical ingredients in Volume 1, Introduction, Part A.*

Chemical Class: Biological Products

Technical/Other Names:
Ruta Graveolens (Rue) Extract (U.S.)
Ruta Graveolens (Rue) Oil (U.S.)

RUTA GRAVEOLENS (RUE) EXTRACT

CAS No.	EINECS No.
84929-47-5	284-531-5

Definition: Ruta Graveolens (Rue) Extract is an extract of the leaves, roots and stems of the rue, *Ruta graveolens. See "Regulatory and Ingredient Use Information," regarding the labeling names for botanical ingredients in Volume 1, Introduction, Part A.*

Chemical Class: Biological Products

Function: Not Reported

Technical/Other Names:
Extract of Rue
Extract of Ruta Graveolens
Rue Extract
Ruta Graveolens (EU)
Ruta Graveolens Extract

Trade Name Mixtures:
Extrait Hydroglycolique de Rue Officinale (Greentech)
Herbasol Extract Rue (Cosmetochem) (Cosmetochem International Ltd.)
VT-070 Extract of Rue (Vege-Tech)

RUTA GRAVEOLENS (RUE) OIL

CAS No.: 8014-29-7

Definition: Ruta Graveolens (Rue) Oil is the volatile oil distilled from the herb *Ruta graveolens. See "Regulatory and Ingredient Use Information," regarding the labeling names for botanical ingredients in Volume 1, Introduction, Part A.*

Information Sources: 21CFR184.1699, FCC, HP, MAR, MI-13(6872), RIFM, TSCA

Chemical Class: Essential Oils

Function: Fragrance Ingredient

Technical/Other Names:
Oil of Rue
Oils, Rue
Rue Oil
Rue oil (Ruta graveolens L.) (RIFM)
Rue (Ruta graveolens L.) (RIFM)
Ruta Graveolens (EU)
Ruta Graveolens Oil

RUTIN

CAS Nos.	EINECS No.
153-18-4	205-814-1
130603-71-3	

Empirical Formula:
$C_{27}H_{30}O_{16}$

Definition: Rutin is the organic compound that conforms to the formula:

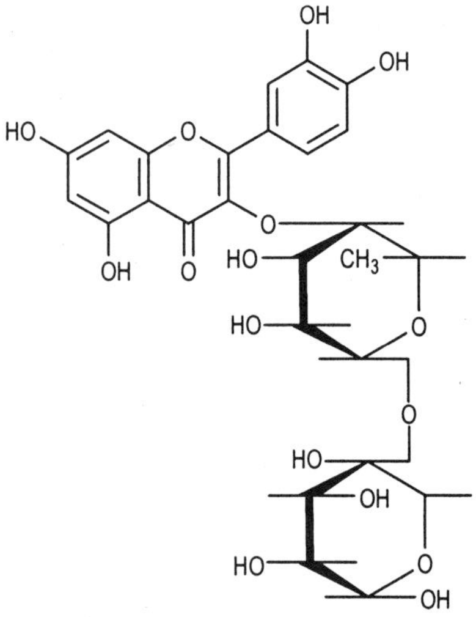

Information Sources: 21CFR330.12, JAN, JCIC, JCLS, JSQI, MI-13(8383), TSCA

Chemical Classes: Carbohydrates; Heterocyclic Compounds; Phenols; Polyols

Functions: Antioxidant; Hair Conditioning Agent; Skin-Conditioning Agent - Miscellaneous

Reported Product Category: Bath Soaps and Detergents

Technical/Other Names:
Eldrin
4H-1-Benzopyran-4-one, 3-[[6-O-(6-Deoxy-α-L-Mannopyranosyl)-β-D-Gluco-pyranosyl]oxy]-2-(3,4-Dihydroxyphenyl)-5,7-Dihydroxy-
Ilixanthin
Myrticolorin
Oxyritin
3,3',4',5,7-Pentahydroxyflavone 3-Rutinoside
Quercetin 3-O-Rutinoside
Rutoside

Trade Name Mixture:
PL-Troxe (Va Ma Farmacosmetica S.R.L)

RYOKU-CHA EKISU

JPN Translation:
チャエキス

Definition: Ryoku-Cha Ekisu is an extract of the green tea leaves of *Camellia sinensis*. See *"Regulatory and Ingredient Use Information,"* regarding use of Japan Trivial names in Volume 1, Introduction, Part A.

Chemical Class: Biological Products

Functions: Antioxidant; Skin-Conditioning Agent - Miscellaneous

Technical/Other Name:
Camellia Sinensis Leaf Extract (U.S.)

RYOKUSO EKISU

JPN Translation:
緑藻エキス

Definition: Ryokuso Ekisu is an extact of *Chlorophyta spp.* See *"Regulatory and Ingredient Use Information,"* regarding use of Japan Trivial names in Volume 1, Introduction, Part A.

Chemical Class: Biological Products

Function: Skin-Conditioning Agent - Miscellaneous

RYOKUSOU EKISU

Definition: See *"Regulatory and Ingredient Use Information,"* regarding use of Japan Trivial names in Volume 1, Introduction, Part A.

Information Source: JCLS

Function: Not Reported

- S -

SABBATIA ANGULARIS

Definition: *See "Regulatory and Ingredient Use Information," regarding EU labeling names for botanical ingredients in Volume 1, Introduction, Part A.*

Chemical Class: Biological Products

Technical/Other Name:
Sabbatia Angularis Extract (U.S.)

SABBATIA ANGULARIS EXTRACT

Definition: Sabbatia Angularis Extract is an extract of the American centaury, *Sabbatia angularis*. See "Regulatory and Ingredient Use Information," regarding the labeling names for botanical ingredients in Volume 1, Introduction, Part A.

Chemical Class: Biological Products

Function: Not Reported

Technical/Other Names:
American Centaury Extract
American Centaury (Sabbatia Angularis) Extract
Extract of American Centaury
Extract of Sabbatia Angularis
Sabbatia Angularis (EU)

Trade Name Mixtures:
Actiphyte of Centaury BG50 (Active Organics)
Actiphyte of Centaury GL50 (Active Organics)
Actiphyte of Centaury Lipo S (Active Organics)
Actiphyte of Centaury PG50 (Active Organics)

SACCHARATED LIME

CAS Nos.
5793-88-4
8002-17-3

EINECS No.
227-334-1

Definition: Saccharated Lime is the reaction product from the oxidation of gluconic acid followed by neutralization with lime.

Information Sources: CTFA D, MI-13 (1705)

Chemical Class: Organic Salts

Function: Emulsion Stabilizer

Technical/Other Names:
Calcium Saccharate
Glucaric Acid, Calcium Salt

SACCHARIDE HYDROLYSATE

CAS No.
8013-17-0

EINECS No.
232-393-1

Definition: Saccharide Hydrolysate is an invert sugar derived by the hydrolysis of sucrose by acid, enzyme, or other method of hydrolysis. It is characterized by a content of fructose and glucose.

Information Source: MI-13(5027)

Chemical Class: Carbohydrates

Function: Skin-Conditioning Agent - Humectant

Technical/Other Names:
Insubeta
Invertose
Invert Sugar
Nulomoline
Travert

Trade Name:
Pepha-Hydrate (Pentapharm/Centerchem)

Trade Name Mixtures:
Lipoderma AA (Lipo)
Unimoist U-125 (Induchem)

SACCHARIDE ISOMERATE

JPN Translation:
異性化糖

Definition: Saccharide Isomerate is a carbohydrate complex formed from a base catalyzed rearrangement of a mixture of saccharides.

Information Sources: JCIC, JCLS, JSQI

Chemical Class: Carbohydrates

Function: Skin-Conditioning Agent - Humectant

Reported Product Categories: Moisturizing Preparations; Eyebrow Pencils; Suntan Gels, Creams, and Liquids; Bath Oils, Tablets, and Salts; Bath Preparations, Misc.; Body and Hand Preparations (Excluding Shaving Preparations); Cleansing Products (Cold Creams, Cleansing Lotions, Liquids and Pads)

Trade Name:
Pentavitin (Pentapharm/Centerchem)

SACCHARIN

CAS No.
81-07-2

EINECS No.
201-321-0

JPN Translation:
サッカリン

Empirical Formula:
$C_7H_5NO_3S$

Definition: Saccharin is the organic compound that conforms to the formula:

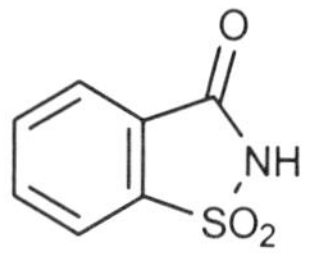

Information Sources: ARG, BP, BPC, BRA, 21CFR145.116, 21CFR145.126, 21CFR145.131, 21CFR145.136, 21CFR145.171, 21CFR145.181, 21CFR150.141, 21CFR150.161, 21CFR180.37, 21CFR310.545, DA, EGY, FCC, IND, ITA, JCLS, JSCI, MAR, MI-13 (8390), NF XIX, PF, PN, POR, RIFM, ROM, TSCA, USAN, USD

Chemical Classes: Amides; Heterocyclic Compounds

Functions: Flavoring Agent; Fragrance Ingredient

Reported Product Categories: Moisturizing Preparations; Mouthwashes and Breath Fresheners (Liquids and Sprays); Dentifrices (Aerosol, Liquid, Pastes and Powders); Lipsticks

Technical/Other Names:
1,2-Benzisothiazol-3(2H)-one, 1,1-Dioxide
o-Benzoic Acid Sulfimide
1,1-Dioxide-1,2-Benzisothiazol-3(2H)-one
Garantose
Saccharin (RIFM)
Saccharine
o-Sulfobenzimide

SACCHAROMYCES/BARLEY SEED FERMENT FILTRATE

Definition: Saccharomyces/Barley Seed Ferment Filtrate is a filtrate of the product obtained by the fermentation of barley seeds by the organism *Saccharomyces*.

Chemical Class: Biological Products

Function: Skin-Conditioning Agent - Humectant

Trade Name Mixture:
Barley Culture Extract (Sansho Seiyaku)

SACCHAROMYCES/CALCIUM FERMENT

Definition: Saccharomyces/Calcium Ferment is an extract of a fermentation product of saccharomyces in the presence of calcium ions.

Chemical Class: Biological Products

Function: Not Reported

Trade Name:
Biomin Ca/P/C (Arch Personal Care Products)

Trade Name Mixtures:
ACB Bio-Chelate Calcium (Active Concepts)
Acqua Biomin Calcium Y3 (Arch Personal Care Products)

SACCHAROMYCES CEREVISIAE EXTRACT

CAS No.	EINECS No.
84604-16-0	283-294-5

Definition: Saccharomyces Cerevisiae Extract is an extract of the yeast cells of *Saccharomyces cerevisiae*.

Chemical Class: Biological Products

Function: Not Reported

Technical/Other Name:
Extract of Saccharomyces Cerevisiae

Trade Name:
Actiflow (Silab)

Trade Name Mixtures:
Barm Extract HS 2963 G (Grau)
Chronosphere TRF (Arch Personal Care Products)
EASHAVE (Pentapharm/Centerchem)
Iricalmin (Pentapharm/Centerchem)
Pro-D-Sine (Sederma)
Pseudofilaggrin (Arch Personal Care Products)
Red Wine Yeast Extract (FAO-II-Red) (Kyowa Hakko Kogyo)
Wine Yeast Extract (FAO) (Kyowa Hakko Kogyo)
Wine Yeast Extract (FAO-II) (Kyowa Hakko Kogyo)
Wine Yeast Extract (FAW) (Kyowa Hakko Kogyo)
Yeast Extract BG (Maruzen Pharmaceuticals Co., Ltd.)
Yeast Extract BGN (Maruzen Pharmaceuticals Co., Ltd.)

SACCHAROMYCES/COPPER FERMENT

Definition: Saccharomyces/Copper Ferment is an extract of a fermentation product of saccharomyces in the presence of copper ions.

Chemical Class: Biological Products

Function: Not Reported

Trade Name Mixtures:
ACB Bio-Chelate (Active Concepts)
ACB Bio-Chelate 5 (Active Concepts)
ACB Bio-Chelate Copper (Active Concepts)
ACB Bio-Chelate Sil (Active Concepts)
ACB Bio-Chelate (Zn,Cu,Mn,Fe,Si,K) (Active Concepts)

ACB Oligomin 5 (Active Concepts)
Acqua Biomin Copper Y3 (Arch Personal Care Products)
Biomin Acquacinque Liquid (Arch Personal Care Products)
Biomin Cu/P/C Liquid (Arch Personal Care Products)
Biomin 3 PF (Arch Personal Care Products)

SACCHAROMYCES/COPPER FERMENT LYSATE FILTRATE

Definition: Saccharomyces/Copper Ferment Lysate Filtrate is the filtration product of a lysate of the fermentation of *Saccharomyces* in the presence of copper ions.

Chemical Class: Biological Products

Function: Skin-Conditioning Agent - Miscellaneous

Trade Name Mixture:
Biomin Triplex Y3 (Arch Personal Care Products)

SACCHAROMYCES/CUCUMIS MELO FRUIT/FRAGARIA CHILOENSIS FRUIT/VITIS VINIFERA FRUIT/AKEBIA QUINATA FRUIT/PYRUS MALUS FRUIT/FICUS CARICA FRUIT/MUSA PARADISIACA FRUIT SUCROSE/FERMENT FILTRATE

Definition: Saccharomyces/Cucumis Melo Fruit/Fragaria Chiloensis Fruit/Vitis Vinifera Fruit/Akebia Quinata Fruit/Pyrus Malus Fruit/Ficus Carica Fruit/Musa Paradisiaca Fruit Sucrose/Ferment Filtrate is a filtrate of the fermentation product of a mixture of sucrose and the fruits of *Cucumis melo*, *Fragaria chiloensis*, *Vitis vinifera*, *Akebia quinata*, *Pyrus malus*, *Ficus carica*, *Musa paradisia* by the microorganism, *Saccharomyces*.

Chemical Class: Biological Products

Function: Skin-Conditioning Agent - Humectant

Trade Name Mixture:
MS Fruits Ferment OM (Maruzen Pharmaceuticals Co., Ltd.)

SACCHAROMYCES FERMENT

Definition: Saccharomyces Ferment is a product obtained by the fermentation of *Saccharomyces*.

Chemical Class: Biological Products

Function: Not Reported

Trade Names:
Intradermin (Pentapharm/Centerchem)
Vegetable Protein Extract (Carrubba)

SACCHAROMYCES FERMENT FILTRATE

Definition: Saccharomyces Ferment Filtrate is a filtrate of the fermentation product of *Saccharomyces*.

Chemical Class: Biological Products

Function: Skin-Conditioning Agent - Humectant

Trade Name Mixture:
EPPC (Arch Personal Care Products)

SACCHAROMYCES FERMENT LYSATE FILTRATE

Definition: Saccharomyces Ferment Lysate Filtrate is the filtration product of a lysate of the fermentation of *Saccharomyces*.

Chemical Class: Biological Products

Function: Skin-Conditioning Agent - Miscellaneous

Trade Name Mixture:
Biodynes EMPP (Arch Personal Care Products)

SACCHAROMYCES/FLUORINE FERMENT

Definition: Saccharomyces/Fluorine Ferment is an extract of a fermentation product of saccharomyces in the presence of fluorine ions.

Chemical Class: Biological Products

Function: Not Reported

Trade Name:
Biomin F/P/C (Arch Personal Care Products)

SACCHAROMYCES/GERMANIUM FERMENT

Definition: Saccharomyces/Germanium Ferment is an extract of a fermentation product of saccharomyces in the presence of germanium ions.

Chemical Class: Biological Products

Function: Not Reported

Trade Name Mixture:
Acqua Biomin Germanium Y3 (Arch Personal Care Products)

SACCHAROMYCES/GOLD FERMENT LYSATE FILTRATE

Definition: Saccharomyces/Gold Ferment Lysate Filtrate is the filtration product of a lysate of the fermentation of *Saccharomyces* in the presence of gold ions.

Information Source: EEC(II-296)

Chemical Class: Biological Products

Function: Skin-Conditioning Agent - Miscellaneous

Trade Name:
Acquabiomin Gold Y3 (Arch Personal Care Products)

Trade Name Mixture:
Biomin Triplex Y3 (Arch Personal Care Products)

SACCHAROMYCES/GRAPE FERMENT EXTRACT

Definition: Saccharomyces/Grape Ferment Extract is an extract of the product obtained from the fermentation of grapes by the microorganism *Saccharomyces*.

Chemical Class: Biological Products

Function: Skin-Conditioning Agent - Miscellaneous

Trade Name Mixture:
Champagne Truffle Complex (Greentech)

SACCHAROMYCES/IRON FERMENT

Definition: Saccharomyces/Iron Ferment is an extract of a fermentation proudct of saccharomyces in the presence of iron ions.

Chemical Class: Biological Products

Function: Not Reported

Trade Name Mixtures:
ACB Bio-Chelate (Active Concepts)
ACB Bio-Chelate 5 (Active Concepts)
ACB Bio-Chelate Sil (Active Concepts)
ACB Bio-Chelate (Zn,Cu,Mn,Fe,Si,K) (Active Concepts)
ACB Oligomin 5 (Active Concepts)
Acqua Biomin Iron Y3 (Arch Personal Care Products)
Biomin Acquacinque Liquid (Arch Personal Care Products)

SACCHAROMYCES/LAMINARIA SACCHARINA FERMENT

Definition: Saccharomyces/Laminaria Saccharina Ferment is the product obtained by the fermentation of the algae *Laminaria saccharina*, by the microorganism *Saccharomyces*.

Chemical Class: Biological Products

Function: Skin-Conditioning Agent - Miscellaneous

Trade Name Mixture:
Laminaria Saccharina 2/033600 (Symrise)

SACCHAROMYCES LYSATE

Definition: Saccharomyces Lysate is the end product of the controlled lysis of various species of *Saccharomyces*.

Chemical Class: Biological Products

Function: Not Reported

Trade Name:
OXY 229-BT (Pentapharm/Centerchem)

SACCHAROMYCES LYSATE EXTRACT

Definition: Saccharomyces Lysate Extract is an extract of Saccharomyces Lysate (q.v.).

Chemical Class: Biological Products

Functions: Skin-Conditioning Agent - Humectant; Skin-Conditioning Agent - Miscellaneous

Reported Product Categories: Moisturizing Preparations; Night Skin Care Preparations

Technical/Other Name:
Extract of Saccharomyces Lysate

Trade Names:
AC Actine Dermal Respiratory Factor Powder (Active Concepts)
Biodynes TRF Powder (Arch Personal Care Products)
Cytocatalyzer (Bio Dell)
Oligocarbopeptides (Arch Personal Care Products)

Trade Name Mixtures:
AC Dermal Respiratory Factor (Active Concepts)
AC Dermal Respiratory Factor 5% (Active Concepts)
AC Dermal Respiratory Factor HSP (Active Concepts)
AC Dermal Respiratory Factor Light (Active Concepts)
AC Dermal Respiratory Factor U5 (Active Concepts)
AC Liposome DRF (Active Concepts)
Biodynes TRF Improved 25 (Arch Personal Care Products)
Biodynes TRF 25% Solution (Arch Personal Care Products)
Biodynes TRF Ultra 5 (Arch Personal Care Products)

Brookosome TRF (Arch Personal Care Products)

SACCHAROMYCES/MAGNESIUM FERMENT

Definition: Saccharomyces/Magnesium Ferment is an extract of a fermentation product of saccharomyces in the presence of magnesium ions.

Chemical Class: Biological Products

Function: Not Reported

Trade Name Mixtures:
ACB Bio-Chelate Magnesium (Active Concepts)
ACB Oligomin 5 (Active Concepts)
Acqua Biomin Magnesium Y3 (Arch Personal Care Products)
Biomin Acquacinque Liquid (Arch Personal Care Products)
Biomin 3 PF (Arch Personal Care Products)

SACCHAROMYCES/MAGNESIUM FERMENT HYDROLYSATE

Definition: Saccharomyces/Magnesium Ferment Hydrolysate is the product formed from the hydrolysis of Saccharomyces/Magnesium Ferment (q.v.).

Chemical Class: Biological Products

Function: Not Reported

Trade Name Mixture:
Biophos 35 (Arch Personal Care Products)

SACCHAROMYCES/MAGNESIUM FERMENT LYSATE FILTRATE

Definition: Saccharomyces/Magnesium Ferment Lysate Filtrate is the filtration product of a lysate of the fermentation of *Saccharomyces* in the presence of magnesium ions.

Chemical Class: Biological Products

Function: Skin-Conditioning Agent - Miscellaneous

Trade Name Mixture:
Biomin Triplex Y3 (Arch Personal Care Products)

SACCHAROMYCES/MANGANESE FERMENT

Definition: Saccharomyces/Manganese Ferment is an extract of a fermentation product of saccharomyces in the presence of manganese ions.

Chemical Class: Biological Products

Function: Not Reported

Trade Name Mixtures:
ACB Bio-Chelate (Active Concepts)
ACB Bio-Chelate 5 (Active Concepts)
ACB Bio-Chelate Sil (Active Concepts)
ACB Bio-Chelate (Zn,Cu,Mn,Fe,Si,K) (Active Concepts)
Acqua Biomin Manganese Y3 (Arch Personal Care Products)

SACCHAROMYCES/MOTHER OF PEARL FERMENT LYSATE FILTRATE

Definition: Saccharomyces/Mother of Pearl Ferment Lysate Filtrate is the filtration product of a lysate of the fermentation of Mother of Pearl (q.v.) by *Saccharomyces*.

Chemical Class: Biological Products

Function: Skin-Conditioning Agent - Miscellaneous

Trade Name:
Acquabiomin Mother of Pearl Y3 (Arch Personal Care Products)

SACCHAROMYCES/OPAL FERMENT LYSATE FILTRATE

Definition: Saccharomyces/Opal Ferment Lysate Filtrate is the filtration product of a lysate of the fermentation of opal by *Saccharomyces*.

Chemical Class: Biological Products

Function: Skin-Conditioning Agent - Miscellaneous

Trade Name:
Aquabiomin Opal Y3 (Arch Personal Care Products)

SACCHAROMYCES POLYPEPTIDES

Definition: Saccharomyces Polypeptides is the protein fraction isolated from Saccharomyces Lysate (q.v.).

Chemical Class: Proteins

Function: Not Reported

Trade Name:
Repitelin (Pentapharm/Centerchem)

SACCHAROMYCES/POTASSIUM FERMENT

Definition: Saccharomyces/Potassium Ferment is an extract of a fermentation

product of saccharomyces in the presence of potassium ions.

Chemical Class: Biological Products

Function: Not Reported

Trade Name:
Biomin K/P/C (Arch Personal Care Products)

Trade Name Mixtures:
ACB Bio-Chelate (Active Concepts)
ACB Bio-Chelate 5 (Active Concepts)
ACB Bio-Chelate Sil (Active Concepts)
ACB Bio-Chelate (Zn,Cu,Mn,Fe,Si,K) (Active Concepts)

SACCHAROMYCES/POTASSIUM FERMENT HYDROLYSATE

Definition: Saccharomyces/Potassium Ferment Hydrolysate is the product formed from the hydrolysis of Saccharomyces/Potassium Ferment (q.v.).

Chemical Class: Biological Products

Function: Not Reported

Trade Name Mixture:
Biophos 35 (Arch Personal Care Products)

SACCHAROMYCES/POTATO EXTRACT FERMENT FILTRATE

Definition: Saccharomyces/Potato Extract Ferment Filtrate is a filtrate of the product obtained by the fermentation of potato extract by the organism *Saccharomyces*.

Chemical Class: Biological Products

Function: Skin-Conditioning Agent - Humectant

SACCHAROMYCES/PRUNUS EXTRACT FERMENT FILTRATE

Definition: Saccharomyces/Prunus Extract Ferment Filtrate is a filtrate of the product obtained by the fermentation of prunus extract by the organism *Saccharomyces*.

Chemical Class: Biological Products

Function: Humectant

Trade Name:
Fermented Plum Extract (Wamiles)

SACCHAROMYCES/RICE BRAN FERMENT

Definition: Saccharomyces/Rice Bran Ferment is a product obtained by the fer-

mentation of Oryza Sativa (Rice) Bran (q.v.) by the organism *Saccharomyces*.

Information Source: JCLS

Chemical Class: Biological Products

Function: Skin-Conditioning Agent - Humectant

Trade Name:
Oryza Liquid S (Ichimaru Pharcos)

SACCHAROMYCES/SEA SALT FERMENT

Definition: Saccharomyces/Sea Salt Ferment is an extract of the fermentation product of saccharomyces in the presence of Sea Salt (q.v.).

Chemical Class: Biological Products

Function: Not Reported

Trade Name:
Biomin Marine 10% Solution (Arch Personal Care Products)

SACCHAROMYCES/SELENIUM FERMENT

Definition: Saccharomyces/Selenium Ferment is an extract of a fermentation product of saccharomyces in the presence of selenium ions.

Information Source: EEC(II-297)

Chemical Class: Biological Products

Function: Not Reported

Trade Name:
ACB Bio-Chelate Selenium (Active Concepts)

Trade Name Mixture:
Acqua Biomin Selenium Y3 (Arch Personal Care Products)

SACCHAROMYCES/SILICON FERMENT

Definition: Saccharomyces/Silicon Ferment is an extract of a fermentation product of saccharomyces in the presence of silicon.

Chemical Class: Biological Products

Function: Not Reported

Trade Name Mixtures:
ACB Bio-Chelate (Active Concepts)
ACB Bio-Chelate 5 (Active Concepts)
ACB Bio-Chelate Sil (Active Concepts)
ACB Bio-Chelate (Zn,Cu,Mn,Fe,Si,K) (Active Concepts)
ACB Oligomin 5 (Active Concepts)
Acqua Biomin Silicon Y3 (Arch Personal Care Products)
Biomin Acquacinque Liquid (Arch Personal Care Products)

SACCHAROMYCES/SOY PROTEIN FERMENT

Definition: Saccharomyces/Soy Protein Ferment is a product obtained by the fermentation of soy protein by the organism *Saccharomyces*.

Chemical Class: Biological Products

Function: Skin-Conditioning Agent - Miscellaneous

Trade Name:
TCN NA (Arval)

SACCHAROMYCES/TURQUOISE FERMENT LYSATE FILTRATE

Definition: Saccharomyces/Turquoise Ferment Lysate Filtrate is the filtration product of a lysate of the fermentation of turquoise by *Saccharomyces*.

Chemical Class: Biological Products

Function: Skin-Conditioning Agent - Miscellaneous

Trade Name:
Aquabiomin Turquoise Y3 (Arch Personal Care Products)

SACCHAROMYCES/XYLINUM BLACK TEA FERMENT

Definition: Saccharomyces/Xylinum Black Tea Ferment is the product obtained by the fermentation of black tea by *Saccharomyces* and *Xylinium*.

Chemical Class: Biological Products

Function: Skin-Conditioning Agent - Miscellaneous

Trade Name:
Kombucha Ferment (Sederma)

SACCHAROMYCES/ZINC FERMENT

Definition: Saccharomyces/Zinc Ferment is an extract of a fermentation product of saccharomyces in the presence of zinc ions.

Chemical Class: Biological Products

Function: Not Reported

Technical/Other Name:
Zinc Yeast Derivative

Trade Name:
Biomin Z/P/C (Arch Personal Care Products)

Trade Name Mixtures:
ACB Bio-Chelate (Active Concepts)

ACB Bio-Chelate 5 (Active Concepts)
ACB Bio-Chelate Sil (Active Concepts)
ACB Bio-Chelate (Zn,Cu,Mn,Fe,Si,K) (Active Concepts)
ACB Oligomin 5 (Active Concepts)
Acqua Biomin Zinc Y3 (Arch Personal Care Products)
Biomin Acquacinque Liquid (Arch Personal Care Products)
Biomin 3 PF (Arch Personal Care Products)

SACCHAROMYCES/ZINC/IRON/ GERMANIUM/COPPER/MAGNESIUM/ SILICON FERMENT

Definition: Saccharomyces/Zinc/Iron/Germanium/Copper/Magnesium/Silicon Ferment is an extract of a fermentation product of *Saccharomyces* in the presence of silicon and zinc, iron, germanium, copper, magnesium ions.

Chemical Class: Biological Products

Function: Not Reported

Trade Name:
BIOMIN 6 MINERALS COMPLEX (Arch Personal Care Products)

SACCHAROMYCES/ZINC/MAGNESIUM/ CALCIUM/GERMANIUM/SELENIUM FERMENT

Definition: Saccharomyces/Zinc/Magnesium/Calcium/Germanium/Selenium Ferment is an extract of the fermentation product of *Saccharomyces* in the presence of zinc, magnesium, calcium, germanium, and selenium ions.

Information Source: EEC(II-297)

Chemical Class: Biological Products

Function: Not Reported

Trade Name Mixture:
Acqua Biomin Zn/Mg/Ca/Ge/Se Y3 (Arch Personal Care Products)

SACCHAROMYCOPSIS FERMENT FILTRATE

Definition: Saccharomycopsis Ferment Filtrate is a filtrate of the fermentation product of the yeast, *Saccharomycopsis*.

Chemical Class: Biological Products

Function: Skin-Conditioning Agent - Humectant

Trade Name:
Pitera 2X, Pitera 4X (Procter & Gamble Far East)

SACCHARUM OFFICINARUM FERMENT EXTRACT

CAS No.	EINECS No.
91770-72-8	294-859-0

Definition: Saccharum Officinarum Ferment Extract is an extract of the fermentation product of *Saccharum officinarum*.

Chemical Class: Biological Products

Function: Not Reported

Technical/Other Names:
Extract of Sugar Cane (Saccharium Officinarium) Ferment
Sugarcane, Fermented, Extract
Sugar Cane (Saccharium Officinarium) Ferment Extract

Trade Name:
Rum Concentrate (Charabot)

SACCHARUM OFFICINARUM (SUGAR CANE) EXTRACT

CAS No.	EINECS No.
91722-22-4	294-424-5

Definition: Saccharum Officinarum (Sugar Cane) Extract is an extract of the sugar cane, *Saccharum officinarum*. *See "Regulatory and Ingredient Use Information," regarding the labeling names for botanical ingredients in Volume 1, Introduction, Part A.*

Chemical Class: Biological Products

Function: Skin-Conditioning Agent - Miscellaneous

Reported Product Category: Skin Care Preparations, Misc.

Technical/Other Names:
Extract of Saccharum Officinarum
Extract of Sugar Cane
Saccharum Officinarum Extract
Sugar Cane Extract
Sugar Cane (Saccharum Officinarum) Extract

Trade Name:
Sugar Cane Extract MSX-245 (Shiseido Company)

Trade Name Mixtures:
Actiphyte Molasses (Active Organics)
Actiphyte Molasses BG50P (Active Organics)
Actiphyte Sugar Cane (Active Organics)
Actiphyte Sugar Cane AQ (Active Organics)
Actiphyte Sugar Cane BG50P (Active Organics)
Actiphyte Sugar Cane GL (Active Organics)
A.H.A. 40 (Ennagram)

Cane Sugar Extract HS 3711 G (Grau)
Hydroxylide QNST (Coletica SA)
Molasses Liquid S (Taiyo Kagaku)
Multifruit BSC (Arch Personal Care Products)
Multifruit MFA (Arch Personal Care Products)
Multifruit PCA (Arch Personal Care Products)
Saccharum Officinarum (Sugar Canne) (IES LABO)
Sugar Cane Extract MSX-245 (Shiseido Company)
Sugar Cane HS (Alban Muller)
VT-205 Extract of Sugar Cane (Vege-Tech)

SAFFLOWER ACID

JPN Translation:
サフラワー脂肪酸

Definition: Safflower Acid is a mixture of fatty acids derived from safflower oil.

Information Source: JCIC

Chemical Class: Fatty Acids

Function: Surfactant - Cleansing Agent

Technical/Other Name:
Safflower Oil Fatty Acid

SAFFLOWERAMIDOPROPYL ETHYLDIMONIUM ETHOSULFATE

CAS No.: 113492-04-9

Definition: Saffloweramidopropyl Ethyldimonium Ethosulfate is the quaternary ammonium salt that conforms to the formula:

$$\left[RC-NH(CH_2)_3-N\begin{matrix}CH_3\\|\\|\\CH_3\end{matrix}-CH_2CH_3 \right]^+ \quad CH_3CH_2\underset{|}{\overset{}{O}}SO_3^-$$

where RCO- represents the fatty acids derived from safflower oil.

Chemical Class: Quaternary Ammonium Compounds

Functions: Antistatic Agent; Hair Conditioning Agent

Technical/Other Names:
1-Propanaminium, 3-Amino-N-Ethyl-N,N-Dimethyl-, N-Safflower Oil Acyl Derivs., Ethyl Sulfates
Saffloweramidopropyl Ethyldimethylammonium Ethylsulfate

Trade Name:
Schercoquat FOAS (Scher)

SAFFLOWER GLYCERIDE

JPN Translation:
サフラワー脂肪酸グリセリル

Definition: Safflower Glyceride is the monoglyceride derived from refined safflower oil.

Information Sources: JCIC, JCLS, JSQI

Chemical Class: Glyceryl Esters and Derivatives

Function: Surfactant - Emulsifying Agent

Technical/Other Names:
Glycerides, Safflower Mono-
Safflower Oil Fatty Acid Glyceryl Ester
Safflower Oil Monoglyceride

SAFFLOWER SEED OIL PEG-8 ESTERS

Definition: Safflower Seed Oil PEG-8 Esters is the product obtained by the transesterification of Carthamus Tinctorius (Safflower) Seed Oil (q.v.) and PEG-8 (q.v.).

Chemical Class: Glyceryl Esters and Derivatives

Functions: Skin-Conditioning Agent - Emollient; Surfactant - Emulsifying Agent

Trade Name:
Viatenza Safflower PE8 (Aldivia)

SAFFLOWER SEED OIL POLYGLYCERYL-6 ESTERS

Definition: Safflower Seed Oil Polyglyceryl-6 Esters is the product obtained by the transesterification of Carthamus Tinctorius (Sunflower) Seed Oil (q.v.) and Polyglycerin-6 (q.v.).

Chemical Class: Glyceryl Esters and Derivatives

Functions: Skin-Conditioning Agent - Emollient; Surfactant - Emulsifying Agent

Trade Name:
Viatenza Safflower PO6 Cx (Aldivia)

SAISHIN EKISU

Definition: See "Regulatory and Ingredient Use Information," regarding use of Japan Trivial names in Volume 1, Introduction, Part A.

Information Source: JCLS

Function: Not Reported

SAISIN EKISU

JPN Translation:
サイシンエキス

Definition: Saisin Ekisu is an extract of the roots of *Asiasarum sieboldii* or of *Asiasarum heterotropoides. See "Regulatory and Ingredient Use Information," regarding use of Japan Trivial names in Volume 1, Introduction, Part A.*

Chemical Class: Biological Products

Function: Not Reported

Technical/Other Names:
Asarum Sieboldi (EU)
Asarum Sieboldi Root Extract (U.S.)

SALICORNIA BIGELOVII EXTRACT

Definition: Salicornia Bigelovii Extract is an extract of *Salicornia bigelovii. See "Regulatory and Ingredient Use Information," regarding the labeling names for botanical ingredients in Volume 1, Introduction, Part A.*

Chemical Class: Biological Products

Function: Not Reported

Technical/Other Name:
Extract of Salicornia Bigelovii

Trade Name:
Seaphire Extract (Seaphire)

SALICORNIA HERBACEA EXTRACT

Definition: Salicornia Herbacea Extract is an extract of *Salicornia herbacea. See "Regulatory and Ingredient Use Information," regarding the labeling names for botanical ingredients in Volume 1, Introduction, Part A.*

Chemical Class: Biological Products

Function: Not Reported

Technical/Other Name:
Extract of Salicornia Herbacea

Trade Name Mixtures:
Cire De Salicorne (Codif)
Extrait de Salicorne MPE PG10 (Yves Rocher)

SALICYLAMIDE

CAS No.	EINECS No.
65-45-2	200-609-3

Empirical Formula:
$C_7H_7NO_2$

Definition: Salicylamide is the aromatic amide that conforms to the formula:

Information Sources: BAN, 21CFR310.545, CZE, DDR, HUN, INN, JAN, KOR, MAR, MI-13(8407), ROM, TSCA, USAN, USD, USP XXIV

Chemical Classes: Amides; Phenols

Function: Not Reported

Technical/Other Names:
Benzamide, 2-Hydroxy-
2-Carboxamidophenol
2-Hydroxybenzamide
Salamide
Salicylic Acid Amide

SALICYLIC ACID

CAS No.	EINECS No.
69-72-7	200-712-3

JPN Translation:
サリチル酸

Empirical Formula:
$C_7H_6O_3$

Definition: Salicylic Acid is the aromatic acid that conforms to the formula:

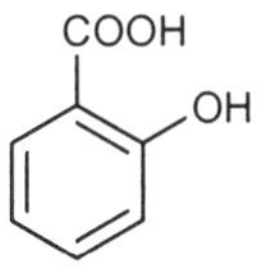

In the United States, Salicylic Acid may be used as an active ingredient in OTC drug products. When used as an active drug ingredient, the established name is *Salicylic Acid. See "Regulatory and Ingredient Use Information," regarding the labeling names for U.S. OTC Drug Ingredients in Volume 1, Introduction, Part A.*

Information Sources: ARG, AUS, BP, BPC, BRA, 21CFR175.105, 21CFR175.300, 21CFR177.2600, 21CFR358.510, 21CFR358.710, 21CFR533.310, CIR: [SQ], CTFA S, CZE, DA, DDR, EEC(VI/1-3), EGY, FIN, HP, HUN, IND, ITA, JAN, JCLS, JSCI, MAR, MEX, MHLW-331/3, MI-13(8411), NED, OTC-I-AK, OTC-I-CR, OTC-I-DP, OTC-I-WR, PF, PN, POR, RIFM, ROM, TSCA, USAN, USD, USP XXIV, USSR, WHO, YUG

Chemical Classes: Carboxylic Acids; Phenols

Functions: Antiacne Agent; Antidandruff Agent; Corn/Callus/Wart Remover; Denaturant; Exfoliant; Fragrance Ingredient; Hair Conditioning Agent; Skin-Conditioning Agent - Miscellaneous

Reported Product Categories: Bath Oils, Tablets, and Salts; Cleansing Products (Cold Creams, Cleansing Lotions, Liquids and Pads); Moisturizing Preparations; Shampoos (Non-coloring); Bath Preparations, Misc.; Body and Hand Preparations (Excluding Shaving Preparations); Tonics, Dressings, and Other Hair Grooming Aids; Skin Care Preparations, Misc.; Paste Masks (Mud Packs); Skin Fresheners; Hair Conditioners; Face and Neck Preparations (Excluding Shaving Preparations); Hair Coloring Preparations, Misc.; Foot Powders and Sprays; Foundations; Bath Soaps and Detergents; Blushers (All types); Eye Lotions; Face Powders; Hair Dyes and Colors (All Types Requiring Caution Statements and Patch Tests); Hair Preparations (Non-coloring), Misc.; Hair Straighteners; Indoor Tanning Preparations; Lipsticks; Makeup Bases; Makeup Fixatives; Makeup Preparations (Not eye), Misc.; Nail Creams and Lotions; Personal Cleanliness Products, Misc.

Technical/Other Names:
Benzoic Acid, 2-Hydroxy-
o-Carboxyphenol
2-Hydroxybenzoic acid (RIFM)
o-Hydroxybenzoic Acid
Phenol-2-Carboxylic Acid

Trade Names:
AEC Salicylic Acid (A & E Connock)
Unichem SALAC (Universal Preserv-A-Chem)

Trade Name Mixtures:
AC PE Sal (Active Concepts)
Antidandruff Agent Special (Crodarom)
Beta Hydroxy Acid (BHA) Willow Bark (Cosmetochem) (Cosmetochem International Ltd.)
Catezomes SA-20 Japan (Collaborative Labs)
Concentrate for Dandruff (Crodarom)
Geogard 361 Preservative (Lonza Inc./Lonza Ltd.)
Glucosamine Salicylate (Bioderm Research)
Glycosan Salicilico (Chemyunion)
Beta-Hydroxyde PVSA (Coletica SA)
LIPO CD-SA (Lipo)
Liponyl N30SA (Lipo)
Molecularsource SA (C.I.T.)
Niacin Salicylic Acid (Bioderm Research)
Pilinhib Veg (Laboratoires Serobiologiques)
Protaderm HA (Protameen)
Salicylic Acid Liposome (Active Concepts)
Salicysome (Coletica SA)
Sveltine (Coletica SA)
Ultrawhite (Coletica SA)

SALICYLOYL PHYTOSPHINGOSINE

Empirical Formula:
$C_{25}H_{43}NO_5$

Definition: Salicyloyl Phytosphingosine is the organic compound that conforms to the formula:

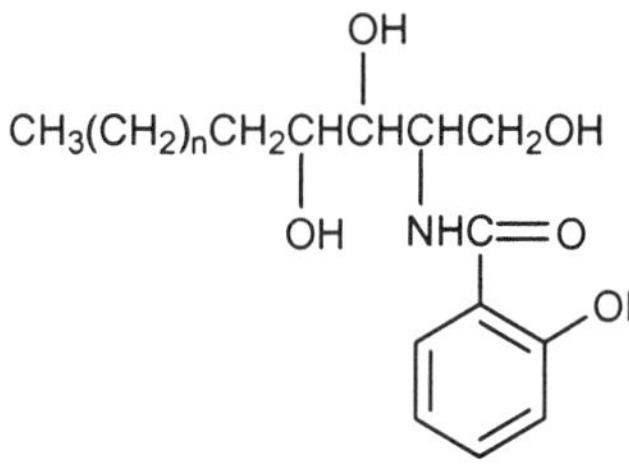

Chemical Classes: Alcohols; Amides; Phenols

Function: Skin-Conditioning Agent - Miscellaneous

Trade Name:
Phytosphingosine SLC (Degussa Care Specialties)

SALICYLYL BEESWAX

Definition: Salicylyl Beeswax is the ester of Salicylic Acid (q.v.) and Beeswax (q.v.).

Chemical Class: Esters

Functions: Skin-Conditioning Agent - Emollient; Skin-Conditioning Agent - Humectant

SALIX ALBA (WILLOW) BARK EXTRACT

CAS No.	EINECS No.
84082-82-6	282-029-0

Definition: Salix Alba (Willow) Bark Extract is an extract of the bark of the white willow, *Salix alba. See "Regulatory and Ingredient Use Information," regarding the labeling names for botanical ingredients in Volume 1, Introduction, Part A.*

Chemical Class: Biological Products

Functions: Hair Conditioning Agent; Skin-Conditioning Agent - Occlusive

Reported Product Category: Shampoos (Non-coloring)

Technical/Other Names:
Extract of Salix Alba Bark
Extract of Willow Bark
Salix Alba Bark Extract
Willow Bark Extract

Trade Names:
ABC White Willow Bark Extract 20% (Active Concepts)
ABS White Willow Bark Extract (Active Concepts)
ABS White Willow Bark Extract Powder (Active Concepts)
Willow bark extract, Premier (Premier Specialties)

Trade Name Mixtures:
Actiphyte of Willow Bark (Active Organics)

Actiphyte of Willow Bark BG50 (Active
  Organics)
Actiphyte of Willow Bark GL50 (Active
  Organics)
Actiphyte of Willow Bark Lipo S (Active
  Organics)
Actiphyte of Willow Bark PG50 (Active
  Organics)
Beta Hydroxy Acid (BHA) Willow Bark
  (Cosmetochem) (Cosmetochem
  International Ltd.)
Clercicyne L (Greentech)
Concentrate for Greasy Hair (Crodarom)
Glycolysat Of Willow (CEP (Solabia))
Herbalcomplex 2 special (Crodarom)
Herbal-Complex 4 special (Crodarom)
Herbaliquid Willow Special (Crodarom)
Herbasol Extract Willow Bark
  (Cosmetochem) (Cosmetochem
  International Ltd.)
Lipoplastidine Salix (Vevy)
Megasol Complex (Vevy)
Natupure White Willow (E.U.K)
Pilinhib Veg (Laboratoires Serobiologiques)
Salix Alba (Willow) Bark Extract ies (IES
  LABO)
Willow Extract HG (Provital/Centerchem)
Willow Extract HS 2377 G (Grau)

SALIX ALBA (WILLOW) BARK WATER

Definition: Salix Alba (Willow) Bark Water
is an aqueous solution containing volatile
oils obtained by the distillation of the bark
of *Salix alba. See "Regulatory and Ingredient Use Information," regarding the
labeling names for botanical ingredients in
Volume 1, Introduction, Part A.*

Chemical Class: Biological Products

Function: Not Reported

Technical/Other Names:
  Salix Alba Distillate
  Willow Bark Distillate
  Willow Distillate

Trade Name:
  Tricosolfan (Vevy)

SALIX ALBA (WILLOW) FLOWER
EXTRACT

Definition: Salix Alba (Willow) Flower
Extract is an extract of the flowers of the
white willow, *Salix alba. See "Regulatory
and Ingredient Use Information," regarding
the labeling names for botanical ingredients
in Volume 1, Introduction, Part A.*

Chemical Class: Biological Products

Function: Not Reported

Technical/Other Names:
  Extract of Salix Alba
  Extract of Willow Flower
  Salix Alba Extract
  Willow Flower Extract

Trade Name Mixture:
  Extrait De Chatons De Saule MP PG 40
  (Yves Rocher)

SALIX ALBA (WILLOW) LEAF EXTRACT

CAS No.	EINECS No.
84082-82-6	282-029-0

Definition: Salix Alba (Willow) Leaf
Extract is an extract of the leaves of the
white willow, *Salix alba. See "Regulatory
and Ingredient Use Information," regarding
the labeling names for botanical ingredients
in Volume 1, Introduction, Part A.*

Chemical Class: Biological Products

Function: Not Reported

Reported Product Category: Shampoos
(Non-coloring)

Technical/Other Names:
  Extract of Salix Alba Leaf
  Extract of Willow
  Extract of Willow Leaf
  Salix Leaf Extract
  Willow Leaf Extract

Trade Name Mixtures:
  Astressyl (Silab)
  Herbasol-Extract Willow Leaf
    (Cosmetochem)
  Vegetol White Willow 4151 Oily (Gattefosse
    s.a.)
  VT-069 Extract of White Willow (Vege-
    Tech)

SALIX NIGRA (WILLOW) BARK EXTRACT

Definition: Salix Nigra (Willow) Bark
Extract is an extract of the bark of the
willow, *Salix nigra. See "Regulatory and
Ingredient Use Information," regarding the
labeling names for botanical ingredients in
Volume 1, Introduction, Part A.*

Chemical Class: Biological Products

Function: Not Reported

Technical/Other Names:
  Black Willow Extract
  Willow Bark Extract
  Willow Extract

Trade Name:
  ABS Willow Bark Extract Powder (Active
  Concepts)

Trade Name Mixtures:
  ACB Willow Bark Extract 20% (Active
    Concepts)
  Brookosome Willow Bark (Arch Personal
    Care Products)
  Enna 216 Breast Development Extract
    (Ennagram)
  NAB Willow Bark Extract (Arch Personal
    Care Products)

SALMON EGG EXTRACT

CAS No.	EINECS No.
94944-92-0	305-671-6

Definition: Salmon Egg Extract is an
extract of salmon eggs. *See "Regulatory
and Ingredient Use Information," regarding
use of EU Trivial names in Volume 1, Introduction, Part A.*

Chemical Class: Biological Products

Function: Not Reported

Technical/Other Name:
  Extract of Salmon Eggs

Trade Names:
  Marine Gamma Nucleotides (Vincience)
  Marine Vitellines (Vincience)

Trade Name Mixtures:
  Hydroplastidine Salmo Ovum (Vevy)
  Lipoplastidine Salmo Ovum (Vevy)
  Marine Phospholipids Deodorized (MMP)
  Vittelines Marines (Vincience)

SALMON OIL

CAS No.	EINECS No.
68991-43-5	273-641-9

Definition: Salmon Oil is the oil expressed
from the fish, salmon, of the genus
*Oncorhynchus. See "Regulatory and Ingre-
dient Use Information," regarding use of EU
Trivial names in Volume 1, Introduction,
Part A.*

Information Source: TSCA

Chemical Class: Fats and Oils

Function: Skin-Conditioning Agent -
Occlusive

Technical/Other Name:
  Oils, Salmon

SALNACEDIN

CAS No.: 87573-01-1

Empirical Formula:
  $C_{12}H_{13}NO_5S$

The inclusion of any compound in the *Dictionary and Handbook* does not indicate that use of that substance as a cosmetic ingredient complies with
the laws and regulations governing such use in the United States or any other country.

Definition: Salnacedin is the organic compound that conforms to the formula:

Information Sources: INN, USAN

Chemical Classes: Amides; Amino Acids; Phenols; Thio Compounds

Functions: Antioxidant; Skin-Conditioning Agent - Miscellaneous

Technical/Other Name:
L-Cysteine, N-Acetyl-, 2-Hydroxybenzoate (Ester)

Trade Name:
N-Acetyl-S-(2-Hydroxybenzoyl)-L-Cysteine / N-Acetyl-L-Cysteine-2-Hydroxybenzoate (Ester) (APR)

SALT MINE MUD

Definition: Salt Mine Mud is the sediment obtained from salt mines.

Chemical Class: Inorganics

Functions: Abrasive; Absorbent; Cosmetic Astringent; Skin-Conditioning Agent - Miscellaneous

Trade Name:
Misola, Berchtesgadener Salzbergschlick, Laist (Sudsalz)

SALVIA HISPANICA SEED EXTRACT

Definition: Salvia Hispanica Seed Extract is an extract of the seeds of *Salvia hispanica*. See *"Regulatory and Ingredient Use Information,"* regarding the labeling names for botanical ingredients in Volume 1, Introduction, Part A.

Chemical Class: Biological Products

Function: Not Reported

Technical/Other Name:
Extract of Salvia Hispanica Seed

Trade Name Mixture:
Chia Extract (Carrubba)

SALVIA HISPANICA SEED OIL

Definition: Salvia Hispanica Seed Oil is the oil expressed from the seed of *Salvia hispanica*. See *"Regulatory and Ingredient Use Information,"* regarding the labeling names for botanical ingredients in Volume 1, Introduction, Part A.

Chemical Class: Fats and Oils

Function: Skin-Conditioning Agent - Occlusive

Technical/Other Names:
Chia Oil
Chia (Salvia Hispanica) Oil

Trade Names:
AEC Chia Oil (A & E Connock)
Chia Seed Oil (Desert Whale)

SALVIA LAVANDULAEFOLIA OIL

Definition: Salvia Lavandulaefolia Oil is the volatile oil obtained from the stems and leaves of *Salvia lavandulaefolia*. See *"Regulatory and Ingredient Use Information,"* regarding the labeling names for botanical ingredients in Volume 1, Introduction, Part A.

Chemical Class: Essential Oils

Function: Fragrance Ingredient

Trade Name:
Sage Oil Spanish (Quest International)

SALVIA MILTIORRHIZA EXTRACT

CAS No.	EINECS No.
90106-50-6	290-273-4

Definition: Salvia Miltiorrhiza Extract is an extract of the roots, flowers and leaves of *Salvia miltiorrhiza*. See *"Regulatory and Ingredient Use Information,"* regarding the labeling names for botanical ingredients in Volume 1, Introduction, Part A.

Chemical Class: Biological Products

Function: Skin-Conditioning Agent - Miscellaneous

Technical/Other Name:
Extract of Salvia Miltiorrhiza

Trade Name:
Phytoselect Miltiorhiza (Indena SpA)

Trade Name Mixtures:
Campo Tan Shen Extract (Campo)
China Extract Tan Shen (E.U.K)
Chine Extract Salva Miltiorhiza (Ennagram)
Tanjin Liquid E (Ichimaru Pharcos)
YSK Magic 3 (Phyto-Technologies)

SALVIA OFFICINALIS

Definition: *See "Regulatory and Ingredient Use Information," regarding EU labeling names for botanical ingredients in Volume 1, Introduction, Part A.*

Chemical Class: Biological Products

Technical/Other Names:
Salvia Officinalis (Sage) Leaf (U.S.)
Salvia Officinalis (Sage) Leaf Extract (U.S.)
Salvia Officinalis (Sage) Leaf Water (U.S.)
Salvia Officinalis (Sage) Oil (U.S.)
Salvia Officinalis (Sage) Water (U.S.)

Trade Name Mixture:
Activated Botanicals Si-LIP AB 100 (Norjin)

SALVIA OFFICINALIS (SAGE) LEAF

JPN Translation:
セージ

Definition: Salvia Officinalis (Sage) Leaf is a plant material derived from the dried, crushed leaves of the sage, *Salvia officinalis*. See *"Regulatory and Ingredient Use Information,"* regarding the labeling names for botanical ingredients in Volume 1, Introduction, Part A.

Information Sources: 21CFR101.22, 21CFR182.10, 21CFR182.20, 21CFR310.545, 21CFR501.22, 21CFR582.10, 21CFR582.20, JCIC, JCLS, JSQI

Chemical Class: Biological Products

Function: Not Reported

Reported Product Category: Bath Soaps and Detergents

Technical/Other Names:
Sage
Sage Leaf
Sage Powder
Salvia Officinalis (EU)

Trade Names:
AEC Sage Leaf Powder (A & E Connock)
Sage (Salvia Officinalis) Leaves (Lebermuth)

SALVIA OFFICINALIS (SAGE) LEAF EXTRACT

CAS No.	EINECS No.
84082-79-1	282-025-9

JPN Translation:
セージエキス

Definition: Salvia Officinalis (Sage) Leaf Extract is an extract of the leaves of the sage, *Salvia officinalis*. See *"Regulatory and Ingredient Use Information,"* regarding the labeling names for botanical ingredients in Volume 1, Introduction, Part A.

Information Sources: 21CFR182.20, DDR, HP, JCIC, JCLS, JSQI

Chemical Class: Biological Products

Functions: Oral Care Agent; Skin-Conditioning Agent - Miscellaneous

Reported Product Categories: Skin Care Preparations, Misc.; Tonics, Dressings, and Other Hair Grooming Aids; Hair Conditioners; Bath Oils, Tablets, and Salts; Bath Preparations, Misc.; Body and Hand Preparations (Excluding Shaving Preparations); Cleansing Products (Cold Creams, Cleansing Lotions, Liquids and Pads); Bath Capsules; Face and Neck Preparations (Excluding Shaving Preparations); Paste Masks (Mud Packs); Bath Soaps and Detergents; Night Skin Care Preparations; Skin Fresheners; Aftershave Lotions; Deodorants (Underarm); Hair Preparations (Non-coloring), Misc.; Moisturizing Preparations; Personal Cleanliness Products, Misc.; Shampoos (Non-coloring)

Technical/Other Names:
Extract of Sage
Extract of Salvia Officinalis
Sage Extract
Sage Leaf Extract
Salvia Extract
Salvia Officinalis (EU)
Salvia Triloba Extract

Trade Names:
Flavex Sage Triloba CO2-se extract, Type 063.003 (Flavex)
Phytelene of Sauge EN 101 powder (Indena SA)
Phytogreen 55 of Sage EP 485 Powder (Phytochim)
Sage Ecoconcentrate (Robertet, Inc.)
Sage Ecoconcentrate Natural (Robertet S.A.)

Trade Name Mixtures:
Acneous Phytogreen Complex GXH 267 (Phytochim)
Actiphyte of Sage BG50 (Active Organics)
Actiphyte of Sage GL50 (Active Organics)
Actiphyte of Sage Lipo S (Active Organics)
Actiphyte of Sage PG50 (Active Organics)
Actiphyte Red Sage (Active Organics)
Aqueous Spray - Dried G Mix (Indena SA)
Aromaphyte of Sage (Active Organics)
Bio-Chelated Derma-Plex I (Bio-Botanica)
Bio-Chelated Neutral Henna Plus I (Bio-Botanica)
275 Blend for Deodorant HS (Alban Muller)
675 Blend for Deodorant LS (Alban Muller)
226 Capilotonique HS (Alban Muller)
245 Capilotonique HS (Alban Muller)
Complex Salvia (Fabriquimica)
Cremogen AF (PN 736567) (Haarmann & Reimer GmbH)
Cremogen MZ (PN 739 032) (Haarmann & Reimer GmbH)
Cremogen MZ/N (PN 755 321) (Haarmann & Reimer GmbH)
Cremogen M-82 (PN 730 337) (Haarmann & Reimer GmbH)

Cremogen Sage (PN 739 016) (Haarmann & Reimer GmbH)
Enna 212 Bust Firmness (Ennagram)
Extract of Sage/Indian Cress (CEP (Solabia))
Extrait De Sauge PP PG40 (Yves Rocher)
Extrapone #1 GW 2/031010 (Symrise)
Extrapone #4 GW 2/031040 (Symrise)
Extrapone #4 Herbs 2/032495 (Symrise)
Extrapone #7 Herbs 2/032535 (Symrise)
Extrapone #1 Special 2/032451 (Symrise)
Extrapone #3 Special 2/034481 (Symrise)
Extrapone #3 Special 2/789490 (Symrise)
Extrapone #5 Special 2/032501 (Symrise)
Extrapone Sage GW 2/031770 (Symrise)
Extrapone Sage Special 2/033291 (Symrise)
Extrapone Seven Herbs Special 2/032527 (Symrise)
Extrapone 1 Special 2/789470 (Symrise)
Extrapone 3 Special 2/789490 (Symrise)
Extrapone 4 Special 2/788400 (Symrise)
Extrapone 5 Special 2/789500 (Symrise)
Extrapone 2 Special J (2/032473) (Symrise)
Extrapone 3 Special New 2/034484 (Symrise)
Extrapone 2 Special 2/789480 (same as 2/032471) (Symrise)
Glycolysat of Sage (CEP (Solabia))
Herbalcomplex 1 Special (Crodarom)
Herbalcomplex 2 special (Crodarom)
Herbalcomplex 3 Special (Crodarom)
Herbal Extract for Normal Hair (Crodarom)
Herbal Extract Glycolic - Article 251172 (Plantextrakt)
Herbaliquid Sage Special (Crodarom)
Herbal Vinegar (Provital/Centerchem)
Herbasec Sage (Cosmetochem) (Cosmetochem International Ltd.)
Herbasol Complex "Herbes de Provence" (Cosmetochem) (Cosmetochem International Ltd.)
Herbasol Extract Oil Soluble Sage (Cosmetochem) (Cosmetochem International Ltd.)
Herbasol-Extract Sage (Cosmetochem)
Hydrumine Sage (Exsymol)
Nail Regenerative Complex (Alban Muller)
Natuscreen (Pronectar)
Nutriplant (Bio-Botanica)
OILY SKIN (Greentech S.A)
Oily Skins Complex 264 (Ennagram)
Oily Skins Phytogreen Complex GXH 264 (Phytochim)
Oleat of Sage (CEP (Solabia))
Pharcolex BX47 (Ichimaru Pharcos)
Pharcolex BX52 (Ichimaru Pharcos)
Phytelene Complex EGX 247 (Indena SA)
Phytelene Complex EGX 252 (Indena SA)
Phytelene of Sage EG 010 liquid (Indena SA)
Phytelene of Sage EN 101 powder (Indena SA)

Phytelene of Sauge EG 010 Liquid (Indena SA)
Phytogreen 55 of Sage EXH 613 Liquid (Phytochim)
Polyplant Anti-Inflammation (Provital/Centerchem)
Polyplant Anti-Wrinkles (Provital/Centerchem)
Polyplant Astringent (Provital/Centerchem)
Polyplant Hair (Provital/Centerchem)
Polyplant Oily Skin (Provital/Centerchem)
Polyplant Sedative (Provital/Centerchem)
Polyplant Skin Purifying (Provital/Centerchem)
Polyplant 5 Special (Provital/Centerchem)
Polyplant Stimulant (Provital/Centerchem)
Premier Sago 10% Extract (Premier Specialties)
Prodhy Extract Suage (Prod'Hyg)
Sage Cl 2/033294 (Symrise)
Sage Extract (Maruzen Pharmaceuticals Co., Ltd.)
Sage Extract BG (Crodarom)
Sage Extract BG (Maruzen Pharmaceuticals Co., Ltd.)
Sage Extract BG-J (Maruzen Pharmaceuticals Co., Ltd.)
Sage Extract HG (Provital/Centerchem)
Sage Extract HS 2428 G (Grau)
Sage Extract LA (Maruzen Pharmaceuticals Co., Ltd.)
Sage Extract PG (Rahn)
Sage Extract Powder-S (Maruzen Pharmaceuticals Co., Ltd.)
Sage Extract SQ (Maruzen Pharmaceuticals Co., Ltd.)
Sage HPG Titrated (Alban Muller)
Sage HS (Alban Muller)
Sage Leaf Liquid B (Ichimaru Pharcos)
Sage Leaf Liquid E (Ichimaru Pharcos)
Sage LS (Alban Muller)
Sage Phytexcell (Crodarom)
Sage Tincture (Rahn)
Salvia Liquid (Ichimaru Pharcos)
Sederma Sage (Sederma)
280 Stimulant HS (Alban Muller)
680 Stimulant LS (Alban Muller)
Teen Age Skin Disorders Complex 267 (Ennagram)
Toothpaste Complex MU 3319 (Greentech S.A)
2305 Vege-Plex Body Complex (Vege-Tech)
2330 Vege-Plex Body Complex (Vege-Tech)
2350 Vege-Plex Body Complex (Vege-Tech)
2270 Vege-Plex Hair Complex (Vege-Tech)
2410 Vege-Plex Skin Complex (Vege-Tech)
2530 Vege-Plex Skin Complex (Vege-Tech)
2610 Vege-Plex Skin Complex (Vege-Tech)

Vegetol Sage GR 377 Hydro (Gattefosse s.a.)
Vegetol Sage MCF 776 Hydro (Gattefosse s.a.)
Vegetol Sage 4138 Oily (Gattefosse s.a.)
Vegetol Sp GR 051 Hydro (Gattefosse s.a.)
Viazest Sage OS (Aldivia)
VT-040 Extract of Sage (Vege-Tech)

SALVIA OFFICINALIS (SAGE) LEAF WATER

JPN Translation:
セージ水

Definition: Salvia Officinalis (Sage) Leaf Water is an aqueous solution of the steam distillate obtained from the leaves of *Salvia officinalis. See "Regulatory and Ingredient Use Information," regarding the labeling names for botanical ingredients in Volume 1, Introduction, Part A.*

Information Sources: JCIC, JCLS

Chemical Class: Biological Products

Functions: Fragrance Ingredient; Skin-Conditioning Agent - Miscellaneous

Technical/Other Names:
Sage Leaf Water
Sage Water
Salvia Officinalis (EU)
Water, Sage

Trade Names:
Extrait Originel Sauge (Gattefosse s.a.)
Vegebios of Sage (CEP (Solabia))

Trade Name Mixtures:
Eau de Aroma Salvia (Ichimaru Pharcos)
Sage Foot Spray (Alban Muller)

SALVIA OFFICINALIS (SAGE) OIL

CAS Nos.: 8022-56-8; 84776-73-8

JPN Translation:
セージ油

Definition: Salvia Officinalis (Sage) Oil is the essential oil derived from the herbal plant, *Salvia officinalis. See "Regulatory and Ingredient Use Information," regarding the labeling names for botanical ingredients in Volume 1, Introduction, Part A.*

Information Sources: 21CFR182.20, DDR, FCC, JCIC, JCLS, NED, RIFM, SNPF, TSCA

Chemical Class: Essential Oils

Function: Fragrance Ingredient

Reported Product Categories: Skin Care Preparations, Misc.; Cleansing Products

(Cold Creams, Cleansing Lotions, Liquids and Pads); Bath Preparations, Misc.; Body and Hand Preparations (Excluding Shaving Preparations); Hair Conditioners; Tonics, Dressings, and Other Hair Grooming Aids; Bath Oils, Tablets, and Salts; Hair Preparations (Non-coloring), Misc.; Moisturizing Preparations; Paste Masks (Mud Packs); Shampoos (Non-coloring)

Technical/Other Names:
Oil of Sage
Sage Dalmatian oil (Salvia officinalis L.) (RIFM)
Sage Oil
Sage oil (Salvia officinalis L.) (RIFM)
Sage oil, Spanish (Salvia lavandulaefolia Vahl.) (RIFM)
Sage oleoresin (Salvia officinalis L.) (RIFM)
Sage (Salvia officinalis L.) (RIFM)
Salvia Officinalis (EU)
Salvia Oil

Trade Names:
AEC Sage Oil (A & E Connock)
Sauge Oil Officinale (Floressence)

Trade Name Mixtures:
ANTI-PERSPIRANT (Greentech S.A)
Aromaphyte of Sage (Active Organics)
Covazen Detox (LCW)

SALVIA OFFICINALIS (SAGE) WATER

Definition: Salvia Officinalis (Sage) Water is an aqueous solution of the steam distillate obtained from *Salvia officinalis. See "Regulatory and Ingredient Use Information," regarding the labeling names for botanical ingredients in Volume 1, Introduction, Part A.*

Chemical Class: Biological Products

Function: Not Reported

Technical/Other Name:
Salvia Officinalis (EU)

Trade Name:
Clary Sage Hydroflorate (Bayliss Ranch)

SALVIA SCLAREA (CLARY) EXTRACT

Definition: Salvia Sclarea (Clary) Extract is an extract of the clary, *Salvia sclarea. See "Regulatory and Ingredient Use Information," regarding the labeling names for botanical ingredients in Volume 1, Introduction, Part A.*

Chemical Class: Biological Products

Function: Not Reported

Technical/Other Names:
Clary Extract

Extract of Clary
Extract of Salvia Sclarea
Salvia Sclarea Extract

Trade Name:
Abs Clary Sage (Charabot)

Trade Name Mixtures:
Actiphyte of Clary Sage BG50 (Active Organics)
Actiphyte of Clary Sage GL50 (Active Organics)
Actiphyte of Clary Sage Lipo S (Active Organics)
Actiphyte of Clary Sage PG50 (Active Organics)
ARP 101 (Greentech)
Clary Sage Extract (Cosmetic Developments)
Herbasol Extract Clary Sage (Cosmetochem) (Cosmetochem International Ltd.)
Kapillarine (Greentech)
VT-148 Extract of Clary (Vege-Tech)

SALVIA SCLAREA (CLARY) OIL

CAS No.: 8016-63-5

Definition: Salvia Sclarea (Clary) Oil is a volatile oil obtained from *Salvia sclarea. See "Regulatory and Ingredient Use Information," regarding the labeling names for botanical ingredients in Volume 1, Introduction, Part A.*

Information Sources: 21CFR182.20, RIFM

Chemical Class: Essential Oils

Function: Fragrance Ingredient

Reported Product Categories: Hair Conditioners; Moisturizing Preparations

Technical/Other Names:
Clary Oil
Clary oil (Salvia sclarea L.) (RIFM)
Clary (Salvia sclarea L.) (RIFM)
Oil, Essential, Clary Sage
Oil of Clary
Sage clary absolute (RIFM)
Sage clary concrete (RIFM)

Trade Names:
AEC Clary Sage Oil (A & E Connock)
Clary Sage Essential Oil (Alban Muller)
Custosense Sage (clary sage oil) (Custom Ingredients)
Huile de Sauge Sclaree (Bertin)

Trade Name Mixtures:
Clary Sage CL Forte 2/033010 (Symrise)
Essentiaderm n.3 (Universal Flavors)
Essentiaderm n.4 (Universal Flavors)
Essentiaderm N.14 (Universal Flavors)

SALVIA SCLAREA (CLARY) WAX

Definition: Salvia Sclarea (Clary) Wax is a wax obtained from the aerial parts of *Salvia sclarea. See "Regulatory and Ingredient Use Information," regarding the labeling names for botanical ingredients in Volume 1, Introduction, Part A.*

Chemical Class: Waxes

Function: Not Reported

Technical/Other Names:
Clary Wax
Salvia Sclarea Wax
Wax, Clary (Salvia Sclarea)

Trade Names:
AEC Cire Essentielle De Sauge Sclaree (A & E Connock)
AEC Clary Sage Oil (Bertin)

SAMBUCUS CANADENSIS EXTRACT

Definition: Sambucus Canadensis Extract is an extract of the common elder, *Sambucus canadensis. See "Regulatory and Ingredient Use Information," regarding the labeling names for botanical ingredients in Volume 1, Introduction, Part A.*

Chemical Class: Biological Products

Function: Not Reported

SAMBUCUS NIGRA FLOWER EXTRACT

CAS No.
84603-58-7

EINECS No.
283-259-4

JPN Translation:
セイヨウニワトコエキス

Definition: Sambucus Nigra Flower Extract is an extract of the flowers of the elder, *Sambucus nigra. See "Regulatory and Ingredient Use Information," regarding the labeling names for botanical ingredients in Volume 1, Introduction, Part A.*

Information Sources: 21CFR172.510, 21CFR182.20, HP, JCIC, JCLS, JSQI, POL

Chemical Class: Biological Products

Functions: Skin-Conditioning Agent - Miscellaneous; Skin-Conditioning Agent - Occlusive

Technical/Other Names:
Elder Flower Extract
Extract of Sambucus
Extract of Sambucus Nigra Flowers
Sambucus Extract

Trade Name:
Hydroessential Sambucus (Vevy)

Trade Name Mixtures:
Actiphyte of Elderflower BG50 (Active Organics)
Actiphyte of Elderflower GL50 (Active Organics)
Actiphyte of Elderflower Lipo S (Active Organics)
Actiphyte of Elderflower PG50 (Active Organics)
265 Babyderme HS (Alban Muller)
665 Babyderme LS (Alban Muller)
Bio-Chelated Derma-Plex I (Bio-Botanica)
615 Blend for Delicate Skin LS (Alban Muller)
Coobato Sambuco (Universal Flavors)
Elder Flower Extract HS 2440 G (Grau)
Elder HS (Alban Muller)
Elder Liquid B (Ichimaru Pharcos)
Elder Liquid E (Ichimaru Pharcos)
Elder LS (Alban Muller)
Elder Tincture (Rahn)
Elder Tree Extract HG (Provital/ Centerchem)
235 Emollient HS (Alban Muller)
635 Emollient LS (Alban Muller)
Extrait De Sureau PPE PG40 (Yves Rocher)
Firmiderm LS 9120 (Laboratoires Sero-biologiques)
Herbaliquid Sambucus Special (Crodarom)
Herbasol-Extract Elder (Cosmetochem)
Herbasol Extract Oil Soluble Elder (Cosmetochem) (Cosmetochem International Ltd.)
Hydroessential Balsamic-1 (Vevy)
Hydroplastidine Sambucus (Vevy)
Moisturizing Complex 266 (Ennagram)
Moisturizing Phytogreen Complex GXH 266 (Phytochim)
Novaplant Sambucus Extract (Crodarom)
Pharcolex BX51 (Ichimaru Pharcos)
Phytelene Complex EGX 251 (Indena SA)
Phytelene of Elder Tree EG 050 liquid (Indena SA)
Phytelene of Sureau EG 050 liquid (Indena SA)
Phytogreen 55 of Elder Tree EXH 623 Liquid (Phytochim)
Phytogreen 55 of Sureau EXH 623 Liquid (Phytochim)
Phytotal SL (Phybiotex/Sederma)
Polyplant Anti-Inflammation (Provital/ Centerchem)
Polyplant Moisturizing (Provital/ Centerchem)
Prodhy Extract Sureau (Prod'Hyg)
PROTECTIVE MOISTURIZER (Greentech S.A)
Sambucus Extract (Maruzen Pharmaceuticals Co., Ltd.)
Sambucus Extract BG (Maruzen Pharmaceuticals Co., Ltd.)

Sambucus Extract LA (Maruzen Pharmaceuticals Co., Ltd.)
Sambucus Nigra Fruit Extract ies (IES LABO)
Vegebois of Elder (CEP (Solabia))
2300 Vege-Plex Body Complex (Vege-Tech)
2340 Vege-Plex Body Complex (Vege-Tech)
2350 Vege-Plex Body Complex (Vege-Tech)
2100 Vege-Plex Hair Complex (Vege-Tech)
2400 Vege-Plex Skin Complex (Vege-Tech)
2410 Vege-Plex Skin Complex (Vege-Tech)
2500 Vege-Plex Skin Complex (Vege-Tech)
2525 Vege-Plex Skin Complex (Vege-Tech)
2540 Vege-Plex Skin Complex (Vege-Tech)
2550 Vege-Plex Skin Complex (Vege-Tech)
2560 Vege-Plex Skin Complex (Vege-Tech)
2570 Vege-Plex Skin Complex (Vege-Tech)
2600 Vege-Plex Skin Complex (Vege-Tech)
Vegetol Elder MCF 1238 Hydro (Gattefosse s.a.)
Vegetol Elder 4144 Oily (Gattefosse s.a.)
VT-210 Extract of Elderberry (Vege-Tech)
VT-041 Extract of Sambucus (Vege-Tech)

SAMBUCUS NIGRA FLOWER JUICE

Definition: Sambucus Nigra Flower Juice is the juice expressed from the flower of *Sambucus nigra. See "Regulatory and Ingredient Use Information," regarding the labeling names for botanical ingredients in Volume 1, Introduction, Part A.*

Chemical Class: Biological Products

Function: Skin-Conditioning Agent - Miscellaneous

Trade Name:
Authenticals of Elder Flowers (CEP (Solabia))

SAMBUCUS NIGRA FLOWER POWDER

Definition: Sambucus Nigra Flower Powder is the dried crushed flowers of the sambucus, *Sambucus nigra. See "Regulatory and Ingredient Use Information," regarding the labeling names for botanical ingredients in Volume 1, Introduction, Part A.*

Information Sources: 21CFR172.510, 21CFR182.10, 21CFR182.20, 21CFR582.10, 21CFR582.20, DA, HP, MI-13(8427), POL, POR, ROM, USSR, YUG

Chemical Class: Biological Products

Function: Not Reported

Technical/Other Names:
Elder Flowers
Sambucus
Sambucus (Sambucus Nigra)

Trade Name Mixture:
Glycolysat of Elder (FL/BE) (CEP (Solabia))

SAMBUCUS NIGRA FLOWER WATER

Definition: Sambucus Nigra Flower Water is an aqueous solution of the steam distillate obtained from the flowers of *Sambucus nigra. See "Regulatory and Ingredient Use Information," regarding the labeling names for botanical ingredients in Volume 1, Introduction, Part A.*

Chemical Class: Biological Products

Function: Not Reported

Technical/Other Name:
Water, Sambucus Nigra

Trade Name:
AEC Elderflower Water (A & E Connock)

Trade Name Mixture:
Elder and Rose Water (Indena SA)

SAMBUCUS NIGRA FRUIT EXTRACT

Definition: Sambucus Nigra Fruit Extract is an extract of the berries of the elder, *Sambucus nigra. See "Regulatory and Ingredient Use Information," regarding the labeling names for botanical ingredients in Volume 1, Introduction, Part A.*

Chemical Class: Biological Products

Function: Not Reported

Trade Name:
AEC Elderberry Extract Powder (A & E Connock)

Trade Name Mixtures:
Actiphyte of Elderberry BG50 (Active Organics)
Actiphyte of Elderberry GL50 (Active Organics)
Actiphyte of Elderberry Lipo S (Active Organics)
Actiphyte of Elderberry PG50 (Active Organics)
Elderberry Extract (Bell Flavors)

Glycolysat of Elder (FL/BE) (CEP (Solabia))
Natupure Elder (E.U.K)

SAMBUCUS NIGRA FRUIT JUICE

Definition: Sambucus Nigra Fruit Juice is the liquid expressed from the fruit of the elderberry, *Sambucus nigra. See "Regulatory and Ingredient Use Information," regarding the labeling names for botanical ingredients in Volume 1, Introduction, Part A.*

Chemical Class: Biological Products

Functions: Cosmetic Astringent; Skin-Conditioning Agent - Miscellaneous

Technical/Other Names:
Elderberry (Sambucus Nigra) Juice
Juice, Elderberry
Juice, Sambucus Nigra
Sambucus Nigra Juice

Trade Name:
AEC Elderberry Conc (A & E Connock)

Trade Name Mixtures:
Fruitapone Elder B 2/036220 (Symrise)
Fruitapone Elder GT 2/037220 (Symrise)
Holunder P-AC-5 (Heidelberger)

SAMBUCUS NIGRA OIL

CAS No.: 68916-55-2

Definition: Sambucus Nigra Oil is the volatile oil obtained from *Sambucus nigra* and other species of *Sambucus. See "Regulatory and Ingredient Use Information," regarding the labeling names for botanical ingredients in Volume 1, Introduction, Part A.*

Information Sources: 21CFR172.510, 21CFR182.20, RIFM, TSCA

Chemical Class: Essential Oils

Function: Fragrance Ingredient

Technical/Other Names:
Oil of Sambucus
Sambucus Oil
Sureau absolute (RIFM)

SAMBUCUS NIGRA WAX

Definition: Sambucus Nigra Wax is a wax obtained from the flowers and leaves of *Sambucus nigra. See "Regulatory and Ingredient Use Information," regarding the labeling names for botanical ingredients in Volume 1, Introduction, Part A.*

Chemical Class: Essential Oils

Function: Fragrance Ingredient

Technical/Other Name:
Waxes, Sambucus Nigra

Trade Names:
AEC Cire Essentielle De Fleurs De Sureau (A & E Connock)
Cire Essentielle de fleurs de Sureau (Bertin)

SAND

Definition: Sand is loose, granular particles of worn or disintegrated rock.

Chemical Class: Inorganics

Function: Abrasive

Trade Names:
Sand Power (Greentech)
Seesand getr. (dried) (Dansk System Mortel)

SANGUINARIA CANADENSIS

Definition: *See "Regulatory and Ingredient Use Information," regarding EU labeling names for botanical ingredients in Volume 1, Introduction, Part A.*

Chemical Class: Biological Products

Technical/Other Names:
Sanguinaria Canadensis Extract (U.S.)
Sanguinaria Canadensis Powder (U.S.)

SANGUINARIA CANADENSIS EXTRACT

CAS No.	EINECS No.
84929-48-6	284-532-0

Definition: Sanguinaria Canadensis Extract is an extract of the rhizomes and roots of the sanguinaria, *Sanguinaria canadensis. See "Regulatory and Ingredient Use Information," regarding the labeling names for botanical ingredients in Volume 1, Introduction, Part A.*

Information Source: HP

Chemical Class: Biological Products

Function: Not Reported

Technical/Other Names:
Bloodroot Extract
Extract of Sanguinaria
Extract of Sanguinaria Canadensis
Sanguinaria Canadensis (EU)
Sanguinaria Extract

Trade Name Mixtures:
Actiphyte of Bloodroot BG50 (Active Organics)

Actiphyte of Bloodroot GL50 (Active
Organics)
Actiphyte of Bloodroot Lipo S (Active
Organics)
Actiphyte of Bloodroot PG50 (Active
Organics)
Actiphyte of Sanguinaria BG50 (Active
Organics)
Actiphyte of Sanguinaria GL50 (Active
Organics)
Actiphyte of Sanguinaria Lipo S (Active
Organics)
Actiphyte of Sanguinaria PG50 (Active
Organics)
Phytelene of Bloodroot EG 531 Liquid
(Indena SA)
Sanguinaria Root Extract HS 2778 G
(Grau)
Vivaderm (Bio-Botanica)
VT-166 Extract of Sanguinaria (Vege-Tech)

SANGUINARIA CANADENSIS POWDER

Definition: Sanguinaria Canadensis
Powder is a plant material derived from the
dried rhizomes and roots of the
sanguinaria, *Sanguinaria canadensis. See
"Regulatory and Ingredient Use
Information," regarding the labeling names
for botanical ingredients in Volume 1, Introduction, Part A.*

Information Sources: HP, MI-13(8432)

Chemical Class: Biological Products

Function: Not Reported

Technical/Other Names:
Sanguinaria
Sanguinaria Canadensis (EU)
Sanguinaria (Sanguinaria Canadensis)

SANGUISORBA OFFICINALIS ROOT EXTRACT

JPN Translation:
ワレモコウエキス

Definition: Sanguisorba Officinalis Root
Extract is an extract of the rhizomes of
*Sanguisorba officinalis. See "Regulatory
and Ingredient Use Information," regarding
the labeling names for botanical ingredients
in Volume 1, Introduction, Part A.*

Chemical Class: Biological Products

Function: Not Reported

Technical/Other Name:
Extract of Sanguisorba Officinalis

Trade Name:
Burnet Extract Powder (Maruzen
Pharmaceuticals Co., Ltd.)

Trade Name Mixtures:
Actiphyte of Burnet Root (Active Organics)
Burnet Extract (Maruzen Pharmaceuticals
Co., Ltd.)
Burnet Extract (Nikko)
Burnet Extract AL-J (Maruzen
Pharmaceuticals Co., Ltd.)
Burnet Extract AL-R (Maruzen
Pharmaceuticals Co., Ltd.)
Burnet Extract BG (Maruzen
Pharmaceuticals Co., Ltd.)
Burnet Extract BG-R (Maruzen
Pharmaceuticals Co., Ltd.)
Burnet Extract LA (Maruzen
Pharmaceuticals Co., Ltd.)
Burnet Extract-R (Maruzen
Pharmaceuticals Co., Ltd.)

SANICULA EUROPAEA EXTRACT

CAS No.	EINECS No.
90106-53-9	290-276-0

Definition: Sanicula Europaea Extract is
an extract of the herb, *Sanicula europaea.
See "Regulatory and Ingredient Use Information," regarding the labeling names for
botanical ingredients in Volume 1, Introduction, Part A.*

Chemical Class: Biological Products

Function: Not Reported

Technical/Other Name:
Extract of Sanicula Europaea

Trade Name Mixture:
Sanicle-Wort-Extract HS 3139 G (Grau)

SANSHOU EKISU

JPN Translation:
サンショウエキス

Definition: Sanshou Ekisu is an extract of
the pericarp of the fruit of *Zanthoxylum
piperitum* or other related species of *Zanthoxylum. See "Regulatory and Ingredient
Use Information," regarding use of Japan
Trivial names in Volume 1, Introduction,
Part A.*

Chemical Class: Biological Products

Function: Skin-Conditioning Agent - Miscellaneous

Technical/Other Names:
Zanthoxylum Piperitum (EU)
Zanthoxylum Piperitum Peel Extract (U.S.)

SANTALUM ALBUM (SANDALWOOD)

Definition: Santalum Album (Sandalwood)
is a plant material derived from the

sandalwood, *Santalum album. See
"Regulatory and Ingredient Use
Information," regarding the labeling names
for botanical ingredients in Volume 1, Introduction, Part A.*

Information Sources: 21CFR172.510, HP

Chemical Class: Biological Products

Function: Not Reported

Technical/Other Names:
Sandalwood
Sandalwood Resin

SANTALUM ALBUM (SANDALWOOD) EXTRACT

CAS No.	EINECS No.
84787-70-2	284-111-1

Definition: Santalum Album (Sandalwood)
Extract is an extract of the whole plant,
*Santalum album. See "Regulatory and
Ingredient Use Information," regarding the
labeling names for botanical ingredients in
Volume 1, Introduction, Part A.*

Chemical Class: Biological Products

Function: Skin-Conditioning Agent - Miscellaneous

Technical/Other Names:
Extract of Sandalwood
Extract of Santalum Album
Sandalwood Extract

Trade Name Mixtures:
Bois II (Barnet)
Sandalwood Extract (Kelisema Italy)

SANTALUM ALBUM (SANDALWOOD) OIL

CAS No.: 8006-87-9

Definition: Santalum Album (Sandalwood)
Oil is the volatile oil obtained from the
heartwood of *Santalum album. See
"Regulatory and Ingredient Use
Information," regarding the labeling names
for botanical ingredients in Volume 1, Introduction, Part A.*

Information Sources: 21CFR172.510,
EGY, FCC, HP, MAR, MI-13(6873), POR,
RIFM, TSCA

Chemical Class: Essential Oils

Function: Fragrance Ingredient

Reported Product Categories: Bath Capsules; Bath Oils, Tablets, and Salts; Cleansing Products (Cold Creams, Cleansing
Lotions, Liquids and Pads); Face and Neck
Preparations (Excluding Shaving Preparations); Skin Care Preparations, Misc.; Mois-

turizing Preparations; Bath Preparations, Misc.; Body and Hand Preparations (Excluding Shaving Preparations); Paste Masks (Mud Packs)

Technical/Other Names:
Oils, Sandalwood
Sandalwood Oil
Sandalwood yellow oil (Santalum album L.) (RIFM)
Santalum Album Oil

Trade Names:
AEC Sandalwood Oil (A & E Connock)
Australian Sandalwood Oil (Southern Cross Botanicals)
Custosense Sandalwood (sandalwood oil) (Custom Ingredients)
Sandalwood Oil (Chauvet)

Trade Name Mixtures:
Aromaphyte of Sandalwood (Active Organics)
Covazen Relax (LCW)
Essentiaderm Capillare N.18 (Universal Flavors)
Essentiaderm n.3 (Universal Flavors)
Essentiaderm N.19 (Universal Flavors)
Hexatrate (Vevy)
Hexatrate Al-Free (Vevy)

SANTALUM ALBUM (SANDALWOOD) SEED OIL

Definition: Santalum Album (Sandalwood) Seed Oil is the fixed oil obtained from the seeds of *Santalum album. See "Regulatory and Ingredient Use Information," regarding the labeling names for botanical ingredients in Volume 1, Introduction, Part A.*

Chemical Class: Fats and Oils

Function: Skin-Conditioning Agent - Occlusive

Technical/Other Names:
Oils, Sandalwood Seed
Sandalwood Seed Oil
Santalum Album Seed Oil

Trade Name:
Ximenoil S (Indena SpA)

SANTALUM ALBUM (SANDALWOOD) WOOD EXTRACT

CAS No. 84787-70-2   **EINECS No.** 284-111-1

JPN Translation:
ビャクダンエキス

Definition: Santalum Album (Sandalwood) Wood Extract is an extract of the wood of the sandalwood, *Santalum album. See "Regulatory and Ingredient Use*

Information," regarding the labeling names for botanical ingredients in Volume 1, Introduction, Part A.

Information Source: 21CFR172.510

Chemical Class: Biological Products

Function: Skin-Conditioning Agent - Occlusive

Technical/Other Names:
Extract of Sandalwood
Extract of Santalum Album
Sandalwood Extract
Sandalwood Wood Extract

Trade Name:
Hydroessential Santalum (Vevy)

Trade Name Mixtures:
Actiphyte of Sandalwood BG50 (Active Organics)
Actiphyte of Sandalwood GL50 (Active Organics)
Actiphyte of Sandalwood Lipo S (Active Organics)
Actiphyte of Sandalwood PG50 (Active Organics)
Aromaphyte of Sandalwood (Active Organics)
Extrait de Santal PPE 100 (Yves Rocher)
Extrapone Sandalwood 2/032161 (Symrise)
Herbasol Extract Sandalwood Red (Cosmetochem) (Cosmetochem International Ltd.)
Sandalwood Liquid B (Ichimaru Pharcos)
Sandalwood Liquid E (Ichimaru Pharcos)

SANTALUM ALBUM SEED EXTRACT

Definition: Santalum Album Seed Extract is an extract of the seeds of *Santalum album. See "Regulatory and Ingredient Use Information," regarding the labeling names for botanical ingredients in Volume 1, Introduction, Part A.*

Chemical Class: Biological Products

Function: Skin-Conditioning Agent - Miscellaneous

Technical/Other Name:
Extract of Santalum Album Seed

Trade Name:
Premier Santalum Album Seed Extract (Premier Specialties)

SANTOLINA CHAMAECYPARISSUS EXTRACT

CAS No. 84961-58-0   **EINECS No.** 284-647-6

Definition: Santolina Chamaecyparissus Extract is an extract of *Santolina chamaecyparissus. See "Regulatory and Ingredient Use Information," regarding the labeling names for botanical ingredients in Volume 1, Introduction, Part A.*

Chemical Class: Biological Products

Function: Not Reported

Technical/Other Name:
Extract of Santolina Chamaecyparissus

Trade Name Mixture:
Extrait De Santoline MP PG 40 (Yves Rocher)

SAPINDUS MUKUROSSI

Definition: *See "Regulatory and Ingredient Use Information," regarding EU labeling names for botanical ingredients in Volume 1, Introduction, Part A.*

Chemical Class: Biological Products

Technical/Other Names:
Sapindus Mukurossi Fruit Extract (U.S.)
Sapindus Mukurossi Peel Extract (U.S.)

SAPINDUS MUKUROSSI FRUIT EXTRACT

Definition: Sapindus Mukurossi Fruit Extract is an extract of the fruit of the soapberry, *Sapindus mukurossi. See "Regulatory and Ingredient Use Information," regarding the labeling names for botanical ingredients in Volume 1, Introduction, Part A.*

Chemical Class: Biological Products

Function: Not Reported

Technical/Other Names:
Extract of Sapindus Mukurossi Fruit
Extract of Soapberry Fruit
Sapindus Mukurossi (EU)
Soapberry Extract
Soapberry Fruit Extract
Soapberry (Sapindus Mukurossi) Extract

Trade Name:
Mukurossi Extract Powder (Maruzen Pharmaceuticals Co., Ltd.)

Trade Name Mixtures:
Crodarom Jaboncillo A (Croda, Inc.)
Soap Nut Extract HS 2754 G (Grau)

SAPINDUS MUKUROSSI PEEL EXTRACT

JPN Translation:
ムクロジエキス

Definition: Sapindus Mukurossi Peel Extract is an extract of the peel of the soapberry, *Sapindus mukurossi. See "Regulatory and Ingredient Use Information," regarding the labeling names for botanical ingredients in Volume 1, Introduction, Part A.*

Chemical Class: Biological Products

Function: Not Reported

Technical/Other Names:
Extract of Sapindus Mukurossi Peel
Extract of Soapberry Peel
Sapindus Mukurossi (EU)
Soapberry Peel Extract
Soapberry (Sapindus Mukurossi) Peel Extract

Trade Name:
Mukurossi Extract Powder (Maruzen Pharmaceuticals Co., Ltd.)

SAPINDUS TRIFOLIATUS FRUIT EXTRACT

Definition: Sapindus Trifoliatus Fruit Extract is an extract of the fruit of *Sapindus trifoliatus. See "Regulatory and Ingredient Use Information," regarding the labeling names for botanical ingredients in Volume 1, Introduction, Part A.*

Chemical Class: Biological Products

Function: Not Reported

Trade Name:
Sapindin (Sabinsa)

Trade Name Mixture:
Aritha Extract (Carlisle)

SAPONARIA OFFICINALIS EXTRACT

CAS No.   **EINECS No.**
84775-97-3   283-921-2

JPN Translation:
サボンソウエキス

Definition: Saponaria Officinalis Extract is an extract of the leaves and roots of the saponaria, *Saponaria officinalis. See "Regulatory and Ingredient Use Information," regarding the labeling names for botanical ingredients in Volume 1, Introduction, Part A.*

Information Sources: JCIC, JCLS, JSQI

Chemical Class: Biological Products

Function: Skin-Conditioning Agent - Miscellaneous

Reported Product Categories: Shampoos (Non-coloring); Bath Oils, Tablets, and Salts; Cleansing Products (Cold Creams, Cleansing Lotions, Liquids and Pads); Fragrance Preparations, Misc.; Personal Cleanliness Products, Misc.

Technical/Other Names:
Extract of Saponaria
Extract of Saponaria Officinalis
Saponaria Extract

Trade Name Mixtures:
Acneous Phytogreen Complex GXH 267 (Phytochim)
Actiphyte of Soapwort BG50 (Active Organics)
Actiphyte of Soapwort GL50 (Active Organics)
Actiphyte of Soapwort Lipo S (Active Organics)
Actiphyte of Soapwort PG50 (Active Organics)
Common Soapwort Extract HG (Provital/Centerchem)
Extrait de Saponaire MPE 100 (Yves Rocher)
Glycolysat of Saponaria (CEP (Solabia))
Herbaliquid Soap Wort special (Crodarom)
Herbasol Extract Soap Wort (Cosmetochem) (Cosmetochem International Ltd.)
Oily Skins Complex 264 (Ennagram)
Oily Skins Phytogreen Complex GXH 264 (Phytochim)
Phytelene Complex EGX 247 (Indena SA)
Phytelene Complex EGX 252 (Indena SA)
Phytelene of Saponaire EG 048 liquid (Indena SA)
Phytelene of Soapwort EG 048 liquid (Indena SA)
Phytogreen 55 of Saponaire EXH 622 Liquid (Phytochim)
Phytogreen 55 of Soapwort EXH 622 Liquid (Phytochim)
Polyplant Oily Skin (Provital/Centerchem)
Prodhy Extract Saponaire (Prod'Hyg)
Saponaria Extract BG (Maruzen Pharmaceuticals Co., Ltd.)
Soap Wort Extract HS 2577 G (Grau)
Soapwort HS (Alban Muller)
Soapwort Phytexcell (Crodarom)
Teen Age Skin Disorders Complex 267 (Ennagram)
2340 Vege-Plex Body Complex (Vege-Tech)
2410 Vege-Plex Skin Complex (Vege-Tech)
2520 Vege-Plex Skin Complex (Vege-Tech)
Vegetol Saponaria LC 386 Hydro (Gattefosse s.a.)
VT-147 Extract of Soapwort (Vege-Tech)

SAPONARIA OFFICINALIS LEAF EXTRACT

Definition: Saponaria Officinalis Leaf Extract is an extract of the leaves of *Saponaria officinalis. See "Regulatory and Ingredient Use Information," regarding the labeling names for botanical ingredients in Volume 1, Introduction, Part A.*

Chemical Class: Biological Products

Function: Cosmetic Biocide

Technical/Other Name:
Extract of Saponaria Officinalis Leaf

Trade Name Mixtures:
Pharcolex BX47 (Ichimaru Pharcos)
Pharcolex BX52 (Ichimaru Pharcos)
Saponaria Extract BG-J (Maruzen Pharmaceuticals Co., Ltd.)
Soapwort Liquid B (Ichimaru Pharcos)

SAPONINS

CAS Nos.   **EINECS No.**
8047-15-2   232-462-6
11006-75-0
72231-29-9

Definition: Saponins are a class of water soluble high molecular weight glycosidal substances naturally occurring in a wide variety of plants.

Information Source: MI-13(8442)

Chemical Class: Biological Products

Functions: Surfactant - Cleansing Agent; Surfactant - Emulsifying Agent

Technical/Other Names:
Sapogenins, Glycosides
Saponosides

Trade Names:
Bio-Saponins (Bio-Botanica)
Botanessentials SAP-50 (Botanigenics)
Phytoselect Soya (Indena SpA)
Saponin Purified (Cosmetochem) (Cosmetochem International Ltd.)

Trade Name Mixture:
Tensami 10/06 (Alban Muller)

SAPPHIRE POWDER

CAS No.: 1317-82-4

Definition: Sapphire Powder is a ground native gem consisting chiefly of aluminum oxide

Information Source: MI-13(8444)

Chemical Class: Inorganics

The inclusion of any compound in the *Dictionary and Handbook* does not indicate that use of that substance as a cosmetic ingredient complies with the laws and regulations governing such use in the United States or any other country.

Function: Opacifying Agent

SARCOSINE

CAS No.	EINECS No.
107-97-1	203-538-6

Empirical Formula:
$C_3H_7NO_2$

Definition: Sarcosine is the organic compound that conforms to the formula:

CH_3NHCH_2COOH

Information Source: MI-13(8450)

Chemical Class: Amino Acids

Function: Skin-Conditioning Agent - Miscellaneous

Technical/Other Names:
N-Methylaminoacetic Acid
N-Methylaminoethanoic Acid
N-Methyl Glycine

Trade Name:
Metil Glicina (I.C.I.M.)

Trade Name Mixtures:
Sepicalm S (SEPPIC)
Sepicontrol A5 (SEPPIC)

SARGASSUM FILIPENDULA EXTRACT

JPN Translation:
サルガッスムフィリベンデュラエキス

Definition: Sargassum Filipendula Extract is an extract of the brown alga, *Sargassum filipendula. See "Regulatory and Ingredient Use Information," regarding the labeling names for botanical ingredients in Volume 1, Introduction, Part A.*

Chemical Class: Biological Products

Function: Not Reported

Reported Product Categories: Hair Dyes and Colors (All Types Requiring Caution Statements and Patch Tests); Bath Capsules; Face and Neck Preparations (Excluding Shaving Preparations); Moisturizing Preparations; Bath Preparations, Misc.; Body and Hand Preparations (Excluding Shaving Preparations); Skin Care Preparations, Misc.; Shampoos (Non-coloring); Bath Oils, Tablets, and Salts; Cleansing Products (Cold Creams, Cleansing Lotions, Liquids and Pads); Hair Conditioners; Paste Masks (Mud Packs); Mascara; Bath Soaps and Detergents; Skin Fresheners; Tonics, Dressings, and Other Hair Grooming Aids; Eyebrow Pencils; Suntan Gels, Creams, and Liquids; Personal Cleanliness Products, Misc.; Aftershave Lotions; Baby Shampoos; Bubble Baths; Fragrance Preparations, Misc.; Hair Rinses (Non-coloring)

Technical/Other Names:
Algae Extract
Extract of Sargassum Filipendula

Trade Name Mixtures:
Ormagel AC-400 (Assessa-Industria)
Ormagel SH (Assessa-Industria)
Ormagel SHE (Assessa-Industria)
Ormagel XPU (Assessa-Industria)
Ormagel XPX (Assessa-Industria)
Seaweederm A-525 (Assessa-Industria)
Seaweedex (Assessa-Industria)

SARGASSUM FUSIFORME EXTRACT

JPN Translation:
サルガッスムフシフォルムエキス

Definition: Sargassum Fusiforme Extract is an extract of the brown alga, *Sargassum fusiforme. See "Regulatory and Ingredient Use Information," regarding the labeling names for botanical ingredients in Volume 1, Introduction, Part A.*

Chemical Class: Biological Products

Function: Not Reported

Technical/Other Name:
Extract of Sargassum Fusiforme

Trade Name Mixture:
Sinominceur (I.D. bio)

SARGASSUM MUTICUM EXTRACT

JPN Translation:
サルガッスムムティカムエキス

Definition: Sargassum Muticum Extract is an extract of the alga *Sargassum muticum. See "Regulatory and Ingredient Use Information," regarding the labeling names for botanical ingredients in Volume 1, Introduction, Part A.*

Chemical Class: Biological Products

Function: Not Reported

Technical/Other Name:
Extract of Sargassum Muticum

Trade Name:
Extract of Sargassum muticum Phyactyl (GELYMA)

SARGASSUM PALLIDUM EXTRACT

Definition: Sargassum Pallidum Extract is an extract of the plant, *Sargassum pallidum. See "Regulatory and Ingredient Use Information," regarding the labeling* names for botanical ingredients in Volume 1, Introduction, Part A. See Reported Ingredient Functions-The Cosmetic Drug Distinction, in Regulatory and Ingredient Use Information, Volume I, Part A.*

Chemical Class: Biological Products

Functions: Antifungal Agent; Antioxidant

Technical/Other Name:
Extract of Sargassum Pallidium

Trade Name Mixture:
Gulfweed Extract 101873 (Fragrance Oils Int. Ltd.)

SARGASSUM VULGARE EXTRACT

Definition: Sargassum Vulgare Extract is an extract of the seaweed, *Sargassum vulgare. See "Regulatory and Ingredient Use Information," regarding the labeling names for botanical ingredients in Volume 1, Introduction, Part A.*

Chemical Class: Biological Products

Function: Skin-Conditioning Agent - Miscellaneous

Technical/Other Name:
Extract of Sargassum Vulgare

Trade Name Mixture:
Quidgel BRM. (Assessa-Industria)

SAROTHAMNUS SCOPARIUS EXTRACT

CAS No.	EINECS No.
84696-48-0	283-653-6

Definition: Sarothamnus Scoparius Extract is an extract of the entire plant, *Sarothamnus scoparius. See "Regulatory and Ingredient Use Information," regarding the labeling names for botanical ingredients in Volume 1, Introduction, Part A.*

Chemical Class: Biological Products

Function: Antioxidant

Technical/Other Name:
Extract of Sarothamnus Scoparius

Trade Name Mixtures:
Melscreen Black (Chemyunion)
Melscreen Black EX BG (Chemyunion)

SASA KURILENSIS WATER

Definition: Sasa Kurilensis Water is an aqueous solution of the steam distillate obtained from the leaves and stems of *Sasa kurilensis. See "Regulatory and*

Ingredient Use Information," *regarding the labeling names for botanical ingredients in Volume 1, Introduction, Part A.*

Chemical Class: Biological Products

Function: Skin-Conditioning Agent - Miscellaneous

Trade Name:
Chishima-zasa Water (Shinwa Kasei Corporation)

SASA VEITCHII EXTRACT

JPN Translation:
クマザサエキス

Definition: Sasa Veitchii Extract is an extract of the leaves of *Sasa veitchii*. See *"Regulatory and Ingredient Use Information," regarding the labeling names for botanical ingredients in Volume 1, Introduction, Part A.*

Chemical Class: Biological Products

Function: Not Reported

Technical/Other Name:
Extract of Sasa Veitchii

Trade Name Mixtures:
Kumazasa Liquid E (Ichimaru Pharcos)
Sasa Veitchii Extract BG (Maruzen Pharmaceuticals Co., Ltd.)
Sasa Veitchii Extract Powder-S (Maruzen Pharmaceuticals Co., Ltd.)

SASSAFRAS OFFICINALE

Definition: *See "Regulatory and Ingredient Use Information," regarding EU labeling names for botanical ingredients in Volume 1, Introduction, Part A.*

Chemical Class: Biological Products

Technical/Other Names:
Sassafras Officinale Extract (U.S.)
Sassafras Officinale Root Oil (U.S.)

SASSAFRAS OFFICINALE EXTRACT

Definition: Sassafras Officinale Extract is an extract of the bark and roots of the sassafras, *Sassafras officinale*. See *"Regulatory and Ingredient Use Information," regarding the labeling names for botanical ingredients in Volume 1, Introduction, Part A.*

Chemical Class: Biological Products

Function: Not Reported

Technical/Other Names:
Extract of Sassafras

Extract of Sassafras Officinale
Sassafras Albidum Extract
Sassafras Extract
Sassafras Officinale (EU)

Trade Name Mixtures:
Actiphyte of Sassafras BG50 (Active Organics)
Actiphyte of Sassafras GL50 (Active Organics)
Actiphyte of Sassafras Lipo S (Active Organics)
Actiphyte of Sassafras PG50 (Active Organics)
VT-109 Extract of Sassafrass (Vege-Tech)

SASSAFRAS OFFICINALE ROOT OIL

CAS No.: 8006-80-2

Definition: Sassafras Officinale Root Oil is the volatile oil obtained from the root of *Sassafras officinale*. It contains approximately 80% safrol. See *"Regulatory and Ingredient Use Information," regarding the labeling names for botanical ingredients in Volume 1, Introduction, Part A.*

Information Sources: BRA, 21CFR172.510, 27CFR21.65, 27CFR21.151, MAR, MI-13(6874), PF, POR, RIFM, TSCA

Chemical Class: Essential Oils

Function: Fragrance Ingredient

Technical/Other Names:
Oil of Sassafras
Oils, Sassafras
Sassafras Officinale (EU)
Sassafras Oil
Sassafras oil (Sassafras albidum (Nutt.) Nees) (RIFM)

SATUREIA HORTENSIS EXTRACT

CAS No.	EINECS No.
84775-98-4	283-922-8

Definition: Satureia Hortensis Extract is an extract of the savory, *Satureia hortensis*. See *"Regulatory and Ingredient Use Information," regarding the labeling names for botanical ingredients in Volume 1, Introduction, Part A.*

Information Sources: 21CFR182.20, RIFM

Chemical Class: Biological Products

Functions: Fragrance Ingredient; Skin-Conditioning Agent - Miscellaneous

Technical/Other Names:
Extract of Satureia Hortensis

Extract of Savory
Savory Extract
Savory (Satureia Hortensis) Extract
Savory, summer, oleoresin (Satureja hortensis L.) (RIFM)

Trade Name Mixtures:
Herbasol Extract Savory (Wort) (Cosmetochem) (Cosmetochem International Ltd.)
Oral Mucous Protection Complex MU 3776 (Greentech S.A)
Phytelene of Savory EG 502 liquid (Indena SA)
Phytogreen 55 of Savory EXH 705 Liquid (Phytochim)
Satureia Hortensis Extract ies (IES LABO)
Savory HS (Alban Muller)
Vegebios of Savory (CEP (Solabia))
VT-181 Extract of Savory Leaves (Vege-Tech)

SATUREIA HORTENSIS LEAF EXTRACT

Definition: Satureia Hortensis Leaf Extract is an extract of the leaves of *Satureia hortensis*. See *"Regulatory and Ingredient Use Information," regarding the labeling names for botanical ingredients in Volume 1, Introduction, Part A.*

Chemical Class: Biological Products

Function: Skin-Conditioning Agent - Miscellaneous

Technical/Other Name:
Extract of Satureia Hortensis Leaf

SAUSSUREA INVOLUCRATA EXTRACT

Definition: Saussurea Involucrata Extract is an extract of the whole plant, *Saussurea involucrata*. See *"Regulatory and Ingredient Use Information," regarding the labeling names for botanical ingredients in Volume 1, Introduction, Part A.*

Function: Skin-Conditioning Agent - Humectant

Technical/Other Name:
Extract of Saussurea Involucrata

Trade Name Mixture:
Saussurea Involucrata Extract BG (Maruzen Pharmaceuticals Co., Ltd.)

SAUSSUREA LAPPA ROOT EXTRACT

Definition: Saussurea Lappa Root Extract is an extract of the roots of *Saussurea lappa*. See *"Regulatory and Ingredient Use Information," regarding the labeling names*

for botanical ingredients in Volume 1, Introduction, Part A.

Chemical Class: Biological Products

Function: Not Reported

Technical/Other Name:
Extract of Saussurea Lappa

Trade Name Mixture:
Mokkou Liquid E (Ichimaru Pharcos)

SAXIFRAGA SARMENTOSA EXTRACT

JPN Translation:
ユキノシタエキス

Definition: Saxifraga Sarmentosa Extract is an extract of the herb of the strawberry begonia, *Saxifraga sarmentosa*. See *"Regulatory and Ingredient Use Information,"* regarding the labeling names for botanical ingredients in Volume 1, Introduction, Part A.

Information Sources: JCIC, JCLS

Chemical Class: Biological Products

Function: Not Reported

Technical/Other Names:
Extract of Saxifraga Sarmentosa
Extract of Saxifraga Stolonifera
Saxifraga Stolonifera Extract
Saxifrage Extract

Trade Name Mixtures:
Biowhite (Coletica SA)
Clarisome (Coletica SA)
Pharcolex PSP (Ichimaru Pharcos)
Phytoclar (Coletica SA)
Phytoclar II (Coletica SA)
Saxifrage Extract (Maruzen
    Pharmaceuticals Co., Ltd.)
Saxifrage Extract BG (Maruzen
    Pharmaceuticals Co., Ltd.)
Ultrawhite (Coletica SA)
Yukinoshita Liquid MB (Ichimaru Pharcos)

SAXIFRAGA STOLONIFERA LEAF POWDER

Definition: Saxifraga Stolonifera Leaf Powder is a powder obtained from the leaves of *Saxifraga stolonifera*. See *"Regulatory and Ingredient Use Information,"* regarding the labeling names for botanical ingredients in Volume 1, Introduction, Part A.

Chemical Class: Biological Products

Function: Skin-Conditioning Agent - Miscellaneous

Trade Name Mixture:
DryLeaf CGS (Shiseido Company)

SCABIOSA ARVENSIS

Definition: See *"Regulatory and Ingredient Use Information,"* regarding EU labeling names for botanical ingredients in Volume 1, Introduction, Part A.

Chemical Class: Biological Products

Technical/Other Name:
Scabiosa Arvensis Extract (U.S.)

SCABIOSA ARVENSIS EXTRACT

CAS No.	EINECS No.
90046-08-5	289-975-3

Definition: Scabiosa Arvensis Extract is an extract of the herb of the scabiosa, *Scabiosa arvensis*. See *"Regulatory and Ingredient Use Information,"* regarding the labeling names for botanical ingredients in Volume 1, Introduction, Part A.

Chemical Class: Biological Products

Function: Not Reported

Technical/Other Names:
Extract of Scabiosa
Extract of Scabiosa Arvensis
Knautia Arvensis Extract
Scabiosa Arvensis (EU)
Scabiosa Extract

Trade Name Mixture:
Scabiosa Herb Extract HS 2680 G (Grau)

SCHINUS MOLLE OIL

CAS No.: 68917-52-2

Definition: Schinus Molle Oil is the oil expressed from the fruit of *Schinus molle*. See *"Regulatory and Ingredient Use Information,"* regarding the labeling names for botanical ingredients in Volume 1, Introduction, Part A.

Information Source: RIFM

Chemical Class: Fats and Oils

Functions: Fragrance Ingredient; Skin-Conditioning Agent - Miscellaneous

Technical/Other Names:
Oil, Schinus Molle
Schinus molle oil (Schinus molle L.) (RIFM)

SCHINUS TEREBINTHIFOLIUS SEED EXTRACT

Definition: Schinus Terebinthifolius Seed Extract is an extract of the seeds of *Schinus terebinthifolius*. See *"Regulatory and Ingredient Use Information,"* regarding

the labeling names for botanical ingredients in Volume 1, Introduction, Part A.

Chemical Class: Biological Products

Functions: Cosmetic Astringent; Skin-Conditioning Agent - Miscellaneous

Trade Name Mixture:
Actiphyte of Pink Peppercorn (Active Organics)

SCHIZANDRA CHINENSIS FRUIT EXTRACT

Definition: Schizandra Chinensis Fruit Extract is an extract of the fruit of *Schizandra chinensis*. See *"Regulatory and Ingredient Use Information,"* regarding the labeling names for botanical ingredients in Volume 1, Introduction, Part A.

Chemical Class: Biological Products

Function: Not Reported

Technical/Other Name:
Extract of Schizandra Chinensis

Trade Name Mixtures:
Actiphyte of Schisandra BG50 (Active
    Organics)
Actiphyte of Schisandra GL50 (Active
    Organics)
Actiphyte of Schisandra Lipo S (Active
    Organics)
Actiphyte of Schisandra PG50 (Active
    Organics)
Campo Wu Wei Zi Extract (Campo)

SCHIZONEPETA TENUIFOLIA EXTRACT

Definition: Schizonepeta Tenuifolia Extract is an extract of the whole plant, *Schizonepeta tenuifolia*. See *"Regulatory and Ingredient Use Information,"* regarding the labeling names for botanical ingredients in Volume 1, Introduction, Part A.

Chemical Class: Biological Products

Function: Not Reported

Technical/Other Name:
Extract of Schizonepeta Tenuifolia

Trade Name:
Macela dry extract special 2.5% NTR
    (Centroflora Group)

Trade Name Mixture:
Keigai Liquid E (Ichimaru Pharcos)

SCHIZOPHYLLAN

CAS No.: 9050-67-3

Definition: Schizophyllan is a poly-saccharide produced by the fungus, *Schizophyllum commune*. It consists of three β-(1->3)-linked D-glucopyranose residues, to one of which is attached a single β-(1->6)-linked D-glucopyranosyl side chain.

Information Source: MI-13(8630)

Chemical Class: Carbohydrates

Function: Humectant

Technical/Other Name:
Sizofiran

Trade Name:
Schizophyllan (CPN)

SCHLEICHERA TRIJUGA SEED OIL

Definition: Schleichera Trijuga Seed Oil is the fixed oil expressed from the seeds of *Schleichera trijuga*. See "Regulatory and Ingredient Use Information," regarding the labeling names for botanical ingredients in Volume 1, Introduction, Part A.

Chemical Class: Fats and Oils

Functions: Hair Conditioning Agent; Skin-Conditioning Agent - Occlusive

Trade Name:
Huile De Macassar (Plantes et Industrie)

SCIADOPITYS VERTICILLATA ROOT EXTRACT

Definition: Sciadopitys Verticillata Root Extract is an extract of the roots of *Sciadopitys verticillata*. See "Regulatory and Ingredient Use Information," regarding the labeling names for botanical ingredients in Volume 1, Introduction, Part A. See "Regulatory and Ingredient Use Information," regarding the labeling names for U.S. OTC Drug Ingredients in Volume 1, Introduction, Part A. See Reported Ingredient Functions-The Cosmetic Drug Distinction, in Regulatory and Ingredient Use Information, Volume I, Part A.

Chemical Class: Biological Products

Functions: Antidandruff Agent; Antifungal Agent; Antistatic Agent; Skin-Conditioning Agent - Miscellaneous

Technical/Other Name:
Extract of Sciadopityl Verticillata Root

Trade Name Mixture:
Kumsong Extract (EUROCOSTECH)

SCLAREOLIDE

CAS No.
564-20-5

EINECS No.
209-269-0

Empirical Formula:
$C_{16}H_{26}O_2$

Definition: Sclareolide is the organic compound that conforms to the formula:

Information Sources: RIFM, TSCA

Chemical Classes: Ethers; Ketones

Functions: Fragrance Ingredient; Skin-Conditioning Agent - Miscellaneous

Technical/Other Names:
Naphtho[2,1-b]Furan-2(1H)-one, Decahydro-3a,6,6,9a-Tetramethyl-, [3aR-(3aα,5aβ,9aα,9bβ)]-
Norambreinolide
Sclareolide (RIFM)

Trade Name:
Fermented Clary Sage Extract (MMP)

SCLEROCARYA BIRREA FRUIT EXTRACT

Definition: Sclerocarya Birrea Fruit Extract is the extract of the fruit of *Sclerocarya birrea*. See "Regulatory and Ingredient Use Information," regarding the labeling names for botanical ingredients in Volume 1, Introduction, Part A.

Chemical Class: Biological Products

Function: Skin-Conditioning Agent - Miscellaneous

Technical/Other Name:
Extract of Sclerocarya Birrea Fruit

Trade Name Mixture:
Marula (Sclerocarya Birrea) Fruit Extract (Hideaway Group)

SCLEROCARYA BIRREA LEAF EXTRACT

Definition: Sclerocarya Birrea Leaf Extract is an extract of the leaf of *Sclerocarya Birrea*. See "Regulatory and Ingredient Use Information," regarding the labeling names for botanical ingredients in Volume 1, Introduction, Part A.

Chemical Class: Biological Products

Function: Skin-Conditioning Agent - Miscellaneous

SCLEROCARYA BIRREA OIL

Definition: Sclerocarya Birrea Oil is the oil expressed from the seeds of *Sclerocarya*

birrea. See "Regulatory and Ingredient Use Information," regarding the labeling names for botanical ingredients in Volume 1, Introduction, Part A.

Chemical Class: Fats and Oils

Functions: Hair Conditioning Agent; Skin-Conditioning Agent - Emollient

Technical/Other Name:
Oils, Sclerocarya Birrea

Trade Names:
AEC Marula Oil (A & E Connock)
Marula Oil (Aldivia)
Marula Oil (Statfold Seed Oils) (Arch Personal Care Products)
Marula Oil (Statfold Seed Oils) (Statfold Seed Oils')

SCLEROTIUM GUM

CAS No.: 39464-87-4

Definition: Sclerotium Gum is the poly-saccharide gum produced by the bacterium *Sclerotium rolfssii*. It is composed of glucose monomers.

Chemical Classes: Biological Polymers and their Derivatives; Gums, Hydrophilic Colloids and Derivatives

Functions: Emulsion Stabilizer; Skin-Conditioning Agent - Miscellaneous; Viscosity Increasing Agent - Aqueous

Reported Product Categories: Hair Dyes and Colors (All Types Requiring Caution Statements and Patch Tests); Tonics, Dressings, and Other Hair Grooming Aids

Technical/Other Names:
Gum, Sclerotium
Scleroglucan
Sclerogum

Trade Names:
AEC Sclerotium Gum (A & E Connock)
Clearogel CS 11 D (MMP)
TINOCARE GL (Ciba Specialty Chemicals)

Trade Name Mixtures:
Acacia Phytolait (Alban Muller)
Almond Phytolait (Alban Muller)
Blond Pea Phytolait (Alban Muller)
Cocoa Phytolait (Alban Muller)
Frangipany Phytolait (Alban Muller)
Green Tea Phytolait (Alban Muller)
Hydriame (Lanatech)
Jasmine Phytolait (Alban Muller)
Lotus Phytolait (Alban Muller)
Lupine Phytolait (Alban Muller)
Oat Phytolait (Alban Muller)
Palm Phytolait (Alban Muller)
Rice Phytolait (Alban Muller)
Soya Bean Phytolait (Alban Muller)
Vegetensor (Alban Muller)
Wheat Phytolait (Alban Muller)

SCORDININE

CAS No.: 37317-75-2

JPN Translation:
スコルジニン

Definition: Scordinine is the organic compound that conforms to the formula:

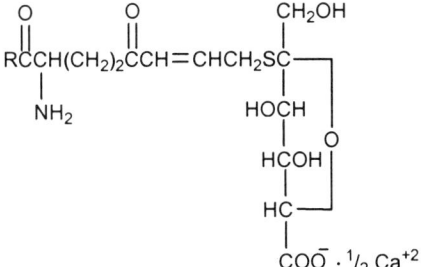

Information Source: JCIC

Chemical Classes: Amines; Heterocyclic Compounds; Thio Compounds

Function: Skin-Conditioning Agent - Miscellaneous

SCROPHULARIA NODOSA EXTRACT

CAS No.	EINECS No.
90106-65-3	290-289-1

Definition: Scrophularia Nodosa Extract is an extract of the figwort, *Scrophularia nodosa*. See *"Regulatory and Ingredient Use Information,"* regarding the labeling names for botanical ingredients in Volume 1, Introduction, Part A.

Chemical Class: Biological Products

Function: Not Reported

Technical/Other Names:
Extract of Figwort
Extract of Scrophularia Nodosa
Figwort Extract
Figwort (Scrophularia Nodosa) Extract

Trade Name Mixtures:
Common Fig Wort HS 2581 G (Grau)
Herbasol Extract Fig Wort (Cosmetochem) (Cosmetochem International Ltd.)
Scrophularia Nodosa Extract ies (IES LABO)

SCUTELLARIA BAICALENSIS ROOT EXTRACT

CAS No.	EINECS No.
94279-99-9	304-845-9

JPN Translation:
オウゴンエキス

Definition: Scutellaria Baicalensis Root Extract is an extract of the roots of the scutellaria, *Scutellaria baicalensis*. See *"Regulatory and Ingredient Use Information,"* regarding the labeling names for botanical ingredients in Volume 1, Introduction, Part A.

Information Sources: JCIC, JCLS, JSQI

Chemical Class: Biological Products

Function: Skin-Conditioning Agent - Humectant

Reported Product Categories: Bath Capsules; Moisturizing Preparations

Technical/Other Names:
Extract of Scutellaria Baicalensis Root
Extract of Scutellaria Root
Ohgon
Scutellaria Root Extract

Trade Names:
Ougon Extract Powder (Ichimaru Pharcos)
Scutellaria Root Extract Powder (Maruzen Pharmaceuticals Co., Ltd.)

Trade Name Mixtures:
Biowhite (Coletica SA)
Clarisome (Coletica SA)
Ougon Liquid B (Ichimaru Pharcos)
Ougon Liquid E (Ichimaru Pharcos)
Ougon Liquid SE (Ichimaru Pharcos)
Phytoclar II (Coletica SA)
Phytolight (Coletica SA)
Phytolight BG (Coletica SA)
Scutellaria Root Extract (Bioland)
Scutellaria Root Extract (Maruzen Pharmaceuticals Co., Ltd.)
Scutellaria Root Extract BG (Maruzen Pharmaceuticals Co., Ltd.)
Scutellaria Root Extract BG-J (Maruzen Pharmaceuticals Co., Ltd.)
Scutellaria Root Extract-J (Maruzen Pharmaceuticals Co., Ltd.)
Scutellaria Root Extract SQ (Maruzen Pharmaceuticals Co., Ltd.)
Turn White Complex (Ennagram)
Ultrawhite (Coletica SA)
Vegewhite (LCW)

SCUTELLARIA GALERICULATA EXTRACT

CAS No.	EINECS No.
90106-66-4	290-290-7

Definition: Scutellaria Galericulata Extract is an extract of the skullcap, *Scutellaria galericulata*. See *"Regulatory and Ingredient Use Information,"* regarding the labeling names for botanical ingredients in Volume 1, Introduction, Part A.

Chemical Class: Biological Products

Function: Not Reported

Technical/Other Names:
Extract of Scutellaria Galericulata
Extract of Skullcap
Skullcap Extract
Skullcap (Scutellaria Galericulata) Extract

Trade Name Mixtures:
Actiphyte of Skullcap BG50 (Active Organics)
Actiphyte of Skullcap GL50 (Active Organics)
Actiphyte of Skullcap Lipo S (Active Organics)
Actiphyte of Skullcap PG50 (Active Organics)
Phytotal RS (Phybiotex/Sederma)
VT-228 Extract of Skullcap (Vege-Tech)

SCUTELLARIA LATERIFLORA FLOWER POWDER

Definition: Scutellaria Lateriflora Flower Powder is the powder obtained from the drived flowers of *Scutellaria lateriflora*. See *"Regulatory and Ingredient Use Information,"* regarding the labeling names for botanical ingredients in Volume 1, Introduction, Part A.

Chemical Class: Biological Products

Function: Skin-Conditioning Agent - Miscellaneous

Trade Name:
Scullcap Powder (Aveda)

SD ALCOHOL 1

CAS No.: 8048-19-9

Definition: SD Alcohol 1 is ethyl alcohol denatured with methyl alcohol and one of the following: denatonium benzoate, MIBK, mixed isomers of nitropropane, or methyl n-butyl ketone, in accordance with 27CFR21. See *"Regulatory and Ingredient Use Information,"* regarding the labeling names for denatured alcohol in Volume 1, Introduction, Part A.

Information Sources: 27CFR20.11, 27CFR21.32, MI-13(3796)

Chemical Class: Alcohols

Functions: Cosmetic Astringent; Solvent; Viscosity Decreasing Agent

Technical/Other Name:
Alcohol Denat.

SD ALCOHOL 3-A

Definition: SD Alcohol 3-A is ethyl alcohol denatured with methyl alcohol in accordance with 27CFR21. *See "Regulatory and Ingredient Use Information," regarding the labeling names for denatured alcohol in Volume 1, Introduction, Part A.*

Information Sources: 27CFR20.11, 27CFR21.35, MI-13(3796)

Chemical Class: Alcohols

Functions: Cosmetic Astringent; Solvent; Viscosity Decreasing Agent

Reported Product Categories: Bubble Baths; Cleansing Products (Cold Creams, Cleansing Lotions, Liquids and Pads); Shampoos (Non-coloring); Skin Fresheners

Technical/Other Name:
Alcohol Denat.

Trade Name:
Eastman SDA-3A (Eastman Chemical)

SD ALCOHOL 3-B

Definition: SD Alcohol 3-B is ethyl alcohol denatured with pine tar in accordance with 27CFR21. *See "Regulatory and Ingredient Use Information," regarding the labeling names for denatured alcohol in Volume 1, Introduction, Part A.*

Information Sources: 27CFR20.11, 27CFR21.36

Chemical Class: Alcohols

Functions: Cosmetic Astringent; Solvent; Viscosity Decreasing Agent

Technical/Other Name:
Alcohol Denat.

SD ALCOHOL 3-C

Definition: SD Alcohol 3-C is ethyl alcohol denatured with isopropyl alcohol in accordance with 27CFR21. *See "Regulatory and Ingredient Use Information," regarding the labeling names for denatured alcohol in Volume 1, Introduction, Part A.*

Information Sources: 27CFR21.37, 27CFR21.161

Chemical Class: Alcohols

Functions: Cosmetic Astringent; Solvent; Viscosity Decreasing Agent

Technical/Other Name:
Alcohol Denat.

Trade Name:
Eastman SDA-3C (Eastman Chemical)

SD ALCOHOL 23-A

Definition: SD Alcohol 23-A is ethyl alcohol denatured with acetone in accordance with 27CFR21. *See "Regulatory and Ingredient Use Information," regarding the labeling names for denatured alcohol in Volume 1, Introduction, Part A.*

Information Sources: 27CFR20.11, 27CFR21.47, MI-13(3796)

Chemical Class: Alcohols

Functions: Cosmetic Astringent; Solvent; Viscosity Decreasing Agent

Reported Product Category: Skin Care Preparations, Misc.

Technical/Other Name:
Alcohol Denat.

SD ALCOHOL 23-F

Definition: SD Alcohol 23-F is ethyl alcohol denatured with salicylic acid, resorcinol, and bergamot oil or bay oil (myrica oil) in accordance with 27CFR21. *See "Regulatory and Ingredient Use Information," regarding the labeling names for denatured alcohol in Volume 1, Introduction, Part A.*

Information Sources: 27CFR20.11, 27CFR21.48

Chemical Class: Alcohols

Functions: Cosmetic Astringent; Solvent; Viscosity Decreasing Agent

Technical/Other Name:
Alcohol Denat.

SD ALCOHOL 23-H

Definition: SD Alcohol 23-H is ethyl alcohol denatured with acetone and MIBK in accordance with 27CFR21. *See "Regulatory and Ingredient Use Information," regarding the labeling names for denatured alcohol in Volume 1, Introduction, Part A.*

Information Sources: 27CFR20.11, 27CFR21.49

Chemical Class: Alcohols

Functions: Cosmetic Astringent; Solvent; Viscosity Decreasing Agent

Technical/Other Name:
Alcohol Denat.

SD ALCOHOL 27-B

Definition: SD Alcohol 27-B is ethyl alcohol denatured with lavender oil and

medicinal soft soap (green soap) in accordance with 27CFR21. *See "Regulatory and Ingredient Use Information," regarding the labeling names for denatured alcohol in Volume 1, Introduction, Part A.*

Information Sources: 27CFR20.11, 27CFR21.54

Chemical Class: Alcohols

Functions: Cosmetic Astringent; Solvent; Viscosity Decreasing Agent

Reported Product Category: Body and Hand Preparations (Excluding Shaving Preparations)

Technical/Other Name:
Alcohol Denat.

SD ALCOHOL 30

Definition: SD Alcohol 30 is ethyl alcohol denatured with methyl alcohol in accordance with 27CFR21. *See "Regulatory and Ingredient Use Information," regarding the labeling names for denatured alcohol in Volume 1, Introduction, Part A.*

Information Sources: 27CFR20.11, 27CFR21.57, MI-13(3796)

Chemical Class: Alcohols

Functions: Cosmetic Astringent; Solvent; Viscosity Decreasing Agent

Technical/Other Name:
Alcohol Denat.

SD ALCOHOL 31-A

Definition: SD Alcohol 31-A is ethyl alcohol denatured with glycerin and hard soap in accordance with 27CFR21. *See "Regulatory and Ingredient Use Information," regarding the labeling names for denatured alcohol in Volume 1, Introduction, Part A.*

Information Sources: 27CFR20.11, 27CFR21.58

Chemical Class: Alcohols

Functions: Cosmetic Astringent; Solvent; Viscosity Decreasing Agent

Technical/Other Name:
Alcohol Denat.

SD ALCOHOL 36

Definition: SD Alcohol 36 is ethyl alcohol denatured with ammonium hydroxide and sodium hydroxide in accordance with 27CFR21. *See "Regulatory and Ingredient Use Information," regarding the labeling names for denatured alcohol in Volume 1, Introduction, Part A.*

Information Sources: 27CFR20.11, 27CFR21.63

Chemical Class: Alcohols

Functions: Cosmetic Astringent; Solvent; Viscosity Decreasing Agent

Technical/Other Name:
Alcohol Denat.

SD ALCOHOL 37

Definition: SD Alcohol 37 is ethyl alcohol denatured with eucalyptol, thymol and menthol in accordance with 27CFR21. *See "Regulatory and Ingredient Use Information," regarding the labeling names for denatured alcohol in Volume 1, Introduction, Part A.*

Information Sources: 27CFR20.11, 27CFR21.64

Chemical Class: Alcohols

Functions: Cosmetic Astringent; Solvent; Viscosity Decreasing Agent

Technical/Other Name:
Alcohol Denat.

SD ALCOHOL 38-B

Definition: SD Alcohol 38-B is ethyl alcohol denatured with one or more of the following: anethol, anise oil, bay oil (myrcia oil), benzaldehyde, bergamot oil, bitter almond oil, camphor, cedar leaf oil, chlorothymol, cinnamic aldehyde, cinnamon oil (cassia oil), citronella oil (natural), clove oil, coal tar, eucalyptol, eucalyptus oil, eugenol, guaiacol, lavender oil, menthol, methyl salicylate, mustard oil, volatile (allyl isothiocyanate), peppermint oil, phenol, phenyl salicylate (salol), pine oil, pine needle oil, dwarf, rosemary oil, safrole, sassafras oil, spearmint oil, spearmint oil (terpeneless), spike lavender oil, natural, storax, thyme oil, thymol, tolu balsam, turpentine oil, or other approved essential oils in accordance with 27CFR21. *See "Regulatory and Ingredient Use Information," regarding the labeling names for denatured alcohol in Volume 1, Introduction, Part A.*

Information Sources: 27CFR20.11, 27CFR21.65

Chemical Class: Alcohols

Functions: Cosmetic Astringent; Solvent; Viscosity Decreasing Agent

Reported Product Categories: Mouthwashes and Breath Fresheners (Liquids and Sprays); Cleansing Products (Cold Creams, Cleansing Lotions, Liquids and Pads); Oral Hygiene Products, Misc.; Skin Care Preparations, Misc.

Technical/Other Name:
Alcohol Denat.

Trade Name Mixture:
Premier Caribbean Denatured Rum (Premier Specialties)

SD ALCOHOL 38-C

Definition: SD Alcohol 38-C is ethyl alcohol denatured with menthol and formaldehyde solution in accordance with 27CFR21. *See "Regulatory and Ingredient Use Information," regarding the labeling names for denatured alcohol in Volume 1, Introduction, Part A.*

Information Sources: 27CFR20.11, 27CFR21.66

Chemical Class: Alcohols

Functions: Cosmetic Astringent; Solvent; Viscosity Decreasing Agent

Technical/Other Name:
Alcohol Denat.

SD ALCOHOL 38-D

Definition: SD Alcohol 38-D is ethyl alcohol denatured with menthol and formaldehyde solution in accordance with 27CFR21. *See "Regulatory and Ingredient Use Information," regarding the labeling names for denatured alcohol in Volume 1, Introduction, Part A.*

Information Sources: 27CFR20.11, 27CFR21.67

Chemical Class: Alcohols

Functions: Cosmetic Astringent; Solvent; Viscosity Decreasing Agent

Reported Product Category: Mouthwashes and Breath Fresheners (Liquids and Sprays)

Technical/Other Name:
Alcohol Denat.

SD ALCOHOL 38-F

Definition: SD Alcohol 38-F is ethyl alcohol denatured with 1) thymol,

chlorathymol, menthol, and either boric acid or polysorbate-80; 2) any two or more denaturants listed under Formula 38-B, and either boric acid or polysorbate-80; 3) any two or more denaturants listed under Formula 38-B, and zinc chloride and hydrochloric acid in accordance with 27CFR21. *See "Regulatory and Ingredient Use Information," regarding the labeling names for denatured alcohol in Volume 1, Introduction, Part A.*

Information Sources: 27CFR20.11, 27CFR21.68

Chemical Class: Alcohols

Functions: Cosmetic Astringent; Solvent; Viscosity Decreasing Agent

Reported Product Category: Mouthwashes and Breath Fresheners (Liquids and Sprays)

Technical/Other Name:
Alcohol Denat.

SD ALCOHOL 39

Definition: SD Alcohol 39 is ethyl alcohol denatured with sodium salicylate or salicylic acid, and fluid extract of quassin, and t-butyl alcohol in accordance with 27CFR21. *See "Regulatory and Ingredient Use Information," regarding the labeling names for denatured alcohol in Volume 1, Introduction, Part A.*

Information Sources: 27CFR20.11, 27CFR21.69

Chemical Class: Alcohols

Functions: Cosmetic Astringent; Solvent; Viscosity Decreasing Agent

Technical/Other Name:
Alcohol Denat.

SD ALCOHOL 39-A

Definition: SD Alcohol 39-A is ethyl alcohol denatured with t-butyl alcohol and one of the following: quinine, quinine bisulfate, quinine dihydrochloride, cinchonidine, or cinchonidine sulfate or their salts in accordance with 27CFR21. *See "Regulatory and Ingredient Use Information," regarding the labeling names for denatured alcohol in Volume 1, Introduction, Part A.*

Information Sources: 27CFR20.11, 27CFR21.70

Chemical Class: Alcohols

Functions: Cosmetic Astringent; Solvent; Viscosity Decreasing Agent

Technical/Other Name:
Alcohol Denat.

SD ALCOHOL 39-B

Definition: SD Alcohol 39-B is ethyl alcohol denatured with t-butyl alcohol and diethyl phthalate in accordance with 27CFR21. *See "Regulatory and Ingredient Use Information," regarding the labeling names for denatured alcohol in Volume 1, Introduction, Part A.*

Information Sources: 27CFR20.11, 27CFR21.71

Chemical Class: Alcohols

Functions: Cosmetic Astringent; Solvent; Viscosity Decreasing Agent

Reported Product Categories: Cleansing Products (Cold Creams, Cleansing Lotions, Liquids and Pads); Foot Powders and Sprays; Tonics, Dressings, and Other Hair Grooming Aids

Technical/Other Name:
Alcohol Denat.

SD ALCOHOL 39-C

Definition: SD Alcohol 39-C is ethyl alcohol denatured with diethyl phthalate in accordance with 27CFR21. *See "Regulatory and Ingredient Use Information," regarding the labeling names for denatured alcohol in Volume 1, Introduction, Part A.*

Information Sources: 27CFR20.11, 27CFR21.72, CTFA D, JCIC, JCLS, MI-13 (3796)

Chemical Class: Alcohols

Functions: Cosmetic Astringent; Solvent; Viscosity Decreasing Agent

Reported Product Categories: Colognes and Toilet Waters; Hair Dyes and Colors (All Types Requiring Caution Statements and Patch Tests); Aftershave Lotions; Perfumes; Deodorants (Underarm); Fragrance Preparations, Misc.; Face and Neck Preparations (Excluding Shaving Preparations); Tonics, Dressings, and Other Hair Grooming Aids; Skin Care Preparations, Misc.; Body and Hand Preparations (Excluding Shaving Preparations); Cleansing Products (Cold Creams, Cleansing Lotions, Liquids and Pads); Hair Sprays (Aerosol Fixatives); Mascara; Hair Preparations (Non-coloring), Misc.; Skin Fresheners; Shaving Preparations, Misc.; Indoor Tanning Preparations; Paste Masks (Mud Packs); Bath Oils, Tablets, and Salts; Bath Preparations, Misc.; Bath Soaps and Detergents; Eye Makeup Preparations, Misc.; Foot Powders and Sprays; Foundations; Hair Conditioners; Makeup Preparations (Not eye), Misc.; Personal Cleanliness Products, Misc.; Shampoos (Non-coloring)

Technical/Other Names:
Alcohol Denat.
Denatured Alcohol, Not Designated by the Government

Trade Name Mixtures:
Almondermin LS (Laboratoires Sero-biologiques)
Arnica Distillate 2/378370 (Symrise)
Birch Distillate 2/384280 (Symrise)
Chamomille Distillate 2/380930 (Symrise)
Dragoplant Witch Hazel 2/034020 (Symrise)
Elespher Almondermin (Laboratoires Sero-biologiques)
Extrapone 2 Special J (2/032473) (Symrise)
Extrapone 2 Special 2/789480 (same as 2/032471) (Symrise)
Extrapone Witch Hazel 2/032893 (Symrise)
Extrapone Witch Hazel Dist. Colorless Special 2/032891 (Symrise)
Hydroviton 2/059353 (Symrise)
Linden Blossom Distillate 2/382920 (Symrise)
Myrrh Extract HA (Provital/Centerchem)
Pollen Extract HG (Provital/Centerchem)
Pronalen A/C HSC (Provital/Centerchem)
Pronalen Anti-Cellulite HSC (Provital/Centerchem)
Pronalen Anti-Fatigue HSC (Provital/Centerchem)
Pronalen Capsicum HSC (Provital/Centerchem)
Pronalen Fruit Acid AHA-5 (Provital/Centerchem)
Pronalen Fruit Acid AHA-20 (Provital/Centerchem)
Pronalen Fruit Acid AHA-50 (Provital/Centerchem)
Pronalen Origanum HSC (Provital/Centerchem)
Pronalen Ruscus HSC (Provital/Centerchem)
Pronalen Silymarin HSC (Provital/Centerchem)

SD ALCOHOL 39-D

Definition: SD Alcohol 39-D is ethyl alcohol denatured with bay oil (myrica oil) and either quinine sulfate, or sodium salicylate in accordance with 27CFR21. *See "Regulatory and Ingredient Use Information," regarding the labeling names for denatured alcohol in Volume 1, Introduction, Part A.*

Information Sources: 27CFR20.11, 27CFR21.73

Chemical Class: Alcohols

Functions: Cosmetic Astringent; Solvent; Viscosity Decreasing Agent

Technical/Other Name:
Alcohol Denat.

SD ALCOHOL 40

CAS No.: 61116-08-3

Definition: SD Alcohol 40 is ethyl alcohol denatured with t-butyl alcohol and any combination of one or more of the following: brucine (alkaloid), brucine sulfate, or quassin in accordance with 27CFR21. *See "Regulatory and Ingredient Use Information," regarding the labeling names for denatured alcohol in Volume 1, Introduction, Part A.*

Information Sources: 27CFR20.11, 27CFR21.74

Chemical Class: Alcohols

Functions: Cosmetic Astringent; Solvent; Viscosity Decreasing Agent

Reported Product Categories: Hair Dyes and Colors (All Types Requiring Caution Statements and Patch Tests); Hair Sprays (Aerosol Fixatives); Tonics, Dressings, and Other Hair Grooming Aids; Hair Preparations (Non-coloring), Misc.; Colognes and Toilet Waters; Aftershave Lotions; Cleansing Products (Cold Creams, Cleansing Lotions, Liquids and Pads); Skin Fresheners; Deodorants (Underarm); Hair Conditioners; Body and Hand Preparations (Excluding Shaving Preparations); Skin Care Preparations, Misc.; Bath Soaps and Detergents; Mascara; Personal Cleanliness Products, Misc.; Makeup Preparations (Not eye), Misc.; Hair Coloring Preparations, Misc.; Paste Masks (Mud Packs); Hair Bleaches; Perfumes; Suntan Gels, Creams, and Liquids; Bath Preparations, Misc.; Moisturizing Preparations; Nail Polish and Enamels; Permanent Waves; Shaving Preparations, Misc.; Nail Polish and Enamel Removers; Preshave Lotions (All types); Shampoos (Non-coloring); Baby Products, Misc.; Eye Shadows; Face and Neck Preparations (Excluding Shaving Preparations); Foundations; Hair Wave Sets; Leg and Body Paints; Makeup Bases; Bubble Baths; Douches; Foot Powders and Sprays; Fragrance Preparations, Misc.; Hair Straighteners; Manicuring Preparations, Misc.; Night Skin Care Preparations

Technical/Other Name:
Alcohol Denat.

Trade Name Mixtures:
Bentone Gel MIO-A40 (ELE)
Bentone Gel VS-5 (ELE)
Chinese Gardenia Extract (Cardre)
Chinese Licorice Extract (Cardre)
Jack-in-the-Pulpit Extract (Cardre)
Notoginseng Extract (Cardre)
Sickle Sienna Extract (Cardre)

SD ALCOHOL 40-A

Definition: SD Alcohol 40-A is ethyl alcohol denatured with t-butyl alcohol and sucrose octaacetate in accordance with 27CFR21. *See "Regulatory and Ingredient Use Information," regarding the labeling names for denatured alcohol in Volume 1, Introduction, Part A.*

Information Sources: 27CFR20.11, 27CFR21.75

Chemical Class: Alcohols

Functions: Cosmetic Astringent; Solvent; Viscosity Decreasing Agent

Reported Product Categories: Cleansing Products (Cold Creams, Cleansing Lotions, Liquids and Pads); Deodorants (Underarm); Fragrance Preparations, Misc.; Skin Care Preparations, Misc.; Aftershave Lotions; Colognes and Toilet Waters; Hair Sprays (Aerosol Fixatives); Moisturizing Preparations; Nail Polish and Enamels; Perfumes

Technical/Other Name:
Alcohol Denat.

Trade Name Mixtures:
Actiphyte Bergamot AL (Active Organics)
Actiphyte Camphor AL (Active Organics)
Actiphyte Grape Seed AL (Active Organics)
Actiphyte Horsetail AL (Active Organics)
Actiphyte Kava Kava AL (Active Organics)
Actiphyte Lemon Bioflavonoids AL (Active Organics)
Actiphyte Myrtle AL (Active Organics)
Actiphyte Orchid AL (Active Organics)
Actiphyte Passionflower AL (Active Organics)
Actiphyte Peach Kernel AL (Active Organics)
Actiphyte Pyrethrum AL (Active Organics)
Actiphyte Sea Buckthorn AL (Active Organics)

SD ALCOHOL 40-B

Definition: SD Alcohol 40-B is ethyl alcohol denatured with denatonium benzoate and t-butyl alcohol in accordance with 27CFR21. *See "Regulatory and Ingredient Use Information," regarding the labeling names for denatured alcohol in Volume 1, Introduction, Part A.*

Information Sources: 27CFR20.11, 27CFR21.76

Chemical Class: Alcohols

Functions: Cosmetic Astringent; Solvent; Viscosity Decreasing Agent

Reported Product Categories: Colognes and Toilet Waters; Hair Preparations (Non-coloring), Misc.; Hair Sprays (Aerosol Fixatives); Perfumes; Aftershave Lotions; Bath Preparations, Misc.; Fragrance Preparations, Misc.; Deodorants (Underarm); Moisturizing Preparations; Hair Conditioners; Paste Masks (Mud Packs); Skin Fresheners; Skin Care Preparations, Misc.; Bubble Baths; Tonics, Dressings, and Other Hair Grooming Aids; Bath Soaps and Detergents; Body and Hand Preparations (Excluding Shaving Preparations); Shampoos (Non-coloring); Bath Oils, Tablets, and Salts; Hair Wave Sets; Cleansing Products (Cold Creams, Cleansing Lotions, Liquids and Pads); Foot Powders and Sprays; Basecoats and Undercoats; Manicuring Preparations, Misc.; Personal Cleanliness Products, Misc.; Powders (Dusting and Talcum, Excluding Aftershave Talcs); Nail Polish and Enamels; Eye Lotions; Eye Makeup Preparations, Misc.; Eyebrow Pencils; Eyeliners; Face and Neck Preparations (Excluding Shaving Preparations); Foundations; Hair Coloring Preparations, Misc.; Hair Dyes and Colors (All Types Requiring Caution Statements and Patch Tests); Hair Rinses (Non-coloring); Indoor Tanning Preparations; Makeup Bases; Makeup Fixatives; Mascara; Nail Creams and Lotions; Night Skin Care Preparations; Preshave Lotions (All types); Shaving Cream (Aerosol, Brushless and Lather); Suntan Gels, Creams, and Liquids; Suntan Preparations, Misc.; Blushers (All types); Eye Makeup Removers; Hair Color Sprays (Aerosol); Hair Lighteners with Color; Makeup Preparations (Not eye), Misc.; Shaving Preparations, Misc.

Technical/Other Name:
Alcohol Denat.

SD ALCOHOL 40-C

Definition: SD Alcohol 40-C is ethyl alcohol denatured with t-butyl alcohol in accordance with 27CFR21. *See "Regulatory and Ingredient Use Information," regarding the labeling names for denatured alcohol in Volume 1, Introduction, Part A.*

Information Sources: 27CFR20.11, 27CFR21.77, JCIC, JCLS

Chemical Class: Alcohols

Functions: Cosmetic Astringent; Solvent; Viscosity Decreasing Agent

Technical/Other Names:
Alcohol Denat.
Denatured Alcohol, Not Designated by the Government

SD ALCOHOL 46

Definition: SD Alcohol 46 is ethyl alcohol denatured with phenol and methyl salicylate in accordance with 27CFR21. *See "Regulatory and Ingredient Use Information," regarding the labeling names for denatured alcohol in Volume 1, Introduction, Part A.*

Information Sources: 27CFR20.11, 27CFR21.81

Chemical Class: Alcohols

Functions: Cosmetic Astringent; Solvent; Viscosity Decreasing Agent

Technical/Other Name:
Alcohol Denat.

SEA CLAY EXTRACT

Definition: Sea Clay Extract is an extract of clay from the sea.

Chemical Class: Quaternary Ammonium Compounds

Functions: Hair Conditioning Agent; Skin-Conditioning Agent - Miscellaneous

Technical/Other Name:
Extract of Sea Clay

Trade Name:
Racine MCE (Racine Kagaku Co., Ltd)

SEA SALT

CAS No.	EINECS No.
7647-14-5	231-598-3

JPN Translation:
海塩

Definition: Sea Salt is a mixture of inorganic salts derived from sea water or from inland bodies of salt water. *See "Regulatory and Ingredient Use Information," regarding use of EU Trivial names in Volume 1, Introduction, Part A.*

Information Sources: JCIC, JCLS

Chemical Class: Inorganic Salts

Functions: Abrasive; Skin-Conditioning Agent - Humectant

Reported Product Categories: Shampoos (Non-coloring); Bath Soaps and Detergents; Bubble Baths; Bath Oils, Tablets, and Salts; Cleansing Products (Cold Creams, Cleansing Lotions, Liquids and Pads); Personal Cleanliness Products, Misc.; Hair Dyes and Colors (All Types Requiring Caution Statements and Patch Tests); Eye Makeup Removers; Hair Conditioners; Skin Care Preparations, Misc.; Body and Hand Preparations (Excluding Shaving Preparations); Moisturizing Preparations; Bath Preparations, Misc.; Hair Preparations (Non-coloring), Misc.; Baby Products, Misc.; Baby Shampoos; Bath Capsules; Foundations; Tonics, Dressings, and Other Hair Grooming Aids; Deodorants (Underarm); Face and Neck Preparations (Excluding Shaving Preparations); Skin Fresheners; Makeup Bases; Paste Masks (Mud Packs); Eye Lotions; Eyeliners; Fragrance Preparations, Misc.; Night Skin Care Preparations; Eyebrow Pencils; Hair Rinses (Non-coloring); Hair Shampoos (Coloring); Hair Sprays (Aerosol Fixatives); Manicuring Preparations, Misc.; Suntan Gels, Creams, and Liquids

Trade Names:
AEC Dead Sea Salt (A & E Connock)
Afrosalt (Vevy)
Atoligomer (Codif)
Atomized Sea Water (SECMA)
Dead Sea Bath Crystals (Dead Sea Labs)
Great Mineral (Green Heart)
Hydroton-Jordan-Biodefender-Dermosal (Termal)
Medizinisches Badesalz aus dem Toten Meer (Dagro)
Medizinisches Badesalz aus dem Toten Meer (Fette)
Miyabi Spa Extract 2 (Nonogawa)
Multimineral (mg + alpha) (Green Heart)

Trade Name Mixtures:
AEC Cosflor Blend 017 Moisture Factor WSS (A & E Connock)
AEC Moisture Factor HV (A & E Connock)
Marine Oligo Elements (Cosmetochem) (Cosmetochem International Ltd.)

SEA SILT

Definition: Sea Silt is sediment from the sea.

Chemical Class: Inorganics

Function: Skin-Conditioning Agent - Miscellaneous

Technical/Other Name:
Maris Limus

Trade Names:
Limon Marin (Setalg)
Natural Oceanic Clay (Ironwood Clay Company)

Trade Name Mixture:
Sediment Marin (Setalg)

SEA SILT EXTRACT

Definition: Sea Silt Extract is an extract of Sea Silt (q.v.). *See "Regulatory and Ingredient Use Information," regarding use of EU Trivial names in Volume 1, Introduction, Part A.*

Chemical Class: Inorganics

Function: Not Reported

Technical/Other Name:
Extract of Sea Silt

Trade Name Mixture:
Oligoceane (Sederma)

SEA URCHIN EXTRACT

Definition: Sea Urchin Extract is an extract of sea urchins.

Chemical Class: Biological Products

Function: Skin-Conditioning Agent - Miscellaneous

Technical/Other Name:
Extract of Sea Urchin

Trade Name Mixtures:
Glycolysat of Sea Urchins (CEP (Solabia))
Urchin HS (Alban Muller)

SEA WATER

Definition: Sea Water is water obtained from the sea or from inland bodies of salt water. *See "Regulatory and Ingredient Use Information," regarding use of EU Trivial names in Volume 1, Introduction, Part A.*

Chemical Class: Inorganics

Functions: Humectant; Skin-Conditioning Agent - Humectant; Solvent

Trade Names:
Dead Sea Mineral Water (Dead Sea Labs)
Deep Water (Maruzen Pharmaceuticals Co., Ltd.)
Earth Marine Water (Codif)
Electrodialyzed deep sea water (Nonogawa)
Inland Sea Water (Green Heart)
Miansoo (Dong Bang Beauty)
Milieu Marin (Eclosarium)
Muroto Deep Sea Water (Shu Uemura)
Oligomarine (Noevir)
PN Sea Water Base (Toyo Kasei)
Reduced-salt Sea Water (The Environmental Control)
Spring Sea Water (Somaig)

Trade Name Mixtures:
Complexe Algomarin (Codif)
Hydrocomplex 2 (Somaig)
Hydrocomplex 3 (Somaig)
Pheohydrane (Codif)

SEA WHIP EXTRACT

Definition: Sea Whip Extract is an extract of the marine invertebrate, *Pseudopterogorgia elisabethae.*

Chemical Class: Biological Products

Function: Not Reported

Technical/Other Name:
Extract of Sea Whip

Trade Name Mixtures:
Gorgonian Extract BG (Lipo)
Gorgonian Extract GC (Lipo)

SEBACIC ACID

CAS No.	EINECS No.
111-20-6	203-845-5

Empirical Formula:
$C_{10}H_{18}O_4$

Definition: Sebacic Acid is the organic dicarboxylic acid that conforms to the formula:

$$HOOC(CH_2)_8COOH$$

Information Sources: 21CFR175.105, MI-13(8490), TSCA

Chemical Class: Carboxylic Acids

Function: pH Adjuster

Technical/Other Names:
Decanedioic Acid
1,8-Octanedicarboxylic Acid

Trade Name Mixtures:
Acnacidol BG (Vincience)
Acnacidol PG (Vincience)

SECALE CEREALE (RYE) PHYTO PLACENTA CULTURE EXTRACT FILTRATE

Definition: Secale Cereale (Rye) Phyto Placenta Culture Extract Filtrate is a filtrate of an extract of a suspension prepared from the growth media of a rye, *Secale* cereale, placenta cell culture. *See "Regulatory and Ingredient Use Information," regarding the labeling names for botanical ingredients in Volume 1, Introduction, Part A.*

Chemical Class: Biological Products

Function: Not Reported

Trade Name:
Seryel (Bio Dell)

SECALE CEREALE (RYE) SEED EXTRACT

Definition: Secale Cereale (Rye) Seed Extract is an extract of the seeds of rye, *Secale cereale. See "Regulatory and Ingredient Use Information," regarding the labeling names for botanical ingredients in Volume 1, Introduction, Part A.*

Chemical Class: Biological Products

Function: Not Reported

Technical/Other Names:
Extract of Secale Cereale
Rye Extract
Rye Seed Extract

Trade Name Mixtures:
4-Cereals Milk (Cosmetochem)
(Cosmetochem International Ltd.)
Glycolysat of 7 Cereals (CEP (Solabia))
Herbasol Extract Rye (Cosmetochem)
(Cosmetochem International Ltd.)
HSP-Balance LS (Laboratoires Sero-biologiques)
Questex Malt Wine Concentrate (Quest International)
Rye-Extract HS 3461 G (Grau)

SECALE CEREALE (RYE) SEED FLOUR

JPN Translation:
ライムギ

Definition: Secale Cereale (Rye) Seed Flour is a powder prepared by grinding of rye seeds, *Secale cereale. See "Regulatory and Ingredient Use Information," regarding the labeling names for botanical ingredients in Volume 1, Introduction, Part A.*

Information Sources: JCIC, JCLS

Chemical Class: Biological Products

Functions: Abrasive; Bulking Agent

Technical/Other Names:
Flour, Rye
Rye Flour
Rye Seed Flour

SECHIUM EDULE FRUIT EXTRACT

Definition: Sechium Edule Fruit Extract is an extract of the fruit of *Sechium edule.*

See "Regulatory and Ingredient Use Information," regarding the labeling names for botanical ingredients in Volume 1, Introduction, Part A.

Chemical Class: Biological Products

Function: Skin-Conditioning Agent - Miscellaneous

Trade Name Mixtures:
Hydraxine (Biophysis)
Phytelene of Christophine BG 739 (Indena SA)
Phytelene of Christophine EG 739 (Indena SA)

SECHIUM EDULE FRUIT JUICE

Definition: Sechium Edule Fruit Juice is the juice expressed from the fruit of *Sechium edule. See "Regulatory and Ingredient Use Information," regarding the labeling names for botanical ingredients in Volume 1, Introduction, Part A.*

Chemical Class: Biological Products

Function: Skin-Conditioning Agent - Emollient

Technical/Other Name:
Juice, Sechium Edule Fruit

Trade Name Mixture:
Hydraxine (Biophysis)

SEDUM ACRE EXTRACT

CAS No.	EINECS No.
90106-69-7	290-293-3

Definition: Sedum Acre Extract is an extract of the stonecrop, *Sedum acre. See "Regulatory and Ingredient Use Information," regarding the labeling names for botanical ingredients in Volume 1, Introduction, Part A.*

Chemical Class: Biological Products

Function: Not Reported

Technical/Other Name:
Extract of Sedum Acre

Trade Name Mixture:
Sedum Extract (Weleda)

SEDUM PURPUREUM EXTRACT

Definition: Sedum Purpureum Extract is an extract of the plant, *Sedum purpureum. See "Regulatory and Ingredient Use Information," regarding the labeling names for*

botanical ingredients in Volume 1, Introduction, Part A.

Chemical Class: Biological Products

Function: Skin-Conditioning Agent - Miscellaneous

Technical/Other Name:
Extract of Sedum Purpureum

Trade Name Mixture:
Sedum Purpureum Extract (Weleda)

SEDUM ROSEA ROOT EXTRACT

Definition: Sedum Rosea Root Extract is an extract of the roots of *Sedum rosea. See "Regulatory and Ingredient Use Information," regarding the labeling names for botanical ingredients in Volume 1, Introduction, Part A.*

Chemical Class: Biological Products

Functions: Antioxidant; Cosmetic Astringent; Skin-Conditioning Agent - Miscellaneous

Technical/Other Name:
Extract of Sedum Rosea Root

Trade Name Mixture:
Draco Rhodiola Rosea Full Spectrum Standardized Extract (Draco)

SEKKEN-K

JPN Translation:
カリ石ケン

Definition: *See "Regulatory and Ingredient Use Information," regarding use of Japan Trivial names in Volume 1, Introduction, Part A.*

Information Source: JCLS

Function: Not Reported

SEKKEN-NA/K

JPN Translation:
ラウロイルメチルアラニン

Definition: *See "Regulatory and Ingredient Use Information," regarding use of Japan Trivial names in Volume 1, Introduction, Part A.*

Information Source: JCLS

Function: Not Reported

SEKKEN SOJI

JPN Translation:
石ケン素地

Definition: *See "Regulatory and Ingredient Use Information," regarding use of Japan Trivial names in Volume 1, Introduction, Part A.*

Information Source: JCLS

Function: Not Reported

SEKKEN SOJI-K

JPN Translation:
カリ石ケン素地

Definition: *See "Regulatory and Ingredient Use Information," regarding use of Japan Trivial names in Volume 1, Introduction, Part A.*

Information Source: JCLS

Function: Not Reported

SELENIUM ASPARTATE

Empirical Formula:
$C_4H_7NO_4 \cdot xSe$

Definition: Selenium Aspartate is the selenium salt of Aspartic Acid (q.v.).

Information Sources: EEC(II-297), MHLW-331/1

Chemical Class: Amino Acids

Function: Skin-Conditioning Agent - Occlusive

Trade Name:
Oligoidyne Selenium (Vevy)

SELENIUM SULFIDE

CAS No. 7488-56-4
EINECS No. 231-303-8

Empirical Formula:
S_2Se

Definition: Selenium Sulfide is the inorganic salt that conforms to the formula:

$$SeS_2$$

In the United States, Selenium Sulfide may be used as an active ingredient in OTC drug products. When used as an active drug ingredient, the established name is *Selenium Sulfide. See "Regulatory and Ingredient Use Information," regarding the labeling names for U.S. OTC Drug Ingredients in Volume 1, Introduction, Part A.*

Information Sources: 21CFR358.710, EEC(III/1-49), MHLW-331/1, OTC-I-DP, TSCA, USAN, USP XXIV

Chemical Class: Inorganic Salts

Functions: Antidandruff Agent; Hair Conditioning Agent

Technical/Other Name:
Selenium Disulfide

SEMIAQVILEGIA ADOXOIDES ROOT EXTRACT

Definition: Semiaqvilegia Adoxoides Root Extract is an extract of the roots of *Semiaqvilegia adoxoides. See "Regulatory and Ingredient Use Information," regarding the labeling names for botanical ingredients in Volume 1, Introduction, Part A.*

Chemical Class: Biological Products

Function: Not Reported

Technical/Other Name:
Extract of Semiaqvilegia Adoxoides

Trade Name Mixture:
Sinopurete (I.D. bio)

SEMPERVIVUM TECTORUM EXTRACT

CAS No. 85117-14-2
EINECS No. 285-565-3

Definition: Sempervivum Tectorum Extract is an extract of the entire plant of the houseleek, *Sempervivum tectorum. See "Regulatory and Ingredient Use Information," regarding the labeling names for botanical ingredients in Volume 1, Introduction, Part A.*

Chemical Class: Biological Products

Function: Skin-Conditioning Agent - Miscellaneous

Technical/Other Names:
Extract of Houseleek
Extract of Sempervivum Tectorum
Houseleek Extract
Houseleek (Sempervivum Tectorum) Extract

Trade Name Mixture:
House-Leek HS (Alban Muller)

SENECIO VULGARIS EXTRACT

CAS No. 84650-58-8
EINECS No. 283-517-6

Definition: Senecio Vulgaris Extract is an extract of the groundsel, *Senecio vulgaris. See "Regulatory and Ingredient Use Information," regarding the labeling names for*

botanical ingredients in Volume 1, Introduction, Part A.

Chemical Class: Biological Products

Function: Not Reported

Technical/Other Names:
Extract of Groundsel
Extract of Senecio Vulgaris
Groundsel Extract
Groundsel (Senecio Vulgaris) Extract

Trade Name Mixture:
Herbasol Extract Groundsel (Cosmetochem) (Cosmetochem International Ltd.)

SEQUOIADENDRON GIGANTEA BUD EXTRACT

Definition: Sequoiadendron Gigantea Bud Extract is an extract of the buds of *Sequoiadendron gigantea. See "Regulatory and Ingredient Use Information," regarding the labeling names for botanical ingredients in Volume 1, Introduction, Part A. See Reported Ingredient Functions-The Cosmetic Drug Distinction, in Regulatory and Ingredient Use Information, Volume I, Part A.*

Chemical Class: Biological Products

Function: Skin Protectant

Technical/Other Name:
Extract of Sequoiadendron Gigantea Bud

Trade Name Mixture:
Cryo-Bourgeon De Sequoia (Greentech)

SEQUOIADENDRON GIGANTEA STEM EXTRACT

Definition: Sequoiadendron Gigantea Stem Extract is an extract of the stems of *Sequoiadendron gigantea. See "Regulatory and Ingredient Use Information," regarding the labeling names for botanical ingredients in Volume 1, Introduction, Part A.*

Chemical Class: Biological Products

Function: Skin-Conditioning Agent - Miscellaneous

Technical/Other Name:
Extract of Sequoiadendron Gigantea Stem

Trade Name Mixtures:
Sequoia Extract (Greentech)
Sequoia Extract CLA (C3D)

SEQUOIA SEMPERVIRENS STEM EXTRACT

Definition: Sequoia Sempervirens Stem Extract is an extract of the stems of

Sequoia sempervirens. See "Regulatory and Ingredient Use Information," regarding the labeling names for botanical ingredients in Volume 1, Introduction, Part A.

Chemical Class: Biological Products

Function: Skin-Conditioning Agent - Miscellaneous

Technical/Other Name:
Extract of Sequoia Sempervirens Stem

Trade Name Mixtures:
Sequoia Extract (Greentech)
Sequoia Extract (I.N.E.A)

SERENOA SERRULATA FRUIT EXTRACT

CAS No.	EINECS No.
84604-15-9	283-292-4

Definition: Serenoa Serrulata Fruit Extract is an extract of the fruit of the saw palmetto, *Serenoa serrulata. See "Regulatory and Ingredient Use Information," regarding the labeling names for botanical ingredients in Volume 1, Introduction, Part A.*

Chemical Class: Biological Products

Functions: Lytic Agent; Skin-Conditioning Agent - Miscellaneous

Technical/Other Names:
Extract of Saw Palmetto
Extract of Serenoa Serrulata
Sabal Serulata Extract
Saw Palmetto Extract
Saw Palmetto (Serenoa Serrulata) Extract

Trade Name:
Saw Palmetto Lipidic Sterolic Extract (Euromed)

Trade Name Mixtures:
AC Saw Palmetto Liposome (Active Concepts)
Actiphyte Saw Palmetto (Active Organics)
Actiphyte Saw Palmetto AQ (Active Organics)
Actiphyte Saw Palmetto BG50P (Active Organics)
Actiphyte Saw Palmetto Lipo S (Active Organics)
Activated Botanicals Estroherb Complex AB 106 (Norjin)
ARP 100 (Greentech)
ARP 101 (Greentech)
Herbasol Extract Sabal (Saw Palmetto) (Cosmetochem) (Cosmetochem International Ltd.)
Phytosterol Complex Concentrate SI 1006 (Norjin)
REGU-SEB (Pentapharm/Centerchem)
Saw Palmetto Berries Extract HS 2792 G (Grau)
Saw Palmetto HPG Titrated (Alban Muller)

Saw Palmetto HS (Alban Muller)
Saw Palmetto LS (Alban Muller)
Saw Palmetto Tincture (Rahn)
Vegeles SR (Laboratoires Serobiologiques)
VT-287 Extract of Saw Palmentto (Vege-Tech)

SERICIN

CAS Nos.: 60650-88-6; 60650-89-7

Definition: Sericin is a protein isolated from the silk produced by the silk worm, *Bombyx mori.*

Chemical Class: Protein Derivatives

Functions: Hair Conditioning Agent; Skin-Conditioning Agent - Miscellaneous

Trade Names:
AEC Sericin (A & E Connock)
Pharconix Sericin P (Ichimaru Pharcos)
Sericina (Sinerga)
Sericin Pentapharm (Pentapharm/Centerchem)

Trade Name Mixtures:
Pharconix Sericin B (Ichimaru Pharcos)
Pharconix Sericin E (Ichimaru Pharcos)
Unicerin C-30 (Induchem)

SERICOSIDE

CAS No.: 55306-04-2

Definition: Sericoside is the organic compound that conforms to the formula:

where R represents Glucose (q.v.).

Chemical Classes: Esters; Sterols

Function: Skin-Conditioning Agent - Miscellaneous

Technical/Other Name:
Olean-12-en-28-oic Acid, 2,3,19,23-Tetrahydroxy, Glucopyranosyl Ester

Trade Name:
Sericoside Vinyals (Vinyals)

SERINE

CAS No.	EINECS No.
56-45-1 (L-Form)	200-274-3

JPN Translation:
セリン

Empirical Formula:
$C_3H_7NO_3$

Definition: Serine is the amino acid that conforms to the formula:

$$HOCH_2\underset{\underset{NH_2}{|}}{C}HCOOH$$

Information Sources: 21CFR172.320, 21CFR582.5701, CTFA D, FCC, INN, JCLS, JSCI, MI-13(8534), RIFM, TSCA, USAN, USP XXIV

Chemical Class: Amino Acids

Functions: Fragrance Ingredient; Hair Conditioning Agent; Skin-Conditioning Agent - Miscellaneous

Reported Product Categories: Moisturizing Preparations; Hair Conditioners; Skin Care Preparations, Misc.; Bath Capsules; Bath Oils, Tablets, and Salts; Cleansing Products (Cold Creams, Cleansing Lotions, Liquids and Pads); Face and Neck Preparations (Excluding Shaving Preparations); Hair Wave Sets; Permanent Waves; Bath Preparations, Misc.; Body and Hand Preparations (Excluding Shaving Preparations)

Technical/Other Names:
Alanine, 3-Hydroxy
2-Amino-3-Hydroxypropionic Acid
Serine (RIFM)
DL-Serine
L-Serine

Trade Name:
AEC Serine (A & E Connock)

Trade Name Mixtures:
B.H.A. Extract (Ennagram)
B.H.A Vegetable Extract (Phytochim)
Crodarom Hygroderm (Croda, Inc.)
Essential Vital Elements (Dipta)
Essential Vital Elements - S (Dipta)
Facteur Hydratant PH (Prod'Hyg)
Fluxhydran (Laboratoires Serobiologiques)
Hydeoviton 5,5 N 2/059359 (Symrise)
Hydro-Diffuser Microreservoir (Sederma)
Hydroviton 2/059353 (Symrise)
Hydroviton 24 (2/059351) (Symrise)
Beta-Hydroxyde AMSE (Coletica SA)
Moisturising Factor Hydrogerm (Crodarom)
Moisturizing Factor Hygro-Complex ARO 5272 (Crodarom)
Moisturizing Phytoamine Biocomplex (Alban Muller)
Osmhydran (Laboratoires Serobiologiques)
Osmhydran LS 8453 (Laboratoires Serobiologiques)
Prodew 400 (Ajinomoto)
Sel-Silk (Seltzer)
Sel-Smooth (Seltzer)
Slimming Phytoamine Biocomplex (Alban Muller)

Slimming Phytoamine Biocomplex (Alban
Muller)
TINODERM NMF (Ciba Specialty
Chemicals)

SERUM ALBUMIN

CAS No.	EINECS No.
9048-46-8	232-936-2

JPN Translation:
ウシ血清アルブミン

Definition: Serum Albumin is the major
protein component of blood plasma.

Information Sources: JCIC, JCLS, MI-13
(8542), TSCA, USP XXIV

Chemical Class: Proteins

Functions: Film Former; Hair Conditioning
Agent; Skin-Conditioning Agent - Miscellaneous

Reported Product Categories: Skin Care
Preparations, Misc.; Bath Capsules; Face
and Neck Preparations (Excluding Shaving
Preparations)

Technical/Other Names:
Bovine Serum Albumin Powder
Bovine Serum Albumin Solution
Serum Proteins

Trade Names:
Bovine Serum Albumin Solution (Intergen)
Serum Albumine Bovine (I.D. bio)

Trade Name Mixture:
BioCare SA (Amerchol)

SERUM PROTEIN

Definition: Serum Protein is the protein or
protein fraction obtained from blood plasma.

Chemical Class: Proteins

Functions: Hair Conditioning Agent; Skin-
Conditioning Agent - Miscellaneous

Reported Product Categories: Skin Care
Preparations, Misc.; Moisturizing Prepara-
tions; Bath Capsules; Body and Hand Prepa-
rations (Excluding Shaving Preparations);
Face and Neck Preparations (Excluding
Shaving Preparations)

Trade Names:
Eleseryl SH (Laboratoires Serobiologiques)
SR-71 (Bottger)

Trade Name Mixture:
PRE Complex (Atrium)

SESAME AMINO ACIDS

Definition: Sesame Amino Acids are the
amino acids obtained from the complete
hydrolysis of sesame flour.

Chemical Class: Amino Acids

Function: Skin-Conditioning Agent -
Humectant

SESAMIDE DEA

CAS No.: 124046-35-1

Definition: Sesamide DEA is a mixture of
diethanolamides of the fatty acids derived
from Sesamum Indicum (Sesame) Oil (q.v.).
It conforms generally to the formula:

$$RC \overset{O}{\underset{\|}{}}—N(CH_2CH_2OH)_2$$

where RCO- represents the fatty acids
derived from sesame oil.

Information Source: EEC(III/1-60)

Chemical Class: Alkanolamides

Functions: Surfactant - Foam Booster;
Viscosity Increasing Agent - Aqueous

Technical/Other Names:
Amides, Sesame Oil, N,N-Bis(2-Hydroxy-
ethyl)-
N,N-Bis(2-Hydroxyethyl)Sesame Oil Amide
Diethanolamine Sesame Oil Amides
Sesame Fatty Acid Diethanolamide

SESAMIDOPROPYLAMINE OXIDE

Definition: Sesamidopropylamine Oxide is
the tertiary amine oxide that conforms to the
formula:

$$RC \overset{O}{\underset{\|}{}}—NH(CH_2)_3—\overset{CH_3}{\underset{CH_3}{N}}\rightarrow O$$

where RCO- represents the fatty acids
derived from sesame oil.

Chemical Class: Amine Oxides

Functions: Hair Conditioning Agent;
Surfactant - Cleansing Agent; Surfactant -
Foam Booster; Surfactant - Hydrotrope

Technical/Other Names:
Amides, Sesame, N-[3-(Dimethylamino)-
Propyl], N-Oxide
N-[3-(Dimethylamino)Propyl]Sesame
Amides-N-Oxide
Sesame Amides, N-[3-(Dimethylamino)-
Propyl], N-Oxide

SESAMIDOPROPYL BETAINE

Definition: Sesamidopropyl Betaine is the
zwitterion (inner salt) that conforms to the
formula:

$$RC \overset{O}{\underset{\|}{}}—NH(CH_2)_3—\overset{CH_3}{\underset{CH_3}{\overset{+}{N}}}—CH_2COO^-$$

where RCO- represents the fatty acids
derived from sesame oil.

Chemical Class: Betaines

Functions: Antistatic Agent; Hair Conditioning Agent; Skin-Conditioning Agent - Miscellaneous; Surfactant - Cleansing Agent;
Surfactant - Foam Booster; Viscosity
Increasing Agent - Aqueous

Technical/Other Names:
N-(Carboxymethyl)-N,N-Dimethyl-3-[(1-
Oxosesame)Amino]-1-Propanaminium
Hydroxide, Inner Salt
1-Propanaminium, N-(Carboxymethyl)-N,N-
Dimethyl-3-[(1-Oxosesame)Amino]-,
Hydroxide, Inner Salt
Quaternary Ammonium Compounds,
(Carboxymethyl)(3-Sesameamidopropyl)
Dimethyl, Hydroxide, Inner Salt
Sesame Amide Propylbetaine
Sesamidopropyl Dimethyl Glycine

SESAMIDOPROPYL DIMETHYLAMINE

Definition: Sesamidopropyl Dimethylamine
is the amidoamine that conforms to the
formula:

$$RC \overset{O}{\underset{\|}{}}—NH(CH_2)_3N\overset{CH_3}{\underset{CH_3}{}}$$

where RCO- represents the fatty acids
derived from sesame oil.

Chemical Class: Amines

Function: Antistatic Agent

Technical/Other Names:
Amides, Sesame, N-[3-(Dimethylamino)-
Propyl]-
N-[3-(Dimethylamino)Propyl]Sesame
Amides

SESAMUM INDICUM (SESAME) OIL UNSAPONIFIABLES

Definition: Sesamum Indicum (Sesame)
Oil Unsaponifiables is the fraction of
sesame oil which is not saponified in the
refining recovery of sesame oil fatty acids.
*See "Regulatory and Ingredient Use Infor-
mation," regarding the labeling names for
botanical ingredients in Volume 1, Introduction, Part A.*

Chemical Class: Unsaponifiables

Functions: Hair Conditioning Agent; Skin-Conditioning Agent - Miscellaneous

Technical/Other Names:
Sesame Oil Unsaponifiables
Sesamum Indicum Unsaponifiables
Unsaponifiables, Sesame Oil
Unsaponifiables, Sesamum Indicum

Trade Names:
Sesaline (Expanscience)
Unsaponifiable of Sesame Oil (Expanscience)

Trade Name Mixtures:
Hierogaline (Expanscience)
Hierogaline Gel (Expanscience)
Sesaline Gel II (Expanscience)

SESAMUM INDICUM (SESAME) SEED

Definition: Sesamum Indicum (Sesame) Seed is the dried seeds of *Sesamum indicum. See "Regulatory and Ingredient Use Information," regarding the labeling names for botanical ingredients in Volume 1, Introduction, Part A.*

Chemical Class: Biological Products

Function: Not Reported

Technical/Other Names:
Seeds, Sesame
Seeds, Sesamum Indicum
Sesame Seed
Sesamum Indicum Seeds

SESAMUM INDICUM (SESAME) SEED EXTRACT

Definition: Sesamum Indicum (Sesame) Seed Extract is an extract of the seeds of the sesame, *Sesamum indicum. See "Regulatory and Ingredient Use Information," regarding the labeling names for botanical ingredients in Volume 1, Introduction, Part A.*

Chemical Class: Biological Products

Function: Not Reported

Technical/Other Names:
Extract of Sesame
Extract of Sesamum Indicum
Sesame Extract
Sesame Seed Extract
Sesamum Indicum Extract

Trade Name Mixtures:
Actiphyte of Sesame BG50 (Active Organics)
Actiphyte of Sesame GL50 (Active Organics)
Actiphyte of Sesame Lipo S (Active Organics)
Actiphyte of Sesame PG50 (Active Organics)
Black Sesame Seed Extract (Carrubba)
Herbasol Extract Sesame Seed (Cosmetochem) (Cosmetochem International Ltd.)
Lait Sesame MEE 100 (Yves Rocher)
REGU-SEB (Pentapharm/Centerchem)

SESAMUM INDICUM (SESAME) SEED OIL

CAS No. 8008-74-0

EINECS No. 232-370-6

JPN Translation:
ゴマ油

Definition: Sesamum Indicum (Sesame) Seed Oil is the oil obtained from the seeds of *Sesamum indicum. See "Regulatory and Ingredient Use Information," regarding the labeling names for botanical ingredients in Volume 1, Introduction, Part A.*

Information Sources: ARG, AUS, BEL, BP 1971, BPC 1971, 21CFR175.105, 21CFR175.300, 21CFR176.200, 21CFR176.210, CIR: [S] JACT-12(3)1993, CTFA S, EGY, FI, IND, ITA, JAN, JCLS, JSCI, MAR, MEX, MI-13(8543), NF XIX, RIFM, USAN, USD, WHO

Chemical Class: Fats and Oils

Functions: Fragrance Ingredient; Skin-Conditioning Agent - Occlusive

Reported Product Categories: Lipsticks; Bath Preparations, Misc.; Body and Hand Preparations (Excluding Shaving Preparations); Moisturizing Preparations; Skin Care Preparations, Misc.; Foundations; Night Skin Care Preparations; Bath Oils, Tablets, and Salts; Cleansing Products (Cold Creams, Cleansing Lotions, Liquids and Pads); Eyebrow Pencils; Suntan Gels, Creams, and Liquids; Bath Capsules; Face and Neck Preparations (Excluding Shaving Preparations); Makeup Bases; Blushers (All types); Eye Makeup Preparations, Misc.; Tonics, Dressings, and Other Hair Grooming Aids; Fragrance Preparations, Misc.; Hair Preparations (Non-coloring), Misc.; Shampoos (Non-coloring); Bath Soaps and Detergents; Cuticle Softeners; Hair Conditioners; Manicuring Preparations, Misc.; Suntan Preparations, Misc.

Technical/Other Names:
Gingilli Oil
Oils, Sesame
Sesame Oil
Sesame Seed Oil
Sesame seed oil (RIFM)

Trade Names:
AEC Sesame Oil (A & E Connock)
Certified Organic Sesame Oil (Formula One Sciences)
Cropure Sesame (Croda Chemicals)
Cropure Sesame (Croda, Inc.)
EmCon SES (Fanning)
Huile de Sesame Vierge (Bertin)
Jeen Sesame Seed Oil USP (Jeen)
Lipovol SES (Lipo)
Phytol SES (sesame oil) (Custom Ingredients)
Sesame Oil (Dekker)
Sesame Oil (Expanscience)
Sesame Oil Extract (Expanscience)
Sesame seed oil (RIFM) (Desert Whale)
Sesame Seed Oil, Premier (Premier Specialties)
Sesame (Sesamum Indicum) Oil PC (Protameen)
Sesamum Indicum Seed Oil ies (IES LABO)
Tri-OL SES (Tri-K)
Uniderm SSME (Universal Preserv-A-Chem)

Trade Name Mixtures:
Algae Comp-C (Bio-Botanica)
Bee's Milk (Koster Keunen)
Crodarom Rhatania O (Croda, Inc.)
Extrapone Cereals GW 2/031300 (Symrise)
Extrapone Sesame GW 2/031304 (Symrise)
LNST 98 (Lanatech)
Oleo-Coll LP (Arch Personal Care Products)
Oleo-Coll LP/LF (Arch Personal Care Products)
Proto-Lan 8 (Maybrook)
Rhatany Root Extract, Oil Soluble (Crodarom)
SESALINE GEL (Expanscience)
Vedacalm (Libiol)

SESAMUM INDICUM (SESAME) SEED POWDER

Definition: Sesamum Indicum (Sesame) Seed Powder is the powder obtained from the ground seeds of *Sesamum indicum. See "Regulatory and Ingredient Use Information," regarding the labeling names for botanical ingredients in Volume 1, Introduction, Part A.*

Chemical Class: Biological Products

Function: Skin-Conditioning Agent - Miscellaneous

Trade Name:
Sesame (Sunka)

SESAMUM INDICUM (SESAME) SPROUT EXTRACT

Definition: Sesamum Indicum (Sesame) Sprout Extract is an extract of the sprouts of *Sesamum indicum*. See "Regulatory and Ingredient Use Information," regarding the labeling names for botanical ingredients in Volume 1, Introduction, Part A.

Chemical Class: Biological Products

Function: Skin-Conditioning Agent - Humectant

Technical/Other Names:
Extract of Sesamum Indicum Sprouts
Sesame Sprout Extract

Trade Name Mixture:
Sesame Sprouts Extract (Nisshin OilliO)

SHAKUYAKU

JPN Translation:
シャクヤク

Definition: Shakuyaku is the root of *Paeonia lactiflora* or other related species of the family *Paeoniaceae*. See "Regulatory and Ingredient Use Information," regarding use of Japan Trivial names in Volume 1, Introduction, Part A.

Chemical Class: Biological Products

Function: Skin-Conditioning Agent - Miscellaneous

Technical/Other Name:
Paeonia Lactiflora Root (U.S.)

SHAKUYAKU EKISU

JPN Translation:
シャクヤクエキス

Definition: Shakuyaku Ekisu is an extract of Shakuyaku (q.v.). See "Regulatory and Ingredient Use Information," regarding use of Japan Trivial names in Volume 1, Introduction, Part A.

Chemical Class: Biological Products

Function: Skin-Conditioning Agent - Miscellaneous

Technical/Other Name:
Paeonia Lactiflora Root Extract (U.S.)

SHALE EXTRACT

Definition: Shale Extract is an extract of shale.

Chemical Class: Inorganics

Function: Skin-Conditioning Agent - Emollient

Trade Name:
Shale Extract Solution (Wamiles)

SHARK LIVER OIL

CAS No.	EINECS No.
68990-63-6	273-616-2

JPN Translation:
サメ肝油

Definition: Shark Liver Oil is the oil expressed from the fresh livers of several species of shark, including *Galeorhinus zyopterus* and *Hypoprion brevirostris*. In the United States, Shark Liver Oil may be used as an active ingredient in OTC drug products. When used as an active drug ingredient, the established name is *Shark Liver Oil*. See "Regulatory and Ingredient Use Information," regarding use of EU Trivial names in Volume 1, Introduction, Part A.

Information Sources: 21CFR175.105, 21CFR176.210, 21CFR177.2800, 21CFR346.14, IND, JCIC, JCLS, MAR, MI-13(8552), TSCA

Chemical Class: Fats and Oils

Functions: Skin-Conditioning Agent - Occlusive; Solvent

Technical/Other Name:
Oils, Shark Liver

SHELLAC

CAS No.	EINECS No.
9000-59-3	232-549-9

JPN Translation:
セラック

Definition: Shellac is the resinous secretion of the insect *Laccifer (Tachardia) lacca*.

Information Sources: BPC 1963, 21CFR73.1, 21CFR175.105, 21CFR175.300, 21CFR175.380, 21CFR175.390, 21CFR182.99, 27CFR21.32, 27CFR21.47, 27CFR21.126, 27CFR21.141, 27CFR21.151, CIR: [SQ] JACT-5(5)1986, CTFA D, JCLS, JSCI, MAR, MI-13(8553), NF XIX, TSCA, USAN

Chemical Class: Biological Products

Functions: Binder; Film Former; Hair Fixative

Reported Product Categories: Mascara; Eyeliners

Technical/Other Name:
Shellac Orange S-40

Trade Names:
Certified Refined Bleached Shellac (Mantrose-Bradshaw-Zinnser)
MHP 101 SB (MHP Shellac)
Orange Flake Shellac (Mantrose-Bradshaw-Zinnser)

SHELLAC WAX

CAS No.	EINECS No.
97766-50-2	307-913-6

Definition: Shellac Wax is a waxy fraction of bleached shellac obtained by physical means. See "Regulatory and Ingredient Use Information," regarding use of EU Trivial names in Volume 1, Introduction, Part A.

Chemical Class: Waxes

Functions: Binder; Hair Conditioning Agent; Skin-Conditioning Agent - Occlusive

Technical/Other Name:
Waxes, Shellac

Trade Names:
Bleached Refined Shellac Wax (Ross)
Bleached Refined Shellac Wax Cosmetic Grade (Ross)
Hansonwax Chirashine JH-7302L (Hansotech)
MHP Shellac Wax IW (MHP Shellac)
Refined Shellac Wax (Ross)
Shellae Wax 7302 L (Kahl)

SHOREA ROBUSTA LEAF EXTRACT

Definition: Shorea Robusta Leaf Extract is an extract of the leaves of *Shorea robusta*. See "Regulatory and Ingredient Use Information," regarding the labeling names for botanical ingredients in Volume 1, Introduction, Part A.

Chemical Class: Biological Products

Function: Not Reported

Technical/Other Names:
Damar (Shorea Robusta) Extract
Extract of Damar (Shorea Robusta)
Extract of Shorea Robusta

Trade Name Mixture:
Campo Sala Siddha (Campo)

SHOREA ROBUSTA RESIN

CAS No.	EINECS No.
9000-16-2	232-528-4

Definition: Shorea Robusta Resin is a resinous exudate from *Shorea robusta*. See "Regulatory and Ingredient Use Information," regarding the labeling names for botanical ingredients in Volume 1, Introduction, Part A.

Information Sources: AUS, 21CFR73.1, 21CFR175.105, 21CFR175.300, 21CFR177.1200, 21CFR177.1400, CTFA D, MI-13(2834), TSCA

Chemical Class: Biological Products

Function: Not Reported

Technical/Other Names:
Damar
Damar Gum
Damar (Shorea Robusta)
Gum Damar
Gum, Shorea Robusta
Shorea Robusta Gum

Trade Name Mixture:
7304 Candelilla Substitute (Kahl)

SHOREA ROBUSTA SEED BUTTER

Definition: Shorea Robusta Seed Butter is the fat obtained from the seeds of *Shorea robusta*. See *"Regulatory and Ingredient Use Information," regarding the labeling names for botanical ingredients in Volume 1, Introduction, Part A.*

Chemical Class: Fats and Oils

Function: Skin-Conditioning Agent - Occlusive

Trade Names:
AEC Shorea Robusta Butter (A & E Connock)
Dub Shorea (Stearinerie Dubois Fils)
ESP Sal Butter (Earth Supplied Products)
Sal Butter (Rahn)
Sal Butter - Ultra Refined (Biochemicals Int'l)

SHOREA STENOPTERA

Definition: *See "Regulatory and Ingredient Use Information," regarding EU labeling names for botanical ingredients in Volume 1, Introduction, Part A.*

Chemical Class: Biological Products

Technical/Other Name:
Shorea Stenoptera Butter (U.S.)

SHOREA STENOPTERA BUTTER

Definition: Shorea Stenoptera Butter is a fat obtained from *Shorea stenoptera*. See *"Regulatory and Ingredient Use Information," regarding the labeling names for botanical ingredients in Volume 1, Introduction, Part A.*

Chemical Class: Biological Products

Function: Skin-Conditioning Agent - Occlusive

Technical/Other Names:
Butter, Shorea Stenoptera

Shorea Butter
Shorea Stenoptera (EU)

Trade Names:
AEC Illipe Butter (A & E Connock)
Cegesoft SH (Cognis Deutschland)
Jarcreme SH (Jarchem)
Lipex 106 (Karlshamns AB)
Shorea Butter (Greeff)

SHRIMP EXTRACT

CAS No. 97766-35-3   **EINECS No.** 307-897-0

Definition: Shrimp Extract is an extract of various species of shrimp.

Chemical Class: Biological Products

Function: Skin-Conditioning Agent - Miscellaneous

Technical/Other Name:
Extract of Shrimp

Trade Name Mixture:
Glycolysat of Shrimp (CEP (Solabia))

SIALYLLACTOSE

CAS No.: 35890-38-1

Definition: Sialyllactose is the product obtained by the reaction of sialic acid with Lactose (q.v.).

Chemical Classes: Carbohydrates; Esters

Function: Skin-Conditioning Agent - Miscellaneous

Technical/Other Names:
N-Acetylneuraminoyllactose
D-Glucose, O-(N-Acetyl-α-Neuraminosyl)-(2.fwdarw. 3)-O-β-D-Galactopyranosyl-(1.fwdarw.4)-Neuraminyllactose

Trade Name:
Cylac 23 (Actives International)

SIGESBECKIA ORIENTALIS EXTRACT

Definition: Sigesbeckia Orientalis Extract is an extract of *Sigesbeckia orientalis*. See *"Regulatory and Ingredient Use Information," regarding the labeling names for botanical ingredients in Volume 1, Introduction, Part A.*

Chemical Class: Biological Products

Function: Skin-Conditioning Agent - Miscellaneous

Reported Product Category: Hair Dyes and Colors (All Types Requiring Caution Statements and Patch Tests)

Technical/Other Name:
Extract of Sigesbeckia Orientalis

Trade Name Mixtures:
Siegesbeckia (Sederma)
Slimming Phytoamine Biocomplex (Alban Muller)

SILANEDIOL SALICYLATE

Definition: Silanediol Salicylate is the product obtained by the hydrolysis of dimethylsilylsalicylate.

Chemical Class: Siloxanes and Silanes

Function: Skin-Conditioning Agent - Miscellaneous

Trade Name:
D.S.B. C (Exsymol)

Trade Name Mixture:
Capillisil Haute Concentration (Exsymol)

SILANETRIOL

Empirical Formula:
CH_6O_3Si

Definition: Silanetriol is the siloxane compound that conforms to the formula:

$$CH_3Si(OH)_3$$

Chemical Class: Siloxanes and Silanes

Function: Not Reported

Technical/Other Names:
Trihydroxymethylsilane
Trihydroxysilane

Trade Name Mixtures:
Organo Silanetriol (Carilene)
Organo Silanetriol (Synthesa)

SILANETRIOL ARGINATE

Definition: Silanetriol Arginate is the product obtained by the reaction of Silanetriol (q.v.) and Arginine (q.v.).

Chemical Classes: Amino Acids; Siloxanes and Silanes

Function: Skin-Conditioning Agent - Miscellaneous

Trade Name:
Argisil C (Exsymol)

SILANETRIOL GLUTAMATE

Definition: Silanetriol Glutamate is the product oftained by the reaction of Silanetriol (q.v.) and Glutamic Acid (q.v.).

Chemical Classes: Amino Acids; Siloxanes and Silanes

Function: Skin-Conditioning Agent - Miscellaneous

Trade Name:
Glutasil C (Exsymol)

SILANETRIOL LYSINATE

Definition: Silanetriol Lysinate is the product obtained by the reaction of Silanetriol (q.v.) and Lysine (q.v.).

Chemical Classes: Amino Acids; Siloxanes and Silanes

Function: Skin-Conditioning Agent - Miscellaneous

Trade Name:
Silysin C (Exsymol)

SILANETRIOL TREHALOSE ETHER

Definition: Silanetriol Trehalose Ether is the product obtained by the reaction of Silanetriol (q.v.) and Trehalose (q.v.).

Chemical Classes: Carbohydrates; Ethers; Siloxanes and Silanes

Function: Skin-Conditioning Agent - Miscellaneous

Trade Name:
GPS (Exsymol)

SILICA

CAS Nos.
7631-86-9
60676-86-0
112945-52-5

EINECS No.
262-373-8

JPN Translation:
シリカ

Empirical Formula:
O_2Si

Definition: Silica is the inorganic oxide that conforms to the formula:

$$SiO_2$$

Information Sources: 21CFR73.1, 21CFR172.230, 21CFR172.480, 21CFR172.864, 21CFR173.165, 21CFR173.340, 21CFR175.105, 21CFR175.300, 21CFR175.350, 21CFR175.390, 21CFR176.170, 21CFR176.200, 21CFR176.210, 21CFR177.1200, 21CFR177.1460, 21CFR177.2420, 21CFR177.2600, 21CFR178.3297, 21CFR178.3620, 21CFR182.90, 21CFR182.1711, 21CFR331.11, 21CFR436.316, 21CFR582.1711, CTFA S, DDR, HUN, ITA, JCIC, JCLS, JSCI, MAR, MI-13(8567), NF XIX, TSCA, USAN

Chemical Classes: Inorganics; Siloxanes and Silanes

Functions: Abrasive; Absorbent; Anticaking Agent; Bulking Agent; Opacifying Agent; Suspending Agent - Nonsurfactant

Reported Product Categories: Eye Shadows; Lipsticks; Face Powders; Foundations; Blushers (All types); Powders (Dusting and Talcum, Excluding Aftershave Talcs); Makeup Preparations (Not eye), Misc.; Mascara; Hair Dyes and Colors (All Types Requiring Caution Statements and Patch Tests); Hair Bleaches; Personal Cleanliness Products, Misc.; Eye Makeup Preparations, Misc.; Eyebrow Pencils; Makeup Bases; Skin Care Preparations, Misc.; Nail Polish and Enamels; Bath Oils, Tablets, and Salts; Bath Preparations, Misc.; Body and Hand Preparations (Excluding Shaving Preparations); Moisturizing Preparations; Deodorants (Underarm); Eyeliners; Bath Capsules; Paste Masks (Mud Packs); Perfumes; Fragrance Preparations, Misc.; Hair Coloring Preparations, Misc.; Basecoats and Undercoats; Face and Neck Preparations (Excluding Shaving Preparations); Foot Powders and Sprays; Hair Preparations (Non-coloring), Misc.; Cleansing Products (Cold Creams, Cleansing Lotions, Liquids and Pads); Dentifrices (Aerosol, Liquid, Pastes and Powders); Hair Lighteners with Color; Manicuring Preparations, Misc.; Night Skin Care Preparations; Skin Fresheners; Tonics, Dressings, and Other Hair Grooming Aids; Bubble Baths; Eye Makeup Removers; Hair Conditioners; Hair Straighteners; Permanent Waves; Rouges; Suntan Gels, Creams, and Liquids

Technical/Other Names:
Amorphous Silica
Amorphous Silicon Oxide Hydrate
Silica, Amorphous
Silicic Anhydride
Silicon Dioxide
Silicon Dioxide, Fumed
Spheron P-1000
Spheron PL-700

Trade Names:
Aerosil 130 (Degussa AG)
Aerosil 200 (Degussa AG)
Aerosil 255 (Degussa AG)
Aerosil 300 (Degussa AG)
Aerosil 380 (Degussa AG)
Aerosil 380 S (Degussa AG)
B-6C (Suzuki Yushi)
CAB-O-SIL EH-5 (Cabot)
CAB-O-SIL Fumed Silica (Cabot)
CAB-O-SIL HS-5 (Cabot)
CAB-O-SIL LM-130 (Cabot)
CAB-O-SIL MS-55 (Cabot)
CAB-O-SIL M-5 (Cabot)
E-6C (Suzuki Yushi)
EP 10TP (Crosfield Co.)
ESP Cosmetic Grade Sand (Earth Supplied Products)
Fossil Flour MBK (MBK)
Greensil K (Greentech)
MSS-500 (Kobo)
Neosil CBT50 (Ineos Silicas)
Neosil CBT60 (Ineos Silicas)
Neosil CBT70 (Ineos Silicas)
Neosil CBT60S (Ineos Silicas)
Neosil CL2000 (Ineos Silicas)
Neosil CT11 (Crosfield Co.)
Neosil PC10 (Ineos Silicas)
Neosil PC50S (Crosfield Co.)
Ronasphere (Merck KGaA/EMD Chemicals Inc.)
SB-705 (US Cosmetics)
Silotrat-1 (Vevy)
Sipernat 22 (Degussa AG)
Sipernat 50 (Degussa AG)
Sipernat 22 LS (Degussa AG)
Sipernat 22 S (Degussa AG)
Sipernat 50 S (Degussa AG)
Sorbosil AC33 (Crosfield Co.)
Sorbosil AC 35 (Crosfield Co.)
Sorbosil AC 37 (Crosfield Co.)
Sorbosil AC 39 (Crosfield Co.)
Sorbosil AC77 (Crosfield Co.)
Sorbosil BFG10 (Ineos Silicas)
Sorbosil BFG50 (Crosfield Co.)
Sorbosil TC15 (Crosfield Co.)
Spherica (Ikeda)
Spheriglass (Potters-Ballotini)
Spheron L-1500 (Presperse)
Spheron N-2000 (Presperse)
Spheron P-1500 (Presperse)
Wacker HDK H 30 (Wacker-Chemie)
Wacker HDK N 20 (Wacker-Chemie)
Wacker HDK P 100 H (Wacker Silicones)
Wacker HDK N 20P (Wacker-Chemie)
Wacker HDK N 25P (Wacker-Chemie)
Wacker HDK S 13 (Wacker-Chemie)
Wacker HDK T 30 (Wacker-Chemie)
Wacker HDK V 15 (Wacker-Chemie)
Wacker HDK V 15 P (Wacker-Chemie)
Zelec Sil (DuPont de Nemours)

Trade Name Mixtures:
AC Nano Vector (Active Concepts)
Aerosil R 202 (Degussa AG)
Aerosil RY 200 (Degussa AG)
Aerosil RY 200 S (Degussa AG)
Aerosil US 202 (Degussa AG)
AF75 (GE Silicones)
AF9000 (GE Silicones)
Albacan (Bio-Botanica)
Amilon (Ikeda)
Apacider-AW (Sangi Co., Ltd.)
Biju BTF-XD (Engelhard Corp.)

Biju Ultra UTF-XD (Engelhard Corp.)
BK-50H (Suzuki Yushi)
Cashmir K-II (Presperse)
Ceriguard D-620 (Nippon Denko)
Ceriguard W-510 (Nippon Denko)
Corneosine (Solabia)
Cosmacol PLG (Sasol Italy)
COSMO S-40SB (Catalysts & Chemicals)
Dichrona Spendid BY (Merck KGaA/EMD Chemicals Inc.)
Dichrona Splendid BR (Merck KGaA/EMD Chemicals Inc.)
Dry Vitamin E 50% SD (Roche)
Elesponge AHA LS 8911 B (Laboratoires Serobiologiques)
Elesponge Melhydran (Laboratoires Sero-biologiques)
Elestab 305 (Laboratoires Serobiologiques)
Eusolex UV-Pearls OMC (Merck KGaA/ EMD Chemicals Inc.)
Facemat (LCW)
FC-45KY (Suzuki Yushi)
Fiberlon Y2 (LCW)
Fiberlon Y10 (LCW)
Hemp Gel 3500 (Natunola)
Hemp Gel 3600 (Natunola)
HP-01 (Suzuki Yushi)
Isocell Slim (Lucas Meyer)
Kalixide AS (Vevy)
Lipase MC (Suzuki Yushi)
Lipogelag (Vevy)
LTSG-flake (Nippon Sheet Glass)
Matsumoto Microsphere S-100 (Tomen America)
Matsumoto Microsphere S-101 (Tomen America)
Matsumoto Microsphere S-102 (Tomen America)
Micronasphere M (Merck KGaA/EMD Chemicals Inc.)
Mirasun TIW 60 (Rhodia)
Morphotex (Kishimoto)
Morphotex Blue (Kishimoto)
Morphotex Red (Kishimoto)
Morphotex Yellow (Kishimoto)
MT-OSX (Tayca)
MT-OSX2 (Tayca)
Nanospheres 100 Lipophilic (Exsymol)
Natunola Castor 1023 (Natunola)
Natunola CWAX 5611 (Natunola)
Natunola Macrice 1501 (Natunola)
Natunola Maple 2701 (Natunola)
Natunola Sunflower 1102 (Natunola)
Neosil CBT71 (Ineos Silicas)
Neosil CBT72 (Ineos Silicas)
Neosil PC20S (Ineos Silicas)
Neosil PC51S (Crosfield Co.)
Neosil PC52S (Crosfield Co.)
Neosil PC53S (Crosfield Co.)
Neosil PC54S (Ineos Silicas)
Oleavine LS (Laboratoires Serobiologiques)
OMC-BMDBM (Sol Gel)
PC-Ball R-50H (Suzuki Yushi)

PC Ball W-50H (Suzuki Yushi)
PCL-Siccum 2/066215 (Symrise)
Photogenica-1 (Ikeda)
Photogenica-2 (Ikeda)
Photogenica-3 (Ikeda)
Photogenica-4 (Ikeda)
Phototan (Laboratoires Serobiologiques)
Plastic Powder CS-400 (Toshiki Pigment Co., Ltd)
Plastic Powder D-400 (Toshiki Pigment Co., Ltd)
Plastic Powder D-800 (Toshiki Pigment Co., Ltd)
Propolis Extract Powder Pan 80 (Paninkret)
Propolis Extract Powder Pan 90 (Paninkret)
PTSG-flake (Nippon Sheet Glass)
Pulaea SQE-10C (Suzuki Yushi)
Pulzea SEQ-10Z (Suzuki Yushi)
Pulzea SQE-10T (Suzuki Yushi)
Ronasphere LDP (Merck KGaA/EMD Chemicals Inc.)
SAG-Antifoam 730 (Dow Chemical)
SA-SB-300 (7%) (US Cosmetics)
Sepipress M (SEPPIC)
Sicopearl Fantastico Gold (BASF)
Sicopearl Fantastico Green (BASF)
Sicopearl Fantastico Pink (BASF)
Sicopearl Fantastico Ruby (BASF)
SilPower (LG Cosmetic)
SilPower Plus (LG Cosmetic)
Sipernat D 11 (Degussa AG)
SM-1000 (Presperse)
SM-2000 (Presperse)
Softshade SZ-5 (Ajinomoto)
Sorbosil BFG51 (Crosfield Co.)
Sorbosil BFG52 (Crosfield Co.)
Sorbosil BFG53 (Crosfield Co.)
Sorbosil BFG54 (Ineos Silicas)
Spectraveil AQ (Uniqema, Belgium)
Spherica HA (Ikeda)
Sphingoceryl Powder VEG (Laboratoires Serobiologiques)
SP-29 UVS (Presperse)
ST-E01 (Suzuki Yushi)
Sun Hemp Gel 5401 (Natunola)
Sunveil (Ikeda)
Sunveil PW-1010 (Ikeda)
Sunveil PW-6010 (Ikeda)
Sunveil PW-6030 (Ikeda)
Timiron Arctic Silver (Merck KGaA/EMD Chemicals Inc.)
Timiron Splendid Blue (Merck KGaA/EMD Chemicals Inc.)
Timiron Splendid Copper (Merck KGaA/ EMD Chemicals Inc.)
Timiron Splendid Gold (Merck KGaA/EMD Chemicals Inc.)
Timiron Splendid Green (Merck KGaA/ EMD Chemicals Inc.)
Timiron Splendid Red (Merck KGaA/EMD Chemicals Inc.)
Timiron Splendid Violet (Merck KGaA/EMD Chemicals Inc.)
Tioveil AQ-G (Uniqema, Belgium)

Tioveil AQ-N (Uniqema, Belgium)
Tioveil AQ-P (Uniqema, Belgium)
Tioveil EUT (Uniqema, Belgium)
Tioveil FCM (Uniqema, Belgium)
Tioveil FIN (Uniqema, Belgium)
Tioveil GCM (Uniqema, Belgium)
Tioveil IPM (Uniqema, Belgium)
Tioveil MOTG (Uniqema, Belgium)
Tioveil OP (Uniqema, Belgium)
Tioveil TG (Uniqema, Belgium)
Tioveil TGOP (Uniqema, Belgium)
Ti-Sphere AB-15155A (Presperse)
Toshiki BINS-1 (Nikko)
Toshiki BINS-2 (Nikko)
Toshiki BINS-3 (Nikko)
Toshiki BINS-4 (Nikko)
TSG-flake (Nippon Sheet Glass)
TSK-5 (Ishihara Sangyo Kaisha)
TTO-51(A) (Ishihara Sangyo Kaisha)
TTO-51(C) (Ishihara Sangyo Kaisha)
Unipure Blue LC 686 (LCW)
UV Pearls OMC (Sol Gel)
Vegelatum C800 (Natunola)
Vegelatum Clear (Natunola)
Vegelatum Clear RM (Natunola)
Vegelatum Compact 9 (Natunola)
Vegelatum Equiline (Natunola)
Vegelatum Equiline EU103 (Natunola)
Vegelatum Equiline SF (Natunola)
Velvet Veil 310 (Presperse)
Velvetveil A (Ikeda)
Velvetveil X (Ikeda)
Velvetveil Y (Ikeda)
Velvetveil Z (Ikeda)
Visionaire Bright Cinnamon (Eckart) (Eckart America)
Visionaire Bright Honey (Eckart) (Eckart America)
Visionaire Bright Natural Gold (Eckart) (Eckart America)
Visionaire Bright Silver Sea (Eckart) (Eckart America)
Visionaire Bright Sunflower Gold (Eckart) (Eckart America)
Vulcanyl CG (ARCO)
White Charcoal MB-15153 (Ikeda)
Xirona Indian Summer (Merck KGaA/EMD Chemicals Inc.)
Xirona Magic Mauve (Merck KGaA/EMD Chemicals Inc.)

SILICA DIMETHICONE SILYLATE

Definition: Silica Dimethicone Silylate is a hydrophobic silica derivative in which the surface of the fumed silica has been modified by the addition of dimethicone.

Chemical Class: Siloxanes and Silanes

Functions: Absorbent; Anticaking Agent; Antifoaming Agent; Suspending Agent - Nonsurfactant; Viscosity Increasing Agent - Nonaqueous

Trade Name:
CAB-O-SIL TS-720 (Cabot)

SILICA DIMETHYL SILYLATE

Definition: Silica Dimethyl Silylate is a silica derivative in which the surface of the fumed silica has been modified by the addition of dimethyl silyl groups.

Information Sources: JCIC, JCLS

Chemical Class: Siloxanes and Silanes

Functions: Anticaking Agent; Bulking Agent; Slip Modifier; Suspending Agent - Nonsurfactant; Viscosity Increasing Agent - Nonaqueous

Reported Product Categories: Lipsticks; Makeup Preparations (Not eye), Misc.

Technical/Other Name:
Dimethylsilyl Silicic Anhydride

Trade Names:
Aerosil R 972 (Degussa AG)
Aerosil R 974 (Degussa AG)
Aerosil R 976 (Degussa AG)
Aerosil R 976 S (Degussa AG)
CAB-O-SIL TS-610 (Cabot)
Covasilic 15 (LCW)
Wacker HDK H15 (Wacker-Chemie)
Wacker HDK H18 (Wacker-Chemie)
Wacker HDK H20 (Wacker-Chemie)

Trade Name Mixtures:
Amino Acid Microspheres (Coletica SA)
Aquafix Marin (Coletica SA)
Aquafix Vegetal (Coletica SA)
Color Marine Filling Spheres (Coletica SA)
Color Marine Vitamine CPMG Spheres (Coletica SA)
Color Retinol Thalasphere (Coletica SA)
Color Vegetal Filling Spheres (Coletica SA)
Eusolex T-Olio F (Merck KGaA)
Flavenger (Coletica SA)
Phenolines (Coletica SA)
Phenolines The Vert (Coletica SA)
Phytokine Color (Coletica SA)

SILICA SILYLATE

JPN Translation:
シリル化シリカ

Definition: Silica Silylate is a hydrophobic silica derivative where some of the hydroxyl groups on the surface of the fumed silica have been replaced by trimethylsiloxyl groups.

Chemical Class: Siloxanes and Silanes

Functions: Antifoaming Agent; Bulking Agent; Skin-Conditioning Agent - Emollient; Suspending Agent - Nonsurfactant

Reported Product Categories: Eyeliners; Eye Shadows; Eyebrow Pencils; Lipsticks

Trade Names:
Aerosil R 812 (Degussa AG)
Aerosil RX 300 (Degussa AG)
CAB-O-SIL TS-530 (Cabot)
Sipernat D 17 (Degussa AG)
Wacker HDK H2000 (Wacker-Chemie)

Trade Name Mixture:
Lipmat (LCW)

SILICON CARBIDE

CAS No.   **EINECS No.**
409-21-2   206-991-8

JPN Translation:
炭化ケイ素

Definition: Silicon Carbide is processed by heating coke and sand and consists chiefly of:

$$SiC$$

Information Sources: JCLS, MI-13(8566)

Chemical Class: Siloxanes and Silanes

Function: Not Reported

SILICONE QUATERNIUM-1

Definition: Silicone Quaternium-1 is the siloxane-derived quaternary ammonium salt that conforms generally to the formula:

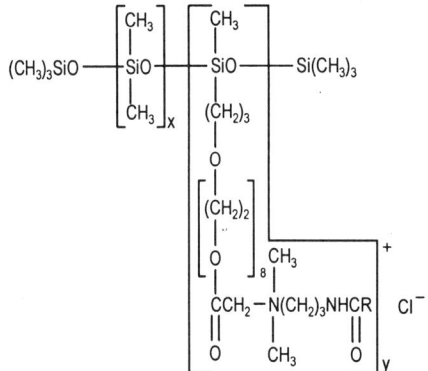

where RCO- represents the fatty acids derived from coconut oil.

Chemical Classes: Quaternary Ammonium Compounds; Siloxanes and Silanes

Function: Hair Conditioning Agent

SILICONE QUATERNIUM-2

Definition: Silicone Quaternium-2 is the siloxane-derived quaternary ammonium salt that conforms generally to the formula:

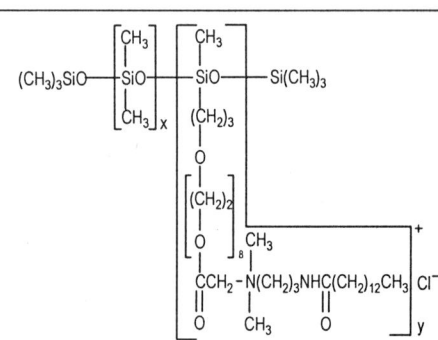

Chemical Classes: Quaternary Ammonium Compounds; Siloxanes and Silanes

Function: Hair Conditioning Agent

SILICONE QUATERNIUM-2 PANTHENOL SUCCINATE

Definition: Silicone Quaternium-2 Panthenol Succinate is the reaction product of Silicone Quaternium-2 (q.v.) and panthenyl succinate. It conforms to the formula:

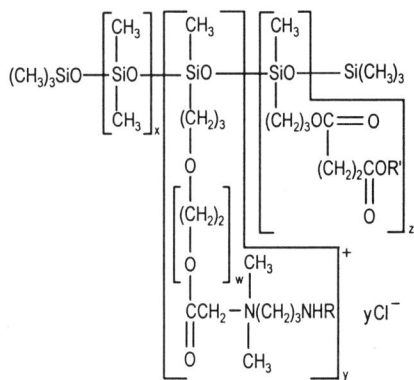

where R represents the panthenyl succinate moiety.

Chemical Classes: Quaternary Ammonium Compounds; Siloxanes and Silanes; Synthetic Polymers

Functions: Antistatic Agent; Hair Conditioning Agent

Trade Name:
Biosil Basics SPQ (Biosil Technologies, Inc.)

SILICONE QUATERNIUM-3

CAS No.: 137145-37-0

Definition: Silicone Quaternium-3 is the polymeric quaternary ammonium salt that conforms generally to the formula:

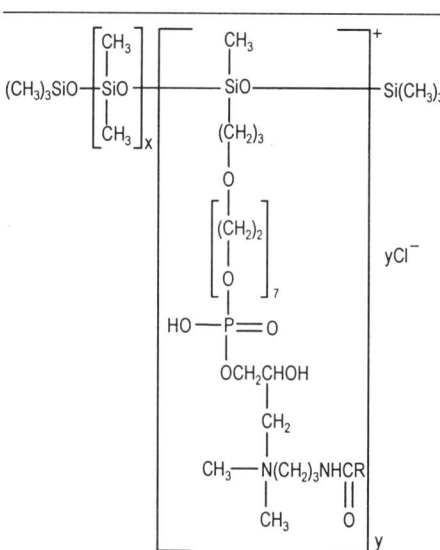

where RCO- represents the myristoyl moiety.

Chemical Classes: Quaternary Ammonium Compounds; Siloxanes and Silanes

Function: Hair Conditioning Agent

Technical/Other Name:
Siloxanes and Silicones, Dimethyl, 3-Hydroxypropylmethyl, Ethers with Poly-ethylene Glycol 3-[Dimethyl[3-[(Oxotetra-decyl)Amino]Propyl]Ammonio]-2-Hydroxypropyl Hydrogen Phosphate, Hydroxides, Inner Salts, Chloride

Trade Name:
Pecosil 14 PQ (Phoenix)

SILICONE QUATERNIUM-4

Definition: Silicone Quaternium-4 is the polymeric quaternary ammonium salt that conforms generally to the formula:

where RCO- represents the dilinoleoyl moiety.

Chemical Classes: Quaternary Ammonium Compounds; Siloxanes and Silanes

Function: Hair Conditioning Agent

Trade Name:
Pecosil 36 PQ (Phoenix)

SILICONE QUATERNIUM-5

Definition: Silicone Quaternium-5 is the polymeric quaternary ammonium salt that conforms generally to the formula:

where R represents the myristyl moiety.

Chemical Classes: Quaternary Ammonium Compounds; Siloxanes and Silanes

Function: Hair Conditioning Agent

Trade Name:
Pecosil SMQ-40 (Phoenix)

SILICONE QUATERNIUM-6

Definition: Silicone Quaternium-6 is the polymeric quaternary ammonium salt that conforms generally to the formula:

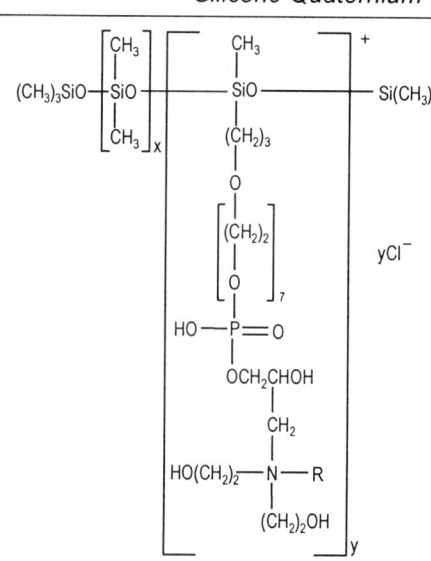

where R represents the coconut fatty moiety.

Chemical Classes: Quaternary Ammonium Compounds; Siloxanes and Silanes

Function: Hair Conditioning Agent

Trade Name:
Pecosil SPB-1240 (Phoenix)

SILICONE QUATERNIUM-7

Definition: Silicone Quaternium-7 is the polymeric quaternary ammonium salt that conforms generally to the formula:

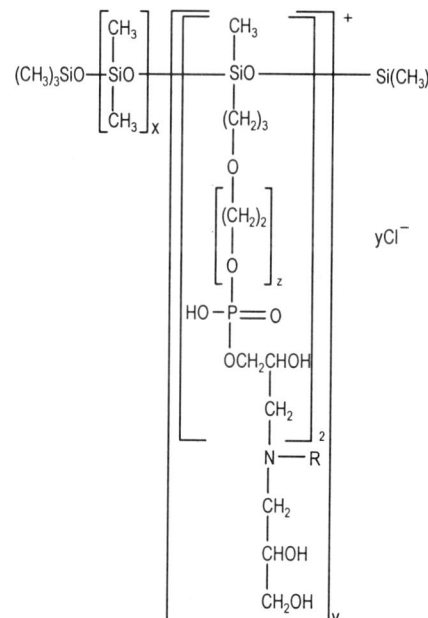

where R represents the hydrolyzed wheat protein moiety.

Chemical Classes: Quaternary Ammonium Compounds; Siloxanes and Silanes

Function: Hair Conditioning Agent

Trade Name:
Pecosil SWPQ-40 (Phoenix)

SILICONE QUATERNIUM-8

Definition: Silicone Quaternium-8 is the polymeric quaternary ammonium salt that conforms generally to the formula:

$$(CH_3)_3SiO - \begin{bmatrix} CH_3 \\ | \\ SiO \\ | \\ CH_3 \end{bmatrix}_x \begin{bmatrix} CH_3 \\ | \\ SiO \\ | \\ (CH_2)_3 \\ | \\ O \\ | \\ \begin{bmatrix} (CH_2)_2 \\ | \\ O \end{bmatrix}_z \\ | \quad CH_3 \quad + \\ CCH_2 - N(CH_2)_3NHCR \\ \| \quad | \quad \| \\ O \quad CH_3 \quad O \end{bmatrix}_y yCl^- \quad - Si(CH_3)_3$$

where RCO- represents the dilinoleoyl moiety.

Chemical Classes: Quaternary Ammonium Compounds; Siloxanes and Silanes

Function: Hair Conditioning Agent

Trade Names:
Silquat AD (Siltech)
Ultrasil Q-8 (Noveon)
Ultrasil Q-Plus Silicone (Noveon)

SILICONE QUATERNIUM-9

Definition: Silicone Quaternium-9 is the polymeric quaternary ammonium salt that conforms generally to the formula:

$$(CH_3)_3SiO - \begin{bmatrix} CH_3 \\ | \\ SiO \\ | \\ CH_3 \end{bmatrix}_x \begin{bmatrix} CH_3 \\ | \\ SiO \\ | \\ (CH_2)_3 \\ | \\ O \\ | \\ \begin{bmatrix} (CH_2)_2 \\ | \\ O \end{bmatrix}_7 \\ | \\ HO-P=O \\ | \\ OCH_2CHOH \\ | \\ CH_2 \\ | \\ CH_3-N(CH_2)_3NHCR \\ | \quad \| \\ CH_3 \quad O \end{bmatrix}_y^+ yCl^- - Si(CH_3)_3$$

where RCO- represents the cocoyl moiety.

Chemical Classes: Quaternary Ammonium Compounds; Siloxanes and Silanes

Function: Hair Conditioning Agent

Trade Name:
Pecosil CAP-1240 (Phoenix)

SILICONE QUATERNIUM-10

Definition: Silicone Quaternium-10 is the polymeric quaternary ammonium salt that conforms generally to the formula:

$$(CH_3)_3SiO - \begin{bmatrix} CH_3 \\ | \\ SiO \\ | \\ CH_3 \end{bmatrix}_x \begin{bmatrix} CH_3 \\ | \\ SiO \\ | \\ (CH_2)_3 \\ | \\ O \\ | \\ \begin{bmatrix} (CH_2)_2 \\ | \\ O \end{bmatrix}_z \\ | \quad CH_3 \quad + \\ CCH_2 - NR \\ \| \quad | \\ O \quad CH_3 \end{bmatrix}_y yCl^- - Si(CH_3)_3$$

where R represents the myristyl moiety.

Chemical Classes: Quaternary Ammonium Compounds; Siloxanes and Silanes

Function: Hair Conditioning Agent

Trade Name:
Pecosil SM-40 (Phoenix)

SILICONE QUATERNIUM-11

Definition: Silicone Quaternium-11 is the polymeric quaternary ammonium salt that conforms generally to the formula:

$$(CH_3)_3SiO - \begin{bmatrix} CH_3 \\ | \\ SiO \\ | \\ CH_3 \end{bmatrix}_x \begin{bmatrix} CH_3 \\ | \\ SiO \\ | \\ (CH_2)_3 \\ | \\ O \\ | \\ \begin{bmatrix} (CH_2)_2 \\ | \\ O \end{bmatrix}_z \\ | \quad CH_2CH_2OH \quad + \\ CCH_2 - NC_{12}H_{25} \\ \| \quad | \\ O \quad CH_2CH_2OH \end{bmatrix}_y yCl^- - Si(CH_3)_3$$

Chemical Classes: Quaternary Ammonium Compounds; Siloxanes and Silanes; Synthetic Polymers

Function: Hair Conditioning Agent

Trade Name:
Pecosil SB-1240 (Phoenix)

SILICONE QUATERNIUM-12

Definition: Silicone Quaternium-12 is the polymeric quaternary ammonium salt that conforms generally to the formula:

$$(CH_3)_3SiO - \begin{bmatrix} CH_3 \\ | \\ SiO \\ | \\ CH_3 \end{bmatrix}_x \begin{bmatrix} CH_3 \\ | \\ SiO \\ | \\ (CH_2)_3 \\ | \\ O \\ | \\ \begin{bmatrix} (CH_2)_2 \\ | \\ O \end{bmatrix}_z \\ | \quad CH_3 \quad + \\ CCH_2 - N(CH_2)_3NHCR \\ \| \quad | \quad \| \\ O \quad CH_3 \quad O \end{bmatrix}_y yCl^- - Si(CH_3)_3$$

where RCO represents the cocoyl moiety.

Chemical Classes: Quaternary Ammonium Compounds; Siloxanes and Silanes; Synthetic Polymers

Function: Hair Conditioning Agent

Trade Name:
Pecosil CA-1240 (Phoenix)

SILICONE QUATERNIUM-15

Definition: Silicone Quaternium-15 is the polymeric quaternary ammonium salt that conforms generally to the formula:

$$\begin{matrix} OH \\ | \\ CH_2CHO \end{matrix} - \begin{bmatrix} CH_3 \\ | \\ SiO \\ | \\ CH_3 \end{bmatrix}_x \begin{bmatrix} CH_3 \\ | \\ SiO \\ | \\ (CH_2)_3 \\ | \\ CH_3NCH_3 \\ | \\ R \end{bmatrix}_y^+ \begin{bmatrix} CH_3 \\ | \\ SiO \\ | \\ CH_3 \end{bmatrix}_z - OCHCH_2 \begin{matrix} OH \\ | \\ | \\ OH \end{matrix} \cdot CH_3COO^-$$

where R represents the coconut alkyl grouping.

Chemical Classes: Quaternary Ammonium Compounds; Siloxanes and Silanes; Synthetic Polymers

Functions: Antistatic Agent; Hair Conditioning Agent

Trade Name:
Hansa SQ 2010 VP (Hansa)

SILICONE QUATERNIUM-16

Definition: Silicone Quaternium-16 is the polymeric quaternary ammonium salt that conforms generally to the formula:

The inclusion of any compound in the *Dictionary and Handbook* does not indicate that use of that substance as a cosmetic ingredient complies with the laws and regulations governing such use in the United States or any other country.

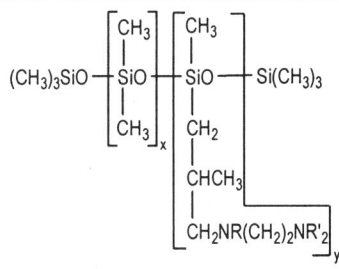

where R represents a 2-hydroxypropyl group, or H, and R' represents a C1-18 alkyl group.

Chemical Classes: Quaternary Ammonium Compounds; Siloxanes and Silanes; Synthetic Polymers

Functions: Hair Conditioning Agent; Skin-Conditioning Agent - Miscellaneous

SILK

Definition: Silk is the fibrous protein obtained from cocoons of the silk worm. *See "Regulatory and Ingredient Use Information," regarding use of EU Trivial names in Volume 1, Introduction, Part A.*

Chemical Class: Proteins

Function: Bulking Agent

Reported Product Categories: Face Powders; Powders (Dusting and Talcum, Excluding Aftershave Talcs); Blushers (All types); Lipsticks; Mascara; Foundations; Bath Preparations, Misc.; Bath Capsules; Eye Shadows; Hair Preparations (Non-coloring), Misc.; Makeup Preparations (Not eye), Misc.

Trade Names:
AC Silk Powder (Active Concepts)
Fibro-Silk Powder (Arch Personal Care Products)
Silkall 100 (Ikeda)
Silk Powder (U.S.) (Proteina)

Trade Name Mixtures:
Silkall CA (Ikeda)
Silkall TI (Ikeda)
Silkall TL (Ikeda)
Silkall ZN (Ikeda)

SILK AMINO ACIDS

Definition: Silk Amino Acids is the mixture of amino acids resulting from the complete hydrolysis of silk.

Chemical Class: Amino Acids

Functions: Hair Conditioning Agent; Skin-Conditioning Agent - Miscellaneous

Reported Product Categories: Shampoos (Non-coloring); Tonics, Dressings, and Other Hair Grooming Aids; Hair Conditioners; Hair Preparations (Non-coloring), Misc.; Moisturizing Preparations; Fragrance Preparations, Misc.; Hair Sprays (Aerosol Fixatives); Bath Soaps and Detergents; Hair Wave Sets

Trade Names:
AC Silk Amino Acids (Active Concepts)
AC Silk Amino Acids LS (Active Concepts)
Amino-Silk SF (Maybrook)
Crosilk Liquid (Croda Chemicals)
Crosilk Liquid (Croda, Inc.)
SILKA3 (Phytocos)
Silkgen G AA (Ichimaru Pharcos)
Solu-Silk 25 (Arch Personal Care Products)
Solu-Silk SF (Arch Personal Care Products)
Tri-Tein Silk AA (Tri-K)

Trade Name Mixtures:
Collamino Complex S (Arch Personal Care Products)
Silk Water (Bioland)

SILK EXTRACT

JPN Translation:
シルクエキス

Definition: Silk Extract is an extract of silk fiber.

Information Source: JCLS

Chemical Class: Biological Products

Function: Skin-Conditioning Agent - Miscellaneous

SILK POWDER

CAS No.: 9009-99-8

JPN Translation:
シルク

Definition: Silk Powder is finely pulverized silk. *See "Regulatory and Ingredient Use Information," regarding use of EU Trivial names in Volume 1, Introduction, Part A.*

Information Sources: JCIC, JCLS, JSQI

Chemical Class: Biological Products

Functions: Bulking Agent; Skin-Conditioning Agent - Miscellaneous; Slip Modifier

Reported Product Categories: Face Powders; Powders (Dusting and Talcum, Excluding Aftershave Talcs); Blushers (All types); Lipsticks; Mascara; Foundations; Shampoos (Non-coloring); Bath Preparations, Misc.; Body and Hand Preparations (Excluding Shaving Preparations); Hair Conditioners; Moisturizing Preparations; Bath Capsules; Eye Shadows; Face and Neck Preparations (Excluding Shaving Prep-

arations); Fragrance Preparations, Misc.; Makeup Preparations (Not eye), Misc.

Technical/Other Name:
Powdered Silk

Trade Names:
AEC Silk Powder (A & E Connock)
BHA Enna HW (Ennagram)
BHA Enna LW (Ennagram)
BHA Enna MW (Ennagram)
Crosilk Powder (Croda Chemicals)
Crosilk Powder (Croda, Inc.)
Silkgen G Powder (Ichimaru Pharcos)

SILKWORM CHRYSALIS EXTRACT

Definition: Silkworm Chrysalis Extract is an extract of the chrysalis obtained from the cocoon of the silkworm. *See Reported Ingredient Functions-The Cosmetic Drug Distinction, in Regulatory and Ingredient Use Information, Volume I, Part A.*

Chemical Class: Biological Products

Functions: Skin-Conditioning Agent - Emollient; Skin Protectant

Technical/Other Name:
Extract of Silkworm Chrysalis

Trade Name:
Chrysalis Oil (Sogo Printing Industrial Art)

SILKWORM COCOON EXTRACT

Definition: Silkworm Cocoon Extract is an extract of the cocoon of the silkworm.

Chemical Class: Biological Products

Function: Skin-Conditioning Agent - Humectant

Technical/Other Name:
Extract of Silkworm Cocoon

Trade Name:
Cocoon Extract (Sogo Printing Industrial Art)

SILK WORM EXTRACT

CAS No.	EINECS No.
91079-16-2	293-402-2

Definition: Silk Worm Extract is an extract obtained from crushed silk worms. *See "Regulatory and Ingredient Use Information," regarding use of EU Trivial names in Volume 1, Introduction, Part A.*

Information Source: JSQI

Chemical Class: Biological Products

Function: Skin-Conditioning Agent - Miscellaneous

Technical/Other Name:
Bombyx Mori Extract

Trade Name Mixtures:
Magilyne (Greentech)
Magnalys (Greentech)
Mimosoie (Alban Muller)
Protelin L (Greentech)
Sansha Extract (Maruzen Pharmaceuticals Co., Ltd.)
Sansha Extract B (Maruzen Pharmaceuticals Co., Ltd.)

SILK WORM LIPIDS

Definition: Silk Worm Lipids are the lipids obtained from crushed silk worms.

Chemical Class: Fats and Oils

Function: Skin-Conditioning Agent - Miscellaneous

Trade Names:
Ceralys (Greentech)
Silwome Pupa's Oil (Silk More)

SILOXANETRIOL ALGINATE

Definition: Siloxanetriol Alginate is the ester of siloxanetriol and Alginic Acid (q.v.).

Chemical Classes: Siloxanes and Silanes; Synthetic Polymers

Function: Skin-Conditioning Agent - Miscellaneous

Trade Name Mixture:
Cafeisilane C (Exsymol)

SILOXANETRIOL PHYTATE

Definition: Siloxanetriol Phytate is the ester of siloxanetriol and Phytic Acid (q.v.).

Chemical Classes: Esters; Phosphorus Compounds; Siloxanes and Silanes

Function: Skin-Conditioning Agent - Miscellaneous

Trade Name:
Inosil (Exsymol)

SILT

Definition: Silt is a sediment from inland bodies of water. *See Reported Ingredient Functions-The Cosmetic Drug Distinction, in Regulatory and Ingredient Use Information, Volume I, Part A.*

Chemical Class: Inorganics

Functions: Absorbent; Antiacne Agent; Cosmetic Astringent; Skin-Conditioning Agent - Humectant

Trade Names:
AEC Dead Sea Mud (A & E Connock)
Dead Sea Mineral Mud (Dead Sea Labs)
Heilmoor Reichenau (Thermal Mud) (Kelisema Italy)
Liposilt Black (Lipo)
Liposilt Green (Lipo)
Pelavie Black Silt (C.I.T.)
Pelavie Blue (C.I.T.)
Pelavie Green Silt (C.I.T.)

Trade Name Mixtures:
Pelabalm (C.I.T.)
Pelabath (C.I.T.)

SILT EXTRACT

Definition: Silt Extract is an extract of Silt (q.v.).

Chemical Class: Inorganics

Function: Not Reported

Technical/Other Name:
Extract of Silt

Trade Name:
Moor Extract (AH Ehrlich GmbH)

SILVER

CAS No. 7440-22-4   **EINECS No.** 231-131-3

Empirical Formula:
Ag

Definition: Silver is a metallic element. *See "Regulatory and Ingredient Use Information," for Colorants in Volume 1, Introduction, Part A. To identify the colorant meeting the requirements for labeling purposes in the US, the INCI Name Silver must be used. To identify the colorant allowed for use in the European Union (EU), the INCI Name CI 77820 must be used, except for hair dye products.*

Information Sources: AUS, 21CFR73.2500, 21CFR165.110, 21CFR175.300, 21CFR176.170, 21CFR176.300, 21CFR310.545, 21CFR440.19, 27CFR21.127, CI 77820, HP, HUN, M3, MAR, MI-13(8577), TSCA

Chemical Classes: Color Additives - Exempt from Batch Certification by the U.S. Food and Drug Administration; Elements; Inorganics

Function: Colorant

Reported Product Category: Nail Polish and Enamels

Technical/Other Name:
CI 77820

Trade Name Mixtures:
AC 305 (Active Concepts)
Apacider-AW (Sangi Co., Ltd.)
Apacider-C (Sangi Co., Ltd.)
Colloid MAG (Grant)
Cosmetallic Scintillating Silver (Engelhard Corp.)
Metashine GP (Nippon Sheet Glass)
Metashine PS (Nippon Sheet Glass)
SilPower (LG Cosmetic)
SilPower Plus (LG Cosmetic)

SILVER ACETYLMETHIONATE

CAS No.: 105883-46-3

Definition: Silver Acetylmethionate is the silver salt of N-acetylmethionine.

Chemical Classes: Amino Acids; Thio Compounds

Function: Not Reported

SILVER BOROSILICATE

Definition: Silver Borosilicate is a synthetic product formed by the fusion of boron oxide, Silica (q.v.), sodium oxide, and silver oxide.

Chemical Class: Inorganics

Functions: Preservative; Skin-Conditioning Agent - Miscellaneous

Trade Name:
Ionpure Type A (US Cosmetics)

SILVER CHLORIDE

CAS No. 7783-90-6   **EINECS No.** 232-033-3

Empirical Formula:
AgCl

Definition: Silver Chloride is the inorganic salt that conforms to the formula:

AgCl

Information Sources: MI-13(8582), TSCA

Chemical Class: Inorganic Salts

Function: Not Reported

Trade Name Mixtures:
AC 304 (Active Concepts)
JMAC (Microbial Systems)
JMAC TD (Microbial Systems)

JM ActiCare (Microbial Systems)
JM ActiCare BG (Microbial Systems)
JM ActiCare Plus (Microbial Systems)

SILVER COPPER ZEOLITE

Definition: Silver Copper Zeolite is the product obtained by the reaction of Zeolite (q.v.) with silver nitrate and cupric nitrate.

Chemical Class: Inorganics

Functions: Absorbent; Deodorant Agent

Trade Name:
Bactekiller - AC (Kanebo)

SILVER MAGNESIUM ALUMINUM PHOSPHATE

Definition: Silver Magnesium Aluminum Phosphate is a synthetic product formed by the fusion of phosphorous pentoxide, Magnesium Oxide (q.v.), Alumina (q.v.) and silver oxide.

Chemical Class: Inorganics

Function: Preservative

Trade Name:
Ionpure Type B (US Cosmetics)

SILVER NITRATE

CAS No.	EINECS No.
7761-88-8	231-853-9

Empirical Formula:
$Ag \cdot HNO_3$

Definition: Silver Nitrate is the inorganic salt that conforms to the formula:

$$AgNO_3$$

Information Sources: AUS, BP, BPC, BRA, 21CFR176.300, CZE, DA, DDR, EEC (III/1-48), EGY, EP, FIN, HP, HUN, IND, ITA, JAN, MAR, MEX, MI-13(8591), MI-13 (8592), PN, POR, ROM, TSCA, USAN, USD, USP XXIV, WHO, YUG

Chemical Class: Inorganic Salts

Function: Not Reported

Technical/Other Name:
Nitric Acid, Silver(1+) Salt

SILVER OXIDE

CAS No.	EINECS No.
20667-12-3	243-957-1

Definition: Silver Oxide is the inorganic oxide that conforms to the formula:

$$Ag_2O$$

Information Sources: MI-13(8595), MI-13 (8596)

Chemical Class: Inorganics

Function: Cosmetic Biocide

Trade Name Mixture:
Ionpure Type H (US Cosmetics)

SILVER SULFATE

CAS No.	EINECS No.
10294-26-5	233-653-7

Empirical Formula:
$Ag \cdot \frac{1}{2}H_2O_4S$

Definition: Silver Sulfate is the inorganic salt that conforms to the formula:

$$Ag_2SO_4$$

Information Sources: MI-13(8603), TSCA

Chemical Class: Inorganic Salts

Function: Not Reported

Technical/Other Name:
Sulfuric Acid, Disilver(1+) Salt

SILYBUM MARIANUM ETHYL ESTER

Definition: Silybum Marianum Ethyl Ester is the ethyl ester of the fatty acids derived from *Silybum marianum* oil.

Chemical Class: Esters

Function: Skin-Conditioning Agent - Miscellaneous

Trade Name:
Lipolami Milk Thistle (Alban Muller)

SILYBUM MARIANUM EXTRACT

CAS No.	EINECS No.
84604-20-6	283-298-7

Definition: Silybum Marianum Extract is an extract of the lady's thistle, *Silybum marianum*. See *"Regulatory and Ingredient Use Information,"* regarding the labeling names for botanical ingredients in Volume 1, Introduction, Part A.

Chemical Class: Biological Products

Function: Skin-Conditioning Agent - Miscellaneous

Technical/Other Names:
Extract of Lady's Thistle
Extract of Silybum Marianum
Lady's Thistle Extract
Lady's Thistle (Silybum Marianum) Extract

Trade Name:
Milk Thistle Purified Spray Dried Extract (Alban Muller)

Trade Name Mixtures:
Actiphyte Milk Thistle (Active Organics)
Actiphyte of Thistle BG50 (Active Organics)
Actiphyte of Thistle GL50 (Active Organics)
Actiphyte of Thistle Lipo S (Active Organics)
Actiphyte of Thistle PG50 (Active Organics)
Herbasec Milk Thistle (St Mary's Thistle) (Cosmetochem) (Cosmetochem International Ltd.)
Herbasol Extract Milk Thistle (St Mary's Thistle) (Cosmetochem) (Cosmetochem International Ltd.)
Milk Thistle HPG Titrated (Alban Muller)
Milk Thistle HS (Alban Muller)
Milk Thistle LS (Alban Muller)
Pronalen Refirming BG (Provital/ Centerchem)
Pronalen Refirming HSC (Provital/ Centerchem)
Silymarin Phytosome (Indena SpA)
St. Mary's Thistle Extract HS 2693 G (Grau)

SILYBUM MARIANUM FRUIT EXTRACT

JPN Translation:
マリアアザミエキス

Definition: Silybum Marianum Fruit Extract is an extract of the fruit of lady's thistle, *Silybum marianum*. See *"Regulatory and Ingredient Use Information,"* regarding the labeling names for botanical ingredients in Volume 1, Introduction, Part A.

Information Source: MI-13(8607)

Chemical Class: Biological Products

Function: Skin-Conditioning Agent - Miscellaneous

Technical/Other Names:
Extract of Lady's Thistle Fruit
Extract of Silybum Marianum Fruit
Lady's Thistle Fruit Extract
Lady's Thistle (Silybum Marianum) Fruit Extract

Trade Names:
Milk Thistle Dry Extract / Silymarin (Euromed)
Pronalen Silymarin SPE (Provital/ Centerchem)
Silymarin (Euromed)
Silymarin (MMP)

Trade Name Mixtures:
Pronalen EAP (Provital/Centerchem)
Pronalen Sensitive Skin (Provital/ Centerchem)
Pronalen Silymarin HSC (Provital/ Centerchem)

St. Mary's Thistle Extract HS 3687 G
(Grau)

SILYBUM MARIANUM SEED EXTRACT

Definition: Silybum Marianum Seed
Extract is an extract of the seeds of
*Silybum marianum. See "Regulatory and
Ingredient Use Information," regarding the
labeling names for botanical ingredients in
Volume 1, Introduction, Part A.*

Chemical Class: Biological Products

Function: Skin-Conditioning Agent - Miscellaneous

Trade Name Mixture:
Premier Milk Thistle Extract BG (Premier
Specialties)

SILYBUM MARIANUM SEED OIL

Definition: Silybum Marianum Seed Oil is
the fixed oil expressed from the seeds of
*Silybum marianum. See "Regulatory and
Ingredient Use Information," regarding the
labeling names for botanical ingredients in
Volume 1, Introduction, Part A.*

Chemical Class: Biological Products

Function: Skin-Conditioning Agent - Miscellaneous

Technical/Other Names:
Lady's Thistle Oil
Lady's Thistle (Silybum Marianum) Oil
Oils, Lady's Thistle (Silybum Marianum)

Trade Name:
Milk Thistle Oil (Alban Muller)

Trade Name Mixture:
Omegaceane (GELYMA)

SIMETHICONE

CAS No.: 8050-81-5

JPN Translation:
シメチコン

Definition: Simethicone is a mixture of
Dimethicone (q.v.) with an average chain
length of 200 to 350 dimethylsiloxane units
and hydrated silica.

Information Sources: 21CFR310.545,
21CFR332.10, JCLS, JSCI, MAR, MI-13
(3241), OTC-I-AA, USAN, USD, USP XXIV

Chemical Class: Siloxanes and Silanes

Function: Antifoaming Agent

Reported Product Categories: Mascara;
Lipsticks; Hair Dyes and Colors (All Types
Requiring Caution Statements and Patch
Tests); Hair Conditioners; Permanent Waves;
Hair Bleaches; Eye Makeup Preparations,
Misc.; Moisturizing Preparations; Hair
Coloring Preparations, Misc.; Eyeliners;
Makeup Bases; Tonics, Dressings, and Other
Hair Grooming Aids; Foundations; Makeup
Preparations (Not eye), Misc.; Night Skin
Care Preparations; Bath Capsules; Bath Oils,
Tablets, and Salts; Cleansing Products (Cold
Creams, Cleansing Lotions, Liquids and
Pads); Face and Neck Preparations
(Excluding Shaving Preparations); Bath
Preparations, Misc.; Body and Hand Prepa-
rations (Excluding Shaving Preparations);
Hair Preparations (Non-coloring), Misc.; Hair
Wave Sets; Personal Cleanliness Products,
Misc.; Skin Care Preparations, Misc.

Technical/Other Names:
Silicone Antifoam Emulsions SE 6 and SE
9
Silicone Resin

Trade Names:
AEC Simethicone (A & E Connock)
Antifoam A Compound (Dow Corning)
Antifoam AF Emulsion (Dow Corning)
Antifoam C Emulsion (Dow Corning)
Dow Corning Medical Antifoam A
Compound (Dow Corning)
Dow Corning Medical Antifoam AF
Emulsion (Dow Corning)
Dow Corning Medical Antifoam C Emulsion
(Dow Corning)
Dow Corning 1510-US Emulsion (Dow
Corning)
KS66 (Shin Etsu)
Mirasil SM (Rhodia)
SENTRY Simethicone USP (OSi Special-
ties)
Silicon-Antifoam Emulsion SE 57 (Wacker-
Chemie)
Silicone Antifoam Agent S 184 (Wacker-
Chemie)
Silicone Antifoam Emulsion SE 2 (Wacker-
Chemie)
Silicone Antifoam Emulsion SE 6 (Wacker-
Chemie)
Silicone Antifoam Emulsion SE 9 (Wacker-
Chemie)
Silicone Antifoam Emulsion SLE (Wacker-
Chemie)
Unisil DF-200S (Universal Preserv-A-
Chem)
Unisil DF-200SP (Universal Preserv-A-
Chem)
Unisil SF (Universal Preserv-A-Chem)

Trade Name Mixtures:
Bellsilk CA (Ichimaru Pharcos)
Bellsilk TI (Ichimaru Pharcos)
Bellsilk TI-UV (Ichimaru Pharcos)
Bellsilk TL (Ichimaru Pharcos)
Bellsilk TL-LT (Ichimaru Pharcos)
Bellsilk ZN (Ichimaru Pharcos)
Bellsilk ZN-CG (Ichimaru Pharcos)
Bellsilk ZN-UV (Ichimaru Pharcos)
Bellsilk ZN-VC (Ichimaru Pharcos)
Bellsilk ZN-VE (Ichimaru Pharcos)
Eusolex T-45D (Merck KGaA/EMD
Chemicals Inc.)
Eusolex T-ECO (Merck KGaA)
Eusolex T-ECO (Merck KGaA/EMD
Chemicals Inc.)
Eusolex T-Olio P (Merck KGaA)
Eusolex T (Merck KGaA)
Eusolex T-2000 (Merck KGaA/EMD
Chemicals Inc.)

SIMMONDSIA CHINENSIS (JOJOBA) BUTTER

Definition: Simmondsia Chinensis
(Jojoba) Butter is the material obtained by
the isomerization of Simmondsia Chinensis
(Jojoba) Oil (q.v.). *See "Regulatory and
Ingredient Use Information," regarding the
labeling names for botanical ingredients in
Volume 1, Introduction, Part A.*

Chemical Class: Waxes

Functions: Hair Conditioning Agent; Skin-
Conditioning Agent - Occlusive

Technical/Other Name:
Jojoba Butter

Trade Names:
ISO-Jojoba 35 (Desert Whale)
ISO-Jojoba 50 (Desert Whale)

SIMMONDSIA CHINENSIS (JOJOBA) LEAF EXTRACT

Definition: Simmondsia Chinensis
(Jojoba) Leaf Extract is an extract of the
leaves of *Simmondsia chinensis. See
"Regulatory and Ingredient Use
Information," regarding the labeling names
for botanical ingredients in Volume 1, Introduction, Part A.*

Chemical Class: Biological Products

Function: Skin-Conditioning Agent -
Humectant

Technical/Other Name:
Extract of Simmondsia Chinensis (Jojoba)
Leaf

Trade Name Mixtures:
Jojoba Leaf Extract BG-50 (C&F Koei
Phyto Corp.)
Jojoba Leaf Extract BG-50 (Koei
Perfumery)

SIMMONDSIA CHINENSIS (JOJOBA) SEED EXTRACT

CAS No.	EINECS No.
90045-98-0	289-964-3

Definition: Simmondsia Chinensis (Jojoba) Seed Extract is an extract of the nuts of the jojoba, *Simmondsia chinensis. See "Regulatory and Ingredient Use Information," regarding the labeling names for botanical ingredients in Volume 1, Introduction, Part A.*

Chemical Class: Biological Products

Function: Not Reported

Reported Product Categories: Shampoos (Non-coloring); Hair Conditioners

Technical/Other Names:
Buxus Chinensis Extract
Extract of Jojoba
Extract of Simmondsia Chinensis
Jojoba Extract
Jojoba Seed Extract

Trade Name:
Jojoba White - XT (Desert Whale)

Trade Name Mixtures:
Actiphyte Jojoba Meal (Active Organics)
Actiphyte Jojoba Meal AQ (Active Organics)
Actiphyte Jojoba Meal BG50P (Active Organics)
Actiphyte Jojoba Meal GL (Active Organics)
Actiphyte Jojoba Meal Lipo S (Active Organics)
Actiphyte Jojoba Meal Lipo SB (Active Organics)
Extract of Jojoba Comp. (Arda Natura)
Extrapone Jojoba Special (2/032990) (Symrise)
Herbasol Extract Jojoba (Cosmetochem) (Cosmetochem International Ltd.)
Jojoba Extract HS 3496 G (Grau)
Phytelene of Jojoba EG 541 Liquid (Indena SA)
Phytogreen 55 of Jojoba EXH Liquid (Phytochim)
Prodhy Extract Jojoba (Prod'Hyg)
VT-112 Extract of Jojoba (Vege-Tech)

SIMMONDSIA CHINENSIS (JOJOBA) SEED OIL

CAS No.: 61789-91-1

JPN Translation:
ホホバ油

Definition: Simmondsia Chinensis (Jojoba) Seed Oil is the fixed oil expressed or extracted from seeds of the desert shrub, Jojoba, *Simmondsia chinensis. See "Regulatory and Ingredient Use Information," regarding the labeling names for botanical ingredients in Volume 1, Introduction, Part A.*

Information Sources: CIR: [S] JACT-11(1)-1992, CTFA S, JCIC, JCLS, JSQI, MI-13 (5285)

Chemical Class: Esters

Functions: Hair Conditioning Agent; Skin-Conditioning Agent - Occlusive

Reported Product Categories: Hair Conditioners; Moisturizing Preparations; Foundations; Lipsticks; Bath Preparations, Misc.; Body and Hand Preparations (Excluding Shaving Preparations); Skin Care Preparations, Misc.; Hair Dyes and Colors (All Types Requiring Caution Statements and Patch Tests); Bath Capsules; Shampoos (Non-coloring); Tonics, Dressings, and Other Hair Grooming Aids; Face and Neck Preparations (Excluding Shaving Preparations); Makeup Preparations (Not eye), Misc.; Bath Oils, Tablets, and Salts; Night Skin Care Preparations; Cleansing Products (Cold Creams, Cleansing Lotions, Liquids and Pads); Eye Makeup Preparations, Misc.; Blushers (All types); Hair Preparations (Non-coloring), Misc.; Hair Sprays (Aerosol Fixatives); Bath Soaps and Detergents; Eyeliners; Paste Masks (Mud Packs); Eye Shadows; Eyebrow Pencils; Face Powders; Hair Rinses (Non-coloring); Makeup Bases; Manicuring Preparations, Misc.; Suntan Gels, Creams, and Liquids; Suntan Preparations, Misc.

Technical/Other Names:
Buxus Chinensis Oil
Jojoba Oil
Jojoba Seed Oil
Oils, Jojoba

Trade Names:
AEC Jojoba Oil Pure (A & E Connock)
AEC Jojoba Oil Refined (A & E Connock)
CoJoba (Costec)
Cojoba Clear (Costec)
Floraesters Jojoba Oil (Floratech)
Huile de Jojoba Vierge (Bertin)
Jeen Jojoba Oil (Jeen)
Jojoba Oil (Active Concepts)
Jojoba Oil (Dekker)
Jojoba Oil (Purcell Jojoba)
Jojoba Oil Colorless (Desert Whale)
Jojoba Oil Colorless Organic (Desert Whale)
Jojobaoil "COS" (Cosmetochem) (Cosmetochem International Ltd.)
Jojoba Oil Golden (Desert Whale)
Jojoba Oil Golden Organic (Desert Whale)
Jojoba Oil Sonoran (Desert Whale)
Lipovol J (Lipo)
Lipovol J Clear (Lipo)
Nikkol Jojoba Oil E (Nikko)
Nikkol Jojoba Oil S (Nikko)
Phytol JOJO (jojoba oil) (Custom Ingredients)
PNJ Deodorized (Purcell Jojoba)
PNJ Golden (Purcell Jojoba)
PNJ Organic Certified (Purcell Jojoba)
Refined Jojoba Oil (Ross)
Simmondsia Chinensis Seed Oil ies (IES LABO)

Trade Name Mixtures:
Actiphyte Evening Primrose Lipo J (Active Organics)
AEC Jojoba Oil Microencapsulated (A & E Connock)
Biosil Basics HMC - Hair Moisture Complex (Biosil Technologies, Inc.)
Biosil Basics HMC-1 Hair Moisture Complex (Biosil Technologies, Inc.)
Biosil Basics HMV- Hair Moisture Complex (Biosil Technologies, Inc.)
Biosil Basics HMW - Hair Moisture Complex (Biosil Technologies, Inc.)
Biosil Basics SMC (Biosil Technologies, Inc.)
Dragobotania 2/H00005 (Symrise)
Extrapone Jojoba/Rose Blend GW 2/031326 (Symrise)
Gelhyperm (Jojoba Oil) (Novoselect)
Gilugel JOB (Giulini/Giulini Chemie)
Hot Dots - Blue (Charm Girl)
Hot Dots - Green (Charm Girl)
Hot Dots - Red (Charm Girl)
Jojoba Glaze (Desert Whale)
Jojoba Oil (Ross)
Jojobaoil "W" Watersoluble (Cosmetochem) (Cosmetochem International Ltd.)
Jojobasomes (Desert Whale)
Jonat AS (Dr. Gerhard Steidl)
Mulberry Extract-JO (Maruzen Pharmaceuticals Co., Ltd.)
Tixogel JOG 1583 (Sud-Chemie, United Catalysts)
Versagel MJ (Penreco)

SIMMONDSIA CHINENSIS (JOJOBA) SEED POWDER

Definition: Simmondsia Chinensis (Jojoba) Seed Powder is a powder of the ground seeds of the jojoba, *Simmondsia chinensis. See "Regulatory and Ingredient Use Information," regarding the labeling names for botanical ingredients in Volume 1, Introduction, Part A.*

Chemical Class: Biological Products

Function: Skin-Conditioning Agent - Miscellaneous

Technical/Other Names:
Buxus Chinensis Powder
Jojoba Powder
Jojoba Seed Powder
Powdered Jojoba Seed

Trade Names:
AEC Jojoba Meal (A & E Connock)
Floragrains JSP (Floratech)
Jojoba Abrasive "GBU" (Cosmetochem) (Cosmetochem International Ltd.)
Jojoba Meal (Desert Whale)

Trade Name Mixture:
AEC Jojoba Wax Meal Beads (A & E Connock)

SIMMONDSIA CHINENSIS (JOJOBA) SEED WAX

CAS No.: 61789-91-1

Definition: Simmondsia Chinensis (Jojoba) Seed Wax is the wax obtained from the seed of the jojoba, *Simmondsia Chinensis. See "Regulatory and Ingredient Use Information," regarding the labeling names for botanical ingredients in Volume 1, Introduction, Part A.*

Information Source: CIR: [S] JACT-11(1)-1992

Chemical Class: Waxes

Functions: Hair Conditioning Agent; Skin-Conditioning Agent - Occlusive; Viscosity Increasing Agent - Nonaqueous

Reported Product Categories: Hair Conditioners; Moisturizing Preparations; Foundations; Lipsticks; Bath Preparations, Misc.; Body and Hand Preparations (Excluding Shaving Preparations); Skin Care Preparations, Misc.; Hair Dyes and Colors (All Types Requiring Caution Statements and Patch Tests); Bath Capsules; Shampoos (Non-coloring); Tonics, Dressings, and Other Hair Grooming Aids; Face and Neck Preparations (Excluding Shaving Preparations); Makeup Preparations (Not eye), Misc.; Bath Oils, Tablets, and Salts; Night Skin Care Preparations; Cleansing Products (Cold Creams, Cleansing Lotions, Liquids and Pads); Eye Makeup Preparations, Misc.; Blushers (All types); Hair Preparations (Non-coloring), Misc.; Hair Sprays (Aerosol Fixatives); Bath Soaps and Detergents; Eyeliners; Paste Masks (Mud Packs); Eye Shadows; Eyebrow Pencils; Face Powders; Manicuring Preparations, Misc.; Suntan Gels, Creams, and Liquids; Suntan Preparations, Misc.

Technical/Other Name:
Waxes, Jojoba

SINANOKI EKISU

JPN Translation:
シナノキエキス

Definition: Sinanoki Ekisu is an extract of the flowers and leaves of *Tilia vulgaris, Tilia cordata,* or *Tilia platyphyllos. See "Regulatory and Ingredient Use Information," regarding use of Japan Trivial names in Volume 1, Introduction, Part A.*

Chemical Class: Biological Products

Function: Skin-Conditioning Agent - Miscellaneous

Technical/Other Names:
Tilia Cordata Flower Extract (U.S.)
Tilia Platyphyllos Flower Extract (U.S.)
Tilia Vulgaris Flower Extract (U.S.)

SIRAKABA EKISU

JPN Translation:
シラカバエキス

Definition: Sirakaba Ekisu is an extract of the leaves and bark of *Betula alba* or other species of the family, *Betulaceae. See "Regulatory and Ingredient Use Information," regarding use of Japan Trivial names in Volume 1, Introduction, Part A.*

Chemical Class: Biological Products

Function: Skin-Conditioning Agent - Miscellaneous

Technical/Other Name:
Betula Alba Extract (U.S.)

SIRAKABA HA EKISU

JPN Translation:
シラカバ葉エキス

Definition: Sirakaba Ha Ekisu is an extract of the leaves of *Betula alba* or other related species of the genus *Betula. See "Regulatory and Ingredient Use Information," regarding use of Japan Trivial names in Volume 1, Introduction, Part A.*

Chemical Class: Biological Products

Function: Skin-Conditioning Agent - Miscellaneous

Technical/Other Name:
Betula Alba Leaf Extract (U.S.)

SIRAKABA JYUHI EKISU

JPN Translation:
シラカバ樹皮エキス

Definition: Sirakaba Jyuhi Ekisu is an extract of the bark of *Betula alba* or other related species of the genus *Betula. See "Regulatory and Ingredient Use Information," regarding use of Japan Trivial names in Volume 1, Introduction, Part A.*

Chemical Class: Biological Products

Function: Skin-Conditioning Agent - Miscellaneous

Technical/Other Name:
Betula Alba Bark Extract (U.S.)

SISYMBRIUM IRIO SEED OIL

Definition: Sisymbrium Irio Seed Oil is the fixed oil obtained from the seeds of *Sisymbrium irio. See "Regulatory and Ingredient Use Information," regarding the labeling names for botanical ingredients in Volume 1, Introduction, Part A.*

Chemical Class: Essential Oils

Function: Fragrance Ingredient

Technical/Other Name:
Oils, Sisymbrium Irio

Trade Name:
AEC Sisymbrium Irio Oil (A & E Connock)

BETA-SITOSTEROL

CAS No.	EINECS No.
83-46-5	201-480-6

JPN Translation:
シトステロール

Empirical Formula:
$C_{29}H_{50}O$

Definition: Beta-Sitosterol is the sterol that conforms to the formula:

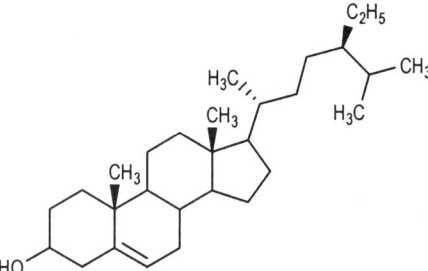

Information Sources: JCIC, JCLS, JSQI, MI-13(8627), MI-13(8628), RIFM

Chemical Class: Sterols

Functions: Fragrance Ingredient; Skin-Conditioning Agent - Miscellaneous

Reported Product Category: Moisturizing Preparations

Technical/Other Names:
24-α-Ethylcholesterol
β-Sitosterol
beta-Sitosterol (RIFM)
(3β)-Stigmast-5-en-3-ol

Trade Name Mixtures:
ALPHAFLOW N.I (Creations Couleurs)
Covasterol (LCW)
Escin/B-Sitosterol Phytosome (Indena SpA)
Phytosterol Complex Concentrate SI 1006 (Norjin)
Ultraspheres-5409 (Lipoid)

BETA-SITOSTERYL ACETATE

Empirical Formula:
$C_{31}H_{52}O_2$

Definition: Beta-Sitosteryl Acetate is the ester of Beta-Sitosterol (q.v.) and Acetic Acid (q.v.).

Information Source: MI-13(8627)

Chemical Classes: Esters; Sterols

Function: Skin-Conditioning Agent - Miscellaneous

SKIN LIPIDS

Definition: Skin Lipids is a mixture of lipids derived from animal skin. *See "Regulatory and Ingredient Use Information," regarding use of EU Trivial names in Volume 1, Introduction, Part A.*

Chemical Class: Fats and Oils

Function: Skin-Conditioning Agent - Miscellaneous

Technical/Other Names:
Animal Skin Lipids
Lipids, Animal Skin

S-LACTOYLGLUTATHIONE

CAS No.: 54398-03-7

Definition: S-Lactoylglutathione is the organic compound that conforms to the formula:

$$\underset{\underset{\underset{\underset{O\ \ OH}{||\ \ |}}{CH_2SC-CHCH_3}}{|}}{\overset{\overset{NH_2}{|}}{HOOCCH(CH_2)_2C}\overset{\overset{O}{||}}{-}NHCHC\overset{\overset{O}{||}}{-}NHCH_2COOH}$$

Chemical Classes: Amides; Carboxylic Acids; Thio Compounds

Function: Skin-Conditioning Agent - Miscellaneous

Technical/Other Name:
Glycine, N-[S-1,2-Dioxopropyl)-N-L-gamma-Glutamyl-L-Cysteinyl-

Trade Name:
SLG (Nonogawa)

SMILAX ARISTOLOCHIAEFOLIA ROOT EXTRACT

Definition: Smilax Aristolochiaefolia Root Extract is an extract of the roots of the sarsaparilla, *Smilax aristolochiaefolia. See "Regulatory and Ingredient Use Information," regarding the labeling names for botanical ingredients in Volume 1, Introduction, Part A.*

Information Source: 21CFR172.510

Chemical Class: Biological Products

Function: Skin-Conditioning Agent - Miscellaneous

Technical/Other Names:
Extract of Sarsaparilla
Extract of Smilax Aristolochiaefolia
Sarsaparilla Extract
Sarsaparilla (Smilax Aristolochiaefolia) Extract

Trade Name Mixtures:
Actiphyte of Sarsaparilla BG50 (Active Organics)
Actiphyte of Sarsaparilla GL50 (Active Organics)
Actiphyte of Sarsaparilla Lipo S (Active Organics)
Actiphyte of Sarsaparilla PG50 (Active Organics)
Activated Botanicals Estroherb Complex AB 106 (Norjin)
Herbasol-Extract Sarsaparilla (Cosmetochem)
Sarsaparilla HBGSC (Alban Muller)
Sarsaparilla HS (Alban Muller)

SMILAX GLABRA ROOT EXTRACT

Definition: Smilax Glabra Root Extract is an extract of tubers of *Smilax glabra. See "Regulatory and Ingredient Use Information," regarding the labeling names for botanical ingredients in Volume 1, Introduction, Part A.*

Chemical Class: Biological Products

Function: Not Reported

Technical/Other Name:
Extract of Smilax Glabra

Trade Name Mixture:
Sankirai Liquid E (Ichimaru Pharcos)

SMILAX MEDICA ROOT EXTRACT

Definition: Smilax Medica Root Extract is an extract of the roots of *Smilax medica. See "Regulatory and Ingredient Use Information," regarding the labeling names for botanical ingredients in Volume 1, Introduction, Part A.*

Chemical Class: Biological Products

Function: Not Reported

Technical/Other Name:
Extract of Smilax Medica Root

Trade Name:
Cosflor Sarsparilla HG-2 (A & E Connock)

SMILAX UTILIS ROOT EXTRACT

Definition: Smilax Utilis Root Extract is an extract of the dried roots of *Smilax utilis. See "Regulatory and Ingredient Use Information," regarding the labeling names for botanical ingredients in Volume 1, Introduction, Part A.*

Chemical Class: Biological Products

Function: Not Reported

Technical/Other Names:
Extract of Sarsaparilla
Extract of Smilax Utilis
Sarsaparilla Extract
Sarsaparilla (Smilax Utilis) Extract

Trade Name Mixtures:
Herbal-Complex 4 special (Crodarom)
Sarsaparilla Extract HS 2995 G (Grau)
Sarsaparilla Extract HS 3334 G (Grau)

SMITHSONITE

CAS Nos.: 3486-35-9; 14476-25-6

Definition: Smithsonite is a mineral consisting chiefly of zinc carbonate.

Information Source: MI-13(10184)

Chemical Class: Inorganics

Function: Cosmetic Astringent

SMITHSONITE EXTRACT

Definition: Smithsonite Extract is an extract of *Smithsonite q.v..*

Chemical Class: Inorganics

Functions: Antioxidant; Skin-Conditioning Agent - Miscellaneous

Trade Name:
Zin'Cite (Libiol)

SODIUM ACETATE

CAS No.	EINECS No.
127-09-3	204-823-8

JPN Translation:
酢酸 Na

Empirical Formula:
$C_2H_4O_2 \cdot Na$

Definition: Sodium Acetate is the sodium salt of acetic acid that conforms to the formula:

$$CH_3COONa$$

Information Sources: BP, BPC, BRA, 21CFR173.310, 21CFR182.70,

21CFR184.1721, DDR, EGY, FCC, HUN,
JAN, JCLS, MAR, MI-13(8642), NED, RIFM,
TSCA, USAN, USD, USP XXIV

Chemical Class: Organic Salts

Functions: Buffering Agent; Fragrance
Ingredient

Reported Product Category: Hair Dyes
and Colors (All Types Requiring Caution
Statements and Patch Tests)

Technical/Other Names:
Acetic Acid, Sodium Salt
Sodium acetate (RIFM)

Trade Name Mixtures:
Essential Vital Elements (Dipta)
Essential Vital Elements - S (Dipta)

SODIUM ACETYLATED HYALURONATE

Definition: Sodium Acetylated Hyaluronate
is the acetyl ester of Sodium Hyaluronate
(q.v.).

Chemical Classes: Biological Polymers and
their Derivatives; Esters

Function: Humectant

Reported Product Categories: Lipsticks;
Eye Shadows; Foundations; Face Powders;
Eye Makeup Preparations, Misc.; Rouges;
Bath Capsules; Bath Oils, Tablets, and Salts;
Cleansing Products (Cold Creams, Cleansing
Lotions, Liquids and Pads); Eyeliners;
Makeup Preparations (Not eye), Misc.;
Mascara; Blushers (All types)

Trade Name:
Sodium Acetylhyaluronate (Shiseido
Company)

SODIUM ACRYLATE/ ACRYLOYLDIMETHYL TAURATE COPOLYMER

Definition: Sodium Acrylate/
Acryloyldimethyl Taurate Copolymer is a
copolymer of sodium acrylate and sodium
acryloyldimethyl taurate monomers.

Chemical Class: Synthetic Polymers

Functions: Anticaking Agent; Emulsion
Stabilizer; Opacifying Agent; Suspending
Agent - Nonsurfactant; Viscosity Increasing
Agent - Aqueous

Trade Name Mixtures:
AEC Sodium Acrylate/Acryloyldimethyl
Taurate Copolymer (&) Isohexadecane
(&) Polysorbate 80 (A & E Connock)
Simulgel EG (SEPPIC)
Simulgel EPG (SEPPIC)

SODIUM ACRYLATES/ACROLEIN COPOLYMER

Definition: Sodium Acrylates/Acrolein
Copolymer is a polymer consisting of sodium
acrylate and acrolein monomers.

Information Source: CIR: [SQ] IJT 21
(SUPPL. 3) 2002

Chemical Class: Synthetic Polymers

Functions: Binder; Film Former; Viscosity
Increasing Agent - Aqueous

SODIUM ACRYLATES/ACRYLO-NITROGENS COPOLYMER

CAS No.: 182576-39-2

Definition: Sodium Acrylates/Acrylo-
nitrogens Copolymer is the polymer formed
by the controlled hydrolysis of polyacrylonitrile neutralized by sodium.

Chemical Class: Synthetic Polymers

Functions: Binder; Film Former; Skin-
Conditioning Agent - Miscellaneous; Viscosity
Increasing Agent - Aqueous

Technical/Other Name:
β-Alanine, Reaction Products with Polyacrylonitrile and Sodium Thiocyanate

Trade Name:
HYPAN AA90 (Hymedix)

SODIUM ACRYLATES/C10-30 ALKYL ACRYLATES CROSSPOLYMER

Definition: Sodium Acrylates/C10-30 Alkyl
Acrylates Crosspolymer is the sodium salt of
Acrylates/C10-30 Alkyl Acrylate Cross-
polymer (q.v.).

Chemical Class: Synthetic Polymers

Function: Film Former

SODIUM ACRYLATES COPOLYMER

Definition: Sodium Acrylates Copolymer is
the sodium salt of a polymer consisting of
acrylic acid, methacrylic acid or one of their
simple esters.

Information Source: CIR: [SQ] IJT 21
(SUPPL. 3) 2002

Chemical Class: Synthetic Polymers

Functions: Binder; Film Former; Viscosity
Increasing Agent - Aqueous

Reported Product Category: Hair Dyes
and Colors (All Types Requiring Caution
Statements and Patch Tests)

Trade Name Mixtures:
DP 705-9339 (Ciba Specialty Chemicals)
Fancorgel A (Fanning)
Heliogel (Advanced Beauty)
Luvigel EM (BASF)
Salcare AST (Ciba Specialty Chemicals)
Salcare SC91 (Ciba Specialty Chemicals)

SODIUM ACRYLATES CROSSPOLYMER

JPN Translation:
アクリル酸 Na クロスポリマー

Definition: Sodium Acrylates Crosspolymer
is the sodium salt of a copolymer of acrylic
acid, methacrylic acid or one or more of its
simple esters crosslinked with ethylene glycol
diglycidyl ether.

Chemical Class: Synthetic Polymers

Functions: Absorbent; Viscosity Increasing
Agent - Aqueous

Trade Name Mixture:
Sofcare S-SP-D5 (Kao Corp.)

SODIUM ACRYLATES/ETHYLHEXYL ACRYLATE COPOLYMER

JPN Translation:
アクリル酸アルキルコポリマーNa

Definition: Sodium Acrylates/Ethylhexyl
Acrylate Copolymer is a copolymer of
ethylhexyl acrylate and the sodium salt of one
or more monomers consisting of acrylic acid,
methacrylic acid or one of their simple esters.

Information Source: JCLS

Chemical Class: Synthetic Polymers

Function: Film Former

SODIUM ACRYLATE/SODIUM ACRYLOYLDIMETHYL TAURATE COPOLYMER

Definition: Sodium Acrylate/Sodium
Acryloyldimethyl Taurate Copolymer is a
copolymer of sodium acrylate and sodium
acryloyldimethyl taurate monomers.

Chemical Class: Synthetic Polymers

Functions: Anticaking Agent; Emulsion
Stabilizer; Opacifying Agent; Suspending
Agent - Nonsurfactant; Viscosity Increasing
Agent - Nonaqueous

Trade Name Mixture:
Flocare ET 30 (SNF)

SODIUM ACRYLATES/VINYL ISODECA-NOATE CROSSPOLYMER

Definition: Sodium Acrylates/Vinyl Isodecanoate Crosspolymer is the sodium salt of

Acrylates/Vinyl Isodecanoate Crosspolymer (q.v.).

Chemical Class: Synthetic Polymers

Functions: Emulsion Stabilizer; Suspending Agent - Nonsurfactant; Viscosity Increasing Agent - Aqueous

Trade Name:
PNC 30 (3V Sigma S.P.A.)

SODIUM ACRYLATE/VINYL ALCOHOL COPOLYMER

CAS Nos.: 27599-56-0; 58374-38-2

Empirical Formula:
$(C_3H_4O_2 \cdot C_2H_4O)_x \cdot xNa$

Definition: Sodium Acrylate/Vinyl Alcohol Copolymer is a polymer of sodium acrylate and vinyl alcohol monomers.

Chemical Class: Synthetic Polymers

Functions: Binder; Emulsion Stabilizer; Film Former; Viscosity Increasing Agent - Aqueous

Technical/Other Names:
Acrylic Acid, Polymer with Vinyl Alcohol, Sodium Salt
2-Propenoic Acid, Polymer with Ethenol, Sodium Salt
2-Propenoic Acid, Sodium Salt, Polymer with Ethenol
Vinyl Alcohol, Polymer with Acrylic Acid, Sodium Salt

SODIUM ACRYLIC ACID/MA COPOLYMER

CAS No.: 52255-49-9

Definition: Sodium Acrylic Acid/MA Copolymer is the sodium salt of a copolymer consisting of maleic anhydride and acrylic acid.

Chemical Class: Synthetic Polymers

Function: Suspending Agent - Nonsurfactant

Technical/Other Name:
2-Propenoic Acid, Polymer with 2,5-Furandione, Sodium Salt

Trade Name:
Sokalan CP5 (BASF)

SODIUM ALGIN SULFATE

CAS No.: 9010-06-4

Definition: Sodium Algin Sulfate is the sulfate ester of Algin (q.v.).

Information Source: JCLS

Chemical Classes: Gums, Hydrophilic Colloids and Derivatives; Sulfuric Acid Esters

Function: Skin-Conditioning Agent - Humectant

Technical/Other Names:
Alginic Acid, Hydrogen Sulfate, Sodium Salt
Hepinoid

Trade Name:
AC Alginate (Meito Sangyo)

SODIUM ALLANTOIN PCA

JPN Translation:
PCA - Na アラントイン

Definition: Sodium Allantoin PCA is the complex formed between Allantoin (q.v.) and Sodium PCA (q.v.).

Information Source: JCIC

Chemical Classes: Heterocyclic Compounds; Organic Salts

Function: Skin-Conditioning Agent - Miscellaneous

Technical/Other Name:
Allantoin Sodium DL-Pyrrolidone Carboxylate

SODIUM ALUM

CAS Nos.	EINECS No.
10024-42-7	
10102-71-3	233-277-3

Empirical Formula:
$Al \cdot 2H_2O_4S \cdot Na$

Definition: Sodium Alum is the inorganic salt that conforms to the formula:

$$NaAl(SO_4)_2 \quad \cdot \quad 12H_2O$$

Information Sources: 21CFR175.105, 21CFR178.3120, 21CFR182.90, 21CFR182.1131, FCC, MI-13(362), TSCA

Chemical Class: Inorganic Salts

Function: Cosmetic Astringent

Technical/Other Names:
Aluminum Sodium Sulfate
Sodium Aluminum Sulfate
Sulfuric Acid, Aluminum Sodium Salt (2:1:1)

SODIUM ALUMINATE

CAS No.	EINECS No.
1302-42-7	215-100-1

Empirical Formula:
$AlO_2 \cdot Na$

Definition: Sodium Aluminate is the inorganic compound that conforms to the formula:

$$AlNaO_2$$

Information Sources: 21CFR173.310, 21CFR182.90, MI-13(8645)

Chemical Class: Inorganic Salts

Functions: Buffering Agent; Corrosion Inhibitor; pH Adjuster

Technical/Other Names:
Aluminate, Sodium
Aluminum Sodium Oxide

SODIUM ALUMINUM CHLOROHYDROXY LACTATE

CAS No.: 8038-93-5

Definition: Sodium Aluminum Chloro-hydroxy Lactate is the sodium salt of a complex of lactic acid and aluminum chloro-hydrate.

Information Source: CTFA S

Chemical Class: Organic Salts

Function: Cosmetic Astringent

Technical/Other Name:
Sodium Aluminum Chlorhydroxy Lactate

Trade Name:
Chloracel (Reheis)

SODIUM ALUMINUM LACTATE

CAS No.	EINECS No.
68953-69-5	273-223-6

Definition: Sodium Aluminum Lactate is a complex salt of sodium and aluminum lactates. It conforms generally to the formula:

$$Na_2HAl(OOCCHOHCH_3)_2(OH)_6$$

Information Sources: CTFA S, TSCA

Chemical Class: Organic Salts

Functions: Buffering Agent; Cosmetic Astringent

Technical/Other Name:
Aluminum, Lactate Sodium Complexes, Basic

SODIUM ASCORBATE

CAS No.	EINECS No.
134-03-2	205-126-1

JPN Translation:
アスコルビン酸 Na

Empirical Formula:
$C_6H_8O_6 \cdot Na$

Definition: Sodium Ascorbate is the sodium salt of ascorbic acid that conforms to the formula:

Information Sources: 21CFR182.3731, CIR: [S], FCC, INN, JCLS, JSCI, MAR, MI-12(8723), TSCA, USAN, USP XXIV

Chemical Classes: Heterocyclic Compounds; Organic Salts

Function: Antioxidant

Reported Product Categories: Hair Dyes and Colors (All Types Requiring Caution Statements and Patch Tests); Lipsticks

Technical/Other Names:
L-Ascorbic Acid, Monosodium Salt
Vitamin C Sodium

Trade Names:
Sodium Ascorbate Fine Powder (Roche)
Sodium Ascorbate Type AG (Roche)

Trade Name Mixtures:
Isocell Citrus (Lucas Meyer)
RetiSTAR (BASF Corporation)
RonaCare ASC III (Merck KGaA)
RonaCare ASC III (Merck KGaA/EMD Chemicals Inc.)
RonaCare VTA (Merck KGaA)

SODIUM ASCORBYL/CHOLESTERYL PHOSPHATE

CAS No.: 185018-43-3

Definition: Sodium Ascorbyl/Cholesteryl Phosphate is the sodium salt of a complex mixture of esters of Ascorbic Acid (q.v.) and Cholesterol (q.v.) with phosphoric acid.

Chemical Class: Phosphorus Compounds

Functions: Antioxidant; Skin-Conditioning Agent - Miscellaneous

Technical/Other Names:
L-Ascorbic Acid, 2-[(3β-Cholest-5-en-3-yl Hydrogen Phosphate], Monosodium Salt
AVC-10

SODIUM ASCORBYL PHOSPHATE

CAS No.: 66170-10-3

Empirical Formula:
$C_6H_6O_9P \cdot 3Na$

Definition: Sodium Ascorbyl Phosphate is the organic compound that conforms to the formula:

Information Source: CIR: [S]

Chemical Classes: Heterocyclic Compounds; Organic Salts; Phosphorus Compounds

Function: Antioxidant

Reported Product Categories: Body and Hand Preparations (Excluding Shaving Preparations); Eye Lotions; Hair Sprays (Aerosol Fixatives); Moisturizing Preparations; Night Skin Care Preparations

Technical/Other Name:
L-Ascorbic Acid, 2-(Dihydrogen Phosphate), Trisodium Salt

Trade Names:
Sodium Ascorbyl Phosphate (BASF)
Sodium L-Ascorbyl-2-Phosphate (Kyowa Hakko Kogyo)
Sodium L-Ascorbyl-2-Phosphate, SAP Ascorbyl Phosphate Sodium, ASP (Showa Denko)
Stay-C 50 (Hoffmann-La Roche)
Stay-C 50 (Roche)
Stay-C 50 (Roche Vitamins)
TINODERM C (Ciba Specialty Chemicals)

Trade Name Mixture:
Ultraspheres 8028 (Lipoid)

SODIUM ASPARTATE

CAS Nos.	EINECS Nos.
3792-50-5	
5598-53-8 (L-Form)	227-012-0
17090-93-6	241-155-6

JPN Translation:
アスパラギン酸 Na

Empirical Formula:
$C_4H_7NO_4 \cdot Na$

Definition: Sodium Aspartate is the sodium salt of aspartic acid that conforms to the formula:

$$HOOCCH_2CHCOONa$$
$$|$$
$$NH_2$$

Information Sources: 21CFR172.320, JCLS, JSCI

Chemical Class: Amino Acids

Function: Skin-Conditioning Agent - Humectant

Technical/Other Names:
Aspartic Acid, Sodium Salt
Monosodium L-Aspartate

Trade Name:
Monosodium L-Aspartate (Kyowa Hakko Kogyo)

SODIUM ASTROCARYUM MURUMURUATE

Definition: Sodium Astrocaryum Murumuruate is the sodium salt of the fatty acids derived from Astrocaryum Murumuru Butter (q.v.).

Chemical Class: Soaps

Functions: Skin-Conditioning Agent - Emollient; Surfactant - Cleansing Agent

Trade Name:
Chemysoap (Chemyunion)

SODIUM BABASSUATE

Definition: Sodium Babassuate is the sodium salt of the fatty acids derived from Orbignya oleifera (babassu) oil.

Chemical Class: Soaps

Functions: Surfactant - Cleansing Agent; Surfactant - Emulsifying Agent

SODIUM BABASSU SULFATE

Definition: Sodium Babassu Sulfate is the sodium salt of the sulfate ester of the fatty alcohols derived from babassu oil.

Chemical Class: Alkyl Sulfates

Function: Surfactant - Cleansing Agent

SODIUM BEESWAX

Definition: Sodium Beeswax is the sodium salt of the fatty acids derived from Beeswax (q.v.).

Chemical Class: Soaps

Function: Surfactant - Emulsifying Agent

Trade Name:
Beeswax Saponified (Weleda)

SODIUM BEHENATE

CAS No.
5331-77-1

EINECS No.
226-234-5

Definition: Sodium Behenate is the sodium salt of Behenic Acid (q.v.).

Chemical Class: Soaps

Function: Surfactant - Cleansing Agent

Technical/Other Name:
Behenic Acid, Sodium Salt

SODIUM BEHENOYL LACTYLATE

Empirical Formula:
$C_{28}H_{52}O_6 \cdot Na$

Definition: Sodium Behenoyl Lactylate is the sodium salt of the behenic acid ester of lactyl lactate. It conforms to the formula:

$$CH_3(CH_2)_{20}\overset{O}{\overset{\|}{C}}-O\overset{|}{\underset{CH_3}{C}}HC-O\overset{|}{\underset{CH_3}{C}}HCOONa$$

Chemical Classes: Esters; Organic Salts

Function: Surfactant - Emulsifying Agent

Trade Name:
Pationic SBL (RITA)

SODIUM BENZOATE

CAS No.
532-32-1

EINECS No.
208-534-8

JPN Translation:
安息香酸 Na

Empirical Formula:
$C_7H_6O_2 \cdot Na$

Definition: Sodium Benzoate is the sodium salt of benzoic acid that conforms to the formula:

COONa

Information Sources: ARG, AUS, BP, BPC, BRA, 21CFR146.152, 21CFR146.154, 21CFR150.141, 21CFR150.161, 21CFR166.40, 21CFR166.110, 21CFR181.22, 21CFR181.23, 21CFR184.1733, CIR: [SQ] IJT-20(SUPPL. 3)2001, CTFA S, CZE, DA, DDR, EEC(VI/1-1), EGY, FCC, FIN, HUN, IND, ITA, JAN, JCLS, JSCI, MAR, MEX, MHLW-331/3, MI-13(8654), NF XVIII, PN, POR, RIFM, ROM, SNPF, TSCA, USAN, USD, USSR, YUG

Chemical Class: Organic Salts

Functions: Fragrance Ingredient; Preservative

Reported Product Categories: Shampoos (Non-coloring); Hair Sprays (Aerosol Fixatives); Aftershave Lotions; Baby Shampoos; Bath Oils, Tablets, and Salts; Skin Care Preparations, Misc.; Cleansing Products (Cold Creams, Cleansing Lotions, Liquids and Pads); Moisturizing Preparations; Dentifrices (Aerosol, Liquid, Pastes and Powders); Bath Preparations, Misc.; Body and Hand Preparations (Excluding Shaving Preparations); Shaving Cream (Aerosol, Brushless and Lather); Eye Makeup Removers; Hair Conditioners; Night Skin Care Preparations; Bath Capsules; Bath Soaps and Detergents; Eye Makeup Preparations, Misc.; Eye Shadows; Tonics, Dressings, and Other Hair Grooming Aids; Bubble Baths; Face and Neck Preparations (Excluding Shaving Preparations); Hair Rinses (Non-coloring); Personal Cleanliness Products, Misc.; Eyeliners

Technical/Other Names:
Benzoic Acid, Sodium Salt
Sodium benzoate (RIFM)

Trade Name:
Unisept SB (Universal Preserv-A-Chem)

Trade Name Mixtures:
AEC Cosflor Blend 017 Moisture Factor WSS (A & E Connock)
AEC Moisture Factor HV (A & E Connock)
Lactil (Degussa Care Specialties)
Nipacombin A (Clariant)
Nipacombin A (Clariant GmbH, Personal Care)
Solusol 85% (American Cyanamid/Fine Chemicals)

SODIUM BENZOTRIAZOLYL BUTYLPHENOL SULFONATE

CAS No.
92484-48-5

EINECS No.
403-080-9

Empirical Formula:
$C_{16}H_{17}N_3O_5S \cdot Na$

Definition: Sodium Benzotriazolyl Butylphenol Sulfonate is the organic compound that conforms to the formula:

Chemical Classes: Glyceryl Esters and Derivatives; Sulfonic Acids

Function: Ultraviolet Light Absorber

Technical/Other Names:
Benzenesulfonic Acid, 3-(2H-Benzotriazol-2-yl)-4-Hydroxy-5-(1-Methylpropyl)-, Monosodium Salt
Benztriazol UV Absorber BUK 4499
Sodium 3-(2H-Benzotriazol-2-yl)-5-sec-Butyl-4-Hydroxybenzenesulfonate

Trade Name:
Tinogard HS (Ciba Specialty Chemicals)

Trade Name Mixture:
Cibafast H LIQ. (Ciba Specialty Chemicals)

SODIUM BICARBONATE

CAS No.
144-55-8

EINECS No.
205-633-8

JPN Translation:
炭酸水素 Na

Empirical Formula:
$CH_2O_3 \cdot Na$

Definition: Sodium Bicarbonate is the inorganic salt that conforms to the formula:

$$NaHCO_3$$

In the United States, Sodium Bicarbonate may be used as an active ingredient in OTC drug products. When used as an active drug ingredient, the established name is *Sodium Bicarbonate. See "Regulatory and Ingredient Use Information," regarding the labeling names for U.S. OTC Drug Ingredients in Volume 1, Introduction, Part A.*

Information Sources: ARG, AUS, BP, BPC, BRA, 21CFR137.180, 21CFR137.270, 21CFR163.110, 21CFR173.385, 21CFR178.1010, 21CFR184.1736, CIR: [S] JACT-6(1)1987, CZE, DA, DDR, EGY, EP, FCC, FIN, HUN, IND, ITA, JAN, JCIC, JCLS, JSCI, MAR, MEX, MI-13(8655), OTC-I-AA, OTC-I-OH, OTC-I-SK, PN, POR, ROM, TSCA, USAN, USD, USP XXIV, YUG

Chemical Class: Inorganic Salts

Functions: Abrasive; Buffering Agent; Deodorant Agent; Oral Care Agent; Oral Health Care Drug; pH Adjuster; Skin Protectant

Reported Product Categories: Mascara; Hair Dyes and Colors (All Types Requiring Caution Statements and Patch Tests); Dentifrices (Aerosol, Liquid, Pastes and Powders); Powders (Dusting and Talcum, Excluding Aftershave Talcs); Bath Oils, Tablets, and Salts; Eyeliners; Foot Powders and Sprays; Skin Care Preparations, Misc.;

Bath Preparations, Misc.; Mouthwashes and Breath Fresheners (Liquids and Sprays); Permanent Waves; Personal Cleanliness Products, Misc.

Technical/Other Names:
Baking Soda
Bicarbonate of Soda
Carbonic Acid, Monosodium Salt

Trade Names:
sodium hydrogen carbonate (Unikem)
Unichem BICARB-S (Universal Preserv-A-Chem)

Trade Name Mixtures:
Essential Vital Elements (Dipta)
Essential Vital Elements - S (Dipta)
Lipothix 100-B (Lipo)

SODIUM BISCHLOROPHENYL SULFAMINE

CAS No.: 58727-01-8

Empirical Formula:
$C_{38}H_{28}Cl_4N_4O_{14}S_6 \cdot 2Na$

Definition: Sodium Bischlorophenyl Sulfamine is the organic compound that conforms generally to the formula:

where R represents H or the 3,4-dichloro-phenyl sulfonyl group.

Chemical Classes: Amines; Halogen Compounds; Sulfonic Acids

Function: Skin-Conditioning Agent - Miscellaneous

Technical/Other Names:
Benzenesulfonic Acid, 2,2'-(1,2-Ethenediyl)Bis[5-[[[4-[[(3,4-Dichloro-phenyl)Sulfonyl]Amino]Phenyl]-Sulfonyl]Amino]-, Disodium Salt
2,2'-(1,2-Ethenediyl)Bis[5-[[[4-[[(3,4-Dichlorophenyl)Sulfonyl]Amino]Phenyl]-Sulfonyl]Amino]Benzenesulfonic Acid, Disodium Salt

Trade Name:
Eucoriol (Stockhausen, Inc.)

SODIUM BISGLYCOL RICINOSULFO-SUCCINATE

Empirical Formula:
$C_{26}H_{46}O_{12}S \cdot Na$

Definition: Sodium Bisglycol Ricinosulfo-succinate is the organic compound which conforms generally to the formula:

Chemical Class: Sulfosuccinates and Sulfosuccinamates

Functions: Surfactant - Cleansing Agent; Surfactant - Solubilizing Agent

SODIUM BISULFATE

CAS No.	EINECS No.
7681-38-1	231-665-7

Empirical Formula:
$H_2O_4S \cdot Na$

Definition: Sodium Bisulfate is the inorganic salt that conforms to the formula:

$$NaHSO_4$$

Information Sources: 21CFR175.105, FCC, MI-13(8658), NFJ, TSCA

Chemical Class: Inorganic Salts

Function: pH Adjuster

Reported Product Category: Bubble Baths

Technical/Other Names:
Monobasic Sodium Sulfate
Sodium Hydrogen Sulfate
Sulfuric Acid, Monosodium Salt

SODIUM BISULFITE

CAS No.	EINECS No.
7631-90-5	231-548-0

JPN Translation:
亜硫酸水素 Na

Empirical Formula:
$H_2O_3S \cdot Na$

Definition: Sodium Bisulfite is the inorganic salt that conforms to the formula:

$$NaHSO_3$$

Information Sources: BRA, 21CFR161.173, 21CFR173.310, 21CFR182.3739, CIR: [S], FCC, JAN, JCLS, JSCI, MI-13(8660), NF XV, POR, TSCA, USD, USP XXIV

Chemical Class: Inorganic Salts

Functions: Antioxidant; Hair-Waving/Straightening Agent; Reducing Agent

Reported Product Categories: Hair Dyes and Colors (All Types Requiring Caution Statements and Patch Tests); Skin Care Preparations, Misc.; Face and Neck Preparations (Excluding Shaving Preparations); Paste Masks (Mud Packs); Tonics, Dressings, and Other Hair Grooming Aids

Technical/Other Names:
Sodium Acid Sulfite
Sulfurous Acid, Monosodium Salt

Trade Name:
Uantox SBS (Universal Preserv-A-Chem)

SODIUM BORAGEAMIDOPROPYL PG-DIMONIUM CHLORIDE PHOSPHATE

Definition: Sodium Borageamidopropyl PG-Dimonium Chloride Phosphate is the amphoteric organic compound that conforms generally to the formula:

where RCO- represents the fatty acids derived from borage seed oil.

Chemical Classes: Phosphorus Compounds; Quaternary Ammonium Compounds

Functions: Antistatic Agent; Surfactant - Cleansing Agent; Surfactant - Foam Booster

Trade Name:
Phospholipid GLA (Uniqema)

SODIUM BORATE

CAS Nos.	EINECS No.
1303-96-4 (Hydrous)	
1330-43-4	215-540-4

JPN Translation:
ホウ酸 Na

Empirical Formula:
$B_4H_2O_7 \cdot 10H_2O \cdot 2Na$

Definition: Sodium Borate is the inorganic salt that conforms generally to the formula:

$$Na_2B_4O_7$$

Information Sources: ARG, AUS, BP, BRA, 21CFR175.105, 21CFR175.210, 21CFR176.180, 21CFR177.2800, 21CFR181.22, 21CFR181.30, CIR: [SQ]

JACT-2(7)1983, CTFA S, CZE, DA, DDR, EEC(III/1-1), EGY, EP, FIN, HP, HUN, IND, ITA, JAN, JCLS, JSCI, MAR, MEX, MHLW-331/2, MI-13(8662), NF XIX, PN, POR, ROM, SNPF, USAN, USD, WHO

Chemical Class: Inorganic Salts

Function: pH Adjuster

Reported Product Categories: Bath Oils, Tablets, and Salts; Cleansing Products (Cold Creams, Cleansing Lotions, Liquids and Pads); Bath Preparations, Misc.; Body and Hand Preparations (Excluding Shaving Preparations); Moisturizing Preparations; Skin Care Preparations, Misc.; Night Skin Care Preparations; Makeup Bases; Mascara; Bath Capsules; Face and Neck Preparations (Excluding Shaving Preparations); Hair Dyes and Colors (All Types Requiring Caution Statements and Patch Tests); Shaving Cream (Aerosol, Brushless and Lather); Tonics, Dressings, and Other Hair Grooming Aids; Paste Masks (Mud Packs); Personal Cleanliness Products, Misc.; Eyebrow Pencils; Permanent Waves; Suntan Gels, Creams, and Liquids; Eyeliners; Skin Fresheners; Bath Soaps and Detergents; Dentifrices (Aerosol, Liquid, Pastes and Powders); Eye Shadows; Foundations; Hair Coloring Preparations, Misc.; Hair Conditioners; Hair Preparations (Non-coloring), Misc.; Suntan Preparations, Misc.

Technical/Other Names:
Borax
Borax Granular
Boric Acid, Disodium Salt
Puffed Borax
Sodium Tetraborate

Trade Name:
Di-Sodium Tetraborate (Merck KGaA)

Trade Name Mixture:
Phenolines The Vert (Coletica SA)

SODIUM BROMATE

CAS No.	EINECS No.
7789-38-0	232-160-4

JPN Translation:
臭素酸 Na

Empirical Formula:
$BrHO_3 \cdot Na$

Definition: Sodium Bromate is the inorganic salt that conforms to the formula:

$$NaBrO_3$$

Information Sources: CIR: [SQ] JACT-13 (5)1994, CTFA S, CZE, JCLS, JSCI, MI-13 (8665), TSCA

Chemical Class: Inorganic Salts

Function: Oxidizing Agent

Reported Product Categories: Permanent Waves; Hair Preparations (Non-coloring), Misc.; Hair Rinses (Non-coloring); Hair Straighteners

Technical/Other Name:
Bromic Acid, Sodium Salt

SODIUM BUTOXYETHOXY ACETATE

CAS No.	EINECS No.
67990-17-4	268-040-3

Empirical Formula:
$C_8H_{16}O_4 \cdot Na$

Definition: Sodium Butoxyethoxy Acetate is the organic salt that conforms to the formula:

$$CH_3(CH_2)_3OCH_2CH_2OCH_2COONa$$

Information Source: TSCA

Chemical Classes: Ethers; Organic Salts

Function: Not Reported

Technical/Other Name:
Acetic Acid, (2-Butoxyethoxy)-, Sodium Salt

SODIUM BUTOXYNOL-12 SULFATE

Definition: Sodium Butoxynol-12 Sulfate is the sodium salt of the sulfuric acid ester of an alkoxylated phenol that conforms generally to the formula:

where n has an average value of 12.

Chemical Class: Alkyl Ether Sulfates

Function: Surfactant - Cleansing Agent

Technical/Other Names:
PEG-12 Butyl Phenyl Ether Sulfate, Sodium Salt
Polyethylene Glycol 600 Butyl Phenyl Ether Sulfate, Sodium Salt
Polyoxyethylene (12) Butyl Phenyl Ether Sulfate, Sodium Salt

Trade Name:
Akyposal TBP 120 (Kao GmbH)

SODIUM BUTYL ESTER OF PVM/MA COPOLYMER

Definition: Sodium Butyl Ester of PVM/MA Copolymer is the sodium salt of Butyl Ester of PVM/MA Copolymer (q.v.).

Chemical Class: Synthetic Polymers

Functions: Binder; Film Former; Hair Fixative

SODIUM BUTYLPARABEN

CAS No.	EINECS No.
36457-20-2	253-049-7

Empirical Formula:
$C_{11}H_{14}O_3 \cdot Na$

Definition: Sodium Butylparaben is the sodium salt of Butylparaben (q.v.) that conforms to the formula:

Information Sources: EEC(VI/1-12), MHLW-331/3

Chemical Classes: Esters; Organic Salts; Phenols

Function: Preservative

Reported Product Categories: Blushers (All types); Colognes and Toilet Waters; Makeup Preparations (Not eye), Misc.

Technical/Other Names:
Benzoic Acid, 4-Hydroxy-, Butyl Ester, Sodium Salt
Butylparaben, Sodium Salt
4-Hydroxybenzoic Acid, Butyl Ester, Sodium Salt

Trade Names:
Nipabutyl Sodium (Clariant)
Nipabutyl Sodium (Clariant GmbH, Personal Care)

Trade Name Mixtures:
Nipacide A Sodium (Clariant)
Nipacide A Sodium (Clariant GmbH, Personal Care)
Nipacombin SK (Clariant)
Nipacombin SK (Clariant GmbH, Personal Care)
Nipastat Sodium (Clariant)
Nipastat Sodium (Clariant GmbH, Personal Care)

SODIUM C13-17 ALKANE SULFONATE

Definition: Sodium C13-17 Alkane Sulfonate is the sodium salt of a sulfonated alkane with 13 to 17 carbon atoms in the alkyl chain.

Chemical Class: Sulfonic Acids

Function: Surfactant - Cleansing Agent

Trade Names:
Marlon PS 30 (Sasol GmbH - Marl)
Marlon PS 60 (Sasol GmbH - Marl)
Marlon PS 65 (Sasol GmbH - Marl)
Marlon PS 60W (Sasol GmbH - Marl)

SODIUM C14-18 ALKANE SULFONATE

Definition: Sodium C14-18 Alkane Sulfonate is the sodium salt of a sulfonated alkane with 14 to 18 carbons in the alkyl chain.

Information Sources: JCIC, JCLS

Chemical Class: Sulfonic Acids

Function: Surfactant - Cleansing Agent

Technical/Other Name:
Sodium Alkanesulfonate

SODIUM C12-15 ALKOXYPROPYL IMINODIPROPIONATE

Definition: Sodium C12-15 Alkoxypropyl Iminodipropionate is the partial sodium salt of a substituted propionic acid. It conforms generally to the formula:

$$RO(CH_2)_3N \begin{cases} CH_2CH_2COONa \\ CH_2CH_2COONa \end{cases}$$

where R represents the 12-15 carbon alkyl chain.

Chemical Class: Alkyl-Substituted Amino Acids

Functions: Hair Conditioning Agent; Surfactant - Cleansing Agent

Trade Name:
Amphoteric N (Tomah)

Trade Name Mixtures:
Biosil Basics Cetylsil NS (Biosil Technologies, Inc.)
Silteric Coco Complex (Siltech)

SODIUM C8-10 ALKYL SULFATE

CAS No.
85338-42-7

EINECS No.
286-718-7

Definition: Sodium C8-10 Alkyl Sulfate is the sodium salt of the sulfate of a mixture of synthetic fatty alcohols with 8 to 10 carbons in the alkyl chain.

Information Source: JCLS

Chemical Class: Alkyl Sulfates

Function: Surfactant - Foam Booster

Trade Name:
Empicol LB 40 (Albright & Wilson UK)

SODIUM C11-15 ALKYL SULFATE

JPN Translation:
アルキル（C11，13，15）硫酸 Na

Definition: Sodium C11-15 Alkyl Sulfate is the sodium salt of the sulfate of a mixture of synthetic alcohols with 11 to 15 carbons in the alkyl chain.

Information Source: JCLS

Chemical Class: Alkyl Sulfates

Function: Surfactant - Cleansing Agent

SODIUM C12-13 ALKYL SULFATE

JPN Translation:
アルキル（C12，13）硫酸 Na

Definition: Sodium C12-13 Alkyl Sulfate is the sodium salt of the sulfate of C12-13 Alcohols (q.v.).

Information Source: JCLS

Chemical Class: Alkyl Sulfates

Function: Surfactant - Cleansing Agent

SODIUM C12-15 ALKYL SULFATE

Definition: Sodium C12-15 Alkyl Sulfate is the sodium salt of the sulfate of C12-15 Alcohols (q.v.).

Information Source: JCLS

Chemical Class: Alkyl Sulfates

Function: Surfactant - Cleansing Agent

Reported Product Categories: Shampoos (Non-coloring); Skin Care Preparations, Misc.

Technical/Other Name:
Sodium C12-15 Alcohols Sulfate

Trade Name:
Unipol WAQ-115 (Universal Preserv-A-Chem)

SODIUM C12-18 ALKYL SULFATE

Definition: Sodium C12-18 Alkyl Sulfate is the sodium salt of the sulfate of a mixture of synthetic alcohols with 12 to 18 carbons in the alkyl chain.

Information Source: JCLS

Chemical Class: Alkyl Sulfates

Function: Surfactant - Cleansing Agent

Technical/Other Names:
Sodium C12-18 Alcohols Sulfate
Sulfuric Acid, Mono-C12-18-Alkyl Esters, Sodium Salts

Trade Names:
Stokopol LO (Stockhausen, Inc.)
Sulfetal TC 50 (Zschimmer & Schwarz)
Tensopol ACL79 (Manro)
Tensopol DCX (Manro)
Tensopol DDX (Manro)
Tensopol PCL94 (Manro)

SODIUM C16-20 ALKYL SULFATE

Definition: Sodium C16-20 Alkyl Sulfate is the sodium salt of the sulfate of a mixture of synthetic alcohols with 16 to 20 carbons in the alkyl chain.

Information Source: JCLS

Chemical Class: Alkyl Sulfates

Function: Surfactant - Cleansing Agent

Technical/Other Name:
Sodium C16-20 Alcohols Sulfate

SODIUM C9-22 ALKYL SEC SULFONATE

Definition: Sodium C9-22 Alkyl Sec Sulfonate is the sodium salt of secondary sulfonated C9-22 alkanes.

Chemical Class: Sulfonic Acids

Function: Surfactant - Cleansing Agent

Trade Name:
Mersolat H 95 (Bayer AG)

SODIUM C14-17 ALKYL SEC SULFONATE

CAS No.
68037-49-0

EINECS No.
268-213-3

JPN Translation:
アルキル（C14 - 18）スルホン酸 Na

Definition: Sodium C14-17 Alkyl Sec Sulfonate is the sodium salt of secondary sulfonated C14-17 alkanes.

Chemical Class: Sulfonic Acids

Function: Surfactant - Cleansing Agent

Technical/Other Name:
Sodium C14-17 Alcohol Sulfonate

Trade Names:
Hostapur SAS60 (Clariant)

Hostapur SAS60 (Clariant GmbH, Personal Care)
Hostapur SAS93 (Clariant)
Hostapur SAS93 (Clariant GmbH, Personal Care)
Hostapur SAS30N (Clariant)
Hostapur SAS30N (Clariant GmbH, Personal Care)

SODIUM CAPRATE

CAS No.   **EINECS No.**
1002-62-6   213-688-4

Definition: Sodium Caprate is the sodium salt of Capric Acid (q.v.).

Chemical Class: Soaps

Function: Surfactant - Cleansing Agent

Technical/Other Names:
Capric Acid, Sodium Salt
Decanoic Acid, Sodium Salt
Sodium Decanoate

SODIUM CAPROAMPHOACETATE

Empirical Formula:
$C_{16}H_{32}N_2O_4 \cdot Na$

Definition: Sodium Caproamphoacetate is the amphoteric organic compound that conforms generally to the formula:

$$CH_3(CH_2)_8\overset{\displaystyle O}{\overset{\|}{C}}-NH(CH_2)_2N\overset{\displaystyle CH_2CH_2OH}{\underset{}{|}}CH_2COONa$$

Information Source: TSCA

Chemical Class: Alkylamido Alkylamines

Functions: Hair Conditioning Agent; Surfactant - Cleansing Agent; Surfactant - Foam Booster; Surfactant - Hydrotrope

Technical/Other Names:
Caproamphoacetate
Caproamphoglycinate

SODIUM CAPROAMPHO-HYDROXYPROPYLSULFONATE

Empirical Formula:
$C_{17}H_{36}N_2O_6S \cdot Na$

Definition: Sodium Caproamphohydroxypropylsulfonate is the amphoteric organic compound that conforms generally to the formula:

$$CH_3(CH_2)_8\overset{\displaystyle O}{\overset{\|}{C}}-NH(CH_2)_2N\overset{\displaystyle CH_2CH_2OH}{\underset{}{|}}CH_2\overset{\displaystyle}{\underset{OH}{\underset{|}{C}}HCH_2SO_3Na}$$

Chemical Classes: Alkylamido Alkylamines; Sulfonic Acids

Functions: Hair Conditioning Agent; Surfactant - Cleansing Agent; Surfactant - Foam Booster; Surfactant - Hydrotrope

Technical/Other Names:
Caproamphohydroxypropylsulfonate
Caproamphopropylsulfonate

SODIUM CAPROAMPHOPROPIONATE

Definition: Sodium Caproamphopropionate is the amphoteric organic compound that conforms generally to the formula:

$$CH_3(CH_2)_8\overset{\displaystyle O}{\overset{\|}{C}}-NH(CH_2)_2N\overset{\displaystyle CH_2CH_2OH}{\underset{}{|}}CH_2CH_2COONa$$

Chemical Class: Alkylamido Alkylamines

Functions: Hair Conditioning Agent; Surfactant - Cleansing Agent; Surfactant - Foam Booster; Surfactant - Hydrotrope

SODIUM CAPROYL LACTYLATE

CAS No.: 42566-88-1

Empirical Formula:
$C_{16}H_{28}O_6 \cdot Na$

Definition: Sodium Caproyl Lactylate is the sodium salt of the capryl ester of lactyl lactate. It conforms generally to the formula:

$$CH_3(CH_2)_8\overset{\displaystyle O}{\overset{\|}{C}}-O\overset{\displaystyle}{\underset{CH_3}{\underset{|}{C}}H}-\overset{\displaystyle O}{\overset{\|}{C}}-O\overset{\displaystyle}{\underset{CH_3}{\underset{|}{C}}H}COONa$$

Chemical Classes: Esters; Organic Salts

Function: Surfactant - Emulsifying Agent

Trade Names:
Dermosoft SCL (Straetmans)
Pationic 122A (RITA)

SODIUM CAPRYLATE

CAS No.   **EINECS No.**
1984-06-1   217-850-5

Empirical Formula:
$C_8H_{16}O_2 \cdot Na$

Definition: Sodium Caprylate is the sodium salt of caprylic acid that conforms to the formula:

$$CH_3(CH_2)_6COONa$$

Information Sources: 21CFR172.863, 21CFR175.105, 21CFR175.320,

21CFR176.170, 21CFR176.200, 21CFR176.210, 21CFR177.1200, 21CFR177.2260, 21CFR177.2600, 21CFR177.2800, 21CFR178.3910, CTFA D, MAR, TSCA

Chemical Class: Soaps

Functions: Surfactant - Cleansing Agent; Surfactant - Emulsifying Agent

Technical/Other Names:
Octanoic Acid, Sodium Salt
Sodium n-Octanoate

SODIUM CAPRYLETH-2 CARBOXYLATE

Definition: Sodium Capryleth-2 Carboxylate is the sodium salt of the carboxylic acid derived from ethoxylated caprylyl alcohol. It conforms generally to the formula:

$$CH_3(CH_2)_7(OCH_2CH_2)_nOCH_2COONa$$

where n has an average value of 1.

Chemical Class: Organic Salts

Function: Surfactant - Cleansing Agent

Technical/Other Names:
PEG-2 Capryl Ether Carboxylic Acid, Sodium Salt
Polyethylene Glycol 100 Capryl Ether Carboxylic Acid, Sodium Salt
Polyoxyethylene (2) Capryl Ether Carboxylic Acid, Sodium Salt
Sodium Polyethylene Glycol 100 Capryl Ether Carboxylate
Sodium Polyoxyethylene (2) Capryl Ether Carboxylate

Trade Name Mixture:
Akypo OCD 10 NV (Kao GmbH)

SODIUM CAPRYLETH-9 CARBOXYLATE

CAS No.: 126646-15-9

Empirical Formula:
$C_{26}H_{52}O_{11} \cdot Na$

Definition: Sodium Capryleth-9 Carboxylate is the sodium salt of the carboxylic acid derived from ethoxylated caprylyl alcohol that conforms generally to the formula:

$$CH_3(CH_2)_7(OCH_2CH_2)_nOCH_2COONa$$

where n has an average value of 8.

Chemical Class: Organic Salts

Functions: Surfactant - Cleansing Agent; Surfactant - Emulsifying Agent

Technical/Other Names:
3,6,9,12,15,18,21,24,27-Nonaoxapentatricontanoic Acid, Sodium Salt

---

PEG-9 Capryl Ether Carboxylic Acid,
Sodium Salt
Polyethylene Glycol 450 Capryl Ether
Carboxylic Acid, Sodium Salt
Polyoxyethylene (9) Capryl Ether
Carboxylic Acid, Sodium Salt
Sodium 3,6,9,12,15,18,21,24,27-
Nonaoxapentatricontanoate
Sodium Polyethylene Glycol 450 Capryl
Ether Carboxylate
Sodium Polyoxyethylene (9) Capryl Ether
Carboxylate

Trade Name Mixture:
Akypo LF 4 N (Kao GmbH)

SODIUM CAPRYLOAMPHOACETATE

Empirical Formula:
$C_{14}H_{28}N_2O_4 \cdot Na$

Definition: Sodium Capryloamphoacetate is
the amphoteric organic compound that
conforms generally to the formula:

$$CH_3(CH_2)_6\overset{\overset{\displaystyle O}{\|}}{C}-NH(CH_2)_2N\overset{\overset{\displaystyle CH_2CH_2OH}{|}}{C}H_2COONa$$

Information Source: TSCA

Chemical Class: Alkylamido Alkylamines

Functions: Hair Conditioning Agent;
Surfactant - Cleansing Agent; Surfactant -
Foam Booster; Surfactant - Hydrotrope

Technical/Other Names:
Capryloamphoacetate
Capryloamphoglycinate

Trade Name:
Zoharteric LF-8 (Zohar)

SODIUM CAPRYLOAMPHOHYDROXY-PROPYLSULFONATE

Empirical Formula:
$C_{15}H_{32}N_2O_6S \cdot Na$

Definition: Sodium Capryloamphohydroxypropylsulfonate is the amphoteric organic
compound that conforms generally to the
formula:

$$CH_3(CH_2)_6\overset{\overset{\displaystyle O}{\|}}{C}-NH(CH_2)_2N\overset{\overset{\displaystyle CH_2CH_2OH}{|}}{C}H_2\underset{\underset{\displaystyle OH}{|}}{C}HCH_2SO_3Na$$

Chemical Classes: Alkylamido Alkylamines;
Sulfonic Acids

Functions: Hair Conditioning Agent;
Surfactant - Cleansing Agent; Surfactant -
Foam Booster; Surfactant - Hydrotrope

Technical/Other Names:
Capryloamphohydroxypropylsulfonate
Capryloamphopropylsulfonate

Trade Name:
Mackam JS (McIntyre)

SODIUM CAPRYLOAMPHOPROPIONATE

Empirical Formula:
$C_{15}H_{30}N_2O_4 \cdot Na$

Definition: Sodium Capryloamphopropionate is the amphoteric organic
compound that conforms generally to the
formula:

$$CH_3(CH_2)_6\overset{\overset{\displaystyle O}{\|}}{C}-NH(CH_2)_2N\overset{\overset{\displaystyle CH_2CH_2OH}{|}}{C}H_2CH_2COONa$$

Chemical Class: Alkylamido Alkylamines

Functions: Hair Conditioning Agent;
Surfactant - Cleansing Agent; Surfactant -
Foam Booster

Technical/Other Name:
Capryloamphopropionate

Trade Name:
Monateric CyNa-50 (Uniqema)

SODIUM CAPRYLOYL GLUTAMATE

Empirical Formula:
$C_{13}H_{23}NO_5 \cdot Na$

Definition: Sodium Capryloyl Glutamate is
the substituted amino acid that conforms to
the formula:

$$CH_3(CH_2)_6\overset{\overset{\displaystyle O}{\|}}{C}-NH\underset{\underset{\displaystyle COOH}{|}}{C}H(CH_2)_2COONa$$

Chemical Class: Amino Acids

Functions: Deodorant Agent; Surfactant -
Cleansing Agent

Trade Name:
Protelan AG 80 (Zschimmer & Schwarz
Italiana)

SODIUM CAPRYLOYL HYDROLYZED WHEAT PROTEIN

Definition: Sodium Capryloyl Hydrolyzed
Wheat Protein is the sodium salt of the
condensation product of caprylic acid
chloride and Hydrolyzed Wheat Protein (q.v.).

Chemical Class: Protein Derivatives

Functions: Hair Conditioning Agent; Skin-
Conditioning Agent - Miscellaneous;
Surfactant - Cleansing Agent

Trade Name:
Protelan W 8 (Zschimmer & Schwarz
Italiana)

SODIUM CAPRYLYL SULFONATE

CAS No.	EINECS No.
5324-84-5	226-195-4

Empirical Formula:
$C_8H_{18}O_3S \cdot Na$

Definition: Sodium Caprylyl Sulfonate is the
sodium salt of caprylyl sulfonate that
conforms to the formula:

$$CH_3(CH_2)_7SO_3Na$$

Chemical Class: Sulfonic Acids

Function: Surfactant - Cleansing Agent

Technical/Other Names:
1-Octanesulfonic Acid, Sodium Salt
Sodium n-Octylsulfonate

Trade Name:
BIO-TERGE PAS-8S (Stepan)

SODIUM CARBOMER

CAS No.: 73298-57-4

Definition: Sodium Carbomer is the sodium
salt of Carbomer (q.v.).

Chemical Classes: Organic Salts; Synthetic
Polymers

Functions: Emulsion Stabilizer; Film
Former; Viscosity Increasing Agent -
Aqueous

Technical/Other Name:
Carbomer, Sodium Salt

Trade Names:
PNC 400 (3V Inc.)
PNC 410 (3V Inc.)
PNC 430 (3V Inc.)

Trade Name Mixture:
Elespher Dermosaccharides HC (Laboratoires Serobiologiques)

SODIUM CARBONATE

CAS No.	EINECS No.
497-19-8	207-838-8

JPN Translation:
炭酸 Na

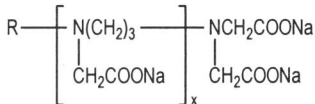

Empirical Formula:
CH$_2$O$_3$ • 2Na

Definition: Sodium Carbonate is the inorganic salt that conforms to the formula:

Na$_2$CO$_3$

Information Sources: ARG, AUS, BPC, BRA, 21CFR173.310, 21CFR184.1742, CIR: [S] JACT-6(1)1987, CTFA D, DA, DDR, EGY, FCC, FIN, HP, HUN, IND, ITA, JCLS, JSCI, MAR, MI-13(8668), NF XIX, OTC-I-AA, PF, PN, POR, ROM, TSCA, USAN, USD, YUG

Chemical Class: Inorganic Salts

Function: pH Adjuster

Reported Product Categories: Hair Dyes and Colors (All Types Requiring Caution Statements and Patch Tests); Bath Oils, Tablets, and Salts; Moisturizing Preparations; Face Powders; Eye Makeup Preparations, Misc.; Lipsticks

Technical/Other Names:
Bisodium Carbonate
Carbonic Acid, Disodium Salt
Soda Ash

Trade Name:
Kristall-Soda Sodium Carbonate, Decahydrate (B. Mueller KG)

SODIUM CARBONATE PEROXIDE

CAS No.	EINECS No.
15630-89-4	239-707-6

Empirical Formula:
CH$_2$O$_3$ • $^3/_2$H$_2$O$_2$ • 2Na

Definition: Sodium Carbonate Peroxide is the inorganic salt that conforms to the formula:

2Na$_2$CO$_3$ • 3H$_2$O$_2$

Information Sources: CTFA D, EEC(III/1-12), TSCA

Chemical Class: Inorganic Salts

Function: Oxidizing Agent

Technical/Other Names:
Carbonic Acid, Disodium Salt, Compd. with Hydrogen Peroxide (2:3)
Peroxy Sodium Carbonate
Sodium Percarbonate

SODIUM CARBOXYDECYL PEG-8 DIMETHICONE

Definition: Sodium Carboxydecyl PEG-8 Dimethicone is the silicone polymer that conforms generally to the formula:

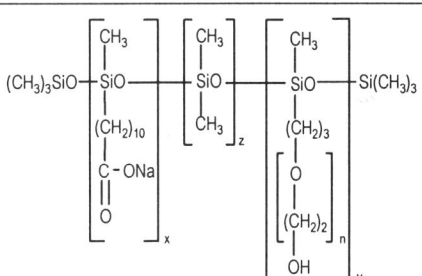

Chemical Classes: Organic Salts; Siloxanes and Silanes

Functions: Skin-Conditioning Agent - Miscellaneous; Surfactant - Emulsifying Agent

Trade Name:
Silube CU-1 (Siltech)

SODIUM CARBOXYETHYL TALLOW POLYPROPYLAMINE

Definition: Sodium Carboxyethyl Tallow Polypropylamine is the organic compound that conforms to the formula:

R—[—N(CH$_2$)$_3$ | (CH$_2$)$_2$COONa—N(CH$_2$)$_2$COONa | (CH$_2$)$_2$COONa—]$_x$

where R represent the alkyl groups derived from Tallow (q.v.).

Chemical Classes: Alkyl-Substituted Amino Acids; Amines; Organic Salts

Function: Antistatic Agent

SODIUM CARBOXYMETHYL CHITIN

Definition: Sodium Carboxymethyl Chitin is the sodium salt of a Carboxymethyl Chitin (q.v.).

Chemical Classes: Biological Polymers and their Derivatives; Gums, Hydrophilic Colloids and Derivatives

Functions: Film Former; Viscosity Increasing Agent - Aqueous

Technical/Other Name:
Chitin, Carboxymethyl, Sodium Salt

Trade Name:
Atomergic Water Soluble Chitin (Atomergic Chemetals)

SODIUM CARBOXYMETHYL COCOPOLY-PROPYLAMINE

Definition: Sodium Carboxymethyl Cocopolypropylamine is the organic compound that conforms generally to the formula:

R—[—N(CH$_2$)$_3$ | CH$_2$COONa—NCH$_2$COONa | CH$_2$COONa—]$_x$

where R represents alkyl groups derived from coconut oil.

Chemical Classes: Alkyl-Substituted Amino Acids; Amines; Organic Salts

Function: Antistatic Agent

Trade Name:
Ampholak 7CX/C (Akzo Nobel Surface AB)

SODIUM CARBOXYMETHYL DEXTRAN

CAS No.: 39422-83-8

JPN Translation:
カルボキシメチルデキストラン Na

Definition: Sodium Carboxymethyl Dextran is the sodium salt of a Carboxymethyl Dextran (q.v.). It conforms generally to the formula:

Information Sources: JCIC, JCLS

Chemical Class: Gums, Hydrophilic Colloids and Derivatives

Functions: Binder; Emulsion Stabilizer; Viscosity Increasing Agent - Aqueous

Trade Name:
CMD (Meito Sangyo)

SODIUM CARBOXYMETHYL BETA-GLUCAN

CAS No.: 9050-93-5

Definition: Sodium Carboxymethyl Beta-Glucan is the sodium salt of a carboxymethyl ether of Beta-Glucan (q.v.).

Chemical Classes: Carbohydrates; Gums, Hydrophilic Colloids and Derivatives

Functions: Binder; Viscosity Increasing Agent - Aqueous

Technical/Other Name:
β-D-Glucan, (1.fwdarw. 3)-, Carboxymethyl Ether, Sodium Salt

Trade Names:
Carboxymethylglucan (CPN)
CM-Glucan (Mibelle AG)
Cromoist CM Glucan (Croda, Inc.)
GluCare S (Degussa Care Specialties)
GluCare S 2% (Degussa Care Specialties)

Trade Name Mixture:
Cosmocair CG (Degussa Care Specialties)

SODIUM CARBOXYMETHYL LAURYL GLUCOSIDE

Definition: Sodium Carboxymethyl Lauryl Glucoside is the sodium carboxymethyl ether of Lauryl Glucoside (q.v.).

Chemical Classes: Organic Salts; Polyols

Function: Surfactant - Cleansing Agent

Trade Name:
Plantapon LGC Sorb (Cognis Deutschland)

SODIUM CARBOXYMETHYL OLEYL POLYPROPYLAMINE

Empirical Formula:
$C_{37}H_{68}N_4O_{10} \cdot 5Na$

Definition: Sodium Carboxymethyl Oleyl Polypropylamine is the organic compound that conforms generally to the formula:

$$R-\left[N(CH_2)_3 \begin{matrix} -NCH_2COONa \\ | \\ CH_2COONa \end{matrix} \begin{matrix} \\ \\ CH_2COONa \end{matrix}\right]_x$$

where x has an average value of 3 and R represents the oleyl moiety.

Chemical Classes: Alkyl-Substituted Amino Acids; Organic Salts

Function: Antistatic Agent

Trade Name:
Ampholak XO7/C (Akzo Nobel Rheology)

SODIUM CARBOXYMETHYL STARCH

CAS No.: 9063-38-1

JPN Translation:
デンプングリコール酸 Na

Definition: Sodium Carboxymethyl Starch is the sodium salt of a carboxymethyl derivative of starch.

Information Source: JCLS

Chemical Classes: Gums, Hydrophilic Colloids and Derivatives; Organic Salts

Functions: Binder; Emulsion Stabilizer; Film Former; Viscosity Increasing Agent - Aqueous

Technical/Other Names:
Sodium Starch Glycolate
Starch, Carboxymethyl Ether, Sodium Salt

Trade Names:
Naturally Thik CS (International Additive)
Primojel (Avebe)
Ultramyl (Crinos)

SODIUM CARBOXYMETHYL TALLOW POLYPROPYLAMINE

Definition: Sodium Carboxymethyl Tallow Polypropylamine is the organic compound that conforms generally to the formula:

$$R-\left[N(CH_2)_3 \begin{matrix} -NCH_2COONa \\ | \\ CH_2COONa \end{matrix} \begin{matrix} \\ \\ CH_2COONa \end{matrix}\right]_x$$

where R represents the alkyl groups derived from Tallow (q.v.) and n has a value from 1 to 4.

Chemical Classes: Alkyl-Substituted Amino Acids; Amines; Organic Salts

Function: Antistatic Agent

Trade Names:
Ampholak 7TX/C (Akzo Nobel Rheology)
Ampholak 7TX-T/C (Akzo Nobel Rheology)

SODIUM CARRAGEENAN

CAS Nos.: 9061-82-9; 60616-95-7

Definition: Sodium Carrageenan is the sodium salt of Carrageenan (q.v.)

Information Sources: 21CFR136.110, 21CFR136.115, 21CFR136.130, 21CFR136.160, 21CFR136.180, 21CFR139.121, 21CFR139.122, 21CFR150.141, 21CFR150.161, 21CFR172.623, 21CFR172.626, 21CFR176.170, CTFA D, TSCA

Chemical Class: Gums, Hydrophilic Colloids and Derivatives

Functions: Binder; Emulsion Stabilizer; Film Former; Viscosity Increasing Agent - Aqueous

Reported Product Category: Dentifrices (Aerosol, Liquid, Pastes and Powders)

Technical/Other Names:
Carrageenan, Sodium Salt
Sodium Carrageenate

SODIUM CASEINATE

CAS No.: 9005-46-3

JPN Translation:
カゼイン Na

Definition: Sodium Caseinate is the sodium salt of Casein (q.v.).

Information Sources: 21CFR135.110, 21CFR135.140, 21CFR166.110, 21CFR182.1748, JCIC, JCLS, JSQI, NED, TSCA

Chemical Class: Protein Derivatives

Functions: Hair Conditioning Agent; Skin-Conditioning Agent - Miscellaneous

Technical/Other Name:
Casein, Sodium Salt

Trade Name:
Alanate 110 (New Zealand)

Trade Name Mixtures:
AEC Coconut Cream Powder (A & E Connock)
Lactofil Foam (Gattefosse s.a.)

SODIUM CASTORATE

CAS Nos.	EINECS Nos.
8013-06-7	232-390-5
96690-37-8	306-231-6

JPN Translation:
ヒマシ脂肪酸 Na

Definition: Sodium Castorate is the sodium salt of the fatty acids derived from Ricinus Communis (Castor) Oil (q.v.).

Information Sources: 21CFR176.200, 21CFR176.210, 21CFR177.2600, 21CFR177.2800, 21CFR178.3910, JCIC, JCLS, JSQI, TSCA

Chemical Class: Soaps

Functions: Surfactant - Cleansing Agent; Surfactant - Emulsifying Agent

Reported Product Categories: Bath Oils, Tablets, and Salts; Cleansing Products (Cold Creams, Cleansing Lotions, Liquids and Pads)

Technical/Other Names:
Castor Oil, Sodium Salt
Sodium Castor Oil Fatty Acid Solution (30%)

SODIUM CELLULOSE SULFATE

CAS No.: 9005-22-5

Definition: Sodium Cellulose Sulfate is the sodium salt of sulfated Cellulose (q.v.).

Information Source: TSCA

Chemical Classes: Gums, Hydrophilic Colloids and Derivatives; Sulfuric Acid Esters

Functions: Binder; Emulsion Stabilizer; Viscosity Increasing Agent - Aqueous

Technical/Other Name:
Cellulose, Sulfate, Sodium Salt

SODIUM CETEARYL SULFATE

CAS No.: 59186-41-3

JPN Translation:
セテアリル硫酸 Na

Definition: Sodium Cetearyl Sulfate is the sodium salt of a mixture of cetyl and stearyl sulfate. It conforms generally to the formula:

$$CH_3(CH_2)_nCH_2OSO_3Na$$

where n has a value of 14 and 16.

Information Sources: AUS, BEL, CIR: [S] JACT-11(1)1992, CTFA D, JCIC, JCLS, JSQI, MAR

Chemical Class: Alkyl Sulfates

Function: Surfactant - Cleansing Agent

Reported Product Categories: Moisturizing Preparations; Bath Oils, Tablets, and Salts; Cleansing Products (Cold Creams, Cleansing Lotions, Liquids and Pads); Hair Dyes and Colors (All Types Requiring Caution Statements and Patch Tests); Bath Preparations, Misc.; Body and Hand Preparations (Excluding Shaving Preparations); Paste Masks (Mud Packs); Skin Care Preparations, Misc.; Night Skin Care Preparations; Suntan Gels, Creams, and Liquids

Technical/Other Names:
Sodium Cetostearyl Sulfate
Sodium Cetyl/Stearyl Sulfate

Trade Names:
Lanette E (Cognis Care Chemicals/NJ)
Lanette E (Cognis Care Chemicals/PA)
Lanette E (Cognis Deutschland)
Sabosol CSSP (Sabo)

Trade Name Mixtures:
Cutina LE (Cognis Deutschland)
Emulgade F (Cognis Care Chemicals/NJ)
Emulgade F (Cognis Care Chemicals/PA)
Emulgade F (Cognis Deutschland)
Galenol 1618 CS (Sasol GmbH - Hamburg)
Galenol 1618 KS (Sasol GmbH - Hamburg)
Lanette N (Cognis Care Chemicals/NJ)
Lanette N (Cognis Care Chemicals/PA)
Lanette N (Cognis Deutschland)
Lanette SX (Cognis Care Chemicals/NJ)
Lanette SX (Cognis Care Chemicals/PA)
Lanette SX (Cognis Deutschland)
Sabowax N (Sabo)
Sabowax NH (Sabo)
Unimulgade-F (Universal Preserv-A-Chem)

SODIUM CETETH-13 CARBOXYLATE

CAS No.: 33939-65-0 (Generic)

Definition: Sodium Ceteth-13 Carboxylate is the sodium salt of the carboxylic acid derived from ethoxylated cetyl alcohol. It conforms generally to the formula:

$$CH_3(CH_2)_{15}(OCH_2CH_2)_nOCH_2COONa$$

where n has an average value of 12.

Chemical Class: Organic Salts

Functions: Surfactant - Cleansing Agent; Surfactant - Emulsifying Agent

Technical/Other Names:
Ceteth-13 Carboxylic Acid, Sodium Salt
PEG-13 Cetyl Ether Carboxylic Acid, Sodium Salt
Polyethylene Glycol (13) Cetyl Ether Carboxylic Acid, Sodium Salt
Polyoxyethylene (13) Cetyl Ether Carboxylic Acid, Sodium Salt

Trade Names:
Sandopan KST (Clariant)
Sandopan KST (Clariant GmbH, Personal Care)

SODIUM CETETH-4 PHOSPHATE

Definition: Sodium Ceteth-4 Phosphate is the sodium salt of esters of Ceteth-4 (q.v.) and Phosphoric Acid (q.v.).

Information Source: JCLS

Chemical Class: Esters

Function: Not Reported

SODIUM CETYL SULFATE

CAS No. 1120-01-0   **EINECS No.** 214-292-4

JPN Translation:
セチル硫酸 Na

Empirical Formula:
$C_{16}H_{34}O_4S \cdot Na$

Definition: Sodium Cetyl Sulfate is the sodium salt of cetyl sulfate that conforms to the formula:

$$CH_3(CH_2)_{15}OSO_3Na$$

Information Sources: 21CFR177.1200, 21CFR177.1210, 21CFR177.2800, CTFA D, CZE, JCLS, JSCI, ROM, TSCA

Chemical Class: Alkyl Sulfates

Function: Surfactant - Cleansing Agent

Reported Product Categories: Bath Oils, Tablets, and Salts; Personal Cleanliness Products, Misc.

Technical/Other Names:
1-Hexadecanol, Hydrogen Sulfate, Sodium Salt
Sodium Hexadecyl Sulfate
Sodium Palmityl Sulfate

Trade Name:
Nikkol SCS (Nikko)

Trade Name Mixtures:
Standapol CS Paste (Cognis Care Chemicals/NJ)
Standapol CS Paste (Cognis Care Chemicals/PA)
Texapon CS Paste (Cognis Deutschland)
Unipol CS-50 (Universal Preserv-A-Chem)
Unipol CS PASTE (Universal Preserv-A-Chem)

SODIUM CHITOSAN METHYLENE PHOSPHONATE

Definition: Sodium Chitosan Methylene Phosphonate is the product obtained by the phosphonomethylation of Chitosan (q.v.) in the presence of sodium hydroxide.

Chemical Classes: Biological Polymers and their Derivatives; Phosphorus Compounds

Function: Chelating Agent

Trade Name:
P-Chitosan (Ciba Specialty Chemicals)

SODIUM CHLORATE

CAS No. 7775-09-9   **EINECS No.** 231-887-4

Empirical Formula:
$HClO_3 \cdot Na$

Definition: Sodium Chlorate is the inorganic salt that conforms to the formula:

$$NaClO_3$$

Information Sources: EEC(III/1-6), JAN, MI-13(8670), TSCA, USAN, USP XXIV

Chemical Class: Inorganic Salts

Function: Oxidizing Agent

Technical/Other Name:
Chloric Acid, Sodium Salt

Trade Name:
Stabilized Chlorine Dioxide (Zinnellé)

SODIUM CHLORIDE

CAS No. 7647-14-5   **EINECS No.** 231-598-3

JPN Translation:
塩化 Na

Empirical Formula:
ClNa

Definition: Sodium Chloride is the inorganic salt that conforms to the formula:

NaCl

Information Sources: ARG, AUS, BP, BPC, 21CFR73.85, 21CFR74.302, 21CFR74.3045, 21CFR182.1, 21CFR182.70, 21CFR182.90, CZE, DA, DDR, EGY, EP, FCC, FIN, HP, HUN, IND, ITA, JAN, JCLS, JSCI, MAR, MEX, MI-13 (8671), OTC-I-OP, PF, PN, POR, ROM, TSCA, USAN, USD, USP XXIV, USSR, WHO, YUG

Chemical Class: Inorganic Salts

Functions: Flavoring Agent; Oral Care Agent; Viscosity Increasing Agent - Aqueous

Reported Product Categories: Shampoos (Non-coloring); Bath Soaps and Detergents; Bubble Baths; Bath Oils, Tablets, and Salts; Cleansing Products (Cold Creams, Cleansing Lotions, Liquids and Pads); Personal Cleanliness Products, Misc.; Hair Dyes and Colors (All Types Requiring Caution Statements and Patch Tests); Eye Makeup Removers; Hair Conditioners; Skin Care Preparations, Misc.; Body and Hand Preparations (Excluding Shaving Preparations); Moisturizing Preparations; Bath Preparations, Misc.; Hair Preparations (Non-coloring), Misc.; Baby Products, Misc.; Baby Shampoos; Bath Capsules; Foundations; Tonics, Dressings, and Other Hair Grooming Aids; Deodorants (Underarm); Face and Neck Preparations (Excluding Shaving Preparations); Skin Fresheners; Makeup Bases; Paste Masks (Mud Packs); Permanent Waves; Eye Lotions; Eyeliners; Fragrance Preparations, Misc.; Night Skin Care Preparations; Eyebrow Pencils; Hair Rinses (Non-coloring); Hair Shampoos (Coloring); Hair Sprays (Aerosol Fixatives); Indoor Tanning Preparations; Manicuring Preparations, Misc.; Suntan Gels, Creams, and Liquids

Technical/Other Names:
Rock Salt
Salt

Trade Name:
Salt (Cargill)

Trade Name Mixtures:
Aminogluten MG (Croda Chemicals)
Anti-Rheumatic Phytospa NaCl (Alban Muller)
Cerealmilk Premium (Chemyunion)
Complexe Hydroxy-Salicylique Color (Coletica SA)
Crotein HKP Powder (Croda Chemicals)
Crotein HKP Powder (Croda, Inc.)
Dacriosalt (Vevy)
Elastase Inhibitor-3 (Arval)
Essential Vital Elements (Dipta)
Essential Vital Elements - S (Dipta)
Extraliss NAT (Coletica SA)
Hydeoviton 5,5 N 2/059359 (Symrise)
Hydroviton 2/059353 (Symrise)
Hydroviton 24 (2/059351) (Symrise)
Hydroxylide QNST (Coletica SA)
Keramino SD (Arch Personal Care Products)
Natunola Flax Extract 1621 (Natunola)
Relaxing Phytospa NaCl (Alban Muller)
Saltworks brine Bad Rothenfelde (Ultra-Pharm)
Seanamine BD (Laboratoires Sero-biologiques)
Slimming/Tonic Phytospa NaCl (Alban Muller)
Stimulating Phytospa NaCl (Alban Muller)

SODIUM CHLORITE

CAS No. 7758-19-2

EINECS No. 231-836-6

Empirical Formula:
$ClO_2 \cdot Na$

Definition: Sodium Chlorite is the inorganic compound that conforms to the formula:

$NaClO_2$

Information Sources: 21CFR173.325, 21CFR186.1750, MI-13(8672), TSCA

Chemical Class: Inorganics

Functions: Oral Care Agent; Oxidizing Agent

Technical/Other Name:
Chlorous Acid, Sodium Salt

Trade Names:
Closys II (Rowpar)
Stabilized Chlorine Dioxide (Bio-Cide)

SODIUM p-CHLORO-m-CRESOL

CAS No. 15733-22-9

EINECS No. 239-825-8

Empirical Formula:
$C_7H_7ClO \cdot Na$

Definition: Sodium p-Chloro-m-Cresol is the organic compound that conforms to the formula:

Information Source: CIR: [SQ]

Chemical Classes: Halogen Compounds; Organic Salts; Phenols

Function: Preservative

Technical/Other Names:
p-Chloro-m-Cresol, Sodium Salt
3-Methyl-4-Chlorophenol, Sodium Salt
Sodium 3-Methyl-4-Chlorophenolate

Trade Name:
Preventol CMK Sodium Salt (Bayer AG)

SODIUM CHOLESTERYL SULFATE

CAS No.: 2864-50-8

Empirical Formula:
$C_{27}H_{46}O_4S \cdot Na$

Definition: Sodium Cholesteryl Sulfate is the sodium salt of the sulfuric acid ester of Cholesterol (q.v.).

Information Source: JCLS

Chemical Classes: Sterols; Sulfuric Acid Esters

Function: Skin-Conditioning Agent - Miscellaneous

Technical/Other Name:
Cholest-5-en-3-ol, (3β)-, Hydrogen Sulfate, Sodium Salt

Trade Name:
Sodium Cholesterol Sulfate (SPCI)

SODIUM CHONDROITIN SULFATE

CAS Nos. 9007-28-7 9082-07-9

EINECS No. 232-696-9

JPN Translation:
コンドロイチン硫酸 Na

Definition: Sodium Chondroitin Sulfate is a derivative of a natural mucopolysaccharide.

Information Sources: 21CFR701.3, JCLS, JSCI, MI-13(2235), TSCA

Chemical Classes: Biological Polymers and their Derivatives; Carbohydrates; Sulfuric Acid Esters

Functions: Hair Conditioning Agent; Skin-Conditioning Agent - Miscellaneous

Reported Product Categories: Moisturizing Preparations; Skin Care Preparations, Misc.; Bath Preparations, Misc.; Paste Masks (Mud Packs); Body and Hand Preparations (Excluding Shaving Preparations)

Trade Names:
AEC Chondroitin Sulfate (A & E Connock)

Chondroitin Sulfate (Bioiberica)
Chondroitin Sulfate (Freeman)

Trade Name Mixtures:
AE-957 (Atrium)
Amino Acid Microspheres (Coletica SA)
Aquafix Marin (Coletica SA)
Azelosome (Coletica SA)
Capigen (Sederma)
Catiosphere Color (Coletica SA)
Clarisome (Coletica SA)
Color Marine Filling Spheres (Coletica SA)
Color Marine Vitamine CPMG Spheres (Coletica SA)
Color Retinol Thalasphere (Coletica SA)
Cromoist CS (Croda Chemicals)
Cromoist CS Powder (Croda Chemicals)
Deperoxidium Marin (Coletica SA)
Elastase Inhibitor-3 (Arval)
Glycacid CMMA (Coletica SA)
Heparinoid (Bioiberica)
Hydromarine (Sederma)
Lipoderma CS (Lipo)
Molecularsource LPC (C.I.T.)
Repairzyme (Coletica SA)
Retinal Microcapsule (Coletica SA)
Retinol Microcapsule (Coletica SA)
Salicysome (Coletica SA)
Slimisome Caffeine (Coletica SA)
Slimisome Carnitine (Coletica SA)
Slimisome Esculoside (Coletica SA)
Sveltine (Coletica SA)
Thalasphere 7-DHC (Coletica SA)
Thioglycans (Esperis)
Unichondrin ATP (Induchem)
Ursolisome (Coletica SA)

SODIUM C8-16 ISOALKYLSUCCINYL LACTOGLOBULIN SULFONATE

Definition: Sodium C8-16 Isoalkylsuccinyl Lactoglobulin Sulfonate is the sodium salt of the sulfonated product of lactoglobulin reacted with an alkyl substituted succinic anhydride.

Chemical Classes: Protein Derivatives; Sulfonic Acids

Functions: Hair Conditioning Agent; Skin-Conditioning Agent - Miscellaneous

Reported Product Category: Face Powders

Trade Names:
AC Biopolymer Lactoglobulin (V1) (Active Concepts)
AC Biopolymer Lactoglobulin (V2) (Active Concepts)
Biopol EA (Arch Personal Care Products)
Bio-Pol OE (Arch Personal Care Products)
Biopol OE/SD (Arch Personal Care Products)

Trade Name Mixtures:
Chronosphere Biopol OE (Arch Personal Care Products)

COULOURMAT WOE BLACK (Creations Couleurs)
COULOURMAT WOE BLUE (Creations Couleurs)
COULOURMAT WOE GREEN (Creations Couleurs)
COULOURMAT WOE GREEN LEAF (Creations Couleurs)
COULOURMAT WOE RED (Creations Couleurs)
COULOURMAT WOE WHITE (Creations Couleurs)
COULOURMAT WOE YELLOW (Creations Couleurs)
HI - 13DC Biopolymer (Hildebrand)
Micapoly WOE Black (Creations Couleurs)
Micapoly WOE Blue (Creations Couleurs)
Micapoly WOE Brown (Creations Couleurs)
Micapoly WOE Geranium (Creations Couleurs)
Micapoly WOE Green 70% (Creations Couleurs)
Micapoly WOE Manganese Violet (Creations Couleurs)
Micapoly WOE Melon (Creations Couleurs)
Micapoly WOE Pink (Creations Couleurs)
Micapoly WOE Red (Creations Couleurs)
Micapoly WOE Violet (Creations Couleurs)
Micapoly WOE White (Creations Couleurs)
Micapoly WOE Yellow (Creations Couleurs)

SODIUM CITRATE

CAS Nos.
68-04-2
6132-04-3

EINECS No.
200-675-3

JPN Translation:
クエン酸 Na

Empirical Formula:
$C_6H_8O_7 \cdot 3Na$

Definition: Sodium Citrate is the sodium salt of citric acid that conforms to the formula:

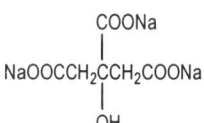

Information Sources: ARG, AUS, BP, BPC, BRA, 21CFR131.111, 21CFR131.160, 21CFR133.144, 21CFR133.169, 21CFR133.173, 21CFR133.179, 21CFR150.141, 21CFR150.161, 21CFR175.300, 21CFR181.22, 21CFR181.29, CZE, DA, DDR, EGY, FCC, FIN, HUN, IND, ITA, JAN, JCLS, JSCI, MAR, MEX, MI-13(8675), PF, PN, POR, RIFM, ROM, TSCA, USAN, USD, USP XXIV, USSR, WHO, YUG

Chemical Class: Organic Salts

Functions: Buffering Agent; Chelating Agent; Fragrance Ingredient; pH Adjuster

Reported Product Categories: Bath Soaps and Detergents; Bath Oils, Tablets, and Salts; Moisturizing Preparations; Cleansing Products (Cold Creams, Cleansing Lotions, Liquids and Pads); Skin Fresheners; Skin Care Preparations, Misc.; Bath Capsules; Permanent Waves; Bubble Baths; Bath Preparations, Misc.; Hair Conditioners; Paste Masks (Mud Packs); Body and Hand Preparations (Excluding Shaving Preparations); Face and Neck Preparations (Excluding Shaving Preparations); Hair Dyes and Colors (All Types Requiring Caution Statements and Patch Tests); Fragrance Preparations, Misc.; Eyeliners; Indoor Tanning Preparations; Makeup Preparations (Not eye), Misc.; Shampoos (Non-coloring)

Technical/Other Names:
Citric Acid, Trisodium Salt
2-Hydroxy-1,2,3-Propanetricarboxylic Acid, Trisodium Salt
1,2,3-Propanetricarboxylic Acid, 2-Hydroxy-, Trisodium Salt
Sodium citrate (RIFM)
Trisodium Citrate

Trade Names:
Sodium Citrate USP Fine Granular (Roche)
Trisodium Citrate (Jungbunzlauer)
Trisodium Citrate Dihydrate (Jungbunzlauer)

Trade Name Mixtures:
Cellryel (Bio Dell)
Collagen S (PF) (Nitta Gelatin)
Collagen S-03 (PF) (Nitta Gelatin)
Collagen S-06 (PF) (Nitta Gelatin)
Connexan LS (Laboratoires Sero-biologiques)
Dermawhite HS (Laboratoires Sero-biologiques)
Dermawhite NF (Laboratoires Sero-biologiques)
Elesponge AHA LS 8911 B (Laboratoires Serobiologiques)
Follicusan (CLR)
Lipobrite BG (Lipo)
Lipobrite PG (Lipo)
Nautigene (Coletica SA)
NIKKOL Aquasome EC-30 (Nikko)
Rovisome AHA - Citric Acid (Rovi)
Uninontan U-34 (Induchem)
Vegeles AHA 8702 (Laboratoires Sero-biologiques)
Vegeles AHA LS 8763 (Laboratoires Sero-biologiques)

SODIUM COCAMIDOPROPYL PG-DIMONIUM CHLORIDE PHOSPHATE

Definition: Sodium Cocamidopropyl PG-Dimonium Chloride Phosphate is the quaternary ammonium salt that conforms generally to the formula:

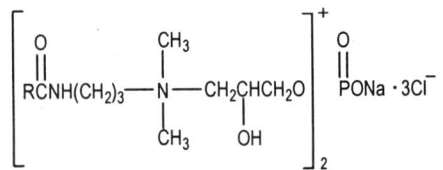

where RCO- represents the coconut acid moiety.

Chemical Classes: Phosphorus Compounds; Quaternary Ammonium Compounds

Functions: Hair Conditioning Agent; Surfactant - Cleansing Agent; Surfactant - Foam Booster

Trade Name:
Phoenotaine C-35 (Phoenix)

SODIUM COCAMINOPROPIONATE

CAS Nos.	EINECS No.
8033-69-0	
12676-37-8	
68608-68-4	271-795-1

Definition: Sodium Cocaminopropionate is the sodium salt of a substituted propionic acid. It conforms generally to the formula:

$$RNHCH_2CH_2COONa$$

where R represents the coconut fatty radical.

Information Sources: CTFA D, TSCA

Chemical Class: Alkyl-Substituted Amino Acids

Functions: Hair Conditioning Agent; Surfactant - Cleansing Agent

Technical/Other Names:
β-Alanine, N-Coco Alkyl Derivs., Sodium Salts
Sodium N-Coco-β-Aminopropionate

SODIUM COCETH SULFATE

Definition: Sodium Coceth Sulfate is the sodium salt of the sulfate ester of the polyethylene glycol ether of coconut alcohol that conforms generally to the formula:

$$R(OCH_2CH_2)_nOSO_3Na$$

where R represents the alkyl groups derived from coconut oil and n has an average value of 1 to 4.

Chemical Class: Alkyl Ether Sulfates

Functions: Surfactant - Cleansing Agent; Surfactant - Emulsifying Agent

Technical/Other Names:
Sodium Polyethylene Glycol (1-4) Coconut Ether Sulfate

Sodium Polyoxyethylene (1-4) Coconut Ether Sulfate

Trade Names:
Zetesol 270/C (Zschimmer & Schwarz Italiana)
Zetesol LES 2/C (Zschimmer & Schwarz Italiana)

Trade Name Mixtures:
Base SP 100 (LCW)
Oronal LCG (SEPPIC)

SODIUM COCETH-30 SULFATE

CAS No.: 68891-38-3

Definition: Sodium Coceth-30 Sulfate is the sodium salt of the sulfuric acid ester of the polyethylene glycol ether of coconut alcohol that conforms generally to the formula:

$$R(OCH_2CH_2)_nOSO_3Na$$

where R represents the alkyl groups derived from coconut oil and n has an average value of 30.

Chemical Class: Alkyl Ether Sulfates

Functions: Surfactant - Cleansing Agent; Surfactant - Solubilizing Agent

SODIUM COCOABUTTER-AMPHOACETATE

Definition: Sodium Cocoabutteramphoacetate is the amphoteric organic compound that conforms generally to the formula:

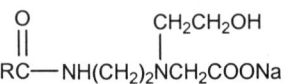

where RCO- represents the fatty acids derived from cocoa butter.

Chemical Class: Alkylamido Alkylamines

Functions: Hair Conditioning Agent; Surfactant - Cleansing Agent; Surfactant - Foam Booster

Trade Name:
Vamasoft Cocoa (Vama Farmacosmetica)

Trade Name Mixture:
Dermocare CB35 (Fratelli Ricci)

SODIUM COCOAMPHOACETATE

JPN Translation:
ココアンホ酢酸 Na

Definition: Sodium Cocoamphoacetate is the amphoteric organic compound that conforms generally to the formula:

where RCO- represents the fatty acids derived from coconut oil.

Information Sources: CIR: [S] JACT-9(2)-1990, CTFA D, TSCA

Chemical Class: Alkylamido Alkylamines

Functions: Hair Conditioning Agent; Surfactant - Cleansing Agent; Surfactant - Foam Booster

Reported Product Categories: Personal Cleanliness Products, Misc.; Shampoos (Non-coloring); Bubble Baths; Hair Conditioners; Hair Dyes and Colors (All Types Requiring Caution Statements and Patch Tests)

Technical/Other Names:
Cocoamphoacetate
Cocoamphoglycinate

Trade Names:
AMPHOSOL 1C (Stepan)
Amphoterge W (Lonza Inc./Lonza Ltd.)
Colateric 1C (Colonial Chemical Inc)
Dehyton MC (Cognis Care Chemicals/NJ)
Dehyton MC (Cognis Deutschland)
Empigen CDR60 (Albright & Wilson UK)
Empigen CDR90 (Albright & Wilson UK)
Ethteric CA (Ethox)
Ethteric C-P (Ethox)
Euroglyc AM (EOC Surfactants)
Jeeteric CM-36S (Jeen)
Mackam 1C (McIntyre)
Mackam HPC-32 (McIntyre)
Miranol Ultra C 32 (Rhodia)
Miranol Ultra C 37 (Rhodia)
Monateric CM-36S (Uniqema)
Nikkol AM-101 (Nikko)
Rewoteric AM C (Degussa Care Specialties)
Schercoteric MS (Scher)
Surfaron N 4112 N 45% (Synthron)
Zoharteric M (Zohar)

Trade Name Mixtures:
Collagen-IMZ-Complex (Kelisema Italy)
Miracare AB-33 (Rhodia)
Mirasheen CP-920 (Rhodia)
Mirasheen CP-820/G (Rhodia)
Zoharteric M-2 (Zohar)
Zoharteric M-3 (Zohar)

SODIUM COCOAMPHOHYDROXY-PROPYLSULFONATE

Definition: Sodium Cocoamphohydroxypropylsulfonate is the amphoteric organic compound that conforms generally to the formula:

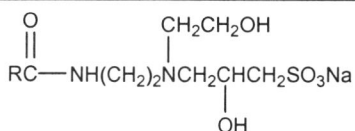

where RCO- represents the fatty acids derived from coconut oil.

Information Source: CTFA D

Chemical Classes: Alkylamido Alkylamines; Sulfonic Acids

Functions: Hair Conditioning Agent; Surfactant - Cleansing Agent; Surfactant - Foam Booster; Surfactant - Hydrotrope

Technical/Other Names:
Cocoamphohydroxypropylsulfonate
Cocoamphopropylsulfonate

Trade Names:
Amphoterge SB (Lonza Inc./Lonza Ltd.)
Ethteric CHS (Ethox)
Schercoteric MS-EP (Scher)

SODIUM COCOAMPHOPROPIONATE

JPN Translation:
ココアンホプロピオン酸 Na

Definition: Sodium Cocoamphopropionate is the amphoteric organic compound that conforms generally to the formula:

$$RC\overset{\displaystyle O}{\overset{\|}{-}}NH(CH_2)_2NCH_2CH_2COONa$$

$$CH_2CH_2OH$$

where RCO- represents the fatty acids derived from coconut oil.

Information Sources: CIR: [S] JACT-9(2)-1990, CTFA D, JCIC, JCLS, JSQI

Chemical Class: Alkylamido Alkylamines

Functions: Hair Conditioning Agent; Surfactant - Cleansing Agent; Surfactant - Foam Booster; Surfactant - Hydrotrope

Technical/Other Names:
Cocoamphopropionate
Sodium N-Cocoyl-N-carboxyethyl-N-hydroxyethyl Ethylenediamine

Trade Names:
Amphosol CSF (Stepan)
Amphoterge K (Lonza Inc./Lonza Ltd.)
Colateric CA-35 (Colonial Chemical Inc)
Mackam CSF (McIntyre)
Mackam CSF-CG (McIntyre)
Monateric CA-35 (Uniqema)
Rewoteric AM KSF 40 (Degussa Care Specialties)

Trade Name Mixture:
Lowenol 4558 (Lowenstein)

SODIUM COCOATE

CAS No.	EINECS No.
61789-31-9	263-050-4

JPN Translation:
ヤシ脂肪酸 Na

Definition: Sodium Cocoate is the sodium salt of Coconut Acid (q.v.).

Information Sources: 21CFR175.105, 21CFR175.320, 21CFR176.170, 21CFR176.200, 21CFR176.210, 21CFR177.1200, 21CFR177.2260, 21CFR177.2600, 21CFR177.2800, 21CFR178.3910, JCIC, JCLS, TSCA

Chemical Class: Soaps

Functions: Surfactant - Cleansing Agent; Surfactant - Emulsifying Agent

Reported Product Categories: Bath Soaps and Detergents; Bath Oils, Tablets, and Salts; Cleansing Products (Cold Creams, Cleansing Lotions, Liquids and Pads); Personal Cleanliness Products, Misc.; Baby Products, Misc.

Technical/Other Name:
Fatty Acids, Coconut Oil, Sodium Salts

Trade Names:
AEC Sodium Cocoate (A & E Connock)
Norfox Coco Powder (Norman, Fox & Co.)

Trade Name Mixtures:
Emulgade CL Special (Cognis Deutschland)
Vegetable Soapbase OC (Weleda)
Vegetable Soapbase PCO (Weleda)

SODIUM COCO/BABASSU/ANDIROBA SULFATE

CAS No.: 91078-92-1

Definition: Sodium Coco/Babassu/Andiroba Sulfate is the organic compound that conforms generally to the formula:

$$ROSO_3Na$$

where R represents the alkyl groups derived from a blend of coconut, babassu and andiroba oils.

Chemical Class: Sulfuric Acid Esters

Function: Surfactant - Cleansing Agent

SODIUM COCO/BABASSU SULFATE

Definition: Sodium Coco/Babassu Sulfate is the sodium salt of the sulfate ester of the mixed fatty alcohols derived from coconut and babassu oils. It conforms generally to the formula:

$$ROSO_3Na$$

where represents the alkyl groups derived from coconut and babassu oil.

Chemical Class: Alkyl Sulfates

Functions: Surfactant - Cleansing Agent; Surfactant - Emulsifying Agent

SODIUM COCO-GLUCOSIDE TARTRATE

Definition: Sodium Coco-Glucoside Tartrate is the sodium salt of the ester of tartaric acid and Coco-Glucoside (q.v.).

Chemical Classes: Carbohydrates; Organic Salts

Functions: Surfactant - Cleansing Agent; Surfactant - Emulsifying Agent

Trade Name:
Eucarol AGE/ET (Cesalpinia)

SODIUM COCOGLYCERYL ETHER SULFONATE

Definition: Sodium Cocoglyceryl Ether Sulfonate is the compound that conforms generally to the formula:

$$ROCH_2CHCH_2SO_3Na$$
$$OH$$

where R represents the coconut alcohol radical.

Chemical Class: Sulfonic Acids

Function: Surfactant - Cleansing Agent

Reported Product Category: Bath Soaps and Detergents

SODIUM COCO/HYDROGENATED TALLOW SULFATE

Definition: Sodium Coco/Hydrogenated Tallow Sulfate is the sodium salt of the sulfate ester of the mixed fatty alcohols derived from coconut oil and hydrogenated tallow. It conforms generally to the formula:

$$ROSO_3Na$$

where R represents the alkyl groups derived from coconut oil and hydrogenated tallow.

Chemical Class: Alkyl Sulfates

Functions: Surfactant - Cleansing Agent; Surfactant - Emulsifying Agent

SODIUM COCOMONOGLYCERIDE SULFATE

CAS No.	EINECS No.
61789-04-6	263-026-3

JPN Translation:
ココグリセリル硫酸 Na

Definition: Sodium Cocomonoglyceride Sulfate is the compound that conforms generally to the formula:

$$RC \overset{\overset{\displaystyle O}{\|}}{—} OCH_2CHCH_2OSO_3Na$$
$$\underset{OH}{|}$$

where RCO- represents the fatty acids derived from coconut oil.

Information Sources: JSQI, TSCA

Chemical Classes: Glyceryl Esters and Derivatives; Sulfuric Acid Esters

Function: Surfactant - Cleansing Agent

Technical/Other Names:
Glycerides, Coconut Oil Mono-, Sulfated, Sodium Salts
Sodium Coconut Monoglyceride Sulfate

Trade Names:
Nikkol SGC-80N (Nikko)
Plantapon CMGS (Cognis Care Chemicals/NJ)
POEM-LS-90 (Riken Vitamin Oil)

Trade Name Mixture:
Plantapon CL 30 (Cognis Care Chemicals/NJ)

SODIUM COCOMONOGLYCERIDE SULFONATE

Definition: Sodium Cocomonoglyceride Sulfonate is the compound that conforms generally to the formula:

$$RC \overset{\overset{\displaystyle O}{\|}}{—} OCH_2CHCH_2SO_3Na$$
$$\underset{OH}{|}$$

where RCO- represents the fatty acids derived from coconut oil.

Chemical Classes: Glyceryl Esters and Derivatives; Sulfonic Acids

Function: Surfactant - Cleansing Agent

SODIUM COCO PG-DIMONIUM CHLORIDE PHOSPHATE

Definition: Sodium Coco PG-Dimonium Chloride Phosphate is the amphoteric organic compound that conforms generally to the formula:

$$\left[R—\overset{\overset{\displaystyle CH_3}{|}}{\underset{\underset{\displaystyle CH_3}{|}}{N}}—CH_2CHCH_2O—\overset{\overset{\displaystyle O}{\|}}{\underset{\underset{\displaystyle OH}{|}}{P}}—ONa \right]^{+} \quad Cl^{-}$$
$$\quad\quad\quad\quad\quad\quad \underset{OH}{}$$

where R represents the coco alkyl moiety.

Chemical Classes: Phosphorus Compounds; Quaternary Ammonium Compounds

Functions: Antistatic Agent; Surfactant - Cleansing Agent; Surfactant - Foam Booster

Trade Names:
Colalipid DCCA (Colonial Chemical Inc)
Phospholipid CDM (Uniqema)

SODIUM COCO-SULFATE

Definition: Sodium Coco-Sulfate is the sodium salt of the sulfate ester of coconut alcohol that conforms generally to the formula:

$$ROSO_3Na$$

where R represents the alkyl groups derived from coconut oil.

Chemical Class: Alkyl Sulfates

Function: Surfactant - Cleansing Agent

Technical/Other Name:
Sulfuric Acid, Monococoyl Ester, Sodium Salt

Trade Names:
Stepanol CFAS-70 (Stepan)
Stepanol DCFAS-F (Stepan)
Stepanol DCFAS-N (Stepan)
Stepanol DCFAS-P (Stepan)
Sulfetal C 38 (Zschimmer & Schwarz)
Sulfopon HC (Cognis Care Chemicals/NJ)
Sulfopon HC Granulat (Cognis Deutschland)

Trade Name Mixtures:
Zetesap 813 A (Zschimmer & Schwarz)
Zetesap 813 P (Zschimmer & Schwarz)

SODIUM COCOYL AMINO ACIDS

Definition: Sodium Cocoyl Amino Acids is the sodium salt of a mixture of amino acids acylated by cocoyl chloride.

Chemical Class: Amino Acids

Function: Surfactant - Cleansing Agent

Trade Name Mixtures:
Sepicalm S (SEPPIC)
Sepicap MP (SEPPIC)

SODIUM COCOYL APPLE AMINO ACIDS

Definition: Sodium Cocoyl Apple Amino Acids is the sodium salt of the condensation product of coconut acid chloride and the amino acids isolated from apple juice.

Chemical Class: Amino Acids

Functions: Hair Conditioning Agent; Skin-Conditioning Agent - Miscellaneous; Surfactant - Foam Booster

Trade Name:
Proteol APL (SEPPIC)

SODIUM COCOYL COLLAGEN AMINO ACIDS

Definition: Sodium Cocoyl Collagen Amino Acids is the sodium salt of the condensation product of coconut acid chloride and Collagen Amino Acids (q.v.).

Chemical Class: Amino Acids

Functions: Hair Conditioning Agent; Surfactant - Cleansing Agent

Reported Product Category: Hair Conditioners

Trade Name Mixture:
Foam-Coll 5W (Arch Personal Care Products)

SODIUM COCOYL GLUTAMATE

CAS No.	EINECS No.
68187-32-6	269-087-2

JPN Translation:
ココイルグルタミン酸 Na

Definition: Sodium Cocoyl Glutamate is the sodium salt of the Cocoyl Glutamic Acid (q.v.). It conforms generally to the formula:

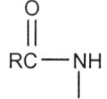

$$RC \overset{\overset{\displaystyle O}{\|}}{—} NH$$
$$\quad\quad\quad \underset{|}{}$$
$$HOOCCH_2CH_2CHCOONa$$

where RCO- represents the fatty acids derived from coconut oil.

Information Sources: JCIC, JCLS, JSQI, TSCA

Chemical Class: Amino Acids

Function: Surfactant - Cleansing Agent

Technical/Other Names:
L-Glutamic Acid, N-Coco Acyl Derivs., Monosodium Salts
Sodium N-Cocoyl-L-glutamate

Trade Names:
Aminosurfact ACDS-L (Asahi Denka Kogyo)
Aminosurfact ACMS (Asahi Denka Kogyo)
Amisoft CS-11 (Ajinomoto)
Hostapon CCG (Clariant)
Hostapon CCG (Clariant GmbH, Personal Care)

Hostapon KCG (Clariant)
Hostapon KCG (Clariant GmbH, Personal Care)
Protelan AGL 95/C (Zschimmer & Schwarz Italiana)

Trade Name Mixtures:
Aminosurfact ACDP-L (Asahi Denka Kogyo)
Amisoft CS-22 (Ajinomoto)
Amisoft GS-11 (Ajinomoto)
Amisoft GS-11P (Ajinomoto)
Elespher Vitaplex Hydro (Laboratoires Serobiologiques)
Emulvama AGW (Vama Farmacosmetica)
Softinat (Vama Farmacosmetica)

SODIUM COCOYL GLYCINATE

CAS No.	EINECS No.
90387-74-9	291-350-5

Definition: Sodium Cocoyl Glycinate is the organic compound that conforms generally to the formula:

$$RC \overset{\displaystyle O}{\overset{\displaystyle \|}{}}-NHCH_2COONa$$

where RCO- represents the cocoyl moiety.

Chemical Classes: Amides; Amino Acids; Organic Salts

Functions: Hair Conditioning Agent; Skin-Conditioning Agent - Miscellaneous; Surfactant - Cleansing Agent

Technical/Other Name:
Glycine, N-Coco Acyl Derivs., Sodium Salts

Trade Names:
Amilite GCS-12 (Ajinomoto)
Amilite GCS-11F (Ajinomoto)

SODIUM COCOYL/HYDROGENATED TALLOW GLUTAMATE

JPN Translation:
（ヤシ脂肪酸 / 水添牛脂脂肪酸）グルタミン酸 Na

Definition: Sodium Cocoyl/Hydrogenated Tallow Glutamate is the organic compound that conforms generally to the formula:

$$HOOCCH_2CH_2CHCOONa$$
$$|$$
$$HN-CR$$
$$\overset{\displaystyle \|}{\underset{\displaystyle O}{}}$$

where RCO- represents a mixture of fatty acids derived from coconut oil and hydrogenated tallow.

Chemical Class: Amino Acids

Function: Surfactant - Cleansing Agent

Trade Name:
Aminosurfact AGMS (Asahi Denka Kogyo)

SODIUM COCOYL HYDROLYZED COLLAGEN

CAS No.: 68188-38-5

JPN Translation:
ココイル加水分解コラーゲン Na

Definition: Sodium Cocoyl Hydrolyzed Collagen is the sodium salt of the condensation product of coconut acid chloride and Hydrolyzed Collagen (q.v.).

Information Sources: JCIC, JCLS, JSQI, TSCA

Chemical Class: Protein Derivatives

Functions: Hair Conditioning Agent; Skin-Conditioning Agent - Miscellaneous; Surfactant - Cleansing Agent

Technical/Other Names:
Acid Chlorides, Coco, Reaction Products with Protein Hydrolyzates, Sodium Salts
Sodium Coco-Hydrolyzed Animal Protein
Sodium Cocoyl Hydrolyzed Animal Protein

Trade Names:
Bio-Pol NCHAP (Arch Personal Care Products)
Foam-Coll SK (Arch Personal Care Products)
Geliderm 3000 P (DGF Stoess)
Geliderm 3000 S (DGF Stoess)
May-Tein SK (Maybrook)
Promois ECS (Seiwa Kasei)
Promois ECS-K (Seiwa Kasei)
Promois EUCS (Seiwa Kasei)

SODIUM COCOYL HYDROLYZED KERATIN

Definition: Sodium Cocoyl Hydrolyzed Keratin is the sodium salt of the condensation product of coconut acid chloride and Hydrolyzed Keratin (q.v.).

Chemical Class: Protein Derivatives

Functions: Hair Conditioning Agent; Skin-Conditioning Agent - Miscellaneous; Surfactant - Cleansing Agent

Trade Names:
May-Tein KT (Maybrook)
Proteol K25 (SEPPIC)

SODIUM COCOYL HYDROLYZED RICE PROTEIN

Definition: Sodium Cocoyl Hydrolyzed Rice Protein is the sodium salt of the condensation product of coconut acid chloride and Hydrolyzed Rice Protein (q.v.)

Chemical Class: Protein Derivatives

Functions: Hair Conditioning Agent; Skin-Conditioning Agent - Miscellaneous; Surfactant - Cleansing Agent

Trade Name:
Protelan R (Zschimmer & Schwarz Italiana)

SODIUM COCOYL HYDROLYZED SOY PROTEIN

Definition: Sodium Cocoyl Hydrolyzed Soy Protein is the sodium salt of the condensation product of coconut acid chloride and Hydrolyzed Soy Protein (q.v.).

Chemical Class: Protein Derivatives

Functions: Hair Conditioning Agent; Skin-Conditioning Agent - Miscellaneous; Surfactant - Cleansing Agent

Reported Product Category: Shampoos (Non-coloring)

Trade Names:
Foam-Soy C (Arch Personal Care Products)
Proteol VS 22 (SEPPIC)

Trade Name Mixture:
Supro-Tein S (Maybrook)

SODIUM COCOYL HYDROLYZED SWEET ALMOND PROTEIN

Definition: Sodium Cocoyl Hydrolyzed Sweet Almond Protein is the sodium salt of the condensation product of coconut acid chloride and Hydrolyzed Sweet Almond Protein (q.v.).

Chemical Class: Protein Derivatives

Functions: Hair Conditioning Agent; Skin-Conditioning Agent - Miscellaneous; Surfactant - Cleansing Agent

Trade Name:
Coccomandel (Sinerga)

SODIUM COCOYL HYDROLYZED WHEAT PROTEIN

Definition: Sodium Cocoyl Hydrolyzed Wheat Protein is the sodium salt of the condensation product of coconut acid chloride and Hydrolyzed Wheat Protein (q.v.).

Chemical Class: Protein Derivatives

Functions: Hair Conditioning Agent; Skin-Conditioning Agent - Miscellaneous; Surfactant - Cleansing Agent

Trade Names:
AC Foaming Wheat (Active Concepts)
Coccopolipeptide Di Grano (Sinerga)
Gluadin WK (Cognis Care Chemicals/PA)
Gluadin WK (Cognis Deutschland)
Protelan VE/K (Zschimmer & Schwarz)

Trade Name Mixture:
Emulvama AGW (Vama Farmacosmetica)

SODIUM COCOYL HYDROLYZED WHEAT PROTEIN GLUTAMATE

Definition: Sodium Cocoyl Hydrolyzed Wheat Protein Glutamate is the reaction product of Sodium Cocoyl Hydrolyzed Wheat Protein (q.v.) and Glutamic Acid (q.v.).

Chemical Class: Protein Derivatives

Functions: Surfactant - Cleansing Agent; Surfactant - Foam Booster

Trade Name:
Plantapon S (Cognis Deutschland)

SODIUM COCOYL ISETHIONATE

CAS Nos.	EINECS No.
58969-27-0	
61789-32-0	263-052-5

JPN Translation:
ココイルイセチオン酸 Na

Definition: Sodium Cocoyl Isethionate is the sodium salt of the coconut fatty acid ester of isethionic acid. It conforms generally to the formula:

$$RC\overset{\displaystyle O}{\overset{\|}{-\!\!-}}OCH_2CH_2SO_3Na$$

where RCO- represents the fatty acids derived from coconut oil.

Information Sources: CIR: [SQ] JACT-12 (5)1993, CTFA S, JCIC, JCLS, JSQI, TSCA

Chemical Class: Isethionates

Function: Surfactant - Cleansing Agent

Reported Product Categories: Bath Soaps and Detergents; Bath Oils, Tablets, and Salts; Cleansing Products (Cold Creams, Cleansing Lotions, Liquids and Pads); Hair Preparations (Non-coloring), Misc.; Shampoos (Non-coloring); Tonics, Dressings, and Other Hair Grooming Aids; Hair Bleaches; Baby Products, Misc.

Technical/Other Names:
Fatty Acids, Coconut Oil, Sulfoethyl Esters, Sodium Salts
Sodium Cocoyl Ethyl Ester Sulfonate

Trade Names:
AEC Sodium Cocoyl Isethionate (A & E Connock)

Elfan AT 84 (Akzo Nobel Surface AB)
Elfan AT 84 G (Akzo Nobel Surface AB)
Elfan AT 90G (Akzo Nobel Surface AB)
Ethox SCI (Ethox)
Geropon AS-200 (Rhodia)
Hostapon SCI-65 (Clariant)
Hostapon SCI-65 (Clariant GmbH, Personal Care)
Hostapon SCI-85 (Clariant)
Hostapon SCI-85 (Clariant GmbH, Personal Care)
Hostapon SCI-78 C (Clariant GmbH, Personal Care)
Hostapon SCI-78 P (Clariant GmbH, Personal Care)
Jeepon AC-78 (Jeen)
Jordapon CI Prill (BASF)
Protapon AC-85 (Protameen)
Tauranol I-78 (Finetex)
Tauranol I-78-6 (Finetex)
Tauranol I-78E (Finetex)
Tauranol I-78E Flakes (Finetex)
Tauranol I-78 Flake (Finetex)

Trade Name Mixtures:
Hostapon SCI-65 C (Clariant GmbH, Personal Care)
Hostapon SCI-40 L (Clariant GmbH, Personal Care)
Jordapon CI 65 (BASF)
Miracare UM-140 (Rhodia)
Zetesap 5165 (Zschimmer & Schwarz)

SODIUM COCOYL LACTYLATE

Definition: Sodium Cocoyl Lactylate is the sodium salt of the coconut acid ester of lactyl lactate. It conforms to the formula:

$$RC\overset{\displaystyle O}{\overset{\|}{-}}OCHC\overset{\displaystyle O}{\overset{\|}{-}}OCHCOONa$$
$$\underset{CH_3}{|} \qquad \underset{CH_3}{|}$$

where RCO- represents the fatty acids derived from coconut oil.

Chemical Classes: Esters; Organic Salts

Function: Surfactant - Emulsifying Agent

Reported Product Category: Bath Soaps and Detergents

Trade Name:
Pationic SCL (RITA)

SODIUM COCOYL METHYL BETA-ALANINE

JPN Translation:
ココイルメチルアラニン Na

Definition: Sodium Cocoyl Methyl Beta-Alanine is the sodium salt of Cocoyl Methyl Beta-Alanine(q.v.).

Information Source: JCLS

Chemical Classes: Amino Acids; Organic Salts

Function: Skin-Conditioning Agent - Miscellaneous

SODIUM COCOYL METHYLAMINOPROPIONATE

CAS No.: 79121-83-8

Definition: Sodium Cocoyl Methylaminopropionate is the organic compound that conforms to the formula:

$$RC\overset{\displaystyle O}{\overset{\|}{-}}N(CH_2)_2COONa$$
$$\underset{CH_3}{|}$$

where RCO- represents the cocoyl moiety.

Chemical Class: Amines

Function: Surfactant - Cleansing Agent

Trade Name:
Alanon ACE (Kawaken)

SODIUM COCOYL OAT AMINO ACIDS

Definition: Sodium Cocoyl Oat Amino Acids is the sodium salt of the condensation product of coconut acid chloride and the amino acids derived from Avena Sativa (Oat) Protein (q.v.).

Chemical Class: Amino Acids

Functions: Hair Conditioning Agent; Skin-Conditioning Agent - Miscellaneous; Surfactant - Cleansing Agent

Trade Name:
Lauroat Di Avena (Sinerga)

SODIUM COCOYL SARCOSINATE

CAS No.	EINECS No.
61791-59-1	263-193-2

JPN Translation:
ココイルサルコシン Na

Definition: Sodium Cocoyl Sarcosinate is the sodium salt of Cocoyl Sarcosine (q.v.).

Information Sources: CIR: [SQ] IJT-20 (SUPPL. 1)2001, JCIC, JCLS, JSQI, SNPF, TSCA

Chemical Class: Sarcosinates and Sarcosine Derivatives

Functions: Hair Conditioning Agent; Surfactant - Cleansing Agent

Reported Product Categories: Bath Oils, Tablets, and Salts; Cleansing Products (Cold Creams, Cleansing Lotions, Liquids and Pads); Shampoos (Non-coloring); Bath Preparations, Misc.

Technical/Other Names:
Amides, Coconut Oil, with Sarcosine, Sodium Salts
Sodium N-Cocoyl Sarcosinate

Trade Names:
Crodasinic CS (Croda Chemicals)
Hamposyl C-30 (Amerchol)
Neoscoap SCN-35 (Toho)
Nikkol Sarcosinate CN-30 (Nikko)
Protelan LS 9011/C (Zschimmer & Schwarz Italiana)
Vanseal NACS-30 (Vanderbilt)

Trade Name Mixture:
LOWENOL T-1106-A (Lowenstein)

SODIUM COCOYL TAURATE

JPN Translation:
ココイルタウリン Na

Definition: Sodium Cocoyl Taurate is the organic salt that conforms to the formula:

$$RC-NHCH_2CH_2SO_3Na$$
(with $\overset{O}{\overset{\|}{}}$ on the RC carbonyl)

where RCO- represents the coconut acid radical.

Chemical Class: Sulfonic Acids

Function: Surfactant - Cleansing Agent

Trade Name:
Neoscoap CTS (Toho)

SODIUM COCOYL WHEAT AMINO ACIDS

Definition: Sodium Cocoyl Wheat Amino Acids is the sodium salt of the condensation product of coconut acid chloride and the amino acids derived from Triticum Vulgare (Wheat) Protein (q.v.).

Chemical Class: Amino Acids

Functions: Hair Conditioning Agent; Skin-Conditioning Agent - Miscellaneous; Surfactant - Cleansing Agent

Trade Names:
Coccoaminoacido Di Grano (Sinerga)
Protelan LWA/C (Zschimmer & Schwarz Italiana)

SODIUM C4-12 OLEFIN/MALEIC ACID COPOLYMER

Definition: Sodium C4-12 Olefin/Maleic Acid Copolymer is a sodium salt of a polymer synthesized from C4-12 olefins and maleic anhydride.

Chemical Class: Synthetic Polymers

Functions: Binder; Emulsion Stabilizer; Film Former; Suspending Agent - Nonsurfactant

Reported Product Category: Mascara

Trade Name:
Demol EP (Kao Corp.)

SODIUM C12-14 OLEFIN SULFONATE

Definition: Sodium C12-14 Olefin Sulfonate is a mixture of long chain sulfonate salts prepared by sulfonation of C12-14 alpha olefins. It consists chiefly of sodium alkene sulfonates and sodium hydroxyalkane sulfonates.

Information Sources: 21CFR175.105, CIR: [SQ] IJT-17(Suppl. 5)1998

Chemical Class: Sulfonic Acids

Function: Surfactant - Cleansing Agent

Trade Name Mixture:
Serdet VLC 1200 (Sasol Servo)

SODIUM C14-16 OLEFIN SULFONATE

CAS No. 68439-57-6
EINECS No. 270-407-8

JPN Translation:
オレフィン（C14 - 16）スルホン酸 Na

Definition: Sodium C14-16 Olefin Sulfonate is a mixture of long chain sulfonate salts prepared by sulfonation of C14-16 alpha olefins. It consists chiefly of sodium alkene sulfonates and sodium hydroxyalkane sulfonates.

Information Sources: 21CFR175.105, CIR: [SQ] IJT-17(Suppl. 5)1998, JCLS, JSCI, TSCA

Chemical Class: Sulfonic Acids

Function: Surfactant - Cleansing Agent

Reported Product Categories: Bath Soaps and Detergents; Shampoos (Non-coloring); Bubble Baths; Bath Oils, Tablets, and Salts; Cleansing Products (Cold Creams, Cleansing Lotions, Liquids and Pads); Bath Preparations, Misc.; Hair Dyes and Colors (All Types Requiring Caution Statements and Patch Tests); Personal Cleanliness Products, Misc.

Technical/Other Name:
Sodium Tetradecenesulfonate

Trade Names:
AEC Sodium C14-16 Olefin Sulfonate (A & E Connock)
BIO-TERGE AS-40 (Stepan)
Bio-Terge AS-90 Beads (Stepan)
BIO-TERGE AS-40 CG (Stepan)
BIO-TERGE AS-40 CG-P (Stepan)
Calsoft AOS-40 (Pilot)
Hostapur OSB (Clariant)
Hostapur OSB (Clariant GmbH, Personal Care)
Jeenate AOS-40 (Jeen)
Nansa LSS480 (Albright & Wilson UK)
Nansa LSS38/AS (Albright & Wilson UK)
Nansa LSS 38A/ZI (Albright & Wilson Asia)
Nansa LSS 92A/ZI (Albright & Wilson Asia)
Nansa LSS480/B (Albright & Wilson UK)
Nansa LSS 490/H (Albright & Wilson UK)
Nikkol OS-14 (Nikko)
Norfox Alpha XL (Norman, Fox & Co.)
Rhodacal A-246 L (Rhodia)
Rhodacal LSS-40 (Rhodia)
Rhodacal LSS-92 (Rhodia)

Trade Name Mixtures:
BIO-TERGE 804 (Stepan)
Bio-Terge 804 M (Stepan)
Dow Corning 2-1998 Anionic Emulsion (Dow Corning)
Gluplex OS (Kelisema Italy)
Tego Pearl S 33 (Degussa Care Specialties)

SODIUM C14-18 OLEFIN SULFONATE

Definition: Sodium C14-18 Olefin Sulfonate is a mixture of long chain sulfonate salts prepared by sulfonation of C14-18 alpha olefins. It consists chiefly of sodium alkene sulfonates and sodium hydroxyalkane sulfonates.

Information Sources: 21CFR175.105, CIR: [SQ] IJT-17(Suppl. 5)1998

Chemical Class: Sulfonic Acids

Function: Surfactant - Cleansing Agent

Trade Names:
Colonial AOS-40 (Colonial Chemical Inc)
Lowenol O-11016 (Lowenstein)

SODIUM C16-18 OLEFIN SULFONATE

Definition: Sodium C16-18 Olefin Sulfonate is a mixture of long chain sulfonate salts prepared by sulfonation of C16-18 alpha olefins. It consists chiefly of sodium alkene sulfonates and sodium hydroxyalkane sulfonates.

Information Sources: 21CFR175.105, CIR: [SQ] IJT-17(Suppl. 5)1998

Chemical Class: Sulfonic Acids

Function: Surfactant - Cleansing Agent

SODIUM CORNAMPHOPROPIONATE

Definition: Sodium Cornamphopropionate is the amphoteric organic compound that conforms generally to the formula:

$$\underset{RC-NH(CH_2)_2NCH_2CH_2COONa}{\overset{\overset{\displaystyle O}{\|}\qquad\qquad \overset{\displaystyle CH_2CH_2OH}{|}}{}}$$

where RCO- represents the fatty acids derived from corn oil.

Chemical Class: Alkylamido Alkylamines

Functions: Hair Conditioning Agent; Surfactant - Cleansing Agent; Surfactant - Foam Booster

SODIUM C13-15 PARETH-8 BUTYL PHOS-PHATE

Definition: Sodium C13-15 Pareth-8 Butyl Phosphate is the sodium salt of a mixture of phosphoric acid esters of C13-15 pareth-8 and butyl alcohol.

Chemical Class: Phosphorus Compounds

Function: Surfactant - Cleansing Agent

Reported Product Category: Hair Sprays (Aerosol Fixatives)

Trade Name Mixtures:
Elfugin AKT 300 Liquid (Clariant)
Elfugin AKT 300 Liquid (Clariant GmbH, Personal Care)

SODIUM C9-11 PARETH-6 CARBOXYLATE

Definition: Sodium C9-11 Pareth-6 Carboxylate is the sodium salt of C9-11 Pareth-6 Carboxylic Acid (q.v) that conforms generally to the formula:

$$R(OCH_2CH_2)_nOCH_2COONa$$

where R represents the C9-11 alkyl group and n has an average value of 5.

Chemical Class: Organic Salts

Function: Surfactant - Cleansing Agent

Technical/Other Names:
PEG-6 C9-11 Alkyl Ether Carboxylic Acid, Sodium Salt
Polyethylene Glycol 300 C9-11 Alkyl Ether Carboxylic Acid, Sodium Salt
Polyoxyethylene (6) C9-11 Alkyl Ether Carboxylic Acid, Sodium Salt

Trade Name:
Neodox 91-5, Sodium Salt (Shell)

SODIUM C11-15 PARETH-7 CARBOXYLATE

CAS No.: 68603-23-6 (Generic)

Definition: Sodium C11-15 Pareth-7 Carboxylate is the sodium salt of C11-15 Pareth-7 Carboxylic Acid (q.v.). It conforms generally to the formula:

$$R(OCH_2CH_2)_nOCH_2COONa$$

where R represents the C11-15 alkyl group and n has an average value of 6.

Chemical Class: Organic Salts

Function: Surfactant - Cleansing Agent

Technical/Other Names:
PEG-7 C11-15 Alkyl Ether Carboxylic Acid, Sodium Salt
Polyethylene Glycol (7) C11-15 Alkyl Ether Carboxylic Acid, Sodium Salt
Polyoxyethylene (7) C11-15 Alkyl Ether Carboxylic Acid, Sodium Salt
Sodium Pareth-15-7 Carboxylate

SODIUM C12-13 PARETH-5 CARBOXYLATE

Definition: Sodium C12-13 Pareth-5 Carboxylate is the sodium salt of C12-13 Pareth-5 Carboxylic Acid (q.v.) that conforms generally to the formula:

$$R(OCH_2CH_2)_nOCH_2COONa$$

where R represents the C12-13 alkyl group and n has an averagea value of 4.

Chemical Class: Organic Salts

Function: Surfactant - Cleansing Agent

Technical/Other Names:
PEG-5 C12-13 Alkyl Ether Carboxylic Acid, Sodium Salt
Polyethylene Glycol (5) C12-13 Alkyl Ether Carboxylic Acid, Sodium Salt
Polyoxyethylene (5) C12-13 Alkyl Ether Carboxylic Acid, Sodium Salt

Trade Name:
Neodox 23-4, Sodium Salt (Shell)

SODIUM C12-13 PARETH-8 CARBOXYLATE

Definition: Sodium C12-13 Pareth-8 Carboxylate is the sodium salt of C12-13 Pareth-8 Carboxylic Acid (q.v.). It conforms generally to the formula:

$$R(OCH_2CH_2)_nOCH_2COONa$$

where R represents the C12-13 alkyl group and n has an average value of 7.

Chemical Class: Organic Salts

Function: Surfactant - Cleansing Agent

Technical/Other Names:
Polyethylene Glycol (8) C12-13 Alkyl Ether Carboxylic Acid, Sodium Salt
Polyoxyethylene (8) C12-13 Alkyl Ether Carboxylic Acid, Sodium Salt

Trade Names:
Neodox 23-7, Sodium Salt (Shell)
Surfine WNT Gel (Finetex)
Surfine WNT Liquid (Finetex)
Surfine WNT-LS (Finetex)

SODIUM C12-13 PARETH-12 CARBOXYLATE

Definition: Sodium C12-13 Pareth-12 Carboxylate is the sodium salt of C12-13 Pareth-12 Carboxylic Acid (q.v.) that conforms generally to the formula:

$$R(OCH_2CH_2)_nOCH_2COONa$$

where R represents the C12-13 alkyl group and n has an average value of 11.

Chemical Class: Organic Salts

Function: Surfactant - Cleansing Agent

Technical/Other Names:
PEG-12 C12-13 Alkyl Ether Carboxylic Acid, Sodium Salt
Polyethylene Glycol 600 C12-13 Alkyl Ether Carboxylic Acid, Sodium Salt
Polyoxyethylene (12) C12-13 Alkyl Ether Carboxylic Acid, Sodium Salt

Trade Name:
Neodox 23-11, Sodium Salt (Shell)

SODIUM C12-15 PARETH-6 CARBOXYLATE

CAS No.: 70632-06-3 (Generic)

Definition: Sodium C12-15 Pareth-6 Carboxylate is the sodium salt of the organic acid that conforms generally to the formula:

$$R(OCH_2CH_2)_nOCH_2COONa$$

where R represents the C12-15 alkyl group.

Chemical Class: Organic Salts

Function: Surfactant - Cleansing Agent

Technical/Other Names:
PEG-6 C12-15 Alkyl Ether Carboxylic Acid, Sodium Salt
Polyethylene Glycol 300 C12-15 Alkyl Ether Carboxylic Acid, Sodium Salt
Polyoxyethylene (6) C12-15 Alkyl Ether Carboxylic Acid, Sodium Salt

SODIUM C12-15 PARETH-7 CARBOXYLATE

CAS No.: 70632-06-3 (Generic)

Definition: Sodium C12-15 Pareth-7 Carboxylate is the sodium salt of C12-15 Pareth-7 Carboxylic Acid (q.v.). It conforms generally to the formula:

$$R(OCH_2CH_2)_nOCH_2COONa$$

where R represents the C12-15 alkyl group.

Chemical Class: Organic Salts

Function: Surfactant - Cleansing Agent

Technical/Other Names:
PEG-7 C12-15 Alkyl Ether Carboxylic Acid, Sodium Salt
Polyethylene Glycol (7) C12-15 Alkyl Ether Carboxylic Acid, Sodium Salt
Polyoxyethylene (7) C12-15 Alkyl Ether Carboxylic Acid, Sodium Salt
Sodium Pareth-25-7 Carboxylate

Trade Name:
Neodox 25-6, Sodium Salt (Shell)

SODIUM C12-15 PARETH-8 CARBOXYLATE

Definition: Sodium C12-15 Pareth-8 Carboxylate is the sodium salt of the organic acid that conforms generally to the formula:

$$R(OCH_2CH_2)_nOCH_2COONa$$

where R represents the C12-15 alkyl group and n has an average value of 7.

Chemical Class: Organic Salts

Function: Surfactant - Cleansing Agent

Technical/Other Names:
Polyethylene Glycol (8) C12-15 Alkyl Ether Carboxylic Acid, Sodium Salt
Polyoxyethylene (8) C12-15 Alkyl Ether Carboxylic Acid, Sodium Salt

Trade Name:
Surfine WLG (Finetex)

SODIUM C14-15 PARETH-8 CARBOXYLATE

Definition: Sodium C14-15 Pareth-8 Carboxylate is the sodium salt of C14-15 Pareth-8 Carboxylic Acid (q.v.) that conforms generally to the formula:

$$R(OCH_2CH_2)_nOCH_2COONa$$

where R represents the C14-15 alkyl group and n has an average value of 7.

Chemical Class: Organic Salts

Function: Surfactant - Cleansing Agent

Technical/Other Names:
PEG-8 C14-15 Alkyl Ether Carboxylic Acid, Sodium Salt

Polyethylene Glycol 400 C14-15 Alkyl Ether Carboxylic Acid, Sodium Salt
Polyoxyethylene (8) C14-15 Alkyl Ether Carboxylic Acid, Sodium Salt

Trade Name:
Neodox 45-7, Sodium Salt (Shell)

SODIUM C12-14 SEC-PARETH-8 CARBOXYLATE

Definition: Sodium C12-14 Sec-Pareth-8 Carboxylate is the sodium salt of the polyethylene glycol ether of a mixture of carboxylated synthetic secondary C12-14 alcohols with an average ethoxylation value of 8.

Chemical Classes: Alkoxylated Carboxylic Acids; Organic Salts

Function: Surfactant - Cleansing Agent

Trade Name:
SOFTANOL 70EC (Nippon Shokubai)

SODIUM C14-15 PARETH-PG SULFONATE

Definition: Sodium C14-15 Pareth-PG Sulfonate is the sodium salt of an ethoxylated alkyl glyceryl ether. It conforms generally to the formula:

$$R(OCH_2CH_2)_nOCH_2\underset{\underset{OH}{|}}{CH}CH_2SO_3Na$$

where R represents the C14-15 fatty alcohol grouping and n has an average value between 1 to 4.

Chemical Classes: Alkoxylated Alcohols; Sulfonic Acids

Function: Surfactant - Cleansing Agent

Trade Name:
Nupore (Shell)

SODIUM C12-13 PARETH-2 PHOSPHATE

Definition: Sodium C12-13 Pareth-2 Phosphate is the sodium salt of a complex mixture of esters of phosphoric acid and C12-13 Pareth-2 (q.v.).

Chemical Class: Phosphorus Compounds

Function: Surfactant - Cleansing Agent

Trade Name:
Estamit ME-502NA (Kao Corp.)

SODIUM C13-15 PARETH-8 PHOSPHATE

Definition: Sodium C13-15 Pareth-8 Phosphate is the sodium salt of a mixture of esters

of phosphoric acid and a mixture of synthetic C13-15 alcohols with an average of 8 moles of ethoxylation.

Chemical Class: Phosphorus Compounds

Function: Surfactant - Cleansing Agent

Reported Product Category: Hair Sprays (Aerosol Fixatives)

Trade Name Mixtures:
Elfugin AKT 300 Liquid (Clariant)
Elfugin AKT 300 Liquid (Clariant GmbH, Personal Care)

SODIUM C9-15 PARETH-3 SULFATE

Definition: Sodium C9-15 Pareth-3 Sulfate is the sodium salt of a sulfated polyethylene glycol ether of a mixture of synthetic C9-15 fatty alcohols. It conforms generally to the formula:

$$R(OCH_2CH_2)_nOSO_3Na$$

where R represents the C9-15 alkyl group and n has an average value of 3.

Chemical Class: Alkyl Ether Sulfates

Function: Surfactant - Cleansing Agent

Trade Name:
Empimin KSN27/XW (Albright & Wilson UK)

SODIUM C10-15 PARETH SULFATE

JPN Translations:
(C11 , 13 , 15) パレス - 1 硫酸 Na
(C11 - 15) パレス - 3 硫酸 Na

Definition: Sodium C10-15 Pareth Sulfate is the sodium salt of a sulfated polyethylene glycol ether of a mixture of synthetic C10-15 alcohols. It conforms generally to the formula:

$$R(OCH_2CH_2)_nOSO_3Na$$

where R represents the C10-15 alkyl group and n has an average value between 1 and 4.

Information Source: JCLS

Chemical Class: Alkyl Ether Sulfates

Function: Surfactant - Cleansing Agent

SODIUM C12-13 PARETH SULFATE

JPN Translation:
綿実油

Definition: Sodium C12-13 Pareth Sulfate is the sodium salt of a sulfated polyethylene glycol ether of a mixture of synthetic C12-13 fatty alcohols. It conforms generally to the formula:

$$R(OCH_2CH_2)_nOSO_3Na$$

where R represents the C12-13 alkyl group and n has an average value between 1 and 4.

Information Sources: JCIC, JCLS, JSQI

Chemical Class: Alkyl Ether Sulfates

Function: Surfactant - Cleansing Agent

Technical/Other Names:
Sodium Pareth-23 Sulfate
Sodium Polyoxyethylene Alkyl (12,13) Ether Sulfate (3E.O.)

Trade Names:
Alscoap DA-3S (Toho)
Alscoap DA-330S (Toho)
Empimin KESH 70 (Albright & Wilson UK)
Empimin KSL 68 (Albright & Wilson UK)
Empimin KSN27/LA (Albright & Wilson UK)
Empimin KSN70/LA (Albright & Wilson UK)

SODIUM C12-14 PARETH-3 SULFATE

JPN Translation:
（C12 - 14）バレス - 3 硫酸 Na

Definition: Sodium C12-14 Pareth-3 Sulfate is the sodium salt of a sulfate ester of C12-14 Pareth-3 (q.v.).

Information Source: JCLS

Chemical Class: Alkyl Ether Sulfates

Function: Surfactant - Emulsifying Agent

SODIUM C12-15 PARETH SULFATE

CAS No.	EINECS No.
91648-56-5	293-918-8

Definition: Sodium C12-15 Pareth Sulfate is the sodium salt of a sulfated polyethylene glycol ether of a mixture of synthetic C12-15 fatty alcohols. It conforms generally to the formula:

$$R(OCH_2CH_2)_nOSO_3Na$$

where R represents the C12-15 alkyl group and n has an average value between 1 and 4.

Information Sources: JCIC, JCLS, JSQI

Chemical Class: Alkyl Ether Sulfates

Function: Surfactant - Cleansing Agent

Technical/Other Names:
Sodium Pareth-25 Sulfate
Sodium Polyoxyethylene Alkyl (12,15) Ether Sulfate (3E.O.)
Sulfuric Acid, Mono[2-[2-[2-(C12-15 Alkyloxy)Ethoxy]Ethoxy]Ethyl] Esters, Sodium Salts

Trade Names:
Empicol ESB 3/X (Albright & Wilson UK)
Empicol ESB70/X (Albright & Wilson UK)
Empimin KSN70/L (Albright & Wilson UK)
Nikkol NES-203-27 (Nikko)
Rhodasurf L-790 (Rhodia)
Rhodasurf LA Series (Rhodia)
Zetesol AO 328 (Zschimmer & Schwarz)
Zoharpon ETA 270 (Zohar)
Zoharpon ETA 603 (Zohar)
Zoharpon ETA 700 (Zohar)

SODIUM C12-15 PARETH-3 SULFATE

JPN Translation:
（C12 - 15）バレス - 3 硫酸 Na

Definition: Sodium C12-15 Pareth-3 Sulfate is the sodium salt of a sulfated polyethylene glycol ether of a mixture of synthetic C12-15 alcohols. It conforms generally to the formula:

$$R(OCH_2CH_2)_nOSO_3Na$$

where R represents the C12-15 alkyl group and n has an average value of 3.

Information Source: JCLS

Chemical Class: Alkyl Ether Sulfates

Function: Surfactant - Cleansing Agent

SODIUM C13-15 PARETH-3 SULFATE

Definition: Sodium C13-15 Pareth-3 Sulfate is the sodium salt of a sulfated polyethylene glycol ether of a mixture of synthetic C13-15 alcohols. It conforms generally to the formula:

$$R(OCH_2CH_2)_nOSO_3Na$$

where R represents the C13-15 alkyl group and n has an average value of 3.

Information Source: JCLS

Chemical Class: Alkyl Ether Sulfates

Function: Surfactant - Cleansing Agent

Trade Names:
Empimin KSN27/X (Albright & Wilson UK)
Empimin KSN70/X (Albright & Wilson UK)

SODIUM C12-14 SEC-PARETH-3 SULFATE

CAS No.: 125736-54-1

Definition: Sodium C12-14 Sec-Pareth-3 Sulfate is the sodium salt of a sulfated mixture of synthetic secondary C12-14 alcohols with an average ethoxylation value of 3.

Chemical Classes: Alkoxylated Alcohols; Sulfuric Acid Esters

Function: Surfactant - Cleansing Agent

Trade Name:
SOFTANOL 30S (Nippon Shokubai)

SODIUM C12-15 PARETH-3 SULFONATE

Definition: Sodium C12-15 Pareth-3 Sulfonate is the sodium sulfonate of an ethoxylated synthetic fatty alcohol. It conforms generally to the formula:

$$R(OCH_2CH_2)_nSO_3Na$$

where R represents the C12-15 fatty alcohols and n has an average value of 3.

Chemical Class: Sulfonic Acids

Function: Surfactant - Cleansing Agent

SODIUM C12-15 PARETH-7 SULFONATE

Definition: Sodium C12-15 Pareth-7 Sulfonate is the sodium sulfonate of an ethoxylated synthetic fatty alcohol. It conforms generally to the formula:

$$R(OCH_2CH_2)_nSO_3Na$$

where R represents the C12-15 fatty alcohol moiety and n has an average value of 7.

Chemical Class: Sulfonic Acids

Function: Surfactant - Cleansing Agent

SODIUM C12-15 PARETH-15 SULFONATE

Definition: Sodium C12-15 Pareth-15 Sulfonate is the sodium sulfonate of an ethoxylated synthetic fatty alcohol. It conforms generally to the formula:

$$R(OCH_2CH_2)_nSO_3Na$$

where R represents C12-15 fatty alcohols and n has an average value of 15.

Chemical Class: Sulfonic Acids

Function: Surfactant - Cleansing Agent

Reported Product Category: Hair Straighteners

Trade Names:
AEC Sodium C12-15 Pareth-15 Sulphonate (A & E Connock)
Avanel S-150 CGN (BASF)

SODIUM CUMENESULFONATE

CAS Nos.	EINECS Nos.
28348-53-0	248-983-7
32073-22-6	250-913-5

Empirical Formula:
$C_9H_{12}O_3S \cdot Na$

Definition: Sodium Cumenesulfonate is the substituted aromatic compound that conforms generally to the formula:

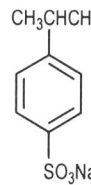

Chemical Class: Alkyl Aryl Sulfonates

Function: Surfactant - Hydrotrope

Technical/Other Names:
Benzene, (1-Methylethyl)-, Monosulfo Deriv., Sodium Salt
(1-Methylethyl)Benzene, Monosulfo Deriv., Sodium Salt
Sodium Monoisopropylbenzenesulfonate

Trade Names:
Eltesol SC40 (Albright & Wilson UK)
Eltesol SC93 (Albright & Wilson UK)
Eltesol SC Pellets (Albright & Wilson UK)
STEPANATE SCS (Stepan)

Trade Name Mixture:
KNa-Cumosulfonat (Sasol GmbH - Marl)

SODIUM CYCLAMATE

CAS No.: 139-05-9

Empirical Formula:
$C_6H_{12}NO_3S \cdot Na$

Definition: Sodium Cyclamate is the organic compound that conforms to the formula:

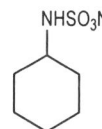

Chemical Class: Organic Salts

Function: Flavoring Agent

Technical/Other Names:
Sodium Cyclohexanesulfamate
Sodium Cyclohexylaminesulfonate

Trade Names:
PA Cyclamate (Productos Additivos)
Sodium Cyclamate Brasfanta (Brasfanta)

SODIUM CYCLODEXTRIN SULFATE

CAS No.: 37191-69-8

Definition: Sodium Cyclodextrin Sulfate is the sodium salt of sulfated Cyclodextrin (q.v.).

Chemical Classes: Carbohydrates; Organic Salts

Functions: Emulsion Stabilizer; Viscosity Increasing Agent - Aqueous

Trade Name:
Cavitron Cyclodextrin - Sulfated (Cerestar USA)

SODIUM CYCLOPENTANE CARBOXYLATE

CAS No.: 17273-89-1

Empirical Formula:
$C_6H_{10}O_2 \cdot Na$

Definition: Sodium Cyclopentane Carboxylate is the sodium salt of cyclopentane carboxylic acid that conforms to the formula:

COONa

Chemical Class: Organic Salts

Function: Not Reported

Technical/Other Name:
Cyclopentane Carboxylic Acid, Sodium Salt

SODIUM DECETH-2 CARBOXYLATE

CAS No.: 38815-93-9 (Generic)

Empirical Formula:
$C_{14}H_{28}O_4 \cdot Na$

Definition: Sodium Deceth-2 Carboxylate is the sodium salt of the organic acid that conforms generally to the formula:

$$CH_3(CH_2)_9OCH_2CH_2OCH_2COONa$$

Chemical Class: Organic Salts

Function: Surfactant - Cleansing Agent

Technical/Other Names:
PEG-2 Decyl Ether Carboxylic Acid, Sodium Salt
Polyethylene Glycol 100 Decyl Ether Carboxylic Acid, Sodium Salt
Polyoxyethylene (2) Decyl Ether Carboxylic Acid, Sodium Salt
Sodium Polyethylene Glycol 100 Decyl Ether Carboxylate
Sodium Polyoxyethylene (2) Decyl Ether Carboxylate

Trade Name Mixture:
Akypo OCD 10 NV (Kao GmbH)

SODIUM DECETH SULFATE

Definition: Sodium Deceth Sulfate is the sodium salt of sulfated ethoxylated decyl alcohol that conforms generally to the formula:

$$CH_3(CH_2)_9(OCH_2CH_2)_nOSO_3Na$$

where n has an average value between 1 and 4.

Chemical Class: Alkyl Ether Sulfates

Function: Surfactant - Cleansing Agent

Technical/Other Name:
Sodium Decyl Ether Sulfate

SODIUM DECYLBENZENESULFONATE

CAS No.	EINECS No.
1322-98-1	215-347-5

Empirical Formula:
$C_{16}H_{26}O_3S \cdot Na$

Definition: Sodium Decylbenzenesulfonate is the substituted aromatic compound that conforms generally to the formula:

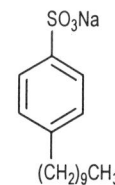

Information Sources: 21CFR172.210, 21CFR175.105, 21CFR175.300, 21CFR176.170, 21CFR176.210, 21CFR177.1210, 21CFR177.2600, 21CFR178.3130, 21CFR178.3400, CIR: [S] JACT-12(3)1993

Chemical Class: Alkyl Aryl Sulfonates

Function: Surfactant - Cleansing Agent

Technical/Other Names:
Benzenesulfonic Acid, Decyl-, Sodium Salt
Decylbenzenesulfonic Acid, Sodium Salt

SODIUM DECYL SULFATE

CAS No.	EINECS No.
142-87-0	205-568-5

Empirical Formula:
$C_{10}H_{22}O_4S \cdot Na$

Definition: Sodium Decyl Sulfate is the sodium salt of decyl sulfate that conforms to the formula:

$$CH_3(CH_2)_9OSO_3Na$$

Information Source: JCLS

Chemical Class: Alkyl Sulfates

Function: Surfactant - Foam Booster

Technical/Other Name:
Sulfuric Acid, Monodecyl Ester, Sodium Salt

Trade Name:
 Empicol 0758 (Albright & Wilson UK)

SODIUM DEHYDROACETATE

CAS No.   **EINECS No.**
4418-26-2   224-580-1

JPN Translation:
 デヒドロ酢酸 Na

Empirical Formula:
 $C_8H_8O_4 \cdot Na$

Definition: Sodium Dehydroacetate is the heterocyclic compound that conforms generally to the formula:

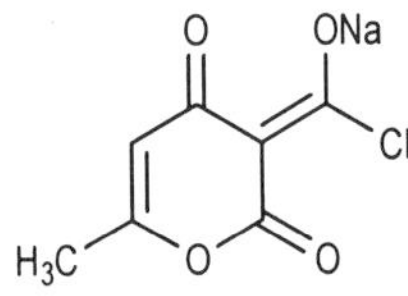

Information Sources: 21CFR172.130, 21CFR175.105, CIR: [S] JACT-4(3)1985, CTFA S, EEC(VI/1-13), FCC, JCLS, JSCI, MHLW-331/3, NF XIX, SNPF, TSCA, USAN

Chemical Classes: Heterocyclic Compounds; Organic Salts

Function: Preservative

Reported Product Categories: Eye Shadows; Mascara; Moisturizing Preparations; Face Powders; Skin Care Preparations, Misc.; Bath Preparations, Misc.; Body and Hand Preparations (Excluding Shaving Preparations); Blushers (All types); Bath Oils, Tablets, and Salts; Cleansing Products (Cold Creams, Cleansing Lotions, Liquids and Pads); Eye Makeup Preparations, Misc.; Foundations; Eyeliners; Makeup Preparations (Not eye), Misc.; Paste Masks (Mud Packs); Bath Capsules; Makeup Bases; Eye Lotions; Face and Neck Preparations (Excluding Shaving Preparations); Hair Preparations (Non-coloring), Misc.; Night Skin Care Preparations; Shaving Cream (Aerosol, Brushless and Lather); Nail Creams and Lotions; Powders (Dusting and Talcum, Excluding Aftershave Talcs); Suntan Gels, Creams, and Liquids; Suntan Preparations, Misc.

Technical/Other Names:
 3-Acetyl-6-Methyl-2H-Pyran-2,4(3H)-Dione, Ion(1-), Sodium Salt
 2H-Pyran-2,4(3H)-dione, 3-Acetyl-6-Methyl-, Ion(1-), Sodium

Trade Names:
 Tristat SDHA (Tri-K)
 Unisept DSA (Universal Preserv-A-Chem)

Trade Name Mixture:
 Fondix G Bis (Gattefosse s.a.)

SODIUM DERMATAN SULFATE

Definition: Sodium Dermatan Sulfate is the sodium salt of the natural mucopolysaccharide, N-acetyl-D-galactosamine 4-sulfate.

Chemical Classes: Biological Polymers and their Derivatives; Carbohydrates

Function: Skin-Conditioning Agent - Miscellaneous

Trade Name:
 Dermatan Sulfate (Bioiberica)

Trade Name Mixture:
 Heparinoid (Bioiberica)

SODIUM DEXTRAN SULFATE

CAS No.: 9011-18-1

JPN Translation:
 デキストラン硫酸 Na

Definition: Sodium Dextran Sulfate is the sodium salt of the sulfuric acid ester of Dextran (q.v.).

Information Sources: JAN, JCIC, JCLS, JSQI, MI-13(2969)

Chemical Class: Biological Polymers and their Derivatives

Function: Suspending Agent - Nonsurfactant

Technical/Other Names:
 Dextran, Hydrogen Sulfate, Sodium Salt
 Dextran Sulfuric Acid Ester Sodium Salt

Trade Name:
 Dextran Sulfate Sodium (Meito Sangyo)

SODIUM DEXTRIN OCTENYLSUCCINATE

Definition: Sodium Dextrin Octenylsuccinate is the sodium salt of the reaction product of octenylsuccinic anhydride with Dextrin (q.v.).

Chemical Classes: Biological Polymers and their Derivatives; Carbohydrates

Functions: Emulsion Stabilizer; Hair Conditioning Agent; Humectant; Skin-Conditioning Agent - Emollient; Surfactant - Emulsifying Agent

Trade Names:
 Naturalnisk NR (Nippon Starch)
 Niskpowder N (Nippon Starch)

SODIUM DICARBOXYETHYLCOCO PHOS-PHOETHYL IMIDAZOLINE

Definition: Sodium Dicarboxyethylcoco Phosphoethyl Imidazoline is the substituted imidazoline compound that conforms generally to the formula:

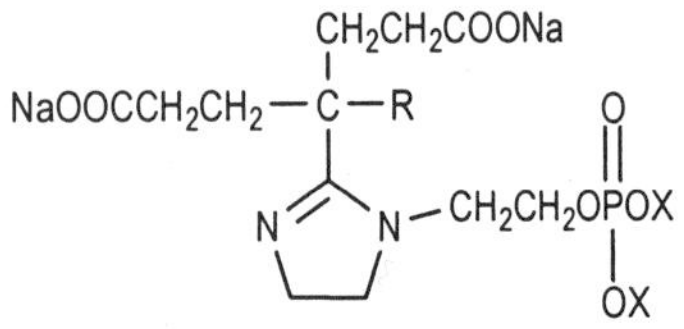

where R respresents the alkyl groups derived from coconut acid and X may be either Na or H.

Chemical Classes: Imidazoline Compounds; Phosphorus Compounds

Functions: Antistatic Agent; Hair Conditioning Agent

Trade Name:
 Phosphoteric T-C6 (Uniqema)

SODIUM DICETEARETH-10 PHOSPHATE

Definition: Sodium Diceteareth-10 Phosphate is the sodium salt of a complex mixture of phosphate diesters of Ceteareth-10 (q.v.).

Chemical Class: Phosphorus Compounds

Function: Surfactant - Emulsifying Agent

SODIUM DICOCOYLETHYLENEDIAMINE PEG-15 SULFATE

Definition: Sodium Dicocoylethylenediamine PEG-15 Sulfate is the organic compound that conforms generally to the formula:

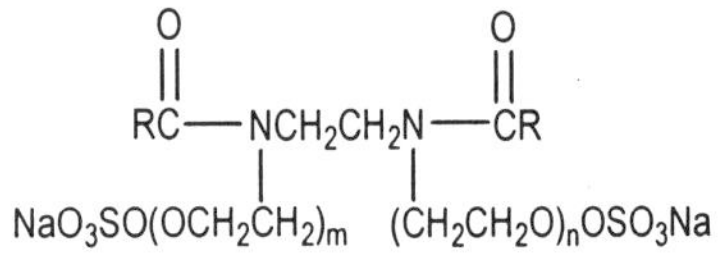

where m + n has an average value of 15 and RCO- represents the coconut acid radical.

Chemical Class: Sulfuric Acid Esters

Functions: Hair Conditioning Agent; Surfactant - Cleansing Agent; Surfactant - Foam Booster

Trade Name Mixtures:
 Ceralution C (Sasol GmbH - Marl)
 Ceralution F (Sasol GmbH - Marl)
 Ceralution H (Sasol GmbH - Marl)

SODIUM DIETHYLAMINOPROPYL COCO-ASPARTAMIDE

Definition: Sodium Diethylaminopropyl Cocoaspartamide is the amidoamine that conforms generally to the formula:

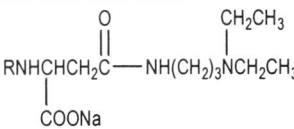

where R represents the alkyl groups derived from coconut oil.

Chemical Classes: Alkyl-Substituted Amino Acids; Amides

Function: Hair Conditioning Agent

Reported Product Category: Hair Dyes and Colors (All Types Requiring Caution Statements and Patch Tests)

Technical/Other Name:
Sodium Monodiethylaminopropyl Coco-aspartamide

Trade Name:
Chimexane HB (Chimex)

SODIUM DIETHYLENETRIAMINE PENTAMETHYLENE PHOSPHONATE

CAS No.   **EINECS No.**
22042-96-2   244-751-4

Empirical Formula:
$C_9H_{23}N_3O_{15}P_5Na_5$

Definition: Sodium Diethylenetriamine Pentamethylene Phosphonate is the substituted amine that conforms generally to the formula:

Information Source: TSCA

Chemical Classes: Amines; Phosphorus Compounds

Function: Chelating Agent

Technical/Other Names:
Phosphonic Acid, [[(Phosphonomethyl)-imino]bis[2,1-Ethanediylnitrilobis (Methylene)]]Tetrakis-, Sodium Salt
[[(Phosphonomethyl)imino]bis [(Ethylenenitrilo)bis(Methylene)]]-Tetrakisphosphonic Acid, Sodium Salt

Trade Names:
Dequest 2066 (Solutia)
Sequion 40 Na 32 (Bozzetto)

SODIUM DIHYDROXYCETYL PHOSPHATE

CAS No.   **EINECS No.**
94277-32-4   304-615-8

Definition: Sodium Dihydroxycetyl Phosphate is the sodium salt of a complex mixture of phosphate esters of dihydroxycetyl alcohol.

Chemical Class: Phosphorus Compounds

Function: Surfactant - Emulsifying Agent

Trade Name:
Dragophos S 2/918501 (Symrise)

SODIUM DIHYDROXYETHYLGLYCINATE

CAS Nos.   **EINECS No.**
139-41-3   205-360-4
17123-43-2
60168-81-2

Empirical Formula:
$C_6H_{13}NO_4 \cdot Na$

Definition: Sodium Dihydroxyethylglycinate is the salt of the substituted amino acid that conforms to the formula:

$$HOCH_2CH_2 \diagdown \atop HOCH_2CH_2 \diagup NCH_2COONa$$

Information Sources: CTFA D, TSCA

Chemical Class: Alkyl-Substituted Amino Acids

Function: Chelating Agent

Technical/Other Names:
N,N-bis-2-Hydroxyethylaminoacetic Acid Sodium Salt
N,N-Bis(2-Hydroxyethyl)Glycine, Monosodium Salt
Glycine, N,N-Bis(2-Hydroxyethyl)-, Monosodium Salt
Sodium N,N-Bis-2-Hydroxyethyl Glycinate
Sodium Dihydroxyglycinate

Trade Name:
Hampshire DEG (Amerchol)

Trade Name Mixture:
Kelene 77 (Lowenstein)

SODIUM DILAURETH-7 CITRATE

Empirical Formula:
$C_{55}H_{113}O_{21} \cdot Na$

Definition: Sodium Dilaureth-7 Citrate is the sodium salt of the diester of Citric Acid (q.v.) and Laureth-7 (q.v.).

Chemical Classes: Esters; Organic Salts

Function: Surfactant - Cleansing Agent

Trade Name:
Eucarol D (Cesalpinia)

SODIUM DILAURETH-10 PHOSPHATE

Definition: Sodium Dilaureth-10 Phosphate is the sodium salt of a complex mixture of phosphate diesters of Laureth-10 (q.v.).

Chemical Class: Phosphorus Compounds

Functions: Surfactant - Cleansing Agent; Surfactant - Solubilizing Agent

Trade Name:
NIKKOL DLP-10 (Nikko)

SODIUM DILINOLEAMIDOPROPYL PG-DIMONIUM CHLORIDE PHOSPHATE

Definition: Sodium Dilinoleamidopropyl PG-Dimonium Chloride Phosphate is the quaternary ammonium salt that conforms generally to the formula:

where RCO- represents the dilinoleic acid moiety.

Chemical Classes: Phosphorus Compounds; Quaternary Ammonium Compounds

Functions: Hair Conditioning Agent; Surfactant - Cleansing Agent; Surfactant - Foam Booster

Trade Name:
Phoenotaine D-35 (Phoenix)

SODIUM DILINOLEATE

CAS No.: 67701-20-6

Definition: Sodium Dilinoleate is the sodium salt of Dilinoleic Acid (q.v.).

Chemical Class: Soaps

Function: Surfactant - Cleansing Agent

Technical/Other Name:
Fatty Acids, C18-Unsatd., Dimers, Sodium Salts

Trade Name:
Sodium Dimerate (Uniqema/Belgium)

SODIUM DIMALTODEXTRIN PHOSPHATE

Definition: Sodium Dimaltodextrin Phosphate is the sodium salt of a complex mixture of diesters of Maltodextrin (q.v.) and phosphoric acid.

Chemical Classes: Carbohydrates; Phosphorus Compounds

Function: Suspending Agent - Nonsurfactant

Trade Name:
Kappaline-P (Kappa Biotech)

SODIUM DIMETHICONE PEG-7 ACETYL METHYLTAURATE

Definition: Sodium Dimethicone PEG-7 Acetyl Methyltaurate is the siloxane polymer that conforms generally to the formula:

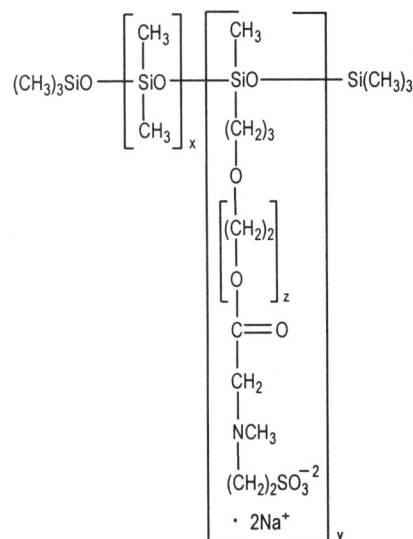

where z has an average value of 7.

Chemical Class: Siloxanes and Silanes

Functions: Hair Conditioning Agent; Skin-Conditioning Agent - Miscellaneous

Trade Name:
Pecosil DCT (Phoenix)

SODIUM DIOLETH-8 PHOSPHATE

Definition: Sodium Dioleth-8 Phosphate is the sodium salt of a complex mixture of phosphate diesters of Oleth-8 (q.v.).

Chemical Class: Phosphorus Compounds

Function: Surfactant - Cleansing Agent

Trade Name:
Nikkol DOP-8N (Nikko)

SODIUM DNA

JPN Translation:
DNA - Na

Definition: Sodium DNA is the sodium salt of DNA (q.v.).

Information Sources: JCIC, JCLS, JSQI

Chemical Classes: Biological Polymers and their Derivatives; Phosphorus Compounds

Function: Skin-Conditioning Agent - Miscellaneous

Reported Product Category: Moisturizing Preparations

Technical/Other Names:
DNA, Sodium Salt
Sodium Deoxyribonucleic Acid

Trade Names:
AEC Sodium DNA (A & E Connock)
BIODNA (Chemyunion)
HPDR: Highly Polymerized Desoxyribonucleic Acid (Sodium Salt) (Javenech)
Nucleic Acids (Sodium Form) (Bioiberica)

Trade Name Mixture:
HPDR: Highly Polymerized Desoxyribonucleic Acid (Mixed Sodium, Calcium, and Magnesium Salt) (Javenech)

SODIUM DODECYLBENZENESULFONATE

CAS No.
25155-30-0

EINECS No.
246-680-4

JPN Translation:
ドデシルベンゼンスルホン酸 Na

Empirical Formula:
$C_{18}H_{30}O_3S \cdot Na$

Definition: Sodium Dodecylbenzene-sulfonate is the substituted aromatic compound that conforms generally to the formula:

$$SO_3Na$$

$(CH_2)_{11}CH_3$

Information Sources: 21CFR173.315, 21CFR175.105, 21CFR175.300, 21CFR175.320, 21CFR176.210, 21CFR177.1010, 21CFR177.1200, 21CFR177.1630, 21CFR177.2600, 21CFR177.2800, 21CFR178.3120, 21CFR178.3130, 21CFR178.3400, CIR: [S] JACT-12(3)1993, CTFA S, JCIC, JCLS, JSQI, MI-13(8686), TSCA

Chemical Class: Alkyl Aryl Sulfonates

Function: Surfactant - Cleansing Agent

Reported Product Categories: Bath Soaps and Detergents; Eyeliners

Technical/Other Names:
Benzenesulfonic Acid, Dodecyl-, Sodium Salt
Dodecylbenzenesulfonic Acid, Sodium Salt
Sodium Lauryl Benzene Sulfonate
Sodium Lauryl Phenyl Sulfonate

Trade Names:
BIO-SOFT D-40 (Stepan)
BIO-SOFT D-62 (Stepan)
Calsoft F-90 (Pilot)
Calsoft L-40 (Pilot)
Calsoft L-60 (Pilot)
Hetsulf Acid (Heterene)
Maranil A 55 (Cognis Care Chemicals/NJ)
Maranil Paste A 55 (Cognis Deutschland)
Marlon A 350 (Sasol GmbH - Marl)
Marlon A 360 (Sasol GmbH - Marl)
Marlon A 365 (Sasol GmbH - Marl)
Marlon A375 (Sasol GmbH - Marl)
NACCONOL 40G (Stepan)
NACCONOL 90G (Stepan)
Nansa HS80/S (Albright & Wilson UK)
Nansa HS85/S (Albright & Wilson UK)
Norfox 90 (Norman, Fox & Co.)
Rhodacal DS-10 (Rhodia)
Rhodacal SS-45 (Rhodia)
Rhodacal SS-60 (Rhodia)
Toshiki TiO2 T-144C (Nikko)

Trade Name Mixtures:
AEC Dimethiconol & Sodium Dodecyl-benzenesulfonate (A & E Connock)
Akypogene VSM-N (Kao GmbH)
B C Dimethiconol Emulsion 95 (Basildon)
BIO-SOFT LD-95 (Stepan)
Lytron 614 (Rohm and Haas)
Silsoft E-623 (OSi Specialties)
SM2725 (GE Silicones)
SM2765 (GE Silicones)
Wacker-Belsil DM 3112 VP (Wacker-Chemie)

SODIUM DVB/ACRYLATES COPOLYMER

Definition: Sodium DVB/Acrylates Copolymer is the sodium salt of a polymer of divinyl benzene and one or more monomers consisting of acrylic acid, methacrylic acid or their simple esters.

Chemical Class: Synthetic Polymers

Function: Film Former

SODIUM EDTMP

CAS No.
22036-77-7

EINECS No.
244-742-5

Empirical Formula:
$C_6H_{20}N_2O_{12}P_4 \cdot xNa$

Definition: Sodium EDTMP is the substituted diamine that conforms to the formula:

$$H_2O_3PCH_2 \qquad CH_2PO_3H_2$$
$$NCH_2CH_2N \qquad \cdot Na$$
$$H_2O_3PCH_2 \qquad CH_2PO_3H_2$$

Chemical Classes: Alkyl-Substituted Amino Acids; Amines; Phosphorus Compounds

Function: Chelating Agent

Technical/Other Names:
[Ethylenebis[Nitrilobis(Methylene)]]-Tetrakisphosphonic Acid, Sodium Salt
Sodium Ethylenediamine Tetramethylene Phosphonate

Trade Name:
Sequion D 30 (Bozzetto)

SODIUM EMUAMIDOPROPYL PG-DIMONIUM CHLORIDE PHOSPHATE

Definition: Sodium Emuamidopropyl PG-Dimonium Chloride Phosphate is the quaternary ammonium salt that conforms generally to the formula:

$$\left[RCNH(CH_2)_3 - \overset{\overset{\displaystyle CH_3}{|}}{\underset{\underset{\displaystyle CH_3}{|}}{N}} - CH_2\overset{\overset{\displaystyle O}{\|}}{\underset{\underset{\displaystyle OH}{|}}{C}HCH_2O} \ \overset{\overset{\displaystyle O}{\|}}{PONa} \right]^{+} \cdot 3Cl^{-}$$

where RCO- represents the fatty acids derived from Emu Oil (q.v.).

Chemical Class: Quaternary Ammonium Compounds

Functions: Antistatic Agent; Hair Conditioning Agent

SODIUM ERYTHORBATE

CAS No. 6381-77-7

EINECS No. 228-973-9

JPN Translation:
エリソルビン酸 Na

Empirical Formula:
$C_6H_8O_6 \cdot Na$

Definition: Sodium Erythorbate is the sodium salt of Erythorbic Acid (q.v.) that conforms to the formula:

Information Sources: CIR: [S] IJT-18 (SUPPL. 3)1999, FCC, JCLS, MAR, MI-13 (5143)

Chemical Classes: Heterocyclic Compounds; Organic Salts

Function: Antioxidant

Reported Product Category: Hair Dyes and Colors (All Types Requiring Caution Statements and Patch Tests)

Technical/Other Names:
Araboascorbic Acid, Monosodium Salt
Erythorbic Acid Sodium Salt
D-Erythro-Hex-2-Enonic Acid, γ-Lactone, Monosodium Salt
Sodium Isoascorbate

Trade Name:
Uantox SEBATE (Universal Preserv-A-Chem)

SODIUM ETHYL ESTER OF PVM/MA COPOLYMER

Definition: Sodium Ethyl Ester of PVM/MA Copolymer is the sodium salt of Ethyl Ester of PVM/MA Copolymer (q.v.).

Chemical Class: Synthetic Polymers

Functions: Binder; Film Former; Hair Fixative

SODIUM ETHYLHEXYL SULFATE

CAS No. 126-92-1

EINECS No. 204-812-8

Empirical Formula:
$C_8H_{18}O_4S \cdot Na$

Definition: Sodium Ethylhexyl Sulfate is the sodium salt of 2-ethylhexyl sulfate that conforms to the formula:

$$\underset{\underset{\displaystyle CH_3CH_2}{|}}{CH_3(CH_2)_3CHCH_2OSO_3Na}$$

Information Sources: 21CFR173.315, 21CFR175.105, 21CFR176.170, INN, JCLS, TSCA, USAN

Chemical Class: Alkyl Sulfates

Function: Surfactant - Hydrotrope

Technical/Other Names:
Sodium 2-Ethylhexyl Sulfate
Sodium Octyl Sulfate
Sulfuric Acid, Mono(2-Ethylhexyl) Ester, Sodium Salt

Trade Names:
DeSULF SEH-40 (DeForest)
Rhodapon OLS (Rhodia)

SODIUM ETHYLPARABEN

CAS No. 35285-68-8

EINECS No. 252-487-6

Empirical Formula:
$C_9H_{10}O_3 \cdot Na$

Definition: Sodium Ethylparaben is the sodium salt of Ethylparaben (q.v.) that conforms to the formula:

Information Sources: EEC(VI/1-12), MHLW-331/3

Chemical Classes: Esters; Organic Salts; Phenols

Function: Preservative

Reported Product Categories: Blushers (All types); Colognes and Toilet Waters; Makeup Preparations (Not eye), Misc.

Technical/Other Names:
Benzoic Acid, 4-Hydroxy-, Ethyl Ester, Sodium Salt
Ethyl p-Hydroxybenzoate Sodium Salt
Ethylparaben, Sodium Salt
4-Hydroxybenzoic Acid, Ethyl Ester, Sodium Salt

Trade Names:
Ethyl 4-hydroxybenzoate Sodium Salt (Merck KGaA)
Nipagin A Sodium (Clariant)
Nipagin A Sodium (Clariant GmbH, Personal Care)

Trade Name Mixtures:
Nipacombin A (Clariant)
Nipacombin A (Clariant GmbH, Personal Care)
Nipasept Sodium (Clariant)
Nipasept Sodium (Clariant GmbH, Personal Care)
Nipastat Sodium (Clariant)
Nipastat Sodium (Clariant GmbH, Personal Care)

SODIUM ETHYL 2-SULFOLAURATE

CAS No.: 7381-01-3

Empirical Formula:
$C_{14}H_{27}O_5S \cdot Na$

Definition: Sodium Ethyl 2-Sulfolaurate is the organic compound that conforms to the formula:

$$\underset{\underset{\displaystyle SO_3Na}{|}}{CH_3(CH_2)_9\overset{\overset{\displaystyle O}{\|}}{C}HC - OCH_2CH_3}$$

Chemical Classes: Esters; Sulfonic Acids

Function: Surfactant - Cleansing Agent

Reported Product Category: Bath Preparations, Misc.

Technical/Other Names:
Dodecanoic Acid, 2-Sulfoethyl Ester, Sodium Salt
Sodium 2-Sulfoethyldodecanoate

SODIUM FLUORIDE

CAS No. 7681-49-4
EINECS No. 231-667-8

Empirical Formula:
FNa

Definition: Sodium Fluoride is the inorganic salt that conforms to the formula:

NaF

In the United States, Sodium Fluoride may be used as an active ingredient in OTC drug products. When used as an active drug ingredient, the established name is *Sodium Fluoride. See "Regulatory and Ingredient Use Information," regarding the labeling names for U.S. OTC Drug Ingredients in Volume 1, Introduction, Part A.*

Information Sources: AUS, BP, BPC, 21CFR175.105, 21CFR177.2800, 21CFR355.10, CZE, DDR, EEC(III/1-31), HP, HUN, JAN, MAR, MI-13(8691), OTC-I-AC, PN, ROM, TSCA, USAN, USD, USP XXIV

Chemical Class: Inorganic Salts

Functions: Anticaries Agent; Oral Care Agent

Reported Product Category: Dentifrices (Aerosol, Liquid, Pastes and Powders)

Trade Name:
RonaCare NaF (Merck KGaA)

SODIUM FLUOROSILICATE

CAS No. 16893-85-9
EINECS No. 240-934-8

Empirical Formula:
F_6H_2Si

Definition: Sodium Fluorosilicate is the inorganic salt that conforms to the formula:

Na_2SiF_6

Information Sources: EEC(III/1-40), MI-13 (8698), TSCA

Chemical Class: Inorganic Salts

Function: Oral Care Agent

Technical/Other Names:
Silicate, Hexafluoro-, Disodium
Sodium Hexafluorosilicate

SODIUM FORMATE

CAS No. 141-53-7
EINECS No. 205-488-0

Empirical Formula:
CH_2O_2 • Na

Definition: Sodium Formate is the organic salt that conforms to the formula:

HCOONa

Information Sources: 21CFR186.1756, EEC(VI/1-14), MI-13(8694)

Chemical Class: Organic Salts

Function: Preservative

Reported Product Category: Shampoos (Non-coloring)

Technical/Other Name:
Formic Acid, Sodium Salt

SODIUM FRUCTOBORATE

Definition: Sodium Fructoborate is the sodium salt of the reaction product of Boric Acid (q.v.) and Fructose (q.v.).

Chemical Classes: Carbohydrates; Organic Salts

Function: Skin-Conditioning Agent - Humectant

Trade Name:
Fruit-BX (VDF FutureCeuticals)

SODIUM FUMARATE

CAS Nos. 5873-57-4 / 7704-73-6
EINECS Nos. 227-535-4 / 231-725-2

JPN Translation:
フマル酸 Na

Empirical Formula:
$C_4H_3O_4$ • Na

Definition: Sodium Fumarate is the sodium salt of Fumaric Acid (q.v.). It conforms to the formula:

HOOCCH＝CHCOONa

Chemical Class: Organic Salts

Functions: Buffering Agent; pH Adjuster

Technical/Other Names:
2-Butenedioic Acid, Monosodium Salt

Fumaric Acid, Monosodium Salt
Monosodium Fumarate

SODIUM GLUCEPTATE

CAS Nos. 13007-85-7 / 31138-65-5
EINECS Nos. 235-849-8 / 250-480-2

Empirical Formula:
$C_7H_{14}O_8$ • Na

Definition: Sodium Gluceptate is the organic salt that conforms to the formula:

$$HOCH_2CH - \overset{\overset{\displaystyle OH}{|}}{CH}CHCH - CHCOONa$$
$$\quad\quad\quad\; \underset{OH\;\;\;OH\;\;\;OH\;\;\;OH}{}$$

Information Sources: MI-13(4468), USAN

Chemical Class: Organic Salts

Function: Chelating Agent

Technical/Other Names:
Gluceptate, Sodium
D-Glycero-D-gulo-Heptonic Acid, Monosodium Salt
Sodium Glucoheptonate

Trade Name:
Sodium Heptonate Dihydrate 300 (Croda Chemicals)

SODIUM GLUCONATE

CAS No. 527-07-1
EINECS No. 208-407-7

JPN Translation:
グルコン酸 Na

Empirical Formula:
$C_6H_{12}O_7$ • Na

Definition: Sodium Gluconate is the sodium salt of gluconic acid. It conforms to the formula:

$$HOCH_2CH - CHCHCHCOONa$$
$$\quad\quad\quad \underset{OH\;\;\;OH\;\;\;OH}{}$$

Information Sources: 21CFR182.6757, JCIC, JCLS, JSQI, MI-13(8695), TSCA, USAN, USP XXIV

Chemical Class: Organic Salts

Functions: Chelating Agent; Skin-Conditioning Agent - Miscellaneous

Technical/Other Name:
D-Gluconic Acid, Monosodium Salt

Trade Names:
Givobio GNa (SEPPIC)

Gluconal NG-C (Glucona America)
Jungbunzlauer Sodium Gluconate
(Jungbunzlauer)

Trade Name Mixture:
Dermawhite NF (Laboratoires Sero-
biologiques)

SODIUM GLUCURONATE

CAS Nos.   **EINECS No.**
4934-42-3
14984-34-0   239-065-7

Empirical Formula:
$C_6H_{10}O_7 \cdot Na$

Definition: Sodium Glucuronate is the
sodium salt of Glucuronic Acid (q.v.).

Information Source: JAN

Chemical Classes: Carbohydrates; Organic
Salts; Polyols

Functions: Humectant; Skin-Conditioning
Agent - Humectant

Technical/Other Names:
Glucuronic Acid, Sodium Salt
Glycopyranuronic Acid, Monosodium Salt

Trade Name Mixture:
Seanamine BD (Laboratoires Sero-
biologiques)

SODIUM GLUTAMATE

CAS Nos.   **EINECS Nos.**
142-47-2   205-538-1
6106-04-3
16177-21-2 (L-Form)   240-313-1
32221-81-1 (dl-alpha)

JPN Translation:
グルタミン酸 Na

Empirical Formula:
$C_5H_{10}NO_4 \cdot Na$

Definition: Sodium Glutamate is the
monosodium salt of the L-form of glutamic
acid. It conforms to the formula:

HOOCCH₂CH₂CHCOONa
|
NH₂

Information Sources: 21CFR145.131,
21CFR155.120, 21CFR155.130,
21CFR155.170, 21CFR155.200,
21CFR158.170, 21CFR161.190,
21CFR169.115, 21CFR169.140,
21CFR169.150, 21CFR172.320,
21CFR182.1, JCLS, MI-13(6278), NF XVIII,
RIFM, TSCA, USAN

Chemical Class: Amino Acids

Functions: Fragrance Ingredient; Hair
Conditioning Agent; Skin-Conditioning Agent
- Miscellaneous

Reported Product Categories: Foundations; Shampoos (Non-coloring); Makeup
Fixatives; Permanent Waves

Technical/Other Names:
Glutamic Acid, Monosodium Salt
Monosodium glutamate (RIFM)
Monosodium L-Glutamate
Monosodium L-Glutamate Monohydrate

Trade Names:
AEC Sodium Glutamate (A & E Connock)
Monosodium L-Glutamate (Ajinomoto)

Trade Name Mixture:
Moisturizing Factor Hygro-Complex ARO
5272 (Crodarom)

SODIUM GLYCERETH-1 POLYPHOSPHATE

Definition: Sodium Glycereth-1
Polyphosphate is the sodium salt of a
complex mixture of esters of tetraphosphoric
acid with a 1 mole ethoxylate of glycerin.

Chemical Class: Phosphorus Compounds

Functions: Chelating Agent; Suspending
Agent - Nonsurfactant

SODIUM GLYCEROPHOSPHATE

CAS Nos.   **EINECS Nos.**
1334-74-3
1555-56-2
17603-42-8   241-577-0
39951-36-5   254-713-9
89923-83-1

Definition: Sodium Glycerophosphate is a
mixture of sodium salts of α and β
glycerophosphoric acid. It conforms generally
to the formula:

$$\begin{bmatrix} CH_2OPO_3H \\ | \\ CHOH \\ | \\ CH_2OH \end{bmatrix}^{-} Na^{+} \qquad \begin{bmatrix} CH_2OH \\ | \\ CHOPO_3 \\ | \\ CH_2OH \end{bmatrix}^{-2} 2Na^{+}$$

a form   b form

Information Source: MI-13(8696)

Chemical Classes: Glyceryl Esters and
Derivatives; Organic Salts; Phosphorus
Compounds

Function: Oral Care Agent

Technical/Other Names:
Glycerol, 1-(Dihydrogen Phosphate),
Sodium Salt
Sodium 3-Phosphoglycerate

Trade Name:
Givobio GPNa (SEPPIC)

SODIUM GLYCERYL OLEATE PHOS-PHATE

Definition: Sodium Glyceryl Oleate Phosphate is the sodium salt of a complex mixture
of phosphate esters of glyceryl monooleate.

Chemical Class: Phosphorus Compounds

Function: Surfactant - Cleansing Agent

SODIUM GRAPESEEDAMIDOPROPYL PG-DIMONIUM CHLORIDE PHOSPHATE

Definition: Sodium Grapeseedamidopropyl
PG-Dimonium Chloride Phosphate is the
quaternary ammonium salt that conforms
generally to the formula:

$$\begin{bmatrix} & O & & CH_3 & & & O & \\ & || & & | & & & || & \\ RCNH(CH_2)_3 & - & N & - & CH_2CHCH_2O & & PONa \\ & & & | & & | & & \\ & & & CH_3 & & OH & & \end{bmatrix}_2^{+} \cdot 2Cl^{-}$$

where RCO- represents the fatty acids
derived from grapeseed oil.

Chemical Class: Quaternary Ammonium
Compounds

Functions: Hair Conditioning Agent; Skin-
Conditioning Agent - Miscellaneous;
Surfactant - Cleansing Agent

Trade Name:
Colalipid GS (Colonial Chemical Inc)

SODIUM GUAIAZULENE SULFONATE

CAS No.   **EINECS No.**
6223-35-4   228-309-8

JPN Translation:
グアイアズレンスルホン酸 Na

Empirical Formula:
$C_{15}H_{18}O_3S \cdot Na$

Definition: Sodium Guaiazulene Sulfonate
is the organic compound that conforms to the
formula:

Information Sources: INN, JAN, JCIC, JCLS, JSQI

Chemical Class: Sulfonic Acids

Function: Not Reported

Technical/Other Names:
1-Azulenesulfonic Acid, 3,8-Dimethyl-5-(1-Methylethyl)-, Sodium Salt
3,8-Dimethyl-5-(1-Methylethyl)-1-Azulenesulfonic Acid, Sodium Salt
7-Isopropyl-1,4-dimethyl-3-sodiumsulfonate Azulene
Sodium 3,8-Dimethyl-5-(1-Methylethyl)-1-Azulenesulfonate
Sodium Guiazulene Sulfonate
Sodium 7-Isopropyl-1,4-Dimethylazulene-3-Sulfonate

SODIUM GUANOSINE CYCLIC MONO-PHOSPHATE

CAS No.	EINECS No.
40732-48-7	255-056-0

Empirical Formula:
$C_{10}H_{12}N_5O_7P \cdot Na$

Definition: Sodium Guanosine Cyclic Monophosphate is the heterocyclic compound that conforms to the formula:

Information Source: MI-13(2738)

Chemical Classes: Amines; Carbohydrates; Heterocyclic Compounds; Phosphorus Compounds

Function: Skin-Conditioning Agent - Miscellaneous

Technical/Other Names:
Cyclic GMP, Sodium Salt
Guanosine, Cyclic 3', 5'-(Hydrogen Phosphate), Monosodium Salt
Guanosine 3, 5'-Monophosphate, Sodium Salt

SODIUM HEPARIN

CAS No.: 9041-08-1 (Generic)

Definition: Sodium Heparin is the sodium salt of Heparin (q.v.).

Information Sources: BAN, INN, JAN, MI-13(4670), USAN, USP XXIV

Chemical Class: Biological Polymers and their Derivatives

Function: Skin-Conditioning Agent - Miscellaneous

Trade Name Mixture:
Heparinoid (Bioiberica)

SODIUM HEXAMETAPHOSPHATE

CAS Nos.	EINECS Nos.
10124-56-8	233-343-1
10361-03-2	
68915-31-1	272-808-3

Empirical Formula:
$H_6O_{18}P_6 \cdot 6Na$

Definition: Sodium Hexametaphosphate is the inorganic salt that conforms generally to the formula:

$$(NaPO_3)_6$$

Information Sources: 21CFR173.310, 21CFR182.90, 21CFR182.6760, CIR: [SQ] IJT-20(SUPPL. 3)2001, CTFA D, FCC, MI-13(8741), RIFM, TSCA

Chemical Classes: Inorganic Salts; Phosphorus Compounds

Functions: Chelating Agent; Corrosion Inhibitor; Fragrance Ingredient

Reported Product Categories: Bath Capsules; Foundations; Moisturizing Preparations; Mascara; Bath Preparations, Misc.; Body and Hand Preparations (Excluding Shaving Preparations); Skin Care Preparations, Misc.; Bath Oils, Tablets, and Salts; Eye Makeup Preparations, Misc.; Eyeliners

Technical/Other Names:
Graham's Salt
Metaphosphoric Acid, Hexasodium Salt
Sodium hexametaphosphate (RIFM)

Trade Names:
Calgon (ONDEO Nalco)
Calgon (ONDEO Nalco Europe)
Vitraphos (Rhodia)

SODIUM HEXETH-4 CARBOXYLATE

CAS No.: 126646-16-0

Empirical Formula:
$C_{14}H_{28}O_6 \cdot Na$

Definition: Sodium Hexeth-4 Carboxylate is the sodium salt of the carboxylic acid derived from an ethoxylated hexyl alcohol that conforms generally to the formula:

$$CH_3(CH_2)_5(OCH_2CH_2)_nOCH_2COONa$$

where n has an average value of 3.

Chemical Class: Organic Salts

Functions: Surfactant - Cleansing Agent; Surfactant - Hydrotrope

Technical/Other Names:
PEG-4 Hexyl Ether Carboxylic Acid, Sodium Salt
Polyethylene Glycol 200 Hexyl Ether Carboxylic Acid, Sodium Salt
Polyoxyethylene (4) Hexyl Ether Carboxylic Acid, Sodium Salt
Sodium Polyethylene Glycol 200 Hexyl Ether Carboxylate
Sodium Polyoxyethylene (4) Hexyl Ether Carboxylate
Sodium 3,6,9,12-Tetraoxaoctadecanoate
3,6,9,12-Tetraoxaoctadecanoic Acid, Sodium Salt

Trade Name Mixture:
Akypo LF 4 N (Kao GmbH)

SODIUM HEXYLDIPHENYL ETHER SULFONATE

CAS No.: 147732-60-3

Definition: Sodium Hexyldiphenyl Ether Sulfonate is the mixture organic compounds that conforms generally to the formula:

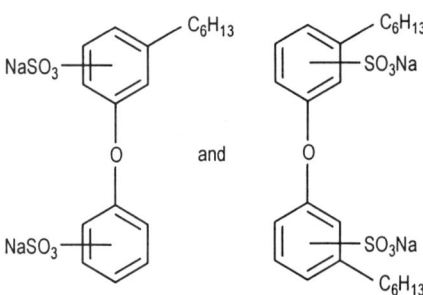

Information Source: TSCA

Chemical Classes: Ethers; Sulfonic Acids

Function: Surfactant - Hydrotrope

Technical/Other Name:
Benzene, 1,1'-Oxybis-, Sec-Hexyl Derivs., Sulfonated , Sodium Salts

Trade Name:
DOWFAX Hydrotrope Surfactant (Dow Chemical)

SODIUM HINOKITIOL

CAS No.: 17387-01-8

Definition: Sodium Hinokitiol is the sodium salt of Hinokitiol (q.v.).

Chemical Classes: Alcohols; Ketones; Organic Salts

Function: Cosmetic Biocide

Technical/Other Names:
2,4,6-Cycloheptatrien-1-one, 2-Hydroxy-4-(1-Methylethyl)-, Sodium Salt
Hinokitiol Sodium Salt

Trade Name:
Hinokitiol Sodium Salt (Osaka Organic)

SODIUM HYALURONATE

CAS No.: 9067-32-7

JPN Translation:
ヒアルロン酸 Na

Definition: Sodium Hyaluronate is the sodium salt of Hyaluronic Acid (q.v.).

Information Sources: JAN, JCIC, JCLS, JSQI, USAN

Chemical Class: Biological Polymers and their Derivatives

Function: Skin-Conditioning Agent - Miscellaneous

Reported Product Categories: Moisturizing Preparations; Lipsticks; Foundations; Skin Care Preparations, Misc.; Bath Capsules; Bath Oils, Tablets, and Salts; Makeup Bases; Makeup Preparations (Not eye), Misc.; Cleansing Products (Cold Creams, Cleansing Lotions, Liquids and Pads); Bath Preparations, Misc.; Body and Hand Preparations (Excluding Shaving Preparations); Eye Makeup Preparations, Misc.; Blushers (All types); Face and Neck Preparations (Excluding Shaving Preparations); Night Skin Care Preparations; Paste Masks (Mud Packs); Mascara; Skin Fresheners; Face Powders; Eye Lotions; Aftershave Lotions; Baby Shampoos; Eyeliners; Hair Conditioners; Eyebrow Pencils; Makeup Fixatives; Shampoos (Non-coloring); Suntan Gels, Creams, and Liquids; Eye Shadows; Suntan Preparations, Misc.

Technical/Other Names:
Hyaluronic Acid, Sodium Salt
Sodium Hyaluronate (1)
Sodium Hyaluronate (2)
Sodium Hyaluronate (3)
Sodium Hyaluronate Solution

Trade Names:
Actimoist (Active Organics)
AEC Sodium Hyaluronate (A & E Connock)
Avian Sodium Hyaluronate Powder (Intergen)
Avian Sodium Hyaluronate Solution (Intergen)
Bio-HE (Pacific)

Dekluron (Dekker)
HTL MYP Hyaluronic Acid (Lipo)
Hyaluronate Na F93 (HTL)
Hyaluronate Na F100 (HTL)
Hyaluronate Na 1,0% Gel (HTL)
Hyaluronate Na P85 (HTL)
Hyaluronate NA P90 (HTL)
Hyaluronate Na P93 (HTL)
Hyaluronate Na P100 (HTL)
Hyaluronic Acid-BT (Pentapharm/Centerchem)
Hyaluronic Acid FCH-150 (Lipo)
Hyaluronic Acid FCH-200 (Lipo)
Hyaluronic Acid (Na) (Ichimaru Pharcos)
Hyaluronic Acid, Sodium Salt (CPN)
Hyaluronic Acid, Sodium Salt (Pentapharm/Centerchem)
Hyaluronsan HA-L510 (Q.P.)
Hyaluronsan HA-M5070 (Q.P.)
Hyaluronsan HA-Q (Q.P.)
Hyaluronsan HA-QSS (Q.P.)
Hyaluronsan Solution HA-Q1 (Q.P.)
Hyaluronsan Solution HA-Q1P (Q.P.)
Hyasol (Pentapharm/Centerchem)
Hyasol-BT (Pentapharm/Centerchem)
Jaluronid (Esperis)
LMW Hyaluronic Acid Na Salt (CPN)
Nikkol Sodium Hyaluronate (Nikko)
OriStar SH (Orient Stars)
RITA HA C-1-C (RITA)
RITA HA C-1-P (RITA)
Saccaluronate CW (LCW)
Sodium Hyaluronate HA-Q (Ikeda)

Trade Name Mixtures:
Actiglide (Active Organics)
Advanced Moisture Complex (Collaborative Labs)
Aragoline (Siera)
Atecoron (GfN)
Bellsilk HA (Ichimaru Pharcos)
Biocrystal (Siera)
Brookosome H (Arch Personal Care Products)
Chronosphere FHC/HA Blend (Arch Personal Care Products)
Chronosphere Hyaluronic (Arch Personal Care Products)
Collagen-Hyaluronic Acid-Jelly (Labopharma)
Desaron (GfN)
EASHAVE (Pentapharm/Centerchem)
Essential Vital Elements - S (Dipta)
Gelhyperm (Avocado oil) (Novoselect)
Gelhyperm (Jojoba Oil) (Novoselect)
Gelhyperm (Macadamia nut oil) (Novoselect)
Gelhyperm (Seabukthorn Oil) (Novoselect)
Gelhyperm (Wheat Germ Oil) (Novoselect)
HA-Sol 2% (Collaborative Labs)
Hyaluronic Acid 1% Solution (Lipo)
Hydralphatine 3P (Lanatech)
Hydroxan (Lanatech)
Hydroxan BG (Lanatech)

Hydroxan CH (Lanatech)
Iricalmin (Pentapharm/Centerchem)
Polysol HQ (Polygon)
Quiditat NwH (Assessa-Industria)
Ritacomplex DF 15 (RITA)
Ritacomplex DF 26 (RITA)
Rovisome H A (Rovi)
Saccaluronate CC (LCW)
Saccaluronate LC (LCW)
Spherica HA (Ikeda)
Thioglycans (Esperis)
Toshiki BINS-3 (Nikko)

SODIUM HYALURONATE CROSS-POLYMER

Definition: Sodium Hyaluronate Cross-polymer is a polymer of Hyaluronic Acid (q.v.) crosslinked with vinyl sulfone.

Chemical Class: Synthetic Polymers

Functions: Skin-Conditioning Agent - Humectant; Skin-Conditioning Agent - Miscellaneous

SODIUM HYALURONATE DIMETHYL-SILANOL

Definition: Sodium Hyaluronate Dimethylsilanol is the sodium salt of an ester of Hyaluronic Acid (q.v.) and dimethylsilanol.

Chemical Classes: Biological Polymers and their Derivatives; Siloxanes and Silanes

Function: Skin-Conditioning Agent - Miscellaneous

Trade Name Mixture:
Recelderm - 503 (Bioland)

SODIUM HYDROGENATED COCOATE

Definition: Sodium Hydrogenated Cocoate is the sodium salt of Hydrogenated Coconut Acid (q.v.)

Chemical Class: Soaps

Function: Surfactant - Cleansing Agent

SODIUM HYDROGENATED PALMATE

Definition: Sodium Hydrogenated Palmate is the sodium salt of Hydrogenated Palm Acid (q.v.).

Chemical Class: Soaps

Function: Surfactant - Cleansing Agent

SODIUM HYDROGENATED TALLOWATE

Definition: Sodium Hydrogenated Tallowate is the sodium salt of Hydrogenated Tallow Acid (q.v.).

Chemical Class: Soaps

Function: Surfactant - Cleansing Agent

SODIUM HYDROGENATED TALLOWOYL GLUTAMATE

JPN Translation:
水添タロウグルタミン酸 Na

Definition: Sodium Hydrogenated Tallowoyl Glutamate is the sodium salt of the hydrogenated tallow acid amide of glutamic acid. It conforms generally to the formula:

$$RC-NHCH(CH_2)_2COOH$$

with carbonyl O above RC and COONa below the CH

where RCO- represents the fatty acids derived from hydrogenated tallow.

Information Sources: JCIC, JCLS, JSQI

Chemical Class: Amino Acids

Function: Surfactant - Cleansing Agent

Reported Product Category: Indoor Tanning Preparations

Technical/Other Names:
Sodium Hydrogenated Tallow Glutamate
Sodium Hydrogenated Tallow-L-glutamate

Trade Name:
Amisoft HS-11 (Ajinomoto)

Trade Name Mixtures:
Amisoft GS-11 (Ajinomoto)
Toshiki Sericite JA-A3 (Nikko)

SODIUM HYDROLYZED CASEIN

JPN Translation:
加水分解カゼイン Na

Definition: Sodium Hydrolyzed Casein is the sodium salt of Hydrolyzed Casein (q.v.).

Chemical Classes: Organic Salts; Protein Derivatives

Functions: Hair Conditioning Agent; Skin-Conditioning Agent - Miscellaneous

Technical/Other Name:
Hydrolyzed Casein, Sodium Salt

SODIUM HYDROSULFITE

CAS Nos. **EINECS Nos.**
7631-94-9   231-550-1
7775-14-6   231-890-0

Empirical Formula:
$H_2O_4S_2 \cdot 2Na$

Definition: Sodium Hydrosulfite is the inorganic salt that conforms to the formula:

$$Na_2S_2O_4$$

Information Sources: 21CFR177.2800, 21CFR182.90, CTFA D, MI-13(8685), MI-13 (8700), TSCA

Chemical Class: Inorganic Salts

Function: Reducing Agent

Reported Product Category: Hair Dyes and Colors (All Types Requiring Caution Statements and Patch Tests)

Technical/Other Names:
Disodium Dithionate
Dithionous Acid, Disodium Salt
Sodium Dithionate
Sodium Hyposulfate

SODIUM HYDROXIDE

CAS No.   **EINECS No.**
1310-73-2   215-185-5

JPN Translation:
水酸化 Na

Empirical Formula:
HNaO

Definition: Sodium Hydroxide is the inorganic base that conforms to the formula:

$$NaOH$$

Information Sources: ARG, AUS, BP, BPC, 21CFR114.90, 21CFR163.110, 21CFR172.560, 21CFR172.814, 21CFR172.892, 21CFR173.310, 21CFR184.1763, 27CFR21.101, DA, DDR, EEC(III/1-15a), EGY, FCC, FIN, HP, HUN, IND, JCLS, JSCI, MAR, MEX, MI-13(8701), NF XIX, PN, POR, TSCA, USAN, USD, YUG

Chemical Class: Inorganic Bases

Functions: Denaturant; pH Adjuster

Reported Product Categories: Moisturizing Preparations; Bath Oils, Tablets, and Salts; Bath Soaps and Detergents; Deodorants (Underarm); Cleansing Products (Cold Creams, Cleansing Lotions, Liquids and Pads); Hair Dyes and Colors (All Types Requiring Caution Statements and Patch Tests); Hair Straighteners; Shampoos (Non-coloring); Bath Preparations, Misc.; Body and Hand Preparations (Excluding Shaving Preparations); Bath Capsules; Hair Rinses (Coloring); Hair Sprays (Aerosol Fixatives); Skin Care Preparations, Misc.; Bubble Baths; Face and Neck Preparations (Excluding Shaving Preparations); Permanent Waves; Tonics, Dressings, and Other Hair Grooming Aids; Eyeliners; Personal Cleanliness Products, Misc.; Aftershave Lotions; Baby Shampoos; Depilatories; Hair Conditioners; Shaving Cream (Aerosol, Brushless and Lather); Baby Lotions, Oils, Powders and Creams; Eye Shadows; Suntan Preparations, Misc.; Eyebrow Pencils; Suntan Gels, Creams, and Liquids; Night Skin Care Preparations; Shaving Preparations, Misc.; Skin Fresheners; Baby Products, Misc.; Cuticle Softeners; Dentifrices (Aerosol, Liquid, Pastes and Powders); Eye Makeup Removers; Hair Preparations (Non-coloring), Misc.; Indoor Tanning Preparations

Technical/Other Names:
Caustic Soda
Sodium Hydroxide Solution

Trade Name:
Unichem SOHYD (Universal Preserv-A-Chem)

Trade Name Mixtures:
Catezomes SA-20 Japan (Collaborative Labs)
Elespher Almondermin (Laboratoires Sero-biologiques)
Noster A (Ennagram)
Noster K (Ennagram)
Noster MX (Ennagram)
Noster S (Ennagram)
Proxel CG 10 (Zeneca)
Proxel CG 20 (Zeneca)
Rovisome AHA - Glycolic Acid (Rovi)

SODIUM HYDROXYLAURYLDIMONIUM ETHYL PHOSPHATE

Definition: Sodium Hydroxylauryldimonium Ethyl Phosphate is the organic compound that conforms to the formula:

$$CH_3(CH_2)_9CHCH_2-N^+-CH_2CH_2PO_3Na$$

with OH below the CHCH2, and two CH3 groups above and below the N

Information Source: JCIC

Chemical Classes: Amines; Phosphorus Compounds

Functions: Surfactant - Cleansing Agent; Surfactant - Emulsifying Agent

SODIUM HYDROXYMETHANE SULFONATE

CAS No.   **EINECS No.**
870-72-4   212-800-9

Empirical Formula:
$CH_4O_4S \cdot Na$

Definition: Sodium Hydroxymethane Sulfonate is the compound that conforms to the formula:

$HOCH_2SO_3Na$

Information Sources: MI-13(4260), TSCA

Chemical Class: Organic Salts

Functions: Preservative; Reducing Agent

Technical/Other Names:
Hydroxymethanesulfonic Acid,
 Monosodium Salt
Methanesulfonic Acid, Hydroxy-,
 Monosodium Salt
Sodium Formaldehyde Bisulfite
Sodium Hydroxymethylsulfonate

SODIUM HYDROXYMETHYLGLYCINATE

CAS No.   **EINECS No.**
70161-44-3   274-357-8

Empirical Formula:
$C_3H_7NO_3$ • Na

Definition: Sodium Hydroxymethylglycinate is the sodium salt of the substituted amino acid that conforms to the formula:

$HOCH_2NHCH_2COONa$

Information Sources: EEC(VI/1-51), TSCA

Chemical Class: Alkyl-Substituted Amino Acids

Functions: Hair Conditioning Agent; Preservative

Reported Product Categories: Tonics, Dressings, and Other Hair Grooming Aids; Bubble Baths; Hair Conditioners; Shampoos (Non-coloring)

Technical/Other Names:
Glycine, N-(Hydroxymethyl)-. Monosodium
 Salt
Glycine, N-(Hydroxymethyl)-, Sodium Salt
N-(Hydroxymethyl)Glycine, Sodium Salt

Trade Names:
Gramcide III (Sinerga)
Nipaguard SMG (Clariant)
Nipaguard SMG (Clariant GmbH, Personal
 Care)
Suttocide A (Sutton)

SODIUM HYDROXYPROPYL STARCH PHOSPHATE

Definition: Sodium Hydroxypropyl Starch Phosphate is the sodium salt of a 2-hydroxy-propyl ether Distarch Phosphate (q.v.).

Chemical Classes: Carbohydrates; Ethers

Functions: Abrasive; Bulking Agent; Viscosity Increasing Agent - Aqueous

Trade Name:
Pure-Gel (Grain Processing)

SODIUM HYDROXYSTEARATE

CAS No.   **EINECS No.**
13329-67-4   236-374-9

Empirical Formula:
$C_{18}H_{36}O_3$ • Na

Definition: Sodium Hydroxystearate is the sodium salt of Hydroxystearic Acid (q.v.).

Chemical Class: Soaps

Function: Surfactant - Cleansing Agent

Technical/Other Names:
12-Hydroxystearic Acid, Sodium Salt
Octadecanoic Acic, 12-Hydroxy-,
 Monosodium Salt
Sodium 12-Hydroxyoctadecanoate

Trade Names:
Casid HSA-Na (CasChem)
Sodium 12-Hydroxystearate (Uniqema/
 Belgium)

SODIUM HYPOCHLORITE

CAS No.   **EINECS No.**
7681-52-9   231-668-3

Empirical Formula:
ClHO • Na

Definition: Sodium Hypochlorite is the sodium salt of hypochlorous acid that conforms to the formula:

NaClOH

Information Source: MI-13(8702)

Chemical Class: Inorganic Acids

Function: Oxidizing Agent

Technical/Other Name:
Hypochlorous Acid, Sodium Salt

Trade Name:
Sodium Hypochlorite 50 High Purity
 (Atofina)

SODIUM IODATE

CAS No.   **EINECS No.**
7681-55-2   231-672-5

Empirical Formula:
HIO_3 • Na

Definition: Sodium Iodate is the inorganic salt that conforms to the formula:

$NaIO_3$

Information Sources: CIR: [I] JACT-14(3)-1995, EEC(VI/1-10), MI-13(8704), TSCA

Chemical Class: Inorganic Salts

Function: Oxidizing Agent

Technical/Other Name:
Iodic Acid, Sodium Salt

SODIUM IODIDE

CAS No.   **EINECS No.**
7681-82-5   231-679-3

Empirical Formula:
INa

Definition: Sodium Iodide is the inorganic salt that conforms to the formula:

NaI

Information Sources: AUS, BP, BPC, BRA, 21CFR178.1010, CZE, DA, DDR, EGY, EP, FIN, HUN, IND, ITA, JAN, MAR, MEX, MI-13(8705), 9262, PF, PN, POR, ROM, TSCA, USAN, USD, USP XXIII, USSR, WHO, YUG

Chemical Class: Inorganic Salts

Function: Not Reported

Technical/Other Name:
Ioduril

SODIUM ISETHIONATE

CAS No.   **EINECS No.**
1562-00-1   216-343-6

JPN Translation:
イセチオン酸 Na

Empirical Formula:
$C_2H_6O_4S$ • Na

Definition: Sodium Isethionate is the organic salt that conforms to the formula:

$HOCH_2CH_2SO_3Na$

Information Sources: CTFA D, JCIC, JCLS, JSQI, TSCA

Chemical Class: Organic Salts

Function: Not Reported

Reported Product Categories: Bath Soaps and Detergents; Bath Oils, Tablets, and Salts; Cleansing Products (Cold Creams, Cleansing Lotions, Liquids and Pads)

Technical/Other Names:
Ethanesulfonic Acid, 2-Hydroxy-,
 Monosodium Salt
2-Hydroxyethanesulfonic Acid, Sodium Salt
Sodium 2-Hydroxyethanesulfonic Acid

Trade Names:
Hostapon SI (Clariant)

Hostapon SI (Clariant GmbH, Personal Care)

SODIUM ISOBUTYLPARABEN

Definition: Sodium Isobutylparaben is the sodium salt of Isobutylparaben (q.v.).

Information Source: MHLW-331/3

Chemical Classes: Esters; Organic Salts; Phenols

Function: Preservative

Reported Product Category: Blushers (All types)

Trade Name Mixtures:
Nipacide A Sodium (Clariant)
Nipacide A Sodium (Clariant GmbH, Personal Care)
Nipastat Sodium (Clariant)
Nipastat Sodium (Clariant GmbH, Personal Care)

SODIUM ISOFERULATE

CAS No.: 110993-57-2

Empirical Formula:
$C_{10}H_{10}O_4Na$

Definition: Sodium Isoferulate is the organic salt that conforms to the formula:

Chemical Class: Organic Salts

Function: Ultraviolet Light Absorber

Technical/Other Name:
2-Propenoic Acid, 3-(3-Hydroxy-4-methoxyphenyl)-, Monosodium Salt

Trade Name:
Sodium Isoferulate (Takasago)

SODIUM ISOOCTYLENE/MA COPOLYMER

Definition: Sodium Isooctylene/MA Copolymer is the sodium salt of a copolymer of isooctylene and maleic anhydride monomers.

Chemical Class: Synthetic Polymers

Functions: Film Former; Suspending Agent - Nonsurfactant; Viscosity Increasing Agent - Aqueous

SODIUM ISOPROPYLPARABEN

Empirical Formula:
$C_{10}H_{12}O_3$ • Na

Definition: Sodium Isopropylparaben is the sodium salt of Isopropylparaben (q.v.).

Information Sources: EEC(VI/1-12), MHLW-331/3

Chemical Classes: Esters; Organic Salts; Phenols

Function: Preservative

Technical/Other Name:
Benzoic Acid, 4-Hydroxy, 1-Methylethyl Ester, Sodium Salt

SODIUM ISOSTEARATE

CAS No.	EINECS No.
64248-79-9	264-754-4

Definition: Sodium Isostearate is the sodium salt of Isostearic Acid (q.v.).

Chemical Class: Soaps

Functions: Surfactant - Cleansing Agent; Surfactant - Emulsifying Agent

Technical/Other Name:
Isooctadecanoic Acid, Sodium Salt

Trade Name:
PRISAVON NAIS (Uniqema Europe)

SODIUM ISOSTEARETH-6 CARBOXYLATE

Empirical Formula:
$C_{30}H_{60}O_8$ • Na

Definition: Sodium Isosteareth-6 Carboxylate is the sodium salt of Isosteareth-6 Carboxylic Acid (q.v.). It conforms generally to the formula:

$$C_{18}H_{37}(OCH_2CH_2)_nOCH_2COONa$$

where n has an average value of 5.

Chemical Class: Organic Salts

Function: Surfactant - Cleansing Agent

Technical/Other Names:
Isosteareth-6 Carboxylic Acid, Sodium Salt
PEG-6 Isostearyl Ether Carboxylic Acid, Sodium Salt
Polyethylene Glycol 300 Isostearyl Ether Carboxylic Acid, Sodium Salt
Polyoxyethylene (6) Isostearyl Ether Carboxylic Acid, Sodium Salt

SODIUM ISOSTEARETH-11 CARBOXYLATE

Empirical Formula:
$C_{40}H_{80}O_{13}$ • Na

Definition: Sodium Isosteareth-11 Carboxylate is the sodium salt of Isosteareth-11 Carboxylic Acid (q.v.). It conforms generally to the formula:

$$C_{18}H_{37}(OCH_2CH_2)_nOCH_2COONa$$

where n has an average value of 10.

Chemical Class: Organic Salts

Function: Surfactant - Cleansing Agent

Technical/Other Names:
Isosteareth-11 Carboxylic Acid, Sodium Salt
PEG-11 Isostearyl Ether Carboxylic Acid, Sodium Salt
Polyethylene Glycol (11) Isostearyl Ether Carboxylic Acid, Sodium Salt
Polyoxyethylene (11) Isostearyl Ether Carboxylic Acid, Sodium Salt

SODIUM ISOSTEAROAMPHOACETATE

Empirical Formula:
$C_{24}H_{48}N_2O_4$ • Na

Definition: Sodium Isostearoamphoacetate is the amphoteric organic compound that conforms generally to the formula:

Chemical Class: Alkylamido Alkylamines

Functions: Hair Conditioning Agent; Surfactant - Cleansing Agent; Surfactant - Foam Booster

Technical/Other Names:
Isostearoamphoacetate
Isostearoamphoglycinate

SODIUM ISOSTEAROAMPHO-PROPIONATE

Empirical Formula:
$C_{25}H_{50}N_2O_4$ • Na

Definition: Sodium Isostearoamphopropionate is the amphoteric organic compound that conforms generally to the formula:

Information Source: TSCA

Chemical Class: Alkylamido Alkylamines

Functions: Hair Conditioning Agent; Surfactant - Cleansing Agent; Surfactant - Foam Booster; Surfactant - Hydrotrope

Reported Product Category: Shampoos (Non-coloring)

Technical/Other Name:
Isostearoamphopropionate

Trade Names:
Monateric ISA-35 (Uniqema)
Schercoteric I-AA (Scher)

SODIUM ISOSTEAROYL LACTATE

Empirical Formula:
$C_{21}H_{39}O_4 \cdot Na$

Definition: Sodium Isostearoyl Lactate is the organic compound that conforms generally to the formula:

$$C_{17}H_{35}C\overset{\displaystyle O}{\overset{\|}{}} - OCHCOONa$$
$$| $$
$$CH_3$$

Chemical Class: Organic Salts

Function: Surfactant - Emulsifying Agent

SODIUM ISOSTEAROYL LACTYLATE

CAS No. 66988-04-3
EINECS No. 266-533-8

JPN Translation:
イソステアロイル乳酸 Na

Empirical Formula:
$C_{24}H_{44}O_6 \cdot Na$

Definition: Sodium Isostearoyl Lactylate is the sodium salt of the isostearic acid ester of lactyl lactate. It conforms to the formula:

$$C_{17}H_{35}C\overset{\displaystyle O}{\overset{\|}{}} - OCHC\overset{\displaystyle O}{\overset{\|}{}} - OCHCOONa$$
$$| \qquad\qquad |$$
$$CH_3 \qquad\quad CH_3$$

Information Source: CTFA D

Chemical Classes: Esters; Organic Salts

Function: Surfactant - Emulsifying Agent

Reported Product Categories: Shampoos (Non-coloring); Bath Soaps and Detergents

Technical/Other Name:
Isooctadecanoic Acid, 2-(1-Carboxyethoxy)-1-Methyl-2-Oxoethyl Ester, Sodium Salt

Trade Names:
Pationic ISL (RITA)
PRIAZUL 2133 (Uniqema Europe)
Protaquat SIL (Protameen)

SODIUM LACTATE

CAS Nos. 72-17-3
867-56-1
EINECS Nos. 200-772-0
212-762-3

JPN Translation:
乳酸 Na

Empirical Formula:
$C_3H_6O_3 \cdot Na$

Definition: Sodium Lactate is the sodium salt of lactic acid that conforms to the formula:

$$CH_3CHCOONa$$
$$|$$
$$OH$$

Information Sources: 21CFR184.1768, CIR: [SQ] IJT-17(Suppl. 1)1998, CTFA D, JAN, JCLS, JSCI, MAR, MI-13(8709), PN, TSCA, USAN, USP XXIV

Chemical Class: Organic Salts

Functions: Buffering Agent; Exfoliant; Skin-Conditioning Agent - Humectant

Reported Product Categories: Moisturizing Preparations; Skin Care Preparations, Misc.; Bath Oils, Tablets, and Salts; Cleansing Products (Cold Creams, Cleansing Lotions, Liquids and Pads); Bath Preparations, Misc.; Body and Hand Preparations (Excluding Shaving Preparations); Hair Conditioners; Bath Capsules; Face and Neck Preparations (Excluding Shaving Preparations); Night Skin Care Preparations; Shampoos (Non-coloring); Tonics, Dressings, and Other Hair Grooming Aids; Aftershave Lotions; Baby Shampoos; Paste Masks (Mud Packs); Permanent Waves; Skin Fresheners; Foundations

Technical/Other Names:
2-Hydroxypropanoic Acid, Monosodium Salt
Propanoic Acid, 2-Hydroxy-, Monosodium Salt
Sodium α-Hydroxypropionate
Sodium Lactate Solution

Trade Names:
PURASAL S/COS (Purac)
Sodium Lactate Solution About 50% (Merck KGaA)

Trade Name Mixtures:
Aquaderm (Crodarom)
Crodarom Hygroderm (Croda, Inc.)
Elesponge AHA LS 8911 B (Laboratoires Serobiologiques)
Facteur Hydratant PH (Prod'Hyg)
Hidroderm (Fabriquimica)
Hydeoviton 5,5 N 2/059359 (Symrise)
Hydrocos (Cosmetochem)
Hydrocos "P" (Cosmetochem) (Cosmetochem International Ltd.)
Hydroviton 2/059353 (Symrise)
Hydroviton 24 (2/059351) (Symrise)
Lactil (Degussa Care Specialties)
Moisturising Factor Hydrogerm (Crodarom)
Moisturizing Factor Hygro-Complex ARO 5272 (Crodarom)

Polymannocym LAC (Polygon)
Polymoist Marine (Cognis Deutschland)
Prodew 100 (Ajinomoto)
Prodew 200 (Ajinomoto)
Prodew 300 (Ajinomoto)
PURAC BF/S (Purac)
Rovisome AHA - Lactic Acid (Rovi)
Umordant P (Cosmetochem) (Cosmetochem International Ltd.)
Vegeles AHA 8702 (Laboratoires Sero-biologiques)
Vegeles AHA LS 8763 (Laboratoires Sero-biologiques)

SODIUM LACTATE METHYLSILANOL

Definition: Sodium Lactate Methylsilanol is a complex of Sodium Lactate (q.v.) and monomethylsilanol.

Chemical Classes: Organic Salts; Siloxanes and Silanes

Function: Skin-Conditioning Agent - Miscellaneous

Reported Product Category: Foundations

Trade Name:
Lasilium (Exsymol)

Trade Name Mixture:
Horsetail AMI Organic Silicon (Alban Muller)

SODIUM LANETH SULFATE

CAS No.: 68919-23-3 (Generic)

Definition: Sodium Laneth Sulfate is the sodium salt of sulfated ethoxylated lanolin alcohol that conforms generally to the formula:

$$R(OCH_2CH_2)_nOSO_3Na$$

where R represents the lanolin alcohol radical and n has a value between 1 and 4.

Information Sources: CTFA D, TSCA

Chemical Classes: Alkyl Ether Sulfates; Lanolin and Lanolin Derivatives

Functions: Skin-Conditioning Agent - Miscellaneous; Surfactant - Cleansing Agent

Technical/Other Name:
Sodium Polyoxyethylene Lanolin Ether Sulfate

SODIUM LANOLATE

Definition: Sodium Lanolate is the sodium salt of Lanolin Acid (q.v.).

Chemical Class: Soaps

Function: Surfactant - Cleansing Agent

Technical/Other Name:
Lanolin Acid, Sodium Salt

SODIUM LARDATE

CAS No.: 68605-06-1

Definition: Sodium Lardate is the sodium salt of the fatty acids derived from Lard (q.v.).

Chemical Class: Soaps

Functions: Surfactant - Cleansing Agent; Surfactant - Emulsifying Agent; Surfactant - Foam Booster

Trade Name Mixture:
Sapo Durus (Block Drug)

SODIUM LAURAMIDO DIACETATE

JPN Translation:
ラウラミノジ酢酸 Na

Empirical Formula:
$C_{16}H_{29}NO_5 \cdot Na$

Definition: Sodium Lauramido Diacetate is the organic compound that conforms to the formula:

$$CH_3(CH_2)_{10}C-NCH_2COONa$$

(with O double-bonded to C, and CH_2COOH branch on N)

Information Source: JCLS

Chemical Classes: Amides; Carboxylic Acids

Function: Surfactant - Cleansing Agent

SODIUM LAURAMINOPROPIONATE

CAS No. 3546-96-1   **EINECS No.** 222-597-9

JPN Translation:
ラウラミノプロピオン酸 Na

Empirical Formula:
$C_{15}H_{31}NO_2 \cdot Na$

Definition: Sodium Lauraminopropionate is the sodium salt of a substituted propionic acid. It conforms generally to the formula:

$$CH_3(CH_2)_{11}NH(CH_2)_2COONa$$

Information Sources: CIR: [l] IJT-16 (Suppl. 1)1997, JCLS, JSCI

Chemical Class: Alkyl-Substituted Amino Acids

Functions: Antistatic Agent; Hair Conditioning Agent; Surfactant - Cleansing Agent; Surfactant - Foam Booster

Reported Product Category: Skin Fresheners

Technical/Other Names:
β-Alanine,N-Dodecyl-, Monosodium Salt
N-Dodecyl-β-Alanine, Monosodium Salt
Sodium β-Laurylaminopropionate

Trade Name Mixture:
Proticute C Alpha (Ichimaru Pharcos)

SODIUM LAURATE

CAS No. 629-25-4   **EINECS No.** 211-082-4

Empirical Formula:
$C_{12}H_{24}O_2 \cdot Na$

Definition: Sodium Laurate is the sodium salt of lauric acid that conforms generally to the formula:

$$CH_3(CH_2)_{10}COONa$$

Information Sources: 21CFR172.863, 21CFR175.105, 21CFR175.320, 21CFR176.170, 21CFR176.200, 21CFR176.210, 21CFR177.1200, 21CFR177.2260, 21CFR177.2600, 21CFR177.2800, 21CFR178.3910, TSCA

Chemical Class: Soaps

Functions: Surfactant - Cleansing Agent; Surfactant - Emulsifying Agent

Reported Product Categories: Bath Soaps and Detergents; Moisturizing Preparations; Bath Oils, Tablets, and Salts; Cleansing Products (Cold Creams, Cleansing Lotions, Liquids and Pads)

Technical/Other Names:
Dodecanoic Acid, Sodium Salt
Sodium Dodecanoate

SODIUM LAURETH-3 CARBOXYLATE

CAS No.: 33939-64-9 (Generic)

JPN Translation:
ラウレス-3 カルボン酸 Na

Empirical Formula:
$C_{18}H_{36}O_5 \cdot Na$

Definition: Sodium Laureth-3 Carboxylate is the sodium salt of the carboxylic acid derived from Laureth-3 (q.v.). It conforms generally to the formula:

$$CH_3(CH_2)_{11}(OCH_2CH_2)_nOCH_2COONa$$

where n has an average value of 2.

Chemical Class: Organic Salts

Function: Surfactant - Cleansing Agent

SODIUM LAURETH-4 CARBOXYLATE

CAS Nos.: 33939-64-9 (Generic); 38975-04-1

JPN Translation:
ラウレス-4 カルボン酸 Na

Empirical Formula:
$C_{20}H_{40}O_6 \cdot Na$

Definition: Sodium Laureth-4 Carboxylate is the sodium salt of the carboxylic acid derived from Laureth-4 (q.v.). It conforms generally to the formula:

$$CH_3(CH_2)_{11}(OCH_2CH_2)_nOCH_2COONa$$

where n has an average value of 3.

Chemical Class: Organic Salts

Function: Surfactant - Cleansing Agent

Technical/Other Names:
PEG-4 Lauryl Ether Carboxylic Acid, Sodium Salt
Polyethylene Glycol 200 Lauryl Ether Carboxylic Acid, Sodium Salt
Polyoxyethylene (4) Lauryl Ether Carboxylic Acid, Sodium Salt
Sodium Polyethylene Glycol 200 Lauryl Ether Carboxylate
Sodium Polyoxyethylene (4) Lauryl Ether Carboxylate
Sodium 3,6,9,12-Tetraoxatetracosanoate
3,6,9,12-Tetraoxatetracosanoic Acid, Sodium Salt

Trade Names:
Akypo NTS (Kao GmbH)
Empicol CBCS (Albright & Wilson UK)
Neoscoap LECN (Toho)

SODIUM LAURETH-5 CARBOXYLATE

CAS Nos.: 33939-64-9 (Generic); 38975-03-0

JPN Translation:
ラウレス-5 カルボン酸 Na

Empirical Formula:
$C_{22}H_{44}O_7 \cdot Na$

Definition: Sodium Laureth-5 Carboxylate is the sodium salt of the carboxylic acid derived from Laureth-5 (q.v.). It conforms generally to the formula:

$$CH_3(CH_2)_{11}(OCH_2CH_2)_nOCH_2COONa$$

where n has an average value of 4.

Chemical Class: Organic Salts

Function: Surfactant - Cleansing Agent

Technical/Other Names:
Laureth-5 Carboxylic Acid, Sodium salt

PEG-5 Lauryl Ether Carboxylic Acid,
Sodium Salt
3,6,9,12,15-Pentaoxaheptacosanoic Acid,
Sodium Salt
Polyethylene Glycol (5) Lauryl Ether
Carboxylic Acid, Sodium Salt
Polyoxyethylene (5) Lauryl Ether
Carboxylic Acid, Sodium Salt
Sodium 3,6,9,12,15-Pentaoxa-
heptacosanoate
Sodium Polyethylene Glycol (5) Lauryl
Ether Carboxylate
Sodium Polyoxyethylene (5) Lauryl Ether
Carboxylate

Trade Name:
Empicol CED 5 S (Albright & Wilson UK)

SODIUM LAURETH-6 CARBOXYLATE

CAS No.: 33939-64-9 (Generic)

JPN Translation:
ラウレス-6カルボン酸Na

Empirical Formula:
$C_{24}H_{48}O_8 \cdot Na$

Definition: Sodium Laureth-6 Carboxylate is the sodium salt of the carboxylic acid derived from Laureth-6 (q.v.). It conforms generally to the formula:

$$CH_3(CH_2)_{11}(OCH_2CH_2)_nOCH_2COONa$$

where n has an average value of 5.

Chemical Class: Organic Salts

Function: Surfactant - Cleansing Agent

Technical/Other Names:
Laureth-6 Carboxylic Acid, Sodium Salt
PEG-6 Lauryl Ether Carboxylic Acid,
Sodium Salt
Polyethylene Glycol 300 Lauryl Ether
Carboxylic Acid, Sodium Salt
Polyoxyethylene (6) Lauryl Ether
Carboxylic Acid, Sodium Salt
Sodium Polyethylene Glycol 300 Lauryl
Ether Carboxylate
Sodium Polyoxyethylene (6) Lauryl Ether
Carboxylate

Trade Name:
Akypo Soft 45 NV (Kao GmbH)

Trade Name Mixture:
Akypogene KTS (Kao GmbH)

SODIUM LAURETH-8 CARBOXYLATE

CAS No.: 33939-64-9 (Generic)

Definition: Sodium Laureth-8 Carboxylate is the sodium salt of Laureth-8 Carboxylic Acid (q.v.) that conforms generally to the formula:

$$CH_3(CH_2)_{11}(OCH_2CH_2)_nOCH_2COONa$$

where n has an average value of 7.

Chemical Class: Organic Salts

Function: Surfactant - Cleansing Agent

Technical/Other Names:
Polyethylene Glycol (8) Lauryl Ether
Carboxylic Acid, Sodium Salt
Polyoxyethylene (8) Lauryl Ether
Carboxylic Acid, Sodium Salt

Trade Names:
Sandosan LNS (Clariant)
Sandosan LNS (Clariant GmbH, Personal
Care)

SODIUM LAURETH-11 CARBOXYLATE

CAS No.: 33939-64-9 (Generic)

JPN Translation:
ラウレス-11カルボン酸Na

Empirical Formula:
$C_{34}H_{68}O_{13} \cdot Na$

Definition: Sodium Laureth-11 Carboxylate is the sodium salt of the carboxylic acid derived from Laureth-11 (q.v.). It conforms generally to the formula:

$$CH_3(CH_2)_{11}(OCH_2CH_2)_nOCH_2COONa$$

where n has an average value of 10.

Chemical Class: Organic Salts

Function: Surfactant - Cleansing Agent

Reported Product Category: Shampoos (Non-coloring)

Technical/Other Names:
Laureth-11 Carboxylic Acid, Sodium Salt
PEG-11 Lauryl Ether Carboxylic Acid,
Sodium Salt
Polyethylene Glycol (11) Lauryl Ether
Carboxylic Acid, Sodium Salt
Polyoxyethylene (11) Lauryl Ether
Carboxylic Acid, Sodium Salt
Sodium Polyethylene Glycol (11) Lauryl
Ether Carboxylate
Sodium Polyoxyethylene (11) Lauryl Ether
Carboxylate

Trade Names:
Akypo RLM 100 NV (Kao GmbH)
Akypo Soft 100 NV (Kao GmbH)
Empicol CBJS (Albright & Wilson UK)
Marlinat CM 105/80 (Sasol GmbH - Marl)

Trade Name Mixture:
Akypo Soft 100 BVC (Kao GmbH)

SODIUM LAURETH-12 CARBOXYLATE

CAS No.: 33939-64-9 (Generic)

Definition: Sodium Laureth-12 Carboxylate is the sodium salt of Laureth-12 Carboxylic Acid (q.v.) that conforms generally to the formula:

$$CH_3(CH_2)_{11}(OCH_2CH_2)_nOCH_2COONa$$

where n has an average value of 11.

Chemical Class: Organic Salts

Function: Surfactant - Cleansing Agent

Technical/Other Names:
Laureth-12 Carboxylic Acid, Sodium Salt
Polyethylene Glycol (12) Lauryl Ether
Carboxylic Acid, Sodium Salt
Polyoxyethylene (12) Lauryl Ether
Carboxylic Acid, Sodium Salt

Trade Names:
Sandosan LNCS (Clariant)
Sandosan LNCS (Clariant GmbH, Personal
Care)

SODIUM LAURETH-13 CARBOXYLATE

CAS No.: 33939-64-9 (Generic)

JPN Translation:
ラウレス-13カルボン酸Na

Definition: Sodium Laureth-13 Carboxylate is the sodium salt of the carboxylic acid derived from Laureth-13 (q.v.). It conforms generally to the formula:

$$CH_3(CH_2)_{11}(OCH_2CH_2)_nOCH_2COONa$$

Chemical Class: Organic Salts

Function: Surfactant - Cleansing Agent

Reported Product Categories: Hair Dyes and Colors (All Types Requiring Caution Statements and Patch Tests); Baby Shampoos; Shampoos (Non-coloring); Bath Oils, Tablets, and Salts; Cleansing Products (Cold Creams, Cleansing Lotions, Liquids and Pads); Tonics, Dressings, and Other Hair Grooming Aids; Baby Products, Misc.

Technical/Other Names:
Laureth-13 Carboxylic Acid, Sodium Salt
PEG-13 Lauryl Ether Carboxylic Acid,
Sodium Salt
Polyethylene Glycol (13) Lauryl Ether
Carboxylic Acid, Sodium Salt
Polyoxyethylene (13) Lauryl Ether
Carboxylic Acid, Sodium Salt

Trade Names:
Ethcarb LTA (Ethox)
Miranate LEC (Rhodia)
Miranate LEC-80 (Rhodia)
Sandopan LS-24 (Clariant)
Sandopan LS-24 (Clariant GmbH, Personal
Care)

Trade Name Mixtures:
Miracare BC-10 (Rhodia)

Miracare BC-20 (Rhodia)
Miracare BC-27 (Rhodia)
Miracare MS-2 (Rhodia)
Miracare MS-4 (Rhodia)

SODIUM LAURETH-14 CARBOXYLATE

CAS No.: 33939-64-9 (Generic)

JPN Translation:
ラウレス-１４カルボン酸 Na

Definition: Sodium Laureth-14 Carboxylate is the sodium salt of the carboxylic acid derived from Laureth-14 (q.v.). It conforms generally to the formula:

$$CH_3(CH_2)_{11}(OCH_2CH_2)_nOCH_2COONa$$

where n has an average value of 13.

Chemical Class: Organic Salts

Function: Surfactant - Cleansing Agent

Technical/Other Names:
Laureth-14 Carboxylic Acid, Sodium Salt
PEG-14 Lauryl Ether Carboxylic Acid, Sodium Salt
Polyethylene Glycol (14) Lauryl Ether Carboxylic Acid, Sodium Salt
Polyoxyethylene (14) Lauryl Ether Carboxylic Acid, Sodium Salt
Sodium Polyethylene Glycol (14) Lauryl Ether Carboxylate
Sodium Polyoxyethylene (14) Lauryl Ether Carboxylate

Trade Name:
Akypo Soft 130 NV (Kao GmbH)

SODIUM LAURETH-16 CARBOXYLATE

Definition: Sodium Laureth-16 Carboxylate is the sodium salt of the carboxylic acid derived from Laureth-16 (q.v.). It conforms to the formula:

$$CH_3(CH_2)_{11}(OCH_2CH_2)_nOCH_2COONa$$

where n has an average value of 15.

Information Source: JCLS

Chemical Class: Organic Salts

Function: Surfactant - Cleansing Agent

SODIUM LAURETH-17 CARBOXYLATE

CAS No.: 33939-64-9 (Generic)

JPN Translation:
ラウレス-１７カルボン酸 Na

Definition: Sodium Laureth-17 Carboxylate is the sodium salt of the carboxylic acid derived from Laureth-17. It conforms generally to the formula:

$$CH_3(CH_2)_{11}(OCH_2CH_2)_nOCH_2COONa$$

where n has an average value of 16.

Chemical Class: Organic Salts

Function: Surfactant - Cleansing Agent

Technical/Other Names:
Laureth-17 Carboxylic Acid, Sodium Salt
PEG-17 Lauryl Ether Carboxylic Acid, Sodium Salt
Polyethylene Glycol (17) Lauryl Ether Carboxylic Acid, Sodium Salt
Polyoxyethylene (17) Lauryl Ether Carboxylic Acid, Sodium Salt
Sodium Polyethylene Glycol (17) Lauryl Ether Carboxylate
Sodium Polyoxyethylene (17) Lauryl Ether Carboxylate

Trade Names:
Akypo RLM 160 N (Kao GmbH)
Akypo Soft 160 NV (Kao GmbH)

SODIUM LAURETH-2 PHOSPHATE

CAS No.: 42612-52-2 (Generic)

Definition: Sodium Laureth-2 Phosphate is the sodium salt of a complex mixture of phosphate esters of Laureth-2 (q.v.).

Chemical Classes: Alkoxylated Alcohols; Phosphorus Compounds

Function: Surfactant - Emulsifying Agent

Technical/Other Name:
Sodium Polyoxyethylene (2) Lauryl Ether Phosphate

Trade Name:
T-5076 (Nicca Chemical)

SODIUM LAURETH-4 PHOSPHATE

CAS No.: 42612-52-2 (Generic)

JPN Translation:
ラウレス - 4 リン酸 Na

Definition: Sodium Laureth-4 Phosphate is the sodium salt of a complex mixture of phosphate esters of Laureth-4 (q.v.).

Information Sources: CTFA D, JCIC, JCLS, JSCI, SNPF, TSCA

Chemical Class: Phosphorus Compounds

Function: Surfactant - Emulsifying Agent

Technical/Other Names:
Partially Neutralized Sodium Polyoxyethylene Lauryl Ether Phosphate
Sodium Polyethylene Glycol 200 Lauryl Ether Phosphate
Sodium Polyoxyethylene Lauryl Ether Phosphate

Sodium Polyoxyethylene (4) Lauryl Ether Phosphate

SODIUM LAURETH SULFATE

CAS Nos.	EINECS No.
1335-72-4	
3088-31-1	221-416-0
9004-82-4 (Generic)	
68585-34-2 (Generic)	
68891-38-3 (Generic)	
91648-56-5	

JPN Translation:
ラウレス硫酸 Na

Definition: Sodium Laureth Sulfate is the sodium salt of sulfated ethoxylated lauryl alcohol that conforms generally to the formula:

$$CH_3(CH_2)_{11}(OCH_2CH_2)_nOSO_3Na$$

where n averages between 1 and 4.

Information Sources: CIR: [S] JACT-2(5)-1983, CTFA S, JCLS, JSCI, SNPF, TSCA

Chemical Classes: Alkyl Ether Sulfates; Sulfonic Acids

Functions: Surfactant - Cleansing Agent; Surfactant - Emulsifying Agent

Reported Product Categories: Shampoos (Non-coloring); Bath Oils, Tablets, and Salts; Bath Soaps and Detergents; Bubble Baths; Cleansing Products (Cold Creams, Cleansing Lotions, Liquids and Pads); Personal Cleanliness Products, Misc.; Hair Dyes and Colors (All Types Requiring Caution Statements and Patch Tests); Eye Makeup Removers; Hair Shampoos (Coloring); Skin Care Preparations, Misc.; Baby Shampoos; Face and Neck Preparations (Excluding Shaving Preparations); Bath Capsules; Mascara; Baby Products, Misc.; Bath Preparations, Misc.; Body and Hand Preparations (Excluding Shaving Preparations); Douches; Fragrance Preparations, Misc.; Hair Preparations (Non-coloring), Misc.; Moisturizing Preparations; Paste Masks (Mud Packs); Tonics, Dressings, and Other Hair Grooming Aids; Foot Powders and Sprays; Hair Conditioners

Technical/Other Names:
Dodecyl Sodium Sulfate
PEG-(1-4) Lauryl Ether Sulfate, Sodium Salt
Polyethylene Glycol (1-4) Lauryl Ether Sulfate, Sodium Salt
Poly(Oxy-1,2-Ethanediyl), α-Sulfo-ω (Dodecyloxy)-, Sodium Salt
Polyoxyethylene (1-4) Lauryl Ether Sulfate, Sodium Salt
Sodium PEG Lauryl Ether Sulfate
Sodium Polyoxyethylene Lauryl Ether Sulfate
Sodium Polyoxyethylene Lauryl Sulfate

Trade Names:

Akyposal EO 20 (Kao GmbH)
Alkyl Ether Sulphate (Kjemi)
Alscoap AT-370 (Toho)
Alscoap Tap-30 (Toho)
Alscoap TAP-230 (Toho)
Alscoap TH-330 (Toho)
Calfoam ES-301 (Pilot)
Calfoam ES-302 (Pilot)
Calfoam ES-303 (Pilot)
Calfoam ES-603 (Pilot)
Calfoam ES-702 (Pilot)
Calfoam ES-703 (Pilot)
Colonial SLES-1 (Colonial Chemical Inc)
Colonial SLES-2 (Colonial Chemical Inc)
Colonial SLES-3 (Colonial Chemical Inc)
Colonial SLES-70 (Colonial Chemical Inc)
DeSULF SLES-301 (DeForest)
DeSULF SLES-302 (DeForest)
DeSULF SLES-303 (DeForest)
DeSULF SLES-603 (DeForest)
Empicol 0251/70 (Albright & Wilson UK)
Empicol ESA (Albright & Wilson UK)
Empicol ESB70 (Albright & Wilson UK)
Empicol ESB3/M (Albright & Wilson UK)
Empicol ESB2/SP (Albright & Wilson UK)
Empicol ESB 70/SP (Albright & Wilson UK)
Empicol ESB3/ZA (Albright & Wilson Asia)
Empicol ESB70/ZA (Albright & Wilson Asia)
Empicol ESC3 (Albright & Wilson UK)
Empicol ESC70 (Albright & Wilson UK)
Empicol ESC3/ZA (Albright & Wilson Asia)
Empicol ESC70/ZA (Albright & Wilson Asia)
Empimin 3750 (Albright & Wilson Asia)
Empimin 3753 (Albright & Wilson Asia)
Empimin KSN27 (Albright & Wilson UK)
Genapol LRO Liquid (Clariant)
Genapol LRO Liquid (Clariant GmbH, Personal Care)
Genapol LRO Paste (Clariant)
Genapol LRO Paste (Clariant GmbH, Personal Care)
Jeelate ES-1 (Jeen)
Jeelate ES-2 (Jeen)
Jeelate ES-3 (Jeen)
Jeelate ES-270 (Jeen)
Jeelate SLES-60 (Jeen)
Mackol 70NS (McIntyre)
Manro BEC28 (Manro)
Manro BEC 70 (Manro)
Manro NEC28 (Manro)
Manro NEC 70 (Manro)
Manro NEC70N2 (Manro)
Marlinat 242/28 (Sasol GmbH - Marl)
Marlinat 242/70 (Sasol GmbH - Marl)
Nikkol SBL-2N-27 (Nikko)
Nikkol SBL-3N-27 (Nikko)
Norfox SLES-01 (Norman, Fox & Co.)
Norfox SLES-02 (Norman, Fox & Co.)
Norfox SLES-03 (Norman, Fox & Co.)
Norfox SLES-60 (Norman, Fox & Co.)
Oronal BLD (SEPPIC)
Protachem ES-1 (Protameen)
Protachem ES-2 (Protameen)

Rhodapex ES-2 (Rhodia)
Rhodapex ESA/A2 (Rhodia)
Rhodapex ESB-70 (Rhodia)
Rhodapex ESC Series (Rhodia)
Rhodapex ES STD (Rhodia)
Rhodapex ESY (Rhodia)
Rhodapex ESY/CG (Rhodia)
Rhodapex KSN-60/A (Rhodia)
Rhodapex PS 603 (Rhodia)
Sabosol EMS (Sabo)
Sabosol ES 51 (Sabo)
Safol 23 E2S-70 (Sasol GmbH - Marl)
Standapol ES-1 (Cognis Care Chemicals/NJ)
Standapol ES-1 (Cognis Care Chemicals/PA)
Standapol ES-2 (Cognis Care Chemicals/NJ)
Standapol ES-2 (Cognis Care Chemicals/PA)
Standapol ES-3 (Cognis Care Chemicals/NJ)
Standapol ES-3 (Cognis Care Chemicals/PA)
Standapol ES-250 (Cognis Care Chemicals/NJ)
Standapol ES-250 (Cognis Care Chemicals/PA)
Standapol ES-350 (Cognis Care Chemicals/NJ)
Standapol ES-350 (Cognis Care Chemicals/PA)
STEOL CS-130 (Stepan)
STEOL CS-230 (Stepan)
STEOL CS-270 (Stepan)
STEOL CS-330 (Stepan)
STEOL CS-370 (Stepan)
STEOL CS-460 (Stepan)
STEOL 4N (Stepan)
Sulfochem ES-1 (Chemron)
Sulfochem ES-2 (Chemron)
Sulfochem ES-3 (Chemron)
Sulfochem ES-60 (Chemron)
Sulfochem ES-70 (Chemron)
Tensagex EOC628 (Manro)
Tensagex EOC670 (Manro)
Texapon N-70 (Cognis Deutschland)
Texapon N 702 (Cognis Deutschland)
Texapon NC-70 (Cognis Deutschland)
Texapon NC-70LS (Cognis Deutschland)
Texapon NSO (Cognis Care Chemicals/NJ)
Texapon NSO (Cognis Deutschland)
Texapon NSO BZ (Cognis Deutschland)
Texapon NSO IS (Cognis Deutschland)
Texapon NSO KC (Cognis Deutschland)
Texapon NSO UP (Cognis Deutschland)
Texapon SPN 70 (Cognis Care Chemicals/NJ)
Unipol ES-1 (Universal Preserv-A-Chem)
Unipol ES-2 (Universal Preserv-A-Chem)
Unipol ES-3 (Universal Preserv-A-Chem)
Unisan E70 (Albright & Wilson Asia)
Zetesol 270 (Zschimmer & Schwarz)

Zetesol LES 2 (Zschimmer & Schwarz Italiana)
Zetesol LES 2/FK (Zschimmer & Schwarz Italiana)
Zetesol 270/N (Zschimmer & Schwarz)
Zetesol 270/N (Zschimmer & Schwarz Italiana)
Zetesol NL (Zschimmer & Schwarz)
Zetesol NL-2 (Zschimmer & Schwarz)
Zoharpon ETA 27 (Zohar)
Zoharpon ETA 70 (Zohar)
Zoharpon ETA 273 (Zohar)
Zoharpon ETA 703 (Zohar)

Trade Name Mixtures:

Abex VA 50 (Kao GmbH)
AEC Glycol Distearate (&) Sodium Laureth Sulfate (&) Cocamide MEA (&) Laureth-10 (A & E Connock)
Base Nacrante 6030CP (SEPPIC)
BIO-SOFT LD-95 (Stepan)
BIO-TERGE 804 (Stepan)
Bio-Terge 804 M (Stepan)
CalBlend Clear (Pilot)
CalBlend GEL (Pilot)
CalBlend Pearl (Pilot)
Cosmacol P-50 (Sasol Italy)
DeCONC C-30 WOA (DeForest)
DeCONC HS-30 (DeForest)
DeCONC SC-35-DF (DeForest)
DeCONC SCE-40 (DeForest)
Dermulsene RA 405 (Les Derives Resiniques)
Empicol BSD (Albright & Wilson UK)
Empicol BSD 52 (Albright & Wilson UK)
Euperlan PK-771 (Cognis Care Chemicals/NJ)
Euperlan PK-771 (Cognis Care Chemicals/PA)
Euperlan PK-810 (Cognis Care Chemicals/NJ)
Euperlan PK-810 (Cognis Care Chemicals/PA)
Euperlan PK-900 BENZ-W (Cognis Care Chemicals/NJ)
Euperlan PK-900 BENZ-W (Cognis Care Chemicals/PA)
Euperlan PK-900 BENZ-W (Cognis Deutschland)
Euromix MEA (EOC Surfactants)
EuroNac AN10 (EOC Surfactants)
Genapol PGM (Clariant)
Genapol PGM (Clariant GmbH, Personal Care)
Genapol TSM (Clariant)
Genapol TSM (Clariant GmbH, Personal Care)
Gluplex LES (Kelisema Italy)
Homulgator 910 G Extra (Grau)
Jeechem CASS (Jeen)
Jeeteric CDL (Jeen)
Kelifluo 1500 (Kelisema Italy)
Kelifluo 3200 (Kelisema Italy)
Kelifluo 6250 (Kelisema Italy)

KM-901 (Shin-Etsu Chemical Co.)
KM-902 (Shin-Etsu Chemical Co.)
KM-903 (Shin-Etsu Chemical Co.)
KM-906 (Shin-Etsu Chemical Co.)
KM-910 (Shin-Etsu Chemical Co.)
KM-906A (Shin-Etsu Chemical Co.)
KM-902C (Shin-Etsu Chemical Co.)
Krim CH 25 (Fabriquimica)
LP110 (Phytocos)
Miracare 2MCAS (Rhodia)
Miracare MP 35 (Rhodia)
Miracare MPC (Rhodia)
Miracare XGD/A (Rhodia)
Mirasheen 207 (Rhodia)
Mirasheen CP-920 (Rhodia)
Mirasheen CP-820/G (Rhodia)
Natunola Flax Extract 1621 (Natunola)
Perlglanzmittel GM 4175 (Zschimmer & Schwarz)
Plantapon 611 C (Cognis Deutschland)
Plantaren LSC (Cognis Care Chemicals/ NJ)
Plantaren LSC (Cognis Care Chemicals/ PA)
Plantaren PS 10 (Cognis Deutschland)
Plantaren PS-200 (Cognis Care Chemicals/ NJ)
Plantaren PS-200 (Cognis Care Chemicals/ PA)
Plantaren PS-400 (Cognis Care Chemicals/ NJ)
Plantaren PS-400 (Cognis Care Chemicals/ PA)
Protachem Shampoo Concentrate (Protameen)
Proteric CDL (Protameen)
Quickpearl II (Chemron)
Rewopol HM 28 (Degussa Care Specialties)
Rewopol HM 80 (Degussa Care Specialties)
Rewopol PGK 2000 (Degussa Care Specialties)
Rewopol SB CS 50 (Degussa Care Specialties)
Rhodapex BSD-FL/A2 (Rhodia)
Saboperl 450 (Sabo)
Sabosol 8882 (Sabo)
Sabosol BBM (Sabo)
Sabosol CST (Sabo)
Sabosol EM (Sabo)
Sabosol HO (Sabo)
Sabosol SDP (Sabo)
Schercoteric MS-2ES Modified (Scher)
Sinnoflor N 28 (Cognis France)
Standapol 7092 (Cognis Care Chemicals/ NJ)
Standapol 7092 (Cognis Care Chemicals/ PA)
Standapol Pearl Conc. 7130 (Cognis Care Chemicals/NJ)
Standapol Pearl Conc. 7130 (Cognis Care Chemicals/PA)
Stepan Pearl 2 (Stepan)

STEPAN PEARL 4 (Stepan)
Tensoplex Calendula (Fabriquimica)
Tensoplex Centella (Fabriquimica)
Tensoplex Henna (Fabriquimica)
Tensoplex Hiedra (Fabriquimica)
Tensoplex Ortiga (Fabriquimica)
Texapon ASV (Cognis Deutschland)
Texapon ASV-50 (Cognis Care Chemicals/ NJ)
Texapon ASV-50 (Cognis Care Chemicals/ PA)
Texapon ASV-50 (Cognis Deutschland)
Texapon ASV-70 Special (Cognis Deutschland)
Texapon EVR K 400 (Cognis Deutschland)
Unette-W (Universal Preserv-A-Chem)
Unipol 7092 (Universal Preserv-A-Chem)
Zohar EGDS 771 (Zohar)
Zoharpon EGMS-771 (Zohar)
Zoharteric D-2 (Zohar)
Zoharteric D-3 (Zohar)
Zoharteric DOT (Zohar)
Zoharteric M-2 (Zohar)
Zoharteric M-3 (Zohar)

SODIUM LAURETH-5 SULFATE

CAS No.: 9004-82-4 (Generic)

JPN Translation:
ラウレス - 5 硫酸 Na

Empirical Formula:
$C_{22}H_{46}O_9S \cdot Na$

Definition: Sodium Laureth-5 Sulfate is the sodium salt of the sulfate ester of the polyethylene glycol ether of lauryl alcohol that conforms generally to the formula:

$$CH_3(CH_2)_{11}(OCH_2CH_2)_nOSO_3Na$$

where n has an average value of 5.

Information Source: TSCA

Chemical Class: Alkyl Ether Sulfates

Function: Surfactant - Cleansing Agent

Reported Product Categories: Shampoos (Non-coloring); Bath Oils, Tablets, and Salts; Bath Soaps and Detergents; Bubble Baths; Cleansing Products (Cold Creams, Cleansing Lotions, Liquids and Pads); Personal Cleanliness Products, Misc.; Hair Dyes and Colors (All Types Requiring Caution Statements and Patch Tests); Eye Makeup Removers; Hair Shampoos (Coloring); Skin Care Preparations, Misc.; Baby Shampoos; Face and Neck Preparations (Excluding Shaving Preparations); Bath Capsules; Mascara; Baby Products, Misc.; Bath Preparations, Misc.; Body and Hand Preparations (Excluding Shaving Preparations); Fragrance Preparations, Misc.; Hair Preparations (Non-coloring), Misc.; Moisturizing Preparations; Paste Masks (Mud Packs); Tonics,

Dressings, and Other Hair Grooming Aids; Foot Powders and Sprays; Hair Conditioners

Technical/Other Names:
PEG-5 Lauryl Ether Sulfate, Sodium Salt
Polyethylene Glycol (5) Lauryl Ether Sulfate, Sodium Salt
Polyoxyethylene (5) Lauryl Ether Sulfate, Sodium Salt
Sodium Polyethylene Glycol (5) Lauryl Ether Sulfate
Sodium Polyoxyethylene (5) Lauryl Ether Sulfate

SODIUM LAURETH-7 SULFATE

CAS No.: 9004-82-4 (Generic)

JPN Translation:
ラウレス - 7 硫酸 Na

Empirical Formula:
$C_{26}H_{54}O_{11}S \cdot Na$

Definition: Sodium Laureth-7 Sulfate is the sodium salt of the sulfate ester of the polyethylene glycol ether of lauryl alcohol that conforms generally to the formula:

$$CH_3(CH_2)_{11}(OCH_2CH_2)_nOSO_3Na$$

where n has an average value of 7.

Information Source: TSCA

Chemical Class: Alkyl Ether Sulfates

Function: Surfactant - Cleansing Agent

Reported Product Categories: Shampoos (Non-coloring); Bath Oils, Tablets, and Salts; Bath Soaps and Detergents; Bubble Baths; Cleansing Products (Cold Creams, Cleansing Lotions, Liquids and Pads); Personal Cleanliness Products, Misc.; Hair Dyes and Colors (All Types Requiring Caution Statements and Patch Tests); Eye Makeup Removers; Hair Shampoos (Coloring); Skin Care Preparations, Misc.; Baby Shampoos; Face and Neck Preparations (Excluding Shaving Preparations); Bath Capsules; Mascara; Baby Products, Misc.; Bath Preparations, Misc.; Body and Hand Preparations (Excluding Shaving Preparations); Fragrance Preparations, Misc.; Hair Preparations (Non-coloring), Misc.; Moisturizing Preparations; Paste Masks (Mud Packs); Tonics, Dressings, and Other Hair Grooming Aids; Foot Powders and Sprays; Hair Conditioners

Technical/Other Names:
PEG-7 Lauryl Ether Sulfate, Sodium Salt
Polyethylene Glycol (7) Lauryl Ether Sulfate, Sodium Salt
Polyoxyethylene (7) Lauryl Ether Sulfate, Sodium Salt
Sodium Polyethylene Glycol (7) Lauryl Ether Sulfate
Sodium Polyoxyethylene (7) Lauryl Ether Sulfate

SODIUM LAURETH-8 SULFATE

CAS No.: 9004-82-4 (Generic)

JPN Translation:
ラウレス - 8 硫酸 Na

Empirical Formula:
$C_{28}H_{58}O_{13}S \cdot Na$

Definition: Sodium Laureth-8 Sulfate is the sodium salt of the sulfate ester of the poly-ethylene glycol ether of lauryl alcohol that conforms generally to the formula:

$$CH_3(CH_2)_{11}(OCH_2CH_2)_nOSO_3Na$$

where n has an average value of 8.

Chemical Class: Alkyl Ether Sulfates

Function: Surfactant - Cleansing Agent

Reported Product Categories: Bath Soaps and Detergents; Bubble Baths; Personal Cleanliness Products, Misc.; Hair Dyes and Colors (All Types Requiring Caution Statements and Patch Tests); Eye Makeup Removers; Hair Shampoos (Coloring); Skin Care Preparations, Misc.; Baby Shampoos; Face and Neck Preparations (Excluding Shaving Preparations); Bath Capsules; Mascara; Bath Oils, Tablets, and Salts; Cleansing Products (Cold Creams, Cleansing Lotions, Liquids and Pads); Baby Products, Misc.; Bath Preparations, Misc.; Body and Hand Preparations (Excluding Shaving Preparations); Fragrance Preparations, Misc.; Hair Preparations (Non-coloring), Misc.; Moisturizing Preparations; Paste Masks (Mud Packs); Shampoos (Non-coloring); Tonics, Dressings, and Other Hair Grooming Aids; Foot Powders and Sprays; Hair Conditioners

Technical/Other Names:
Laureth-8 Carboxylic Acid, Sodium Salt
PEG-8 Lauryl Ether Sulfate, Sodium Salt
Polyethylene Glycol 400 Lauryl Ether Sulfate, Sodium Salt
Polyoxyethylene (8) Lauryl Ether Sulfate, Sodium Salt
Sodium Polyethylene Glycol 400 Lauryl Ether Sulfate
Sodium Polyoxyethylene (8) Lauryl Ether Sulfate

Trade Name Mixtures:
Plantaren PS-400 (Cognis Care Chemicals/ NJ)
Plantaren PS-400 (Cognis Care Chemicals/ PA)
Texapon ASV (Cognis Deutschland)
Texapon ASV-50 (Cognis Care Chemicals/ NJ)
Texapon ASV-50 (Cognis Care Chemicals/ PA)
Texapon ASV-50 (Cognis Deutschland)
Texapon ASV-70 Special (Cognis Deutschland)

SODIUM LAURETH-12 SULFATE

CAS Nos.: 9004-82-4 (Generic); 66161-57-7

JPN Translation:
ラウレス - 12 硫酸 Na

Empirical Formula:
$C_{36}H_{74}O_{16}S \cdot Na$

Definition: Sodium Laureth-12 Sulfate is the sodium salt of the sulfate ester of the poly-ethylene glycol ether of lauryl alcohol that conforms generally to the formula:

$$CH_3(CH_2)_{11}(OCH_2CH_2)_nOSO_3Na$$

where n has an average value of 12.

Information Sources: CTFA D, TSCA

Chemical Class: Alkyl Ether Sulfates

Function: Surfactant - Cleansing Agent

Reported Product Categories: Shampoos (Non-coloring); Bath Oils, Tablets, and Salts; Bath Soaps and Detergents; Bubble Baths; Cleansing Products (Cold Creams, Cleansing Lotions, Liquids and Pads); Personal Cleanliness Products, Misc.; Hair Dyes and Colors (All Types Requiring Caution Statements and Patch Tests); Eye Makeup Removers; Hair Shampoos (Coloring); Skin Care Preparations, Misc.; Baby Shampoos; Face and Neck Preparations (Excluding Shaving Preparations); Bath Capsules; Mascara; Baby Products, Misc.; Bath Preparations, Misc.; Body and Hand Preparations (Excluding Shaving Preparations); Fragrance Preparations, Misc.; Hair Preparations (Non-coloring), Misc.; Moisturizing Preparations; Paste Masks (Mud Packs); Tonics, Dressings, and Other Hair Grooming Aids; Foot Powders and Sprays; Hair Conditioners

Technical/Other Names:
3,6,9,12,15,18,21,24,27,30,33,36-Dodecaoxaoctatetracosodium Salt
PEG-12 Lauryl Ether Sulfate, Sodium Salt
Polyethylene Glycol 600 Lauryl Ether Sulfate, Sodium Salt
Polyoxyethylene (12) Lauryl Ether Sulfate, Sodium Salt
Sodium Polyethylene Glycol 600 Lauryl Ether Sulfate
Sodium Polyoxyethylene (12) Lauryl Ether Sulfate

Trade Name:
Unipol 125-E (Universal Preserv-A-Chem)

Trade Name Mixtures:
Syntran 5190 (Interpolymer)
Syntran 5760 (Interpolymer)

SODIUM LAURETH-40 SULFATE

CAS No.: 9004-82-4 (Generic)

Definition: Sodium Laureth-40 Sulfate is the sodium salt of the sulfuric acid ester of Laureth-40 (q.v.). It conforms generally to the formula:

$$CH_3(CH_2)_{11}(OCH_2CH_2)_nOSO_3Na$$

where n has an average value of 40.

Chemical Class: Alkyl Ether Sulfates

Function: Surfactant - Cleansing Agent

Reported Product Categories: Shampoos (Non-coloring); Bath Oils, Tablets, and Salts; Bubble Baths; Hair Dyes and Colors (All Types Requiring Caution Statements and Patch Tests); Eye Makeup Removers; Hair Shampoos (Coloring); Baby Shampoos; Bath Capsules; Mascara; Baby Products, Misc.; Bath Preparations, Misc.; Fragrance Preparations, Misc.; Hair Preparations (Non-coloring), Misc.; Tonics, Dressings, and Other Hair Grooming Aids; Hair Conditioners

Technical/Other Names:
PEG-40 Lauryl Ether Sulfate, Sodium Salt
Polyethylene Glycol (40) Lauryl Ether, Sodium Salt
Polyoxyethylene (40) Lauryl Ether, Sodium Salt
Sodium Polyethylene Glycol (40) Lauryl Ether Sulfate
Sodium Polyoxyethylene (40) Lauryl Ether Sulfate

SODIUM LAURETH-7 TARTRATE

CAS No.: 141250-42-2

Empirical Formula:
$C_{30}H_{59}O_{13} \cdot Na$

Definition: Sodium Laureth-7 Tartrate is the sodium salt of the ester of tartaric acid with Laureth-7 (q.v.).

Chemical Classes: Esters; Organic Salts

Function: Surfactant - Cleansing Agent

Technical/Other Names:
PEG-7 Lauryl Ether Tartrate, Sodium Salt
Polyethylene Glycol (7) Lauryl Ether Tartrate, Sodium Salt
Polyoxyethylene (7) Lauryl Ether Tartrate, Sodium Salt
Sodium Polyethylene Glycol (7) Lauryl Ether Tartrate
Sodium Polyoxyethylene (7) Lauryl Ether Tartrate

Trade Name:
Eucarol TA (Cesalpinia)

SODIUM LAURIMINODIPROPIONATE

CAS Nos.	EINECS Nos.
14960-06-6	239-032-7
26256-79-1	247-552-0

JPN Translation:
ラウリミノジプロピオン酸 Na

Empirical Formula:
$C_{18}H_{35}NO_4 \cdot Na$

Definition: Sodium Lauriminodipropionate is the partial sodium salt of a substituted propionic acid. It conforms generally to the formula:

$$CH_3(CH_2)_{11}N \begin{array}{c} CH_2CH_2COONa \\ \\ CH_2CH_2COOH \end{array}$$

Information Sources: CIR: [I] IJT-16 (Suppl. 1)1997, CTFA D, JCIC, JCLS, JSQI

Chemical Class: Alkyl-Substituted Amino Acids

Functions: Antistatic Agent; Hair Conditioning Agent; Surfactant - Cleansing Agent; Surfactant - Foam Booster

Reported Product Categories: Hair Conditioners; Bath Soaps and Detergents; Hair Shampoos (Coloring); Bath Oils, Tablets, and Salts; Cleansing Products (Cold Creams, Cleansing Lotions, Liquids and Pads); Shampoos (Non-coloring); Skin Fresheners

Technical/Other Names:
β-Alanine, N-(2-Carboxyethyl)-N-Dodecyl-, Monosodium Salt
N-(2-Carboxyethyl)-N-Dodecyl-β-Alanine, Monosodium Salt
Sodium Laurylaminodipropionate Solution (30%)
Sodium (Laurylimino)Dipropionate
Sodium N-Lauryl-β-Iminodipropionate

Trade Names:
Amphosol 160C-30 (Stepan)
Deriphat 160-C (Cognis Care Chemicals/NJ)
Deriphat 160-C (Cognis Care Chemicals/PA)
DeTERIC LP (DeForest)
Ethtaine LIDP (Ethox)
Mackam 160C-30 (McIntyre)
Unitex 610-L (Universal Preserv-A-Chem)

SODIUM LAUROAMPHOACETATE

CAS No.: 156028-14-7

JPN Translation:
ラウロアンホ酢酸 Na

Empirical Formula:
$C_{18}H_{36}N_2O_4 \cdot Na$

Definition: Sodium Lauroamphoacetate is the amphoteric organic compound that conforms generally to the formula:

$$CH_3(CH_2)_{10}C \begin{array}{c} O \\ \| \end{array} NHCH_2CH_2NCH_2COONa \begin{array}{c} CH_2CH_2OH \\ | \end{array}$$

Information Source: TSCA

Chemical Class: Alkylamido Alkylamines

Functions: Hair Conditioning Agent; Surfactant - Cleansing Agent; Surfactant - Foam Booster

Reported Product Categories: Bath Soaps and Detergents; Shampoos (Non-coloring); Bath Oils, Tablets, and Salts; Cleansing Products (Cold Creams, Cleansing Lotions, Liquids and Pads)

Technical/Other Names:
Lauroamphoacetate
Lauroamphoglycinate

Trade Names:
AMPHOSOL 1L (Stepan)
Colateric LAA-30 (Colonial Chemical Inc)
Colateric SLAA (Colonial Chemical Inc)
Empigen CDL 30/J (Albright & Wilson UK)
Empigen CDL 30/J/35 (Albright & Wilson UK)
Empigen CDL 60/P (Albright & Wilson UK)
Ethteric LA (Ethox)
Ethteric LMA (Ethox)
Ethteric L-P (Ethox)
Genagen LAA (Clariant)
Genagen LAA (Clariant GmbH, Personal Care)
Genagen LDA (Clariant)
Genagen LDA (Clariant GmbH, Personal Care)
Jeechem LMM-30 (Jeen)
Mackam HPL-28 (McIntyre)
Mackam 1L (McIntyre)
Miranol HMA (Rhodia)
Miranol Ultra L-32 (Rhodia)
Monateric LM-M30 (Uniqema)
Proteric LM-M30 (Protameen)

Trade Name Mixtures:
Miracare BC-10 (Rhodia)
Miracare BC-20 (Rhodia)
Miracare BC-27 (Rhodia)
Miracare MHT (Rhodia)
Miracare UM-140 (Rhodia)
Monateric 985A (Uniqema)
Proteric 1095 (Protameen)

SODIUM LAUROAMPHOHYDROXY-PROPYLSULFONATE

Empirical Formula:
$C_{19}H_{40}N_2O_6S \cdot Na$

Definition: Sodium Lauroamphohydroxypropylsulfonate is the amphoteric organic compound that conforms generally to the formula:

$$CH_3(CH_2)_{10}C \begin{array}{c} O \\ \| \end{array} NHCH_2CH_2NCH_2CHCH_2SO_3Na \begin{array}{c} CH_2CH_2OH \\ | \\ OH \end{array}$$

Information Sources: CTFA D, TSCA

Chemical Classes: Alkylamido Alkylamines; Sulfonic Acids

Functions: Hair Conditioning Agent; Surfactant - Cleansing Agent; Surfactant - Foam Booster; Surfactant - Hydrotrope

Technical/Other Names:
Lauroamphohydroxypropylsulfonate
Lauroamphopropylsulfonate

SODIUM LAUROAMPHO PG-ACETATE PHOSPHATE

CAS No.: 193888-44-7

JPN Translation:
ラウロアンホ PG 酢酸リン酸 Na

Empirical Formula:
$C_{21}H_{42}N_2O_9 \cdot Na$

Definition: Sodium Lauroampho PG-Acetate Phosphate is the zwitterion (inner salt) that conforms generally to the formula:

$$CH_3(CH_2)_{10}C \begin{array}{c} O \\ \| \end{array} NH(CH_2)_2NCH_2CHCH_2OPONa \begin{array}{c} CH_2COO^- \\ |+ \\ HOCH_2CH_2 \ OH \end{array} \begin{array}{c} O \\ \| \\ O \end{array}$$

Information Sources: JCIC, JCLS

Chemical Classes: Alkylamido Alkylamines; Phosphorus Compounds

Functions: Hair Conditioning Agent; Surfactant - Cleansing Agent; Surfactant - Foam Booster; Surfactant - Hydrotrope

Technical/Other Names:
Sodium Lauroamphoglycinohydroxypropyl Phosphate
Sodium Lauroylamidoethyl Hydroxyethyl Carboxymethylbetaine Hydroxypropyl Phosphate Solution

SODIUM LAUROAMPHOPROPIONATE

Empirical Formula:
$C_{19}H_{38}N_2O_4 \cdot Na$

Definition: Sodium Lauroamphopropionate is the amphoteric organic compound that conforms generally to the formula:

$$CH_3(CH_2)_{10}C \begin{array}{c} O \\ \| \end{array} NHCH_2CH_2NCH_2CH_2COONa \begin{array}{c} CH_2CH_2OH \\ | \end{array}$$

Chemical Class: Alkylamido Alkylamines

Functions: Hair Conditioning Agent; Surfactant - Cleansing Agent; Surfactant - Foam Booster

SODIUM LAUROYL ASPARTATE

CAS No.: 41489-18-3

JPN Translation:
ラウロイルアスパラギン酸 Na

Empirical Formula:
$C_{16}H_{29}NO_5 \cdot Na$

Definition: Sodium Lauroyl Aspartate is the organic compound that conforms to the formula:

$$CH_3(CH_2)_{10}C(=O)-NHCHCOOH$$
$$| \quad CH_2COONa$$

Information Source: JCLS

Chemical Classes: Amino Acids; Organic Salts

Functions: Hair Conditioning Agent; Surfactant - Cleansing Agent

Technical/Other Names:
L-Aspartic Acid, N-(1-Oxododecyl)-, Monosodium Salt
N-(1-Oxododecyl)-L-Aspartic Acid, Monosodium Salt

Trade Name:
Monosodium N-Lauroyl-L-Aspartate (Mitsubishi Petrochemical)

SODIUM LAUROYL COLLAGEN AMINO ACIDS

Definition: Sodium Lauroyl Collagen Amino Acids is the sodium salt of the condensation product of lauric acid chloride and Collagen Amino Acids (q.v.).

Chemical Class: Amino Acids

Functions: Hair Conditioning Agent; Surfactant - Cleansing Agent

Technical/Other Name:
Sodium Lauroyl Animal Collagen Amino Acids

SODIUM LAUROYL ETHYLENEDIAMINE TRIACETATE

CAS No.: 206886-68-2

Definition: Sodium Lauroyl Ethylenediamine Triacetate is the organic compound that conforms to the formula:

$$NaOOCCH_2 \diagdown \qquad \diagup C(=O)(CH_2)_{10}CH_3$$
$$NCH_2CH_2N$$
$$NaOOCCH_2 \diagup \qquad \diagdown CCH_2COOH$$

Chemical Classes: Amides; Amines

Functions: Antistatic Agent; Chelating Agent; Hair Conditioning Agent; Oral Care Agent; Surfactant - Foam Booster

Technical/Other Name:
Glycine, N-[2-[Bis(Carboxymethyl)Amino]-Ethyl]-N-(1-Oxododecyl)-, Sodium Salt

Trade Name:
Hampshire LED3A Sodium Salt Solution (Amerchol)

SODIUM LAUROYL GLUTAMATE

CAS Nos.
29923-31-7 (L-Form)
29923-34-0 (dl-alpha)
42926-22-7 (L-Form)
98984-78-2

EINECS No.
249-958-3

JPN Translation:
ラウロイルグルタミン酸 Na

Empirical Formula:
$C_{17}H_{31}NO_5 \cdot Na$

Definition: Sodium Lauroyl Glutamate is the sodium salt of the lauric acid amide of glutamic acid. It conforms generally to the formula:

$$CH_3(CH_2)_{10}C(=O)-NHCHCOONa$$
$$| \quad CH_2CH_2COOH$$

Information Sources: JCLS, JSCI, TSCA

Chemical Class: Amino Acids

Function: Hair Conditioning Agent

Reported Product Categories: Bath Oils, Tablets, and Salts; Cleansing Products (Cold Creams, Cleansing Lotions, Liquids and Pads)

Technical/Other Names:
N-Dodecyl-L-Glutamic Acid, Monosodium Salt
L-Glutamic Acid, N-Dodecyl-, Monosodium Salt
Glutamic Acid, N-(1-Oxododecyl)-, Monosodium Salt
N-(1-Oxododecyl)Glutamic Acid, Monosodium Salt
Sodium N-Lauroyl-L-glutamate

Trade Names:
Aminosurfact ALMS (Asahi Denka Kogyo)
Amisoft LS-11 (Ajinomoto)
Hostapon CLG (Clariant)
Hostapon CLG (Clariant GmbH, Personal Care)
Protelan AGL 95 (Zschimmer & Schwarz)
Protelan AGL 95 (Zschimmer & Schwarz Italiana)

Trade Name Mixture:
Amisoft LS-22 (Ajinomoto)

SODIUM LAUROYL GLYCINE PROPIONATE

Empirical Formula:
$C_{17}H_{30}NO_5Na$

Definition: Sodium Lauroyl Glycine Propionate is the substituted amino acid that conforms to the formula:

$$CH_3(CH_2)_{10}C(=O)-NCH_2CH_2COOH$$
$$| \quad CH_2COOH$$

Chemical Class: Alkyl-Substituted Amino Acids

Function: Surfactant - Cleansing Agent

SODIUM LAUROYL HYDROLYZED COLLAGEN

CAS No.: 68989-51-5

JPN Translation:
ラウロイル加水分解コラーゲン Na

Definition: Sodium Lauroyl Hydrolyzed Collagen is the sodium salt of the condensation product of lauric acid chloride and Hydrolyzed Collagen (q.v.).

Information Sources: JCIC, JCLS, JSQI

Chemical Class: Protein Derivatives

Functions: Hair Conditioning Agent; Skin-Conditioning Agent - Miscellaneous; Surfactant - Cleansing Agent

Technical/Other Names:
Collagens, Lauroyl Derivs., Sodium Salts
Proteins, Hydrolysates, Reaction Products with Lauroyl Chloride, Compds. with Sodium
Sodium Lauroyl Hydrolyzed Animal Protein

Trade Names:
Promois ELS (Seiwa Kasei)
Promois EULS (Seiwa Kasei)

SODIUM LAUROYL HYDROLYZED SILK

JPN Translation:
ラウロイル加水分解シルク Na

Definition: Sodium Lauroyl Hydrolyzed Silk is the sodium salt of the condensation product of lauric acid chloride and Hydrolyzed Silk (q.v.).

Information Sources: JCIC, JCLS

Chemical Class: Protein Derivatives

Functions: Hair Conditioning Agent; Skin-Conditioning Agent - Miscellaneous; Surfactant - Cleansing Agent

Technical/Other Name:
Sodium Lauroyl Hydrolyzed Silk Solution

Trade Names:
Promois EFLS (Seiwa Kasei)
Promois EFLS-F (Seiwa Kasei)

SODIUM LAUROYL ISETHIONATE

CAS No. 7381-01-3   **EINECS No.** 230-949-8

Empirical Formula:
$C_{14}H_{28}O_5S \cdot Na$

Definition: Sodium Lauroyl Isethionate is the sodium salt of the lauric acid ester of isethionic acid. It conforms generally to the formula:

$$CH_3(CH_2)_{10}\overset{\overset{O}{\|}}{C}-OCH_2CH_2SO_3Na$$

Information Source: TSCA

Chemical Class: Isethionates

Function: Surfactant - Cleansing Agent

Reported Product Category: Bath Preparations, Misc.

Technical/Other Names:
Ethanesulfonic Acid, 2-Hydroxy-, Laurate, Sodium Salt
2-Sulfoethyl Laurate Sodium Salt

SODIUM LAUROYL LACTYLATE

CAS No. 13557-75-0   **EINECS No.** 236-942-6

JPN Translation:
ラウロイル乳酸 Na

Empirical Formula:
$C_{18}H_{32}O_6 \cdot Na$

Definition: Sodium Lauroyl Lactylate is the sodium salt of the lauric acid ester of lactyl lactate. It conforms to the formula:

$$CH_3(CH_2)_{10}\overset{\overset{O}{\|}}{C}-O\underset{\underset{CH_3}{|}}{C}H-O\underset{\underset{CH_3}{|}}{C}HCOONa$$

Chemical Classes: Esters; Organic Salts

Function: Surfactant - Emulsifying Agent

Reported Product Categories: Bath Soaps and Detergents; Bath Oils, Tablets, and Salts

Technical/Other Names:
Dodecanoic Acid, 2-(1-Carboxyethoxy)-1-Methyl-2-Oxoethyl Ester, Sodium Salt
Soduim Lauroyl Dilactate

Trade Names:
Dermasoft SLL (Straetmans)
Dermosoft SLL (Straetmans)
Pationic 138C (RITA)

Trade Name Mixtures:
Ceralution C (Sasol GmbH - Marl)
Ceralution F (Sasol GmbH - Marl)
SK - Influx (Degussa Care Specialties)

SODIUM LAUROYL METHYLAMINO-PROPIONATE

CAS No. 21539-58-2   **EINECS No.** 244-429-3

JPN Translation:
ラウロイルメチルアラニン Na

Empirical Formula:
$C_{16}H_{31}NO_3 \cdot Na$

Definition: Sodium Lauroyl Methylamino-propionate is the sodium salt of the lauric acid amide of N-methyl beta-alanine. It conforms to the formula:

$$CH_3(CH_2)_{10}\overset{\overset{O}{\|}}{C}-\underset{\underset{CH_3}{|}}{N}CH_2CH_2\overset{\overset{O}{\|}}{C}-ONa$$

Information Sources: JCIC, JCLS, JSQI

Chemical Class: Alkyl-Substituted Amino Acids

Function: Surfactant - Cleansing Agent

Technical/Other Names:
Sodium N-Lauroyl N-Methyl beta-Alanine
Sodium N-Lauroyl-N-methyl-β-amino-propionate Solution

Trade Names:
Alanon ALE (Kawaken)
Nikkol Alaninate LN-30 (Nikko)

SODIUM LAUROYL/MYRISTOYL ASPARTATE

JPN Translation:
アシル (C12 , 14) アスパラギン酸 Na

Definition: Sodium Lauroyl/Myristoyl Aspartate is the sodium salt of the substituted amino acid that conforms generally to the formula:

$$RC\overset{\overset{O}{\|}}{-}NH\underset{\underset{CH_2COOH}{|}}{C}HCOONa$$

where RCO- represents the lauroyl/myristoyl grouping.

Chemical Classes: Amino Acids; Organic Salts

Functions: Hair Conditioning Agent; Surfactant - Cleansing Agent

Trade Name:
Asparack LM-NS (Mitsubishi Petrochemical)

SODIUM LAUROYL OAT AMINO ACIDS

Definition: Sodium Lauroyl Oat Amino Acids is the sodium salt of the condensation product of lauric acid chloride with the amino acids derived from Avena Sativa (Oat) Protein (q.v.).

Chemical Class: Amino Acids

Functions: Hair Conditioning Agent; Skin-Conditioning Agent - Miscellaneous; Surfactant - Cleansing Agent

Reported Product Category: Shampoos (Non-coloring)

Trade Names:
Lauroat (Sinerga)
Proteol O.A.T. (SEPPIC)

SODIUM LAUROYL SARCOSINATE

CAS No. 137-16-6   **EINECS No.** 205-281-5

JPN Translation:
ラウロイルサルコシン Na

Empirical Formula:
$C_{15}H_{29}NO_3 \cdot Na$

Definition: Sodium Lauroyl Sarcosinate is the sodium salt of Lauroyl Sarcosine (q.v.). It conforms generally to the formula:

$$CH_3(CH_2)_{10}\overset{\overset{O}{\|}}{C}-\underset{\underset{CH_3}{|}}{N}CH_2COONa$$

Information Sources: 21CFR175.105, 21CFR177.1200, CIR: [SQ] IJT-20(SUPPL. 1)2001, CTFA D, JCLS, JSCI, MHLW-331/2, MI-13(4383), SNPF, TSCA

Chemical Class: Sarcosinates and Sarcosine Derivatives

Functions: Hair Conditioning Agent; Surfactant - Cleansing Agent

Reported Product Categories: Shampoos (Non-coloring); Bath Soaps and Detergents; Bath Oils, Tablets, and Salts; Bath Preparations, Misc.; Cleansing Products (Cold Creams, Cleansing Lotions, Liquids and Pads); Hair Shampoos (Coloring); Foundations; Personal Cleanliness Products, Misc.

Technical/Other Names:
N-Dodecanoylsarcosine Sodium Salt
Glycine, N-Methyl-N-(1-Oxododecyl)-, Sodium Salt
N-Methyl-N-(1-Oxododecyl)Glycine, Sodium Salt
Sodium N-Lauroyl Sarcosinate

Trade Names:
Crodasinic LS30 (Croda Chemicals)
Crodasinic LS-30 (Croda, Inc.)
Crodasinic LS35 (Croda Chemicals)
Crodasinic LS95 (Croda Chemicals)
Crodasinic LS-95 (Croda, Inc.)

Hamposyl L-30 (Amerchol)
Hamposyl L-95 (Amerchol)
Lowenol L-95 (Lowenstein)
MAPROSYL 30 (Stepan)
Medialan LD (Clariant)
Medialan LD (Clariant GmbH, Personal
Care)
Neoscoap SLN-100 (Toho)
Nikkol Sarcosinate LN (Nikko)
Nikkol Sarcosinate LN-30 (Nikko)
Oramix L 30 (SEPPIC)
Protelan LS 9011 (Zschimmer & Schwarz)
Protelan LS 9011 (Zschimmer & Schwarz
Italiana)
Sabosol L 30 (Sabo)
Soypon SLE (Kawaken)
Vanseal NALS-30 (Vanderbilt)
Vanseal NALS-95 (Vanderbilt)
Zoharsyl L-30 (Zohar)

Trade Name Mixture:
Afron-LS (Vevy)

SODIUM LAUROYL SILK AMINO ACIDS

Definition: Sodium Lauroyl Silk Amino Acids
is the sodium salt of the condensation
product of lauric acid chloride and Silk Amino
Acids (q.v.).

Chemical Class: Amino Acids

Functions: Hair Conditioning Agent; Skin-
Conditioning Agent - Miscellaneous;
Surfactant - Cleansing Agent

Trade Names:
C12 Soie Na (Phytocos)
Promois EFLS-C (Seiwa Kasei)

SODIUM LAUROYL TAURATE

CAS No.	EINECS No.
70609-66-4	274-695-6

Empirical Formula:
$C_{14}H_{29}NO_4S \cdot Na$

Definition: Sodium Lauroyl Taurate is the
organic salt that conforms generally to the
formula:

$$CH_3(CH_2)_{10}\overset{\displaystyle O}{\overset{\|}{C}} - NHCH_2CH_2SO_3Na$$

Chemical Class: Sulfonic Acids

Function: Surfactant - Cleansing Agent

Technical/Other Names:
Ethanesulfonic Acid, 2-[(1-Oxododecyl)-
Amino]-, Sodium Salt
2-[(1-Oxododecyl)Amino]Ethanesulfonic
Acid, Sodium Salt
Sodium 2-[(1-Oxododecyl)Amino]Ethane-
sulfonate

SODIUM LAUROYL WHEAT AMINO ACIDS

Definition: Sodium Lauroyl Wheat Amino
Acids is the sodium salt of the condensation
product of lauric acid chloride and Wheat
Amino Acids (q.v.).

Chemical Class: Amino Acids

Functions: Hair Conditioning Agent; Skin-
Conditioning Agent - Miscellaneous;
Surfactant - Cleansing Agent

Trade Names:
Protelan LWA (Zschimmer & Schwarz
Italiana)
Proteol LW 30 (SEPPIC)

SODIUM LAURYL DIETHYLENE-DIAMINOGLYCINATE

JPN Translation:
ラウリルジアミノエチルグリシン Na

Definition: Sodium Lauryl Diethylene-
diaminoglycinate is the organic compound
that conforms to the formula:

$$C_{12}H_{25} - \overset{\displaystyle C_2H_5NH_2}{\underset{\displaystyle C_2H_5NH_2}{\overset{|}{\underset{|}{N^+}}}} - CH_2COONa$$

Information Sources: JCIC, MHLW-331/3

Chemical Classes: Amines; Amino Acids

Functions: Preservative; Surfactant -
Cleansing Agent; Surfactant - Emulsifying
Agent

SODIUM LAURYL GLUCOSIDEOXY-ACETATE

CAS No.: 383178-66-3

Definition: Sodium Lauryl Glucosideoxy-
acetate is the organic compound that
conforms generally to the formula:

where R represents the lauryl moiety.

Chemical Classes: Carbohydrates; Ethers;
Organic Salts

Function: Not Reported

Technical/Other Name:
D-Glucopyranose, Oligomeric, C10-16
Alkyl Glycosides, Carboxymethyl Ethers,
Sodium Salts

Trade Name Mixture:
Plantapon LGC (Cognis Deutschland)

SODIUM LAURYL GLYCOL CARBOXYLATE

Definition: Sodium Lauryl Glycol
Carboxylate is the product obtained by the
reaction of Lauryl Glycol (q.v.) and sodium
monochloroacetate.

Chemical Classes: Ethers; Organic Salts

Function: Surfactant - Cleansing Agent

Trade Name:
Beaulight SHAA (Sanyo)

SODIUM LAURYL HYDROXYACETAMIDE SULFATE

Definition: Sodium Lauryl
Hydroxyacetamide Sulfate is the sodium salt
of a laurylacetamide derivative. It conforms to
the formula:

$$NaO_3SOCH_2\overset{\displaystyle O}{\overset{\|}{C}} - NH(CH_2)_{11}CH_3$$

Information Source: JCLS

Chemical Class: Alkyl Sulfates

Function: Not Reported

SODIUM LAURYL PHOSPHATE

JPN Translation:
ラウリルリン酸 Na

Definition: Sodium Lauryl Phosphate is the
sodium salt of a complex mixture of esters of
lauryl alcohol and phosphoric acid.

Information Sources: JCIC, JCLS, JSQI

Chemical Class: Phosphorus Compounds

Functions: Surfactant - Cleansing Agent;
Surfactant - Emulsifying Agent

Technical/Other Names:
Dodecyl Disodium Phosphate
Phosphoric Acid, Dodecyl Ester, Sodium
Salt
Sodium Lauryl Phosphate (1)
Sodium Lauryl Phosphate (2)

Trade Name:
Nikkol Phosten HLP-N (Nikko)

SODIUM LAURYL SULFATE

CAS Nos.	EINECS Nos.
151-21-3	205-788-1

68585-47-7 (Generic)
68955-19-1
73296-89-6   277-362-3

JPN Translation:
ラウリル硫酸 Na

Empirical Formula:
$C_{12}H_{26}O_4S \cdot Na$

Definition: Sodium Lauryl Sulfate is the sodium salt of lauryl sulfate that conforms to the formula:

$$CH_3(CH_2)_{11}OSO_3Na$$

Information Sources: BP, BPC, BRA, 21CFR172.210, 21CFR172.822, 21CFR175.105, 21CFR175.300, 21CFR175.320, 21CFR176.170, 21CFR176.210, 21CFR177.1200, 21CFR177.1210, 21CFR177.1630, 21CFR177.2600, 21CFR177.2800, 21CFR178.1010, 21CFR178.3400, CIR: [SQ] JACT-2(7)1983, CTFA S, CZE, FCC, HUN, IND, ITA, JAN, JCLS, JSCI, MAR, MI-13(8710), NF XIX, ROM, SNPF, TSCA, USAN, USD

Chemical Class: Alkyl Sulfates

Functions: Denaturant; Surfactant - Cleansing Agent

Reported Product Categories: Hair Dyes and Colors (All Types Requiring Caution Statements and Patch Tests); Shampoos (Non-coloring); Bath Soaps and Detergents; Bath Oils, Tablets, and Salts; Cleansing Products (Cold Creams, Cleansing Lotions, Liquids and Pads); Dentifrices (Aerosol, Liquid, Pastes and Powders); Shaving Cream (Aerosol, Brushless and Lather); Bath Preparations, Misc.; Body and Hand Preparations (Excluding Shaving Preparations); Moisturizing Preparations; Personal Cleanliness Products, Misc.; Hair Bleaches; Skin Care Preparations, Misc.; Permanent Waves; Hair Straighteners; Mascara; Foundations; Makeup Bases; Eye Makeup Preparations, Misc.; Hair Conditioners; Hair Wave Sets; Bath Capsules; Bubble Baths; Depilatories; Hair Coloring Preparations, Misc.; Hair Preparations (Non-coloring), Misc.; Mouthwashes and Breath Fresheners (Liquids and Sprays); Deodorants (Underarm); Face and Neck Preparations (Excluding Shaving Preparations); Hair Lighteners with Color; Paste Masks (Mud Packs); Eye Shadows; Eyeliners; Hair Shampoos (Coloring); Suntan Preparations, Misc.

Technical/Other Names:
Sodium Dodecyl Sulfate
Sulfuric Acid, Monododecyl Ester, Sodium Salt

Trade Names:
Akyposal NLS (Kao GmbH)

Alscoap LN-40P (Toho)
Alscoap LN-90(P/N) (Toho)
Alscoap TL-30 (Toho)
Calfoam SLS-30 (Pilot)
Colonial SLS (Colonial Chemical Inc)
DeSULF SLS-30-LC (DeForest)
DeSULF SLS-30LS (DeForest)
Empicol 0045 (Albright & Wilson UK)
Empicol 9406 (Albright & Wilson Asia)
Empicol 9418 (Albright & Wilson Asia)
Empicol LX (Albright & Wilson UK)
Empicol LX28 (Albright & Wilson UK)
Empicol LX32 (Albright & Wilson UK)
Empicol LX42 (Albright & Wilson UK)
Empicol LX100 (Albright & Wilson UK)
Empicol LXS95/S (Albright & Wilson UK)
Empicol LXSV/S (Albright & Wilson UK)
Empicol LXV (Albright & Wilson UK)
Empicol LXV100 (Albright & Wilson UK)
Empicol LXV/D/ZA (Albright & Wilson Asia)
Empicol LXV/ZA (Albright & Wilson Asia)
Empicol LX 28/ZA (Albright & Wilson Asia)
Empicol LZ (Albright & Wilson UK)
Empicol LZ/D (Albright & Wilson UK)
Empicol LZV (Albright & Wilson UK)
Empicol LZV/D (Albright & Wilson UK)
Empicol LZV/ZA (Albright & Wilson Asia)
Empicol 0303/VA (Albright & Wilson UK)
Jeelate SLS-30 (Jeen)
Jeelate SLSP-90 (Jeen)
Manro SLS28 (Manro)
Nikkol SLS (Nikko)
Nikkol SLS-30 (Nikko)
Protachem SLP95 (Protameen)
Rhodapon LCP (Rhodia)
Rhodapon LSB (Rhodia)
Rhodapon LS-94/P (Rhodia)
Rhodapon LS-92/RN (Rhodia)
Rhodapon LX-28 (Rhodia)
Rhodapon SB-8208/S (Rhodia)
Sabosol DSP (Sabo)
Sabosol LSSP (Sabo)
Serdet DFK 30 (Sasol Servo)
Standapol WAQ-LC (Cognis Care Chemicals/NJ)
Standapol WAQ-LC (Cognis Care Chemicals/PA)
Standapol WAQ-Special (Cognis Care Chemicals/NJ)
Standapol WAQ-Special (Cognis Care Chemicals/PA)
STEPANOL ME-DRY (Stepan)
STEPANOL WAC (Stepan)
STEPANOL WA-EXTRA (Stepan)
STEPANOL WA-100NF/USP (Stepan)
STEPANOL WA-PASTE (Stepan)
STEPANOL WA-SPECIAL (Stepan)
Sulfetal LS (Zschimmer & Schwarz)
Sulfochem SAC (Chemron)
Sulfochem SLC (Chemron)
Sulfochem SLN-95 (Chemron)
Sulfochem SLP-95 (Chemron)
Sulfochem SLS (Chemron)
Sulfopon 101 Special (Cognis Deutschland)

Tensopol A79 (Manro)
Tensopol A795 (Manro)
Tensopol DP (Manro)
Tensopol DSX (Manro)
Tensopol S30LS (Manro)
Tensopol USP94 (Manro)
Tensopol USP97 (Manro)
Texapon CP-G 95 (Cognis Deutschland)
Texapon K-12G (Cognis Deutschland)
Texapon K 12 P (Cognis Deutschland)
Texapon K 12 P PH (Cognis Deutschland)
Texapon K 400 S&M (Cognis Deutschland)
Texapon K-1296 USP (Cognis Care Chemicals/NJ)
Texapon K-1296 USP (Cognis Care Chemicals/PA)
Texapon V 95 G (Cognis Deutschland)
Texapon VHC Needles (Cognis Care Chemicals/NJ)
Texapon VHC Needles (Cognis Care Chemicals/PA)
Texapon ZHC Needles (Cognis Care Chemicals/NJ)
Texapon ZHC Needles (Cognis Care Chemicals/PA)
Texapon ZHC Powder (Cognis Care Chemicals/PA)
Texapon Z 95 P (Cognis Deutschland)
Unipol WA-AC (Universal Preserv-A-Chem)
Unipol WAC Special (Universal Preserv-A-Chem)
Unipol WAQ-LC (Universal Preserv-A-Chem)
Unipon K12 (Universal Preserv-A-Chem)
Unipon K1296 (Universal Preserv-A-Chem)
Unipon L-100 (Universal Preserv-A-Chem)
Unipon ZHC Needles (Universal Preserv-A-Chem)
Unipon ZHC Powder (Universal Preserv-A-Chem)
Zoharpon LAS (Zohar)
ZOHARPON SDS (Zohar)
Zoharpon SLS (Zohar)

Trade Name Mixtures:
Afron-A (Vevy)
Amisol 638 (Lucas Meyer GmbH)
Ceral 10 (Fabriquimica)
Ceral EF (Fabriquimica)
Ceral EFN (Fabriquimica)
Ceral LE (Fabriquimica)
Cerasynt LP (International Specialty Products)
Cerasynt WM (International Specialty Products)
Collagen-LSS-Complex (Kelisema Italy)
Crodex A (Croda Chemicals)
Custoblend BAC (Anti-Bac Blend) (Custom Ingredients)
Custoblend BAT (Anti-Bac Blend) (Custom Ingredients)
Custoblend UB (Universal Blend) (Custom Ingredients)
DeCONC SC-35-DF (DeForest)
DeCONC SCE-40 (DeForest)

Emulmetik 310 (Lucas Meyer GmbH)
Galenol 1618 DSN (Sasol GmbH - Hamburg)
Galenol 1618 KS (Sasol GmbH - Hamburg)
Gluplex LS (Kelisema Italy)
Jeeteric CDL (Jeen)
Klensoft (Guardian)
Lait Vegetal (Greentech)
Lanette SX (Cognis Care Chemicals/NJ)
Lanette SX (Cognis Care Chemicals/PA)
Lanette SX (Cognis Deutschland)
Lanette W (Cognis Care Chemicals/NJ)
Lanette W (Cognis Care Chemicals/PA)
Lanette W (Cognis Deutschland)
Lexemul AS (Inolex)
Lytron 450 (Omnova Solutions, Inc)
Lytron 621 (Rohm and Haas)
Lytron 631 (Rohm and Haas)
Lytron 651 (Rohm and Haas)
Mearlmaid TR (Engelhard Corp.)
Miracare 2MCA (Rhodia)
Miracare 2MCAS (Rhodia)
Miracare XGD/A (Rhodia)
Oxodry (LCW)
Oxothick (LCW)
Protachem GMS-AS (Protameen)
Proteric CDL (Protameen)
Proticute U Alpha (Ichimaru Pharcos)
Quickpearl I (Chemron)
Rewoteric AM G 30 (Degussa Care Specialties)
Sabowax FLT (Sabo)
Sabowax SX (Sabo)
Schercoteric MS-2 Modified (Scher)
Silsoft E-50 (OSi Specialties)
Standapol CS Paste (Cognis Care Chemicals/NJ)
Standapol CS Paste (Cognis Care Chemicals/PA)
STEPANOL 360 (Stepan)
Syntran 5009 (Interpolymer)
Syntran KL-219C (Interpolymer)
Tensoplex Manzanilla (Fabriquimica)
Texapon CS Paste (Cognis Deutschland)
Texapon EVR K 400 (Cognis Deutschland)
Unihydag WAX-SX (Universal Preserv-A-Chem)
Unipol CS-50 (Universal Preserv-A-Chem)
Unipol CS PASTE (Universal Preserv-A-Chem)
Zoharteric DO (Zohar)

SODIUM LAURYL SULFOACETATE

CAS No.
1847-58-1

EINECS No.
217-431-7

JPN Translation:
ラウリルスルホ酢酸 Na

Empirical Formula:
$C_{14}H_{28}O_5S \cdot Na$

Definition: Sodium Lauryl Sulfoacetate is the organic salt that conforms generally to the formula:

$$NaSO_3CH_2C \overset{\overset{\displaystyle O}{\|}}{} - O(CH_2)_{11}CH_3$$

Information Sources: CIR: [S] JACT-6(3)-1987, CTFA D, JCIC, JCLS, JSQI, SNPF, TSCA

Chemical Class: Sulfonic Acids

Function: Surfactant - Cleansing Agent

Reported Product Categories: Bath Preparations, Misc.; Bubble Baths; Bath Oils, Tablets, and Salts; Cleansing Products (Cold Creams, Cleansing Lotions, Liquids and Pads)

Technical/Other Names:
Acetic Acid, Sulfo-, 1-Dodecyl Ester, Sodium Salt
Dodecyl Sodium Sulfoacetate
Sulfoacetic Acid, 1-Dodecyl Ester, Sodium Salt

Trade Names:
LATHANOL LAL (Stepan)
Nikkol LSA (Nikko)

Trade Name Mixtures:
Lumorol K 5019 (Zschimmer & Schwarz)
STEPAN-MILD LSB (Stepan)
Zetesap 5213 (Zschimmer & Schwarz)

SODIUM LEVULINATE

CAS No.
19856-23-6

EINECS No.
243-378-4

Empirical Formula:
$C_5H_8O_3 \cdot Na$

Definition: Sodium Levulinate is the sodium salt of Levulinic Acid (q.v.).

Chemical Class: Organic Salts

Function: Skin-Conditioning Agent - Miscellaneous

Technical/Other Name:
Pentanoic Acid, 4-oxo-, Sodium Salt

Trade Name:
Dermosoft 700 (Straetmans)

Trade Name Mixtures:
Dermosoft 690 (Straetmans)
Lexgard 690 (Inolex)

SODIUM LIGNOSULFONATE

CAS No.: 8061-51-6

Definition: Sodium Lignosulfonate is the sodium salt of polysulfonated lignin, which is a dark brown polymeric material found in wood.

Information Sources: 21CFR173.310, 21CFR175.105, 21CFR176.170, 21CFR176.210, 21CFR177.1210, CTFA D, TSCA, USAN

Chemical Class: Sulfonic Acids

Function: Surfactant - Suspending Agent

Technical/Other Names:
Lignosulfonic Acid, Sodium Salt
Sodium Polignate
Sulfonated Lignin Sodium Salt

Trade Name:
Darvan #2 (Vanderbilt)

SODIUM LINOLEATE

CAS No.
822-17-3

EINECS No.
212-491-0

Definition: Sodium Linoleate is the sodium salt of Linoleic Acid (q.v.).

Chemical Class: Soaps

Functions: Surfactant - Cleansing Agent; Surfactant - Emulsifying Agent; Viscosity Increasing Agent - Nonaqueous

Technical/Other Names:
Linoleic Acid, Sodium Salt
9,12-Octadecadienoic Acid, Sodium Salt

Trade Name:
PRISAVON L (Uniqema Europe)

SODIUM MA/DIISOBUTYLENE COPOLYMER

JPN Translation:
(マレイン酸 / ジイソブチレン) コポリマーNa

Definition: Sodium MA/Diisobutylene Copolymer is the sodium salt of a copolymer of maleic anhydride and diisobutylene monomers.

Chemical Class: Synthetic Polymers

Function: Film Former

Trade Name:
Acusol 460N Polymer (Rohm and Haas)

SODIUM MAGNESIUM FLUOROSILICATE

Definition: Sodium Magnesium Fluorosilicate is the inorganic salt that conforms generally to the formula:

$$Na_2Mg(SiF_6)_2$$

Chemical Class: Inorganics

Functions: Abrasive; Absorbent; Opacifying Agent; Suspending Agent - Nonsurfactant; Viscosity Increasing Agent - Aqueous

Technical/Other Name: Synthetic Hectorite	**Trade Name Mixture:** Rovisome AHA - Malic Acid (Rovi)	**Trade Name:** K-polymer (Kuraray)

SODIUM MAGNESIUM SILICATE

JPN Translation:
ケイ酸 (Na / Mg)

Definition: Sodium Magnesium Silicate is a synthetic silicate clay with a composition mainly of magnesium and sodium silicate.

Information Sources: CIR: [S] IJT-22 (SUPPL. 1)2003, CTFA D, JCIC, JCLS, JSQI, TSCA

Chemical Class: Inorganic Salts

Functions: Binder; Bulking Agent

Reported Product Categories: Lipsticks; Eye Shadows; Foundations; Face Powders; Blushers (All types); Rouges; Eye Makeup Preparations, Misc.; Makeup Preparations (Not eye), Misc.; Eyeliners; Bath Capsules; Eyebrow Pencils; Face and Neck Preparations (Excluding Shaving Preparations); Body and Hand Preparations (Excluding Shaving Preparations); Makeup Bases; Mascara; Moisturizing Preparations

Technical/Other Name:
Synthetic Sodium Magnesium Silicate

Trade Names:
Ionite (Mizusawa)
Optigel SH (Sud-Chemie, United Catalysts)

Trade Name Mixtures:
Electorized, Deoxidized and Ionized Water GE-100 (Ikeda)
Optigel GSH - 1330 (Sud-Chemie, United Catalysts)
WHP-CL-1 (US Cosmetics)

SODIUM MALATE

JPN Translation:
リンゴ酸 Na

Empirical Formula:
$C_4H_6O_4 \cdot Na$

Definition: Sodium Malate is the sodium salt of Malic Acid (q.v.). It conforms to the formula:

$$NaOOCCHCH_2COOH$$
$$|$$
$$OH$$

Information Source: CIR: [SQ] IJT-20 (SUPPL. 1)2001

Chemical Class: Organic Salts

Function: Skin-Conditioning Agent - Humectant

Technical/Other Names:
Butanedioic Acid, Hydroxy-, Monosodium Salt
Malic Acid, Monosodium Salt

SODIUM MANNOSE PHOSPHATE

CAS No.: 70442-25-0

Definition: Sodium Mannose Phosphate is the sodium salt of a complex mixture of esters of phosphoric acid and Mannose (q.v.).

Chemical Classes: Carbohydrates; Phosphorus Compounds

Functions: Skin-Conditioning Agent - Humectant; Skin-Conditioning Agent - Miscellaneous

Technical/Other Name:
D-Mannose, 6-(Dihydrogen Phosphate), Monosodium Salt

Trade Name:
Manophos-6 (Independent Chemical)

SODIUM MANNURONATE METHYL-SILANOL

CAS No.: 23732-95-8

Definition: Sodium Mannuronate Methyl-silanol is a complex of sodium mannuronate and monomethylsilanol.

Chemical Classes: Organic Salts; Siloxanes and Silanes

Function: Skin-Conditioning Agent - Miscellaneous

Reported Product Categories: Skin Care Preparations, Misc.; Body and Hand Preparations (Excluding Shaving Preparations)

Trade Name:
Algisium (Exsymol)

SODIUM MA/VINYL ALCOHOL COPOLYMER

CAS No.: 139871-83-3

Definition: Sodium MA/Vinyl Alcohol Copolymer is the sodium salt of a copolymer of maleic anhydride and vinyl alcohol monomers.

Chemical Class: Synthetic Polymers

Functions: Binder; Film Former

Technical/Other Name:
2-Butenedioic Acid (Z)-, Polymer with Ethenol and Ethenyl Acetate, Sodium Salt

SODIUM/MEA LAURETH-2 SULFOSUCCINATE

Definition: Sodium/MEA Laureth-2 Sulfosuccinate is a mixture of sodium and monoethanolamine salts of the Laureth-2 (q.v.) half ester of sulfosuccinic acid.

Chemical Class: Sulfosuccinates and Sulfosuccinamates

Functions: Surfactant - Cleansing Agent; Surfactant - Emulsifying Agent; Surfactant - Foam Booster; Surfactant - Hydrotrope

SODIUM/MEA-PEG-3 COCAMIDE SULFATE

Definition: Sodium/MEA-PEG-3 Cocamide Sulfate is the mixed sodium and monoethanolamine salts of the sulfate esters of PEG-3 Cocamide (q.v.).

Chemical Class: Alkoxylated Amides

Function: Surfactant - Emulsifying Agent

Trade Name:
Neoscoap CME-30S (Toho)

SODIUM METABISULFITE

CAS Nos.	**EINECS No.**
7681-57-4	231-673-0
7757-74-6	

JPN Translation:
ピロ亜硫酸 Na

Empirical Formula:
$H_2O_5S_2 \cdot 2Na$

Definition: Sodium Metabisulfite is the inorganic salt that conforms to the formula:

$$Na_2S_2O_6$$

Information Sources: 21CFR173.310, 21CFR177.1200, 21CFR182.3766, CIR: [S], JCLS, JSCI, MI-13(8712), NF XIX, TSCA, USAN

Chemical Class: Inorganic Salts

Functions: Antioxidant; Reducing Agent

Reported Product Categories: Hair Dyes and Colors (All Types Requiring Caution Statements and Patch Tests); Hair Color Sprays (Aerosol); Eyeliners; Indoor Tanning Preparations; Deodorants (Underarm); Bath Oils, Tablets, and Salts; Skin Care Preparations, Misc.; Bath Preparations, Misc.; Body and Hand Preparations (Excluding Shaving

Preparations); Cleansing Products (Cold Creams, Cleansing Lotions, Liquids and Pads); Face and Neck Preparations (Excluding Shaving Preparations); Foundations; Hair Wave Sets; Moisturizing Preparations; Night Skin Care Preparations

Technical/Other Names:
Disulfurous Acid, Disodium Salt
Pyrosulfurous Acid, Disodium Salt
Sodium Pyrosulfite

Trade Names:
Covastyle MBS (LCW)
Uantox SMBS (Universal Preserv-A-Chem)

Trade Name Mixtures:
Albacan (Bio-Botanica)
Phytolight BG (Coletica SA)

SODIUM METAPHOSPHATE

CAS Nos.	EINECS Nos.
10361-03-2	233-782-9
50813-16-6	256-779-4

JPN Translation:
メタリン酸 Na

Definition: Sodium Metaphosphate is a linear sodium polyphosphate that conforms generally to the formula:

$$(NaPO_3)_n$$

Information Sources: 21CFR182.6769, 21CFR200.5, 21CFR201.63, 21CFR355.55, CIR: [SQ] IJT-20(SUPPL. 3)2001, CTFA S, JCLS, JSCI, MI-13(5668), MI-13(8715), TSCA

Chemical Classes: Inorganic Salts; Phosphorus Compounds

Functions: Chelating Agent; Oral Care Agent

Technical/Other Names:
IMP
Insoluble Metaphosphate
Insoluble Sodium Metaphosphate
Sodium Metaphosphate, Insoluble

Trade Name Mixture:
Eusolex T-Aqua (Merck KGaA/EMD Chemicals Inc.)

SODIUM METASILICATE

CAS No.	EINECS No.
6834-92-0	229-912-9

Empirical Formula:
$H_2O_3Si \cdot 2Na$

Definition: Sodium Metasilicate is the inorganic salt that conforms to the formula:

$$Na_2SiO_3$$

Information Sources: 21CFR173.310, 21CFR184.1769a, CIR: [SQ], CTFA S, JSQI, MI-13(8716), TSCA

Chemical Class: Inorganic Salts

Functions: Chelating Agent; Corrosion Inhibitor

Reported Product Categories: Hair Dyes and Colors (All Types Requiring Caution Statements and Patch Tests); Hair Bleaches; Shaving Cream (Aerosol, Brushless and Lather); Hair Lighteners with Color; Hair Coloring Preparations, Misc.

Technical/Other Names:
Disodium Silicate
Silicic Acid, Disodium Salt
Sodium Silicon Oxide

Trade Names:
Metso Beads 2048 (PQ Corporation)
Metso Pentabead 20 (PQ Corporation)

SODIUM METHACRYLATE/STYRENE COPOLYMER

CAS No.: 33970-45-5

Definition: Sodium Methacrylate/Styrene Copolymer is a copolymer of sodium methacrylate and styrene monomers.

Chemical Class: Synthetic Polymers

Function: Opacifying Agent

Technical/Other Names:
Benzene, Ethenyl-, Polymer with Sodium 2-Methyl-2-Propenoate
2-Propenoic Acid, 2-Methyl-, Sodium Salt, Polymer with Ethenylbenzene

Trade Name Mixture:
Lytron 651 (Rohm and Haas)

SODIUM METHOXY PPG-2 ACETATE

CAS No.: 165038-56-2

Empirical Formula:
$C_9H_{18}O_5 \cdot Na$

Definition: Sodium Methoxy PPG-2 Acetate is the organic compound that conforms to the formula:

$$CH_3(OCHCH_2)_2OCH_2COONa$$
$$CH_3$$

Chemical Classes: Alkoxylated Alcohols; Organic Salts

Functions: Surfactant - Cleansing Agent; Surfactant - Emulsifying Agent; Surfactant - Hydrotrope

Technical/Other Name:
Acetic Acid, [2-(2-Methoxymethylethoxy)-Methylethoxy]-, Sodium Salt

Trade Name:
Gemtex DPM-101 (Finetex)

SODIUM METHYL COCOYL TAURATE

CAS No.: 12765-39-8

JPN Translation:
ココイルメチルタウリン Na

Definition: Sodium Methyl Cocoyl Taurate is the sodium salt of the coconut fatty acid amide of N-methyltaurine. It conforms generally to the formula:

$$RC \overset{O}{\overset{\|}{-}} NCH_2CH_2SO_3Na$$
$$CH_3$$

where RCO- represents the coconut acid radical.

Information Sources: CTFA D, JCIC, JCLS, JSQI, SNPF, TSCA

Chemical Class: Sulfonic Acids

Function: Surfactant - Cleansing Agent

Reported Product Categories: Bath Oils, Tablets, and Salts; Hair Dyes and Colors (All Types Requiring Caution Statements and Patch Tests); Cleansing Products (Cold Creams, Cleansing Lotions, Liquids and Pads); Bath Preparations, Misc.; Bath Soaps and Detergents; Shampoos (Non-coloring); Hair Preparations (Non-coloring), Misc.

Technical/Other Names:
Amides, Coconut Oil, with N-Methyltaurine, Sodium Salts
Sodium N-Cocoyl-N-Methyl Taurate
Sodium N-Methyl-N-Cocoyl Taurate

Trade Names:
Adinol CT (Croda Chemicals)
Adinol CT-95 (Croda, Inc.)
Geropon TC-42 (Rhodia)
Geropon TC 270 (Rhodia)
Hostapon CT (Clariant)
Hostapon CT (Clariant GmbH, Personal Care)
Jeepon 24A (Jeen)
Jeepon 30-A (Jeen)
Neoscoap CN-30 (Toho)
Neoscoap CN-30-SF (Toho)
Nikkol CMT-30 (Nikko)
Nikkol CMT-30T (Nikko)
Protapon 24A (Protameen)
Protapon 30A (Protameen)
Somepon T 25 (SEPPIC)
Tauranol WS (Finetex)
Tauranol WS Conc. (Finetex)
Tauranol WS H.P. (Finetex)
Tauranol WS P (Finetex)

Trade Name Mixtures:
Miracare UM-140 (Rhodia)
Natrlfine T-1 (Finetex)

SODIUM METHYLESCULETIN ACETATE

CAS No.	EINECS No.
95873-69-1	306-060-7

Empirical Formula:
$C_{12}H_{10}O_6 \cdot Na$

Definition: Sodium Methylesculetin Acetate is the organic compound that conforms to the formula:

Chemical Classes: Heterocyclic Compounds; Organic Salts; Phenols

Function: Oral Care Agent

Technical/Other Name:
Sodium [(6-Hydroxy-4-Methyl-2-Oxo-2H-1-Benzopyran-7-yl)Oxy]Acetate

Trade Name:
Permethol (Sochibo)

SODIUM METHYL LAUROYL TAURATE

CAS No.	EINECS No.
4337-75-1	224-388-8

JPN Translation:
ラウロイルメチルタウリン Na

Empirical Formula:
$C_{15}H_{31}NO_4S \cdot Na$

Definition: Sodium Methyl Lauroyl Taurate is the sodium salt of the lauric acid amide of N-methyl taurine. It conforms to the formula:

Information Sources: JCLS, JSCI

Chemical Class: Sulfonic Acids

Function: Surfactant - Cleansing Agent

Technical/Other Names:
N-Dodecanoyl-N-Methyltaurine Sodium Salt
Ethanesulfonic Acid, 2-[Methyl(1-Oxododecyl)Amino]-, Sodium Salt
2-[Methyl(1-Oxododecyl)Amino]Ethane-sulfonic Acid, Sodium Salt
Sodium N-Lauroyl-N-Methylaminoethane-sulfonate
Sodium Lauroylmethyl Taurate
Sodium N-Lauroyl Methyl Taurate
Sodium Lauroylmethyl Taurate Solution
Sodium N-Methyl-N-Lauroyl Taurate

Trade Names:
Nikkol LMT (Nikko)
Zoharpon LMT-42 (Zohar)

SODIUM METHYL MYRISTOYL TAURATE

CAS No.	EINECS No.
18469-44-8	242-349-3

JPN Translation:
ミリストイルメチルタウリン Na

Empirical Formula:
$C_{17}H_{35}NO_4S \cdot Na$

Definition: Sodium Methyl Myristoyl Taurate is the sodium salt of the myristic acid amide of N-methyl taurine. It conforms to the formula:

Information Sources: JCIC, JCLS, JSQI

Chemical Class: Sulfonic Acids

Function: Surfactant - Cleansing Agent

Technical/Other Names:
Ethanesulfonic Acid, 2-[Methyl(1-Oxotetra-decyl)Amino]-, Sodium Salt
2-[Methyl(1-Oxotetradecyl)Amino]Ethane-sulfonic Acid, Sodium Salt
Sodium N-Methyl-N-Myristoyl Taurate
Sodium N-Myristoyl-N-Methyl-2-Aminoethanesulfonate
Sodium Myristoylmethyl Taurate
Sodium N-Myristoyl Methyl Taurate

Trade Name:
Nikkol MMT (Nikko)

SODIUM METHYLNAPHTHALENESULFO-NATE

CAS No.	EINECS No.
26264-58-4	247-564-6

Definition: Sodium Methylnaphthalene-sulfonate is a mixture of mono- and dimethyl substituted naphthalene sulfonates that conforms generally to the formula:

where n has a value of 1 or 2.

Information Sources: 21CFR172.824, 21CFR173.315, 21CFR175.105, 21CFR176.170, 21CFR176.180, 21CFR176.210, TSCA

Chemical Class: Alkyl Aryl Sulfonates

Functions: Surfactant - Hydrotrope; Surfactant - Suspending Agent

SODIUM METHYL OLEOYL TAURATE

CAS Nos.	EINECS Nos.
137-20-2	205-285-7
7308-16-9	230-762-1

JPN Translation:
オレオイルメチルタウリン Na

Empirical Formula:
$C_{21}H_{41}NO_4S \cdot Na$

Definition: Sodium Methyl Oleoyl Taurate is the sodium salt of the oleic acid amide of N-methyl taurine. It conforms generally to the formula:

Information Sources: 21CFR176.170, 21CFR176.180, 21CFR176.210, JCIC, JCLS, JSQI, SNPF, TSCA

Chemical Class: Sulfonic Acids

Function: Surfactant - Cleansing Agent

Technical/Other Names:
Ethanesulfonic Acid, 2-[Methyl(1-Oxo-9-Octadecenyl)Amino]-, Sodium Salt
2-[Methyl(1-Oxo-9-Octadecenyl)Amino]-Ethanesulfonic Acid, Sodium Salt
Sodium N-Methyl-N-Oleoyl Taurate
Sodium Oleoyl Methyl Taurate
Sodium N-Oleoyl-N-Methyl Taurate
Taurine, N-Methyl-N-Oleoyl-, Sodium Salt

Trade Names:
Adinol OT (Croda Chemicals)
Geropon T-77 (Rhodia)
Hostapon T (Clariant)
Hostapon T (Clariant GmbH, Personal Care)
Jeepon T-33 (Jeen)
Protapon T-33 (Protameen)
Tauranol MS (Finetex)

Trade Name Mixture:
Tauranol ML (Finetex)

SODIUM METHYL PALMITOYL TAURATE

CAS No.	EINECS No.
3737-55-1	223-114-4

JPN Translation:
パルミトイルメチルタウリン Na

Empirical Formula:
$C_{19}H_{39}NO_4S \cdot Na$

Definition: Sodium Methyl Palmitoyl Taurate is the sodium salt of the palmitic acid amide of N-methyl taurine. It conforms to the formula:

Information Sources: JCIC, JCLS

Chemical Class: Sulfonic Acids

Function: Surfactant - Cleansing Agent

Technical/Other Names:
Ethanesulfonic Acid, 2-[Methyl(1-Oxohexa-decyl)Amino]-, Sodium Salt
2-[Methyl(1-Oxohexadecyl)Amino]Ethane-sulfonic Acid, Sodium Salt
Sodium N-Methyl-N-Palmitoyl Taurate
Sodium N-Palmitoyl Methyl Taurate
Taurine, N-Methyl-N-Palmitoyl-, Sodium Salt

Trade Name:
Nikkol PMT (Nikko)

SODIUM METHYLPARABEN

CAS No. 5026-62-0

EINECS No. 225-714-1

JPN Translation:
メチルパラベン Na

Empirical Formula:
$C_8H_8O_3$ • Na

Definition: Sodium Methylparaben is the sodium salt of Methylparaben (q.v.) that conforms to the formula:

Information Sources: EEC(VI/1-12), JCIC, JCLS, MHLW-331/3, NF XIX, USAN

Chemical Classes: Esters; Organic Salts; Phenols

Function: Preservative

Reported Product Categories: Bubble Baths; Bath Capsules; Bath Soaps and Detergents; Moisturizing Preparations; Face and Neck Preparations (Excluding Shaving Preparations); Bath Preparations, Misc.; Body and Hand Preparations (Excluding Shaving Preparations); Shampoos (Non-coloring); Skin Care Preparations, Misc.; Skin Fresheners; Blushers (All types); Cleansing Products (Cold Creams, Cleansing Lotions, Liquids and Pads); Colognes and Toilet Waters; Eye Shadows; Eyebrow Pencils; Makeup Preparations (Not eye), Misc.; Suntan Gels, Creams, and Liquids

Technical/Other Names:
Benzoic Acid, 4-Hydroxy-, Methyl Ester, Sodium Salt
Methylparaben, Sodium Salt
Sodium 4-Carbomethoxyphenolate
Sodium p-Methoxycarbonylphenoxide
Sodium Methyl p-Hydroxybenzoate

Trade Names:
Nipagin M Sodium (Clariant)
Nipagin M Sodium (Clariant GmbH, Personal Care)
NS 3500S (Nutri-Shield)
Paridol MNA (Dekker)
Solbrol M Sodium Salt (Bayer AG)

Trade Name Mixtures:
Conservateur TW (LCW)
Elestab 48 (Laboratoires Serobiologiques)
Eusolex T-Aqua (Merck KGaA/EMD Chemicals Inc.)
Fondix G Bis (Gattefosse s.a.)
Nipacide A Sodium (Clariant)
Nipacide A Sodium (Clariant GmbH, Personal Care)
Nipacombin A (Clariant)
Nipacombin A (Clariant GmbH, Personal Care)
Nipasept Sodium (Clariant)
Nipasept Sodium (Clariant GmbH, Personal Care)
Nipastat Sodium (Clariant)
Nipastat Sodium (Clariant GmbH, Personal Care)

SODIUM METHYL STEAROYL TAURATE

CAS No. 149-39-3

EINECS No. 205-738-9

JPN Translation:
ステアロイルメチルタウリン Na

Empirical Formula:
$C_{21}H_{43}NO_4S$ • Na

Definition: Sodium Methyl Stearoyl Taurate is the sodium salt of the stearic acid amide of N-methyl taurine. It conforms generally to the formula:

Information Sources: JCIC, JCLS, JSQI

Chemical Class: Sulfonic Acids

Function: Surfactant - Cleansing Agent

Technical/Other Names:
Ethanesulfonic Acid, 2-[Methyl(1-Oxoocta-decyl)Amino]-, Sodium Salt
2-[Methyl(1-Oxooctadecyl)Amino]Ethane-sulfonic Acid, Sodium Salt
Sodium N-Methyl-N-Stearoyl Taurate
Sodium N-Stearoyl Methyl Taurate
Sodium N-Stearoyl-N-Methyl Taurate

Trade Name:
Nikkol SMT (Nikko)

Trade Name Mixture:
Nikkol Aquasome AE Conc (Nikko)

SODIUM METHYL 2-SULFOLAURATE

Empirical Formula:
$C_{13}H_{26}O_5S$ • Na

Definition: Sodium Methyl 2-Sulfolaurate is the organic compound that conforms to the formula:

Chemical Classes: Esters; Sulfonic Acids

Function: Surfactant - Cleansing Agent

Reported Product Category: Bath Preparations, Misc.

Trade Name Mixtures:
Alpha-Step BSS-45 (Stepan)
ALPHA-STEP MC-48 (Stepan)
ALPHA-STEP ML-40 (Stepan)
Alpha-Step PC-48 (Stepan)

SODIUM METHYLTAURATE

CAS No. 4316-74-9

EINECS No. 224-339-0

Empirical Formula:
$C_3H_9NO_3S$ • Na

Definition: Sodium Methyltaurate is the organic compound that conforms to the formula:

$$CH_3NHCH_2CH_2SO_3Na$$

Chemical Class: Sulfonic Acids

Function: Skin-Conditioning Agent - Miscellaneous

Technical/Other Names:
Ethanesulfonic Acid, 2-(Methylamino)-, Monosodium Salt
Sodium N-Methyl Taurinate

Trade Name:
MTS-50 (Clariant (Japan))

SODIUM METHYLTAURINE COCOYL METHYLTAURATE

Definition: Sodium Methyltaurine Cocoyl Methyltaurate is the organic salt that conforms to the formula:

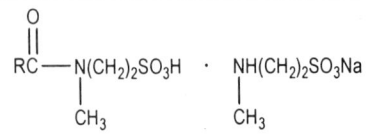

where RCO represents the cocoyl group.

Chemical Classes: Amides; Sulfonic Acids

Functions: Surfactant - Cleansing Agent; Surfactant - Emulsifying Agent

Technical/Other Name:
Amides, Coconut Oil, with N-Methyltaurine, Sodium N-Methyltaurine Salt

Trade Name:
Diapon K-1T2 (NOF)

SODIUM MILKAMIDOPROPYL PG-DIMONIUM CHLORIDE PHOSPHATE

Definition: Sodium Milkamidopropyl PG-Dimonium Chloride Phosphate is the quaternary ammonium salt that conforms generally to the formula:

$$\left[RCNH(CH_2)_3 \overset{\overset{\displaystyle CH_3}{|}}{\underset{\underset{\displaystyle CH_3}{|}}{N}} - CH_2\underset{\underset{\displaystyle OH}{|}}{CH}CH_2O \overset{\overset{\displaystyle O}{\|}}{P}ONa \right]_3^+ \cdot 3Cl^-$$

where RCO- represents the fatty acids derived from milk.

Chemical Class: Quaternary Ammonium Compounds

Functions: Antistatic Agent; Hair Conditioning Agent

Trade Name:
Monalac MPL (Uniqema)

SODIUM MOLYBDATE

CAS Nos.
7631-95-0
12680-49-8

EINECS No.
231-551-7

Empirical Formula:
$MoO_4 \cdot 2Na$

Definition: Sodium Molybdate is the sodium salt of molybdic acid. It conforms to the formula:

$$Na_2M_8O_4$$

Information Source: MI-13(8718)

Chemical Class: Inorganic Salts

Function: Corrosion Inhibitor

Technical/Other Names:
Disodium Molybdate

Molybdic Acid, Sodium Salt
Sodium Molybdenum Oxide

SODIUM MONOFLUOROPHOSPHATE

CAS No.
7631-97-2

EINECS No.
231-552-2

Empirical Formula:
$FH_2O_3P \cdot 2Na$

Definition: Sodium Monofluorophosphate is the inorganic salt that conforms to the formula:

$$Na_2PO_3F$$

In the United States, Sodium Monofluorophosphate may be used as an active ingredient in OTC drug products. When used as an active drug ingredient, the established name is *Sodium Monofluorophosphate. See "Regulatory and Ingredient Use Information," regarding the labeling names for U.S. OTC Drug Ingredients in Volume 1, Introduction, Part A.*

Information Sources: 21CFR355.10, EEC (III/1-27), OTC-I-AC, TSCA, USAN, USP XXIV

Chemical Classes: Inorganic Salts; Phosphorus Compounds

Functions: Anticaries Agent; Oral Care Agent

Reported Product Category: Dentifrices (Aerosol, Liquid, Pastes and Powders)

Technical/Other Names:
MFP
Phosphorofluoridic Acid, Sodium Salt
Sodium Phosphorofluoridate

Trade Name:
MFP (Ozark)

SODIUM MYRETH SULFATE

CAS No.
25446-80-4

EINECS No.
246-986-8

JPN Translations:
ミレス - 3 硫酸 Na
ミレス - 4 硫酸 Na

Definition: Sodium Myreth Sulfate is the sodium salt of sulfated ethoxylated myristyl alcohol that conforms generally to the formula:

$$CH_3(CH_2)_{13}(OCH_2CH_2)_nOSO_3Na$$

where n has a value between 1 and 4.

Information Sources: CIR: [S] JACT-11(1)-1992, CTFA S, JCIC, JCLS, JSQI, SNPF

Chemical Classes: Alkyl Ether Sulfates; Sulfonic Acids

Functions: Surfactant - Cleansing Agent; Surfactant - Emulsifying Agent

Reported Product Categories: Shampoos (Non-coloring); Bubble Baths; Cleansing Products (Cold Creams, Cleansing Lotions, Liquids and Pads); Bath Preparations, Misc.

Technical/Other Names:
PEG-(1-4) Myristyl Ether Sulfate, Sodium Salt
Polyethylene Glycol (1-4) Myristyl Ether Sulfate, Sodium Salt
Sodium Myristyl Ether Sulfate
Sodium Polyoxyethylene Myristyl Ether Sulfate (3E.O.)
Sodium Polyoxyethylene Myristyl Ether Sulfate Solution

Trade Names:
DeSULF SMES-603 (DeForest)
Standapol ES-40 (Cognis Care Chemicals/NJ)
Standapol ES-40 (Cognis Care Chemicals/PA)
Sulfochem ME-60 (Chemron)
Texapon K 14S Spezial (Cognis Deutschland)
Texapon K 14 S Spezial 70% (Cognis Deutschland)
Unipol ES-40 (Universal Preserv-A-Chem)
Zetesol 470 (Zschimmer & Schwarz)

Trade Name Mixture:
Homulgator 910 G Extra (Grau)

SODIUM MYRISTATE

CAS No.
822-12-8

EINECS No.
212-487-9

Empirical Formula:
$C_{14}H_{28}O_2 \cdot Na$

Definition: Sodium Myristate is the sodium salt of myristic acid that conforms to the formula:

$$CH_3(CH_2)_{12}COONa$$

Information Sources: 21CFR172.863, 21CFR175.105, 21CFR175.320, 21CFR176.170, 21CFR176.200, 21CFR176.210, 21CFR177.1200, 21CFR177.2260, 21CFR177.2600, 21CFR177.2800, 21CFR178.3910, TSCA

Chemical Class: Soaps

Functions: Surfactant - Cleansing Agent; Surfactant - Emulsifying Agent

Reported Product Categories: Moisturizing Preparations; Bath Oils, Tablets, and Salts

Technical/Other Names:
Sodium Tetradecanoate
Tetradecanoic Acid, Sodium Salt

SODIUM MYRISTOAMPHOACETATE

CAS No. 68647-45-0
EINECS No. 271-950-3

Empirical Formula:
$C_{20}H_{40}N_2O_4 \cdot Na$

Definition: Sodium Myristoamphoacetate is the amphoteric organic compound that conforms generally to the formula:

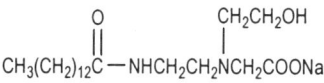

Information Source: TSCA

Chemical Class: Alkylamido Alkylamines

Functions: Hair Conditioning Agent; Surfactant - Cleansing Agent; Surfactant - Foam Booster

Technical/Other Names:
Myristoamphoacetate
Myristoamphoglycinate

SODIUM MYRISTOYL GLUTAMATE

CAS Nos. 38517-37-2
38754-83-5 (dl-alpha)
71368-20-2
EINECS No. 253-981-4

JPN Translation:
ミリストイルグルタミン酸 Na

Empirical Formula:
$C_{19}H_{35}NO_5 \cdot Na$

Definition: Sodium Myristoyl Glutamate is the sodium salt of the myristic acid amide of glutamic acid. It conforms generally to the formula:

$$CH_3(CH_2)_{12}C \overset{O}{\overset{\|}{-}} NHCHCOONa$$
$$CH_2CH_2COOH$$

Information Sources: JCLS, JSCI

Chemical Class: Amino Acids

Function: Surfactant - Cleansing Agent

Technical/Other Names:
Glutamic Acid, N-(1-Oxotetradecyl)-, Monosodium Salt
N-(1-Oxotetradecyl)Glutamic Acid, Monosodium Salt
Sodium N-Myristoyl-L-glutamate

Trade Names:
Aminosurfact AMMS (Asahi Denka Kogyo)
Amisoft MS-ll (Ajinomoto)

Trade Name Mixture:
VAI-C47051-10 (US Cosmetics)

SODIUM MYRISTOYL HYDROLYZED COLLAGEN

Definition: Sodium Myristoyl Hydrolyzed Collagen is the sodium salt of the conden-

sation product of myristic acid chloride and Hydrolyzed Collagen (q.v.).

Chemical Class: Protein Derivatives

Functions: Hair Conditioning Agent; Skin-Conditioning Agent - Miscellaneous; Surfactant - Cleansing Agent

Technical/Other Name:
Proteins, Hydrolysates, Reaction Products with Myristoyl Chloride, Compds. with Sodium

Trade Names:
Promois EMS (Seiwa Kasei)
Promois EUMS (Seiwa Kasei)

SODIUM MYRISTOYL ISETHIONATE

CAS No.: 37747-10-7

Empirical Formula:
$C_{16}H_{32}O_5S \cdot Na$

Definition: Sodium Myristoyl Isethionate is the organic salt that conforms generally to the formula:

$$CH_3(CH_2)_{12}C \overset{O}{\overset{\|}{-}} OCH_2CH_2SO_3Na$$

Information Source: TSCA

Chemical Class: Isethionates

Functions: Hair Conditioning Agent; Surfactant - Cleansing Agent

Technical/Other Names:
Sodium 2-Sulfoethyl Myristate
2-Sulfoethyl Tetradecanoate, Sodium Salt
Tetradecanoic Acid, 2-Sulfoethyl Ester, Sodium Salt

SODIUM MYRISTOYL METHYL BETA-ALANINE

Definition: Sodium Myristoyl Methyl Beta-Alanine is the sodium salt of Myristoyl Methyl Beta-Alanine (q.v.). It conforms to the formula:

$$CH_3(CH_2)_{12}C \overset{O}{\overset{\|}{-}} \overset{CH_3}{\underset{CH_3}{\overset{|}{N}}CHCOONa}$$

Chemical Classes: Amino Acids; Organic Salts

Function: Skin-Conditioning Agent - Miscellaneous

SODIUM MYRISTOYL SARCOSINATE

CAS No. 30364-51-3
EINECS No. 250-151-3

JPN Translation:
ミリストイルサルコシン Na

Empirical Formula:
$C_{17}H_{33}NO_3 \cdot Na$

Definition: Sodium Myristoyl Sarcosinate is the sodium salt of Myristoyl Sarcosine (q.v.). It conforms generally to the formula:

$$CH_3(CH_2)_{12}C \overset{O}{\overset{\|}{-}} \overset{}{\underset{CH_3}{\overset{|}{N}}CH_2COONa}$$

Information Sources: CIR: [SQ] IJT-20 (SUPPL. 1)2001, JCIC, JCLS, JSQI, TSCA

Chemical Class: Sarcosinates and Sarcosine Derivatives

Functions: Hair Conditioning Agent; Surfactant - Cleansing Agent

Reported Product Categories: Bath Oils, Tablets, and Salts; Cleansing Products (Cold Creams, Cleansing Lotions, Liquids and Pads)

Technical/Other Names:
Glycine, N-Methyl-N-(1-Oxotetradecyl)-, Sodium Salt
N-Methyl-N-(1-Oxotetradecyl)Glycine, Sodium Salt
Sodium Myristoylmethyl Glycinate

Trade Names:
Crodasinic MS (Croda Chemicals)
Hamposyl M-30 (Amerchol)
Nikkol Sarcosinate MN (Nikko)

SODIUM MYRISTYL SULFATE

CAS No. 1191-50-0
EINECS No. 214-737-2

JPN Translation:
ミリスチル硫酸 Na

Empirical Formula:
$C_{14}H_{30}O_4S \cdot Na$

Definition: Sodium Myristyl Sulfate is the sodium salt of myristyl sulfate that conforms to the formula:

$$CH_3(CH_2)_{13}OSO_3Na$$

Information Sources: 21CFR175.105, 21CFR177.1210, 21CFR177.2800, JCIC, JCLS, JSQI, MI-13(8764), TSCA

Chemical Class: Alkyl Sulfates

Function: Surfactant - Cleansing Agent

Reported Product Category: Bubble Baths

Technical/Other Names:
Sodium Tetradecyl Sulfate
Sulfuric Acid, Monotetradecyl Ester, Sodium Salt
1-Tetradecanol, Hydrogen Sulfate, Sodium Salt

Trade Name:
Nikkol SMS (Nikko)

Trade Name Mixture:
Texapon CS Paste (Cognis Deutschland)

SODIUM NAPHTHALENESULFONATE

CAS Nos.
532-02-5
1321-69-3

EINECS Nos.
208-523-8
215-323-4

Empirical Formula:
$C_{10}H_8O_3S \cdot Na$

Definition: Sodium Naphthalenesulfonate is the sodium salt of naphthalene sulfonic acid that conforms to the formula:

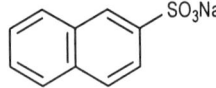

Information Sources: CIR: [SQ], TSCA

Chemical Class: Sulfonic Acids

Function: Surfactant - Hydrotrope

Technical/Other Name:
2-Naphthalenesulfonic Acid, Sodium Salt

Trade Name:
NSA (Kao Corp.)

SODIUM NAPHTHOL SULFONATE

CAS No.
6099-57-6

EINECS No.
228-050-0

Empirical Formula:
$C_{10}H_7O_4S \cdot Na$

Definition: Sodium Naphthol Sulfonate is the organic compound that conforms to the formula:

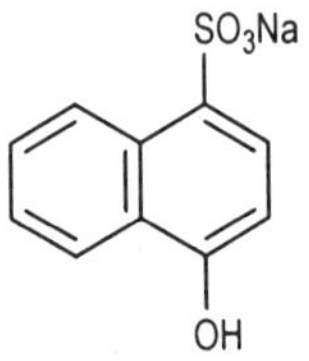

Information Source: TSCA

Chemical Classes: Organic Salts; Phenols; Sulfonic Acids

Function: Absorbent

Technical/Other Names:
1-Naphthalenesulfonic Acid, 4-Hydroxy-, Monsodium Salt
Sodium 4-Hydroxynaphthalene-1-Sulphonate

Trade Name:
Neville Winter Acid (IN.CHI.CO)

SODIUM NITRATE

CAS No.
7631-99-4

EINECS No.
231-554-3

JPN Translation:
硝酸 Na

Empirical Formula:
$NaNO_3$

Definition: Sodium Nitrate is the sodium salt of nitric acid that conforms to the formula:

Information Source: MI-13(8720)

Chemical Class: Inorganic Salts

Function: Not Reported

Technical/Other Name:
Nitric Acid, Sodium Salt

SODIUM NITRITE

CAS No.
7632-00-0

EINECS No.
231-555-9

Empirical Formula:
$HNO_2 \cdot Na$

Definition: Sodium Nitrite is the inorganic salt that conforms to the formula:

$NaNO_2$

Information Sources: EEC(III/1-17), JCLS, MI-13(8721), TSCA, USAN, USP XXIV

Chemical Class: Inorganic Salts

Function: Corrosion Inhibitor

Technical/Other Name:
Nitrous Acid, Sodium Salt

SODIUM m-NITROBENZENESULFONATE

CAS No.
127-68-4

EINECS No.
204-857-3

Empirical Formula:
$C_6H_5NO_5S \cdot Na$

Definition: Sodium m-Nitrobenzene-sulfonate is the substituted aromatic compound that conforms to the formula:

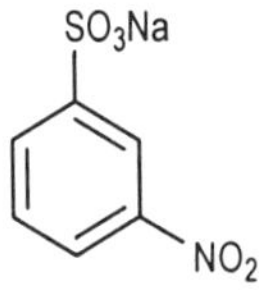

Information Sources: CIR: [I] JACT-15(4)-1996, TSCA

Chemical Class: Sulfonic Acids

Function: Not Reported

Reported Product Category: Hair Dyes and Colors (All Types Requiring Caution Statements and Patch Tests)

Technical/Other Names:
Benzenesulfonic Acid, 3-Nitro-, Sodium Salt
3-Nitrobenzenesulfonic Acid, Sodium Salt
Sodium 3-Nitrophenylsulfonate

SODIUM 5-NITROGUAIACOLATE

JPN Translation:
ニトログアヤコール Na

Empirical Formula:
$C_7H_7NO_4 \cdot Na$

Definition: Sodium 5-Nitroguaiacolate is the organic compound that conforms to the formula:

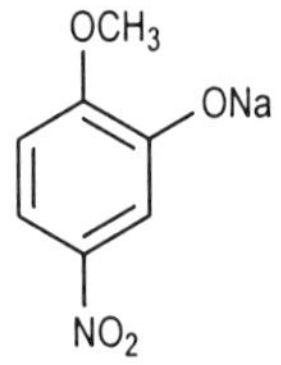

Chemical Classes: Ethers; Organic Salts

Function: Colorant

Technical/Other Names:
2-Methoxy-5-Nitrophenol, Sodium Salt
5-Nitroguaiacol, Sodium Salt

Trade Name:
5-NGSS (Mitajiri Chemical)

SODIUM NONOXYNOL-6 PHOSPHATE

CAS No.
12068-19-8

EINECS No.
235-093-9

JPN Translation:
ノノキシノール - 6 リン酸 Na

Definition: Sodium Nonoxynol-6 Phosphate is the sodium salt of a complex mixture of esters of phosphoric acid and Nonoxynol-6 (q.v.).

Information Sources: JCLS, TSCA

Chemical Class: Phosphorus Compounds

Functions: Surfactant - Cleansing Agent; Surfactant - Hydrotrope

Technical/Other Names:
PEG-6 Nonyl Phenyl Ether Phosphate, Sodium Salt
Polyethylene Glycol 300 Nonyl Phenyl Ether Phosphate, Sodium Salt
Polyoxyethylene (6) Nonyl Phenyl Ether Phosphate, Sodium Salt

Trade Names:
Phosphanol LO-529 (Toho)
Surfagene FAD 106 (Kao GmbH)

SODIUM NONOXYNOL-9 PHOSPHATE

JPN Translation:
ノノキシノール - 9 リン酸 Na

Definition: Sodium Nonoxynol-9 Phosphate is a complex mixture of esters of phosphoric acid and Nonoxynol-9 (q.v.).

Information Source: JCLS

Chemical Class: Phosphorus Compounds

Functions: Surfactant - Cleansing Agent; Surfactant - Hydrotrope

Reported Product Categories: Baby Products, Misc.; Personal Cleanliness Products, Misc.

Technical/Other Names:
PEG-9 Nonyl Phenyl Ether Phosphate, Sodium Salt
Polyethylene Glycol 450 Nonyl Phenyl Ether Phosphate, Sodiumalt
Polyoxyethylene (9) Nonyl Phenyl Ether Phosphate, Sodium Sal

SODIUM NONOXYNOL-1 SULFATE

CAS No.: 9014-90-8 (Generic)

Empirical Formula:
$C_{17}H_{28}O_5S \cdot Na$

Definition: Sodium Nonoxynol-1 Sulfate is the organic compound that conforms generally to the formula:

$$C_9H_{19}C_6H_4OCH_2CH_2OSO_3Na$$

Information Sources: 21CFR175.105, CTFA D, TSCA

Chemical Class: Alkyl Ether Sulfates

Function: Surfactant - Cleansing Agent

Technical/Other Names:
PEG-1 Nonyl Phenyl Ether Sulfate, Sodium Salt
Polyethylene Glycol (1) Nonyl Phenyl Ether Sulfate, Sodium Salt
Polyoxyethylene (1) Nonyl Phenyl Ether Sulfate, Sodium Salt
Sodium Nonylphenyl Ether Sulfate

SODIUM NONOXYNOL-3 SULFATE

CAS No.: 9014-90-8 (Generic)

Empirical Formula:
$C_{21}H_{36}O_7S \cdot Na$

Definition: Sodium Nonoxynol-3 Sulfate is the sodium salt of the sulfuric acid ester of Nonoxynol-3 (q.v.) that conforms generally to the formula:

$$C_9H_{19}C_6H_4(OCH_2CH_2)_nOSO_3Na$$

where n has an average value of 3.

Information Source: 21CFR175.105

Chemical Class: Alkyl Ether Sulfates

Function: Surfactant - Cleansing Agent

Technical/Other Names:
PEG-3 Nonyl Phenyl Ether Sulfate, Sodium Salt
Polyethylene Glycol (3) Nonyl Phenyl Ether Sulfate, Sodium Salt
Polyoxyethylene (3) Nonyl Phenyl Ether Sulfate, Sodium Salt

SODIUM NONOXYNOL-4 SULFATE

CAS No.: 9014-90-8 (Generic)

JPN Translation:
ノノキシノール - 4 硫酸 Na

Empirical Formula:
$C_{23}H_{40}O_8S \cdot Na$

Definition: Sodium Nonoxynol-4 Sulfate is the sodium salt of the sulfuric acid ester of Nonoxynol-4 that conforms generally to the formula:

$$C_9H_{19}C_6H_4(OCH_2CH_2)_nOSO_3Na$$

where n has an average value of 4.

Information Sources: 21CFR175.105, 21CFR178.3400, JCIC, JCLS, TSCA

Chemical Class: Alkyl Ether Sulfates

Function: Surfactant - Cleansing Agent

Technical/Other Names:
PEG-4 Nonyl Phenyl Ether Sulfate, Sodium Salt
Polyethylene Glycol 200 Nonyl Phenyl Ether Sulfate, Sodium Salt
Polyoxyethylene (4) Nonyl Phenyl Ether Sulfate, Sodium Salt
Sodium Polyoxyethylene Nonyl Phenyl Ether Sulfate Solution

Trade Names:
Nikkol SNP-4N (Nikko)
Rhodapex CO-433 (Rhodia)

Trade Name Mixture:
OPD-140 (Importaciones y Suministros)

SODIUM NONOXYNOL-6 SULFATE

CAS No.: 9014-90-8 (Generic)

Empirical Formula:
$C_{27}H_{48}O_{10}S \cdot Na$

Definition: Sodium Nonoxynol-6 Sulfate is the sodium salt of the sulfuric acid ester of Nonoxynol-6 (q.v.) that conforms generally to the formula:

$$C_9H_{19}C_6H_4(OCH_2CH_2)_nOSO_3Na$$

where n has an average value of 6.

Information Source: 21CFR175.105

Chemical Class: Alkyl Ether Sulfates

Function: Surfactant - Cleansing Agent

Technical/Other Names:
PEG-6 Nonyl Phenyl Ether Sulfate, Sodium Salt
Polyethylene Glycol 300 Nonyl Phenyl Ether Sulfate, Sodium Salt
Polyoxyethylene (6) Nonyl Phenyl Ether Sulfate, Sodium Salt

SODIUM NONOXYNOL-8 SULFATE

CAS No.: 9014-90-8 (Generic)

Empirical Formula:
$C_{31}H_{56}O_{12}S \cdot Na$

Definition: Sodium Nonoxynol-8 Sulfate is the sodium salt of the sulfuric acid ester of Nonoxynol-8 (q.v.) that conforms generally to the formula:

$$C_9H_{19}C_6H_4(OCH_2CH_2)_nOSO_3Na$$

where n has an average value of 8.

Information Source: 21CFR175.105

Chemical Class: Alkyl Ether Sulfates

Function: Surfactant - Cleansing Agent

Technical/Other Names:
PEG-8 Nonyl Phenyl Ether Sulfate, Sodium Salt
Polyethylene Glycol 400 Nonyl Phenyl Ether Sulfate, Sodium Salt
Polyoxyethylene (8) Nonyl Phenyl Ether Sulfate, Sodium Salt

SODIUM NONOXYNOL-10 SULFATE

CAS No.: 9014-90-8 (Generic)

Empirical Formula:
$C_{35}H_{64}O_{14}S \cdot Na$

Definition: Sodium Nonoxynol-10 Sulfate is the sodium salt of the sulfuric acid ester of Nonoxynol-10 (q.v.) that conforms generally to the formula:

$$C_9H_{19}C_6H_4(OCH_2CH_2)_nOSO_3Na$$

where n has an average value of 10.

Information Source: 21CFR175.105

Chemical Class: Alkyl Ether Sulfates

Function: Surfactant - Cleansing Agent

Technical/Other Names:
PEG-10 Nonyl Phenyl Ether Sulfate, Sodium Salt

Polyethylene Glycol 500 Nonyl Phenyl Ether Sulfate, Sodium Salt
Polyoxyethylene (10) Nonyl Phenyl Ether Sulfate, Sodium Salt

SODIUM NONOXYNOL-25 SULFATE

CAS No.: 9014-90-8 (Generic)

Definition: Sodium Nonoxynol-25 Sulfate is the sodium salt of the sulfuric acid ester of Nonoxynol-25 that conforms generally to the formula:

$$C_9H_{19}C_6H_4(OCH_2CH_2)_nOSO_3Na$$

where n has an average value of 25.

Information Source: 21CFR175.105

Chemical Class: Alkyl Ether Sulfates

Function: Surfactant - Cleansing Agent

Technical/Other Names:
PEG-25 Nonyl Phenyl Ether Sulfate, Sodium Salt
Polyethylene Glycol (25) Nonyl Phenyl Ether Sulfate, Sodium Salt
Polyoxyethylene (25) Nonyl Phenyl Ether Sulfate, Sodium Salt

SODIUM OCTOXYNOL-2 ETHANE SULFONATE

CAS Nos.	EINECS Nos.
2917-94-4	220-851-3
55837-16-6	
67923-87-9	267-791-4

JPN Translation:
オクトキシノール - 2 エタンスルホン酸 Na

Empirical Formula:
$C_{20}H_{34}O_6S \cdot Na$

Definition: Sodium Octoxynol-2 Ethane Sulfonate is the organic compound that conforms to the formula:

$$C_8H_{17}C_6H_4O(CH_2CH_2O)_2CH_2CH_2SO_3Na$$

Information Sources: 21CFR176.180, CIR: [SQ], INN, JCIC, JCLS, JSQI, USAN

Chemical Class: Sulfonic Acids

Function: Surfactant - Cleansing Agent

Technical/Other Names:
Entsufon
2-[2-[2-(Octylphenoxy)Ethoxy]Ethoxy]-Ethanesulfonic Acid, Sodium Salt
Sodium Octoxynol-3 Sulfonate
Sodium Octylphenoxy Diethoxyethyl Sulfonate

Trade Name:
TRITON X-200 Surfactant (Dow Chemical)

SODIUM OCTOXYNOL-2 SULFATE

JPN Translation:
オクトキシノール - 2 硫酸 Na

Definition: Sodium Octoxynol-2 Sulfate is the sodium salt of the sulfuric acid ester of Octoxynol-2 that conforms generally to the formula:

$$C_8H_{17}C_6H_4(OCH_2CH_2)_nOSO_3Na$$

where n has an average value of 2.

Information Source: CIR: [SQ]

Chemical Class: Alkyl Ether Sulfates

Function: Surfactant - Cleansing Agent

Technical/Other Names:
PEG-2 Octyl Phenyl Ether Sulfate, Sodium Salt
Polyethylene Glycol (2) Octyl Phenyl Ether Sulfate, Sodium Salt
Polyoxyethylene (2) Octyl Phenyl Ether Sulfate, Sodium Salt

SODIUM OCTOXYNOL-6 SULFATE

Empirical Formula:
$C_{26}H_{46}O_{10}S \cdot Na$

Definition: Sodium Octoxynol-6 Sulfate is the sodium salt of the sulfuric acid ester of Octoxynol-6 that conforms generally to the formula:

$$C_8H_{17}C_6H_4(OCH_2CH_2)_nOSO_3Na$$

where n has an average value of 6.

Information Source: CIR: [SQ]

Chemical Class: Alkyl Ether Sulfates

Function: Surfactant - Cleansing Agent

Technical/Other Names:
PEG-6 Octyl Phenyl Ether Sulfate, Sodium Salt
Polyethylene Glycol 300 Octyl Phenyl Ether Sulfate, Sodium Salt
Polyoxyethylene (6) Octyl Phenyl Ether Sulfate, Sodium Salt

Trade Name:
Akyposal BD (Kao GmbH)

SODIUM OCTOXYNOL-9 SULFATE

Empirical Formula:
$C_{32}H_{58}O_{13}S \cdot Na$

Definition: Sodium Octoxynol-9 Sulfate is the sodium salt of the sulfuric acid ester of Octoxynol-9 (q.v.) that conforms generally to the formula:

$$C_8H_{17}C_6H_4(OCH_2CH_2)_nOSO_3Na$$

where n has an average value of 9.

Information Source: CIR: [S]

Chemical Class: Alkyl Ether Sulfates

Function: Surfactant - Cleansing Agent

Technical/Other Names:
PEG-9 Octyl Phenyl Ether Sulfate, Sodium Salt
Polyethylene Glycol 450 Octyl Phenyl Ether Sulfate, Sodium Salt
Polyoxyethylene (9) Octyl Phenyl Ether Sulfate, Sodium Salt

SODIUM OLEAMIDOPROPYL PG-DIMONIUM CHLORIDE PHOSPHATE

Definition: Sodium Oleamidopropyl PG-Dimonium Chloride Phosphate is the quaternary ammonium salt that conforms generally to the formula:

where RCO- represents the oleic acid radical.

Chemical Classes: Phosphorus Compounds; Quaternary Ammonium Compounds

Functions: Antistatic Agent; Hair Conditioning Agent

SODIUM OLEANOLATE

Empirical Formula:
$C_{30}H_{48}O_3 \cdot Na$

Definition: Sodium Oleanolate is the organic compound that conforms to the formula:

Chemical Class: Organic Salts

Function: Skin-Conditioning Agent - Miscellaneous

Trade Name Mixture:
Ursolic Acid Sodium Salt (Boehringer)

The inclusion of any compound in the *Dictionary and Handbook* does not indicate that use of that substance as a cosmetic ingredient complies with the laws and regulations governing such use in the United States or any other country.

SODIUM OLEATE

CAS Nos.
143-19-1
16558-02-4

EINECS No.
205-591-0

JPN Translation:
オレイン酸 Na

Empirical Formula:
$C_{18}H_{34}O_2$ • Na

Definition: Sodium Oleate is the sodium salt of oleic acid that conforms generally to the formula:

$$CH_3(CH_2)_7CH=CH(CH_2)_7COONa$$

Information Sources: 21CFR172.863, 21CFR175.105, 21CFR175.300, 21CFR175.320, 21CFR176.170, 21CFR176.200, 21CFR176.210, 21CFR177.1200, 21CFR177.2260, 21CFR177.2600, 21CFR177.2800, 21CFR178.3910, 21CFR186.1770, 21CFR310.545, JCIC, JCLS, JSQI, MAR, MI-13(6898), TSCA

Chemical Class: Soaps

Functions: Surfactant - Cleansing Agent; Surfactant - Emulsifying Agent; Viscosity Increasing Agent - Aqueous

Reported Product Categories: Eyeliners; Baby Lotions, Oils, Powders and Creams; Bath Oils, Tablets, and Salts; Bath Preparations, Misc.; Body and Hand Preparations (Excluding Shaving Preparations); Cleansing Products (Cold Creams, Cleansing Lotions, Liquids and Pads)

Technical/Other Names:
9-Octadecenoic Acid, Sodium Salt
Sodium 9-Octadecenoate

Trade Names:
AEC Sodium Oleate (A & E Connock)
Norfox Oleic Flakes (Norman, Fox & Co.)

SODIUM OLEOAMPHOACETATE

Empirical Formula:
$C_{24}H_{46}N_2O_4$ • Na

Definition: Sodium Oleoamphoacetate is the amphoteric organic compound that conforms generally to the formula:

Chemical Class: Alkylamido Alkylamines

Functions: Hair Conditioning Agent; Surfactant - Cleansing Agent; Surfactant - Foam Booster; Surfactant - Hydrotrope

Technical/Other Names:
Oleoamphoacetate
Oleoamphoglycinate

SODIUM OLEOAMPHOHYDROXY-PROPYLSULFONATE

Empirical Formula:
$C_{25}H_{50}N_2O_6S$ • Na

Definition: Sodium Oleoamphohydroxypropylsulfonate is the amphoteric organic compound that conforms generally to the formula:

Chemical Classes: Alkylamido Alkylamines; Sulfonic Acids

Functions: Hair Conditioning Agent; Surfactant - Cleansing Agent; Surfactant - Foam Booster; Surfactant - Hydrotrope

Technical/Other Names:
Oleoamphohydroxypropylsulfonate
Oleoamphopropylsulfonate

Trade Names:
Mackam OS (McIntyre)
Sandopan TFLA DBL (Clariant)
Sandopan TFLA DBL (Clariant GmbH, Personal Care)

SODIUM OLEOAMPHOPROPIONATE

Empirical Formula:
$C_{25}H_{48}N_2O_4$ • Na

Definition: Sodium Oleoamphopropionate is the amphoteric organic compound that conforms generally to the formula:

Information Source: TSCA

Chemical Class: Alkylamido Alkylamines

Functions: Hair Conditioning Agent; Surfactant - Cleansing Agent; Surfactant - Foam Booster

Technical/Other Name:
Oleoamphopropionate

Trade Names:
Mackam OSF (McIntyre)
Schercoteric O-AA (Scher)

SODIUM OLEOYL HYDROLYZED COLLAGEN

Definition: Sodium Oleoyl Hydrolyzed Collagen is the sodium salt of the condensation product of oleic acid chloride and Hydrolyzed Collagen (q.v.).

Chemical Class: Protein Derivatives

Functions: Hair Conditioning Agent; Skin-Conditioning Agent - Miscellaneous; Surfactant - Cleansing Agent

Technical/Other Names:
Proteins, Hydrolysates, Reaction Products with Oleoyl Chloride, Compds. with Sodium
Sodium Oleoyl Hydrolyzed Animal Protein

Trade Name:
Promois EOS (Seiwa Kasei)

SODIUM OLEOYL ISETHIONATE

CAS No.
142-15-4

EINECS No.
205-522-4

Empirical Formula:
$C_{20}H_{38}O_5S$ • Na

Definition: Sodium Oleoyl Isethionate is the organic salt that conforms generally to the formula:

Information Source: TSCA

Chemical Class: Isethionates

Functions: Hair Conditioning Agent; Surfactant - Cleansing Agent

Technical/Other Names:
9-Octadecenoic Acid, 2-Sulfoethyl Ester, Sodium Salt
2-Sulfoethyl 9-Octadecenoate, Sodium Salt

SODIUM OLEOYL LACTYLATE

Empirical Formula:
$C_{24}H_{42}O_6$ • Na

Definition: Sodium Oleoyl Lactylate is the sodium salt of the oleic acid ester of lactyl lactate. It conforms to the formula:

Chemical Classes: Esters; Organic Salts

Function: Surfactant - Emulsifying Agent

Technical/Other Name:
9-Octadecenoic Acid, 2-(1-Carboxyethoxy)-1-Methyl-2-Oxoethyl Ester, Sodium Salt

SODIUM OLETH-7 PHOSPHATE

CAS Nos.: 57486-09-6 (Generic); 68936-83-4 (Generic); 80145-09-1 (Generic)

JPN Translation:
オレス - 7 リン酸 Na

Definition: Sodium Oleth-7 Phosphate is the sodium salt of the phosphate ester of Oleth-7 (q.v.).

Information Sources: CTFA D, JCIC, JCLS

Chemical Class: Phosphorus Compounds

Function: Surfactant - Emulsifying Agent

Technical/Other Names:
Partially Neutralized Sodium Polyoxyethylene Oleyl Ether Phosphate
Sodium Polyoxyethylene (7) Oleyl Ether Phosphate

Trade Names:
Phosphanol 701 (Toho)
Phosphanol GB-520 (Toho)
Phosphanol RD-720N (Toho)

SODIUM OLETH-8 PHOSPHATE

CAS Nos.: 57486-09-6 (Generic); 68936-83-4 (Generic); 80145-09-1 (Generic)

JPN Translation:
オレス - 8 リン酸 Na

Definition: Sodium Oleth-8 Phosphate is the sodium salt of the phosphate ester of Oleth-8 (q.v.).

Information Source: JCLS

Chemical Class: Phosphorus Compounds

Function: Surfactant - Emulsifying Agent

SODIUM OLETH SULFATE

CAS No.: 27233-34-7

Definition: Sodium Oleth Sulfate is the sodium salt of the sulfate ester of the polyethylene glycol ether of oleyl alcohol that conforms generally to the formula:

$$CH_3(CH_2)_7CH = CH(CH_2)_7CH_2(OCH_2CH_2)_nOSO_3Na$$

where n has an average value between 1 and 4.

Chemical Class: Alkyl Ether Sulfates

Function: Surfactant - Cleansing Agent

Reported Product Categories: Eye Makeup Removers; Bath Oils, Tablets, and Salts; Cleansing Products (Cold Creams, Cleansing Lotions, Liquids and Pads); Shampoos (Non-coloring)

Technical/Other Names:
Sodium Polyethylene Glycol (1-4) Oleyl Ether Sulfate

Sodium Polyoxyethylene (1-4) Oleyl Ether Sulfate

Trade Name Mixtures:
LP110 (Phytocos)
Plantaren PS-400 (Cognis Care Chemicals/NJ)
Plantaren PS-400 (Cognis Care Chemicals/PA)
Texapon ASV (Cognis Deutschland)
Texapon ASV-50 (Cognis Care Chemicals/NJ)
Texapon ASV-50 (Cognis Care Chemicals/PA)
Texapon ASV-50 (Cognis Deutschland)
Texapon ASV-70 Special (Cognis Deutschland)

SODIUM OLEYL SULFATE

CAS Nos.	**EINECS Nos.**
1847-55-8	217-430-1
16979-51-4	241-058-9

JPN Translation:
オレイル硫酸 Na

Empirical Formula:
$C_{18}H_{36}O_4S \cdot Na$

Definition: Sodium Oleyl Sulfate is the sodium salt of oleyl sulfate that conforms to the formula:

$$CH_3(CH_2)_7CH = CH(CH_2)_7CH_2OSO_3Na$$

Information Sources: 21CFR175.105, 21CFR176.170, 21CFR176.200, 21CFR177.1200, JCIC, JCLS, JSQI

Chemical Class: Alkyl Sulfates

Function: Surfactant - Cleansing Agent

Reported Product Categories: Moisturizing Preparations; Eye Shadows; Body and Hand Preparations (Excluding Shaving Preparations); Skin Care Preparations, Misc.; Foundations; Cleansing Products (Cold Creams, Cleansing Lotions, Liquids and Pads); Mascara; Night Skin Care Preparations; Makeup Bases; Eyeliners; Eye Makeup Preparations, Misc.; Paste Masks (Mud Packs); Bath Soaps and Detergents; Face and Neck Preparations (Excluding Shaving Preparations); Hair Dyes and Colors (All Types Requiring Caution Statements and Patch Tests); Eye Makeup Removers; Face Powders; Shampoos (Non-coloring); Skin Fresheners; Aftershave Lotions; Indoor Tanning Preparations; Hair Conditioners; Makeup Preparations (Not eye), Misc.; Cuticle Softeners; Fragrance Preparations, Misc.; Personal Cleanliness Products, Misc.; Bath Preparations, Misc.; Blushers (All types); Eye Lotions; Shaving Preparations, Misc.; Suntan Preparations, Misc.; Hair Wave Sets; Lipsticks; Nail Creams and Lotions;

Shaving Cream (Aerosol, Brushless and Lather); Tonics, Dressings, and Other Hair Grooming Aids

Technical/Other Name:
9-Octadecen-1-ol, Hydrogen Sulfate, Sodium Salt

SODIUM OLIVAMIDOPROPYL PG-DIMONIUM CHLORIDE PHOSPHATE

Definition: Sodium Olivamidopropyl PG-Dimonium Chloride Phosphate is the quaternary ammonium salt that conforms to the formula:

where RCO- represents the fatty acids derived from olive oil.

Chemical Classes: Phosphorus Compounds; Quaternary Ammonium Compounds

Functions: Antistatic Agent; Hair Conditioning Agent

Trade Name:
Colalipid OL (Colonial Chemical Inc)

SODIUM OLIVAMPHOACETATE

Definition: Sodium Olivamphoacetate is the amphoteric organic compound that conforms generally to the formula:

where RCO- represents the fatty acids derived from olive oil.

Chemical Class: Alkylamido Alkylamines

Functions: Hair Conditioning Agent; Surfactant - Cleansing Agent; Surfactant - Foam Booster

Trade Names:
Resassol AGO (Res Pharma)
Vamasoft Olive (Vama Farmacosmetica)

Trade Name Mixture:
Anfolive (Naturactiva)

SODIUM OLIVATE

CAS No.: 61789-88-6

Definition: Sodium Olivate is the sodium salt of the fatty acids derived from Olea Europaea (Olive) Oil (q.v.).

Chemical Class: Soaps

Functions: Surfactant - Cleansing Agent; Surfactant - Emulsifying Agent; Viscosity Increasing Agent - Nonaqueous

Technical/Other Name:
Fatty Acids, Olive Oil, Sodium Salts

Trade Name:
PRISAVON 1876 (Uniqema Europe)

Trade Name Mixtures:
Sapo Durus (Block Drug)
Vegetable Soapbase OC (Weleda)

SODIUM OLIVOYL GLUTAMATE

Definition: Sodium Olivoyl Glutamate is the sodium salt of olivoyl glutamic acid. It conforms generally to the formula:

$$RC-NH$$
$$HOOCCH_2CH_2CHCOONa$$

where RCO- represents the fatty acids derived from olive oil.

Information Source: JCLS

Chemical Class: Amino Acids

Function: Surfactant - Cleansing Agent

Technical/Other Name:
Sodium N-Olivoyl L-Glutamate

Trade Name:
Olivoil Glutammate (Keminova Italiana)

SODIUM OXALATE

CAS No.	EINECS No.
62-76-0	200-550-3

JPN Translation:
シュウ酸 Na

Empirical Formula:
$C_2H_2O_4 \cdot 2Na$

Definition: Sodium Oxalate is the sodium salt of oxalic acid that conforms to the formula:

$$NaOOC-COONa$$

Information Sources: EEC(III/1-3), JCLS, JSCI, MI-13(8723), TSCA

Chemical Class: Organic Salts

Function: Corrosion Inhibitor

Technical/Other Names:
Ethanedioic Acid, Disodium Salt
Oxalic Acid, Disodium Salt

SODIUM OXIDE

CAS No.	EINECS No.
1313-59-3	215-208-9

Definition: Sodium Oxide is the inorganic oxide that conforms to the formula:

$$Na_2O$$

Information Source: MI-13(8724)

Chemical Class: Inorganics

Function: pH Adjuster

Trade Name Mixtures:
SilPower (LG Cosmetic)
SilPower Plus (LG Cosmetic)

SODIUM PALMAMPHOACETATE

Definition: Sodium Palmamphoacetate is the amphoteric organic compound that conforms generally to the formula:

$$RC-NHCH_2CH_2NCH_2COONa$$
$$CH_2CH_2OH$$

where RCO- represents the fatty acids derived from palm oil.

Chemical Class: Alkylamido Alkylamines

Functions: Hair Conditioning Agent; Surfactant - Cleansing Agent; Surfactant - Foam Booster

Trade Name:
Resassol AGP (Res Pharma)

SODIUM PALMATE

CAS No.	EINECS No.
61790-79-2	263-162-3

Definition: Sodium Palmate is the sodium salt of the acids derived from Elaeis Guineensis (Palm) Oil (q.v.).

Chemical Class: Soaps

Functions: Surfactant - Cleansing Agent; Surfactant - Emulsifying Agent; Viscosity Increasing Agent - Nonaqueous

Reported Product Categories: Bath Oils, Tablets, and Salts; Bath Soaps and Detergents

Technical/Other Name:
Fatty Acids, Palm Oil, Sodium Salts

Trade Names:
PRISAVON 1815 (Uniqema Europe)
Sodium Palm Stearinate (Dial)

Trade Name Mixture:
Vegetable Soapbase PCO (Weleda)

SODIUM PALM GLYCERIDE SULFONATE

Definition: Sodium Palm Glyceride Sulfonate is the compound that conforms generally to the formula:

$$RC-OCH_2CHCH_2SO_3Na$$
$$OH$$

where RCO- represents the fatty acids dervied from palm oil.

Chemical Classes: Glyceryl Esters and Derivatives; Sulfonic Acids

Function: Surfactant - Cleansing Agent

Trade Names:
MGS-Palm (LG Cosmetic)
MGS-PO (LG Cosmetic)

SODIUM PALMITATE

CAS No.	EINECS No.
408-35-5	206-988-1

JPN Translation:
パルミチン酸 Na

Empirical Formula:
$C_{16}H_{32}O_2 \cdot Na$

Definition: Sodium Palmitate is the sodium salt of palmitic acid that conforms generally to the formula:

$$CH_3(CH_2)_{14}COONa$$

Information Sources: 21CFR172.863, 21CFR175.105, 21CFR175.320, 21CFR176.170, 21CFR176.200, 21CFR176.210, 21CFR177.1200, 21CFR177.2600, 21CFR177.2800, 21CFR178.3910, 21CFR186.1771, JCIC, JCLS, TSCA

Chemical Class: Soaps

Functions: Surfactant - Cleansing Agent; Surfactant - Emulsifying Agent; Viscosity Increasing Agent - Aqueous

Reported Product Categories: Bath Oils, Tablets, and Salts; Bath Soaps and Detergents; Cleansing Products (Cold Creams, Cleansing Lotions, Liquids and Pads); Moisturizing Preparations

Technical/Other Names:
Hexadecanoic Acid, Sodium Salt
Sodium Hexadecanoate
Sodium Pentadecanecarboxylate

SODIUM PALMITOYL CHONDROITIN SULFATE

Definition: Sodium Palmitoyl Chondroitin Sulfate is the condensation product of palmitic acid chloride and Sodium Chondroitin Sulfate (q.v.).

Chemical Classes: Biological Polymers and their Derivatives; Carbohydrates; Sulfuric Acid Esters

Functions: Hair Conditioning Agent; Skin-Conditioning Agent - Miscellaneous

SODIUM PALMITOYL HYDROLYZED COLLAGEN

Definition: Sodium Palmitoyl Hydrolyzed Collagen is the sodium salt of the condensation product of palmitic acid chloride and Hydrolyzed Collagen (q.v.).

Chemical Class: Protein Derivatives

Functions: Hair Conditioning Agent; Skin-Conditioning Agent - Miscellaneous; Surfactant - Cleansing Agent

Technical/Other Name:
Proteins, Hydrolysates, Reaction with Palmitoyl Chloride, Compds. with Sodium

SODIUM PALMITOYL HYDROLYZED WHEAT PROTEIN

Definition: Sodium Palmitoyl Hydrolyzed Wheat Protein is the sodium salt of the condensation product of palmitic acid chloride and Hydrolyzed Wheat Protein (q.v.).

Chemical Class: Protein Derivatives

Functions: Hair Conditioning Agent; Skin-Conditioning Agent - Miscellaneous; Surfactant - Cleansing Agent

Technical/Other Name:
Proteins, Hydrolysates, Reaction Products with Palmitoyl Chloride, Compds. with Sodium

SODIUM PALMITOYL PROLINE

CAS No.	EINECS No.
58725-33-0	261-406-3

Empirical Formula:
$C_{21}H_{38}NO_3 \cdot Na$

Definition: Sodium Palmitoyl Proline is the substituted amino acid that conforms to the formula:

Chemical Classes: Amides; Amino Acids

Function: Skin-Conditioning Agent - Miscellaneous

Technical/Other Names:
N-Palmitoyl-L-Proline Sodium Salt
Sodium 5-Oxo-1-Palmitoyl-L-Prolinate

Trade Name Mixture:
Sepicalm VG (SEPPIC)

SODIUM PALMITOYL SARCOSINATE

CAS No.	EINECS No.
4028-10-8	223-705-7

Empirical Formula:
$C_{19}H_{37}NO_3 \cdot Na$

Definition: Sodium Palmitoyl Sarcosinate is the sodium salt of palmitoyl sarcosine that conforms to the formula:

Chemical Class: Sarcosinates and Sarcosine Derivatives

Functions: Hair Conditioning Agent; Surfactant - Cleansing Agent

Technical/Other Names:
Glycine, N-Methyl-N-(1-Oxohexadecyl)-, Sodium Salt
Sarcosine, N-Palmitoyl-, Sodium Salt
Sodium N-Methyl-N-(1-Oxohexadaecyl) Aminoacetate

Trade Name:
Nikkol Sarcosinate PN (Nikko)

Trade Name Mixture:
Sepifeel One (SEPPIC)

SODIUM PALM KERNELATE

CAS No.	EINECS No.
61789-89-7	263-097-0

Definition: Sodium Palm Kernelate is the sodium salt of the acids derived from palm kernel oil.

Chemical Class: Soaps

Functions: Surfactant - Cleansing Agent; Surfactant - Emulsifying Agent; Viscosity Increasing Agent - Aqueous

Reported Product Categories: Bath Soaps and Detergents; Bath Oils, Tablets, and Salts; Cleansing Products (Cold Creams, Cleansing Lotions, Liquids and Pads)

Technical/Other Name:
Palm Kernel Acids, Sodium Salt

Trade Name Mixture:
PRISAVON 9271 (Uniqema Europe)

SODIUM PALMOYL GLUTAMATE

Definition: Sodium Palmoyl Glutamate is the sodium salt of palmoyl glutamic acid. It conforms generally to the formula:

where RCO- represents the palmoyl radical.

Chemical Class: Amino Acids

Function: Surfactant - Cleansing Agent

Trade Name:
AMISOFT GS-11PF (Ajinomoto)

SODIUM PANTETHEINE SULFONATE

JPN Translation:
パンテチンスルホン酸 Na

Empirical Formula:
$C_{11}H_{21}N_2O_7S_2 \cdot Na$

Definition: Sodium Pantetheine Sulfonate is the bunte salt of pantetheine. It conforms to the formula:

Information Sources: JCIC, JCLS

Chemical Classes: Amides; Esters; Organic Salts; Sulfonic Acids; Thio Compounds

Function: Skin-Conditioning Agent - Miscellaneous

Trade Name:
Sodium D-Pantetheine-S-Sulfonate Solution (Ikeda)

SODIUM PANTOTHENATE

CAS No.	EINECS No.
867-81-2	212-768-6

JPN Translation:
パントテン酸 Na

Definition: Sodium Pantothenate is the sodium salt of Pantothenic Acid (q.v.).

Chemical Class: Organic Salts

Function: Hair Conditioning Agent

Technical/Other Names:
β-Alanine, N-2,4-Dihydroxy-3,3-Dimethyl-1-Oxobutyl)-, Monosodium Salt
Pantothenic Acid, Sodium Salt

SODIUM PARABEN

CAS Nos.	EINECS No.
114-63-6	204-051-1
85080-04-2	

Empirical Formula:
$C_7H_8O_3$ • Na

Definition: Sodium Paraben is the organic salt that conforms to the formula:

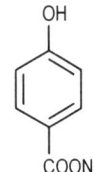

Information Sources: EEC(VI/1-12), MHLW-331/3, TSCA

Chemical Classes: Organic Salts; Phenols

Function: Preservative

Reported Product Categories: Douches; Personal Cleanliness Products, Misc.

Technical/Other Names:
Benzoic Acid, 4-Hydroxy-, Sodium Salt
4-Hydroxybenzoic Acid, Sodium Salt
p-Hydroxybenzoic Acid, Sodium Salt
Sodium p-Hydroxybenzoate

SODIUM PCA

CAS No.
28874-51-3

EINECS No.
249-277-1

JPN Translation:
PCA - Na

Empirical Formula:
$C_5H_7NO_3$ • Na

Definition: Sodium PCA is the sodium salt of PCA (q.v.) that conforms to the formula:

Information Sources: CIR: [SQ] IJT-18 (SUPPL. 2)1999, JCLS, JSCI, TSCA

Chemical Classes: Amides; Heterocyclic Compounds; Organic Salts

Functions: Hair Conditioning Agent; Humectant; Skin-Conditioning Agent - Humectant

Reported Product Categories: Moisturizing Preparations; Bath Preparations, Misc.; Body and Hand Preparations (Excluding Shaving Preparations); Skin Care Preparations, Misc.; Bath Oils, Tablets, and Salts; Cleansing Products (Cold Creams, Cleansing Lotions, Liquids and Pads); Shampoos (Non-coloring); Hair Conditioners; Hair Sprays (Aerosol Fixatives); Night Skin Care Preparations; Tonics, Dressings, and Other Hair Grooming Aids; Bath Capsules; Face and Neck Preparations (Excluding Shaving Preparations); Skin Fresheners; Paste Masks (Mud Packs); Aftershave Lotions; Baby Shampoos; Eyebrow Pencils; Suntan Gels, Creams, and Liquids; Bath Soaps and Detergents; Eyeliners; Indoor Tanning Preparations; Permanent Waves; Eye Makeup Preparations, Misc.; Eye Makeup Removers; Hair Preparations (Non-coloring), Misc.; Hair Rinses (Non-coloring); Suntan Preparations, Misc.

Technical/Other Names:
5-Oxo-DL-Proline, Monosodium Salt
PCA Soda
DL-Proline, 5-Oxo-, Monosodium Salt
Sodium Pyroglutamate
Sodium DL-2-Pyrrolidone-5-Carboxylate
Sodium DL-Pyrrolidonecarboxylate Solution

Trade Names:
AEC Sodium PCA (A & E Connock)
Ajidew N-50 (Ajinomoto)
Dermidrol (Esperis)
Nalidone (UCIB (Solabia))
Ritamectant PCA (RITA)

Trade Name Mixtures:
Advanced Moisture Complex (Collaborative Labs)
AEC Cosflor Blend 017 Moisture Factor WSS (A & E Connock)
AEC Moisture Factor HV (A & E Connock)
Ajidew SP-100 (Ajinomoto)
Aquaderm (Crodarom)
Elesponge AHA LS 8911 B (Laboratoires Serobiologiques)
Hair Care Phytoamine Biocomplex SP Lotion (Alban Muller)
Hidroderm (Fabriquimica)
Hydrolyzed NMF (Proalan)
Hydroveg VV (Variati)
Lactil (Degussa Care Specialties)
Moisturizing Liposomes (Collaborative Labs)
Multifruit PCA (Arch Personal Care Products)
Physiogenyl (UCIB (Solabia))
Prodew 100 (Ajinomoto)
Prodew 200 (Ajinomoto)
Prodew 300 (Ajinomoto)
Prodew 400 (Ajinomoto)
Ritaderm (RITA)
Umordant P (Cosmetochem) (Cosmetochem International Ltd.)
Vegeles AHA 8702 (Laboratoires Serobiologiques)

SODIUM PCA METHYLSILANOL

Definition: Sodium PCA Methylsilanol is a complex of Sodium PCA (q.v.) and mono-methylsilanol .

Chemical Classes: Amides; Heterocyclic Compounds; Organic Salts; Siloxanes and Silanes

Function: Skin-Conditioning Agent - Miscellaneous

Trade Name:
Silhydrate (Exsymol)

SODIUM PEANUTAMPHOACETATE

Definition: Sodium Peanutamphoacetate is the amphoteric organic compound that conforms generally to the formula:

$$RC\overset{\displaystyle O}{\overset{\|}{—}}NH(CH_2)_2NCH_2COONa \quad | \quad CH_2CH_2OH$$

where RCO- represents the fatty acids derived from peanut oil.

Chemical Class: Alkylamido Alkylamines

Functions: Hair Conditioning Agent; Surfactant - Cleansing Agent; Surfactant - Foam Booster

Trade Name:
Vamasoft Peanut (Vama Farmacosmetica)

Trade Name Mixture:
Dermocare PO 35 (Fratelli Ricci)

SODIUM PEANUTATE

Definition: Sodium Peanutate is the sodium salt of the fatty acids derived from Arachis Hypogaea (Peanut) Oil (q.v.).

Chemical Class: Soaps

Functions: Surfactant - Cleansing Agent; Surfactant - Emulsifying Agent; Viscosity Increasing Agent - Nonaqueous

SODIUM PEG-6 COCAMIDE CARBOXYLATE

Definition: Sodium PEG-6 Cocamide Carboxylate is the sodium salt of the organic acid derived from PEG-6 Cocamide (q.v.). It conforms generally to the formula:

$$RC\overset{\displaystyle O}{\overset{\|}{—}}NH(CH_2CH_2O)_nCH_2COONa$$

where RCO- represents the fatty acids derived from coconut oil and n has an average value of 5.

Chemical Classes: Amides; Organic Salts

Function: Surfactant - Cleansing Agent

Reported Product Category: Shampoos (Non-coloring)

Technical/Other Names:
PEG-6 Cocamide Ether Carboxylic Acid, Sodium Salt
Polyethylene Glycol 300 Cocamide Ether Carboxylic Acid, Sodium Salt
Polyoxyethylene (6) Cocamide Ether Carboxylic Acid, Sodium Salt

Trade Name:
Akypo Soft KA 250 BVC (Kao GmbH)

SODIUM PEG-8 COCAMIDE CARBOXYLATE

Definition: Sodium PEG-8 Cocamide Carboxylate is the sodium salt of the organic acid derived from PEG-8 Cocamide. It conforms generally to the formula:

$$RC\!\!\overset{\displaystyle O}{\overset{\|}{}}\!\!-\!\!NH(CH_2CH_2O)_nCH_2COONa$$

where RCO- represents the fatty acids derived from coconut oil and n has an average value of 7.

Chemical Classes: Amides; Organic Salts

Function: Surfactant - Cleansing Agent

Technical/Other Names:
PEG-8 Cocamide Ether Carboxylic Acid, Sodium Salt
Polyethylene Glycol 400 Cocamide Ether Carboxylic Acid, Sodium Salt
Polyoxyethylene (8) Cocamide Ether Carboxylic Acid, Sodium Salt

SODIUM PEG-4 COCAMIDE SULFATE

JPN Translation:
PEG - 3 ヤシ脂肪酸アミド MEA 硫酸 Na

Definition: Sodium PEG-4 Cocamide Sulfate is the organic compound that conforms generally to the formula:

$$RC\!\!\overset{\displaystyle O}{\overset{\|}{}}\!\!-\!\!NH(CH_2CH_2O)_nOSO_3Na$$

where RCO- represents the coconut fatty acid radical and n has an average value of 4.

Chemical Classes: Organic Salts; Sulfuric Acid Esters

Functions: Surfactant - Cleansing Agent; Surfactant - Foam Booster

Trade Names:
Neoscoap CME-30S (Toho)
Nissan Sunamide C-3 (NOF)

SODIUM PEG-50 HYDROGENATED CASTOR OIL SUCCINATE

Definition: Sodium PEG-50 Hydrogenated Castor Oil Succinate is the sodium salt of PEG-50 Hydrogenated Castor Oil Succinate (q.v.).

Chemical Classes: Alkoxylated Alcohols; Organic Salts

Function: Not Reported

SODIUM PEG-3 LAURAMIDE CARBOXYLATE

Empirical Formula:
$C_{16}H_{35}NO_5 \cdot Na$

Definition: Sodium PEG-3 Lauramide Carboxylate is the sodium salt of the organic acid derived from PEG-3 Lauramide (q.v.). It conforms generally to the formula:

$$CH_3(CH_2)_{10}C\overset{\displaystyle O}{\overset{\|}{}}\!\!-\!\!NH(CH_2CH_2O)_nCOONa$$

where n has an average value of 2.

Chemical Classes: Amides; Organic Salts

Function: Surfactant - Cleansing Agent

Technical/Other Names:
PEG-3 Lauramide Ether Carboxylic Acid, Sodium Salt
Polyethylene Glycol (3) Lauramide Ether Carboxylic Acid, Sodium Salt
Polyoxyethylene (3) Lauramide Ether Carboxylic Acid, Sodium Salt

SODIUM PEG-4 LAURAMIDE CARBOXYLATE

Definition: Sodium PEG-4 Lauramide Carboxylate is the sodium salt of the organic acid derived from PEG-4 Lauramide. It conforms generally to the formula:

$$CH_3(CH_2)_{10}C\overset{\displaystyle O}{\overset{\|}{}}\!\!-\!\!NH(CH_2CH_2O)_nCOONa$$

where n has an average value of 3.

Chemical Classes: Amides; Organic Salts

Function: Surfactant - Cleansing Agent

Technical/Other Names:
PEG-4 Lauramide Ether Carboxylic Acid, Sodium Salt
Polyethylene Glycol 200 Lauramide Ether Carboxylic Acid, Sodium Salt
Polyoxyethylene (4) Lauramide Ether Carboxylic Acid, Sodium Salt

SODIUM PEG-7 OLIVE OIL CARBOXYLATE

Definition: Sodium PEG-7 Olive Oil Carboxylate is the reaction product of olive oil peg-7 esters and sodium monochloroacetate.

Chemical Classes: Alkoxylated Carboxylic Acids; Organic Salts

Functions: Surfactant - Emulsifying Agent; Surfactant - Foam Booster; Surfactant - Hydrotrope

Trade Name:
Olivem 400 (B & T)

SODIUM PEG-8 PALM GLYCERIDES CARBOXYLATE

Definition: Sodium PEG-8 Palm Glycerides Carboxylate is the organic compound that conforms to the formula:

$$R(OCH_2CH_2)_nOCH_2COONa$$

where R represents Palm Glycerides (q.v.) and n has an average value of 8.

Chemical Classes: Alkoxylated Carboxylic Acids; Ethers; Organic Salts

Functions: Surfactant - Cleansing Agent; Surfactant - Foam Booster

Trade Name:
Resassol CBP (Res Pharma)

SODIUM PERBORATE

CAS Nos.	EINECS Nos.
7632-04-4	231-556-4
11138-47-9	234-390-0

Empirical Formula:
$BNaO_3$

Definition: Sodium Perborate is the inorganic salt that conforms to the formula:

$$NaBO_3$$

In the United States, Sodium Perborate may be used as an active ingredient in OTC drug products. When used as an active drug ingredient, the established name for Sodium Perborate is *Sodium Perborate. See "Regulatory and Ingredient Use Information," regarding the labeling names for U.S. OTC Drug Ingredients in Volume 1, Introduction, Part A.*

Information Sources: BPC, BRA, 21CFR175.105, CTFA S, EEC(III/1-1), IND, MAR, MHLW-331/1, MI-13(8725), NED, NFJ, OTC-I-OH, OTC-I-OI, PF, POR, TSCA, USAN, USD, YUG

Chemical Class: Inorganic Salts

Functions: Oral Health Care Drug; Oxidizing Agent

Technical/Other Names:
Perboric Acid, Sodium Salt
Sodium Peroxoborate

Trade Names:
AEC Sodium Perborate (A & E Connock)
Sodium Perborate Trihydrate (Merck KGaA)

SODIUM PERSULFATE

CAS No.   **EINECS No.**
7775-27-1   231-892-1

Empirical Formula:
$H_2O_8S_2 \cdot 2Na$

Definition: Sodium Persulfate is the inorganic salt that conforms to the formula:

$$Na_2S_2O_8$$

Information Sources: 21CFR175.105, 21CFR176.170, 21CFR177.1210, CIR: [SQ] IJT-20(SUPPL. 3)2001, MI-13(8729), TSCA

Chemical Class: Inorganic Salts

Function: Oxidizing Agent

Reported Product Categories: Hair Bleaches; Hair Lighteners with Color

Technical/Other Names:
Peroxydisulfuric Acid, Disodium Salt
Sodium Peroxydisulfate

SODIUM PG-PROPYLDIMETHICONE THIOSULFATE COPOLYMER

Definition: Sodium PG-Propyldimethicone Thiosulfate Copolymer is the siloxane polymer that conforms generally to the formula:

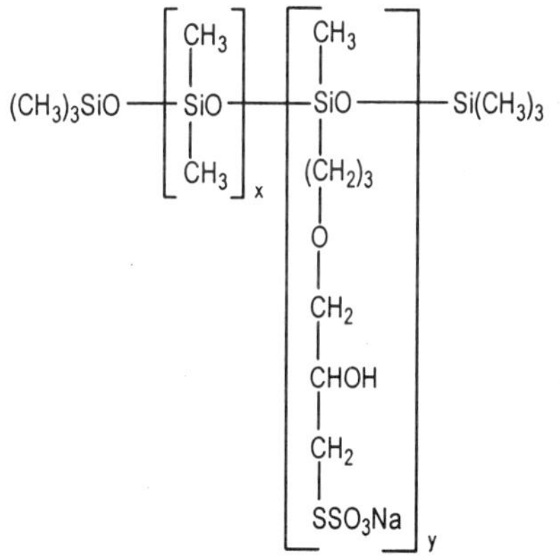

Chemical Classes: Siloxanes and Silanes; Thio Compounds

Functions: Film Former; Hair Conditioning Agent

SODIUM PG-PROPYL THIOSULFATE DIMETHICONE

Definition: Sodium PG-Propyl Thiosulfate Dimethicone is Dimethicone (q.v.) end-

blocked with sodium propylhydroxypropyl thiosulfate. It conforms generally to the formula:

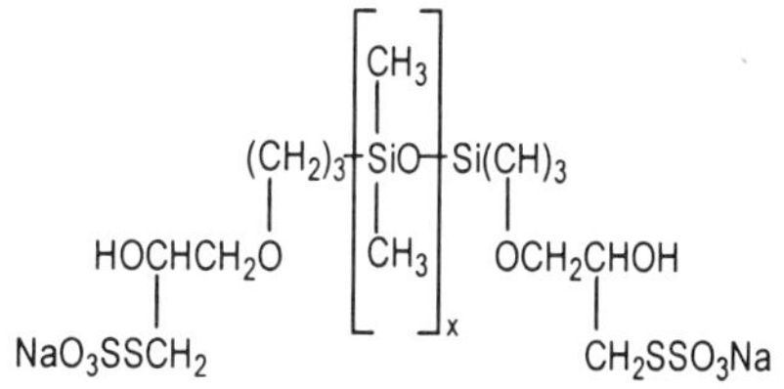

Chemical Classes: Siloxanes and Silanes; Thio Compounds

Function: Hair Conditioning Agent

SODIUM PG-SULFONATE

Empirical Formula:
$C_3H_7O_5S \cdot Na$

Definition: Sodium PG-Sulfonate is the organic compound that conforms to the formula:

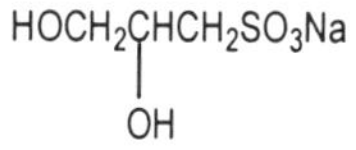

$$HOCH_2\underset{\underset{OH}{|}}{C}HCH_2SO_3Na$$

Chemical Classes: Organic Salts; Polyols; Sulfonic Acids

Function: Skin-Conditioning Agent - Humectant

Technical/Other Name:
2,3-Dihydroxy-1-Propanesulfonic Acid, Sodium Salt

Trade Name:
SDHS (LG Cosmetic)

SODIUM PHENOLSULFONATE

CAS No.   **EINECS No.**
1300-51-2   215-087-2

JPN Translation:
フェノールスルホン酸 Na

Empirical Formula:
$C_6H_6O_4S \cdot Na$

Definition: Sodium Phenolsulfonate is the organic salt that conforms to the formula:

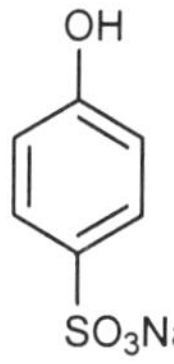

Information Sources: JCIC, JCLS, MI-13 (8731), TSCA

Chemical Classes: Phenols; Sulfonic Acids

Functions: Cosmetic Biocide; Deodorant Agent; Preservative

Technical/Other Names:
Benzenesulfonic Acid, Hydroxy-, Monosodium Salt
Hydroxybenzenesulfonic Acid, Monosodium Salt
Sodium Paraphenolsulfonate (Dihydrate)
Sodium Sulfocarbolate

Trade Name:
Sodium paraphenolsulfonate (Matsumoto Trading)

SODIUM PHENOXIDE

CAS No.   **EINECS No.**
139-02-6   205-347-3

Empirical Formula:
$C_6H_6O \cdot Na$

Definition: Sodium Phenoxide is the aromatic compound that conforms to the formula:

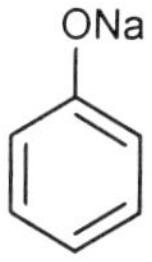

See Reported Ingredient Functions-The Cosmetic Drug Distinction, in Regulatory and Ingredient Use Information, Volume I, Part A.

Information Sources: EEC(III/1-19), MI-13 (8732), OTC-I-EA, OTC-I-OD, USAN

Chemical Classes: Organic Salts; Phenols

Functions: Cosmetic Biocide; External Analgesic; Preservative

Technical/Other Names:
Phenol, Sodium Salt
Sodium Carbolate
Sodium Phenate

SODIUM PHENYLBENZIMIDAZOLE SULFONATE

CAS No.: 5997-53-5

Empirical Formula:
$C_{13}H_{10}N_2O_3S \cdot Na$

Definition: Sodium Phenylbenzimidazole Sulfonate is the organic salt that conforms to the formula:

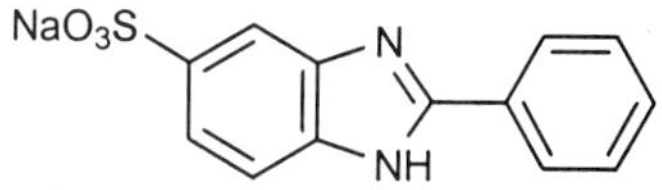

Information Source: EEC(VII/1-6)

Chemical Classes: Heterocyclic Compounds; Organic Salts; Sulfonic Acids

Function: Ultraviolet Light Absorber

Technical/Other Name:
2-Phenylbenzimidazole-5-Sulfonic Acid, Sodium Salt

Trade Name:
Phenylbenzimidazole Sulfonic Acid Sodium Salt (Merck KGaA)

SODIUM o-PHENYLPHENATE

CAS No. 132-27-4

EINECS No. 205-055-6

JPN Translation:
フエニルフエノール Na

Empirical Formula:
$C_{12}H_{10}O \cdot Na$

Definition: Sodium o-Phenylphenate is the sodium salt of o-phenylphenol that conforms to the formula:

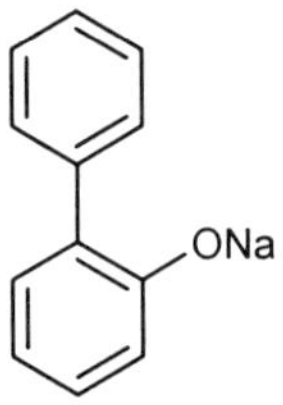

Information Sources: 21CFR175.105, 21CFR175.300, 21CFR176.170, 21CFR176.210, 21CFR177.1210, 21CFR178.3120, CTFA D, EEC(VI/1-7), JCIC, JCLS, MAR, MHLW-331/3, MI-13 (7388), TSCA

Chemical Classes: Organic Salts; Phenols

Functions: Cosmetic Biocide; Preservative

Reported Product Category: Hair Dyes and Colors (All Types Requiring Caution Statements and Patch Tests)

Technical/Other Names:
(1,1'-Biphenyl)-2-ol, Sodium Salt
2-Biphenylol, Sodium Salt
2-Hydroxydiphenyl Sodium
o-Phenyl Phenol Sodium Salt
Sodium o-Phenylphenoxide

Trade Names:
DOWICIDE A (Dow Chemical)
Preventol ON Extra (Bayer AG)

SODIUM PHOSPHATE

CAS Nos. 7558-80-7
7632-05-5

EINECS Nos. 231-449-2
231-558-5

JPN Translation:
リン酸 Na

Empirical Formula:
$H_3O_4P \cdot Na$

Definition: Sodium Phosphate is the inorganic salt that conforms to the formula:

$$NaH_2PO_4$$

Information Sources: ARG, AUS, BP, BRA, 21CFR133.169, 21CFR133.173, 21CFR133.179, 21CFR150.141, 21CFR150.161, 21CFR160.110, 21CFR163.123, 21CFR163.130, 21CFR163.135, 21CFR163.140, 21CFR163.145, 21CFR163.150, 21CFR163.153, 21CFR163.155, 21CFR172.892, 21CFR173.310, 21CFR182.1778, 21CFR182.6085, 21CFR182.6778, 21CFR182.8778, DA, DDR, EGY, FCC, HP, HUN, IND, JCIC, JCLS, JSCI, MAR, MEX, MI-13(8734), NED, OTC-I-LX, PN, TSCA, USAN, USD, USP XXIV, YUG

Chemical Classes: Inorganic Salts; Phosphorus Compounds

Function: Buffering Agent

Reported Product Categories: Shampoos (Non-coloring); Dentifrices (Aerosol, Liquid, Pastes and Powders); Cleansing Products (Cold Creams, Cleansing Lotions, Liquids and Pads); Moisturizing Preparations; Bath Preparations, Misc.; Body and Hand Preparations (Excluding Shaving Preparations); Hair Conditioners; Permanent Waves; Skin Fresheners

Technical/Other Names:
Monobasic Sodium Phosphate
Monobasic Sodium Phosphate (Monohydrate)
Monosodium Phosphate
Phosphoric Acid, Monosodium Salt
Sodium Biphosphate
Sodium Dihydrogen Phosphate
Sodium Phosphate, Monobasic

Trade Name Mixtures:
Bovine Fibronectin Solution (Intergen)
Hydrocos "P" (Cosmetochem) (Cosmetochem International Ltd.)
Umordant P (Cosmetochem) (Cosmetochem International Ltd.)

SODIUM PHOSPHONO-PYRIDOXYLIDENERHODANINE

Definition: Sodium Phosphono-Pyridoxylidenerhodanine is the organic compound that conforms to the formula:

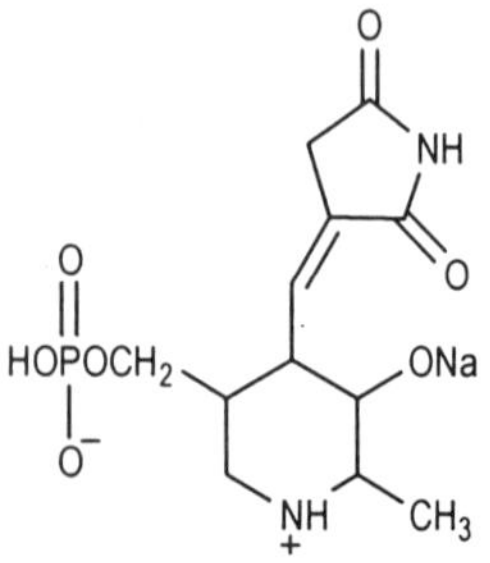

Chemical Classes: Heterocyclic Compounds; Phosphorus Compounds; Thio Compounds

Function: Antioxidant

Trade Name:
B6Pr Sodium Hydrate (Pharma Cosmetix)

SODIUM PHYTATE

CAS Nos. 14306-25-3
34367-89-0

EINECS No. 238-242-6

Definition: Sodium Phytate is the complex sodium salt of Phytic Acid (q.v.).

Information Source: USAN

Chemical Classes: Esters; Organic Salts; Phosphorus Compounds

Functions: Chelating Agent; Oral Care Agent

Technical/Other Names:
Myo-Inositol, Hexakis(Dihydrogen Phosphate), Hexasodium Salt
Phytic Acid, Sodium Salt

Trade Name Mixture:
Dermofeel PA-3 (Straetmans)

SODIUM PICRAMATE

CAS No. 831-52-7

EINECS No. 212-603-8

Empirical Formula:
$C_6H_5N_3O_5 \cdot Na$

Definition: Sodium Picramate is the sodium salt of Picramic Acid (q.v.) that conforms to the formula:

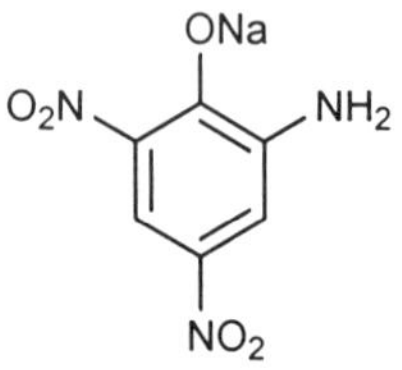

See "Regulatory and Ingredient Use Information," for Colorants in Volume 1, Introduction, Part A.

The inclusion of any compound in the *Dictionary and Handbook* does not indicate that use of that substance as a cosmetic ingredient complies with the laws and regulations governing such use in the United States or any other country.

Information Sources: CIR: [SQ] JACT-11 (4)1992, TSCA

Chemical Classes: Amines; Color Additives - Hair; Organic Salts

Function: Hair Colorant

Reported Product Categories: Hair Dyes and Colors (All Types Requiring Caution Statements and Patch Tests); Hair Tints

Technical/Other Names:
2-Amino-4,6-Dinitrophenol, Sodium Salt
Phenol, 2-Amino-4,6-Dinitro-, Sodium Salt
Picramic Acid, Sodium Salt

Trade Name:
Rodol 4R (Lowenstein)

SODIUM POLYACRYLATE

CAS Nos.: 9003-04-7; 25549-84-2

JPN Translation:
ポリアクリル酸 Na

Empirical Formula:
$(C_3H_4O_2)_x \cdot xNa$

Definition: Sodium Polyacrylate is the sodium salt of Polyacrylic Acid (q.v.).

Information Sources: 21CFR173.73, CIR: [SQ] IJT 21(SUPPL. 3) 2002, JCLS, JSCI, TSCA

Chemical Class: Synthetic Polymers

Functions: Absorbent; Emulsion Stabilizer; Film Former; Hair Fixative; Skin-Conditioning Agent - Emollient; Viscosity Controlling Agent; Viscosity Increasing Agent - Aqueous

Reported Product Categories: Skin Care Preparations, Misc.; Hair Dyes and Colors (All Types Requiring Caution Statements and Patch Tests)

Technical/Other Names:
Acrylic Acid Homopolymer Sodium Salt
Polyacrylic Acid, Sodium Salt
2-Propenoic Acid, Homopolymer, Sodium Salt

Trade Names:
Aronvis (Nihon Junyaku)
Covacryl AC (LCW)
Covacryl RH (LCW)
Flocare CGEL 100 (SNF)
FLOCARE DP/PSD 100 (SNF)
Flocare G300 (SNF)
Flocare G800 (SNF)
Hysorb 8200 (BASF)
Octacare RM100 (Associated Octel)
Octacare X100 (Associated Octel)
Octacare X110 (Associated Octel)
RapiThix A-100 (International Specialty Products)
Rheogic 250H (Nihon Junyaku)

Trade Name Mixtures:
Cassava II Extract (Sederma)
Ceramide A2 (Sederma)
Cosmedia SPL (Cognis France)
Creagel NP (C.I.T.)
Flocare ET 75 (SNF)
Flocare ET 76 (SNF)
Flocare ET 100 (SNF)
Osmohair (Sederma)
PALI-M-102 (US Cosmetics)
PALI-S-100 (US Cosmetics)
PALI-TA-13R (US Cosmetics)
RapiThix A-60 (International Specialty Products)
Rheocare ATH (Cosmetic Rheologies)
Sepigel 501 (SEPPIC)
Sepigel 502 (SEPPIC)
Tioveil AQ-G (Uniqema, Belgium)
Tioveil AQ-N (Uniqema, Belgium)
Tioveil AQ-P (Uniqema, Belgium)
Zilgel NP (Zile)
Zilgel Oil (Zile)
Zilgel SM (Zile)

SODIUM POLYACRYLATE STARCH

JPN Translation:
アクリル酸 Na グラフトデンプン

Definition: Sodium Polyacrylate Starch is a polymer of sodium acrylate grafted with starch.

Chemical Classes: Biological Polymers and their Derivatives; Gums, Hydrophilic Colloids and Derivatives; Synthetic Polymers

Functions: Absorbent; Binder; Emulsion Stabilizer; Viscosity Increasing Agent - Aqueous

Trade Names:
Sanfresh ST-100C (Sanyo Chemical)
Sanfresh ST-100MC (Sanyo Chemical)
Sanwet IM-300MC (Sanyo Chemical)

SODIUM POLYACRYLOYLDIMETHYL TAURATE

Definition: Sodium Polyacryloyldimethyl Taurate is the polymer that conforms generally to the formula:

Chemical Classes: Amides; Sulfonic Acids; Synthetic Polymers

Functions: Emulsion Stabilizer; Viscosity Increasing Agent - Aqueous

Trade Name Mixture:
Simulgel 800 (SEPPIC)

SODIUM POLYASPARTATE

CAS No.: 94525-01-6

JPN Translation:
ポリアスパラギン酸 Na

Definition: Sodium Polyaspartate is the sodium salt of a polymer of Aspartic Acid (q.v.).

Information Source: TSCA

Chemical Classes: Amino Acids; Organic Salts; Synthetic Polymers

Functions: Hair Conditioning Agent; Skin-Conditioning Agent - Humectant

Reported Product Category: Moisturizing Preparations

Technical/Other Name:
DL-Aspartic Acid, Homopolymer, Sodium Salt

Trade Name:
Aquadew SPA - 30 (Ajinomoto)

Trade Name Mixtures:
HydroSpers-B-335198 (US Cosmetics)
HydroSpers-BR-33115 (US Cosmetics)
HydroSpers-R-338075 (US Cosmetics)
HydroSpers-UB-431810 (US Cosmetics)
HydroSpers-Y-338073 (US Cosmetics)

SODIUM POLYDIMETHYL-GLYCINOPHENOLSULFONATE

Empirical Formula:
$(C_{10}H_{11}NO_5 \cdot 2Na)_x$

Definition: Sodium Polydimethyl-glycinophenolsulfonate is the polymer that conforms to the formula:

Chemical Classes: Alkyl-Substituted Amino Acids; Phenols; Sulfonic Acids

Function: Chelating Agent

SODIUM POLYGAMMA-GLUTAMATE

Definition: Sodium Polygamma-Glutamate is the organic compound that conforms generally to the formula:

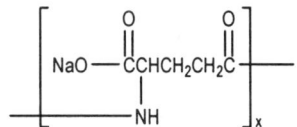

Chemical Classes: Amides; Synthetic Polymers

Functions: Emulsion Stabilizer; Hair Fixative; Viscosity Increasing Agent - Aqueous

Technical/Other Name:
Gamma-Glutamic Acid, Homopolymer, Sodium Salt

Trade Names:
Gelprotein A-8000 (Idemitsu Technofine Co., Ltd)
Gelprotein A - 8001 (Idemitsu Technofine Co., Ltd)
Gelprotein A - 8002 (Idemitsu Technofine Co., Ltd)

SODIUM POLYGLUTAMATE

CAS No.: 28829-38-1

Empirical Formula:
$(C_5H_7NO_3)_x \cdot xNa$

Definition: Sodium Polyglutamate is the organic compound that conforms generally to the formula:

Chemical Classes: Amides; Synthetic Polymers

Functions: Hair Conditioning Agent; Skin-Conditioning Agent - Miscellaneous

Reported Product Category: Moisturizing Preparations

Technical/Other Names:
Glutamic Acid Homopolymer Sodium Salt
Poly(glutamic Acid) Sodium Salt

Trade Name:
Ajicoat SPG (Ajinomoto)

SODIUM POLYMETHACRYLATE

CAS No.: 25086-62-8

Empirical Formula:
$(C_4H_6O_2)_x \cdot xNa$

Definition: Sodium Polymethacrylate is the polymer that conforms generally to the formula:

Information Sources: 21CFR173.310, 21CFR175.105, TSCA

Chemical Class: Synthetic Polymers

Functions: Binder; Emulsion Stabilizer; Film Former; Viscosity Increasing Agent - Aqueous

Reported Product Categories: Mascara; Foundations

Technical/Other Names:
Methacrylic Acid Homopolymer Sodium Salt
2-Propenoic Acid, 2-Methyl-, Homopolymer, Sodium Salt

Trade Names:
Darvan #7 (Vanderbilt)
Daxad 30 (Amerchol)

SODIUM POLYNAPHTHALENE-SULFONATE

CAS No.: 9084-06-4

Empirical Formula:
$(C_{10}H_8O_3S \cdot CH_2O)_x \cdot xNa$

Definition: Sodium Polynaphthalene-sulfonate is the sodium salt of the product obtained by the condensation polymerization of naphthalene sulfonic acid and formaldehyde.

Information Sources: 21CFR175.105, 21CFR176.170, 21CFR176.180, 21CFR177.1200, 21CFR177.1210, 21CFR177.1550, 21CFR177.1650, 21CFR177.2600, 21CFR178.3910, CIR: [SQ], TSCA

Chemical Classes: Sulfonic Acids; Synthetic Polymers

Functions: Emulsion Stabilizer; Surfactant - Hydrotrope; Surfactant - Suspending Agent

Reported Product Categories: Makeup Bases; Mascara; Foundations; Makeup Preparations (Not eye), Misc.; Eye Makeup Preparations, Misc.; Blushers (All types)

Trade Names:
Darvan #1 (Vanderbilt)
Daxad 11 (Amerchol)

SODIUM POLYPHOSPHATE

CAS No.	EINECS No.
68915-31-1	272-808-3

Definition: Sodium Polyphosphate is a mixture of the sodium salts of polyphosphoric acid.

Information Source: TSCA

Chemical Classes: Inorganic Salts; Phosphorus Compounds

Function: Chelating Agent

Reported Product Categories: Bath Capsules; Foundations; Moisturizing Preparations; Mascara; Bath Preparations, Misc.; Body and Hand Preparations (Excluding Shaving Preparations); Skin Care Preparations, Misc.; Bath Oils, Tablets, and Salts; Eyeliners

Technical/Other Names:
Polyphosphoric Acids, Sodium Salts
Sodium hexametaphosphate (RIFM)

Trade Name:
Glass H (FMC/Pharmaceutical Division)

SODIUM POLYSTYRENE SULFONATE

CAS Nos.: 9003-59-2; 9080-79-9; 62744-35-8

Empirical Formula:
$(C_8H_8O_3S)_x \cdot xNa$

Definition: Sodium Polystyrene Sulfonate is the polymer that conforms generally to the formula:

Information Sources: 21CFR175.105, JAN, MAR, MI-13(8742), TSCA, USAN, USD, USP XXIV

Chemical Classes: Sulfonic Acids; Synthetic Polymers

Functions: Emulsion Stabilizer; Film Former; Surfactant - Suspending Agent; Viscosity Increasing Agent - Aqueous

Reported Product Categories: Shampoos (Non-coloring); Tonics, Dressings, and Other Hair Grooming Aids

Technical/Other Names:
Benzenesulfonic Acid, Ethenyl-, Homopolymer, Sodium Salt

Benzenesulfonic Acid, Ethenyl-, Sodium
Salt, Homopolymer
Ethenylbenzenesulfonic Acid, Homo-
polymer, Sodium Salt
Ethenylbenzenesulfonic Acid, Sodium Salt,
Homopolymer
Polyethenylbenzenesulfonic Acid, Sodium
Salt
Styrenesulfonic Acid Homopolymer Sodium
Salt

Trade Names:
Flexan 130 (National Starch)
Flexan II (National Starch)

SODIUM POTASSIUM ALUMINUM SILICATE

CAS No.: 66402-68-4

Definition: Sodium Potassium Aluminum
Silicate is a complex silicate refined from
naturally occurring minerals.

Chemical Class: Inorganics

Function: Bulking Agent

Trade Names:
Kiyouseki (Picaso Cosmetic Laboratory)
3M Cosmetic Microspheres CM-210, CM-
410, CM-610 (3M)

SODIUM PROPIONATE

CAS No.	EINECS No.
137-40-6	205-290-4

JPN Translation:
プロピオン酸 Na

Empirical Formula:
$C_3H_6O_2 \cdot Na$

Definition: Sodium Propionate is the sodium
salt of propionic acid that conforms to the
formula:

$$CH_3CH_2COONa$$

Information Sources: 21CFR133.123,
21CFR133.124, 21CFR133.169,
21CFR133.173, 21CFR133.179,
21CFR150.141, 21CFR150.161,
21CFR179.45, 21CFR181.22,
21CFR181.23, 21CFR184.1784, EEC(VI/1-
2), FCC, ITA, JCLS, KOR, MAR, MI-13
(8743), NF XIX, TSCA, USAN, USD

Chemical Class: Organic Salts

Function: Preservative

Technical/Other Names:
Propanoic Acid, Sodium Salt
Sodium Ethanecarboxylate

Trade Name:
Unistat SOPRO (Universal Preserv-A-
Chem)

SODIUM PROPOXYHYDROXYPROPYL THIOSULFATE SILICA

Definition: Sodium Propoxyhydroxypropyl
Thiosulfate Silica is a substituted silica
prepared from the hydrolysis of an
alkylthiosulfate substituted trimethoxysilane
in the presence of Silica (q.v.). It conforms to
the formula:

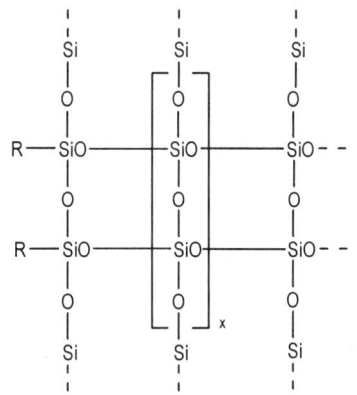

where R is $(CH_2)_3OCH_2CHOHCH_2S_2O_3Na$.

Chemical Class: Siloxanes and Silanes

Function: Not Reported

Trade Name:
Nanospheres NCK Hydrophilic (Exsymol)

SODIUM PROPOXY PPG-2 ACETATE

CAS No.: 165038-54-0

Empirical Formula:
$C_{11}H_{22}O_5 \cdot Na$

Definition: Sodium Propoxy PPG-2 Acetate
is the organic compound that conforms to the
formula:

$$CH_3CH_2CH_2(OCHCH_2)_2OCH_2COONa$$
$$|$$
$$CH_3$$

Chemical Classes: Alkoxylated Alcohols;
Organic Salts

Functions: Surfactant - Cleansing Agent;
Surfactant - Emulsifying Agent; Surfactant -
Hydrotrope; Surfactant - Solubilizing Agent

Technical/Other Name:
Acetic Acid, [1-Methyl-2-(1-Methyl-2-
Propoxyethoxy)Ethoxy]-, Sodium Salt

Trade Name:
Gemtex DPNP-101 (Finetex)

SODIUM PROPYLPARABEN

CAS No.	EINECS No.
35285-69-9	252-488-1

Empirical Formula:
$C_{10}H_{12}O_3 \cdot Na$

Definition: Sodium Propylparaben is the
sodium salt of Propylparaben (q.v.) that
conforms to the formula:

Information Sources: EEC(VI/1-12),
MHLW-331/3, NF XIX, USAN

Chemical Classes: Esters; Organic Salts;
Phenols

Function: Preservative

Reported Product Categories: Blushers (All
types); Colognes and Toilet Waters; Eyebrow
Pencils; Makeup Preparations (Not eye),
Misc.

Technical/Other Names:
Benzoic Acid, 4-Hydroxy-, Propyl Ester,
Sodium Salt
4-Hydroxybenzoic Acid, Propyl Ester,
Sodium Salt
Propyl p-Hydroxybenzoate Sodium Salt
Propylparaben, Sodium Salt

Trade Names:
Nipasol M Sodium (Clariant)
Nipasol M Sodium (Clariant GmbH, Per-
sonal Care)
Paridol PNA (Dekker)
Propyl 4-Hydroxybenzoate Sodium Salt
(Merck KGaA)

Trade Name Mixtures:
Conservateur TW (LCW)
Nipacombin A (Clariant)
Nipacombin A (Clariant GmbH, Personal
Care)
Nipacombin SK (Clariant)
Nipacombin SK (Clariant GmbH, Personal
Care)
Nipasept Sodium (Clariant)
Nipasept Sodium (Clariant GmbH, Per-
sonal Care)
Nipastat Sodium (Clariant)
Nipastat Sodium (Clariant GmbH, Personal
Care)

SODIUM PVM/MA/DECADIENE CROSS-POLYMER

Definition: Sodium PVM/MA/Decadiene
Crosspolymer is a copolymer of methyl vinyl
ether and maleic anhydride crosslinked with
1,9-decadiene.

Chemical Class: Synthetic Polymers

Functions: Film Former; Viscosity Increasing Agent - Nonaqueous

SODIUM PYRITHIONE

CAS No.
3811-73-2

EINECS No.
223-296-5

Empirical Formula:
$C_5H_5NOS \cdot Na$

Definition: Sodium Pyrithione is the organic compound that conforms to the formula:

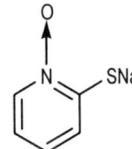

Information Sources: CTFA D, EEC(II-369), TSCA, USAN

Chemical Classes: Heterocyclic Compounds; Organic Salts; Thio Compounds

Function: Preservative

Technical/Other Names:
1-Oxo-2-Pyridinethiol Sodium Salt
2-Pyridinethiol, 1-Oxide, Sodium Salt
Pyrithione Sodium
Sodium 2-Mercaptopyridine 1-Oxide
Sodium (2-Pyridylthio)-N-Oxide

Trade Names:
Natrium-Pyrion (Rutgers Organics)
Sodium OMADINE (Arch Chemical)

SODIUM PYRUVATE

CAS No.
113-24-6

EINECS No.
204-024-4

Empirical Formula:
$C_3H_4O_3 \cdot Na$

Definition: Sodium Pyruvate is the organic salt that conforms to the formula:

$$CH_3C-C-ONa$$

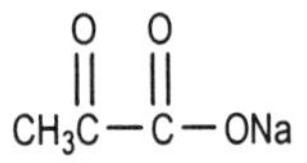

Information Source: TSCA

Chemical Class: Organic Salts

Function: Skin-Conditioning Agent - Miscellaneous

Technical/Other Names:
2-Oxopropanoic Acid, Sodium Salt
Propanoic Acid, 2-Oxo-, Sodium Salt
Sodium α-Ketopropionate
Sodium 2-Oxopropanoate

Trade Name:
Pyruvic Acid, Sodium Salt (Merck KGaA)

SODIUM RAPESEEDATE

Definition: Sodium Rapeseedate is the sodium salt of the fatty acids derived from Brassica Campestris (Rapeseed) Oil (q.v.).

Chemical Class: Soaps

Functions: Surfactant - Cleansing Agent; Surfactant - Emulsifying Agent; Viscosity Increasing Agent - Nonaqueous

Technical/Other Name:
Fatty Acids, Rapeseed Oil, Sodium Salt

Trade Name:
PRISAVON RP (Uniqema Europe)

Trade Name Mixture:
PRISAVON 9271 (Uniqema Europe)

SODIUM RIBOFLAVIN PHOSPHATE

CAS No.
130-40-5

EINECS No.
204-988-6

JPN Translation:
リボフラビンリン酸 Na

Empirical Formula:
$C_{17}H_{21}N_4O_9P \cdot Na$

Definition: Sodium Riboflavin Phosphate is the ester of Riboflavin (q.v.) and Sodium Phosphate (q.v.). It conforms generally to the formula:

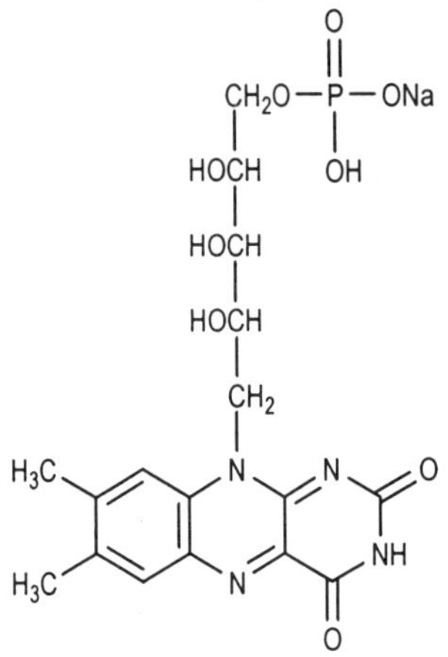

Information Sources: JCLS, JP

Chemical Classes: Heterocyclic Compounds; Organic Salts; Phosphorus Compounds; Polyols

Function: Skin-Conditioning Agent - Miscellaneous

Technical/Other Names:
Riboflavine Sodium Phosphate
Riboflavin Sodium Phosphate

Trade Name:
Riboflavin-5'-Phosphate Sodium (Roche)

Trade Name Mixtures:
Amaryl Hydro (Laboratoires Sero-biologiques)
Elespher Vitaplex Hydro (Laboratoires Serobiologiques)

SODIUM RICINOLEATE

CAS No.
5323-95-5

EINECS No.
226-191-2

Empirical Formula:
$C_{18}H_{34}O_3 \cdot Na$

Definition: Sodium Ricinoleate is the sodium salt of Ricinoleic Acid (q.v.).

Information Sources: 21CFR175.300, MI-13(8295), TSCA

Chemical Class: Soaps

Functions: Surfactant - Cleansing Agent; Surfactant - Emulsifying Agent

Reported Product Category: Bath Soaps and Detergents

Technical/Other Names:
12-Hydroxy-9-Octadecenoic Acid, Sodium Salt
9-Octadecenoic Acid, 12-Hydroxy-, Sodium Salt
Sodium Ricinate

SODIUM RICINOLEOAMPHOACETATE

Empirical Formula:
$C_{24}H_{46}N_2O_5 \cdot Na$

Definition: Sodium Ricinoleoamphoacetate is the amphoteric organic compound that conforms generally to the formula:

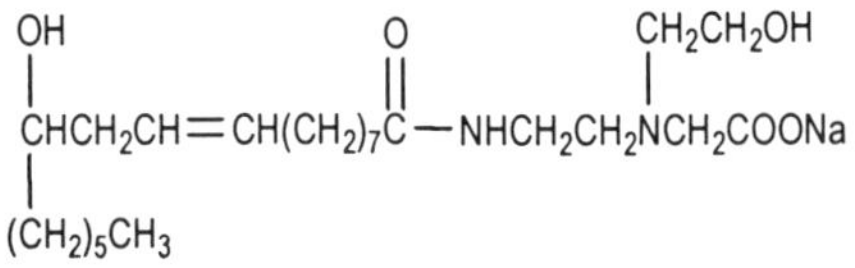

Chemical Class: Alkylamido Alkylamines

Functions: Hair Conditioning Agent; Surfactant - Cleansing Agent; Surfactant - Foam Booster

SODIUM RNA

JPN Translation:
RNA - Na

Definition: Sodium RNA is the sodium salt of RNA (q.v.).

Chemical Classes: Biological Polymers and their Derivatives; Phosphorus Compounds

Function: Skin-Conditioning Agent - Miscellaneous

Reported Product Categories: Makeup Bases; Skin Care Preparations, Misc.

Technical/Other Names:
RNA, Sodium Salt
Sodium Ribonucleic Acid

Trade Name:
AEC Sodium RNA (A & E Connock)

SODIUM ROSINATE

JPN Translation:
ロジン酸 Na

Definition: Sodium Rosinate is the sodium salt of the acids derived from Rosin (q.v.).

Chemical Class: Soaps

Functions: Surfactant - Cleansing Agent; Viscosity Increasing Agent - Nonaqueous

Technical/Other Names:
Fatty Acids, Rosin, Sodium Salt
Rosin Acid, Sodium Salt

Trade Name:
Colophonium, Sodium Salt (Balsamharz-Natriumsalz) (Walter Rau)

SODIUM SACCHARIN

CAS Nos.   **EINECS No.**
128-44-9   204-886-1
6155-57-3

JPN Translation:
サッカリン Na

Empirical Formula:
$C_7H_5NO_3S \cdot 2H_2O \cdot Na$

Definition: Sodium Saccharin is the organic compound that conforms to the formula:

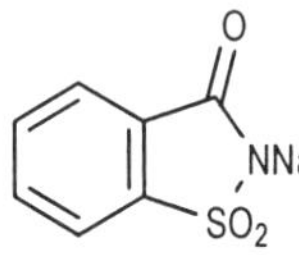

Information Sources: ARG, AUS, BP, BPC, 21CFR145.116, 21CFR145.126, 21CFR145.131, 21CFR145.136, 21CFR145.171, 21CFR145.181, 21CFR150.141, 21CFR150.161, 21CFR180.37, DA, DDR, EGY, FCC, HUN, JAN, JCLS, JSCI, MAR, MEX, MI-13(8390), PN, RIFM, TSCA, USAN, USD, USP XXIV, YUG

Chemical Classes: Heterocyclic Compounds; Organic Salts

Functions: Flavoring Agent; Fragrance Ingredient

Reported Product Categories: Mouthwashes and Breath Fresheners (Liquids and Sprays); Dentifrices (Aerosol, Liquid, Pastes and Powders); Lipsticks

Technical/Other Names:
1,2-Benzisothiazol-3(2H)-one, 1,1-Dioxide, Sodium Salt, Dihydrate
o-Benzoylsulfimide Sodium Salt
Crystallose
1,1-Dioxide-1,2-Benzisothiazol-3(2H)-one, Sodium Salt, Dihydrate
Saccharine, sodium salt (RIFM)
Saccharin Sodium
Soduim Saccharine

Trade Name:
Unisweet SOSAC (Universal Preserv-A-Chem)

SODIUM SAFFLOWERATE

Definition: Sodium Safflowerate is the sodium salt of the fatty acids derived from Carthamus Tinctorius (Safflower) Seed Oil (q.v.).

Chemical Class: Soaps

Function: Surfactant - Cleansing Agent

SODIUM SALICYLATE

CAS No.   **EINECS No.**
54-21-7   200-198-0

JPN Translation:
サリチル酸 Na

Empirical Formula:
$C_7H_6O_3 \cdot Na$

Definition: Sodium Salicylate is the sodium salt of salicylic acid that conforms to the formula:

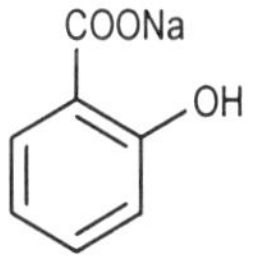

Information Sources: ARG, AUS, BP, BPC, BRA, 21CFR175.105, CIR: [SQ], CZE, DA, DDR, EEC(VI/1-3), EGY, FIN, HP, HUN, IND, ITA, JAN, JCLS, JSCI, MEX, MHLW-331/3, MI-12(8819), OTC-I-IA, OTC-I-MD, PF, PN, POR, ROM, TSCA, USAN, USD, USP XXIV, USSR, WHO, YUG

Chemical Classes: Organic Salts; Phenols

Functions: Denaturant; Preservative

Reported Product Categories: Moisturizing Preparations; Mouthwashes and Breath Fresheners (Liquids and Sprays); Tonics, Dressings, and Other Hair Grooming Aids

Technical/Other Names:
Benzoic Acid, 2-Hydroxy-, Monosodium Salt
2-Hydroxybenzoic Acid, Monosodium Salt
Sodium o-Hydroxybenzoate

Trade Name Mixtures:
Complexe Hydroxy-Salicylique Color (Coletica SA)
Gatuline Equalizing (Gattefosse s.a.)
Isocell Slim (Lucas Meyer)

SODIUM SARCOSINATE

CAS No.   **EINECS No.**
4316-73-8   224-338-5

Empirical Formula:
$C_3H_7NO_2 \cdot Na$

Definition: Sodium Sarcosinate is the sodium salt of sarcosine. It conforms to the formula:

$$CH_3NHCH_2COONa$$

Information Source: TSCA

Chemical Class: Sarcosinates and Sarcosine Derivatives

Function: Not Reported

Technical/Other Names:
Glycine, N-Methyl-, Sodium Salt
N-Methylglycine, Sodium Salt
Sodium Methylaminoacetate

SODIUM SCYMNOL SULFATE

CAS No.: 119068-78-9

Empirical Formula:
$C_{27}H_{47}O_9S \cdot Na$

Definition: Sodium Scymnol Sulfate is the organic compound that conforms to the formula:

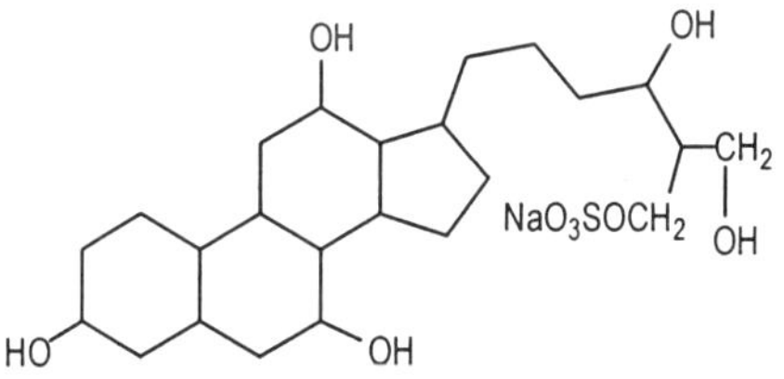

Chemical Classes: Alkyl Sulfates; Sterols

Function: Skin-Conditioning Agent - Miscellaneous

Technical/Other Names:
Cholestane-3,7,12,24,26,27-hexol, 26-(Hydrogen Sulfate), Monosodium Salt
24R-(+) 3a, 7a, 12a, 24, 26-Pentahydroxy-coprastane-27-Sodium Sulfate Ester

Trade Name:
Isolutrol (McFarlane)

SODIUM SESAMPHOACETATE

Definition: Sodium Sesamphoacetate is the amphoteric organic compound that conforms generally to the formula:

$$RC{-}NHCH_2CH_2NCH_2COONa$$

(with O double-bonded to RC and a CH_2CH_2OH group on the nitrogen)

where RCO- represents the fatty acids derived from sesame oil.

Chemical Class: Alkylamido Alkylamines

Functions: Hair Conditioning Agent; Surfactant - Cleansing Agent; Surfactant - Foam Booster

Trade Name:
Vamasoft Sesame (Vama Farmacosmetica)

SODIUM SESQUICARBONATE

CAS No. 533-96-0
EINECS No. 208-580-9

JPN Translation:
セスキ炭酸 Na

Empirical Formula:
$CH_2O_3 \cdot {}^3/_2 Na$

Definition: Sodium Sesquicarbonate is the inorganic salt that conforms generally to the formula:

$$Na_2CO_3 \quad \cdot \quad NaHCO_3$$

Information Sources: 21CFR184.1792, CIR: [S] JACT-6(1)1987, CTFA S, FCC, JCIC, JCLS, MI-13(8749), TSCA

Chemical Class: Inorganic Salts

Function: pH Adjuster

Reported Product Categories: Bath Oils, Tablets, and Salts; Bubble Baths; Bath Preparations, Misc.

Technical/Other Name:
Carbonic Acid, Sodium Salt (2:3)

SODIUM SHALE OIL SULFONATE

Definition: Sodium Shale Oil Sulfonate is the sodium salt of sulfonated shale oil.

Chemical Class: Sulfonic Acids

Function: Cosmetic Biocide

Trade Name:
Ichthyol Pale (Ichthyol)

SODIUM SILICATE

CAS No. 1344-09-8
EINECS No. 215-687-4

JPN Translation:
ケイ酸 Na

Definition: Sodium Silicate is a sodium salt of silicic acid.

Information Sources: 21CFR173.310, 21CFR177.1200, 21CFR182.70, 21CFR182.90, CIR: [SQ], CTFA S, JCLS, JSCI, MAR, MI-13(8750), MI-13(8751), TSCA

Chemical Class: Inorganic Salts

Functions: Buffering Agent; Corrosion Inhibitor; pH Adjuster

Reported Product Categories: Hair Bleaches; Shaving Cream (Aerosol, Brushless and Lather); Bath Soaps and Detergents; Bubble Baths; Depilatories; Personal Cleanliness Products, Misc.; Hair Coloring Preparations, Misc.; Hair Dyes and Colors (All Types Requiring Caution Statements and Patch Tests); Skin Care Preparations, Misc.

Technical/Other Names:
Silicic Acid, Sodium Salt
Silicon Sodium Oxide

Trade Names:
Britesil (PQ Corporation)
'N' Silicate (PQ Corporation)
'O' Silicate (PQ Corporation)

SODIUM SILICOALUMINATE

CAS No. 1344-00-9
EINECS No. 215-684-8

Definition: Sodium Silicoaluminate is a series of hydrated sodium aluminum silicates.

Information Sources: 21CFR133.146, 21CFR160.105, 21CFR160.185, 21CFR182.2727, 21CFR582.2727, FCC, TSCA

Chemical Class: Inorganic Salts

Functions: Abrasive; Viscosity Increasing Agent - Aqueous

Reported Product Category: Bath Soaps and Detergents

Technical/Other Names:
Aluminosilicic Acid, Sodium Salt
Aluminum Silicon Sodium Oxide
Aluminum Sodium Silicate
Silic Acid, Aluminum Sodium Salt
Sodium Aluminosilicate

Trade Names:
Alusil ET (Crosfield Co.)
Alusil SD533 (Crosfield Co.)
Valfor Zeolite Na-A (PQ Corporation)
Zeodent 012 (Huber)
Zeolex 7 (Huber)
Zeolex 35 (Huber)
Zeolex 7A (J.M Huber)
Zeolex 23A (Huber)

SODIUM BETA-SITOSTERYL SULFATE

Definition: Sodium Beta-Sitosteryl Sulfate is the sodium salt of sulfated Beta-Sitosterol (q.v.).

Chemical Classes: Sterols; Sulfuric Acid Esters

Function: Skin-Conditioning Agent - Miscellaneous

Trade Name:
Phytocohesine (Vincience)

SODIUM SORBATE

CAS No. 7757-81-5
EINECS No. 231-819-3

Empirical Formula:
$C_6H_8O_2 \cdot Na$

Definition: Sodium Sorbate is the sodium salt of Sorbic Acid (q.v.) that conforms to the formula:

$$CH_3CH{=}CHCH{=}CHCOONa$$

Information Sources: EEC(VI/1-4), MHLW-331/3

Chemical Class: Organic Salts

Function: Preservative

Technical/Other Names:
2,4-Hexadienoic Acid, Sodium Salt
Sodium 2,4-Hexadienoate

SODIUM SOYATE

Definition: Sodium Soyate is the sodium salt of Soy Acid (q.v.).

Chemical Class: Soaps

Functions: Surfactant - Cleansing Agent; Surfactant - Emulsifying Agent; Viscosity Increasing Agent - Nonaqueous

Technical/Other Name:
Soya Acid, Sodium Salt

SODIUM SOY HYDROLYZED COLLAGEN

CAS No.: 68188-31-8

Definition: Sodium Soy Hydrolyzed Collagen is the sodium salt of the condensation product of soya acid chloride and Hydrolyzed Collagen (q.v.).

International Cosmetic Ingredient Dictionary and Handbook

Information Source: TSCA

Chemical Class: Protein Derivatives

Functions: Hair Conditioning Agent; Skin-Conditioning Agent - Miscellaneous; Surfactant - Cleansing Agent

Technical/Other Names:
Acid Chlorides, Soy, Reaction Products with Protein Hydrolyzates, Sodium Salts
Sodium Soya Hydrolyzed Animal Protein
Sodium Soya Hydrolyzed Collagen
Sodium Soy Hydrolyzed Animal Protein

SODIUM STANNATE

CAS No.	EINECS No.
12058-66-1	235-030-5

Empirical Formula:
$H_2O_3Sn \cdot 2Na$

Definition: Sodium Stannate is the inorganic salt that conforms to the formula:

$$Na_2SnO_3 \quad \cdot \quad 3H_2O$$

Information Sources: CTFA D, JSQI, MI-13(8752), TSCA

Chemical Class: Inorganic Salts

Function: Not Reported

Reported Product Categories: Hair Coloring Preparations, Misc.; Hair Dyes and Colors (All Types Requiring Caution Statements and Patch Tests); Permanent Waves

Technical/Other Name:
Sodium Tin Oxide

SODIUM STARCH OCTENYLSUCCINATE

CAS Nos.: 52906-93-1; 70714-61-3

JPN Translation:
オクテニルコハク酸トウモロコシデンプン Na

Definition: Sodium Starch Octenylsuccinate is the sodium salt of the reaction product of octenylsuccinic anhydride with Corn Starch (q.v.).

Information Sources: JCIC, JCLS

Chemical Classes: Biological Polymers and their Derivatives; Carbohydrates

Functions: Absorbent; Emulsion Stabilizer; Viscosity Increasing Agent - Aqueous

Technical/Other Name:
Sodium Corn Starch Octenylsuccinate

Trade Names:
Amycol Nyuka (Nippon Starch)
Capsul (National Starch)
C* EmCap - Instant 12639 (Cerestar USA)
Cerestar EmCap 06375 (Cerestar USA)
Cerestar EmCap 06376 (Cerestar USA)
Cerestar EmCap 06377 (Cerestar USA)
Cerestar EmCap 12633 (Cerestar USA)
Cerestar EmCap 12634 (Cerestar USA)
Cerestar EmCap 12635 (Cerestar USA)
Cerestar EmCap Instant 12639 (Cerestar USA)
Cerestar EmTex 06369 (Cerestar USA)
Cerestar EmTex 12638 (Cerestar USA)
Nyuka (Nippon Starch)
Nyuka W (Nippon Starch)
Oil Q S (Nippon Starch)

SODIUM STEARATE

CAS No.	EINECS No.
822-16-2	212-490-5

JPN Translation:
ステアリン酸 Na

Empirical Formula:
$C_{18}H_{36}O_2 \cdot Na$

Definition: Sodium Stearate is the sodium salt of stearic acid that conforms generally to the formula:

$$CH_3(CH_2)_{16}COONa$$

Information Sources: 21CFR172.615, 21CFR172.863, 21CFR175.105, 21CFR175.300, 21CFR175.320, 21CFR176.170, 21CFR176.200, 21CFR176.210, 21CFR177.1200, 21CFR177.2260, 21CFR177.2600, 21CFR177.2800, 21CFR178.3910, 21CFR179.45, 21CFR181.22, 21CFR181.29, CIR: [S] JACT-1(2)1982, CTFA S, ITA, JCIC, JCLS, JSQI, MAR, MI-13(8753), NF XV, NF XIX, POR, SNPF, TSCA, USAN

Chemical Class: Soaps

Functions: Surfactant - Cleansing Agent; Surfactant - Emulsifying Agent; Viscosity Increasing Agent - Aqueous

Reported Product Categories: Deodorants (Underarm); Bath Soaps and Detergents; Hair Bleaches; Bath Preparations, Misc.; Body and Hand Preparations (Excluding Shaving Preparations); Bath Oils, Tablets, and Salts; Hair Lighteners with Color; Moisturizing Preparations; Colognes and Toilet Waters; Baby Lotions, Oils, Powders and Creams; Cleansing Products (Cold Creams, Cleansing Lotions, Liquids and Pads); Eyeliners; Shaving Cream (Aerosol, Brushless and Lather); Skin Care Preparations, Misc.

Technical/Other Names:
Octadecanoic Acid, Sodium Salt
Sodium Octadecanoate

Trade Names:
AEC Sodium Stearate (A & E Connock)
Jeechem Sodium Stearate (Jeen)
RTD Sodium Stearate OP-100 (RTD Chemicals)
RTD Sodium Stearate OP-200 (RTD Chemicals)
Unichem SS (Universal Preserv-A-Chem)

Trade Name Mixtures:
Cerealmilk Premium (Chemyunion)
Complexe Hydroxy-Salicylique Color (Coletica SA)
Customulse GMS SE (glyceryl stearate, SE) (Custom Ingredients)
Elastase Inhibitor-3 (Arval)
Emulgade CL Special (Cognis Deutschland)
Hydroxylide QNST (Coletica SA)

SODIUM STEARETH-4 PHOSPHATE

CAS Nos.: 57829-61-5; 87666-99-7

Definition: Sodium Steareth-4 Phosphate is the sodium salt of a complex mixture of phosphate esters of Steareth-4 (q.v.).

Chemical Class: Phosphorus Compounds

Function: Surfactant - Emulsifying Agent

SODIUM STEAROAMPHOACETATE

CAS No.	EINECS No.
30473-39-3	250-215-0

Empirical Formula:
$C_{24}H_{48}N_2O_4 \cdot Na$

Definition: Sodium Stearoamphoacetate is the amphoteric organic compound that conforms generally to the formula:

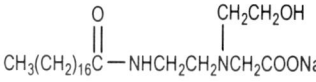

Chemical Class: Alkylamido Alkylamines

Functions: Hair Conditioning Agent; Surfactant - Cleansing Agent; Surfactant - Foam Booster

Technical/Other Names:
Stearoamphoacetate
Stearoamphoglycinate

SODIUM STEAROAMPHOHYDROXY-PROPYLSULFONATE

Empirical Formula:
$C_{25}H_{52}N_2O_6S \cdot Na$

Definition: Sodium Stearoamphohydroxypropylsulfonate is the amphoteric organic compound that conforms generally to the formula:

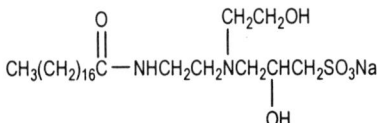

Chemical Classes: Alkylamido Alkylamines; Sulfonic Acids

Functions: Hair Conditioning Agent; Surfactant - Cleansing Agent; Surfactant - Foam Booster; Surfactant - Hydrotrope

Technical/Other Names:
Stearamphopropylsulfonate
Stearoamphohydroxypropylsulfonate

SODIUM STEAROAMPHOPROPIONATE

Empirical Formula:
$C_{25}H_{50}N_2O_4 \cdot Na$

Definition: Sodium Stearoamphopropionate is the amphoteric organic compound that conforms generally to the formula:

$$CH_3(CH_2)_{16}C \overset{O}{\overset{\|}{-}} NHCH_2CH_2NCH_2CH_2COONa$$
$$\overset{|}{CH_2CH_2OH}$$

Chemical Class: Alkylamido Alkylamines

Functions: Hair Conditioning Agent; Surfactant - Cleansing Agent; Surfactant - Foam Booster

SODIUM STEAROXY PG-HYDROXY-ETHYLCELLULOSE SULFONATE

Definition: Sodium Stearoxy PG-Hydroxy-ethylcellulose Sulfonate is the reaction product of Hydroxyethylcellulose (q.v.) with sodium 3-chloro-2-hydroxypropanesulfonate and stearyl glycidyl ether.

Chemical Classes: Gums, Hydrophilic Colloids and Derivatives; Organic Salts

Function: Viscosity Increasing Agent - Aqueous

SODIUM STEAROYL CASEIN

Definition: Sodium Stearoyl Casein is the sodium salt of the condensation product of stearic acid chloride and Casein (q.v.).

Chemical Class: Protein Derivatives

Functions: Hair Conditioning Agent; Skin-Conditioning Agent - Miscellaneous; Surfactant - Cleansing Agent

Trade Name:
Lifidrem CAST (Coletica SA)

SODIUM STEAROYL CHONDROITIN SULFATE

Definition: Sodium Stearoyl Chondroitin Sulfate is the condensation product of stearic acid chloride and Sodium Chondroitin Sulfate (q.v.).

Chemical Classes: Biological Polymers and their Derivatives; Carbohydrates; Sulfuric Acid Esters

Functions: Hair Conditioning Agent; Skin-Conditioning Agent - Miscellaneous

Trade Name:
Lifidrem CMST (Coletica SA)

SODIUM STEAROYL DNA

Definition: Sodium Stearoyl DNA is the sodium salt of the condensation product of stearic acid chloride and DNA (q.v.).

Chemical Classes: Biological Polymers and their Derivatives; Phosphorus Compounds

Function: Skin-Conditioning Agent - Miscellaneous

Trade Name:
Lifidrem DNST (Coletica SA)

SODIUM STEAROYL GLUTAMATE

CAS Nos.
38517-23-6
79811-24-8 (L-Form)

EINECS No.
253-980-9

JPN Translation:
ステアロイルグルタミン酸 Na

Empirical Formula:
$C_{23}H_{43}NO_5 \cdot Na$

Definition: Sodium Stearoyl Glutamate is the organic compound that conforms to the formula:

$$CH_3(CH_2)_{16}C \overset{O}{\overset{\|}{-}} NHCHCOONa$$
$$\overset{|}{CH_2CH_2COOH}$$

Information Sources: JCLS, JSCI, TSCA

Chemical Class: Amino Acids

Functions: Hair Conditioning Agent; Skin-Conditioning Agent - Miscellaneous; Surfactant - Cleansing Agent

Technical/Other Names:
L-Glutamic Acid, N-(1-Oxooctadecyl)-, Monosodium Salt
N-(1-Oxooctadecyl)-L-Glutamic Acid, Monosodium Salt
Sodium N-(1-Oxooctadecyl)-L-Glutamate
Sodium N-Stearoyl L-Glutamate

Trade Name:
Amisoft HS-11P (Ajinomoto)

Trade Name Mixture:
Amisoft GS-11P (Ajinomoto)

SODIUM STEAROYL HYALURONATE

Definition: Sodium Stearoyl Hyaluronate is the sodium salt of the condensation product of stearic acid chloride and Hyaluronic Acid (q.v.).

Chemical Classes: Biological Polymers and their Derivatives; Carbohydrates; Organic Salts

Functions: Hair Conditioning Agent; Skin-Conditioning Agent - Miscellaneous; Surfactant - Cleansing Agent

Trade Names:
Lifiderm AHST (Coletica SA)
Lifidrem HAST (Coletica SA)
Lifidrem HASV (Coletica SA)

SODIUM STEAROYL HYDROLYZED COLLAGEN

JPN Translation:
ステアロイル加水分解コラーゲン Na

Definition: Sodium Stearoyl Hydrolyzed Collagen is the sodium salt of the condensation product of stearic acid chloride with Hydrolyzed Collagen (q.v.).

Information Sources: JCIC, JCLS

Chemical Class: Protein Derivatives

Functions: Hair Conditioning Agent; Skin-Conditioning Agent - Miscellaneous; Surfactant - Cleansing Agent

Technical/Other Name:
Proteins, Hydrolysates, Reaction Products with Stearoyl Chloride, Compounds, with Sodium

Trade Names:
Lifidrem COST (Coletica SA)
Promois ESS (Seiwa Kasei)

Trade Name Mixture:
Lifidrem COSTLJVE (Coletica SA)

SODIUM STEAROYL HYDROLYZED CORN PROTEIN

Definition: Sodium Stearoyl Hydrolyzed Corn Protein is the sodium salt of the condensation product of stearic acid chloride and Hydrolyzed Corn Protein (q.v.).

Chemical Class: Protein Derivatives

Functions: Hair Conditioning Agent; Skin-Conditioning Agent - Miscellaneous; Surfactant - Cleansing Agent

Trade Name Mixture:
Lifidrem PVST (Coletica SA)

SODIUM STEAROYL HYDROLYZED SILK

Definition: Sodium Stearoyl Hydrolyzed Silk is the sodium salt of the condensation product of stearic acid chloride and Hydrolyzed Silk (q.v.).

Chemical Class: Protein Derivatives

Functions: Hair Conditioning Agent; Skin-Conditioning Agent - Miscellaneous; Surfactant - Cleansing Agent

Trade Name:
Lifidrem SOST (Coletica SA)

SODIUM STEAROYL HYDROLYZED SOY PROTEIN

Definition: Sodium Stearoyl Hydrolyzed Soy Protein is the sodium salt of the condensation product of stearic acid chloride and Hydrolyzed Soy Protein (q.v.).

Chemical Class: Protein Derivatives

Functions: Hair Conditioning Agent; Skin-Conditioning Agent - Miscellaneous; Surfactant - Cleansing Agent

Trade Name Mixture:
Lifidrem PVST (Coletica SA)

SODIUM STEAROYL HYDROLYZED SWEET ALMOND PROTEIN

Definition: Sodium Stearoyl Hydrolyzed Sweet Almond Protein is the sodium salt of the condensation product of stearic acid chloride and Hydrolyzed Sweet Almond Protein (q.v.).

Chemical Class: Protein Derivatives

Functions: Antistatic Agent; Skin-Conditioning Agent - Emollient; Surfactant - Emulsifying Agent

Trade Name:
Lifidrem AMST (Coletica SA)

SODIUM STEAROYL HYDROLYZED WHEAT PROTEIN

Definition: Sodium Stearoyl Hydrolyzed Wheat Protein is the sodium salt of the condensation product of stearic acid chloride and Hydrolyzed Wheat Protein (q.v.).

Chemical Class: Protein Derivatives

Functions: Hair Conditioning Agent; Skin-Conditioning Agent - Miscellaneous; Surfactant - Cleansing Agent

Technical/Other Name:
Proteins, Wheat, Hydrolysates, Reaction Products with Stearoyl Chloride, Compounds with Sodium

Trade Names:
Lifidrem BLST (Coletica SA)
Lifidrem VGST (Coletica SA)

SODIUM STEAROYL LACTALBUMIN

Definition: Sodium Stearoyl Lactalbumin is the sodium salt of the condensation product of stearic acid chloride with milk albumin.

Chemical Class: Proteins

Functions: Hair Conditioning Agent; Skin-Conditioning Agent - Miscellaneous; Surfactant - Cleansing Agent

Trade Name:
Lifidrem LAST (Coletica SA)

SODIUM STEAROYL LACTYLATE

CAS Nos.	EINECS Nos.
18200-72-1	242-090-6
25383-99-7	246-929-7

JPN Translation:
ステアロイル乳酸 Na

Empirical Formula:
$C_{24}H_{44}O_6 \cdot Na$

Definition: Sodium Stearoyl Lactylate is the sodium salt of the stearic acid ester of lactyl lactate. It conforms to the formula:

$$CH_3(CH_2)_{16}\overset{O}{\overset{\|}{C}}-O\overset{}{\underset{CH_3}{C}}H-O\overset{O}{\overset{\|}{C}}H\overset{}{\underset{CH_3}{C}}HCOONa$$

Information Sources: 21CFR172.846, 21CFR177.1200, FCC, JCIC, JCLS, TSCA

Chemical Classes: Esters; Organic Salts

Function: Surfactant - Emulsifying Agent

Reported Product Categories: Bath Capsules; Face and Neck Preparations (Excluding Shaving Preparations)

Technical/Other Names:
Octadecanoic Acid, 2-(1-Carboxyethoxy)-1-Methyl-2-Oxoethyl Ester, Sodium Salt

Sodium Stearoyl Lactate
Sodium 2-(Stearoyloxy) Propionate
Sodium Stearyl-2-Lactylate

Trade Names:
AEC Sodium Stearoyl Lactylate (A & E Connock)
Akoline SL (Karlshamns AB)
Dermofeel SL (Straetmans)
Pationic SSL (RITA)
PRIAZUL 2134 (Uniqema Europe)
Radiamuls 2990 (Oleon NV)
Rylo SL 18 (Danisco)

Trade Name Mixture:
Nikkol Nikkomulese 41 (Nikko)

SODIUM STEAROYL OAT PROTEIN

Definition: Sodium Stearoyl Oat Protein is the sodium salt of the condensation product of stearic acid chloride and Avena Sativa (Oat) Protein (q.v.).

Chemical Class: Protein Derivatives

Functions: Hair Conditioning Agent; Skin-Conditioning Agent - Miscellaneous; Surfactant - Cleansing Agent

Trade Name:
Lifidrem AVST (Coletica SA)

SODIUM STEAROYL PEA PROTEIN

Definition: Sodium Stearoyl Pea Protein is the sodium salt of the condensation product of stearic acid chloride and pea protein.

Chemical Class: Protein Derivatives

Functions: Hair Conditioning Agent; Skin-Conditioning Agent - Miscellaneous; Surfactant - Cleansing Agent

Trade Name:
Lifidrem PPST (Coletica SA)

SODIUM STEAROYL SOY PROTEIN

Definition: Sodium Stearoyl Soy Protein is the sodium salt of the condensation product stearic acid chloride and Glycine Soja (Soy) Protein (q.v.).

Chemical Class: Protein Derivatives

Functions: Hair Conditioning Agent; Skin-Conditioning Agent - Miscellaneous; Surfactant - Cleansing Agent

Trade Name:
Lifidrem SJST (Coletica SA)

SODIUM STEARYL DIMETHYL GLYCINE

JPN Translation:
ステアリルジメチルベタイン Na

Definition: Sodium Stearyl Dimethyl Glycine is the organic compound that conforms to the formula:

$$C_{18}H_{37} - \overset{\overset{\displaystyle CH_3}{|}}{\underset{\underset{\displaystyle CH_3}{|}}{N^+}} - CH_2COONa$$

Information Source: JCIC

Chemical Class: Amino Acids

Functions: Surfactant - Cleansing Agent; Surfactant - Emulsifying Agent

SODIUM STEARYL PHTHALAMATE

CAS No.: 86432-23-7

Empirical Formula:
$C_{26}H_{43}NO_3 \cdot Na$

Definition: Sodium Stearyl Phthalamate is the organic compound that conforms to the formula:

$$\text{(structure: benzene ring with } C(=O)-NH(CH_2)_{17}CH_3 \text{ and } COONa \text{ substituents)}$$

Chemical Classes: Amides; Organic Salts

Function: Surfactant - Emulsifying Agent

Technical/Other Names:
Benzoic Acid, 2-[(Octadecylamino)-Carbonyl]-, Monosodium Salt
Sodium 2-[(Octadecylamino)Carbonyl]-Benzoate

Trade Name:
STEPAN-MILD RM1 (Stepan)

SODIUM STEARYL SULFATE

CAS No.	EINECS No.
1120-04-3	214-295-0

JPN Translation:
ステアリル硫酸 Na

Empirical Formula:
$C_{18}H_{38}O_4S \cdot Na$

Definition: Sodium Stearyl Sulfate is the sodium salt of stearyl sulfate that conforms to the formula:

$$CH_3(CH_2)_{17}OSO_3Na$$

Information Sources: JCIC, JCLS

Chemical Class: Alkyl Sulfates

Functions: Surfactant - Cleansing Agent; Surfactant - Emulsifying Agent

Technical/Other Names:
Sodium Octadecyl Sulfate
Sulfuric Acid, Monooctadecyl Ester, Sodium Salt

Trade Name Mixture:
Texapon CS Paste (Cognis Deutschland)

SODIUM STYRENE/ACRYLATES COPOLYMER

CAS No.: 9010-92-8

Definition: Sodium Styrene/Acrylates Copolymer is the sodium salt of a polymer of styrene and a monomer consisting of acrylic acid, methacrylic acid or one of their simple esters.

Information Source: CIR: [SQ] IJT 21 (SUPPL. 3) 2002

Chemical Class: Synthetic Polymers

Functions: Film Former; Viscosity Increasing Agent - Aqueous

Trade Names:
Europacif 2146 (EOC Surfactants)
Lytron 180 (Omnova Solutions, Inc)
Lytron 653 (Rohm and Haas)

Trade Name Mixtures:
Isocell Care (Lucas Meyer)
Isocell Life (Lucas Meyer)
Lytron 631 (Rohm and Haas)

SODIUM STYRENE/ACRYLATES/DIVINYL-BENZENE COPOLYMER

Definition: Sodium Styrene/Acrylates/Divinylbenzene Copolymer is the sodium salt of a polymer of styrene, divinylbenzene and two or more monomers consisting of acrylic acid, methacrylic acid or their simple esters.

Chemical Class: Synthetic Polymers

Functions: Opacifying Agent; Oral Care Agent

Reported Product Category: Hair Dyes and Colors (All Types Requiring Caution Statements and Patch Tests)

Trade Name:
Lytron 170 (Omnova Solutions, Inc)

Trade Name Mixture:
Lytron 295 (Rohm and Haas)

SODIUM STYRENE/ACRYLATES/ETHYLHEXYL ACRYLATE/LAURYL ACRYLATE COPOLYMER

JPN Translation:
（スチレン / アクリル酸アルキル）コポリマーNa

Definition: Sodium Styrene/Acrylates/Ethylhexyl Acrylate/Lauryl Acrylate Copolymer is the sodium salt of Styrene/Acrylates/Ethylhexyl Acrylate/Lauryl Acrylate Copolymer (q.v.).

Information Source: JCLS

Chemical Class: Synthetic Polymers

Function: Film Former

SODIUM STYRENE/ACRYLATES/PEG-10 DIMALEATE COPOLYMER

Definition: Sodium Styrene/Acrylates/PEG-10 Dimaleate Copolymer is the sodium salt of a polymer of styrene, PEG-10 dimaleate and a monomer consisting of acrylic acid, methacrylic acid or one of their simple esters.

Chemical Class: Synthetic Polymers

Function: Opacifying Agent

Reported Product Category: Bath Soaps and Detergents

Technical/Other Name:
Sodium Styrene/Acrylate/PEG-10 Dimaleate Copolymer

Trade Name Mixture:
Lytron 300 (Rohm and Haas)

SODIUM STYRENE/PEG-10 MALEATE/NONOXYNOL-10 MALEATE/ACRYLATES COPOLYMER

Definition: Sodium Styrene/PEG-10 Maleate/Nonoxynol-10 Maleate/Acrylates Copolymer is the sodium salt of a polymer of styrene, PEG-10 maleate, nonoxynol-10 maleate and a monomer consisting of acrylic acid, methacrylic acid or one of their simple esters.

Chemical Class: Synthetic Polymers

Function: Opacifying Agent

Technical/Other Name:
Sodium Styrene/PEG-10 Maleate/Nonoxynol-10 Maleate/Acrylate Copolymer

Trade Name Mixture:
Lytron 305 (Rohm and Haas)

SODIUM SUCCINATE

CAS No.	EINECS No.
2922-54-5	220-871-2

JPN Translation:
コハク酸 Na

Empirical Formula:
$C_4H_5O_4 \cdot Na$

Definition: Sodium Succinate is the sodium salt of succinic acid. It conforms to the formula:

$$HOOCCH_2CH_2COONa$$

Chemical Class: Organic Salts

Functions: Buffering Agent; pH Adjuster

Technical/Other Names:
Butanedioic Acid, Monosodium Salt
Monosodium Succinate

Trade Name Mixture:
Cellryel (Bio Dell)

SODIUM SUCCINOYL GELATIN

Definition: Sodium Succinoyl Gelatin is the sodium salt of the condensation product of succinic acid anhhydride with gelatin.

Chemical Classes: Organic Salts; Protein Derivatives

Function: Skin-Conditioning Agent - Miscellaneous

Trade Name:
Succinated Gelatin (DGF Stoess AG)

SODIUM SUCROSE OCTASULFATE

CAS No.: 74135-10-7

Definition: Sodium Sucrose Octasulfate is the sodium salt of the octaester of sulfuric acid and Sucrose (q.v.).

Chemical Classes: Carbohydrates; Organic Salts; Sulfuric Acid Esters

Function: Skin-Conditioning Agent - Miscellaneous

Technical/Other Name:
α-D-Glucopyranoside, 1,3,4,6-Tetra-O-Sulfo-β-D-Fructofuranosyl, tetrakis (Hydrogen Sulfate), Octasodium Salt

Trade Name:
Sucrose Sulfate (Egalet)

SODIUM SULFANILATE

CAS No.	**EINECS No.**
515-74-2	208-208-5

Empirical Formula:
$C_6H_7NO_3S \cdot Na$

Definition: Sodium Sulfanilate is the sodium salt of sulfanilic acid that conforms to the formula:

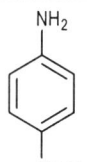

Information Sources: MI-13(9012), TSCA

Chemical Classes: Amines; Sulfonic Acids

Function: Hair Colorant

Technical/Other Names:
4-Aminobenzenesulfonic Acid, Monosodium Salt
Benzenesulfonic Acid, 4-Amino-, Monosodium Salt
Sodium Aniline Sulfonate
Sulfanilic Acid, Sodium Salt

SODIUM SULFATE

CAS Nos.	**EINECS No.**
7727-73-3 (Decahydrate)	
7757-82-6	231-820-9

JPN Translation:
硫酸 Na

Empirical Formula:
$H_2O_4S \cdot 2Na$

Definition: Sodium Sulfate is the inorganic salt that conforms to the formula:

$$Na_2SO_4$$

Information Sources: ARG, AUS, BP, BPC, BRA, 21CFR172.615, 21CFR172.886, 21CFR173.310, 21CFR177.1200, 21CFR186.1797, CIR: [SQ] IJT-19 (SUPPL.1)2000, CTFA S, CZE, DA, DDR, EGY, EP, FCC, FIN, HP, HUN, IND, ITA, JAN, JCIC, JCLS, JSCI, JSQI, KOR, MAR, MEX, MI-13(8755), PF, PN, POR, ROM, TSCA, USAN, USD, USP XXIV, YUG

Chemical Class: Inorganic Salts

Function: Viscosity Increasing Agent - Aqueous

Reported Product Categories: Bubble Baths; Bath Soaps and Detergents; Moisturizing Preparations; Bath Oils, Tablets, and Salts; Bath Preparations, Misc.; Body and Hand Preparations (Excluding Shaving Preparations); Cleansing Products (Cold Creams, Cleansing Lotions, Liquids and Pads); Shampoos (Non-coloring); Skin Care Preparations, Misc.

Technical/Other Names:
Disodium Sulfate
Exsiccated Sodium Sulfate
Sodium Sulfate, Anhydrous
Sulfuric Acid, Disodium Salt

Trade Name Mixtures:
Algae Extract Powder-S (Maruzen Pharmaceuticals Co., Ltd.)
Aloe Extract (Kidachi) Powder-S (Maruzen Pharmaceuticals Co., Ltd.)
Aloe Extract Powder-S (Maruzen Pharmaceuticals Co., Ltd.)
Apricot Kernel Extract Powder-S (Maruzen Pharmaceuticals Co., Ltd.)
Artemisia Capillaris Extract Powder-S (Maruzen Pharmaceuticals Co., Ltd.)
Balm Mint Extract Powder-S (Maruzen Pharmaceuticals Co., Ltd.)
Bathgranue Aloe Vera (Ichimaru Pharcos)
Bathgranue Benibana (Ichimaru Pharcos)
Bathgranue Biwa Leaf (Ichimaru Pharcos)
Bathgranue Chinpi (Ichimaru Pharcos)
Bathgranue Eucariptus (Ichimaru Pharcos)
Bathgranue Garlic (Ichimaru Pharcos)
Bathgranue Hetima (Ichimaru Pharcos)
Bathgranue Hibamata (Ichimaru Pharcos)
Bathgranue Kakkon (Ichimaru Pharcos)
Bathgranue Kamitsure (Ichimaru Pharcos)
Bathgranue Oubaku (Ichimaru Pharcos)
Bathgranue Ouren (Ichimaru Pharcos)
Bathgranue Peach Leaf (Ichimaru Pharcos)
Bathgranue Peppermint (Ichimaru Pharcos)
Bathgranue Senkyu (Ichimaru Pharcos)
Bathgranue Shakuyaku (Ichimaru Pharcos)
Bathgranue Shiso (Ichimaru Pharcos)
Bathgranue Shoubu (Ichimaru Pharcos)
Bathgranue Silk (Ichimaru Pharcos)
Bathgranue Suikazura (Ichimaru Pharcos)
Bathgranue Touhi (Ichimaru Pharcos)
Bathgranue Touki (Ichimaru Pharcos)
Bathgranue Yomogi (Ichimaru Pharcos)
Beefsteak Plant Extract Powder-S (Maruzen Pharmaceuticals Co., Ltd.)
Birch Bark Extract Powder-S (Maruzen Pharmaceuticals Co., Ltd.)
Bitter Orange Peel Extract Powder-S (Maruzen Pharmaceuticals Co., Ltd.)
Chamomile Extract Powder-S (Maruzen Pharmaceuticals Co., Ltd.)
Citrus Unshiu Extract Powder-S (Maruzen Pharmaceuticals Co., Ltd.)
Cnidium Extract Powder-S (Maruzen Pharmaceuticals Co., Ltd.)
Coix Seed Extract Powder-S (Maruzen Pharmaceuticals Co., Ltd.)
Fennel Extract Powder-S (Maruzen Pharmaceuticals Co., Ltd.)
Ganoderma Extract Powder-S (Maruzen Pharmaceuticals Co., Ltd.)
Gentian Extract Powder-S (Maruzen Pharmaceuticals Co., Ltd.)
Ginger Extract Powder-S (Maruzen Pharmaceuticals Co., Ltd.)
Ginseng Extract Powder-S (Maruzen Pharmaceuticals Co., Ltd.)
Honeysuckle Extract Powder-S (Maruzen Pharmaceuticals Co., Ltd.)
Houttuynia Extract Powder-S (Maruzen Pharmaceuticals Co., Ltd.)

Japanese Angelica Extract Powder-S
(Maruzen Pharmaceuticals Co., Ltd.)
Linden Extract Powder-S (Maruzen
Pharmaceuticals Co., Ltd.)
Loquat Leaf Powder-S (Maruzen
Pharmaceuticals Co., Ltd.)
Luffa Extract Powder-S (Maruzen
Pharmaceuticals Co., Ltd.)
Mugwort Extract Powder-S (Maruzen
Pharmaceuticals Co., Ltd.)
Mulberry Extract Powder-S (Maruzen
Pharmaceuticals Co., Ltd.)
Peach Leaf Extract Powder-S (Maruzen
Pharmaceuticals Co., Ltd.)
Peony Root Extract Powder-S (Maruzen
Pharmaceuticals Co., Ltd.)
Peppermint Extract Powder-S (Maruzen
Pharmaceuticals Co., Ltd.)
Phellodendron Extract Powder-S (Maruzen
Pharmaceuticals Co., Ltd.)
Rehmannia Extract Powder-S (Maruzen
Pharmaceuticals Co., Ltd.)
Royal Jelly Extract Powder-S (Maruzen
Pharmaceuticals Co., Ltd.)
Sage Extract Powder-S (Maruzen
Pharmaceuticals Co., Ltd.)
Sasa Veitchii Extract Powder-S (Maruzen
Pharmaceuticals Co., Ltd.)
Sweet Flag Extract Powder-S (Maruzen
Pharmaceuticals Co., Ltd.)
Turmeric Extract Powder-S (Maruzen
Pharmaceuticals Co., Ltd.)
Wild Rose Extract Powder-S (Maruzen
Pharmaceuticals Co., Ltd.)
Yuzu Extract Powder-S (Maruzen
Pharmaceuticals Co., Ltd.)

SODIUM SULFIDE

CAS No.	EINECS No.
1313-82-2	215-211-5

Empirical Formula:
Na_2S

Definition: Sodium Sulfide is the inorganic salt that conforms to the formula:

$$Na_2S$$

Information Sources: EEC(III/1-23), MI-13 (8756), TSCA

Chemical Class: Inorganic Salts

Function: Depilating Agent

SODIUM SULFITE

CAS No.	EINECS No.
7757-83-7	231-821-4

JPN Translation:
亜硫酸 Na

Empirical Formula:
$H_2O_3S \cdot 2Na$

Definition: Sodium Sulfite is the inorganic salt that conforms to the formula:

$$Na_2SO_3$$

Information Sources: BPC, BRA, 21CFR173.310, 21CFR177.1200, 21CFR182.3798, CIR: [S], CTFA S, EEC(VI/1-9), FCC, HP, JCLS, JSCI, MAR, MI-13 (8757), TSCA

Chemical Class: Inorganic Salts

Functions: Antioxidant; Hair-Waving/Straightening Agent; Reducing Agent

Reported Product Categories: Hair Dyes and Colors (All Types Requiring Caution Statements and Patch Tests); Hair Tints; Shampoos (Non-coloring); Hair Coloring Preparations, Misc.; Personal Cleanliness Products, Misc.; Skin Care Preparations, Misc.

Technical/Other Names:
Anhydrous Sodium Sulfite
Sulfurous Acid, Disodium Salt

Trade Name Mixtures:
Phenolines The Vert (Coletica SA)
Phytolight BG (Coletica SA)

SODIUM SUNFLOWERAMIDOPROPYL PG-DIMONIUM CHLORIDE PHOSPHATE

Definition: Sodium Sunfloweramidopropyl PG-Dimonium Chloride Phosphate is the quaternary ammonium salt that conforms generally to the formula:

where RCO- represents the fatty acids derived from sunflower seed oil.

Chemical Class: Quaternary Ammonium Compounds

Functions: Antistatic Agent; Hair Conditioning Agent

SODIUM SUNFLOWERSEED-AMPHOACETATE

Definition: Sodium Sunflowerseed-amphoacetate is the amphoteric organic compound that conforms generally to the formula:

where RCO- represents the fatty acids derived from sunflower seed oil.

Chemical Class: Alkylamido Alkylamines

Functions: Hair Conditioning Agent; Surfactant - Cleansing Agent; Surfactant - Foam Booster

Trade Name:
Vamasoft Sunfower (Vama Farmacosmetica)

Trade Name Mixtures:
Dermocare S 35 (Fratelli Ricci)
Softinat (Vama Farmacosmetica)

SODIUM SURFACTIN

Definition: Sodium Surfactin is a peptidelipid composed of amino acids and fatty acids and is produced by the fermentation of *Bacillus subtilis*.

Chemical Class: Biological Products

Functions: Surfactant - Cleansing Agent; Surfactant - Emulsifying Agent; Surfactant - Solubilizing Agent; Surfactant - Suspending Agent

Technical/Other Name:
Extract of Bacillus Ferment

Trade Name:
Aminofect (Showa Denko)

SODIUM SWEETALMOND-AMPHOACETATE

Definition: Sodium Sweetalmond-amphoacetate is the amphoteric organic compound that conforms generally to the formula:

where RCO- represents the fatty acids derived from sweet almond oil.

Chemical Class: Alkylamido Alkylamines

Functions: Hair Conditioning Agent; Surfactant - Cleansing Agent; Surfactant - Foam Booster

Trade Name:
Vamasoft Sweet Almond (Vama Farmacosmetica)

Trade Name Mixture:
Dermocare SA 35 (Fratelli Ricci)

SODIUM TALLAMPHOPROPIONATE

CAS No.: 68991-88-8

Definition: Sodium Tallamphopropionate is the amphoteric organic compound that conforms generally to the formula:

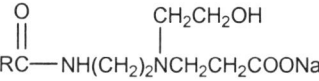

where RCO- represents the fatty acids derived from tall oil.

Information Sources: JCLS, TSCA

Chemical Class: Alkylamido Alkylamines

Functions: Hair Conditioning Agent; Surfactant - Cleansing Agent; Surfactant - Foam Booster

SODIUM TALLATE

CAS No. 61790-45-2   **EINECS No.** 263-137-7

Definition: Sodium Tallate is the sodium salt of Tall Oil Acid (q.v.).

Chemical Class: Soaps

Functions: Surfactant - Cleansing Agent; Surfactant - Emulsifying Agent

SODIUM TALLOWAMPHOACETATE

CAS No.: 124046-48-6

Definition: Sodium Tallowamphoacetate is the amphoteric organic compound that conforms generally to the formula:

$$\underset{RC}{\overset{O}{\|}}\text{—NH(CH}_2)_2\text{NCH}_2\text{COONa}$$
with $\text{CH}_2\text{CH}_2\text{OH}$ branch

where RCO- represents the fatty acids derived from tallow.

Chemical Class: Alkylamido Alkylamines

Functions: Hair Conditioning Agent; Surfactant - Cleansing Agent; Surfactant - Foam Booster

Technical/Other Names:
Glycine, N-2-Aminoethyl)-N-(2-Hydroxyethyl)-, N-Tallow Acyl Derivs., Monosodium Salts
Tallowamphoacetate
Tallowamphoglycinate

SODIUM TALLOWATE

CAS No. 8052-48-0   **EINECS No.** 232-491-4

Definition: Sodium Tallowate is the sodium salt of Tallow Acid (q.v.).

Information Sources: 21CFR175.105, 21CFR175.320, 21CFR176.170, 21CFR176.200, 21CFR177.2600, 21CFR177.2800, 21CFR178.3910, TSCA

Chemical Class: Soaps

Functions: Surfactant - Cleansing Agent; Surfactant - Foam Booster; Viscosity Increasing Agent - Aqueous

Reported Product Categories: Bath Soaps and Detergents; Bath Oils, Tablets, and Salts; Cleansing Products (Cold Creams, Cleansing Lotions, Liquids and Pads); Personal Cleanliness Products, Misc.

Technical/Other Name:
Tallow, Sodium Salt

Trade Name:
Norfox XXX Granules (Norman, Fox & Co.)

SODIUM TALLOW SULFATE

CAS Nos. 8052-50-4 68140-10-3   **EINECS Nos.** 232-494-0 268-773-9

Definition: Sodium Tallow Sulfate is a mixture of sodium alkyl sulfates that conforms generally to the formula:

$$ROSO_3Na$$

where R represents the alkyl groups derived from tallow.

Information Sources: 21CFR175.105, 21CFR176.170, 21CFR176.210, 21CFR177.2800

Chemical Class: Alkyl Sulfates

Function: Surfactant - Cleansing Agent

Technical/Other Names:
Sodium Tallow Alcohol Sulfate
Sulfuric Acid, Monotallow Alkyl Esters, Sodium Salts
Tallow, Sulfated, Sodium Salt

SODIUM TAURATE

Empirical Formula:
$C_2H_7NSO_3Na$

Definition: Sodium Taurate is the organic salt that conforms to the formula:

$$NH_2CH_2CH_2SO_3Na$$

Chemical Class: Sulfonic Acids

Functions: Surfactant - Cleansing Agent; Surfactant - Foam Booster

Technical/Other Name:
2-Aminoethanesulfonic Acid, Sodium Salt

SODIUM TAURIDE ACRYLATES/ACRYLIC ACID/ACRYLONITROGENS COPOLYMER

Definition: Sodium Tauride Acrylates/Acrylic Acid/Acrylonitrogens Copolymer is the polymer formed by the controlled hydrolysis of polyacrylonitrile in the presence of sodium taurate.

Chemical Class: Synthetic Polymers

Functions: Film Former; Viscosity Increasing Agent - Aqueous

SODIUM TAURINE COCOYL METHYLTAURATE

Definition: Sodium Taurine Cocoyl Methyltaurate is the organic salt that conforms to the formula:

$$\underset{RC}{\overset{O}{\|}}\text{—N(CH}_2)_2\text{SO}_3\text{H} \cdot \text{NH}_2\text{CH}_2\text{SO}_3\text{Na}$$
with CH_3 branch

where RCO represents the cocoyl group.

Chemical Classes: Amides; Sulfonic Acids

Functions: Surfactant - Cleansing Agent; Surfactant - Emulsifying Agent

Technical/Other Name:
Amides, Coconut Oil, with N-Methyltaurine, Sodium Taurine Salt

Trade Name:
Diapon K-0T2 (NOF)

SODIUM TAURINE LAURATE

Definition: Sodium Taurine Laurate is the organic salt that conforms to the formula:

$$C_{11}H_{23}COOH \cdot NH_2CH_2CH_2SO_3Na$$

Chemical Classes: Organic Salts; Sulfonic Acids

Function: Surfactant - Cleansing Agent

Technical/Other Name:
Dodecanoic Acid, 2-Aminoethanesulfonic, Sodium Salt

Trade Name:
LT-2 (NOF)

SODIUM/TEA-LAUROYL COLLAGEN AMINO ACIDS

Definition: Sodium/TEA-Lauroyl Collagen Amino Acids is a mixture of sodium and triethanolamine salts of the condensation product of lauric acid chloride and Collagen Amino Acids (q.v.).

Chemical Class: Amino Acids

Functions: Hair Conditioning Agent; Surfactant - Cleansing Agent

Technical/Other Name:
Proteins, Hydrolysates, Reaction Products with Lauroyl Chloride, Compds. with Sodium and Triethanolamine

Trade Name:
Proteol LCO (SEPPIC)

SODIUM/TEA-LAUROYL HYDROLYZED COLLAGEN

Definition: Sodium/TEA-Lauroyl Hydrolyzed Collagen is a mixed sodium and triethanolamine salt of the condensation product of lauric acid chloride and Hydrolyzed Collagen (q.v.).

Chemical Class: Protein Derivatives

Functions: Hair Conditioning Agent; Skin-Conditioning Agent - Miscellaneous; Surfactant - Cleansing Agent

Technical/Other Name:
Sodium/TEA-Lauroyl Hydrolyzed Animal Protein

Trade Name:
Foam-Keratin LK (Arch Personal Care Products)

SODIUM/TEA-LAUROYL HYDROLYZED KERATIN

Definition: Sodium/TEA-Lauroyl Hydrolyzed Keratin is a mixture of sodium and triethanolamine salts of the condensation product of lauric acid chloride and Hydrolyzed Keratin (q.v.).

Chemical Class: Protein Derivatives

Functions: Hair Conditioning Agent; Skin-Conditioning Agent - Miscellaneous; Surfactant - Cleansing Agent

Trade Name:
May-Tein KTS (Maybrook)

SODIUM/TEA-LAUROYL KERATIN AMINO ACIDS

Definition: Sodium/TEA-Lauroyl Keratin Amino Acids is a mixture of sodium and triethanolamine salts of the condensation product of lauric acid chloride and Keratin Amino Acids (q.v.).

Chemical Class: Amino Acids

Functions: Hair Conditioning Agent; Surfactant - Cleansing Agent

SODIUM/TEA-UNDECYLENOYL ALGINATE

Definition: Sodium/TEA-Undecylenoyl Alginate is the mixed sodium and triethanolamine salt of the condensation product of undecylenic acid chloride and Alginic Acid (q.v.).

Chemical Classes: Gums, Hydrophilic Colloids and Derivatives; Organic Salts

Functions: Emulsion Stabilizer; Hair Conditioning Agent; Skin-Conditioning Agent - Miscellaneous

Trade Name:
Lifidrem XLUN (Coletica SA)

SODIUM/TEA-UNDECYLENOYL CARRAGEENAN

Definition: Sodium/TEA-Undecylenoyl Carrageenan is the mixed sodium and triethanolamine salt of the condensation product of undecylenic acid chloride and Carrageenan (q.v.).

Chemical Classes: Gums, Hydrophilic Colloids and Derivatives; Organic Salts

Functions: Emulsion Stabilizer; Hair Conditioning Agent; Skin-Conditioning Agent - Miscellaneous

Trade Name:
Lifidrem ARUN (Coletica SA)

SODIUM/TEA-UNDECYLENOYL COLLAGEN AMINO ACIDS

Definition: Sodium/TEA-Undecylenoyl Collagen Amino Acids is a mixture of sodium and triethanolamine salts of the condensation product of undecylenic acid chloride and Collagen Amino Acids (q.v.).

Chemical Class: Amino Acids

Functions: Hair Conditioning Agent; Surfactant - Cleansing Agent

SODIUM/TEA-UNDECYLENOYL HYDROLYZED COLLAGEN

Definition: Sodium/TEA-Undecylenoyl Hydrolyzed Collagen is a mixed sodium and triethanolamine salt of the condensation product of undecylenic acid chloride and Hydrolyzed Collagen (q.v.).

Chemical Class: Protein Derivatives

Functions: Hair Conditioning Agent; Skin-Conditioning Agent - Miscellaneous; Surfactant - Cleansing Agent

Technical/Other Name:
Sodium/TEA-Undecylenoyl Hydrolyzed Animal Protein

SODIUM/TEA-UNDECYLENOYL HYDROLYZED CORN PROTEIN

Definition: Sodium/TEA-Undecylenoyl Hydrolyzed Corn Protein is the mixed sodium and triethanolamine salt of the condensation product of undecylenic acid chloride and Hydrolyzed Corn Protein (q.v.).

Chemical Class: Protein Derivatives

Functions: Hair Conditioning Agent; Skin-Conditioning Agent - Miscellaneous; Surfactant - Cleansing Agent

Trade Name Mixture:
Lifidrem PVUN (Coletica SA)

SODIUM/TEA-UNDECYLENOYL HYDROLYZED SOY PROTEIN

Definition: Sodium/TEA-Undecylenoyl Hydrolyzed Soy Protein is the mixed sodium and triethanolamine salt of the condensation product of undecylenic acid chloride and Hydrolyzed Soy Protein (q.v.).

Chemical Class: Protein Derivatives

Functions: Hair Conditioning Agent; Skin-Conditioning Agent - Miscellaneous; Surfactant - Cleansing Agent

Trade Name Mixture:
Lifidrem PVUN (Coletica SA)

SODIUM/TEA-UNDECYLENOYL HYDROLYZED WHEAT PROTEIN

Definition: Sodium/TEA-Undecylenoyl Hydrolyzed Wheat Protein is the mixed sodium and triethanolamine salt of the condensation product of undecylenic acid chloride and Hydrolyzed Wheat Protein (q.v.).

Chemical Class: Protein Derivatives

Functions: Hair Conditioning Agent; Skin-Conditioning Agent - Miscellaneous; Surfactant - Cleansing Agent

Trade Name:
Lifidrem BLUN (Coletica SA)

SODIUM THIOCYANATE

CAS No.	EINECS No.
540-72-7	208-754-4

Empirical Formula:
CHNS • Na

Definition: Sodium Thiocyanate is the inorganic salt that conforms to the formula:

NaSCN

Information Sources: MI-13(9401), TSCA

Chemical Class: Inorganic Salts

Function: Not Reported

Technical/Other Names:
Sodium Rhodanate
Sodium Thiocyanide
Thiocyanic Acid, Sodium Salt

SODIUM THIOGLYCOLATE

CAS No. 367-51-1 **EINECS No.** 206-696-4

Empirical Formula:
$C_2H_4O_2S \cdot Na$

Definition: Sodium Thioglycolate is the organic salt that conforms to the formula:

$HSCH_2COONa$

Information Sources: EEC(III/1-2a), MI-13 (8767), TSCA

Chemical Classes: Organic Salts; Thio Compounds

Functions: Antioxidant; Depilating Agent; Hair-Waving/Straightening Agent; Reducing Agent

Reported Product Category: Depilatories

Technical/Other Names:
Mercaptoacetic Acid, Sodium Salt
Sodium 2-Mercaptoethanoate
Sodium Thioglycollate

SODIUM THIOSULFATE

CAS Nos. 7772-98-7 10102-17-7 **EINECS No.** 231-867-5

JPN Translation:
チオ硫酸 Na

Empirical Formula:
$Na_2S_2O_3$

Definition: Sodium Thiosulfate is the inorganic salt that conforms to the formula:

$Na_2S_2O_3$

Information Sources: JAN, JCIC, JCLS, JSCI, MI-13(8769), TSCA, USAN, USP XXIV

Chemical Class: Inorganic Salts

Function: Not Reported

Technical/Other Names:
Sodium Hyposulfite

Sodium Oxide Sulfide
Sodium Thiosulfate, Anhydrous
Thiosulfuric Acid, Disodium Salt

Trade Name:
Sodium Thiosulfate Pentahydrate (William Blythe)

SODIUM TOCOPHERYL PHOSPHATE

Definition: Sodium Tocopheryl Phosphate is the sodium salt of a complex mixture of esters of phosphoric acid and Tocopherol (q.v.).

Chemical Classes: Heterocyclic Compounds; Phosphorus Compounds

Functions: Antioxidant; Emulsion Stabilizer; Reducing Agent; Skin-Conditioning Agent - Miscellaneous; Surfactant - Emulsifying Agent; Viscosity Increasing Agent - Aqueous

Trade Name:
Sodium Vitamin E Phosphate (Showa Denko)

SODIUM TOLUENESULFONATE

CAS Nos. 657-84-1 12068-03-0 **EINECS Nos.** 211-522-5 235-088-1

Empirical Formula:
$C_7H_8O_3S \cdot Na$

Definition: Sodium Toluenesulfonate is the substituted aromatic compound that conforms generally to the formula:

Information Sources: 21CFR178.1010, TSCA

Chemical Class: Alkyl Aryl Sulfonates

Function: Surfactant - Hydrotrope

Reported Product Category: Cleansing Products (Cold Creams, Cleansing Lotions, Liquids and Pads)

Technical/Other Names:
Benzenesulfonic Acid, Methyl-, Sodium Salt
Methylbenzenesulfonic Acid, Sodium Salt
Sodium p-Tolyl Sulfonate
Sodium Tosylate

Trade Names:
Eltesol ST 40 (Albright & Wilson UK)
Eltesol ST90 (Albright & Wilson UK)

SODIUM TREHALOSE SULFATE

JPN Translation:
トレハロース硫酸 Na

Definition: Sodium Trehalose Sulfate is the sodium salt of sulfated Trehalose (q.v.).

Information Source: JCLS

Chemical Classes: Carbohydrates; Sulfonic Acids

Function: Skin-Conditioning Agent - Humectant

Trade Name:
Trehalose S (Nihon Shokuhin Kako)

SODIUM TRIDECETH-3 CARBOXYLATE

CAS Nos.: 61757-59-3 (Generic); 68891-17-8 (Generic)

JPN Translation:
トリデセス - 3 カルボン酸 Na

Empirical Formula:
$C_{19}H_{38}O_5 \cdot Na$

Definition: Sodium Trideceth-3 Carboxylate is the sodium salt of Trideceth-3 Carboxylic Acid (q.v.).

Information Source: JCLS

Chemical Class: Organic Salts

Function: Surfactant - Cleansing Agent

Technical/Other Names:
PEG-3 Tridecyl Ether Carboxylic Acid, Sodium Salt
Polyethylene Glycol (3) Tridecyl Ether Carboxylic Acid, Sodium Salt
Polyoxyethylene (3) Tridecyl Ether Carboxylic Acid, Sodium Salt
Sodium Polyethylene Glycol (3) Tridecyl Ether Carboxylate
Sodium Polyoxyethylene (3) Tridecyl Ether Carboxylate

Trade Name Mixture:
Toshiki TWB-6033 (Nikko)

SODIUM TRIDECETH-4 CARBOXYLATE

CAS Nos.: 61757-59-3 (Generic); 68891-17-8 (Generic)

Definition: Sodium Trideceth-4 Carboxylate is the sodium salt of Trideceth-4 Carboxylic Acid (q.v.).

Chemical Class: Organic Salts

Function: Surfactant - Cleansing Agent

Technical/Other Names:
Sodium Polyethylene Glycol 200 Tridecyl Ether Carboxylate

Sodium Polyoxyethylene (4) Tridecyl Ether Carboxylate

Trade Names:
Nikkol ECTD-3NEX (Nikko)
Nikkol ECT-3NEX (Nikko)

SODIUM TRIDECETH-6 CARBOXYLATE

CAS Nos.: 61757-59-3 (Generic); 68891-17-8 (Generic)

JPN Translation:
トリデセス - 6 カルボン酸 Na

Empirical Formula:
$C_{25}H_{50}O_8$ • Na

Definition: Sodium Trideceth-6 Carboxylate is the sodium salt of Trideceth-6 Carboxylic Acid (q.v.).

Information Source: JCLS

Chemical Class: Organic Salts

Function: Surfactant - Cleansing Agent

Technical/Other Names:
PEG-6 Tridecyl Ether Carboxylic Acid, Sodium Salt
Polyethylene Glycol 300 Tridecyl Ether Carboxylic Acid, Sodium Salt
Polyoxyethylene (6) Tridecyl Ether Carboxylic Acid, Sodium Salt
Sodium Polyethylene Glycol 300 Tridecyl Ether Carboxylate
Sodium Polyoxyethylene (6) Tridecyl Ether Carboxylate

SODIUM TRIDECETH-7 CARBOXYLATE

CAS Nos.: 61757-59-3 (Generic); 68891-17-8 (Generic)

Empirical Formula:
$C_{27}H_{54}O_9$ • Na

Definition: Sodium Trideceth-7 Carboxylate is the sodium salt of Trideceth-7 Carboxylic Acid (q.v.).

Information Sources: JCLS, JSQI

Chemical Class: Organic Salts

Function: Surfactant - Cleansing Agent

Reported Product Category: Shampoos (Non-coloring)

Technical/Other Names:
PEG-7 Tridecyl Ether Carboxylic Acid, Sodium Salt
Polyethylene Glycol (7) Tridecyl Ether Carboxylic Acid, Sodium Salt
Polyoxyethylene (7) Tridecyl Ether Carboxylic Acid, Sodium Salt

Trade Names:
Nikkol ECTD-6NEX (Nikko)

Sandopan DTC (Clariant)
Sandopan DTC (Clariant GmbH, Personal Care)

SODIUM TRIDECETH-8 CARBOXYLATE

CAS Nos.: 61757-59-3 (Generic); 68891-17-8 (Generic)

Empirical Formula:
$C_{29}H_{58}O_{10}$ • Na

Definition: Sodium Trideceth-8 Carboxylate is the sodium salt of the carboxylic acid derived from Trideceth-8 (q.v.). It conforms generally to the formula:

$$CH_3(CH_2)_{12}(OCH_2CH_2)_nOCH_2COONa$$

Information Source: JCLS

Chemical Class: Organic Salts

Function: Surfactant - Cleansing Agent

Technical/Other Names:
PEG-8 Tridecyl Ether Carboxylic Acid, Sodium Salt
Polyethylene Glycol 400 Tridecyl Ether Carboxylic Acid, Sodium Salt
Polyoxyethylene (8) Tridecyl Ether Carboxylic Acid, Sodium Salt

SODIUM TRIDECETH-12 CARBOXYLATE

CAS Nos.: 61757-59-3 (Generic); 68891-17-8 (Generic)

Definition: Sodium Trideceth-12 Carboxylate is the sodium salt of the carboxylic acid derived from Trideceth-12 (q.v.). It conforms generally to the formula:

$$CH_3(CH_2)_{12}(OCH_2CH_2)_nOCH_2COONa$$

Information Source: JCLS

Chemical Class: Organic Salts

Function: Surfactant - Cleansing Agent

Technical/Other Names:
PEG-12 Tridecyl Ether Carboxylic Acid, Sodium Salt
Polyethylene Glycol 600 Tridecyl Ether Carboxylic Acid, Sodium Salt

SODIUM TRIDECETH-15 CARBOXYLATE

CAS Nos.: 61757-59-3 (Generic); 68891-17-8 (Generic)

Definition: Sodium Trideceth-15 Carboxylate is the sodium salt of Trideceth-15 Carboxylic Acid (q.v.).

Chemical Class: Organic Salts

Function: Surfactant - Cleansing Agent

Technical/Other Names:
Sodium Polyethylene Glycol (15) Tridecyl Ether Carboxylate
Sodium Polyoxyethylene (15) Tridecyl Ether Carboxylate

SODIUM TRIDECETH-19 CARBOXYLATE

CAS Nos.: 61757-59-3 (Generic); 68891-17-8 (Generic)

Definition: Sodium Trideceth-19 Carboxylate is the sodium salt of Trideceth-19 Carboxylic Acid (q.v.).

Chemical Class: Organic Salts

Function: Surfactant - Cleansing Agent

Technical/Other Names:
Sodium Polyethylene Glycol (19) Tridecyl Ether Carboxylate
Sodium Polyoxyethylene (19) Tridecyl Ether Carboxylate

SODIUM TRIDECETH SULFATE

CAS No.
25446-78-0 (n=3)

EINECS No.
246-985-2

JPN Translation:
ポリオキシエチレントリデシル硫酸 Na

Definition: Sodium Trideceth Sulfate is the sodium salt of sulfated ethoxylated Tridecyl Alcohol (q.v.) that conforms generally to the formula:

$$CH_3(CH_2)_{12}(OCH_2CH_2)_nOSO_3Na$$

where n has a value between 1 and 4.

Information Sources: 21CFR175.105, JCLS, TSCA

Chemical Class: Alkyl Ether Sulfates

Functions: Surfactant - Cleansing Agent; Surfactant - Emulsifying Agent

Reported Product Categories: Bath Oils, Tablets, and Salts; Cleansing Products (Cold Creams, Cleansing Lotions, Liquids and Pads); Baby Shampoos; Shampoos (Non-coloring); Eye Makeup Removers; Baby Products, Misc.; Bath Soaps and Detergents

Technical/Other Names:
Sodium Polyoxyethylene Tridecyl Sulfate
Sodium Tridecyl Ether Sulfate

Trade Names:
CEDEPAL TD-407 (Stepan)
CEDEPAL TD-484 (Stepan)
CEDEPAL TD-403 MFLD (Stepan)
DeSulf STDES-30 (DeForest)
Genapol XRO (Clariant)

The inclusion of any compound in the *Dictionary and Handbook* does not indicate that use of that substance as a cosmetic ingredient complies with the laws and regulations governing such use in the United States or any other country.

Genapol XRO (Clariant GmbH, Personal
  Care)
Rhodapex EST-30 (Rhodia)
Rhodapex EST-65 (Rhodia)
Sulfochem TD-3 (Chemron)

Trade Name Mixtures:
Custoblend BSC-50 (Baby Shampoo)
  (Custom Ingredients)
DeCONC BSC-50 (DeForest)
Jeeteric CDTD (Jeen)
Miracare BC-10 (Rhodia)
Miracare BC-20 (Rhodia)
Miracare BC-27 (Rhodia)
Miracare BT (Rhodia)
Miracare MHT (Rhodia)
Miracare 2MHT (Rhodia)
Miracare MS-2 (Rhodia)
Miracare MS-4 (Rhodia)
Miranol BM Conc. (Rhodia)
Miranol BT (Rhodia)
Monateric 985A (Uniqema)
Proteric 1095 (Protameen)
Proteric CDTD (Protameen)

SODIUM TRIDECYLBENZENESULFONATE

CAS No.   **EINECS No.**
26248-24-8   247-536-3

Empirical Formula:
  $C_{19}H_{32}O_3S \cdot Na$

Definition: Sodium Tridecylbenzene-
sulfonate is the substituted aromatic
compound that conforms generally to the
formula:

SO_3Na

(CH_2)_{12}CH_3

Information Sources: 21CFR175.105,
21CFR176.210, 21CFR178.3130,
21CFR178.3400

Chemical Class: Alkyl Aryl Sulfonates

Function: Surfactant - Cleansing Agent

Technical/Other Names:
  Benzenesulfonic Acid, Tridecyl-, Sodium
    Salt
  Tridecylbenzenesulfonic Acid, Sodium Salt

SODIUM TRIDECYL SULFATE

CAS No.   **EINECS No.**
3026-63-9   221-188-2

Empirical Formula:
  $C_{13}H_{28}O_4S \cdot Na$

Definition: Sodium Tridecyl Sulfate is the
sodium salt of tridecyl sulfate that conforms
to the formula:

$$CH_3(CH_2)_{12}OSO_3Na$$

Information Sources: 21CFR177.1210,
JCLS, TSCA

Chemical Class: Alkyl Sulfates

Functions: Surfactant - Cleansing Agent;
Surfactant - Emulsifying Agent

Technical/Other Name:
  1-Tridecanol, Hydrogen Sulfate, Sodium
    Salt

Trade Name:
  Rhodapon TDS (Rhodia)

SODIUM TRIMETAPHOSPHATE

CAS No.   **EINECS No.**
7785-84-4   232-088-3

Empirical Formula:
  $H_3O_9P_3 \cdot 3Na$

Definition: Sodium Trimetaphosphate is the
inorganic salt that conforms to the formula:

$$(NaPO_3)_3$$

Information Sources: CIR: [SQ] IJT-20
(SUPPL. 3)2001, MI-13(8770), TSCA,
USAN

Chemical Class: Inorganic Salts

Functions: Buffering Agent; Chelating
Agent; pH Adjuster

Technical/Other Name:
  Metaphosphoric Acid, Trisodium Salt

SODIUM TRIMETHYLPENTENE/MA COPOLYMER

CAS No.: 37199-81-8

Definition: Sodium Trimethylpentene/MA
Copolymer is the soidum salt of a copolymer
of maleic anhydride and 2,4,4-trimethyl-
pentene monomers.

Chemical Class: Synthetic Polymers

Function: Viscosity Controlling Agent

Technical/Other Name:
  2,5-Furandione, Polymer with 2,4,4-
    Trimethylpentene, Sodium Salt

SODIUM UNDECETH-5 CARBOXYLATE

Definition: Sodium Undeceth-5 Carboxylate
is the sodium salt of Undeceth-5 Carboxylic
Acid (q.v.) that conforms generally to the
formula:

$$CH_3(CH_2)_{10}(OCH_2CH_2)_nOCH_2COONa$$

where n has an average value of 4.

Chemical Class: Organic Salts

Function: Surfactant - Cleansing Agent

Technical/Other Names:
  PEG-5 Undecyl Ether Carboxylic Acid,
    Sodium Salt
  Polyethylene Glycol (5) Undecyl Ether
    Carboxylic Acid, Sodium Salt
  Polyoxyethylene (5) Undecyl Ether
    Carboxylic Acid, Sodium Salt

Trade Name:
  Neodox 1-4 Sodium Salt (Shell)

SODIUM UNDECYLENATE

CAS No.   **EINECS No.**
3398-33-2   222-264-8

Empirical Formula:
  $C_{11}H_{20}O_2 \cdot Na$

Definition: Sodium Undecylenate is the
sodium salt of undecylenic acid that conforms
generally to the formula:

$$CH_2{=}CH(CH_2)_8COONa$$

Information Source: EEC(VI/1-18)

Chemical Class: Soaps

Functions: Surfactant - Cleansing Agent;
Surfactant - Emulsifying Agent

Technical/Other Name:
  10-Undecenoic Acid, Sodium Salt

Trade Name:
  Undenat (Vevy)

SODIUM UNDECYLENOAMPHOACETATE

Empirical Formula:
  $C_{17}H_{32}N_2O_4 \cdot Na$

Definition: Sodium Undecylenoamphoacetate is the amphoteric organic compound
that conforms generally to the formula:

$$CH_2{=}CH(CH_2)_8\overset{\displaystyle O}{\overset{\displaystyle \|}{C}}-NHCH_2CH_2\overset{\displaystyle CH_2CH_2OH}{\overset{\displaystyle |}{N}}CH_2COONa$$

Chemical Class: Alkylamido Alkylamines

Functions: Hair Conditioning Agent;
Surfactant - Cleansing Agent; Surfactant -
Foam Booster

Technical/Other Names:
  Undecylenoamphoacetate
  Undecylenoamphoglycinate

SODIUM UNDECYLENOAMPHO-PROPIONATE

Empirical Formula:
$C_{18}H_{34}N_2O_4 \cdot Na$

Definition: Sodium Undecylenoamphopropionate is the amphoteric organic compound that conforms generally to the formula:

$$CH_2=CH(CH_2)_8\overset{\displaystyle O}{\overset{\|}{C}}-NHCH_2CH_2N\overset{\displaystyle CH_2CH_2OH}{\underset{CH_2CH_2COONa}{|}}$$

Chemical Class: Alkylamido Alkylamines

Functions: Hair Conditioning Agent; Surfactant - Cleansing Agent; Surfactant - Foam Booster

SODIUM UNDECYLENOYL GLUTAMATE

Definition: Sodium Undecylenoyl Glutamate is the substituted amino acid that conforms generally to the formula:

$$CH_2=CH(CH_2)_8\overset{\displaystyle O}{\overset{\|}{C}}-NHCH\overset{\displaystyle (CH_2)_2COONa}{\underset{COOH}{|}}$$

Chemical Class: Amino Acids

Functions: Hair Conditioning Agent; Skin-Conditioning Agent - Miscellaneous; Surfactant - Cleansing Agent

Trade Name:
Protelan AG 11/N (Zschimmer & Schwarz Italiana)

SODIUM UROCANATE

CAS No.: 6159-49-5

Empirical Formula:
$C_6H_6N_2O_2 \cdot Na$

Definition: Sodium Urocanate is the sodium salt of Urocanic Acid (q.v.). It conforms to the formula:

CH=CHCOONa (imidazole ring structure)

Chemical Classes: Amino Acids; Heterocyclic Compounds

Function: Ultraviolet Light Absorber

Technical/Other Names:
3-(1H-Imidazol-4-yl)-2-Propenoic Acid, Sodium Salt
2-Propenoic Acid, 3-(1H-Imidazol-4-yl)-, Sodium Salt

SODIUM URSOLATE

CAS No.: 220435-39-2

Empirical Formula:
$C_{30}H_{48}O_3 \cdot Na$

Definition: Sodium Ursolate is the sodium salt of Ursolic Acid (q.v.).

Chemical Class: Organic Salts

Function: Skin-Conditioning Agent - Miscellaneous

Technical/Other Name:
Urs-12-3en-28-oic Acid, 3-Hydroxy-, Monosodium Salt, (3-Beta)-

Trade Name:
Dubosia Ursolic Acid, Sodium Salt, Premier (Premier Specialties)

Trade Name Mixture:
Ursolic Acid Sodium Salt (Boehringer)

SODIUM USNATE

CAS No.: 34769-44-3

Empirical Formula:
$C_{18}H_{16}O_7 \cdot Na$

Definition: Sodium Usnate is the sodium salt of Usnic Acid (q.v.).

Chemical Classes: Heterocyclic Compounds; Organic Salts

Function: Cosmetic Biocide

Technical/Other Name:
1,3(2H,9bH)-Dibenzofurandione, 2,6-Diacetyl-7,9-Dihydroxy--8,9b-Dimethyl-, Monosodium Salt

Trade Names:
AEC Sodium Usnate (A & E Connock)
Evosina 100% (Variati)

Trade Name Mixtures:
Evosina NA2GP (Variati)
Evosina SBS (Variati)

SODIUM WHEAT GERMAMPHOACETATE

Definition: Sodium Wheat Germamphoacetate is the amphoteric organic compound that conforms to the formula:

$$RC\overset{\displaystyle O}{\overset{\|}{}}-NH(CH_2)_2N\overset{\displaystyle CH_2CH_2OH}{\underset{CH_2COONa}{|}}$$

where RCO- represents the fatty acids derived from wheat germ oil.

Chemical Class: Alkylamido Alkylamines

Functions: Hair Conditioning Agent; Surfactant - Cleansing Agent; Surfactant - Foam Booster

SODIUM XYLENESULFONATE

CAS No.	EINECS No.
1300-72-7	215-090-9

Empirical Formula:
$C_8H_{10}O_3S \cdot Na$

Definition: Sodium Xylenesulfonate is the sodium salt of ring sulfonated mixed xylene isomers that conforms generally to the formula:

$$(CH_3)_2C_6H_3SO_3Na$$

Information Sources: 21CFR175.105, 21CFR176.180, 21CFR178.1010, TSCA

Chemical Class: Alkyl Aryl Sulfonates

Function: Surfactant - Hydrotrope

Reported Product Categories: Shampoos (Non-coloring); Personal Cleanliness Products, Misc.

Technical/Other Names:
Benzenesulfonic Acid, Dimethyl-, Sodium Salt
Dimethylbenzenesulfonic Acid, Sodium Salt
Sodium Dimethylbenzenesulfonate
Xylenesulfonic Acid, Sodium Salt

Trade Names:
Eltesol SX30 (Albright & Wilson UK)
Eltesol SX40 (Albright & Wilson UK)
Eltesol SX93 (Albright & Wilson UK)
Eltesol SX Pellets (Albright & Wilson UK)
Norfox SXS-40 (Norman, Fox & Co.)
Pilot SXS-40 (Pilot)
STEPANATE SXS (Stepan)

Trade Name Mixtures:
Custoblend DTS (Conditioning Blend) (Custom Ingredients)
Custopearl 1000 (Custom Ingredients)
Miracare UM-140 (Rhodia)
Mirasheen 207 (Rhodia)

SODIUM ZINC CETYL PHOSPHATE

Definition: Sodium Zinc Cetyl Phosphate is the organic compound that conforms to the formula:

$$CH_3(CH_2)_{15}O-\overset{\displaystyle O}{\overset{\|}{P}}\overset{}{\underset{ONa}{|}}-O-Zn-O-\overset{\displaystyle O}{\overset{\|}{P}}\overset{}{\underset{OH}{|}}-O(CH_2)_{15}CH_3$$

Chemical Classes: Organic Salts; Phosphorus Compounds

Function: Colorant

SOLANUM DULCAMARA

Definition: *See "Regulatory and Ingredient Use Information," regarding EU labeling*

---

names for botanical ingredients in Volume 1, Introduction, Part A.

Chemical Class: Biological Products

Technical/Other Name:
Solanum Dulcamara Stem Extract (U.S.)

SOLANUM DULCAMARA STEM EXTRACT

CAS No.	EINECS No.
84696-50-4	283-655-7

Definition: Solanum Dulcamara Stem Extract is an extract of the stems of the dulcamara, *Solanum dulcamara*. See *"Regulatory and Ingredient Use Information,"* regarding the labeling names for botanical ingredients in Volume 1, Introduction, Part A.

Chemical Class: Biological Products

Function: Not Reported

Technical/Other Names:
Bittersweet Extract
Dulcamara Extract
Dulcamara (Solanum Dulcamara) Extract
Extract of Bittersweet
Extract of Dulcamara
Extract of Solanum Dulcamara
Solanum Dulcamara (EU)

Trade Name Mixtures:
Actiphyte Bittersweet (Active Organics)
Bittersweet HS (Alban Muller)
Bittersweet Stalks Extract HS 3578 G
(Grau)

SOLANUM LYCOCARPUM FRUIT EXTRACT

Definition: Solanum Lycocarpum Fruit Extract is an extract of the fruit of *Solanum lycocarpum*. See *"Regulatory and Ingredient Use Information,"* regarding the labeling names for botanical ingredients in Volume 1, Introduction, Part A.

Chemical Class: Biological Products

Function: Not Reported

Trade Name Mixture:
Fruta De Lobo (Sederma)

SOLANUM LYCOPERSICUM (TOMATO) EXTRACT

CAS No.	EINECS No.
90131-63-8	290-375-9

JPN Translation:
トマトエキス

Definition: Solanum Lycopersicum (Tomato) Extract is an extract of the leaves, stems and fruit of the tomato, *Solanum lycopersicum*. See *"Regulatory and Ingredient Use Information,"* regarding the labeling names for botanical ingredients in Volume 1, Introduction, Part A.

Information Sources: JCIC, JCLS, JSQI

Chemical Class: Biological Products

Function: Skin-Conditioning Agent - Miscellaneous

Technical/Other Names:
Extract of Solanum Lycopersicum
Extract of Tomato
Lycopersicon Esculentum Extract
Lycopersicum Esculentum Extract
Tomato Extract

Trade Names:
Phytelene of Tomato EN 310 powder
(Indena SA)
Phytogreen of Tomato EP 505 Powder
(Phytochim)
Tomato Extract (Quest International)

Trade Name Mixtures:
Acifructol Complex P 63 (Gattefosse s.a.)
Acifructol Tomato P 62 (Gattefosse s.a.)
Actiphyte of Tomato BG50 (Active
Organics)
Actiphyte of Tomato GL50 (Active
Organics)
Actiphyte of Tomato Lipo S (Active
Organics)
Actiphyte of Tomato PG50 (Active
Organics)
BMX Complex (Barnet)
Extrait De Tomate (Silab)
Herbasec Tomato KBA (Cosmetochem)
(Cosmetochem International Ltd.)
Herbasol Extract Tomato (Cosmetochem)
(Cosmetochem International Ltd.)
Lipoplastidine Solanum Lycopersicum
(Vevy)
Lycopersidine (Exsymol)
Lyco-Sol (Libiol)
Phytami Tomato (Alban Muller)
Phytostimulines of Tomato (CEP (Solabia))
Polyplant Anti-Acne (Provital/Centerchem)
Tomato Extract BG (Maruzen
Pharmaceuticals Co., Ltd.)
Tomato HS (Alban Muller)
Tomato Liquid (Ichimaru Pharcos)
Vegebios of Tomato (CEP (Solabia))
Vegepone Tomato 2/031307 (Symrise)
Vegetol Tomato GR 207 Hydro (Gattefosse
s.a.)

SOLANUM LYCOPERSICUM (TOMATO) FRUIT JUICE

JPN Translation:
トマト果汁

Definition: Solanum Lycopersicum (Tomato) Fruit Juice is the juice expressed from the fruit of *Solanum lycopersicum*. See *"Regulatory and Ingredient Use Information,"* regarding the labeling names for botanical ingredients in Volume 1, Introduction, Part A.

Chemical Class: Biological Products

Function: Skin-Conditioning Agent - Miscellaneous

Technical/Other Names:
Juice, Solanum Lycopersicum
Juice, Tomato
Lycopersicon Esculentum Juice
Lycopersicum Esculentum Juice
Solanum Lycopersicum Juice
Tomato Fruit Juice
Tomato Juice

Trade Names:
AEC Tomato Distillate Aqueous Natural (A
& E Connock)
Authenticals of Tomato (CEP (Solabia))

SOLANUM LYCOPERSICUM (TOMATO) FRUIT LIPIDS

Definition: Solanum Lycopersicum (Tomato) Fruit Lipids are the lipids extracted from the fruit of *Solanum lycopersicum*. See *"Regulatory and Ingredient Use Information,"* regarding the labeling names for botanical ingredients in Volume 1, Introduction, Part A.

Chemical Class: Biological Products

Function: Skin-Conditioning Agent - Miscellaneous

Trade Name:
Lyc-O-Mato 5-15% (LycoRed USA)

SOLANUM LYCOPERSICUM (TOMATO) FRUIT OIL

Definition: Solanum Lycopersicum (Tomato) Fruit Oil is the oil extracted from the fruit of the tomato, *Solanum lycopersicum*. See *"Regulatory and Ingredient Use Information,"* regarding the labeling names for botanical ingredients in Volume 1, Introduction, Part A.

Chemical Class: Fats and Oils

Function: Skin-Conditioning Agent - Occlusive

Trade Name:
Lycomax 7% (Carotech)

Trade Name Mixtures:
Lycomax 5% (Carotech)
Lycomax 6% (Carotech)

SOLANUM LYCOPERSICUM (TOMATO) FRUIT WATER

Definition: Solanum Lycopersicum (Tomato) Fruit Water is an aqueous solution of the steam distillate obtained from the fruit of the tomato, *Solanum lycopersicum*. See "Regulatory and Ingredient Use Information," regarding the labeling names for botanical ingredients in Volume 1, Introduction, Part A.

Chemical Class: Biological Products

Function: Fragrance Ingredient

Technical/Other Names:
Lycopersicon Esculentum Water
Lycopersicum Esculentum Water
Tomato Fruit Water

Trade Name:
Extrait Originel Tomate (Gattefosse s.a.)

SOLANUM LYCOPERSICUM (TOMATO) SEED OIL

Definition: Solanum Lycopersicum (Tomato) Seed Oil is the fixed oil obtained from the seeds of the tomato, *Solanum lycopersicum*. See "Regulatory and Ingredient Use Information," regarding the labeling names for botanical ingredients in Volume 1, Introduction, Part A.

Chemical Class: Fats and Oils

Function: Skin-Conditioning Agent - Occlusive

Technical/Other Names:
Lycopersicon Esculentum Seed Oil
Lycopersicum Esculentum Oil
Tomato Oil
Tomato Seed Oil

Trade Names:
Huile de Graines de Tomates (Bertin)
Oleoresin From Tomatoes (Inocosm)
Tomato Seed Oil (Nestle World Trade)

SOLANUM MELONGENA (EGGPLANT) FRUIT EXTRACT

CAS No.	EINECS No.
84012-19-1	281-665-6

Definition: Solanum Melongena (Eggplant) Fruit Extract is an extract of the fruit of the eggplant, *Solanum melongena*. See "Regulatory and Ingredient Use Information," regarding the labeling names for botanical ingredients in Volume 1, Introduction, Part A.

Chemical Class: Biological Products

Function: Skin-Conditioning Agent - Miscellaneous

Technical/Other Names:
Eggplant Extract
Eggplant Fruit Extract
Extract of Eggplant
Extract of Solanum Melongena
Solanum Esculentum Extract
Solanum Melongena Extract

Trade Name Mixtures:
Actiphyte of Eggplant BG50 (Active Organics)
Actiphyte of Eggplant GL50 (Active Organics)
Actiphyte of Eggplant Lipo S (Active Organics)
Actiphyte of Eggplant PG50 (Active Organics)
Aubergine Extract HS 2477 G (Grau)
Eggplant Fruit Extract (Libiol)
Eggplant HS (Alban Muller)
Phytami Eggplant (Alban Muller)

SOLANUM MURICATUM EXTRACT

Definition: Solanum Muricatum Extract is the extract obtained from the fruit of *Solanum muricatum*. See "Regulatory and Ingredient Use Information," regarding the labeling names for botanical ingredients in Volume 1, Introduction, Part A.

Chemical Class: Biological Products

Function: Not Reported

Technical/Other Name:
Extract of Solanum Muricatum

Trade Name Mixture:
Cosflor Pepino HGS (A & E Connock)

SOLANUM TUBEROSUM (POTATO) EXTRACT

CAS No.	EINECS No.
90083-08-2	290-202-7

Definition: Solanum Tuberosum (Potato) Extract is an extract of the pulp of the potato, *Solanum tuberosum*. See "Regulatory and Ingredient Use Information," regarding the labeling names for botanical ingredients in Volume 1, Introduction, Part A.

Chemical Class: Biological Products

Function: Skin-Conditioning Agent - Miscellaneous

Technical/Other Names:
Extract of Potato
Extract of Solanum Tuberosum Pulp
Potato Extract
Potato Pulp Extract
Solanum Tuberosum Extract

Trade Name Mixtures:
Actiphyte of Potato (Active Organics)
Herbasol Extract Potato (Cosmetochem) (Cosmetochem International Ltd.)
Potato Extract HS 3327 G (Grau)
Potato HS (Alban Muller)
VT-292 Extract of Potato (Vege-Tech)

SOLANUM TUBEROSUM (POTATO) PEEL EXTRACT

Definition: Solanum Tuberosum (Potato) Peel Extract is an extract of the peel of the potato, *Solanum tuberosum*. See "Regulatory and Ingredient Use Information," regarding the labeling names for botanical ingredients in Volume 1, Introduction, Part A.

Chemical Class: Biological Products

Function: Not Reported

Technical/Other Names:
Extract of Potato Peel
Extract of Solanum Tuberosum Peel
Potato Peel Extract
Solanum Tuberosum Peel Extract

Trade Name Mixtures:
VT-291 Extract of Potato Peel (Vege-Tech)
VT-292 Extract of Potato Peel (Vege-Tech)

SOLANUM TUBEROSUM (POTATO) STARCH

CAS No.	EINECS No.
9005-25-8	232-679-6

JPN Translation:
バレイショデンプン

Definition: Solanum Tuberosum (Potato) Starch is a polysaccharide obtained from the potato, *Solanum tuberosum*. See "Regulatory and Ingredient Use Information," regarding the labeling names for botanical ingredients in Volume 1, Introduction, Part A.

Information Sources: AUS, BEL, BP, BPC, BRA, 21CFR175.105, 21CFR178.3520, 21CFR182.70, DA, DDR, EGY, EP, FIN, HUN, IND, ITA, JAN, JCLS, JSCI, MAR, MEX, MI-13(8877), NF XV, NF XIX, PN, POL, ROM, SNPF, TSCA, USAN, USSR

Chemical Class: Biological Products

Functions: Absorbent; Binder; Bulking Agent; Viscosity Increasing Agent - Aqueous

Technical/Other Names:
Potato Starch

Solanum Tuberosum Starch
Starch, Potato
Starch, Solanum Tuberosum

Trade Names:
Amycol HF (Nippon Starch)
Graflow F (Nippon Starch)
JP Potato Starch ST-P (Nippon Starch)
ST Starch P (Nippon Starch)

Trade Name Mixture:
Vegewhite (LCW)

SOLIDAGO ODORA (GOLDENROD) EXTRACT

Definition: Solidago Odora (Goldenrod) Extract is an extract of the goldenrod, *Solidago odora*. See *"Regulatory and Ingredient Use Information," regarding the labeling names for botanical ingredients in Volume 1, Introduction, Part A.*

Chemical Class: Biological Products

Function: Not Reported

Technical/Other Names:
Extract of Goldenrod
Extract of Solidago Odora
Goldenrod Extract
Solidago Odora Extract

Trade Name Mixtures:
Actiphyte Goldenrod (Active Organics)
Phytelene of Golden Rod EG 474 liquid (Indena SA)
Phytogreen 55 of Golden Rod EXH 688 Liquid (Phytochim)

SOLIDAGO VIRGAUREA (GOLDENROD) EXTRACT

CAS No.: 85117-06-2

Definition: Solidago Virgaurea (Goldenrod) Extract is an extract of the goldenrod, *Solidago virgaurea*. See *"Regulatory and Ingredient Use Information," regarding the labeling names for botanical ingredients in Volume 1, Introduction, Part A.*

Chemical Class: Biological Products

Function: Skin-Conditioning Agent - Miscellaneous

Technical/Other Names:
Extract of Goldenrod
Extract of Solidago Virgaurea
Goldenrod Extract
Solidago Virgaurea Extract

Trade Name Mixtures:
Golden Rod HPG Titrated (Alban Muller)
Golden Rod HS (Alban Muller)

SOLUBLE COLLAGEN

JPN Translation:
水溶性コラーゲン

Definition: Soluble Collagen is a nonhydrolyzed, native protein derived from the connective tissue of young animals. It consists essentially of a mixture of the precursors of mature collagen. It has a triple helical structure and is predominantly not cross-linked.

Information Sources: JCIC, JCLS, JSQI

Chemical Class: Protein Derivatives

Functions: Hair Conditioning Agent; Skin-Conditioning Agent - Miscellaneous

Reported Product Categories: Moisturizing Preparations; Bath Preparations, Misc.; Body and Hand Preparations (Excluding Shaving Preparations); Skin Care Preparations, Misc.; Foundations; Bath Capsules; Bath Oils, Tablets, and Salts; Cleansing Products (Cold Creams, Cleansing Lotions, Liquids and Pads); Face and Neck Preparations (Excluding Shaving Preparations); Bubble Baths; Night Skin Care Preparations; Paste Masks (Mud Packs); Aftershave Lotions; Baby Shampoos; Eye Makeup Preparations, Misc.; Eye Shadows; Hair Conditioners; Mascara; Suntan Preparations, Misc.

Technical/Other Names:
Soluble Animal Collagen
Water-soluble Collagen

Trade Names:
AC Marine Collagen (Active Concepts)
AC Soluble Collagen (Active Concepts)
Alfomarine-CL (Technoble)
Aquagene (Coletica SA)
Clearcol (Croda Chemicals)
Clearcol (Croda, Inc.)
COLLAGEN BP (Nitta Gelatin)
COLLAGEN BP-03 (Nitta Gelatin)
COLLAGEN BP(PF) (Nitta Gelatin)
COLLAGEN BP-03(PF) (Nitta Gelatin)
Collagen CLR (CLR)
Collagen Complex (Maybrook)
Collagen HEYL (Nikko Rica)
Collagen Nativ, 1% (Crodarom)
Collagen Native Extra 1% (Maybrook)
Collagen P (Nitta Gelatin)
COLLAGEN P(PF) (Nitta Gelatin)
COLLAGEN P-03(PF) (Nitta Gelatin)
COLLAGEN ST-03 (Nitta Gelatin)
Collaplex 0.3 (GfN)
Collaplex 1.0 (GfN)
Collasol (Croda Chemicals)
Collasol (Croda, Inc.)
Collasol M (Croda, Inc.)
Collegen HEYL Clear (Nikko Rica)
Etigene (Coletica SA)
Grancol-1 (Grant)
Grancol SP-01 (Grant)
Ichtyocollagene (Sederma)

Maricol CLR (CLR)
Maricol LO (CLR)
Marine Colladerm (Vincience)
Marine Collagen (Katakura)
Marine Native Collagen (Vincience)
Marinepure Collagen (Katakura)
Native-Collagen pur COS (Cosmetochem)
Native Soluble Collagen (Esperis)
Natural Soluble Collagen (Pentapharm/ Centerchem)
Neptigene I (Coletica SA)
Neptigene II (Coletica)
Neptuline C (Gattefosse s.a.)
Nerecoll (Coletica SA)
Oceagen LS (Laboratoires Serobiologiques)
Pancogene Marin (Gattefosse s.a.)
Seagem Collagen (Katakura)
Soluble Collagen (Proteina)
Solu-Col Complex VY (Arch Personal Care Products)
Solu-Coll (Arch Personal Care Products)
Solu-Coll C (Arch Personal Care Products)
Solu-Coll CLR (Arch Personal Care Products)
Solu-Coll Native (Arch Personal Care Products)
Solu-Mar Native (Arch Personal Care Products)

Trade Name Mixtures:
Brookosome SC (Arch Personal Care Products)
Collagel S 100 (Vincience)
Collagen-CCK-Complex (Kelisema Italy)
Collagen-Hyaluronic Acid-Jelly (Labopharma)
Collagen-IMZ-Complex (Kelisema Italy)
Collagen-LSS-Complex (Kelisema Italy)
Collagen S (PF) (Nitta Gelatin)
Collagen S-03 (PF) (Nitta Gelatin)
Collagen S-06 (PF) (Nitta Gelatin)
Collagen SP-03 (PF) (Nitta Gelatin)
Colla-Tein Collagen Mask (Maybrook)
Desaron (GfN)
Elastin PG 2000 (GfN)
Marine Plasma Extract (Arch Personal Care Products)
Marine Plasma Extract III (Arch Personal Care Products)
Ocean Collagen B-03 (Air Water)
Ocean Collagen B-05 (Air Water)
Soluble Collagen With Elastin (Proteina)
Solu-Coll Complex (Arch Personal Care Products)
Thermoplex (GfN)

SOLUBLE ELASTIN

JPN Translation:
水溶性エラスチン

Definition: Soluble Elastin a water soluble nonhydrolyzed, native protein derived from Elastin (q.v.).

Information Source: JCLS

Chemical Class: Proteins

Function: Skin-Conditioning Agent - Miscellaneous

SOLUBLE PROTEOGLYCAN

JPN Translation:
水溶性プロテオグリカン

Definition: Soluble Proteoglycan is a solubilized Glycoproteins (q.v.) having a very high carbohydrate content.

Chemical Classes: Biological Polymers and their Derivatives; Carbohydrates; Proteins

Functions: Hair Conditioning Agent; Skin-Conditioning Agent - Miscellaneous

Trade Name:
Proteodermin (CLR)

SOLVENT BLACK 3

CAS No. 4197-25-5
EINECS No. 224-087-1

Empirical Formula:
$C_{29}H_{24}N_6$

Definition: Solvent Black 3 is classed chemically as a disazo color. It conforms to the formula:

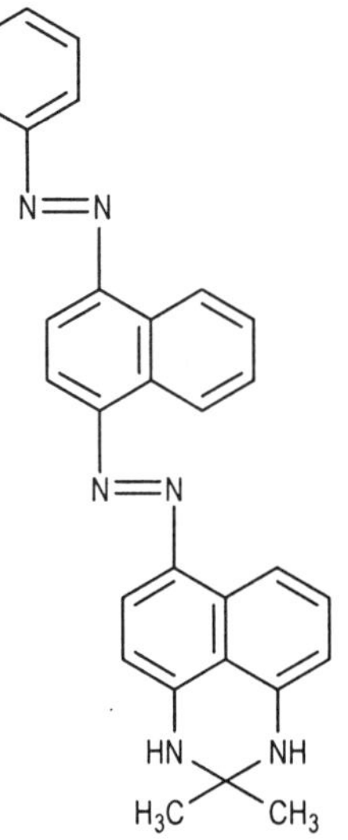

See "Regulatory and Ingredient Use Information," for Colorants in Volume 1, Introduction, Part A.

Information Sources: CI 26150, MI-13 (8970), TSCA

Chemical Class: Color Additives - Hair

Function: Hair Colorant

Technical/Other Names:
CI 26150
2,3-Dihydro-2,2-Dimethyl-6-[[4-(Phenylazo)-1-Naphthalenyl]Azo]-1H-Pyrimidine
1H-Pyrimidine, 2,3-Dihydro-2,2-Dimethyl-6-[[4-(Phenylazo)-1-Naphthalenyl]Azo]-
Sudan Black B

SOLVENT BLACK 5

CAS No.: 11099-03-9

Definition: Solvent Black 5 is the azine color that is obtained when nitrophenol, nitrobenzene, aniline, and aniline hydrochloride are heated at elevated temperatures and pressure in an inert atmosphere. *See "Regulatory and Ingredient Use Information," for Colorants in Volume 1, Introduction, Part A.*

Chemical Class: Color Additives - Hair

Function: Hair Colorant

Technical/Other Name:
CI 50415

Trade Name:
Lowasol Black 5 (Lowenstein)

SOLVENT BLUE 35

CAS Nos.
12769-17-4
17354-14-2
EINECS No.

241-379-4

Empirical Formula:
$C_{22}H_{26}N_2O_2$

Definition: Solvent Blue 35 is classed chemically as an anthraquinone color. It conforms to the formula:

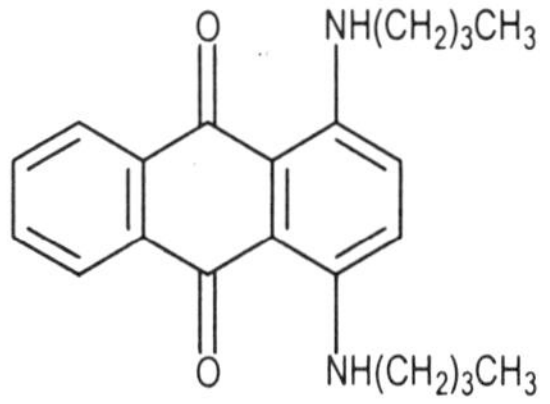

See "Regulatory and Ingredient Use Information," for Colorants in Volume 1, Introduction, Part A.

Information Sources: CI 61554, EEC(II-389), TSCA

Chemical Class: Color Additives - Hair

Function: Hair Colorant

Technical/Other Names:
9,10-Anthracenedione, 1,4-bis (Butylamino)-
9,10-Anthracenedione, 1,4-Di(Butylamino)-
Anthraquinone, 1,4-Bis(Butylamino)-
1,4-Bis(Butylamino)Anthraquinone
CI 61554
1,4-Di(Butylamino)-9,10-Anthracenedione
Sudan Blue 2

SOLVENT GREEN 3

CAS No. 128-80-3
EINECS No. 204-909-5

Empirical Formula:
$C_{28}H_{22}N_2O_2$

Definition: Solvent Green 3 is classed chemically as an anthraquinone color. It conforms to the formula:

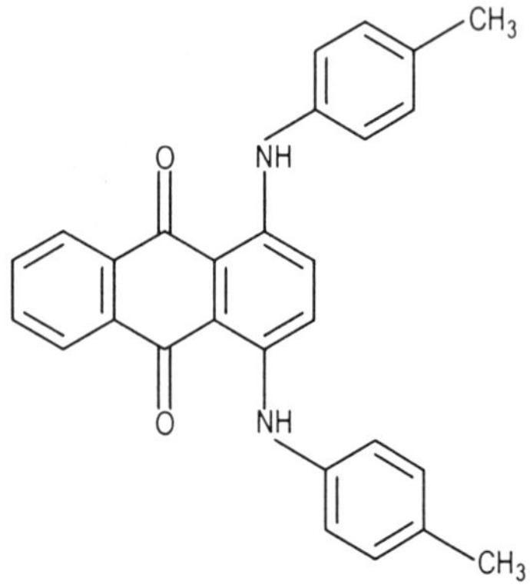

See "Regulatory and Ingredient Use Information," for Colorants in Volume 1, Introduction, Part A. To identify the certified colorant for labeling purposes in the US, the INCI Name Green 6 must be used. To identify the colorant allowed for use in the European Union (EU), the INCI Name CI 61565 must be used, except for hair dye products. The INCI Name for batches of this colorant that have not been certified is Solvent Green 3. To identify the colorant allowed for use in Japan, the INCI name Midori202 must be used.

Information Sources: CI 61565, M3, MI-13(8157), TSCA

Chemical Class: Color Additives - Miscellaneous

Function: Colorant

Reported Product Categories: Tonics, Dressings, and Other Hair Grooming Aids; Bath Oils, Tablets, and Salts; Hair Conditioners; Deodorants (Underarm); Perfumes; Hair Preparations (Non-coloring), Misc.

Technical/Other Names:
9,10-Anthracenedione, 1,4-Bis[(4-Methylphenyl)Amino]-
1,4-bis(4'-Methylanilino)Anthraquinone
1,4-Bis[(4-Methylphenyl)Amino-9,10-Anthracenedione

Ceres Green BB
CI 61565
1,4-Di-p-Toluidinoanthraquinone
Green No. 202
Quinizarin Green SS

SOLVENT GREEN 7

CAS No. 6358-69-6   **EINECS No.** 228-783-6

Empirical Formula:
$C_{16}H_{10}O_{10}S_3 \cdot 3Na$

Definition: Solvent Green 7 is classed chemically as a pyrene color. It conforms to the formula:

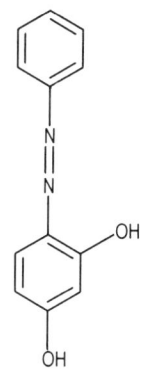

See "Regulatory and Ingredient Use Information," for Colorants in Volume 1, Introduction, Part A. To identify the certified colorant for labeling purposes in the US, the INCI Name Green 8 must be used. To identify the colorant allowed for use in the European Union (EU), the INCI Name CI 59040 must be used, except for hair dye products. The INCI Name for batches of this colorant that have not been certified is Solvent Green 7. To identify the colorant allowed for use in Japan, the INCI name Midori204 must be used.

Information Sources: CI 59040, M3, TSCA

Chemical Class: Color Additives - Miscellaneous

Function: Colorant

Reported Product Categories: Shampoos (Non-coloring); Bath Soaps and Detergents

Technical/Other Names:
CI 59040
Green No. 204
8-Hydroxy-1,3,6-Pyrenetrisulfonic Acid, Trisodium Salt
Pyranine
1,3,6-Pyrenetrisulfonic Acid, 8-Hydroxy-, Trisodium Salt
Trisodium 8-Hydroxypyrene-1,3,6-Trisulfonate

SOLVENT ORANGE 1

CAS No. 2051-85-6   **EINECS No.** 218-131-9

Empirical Formula:
$C_{12}H_{10}N_2O_2$

Definition: Solvent Orange 1 is classed chemically as a monoazo color. It conforms to the formula:

See "Regulatory and Ingredient Use Information," for Colorants in Volume 1, Introduction, Part A. To identify the colorant allowed for use in the European Union (EU), the INCI Name CI 11920 must be used, except for hair dye products.

Information Sources: CI 11920, M3, TSCA

Chemical Class: Color Additives - Miscellaneous

Function: Colorant

Technical/Other Names:
1,3-Benzenediol, 4-(Phenylazo)-
CI 11920
2,4-Dibenzeneazoresorcinol
2,4-Dihydroxyazobenzene
Food Orange 3
4-(Phenylazo)-1,3-Benzenediol
4-Phenylazoresorcinol
Sudan Orange G

SOLVENT RED 1

CAS No. 1229-55-6   **EINECS No.** 214-968-9

Empirical Formula:
$C_{17}H_{14}N_2O_2$

Definition: Solvent Red 1 is classed chemically as a monoazo color. It conforms to the formula:

See "Regulatory and Ingredient Use Information," for Colorants in Volume 1, Introduction, Part A. To identify the colorant allowed for use in the European Union (EU), the INCI Name CI 12150 must be used, except for hair dye products.

Information Sources: CI 12150, M3, TSCA

Chemical Class: Color Additives - Miscellaneous

Function: Colorant

Technical/Other Names:
Anisol-2-Azo-beta-Naphthol
CI 12150
1-[(2-Methoxyphenyl)Azo]-2-Naphthalenol
2-Naphthalenol, 1-[(2-Methoxyphenyl)Azo]-
Sudan Red G

SOLVENT RED 3

CAS No. 6535-42-8   **EINECS No.** 229-439-8

Empirical Formula:
$C_{18}H_{16}N_2O_2$

Definition: Solvent Red 3 is classed chemically as a monoazo color. It conforms to the formula:

See "Regulatory and Ingredient Use Information," for Colorants in Volume 1, Introduction, Part A. To identify the colorant allowed for use in the European Union (EU), the INCI Name CI 12010 must be used, except for hair dye products.

Information Sources: CI 12010, M3, TSCA

Chemical Class: Color Additives - Miscellaneous

Function: Colorant

Technical/Other Names:
Ceres Brown B

CI 12010
4-[(4-Ethoxyphenyl)Azo]-1-Naphthalenol
1-Naphthalenol, 4-[(4-Ethoxyphenyl)Azo]-
Sudan Brown B

SOLVENT RED 23

CAS No.
85-86-9

EINECS No.
201-638-4

Empirical Formula:
$C_{22}H_{16}N_4O$

Definition: Solvent Red 23 is classed chemically as a disazo color. It conforms to the formula:

See "Regulatory and Ingredient Use Information," for Colorants in Volume 1, Introduction, Part A. To identify the certified colorant for labeling purposes in the US, the INCI Name Red 17 must be used. To identify the colorant allowed for use in the European Union (EU), the INCI Name CI 26100 must be used, except for hair dye products. The INCI Name for batches of this colorant that have not been certified is Solvent Red 23. To identify the colorant allowed for use in Japan, the INCI name Aka225 must be used.

Information Sources: CI 26100, M3, MI-13(8969), TSCA

Chemical Class: Color Additives - Miscellaneous

Function: Colorant

Reported Product Categories: Tonics, Dressings, and Other Hair Grooming Aids; Nail Polish and Enamel Removers; Bath Oils, Tablets, and Salts; Perfumes; Bath Soaps and Detergents; Cuticle Softeners; Eyebrow Pencils; Hair Conditioners; Hair Preparations (Non-coloring), Misc.; Moisturizing Preparations; Suntan Gels, Creams, and Liquids

Technical/Other Names:
CI 26100

2-Naphthalenol, 1-[[4-(Phenylazo)-Phenyl]Azo]-
1-[[4-(Phenylazo)Phenyl]Azo]-2-Naphthalenol
Red No. 225
Sudan III
Sudan Red BK
Tetrazobenzene-beta-Naphthol

SOLVENT RED 24

CAS No.
85-83-6

EINECS No.
201-635-8

Empirical Formula:
$C_{24}H_{20}N_4O$

Definition: Solvent Red 24 is classed chemically as a disazo color. It conforms to the formula:

See "Regulatory and Ingredient Use Information," for Colorants in Volume 1, Introduction, Part A. To identify the colorant allowed for use in Japan, the INCI name Aka501 must be used.

Information Sources: CI 26105, EEC(II-379), MI-13(8469), TSCA

Chemical Class: Color Additives - Miscellaneous

Function: Colorant

Technical/Other Names:
CI 26105
2',3-Dimethyl-4-(2-Hydroxynaphthylazo) Azobenzene
1-[[2-Methyl-4-[(2-Methylphenyl)Azo]-Phenyl]Azo]-2-Naphthalenol
2-Naphthalenol, 1-[[2-Methyl-4-[(2-Methylphenyl)Azo]Phenyl]Azo]-
2-Naphthol, 1-(4-o-Tolylazo-o-Tolylazo)-
Red No. 501
Scarlet Red

SOLVENT RED 43

CAS No.
15086-94-9

EINECS No.
239-138-3

Empirical Formula:
$C_{20}H_8Br_4O_5$

Definition: Solvent Red 43 is classed chemically as a fluoran color. It conforms to the formula:

See "Regulatory and Ingredient Use Information," for Colorants in Volume 1, Introduction, Part A. To identify the certified colorant for labeling purposes in the US, the INCI Name Red 21 must be used. To identify the certified colorant (sodium salt) for labeling purposes in the US, the INCI Name Red 22 must be used. To identify the colorant allowed for use in the European Union (EU), the INCI Name CI 45380 must be used, except for hair dye products. The INCI Name for batches of this colorant that have not been certified is Solvent Red 43. To identify the colorant allowed for use in Japan, the INCI name Aka223 must be used.

Information Sources: CI 45380, CI 45380:2, M3, TSCA

Chemical Classes: Color Additives - Miscellaneous; Halogen Compounds

Function: Colorant

Reported Product Categories: Lipsticks; Makeup Preparations (Not eye), Misc.; Rouges

Technical/Other Names:
Bromofluoresceic Acid
CI 45380:2
Eosin
Eosine G
Red No. 223
Spiro[Isobenzofuran-1(3H),9'-[9H]-Xanthen]-3-one, 2',4',5',7'-Tetrabromo-3', 6'-Dihydroxy-
2',4',5',7'-Tetrabromo-3',6'-Dihydroxyspiro [Isobenzofuran-1(3H),9'-[9H]Xanthen]-3-one
Tetrabromofluorescein

SOLVENT RED 48

CAS No.
13473-26-2

EINECS No.
236-747-6

Empirical Formula:
$C_{20}H_4Br_4Cl_4O_5$

Definition: Solvent Red 48 is classed chemically as a fluoran color. It conforms to the formula:

See "Regulatory and Ingredient Use Information," for Colorants in Volume 1, Introduction, Part A. To identify the certified colorant for labeling purposes in the US, the INCI Name Red 27 must be used. To identify the certified colorant (sodium salt) for labeling purposes in the US, the INCI Name Red 28 must be used. To identify the colorant allowed for use in the European Union (EU), the INCI Name CI 45410 must be used, except for hair dye products. The INCI Name for batches of this colorant that have not been certified is Solvent Red 48. To identify the colorant allowed for use in Japan, the INCI name Aka104(1), Aka218, or Aka231 must be used.

Information Sources: CI 45410, CI 45410:1, M3, TSCA

Chemical Classes: Color Additives - Miscellaneous; Halogen Compounds

Function: Colorant

Reported Product Category: Lipsticks

Technical/Other Names:
Benzoic Acid, 2,3,4,5-Tetrachloro-6-(2,4,5,
7-Tetrabromo-6-Hydroxy-3-Oxo-3H-
Xanthen-9-yl
CI 45410
Red No. 218
Spiro[Isobenzofuran-1(3H),9'-[9H]-
Xanthen]-3-one, 2',4',5',7'-Tetrabromo-4,
5,6,7-Tetrachloro-3',6'-Dihydroxy-
2',4',5',7'-Tetrabromo-4,5,6,7-Tetrachloro-
3',6'-Dihydroxyspiro[Isobenzofuran-1(3H),
9'-[9H]Xanthen]-3-one
Tetrachlorotetrabromofluorescein

SOLVENT RED 49:1

CAS No.	EINECS No.
6373-07-5	228-908-4

Empirical Formula:
$C_{28}H_{31}N_2O_3 \cdot C_{18}H_{35}O_2$

Definition: Solvent Red 49:1 is classed chemically as a xanthene color. It conforms to the formula:

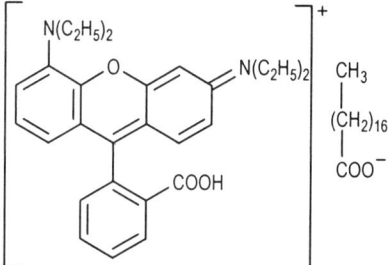

This color is the stearate of Basic Violet 10. *See "Regulatory and Ingredient Use Information," for Colorants in Volume 1, Introduction, Part A. To identify the colorant allowed for use in Japan, the INCI name Aka215 must be used.*

Information Sources: CI 45170:1, EEC(II-398), TSCA

Chemical Class: Color Additives - Miscellaneous

Function: Colorant

Technical/Other Names:
Ammonium, (9-(o-Carboxyphenyl)-6(Di-
ethylamino)-3H-xanthen-3-ylidene)-
diethyl-, Stearate
N-[9-(2-Carboxyphenyl)-6-(Diethylamino)-
3H-Xanthen-3-ylidene]-N-Ethylethana-
minium Octadecanoate
CI 45170:1
Ethanaminium, N-[9-(2-Carboxyphenyl)-
6-(Diethylamino)-3H-Xanthen-3-ylidene]-
N-Ethyl-, Octadecanoate
Red 215
Rhodamine B-Stearate
Xanthylium, 9-(2-Carboxyphenyl)-3,6-bis
(Diethylamino)-, Octadecanoate

SOLVENT RED 72

CAS Nos.	EINECS Nos.
596-03-2	209-876-0
4372-02-5	224-468-2

Empirical Formula:
$C_{20}H_{10}Br_2O_5 \cdot 2Na$

Definition: Solvent Red 72 is classed chemically as a xanthene color. It conforms to the formula:

See "Regulatory and Ingredient Use Information," for Colorants in Volume 1, Introduction, Part A. To identify the certified colorant for labeling purposes in the US, the INCI Name Orange 5 must be used. To identify the colorant allowed for use in the European Union (EU), the INCI Name CI 45370 must be used, except for hair dye products. The INCI Name for batches of this colorant that have not been certified is Solvent Red 72. To identify the colorant allowed for use in Japan, the INCI name Daidai201 must be used.

Information Sources: 21CFR701.3, CI 45370, M3, MHLW Ord. No. 30, MI-13 (3046), TSCA

Chemical Classes: Color Additives - Miscellaneous; Halogen Compounds

Function: Colorant

Reported Product Categories: Lipsticks; Makeup Preparations (Not eye), Misc.; Blushers (All types); Rouges

Technical/Other Names:
Acid Orange 11
CI 45370
4',5'-Dibromo-3',6'-Dihydroxy-Spiro[Isoben-
zofuran-1(3H),9'-[9H]Xanthen]-3-One,
Disodium Salt
Dibromofluorescein
Orange No. 201
Spiro[Isobenzofuran-1(3H),9'-[9H]-
Xanthen]-3-one, 4',5'-Dibromo-3',6'-
Dihydroxy-, Disodium Salt

SOLVENT RED 73

CAS Nos.	EINECS No.
518-40-1	
38577-97-8	245-010-7

Empirical Formula:
$C_{20}H_{10}I_2O_5$

Definition: Solvent Red 73 is classed chemically as a fluoran color. It conforms to the formula:

See "Regulatory and Ingredient Use Information," for Colorants in Volume 1, Introduction, Part A. To identify the certified colorant for labeling purposes in the US, the INCI Name Orange 10 must be used. To identify the certified colorant (sodium

---

The inclusion of any compound in the *Dictionary and Handbook* does not indicate that use of that substance as a cosmetic ingredient complies with the laws and regulations governing such use in the United States or any other country.

salt) for labeling purposes in the US, the INCI Name Orange 11 must be used. To identify the colorant allowed for use in the European Union (EU), the INCI Name CI 45425 must be used, except for hair dye products. The INCI Name for batches of this colorant that have not been certified is Solvent Red 73. To identify the colorant allowed for use in Japan, the INCI name Daidai206 or Daidai207 must be used.

Information Sources: CI 45425:1, M3, MI-13(3211), TSCA

Chemical Classes: Color Additives - Miscellaneous; Halogen Compounds

Function: Colorant

Technical/Other Names:
Benzoic Acid, 2-(6-Hydroxy-4,5-Diiodo-3-oxo-3H-Xanthen-9-yl)-
CI 45425:1
3',6'-Dihydroxy-4',5'-Diiodospiro[Isobenzofuran-1(3H),9'-[9H]Xanthen]-3-one
4,5-Diiodo-3,6-Fluorandiol
Diiodofluorescein
Hydroxydiiodo-o-Carboxyphenylfluorone
Orange No. 206
Spiro[Isobenzofuran-1(3H),9'-[9H]-Xanthen]-3-one, 3',6'-Dihydroxy-4',5'-Diiodo-

SOLVENT VIOLET 13

CAS No.	EINECS No.
81-48-1	201-353-5

Empirical Formula:
$C_{21}H_{15}NO_3$

Definition: Solvent Violet 13 is classed chemically as an anthraquinone color. It conforms to the formula:

See "Regulatory and Ingredient Use Information," for Colorants in Volume 1, Introduction, Part A. To identify the certified colorant for labeling purposes in the US, the INCI Name Violet 2 must be used. To identify the colorant allowed for use in the European Union (EU), the INCI Name CI 60725 must be used, except for hair dye products. The INCI Name for batches of this colorant that have not been certified is

Solvent Violet 13. To identify the colorant allowed for use in Japan, the INCI name Murasaki201 must be used.

Information Sources: CI 60725, M3, TSCA

Chemical Class: Color Additives - Miscellaneous

Function: Colorant

Reported Product Categories: Hair Dyes and Colors (All Types Requiring Caution Statements and Patch Tests); Colognes and Toilet Waters; Nail Polish and Enamels; Tonics, Dressings, and Other Hair Grooming Aids; Perfumes; Basecoats and Undercoats; Bath Oils, Tablets, and Salts; Hair Bleaches; Aftershave Lotions; Baby Shampoos; Deodorants (Underarm); Fragrance Preparations, Misc.; Manicuring Preparations, Misc.; Shampoos (Non-coloring); Bubble Baths; Nail Polish and Enamel Removers; Skin Care Preparations, Misc.

Technical/Other Names:
9,10-Anthracenedione, 1-Hydroxy-4-[(4-Methylphenyl)Amino]-
Anthraquinone, 1-Hydroxy-4-p-toluidino-
CI 60725
Disperse Blue 72
1-Hydroxy-4-(4-Methylanilino)Anthraquinone
1-Hydroxy-4-[(4-Methylphenyl)Amino]-9,10-Anthracenedione
1-Hydroxy-4-(p-Toluidino)Anthraquinone
Violet No. 201

SOLVENT YELLOW 18

CAS No.	EINECS No.
6407-78-9	229-043-5

Empirical Formula:
$C_{18}H_{18}N_4O$

Definition: Solvent Yellow 18 is classed chemically as a monoazo color. It conforms to the formula:

See "Regulatory and Ingredient Use Information," for Colorants in Volume 1, Introduction, Part A.

Information Sources: 21CFR73.3122, CI 12740, TSCA

Chemical Class: Color Additives - Miscellaneous

Function: Colorant

Technical/Other Names:
CI 12740
4-[(2,4-Dimethylphenyl)Azo]-2,4-Dihydro-5-Methyl-2-Phenyl-3H-Pyrazol-3-one
Food Yellow 12
3H-Pyrazol-3-one, 4-[(2,4-Dimethylphenyl)Azo]-2,4-Dihydro-5-Methyl-2-Phenyl-

SOLVENT YELLOW 29

CAS No.	EINECS No.
6706-82-7	229-754-0

Empirical Formula:
$C_{44}H_{52}N_4O_2$

Definition: Solvent Yellow 29 is classed chemically as a disazo color. It conforms to the formula:

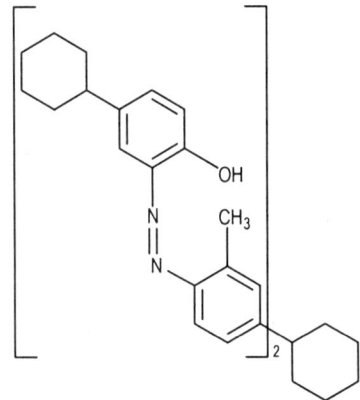

See "Regulatory and Ingredient Use Information," for Colorants in Volume 1, Introduction, Part A. To identify the colorant allowed for use in the European Union (EU), the INCI Name CI 21230 must be used, except for hair dye products.

Information Sources: CI 21230, M3, TSCA

Chemical Class: Color Additives - Miscellaneous

Function: Colorant

Technical/Other Names:
CI 21230
2,2'-[Cyclohexylidenebis[(2-Methyl-4,1-Phenylene)Azo]]Bis(4-Cyclohexylphenol]
Phenol, 2,2'-[Cyclohexylidenebis[(2-Methyl-4,1-Phenylene)Azo]]Bis[4-Cyclohexyl-
Sudan Yellow GRN

SOLVENT YELLOW 33

CAS No.	EINECS No.
8003-22-3	232-318-2

Empirical Formula:
$C_{18}H_{11}NO_2$

Definition: Solvent Yellow 33 is classed chemically as a quinoline color. It conforms to the formula:

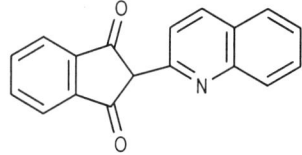

See "Regulatory and Ingredient Use Information," for Colorants in Volume 1, Introduction, Part A. To identify the certified colorant for labeling purposes in the US, the INCI Name Yellow 11 must be used. To identify the colorant allowed for use in the European Union (EU), the INCI Name CI 47000 must be used, except for hair dye products. The INCI Name for batches of this colorant that have not been certified is Solvent Yellow 33. To identify the colorant allowed for use in Japan, the INCI name Ki204 must be used.

Information Sources: CI 47000, M3, MI-13(8164), TSCA

Chemical Class: Color Additives - Miscellaneous

Function: Colorant

Reported Product Categories: Tonics, Dressings, and Other Hair Grooming Aids; Hair Conditioners; Nail Polish and Enamel Removers; Bath Oils, Tablets, and Salts; Moisturizing Preparations; Perfumes; Basecoats and Undercoats; Bath Preparations, Misc.; Body and Hand Preparations (Excluding Shaving Preparations); Bubble Baths; Depilatories; Hair Preparations (Non-coloring), Misc.; Shampoos (Non-coloring)

Technical/Other Names:
CI 47000
1H-Indene-1,3-(2H)-dione, 2-(2-Quinolinyl)-
Quinoline Yellow SS
2-(2-Quinolyl)-1,3-Indandione
Quinophthalone
Yellow No. 204

SOLVENT YELLOW 44

CAS No.
2478-20-8

EINECS No.
219-607-9

Empirical Formula:
$C_{20}H_{16}N_2O_2$

Definition: Solvent Yellow 44 is classed chemically as an aminoketone color. It conforms to the formula:

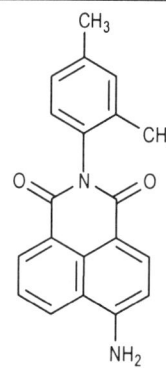

See "Regulatory and Ingredient Use Information," for Colorants in Volume 1, Introduction, Part A.

Information Sources: CI 56200, TSCA

Chemical Class: Color Additives - Miscellaneous

Function: Colorant

Technical/Other Names:
6-Amino-2-(2,4-Dimethylphenyl)-1H-Benz[de]Isoquinoline-1,3[2H]-Dione
CI 56200
Disperse Yellow 11
1H-Benz[de]Isoquinoline-1,3[2H]-Dione, 6-Amino-2-(2,4-Dimethylphenyl)-
Naphthalimide, 4-Amino-N-2,4-xylyl-

SOLVENT YELLOW 85

CAS No.
1742-95-6

EINECS No.
217-110-1

Empirical Formula:
$C_{12}H_8N_2O_2$

Definition: Solvent Yellow 85 is the colorant that conforms to the formula:

See "Regulatory and Ingredient Use Information," for Colorants in Volume 1, Introduction, Part A.

Chemical Class: Color Additives - Hair

Function: Hair Colorant

Technical/Other Names:
4-Aminonaphthalene-1,8-Dicarboximide
1H-Benz[de]Isoquinoline-1,3(2H)-Dione, 6-Amino-

Trade Name Mixtures:
A-11 Apollo Red (Swada)

T-11 Apollo Red (Swada)
A-7 Solar Yellow (Swada)
E-3 Laser Red (Swada)
FEX-21 Strong Magenta (Swada)
Fluorescent Pigments (Day Light) - Yellow (Aron Universal)
T-3 Laser Red (Swada)
T-2 Nova Red (Swada)
610 Red 11 (Sterling)
710 Red 3 (Sterling)
810 Red 2 (Sterling)
810 Red 3 (Sterling)
810 Red 11 (Sterling)
T-7 Solar Yellow (Swada)
650 Strong Magenta 21 (Sterling)
610 Yellow 7 (Sterling)
810 Yellow 7 (Sterling)

SOLVENT YELLOW 172

CAS No.
68427-35-0

EINECS No.
270-393-3

Empirical Formula:
$C_{20}H_{19}N_3O_5S$

Definition: Solvent Yellow 172 is classed chemically as a coumarin color. It conforms to the formula:

See "Regulatory and Ingredient Use Information," for Colorants in Volume 1, Introduction, Part A.

Chemical Class: Color Additives - Hair

Function: Hair Colorant

Technical/Other Name:
5-Benzoxazolesulfonamide, 2-(7-(Diethylamino)-2-oxo-2H-1-Benzopyran-3-yl

Trade Name Mixtures:
A-11 Apollo Red (Swada)
A-6 Arc Chrome (Swada)
A-5 Blaze (Swada)
A-4 Flame Orange (Swada)
A-3 Laser Red (Swada)
A-27 Lunar Yellow (Swada)
T-6 Arc Chrome (Swada)
A-8 Stellar Green (Swada)
T-5 Blaze (Swada)
E-6 Arc Chrome (Swada)
E-5 Blaze (Swada)
EBT-31 Yellow (Swada)
E-4 Flame Orange (Swada)
E-27 Lunar Yellow (Swada)
E-8 Stellar Green (Swada)
FEX-6 Arc Chrome (Swada)
FEX-1 Astral Pink (Swada)

FEX-5 Blaze (Swada)
FEX-15 Fire Red (Swada)
FEX-4 Flame Orange (Swada)
FEX-27 Lunar Yellow (Swada)
FEX-8 Stellar Green (Swada)
T-15 Fire Red (Swada)
T-4 Flame Orange (Swada)
Flare 610 Red 3 (Sterling)
Fluorescent Pigments (Day Light) - Green (Aron Universal)
Fluorescent Pigments (Day Light) - Yellow (Aron Universal)
FTX Astral Pink 1 (Swada)
FTX Blaze 5 (Swada)
FTX Flame Orange 4 (Swada)
FTX Laser Red 3 (Swada)
FTX Lunar Yellow 27 (Swada)
FTX Stellar Green 8 (Swada)
210 Green 8 (Sterling)
610 Green 8 (Sterling)
650 Green 8 (Sterling)
710 Green 8 (Sterling)
810 Green 8 (Sterling)
911 Green 8 (Sterling)
915 Green 8 (Sterling)
HMP-5 Blaze (Swada)
HMP-4 Flame Orange (Swada)
HMP-3 Laser Red (Swada)
HMP-27 Lunar Yellow (Swada)
HMP-2 Nova Red (Swada)
HMP-8 Stellar Green (Swada)
LMP-6 Arc Chrome (Swada)
LMP-5 Blaze (Swada)
LMP-4 Flame Orange (Swada)
LMP-3 Laser Red (Swada)
LMP-27 Lunar Yellow (Swada)
LMP-8 Stellar Green (Swada)
LPF ARC Chrome 6 (Swada)
LPF Blaze 5 (Swada)
LPF Flame Orange 4 (Swada)
LPF Laser Red 3 (Swada)
LPF Lunar Yellow 27 (Swada)
LPF Stellar Green 8 (Swada)
T-27 Lunar Yellow (Swada)
210 Orange 4 (Sterling)
210 Orange 5 (Sterling)
210 Orange 6 (Sterling)
610 Orange 4 (Sterling)
610 Orange 5 (Sterling)
610 Orange 6 (Sterling)
650 Orange 4 (Sterling)
650 Orange 5 (Sterling)
650 Orange 6 (Sterling)
710 Orange 4 (Sterling)
710 Orange 5 (Sterling)
710 Orange 6 (Sterling)
810 Orange 4 (Sterling)
810 Orange 5 (Sterling)
810 Orange 6 (Sterling)
911 Orange 4 (Sterling)
911 Orange 5 (Sterling)
915 Orange 4 (Sterling)
915 Orange 5 (Sterling)
915 Orange 6 (Sterling)

210 Pink 1 (Sterling)
650 Pink 1 (Sterling)
210 Red 3 (Sterling)
610 Red 11 (Sterling)
650 Red 15 (Sterling)
810 Red 15 (Sterling)
911 Red 2 (Sterling)
911 Red 15 (Sterling)
915 Red 3 (Sterling)
RTS-6 Arc Chrome (Swada)
RTS-1 Astral Pink (Swada)
RTS-5 Blaze (Swada)
RTS-4 Flame Orange (Swada)
RTS-3 Laser Red (Swada)
RTS-27 Lunar Yellow (Swada)
RTS-8 Stellar Green (Swada)
850 Series Green 8 (Sterling)
916 Series Green 8 (Sterling)
850 Series Orange 4 (Sterling)
850 Series Orange 5 (Sterling)
916 Series Orange 4 (Sterling)
916 Series Orange 5 (Sterling)
916 Series Orange 6 (Sterling)
850 Series Pink 1 (Sterling)
850 Series Red 3 (Sterling)
916 Series Red 3 (Sterling)
850 Series Yellow 27 (Sterling)
916 Series Yellow 27 (Sterling)
T-8 Stellar Green (Swada)
920 Strong Red 23 (Sterling)
920 Strong Yellow 29 (Sterling)
XSP-23 Strong Red (Swada)
XSP-29 Strong Yellow (Swada)
210 Yellow 27 (Sterling)
410 Yellow 31 (Sterling)
610 Yellow 27 (Sterling)
650 Yellow 27 (Sterling)
710 Yellow 27 (Sterling)
810 Yellow 27 (Sterling)
915 Yellow 27 (Sterling)

SONCHUS OLERACEUS EXTRACT

Definition: Sonchus Oleraceus Extract is an extract of the juice of the thistle, *Sonchus oleraceus*. See "Regulatory and Ingredient Use Information," regarding the labeling names for botanical ingredients in Volume 1, Introduction, Part A.

Information Source: JCLS

Chemical Class: Biological Products

Function: Not Reported

Technical/Other Name:
Extract of Sonchus Oleraceus

Trade Name Mixture:
VT-225 Extract of Milk Thistle (Vege-Tech)

SOPHORA ANGUSTIFOLIA ROOT EXTRACT

JPN Translation:
ラウロアンホジ酢酸 2Na / ラウロイルサルコシン

Definition: Sophora Angustifolia Root Extract is an extract of the roots of *Sophora angustifolia*. See "Regulatory and Ingredient Use Information," regarding the labeling names for botanical ingredients in Volume 1, Introduction, Part A.

Information Source: JCLS

Chemical Class: Biological Products

Function: Not Reported

Technical/Other Names:
Sophora Angustifolia Extract
Sophora Flavescens Root Extract

Trade Names:
Sophora Extract Powder (Maruzen Pharmaceuticals Co., Ltd.)
Sophora Extract Powder-S (Maruzen Pharmaceuticals Co., Ltd.)
Sophora powder (Bioland)

Trade Name Mixtures:
Kurara Liquid (Ichimaru Pharcos)
Kurara Liquid B (Ichimaru Pharcos)
Pharcolex MSTC (Ichimaru Pharcos)
Sophora Extract (Bioland)
Sophora Extract (Maruzen Pharmaceuticals Co., Ltd.)
Sophora Extract AL (Maruzen Pharmaceuticals Co., Ltd.)
Sophora Extract AL-J (Maruzen Pharmaceuticals Co., Ltd.)
Sophora Extract BG (Maruzen Pharmaceuticals Co., Ltd.)
Sophora Extract BG-J (Maruzen Pharmaceuticals Co., Ltd.)
Sophora Extract SQ (Maruzen Pharmaceuticals Co., Ltd.)
Sophora Flavescentis Root Extract (Draco)
Sophora Liquid (Maruzen Pharmaceuticals Co., Ltd.)
Synerlight (Libiol)

SOPHORA JAPONICA BUD EXTRACT

CAS No.	EINECS No.
90131-19-4	290-329-8

Definition: Sophora Japonica Bud Extract is an extract of the buds of *Sophora japonica*. See "Regulatory and Ingredient Use Information," regarding the labeling names for botanical ingredients in Volume 1, Introduction, Part A.

Chemical Class: Biological Products

Function: Not Reported

Technical/Other Name:
Extract of Sophora Japonica Bud

SOPHORA JAPONICA FLOWER EXTRACT

Definition: Sophora Japonica Flower Extract is an extract of the flowers of the

Chinese scholar tree or the Japanese pagoda tree, *Sophora japonica. See "Regulatory and Ingredient Use Information," regarding the labeling names for botanical ingredients in Volume 1, Introduction, Part A.*

Chemical Class: Biological Products

Function: Not Reported

Technical/Other Name:
Extract of Sophora Japonica Flower

Trade Name Mixtures:
Extrait De Sophora MBE BG 40 (Yves Rocher)
Herbasol Extract Japanese Pagoda (Cosmetochem) (Cosmetochem International Ltd.)
Sophora Japonica Extract HS 2665 G (Grau)
Sophora Japonica Flower Extract ies (IES LABO)

SOPHORA JAPONICA LEAF EXTRACT

Definition: Sophora Japonica Leaf Extract is an extract of the leaves of the Chinese scholar tree or the Japanese pagoda tree, *Sophora japonica. See "Regulatory and Ingredient Use Information," regarding the labeling names for botanical ingredients in Volume 1, Introduction, Part A.*

Chemical Class: Biological Products

Function: Not Reported

Technical/Other Names:
Extract of Sophora Japonica
Extract of Sophora Japonica Leaves

Trade Name Mixtures:
Sophorine (Solabia)
Vegewhite (LCW)

SOPHORA JAPONICA ROOT EXTRACT

Definition: Sophora Japonica Root Extract is an extract of the roots of *Sophora japonica. See "Regulatory and Ingredient Use Information," regarding the labeling names for botanical ingredients in Volume 1, Introduction, Part A.*

Information Source: JCLS

Chemical Class: Biological Products

Function: Not Reported

Technical/Other Name:
Extract of Sophora Japonica Root

Trade Name Mixture:
Sophora Extract (Koei Perfumery)

SORBETH-6

Empirical Formula:
$C_{18}H_{38}O_{12}$

Definition: Sorbeth-6 is the polyethylene glycol ether of Sorbitol (q.v.) with an average of 6 moles of ethylene oxide.

Chemical Classes: Alkoxylated Alcohols; Polyols

Functions: Humectant; Solvent; Viscosity Decreasing Agent

Technical/Other Names:
PEG-6 Sorbitol Ether
Polyethylene Glycol 300 Sorbitol Ether
Polyoxyethylene (6) Sorbitol Ether

SORBETH-20

Definition: Sorbeth-20 is polyethylene glycol ether of Sorbitol (q.v.) with an average of 20 moles of ethylene oxide.

Chemical Classes: Alkoxylated Alcohols; Polyols

Functions: Humectant; Solvent; Viscosity Decreasing Agent

Technical/Other Names:
PEG-20 Sorbitol Ether
Polyethylene Glycol 1000 Sorbitol Ether
Polyoxyethylene (20) Sorbitol Ether

SORBETH-30

Definition: Sorbeth-30 is the polyethylene glycol ether of Sorbitol (q.v.) with an average of 30 moles of ethylene oxide.

Chemical Classes: Alkoxylated Alcohols; Polyols

Functions: Humectant; Solvent; Viscosity Decreasing Agent

Technical/Other Names:
PEG-30 Sorbitol Ether
Polyethylene Glycol (30) Sorbitol Ether
Polyoxyethylene (30) Sorbitol Ether

Trade Name:
Atlas G-2330 (Uniqema Americas)

SORBETH-40

Definition: Sorbeth-40 is the polyethylene glycol ether of Sorbitol (q.v.) with an average of 40 moles of ethylene oxide.

Chemical Classes: Alkoxylated Alcohols; Polyols

Functions: Humectant; Solvent; Viscosity Decreasing Agent

Technical/Other Names:
PEG-40 Sorbitol Ether
Polyethylene Glycol 2000 Sorbitol
Polyoxyethylene (40) Sorbitol

SORBETH-2 BEESWAX

Definition: Sorbeth-2 Beeswax is an ethoxylated sorbitan derivative of beeswax with an average of 2 moles of ethylene oxide.

Chemical Class: Sorbitan Derivatives

Function: Surfactant - Emulsifying Agent

Technical/Other Names:
PEG-2 Sorbitan Beeswax
Polyethylene Glycol (100) Sorbitan Beeswax
Polyoxyethylene (2) Sorbitan Beeswax

Trade Name:
Nikkol BX-2V (Nikko)

SORBETH-6 BEESWAX

CAS No.: 8051-15-8

JPN Translation:
PEG - 6 ソルビットミツロウ

Definition: Sorbeth-6 Beeswax is an ethoxylated sorbitol derivative of Beeswax (q.v.) with an average of 6 moles of ethylene oxide.

Information Sources: CIR: [SQ] IJT-20 (SUPPL. 4)2001, CTFA D, JCLS

Chemical Class: Sorbitan Derivatives

Function: Surfactant - Emulsifying Agent

Technical/Other Names:
PEG-6 Sorbitan Beeswax
Polyethylene Glycol 300 Sorbitan Beeswax
Polyoxyethylene (6) Sorbitol Beeswax

Trade Name:
Nikkol GBW-25 (Nikko)

SORBETH-8 BEESWAX

JPN Translation:
PEG - 8 ソルビットミツロウ

Definition: Sorbeth-8 Beeswax is an ethoxylated sorbitan derivative of Beeswax (q.v.) with an average of 8 moles of ethylene oxide.

Information Sources: CIR: [SQ] IJT-20 (SUPPL. 4)2001, JCLS

Chemical Class: Sorbitan Derivatives

Function: Surfactant - Emulsifying Agent

Technical/Other Names:
PEG-8 Sorbitan Beeswax

Polyethylene Glycol 400 Sorbitan Beeswax
Polyoxyethylene (8) Sorbitol Beeswax

Trade Name:
Nikkol GBW-8 (Nikko)

SORBETH-20 BEESWAX

CAS No.: 8051-73-8

JPN Translation:
PEG - 20 ソルビットミツロウ

Definition: Sorbeth-20 Beeswax is an ethoxylated sorbitan derivative of Beeswax (q.v.) with an average of 20 moles of ethylene oxide.

Information Sources: CIR: [SQ] IJT-20 (SUPPL. 4)2001, CTFA D, JCLS

Chemical Class: Sorbitan Derivatives

Functions: Surfactant - Emulsifying Agent; Surfactant - Solubilizing Agent

Reported Product Categories: Mascara; Lipsticks; Eye Makeup Preparations, Misc.; Blushers (All types); Eyebrow Pencils; Eyeliners; Foot Powders and Sprays; Makeup Fixatives; Makeup Preparations (Not eye), Misc.

Technical/Other Names:
PEG-20 Sorbitan Beeswax
Polyethylene Glycol 1000 Sorbitan Beeswax
Polyoxyethylene (20) Sorbitol Beeswax

Trade Names:
Atlas G-1726 (Uniqema Americas)
Nikkol GBW-125 (Nikko)

SORBETH-2 COCOATE

Definition: Sorbeth-2 Cocoate is the ester of the fatty acids derived from Cocos Nucifera (Coconut) Oil (q.v.) and a polyethylene glycol ether of Sorbitol (q.v.) containing an average of 2 moles of ethylene oxide.

Chemical Classes: Alkoxylated Alcohols; Esters

Function: Surfactant - Emulsifying Agent

SORBETH-40 HEXAOLEATE

Definition: Sorbeth-40 Hexaoleate is the oleic acid hexaester of ethoxylated sorbitol with an average of 40 moles of ethylene oxide.

Information Sources: 21CFR175.300, 21CFR176.210, CIR: [S] IJT-19(SUPPL.2)-2000

Chemical Classes: Alkoxylated Alcohols; Esters

Function: Surfactant - Emulsifying Agent

Technical/Other Names:
Polyethylene Glycol 2000 Sorbitol Hexaoleate
Polyoxyethylene (40) Sorbitol Hexaoleate

Trade Names:
Atlas G-1086 (Uniqema Americas)
Sympatens-SHO/400 (Kolb)

SORBETH-50 HEXAOLEATE

Definition: Sorbeth-50 Hexaoleate is the oleic acid hexaester of ethoxylated sorbitol with an average of 50 moles of ethylene oxide.

Information Sources: 21CFR175.300, CIR: [S] IJT-19(SUPPL. 2) 2000

Chemical Classes: Alkoxylated Alcohols; Esters

Function: Surfactant - Emulsifying Agent

Technical/Other Names:
Polyethylene Glycol (50) Sorbitol Hexaoleate
Polyoxyethylene (50) Sorbitol Hexaoleate

Trade Names:
Atlas G-1096 (Uniqema Americas)
Ethox 3095 (Ethox)

SORBETH-6 HEXASTEARATE

CAS No.: 66828-20-4

JPN Translation:
ヘキサステアリン酸ソルベス - 6

Definition: Sorbeth-6 Hexastearate is the hexaester of ethoxylated sorbitol and stearic acid.

Chemical Class: Esters

Function: Surfactant - Emulsifying Agent

Technical/Other Name:
Sorbitol Hexaethoxylate, Stearate

Trade Name:
Nikkol GS-6 (Nikko)

SORBETH-3 ISOSTEARATE

JPN Translation:
イソステアリン酸ソルベス - 3

Empirical Formula:
$C_{30}H_{60}O_{10}$

Definition: Sorbeth-3 Isostearate is the ester of Isostearic Acid (q.v.) and a poly-

ethylene glycol ether of Sorbitol (q.v.) containing an average of 3 moles of ethylene oxide.

Chemical Classes: Alkoxylated Alcohols; Esters

Function: Surfactant - Emulsifying Agent

SORBETH-6 LAURATE

JPN Translation:
ラウリン酸 PEG ソルビット

Definition: Sorbeth-6 Laurate is the ester of lauric acid and a polyethylene glycol ether of Sorbitol (q.v.) containing an average of 6 moles of ethylene oxide.

Chemical Classes: Alkoxylated Alcohols; Esters

Functions: Surfactant - Emulsifying Agent; Surfactant - Solubilizing Agent

Technical/Other Names:
Polyethylene Glycol (6) Sorbitol
Polyoxyethylene (6) Sorbitol

Trade Name:
Nikkol GL-1 (Nikko)

SORBETH-20 PENTAISOSTEARATE

Definition: Sorbeth-20 Pentaisostearate is the pentaester of Isostearic Acid (q.v.) and a polyethylene glycol ether of Sorbitol (q.v.) containing an average of 20 moles of ethylene oxide.

Chemical Classes: Alkoxylated Alcohols; Esters

Function: Surfactant - Emulsifying Agent

Technical/Other Names:
PEG-20 Sorbitol Pentaisostearate
Polyethylene Glycol 1000 Sorbitol Pentaisostearate
Polyoxyethylene (20) Sorbitol Pentaisostearate

Trade Name:
Emalex PESIS-520 (Nihon Emulsion)

SORBETH-30 PENTAISOSTEARATE

Definition: Sorbeth-30 Pentaisostearate is the pentaester of Isostearic Acid (q.v.) and a polyethylene glycol ether of Sorbitol (q.v.) containing an average of 30 moles of ethylene oxide.

Chemical Classes: Alkoxylated Alcohols; Esters

Function: Surfactant - Emulsifying Agent

Technical/Other Names:
PEG-30 Sorbitol Pentaisostearate
Polyethylene Glycol (30) Pentaisostearate
Polyethylene Glycol (30) Sorbitol
Pentaisostearate

Trade Name:
Emalex PESIS-530 (Nihon Emulsion)

SORBETH-40 PENTAISOSTEARATE

Definition: Sorbeth-40 Pentaisostearate is the pentaester of Isostearic Acid (q.v.) and a polyethylene glycol ether of Sorbitol (q.v.) containing an average of 40 moles of ethylene oxide.

Chemical Classes: Alkoxylated Alcohols; Esters

Function: Surfactant - Emulsifying Agent

Technical/Other Names:
PEG-40 Sorbitol Pentaisostearate
Polyethylene Glycol 2000 Sorbitol
Pentaisostearate
Polyoxyethylene (40) Pentaisostearate

Trade Name:
Emalex Pesis-540 (Nihon Emulsion)

SORBETH-50 PENTAISOSTEARATE

Definition: Sorbeth-50 Pentaisostearate is the pentaester of Isostearic Acid (q.v.) and a polyethylene glycol ether of Sorbitol (q.v.) containing an average of 50 moles of ethylene oxide.

Chemical Classes: Alkoxylated Alcohols; Esters

Function: Surfactant - Emulsifying Agent

Technical/Other Names:
PEG-50 Sorbitol Pentaisostearate
Polyethylene Glycol (50) Sorbitol
Pentaisostearate
Polyoxyethylene (50) Sorbitol
Pentaisostearate

Trade Name:
Emalex PESIS-550 (Nihon Emulsion)

SORBETH-40 PENTAOLEATE

JPN Translation:
ペンタオレイン酸 PEG - 40 ソルビット

Definition: Sorbeth-40 Pentaoleate is the pentaester of Oleic Acid (q.v.) and a poly-ethylene glycol ether of Sorbitol (q.v.) containing an average of 40 moles of ethylene oxide.

Chemical Classes: Alkoxylated Alcohols; Esters

Function: Surfactant - Emulsifying Agent

Technical/Other Names:
PEG-40 Sorbitol Pentaoleate
Polyethylene Glycol 2000 Sorbitol
Pentaoleate
Polyoxyethylene (40) Sorbitol Pentaoleate

Trade Name:
Emalex PESO-540 (Nihon Emulsion)

SORBETH-20 TETRAISOSTEARATE

Definition: Sorbeth-20 Tetraisostearate is the tetraester of Isostearic Acid (q.v.) and a polyethylene glycol ether of Sorbitol (q.v.) containing an average of 20 moles of ethylene oxide.

Chemical Classes: Alkoxylated Alcohols; Esters

Function: Surfactant - Emulsifying Agent

Technical/Other Names:
PEG-20 Sorbitol Tetraisostearate
Polyethylene Glycol 1000 Sorbitol Tetra-
isostearate
Polyoxyethylene (20) Sorbitol Tetra-
isostearate

Trade Name:
Emalex PESIS-420 (Nihon Emulsion)

SORBETH-30 TETRAISOSTEARATE

Definition: Sorbeth-30 Tetraisostearate is the tetraester of Isostearic Acid (q.v.) and a polyethylene glycol ether of Sorbitol (q.v.) containing an average of 30 moles of ethylene oxide.

Chemical Classes: Alkoxylated Alcohols; Esters

Function: Surfactant - Emulsifying Agent

Technical/Other Names:
PEG-30 Sorbitol Tetraisostearate
Polyethylene Glycol (30) Sorbitol Tetra-
isostearate
Polyoxyethylene (30) Tetraisostearate

Trade Name:
Emalex PESIS-430 (Nihon Emulsion)

SORBETH-40 TETRAISOSTEARATE

Definition: Sorbeth-40 Tetraisostearate is the tetraester of Isostearic Acid (q.v.) and a polyethylene glycol ether of Sorbitol (q.v.) containing an average of 40 moles of ethylene oxide.

Chemical Classes: Alkoxylated Alcohols; Esters

Function: Surfactant - Emulsifying Agent

Technical/Other Names:
PEG-40 Sorbitol Tetraisostearate
Polyethylene Glycol 2000 Sorbitol Tetra-
isostearate
Polyoxyethylene (40) Sorbitol Tetra-
isostearate

Trade Name:
Emalex PESIS-440 (Nihon Emulsion)

SORBETH-50 TETRAISOSTEARATE

Definition: Sorbeth-50 Tetraisostearate is the tetraester of Isostearic Acid (q.v.) and a polyethylene glycol ether of Sorbitol (q.v.) containing an average of 50 moles of ethylene oxide.

Chemical Classes: Alkoxylated Alcohols; Esters

Function: Surfactant - Emulsifying Agent

Technical/Other Names:
PEG-50 Sorbitol Tetraisostearate
Polyethylene Glycol (50) Sorbitol Tetra-
isostearate
Polyoxyethylene (50) Sorbitol Tetra-
isostearate

Trade Name:
Emalex PESIS-450 (Nihon Emulsion)

SORBETH-4 TETRAOLEATE

Definition: Sorbeth-4 Tetraoleate is the tetraester of oleic acid and ethoxylated sorbitol containing an average of 4 moles of ethylene oxide.

Chemical Classes: Alkoxylated Alcohols; Esters

Function: Surfactant - Emulsifying Agent

Trade Name:
Emalex EG-2854-OL (Nihon Emulsion)

SORBETH-6 TETRAOLEATE

Definition: Sorbeth-6 Tetraoleate is the oleic acid tetraester of ethoxylated sorbitol with an average of 6 moles of ethylene oxide.

Chemical Classes: Alkoxylated Alcohols; Esters

Function: Surfactant - Emulsifying Agent

Technical/Other Names:
Polyethylene Glycol 300 Sorbitol Tetra-
oleate
Polyoxyethylene (6) Sorbitol Tetraoleate

Trade Name:
Nikkol GO-4 (Nikko)

SORBETH-30 TETRAOLEATE

JPN Translation:
テトラオレイン酸ソルベス - 30

Definition: Sorbeth-30 Tetraoleate is the tetraester of Oleic Acid (q.v.) and a polyethylene glycol ether of Sorbitol (q.v.) containing an average of 30 moles of ethylene oxide.

Chemical Classes: Alkoxylated Alcohols; Esters

Function: Surfactant - Emulsifying Agent

Technical/Other Names:
PEG-30 Sorbitol Tetraoleate
Polyethylene Glycol (30) Sorbitol Tetraoleate
Polyoxyethylene (30) Sorbitol Tetraoleate

Trade Names:
Nikkol GO-430 (Nikko)
Nikkol GO-430N (Nikko)

SORBETH-40 TETRAOLEATE

JPN Translation:
テトラオレイン酸ソルベス - 40

Definition: Sorbeth-40 Tetraoleate is the tetraester of Oleic Acid (q.v.) and a polyethylene glycol ether of Sorbitol (q.v.) with an average of 40 moles of ethylene oxide.

Chemical Classes: Alkoxylated Alcohols; Esters

Function: Surfactant - Emulsifying Agent

Technical/Other Names:
PEG-40 Sorbitol Tetraoleate
Polyethylene Glycol 2000 Sorbitol Tetraoleate
Polyoxyethylene (40) Sorbitol Tetraoleate

Trade Names:
Emalex PESO-440 (Nihon Emulsion)
Nikkol GO-440 (Nikko)

SORBETH-60 TETRAOLEATE

JPN Translation:
テトラオレイン酸ソルベス - 60

Definition: Sorbeth-60 Tetraoleate is the tetraester of Oleic Acid (q.v.) and a polyethylene glycol ether of Sorbitol (q.v.) with an average of 60 moles of ethylene oxide.

Chemical Classes: Alkoxylated Alcohols; Esters

Function: Surfactant - Emulsifying Agent

Technical/Other Names:
PEG-60 Sorbitol Tetraoleate
Polyethylene Glycol 3000 Sorbitol Tetraoleate
Polyoxyethylene (60) Sorbitol Tetraoleate

Trade Name:
Nikkol GO-460 (Nikko)

SORBETH-30 TETRAOLEATE LAURATE

Definition: Sorbeth-30 Tetraoleate Laurate is the oleic acid tetraester and lauric acid ester of ethoxylated sorbitol with an average of 30 moles of ethylene oxide.

Information Sources: 21CFR175.300, CIR: [S] IJT-19(SUPPL. 2)2000

Chemical Classes: Alkoxylated Alcohols; Esters

Function: Surfactant - Emulsifying Agent

Technical/Other Names:
Polyethylene Glycol (30) Sorbitol Tetraoleate Laurate
Polyoxyethylene (30) Sorbitol Tetraoleate Laurate

Trade Name:
Atlas G-1144 (Uniqema Americas)

SORBETH-60 TETRASTEARATE

JPN Translation:
テトラステアリン酸ソルベス - 60

Definition: Sorbeth-60 Tetrastearate is the stearic acid tetraester of ethoxylated sorbitol with an average of 60 moles of ethylene oxide.

Information Sources: CIR: [S] IJT-19 (SUPPL.2)2000, JCIC, JCLS

Chemical Classes: Alkoxylated Alcohols; Esters

Function: Surfactant - Emulsifying Agent

Technical/Other Names:
Polyethylene Glycol (60) Sorbitol Tetrastearate
Polyoxyethylene (60) Sorbitol Tetrastearate
Polyoxyethylene Sorbitol Tetrastearate (60E.O.)

Trade Name:
Nikkol GS-460 (Nikko)

SORBETH-3 TRISTEARATE

Definition: Sorbeth-3 Tristearate is the triester of stearic acid and ethoxylated sorbitol containing an average of 3 moles of ethylene oxide.

Chemical Classes: Alkoxylated Alcohols; Esters

Function: Surfactant - Emulsifying Agent

Trade Name:
Emalex EG-2854-ST (Nihon Emulsion)

SORBETH-160 TRISTEARATE

JPN Translation:
トリイソステアリン

Definition: Sorbeth-160 Tristearate is the triester of Stearic Acid (q.v.) and a polyethylene glycol ether of Sorbitol (q.v.) with an average of 160 moles of ethylene oxide.

Information Source: JCLS

Chemical Class: Esters

Function: Not Reported

SORBIC ACID

CAS No.	EINECS No.
110-44-1	203-768-7

JPN Translation:
ソルビン酸

Empirical Formula:
$C_6H_8O_2$

Definition: Sorbic Acid is the organic acid that conforms generally to the formula:

$$CH_3CH=CHCH=CHCOOH$$

Information Sources: BPC, 21CFR133, 21CFR146.152, 21CFR146.154, 21CFR150.141, 21CFR150.161, 21CFR166.110, 21CFR181.22, 21CFR181.23, 21CFR182.3089, CIR: [S] JACT-7(6)1988, CTFA S, CZE, EEC(VI/1-4), FCC, JCLS, JSCI, MAR, MHLW-331/3, MI-13(8793), NF XIX, RIFM, TSCA, USAN

Chemical Class: Carboxylic Acids

Functions: Fragrance Ingredient; Preservative

Reported Product Categories: Bath Preparations, Misc.; Body and Hand Preparations (Excluding Shaving Preparations); Bath Oils, Tablets, and Salts; Lipsticks; Cleansing Products (Cold Creams, Cleansing Lotions, Liquids and Pads); Skin Care Preparations, Misc.; Eye Shadows; Face Powders; Foundations; Paste Masks (Mud Packs); Bubble Baths; Eyeliners; Indoor Tanning Preparations; Moisturizing Preparations; Bath Capsules; Hair Conditioners; Night Skin Care Preparations; Makeup Preparations (Not eye), Misc.; Blushers (All types); Personal Cleanliness Products, Misc.; Tonics, Dressings, and Other Hair Grooming Aids; Eye Makeup Preparations, Misc.; Eye Makeup Removers; Face and Neck Preparations (Excluding Shaving Preparations); Baby Products, Misc.; Eyebrow Pencils; Mascara; Shampoos (Non-coloring); Suntan Gels, Creams, and Liquids; Suntan Preparations, Misc.; Eye Lotions; Manicuring Preparations, Misc.; Skin Fresheners

Technical/Other Names:
2,4-Hexadienoic Acid
2,4-Hexadienoic acid, (E,E)- (RIFM)
1,3-Pentadiene-1-Carboxylic Acid

Trade Names:
Jeen Sorbic Acid (Jeen)
Nutrinova Sorbic Acid (Nutrinova)
Sorbic Acid PC (Protameen)
Tristat (Tri-K)
Unistat SBA (Universal Preserv-A-Chem)

Trade Name Mixtures:
Dragocid Forte 2/027045 (Symrise)
Elestab 4150 Lipo (Laboratoires Sero-biologiques)
Fondix G Bis (Gattefosse s.a.)
Germazide WS (Collaborative Labs)
Neo Dragocide Powder 2/060100 (Symrise)

SORBITAN CAPRYLATE

Empirical Formula:
$C_{14}H_{26}O_6$

Definition: Sorbitan Caprylate is the monoester of caprylic acid and hexitol anhydrides derived from sorbitol. It conforms generally to the formula:

Information Source: CIR: [S] IJT-21 (SUPPL. 1)2002

Chemical Class: Sorbitan Derivatives

Function: Surfactant - Emulsifying Agent

Technical/Other Name:
Sorbitan, Octanoate

Trade Names:
Antistatique WL 879 (Gattefosse s.a.)
Sorbon S-10 (Toho)

SORBITAN COCOATE

CAS No.	EINECS No.
68154-36-9	268-910-2

JPN Translation:
ヤシ脂肪酸ソルビタン

Definition: Sorbitan Cocoate is the monoester of coconut acid and hexitol anhydrides derived from sorbitol. It conforms generally to the formula:

where RCO- represents the fatty acids derived from coconut oil.

Information Sources: CIR: [S] IJT-21 (SUPPL. 1)2002, JCIC, JCLS, JSQI, TSCA

Chemical Class: Sorbitan Derivatives

Function: Surfactant - Emulsifying Agent

Technical/Other Names:
Anhydrosorbitol Monococoate
Fatty Acids, Coco, Monoesters with Sorbitan
Sorbitan Monococoate

SORBITAN DIISOSTEARATE

CAS No.	EINECS No.
68238-87-9	269-410-7

Empirical Formula:
$C_{42}H_{80}O_7$

Definition: Sorbitan Diisostearate is the diester of Isostearic Acid (q.v.) and hexitol anhydrides derived from sorbitol.

Information Source: CIR: [S] IJT-21 (SUPPL. 1)2002

Chemical Class: Sorbitan Derivatives

Function: Surfactant - Emulsifying Agent

Technical/Other Names:
Anhydrohexitol Diisostearate
Sorbitan Diisooctadecanoate

SORBITAN DIOLEATE

CAS No.	EINECS No.
29116-98-1	249-448-0

Empirical Formula:
$C_{42}H_{76}O_7$

Definition: Sorbitan Dioleate is the diester of oleic acid and hexitol anhydrides derived from sorbitol.

Information Sources: 21CFR175.320, CIR: [S] IJT-21(SUPPL. 1)2002, CTFA D

Chemical Class: Sorbitan Derivatives

Function: Surfactant - Emulsifying Agent

Technical/Other Names:
Anhydrosorbitol Dioleate
Oleic Acid, Diester with Sorbitan
Sorbide Dioleate
Sorbitan, Di-9-Octadecenoate

Trade Name:
AEC Sorbitan Dioleate (A & E Connock)

SORBITAN DISTEARATE

CAS No.: 36521-89-8

JPN Translation:
ジステアリン酸ソルビタン

Empirical Formula:
$C_{42}H_{88}O_7$

Definition: Sorbitan Distearate is the diester of stearic acid and the hexitol anhydrides derived from sorbitol.

Information Sources: 21CFR175.320, CIR: [S] IJT-21(SUPPL. 1)2002, JCIC, JCLS

Chemical Class: Sorbitan Derivatives

Function: Surfactant - Emulsifying Agent

Technical/Other Names:
Anhydrosorbitol Distearate
Sorbitan Dioctadecanoate

Trade Names:
Sorbon S-66 (Toho)
Sorbon S-66F (Toho)

SORBITAN FATTY ACID ESTER

JPN Translation:
加水分解トサカ

Definition: *See "Regulatory and Ingredient Use Information," regarding use of Japan Trivial names in Volume 1, Introduction, Part A.*

Information Source: JCLS

Chemical Classes: Esters; Sorbitan Derivatives

Function: Not Reported

SORBITAN ISOSTEARATE

CAS Nos.	EINECS No.
54392-26-6	
71902-01-7	276-171-2

JPN Translation:
イソステアリン酸ソルビタン

Empirical Formula:
$C_{24}H_{46}O_6$

Definition: Sorbitan Isostearate is the monoester of Isostearic Acid (q.v.) and hexitol anhydrides derived from sorbitol.

Information Sources: CIR: [S] IJT-21 (SUPPL. 1)2002, JCIC, JCLS, JSCI

Chemical Class: Sorbitan Derivatives

Function: Surfactant - Emulsifying Agent

Reported Product Categories: Eye Shadows; Blushers (All types); Makeup Bases; Eye Makeup Preparations, Misc.; Baby Lotions, Oils, Powders and Creams; Foundations; Eyebrow Pencils; Eyeliners; Moisturizing Preparations; Suntan Gels, Creams, and Liquids

Technical/Other Names:
1,4-Anhydro-D-Glucitol, 6-Isooctadecanoate
Anhydrosorbitol Monoisostearate
D-Glucitol, 1,4-Anhydro-, 6-Isooctadecanoate
Sorbitan, Monoisooctadecanoate
Sorbitan Monoisostearate

Trade Names:
AEC Sorbitan Isostearate (A & E Connock)
Arlacel 987 (Uniqema Americas)
Crill 6 (Croda Chemicals)
Crill 6 (Croda, Inc.)
Kosteran-I/1 (Kolb)
Megatan 20 (Megachem)
Montane 70 (SEPPIC)
Nikkol SI-10R (Nikko)
Nikkol SI-10T (Nikko)
PRISORINE 3768 (Uniqema Europe)

SORBITAN LAURATE

CAS Nos.	EINECS Nos.
1337-30-0	
1338-39-2	215-663-3
5959-89-7	227-729-9

JPN Translation:
ラウリン酸ソルビタン

Empirical Formula:
$C_{18}H_{34}O_6$

Definition: Sorbitan Laurate is the monoester of lauric acid and hexitol anhydrides derived from sorbitol. It conforms generally to the formula:

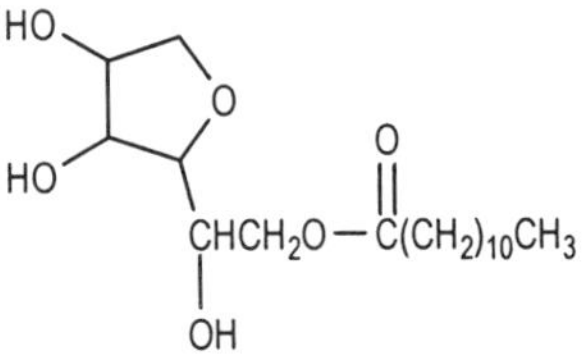

Information Sources: BAN, BPC, 21CFR175.320, 21CFR178.3400, CIR: [S]

JACT-4(3)1985, CTFA D, INN, JCLS, JSCI, MAR, NF XIX, TSCA, USAN

Chemical Class: Sorbitan Derivatives

Function: Surfactant - Emulsifying Agent

Reported Product Categories: Foundations; Lipsticks; Moisturizing Preparations; Bath Preparations, Misc.; Body and Hand Preparations (Excluding Shaving Preparations); Fragrance Preparations, Misc.; Makeup Bases; Bath Oils, Tablets, and Salts; Cleansing Products (Cold Creams, Cleansing Lotions, Liquids and Pads); Mascara; Skin Care Preparations, Misc.; Eyeliners; Paste Masks (Mud Packs); Shampoos (Non-coloring); Makeup Fixatives; Makeup Preparations (Not eye), Misc.; Suntan Gels, Creams, and Liquids

Technical/Other Names:
1,4-Anhydro-D-Glucitol, 6-Dodecanoate
Anhydrosorbitol Monolaurate
D-Glucitol, 1,4-Anhydro-, 6-Dodecanoate
Sorbitan, Monododecanoate
Sorbitan Monolaurate

Trade Names:
AEC Sorbitan Laurate (A & E Connock)
Alkamuls SML (Rhodia)
Arlacel 20 (Uniqema Americas)
Crill 1 (Croda Chemicals)
DeMULS SML (DeForest)
Extan-LT (Lanaetex)
Glycomul L (Lonza Inc./Lonza Ltd.)
Ixolene 2 (Vevy)
Jeechem SML (Jeen)
Kosteran-L/1 (Kolb)
Liposorb L (Lipo)
Lumisorb SML (Lambent)
Montane 20 (SEPPIC)
Nikkol SL-10 (Nikko)
Protachem SML (Protameen)
Radiasurf 7125 (Oleon NV)
Sabosorb ML (Sabo)
Sorbax SML (Chemax)
Sorbirol L (Cesalpinia)
Sorbon S-20 (Toho)
Span 20 (Uniqema Americas)
Tego SML (Degussa Care Specialties)

Trade Name Mixtures:
Amisol 4135 (Lucas Meyer GmbH)
Tego Pearl S 33 (Degussa Care Specialties)

SORBITAN OLEATE

CAS Nos.	EINECS No.
1338-43-8	215-665-4
37318-79-9	

JPN Translation:
オレイン酸ソルビタン

Empirical Formula:
$C_{24}H_{44}O_6$

Definition: Sorbitan Oleate is the monoester of oleic acid and hexitol anhydrides derived from sorbitol. It conforms generally to the formula:

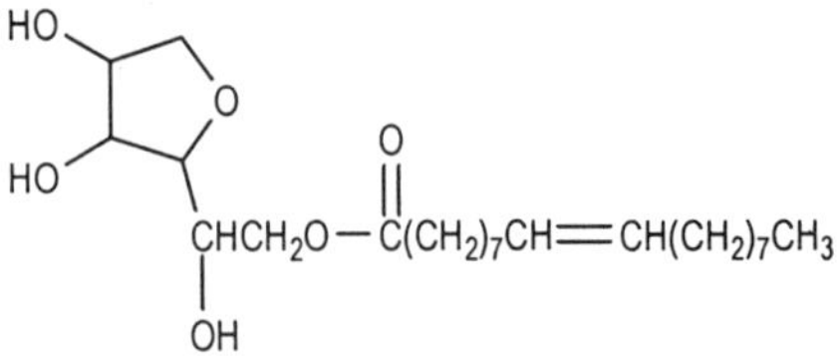

Information Sources: BAN, BPC, 21CFR73.1001, 21CFR173.75, 21CFR175.105, 21CFR175.320, 21CFR178.3400, CIR: [S] JACT-4(3)1985, CTFA D, DDR, INN, JCLS, JSCI, MAR, NF XIX, RIFM, TSCA, USAN

Chemical Class: Sorbitan Derivatives

Functions: Fragrance Ingredient; Surfactant - Emulsifying Agent

Reported Product Categories: Moisturizing Preparations; Foundations; Permanent Waves; Bath Oils, Tablets, and Salts; Bath Preparations, Misc.; Body and Hand Preparations (Excluding Shaving Preparations); Cleansing Products (Cold Creams, Cleansing Lotions, Liquids and Pads); Fragrance Preparations, Misc.; Eye Shadows; Makeup Bases; Night Skin Care Preparations; Skin Fresheners

Technical/Other Names:
Alkamuls S80
1,4-Anhydro-D-Glucitol, 6-(9-Octadecenoate)
Anhydrosorbitol Monooleate
D-Glucitol, 1,4-Anhydro-, 6-(9-Octadecenoate)
Sorbitan, Mono-9-Octadecenoate
Sorbitan monooleate (RIFM)

Trade Names:
AEC Sorbitan Oleate (A & E Connock)
Alkamuls SMO (Rhodia)
Arlacel 80 (Uniqema Americas)
Crill 4 (Croda Chemicals)
Crill 50 (Croda Chemicals)
Customulse O (sorbitan oleate) (Custom Ingredients)
DeMULS SMO (DeForest)
Extan-OT (Lanaetex)
Ixolene 8 (Vevy)
Jeechem SMO (Jeen)
Kosteran-O/1 (Kolb)
Liposorb O (Lipo)
Lumisorb SMO (Lambent)
Montane 80 (SEPPIC)
Nikkol SO-10 (Nikko)
Nikkol SO-10R (Nikko)
Protachem SMO (Protameen)

Radiasurf 7155 (Oleon NV)
Radiasurf 7755 (Oleon NV)
Rylo SO 17 (Danisco)
Sabosorb MO (Sabo)
Sorbac 80-VDO (Specialty Industrial)
Sorbax SMO (Chemax)
Sorbirol O (Cesalpinia)
Sorbon S-80 (Toho)
Span 80 (Uniqema Americas)
Sunsoft 81S (Taiyo Kagaku)
Tego SMO V (Degussa Care Specialties)

Trade Name Mixtures:
Base W/O 126 (LCW)
Eusolex T-Olio P (Merck KGaA)
Liant TW 406 (LCW)
Liant TW 729 (LCW)
Mascawax 012 (LCW)
Montane 481 (SEPPIC)
Oleo-Coll A240 (Arch Personal Care
   Products)
Oleo-Coll A240-20 (Arch Personal Care
   Products)
Proto-Lan KT (Maybrook)
Sabowax FL 81 (Sabo)
Simulgel 800 (SEPPIC)
Vitaphyle ACE (LCW)

SORBITAN OLIVATE

Definition: Sorbitan Olivate is the monoester of the fatty acids derived from olive oil and hexitol anhydrides derived from sorbitol. It conforms generally to the formula:

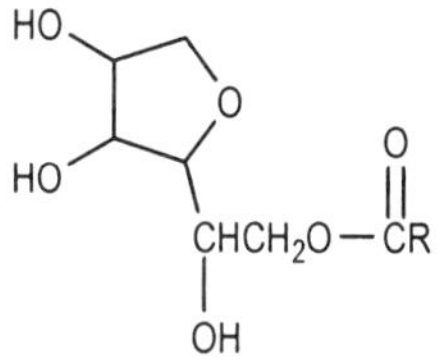

where RCO- represents the fatty acids derived from olive oil.

Information Source: CIR: [S] IJT-21 (SUPPL. 1)2002

Chemical Class: Sorbitan Derivatives

Function: Surfactant - Emulsifying Agent

Technical/Other Names:
Anhydrosorbitol Monoolivate
Fatty Acids, Olive, Monoesters with
   Sorbitan
Sorbitan Monoolivate

Trade Name:
Olivem 900 (B & T)

Trade Name Mixture:
Olivem 1000 (B & T)

SORBITAN PALMITATE

CAS Nos.	EINECS No.
5050-91-9	
26266-57-9	247-568-8

JPN Translation:
パルミチン酸ソルビタン

Empirical Formula:
$C_{22}H_{42}O_6$

Definition: Sorbitan Palmitate is the monoester of palmitic acid and hexitol anhydrides derived from sorbitol. It conforms generally to the formula:

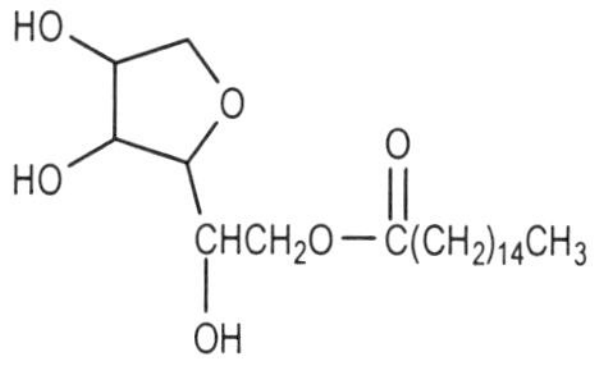

Information Sources: BAN, 21CFR175.320, 21CFR178.3400, CIR: [S] JACT-4(3)1985, CTFA D, INN, JCLS, JSCI, MAR, NF XIX, TSCA, USAN

Chemical Class: Sorbitan Derivatives

Function: Surfactant - Emulsifying Agent

Reported Product Categories: Eye Makeup Preparations, Misc.; Makeup Preparations (Not eye), Misc.; Bath Oils, Tablets, and Salts; Cleansing Products (Cold Creams, Cleansing Lotions, Liquids and Pads); Eyebrow Pencils; Indoor Tanning Preparations; Eyeliners; Lipsticks; Moisturizing Preparations; Paste Masks (Mud Packs)

Technical/Other Names:
1,4-Anhydro-D-Glucitol, 6-Hexadecanoate
D-Glucitol, 1,4-Anhydro-, 6-Hexadecanoate
Sorbitan, Monohexadecanoate
Sorbitan Monopalmitate

Trade Names:
AEC Sorbitan Palmitate (A & E Connock)
Arlacel 40 (Uniqema Americas)
Armotan MP (Akzo Nobel Surface AB)
Crill 2 (Croda Chemicals)
Extan-PT (Lanaetex)
Ixolene 4 (Vevy)
Jeechem SMP (Jeen)
Kosteran-P/1 (Kolb)
Liposorb P (Lipo)
Lonzest SMP (Lonza Inc./Lonza Ltd.)
Montane 40 (SEPPIC)
Nikkol SP-10 (Nikko)
Protachem SMP (Protameen)
Radiasurf 7135 (Oleon NV)
Sabosorb MP (Sabo)
Sorbac 40 (Specialty Industrial)
Sorbax SMP (Chemax)
Sorbirol P (Cesalpinia)
Sorbon S-40 (Toho)
Span 40 (Uniqema Americas)

SORBITAN SESQUIISOSTEARATE

CAS No.: 71812-38-9

JPN Translation:
セスキイソステアリン酸ソルビタン

Definition: Sorbitan Sesquiisostearate is a mixture of mono and diesters of Isostearic Acid (q.v.) and hexitol anhydrides derived from sorbitol.

Information Sources: CIR: [S] IJT-21 (SUPPL. 1)2002, JCIC, JCLS

Chemical Class: Sorbitan Derivatives

Function: Surfactant - Emulsifying Agent

Reported Product Categories: Eye Shadows; Face Powders; Foundations; Eyeliners; Blushers (All types); Eye Makeup Preparations, Misc.; Eyebrow Pencils; Makeup Preparations (Not eye), Misc.; Makeup Fixatives; Depilatories

Technical/Other Name:
Sorbitan, Isooctadecanoate (2:3)

Trade Names:
AEC Sorbitan Sesquiisostearate (A & E
   Connock)
Cosmol 182 (Nisshin OilliO)
Nikkol SI-15R (Nikko)
Salacos 182 (Nisshin OilliO)

SORBITAN SESQUIOLEATE

CAS No.	EINECS No.
8007-43-0	232-360-1

JPN Translation:
セスキオレイン酸ソルビタン

Definition: Sorbitan Sesquioleate is a mixture of mono and diesters of oleic acid and hexitol anhydrides derived from sorbitol.

Information Sources: BAN, 21CFR175.320, CIR: [S] JACT-4(3)1985, CTFA D, INN, JCLS, JSCI, MAR, NF XIX, TSCA, USAN

Chemical Class: Sorbitan Derivatives

Function: Surfactant - Emulsifying Agent

Reported Product Categories: Foundations; Mascara; Lipsticks; Moisturizing Preparations; Face Powders; Makeup Preparations (Not eye), Misc.; Skin Care Preparations, Misc.; Bath Oils, Tablets, and Salts; Cleansing Products (Cold Creams, Cleansing Lotions, Liquids and Pads); Night Skin Care Preparations; Eye Shadows; Eyebrow Pencils; Suntan Gels, Creams, and Liquids; Bath Preparations, Misc.; Body and Hand Preparations (Excluding Shaving Preparations); Bath Capsules; Eye Makeup Preparations, Misc.; Eyeliners; Blushers (All types); Face and Neck Preparations (Excluding Shaving Preparations); Hair Conditioners

Technical/Other Names:
Anhydrohexitol Sesquioleate

Anhydrosorbitol Sesquioleate
Sorbitan, 9-Octadecenoate (2:3)

Trade Names:
AEC Sorbitan Sesquioleate (A & E Connock)
Arlacel 83 (Uniqema Americas)
Arlacel C (Uniqema Americas)
Cosmol 82 (Nisshin OilliO)
Crill 43 (Croda Chemicals)
Extan-SOT (Lanaetex)
Jeechem SOC (Jeen)
Kosteran-SQ/O (Kolb)
Liposorb SQO (Lipo)
Lonzest SOC (Lonza Inc./Lonza Ltd.)
Lumisorb SSO (Lambent)
Montane 83 (SEPPIC)
Nikkol SO-15 (Nikko)
Nikkol SO-15EX (Nikko)
Nikkol SO-15R (Nikko)
Nikkol SO-15TX (Nikko)
Protachem SOC (Protameen)
Sabosorb SQ (Sabo)
Salacos 82 (Nisshin OilliO)
Sorbon S-83L (Toho)

Trade Name Mixtures:
Dehymuls E (Cognis Care Chemicals/NJ)
Dehymuls K (Cognis Deutschland)
Dehymuls SBL (Cognis Care Chemicals/NJ)
Forlan L (RITA)
Forlan LM (RITA)
Lanaetex-H (Lanaetex)
Neo-PCL w/o s.e. 2/066255 (Symrise)
PCL SE w/o 2/066255 (Symrise)
Umulse-E (Universal Preserv-A-Chem)

SORBITAN SESQUISTEARATE

CAS No. 51938-44-4
EINECS No. 257-529-7

JPN Translation:
セスキステアリン酸ソルビタン

Definition: Sorbitan Sesquistearate is a mixture of mono and diesters of stearic acid and hexitol anhydrides derived from sorbitol.

Information Sources: 21CFR175.320, CIR: [S] IJT-21(SUPPL. 1)2002, JCLS, JSCI

Chemical Class: Sorbitan Derivatives

Function: Surfactant - Emulsifying Agent

Technical/Other Names:
Anhydrosorbitol Sesquistearate
Sorbitan Octadecanoate (2:3)

Trade Name:
Nikkol SS-15 (Nikko)

SORBITAN STEARATE

CAS Nos. 1338-41-6
EINECS No. 215-664-9

5093-91-4
56451-84-4

JPN Translation:
ステアリン酸ソルビタン

Empirical Formula:
$C_{24}H_{46}O_6$

Definition: Sorbitan Stearate is the monoester of stearic acid and hexitol anhydrides derived from sorbitol. It conforms generally to the formula:

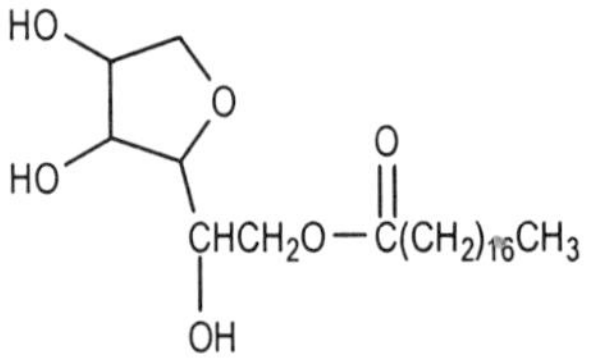

Information Sources: BAN, BPC, 21CFR73.1001, 21CFR163.123, 21CFR163.130, 21CFR163.135, 21CFR163.140, 21CFR163.145, 21CFR163.150, 21CFR163.153, 21CFR163.155, 21CFR172.515, 21CFR172.842, 21CFR173.340, 21CFR175.105, 21CFR175.320, 21CFR178.3400, 21CFR573.960, CIR: [S] JACT-4(3)1985, CTFA D, FCC, INN, JCLS, JSCI, MAR, NF XIX, RIFM, TSCA, USAN

Chemical Class: Sorbitan Derivatives

Functions: Fragrance Ingredient; Surfactant - Emulsifying Agent

Reported Product Categories: Moisturizing Preparations; Body and Hand Preparations (Excluding Shaving Preparations); Bath Preparations, Misc.; Skin Care Preparations, Misc.; Bath Oils, Tablets, and Salts; Cleansing Products (Cold Creams, Cleansing Lotions, Liquids and Pads); Mascara; Bath Capsules; Face and Neck Preparations (Excluding Shaving Preparations); Night Skin Care Preparations; Fragrance Preparations, Misc.; Paste Masks (Mud Packs); Deodorants (Underarm); Foundations; Makeup Bases; Makeup Preparations (Not eye), Misc.; Baby Lotions, Oils, Powders and Creams; Eye Shadows; Eyeliners; Hair Conditioners; Indoor Tanning Preparations; Tonics, Dressings, and Other Hair Grooming Aids; Cuticle Softeners; Eye Makeup Preparations, Misc.; Eye Makeup Removers; Eyebrow Pencils; Foot Powders and Sprays; Suntan Gels, Creams, and Liquids

Technical/Other Names:
1,4-Anhydro-D-Glucitol, 6-Octadecanoate
Anhydrosorbitol Monostearate
D-Glucitol, 1,4-Anhydro-, 6-Octadecanoate
Sorbitan, Monooctadecanoate
Sorbitan monostearate (RIFM)

Trade Names:
AEC Sorbitan Stearate (A & E Connock)

Alkamuls SMS (Rhodia)
Arlacel 60 (Uniqema Americas)
Crill 3 (Croda Chemicals)
Crill 3 (Croda, Inc.)
DeMULS SMS (DeForest)
ESTOL 1963 (Uniqema Europe)
Extan-ST (Lanaetex)
Ixolene 6 (Vevy)
Jeechem SMS (Jeen)
Kosteran-S/1 (Kolb)
Liposorb S (Lipo)
Lonzest SMS (Lonza Inc./Lonza Ltd.)
Lumisorb SMS (Lambent)
Montane 60 (SEPPIC)
Nikkol SS-10 (Nikko)
Nikkol SS-10M (Nikko)
Protachem SMS (Protameen)
Radiasurf 7145 (Oleon NV)
Rylo SO 18 (Danisco)
Sabosorb MS (Sabo)
Sorbac 60-V (Specialty Industrial)
Sorbax SMS (Chemax)
Sorbirol S (Cesalpinia)
Sorbon S-60 (Toho)
Span 60 (Uniqema Americas)
Sunsoft 61S (Taiyo Kagaku)
Tego SMS (Degussa Care Specialties)

Trade Name Mixtures:
AF75 (GE Silicones)
Arlatone 2121 (Uniqema Americas)
Covawax 501 (LCW)
Micapoly MSL Red (Creations Couleurs)
Micapoly MSL White (Creations Couleurs)
Micapoly MSL Yellow (Creations Couleurs)
Nikkol MGS-C (Nikko)

SORBITAN TRIISOSTEARATE

CAS No. 54392-27-7
EINECS No. 259-141-3

Empirical Formula:
$C_{60}H_{114}O_8$

Definition: Sorbitan Triisostearate is the triester of Isostearic Acid (q.v.) and hexitol anhydrides derived from sorbitol.

Information Sources: CIR: [S] IJT-21 (SUPPL. 1)2002, TSCA

Chemical Class: Sorbitan Derivatives

Function: Surfactant - Emulsifying Agent

Technical/Other Names:
Anhydrosorbitol Triisostearate
Sorbitan, Triisooctadecanoate

Trade Names:
AEC Sorbitan Triisostearate (A & E Connock)
Crill 65 (Croda Chemicals)

SORBITAN TRIOLEATE

CAS No. 26266-58-0
EINECS No. 247-569-3

JPN Translation:
トリオレイン酸ソルビタン

Empirical Formula:
$C_{60}H_{108}O_8$

Definition: Sorbitan Trioleate is the triester of oleic acid and hexitol anhydrides derived from sorbitol.

Information Sources: BAN, 21CFR175.320, 21CFR178.3400, CIR: [S] JACT-4(3)1985, INN, JCLS, JSCI, MAR, NF XIX, TSCA, USAN

Chemical Class: Sorbitan Derivatives

Function: Surfactant - Emulsifying Agent

Reported Product Categories: Blushers (All types); Cleansing Products (Cold Creams, Cleansing Lotions, Liquids and Pads); Tonics, Dressings, and Other Hair Grooming Aids

Technical/Other Names:
Anhydrosorbitol Trioleate
Sorbitan, Tri-9-Octadecenoate

Trade Names:
AEC Sorbitan Trioleate (A & E Connock)
Arlacel 85 (Uniqema Americas)
Crill 45 (Croda Chemicals)
Dehymuls STO (Cognis France)
Jeechem STO (Jeen)
Kosteran-O/3 (Kolb)
Liposorb TO (Lipo)
Lonzest STO (Lonza Inc./Lonza Ltd.)
Montane 85 (SEPPIC)
Nikkol SO-30 (Nikko)
Nikkol SO-30R (Nikko)
Protachem STO (Protameen)
Sabosorb TO (Sabo)
Sorbax STO (Chemax)
Sorbon S-85 (Toho)
Span 85 (Uniqema Americas)
Tego STO V (Degussa Care Specialties)

SORBITAN TRISTEARATE

CAS No.	EINECS No.
26658-19-5	247-891-4

JPN Translation:
トリステアリン酸ソルビタン

Empirical Formula:
$C_{60}H_{114}O_8$

Definition: Sorbitan Tristearate is the triester of stearic acid and hexitol anhydrides derived from sorbitol.

Information Sources: BAN, 21CFR175.320, 21CFR178.3400, CIR: [S] JACT-4(3)1985, CTFA D, INN, JCLS, JSCI, MAR, TSCA, USAN

Chemical Class: Sorbitan Derivatives

Function: Surfactant - Emulsifying Agent

Reported Product Categories: Moisturizing Preparations; Body and Hand Preparations (Excluding Shaving Preparations)

Technical/Other Names:
Anhydrosorbitol Tristearate
Sorbitan, Trioctadecanoate

Trade Names:
AEC Sorbitan Tristearate (A & E Connock)
Crill 35 (Croda Chemicals)
Crill 41 (Croda Chemicals)
Jeechem STS (Jeen)
Kosteran-S/3 (Kolb)
Liposorb TS (Lipo)
Lonzest STS (Lonza Inc./Lonza Ltd.)
Lumisorb STS (Lambent)
Montane 65 (SEPPIC)
Nikkol SS-30 (Nikko)
Protachem STS (Protameen)
Rylo SO 38 (Danisco)
Sabosorb TS (Sabo)
Sorbac 65-K (Specialty Industrial)
Sorbax STS (Chemax)
Span 65 (Uniqema Americas)

Trade Name Mixtures:
Suncaps 664 (Particle Sciences)
Suncaps 903 (Particle Sciences)

SORBITAN UNDECYLENATE

CAS No.	EINECS No.
93963-92-9	300-837-4

Empirical Formula:
$C_{17}H_{30}O_6$

Definition: Sorbitan Undecylenate is the monoester of undecylenic acid and the hexitol anhydrides derived from soribtol. It conforms to the formula:

Chemical Class: Sorbitan Derivatives

Function: Surfactant - Emulsifying Agent

Technical/Other Name:
Sorbitan, Mono-1-Undecenoate

Trade Name:
DUB MUSO (Stearinerie Dubois Fils)

SORBITOL

CAS No.	EINECS No.
50-70-4	200-061-5

JPN Translation:
ソルビトール

Empirical Formula:
$C_6H_{14}O_6$

Definition: Sorbitol is the hexahydric alcohol that conforms to the formula:

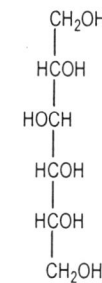

Information Sources: BP, BPC, 21CFR73.2125, 21CFR172.820, 21CFR175.300, 21CFR175.320, 21CFR176.210, 21CFR177.2420, 21CFR182.90, 21CFR184.1835, 21CFR310.545, 21CFR582.5835, CTFA S, DA, DDR, FCC, HUN, ITA, JAN, JCLS, JP, JSCI, KOR, MAR, MI-13(8797), NF XIX, OTC-I-LX, RIFM, TSCA, USAN, USD, USP XX, USP XXIV

Chemical Class: Polyols

Functions: Flavoring Agent; Fragrance Ingredient; Humectant; Skin-Conditioning Agent - Humectant

Reported Product Categories: Bath Oils, Tablets, and Salts; Hair Dyes and Colors (All Types Requiring Caution Statements and Patch Tests); Moisturizing Preparations; Cleansing Products (Cold Creams, Cleansing Lotions, Liquids and Pads); Bath Preparations, Misc.; Body and Hand Preparations (Excluding Shaving Preparations); Skin Care Preparations, Misc.; Tonics, Dressings, and Other Hair Grooming Aids; Night Skin Care Preparations; Bath Capsules; Bath Soaps and Detergents; Face and Neck Preparations (Excluding Shaving Preparations); Hair Conditioners; Paste Masks (Mud Packs); Hair Sprays (Aerosol Fixatives); Dentifrices (Aerosol, Liquid, Pastes and Powders); Shaving Preparations, Misc.; Shaving Cream (Aerosol, Brushless and Lather); Eyebrow Pencils; Suntan Gels, Creams, and Liquids; Shampoos (Non-coloring); Deodorants (Underarm); Hair Bleaches; Mouthwashes and Breath Fresheners (Liquids and Sprays); Skin Fresheners; Aftershave Lotions; Baby Shampoos; Eyeliners; Indoor Tanning Preparations; Baby Lotions, Oils, Powders and Creams; Eye Makeup Preparations, Misc.; Hair Preparations (Non-coloring), Misc.; Sachets; Suntan Preparations, Misc.; Fragrance Preparations, Misc.; Hair Wave Sets; Permanent Waves; Personal Cleanliness Products, Misc.; Foundations

Technical/Other Names:
D-Glucitol
d-Sorbitol (RIFM)
Sorbitol Solution

Trade Names:
A-642 (SPI Polyols)
A-625 non-crystallizing sorbitol solution (SPI Polyols)
Arlex Polyol (SPI Polyols)
A-642 Sorbitol (SPI Polyols)
Karion FP (Merck KGaA)
Karion FP Liquid (Merck KGaA)
Karion P (Merck KGaA)
Liponic 70 NC (Lipo)
Neosorb (Roquette)
Sorbitol F Liquid (Karion F Liquid) (Merck KGaA)
Sorbitol Sol'n 70% (Lipo)
Sorbitol USP Powder (Lipo)
Sorbo 70% Sorbitol Solution (SPI Polyols)
Unisweet 70 (Universal Preserv-A-Chem)
Unisweet CONC (Universal Preserv-A-Chem)

Trade Name Mixtures:
Activator Omega MO Type B (Derma-Search)
AEC Cream of Almonds (A & E Connock)
Beta Hydroxy Acid (BHA) Willow Bark (Cosmetochem) (Cosmetochem International Ltd.)
Biosil Basics HMC - Hair Moisture Complex (Biosil Technologies, Inc.)
Biosil Basics HMC-1 Hair Moisture Complex (Biosil Technologies, Inc.)
Biosil Basics HMV- Hair Moisture Complex (Biosil Technologies, Inc.)
Biosil Basics HMW - Hair Moisture Complex (Biosil Technologies, Inc.)
Biosil Basics SMC (Biosil Technologies, Inc.)
Bleu Covasorb W 6783 A (LCW)
Carotenic 1% (Nikken Sohonsha)
Cellryel (Bio Dell)
Omega-CH-Activator-A (GfN)
Omega-CH-Activator (GfN)
Chamomile Extract 21006 (Fragrance Oils Int. Ltd.)
Crodarom Hygroderm (Croda, Inc.)
Drieline (Lanatech Paris)
Ester-C Topical Concentrate (Inter-Cal)
Eucalyptus Softcream (CEP (Solabia))
Facteur Hydratant PH (Prod'Hyg)
Foam-Coll 5 (Arch Personal Care Products)
Foam-Coll 5W (Arch Personal Care Products)
Glucolysan (Vincience)
Green Tea Extract 101266 (Fragrance Oils Int. Ltd.)
Gulfweed Extract 101873 (Fragrance Oils Int. Ltd.)
Herbasol Complex "Herbes de Provence" (Cosmetochem) (Cosmetochem International Ltd.)
Herbasol Complex 7-Herbs (Cosmetochem) (Cosmetochem International Ltd.)
Herbasol Complex "Sedative/Relaxing" (Cosmetochem) (Cosmetochem International Ltd.)
Herbasol Extract Myrrh (Cosmetochem) (Cosmetochem International Ltd.)
Hydeoviton 5,5 N 2/059359 (Symrise)
Hydralphatine Asiatique (Lanatech)
Hydralphatine 3P (Lanatech)
Hydrocos (Cosmetochem)
Hydrocos "P" (Cosmetochem) (Cosmetochem International Ltd.)
Hydrofacteur LC (LCW)
Hydroveg VV (Variati)
Hydroviton 2/059353 (Symrise)
Hydroviton 24 (2/059351) (Symrise)
Hydroxan (Lanatech)
Hydroxan BG (Lanatech)
Hydroxan CH (Lanatech)
Lime Softcream (CEP (Solabia))
Liprot CTS (Fabriquimica)
Lowenol Conditioner 288 (Lowenstein)
Lowenol Conditioner C-727 (Lowenstein)
Microderm AE (Zehentmayer)
Moisturising Factor Hydrogerm (Crodarom)
Monteine V (SEPPIC)
Murumuru Extract 136286 (Fragrance Oils Int. Ltd.)
Nectar de Kniphofia (Pronectar)
Nikkol Aquasome LA (Nikko)
Ormagel AC-400 (Assessa-Industria)
Ormagel SH (Assessa-Industria)
Ormagel XPU (Assessa-Industria)
P.A. Reviviscence PW (Laboratoires Sero-biologiques)
Peppermint Softcream (CEP (Solabia))
Phototan (Laboratoires Serobiologiques)
Phototan 2400 B (Laboratoires Sero-biologiques)
Plantazyme Phase I (Coletica SA)
Prodew 100 (Ajinomoto)
Prodew 200 (Ajinomoto)
Prodew 300 (Ajinomoto)
Prodew 400 (Ajinomoto)
Pro-Lan V (Lanaetex)
Promois BE-1 (Seiwa Kasei)
PW 2000(Phyto-Whitening) (Proimex GmbH)
Quiditat NwA (Assessa-Industria)
Quiditat NwH (Assessa-Industria)
Quiditat SR (Assessa-Industria)
Rouge Covasorb W 3784-A (LCW)
Scavenzyme (Coletica SA)
Seanamine BD (Laboratoires Sero-biologiques)
Seanamine SU (Laboratoires Sero-biologiques)
Seaweederm A-525 (Assessa-Industria)
Sorbaine (Vincience)
Superpro 5A (Inolex)
Supro-Tein S (Maybrook)
Supro-Tein V (Maybrook)

SORBITYL ACETATE

CAS No.	EINECS No.
39346-74-2	254-423-2

Empirical Formula:
$C_8H_{16}O_7$

Definition: Sorbityl Acetate is the ester of Sorbitol (q.v.) and acetic acid.

Chemical Classes: Esters; Polyols

Function: Not Reported

Technical/Other Names:
Acetic Acid, Sorbitol Ester
D-Glucitol, Acetate

SORBITYL FURFURAL

Empirical Formula:
$C_{10}H_{16}O_7$

Definition: Sorbityl Furfural is the cyclic acetal formed between Sorbitol (q.v.) and Furfural (q.v.). It conforms to the formula:

Chemical Classes: Heterocyclic Compounds; Polyols

Function: Antioxidant

Trade Name:
AR-GB11 (Laboratori Fitocosmesi E Farmaceutici)

SORBITYL SILANEDIOL

Definition: Sorbityl Silanediol is the organic compound that conforms to the formula:

Chemical Classes: Polyols; Siloxanes and Silanes

Functions: Humectant; Skin-Conditioning Agent - Humectant

Trade Name:
Dimethyl Silanol Sorbitol (Exsymol)

SORBUS AUCUPARIA FRUIT EXTRACT

CAS No. 84776-90-9 **EINECS No.** 284-017-0

Definition: Sorbus Aucuparia Fruit Extract is an extract of the berries of the mountain ash, *Sorbus aucuparia*. See *"Regulatory and Ingredient Use Information," regarding the labeling names for botanical ingredients in Volume 1, Introduction, Part A.*

Chemical Class: Biological Products

Function: Not Reported

Technical/Other Names:
Extract of Mountain Ash
Extract of Sorbus Aucuparia
Mountain Ash Extract
Mountain Ash (Sorbus Aucuparia) Extract

Trade Name Mixtures:
Herbasol-Extract Mountain Ash (Cosmetochem)
Mountain Ash Fruit Extract HS 2655 G (Grau)
Sorbus Aucuparia Fruit Extract ies (IES LABO)
Vegebois of Sorb (CEP (Solabia))

SORGHUM VULGARE EXTRACT

Definition: Sorghum Vulgare Extract is an extract of the seeds, skin and stalks of *Sorghum vulgare*. See *"Regulatory and Ingredient Use Information," regarding the labeling names for botanical ingredients in Volume 1, Introduction, Part A. See "Regulatory and Ingredient Use Information," for Colorants in Volume 1, Introduction, Part A.*

Chemical Class: Biological Products

Function: Colorant

Technical/Other Name:
Extract of Sorghum Vulgare

Trade Name:
Premier Sorghum Extract (Premier Specialties)

SOUHAKUHI EKISU

JPN Translation:
ソウハクヒエキス

Definition: Souhakuhi Ekisu is an extract of the root bark of *Morus alba* or other related species of the family, *Moraceae*. See *"Regulatory and Ingredient Use Information," regarding use of Japan Trivial names in Volume 1, Introduction, Part A.*

Chemical Class: Biological Products

Function: Skin-Conditioning Agent - Miscellaneous

Technical/Other Name:
Morus Alba Root Extract (U.S.)

SOY ACID

CAS No. 68308-53-2 **EINECS No.** 269-657-0

Definition: Soy Acid is the mixture of fatty acids derived from Glycine Soja (Soybean) Oil (q.v.).

Information Sources: 21CFR175.105, 21CFR177.2800, 21CFR178.3570, TSCA

Chemical Class: Fatty Acids

Functions: Surfactant - Cleansing Agent; Surfactant - Emulsifying Agent

Surfactant-Cleansing Agent is included as a function for the soap form of Soy Acid.

Technical/Other Names:
Acids, Soy
Fatty Acids, Soy
Glycine Soja (Soy) Acid
Soya Acid

Trade Name:
S-210 (Procter & Gamble)

SOYAMIDE DEA

CAS No. 68425-47-8 **EINECS No.** 270-355-6

Definition: Soyamide DEA is a mixture of ethanolamides of Soy Acid (q.v.). It conforms generally to the formula:

$$RC{-}N(CH_2CH_2OH)_2$$

with carbonyl oxygen (O) double-bonded to the RC group.

where RCO- represents the fatty acids derived from soy.

Information Sources: 21CFR172.710, 21CFR175.105, 21CFR176.210, EEC(III/1-60), TSCA

Chemical Class: Alkanolamides

Functions: Surfactant - Foam Booster; Viscosity Increasing Agent - Aqueous

Reported Product Categories: Shampoos (Non-coloring); Hair Sprays (Aerosol Fixatives); Tonics, Dressings, and Other Hair Grooming Aids; Permanent Waves; Hair Conditioners

Technical/Other Names:
Amides, Soy, N,N-Bis(Hydroxyethyl)-
N,N-Bis(Hydroxyethyl)Soy Amides
Soy Amides, N,N-Bis(Hydroxyethyl)-

Trade Names:
Amidex S (Chemron)
Eur-Amid Soy (EOC Surfactants)
Jeemide SS-100 (Jeen)
Mackamide S (McIntyre)
Purton SFD (Zschimmer & Schwarz)
Schercomid SLS (Scher)
Serdolamide PQF 74 (Sasol Servo)

Trade Name Mixtures:
Mackamide SD (McIntyre)
Upamide SS-10 (Universal Preserv-A-Chem)

SOYAMIDOETHYLDIMONIUM/TRIMONIUM HYDROXYPROPYL HYDROLYZED WHEAT PROTEIN

Definition: Soyamidoethyldimonium/ Trimonium Hydroxypropyl Hydrolyzed Wheat Protein is the quaternary ammonium salt produced by the reaction of equimolar quantities of chlorohydroxypropyl trimethylammonium chloride and dimethylsoya ammonium chloride with hydrolyzed wheat protein.

Chemical Class: Quaternary Ammonium Compounds

Functions: Antistatic Agent; Hair Conditioning Agent

SOYAMIDOPROPALKONIUM CHLORIDE

Definition: Soyamidopropalkonium Chloride is the quaternary ammonium salt that conforms generally to the formula:

where RCO- represents the fatty acids derived from soy.

Chemical Class: Quaternary Ammonium Compounds

Function: Antistatic Agent

Trade Name:
Quatrex S (Chemron)

SOYAMIDOPROPYLAMINE OXIDE

Definition: Soyamidopropylamine Oxide is the tertiary amine oxide that conforms generally to the formula:

---

where RCO- represents Soy Acid (q.v.).

Chemical Class: Amine Oxides

Functions: Surfactant - Cleansing Agent; Surfactant - Foam Booster; Surfactant - Hydrotrope

Trade Name:
Chemoxide SO (Chemron)

SOYAMIDOPROPYL BETAINE

Definition: Soyamidopropyl Betaine is the zwitterion (inner salt) that conforms generally to the formula:

$$RC-NH(CH_2)_3-\overset{CH_3}{\underset{CH_3}{\overset{O}{\overset{\|}{}}\overset{+}{N}}}-CH_2COO^-$$

where RCO- represents the fatty acids derived from soy.

Chemical Class: Betaines

Functions: Antistatic Agent; Hair Conditioning Agent; Skin-Conditioning Agent - Miscellaneous; Surfactant - Cleansing Agent; Surfactant - Foam Booster; Viscosity Increasing Agent - Aqueous

Technical/Other Names:
N-(Carboxymethyl)-N,N-Dimethyl-3-[(1-Oxosoy)Amino]-1-Propanaminium Hydroxide, Inner Salt
1-Propanaminium, N-(Carboxymethyl)-N,N-Dimethyl-3-[(1-Oxosoy)Amino]-, Hydroxide, Inner Salt
Quaternary Ammonium Compounds, (Carboxymethyl)(3-Soyamidopropyl)Dimethyl, Hydroxide, Inner Salt
Soy Amide Propylbetaine
Soyamidopropyl Dimethyl Glycine

Trade Name:
Chembetaine S (Chemron)

SOYAMIDOPROPYL DIMETHYLAMINE

CAS No.: 68188-30-7

Definition: Soyamidopropyl Dimethylamine is the amidoamine that conforms generally to the formula:

$$RC-NH(CH_2)_3N\overset{CH_3}{\underset{CH_3}{\overset{O}{\overset{\|}{}}}}$$

where RCO- represents the fatty acids derived from soy.

Chemical Class: Amines

Function: Antistatic Agent

Technical/Other Names:
Amides, Soy, N-[3-(Dimethylamino)Propyl]-
Dimethylaminopropyl Soyamide
N-[3-(Dimethylamino)Propyl] Soy Amides

Trade Name:
Mackine 901 (McIntyre)

SOYAMIDOPROPYL ETHYLDIMONIUM ETHOSULFATE

CAS No.	EINECS No.
90529-57-0	291-990-5

Definition: Soyamidopropyl Ethyldimonium Ethosulfate is the quaternary ammonium salt that conforms generally to the formula:

$$\left[RC-NH(CH_2)_3-\overset{CH_3}{\underset{CH_3}{\overset{O}{\overset{\|}{}}}\overset{+}{N}}-CH_2CH_3 \right]^+ \quad CH_3CH_2\overset{}{\underset{}{OSO_3^-}}$$

where RCO- represents the fatty acids derived from soy.

Chemical Class: Quaternary Ammonium Compounds

Function: Antistatic Agent

Trade Names:
Mackernium SODES (McIntyre)
Schercoquat SOAS (Scher)

SOYAMINE

CAS No.	EINECS No.
61790-18-9	263-112-0

Definition: Soyamine is the primary aliphatic amine derived from Soy Acid (q.v.). It conforms generally to the formula:

$$RNH_2$$

where R represents the alkyl groups derived soy.

Information Source: TSCA

Chemical Class: Amines

Function: Antistatic Agent

Technical/Other Name:
Amines, Soy Alkyl

Trade Names:
Armeen S (Akzo Nobel)
Armeen SD (Akzo Nobel)

SOY AMINO ACIDS

Definition: Soy Amino Acids is a mixture of amino acids derived from the complete hydrolysis of soy protein.

Chemical Class: Amino Acids

Functions: Hair Conditioning Agent; Skin-Conditioning Agent - Miscellaneous

Trade Names:
AquaPro II SAA (MGP)
Soypeptide AC (Sinerga)
Soy-Tein AA (Maybrook)

SOYAMINOPROPYLAMINE

Definition: Soyaminopropylamine is the substituted amine that conforms generally to the formula:

$$RNH(CH_2)_3NH_2$$

where R represents the alkyl groups derived from soy.

Chemical Class: Amines

Function: Antistatic Agent

SOYBEAN OIL PEG-36 ESTERS

Definition: Soybean Oil PEG-36 Esters is a polyethylene glycol derivative of the mono- and diglycerides from Glycine Soja (Soybean) Oil (q.v.) with an average of 36 moles of ethylene oxide.

Chemical Classes: Alkoxylated Alcohols; Glyceryl Esters and Derivatives

Functions: Skin-Conditioning Agent - Emollient; Surfactant - Emulsifying Agent

Technical/Other Names:
Polyethylene Glycol (36) Soy Glycerides
Polyoxyethylene (36) Soy Glycerides

SOYBEAN PALMITATE

Definition: Soybean Palmitate is the product obtained by the reaction of crushed soybeans with palmitic acid chloride.

Chemical Classes: Amides; Carbohydrates; Esters; Proteins

Function: Skin-Conditioning Agent - Miscellaneous

SOY DIHYDROXYPROPYLDIMONIUM GLUCOSIDE

Definition: Soy Dihydroxypropyldimonium Glucoside is the quaternary ammonium chloride that conforms generally to the formula:

$$\left[R-\overset{CH_3}{\underset{CH_3}{\overset{|}{N}}}-CH_2\overset{}{\underset{OH}{CH}}CH_2OH \quad R' \right]^+$$

where R represents the alkyl groups derived from hydrogenated soy and R' represents glucose.

Chemical Class: Quaternary Ammonium Compounds

Function: Hair Conditioning Agent

Trade Name Mixture:
  Brocose Q (Arch Personal Care Products)

SOYDIMONIUM HYDROXYPROPYL HYDROLYZED WHEAT PROTEIN

Definition: Soydimonium Hydroxypropyl Hydrolyzed Wheat Protein is the quaternary ammonium chloride that conforms generally to the formula:

where R represents the alkyl groups derived from soy and R' represents the hydrolyzed wheat protein moiety.

Chemical Class: Quaternary Ammonium Compounds

Function: Hair Conditioning Agent

Trade Name:
  Quat-Wheat SDMA-30 (Arch Personal Care Products)

SOYETHYLDIMONIUM ETHOSULFATE

CAS No.	EINECS No.
68308-67-8	269-663-3

Definition: Soyethyldimonium Ethosulfate is the quaternary ammonium compound that conforms to the formula:

where R represents the alkyl groups derived from soy.

Chemical Class: Quaternary Ammonium Compounds

Functions: Antistatic Agent; Hair Conditioning Agent

SOYETHYL MORPHOLINIUM ETHOSULFATE

CAS No.	EINECS No.
61791-34-2	263-167-0

Definition: Soyethyl Morpholinium Ethosulfate is the quaternary ammonium salt that conforms generally to theformula:

where R represents the alkyl groups derived from soy.

Information Sources: CTFA D, TSCA

Chemical Classes: Heterocyclic Compounds; Quaternary Ammonium Compounds

Function: Antistatic Agent

Reported Product Categories: Permanent Waves; Hair Conditioners

Technical/Other Names:
  Morpholinium Compounds, N-Ethyl-N-Soy Alkyl, Ethyl Sulfates
  Quaternium-2
  N-Soy-N-Ethyl Morpholinium Ethosulfate

Trade Name:
  Forestall (Uniqema Americas)

SOY HYDROXYETHYL IMIDAZOLINE

CAS No.	EINECS No.
70024-77-0	274-267-9

Definition: Soy Hydroxyethyl Imidazoline is the heterocyclic compound that conforms generally to the formula:

where R is derived from the alkyl groups derived from soy.

Chemical Class: Imidazoline Compounds

Functions: Antistatic Agent; Hair Conditioning Agent

Technical/Other Names:
  Soya Hydroxyethyl Imidazoline
  Soya Imidazoline
  Soy Imidazoline

SOY ISOFLAVONES

Definition: Soy Isoflavones is a mixture of isoflavones derived from Soybean (Glycine Soja) Extract (q.v.). Its consists chiefly of genistein and daidzein.

Chemical Class: Biological Products

Function: Skin-Conditioning Agent - Miscellaneous

Trade Names:
  Novasoy (Archer Daniels Midland)
  Solgen 40 (Solbar Plant)

Trade Name Mixtures:
  Iso-SlimComplex (Mibelle AG)
  Lipobelle Soyaglycone (Mibelle AG)

SOYMILK

Definition: Soymilk is an emulsion containing the water-soluble proteins and carbohydrates, and most of the oil derived from the whole soybean, *Glycine soja*.

Chemical Class: Biological Products

Function: Skin-Conditioning Agent - Miscellaneous

Trade Name:
  ESP Soy Milk Powder (Earth Supplied Products)

Trade Name Mixture:
  Soy Milk Protein Extract (Carrubba)

SOYOU EKISU

JPN Translation:
  シソエキス

Definition: Soyou Ekisu is an extract of the leaves and twigs of *Perilla ocymoides* or other related species of the family, *Labiatae*. See "Regulatory and Ingredient Use Information," regarding use of Japan Trivial names in Volume 1, Introduction, Part A.

Chemical Class: Biological Products

Function: Skin-Conditioning Agent - Miscellaneous

Technical/Other Name:
  Perilla Ocymoides Leaf Extract (U.S.)

SOY STEROL ACETATE

Definition: Soy Sterol Acetate is the acetic acid ester of soy sterol.

Chemical Classes: Esters; Sterols

Function: Skin-Conditioning Agent - Occlusive

SOYTRIMONIUM CHLORIDE

CAS No.	EINECS No.
61790-41-8	263-134-0

Definition: Soytrimonium Chloride is the quaternary ammonium salt that conforms generally to the formula:

$$\left[\begin{array}{c} CH_3 \\ | \\ R-N^+-CH_3 \\ | \\ CH_3 \end{array} \right] Cl^-$$

where R represents the alkyl groups derived from soy.

Information Sources: JCLS, TSCA

Chemical Class: Quaternary Ammonium Compounds

Functions: Antistatic Agent; Hair Conditioning Agent

Reported Product Categories: Hair Dyes and Colors (All Types Requiring Caution Statements and Patch Tests); Hair Shampoos (Coloring); Hair Conditioners

Technical/Other Names:
Quaternary Ammonium Compounds, Trimethylsoy Alkyl, Chlorides
Quaternium-9
N-(Soy Alkyl)-N,N,N-Trimethyl Ammonium Chloride
Soyatrimonium Chloride
Soy Trimethyl Ammonium Chloride

Trade Name:
Arquad S-60 PG (Akzo Nobel)

Trade Name Mixture:
Arquad S-50 (Akzo Nobel)

SPARTIUM JUNCEUM FLOWER EXTRACT

Definition: Spartium Junceum Flower Extract is an extract of the flowers of *Spartium junceum. See "Regulatory and Ingredient Use Information," regarding the labeling names for botanical ingredients in Volume 1, Introduction, Part A.*

Chemical Class: Biological Products

Function: Fragrance Ingredient

Technical/Other Names:
Extract of Spartium Junceum Flower
Genet extract (Spartium junceum L.) (RIFM)

Trade Name:
Genet Extract (Charabot)

SPENT GRAIN WAX

Definition: Spent Grain Wax is a product derived from the extraction of the residual dry spent grains obtained from the malting process in beer production. *See*

"Regulatory and Ingredient Use Information," regarding use of EU Trivial names in Volume 1, Introduction, Part A.

Chemical Class: Waxes

Functions: Skin-Conditioning Agent - Miscellaneous; Viscosity Increasing Agent - Nonaqueous

Technical/Other Name:
Waxes, Spent Grain

Trade Name:
Stimu-Tex (Pentapharm/Centerchem)

Trade Name Mixture:
Stimu-Tex AS (Pentapharm/Centerchem)

SPERGULARIA RUBRA EXTRACT

CAS No.	EINECS No.
84776-02-3	283-927-5

Definition: Spergularia Rubra Extract is an extract of the whole plant, *Spergularia rubra. See "Regulatory and Ingredient Use Information," regarding the labeling names for botanical ingredients in Volume 1, Introduction, Part A.*

Chemical Class: Biological Products

Function: Not Reported

Technical/Other Name:
Extract of Spergularia Rubra

Trade Name Mixture:
Extrait De Sabline PPE PG 40 (Yves Rocher)

SPERMIDINE HCl

CAS No.	EINECS No.
334-50-9	206-379-0

Empirical Formula:
$C_7H_{19}N_3 \cdot 3HCl$

Definition: Spermidine HCL is the amine salt that conforms to the formula:

$$NH_2(CH_2)_4NH(CH_2)_3NH_2 \quad \cdot \quad 3HCl$$

Information Source: MI-13(8816)

Chemical Class: Amines

Function: Hair Conditioning Agent

Technical/Other Name:
1,4-Butanediamine, N-(3-Aminopropyl)-, Trihydrochloride

Trade Name:
Spermidine HCL (Giuliani spa)

SPHACELARIA SCOPARIA EXTRACT

Definition: Sphacelaria Scoparia Extract is an extract of the alga, *Sphacelaria*

scoparia. See Reported Ingredient Functions-The Cosmetic Drug Distinction, in Regulatory and Ingredient Use Information, Volume I, Part A.

Chemical Class: Biological Products

Function: Corn/Callus/Wart Remover

Technical/Other Name:
Extract of Sphacelaria Scoparia

Trade Name Mixture:
Scopariane (Codif)

SPHAGNUM SQUARROSUM EXTRACT

CAS No.	EINECS No.
90131-22-9	290-332-4

Definition: Sphagnum Squarrosum Extract is an extract of the peat of *Sphagnum squarrosum. See "Regulatory and Ingredient Use Information," regarding the labeling names for botanical ingredients in Volume 1, Introduction, Part A.*

Chemical Class: Biological Products

Function: Skin-Conditioning Agent - Miscellaneous

Technical/Other Name:
Extract of Sphagnum Squarrosum

Trade Name Mixture:
Moor Extract HS 3665 G (Grau)

SPHINGANINE

CAS Nos.	EINECS No.
764-22-7	
3102-56-5	212-116-0

Empirical Formula:
$C_{18}H_{39}NO_2$

Definition: Sphinganine is the organic compound that conforms to the formula:

$$\begin{array}{c} NH_2 \\ | \\ HOCH_2CHCH(CH_2)_{14}CH_3 \\ | \\ OH \end{array}$$

Chemical Classes: Alcohols; Amines

Functions: Hair Conditioning Agent; Skin-Conditioning Agent - Miscellaneous

Technical/Other Names:
Dihydrosphingosine
D-Erythro-1,3-Dihydroxy-2-Aminooctadecane
1,3-Octadecanediol, 2-Amino-, (2S,3R)-

Trade Name Mixture:
Sphinganine S (Sederma)

SPHINGOBACTERIUM FERMENT EXTRACT

Definition: Sphingobacterium Ferment Extract is an extract of the fermentation product of *Sphingobacterium*.

Chemical Class: Biological Products

Function: Skin-Conditioning Agent - Emollient

Trade Name:
Club Sphingo-Lipo (Club Cosmetics)

SPHINGOLIPIDS

CAS No.	EINECS No.
85116-74-1	285-526-0

Definition: Sphingolipids are complex lipids which contain sphingosine or a related base, a polar head group and a long saturated or monounsaturated fatty acid connected to the backbone at its amino group.

Chemical Classes: Fats and Oils; Phosphorus Compounds

Function: Skin-Conditioning Agent - Miscellaneous

Reported Product Categories: Moisturizing Preparations; Bath Capsules; Face and Neck Preparations (Excluding Shaving Preparations); Bath Oils, Tablets, and Salts; Cleansing Products (Cold Creams, Cleansing Lotions, Liquids and Pads); Shampoos (Noncoloring); Eye Makeup Preparations, Misc.; Hair Conditioners; Hair Preparations (Noncoloring), Misc.

Trade Names:
Club Sphingo-Lipo C200 (Club Cosmetics)
Milk Sphingomyelins (LTP)

Trade Name Mixtures:
ABS Plant Sil Blend (REU) (Active Concepts)
A.F.R./Veg. (Laboratoires Serobiologiques)
Alpha Lipid 100 (Lucas Meyer)
Ceramide Complex CLR (P) (CLR)
Eticerav (Coletica SA)
Glycoderm P (CLR)
Glycosome (Pentapharm/Centerchem)
Liposomes Trichogen Veg (Laboratoires Serobiologiques)
Sphingosome AL/VEG (Laboratoires Serobiologiques)

SPHINGOMONAS FERMENT EXTRACT

Definition: Sphingomonas Ferment Extract is an extract of the fermentation product of *Sphingomonas*.

Chemical Class: Biological Products

Function: Not Reported

Technical/Other Name:
Extract of Sphingomonas Ferment

Trade Names:
Bioceramide (Pronova Biopolymer Inc.)
Ferment Ceramide (Kibun)

SPILANTHES ACMELLA FLOWER EXTRACT

Definition: Spilanthes Acmella Flower Extract is an extract of the flowers of *Spilanthes acmella*. *See "Regulatory and Ingredient Use Information," regarding the labeling names for botanical ingredients in Volume 1, Introduction, Part A.*

Chemical Class: Biological Products

Functions: Hair Conditioning Agent; Skin-Conditioning Agent - Miscellaneous

Technical/Other Name:
Extract of Spilanthes Acmella

Trade Name Mixtures:
Jambu Oleoresin (Takasago)
Para Cress Extract HS 3738 G (Grau)

SPINACIA OLERACEA (SPINACH)

Definition: Spinacia Oleracea (Spinach) is a plant material derived from *Spinacia oleracea*. *See "Regulatory and Ingredient Use Information," regarding the labeling names for botanical ingredients in Volume 1, Introduction, Part A.*

Chemical Class: Biological Products

Function: Not Reported

Technical/Other Name:
Spinach

Trade Name:
Spinach (Spinacia Oleracea) (Vege-Tech)

SPINACIA OLERACEA (SPINACH) LEAF EXTRACT

CAS No.	EINECS No.
90131-25-2	290-336-6

Definition: Spinacia Oleracea (Spinach) Leaf Extract is an extract of the leaves of the spinach, *Spinacia oleracea*. *See "Regulatory and Ingredient Use Information," regarding the labeling names for botanical ingredients in Volume 1, Introduction, Part A.*

Chemical Class: Biological Products

Function: Skin-Conditioning Agent - Miscellaneous

Technical/Other Names:
Extract of Spinach
Extract of Spinacia Oleracea
Spinach Extract
Spinach Leaf Extract
Spinacia Oleracea Extract

Trade Name Mixtures:
Actiphyte of Spinach BG50 (Active Organics)
Actiphyte of Spinach GL50 (Active Organics)
Actiphyte of Spinach Lipo S (Active Organics)
Actiphyte of Spinach PG50 (Active Organics)
Glycolysat of Spinach (CEP (Solabia))
Lipoplastidine Spinacia (Vevy)
Oleophylle-10 (Vevy)
Spinach Extract HS 3538 G (Grau)
Spinach HS (Alban Muller)
Vegebios of Spinach (CEP (Solabia))

SPINACIA OLERACEA (SPINACH) LEAF POWDER

Definition: Spinacia Oleracea (Spinach) Leaf Powder is the powder obtained from the dehydrated, crushed leaves of *Spinacia oleracea*. *See "Regulatory and Ingredient Use Information," regarding the labeling names for botanical ingredients in Volume 1, Introduction, Part A.*

Chemical Class: Biological Products

Function: Skin-Conditioning Agent - Miscellaneous

Trade Name:
Premier Spinach Powder BC (Premier Specialties)

SPINAL CORD EXTRACT

Definition: Spinal Cord Extract is an extract obtained from animal spinal cords.

Information Source: EEC(II-419)

Chemical Class: Biological Products

Function: Not Reported

Technical/Other Name:
Extract of Spinal Cords

Trade Name Mixture:
Horse Spinalcord Extract in oil (Paninkret)

SPINAL CORD LIPIDS

Definition: Spinal Cord Lipids are the lipids derived from spinal cord tissue.

Information Source: EEC(II-419)

Chemical Class: Fats and Oils

Function: Skin-Conditioning Agent - Emollient

SPINAL LIPID EXTRACT

Definition: Spinal Lipid Extract is an extract of animal spinal lipids.

Information Source: EEC(II-419)

Chemical Class: Biological Products

Function: Not Reported

Technical/Other Name:
Extract of Spinal Lipids

SPIRAEA ULMARIA EXTRACT

JPN Translation:
シモツケソウエキス

Definition: Spiraea Ulmaria Extract is an extract of the whole plant of *Spiraea ulmaria.See "Regulatory and Ingredient Use Information," regarding the labeling names for botanical ingredients in Volume 1, Introduction, Part A.*

Chemical Class: Biological Products

Function: Skin-Conditioning Agent - Miscellaneous

Trade Name Mixtures:
ACB Meadowsweet Extract (Active Concepts)
ACB Meadowsweet Extract 20% (Active Concepts)
Actiphyte of Queen of the Meadow BG50 (Active Organics)
Actiphyte of Queen of the Meadow GL50 (Active Organics)
Actiphyte of Queen of the Meadow Lipo S (Active Organics)
Actiphyte of Queen of the Meadow PG50 (Active Organics)
Herbasol-Extract Meadowsweet (Cosmetochem)
Sebonormine (Silab)
Vegebios of Meadowsweet (CEP (Solabia))

SPIRAEA ULMARIA FLOWER EXTRACT

CAS No. 84775-57-5

EINECS No. 283-886-3

Definition: Spiraea Ulmaria Flower Extract is an extract of the flower of *Spiraea ulmaria. See "Regulatory and Ingredient Use Information," regarding the labeling*

names for botanical ingredients in Volume 1, Introduction, Part A.

Information Sources: JCIC, JCLS

Chemical Class: Biological Products

Function: Skin-Conditioning Agent - Miscellaneous

Reported Product Categories: Bath Preparations, Misc.; Body and Hand Preparations (Excluding Shaving Preparations); Skin Care Preparations, Misc.

Technical/Other Names:
Drupwort Extract
Extract of Drupwort
Extract of Meadowsweet
Extract of Spiraea
Extract of Spiraea Ulmaria
Filipendula Extract
Filipendula Ulmaria Extract
Meadowsweet Extract
Meadowsweet (Spiraea Ulmaria) Extract
Queen of the Meadow Extract
Spiraea Extract

Trade Name:
Meadow - Sweet Supextrat (Indena SA)

Trade Name Mixtures:
Actiphyte of Meadowsweet BG50 (Active Organics)
Actiphyte of Meadowsweet GL50 (Active Organics)
Actiphyte of Meadowsweet Lipo S (Active Organics)
Actiphyte of Meadowsweet PG50 (Active Organics)
Actiplex 1072 Skin Renewal Complex (Active Organics)
Anti-Inflammatory Phytoamine Biocomplex (Alban Muller)
Anti-Rheumatic Phytospa (Alban Muller)
Anti-Rheumatic Phytospa NaCl (Alban Muller)
255 Blend for Slenderizing Products HS (Alban Muller)
655 Blend for Slenderizing Products LS (Alban Muller)
316 Blend for Slimming Products HS (Alban Muller)
616 Blend for Slimming Products LS (Alban Muller)
Blend 3250 HS (Alban Muller)
Enna 201 Slimming Vegetal Extract (Ennagram)
Extrait d'Ulmaire LPS (Phytocos)
IPF Meadowsweet (Solabia)
Lipoplastidine Spiraea (Vevy)
Meadowsweet Extract HG (Provital/Centerchem)
Meadow Sweet Extract HS 2473 G (Grau)
Meadow Sweet HPG Titrated (Alban Muller)
Meadow Sweet HS (Alban Muller)

Meadowsweet Liquid B (Ichimaru Pharcos)
Meadow Sweet LS (Alban Muller)
Meadowsweet Phytexcell (Crodarom)
Meadowsweet Tincture (Rahn)
Natupure Meadowsweet (E.U.K)
Pharcolex BX50 (Ichimaru Pharcos)
Phytelene Complex EGX 250 (Indena SA)
Phytelene of Queen Meadow EG 213 liquid (Indena SA)
Phytelene of Ulmaire EG 213 liquid (Indena SA)
Phytogreen 55 of Queen Meadow EXH 639 Liquid (Phytochim)
Phytogreen 55 of Ulmarie EXH 639 Liquid (Phytochim)
Prodhy Extract Ulmaire (Prod'Hyg)
Queen of the Prairie Extract (Arch Personal Care Products)
278 Relaxant HS (Alban Muller)
678 Relaxant LS (Alban Muller)
328 Rhumacalm HS (Alban Muller)
SLIMMING (Greentech S.A)
Slimming Complex 265 (Ennagram)
Slimming Phytogreen Complex GXH 265 (Phytochim)
Soothing Milk (CEP (Solabia))
Vegetol Ulmary LC 383 Hydro (Gattefosse s.a.)
VT-053 Extract of Meadowsweet (Vege-Tech)

SPIRAEA ULMARIA (MEADOWSWEET) LEAF EXTRACT

Definition: Spiraea Ulmaria (Meadowsweet) Leaf Extract is an extract of the leaves of *Spiraea ulmaria. See "Regulatory and Ingredient Use Information," regarding the labeling names for botanical ingredients in Volume 1, Introduction, Part A.*

Chemical Class: Biological Products

Function: Skin-Conditioning Agent - Miscellaneous

Technical/Other Names:
Extract of Spiraea Ulmaria (Meadowsweet) Leaf
Meadowsweet Leaf Extract

Trade Name Mixture:
CYTOBIOL ULMAIRE (Libiol)

SPIRODELA POLYRRHIZA EXTRACT

Definition: Spirodela Polyrrhiza Extract is an extract of the plant *Spirodela polyrrhiza. See "Regulatory and Ingredient Use Information," regarding the labeling names for botanical ingredients in Volume 1, Introduction, Part A.*

Chemical Class: Biological Products

Function: Not Reported

Technical/Other Names:
Duckweed Extract
Extract of Duckweed
Extract of Spirodela Polyrrhiza

Trade Name Mixture:
Sinominceur (I.D. bio)

SPIRULINA AMINO ACIDS

Definition: Spirulina Amino Acids is the mixture of amino acids derived from the complete hydrolysis of the protein obtained from the aquatic plant, *Spirulina platensis.*

Chemical Class: Amino Acids

Functions: Hair Conditioning Agent; Skin-Conditioning Agent - Humectant

SPIRULINA MAXIMA EXTRACT

JPN Translation:
スピルリナマキシマエキス

Definition: Spirulina Maxima Extract is an extract of the fronds of the spirulina, *Spirulina maxima. See "Regulatory and Ingredient Use Information," regarding the labeling names for botanical ingredients in Volume 1, Introduction, Part A.*

Chemical Class: Biological Products

Function: Skin-Conditioning Agent - Miscellaneous

Reported Product Category: Skin Care Preparations, Misc.

Technical/Other Names:
Extract of Spirulina
Extract of Spirulina Maxima
Spirulina Extract

Trade Name:
Biomin Selenium (Greentech)

Trade Name Mixtures:
Actiphyte of Spirulina BG50 (Active Organics)
Actiphyte of Spirulina GL50 (Active Organics)
Actiphyte of Spirulina Lipo S (Active Organics)
Actiphyte of Spirulina PG50 (Active Organics)
Extrait de Spiruline MPE 100 (Yves Rocher)
Extrait Hydroglycolique d'Algue Bleue (Greentech)
Floraceutical Spirulina Extract-Standardized (Bio-Botanica)
Phytelene of Spirulina EG 718 Liquid (Indena SA)
Phytelenes of Spirulina EG 718 Liquid (Indena SA)

Phytogreen 55 of Spirulina EXH 747 Liquid (Phytochim)
Protulines (Exsymol)
Spirulina Extract (CEP (Solabia))
Spirulina Extract HS 3511 G (Grau)
Spirulina HPG Titrated (Alban Muller)
Spirulina Maxima Extract ies (IES LABO)
Vege Plex VP-1297.050WB Sea Plex in Butylene Glycol (Vege-Tech)
VT-063 Extract of Spirulina (Vege-Tech)

SPIRULINA PLATENSIS EXTRACT

JPN Translation:
スピルリナプラテンシスエキス

Definition: Spirulina Platensis Extract is an extract of the alga, *Spirulina platensis. See "Regulatory and Ingredient Use Information," regarding the labeling names for botanical ingredients in Volume 1, Introduction, Part A.*

Chemical Class: Biological Products

Function: Not Reported

Reported Product Categories: Bath Capsules; Face and Neck Preparations (Excluding Shaving Preparations)

Trade Name Mixtures:
Algemara (Bottger)
Blue Algae (SECMA)
Japanische Blaualge P-AC-7 (Heidelberger)
Spirox (Collaborative Labs)
Spirulina Extract Powder (MicroAlgae)
Spirulina Extract Powder (MicroAlgae)

SPIRULINA PLATENSIS POWDER

Definition: Spirulina Platensis Powder is a powder of the common alga, *Spirulina platensis. See "Regulatory and Ingredient Use Information," regarding the labeling names for botanical ingredients in Volume 1, Introduction, Part A. See Reported Ingredient Functions-The Cosmetic Drug Distinction, in Regulatory and Ingredient Use Information, Volume I, Part A.*

Chemical Class: Biological Products

Functions: Skin Protectant; Sunscreen Agent; Ultraviolet Light Absorber

Technical/Other Name:
Powdered Spirulina Platensis

Trade Name:
Spirulina Pacifica (Cyanotech)

SPIRULINA SUBSALSA EXTRACT

Definition: Spirulina Subsalsa Extract is an extract of the alga, *Spirulina subsalsa.*

See "Regulatory and Ingredient Use Information," regarding the labeling names for botanical ingredients in Volume 1, Introduction, Part A.

Chemical Class: Biological Products

Function: Not Reported

SPLEEN EXTRACT

CAS No.: 84540-14-7

JPN Translation:
ウシ脾臓エキス

Definition: Spleen Extract is an extract of bovine spleens.

Information Sources: EEC(II-419), JCIC, JCLS

Chemical Class: Biological Products

Function: Not Reported

Technical/Other Names:
Bovine Spleen Extract (1)
Bovine Spleen Extract (2)

Trade Name Mixtures:
Brookosome SE (Arch Personal Care Products)
Mesopol II (Arch Personal Care Products)
Nutrex RT (Fabriquimica)

SPLEEN HYDROLYSATE

JPN Translation:
加水分解脾臓エキス

Definition: Spleen Hydrolysate is an hydrolysate of animal spleen derived by acid, enzyme or other method of hydrolysis.

Information Source: EEC(II-419)

Chemical Class: Biological Products

Function: Not Reported

Technical/Other Name:
Hydrolyzed Spleen

SPONDIAS AMARA FRUIT EXTRACT

Definition: Spondias Amara Fruit Extract is an extract of the whole fruit of *Spondias amara. See "Regulatory and Ingredient Use Information," regarding the labeling names for botanical ingredients in Volume 1, Introduction, Part A.*

Chemical Class: Biological Products

Function: Not Reported

Technical/Other Name:
Extract of Spondias Amara

Trade Name Mixture:
Campo An Mo Le Extract (Campo)

SPONGE

Definition: Sponge is the dried, ground skeleton of the freshwater sponge, *Spongia fluviatilis*.

Chemical Class: Biological Products

Function: Proprietary

SPOTTED DOGFISH SKIN EXTRACT

Definition: Spotted Dogfish Skin Extract is an extract of the skin of *Scyliorhinus canicula*.

Chemical Class: Biological Products

Function: Not Reported

Trade Name Mixtures:
Collagenna (Ennagram)
Elastenna (Ennagram)
Managen (Phytochim)
Manastin (Phytochim)

SQUALANE

CAS No.	EINECS No.
111-01-3	203-825-6

JPN Translation:
スクワラン

Empirical Formula:
$C_{30}H_{62}$

Definition: Squalane is the saturated branched chain hydrocarbon, obtained by hydrogenation of shark liver oil or other natural oils, that conforms to the formula:

$$CH_3-CH(CH_2)_3CH(CH_2)_3CH(CH_2)_4CH(CH_2)_3CH(CH_2)_3CH-CH_3$$

(with CH_3 branches)

Information Sources: 21CFR524.2140, CIR: [S] JACT-1(2)1982, CTFA D, JCIC, JCLS, JSCI, MI-13(8846), NF XIX, USAN

Chemical Class: Hydrocarbons

Functions: Hair Conditioning Agent; Skin-Conditioning Agent - Occlusive

Reported Product Categories: Moisturizing Preparations; Lipsticks; Skin Care Preparations, Misc.; Bath Preparations, Misc.; Body and Hand Preparations (Excluding Shaving Preparations); Foundations; Face Powders; Night Skin Care Preparations; Bath Capsules; Bath Oils, Tablets, and Salts; Cleansing Products (Cold Creams, Cleansing Lotions, Liquids and Pads); Makeup Preparations (Not eye), Misc.; Face and Neck Preparations (Excluding Shaving Preparations); Paste Masks (Mud Packs); Blushers (All types); Makeup Bases; Eye Makeup Preparations, Misc.; Hair Conditioners; Fragrance Preparations, Misc.; Eyebrow Pencils; Suntan Gels, Creams, and Liquids; Eye Lotions; Hair Straighteners; Manicuring Preparations, Misc.; Aftershave Lotions; Baby Shampoos; Eye Shadows; Powders (Dusting and Talcum, Excluding Aftershave Talcs); Suntan Preparations, Misc.

Technical/Other Names:
2,6,10,15,19,23-Hexamethyltetracosane
Vegetable Squalane

Trade Names:
AEC Squalane (A & E Connock)
C.I.T. Vegetable Squalane (C.I.T.)
Cosbiol (Laserson)
Dermane (Universal Preserv-A-Chem)
Exolive (Caro'iline)
Fanoliv Squalane (Fanning)
Fitoderm (Cognis Deutschland)
Fitoderm (Cognis Iberia/Centerchem)
Keteol N (Prod'Hyg)
Keteol V (Prod'Hyg)
Nikkol Olive Squalane (Nikko)
Nikkol Squalane (Nikko)
Phytosqualan (Sophim)
Phytosqual Hydrogen (Vevy)
PRIPURE 3759 (Uniqema Europe)
Robane (Robeco)
Salacos RS (Nisshin OilliO)
Salacos RSP (Nisshin OilliO)
Sophiderm (Sophim)
Trilane (Tri-K)
Vegelane (Naturactiva)

Trade Name Mixtures:
ABS Chamomile Extract OSQ (Active Concepts)
ABS Chamomile Extract SQ (Active Concepts)
ABS Edelweiss Extract OS (Active Concepts)
Aloe Extract SQ (Maruzen Pharmaceuticals Co., Ltd.)
Artemisia Capillaris Extract SQ (Maruzen Pharmaceuticals Co., Ltd.)
Beefsteak Plant Extract SQ (Maruzen Pharmaceuticals Co., Ltd.)
Birch Bark Extract SQ (Maruzen Pharmaceuticals Co., Ltd.)
Brookosome S (Arch Personal Care Products)
Burdock Extract SQ (Maruzen Pharmaceuticals Co., Ltd.)
Chamomile Extract KSQ (Maruzen Pharmaceuticals Co., Ltd.)
Chamomile Extract SQ (Maruzen Pharmaceuticals Co., Ltd.)
Chamomile Extract SQ-J (Maruzen Pharmaceuticals Co., Ltd.)
C.I.T. Alphaflo 30 VEG (C.I.T.)
Citrus Unshiu Extract SQ (Maruzen Pharmaceuticals Co., Ltd.)
Cnidium Extract SQ (Maruzen Pharmaceuticals Co., Ltd.)
Crataegus Fruit Extract SQ (Maruzen Pharmaceuticals Co., Ltd.)
Dermasome-S (ChemMark)
DS-SBS (Doosan)
Essential Vital Elements - S (Dipta)
Ganoderma Extract SQ (Maruzen Pharmaceuticals Co., Ltd.)
Ginger Extract SQ (Maruzen Pharmaceuticals Co., Ltd.)
Ginseng Extract SQ (Maruzen Pharmaceuticals Co., Ltd.)
Hoelen Extract SQ (Maruzen Pharmaceuticals Co., Ltd.)
Hops Extract SQ (Maruzen Pharmaceuticals Co., Ltd.)
Horse Chestnut Extract SQ (Maruzen Pharmaceuticals Co., Ltd.)
Houttuynia Extract SQ (Maruzen Pharmaceuticals Co., Ltd.)
Hypdricum Extract SQ (Maruzen Pharmaceuticals Co., Ltd.)
Ivy Extract-SQ (Maruzen Pharmaceuticals Co., Ltd.)
Japanese Angelica Extract SQ (Maruzen Pharmaceuticals Co., Ltd.)
Juniper Extract SQ (Maruzen Pharmaceuticals Co., Ltd.)
KSG-34 (Shin-Etsu Chemical Co.)
KSG-44 (Shin-Etsu Chemical Co.)
KSG-340 (Shin-Etsu Chemical Co.)
KSG-840 (Shin-Etsu Chemical Co.)
Lavender Extract SQ (Maruzen Pharmaceuticals Co., Ltd.)
Lentisterol (Bioland)
Lipogard (Pentapharm/Centerchem)
Loquat Leaf Extract SQ (Maruzen Pharmaceuticals Co., Ltd.)
Mallow Extract SQ (Maruzen Pharmaceuticals Co., Ltd.)
Marigold Extract-SQ (Maruzen Pharmaceuticals Co., Ltd.)
Mugwort Extract SQ (Maruzen Pharmaceuticals Co., Ltd.)
Mulberry Extract SQ (Maruzen Pharmaceuticals Co., Ltd.)
Nikkol Nikkosome OS (Nikko)
Peach Leaf Extract SQ (Maruzen Pharmaceuticals Co., Ltd.)
Phellodendron Extract SQ (Maruzen Pharmaceuticals Co., Ltd.)
Rosemary Extract SQ (Maruzen Pharmaceuticals Co., Ltd.)
Royal Jelly Extract SQ (Maruzen Pharmaceuticals Co., Ltd.)
Sage Extract SQ (Maruzen Pharmaceuticals Co., Ltd.)
Scutellaria Root Extract SQ (Maruzen Pharmaceuticals Co., Ltd.)
Simulgel NS (SEPPIC)
Solid Vegetable Squalane (Cognis Iberia/Centerchem)

Sophora Extract SQ (Maruzen
  Pharmaceuticals Co., Ltd.)
Sweet Flag Extract SQ (Maruzen
  Pharmaceuticals Co., Ltd.)
Toshiki BINS-1 (Nikko)

SQUALENE

CAS Nos.   **EINECS No.**
111-02-4   203-826-1
7683-64-9

JPN Translation:
スクワレン

Empirical Formula:
  $C_{30}H_{50}$

Definition: Squalene is an unsaturated
branched chain isoprenoid hydrocarbon
found in large quantities in shark liver oil. It
conforms to the formula:

Information Sources: CIR: [S] JACT-1(2)-
1982, JCIC, JCLS, JSQI, MI-13(8847),
TSCA

Chemical Class: Hydrocarbons

Functions: Hair Conditioning Agent; Skin-
Conditioning Agent - Emollient

Reported Product Categories: Moisturizing
Preparations; Bath Capsules; Bath Prepara-
tions, Misc.; Body and Hand Preparations
(Excluding Shaving Preparations); Eye
Shadows; Face and Neck Preparations
(Excluding Shaving Preparations); Lipsticks;
Rouges; Foundations; Bath Oils, Tablets, and
Salts; Cleansing Products (Cold Creams,
Cleansing Lotions, Liquids and Pads); Skin
Care Preparations, Misc.

Technical/Other Names:
  2,6,10,15,19,23-Hexamethyl-2,6,10,14,18,
    22-Tetracosahexaene
  2,6,10,14,18,22-Tetracosahexaene, 2,6,10,
    15,19,23-Hexamethyl-,

Trade Names:
  AEC Olive Oil Unsaponifiables (A & E
    Connock)
  AEC Squalene (A & E Connock)
  Dermene (Universal Preserv-A-Chem)
  Escualsin (Fabriquimica)
  Nikkol Squalene EX (Nikko)
  Phytosqual (Vevy)
  Phytosqualene (Sophim)
  Supraene (Robeco)

Trade Name Mixtures:
  Biophytosebum (Sophim)
  Nutriene Tocotrienols (Eastman Chemical)
  Tocomin 30% (Carotech)
  Tocomin 50% (Carotech)

STACHYS OFFICINALIS EXTRACT

Definition: Stachys Officinalis Extract is
an extract of the aerial parts of the betony,
*Stachys officinalis. See "Regulatory and
Ingredient Use Information," regarding the
labeling names for botanical ingredients in
Volume 1, Introduction, Part A.*

Chemical Class: Biological Products

Function: Not Reported

Technical/Other Names:
  Betony Extract
  Extract of Betony
  Extract of Stachys Officinalis

Trade Name Mixture:
  VT-239 Extract of Wood Betony (Vege-
  Tech)

STANNOUS CHLORIDE

CAS No.   **EINECS No.**
7772-99-8   231-868-0

Empirical Formula:
  Cl_2Sn

Definition: Stannous Chloride is the
inorganic salt that conforms to the formula:

$$SnCl_2$$

Information Sources: 21CFR155.200,
21CFR172.180, 21CFR175.300,
21CFR177.2600, 21CFR184.1845, DDR,
FCC, MI-13(8861), TSCA, USAN

Chemical Class: Inorganic Salts

Function: Not Reported

Technical/Other Name:
  Tin Dichloride

STANNOUS FLUORIDE

CAS No.   **EINECS No.**
7783-47-3   231-999-3

Empirical Formula:
  F_2Sn

Definition: Stannous Fluoride is the
inorganic salt that conforms to the formula:

$$SnF_2$$

In the United States, Stannous Fluoride
may be used as an active ingredient in

OTC drug products. When used as an
active drug ingredient, the established
name is *Stannous Fluoride. See
"Regulatory and Ingredient Use
Information," regarding the labeling names
for U.S. OTC Drug Ingredients in Volume 1,
Introduction, Part A.*

Information Sources: 21CFR355.10, EEC
(III/1-35), MAR, MI-13(8862), OTC-I-AC,
TSCA, USAN, USD, USP XXIV

Chemical Class: Inorganic Salts

Functions: Anticaries Agent; Oral Care
Agent

Technical/Other Name:
  Tin Difluoride

STANNOUS PYROPHOSPHATE

CAS No.   **EINECS No.**
15578-26-4   239-635-5

Empirical Formula:
  $H_4O_7P_2 \cdot 2Sn$

Definition: Stannous Pyrophosphate is the
inorganic salt that conforms to the formula:

$$Sn_2P_2O_7$$

Information Sources: CTFA D, MI-13
(8867), TSCA, USAN

Chemical Class: Inorganic Salts

Function: Oral Care Agent

Technical/Other Names:
  Diphosphoric Acid, Tin Salt (1:2)
  Tin (II) Pyrophosphate

STARCH ACETATE

CAS No.: 9045-28-7

Definition: Starch Acetate is the product
obtained by the reaction of acetic acid with
starch.

Chemical Class: Carbohydrates

Functions: Hair Conditioning Agent; Skin-
Conditioning Agent - Emollient

STARCH/ACRYLATES/ACRYLAMIDE COPOLYMER

Definition: Starch/Acrylates/Acrylamide
Copolymer is a polymer of starch, acrylamide
and a monomer consisting of acrylic acid,
methacrylic acid or one of their simple esters.

Chemical Classes: Biological Polymers and
their Derivatives; Synthetic Polymers

Functions: Film Former; Viscosity Increasing Agent - Aqueous

Trade Name:
Water-Lock (Grain Processing)

STARCH DIETHYLAMINOETHYL ETHER

CAS No.: 9041-94-5

Definition: Starch Diethylaminoethyl Ether is the product obtained by conversion of some hydroxyl groups in starch to diethylaminoethyl ether groups.

Information Source: TSCA

Chemical Classes: Amines; Biological Polymers and their Derivatives; Carbohydrates

Functions: Film Former; Skin-Conditioning Agent - Miscellaneous

Reported Product Categories: Bath Preparations, Misc.; Body and Hand Preparations (Excluding Shaving Preparations)

Technical/Other Name:
Starach, 2-(Diethylamino)Ethyl Ether

STARCH HYDROXYPROPYLTRIMONIUM CHLORIDE

CAS No.: 56780-58-6

Definition: Starch Hydroxypropyltrimonium Chloride is the quaternary ammonium compound formed by the reaction of starch with 2,3-epoxypropyltrimethylammonium chloride.

Chemical Classes: Carbohydrates; Quaternary Ammonium Compounds

Functions: Antistatic Agent; Emulsion Stabilizer; Hair Conditioning Agent; Suspending Agent - Nonsurfactant; Viscosity Increasing Agent - Aqueous

Technical/Other Name:
Starch, 2-Hydroxy-3-(Trimethylammonio)-Propyl Ether, Chloride

Trade Names:
Excell (Nippon Starch)
Sensomer CI-50 (ONDEO Nalco)
Sensomer CI-50 (ONDEO Nalco Europe)
Starch hydroxypropyl trimonium chloride (Avebe)

STARCH TALLOWATE

JPN Translation:
牛脂脂肪酸デンプン

Definition: Starch Tallowate is the ester of starch with the fatty acids derived from Tallow (q.v.).

Information Source: JCIC

Chemical Classes: Carbohydrates; Esters

Function: Skin-Conditioning Agent - Emollient

Technical/Other Name:
Starch Tallow Fatty Acid Ester

STEAPYRIUM CHLORIDE

CAS Nos.	EINECS No.
1341-08-8	
14492-68-3	238-501-3
42566-92-7	

JPN Translation:
ステアロイルコラミノホルミルメチルピリジニウムクロリド

Empirical Formula:
$C_{27}H_{47}N_2O_3 \cdot Cl$

Definition: Steapyrium Chloride is the quaternary ammonium salt that conforms generally to the formula:

Information Sources: CIR: [S] JACT-10(1)-1991, CTFA D, JCIC, JCLS, JSQI

Chemical Class: Quaternary Ammonium Compounds

Function: Antistatic Agent

Reported Product Categories: Hair Conditioners; Moisturizing Preparations; Shampoos (Non-coloring); Body and Hand Preparations (Excluding Shaving Preparations)

Technical/Other Names:
1-[[(2-Hydroxyethyl)Carbamoyl]Methyl]-Pyridinium Chloride, Stearate
1-[2-Oxo-2-[[2-[(1-Oxooctadecyl)Oxy]-Ethyl]Amino]Ethyl]Pyridinium Chloride
Pyridinium, 1-[[(2-Hydroxyethyl)-Carbamoyl]Methyl]-, Chloride, Stearate
Pyridinium, 1-[2-Oxo-2-[[2-[(1-Oxooctadecyl)Oxy]Ethyl]Amino]Ethyl], Chloride
Quaternium-7
N-(Stearoyl Colamino Formyl Methyl) Pyridinium Chloride

Trade Name:
Catemol WPC (Phoenix)

STEARALKONIUM BENTONITE

CAS No.: 130501-87-0

Definition: Stearalkonium Bentonite is a reaction product of Bentonite (q.v.) and Stearalkonium Chloride (q.v.).

Chemical Classes: Inorganics; Quaternary Ammonium Compounds

Function: Suspending Agent - Nonsurfactant

Reported Product Categories: Nail Polish and Enamels; Manicuring Preparations, Misc.; Basecoats and Undercoats; Lipsticks

Trade Names:
Claytone AF (Southern Clay)
Garamite VT (Southern Clay)
Tixogel LG (Sud-Chemie, United Catalysts)
Tixogel MP - 250 (Sud-Chemie, United Catalysts)
Tixogel VZ (Sud-Chemie, United Catalysts)
Tixogel VZ - V (Sud-Chemie, United Catalysts)
Viscogel B3 (Bentec)
Viscogel B4 (Bentec)
Viscogel B7 (Bentec)
Viscogel B8 (Bentec)
Viscogel ED (Bentec)
Viscogel GM (Bentec)
Viscogel S4 (Bentec)
Viscogel SD (Bentec)

Trade Name Mixtures:
Miglyol Gel T (Sasol GmbH - Witten)
Tixogel CCT 6030 (Sud-Chemie, United Catalysts)
Tixogel FTN (Sud-Chemie, United Catalysts)
Tixogel FTN 1564 (Sud-Chemie, United Catalysts)
Tixogel IPM (Sud-Chemie, United Catalysts)
Tixogel LAN (Sud-Chemie, United Catalysts)

STEARALKONIUM CHLORIDE

CAS No.	EINECS No.
122-19-0	204-527-9

JPN Translation:
ステアラルコニウムクロリド

Empirical Formula:
$C_{27}H_{50}N \cdot Cl$

Definition: Stearalkonium Chloride is the quaternary ammonium salt that conforms generally to the formula:

Information Sources: 21CFR172.165, 21CFR173.320, 21CFR175.105, CIR: [S] JACT-1(2)1982, CTFA S, JCLS, JSCI, TSCA

Chemical Class: Quaternary Ammonium Compounds

Function: Antistatic Agent

Reported Product Categories: Hair Conditioners; Hair Dyes and Colors (All Types Requiring Caution Statements and Patch Tests); Tonics, Dressings, and Other Hair Grooming Aids; Hair Rinses (Coloring); Hair Wave Sets; Hair Rinses (Non-coloring); Moisturizing Preparations; Permanent Waves; Hair Preparations (Non-coloring), Misc.; Hair Sprays (Aerosol Fixatives)

Technical/Other Names:
Benzenemethanaminium, N,N-Dimethyl-N-Octadecyl-, Chloride
Benzyl Dimethyl Stearyl Ammonium Chloride
N,N-Dimethyl-N-Octadecylbenzene-methanaminium Chloride
Stearyl Dimethyl Benzyl Ammonium Chloride

Trade Names:
AEC Stearalkonium Chloride (A & E Connock)
AMMONYX 4 (Stepan)
AMMONYX 485 (Stepan)
AMMONYX 4002 (Stepan)
AMMONYX 4B (Stepan)
Carsoquat SDQ-25 (Lonza Inc./Lonza Ltd.)
Carsoquat SDQ-85 (Lonza Inc./Lonza Ltd.)
Custom SDQ-85 (stearalkonium chloride) (Custom Ingredients)
HOE S 4131 (Clariant)
HOE S 4131 (Clariant GmbH, Personal Care)
Incroquat S-85 (Croda, Inc.)
Incroquat SDQ-25 (Croda Chemicals)
Incroquat SDQ-25 (Croda, Inc.)
Jeequat SDQ-85 (Jeen)
Mackernium 25-NA (McIntyre)
Mackernium SDC-25 (McIntyre)
Mackernium SDC-85 (McIntyre)
Maquat SC18-25% (Mason)
Maquat SC 18-85% (Mason)
Stedbac (Zecland)
Sumquat 6210 (Zeeland)
Unisoft SAC (Universal Preserv-A-Chem)

Trade Name Mixtures:
Catinal OB-80E (Toho)
CoChem SCS (Costec)
Conditioner Base (Croda Chemicals)
Incroquat CR Concentrate (Croda, Inc.)
Maquat SC 1632 (Mason)
Protaquat ASP (Protameen)
Quatrex CRC (Chemron)
Standamul Conc. 1002 (Cognis Care Chemicals/NJ)
Standamul Conc. 1002 (Cognis Care Chemicals/PA)

Unimul-1002 Conc. (Universal Preserv-A-Chem)
Uniquart H (Universal Preserv-A-Chem)

STEARALKONIUM DIMETHICONE PEG-8 PHTHALATE

Definition: Stearalkonium Dimethicone PEG-8 Phthalate is a salt of stearyl dimethyl benzylamine and Dimethicone PEG-8 Phthalate (q.v.).

Chemical Classes: Organic Salts; Quaternary Ammonium Compounds; Siloxanes and Silanes

Function: Hair Conditioning Agent

STEARALKONIUM HECTORITE

CAS Nos.: 12691-60-0; 94891-33-5

JPN Translation:
ステアラルコニウムヘクトライト

Definition: Stearalkonium Hectorite is a reaction product of Hectorite (q.v.) and Stearalkonium Chloride (q.v.).

Information Sources: 21CFR175.300, CIR: [S] IJT-19(SUPPL. 2)2000, CTFA S, JCIC, JCLS

Chemical Classes: Inorganics; Quaternary Ammonium Compounds

Function: Suspending Agent - Nonsurfactant

Reported Product Categories: Nail Polish and Enamels; Lipsticks; Basecoats and Undercoats; Manicuring Preparations, Misc.; Eyeliners; Eye Makeup Preparations, Misc.; Makeup Preparations (Not eye), Misc.; Eyebrow Pencils; Personal Cleanliness Products, Misc.; Moisturizing Preparations

Technical/Other Name:
Benzyldimethylstearylammonium Hectorite

Trade Names:
Bentone 27 (ELE)
Lucentite SSN (Co-Op/Kobo)

Trade Name Mixtures:
Bentone Gel CAO (ELE)
Bentone Gel IPM (ELE)
Bentone Gel LOI (ELE)
Bentone Gel M-20 (ELE)
Bentone Gel TN (ELE)
Biju BNT (Engelhard Corp.)
Biju BTF-WD (Engelhard Corp.)
Biju BTF-XD (Engelhard Corp.)
BIJU BXD (Engelhard Corp.)
Biju Ultra UNT (Engelhard Corp.)
Biju Ultra UTF-WD (Engelhard Corp.)
Biju Ultra UTF-XD (Engelhard Corp.)
BIJU Ultra UXD (Engelhard Corp.)

Miglyol Gel B (Sasol GmbH - Witten)
Miglyol 840 Gel B (Sasol GmbH - Witten)
Nailsyn II C2X (Merck KGaA/EMD Chemicals Inc.)
Nailsyn II Platinum 25 (Merck KGaA/EMD Chemicals Inc.)
Simagel C (Biophil)
Simagel IM (Biophil)
Simagel NO (Biophil)
Softisan Gel (Sasol GmbH - Witten)

STEARAMIDE

CAS No.	EINECS No.
124-26-5	204-693-2

JPN Translation:
ステアラミド

Empirical Formula:
$C_{18}H_{37}NO$

Definition: Stearamide is the aliphatic amide that conforms generally to the formula:

$$CH_3(CH_2)_{16}\overset{\overset{\displaystyle O}{\|}}{C}-NH_2$$

Information Sources: 21CFR175.105, 21CFR177.1210, 21CFR178.3860, 21CFR178.3910, 21CFR179.45, 21CFR181.22, 21CFR181.28, JCIC, JCLS, JSQI, TSCA

Chemical Class: Amides

Functions: Opacifying Agent; Viscosity Increasing Agent - Nonaqueous

Reported Product Category: Hair Dyes and Colors (All Types Requiring Caution Statements and Patch Tests)

Technical/Other Names:
Octadecanamide
Stearic Acid Amide

Trade Name:
Armid 18 (Akzo Nobel)

STEARAMIDE AMP

CAS No.: 36284-86-3

Empirical Formula:
$C_{22}H_{45}NO_2$

Definition: Stearamide AMP is the organic compound that conforms to the formula:

$$CH_3(CH_2)_{16}\overset{\overset{\displaystyle O}{\|}}{C}-NHC\overset{\displaystyle CH_3}{\underset{\displaystyle CH_3}{|}}CH_2OH$$

Chemical Class: Alkanolamides

Functions: Surfactant - Foam Booster; Viscosity Increasing Agent - Aqueous

Reported Product Categories: Moisturizing Preparations; Bath Preparations, Misc.; Body and Hand Preparations (Excluding Shaving Preparations)

Technical/Other Names:
N-(2-Hydroxy-1,1-Dimethylethyl)-Octadecanamide
Octadecanamide, N-(2-Hydroxy-1,1-Dimethylethyl)-

Trade Name Mixtures:
Cerasynt IP (International Specialty Products)
Crodapearl EM04807 (Croda Chemicals)
Jeechem SAS (Jeen)
Polytex 10M (Lipo)
Protachem SAS (Protameen)
Ritasynt IP (RITA)
STEPAN EGAS (Stepan)

STEARAMIDE DEA

CAS No.	EINECS No.
93-82-3	202-280-1

JPN Translation:
ステアラミド DEA

Empirical Formula:
$C_{22}H_{45}NO_3$

Definition: Stearamide DEA is a mixture of ethanolamides of stearic acid. It conforms generally to the formula:

$$CH_3(CH_2)_{16}C\overset{O}{\overset{\|}{-}}N(CH_2CH_2OH)_2$$

Information Sources: 21CFR175.105, 21CFR176.180, 21CFR177.2260, 21CFR177.2800, CIR: [SQ], CTFA D, EEC (III/1-60), JCLS, JSCI, TSCA

Chemical Class: Alkanolamides

Functions: Surfactant - Foam Booster; Viscosity Increasing Agent - Aqueous

Reported Product Categories: Bath Preparations, Misc.; Hair Conditioners; Body and Hand Preparations (Excluding Shaving Preparations); Foundations; Shampoos (Non-coloring); Bubble Baths; Tonics, Dressings, and Other Hair Grooming Aids

Technical/Other Names:
N,N-Bis(2-Hydroxyethyl)Octadecanamide
N,N-Bis(2-Hydroxyethyl)Stearamide
Diethanolamine Stearic Acid Amide
Octadecanamide, N,N-Bis(2-Hydroxyethyl)-
Stearic Acid Diethanolamide
Stearoyl Diethanolamide

Trade Names:
Colamid 286 (Colonial Chemical Inc)
Ethox 2984 (Ethox)
Hetamide DS (Heterene)

Jeemide N-1918 (Jeen)
Jeemide S (Jeen)
Lipamide S (Lipo)
Lipamide S Pastilles (Lipo)
Mulsor OC (Fabriquimica)
Olamida ED (Fabriquimica)

STEARAMIDE DEA-DISTEARATE

Empirical Formula:
$C_{58}H_{113}NO_5$

Definition: Stearamide DEA-Distearate is the substituted ethanolamide that conforms generally to the formula:

$$CH_3(CH_2)_{16}C\overset{O}{\overset{\|}{-}}N\begin{matrix}(CH_2)_2O-C(CH_2)_{16}CH_3 \\ (CH_2)_2-C(CH_2)_{16}CH_3\end{matrix}$$

Chemical Classes: Alkanolamides; Esters

Functions: Opacifying Agent; Surfactant - Foam Booster; Viscosity Increasing Agent - Aqueous; Viscosity Increasing Agent - Nonaqueous

Technical/Other Name:
Stearic Diethanolamide Distearate

Trade Name:
AEC Stearamide DEA-Distearate (A & E Connock)

STEARAMIDE DIBA-STEARATE

CAS No.: 60209-70-3

Empirical Formula:
$C_{40}H_{79}NO_4$

Definition: Stearamide DIBA-Stearate is the substituted dihydroxyisobutylamine that conforms generally to the formula:

$$CH_3(CH_2)_{16}C\overset{O}{\overset{\|}{-}}NHCH_2CHCH_2O-\overset{O}{\overset{\|}{C}}(CH_2)_{16}CH_3$$
$$|$$
$$CH_2OH$$

Information Sources: CIR: [I] IJT-20 (SUPPL. 3)2001, JSQI

Chemical Classes: Alkanolamides; Esters

Functions: Opacifying Agent; Surfactant - Foam Booster; Viscosity Increasing Agent - Aqueous; Viscosity Increasing Agent - Nonaqueous

Reported Product Category: Body and Hand Preparations (Excluding Shaving Preparations)

Technical/Other Name:
Octadecanoic Acid, 3-Hydroxy-2-Methyl-2-((1-Oxooctadecyl)Amino)Propyl Ester

Trade Name:
Paramul SAS (Bernel)

STEARAMIDE MEA

CAS No.	EINECS No.
111-57-9	203-883-2

JPN Translation:
ステアラミド MEA

Empirical Formula:
$C_{20}H_{41}NO_2$

Definition: Stearamide MEA is a mixture of ethanolamides of stearic acid. It conforms generally to the formula:

$$CH_3(CH_2)_{16}C\overset{O}{\overset{\|}{-}}NHCH_2CH_2OH$$

Information Sources: CIR: [SQ], CTFA D, JCLS, JSCI, TSCA

Chemical Class: Alkanolamides

Functions: Surfactant - Foam Booster; Viscosity Increasing Agent - Aqueous

Reported Product Categories: Hair Dyes and Colors (All Types Requiring Caution Statements and Patch Tests); Hair Conditioners

Technical/Other Names:
N-(2-Hydroxyethyl)Octadecanamide
N-(2-Hydroxyethyl)Stearamide
Monoethanolamine Stearic Acid Amide
Octadecanamide, N-(2-Hydroxyethyl)-
Stearic Acid Monoethanolamide
Stearoyl Monoethanolamide

Trade Names:
Colamid SA (Colonial Chemical Inc)
Hetamide MS (Heterene)
Mackamide SMA (McIntyre)
Monamid S (Uniqema)
Olamida SM (Fabriquimica)
Rewomid S 280 (Degussa Care Specialties)
Upamide SME-M (Universal Preserv-A-Chem)

Trade Name Mixture:
Rewopol PGK 2000 (Degussa Care Specialties)

STEARAMIDE MEA-STEARATE

CAS No.	EINECS No.
14351-40-7	238-310-5

JPN Translation:
ステアリン酸ステアラミド MEA

Empirical Formula:
$C_{38}H_{75}NO_3$

Definition: Stearamide MEA-Stearate is the substituted ethanolamide that conforms generally to the formula:

$$CH_3(CH_2)_{16}\overset{O}{\overset{\|}{C}}-NH(CH_2)_2O-\overset{O}{\overset{\|}{C}}(CH_2)_{16}CH_3$$

Information Sources: JCIC, JCLS, JSQI

Chemical Classes: Alkanolamides; Esters

Functions: Opacifying Agent; Surfactant - Foam Booster; Viscosity Increasing Agent - Aqueous; Viscosity Increasing Agent - Nonaqueous

Reported Product Categories: Bubble Baths; Moisturizing Preparations

Technical/Other Names:
Octadecanoic Acid, 2-[(1-Oxooctadecyl)-Amino]Ethyl Ester
2-[(1-Oxooctadecyl)Amino]Ethyl Octadeca-noate
Stearic Monoethanolamide Stearate
Stearoyl Monoethanolamide Stearate

Trade Names:
AEC Stearamide MEA-Stearate (A & E Connock)
Cerasynt D (International Specialty Products)
Varamide MAS (Degussa Care Specialties)

STEARAMIDE MIPA

CAS No. 35627-96-4
EINECS No. 252-648-0

Empirical Formula:
$C_{21}H_{43}NO_2$

Definition: Stearamide MIPA is a mixture of isopropanolamides of stearic acid. It conforms generally to the formula:

$$CH_3(CH_2)_{16}\overset{O}{\overset{\|}{C}}-NHCH_2\underset{\underset{CH_3}{|}}{C}HOH$$

Information Sources: 21CFR176.210, CTFA D, TSCA

Chemical Class: Alkanolamides

Functions: Surfactant - Foam Booster; Viscosity Increasing Agent - Aqueous

Technical/Other Names:
N-(2-Hydroxypropyl)Octadecanamide
Monoisopropanolamine Stearic Acid Amide
Octadecanamide, N-(2-Hydroxypropyl)-
Stearoyl Isopropanolamide
Stearoyl Monoisopropanolamide

STEARAMIDODIHYDROXYISOBUTYL STEARATE

Definition: Stearamidodihydroxyisobutyl Stearate is the organic compound that conforms to the formula:

$$CH_3(CH_2)_{16}\overset{O}{\overset{\|}{C}}-NH\underset{\underset{CH_2OH}{|}}{\overset{\overset{CH_3}{|}}{C}}CH_2O-\overset{O}{\overset{\|}{C}}(CH_2)_{16}CH_3$$

Information Sources: JCIC, JCLS

Chemical Classes: Amides; Esters

Function: Not Reported

STEARAMIDOETHYL DIETHANOLAMINE

Empirical Formula:
$C_{24}H_{50}N_2O_3$

Definition: Stearamidoethyl Diethanolamine is the amidoamine that conforms to the formula:

$$CH_3(CH_2)_{10}\overset{O}{\overset{\|}{C}}-NH(CH_2)_2N(CH_2CH_2OH)_2$$

Chemical Class: Amines

Function: Antistatic Agent

STEARAMIDOETHYL DIETHYLAMINE

CAS No. 16889-14-8
EINECS No. 240-924-3

JPN Translation:
ステアラミドエチルジエチルアミン

Empirical Formula:
$C_{24}H_{50}N_2O$

Definition: Stearamidoethyl Diethylamine is the amidoamine that conforms generally to the formula:

$$CH_3(CH_2)_{16}\overset{O}{\overset{\|}{C}}-NH(CH_2)_3N\overset{\diagup CH_2CH_3}{\diagdown CH_2CH_3}$$

Information Sources: CTFA D, JCIC, JCLS, JSQI, TSCA

Chemical Class: Amines

Functions: Antistatic Agent; Hair Condition-ing Agent

Reported Product Categories: Bath Oils, Tablets, and Salts; Cleansing Products (Cold Creams, Cleansing Lotions, Liquids and Pads)

Technical/Other Names:
N-[2-Diethylamino)Ethyl]Octadecanamide
Diethylaminoethyl Stearamide
Octadecanamide, N-[2-(Diethylamino)-Ethyl]-

Trade Names:
Lexamine 22 (Inolex)
Nikkol Amidoamine S (Nikko)

Trade Name Mixture:
Lexemul AR (Inolex)

STEARAMIDOETHYL DIETHYLAMINE PHOSPHATE

CAS No. 68133-34-6
EINECS No. 268-677-7

Definition: Stearamidoethyl Diethylamine Phosphate is a complex mixture of phosphoric acid and Stearamidoethyl Diethylamine (q.v.).

Information Source: TSCA

Chemical Classes: Amines; Phosphorus Compounds

Functions: Antistatic Agent; Hair Condition-ing Agent

Technical/Other Names:
N-[2-(Diethylamino)Ethyl]Octadecanamide Phosphate (1:1)
Diethylaminoethyl Stearamide Phosphate
Octadecanamide, N-[2-(Diethylamino)-Ethyl]-, Phosphate (1:1)

STEARAMIDOETHYL ETHANOLAMINE

CAS No. 141-21-9
EINECS No. 205-469-7

Empirical Formula:
$C_{22}H_{46}N_2O_2$

Definition: Stearamidoethyl Ethanolamine is the amidoamine that conforms to the formula:

$$CH_3(CH_2)_{16}\overset{O}{\overset{\|}{C}}-NH(CH_2)_2NHCH_2CH_2OH$$

Information Source: TSCA

Chemical Class: Amines

Function: Antistatic Agent

Technical/Other Names:
Ethanolaminoethyl Stearamide
N-[2-[(2-Hydroxyethyl)Amino]Ethyl]-Octadecanamide
Octadecanamide, N-[2-[(2-Hydroxyethyl)-Amino]Ethyl]-

Trade Name:
Catemol 18SA (Phoenix)

STEARAMIDOETHYL ETHANOLAMINE PHOSPHATE

Definition: Stearamidoethyl Ethanolamine Phosphate is the phosphoric acid salt of Stearamidoethyl Ethanolamine (q.v.).

Chemical Classes: Amines; Phosphorus Compounds

Function: Antistatic Agent

STEARAMIDOPROPALKONIUM CHLORIDE

CAS No. 65694-10-2

EINECS No. 265-880-2

Empirical Formula:
$C_{30}H_{55}N_2O \cdot Cl$

Definition: Stearamidopropalkonium Chloride is the quaternary ammonium salt that conforms generally to the formula:

$$CH_3(CH_2)_{16}\overset{\overset{O}{\|}}{C}-NH(CH_2)_3-\overset{\overset{CH_3}{|}}{\underset{\underset{CH_2}{|}}{N^+}}-CH_3 \quad Cl^-$$

Information Source: TSCA

Chemical Class: Quaternary Ammonium Compounds

Function: Antistatic Agent

Technical/Other Names:
Benzenemethanaminium, N,N-Dimethyl-N-[3-[(1-Oxooctadecyl)Amino]Propyl]-, Chloride
N,N-Dimethyl-N-[3-[(1-Oxooctadecyl)-Amino]Propyl]Benzenemethanaminium Chloride

STEARAMIDOPROPYLAMINE OXIDE

CAS No. 25066-20-0

EINECS No. 246-598-9

Empirical Formula:
$C_{23}H_{48}N_2O_2$

Definition: Stearamidopropylamine Oxide is the amine oxide that conforms to the formula:

$$CH_3(CH_2)_{16}\overset{\overset{O}{\|}}{C}-NH(CH_2)_3\overset{\overset{CH_3}{|}}{\underset{\underset{CH_3}{|}}{N}}\longrightarrow O$$

Chemical Class: Amine Oxides

Functions: Hair Conditioning Agent; Surfactant - Cleansing Agent; Surfactant - Foam Booster; Surfactant - Hydrotrope

Technical/Other Names:
Amides, Stearic, N-[3-(Dimethylamino)-Propyl], N-Oxide
N-[3-(Dimethylamino)Propyl]-Octadecanamide-N-Oxide
Octadecanamide, N-[3-(Dimethylamino)-Propyl]-N-Oxide

Trade Name:
Mackamine SAO (McIntyre)

STEARAMIDOPROPYL BETAINE

CAS No. 6179-44-8

EINECS No. 228-227-2

Empirical Formula:
$C_{25}H_{50}N_2O_3$

Definition: Stearamidopropyl Betaine is the zwitterion (inner salt) that conforms generally to the formula:

$$CH_3(CH_2)_{16}\overset{\overset{O}{\|}}{C}-NH(CH_2)_3\overset{\overset{CH_3}{|}}{\underset{\underset{CH_3}{|}}{N^+}}-CH_2COO^-$$

Information Source: TSCA

Chemical Class: Betaines

Functions: Antistatic Agent; Hair Conditioning Agent; Skin-Conditioning Agent - Miscellaneous; Surfactant - Cleansing Agent; Surfactant - Foam Booster; Viscosity Increasing Agent - Aqueous

Technical/Other Names:
N-(Carboxymethyl)-N,N-Dimethyl-3-[(1-Oxooctadecyl)Amino]-1-Propanaminium Hydroxide, Inner Salt
1-Propanaminium, N-(Carboxymethyl)-N,N-Dimethyl-3-[(1-Oxooctadecyl)Amino]-, Hydroxide, Inner Salt
Stearoyl Amide Propyl Dimethyl Glycine

STEARAMIDOPROPYL CETEARYL DIMONIUM TOSYLATE

CAS No.: 97616-63-2

Definition: Stearamidopropyl Cetearyl Dimonium Tosylate is the quaternary ammonium salt that conforms to the formula:

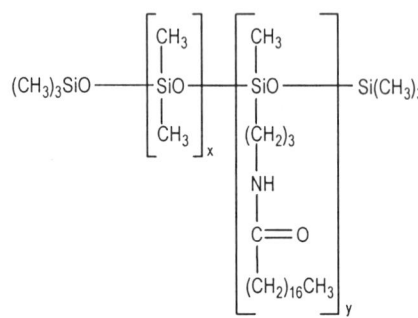

where R represents a blend of cetyl and stearyl alkyl groups.

Chemical Class: Quaternary Ammonium Compounds

Function: Antistatic Agent

Technical/Other Name:
Stearamidopropyl Dimethyl Cetearyl Ammonium Tosylate

Trade Name Mixture:
Ceraphyl 85 (International Specialty Products)

STEARAMIDOPROPYL DIMETHICONE

Definition: Stearamidopropyl Dimethicone is the siloxane polymer that conforms to the formula:

$$(CH_3)_3SiO-\left[\overset{\overset{CH_3}{|}}{\underset{\underset{CH_3}{|}}{SiO}}\right]_x\left[\overset{\overset{CH_3}{|}}{\underset{\underset{(CH_2)_{16}CH_3}{|}}{\underset{|}{\underset{C=O}{\underset{|}{\underset{NH}{\underset{|}{(CH_2)_3}}}}}}}{SiO}}\right]_y-Si(CH_3)_3$$

Information Source: CIR: [S]

Chemical Classes: Siloxanes and Silanes; Synthetic Polymers

Functions: Corrosion Inhibitor; Film Former

Trade Names:
GP-7104 (Genesee)
GP-7105 (Genesee)

STEARAMIDOPROPYL DIMETHYLAMINE

CAS Nos. 7651-02-7 20182-63-2

EINECS No. 231-609-1

JPN Translation:
ステアラミドプロピルジメチルアミン

Empirical Formula:
$C_{23}H_{48}N_2O$

Definition: Stearamidopropyl Dimethylamine is the amidoamine that conforms generally to the formula:

$$CH_3(CH_2)_{16}\overset{\overset{O}{\|}}{C}-NH(CH_2)_3N\overset{CH_3}{\underset{CH_3}{<}}$$

Information Sources: CTFA D, TSCA

Chemical Class: Amines

Functions: Antistatic Agent; Hair Conditioning Agent

Reported Product Categories: Hair Conditioners; Hair Coloring Preparations, Misc.; Hair Bleaches; Hair Preparations (Non-coloring), Misc.; Hair Rinses (Non-coloring); Skin Care Preparations, Misc.

Technical/Other Names:
N-[3-(Dimethylamino)Propyl]-Octadecanamide
Dimethylaminopropyl Stearamide
Octadecanamide, N-[3-(Dimethylamino)-Propyl]-

Trade Names:
Chemidex S (Chemron)
Incromine SB (Croda Chemicals)
Incromine SB (Croda, Inc.)
Jeechem S-13 (Jeen)
Lexamine S-13 (Inolex)
Lipamine SPA (Lipo)
Mackine 301 (McIntyre)
Schercodine S (Scher)
Tego Amid S 18 (Degussa Care Specialties)
Unizeen SA (Universal Preserv-A-Chem)

Trade Name Mixtures:
Lexate CRC (Inolex)
Solarcat (Collaborative Labs)
Stepanquat DC1 (Stepan)
Varisoft CRC (Degussa Care Specialties)

STEARAMIDOPROPYL DIMETHYLAMINE LACTATE

CAS Nos.	EINECS No.
55819-53-9	259-837-7
133681-90-0	

Empirical Formula:
$C_{23}H_{48}N_2O \cdot C_3H_6O_3$

Definition: Stearamidopropyl Dimethylamine Lactate is the lactic acid salt of Stearamidopropyl Dimethylamine (q.v.).

Information Source: TSCA

Chemical Class: Amines

Functions: Antistatic Agent; Hair Conditioning Agent

Reported Product Categories: Hair Dyes and Colors (All Types Requiring Caution Statements and Patch Tests); Hair Conditioners

Technical/Other Name:
Propanoic Acid, 2-Hydroxy-, Compd. with N-[3-(Dimethylamino)Propyl]-Octadecanamide

Trade Names:
Hetamine 5L25 (Heterene)
Incromate SDL (Croda Chemicals)
Incromate SDL (Croda, Inc.)
Jeechem SDM (Jeen)
Mackalene 316 (McIntyre)
Protachem SDM (Protameen)

Trade Name Mixture:
Nikkol Nikkomulese 61H (Nikko)

STEARAMIDOPROPYL DIMETHYLAMINE STEARATE

Empirical Formula:
$C_{23}H_{48}N_2O \cdot C_{18}H_{36}O_2$

Definition: Stearamidopropyl Dimethylamine Stearate is the stearic acid salt of Stearamidopropyl Dimethylamine (q.v.). It conforms to the formula:

Chemical Class: Amines

Functions: Antistatic Agent; Hair Conditioning Agent

Trade Name:
Catemol S180-S (Phoenix)

STEARAMIDOPROPYL ETHYLDIMONIUM ETHOSULFATE

CAS No.	EINECS No.
67846-16-6	267-360-0

Empirical Formula:
$C_{25}H_{53}N_2O \cdot C_2H_5O_4S$

Definition: Stearamidopropyl Ethyldimonium Ethosulfate is the quaternary ammonium salt that conforms to the formula:

Chemical Class: Quaternary Ammonium Compounds

Functions: Antistatic Agent; Hair Conditioning Agent

Technical/Other Names:
N-Ethyl-N,N-Dimethyl-3-[(1-Oxooctadecyl)Amino]-1-Propanaminium Ethyl Sulfate
1-Propanaminium, N-Ethyl-N,N-Dimethyl-3-[(1-Oxooctadecyl)Amino]-, Ethyl Sulfate

Trade Name:
Schercoquat SAS (Scher)

STEARAMIDOPROPYL MORPHOLINE

CAS No.: 55852-13-6

Empirical Formula:
$C_{25}H_{50}N_2O_2$

Definition: Stearamidopropyl Morpholine is the amidoamine that conforms generally to the formula:

Information Source: TSCA

Chemical Classes: Amines; Heterocyclic Compounds

Function: Antistatic Agent

Technical/Other Names:
N-[3-(4-Morpholinyl)Propyl]-Octadecanamide
Octadecanamide, N-[3-(4-Morpholinyl)-Propyl]-

Trade Name:
Mackine 321 (McIntyre)

STEARAMIDOPROPYL MORPHOLINE LACTATE

CAS Nos.	EINECS No.
55852-14-7	259-860-2
133681-88-6	

Empirical Formula:
$C_{25}H_{50}N_2O_2 \cdot C_3H_6O_3$

Definition: Stearamidopropyl Morpholine Lactate is the lactic acid salt of Stearamidopropyl Morpholine (q.v.).

Information Source: TSCA

Chemical Classes: Amines; Heterocyclic Compounds

Function: Antistatic Agent

Technical/Other Name:
Propanoic Acid, 2-Hydroxy-, Compd. with N-[3-(4-Morpholinyl)Propyl]-Octadecanamide

Trade Name:
Mackalene 326 (McIntyre)

STEARAMIDOPROPYL PG-DIMONIUM CHLORIDE PHOSPHATE

Definition: Stearamidopropyl PG-Dimonium Chloride Phosphate is the quaternary ammonium salt that conforms generally to the formula:

where RCO- represents the stearoyl moiety.

Chemical Classes: Phosphorus Compounds; Quaternary Ammonium Compounds

Functions: Antistatic Agent; Hair Conditioning Agent

Reported Product Category: Bath Soaps and Detergents

Trade Name:
Colalipid ST (Colonial Chemical Inc)

Trade Name Mixture:
Phospholipid SV (Uniqema)

STEARAMIDOPROPYL PYRROLIDONYL-METHYL DIMONIUM CHLORIDE

Empirical Formula:
$C_{28}H_{56}N_2O_2 \cdot Cl$

Definition: Stearamidopropyl Pyrrolidonyl-methyl Dimonium Chloride is the quaternary ammonium salt that conforms to the formula:

Chemical Classes: Heterocyclic Compounds; Quaternary Ammonium Compounds

Functions: Antistatic Agent; Hair Conditioning Agent

Technical/Other Name:
N,N-Dimethyl-N-[(2-Pyrrolidonyl)Methyl]-N-(3-Stearamidopropyl)Ammonium Chloride

STEARAMIDOPROPYL TRIMONIUM METHOSULFATE

CAS No.	EINECS No.
19277-88-4	242-930-1

Empirical Formula:
$C_{24}H_{51}N_2O \cdot CH_3O_4S$

Definition: Stearamidopropyl Trimonium Methosulfate is the quaternary ammonium salt that conforms to the formula:

Chemical Class: Quaternary Ammonium Compounds

Functions: Antistatic Agent; Hair Conditioning Agent

Technical/Other Names:
1-Propanaminium, N,N,N-Trimethyl-3-[(1-Oxooctadecyl)Amino]-, Methyl Sulfate
N,N,N-Trimethyl-3-[(1-Oxooctadecyl)-Amino]-1-Propanaminium Methyl Sulfate

Trade Name:
CATIGENE SA-70 (Stepan)

STEARAMINE

CAS No.	EINECS No.
124-30-1	204-695-3

Empirical Formula:
$C_{18}H_{39}N$

Definition: Stearamine is the aliphatic amine that conforms generally to the formula:

$$CH_3(CH_2)_{17}NH_2$$

Information Sources: 21CFR173.310, CIR: [I] JACT-14(3)1995, TSCA

Chemical Class: Amines

Function: Antistatic Agent

Reported Product Category: Shampoos (Non-coloring)

Technical/Other Names:
1-Octadecanamine
Octadecylamine
Stearyl Amine

Trade Names:
Armeen 18 (Akzo Nobel)
Armeen 18D (Akzo Nobel)

STEARAMINE OXIDE

CAS No.	EINECS No.
2571-88-2	219-919-5

JPN Translation:
ステアラミンオキシド

Empirical Formula:
$C_{20}H_{43}NO$

Definition: Stearamine Oxide is the tertiary amine oxide that conforms generally to the formula:

Information Sources: CIR: [SQ] JACT-13 (5)1994, CTFA S, JCIC, JCLS, JSQI, TSCA

Chemical Class: Amine Oxides

Functions: Hair Conditioning Agent; Surfactant - Cleansing Agent; Surfactant - Foam Booster; Surfactant - Hydrotrope

Reported Product Categories: Hair Conditioners; Shampoos (Non-coloring); Aftershave Lotions; Hair Coloring Preparations, Misc.; Skin Care Preparations, Misc.

Technical/Other Names:
N,N-Dimethyl-1-Octadecanamine-N-Oxide
1-Octadecanamine, N,N-Dimethyl-, N-Oxide
Stearamine Oxide Solution
Stearyl Dimethylamine Oxide

Trade Names:
AMMONYX SO (Stepan)
Barlox 18S (Lonza Inc./Lonza Ltd.)
Incromine Oxide S (Croda, Inc.)
Mackamine SO (McIntyre)
Schercamox DMS (Scher)

STEARDIMONIUM HYDROXYPROPYL HYDROLYZED CASEIN

Definition: Steardimonium Hydroxypropyl Hydrolyzed Casein is the quaternary ammonium chloride that conforms to the formula:

where R represents the hydrolyzed casein moiety.

Chemical Classes: Protein Derivatives; Quaternary Ammonium Compounds

Functions: Antistatic Agent; Hair Conditioning Agent; Skin-Conditioning Agent - Miscellaneous

Technical/Other Name:
Stearyldimonium Hydroxypropyl Hydrolyzed Casein

Trade Name:
Promois Milk-SAQ (Seiwa Kasei)

STEARDIMONIUM HYDROXYPROPYL HYDROLYZED COLLAGEN

JPN Translation:
ステアルジモニウムヒドロキシプロピル加水分解コラーゲン

Definition: Steardimonium Hydroxypropyl Hydrolyzed Collagen is the quaternary

---

The inclusion of any compound in the *Dictionary and Handbook* does not indicate that use of that substance as a cosmetic ingredient complies with the laws and regulations governing such use in the United States or any other country.

ammonium chloride that conforms generally to the formula:

$$\left[CH_3(CH_2)_{17} - \overset{\overset{\displaystyle CH_3}{|}}{\underset{\underset{\displaystyle CH_3}{|}}{N}} - CH_2\underset{\underset{\displaystyle OH}{|}}{C}HCH_2R \right]^+ \quad Cl^-$$

where R represents the hydrolyzed collagen moiety.

Information Sources: JCIC, JCLS

Chemical Classes: Protein Derivatives; Quaternary Ammonium Compounds

Functions: Antistatic Agent; Hair Conditioning Agent; Skin-Conditioning Agent - Miscellaneous

Reported Product Category: Tonics, Dressings, and Other Hair Grooming Aids

Technical/Other Names:
N-[2-Hydroxy-3-(stearyldimethylammonio)propyl] Hydrolyzed Collagen
Stearyldimonium Hydroxypropyl Hydrolyzed Collagen

Trade Names:
Croquat S (Croda Chemicals)
Croquat S (Croda, Inc.)
Promois W-32RHSAQ (Seiwa Kasei)
Promois W-32SAQ (Seiwa Kasei)
Promois W-42SAQ (Seiwa Kasei)
Quat-Coll SDMA-40 (Arch Personal Care Products)

STEARDIMONIUM HYDROXYPROPYL HYDROLYZED JOJOBA PROTEIN

CAS No.: 333338-12-8

Definition: Steardimonium Hydroxypropyl Hydrolyzed Jojoba Protein is the quaternary ammonium salt that conforms generally to the formula:

$$\left[CH_3(CH_2)_{17} - \overset{\overset{\displaystyle CH_3}{|}}{\underset{\underset{\displaystyle CH_3}{|}}{N}} - CH_2\underset{\underset{\displaystyle OH}{|}}{C}HCH_2 - R \right]^+ \quad Cl^-$$

where R represents the Hydrolyzed Jojoba Protein (q.v.) moiety.

Chemical Classes: Protein Derivatives; Quaternary Ammonium Compounds

Functions: Antistatic Agent; Hair Conditioning Agent

Technical/Other Name:
Protein Hydrolyzates, Jojoba, [3-(Dimethyloctadecylammonio)-2-Hydroxypropyl], Chlorides

Trade Name:
Jojoba Quat - SH (Desert Whale)

STEARDIMONIUM HYDROXYPROPYL HYDROLYZED KERATIN

JPN Translation:
ステアルジモニウムヒドロキシプロピル加水分解ケラチン

Definition: Steardimonium Hydroxypropyl Hydrolyzed Keratin is the quaternary ammonium chloride that conforms generally to the formula:

$$\left[CH_3(CH_2)_{17} - \overset{\overset{\displaystyle CH_3}{|}}{\underset{\underset{\displaystyle CH_3}{|}}{N}} - CH_2\underset{\underset{\displaystyle OH}{|}}{C}HCH_2R \right]^+ \quad Cl^-$$

where R represents the hydrolyzed keratin moiety.

Information Sources: JCIC, JCLS

Chemical Classes: Protein Derivatives; Quaternary Ammonium Compounds

Functions: Antistatic Agent; Hair Conditioning Agent; Skin-Conditioning Agent - Miscellaneous

Technical/Other Names:
N-[2-Hydroxy-3-(stearyldimethylammonio)propyl] Hydrolyzed Keratin Chloride
Stearyldimonium Hydroxypropyl Hydrolyzed Keratin

Trade Name:
Promois WK-HSAQ (Seiwa Kasei)

STEARDIMONIUM HYDROXYPROPYL HYDROLYZED RICE PROTEIN

Definition: Steardimonium Hydroxypropyl Hydrolyzed Rice Protein is the quaternary ammonium chloride that conforms generally to the formula:

$$\left[CH_3(CH_2)_{17} - \overset{\overset{\displaystyle CH_3}{|}}{\underset{\underset{\displaystyle CH_3}{|}}{N}} - CH_2\underset{\underset{\displaystyle OH}{|}}{C}HCH_2R \right]^+ \quad Cl^-$$

where R represents the hydrolyzed rice protein moiety.

Chemical Classes: Protein Derivatives; Quaternary Ammonium Compounds

Functions: Antistatic Agent; Hair Conditioning Agent; Skin-Conditioning Agent - Miscellaneous

Technical/Other Name:
Stearyldimonium Hydroxypropyl Hydrolyzed Rice Protein

Trade Name:
Quat-Rice SDMA-25 (Arch Personal Care Products)

STEARDIMONIUM HYDROXYPROPYL HYDROLYZED SILK

JPN Translation:
ステアルジモニウムヒドロキシプロピル加水分解シルク

Definition: Steardimonium Hydroxypropyl Hydrolyzed Silk is the quaternary ammonium chloride that conforms generally to the formula:

$$\left[CH_3(CH_2)_{17} - \overset{\overset{\displaystyle CH_3}{|}}{\underset{\underset{\displaystyle CH_3}{|}}{N}} - CH_2\underset{\underset{\displaystyle OH}{|}}{C}HCH_2R \right]^+ \quad Cl^-$$

where R represents the hydrolyzed silk moiety.

Information Sources: JCIC, JCLS

Chemical Classes: Protein Derivatives; Quaternary Ammonium Compounds

Functions: Antistatic Agent; Hair Conditioning Agent; Skin-Conditioning Agent - Miscellaneous

Technical/Other Names:
N-[2-Hydroxy-3-(stearyldimethylammonio)propyl] Hydrolyzed Silk
Stearyldimonium Hydroxypropyl Hydrolyzed Silk

Trade Name:
Promois Silk-SAQ (Seiwa Kasei)

STEARDIMONIUM HYDROXYPROPYL HYDROLYZED SOY PROTEIN

JPN Translation:
カチオン化加水分解ダイズタンパク - 1

Definition: Steardimonium Hydroxypropyl Hydrolyzed Soy Protein is the quaternary ammonium chloride that conforms generally to the formula:

$$\left[CH_3(CH_2)_{17} - \overset{\overset{\displaystyle CH_3}{|}}{\underset{\underset{\displaystyle CH_3}{|}}{N}} - CH_2\underset{\underset{\displaystyle OH}{|}}{C}HCH_2R \right]^+ \quad Cl^-$$

where R represents the hydrolyzed soy protein moiety.

Information Source: JCLS

Chemical Classes: Protein Derivatives; Quaternary Ammonium Compounds

Functions: Antistatic Agent; Hair Conditioning Agent; Skin-Conditioning Agent - Miscellaneous

Trade Names:
Aqua Pro II QSS (MGP)
Promois WS-SAQ (Seiwa Kasei)

STEARDIMONIUM HYDROXYPROPYL HYDROLYZED VEGETABLE PROTEIN

Definition: Steardimonium Hydroxypropyl Hydrolyzed Vegetable Protein is the quaternary ammonium chloride that conforms generally to the formula:

$$\left[CH_3(CH_2)_{17} - \underset{\underset{CH_3}{|}}{\overset{\overset{CH_3}{|}}{N}} - CH_2\underset{\underset{OH}{|}}{CH}CH_2R \right]^+ \quad Cl^-$$

where R represents the hydrolyzed vegetable protein moiety.

Chemical Classes: Protein Derivatives; Quaternary Ammonium Compounds

Functions: Antistatic Agent; Hair Conditioning Agent; Skin-Conditioning Agent - Miscellaneous

Technical/Other Name:
Stearyldimonium Hydroxypropyl Hydrolyzed Vegetable Protein

Trade Name:
Quat-Veg SDA-30 (Arch Personal Care Products)

STEARDIMONIUM HYDROXYPROPYL HYDROLYZED WHEAT PROTEIN

CAS No.: 130381-05-4

JPN Translation:
カチオン化加水分解コムギタンパク - 1

Definition: Steardimonium Hydroxypropyl Hydrolyzed Wheat Protein is the quaternary ammonium chloride that conforms generally to the formula:

$$\left[CH_3(CH_2)_{17} - \underset{\underset{CH_3}{|}}{\overset{\overset{CH_3}{|}}{N}} - CH_2\underset{\underset{OH}{|}}{CH}CH_2R \right]^+ \quad Cl^-$$

where R represents the hydrolyzed wheat protein moiety.

Chemical Classes: Protein Derivatives; Quaternary Ammonium Compounds

Functions: Antistatic Agent; Hair Conditioning Agent; Skin-Conditioning Agent - Miscellaneous

Technical/Other Names:
Protein Hydrolysates, Wheat Germ, [3-(Dimethyloctadecylammonio)-2-Hydroxypropyl], Chloride
Protein Hydrolysates, Wheat Germ, [3-(Dimethyloctadecylammonio)-2-Hydroxypropyl], Chlorides
Stearyldimonium Hydroxypropyl Hydrolyzed Wheat Protein

Trade Names:
Aqua Pro II QWS (MGP)
Hydrotriticum QS (Croda Chemicals)
Hydrotriticum QS (Croda, Inc.)
Promois WG-SAQ (Seiwa Kasei)
Quat-Wheat SDMA-25 (Arch Personal Care Products)

STEARDIMONIUM HYDROXYPROPYL PANTHENYL PEG-7 DIMETHICONE PHOSPHATE CHLORIDE

CAS No.: 220714-77-2

Definition: Steardimonium Hydroxypropyl Panthenyl PEG-7 Dimethicone Phosphate Chloride is the siloxane polymer that conforms generally to the formula:

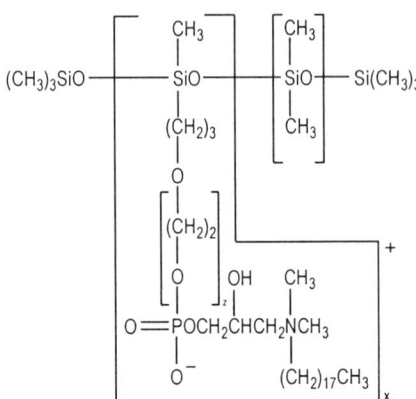

· xCl⁻

where z has an average value of 7.

Chemical Classes: Phosphorus Compounds; Siloxanes and Silanes; Synthetic Polymers

Function: Hair Conditioning Agent

Trade Names:
Pecosil PAN-150 (Phoenix)
Pecosil PAN-418 (Phoenix)

STEARDIMONIUM HYDROXYPROPYL PEG-7 DIMETHICONE PHOSPHATE CHLORIDE

CAS No.: 220714-63-6

Definition: Steardimonium Hydroxypropyl PEG-7 Dimethicone Phosphate Chloride is the siloxane polymer that conforms generally to the formula:

where z has an average value of 7.

Chemical Classes: Phosphorus Compounds; Siloxanes and Silanes; Synthetic Polymers

Functions: Hair Conditioning Agent; Skin-Conditioning Agent - Emollient

Technical/Other Names:
Siloxanes and Silicones, di-Me, 3-Hydroxypropyl Me, Ethers with Polyethylene Glycol Mono[3-[(2,4-Dihydroxy-3,3-Dimethyl-1-Oxobutyl)Amino]Propyl 3-(Dimethyloctadecylammonio)-2-Hydroxypropylphosphate], Chlorides
Siloxanes and Silicones, Di-Me, 3-Hydroxypropyl Me, Ethers with Polyethylene Glycol Mono[3-(Dimethyloctadecylammonio)-2-Hydroxypropyl Hydrogen Phosphate], Inner Salts

Trade Name:
Pecosil PSQ-418 (Phoenix)

STEARETH-2

CAS Nos.: 9005-00-9 (Generic); 16057-43-5

JPN Translation:
ステアレス - 2

Empirical Formula:
$C_{22}H_{46}O_3$

Definition: Steareth-2 is the polyethylene glycol ether of Stearyl Alcohol (q.v.) that conforms to the formula:

$$CH_3(CH_2)_{17}(OCH_2CH_2)_nOH$$

where n has an average value of 2.

Information Sources: CIR: [S] JACT-7(6)-1988, CTFA S, JCLS, MI-13(7659), SNPF, TSCA

Chemical Class: Alkoxylated Alcohols

Function: Surfactant - Emulsifying Agent

Reported Product Categories: Bath Preparations, Misc.; Body and Hand Preparations (Excluding Shaving Preparations); Moisturizing Preparations; Personal Cleanliness Products, Misc.; Foundations; Skin Care Preparations, Misc.; Aftershave Lotions; Baby Shampoos; Bath Capsules; Bath Oils, Tablets, and Salts; Cleansing Products (Cold Creams, Cleansing Lotions, Liquids and Pads); Eyeliners; Indoor Tanning Preparations; Face and Neck Preparations (Excluding Shaving Preparations); Night Skin Care Preparations; Makeup Bases; Mascara; Eye Makeup Preparations, Misc.; Hair Conditioners; Hair Preparations (Noncoloring), Misc.; Makeup Preparations (Not eye), Misc.; Paste Masks (Mud Packs); Baby Lotions, Oils, Powders and Creams; Eye Makeup Removers; Foot Powders and Sprays; Fragrance Preparations, Misc.; Hair Straighteners; Suntan Gels, Creams, and Liquids; Suntan Preparations, Misc.; Tonics, Dressings, and Other Hair Grooming Aids

Technical/Other Names:
Ethanol, 2-[2-Octadecyloxy)Ethoxy]-
2-[2-(Octadecyloxy)Ethoxy]Ethanol
PEG-2 Stearyl Ether
Polyethylene Glycol (100) Stearyl Ether
Polyoxyethylene (2) Stearyl Ether

Trade Names:
Brij 72 (Uniqema Americas)
Chemonic S-2 (Chemron)
Customol SA-2 (Steareth-2) (Custom Ingredients)
Ethal SA-2 (Ethox)
Genapol HS 020 (Clariant)
Genapol HS 020 (Clariant GmbH, Personal Care)
Hetoxol STA-2 (Heterene)
Jeecol SA-2 (Jeen)
Lanycol-72 (Lanaetex)
Lipocol S-2 (Lipo)
Lumulse S-2 (Lambent)
Nikkol BS-2 (Nikko)
Procol SA-2 (Protameen)
Sipol SAL-2 (Specialty Industrial)
Sympatens-AS/020 (Kolb)
Tego Alkanol S 2 Pellets (Degussa Care Specialties)
Unicol SA-2 (Universal Preserv-A-Chem)

Volpo S2 (Croda Chemicals)
Volpo S-2 (Croda, Inc.)
Volpo S2A (Croda Chemicals)

Trade Name Mixtures:
Emulgade NLB (Cognis Deutschland)
Nano-emulsion Concentrate (Active Concepts)
Nano-emulsion Concentrate Sun (Active Concepts)

STEARETH-3

CAS Nos.: 4439-32-1; 9005-00-9 (Generic)

JPN Translation:
ステアレス - 3

Empirical Formula:
$C_{24}H_{50}O_4$

Definition: Steareth-3 is the polyethylene glycol ether of Stearyl Alcohol (q.v.) that conforms generally to the formula:

$$CH_3(CH_2)_{17}(OCH_2CH_2)_nOH$$

where n has an average value of 3.

Information Sources: JCLS, MI-13(7659)

Chemical Class: Alkoxylated Alcohols

Function: Surfactant - Emulsifying Agent

Technical/Other Names:
Ethanol, 2-[2-[2-(Octadecyloxy)Ethoxy]-Ethoxy]-
2-[2-[2-(Octadecyloxy)Ethoxy]Ethoxy]-Ethanol
PEG-3 Stearyl Ether
Polyethylene Glycol (3) Stearyl Ether
Polyoxyethylene (3) Stearyl Ether

Trade Name Mixture:
Isoxal 5 (Vevy)

STEARETH-4

CAS Nos.: 9005-00-9 (Generic); 59970-10-4

JPN Translation:
ステアレス - 4

Empirical Formula:
$C_{26}H_{54}O_5$

Definition: Steareth-4 is the polyethylene glycol ether of Stearyl Alcohol (q.v.) that conforms generally to the formula:

$$CH_3(CH_2)_{17}(OCH_2CH_2)_nOH$$

where n has an average value of 4.

Information Sources: CIR: [S] JACT-7(6)-1988, JCLS, MI-13(7659), TSCA

Chemical Class: Alkoxylated Alcohols

Function: Surfactant - Emulsifying Agent

Reported Product Categories: Bath Preparations, Misc.; Bath Oils, Tablets, and Salts; Cleansing Products (Cold Creams, Cleansing Lotions, Liquids and Pads)

Technical/Other Names:
PEG-4 Stearyl Ether
Polyethylene Glycol 200 Stearyl Ether
Polyoxyethylene (4) Stearyl Ether
3,6,9,12-Tetraoxatriacontan-1-ol

Trade Names:
Jeecol SA-4 (Jeen)
Nikkol BS-4 (Nikko)
Pegnol S-4D (Toho)
Procol SA-4 (Protameen)

Trade Name Mixture:
Tego Pearl N 100 (Degussa Care Specialties)

STEARETH-5

CAS Nos.: 9005-00-9 (Generic); 71093-13-5

JPN Translation:
ステアレス - 5

Empirical Formula:
$C_{28}H_{58}O_6$

Definition: Steareth-5 is the polyethylene glycol ether of Stearyl Alcohol (q.v.) that conforms generally to the formula:

$$CH_3(CH_2)_{17}(OCH_2CH_2)_xOH$$

where n has an average value of 5.

Information Sources: JCLS, MI-13(7659)

Chemical Class: Alkoxylated Alcohols

Function: Surfactant - Emulsifying Agent

Technical/Other Names:
PEG-5 Stearyl Ether
3,6,9,12,15-Pentaoxatritriacontan-1-ol
Polyethylene Glycol (5) Stearyl Ether
Polyoxyethylene (5) Stearyl Ether

Trade Name:
Jeecol SA-5 (Jeen)

Trade Name Mixture:
Isoxal 12 (Vevy)

STEARETH-6

CAS Nos.: 2420-29-3; 9005-00-9 (Generic)

JPN Translation:
ステアレス - 6

Empirical Formula:
$C_{30}H_{62}O_7$

Definition: Steareth-6 is the polyethylene glycol ether of Stearyl Alcohol (q.v.) that conforms generally to the formula:

$$CH_3(CH_2)_{17}(OCH_2CH_2)_nOH$$

where n has an average value of 6.

Information Sources: CIR: [S] JACT-7(6)-1988, JCLS, MI-13(7659), TSCA

Chemical Class: Alkoxylated Alcohols

Function: Surfactant - Emulsifying Agent

Technical/Other Names:
3,6,9,12,15,18-Hexaoxahexatriacontan-1-ol
PEG-6 Stearyl Ether
Polyethylene Glycol 300 Stearyl Ether
Polyoxyethylene (6) Stearyl Ether

Trade Name:
Emalex 606 (Nihon Emulsion)

STEARETH-7

CAS Nos.: 9005-00-9 (Generic); 66146-84-7

JPN Translation:
ステアレス - 7

Empirical Formula:
$C_{32}H_{66}O_8$

Definition: Steareth-7 is the polyethylene glycol ether of Stearyl Alcohol (q.v.) that conforms generally to the formula:

$$CH_3(CH_2)_{17}(OCH_2CH_2)_nOH$$

where n has an average value of 7.

Information Sources: CIR: [S] JACT-7(6)-1988, JCLS, MI-13(7659), TSCA

Chemical Class: Alkoxylated Alcohols

Function: Surfactant - Emulsifying Agent

Reported Product Category: Body and Hand Preparations (Excluding Shaving Preparations)

Technical/Other Names:
3,6,9,12,15,18,21-Heptaoxanonatriacontan-1-ol
PEG-7 Stearyl Ether
Polyethylene Glycol (7) Stearyl Ether
Polyoxyethylene (7) Stearyl Ether

Trade Name Mixture:
Neo PCL SE o/w 2/066280 (Symrise)

STEARETH-8

CAS No.: 9005-00-9 (Generic)

JPN Translation:
ステアレス - 8

Empirical Formula:
$C_{34}H_{70}O_9$

Definition: Steareth-8 is the polyethylene glycol ether of stearyl alcohol that conforms generally to the formula:

$$CH_3(CH_2)_{17}(OCH_2CH_2)_nOH$$

where n has an average value of 8.

Information Sources: JCLS, TSCA

Chemical Class: Alkoxylated Alcohols

Function: Surfactant - Emulsifying Agent

Technical/Other Names:
PEG-8 Stearyl Ether
Polyethylene Glycol 400 Stearyl Ether
Polyoxyethylene (8) Stearyl Ether

STEARETH-10

CAS Nos.: 9005-00-9 (Generic); 13149-86-5

JPN Translation:
ステアレス - 10

Empirical Formula:
$C_{38}H_{78}O_{11}$

Definition: Steareth-10 is the polyethylene glycol ether of Stearyl Alcohol (q.v.) that conforms to the formula:

$$CH_3(CH_2)_{17}(OCH_2CH_2)_nOH$$

where n has an average value of 10.

Information Sources: 21CFR177.2800, CIR: [S] JACT-7(6)1988, CTFA S, JCLS, MI-13(7659), SNPF, TSCA

Chemical Class: Alkoxylated Alcohols

Function: Surfactant - Emulsifying Agent

Reported Product Categories: Moisturizing Preparations; Suntan Gels, Creams, and Liquids; Bath Capsules; Face and Neck Preparations (Excluding Shaving Preparations); Hair Preparations (Non-coloring), Misc.

Technical/Other Names:
3,6,9,12,15,18,21,24,27,30-Decaoxaoctatetracontan-1-ol
PEG-10 Stearyl Ether
Polyethylene Glycol 500 Stearyl Ether
Polyoxyethylene (10) Stearyl Ether

Trade Names:
Brij 76 (Uniqema Americas)
Chemonic S-10 (Chemron)
Ethal SA-10 (Ethox)
Hetoxol STA-10 (Heterene)
Jeecol SA-10 (Jeen)
Lipocol S-10 (Lipo)
Procol SA-10 (Protameen)
Sympatens-AS/100 (Kolb)
Unicol SA-10 (Universal Preserv-A-Chem)
Volpo S10 (Croda Chemicals)
Volpo S-10 (Croda, Inc.)

Trade Name Mixtures:
Cosmowax J (Croda Chemicals)
Isoxal 11 (Vevy)
Neo PCL SE o/w 2/066280 (Symrise)

Phoenoxol J (Phoenix)
Ritapro 100 (RITA)
Unicol 123 (Universal Preserv-A-Chem)

STEARETH-11

CAS No.: 9005-00-9 (Generic)

JPN Translation:
ステアレス - 11

Definition: Steareth-11 is the polyethylene glycol ether of Stearyl Alcohol (q.v.) that conforms generally to the formula:

$$CH_3(CH_2)_{17}(OCH_2CH_2)_nOH$$

where n has an average value of 11.

Information Sources: 21CFR177.2800, CIR: [S] JACT-7(6)1988, JCLS, MI-13 (7659), TSCA

Chemical Class: Alkoxylated Alcohols

Function: Surfactant - Emulsifying Agent

Technical/Other Names:
PEG-11 Stearyl Ether
Polyethylene Glycol (11) Stearyl Ether
Polyoxyethylene (11) Stearyl Ether

STEARETH-13

CAS No.: 9005-00-9 (Generic)

JPN Translation:
ステアレス - 13

Definition: Steareth-13 is the polyethylene glycol ether of Stearyl Alcohol (q.v.) that conforms generally to the formula:

$$CH_3(CH_2)_{17}(OCH_2CH_2)_nOH$$

where n has an average value of 13.

Information Sources: 21CFR177.2800, CIR: [S] JACT-7(6)1988, MI-13 (7659), TSCA

Chemical Class: Alkoxylated Alcohols

Function: Surfactant - Emulsifying Agent

Technical/Other Names:
PEG-13 Stearyl Ether
Polyethylene Glycol (13) Stearyl Ether
Polyoxyethylene (13) Stearyl Ether

Trade Name:
Unicol SA-13 (Universal Preserv-A-Chem)

STEARETH-14

CAS No.: 9005-00-9 (Generic)

JPN Translation:
ステアレス - 14

Definition: Steareth-14 is the polyethylene glycol ether of Stearyl Alcohol (q.v.) that conforms generally to the formula:

$$CH_3(CH_2)_{17}(OCH_2CH_2)_nOH$$

where n has an average value of 14.

Information Source: JCLS

Chemical Class: Alkoxylated Alcohols

Function: Surfactant - Emulsifying Agent

Technical/Other Names:
PEG-14 Stearyl Ether
Polyethylene Glycol (14) Stearyl Ether
Polyoxyethylene (14) Stearyl Ether

STEARETH-15

CAS No.: 9005-00-9 (Generic)

JPN Translation:
ステアレス - 15

Definition: Steareth-15 is the polyethylene glycol ether of Stearyl Alcohol (q.v.) that conforms generally to the formula:

$$CH_3(CH_2)_{17}(OCH_2CH_2)_nOH$$

where n has an average value of 15.

Information Sources: 21CFR177.2800, CIR: [S] JACT-7(6)1988, JCLS, MI-13 (7659), TSCA

Chemical Class: Alkoxylated Alcohols

Functions: Surfactant - Cleansing Agent; Surfactant - Emulsifying Agent

Technical/Other Names:
PEG-15 Stearyl Ether
Polyethylene Glycol (15) Stearyl Ether
Polyoxyethylene (15) Stearyl Ether

Trade Name:
Unicol SA-15 (Universal Preserv-A-Chem)

STEARETH-16

CAS No.: 9005-00-9 (Generic)

JPN Translation:
ステアレス - 16

Definition: Steareth-16 is the polyethylene glycol ether of Stearyl Alcohol (q.v.) that conforms generally to the formula:

$$CH_3(CH_2)_{17}(OCH_2CH_2)_nOH$$

where n has an average value of 16.

Information Sources: 21CFR177.2800, JCLS, MI-13(7659)

Chemical Class: Alkoxylated Alcohols

Functions: Surfactant - Cleansing Agent; Surfactant - Emulsifying Agent

Reported Product Categories: Moisturizing Preparations; Tonics, Dressings, and Other Hair Grooming Aids

Technical/Other Names:
PEG-16 Stearyl Ether
Polyethylene Glycol (16) Stearyl Ether
Polyoxyethylene (16) Stearyl Ether

Trade Name Mixture:
Solulan 16 (Amerchol)

STEARETH-20

CAS No.: 9005-00-9 (Generic)

JPN Translation:
ステアレス - 20

Definition: Steareth-20 is the polyethylene glycol ether of Stearyl Alcohol (q.v.) that conforms generally to the formula:

$$CH_3(CH_2)_{17}(OCH_2CH_2)_nOH$$

where n has an average value of 20.

Information Sources: 21CFR177.2800, CIR: [S] JACT-7(6)1988, CTFA S, JCLS, MI-13(7659), SNPF, TSCA

Chemical Class: Alkoxylated Alcohols

Functions: Surfactant - Cleansing Agent; Surfactant - Emulsifying Agent; Surfactant - Solubilizing Agent

Reported Product Categories: Bath Preparations, Misc.; Body and Hand Preparations (Excluding Shaving Preparations); Moisturizing Preparations; Makeup Bases; Personal Cleanliness Products, Misc.; Eyeliners; Indoor Tanning Preparations; Hair Conditioners; Skin Care Preparations, Misc.; Bath Oils, Tablets, and Salts; Cleansing Products (Cold Creams, Cleansing Lotions, Liquids and Pads); Eye Makeup Preparations, Misc.; Mascara; Tonics, Dressings, and Other Hair Grooming Aids; Bath Capsules; Deodorants (Underarm); Hair Preparations (Non-coloring), Misc.; Paste Masks (Mud Packs); Suntan Gels, Creams, and Liquids; Face and Neck Preparations (Excluding Shaving Preparations); Foot Powders and Sprays; Foundations; Fragrance Preparations, Misc.; Hair Straighteners; Permanent Waves; Suntan Preparations, Misc.

Technical/Other Names:
PEG-20 Stearyl Ether
Polyethylene Glycol 1000 Stearyl Ether
Polyoxyethylene (20) Stearyl Ether

Trade Names:
AEC Steareth-20 (A & E Connock)
Brij 78 (Uniqema Americas)
Chemonic S-20 (Chemron)
Customol SA-20 (Steareth-20) (Custom Ingredients)
Ethal SA-20 (Ethox)
Genapol HS 200 (Clariant)
Genapol HS 200 (Clariant GmbH, Personal Care)
Hetoxol STA-20 (Heterene)
Jeecol SA-20 (Jeen)
Lanycol-78 (Lanaetex)
Lipocol S-20 (Lipo)
Lumulse S-20 (Lambent)
Nikkol BS-20 (Nikko)
Procol SA-20 (Protameen)
Simulsol 78 (SEPPIC)
Sipol SAL-20 (Specialty Industrial)
Sympatens-AS/200 (Kolb)
Tego Alkanol S 20 P (Degussa Care Specialties)
Unicol SA-20 (Universal Preserv-A-Chem)
Volpo S20 (Croda Chemicals)
Volpo S-20 (Croda, Inc.)

Trade Name Mixtures:
Brookswax R (Arch Personal Care Products)
Cosmowax J (Croda Chemicals)
Emulcire 61 WL 2659 (Gattefosse s.a.)
Emulium Delta (Gattefosse s.a.)
Jeecol P (Jeen)
Lipowax PR (Lipo)
Lipowax R-2 (Lipo)
Procol P (Protameen)
Relaxer Concentrate No. 1 (Arch Personal Care Products)
Relaxer Concentrate #2 (Arch Personal Care Products)
Relaxer Concentrate #3 (Arch Personal Care Products)
Ritachol 1000 (RITA)
Ritapro 100 (RITA)
Simulsol SPK (SEPPIC)
Tefose 2000 (Gattefosse s.a.)
Tefose 2561 (Gattefosse s.a.)
Teinowax (Lanaetex)
Unicol CPS (Universal Preserv-A-Chem)

STEARETH-21

CAS No.: 9005-00-9 (Generic)

JPN Translation:
ステアレス - 21

Definition: Steareth-21 is the polyethylene glycol ether of Stearyl Alcohol (q.v.) that conforms generally to the formula:

$$CH_3(CH_2)_{17}(OCH_2CH_2)_nOH$$

where n has an average value of 21.

Information Sources: 21CFR177.2800, JCLS, MI-13(7659)

Chemical Class: Alkoxylated Alcohols

Functions: Surfactant - Cleansing Agent; Surfactant - Emulsifying Agent; Surfactant - Solubilizing Agent

---

Reported Product Categories: Hair Dyes and Colors (All Types Requiring Caution Statements and Patch Tests); Moisturizing Preparations; Hair Conditioners; Bath Capsules; Hair Rinses (Coloring); Skin Care Preparations, Misc.; Face and Neck Preparations (Excluding Shaving Preparations); Night Skin Care Preparations; Bath Oils, Tablets, and Salts; Cleansing Products (Cold Creams, Cleansing Lotions, Liquids and Pads); Hair Bleaches; Indoor Tanning Preparations; Lipsticks; Makeup Bases

Technical/Other Names:
PEG-21 Stearyl Ether
Polyethylene Glycol (21) Stearyl Ether
Polyoxyethylene (21) Stearyl Ether

Trade Names:
Brij 721 (Uniqema Americas)
Cromul EM1207 (Croda Chemicals)
Lipocol S-21 (Lipo)
Procol SA-21 (Protameen)

STEARETH-25

CAS No.: 9005-00-9 (Generic)

JPN Translation:
ステアレス - 25

Definition: Steareth-25 is the polyethylene glycol ether of Stearyl Alcohol (q.v.) that conforms generally to the formula:

$$CH_3(CH_2)_{17}(OCH_2CH_2)_nOH$$

where n has an average value of 25.

Information Sources: 21CFR177.2800, JCLS, MI-13(7659)

Chemical Class: Alkoxylated Alcohols

Functions: Surfactant - Cleansing Agent; Surfactant - Solubilizing Agent

Technical/Other Names:
PEG-25 Stearyl Ether
Polyethylene Glycol (25) Stearyl Ether
Polyoxyethylene (25) Stearyl Ether

Trade Name Mixture:
Tego Care 150 (Degussa Care Specialties)

STEARETH-27

CAS No.: 9005-00-9 (Generic)

JPN Translation:
ステアレス - 27

Definition: Steareth-27 is the polyethylene glycol ether of Stearyl Alcohol (q.v.) that conforms generally to the formula:

$$CH_3(CH_2)_{17}(OCH_2CH_2)_nOH$$

where n has an average value of 27.

Information Sources: 21CFR177.2800, JCLS, MI-13(7659), TSCA

Chemical Class: Alkoxylated Alcohols

Functions: Surfactant - Cleansing Agent; Surfactant - Solubilizing Agent

Technical/Other Names:
PEG-27 Stearyl Ether
Polyethylene Glycol (27) Stearyl Ether
Polyoxyethylene (27) Stearyl Ether

STEARETH-30

CAS No.: 9005-00-9 (Generic)

JPN Translation:
ステアレス - 30

Definition: Steareth-30 is the polyethylene glycol ether of Stearyl Alcohol (q.v.) that conforms generally to the formula:

$$CH_3(CH_2)_{17}(OCH_2CH_2)_nOH$$

where n has an average value of 30.

Information Sources: 21CFR177.2800, JCLS, MI-13(7659), TSCA

Chemical Class: Alkoxylated Alcohols

Functions: Surfactant - Cleansing Agent; Surfactant - Solubilizing Agent

Technical/Other Names:
PEG-30 Stearyl Ether
Polyethylene Glycol (30) Stearyl Ether
Polyoxyethylene (30) Stearyl Ether

STEARETH-40

CAS No.: 9005-00-9 (Generic)

JPN Translation:
ステアレス - 40

Definition: Steareth-40 is the polyethylene glycol ether of Stearyl Alcohol (q.v.) that conforms generally to the formula:

$$CH_3(CH_2)_{17}(OCH_2CH_2)_nOH$$

where n has an average value of 40.

Information Sources: 21CFR177.2800, JCLS, MI-13(7659), TSCA

Chemical Class: Alkoxylated Alcohols

Functions: Surfactant - Cleansing Agent; Surfactant - Solubilizing Agent

Technical/Other Names:
PEG-40 Stearyl Ether
Polyethylene Glycol 2000 Stearyl Ether
Polyoxyethylene (40) Stearyl Ether

Trade Names:
Lumulse S-40 (Lambent)
Unicol SA-40 (Universal Preserv-A-Chem)

STEARETH-50

CAS No.: 9005-00-9 (Generic)

JPN Translation:
ステアレス - 50

Definition: Steareth-50 is the polyethylene glycol ether of Stearyl Alcohol (q.v.) that conforms to the formula:

$$CH_3(CH_2)_{17}(OCH_2CH_2)_nOH$$

where n has an average value of 50.

Information Sources: 21CFR177.2800, JCLS, MI-13(7659), TSCA

Chemical Class: Alkoxylated Alcohols

Functions: Surfactant - Cleansing Agent; Surfactant - Solubilizing Agent

Technical/Other Names:
PEG-50 Stearyl Ether
Polyethylene Glycol (50) Stearyl Ether
Polyoxyethylene (50) Stearyl Ether

STEARETH-80

CAS No.: 9005-00-9

JPN Translation:
ステアレス - 80

Definition: Steareth-80 is the polyethylene glycol ether of stearyl alcohol that conforms to the formula:

$$CH_3(CH_2)_{17}(OCH_2CH_2)_nOH$$

where n has an average value of 80.

Information Source: JCLS

Chemical Class: Alkoxylated Alcohols

Functions: Surfactant - Cleansing Agent; Surfactant - Solubilizing Agent

Technical/Other Names:
Polyethylene Glycol (80) Stearyl Ether
Polyoxyethylene (80) Stearyl Ether

Trade Name:
DeTHOX SA-80 (DeForest)

STEARETH-100

CAS No.: 9005-00-9 (Generic)

JPN Translation:
ステアレス - 100

Definition: Steareth-100 is the polyethylene glycol ether of Stearyl Alcohol (q.v.) that conforms generally to the formula:

$$CH_3(CH_2)_{17}(OCH_2CH_2)_nOH$$

where n has an average value of 100.

Information Sources: JCLS, MI-13(7659), TSCA

Chemical Class: Alkoxylated Alcohols

Functions: Surfactant - Cleansing Agent; Surfactant - Solubilizing Agent

Reported Product Categories: Moisturizing Preparations; Deodorants (Underarm); Hair Conditioners; Personal Cleanliness Products, Misc.

Technical/Other Names:
PEG-100 Stearyl Ether
Polyethylene Glycol (100) Stearyl Ether
Polyoxyethylene (100) Stearyl Ether

Trade Names:
Brij 700 (Uniqema Americas)
Jeecol SA-100 (Jeen)
Lanycol-700 (Lanaetex)
Sympatens-AS/1000 (Kolb)
Volpo S100 (Croda Chemicals)

Trade Name Mixtures:
Suncaps 664 (Particle Sciences)
Suncaps 903 (Particle Sciences)

STEARETH-10 ALLYL ETHER/ACRYLATES COPOLYMER

CAS No.: 109292-17-3

Definition: Steareth-10 Allyl Ether/Acrylates Copolymer is a copolymer of the allyl ether of Steareth-10 (q.v.) and one or more monomers consisting of acrylic acid, methacrylic acid or one of their simple esters.

Information Sources: CIR: [SQ] IJT 21 (SUPPL. 3) 2002, TSCA

Chemical Class: Synthetic Polymers

Functions: Film Former; Viscosity Increasing Agent - Nonaqueous

Reported Product Category: Hair Dyes and Colors (All Types Requiring Caution Statements and Patch Tests)

Trade Name:
Salcare SC80 (Ciba Specialty Chemicals)

STEARETH-60 CETYL ETHER

Definition: Steareth-60 Cetyl Ether is the organic compound that conforms generally to the formula:

$$CH_3(CH_2)_{16}CH_2(OCH_2CH_2)nOCH_2(CH_2)_{14}CH_3$$

where n has an average value of 60.

Chemical Classes: Alkoxylated Alcohols; Ethers

Functions: Viscosity Increasing Agent - Aqueous; Viscosity Increasing Agent - Nonaqueous

Trade Name:
NJbon ECS-600 (New Japan Chemical)

STEARETH-100/PEG-136/HDI COPOLYMER

CAS No.: 103777-69-1

Definition: Steareth-100/PEG-136/HDI Copolymer is a copolymer of steareth-100, PEG-136 and hexamethylene diisocyanate monomers.

Chemical Class: Synthetic Polymers

Function: Viscosity Increasing Agent - Aqueous

Technical/Other Name:
Poly(Oxy-1,2-Ethanediyl), α-Hydro-θ-Hydroxy, Polymer with 1,6-Diisocyanatohexane and α-Octadecyl-θ-Hydroxypoly(Oxy-1,2- ethanediyl)

Trade Name:
Ser-Ad FX 1100 (Sasol Servo)

STEARETH-2 PHOSPHATE

CAS No.: 62362-49-6

JPN Translation:
ステアレス - 2 リン酸

Definition: Steareth-2 Phosphate is a complex mixture of esters of phosphoric acid and Steareth-2 (q.v.).

Information Sources: JCLS, JSCI

Chemical Class: Phosphorus Compounds

Function: Surfactant - Cleansing Agent

Technical/Other Names:
Poly(Oxy-1,2-Ethanediyl), α-Octadecyl-ω-Hydroxy-, Phosphate
Polyoxyethylene Stearylether Phosphate

Trade Name:
Crodafos S2A (Croda Chemicals)

STEARETH-3 PHOSPHATE

CAS No.: 62362-49-6 (Generic)

Definition: Steareth-3 Phosphate is a complex mixture of esters of Steareth-3 (q.v.) and phosphoric acid.

Information Source: TSCA

Chemical Class: Phosphorus Compounds

Function: Surfactant - Cleansing Agent

Technical/Other Names:
Polyethylene Glycol (3) Stearyl Ether Phosphate
Polyoxyethylene (3) Stearyl Ether Phosphate

Trade Name:
Phosphanol RL-310 (Toho)

STEARETH-4 STEARATE

JPN Translation:
ステアリン酸ステアレス - 4

Definition: Steareth-4 Stearate is the ester of Steareth-4 (q.v.) and Stearic Acid (q.v.). It conforms generally to the formula:

$$CH_3(CH_2)_{16}CH_3(OCH_2CH_2)_nO(CH_2)_{17}CH_3$$

Information Source: JCLS

Chemical Class: Esters

Function: Skin-Conditioning Agent - Miscellaneous

STEARETH-5 STEARATE

CAS No.: 85066-57-5

JPN Translation:
ステアリン酸ステアレス - 5

Empirical Formula:
$C_{46}H_{92}O_7$

Definition: Steareth-5 Stearate is the ester of Steareth-5 (q.v.) and stearic acid. It conforms generally to the formula:

$$CH_3(CH_2)_{16}CH_3(OCH_2CH_2)_nO(CH_2)_{17}CH_3$$

where n has an average value of 5.

Information Sources: JCIC, JCLS

Chemical Class: Esters

Function: Skin-Conditioning Agent - Occlusive

Technical/Other Names:
Polyethylene Glycol (5) Stearyl Ether Stearate
Poly(Oxy-1,2-Ethanediyl), α-(1-Oxooctadecyl)-ω-(Octadecyloxy)-
Polyoxyethylene Stearyl Ether Stearate
Polyoxyethylene (5) Stearyl Ether Stearate

STEARETH-9 STEARATE

JPN Translation:
ステアリン酸ステアレス - 9

Definition: Steareth-9 Stearate is the ester of polyoxyethylene stearyl ether and Stearic Acid (q.v.). It conforms generally to the formula:

$$CH_3(CH_2)_{16}\overset{\displaystyle O}{\overset{\displaystyle \|}{C}}H_3(OCH_2CH_2)_nO(CH_2)_{17}CH_3$$

where n has the average value of 9.

Information Source: JCLS

Chemical Class: Esters

Function: Skin-Conditioning Agent - Emollient

STEARETH-12 STEARATE

Definition: Steareth-12 Stearate is the ester of Steareth-12 (q.v.) and stearic acid. It conforms generally to the formula:

$$CH_3(CH_2)_{16}\overset{\displaystyle O}{\overset{\displaystyle \|}{C}}H_3(OCH_2CH_2)_nO(CH_2)_{17}CH_3$$

where n has an average value of 12.

Chemical Class: Esters

Function: Skin-Conditioning Agent - Occlusive

Technical/Other Names:
Polyethylene Glycol (12) Stearyl Ether Stearate
Polyoxyethylene (12) Stearyl Ether Stearate

Trade Name:
Emalex SWS-12 (Nihon Emulsion)

STEARIC ACID

CAS No. 57-11-4   **EINECS No.** 200-313-4

JPN Translation:
ステアリン酸

Empirical Formula:
$C_{18}H_{36}O_2$

Definition: Stearic Acid is the fatty acid that conforms generally to the formula:

$$CH_3(CH_2)_{16}COOH$$

Information Sources: ARG, AUS, BPC, BRA, 21CFR172.210, 21CFR172.615, 21CFR172.860, 21CFR175.105, 21CFR175.300, 21CFR175.320, 21CFR176.170, 21CFR176.200, 21CFR176.210, 21CFR177.1010, 21CFR177.1200, 21CFR177.2260, 21CFR177.2600, 21CFR177.2800, 21CFR178.3570, 21CFR178.3910, 21CFR184.1090, CIR: [S] JACT-6(3)1987, CTFA S, DDR, EGY, FCC, IND, ITA, JAN, JCLS, JSCI, MEX, MI-13(8882), NED, NF XIX, PF, PN, POR, RIFM, ROM, SNPF, TSCA, USAN, USD, YUG

Chemical Class: Fatty Acids

Functions: Fragrance Ingredient; Surfactant - Cleansing Agent; Surfactant - Emulsifying Agent

Surfactant-Cleansing Agent is included as a function for the soap form of Stearic Acid.

Reported Product Categories: Moisturizing Preparations; Bath Preparations, Misc.; Body and Hand Preparations (Excluding Shaving Preparations); Hair Dyes and Colors (All Types Requiring Caution Statements and Patch Tests); Bath Oils, Tablets, and Salts; Cleansing Products (Cold Creams, Cleansing Lotions, Liquids and Pads); Skin Care Preparations, Misc.; Foundations; Mascara; Shaving Cream (Aerosol, Brushless and Lather); Bath Capsules; Face and Neck Preparations (Excluding Shaving Preparations); Night Skin Care Preparations; Paste Masks (Mud Packs); Makeup Bases; Eyebrow Pencils; Suntan Gels, Creams, and Liquids; Bath Soaps and Detergents; Lipsticks; Eye Makeup Preparations, Misc.; Fragrance Preparations, Misc.; Makeup Preparations (Not eye), Misc.; Deodorants (Underarm); Eye Shadows; Suntan Preparations, Misc.; Baby Lotions, Oils, Powders and Creams; Baby Shampoos; Hair Coloring Preparations, Misc.; Shampoos (Non-coloring); Aftershave Lotions; Eyeliners; Indoor Tanning Preparations; Cuticle Softeners; Hair Straighteners; Baby Products, Misc.; Hair Conditioners; Face Powders; Personal Cleanliness Products, Misc.; Foot Powders and Sprays; Nail Creams and Lotions; Blushers (All types); Eye Lotions; Makeup Fixatives; Sachets; Shaving Preparations, Misc.; Skin Fresheners; Tonics, Dressings, and Other Hair Grooming Aids; Eye Makeup Removers; Rouges

Technical/Other Names:
n-Octadecanoic Acid
Stearic acid (RIFM)

Trade Names:
AEC Stearic Acid (A & E Connock)
Emersol 120 (Cognis Corp.)
Emersol 132 (Cognis Corp.)
Emersol 150 (Cognis Corp.)
Glycon DP (Lonza Inc./Lonza Ltd.)
Glycon P-45 (Lonza Inc./Lonza Ltd.)
Glycon S-65 (Lonza Inc./Lonza Ltd.)
Glycon S-70 (Lonza Inc./Lonza Ltd.)
Glycon S-90 (Lonza Inc./Lonza Ltd.)
Glycon TP (Lonza Inc./Lonza Ltd.)
Jeen Stearic Acid TP NF (Jeen)
Kortacid 1895 (Akzo Nobel)
Kortacid 1895 (Akzo Nobel Surface AB)
Kortacid PH05-C (Akzo Nobel)
Megalube 2621 (Megachem)
PRIFRAC 2981 (Uniqema Europe)
PRISTERENE 4900 (Uniqema Europe)
PRISTERENE 4911 (Uniqema Europe)
PRISTERENE 4916 (Uniqema Europe)
PRISTERENE 4917 (Uniqema Europe)
PRISTERENE 4922 (Uniqema Europe)
PRISTERENE 4960 (Uniqema Europe)
PRISTERENE 9551 (Uniqema Europe)
PRISTERENE 9552 (Uniqema Europe)
PRISTERENE 9554 (Uniqema Europe)
PRISTERENE 9555 (Uniqema Europe)
PRISTERENE 9559 (Uniqema Europe)
Radiacid 0152 (Oleon NV)
Radiacid 0417 (Oleon NV.)
Radiacid 0422-0427-0416 (Oleon NV)
Stearic Acid (Lipo)
Stearic Acid PC (Protameen)
Ultrapure SA (Ultra Chemical)
Unifat 132 (Universal Preserv-A-Chem)

Trade Name Mixtures:
B-122 (Guardian)
Base 4978 (Gattefosse s.a.)
Base RW 135 (LCW)
Base RW 136 (LCW)
Basis LP-20H (Nisshin OilliO)
Beeswax Substitute 81-1104 (Ross)
Beeswax Substitute 628/5 (Ross)
B-Wax White (Ross)
Capispheres (Barnet)
Cetasal (Gattefosse s.a.)
Chelonine (JUVEX)
Covacrem LP (LCW)
Covacrem MK (LCW)
Crodafos CP-50 (Croda, Inc.)
Cutina FS 45 (Cognis Deutschland)
DCH45TS (Kobo)
DID50TS (Kobo)
DS-CERIX (Doosan)
DS-SBS (Doosan)
Emulgade CL Special (Cognis Deutschland)
Essential Vital Elements - S (Dipta)
Eusolex T-Olio P (Merck KGaA)
Eusolex T-S (Merck KGaA)
Eusolex T-S (Merck KGaA/EMD Chemicals Inc.)
Hostapon SCI-65 C (Clariant GmbH, Personal Care)
IN65TS (Kobo)
Isobeeswax 6110 (Kahl)
Isobeeswax SP 154 (Strahl & Pitsch)
Isobeeswax - STRALPITZ (Strahl & Pitsch)
Jordapon CI 65 (BASF)
Lubraslide (Guardian)
Macamat Wax (LCW)
Mascawax 012 (LCW)
Micromac Wax (LCW)
Micro Titanium Dioxide MT-01 (Tayca)
Micro Titanium Dioxide MT-100T (Tri-K)
Micro Titanium Dioxide MT-100TV (Tayca)
Micro Titanium Dioxide MT-100Z (Tayca)
Montane 481 (SEPPIC)
MT-100Z (Tri-K)
MT-100ZH (Tri-K)
Natrlfine 137-T (Finetex)
Natrlfine TP-T (Finetex)

Nikkol Nikkolipid 81S (Nikko)
Nikkol Nikkomulese GT (Nikko)
N.S.L.E. (Sederma)
OP60S4 (Kobo)
Oxowax (LCW)
Phytosphingosine ETP (Degussa Goldschmidt AG)
ProLipid 141 (International Specialty Products)
Questamix H (Quest International)
Ross Synthetic Candelilla Wax (Ross)
Rosswax 2660 (Ross)
Sabowax FL 81 (Sabo)
SA-TTO-S-4 (10%) (US Cosmetics)
SA-TTO-S-4/D5 (50%) (US Cosmetics)
SA-TTO-S-4/ININ (55%) (US Cosmetics)
Setacire (Phytocos)
Solarcat (Collaborative Labs)
SPD-T5 (Shin-Etsu Chemical Co.)
SPD - TIS (Shin-Etsu Chemical Co.)
SPD-T1V (Shin-Etsu Chemical Co.)
Stearina COS (Sabo)
STEPAN DGS SE and Stearic Acid (Stepan)
ST-485SA15 (Titan Kogyo)
Synthetic Beeswax 6103 (Kahl)
Toshiki Nylon IVCF-1 (Nikko)
TTO-51(C) (Ishihara Sangyo Kaisha)
TTO-55(C) (Ishihara Sangyo Kaisha)
TTO-S-2 (Ishihara Sangyo Kaisha)
TTO-S-4 (Ishihara Sangyo Kaisha)
TTO-S-6 (Ishihara Sangyo Kaisha)
TTO-V-4 (Ishihara Sangyo Kaisha)
Ultrapure Micronized Titanium Dioxide (Ultra Chemical)
UV-Titan M160 (Kemira Pigments OY)
Vitamin F Oilsoluble (Cosmetochem) (Cosmetochem International Ltd.)

STEARIC HYDRAZIDE

CAS No.	EINECS No.
4130-54-5	223-946-8

Empirical Formula:
$C_{18}H_{38}N_2O$

Definition: Stearic Hydrazide is the organic compound that conforms generally to the formula:

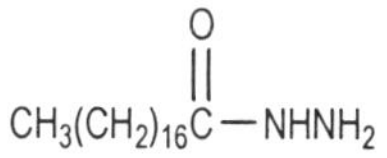
O
CH3(CH2)16C—NHNH2

Information Sources: CIR: [I] JACT-10(1)-1991, EEC(II-200), TSCA

Chemical Class: Amides

Function: Not Reported

Technical/Other Name:
Octadecanoic Acid, Hydrazide

STEARONE

CAS No.	EINECS No.
504-53-0	207-993-1

JPN Translation:
ステアロン

Empirical Formula:
$C_{35}H_{70}O$

Definition: Stearone is the aliphatic ketone that conforms to the formula:

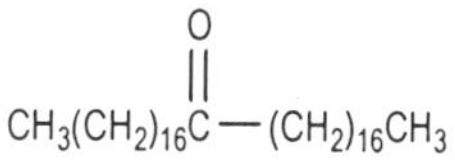
O
CH3(CH2)16C—(CH2)16CH3

Information Sources: JCIC, JCLS, TSCA

Chemical Class: Ketones

Function: Viscosity Increasing Agent - Nonaqueous

Technical/Other Names:
Diheptadecyl Ketone
18-Pentatriacontanone

STEAROXY DIMETHICONE

CAS No.: 68554-53-0

Definition: Stearoxy Dimethicone is a polymer of dimethylpolysiloxane endblocked with stearoxy groups.

Information Sources: CIR: [S], JCIC, JCLS, JSQI, TSCA

Chemical Class: Siloxanes and Silanes

Function: Skin-Conditioning Agent - Emollient

Reported Product Categories: Foundations; Bath Preparations, Misc.; Body and Hand Preparations (Excluding Shaving Preparations); Moisturizing Preparations; Skin Care Preparations, Misc.; Face and Neck Preparations (Excluding Shaving Preparations); Cleansing Products (Cold Creams, Cleansing Lotions, Liquids and Pads); Eye Lotions; Eye Shadows; Hair Sprays (Aerosol Fixatives); Lipsticks; Makeup Bases; Tonics, Dressings, and Other Hair Grooming Aids

Technical/Other Names:
Dimethylsiloxane•Methylstearoxysiloxane Copolymer
Dimethyl Siloxy Stearoxy Siloxane Polymer
Poly(Dimethylsiloxy) Stearoxysiloxane
Siloxanes and Silicones, Dimethyl, (Octadecyloxy)- Terminated
Stearoxymethylpolysiloxane

Trade Name:
Abil Wax 2434 (Degussa Care Specialties)

Trade Name Mixture:
Wacker Belsil SDM 6022 (Wacker-Chemie)

STEAROXYMETHICONE/DIMETHICONE COPOLYMER

JPN Translations:
(ステアロキシメチコン/ジメチコン)コポリマー
ステアロキシメチルポリシロキサン

Definition: Stearoxymethicone/Dimethicone Copolymer is the siloxane polymer that conforms generally to the formula:

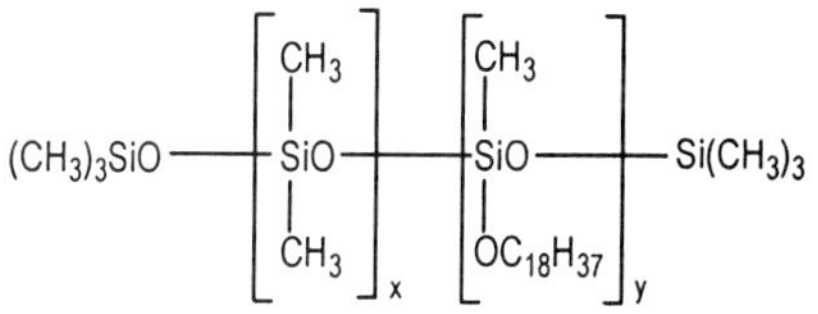
CH3
CH3
(CH3)3SiO
SiO
SiO
Si(CH3)3
CH3
OC18H37
x
y

Chemical Class: Siloxanes and Silanes

Functions: Skin-Conditioning Agent - Occlusive; Viscosity Increasing Agent - Nonaqueous

Reported Product Category: Lipsticks

Trade Names:
Gransil SRP (Grant)
KF7002 (Shin Etsu)

STEAROXYPROPYL DIMETHYLAMINE

CAS No.	EINECS No.
17517-01-0	241-516-8

Empirical Formula:
$C_{23}H_{49}NO$

Definition: Stearoxypropyl Dimethylamine is the organic compound that conforms to the formula:

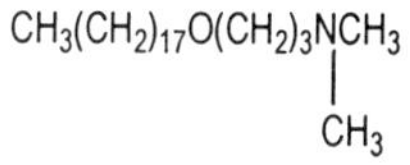
CH3(CH2)17O(CH2)3NCH3
CH3

Chemical Classes: Amines; Ethers

Function: Hair Conditioning Agent

Technical/Other Name:
1-Propanamine, N,N-Dimethyl-3-(Octadecyloxy)-

Trade Name:
Farmin DM E-80 (Kao Corp.)

STEAROXYPROPYLTRIMONIUM CHLORIDE

CAS No.: 23328-71-4

Empirical Formula:
$C_{24}H_{52}NO \cdot Cl$

Definition: Stearoxypropyltrimonium Chloride is the quaternary ammonium compound that conforms to the formula:

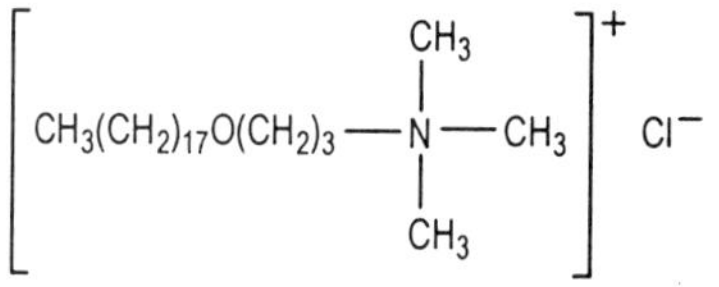
CH3
CH3(CH2)17O(CH2)3—N—CH3
CH3
+
Cl−

Chemical Class: Quaternary Ammonium Compounds

Functions: Hair Conditioning Agent; Surfactant - Cleansing Agent

Technical/Other Name:
Ammonium, Trimethyl[3-Octadecyloxy)-Propyl]-, Chloride

Trade Name:
Quartamin E-80K (Kao Corp.)

STEAROXYTRIMETHYLSILANE

CAS No.	EINECS No.
18748-98-6	242-554-8

JPN Translation:
ステアロキシトリメチルシラン

Empirical Formula:
$C_{21}H_{46}OSi$

Definition: Stearoxytrimethylsilane is the organo-silicon compound that conforms to the formula:

$$CH_3(CH_2)_{17}OSi(CH_3)_3$$

Information Sources: JCIC, JCLS

Chemical Class: Siloxanes and Silanes

Function: Skin-Conditioning Agent - Emollient

Reported Product Categories: Eyeliners; Indoor Tanning Preparations; Moisturizing Preparations

Technical/Other Names:
Silane, Trimethyl(Octadecyloxy)-Trimethyl(Octadecyloxy)Silane

Trade Names:
SilCare Silicone 1M71 Stearoxytrimethyl-silane (Clariant)
SilCare Silicone 1M71 Stearoxytrimethyl-silane (Clariant GmbH, Personal Care)

Trade Name Mixtures:
Dow Corning 580 Wax (Dow Corning)
MagiSil 451 (Diow)
MagiSil 509 (Diow)
MagiSil 451S (Diow)
MagiSil 451SS (Diow)
MagiSil 451SW (Diow)
StrataGel (Diow)
StrataGel 171 (Diow)

N-STEAROYL-DIHYDROSPHINGOSINE

Definition: N-Stearoyl-Dihydrosphingosine is the sphingolipid that conforms to the formula:

Chemical Classes: Alcohols; Amides

Function: Skin-Conditioning Agent - Miscellaneous

STEAROYL DIHYDROXY ISOBUTYLAMIDE STEARATE

JPN Translation:
ステアリン酸ステアロイルジヒドロキシイ
ソブチルアミド

Definition: Stearoyl Dihydroxy Isobutyl-amide Stearate is an ester of Stearic Acid (q.v.) and dihydroxyisobutyl stearate.

Information Source: JCLS

Chemical Class: Esters

Function: Not Reported

STEAROYL EPOXY RESIN

JPN Translations:
エポキシエステル - 3
エポキシエステル - 4

Definition: Stearoyl Epoxy Resin is the stearic acid ester of 4,4'-Isopropylidene-diphenol/Epichlorohydrin Copolymer (q.v.), also known as epoxy resin.

Information Sources: JCIC, JCLS

Chemical Class: Synthetic Polymers

Function: Film Former

Technical/Other Name:
Bisphenol A Type Epoxy Resin Stearate (2)

STEAROYL GLUTAMIC ACID

CAS No.	EINECS No.
3397-16-8	222-252-2

JPN Translation:
ステアロイルグルタミン酸

Empirical Formula:
$C_{23}H_{43}NO_5$

Definition: Stearoyl Glutamic Acid is the substituted amino acid that conforms to the formula:

$$HOOCCH_2CH_2CHCOOH$$

Information Sources: JCIC, JCLS

Chemical Classes: Amides; Amino Acids

Functions: Hair Conditioning Agent; Skin-Conditioning Agent - Miscellaneous; Surfactant - Cleansing Agent

Technical/Other Names:
L-Glutamic Acid, N-(1-Oxooctadecyl)-
Glutamic Acid, N-Stearoyl-
N-Stearoyl-L-Glutamic Acid

STEAROYL INULIN

Definition: Stearoyl Inulin is the condensation product of stearic acid chloride with the carbohydrate, Inulin (q.v.).

Chemical Class: Carbohydrates

Functions: Skin-Conditioning Agent - Emollient; Surfactant - Emulsifying Agent

Trade Names:
Lifidrem INST (Coletica SA)
Rheopearl INS (Chiba)

STEAROYL LACTYLIC ACID

CAS No.	EINECS No.
14440-80-3	238-418-2

Empirical Formula:
$C_{24}H_{44}O_6$

Definition: Stearoyl Lactylic Acid is the ester of stearic acid and lactyl lactate. It conforms to the formula:

Information Source: 21CFR176.170

Chemical Classes: Carboxylic Acids; Esters

The inclusion of any compound in the *Dictionary and Handbook* does not indicate that use of that substance as a cosmetic ingredient complies with the laws and regulations governing such use in the United States or any other country.

Function: Surfactant - Emulsifying Agent

STEAROYL LEUCINE

JPN Translation:
ステアロイルロイシン

Empirical Formula:
$C_{24}H_{47}NO_3$

Definition: Stearoyl Leucine is the stearoyl derivative of leucine that conforms to the formula:

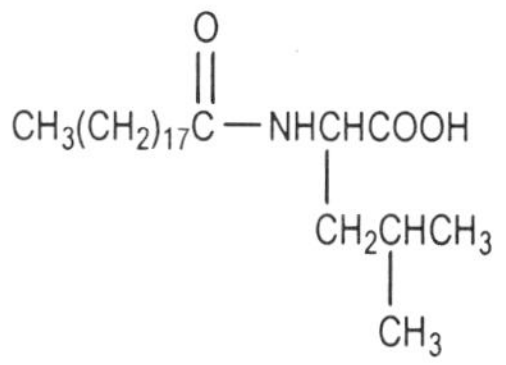

Chemical Class: Amino Acids

Functions: Hair Conditioning Agent; Skin-Conditioning Agent - Miscellaneous; Surfactant - Emulsifying Agent

Technical/Other Name:
N-Stearoyl L-Leucine

STEAROYL PG-TRIMONIUM CHLORIDE

Empirical Formula:
$C_{24}H_{50}NO_3 \cdot Cl$

Definition: Stearoyl PG-Trimonium Chloride is the quaternary ammonium salt that conforms to the formula:

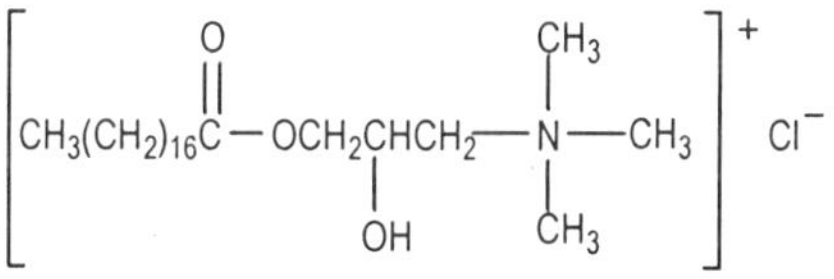

Chemical Class: Quaternary Ammonium Compounds

Functions: Antistatic Agent; Hair Conditioning Agent

Technical/Other Name:
Stearoyl Propylene Glycol Trimethylammonium Chloride

Trade Name Mixture:
Akypoquat 40 (Kao GmbH)

STEAROYL SARCOSINE

CAS No.
142-48-3

EINECS No.
205-539-7

Empirical Formula:
$C_{21}H_{41}NO_3$

Definition: Stearoyl Sarcosine is the N-stearoyl derivative of N-methylglycine that conforms generally to the formula:

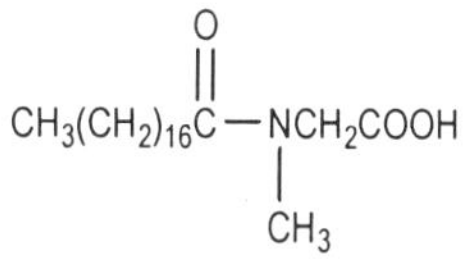

Information Sources: 21CFR177.1200, CIR: [SQ] IJT-20(SUPPL. 1)2001, TSCA

Chemical Class: Sarcosinates and Sarcosine Derivatives

Functions: Hair Conditioning Agent; Surfactant - Cleansing Agent

Reported Product Category: Shaving Preparations, Misc.

Technical/Other Names:
Glycine, N-Methyl-N-(1-Oxooctadecyl)-
N-Methyl-N-(1-Oxooctadecyl)Glycine
Stearoyl N-Methylaminoacetic Acid
Stearoyl N-Methylglycine

Trade Names:
Crodasinic S (Croda Chemicals)
Hamposyl S (Amerchol)

STEARTRIMONIUM BROMIDE

JPN Translation:
ステアリルトリモニウムブロミド

Empirical Formula:
$C_{21}H_{46}N \cdot Br$

Definition: Steartrimonium Bromide is the quaternary ammonium salt that conforms to the formula:

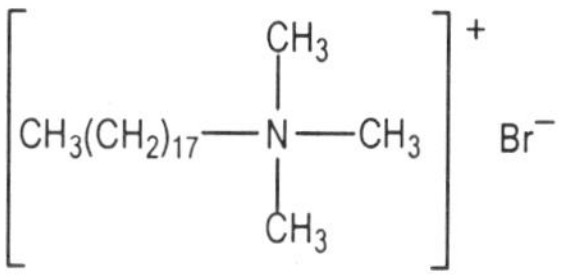

Chemical Class: Quaternary Ammonium Compounds

Functions: Antistatic Agent; Hair Conditioning Agent

Trade Name Mixture:
Catinal STB-70 (Toho)

STEARTRIMONIUM CHLORIDE

CAS No.
112-03-8

EINECS No.
203-929-1

JPN Translation:
ステアルトリモニウムクロリド

Empirical Formula:
$C_{21}H_{46}N \cdot Cl$

Definition: Steartrimonium Chloride is the quaternary ammonium salt that conforms to the formula:

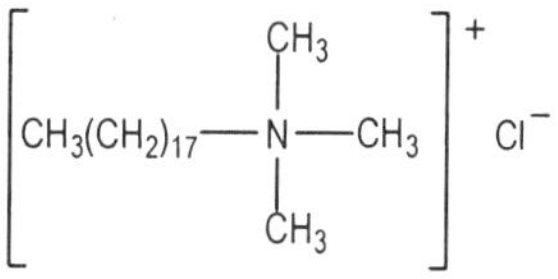

Information Sources: CIR: [SQ] IJT-16(3)-1997, JCIC, JCLS, JSCI, JSQI, TSCA

Chemical Class: Quaternary Ammonium Compounds

Functions: Antistatic Agent; Hair Conditioning Agent

Reported Product Category: Hair Conditioners

Technical/Other Names:
1-Octadecanaminium, N,N,N-Trimethyl-, Chloride
Quaternium-10
Stearyl Trimethyl Ammonium Chloride
Stearyl Trimethyl Ammonium Chloride Solution
N,N,N-Trimethyl-1-Octadecanaminium Chloride

Trade Names:
Genamin STAC (Clariant)
Genamin STAC (Clariant GmbH, Personal Care)
Nikkol CA-2450 (Nikko)
Nikkol CA-2465 (Nikko)

Trade Name Mixture:
Arquad 18-50 (Akzo Nobel)

STEARTRIMONIUM HYDROXYETHYL HYDROLYZED COLLAGEN

Definition: Steartrimonium Hydroxyethyl Hydrolyzed Collagen is the quaternary ammonium chloride formed by the reaction of a terminally epoxide substituted stearyl trimethyl ammonium chloride with Hydrolyzed Collagen (q.v.).

Chemical Classes: Protein Derivatives; Quaternary Ammonium Compounds

Functions: Antistatic Agent; Hair Conditioning Agent; Skin-Conditioning Agent - Miscellaneous

Technical/Other Names:
Stearyltrimonium Hydrolyzed Animal Protein
Stearyltrimonium Hydroxyethyl Hydrolyzed Collagen

Trade Names:
Pran QC (Fabriquimica)
Quat Coll QS-30 (Arch Personal Care Products)
Quat-Pro S (Maybrook)
Quat-Pro S30 (Maybrook)

STEARTRIMONIUM METHOSULFATE

CAS No.: 18684-11-2

Empirical Formula:
$C_{21}H_{46}N \cdot CH_3O_4S$

Definition: Steartrimonium Methosulfate is the quaternary ammonium salt that conforms to the formula:

$$\left[CH_3(CH_2)_{17} - \overset{\overset{\displaystyle CH_3}{|}}{\underset{\underset{\displaystyle CH_3}{|}}{N}} - CH_3 \right]^{+} \quad CH_3OSO_3^{-}$$

Chemical Class: Quaternary Ammonium Compounds

Functions: Antistatic Agent; Hair Conditioning Agent

Technical/Other Names:
1-Octadecanaminium, N,N,N-Trimethyl-, Methyl Sulfate
Stearyl Trimethylammonium Methyl Sulfate
Stearyltrimonium Methosulfate
N,N,N-Trimethyl-1-Octadecanaminium Methyl Sulfate

Trade Names:
CATIGENE ST-70 (Stepan)
Empigen CM (Albright & Wilson UK)

STEARTRIMONIUM SACCHARINATE

JPN Translation:
ステアリルトリモニウムサッカリン

Empirical Formula:
$C_{28}H_{50}N_2O_3S$

Definition: Steartrimonium Saccharinate is the quaternary ammonium salt that conforms to the formula:

$$\left[CH_3(CH_2)_{17} - \overset{\overset{\displaystyle CH_3}{|}}{\underset{\underset{\displaystyle CH_3}{|}}{N}} - CH_3 \right]^{+} \left[O_2S \overset{\displaystyle N}{\diagdown} C=O \right]^{-}$$

Chemical Classes: Heterocyclic Compounds; Quaternary Ammonium Compounds

Functions: Antistatic Agent; Hair Conditioning Agent

STEARYL ACETATE

CAS No.
822-23-1

EINECS No.
212-493-1

Empirical Formula:
$C_{20}H_{40}O_2$

Definition: Stearyl Acetate is the ester of stearyl alcohol and acetic acid. It conforms to the formula:

$$CH_3\overset{\overset{\displaystyle O}{\|}}{C} - O(CH_2)_{17}CH_3$$

Information Source: TSCA

Chemical Class: Esters

Function: Skin-Conditioning Agent - Emollient

Technical/Other Name:
Acetic Acid, Octadecyl Ester

Trade Name:
AEC Stearyl Acetate (A & E Connock)

STEARYL ACETYL GLUTAMATE

Empirical Formula:
$C_{25}H_{47}NO_5$

Definition: Stearyl Acetyl Glutamate is the substituted amino acid that conforms to the formula:

$$CH_3\overset{\overset{\displaystyle O}{\|}}{C} - NHCHCH_2CH_2\overset{\overset{\displaystyle O}{\|}}{C} - O(CH_2)_{17}CH_3$$
$$\underset{\displaystyle COOH}{|}$$

Information Sources: JCIC, JCLS

Chemical Classes: Amides; Amino Acids; Esters

Function: Skin-Conditioning Agent - Miscellaneous

Technical/Other Name:
Stearyl N-Acetyl-L-Glutamate

STEARYL ACETYL GLUTAMINATE

JPN Translation:
アセチルグルタミンステアリル

Definition: Stearyl Acetyl Glutaminate is the ester of stearyl alcohol and Acetyl Glutamine (q.v.).

Chemical Classes: Amino Acids; Esters

Functions: Humectant; Skin-Conditioning Agent - Emollient

Technical/Other Name:
N-Acetyl Glutamine, Stearyl Ester

Trade Name:
NAGS (Kyowa Hakko Kogyo)

STEARYL ALCOHOL

CAS No.
112-92-5

EINECS No.
204-017-6

JPN Translation:
ステアリルアルコール

Empirical Formula:
$C_{18}H_{38}O$

Definition: Stearyl Alcohol is the fatty alcohol that conforms generally to the formula:

$$CH_3(CH_2)_{17}OH$$

Information Sources: AUS, 21CFR172.755, 21CFR172.864, 21CFR175.105, 21CFR175.300, 21CFR176.200, 21CFR176.210, 21CFR177.1010, 21CFR177.1200, 21CFR177.2800, 21CFR178.3480, 21CFR178.3910, CIR: [S] JACT-4(5)1985, CTFA S, DDR, ITA, JAN, JCLS, JSCI, KOR, MI-13(8883), NF XIX, RIFM, SNPF, TSCA, USAN, USD

Chemical Classes: Fatty Alcohols; Sulfonic Acids

Functions: Emulsion Stabilizer; Fragrance Ingredient; Surfactant - Emulsifying Agent; Surfactant - Foam Booster; Viscosity Increasing Agent - Aqueous; Viscosity Increasing Agent - Nonaqueous

Reported Product Categories: Hair Dyes and Colors (All Types Requiring Caution Statements and Patch Tests); Hair Conditioners; Moisturizing Preparations; Body and Hand Preparations (Excluding Shaving Preparations); Bath Preparations, Misc.; Personal Cleanliness Products, Misc.; Bath Oils, Tablets, and Salts; Cleansing Products (Cold Creams, Cleansing Lotions, Liquids and Pads); Foundations; Skin Care Preparations, Misc.; Hair Bleaches; Bath Capsules; Shampoos (Non-coloring); Face and Neck Preparations (Excluding Shaving Preparations); Eyeliners; Indoor Tanning Preparations; Night Skin Care Preparations; Makeup Bases; Paste Masks (Mud Packs); Baby Lotions, Oils, Powders and Creams; Eye Makeup Preparations, Misc.; Tonics, Dressings, and Other Hair Grooming Aids; Deodorants (Underarm); Fragrance Preparations, Misc.; Hair Straighteners; Shaving Cream (Aerosol, Brushless and Lather); Eye Shadows; Makeup Preparations (Not eye), Misc.; Aftershave Lotions; Baby Shampoos; Eye Lotions; Mascara; Hair Rinses (Non-coloring); Permanent Waves; Suntan Preparations, Misc.; Bath Soaps and Detergents; Eyebrow Pencils; Foot Powders and Sprays; Hair Coloring Preparations, Misc.; Hair Preparations (Non-coloring), Misc.; Makeup Fixatives; Suntan Gels, Creams, and Liquids

Technical/Other Name:
1-Octadecanol (RIFM)

Trade Names:
AEC Stearyl Alcohol (A & E Connock)

Alfol 18 Alcohol (Sasol North America)
Cachalot S-56 (Michel)
CO-1895 (Procter & Gamble)
CoChem SA (Costec)
Cochem SAN (Costec)
Crodacol S95 (Croda Chemicals)
Crodacol S-95 (Croda, Inc.)
Custom Stearyl (stearyl alcohol) (Custom Ingredients)
Hainol 18SS (Kokyu Alcohol)
Hyfatol 18-95 (Aarhus)
Hyfatol 18-98 (Aarhus)
Lanette 18 (Cognis Care Chemicals/NJ)
Lanette 18 (Cognis Care Chemicals/PA)
Lanette 18 (Cognis Deutschland)
Lanol S (SEPPIC)
Lipocol S (Lipo)
Lipocol S-DEO (Lipo)
Nacol 18-94 Alcohol (Sasol GmbH - Hamburg)
Nacol 18-98 Alcohol (Sasol GmbH - Hamburg)
Nacol 18-99 Alcohol (Sasol GmbH - Hamburg)
Nikkol Stearyl Alcohol (Nikko)
RITA SA (RITA)
Sabonal C 18 95 (Sabo)
Stearyl Alcohol 98/F (Zschimmer & Schwarz Italiana)
Stearyl Alcohol NF (Jeen)
Stearyl Alcohol NX (Kokyu Alcohol)
Stearyl Alcohol 98/P (Zschimmer & Schwarz Italiana)
Stearyl Alcohol PC (Protameen)
Stearyl Alcohol S (Kokyu Alcohol)
Stearyl Alcohol SP (Kokyu Alcohol)
Tego Alkanol 18 (Degussa Care Specialties)
Ultrapure S (Ultra Chemical)
Unihydag WAX-18 (Universal Preserv-A-Chem)

Trade Name Mixtures:
Almond Milk (Cosmetochem) (Cosmetochem International Ltd.)
Aloe Vera Milk (Cosmetochem) (Cosmetochem International Ltd.)
Apricot Milk (Cosmetochem) (Cosmetochem International Ltd.)
Apricot Milkl (Cosmetochem) (Cosmetochem International Ltd.)
Atlas G-1875 (Uniqema Americas)
Avocado Milk (Cosmetochem) (Cosmetochem International Ltd.)
Bamboo Milk (Cosmetochem) (Cosmetochem International Ltd.)
Brookswax G (Arch Personal Care Products)
Cactus Milk (Cosmetochem) (Cosmetochem International Ltd.)
Ceral G (Fabriquimica)
Cerasynt LP (International Specialty Products)

Cerasynt WM (International Specialty Products)
4-Cereals Milk (Cosmetochem) (Cosmetochem International Ltd.)
Cetearyl Alcohol 30/F (Zschimmer & Schwarz Italiana)
Cetearyl Alcohol 50/F (Zschimmer & Schwarz Italiana)
Cetearyl Alcohol 70/F (Zschimmer & Schwarz Italiana)
Cocoa Milk (Cosmetochem) (Cosmetochem International Ltd.)
Cocos Milk (Cosmetochem) (Cosmetochem International Ltd.)
Cotton Milk (Cosmetochem) (Cosmetochem International Ltd.)
Cremophor A6 (BASF)
Dispersen-G (Lanaetex)
Dow Corning 580 Wax (Dow Corning)
DS-SBS (Doosan)
Emulgade NLB (Cognis Deutschland)
Emulsifying Wax EM0010 (Croda Chemicals)
Fig Milk (Cosmetochem) (Cosmetochem International Ltd.)
Forlan L (RITA)
Forlan LM (RITA)
Hazelnut Milk (Cosmetochem) (Cosmetochem International Ltd.)
Homulgator 920 G (Grau)
Homulgator 1330 G (Grau)
Homulgator 910 G Extra (Grau)
Hydro Myristenol 14082 2/014082 (Symrise)
Incroquat BBO-35 (Croda, Inc.)
Incroquat BES-35 S (Croda, Inc.)
Jeecol ST-20-G (Jeen)
Lexemul BEO (Inolex)
Lipowax G (Lipo)
Lowenol Emulsion 79 (Lowenstein)
Lowenol Emulsion 270 (Lowenstein)
Macadamia Milk (Cosmetochem) (Cosmetochem International Ltd.)
Mango Milk (Cosmetochem) (Cosmetochem International Ltd.)
Miracare CT-100 (Rhodia)
Neo PCL SE o/w 2/066280 (Symrise)
Oat Milk (Cosmetochem) (Cosmetochem International Ltd.)
Palm Milk (Cosmetochem) (Cosmetochem International Ltd.)
Peach Milk (Cosmetochem) (Cosmetochem International Ltd.)
Procol ST-20-G (Protameen)
Promulgen G (Amerchol)
Prozymex HBT (Laboratoires Sero-biologiques)
Ritapro 200 (RITA)
Sabowax CS 6 (Sabo)
Simulsol SPK (SEPPIC)
Soya Milk (Cosmetochem) (Cosmetochem International Ltd.)
4-Spices Milk (Cosmetochem) (Cosmetochem International Ltd.)

Stepanquat DC1 (Stepan)
Tego Care 150 (Degussa Care Specialties)
Tewax TC 1 (Cesalpinia)
Unieucerin (Chemyunion)

STEARYL AMINOPROPYL METHICONE

CAS No.: 110720-64-4

Definition: Stearyl Aminopropyl Methicone is the siloxane polymer that conforms generally to the formula:

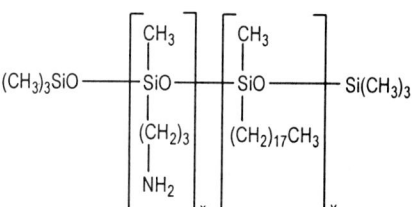

Chemical Classes: Siloxanes and Silanes; Synthetic Polymers

Function: Hair Conditioning Agent

Trade Name:
EXP-61 (Genesee)

STEARYL BEESWAX

Definition: Stearyl Beeswax is the ester of Stearyl Alcohol (q.v.) and Beeswax Acid (q.v.).

Chemical Class: Esters

Function: Skin-Conditioning Agent - Occlusive

Trade Name Mixtures:
BW Ester BW67 (Koster Keunen Holland)
Cera Bellina ST (Dekker)

STEARYL BEHENATE

CAS No.	EINECS No.
24271-12-3	246-115-1

Empirical Formula:
$C_{40}H_{80}O_2$

Definition: Stearyl Behenate is the ester of Stearyl Alcohol (q.v.) and Behenic Acid (q.v.). It conforms to the formula:

$$CH_3(CH_2)_{20}\overset{\displaystyle O}{\overset{\|}{C}} - O(CH_2)_{17}CH_3$$

Chemical Class: Esters

Function: Skin-Conditioning Agent - Occlusive

Technical/Other Name:
Docosanoic Acid, Octadecyl Ester

Trade Names:
  Pelemol SB (Phoenix)
  Purester 40 (Strahl & Pitsch)

STEARYL BENZOATE

CAS No.   **EINECS No.**
10578-34-4   234-169-9

Empirical Formula:
  $C_{25}H_{42}O_2$

Definition: Stearyl Benzoate is the ester of stearyl alcohol and benzoic acid that conforms to the formula:

Chemical Class: Esters

Functions: Skin-Conditioning Agent - Emollient; Solvent

Technical/Other Names:
  Benzoic Acid, Octadecyl Ester
  Benzoic Acid, Stearyl Ester
  Octadecyl Benzoate

Trade Names:
  DUB PG (Stearinerie Dubois Fils)
  Finsolv 116 (Finetex)

STEARYL BETAINE

CAS No.   **EINECS No.**
820-66-6   212-470-6

JPN Translation:
  ステアリルベタイン

Empirical Formula:
  $C_{22}H_{45}NO_2$

Definition: Stearyl Betaine is the zwitterion (inner salt) that conforms to the formula:

Information Sources: JCIC, JCLS, JSQI, TSCA

Chemical Class: Betaines

Functions: Antistatic Agent; Hair Conditioning Agent; Skin-Conditioning Agent - Miscellaneous; Surfactant - Cleansing Agent; Surfactant - Foam Booster; Viscosity Increasing Agent - Aqueous

Technical/Other Names:
  N-(Carboxymethyl)-N,N-Dimethyl-1-Octadecanaminium Hydroxide, Inner Salt

1-Octadecanaminium, N-(Carboxymethyl)-N,N-Dimethyl-, Hydroxide, Inner Salt
Stearyl Dimethyl Glycine

STEARYL CAPRYLATE

CAS No.   **EINECS No.**
18312-31-7   242-200-2

Empirical Formula:
  $C_{26}H_{52}O_2$

Definition: Stearyl Caprylate is the ester of stearyl alcohol and caprylic acid. It conforms to the formula:

Chemical Class: Esters

Function: Skin-Conditioning Agent - Occlusive

Technical/Other Name:
  Octanoic Acid, Octadecyl Ester

Trade Names:
  AEC Stearyl Caprylate (A & E Connock)
  Pelemol SNO (Phoenix)

Trade Name Mixtures:
  AEC Stearyl Heptanoate (&) Stearyl
    Caprylate (A & E Connock)
  DUB Solide (Stearinerie Dubois Fils)
  Neo PCL SE o/w 2/066280 (Symrise)
  Neo-PCL w/o s.e. 2/066255 (Symrise)
  PCL SE w/o 2/066255 (Symrise)
  PCL-Siccum 2/066215 (Symrise)
  PCL Solid 2/066220 (Symrise)

STEARYL CEROTATE/CARNAUBATE

JPN Translation:
  銅クロロフィル

Definition: Stearyl Cerotate/Carnaubate is the ester of Stearic Acid (q.v.) and a mixture of Cerotic Acid (q.v.) and a fatty acid obtained from carnauba wax.

Information Source: JCLS

Chemical Class: Esters

Function: Skin-Conditioning Agent - Miscellaneous

STEARYL CITRATE

CAS Nos.   **EINECS No.**
1323-66-6
1337-33-3   215-654-4

Empirical Formula:
  $C_{24}H_{44}O_7$

Definition: Stearyl Citrate is the ester of stearyl alcohol and citric acid. It conforms to the formula:

Information Sources: 21CFR166.40, 21CFR166.110, 21CFR175.300, 21CFR178.3910, 21CFR181.22, 21CFR181.27, TSCA

Chemical Class: Esters

Function: Skin-Conditioning Agent - Emollient

Technical/Other Names:
  Citric Acid, Octadecyl Ester
  2-Hydroxy-1,2,3-Propanetricarboxylic Acid, Monooctadecyl Ester
  Octadecyl Citrate
  1,2,3-Propanetricarboxylic Acid, 2-Hydroxy-monooctadecyl Ester

STEARYL DIMETHICONE

CAS No.: 67762-83-8

Definition: Stearyl Dimethicone is the siloxane polymer that conforms generally to the formula:

Information Source: CIR: [S]

Chemical Class: Siloxanes and Silanes

Function: Skin-Conditioning Agent - Occlusive

Reported Product Categories: Blushers (All types); Eye Shadows; Foundations; Lipsticks; Makeup Bases; Makeup Fixatives; Mascara

Technical/Other Name:
  Siloxanes and Silicones, di-Me, Me Stearyl

Trade Names:
  Abil Wax 9800 (Degussa Care Specialties)
  Dow Corning 2503 Cosmetic Wax (Dow Corning)
  SilCare Silicone 41M65 Stearyl Dimethicone (Clariant)
  SilCare Silicone 41M65 Stearyl Dimethicone (Clariant GmbH, Personal Care)
  Silsoft W-18 (OSi Specialties)
  Silwax L-118 (Siltech LLC)
  Wacker -Belsil SDM 5055 (Wacker-Chemie)

The inclusion of any compound in the *Dictionary and Handbook* does not indicate that use of that substance as a cosmetic ingredient complies with the laws and regulations governing such use in the United States or any other country.

STEARYL ERUCAMIDE

CAS Nos.
10094-45-8
96810-35-4

EINECS No.
233-226-5

Empirical Formula:
$C_{40}H_{79}NO$

Definition: Stearyl Erucamide is the substituted aliphatic amide that conforms to the formula:

$$CH_3(CH_2)_7CH=CH(CH_2)_{11}C(=O)-NH(CH_2)_{17}CH_3$$

Information Source: 21CFR178.3860

Chemical Class: Amides

Function: Viscosity Increasing Agent - Nonaqueous

Technical/Other Names:
13-Docosenamide, N-Octadecyl-
N-Octadecyl-13-Docosenamide

STEARYL ERUCATE

CAS Nos.
86601-84-5
96810-34-3

EINECS No.
289-256-4

Empirical Formula:
$C_{40}H_{78}O_2$

Definition: Stearyl Erucate is the ester of stearyl alcohol and erucic acid. It conforms generally to the formula:

$$CH_3(CH_2)_7CH=CH(CH_2)_{11}C(=O)-O(CH_2)_{17}CH_3$$

Chemical Class: Esters

Function: Viscosity Increasing Agent - Nonaqueous

Technical/Other Names:
13-Docosenoic Acid, Octadecyl Ester
Octadecyl 13-Docosenoate

Trade Name:
Schercemol SE (Scher)

STEARYL ETHYLHEXANOATE

CAS No.
59130-70-0

EINECS No.
261-620-7

JPN Translation:
オクタン酸ステアリル

Empirical Formula:
$C_{26}H_{52}O_2$

Definition: Stearyl Ethylhexanoate is the ester of stearyl alcohol and 2-ethylhexanoic acid. It conforms to the formula:

$$CH_3(CH_2)_3CHC(=O)-O(CH_2)_{17}CH_3$$
$$|$$
$$CH_2CH_3$$

Information Sources: JCIC, JCLS, JSQI

Chemical Class: Esters

Function: Skin-Conditioning Agent - Occlusive

Reported Product Categories: Body and Hand Preparations (Excluding Shaving Preparations); Fragrance Preparations, Misc.; Hair Sprays (Aerosol Fixatives); Personal Cleanliness Products, Misc.

Technical/Other Names:
2-Ethylhexanoic Acid, Octadecyl Ester
Hexanoic Acid, 2-Ethyl-, Octadecyl Ester
Octadecyl 2-Ethyl Hexanoate
Stearyl 2-Ethylhexanoate
Stearyl Octanoate

Trade Names:
AEC Stearyl Ethylhexanoate (A & E Connock)
KAK 817 (Kokyu Alcohol)
Pelemol 188 (Phoenix)
SEH (Kokyu Alcohol)

STEARYL ETHYLHEXYLDIMONIUM CHLORIDE

Empirical Formula:
$C_{28}H_{60}N \cdot Cl$

Definition: Stearyl Ethylhexyldimonium Chloride is the quaternary organic compound that conforms to the formula:

$$\left[CH_3(CH_2)_{17}-\overset{\overset{\displaystyle CH_3}{|}}{\underset{\underset{\displaystyle CH_3}{|}}{N}}-CH_2CH(CH_2)_3CH_3 \right]^+ Cl^-$$
$$CH_2CH_3$$

Chemical Class: Quaternary Ammonium Compounds

Functions: Antistatic Agent; Hair Conditioning Agent

Technical/Other Names:
Hydrogenated Tallow Octyldimonium Chloride
Quaternary Ammonium Compounds, (Stearyl, 2-Ethylhexyl)Dimethyl, Chlorides
Stearyl Octyldimonium Chloride

STEARYL ETHYLHEXYLDIMONIUM METHOSULFATE

Empirical Formula:
$C_{28}H_{60}N \cdot CH_3O_4S$

Definition: Stearyl Ethylhexyldimonium Methosulfate is the quaternary ammonium salt that conforms to the formula:

$$\left[CH_3(CH_2)_{17}-\overset{\overset{\displaystyle CH_3}{|}}{\underset{\underset{\displaystyle CH_3}{|}}{N}}-CH_2CH(CH_2)_3CH_3 \right]^+ \begin{matrix} CH_3 \\ OSO_3^- \end{matrix}$$
$$CH_2CH_3$$

Chemical Class: Quaternary Ammonium Compounds

Functions: Antistatic Agent; Hair Conditioning Agent

Reported Product Category: Hair Conditioners

Technical/Other Names:
Hydrogenated Tallow Octyldimonium Methosulfate
Quaternary Ammonium Compounds, (Stearyl, 2-Ethylhexyl)Dimethyl, Ethyl Sulfates
Stearyl Octyldimonium Methosulfate

Trade Name:
Arquad HTL8-MS (Akzo Nobel)

STEARYLGLUCONAMIDE DILAURATE

Definition: Stearylgluconamide Dilaurate is the organic compound that conforms generally to the formula:

$$CH_3(CH_2)_{17}NH-C(=O)-CHCHCHCHCH_2OR$$
$$OR \quad OR \quad OR \quad OR$$

where R represents either H or the lauryl grouping.

Chemical Classes: Amides; Esters

Functions: Hair Conditioning Agent; Skin-Conditioning Agent - Occlusive

Trade Name:
SGA-12,12 (Nippon Chemical)

STEARYL GLYCOL

CAS No.
20294-76-2

EINECS No.
243-711-3

Empirical Formula:
$C_{18}H_{38}O_2$

Definition: Stearyl Glycol is the diol that conforms to the formula:

$$CH_3(CH_2)_{15}CHCH_2OH$$
$$|$$
$$OH$$

Chemical Class: Alcohols

Functions: Emulsion Stabilizer; Skin-Conditioning Agent - Emollient; Viscosity Increasing Agent - Nonaqueous

Technical/Other Names:
1,2-Dihydroxyoctadecane
1,2-Octadecanediol

STEARYL GLYCOL ISOSTEARATE

Empirical Formula:
$C_{36}H_{72}O_3$

Definition: Stearyl Glycol Isostearate is the ester of Stearyl Glycol (q.v.) and Isostearic Acid (q.v.).

Chemical Class: Esters

Functions: Skin-Conditioning Agent - Emollient; Viscosity Increasing Agent - Nonaqueous

STEARYL GLYCYRRHETINATE

CAS No.: 13832-70-7

JPN Translation:
グリチルレチン酸ステアリル

Empirical Formula:
$C_{48}H_{82}O_4$

Definition: Stearyl Glycyrrhetinate is the ester of stearyl alcohol and Glycyrrhetinic Acid (q.v.). It conforms to the formula:

Information Sources: JCLS, JSCI

Chemical Class: Esters

Function: Flavoring Agent

Reported Product Categories: Moisturizing Preparations; Bath Oils, Tablets, and Salts; Lipsticks; Nail Polish and Enamels; Cleansing Products (Cold Creams, Cleansing Lotions, Liquids and Pads); Skin Care Preparations, Misc.

Technical/Other Names:
Octadecyl Glycyrrhetinate
Octadecyl 3-Hydroxy-11-Oxoolean-12-En-29-Oate
Olean-12-En-29-Oic Acid, 3-Hydroxy-11-Oxo-, Octadecyl Ester

Trade Names:
Co-Grhetinol (Maruzen Pharmaceuticals Co., Ltd.)
Magnasoothe (MAFCO)
Plantactiv GLAS 18 (Cognis Deutschland)
Stearyl Glycyrrhetinate Vinyals (Vinyals)
ST-Glycyrrhetinate (Ichimaru Pharcos)

Trade Name Mixture:
Bellsilk ZN-CG (Ichimaru Pharcos)

STEARYL HDI/PEG-50 COPOLYMER

Definition: Stearyl HDI/PEG-50 Copolymer is a copolymer of hexylmethylene diisocyanate, PEG-50 and stearyl alcohol monomers.

Chemical Class: Synthetic Polymers

Function: Film Former

Trade Name:
Borchigel LW 44 (Borchers)

STEARYL HEPTANOATE

CAS No.	EINECS No.
66009-41-4	266-065-4

JPN Translation:
ヘプタン酸ステアリル

Empirical Formula:
$C_{25}H_{50}O_2$

Definition: Stearyl Heptanoate is the ester of stearyl alcohol and heptanoic acid. It conforms to the formula:

Information Sources: CIR: [S] JACT-14(6)-1995, CTFA D, JCIC, JCLS

Chemical Class: Esters

Function: Skin-Conditioning Agent - Occlusive

Reported Product Categories: Bath Preparations, Misc.; Body and Hand Preparations (Excluding Shaving Preparations); Eyeliners; Skin Care Preparations, Misc.; Makeup Preparations (Not eye), Misc.; Bath Capsules; Eye Makeup Preparations, Misc.; Face and Neck Preparations (Excluding Shaving Preparations); Night Skin Care Preparations; Lipsticks; Moisturizing Preparations; Paste Masks (Mud Packs); Eye Shadows

Technical/Other Name:
Heptanoic Acid, Octadecyl Ester

Trade Names:
AEC Stearyl Heptanoate (A & E Connock)
Crodamol W (Croda Chemicals)
Crodamol W (Croda, Inc.)
Pelemol 187 (Phoenix)
Prodhyphore ST (Prod'Hyg)
Tegosoft SH (Degussa Care Specialties)

Trade Name Mixtures:
AEC Stearyl Heptanoate (&) Stearyl Caprylate (A & E Connock)
DUB Solide (Stearinerie Dubois Fils)
Emulzome (Exsymol)
Neo PCL SE o/w 2/066280 (Symrise)
Neo-PCL w/o s.e. 2/066255 (Symrise)
PCL SE w/o 2/066255 (Symrise)
PCL-Siccum 2/066215 (Symrise)
PCL Solid 2/066220 (Symrise)

STEARYL HYDROXYETHYL IMIDAZOLINE

CAS No.	EINECS No.
95-19-2	202-397-8

Empirical Formula:
$C_{22}H_{44}N_2O$

Definition: Stearyl Hydroxyethyl Imidazoline is the heterocyclic compound that conforms to the formula:

Information Source: TSCA

Chemical Class: Imidazoline Compounds

Functions: Antistatic Agent; Hair Conditioning Agent

Reported Product Category: Shampoos (Non-coloring)

Technical/Other Names:
2-Heptadecyl-4,5-Dihydro-1H-Imidazole
1H-Imidazole-1-Ethanol, 2-Heptadecyl-4,5-Dihydro-
Stearyl Imidazoline

STEARYL HYDROXYETHYLIMIDONIUM CHLORIDE

Empirical Formula:
$C_{24}H_{49}N_2O_2 \cdot Cl$

Definition: Stearyl Hydroxyethylimidonium Chloride is the quaternary ammonium salt that conforms generally to the formula:

The inclusion of any compound in the *Dictionary and Handbook* does not indicate that use of that substance as a cosmetic ingredient complies with the laws and regulations governing such use in the United States or any other country.

Chemical Classes: Imidazoline Compounds; Quaternary Ammonium Compounds

Functions: Antistatic Agent; Hair Conditioning Agent

Technical/Other Names:
1,1-Bis(2-Hydroxyethyl)-4,5-Dihydro-2-Heptadecylimidazolium Chloride
Imidazolium, 1,1-Bis(2-Hydroxyethyl)-4,5-Dihydro-2-Heptadecyl-, Chloride
Quaternium-46

STEARYL LACTATE

CAS No.	EINECS No.
35230-14-9	252-447-8

Empirical Formula:
$C_{21}H_{42}O_3$

Definition: Stearyl Lactate is the ester of stearyl alcohol and lactic acid. It conforms generally to the formula:

$$CH_3CHC-O(CH_2)_{17}CH_3$$

Chemical Class: Esters

Function: Skin-Conditioning Agent - Emollient

Technical/Other Names:
2-Hydroxypropanoic Acid, Octadecyl Ester
Octadecyl 2-Hydroxypropanoate
Octadecyl Lactate
Propanoic Acid, 2-Hydroxy-, Octadecyl Ester

Trade Names:
AEC Stearyl Lactate (A & E Connock)
Lactabase C18 (Prod'Hyg)

STEARYL LINOLEATE

CAS No.	EINECS No.
17673-53-9	241-653-3

Empirical Formula:
$C_{36}H_{68}O_2$

Definition: Stearyl Linoleate is the ester of stearyl alcohol and linoleic acid that conforms to the formula:

$$CHCH_2CH=CH(CH_2)_7C-O(CH_2)_{17}CH_3$$

Chemical Class: Esters

Functions: Skin-Conditioning Agent - Occlusive; Viscosity Increasing Agent - Nonaqueous

Technical/Other Name:
9,12-Octadecadienoic Acid, Octadecyl Ester

STEARYL METHACRYLATE/ PERFLUOROOCTYLETHYL METHACRYLATE COPOLYMER

Definition: Stearyl Methacrylate/ Perfluorooctylethyl Methacrylate Copolymer is a copolymer of stearyl methacrylate and perfluorooctylethyl methacrylate monomers.

Chemical Class: Synthetic Polymers

Functions: Film Former; Viscosity Increasing Agent - Nonaqueous

STEARYL METHICONE

Definition: Stearyl Methicone is the siloxane polymer that conforms to the formula:

$$(CH_3)_3SiO - SiO - Si(CH_3)_3$$

Information Source: CIR: [S]

Chemical Class: Siloxanes and Silanes

Function: Skin-Conditioning Agent - Occlusive

Trade Names:
SilCare Silicone 41M30 Stearyl Methicone (Clariant)
SilCare Silicone 41M30 Stearyl Methicone (Clariant GmbH, Personal Care)
Wacker-Belsil SM 6018 (Wacker-Chemie)

Trade Name Mixtures:
SilCare Silicone 51M30 Trimethylsiloxysilicate in Stearyl Methicone (Clariant)
SilCare Silicone 51M30 Trimethylsiloxysilicate in Stearyl Methicone (Clariant GmbH, Personal Care)
Silicone Fluid Emulsion E 32 (Wacker-Chemie)

STEARYL PALMITATE

CAS No.: 2598-99-4

Definition: Stearyl Palmitate is the ester of Stearyl Alcohol (q.v.) and Palmitic Acid (q.v.).

Chemical Class: Esters

Functions: Binder; Emulsion Stabilizer; Hair Conditioning Agent; Humectant; Opacifying Agent; Skin-Conditioning Agent - Miscellaneous

Technical/Other Name:
Hexadecanoic Acid, Octadecyl Ester

Trade Names:
Purester 34 (Strahl & Pitsch)
Stearyl Palmitate (Procter & Gamble)

STEARYL PG-DIMONIUM CHLORIDE PHOSPHATE

Definition: Stearyl PG-Dimonium Chloride Phosphate is the quaternary ammonium salt that conforms to the formula:

Chemical Classes: Phosphorus Compounds; Quaternary Ammonium Compounds

Functions: Antistatic Agent; Hair Conditioning Agent

STEARYL PHOSPHATE

CAS No.	EINECS No.
2958-09-0	220-983-1

JPN Translation:
リン酸ステアリル

Definition: Stearyl Phosphate is a mixture of mono- and diesters of stearyl alcohol and phosphoric acid.

Information Source: JCLS

Chemical Class: Phosphorus Compounds

Function: Surfactant - Emulsifying Agent

Technical/Other Names:
Dihydrogen Stearylphosphate
Octadecyl Dihydrogen Phosphate
Octadecyl Phosphate
Phosphoric Acid, Monooctadecyl Ester

Trade Names:
Arlatone MAP 180 (Uniqema Americas)
Hostaphat CS 120 (Clariant)
Hostaphat CS 120 (Clariant GmbH, Personal Care)

STEARYL STEARATE

CAS No.	EINECS No.
2778-96-3	220-476-5

JPN Translation:
ステアリン酸ステアリル

Empirical Formula:
$C_{36}H_{72}O_2$

Definition: Stearyl Stearate is the ester of stearyl alcohol and stearic acid. It conforms to the formula:

$$CH_3(CH_2)_{16}C - O(CH_2)_{17}CH_3$$

with O double-bonded to C.

Information Sources: 21CFR178.3910, JCIC, JCLS, JSQI, TSCA

Chemical Class: Esters

Functions: Skin-Conditioning Agent - Occlusive; Viscosity Increasing Agent - Nonaqueous

Reported Product Categories: Hair Conditioners; Moisturizing Preparations; Body and Hand Preparations (Excluding Shaving Preparations)

Technical/Other Name:
Octadecanoic Acid, Octadecyl Ester

Trade Names:
AEC Stearyl Stearate (A & E Connock)
Cetinol EE (Fabriquimica)
ESTOL 3706 (Uniqema Europe)
Hetester 412 (Bernel)
Jeechem SS (Jeen)
Liponate SS (Lipo)
Megalube G4 (Megachem)
Radia 7501 (Oleon NV)
Ritachol SS (RITA)

Trade Name Mixtures:
Cerenat N11 (Chemtec Leuna)
Lexate TA (Inolex)
Lipocerite Standard (Vevy)

STEARYL STEAROYL STEARATE

CAS No.: 54684-78-5

JPN Translation:
ステアロイルオキシステアリン酸ステアリル

Empirical Formula:
$C_{54}H_{106}O_4$

Definition: Stearyl Stearoyl Stearate is the ester that conforms generally to the formula:

$$CH_3(CH_2)_{16}C - OCH(CH_2)_{10}C - O(CH_2)_{17}CH_3$$
$$(CH_2)_5CH_3$$

with O double-bonded to both C positions.

Chemical Class: Esters

Functions: Skin-Conditioning Agent - Occlusive; Viscosity Increasing Agent - Nonaqueous

Technical/Other Names:
Octadecanoic Acid, 12-[(1-Oxooctadaecyl)Oxy]-, Octadecyl Ester
12-[(1-Oxooctadecyl)Oxy]Octadecanoic Acid, Octadecyl Ester

Trade Name:
Hetester SSS (Bernel)

STEARYLVINYL ETHER/MA COPOLYMER

CAS No.: 28214-64-4

Definition: Stearylvinyl Ether/MA Copolymer is a polymer of stearylvinyl ether and maleic anhydride.

Chemical Classes: Ethers; Synthetic Polymers

Functions: Emulsion Stabilizer; Film Former; Suspending Agent - Nonsurfactant

Technical/Other Names:
1-(Ethenyloxy)Octadecane, Polymer with 2, 5-Furandione
2,5-Furandione, Polymer with 1-(Ethenyloxy)Octadecane
Stearylvinyl Ether/Maleic Anhydride Copolymer

Trade Name:
Gantrez AN-8194 (International Specialty Products)

STELLARIA MEDIA (CHICKWEED) EXTRACT

CAS No.	EINECS No.
90131-34-3	290-345-5

Definition: Stellaria Media (Chickweed) Extract is an extract of the herb of the chickweed, *Stellaria media*. See "Regulatory and Ingredient Use Information," regarding the labeling names for botanical ingredients in Volume 1, Introduction, Part A.

Chemical Class: Biological Products

Function: Not Reported

Technical/Other Names:
Chickweed Extract
Extract of Chickweed
Extract of Stellaria Media
Stellaria Media Extract

Trade Name Mixtures:
Actiphyte of Chickweed BG50 (Active Organics)
Actiphyte of Chickweed GL50 (Active Organics)
Actiphyte of Chickweed Lipo S (Active Organics)
Actiphyte of Chickweed PG50 (Active Organics)
Herbasol Extract Chickweed (Cosmetochem) (Cosmetochem International Ltd.)
VT-141 Extract of Chickweed (Vege-Tech)

STEMMACANTHA CARTHAMOIDES ROOT EXTRACT

Definition: Stemmacantha Carthamoides Root Extract is an extract of the roots of *Stemmacantha carthamoides*. See "Regulatory and Ingredient Use Information," regarding the labeling names for botanical ingredients in Volume 1, Introduction, Part A. See Reported Ingredient Functions-The Cosmetic Drug Distinction, in Regulatory and Ingredient Use Information, Volume I, Part A.

Chemical Class: Biological Products

Functions: Deodorant Agent; Skin Protectant

Technical/Other Name:
Extract of Stemmacantha Carthamoides Root

Trade Name Mixture:
Maralen (Hlavin Industries)

STEPHANIA CEPHARANTHA ROOT EXTRACT

Definition: Stephania Cepharantha Root Extract is an extract of the roots of *Stephania Cepharantha*. See "Regulatory and Ingredient Use Information," regarding the labeling names for botanical ingredients in Volume 1, Introduction, Part A.

Chemical Class: Biological Products

Function: Skin-Conditioning Agent - Miscellaneous

Technical/Other Name:
Extract of Stephania Cepharantha Root

Trade Names:
Biscotin (Technoble)
Cepharantine (Maruzen Pharmaceuticals Co., Ltd.)

Trade Name Mixture:
Cepharanthin (Kaken)

STEPHANIA TETRANDA ROOT EXTRACT

Definition: Stephania Tetranda Root Extract is an extract of the roots of *Stephania tetranda*. See "Regulatory and Ingredient Use Information," regarding the labeling names for botanical ingredients in Volume 1, Introduction, Part A.

Chemical Class: Biological Products

Function: Not Reported

Technical/Other Name:
Extract of Stephania Tetranda

STERCULIA PLANTIFOLIA EXTRACT

Definition: Sterculia Plantifolia Extract is an extract of the seeds, leaves and flowers of *Sterculia plantifolia*. See *"Regulatory and Ingredient Use Information,"* regarding the labeling names for botanical ingredients in Volume 1, Introduction, Part A.

Chemical Class: Biological Products

Function: Not Reported

Technical/Other Name:
Extract of Sterculia Plantifolia

Trade Name Mixture:
Campo Wu Tung Extract (Campo)

STERCULIA URENS GUM

CAS No.   **EINECS No.**
9000-36-6   232-539-4

JPN Translation:
カラヤガム

Definition: Sterculia Urens Gum is a dried exudate from the tree, *Sterculia urens*. See *"Regulatory and Ingredient Use Information,"* regarding the labeling names for botanical ingredients in Volume 1, Introduction, Part A.

Information Sources: BPC, BRA, 21CFR133.133, 21CFR133.134, 21CFR133.162, 21CFR133.178, 21CFR133.179, 21CFR150.141, 21CFR150.161, 21CFR184.1349, FCC, JCIC, JCLS, JSQI, MAR, MI-13(5302), OTC-I-LX, RIFM, TSCA

Chemical Class: Gums, Hydrophilic Colloids and Derivatives

Functions: Adhesive; Binder; Emulsion Stabilizer; Fragrance Ingredient; Hair Fixative; Viscosity Increasing Agent - Aqueous

Technical/Other Names:
Gum Karaya
Gum, Sterculia Urens
Indian Tragacanth Gum
Karaya Gum
Karaya gum (Sterculia urens Roxb.) (RIFM)
Karaya (Sterculia Urens) Gum

STEVIA REBAUDIANA EXTRACT

CAS No.   **EINECS No.**
91722-21-3   294-422-4

JPN Translation:
ステビアエキス

Definition: Stevia Rebaudiana Extract is an extract of the aerial parts of *Stevia rebaudiana*. See *"Regulatory and Ingredient Use Information,"* regarding the labeling names for botanical ingredients in Volume 1, Introduction, Part A.

Chemical Class: Biological Products

Function: Flavoring Agent

Technical/Other Name:
Extract of Stevia Rebaudiana

Trade Name:
Marumilon Pure (Maruzen Pharmaceuticals Co., Ltd.)

Trade Name Mixtures:
Kahelift (Silab)
Stevia Liquid B (Ichimaru Pharcos)

STEVIOSIDE

CAS No.   **EINECS No.**
57817-89-7   260-975-5

Empirical Formula:
$C_{38}H_{60}O_{18}$

Definition: Stevioside is the carbohydrate isolated from *Eupatorium rebaudianum*.

Information Source: MI-13(8888)

Chemical Class: Carbohydrates

Function: Flavoring Agent

Technical/Other Names:
13-[(2-O-β-D-Glucopyranosyl-α-D-Gluco-pyranosyl)Oxy]Kaur-16-en-18-oic Acid, β-D-Glucopyranosyl Ester
Kaur-16-en-18-oic Acid, 13-[2-O-β-D-Glucopyranosyl-α-D-Glucoopyranosyl)-Oxy]-, β-D-Glucopyranosyl Ester
Kaur-16-en-18-oic Acid, 13-[(2-O-β-D-Glucopyranosyl-α-D-Glucopyranosyl)-Oxy-, β-D-Glucopyranosyl Ester, (4α)-

Trade Names:
AEC Stevioside (A & E Connock)
Stevita Cristal (Steviafarma)

STIGMASTANOL MALTOSIDE

Empirical Formula:
$C_{41}H_{72}O_{11}$

Definition: Stigmastanol Maltoside is the ether formed by the reaction of stigmastanol and Maltose (q.v.). It conforms to the formula:

where R represents the maltose grouping.

Chemical Class: Sterols

Function: Skin-Conditioning Agent - Emollient

Trade Name:
Stigma M-P (Nikko)

STILLINGIA SYLVATICA

Definition: *See "Regulatory and Ingredient Use Information,"* regarding EU labeling names for botanical ingredients in Volume 1, Introduction, Part A.

Chemical Class: Biological Products

Technical/Other Name:
Stillingia Sylvatica Root (U.S.)

STILLINGIA SYLVATICA ROOT

Definition: Stillingia Sylvatica Root is a plant material derived from the dried roots of the stillingia, *Stillingia sylvatica*. See *"Regulatory and Ingredient Use Information,"* regarding the labeling names for botanical ingredients in Volume 1, Introduction, Part A.

Information Sources: HP, MI-13(8896)

Chemical Class: Biological Products

Function: Not Reported

Technical/Other Names:
Stillingia
Stillingia Sylvatica (EU)

STIPA TENACISSIMA WAX

Definition: Stipa Tenacissima Wax is a wax obtained from the whole plant of *Stipa tenacissima*. See *"Regulatory and Ingredient Use Information,"* regarding the labeling names for botanical ingredients in Volume 1, Introduction, Part A.

Chemical Class: Waxes

Function: Not Reported

Technical/Other Name:
Waxes, Stipa Tenacissima

Trade Names:
AEC Cire Essentielle D'Alpha Du Desert (A & E Connock)
Cire Essentielle d'Alpha du Desert (Bertin)

STOMACH EXTRACT

CAS No.   **EINECS No.**
84540-15-8   283-106-1

Definition: Stomach Extract is an extract of bovine stomachs.

Chemical Class: Biological Products

Function: Skin-Conditioning Agent - Humectant

Technical/Other Name:
Extract of Stomach

STREPTOCOCCUS LACTIS EXTRACT

Definition: Streptococcus Lactis Extract is an extract of the bacterial culture, *Streptococcus lactis.*

Chemical Class: Biological Products

Function: Not Reported

Technical/Other Name:
Extract of Streptococcus Lactis

STREPTOCOCCUS THERMOPHILUS FERMENT

Definition: Streptococcus Thermophilus Ferment is the fermentation product of *Streptococcus thermophilus.*

Chemical Class: Biological Products

Function: Skin-Conditioning Agent - Humectant

Trade Name:
Streptococcus Ferment D2 (Nonogawa)

Trade Name Mixture:
Streptococcus Thermophilus Ferment D2 (Nonogawa)

STREPTOCOCCUS THERMOPHILUS/MILK FERMENT LYSATE

Definition: Streptococcus Thermophilus/Milk Ferment Lysate is the fermentation product of milk by the microorganism *Streptococcus thermophilus* with subsequent lysing of the microorganism's cells.

Chemical Class: Biological Products

Function: Skin-Conditioning Agent - Miscellaneous

Trade Name:
Streptococcus Thermophilus (Avantgarde)

STRONTIUM ACETATE

CAS Nos.	EINECS No.
543-94-2	208-853-2
14692-29-6	

Empirical Formula:
$C_2H_4O_2 \cdot \frac{1}{2}Sr$

Definition: Strontium Acetate is the organic salt that conforms to the formula:

$$(CH_3COO)_2Sr$$

Information Sources: EEC(III/1-58), MHLW-331/1, MI-13(8916)

Chemical Class: Organic Salts

Function: Oral Care Agent

Technical/Other Names:
Acetic Acid, Strontium Salt
Strontium Acetate Hemihydrate

STRONTIUM CHLORIDE

CAS No.	EINECS No.
10476-85-4	233-971-6

Empirical Formula:
$H_2Cl_2 \cdot Sr$

Definition: Strontium Chloride is the inorganic salt that conforms to the formula:

$$SrCl_2$$

Information Sources: EEC(III/1-57), MHLW-331/1, MI-13(8920), TSCA

Chemical Class: Inorganic Salts

Functions: Oral Care Agent; Skin-Conditioning Agent - Miscellaneous

Technical/Other Name:
Strontium Dichloride

Trade Name:
COSMEDERM-7 (Cosmederm)

STRONTIUM CHLORIDE HEXAHYDRATE

CAS No.: 10025-70-4

Empirical Formula:
$Cl_2Sr \cdot 6H_2O$

Definition: Strontium Chloride Hexahydrate is the inorganic salt that conforms to the formula:

$$SrCl_2 \quad \cdot \quad 6H_2O$$

Information Sources: EEC(III/1-57), MHLW-331/1, MI-13(8920), TSCA

Chemical Class: Inorganic Salts

Functions: Oral Care Agent; Skin-Conditioning Agent - Miscellaneous

Trade Name:
COSMEDERM-7 (Cosmederm)

STRONTIUM HYDROXIDE

CAS Nos.	EINECS No.
1311-10-0 (Octahydrate)	
18480-07-4	242-367-1

Empirical Formula:
H_2O_2Sr

Definition: Strontium Hydroxide is the inorganic compound that conforms to the formula:

$$Sr(OH)_2$$

Information Sources: CTFA D, EEC(III/1-63), MHLW-331/1, MI-13(8924), TSCA

Chemical Class: Inorganic Bases

Function: pH Adjuster

Reported Product Categories: Bubble Baths; Depilatories

Technical/Other Names:
Strontium Hydrate
Strontium Hydroxide, Octahydrate

Trade Name:
Strontium Hydroxide, Octahydrate (Merck KGaA)

STRONTIUM NITRATE

CAS No.	EINECS No.
10042-76-9	233-131-9

Empirical Formula:
$Sr(NO_3)_2$

Definition: Strontium Nitrate is the inorganic salt that conforms to the formula:

$$Sr(NO_3)_2$$

Information Sources: EEC(II-403), MHLW-331/1, MI-13(8926), TSCA

Chemical Class: Inorganics

Functions: Oral Care Agent; Skin-Conditioning Agent - Miscellaneous

Technical/Other Name:
Nitric Acid, Strontium Salt

Trade Name:
COSMEDERM-7 (Cosmederm)

STRONTIUM PEROXIDE

CAS No.	EINECS No.
1314-18-7	215-224-6

Empirical Formula:
O_2Sr

Definition: Strontium Peroxide is the inorganic oxide that conforms to the formula:

$$SrO_2$$

Information Sources: EEC(III/1-64), MHLW-331/1, MI-13(8929), TSCA

Chemical Class: Inorganics

Function: Oxidizing Agent

Technical/Other Name:
Strontium Dioxide

STRONTIUM SULFIDE

CAS No.	EINECS No.
1314-96-1	215-249-2

Empirical Formula:
SSr

Definition: Strontium Sulfide is the inorganic salt that conforms to the formula:

SrS

Information Sources: EEC(III/1-23), MHLW-331/1, MI-13(8932), TSCA

Chemical Class: Inorganic Salts

Function: Depilating Agent

STRONTIUM THIOGLYCOLATE

CAS No.	EINECS No.
38337-95-0	253-888-9

Empirical Formula:
$C_2H_4O_2S \cdot \frac{1}{2}Sr$

Definition: Strontium Thioglycolate is the strontium salt of thioglycolic acid that conforms to the formula:

$(HSCH_2COO)_2Sr$

Information Sources: EEC(III/1-2a), MHLW-331/1

Chemical Classes: Organic Salts; Thio Compounds

Functions: Depilating Agent; Hair-Waving/Straightening Agent; Reducing Agent

Technical/Other Names:
Acetic Acid, Mercapto-, Strontium Salt
Strontium Mercaptoacetate
Strontium Thioglycollate
Thioglycolic Acid, Strontium Salt

STRYPHNODENDRON ADSTRINGENS BARK EXTRACT

Definition: Stryphnodendron Adstringens Bark Extract is an extract of the bark of *Stryphnodendron adstringens*. See *"Regulatory and Ingredient Use Information,"* regarding the labeling names for botanical ingredients in Volume 1, Introduction, Part A.

Chemical Class: Biological Products

Function: Not Reported

Technical/Other Names:
Extract of Stryphnodendron Adstringens Bark
Stryyphnodendron Barbadetimam Extract

Trade Name:
Barbitamao (Greentech)

STYRAX BENZOIN GUM

CAS Nos.	EINECS Nos.
9000-05-9	232-523-7
9000-72-0	232-556-7

Definition: Styrax Benzoin Gum is a balsamic resin obtained from *Styrax benzoin*. See *"Regulatory and Ingredient Use Information,"* regarding the labeling names for botanical ingredients in Volume 1, Introduction, Part A.

Information Sources: AUS, BEL, BPC, BRA, 21CFR73.1, 21CFR172.510, 21CFR172.515, 21CFR310.545, EGY, FI, FIN, HP, HUN, IND, JAN, MAR, MEX, MI-13 (1094), MI-13(4594), PF, POR, RIFM, ROM, SNPF, TSCA, USAN, USD, USP XXIV

Chemical Class: Gums, Hydrophilic Colloids and Derivatives

Functions: Adhesive; Film Former; Fragrance Ingredient; Skin-Conditioning Agent - Miscellaneous

Reported Product Categories: Bath Preparations, Misc.; Body and Hand Preparations (Excluding Shaving Preparations)

Technical/Other Names:
Benzoin (RIFM)
Benzoin
Benzoin Gum
Benzoin gum, Siam (RIFM)
Benzoin resinoid (RIFM)
Benzoin (Styrax Benzoin) Gum
Gum Benzoin
Gum Benzoin Siam
Gum, Styrax Benzoin
Gum Sumatra
Indochina Benzoin
Resins, Benzoin
Siam Benzoin
Sumatra Benzoin

Trade Name:
AEC Benzoin Powder (A & E Connock)

Trade Name Mixtures:
Phytelene of Gum Benzoin EG 500 liquid (Indena SA)
Phytogreen 55 of Gum Benzoin EXH 704 Liquid (Phytochim)
Tropical Resins EA (Alban Muller)
Tropical Resins H (Alban Muller)
Tropical Resins HS (Alban Muller)
Tropical Resins LS (Alban Muller)

STYRAX BENZOIN RESIN EXTRACT

CAS Nos.	EINECS No.
84696-18-4	
84929-79-3	284-557-7

JPN Translation:
アンソッコウエキス

Definition: Styrax Benzoin Resin Extract is an extract of the balsamic resin of *Styrax benzoin*. See *"Regulatory and Ingredient Use Information,"* regarding the labeling names for botanical ingredients in Volume 1, Introduction, Part A.

Information Sources: 21CFR172.510, JCIC, JCLS, RIFM

Chemical Class: Biological Products

Functions: Fragrance Ingredient; Not Reported

Technical/Other Names:
Benzoin Extract
Benzoin (Styrax Benzoin) Extract
Extract of Benzoin
Extract of Styrax Benzoin
Styrax benzoin, ext. (RIFM)
Styrax Extract
Styrax Resin Extract

Trade Name Mixtures:
Actiphyte of Benzoin BG50 (Active Organics)
Actiphyte of Benzoin GL50 (Active Organics)
Actiphyte of Benzoin Lipo S (Active Organics)
Actiphyte of Benzoin PG50 (Active Organics)
Styrax Benzoin Resin Extract ies (IES LABO)
VT-275 Extract of Benzoin Extract (Vege-Tech)

STYRAX TONKINENSIS RESIN EXTRACT

Definition: Styrax Tonkinensis Resin Extract is an extract of the resin of *Styrax tonkinensis*. See *"Regulatory and Ingredient Use Information,"* regarding the labeling names for botanical ingredients in Volume 1, Introduction, Part A.

Chemical Class: Biological Products

Function: Skin-Conditioning Agent - Miscellaneous

Technical/Other Name:
Extract of Styrax Tonkinensis

Trade Name Mixtures:
Benzoin HS (Alban Muller)
285 Stimulant HS (Alban Muller)

STYRENE/ACRYLAMIDE COPOLYMER

CAS No.: 24981-13-3

JPN Translation:
（スチレン / アクリルアミド）コポリマー

Styrene/Acrylamide Copolymer (Cont.)

Empirical Formula: $(C_8H_8 \cdot C_3H_5NO)_x$

Definition: Styrene/Acrylamide Copolymer is a polymer formed from styrene and acrylamide monomers.

Information Sources: JCIC, JCLS, TSCA

Chemical Class: Synthetic Polymers

Function: Opacifying Agent

Reported Product Categories: Hair Preparations (Non-coloring), Misc.; Permanent Waves

Technical/Other Names:
Acrylamide•Styrene Copolymer
Ethenylbenzene, Polymer with 2-Propenamide
2-Propenamide, Polymer with Ethenylbenzene

Trade Names:
Acusol OP303P Opacifier (Rohm and Haas)
Esi-Cryl 12 (ESI)

Trade Name Mixtures:
Lytron 308 (Rohm and Haas)
Lytron 318 (Rohm and Haas)
Syntran 5905 (Interpolymer)

STYRENE/ACRYLATES/ACRYLONITRILE COPOLYMER

JPN Translation:
(スチレン / アクリル酸ブチル / アクリロニトリル) コポリマー

Definition: Styrene/Acrylates/Acrylonitrile Copolymer is a polymer of styrene, acrylonitrile and a monomer consisting of acrylic acid, methacrylic acid or one of their simple esters.

Information Sources: 21CFR177.1010, JCIC, JCLS

Chemical Class: Synthetic Polymers

Functions: Film Former; Opacifying Agent; Suspending Agent - Nonsurfactant

Reported Product Category: Nail Polish and Enamels

Technical/Other Name:
Styrene/Acrylate/Acrylonitrile Copolymer

Trade Name:
Neo Cryl B-1000 (ICI Resins)

STYRENE/ACRYLATES/AMMONIUM METHACRYLATE COPOLYMER

Definition: Styrene/Acrylates/Ammonium Methacrylate Copolymer is a polymer of styrene, ammonium methacrylate and a monomer consisting of acrylic acid, methacrylic acid or one of their simple esters.

Information Sources: 21CFR177.1010, CIR: [SQ] IJT 21(SUPPL. 3) 2002

Chemical Class: Synthetic Polymers

Functions: Film Former; Suspending Agent - Nonsurfactant

Technical/Other Name:
Styrene/Acrylate/Ammonium Methacrylate Copolymer

Trade Name Mixtures:
Syntran 5009 (Interpolymer)
Syntran 5760 (Interpolymer)

STYRENE/ACRYLATES COPOLYMER

CAS No.: 9010-92-8

Definition: Styrene/Acrylates Copolymer is a polymer of styrene and a monomer consisting of acrylic acid, methacrylic acid or one of their simple esters.

Information Sources: 21CFR175.300, 21CFR175.320, 21CFR176.170, 21CFR177.1010, 21CFR177.1830, CIR: [SQ] IJT 21(SUPPL. 3) 2002, JCLS

Chemical Class: Synthetic Polymers

Function: Film Former

Reported Product Categories: Hair Dyes and Colors (All Types Requiring Caution Statements and Patch Tests); Nail Polish and Enamels; Permanent Waves; Personal Cleanliness Products, Misc.; Bath Capsules; Face and Neck Preparations (Excluding Shaving Preparations); Eyeliners

Technical/Other Names:
2-Propenoic Acid, Butyl Ester, Polymer with Ethylbenzene
Styrene/Acrylate Copolymer

Trade Names:
Acronal 290 D (BASF)
Acronal 296 D (BASF)
Acusol OP301 Emulsion (Rohm and Haas)
Acusol OP304 Emulsion (Rohm and Haas)
Acusol OP302P Opacifier (Rohm and Haas)
Custom OP 300M (Custom Ingredients)
Daitosol SPA (LCW)
Esi-Cryl 11 (ESI)
Joncryl 50 (Johnson Polymer)
Joncryl 52 (Johnson Polymer)
Joncryl 56 (Johnson Polymer)
Joncryl 57 (Johnson Polymer)
Joncryl 58 (Johnson Polymer)
Joncryl 59 (Johnson Polymer)
Joncryl 60 (Johnson Polymer)
Joncryl 61 (Johnson Polymer)
Joncryl 62 (Johnson Polymer)
Joncryl 63 (Johnson Polymer)
Joncryl 67 (Johnson Polymer)
Joncryl 70 (Johnson Polymer)
Joncryl 73 (Johnson Polymer)
Joncryl 74 (Johnson Polymer)
Joncryl 77 (Johnson Polymer)
Joncryl 89 (Johnson Polymer)
Joncryl 95 (Johnson Polymer)
Joncryl 98 (Johnson Polymer)
Joncryl 537 (Johnson Polymer)
Joncryl 538 (Johnson Polymer)
Joncryl 585 (Johnson Polymer)
Joncryl 586 (Johnson Polymer)
Joncryl 611 (Johnson Polymer)
Joncryl 617 (Johnson Polymer)
Joncryl 624 (Johnson Polymer)
Joncryl 631 (Johnson Polymer)
Joncryl 660 (Johnson Polymer)
Joncryl 674 (Johnson Polymer)
Joncryl 678 (Johnson Polymer)
Joncryl 680 (Johnson Polymer)
Joncryl 682 (Johnson Polymer)
Joncryl 690 (Johnson Polymer)
Joncryl 693 (Johnson Polymer)
Joncryl 817 (Johnson Polymer)
Joncryl 819 (Johnson Polymer)
Joncryl 1536 (Johnson Polymer)
Joncryl 1601 (Johnson Polymer)
Joncryl 1645 (Johnson Polymer)
Joncryl 1661 (Johnson Polymer)
Joncryl 1663 (Johnson Polymer)
Joncryl 1674 (Johnson Polymer)
Joncryl 1695 (Johnson Polymer)
Joncryl 2153 (Johnson Polymer)
Joncryl 2161 (Johnson Polymer)
Joncryl 2640 (Johnson Polymer)
Joncryl 2660 (Johnson Polymer)
Joncryl 8055 (Johnson Polymer)
Joncryl 8281 (Johnson Polymer)
Joncryl ECO 94 (Johnson Polymer)
Joncryl ECO 675 (Johnson Polymer)
Joncryl ECO 684 (Johnson Polymer)
Joncryl ECO 694 (Johnson Polymer)
Joncryl ECO 2117 (Johnson Polymer)
Joncryl ECO 2124 (Johnson Polymer)
Joncryl ECO 2177 (Johnson Polymer)
Joncryl ECO 2189 (Johnson Polymer)
Joncryl HPD 71 (Johnson Polymer)
Joncryl HPD 671 (Johnson Polymer)
Lytron 180AV (Omnova Solutions, Inc)
Modarez O AMS 1982 (Synthron)
Nanospheres 100 SA Hydrophilic (Exsymol)
SCX 86 (Johnson Polymer)
SCX 686 (Johnson Polymer)
SCX 804 (Johnson Polymer)
SCX 806 (Johnson Polymer)
SCX 815 (Johnson Polymer)
SCX 820 (Johnson Polymer)
SCX 821 (Johnson Polymer)
SCX 822 (Johnson Polymer)
SCX 835 (Johnson Polymer)
SCX 839 (Johnson Polymer)

SCX 880 (Johnson Polymer)
SCX 2155 (Johnson Polymer)
SCX 2610 (Johnson Polymer)
SCX 2641 (Johnson Polymer)
SCX 8211 (Johnson Polymer)
SCX 8280 (Johnson Polymer)
SCX 8320 (Johnson Polymer)
SunSpheres LCG Polymer (Rohm and Haas)
SunSpheres PGL Polymer (Rohm and Haas)
SunSpheres Polymer (Rohm and Haas)
Syntran 5904 (Interpolymer)
Yodosol GH840 (National Starch)
Yodosol GH4l (National Starch)

Trade Name Mixtures:
Colourspheres FD&C Green No. 3 HL 25% (Creations Couleurs)
Colourspheres Alizuridine HL 25% (Creations Couleurs)
Colourspheres Black HB 25% (Creations Couleurs)
Colourspheres Black HB 70% (Creations Couleurs)
Colourspheres Black HL 25% (Creations Couleurs)
Colourspheres Black HL 70% (Creations Couleurs)
Colourspheres Blue HB 70% (Creations Couleurs)
Colourspheres Blue HL 25% (Creations Couleurs)
Colourspheres Blue HL 70% (Creations Couleurs)
Colourspheres Blue Sky Dye HL 25% (Creations Couleurs)
Colourspheres Bright Yellow HL 25% (Creations Couleurs)
Colourspheres Bright Yellow Lake HL 25% (Creations Couleurs)
Colourspheres Coral Dye HL 25% (Creations Couleurs)
Colourspheres Electric Pink Lake HL 25% (Creations Couleurs)
Colourspheres Eosin Dye HL 25% (Creations Couleurs)
Colourspheres Geranium HB 25% (Creations Couleurs)
Colourspheres Gold Grass Dye HL 25% (Creations Couleurs)
Colourspheres Green HL 25% (Creations Couleurs)
Colourspheres Harvest Dye HL 25% (Creations Couleurs)
Colourspheres Helindone Lake HL 25% (Creations Couleurs)
Colourspheres Magenta Dye HL 25% (Creations Couleurs)
Colourspheres Magenta HB 25% (Creations Couleurs)
Colourspheres Magenta Lake HL 25% (Creations Couleurs)
Colourspheres Mandarin Dye HL 25% (Creations Couleurs)
Colourspheres Melon Dye HL 25% (Creations Couleurs)
Colourspheres Melon Lake HL 25% (Creations Couleurs)
Colourspheres Ocean Blue Dye HL 25% (Creations Couleurs)
Colourspheres Ocean Blue HL 25% (Creations Couleurs)
Colourspheres Orange Dye HL 25% (Creations Couleurs)
Colourspheres Orange Lake HL 25% (Creations Couleurs)
Colourspheres Pinky Dye HL 25% (Creations Couleurs)
Colourspheres Poppy Lake HL 25% (Creations Couleurs)
Colourspheres Prussian Blue HL 25% (Creations Couleurs)
Colourspheres Red HB 25% (Creations Couleurs)
Colourspheres Red HB 70% (Creations Couleurs)
Colourspheres Red HL 25% (Creations Couleurs)
Colourspheres Red HL 70% (Creations Couleurs)
Colourspheres Safran Dye HL 25% (Creations Couleurs)
Colourspheres Safran HB 25% (Creations Couleurs)
Colourspheres Safran HL 25% (Creations Couleurs)
Colourspheres White R HB 25% (Creations Couleurs)
Colourspheres White R HB 70% (Creations Couleurs)
Colourspheres White R HL 25% (Creations Couleurs)
Colourspheres White R HL 70% (Creations Couleurs)
Colourspheres Wood Dye HL 25% (Creations Couleurs)
Colourspheres Yellow Grass Dye HL 25% (Creations Couleurs)
Colourspheres Yellow HB 25% (Creations Couleurs)
Colourspheres Yellow HB 70% (Creations Couleurs)
Colourspheres Yellow HL 25% (Creations Couleurs)
Colourspheres Yellow HL 70% (Creations Couleurs)
CosmoTurb SB (CosmoCare)
Eospoly UV Cristal HB 50% (Creations Couleurs)
Eospoly UV Cristal HL 50% (Creations Couleurs)
Eospoly UV Shadow HL 12,5% (Creations Couleurs)
Eospoly UV Zinc HB 50% (Creations Couleurs)
Eospoly UV Zinc HL 50% (Creations Couleurs)
Lait Vegetal (Greentech)
Lytron 450 (Omnova Solutions, Inc)
Lytron 614 (Rohm and Haas)
Lytron 621 (Rohm and Haas)
Nanosource CL (C.I.T.)
Nanosource GB (C.I.T.)
Nanosource LI (C.I.T.)
Nanosource P (C.I.T.)
Nanosource PA (C.I.T.)
Nanosource PN (C.I.T.)
Nanosource PO (C.I.T.)
Nanosource TL (C.I.T.)
Nanosource UQ (C.I.T.)
OPD-140 (Importaciones y Suministros)

STYRENE/ACRYLATES COPOLYMER/ POLYURETHANE

Definition: Styrene/Acrylates Copolymer/ Polyurethane is a copolymer of Styrene/ Acrylates Copolymer (q.v.) , with cyclohexanedimethanol, benzoic acid, phthalic anhydride and isophoronediisocyanate.

Chemical Class: Synthetic Polymers

Functions: Binder; Film Former; Plasticizer

Trade Name Mixture:
KFilm 2001 (Kane)

STYRENE/ACRYLATES/DIMETHICONE ACRYLATE CROSSPOLYMER

Definition: Styrene/Acrylates/Dimethicone Acrylate Crosspolymer is a copolymer of styrene, dimethicone acrylate and one or more monomers of acrylic acid, methacrylic acid or one of their simple esters crosslinked with divinylbenzene.

Chemical Classes: Siloxanes and Silanes; Synthetic Polymers

Function: Skin-Conditioning Agent - Miscellaneous

STYRENE/ACRYLATES/DIMETHICONE COPOLYMER

Definition: Styrene/Acrylates/Dimethicone Copolymer is a polymer of styrene, dimethicone and a monomer consisting of acrylic acid, methacrylic acid or one of their simple esters.

Chemical Classes: Siloxanes and Silanes; Synthetic Polymers

Function: Skin-Conditioning Agent - Miscellaneous

STYRENE/ACRYLATES/ETHYLHEXYL ACRYLATE/LAURYL ACRYLATE COPOLYMER

JPN Translation:
（スチレン / アクリル酸アルキル）コポリマー

Definition: Styrene/Acrylates/Ethylhexyl Acrylate/Lauryl Acrylate Copolymer is a coplymer of styrene, acrylates, ethylhexyl acrylate and lauryl acrylate.

Information Source: JCLS

Chemical Class: Synthetic Polymers

Function: Film Former

STYRENE/ALLYL BENZOATE COPOLYMER

Empirical Formula:
$(C_8H_8)_x(C_{10}H_{12}O_2)_y$

Definition: Styrene/Allyl Benzoate Copolymer is the copolymer of allyl benzoate and styrene monomers.

Chemical Class: Synthetic Polymers

Function: Film Former

STYRENE/BUTADIENE COPOLYMER

CAS No.: 9003-55-8

Definition: Styrene/Butadiene Copolymer is a copolymer of styrene and butadiene monomers.

Chemical Class: Synthetic Polymers

Function: Opacifying Agent

Technical/Other Name:
Benzene, Ethenyl-, Polymer with 1,3-Butadiene

Trade Names:
Baystal S 61 (Polymerlatex)
Opacifier 43.009 (Augusto Bellinvia)

STYRENE/DVB CROSSPOLYMER

CAS No.: 9003-70-7

Empirical Formula:
$(C_8H_8)_x(C_{10}H_{10})_y$

Definition: Styrene/DVB Crosspolymer is the crosspolymer of styrene and divinylbenzene monomers.

Information Source: 21CFR177.2710

Chemical Class: Synthetic Polymers

Function: Film Former

Technical/Other Names:
Benzene, Diethenyl-, Polymer with Ethenylbenzene
Divinylbenzene-Styrene Copolymer
Styrene/DVB Copolymer

Trade Name:
Toshiki Plastic Powder FP-SQ (Nikko)

Trade Name Mixture:
Microsponge 5645 Mineral Oil (EDT, Inc.)

STYRENE/ISOPRENE COPOLYMER

CAS No.: 25038-32-8

Definition: Styrene/Isoprene Copolymer is a copolymer of styrene and isoprene monomers.

Information Source: TSCA

Chemical Class: Synthetic Polymers

Functions: Film Former; Opacifying Agent

Technical/Other Names:
Benzene, Ethenyl, Polymer with 2-Methyl-1,3-Butadiene
Styrene, Polymer with Isoprene

Trade Names:
Kraton (Coloplast Consumer)
Kraton D-1161NS (Shell)

STYRENE/MA COPOLYMER

CAS No.: 9011-13-6

Empirical Formula:
$(C_8H_8 \cdot C_4H_2O_3)_x$

Definition: Styrene/MA Copolymer is a polymer of styrene and maleic anhydride monomers.

Information Sources: 21CFR176.170, 21CFR177.1200, 21CFR177.1210, 21CFR177.1820, JCLS, TSCA

Chemical Class: Synthetic Polymers

Functions: Binder; Emulsion Stabilizer; Film Former; Suspending Agent - Nonsurfactant

Technical/Other Names:
2,5-Furandione, Polymer with Ethenylbenzene
Styrene/Maleic Anhydride Copolymer

STYRENE/METHACRYLAMIDE/ ACRYLATES COPOLYMER

CAS No.: 28632-83-9

Definition: Styrene/Methacrylamide/Acrylates Copolymer is a copolymer of styrene, methacrylamide and one or more monomers of acrylic acid, methacrylic acid or one of their simple esters.

Chemical Class: Synthetic Polymers

Functions: Film Former; Viscosity Increasing Agent - Nonaqueous

Technical/Other Names:
Acrylic Acid-Methacrylamide-Styrene Copolymer
2-Propenoic Acid, Polymer with Ethenylbenzene and 2-Methyl-2-Propenamide

Trade Names:
Yodosol GH52 (National Starch)
Yodosol GH52-OP (National Starch)

STYRENE/METHYLSTYRENE COPOLYMER

CAS Nos.: 9011-11-4; 37218-15-8

Definition: Styrene/Methylstyrene Copolymer is a copolymer of styrene and methyl styrene monomers.

Information Source: TSCA

Chemical Class: Synthetic Polymers

Functions: Binder; Depilating Agent

Technical/Other Names:
Benzene, Ethenyl-, Polymer with (1-Methylethenyl)Benzene
Ethenylbenzene, Copolymer with (1-Methylethenyl)Benzene
Styrene-Vinyltoluene Copolymer

Trade Names:
Norsolene W 90 (Cray Valley)
Norsolene W 90 (Cray Valley Ltd.)
Norsolene W 100 (Cray Valley)
Norsolene W 110 (Cray Valley)

STYRENE/METHYLSTYRENE/INDENE COPOLYMER

JPN Translation:
（スチレン / ビニルトルエン / メチルスチレン / インデン）コポリマー

Definition: Styrene/Methylstyrene/Indene Copolymer is the copolymer of styrene, methylstyrene and indene monomers.

Chemical Class: Synthetic Polymers

Functions: Film Former; Opacifying Agent

STYRENE/STEARYL METHACRYLATE CROSSPOLYMER

CAS No.: 91838-84-5

Definition: Styrene/Stearyl Methacrylate Crosspolymer is a copolymer of styrene and stearyl methacrylate monomers crosslinked with divinylbenzene.

Chemical Class: Synthetic Polymers

Functions: Absorbent; Skin-Conditioning Agent - Miscellaneous

Technical/Other Name:
2-Propenoic Acid, 2-Methyl-, Octadecyl Ester, Polymer with Diethenylbenzene and Ethenylbenzene

STYRENE/VA COPOLYMER

JPN Translation:
（スチレン/VA）コポリマー

Definition: Styrene/VA Copolymer is a copolymer of styrene and vinyl acetate monomers.

Information Sources: JCIC, JCLS

Chemical Class: Synthetic Polymers

Functions: Film Former; Opacifying Agent

Technical/Other Name:
Vinyl Acetate•Styrene Copolymer Emulsion

STYRENE/VP COPOLYMER

CAS No.: 25086-29-7

JPN Translation:
（スチレン / ビニルピロリドン）コポリマー

Empirical Formula:
$(C_8H_8 \cdot C_6H_9NO)_x$

Definition: Styrene/VP Copolymer is a copolymer prepared from vinylpyrrolidone and styrene monomers.

Information Sources: CTFA D, JCIC, JCLS, JSQI, TSCA

Chemical Class: Synthetic Polymers

Function: Film Former

Reported Product Categories: Permanent Waves; Cleansing Products (Cold Creams, Cleansing Lotions, Liquids and Pads); Hair Conditioners; Shampoos (Non-coloring); Bath Oils, Tablets, and Salts; Hair Bleaches; Hair Preparations (Non-coloring), Misc.; Hair Straighteners; Hair Wave Sets; Tonics, Dressings, and Other Hair Grooming Aids

Technical/Other Names:
1-Ethenyl-2-Pyrrolidinone, Polymer with Ethenylbenzene
Polyvinylpyrrolidone•Styrene Copolymer
Polyvinylpyrrolidone•Styrene Copolymer Emulsion
PVP/Styrene Copolymer
2-Pyrrolidinone, 1-Ethenyl-, Polymer with Ethenylbenzene
Styrene/PVP Copolymer
Vinylpyrrolidone/Styrene Copolymer

Trade Names:
Antara 430 (International Specialty Products)
Polectron 430 (International Specialty Products)

SUBTILISIN

CAS No.	EINECS No.
9014-01-1	232-752-2

Definition: Subtilisin is the enzyme obtained by fermentation of *Bacillus subtilis* or *Bacillus licheniformis*.

Chemical Class: Proteins

Functions: Lytic Agent; Skin-Conditioning Agent - Miscellaneous

Technical/Other Names:
Proteinase, Bacillus Subtilis Alkaline
Subtilopeptidase

Trade Names:
Alcalase 1.5 P Type E (Novozymes)
Bioprase Conc (Nagase)

Trade Name Mixtures:
Depil Enzyme (I.R.A.)
Erase (Degussa Care Specialties)
Erase-AP (Degussa Care Specialties)
Erase-NP (Degussa Care Specialties)
Exfocellia (Coletica SA)
Prozymex HBT (Laboratoires Sero-biologiques)

SUCCINIC ACID

CAS No.	EINECS No.
110-15-6	203-740-4

JPN Translation:
コハク酸

Empirical Formula:
$C_4H_6O_4$

Definition: Succinic Acid is the dicarboxylic acid that conforms to the formula:

$$HOOCCH_2CH_2COOH$$

Information Sources: 21CFR184.1091, CTFA D, FCC, JCLS, JSCI, MI-13(8953), NED, RIFM, TSCA

Chemical Class: Carboxylic Acids

Functions: Fragrance Ingredient; pH Adjuster

Reported Product Category: Hair Straighteners

Technical/Other Names:
Butanedioic Acid
Dihydrofumaric Acid
1,2-Ethanedicarboxylic Acid
Succinic acid (RIFM)

Trade Name Mixture:
Shiconix Liquid AB (N) (Ichimaru Pharcos)

SUCCINOGLYCAN

CAS No.: 73667-50-2

Definition: Succinoglycan is a polysaccharide produced by the fermentation of *Agrobacterium tumefaciens*.

Chemical Class: Carbohydrates

Function: Skin-Conditioning Agent - Miscellaneous

Trade Name:
Rheozan (Rhodia)

Trade Name Mixtures:
Mirasil DME-2 (Rhodia)
Mirasil DME-30 (Rhodia)
Mirasil DME-40A (Rhodia)

SUCCINOYL ATELOCOLLAGEN

Definition: Succinoyl Atelocollagen is the product formed by the reaction of succinic anhydride and Atelocollagen (q.v.).

Chemical Classes: Amides; Protein Derivatives

Function: Skin-Conditioning Agent - Miscellaneous

Trade Name:
Marine Collagen AS (Katakura)

SUCCINOYL SERUM ALBUMIN

JPN Translation:
安息香酸パントテニルエチル

Definition: Succinoyl Serum Albumin is the reaction product of Serum Albumin (q.v.) and Succinic Acid (q.v.).

Information Source: JCLS

Chemical Classes: Amides; Protein Derivatives

Function: Skin-Conditioning Agent - Miscellaneous

SUCRALOSE

CAS No.	EINECS No.
56038-13-2	259-952-2

Empirical Formula:
$C_{12}H_{19}Cl_3O_8$

Definition: Sucralose is the organic compound that conforms to the formula:

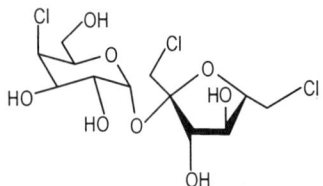

Information Source: MI-13(8965)

Chemical Class: Carbohydrates

Function: Flavoring Agent

Technical/Other Names:
α-D-Galactopyranoside, 1,6-Dichloro-1,6-Dideoxy-β-D-Fructofuranosyl 4-Chloro-4-Deoxy-
1',4,6'-Trichloro-Galacto-Sucrose

Trade Name:
Splenda Brand Sweetner (McNeil Nutritionals)

SUCROSE

CAS No.	EINECS No.
57-50-1	200-334-9

JPN Translation:
スクロース

Empirical Formula:
$C_{12}H_{22}O_{11}$

Definition: Sucrose is the disaccharide that conforms to the formula:

Information Sources: BP, BPC, BRA, 21CFR73.85, 21CFR184.1854, DDR, EP, FIN, HUN, IND, ITA, JAN, JCLS, JP, JSCI, MAR, MEX, MI-13(8966), NF XIX, PN, TSCA, USAN, USD, YUG

Chemical Classes: Carbohydrates; Polyols

Functions: Flavoring Agent; Humectant

Reported Product Categories: Bath Soaps and Detergents; Bath Oils, Tablets, and Salts; Cleansing Products (Cold Creams, Cleansing Lotions, Liquids and Pads); Bath Preparations, Misc.; Eyeliners; Bath Capsules; Face and Neck Preparations (Excluding Shaving Preparations); Foundations; Indoor Tanning Preparations; Moisturizing Preparations; Eye Makeup Preparations, Misc.

Technical/Other Names:
β-D-Fructofuranosyl-α-D-Glucopyranoside
α-D-Glucopyranoside, β-D-Fructofuranosyl-
Purified Sucrose
Saccharose
Sugar

Trade Names:
ESP Demerera Sugar (Earth Supplied Products)
ESP Organic Evaporated Cane Juice Sugar (Earth Supplied Products)
ESP Turbinade Sugar (Earth Supplied Products)
ESP White AA Con Sugar (Earth Supplied Products)
ESP White Special Sanding Sugar (Earth Supplied Products)

Trade Name Mixtures:
Dry Vitamin A Acetate + D3 500/50 (Roche)
Dry Vitamin A Palmitate 500 (Roche)
Elesponge AHA LS 8911 B (Laboratoires Serobiologiques)
Fluxhydran (Laboratoires Serobiologiques)
Hygroplex HHG (CLR)
Kryosome 1708 (Lipoid)
Kryosome 1709 (Lipoid)
Osmhydran (Laboratoires Serobiologiques)
Osmhydran LS 8453 (Laboratoires Serobiologiques)
P.A. Antifroid LS 9224B (Laboratoires Serobiologiques)
ROD Extractive Bifidus B (Youmedica)
Vegeles AHA 8702 (Laboratoires Serobiologiques)
Vegeles AHA LS 8763 (Laboratoires Serobiologiques)
Vitamin A Acetate 500/BASF (BASF Corporation)

SUCROSE ACETATE ISOBUTYRATE

CAS No.	EINECS No.
126-13-6	204-771-6

JPN Translation:
イソ酪酸酢酸スクロース

Empirical Formula:
$C_{40}H_{62}O_{19}$

Definition: Sucrose Acetate Isobutyrate is the mixed ester of sucrose and acetic and isobutyric acids.

Information Sources: 21CFR175.105, JCIC, JCLS, TSCA

Chemical Classes: Carbohydrates; Esters

Function: Plasticizer

Reported Product Categories: Nail Polish and Enamels; Basecoats and Undercoats; Eye Makeup Preparations, Misc.; Manicuring Preparations, Misc.

Technical/Other Names:
α-D-Glucopyranoside, 6-O-Acetyl-1,3,4-Tris-O-(2-Methyl-1-Oxopropyl)-β-D-Fructofuranosyl, 6-Acetate 2,3,4-Tris(2-Methylpropanoate)
Isobutyric Acid, Hexaester with Sucrose Diacetate
Sucrose Acetate Isobutylate

Trade Names:
Eastman SAIB-90 (Eastman Chemical)
Eastman SAIB-100 (Eastman Chemical)
Eastman SAIB-Food Grade (Eastman Chemical)

Trade Name Mixtures:
Eastman SAIB-90EA (Eastman Chemical)
Eastman SAIB-ET10, Food Grade, Kosher (Eastman Chemical)

SUCROSE ACETATE/STEARATE

Definition: Sucrose Acetate/Stearate is the mixed ester of Sucrose (q.v.) with acetic and stearic acids.

Chemical Classes: Carbohydrates; Esters

Function: Skin-Conditioning Agent - Emollient

Reported Product Categories: Eyeliners; Makeup Preparations (Not eye), Misc.

Trade Name:
Sugarwax (Shiseido Company)

SUCROSE BENZOATE

CAS No.	EINECS No.
12738-64-6	235-795-5

JPN Translation:
安息香酸スクロース

Empirical Formula:
$C_{19}H_{26}O_{12}$

Definition: Sucrose Benzoate is the disaccharide ester that conforms generally to the formula:

Information Sources: 21CFR175.105, JCIC, JCLS, TSCA

Chemical Classes: Carbohydrates; Esters

Function: Plasticizer

Reported Product Categories: Nail Polish and Enamels; Basecoats and Undercoats; Manicuring Preparations, Misc.

Technical/Other Names:
β-D-Fructofuranosyl-α-D-Glucopyranoside Benzoate
α-D-Glucopyranoside, β-D-Fructofuranosyl, Benzoate

SUCROSE BENZOATE/SUCROSE ACETATE ISOBUTYRATE/BUTYL BENZYL PHTHALATE COPOLYMER

Definition: Sucrose Benzoate/Sucrose Acetate Isobutyrate/Butyl Benzyl Phthalate Copolymer is the condensation polymer of Sucrose Benzoate (q.v.), Sucrose Acetate Isobutyrate (q.v.) and Butyl Benzyl Phthalate (q.v.) monomers.

Chemical Class: Synthetic Polymers

Functions: Film Former; Suspending Agent - Nonsurfactant

SUCROSE BENZOATE/SUCROSE ACETATE ISOBUTYRATE/BUTYL BENZYL PHTHALATE/METHYL METHACRYLATE COPOLYMER

Definition: Sucrose Benzoate/Sucrose Acetate Isobutyrate/Butyl Benzyl Phthalate/ Methyl Methacrylate Copolymer is the condensation polymer of Sucrose Benzoate (q.v.), Sucrose Acetate Isobutyrate (q.v.) Butyl Benzyl Phthalate (q.v.) and methyl methacrylate monomers.

Chemical Class: Synthetic Polymers

Function: Film Former

SUCROSE BENZOATE/SUCROSE ACETATE ISOBUTYRATE COPOLYMER

Definition: Sucrose Benzoate/Sucrose Acetate Isobutyrate Copolymer is the condensation polymer of Sucrose Benzoate (q.v.) and Sucrose Acetate Isobutyrate (q.v.) monomers.

Chemical Class: Synthetic Polymers

Function: Film Former

Trade Name Mixtures:
Biju BTF-XD (Engelhard Corp.)
Biju Ultra UTF-XD (Engelhard Corp.)

SUCROSE COCOATE

JPN Translation:
ヤシ脂肪酸スクロース

Definition: Sucrose Cocoate is a mixture of sucrose esters of Coconut Acid (q.v.), consisting primarily of the monoesters.

Information Sources: JCIC, JCLS, JSQI

Chemical Classes: Carbohydrates; Esters

Functions: Skin-Conditioning Agent - Emollient; Surfactant - Emulsifying Agent

Trade Names:
Crodesta SL-40 (Croda Chemicals)
Crodesta SL-40 (Croda, Inc.)
Tegosoft LSE 65 K (Degussa Care Specialties)
Tegosoft LSE 65 K Soft (Degussa Care Specialties)

Trade Name Mixture:
Arlatone 2121 (Uniqema Americas)

SUCROSE DILAURATE

CAS No.	EINECS No.
25915-57-5	247-345-5

JPN Translation:
ジラウリン酸スクロース

Empirical Formula:
$C_{36}H_{66}O_{13}$

Definition: Sucrose Dilaurate is the diester of lauric acid and Sucrose (q.v.).

Chemical Classes: Carbohydrates; Esters

Functions: Skin-Conditioning Agent - Emollient; Surfactant - Emulsifying Agent

Trade Name:
Surfhope SE Cosme C-1205 (Mitsubishi-Kagaku)

SUCROSE DISTEARATE

CAS No.	EINECS No.
27195-16-0	248-317-5

JPN Translation:
ジステアリン酸スクロース

Empirical Formula:
$C_{48}H_{90}O_{13}$

Definition: Sucrose Distearate is a mixture of sucrose esters of stearic acid consisting primarily of the diester.

Chemical Classes: Carbohydrates; Esters

Functions: Skin-Conditioning Agent - Emollient; Surfactant - Emulsifying Agent

Technical/Other Name:
α-D-Glucopyranoside, β-D-Fructofuranosyl, Dioctadecanoate

Trade Names:
Crodesta F-10 (Croda Chemicals)
Crodesta F-10 (Croda, Inc.)
Crodesta F20 (Croda Chemicals)
Crodesta F70 (Croda Chemicals)
DUB SE 5S (Stearinerie Dubois Fils)
Surfhope SE Cosme C-1805 (Mitsubishi-Kagaku)
Surfhope SE Cosme C-1807 (Mitsubishi-Kagaku)
Surfhope SE Cosme C-1809 (Mitsubishi-Kagaku)

Trade Name Mixtures:
Crodesta F-110 (Croda Chemicals)
Crodesta F-110 (Croda, Inc.)
Crodesta F-140 (Croda Chemicals)

SUCROSE HEXAERUCATE

Definition: Sucrose Hexaerucate is the hexaester of Sucrose (q.v.) and Erucic Acid (q.v.).

Chemical Classes: Carbohydrates; Esters

Functions: Skin-Conditioning Agent - Emollient; Surfactant - Emulsifying Agent

Trade Name:
Surfhope SE Cosme C-2101 (Mitsubishi-Kagaku)

SUCROSE HEXAOLEATE/ HEXAPALMITATE/HEXASTEARATE

Definition: Sucrose Hexaoleate/ Hexapalmitate/Hexastearate is the hexaester of Sucrose (q.v.) and oleic, palmitic, and stearic acids.

Chemical Classes: Carbohydrates; Esters

Functions: Surfactant - Emulsifying Agent; Surfactant - Suspending Agent

Trade Name:
Surfhope SE Cosme C-POS01 (Mitsubishi-Kagaku)

SUCROSE HEXAPALMITATE

CAS No.	EINECS No.
29130-29-8	249-462-7

Definition: Sucrose Hexapalmitate is a mixture of esters of Palmitic Acid (q.v.) and Sucrose (q.v.), consisting primarily of the hexaester.

Chemical Classes: Carbohydrates; Esters

Functions: Surfactant - Emulsifying Agent; Surfactant - Suspending Agent

Technical/Other Name:
α-D-Glucopyranoside, β-D-Fructofuranosyl, Hexahexadecanoate

Trade Name:
Surfhope SE Cosme C-1601 (Mitsubishi-Kagaku)

SUCROSE LAURATE

CAS Nos.	EINECS Nos.
25339-99-5	246-873-3
37266-93-6	253-432-9

JPN Translation:
ラウリン酸スクロース

Empirical Formula:
$C_{24}H_{44}O_{12}$

Definition: Sucrose Laurate is a mixture of sucrose esters of lauric acid consisting primarily of the monoester.

Chemical Classes: Carbohydrates; Esters

Functions: Skin-Conditioning Agent - Emollient; Surfactant - Emulsifying Agent

Technical/Other Names:
β-D-Fructofuranosyl-α-D-Glucopyranoside, Monododecanoate
α-D-Glucopyranoside, β-D-Fructofuranosyl, Monododecanoate

Trade Names:
Surfhope SE Cosme C-1216 (Mitsubishi-Kagaku)
Surfhope SE Cosme C-1215L (Mitsubishi-Kagaku)

Trade Name Mixture:
Floraglo Lutein 15% with Soltec Coating (Kemin)

SUCROSE MORTIERELLATE

Definition: Sucrose Mortierellate is the ester of Sucrose (q.v.) with the fatty acids derived from Mortierella Oil (q.v.).

Chemical Class: Esters

Functions: Skin-Conditioning Agent - Emollient; Surfactant - Emulsifying Agent

Trade Name:
Crodesta Gamma (Croda Japan)

SUCROSE MYRISTATE

CAS Nos.	EINECS No.
9042-71-1	
27216-47-3	248-340-0

JPN Translation:
ミリスチン酸スクロース

Empirical Formula:
$C_{26}H_{48}O_{12}$

Definition: Sucrose Myristate is the monoester of myristic acid and Sucrose (q.v.).

Chemical Classes: Carbohydrates; Esters

Functions: Skin-Conditioning Agent - Emollient; Surfactant - Emulsifying Agent

Technical/Other Names:
α-D-Glucopyranoside, β-D-Fructofuranosyl, Monotetradecanoate
α-D-Glucopyranoside, β-D-Fructofuranosyl, Tetradecanoate

Trade Name:
Surfhope SE Cosme C-1416 (Mitsubishi-Kagaku)

SUCROSE OCTAACETATE

CAS No.	EINECS No.
126-14-7	204-772-1

Empirical Formula:
$C_{28}H_{38}O_{19}$

Definition: Sucrose Octaacetate is an acetylation product of sucrose. It conforms to the formula:

where R represents the acetic acid radical.

Information Sources: 21CFR172.515, 21CFR175.105, 27CFR21.130, MAR, MI-13 (8967), NF XIX, RIFM, TSCA, USAN

Chemical Classes: Carbohydrates; Esters

Functions: Denaturant; Fragrance Ingredient

Technical/Other Names:
α-D-Glucopyranoside, 1,3,4,6-Tetra-O-Acetyl-β-D-Fructofuranosyl, Tetraacetate
Sucrose octaacetate (RIFM)
1,3,4,6-Tetra-O-Acetyl-β-D-Fructofuranosyl-α-D-Glucopyranoside Tetraacetate

SUCROSE OLEATE

CAS No.	EINECS No.
52683-61-1	258-098-8

JPN Translation:
オレイン酸スクロース

Empirical Formula:
$C_{30}H_{54}O_{12}$

Definition: Sucrose Oleate is the monoester of oleic acid and Sucrose (q.v.).

Chemical Classes: Carbohydrates; Esters

Functions: Skin-Conditioning Agent - Emollient; Surfactant - Emulsifying Agent

Technical/Other Name:
α-D-Glucopyranoside, β-D-Fructofuranosyl, (9Z)-9-Octadecenoate

Trade Names:
Surfhope SE Cosme C-1715 (Mitsubishi-Kagaku)
Surfhope SE Cosme C-1715L (Mitsubishi-Kagaku)

SUCROSE PALMITATE

CAS Nos.	EINECS Nos.
26446-38-8	247-706-7
39300-95-3	254-410-1

JPN Translation:
パルミチン酸スクロース

Empirical Formula:
$C_{28}H_{52}O_{12}$

Definition: Sucrose Palmitate is the monoester of palmitic acid and Sucrose (q.v.).

Chemical Classes: Carbohydrates; Esters

Functions: Skin-Conditioning Agent - Emollient; Surfactant - Emulsifying Agent

Technical/Other Names:
β-D-Fructofuranosyl-α-D-Glucopyranoside Monohexadecanoate
α-D-Glucopyranoside, β-D-Fructofuranosyl, Hexadecanoate
α-D-Glucopyranoside, β-D-Fructofuranosyl, Monohexadecanoate

Trade Names:
DUB SE 15P (Stearinerie Dubois Fils)
Surfhope SE Cosme C-1615 (Mitsubishi-Kagaku)
Surfhope SE Cosme C-1616 (Mitsubishi-Kagaku)

Trade Name Mixtures:
Lyc-O-Mato 2-4%SG NG(LycoRed USA)
Lyc-O-Mato 2-4%SG (LycoRed USA)

SUCROSE PENTAERUCATE

Definition: Sucrose Pentaerucate is the pentaester of Sucrose (q.v.) and Erucic Acid (q.v.).

Chemical Classes: Carbohydrates; Esters

Functions: Skin-Conditioning Agent - Emollient; Surfactant - Emulsifying Agent

Trade Name:
Surfhope SE Cosme C-2102 (Mitsubishi-Kagaku)

SUCROSE POLYBEHENATE

CAS No. 93571-82-5 **EINECS No.** 297-409-1

Definition: Sucrose Polybehenate is a mixture of esters of Behenic Acid (q.v.) and Sucrose (q.v.).

Chemical Classes: Carbohydrates; Esters

Functions: Skin-Conditioning Agent - Emollient; Surfactant - Emulsifying Agent

Trade Name:
SEFA Behenate (Procter & Gamble)

SUCROSE POLYCOTTONSEEDATE

CAS No. 93571-82-5 **EINECS No.** 297-409-1

Definition: Sucrose Polycottonseedate is a mixture of esters of Cottonseed Acid (q.v.) and Sucrose (q.v.).

Chemical Classes: Carbohydrates; Esters

Functions: Skin-Conditioning Agent - Emollient; Surfactant - Emulsifying Agent

Trade Name:
SEFA Cottonate (Procter & Gamble)

Trade Name Mixture:
SEFA Soyate/Cottonate (Procter & Gamble)

SUCROSE POLYLAURATE

JPN Translation:
ポリラウリン酸スクロース

Definition: Sucrose Polylaurate is a mixture of esters of lauric acid and Sucrose (q.v.).

Chemical Classes: Carbohydrates; Esters

Functions: Skin-Conditioning Agent - Emollient; Surfactant - Emulsifying Agent

Trade Name:
Surfhope SE Cosme C-1201 (Mitsubishi-Kagaku)

SUCROSE POLYLINOLEATE

JPN Translation:
ポリリノール酸スクロース

Definition: Sucrose Polylinoleate is a mixture of esters of linoleic acid and Sucrose (q.v.).

Chemical Classes: Carbohydrates; Esters

Functions: Skin-Conditioning Agent - Emollient; Surfactant - Emulsifying Agent

SUCROSE POLYOLEATE

JPN Translation:
ポリオレイン酸スクロース

Definition: Sucrose Polyoleate is a mixture of esters of oleic acid and Sucrose (q.v.).

Chemical Classes: Carbohydrates; Esters

Functions: Skin-Conditioning Agent - Emollient; Surfactant - Emulsifying Agent

Trade Name:
Surfhope SE Cosme C-1701 (Mitsubishi-Kagaku)

SUCROSE POLYPALMATE

JPN Translation:
ポリパーム脂肪酸スクロース

Definition: Sucrose Polypalmate is a mixture of esters of the acids derived from Hydrogenated Palm Oil (q.v.) with Sucrose (q.v.).

Chemical Class: Esters

Functions: Emulsion Stabilizer; Skin-Conditioning Agent - Emollient; Surfactant - Emulsifying Agent

SUCROSE POLYSOYATE

CAS No. 93571-82-5 **EINECS No.** 297-409-1

Definition: Sucrose Polysoyate is a mixture of esters of Soy Acid (q.v.) and Sucrose (q.v.).

Chemical Classes: Carbohydrates; Esters

Functions: Skin-Conditioning Agent - Emollient; Surfactant - Emulsifying Agent

Trade Name:
SEFA Soyate (Procter & Gamble)

Trade Name Mixture:
SEFA Soyate/Cottonate (Procter & Gamble)

SUCROSE POLYSTEARATE

JPN Translation:
ポリステアリン酸スクロース

Definition: Sucrose Polystearate is a mixture of esters of stearic acid and Sucrose (q.v.).

Chemical Classes: Carbohydrates; Esters

Functions: Skin-Conditioning Agent - Emollient; Surfactant - Emulsifying Agent

Trade Names:
Surfhope SE Cosme C-1800 (Mitsubishi-Kagaku)
Surfhope SE Cosme C-1801 (Mitsubishi-Kagaku)
Surfhope SE Cosme C-1802 (Mitsubishi-Kagaku)

SUCROSE RICINOLEATE

CAS No.: 100358-63-2

JPN Translation:
リシノレイン酸スクロース

Empirical Formula:
$C_{30}H_{54}O_{13}$

Definition: Sucrose Ricinoleate is the monoester of ricinoleic acid and Sucrose (q.v.).

Chemical Classes: Carbohydrates; Esters

Functions: Skin-Conditioning Agent - Emollient; Surfactant - Emulsifying Agent

Technical/Other Names:
β-D-Fructofuranosyl-γ-D-Glucopyranoside Monoricinoleate
α-D-Glucopyranoside, β-D-Fructofuranosyl, (R-((Z))-12-Hydroxy-9-Octadecenoate

SUCROSE STEARATE

CAS Nos.	EINECS Nos.
25168-73-4	246-705-9
37318-31-3	253-459-6

JPN Translation:
ステアリン酸スクロース

Empirical Formula:
$C_{30}H_{56}O_{12}$

Definition: Sucrose Stearate is the monoester of stearic acid and Sucrose (q.v.).

Chemical Classes: Carbohydrates; Esters

Functions: Skin-Conditioning Agent - Emollient; Surfactant - Emulsifying Agent

Reported Product Categories: Mascara; Bath Oils, Tablets, and Salts; Bath Capsules; Cleansing Products (Cold Creams, Cleansing Lotions, Liquids and Pads); Moisturizing Preparations

Technical/Other Names:
β-D-Fructofuranosyl-α-D-Glucopyranoside, Monooctadecanoate

α-D-Glucopyranoside, β-D-Fructofuranosyl, Monooctadecanoate

α-D-Glucopyrpanoside, β-D-Fructofuranosyl, Octadecanoate

Trade Names:
Crodesta F-160 (Croda Chemicals)
Crodesta F-160 (Croda, Inc.)
DUB SE 11S (Stearinerie Dubois Fils)
DUB SE 16S (Stearinerie Dubois Fils)
Surfhope SE Cosme C-1811 (Mitsubishi-Kagaku)
Surfhope SE Cosme C-1815 (Mitsubishi-Kagaku)
Surfhope SE Cosme C-1816 (Mitsubishi-Kagaku)
Tegosoft PSE 141 G (Degussa Care Specialties)

Trade Name Mixtures:
Crodesta F-110 (Croda Chemicals)
Crodesta F-110 (Croda, Inc.)
Crodesta F-140 (Croda Chemicals)

SUCROSE TETRAISOSTEARATE

Definition: Sucrose Tetraisostearate is a mixture of esters of Isostearic Acid (q.v.) and Sucrose (q.v.), consisting primarily of the tetraester.

Chemical Classes: Carbohydrates; Esters

Functions: Skin-Conditioning Agent - Emollient; Surfactant - Emulsifying Agent

Technical/Other Name:
β-D-Fructofuranosyl-α-D-Glucopyranoside, Tetraisooctadecanoate

Trade Name:
Crodesta 4I-S (Croda Japan)

SUCROSE TETRASTEARATE TRIACETATE

JPN Translation:
酢酸ステアリン酸スクロース

Definition: Sucrose Tetrastearate Triacetate is a mixture of esters of stearic acid, acetic acid and Sucrose (q.v.).

Chemical Classes: Carbohydrates; Esters

Function: Skin-Conditioning Agent - Emollient

Reported Product Categories: Eyebrow Pencils; Eyeliners; Mascara; Rouges; Makeup Fixatives

SUCROSE TRIBEHENATE

CAS No.: 84798-44-7

JPN Translation:
トリベヘン酸スクロース

Empirical Formula:
$C_{78}H_{148}O_{14}$

Definition: Sucrose Tribehenate is the triester of behenic acid and Sucrose (q.v.).

Chemical Classes: Carbohydrates; Esters

Function: Skin-Conditioning Agent - Emollient

Technical/Other Name:
α-D-Glucopyranoside, β-D-Fructofuranosyl, Tridocosanoate

Trade Name:
Surfhope SE Cosme C-2203 (Mitsubishi-Kagaku)

SUCROSE TRISTEARATE

CAS No.	EINECS No.
27923-63-3	248-731-6

JPN Translation:
トリステアリン酸スクロース

Empirical Formula:
$C_{66}H_{124}O_{14}$

Definition: Sucrose Tristearate is the triester of stearic acid and Sucrose (q.v.).

Chemical Classes: Carbohydrates; Esters

Function: Skin-Conditioning Agent - Emollient

Technical/Other Names:
β-D-Fructofuranosyl-α-D-Glucopyranoside Trioctadecanoate
α-D-Glucopyranoside, β-D-Fructofuranosyl, Trioctadecanoate

Trade Names:
DUB SE 3S (Stearinerie Dubois Fils)
Surfhope SE Cosme C-1803 (Mitsubishi-Kagaku)

SUIKAZURA EKISU

JPN Translation:
スイカズラエキス

Definition: Suikazura Ekisu is an extract of the flowers, leaves, or stems of the honeysuckle, *Lonicera japonica* or other related species of the family, *Caprifoliaceae*. See *"Regulatory and Ingredient Use Information,"* regarding use of Japan Trivial names in Volume 1, Introduction, Part A.

Chemical Class: Biological Products

Function: Skin-Conditioning Agent - Miscellaneous

Technical/Other Names:
Lonicera Japonica (Honeysuckle) Flower Extract (U.S.)
Lonicera Japonica (Honeysuckle) Leaf Extract (U.S.)

SULFATED CASTOR OIL

CAS Nos.	EINECS No.
8002-33-3	232-306-7
101316-48-7	

JPN Translation:
硫酸化ヒマシ油

Definition: Sulfated Castor Oil is the oil that consists primarily of the sodium salt of the sulfated triglyceride of castor oil.

Information Sources: 21CFR175.105, 21CFR176.170, 21CFR176.200, 21CFR177.1200, CTFA D, JCIC, JCLS, JSQI, MAR, MI-13(9891), TSCA

Chemical Classes: Glyceryl Esters and Derivatives; Sulfuric Acid Esters

Function: Surfactant - Cleansing Agent

Reported Product Categories: Hair Dyes and Colors (All Types Requiring Caution Statements and Patch Tests); Hair Conditioners; Shampoos (Non-coloring)

Technical/Other Names:
Castor Oil, Sulfated
Castor Oil, Sulfonated
Sulfonated Castor Oil
Turkey-Red Oil

Trade Names:
AEC Castor Oil Sulphated (A & E Connock)
Norfox Sulfated Castor Oil #75 (Norman, Fox & Co.)
Standapol SCO (Cognis Care Chemicals/NJ)
Standapol SCO (Cognis Care Chemicals/PA)
Tuerkischrotoel 100% (Zschimmer & Schwarz)
Unipol SCO (Universal Preserv-A-Chem)

Trade Name Mixtures:
Amisol HS-2 (Lucas Meyer GmbH)
Amisol HS-3 (Lucas Meyer GmbH)
Amisol HS3US (Lucas Meyer)
Cetasal (Gattefosse s.a.)

SULFATED GLYCERYL OLEATE

Definition: Sulfated Glyceryl Oleate is the product of the sulfation of Glyceryl Oleate (q.v.).

Chemical Classes: Glyceryl Esters and Derivatives; Sulfuric Acid Esters

Function: Surfactant - Cleansing Agent

SULFATED OLIVE OIL

Definition: Sulfated Olive Oil is the product obtained by the sulfation of Olea Europaea (Olive) Oil (q.v.).

Chemical Classes: Glyceryl Esters and Derivatives; Sulfuric Acid Esters

Function: Surfactant - Cleansing Agent

Trade Name:
Sulfolivat (Weleda)

SULFATED PEANUT OIL

CAS No. 73138-79-1   **EINECS No.** 277-298-6

Definition: Sulfated Peanut Oil is the product obtained by the sulfation of Arachis Hypogaea (Peanut) Oil (q.v.).

Information Source: 21CFR175.105

Chemical Classes: Glyceryl Esters and Derivatives; Sulfuric Acid Esters

Function: Surfactant - Cleansing Agent

Technical/Other Names:
Oils, Peanut, Sulfated
Peanut Oil, Sulfated

SULFUR

CAS No. 7704-34-9   **EINECS No.** 231-722-6

Empirical Formula:
S

Definition: Sulfur is a naturally occurring element. In the United States, Sulfur may be used as an active ingredient in OTC drug products. When used as an active drug ingredient, the established name is *Sulfur. See "Regulatory and Ingredient Use Information," regarding the labeling names for U.S. OTC Drug Ingredients in Volume 1, Introduction, Part A.*

Information Sources: ARG, AUS, BP, BPC, 21CFR73.85, 21CFR101.100, 21CFR130.9, 21CFR168.111, 21CFR168.120, 21CFR175.105, 21CFR176.210, 21CFR177.1210, 21CFR177.2210, 21CFR177.2490, 21CFR177.2600, 21CFR177.2800, 21CFR178.1010, 21CFR182.3862, 21CFR184.1095, 21CFR310.201, 21CFR310.503, 21CFR310.545, 21CFR333.310, 21CFR358.710, 21CFR436.515, 21CFR436.545,

21CFR442.40, 21CFR444.42, 21CFR444.42a, 21CFR452.75, 21CFR533.310, 21CFR582.1095, 21CFR582.3862, 21CFR1310.02, CZE, DA, DDR, EGY, FIN, HP, HUN, IND, ITA, JAN, MAR, MI-13(9059), MI-13(9067), NFJ, OTC-I-AK, OTC-I-DP, PN, ROM, TSCA, USAN, USD, USP XXIV, USSR, YUG

Chemical Classes: Elements; Inorganics

Functions: Antiacne Agent; Antidandruff Agent; Hair Conditioning Agent; Skin-Conditioning Agent - Miscellaneous

Reported Product Categories: Hair Conditioners; Hair Coloring Preparations, Misc.; Shampoos (Non-coloring); Tonics, Dressings, and Other Hair Grooming Aids; Paste Masks (Mud Packs); Face Powders; Hair Preparations (Non-coloring), Misc.

Trade Name:
Royal RM98 D Sulfur (Royal)

Trade Name Mixtures:
Azufre Soluble (Fabriquimica)
Biosil Basics HMC - Hair Moisture Complex (Biosil Technologies, Inc.)
Biosil Basics HMC-1 Hair Moisture Complex (Biosil Technologies, Inc.)
Biosil Basics HMV- Hair Moisture Complex (Biosil Technologies, Inc.)
Biosil Basics HMW - Hair Moisture Complex (Biosil Technologies, Inc.)
Biosulphur Fluid (CLR)
Bio-Sulphur Liquid (Crodarom)
Concentrate for Dandruff (Crodarom)
Granule AA (Ichimaru Pharcos)
Neo-Haircomplex Special (Crodarom)
Soufre Soluble M 2996 (Gattefosse s.a.)

SULFURIC ACID

CAS No. 7664-93-9   **EINECS No.** 231-639-5

JPN Translation:
硫酸

Empirical Formula:
H_2O_4S

Definition: Sulfuric Acid is the inorganic acid that conforms to the formula:

$$H_2SO_4$$

Information Sources: ARG, AUS, BPC, BRA, 21CFR172.560, 21CFR172.892, 21CFR173.385, 21CFR178.1010, 21CFR184.1095, 21CFR1310.02, 21CFR1310.04, 21CFR1310.08, DDR, EGY, FCC, FI, HP, IND, JCLS, MAR, MI-13 (9064), NF XIX, PF, PN, POR, TSCA, USAN, YUG

Chemical Class: Inorganic Acids

Function: pH Adjuster

Reported Product Categories: Hair Dyes and Colors (All Types Requiring Caution Statements and Patch Tests); Permanent Waves

SULFURIZED HYDROLYZED CORN PROTEIN

Definition: Sulfurized Hydrolyzed Corn Protein is the product of the reaction of sulfur with Hydrolyzed Corn Protein (q.v.).

Chemical Class: Protein Derivatives

Function: Hair Conditioning Agent

SULFURIZED HYDROLYZED ZEIN

Definition: Sulfurized Hydrolyzed Zein is the product of the reaction of sulfur with Hydrolyzed Zein (q.v.).

Chemical Class: Protein Derivatives

Function: Skin-Conditioning Agent - Miscellaneous

SULFURIZED JOJOBA OIL

Definition: Sulfurized Jojoba Oil is the reaction product obtained from the chemical addition of elemental sulfur to Simmondsia Chinensis (Jojoba) Oil (q.v.).

Chemical Classes: Thio Compounds; Waxes

Function: Skin-Conditioning Agent - Miscellaneous

SULFURIZED TEA-RICINOLEATE

Definition: Sulfurized TEA-Ricinoleate is the product of the reaction of sulfur with the triethanolamine salt of Ricinoleic Acid (q.v.).

Chemical Classes: Soaps; Thio Compounds

Function: Hair Conditioning Agent

Trade Name Mixture:
Sulfoconcentrol (2/380011) (Symrise)

SUNFLOWER SEED ACID

CAS No. 84625-38-7   **EINECS No.** 283-413-0

Definition: Sunflower Seed Acid is a mixture of fatty acids derived from Helianthus Annuus (Sunflower) Seed Oil (q.v.).

Chemical Class: Fatty Acids

Function: Surfactant - Cleansing Agent

Technical/Other Names:
Acids, Sunflower Seed
Helianthus Annuus (Sunflower) Seed Acid

Trade Names:
Extra Oleic-99 (NOF)
Extra Olein 99 (NOF)
Extra OS-85 (NOF)
PRIFAC 8960 (Uniqema Europe)

SUNFLOWERSEEDAMIDOPROPYL DIMETHYLAMINE

Definition: Sunflowerseedamidopropyl Dimethylamine is the amidoamine that conforms to the formula:

$$RC\overset{O}{\underset{\|}{\,}}-NH(CH_2)_3N\overset{CH_3}{\underset{CH_3}{\diagdown}}$$

where RCO- represents the fatty acids derived from sunflowerseed oil.

Chemical Class: Amines

Function: Antistatic Agent

Trade Name:
MACKINE 1201 (McIntyre)

SUNFLOWERSEEDAMIDOPROPYL DIMETHYLAMINE LACTATE

Definition: Sunflowerseedamidopropyl Dimethylamine Lactate is the lactic acid salt of Sunflowerseedamidopropyl Dimethylamine (q.v.). It conforms generally to the formula:

$$RC\overset{O}{\underset{\|}{\,}}-NH(CH_2)_3N\overset{CH_3}{\underset{CH_3}{|}} \cdot HOOCCHCH_3\underset{OH}{|}$$

where RCO- represents the fatty acids derived sunflower seed oil.

Chemical Class: Amines

Function: Hair Conditioning Agent

Technical/Other Name:
Amides, Sunflower Seed Oil,
N-[2-(Dimethylamino)Propyl], Lactates

Trade Name:
MACKALENE 1216 (McIntyre)

SUNFLOWERSEEDAMIDOPROPYL DIMETHYLAMINE MALATE

Definition: Sunflowerseedamidopropyl Dimethylamine Malate is the malic acid salt of Sunflowerseedamidopropyl Dimethylamine (q.v.). It conforms generally to the formula:

$$RC\overset{O}{\underset{\|}{\,}}-NH(CH_2)_3N\overset{CH_3}{\underset{CH_3}{|}} \cdot HOOCCHCH_2COOH\underset{OH}{|}$$

where RCO- represents the fatty acids derived from sunflower seed oil.

Chemical Class: Amines

Function: Hair Conditioning Agent

Technical/Other Name:
Amides, Sunflower Seed Oil,
N-[2-(Dimethylamino)Propyl]-, Malate

Trade Name:
MACKALENE 1218 (McIntyre)

SUNFLOWERSEEDAMIDOPROPYL ETHYLDIMONIUM ETHOSULFATE

Definition: Sunflowerseedamidopropyl Ethyldimonium Ethosulfate is the quaternary ammonium salt that conforms generally to the formula :

$$\left[RC\overset{O}{\underset{\|}{\,}}-NH(CH_2)_3-\overset{CH_3}{\underset{CH_3}{\overset{|}{N}}}-CH_2CH_3 \right]^+ \; CH_3CH_2\underset{OSO_3^-}{}$$

where RCO- represents the fatty acids derived from sunflower seed oil.

Chemical Class: Quaternary Ammonium Compounds

Functions: Antistatic Agent; Hair Conditioning Agent

Trade Name Mixture:
MACKERNIUM SFES (McIntyre)

SUNFLOWERSEEDAMIDOPROPYL HYDROXYETHYLDIMONIUM CHLORIDE

Definition: Sunflowerseedamidopropyl Hydroxyethyldimonium Chloride is the quaternary ammonium salt that conforms generally to the formula:

$$\left[RC\overset{O}{\underset{\|}{\,}}-NH(CH_2)_3-\overset{CH_3}{\underset{CH_3}{\overset{|}{N}}}-CH_2CH_2OH \right]^+ \; Cl^-$$

where RCO- represents the fatty acids derived from sunflower seed oil.

Chemical Class: Quaternary Ammonium Compounds

Functions: Antistatic Agent; Hair Conditioning Agent; Surfactant - Cleansing Agent

Trade Name:
Conditioneze SFQ (International Specialty Products)

SUNFLOWERSEEDAMIDOPROPYL MORPHOLINE LACTATE

Definition: Sunflowerseedamidopropyl Morpholine Lactate is the lactic acid salt of sunflowerseedamidopropyl morpholine.

Chemical Classes: Amines; Heterocyclic Compounds

Functions: Antistatic Agent; Hair Conditioning Agent

Trade Name:
Mackalene 1226 (McIntyre)

SUNFLOWERSEEDAMIDOPROPYL PG-DIMONIUM CHLORIDE PHOSPHATE

Definition: Sunflowerseedamidopropyl PG-Dimonium Chloride Phosphate is the quaternary ammonium salt that conforms generally to the formula:

$$\left[RC\overset{O}{\underset{\|}{\,}}-NH(CH_2)_3-\overset{CH_3}{\underset{CH_3}{\overset{|}{N}}}-CH_2CHCH_2O\underset{OH}{|}-\overset{O}{\underset{\|}{P}} \cdot 3Cl^- \right]^+_3$$

where RCO- represents the fatty acids derived from sunflower seed oil.

Chemical Class: Quaternary Ammonium Compounds

Functions: Antistatic Agent; Hair Conditioning Agent

Trade Name:
Colalipid SUN (Colonial Chemical Inc)

SUNFLOWER SEED OIL GLYCERIDE

Definition: Sunflower Seed Oil Glyceride is the monoglyceride derived from Helianthus Annuus (Sunflower) Seed Oil (q.v.).

Chemical Class: Glyceryl Esters and Derivatives

Functions: Skin-Conditioning Agent - Emollient; Surfactant - Emulsifying Agent

Reported Product Category: Shaving Preparations, Misc.

Technical/Other Name:
Glycerides, Sunflower Seed Mono-

SUNFLOWER SEED OIL GLYCERIDES

Definition: Sunflower Seed Oil Glycerides is a mixture of mono, di and triglycerides derived from Helianthus Annuus (Sunflower) Seed Oil (q.v.).

Chemical Class: Glyceryl Esters and Derivatives

Function: Skin-Conditioning Agent - Emollient

Technical/Other Name:
Glycerides, Sunflower Seed Mono-, Di- and Tri-

SUNFLOWER SEED OIL PEG-8 ESTERS

Definition: Sunflower Seed Oil PEG-8 Esters is the product obtained by the transesterification of Helianthus Annuus (Sunflower) Seed Oil (q.v.) and PEG-8 (q.v.).

Chemical Class: Glyceryl Esters and Derivatives

Functions: Skin-Conditioning Agent - Emollient; Surfactant - Emulsifying Agent

Trade Name:
Viatenza Sunflower PE8 (Aldivia)

SUNSET YELLOW

CAS No.	EINECS No.
2783-94-0	220-491-7

Empirical Formula:
$C_{16}H_{12}N_2O_7S_2 \cdot 2Na$

Definition: Sunset Yellow is classed chemically as a monoazo color. It conforms to the formula:

See "Regulatory and Ingredient Use Information," for Colorants in Volume 1, Introduction, Part A. To identify the certified colorant for labeling purposes in the US, the INCI Name Yellow 6 must be used. To identify the colorant allowed for use in the
European Union (EU), the INCI Name CI 15985 must be used, except for hair dye products. The INCI Name for batches of this colorant that have not been certified is Sunset Yellow. To identify the colorant allowed for use in Japan, the INCI name Ki5 must be used.

Information Sources: CI 15985, M3, MI-13(9091), TSCA

Chemical Class: Color Additives - Miscellaneous

Function: Colorant

Reported Product Categories: Colognes and Toilet Waters; Shampoos (Non-coloring); Hair Dyes and Colors (All Types Requiring Caution Statements and Patch Tests); Hair Conditioners; Bath Preparations, Misc.; Moisturizing Preparations; Body and Hand Preparations (Excluding Shaving Preparations); Bath Oils, Tablets, and Salts; Cleansing Products (Cold Creams, Cleansing Lotions, Liquids and Pads); Bath Soaps and Detergents; Skin Care Preparations, Misc.; Aftershave Lotions; Baby Shampoos; Tonics, Dressings, and Other Hair Grooming Aids; Perfumes; Hair Preparations (Non-coloring), Misc.; Bath Capsules; Face and Neck Preparations (Excluding Shaving Preparations); Fragrance Preparations, Misc.; Bubble Baths; Deodorants (Underarm); Hair Rinses (Coloring); Lipsticks; Night Skin Care Preparations; Paste Masks (Mud Packs); Personal Cleanliness Products, Misc.; Skin Fresheners; Shaving Preparations, Misc.; Hair Wave Sets; Mouthwashes and Breath Fresheners (Liquids and Sprays); Nail Creams and Lotions; Shaving Cream (Aerosol, Brushless and Lather)

Technical/Other Names:
CI 15985
Food Yellow 3, Disodium Salt
6-Hydroxy-5-[(4-Sulfophenyl)Azo]-2-Naphthalenesulfonic Acid, Disodium Salt
Japan Yellow 5
2-Naphthalenesulfonic Acid, 6-Hydroxy-5-[(4-Sulfophenyl)Azo]-, Disodium Salt
Pigment Yellow 104
Sunset Yellow FCF

Trade Name:
Sicovit Yellow Orange 85 E 110 (BASF)

SUNSET YELLOW ALUMINUM LAKE

Definition: Sunset Yellow Aluminum Lake is an insoluble pigment composed of the aluminum salt of Sunset Yellow (q.v.) extended on an appropriate substrate. *See "Regulatory and Ingredient Use Information," for Colorants in Volume 1,*
Introduction, Part A. To identify the certified colorant for labeling purposes in the US, the INCI Name Yellow 6 Lake must be used. The INCI Name for batches of this colorant that have not been certified is Sunset Yellow Aluminum Lake. To identify the colorant allowed for use in the European Union (EU), the INCI Name CI 15985 must be used, except for hair dye products. To identify the colorant allowed for use in Japan, the INCI name Ki5 must be used.

Information Source: CI 15985:1

Chemical Class: Color Additives - Miscellaneous

Function: Colorant

Technical/Other Name:
CI 15985:1

SUPEROXIDE DISMUTASE

CAS No.	EINECS No.
9054-89-1	232-943-0

Definition: Superoxide Dismutase is a mixture of metaloenzymes found in aerobic cells.

Information Source: MI-13(9092)

Chemical Class: Proteins

Functions: Reducing Agent; Skin-Conditioning Agent - Miscellaneous

Reported Product Categories: Moisturizing Preparations; Skin Care Preparations, Misc.

Technical/Other Name:
Dismutase, Superoxide

Trade Names:
ACB SOD Powder (Active Concepts)
Biocell S.O.D. (Arch Personal Care Products)
Dismutin BT (Pentapharm/Centerchem)
Sod C.a.m. (Silab)

Trade Name Mixtures:
AC Colorplex (Active Concepts)
AC SOD Liposome (Active Concepts)
Brookosome SOD (Arch Personal Care Products)
Chronosphere SOD (Arch Personal Care Products)
Enzyami 1 (Alban Muller)
Enzyami 5 (Alban Muller)
Plantazyme Phase I (Coletica SA)

SUTILAINS

CAS Nos.	EINECS Nos.
9001-92-7	232-642-4
12211-28-8	235-390-3

Definition: Sutilains are the enzymes obtained from *Bacillus subtilis*.

Information Source: USP XXIII

Chemical Class: Proteins

Function: Lytic Agent

Technical/Other Names:
Proteinase
Proteinase, Bacillus Subtilis, Sutilain

Trade Name Mixtures:
Keratoline CL (Sederma)
Melaclear 2 (Sederma)

SWEET ALMOND OIL PEG-8 ESTERS

Definition: Sweet Almond Oil PEG-8 Esters is the product obtained by the transesterification of Prunus Amygdalus (Sweet Almond) Oil (q.v.) and PEG-6 (q.v.).

Chemical Class: Glyceryl Esters and Derivatives

Functions: Skin-Conditioning Agent - Emollient; Surfactant - Emulsifying Agent

Trade Name:
Viatenza Almond PE8 (Aldivia)

SWEET ALMOND OIL POLYGLYCERYL-6 ESTERS

Definition: Sweet Almond Oil Polyglyceryl-6 Esters is the product obtained by the transesterification of Prunus Amygdalus Dulcis (Sweet Almond) Oil (q.v.) and Polyglycerin-6 (q.v.).

Chemical Class: Glyceryl Esters and Derivatives

Functions: Skin-Conditioning Agent - Emollient; Surfactant - Emulsifying Agent

Trade Name:
Viatenza Almond PO6 (Aldivia)

SWERTIA CHIRATA EXTRACT

CAS No. 97766-44-4

EINECS No. 307-906-8

Definition: Swertia Chirata Extract is an extract of the flowers, leaves and stems of the swertia, *Swertia chirata. See "Regulatory and Ingredient Use Information," regarding the labeling names for botanical ingredients in Volume 1, Introduction, Part A.*

Information Sources: 21CFR172.510, JCIC, JCLS, JSQI

Chemical Class: Biological Products

Function: Not Reported

Technical/Other Names:
Extract of Swertia
Extract of Swertia Chirata
Swertia Extract
Swertia Herb Extract

Trade Name Mixture:
East Indian Balmony Extract HS 2748 G (Grau)

SWERTIA JAPONICA EXTRACT

CAS No. 94167-11-0

EINECS No. 303-410-0

JPN Translation:
センブリエキス

Definition: Swertia Japonica Extract is an extract of the whole plant of *Swertia japonica. See "Regulatory and Ingredient Use Information," regarding the labeling names for botanical ingredients in Volume 1, Introduction, Part A.*

Chemical Class: Biological Products

Function: Not Reported

Reported Product Category: Tonics, Dressings, and Other Hair Grooming Aids

Technical/Other Name:
Extract of Swertia Japonica

Trade Name:
Swertia Herb Extract SQ (Maruzen Pharmaceuticals Co., Ltd.)

Trade Name Mixtures:
Swertia Herb Extract (Maruzen Pharmaceuticals Co., Ltd.)
Swertia Herb Extract BG (Maruzen Pharmaceuticals Co., Ltd.)
Swertia Herb Liquid (Maruzen Pharmaceuticals Co., Ltd.)
Swertia Herb Liquid ET (Maruzen Pharmaceuticals Co., Ltd.)
Swertia Herb Liquid S (Maruzen Pharmaceuticals Co., Ltd.)
Swertia Herb Liquid SS (Maruzen Pharmaceuticals Co., Ltd.)
Swertianin (Ichimaru Pharcos)
Swertianin P (Ichimaru Pharcos)

SWERTIA JAPONICA POWDER

JPN Translation:
加水分解ゼラチン

Definition: is the powder derived from *Swertia Japonica.*

Information Source: JCLS

Chemical Class: Biological Products

Function: Not Reported

SWERTIA PSEUDOCHINENSIS EXTRACT

JPN Translation:
ムラサキセンブリエキス

Definition: Swertia Pseudochinensis Extract is an extract of *Swertia pseudochinensis. See "Regulatory and Ingredient Use Information," regarding the labeling names for botanical ingredients in Volume 1, Introduction, Part A.*

Chemical Class: Biological Products

Function: Skin-Conditioning Agent - Miscellaneous

Technical/Other Name:
Extract of Swertia Pseudochinensis

Trade Name Mixture:
Purple Swertia Herb Extract (Maruzen Pharmaceuticals Co., Ltd.)

SWIFTLET NEST EXTRACT

Definition: Swiftlet Nest Extract is an extract of the nest of the swiftlet, *Collocalia esculenta, Collocalia vestira,* or *Collocalia inexpectra.*

Chemical Class: Biological Products

Function: Skin-Conditioning Agent - Miscellaneous

Technical/Other Name:
Extract of Bird Nest

Trade Name:
Swiftlet's Nest Extract (Iwase Cosfa)

SYMPHYTUM OFFICINALE

Definition: *See "Regulatory and Ingredient Use Information," regarding EU labeling names for botanical ingredients in Volume 1, Introduction, Part A.*

Chemical Class: Biological Products

Technical/Other Names:
Symphytum Officinale Extract (U.S.)
Symphytum Officinale Leaf Extract (U.S.)
Symphytum Officinale Leaf Powder (U.S.)

SYMPHYTUM OFFICINALE EXTRACT

CAS No. 84696-05-9

EINECS No. 283-625-3

The inclusion of any compound in the *Dictionary and Handbook* does not indicate that use of that substance as a cosmetic ingredient complies with the laws and regulations governing such use in the United States or any other country.

Definition: Symphytum Officinale Extract is an extract of the rhizomes and roots of the comfrey, *Symphytum officinale. See "Regulatory and Ingredient Use Information," regarding the labeling names for botanical ingredients in Volume 1, Introduction, Part A.*

Information Sources: HP, JSQI

Chemical Class: Biological Products

Function: Skin-Conditioning Agent - Miscellaneous

Reported Product Categories: Shampoos (Non-coloring); Tonics, Dressings, and Other Hair Grooming Aids; Skin Care Preparations, Misc.; Bath Preparations, Misc.; Body and Hand Preparations (Excluding Shaving Preparations); Eye Makeup Preparations, Misc.; Bath Soaps and Detergents; Moisturizing Preparations; Personal Cleanliness Products, Misc.; Cleansing Products (Cold Creams, Cleansing Lotions, Liquids and Pads); Foundations; Hair Conditioners; Skin Fresheners

Technical/Other Names:
Comfrey Extract
Comfrey (Symphytum Officinale) Extract
Extract of Comfrey
Extract of Symphytum Officinale
Symphytum Officinale (EU)
Symphytum Officinalis Extract

Trade Name Mixtures:
Actiphyte of Comfrey Root BG50 (Active Organics)
Actiphyte of Comfrey Root GL50 (Active Organics)
Actiphyte of Comfrey Root Lipo S (Active Organics)
Actiphyte of Comfrey Root PG50 (Active Organics)
BBC Moisture Trol (Bio-Botanica)
Bio-Chelated Derma-Plex I (Bio-Botanica)
Bio-Chelated Derma Plex II (Bio-Botanica)
Bio-Chelated Nutra Plant Complex I (Bio-Botanica)
Bio-Chelated Nutra Plant Complex II (Bio-Botanica)
Bio-Chelated Sauna-Derm I (Bio-Botanica)
Bio-Chelated Sauna-Derm II (Bio-Botanica)
Bio-Dandra Plex (Bio-Botanica)
Comfrey Extract (Maruzen Pharmaceuticals Co., Ltd.)
Comfrey Extract BG (Maruzen Pharmaceuticals Co., Ltd.)
Comfrey Extract LA (Maruzen Pharmaceuticals Co., Ltd.)
Comfrey HS (Alban Muller)
Comfrey Phytexcell (Crodarom)
Comfrey Root Extract HS 2446 G (Grau)
Comfrey Root Extract Liquid (Crodarom)
Glycolysat Complex 85-5-10 (CEP (Solabia))
Herbasol Complex GU-61-A (Cosmetochem)
Herbasol Complex GU-61-Standard (Cosmetochem)
Moisturizing Phytoamine Biocomplex (Alban Muller)
Phytelene of Comfrey EG 032 liquid (Indena SA)
Phytogreen 55 of Comfrey EXH 617 Liquid (Phytochim)
Prodhy Extract Consoude (Prod'Hyg)
Rhizodermin (Laboratoires Serobiologiques)
Sederma Comfrey (Sederma)
Symphytum Officinale Extract ies (IES LABO)
Vegebios of Comfrey (CEP (Solabia))
2300 Vege-Plex Body Complex (Vege-Tech)
2303 Vege-Plex Body Complex (Vege-Tech)
2310 Vege-Plex Body Complex (Vege-Tech)
2320 Vege-Plex Body Complex (Vege-Tech)
2325 Vege-Plex Body Complex (Vege-Tech)
2100 Vege-Plex Hair Complex (Vege-Tech)
2225 Vege-Plex Hair Complex (Vege-Tech)
2280 Vege-Plex Hair Complex (Vege-Tech)
2110 Vege-Plex Hair Complex Shampoo (Vege-Tech)
2120 Vege-Plex Hair Complex Shampoo (Vege-Tech)
2410 Vege-Plex Skin Complex (Vege-Tech)
2500 Vege-Plex Skin Complex (Vege-Tech)
2525 Vege-Plex Skin Complex (Vege-Tech)
2540 Vege-Plex Skin Complex (Vege-Tech)
2550 Vege-Plex Skin Complex (Vege-Tech)
2560 Vege-Plex Skin Complex (Vege-Tech)
2600 Vege-Plex Skin Complex (Vege-Tech)
VT-0014.000W Extract of Comfrey (Vege-Tech)

SYMPHYTUM OFFICINALE LEAF EXTRACT

CAS No.	EINECS No.
84696-05-9	283-625-3

JPN Translation:
コンフリーエキス

Definition: Symphytum Officinale Leaf Extract is an extract of the leaves of the comfrey, *Symphytum officinale. See "Regulatory and Ingredient Use Information," regarding the labeling names for botanical ingredients in Volume 1, Introduction, Part A.*

Information Sources: JCIC, JCLS

Chemical Class: Biological Products

Function: Not Reported

Reported Product Categories: Shampoos (Non-coloring); Tonics, Dressings, and Other Hair Grooming Aids; Skin Care Preparations, Misc.; Bath Preparations, Misc.; Body and Hand Preparations (Excluding Shaving Preparations); Eye Makeup Preparations, Misc.; Moisturizing Preparations; Foundations; Hair Conditioners; Skin Fresheners

Technical/Other Names:
Comfrey Extract
Comfrey Leaf Extract
Comfrey (Symphytum Officinale) Leaf Extract
Extract of Comfrey Leaf
Extract of Symphytum Officinale Leaf
Symphytum Officinale (EU)
Symphytum Officinalis Leaf Extract

Trade Name:
Comfrey Leaf Extract (Bio-Botanica)

Trade Name Mixtures:
Actiphyte of Comfrey Leaf BG50 (Active Organics)
Actiphyte of Comfrey Leaf GL50 (Active Organics)
Actiphyte of Comfrey Leaf Lipo S (Active Organics)
Actiphyte of Comfrey Leaf PG50 (Active Organics)
Activated Botanicals AB 110 (Norjin)
Comfrey Leaf Extract (Maruzen Pharmaceuticals Co., Ltd.)
Comfrey Leaf Extract BG (Maruzen Pharmaceuticals Co., Ltd.)
Comfrey Leaf Extract LA (Maruzen Pharmaceuticals Co., Ltd.)
Comfrey Liquid B (Ichimaru Pharcos)
Comfrey Liquid CE (Ichimaru Pharcos)
Comfrey Tincture (Rahn)
Herbasec Comfrey (Cosmetochem) (Cosmetochem International Ltd.)
Herbasol-Extract Comfrey (Cosmetochem)
Phytoblend TIPS (Ichimaru Pharcos)
VT-127 Extract of Comfrey Leaves (Vege-Tech)

SYMPHYTUM OFFICINALE LEAF POWDER

JPN Translation:
コンフリー

Definition: Symphytum Officinale Leaf Powder is a powder of finely ground leaves from the comfrey, *Symphytum officinale. See "Regulatory and Ingredient Use Information," regarding the labeling names for botanical ingredients in Volume 1, Introduction, Part A.*

Information Sources: JCIC, JCLS

Chemical Class: Biological Products

Function: Abrasive

Technical/Other Names:
Comfrey Leaf Powder
Comfrey (Symphytum Officinale) Leaf Powder
Powder, Comfrey Leaves
Powder, Symphytum Officinale
Symphytum Officinale (EU)
Symphytum Officinale Powder

Trade Name Mixture:
DryLeaf CGS (Shiseido Company)

SYNECHOCOCCUS ELONGATUS/ALGAE FERMENT

Definition: Synechococcus Elongatus/Algae Ferment is the product of the fermentation of algae by the organism *Synechococcus elongatus.*

Chemical Class: Biological Products

Function: Skin-Conditioning Agent - Miscellaneous

Trade Name:
Synechococcus 400M AS (Eclosarium)

SYNECHOCOCCUS/MANGANESE FERMENT

Definition: Synechococcus/Manganese Ferment is an extract of a fermentation product of *Synechococcus* in the presence of manganese ions.

Chemical Class: Biological Products

Function: Skin-Conditioning Agent - Miscellaneous

Trade Name Mixture:
Synechococcus 400 M Etoh 20 (Eclosarium)

SYNSEPALUM DULCIFICUM FRUIT EXTRACT

Definition: Synsepalum Dulcificum Fruit Extract is an extract of the fruit of the berry, *Synsepalum dulcificum.* See "Regulatory and Ingredient Use Information," regarding the labeling names for botanical ingredients in Volume 1, Introduction, Part A.

Chemical Class: Biological Products

Function: Skin-Conditioning Agent - Miscellaneous

Technical/Other Name:
Extract of Synsepalum Dulcificum Fruit

Trade Name Mixture:
Miraclefruit Liquid B (Ichimaru Pharcos)

SYNTHETIC BEESWAX

CAS No. 71243-51-1   **EINECS No.** 275-286-5

Definition: Synthetic Beeswax is a wax synthetically derived to be generally indistinguishable from natural beeswax with regard to composition and properties. It consists chiefly of a mixture of esters of even-numbered, straight chain acids and alcohols containing 16-36 carbon atoms.

Information Sources: 21CFR701.3, CIR: [S] JACT-3(3)1984, JCIC, JCLS, JSQI

Chemical Class: Waxes

Functions: Binder; Emulsion Stabilizer; Viscosity Increasing Agent - Nonaqueous

Reported Product Categories: Lipsticks; Eyeliners; Makeup Preparations (Not eye), Misc.; Eyebrow Pencils; Mascara; Foundations; Eye Makeup Preparations, Misc.; Moisturizing Preparations; Baby Lotions, Oils, Powders and Creams; Body and Hand Preparations (Excluding Shaving Preparations); Cleansing Products (Cold Creams, Cleansing Lotions, Liquids and Pads); Eye Shadows; Leg and Body Paints; Tonics, Dressings, and Other Hair Grooming Aids

Technical/Other Names:
Beeswax, Synthetic
Mixed Wax (1)

Trade Names:
Abesin E (Fabriquimica)
Abesin NE (Fabriquimica)
Abesin s 21 (Fabriquimica)
Beeswax Substitue No. 47 (Ross)
Beeswax, Synthetic (Ross)
Beeswax Synthetic Cosmetic Grade (Ross)
Beeswax, Synthetic - STRALPITZ (Strahl & Pitsch)
Botaniwax SBW-120 (Botanigenics)
Hansonwax JH-1540, 1545 (Hansotech)
Kester Wax K80H (Koster Keunen Holland)
LIPOBEE 102 (Lipo)
Syncrowax BB4 (Croda Chemicals)
Syncrowax BB-4 (Croda, Inc.)
Synthetic Beeswax K-82P (Koster Keunen)
Synthetic Beeswax SP-58 (Strahl & Pitsch)
Synthetic Beeswax SP-446 (Strahl & Pitsch)
Synthetic Beeswax SP-755 (Strahl & Pitsch)
Synthetic Beeswax SP-780 (Strahl & Pitsch)
Synthetic Beeswax SP-1290 (Strahl & Pitsch)
Uninatwax-Bee (Universal Preserv-A-Chem)

Trade Name Mixtures:
Base 323 MTC (LCW)
Base RW 135 (LCW)
Base RW 136 (LCW)
Covacrem LP (LCW)
Covalip 22 (LCW)
Covalip 25 (LCW)
Covalip LL 48 (LCW)
Covawax 501 (LCW)
Mascawax 012 (LCW)
Suncaps 664 (Particle Sciences)
Suncaps 903 (Particle Sciences)

SYNTHETIC CANDELILLA WAX

Definition: Synthetic Candelilla Wax is a synthetic wax intended to be generally indistinguishable from natural candelilla wax with regard to chemical composition and physical characteristics.

Chemical Class: Waxes

Functions: Binder; Viscosity Increasing Agent - Nonaqueous

Trade Name:
Synthetic Candelilla (Koster Keunen)

SYNTHETIC CARNAUBA

Definition: Synthetic Carnauba is a synthetic wax intended to be generally indistinguishable from natural carnauba with regard to chemical composition and physical characteristics.

Chemical Class: Waxes

Functions: Binder; Viscosity Increasing Agent - Nonaqueous

SYNTHETIC FLUORPHLOGOPITE

CAS No. 12003-38-2   **EINECS No.** 234-426-5

JPN Translation:
合成金雲母

Definition: Synthetic Fluorphlogopite is a synthetic mineral that conforms generally to the formula:

$$Mg_3K\left[AlF_2O(SiO_3)_3\right]$$

Chemical Class: Inorganics

Functions: Bulking Agent; Viscosity Increasing Agent - Aqueous

Reported Product Categories: Makeup Bases; Foundations; Face Powders; Blushers (All types); Lipsticks

Trade Names:
FNK-100 (Nihon Koken Kogyo)
PDM-FE (Topy Industries)
PDM-FEM2 (Topy Industries)
PDM-HS (Topy Industries)
PDM-9L-20 (Topy Industries)
PDM-10S (Topy Industries)
Synthecite FNK-100 (Presperse)

Trade Name Mixtures:
Prominence BH (Nihon Koken Kogyo)
Prominence GH (Nihon Koken Kogyo)
Prominence Rh (Nihon Koken Kogyo)
Prominence SH (Nihon Koken Kogyo)
Prominence YH (Nihon Koken Kogyo)
Sunshine Fine White (Sun Pigments)
Sunshine Glitter Golden (Sun Pigments)
Sunshine Glitter Red (Sun Pigments)
Sunshine Glitter White (Sun Pigments)
Sunshine Super Blue (Sun Pigments)
Sunshine Super Bronze (Sun Pigments)
Sunshine Super Copper (Sun Pigments)
Sunshine Super Gold (Sun Pigments)
Sunshine Super White (Sun Pigments)
Ultimica BH (Nihon Koken Kogyo)
Ultimica GH (Nihon Koken Kogyo)
Ultimica RH (Nihon Koken Kogyo)
Ultimica RYH (Nihon Koken Kogyo)
Ultimica SH (Nihon Koken Kogyo)
Ultimica YH (Nihon Koken Kogyo)

SYNTHETIC JAPAN WAX

Definition: Synthetic Japan Wax is a synthetic wax intended to be generally indistinguishable from natural Japan Wax (q.v.) with regard to chemical composition and physical properties.

Information Sources: JCIC, JCLS

Chemical Class: Waxes

Functions: Binder; Viscosity Increasing Agent - Nonaqueous

Technical/Other Name:
Mixed Wax (2)

Trade Names:
Synthetic Japan Wax SP-1 (Strahl & Pitsch)
Synthetic Japan Wax SP-3 (Strahl & Pitsch)
Synthetic Japan Wax SP-17 (Strahl & Pitsch)
Synthetic Japan Wax SP-525 (Strahl & Pitsch)
Synthetic Japan Wax SP-760 (Strahl & Pitsch)

SYNTHETIC JOJOBA OIL

Definition: Synthetic Jojoba Oil is a synthetic oil intended to be generally indistinguishable from natural jojoba oil with regard to chemical composition and physical characteristics.

Chemical Class: Waxes

Function: Skin-Conditioning Agent - Occlusive

Reported Product Category: Eyeliners

Trade Name:
Dermol Jojoba E (Fabriquimica)

Trade Name Mixture:
Dermol Jojoba S (Fabriquimica)

SYNTHETIC THYMUS HYDROLYSATE

Definition: Synthetic Thymus Hydrolysate is a mixture of amino acids and peptides compounded to be generally indistinguishable from natural thymus hydrolysate with regard to composition and properties.

Chemical Classes: Amino Acids; Protein Derivatives

Functions: Skin-Conditioning Agent - Humectant; Skin-Conditioning Agent - Miscellaneous

Trade Name:
S*T*P*L (Statistic Thymic Peptide Library) (Thymuskin)

SYNTHETIC WAX

CAS No.: 8002-74-2

JPN Translation:
合成ワックス

Definition: Synthetic Wax is a hydrocarbon wax derived generally through Fischer-Tropsch or ethylene polymerization processes.

Information Sources: 21CFR172.615, 21CFR172.888, 21CFR173.340, 21CFR175.105, 21CFR175.250, 21CFR176.170, 21CFR177.1200, 21CFR178.3720, CIR: [S] JACT-3(3)1984, RIFM, TSCA

Chemical Class: Waxes

Functions: Binder; Emulsion Stabilizer; Fragrance Ingredient; Viscosity Increasing Agent - Nonaqueous

Reported Product Categories: Lipsticks; Makeup Bases; Eyeliners; Skin Care Preparations, Misc.; Bath Oils, Tablets, and Salts; Cleansing Products (Cold Creams, Cleansing Lotions, Liquids and Pads); Moisturizing Preparations; Tonics, Dressings, and Other Hair Grooming Aids; Bath Preparations, Misc.; Body and Hand Preparations (Excluding Shaving Preparations); Personal Cleanliness Products, Misc.; Blushers (All types); Eye Shadows; Foundations; Hair Coloring Preparations, Misc.; Mascara; Bath Soaps and Detergents; Hair Conditioners; Paste Masks (Mud Packs); Night Skin Care Preparations; Rouges; Nail Polish and Enamels; Bath Capsules; Eye Makeup Preparations, Misc.; Face Powders; Makeup Preparations (Not eye), Misc.; Eye Makeup Removers; Eyebrow Pencils; Face and Neck Preparations (Excluding Shaving Preparations); Nail Creams and Lotions; Suntan Gels, Creams, and Liquids; Aftershave Lotions; Baby Lotions, Oils, Powders and Creams; Cuticle Softeners; Suntan Preparations, Misc.

Technical/Other Name:
Paraffin wax (RIFM)

Trade Names:
AEC White Wax (A & E Connock)
CREABASE 60 (Creations Couleurs)
Microease 110S (Micro Powders)
Microease 114S (Micro Powders)
Microease 225S (Presperse)
Paraflint H1 (Moore & Munger)
Rosswax 100 (Ross)
Synthetic Wax 7050 (Kahl)
Synthetic Wax SP-85 (Strahl & Pitsch)
Synthetic Wax SP-170 (Strahl & Pitsch)
Synthetic Wax SP-171 (Strahl & Pitsch)
Synthetic Wax SP-448 (Strahl & Pitsch)
Synthetic Wax SP-677 (Strahl & Pitsch)
Synthetic Wax SP-1124 (Strahl & Pitsch)

Trade Name Mixtures:
Lip Wax PZ80 (Ina Trading)
Microcare 325 (Micro Powders)
Microease 110S (Presperse)
Microease 110 XF (Presperse)
Microsilk 418 (Micro Powders)
Microsilk 418 (Presperse)
2901 Synthetic Carnauba (Kahl)

SYRINGA VULGARIS (LILAC) EXTRACT

CAS No. 90063-50-6 **EINECS No.** 290-008-2

Definition: Syringa Vulgaris (Lilac) Extract is an extract of the lilac, *Syringa vulgaris*. *See "Regulatory and Ingredient Use Information," regarding the labeling names for botanical ingredients in Volume 1, Introduction, Part A.*

Chemical Class: Biological Products

Function: Skin-Conditioning Agent - Occlusive

Technical/Other Names:
Extract of Lilac
Extract of Syringa Vulgaris
Lilac Extract
Syringa Vulgaris Extract

Trade Name:
   Hydroessential Syringa (Vevy)

Trade Name Mixtures:
   Actiphyte Lilac Blossom (Active Organics)
   Herbasol Distillate Lilac (Cosmetochem)
      (Cosmetochem International Ltd.)
   Herbasol Extract Lilac (Cosmetochem)
      (Cosmetochem International Ltd.)
   Lilac Extract (Carrubba)
   VT-302 Extract of Lilac Blossum (Vege-
      Tech)
   VT-209 Extract of Lilac Leaves (Vege-
      Tech)

SYZYGIUM JAMBOS LEAF EXTRACT

Definition: Syzygium Jambos Leaf Extract
is an extract of the leaves of *Syzygium
jambos. See "Regulatory and Ingredient
Use Information," regarding the labeling
names for botanical ingredients in Volume
1, Introduction, Part A.*

Chemical Class: Biological Products

Function: Skin-Conditioning Agent - Miscellaneous

Technical/Other Names:
   Extract of Eugenia Jambos Leaf
   Extract of Syzygium Jambos Leaf
   Rose Apple Leaf Extract BG-30

Trade Name Mixture:
   Rose Apple Leaf Extract BG-30 (C&F Koei
      Phyto Corp.)